AutoCAD 2004
A Problem Solving Approach

Sham Tickoo

Professor
Department of Manufacturing Engineering Technologies
Purdue University Calumet
Hammond, Indiana
U.S.A

CADCIM Technologies

(www.cadcim.com)
U.S.A.

THOMSON

DELMAR LEARNING

autodesk® Press

AutoCAD 2004®: A Problem Solving Approach

Sham Tickoo

Vice President, Technology and Trades SBU:
Alar Elken

Editorial Director:
Sandy Clark

Senior Acquisitions Editor:
James DeVoe

Senior Development Editor:
John Fisher

Editorial Assistant:
Mary Ellen Martino

Marketing Director:
Cynthia Eichelman

Channel Manager:
Fair Huntoon

Marketing Coordinator:
Sarena Douglass

Production Director:
Mary Ellen Black

Production Manager:
Larry Main

Production Editor:
Tom Stover

Art/Design Coordinator:
Mary Beth Vought

Cover Image:
Corbis Corporation

Printed in the United States of America
1 2 3 4 5 XXX 06 05 04 03

For more information contact Delmar Learning
Executive Woods
5 Maxwell Drive, PO Box 8007,
Clifton Park, NY 12065-8007
Or find us on the World Wide Web at
www.delmarlearning.com

Library of Congress Cataloging-in-Publication Data
Tickoo, Sham.
AutoCAD 2004: a problem solving approach / Sham Tickoo.
p. cm.
ISBN 1-4018-5133-9
1.Computer Graphics.
2. AutoCAD. I. Title.
T385 .T5235 2003
620.0042'0285'5369—dc21
2003011156

NOTICE TO THE READER

Publisher does not warrant or guarantee any of the products described herein or perform any independent analysis in connection with any of the product information contained herein. Publisher does not assume, and expressly disclaims, any obligation to obtain and include information other than that provided to it by the manufacturer.

The reader is expressly warned to consider and adopt all safety precautions that might be indicated by the activities herein and to avoid all potential hazards. By following the instructions contained herein, the reader willingly assumes all risks in connection with such instructions.

The publisher makes no representation or warranties of any kind, including but not limited to, the warranties of fitness for particular purpose or merchantability, nor are any such representations implied with respect to the material set forth herein, and the publisher takes no responsibility with respect to such material. The publisher shall not be liable for any special, consequential, or exemplary damages resulting, in whole or part, from the readers' use of, or reliance upon, this material.

Table of Contents

Chapter 5: Editing Sketched Objects-I

Chapter 6: Editing Sketched Objects-II

Chapter 7: Controlling the Drawing Display and Creating Text

Chapter 8: Basic Dimensioning, Geometric Dimensioning and Tolerancing

Chapter 9: Editing Dimensions

Chapter 10: Dimension Styles and Dimensioning System Variables

Chapter 11: Model Space Viewports, Paper Space Viewports, and Layouts

Chapter 12: Plotting Drawings

Chapter 13: Hatching Drawings

Chapter 14: Working with Blocks

AutoCAD Part II

Chapter 15: Defining Block Attributes

Chapter 16: Understanding External References

Chapter 17: Working with Advanced Drawing Options

Chapter 18: Grouping and Advanced Editing of Sketched Objects

Chapter 19: Working With Data Exchange and Object Linking and Embedding

Chapter 20: Technical Drawing with AutoCAD

Chapter 21: Isometric Drawing

Chapter 22: The User Coordinate System

Chapter 23: Getting Started with 3D

Chapter 24: Creating Solid Models

Chapter 25: Editing and Dynamic Viewing of 3D Objects

Chapter 26: Rendering Designs

Chapter 27: Accessing External Database

Chapter 28: AutoCAD on the Internet

AutoCAD Part III (Customizing)

Chapter 29: Template Drawings

Chapter 30: Script Files and Slide Shows

Chapter 31: Creating Linetypes and Hatch Patterns

Chapter 32: Pull-down, Shortcut, and Partial Menus and Customizing Toolbars

Index

Chapters available for free download

The following chapters are available on the author's and publisher's Web site for free download. To download the free chapters, logon to

www.cadcim.com or
http://technology.calumet.purdue.edu/met/tickoo/students/students.htm
www.autodeskpress.com (Follow the **Online Companions** link)

Chapter 33: Working with AutoLISP

Chapter For Free Download

Chapter 34: Visual Basic for Application *Chapter For Free Download*

Chapter 35: Working with Visual LISP *Chapter For Free Download*

Chapter 36: Student Projects *Chapter For Free Download*

Appendices

The followings appendices are available on the author's Web site, **www.cadcim.com** or **http://technology.calumet.purdue.edu/met/tickoo/students/students.htm**

Preface

AutoCAD, developed by Autodesk Inc., is the most popular PC-CAD system available in the market. Over two million people in 80 countries around the world use AutoCAD to generate various kinds of drawings. In 1998 the market share of AutoCAD grew to 78 percent, making it the worldwide standard for generating drawings. Also, AutoCAD's open architecture has allowed third-party developers to write application software that has significantly added to its popularity. For example, the author of this book has developed a software package "**SMLayout**" for sheet metal products that generates flat layout of various geometrical shapes such as transitions, intersections, cones, elbows, and tank heads. Several companies in Canada and the United States are using this software package with AutoCAD to design and manufacture various products. AutoCAD has also provided facilities that allow users to customize AutoCAD to make it more efficient and therefore increase their productivity.

This book contains a detailed explanation of AutoCAD 2004 commands and how to use them to solve drafting and design problems. The book also unravels the customizing power of AutoCAD. Every AutoCAD command and customizing technique is thoroughly explained with examples and illustrations that make it easy to understand their function and application. At the end of each topic, there are examples that illustrate the function of the command and how it can be used in the drawing. When you are done reading this book, you will be able to use AutoCAD commands to make a drawing, create text, make and insert symbols, dimension a drawing, create 3D objects and solid models, write script files, define linetypes and hatch patterns, write your own menus, and write programs in the AutoLISP programming languages.

The book also covers basic drafting and design concepts such as orthographic projections, dimensioning principles, sectioning, auxiliary views, and assembly drawings that provide you with the essential drafting skills you need to solve drawing problems with AutoCAD. In the process, you will discover some new applications of AutoCAD that are unique and might have a significant effect on your drawings. You will also get a better idea of why AutoCAD has become such a popular software package and an international standard in PC-CAD. Please refer to the following table for conventions used in this text.

Convention	**Example**
• Command names are capitalized and bold.	The **MOVE** command
• A key icon appears when you should respond by pressing the ENTER or RETURN key.	Enter

Convention	Examples
• Command sequences are indented. The responses are indicated by boldface. The directions are indicated by italics and the comments are enclosed in parentheses.	Command: **MOVE** Select object: **G** Enter group name: *Enter a group name (the group name is group1)*
• The command selection from the toolbars, menus, and Command prompt are enclosed in a shaded box.	**Toolbar:** Draw > Arc **Menu:** Draw > Arc **Command:** ARC
• AutoCAD 2004 features are indicated by an asterisk symbol at the end of the feature.	Shaded Plotting*

e-resource

This is an educational resource that creates a truly electronic classroom. It is a CD-ROM containing tools and instructional resources that enrich your classroom and make your preparation time shorter. The elements of **e-resource** link directly to the text and tie together to provide a unified instructional system. With **e-resource** you can spend your time teaching, not preparing to teach.
(ISBN: 1-4018-5157-6)

Features contained in e-resource include:

Syllabus: Lesson plans created by chapter. You have the option of using these lesson plans with your own course information.

Chapter Hints: Objectives and teaching hints that provide the basis for a lecture outline that helps you to present concepts and material.

Answers to Review Questions: These solutions enable you to grade and evaluate end-of-chapter tests.

PowerPoint® Presentation: These slides provide the basis for a lecture outline that helps you to present concepts and material. Key points and concepts can be graphically highlighted for student retention.

Exam View Test Bank: Over 800 questions of varying levels of difficulty are provided in true/false and multiple-choice formats so you can assess student comprehension.

AVI Files: AVI files, listed by topic, allow you to view a quick video illustrating and explaining key concepts.

Drawing Files
The drawing files used in examples and exercises.

Author's Web Sites

For Faculty: Please contact the author at **stickoo@calumet.purdue.edu** to access the Web site that contains the following:

1. **PowerPoint presentations**, program listings, and **drawings** used in this textbook.
2. Syllabus, chapter objectives and hints, and questions with answers for every chapter.

For Students: You can download drawing-exercises, tutorials, programs, and special topics by accessing the author's Web site at **www.cadcim.com** or **http://technology.calumet.purdue.edu/met/tickoo/students/students.htm**.

DEDICATION

*To teachers, who make it possible to disseminate knowledge
to enlighten the young and curious minds
of our future generations*

*To students, who are dedicated to learning new technologies
and making the world a better place to live*

THANKS

*To the faculty and students of the MET Department of
Purdue University Calumet for their cooperation*

*To Mohd. Khalid, Ankur Mathur, and Manash Chatterjee
for performing the technical edit and Vivek Kumar and Manoj Bachheti for copy edit*

*To Deepak Maini, Puja Manchanda, Abhinav Rai,
Saket Singhal, Gurpreet Singh, and Santosh Tickoo
of CADCIM/CADSoft Technologies
for their valuable help*

AutoCAD Part I

AutoCAD Part I

Author's Web Sites

For Faculty: Please contact the author at **stickoo@calumet.purdue.edu** or **tickoo@cadcim.com** to access the Web site that contains the following.

1. **PowerPoint presentations, programs, and drawings used in this textbook.**
2. **Syllabus, chapter objectives and hints, and questions with answers for every chapter.**

For Students: You can download drawing-exercises, tutorials, programs, and special topics by accessing author's web site at **www.cadcim.com** or **http://technology.calumet.purdue.edu/met/tickoo/students/students.htm**.

Chapter 1

Introduction to AutoCAD

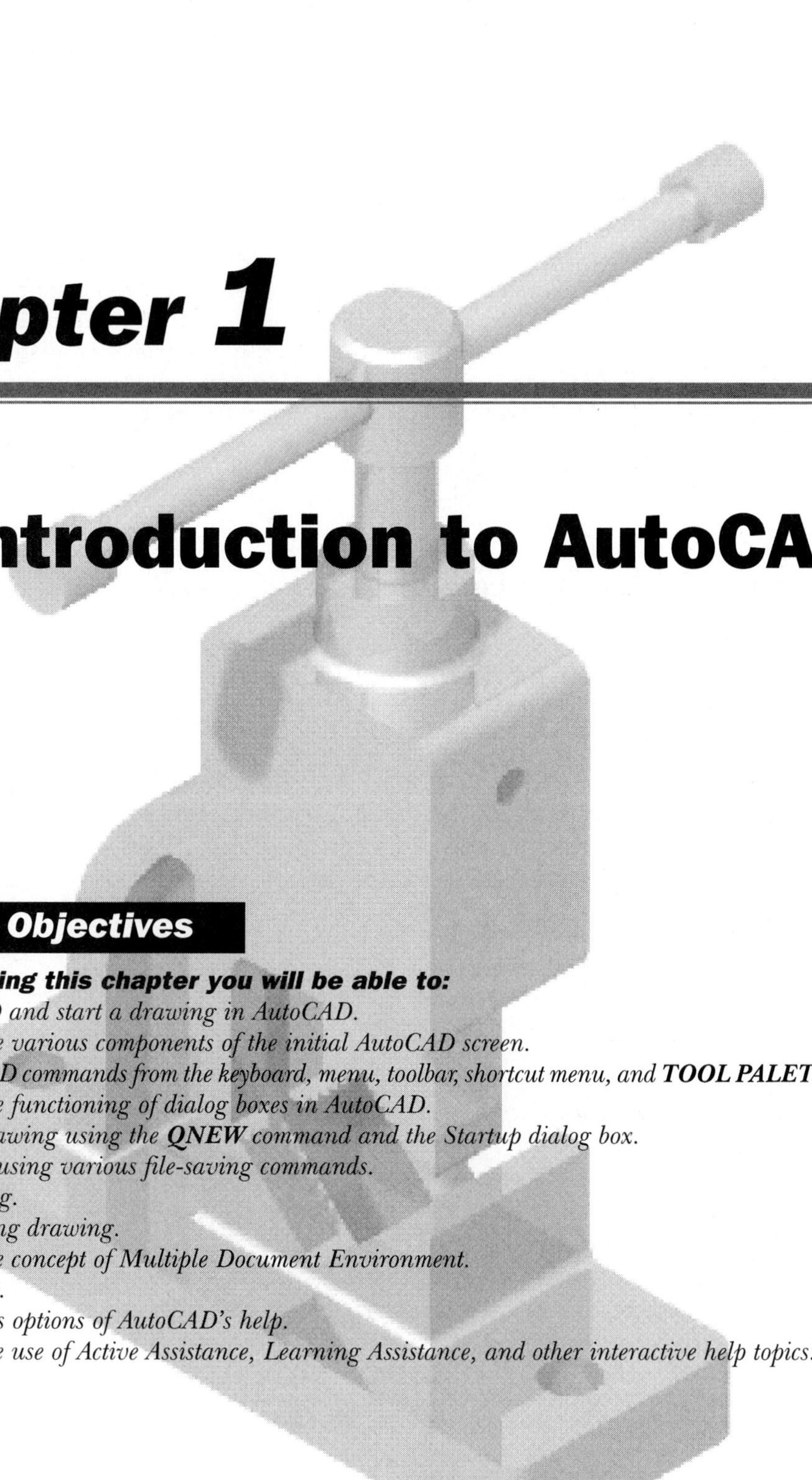

Learning Objectives

After completing this chapter you will be able to:

- *Start AutoCAD and start a drawing in AutoCAD.*
- *Understand the various components of the initial AutoCAD screen.*
- *Invoke AutoCAD commands from the keyboard, menu, toolbar, shortcut menu, and* ***TOOL PALETTES****.*
- *Understand the functioning of dialog boxes in AutoCAD.*
- *Start a new drawing using the* ***QNEW*** *command and the Startup dialog box.*
- *Save the work using various file-saving commands.*
- *Close a drawing.*
- *Open an existing drawing.*
- *Understand the concept of Multiple Document Environment.*
- *Quit AutoCAD.*
- *Use the various options of AutoCAD's help.*
- *Understand the use of Active Assistance, Learning Assistance, and other interactive help topics.*

STARTING AutoCAD

When you turn on your computer, the operating system (Windows 95, Windows NT, Windows 98, Windows XP, and so on) is automatically loaded. This will display the Windows screen with various application icons. You can start AutoCAD by double-clicking on the AutoCAD 2004 icon available on the desktop of your computer. You can also load AutoCAD from the Windows taskbar by choosing the **Start** button at the bottom left corner of the screen (default position) to display the menu. Choose **Programs** to display the program folders. Now, choose the **Autodesk > AutoCAD 2004** folder to display the AutoCAD programs and then choose **AutoCAD 2004** to start AutoCAD, see Figure 1-1.

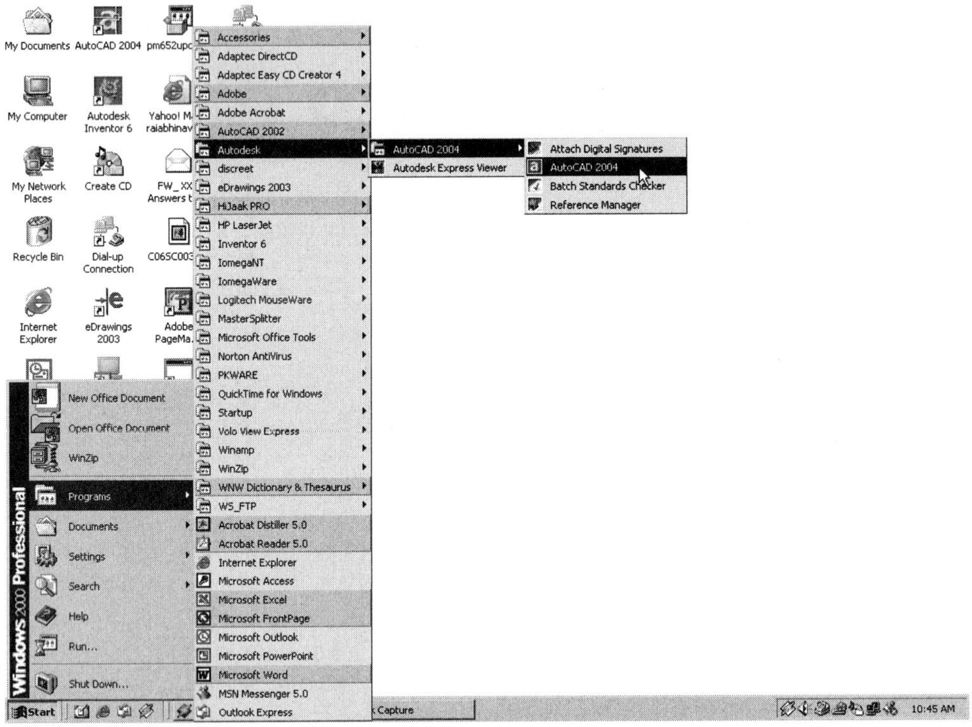

Figure 1-1 *Windows screen with taskbar and application icons*

AutoCAD SCREEN COMPONENTS

The various components of the initial AutoCAD screen are the drawing area, the command window, menu bar, several toolbars, model and layout tabs, and the status bar (Figure 1-2). A title bar that has the AutoCAD symbol and the current drawing name is displayed on top of the screen.

Drawing Area

The drawing area covers the major portion of the screen. Here you can draw the various objects and use the various commands. To draw the objects you need to define the coordinate points, which can be selected by using your pointing device. The position of the pointing

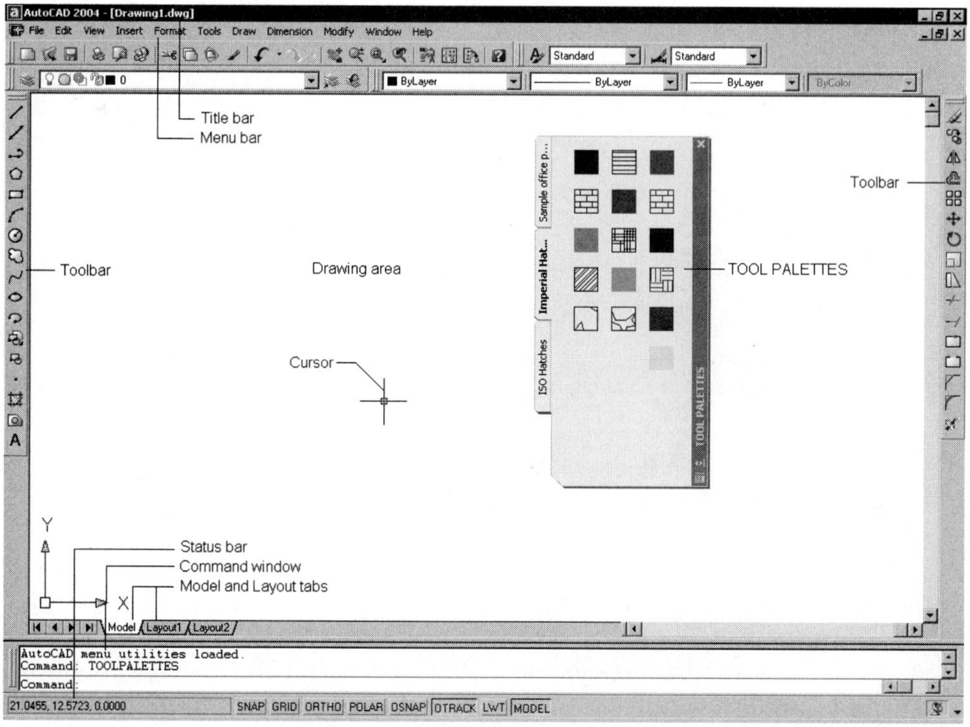

Figure 1-2 *AutoCAD screen components*

device is represented on the screen by the cursor. There is a coordinate system icon at the lower left corner of the drawing area. The window also has the standard Windows buttons such as close, minimize, scroll bar, and so on, available on the top right corner. These buttons have the same functions as for any other standard window.

Command Window

The command window is present at the bottom of the drawing area and has the Command prompt where you can enter the commands. It also displays the subsequent prompt sequences and the messages. You can change the size of the window by placing the cursor on the top edge (double line bar known as the grab bar) and then dragging it. This way you can increase its size to see all the previous commands you have used. By default the command window displays only three lines. You can also press the F2 key to display the **AutoCAD Text window**, which displays the previous commands and prompts.

Tip

You can hide all the toolbars displayed on the screen by either pressing the CTRL+0 keys or by choosing the ***Clean Screen*** *option from the* ***View*** *menu. To turn on the display of the toolbars again press the CTRL+0 keys on the keyboard.*

Status Bar

The status bar is displayed at the bottom of the screen (Figure 1-3). This bar contains some useful information and buttons that will make it easy to change the status of some AutoCAD functions. To change the status, you must choose the buttons that toggle between on and off.

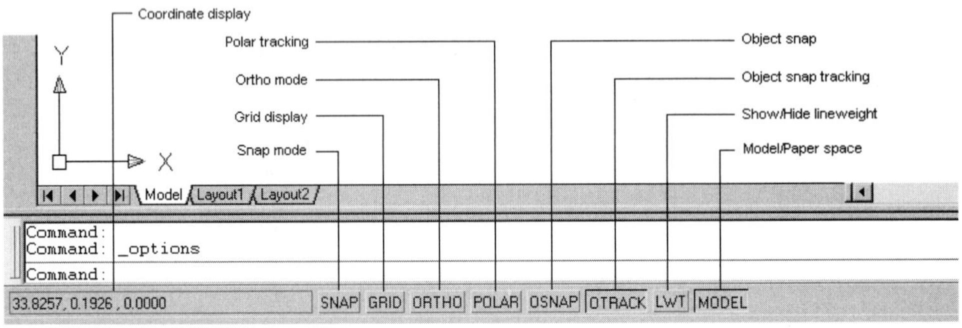

Figure 1-3 *Default status bar display*

Coordinate Display

The coordinates information is displayed in the left corner of the status bar. You can select this coordinate button to toggle between on and off. The **COORDS** system variable controls the type of display of the coordinates. If the value of the **COORDS** variable is set to 0, the coordinate display is static, that is, the coordinate values displayed in the status bar change only when you specify a point. If the value of the **COORDS** variable is set to 1 or 2, the coordinate display is dynamic. When the variable is set to 1, AutoCAD constantly displays the absolute coordinates of the graphics cursor with respect to the UCS origin. The polar coordinates (length<angle) are displayed if you are in an AutoCAD command and the **COORDS** variable is set to 2. You can use the key F6 to turn the coordinate display on or off.

SNAP

The snap mode allows you to move the cursor in fixed increments. If the snap mode is on, the **SNAP** button is displayed as pressed in the status bar; otherwise, it is not displayed. You can also use the function key F9 as a toggle key to turn Snap off or on.

GRID

The grid lines are used as reference lines to draw objects in AutoCAD. If the grid display is on, the **GRID** button is displayed as pressed and the grid lines are displayed on the screen. The function key F7 can be used to turn the grid display on or off.

ORTHO

The ortho mode allows you to draw lines at right angles only. If this mode is on, the **ORTHO** button is pressed in the status bar. You can use the F8 key to turn Ortho on or off.

POLAR

If you turn the polar tracking on, the movement of the cursor is restricted along a path based on the angle set as the polar angle settings. Choosing the **POLAR** button in the status bar turns the polar tracking on. You can also use the function key F10. Remember that turning the polar tracking on, automatically turns off the ortho mode.

OSNAP

When object snap is on, you can use the running object snaps to snap on to a point. If object snap is on, the **OSNAP** button is displayed as pressed in the status bar. You can also use the F3 key to turn the object snap on or off. If **OSNAP** is off, the running object snaps are temporarily disabled. The status of **OSNAP** (off or on) does not prevent you from using immediate mode object snaps.

OTRACK

Choosing the **OTRACK** button turns object snap tracking on or off.

LWT

Choosing this button in the status bar allows you to turn on or off the display of lineweights in the drawing. If the **LWT** button is not pressed, the display of Lineweight is turned off.

MODEL

The **MODEL** button is displayed in the status bar when you are working in the model space to create drawings. You can choose this button to shift to the layouts (paper space) where you can create drawing views. Once you switch to layouts, this button is replaced by the **PAPER** button. You can choose the **PAPER** button to shift back to the model space.

Note

All the buttons in the status bar are discussed in detail in Chapter 4, Working With Drawing Aids.

The menu bar and toolbar are discussed in the following section. The model and layout tabs are discussed in Chapter 11, Model Space Viewports, Paper Space Viewports, and Layouts.

Status Bar Tray Options*

The status bar tray options are displayed at the lower-right corner of the screen. These options are used to access the frequently used commands in AutoCAD.

Communication Center*

The **Communication Center** displays a message and an alert whenever Autodesk provides the latest information regarding software updates and their other products. You can configure the settings of the **Communication Center** by clicking on the **Communication Center** icon. The **Communication Center** dialog box is displayed. Choose the Settings button to display the **Configuration Settings** dialog box. The **Please Select Country** drop-down list available in the **Country** area allows you to select the name of your country. The **Check for New Content** area allows the **Communication Center** to update you daily, weekly, monthly, or on demand on the latest information provided by Autodesk.

Manage Xrefs*

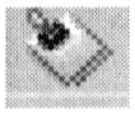

The **Manage Xrefs** icon is displayed whenever an external reference drawing is attached to the selected drawing. This icon displays a message and an alert whenever an xref drawing is required to be reloaded. To know about the detailed information regarding the status of each xref in the drawing and the relation between the various Xrefs, pick the **Manage Xrefs** icon. The **XRef Manager** dialog box is displayed. The Xrefs are discussed in detail in Chapter 16, Understanding External References.

CAD Standards*

The **CAD Standards** icon is displayed whenever a standard drawing is associated with the selected drawing to compare the standards. This icon displays a message and an alert whenever standard violation occurs in the drawing. The drawing can be checked for the standard violation and edited using the **Check Standards** dialog box, which is invoked by clicking on the **CAD Standards** icon.

Validate Digital Signatures*

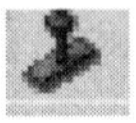

The **Validate Digital Signatures** icon is displayed whenever AutoCAD drawing has a valid digital signature. You can validate the digital signature by clicking on this icon.

INVOKING COMMANDS IN AutoCAD

When you start AutoCAD and you are in the drawing area, you need to invoke AutoCAD commands to perform any operation. For example, if you want to draw a line, first you have to invoke the **LINE** command, and then you define the start point and endpoint of the line. Similarly, if you want to erase objects, you must enter the **ERASE** command, and then select the objects for erasing. AutoCAD has provided the following methods to invoke commands:

Keyboard	**Menu**	**Toolbar**
Shortcut menu	**TOOL PALETTES***	

Keyboard

You can invoke any AutoCAD command at the keyboard by typing the command name at the Command prompt, and then pressing ENTER or the SPACEBAR. Before you enter a command, make sure the Command prompt is displayed as the last line in the command window area. If the Command prompt is not displayed, you must cancel the existing command by pressing ESC (escape) on the keyboard. The following example shows how to invoke the **LINE** command from the keyboard:

Command: **LINE or L** Enter (L is command alias)

Menu

You can also select commands from the menu. The menu bar that displays the menu bar titles is at the top of the screen. As you move the cursor over the menu bar, different titles are highlighted. You can choose the desired item by pressing the pick button of your pointing device. Once the item is selected, the corresponding menu is displayed directly under the title. You can invoke a command from the menu by pressing the pick button of your pointing device. Some of the menu items in the menu display an arrow on the right side, which indicates that the menu item has a cascading menu. The cascading menu provides various options to execute the same AutoCAD command. You can display the cascading menu by choosing the menu item or by just moving the arrow pointer to the right of that item. You can then choose any item in the cascading menu by highlighting the item or command and pressing the pick button of your pointing device. For example, if you want to draw an ellipse using the Center option, choose **Draw** from the menu bar, then choose **Ellipse** from the **Draw** menu, and finally choose **Center** from the cascading menu as shown in Figure 1-4. In this text, this command selection sequence will be referenced as **Draw** > **Ellipse** > **Center**.

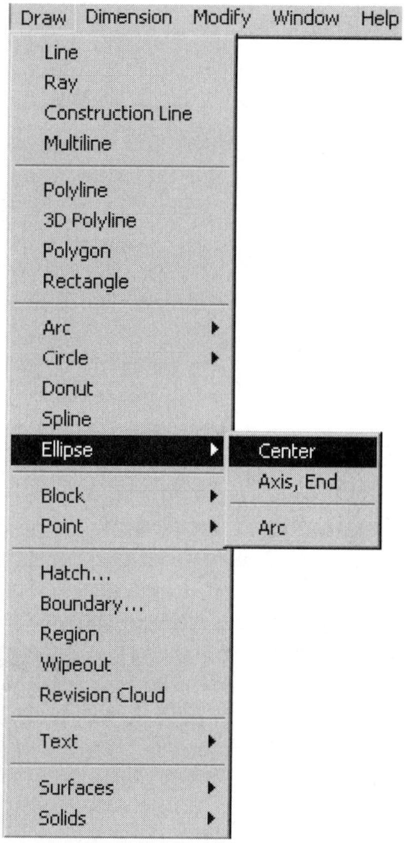

Figure 1-4 *Invoking the* ***ELLIPSE*** *command from the* ***Draw*** *menu*

Note

The menus and toolbars tend to vary in every release of AutoCAD.

Toolbar

In Windows, the toolbar is an easy and convenient way to invoke a command. Each toolbar contains a group of buttons representing various AutoCAD commands. When you move the cursor over the buttons of a toolbar, the button gets lifted and a three-dimensional (3D) box encloses the button on which the cursor is resting. The tooltip (name of the button) is also displayed below the button. Once you locate the desired button, the command associated with that button can be invoked by choosing the button. For example, you can invoke the **LINE** command by choosing the **Line** button from the **Draw** toolbar, see Figure 1-5.

Some of the buttons in a toolbar have a small triangular mark at the lower right corner. This indicates that the button has a flyout attached to it. If you hold the button down, the flyout is displayed. The flyout contains the various options for the command. When you choose a command from the toolbar, the command prompts are displayed in the command window. By default the **Standard**, **Styles**, **Layers**, **Properties**, **Draw**, and **Modify** toolbars are displayed on the screen and are docked to the top and the left side edges of the drawing area.

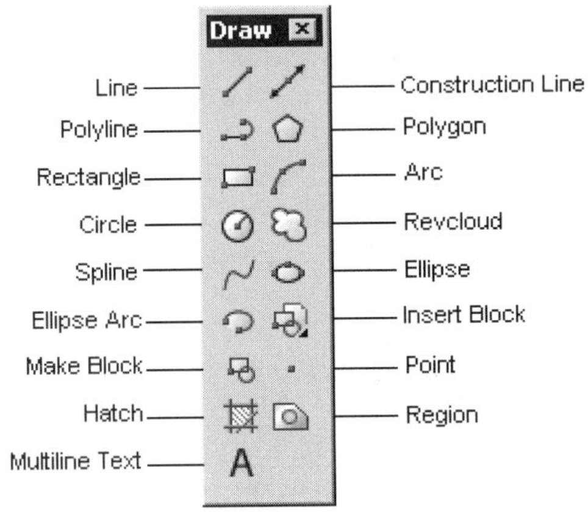

Figure 1-5 *The **Draw** toolbar*

Displaying Toolbars

The various toolbars can be displayed by selecting their respective check boxes in the **Toolbars** tab of the **Customize** dialog box (Figure 1-6). The **Customize** dialog box can be invoked by choosing **View > Toolbars** from the menu bar. You can also display a toolbar from the shortcut menu, which is displayed by right-clicking anywhere on any toolbar on the screen and choosing the name of the toolbar to display from the shortcut menu.

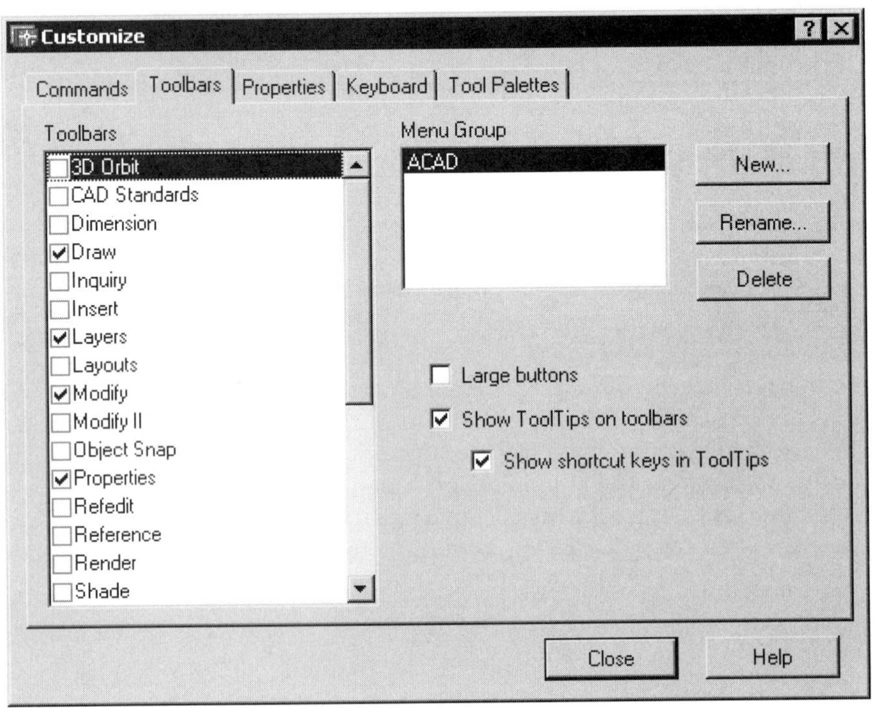

Figure 1-6 *List of toolbars in the **Toolbars** tab of the **Customize** dialog box*

Moving and Reshaping Toolbars

The toolbars can be moved anywhere on the screen by placing the cursor on the title bar area and then dragging it to the desired location. You must hold the pick button down while dragging. While moving the toolbars you can dock them to the top or sides of the screen by dropping them in the docking area. You may also prevent docking by holding the CTRL key when moving the toolbar to a desired location. You can also change the shape of the toolbars by placing the cursor anywhere on the border of the toolbar where it takes the shape of a double arrow (Figure 1-7), and then pulling it in the desired direction (Figure 1-8). You can also customize toolbars to meet your requirements (see Chapter 32, Pull-down, Shortcut, and Partial Menus and Customizing Toolbars).

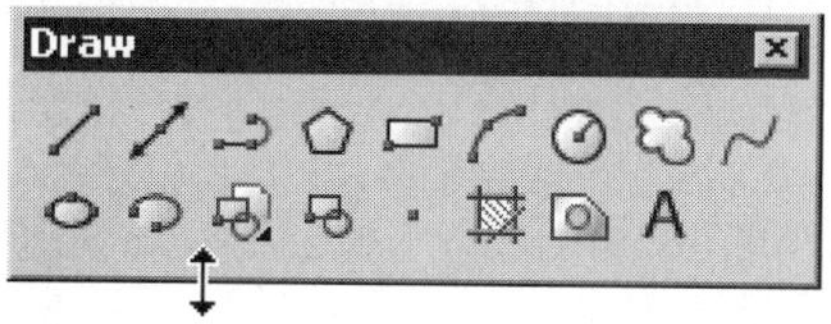

Figure 1-7 *Reshaping the* ***Draw*** *toolbar*

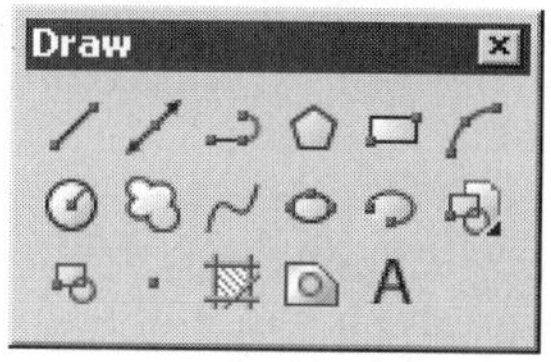

Figure 1-8 ***Draw*** *toolbar reshaped*

Shortcut Menu

AutoCAD has provided shortcut menus as an easy and convenient way of invoking commands. These menus are context-sensitive, which means that the commands present in them are dependent on the place/object for which they are displayed. This menu is invoked by right-clicking and is displayed at the cursor location.

You can right-click anywhere on the drawing area to display the general shortcut menu. It generally contains an option to select the previously invoked command again (Figure 1-9) apart from the common commands for Windows.

If you right-click on the drawing area while a command is in effect, the shortcut menu displayed contains the various options of that particular command. Figure 1-10 shows the shortcut menu when the **POLYLINE** command is active.

If you right-click on the layout tabs, the shortcut menu displayed contains the various options for the layouts (Figure 1-11).

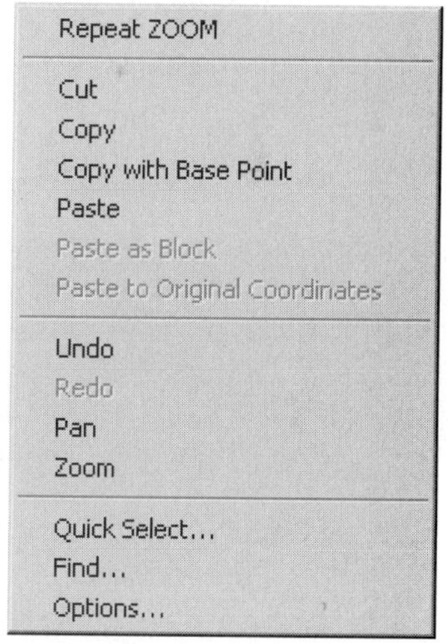

Figure 1-9 *Shortcut menu without an active command*

Figure 1-10** Shortcut menu with the* ***POLYLINE *command active*

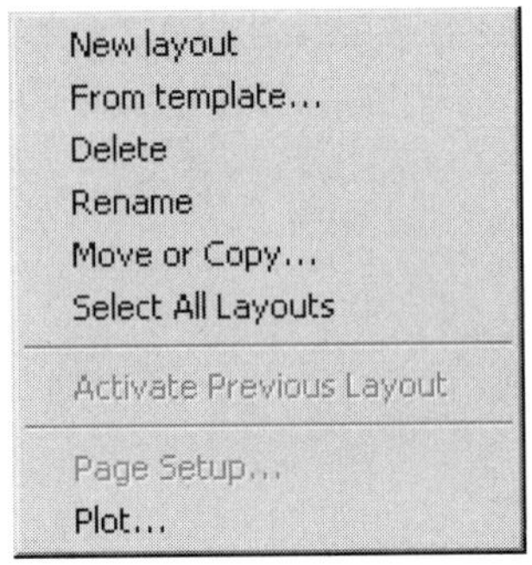

***Figure 1-11** Shortcut menu for the Layout tab*

You can also right-click on the command window to display the shortcut menu. This menu displays the six most recently used commands and some of the window options like Copy and Paste (Figure 1-12). The commands and their prompt entries are displayed in the History window (previous command lines not visible) and can be selectd, copied, and pasted in the command line using the shortcut menu. As you press the up arrow key, the previously entered commands are displayed in the command window. Once the desired command is displayed at the Command prompt you can execute the command by simply pressing the ENTER key. You can also copy and edit any previously invoked command by locating it in the History window and then selecting the lines. Right-click in the command window to display the shortcut menu (Figure 1-12); select copy, and then paste the selected lines in the command line. After the lines are pasted, you can edit them.

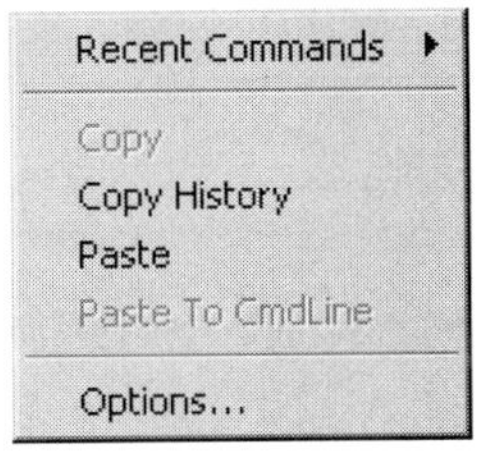

***Figure 1-12** Command line window shortcut menu*

You can right-click on the status bar to display the shortcut menu. This menu contains the options to change the settings of drawing tools (Figure 1-13).

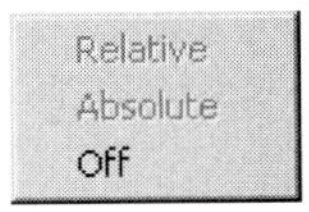

***Figure 1-13** Status bar shortcut menu*

You can also right-click on any of the toolbars to display the shortcut menu from where you can choose any toolbar to be displayed.

Tip

A shortcut menu is available for any situation while working in AutoCAD. You should try to make use of it frequently by right-clicking at various positions.

TOOL PALETTES*

AutoCAD has provided **TOOL PALETTES** (Figure 1-14) as an easy and convenient way of placing and sharing hatch patterns and blocks in the current drawing. By default, AutoCAD displays the Tool Palettes as a window on the right of the drawing area. Also, the **TOOL PALETTES** window can be turned on or off by either choosing the **Tool Palettes** button available on the **Standard** toolbar or by pressing the CTRL+3 keys. The **TOOL PALETTES** are discussed in detail in Chapter 13, Hatching Drawings.

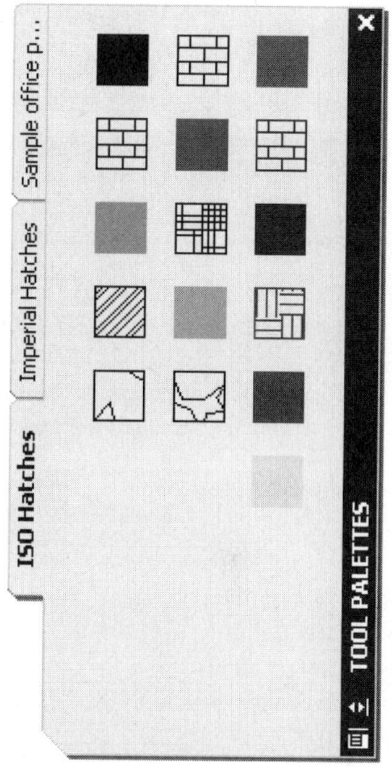

Figure 1-14 TOOL PALETTES window

AutoCAD DIALOG BOXES

There are certain commands which when invoked display a dialog box. A dialog box is a convenient method of user interface. In the menus, the menu item with the ellipses [...] displays the dialog box when you choose that item. For example, **Options** in the **Tools** menu displays the **Options** dialog box. Any dialog box contains a number of parts like the dialog label, radio buttons, text or edit boxes, check boxes, slider bars, image boxes, and command buttons. These components are also referred to as TILES. Some of the components of a dialog box are shown in Figure 1-15.

You can select the desired tile using the pointing device, which is represented by an arrow when a dialog box is invoked. The title bar displays the name of the dialog box. The **tabs** specify the various sections with a group of related options under them. The **check boxes** are toggle buttons for making the particular option available or unavailable. The **drop-down list** displays an item and an arrow on the right which when selected displays a list of items to choose from. You can make a selection in the **radio buttons.** Only one can be selected at a time. The **image box** displays the preview image of the item selected. The **text box** is an area where you can enter a text like a file name. It is also called an **edit box** because you can make any change to the text entered. In some dialog boxes there is the **[...] button**, which displays another related dialog box. There are certain **command buttons** (OK, Cancel, Help) at the bottom of the dialog box. The name implies their functions. The button with a dark border is the default button. The dialog box has a **Help** button for getting help on the various features of the dialog box. There is also a **question mark** (?) button near the top right corner of the dialog box meant for feature-specific help. If you want help on a particular feature of a dialog box, select the ? button. The ? gets attached with the cursor. You can then select the feature to display its help.

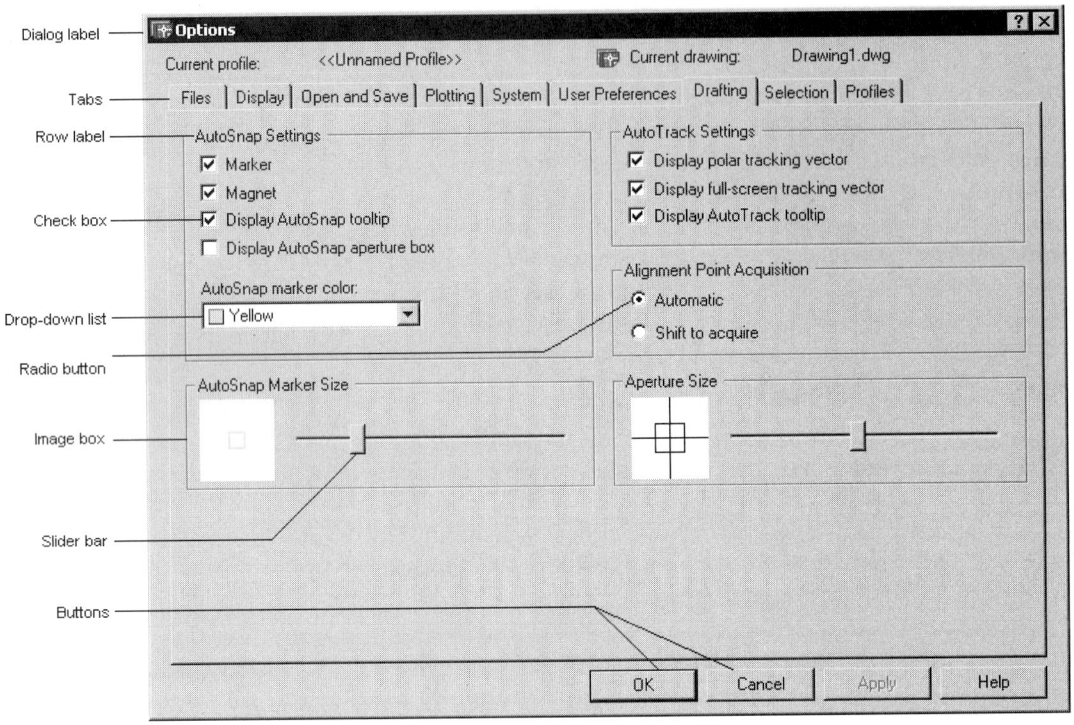

Figure 1-15 Components of a dialog box

STARTING A NEW DRAWING

Toolbar: Standard toolbar > QNew
Menu: File > New
Command: NEW or QNEW

You can open a new drawing using the **QNEW** command. When you invoke the **QNEW** command, by default AutoCAD will display the **Select template** dialog box, as shown in Figure 1-16. The dialog box displays a list of the default templates available in AutoCAD 2004. You can select the desired template to open a new drawing, which will use the settings of the selected template.

By default, when you invoke the **QNEW** command, AutoCAD displays the **Select template** dialog box. You can also open a new drawing using the **Use a Wizard** and **Start from Scratch** options available in the **Create New Drawing** dialog box. By default, the display of this dialog box is turned off. To turn on the display of the **Create New Drawing** dialog box, choose **Tools > Options** from the menu bar. The **Options** dialog box is invoked; choose the **System** tab. Under the **General Options** area, select the **Show Startup dialog box** from the **Startup** drop-down list. Choose **Apply** and then choose the **OK** button. Next, whenever you invoke the **QNEW** command, the **Create New Drawing** dialog box will be displayed as shown in Figure 1-17. The options provided in this dialog box are discussed next.

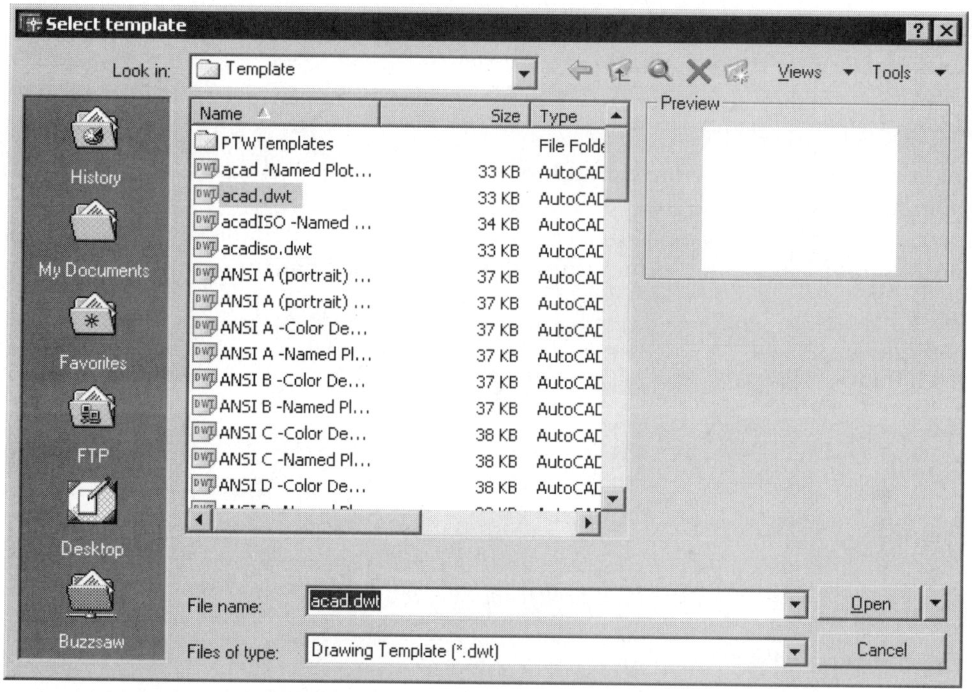

Figure 1-16 *Select template dialog box*

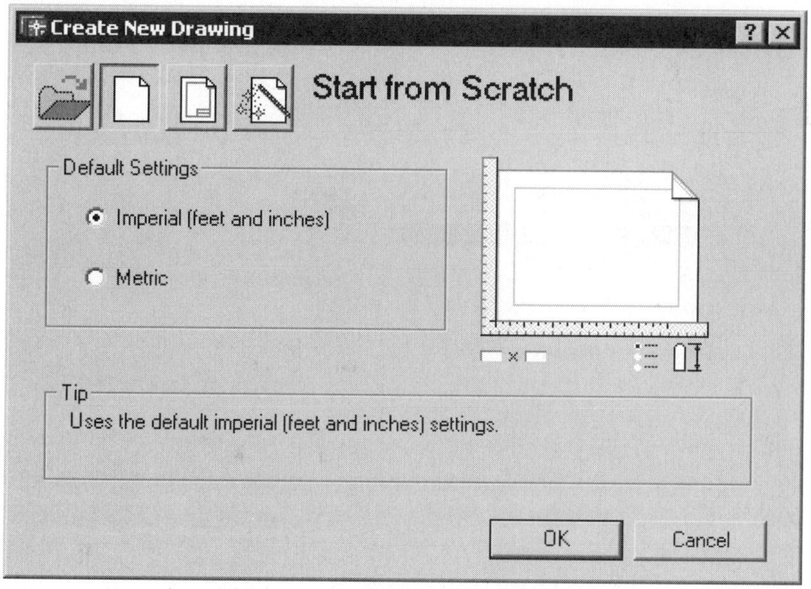

Figure 1-17 Create New Drawing *dialog box*

Open a Drawing

By default this option is not available.

Start from Scratch

When you choose the **Start from Scratch** button (Figure 1-18), AutoCAD provides you with options to start a new drawing that contains the default AutoCAD setup for Imperial (*acad.dwt*) or Metric drawings (*acadiso.dwt*). If you select the Imperial default setting, the limits are 12X9, text height is 0.20, and dimensions and linetype scale factors are 1.

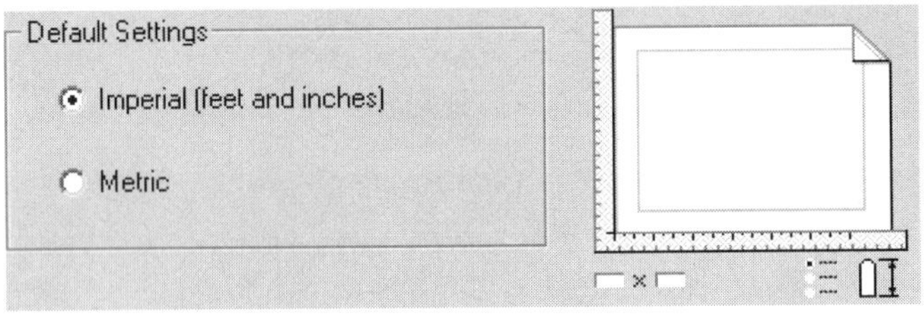

Figure 1-18 *Options that are displayed to start a new drawing when you choose the* ***Start from Scratch*** *button*

Use a Template

When you choose the **Use a Template** button in the **Create New Drawing** dialog box, AutoCAD displays a list of templates supplied with AutoCAD, see Figure 1-19.

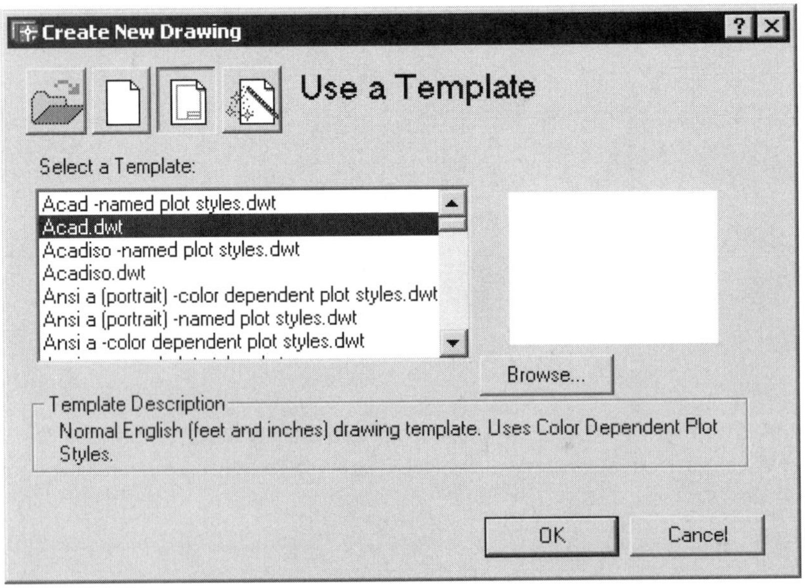

Figure 1-19 *The default templates that are displayed when you choose the* ***Use a Template*** *button*

The default template file is *acad.dwt* or *acadiso.dwt*, depending on the installation. If you use a template file, the new drawing will have the same settings as specified in the template file. All the drawing parameters of the new drawing such as the units, limits, and other settings are already set according to the template file used. The preview of the template file selected is displayed in the dialog box. You can also define your own template files that are customized to your requirements (see Chapter 29, Template Drawings). To differentiate the template files from the drawing files, the template files have *.dwt* extension whereas the drawing files have *.dwg* extension. Any drawing file can be saved as a template file. You can use the **Browse** button to select other template files. When you choose the **Browse** button, the **Select a template file** dialog box is displayed with the **Template** folder open, displaying all the template files.

Use a Wizard

The **Use a Wizard** option allows you to set the initial drawing settings before actually starting a new drawing. When you choose the **Use a Wizard** button, AutoCAD provides you the option of using the **Quick Setup** or **Advanced Setup**, see Figure 1-20.

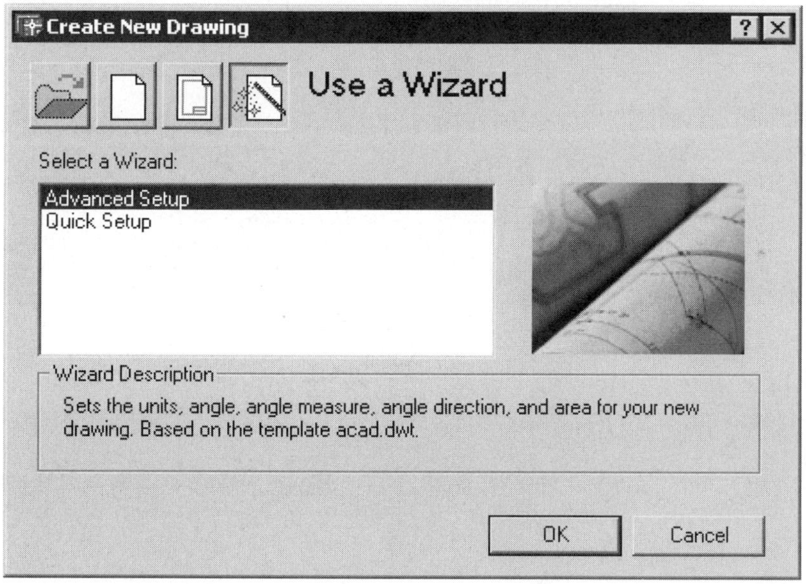

Figure 1-20 *The Wizard options that are displayed when you choose the* ***Use a Wizard*** *button*

In the **Quick Setup**, you can specify the units and the limits of the work area. In the **Advanced Setup**, you can set the units, limits, and the different types of settings for a drawing.

Advanced Setup

This option allows you to preselect various parameters of a new drawing such as the units of linear and angular measurements, type and direction of angular measurements, approximate area desired for the drawing, precision for displaying the units after decimal, and so on.

When you select the **Advanced Setup** wizard option from the **Create New Drawing** dialog box and choose the **OK** button, the **Advanced Setup** dialog box is displayed. The **Units** page is displayed by default as shown in Figure 1-21.

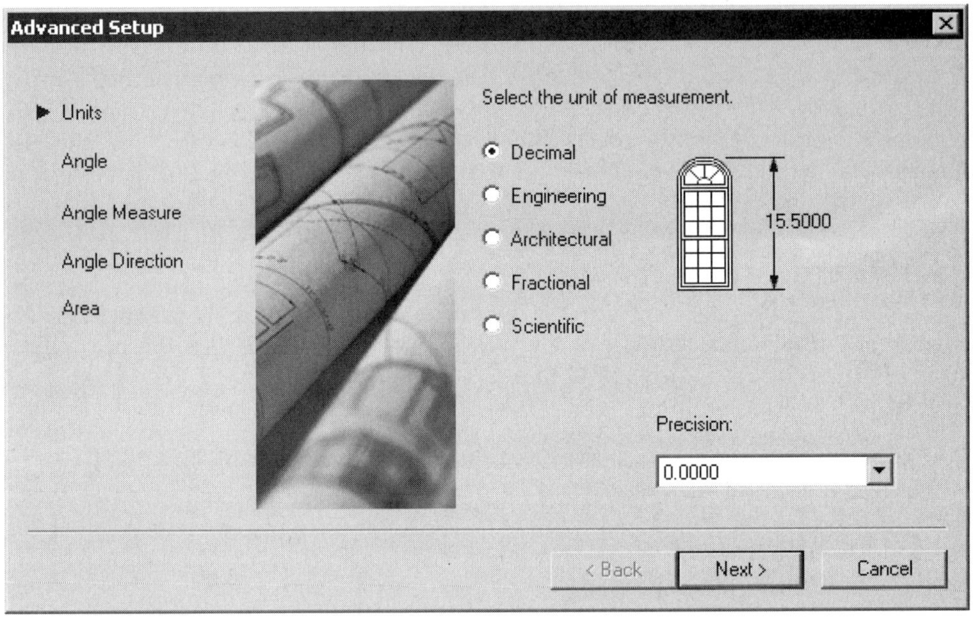

Figure 1-21 *The* ***Units*** *page of the* ***Advanced Setup*** *dialog box*

This page is used to set the units for measurement in the current drawing. You can select the required unit of measurement by selecting its respective radio button. You will notice that the preview image is modified accordingly. The different units of measurement you can choose from are Decimal, Engineering, Architectural, Fractional, and Scientific. You can also set the precision for measurement units by selecting it from the **Precision** drop-down list.

Choose the **Next** button to open the **Angle** page as shown in Figure 1-22. You will notice that an arrow appears on the left of **Angle** in the **Advanced Setup** dialog box. This suggests that this page is current.

This page is used to set the units for angular measurements and the precision for it. The units for angle measurement can be set by selecting its radio button. The units for angular measurements that you can select are Decimal Degrees, Deg/Min/Sec, Grads, Radians, and Surveyor. The preview of the selected angular unit is displayed on the right of the radio buttons. The precision format changes automatically in the **Precision** drop-down list depending on the angle measuring system selected. You can then select the precision from the drop-down list.

The next page is the **Angle Measure** page as shown in Figure 1-23. This page is used to select the direction of the base angle from which the angles will be measured. You can also set your

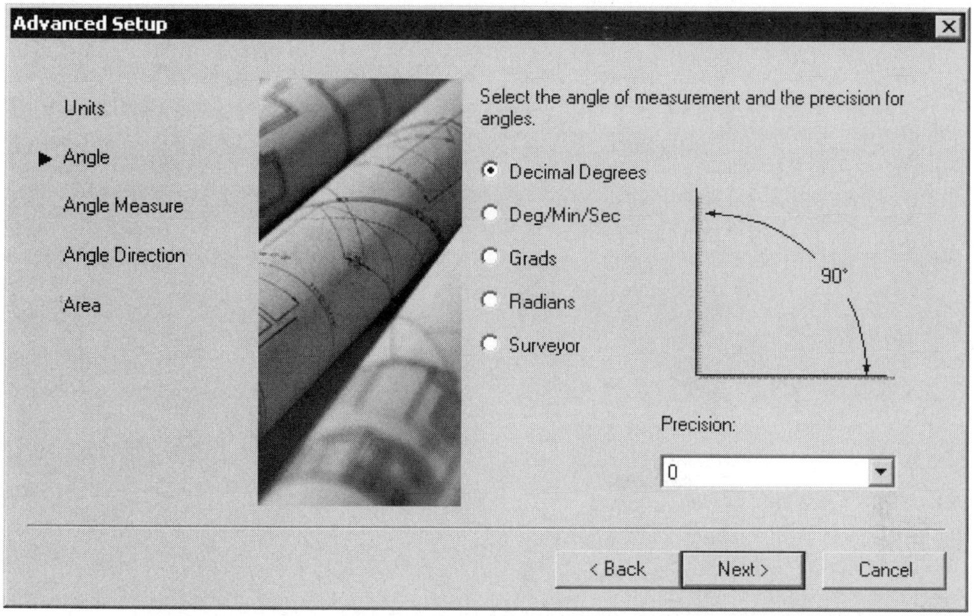

Figure 1-22 *The **Angle** page of the **Advanced Setup** dialog box*

own direction by selecting the **Other** radio button and then entering the value in its edit box. This edit box is available when you select the **Other** radio button.

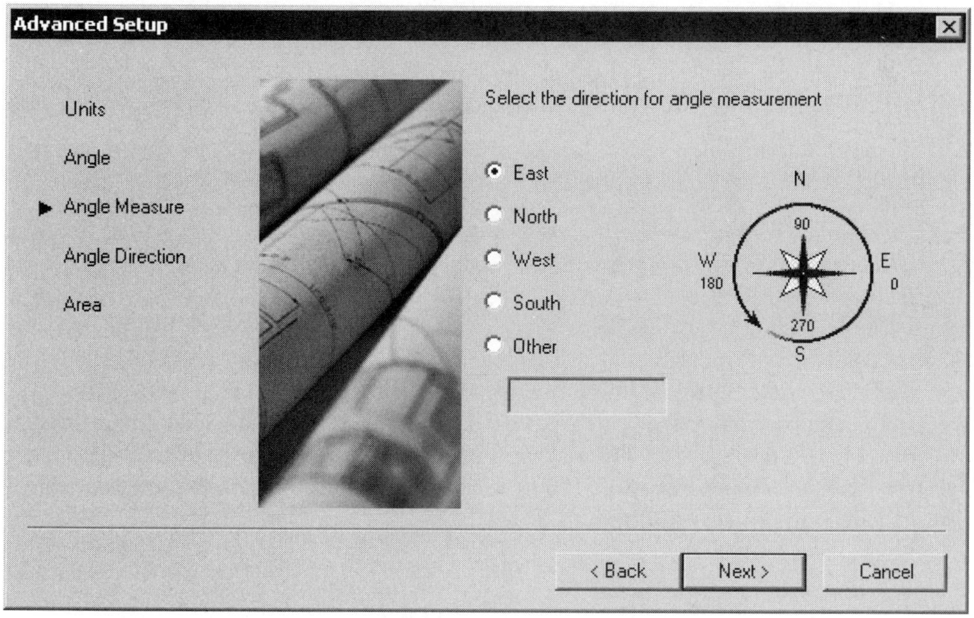

Figure 1-23 *The **Angle Measure** page of the **Advanced Setup** dialog box*

Choose **Next** to display the **Angle Direction** page (Figure 1-24) to set the orientation for angle measurement. By default the angles are positive if measured in a counterclockwise direction. This is because the **Counter-Clockwise** radio button is selected. If you select the **Clockwise** radio button, the angles will be considered positive when measured in clockwise direction.

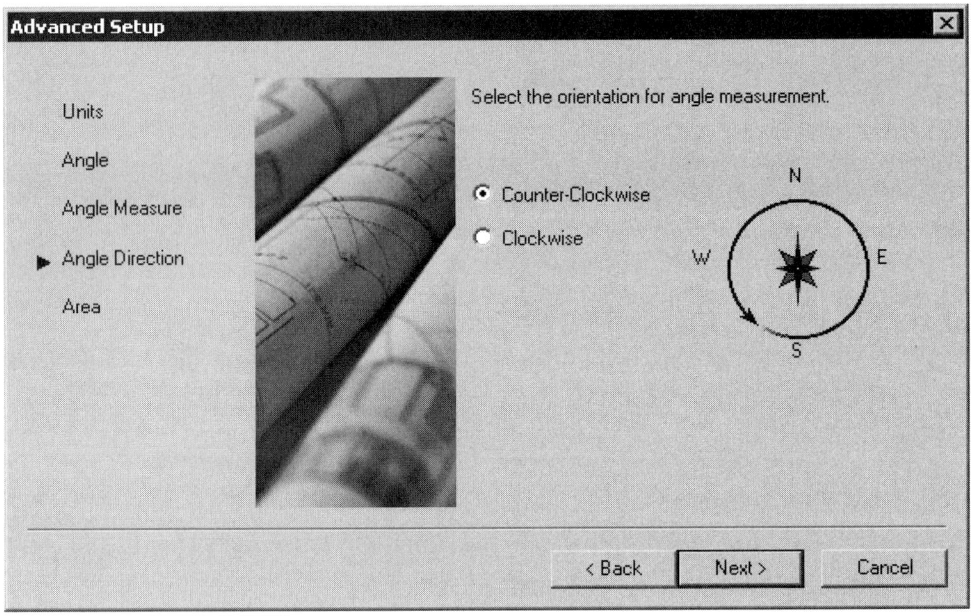

Figure 1-24 *The* ***Angle Direction*** *page of the* ***Advanced Setup*** *dialog box*

To set the limits of the drawing, choose the **Next** button. The **Area** page is displayed as shown in Figure 1-25. You can enter the width and length of the drawing area in the respective edit boxes.

Note

Even after you increase the limits of the drawing, the drawing display area is not increased. You need to invoke the ***ZOOM*** *command and then invoke the* ***All*** *option to increase the drawing display area.*

Quick Setup

When you select **Quick Setup** and choose the **OK** button, the **Quick Setup** dialog box is displayed. This dialog box has just two pages: **Units** and **Area**. The **Units** page opened by default as shown in Figure 1-26. The options in the **Units** page are similar to those in the **Units** page of the **Advanced Setup** dialog box. The only difference is that you cannot set the precision for the units in this dialog box.

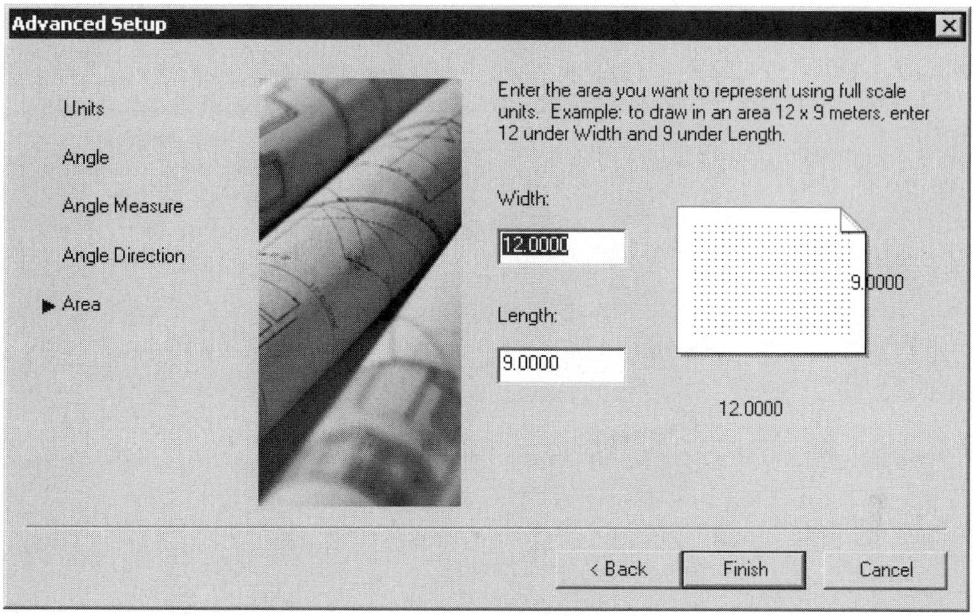

Figure 1-25 *The* ***Area*** *page of the* ***Advanced Setup*** *dialog box*

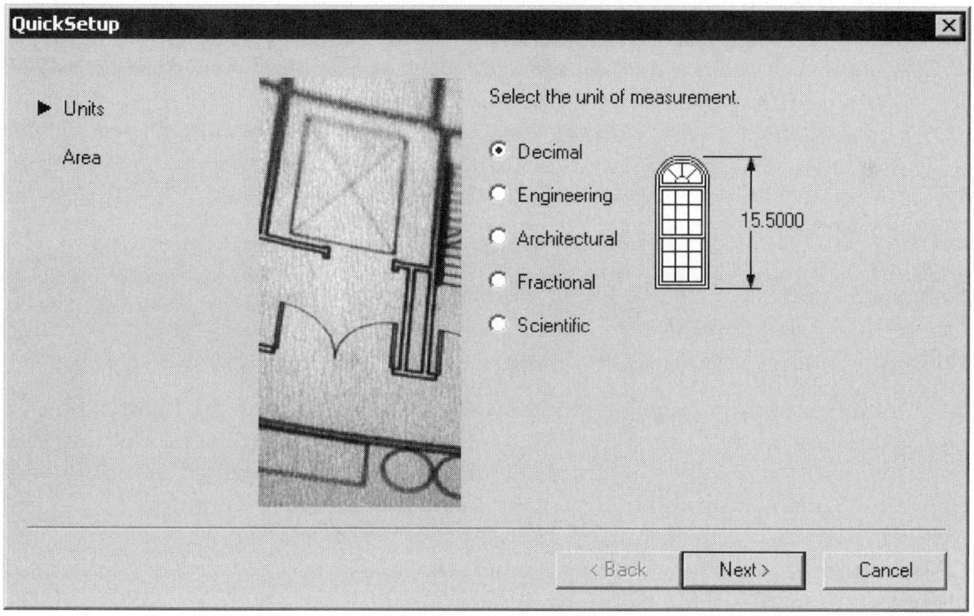

Figure 1-26 *The* ***Units*** *page of the* ***QuickSetup*** *dialog box*

Choose **Next** to display the **Area** page as shown in Figure 1-27. The **Area** page is similar to that of the **Advanced Setup** dialog box where you can set the drawing limits.

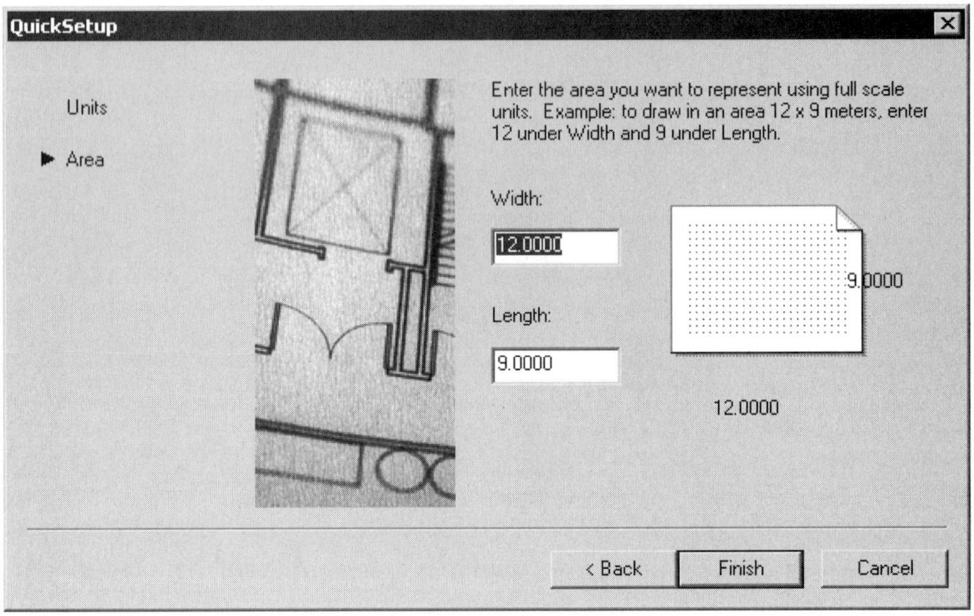

*Figure 1-27 The **Area** page of the **QuickSetup** dialog box*

Tip

*By default, when you open an AutoCAD session, a drawing opens automatically. But you can open a new drawing using various options such as **Start from Scratch** and **Wizards** before entering into the AutoCAD environment using the **Startup** dialog box. As mentioned earlier, the display of the **Startup** dialog box is turned off by default. Refer to the section of **Starting a New Drawing** to know how to turn on the display of this dialog box.*

SAVING YOUR WORK

Toolbar:	Standard > Save
Menu:	File >Save or Save As
Command:	QSAVE , SAVEAS , SAVE

In AutoCAD or any computer system, you must save your work before you exit from the drawing editor or turn the system off. Also, it is recommended that you save your drawings after regular time intervals. In case of a power failure, an editing error, or other problems, all work saved before the problem started will be retained.

AutoCAD has provided the following commands that allow you to save your work on the hard disk of the computer or on the floppy diskette.

QSAVE **SAVEAS** **SAVE**

The **QSAVE**, **SAVEAS**, and **SAVE** commands allow you to save your drawing by writing it to a permanent storage device, such as a hard drive, or on a diskette in any removable drive.

When you choose **Save** from the **File** menu as shown in Figure 1-28, or choose the **Save** button in the **Standard** toolbar, the **QSAVE** command is invoked. If the current drawing is unnamed and you are saving the drawing for the first time in the present session, the **QSAVE** command will prompt you to enter the file name in the **Save Drawing As** dialog box (Figure 1-29). You can enter the name for the drawing and then choose the **Save** button in the dialog box. Once the drawing is saved and you make some changes to it, you can use the **QSAVE** command to save the drawing with the current name without prompting you to enter a file name. This allows you to do a quick save.

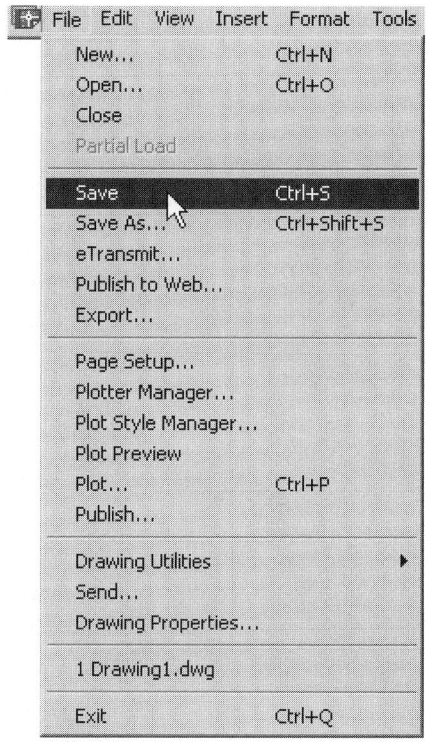

Figure 1-28 *Different Save options in the* ***File*** *menu*

When you invoke the **SAVEAS** command, the **Save Drawing As** dialog box is displayed as shown in Figure 1-29. Even if the drawing has been saved with a file name, this command gives you an option to save it with a different file name. In addition to saving the drawing, it sets the name of the current drawing to the file name you specify, which is displayed in the title bar. This command is used when you want to save a previously saved drawing under a different file name. You can also use this command when you make certain changes to a template and want to save the changed template drawing but leave the original template unchanged.

The **SAVE** command is the most rarely used command and can be invoked only from the command line by entering **SAVE** at the Command prompt. It is similar to the **SAVEAS** command and displays the **Save Drawing As** dialog box always. With this command you can save a previously saved drawing under a different file name, but this command does not set it as the current drawing.

Save Drawing As Dialog Box

The **Save Drawing As** dialog box displays the information related to the drawing files on your system. The various components of the dialog box are described next.

Places list

A column of icons is displayed on the left side of the dialog box. These icons contain the shortcuts to the folders that are frequently used. You can quickly save your drawings in one of these folders. The **History** folder displays the list of most recently saved drawings. You can save your personal drawings in the **My Documents** or the **Favorites** folder. The **FTP** folder

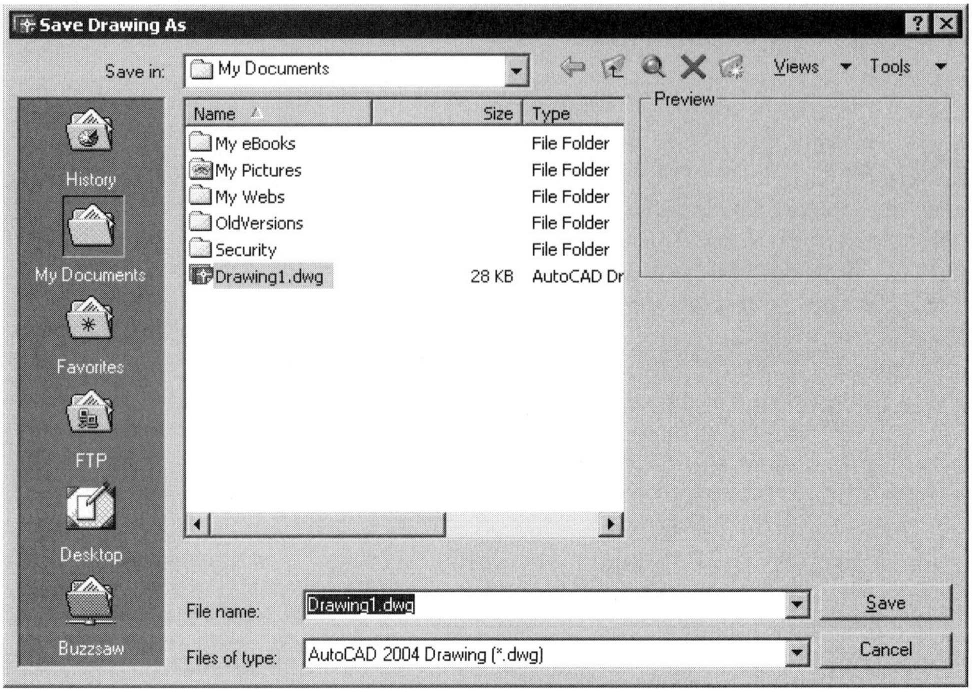

*Figure 1-29 **Save Drawing As** dialog box*

displays the list of the various FTP sites that are available for saving the drawing. By default no FTP sites are shown in the dialog box. To add a FTP site to the dialog box choose the Tools button available on the upper-right corner of the dialog box. When you choose this button a shortcut menu is displayed. Select **Add/Modify FTP Locations**. The **Desktop** folder displays the list of contents available on the desktop. The **Buzzsaw** icons connect you to their respective pages on the Web. You can add any new folder in this list for easy access by simply dragging this folder on to the Places list area and then leaving it. You can rearrange all these folders by dragging them and then placing them at the desired locations. It is also possible to remove the folders when not in frequent use. Right-click on the particular folder and then select **Remove** from the shortcut menu.

File name edit box

To save your work, enter the name of the drawing in the **File Name** edit box. This can be done by typing the file name or selecting it from the drop-down list. If you select the file name you want from the drop-down list, the name you select automatically appears in the **File name** edit box. If you have already assigned a name to the drawing, the current drawing name is taken as default. If the drawing is unnamed, the default name *Drawing1* is displayed in the **File Name** edit box. You can also choose the down arrow at the right of the edit box to display the names of the previously saved drawings and choose a name here.

Files of type drop-down list

The **Files of type** drop-down list (Figure 1-30) is used to specify the drawing format in which

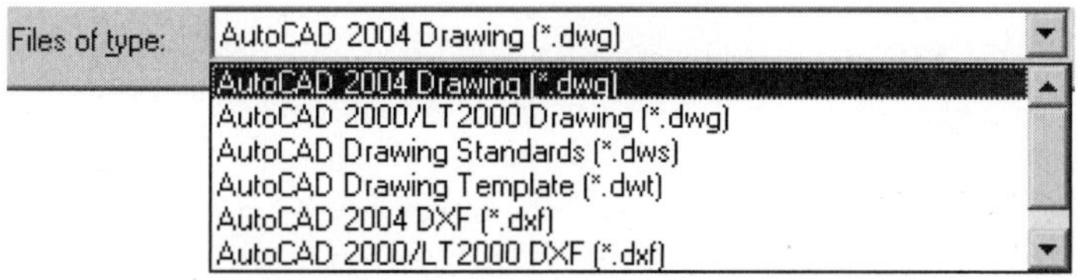

Figure 1-30 Files of type *drop-down list*

you want to save the file. For example, to save the file as AutoCAD 2000 drawing file, select **AutoCAD 2000/LT Drawing (*.dwg)** from the drop-down list.

Save in drop-down list

The current drive and path information is listed in the **Save in** drop-down list, see Figure 1-31. AutoCAD will initially save the drawing in the default directory, but if you want to save the drawing in a different directory, you have to specify the path. For example, if you want to save the present drawing under the file name *house* in the *C1* subdirectory, choose the arrow button in the **Save in** drop-down list to display the drop-down list and select C:. When you select C: all directories in C drive will be listed in the **File** list box. Double-click on 2004 or select 2004 and choose the **Open** button to display its directories. Again double-click on *C1* or select *C1* and choose the **Open** button to display drawing names in the **File** list box. Select *house* from the list, if it is already listed there, or enter it in the **File name** edit box and then choose the **Save** button. Your drawing (*house*) will be saved in the *C1* folder (*C:\2004\C1\house.dwg*). If you want to save the drawing on the A drive, select A: in the **Save in** drop-down list.

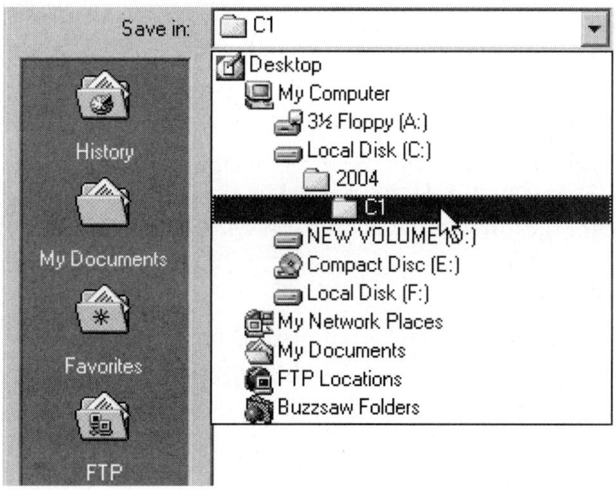

Figure 1-31 Save in *drop-down list*

Tip

The file name you enter to save a drawing should match the contents of the drawing. This helps you to remember the drawing details and makes it easier to refer to them later. Also the file name can be 255 characters long and can contain spaces and punctuation marks.

Note

If you want to save a drawing on the A or B drive, make sure the diskette you are using to save the drawings is formatted.

Views List

The **Views** drop-down list has options for the type of listing of files and displaying the preview images (Figure 1-32).

***Figure 1-32** Views list*

List, Details, Thumbnails, and Preview options

If you choose the **Details** option, it will display detailed information about the files (size, type, date, and time of modification) in the **Files** list box. In the detailed information if you click on the **Name** label, the files are listed with the names in alphabetical order. If you double-click on the **Name** label, the files will be listed in reverse order. Similarly if you click on the **Size** label the files are listed according to the size in ascending order. Double-clicking on the **Size** label will list the files descending by order of size. Similarly you can click on the **Type** label or the **Modified** label to list the files accordingly. If you choose the **List** option, all files present in the current directory will be listed in the **File** list box. If you select the **Preview** option, the list box displays the Preview image box wherein the bitmap image of the file chosen is displayed. If cleared, the Preview box is not displayed. If you select the **Thumbnails** option, the list box displays the preview of the all the drawings along with their names displayed at the bottom of the drawing preview.

Folder Button

If you choose the **Create New Folder** button, AutoCAD creates a new directory under the name **New Folder**. The new folder is displayed in the **File** list box. You can accept the name or change it to your requirement.

Up One Level Button

The **Up one level** button displays the directories that are up by one level. For example, if you are in the *Sample* subdirectory of the *AutoCAD 2004* directory, then choosing the **Up one level** button will take you to the *AutoCAD 2004* directory.

Search the Web

It displays the **Browse the Web** dialog box that enables you to access and store AutoCAD files on the Internet. You can also use ATL+3 keys to browse the Web when this dialog box is available on the screen.

Tools List

The **Tools** drop-down list (Figure 1-33) has an option for adding or modifying the FTP sites. These sites can then be browsed from the FTP shortcut in the **Places** list. The **Add Current Folder to Places** and **Add to Favorites** options add the folder displayed in the **Save in** edit box to the Places list or to the Favorites folder. The **Options** button displays the **Saveas**

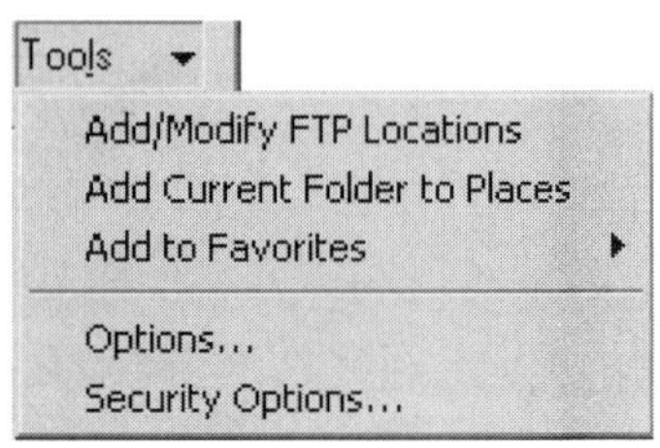

***Figure 1-33** Tools list*

Options dialog box where you can save the proxy images of custom objects. It has the **DWG Options** and **DXF Options** tabs. The **Security Options** button displays the **Security Options** dialog box, which is used to configure the security options of the drawing.

AUTOMATIC TIMED SAVE

AutoCAD allows you to save your work automatically at specific intervals. To change the time intervals you can enter the time intervals in minutes in the **Minutes between saves** text box in the **File Safety Precautions** area available in the **Options** dialog box (**Open and Save** tab). This dialog box can be invoked from the **Tools** menu. Depending on the power supply, hardware, and type of drawings, you should decide on an appropriate time and assign that time to this variable. AutoCAD saves the drawing under the file name *auto.sv$*. The extension of the auto-save file is *.sv$*. You can also change the time interval by using the **SAVETIME** system variable.

CREATION OF BACKUP FILES

If the drawing file already exists and you use **SAVE** or **SAVEAS** commands to update the current drawing, AutoCAD creates a backup file. AutoCAD takes the previous copy of the drawing and changes it from a file type *.dwg* to *.bak*, and the updated drawing is saved as a drawing file with the *.dwg* extension. For example, if the name of the drawing is *myproj.dwg*, AutoCAD will change it to *myproj.bak* and save the current drawing as *myproj.dwg*.

Tip

Although the Automatic save saves your drawing after a certain time interval, you should not completely depend on it because the procedure for converting the sv$ file into a drawing file is cumbersome. Therefore, it is recommended that you save your files regularly using the QSAVE or SAVEAS commands.

CHANGING AUTOMATIC TIMED SAVED AND BACKUP FILES INTO AutoCAD FORMAT

Sometimes you may need to change the automatic timed saved and backup files into the AutoCAD format. To change the backup file into an AutoCAD format, open the folder in which you have saved the backup or the automatic timed saved drawing using **My Computer** or **Windows Explorer**. Choose the **Tools > Folder Options** from the menu bar to invoke the **Folder Options** dialog box. Choose the **View** tab and under the **Advanced settings** area, clear the **Hide file extensions for known file types** text box, if selected. Exit the dialog box. Rename the automatic saved drawing or the backup file with a different name and also change the extension of the drawing from *.sv$* or *.bak* to *.dwg*. After you rename the drawing, you will notice that the icon of the automatic saved drawing or the backup file is replaced by the AutoCAD icon. This indicates that the automatic saved drawing or the backup file is changed into an AutoCAD drawing.

CLOSING A DRAWING

You can use the **CLOSE** command to close the current drawing file without actually quitting AutoCAD. If you choose **Close** from the **File** menu or enter **CLOSE** at the Command prompt, the current drawing file is closed. If you have not saved the drawing after making the last changes to it and you invoke the **CLOSE** command, AutoCAD displays a dialog box that allows you to save the drawing before closing. This box gives you an option to discard the current drawing or changes made to it. It also gives you an option to cancel the command. After closing the drawing you are still in AutoCAD from where you can open a new or an already saved drawing file. You can also use the close button (**X**) of the drawing window to close the drawing.

Note

You can close a drawing in AutoCAD 2004 even if a command is active. But in earlier AutoCAD releases, you were prompted to first exit the current command and then the selected drawing could be closed.

OPENING AN EXISTING DRAWING

You can open an existing drawing file that has been saved previously. There are three methods that can be used to open a drawing file, the **Select File** dialog box, **Startup** dialog box, and by **Dragging and Dropping**.

Opening an Existing Drawing Using the Select File Dialog Box

Toolbar:	Standard > Open
Menu:	File >Open
Command:	OPEN

If you are already in the drawing editor, and you want to open a drawing file you can use the **OPEN** command. The **OPEN** command also displays the **Select File** dialog box, see Figure 1-34. You can select the drawing to open using this dialog box. This dialog box is similar to the standard dialog boxes. You can choose the particular file you want to open from the particular directory. You can change the directory from the **Look in** drop-down list. You can then select the name of the drawing from the list box or you can enter the name of the drawing file you want to open in the **File name** edit box. After selecting the drawing file you can select the **Open** button to open the file. Here you can choose *Drawing1* from the list and then choose the **Open** button to open the drawing.

When you select a file name, its image is displayed in the **Preview** box. If you are not sure about the file name of a particular drawing but know the contents, you can select the file names and look for the particular drawing in the **Preview** box. You can also change the file type by selecting it in the **Files of type** drop-down list. Apart from the *dwg* files, you can open the *dwt* (template) files or the *dxf* files. You have all the standard icons in the **Places** list, which can be used to open the drawing files from different locations. The **Open** button has a

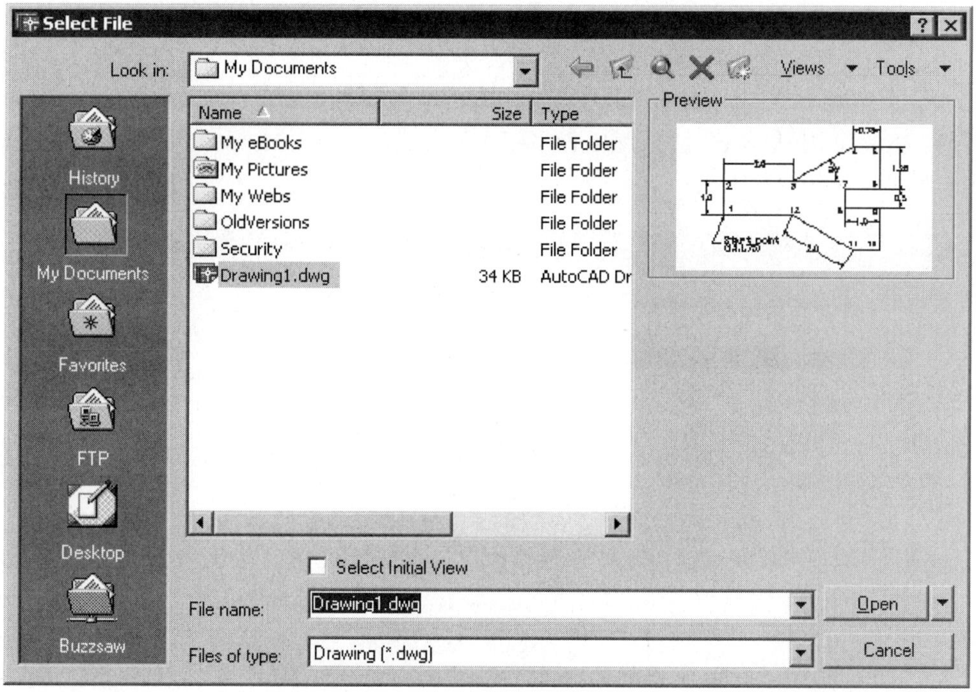

***Figure 1-34 Select File** dialog box*

drop-down list as shown in Figure 1-35. You can choose a method for opening the file using this drop-down list. These methods are discussed next.

***Figure 1-35 Open** drop-down list*

Open Read-Only

If you want to view a drawing without altering it, you must select the **Open Read-Only** option from the drop-down list. In other words, read only protects the drawing file from changes. AutoCAD does not prevent you from editing the drawing, but if you try to save the opened drawing to the original file name, AutoCAD warns you that the drawing file is **write protected**. However, you can save the edited drawing to a file with a different file name using the SAVEAS command. This way you can preserve your drawing.

Partial Open

The **Partial Open** enables you to open only a selected view or a selected layer of a selected drawing. This option can be used to edit small portions of a complicated drawing and then save it with the complete drawing. When you choose the **Partial Open** option from the **Open** drop-down list, the **Partial Open** dialog box (Figure 1-36) is displayed, which contains the different views and layers of the selected drawing. When you select a check box for a layer and then choose the **Open** button, only the objects drawn in that particular layer for the drawing are displayed in a new drawing window. You can make the changes and then save it. For example, in the *C/Program Files/AutoCAD 2004* directory, double-click on the **Sample** folder and then choose *8th floor furniture.dwg* from the list. Now, choose the down arrow available on the right of the **Open** button to display the drop-down list and choose **Partial Open**. All the views and the layers of this drawing are displayed in the **Partial Open** dialog box. Select the check box available on the right of the layer that you want to be opened. When you choose the **Open** button after selecting the layers, the drawing will be opened with only the selected layers.

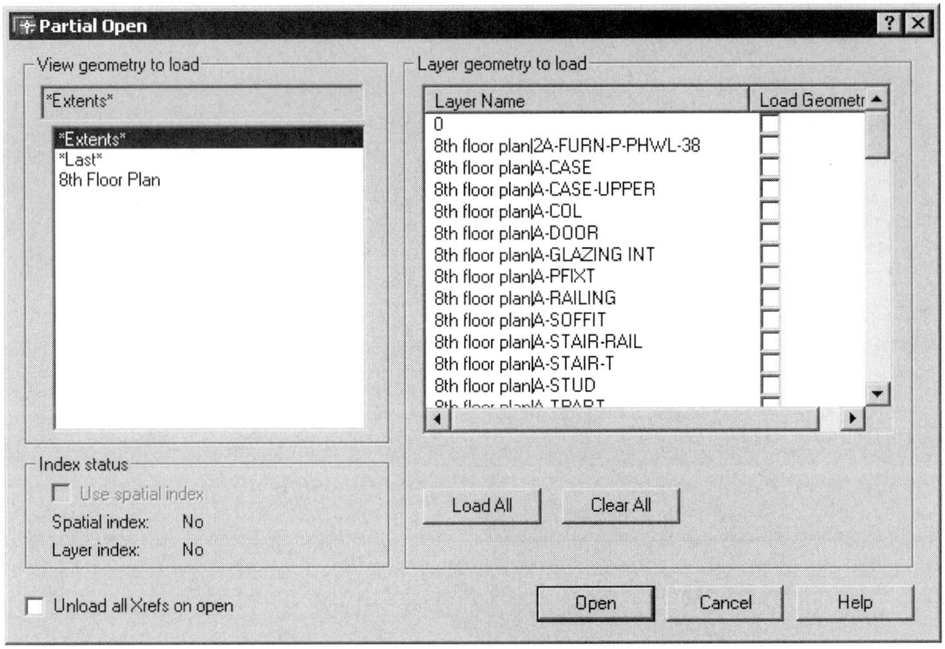

***Figure 1-36 Partial Open** dialog box*

Note

The concept of layers is discussed in Chapter 4, Working with Drawing Aids.

Using the PARTIALOAD Command

Once you have opened a part of a drawing and made the necessary changes, you may want to load additional objects or layers on the existing ones. This can be done by using the **PARTIALOAD** command, which can be invoked by choosing **Partial Load** in the **File** menu, or by entering **PARTIALOAD** at the Command prompt. This command displays the **Partial Load** dialog box, which is similar to the **Partial Open** dialog box. You can choose another layer and the objects drawn in this layer will be added to the partially loaded drawing.

Note

*The **Partial Load** option is not enabled in the **File** menu unless a drawing has been partially opened.*

Loading a drawing partially is a good practice when you are working with objects on a specific layer in a large complicated drawing.

Tip

If a drawing was partially opened and saved previously, it is possible to open it again with the same layers and views. AutoCAD remembers the settings so that while opening a previously partially opened drawing, a dialog box is displayed asking for an option to either fully open it or restore the partially opened drawing.

Note

*In the **Select File** dialog box, the preview of a drawing which was partially opened and then saved is not displayed.*

Select Initial View

A view is defined as the way you look at an object. The **Select Initial View** option allows you to specify the view you want to load initially when AutoCAD loads the drawing. This option will work if the drawing has saved views. This option is generally used while working on a large complicated drawing in which you want to work on a particular portion of the drawing. You can save that particular portion as a view and then select it to open the drawing next time. You can save a desired view by using AutoCAD's **VIEW** command (see "**Creating Views**", Chapter 7). If the drawing has no saved views, selecting this option will load the last view. If you select the **Select Initial View** check box and then the **OK** button, AutoCAD will display the **Select Initial View** dialog box. You can select the view name from this dialog box, and AutoCAD will load the drawing with the selected view displayed.

Tip

*Apart from opening a drawing from the **Startup** dialog box or the **Select File** dialog box, you can also open a drawing from the **File** menu, which displays the four most recently opened drawings, by simply choosing the desired file name.*

It is possible to open an AutoCAD 2000 drawing in AutoCAD 2004. When you save this drawing it is automatically converted and saved as an AutoCAD 2004 drawing file.

Opening an Existing Drawing Using the Startup Dialog Box

If you have configured the settings to show the **Startup** dialog box from the **Options** dialog box, the **Startup** dialog box is displayed every time you start a new AutoCAD session. The first button in this dialog box is the **Open a Drawing** button. When you choose this button, a list of the most recently opened drawings is displayed for you to select from, see Figure 1-37. The **Browse** button displays the **Select File** dialog box, which allows you to browse another file.

Note

*The display of the dialog boxes related to opening and saving drawings is disabled if the **FILEDIA** system variable is set to 0. The initial value of this variable is 1.*

Opening an Existing Drawing Using the Drag and Drop Method

You can also open an existing drawing in AutoCAD by dragging it from the Window Explorer and dropping it into AutoCAD. If you drop the selected drawing in the drawing area, the

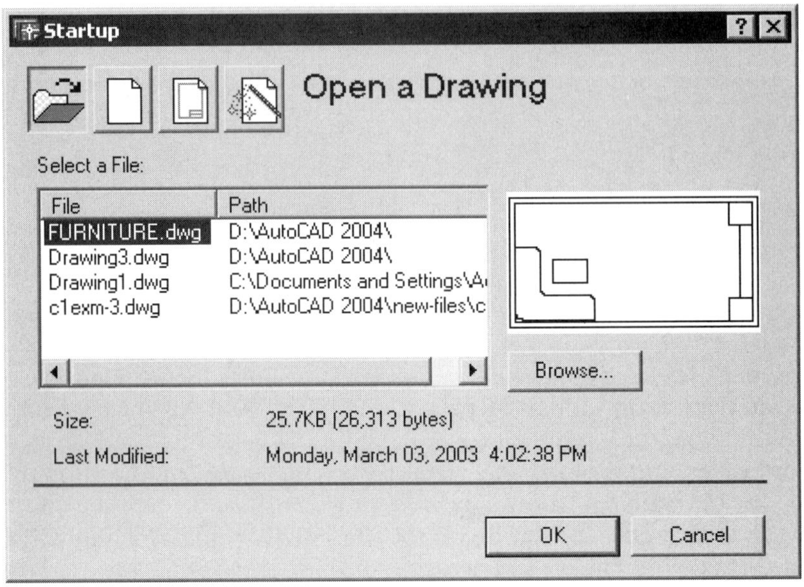

Figure 1-37 *List of recently opened drawings*

drawing will be inserted as a block and as a result it cannot be modified. But, if you drag the drawing from the Window Explorer and drop it anywhere other than the drawing area AutoCAD opens the selected drawing.

MULTIPLE DOCUMENT ENVIRONMENT (MDE)

The MDE is a Windows feature that allows you to open more than one drawing at a time in one AutoCAD session. This feature is very helpful when you want to work on different drawings simultaneously and make changes to them with reference to each other. For example, if you are working on an architectural plan of a building and you want to make changes to it while referring to its elevations, this feature allows you to open all these drawings simultaneously and work on them. Sometimes you may want to incorporate certain features from different old drawings into a new drawing. With the help of MDE you can open all these drawings and the new drawing and then copy the features from the old drawings into the new drawing by simply dragging and dropping. This way it is very convenient to create an assembly by copying the components created in separate drawings. You can also use the **Cut/Copy** and **Paste** instead of drag and drop. These commands are invoked from the shortcut menu, or the **Edit** menu, or you can use the keyboard shortcuts such as CTRL+C, CTRL+V, and so on. It is possible for you to shift from one drawing to another leaving a command incomplete and then start working on another drawing. When you come back to that particular drawing the command resumes from the point you had left it.

Tip

If you want to place an object at a particular location in a drawing, it is better to use the ***Copy/Cut*** *and* ***Paste*** *options rather than drag and drop. AutoCAD allows you to define the coordinates for the insertion point while pasting the object. This makes the placement of objects accurate.*

When you open the first drawing, by default its file name merges with the AutoCAD title bar. When you want to work on more than one drawing simultaneously, choose the **Restore** button present in the right corner of the menu bar to make it floating. This drawing will be placed as a floating window inside the main AutoCAD window. Now if you open another drawing, this window will be placed over the first drawing window but the title bar of the previous window will remain visible. Similarly the rest of the drawings opened are placed over the previous ones with the latest drawing opened being active. The active drawing is specified by the dark blue band in the title bar. You can make any drawing active by simply clicking in it. This type of arrangement is the cascading arrangement and can be changed as per your requirements. If the title bar of a particular drawing is not visible and you want it active, choose the drawing name in the **Window** menu, see Figure 1-38.

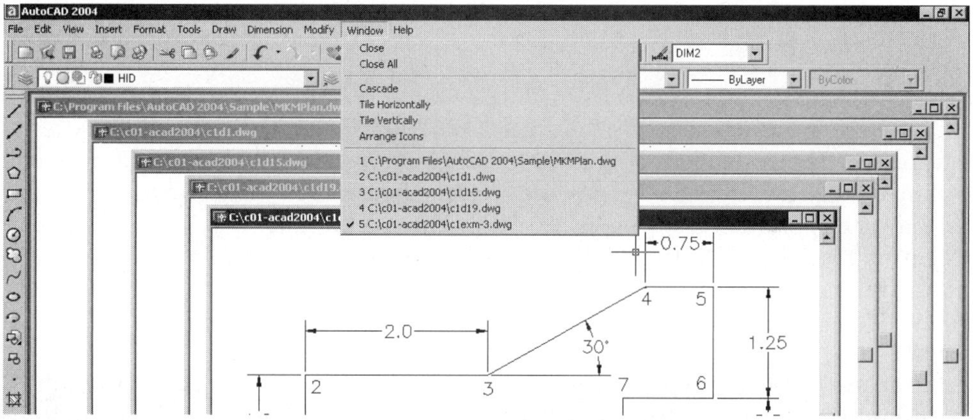

Figure 1-38 *MDE showing different drawings and the* ***Window*** *menu (c1exm-3.dwg active)*

A maximum of nine drawing names are listed in the **Window** menu. If more than nine drawings are opened at a time, **More Windows** is displayed in the list. Choosing this option opens the **Select Window** dialog box where all the drawing names are listed. You can arrange the drawing windows vertically and horizontally by choosing the desired option from the **Window** menu. You can also minimize the drawing files using the **Minimize** button at the top right corner of any window. Here only a small portion of the title bar is visible at the bottom of the window area. You can get the drawings to fill the drawing area using the **Restore** button, which replaces the Minimize button after minimizing. The **Arrange Icons** option in the **Window** menu arranges these minimized title bars at the bottom.

QUITTING AutoCAD

You can exit from the AutoCAD program by using the **EXIT** or **QUIT** command. Even if you have a drawing file open you can use the EXIT command to close it as well as quit the AutoCAD program. In case the drawing has not been saved, it allows you to save the work first through a dialog box. Note that if you choose **No** in this dialog box, all the changed made in the current list till the last save will be lost. You can also use the close button (**X**) of the main AutoCAD window (present in the title bar) to end the AutoCAD session.

AutoCAD'S HELP

Toolbar:	Standard > Help
Menu:	Help > Help
Command:	HELP

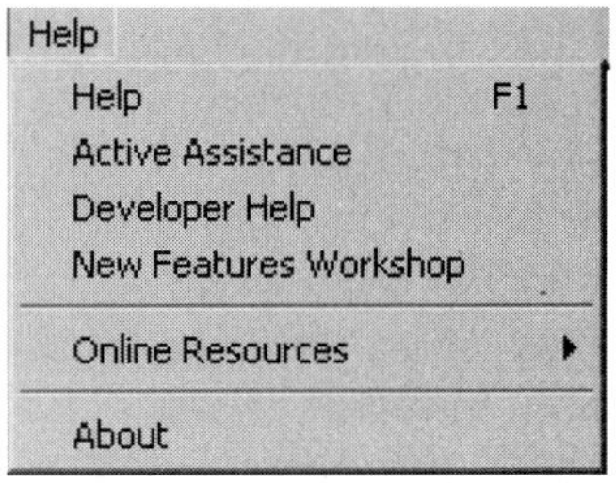

Figure 1-39 Help menu

You can get the online help and documentation on the working of AutoCAD 2004 commands from the **Help** menu, see Figure 1-39, or by pressing the F1 key. The various options available in the **Help** menu are discussed next.

Help

Choosing the **Help** option displays the **AutoCAD 2004 Help: User Documentation** dialog box as shown in Figure 1-40. You can use this dialog box to access help on different topics and commands. It has five tabs: **Contents**, **Index**, **Search**, **Favorites**, and **Ask me**, which display the corresponding help topics. If you are in the middle of a command and require help regarding it, choosing the **Help** button displays information about that particular command in the dialog box.

Contents

This tab displays the help topics, which are organized by categories, pertaining to different sections of AutoCAD such as the User's Guide, Command Reference, and so on. To select a category, double-click on the corresponding book icon or choose the plus sign on the left. The icon becomes an open book with a minus (-) sign and a list of headings associated with that category are displayed. Use the plus sign (+) to further open the headings until you reach the help topic which has a question mark (?) displayed with it. Choose the topic to display information about the selected topic or command in the window present on the right side of the dialog box.

Index

This tab displays the complete index (search keywords) in an alphabetical order. To display information about an item or command, type the item (word) or command name in the edit box. With each letter entered, the listing keeps on changing in the list area, displaying the possible topics. When you enter the word and if AutoCAD finds that word, it is automatically highlighted in the list area. Choose the **Display** button to display information about it.

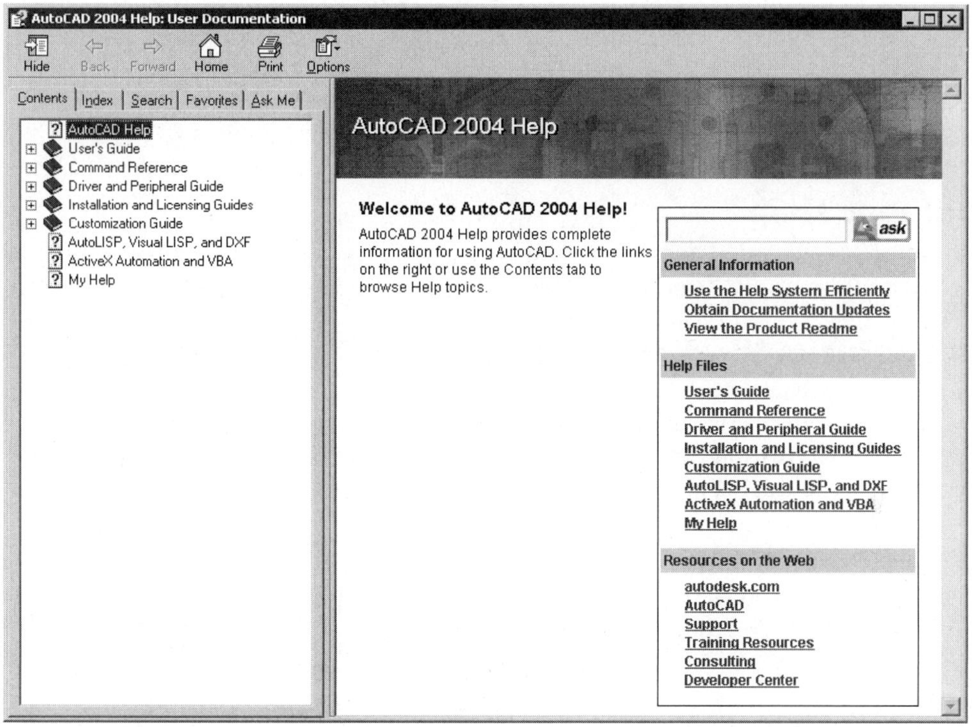

***Figure 1-40 AutoCAD 2004 Help: User Documentation** dialog box (**Contents** tab)*

Search

This tab creates a word list based on all the keywords present in the online hlp files. When you type a word you are looking for and then choose the **List Topics** button, a list of matching words appears in a window below to narrow down your search. This search is dependent on the option you have selected at the bottom of the dialog box where you can search the previous results and also match words similar to those you searched. Use the scroll bar to scroll through the list, select the desired topic, and then choose the **Display** button to display its help.

Favorites

This tab lets you create a list of your own topics that you need to access regularly. The topic which you have chosen in any other tab of Help is displayed in the **Current Topic** box when you choose the **Favorites** tab. Use the **Add** button to add it to your own list. You can use the **Remove** button to remove a topic from your list and the **Display** button to display its help.

Ask Me

When you choose this tab, you are allowed to enter a query in the edit box and then press ENTER. A list of topics related to the question follow. It also shows the book (category) from which it has been selected.

Active Assistance

This option gives you an access to context-sensitive help. This help can be started by choosing the **Active assistance** in the **Help** menu or by entering **ASSIST** at the Command prompt. Whenever you enter a command, an Active Assistance window with information regarding that command is displayed. Its icon is displayed in the main task bar of the Windows at the bottom of the screen. To disable this help, right-click on the Active Assistance icon in the window's bar and choose Exit.

Developer Help

This option gives a detailed help on customizing AutoCAD. You can click on any link provided on the right of the window that is displayed when you choose this option.

New Features Workshop

This option gives you an interactive list of all the new features in AutoCAD 2004. You can choose this option from the **Help** menu that displays the **New Features Workshop** window with a list of various topics (Figure 1-41). When you choose a topic, a description of the feature improvement is displayed.

Online Resources

This utility connects you to the **Product Support**, **Training**, **Customization**, **Autodesk User Group International** Web pages and sites through the Microsoft Internet Explorer.

About

This option gives you information about the Release, Serial Number, Licensed To, and also the legal description about AutoCAD.

ADDITIONAL HELP RESOURCES

1. You can get help for a command while working by pressing the F1 key. The **Help** dialog box containing information about the command is displayed. You can exit the dialog box and continue with the command.

2. You can get help about a certain dialog box by choosing the **Help** button in that dialog box.

3. Some of the dialog boxes have a **question mark** (?) button at the top right corner just adjacent to the **close** button. When you choose this button the ? gets attached to the cursor. You can then drop it on any item in the dialog box to display information about that particular item.

4. Autodesk has provided several resources that you can use to get assistance with your AutoCAD questions. You can also use these resources to get information on Autodesk products, product updates, and other services provided by Autodesk. The following is a list of some of the resources:

   a. Autodesk Web site *http://www.autodesk.com*
   b. AutoCAD 2004 Web site: *http://www.autodesk.com/autocad2004*

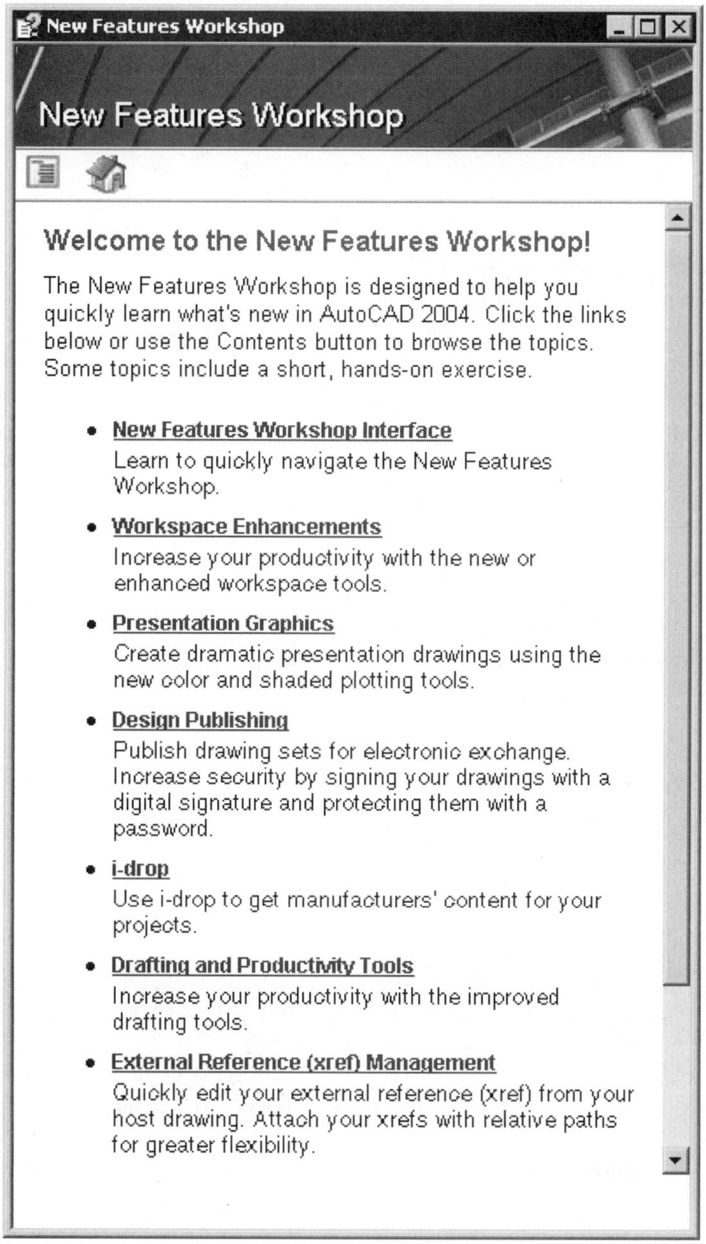

***Figure 1-41 New Features Workshop** window*

c. Autodesk Fax Information System: **(415) 446-1919**
d. AutoCAD Technical Assistance Web site *autodesk.com/support*
e. AutoCAD Discussion Groups Web site *autodesk.com/support/discsgrp/acad.htm*

5. You can also get help by contacting the author, Sham Tickoo, at *stickoo@calumet.purdue.edu*.

6. You can download AutoCAD drawings, programs, and special topics by accessing the authors Web site at *http://technology.calumet.purdue.edu/met/tickoo/students/students.htm*.

Self-Evaluation Test

Answer the following questions and then compare your answers to the correct answers given at the end of this chapter.

1. You can press the F3 key to display the **AutoCAD Text window**, which displays the previous commands and prompts. (T/F)

2. If the value of the **COORDS** variable is set to 1, the coordinate display is static, that is, the coordinate values displayed in the status bar change only when you specify a point. (T/F)

3. If a drawing was partially opened and saved previously, it is not possible to open it again with the same layers and views. (T/F)

4. If the current drawing is unnamed and you are saving the drawing for the first time in the present session, the **QSAVE** command will prompt you to enter the file name in the **Save Drawing As** dialog box. (T/F)

5. A __________ is displayed on top of the AutoCAD screen that has the AutoCAD symbol and the current drawing name.

6. The __________displays a message and an alert whenever Autodesk provides the latest information regarding software updates and their other products.

7. If you want to work on a drawing without altering the original, you must select the __________ option from the **Open** drop-down list in the **Select File** dialog box.

8. The __________ enables you to open only a selected view or a selected layer of a selected drawing.

9. You can use the ________________ command to close the current drawing file without actually quitting AutoCAD.

10. The __________system variable can be used to change the time interval for automatic save.

Review Questions

Answer the following questions.

1. The shortcut menu invoked by right-clicking in the command window displays the six most recently used commands and some of the window options such as **Copy**, **Paste**, and so on. (T/F)

2. It is possible to open an AutoCAD 2002 drawing in AutoCAD 2004. (T/F)

3. The file name you enter to save a drawing in the **Save Drawing As** dialog box file name can be 255 characters long but cannot contain spaces and punctuation marks. (T/F)

4. You can close a drawing in AutoCAD 2004 even if a command is active. (T/F)

5. You can hide all the toolbars displayed on the screen by pressing which of the following keys?

   (a) CTRL+3 (b) CTRL+0
   (c) CTRL+5 (d) CTRL+2

6. The **TOOL PALETTES** window can be turned on or off by pressing which of the following keys?

   (a) CTRL+3 (b) CTRL+0
   (c) CTRL+5 (d) CTRL+2

7. Which of the following commands is used to exit from the AutoCAD program?

   (a) **QUIT** (b) **END**
   (c) **CLOSE** (d) None

8. Which of the following options available in the **Startup** dialog box is used to set the initial drawing settings before actually starting a new drawing?

   (a) **Start from Scratch** (b) **Use a Template**
   (c) **Use a Wizard** (d) None

9. When you choose **Save** from the **File** menu or choose the **Save** button in the **Standard** toolbar, which of the following commands is invoked?

   (a) **SAVE** (b) **LSAVE**
   (c) **QSAVE** (d) **SAVEAS**

10. AutoCAD has provided __________ as an easy and convenient way of placing and sharing hatch patterns and blocks in the current drawing.

11. By default the angles are positive if measured in a __________ direction.

12. You can change the shape of the toolbars by placing the cursor anywhere on the __________ of the toolbar where it takes the shape of a double sided arrow.

13. To differentiate the template files from the drawing files, the template files have __________ extension whereas the drawing files have __________ extension.

14. You can also use __________ and __________ instead of dragging and dropping the objects from one drawing to another while multiple drawings are opened.

15. The __________ tab of the **AutoCAD 2004 Help: User Documentation** dialog box displays the help topics that are organized by categories pertaining to different sections of AutoCAD.

Answers to Self-Evaluation Test

1 - F, **2** - F, **3** - F, **4** - T, **5** - title bar, **6** - **Open Read-Only**, **7** - **Communication Center**, **8** - Partial Open, **9** - **CLOSE**, **10** - **SAVETIME**

Chapter 2

Getting Started with AutoCAD

Learning Objectives

After completing this chapter you will be able to:

- *Draw lines using the* ***LINE*** *command and its options.*
- *Understand various coordinate systems used in AutoCAD.*
- *Use the* ***ERASE*** *commands to clear the drawing area.*
- *Understand the two basic object selection methods: Window and Crossing options.*
- *Draw circles using various options of the* ***CIRCLE*** *command.*
- *Use the* ***ZOOM*** *and* ***PAN*** *display commands.*
- *Set up units using the* ***UNITS*** *command.*
- *Set up and determine limits for a given drawing.*
- *Plot drawings using the basic plotting options.*
- *Use the* ***Options*** *dialog box to specify the settings.*

DRAWING LINES IN AutoCAD

Toolbar:	Draw > Line
Menu:	Draw > Line
Command:	LINE or L

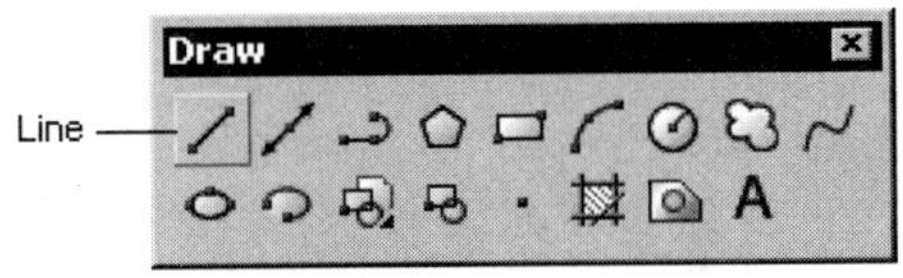

Figure 2-1 *Invoking the **LINE** command from the **Draw** toolbar*

The most fundamental object in a drawing is the line. A line can be drawn between any two points by using AutoCAD's **LINE** command. You can invoke the LINE command by choosing the **Line** button in the **Draw** toolbar (Figure 2-1), or by choosing **Line** from the **Draw** menu (Figure 2-2), or by entering **LINE** at the Command prompt. Once you have invoked the **LINE** command, the next prompt, **Specify first point,** requires you to specify the starting point of the line. You can either select a point using the pointing device or you can enter its coordinates. After the first point is selected, AutoCAD will prompt you to enter the second point at the **Specify next point or [Undo]** prompt. At this point you may continue to select points or terminate the **LINE** command by pressing ENTER, ESC, or the SPACEBAR. You can also right-click to display the shortcut menu from where you can choose the **Enter** or **Cancel** options to exit from the **LINE** command. After terminating the **LINE** command, AutoCAD will again display the Command prompt. The prompt sequence for the drawing, Figure 2-3, is as follows.

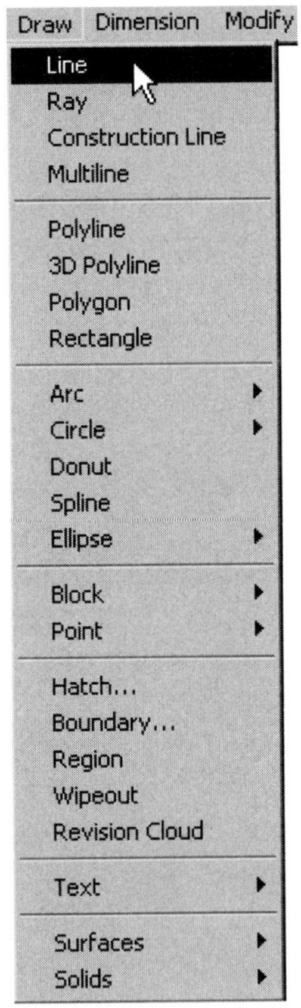

Figure 2-2 *Invoking the **LINE** command from the **Draw** menu*

Command: **LINE** [Enter]
Specify first point: *Move the cursor (mouse) and left-click to specify the first point.*
Specify next point or [Undo]: *Move the cursor and left-click to specify the second point.*
Specify next point or [Undo]: *Specify the third point.*
Specify next point or [Close/Undo]: [Enter] *(Press ENTER to exit the **LINE** command.)*

The **LINE** command has the following three options.

Continue **Close**
Undo

Tip

When you select the points with the cursor, a rubber-band line appears that stretches between the previous point selected and the current position of the cursor. This line is sensitive to the movement of the cursor and helps you to select the direction and the placement of the next point for the line.

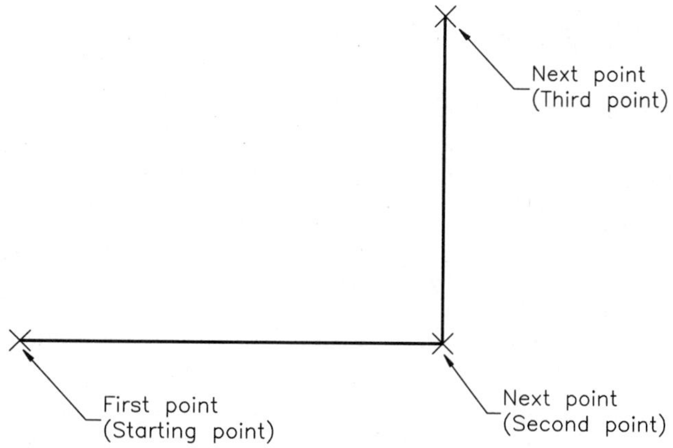

Figure 2-3 *Drawing lines using the* ***LINE*** *command*

Note

To clear the drawing area to gain space to work out the exercises and the examples, choose the ***Erase*** *button from the* ***Modify*** *toolbar or type* ***ERASE*** *at the Command prompt and press ENTER. The screen crosshairs will change into a box called a pick box and AutoCAD will prompt you to select objects. You can select the object by positioning the pick box anywhere on the object and pressing the pick button of the pointing device. Once you have finished selecting the objects, press ENTER to terminate the* ***ERASE*** *command and the objects you selected will be erased. If you enter* ***All*** *at the* ***Select objects:*** *prompt, AutoCAD will erase all objects from the screen. (See "Erasing Objects" discussed later in this chapter.) You can use the* ***U*** (undo) *command to undo the last command by choosing the* ***Undo*** *button from the* ***Standard*** *toolbar.*

Command: **ERASE or E** Enter *(**E** is the command alias of the **ERASE** command.)*
Select objects: *Select objects.* *(Select objects using the pick box.)*
Select objects: Enter

Command: **ERASE** Enter
Select objects: **ALL** Enter
Select objects: Enter
Command: **U** Enter *(The **U** command will undo the last command.)*

The Continue Option

After exiting from the **LINE** command, you may want to draw another line starting from the point where the previous line ended. In such cases you can use the **Continue** option. This option enables you to grab the endpoint of the previous line and to continue drawing the line from that point (Figure 2-4). The following is the prompt sequence for the **Continue** option.

Command: **LINE or L** [Enter] *(**L** is the command alias of the **LINE** command.)*
Specify first point: *Pick first point of the line.*
Specify next point or [Undo]: *Pick second point.*
Specify next point or [Undo]: [Enter]

Command: **LINE** [Enter] *(Or select **Repeat Line** from the shortcut menu.)*
Specify first point: [Enter] *(Press ENTER or right-click to continue the line from the last line.)*
Specify next point or [Undo]: *Pick second point of second line (third point in Figure 2-4).*
Specify next point or [Undo]: [Enter]

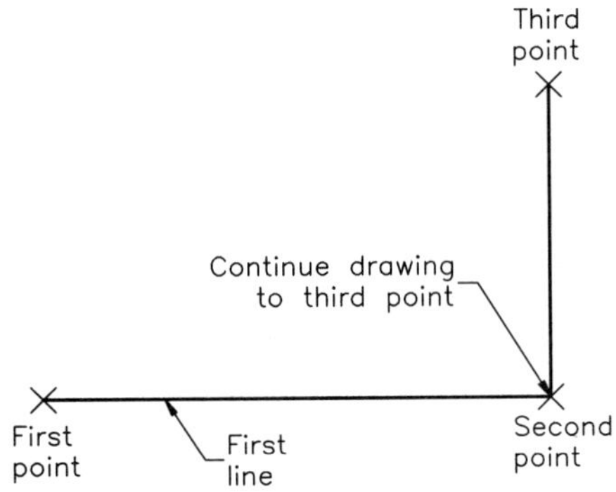

***Figure 2-4** Using the **Continue** option with the **LINE** command*

You can also type the @ symbol to start the line from the **last point**. For example, if you draw a circle and then immediately start the **LINE** command, the @ will snap to the center point of the circle. The **Continue** option snaps to the endpoint of the last line or arc, even if other points have been defined after the line was drawn.

Command: **LINE** [Enter]
Specify first point: *Pick first point of the line.*
Specify next point or [Undo]: *Pick second point.*
Specify next point or [Undo]: [Enter]

Command: **LINE or L** [Enter] *(**L** is the command alias of the **LINE** command)*
Specify first point: @ [Enter] *(Continues drawing the line from the last point.)*
Specify next point or [Undo]: *Pick second point of the second line.*
Specify next point or [Undo]: [Enter]

The Close Option

The **Close** option can be used to join the current point with the initial point of the first line when two or more lines are **drawn in continuation**. For example, this option can be used when an open figure needs one more line to close it and make a polygon (a polygon is a closed figure with at least three sides, for example, a triangle or rectangle). The following is the prompt sequence for the **Close** option (Figure 2-5).

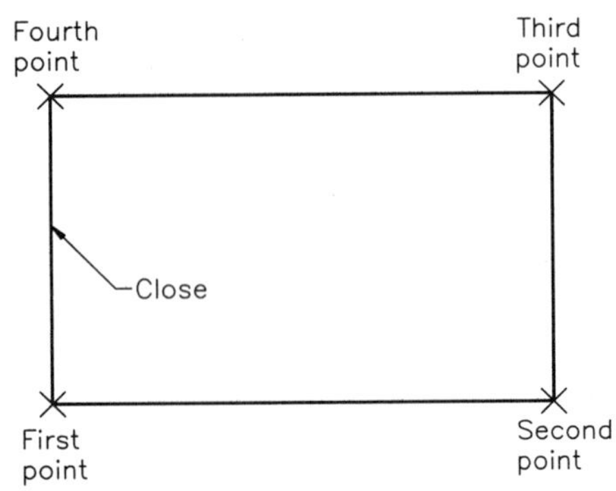

***Figure 2-5** Using the **Close** option with the **LINE** command*

Command: **LINE** [Enter]
Specify first point: *Pick first point.*
Specify next point or [Undo]: *Pick second point.*
Specify next point or [Undo]: *Pick third point.*
Specify next point or [Close/Undo]: *Pick fourth point.*
Specify next point or [Close/Undo]: **C** [Enter] *(Joins the fourth point with the first point.)*

You can also choose the **Close** option from the shortcut menu, which appears when you right-click in the drawing area.

The Undo Option

If you draw a line, and then realize that you made an error, you can remove the line using the **Undo** option. If you need to remove more than one line, you can use this option multiple times and go as far back as you want. In this option, you can type **Undo** (or just **U**) at the **Specify next point or [Undo]:** prompt. You can also right-click to display the shortcut menu, which gives you the **Undo** option. The following example illustrates the use of the **Undo** option (Figure 2-6).

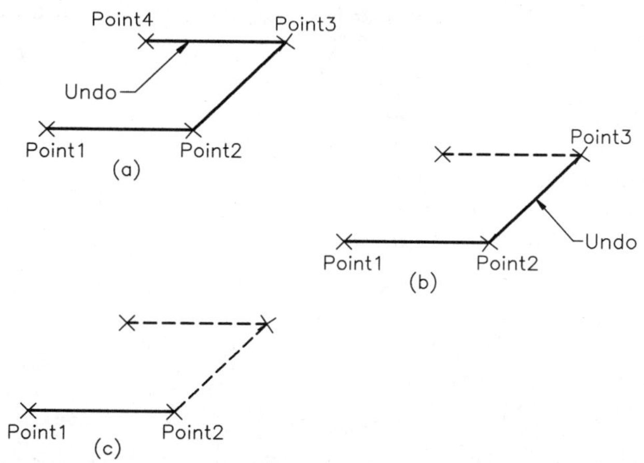

Figure 2-6 *Removing lines using the **Undo** option of the **LINE** command*

Command: **LINE or L** [Enter] *(**L** is the command alias of the **LINE** command)*
Specify first point: *Pick first point (Point 1 in Figure 2-6).*
Specify next point or [Undo]: *Pick second point (Point 2).*
Specify next point or [Undo]: *Pick third point.*
Specify next point or [Close/Undo]: *Pick fourth point.*
Specify next point or [Close/Undo]: **U** [Enter] *(Removes last line from Point 3 to Point 4.)*
Specify next point or [Close/Undo]: **U** [Enter] *(Removes next line from Point 2 to Point 3.)*
Specify next point or [Close/Undo]: [Enter]

Tip

*AutoCAD allows you to enter the command aliases in place of the complete command name. For example, you can enter **L** instead of **LINE** at the Command prompt to invoke the **LINE** command.*

Note

By default, whenever you open a new drawing, you need to modify the drawing display area. To modify the display area type ZOOM at the command prompt and press ENTER. In the command sequence that appears type ALL and press ENTER. The drawing display is modified. You will learn more about the ZOOM command later in this chapter.

COORDINATE SYSTEMS

To specify a point in a plane, take two mutually perpendicular lines as references. The horizontal line is called the ***X* axis**, and the vertical line is called the ***Y* axis.** The point of intersection of these two axes is called the **origin**. The *X* and *Y* axes divide the *XY* plane into four parts, generally known as quadrants. The *X* coordinate measures the horizontal distance from the origin (how far left or right) on the *X* axis. The *Y* coordinate measures the vertical distance from the origin (how far up or down) on the *Y* axis. The origin has the coordinate values of X = 0, Y = 0. The origin is taken as the reference for locating any point in the *XY* plane. The *X* coordinate is positive if measured to the right of the origin and negative if measured to the left of the origin. The *Y* coordinate is positive if measured above the origin and negative if measured below the origin. This method of specifying points is called the **Cartesian coordinate system**, see Figure 2-7. In AutoCAD, the default origin is located at the lower left corner of the graphics area of the screen. AutoCAD uses the following coordinate systems to locate a point in an *XY* plane.

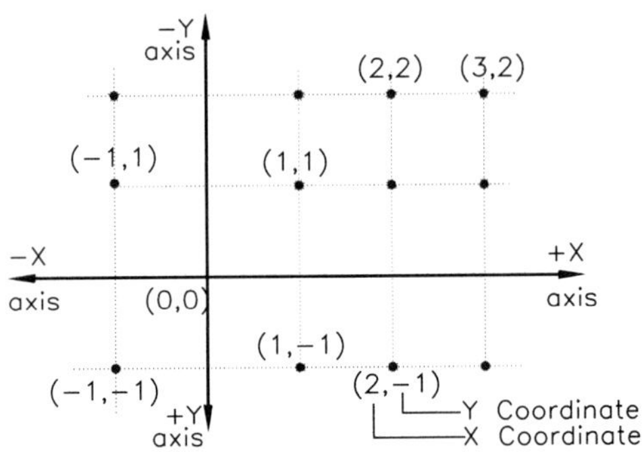

Figure 2-7 *Cartesian coordinate system*

1. **Absolute coordinates**

2. **Relative coordinates**
   a. **Relative rectangular coordinates**
   b. **Relative polar coordinates**

3. **Direct distance entry**

Absolute Coordinate System

In the absolute coordinate system the points are located with respect to the origin (0,0). For example, a point with X = 4 and Y = 3 is measured 4 units horizontally (displacement along the *X* axis) and 3 units vertically (displacement along the *Y* axis) from the origin as shown in Figure 2-8.

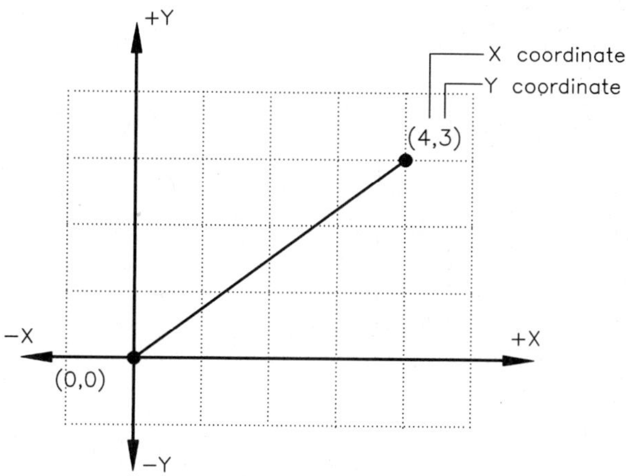

Figure 2-8 *Absolute coordinate system*

In AutoCAD, the absolute coordinates are specified by entering *X* and *Y* coordinates, separated by a comma. The following example illustrates the use of absolute coordinates (Figure 2-9).

Command: **LINE** [Enter]
Specify first point: **1,1** [Enter]
(X = 1 and Y = 1.)
Specify next point or [Undo]: **4,1** [Enter]
(X = 4 and Y = 1.)
Specify next point or [Undo]: **4,3** [Enter]
Specify next point or [Close /Undo]: **1,3** [Enter]
Specify next point or [Close/Undo]: **1,1** [Enter]
Specify next point or [Close/Undo]: [Enter]

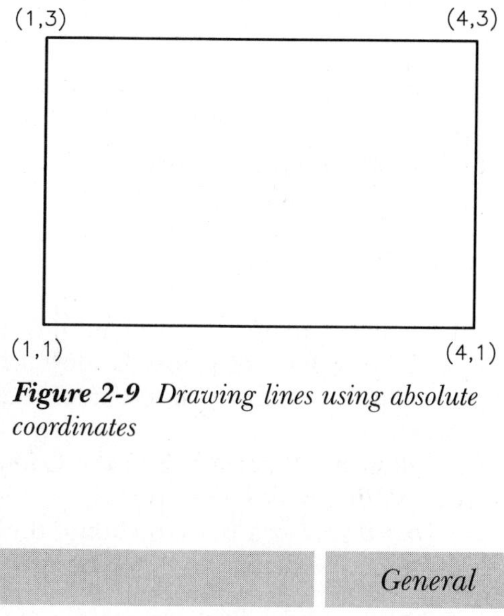

Figure 2-9 *Drawing lines using absolute coordinates*

Example 1 *General*

For Figure 2-10, enter the absolute coordinates of the points in the following table. Then draw the figure using absolute coordinates. Save the drawing under the name *Exam1.dwg*.

Point	Coordinates	Point	Coordinates
1	3,1	5	5,2
2	3,6	6	6,3
3	4,6	7	7,3
4	4,2	8	7,1

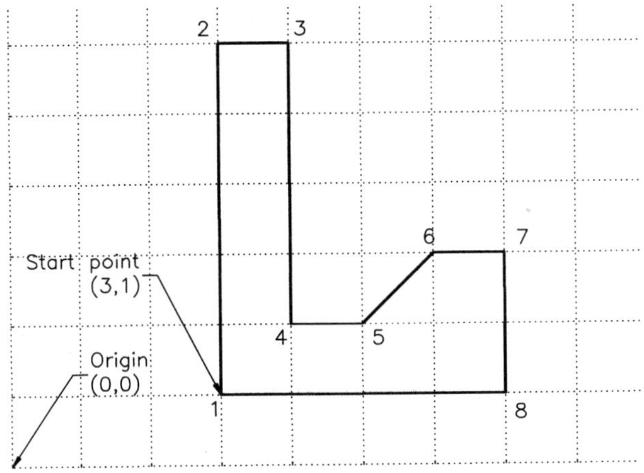

Figure 2-10 *Drawing a figure using absolute coordinates*

Once the coordinates of the points are known you can draw the figure by using AutoCAD's **LINE** command. The prompt sequence is given next.

Command: **LINE** [Enter]
Specify first point: **3,1** [Enter] *(Start point.)*
Specify next point or [Undo]: **3,6** [Enter]
Specify next point or [Undo]: **4,6** [Enter]
Specify next point or [Close/Undo]: **4,2** [Enter]
Specify next point or [Close/Undo]: **5,2** [Enter]
Specify next point or [Close/Undo]: **6,3** [Enter]
Specify next point or [Close/Undo]: **7,3** [Enter]
Specify next point or [Close/Undo]: **7,1** [Enter]
Specify next point or [Close/Undo]: **3,1** [Enter]
Specify next point or [Close/Undo]: [Enter]

Save this drawing. Enter **SAVE** at the Command prompt and then press ENTER. The **Save Drawing As** dialog box is displayed. Enter the name *Exam1* in the **File name** edit box to replace *Drawing1.dwg* and then choose the **Save** button. The drawing will be saved with the given name in the default **My Documents**.

Exercise 1 *General*

For Figure 2-11, enter the absolute coordinates of the points in the following table, and then use these coordinates to draw the same figure. Distance between the dotted lines is 1 unit.

Point	Coordinates	Point	Coordinates
1	2, 1	6	________
2	________	7	________
3	________	8	________
4	________	9	________
5	________		

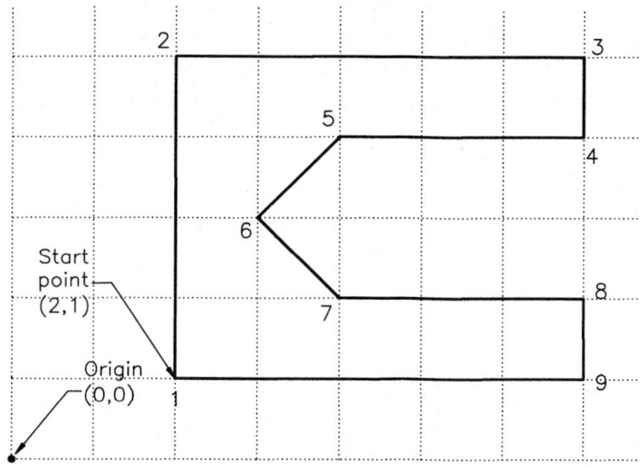

***Figure 2-11** Drawing for Exercise 1*

Relative Coordinate System

There are two types of relative coordinates: the relative rectangular and the relative polar.

Relative Rectangular Coordinates

In the relative rectangular coordinate system, the displacements along the *X* and *Y* axes (DX and DY) are measured with reference to the previous point rather than to the origin. In AutoCAD, the relative coordinate system is designated by the symbol @ and it should precede any relative entry. The following prompt sequence illustrates the use of the relative rectangular coordinate system to draw a rectangle with the lower left corner at point (1,1). The length of the rectangle is 4 units and the width is 3 units (Figure 2-12).

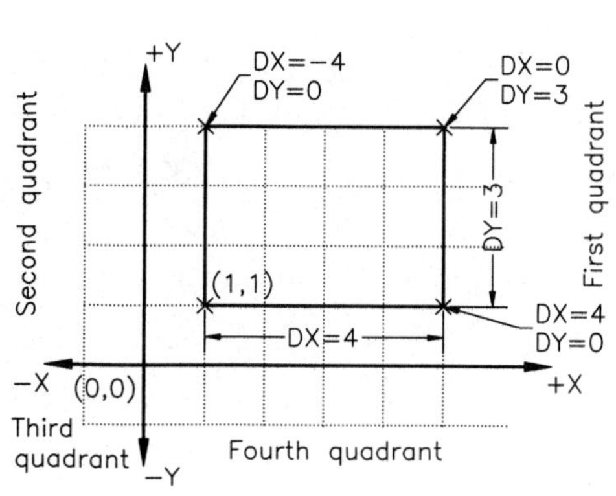

***Figure 2-12** Drawing lines using the relative rectangular coordinates*

Command: **LINE** [Enter]
Specify first point: **1,1** [Enter] *(Start point)*
Specify next point or [Undo]: **@4,0** [Enter] *(Second point DX = 4, DY = 0.)*

Specify next point or [Undo]: **@0,3** [Enter] *(Third point DX = 0, DY = 3.)*
Specify next point or [Close/Undo]: **@-4,0** [Enter] *(Fourth point DX = -4, DY = 0.)*
Specify next point or [Close/Undo]: **@0,-3** [Enter] *(Start point DX = 0, DY = -3.)*
Specify next point or [Close/Undo]: [Enter]

Sign Convention. As just mentioned, in the relative rectangular coordinate system the displacements along the *X* and *Y* axes are measured with respect to the previous point. Imagine a horizontal line and a vertical line passing through the previous point so that you get four quadrants. If the new point is located in the first quadrant, the displacements DX and DY are both positive. If the new point is located in the third quadrant, the displacements DX and DY are both negative. In other words up or right are positive and down or left are negative.

Example 2 — *General*

Draw Figure 2-13 using relative rectangular coordinates of the points given in the table that follows.

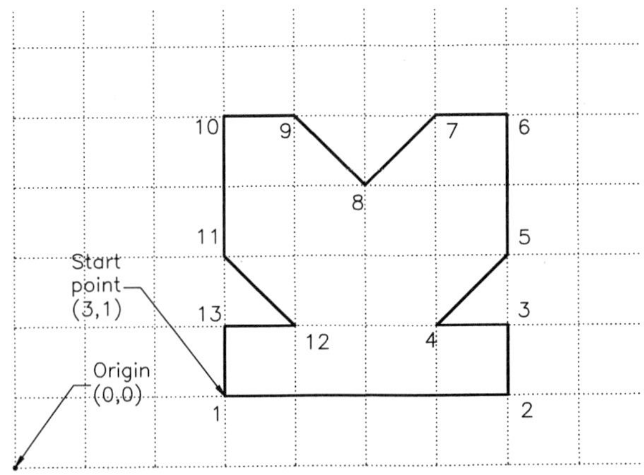

Figure 2-13 *Using relative rectangular coordinates with the* ***LINE*** *command*

Point	Coordinates	Point	Coordinates
1	3,1	8	@-1,-1
2	@4,0	9	@-1,1
3	@0,1	10	@-1,0
4	@-1,0	11	@0,-2
5	@1,1	12	@1,-1
6	@0,2	13	@-1,0
7	@-1,0	14	@0,-1

Once you know the coordinates of the points, you can draw the figure by using AutoCAD's **LINE** command and entering the coordinates of the points.

Command: **LINE** [Enter]
Specify first point: **3,1** [Enter] *(Start point.)*
Specify next point or [Undo]: **@4,0** [Enter]
Specify next point or [Undo]: **@0,1** [Enter]
Specify next point or [Close/Undo]: **@-1,0** [Enter]
Specify next point or [Close/Undo]: **@1,1** [Enter]
Specify next point or [Close/Undo]: **@0,2** [Enter]
Specify next point or [Close/Undo]: **@-1,0** [Enter]
Specify next point or [Close/Undo]: **@-1,-1** [Enter]
Specify next point or [Close/Undo]: **@-1,1** [Enter]
Specify next point or [Close/Undo]: **@-1,0** [Enter]
Specify next point or [Close/Undo]: **@0,-2** [Enter]
Specify next point or [Close/Undo]: **@1,-1** [Enter]
Specify next point or [Close/Undo]: **@-1,0** [Enter]
Specify next point or [Close/Undo]: **@0,-1** [Enter]
Specify next point or [Close/Undo]: [Enter]

Exercise 2 *General*

For Figure 2-14, enter the relative rectangular coordinates of the points in the following table, and then use these coordinates to draw the figure. The distance between the dotted lines is 1 unit.

Point	Coordinates	Point	Coordinates
1	2, 1	12	______
2	______	13	______
3	______	14	______
4	______	15	______
5	______	16	______
6	______	17	______
7	______	18	______
8	______	19	______
9	______	20	______
10	______	21	______
11	______	22	______

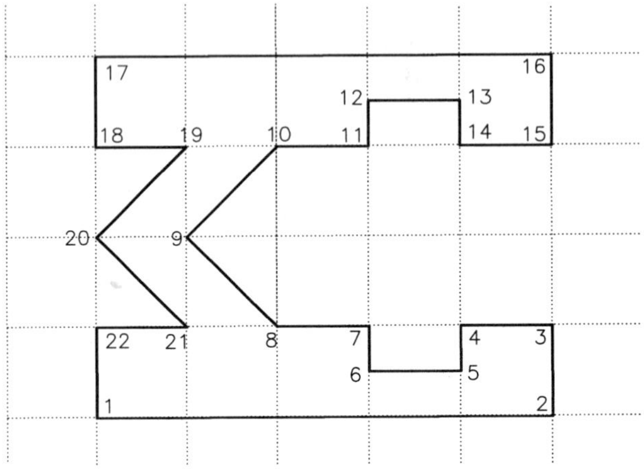

Figure 2-14 Drawing for Exercise 2

Relative Polar Coordinates

In the relative polar coordinate system, a point can be located by defining both the distance of the point from the current point and the angle that the line between the two points makes with the positive *X* axis. The prompt sequence to draw a line from a point at 1,1 to a point at a distance of 5 units from the point (1,1), and at an angle of 30-degree to the *X* axis (Figure 2-15) is given next.

Command: **LINE** Enter
Specify first point: **1,1** Enter
Specify next point or [Undo]: **@5<30** Enter

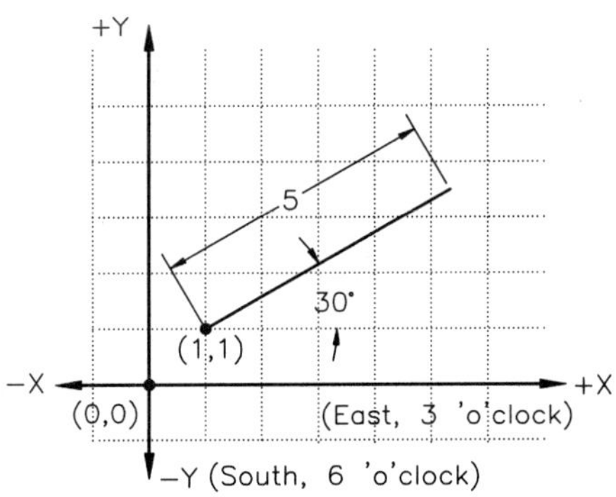

Figure 2-15 *Drawing a line using relative polar coordinates*

Sign Convention. In the relative polar coordinate system, the angle is measured from the horizontal axis (3 'o' clock) as the zero degree baseline. Also, the angle is positive if measured in a counterclockwise direction and negative if measured in a clockwise direction. Here it is assumed that the default setup of angle measurement has not been changed.

Note

You can modify the default settings of angle measurement direction using the ***UNITS*** *command discussed later.*

Example 3 — *General*

For Figure 2-16, enter the relative polar coordinates of each point in the table, and then generate the drawing. Use absolute coordinates for the start point (1.5, 1.75). The dimensions are shown in the drawing. Also, save this drawing as *Exam3.dwg*.

Point	Coordinates	Point	Coordinates
1	1.5,1.75	7	@1.0<180
2	@1.0<90	8	@0.5<270
3	@2.0<0	9	@1.0<0
4	@2.0<30	10	@1.25<270
5	@0.75<0	11	@0.75<180
6	@1.25<-90 (or <270)	12	@2.0<150

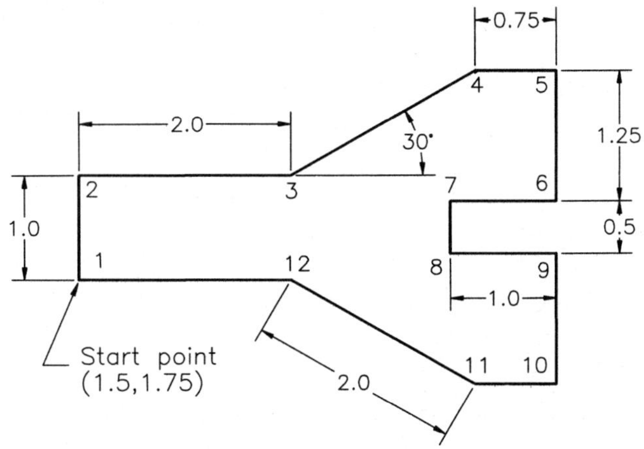

Figure 2-16 *Drawing for Example 3*

Once you know the coordinates of the points, you can generate the drawing by using AutoCAD's **LINE** command and entering the coordinates of the points.

Command: **LINE** [Enter]
Specify first point: **1.5,1.75** [Enter] *(Start point.)*
Specify next point or [Undo]: **@1<90** [Enter]
Specify next point or [Undo]: **@2.0<0** [Enter]
Specify next point or [Close/Undo]: **@2<30** [Enter]
Specify next point or [Close/Undo]: **@0.75<0** [Enter]
Specify next point or [Close/Undo]: **@1.25<-90** [Enter]
Specify next point or [Close/Undo]: **@1.0<180** [Enter]
Specify next point or [Close/Undo]: **@0.5<270** [Enter]
Specify next point or [Close/Undo]: **@1.0<0** [Enter]
Specify next point or [Close/Undo]: **@1.25<270** [Enter]
Specify next point or [Close/Undo]: **@0.75<180** [Enter]
Specify next point or [Close/Undo]: **@2.0<150** [Enter]
Specify next point or [Close/Undo]: **C** [Enter] *(Joins the last point with the first point.)*

Save this drawing by entering **SAVE** at the Command prompt and then press ENTER. The **Save Drawing As** dialog box is displayed. Enter the name *Exam3* in the **File name** edit box to replace *Drawing1.dwg* and then choose the **Save** button. The drawing will be saved with the given name in the default **My Documents** directory.

Exercise 3 *General*

Draw the object shown in Figure 2-17 using the absolute, relative rectangular, and relative polar coordinate systems to locate the points. Do not draw the dimensions; they are for reference only.

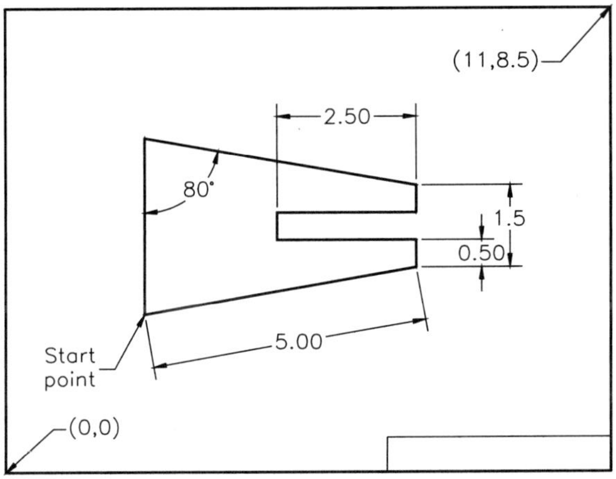

Figure 2-17 *Drawing for Exercise 3*

Direct Distance Entry

You can draw a line by specifying the length of the line and its direction, using Direct Distance Entry (Figure 2-18). The direction is determined by the position of the cursor, and the length of the line is entered from the keyboard. If Ortho is on, you can draw lines along the *X* or *Y* axis by specifying the length of line and positioning the cursor along the ortho direction. You can also use it with other draw commands like **RECTANGLE**. You can also use Direct Distance Entry with polar tracking and **SNAPANG.** For example, if **SNAPANG** is 45-degree and ortho is off, you can draw a line at 45 or 135-degree direction by positioning the cursor and entering the distance from the keyboard. Similarly, if the polar tracking is on, you can position the cursor at the predefined angles and then enter the length of the line from the keyboard.

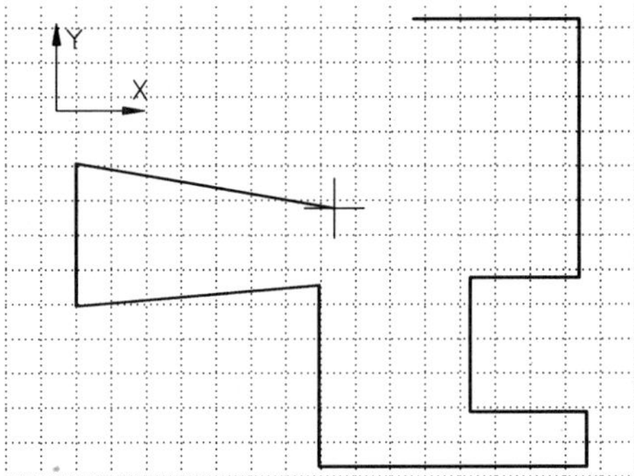

Figure 2-18 *Using Direct Distance Entry to draw lines*

Command: **LINE** [Enter]
Specify first point: *Start point.*
Specify next point or [Undo]: *Position the cursor and then enter distance.*
Specify next point or [Undo]: *Position the cursor and then enter distance.*

Example 4

General

In this example you will draw the object as shown in Figure 2-19, using Direct Distance Entry. The starting point is 2,2.

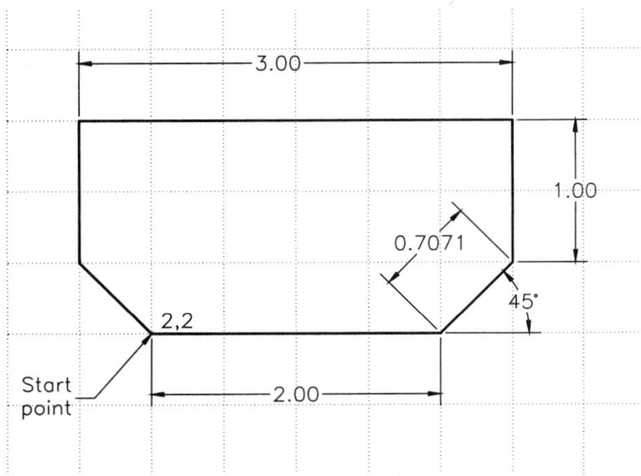

Figure 2-19 *Drawing for Example 4*

In this example, you will use the polar tracking option to draw the lines. The polar tracking option allows you to track lines that are drawn at the specified angles. The default angle that is specified for polar tracking is 90-degree. As a result, you can use the polar tracking to draw lines at an angle that is divisible by 90 such as 90, 180, 270, and 360. This is the reason, you first need to add another angle of 45-degree to the polar tracking that will allow you to track the lines drawn at an angle divisible by 45 such as 45, 90, 135, and so on.

To add 45-degree angle to polar tracking, right-click on the **POLAR** button on the status bar and choose **Settings** from the shortcut menu. Select the **Additional angles** check box in the **Polar Angle Settings** area and then choose the **New** button. Enter 45 in the field that appears and then press ENTER. Choose **OK** to close the dialog box. Now, to turn the polar tracking on, choose the **POLAR** button in the status bar. You can also turn polar tracking on or off while you are in a command. As you move the cursor to draw lines, AutoCAD displays a dotted line when the position of the cursor matches one of the predefined angles for polar tracking. Also, a tooltip is displayed that shows the length of the line and the angle at which it is being drawn. The following is the Command prompt sequence for drawing the object in Figure 2-19.

Command: **LINE**

Specify first point: **2,2**
Specify next point or [Close/Undo]: *Move the cursor horizontally toward the right and when the dotted line and tooltip appear, enter* ***2***.
Specify next point or [Close/Undo]: *Move the cursor at an angle close to 45-degree and when the dotted line and tooptip appear, enter* ***0.7071***.
Specify next point or [Close/Undo]: *Move the cursor vertically upward and when the dotted line and tooltip appear, enter* ***1***.
Specify next point or [Close/Undo]: *Move the cursor horizontally toward the left and when the dotted line and tooltip appear, enter* ***3***.
Specify next point or [Close/Undo]: *Move the cursor vertically downward and when the dotted line and tooltip appear, enter* ***1***.
Specify next point or [Close/Undo]: **C**

Note

You will learn more about the polar tracking in Chapter 4, Working with Drawing Aids.

Exercise 4 — *General*

Use the direct distance entry method to draw a parallelogram. The base of the parallelogram equals 4 units, the side equals 2.25 units, and the angle equals 45-degree. Draw the same parallelogram using absolute, relative, and polar coordinates. Note the differences and the advantage of using direct distance entry.

ERASING OBJECTS

Toolbar:	Modify > Erase
Menu:	Modify > Erase
Command:	ERASE

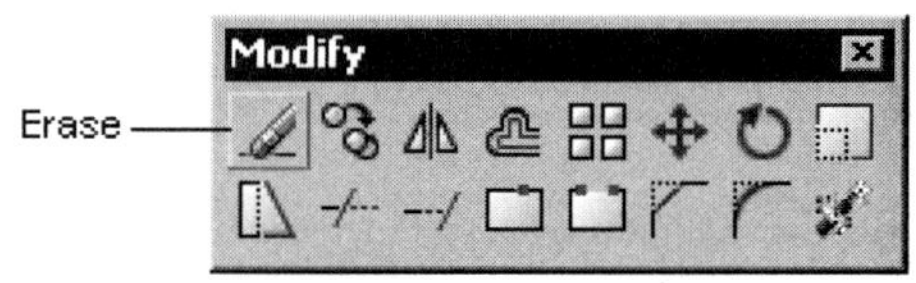

Figure 2-20 *Invoking the* ***ERASE*** *command from the* ***Modify*** *toolbar*

After drawing some objects you may want to erase some of them from the screen. To erase you can use AutoCAD's **ERASE** command (Figure 2-20). This command is used exactly the same way as an eraser is used in manual drafting to remove unwanted information. When you invoke the **ERASE** command, a small box known as the pick box replaces the screen cursor. To erase an object, move the **pick box** so that it touches the object. You can select the object by pressing the pick button of your pointing device (Figure 2-21). AutoCAD confirms the selection by changing the selected obejcts into dashed lines, and the **Select objects:** prompt returns. You can either continue selecting objects or press ENTER to

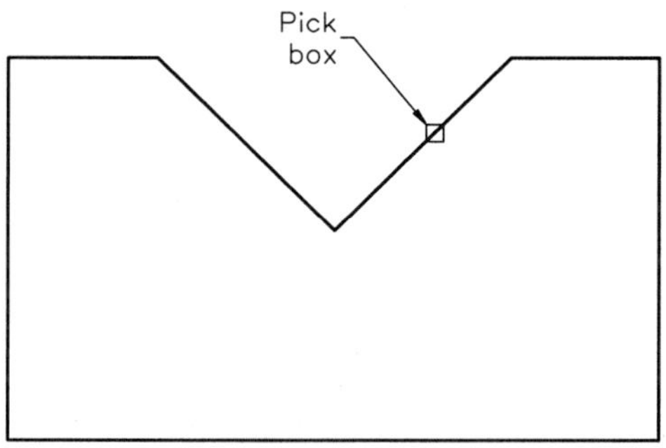

Figure 2-21 *Selecting objects by positioning the pick box at the top of the object and then pressing the pick button on the pointing device*

terminate the object selection process and erase the selected objects. If you are entering the command from the keyboard, you can type **E** or **ERASE**. The prompt sequence is given next.

Command: **ERASE** Enter
Select objects: *Select first object.*
Select objects: *Select second object.*
Select objects: Enter

If you enter All at the **Select objects:** prompt, AutoCAD will erase all objects in the drawing, even if the objects are outside the screen display area.

Command: **ERASE** Enter
Select objects: **All**

You can also first select the objects to be erased from the drawing and then right-click in the drawing area to display the shortcut menu. From this menu, you can choose the **Erase** option.

CANCELING AND UNDOING A COMMAND

If you are in a command and you want to cancel or get out of that command, press the ESC (Escape) key on the keyboard.

Command: **ERASE** Enter
Select objects: *Press ESC (Escape) to cancel the command.*

Similarly, sometimes you unintentionally erase some object from the screen. When you discover such an error, you can correct it by restoring the erased object by means of the **OOPS** command. The **OOPS** command restores objects that have been accidentally erased by the previous **ERASE** command, Figure 2-22. You can also use the **U** (Undo) command to undo the last command.

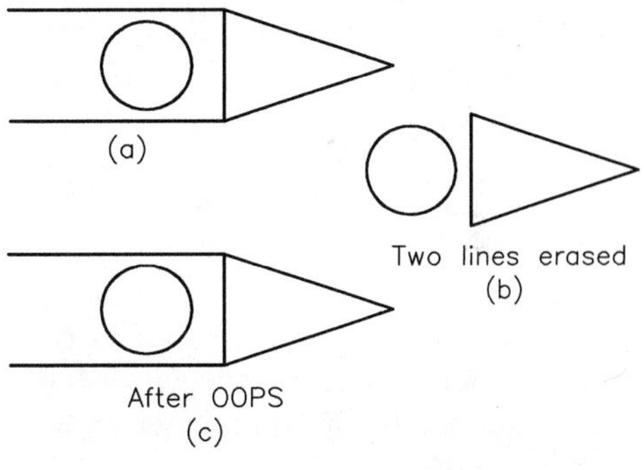

Figure 2-22 *Use of the **OOPS** command*

Command: **OOPS** Enter (*Restores erased objects.*)
Command: **U** Enter (*Undoes the last command.*)

OBJECT SELECTION METHODS

One of the ways to select objects is to select them individually, which can be time-consuming if you have a number of objects to edit. This problem can be solved by creating a selection set that enables you to select several objects at a time. The selection set options can be used with those commands that require object selection, such as **ERASE** and **MOVE**. There are many object selection methods, such as All, Last, and Add. At this point we will explore the two options: **Window** and **Crossing.** The remaining options are discussed in Chapter 5.

The Window Option

This option is used to select an object or group of objects by enclosing them in a box or window. The objects to be selected should be completely enclosed within the window; those objects that lie partially inside the boundaries of the window are not selected. You can select the **Window** option by typing W at the **Select objects:** prompt. You are prompted to select the two opposite corners of the window. After selecting the first corner, you can select the other corner by dragging the cursor to the desired position and specifying the particular point. As you move the cursor, a window is displayed that changes in size as you move the cursor. The objects selected by the **Window** option are displayed as dashed objects (Figure 2-23).

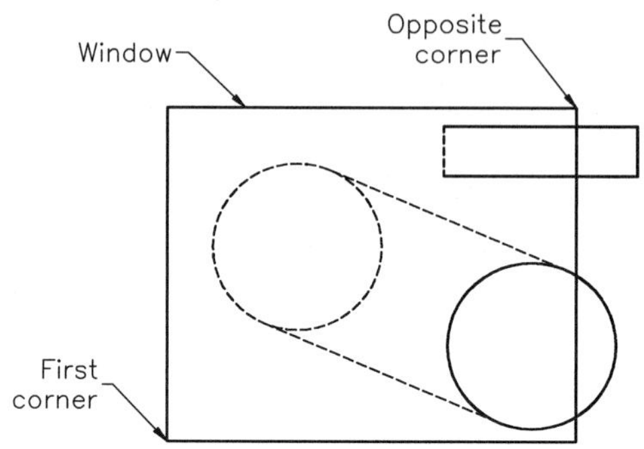

Figure 2-23 *Selecting objects using the* ***Window*** *option*

The prompt sequence for using the **Window** option with the **ERASE** command is given next.

Command: **ERASE** [Enter]
Select objects: **W** [Enter]
Specify first corner: *Select the first corner.*
Specify opposite corner: *Select the second corner.*
Select objects: [Enter]

You can also select the **Window** option by selecting a blank point on the screen at the **Select objects:** prompt. This is automatically taken as the first corner of the window. Dragging the cursor to the right will display a window. After getting all the objects to be selected inside this window, you can specify the other corner with your pointing device. The objects that are completely enclosed within the window will be selected and highlighted. The following is the prompt sequence for automatic window selection with the **ERASE** command.

Command: **ERASE** [Enter]
Select objects: *Select a blank point as the first corner of the window.*
Specify opposite corner: *Drag the cursor to the right to select the other corner of the window.*
Select objects: [Enter]

The Crossing Option

This option is used to select an object or group of objects by creating a box or window around them. The objects to be selected should be touching the window boundaries or completely enclosed within the window. You can invoke the **Crossing** option by entering **C** at the **Select objects:** prompt. After you choose the Crossing option, AutoCAD prompts you to select the first corner at the **Specify first corner:** prompt. Once you have selected the first corner, a box or window made of dashed lines is displayed. By moving the cursor you can change the size of the crossing box, hence putting the objects to be selected within (or touching) the box. Here you can select the other corner. The objects selected by the Crossing option are highlighted by displaying them as dashed objects, Figure 2-24. The following prompt sequence illustrates the use of the Crossing option when you choose the **Erase** button.

Select objects: **C** [Enter]
Specify first corner: *Select the first corner of the crossing window.*
Specify opposite corner: *Select the other corner of the crossing window.*
Select objects: [Enter]

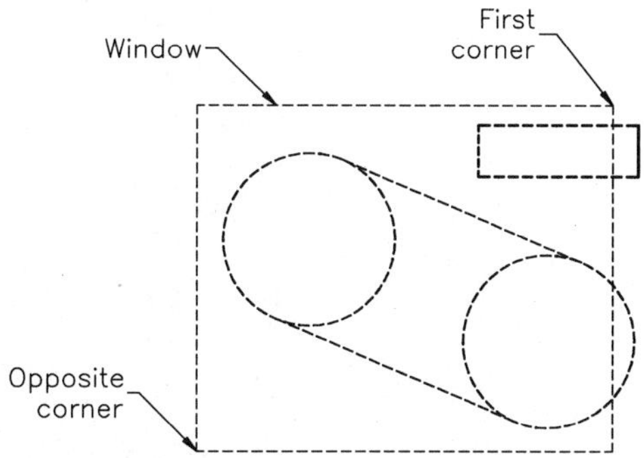

Figure 2-24 *Selecting objects using the Crossing option*

You can also select the **Crossing** option automatically by selecting a blank point on the screen at the **Select objects:** prompt and dragging the cursor to the left. The blank point you selected becomes the first corner of the crossing window and AutoCAD will then prompt you to select the other corner. As you move the cursor, a box or window made of dashed lines is displayed. The objects that are touching or completely enclosed within the window will be selected. The objects selected by the Crossing option are highlighted by being displayed as dashed objects. The prompt sequence for automatic crossing selection when you choose the **Erase** button is given next.

Select objects: *Select a blank point as the first corner of the crossing window.*
Specify opposite corner: *Drag the cursor to the left to select the other corner of the crossing window.*
Select objects: [Enter]

DRAWING CIRCLES

Toolbar:	Draw > Circle
Menu:	Draw > Circle
Command:	CIRCLE

To draw a circle you can use the AutoCAD **CIRCLE** command. You can invoke the **CIRCLE** command from the **Draw** toolbar (Figure 2-25) or from the **Draw** menu (Figure 2-26). The following is the prompt sequence for the **CIRCLE** command.

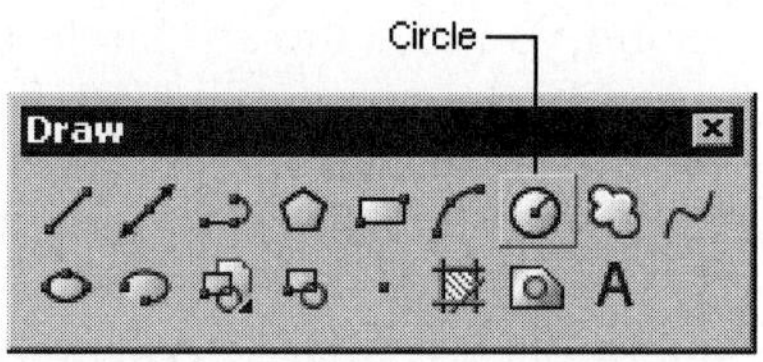

Figure 2-25 *Invoking the **CIRCLE** command from the **Draw** toolbar*

Command: **CIRCLE** [Enter]
Specify center point for circle or [3P/2P/Ttr (tan tan radius)]:

The various options of the **CIRCLE** command are explained in the following sections.

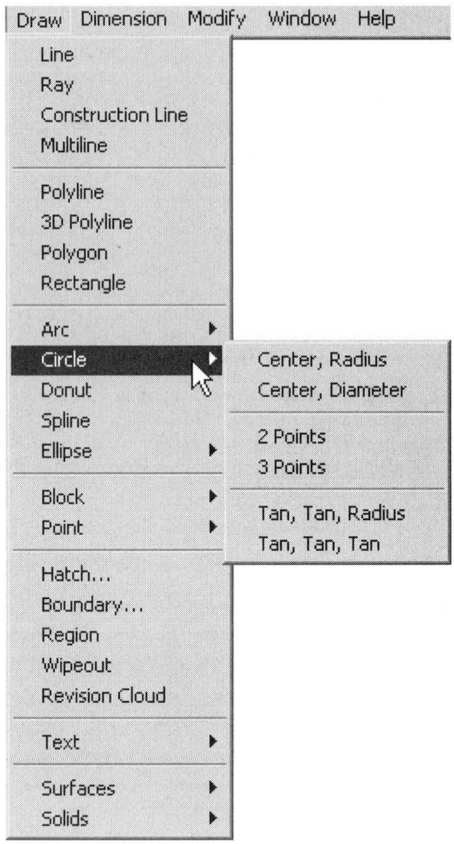

Figure 2-26 *Invoking the* ***CIRCLE*** *command from the* ***Draw*** *menu*

The Center and Radius Option

In this option you can draw a circle by defining the center and the radius of the circle, Figure 2-27. After entering the **CIRCLE** command, AutoCAD will prompt you to enter the center of the circle, which can be selected by specifying a point on the screen or by entering the coordinates of the center point. Next, you will be prompted to enter the radius of the circle. Here you can accept the default value, enter a new value, or select a point on the circumference of the circle to specify the radius. The following is the prompt sequence for drawing a circle with a center at 3,2 and a radius of 1 unit.

Command: **CIRCLE** Enter
Specify center point for circle or [3P/2P/Ttr(tan tan radius): **3,2** Enter
Specify radius of circle or [Diameter]<current>: **1** Enter

Note

You can also set the radius by assigning a value to the ***CIRCLERAD*** *system variable. The value you assign becomes the default value for radius.*

The Center and Diameter Option

In this option you can draw a circle by defining the center and diameter of the circle. After invoking the **CIRCLE** command, AutoCAD prompts you to enter the center of the circle, which can be selected by specifying a point on the screen or by entering the coordinates of the center point. Next, you will be prompted to enter the radius of the circle. At this prompt enter **D**. After this you will be prompted to enter the diameter of the circle. For entering the diameter you can accept the default value, enter a new value, or drag the circle to the desired diameter and select a point. If you use a menu option to select the **CIRCLE** command with the **Diameter** option, the menu automatically enters the **Diameter** option and prompts for the diameter after you specify the center. The following is the prompt sequence for drawing a circle with the center at (2,3) and a diameter of 2 units, Figure 2-28.

Command: **CIRCLE** Enter
Specify center point for circle or [3P/2P/Ttr(tan tan radius): **2,3** Enter
Specify radius of circle or [Diameter]<current>: **D** Enter
Specify diameter of circle <current>: **2** Enter

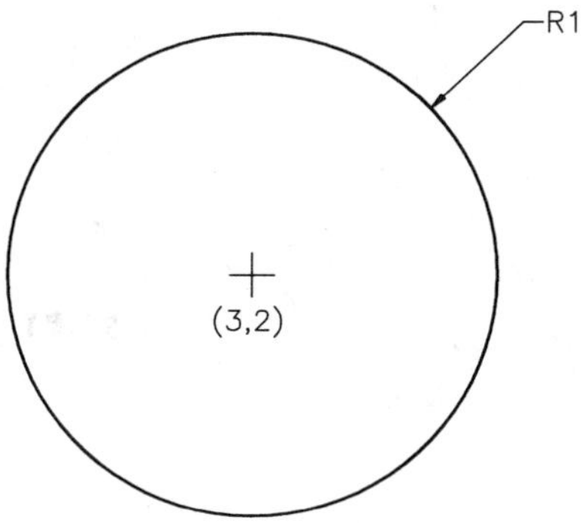

Figure 2-27 *Drawing a circle using the Center and Radius option*

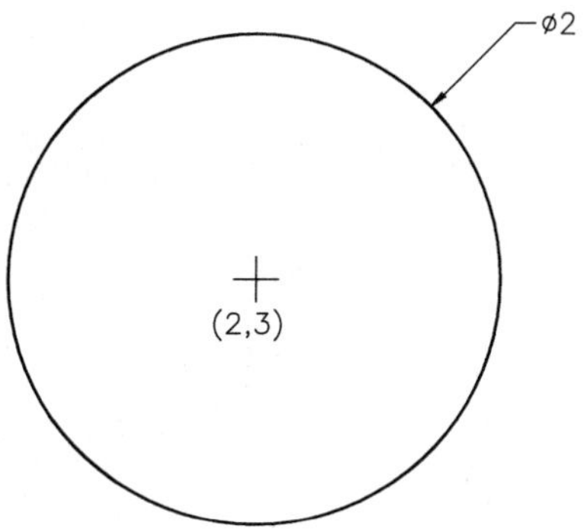

Figure 2-28 *Drawing a circle using the Center and Diameter option*

The Two-Point Option

You can also draw a circle using the **Two-Point** option. In this option AutoCAD lets you draw the circle by specifying the two endpoints of the circle's diameter. For example, if you want to draw a circle that passes through the points (1,1) and (2,1), you can use the **CIRCLE** command with 2P option, as shown in the following example (Figure 2-29).

Command: **CIRCLE** Enter
Specify center point for circle or [3P/2P/Ttr (tan tan radius)]: **2P** Enter
Specify first end point of circle's diameter: **1,1** Enter
Specify second end point of circle's diameter: **2,1** Enter *(You can also use the relative coordinates.)*

The Three-Point Option

For drawing a circle, you can also use the **Three-Point** option by defining three points on the circumference of the circle. The three points may be entered in any order. To draw a circle that passes through the points (3,3), (3,1), and (4,2), Figure 2-30, the prompt sequence is given next.

Command: **CIRCLE** Enter
Specify center point for circle or [3P/2P/Ttr(tan tan radius)]: 3P Enter
Specify first point on circle: **3,3** Enter
Specify second point on circle: **3,1** Enter
Specify third point on circle: **4,2** Enter

You can also use **relative rectangular coordinates** to define the points.

Command: **CIRCLE** Enter
Specify center point for circle or [3P/2P/Ttr(tan tan radius)]: **3P** Enter
Specify first point on circle: **3,3** Enter
Specify second point on circle: **@0,-2** Enter
Specify third point on circle: **@1,1** Enter

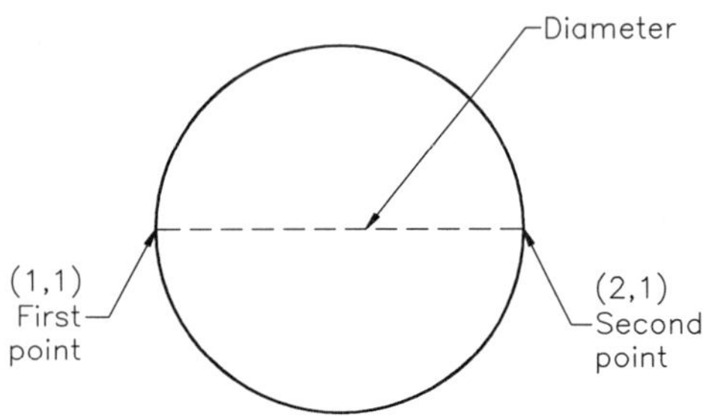

Figure 2-29 *Drawing a circle using the Two-Point option*

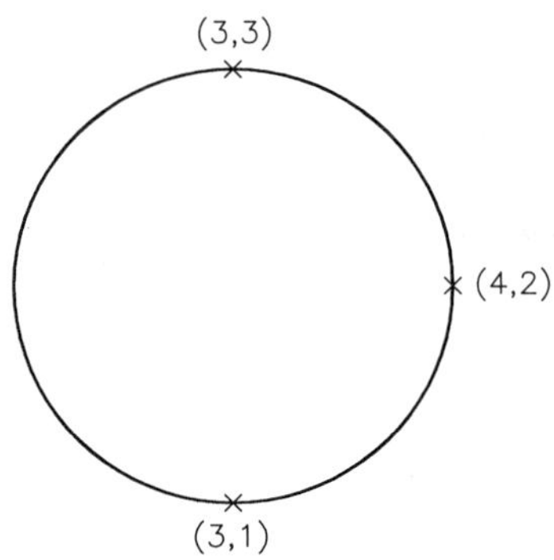

Figure 2-30 *Drawing a circle using the Three-Point option*

The Tangent Tangent Radius Option

A tangent is an object (line, circle, or arc) that contacts the circumference of a circle at only one point. In this option AutoCAD uses the Tangent object snap to locate two tangent points on the selected objects that are to be tangents to the circle. Then you have to specify the radius of the circle. The prompt sequence for drawing a circle using the **Ttr** option is given next.

Command: **CIRCLE** Enter
Specify center point for circle or [3P/2P/Ttr(tan tan radius)]: **T** Enter
Specify point on object for first tangent of circle: *Select first line, circle, or arc.*
Specify point on object for second tangent of circle: *Select second line, circle, or arc.*
Specify radius of circle <current>: **0.75** Enter

In Figures 2-31 through 2-34, the dotted circles represent the circles that are drawn by using the **Ttr** option. The circle actually drawn depends on how you select the objects that are to be tangent to the new circle. The figures show the effect of selecting different points on the objects. If you specify too small or large radius, you may get unexpected results or the "Circle does not exist" prompt.

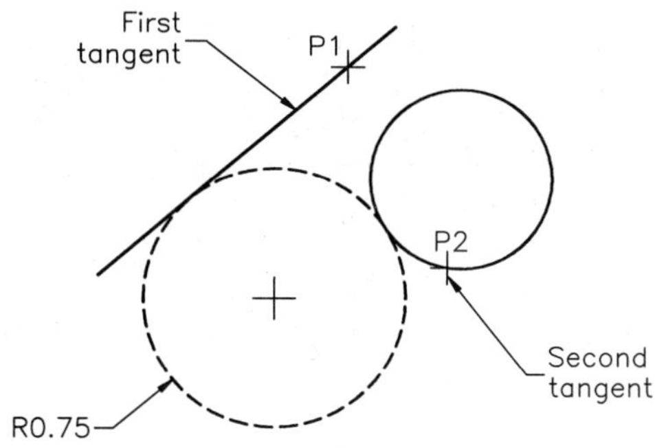

Figure 2-31 *Tangent, tangent, radius (Ttr) option*

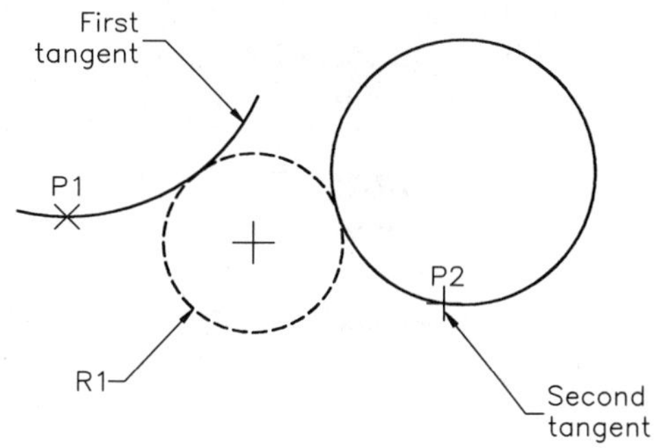

Figure 2-32 *Drawing a circle using the tangent, tangent, radius (Ttr) option*

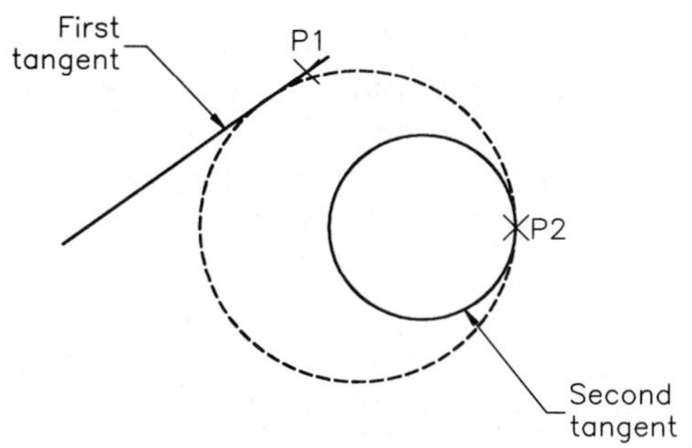

Figure 2-33 *Tangent, tangent, radius option*

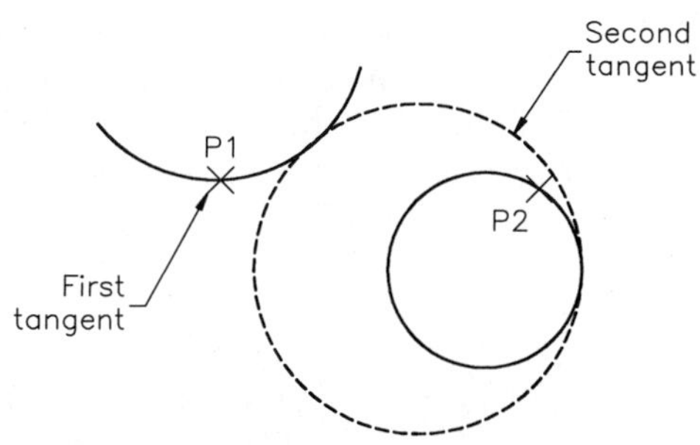

Figure 2-34 *Tangent, tangent, radius (Ttr) option*

The Tangent, Tangent, Tangent Option

You can invoke this option from the menu bar. This option is a modification of the **3P** option. In this option AutoCAD uses the Tangent osnap to locate three points on three selected objects to which the circle is drawn tangent. The following is the prompt sequence for drawing a circle using the **Tan, Tan, Tan** option (Figure 2-35).

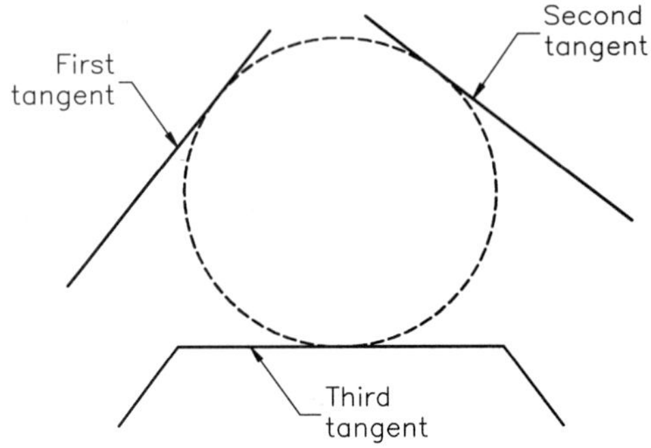

Figure 2-35 *Drawing a circle using the Tan, Tan, Tan option*

Command: **CIRCLE** [Enter]
Specify center point for circle or [3P/2P/Ttr(tan tan radius)]: *Select Tan, Tan, Tan option from the **Draw** menu.*
_ 3P Specify first point on circle: _tan to *Select the first object.*
Specify second point on circle: _tan to *Select the second object.*
Specify third point on circle: _tan to *Select the third object.*

Exercise 5 — *Mechanical*

Draw Figure 2-36 using the various options of the **LINE** and **CIRCLE** commands. Use absolute, relative rectangular, or relative polar coordinates for drawing the triangle. The vertices of the triangle will be used as the center of the circles. The circles can be drawn using the Center and Radius, Center and Diameter, or Tan, Tan, Tan options. (Height of triangle = 4.5 X sin 60 = 3.897.) Do not draw the dimensions; they are for reference only.

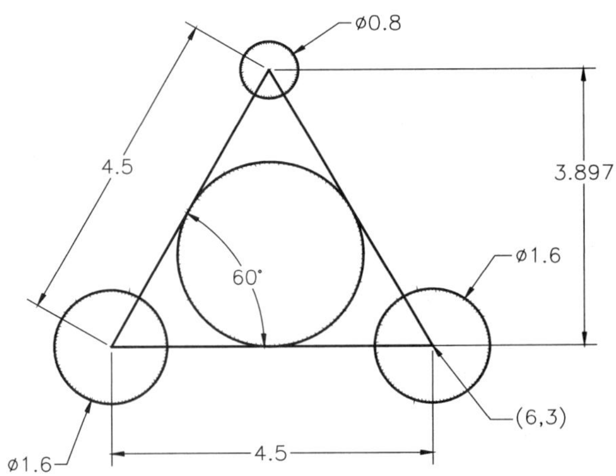

Figure 2-36 *Drawing for Exercise 5*

BASIC DISPLAY COMMANDS

Drawing in AutoCAD is much simpler than manual drafting in many ways. Sometimes while drawing, it is very difficult to see and alter minute details. In AutoCAD, you can overcome this problem by viewing only a specific portion of the drawing. This is done using the **ZOOM** command. This command lets you enlarge or reduce the size of the drawing displayed on the screen. Some of the drawing display commands such as **ZOOM** and **PAN** will be introduced here. A detailed explanation of these commands and other display options appears in Chapter 7 (Controlling the Drawing Display and Creating Text).

Zooming the Drawings

Toolbar:	Zoom toolbar, Standard > Zoom flyout
Menu:	View > Zoom
Command:	ZOOM

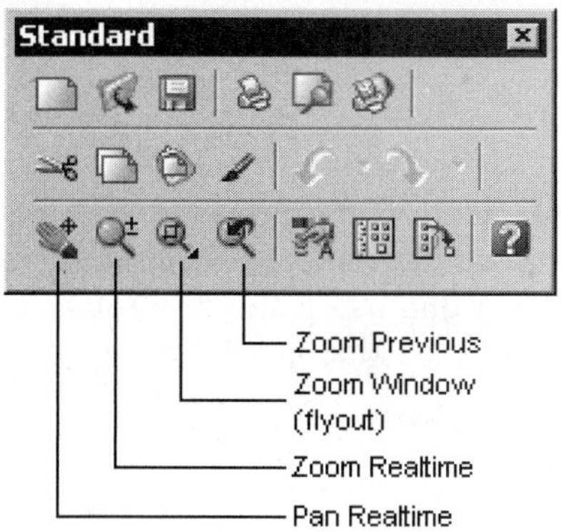

Figure 2-37 Selecting ***ZOOM*** *options from the* ***Standard*** *toolbar*

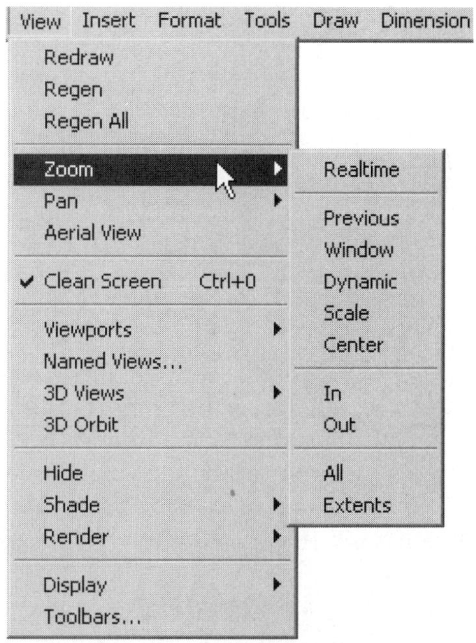

Figure 2-38 Invoking the ***ZOOM*** *command from the* ***View*** *menu*

The **ZOOM** command (Figure 2-37) enlarges or reduces the view of the drawing on the screen, but it does not affect the actual size of the entities. After the **ZOOM** command has been invoked, (Figure 2-38) various options can be used to obtain the desired display. If you use a menu, it issues the appropriate option at the initial **ZOOM** prompt. The following is the prompt sequence of the **ZOOM** command.

Command: **ZOOM** or **Z** Enter
Specify corner of window, enter a scale factor (nX or nXP), or [All/Center/Dynamic/Extents/Previous/Scale/Window]<real time>:

Realtime Zooming

You can use the **Realtime Zoom** to zoom in and zoom out interactively. To zoom in, invoke the command, and then hold the pick button down and move the cursor up. If you want to zoom in further, bring the cursor down, specify a point, and move the cursor up. Similarly, to zoom out, hold the pick button down and move the cursor down. Realtime zoom is the default setting for the **ZOOM** command. At the Command prompt, pressing ENTER after invoking the **ZOOM** command automatically invokes the realtime zoom. To exit the Realtime Zoom, right-click to display the shortcut menu and choose **Exit**. You can also press ESC or the ENTER key to exit the command.

Window Option

This is the most commonly used option of the **ZOOM** command. It lets you specify the area you want to zoom in on by letting you specify two opposite corners of a rectangular window. The center of the specified window becomes the center of the new display screen. The area inside the window is magnified in size to fill the drawing area as completely as possible. The points can be specified by selecting them with the help of the pointing device or by entering their coordinates.

Previous Option

While working on a complex drawing, you may need to zoom in on a portion of the drawing to edit some minute details. When you have completed the editing you may want to return to the previous view. This can be done using the **Previous** option of the **ZOOM** command. AutoCAD remembers the last ten views that can be recalled by using the **Previous** option.

All Option

This option zooms to the drawing limits or the extents, whichever is greater. Whenever you increase the limits (**Quick Setup** in **Use a Wizard** dialog box) the current display is not affected and hence does not show. You need to use the **Zoom All** option to display the limits of the drawing. Sometimes it is possible that the objects are drawn beyond the limits. In such a case the **Zoom All** option zooms to fill the drawn objects in the drawing area irrespective of its limits.

Panning in Realtime

You can use **Pan Realtime** to pan the drawing interactively, by sliding the drawing and placing it at the required position. To pan a drawing, invoke the command and then hold the pick button down and move the cursor in any direction. When you select realtime pan, AutoCAD displays an image of a hand, indicating that you are in PAN mode. You can drag the hand anywhere on the screen to move the drawing. To exit realtime pan, right-click to display the shortcut menu and choose **Exit**. You can also press ESC or the ENTER key to exit the command.

Tip

*You can right-click to display a shortcut menu while the realtime zoom or realtime pan options are active. The **Realtime Zoom**, **Realtime Pan**, **Exit**, and other **ZOOM** command options are available in this shortcut menu.*

SETTING UNITS

In Chapter 1 you have already learned to set units while starting a drawing from the **Startup** dialog box using the **Wizards** option. If you want to change the units while you are already working on a drawing, the **UNITS** command can be used.

Setting Units Using the Drawing Units Dialog Box

Menu:	Format > Units
Command:	UNITS

The **UNITS** command is used to select a format for the units of distance and angle measurement. You can invoke this command using the **Format** menu, see Figure 2-39.

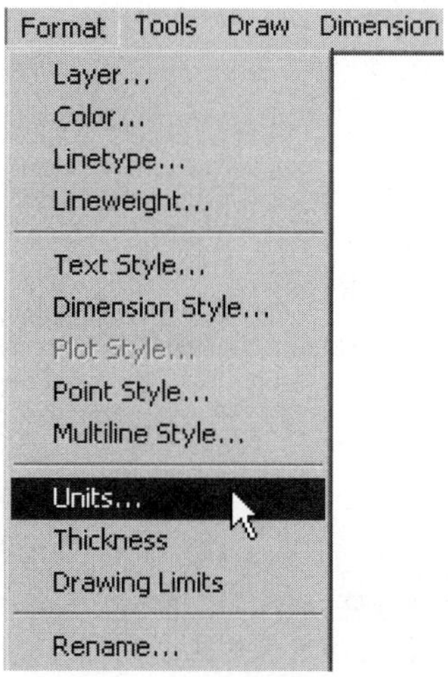

Figure 2-39 *Invoking the **UNITS** command from the **Format** menu*

The **UNITS** command displays the **Drawing Units** dialog box as shown in Figure 2-40. You can then specify the precision for the units and angles from the corresponding **Precision** drop-down list, see Figure 2-41. You can also set the units from the command line by entering **-UNITS** at the Command prompt.

Specifying Units

In the **Drawing Units** dialog box, you can select a desired format of units from the drop-down list displayed when you choose the down arrow to the right of the **Type** edit box. You can select one of the following five formats.

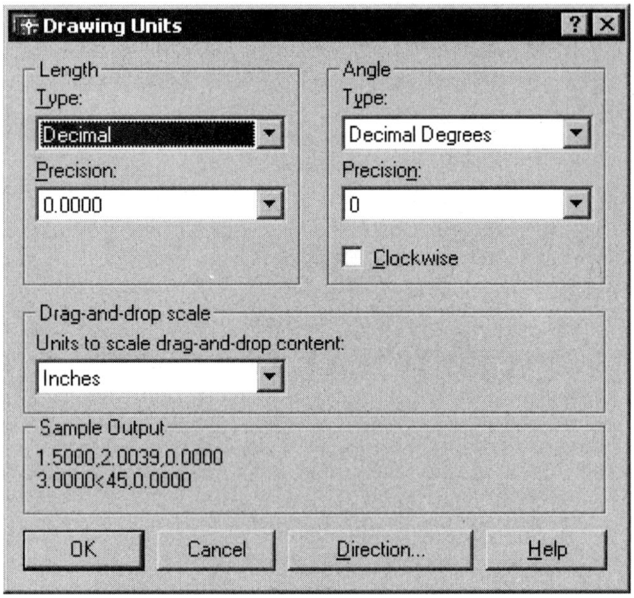

Figure 2-40 ***Drawing Units*** *dialog box*

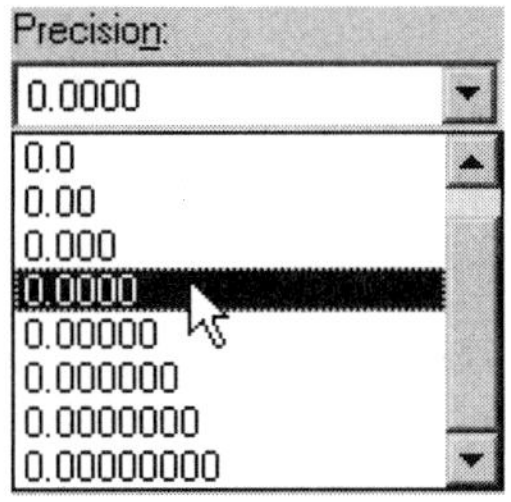

Figure 2-41 *Specifying Precision from the* ***Drawing Units*** *dialog box*

1. Architectural (0'-01/16") 2. Decimal (0.00) 3. Engineering (0'-0.00")
4. Fractional (0 1/16) 5. Scientific (0.00E+01)

If you select the scientific, decimal, or fractional format, you can enter the distances or coordinates in any of these three formats, but not in engineering or architectural units. In the following example, the units are set as decimal, scientific, fractional, and decimal and fractional to enter the coordinates of different points.

Command: **LINE** Enter
Specify from point: **1.75,0.75** Enter (*Decimal.*)
Specify next point or [Undo]: **1.75E+01, 3.5E+00** Enter (*Scientific.*)
Specify next point or [Undo]: **10-3/8,8-3/4** Enter (*Fractional.*)
Specify next point or [Close/Undo]: **0.5,17/4** Enter (*Decimal and fractional.*)

If you choose the engineering or architectural format, you can enter the distances or coordinates in any of the five formats. In the following example, the units are set as architectural; hence, different formats are used to enter the coordinates of points.

Command: **LINE** Enter
Specify first point: **1-3/4,3/4** Enter (*Fractional.*)
Specify next point or [Undo]: **1'1-3/4",3-1/4** Enter (*Architectural.*)
Specify next point or [Undo]: **0'10.375,0'8.75** Enter (*Engineering.*)
Specify next point or [Close/Undo]: **0.5,4-1/4"** Enter (*Decimal and engineering.*)

Note

The inch symbol (") is optional. For example, 1'1-3/4" is the same as 1'1-3/4, and 3/4" is the same as 3/4.

You cannot use the feet (') or inch (") symbols if you have selected scientific, decimal, or fractional unit formats.

Specifying Angle

You can select one of the following five angle measuring systems:

1. Decimal Degrees (0.00)
2. Deg/min/sec (0d00'00")
3. Grads (0.00g)
4. Radians (0.00r)
5. Surveyor's Units (N 0d00'00" E)

If you select any of the first four measuring systems, you can enter the angle in the Decimal, Degrees/minutes/seconds, Grads, or Radians system, but you cannot enter the angle in Surveyor's units. However, if you select Surveyor's units, you can enter the angles in any of the five systems. If you enter an angle value without any indication of a measuring system, it is taken in the current system. To enter the value in another system, use the appropriate suffixes and symbols, such as r (Radians), d (Degrees), g (Grads), or the others shown in the following examples. In the following example, the system of angle measure is Surveyor's units and different systems of angle measure are used to define the angle of the line.

Command: **LINE** Enter
Specify first point: **3,3** Enter
Specify next point or [Undo]: **@3<45.5** Enter *(Decimal degrees.)*
Specify next point or [Undo]: **@3<90d30'45"** Enter *(Degrees/min/sec.)*
Specify next point or [Close/Undo]: **@3<75g** Enter *(Grads.)*
Specify next point or [Close/Undo]: **@3<N45d30'E** Enter *(Surveyor's units.)*

In Surveyor's units you must specify the bearing angle that the line makes with the north-south direction (Figure 2-42). For example, if you want to define an angle of 60-degree with north, in the Surveyor's units the angle will be specified as N60dE. Similarly, you can specify angles such as S50dE, S50dW, and N75dW, as shown in Figure 2-42. You cannot specify an angle that exceeds 90-degree (N120E). The angles can also be specified in radians or grads, for example, 180-degree is equal to **PI** (3.14159)

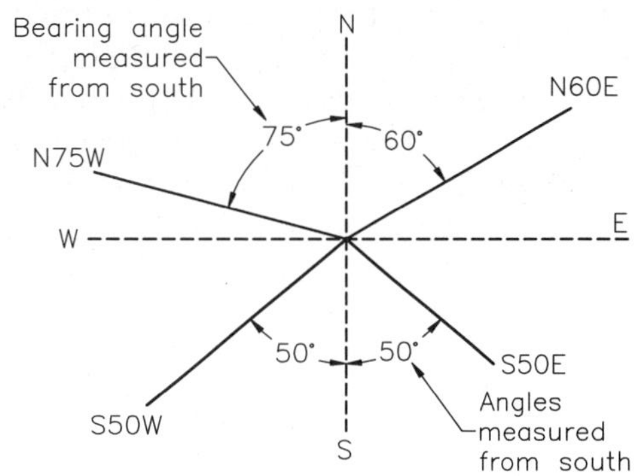

Figure 2-42 *Specifying angle in Surveyor's units*

radians. You can convert degrees into radians or radians into degrees using the following equations:

radians = degrees X 3.14159/180; degrees = radians X 180/3.14159

Grads are generally used in land surveys. There are 400 grads or 360-degree in a circle. A 90-degree angle is equal to 100 grads.

Tip
An example corresponding to the type of unit and angle selected from the ***Length*** *or* ***Angle*** *area of the dialog box can be seen in the* ***Sample Output*** *area of the dialog box.*

In AutoCAD, by default the angles are positive if measured in the counterclockwise direction (Figure 2-43) and the angles are measured from the positive *X* axis, see Figure 2-44. The angles are negative if measured clockwise. If you want the angles to be measured as positive in the clockwise direction, select the **Clockwise** check box. Then the positive angles will be measured in the clockwise direction and the negative angles in the counterclockwise direction.

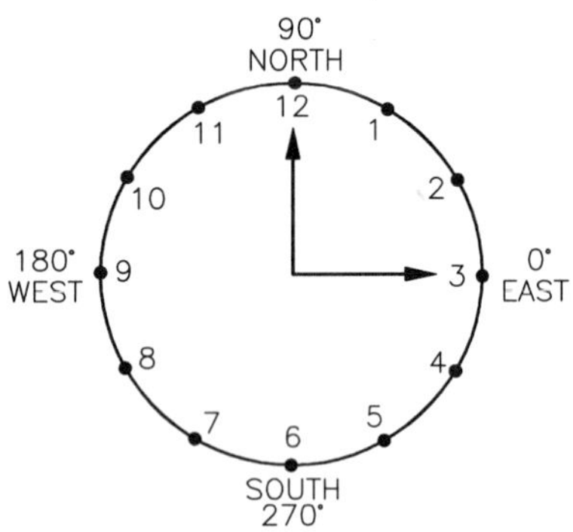

Figure 2-43 *N,S,E,W directions*

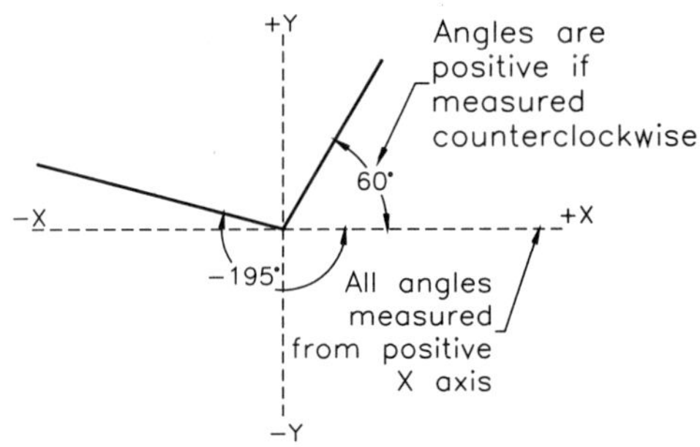

Figure 2-44 *Measuring angles counterclockwise from the positive X axis (default)*

When you choose the **Direction** button in the **Drawing Units** dialog box, the **Direction Control** dialog box appears, which gives you an option of selecting the setting for direction of the base angle, see Figure 2-45. If you select the **Other** option, you can set your own direction for the base angle by either entering a value in the **Angle** edit box or choosing the **Pick an Angle** button to specify the angle on the screen. After selecting an angle you can choose the **OK** button to apply the settings. This will get you back to the **Drawing Units** dialog box.

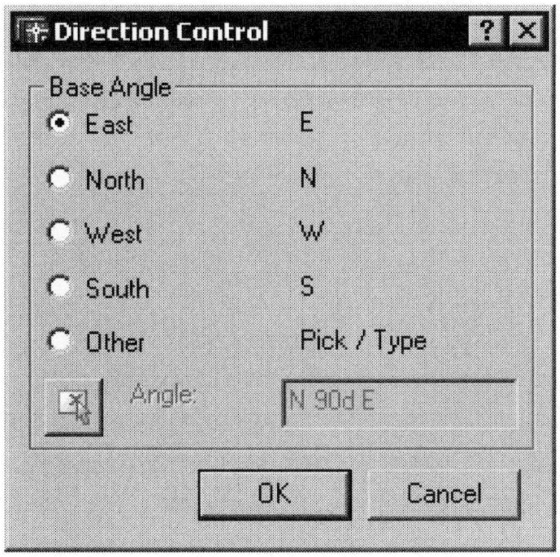

***Figure 2-45** Setting direction from the **Direction Control** dialog box*

You can also set the units of measure while inserting a block or a drawing from the DESIGNCENTER. In the **Drawing Units** dialog box choose any measuring unit from the **Units to scale drag-and-drop content** drop-down list (Figure 2-46). Now, while inserting a block or a drawing from the DESIGNCENTER, AutoCAD inserts the block with the chosen unit. Even if the block was created using a different measuring unit, AutoCAD scales it and inserts it using the specified measuring unit. If you want to insert the block with the original units, then choose **Unitless** from the drop-down list.

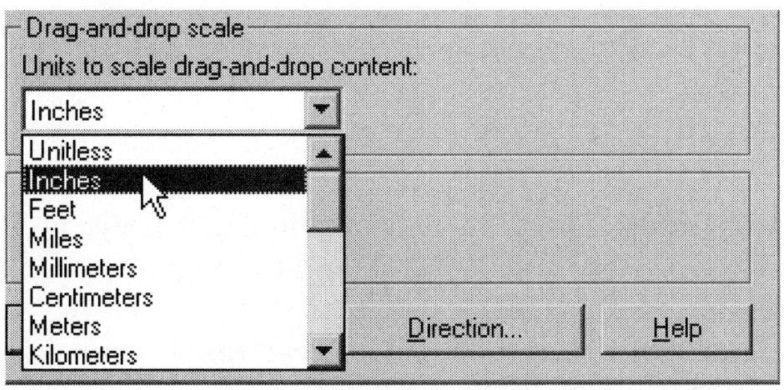

***Figure 2-46** Choosing measuring units for inserting the drawings and blocks using the drag-and-drop method*

Note

The insertion of blocks from the DESIGNCENTER into a drawing is discussed in detail in Chapter 6 (Editing Sketched Objects-II).

The **Sample Output** area in the **Drawing Units** dialog box shows an example of the current format of the units and angles. When you change the type of length and angle measure in the Length and Angle areas of the **Drawing Units** dialog box, the corresponding example is displayed in the **Sample Output** area.

Example 5 *General*

In this example you will set the units for a drawing according to the following specifications and then draw Figure 2-47.

1. Set **UNITS** to fractional, with the denominator of the smallest fraction equal to 32.

2. Set the angular measurement to Surveyor's units, with the number of fractional places for display of angles equal to zero.

3. Set the direction to 90-degree (north) and the direction of measurement of angles to clockwise (angles measured positive in clockwise direction), Figure 2-47.

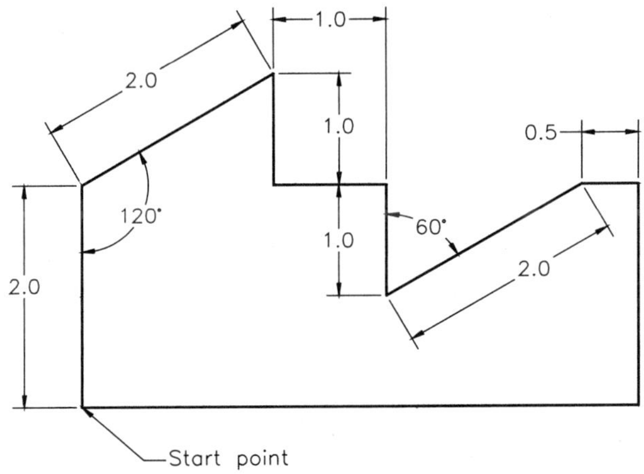

Figure 2-47 *Drawing for Example 5*

1. Invoke the **Drawing Units** dialog box by choosing **Units** from the **Format** menu. You can also invoke the dialog box by entering **UNITS** at the Command prompt.

2. In the **Length** area of the dialog box, select **Fractional** from the **Type** drop-down list. From the **Precision** drop-down list select **0 1/32** (Figure 2-48).

3. In the **Angle** area of the dialog box, select **Surveyor's Units** from the **Type** drop-down list. From the **Precision** drop-down list select **N 0d E** if it is not already selected. Also, select the **Clockwise** check box to set the clockwise angle measurement as positive.

4. Choose the **Direction** button to display the **Direction Control** dialog box. Select the **North** radio button. Choose the **OK** button to exit the **Direction Control** dialog box.

5. Choose the **OK** button to exit the **Drawing Units** dialog box.

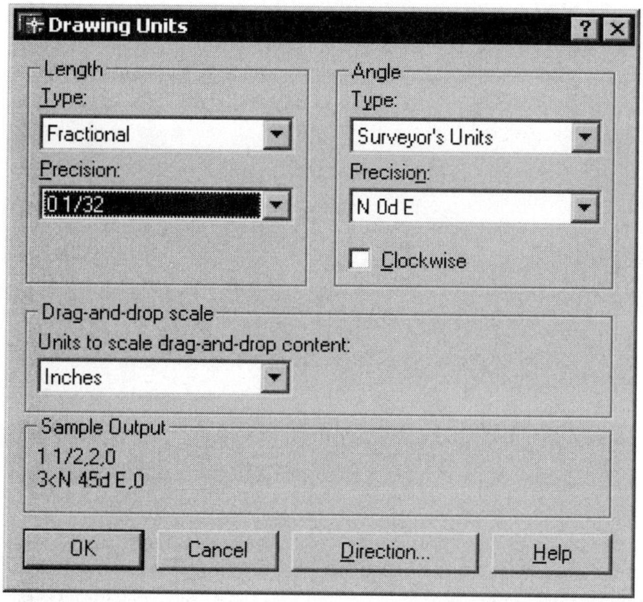

Figure 2-48 *Setting units for Example 5 in the* ***Drawing Units*** *dialog box*

6. With the units set, draw Figure 2-47 using relative polar coordinates. Here the units are fractional and the **angles are measured from north** (90-degree axis). Also, the angles are measured as positive in the clockwise direction and **negative** in the counterclockwise direction (Figure 2-49). Invoke the LINE command and specify the points as follows.

   Command: **LINE** Enter
   Specify first point: **2,2** Enter
   Specify next point or [Undo]: **@2.0<0** Enter
   Specify next point or [Undo]: **@2.0<60** Enter
   Specify next point or [Close/Undo]: **@1<180** Enter
   Specify next point or [Close/Undo]: **@1<90** Enter
   Specify next point or [Close/Undo]: **@1<180** Enter
   Specify next point or [Close/Undo]: **@2.0<60** Enter
   Specify next point or [Close/Undo]: **@0.5<90** Enter
   Specify next point or [Close/Undo]: **@2.0<180** Enter
   Specify next point or [Close/Undo]: **C** Enter

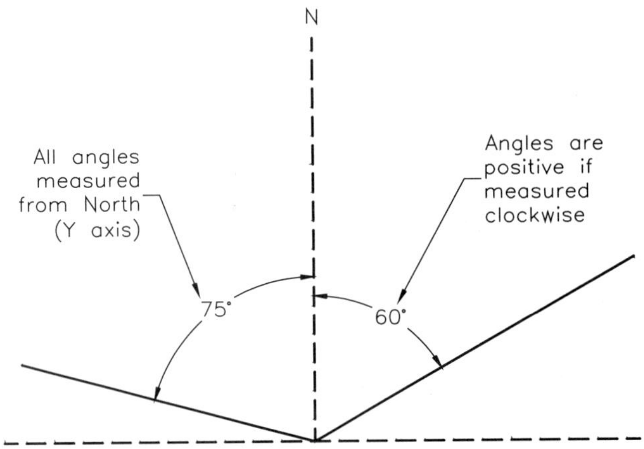

Figure 2-49 Angles measured from north (Y axis)

Forcing Default Angles

When you define the direction by specifying the angle, the output of the angle depends on the following (Figure 2-50):

Angular units
Angle direction
Angle base

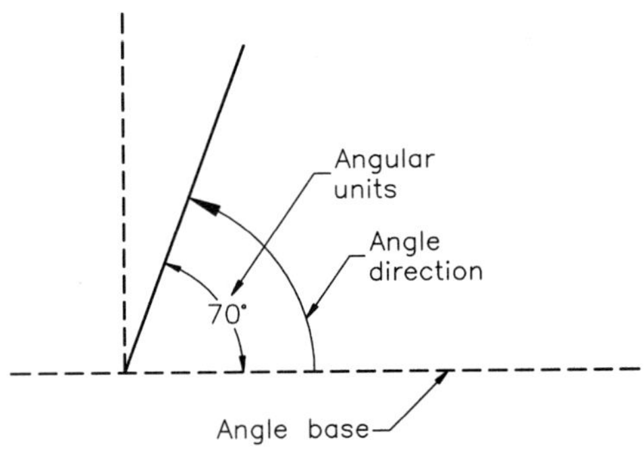

Figure 2-50 *Default angular units, direction, and base*

For example, if you are using the AutoCAD default setting, <70 represents an angle of 70 decimal degrees from the positive *X* axis, measured counterclockwise. The decimal degrees represent angular units, the *X* axis represents the angle base, and counterclockwise represents the angle direction. If you have changed the default settings for measuring angles, it might be confusing to enter the angles. AutoCAD lets you bypass the current settings by entering << or <<< before the angle. If you enter << before the angle, AutoCAD will bypass the current angle settings and use the angle as decimal degrees from the default angle base and default angle direction (angles are referenced from the positive *X* axis and are positive if measured counterclockwise). If you enter <<< in front of the angle, AutoCAD will use current angular units, but it will bypass the current settings for angle base and angle direction and use the default angle setting (angles are referenced from the positive *X* axis and are positive if measured counterclockwise).

Assume, for example, that you have changed the current settings and made the system of angle measure radians with two places of precision: angle base north (the *Y* axis), and the direction clockwise. Now, if you enter <1.04 or <1.04r, all the current settings will be taken into consideration and you will get an angle of 1.04 radians, measured in a clockwise direction from the positive *Y* axis, Figure 2-51(a). If you enter <<60, AutoCAD will bypass the current settings and reference the angle in degrees from the positive *X* axis, measuring 60-degree in a counterclockwise direction, Figure 2-51(b). If you enter <<<1.04r, AutoCAD will use the current angular units, but will bypass the current angle base and angle direction. Hence, the angle will be referenced from the positive *X* axis, and will be measured 1.04 radians in a counterclockwise direction, Figure 2-51(c).

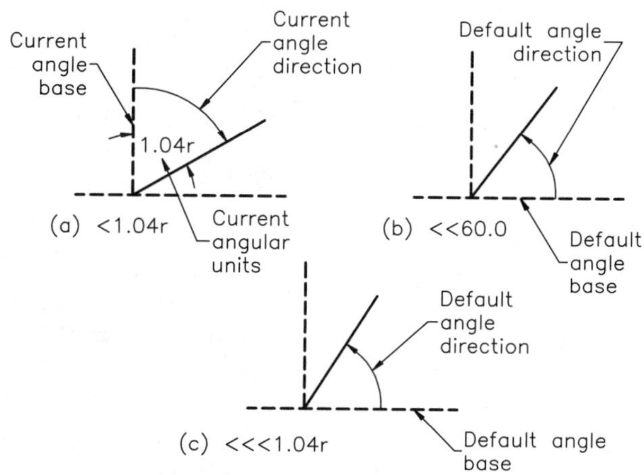

Figure 2-51 *Forcing default angles*

The effect of using the open angle brackets is summarized in the following table.

Angle prefix	Angular units	Angle direction	Angle base	Example
<	Current	Current	Current	<1.04r
<<	Degrees	Counterclockwise	Default	<<60.0
<<<	Current	Counterclockwise	Default	<<<1.04r

You can also override the current angle units by entering the appropriate suffix, such as 60d for degrees.

SETTING THE LIMITS OF THE DRAWING

Menu: Format > Drawing Limits
Command: LIMITS

In AutoCAD, the drawings must be drawn full scale and, therefore, limits are needed to size up a drawing area. The limits of the drawing area are usually determined by the following factors.

1. The actual size of the drawing.
2. The space needed for putting down the dimensions, notes, bill of materials, and other necessary details.
3. The space between various views so that the drawing does not look cluttered.
4. The space for the border and a title block, if any.

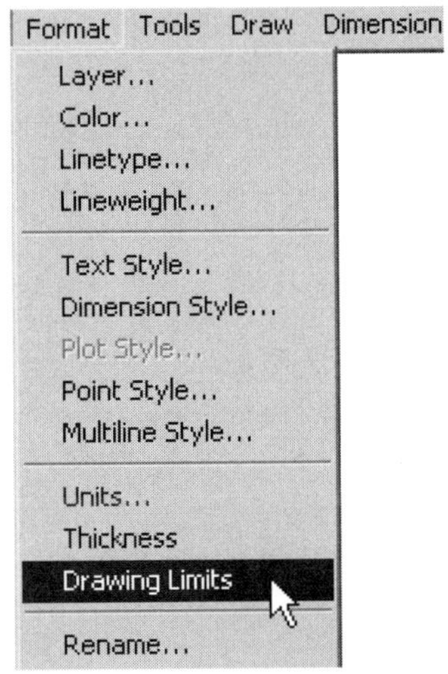

Figure 2-52 *Choosing* ***Drawing Limits*** *from the* ***Format*** *menu*

In Chapter 1 you have already learned to set the limits while starting a drawing from the **Startup** dialog box using the **Wizards** option. If you want to change the limits while you are already working in a drawing, the **LIMITS** command can be used. When you start AutoCAD, the default limits are 12.00,9.00. You can use the **LIMITS** command (Figure 2-52) to set up new limits. The following is the prompt sequence of the **LIMITS** command for setting the limits of 24,18.

Command: **LIMITS** Enter
Reset Model space limits:
Specify lower left corner or [ON/OFF]<current>: **0,0** Enter
Specify upper right corner <current>: **24,18** Enter

At the preceding two prompts you are required to specify the lower left corner and the upper right corner of the sheet. Normally you choose (0,0) as the lower left corner, but you can enter any other point. If the sheet size is 24 X 18, enter (24,18) as the coordinates of the upper right corner.

Tip
Whenever you increase the drawing limits, the display area does not change. You need to use the ***All*** *option of the* ***ZOOM*** *command to display the complete area inside the drawing area.*

Setting Limits

To get a good idea of how to set up limits, it is always better to draw a rough sketch of the drawing to help calculate the area needed. For example, if an object has a front view size of 5 X 5, a side view size of 3 X 5, and a top view size of 5 X 3, the limits should be set so that they can accommodate the drawing and everything associated with it. In Figure 2-53, the space between the front and side views is 4 units and between the front and top views is 3 units. Also, the space between the border and the drawing is 5 units on the left, 5 units on the right, 3 units at the bottom, and 2 units at the top. (The space between the views and between the borderline and the drawing depends on the drawing.)

After you know the sizes of various views and have determined the space required between

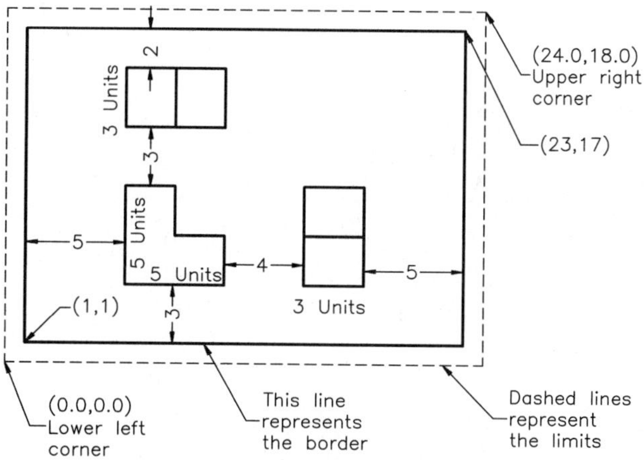

Figure 2-53 *Setting limits in a drawing*

views, between the border and the drawing, and between the borderline and the edges of the paper, you can calculate the space you need as follows.

Space along (*X* axis) = 1 + 5 + 5 + 4 + 3 + 5 + 1 = 24
Space along (*Y* axis) = 1 + 3 + 5 + 3 + 3 + 2 + 1 = 18

Thus, the space or the work area you need for the drawing is 24 X 18. Once you have determined the space you need, select the sheet size that can accommodate your drawing. In the case just explained, you will select a D size (34 X 22) sheet. Therefore, the actual drawing limits are 34,22.

Standard Sheet Sizes

When you make a drawing, you might want to plot the drawing to get a hard copy. Several standard sheet sizes are available to plot your drawing. Although in AutoCAD you can select any work area, it is recommended that you select the work area on the basis of the sheet size you will be using to plot the drawing. The sheet size is the deciding factor for determining the limits (work area), text size **(TEXTSIZE)**, dimensioning scale factor **(DIMSCALE)**, linetype scale factor **(LTSCALE)**, and other drawing-related parameters. The following tables list standard sheet sizes and the corresponding drawing limits for different scale factors.

Standard U.S. Size

Letter size	Sheet size	Limits (1:1)	Limits (1:4)	Limits (1/4"=1')
A	8.5 x 11	8.5,11	34,44	34',44'
B	11 x 17	11,17	44,68	44',68'

C	17 x 22	17,22	68,88	68',88'
D	22 x 34	22,34	88,136	88',136'
E	34 x 44	34,44	136,176	136'x176'

International Size

Letter size	Sheet size	Limits (1:1)	Limits (1:20)
A4	210 x 297	210,297	4200,5940
A3	297 x 420	297,420	5940,8400
A2	420 x 594	420,594	8400,11940
A1	594 x 841	594,841	11940,16820
A0	841 x 1189	841,1189	16820,23780

Limits for Architectural Drawings

Most architectural drawings are drawn at a scale of 1/4" = 1', 1/8" = 1', or 1/16" = 1'. You must set the limits accordingly. The following example illustrates how to calculate the limits in architectural drawings.

Given
Sheet size = 24 X 18
Scale is 1/4" = 1'

Calculate limits
Scale is 1/4" = 1'
or 1/4" = 12"
or 1" = 48"
X limit = 24 X 48
= 1152" or 1152 Units
= 96'
Y limit = 18 X 48
= 864" or 864 Units
= 72'

Thus, the scale factor is 48 and the limits are 1152",864", or 96',72'.

Example 6 — *General*

In this example you will calculate limits and determine an appropriate drawing scale factor for Figure 2-54. The drawing is to be plotted on a 12" X 9" sheet.

The scale factor can be calculated as follows:

Given or known
Overall length of the drawing = 31'
Length of the sheet = 12"

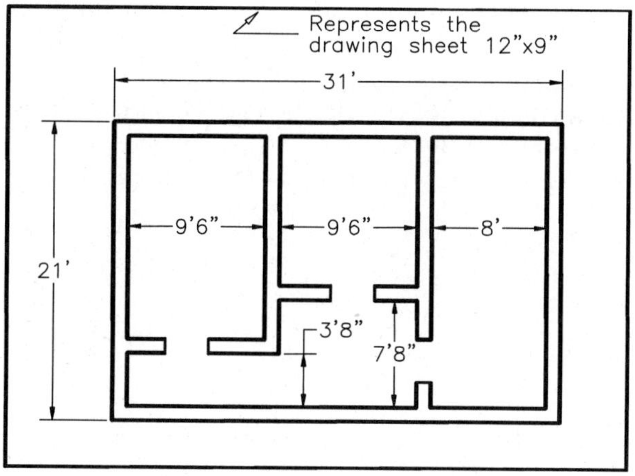

Figure 2-54 Drawing for Example 6

Approximate space between the drawing and the edges of the paper = 2"

Calculate scale factor
To calculate the scale factor, you have to try various scales until you find one that satisfies the given conditions. After some experience you will find this fairly easy to do. For this example, assume a scale factor of 1/4" = 1'.

Scale factor 1/4" = 1'
or 1" = 4'
Thus, a line 31' long will be = 31'/4' = 7.75" on paper. Similarly, a line 21' long = 21'/4' = 5.25".

Approximate space between the drawing and the edges of paper = 2"
Therefore, total length of the sheet = 7.75 + 2 + 2 = 11.75"
Similarly, total width of the sheet = 5.25 + 2 + 2 = 9.25"

Because you selected the scale 1/4" = 1', the drawing will definitely fit on the given sheet of paper (12" x 9"). Therefore, the scale for this drawing is 1/4" = 1'.

Calculate limits
Scale factor = 1" = 48" or 1" = 4'
The length of the sheet is 12"
Therefore, X limit = 12 X 4' = 48'
Also, Y limit = 9 X 4' = 36'

Limits for Metric Drawings

When the drawing units are metric, you must use **standard metric size sheets** or calculate the limits in millimeters (mm). For example, if the sheet size you decide to use is 24 X 18, the limits after conversion to the metric system will be 609.6,457.2 (multiply length and width by 25.4). You can round these numbers to the nearest whole numbers 610,457. Note that metric drawings do not require any special setup, except for the limits. Metric drawings are like any other drawings that use decimal units. As with architectural drawings, you can draw metric drawings to a scale. For example, if the scale is 1:20 you must calculate the limits accordingly. The following example illustrates how to calculate the limits for metric drawings.

Given
Sheet size = 24" X 18"
Scale = 1:20

Calculate limits
Scale is 1:20
Therefore, scale factor = 20
X limit = 24 X 25.4 X 20 = 12192 units
Y limit = 18 X 25.4 X 20 = 9144 units

Thus, the limits are 12192 and 9144.

Exercise 6 *General*

Set the units of the drawing according to the following specifications and then make the drawing shown in Figure 2-55 (leave a space of 3 to 5 units around the drawing for dimensioning and title block). The space between the dotted lines is 1 unit.

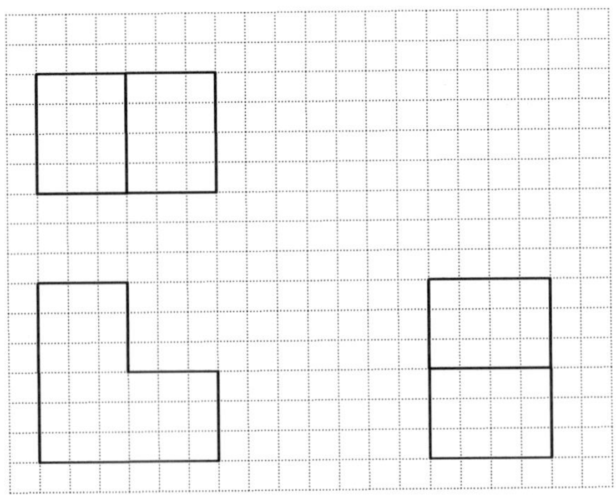

Figure 2-55 *Drawing for Exercise 6*

1. Set **UNITS** to decimal units, with two digits to the right of the decimal point.

2. Set the angular measurement to decimal degrees, with the number of fractional places for display of angles equal to 1.

3. Set the direction to 0-degree (east) and the direction of measurement of angles to counterclockwise (angles measured positive in a counterclockwise direction).

4. Set the limits leaving a space of 3 to 5 units around the drawing for dimensioning and title block.

INTRODUCTION TO PLOTTING DRAWINGS

Toolbar:	Standard toolbar > Plot
Menu:	File > Plot
Command:	PLOT or PRINT

Once you have created a drawing in the current session of AutoCAD, you may need to have its hard copy. This hard copy is very useful in the industry and can be created by plotting and printing it on a sheet of paper. Suppose you have drawn an architectural plan on the computer, and you need to send its hard copy to the site for implementation. Similarly, if you have created a mechanical component, you may need to send its hard copy to the shop floor for manufacturing. Drawings can be plotted by using the **PLOT** command (Figure 2-56), and AutoCAD will display the **Plot** dialog box when you invoke the **PLOT** command.

The values in this dialog box are the ones that were set during the configuring of AutoCAD. If the displayed values conform to your requirements, you can start plotting without making any changes. If necessary, you can make changes in the default values according to your plotting requirements.

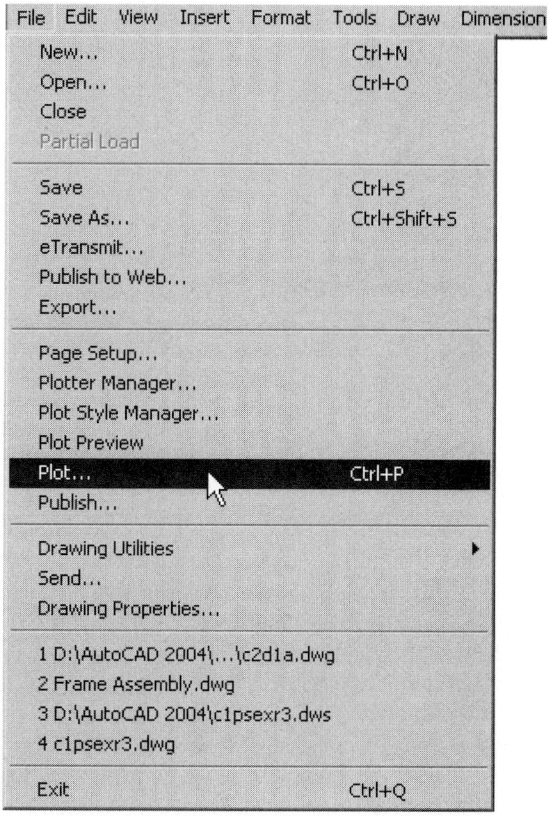

Figure 2-56 *Invoking the* ***PLOT*** *command from the* ***File*** *menu*

Basic Plotting

In this section you will learn how to set up the basic plotting parameters. Later, you will learn about the advance options that allow you to plot according to your plot drawing specifications. Basic plotting involves selecting the correct output device (plotter), specifying the area to plot, selecting paper size, specifying the plot origin, orientation, and the plot scale.

Example 7 *General*

You will plot the drawing shown in Figure 2-57 using the **Window** option to select the area to plot. The drawing was drawn in **Example 3** of this chapter and here it is assumed to be open on the screen. Assume that AutoCAD is configured for two output devices: Default System Printer and **HP Laserjet 2100 Series PS.**

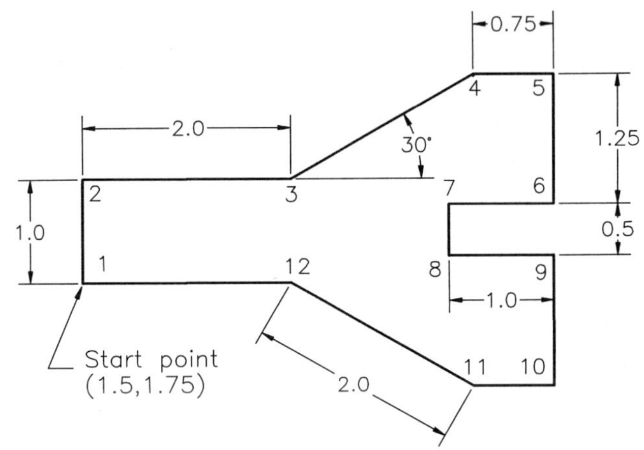

Figure 2-57 *Drawing for Exercise 6*

1. Invoke the **Plot** dialog box from the **Standard** toolbar, or the **File** menu (choose **Plot**), or by entering **PLOT** at the Command prompt. You can also invoke it by choosing **Plot** from the shortcut menu, which is displayed by right-clicking on the **Model/Layout** tabs.

2. Choose the **Plot Device** tab if not already open, Figure 2-58. Under **Plotter configuration** area the name of the default system printer will be displayed (in this example it is **HP Laser jet 2100 Series PS**). You can use any other printer by selecting the name of the device from the drop-down list displayed when you select the arrow button in the **Name** edit box. This displays the list of the currently configured devices. Related information about the selected device is displayed below this box.

Note

The default system printer varies from system to system. It is the printer chosen as the default printer while configuring your system for the output device.

3. Now choose the **Plot Settings** tab and here, choose the **Window** button in the **Plot area**. You can select the object for plotting by creating a window around it on the screen. When you choose the **Window** button, the dialog box will temporarily disappear and the drawing area will appear. Now, select the first and second corners (Points **P1** and **P2**) and specify the plot area (the area you want to plot). Once you have defined the two corners, the **Plot** dialog box will reappear.

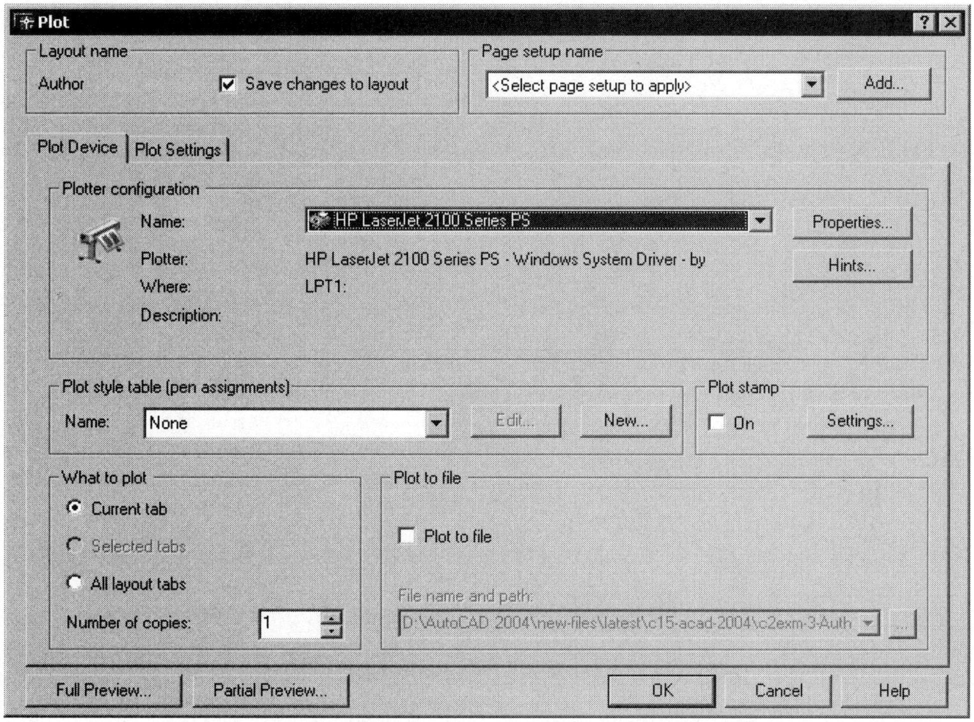

***Figure 2-58 Plot** dialog box with the **Plot Device** tab*

4. To set the size for the plot, you can select a size from the **Paper size** drop-down list in the **Paper size and paper units** area (Figure 2-59), which lists all the plotting sizes that the present plotter can support. You can select any one of the sizes listed in the dialog box or specify a size (width and height) of your own through the **Plotter Manager**. (This option is discussed later in Chapter 12, Plotting Drawings.) Once you select a size, you can also select the orientation by choosing either the **Landscape** or **Portrait** radio buttons under the **Drawing orientation** area. The sections in the **Plot** dialog box pertaining to paper size and orientation are automatically revised to reflect the new paper size and orientation. In this example you will specify Paper size **A4** and Orientation **Portrait**.

5. You can also modify values for **Plot offset**; the default values for X and Y are 0. For this example you can select the **Center the plot** check box to get the drawing in the center of the paper. Similarly, you can enter values for **Plot scale**. Open the drop-down list available in the **Plot scale** area to display the various scale factors. From this list you can select a scale factor you want to use. For example if you select the scale factor 1/4"=1'-0", the edit boxes available below the drop-down list will show 1 mm = 48 units. If you want the drawing to be plotted so that it fits on the specified sheet of paper, select the **Scaled to Fit** option. When you select this option, AutoCAD will determine the scale factor and display the scale factor in the edit boxes. In this example, you will plot the drawing so that it scales to fit the paper. Therefore, select the **Scaled to Fit** option and notice the change in the edit boxes. You can also enter your own values in the edit boxes.

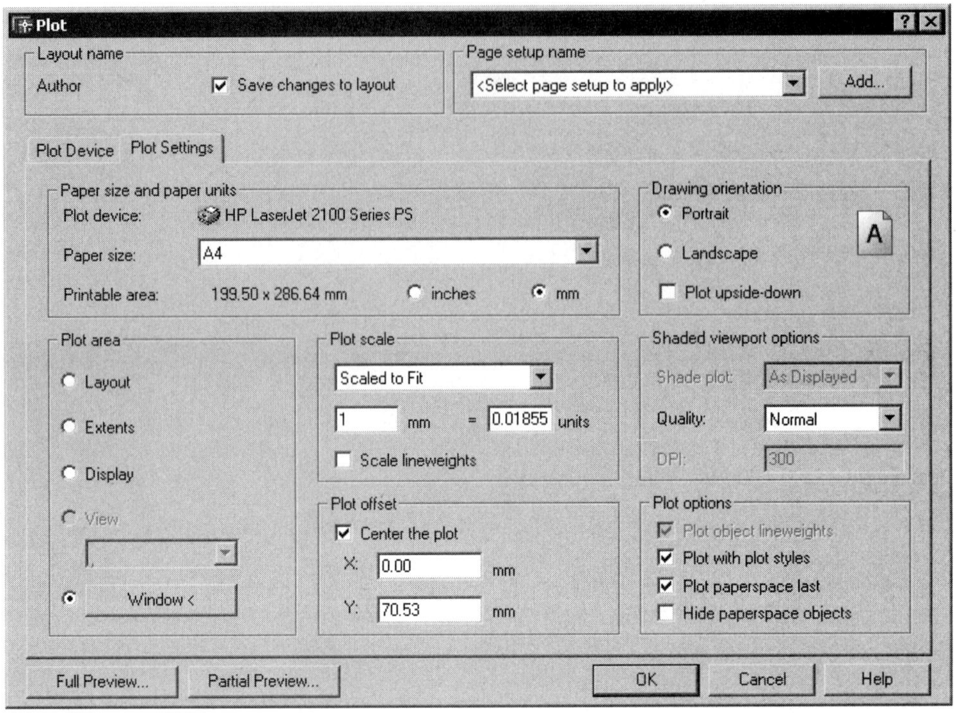

Figure 2-59 ***Plot*** *dialog box with the* ***Plot Settings*** *tab*

6. You can view the plot on the specified paper size before actually plotting it by choosing the Preview buttons in the **Plot** dialog box. This way you can save time and stationery. AutoCAD provides two types of Plot Previews, partial and full. To generate a partial preview of a plot, choose the **Partial Preview** button. The **Partial Plot Preview** dialog box is displayed as shown in Figure 2-60. To display the drawing just as it would be

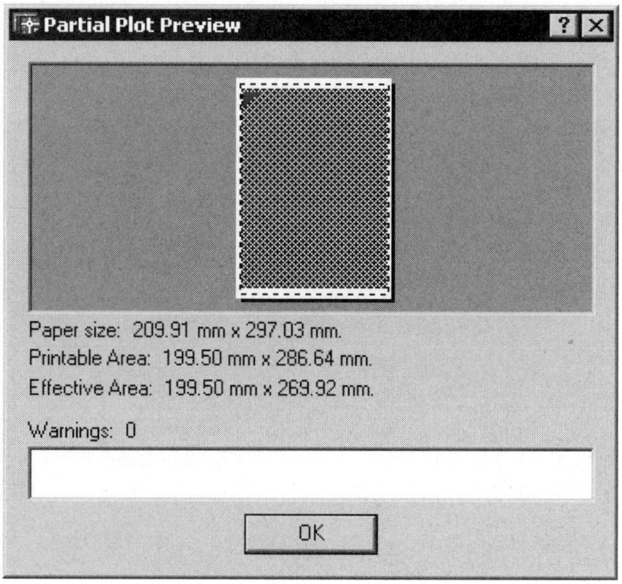

***Figure 2-60 Partial Plot Preview** dialog box*

plotted on the paper, choose the **Full Preview** button. **Full Preview** takes more time than **Partial Preview** because the drawing is regenerated. Once regeneration is complete, the preview image is displayed on the screen, Figure 2-61. Here, in place of the cursor, a realtime zoom icon is displayed. You can hold the pick button of your pointing device and then move it up to zoom into the preview image and move the cursor down to zoom out of the preview image.

7. If the plot preview is satisfactory, you can directly plot your drawing by choosing **Plot** from the shortcut menu, which is displayed by right-clicking. If you want to make some changes in the settings choose **Exit** in the shortcut menu or press the ESC or the ENTER key to get back to the dialog box. You can choose the **OK** button in the dialog box to plot the drawing.

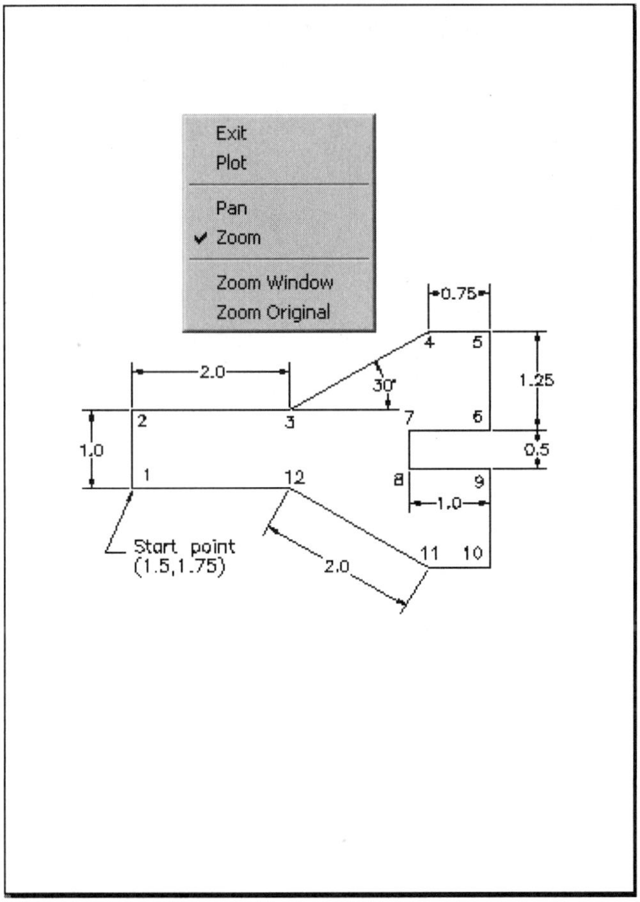

Figure 2-61 Full plot preview with the shortcut menu

MODIFYING AutoCAD SETTINGS USING THE OPTIONS DIALOG BOX

Menu:	Tools > Options
Command:	OPTIONS

You can use the **Options** dialog box to change the settings that affect the drawing environment or the AutoCAD interface. This dialog box is displayed by choosing **Options** from the **Tools** menu, or by entering **OPTIONS** at the command prompt. You can also invoke this command by choosing **Options** from the shortcut menu, which is displayed by right-clicking in the command window or in the drawing area when no command is active or no object is selected, see Figure 2-62.

You can change the AutoCAD default settings and customize them to your requirements

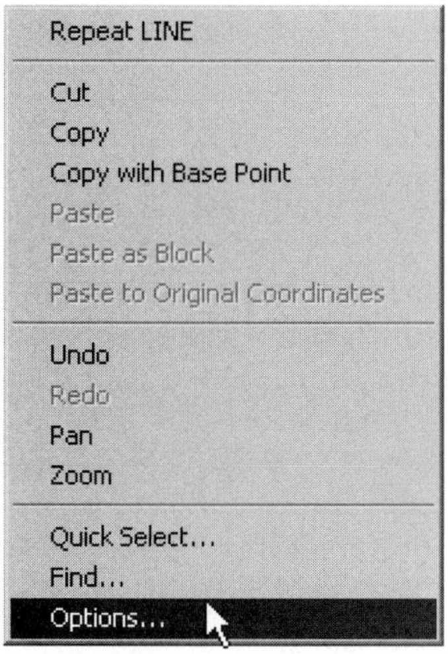

***Figure 2-62** Options in the drawing area shortcut menu*

using the **Options** dialog box (Figure 2-63). For example, you can use this dialog box to turn off the settings to display the shortcut menu by right-clicking or specify the support directories that contain the files you need.

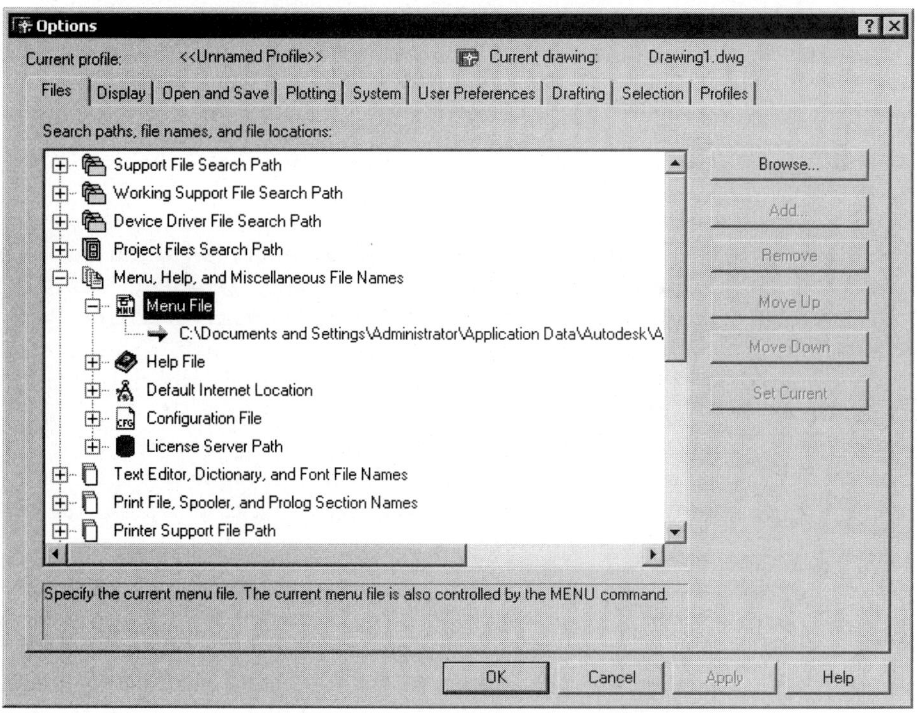

***Figure 2-63 Options** dialog box (**Files** tab)*

The dialog box contains nine tabs that display the sections to change the various environmental aspects. The Current profile: name and the Current drawing: name are displayed on the top, above the tabs. You can save a set of custom settings in a profile to be used later for other drawings. If you do not specify a profile, the current settings are stored in an Unnamed Profile. The various tabs available in the **Options** dialog box are discussed next.

Files

This tab stores the directories in which AutoCAD looks for driver, support, menu, project, template, and other files. It uses three icons: folder, paper stack, and file cabinet. The folder icon is for search path, the paper stack icon is for files, and the file cabinet icon is for a specific folder. Suppose you want to know the path of the menu file. You can select the **Menu, Help, and Miscellaneous File Names** folder and then select the **Menu File** icon to display the path (Figure 2-63). Similarly, you can define a custom hatch pattern file and then add its search path. This way AutoCAD can locate the custom hatch pattern.

Display

This tab controls drawing and window settings like screen menu display and scroll bar. For example, if you want to display the screen menu, select the **Display screen menu** check box in the **Window Elements** area. You can also change the color of the graphics window background, layout window background, command line background, and also the color of the command line text using the **Colors** button in the **Window Elements** area. This tab also allows you to modify the display resolution and display performance. You can also set the smoothness and resolutions of certain objects such as the circle, arc, rendered object, and polyline curve. Here you can toggle on and off the various layout elements such as the layout tabs on the screen, margins, paper background, and so on. You can also toggle on and off the display performance such as the pan and zoom with raster images, apply the solid fills, and so on.

Open and Save

This tab controls the parameters related to opening and saving of files in AutoCAD. You can specify the file type for saving while using the SAVEAS command. The various formats are **AutoCAD 2000/LT2000 Drawing (*.dwg)**, **AutoCAD Drawing Template(*.dwt)**, **AutoCAD 2004 DXF (*.dxf)**, **AutoCAD 2000/LT2000 DXF(*.dxf)**, **AutoCAD R12/LT2 DXF (*.dxf)**, and so on. You can also choose the various file safety precautions such as the Automatic Save feature, or the creation of a backup copy. You can add a password and digital signatures to your drawing while saving using the **Security Options** button in the **File Safety Precautions** area. You can control the display of the digital signature information when a file with a valid digital signature is opened with the help of the **Display digital signature information** check box. You can change the number of recently saved files to be displayed in the **File** menu for opening. You can also set the various parameters for External references and the ObjectARX Applications.

Plotting

The **Plotting** tab controls the parameters related to the plotting of drawings in AutoCAD. You can set the Default output device and also add a new plotter. You can set the general parameters such as the layout or plot device paper size. It is possible to select the spool alert for the system printer and also the OLE plot quality. You can also set the parameters for the plot style such as using the color-dependent plot styles or the named plot styles.

System

This tab contains AutoCAD system settings options such as the 3D graphics display and pointing device settings options where you can choose the pointing device driver. Here you can also set the various system parameters such as the single drawing mode instead of MDE, the display of the **Startup** option while opening a new session of AutoCAD and the **OLE Properties** dialog box, and beep for wrong user input. You also have options to set the parameters for database connectivity.

User Preferences

This tab controls settings that depend on the way the user prefers working on AutoCAD, such as the Right-click Customization where you can change the shortcut menus. You can set the units parameters for the DESIGNCENTER as well as the priorities for various data entry methods. Here it is possible to set the order of object sorting methods and also set the lineweight options.

Drafting

This tab controls settings such as the Autosnap Settings and the Aperture Size. Here you can also set the toggles on and off for the various Autotracking settings.

Selection

This tab controls settings related to the methods of object selection such as the Grips, which enables you to change the various grip colors and the grip size. You can also set the toggles on or off for the various selection modes.

Profiles

This tab saves and restores the system settings. To set the Profile in the **Options** dialog box, choose the **Profile** tab and then choose the **Add to List** button. The **Add Profile** dialog box appears. Enter the Profile name and description. Then choose the **Apply & Close** button. Next, make the new profile current and then specify the settings. The settings are saved in the new profile and can be restored anytime by making the profile current.

Note

*The options in the various tabs of the **Options** dialog box have been discussed throughout the book wherever applicable.*

Tip

*Some options in the various tabs of the **Options** dialog box have a drawing file icon in front of them. For example, the options in the **Display resolution** of the **Display** tab have the drawing file icons. This specifies that these parameters are saved with the current drawing only and therefore affects it. The rest of the options (without the drawing file icon) are saved with the current profile and also affect all the drawings present in that AutoCAD session.*

Example 8 *General*

In the following example you will use the **Options** dialog box to create a profile that contains the specified settings.

1. To invoke the **Options** dialog box, choose **Options** from the **Tools** menu. You can also right-click in the Command window to display the shortcut menu and choose **Options**. Typing **OPTIONS** in the Command line also displays the **Options** dialog box.

2. In the **Options** dialog box choose the **Profiles** tab.

3. Choose the **Add to List** button to display the **Add Profile** dialog box. Enter the name of the new profile (**Myprofile1**) and a description of the new profile, and then choose the **Apply and Close** button to exit.

4. Select **Myprofile1** profile and then choose the **Set Current** button to make **Myprofile1** current. You will notice that the **Current Profile** name above the tabs changes from Unnamed to **Myprofile1**.

5. Choose the **Display** tab and then select the **Display screen menu** check box.

6. Now, choose the **Colors** button. The **Color Options** dialog box is displayed. Change the color of the graphics window background (displayed as **Model tab background** in the Window Element edit box) to white (Figure 2-64). Choose the **Apply & Close** button to return to the **Options** dialog box. Notice the changes in screen color to white.

7. Choose the **Plotting** tab and then select the **Add or Configure Plotters** button to add another printer. When you choose this button, the **Plotters** window is displayed on the screen. Double-click on the **Add-A-Plotter Wizard** icon, which guides you through the process of selecting and adding a plotter. When you choose the **Finish** button in the **Add-Plotter** wizard, an icon for the newly configured plotter appears in the **Plotters** window. Close this box. From the **Use as default output device** drop-down list, select a printer.

8. Choose the **Drafting** tab and change the AutoSnap Marker Size to the maximum. Choose the **Apply** button and then the **OK** button to exit.

9. Notice that the screen menu is displayed on the right side of the screen. Also to see the AutoSnap Marker size, draw a line and then choose the Osnap button in the status bar. Again invoke the LINE command and move the cursor on the previously drawn line; a marker is displayed and the endpoint.

10. From the shortcut menu invoke the **Options** dialog box again or choose **Options** from the **Tools** menu.

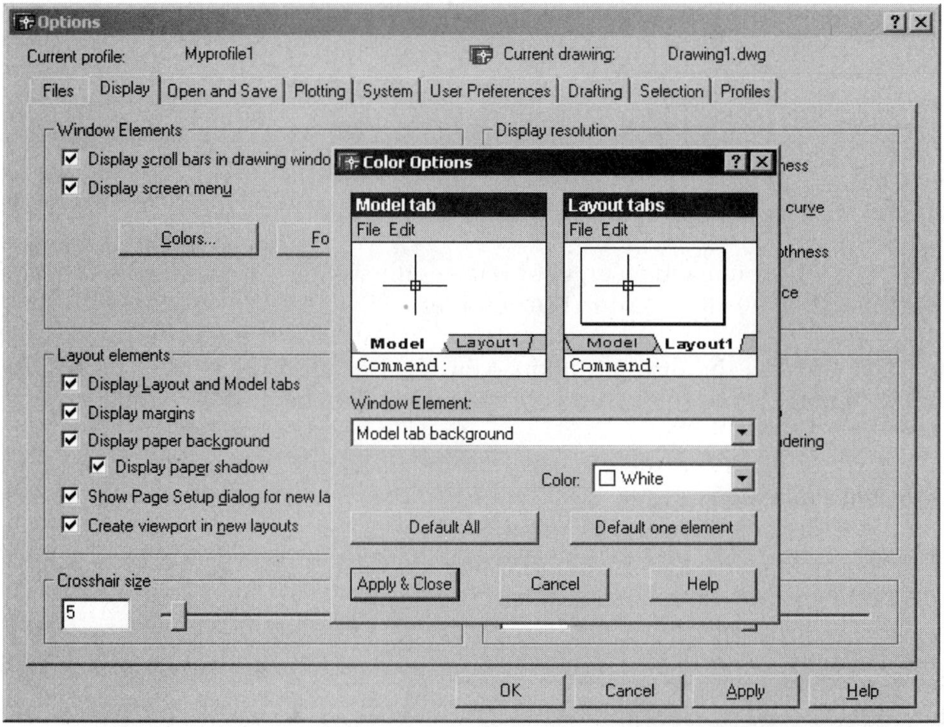

Figure 2-64 ***Color Options*** *dialog box with the* ***Options*** *dialog box*

11. Choose the **Profiles** tab and then select the default profile (<<Unnamed Profile>>). Choose the **Set Current** button to load the default settings. The screen settings will change as specified in the default profile.

12. Select the **Myprofile1** profile and then choose the **Set Current** button to change the settings assigned to **Myprofile1** profile.

Self-Evaluation Test

Answer the following questions and then compare your answers to the correct answers given at the end of this chapter.

1. You can draw a line by specifying the length of the line and its direction, using **Direct distance Entry**. (T/F)

2. Using the **Crossing** method of object selection, the objects to be selected should be completely enclosed within the boundaries of the crossing box. (T/F)

3. The **Three-Point** option of the **CIRCLE** command lets you draw the circle by specifying the two endpoints of the circle's diameter. (T/F)

4. If you choose the engineering or architectural format for units in the **Drawing Units** dialog box, you can enter the distances or coordinates in any of the five formats. (T/F)

5. You can erase a previously drawn line using the __________ option of the **LINE** command.

6. The __________option of the **CIRCLE** command can be used to draw a circle, if you want the circle to be tangent to two previously drawn objects.

7. The __________ command enlarges or reduces the view of the drawing on the screen, but it does not affect the actual size of the entities.

8. After you increase the drawing limits, you need to use the __________ option of the **ZOOM** command to display the complete area inside the drawing area.

9. In __________ units you must specify the bearing angle that the line makes with the north-south direction.

10. For plotting a drawing, __________ takes more time than __________ because the drawing is regenerated.

Review Questions

Answer the following questions.

1. In the **Relative rectangular** coordinate system, the displacements along the *X* and *Y* axes (DX and DY) are measured with reference to the previous point rather than to the origin. (T/F)

2. In AutoCAD, by default the angles are positive if measured in the counterclockwise direction and the angles are measured from the positive *X* axis. (T/F)

3. You can also invoke the **PLOT** command by choosing **Plot** from the shortcut menu, which is displayed by right-clicking on the Command window. (T/F)

4. The **Files** tab of the **Options** dialog box stores the directories in which AutoCAD looks for driver, support, menu, project, template, and other files. (T/F)

6. You cannot terminate the **LINE** command by pressing which of the following key on the keyboard at the **Specify next point or [Close/Undo]:** prompt?

    (a) SPACEBAR (b) BACKSPACE
    (c) ENTER (d) ESC

6. Which of the following options of the **ZOOM** command zooms to the drawing limits or the extents, whichever is greater?

   (a) **Previous** (b) **Window**
   (c) **All** (c) **Realtime**

7. How many formats of units can you choose from in the **Drawing Units** dialog box?

   (a) Three (b) Five
   (c) Six (d) Seven

8. Which of the following input methods cannot be used to invoke the **OPTIONS** command, which displays the **Options** dialog box?

   (a) Menu (b) Toolbar
   (c) Shortcut menu (d) Command prompt

9. When you define the direction by specifying the angle, the output of the angle does not depend on which one of the following factors.

   (a) Angular units (b) Angle value
   (c) Angle direction (d) Angle base

10. The __________ option of the **LINE** command can be used to join the current point with the initial point of the first line when two or more lines are drawn in continuation.

11. The __________ system variable stores the default value of the radius of the circle.

12. When you select any type of unit and angle in the **Length** or **Angle** area of the **Drawing Units** dialog box, the corresponding example is displayed in the __________ area of the dialog box.

13. If you want the drawing to be plotted so that it fits on the specified sheet of paper, select the __________ option in the **Plot** dialog box.

14. The __________ tab in the **Options** dialog box stores the directories in which AutoCAD looks for driver, support, menu, project, template, and other files.

15. You can use the __________ command to change the settings that affect the drawing environment or the AutoCAD interface.

Exercises

Exercise 7

General

Use the following relative rectangular and absolute coordinate values in the **LINE** command to draw the object.

Point	Coordinates	Point	Coordinates
1	3.0, 3.0	5	@3.0,5.0
2	@3,0	6	@3,0
3	@-1.5,3.0	7	@-1.5,-3
4	@-1.5,-3.0	8	@-1.5,3

Exercise 8

General

For Figure 2-65, enter the relative rectangular and relative polar coordinates of the points in the following table, and then use these coordinates to draw the figure. The distance between the dotted lines is 1 unit. Save this drawing as *Exer8.dwg*.

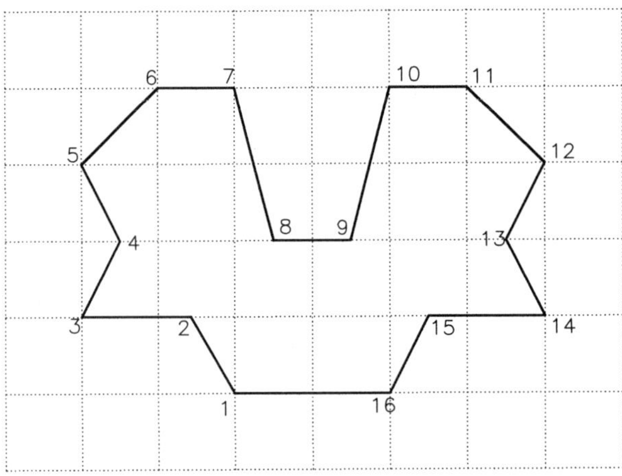

Figure 2-65 *Drawing for Exercise 8*

1	3.0, 1.0	9	______
2	______	10	______
3	______	11	______
4	______	12	______
5	______	13	______
6	______	14	______
7	______	15	______
8	______	16	______

Exercise 9

Mechanical

For Figure 2-66, enter the relative polar coordinates of the points in the following table. Then use these coordinates to draw the figure. Do not draw the dimensions.

Point	Coordinates	Point	Coordinates
1	1.0, 1.0	6	________
2	________	7	________
3	________	8	________
4	________	9	________
5	________		

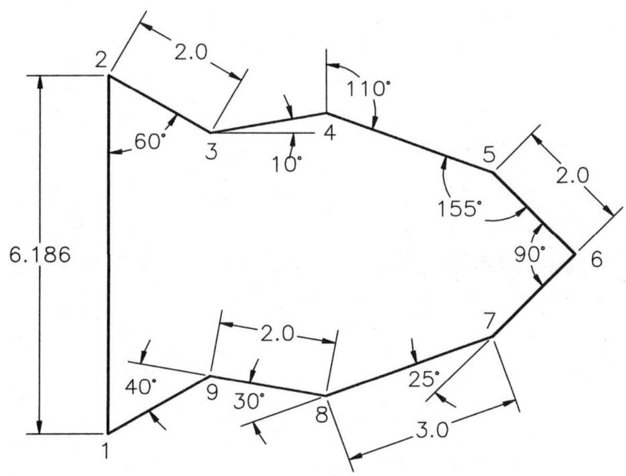

Figure 2-66 *Drawing for Exercise 9*

Exercise 10

Mechanical

Draw Figure 2-67, using the **LINE** and **CIRCLE** commands. The distance between the dotted lines is 1.0 units.

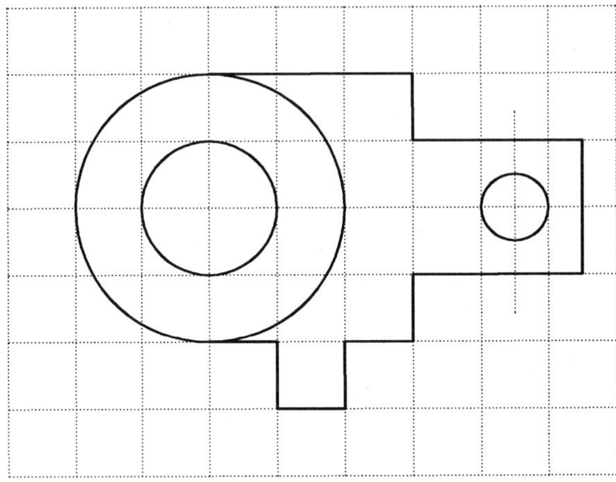

Figure 2-67 *Drawing for Exercise 10*

Exercise 11 *Mechanical*

Draw Figure 2-68, using the **LINE** command and the **Ttr** option of the **CIRCLE** command. Set the units for a drawing according to the following specifications:

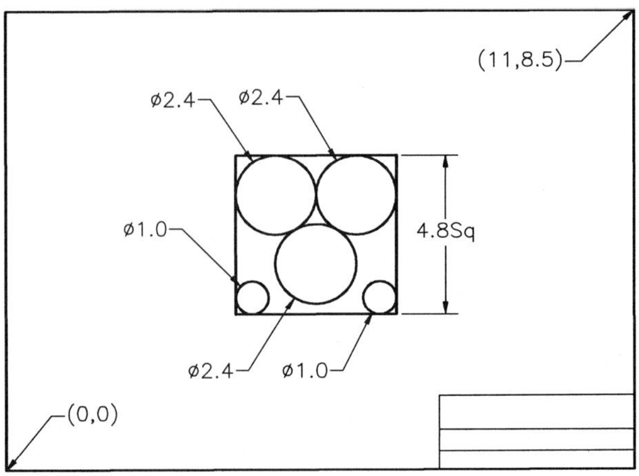

Figure 2-68 *Drawing for Exercise 11*

Exercise 12 *Mechanical*

1. Set the **UNITS** to architectural, with the denominator of the smallest fraction equal to 16.
2. Set the angular measurement to degrees/minutes/seconds, with the number of fractional places for display of angles equal to 0d00'.

3. Set the direction to 0-degree (east) and the direction of measurement of angles to counterclockwise (angles measured positive in a counterclockwise direction).

Based on Figure 2-69, determine and set the limits of the drawing. The scale for this drawing is 1/4" = 1'. Leave enough space around the drawing for dimensioning and title block. (HINT: Scale factor = 48; sheet size required is 12 x 9; therefore, the limits are 12 X 48, 9 X 48 = 576, 432. Use the **ZOOM** command and then select the **All** option to display the new limits.)

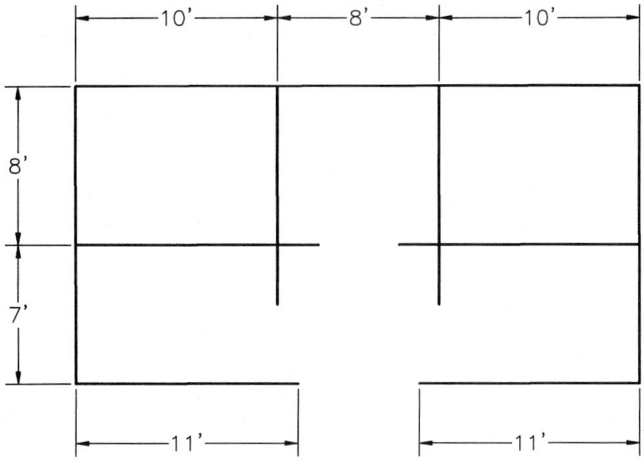

Figure 2-69 *Drawing for Exercise 12*

Exercise 13 *Mechanical*

Draw the object shown in Figure 2-70. The distance between the dotted lines is 10 feet. Determine the limits for this drawing and use the Architectural units with 0'-01/32" precision.

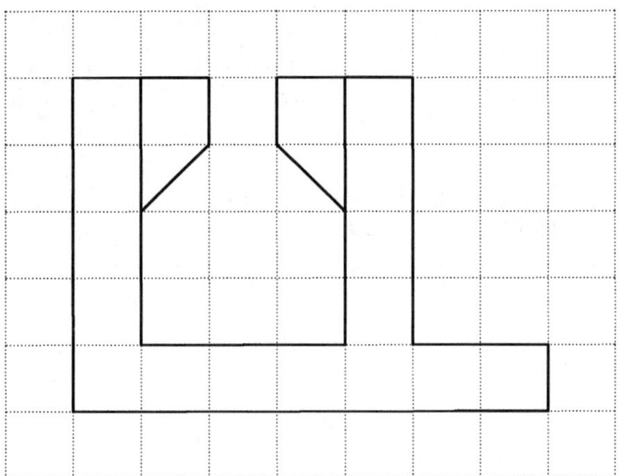

Figure 2-70 *Drawing for Exercise 13*

Exercise 14 *General*

Draw the object shown in Figure 2-71. The distance between the dotted lines is 5 inches. Determine the limits for this drawing and use the Fractional units with 1 1/16 precision.

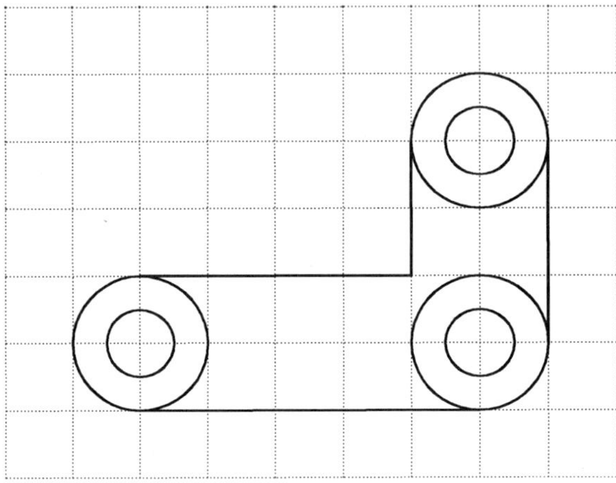

Figure 2-71 *Drawing for Exercise 14*

Exercise 15 *General*

Draw the object shown in Figure 2-72. The distance between the dotted lines is 1 unit. Determine the limits for this drawing and use the Decimal units with 0.00 precision.

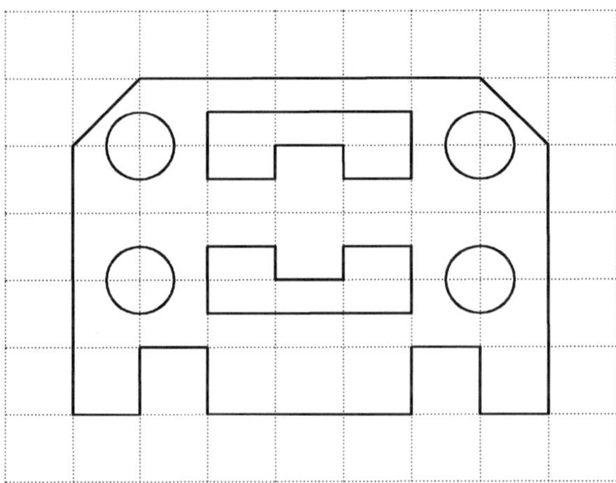

Figure 2-72 *Drawing for Exercise 15*

Exercise 16 *General*

Draw the object shown in Figure 2-73. The distance between the dotted lines is 10 feet. Determine the limits for this drawing and use the Engineering units with 0'0.00" precision.

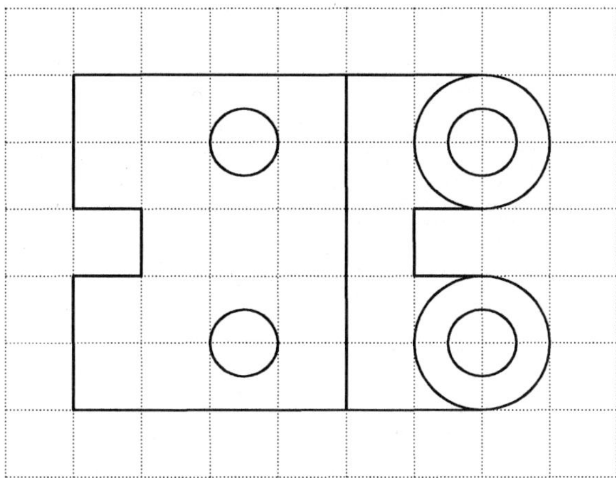

Figure 2-73 *Drawing for Exercise 16*

Problem Solving Exercise 1 *Mechanical*

Draw the object shown in Figure 2-74, using the **LINE** and **CIRCLE** commands. In this exercise only the diameters of the circles are given. To draw the lines and small circles (Dia 0.6), you need to find the coordinate points for the lines and the center points of the circles. For example, if the center of concentric circles is at 5,3.5, then the *X* coordinate of the lower left corner of the rectangle is 5.0 - 2.4 = 2.6.

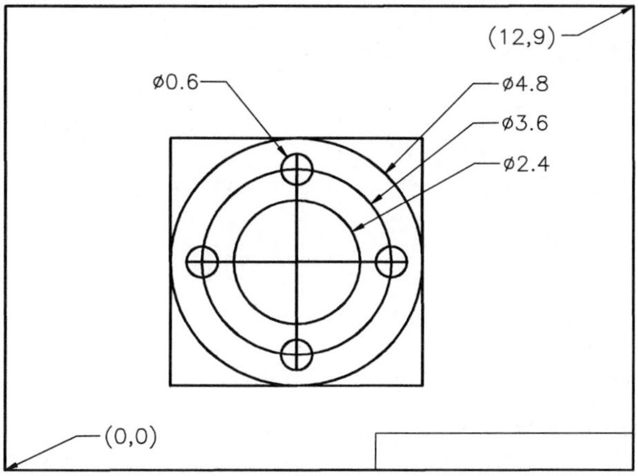

Figure 2-74 *Drawing for Problem Solving Exercise 1*

Problem Solving Exercise 2 *Mechanical*

Draw the object shown in Figure 2-75 using various options of the **CIRCLE** and **LINE** commands. In this exercise you have to find the coordinate points for drawing the lines and circles. Also, you need to determine the best and easiest method to draw the 0.85 diameter circles along the outermost circle.

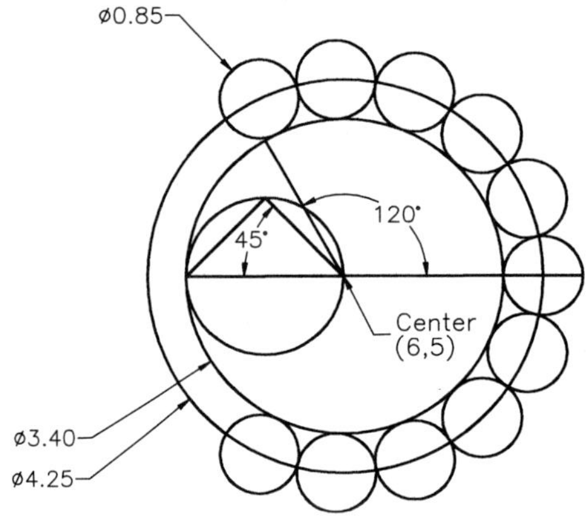

Figure 2-75 *Drawing for Problem Solving Exercise 2*

Problem Solving Exercise 3 *Mechanical*

Draw the drawing in Figure 2-76 using the absolute, relative rectangular, or relative polar coordinate system. Draw according to the dimensions shown in the figure, but do not draw the dimensions.

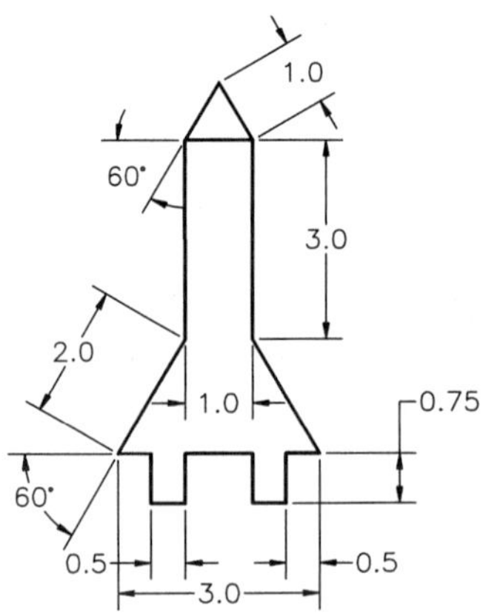

Figure 2-76 *Drawing for Problem Solving Exercise 3*

Answers to Self-Evaluation Test

1 - T, **2** - F, **3** - F, **4** - T, **5** - Undo, **6** - Tangent tangent radius, **7** - **ZOOM**, 8 - All, **9** - Surveyor's, **10** - **Full preview, Partial Preview**

Chapter 3

Drawing Sketches

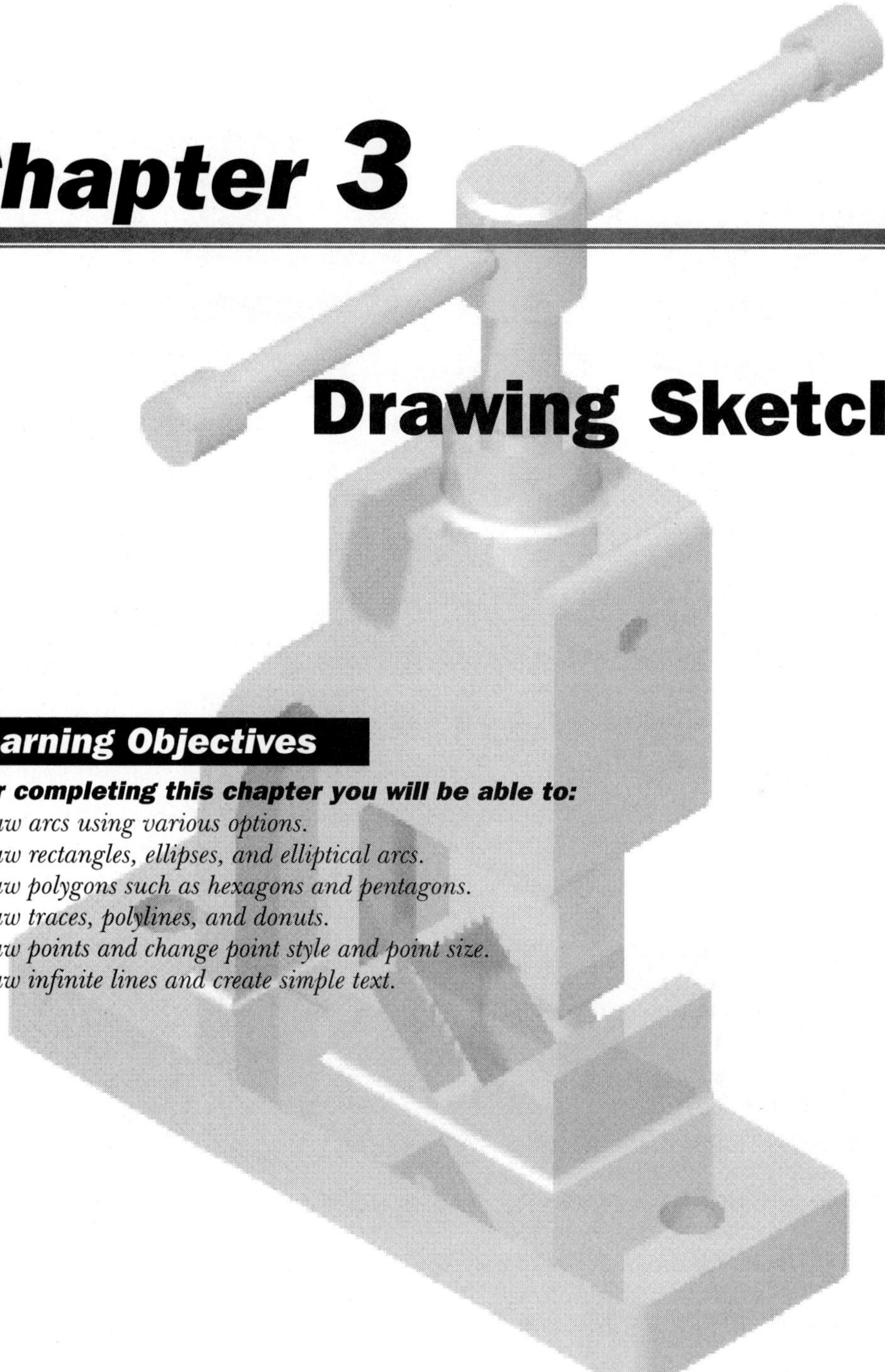

Learning Objectives

After completing this chapter you will be able to:

- *Draw arcs using various options.*
- *Draw rectangles, ellipses, and elliptical arcs.*
- *Draw polygons such as hexagons and pentagons.*
- *Draw traces, polylines, and donuts.*
- *Draw points and change point style and point size.*
- *Draw infinite lines and create simple text.*

DRAWING ARCS

Toolbar:	Draw > Arc
Menu:	Draw > Arc
Command:	ARC or A

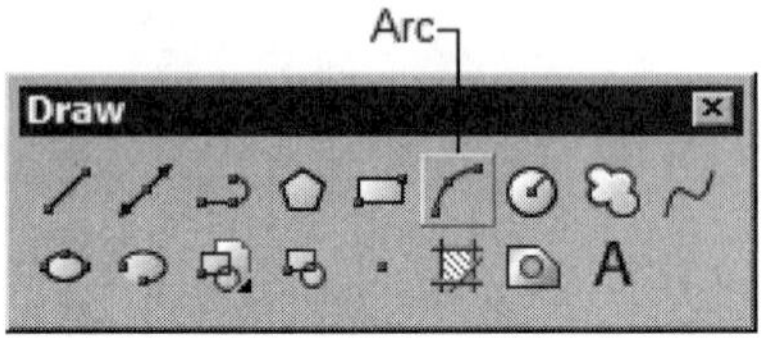

Figure 3-1 *Invoking the* ***ARC*** *command from the* ***Draw*** *toolbar*

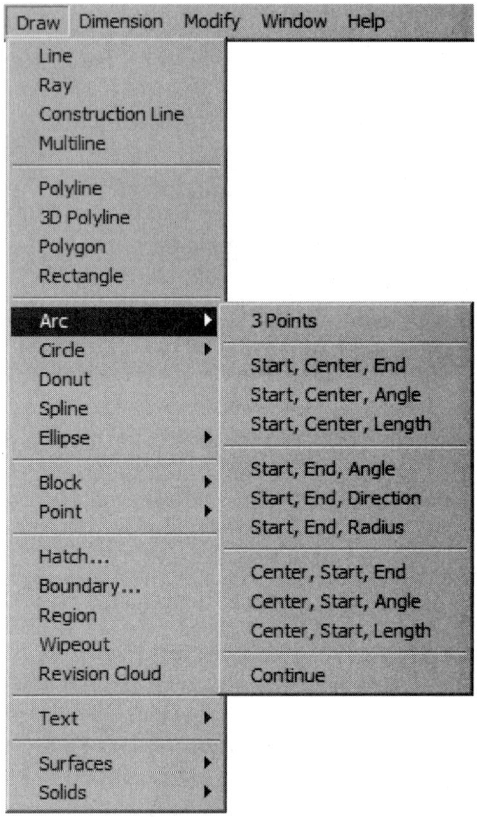

Figure 3-2 *Invoking the* ***ARC*** *command from the* ***Draw*** *menu*

An arc is defined as a part of a circle; it can be drawn using the **ARC** command (Figure 3-1). An arc can be drawn in **11** distinct ways using the options listed under the **ARC** command (Figure 3-2). The default method for drawing an arc is the **3 Points** option. Other options can be invoked by entering the appropriate letter to select an option. The last parameter to be specified in any arc generation is automatically dragged into the relevant location. This is dependent on the **DRAGMODE** variable, which should be set to **Auto** (default).

The 3 Points Option

When you choose the **Arc** button from the **Draw** toolbar, or enter **ARC** at the Command prompt, you automatically enter the **3 Points** option. The 3 Points option requires the start point, the second point, and the endpoint of the arc (Figures 3-3 and Figure 3-4). The arc can be drawn in a clockwise or counterclockwise direction by dragging the arc with the cursor. The following is the prompt sequence to draw an arc with a start point at (2,2), second point at (3,3), and an endpoint at (3,4). (You can also specify the points by moving the cursor and then specifying points on the screen.)

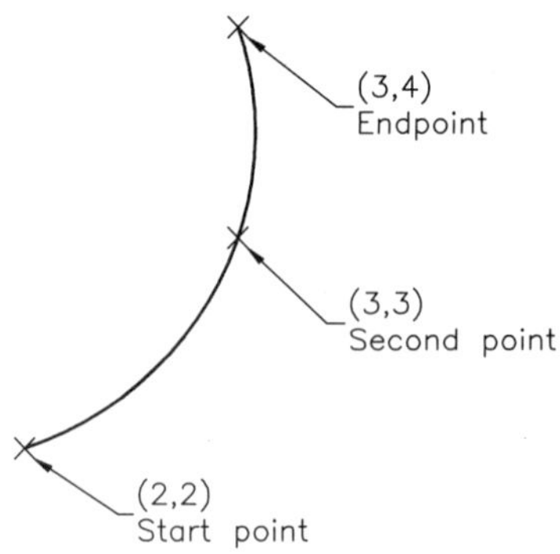

Figure 3-3 *Drawing an arc using the* ***3 Points*** *option*

Command: *Choose the* ***Arc*** *button from the* ***Draw*** *toolbar (****3 Points*** *is the default option).*

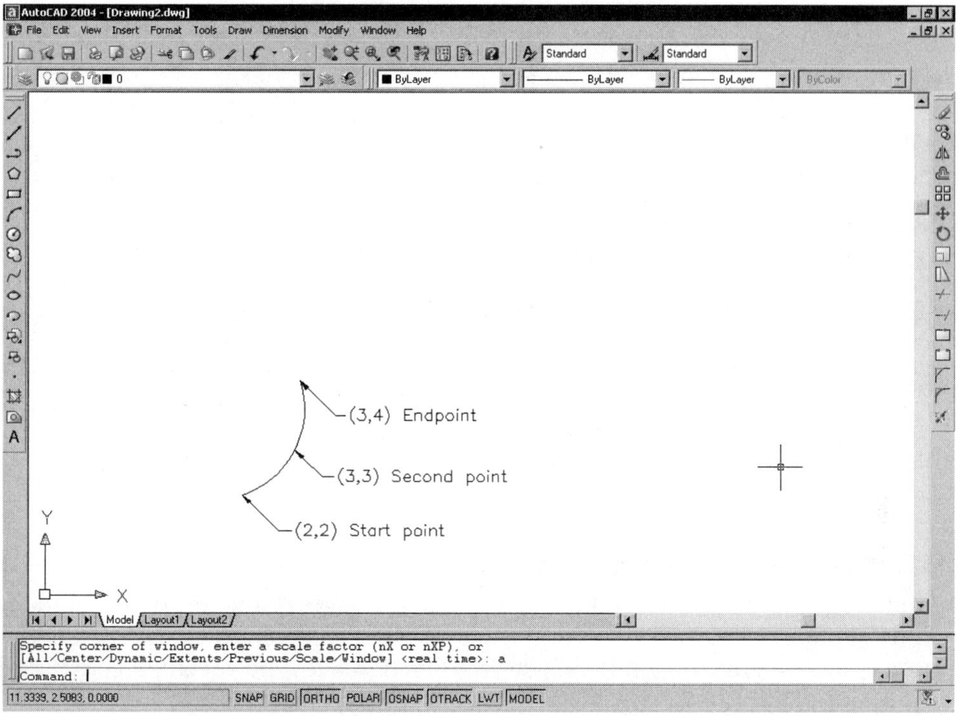

***Figure 3-4** Drawing an arc using the **3 Points** option*

Specify start point of arc or [Center]: **2,2** Enter
Specify second point of arc or [Center/End]: **3,3** Enter
Specify end point of arc: **3,4** Enter

Exercise 1 *General*

Draw several arcs using the **3 Points** option. The points can be selected by entering coordinates or by specifying points on the screen. Also, try to create a circle by drawing two separate arcs and by drawing a single arc. Notice the limitations of the **ARC** command.

The Start, Center, End Option

This option is slightly different from the 3 Points option. In this option, instead of entering the second point, you enter the center of the arc. Choose this option when you know the start point, endpoint, and center point of the arc. The arc is drawn in a counterclockwise direction from the start point to the endpoint around the specified center. The endpoint specified need not be on the arc and is used only to calculate the angle at which the arc ends. The radius of the arc is determined by the distance between the center point and the start point. The prompt sequence for drawing an arc with a start point of (3,2), center point at (2,2), and endpoint of (2,3.5) (Figure 3-5) is as follows.

Command: **ARC** [Enter]
Specify start point of arc or [Center]: **3,2** [Enter]
Specify second point of arc or [Center/End]: **C** [Enter]
Specify center point of arc: **2,2** [Enter]
Specify end point of the arc or [Angle/chord Length]: **2,3.5** [Enter]

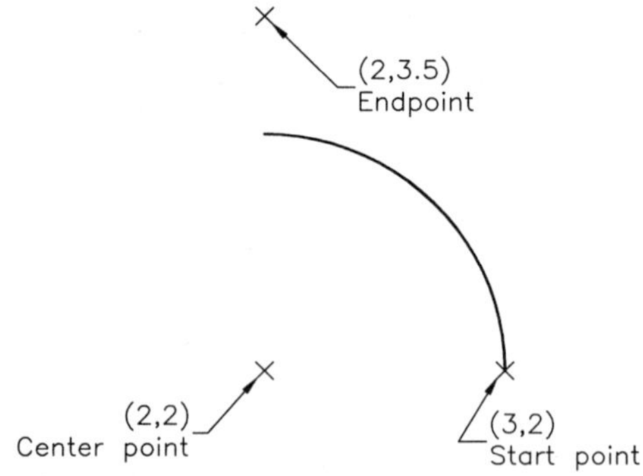

Figure 3-5 Drawing an arc using the ***Start, Center, End*** *option*

Note

If you choose the ***Start, Center, End*** *option from the* ***Draw > Arc*** *menu, after entering the start point, the center option is automatically invoked. This way you bypass the prompt* ***Specify second point of arc or [Center/End]****. You need to simply specify the center point.*

The Start, Center, Angle Option

This option is the best choice if you know the **included angle** of the arc. The included angle is the angle formed by the start point and the endpoint of the arc with the specified center. This option draws an arc in a counterclockwise direction with the specified center and start point spanning the indicated angle (Figure 3-6). If the specified angle is negative, the arc is drawn in a clockwise direction (Figure 3-7).

The prompt sequence for drawing an arc with center at (2,2), a start point of (3,2), and an included angle of 60-degree (Figure 3-6) is given next.

Command: **ARC** [Enter]
Specify start point of arc or [Center]: **3,2** [Enter]
Specify second point of arc or [Center/End]: **C** [Enter]
Specify center point of arc: **2,2** [Enter]
Specify end point of the arc or [Angle/chord Length]: **A** [Enter]
Specify included angle: **60** [Enter]

You can draw arcs with negative angle values in the Start, Center, Included Angle (St,C,Ang) option by entering "-" (negative sign) followed by the angle values of your requirement at the **Specify included angle:** prompt (Figure 3-7). The prompt sequence is given next.

Command: **ARC** [Enter]
Specify start point of arc or [Center]: **4,3** [Enter]
Specify second point of arc or [Center/End]: **C** [Enter]
Specify center point of arc: **3,3** [Enter]
Specify end point of the arc or [Angle/ chord Length]: **A** [Enter]
Specify included angle: **-180** [Enter]

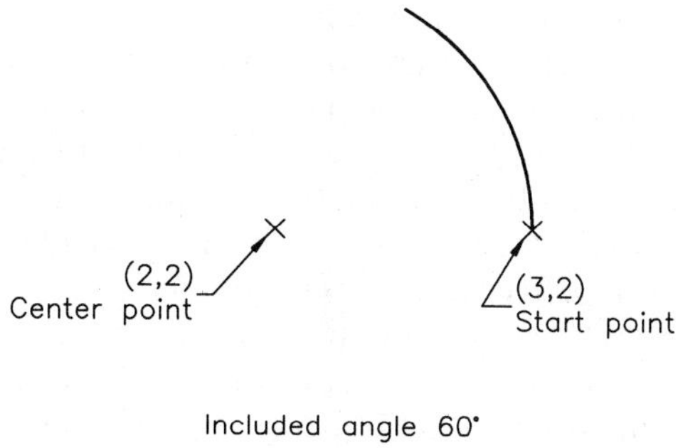

Figure 3-6 *Drawing an arc using the* ***Start, Center, Angle*** *option*

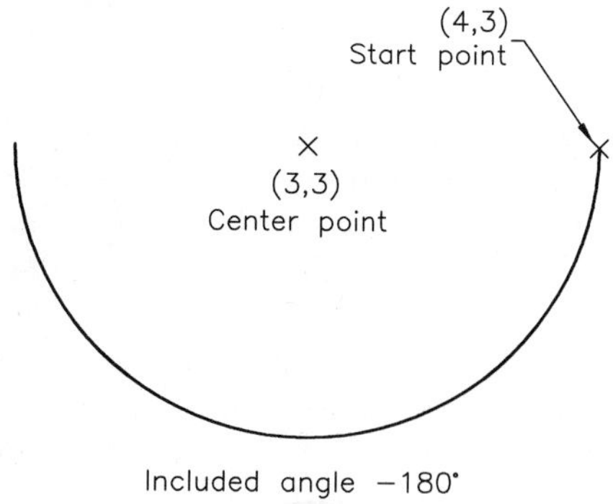

Figure 3-7 *Drawing an arc using a negative angle in the* ***Start, Center, Angle*** *option*

Note

If you choose this option from the ***Draw*** *menu, the center and the angle options are invoked and you simply have to specify the start point, center point, and the angle.*

To invoke the rest of the options, you will use the ***Draw*** *menu.*

Exercise 2

Mechanical

a. Draw an arc using the **St,C,Ang** option. The start point is (6,3), the center point is (3,3), andthe angle is 240-degree.

b. Make the drawing shown in Figure 3-8. The distance between the dotted lines is 1.0 unit. Create the radii by using the arc command options as indicated in the drawing.

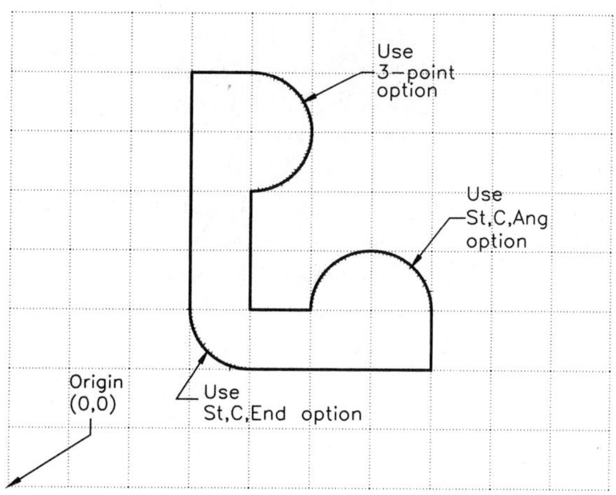

Figure 3-8 *Drawing for Exercise 2(b)*

The Start, Center, Length Option

In this option you are required to specify the start point, center point, and length of chord. A **chord** is defined as the straight line connecting the start point and the endpoint of an arc. The chord length needs to be specified so that AutoCAD can calculate the ending angle. Identical start, center, and chord length specifications can be used to define four different arcs. AutoCAD always draws this type of arc counterclockwise from the start point. Therefore, a positive chord length gives the smallest possible arc with that length. This is known as the minor arc. The minor arc is less than 180-degree. A negative value for chord length results in the largest possible arc, also known as the major arc. The chord length can be determined by using the standard chord length tables or using the mathematical relation (L = 2*Sqrt [h(2r-h)]). For example, an arc of radius 1 unit, with an included angle of 30-degree, has a chord length of 0.51764 units. The prompt sequence for drawing an arc that has a start point of (3,1), center of (2,2), and the chord length of (2) (Figure 3-9) is given next.

Command: *Choose the **Start, Center, Length** option from the **Draw > Arc** menu.*
Specify start point of arc or [Center]: **3,1** Enter
Specify second point of arc or [Center/End]: _c Specify center point of arc: **2,2** Enter
Specify end point of the arc or [Angle/ chord Length]: _l Specify length of chord: **2** Enter

You can draw the major arc by defining the length of the chord as negative (Figure 3-10). In this case the arc with a start point of (3,1), a center point of (2,2), and a negative chord length of (-2) is drawn with the following prompt sequence.

Command: *Choose the **Start, Center, Length** option from the **Draw > Arc** menu.*
Specify start point of arc or [Center]: **3,1** Enter
Specify second point of arc or [Center/End]: _c Specify center point of arc: **2,2** Enter
Specify end point of the arc or [Angle/ chord Length]: _l Specify length of chord: **-2** Enter

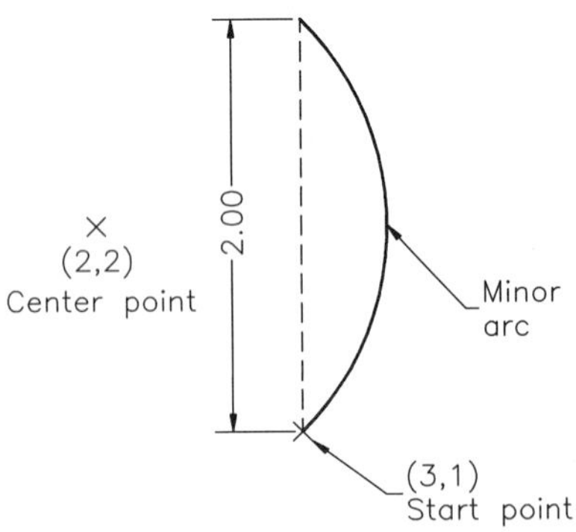

***Figure 3-9** Drawing an arc using the **Start, Center, Length** option*

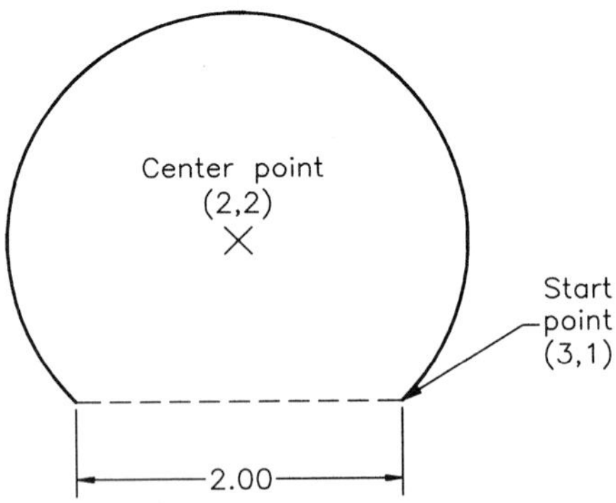

***Figure 3-10** Drawing an arc using a negative chord length in the **Start, Center, Length** option*

Note

The points that you have specified for drawing the arcs using different options are quite close to each other and therefore the arcs created might overlap each other. You can use the ERASE command to erase some of the previously drawn arcs and gain space to draw new arcs.

Exercise 3 — *General*

Draw a minor arc with the center point at (3,4), start point at (4,2), and chord length of 4 units.

The Start, End, Angle Option

With this option you can draw an arc by specifying the start point of the arc, the endpoint, and the included angle. A positive included angle value draws an arc in a counterclockwise direction from the start point to the endpoint, spanning the included angle; a negative included angle value draws the arc in a clockwise direction. The prompt sequence for drawing an arc with a start point of (3,2), endpoint of (2,4), and included angle of 120-degree (Figure 3-11) is given next.

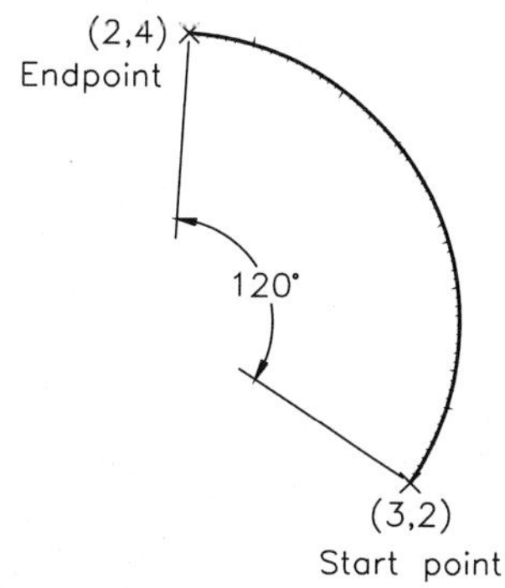

Figure 3-11 *Drawing an arc using the* ***Start, End, Angle*** *option*

Command: *Choose the* ***Start, End, Angle*** *option from the* ***Draw > Arc*** *menu.*
Specify start point of arc or [Center]: **3,2** Enter
Specify second point of arc or [Center/End]: _e Specify end point of arc: **2,4** Enter
Specify center point of arc or [Angle/Direction/Radius]: _a Specify included angle: **120** Enter

The Start, End, Direction Option

In this option you can draw an arc by specifying the start point, endpoint, and starting direction of the arc, in degrees. In other words, the arc starts in the direction you specify (the start of the arc is established tangent to the direction you specify). This option can be used to draw a major or minor arc, in a clockwise or counterclockwise direction. The size and position of the arc are determined by the distance between the start point and endpoint and the direction specified. To illustrate the positive direction option (Figure 3-12), the prompt sequence for an arc having a start point of (4,3), endpoint of (3,5), and direction of 90-degree is given next.

Command: *Choose the* ***Start, End, Direction*** *option from the* ***Draw > Arc*** *menu.*
Specify start point of arc or [Center]: **4,3** Enter
Specify second point of arc or [Center/End]: _e Specify end point of arc: **3,5** Enter
Specify center point of arc or [Angle/Direction/Radius]: _d Specify tangent direction for the start point of arc: **90** Enter

To illustrate the option of using a negative direction degree specification (Figure 3-13), the prompt sequence for an arc having a start point of (4,3), endpoint of (3,4), and direction of -90-degree is given next.

Command: *Choose the* ***Start, End, Direction*** *option from the* ***Draw > Arc*** *menu.*
Specify start point of arc or [Center]: **4,3** Enter
Specify second point of arc or [Center/End]: _e Specify end point of arc: **3,4** Enter
Specify center point of arc or [Angle/Direction/Radius]: _d Specify tangent direction for the start point of arc: **-90** Enter

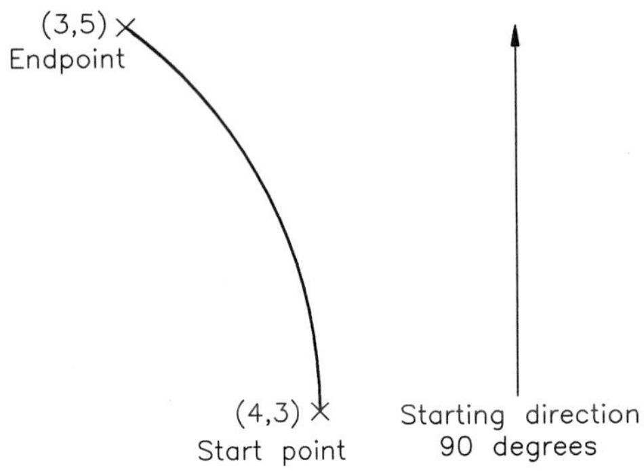

Figure 3-12 *Drawing an arc using the* ***Start, End, Direction*** *option*

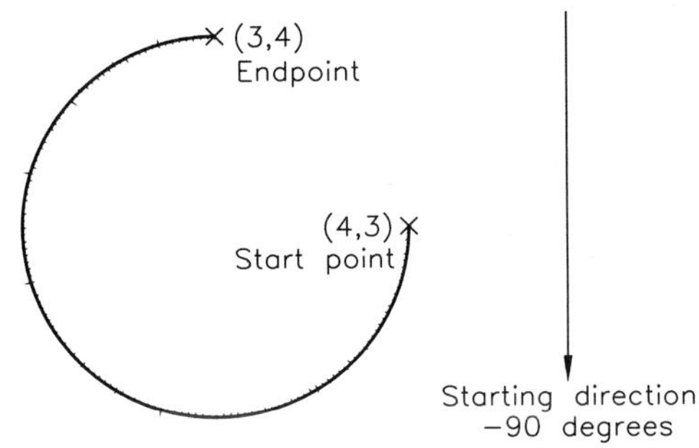

Figure 3-13 *Drawing an arc using a negative direction in the* ***Start, End, Direction*** *option*

Note

With the ***Start, End, Direction*** *option if you do not specify a start point but just press ENTER at the* ***Specify start point of arc or [Center]*** *prompt, the start point and direction of the arc will be taken from the* ***endpoint and ending direction*** *of the previous line or arc drawn on the current screen. You are then required to specify only the endpoint of the arc.*

Exercise 4 *Mechanical*

a. Specify the directions and the coordinates of two arcs in such a way that they form a circular figure.
b. Make the drawing shown in Figure 3-14. Create the curves using the **Arc** command. The distance between the dotted lines is 1.0 unit and the diameter of the circles is 1 unit.

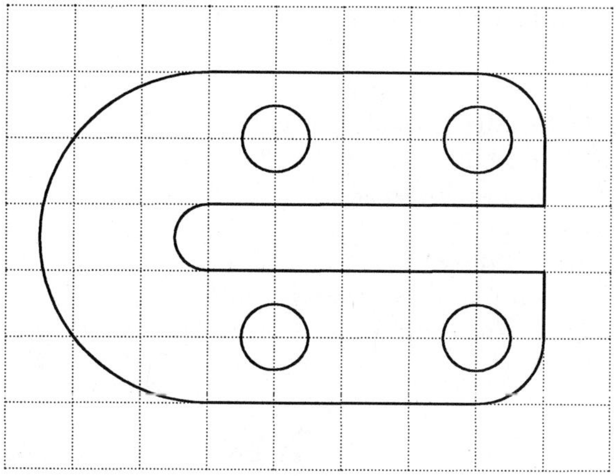

Figure 3-14 Drawing for Exercise 4(b)

The Start, End, Radius Option

This option is used when you know the start point, endpoint, and radius of the arc. The same values for the three variables (start point, endpoint, and radius) can result in four different arcs. AutoCAD resolves this by always drawing this type of arc in a counterclockwise direction from the start point. Therefore, a negative radius value results in a **major arc** (the largest arc between two endpoints), Figure 3-15(a), while a positive radius value results in a **minor arc** (smallest arc between the start point and the endpoint), Figure 3-15(b). The prompt sequence to draw a major arc having a start point of (3,3), endpoint of (2,5), and radius of -2, Figure 3-15(a), is given next.

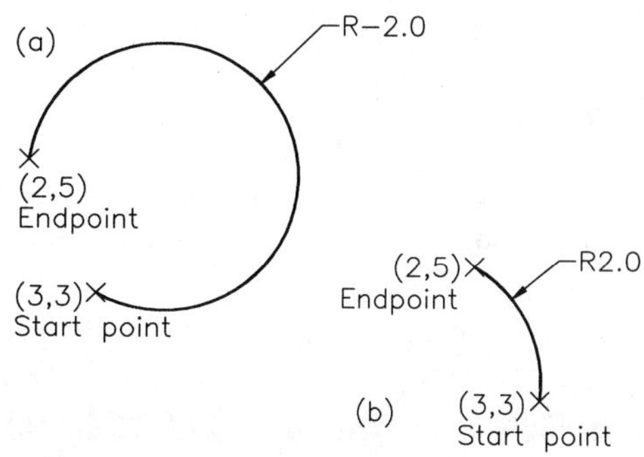

*Figure 3-15 Drawing an arc using the **Start, End, Radius** option*

Command: *Choose the **Start, End, Radius** option from the **Draw > Arc** menu.*
Specify start point of arc or [Center]: **3,3** Enter
Specify second point of arc or [Center/End]: _e Specify end point of arc: **2,5** Enter
Specify center point of arc or [Angle/Direction/Radius]: _r Specify radius of arc: **-2** Enter

The prompt sequence to draw a minor arc having its start point at (3,3), endpoint at (2,5), and radius as 2 as shown in Figure 3-15(b) is given next.

Command: *Choose the **Start, End, Radius** option from the **Draw/Arc** menu.*
Specify start point of arc or [Center]: **3,3** Enter
Specify second point of arc or [Center/End]: _e Specify end point of arc: **2,5** Enter
Specify center point of arc or [Angle/Direction/Radius]: _r Specify radius of arc: **2** Enter

The Center, Start, End Option

The **Center, Start, End** option is a modification of the Start, Center, End option. Use this option whenever it is easier to start drawing an arc by establishing the center first. Here the arc is always drawn in a counterclockwise direction from the start point to the endpoint, around the specified center. The prompt sequence for drawing an arc that has a center point at (3,3), start point at (5,3), and endpoint at (3,5) (Figure 3-16) is given next.

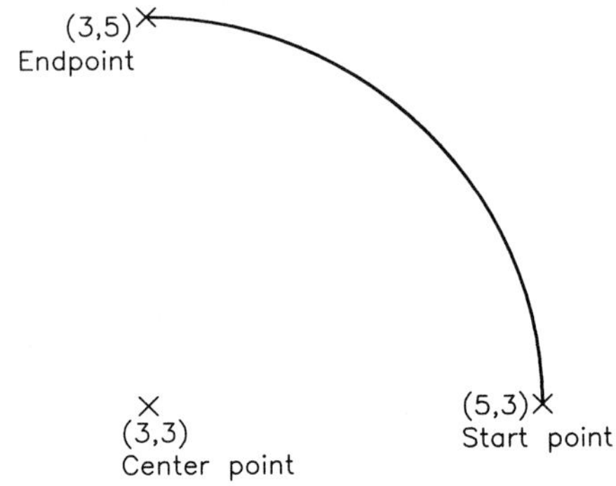

Figure 3-16 *Drawing an arc using the* ***Center, Start, End*** *option*

Command: *Choose the* ***Center, Start, End*** *option from the* ***Draw > Arc*** *menu.*
Specify start point of arc or [Center]: _c
Specify center point of arc: **3,3** Enter
Specify start point of arc: **5,3** Enter
Specify end point of arc or [Angle/chord Length]: **3,5** Enter

The Center, Start, Angle Option

This option is a variation of the **Start, Center, Angle** option. Use this option whenever it is easier to draw an arc by giving the center first. The prompt sequence for drawing an arc that has a center point at (4,5), start point at (5,4), and included angle of 120-degree (Figure 3-17) is given next.

Command: *Choose the* ***Center, Start, Angle*** *option from the* ***Draw > Arc*** *menu.*
Specify start point of arc or [Center]: _c Specify center point of arc: **4,5** Enter
Specify start point of arc: **5,4** Enter
Specify end point of arc or [Angle/chord Length]: _a Specify included angle: **120** Enter

The Center, Start, Length Option

The **Center, Start, Length** option is a modification of the Start, Center, Length option. This option is used whenever it is easier to draw an arc by establishing the center first. The prompt sequence for drawing an arc that has a center point at (2,2), start point at (4,3), and length of chord of 3 (Figure 3-18) is given next.

Command: *Choose the* ***Center, Start, Angle*** *option from the* ***Draw > Arc*** *menu.*
Specify start point of arc or [Center]: _c Specify center point of arc: **2,2** Enter
Specify start point of arc: **4,3** Enter
Specify end point of arc or [Angle/chord Length]: _l Specify length of chord: **3** Enter

Continue Option

With this option you can continue drawing an arc from a previously drawn arc or line. When you select the **Continue** option (Draw menu), then the start point and direction of the arc will be taken from the **endpoint and ending direction** of the previous line or arc drawn on the

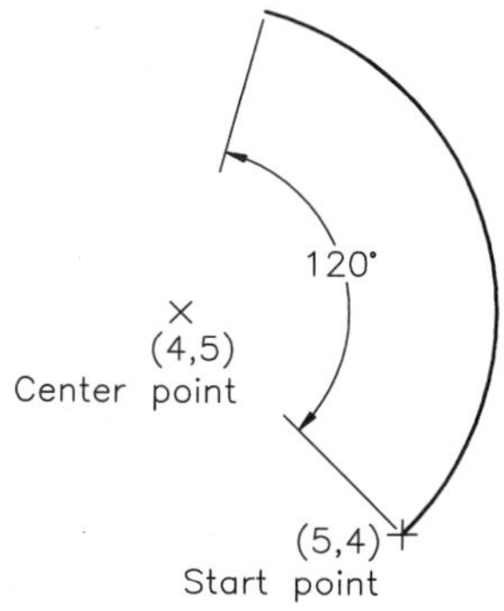

Figure 3-17 *Drawing an arc using the* ***Center, Start, Angle*** *option*

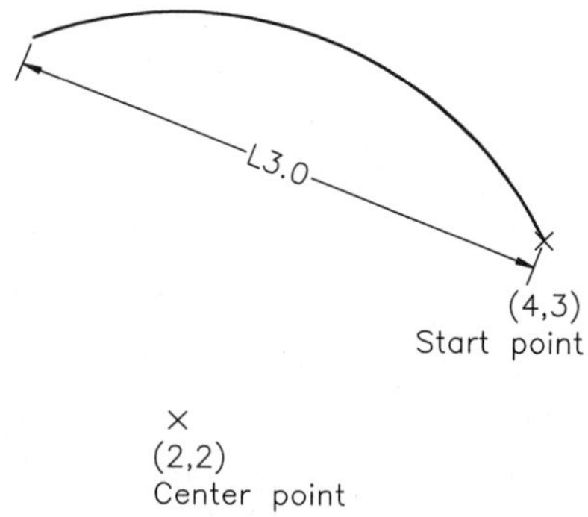

Figure 3-18 *Drawing an arc using the* ***Center, Start, Length*** *option*

current screen. When this option is used to draw arcs, each successive arc is tangent to the previous one. Most often this option is used to draw arcs tangent to a previously drawn line. The prompt sequence to draw an arc tangent to an earlier drawn line using the **Continue** option (Figure 3-19) is given next.

Command: **LINE** Enter
Specify first point: **2,2** Enter
Specify next point or [Undo]: **4,3** Enter
Specify next point or [Undo]: Enter

Command: *Choose the* **Continue** *option from the* ***Draw > Arc*** *menu.*
Specify endpoint of arc: **4,5** Enter

The prompt sequence to draw an arc continued from a previously drawn arc (Figure 3-20) is given next.

Command: **ARC** Enter
Specify start point of arc or [Center]: **2,2** Enter
Specify second point of arc or [Center/End]: **E** Enter
Specify endpoint of arc : **3,4** Enter
Specify center point of arc or [Angle/Direction/Radius]: **R** Enter
Specify radius of arc: **2** Enter

Command: *Choose the* **Continue** *option from the* ***Draw > Arc*** *menu.*
Specify end point of arc: **5,4** Enter

Tip

The ***Continue*** *option can also be invoked by pressing ENTER at the* ***Specify start point of arc or [Center]*** *prompt after entering* ***ARC*** *at the Command prompt. Even when you select any option of the ARC command from the* ***Draw > Arc*** *menu, pressing ENTER at the* ***Specify start point of arc or [Center]*** *prompt invokes the* ***Continue*** *option by terminating the selected option. Here you are asked to specify the endpoint to complete the arc.*

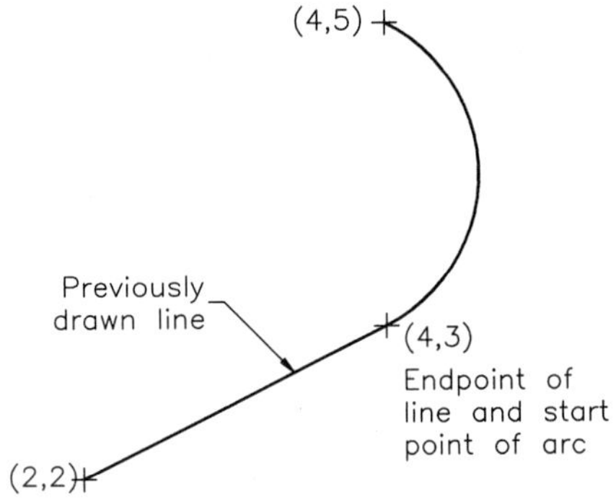

Figure 3-19 Drawing an arc using the ***Continue*** *option*

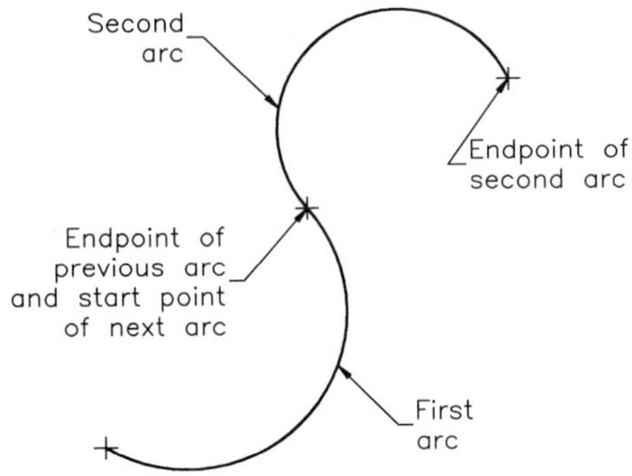

Figure 3-20 Drawing an arc using the ***Continue*** *option*

Continue (LineCont:) Option

This option is used when you want to continue drawing a line from the endpoint of a previously drawn arc. When you use this option, the start point and direction of the line will be taken from the **endpoint and ending direction** of the previous arc. In other words, the line will be tangent to the arc drawn on the current screen. This option is invoked when you press ENTER at the **Specify first point:** prompt of the LINE command. The prompt sequence to draw a line tangent to an earlier drawn arc (Figure 3-21) is given next.

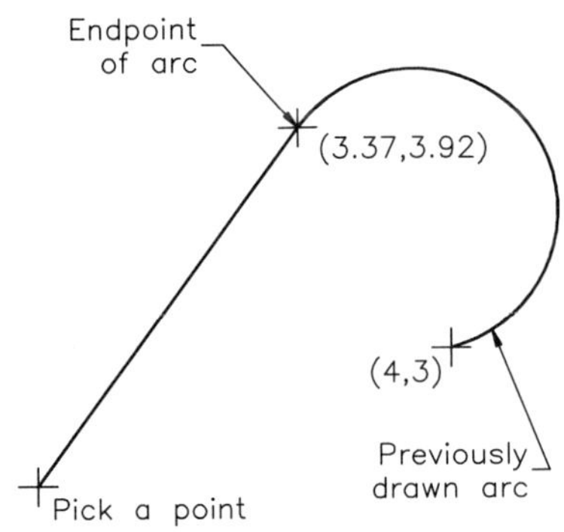

Figure 3-21 Drawing a line from the endpoint of an arc

Command: **ARC** Enter
Specify start point of arc or [Center]: **4,3** Enter
Specify second point of arc or [Center/End]: **E** Enter
Specify end point of arc : **3.37,3.92** Enter
Specify center point of arc or [Angle/Direction/Radius]: *Specify the center point.*

Command: **LINE** Enter
Specify first point: Enter
Length of line: *Enter a value or pick a point.*

Note

You have the ***LinCont*** *option in the* ***DRAW1 > Arc*** *screen menu that can be used to draw a line from the endpoint of a previously drawn arc.*

Exercise 5 General

a. Use the **Center, Start, Angle** and the **Continue** options to draw the figures shown in Figure 3-22.

b. Make the drawing shown in Figure 3-23. The distance between the dotted lines is 1.0 units. Create the radii as indicated in the drawing by using the **ARC** command options.

Figure 3-22 *Drawing for Exercise 5(a)*

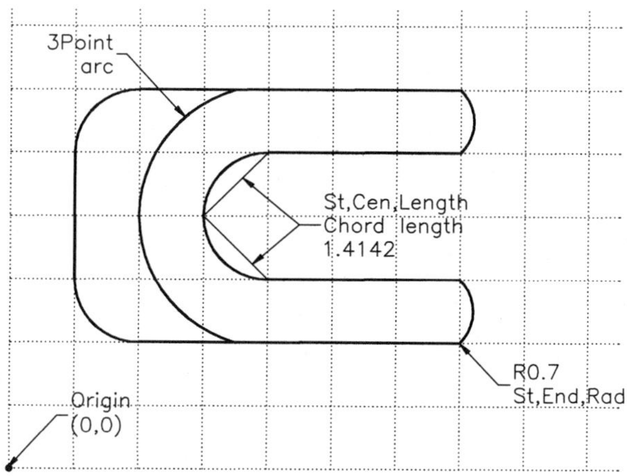

Figure 3-23 *Drawing for Exercise 5(b)*

DRAWING RECTANGLES

Toolbar:	Draw > Rectangle
Menu:	Draw > Rectangle
Command:	RECTANG

A rectangle can be drawn using the **RECTANG** command (Figures 3-24 and Figure 3-25). After invoking the **RECTANG** command, you are prompted to specify the first corner of the rectangle at the **Specify first corner point or [Chamfer/Elevation/Fillet/Thickness/Width]** prompt. Here you can enter the coordinates of the first corner or specify the desired point with the pointing device. The first corner can be any one of the

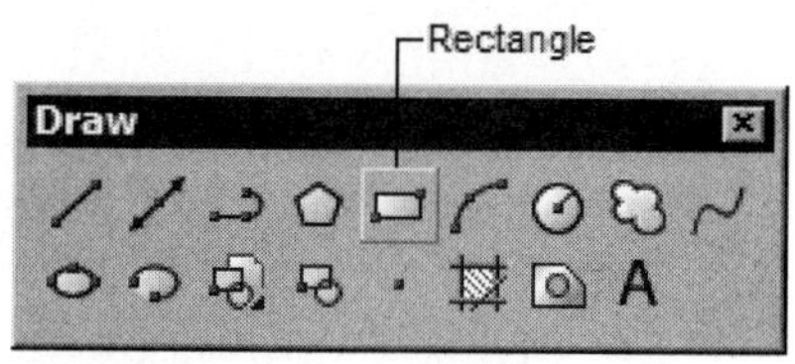

Figure 3-24 *Invoking the* ***RECTANG*** *command from the* ***Draw*** *toolbar*

four corners. Then you are prompted to enter the coordinates or specify the other corner at the **Specify other corner point or [Dimensions]** prompt. This corner is taken as the corner diagonally opposite the first corner. You can specify the coordinates for the other corner or simply drag the cursor to specify it. The prompt sequence for drawing a rectangle with (3,3) as its lower left corner coordinate and (6,5) as its upper right corner (Figure 3-26) is given next.

Command: **RECTANG** Enter
Specify first corner point or [Chamfer/Elevation/Fillet/Thickness/Width]: **3,3** Enter *(Lower left corner location.)*
Specify other corner point or [Dimensions]: **6,5** Enter *(Upper right corner location.)*

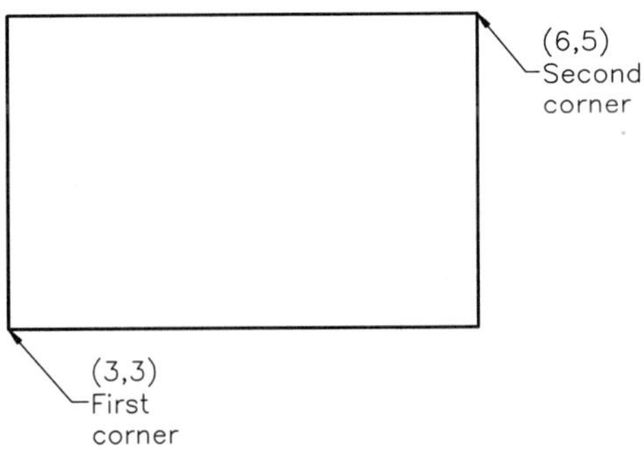

Figure 3-26 *Drawing a rectangle using the* ***RECTANG*** *command*

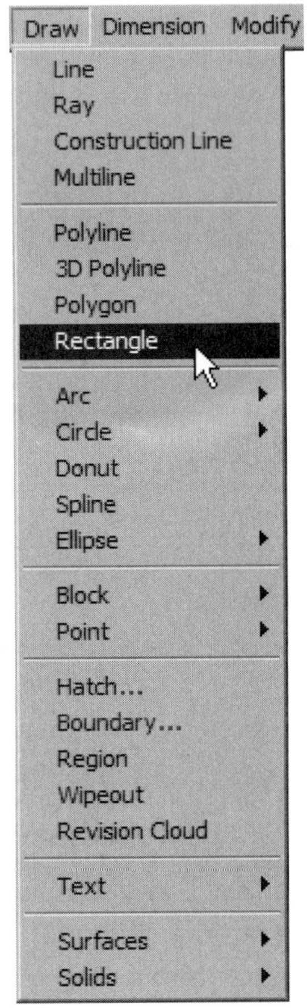

Figure 3-25 *Invoking the* ***RECTANG*** *command from the* ***Draw*** *menu*

After specifying the first corner instead of specifying the diagonally opposite corner you can give the dimensions of the rectangle. This is possible by entering **D** at the **Specify other corner point or [Dimensions]** prompt, which then allows you to enter the length and the width of the rectangle. The prompt sequence for drawing a rectangle with a length of **5** units and a width of **3** units is given next.

Command: **RECTANG** Enter
Specify first corner point or [Chamfer/Elevation/Fillet/Thickness/Width]: **3,3** Enter
Specify other corner point or [Dimensions]: **D** Enter
Specify length for rectangles <0.0000>: **5** Enter
Specify width for rectangles <0.0000>: **3** Enter

Here you are allowed to choose any one of the four locations for placing the rectangle. You can move the cursor to see the four locations of the rectangle. Depending on the location of the cursor, the specified first corner point holds the position of either the lower left corner, or the lower right corner, or the upper right corner, or the upper left corner. After deciding the position you can then click to place the rectangle.

Chamfer

The **Chamfer** option creates a chamfer, which is an angled corner, by specifying the chamfer distances (Figure 3-27). The chamfer is created at all the four corners. You can give two different chamfer values to create an unequal chamfer.

> Command: **RECTANG** Enter
> Specify first corner point or [Chamfer/Elevation/Fillet/Thickness/Width]: **C** Enter
> Specify first chamfer distance for rectangles <0.0000>: *Enter a value.*
> Specify second chamfer distance for rectangles <0.0000>: *Enter a value.*
> Specify first corner point or [Chamfer/Elevation/Fillet/Thickness/Width]: *Select a point as lower left corner location.*
> Specify other corner point or [Dimensions]: *Select a point as upper right corner location.*

Note

The first corner point that you specify need not be the lower left corner location. While selecting the other corner you can select a location such that the first corner point becomes the lower right corner or the upper left corner or the upper right corner.

Fillet

The **Fillet** option allows you to create a filleted rectangle by specifying the fillet radius (Figure 3-28). A fillet is a rounded corner which is created at all the four corners of the rectangle.

> Specify first corner point or [Chamfer/Elevation/Fillet/Thickness/Width]: **F** Enter
> Specify fillet radius for rectangles <0.0000>: *Enter a value.*

Now, if you draw a rectangle it will be filleted provided the length and width of the rectangle are equal to or greater than twice the value of the specified fillet. Otherwise, AutoCAD will draw a rectangle without fillet.

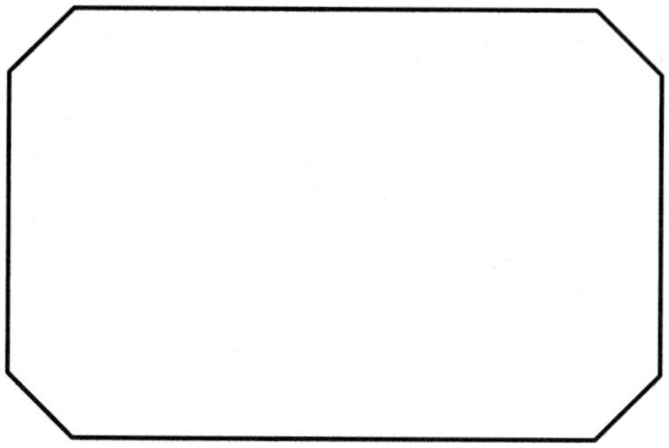

Figure 3-27 *Drawing a rectangle with chamfer*

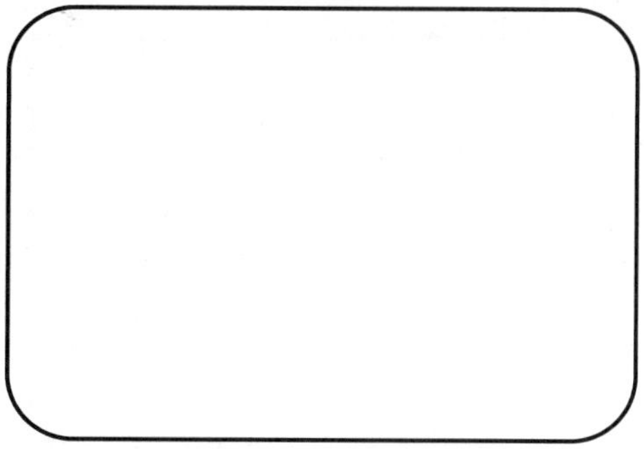

Figure 3-28 *Drawing a rectangle with fillet*

Note

You can draw a rectangle either with chamfers or with fillets. If you specify the chamfer distances first and then specify the fillet radius in the same ***RECTANG*** *command, the rectangle will be drawn with fillets only.*

Width

The **Width** option allows you to create a rectangle whose line segments have some specified width, as shown in Figure 3-29.

Specify first corner point or [Chamfer/Elevation/Fillet/Thickness/Width]: **W** Enter
Specify line width for rectangles <0.0000>: *Enter a value.*

Thickness

The **Thickness** option allows you to draw a rectangle that is extruded in the Z direction by the specified value of thickness. For example, if you draw a rectangle with thickness of 2 units, you will get a rectangular box whose height is 2 units (Figure 3-30). To view the box, choose **View > 3D Views > SE Isometric** from the menu bar.

Specify first corner point or [Chamfer/Elevation/Fillet/Thickness/Width]: **T** Enter
Specify thickness for rectangles <0.0000>: *Enter a value.*

Elevation

The **Elevation** option allows you to draw a rectangle at a specified distance from the *XY* plane along the *Z* axis. For example, if the elevation is 2 units, the rectangle will be drawn two units above the *XY* plane. If the thickness of the rectangle is 1 unit, you will get a rectangular box of 1 unit height located 2 units above the *XY* plane (Figure 3-30).

Chamfer/Elevation/Fillet/Thickness/Width/<First corner>: **E** Enter
Specify elevation for rectangles <0.0000>: *Enter a value.*

To view the objects in 3D space, change the viewpoint by choosing **View > 3D Views > SE Isometric** from the menu bar.

Figure 3-29 *Drawing a rectangle with specified width*

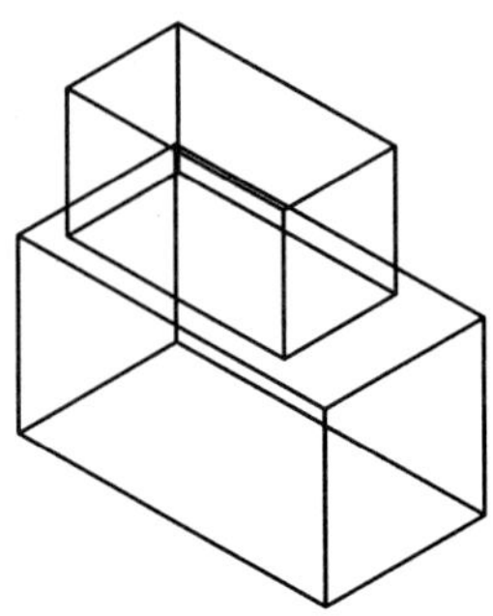

Figure 3-30 *Drawing rectangles with thickness and elevation specified*

Note

*You can change back your viewpoint from SE Isometric to the top view by choosing the **Top View** button in the **View** toolbar or by selecting **Top** in the **View > 3D Views** menu.*

*The value you enter for Fillet, Width, Elevation, and Thickness becomes the current value for the subsequent **RECTANG** command. Therefore, you must reset the values if they are different from the current values. The thickness of rectangle is always controlled by its thickness settings.*

*The rectangle generated on the screen is treated as a single object. Therefore, the individual sides can be edited only after the rectangle has been exploded using the **EXPLODE** command.*

Tip

*You can combine the different options in one **RECTANG** command and then draw the rectangle with the specified characteristics. When you invoke the **RECTANG** command again, the previously set options and their values are displayed before the first prompt. This allows you to change the settings according to the new specifications.*

Exercise 6 — *General*

Draw a rectangle 4 units long, 3 units wide, and with its first corner at (1,1). Draw another rectangle of length 2 units and width 1 unit, and with first corner at 1.5,1.5.

DRAWING ELLIPSES

Toolbar:	Draw > Ellipse
Menu:	Draw > Ellipse
Command:	ELLIPSE

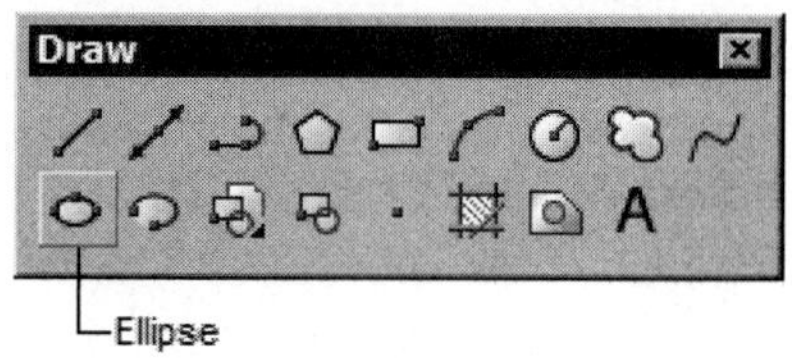

Figure 3-31 *Invoking the **ELLIPSE** command from the **Draw** toolbar*

If a circle is observed from an angle, the shape seen is called an **Ellipse**, which can be created in AutoCAD using the **ELLIPSE** command (Figure 3-31). An ellipse can be created using various options listed within the **ELLIPSE** command. AutoCAD creates a true ellipse, also known as a NURBS-based (Non-Uniform Rational Bezier Spline) ellipse. The true ellipse has a center and quadrant points. If you select it, the grips (small blue squares) will be displayed at the center and the quadrant points of the ellipse. If you move one of the grips located on the perimeter of the ellipse, the major or minor axis will change, which changes the size of the ellipse, as shown in Figure 3-32(d). The creation of a true ellipse is dependent on the **PELLIPSE** system variable, which has a value 0 by default.

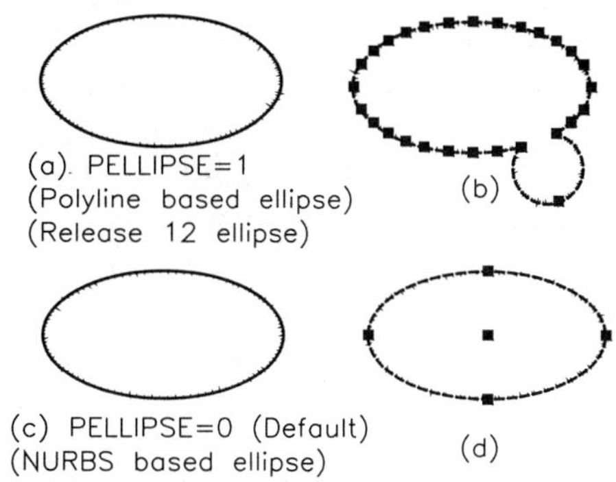

Figure 3-32 *Drawing polyline and NURBS-based ellipses*

Note

*Up to Release 12, ellipses were based on polylines. They were made of multiple polyarcs and as a result, it was difficult to edit an ellipse. For example, if you select a polyline-based ellipse, the grips will be displayed at the endpoints of each polyarc. If you move a vertex point, you get the shape shown in Figure 3-32(b). Also, you cannot snap to the center or the quadrant points of a polyline-based ellipse. In AutoCAD 2004, you can still draw the polyline-based ellipse by setting the value of the **PELLIPSE** system variable to **1**, which is **0** (true ellipse) by default.*

Once you invoke the **ELLIPSE** command, AutoCAD will acknowledge with the prompt **Specify axis endpoint of ellipse or [Arc/Center]** or **Specify axis endpoint of ellipse or [Arc/Center/Isocircle]** (if Isometric snap is on). The response to this prompt depends on the option you want to choose. The various options are explained next.

Note

*The **Isocircle** option is not available by default in the **ELLIPSE** command. To display this option, you have to select the **Isometric snap** radio button in the **Snap and Grid** tab of the **Drafting Settings** dialog box. The Isocircle option will be discussed in Chapter 21.*

*The **Arc** option is not available if you set the value of the **PELLIPSE** system variable to 1 for drawing the polyline-based ellipse.*

Drawing an Ellipse Using the Axis and Endpoint Option

In this option you draw an ellipse by specifying one of its axes and the endpoint of the other axis. To use this option, acknowledge the **Specify axis endpoint of ellipse or [Arc/Center]** prompt by specifying a point, either by using a pointing device or by entering its coordinates. This is the first endpoint of one axis of the ellipse. AutoCAD will then respond with the prompt **Specify other endpoint of axis**. Here, specify the other endpoint of the axis. The angle at which the ellipse is drawn depends on the angle made by these two axis endpoints. Your response to the next prompt determines whether the axis is the **major axis** or the **minor axis**.

The next prompt is **Specify distance to other axis or [Rotation]**. If you specify a distance, it is presumed as half the length of the second axis. You can also specify a point. The distance from this point to the midpoint of the first axis is again taken as half the length of this axis. The ellipse will pass through the selected point only if it is perpendicular to the midpoint of the first axis. To visually analyze the distance between the selected point and the midpoint of the first axis, AutoCAD appends an elastic line to the crosshairs, with one end fixed at the midpoint of the first axis. You can also drag the point, dynamically specifying half of the other axis distance. This helps you to visualize the ellipse. The prompt sequence for drawing an ellipse with one axis endpoint located at (3,3), the other at (6,3), and the distance of the other axis being 1 (Figure 3-33) is given next.

Command: **ELLIPSE** Enter
Specify axis endpoint of ellipse or [Arc/Center]: **3,3** Enter
Specify other endpoint of axis : **6,3** Enter
Specify distance to other axis or [Rotation]: **1** Enter

Another example for drawing an ellipse (Figure 3-34) using this option is illustrated by the following prompt sequence.

Command: **ELLIPSE** Enter
Specify axis endpoint of ellipse or [Arc/Center]: **3,3** Enter
Specify other endpoint of axis : **4,2** Enter
Specify distance to other axis or [Rotation]: **2** Enter

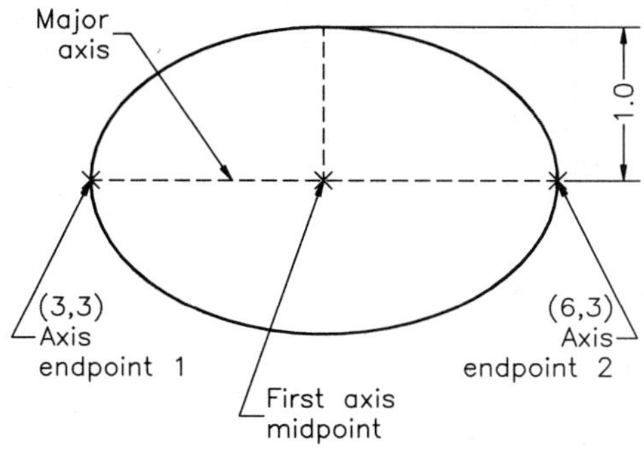

Figure 3-33 *Drawing an ellipse using the* ***Axis and Endpoint*** *option*

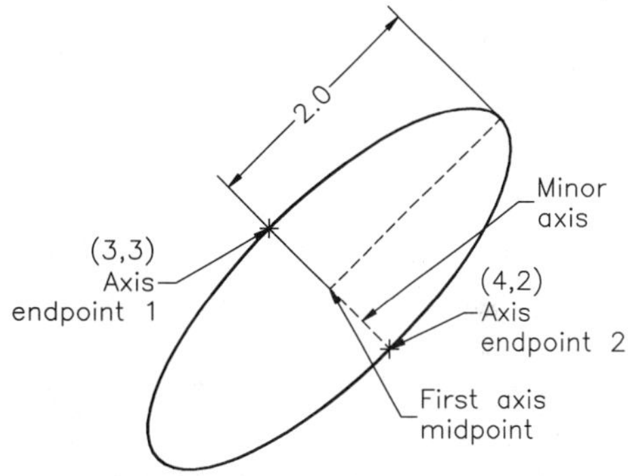

Figure 3-34 *Drawing an ellipse using the* ***Axis and Endpoint*** *option*

If you enter **Rotation** or **R** at the **Specify distance to other axis or [Rotation]** prompt, the first axis specified is automatically taken as the major axis of the ellipse. The next prompt is **Specify rotation around major axis**. The major axis is taken as the diameter line of the circle, and the rotation takes place around this diameter line into the third dimension. The ellipse is formed when AutoCAD projects this rotated circle into the drawing plane. You can enter the rotation angle value in the range of 0 to 89.4-degree only, because an angle value greater than 89.4-degree changes the circle into a line. Instead of entering a definite angle value at the **Specify rotation around major axis** prompt, you can specify a point relative to the midpoint of the major axis. This point can be dragged to specify the ellipse dynamically. The following is the prompt sequence for a rotation of 0-degree around the major axis, as shown in Figure 3-35(a).

Command: **ELLIPSE** Enter
Specify axis endpoint of ellipse or [Arc/Center]: *Select point (P1).*
Specify other endpoint of axis : *Select another point (P2).*
Specify distance to other axis or [Rotation]: **R** Enter
Specify rotation around major axis: **0** Enter

Figure 3-35 also shows rotations of 45-degree, 60-degree, and 89.4-degree.

Chapter 3

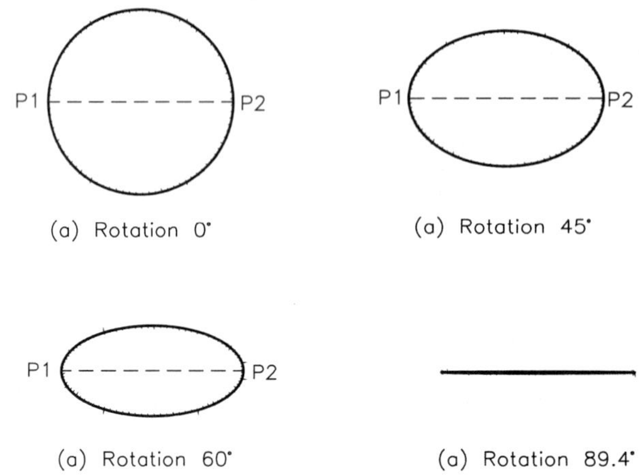

Figure 3-35 *Rotation about the major axis*

Exercise 7 — *General*

Draw an ellipse whose major axis is 4 units and whose rotation around this axis is 60-degree. Draw another ellipse, whose rotation around the major axis is 15-degree.

Drawing Ellipse Using the Center and Two Axes Option

In this option you can construct an ellipse by specifying the center point, the endpoint of one axis, and the length of the other axis. The only difference between this method and the ellipse by axis and endpoint method is that instead of specifying the second endpoint of the first axis, the center of the ellipse is specified. The center of an ellipse is defined as the point of intersection of the major and minor axes. In this option, the first axis need not be the major axis. For example, to draw an ellipse with center at (4,4), axis endpoint at (6,4), and length of the other axis as 2 units (Figure 3-36), the prompt sequence is given next.

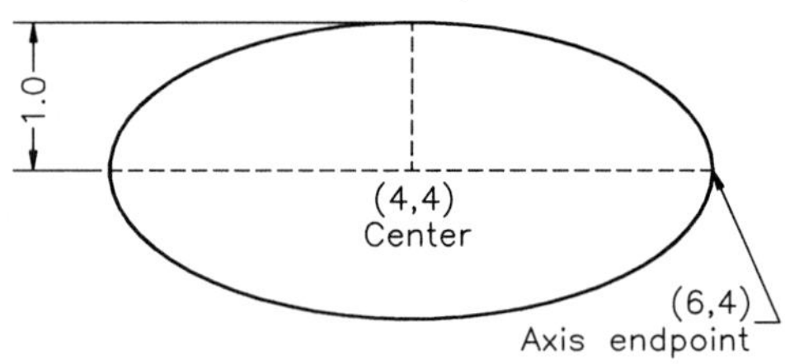

Figure 3-36 *Drawing an ellipse using the* ***Center*** *option*

Command: **ELLIPSE** Enter
Specify axis endpoint of ellipse or [Arc/Center]: **C** Enter
Specify center of ellipse: **4,4** Enter
Specify endpoint of axis: **6,4** Enter
Specify distance to other axis or [Rotation]: **1** Enter

Instead of entering the distance, you can enter Rotation or R at the **Specify distance to other**

axis [Rotation] prompt. This takes the first axis specified as the major axis. The next prompt, **Specify rotation around major axis**, prompts you to enter the rotation angle value. The rotation takes place around the major axis, which is taken as the diameter line of the circle. The rotation angle values should range from 0 to 89.4-degree.

Drawing Elliptical Arcs

Toolbar:	Draw > Ellipse Arc
Menu:	Draw > Ellipse > Arc
Command:	ELLIPSE > Arc

You can use the Arc option of the **ELLIPSE** command to draw an elliptical arc. When you choose the **Ellipse Arc** button from the **Draw** toolbar, the **ELLIPSE** command is invoked with the Arc option selected. AutoCAD will prompt you to enter information about the geometry of the ellipse and the arc limits. You can define the arc limits by using the following options.

1. Start and End angle of the arc.
2. Start and Included angle of the arc.
3. Start and End parameters.

The angles are measured from the first point and in counterclockwise direction if AutoCAD's default setup is not changed. The following example illustrates the use of these three options.

Example 1 *General*

Draw the following elliptical arcs as shown in Figures 3-37 and 3-38.

a. Start angle = -45, end angle = 135
b. Start angle = -45, included angle = 225
c. Start parameter = @1,0, end parameter = @1<225

Specifying Start and End Angle of the Arc [Figure 3-37(a)]

Command: *Choose the **Ellipse Arc** button from the **Draw** toolbar.*
Specify axis endpoint of ellipse or [Arc/Center]: _a
Specify axis endpoint of elliptical arc or [Center]: *Select the first endpoint.*
Specify other endpoint of axis : *Select the second point.*
Specify distance to other axis or [Rotation]: *Select a point or enter a distance.*
Specify start angle or [Parameter]: **-45** Enter
Specify end angle or [Parameter/Included angle]: **135** Enter *(Angle where arc ends.)*

Specifying Start and Included Angle of the Arc [Figure 3-37(b)]

Command: *Choose the **Ellipse Arc** button from the **Draw** toolbar.*
Specify axis endpoint of ellipse or [Arc/Center]: _a

Specify axis endpoint of elliptical arc or [Center]: *Select the first endpoint.*
Specify other endpoint of axis : *Select the second point.*
Specify distance to other axis or [Rotation]: *Select a point or enter a distance.*
Specify start angle or [Parameter]: **-45** Enter
Specify end angle or [Parameter/Included angle]: **I** Enter
Specify included angle for arc<current>: 225 Enter *(Included angle.)*

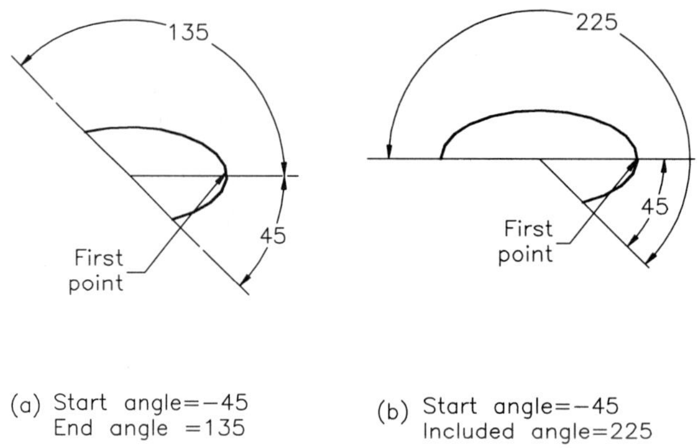

Figure 3-37 *Drawing elliptical arcs*

Specifying Start and End Parameters (Figure 3-38):

Command: *Choose the* ***Ellipse Arc*** *button from the* ***Draw*** *toolbar.*
Specify axis endpoint of ellipse or [Arc/Center]: _a
Specify axis endpoint of elliptical arc or [Center]: *Select the first endpoint.*
Specify other endpoint of axis: *Select the second endpoint.*
Specify distance to other axis or [Rotation]: *Select a point or enter a distance.*
Specify start angle or [Parameter]: **P**
Specify start parameter or [Angle]: **@1,0**
Specify end parameter or [Angle/ Included angle]: **@1<225**

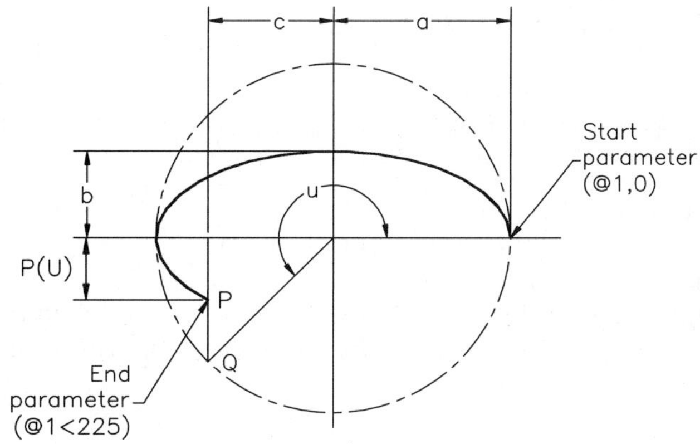

Figure 3-38 *Drawing an elliptical arc by specifying the start and end parameters*

Calculating Parameters for an Elliptical Arc

The start and end parameters of an elliptical arc are determined by specifying a point on the circle whose diameter is equal to the major diameter of the ellipse as shown in Figure 3-38. In this drawing, the major axis of the ellipse is 2.0 and the minor axis is 1.0. The diameter of the circle is 2.0. To determine the start and end parameters of the elliptical arc, you must specify the points on the circle. In the example, the start parameter is @1,0 and the end parameter is @1<225. Once you specify the points on the circle, AutoCAD will project these points on the major axis and determine the endpoint of the elliptical arc. In Figure 3-38, Q is the end parameter of the elliptical arc. AutoCAD projects point Q on the major axis and locates intersection point P, which is the endpoint of the elliptical arc. The coordinates of point P can be calculated by using the following equations.

The equation of an ellipse with center as origin is
x^2/a^2 + y^2/b^2 = 1
In parametric form x = a * cos(u)
y = b * sin(u)
For the example a = 1
b = 0.5
Therefore x = 1 * cos(225) = -0.707
y = 0.5 * sin(225) = -0.353
The coordinates of point P are (-0.707, -0.353) with respect to the center of the ellipse.
Note: v = atan(b/a*tan(u)) = end angle
v = atan(0.5/1*tan(225)) = 206.56o
Also e = 1-b^2/a^2)^.5 = eccentricity
e = 1-.5^2/1^2)^.5 = .866
r = x^2 + y^2)^.5
r = .707^2 + .353^2)^.5 = 0.790

or using the polar equation r = b/(1 - e^2 * cos(v)^2)^.5

r = .5/(1 - .866^2 * cos(206.56)^2)^.5

r = 0.790

Exercise 8 *General*

a. Construct an ellipse with center at (2,3), axis endpoint at (4,6), and the other axis endpoint a distance of 0.75 units from the midpoint of the first axis.

b. Make the drawing as shown in Figure 3-39. The distance between the dotted lines is 1.0 unit. Create the elliptical arcs using the **ELLIPSE** command options.

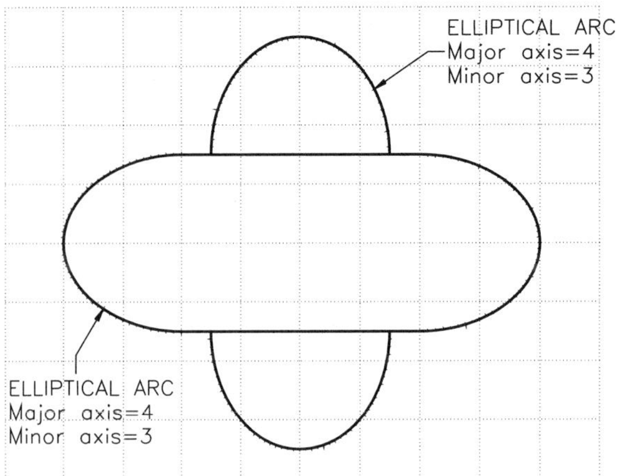

Figure 3-39 *Drawing for Exercise 8(b)*

DRAWING REGULAR POLYGONS

Toolbar:	Draw > Polygon
Menu:	Draw > Polygon
Command:	POLYGON

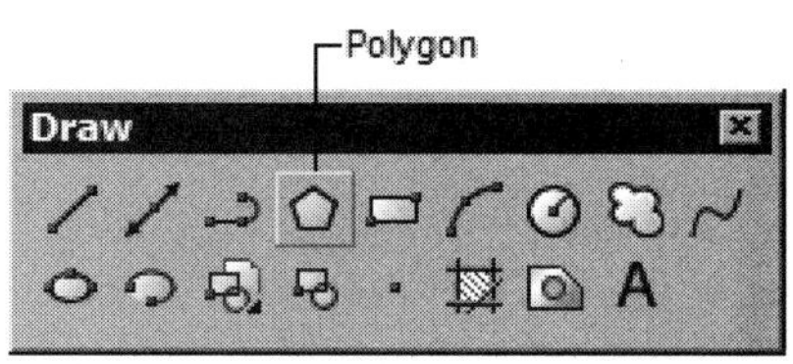

Figure 3-40 *Invoking the **POLYGON** command from the **Draw** toolbar*

A **regular polygon** is a closed geometric figure with equal sides and equal angles. The number of sides varies from **3** to **1024**. For example, a triangle is a three-sided polygon and a pentagon is a five-sided polygon. In AutoCAD, the **POLYGON** command (Figure 3-40) is used to draw regular 2D polygons. The characteristics of a polygon drawn in AutoCAD are those of a closed polyline having 0 width. You can change the width of the polyline forming the polygon. The prompt sequence is given next.

Command: **POLYGON** [Enter]

Enter number of sides <4>:

Once you invoke the **POLYGON** command, it prompts you to enter the number of sides. The number of sides determines the type of polygon (for example, six sides define a hexagon). The default value for the number of sides is **4**. You can change the number of sides to your requirement and then the new value becomes the default. You can also set a different default value for the number of sides by using the **POLYSIDES** system variable.

The Center of Polygon Option

After you specify the number of sides, the next prompt is **Specify center of polygon or [Edge]**, where the default option prompts you to select a point that is taken as the center point of the polygon. The next prompt is **Enter an option [Inscribed in circle/Circumscribed about circle]<I>**. A polygon is said to be **inscribed** when it is drawn inside an imaginary circle and its vertices (corners) touch the circle (Figure 3-41). Likewise, a polygon is **circumscribed** when it is drawn outside the imaginary circle and the sides of the polygon are tangent to the circle (midpoint of each side of the polygon will lie on the circle) (Figure 3-42). If you want to have an inscribed polygon, enter **I** at the prompt. The next prompt issued is **Specify radius of circle**. Here you are required to specify the radius of the circle on which all the vertices of the polygon will lie. Once you specify the radius, a polygon will be generated. If you want to select the circumscribed option, enter **C** at the prompt **Enter an option[Inscribed in circle/Circumscribed about circle]<I>**. After this, enter the radius of the circle. The inscribed or circumscribed circle is not drawn on the screen. The radius of the circle can be dynamically dragged instead of a numerical value entered. The prompt sequence for drawing an inscribed octagon with a center at (4,4) and a radius of 1.5 units (Figure 3-41) is given next.

Command: **POLYGON** Enter
Enter number of sides<4>: **8** Enter
Specify center of polygon or [Edge]: **4,4** Enter
Enter an option[Inscribed in circle/Circumscribed about circle]<I>: **I** Enter
Specify radius of circle: **1.5** Enter

The prompt sequence for drawing a circumscribed pentagon with center at (4,4) and a radius of 1.5 units (Figure 3-42) is given next.

Command: **POLYGON** Enter
Enter number of sides<4>: **5** Enter
Specify center of polygon or [Edge]: **4,4** Enter
Enter an option[Inscribed in circle/Circumscribed about circle]<I>: **C** Enter
Specify radius of circle: **1.5** Enter

Note

If you select a point to specify the radius of an inscribed polygon, one of the vertices is positioned on the selected point. In the case of circumscribed polygons, the midpoint of an edge is placed on the point you have specified. In this manner you can specify the size and rotation of the polygon.

In the case of numerical specification of the radius, the bottom edge of the polygon is rotated by the prevalent snap rotation angle.

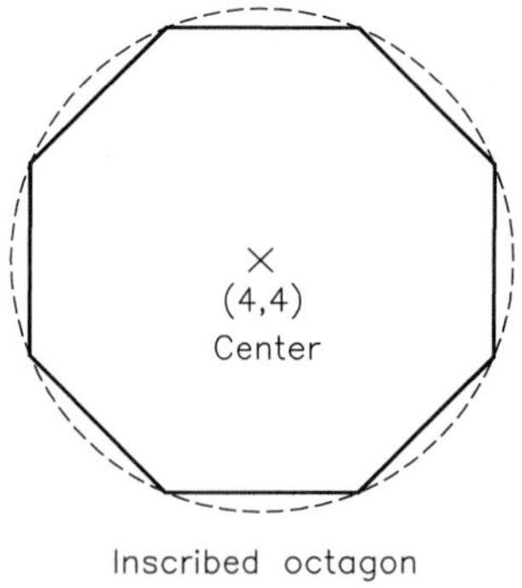

Figure 3-41 Drawing an inscribed polygon using the ***Center of Polygon*** *option*

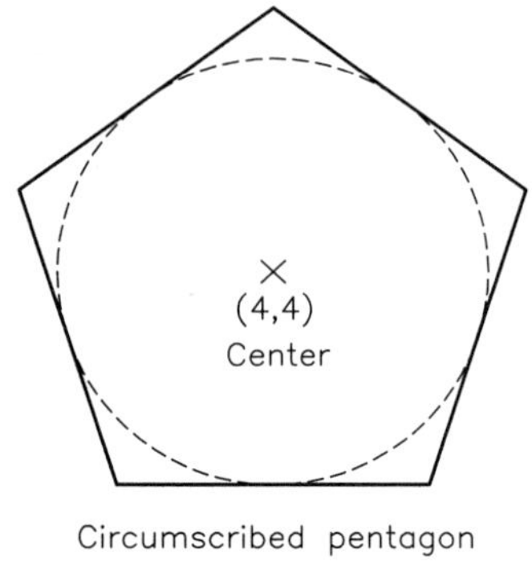

Figure 3-42 Drawing a circumscribed polygon using the ***Center of Polygon*** *option*

Exercise 9 *General*

Draw a circumscribed polygon of eight sides. The polygon should be drawn by the **Center of Polygon** method.

The Edge Option

The other method for drawing a polygon is to select the **Edge** option. This can be done by entering E at the **Specify center of polygon or [Edge]** prompt. The next two prompts issued are **Specify first endpoint of edge** and **Specify second endpoint of edge**. Here you need to specify the two endpoints of an edge of the polygon. The polygon is drawn in a counterclockwise direction, with the two points entered defining its first edge. To draw a hexagon (six-sided polygon) using the Edge option, with the first endpoint of the edge at (2,4) and the second endpoint of the edge at (2,2.5) (Figure 3-43), the prompt sequence is given next.

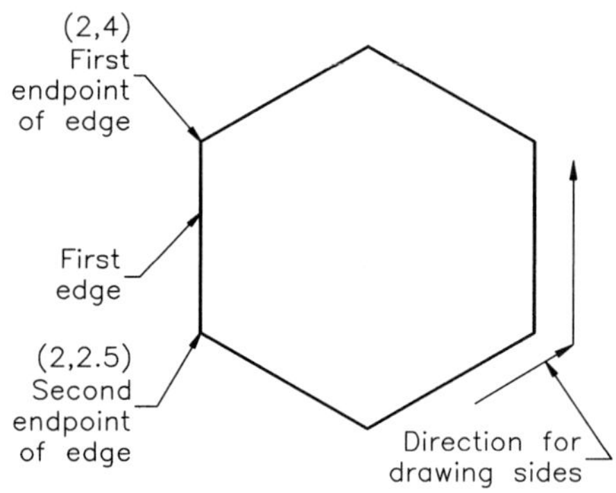

Figure 3-43 Drawing a polygon (hexagon) using the ***Edge*** *option*

Command: **POLYGON** Enter
Enter number of sides<4>: **6** Enter
Specify center of polygon or [Edge]: **E** Enter
Specify first endpoint of edge: **2,4** Enter
Specify second endpoint of edge: **2,2.5** Enter

Exercise 10 *General*

Draw a polygon with ten sides using the **Edge** option and an elliptical arc as shown in Figure 3-44. Let the first endpoint of the edge be at (7,1) and the second endpoint be at (8,2).

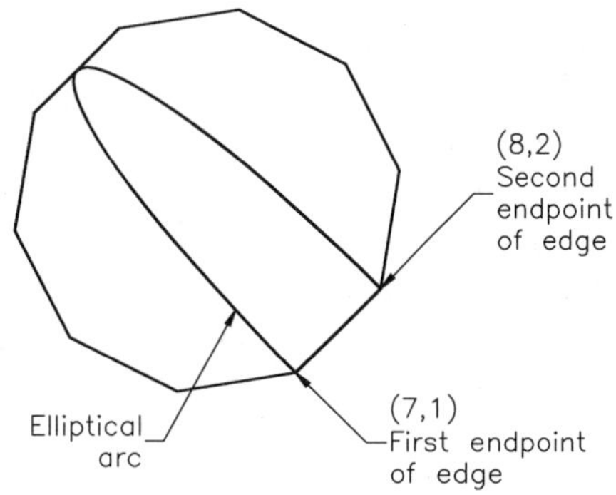

Figure 3-44 *Polygon and elliptical arc for Exercise 10*

DRAWING TRACES

Command: TRACE

TRACES are lines that have a thickness (specified as its width) and are solid-filled if the **FILLMODE** is on (value = 1). If the **FILLMODE** is off (value = 0), only the trace outline is drawn. The fill option can be accessed through the **FILL** command or the **FILLMODE** system variable.

The specifications of a trace are entered just like those of a line. The difference between the **LINE** command and the **TRACE** command is that in the latter the width of the trace is entered before entering the rest of the specifications. To specify the width of the trace, you can enter the distance or select two points. AutoCAD automatically measures the distance between those two points and assumes it as the width of the trace. All of the trace segments drawn by a single **TRACE** command will have the same width. The default value for the width of the trace is the width of the previous trace drawn. This value is stored in the **TRACEWID** system variable, and therefore you can change the default setting for the width of the trace through the **TRACEWID** variable. The endpoints of the trace lie on the center of the trace width. The start and end of traces are always cut square and are made of TRACE segments. The drawing of each segment is suspended until you specify the next segment or you end the trace by pressing ENTER. This is because trace ends are shaped to fit the start of the next segment. The Close and the Undo options do not work with the **TRACE** command. For example, to draw a solid-filled trace (Figure 3-45) with a width of 0.25 units starting from (1,3), to point (3,3), to point (2,1), the prompt sequence is given next.

Command: **FILLMODE** [Enter]
Enter new value for FILLMODE <1>: [Enter]
Command: **TRACE** [Enter]
Specify trace width <0.0500>: **0.25** [Enter]
Specify start point: **1,3** [Enter]

Specify next point: **3,3** [Enter]
Specify next point: **2,1** [Enter]
Specify next point: [Enter]

The prompt sequence to draw a nonsolid trace (Figure 3-46) with a width of 0.25 unit, starting from point (3,3), to point (5,3), to point (5,1), is given next.

Command: **FILLMODE** [Enter]
Enter new value for FILLMODE <1>:**0** [Enter]
Command: **TRACE** [Enter]
Specify trace width <0.0500>: **0.25** [Enter]
Specify start point: **3,3** [Enter]
Specify next point: **5,3** [Enter]
Specify next point: **5,1** [Enter]
Specify next point: [Enter]

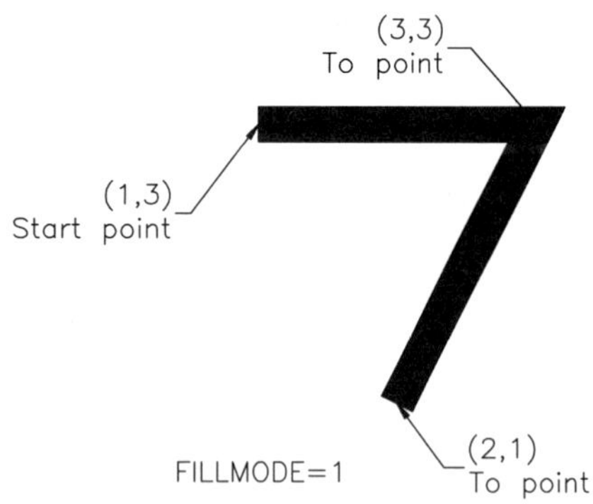

Figure 3-45 *Drawing a solid trace using the* ***TRACE*** *command*

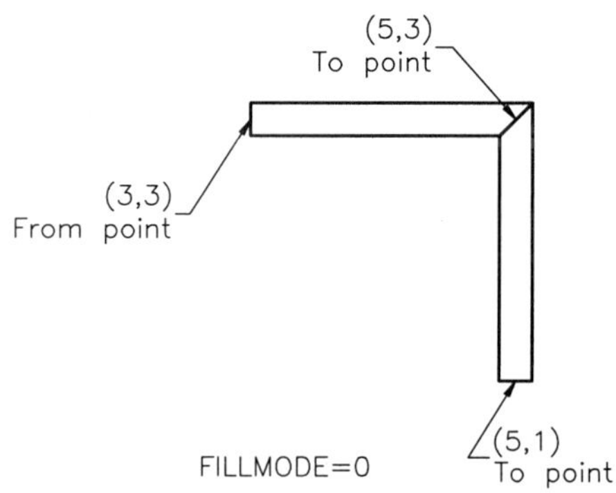

Figure 3-46 *Drawing a nonsolid trace using the* ***TRACE*** *command*

Tip

You should always set the ***FILLMODE*** *off if there are many traces to be drawn because regeneration takes less time for nonsolid traces. You can turn the* ***FILLMODE*** *on when the drawing has taken final shape; then use the* ***REGEN*** *command to fill the traces.*

Exercise 11 *General*

Using the **TRACE** command, draw "Au" on the screen (both as a solid-filled trace and nonsolid trace) as shown in Figure 3-47. Assume the width of the trace = 0.20 units.

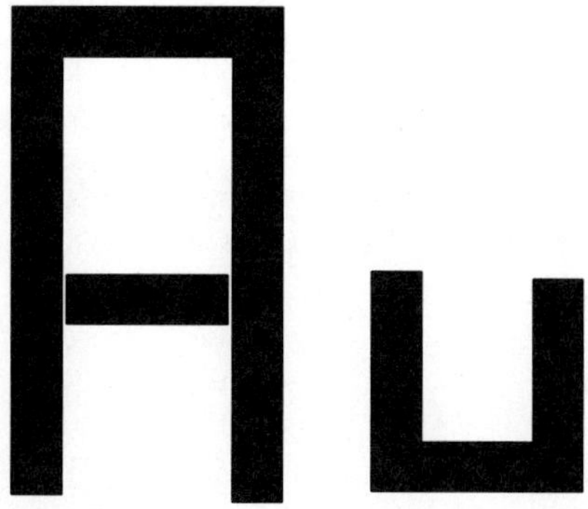

Figure 3-47 *Drawing for Exercise 11*

DRAWING POLYLINES

Toolbar:	Draw > Polyline
Menu:	Draw > Polyline
Command:	PLINE (or PL)

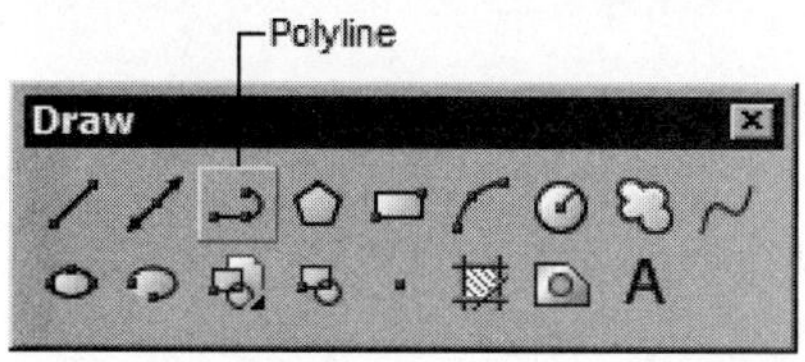

Figure 3-48 *Invoking the **Polyline** command from the **Draw** toolbar*

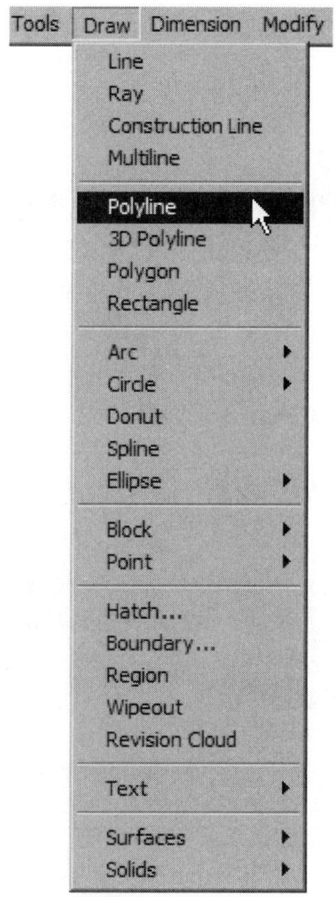

Figure 3-49 *Invoking the **Polyline** command from the **Draw** menu*

A polyline created using the **PLINE** command (Figures 3-48 and 3-49) is a line that can have different characteristics. The term POLYLINE can be broken into two parts: POLY and LINE. POLY means "many". This signifies that a polyline can have many features. Some of the features of polylines are as follows.

1. Polylines, like traces, are thick lines having a desired width.
2. Polylines are very flexible and can be used to draw any shape, such as a filled circle or a doughnut.
3. Polylines can be used to draw objects in any linetype (for example, hidden linetype).
4. Advanced editing commands can be used to edit polylines (for example, the **PEDIT** command).
5. A single polyline object can be formed by joining polylines and polyarcs of different thicknesses.

6. It is easy to determine the area or perimeter of a polyline feature. Also, it is easy to offset when drawing walls.

The **PLINE** command functions fundamentally like the **LINE** command, except that additional options are provided and all the segments of the polyline form a single object. After invoking the **PLINE** command, the next prompt is given next.

> Specify start point: *Specify the starting point or enter its coordinates.*
> Current line width is nn.n.

Current line width is nn.n is displayed automatically, which indicates that the polyline drawn will have nn.n width. If you want the polyline to have a different width, invoke the Width option at the next prompt and then set the polyline width. The next prompt is given next.

> Specify next point or [Arc/Halfwidth/Length/Undo/Width]: *Specify next point or enter an option.*

Depending on your requirements, the options that can be invoked at this prompt are as follows.

Next Point of Line

This option is maintained as the default and is used to specify the next point of the current polyline segment. If additional polyline segments are added to the first polyline, AutoCAD automatically makes the endpoint of the previous polyline segment the start point of the next polyline segment. The prompt sequence is given next.

> Command: **PLINE** Enter
> Specify start point: *Specify the starting point of the polyline.*
> Current line width is 0.0000.
> Specify next point or [Arc/Halfwidth/Length/Undo/Width]: *Specify the endpoint of the first polyline segment.*
> Specify next point or [Arc/Close/Halfwidth/Length/Undo/Width]: *Specify the endpoint of the second polyline segment, or press ENTER to exit the command.*

Width

You can change the current polyline width by entering **W** (width option) at the last prompt. You can also right-click and choose the Width option from the shortcut menu. Then you are prompted for the starting width and the ending width of the polyline.

> Specify next point or [Arc/Halfwidth/Length/Undo/Width]: **W** Enter
> Specify starting width <current>: *Specify the starting width.*
> Specify ending width <starting width>: *Specify the ending width.*

The starting width value is taken as the default value for the ending width. Hence, to have a uniform polyline you need to press ENTER at the **Specify ending width <>** prompt. As in the case of traces, the start point and endpoint of the polyline are located at the center of the line width. To draw a uniform polyline (Figure 3-50) with a width of 0.25 units, start point at (4,5), endpoint at (5,5), and the next endpoint at (3,3), the following is the prompt sequence.

Command: **PLINE** Enter
Specify start point: **4,5** Enter
Current line-width is 0.0000
Specify next point or [Arc/Halfwidth/Length/Undo/Width]: **W** Enter
Specify starting width <current>: **0.25** Enter
Specify ending width <0.25>: Enter
Specify next point or [Arc/Halfwidth/Length/Undo/Width]: **5,5** Enter
Specify next point or [Arc/Close/Halfwidth/Length/Undo/Width]: **3,3** Enter
Specify next point or [Arc/Close/Halfwidth/Length/Undo/Width]: Enter

You can get a tapered polyline by entering two different values at the starting width and the ending width prompts. To draw a tapered polyline (Figure 3-51) with a starting width of 0.5 units and an ending width of 0.15 units, a start point at (2,4), and an endpoint at (5,4), the prompt sequence is given next.

Command: **PLINE** Enter
Specify start point: **2,4** Enter
Current line-width is 0.0000
Specify next point or [Arc/Halfwidth/Length/Undo/Width]: **W** Enter
Specify starting width <0.0000>: **0.50** Enter
Specify ending width <0.50>: **0.15** Enter
Specify next point or [Arc/Halfwidth/Length/Undo/Width]: **5,4** Enter
Specify next point or [Arc/Close/Halfwidth/Length/Undo/Width]: Enter

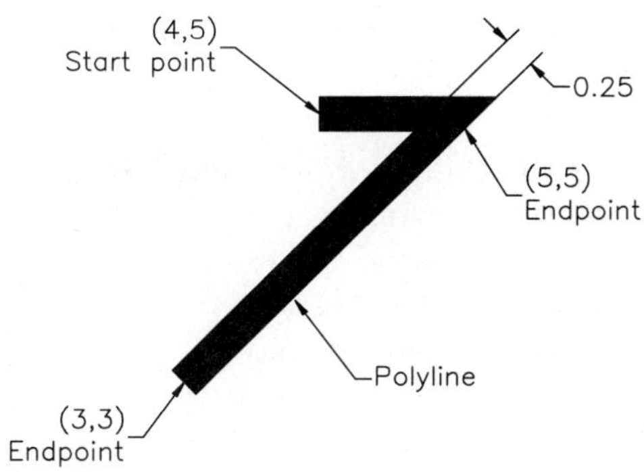

***Figure 3-50** Drawing a uniform polyline using the **PLINE** command*

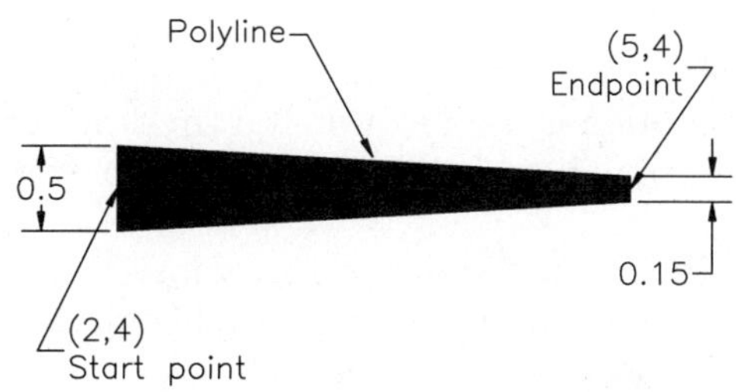

***Figure 3-51** Drawing a tapered polyline using the **PLINE** command*

Halfwidth

With this option you can specify the starting and ending halfwidth of a polyline. This halfwidth distance is equal to half of the actual width of the polyline. This option can be invoked by entering **H** or choosing **Halfwidth** from the shortcut menu at the following prompt.

Specify next point or [Arc/Halfwidth/Length/Undo/Width]: **H** Enter
Specify starting half-width <0.0000>: **0.12** Enter *(Specify desired starting halfwidth.)*
Specify ending half-width <0.1200>: **0.05** Enter *(Specify desired ending halfwidth.)*

Length

This option prompts you to enter the length of a new polyline segment. The new polyline segment will be the length you have entered. It will be drawn at the same angle as the last polyline segment or tangent to the previous polyarc segment. This option can be invoked by entering **L** at the following prompt or by choosing **Length** from the shortcut menu.

Specify next point or [Arc/Close/Halfwidth/Length/Undo/Width]: **L** Enter
Specify length of line: *Specify the desired length of the Pline.*

Undo

This option erases the most recently drawn polyline segment. This option can be invoked by entering **U** at the following prompt.

Specify next point or [Arc/Close/Halfwidth/Length/Undo/Width]: **U** Enter

You can use this option repeatedly until you reach the start point of the first polyline segment. Further use of **Undo** option evokes the message **All segments already undone**.

Close

This option is available when at least one segment of the polyline is drawn. It closes the polyline by drawing a polyline segment from the most recent endpoint to the initial start point. At the same time, it exits from the **PLINE** command. The width of the closing segment can be changed by using the Width/Halfwidth option before invoking the Close option.

Arc

This option is used to switch from drawing polylines to drawing polyarcs, and provides you the options associated with drawing polyarcs. The prompt sequence is given next.

Specify next point or [Arc/Close/Halfwidth/Length/Undo/Width]: **A** Enter
Specify endpoint of arc or [Angle/CEnter/Direction/Halfwidth/Line/Radius/Second pt/Undo/Width]: *Enter an option.*

By default the arc segment is drawn tangent to the previous segment of the polyline. The direction of the previous line, arc, or polyline segment is the default direction for polyarc. The preceding prompt contains options associated with the PLINE Arc. The detailed explanation of each of these options follows.

Angle

This option prompts you to enter the included angle for the arc. If you enter a positive angle, the arc is drawn in a counterclockwise direction from the start point to the endpoint. If the angle specified is negative, the arc is drawn in a clockwise direction. The prompts are given next.

Specify included angle: *Specify the included angle.*
Specify endpoint of arc or [Center/Radius]:

Center refers to the center of the arc segment, Radius refers to the radius of the arc, and Endpoint draws the arc.

CEnter

This option prompts you to specify the center of the arc to be drawn. As mentioned before, usually the arc segment is drawn so that it is tangent to the previous polyline segment; in such cases AutoCAD determines the center of the arc automatically. Hence, the **CEnter** option provides the freedom to choose the center of the arc segment. The CEnter option can be invoked by entering **CE** at the **Specify end point of arc or [Angle/CEnter/Direction/Halfwidth/Line/Radius/Second pt/Undo/Width]** prompt. Once you specify the center point, AutoCAD issues the following prompt:

Specify endpoint of arc or [Angle/Length]:

Angle refers to the included angle, Length refers to the length of the chord, and Endpoint refers to the endpoint of the arc.

Direction

Usually, the arc drawn with the **PLINE** command is tangent to the previous polyline segment. In other words, the starting direction of the arc is the ending direction of the previous segment. The Direction option allows you to specify the tangent direction of your choice for the arc segment to be drawn. You can specify the direction by specifying a point. The prompts are given next.

Specify tangent direction for the start point of arc: *Specify the direction.*
Specify endpoint of arc: *Specify the endpoint of arc.*

Halfwidth

This option is the same as for the PLine and prompts you to specify the starting and ending halfwidth of the arc segment.

Line

This option takes you back to the **Line mode**. You can draw polylines only in this mode.

Radius

This option prompts you to specify radius of the arc segment. The prompt sequence is given next.

Specify radius of arc: *Specify the radius of the arc segment.*
Specify endpoint of arc or [Angle]:

If you specify a point, the arc segment is drawn. If you enter an angle, you will have to specify the angle and the direction of the chord at the **Specify included angle** and **Specify direction of chord for arc<current>** prompts, respectively.

Second pt

This option selects the second point of an arc in the three-point arc option. The prompt sequence is given next.

Specify second point of arc: *Specify the second point on the arc.*
Specify endpoint of arc: *Specify the third point on the arc.*

Undo

This option reverses the changes made in the previously drawn segment.

Width

This option prompts you to enter the width of the arc segment. To draw a tapered arc segment you can enter different values at the starting width and ending width prompts. The prompt sequence is identical to that for the polyline. Also, a specified point on a polyline refers to the midpoint on its width.

Endpoint of arc

This option is maintained as the default and prompts you to specify the endpoint of the current arc segment. The following is the prompt sequence for drawing an arc with start point at (3,3), endpoint at (3,5), starting width of 0.50 units, and ending width of 0.15 units (Figure 3-52).

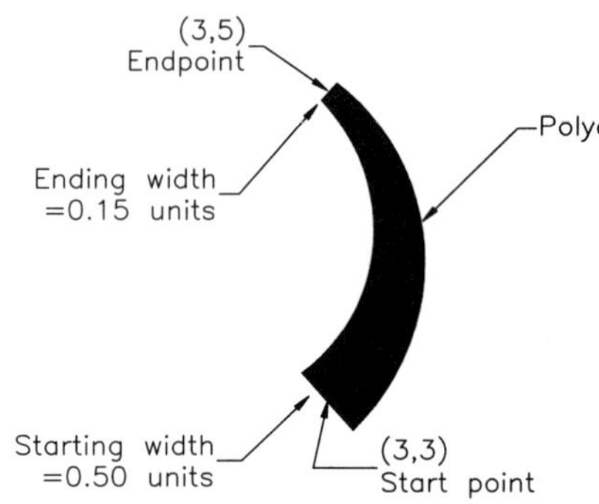

***Figure 3-52** Drawing a Polyarc*

Command: **PLINE** Enter
Specify start point: **3,3** Enter
Current line-width is 0.0000
Specify next point or [Arc/Close/Halfwidth/Length/Undo/Width]: **A** Enter
Specify endpoint of arc or [Angle/CEnter/CLose/Direction/Halfwidth/Line/Radius/Second pt/Undo/Width]: **W** Enter
Specify starting width <current>: **0.50** Enter
Specify ending width <0.50>: **0.15** Enter
Specify endpoint of arc or [Angle/CEnter/CLose/Direction/Halfwidth/Line/Radius/Second pt/Undo/Width]: **3,5** Enter
Specify endpoint of arc or [Angle/CEnter/CLose/Direction/Halfwidth/Line/Radius/Second pt/Undo/Width]: Enter

Tip

*After invoking the **PLINE** command and specifying the start point, you can right-click to display the shortcut menu. You can choose any Pline option directly from the shortcut menu instead of entering the appropriate letters at the Command prompt. Similarly after invoking the **Arc** option of the PLINE command, you can right-click to display the shortcut menu and choose any polyarc option.*

Note

*If **FILL** is On or if **FILLMODE** is 1, the polylines are drawn filled. If you change FILL to OFF or **FILLMODE** to 0, only the outlines are drawn for the new plines and previously drawn plines are also changed from filled to empty. However, note that the change is effective on regeneration. Similarly it works in reverse also.*

*Also, **PLINEGEN** system variable controls the linetype pattern between the vertex points of a 2D polyline. A value of 0 centers the linetype for each polyline segment and 1 makes them continuous.*

Optimized Polylines

The optimized polylines also known as **lightweight polylines** are created when you use the **PLINE** command (an AutoCAD feature since Release 14). The optimized polylines are similar to regular 2D polylines in functionality, but the database format of an optimized polyline is different from that of a 2D polyline. In case of an optimized polyline, the vertices are not stored as separate entities, but as a single object with an array of information. This feature results in reduced object and file size.

If you use the **PEDIT** command (discussed in Chapter 18) on an optimized polyline and use the Spline or Fit options, the polyline automatically becomes a regular 2D polyline. Also, when you load a Release 13 or earlier drawing, AutoCAD automatically converts the polylines into optimized polylines. This is dependent on the **PLINETYPE** system variable. The default value of **2** for this variable automatically converts a 2D polyline into an optimized polyline while opening any old drawing. Commands such as **POLYGON**, **DONUT**, **SKETCH**, **ELLIPSE**, **PEDIT**, **BOUNDARY**, and **RECTANG** create optimized polylines. To see the difference between these two polylines, draw a polyline consisting of several segments and then use the **LIST** command to list the information about the object. Now, use the **PEDIT** command to change this polyline into a splined polyline and then use the **LIST** command again. Notice the difference in the database associated with these polylines.

CONVERT Command

You can use the **CONVERT** command to manually change a 2D polyline to an optimized (lightweight) polyline. You can also use this command to change a Release 13 associative hatch pattern to an AutoCAD 2004 hatch object, with the exception of the solid-fill hatches.

Command: **CONVERT** Enter
Enter type of objects to convert [Hatch/Polyline/All]: *Enter an option.*
Enter object selection preference [Select/All]<All>:

Tip

*The **CONVERT** command is used very rarely to change the polylines because the 2D polylines in any old drawing are automatically converted to lightweight polylines when you open the drawings in AutoCAD 2004. This is dependent on the **PLINETYPE** system variable that has a default value of 2. If you want the old 2D polylines should not get converted to lightweight polylines when you open the drawings, set the value of the **PLINETYPE** variable to 0. This also lets you create 2D polylines in AutoCAD 2004.*

Exercise 12 — *General*

Draw the objects shown in Figures 3-53 and 3-54. Approximate the width of different polylines.

Figure 3-53 *Drawing for Exercise 12*

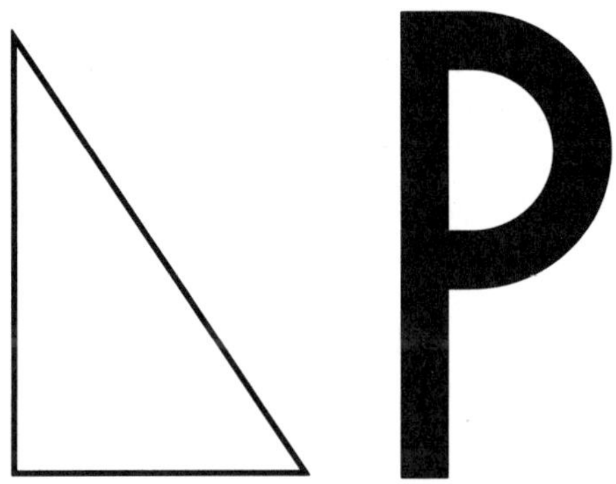

Figure 3-54 *Drawing for Exercise 12*

DRAWING DONUTS

Menu:	Draw > Donut
Command:	DONUT

In AutoCAD the **DONUT** or **DOUGHNUT** command is issued to draw an object that looks like a filled circle ring called a donut. Actually, AutoCAD's donuts are made of two semicircular polyarcs having a certain width. Hence the **DONUT** command allows you to draw a thick circle. The donuts can have any inside and outside diameters. If the FILLMODE is off, the donuts look like circles (if the inside diameter is zero) or concentric circles (if the inside diameter is not zero). After specifying the two diameters, the donut gets attached to the crosshair. You can select a point for the center of the donut anywhere on the screen with the help of a pointing device, and then place the donut. You can place the donuts by clicking your pointing device. The prompt sequence for drawing donuts is given next.

Command: **DONUT** Enter
Specify inside diameter of donut <current>: *Specify the inner diameter of the donut.*
Specify outside diameter of donut <current>: *Specify the outer diameter of the donut.*
Specify center of donut or <exit>: *Specify the center of the donut.*
Specify center of donut or <exit>: *Specify the center of the donut to draw more donuts of previous specifications or give a null response to exit.*

The defaults for the inside and outside diameters are the respective diameters of the most recent donut drawn. The values for the inside and outside diameters are saved in the **DONUTID** and **DONUTOD** system variables. A solid-filled circle is drawn by specifying the inside diameter as zero (FILLMODE is on). Once the diameter specification is completed, the donuts are formed at the crosshairs and can be placed anywhere on the screen. For the location, you can enter the coordinates of the point or specify the point by dragging the center point. Once you have specified the center of the donut, AutoCAD repeats the **Specify center of donut or <exit>** prompt. As you go on specifying the locations for the center point, donuts with the specified diameters are drawn at specified locations. To end the **DONUT** command, give a null response to this prompt by pressing ENTER. Since donuts are circular polylines, the donut can be edited with the **PEDIT** command or any other editing command that can be used to edit polylines.

Example 2 *General*

You will draw an unfilled donut with an inside diameter of 0.75 units, an outside diameter of 2.0 units, and centered at (2,2) (Figure 3-55). You will also draw a filled donut and a solid-filled donut with the specifications given.

The following is the prompt sequence to draw an unfilled donut (Figure 3-55).

Command: **FILLMODE** [Enter]
New value for FILLMODE <1>: **0** [Enter]

Command: **DONUT** [Enter]
Specify inside diameter of donut<0.5000>: **0.75** [Enter]
Specify outside diameter of donut <1.000>: **2** [Enter]
Specify center of donut or <exit>: **2,2** [Enter]
Specify center of donut or <exit>: [Enter]

The following is the prompt sequence for drawing a filled donut with an inside diameter of 0.5 units, outside diameter of 2.0 units, centered at a specified point (Figure 3-56).

Command: **FILLMODE** [Enter]
Enter new value for FILLMODE <0>: **1** [Enter]

Command: **DONUT** [Enter]
Specify inside diameter of donut<0.5000>: **0.50** [Enter]
Specify outside diameter of donut <1.000>: **2** [Enter]
Specify center of donut or <exit>: *Specify a point.*
Specify center of donut or <exit>: [Enter]

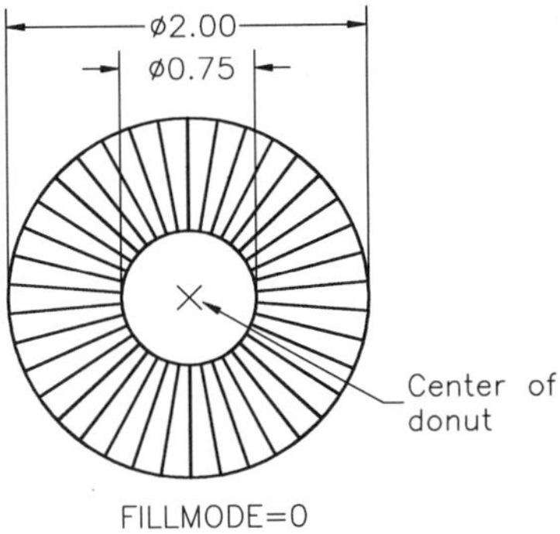

Figure 3-55 *Drawing an unfilled donut using the **DONUT** command*

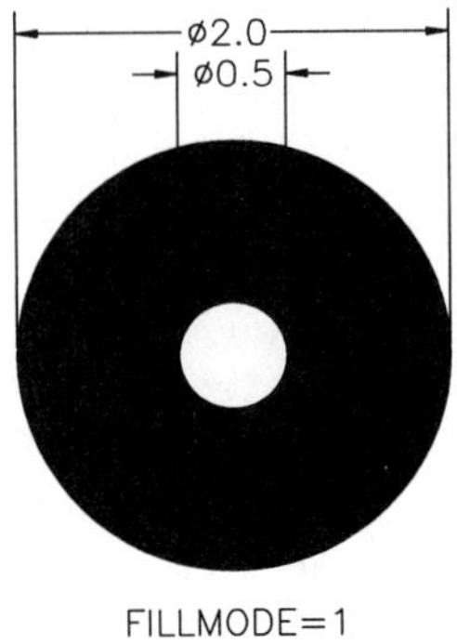

Figure 3-56 *Drawing a filled doughnut using the **DONUT** command*

To draw a solid-filled donut with an outside diameter of 2.0 units (Figure 3-57), the following is the prompt sequence.

Command: **DONUT** [Enter]
Specify inside diameter of donut <0.50>: **0** [Enter]
Specify outside diameter of donut <1.0>: **2** [Enter]
Specify center of donut or <exit>: *Specify a point.*
Specify center of donut or <exit>: [Enter]

Figure 3-57 *Solid-filled donut*

DRAWING POINTS

Toolbar:	Draw > Point
Menu:	Draw > Point
Command:	POINT

The **point** is the basic drawing object. Points are invaluable in building a drawing file. To draw a point anywhere on the screen, AutoCAD provides the **POINT** command (Figures 3-58 and 3-59).

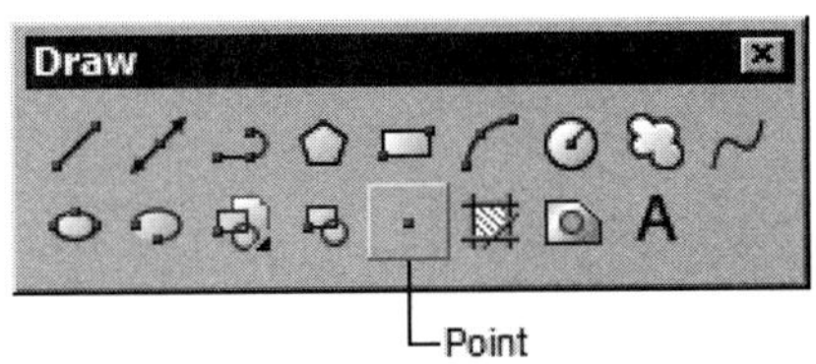

Figure 3-58 *Invoking the **POINT** command from the **Draw** toolbar*

Command: **POINT** [Enter]
Current point modes: PDMODE=n PDSIZE=n.nn
Specify a point: *Specify the location where you want to place the point.*

If you invoke the **POINT** command from the toolbar or the menu (**Multiple Point** option), you can draw as many points as you desire in a single command. In this case you can exit from the **POINT** command by pressing ESC. If you invoke this command by entering **POINT**

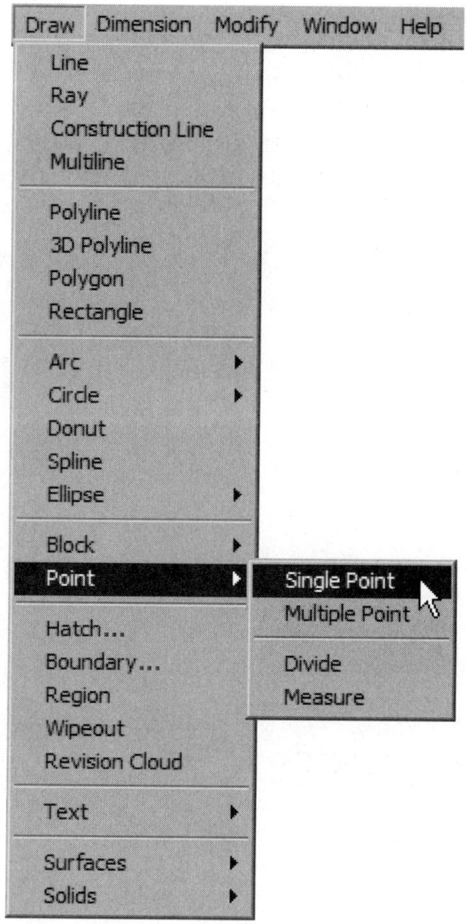

Figure 3-59 *Invoking the **POINT** command from the **Draw** menu*

at the Command prompt or use the **Single Point** option from the menu, you can draw only one point in a single point command.

Note

*It is possible to have a temporary construction marker for the point known as blip. A mark appears on the screen where you place the point. This blip mark can then be cleared once the screen is redrawn using the **REDRAW** command and the point is left on the screen. The visibility of blips can be controlled using the **BLIPMODE** system variable.*

Changing the Point Type

Menu:	Format > Point Style
Command:	DDPTYPE

The point type can be set from the **Point Style** dialog box (Figure 3-60). There are twenty combinations of point types. The **Point Style** dialog box can be accessed from the **Format** menu (Select Format > Point Style) as shown in Figure 3-61. You can choose a point style in this dialog box, which is indicated by highlighting that particular point style. Next, choose the **OK** button. Now all the points will be drawn in the selected style until you change it to a new style. The type of point drawn is stored in the **PDMODE** (Point Display MODE) system variable. You can change the point style by entering a numeric value in the PDMODE variable.

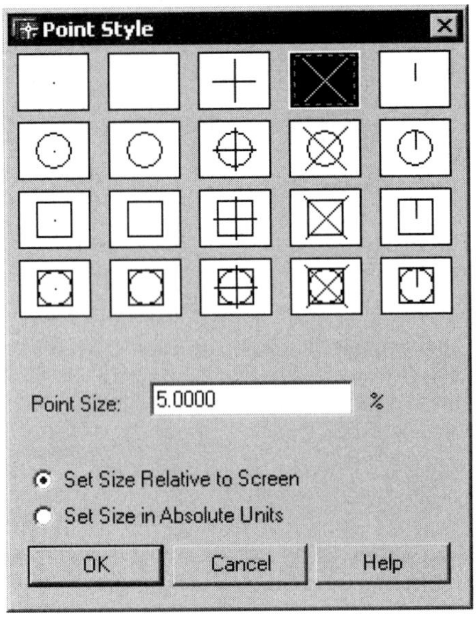

***Figure 3-60 Point Style** dialog box*

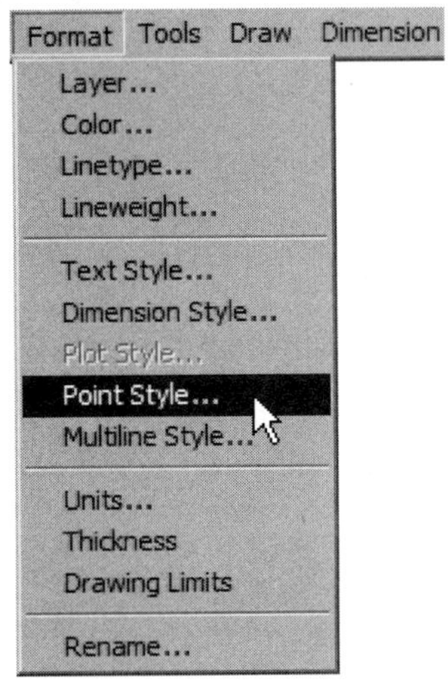

***Figure 3-61** Choosing **Point Style** from the **Format** menu*

The PDMODE values for different point types area as follows.

1. A value of 0 is the default value for the PDMODE variable and generates dot at the specified point.
2. A value of 1 for the PDMODE variable generates nothing at the specified point.
3. A value of 2 for the PDMODE variable generates a plus sign (+) at the specified point.
4. A value of 3 for the PDMODE variable generates a cross mark (X) at the specified point.
5. A value of 4 for the PDMODE variable generates a vertical line in the upward direction from the specified point.
6. When you add **32** to the **PDMODE** values of 0 to 4, a circle is generated around the symbol obtained from the original **PDMODE** value. For example, to draw a point having a cross mark and a circle around it, the value of the **PDMODE** variable will be **3 + 32 = 35.** Similarly, you can generate a square around the symbol with the **PDMODE** of value 0 to 4 by adding **64** to the original **PDMODE** value. For example, to draw a point having a plus sign and a square around it, the value of the **PDMODE** variable will be **2 + 64 = 66**. You can also generate a square and a circle around the symbol with the **PDMODE** of value 0 to 4 by adding **96** to the original **PDMODE** value. For example, to draw a point having a dot mark and a circle and a square around it, the value of the **PDMODE** variable

will be **0 + 96 = 96**. Figure 3-62 shows the **PDMODE** values for different point types area

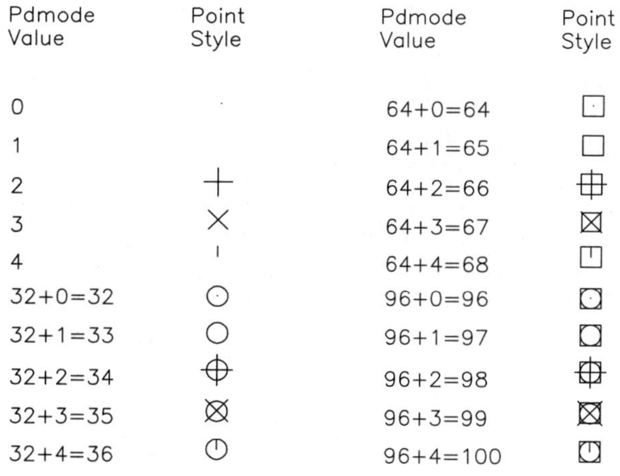

***Figure 3-62** Different point style for **PDMODE** values*

Exercise 13 — *General*

Check what types of points are drawn for each value of the PDMODE variable.

Changing the Point Size

Menu:	Format > Point Style
Command:	DDPTYPE

***Figure 3-63** Changing point size using the PDSIZE variable*

The size of a point can be set from the **Point Style** dialog box (Figure 3-60) by entering the desired point size in the **Point Size** edit box. You can generate the point at a specified percentage of the graphics area height or define an absolute size for the point. An absolute size for the point can be specified by selecting the **Set Size in Absolute Units** radio button in the **Point Style** dialog box and then entering a value in the **Point Size** edit box. The point size can also be set by changing the value of PDSIZE (Figure 3-63). The variable **PDSIZE** governs the size of the point (except for the **PDMODE** values of 0 and 1). You can set the size in absolute units by specifying a positive value for the **PDSIZE** variable. If the **Set Size Relative to Screen** radio button is selected in the **Point Style** dialog box, the size is taken as a percentage of the viewport size. This can also be set by entering a negative value for the **PDSIZE** variable. For example, a setting of 5 makes the point 5 units high; a setting of -5 makes the point 5 percent of the current drawing area.

Exercise 14 *General*

a. Try various combinations of the **PDMODE** and **PDSIZE** variables.
b. Check the difference between the points generated from negative values of **PDSIZE** and the points generated from positive values of **PDSIZE**.

DRAWING INFINITE LINES

The **XLINE** and **RAY** commands can be used to draw construction or projection lines. These are lines that aid in construction or projection and are drawn very lightly when drafting manually. An **xline** (construction line) is a 3D line that extends to infinity on both ends (Figure 3-66). Since the line is infinite in length, it does not have any endpoints. A **ray** is a 3D line that extends to infinity on only one end. The other end of the ray has a finite endpoint. The xlines and rays have zero extents. This means that the extents of the drawing will not change if you use the commands that change the drawing extents, such as the **ZOOM** command with the **All** option. Most of the object snap modes work with both xlines and rays, with some limitations: You cannot use the **Endpoint** object snap with the xline because by definition an xline does not have any endpoints. However, for rays you can use the **Endpoint** snap on one end only. Also, xlines and rays take the properties of the layer in which they are drawn.

Tip
Xlines and rays plot like any other objects in a drawing and hence may create confusion. It is therefore a good idea to create the construction lines in a different layer altogether, such that you can recognize them easily.

XLINE Command

Toolbar:	Draw > Construction Line
Menu:	Draw > Construction Line
Command:	XLINE

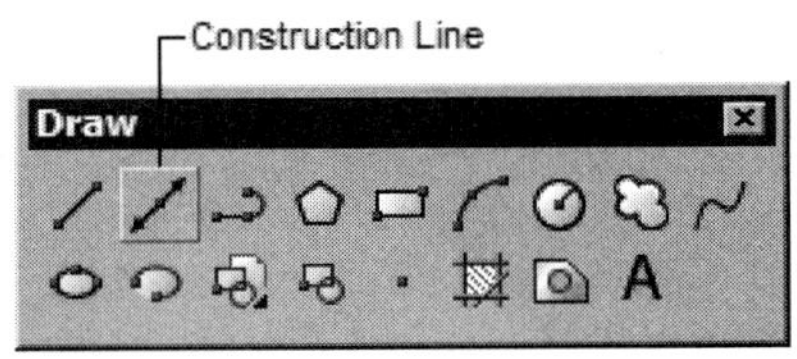

Figure 3-64 *Choosing the **Construction Line** button from the **Draw** toolbar*

When you invoke the **XLINE** command (Figures 3-64 and 3-65) the prompt sequence is as follows.

Command: **XLINE** Enter
Specify a point or [Hor/Ver/Ang/Bisect/Offset]:
Specify an option or select a point through which the xline will pass.

The various options of the command are discussed next.

Point

If you use the default option, AutoCAD will prompt you to select two points through which the xline shall pass at the **Specify a point** and the **Specify through point** prompts. After you select the first point, AutoCAD will dynamically rotate the xline through the specified point as you move the cursor. When you select the second point, an xline will be created that passes through the first and second points (Figure 3-66).

Command: **XLINE** [Enter]
Specify a point or [Hor/Ver/Ang/Bisect/Offset]: *Specify a point.*
Specify through point: *Specify the second point.*

You can continue to select more points to create more xlines. All these xlines will pass through the first point you had selected at the **Specify a point** prompt. This point is also called the root point. Right-click or press ENTER to end the command.

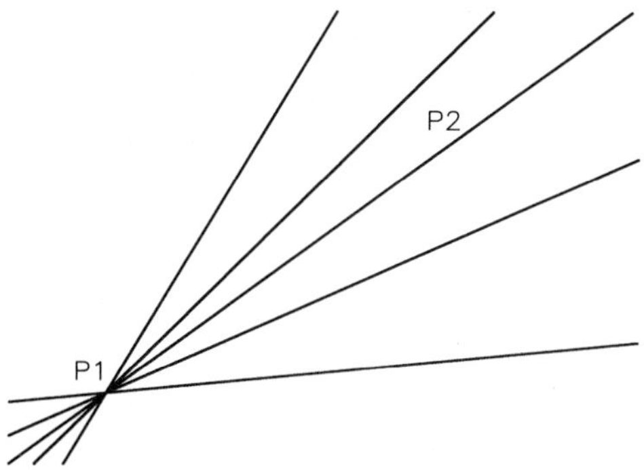

Figure 3-66 *Drawing xlines*

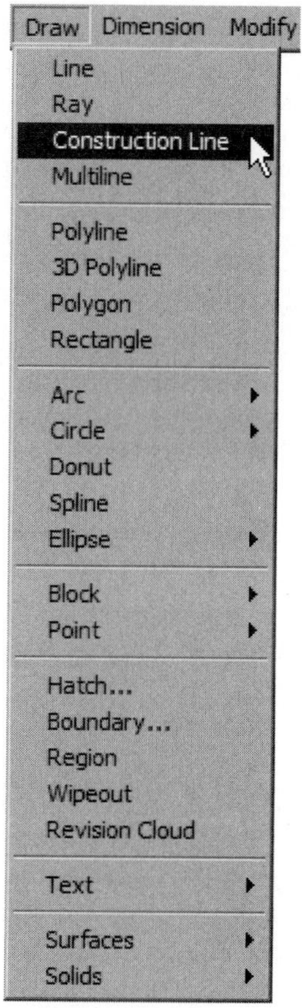

Figure 3-65 *Choosing* ***Construction Line*** *from the* ***Draw*** *menu*

Horizontal

This option will create horizontal xlines of infinite length that pass through the selected points. The xlines will be parallel to the *X* axis of the current UCS (Figure 3-67). As you invoke this option the horizontal xline gets attached to the cursor. You are prompted to select only one point through which the horizontal xline passes. You can continue selecting points to draw horizontal xlines and right-click or press ENTER to end the command.

Vertical

This option will create vertical xlines of infinite length that pass through the selected points. The xlines will be parallel to the *Y* axis of the current UCS (Figure 3-67). As you invoke this option the vertical xline gets attached to the cursor. You are prompted to select only one point through which the vertical xline passes. You can continue selecting points to draw vertical xlines and right-click or press ENTER to end the command.

Angular

This option will create xlines of infinite length that pass through the selected point at a specified angle (in Figure 3-68, the angle specified is 38-degree). The angle can be specified by entering a value at the keyboard. You can also use the reference option by selecting an object and then specifying an angle relative to it. The **Reference** option is useful when the actual angle is not known but the angle relative to an existing object can be specified.

Command: **XLINE** Enter
Specify a point or [Hor/Ver/Ang/Bisect/Offset]: **A** Enter
Enter angle of xline (0) or [Reference]: **R** Enter *(Here you use the **Reference** method for specifying the angle.)*
Select a line object: *Select a line.*
Enter angle of xline <0>: *Enter angle (the angle will be measured counterclockwise with respect to the selected line.)*
Specify through point: *Specify the second point.*

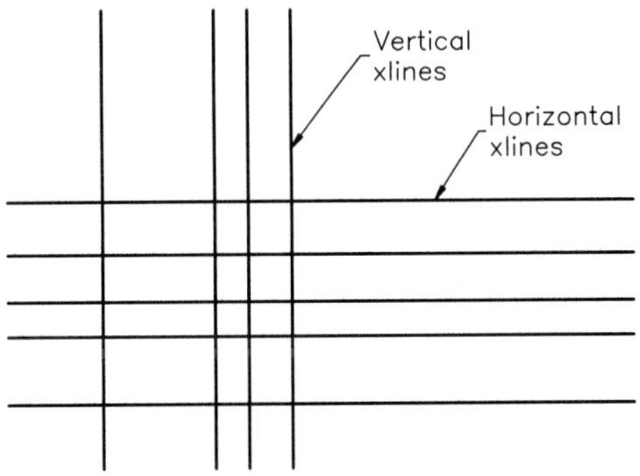

Figure 3-67 *Horizontal and vertical xlines*

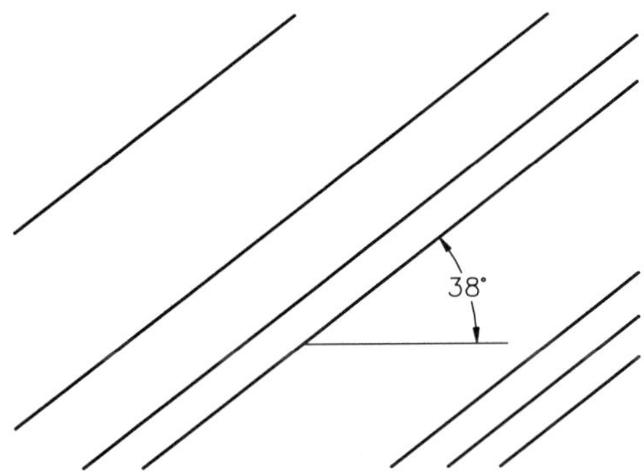

Figure 3-68 *Angular xlines*

Bisect

This option will create an xline that passes through the angle vertex and bisects the angle you specify by selecting two points. The xline created using this option will lie in the plane defined by the selected points. You can use the object snaps to select the points on the existing objects. The following is the prompt sequence for this option (Figure 3-69).

Command: **XLINE** Enter
Specify a point or [Hor/Ver/Ang/Bisect/Offset]: **B** Enter
Specify angle vertex point: *Enter a point (P1).*
Specify angle start point: *Enter a point (P2).*
Specify angle end point: *Enter a point (P3).*
Specify angle end point: *Select more points or press ENTER or right-click to end command.*

Offset

The **Offset** option creates xlines that are parallel to the selected line/xline at a specified offset distance. You can specify the offset distance by entering a numerical value or by selecting two points on the screen. If you select the **Through** option, the offset line will pass through the selected point. This option works like the **OFFSET** editing command. The prompts at the command line are as follows.

Command: **XLINE** Enter

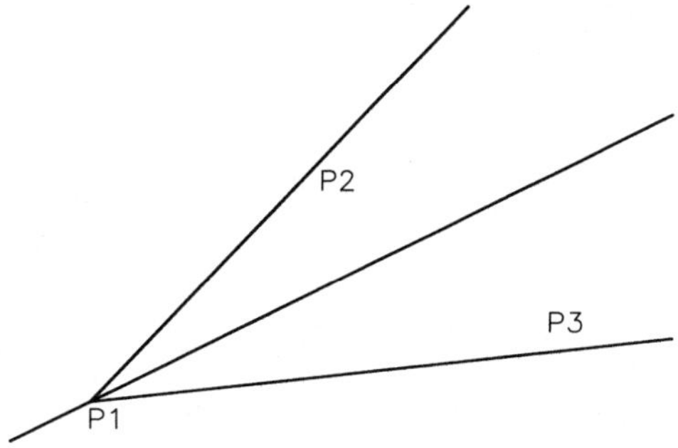

Figure 3-69 *Using the **Bisect** option to draw xlines*

Specify a point or [Hor/Ver/Ang/Bisect/Offset]: **O** Enter
Specify offset distance or [Through] <Through>: *Press ENTER to accept the **Through** option or specify a distance from the selected line object at which the xline shall be drawn.*
Select a line object: *Select the object to which the xline is drawn parallel at a specified distance.*
Specify through point: *Select a point through which the xline should pass.*

If you specify the offset distance, and after you have selected a line object, you are prompted to specify the direction in which the xline is to be offset. You can continue drawing xlines or right-click or press ENTER to end the command.

RAY Command

Menu:	Draw > Ray
Command:	RAY

A ray is a 3D line similar to the xline construction line with the difference being that it extends to infinity only in one direction. It starts from a point you specify and extends to infinity through the specified point. The prompt sequence is give next.

Command: **RAY** Enter
Specify start point: *Select the starting point for the ray.*
Specify through point: *Specify the second point.*

Press ENTER or right-click to exit the command.

Note
When you trim an xline, it gets converted into a ray and when a ray is trimmed at the end that is infinite, it gets converted into a line object.

WRITING SINGLE LINE TEXT

Toolbar:	Text > Single Line Text
Menu:	Draw > Text > Single Line Text
Command:	TEXT

The **TEXT** command lets you write single line text in the drawing. Although you can also write more than one lines of text using this command, but all the lines will be separate text entities. After invoking this command, you have to specify the start point for the text. Then you need to specify the height of the text and also the rotation angle. The characters appear on the screen as you enter them. When you press ENTER after typing a line, the cursor automatically places itself at the start of the next line and repeats the prompt for entering another line. You can end the command by pressing the ENTER key. You can use the BACKSPACE key to edit the text on the screen while you are writing it. The prompt sequence is given next.

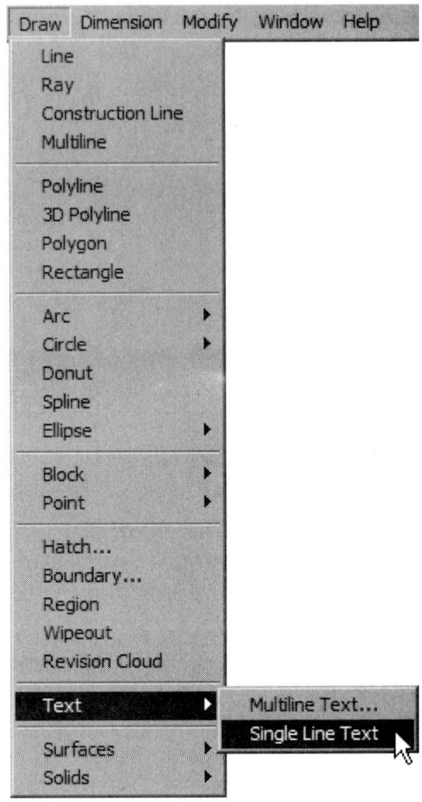

***Figure 3-70** Invoking the **TEXT** command from the **Draw** menu*

Command: **TEXT** Enter
Current text style: "current" Text height: current
Specify start point of text or [Justify/Style]: *Specify the starting point of the text.*
Specify height<current>: *Enter the text height.*
Specify rotation angle of text <0>: Enter
Enter text: *Enter first line of text.*
Enter text: *Enter the second line of text.*
Enter text: Enter

Note

The other commands to enter text are discussed in detail in Chapter 7.

*If you want to move the objects, use the **MOVE** command and then select the objects, specify the base point, and the second point of displacement. The **MOVE** command is discussed in detail in Chapter 5.*

Self-Evaluation Test

Answer the following questions and then compare your answers to the correct answers given at the end of this chapter.

1. A negative value for chord length in the **Start, Center, Length** option of the **ARC** command results in the largest possible arc, also known as the major arc. (T/F)
2. In the **ARC** command, if you do not specify a start point but just press ENTER or choose the **Continue** option, the start point and direction of the arc is taken from the endpoint and ending direction of the previous line or arc drawn on the current screen. (T/F)
3. If the **PELLIPSE** is set to 1, AutoCAD creates a true ellipse, also known as NURBS-based (Non-Uniform Rational Bezier Spline) ellipse. (T/F)
4. The start and end parameters of an elliptical arc are determined by specifying a point on the circle whose diameter is equal to the minor diameter of the ellipse. (T/F)
5. The ________ option of the **RECTANG** command allows you to draw a rectangle at a specified distance from the *XY* plane along the *Z* axis.
6. If the **FILLMODE** is off, only the trace ________ is drawn.
7. You can get a ________ polyline by entering two different values at the starting width and the ending width prompts.
8. In case of a(n) ___________ polyline the vertices are not stored as separate entities, but as a single object with an array of information.
9. If you invoke the **POINT** command from the ___________, you can draw as many points as you desire in a single command.
10. The size of a point is taken as a percentage of the viewport size if you enter a ___________ value for the **PDSIZE** variable.

Review Questions

Answer the following questions.

1. Using the **Start, End, Angle** option of the **ARC** command, a negative included angle value draws the arc in a clockwise direction. (T/F)
2. When the **Continue** option of the **ARC** command is used to draw arcs, each successive arc is perpendicular to the previous one. (T/F)

Chapter 3

3. If you specify the chamfer distances first and then specify the fillet radius in the same **RECTANG** command, the rectangle will be drawn with chamfers only. (T/F)

4. Using the **RECTANG** command, the rectangle drawn is treated as a combination of different objects; therefore individual sides can be edited independently. (T/F)

5. Using the **Start, Center, Length** option of the **ARC** command, a positive chord length generates the smallest possible arc (minor arc) with this length and the arc is always less than

   (a) 90-degree (b) 180-degree
   (c) 270-degree (d) 360-degree

6. Which one of the following options of the **RECTANG** command allows you to draw a rectangle that is extruded in the Z direction by the specified value.

   (a) **Elevation** (b) **Thickness**
   (c) **Extrude** (d) **Width**

7. Which of the following commands draws a line anywhere in 3D space which starts from a point that you specify with the other end extending to infinity.

   (a) **PLINE** (b) **RAY**
   (c) **XLINE** (d) **MLINE**

8. If the old 2D polylines should not get converted to lightweight polylines when opening the drawings in AutoCAD 2004 and also the new polylines drawn should be 2D polylines, the **PLINETYPE** variable should be set to which of the following values?

   (a) 0 (b) 1
   (c) 2 (d) 3

9. Which of the following values should be assigned to the **PDMODE** variable such that a cross mark (X) is generated through the specified point.

   (a) 1 (b) 3
   (c) 5 (d) 7

10. A polygon is said to be __________ when it is drawn inside an imaginary circle and its vertices (corners) touch the circle.

11. If additional polyline segments are added to the first polyline, AutoCAD automatically makes the __________ of the first polyline segment the start point of the next polyline segment.

12. The drawing of each segment of a trace is __________ until you specify the next segment or you end the trace by pressing ENTER.

13. With the **DONUT** command, a solid-filled circle is drawn by specifying the inside diameter as __________ and FILLMODE is on.

14. The visibility of blips can be controlled using the __________ system variable.

15. An absolute size for the point can be specified by entering a __________ value for **PDSIZE** variable.

16. The __________ option of the **XLINE** creates xlines of infinite length that are parallel to the *Y* axis of the current UCS.

17. You can use the __________ key to edit the text on the screen while you are writing it using the **TEXT** command.

Exercises

Exercise 15 *Mechanical*

Draw the sketch shown in Figure 3-71. The distance between the dotted lines is 1.0 unit. Create the radii using appropriate **ARC** command options.

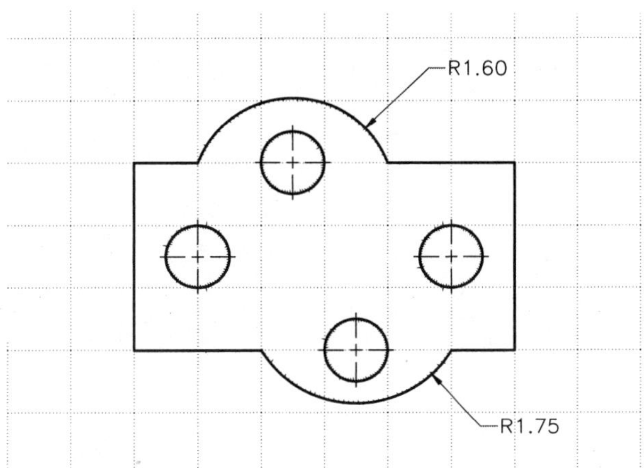

Figure 3-71 *Drawing for Exercise 15*

Exercise 16 *Graphics*

Draw the sketch shown in Figure 3-72. The distance between the dotted lines is 1.0 unit. Create the radii using appropriate **ARC** command options.

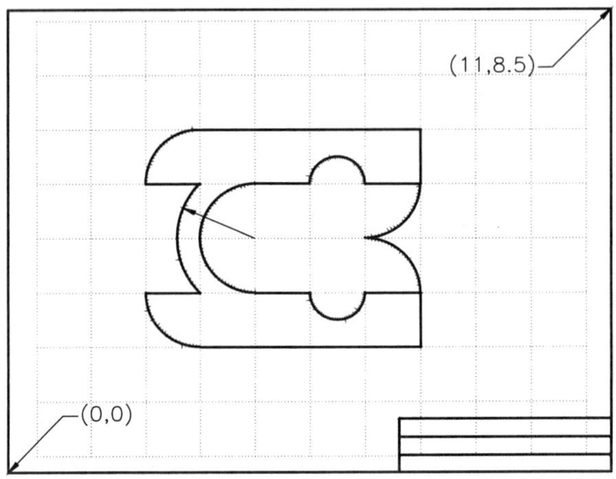

Figure 3-72 *Drawing for Exercise 16*

Exercise 17 *Mechanical*

Draw the sketch shown in Figure 3-73. The distance between the dotted lines is 0.5 unit. Create the ellipses using the **ELLIPSE** command.

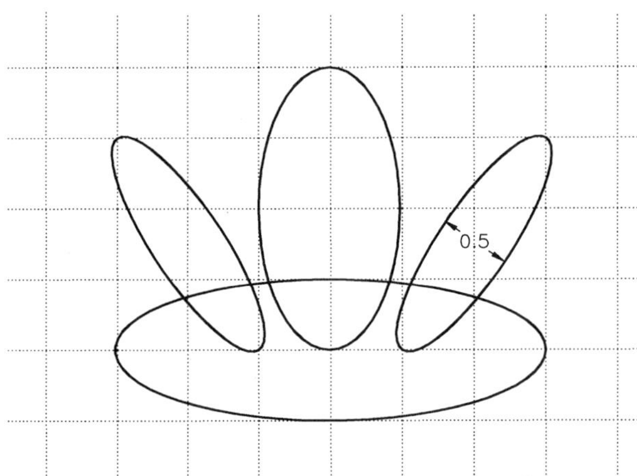

Figure 3-73 *Drawing for Exercise 17*

Exercise 18 *Mechanical*

Draw Figure 3-74 using the **LINE, CIRCLE,** and **ARC** commands. The distance between the dotted lines is 1.0 unit and the diameter of the circles is 1.0 units.

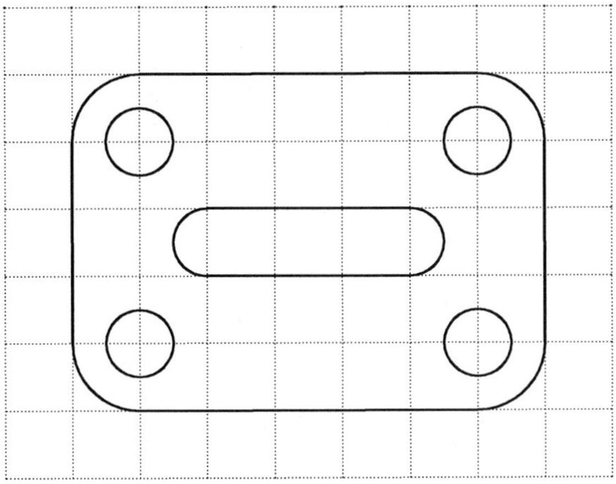

Figure 3-74 *Drawing for Exercise 18*

Exercise 19 *Mechanical*

Draw Figure 3-75 using the **LINE, CIRCLE,** and **ARC** commands or their options. The distance between the grid lines is 1.0 unit and the diameter of the circle is 1.0 unit.

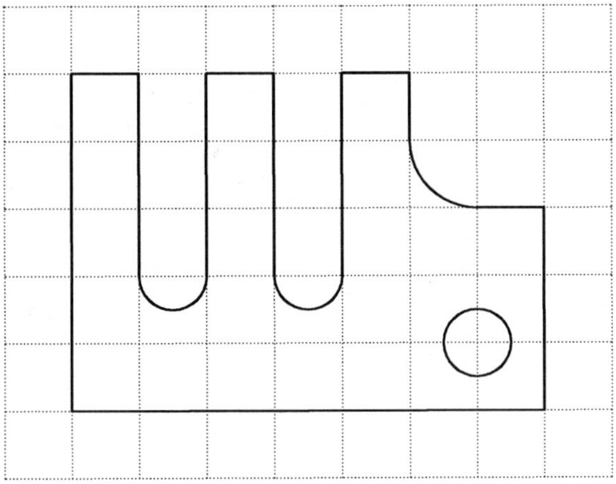

Figure 3-75 *Drawing for Exercise 19*

Problem Solving Exercise 1

Mechanical

Draw the sketch shown in Figure 3-76. Create the radii by using the arc command options indicated in the drawing. (Use the @ symbol to snap to the previous point. Example: Specify start point of arc or [Center]: @)

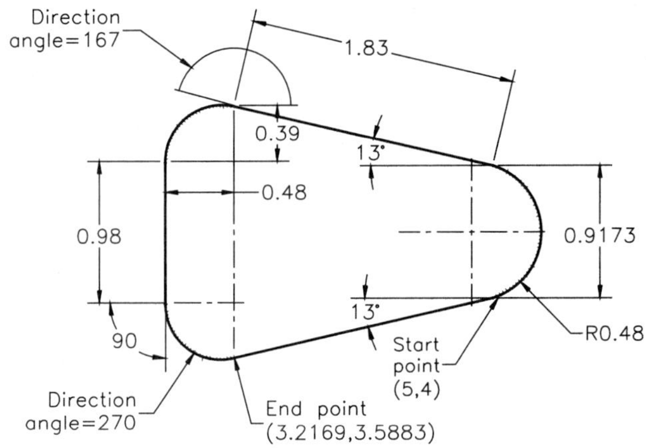

Figure 3-76 *Drawing for Problem Solving Exercise 1*

Problem Solving Exercise 2

Mechanical

Draw the sketch shown in Figure 3-77. Create the radii by using the arc command options. The distance between the dotted lines is 0.5 units.

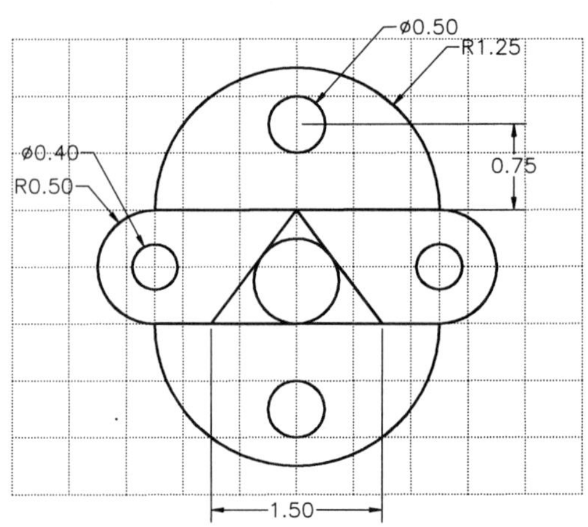

Figure 3-77 *Drawing for Problem Solving Exercise 2*

Problem Solving Exercise 3 *Mechanical*

Draw the sketch shown in Figure 3-78. Create the radii by using the **ARC** command options. The distance between the dotted lines is 1.0 unit.

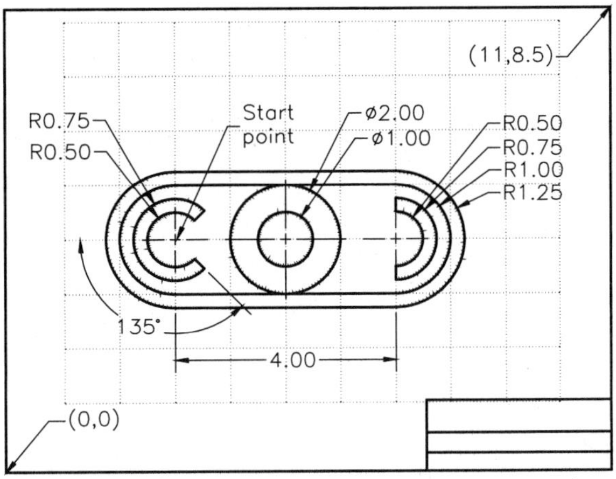

Figure 3-78 *Drawing for Problem Solving Exercise 3*

Problem Solving Exercise 4 *Mechanical*

Draw the sketch shown in Figure 3-79 using the **POLYGON**, **CIRCLE**, and **LINE** commands.

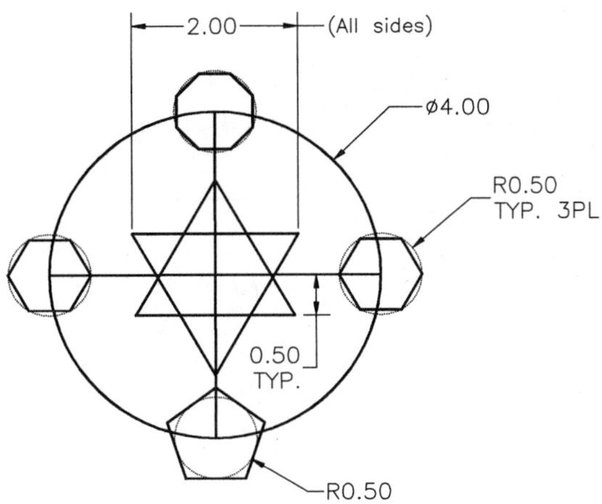

Figure 3-79 *Drawing for Problem Solving Exercise 4*

Problem Solving Exercise 5 *Mechanical*

Draw the sketch shown in Figure 3-80 using the draw commands. Note, Sin30=0.5, Sin60=0.866. The distance between the dotted lines is 1 unit.

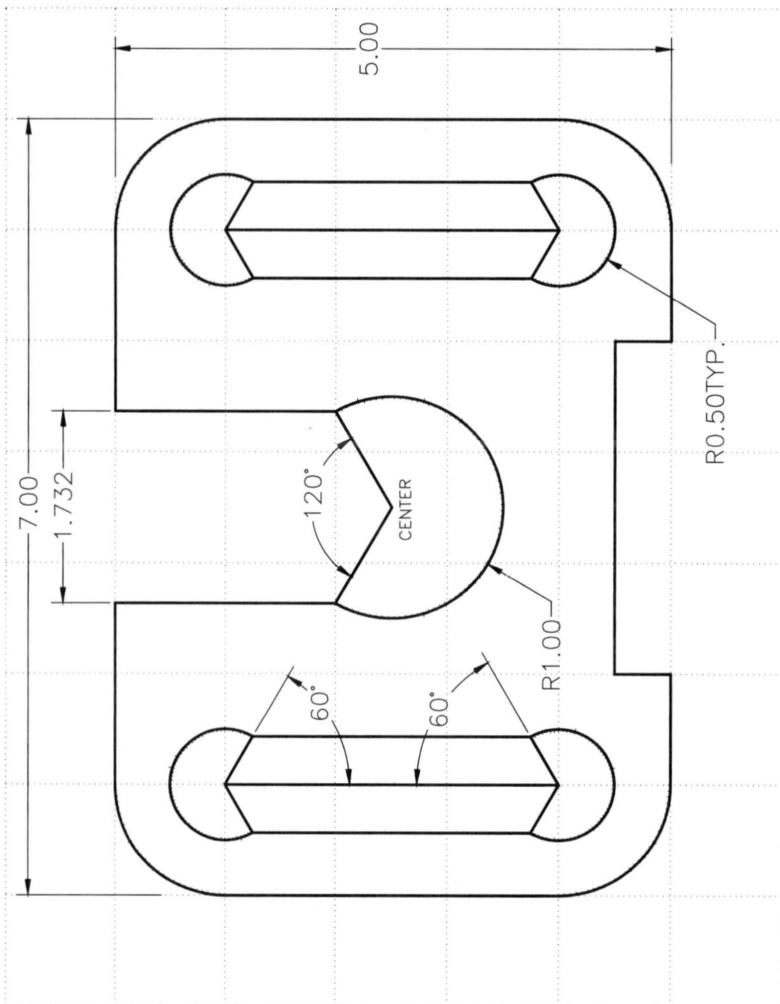

Figure 3-80 *Drawing for Problem Solving Exercise 5*

Answers to Self-Evaluation Test

1 - T, **2** - T, **3** - F, **4** - F, **5** - **Elevation**, **6** - outline, **7** - tapered, 8 - optimized, **9** - toolbar, **10** -negative.

Chapter 4

Working with Drawing Aids

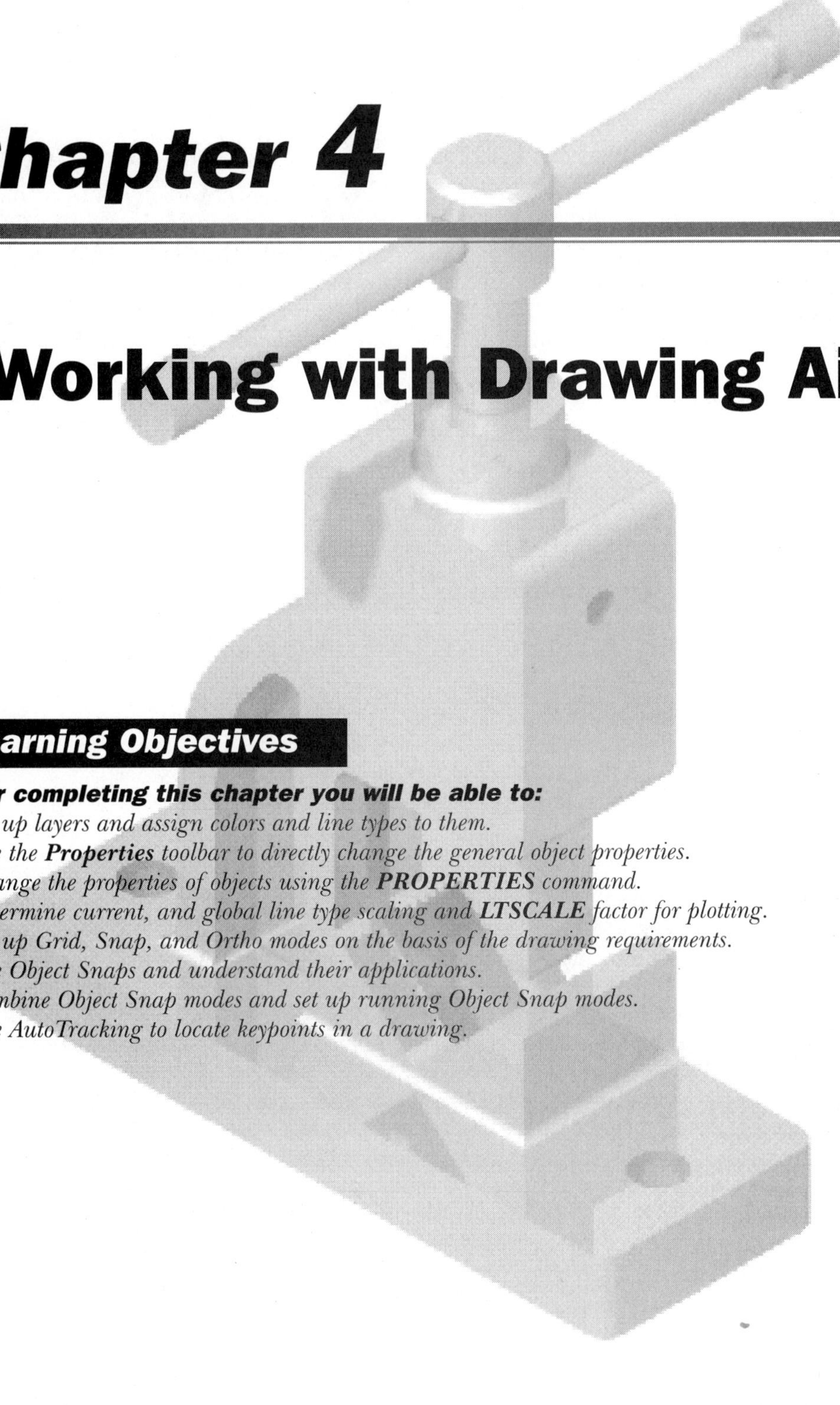

Learning Objectives

After completing this chapter you will be able to:

- *Set up layers and assign colors and line types to them.*
- *Use the* **Properties** *toolbar to directly change the general object properties.*
- *Change the properties of objects using the* **PROPERTIES** *command.*
- *Determine current, and global line type scaling and* **LTSCALE** *factor for plotting.*
- *Set up Grid, Snap, and Ortho modes on the basis of the drawing requirements.*
- *Use Object Snaps and understand their applications.*
- *Combine Object Snap modes and set up running Object Snap modes.*
- *Use AutoTracking to locate keypoints in a drawing.*

In this chapter, you will learn about the drawing setup and the factors that affect the quality and accuracy of a drawing. This chapter contains a detailed description of how to set up layers. You will also learn about some other drawing aids, such as Grid, Snap, and Ortho. These aids will help you to draw accurately and quickly.

UNDERSTANDING THE CONCEPT AND USE OF LAYERS

The concept of layers can be best explained by using the concept of overlays in manual drafting. In manual drafting, different details of the drawing can be drawn on different sheets of paper, or overlays. Each overlay is perfectly aligned with the others. Once all of them are placed on top of each other, you can reproduce the entire drawing. As shown in Figure 4-1, the object lines are drawn in the first overlay and the dimensions in the second. You can place these overlays on top of each other and get a combined look at the drawing.

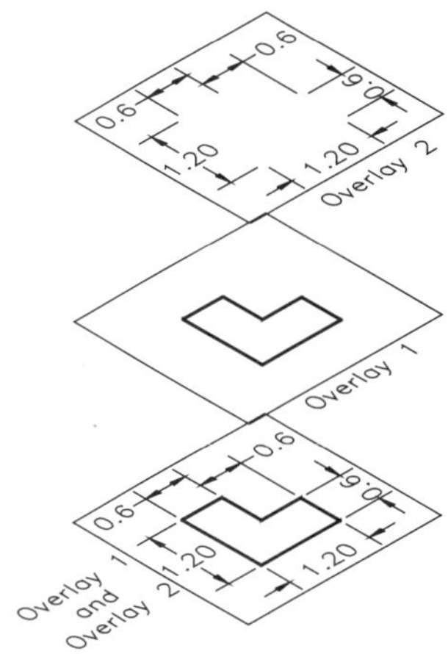

Figure 4-1 *Drawing lines and dimensions in different overlays*

Instead of using overlays in AutoCAD, you use layers. Each layer is assigned a name. You can also assign a color and line type to these layers. For example, in Figure 4-2 the object lines are drawn in the OBJECT layer and the dimensions are drawn in the DIM layer. The object lines will be red because red color has been assigned to the OBJECT layer. Similarly, the dimension lines will be green because the DIM layer has been assigned the green color. You can display all of the layers or you can display the layers individually or in any combination.

Advantages of Layers

1. Each layer can be assigned a different color. Assigning a particular color to a group of objects is very important for plotting. For example, if all object lines are red, at the time of plotting you can assign the red color to a slot (pen) that has the desired tip width (e.g., medium). Similarly, if the dimensions are green, you can assign the green color to another slot (pen) that has a thin tip. By assigning different colors to different layers you can control the width of the lines when the drawing is plotted. You can also make a layer plottable or nonplottable.

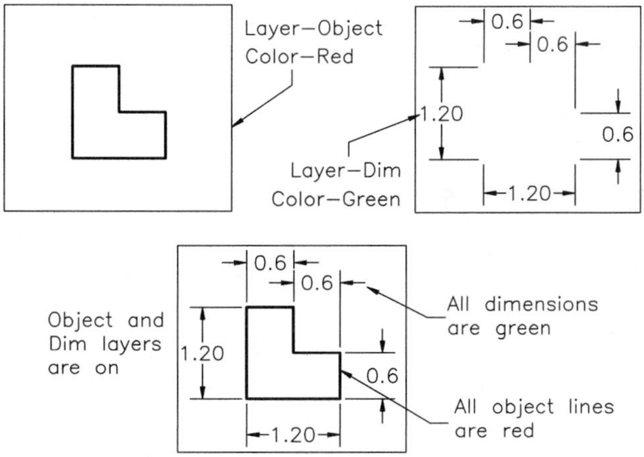

Figure 4-2 *Drawing lines and dimensions in different layers*

2. The layers are useful for some editing operations. For example, if you want to erase all dimensions in a drawing, you can freeze all layers except the dimension layer and then erase all dimensions by using the **Crossing** option to select objects.
3. You can turn a layer off or freeze a layer that you do not want to be displayed or plotted.
4. You can lock a layer, which will prevent the user from accidentally editing the objects in that layer.
5. The colors also help you to distinguish different groups of objects. For example, in architectural drafting, the plans for foundation, floors, plumbing, electrical, and heating systems may all be made in different layers. In electronics drafting and in PCB (printed circuit board), the design of each level of a multilevel circuit board can be drawn on a separate layer. Similarly, in mechanical engineering the main components of an assembly can be made in one layer, other components such as the nuts, bolts, keys, and washers can be made in another layer, and the annotations such as datum symbols and identifiers, texture symbols, Balloons, and Bill of Materials can be made in another layer.

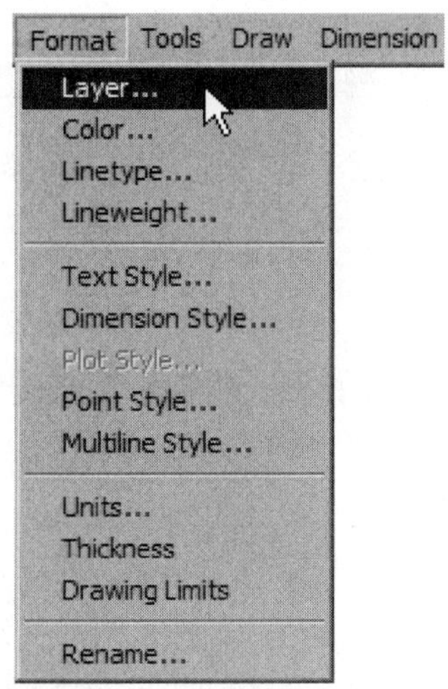

Figure 4-3 *Invoking the **LAYER** command from the **Format** menu*

WORKING WITH LAYERS

Toolbar:	Layers > Layer Properties Manager
Menu:	Format > Layer
Command:	LAYER or LA

You can use the **Format** menu (Figure 4-3) or the **Layers** toolbar (Figure 4-4) to invoke the **Layer Properties Manager** dialog box. Using this dialog box, you can perform the functions associated with layers. For example, you can create new layers, assign colors, assign linetypes, or perform any operation that is shown in the dialog box. You can also perform some layer functions, such as freeze, thaw, lock, unlock, and so on, directly from the **Layer** toolbar. When you invoke the **Layer Properties Manager** dialog box, a default layer with the name **0** is displayed. It is the current layer and any object you draw is created in this layer. There are certain features associated with each layer such as the color, linetype, and lineweight. The layer 0 has a default color of white, linetype continuous, and lineweight default.

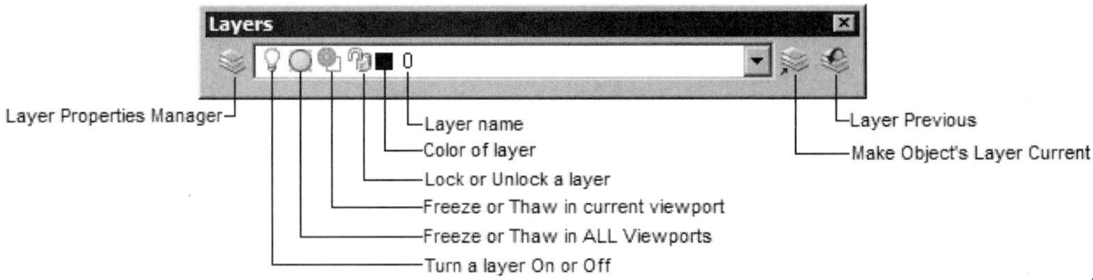

***Figure 4-4 Layers** toolbar*

Creating New Layers

If you want to create new layers, choose the **New** button in the **Layer Properties Manager** dialog box (Figure 4-5). A new layer with name **Layer1** and having the properties of layer 0 is created and listed in the dialog box just below layer 0. If you have more layers apart from the layer 0, the new layer has the properties of the selected layer and is placed below the same. You can change or edit the name by selecting it and then entering a new name. If no layer is selected, the new layer is placed at the end of the layers list and has the properties of the default layer 0. If more than one layer is selected, the new layer is placed below the last highlighted layer and has its properties. Also, right-clicking anywhere in the **Layers** list area of the **Layer Properties Manager** dialog box, displays a shortcut menu that gives you an option to create a new layer. You can right-click a layer whose properties you want to use in the new layer and then select **New Layer** from the shortcut menu.

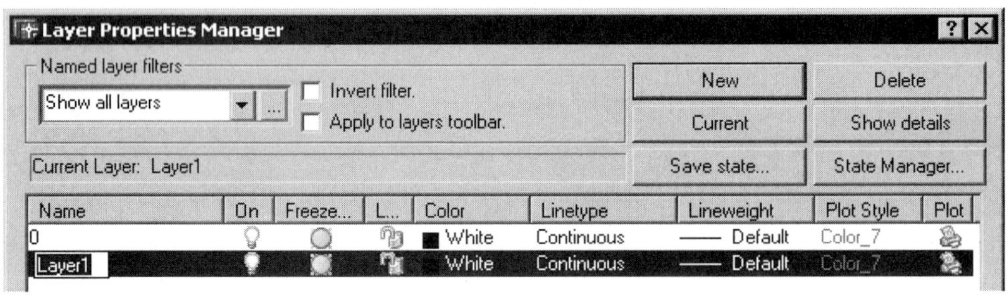

***Figure 4-5 Layer Properties Manager** dialog box with a new layer created*

Layer names

1. A layer name can be up to 255 characters long, including letters (a-z), numbers (0-9), special characters ($ _ -), and spaces. Any combination of lower and uppercase letters can be used while naming a layer. However, characters such as <>;:,'?"=, and so on are not valid characters while naming a layer.

2. The layers should be named to help the user identify the contents of the layer. For example, if the layer name is HATCH, a user can easily recognize the layer and its contents. On the other hand, if the layer name is X261, it is hard to identify the contents of the layer.

3. Layer names should be short, but should also convey the meaning.

Note

All the previous releases of AutoCAD, before AutoCAD 2000, allowed the layer name to be thirty-one characters long. If you give a long name to the layer in AutoCAD 2004 and save the drawing in the previous release, the long name is shortened to thirty-one characters and the illegal characters are replaced by underscores.

The length of the layer name is controlled by the ***EXTNAMES*** *system variable that has a default value 1. If you change it to 0, the layer name is allowed to be up to thirty-one characters long and cannot include the special characters and spaces.*

Tip

If you exchange drawings with or provide drawings to consultants or others, it is very important that you standardize and coordinate layer names and other layer settings.

Making a Layer Current

To draw an object in a particular layer, you need to make it the current layer. Only one layer can be made current and it is the layer where new objects will be drawn. To make a layer current, select the name of the desired layer and then choose the **Current** button in the dialog box. AutoCAD will display the name of the selected layer next to **Current Layer** above the list of layers in the dialog box. Choose **OK** to exit the dialog box. You can also make a layer current by double-clicking on the particular layer from the layer list box of the **Layer Properties Manager** dialog box. Right-clicking a layer in the layer list box displays a shortcut menu that gives you an option (**Make Current**) to make the selected layer current (Figure 4-6). The name and properties of the current layer are also displayed in the **Layers** toolbar. You can also make a layer current by selecting the layer from the **Layer Control** drop-down list in the **Layers** toolbar. You can use the **CLAYER** system variable to make the layer current from the Command prompt. Choosing the **Make Object's Layer Current** button from the **Layers** toolbar prompts you to select the object whose layer you wish to make current. After selecting an object, the layer associated with that object will be made current.

Note

When you select more than one layer at a time using the SHIFT key, the ***Make Current*** *option is not displayed in the shortcut menu. Only one layer can be made current at one time.*

Controlling Display of Layers

You can control the display of the layers by selecting the **Turn a layer On or Off**, **Freeze or thaw in ALL viewports,** and **Lock or Unlock a layer** toggle buttons in the list box of any particular layer.

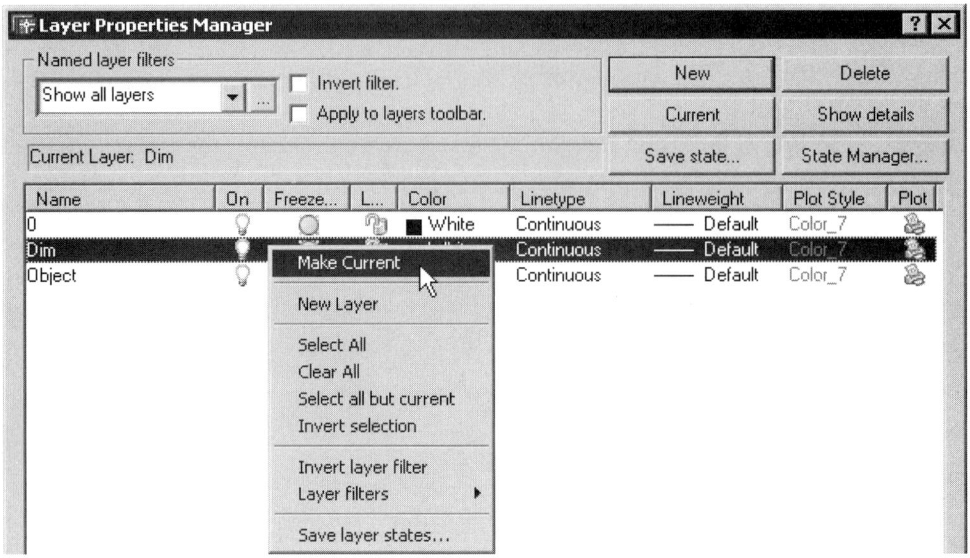

Figure 4-6 ***Layer Properties Manager*** *dialog box with the Layer shortcut menu*

Turn a Layer On or Off

With the **Turn a layer On or Off** toggle icon (light bulb) you can turn the layers on or off. The layers that are turned on are displayed and can be plotted. The layers that are turned off are not displayed and cannot be plotted. You can perform all the operations such as drawing and editing in the layer that has been turned off. You can turn the current layer off, but AutoCAD will display a warning box informing you that the current drawing layer has been turned off. You can also turn the layer on or off by clicking on the On/Off toggle icon from the **Layer** drop-down list in the **Layers** toolbar (Figure 4-7).

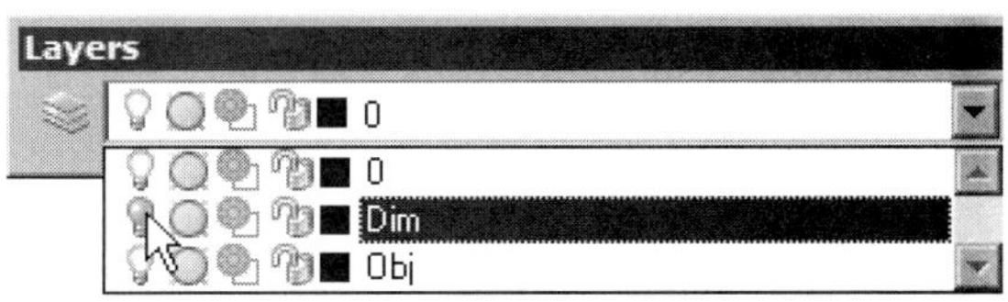

Figure 4-7 *Turning layer* ***Dim*** *Off from the* ***Layers*** *toolbar*

Freeze or Thaw in ALL Viewports

While working on a drawing, if you do not want to see certain layers you can also use the **Freeze or thaw in ALL viewports** toggle icon (sun/snowflakes) to freeze the layers. You can use the **Layers** toolbar or the **Layer Properties Manager** dialog box to freeze or thaw a layer. No modifications can be done in the frozen layer. For example, while editing a drawing you may not want the dimensions to be changed and displayed on the screen. To avoid this, you can freeze the Dim layer in which you are dimensioning the objects. The frozen layers are invisible and cannot be plotted. The Thaw option negates the effect of the Freeze option, and the frozen layers are restored to normal. The difference between the **Off** option and the **Freeze** option is that the frozen layers are not calculated by the computer while regenerating the drawing, and this saves time. The current layer cannot be frozen.

Current or New VP Freeze

When you select a layout in the **Model/Layout** tab (by clicking on Layout1), or you set the **TILEMODE** variable to 0 (see Chapter 11, Model Space Viewports, Paper Space Viewports and Layouts), you can freeze or thaw the selected layers in the active floating viewport by selecting the **Current VP Freeze** icon for the selected layers. Once you are in a floating viewport, the Current and New VP Freeze icons are added in the **Layer Properties Manager** dialog box toward the right side (Figure 4-8). If the icon is not visible you have a scroll bar at the bottom of the layer list box that you can move to display the icons. Also the **Freeze or Thaw in Current viewport** icon in the **Layers** toolbar becomes available once you have viewports. Selecting the **Current VP Freeze** icon makes the selected layers invisible in the active floating viewport only. The frozen layers will still be visible in other viewports. If you want to freeze some layers in the new floating viewports, then select the **New VP Freeze** toggle icon for the selected layers. AutoCAD will freeze the layers in subsequently created new viewports without affecting the viewports that already exist. (Paper space is discussed in Chapter 11.) Also, check the **VPLAYER** command for selectively freezing layers in viewports.

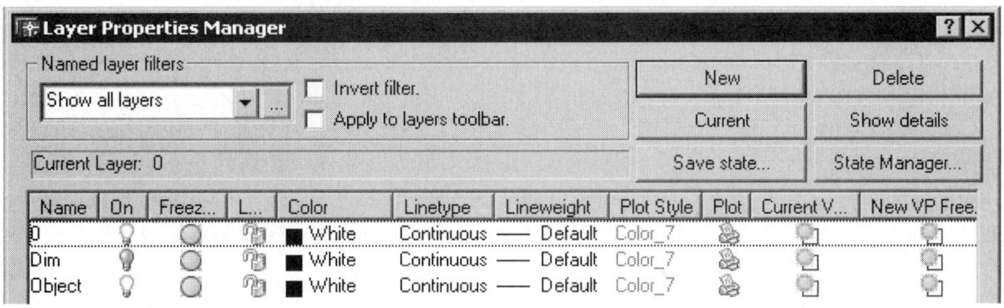

Figure 4-8 ***Layer Properties Manager*** *dialog box with* ***Current*** *and* ***New VP Freeze*** *icons*

Tip

*The widths of the column headings in the **Layer Properties Manager** dialog box can be decreased or increased by positioning the cursor between the column headings on the separator (the cursor turns into a plus sign). Now, hold down the pick button of your pointing device and drag the cursor to the right or left. This way you can make the widths smaller to display all the headings.*

Lock or Unlock a Layer

While working on a drawing, if you do not want to accidentally edit some objects on a particular layer but still need to have them visible, you can use the **Lock/Unlock** toggle icon to lock the layers. When a layer is locked you can still use the objects in the locked layer for Object Snaps and inquiry commands such as **LIST**. You can also make the locked layer the current layer and draw objects on it. The locked layers are plotted. The **Unlock** option negates the **Lock** option and allows you to edit objects on the layers previously locked.

Make a Layer Plottable or Nonplottable

If you do not wish to plot a particular layer, for example, construction lines, you can use the **Make a layer plottable or non-plottable** toggle icon (printer) to make the layer nonplottable.

This icon is available in the **Layer Properties Manager** dialog box. The construction lines will not be plotted if its layer is made nonplottable.

Tip

*It is faster and convenient to use the layer drop-down list in the **Layers** toolbar to make a layer current and control the display features of the layer (On/Off, Freeze/Thaw, Lock/Unlock).*

Assigning Linetype to Layer

To assign a new linetype to a layer, click on the current linetype displayed with a particular layer in the **Layer Properties Manager** dialog box. When you click on the linetype, AutoCAD will display the **Select Linetype** dialog box (Figure 4-9), which displays the linetypes that are defined and loaded on your system. Select the new linetype and then choose the **OK** button. The linetype that you select is assigned to the layer you selected initially. The layers are, by default, assigned continuous linetype and white color if no layer is selected at the time of creating a new layer. Otherwise, the new layer takes the properties of the selected layer.

If you have not loaded the linetypes and are opening the **Select Linetype** dialog box for the first time, only the **Continuous** linetype is displayed. You need to load the linetypes you want and then assign them to the layers. To load the linetypes, choose the **Load** button in the **Select Linetype** dialog box. This displays the **Load or Reload Linetypes** dialog box (Figure 4-10), which displays all linetypes in the *acad.lin* file. In this dialog box you can select individual linetypes or a number of linetypes by holding the SHIFT or CTRL key on the keyboard and then selecting the linetypes. If you right-click, AutoCAD displays the shortcut menu, which you can use to select all linetypes. Then by choosing the **OK** button, the selected linetypes are loaded and therefore displayed in the **Select Linetype** dialog box. Now, select the desired linetype and choose **OK**. The selected linetype is assigned to the selected layer.

Note

*By default, the linetypes in the acad.lin file are displayed in the **Load or Reload Linetypes** dialog box. You can select the linetypes in the acadiso.lin file by choosing the **File** button in the dialog box and then opening the acadiso.lin file from the **Select Linetype Files** dialog box.*

Tip

You can also create your own linetypes. This is discussed in detail in Chapter 31, Creating Linetypes and Hatch Patterns.

Assigning Color to Layer

To assign a color, select the color swatch in a particular layer in the **Layer Properties Manager** dialog box; AutoCAD will display a **Select Color** dialog box. Select the desired color and then choose the **OK** button. The color you selected will be assigned to the selected layer. The number of colors is determined by your graphics card and monitor. Most color systems support eight or more colors. If your system allows it, you may choose a color number between 0 and 255 (256 colors). The following are the first seven standard colors:

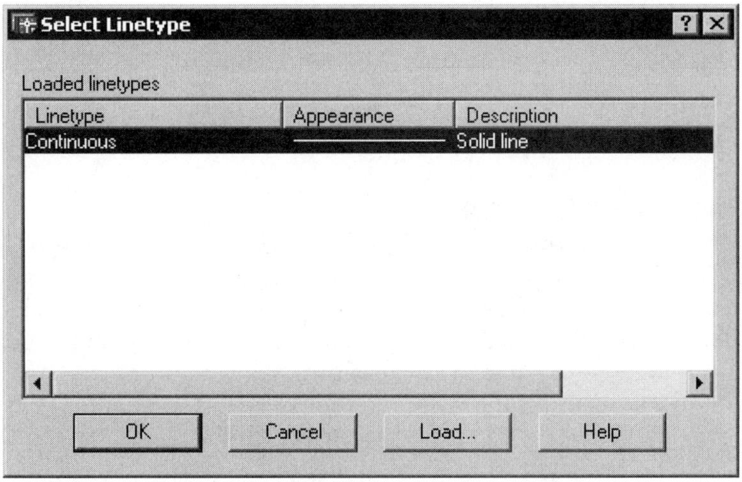

Figure 4-9 ***Select Linetype*** *dialog box*

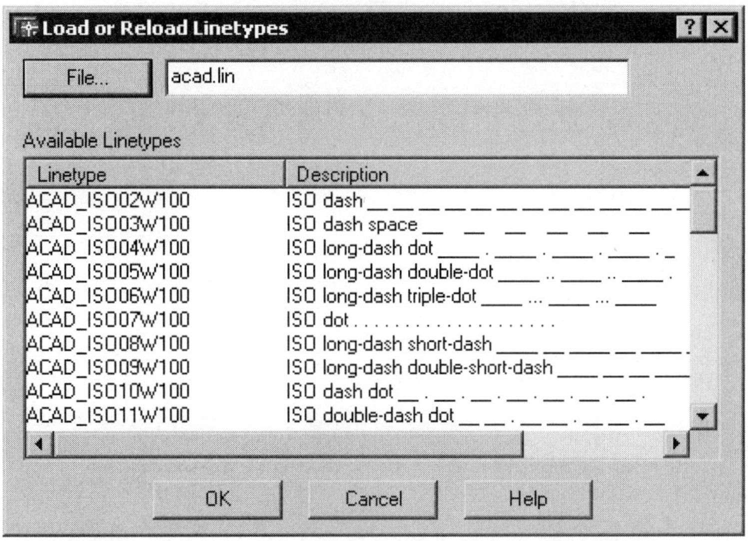

Figure 4-10 ***Load or Reload Linetypes*** *dialog box*

Color number	Color name	Color number	Color name
1	Red	5	Blue
2	Yellow	6	Magenta
3	Green	7	White
4	Cyan		

Note

The use of nonstandard colors may cause compatibility problems if the drawings are used on other systems with different colors. On a light background, the color "white" appears black. Certain colors are hard to see on light backgrounds and others are hard to see on dark backgrounds, and so you may have to use different colors than those specified in some examples and exercises.

Assigning Lineweight to Layer

Lineweight is used to give thickness to the objects in a layer. This thickness is displayed on the screen if the display of the lineweight is on. The lineweight assigned to the objects can also be plotted. If you are making a sectional plan at a certain height, you can assign a layer with a larger value of lineweight to create the objects through which the section is made. Another layer with lesser lineweight can be used to show the objects through which the section does not pass. To assign a lineweight to a layer, select the layer and then click on the lineweight associated with it; the **Lineweight** dialog box (Figure 4-11) appears. Select a lineweight from the **Lineweights** list. Choose **OK** to return to the **Layer Properties Manager** dialog box.

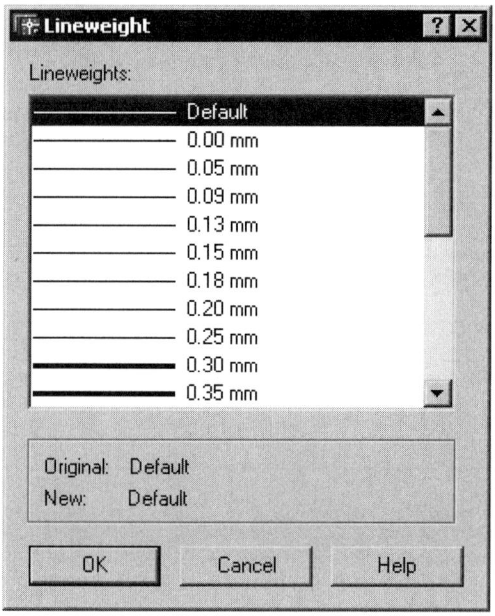

***Figure 4-11 Lineweight** dialog box*

Note

*The **LWEIGHT** command for editing and displaying the lineweight is discussed in detail later in this chapter.*

Assigning Plot Style to Layer

Plot style is a group of property settings such as color, linetype, and lineweight that can be assigned to a layer. The applied plot style affects the drawings while plotting only. The drawing in which you are working should be in named plot style mode (*.stb*) to make the plot style available in the **Layer Properties Manager** dialog box. If the **Plot Style** icon in the dialog box is not available, then you are in color-dependent mode (*.ctb*). To make it available, open the **Options** dialog box from the **Tools** menu and select the **Use named plot style** radio button in the **Plotting** tab. Choose **OK**. After changing the plot style to named plot style dependent, you have to open a new drawing to apply this setting. The default Plot Style is **Normal** in which the color, linetype, and lineweight are BYLAYER. To assign a plot style to a layer, select the layer and then click on its plot style; the **Select Plot Style** dialog box (Figure 4-12) appears where you can select a specific plot style from the **Plot styles** list of available plot styles. Plot styles have to be created before you can use them (see Chapter 12, Plotting Drawings). Choose **OK** to return to the **Layer Properties Manager** dialog box.

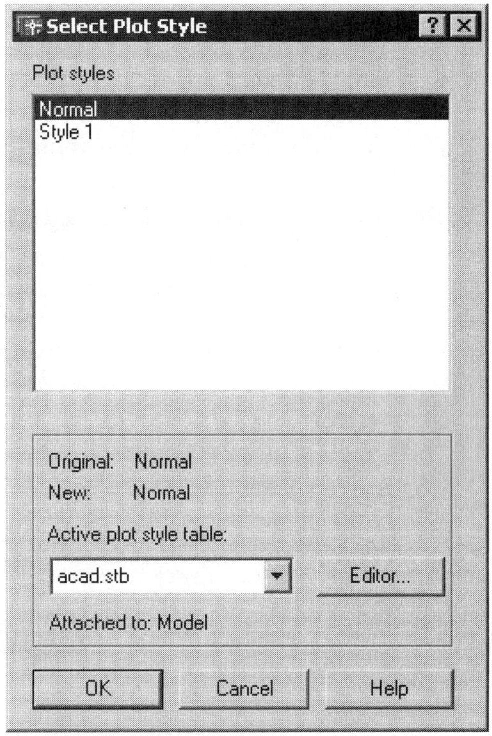

Figure 4-12 ***Select Plot Style*** *dialog box*

Note

After changing the plot style mode (color-dependent or named) in the ***Plotting*** *tab of the* ***Options*** *dialog box you need to open a new drawing to apply the settings.*

Tip

You can also change the plot style mode from the command line by using the ***PSTYLEPOLICY*** *system variable. A value of* ***0*** *sets it to color-dependent mode and a value of* ***1*** *sets it to named mode.*

Deleting Layers

You can delete a layer by selecting the layer and then choosing the **Delete** button in the **Layer Properties Manager** dialog box. To delete a layer it is necessary that the layer should not contain any objects. You cannot delete the Layers 0 and Defpoints, a current layer, and a Xref-dependent layer.

Show Details

You can also control the display of layers by first selecting a layer and then choosing the **Show details** button in the **Layer Properties Manager** dialog box. This displays the layer details area in the lower portion of the **Layer Properties Manager** dialog box. The details area displays the properties associated with the selected layer. Using this box, you can change the name, color, linetype, lineweight, and other properties of the selected layer. Note that you cannot change the name of 0 layer. Once you have chosen the **Show details** button, the **Hide details** button becomes available. This button closes the details area.

Selective Display of Layers

If you have a limited number of layers it is easy to scan through them. However, if you have many layers it is sometimes difficult to search through the layers. To solve this problem you can use the drop-down list in the **Named layer filters** area of the **Layer Properties Manager** dialog box to display the layers selectively. If you want to display only **Xref-dependent** layers, you can select this option in the drop-down list. You can also define your own filter specification by choosing the **[...]** button to display the **Named Layers Filters** dialog box, see Figure 4-13. When loaded, initially this dialog box contains the default specifications. If you want to list only those layers that are red in color, enter red in the **Color** edit box, then give this filter a name in the **Filter name** edit box (Red color), and then choose the **Add** button. This filter gets added to the list of Named layer filters. Now, when you select this named filter from the drop-down list in the Named layer filters area, AutoCAD will display only the layers that

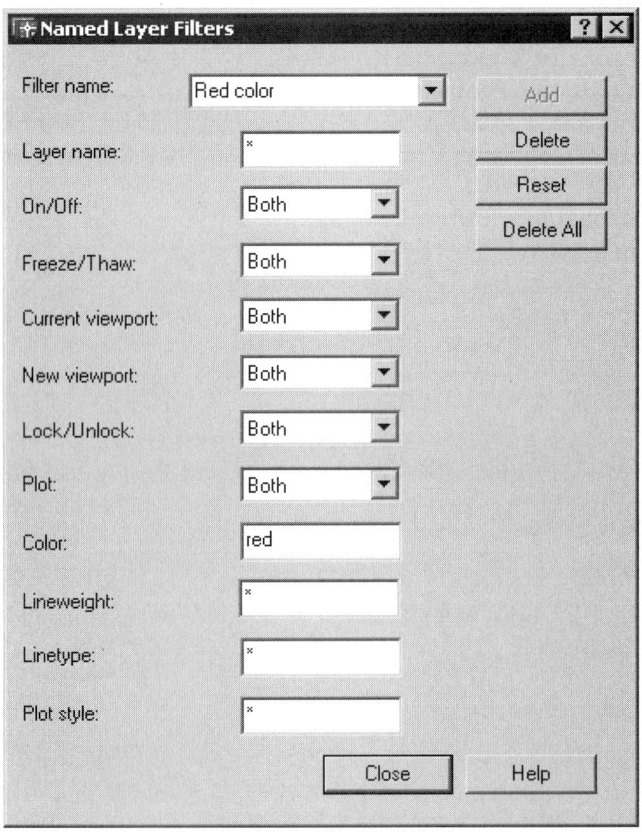

***Figure 4-13 Named Layer Filters** dialog box*

are red in color in the **Layer Properties Manager** dialog box. If you select **Show all layers** in the drop-down list, all layers will be displayed.

In the **Layer Properties Manager** dialog box, when you select the **Invert filter** check box, you invert the filter that you have selected. For example, if you have selected the Show all layers, none of the layers will be displayed. You can right-click in the layers list box of the dialog box to display the shortcut menu. You can then use the **Layer Filter** and **Named Filter** options to selectively display layers. You can also apply the current layer filter to the **Layer Control** list in the **Layers** toolbar by selecting the **Apply to layers toolbar** check box. Choose **OK** in the dialog box. You will notice that only the filtered layers are displayed in the **Layer Control** drop-down list of the **Layers** toolbar. Also if you place your cursor on the arrow of the drop-down list, the tooltip indicates that the filter has been applied. If the current layer is not one among the filtered layers, it will still be displayed in the **Layer Control** list box of the **Layers** toolbar. In the **Layer Properties Manager** dialog box, the current layer is not displayed in the list box if it is not among the filtered layers.

Layer States

You can save and then restore the properties of all the layers in a drawing using the **Save state** and **Restore state** buttons in the **Layer Properties Manager** dialog box. While working on a drawing, at any point in time you can save all the layers with their present properties settings under one name and then restore it anytime later. Choosing the **Save state** button displays the **Save Layer States** dialog box, where you can give the name for saving the layers and you can specify the different states and properties of the layers that you want to save. Choosing the **OK** button saves the checked states and properties of the layers. However, this state is saved only for the current file. If you want to use the current layer state, choose the **State Manager** button to invoke the **Layer States Manager** dialog box. In this dialog box, choose the **Export** button to invoke the **Export layer state** dialog box. Enter the name for the layer state. The layer state is exported with *.las* extension.

You can import the layer state later in any other file using the **Layer States Manager** dialog box. In this dialog box, you are allowed to edit the states and properties of the saved state. You can rename and delete a state.

Tip
*The linetypes will be restored with the layer state in a new drawing file only if those linetypes are already loaded using the **Select Linetype** dialog box.*

Example 1 *Mechanical*

Set up four layers with the following linetypes and colors. Then make the drawing (without dimensions) as shown in Figure 4-14.

Layer name	Color	Linetype	Lineweight
Obj	Red	Continuous	0.031"
Hid	Yellow	Hidden	0.008"
Cen	Green	Center	0.006"

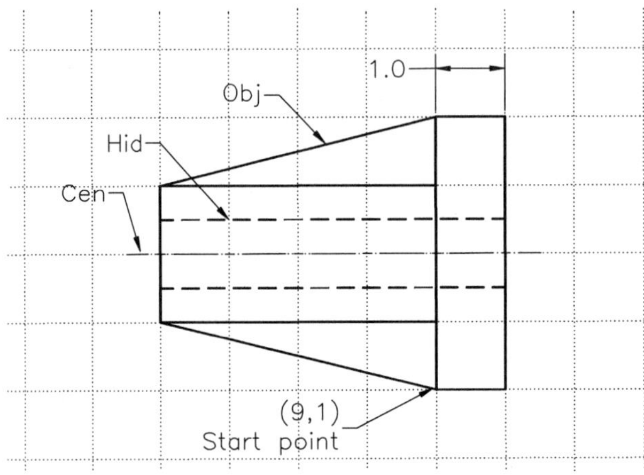

Figure 4-14 *Drawing for Example 1*

In this example, assume that the limits and units are already set. Before drawing the lines, you need to create layers and assign colors, linetypes, and lineweights to these layers. Also, depending on the objects that you want to draw, you need to set that layer as current. In this example you will create the layers using the **Layer Properties Manager** dialog box. You will use the **Layers** toolbar to set the layers current and then draw the figure.

1. As the lineweights specified are in inches, first change the units for lineweight if they are in millimeters. Choose **Lineweight** from the **Format** menu to display the **Lineweight Settings** dialog box. Select the **Inches [in]** radio button in the **Units for Listing** area of the dialog box and then choose the **OK** button.

2. Choose the **Layer Properties Manager** button in the **Layers** toolbar, or choose **Layer** in the **Format** menu, or enter **LAYER** at the Command prompt to display the **Layer Properties Manager** dialog box. The layer **0** with default properties is displayed in the list box.

3. Choose the **New** button. AutoCAD automatically creates a new layer (Layer1) having the default properties and displays it in the list box. Change its name by entering **Obj** in place of Layer1.

4. Now choose the color swatch to display the **Select Color** dialog box. Select color **Red** and then choose **OK**. Red color is assigned to layer **Obj**.

5. Choose the lineweight of the layer to display the **Lineweight** dialog box and choose 0.031" and then choose **OK**. 0.031 lineweight is assigned to layer **Obj**.

6. Again choose the New button. AutoCAD automatically creates a new layer (Layer1) having the properties of layer Obj. Change its name by entering **Hid** in place of Layer1.

7. Choose the color swatch to display the **Select Color** dialog box. Select color **Yellow** and then choose **OK**.

8. Choose the linetype of the layer to display the **Select Linetype** dialog box. If the linetype Hidden is not displayed in the dialog box, choose the **Load** button to display the **Load and Reload** dialog box. Choose **Hidden** from the list and choose the **OK** button. Choose **HIDDEN** in the **Select Linetype** dialog box and then choose **OK**.

9. Choose the lineweight of the layer to display the **Lineweight** dialog box and choose 0.008" and then choose OK. The 0.008 lineweight is assigned to layer **Hid**.

10. Similarly create the new layer **Cen** and assign color Green, linetype CENTER, and lineweight 0.006" to it.

11. Select the **Obj** layer and then choose the **Current** button to make the **Obj** layer current (Figure 4-15). Choose the **OK** button to exit the dialog box.

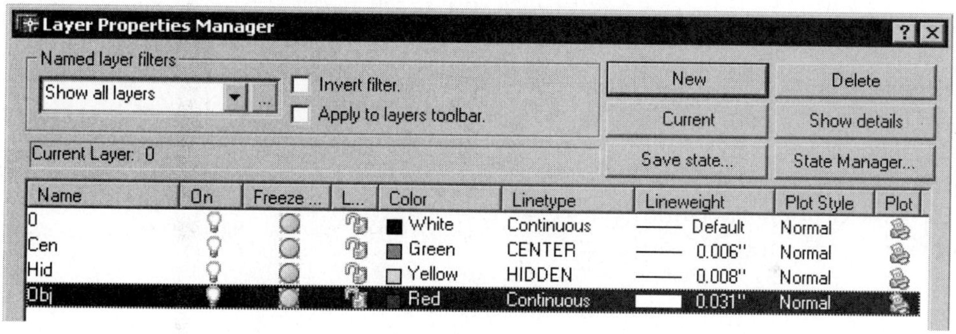

Figure 4-15 *Layers created for Example 1*

12. Using the **LINE** command, draw the object lines in layer **Obj**.

Command: **LINE** Enter
Specify first point: **9,1** Enter
Specify next point or [Undo]: **9,9** Enter
Specify next point or [Undo]: **11,9** Enter
Specify next point or [Close/Undo]: **11,1** Enter
Specify next point or [Close/Undo]: **9,1** Enter
Specify next point or [Close/Undo]: **1,3** Enter
Specify next point or [Close/Undo]: **1,7** Enter
Specify next point or [Close/Undo]: **9,9** Enter
Specify next point or [Close/Undo]: Enter

Command: Enter *(Invokes the LINE command.)*
Specify first point: **1,3** Enter
Specify next point or [Undo]: **9,3** Enter

Specify next point or [Undo]: Enter

Command: Enter *(Invokes the LINE command.)*
Specify first point: **1,7** Enter
Specify next point or [Undo]: **9,7** Enter
Specify next point or [Undo]: Enter

13. Now, make the **Hid** layer current from the **Layers** toolbar. Select the down arrow in the **Layer Control** list box to display the drop-down list (Figure 4-16). Choose the **Hid** layer from the list to make it current.

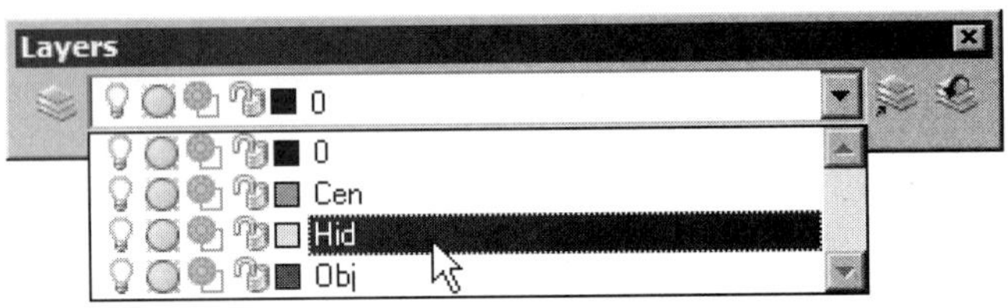

Figure 4-16 *Making Hid layer current from the* ***Layers*** *toolbar*

14. Now draw the following lines using the **LINE** command.

Command: **LINE** Enter
Specify first point: **1,4** Enter
Specify next point or [Undo]: **11,4** Enter
Specify next point or [Undo]: Enter

Command: Enter *(Invokes the LINE command.)*
Specify first point: **1,6** Enter
Specify next point or [Undo]: **11,6** Enter
Specify next point or [Undo]: Enter

15. Again choose the down arrow in the **Layer Control** list box to display the drop-down list. Choose the **Cen** layer from the list to make it current.

16. Now draw the following lines using the **LINE** command.

Command: **LINE** Enter
Specify first point: **0,5** Enter
Specify next point or [Undo]: **12,5** Enter
Specify next point or [Undo]: Enter

Exercise 1 *Mechanical*

Set up layers with the following linetypes and colors. Then make the drawing (without dimensions) as shown in Figure 4-17. The distance between the dotted lines is 1 unit.

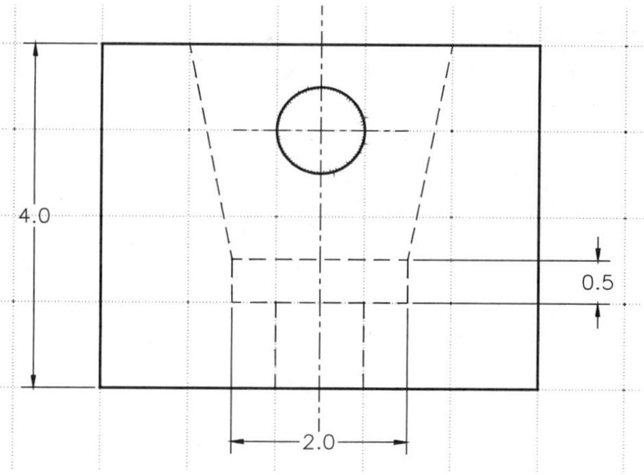

Figure 4-17 Drawing for Exercise 1

Layer name	Color	Linetype
Object	Red	Continuous
Hidden	Yellow	Hidden
Center	Green	Center
Dimension	Blue	Continuous

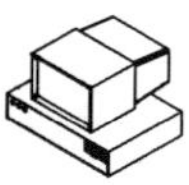

Tip

*Remember that the **Linetype Control**, **Lineweight Control**, **Color Control**, and the **Plot Style Control** list boxes in the **Properties** toolbar should display **ByLayer** as the current properties of the objects. This is to ensure that when you draw an object, it takes those properties from the current layer in which the object is drawn.*

LINETYPE COMMAND

The **LINETYPE** command can be used to load, delete, and make current the linetype which can be used to draw an object. You can invoke this command from the **Format** menu or by entering LINETYPE at the Command prompt. The **Linetype Manager** dialog box (Figure 4-18) is displayed when you invoke this command. By default, the linetypes displayed are ByLayer, ByBlock, and Continuous. Choosing the **Load** button displays the **Load or Reload Linetype** dialog box, which is also displayed from the **Layer Properties Manager** dialog box. The linetypes that you load here will be displayed in the **Select Linetypes** dialog box also. Similarly, if you have loaded some linetypes using the **LAYER** command, they will be displayed in the **Linetype Manager** dialog box. You can make a linetype current by using the **Current** button. All the new objects will be drawn with the current linetype. If you have chosen the current linetype as **ByLayer** then the object will be drawn with the linetype of the layer. You can also make any linetype current other than ByLayer, and the object will be drawn with the particular linetype. The current linetype is also displayed in the **Linetype Control** field of the **Properties** toolbar.

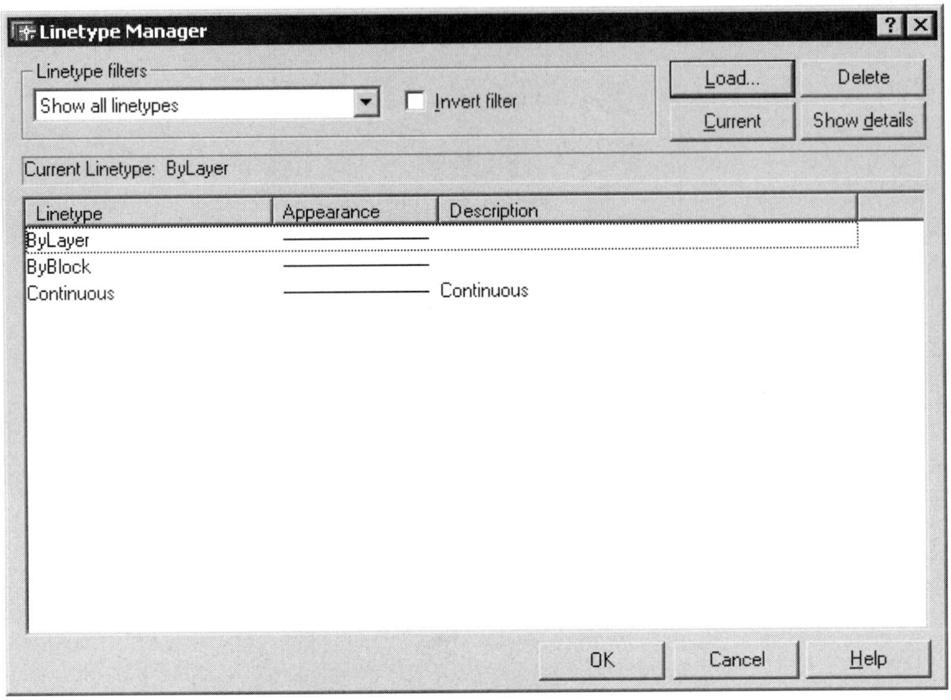

Figure 4-18 ***Linetype Manager*** *dialog box*

LWEIGHT COMMAND

The **LWEIGHT** command can be used to assign a lineweight to an object. You can invoke this command by selecting **Lineweight** from the **Format** menu or by entering **LWEIGHT** at the Command prompt to display the **Lineweight Settings** dialog box (Figure 4-19). You can also use the **Settings** option in the shortcut menu, displayed on right-clicking the **LWT** button on the status bar, to invoke the dialog box. In this dialog box you can choose the

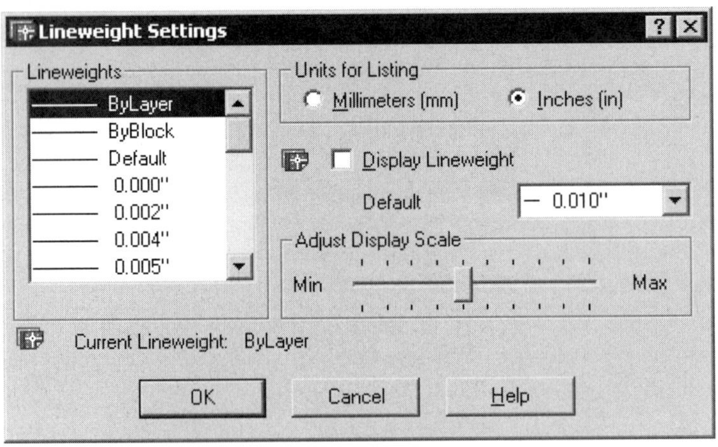

Figure 4-19 ***Lineweight Settings*** *dialog box*

current lineweight for the objects. You can change the units for lineweight and also the display of lineweights for the current drawing. The lineweights are displayed in pixel widths and depending on the lineweight value chosen, the lineweights are displayed if the **Display Lineweight** check box is selected. For a large drawing, this increases the regeneration time and should be cleared. You can also turn the display of lineweights to on or off directly from the status bar by choosing the **LWT** button. The slider bar for **Adjust Display Scale** affects the regeneration time as well. You can keep the slider bar at **Max** for getting a good display of different lineweights on the screen in the Model space; otherwise, keep it at **Min** for faster regeneration.

Tip

For large drawings you should keep the display scale at the minimum value for increasing the performance of AutoCAD. It is also recommended that you clear the ***Display Lineweight*** *check box or simply turn the display of lineweight to off from the status bar.*

OBJECT PROPERTIES

An object when created has certain properties such as color, linetype, lineweight, plot style, and layer associated with it. These properties of an object can be changed and made current using the **Properties** toolbar or the **Properties** palette.

Properties Toolbar*

You can use the **Properties** toolbar to directly change the general properties of the selected objects and also set those properties current.

Color

Select the object whose color you want to change. The current color of the object is displayed in the **Color Control** list box. Display the list of colors by selecting the drop-down arrow (Figure 4-20) and select a new color. The color of the selected object is changed to the new color. If you want to set a different color current, choose a color from the drop-down list without selecting any object. By default the color of an object is **ByLayer** and hence takes the color assigned to the layer in which it is created. From the **Color** drop-down list you can select any other color to assign the color explicitly to the objects drawn. This color is set as

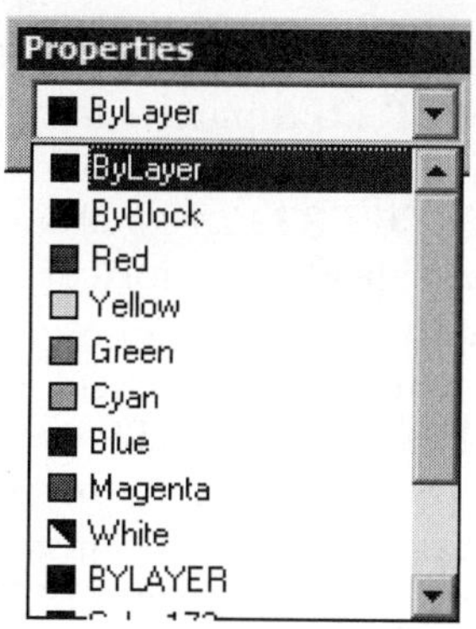

Figure 4-20 *Setting object color from the* ***Properties*** *toolbar*

current and all the new objects will be drawn in this color. If you want to assign a color that is not displayed in the list, choose the **Other** option to display the **Select Color** dialog box. You can select the desired color in this dialog box, and then choose the **OK** button.

Linetype

From the **Linetype Control** drop-down list (Figure 4-21), you can select the new linetype and assign it to the selected object. You can also make a linetype current by selecting it from the list. By default the linetype of an object is **ByLayer**. The linetypes that are loaded from the **LAYER** or the **LINETYPE** command will be listed. If you want to assign a linetype that is not displayed in the list, choose the **Other** option to display the **Linetype Manager** dialog box. You can use the **Load** button to load other linetypes.

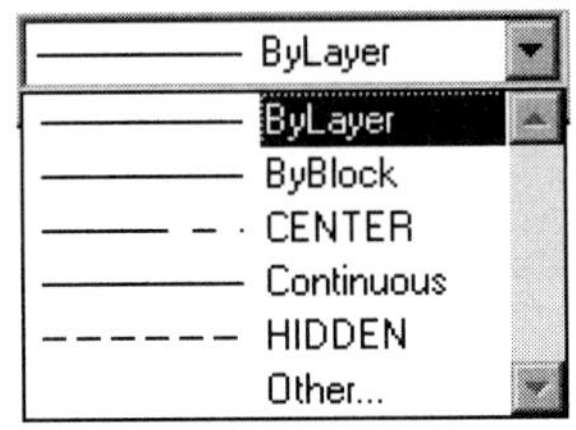

Figure 4-21 *Linetype Control list box*

Lineweight

Similarly, you can select a different lineweight value from the **Lineweight Control** drop-down list in the **Properties** toolbar to assign to the selected object. Also, you can set a new lineweight current by selecting it from the list.

Plot Style

You can select a different plot style for the selected object from the **Plot Style Control** drop-down list. Also, you can set a new plot style current by selecting it from the list.

Note

When you set a linetype, lineweight, or color current, all the objects drawn thereafter will have the current linetype, lineweight, and color. The properties of the current layer are not considered. You need to set current those properties to ***ByLayer*** *again such that the objects assume the linetype, lineweight, color, or plot style of the layer in which they are created.*

PROPERTIES Palette*

Toolbar:	Standard > Properties
Menu:	Tools > Properties
Command:	PROPERTIES, CH, MO

You can also use the **PROPERTIES** palette to change the general properties of the selected objects and also set those properties current. The Properties palette is displayed by invoking the **PROPERTIES** command. It can also be invoked by selecting an object and then right-clicking to display a shortcut menu; choose **Properties** from the shortcut menu. The **PROPERTIES** palette is displayed (Figure 4-22), from where you can change the different properties of the selected object. When you select the objects whose properties you want to change, the **PROPERTIES** palette displays the properties of the selected object. Depending on the object selected, the properties differ. You can also make the different properties current by selecting no object and then selecting the particular properties in the window. Right-clicking in the **PROPERTIES** palette displays a shortcut menu from where you can choose to **Allow Docking** or **Hide** the palette.

When you select **Color** from the **General** list, the color of the selected object is displayed along with a drop-down arrow. By default the color of an object is **ByLayer** and hence takes the color assigned to the layer in which it is created. From the **Color** drop-down list you can select any other color you wish to assign to the selected object. If you want to assign a color that is not displayed in the list, choose the **Other** option to display the **Select Color** dialog box. You can select the desired color in this dialog box, and then choose the **OK** button.

Similarly, you can set current the linetype, lineweight, linetype scale, and other properties for the other objects individually. You can also assign a hyperlink to an object. The **Hyperlink** name and description assigned to the object is displayed. Choosing the **[...]** button, displays the **Insert Hyperlink** dialog box where you can enter the path of the URL or file you want to link the selected object to.

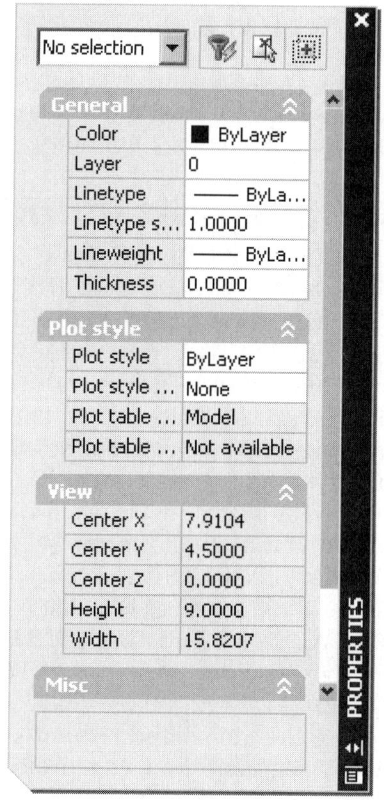

Figure 4-22 ***PROPERTIES*** *palette*

Tip

*You can keep the **PROPERTIES** palette open while working in a drawing. When no object is selected the current settings of the drawing are listed. Whenever you need to change a certain property of an object, select the object so that its properties are listed in the **PROPERTIES** palette and make the required changes.*

Note

*You can use the **CHPROP** command to change the properties of an object from the command line. You can also change the general properties of an object by using the **Properties** option of the **CHANGE** command. Both these commands along with the **Geometry** portion of the **PROPERTIES** palette to edit the objects have been discussed in Chapter 18.*

*The system variables **CECOLOR, CELTYPE, CELWEIGHT** control the current color, linetype, and lineweight of the object. You can set the properties of the objects current using these variables from the command line.*

When you set the property of the object current, it does not consider the property of the layer in which the object will be drawn.

Tip

*It is easier and faster to set the properties of the object current from the **Properties** toolbar. It is better to leave the different properties of objects at **ByLayer** to avoid confusion.*

Exercise 2 *General*

Draw a hexagon on layer OBJ in red color. Let the linetype be hidden. Now, use the **PROPERTIES** command to change the layer to some other existing layer, the color to yellow, and the linetype to continuous.

GLOBAL AND CURRENT LINETYPE SCALING

The **LTSCALE** system variable controls the global scale factor of the lines in a drawing. For example, if **LTSCALE** is set to 2, all lines in the drawing will be affected by a factor of 2. Like **LTSCALE**, the **CELTSCALE** system variable controls the linetype scaling. The difference is that **CELTSCALE** determines the current linetype scaling. For example, if you set **CELTSCALE** to 0.5, all lines drawn after setting the new value for **CELTSCALE** will have the linetype scaling factor of 0.5. The value is retained in the **CELTSCALE** system variable. Line (a) in Figure 4-23 is drawn with a **CELTSCALE** factor of 1; line (b) is drawn with a **CELTSCALE** factor of 0.5. The length of the dash is reduced by a factor of 0.5 when **CELTSCALE** is 0.5. The net scale factor is equal to the product of **CELTSCALE** and **LTSCALE**. Figure 4-23(c) shows a line that is drawn with **LTSCALE** of 2 and **CELTSCALE** of 0.25. The net scale factor = **LTSCALE** X **CELTSCALE** = 2 X 0.25 = 0.5. You can also change the global and current scale factors by entering a desired value in the **Linetype Manager** dialog box. If you choose the **Show details** button, the properties associated with the selected linetype are displayed (Figure 4-24). You can change the values according to your drawing requirements.

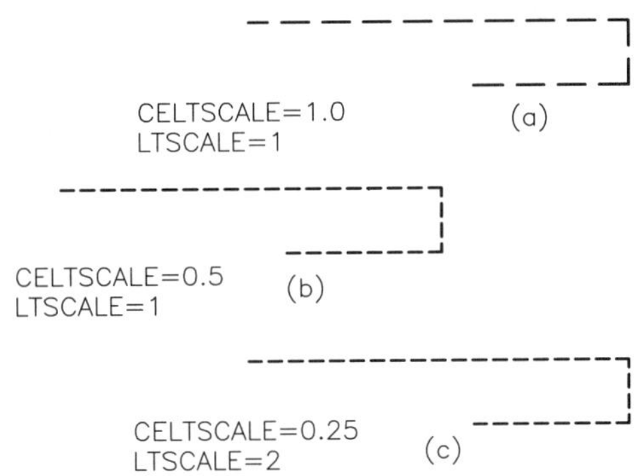

Figure 4-23 *Using* ***CELTSCALE*** *to control current linetype scaling*

Figure 4-24 Details *area of the* ***Linetype Manager*** *dialog box*

LTSCALE FACTOR FOR PLOTTING

The **LTSCALE** factor for plotting depends on the size of the sheet you are using to plot the drawing. For example, if the limits are 48 by 36, the drawing scale is 1:1, and you want to plot the drawing on a 48" by 36" size sheet, then the **LTSCALE** factor is 1. If you check the specification of the Hidden linetype in the *acad.lin* file, the length of each dash is 0.25. Hence, when you plot a drawing with 1:1 scale, the length of each dash in a hidden line is 0.25.

However, if the drawing scale is 1/8" = 1' and you want to plot the drawing on 48" by 36" paper, the **LTSCALE** factor must 8 X 12 = 96. The length of each dash in the hidden line will increase by a factor of 96, because the **LTSCALE** factor is 96. Therefore, the length of each dash will be (0.25 X 96 = 24) units. At the time of plotting, the scale factor for plotting must be 1:96 to plot the 384' by 288' drawing on 48" by 36" paper. Each dash of the hidden line that was 24" long on the drawing will be 24/96 = 0.25" long when plotted. Similarly, if the desired text size on the paper is 1/8", the text height in the drawing must be 1/8 X 96 = 12".

LTSCALE factor for PLOTTING = Drawing Scale

Sometimes your plotter may not be able to plot a 48" by 36" drawing, or you might like to decrease the size of the plot so that the drawing fits within a specified area. To get the correct dash lengths for hidden, center, or other lines, you must adjust the **LTSCALE** factor. For example, if you want to plot the previously mentioned drawing in a 45" by 34" area, the correction factor is:

Correction factor = 48/45
= 1.0666

New LTSCALE factor = **LTSCALE** factor x Correction factor
= 96 x 1.0666
= 102.4

New LTSCALE factor for PLOTTING = Drawing Scale x Correction Factor

Note

If you change the LTSCALE factor, all lines in the drawing are affected by the new ratio.

Changing Linetype Scale Using the PROPERTIES Command

You can also change the linetype scale of an object by using the **PROPERTIES** command. When you invoke this command, AutoCAD will display the **PROPERTIES** palette. All the properties of the selected objects are displayed in the **PROPERTIES** palette. To change the current linetype scale, select the objects and then invoke the **PROPERTIES** command to display the **PROPERTIES** palette (Figure 4-25). In this window, locate the **Linetype scale** edit box in the **General** list and enter the new linetype scale in the edit box. The linetype scale of the selected objects is changed to the new value you entered.

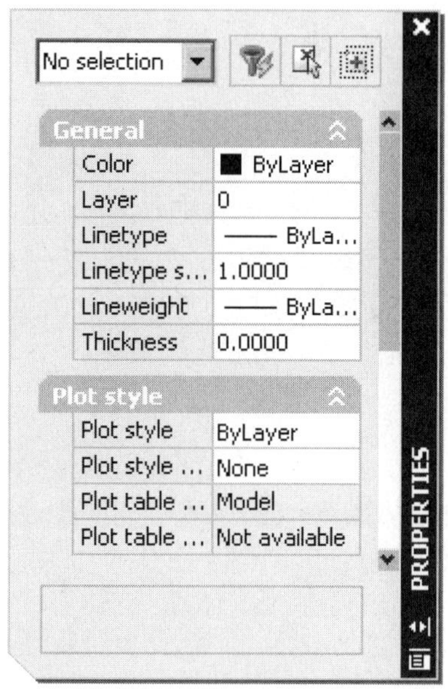

***Figure 4-25 PROPERTIES** palette*

WORKING WITH THE DESIGNCENTER

Toolbar:	Standard > DesignCenter
Menu:	Tools > DesignCenter
Command:	ADCENTER

The **DESIGNCENTER** allows you to reuse and share contents in different drawings. You can use this window to locate the drawing data with the help of search tools and then use it in your drawing. You can insert layers, linetypes, blocks, layouts, external references, and other drawing content in any number of drawings. Hence, if a layer is created once it can be repeatedly used any number of times.

Open the **DESIGNCENTER** window and choose the **Tree View Toggle** button to display the **Tree pane** and the **Palette** side by side (if they are not already displayed). Open the folder in which the drawing is saved containing the layers and linetypes you want to insert. You can use the **Load** and the **Up** buttons to open the folder you want. For example, you want to insert some layers and linetypes from the drawing *c4d1* in the directory *C:\AutoCAD 2004\c04-acad2004*. Open the desired folders and open *c4d1* to display its contents. Choose Layers and all the layers created in *c4d1* are displayed in the Palette. Select **Border** and **CEN** and right-click to display the shortcut menu. Choose **Add Layer[s]** as shown in Figure 4-26. The two layers are added to the current drawing. You can also drag and drop the desired layers into the current drawing. If you open the **Layer control** drop-down list in the **Properties** toolbar, the two layers are listed. Similarly, you can insert the linetypes. You will learn more about **DESIGNCENTER** in Chapter 6.

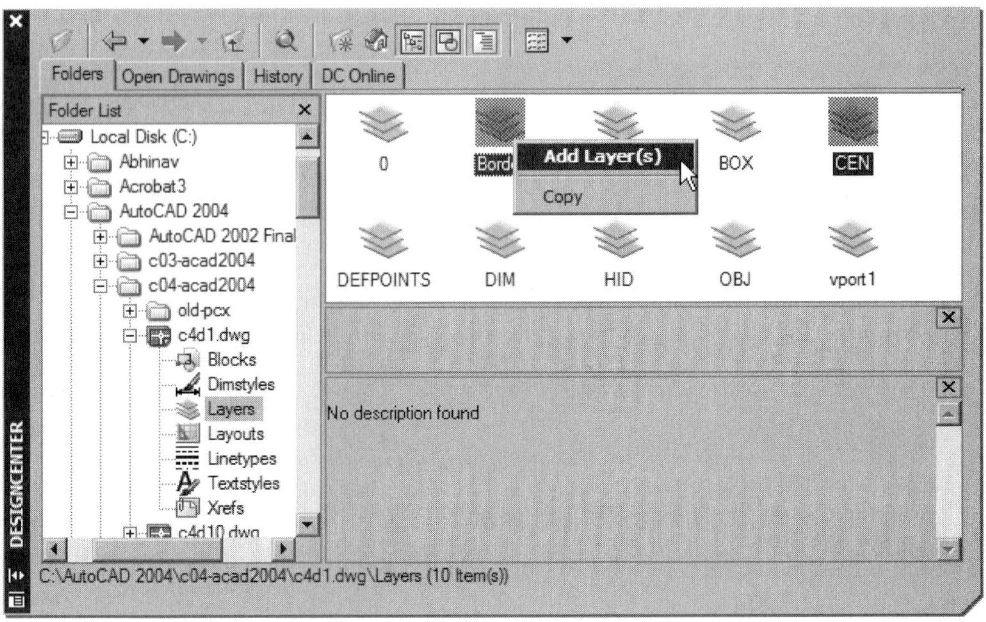

Figure 4-26 ***DESIGNCENTER*** *with the shortcut menu*

DRAFTING SETTINGS DIALOG BOX

Menu:	Tools > Drafting Settings
Command:	DSETTINGS

You can use the **Drafting Settings** dialog box to set drawing modes such as Grid, Snap, Object Snap, Polar, and Object Snap tracking. All these aids help you to draw accurately and also increase the drawing speed. You can right-click on the **SNAP**, **GRID**, **POLAR, OSNAP**, or **OTRACK** buttons on the status bar to display a shortcut menu (Figure 4-27) and choose **Settings** to display the **Drafting Settings** dialog box. This dialog box operates under three main tabs: **Snap and Grid** (Figure 4-28), **Object Snap**, and **Polar Tracking**. When you start AutoCAD you are provided with the default settings of these aids. You can change them according to your requirements by using the **Drafting Settings** dialog box.

Figure 4-27 Settings *in the shortcut menu*

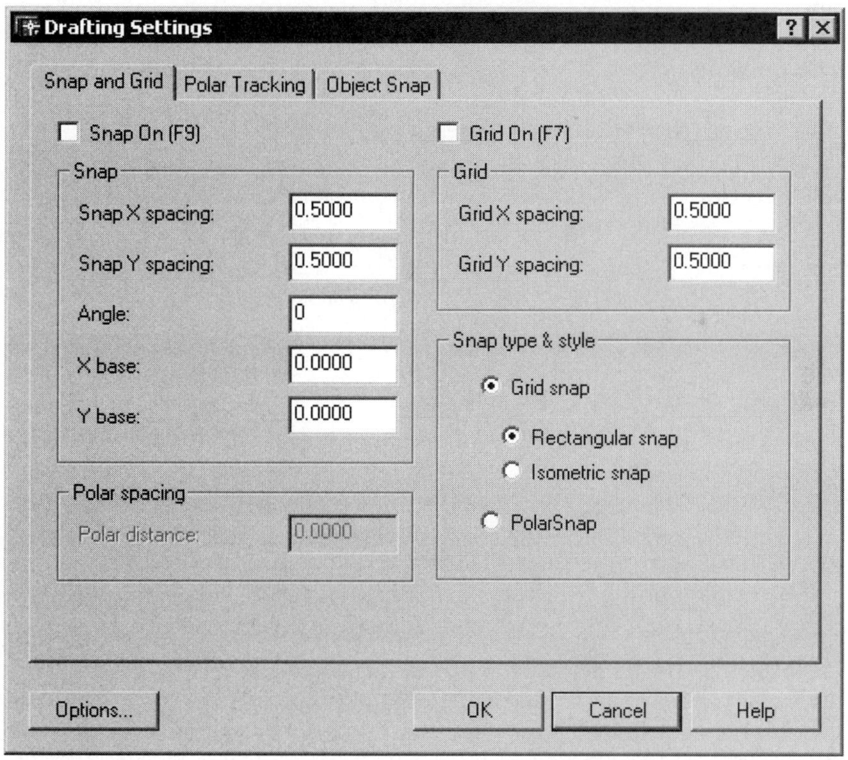

Figure 4-28 Drafting Settings *dialog box (**Snap and Grid** tab)*

Setting Grid

The grid lines are lines of dots on the screen at predefined spacing (Figure 4-29). These dotted lines act as a graph that can be used as reference lines in a drawing. You can change the distance between the grid dots as per your requirement. The grid pattern appears within

Figure 4-29 Grid lines

the drawing limits, which helps to define the working area. The grid also gives you a sense of the size of the drawing objects.

Grid On (F7): Turning the Grid On or Off

You can turn the grid on/off by using the **Grid On** check box in the **Drafting Settings** dialog box (Figure 4-28). You can also turn the grid on or off by choosing the **GRID** button in the status bar, or using the **On** or **Off** option in the shortcut menu when you right-click on the **GRID** button in the status bar, or using the **GRID** command. The function key F7 acts as a toggle key for turning the grid on or off. When the grid is turned on after it has been off, the grid is set to the previous grid spacing.

Grid X Spacing and Grid Y Spacing

The **Grid X spacing** and **Grid Y spacing** edit boxes in the **Drafting Settings** dialog box are used to define a desired grid spacing along the *X* and *Y* axes. For example, to set the grid spacing to 0.5 units, enter 0.5 in the **Grid X spacing** and **Grid Y spacing** edit boxes (Figure 4-30). You can also enter different values for horizontal and vertical grid spacing (Figure 4-31). If you enter only the grid X spacing and then choose the **OK** button in the dialog box, the corresponding Y spacing value is automatically set to match the X spacing value. Therefore, if you want different X and Y spacing values, you need to set the X spacing first, and then set the Y spacing.

Note

Grids are specially effective in drawing when the objects in the drawing are placed at regular interval.

*You can also use the **GRID** command to set the grid from the command line. At the prompt sequence for the **GRID** command you can use the **Aspect** option to assign a different value to the horizontal and vertical grid spacings.*

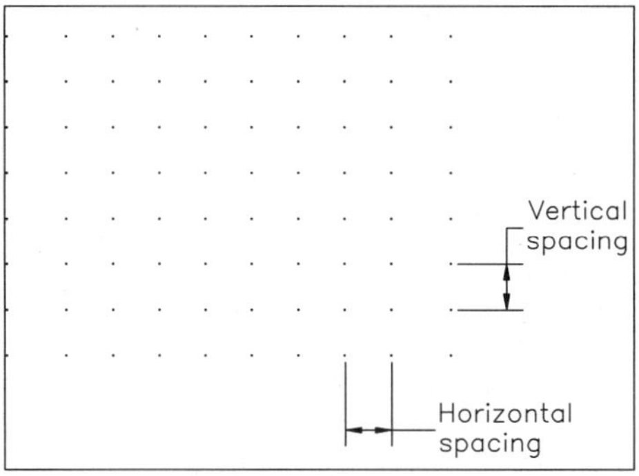

Figure 4-30 Controlling grid spacing

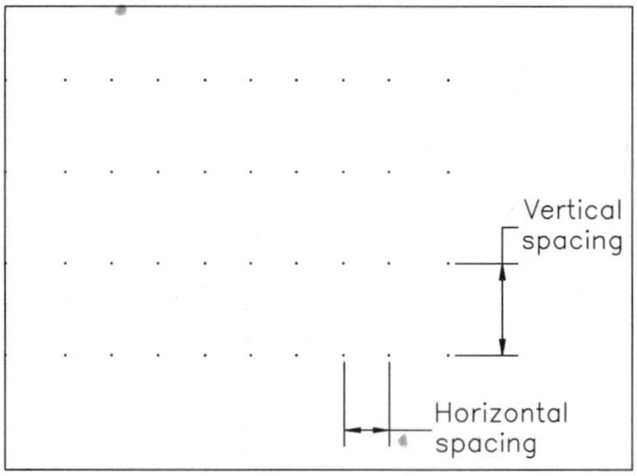

Figure 4-31 Creating unequal grid spacing

Tip

*The grid and the snap grid (discussed in the next section) are independent of each other. However, you can automatically display the grid lines at the same resolution as that of the snap grid by using the **Snap** option of the **GRID** command. When you use this option, AutoCAD will automatically change the grid spacing to zero and display the grid lines at the same resolution as set for Snap. Therefore, in the **Drafting Settings** dialog box if the grid spacing is specified as zero, it automatically adjusts to equal the Snap resolution.*

*In the **GRID** command, to specify the grid spacing as a multiple of the Snap spacing, enter X after the value (2X).*

Setting Snap

The snap is used to set increments for cursor movement. While moving the cursor it is sometimes difficult to position a point accurately. The **SNAP** command allows you to set up an invisible grid (Figure 4-32) that allows the cursor to move in fixed increments from one snap point to another. The snap points are the points where the invisible snap lines intersect. The snap spacing is independent of the grid spacing, and so the two can have equal or different values. You generally set snap to an increment of the grid setting, for example, Snap=2 and Grid=10.

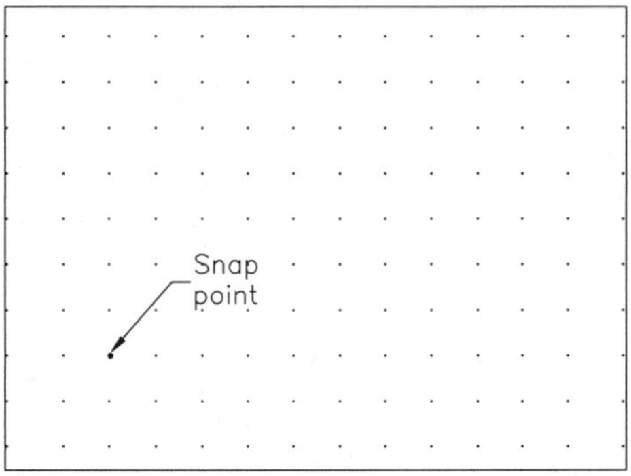

Figure 4-32 Invisible Snap grid

Snap On

You can turn the snap on/off and set the snap spacing from the **Drafting Settings** dialog box (Figure 4-28). As in the case of GRID, the Snap On/Off selection turns the invisible snap grid on or off and moves the cursor by increments. If the SNAP is off, the cursor will not move by increments. It will move freely as you move the pointing device. When you turn the snap off,

AutoCAD remembers the value of the snap, and this value is restored when you turn the snap back on. You can also turn snap on or off by choosing the **Snap** button in the status bar, or from the shortcut menu displayed by right-clicking on the Snap button in the status bar, or by using the function key F9 as a toggle key for turning snap on or off.

Snap Angle

The **Angle** edit box in the **Drafting Settings** dialog box is used to rotate the snap grid through an angle (Figure 4-33). Normally, the snap grid has horizontal and vertical lines, but sometimes you need the snap grid at an angle. For example, when you are drawing an auxiliary view (a drawing view that is at an angle to other views of the drawing), it is more useful to have the snap grid at an angle. If the rotation angle is positive, the grid rotates in a counterclockwise direction; if the rotation angle is negative, it rotates in a clockwise direction.

Figure 4-33 *Snap grid rotated 30-degree*

X Base and Y Base

The **X base** and **Y base** edit boxes can be used to define the snap and grid origin point in the current viewport relative to current UCS. When you specify the snap angle the grid is rotated around the base point whose coordinates are specified as X base and Y base. Hence, you will always have a grid point at this base point. The value is stored in the **SNAPBASE** system variable.

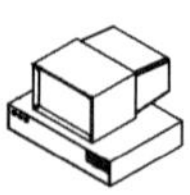

Tip

It is generally preferable to rotate the UCS rather than rotating the snap grid. See Chapter 22 for UCS.

Example 2 *Mechanical*

Draw the auxiliary view of the object whose front view is shown in Figure 4-34. The auxiliary view is shown in Figure 4-35. The upper edge of the object is 2.5 units. The X and Y spacing of the grid is 0.5 units. The thickness of the plate is 2 units; the length of the incline's face is 5 units.

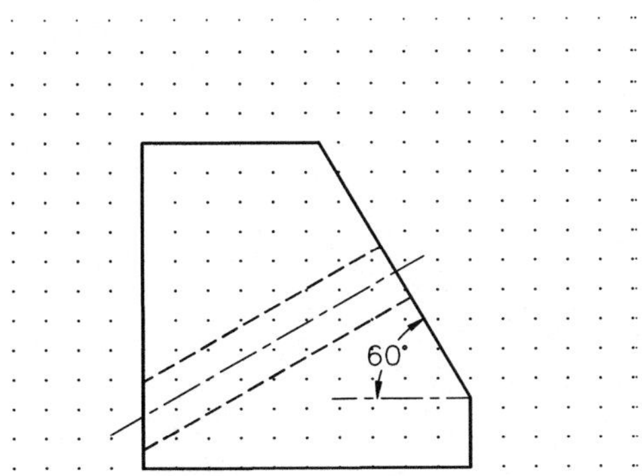

Figure 4-34 Front view of the plate for Example 2

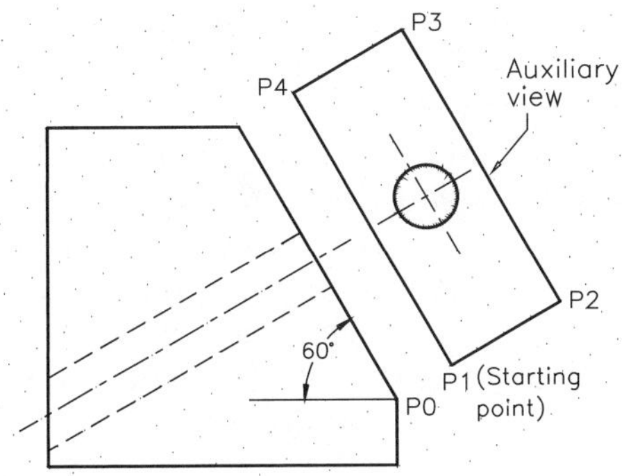

Figure 4-35 Auxiliary view using rotated snap grid

Rotating the Snap Grid

It is assumed that the front view of the plate is already drawn with the coordinates of its starting point as 1,1. Invoke the **Drafting Settings** dialog box and open the **Snap and Grid** tab. In the **Angle** edit box enter **30** and in the **X base** and **Y base** edit boxes enter **6** and **2** respectively as the coordinates of point **P0**. This will rotate the snap grid at an angle of 30-degree through point P0. Now you can draw the auxiliary view easily.

Command: **LINE** Enter
Specify first point: *Select point (P1).*
Specify next point or [Undo]: *Move the cursor 2 units (4 grid points) right and select point (P2).*
Specify next point or [Undo]: *Move the cursor 5 units up and select point (P3).*
Specify next point or [Close/Undo]: *Move the cursor 2 units left and select point (P4).*
Specify next point or [Close/Undo]: **C** Enter

Snap Type and Style

There are two types of Snap types, **Polar** snap and **Grid** snap. **Grid** snap snaps along the grid and is either of **Rectangular** style or **Isometric** style. Rectangular is the default standard style.

Isometric Snap/Grid

You can select the **Isometric snap** radio button in the **Snap type & style** area (Figure 4-36) of the **Drafting Settings** dialog box to set the snap grid to isometric mode. The default is off (standard). The isometric mode is used to make isometric drawings. In isometric drawings, the isometric axes are at angles of 30, 90, and 150-degree. The Isometric snap/grid enables you to display the grid lines along these axes (Figure 4-37). Once you select the **Isometric snap** radio button and choose OK in the dialog box, AutoCAD

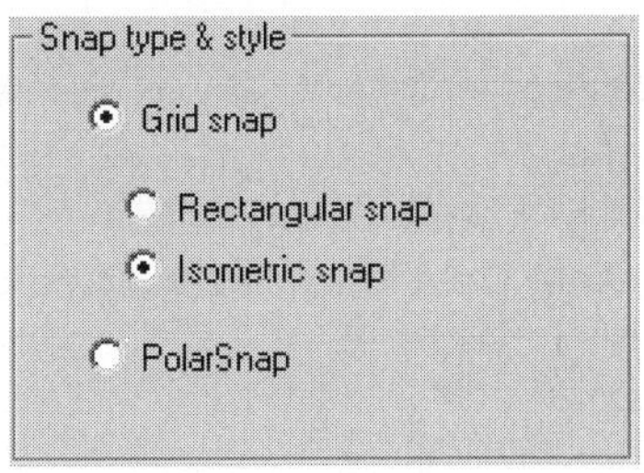

Figure 4-36 Selecting **Isometric snap** in **Snap type & style** area

Chapter 4

automatically changes the cursor to align with the isometric axis. You can adjust the cursor orientation when you are working on the left, top, or right plane of the drawing by using the F5 key or holding down the CTRL key and then pressing the E key to cycle the cursor through different isometric planes (left, right, top). You can change the vertical snap and grid spacing by entering values in the **Snap Y spacing** and **Grid Y spacing** edit boxes. You will notice that the X spacing is not available for this option.

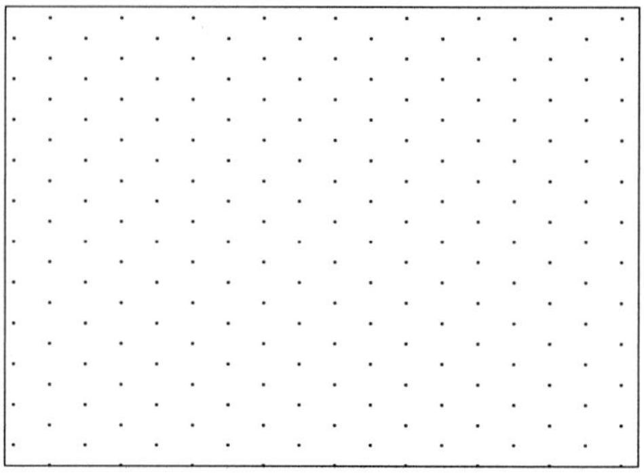

Figure 4-37 *Isometric snap grid*

Polar Snap

You can use polar snap with polar tracking. Here the cursor snaps to points at a specified distance along the Polar alignment angles. You can select the **PolarSnap** radio button in the **Snap style & type** area to set the snap grid to the polar mode. When you choose this option the normal snap options are not available. You can enter a desired distance in the **Polar distance** edit box. If this value is zero, it assumes the same value as the Snap X spacing. The cursor snaps along an imaginary path on the basis of Polar tracking angles, relative to the last point selected or acquired. The angles can be set in the **Polar Tracking** tab of the **Drafting Settings** dialog box (discussed later in the chapter).

For example, select the **PolarSnap** radio button, enter 0.5 in the **Polar distance** edit box, and then choose the **OK** button. Select the **POLAR** button in the status bar for polar tracking. Invoke the **LINE** command and select the starting point anywhere on the screen. Now move the cursor; an imaginary path will be displayed at 0, 90, 180, and 270 angles (polar tracking) with small cross marks displayed on this imaginary line at a distance of 0.5 (polar snap).

Note

The polar snap works along with polar and object tracking. Hence, it is used when the polar tracking is on by selecting the ***Polar*** *button on the status bar, or by selecting the* ***Polar Tracking On*** *check box in the* ***Polar Tracking*** *tab of the* ***Drafting Settings*** *dialog box.*

The snap type is also controlled by the ***SNAPTYPE*** *system variable.*

DRAWING STRAIGHT LINES USING THE ORTHO MODE

You can turn the Ortho mode on or off by choosing the **ORTHO** button in the status bar, or by using the function key F8, or using the **ORTHO** command. The **ORTHO** mode allows you to draw lines at right angles only. Whenever you are using the pointing device to specify the next point, the movement of the rubber-band line connected to the cursor is either horizontal (parallel to the *X* axis) or vertical (parallel to the *Y* axis). If you want to draw a line in the Ortho mode, specify the starting point at the **Specify first point:** prompt. To specify the second point, move the cursor with the pointing device and specify a desired point. The line drawn will be either vertical or horizontal, depending on the direction in which you moved the cursor (Figures 4-38 and Figure 4-39).

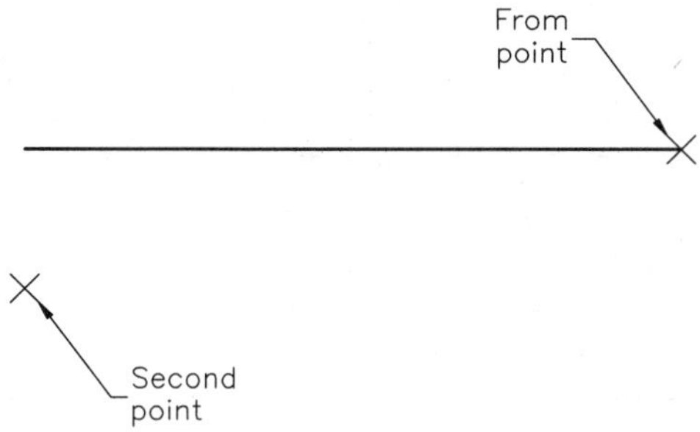

Figure 4-38 *Drawing a horizontal line using the Ortho mode*

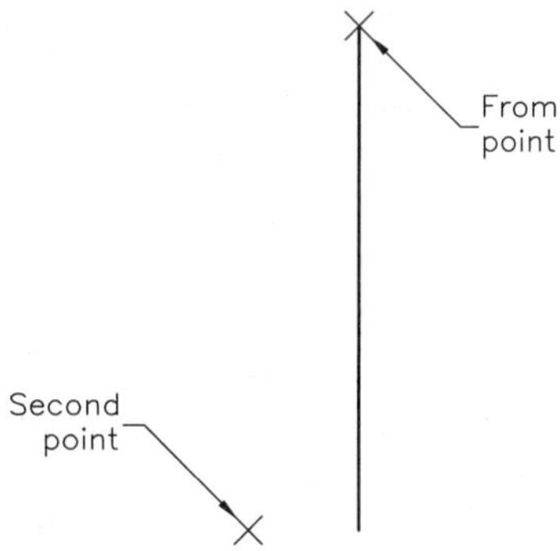

Figure 4-39 *Drawing a vertical line using the Ortho mode*

Tip

You can use the status bar at the bottom of the graphics area to easily and conveniently toggle between on or off for the different drafting functions like the Snap, Grid, and Ortho.

WORKING WITH OBJECT SNAPS

Toolbar:	Object Snap

Object snaps are one of the most useful features of AutoCAD. They improve your performance and the accuracy of your drawing and make drafting much simpler than it normally would be. Object snaps can also be invoked from the shortcut menu (Figure 4-40). This shortcut menu can be accessed by holding down the SHIFT key on the keyboard and then pressing the ENTER button (right-click) on your pointing device. The term object snap refers to the cursor's ability to snap exactly to a geometric point on an object. The advantage of using object snaps is that you do not have to specify an exact point. For example, if you want to place a point at the midpoint of a line, you may not be able to specify the exact point. Using the **MIDpoint** object snap, all you do is move the cursor somewhere on the object. You will notice a marker (in the form of a geometric shape, a triangle for Midpoint) is automatically displayed at the middle point (snap point). You can click to

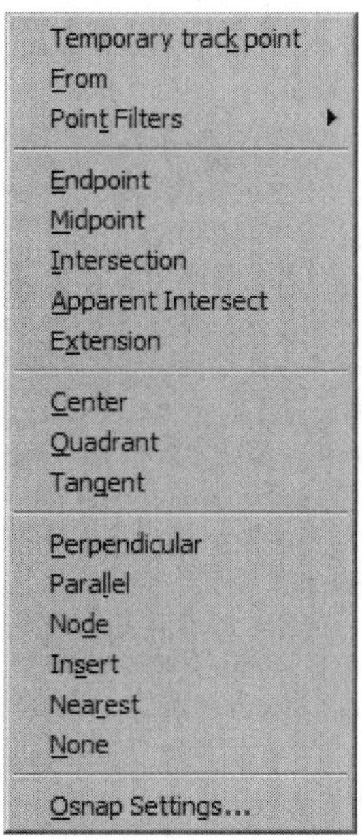

Figure 4-40 *Selecting object snap modes from the shortcut menu*

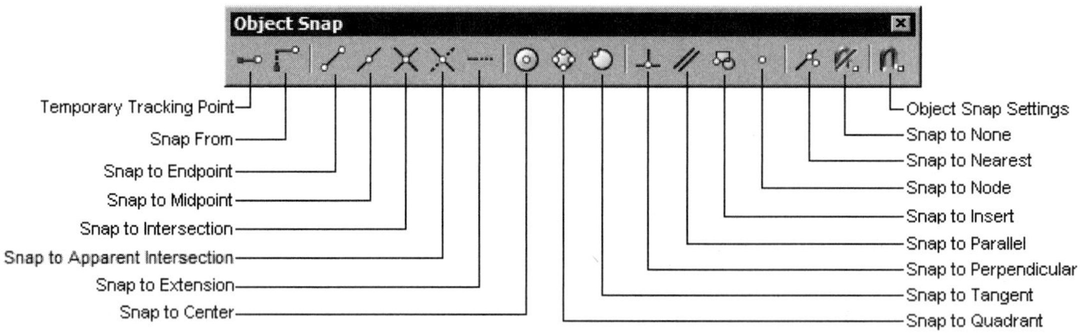

***Figure 4-41 Object Snap** toolbar*

place a point at the position of the **marker**. You also have a **tooltip** with the object snap marker. When you place the cursor on the marker, a tooltip having the name of the object snap will be displayed. The object snaps recognize only the objects that are visible on the screen, which include the objects on locked layers. The objects on layers that are turned off or frozen are not visible, and so they cannot be used for object snaps. The following are the object snap modes available in AutoCAD:

ENDpoint	CENter	PERpendicular	NEArest
MIDpoint	QUAdrant	PARallel	NONe
INTersection	TANgent	INSert	From
APParent Intersection	EXTension	NODe	

You can also invoke an object snap by entering an abbreviation in the command line.

AutoSnap

The AutoSnap feature controls the various characteristics for object snap. As you move the target box over the object, AutoCAD displays the geometric marker corresponding to the shapes shown in the **Object Snap** tab of the **Drafting Settings** dialog box. You can change the different AutoSnap settings such as attaching a target box to the cursor when you invoke any object snap, or change the size and color of the marker. These settings can be changed from the **Options** dialog box (**Drafting** tab). You can invoke the **Options** dialog box from the **Tools** menu (**Tools** > **Options**), or from the shortcut menu when you right-click when no command is active, or entering **OPTIONS** at the Command prompt. You can also invoke it by selecting the **Options** button in the **Drafting Settings** dialog box. When you choose the **Drafting** tab in the **Options** dialog box, the AutoSnap options are displayed (Figure 4-42). You can select the **Marker** check box to toggle the display of marker. You can use the **Magnet** check box to toggle the magnet that snaps the crosshair to the particular point of the object for that object snap. You can use the **Display AutoSnap tooltip** and **Display AutoSnap aperture box** check boxes to toggle the display of the tooltip and the aperture box. Selecting the **Display AutoSnap tooltip** check box shows a flag that gives the name of the object snap that AutoCAD has detected. You can change the size of the marker and the aperture box by moving the **AutoSnap Marker Size** and **Aperture Size** slider bars respectively. You can also change the color of the markers through the **AutoSnap marker color** list box. The size of the

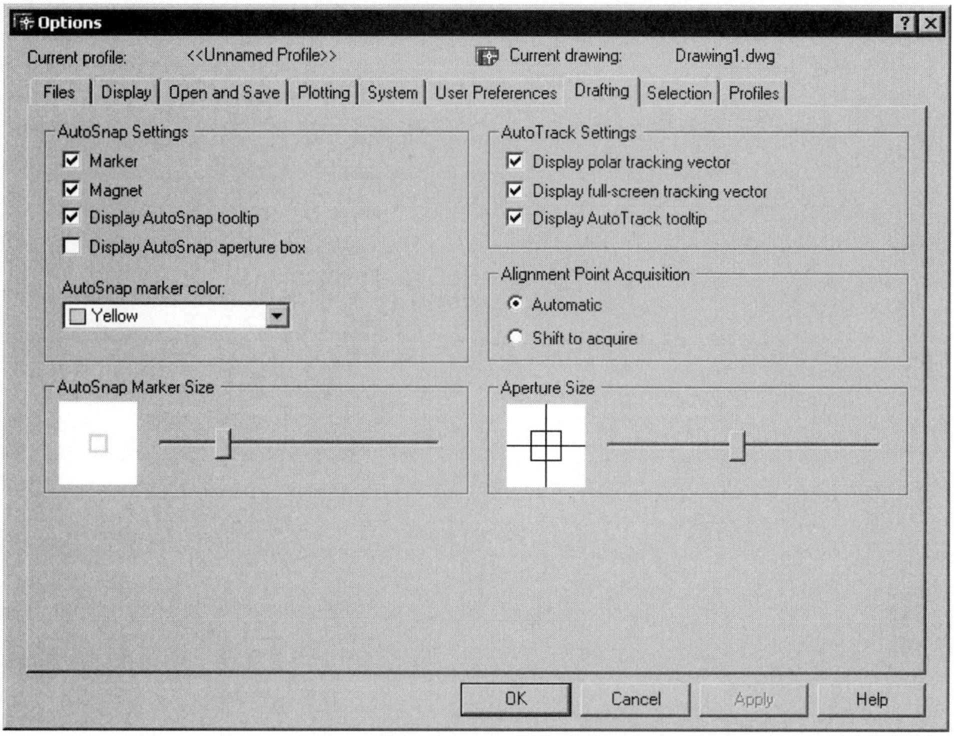

***Figure 4-42 Drafting** tab of the **Options** dialog box*

aperture is measured in pixels, short for picture elements. Picture elements are dots that make up the screen picture. The aperture size can also be changed using the **APERTURE** command. In AutoCAD the default value for the aperture size is 10 pixels. The display of the marker and the tooltip is controlled by the **AUTOSNAP** system variable. The following are the bit values for **AUTOSNAP**.

Bit Values	Function
0	Turns off the Marker, AutoSnap Tooltip, and Magnet
1	Turns on the Marker
2	Turns on the AutoSnapTooltip
4	Turns on the Magnet

Endpoint

The ENDpoint Object Snap mode snaps to the closest endpoint of a line or an arc. To use this Object Snap mode, select the Endpoint button, and move the cursor (crosshairs) anywhere close to the endpoint of the object. The marker will be displayed at the endpoint; click to specify that point. AutoCAD will grab the endpoint of the object. If there are several objects near the cursor crosshairs, AutoCAD will grab the endpoint of the object that is closest to the crosshairs, or if the Magnet is on you can move to grab the desired endpoint. For Figure 4-43, invoke the **LINE** command from the **Draw** toolbar. The following is the prompt sequence:

Chapter 4

Specify first point: *Select the **Snap to Endpoint** button from the **Object Snap** toolbar.*
_endp of *Move the crosshair and select the arc.*
Specify next point or [Undo]: *Select the endpoint of the line.*

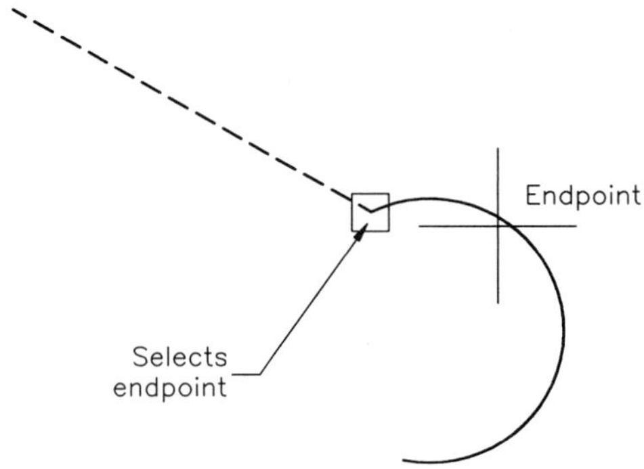

Figure 4-43 *ENDpoint Object Snap mode*

Midpoint

The MIDpoint Object Snap mode snaps to the midpoint of a line or an arc. To use this Object Snap mode, select Midpoint osnap and select the object anywhere. AutoCAD will grab the midpoint of the object. For Figure 4-44, invoke the **LINE** command from the **Draw** toolbar. The following is the prompt sequence.

Specify first point: *Select the starting point of the line.*
Specify next point or [Undo]: *Choose the **Snap to Midpoint** button from the **Object Snap** toolbar.*
_mid of *Move the cursor and select the original line.*

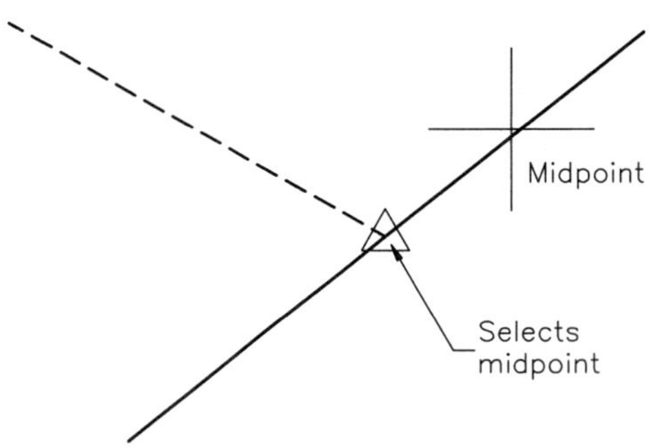

Figure 4-44 *MIDpoint Object Snap mode*

Nearest

The NEArest Object Snap mode selects a point on an object (line, arc, circle, or ellipse) that is visually closest to the graphics cursor (crosshairs). To use this mode, enter the command, and then choose the Nearest object snap. Move the crosshairs near the intended point on the object so as to display the marker at the desired point and then select the object. AutoCAD will grab a point on the line where the marker was displayed. For Figure 4-45, invoke the **LINE** command from the **Draw** toolbar. The following is the prompt sequence:

Specify first point: *Choose the* ***Snap to Nearest*** *button from the* ***Object Snap*** *toolbar.*
_nea to *Select a point near an existing object.*
Specify next point or [Undo]: *Select endpoint of the line.*

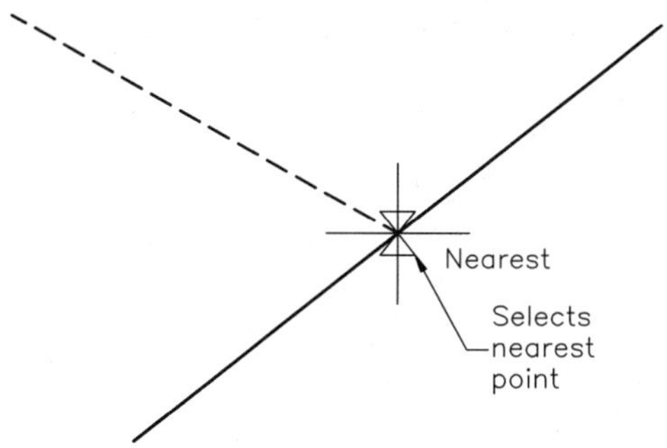

Figure 4-45 *NEArest Object Snap mode*

Center

The CENter Object Snap mode allows you to snap to the center point of an ellipse, circle, or arc. After selecting this option, you must point to the visible part of the circumference of a circle or arc. For Figure 4-46, invoke the **LINE** command from the **Draw** toolbar. The following is the prompt sequence:

Specify first point: *Choose the* ***Snap to Center*** *button from the* ***Object Snap*** *toolbar.*
_cen of *Move the cursor and select the circle.*
Specify next point or [Undo]: *Select the endpoint of the line.*

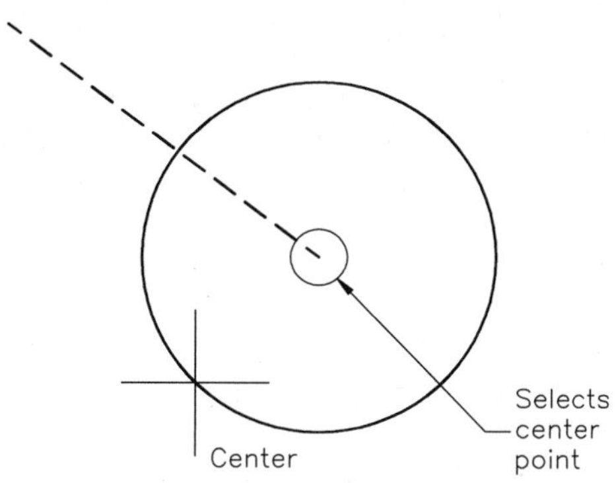

Figure 4-46 *CENter Object Snap mode*

Tangent

The TANgent Object Snap allows you to draw a tangent to or from an existing ellipse, circle, or arc. To use this object snap, place the cursor on the circumference of the circle or arc to select it. For Figure 4-47, invoke the **LINE** command from the **Draw** toolbar. The following is the prompt sequence:

Specify first point: *Select the starting point of the line.*
Specify next point or [Undo]: *Choose the* ***Snap to Tangent*** *button from the* ***Object Snap*** *toolbar.*
_tan to *Move the cursor and select the circle.*
Specify next point or [Undo]: *Select the endpoint of the line (tangent of the circle).*

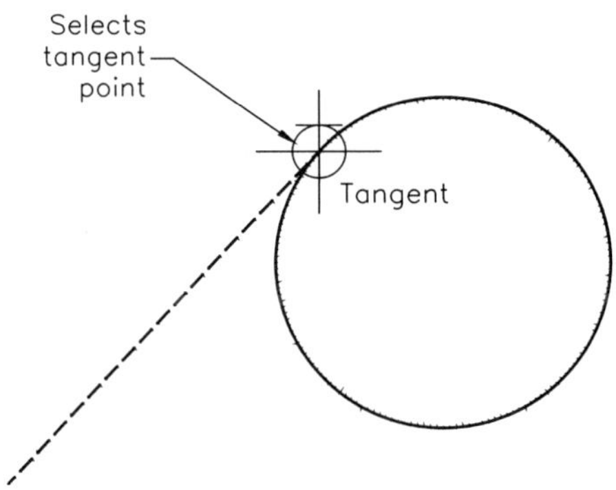

Figure 4-47 *TANgent Object Snap mode*

Note

If the start point of a line is defined using the Tangent Object Snap, the tip shows Deferred Tangent. However, if you end the line using this Object Snap, the tip shows Tangent.

Figure 4-48 shows the use of NEArest, ENDpoint, MIDpoint, and TANgent Object Snap modes.

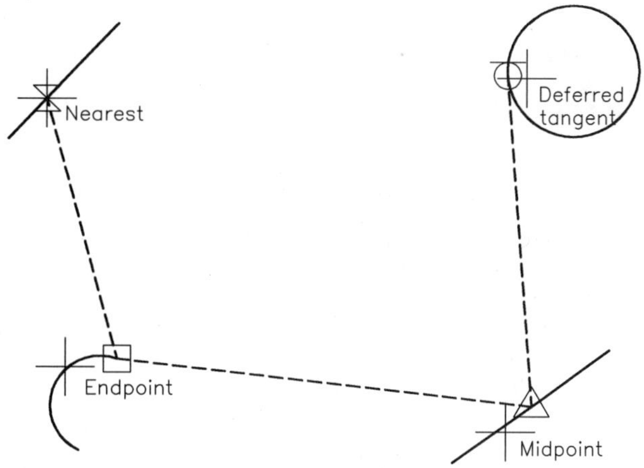

Figure 4-48 *Using the NEArest, ENDpoint, MIDpoint, and TANgent Object Snap modes*

Quadrant

The QUAdrant Object Snap mode is used when you need to snap to a quadrant point of an ellipse, arc, or circle. A circle has four quadrants, and each quadrant subtends an angle of 90-degree. The quadrant points are located at 0-, 90-, 180-, and 270-degree positions. If the circle is inserted as a block (see Chapter 14), that is rotated, the quadrant points are also rotated by the same amount, see Figures 4-49 and 4-50.

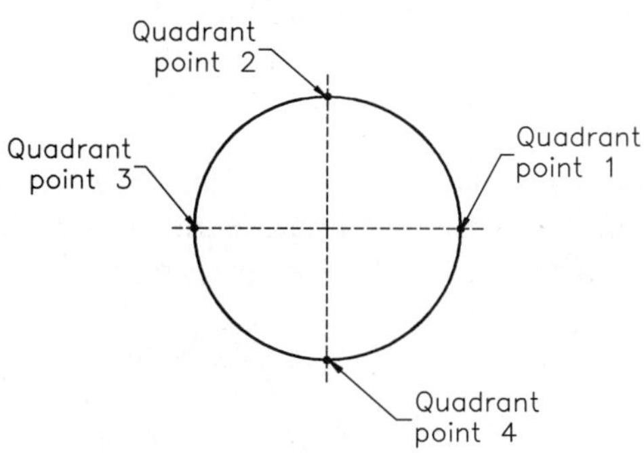

Figure 4-49 *Location of circle quadrants*

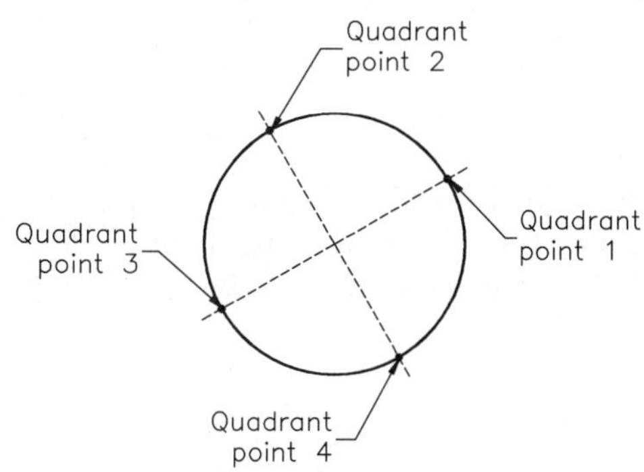

Figure 4-50 *Quadrants in a rotated circle*

To use this object snap, position the cursor on the circle or arc closest to the desired quadrant. The prompt sequence for drawing a line from the third quadrant of a circle as shown in Figure 4-51 is given next.

Specify first point: *Choose the* ***Snap to Quadrant*** *button from the* ***Object Snap*** *toolbar.*
_qua of *Move the cursor close to the third quadrant of the circle and select it.*
Specify next point or [Undo]: *Select the endpoint of the line.*

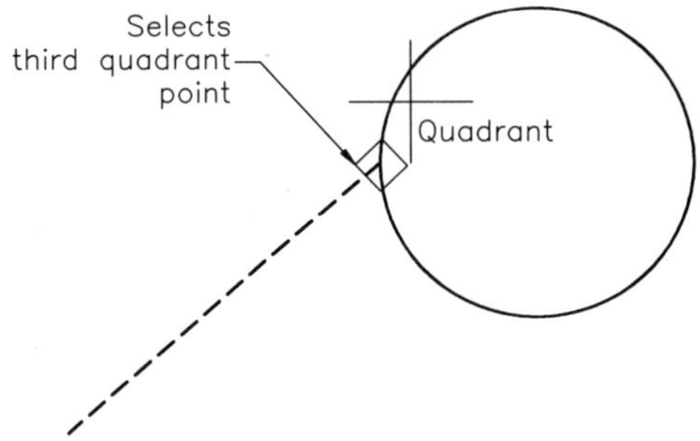

Figure 4-51 *QUAdrant Object Snap mode*

Intersection

The INTersection Object Snap mode is used to snap to a point where two or more lines, circles, ellipses, or arcs intersect. To use this object snap, move the cursor close to the desired intersection so that the intersection is within the target box, and then specify that point. For Figure 4-52, invoke the **LINE** command. The prompt sequence is given next.

Specify first point: *Choose the* ***Snap to Intersection*** *button from the* ***Object Snap*** *toolbar.*
_int of *Position the cursor near the intersection and select it.*
Specify next point or [Undo]: *Select the endpoint of the line.*

After selecting the Intersection object snap, if your cursor is close to an object and not close to an actual intersection, the intersection marker displays ellipses [...] with it. This indicates an extended intersection. If you select this object now, AutoCAD prompts **and** for selection of another object. If your cursor is close to another object, AutoCAD marks the extended intersection point between these two objects. This mode selects extended or visual intersections of lines, arcs, circles, or ellipses (Figure 4-53). The extended intersections are the intersections that do not exist at present, but are imaginary and formed if the line or arc is extended.

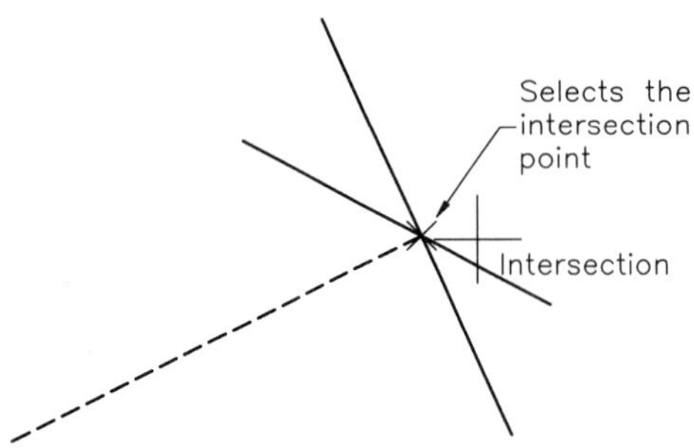

Figure 4-52 *INTersection Object Snap mode*

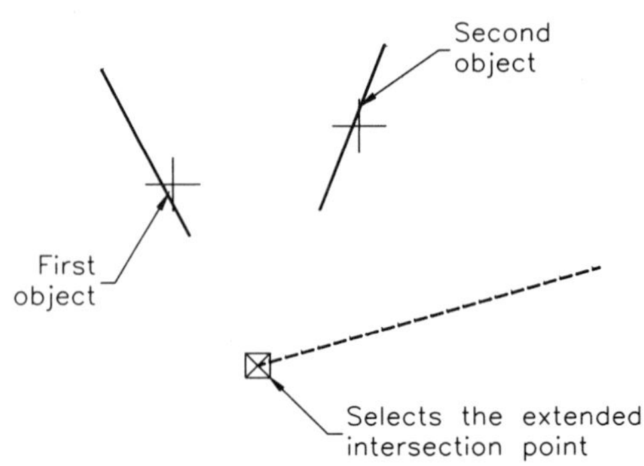

Figure 4-53 *Extended Intersection Object Snap mode*

Apparent Intersection

The APParent Intersection Object Snap mode selects projected or visual intersections of two objects in 3D space. Sometimes, two objects appear to intersect one another in the current view, but in 3D space the two objects do not actually intersect. The Apparent Intersection snap mode selects such visual intersections. This mode works on wireframes in 3D space (See Chapter 23 for wireframe models). If you try to use this object snap mode in 2D, it works like extended intersection.

Perpendicular

The PERpendicular Object Snap mode is used to draw a line perpendicular to or from another line, or normal to or from an arc or circle, or to an ellipse. When you use this mode and select an object, AutoCAD calculates the point on the selected object so that the previously selected point is perpendicular to the line. The object can be selected by positioning the cursor anywhere on the line. First invoke the **LINE** command; the prompt sequence to draw a line perpendicular to a given line (Figure 4-54) is given next.

Specify first point: *Select the starting point of the line.*
Specify next point or [Undo]: *Choose the **Snap to Perpendicular** button from the **Object Snap** toolbar.*
_per to *Select the line on which you want to draw perpendicular.*

When you select the line first, the rubber-band feature of the line is disabled. The line will appear only after the second point is selected. Invoke the **LINE** command. The prompt sequence for drawing a line perpendicular from a given line (Figure 4-55) is given next.

Specify first point: *Choose the **Snap to Perpendicular** button from the **Object Snap** toolbar.*
_per to *Select the line on which you want to draw perpendicular.*
Specify next point or [Undo]: *Select the endpoint of the line.*

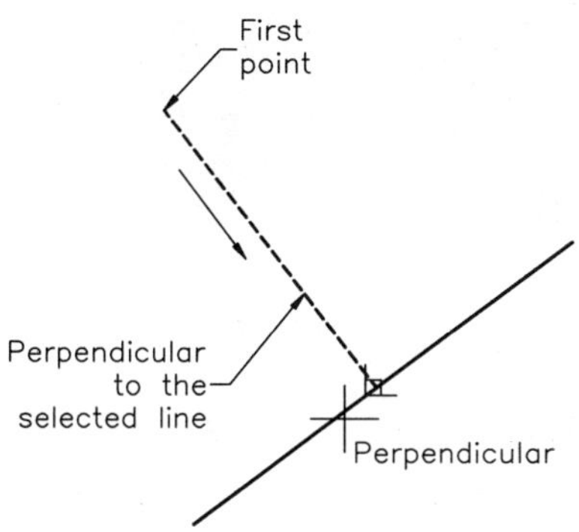

***Figure 4-54** Selecting the start point and then the perpendicular snap*

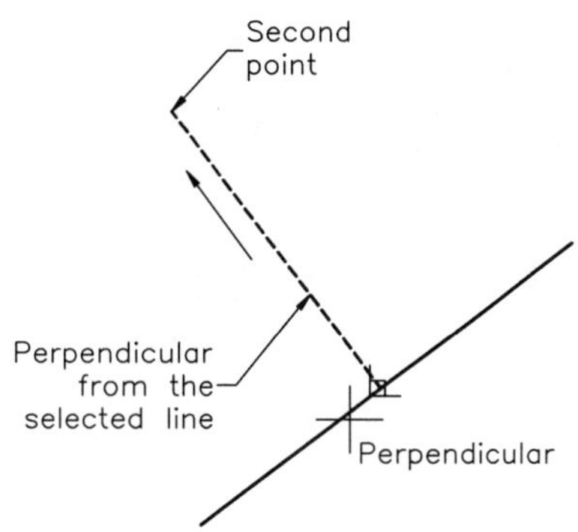

***Figure 4-55** Selecting the perpendicular snap first*

Figure 4-56 shows the use of the various object snap modes.

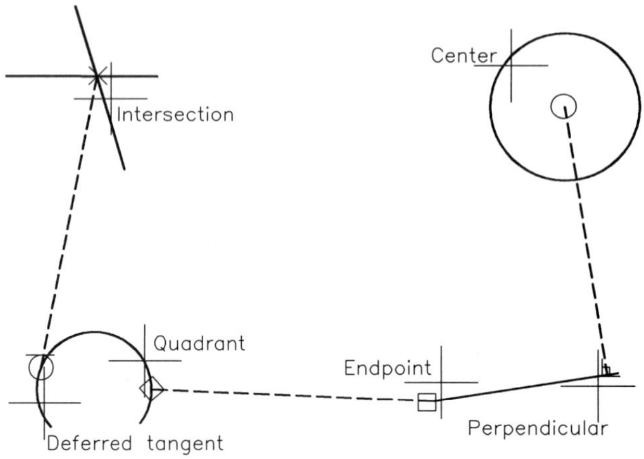

Figure 4-56 *Using various object snap modes to locate points*

Exercise 3 *General*

Make the drawing shown in Figure 4-57. P1 and P2 are the center points of the top and bottom arcs. The space between the dotted lines is 1 unit.

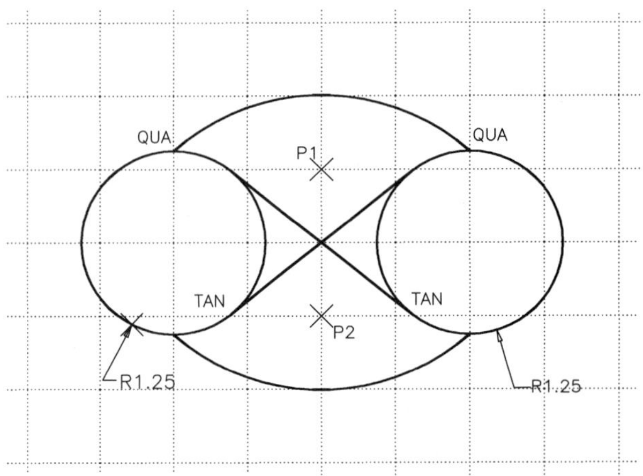

Figure 4-57 *Drawing for Exercise 3*

Node

You can use the NODe Object Snap to snap to a point object drawn using the **POINT** command, or placed using the **DIVIDE** or **MEASURE** commands. In

Figure 4-58, three points have been drawn using the AutoCAD **POINT** command. You can snap to these points by using the NODe snap mode. Invoke the **LINE** command and the prompt sequence is as follows (Figure 4-59):

Specify first point: *Choose the **Snap to Node** button from the **Object Snap** toolbar.*
_nod of *Select point P1.*
Specify next point or [Undo]: *Choose the **Snap to Node** button from the **Object Snap** toolbar.*
_nod of *Select point P2.*
Specify next point or [Undo]: *Choose the **Snap to Node** button from the **Object Snap** toolbar.*
_nod of *Select point P3.*

Figure 4-58 *Point objects*

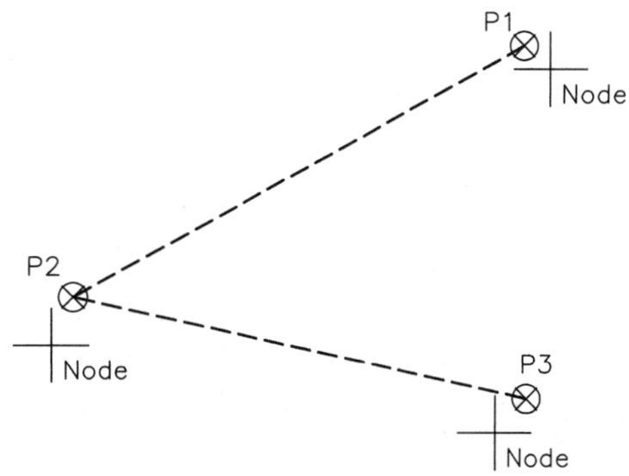

Figure 4-59 *Using NODe object snap*

Insert

The INSert Object Snap mode is used to snap to the insertion point of a text, shape, block, attribute, or attribute definition. In Figure 4-60, the text **WELCOME** is left-justified and the text **AutoCAD** is center-justified. The point with respect to which the text is justified is the insertion point of that text string. If you want to snap to these insertion points or the insertion point of a block, you must use the INSert Object Snap mode. Invoke the **LINE** command and following is the prompt sequence:

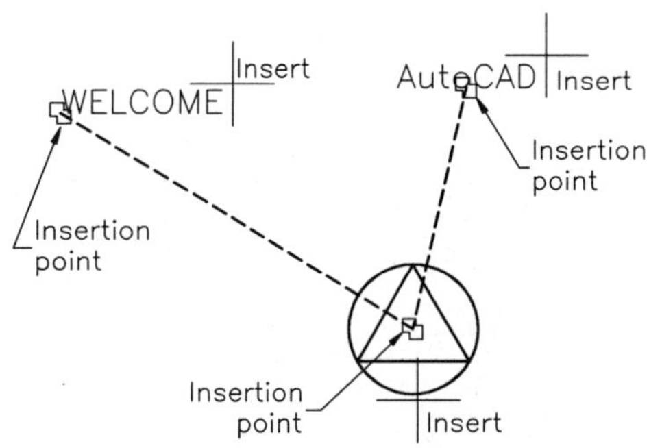

Figure 4-60 *INSert Object Snap mode*

Specify first point: *Choose the **Snap to Insert** button from the **Object Snap** toolbar.*
_ins of *Select WELCOME text.*
Specify next point or [Undo]: *Choose the **Snap to Insert** button from the **Object Snap** toolbar.*
_ins of *Select the block.*
Specify next point or [Undo]: *Choose the **Snap to Insert** button from the **Object Snap** toolbar.*
_ins of *Select AutoCAD text.*

Chapter 4

None

The NONe Object Snap mode turns off any running object snap (see the section "Running Object Snap Mode" that follows) for one point only. The following example illustrates the use of this Object Snap mode:

Invoke the **Drafting Settings** dialog box and select the **Object Snap** tab. Select the **Midpoint** and **Center** check boxes. This sets the object snaps to Mid and Cen. Now, suppose you want to draw a line whose starting point is a point closer to the endpoint of another line. After you invoke the **LINE** command and move the cursor to the desired position on the previous line, it automatically snaps to its midpoint. You can disable this using the NONe object snap by choosing it from the **Object Snap** toolbar. You can select the desired point on the line. Once you have selected a point it continues the command with the specified object snaps.

Parallel

When you need to draw a line parallel to a line or polyline on the screen, you can use the PARallel object snap (Figure 4-61). For example, when you are in the middle of the **LINE** command, and you have to draw a line parallel to the one already on the screen, you can use the **PARallel** object snap as follows:

Command: *Choose **Line** from the **Draw** toolbar.*
Specify first point: *Select a point on the screen.*
Specify next point or [Undo]: *Choose the **Parallel** button from the **Object Snap** toolbar.*
_par to *Specify object to which parallel is to be drawn.*

When you specify the reference object, a parallel sign is displayed. You should briefly pause on that line so that a small plus sign appears on it to indicate that it has been selected. Now, on moving the cursor close to an angle parallel to the line, an imaginary parallel line (construction line) appears on which you can select the next point. A tooltip which has the relative polar coordinates is displayed with the cursor as you move it on the construction line. This helps you to select the next point. This line which has been drawn is parallel to the selected object.

Extension

The EXTension Object Snap gives you an option to locate a point on the extension path of a line or an arc (Figure 4-62). It can also be used with Intersection to determine the point of extended intersection. To use extension, choose the **Line** button from the **Draw** toolbar and then choose the **Snap to Extension** button from the **Object Snap** toolbar. Briefly pause at the end of the line or arc you want to use. A small plus sign (+) appears at the end of the line or arc, indicating that it has been selected. If you moving the cursor along the extension path, a temporary extension path is displayed and the tooltip displays relative polar coordinates from the end of the line. Select a point or enter a distance to begin a line and then select another point to finish the line.

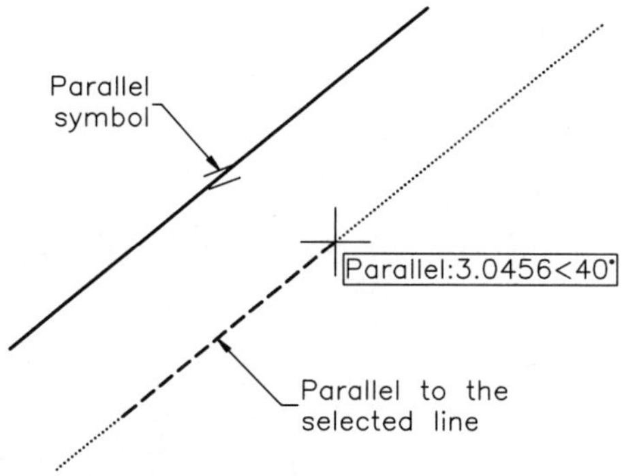

Figure 4-61 Using the PARallel Object Snap mode

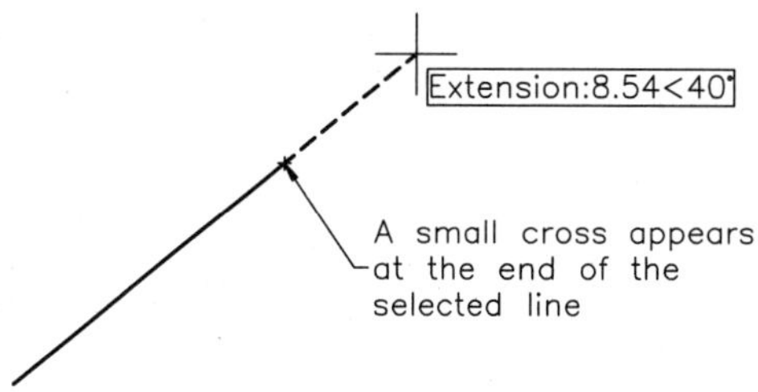

Figure 4-62 Using the EXTension Object Snap mode

Note

While using the PARallel object snap, you can choose more than one line as reference lines that are indicated by plus sign. Depending upon the direction in which you move the cursor, AutoCAD will choose any one of these reference lines. The line that is chosen gets marked by the parallel symbol in place of plus.

Similarly, while using the EXTension object snap, you can choose more than one endpoint as reference.

From

The From Object Snap mode can be used to locate a point relative to a given point (Figure 4-63). For example, if you want to locate a point that is 2.5 units up and 1.5 units right from the endpoint of a given line, you can use the **From** object snap as follows:

Command: **LINE**
Specify first point: *Choose the **Snap From** button from the **Object Snap** toolbar.*
_from Base point: *Choose the **Snap to Endpoint** button from the **Object Snap** toolbar.*
_endp of *Specify the endpoint of the given line*
<Offset>: **@1.5,2.5**

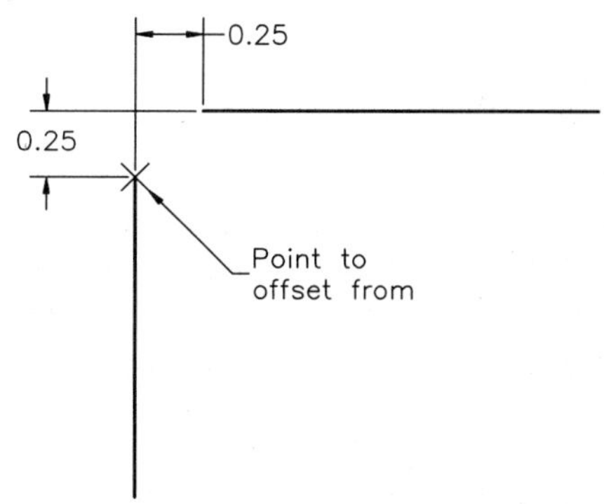

Figure 4-63 Using the From Object Snap mode to locate a point

Note

The From object snap cannot be used as running object snap.

Temporary Tracking Point

The Temporary Tracking Point can be used to locate a point with respect to two different points. When you select a point that is indicated by a plus sign, depending upon the direction in which you move the cursor, an orthogonal imaginary line is displayed either horizontally or vertically. Then you can select another point to display another orthogonal imaginary line in other direction. The desired point will be where these two imaginary lines intersect. You can use this option along with other Object Snap modes to locate a point. For Figure 4-64, invoke the **LINE** command from the **Draw** toolbar and then use the temporary tracking as follows:

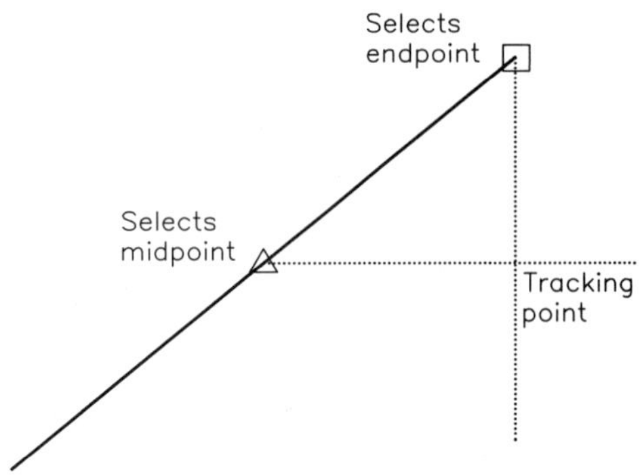

Figure 4-64 *Using temporary tracking*

Specify first point: *Choose the **Temporary Tracking Point** button from the **Object Snap** toolbar.*
Specify temporary OTRACK point: *Choose the **Snap to Midpoint** button from the **Object Snap** toolbar.*
of *Select the midpoint of the line and move the cursor horizontally toward right.*
Specify first point: *Choose the **Temporary Tracking Point** button from the **Object Snap** toolbar.*
Specify temporary OTRACK point: *Choose the **Snap to Endpoint** button from the **Object Snap** toolbar.*
of *Select the upper endpoint of the line and move the cursor vertically down.*

As you move your cursor down both the horizontal and vertical imaginary lines are displayed. Select the point at their intersection point and then draw the line.

Combining Object Snap Modes

You can also combine the snaps from the command line by separating the snap modes with a comma. AutoCAD will search for the specified modes and grab the point on the object that is closest to the point where the object is selected. The prompt sequence for using the Midpoint and Endpoint object snaps is given next.

Command: **LINE** Enter
Specify first point: **MID, END** Enter *(MIDpoint or ENDpoint object snap.)*
to *Select the object.*

Note

In the discussions of object snaps, "line" generally includes xlines, rays, and polyline segments, and "arc" generally includes polyarc segments.

Exercise 4 *General*

Draw the sketch shown in Figure 4-65. The space between the dotted lines is 1 unit.

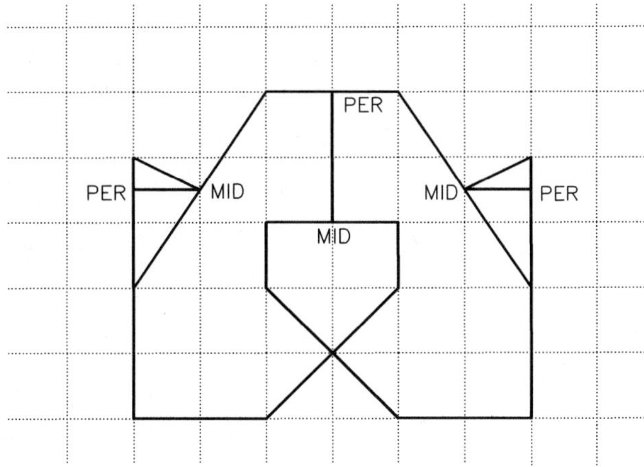

Figure 4-65 *Drawing for Exercise 4*

RUNNING OBJECT SNAP MODE

Toolbar:	Object Snap > Object Snap Settings
Menu:	Tools > Drafting Settings
Command:	OSNAP

In the previous sections you have learned how to use the object snaps to snap to different points of an object. One of the drawbacks of these object snaps is that you have to select them every time you use them, even if it is the same snap mode. This can be solved by using **running object snaps**. The Running Osnap can be invoked from the **Object Snap** tab of the **Drafting Settings** dialog box (Figure 4-66). If you choose the **Object Snap Settings** button from the toolbar, or enter **OSNAP** at the Command line, or you choose **Settings** from the shortcut menu displayed when you right-click on the **OSNAP** button in the status bar, the **Object Snap** tab is displayed in the **Drafting Settings** dialog box. In this tab you can set the running object snap modes by selecting the check boxes next to the snap modes. For example, if you want to set Endpoint as the running Object Snap mode, select the **Endpoint** check box and then the **OK** button in the dialog box.

Once you set the running Object Snap mode, you are automatically in that mode and the marker is displayed when you move the crosshairs over the snap points. If you had selected a combination of modes, AutoCAD selects the mode that is closest to the screen crosshairs. For example, you have selected the **Endpoint**, **Midpoint**, and **Center** check boxes in the dialog box. Now for selection of a point when you move the cursor over a previously drawn line and whenever your cursor is closer to the midpoint of the line the marker is displayed there. You can also move the cursor to the end of the line to invoke the Endpoint object snap. If you place the cursor over the circumference of a previously drawn circle, the Center osnap is invoked. The running Object Snap mode can be turned on or off, without losing the object snap settings, by choosing the **OSNAP** button in the status bar. You can also accomplish this by pressing the function key F3 or CTRL+F keys.

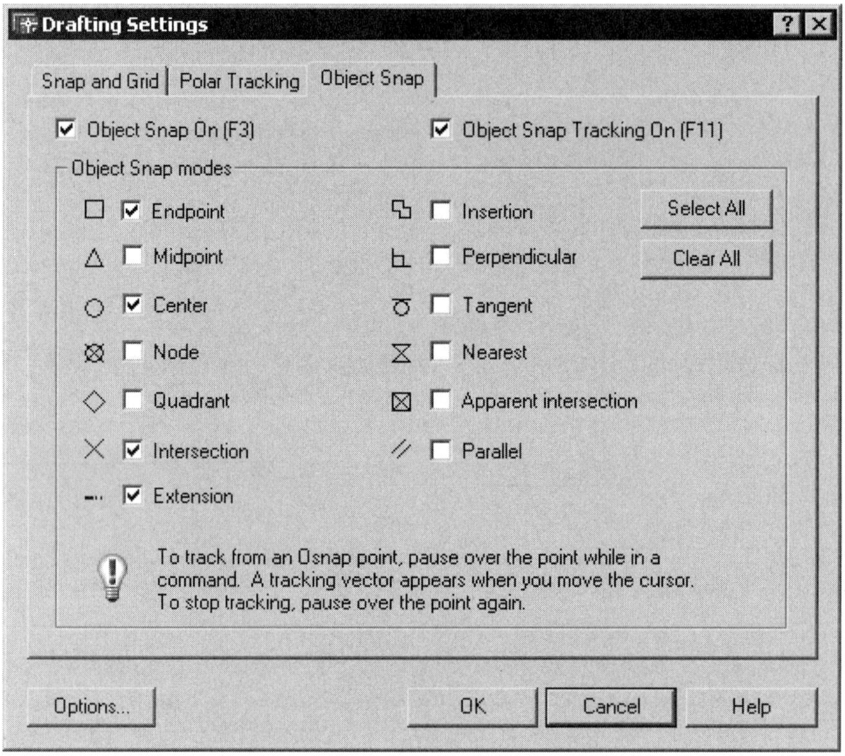

***Figure 4-66 Drafting Settings** dialog box (**Object Snap** tab)*

Note

When you use a temporary object snap and fail to specify a suitable point, AutoCAD issues an Invalid point prompt and reissues the pending command's prompt. However, with a running Object Snap mode, if you fail to specify a suitable point, AutoCAD accepts the point specified without snapping to an object.

*The running osnap works only when it is turned on by using the **Osnap** button in the status bar.*

Overriding the Running Snap

When you select the running object snaps, all other Object Snap modes are ignored unless you select another Object Snap mode. Once you select a different osnap mode, the running OSNAP mode is temporarily overruled. After the operation has been performed, the running OSNAP mode goes into effect again. If you want to discontinue the current running Object Snap modes totally, choose the **Clear all** button in the **Drafting Settings** dialog box. If you want to temporarily disable the running object snap, choose the **OSNAP** button (off position) in the status bar.

If you are overriding the running object snap modes for a point selection and no point is found to satisfy the override Object Snap mode, AutoCAD displays a message to this effect. For example, if you specify an override Object Snap mode of Center and no circle, ellipse, or arc is found at that location, AutoCAD will display the message "No center found for specified point. Point or option keyword required."

Cycling through Snaps

AutoCAD displays the geometric marker corresponding to the shapes shown in the **Object snap settings** tab of the **Drafting Settings** dialog box. You can use the TAB key to cycle through the snaps. For example, if you have a circle with an intersecting rectangle as shown in Figure 4-67 and you want to snap to one of the geometric points on the circle, you can use the TAB key to cycle through the geometric points. The geometric points for a circle are the center point, quadrant points, and the intersecting points with the rectangle. To snap to one of these points, the first thing you need to do is to set the running object snaps (center, quadrant, and intersection object snaps) in the **Drafting Settings** dialog box. After entering a command, when you drag the cursor over the objects, AutoSnap displays a marker and a SnapTip. You can cycle through the snap points available for an object by pressing the TAB key. For example, if you press the TAB key while the aperture box is on the circle and the rectangle (near the lower left intersection point), AutoSnap will display the intersection, center, and quadrant points one by one. When a certain osnap is displayed its pertaining object is highlighted by making it dashed. For example, when the center or quadrant snaps are displayed the circle becomes dashed, and when the intersection snap is displayed both the circle and the rectangle become dashed. By selecting a point you can snap to one of these points.

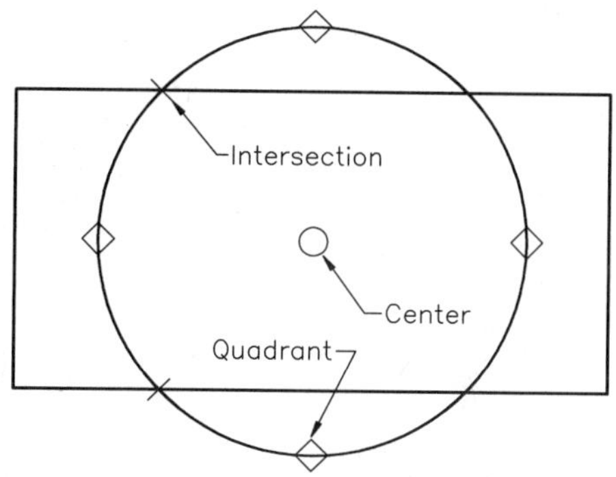

Figure 4-67 *Using TAB key to cycle through the snaps*

Setting the Priority for Coordinate Entry

Sometimes you may want the keyboard entry to take precedence over the running Object Snap modes. This is useful when you want to locate a point that is close to the running osnap. By default, when you specify the coordinates (by using the keyboard) of a point located near a running osnap, AutoCAD ignores the point and snaps to the running osnap. For example, you select the **Intersection** object snap in the **Object Snap** tab of the **Drafting Settings** dialog box. Now if you enter the coordinates of the endpoint of a line very close to the intersection point (Figure 4-68a), the line will snap to the intersection point and ignore the keyboard entry point. You can set the priority between the keyboard entry and object snap through the **User Preferences** tab of the **Options** dialog box (Figure 4-69). Select the **Keyboard Entry** radio button and then choose **OK**. Now if you enter the coordinates of the endpoint of a line very close to the intersection point (Figure 4-68b), the starting point will snap to the coordinates specified and the running osnap (intersection) is ignored. This setting is stored in the **OSNAPCOORD** system variable and you can also set the priority through this variable.

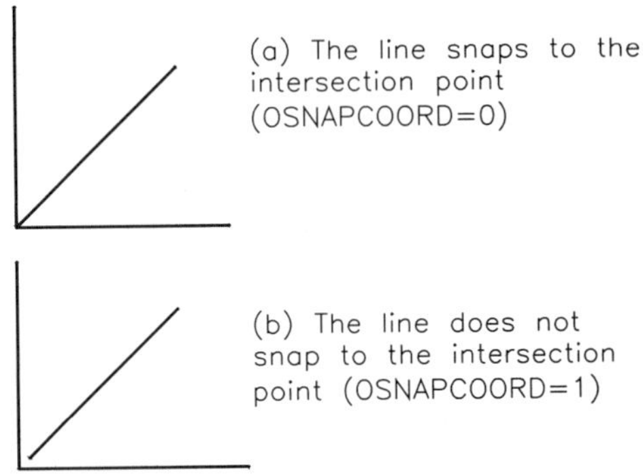

Figure 4-68 *Setting the coordinate entry priority*

A value **0** gives priority to running osnap, value **1** gives priority to keyboard entry, and value **2** gives priority to keyboard entry except scripts.

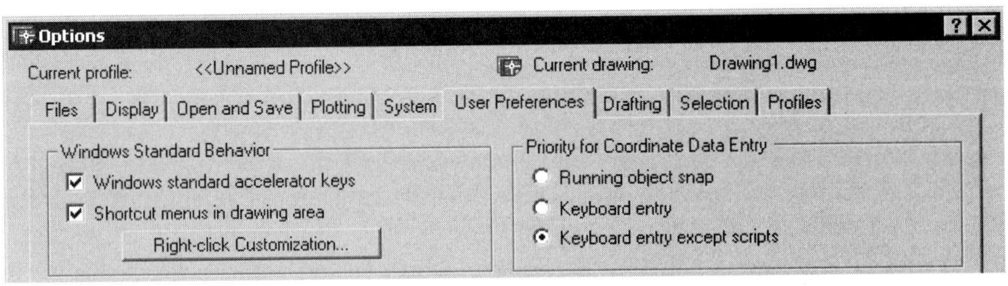

Figure 4-69 ***User Preferences*** *tab of the* ***Options*** *dialog box*

Tip

If you do not want the running osnap to take precedence over the keyboard entry, you can simply disable running osnap temporarily by selecting the ***OSNAP*** *button on the status bar to the off position. This lets you specify a point close to a running osnap. This way you do not have to change the settings for coordinate entry in the* ***Options*** *dialog box.*

USING AUTOTRACKING

When using AutoTracking, the cursor moves along temporary paths to locate key points in a drawing. It can be used to locate points with respect to other points or objects in the drawing. There are two types of AutoTracking options: **Object Snap Tracking** and **Polar Tracking**.

Object Snap Tracking

Object Snap Tracking tracks the movement of the cursor along alignment paths based on Object Snap points (running osnaps) that are selected in the **Object Snap** tab of the **Drafting Settings** dialog box. Selecting the **Object Snap Tracking On (F11)** check box in the **Object Snap** tab of the **Drafting Settings** dialog box, choosing the **OTRACK** button on the status bar, or using the function key F11 sets the object snap tracking on.

The direction of the path is determined by the motion of the cursor or the point you select on an object. For example, if you want to draw a circle whose center is located in line with the center of two existing circles (Figure 4-70), you can use AutoTracking. First select **Center** in the **Object Snap** tab of the **Drafting Settings** dialog box to set Center as the running osnap. Also choose the Osnap button on the status bar to the On position. Now you can activate object tracking by choosing the **OTRACK** button on the status bar. Choose the **Circle** button from the **Draw** toolbar. At the next prompt **Specify center point for circle or [3P/2P/Ttr (tan, tan, radius)]**, pause the cursor at the center of the first circle to attain the point (plus sign) and then move it slightly to get the imaginary horizontal line. Move the cursor toward the second circle; the horizontal path disappears. Now place the cursor briefly at the center of the second circle and move it slightly to get the imaginary vertical line. Move the cursor up along the vertical alignment path and when you are nearly in line with the other circle, the horizontal path is also displayed. Select the intersection of the two alignment paths and it becomes the center of the circle. At the next prompt **Specify radius of circle or [Diameter] <current>**, enter radius or specify a point.

Similarly, you can use AutoTracking in combination with the **midpoint** object snap to locate the center of the rectangle and then draw a circle (Figure 4-71).

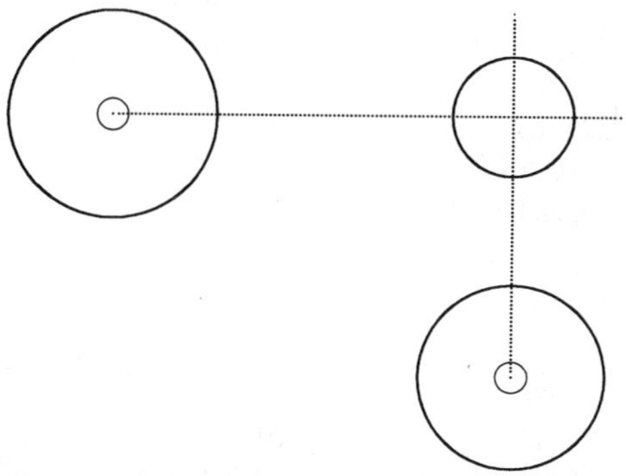

Figure 4-70 *Using AutoTracking to locate a point (center of circle)*

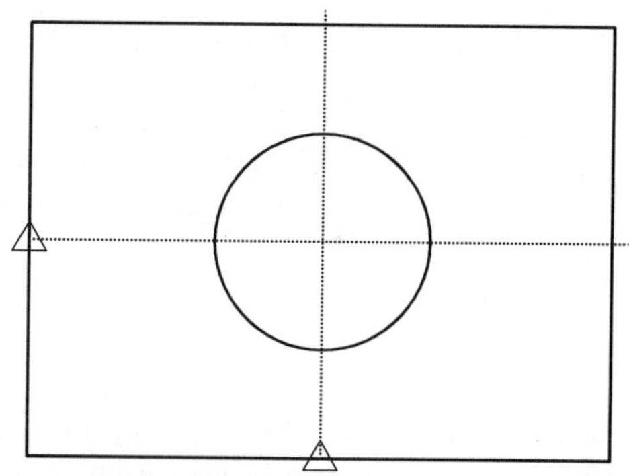

Figure 4-71 *Using AutoTracking to locate a point (midpoint of rectangle)*

Note

Object tracking works only when ***OSNAP*** *is on and some running object snaps have been set.*

Polar Tracking

Polar tracking is used to locate points on an angular alignment path. Polar tracking can be selected by choosing the **POLAR** button on the status bar, by using the function key F10, or by selecting the **Polar Tracking On (F10)** check box in the **Polar Tracking** tab of the **Drafting Settings** dialog box (Figure 4-72). Polar Tracking constrains the movement of the cursor along a path that is based on the polar angle settings. For example, if the **Increment angle** list box value is set to 15-degree in the **Polar Angle Settings** area, the cursor will move along alignment paths that are multiples of 15-degree (0, 15, 30, 45, 60, and so on) and a

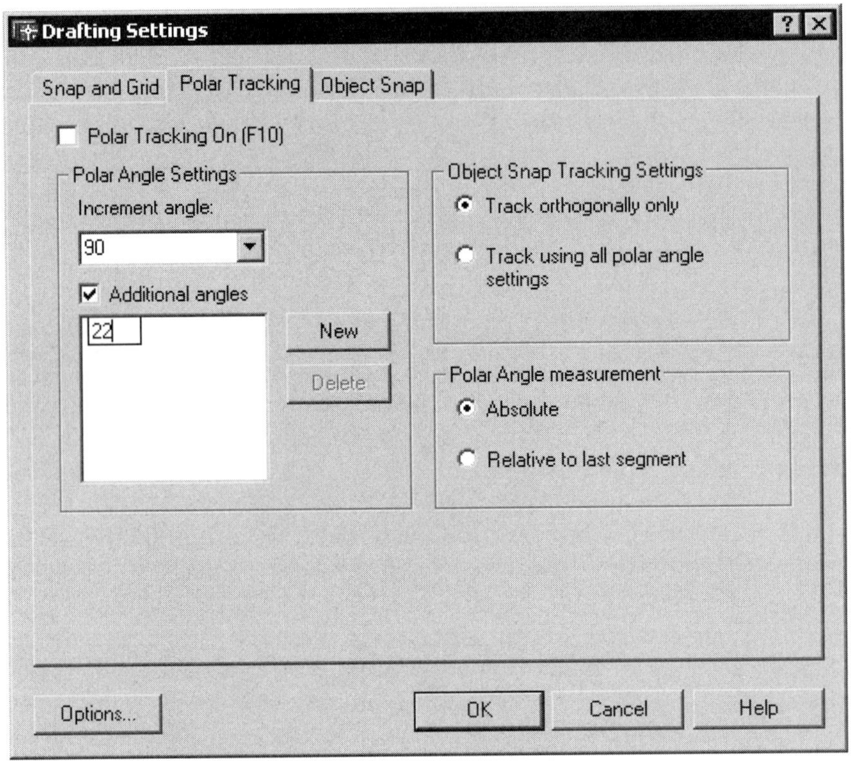

***Figure 4-72 Polar Tracking** tab of the **Drafting Settings** dialog box*

tooltip will display a distance and angle. Selecting the **Additional angles** check box and choosing the **New** button allows you to add additional angle value. The imaginary path will also be displayed at these new angles apart from the increments of the increment angle selected. For example, if the increment angle is set at **15** and you add an additional angle of **22**, the imaginary path will be displayed at 0, 15, 22, 30, 45, and the increments of 15. Polar tracking is on only when the Ortho mode is off.

In the **Drafting Settings** dialog box (**Polar Tracking** tab), you can set the polar tracking to absolute or relative to last segment. If you select the **Absolute** radio button, which is the default, the base angle is taken from 0. If you select the **Relative to last segment** radio button, the base angle for the increments is set to the last segment drawn. You can also use Polar tracking together with Object tracking (Otrack). You can select the **Track using all polar angle settings** radio button in the dialog box.

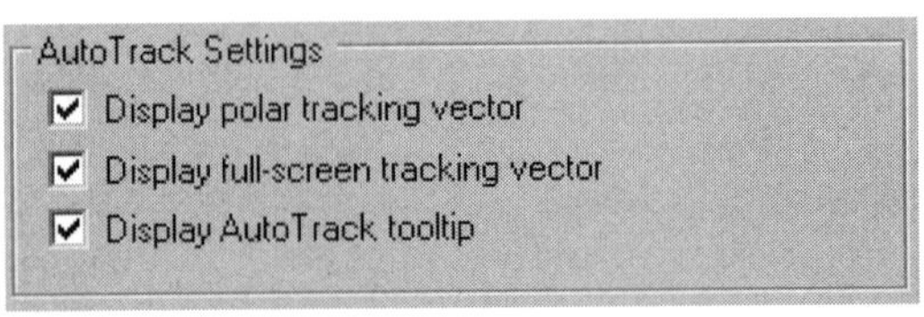

***Figure 4-73 AutoTrack Settings** in the **Options** dialog box (**Drafting** tab)*

AutoTrack Settings

You have different settings while working with autotracking. These can be set in the **Drafting** tab of the **Options** dialog box (Figure 4-73). If you choose the **Options** button in the **Polar**

Tracking tab of the **Drafting Settings** dialog box, the **Options** dialog box with the **Drafting** tab open is displayed. You can use the **Display polar tracking vector** check box to toggle the display of the angle alignment path for Polar tracking. You can also use the **Display full-screen tracking vector** check box to toggle the display of a full-screen construction line for Otrack. You can use the **Display AutoTrack tooltip** check box to toggle the display of tooltips with the paths. You can also use the **TRACKPATH** system variable to set the path display settings.

Tip

*You need the **Options** dialog box quite frequently to change the different drafting settings. When you choose the **Options** button from the **Drafting Settings** dialog box, it directly opens the required tab. After making the changes in the dialog box, choose the **OK** button to get back to the **Drafting Settings** dialog box. The **Options** button is available in all the three tabs of the **Drafting Settings** dialog box.*

FUNCTION AND CONTROL KEYS

You can also use the function and control keys to change the status of coordinate display, Snap, Ortho, Osnap, tablet, screen, isometric planes, running Object Snap, Grid, Polar, and Object tracking. The following is a list of function and control keys.

F1	Help	F7	Grid On/Off (CTRL+G)
F2	Graphics Screen/Text Window	F8	Ortho On/Off (CTRL+L)
F3	Osnap On/Off (CTRL+F)	F9	Snap On/Off (CTRL+B)
F4	Tablet mode On/Off (CTRL+T)	F10	Polar tracking On/Off
F5	Isoplane top/right/left (CTRL+E)	F11	Object Snap tracking
F6	Coordinate display On/Off (CTRL+D)		

Self-Evaluation Test

Answer the following questions and then compare your answers to the correct answers given at the end of this chapter.

1. The layers that are turned off are displayed on the screen but cannot be plotted. (T/F)

2. The drawing in which you are working should be in named plot style mode (*.stb*) to make the plot style available in the **Layer Properties Manager** dialog box. (T/F)

3. The grid pattern appears within the drawing limits, which helps to define the working area. (T/F)

4. If the circle is inserted as a rotated block, the quadrant points are not rotated by the same amount. (T/F)

5. You can also change the plot style mode from the command line by using the ________ system variable.

6. The __________ command enables you to set up an invisible grid that allows the cursor to move in fixed increments from one snap point to another.

7. The __________ snap works along with polar and object tracking only.

8. The __________ Object Snap mode selects projected or visual intersections of two objects in 3D space.

9. The __________ can be used to locate a point with respect to two different points.

10. You can set the priority between the keyboard entry and object snap through the __________ tab of the **Options** dialog box.

Review Questions

Answer the following questions.

1. You cannot enter different values for horizontal and vertical grid spacing. (T/F)

2 You can lock a layer, which will prevent the user from accidentally editing the objects in that layer. (T/F)

3. When a layer is locked, you cannot use the objects in the locked layer for Osnaps. (T/F)

4. The thickness given to the objects in a layer using the Lineweight option is displayed on the screen and is plotted. (T/F)

5. When you select more than one layer using the SHIFT key, which of the following options is not displayed in the **Layer** shortcut menu in the **Layer Properties Manager** dialog box?

   (a) **New Layer** (b) **Select All**
   (c) **Make Current** (d) **Clear All**

6. Which of the following function keys acts as a toggle key for turning the grid on or off?

   (a) F5 (b) F6
   (c) F7 (d) F8

7. Which one of the following object snap mode turns off any running object snap for one point only?

   (a) **NODe** (b) **NONe**
   (c) **From** (d) **NEArest**

8. Which one of the following object snaps cannot be used as running object snap?

(a) **EXTension**	(b) **PARallel**
(c) **From**	(d) **NODe**

9. Which of the following keys can used to cycle through the different running object snaps?

(a) ENTER	(b) SHIFT
(c) CTRL	(d) TAB

10. While working on a drawing, at any point in time you can save all the layers with their present properties' settings under one name and then restore it anytime later using the __________ button in the **Layer Properties Manager** dialog box.

11. You can use the __________ window to locate drawing data with the help of search tools and then use it in your drawing.

12. The grid pattern appears within the drawing __________, which helps to define the working area.

13. The difference between the **Off** option and the **Freeze** option is that the frozen layers are not __________ by the computer while regenerating the drawing.

14. The size of the aperture is measured in __________ , short for picture elements.

15. Using the **Extension** object snap, moving the cursor along the path displays a temporary extension path and the tooltip displays __________ coordinates from the end of the line.

Exercises

Exercise 5 *Mechanical*

Set up layers with the following linetypes and colors. Then make the drawing as shown in Figure 4-74. The distance between the dotted lines is 1.0 unit.

Layer name	Color	Linetype
Object	Red	Continuous
Hidden	Yellow	Hidden
Center	Green	Center

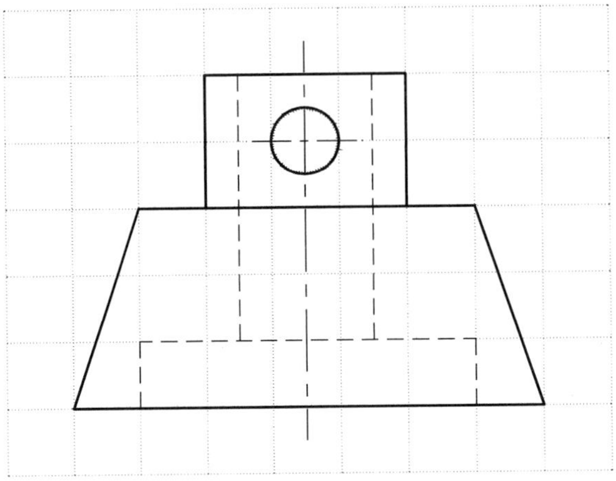

Figure 4-74 Drawing for Exercise 5

Exercise 6 — *Mechanical*

Set up layers, linetypes, and colors as given in Exercise 5. Then make the drawing shown in Figure 4-75. The distance between the dotted lines is 1.0 unit.

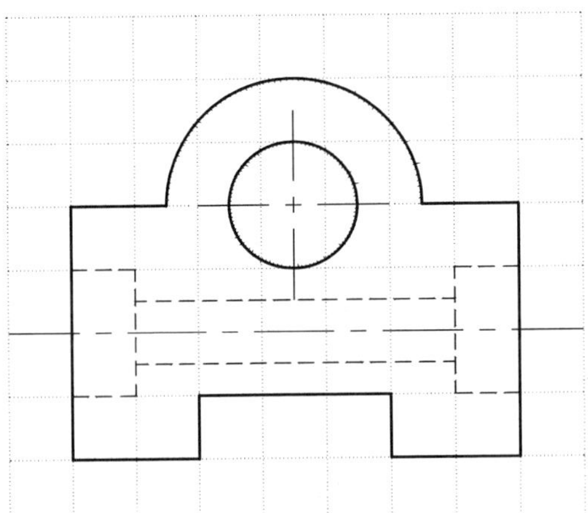

Figure 4-75 Drawing for Exercise 6

Exercise 7 — *Mechanical*

Set up layers, linetypes, and colors and then make the drawing shown in Figure 4-76. The distance between the dotted lines is 1.0 unit.

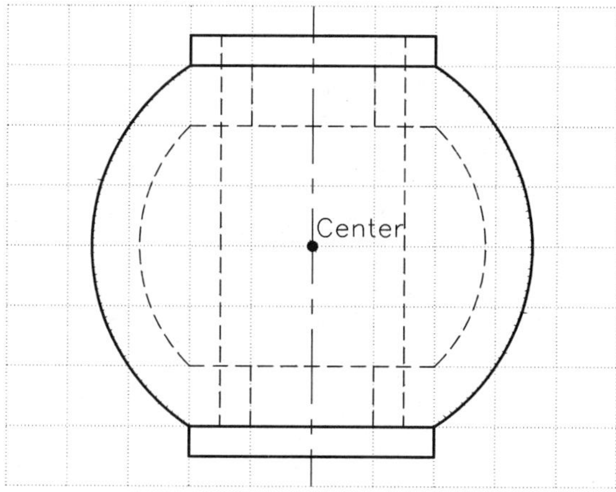

Figure 4-76 Drawing for Exercise 7

Exercise 8

Mechanical

Set up layers, linetypes, and colors and then make the drawing shown in Figure 4-77. Use the object snaps as indicated.

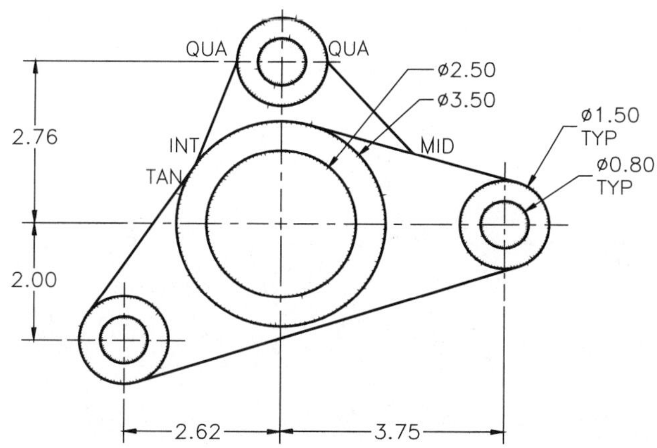

Figure 4-77 Drawing for Exercise 8

Problem Solving Exercise 1

Mechanical

Draw the object shown in Figure 4-78. First draw the lines and then draw the arcs using appropriate ARC command options.

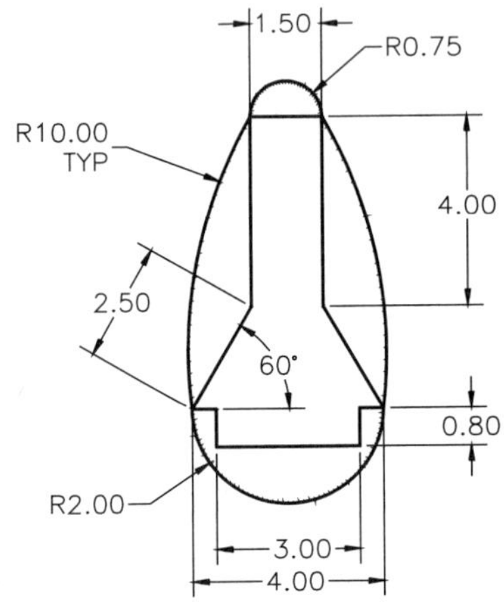

Figure 4-78 Drawing for Problem Solving Exercise 1

Problem Solving Exercise 2 *Mechanical*

Draw the object shown in Figure 4-79. First draw the front view (bottom left) and then the side and top views. Assume the missing dimensions.

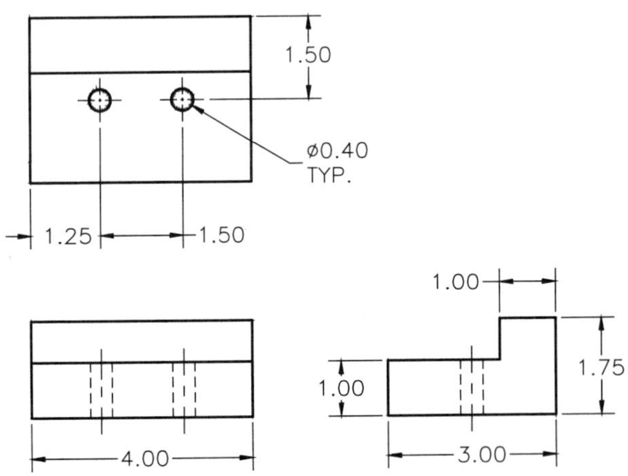

Figure 4-79 Drawing for Problem Solving Exercise 2

Answers to Self-Evaluation Test

1 - F, **2 -** T, **3 -** T, **4 -** F, **5- PSTYLEPOLICY**, **6 -** SNAP, **7 -** Polar , **8 -** Apparent Intersection, **9 -** Temporary Tracking, **10 -** User Preferences

Chapter 5

Editing Sketched Objects-I

Learning Objectives

After completing this chapter you will be able to:

- *Create selection sets using various object selection options.*
- *Move the objects using the **MOVE** command and copy existing objects using the **COPY** command.*
- *Copy objects with base point using the **COPYBASE** command.*
- *Use the **OFFSET** and **BREAK** commands.*
- *Fillet and chamfer objects using the **FILLET** and **CHAMFER** commands.*
- *Cut and extend objects using the **TRIM** and **EXTEND** commands.*
- *Stretch objects using the **STRETCH** command.*
- *Create polar and rectangular arrays using the **ARRAY** command.*
- *Use the **ROTATE** and **MIRROR** commands.*
- *Scale objects using the **SCALE** command.*
- *Lengthen objects such as line, arc, and spline using the **LENGTHEN** command.*
- *Use the **MEASURE** and **DIVIDE** commands.*

CREATING A SELECTION SET

In Chapter 2 (Getting Started with AutoCAD), only two options of the selection set were discussed (Window and Crossing). In this chapter you will learn additional selection set options that can be used to select objects. The following options are explained here.

Last	CPolygon	Add	Undo	Previous
Fence	BOX	SIngle	ALL	Group
AUto	WPolygon	Remove	Multiple	

Note

The default object selection method for most of the commands is pick box to select one entity at a time. If you click in a blank area using a pick box, the window or the crossing option is invoked.

Last

This option is used to select the most recently drawn object that is partially or fully visible in the current display of the screen. This option is the most convenient option if you want to select the most recently drawn object that is visible on the screen. Keep in mind that if the last drawn object is not in the current display, the object in the current display that was drawn last will be selected. Although a selection set is being formed using the **Last** option, only one object is selected. However, you can use the Last option a number of times. You can use the Last selection option with any command that requires selection of objects (e.g., **COPY**, **MOVE**, and **ERASE**). After invoking the particular command, enter LAST or L at the **Select objects** prompt to use this object selection method.

Exercise 1 *General*

Using the **LINE** command, draw Figure 5-1(a). Then use the **ERASE** command with the **Last** option to erase the three most recently drawn lines to obtain Figure 5-1(d).

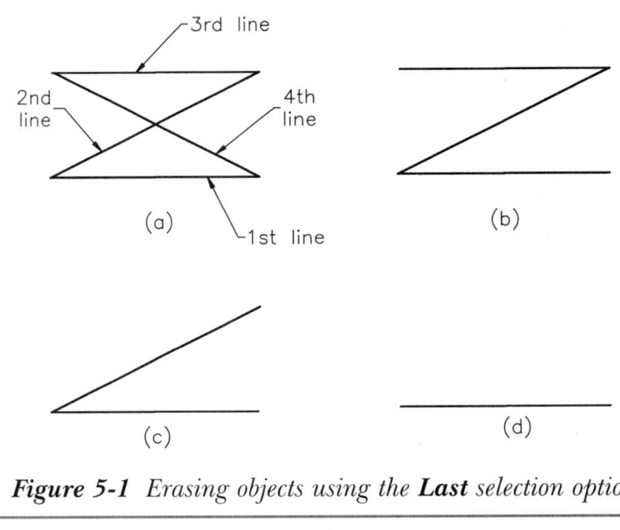

Figure 5-1 *Erasing objects using the* ***Last*** *selection option*

Previous

The **Previous** option automatically selects the objects in the most recently created selection set. To invoke this option you can enter **P** at the **Select objects** prompt. AutoCAD saves the previous selection set and lets you select it again by using this option. In other words, with the help of the **Previous** option you can edit the previous set without reselecting its objects individually. Another advantage of the **Previous** option is that you need not remember the objects if more than one editing operation has to be carried out on the same set of objects. For example, if you want to copy a number of objects and then move them, you can use the **Previous** option to select the same group of objects with the **MOVE** command. The prompt sequence will be as follows.

Command: Choose the **Copy** button.
Select objects: *Select the objects.*
Select objects: Enter
Specify base point or displacement, or [Multiple]: *Specify the base point.*
Specify second point of displacement or <use first point as displacement>: *Specify the point for displacement.*

Command: Choose the **Move** button.
Select objects: **P** Enter
\# found

The **Previous** option does not work with some editing commands, such as **STRETCH**. A previous selection set is cleared by the various deletion operations and the commands associated with them, like **UNDO**. You cannot select the objects in model space and then use the same selection set in paper space, or vice versa. This is because AutoCAD keeps the record of the space (paper space or model space) in which the individual selection set is created.

WPolygon

This option is similar to the **Window** option, except that in this option you can define a window that consists of an irregular polygon. You can specify the selection area by specifying points around the object you want to select (Figure 5-2). Similar to the window method, all the objects to be selected using this method should be completely enclosed within the polygon. The polygon is formed as you specify the points and can take any shape except the one that is self-intersecting. The last segment of the polygon is automatically drawn to close the polygon. The polygon can be created by specifying the coordinates of the points or by specifying the points with the help of a pointing device.

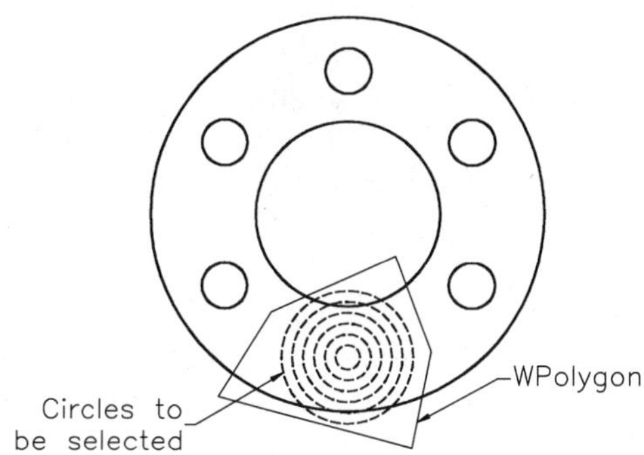

Figure 5-2 *Selecting objects using the **WPolygon** option*

With the **Undo** option, the most recently specified WPolygon point can be undone. To use the **WPolygon** option with object selection commands (like **ERASE**, **MOVE**, **COPY**), first invoke the particular command, and then enter WP at the **Select objects** prompt. The prompt sequence for selecting the objects using the **WPolygon** option is given next.

Command: *Invoke any command such as **Erase, Move, Copy**, and so on.*
Select objects: **WP** Enter
First polygon point: *Specify the first point.*
Specify endpoint of line or [Undo]: *Specify the second point.*
Specify endpoint of line or [Undo]: *Specify the third point.*
Specify endpoint of line or [Undo]: *Specify the fourth point.*
Specify endpoint of line or [Undo]: *Press ENTER after specifying the last point of polygon.*

Exercise 2 *General*

Draw a number of objects on the screen, and then erase some of them using the **WPolygon** option to select the objects you want to erase.

CPolygon

This method of selection is similar to the WPolygon method except that just like a crossing, a **CPolygon** also selects those objects that are not completely enclosed within the polygon, but are touching the polygon boundaries. In other words, the object that is lying partially inside the polygon or is even touching it is also selected in addition to those objects that are completely enclosed within the polygon (Figure 5-3). CPolygon is formed as you specify the points. The points can be specified at the Command line or by specifying points with the pointing device. Just as in the **WPolygon** option, the crossing polygon can take any shape except the one in which it intersects itself. Also, the last segment of the polygon is drawn automatically, and so the CPolygon is closed at all times. The prompt sequence for the **CPolygon** option is given next.

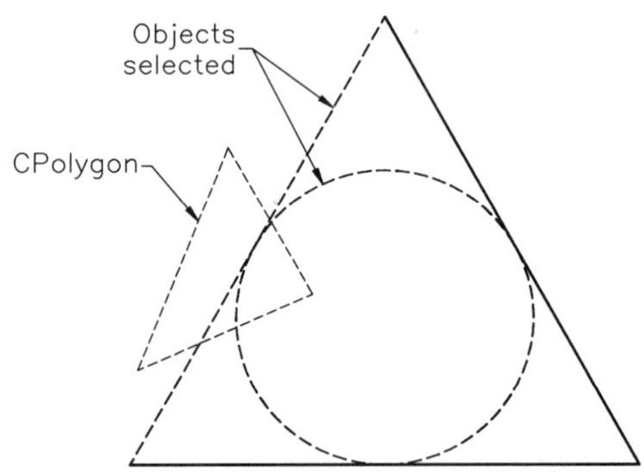

***Figure 5-3** Selecting objects using the **CPolygon** option*

Select objects: **CP** Enter
First polygon point: *Specify the first point.*
Specify endpoint of line or [Undo]: *Specify the second point.*
Specify endpoint of line or [Undo]: *Specify the third point.*
Specify endpoint of line or [Undo]: *Press ENTER after specifying the last point of the polygon.*

Remove

The **Remove** option is used to remove the objects from the selection set (but not from the drawing). After selecting many objects from a drawing by any selection method, there may be a need for removing some of the objects from the selection set. The following prompt sequence displays the use of the **Remove** option for removing the objects from the selection set shown in Figure 5-4.

Command: Choose the **Erase** button.
Select objects: *Select objects (a), (b), and (c).*
Select objects: **R**
Remove objects: *Select object (a).*
1 found, 1 removed, 2 total.
Select objects: Enter

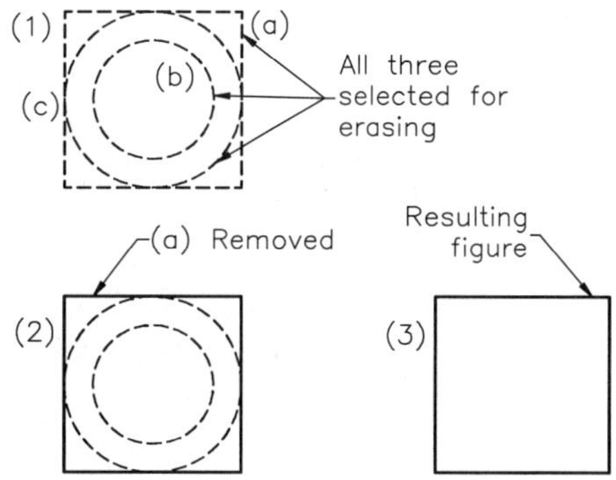

Figure 5-4 *Using the* ***Remove*** *option*

Tip

The objects can also be removed from the selection set using the SHIFT key. For example, pressing the SHIFT key and selecting object (a) with the pointing device will remove it from the selection set. AutoCAD will display the following message; ***1 found, 1 removed, 2 total.***

Add

You can use the **Add** option to add objects to the selection set. When you begin creating a selection set, you are in **Add** mode. After you create a selection set by any selection method, you can add more objects by simply selecting them with the pointing device, when system variable **PICKADD** is set to 1 (default). When the system variable **PICKADD** is set to 0, to add objects to the selection set you will have to press the SHIFT key and then select the objects.

ALL

The **ALL** selection option is used to select all the objects in the current working environment of current drawing. Note that the objects in the "OFF" layers are also selected with the **ALL** option. However, the objects that are in the frozen layers are not selecting using this method. You can use this selection option with any command that requires object selection. After invoking the command, the **ALL** option can be entered at the **Select objects** prompt. Once you enter this option, all the objects drawn on the screen will be highlighted (dashed). For example, if there are four objects on the screen and you want to erase all of them, the prompt sequence is given next.

Command: Choose the **Erase** button.
Select objects: **ALL** Enter
4 found
Select objects: Enter

You can use this option in combination with other selection options. For example, consider there are five objects on the drawing screen and you want to erase three of them. After invoking the **ERASE** command, enter **ALL** at the **Select objects** prompt. Then press the

Chapter 5

SHIFT key and select the two objects you want to remove from the selection set; the remaining three objects are erased.

Fence

In the **Fence** option, a selection set is created by drawing an open polyline fence through the objects to be selected. Any object touched by the fence polyline is selected (Figure 5-5). The selection fence can be created by entering the coordinates at the Command line or by specifying the points with the pointing device. With this option, more flexibility for selection is provided because the fence can intersect itself. The Undo option can be used to undo the most recently selected fence point. Like the other selection options, this option is also used with the commands that need object selection. The prompt sequence for using the **Fence** option is given next.

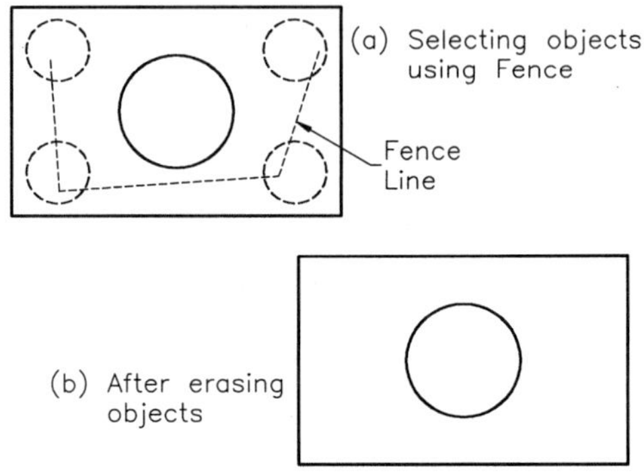

***Figure 5-5** Erasing objects using the **Fence** option*

Select objects: **F** [Enter]
First fence point: *Specify the first point.*
Specify endpoint of line or [Undo]: *Specify the second point.*
Specify endpoint of line or [Undo]: *Specify the third point.*
Specify endpoint of line or [Undo]: *Specify the fourth point.*
Specify endpoint of line or [Undo]: [Enter]

Group

The **Group** option enables you to select a group of objects by their group name. You can create a group and assign a name to it with the help of the **GROUP** command (see Chapter 18). Once a group has been created, you can select the group using the **Group** option for editing purposes. This makes the object selection process easier and faster, since a set of objects is selected by entering just the group name. The prompt sequence is given next.

Command: **MOVE** [Enter]
Select objects: **G** [Enter]
Enter group name: *Enter the name of the predefined group you want to select.*
4 found
Select objects: [Enter]

Exercise 3 *General*

Draw six circles and select all of them to erase using the **ERASE** command with the **ALL** option. Now change the selection set contents by removing alternate circles from the selection set by using the fence so that alternate circles are erased.

BOX

When system variable **PICKAUTO** is set to 1 (default), the **BOX** selection option is used to select objects inside a rectangle. After you enter BOX at the **Select objects** prompt, you are required to specify the two corners of a rectangle at the **Specify first corner** and the **Specify opposite corner** prompts. If you define the Box from right to left, it is equivalent to the Crossing selection option. Hence, it also selects those objects that are touching the rectangle boundaries in addition to those that are completely enclosed within the rectangle. If you define the Box from left to right, this option is equivalent to the Window option and selects only those objects that are completely enclosed within the rectangle. The prompt sequence is given next.

Select objects: **BOX** Enter
Specify first corner: *Specify a point.*
Specify opposite corner: *Specify opposite corner point of the box.*

AUto

The **AUto** option is used to establish automatic selection. You can select a single object, by selecting that object, as well as select a number of objects, by creating a window or a crossing. If you select a single object, it is selected; if you specify a point in the blank area you are automatically in the BOX selection option and the point you have specified becomes the first corner of the box. Auto and Add are default selections.

Multiple

When you enter **M** (Multiplc) at the **Select objects** prompt, you can select multiple objects at a single **Select objects** prompt without the objects being highlighted. Once you give a null response to the **Select objects** prompt, all the selected objects are highlighted together.

Undo

This option removes the most recently selected object from the selection set.

SIngle

When you enter SI (SIngle) at the **Select objects** prompt, the selection takes place in the SIngle selection mode. Once you select an object or a number of objects using a Window or Crossing option, the **Select objects** prompt is not repeated. AutoCAD proceeds with the command for which the selection is made.

Command: *Choose the* ***Erase*** *button.*
Select objects: **SI** Enter
Select objects: *Select the object for erasing. The selected object is erased.*

You can also create a selection set using the **SELECT** command. The prompt sequence is given next.

Command: **SELECT** Enter
Select objects: *Use any selection method.*

EDITING THE SKETCHES

To use AutoCAD effectively, you need to know the editing commands and how to use them. In this section you will learn about the editing commands. These commands can be invoked from the toolbar, menu, or can be entered at the Command prompt. Some of the editing commands such as **ERASE** and **OOPS** have been discussed in Chapter 2 (Getting Started with AutoCAD). The rest of the editing commands will be discussed in this section.

MOVING THE SKETCHED OBJECTS

Toolbar:	Modify > Move
Menu:	Modify > Move
Command:	MOVE

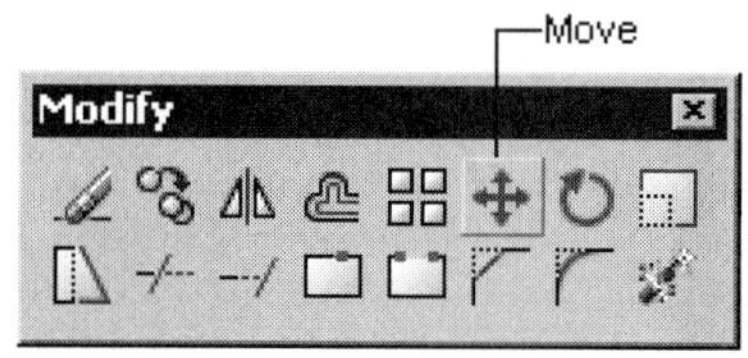

Figure 5-6 *Invoking the* ***MOVE*** *command from the* ***Modify*** *toolbar*

Sometimes, the objects are not located at the position where they actually should be. In these situations you can use the **MOVE** command (Figure 5-6). This command allows you to move one or more objects from their current location to a new location specified by you. This change in the location of the objects does not change their size or orientation. When you invoke this command, you will be prompted to select the objects to be moved. You can use any of the object selection techniques for selecting one or more objects. Once you have selected the objects, you will be prompted to specify the base point. This base point is the reference point with which the object will be picked and moved. It is advisable to select the base point on the object selected to move. Next, you will be prompted to specify the second point of displacement. This is the new location point where you want to move the object. When you specify this point, the selected objects will be moved to this point. Figure 5-7 shows an object moved using the **MOVE** command. The prompt sequence that will be followed when you choose the **Move** button from the **Modify** toolbar is given next.

Select objects: *Select the objects to be moved.*
Select objects: Enter
Specify base point or displacement: *Specify the base point for moving the selected object(s).*
Specify second point of displacement or <use first point as displacement>: *Specify second point or press ENTER to use the first point.*

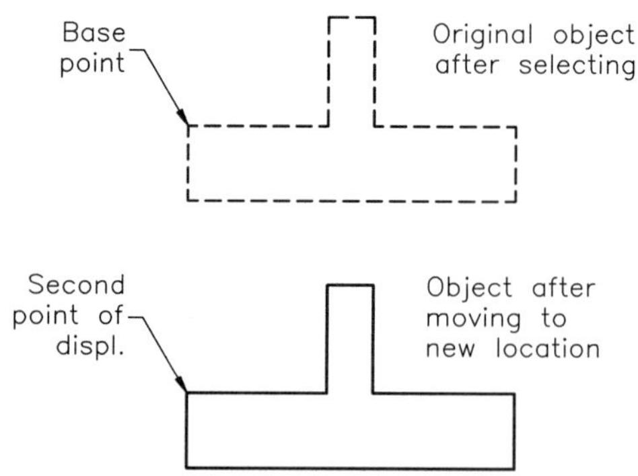

Figure 5-7 *Moving the objects to a new location*

If you press ENTER at the **Specify second point of displacement or <use first point as displacement>** prompt, AutoCAD interprets the first point as the relative value of the displacement in the *X* axis and *Y* axis directions. This value will be added in the *X* and *Y* axis coordinates and the object will be automatically moved to the resultant location. For example, draw a circle with its center at (3,3) and then select the center point of the circle as the base point. Now, at the **Specify second point of**

displacement or <use first point as displacement> prompt, press ENTER. You will notice that the circle is moved such that its center is now placed at 6,6. This is because 3 units (initial coordinates) are added along both the X and Y directions.

COPYING THE SKETCHED OBJECTS

Toolbar:	Modify > Copy Object
Menu:	Modify > Copy
Command:	COPY

The **COPY** command is used to copy an existing object. This command is similar to the **MOVE** command in the sense that it makes copies of the selected objects and places them at specified locations, but the originals are left intact. Also, in this command you need to select the objects and then specify the base point. Then you are required to specify the second point, that is, where you want the copied object to be placed. Figure 5-8 shows the objects copied using this command. The prompt sequence that will be followed when you choose the **Copy Object** button from the **Modify** toolbar is given next.

Select objects: *Select the objects to copy.*
Select objects: Enter
Specify base point or displacement, or [Multiple]: *Specify the base point.*
Specify second point of displacement, or <use first point as displacement>: *Specify a new position on the screen using the pointing device or entering coordinates.*

Creating Multiple Copies

The **Multiple** option of the **COPY** command is used to make multiple copies of the same object (Figure 5-9). To use this option, select the **Multiple** option or enter **M** at the **Specify base point or displacement, or [Multiple]** prompt. Next, you are prompted to enter the base point and then the second point. When you select a point, a copy is placed at this point, and the prompt is automatically repeated until you press ENTER to terminate the **COPY** command. The prompt sequence for the **COPY** command with the **Multiple** option is given next.

Specify base point or displacement, or [Multiple]: **M** Enter
Specify base point: *Specify the base point.*
Specify second point of displacement or <use first point as displacement>: *Specify a point for placement.*
Specify second point of displacement or <use first point as displacement>: *Specify another point for placement.*
Specify second point of displacement or <use first point as displacement>: *Specify another point for placement.*
Specify second point of displacement or <use first point as displacement>: Enter

COPYING THE OBJECT USING THE BASE POINT

Menu:	Edit > Copy with Basepoint
Command:	COPYBASE

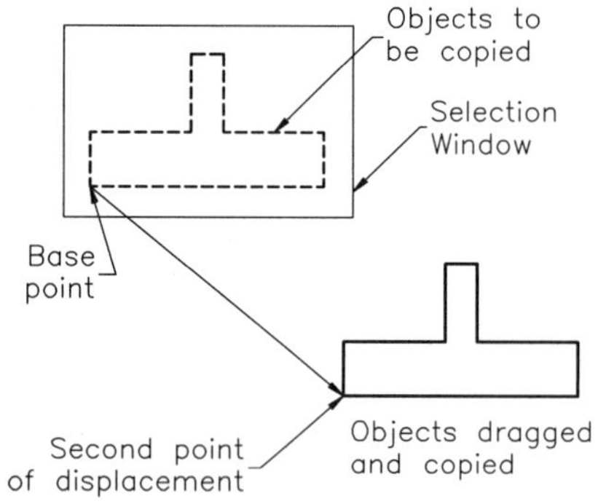

***Figure 5-8** Using the **COPY** command*

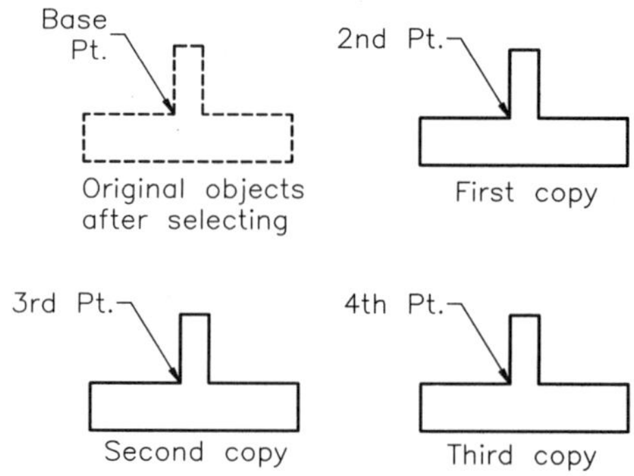

***Figure 5-9** Making multiple copies*

The command **COPYBASE** is used to specify the base point of objects to be copied. This command can also be invoked from the shortcut menu that is displayed upon selecting the object and right-clicking. Choose **Copy with Base Point** to invoke this command, Figure 5-10. This command can be very useful while pasting an object very precisely in the same diagram or into another diagram. When you invoke this command from the shortcut menu, you will be prompted to specify the base point for copying the objects. Unlike the **COPY** command, where the objects are dragged and placed at the desired location, this command copies the selected object on to the Clipboard. Then from the Clipboard, the objects will be placed at the specified location. The Clipboard is defined as a medium for storing the data while transferring the data from one place to the other. Note that the contents of the Clipboard are not visible. The objects copied using this command can be copied from one drawing file to the other or from one working environment to the other. After copying objects with the **COPYBASE** command, you can paste the objects at the desired place with greater accuracy.

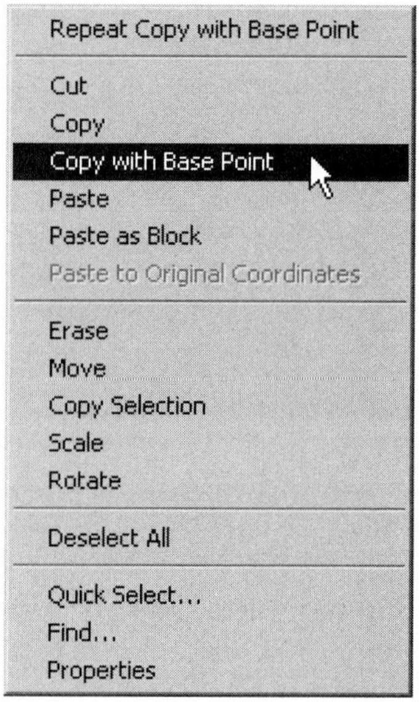

***Figure 5-10** Invoking the **COPYBASE** command from the shortcut menu*

Note

You will learn more about the working environments and the Clipboard in the following chapters.

PASTING THE CONTENTS FROM THE CLIPBOARD

Menu:	Edit > Paste as Block
Command:	PASTEBLOCK

The **PASTEBLOCK** command is used to paste the contents of the Clipboard into a new drawing or in the same drawing at a new location. You can also invoke the **PASTEBLOCK** command from the shortcut menu by right-clicking in the drawing area and choosing **Paste as Block**.

PASTING CONTENTS USING THE ORIGINAL COORDINATES

Menu:	Edit > Paste to Original Coordinates
Command:	PASTEORIG

The **PASTEORIG** command is used to paste the contents of the Clipboard into a new drawing using the coordinates from the original drawing. You can also invoke the **PASTEORIG** command from the shortcut menu by right-clicking in the drawing area and choosing **Paste to Original Coordinates**. The **PASTEORIG** command is available only when the Clipboard contains AutoCAD data from a drawing other than the current drawing.

Exercise 4 — *Mechanical*

In this exercise you will draw the object shown in Figure 5-11. Use the **Copy** command for creating the drawing.

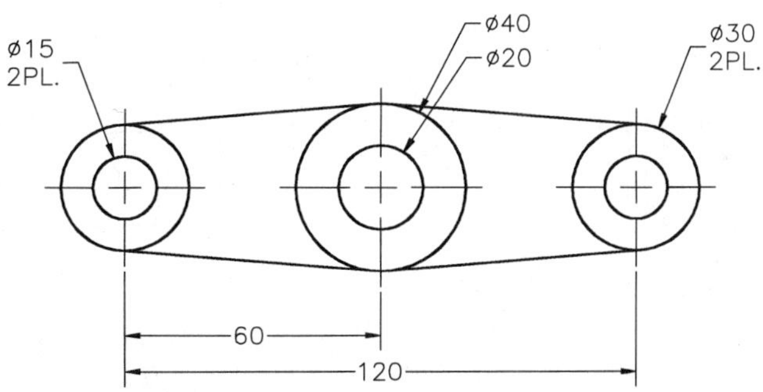

Figure 5-11 *Drawing for Exercise 4*

OFFSETTING THE SKETCHED OBJECTS

Toolbar:	Modify > Offset
Menu:	Modify > Offset
Command:	OFFSET

If you want to draw parallel lines, polylines, concentric circles, arcs, curves, and so on, you can use the **OFFSET** command (Figure 5-12). This command creates another object that is similar to the selected one. Remember that you are allowed to select only one entity at a time to offset. When offsetting an object you can specify the offset distance and the side to offset, or you can specify a distance through which you want to offset the selected object. Depending on the side to offset, you can create smaller or larger circles, ellipses, and arcs. If the offset side is toward the inner side of the perimeter, the arc, ellipse, or circle will be smaller than the original. The prompt sequence that will follow when you choose the **Offset** button from the **Modify** menu is given next.

Specify offset distance or [Through] <current>: *Specify the offset distance or press ENTER to use the **Through** option.*
Select object to offset or <exit>: *Select the object to offset.*

The offset distance can be specified by entering a value or by specifying two points with the pointing device. AutoCAD will measure the distance between these two points and use it as the offset distance. The through option is generally used for creating orthographic views. In this case, you do not have to specify a distance; you simply specify an offset point, see Figure 5-13.

The offset distance is stored in the **OFFSETDIST** system variable. A negative value indicates OFFSET is set to the Through option. You can offset lines, arcs, 2D polylines, xlines, circles, ellipses, elliptical arcs, rays, and planar splines. If you try to offset objects other than these, the message **Cannot offset that object** is displayed.

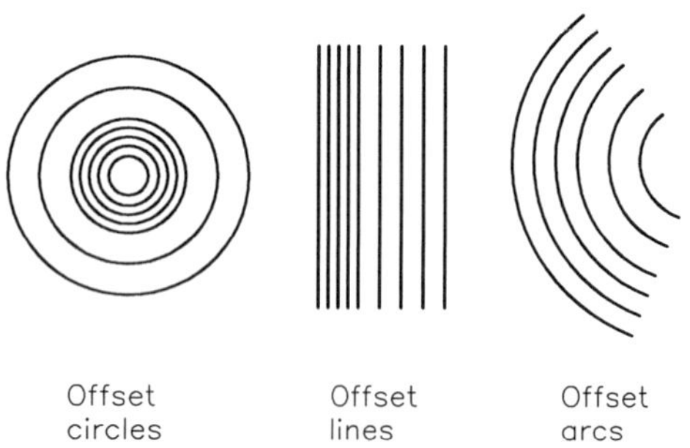

***Figure 5-12** Using the **OFFSET** command multiple times to create multiple offset entities*

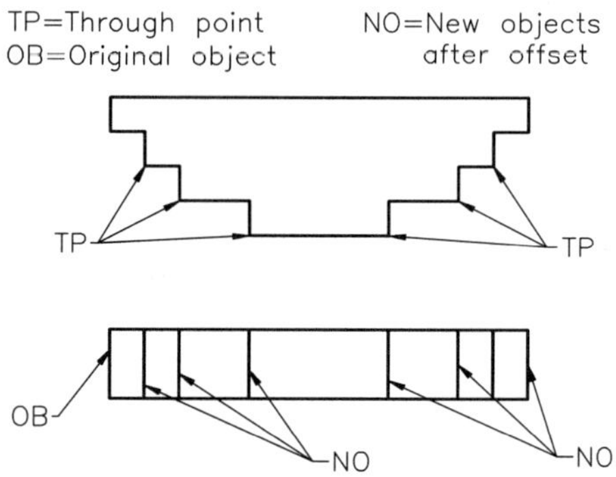

***Figure 5-13** Using the **Through** option*

Exercise 5 *General*

Use the **OFFSET** edit command to draw Figures 5-14 and 5-15.

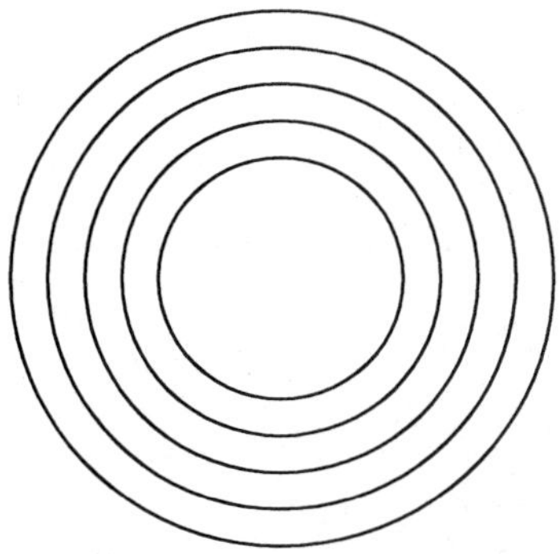

Figure 5-14 Drawing for Exercise 5

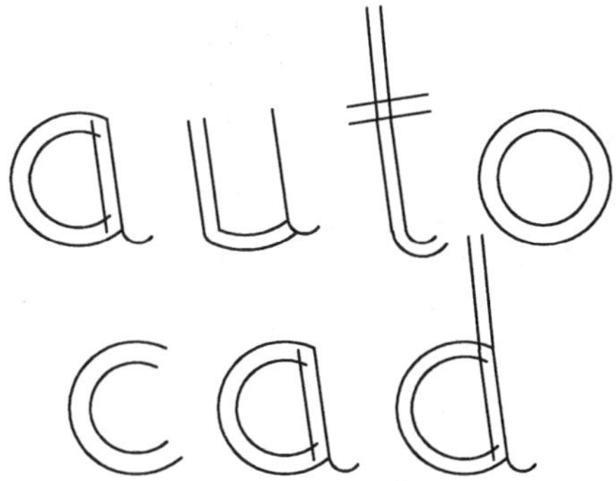

Figure 5-15 Drawing for Exercise 5

ROTATING THE SKETCHED OBJECTS

Toolbar:	Modify > Rotate
Menu:	Modify > Rotate
Command:	ROTATE

While creating designs there are many occasions when you have to rotate an object or a group of objects. You can accomplish this by using the **ROTATE** command. When you invoke this command, AutoCAD will prompt you to select the objects and the base point about which the selected objects will be rotated. You should be careful when selecting the base point as it is easy to get confused if the base point is not located on a known object. After you specify the base point you are required to enter a rotation angle. By default, positive angles produce a counterclockwise rotation and negative angles produce a clockwise rotation (Figure 5-16). The **ROTATE** command can also be invoked from the shortcut menu by selecting the object and clicking the right button in the drawing area and choosing **Rotate**. The prompt sequence that will follow when you choose the **Rotate** button is given next.

Current positive angle in UCS: ANGDIR=*current* ANGBASE=*current*
Select objects: *Select the objects for rotation.*
Select objects: 
Specify base point: *Specify a base point about which the selected objects will be rotated.*
Specify rotation angle or [Reference]: *Enter a positive or negative rotation angle, or specify a point.*

If you need to rotate objects with respect to a known angle, you can do this in two different ways using the **Reference** option. The first way is to specify the known angle as the reference angle, followed by the proposed angle to which the objects will be rotated (Figure 5-17). Here the object is first rotated clockwise from the *X* axis, through the reference angle. Then the object is rotated through the new angle from this reference position in a counterclockwise direction. The prompt sequence is given next.

Current positive angle in UCS: ANGDIR=*current* ANGBASE=*current*
Select objects: *Select the objects for rotation.*

Select objects: Enter
Specify base point: *Specify the base point.*
Specify rotation angle or [Reference]: **R** Enter
Specify the reference angle <0>: *Enter reference angle.*
Specify the new angle: *Enter new angle.*

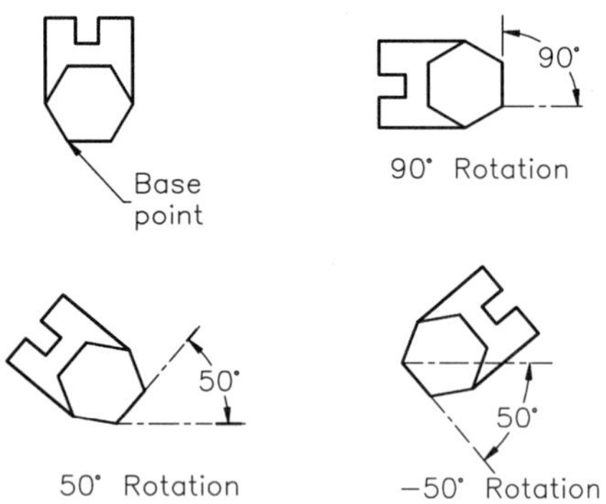

Figure 5-16 *Rotation of objects with different rotation angles*

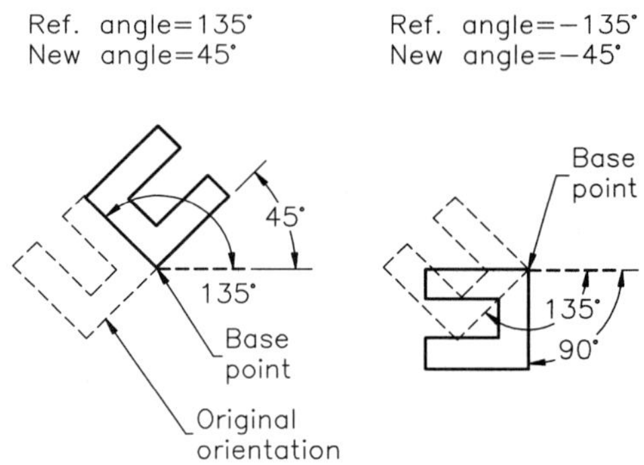

Figure 5-17 *Rotation using the **Reference** Angle option*

The other method is used when the reference angle and the new angle are not known. In this case you can use the edges of the original object and the reference object to specify the original object and reference angle, respectively. Figure 5-18 shows a model created at an unknown angle and a line also at an unknown angle. In this case this line will be used as a reference object for rotating the object. In such cases remember that the base point should be taken on the reference object. This is because you cannot define two points for specifying the new angle. You have to directly enter the angle value or specify only one point. Therefore, the base point will be taken as the first point and the second point can be defined for the new angle. Figure 5-19 shows the model after rotating with reference to the line such that the line and the model are inclined at similar angles. The prompt sequence to rotate the model in Figure 5-18 is given next.

Current positive angle in UCS: ANGDIR=*current* ANGBASE=*current*
Select objects: *Select the object for rotation.*
Select objects: Enter
Specify base point: *Specify the base point as the lower endpoint of the line. See Figure 5-18.*
Specify rotation angle or [Reference]: **R** Enter
Specify the reference angle <0>: *Specify the first point on the edge of the model. See Figure 5-18.*
Specify second point: *Specify the second point on the same edge of the model. See Figure 5-18.*
Specify new angle: *Select the other endpoint of the reference line. See Figure 5-18.*

SCALING THE SKETCHED OBJECTS

Toolbar:	Modify > Scale
Menu:	Modify > Scale
Command:	SCALE

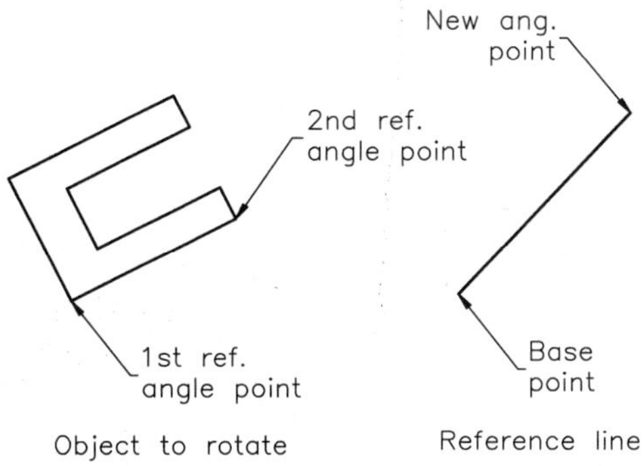

Figure 5-18 *Rotating the objects using a reference line*

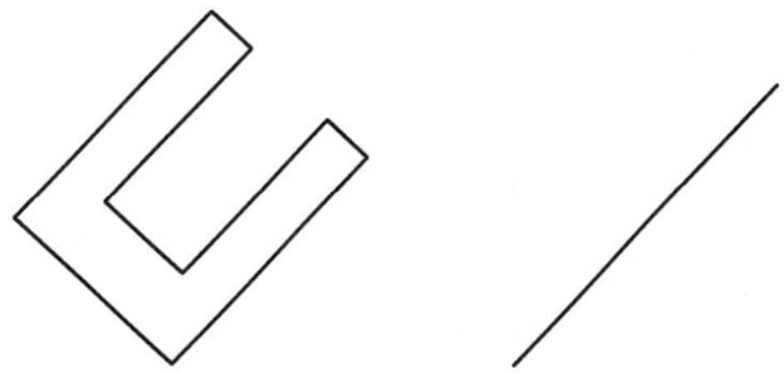

Figure 5-19 *Model after rotating with reference to the line*

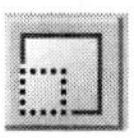

Many times you will need to change the size of objects in a drawing. You can do this with the **SCALE** command. This command dynamically enlarges or shrinks the selected object about a base point, keeping the aspect ratio of the object constant. This means that the size of the object will be increased or reduced equally in the X, Y, and Z directions. The dynamic scaling property allows you to view the object as it is being scaled. This can be viewed by moving the pointing device in the drawing area after selecting the base point. Application of the identical scale factor to the X, Y, and Z dimensions ensures that the shape of the objects being scaled do not change. This is a useful and timesaving editing command because instead of redrawing objects to the required size, you can scale the objects with a single **SCALE** command. Another advantage of this command is that if you have already put the dimensions on the drawing, they will also change accordingly. You can also invoke the **SCALE** command from the shortcut menu by right-clicking in the drawing area and choosing **Scale**. The prompt sequence that will follow when you choose the **Scale** button is given next.

Select objects: *Select objects to be scaled.*
Select objects: Enter
Specify base point: *Specify the base point, preferably a known point.*
Specify scale factor or [Reference]: *Specify the scale factor.*

The base point will not be moved from its position and the selected object(s) will be scaled around the base point as shown in Figures 5-20 and 5-21. To reduce the size of an object, the scale factor should be less than 1 and to increase the size of an object, the scale factor should be greater than 1. You can enter a scale factor or select two points to specify a distance as a factor. When you select two points to specify a distance as a factor, the first point should be on the referenced object.

Sometimes it is time-consuming to calculate the relative scale factor. In such cases you can scale the object by specifying a desired size in relation to the existing size (a known dimension). In other words, you can use a **reference length**. This can be done by entering **R** at the **Specify scale factor or [Reference]** prompt. Then, you either specify two points to specify

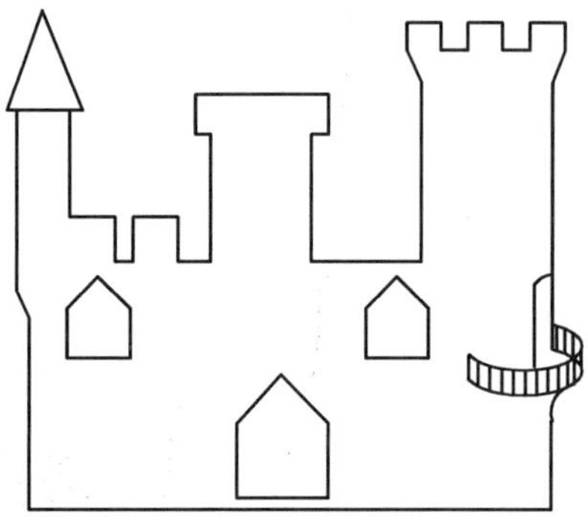

Figure 5-20 *Original object before scaling*

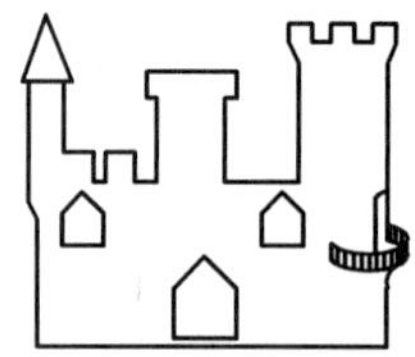

Figure 5-21 *After scaling to 0.5 of actual size*

the length or enter a length. At the next prompt, enter the length relative to the reference length. For example, if a line is **2.5** units long and you want the length of the line to be **1.00** units, instead of calculating the relative scale factor, you can use the **Reference** option. The prompt sequence for using the **Reference** option is given next.

Select objects: *Select the object to scale.*
Select objects: Enter
Specify base point: *Specify the base point.*
Specify scale factor or [Reference]: **R** Enter
Specify reference length <1>: *Specify the reference length.*
Specify new length: *Specify the new length.*

Similar to the **ROTATE** command, here also you can scale one object using the reference of another object. Again, the base point has to be taken on the reference object as you can define only one point for the new length.

Tip

*If you have not used the required drawing units for a drawing, you can use the **Reference** option of the **SCALE** command to correct the error. Select the entire drawing with the help of the All selection option. Specify the **Reference option**, and then select the endpoints of the object whose desired length you know. Specify the new length, and all objects in thé drawing will be rescaled automatically to the desired size.*

FILLETING THE SKETCHES

Toolbar:	Modify > Fillet
Menu:	Modify > Fillet
Command:	FILLET

The edges in the design are generally filleted to reduce the area of stress concentration. The **FILLET** command helps you form round corners between any two entities by allowing you to define two entities that form a sharp vertex. The result is that a smooth round arc is created that connects the two objects. A fillet can also be created between two intersecting parallel lines as well as nonintersecting and nonparallel lines, arcs, polylines, xlines, rays, splines, circles, and true ellipses. The fillet arc created will be tangent to both the selected entities. The default fillet radius is 0.0000. As a result, when you invoke this command, you first need to secify the radius value. The prompt sequence that will follow when you choose the **Fillet** button is given next.

Current Settings: Mode= TRIM, Radius= 0.0000
Select first object or [Polyline/Radius/Trim/mUltiple]:

Creating Fillets Using the Radius Option

The fillet you create depends on the radius distance you specify. The default radius is 0.0000. You can enter a distance or two points. The new radius you enter becomes the default radius and remains in effect until changed. The prompt sequence is given next.

Select first object or [Polyline/Radius/Trim/mUltiple]: **R** Enter
Specify fillet radius <current>: *Enter a fillet radius or press ENTER to accept the current value.*

Note

*The **FILLETRAD** system variable controls and stores the current fillet radius. The default value of this variable is 0.0000.*

You can also fillet 3D objects in different planes. This is discussed in later chapters.

Tip

A fillet with a zero radius creates sharp corners and is used to clean up lines at corners if they overlap or have a gap.

Creating Fillets Using The Select First Object Option

This is the default method to fillet two objects. As the name implies, it prompts for the first object required for filleting. The prompt sequence to use this option is given next.

Current Settings: Mode= TRIM, Radius= modified value
Select first object or [Polyline/Radius/Trim/mUltiple]: *Specify first object.*
Select second object: *Select second object.*

The **FILLET** command can also be used to cap the ends of two parallel lines (Figure 5-22). The cap is a semicircle whose radius is equal to half the distance between the two parallel lines. The cap distance is calculated automatically when you select the two parallel lines for filleting. You can select lines with the Window, Crossing, or Last option, but to avoid unexpected results, select the objects by picking objects individually. Also, selection by picking objects is necessary in the case of arcs and circles that have the possibility of more than one fillet. They are filleted closest to the select points (Figure 5-23).

Creating Fillets Using The Trim Option

When you create a fillet, an arc is created and the selected objects are either trimmed or

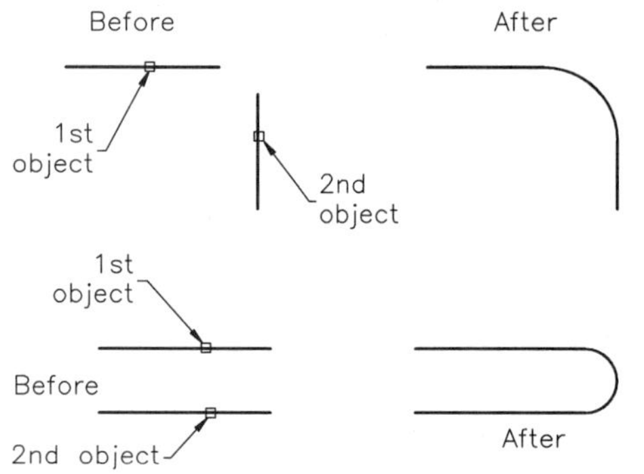

***Figure 5-22** Filleting the parallel and nonparallel lines*

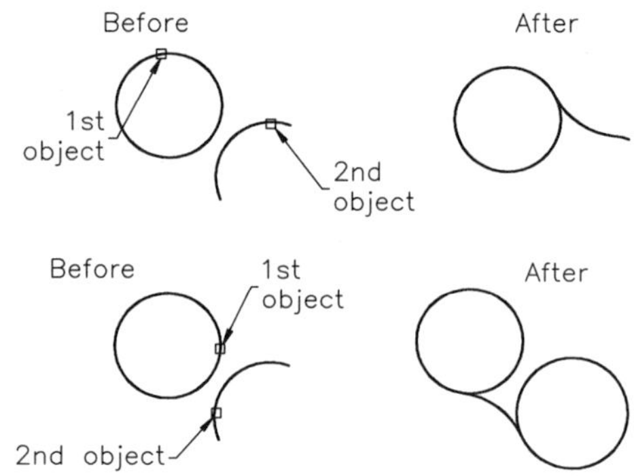

***Figure 5-23** Using the **FILLET** command on circles and arcs*

extended at the fillet endpoint. This is because the **Trim** mode is set to **Trim**. If it is set to **No Trim** they are left intact. Figure 5-24 shows a model filleted with the **Trim** mode set to **Trim** and to **No Trim**. The prompt sequence is given next.

> Select first object or [Polyline/Radius/Trim/mUltiple]: **T** Enter
> Specify Trim mode option [Trim/No trim] <current>: *Enter* **T** *to trim the edges,* **N** *to leave them intact.*

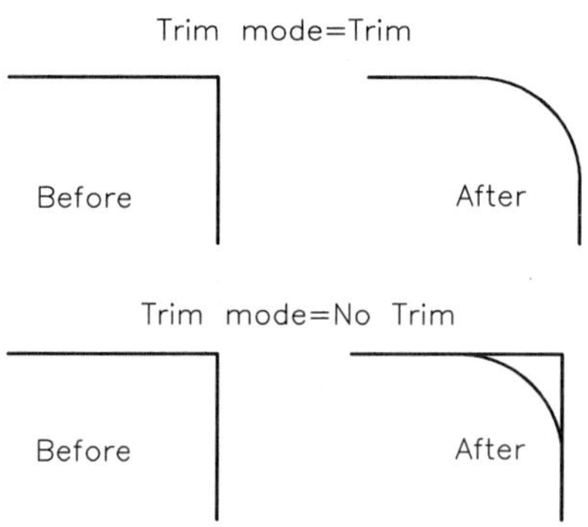

Figure 5-24** Filleting the lines with the **Trim** mode set to **Trim** and **No Trim

Creating Fillets Using The Polyline Option

Using the **Polyline** option you can fillet a number of entities that comprise a single polyline (Figure 5-25). The polylines can be created using the **POLYLINE** or the **RECTANG** command. If the object is created using the **POLYLINE** command, it must be closed using the **Close** option. Otherwise the last corner of the polyline will not be filleted. When you select this option, AutoCAD prompts you to select a polyline. All the vertices of the polyline are filleted when you select the polyline. The fillet radius for all the vertices will be the same. If the selected polyline is not closed, then the last corner is not filleted. The prompt sequence for using this option is given next.

> Current Settings: Mode= *current*, Radius= *current*
> Select first object or [Polyline/Radius/Trim/mUltiple]: **P** Enter
> Select 2D polyline: *Select the polyline.*

Creating Fillets Using The Multiple Option*

When you invoke the **Fillet** command, by default the fillet is created between a single set of entities only. But, with the help of the **Multiple** option you can add fillets to more than one

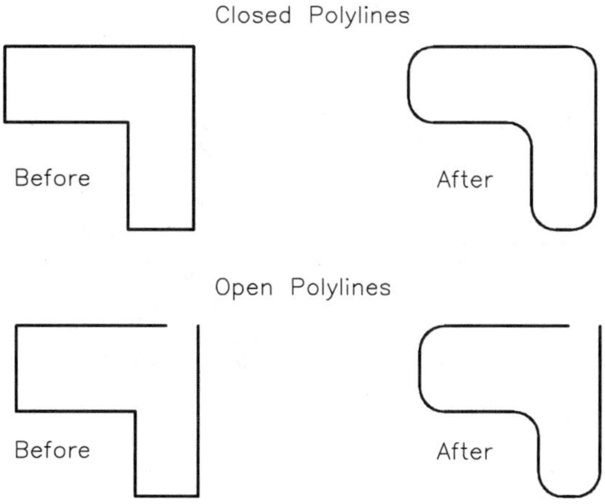

Figure 5-25 *Filleting closed and open polylines*

set of entities. When you select this option, AutoCAD prompts you to select the first object and then the second object. A fillet will be created between the two entities. Next, you are again prompted to select the first object and then the second object. This prompt continues until you press ENTER to terminate the **Fillet** command. The prompt sequence to use this option is given next.

Current Settings: Mode= TRIM, Radius= current
Select first object or [Polyline/Radius/Trim/mUltiple]: **U** Enter
Select first object or [Polyline/Radius/Trim/mUltiple]: *Specify first object of one set.*
Select second object: *Select second object of the same set.*
Select first object or [Polyline/Radius/Trim/mUltiple]: *Specify first object of the other set.*
Select second object: *Select second object of the same set.*
Select first object or [Polyline/Radius/Trim/mUltiple]: *Specify first object of the other set or press* Enter

Filleting Objects with a Different UCS

The fillet command will also fillet objects that are not in the current UCS plane. To create a fillet for these objects, AutoCAD will automatically change the UCS transparently so that it can generate a fillet between the selected objects.

Note

Filleting the objects in different planes is discussed in detail in Chapter 23 (Getting Started with 3D).

Setting the TRIMMODE System Variable

The **TRIMMODE** system variable is a variable that eliminates any size restriction on the **FILLET** command. By setting **TRIMMODE** to **0**, you can create a fillet of any size without actually cutting the existing geometry. Also, there is no restriction on the fillet radius. This means that the fillet radius can be larger than one or both objects that are being filleted. The default value of this variable is 1.

Note

TRIMMODE = 0 *Fillet or chamfer without cutting the existing geometry.*
TRIMMODE = 1 *Extend or trim the geometry.*

*When you enter the **FILLET** command, AutoCAD displays the current* **TRIMMODE** *and the current fillet radius.*

CHAMFERING THE SKETCHES

Toolbar:	Modify > Chamfer
Menu:	Modify > Chamfer
Command:	CHAMFER

Chamfering the sharp corners is another method of reducing the areas of stress concentration in the design. Chamfering is defined as the process in which the sharp edges or corners are beveled. In simple words, it is defined as the taper provided on a surface. A beveled line connects two separate objects to create a chamfer. The size of a chamfer depends on its distance from the corner. If a chamfer is equidistant from the corner in both directions, it is a 45-degree chamfer. A chamfer can be drawn between two lines that may or may not intersect. However, remember that the lines are not parallel because parallel lines cannot be chamfered. This command also works on a single polyline. In AutoCAD, the chamfers can be created using two methods: by defining two distances, or by defining one distance and the chamfer angle. The prompt sequence that will follow when you choose the **Chamfer** button from the **Modify** toolbar is given next.

(TRIM mode) Current chamfer Dist1 = 0.0000, Dist2 = 0.0000
Select first line or [Polyline/Distance/Angle/Trim/Method/mUltiple]:

Creating Chamfer Using The Distance Option

This option is used to enter the chamfer distance. This option can be invoked by entering **D** at the **Select first line or [Polyline/Distance/Angle/Trim/Method/mUltiple]** prompt. Next, enter the first and second chamfer distances. The first distance is the distance of the corner calculated along the edge selected first. Similarly, the second distance is calculated along the edge that is selected last. The new chamfer distances remain in effect until you change them. Instead of entering the distance values, you can specify two points to indicate each distance (Figure 5-26). The prompt sequence is given next.

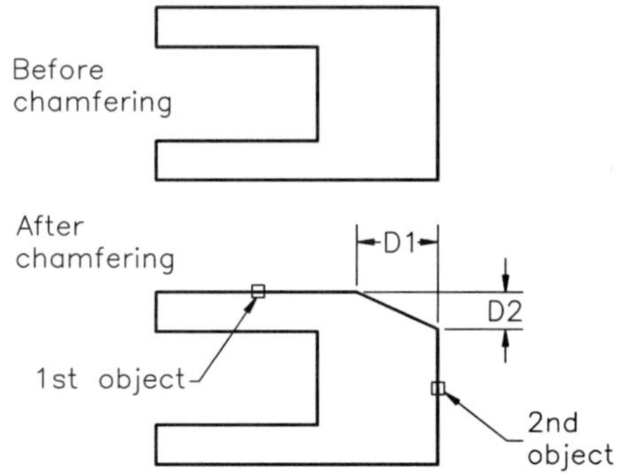

Figure 5-26 *Chamfering the model using the* ***Distance*** *option*

Select first line or [Polyline/Distance/Angle/Trim/Method/mUltiple]: **D** Enter
Specify first chamfer distance <current>: *Enter a distance value or specify two points.*
Specify second chamfer distance <current>: *Enter a distance value or specify two points.*

The first and second chamfer distances are stored in the **CHAMFERA** and **CHAMFERB** system variables. The default value of these variables is 0.0000.

Tip
*If Dist 1 and Dist 2 are set to zero, the **CHAMFER** command will extend or trim the selected lines so that they end at the same point.*

Creating Chamfer Using The Select First Line Option

In this option you need to select two nonparallel objects so that they are joined with a beveled line. The size of the chamfer depends upon the values of the two distances. The default values of the distances are 0.5000 and 0.5000. The prompt sequence is given next.

(TRIM mode) Current chamfer Dist1 = current, Dist2 = current
Select first line or [Polyline/Distance/Angle/Trim/Method/mUltiple]: *Specify the first line.*
Select second line: *Select the second line.*

Creating Chamfer Using The Polyline Option

You can use the **CHAMFER** command to chamfer all corners of a closed or open polyline (Figure 5-27). With a closed polyline, all the corners of the polyline are chamfered to the set distance values. Sometimes the polyline may appear closed. But if the **Close** option was not used to create it, it may not be. In this case, the last corner is not chamfered. The prompt sequence is given next.

(TRIM mode) Current chamfer Dist1 = current, Dist2 = current
Select first line or [Polyline/Distance/Angle/Trim/Method/mUltiple]: **P** Enter
Select 2D polyline: *Select the polyline.*

Creating Chamfer Using The Angle Option

The second method of creating the chamfer is by specifying a distance and the chamfer angle, Figure 5-28. The chamfer angle can be defined using this option. The prompt sequence is given next.

Select first line or [Polyline/Distance/Angle/Trim/Method/mUltiple]: **A** Enter
Specify chamfer length on the first line <current>: *Specify a length.*
Specify chamfer angle from the first line <current>: *Specify an angle.*

Creating Chamfer Using The Trim Option

Depending on this option, the selected objects are either trimmed or extended to the endpoints of the chamfer line or left intact. The prompt sequence is given next.

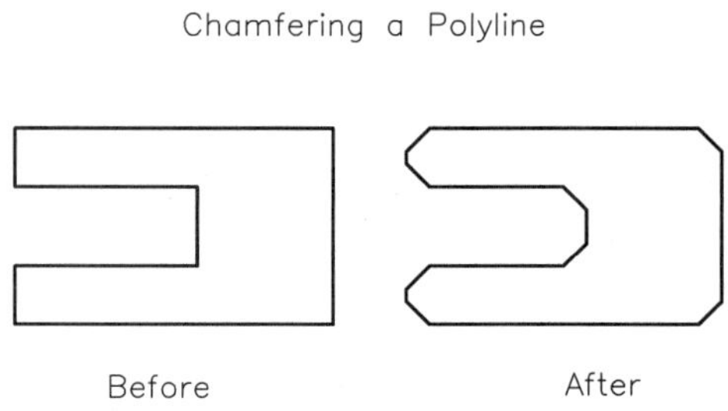

Figure 5-27 Chamfering a polyline

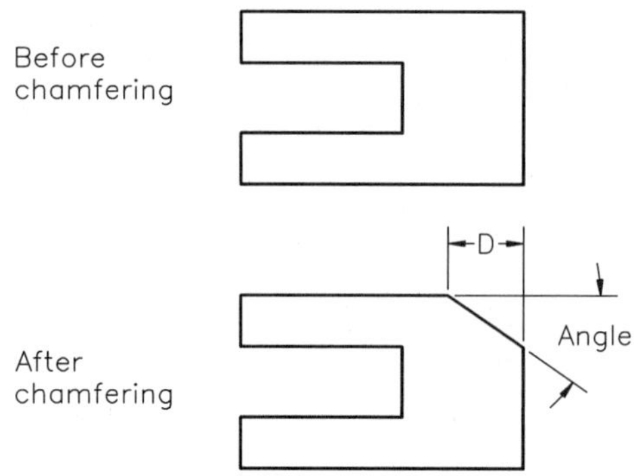

*Figure 5-28 Chamfering using the **Angle** option*

Select first line or [Polyline/Distance/Angle/Trim/Method/mUltiple]: **T** Enter
Enter Trim mode option [Trim/No Trim] <current>:

Creating Chamfer Using The Method Option

This option is used to toggle between the **Distance** method and the **Angle** method for creating the chamfer. The current settings of the selected method will be used for creating the chamfer. The prompt sequence is given next.

Select first line or [Polyline/Distance/Angle/Trim/Method/mUltiple]: **M** Enter
Enter trim method [Distance/Angle] <current>: *Enter* **D** *for Distance option,* **A** *for Angle option.*

Note

*If you set the value of the **TRIMMODE** system variable to **1** (default value), the objects will be trimmed or extended after they are chamfered and filleted. If **TRIMMODE** is set to zero, the objects are left untrimmed.*

Creating Chamfer Using The Multiple Option*

When you invoke the **Chamfer** command, by default the chamfer is created between a single set of entities only. But, with the help of the **Multiple** option you can add chamfers to the multiple sets of entities. When you select this option, AutoCAD prompts you to select the first line and then the second line. The chamfer is added to the two selected lines Next, you are again prompted to select the first line and then the second line. This prompt continues until you press ENTER to terminate the **Chamfer** command. The prompt sequence that follows to create multiple chamfers is given next.

(TRIM mode) Current chamfer Dist1 = current, Dist2= current
Select first line or [Polyline/Distance/Angle/Trim/Method/mUltiple]: **U** Enter
Select first line or [Polyline/Distance/Angle/Trim/Method/mUltiple]: *Specify first line of one set.*
Select second line: *Select second line of the same set.*
Select first line or [Polyline/Distance/Angle/Trim/Method/mUltiple]: *Specify first line of the other set.*

Select second line: *Select second line of the same set.*
Select first line or [Polyline/Distance/Angle/Trim/Method/mUltiple]: *Specify first line of the other set or press* Enter *to terminate the* ***Chamfer*** *command.*

Setting the Chamfering System Variables

The chamfer modes, distances, length, and angle can also be set using the following variables.

CHAMMODE = 0	Distance/Distance (default)
CHAMMODE = 1	Length/Angle
CHAMFERA	Sets first chamfer distance on the first selected line (default = 0.0000)
CHAMFERB	Sets second chamfer distance on the second selected line (default = 0.0000)
CHAMFERC	Sets the chamfer length (default = 0.0000)
CHAMFERD	Sets the chamfer angle from the first line (default = 0)

TRIMMING THE SKETCHED OBJECTS

Toolbar:	Modify > Trim
Menu:	Modify > Trim
Command:	TRIM

When creating a design, there are a number of places where you have to remove the unwanted and extending edges. Breaking individual objects takes time if you are working on a complex design with many objects. In such cases you can use the **TRIM** command. This command trims objects that extend beyond a required point of intersection. When you invoke this command, you will be prompted to select the cutting edges or boundaries. These edges can be lines, polylines, circles, arcs, ellipses, xlines, rays, splines, text, blocks, or even viewports. There can be more than one cutting edge and you can use any selection method to select them. After the cutting edge or edges are selected, you must select each object to be trimmed. An object can be both a cutting edge and an object to trim. You can trim lines, circles, arcs, polylines, splines, ellipses, xlines, and rays. The prompt sequence is given next.

Current settings:Projection=UCS Edge = None
Select cutting edges...
Select objects: *Select the cutting edges.*
Select objects: Enter
Select object to trim or shift-select to extend or [Project/Edge/Undo]:

Select Object to trim Option

Here you have to specify the objects you want to trim and the side from which the object will be trimmed. This prompt is repeated until you press ENTER. This way you can trim several objects with a single **TRIM** command (Figure 5-29). The prompt sequence is given next.

Current settings: Projection= UCS Edge= None
Select cutting edges...

Select objects: *Select the first cutting edge.*
Select objects: *Select the second cutting edge.*
Select objects: Enter
Select object to trim or shift-select to extend or [Project/Edge/Undo]: *Select the first object.*
Select object to trim or shift-select to extend or [Project/Edge/Undo]: *Select the second object.*
Select object to trim or shift-select to extend or [Project/Edge/Undo]: Enter

Shift-select to extend Option

This option is used to switch to the extend mode. It is used to extend the object instead of trimming. In case the object to extend does not intersect with the cutting edge, you can press the SHIFT key and then select the object to extend. The selected edge will be extended taking the cutting edge as the boundary for extension.

Edge Option

This option is used whenever you want to trim those objects that do not intersect the cutting edges, but would intersect if the cutting edges were extended (Figure 5-30). The prompt sequence is given next.

Command: **TRIM** Enter
Current settings: Projection= UCS Edge= None
Select cutting edges...
Select objects: *Select the cutting edge.*
Select objects: Enter
Select object to trim or shift-select to extend or [Project/Edge/Undo]: **E** Enter
Enter an implied edge extension mode [Extend/No extend] <current>: **E** Enter
Select object to trim or shift-select to extend or [Project/Edge/Undo]: *Select object to trim.*

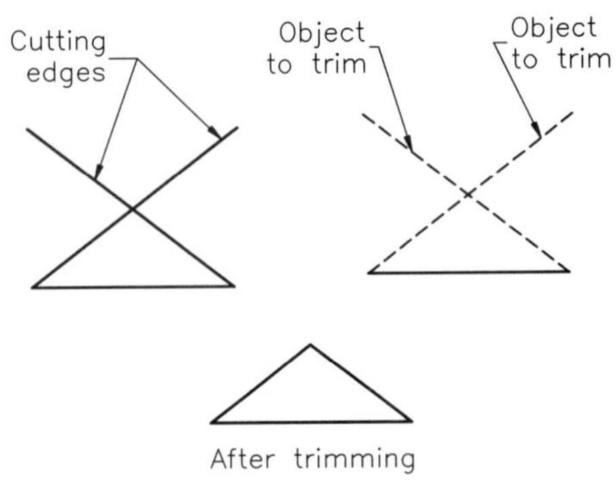

Figure 5-29 *Using the* ***TRIM*** *command*

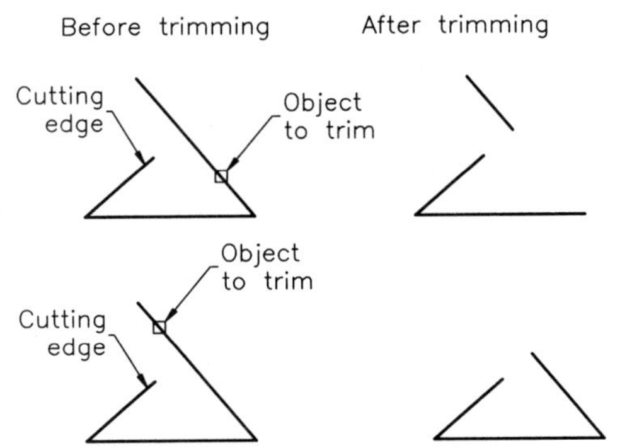

Figure 5-30 *Trimming an object using the* ***Edge*** *option (**Extend**)*

Project Option

In this option you can use the Project mode while trimming objects. The prompt sequence is given next.

Select object to trim or shift-select to extend or [Project/Edge/Undo]: **P** Enter
Enter a projection option [None/Ucs/View] <current>:

The **None** option is used whenever the objects to trim intersect the cutting edges in 3D space. If you want to trim those objects that do not intersect the cutting edges in 3D space, but do visually appear to intersect in a particular UCS or the current view, use the **UCS** or **View** options. The UCS option projects the objects to the *XY* plane of the current UCS, while the **View** option projects the objects to the current view direction (trims to their apparent visual intersections).

Undo Option

If you want to remove the previous change created by the **TRIM** command, enter **U** at the **Select object to trim or [Project/Edge/Undo]:** prompt.

Exercise 6 *Mechanical*

Draw the top illustration in Figure 5-31 and then use the **FILLET**, **CHAMFER**, and **TRIM** commands to obtain the following figure. (Assume the missing dimensions.)

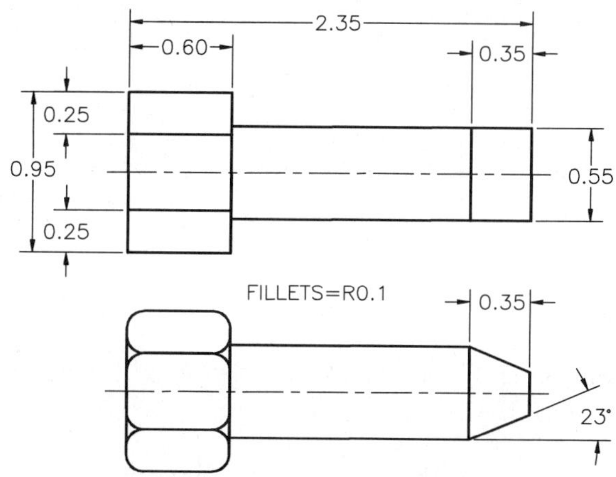

Figure 5-31 *Drawing for Exercise 6*

EXTENDING THE SKETCHED OBJECTS

Toolbar:	Modify > Extend
Menu:	Modify > Extend
Command:	EXTEND

The **EXTEND** command may be considered the opposite of the **TRIM** command. In the **TRIM** command you trim objects; in the **EXTEND** command you can extend lines, polylines, rays, and arcs to meet other objects. This command does not extend closed loops. The command format is similar to that of the **TRIM** command. You are required to select the boundary edges first. The boundary edges are those objects that the selected lines or arcs extend to meet. These edges can be lines, polylines, circles, arcs, ellipses, xlines, rays, splines, text, blocks, or even viewports. The prompt sequence that will follow when you choose the **Extend** button is given next.

Current settings: Projection= UCS, Edge= None
Select boundary edges ...
Select objects: *Select the boundary edges.*
Select objects: Enter
Select object to extend or shift-select to trim or [Project/Edge/Undo]:

Select object to extend Option

Here you have to specify the object you want to extend to the selected boundary (Figure 5-32). This prompt is repeated until you press ENTER. Now you can select a number of objects in a single **EXTEND** command.

Shift-select to trim Option

This option is used to switch to the trim mode in the **EXTEND** command. You can press the SHIFT key and then select the object to be trimmed. In this case the boundary edges will be taken as the cutting edges.

Project Option

In this option you can use the projection mode while trimming objects. The prompt sequence that will follow when you choose the **Extend** button is given next.

Current settings: Projection= UCS Edge= None
Select boundary edges ...
Select objects: *Select the boundary edges.*
Select objects: Enter
Select object to extend or shift-select to trim or [Project/Edge/Undo]: **P** Enter
Enter a projection option [None/UCS/View] <current>:

The **None** option is used whenever the objects to be extended intersect with the boundary edge in 3D space. If you want to extend those objects that do not intersect the boundary edge in 3D space, use the **UCS** or **View** option. The **UCS** option projects the objects to the *XY* plane of the current UCS, while the **View** option projects the objects to the current view.

Edge Option

You can use this option whenever you want to extend objects that do not actually intersect the boundary edge, but would intersect its edge if the boundary edge were extended (Figure 5-33). If you enter **E** at the prompt, the selected object is extended to the implied boundary edge. If you enter **N** at the prompt, only those objects that would actually intersect the real boundary edge are extended (the default). The prompt sequence is given next.

Current settings: Projection= UCS Edge= None
Select boundary edges ...
Select objects: *Select the boundary edge.*
Select objects: Enter
Select object to extend or shift-select to trim or [Project/Edge/Undo]: **E** Enter
Enter an implied extension mode [Extend/No extend] <current>: **E** Enter
Select object to extend or shift-select to trim or [Project/Edge/Undo]: *Select the line to extend.*

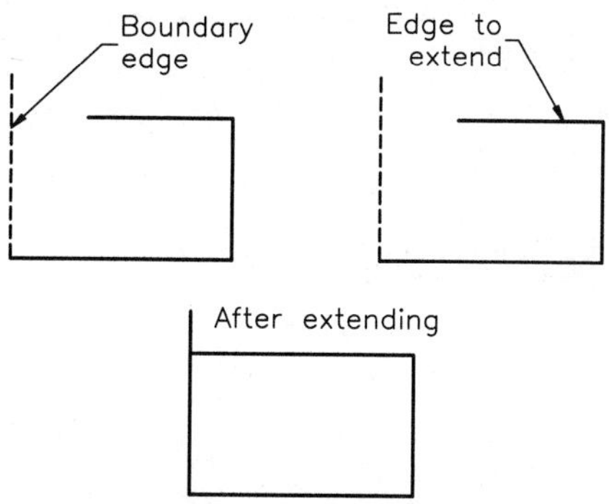

Figure 5-32 *Extending an edge*

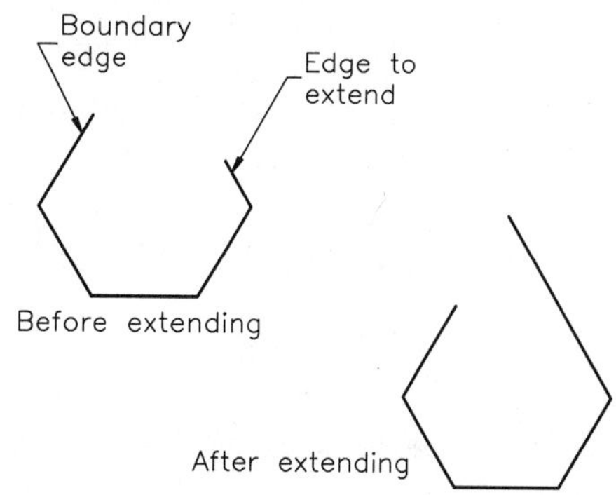

Figure 5-33 *Extending an edge using the* ***Edge*** *option (****Extend****)*

Undo Option

If you want to remove the previous change created by the **EXTEND** command, enter **U** at the **Select object to extend or [Project/Edge/Undo]:** prompt.

Trimming and Extending with Text, Region, or Spline

The **TRIM** and **EXTEND** commands can be used with text, regions, or splines as edges (Figure 5-34). This makes the **TRIM** and **EXTEND** commands two of the most useful editing commands in AutoCAD. The **TRIM** and **EXTEND** commands can also be used with arcs, elliptical arcs, splines, ellipses, 3D Pline, rays, and lines. See Figure 5-35.

The system variables **PROJMODE** and **EDGEMODE** determine how the **TRIM** and **EXTEND** commands are executed. The following are the values that can be assigned to these variables.

Value	PROJMODE	EDGEMODE
0	True 3D mode	Use regular edge without extension (default)
1	Project to current UCS *XY* plane (default)	Extend the edge to natural boundary
2	Project to current view plane	

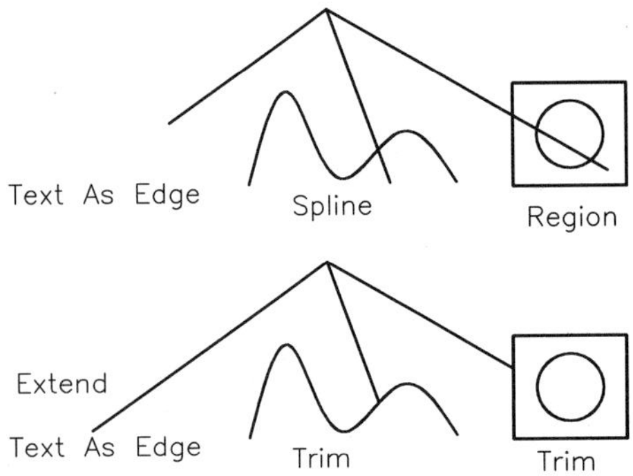

Figure 5-34 *Using the* ***TRIM*** *and* ***EXTEND*** *commands with text, spline, and region*

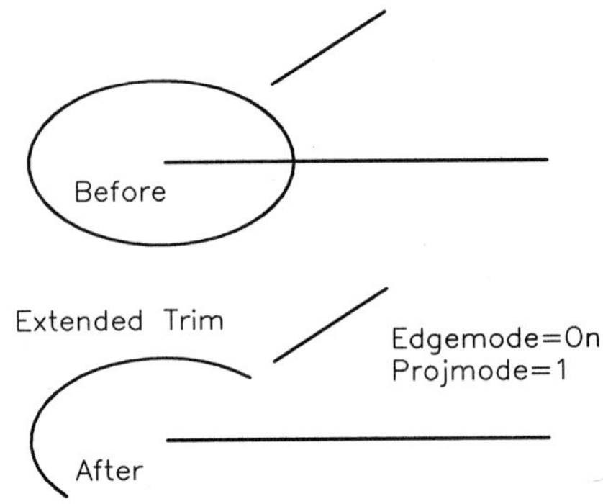

Figure 5-35 *Using the* ***Edge*** *option to do an implied trim*

STRETCHING THE SKETCHED OBJECTS

Toolbar:	Modify > Stretch
Menu:	Modify > Stretch
Command:	STRETCH

This command can be used to stretch objects, altering selected portions of the objects. With this command you can lengthen objects, shorten them, and alter their shapes, see Figure 5-36. You must use a Crossing, or CPolygon selection to specify the objects to stretch. The prompt sequence that will follow when you choose the **Stretch** button is given next.

> Select objects to stretch by crossing-window or crossing-polygon ...
> Select objects: *Select the objects using crossing window or polygon.*
> Select objects: [Enter]

After selecting the objects, you have to specify the point of displacement. You should select only that portion of the object that needs stretching.

> Specify base point or displacement: *Specify the base point.*
> Specify second point of displacement or <use first point as displacement>: *Specify the displacement point or press ENTER to use the first point as the displacement.*

You normally use a Crossing or CPolygon selection with **STRETCH** because if you use a Window, those objects that cross the window are not selected, and the objects selected because they are fully within the window are moved, not stretched. The object selection and stretch specification process of **STRETCH** is a little unusual. You are really specifying two things: first, you are selecting objects. Second, you are specifying the portions of those selected objects to be stretched. You can use a Crossing or CPolygon selection to simultaneously specify both, or you can select objects by any method, and then use any window or crossing specification to specify what parts of those objects to stretch. Objects or portions of selected objects completely within the window or crossing specification are moved. If selected objects cross the window or crossing specification, their defining points within the window or crossing specification are moved, their defining points outside the window or crossing specification remain fixed, and the parts crossing the window or crossing specification are stretched. Only the last window or crossing specification made determines what is stretched or moved. Figure 5-36 illustrates using a crossing selection to simultaneously select the two angled lines and specify that their right ends will be stretched. Alternatively, you can select the lines by any method and then use a Crossing selection (which will not actually select anything) to specify that their right ends will be stretched.

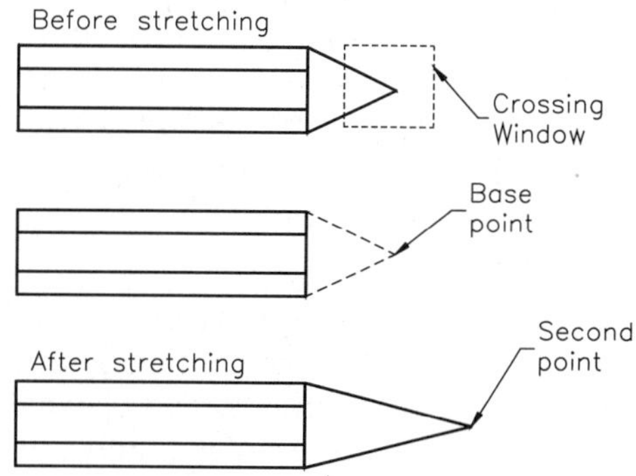

Figure 5-36 *Stretching the entities*

Note

The regions and solids cannot be stretched. When you select them, they will be moved instead of stretching.

LENGTHENING THE SKETCHED OBJECTS

Menu:	Modify > Lengthen
Command:	LENGTHEN

Like the **TRIM** and **EXTEND** commands, the **LENGTHEN** command can be used to extend or shorten lines, polylines, elliptical arcs, and arcs. The **LENGTHEN** command has several options that allow you to change the length of objects by dynamically dragging the object endpoint, entering the delta value, entering the percentage value, or entering the total length of the object. This command also allows the repeated selection of objects for editing. This command has no effect on closed objects such as circles. The selected entity will be increased or decreased from the endpoint closest to the selection point. The prompt sequence that will follow when you invoke the **Lengthen** command is given next.

Select an object or [DElta/Percent/Total/DYnamic]:

Note

*By default, the **Lengthen** button is not added in the **Modify** toolbar. You can add the **Lengthen** button in the **Modify** toolbar by right-clicking on the free space available next to any toolbar to display the shortcut menu. Select **Customize** from the shortcut menu to display the **Customize** dialog box. Select **Modify** from the **Categories** list box to display the various commands available under the **Modify** pull-down menu in the **Commands** list box. Pick the down arrow key and select **Lengthen** from the **Commands** list box using the left mouse button. Drag the **Lengthen** command and drop it on the **Modify** toolbar.*

Select an object Option

This is the default option that returns the current length or the included angle of the selected object. If the object is a line, AutoCAD returns only the length. However, if the selected object is an arc, AutoCAD returns the length and the angle. The same prompt sequence will be displayed after you select the object.

DElta Option

The **DElta** option is used to increase or decrease the length or angle of an object by defining the distance or angle by which the object will be extended. The delta value can be entered by entering a numerical value or by specifying two points. A positive value will increase (Extend) the length of the selected object and a negative value will decrease the length (Trim), see Figure 5-37. The prompt sequence to use this option is given next.

> Select an object or [DElta/Percent/Total/DYnamic]: **DE** Enter
> Enter delta length or [Angle] <current>: *Specify the length or enter A for angle.*
> Enter delta angle <current>: *Specify the delta angle.*
> Select object to change or [Undo]: *Select object to be extended.*
> Select object to change or [Undo]: Enter

Percent

The **Percent** option is used to extend or trim an object by defining the change as a percentage of the original length or the angle, Figure 5-37. The current length of the line is taken as 100 percent. If you enter a value more than 100, the length will increase by that amount. Similarly, if you enter a value less than 100 then the length will decrease by that amount. For example, a positive number of 150 will increase the length by 50 percent and a positive number of 75 will decrease the length by 25 percent of the original value (negative values are not allowed).

Total

The **Total** option is used to extend or trim an object by defining the new total length or angle. For example, if you enter a total length of 1.25, AutoCAD will automatically increase or decrease the length of the object so that the new length of the object is 1.25. The value can be entered by entering a numerical value or by specifying two points. The object is shortened or lengthened with respect to the endpoint that is closest to the selection point. The selection point is determined by where the object was selected. The prompt sequence for using this option is given next.

> Specify total length or [Angle] <current>: *Specify the length or enter* **A** *for angle.*

If you enter a numeric value in the previous prompt, the length of the selected objects is changed accordingly, see Figure 5-38. You can change the angle of an arc by entering **A** and giving the new value.

DYnamic

The **DYnamic** option allows you to dynamically change the length or angle of an object by

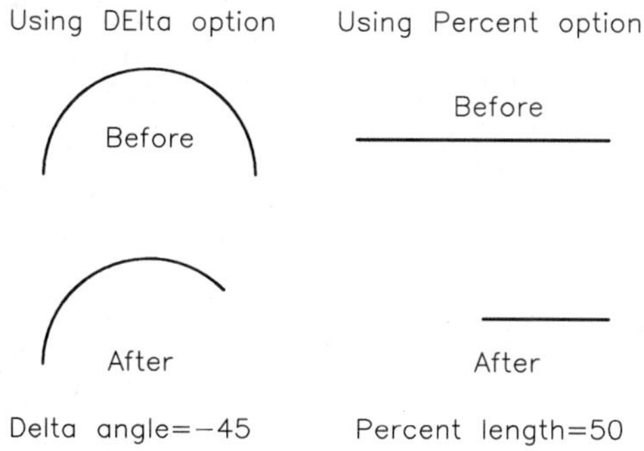

Figure 5-37 *Using the **DElta** and **Percent** options*

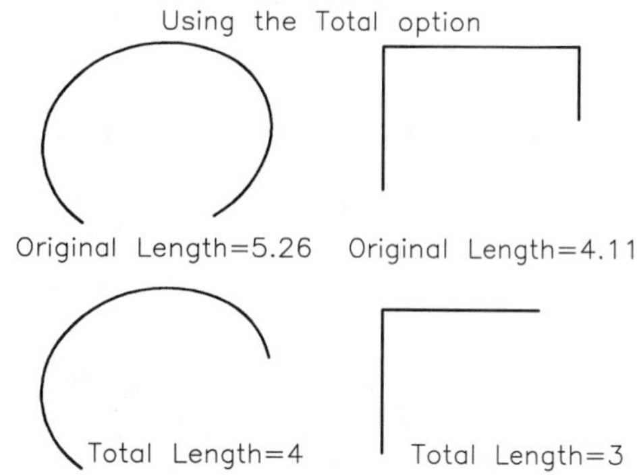

Figure 5-38 *Using the **Total** option*

specifying one of the endpoints and dragging it to a new location. The other end of the object stays fixed and is not affected by dragging. The angle of lines, radius of arcs, and shape of elliptical arcs are unaffected.

ARRAYING THE SKETCHED OBJECTS

Toolbar:	Modify > Array
Menu:	Modify > Array
Command:	ARRAY

Array is defined as the method of creating multiple copies of the selected object and arranging them in a rectangular or circular fashion. In some drawings you may need to specify an object multiple times in a rectangular or circular arrangement. For example, suppose you have to draw six chairs around a table. This job can be accomplished by drawing each chair separately or by using the **COPY** command to make multiple copies of the chair. But it is a very tedious process and also the alignment of the chairs will have to be adjusted. This can be easily done using the **ARRAY** command. All you have to do is to create just one chair and the remaining five will be created and automatically arranged around the table by the **ARRAY** command. This method is more efficient and less time-consuming. This command allows you to make multiple copies of selected objects in a rectangular or polar fashion. Each resulting element of the array can be controlled separately. When you choose the **Array** button, the **Array** dialog box is displayed. You can use this dialog box for creating a rectangular array or a polar array.

Rectangular Array

A rectangular array is formed by making copies of the selected object along the *X* and *Y* axes directions of an imaginary rectangle (along rows and columns). The rectangular array can be created by selecting the **Rectangular Array** radio button in the **Array** dialog box, see Figure 5-39.

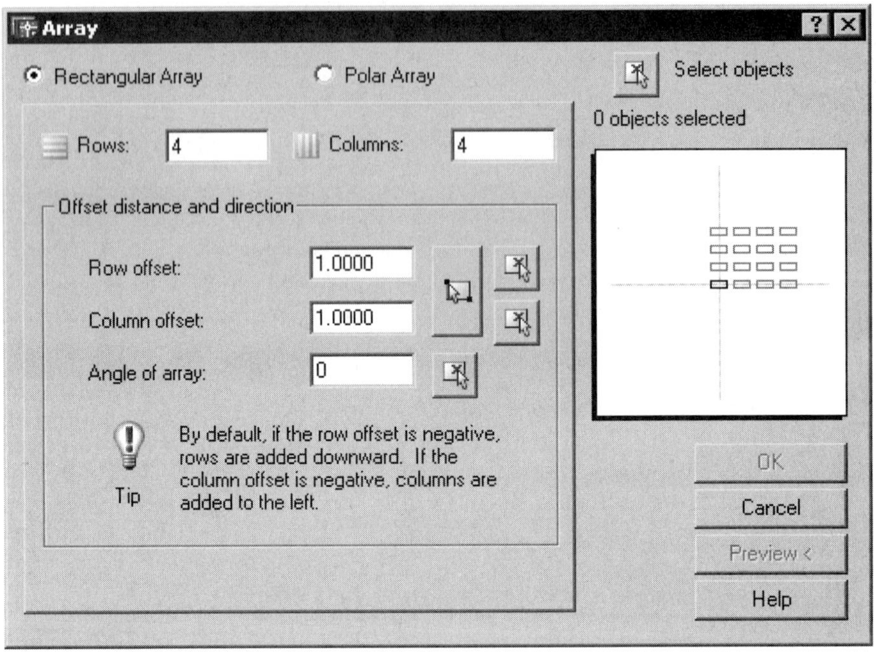

Figure 5-39 ***Rectangular Array*** *options in the* ***Array*** *dialog box*

Rows

This edit box is used to specify the number of rows in the rectangular array. The rows are arranged along the *X* axis of the current UCS.

Columns

This edit box is used to specify the number of columns in the rectangular array. The columns are arranged along the *Y* axis of the current UCS.

Offset distance and direction Area

The options under this area are used to define the distance between the rows and the columns and the angle of the array.

Row offset. This edit box is used to specify the distance between the rows, see Figure 5-40. You can either enter the distance value in this edit box or choose the **Pick Row Offset** button to define the distance. This button is provided on the right of the **Row offset** edit box. When you choose this button, the **Array** dialog box will be temporary closed and you will be prompted to specify two points on the screen to define the row offset distance. A positive value of row offset will create the rows along the positive *X* axis direction. A negative value of row offset will create the rows along the negative *X* axis direction, see Figure 5-41.

Column offset. This edit box is used to specify the distance between the columns, Figure 5-40. You can either enter the distance value in this edit box or choose the **Pick Column Offset**

button to define the distance. This button is provided on the right of the **Column offset** edit box. When you choose this button, the **Array** dialog box will be temporary closed and you will be prompted to specify two points on the screen to define the column offset distance. A positive value of row offset will create the rows along the positive *Y* axis direction. A negative value of row offset will create the rows along the negative *Y* axis direction, see Figure 5-41.

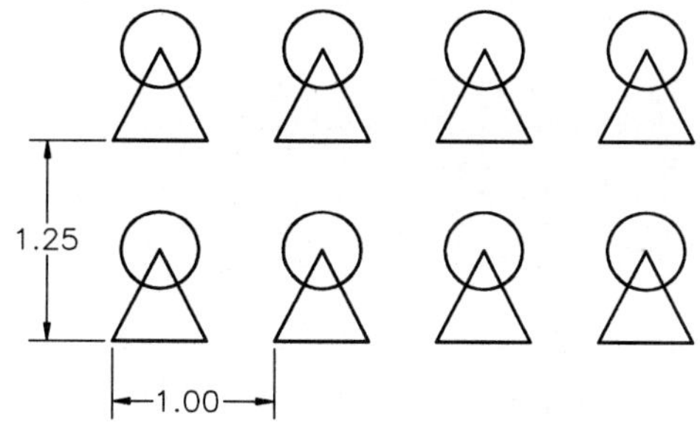

Figure 5-40 *Rectangular array with row and column distance*

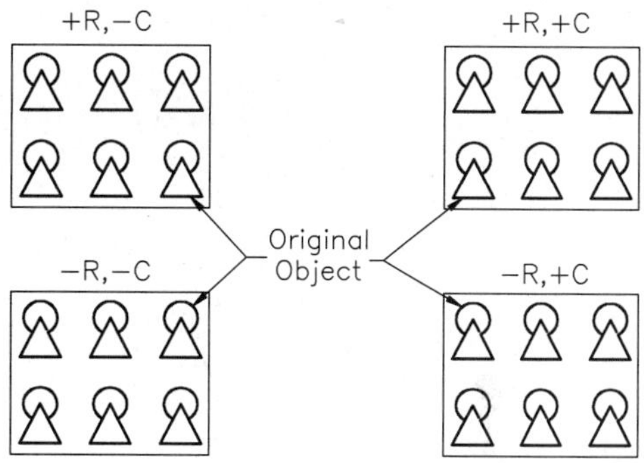

Figure 5-41 *Specifying the direction of arrays*

You can define both the row offset distance and the column offset distance simultaneously by choosing the **Pick Both Offsets** button. When you choose this button, the **Array** dialog box will be temporarily closed and you will be prompted to define a unit cell by specifying two opposite corners. The distance along the *X* axis of the unit cell defines the distance between the rows and the distance along the *Y* axis of the unit cell defines the distance between the columns.

Angle of array. This edit box is used to define the angle of the array. This is the value by which the rows and the columns will be rotated, Figure 5-42. A positive value will rotate them in the counterclockwise direction and a negative value will rotate them in the clockwise direction. You can also define the angle by choosing the **Pick Angle of Array** button and specifying two points on the screen.

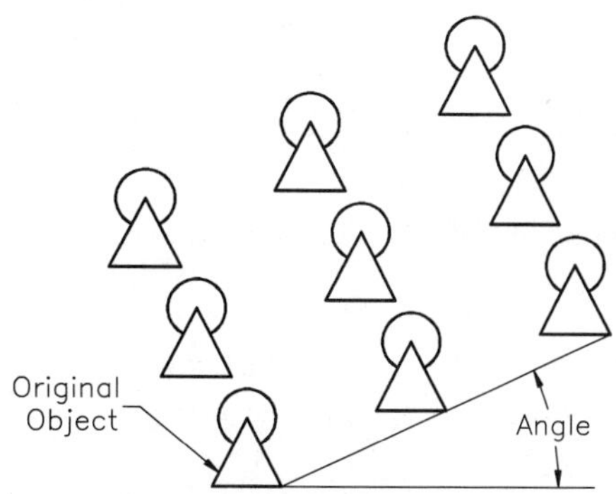

Figure 5-42 *Rotated rectangular array*

The object for array can be selected by choosing the **Select Objects** button. You can preview the creation of array by choosing the **Preview** button. When you choose this button, the **Array** dialog box will be displayed, Figure 5-43. If you are satisfied

Figure 5-43 ***Array*** *dialog box for accepting or modifying the array*

with the array, choose the **Accept** button, and the array will be created. If you are not satisfied with the array, choose the **Modify** button. The **Array** dialog box will be redisplayed on the screen and you can make the necessary changes.

Polar Array

A polar array is an arrangement of objects around a point in a circular fashion. This kind of array is created by selecting the **Polar Array** radio button in the **Array** dialog box, Figure 5-44.

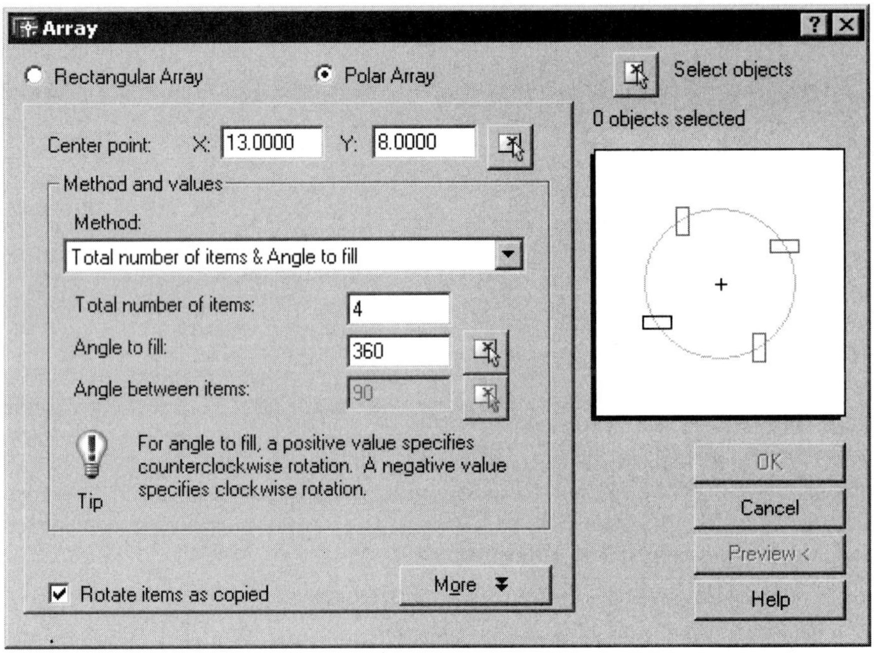

Figure 5-44 ***Polar Array*** *options in the* ***Array*** *dialog box*

Center point

The center point of the array is defined as the point around which the selected items will be arranged. It is considered as the center point of the imaginary circle on whose circumference the items will be placed. The coordinates of the center of the array can be specified in the **X** and **Y** edit boxes. You can either enter the values in these edit boxes or select the center point of the array from the screen. To select the center point of the array from the screen, choose the **Pick Center Point** button provided on the right side of the **Y** edit box. When you choose this button, the **Array** dialog box will be temporarily closed and you will be prompted to select the center of the array. The **Array** dialog box will be redisplayed when you select the center of the array. Figure 5-45 shows a polar array displaying the center of the imaginary circle about which the objects are rotated.

Method and values Area

The options under this area are used to set the parameters related to the method that will be

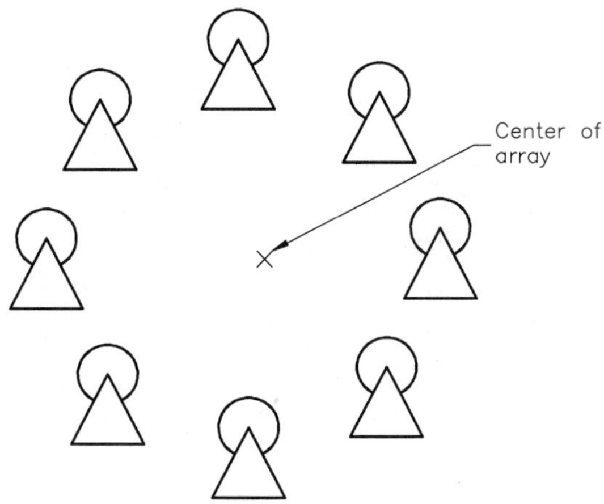

Figure 5-45 Center point of array

employed to create the polar array. The other options under this area will be available based on the method you select for creating the polar array.

Method. This drop-down list provides you with three methods for creating the polar array. The three methods that are available are discussed next.

Total number of items & Angle to fill. This method is used to create a polar array by specifying the total number of items in the array and total included angle between the first and the last item of the array, see Figure 5-46. The number of items and the angle to fill can be specified in the **Total number of items** and **Angle to fill** edit boxes, respectively. You can also specify the angle to fill on the screen. This is done by choosing the **Pick Angle to Fill** button provided on the right side of the **Angle to fill** edit box. The **Array** dialog box will be temporarily closed and you will be prompted to specify the angle to fill.

Total number of items & Angle between the items. This method is used when you want to create a polar array by specifying the total number of items in the array and the included angle between the two adjacent items of the array, see Figure 5-47. The angle between the items is also called the incremental angle. The number of items and the angle between the items can be specified in the **Total number of items** and **Angle between items** edit boxes, respectively. You can also specify the angle between the items on the screen. This is done by choosing the **Pick Angle Between Items** button provided on the right side of the **Angle between items** edit box. The **Array** dialog box will be temporarily closed and you will be prompted to specify the angle between the items.

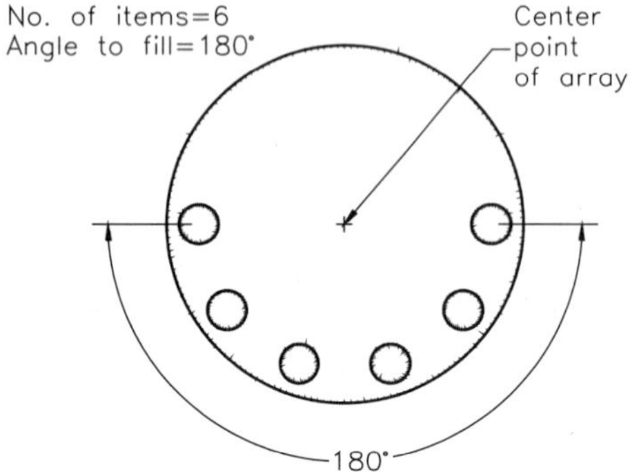

Figure 5-46 *Array created using number of items and angle to fill*

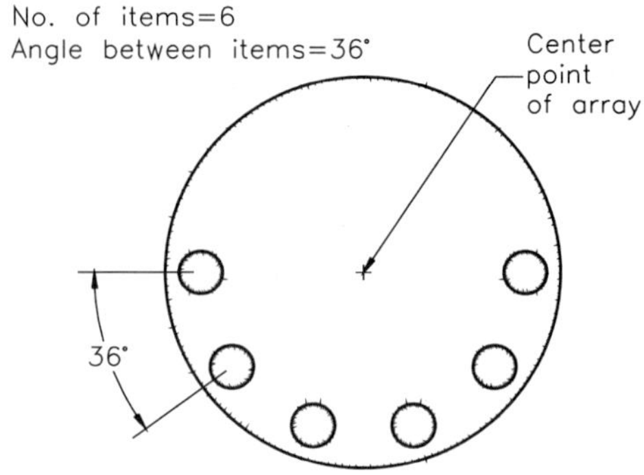

Figure 5-47 *Array created using number of items and angle between items*

Angle to fill & Angle between items. This method is used when you want to specify the angle between the items and total angle to fill, Figure 5-48. In this case the number of items is not specified, but is automatically calculated using the total angle to fill in the array and the angle between the items. The angle to fill and the angle between the items can be entered in the respective edit boxes or can be specified on the screen using the respective buttons.

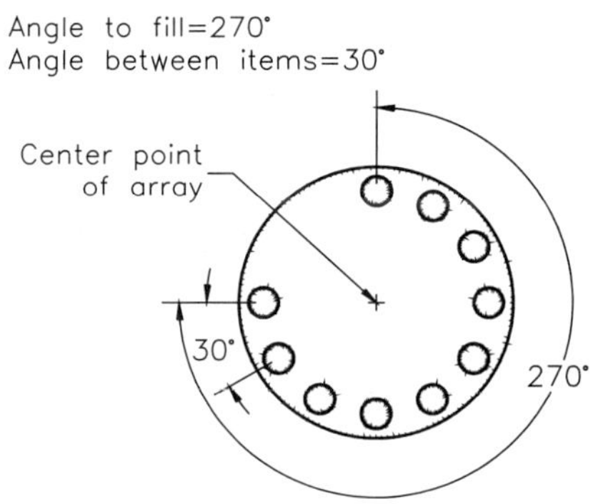

Figure 5-48 *Array created using angle to fill and angle between items*

Rotate items as copied

This check box is selected to rotate the objects as they are copied around the center point. If this check box is cleared, the objects are not rotated as they are copied. This means that the replicated objects remain in the same orientation as the original object. You can have the items rotated around the center point as they are copied by selecting this check box. Here, the same face of each object points toward the pivot point. Figure 5-49 shows polar array with items rotated as copied and Figure 5-50 shows polar array with items not rotated as copied.

More

When this button is chosen, the **Array** dialog box expands providing the options of defining the base point of the objects, see Figure 5-51. The base point will maintain a constant distance from the center of the array when the items are copied in polar fashion. You can use the object's default base points by selecting the **Set to object's default** check box. You can also specify the user-defined points by clearing this check box. Specify the *X* and *Y* coordinate values of the center of the array in the **X** and **Y** edit boxes in this area. You can also specify the base point on the screen by choosing the **Pick Base Point** button.

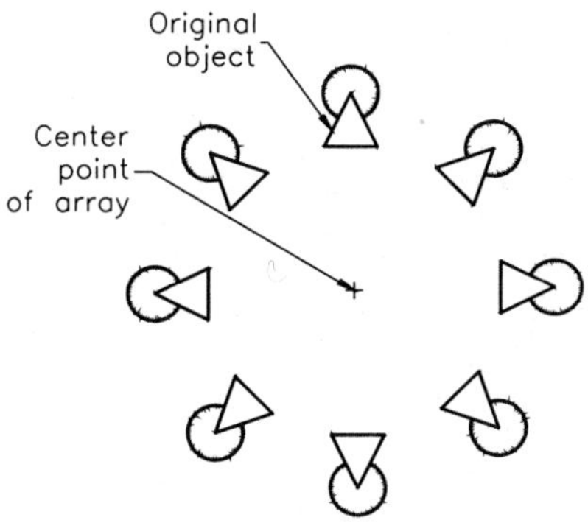

Figure 5-49 *Items rotated as copied*

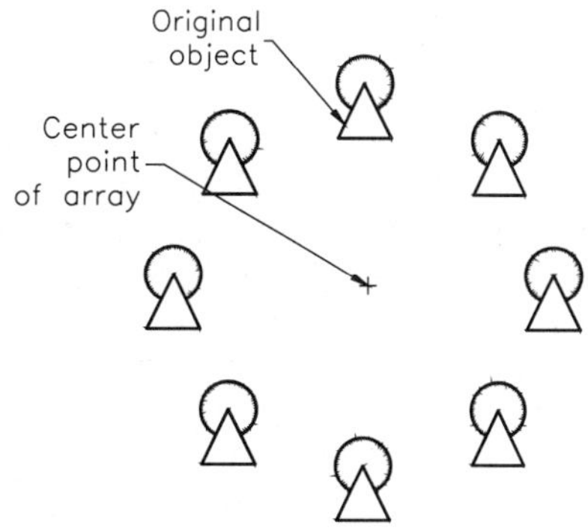

Figure 5-50 *Items not rotated as copied*

You can preview the array using the **Preview** button. The **Array** dialog box will be displayed and you can choose the **Accept** button if you are satisfied with the array or choose the **Modify** button if you want to make any modifications.

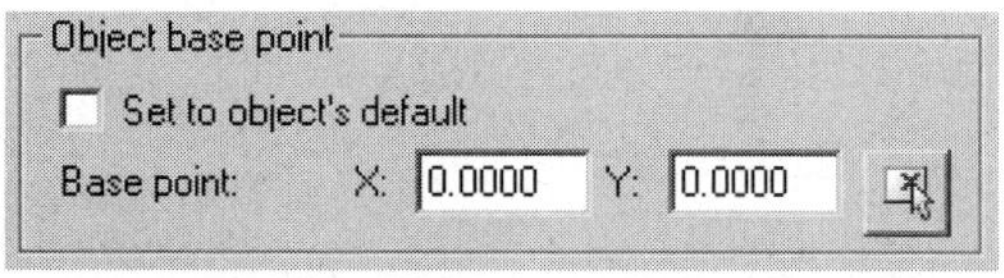

Figure 5-51 *More options of the **Array** dialog box*

Tip

*By default, in polar array the objects are arrayed in counterclockwise direction. But if you change the angle measuring direction from counterclockwise to clockwise direction using the **Units** dialog box then the objects will be arrayed in clockwise direction and not in counterclockwise direction.*

MIRRORING THE SKETCHED OBJECTS

Toolbar:	Modify > Mirror
Menu:	Modify > Mirror
Command:	MIRROR

The **MIRROR** command creates a mirror copy of the selected objects. The objects can be mirrored at any angle. This command is helpful in drawing symmetrical figures. When you invoke this command, AutoCAD will prompt you to select the objects and then the mirror line.

After you select the objects to be mirrored, AutoCAD prompts you to enter the first point of the mirror line and the second point of the mirror line. A mirror line is an imaginary line about which objects are reflected. You can specify the endpoints of the mirror line by specifying the points on the screen or by entering their coordinates. The mirror line can be specified at any angle. After the first point on the mirror line has been selected, AutoCAD displays the selected objects as they would appear after mirroring. Next, you need to specify the second endpoint of the mirror line. Once this is accomplished, AutoCAD prompts you to specify whether you want to retain the original figure (Figure 5-52) or delete it and just keep the mirror image of the figure (Figure 5-53). The prompt sequence when you choose this button is given next.

Select objects: *Select the objects to be mirrored.*
Select objects: [Enter]
Specify first point of mirror line: *Specify the first endpoint.*
Specify second point of mirror line: *Specify the second endpoint.*
Delete source objects ? [Yes/No] <N>: *Enter* Y *for deletion,* N *for retaining the previous objects.*

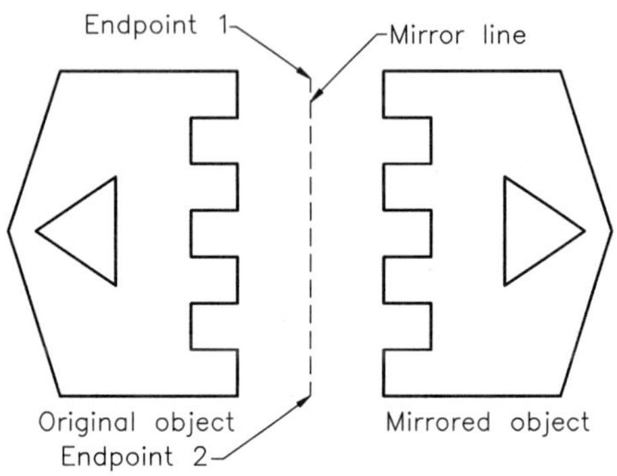

Figure 5-52 *Creating a mirror image of an object using the **MIRROR** command*

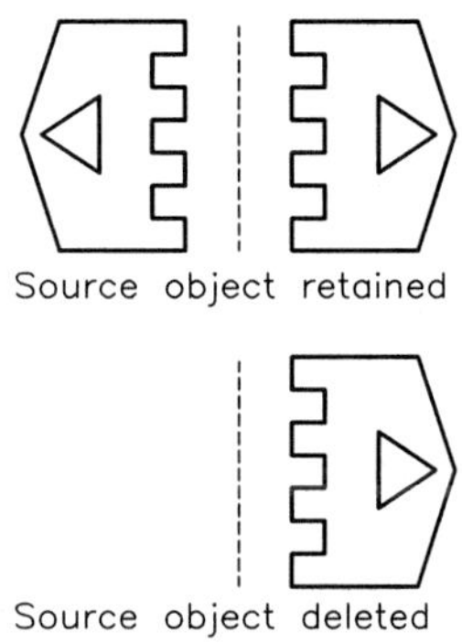

Figure 5-53 *Retaining and deleting old objects after mirroring*

In case you want to mirror the objects at some angle, define the mirror line accordingly. For example, if you want to mirror an object such that the mirrored object is placed at an angle of 90-degree from the original object, define the mirror line at an angle of 45-degree, see Figure 5-54.

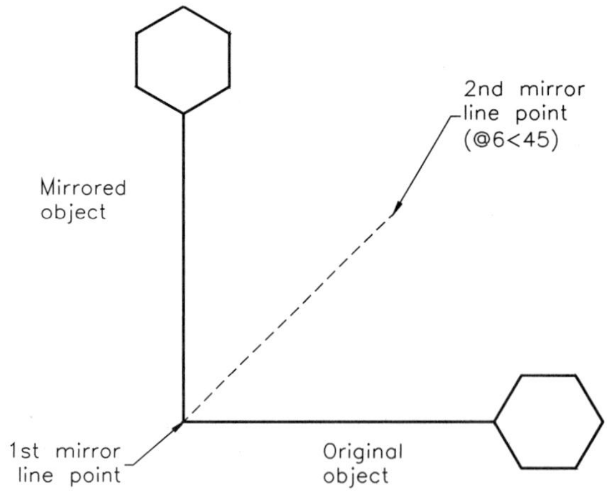

Figure 5-54 *Mirroring the object at an angle*

Text Mirroring

By default, the **MIRROR** command reverses all the objects including texts and dimensions. But, you may not want the text reversed (written backward). In such a situation, you should use the system variable **MIRRTEXT**. The **MIRRTEXT** variable has the following two values (Figure 5-55).

1 = Text is reversed in relation to the original object. This is the default value.
0 = Inhibits the text from being reversed with respect to the original object.

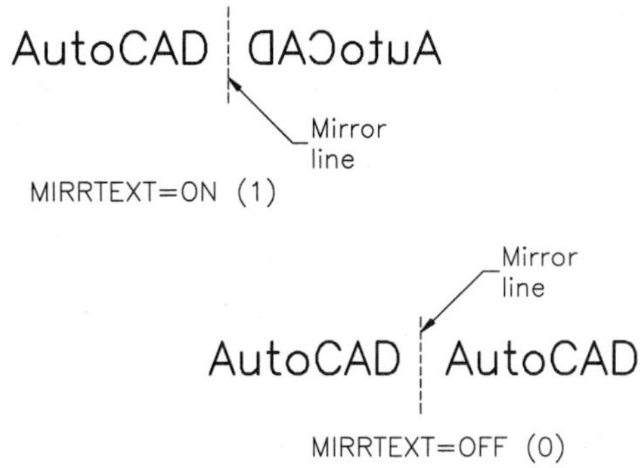

Figure 5-55 Using the ***MIRRTEXT*** *variable for mirroring the text*

Therefore, if you want the existing object to be mirrored, but at the same time you want the text to be readable and not reversed, set the value of the **MIRRTEXT** variable to **0**, and then mirror the objects.

BREAKING THE SKETCHED OBJECTS

Toolbar:	Modify > Break at Point, Break
Menu:	Modify > Break
Command:	BREAK

The **BREAK** command breaks an existing object into two or erases portions of the objects. This command can be used to remove a part of the selected objects or to break objects such as lines, arcs, circles, ellipses, xlines, rays, splines, and polylines. You can break the objects using the following methods.

1 Point Option

This method of breaking the existing entities can be directly invoked by choosing the **Break at Point** button from the **Modify** toolbar. Using this method, you can break the object into two parts. When you choose this button, you will be prompted to select the object to be broken. Once you select the object to be broken, you will be prompted to specify the first break point. You will not be prompted to specify the second break point and the object will be automatically broken at the first point. The prompt sequence that will follow when you choose this button is given next.

Select object: *Select the object to be broken.*
Specify second break point or [First point]: f
Specify first break point: *Specify the point at which the object should be broken.*
Specify second break point: @

Chapter 5

2 Points Option

This method allows you to break an object between two selected points. The portion of the object between the two selected point is removed. The point at which you select the object becomes the first break point and then you are prompted to enter the second break point. The prompt sequence that will follow when you choose this button is given next.

Select object: *Select the object to be broken.*
Specify second break point or [First point]: *Specify the second break point on the object.*

The object is broken between these two points and the in-between portion of the object is removed (Figure 5-56).

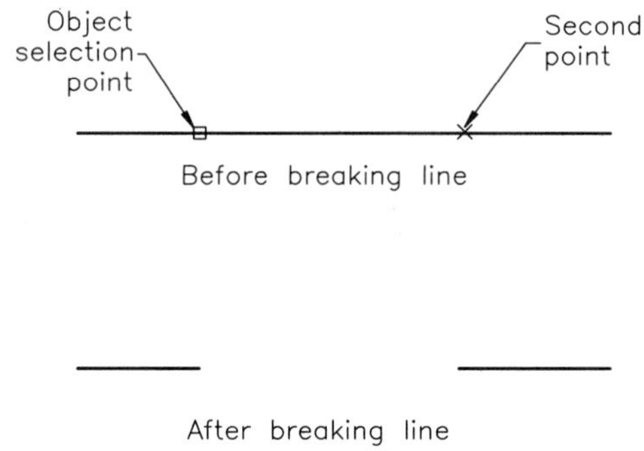

***Figure 5-56** Using the **2 Point** option for breaking the line*

2 Points Select Option

This method is similar to the 2 Points option; the only difference is that instead of making the selection point as the first break point, you are allowed to specify a new first point, see Figure 5-57. The prompt sequence that will follow when you choose the **Break** button is given next.

Select object: *Select the object to be broken.*
Enter second break point or [First point]: **F** Enter
Specify first break point: *Specify a new break point.*
Specify second break point: *Specify second break point on the object.*

If you need to work on arcs or circles, make sure that you work in a counterclockwise direction, or you may end up cutting the wrong part. In this case, the second point should be selected in a counterclockwise direction with respect to the first one (Figure 5-58).

You can use the **2 Points** and **2 Points Select** method to break an object into two without removing a portion in between. This can be achieved by specifying the same point on the object as the first and the second break points. If you specify the first break point on the line and the second break point beyond the end of the line, one complete end starting from the first break point will be removed. The extrusion direction of the selected object need not be parallel to the *Z* axis of the UCS.

Exercise 7 — *General*

Break a line at five different places and then erase the alternate segments. Next, draw a circle and break it into four equal parts.

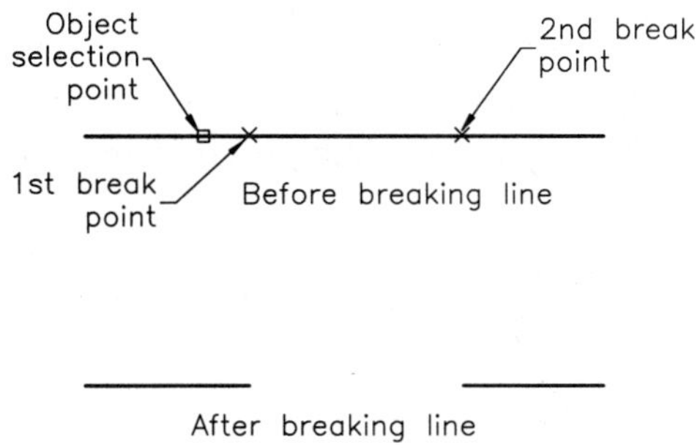

Figure 5-57 Respecifying the first break point for breaking the lines

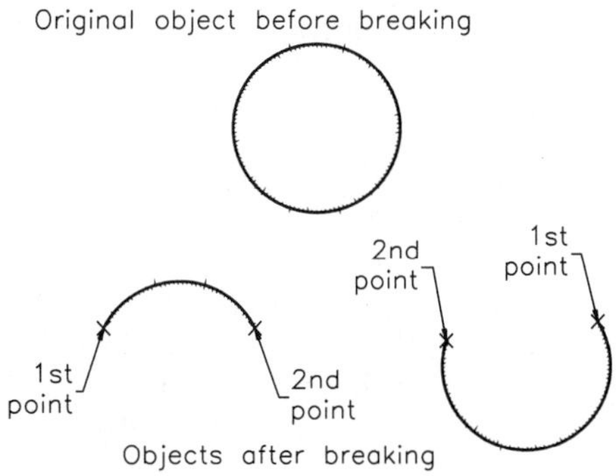

Figure 5-58 Breaking a circle

MEASURING THE SKETCHED OBJECTS

Menu:	Draw > Point > Measure
Command:	MEASURE

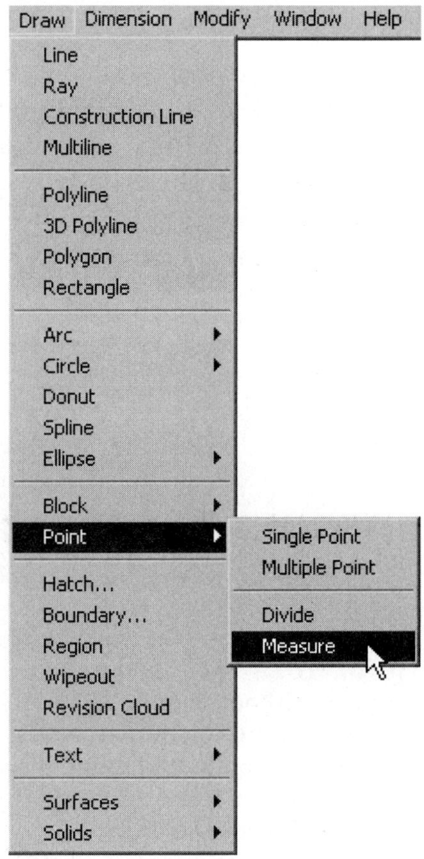

*Figure 5-59 Invoking the **MEASURE** command from the **Draw** menu*

While drawing, you may need to segment an object at fixed distances without actually dividing it. You can use the **MEASURE** command (Figure 5-59) to accomplish this. This command places points or blocks (nodes) on the given object at a specified distance. The shape of these points is determined by the **PDMODE** system variable and the size is determined by the **PDSIZE** variable. The **MEASURE** command starts measuring the object from the endpoint closest to where the object is selected. When a circle is to be measured, an angle from the center is formed that is equal to the Snap rotation angle. This angle becomes the starting point of measurement, and the markers are placed at equal intervals in a counterclockwise direction.

This command goes on placing markers at equal intervals of the specified distance without considering whether the last segment is the same distance or not. Instead of entering a value, you can also select two points that will be taken as the distance. In Figure 5-60 a line and a circle is measured. The Snap rotation angle is zero-degree. The **PDMODE** variable is set to 3 so that X marks are placed as markers. The prompt sequence that will follow when you invoke this option from the **Draw** menu is given next.

Select object to measure: *Select the object to be measured.*
Specify length of segment or [Block]: *Specify the length for measuring the object.*

You can also place blocks as markers (Figure 5-61), but the block must already be defined within the drawing. You can align these blocks with the object to be measured. The prompt sequence is given next.

Select object to measure: *Select the object to be measured.*
Specify length of segment or [Block]: **B** Enter
Enter name of block to insert: *Enter the name of the block.*
Align block with object? [Yes/No] <Y>: *Enter* Y *to align,* N *to not align.*
Specify length of segment: *Enter the measuring distance.*

Note

You will learn more about creating the blocks and inserting them in Chapter 14 (Working with Blocks).

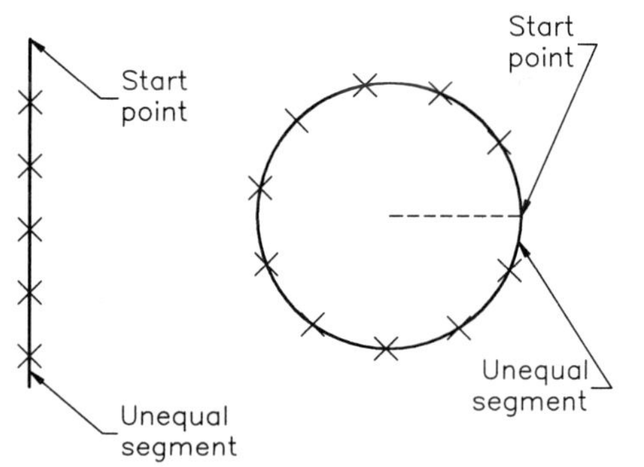

***Figure 5-60** Using the **MEASURE** command*

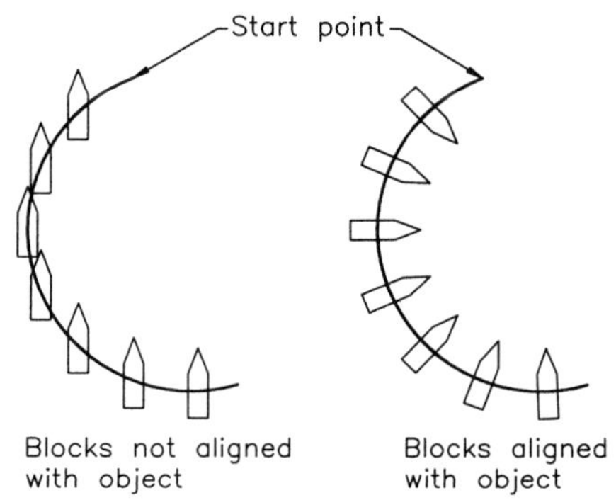

***Figure 5-61** Using blocks with the **MEASURE** command*

DIVIDING THE SKETCHED OBJECTS

Menu:	Draw >Point > Divide
Command:	DIVIDE

The **DIVIDE** command is used to divide an object into a specified number of equal length segments without actually breaking it. This command is similar to the **MEASURE** command except that here you do not have to specify the distance. The **DIVIDE** command calculates the full length of the object and places markers at equal intervals. This makes the last interval equal to the rest of the intervals. If you want a line to be divided, invoke this command and select the object to be divided. After this you enter the number of divisions or segments. The number of divisions entered can range from 2 to 32,767. The prompt sequence that will follow when you invoke this command is given next.

Select object to divide: *Select the object you want to divide.*
Enter number of segments or [Block]: *Specify the number of segments.*

You can also place blocks as markers, but the block must be defined within the drawing. You can align these blocks with the object to be measured. The prompt sequence is given next.

Select object to divide: *Select the object to be measured.*
Enter number of segments or [Block]: **B** Enter
Enter name of block to insert: *Enter the name of the block.*
Align block with object? [Yes/No] <Y>: *Enter Y to align, N to not align.*
Enter number of segments: *Enter the number of segments.*

Figure 5-62 shows a line and a circle divided using this command and Figure 5-63 shows the use of blocks for dividing the selected segment.

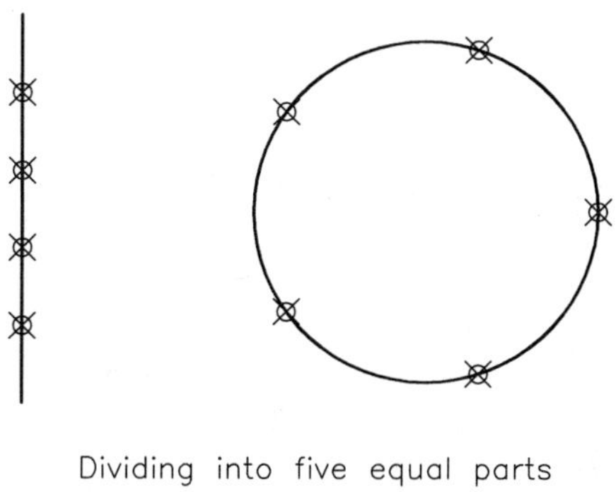

Figure 5-62 *Using the* ***DIVIDE*** *command*

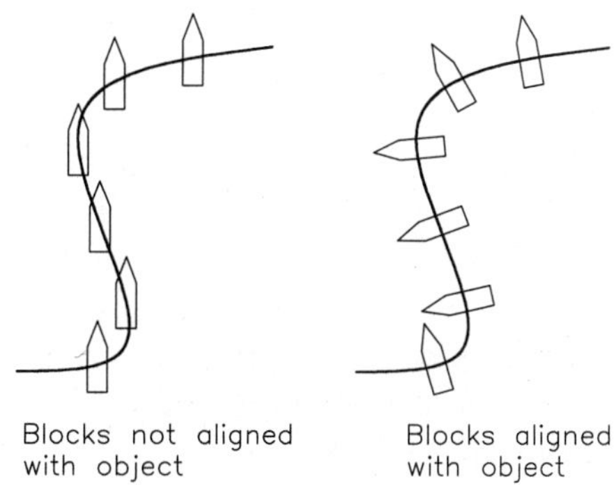

Figure 5-63 *Using blocks with the* ***DIVIDE*** *command*

Note

The size and shape of points placed by the ***DIVIDE*** *and* ***MEASURE*** *commands are controlled by the* ***PDSIZE*** *and* ***PDMODE*** *system variables.*

Self-Evaluation Test

Answer the following questions and then compare your answers to the correct answers given at the end of this chapter.

1. When you shift a group of objects using the **MOVE** command, the size and orientation of these objects are changed. (T/F)

2. The **COPY** command makes copies of the selected object, leaving the original object intact. (T/F)

3. A fillet cannot be created between two parallel and nonintersecting lines. (T/F)

4. With the **BREAK** command, when you select an object, the selection point becomes the first break point. (T/F)

5. Depending on the side to offset, you can create smaller or larger circles, ellipses, and arcs with the __________ command.

6. The __________ command prunes objects that extend beyond a required point of intersection.

7. The offset distance is stored in the __________ system variable.

8. Instead of specifying the scale factor, you can use the __________ option to scale an object with reference of another object.

9. If the __________ system variable is set to a value of 1, the mirrored text is not reversed with respect to the original object. (T/F)

10. There are two types of arrays: __________ and __________.

Review Questions

Answer the following questions.

1. In the case of the **Through** option of the **OFFSET** command, you do not have to specify a distance; you simply have to specify an offset point. (T/F)

2. With the **BREAK** command, you cannot break an object in two without removing a portion in between. (T/F)

3. With the **FILLET** command, the extrusion direction of the selected object must be parallel to the *Z* axis of the UCS. (T/F)

4. A Rectangular array can be rotated. (T/F)

5. Which object selection allows you to create a polygon that selects all the objects that it touches or that lie inside it?

   (a) Last (b) WPolygon
   (c) CPolygon (d) Fence

6. Which command is used to copy an existing object using a base point to another drawing?

   (a) **COPY** (b) **COPYBASE**
   (c) **MOVE** (d) **None**

7. Which command is used to change the size of an existing object with respect to an existing entity?

   (a) **ROTATE** (b) **SCALE**
   (c) **MOVE** (d) **None**

8. Which option of the **LENGTHEN** command modifies the length of the selected entity such that irrespective of the original length, the entity acquires the specified length?

   (a) **DElta** (b) **DYnamic**
   (c) **Percent** (d) **Total**

9. Which selection option is used to select the most recently drawn entity?

   (a) **Last** (b) **Previous**
   (c) **Add** (d) **None**

10. If an entire selected object is within a window or crossing specification, the ________ command works like the **MOVE** command.

11. The ________ option of the **EXTEND** command is used to extend objects to the implied boundary.

12. With the **FILLET** command, using the **Polyline** option, if the selected polyline is not closed, the ________ corner is not filleted.

13. When the chamfer distance is zero, the chamfer created is in the form of a ________.

14. AutoCAD saves the previous selection set and lets you select it again by using the ________ selection option.

15. The ________ selection option is used to select objects by touching the objects to be selected with the selection line.

Exercises

Exercise 8 *General*

Draw the object shown in Figure 5-64. Use the **DIVIDE** command to divide the circle and use the **NODE** object snap to select the points. Assume the dimensions of the drawing.

Figure 5-64 *Drawing for Exercise 8*

Exercise 9 *Mechanical*

Draw the object shown in Figure 5-65 and save the drawing. Assume the missing dimensions.

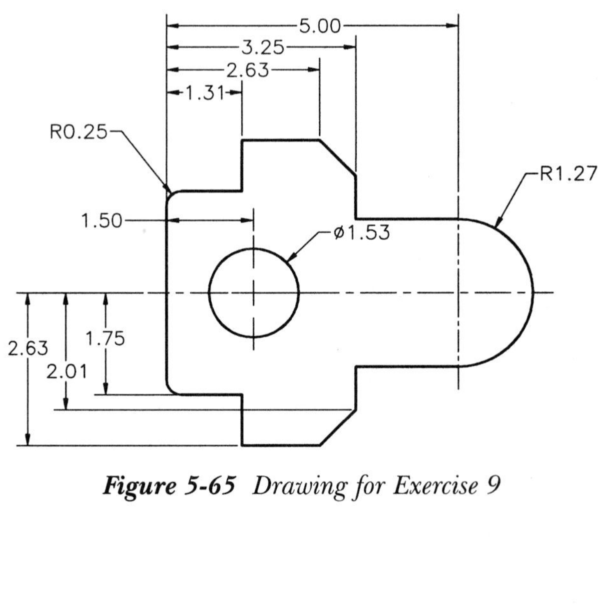

Figure 5-65 *Drawing for Exercise 9*

Exercise 10 Mechanical

Draw the drawing shown in Figure 5-66 and save the drawing.

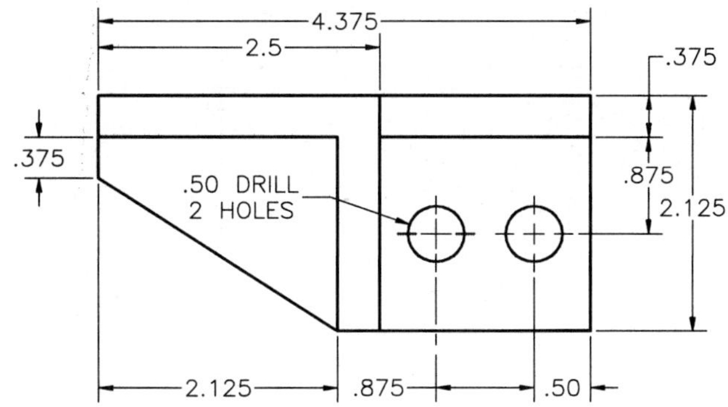

Figure 5-66 *Drawing for Exercise 10*

Exercise 11 Mechanical

Draw the object shown in Figure 5-67 and save the drawing.

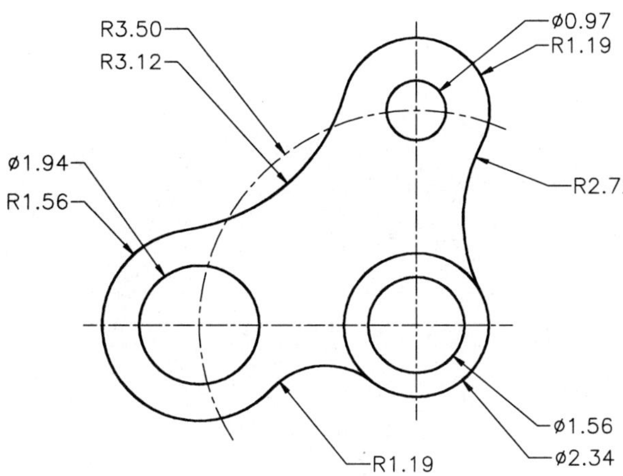

Figure 5-67 *Drawing for Exercise 11*

Exercise 12 *Mechanical*

Draw the object shown in Figure 5-68 and save the drawing. Assume the missing dimensions.

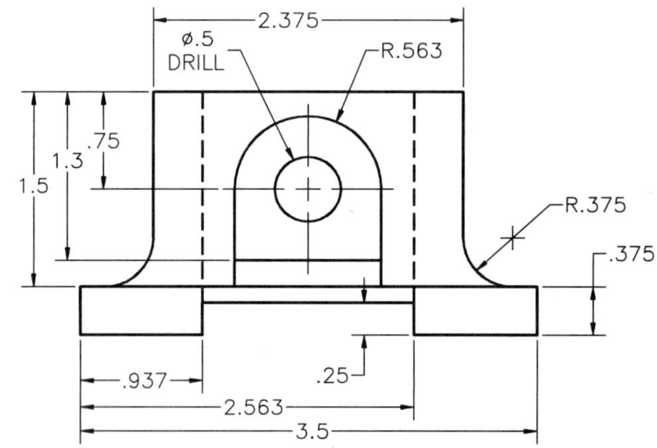

Figure 5-68 *Drawing for Exercise 12*

Problem Solving Exercise 1 *Mechanical*

Draw the object shown in Figure 5-69 and save the drawing. Assume the missing dimensions.

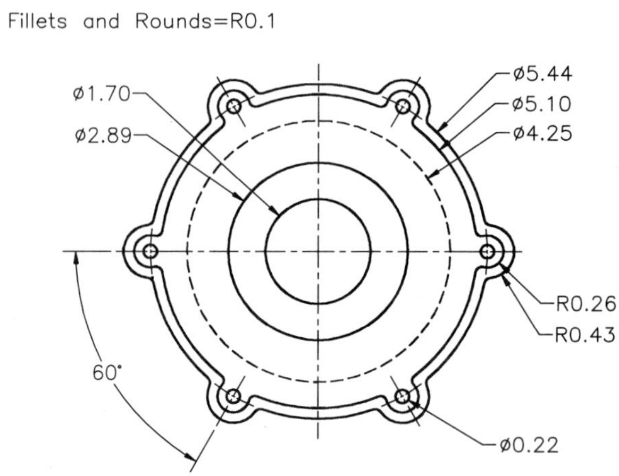

Figure 5-69 *Drawing for Problem Solving Exercise 1*

Problem Solving Exercise 2 *Mechanical*

Draw the object shown in Figure 5-70 and save the drawing. Assume the missing dimensions.

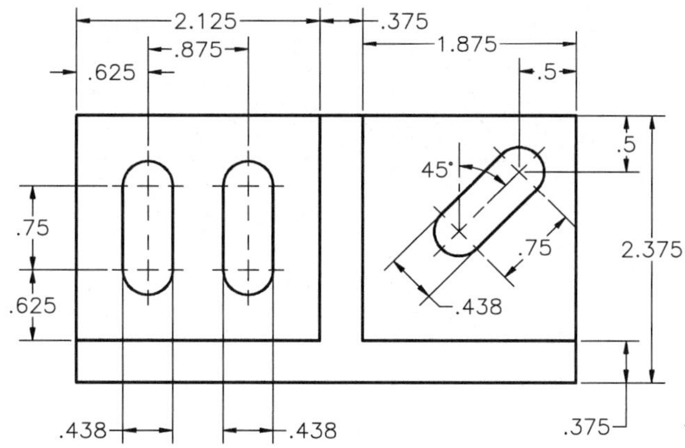

Figure 5-70 *Drawing for Problem Solving Exercise 2*

Problem Solving Exercise 3 *Architectural*

Draw the dining table with chairs shown in Figure 5-71 and save the drawing. Assume the missing dimensions.

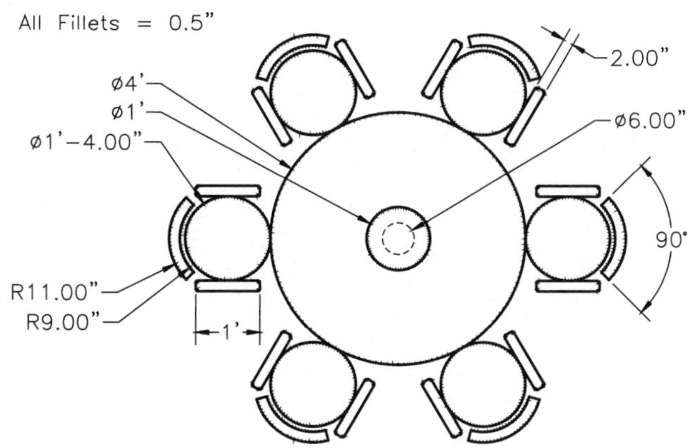

Figure 5-71 *Drawing for Problem Solving Exercise 3*

Problem Solving Exercise 4 *Architectural*

Draw the reception table with chairs shown in Figure 5-72 and save the drawing. The dimensions of the chairs are the same as those in Problem Solving Exercise 3.

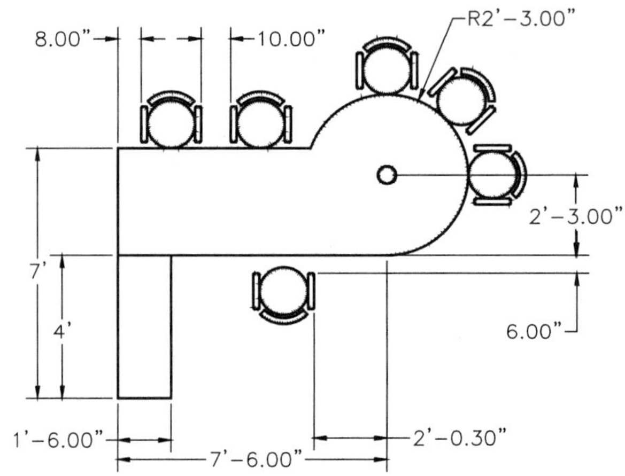

Figure 5-72 *Drawing for Problem Solving Exercise 4*

Problem Solving Exercise 5 *Architectural*

Draw the center table with chairs shown in Figure 5-73 and save the drawing. The dimensions of the chairs are the same as those in Problem Solving Exercise 3.

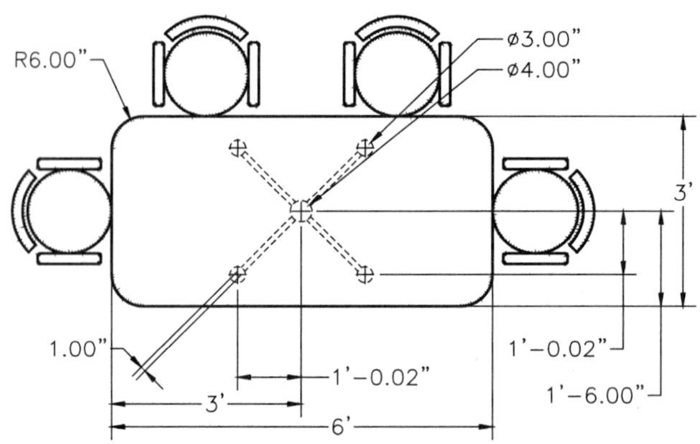

Figure 5-73 *Drawing for Problem Solving Exercise 5*

Problem Solving Exercise 6

Mechanical

Draw the object shown in Figure 5-74 and save the drawing. Refer to the note mentioned in the drawing to create the arc of radius 30.

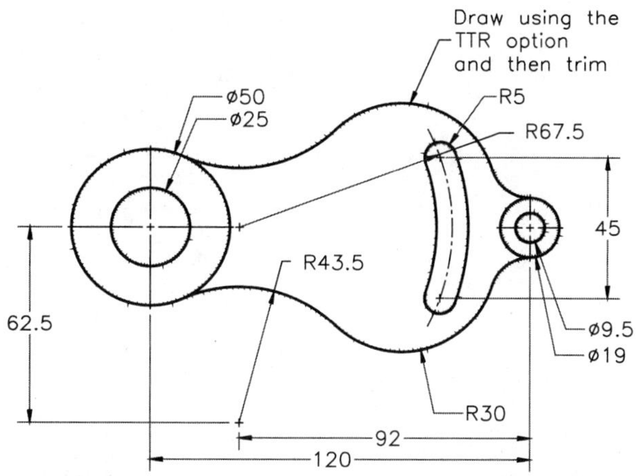

Figure 5-74 *Drawing for Problem Solving Exercise 6*

Problem Solving Exercise 7

Mechanical

Draw the object shown in Figure 5-75 and save the drawing.

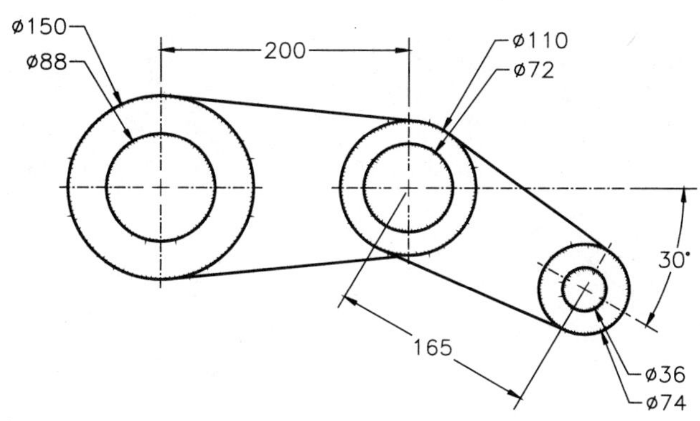

Figure 5-75 *Drawing for Problem Solving Exercise 7*

Problem Solving Exercise 8 — *Mechanical*

Draw the object shown in Figure 5-76 and save the drawing.

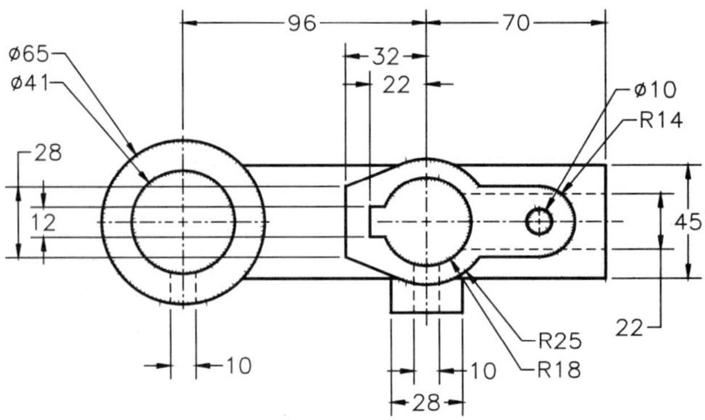

Figure 5-76 *Drawing for Problem Solving Exercise 8*

Answers to Self-Evaluation Test

1 - F, **2** - T, **3** - F, **4** - T, **5** - **OFFSET**, **6** - **TRIM**, **7** - **OFFSETDIST**, **8** - **Reference**, **9** - **MIRRTEXT**, **10** - **Rectangular**, **Polar**

Chapter 6

Editing Sketched Objects-II

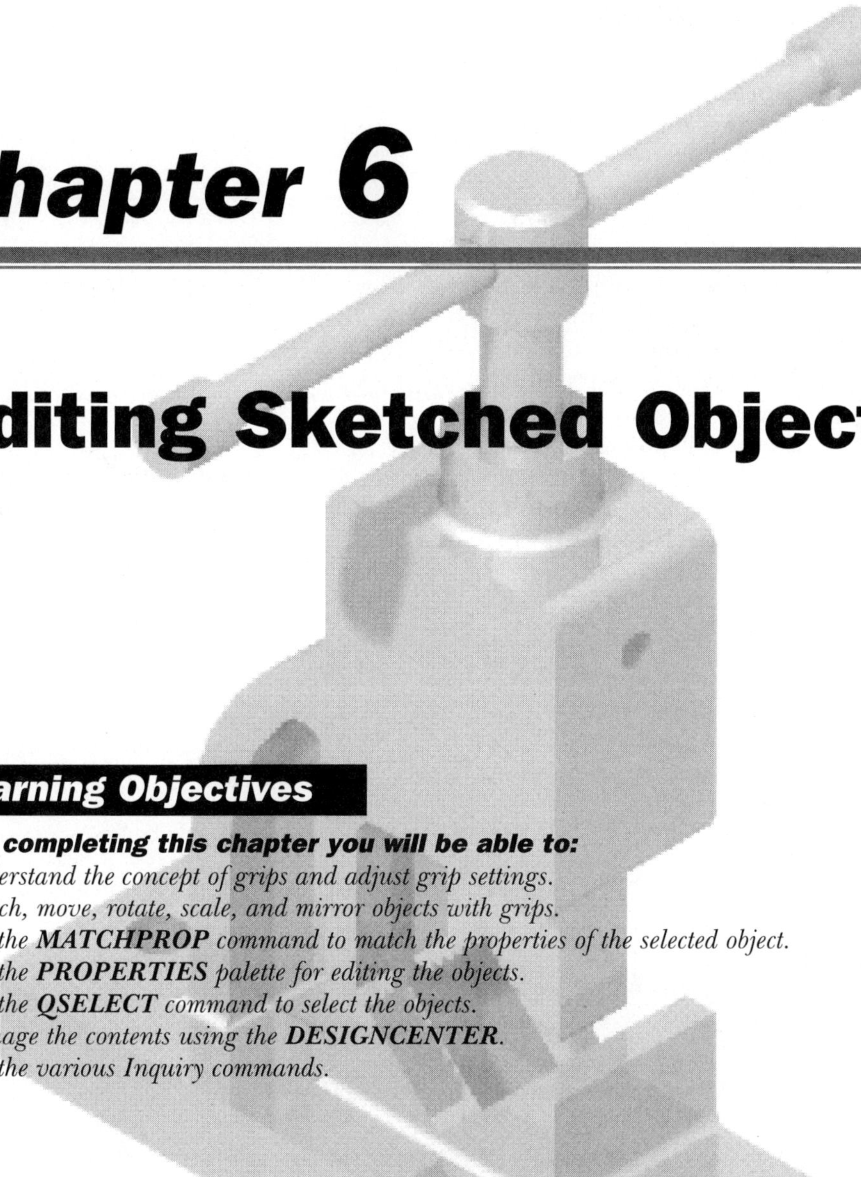

Learning Objectives

After completing this chapter you will be able to:

- *Understand the concept of grips and adjust grip settings.*
- *Stretch, move, rotate, scale, and mirror objects with grips.*
- *Use the **MATCHPROP** command to match the properties of the selected object.*
- *Use the **PROPERTIES** palette for editing the objects.*
- *Use the **QSELECT** command to select the objects.*
- *Manage the contents using the **DESIGNCENTER**.*
- *Use the various Inquiry commands.*

EDITING WITH GRIPS

Grips provide a convenient and quick means of editing objects. With grips you can stretch, move, rotate, scale, and mirror objects, change properties, and load the Web browser. Grips are small squares that are displayed on an object at its definition points when the object is selected. The number of grips depends on the selected object. For example, a line has three grip points, a polyline segment has two, and an arc has three. Similarly, a circle has five grip points and a dimension (vertical) has five. When you select the **Enable grips** and the **Noun/verb selection** check boxes in the **Selection** tab of the **Options** dialog box, a small square (aperture box) at the intersection of the crosshair is displayed (Figure 6-1). The grip location of some of the objects is shown in Figure 6-2.

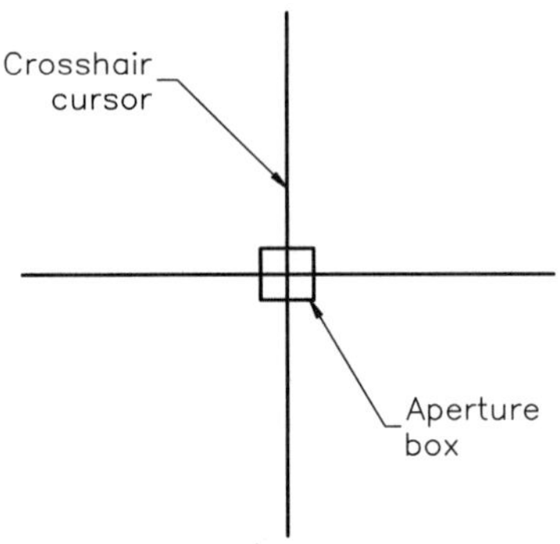

Figure 6-1 *Aperture box at the intersection of crosshair*

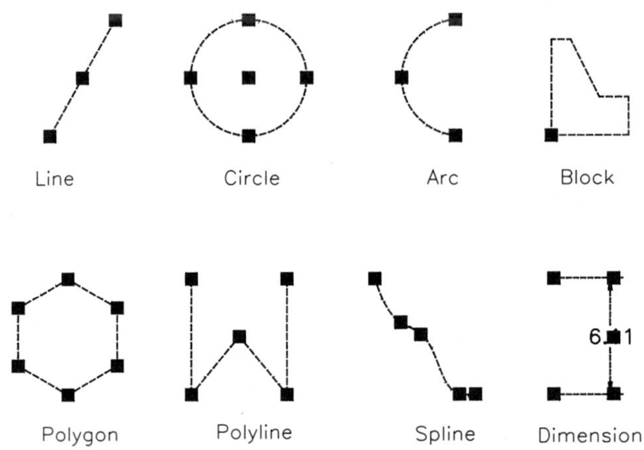

Figure 6-2 *Grip location of various objects*

Note

*AutoCAD also displays a small square (aperture box) at the intersection of crosshairs when the **PICKFIRST** (Noun/Verb Selection) system variable is set to 1 (On).*

TYPES OF GRIPS

Grips can be classified into three types: unselected grips, hover grips, and selected grips. Selected grips are also called hot grips. When you select an object, the grips are displayed at the definition points of the object, and the object is highlighted by displaying it as a dashed line. These grips are called unselected grips (blue). Now, if you move the cursor over the unselected grip, and pause for a second, the grid is displayed in green color. These grips are called hover grips. Next, if you select a grip on this object, the grip becomes a hot grip (filled red square), Figure 6-3. Once the grip is hot, the object can be edited. To cancel the grip, press ESC. If you press ESC once, the hot grip changes to unselected grip. You can also snap to the unselected grip.

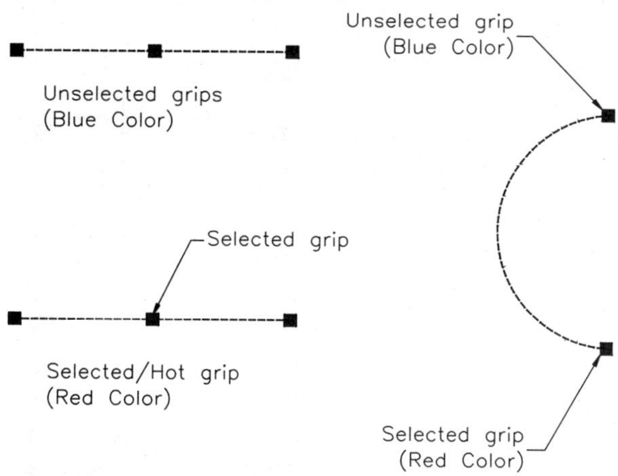

Figure 6-3 *Figure showing the selected and unselected grips*

ADJUSTING GRIP SETTINGS

Menu:	Tools > Options
Command:	OPTIONS

The grip settings can be adjusted using the options under the **Selection** tab of the **Options** dialog box. This dialog box can also be invoked by choosing **Options** from the shortcut menu, see Figure 6-4. The shortcut menu is displayed upon right-clicking in the drawing area.

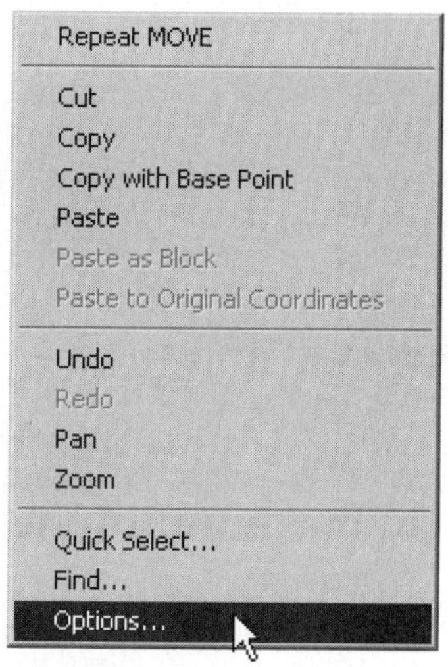

Figure 6-4 *Invoking the **Options** dialog box from the shortcut menu*

The options related to grips that are provided under the **Selection** tab of the **Options** dialog box (Figure 6-5) are discussed next.

Grip Size Area

The **Grip Size** area of the **Selection** tab of the **Options** dialog box consists of a slider bar and a rectangular box that displays the size of the grip. To adjust the size of the grip, move the slider box left or right. The size of the grip can also be adjusted by using the **GRIPSIZE** system variable. The **GRIPSIZE** variable is defined in pixels, and its value can range from 1 to 255 pixels.

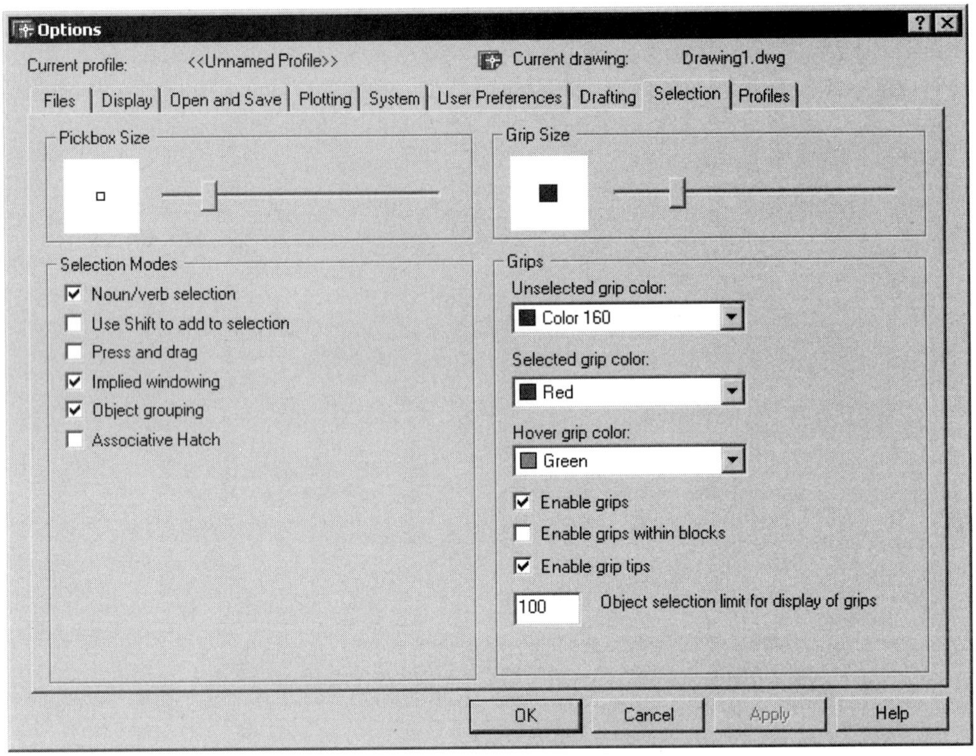

Figure 6-5 ***Selection*** *tab of the* ***Options*** *dialog box*

Grips Area

The Grips area is used to control the display and the color of the grips.

Unselected grip color

This drop-down list is used to set the color of the unselected grip. You can set the color by selecting it from this drop-down list or by selecting the **Select Color** option to display the **Select Color** dialog box. You can select the color for the unselected grip from this dialog box. This color can also be set using the **GRIPCOLOR** system variable.

Selected grip color

This drop-down list is used to set the color of the selected grip. You can set the color by selecting it from this drop-down list or by selecting the **Select Color** option to display the **Select Color** dialog box. You can select the color for the selected grip from this dialog box. This color can also be set using the **GRIPHOT** system variable.

Hover grip color*

This drop-down list is used to set the color of the hover grip. You can set the color by selecting it from this drop-down list or by selecting the **Select Color** option to display the **Select Color** dialog box. You can select the color for the hover grip from this dialog box. This color can also be set using the **GRIPHOVER** system variable.

The Grips area has three check boxes; **Enable grips**, **Enable grips within blocks**, and **Enable grip tips**. The grips can be enabled by selecting the **Enable Grips** check box. They can also be enabled by setting the **GRIPS** system variable to 1. The second check box, **Enable grips within blocks**, enables the grips within a block. If you select this box, AutoCAD will display grips for every object in the block. If you disable the display of grips within a block, the block will have only one grip at its insertion point. You can also enable the grips within a block by setting the value of the **GRIPBLOCK** system variable to 1 (On). If **GRIPBLOCK** is set to 0 (Off), AutoCAD will display only one grip for a block at its insertion point (Figure 6-6). The third check box, **Enable grip tips**, enables you to display the grip tips when the cursor moves over the custom object that supports grip tips. If you disable this check box the grip tips are not displayed when the cursor moves over the custom object. You can also enable the grip tips by setting the value of the **GRIPTIPS** system variable to 0 (Off). If **GRIPTIP** is set to 1 (On), AutoCAD will display the grip tips for the custom object.

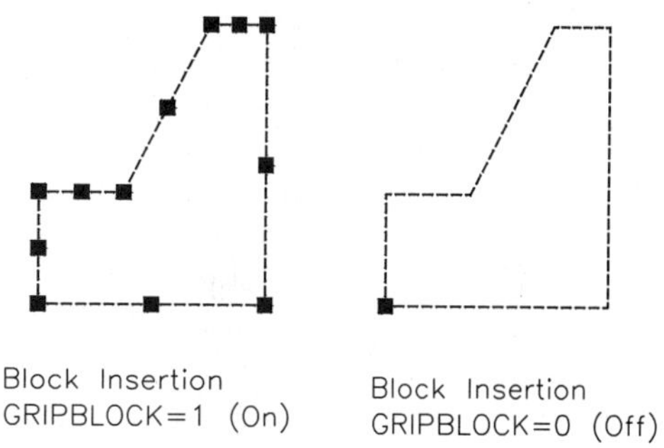

*Figure 6-6 Block insertion with **GRIPBLOCK** set to 1 and to 0*

Note

*If the block has a large number of objects, and if **GRIPBLOCK** is set to 1 (On), AutoCAD will display grips for every object in the block. Therefore, it is recommended that you set the system variable **GRIPBLOCK** to 0 or clear the **Enable grips within blocks** check box in the **Selection** tab of the **Options** dialog box.*

Object selection limit for display of grips*

This text box is used to specify the maximum number of objects that can be selected at a single attempt for the display of grips. If you select objects more than that specified in the text box using a single selection method, grips will not be displayed. Note that this limit is set only for those objects that are selected at a single attempt using any of the **Crossing window**, **Fence**, or the **All** options.

EDITING OBJECTS WITH GRIPS

As mentioned earlier, you can perform different kinds of editing operations using the selected grip. The editing operations are discussed next.

Stretching Objects With Grips (Stretch Mode)

If you select an object, AutoCAD displays unselected grips at the definition points of the object. When you select a grip for editing, you are automatically in the **Stretch** mode. The Stretch mode has a function similar to the **STRETCH** command. When you select a grip, it acts as a base point and is called a base grip. You can also select several grips by holding the SHIFT key down and then selecting the grips. Now, release the SHIFT key and select one of the hot grips to stretch them simultaneously. The geometry between the selected base grips is not altered. You can also make copies of the selected objects or define a new base point. When selecting grips on text objects, blocks, midpoints of lines, centers of circles and ellipses, and point objects in the stretch mode, the selected objects are moved to a new location. The following example illustrates the use of the **Stretch** mode.

1. Use the **PLINE** command to draw a W-shaped figure as shown in Figure 6-7(a).

2. Select the object that you want to stretch [Figure 6-7(a)]. When you select the object, grips will be displayed at the endpoints of each object. A polyline has two grip points. If you use the **LINE** command to draw the object, AutoCAD will display three grips for each object.

3. Hold the SHIFT key down, and select the grips that you want to stretch [Figure 6-7(b)]. The selected grips will become hot grips, and the color of the grip will change from blue to red.

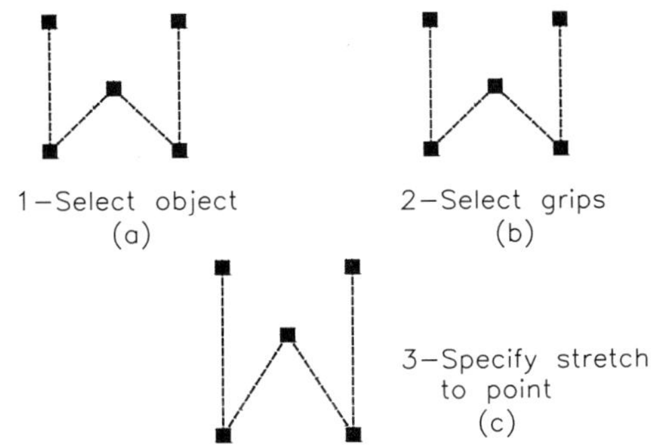

***Figure 6-7** Using the **Stretch** mode to stretch the lines*

Note

You have to make sure that you hold the SHIFT key before selecting even the first grip. You can not hold the SHIFT key and select more grips if the first grip is selected without holding the SHIFT key.

4. Choose one of the selected (hot grip) grips, and specify a point to which you want to stretch the line [Figure 6-7(c)]. When you select a grip, the following prompt is displayed in the Command prompt area.

   STRETCH
   Specify stretch point or [Base point/Copy/Undo/eXit]:

   The Stretch mode has several options: **Base point**, **Copy**, **Undo**, and **eXit**. You can use the Base point option to define the base point and the Copy option to make copies.

5. Select the grip where the two lines intersect. Right-click to display the shortcut menu (Figure 6-8) and choose the **Copy** option. Select the points as shown in Figure 6-9(b). Each time you select a point, AutoCAD will make a copy.

If you press the SHIFT key when specifying the point to which the object is to be stretched, without selecting the copy option, then also AutoCAD allows you to make multiple copies of the selected object. Also, if you press the SHIFT key again when specifying the next point, the cursor snaps to a point whose location is based on the distance between the first two points, that is, the distance between the selected object and the location of the copy of the selected object.

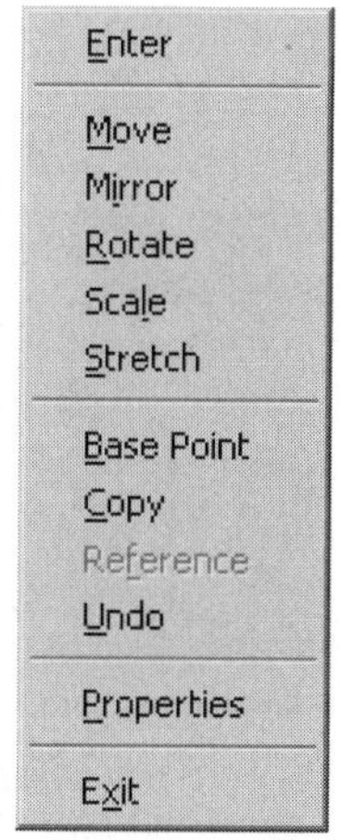

Figure 6-8 *Selecting different Grip options from the shortcut menu*

6. Make a copy of the drawing as shown in Figure 6-9(c). Select the object, and then select the grip where the two lines intersect. When AutoCAD displays the ****STRETCH**** prompt, select the **Base Point** option from the cursor menu or enter B at the Command prompt. Select the bottom left grip as the base point, and then give the displacement point as shown in Figure 6-9(d).

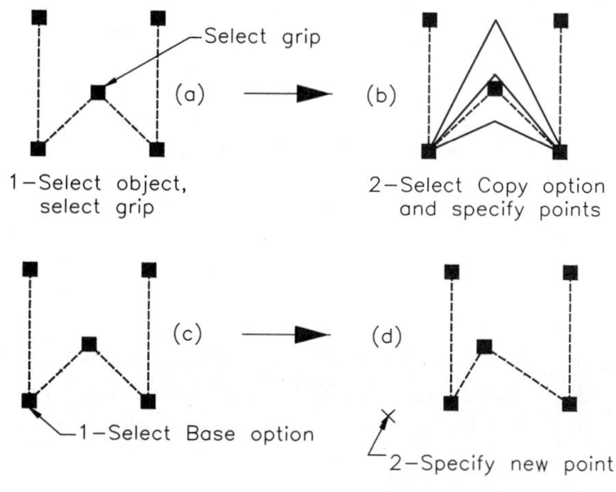

Figure 6-9 *Using the Stretch mode's* ***Copy*** *and* ***Base point*** *options*

7. To terminate the grip editing mode, right-click when the grip is hot to display the shortcut menu and then select **eXit**. You can also enter X at the Command prompt or press ESC to exit.

Note

*You can select an option (**Copy** or **Base Point**) from the shortcut menu that can be invoked by right-clicking your pointing device after selecting a grip. The different modes can also be selected from the shortcut menu. You can also cycle through all the different modes by selecting a grip and pressing the ENTER key or the SPACEBAR.*

Moving Objects with Grips (Move Mode)

The Move mode lets you move the selected objects to a new location. When you move objects, the size of the objects and their angle do not change. You can also use this mode to make copies of the selected objects or to redefine the base point. The following example illustrates the use of the **Move** mode.

1. Use the **LINE** command to draw the shape as shown in Figure 6-10(a). When you select the objects, grips will be displayed at the definition points and the object will be highlighted.

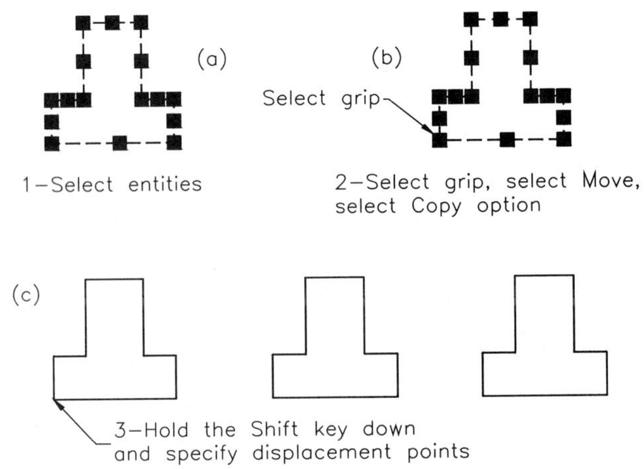

Figure 6-10 *Using the **Move** mode to move and make copies of the selected objects*

2. Select the grip located at the lower left corner, and then select **Move** from the shortcut menu. You can also invoke the Move mode by entering **MOVE** or **MO** at the keyboard or giving a null response by pressing the SPACEBAR or ENTER key. AutoCAD will display the following prompt in the Command: prompt area.

   MOVE
   Specify move point or [Base point/Copy/Undo/eXit]:

3. Hold down the SHIFT key, and then enter the first displacement point. The distance between the first and the second object defines the snap offset for subsequent copies. While holding down the SHIFT key, move the screen crosshair to the next snap point and select the point. AutoCAD will make a copy of the object at this location. If you release the SHIFT key, you can specify any point where you want to place a copy of the object. You can also enter coordinates to specify the displacement.

Rotating Objects with Grips (Rotate Mode)

The **Rotate** mode allows you to rotate objects around the base point without changing their size. The options of Rotate mode can be used to redefine the base point, specify a reference angle, or make multiple copies that are rotated about the specified base point. You can access the Rotate mode by selecting the grip and then selecting **Rotate** from the shortcut menu, or by entering **ROTATE** or **RO** at the keyboard or giving a null response twice by pressing the SPACEBAR or the ENTER key. The following example illustrates the use of **Rotate** mode.

1. Use the **LINE** command to draw the shape as shown in Figure 6-11(a). When you select the objects, grips will be displayed at the definition points and the shape will be highlighted.

2. Select the grip located at the lower left corner and then invoke the Rotate mode. AutoCAD will display the following prompt:

   ROTATE
   Specify rotation angle or [Base point/Copy/Undo/Reference/eXit]:

3. At this prompt, enter the rotation angle. AutoCAD will rotate the selected objects by the specified angle [Figure 6-11(b)].

4. Make a copy of the original drawing as shown in Figure 6-11(c). Select the objects, and then select the grip located at the lower left corner of the object. Invoke the Rotate mode and then select the **Copy** option from the shortcut menu or enter **C** (Copy) at the Command prompt. Enter the rotation angle. AutoCAD will rotate the copy of the object through the specified angle [Figure 6-11(d)].

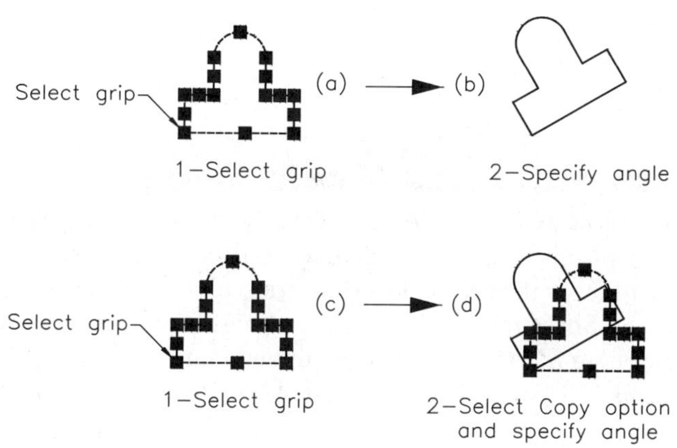

***Figure 6-11** Using the **ROTATE** mode to rotate and make copies of the selected objects*

5. Make another copy of the object as shown in Figure 6-12(a). Select the object, and then select the grip at point (P0). Access the Rotate mode and copy option as described earlier. Select the **Reference** option from the shortcut menu or enter **R** at the following prompt.

**ROTATE (multiple) **
Specify rotation angle or [Base point/Copy/Undo/Reference/eXit]: **R**
Specify reference angle <0>: *Select the grip at (P1).*
Specify second point: *Select the grip at (P2).*
Specify new angle or [Base point/Copy/Undo/Reference/eXit]: **45**

In response to the **Specify reference angle <0>:** prompt, select the grips at points (P1) and (P2) to define the reference angle. When you enter the new angle, AutoCAD will rotate and insert a copy at the specified angle [Figure 6-12(c)]. For example, if the new angle is 45-degree, the selected objects will be rotated about the base point (P0) so that the line P1P2 makes a 45-degree angle with respect to the positive *X* axis.

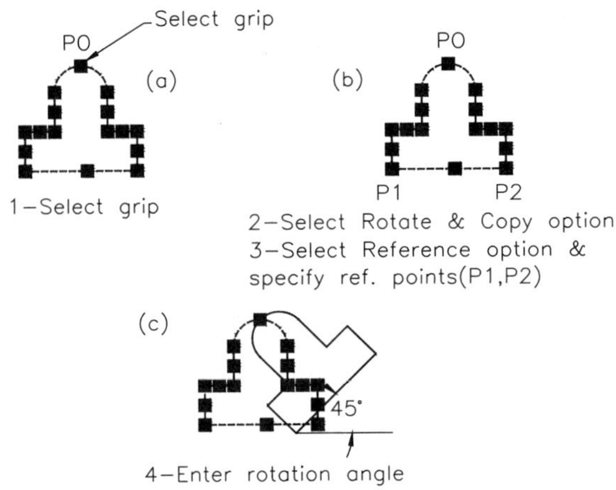

Figure 6-12 *Using the **ROTATE** mode to rotate by giving a reference angle*

Scaling Objects with Grips (Scale Mode)

The **Scale** mode allows you to scale objects with respect to the base point without changing their orientation. The options of **Scale** mode can be used to redefine the base point, specify a reference length, or make multiple copies that are scaled with respect to the specified base point. You can access the **Scale** mode by selecting the grip and then selecting Scale from the shortcut menu, or entering **SCALE** or **SC** on the keyboard, or giving a null response three times by pressing the SPACEBAR or the ENTER key. The following example illustrates the use of the **Scale** mode.

1. Use the **PLINE** command to draw the shape as shown in Figure 6-13(a). When you select the objects, grips will be displayed at the definition points, and the object will be highlighted.

2. Select the grip located at the lower left corner as the base grip, and then invoke the **SCALE** mode. AutoCAD will display the following prompt in the Command prompt area.

SCALE
Specify scale factor or [Base point/Copy/Undo/Reference/eXit]:

3. At this prompt enter the scale factor or move the cursor and select a point to specify a new size. AutoCAD will scale the selected objects by the specified scale factor [Figure 6-13(b)]. If the scale factor is less than 1 (<1), the objects will be scaled down by the specified factor. If the scale factor is greater than 1 (>1), the objects will be scaled up.

4. Make a copy of the original drawing as shown in Figure 6-13(c). Select the objects, and then select the grip located at the lower left corner of the object. Invoke the Scale mode. At the following prompt, enter C (Copy), and then enter B for base point.

**SCALE (multiple) **
Specify scale factor or [Base point/Copy/Undo/Reference/eXit]: **B**

5. At the **Specify base point prompt:**, select the point (P0) as the new base point, and then enter **R** at the following prompt.

**SCALE (multiple) **
Specify scale factor or [Base point/Copy/Undo/Reference/eXit]: **R**
Specify reference length <1.000>: *Select grips at (P1) and (P2).*

After specifying the reference length at the **Specify new length or [Base point/Copy/Reference/eXit]:** prompt, enter the actual length of the line. AutoCAD will scale the objects so that the length of the bottom edge is equal to the specified value [Figure 6-13(c)].

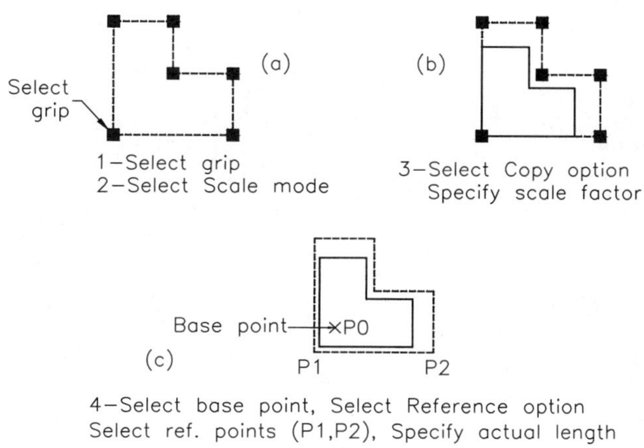

Figure 6-13 *Using the* ***SCALE*** *mode to scale and make copies of selected objects*

Mirroring Objects with Grips (Mirror Mode)

The **Mirror** mode allows you to mirror the objects across the mirror axis without changing the size of the objects. The mirror axis is defined by specifying two points. The first point is the base point, and the second point is the point that you select when AutoCAD prompts for the second point. The options of the Mirror mode can be used to redefine the base point and make a mirror copy of the objects. You can access the **Mirror** mode by selecting a grip and then choosing Mirror from the shortcut menu, or by entering **MIRROR** or **MI** at the keyboard, or giving a null response four times by pressing the SPACEBAR or the ENTER key. The following is the example for the **Mirror** mode.

1. Use the **PLINE** command to draw the shape as shown in Figure 6-14(a). When you select the object, grips will be displayed at the definition points and the object will be highlighted.

2. Select the grip located at the lower right corner (P1), and then invoke the **Mirror** mode. The following prompt is displayed.

   MIRROR
   Specify second point or [Base point/Copy/Undo/eXit]:

3. At this prompt, enter the second point (P2). AutoCAD will mirror the selected objects with line P1P2 as the mirror axis as shown in Figure 6-14(b).

4. Make a copy of the original figure as shown in Figure 6-14(c). Select the object, and then select the grip located at the lower right corner (P1) of the object. Invoke the **Mirror** mode and then select the **Copy** option to make a mirror image while retaining the original object. Alternatively, you can also hold down the SHIFT key and make several mirror copies by specifying the second point.

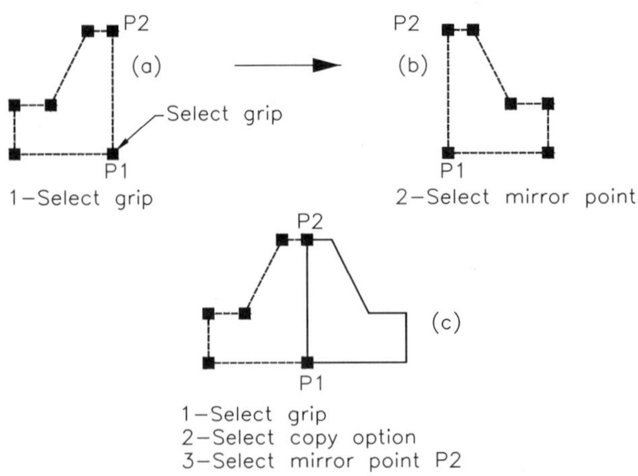

***Figure 6-14** Using the **MIRROR** mode to create a mirror image of selected objects*

5. Select point (P2) in response to the prompt **Specify second point or [Base point/Copy/Undo/eXit]**. AutoCAD will create a mirror image, and the original object will be retained.

Note

*You can use some editing commands such as **ERASE, MOVE, ROTATE, SCALE, MIRROR,** and **COPY** on an object with unselected grips. However, this is possible only if the **PICKFIRST** system variable is set to 1 (On).*

You cannot select an object when the grip is hot.

If you want to remove an object from the selection set displaying grips, press the SHIFT key and then select the particular object. This object, which is removed from the selection set, will now not be highlighted.

LOADING HYPERLINKS

If you have already added hyperlink to the object, you can also use the grips to open a file associated with the hyperlink. For example, the hyperlink could start a word processor, or activate the Web browser and load a Web page that is embedded in the selected object. If you want to launch the Web browser that provides hyperlinks to other Web pages, select the URL-embedded object and then right-click to display the shortcut menu. In the shortcut menu, select the **Hyperlink** option and AutoCAD will automatically load the Web browser. When you move the cursor over or near the object that contains a hyperlink, AutoCAD displays the hyperlink information with the cursor.

EDITING GRIPPED OBJECTS

You can also edit the properties of the gripped objects by using the **Properties** toolbar (Figure 6-15). The gripped objects are created when you select objects without invoking a command. The gripped objects are highlighted and will display grips (rectangular boxes) at their grip points. For example, if you want to change the color of the gripped objects, select the **Color** drop-down list in the **Properties** toolbar and then select a color. The color of the gripped objects will change to the selected color. Similarly, if you want to change the layer, lineweight, or linetype of the gripped objects, select the linetype, lineweight, or layer from the corresponding drop-down lists. If the gripped objects have different colors, linetypes, or lineweights, the Color Control, Linetype Control, and Lineweight Control boxes will appear blank. You can also change the Plot Style of the selected objects.

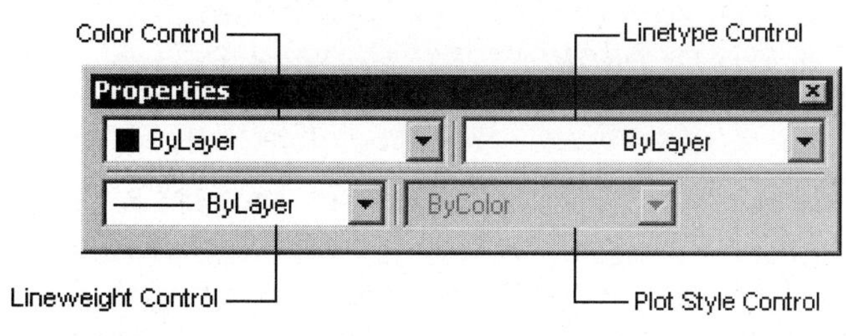

Figure 6-15 *Using the **Properties** toolbar to change properties of the gripped objects*

GRIP SYSTEM VARIABLES

System variable	Default	Setting	Function
GRIPS	1	1=On, 0=Off	Enables or disables Grip mode
GRIPBLOCK	0	1=On, 0=Off	Controls the display of grips in a block
GRIPCOLOR	160	1-255	Specifies the color of unselected grips
GRIPHOT	1	1-255	Specifies the color of selected grips
GRIPSIZE	5	1-255	Specifies the size of the grip box in pixels

CHANGING THE PROPERTIES USING THE PROPERTIES PALETTE

Toolbar:	Standard > Properties
Menu:	Modify > Properties
Command:	PROPERTIES

As mentioned earlier, each object has a number of properties associated to it such as the color, layer, linetype, line weight, and so on. You can modify the properties of an object by using the **PROPERTIES** command. When you invoke this command, AutoCAD will display the **PROPERTIES** palette, see Figure 6-16. The **PROPERTIES** palette can also be displayed when you double-click on the object to be edited. The contents of the **PROPERTIES** palette change according to the objects selected. For example, if you select text entity, the properties related to the text such as its height, justification, style, rotation angle, obliquing factor, and so on will be displayed.

The **PROPERTIES** palette can also be invoked from the shortcut menu displayed when you right-click in the drawing area. Choose the **Properties** option to display the **PROPERTIES** palette. If you select more than one object, the common properties of the selected objects will be displayed in the **PROPERTIES** palette. To change properties of the selected objects, you can click in the cell next to the name of the property and change the values manually or you can choose from the available options in the drop-down list, if one is available. You can cycle through the options by double-clicking in the property cell.

Note

Some of the options of the ***PROPERTIES*** *palette have been explained in Chapter 4. Other options of the* ***PROPERTIES*** *palette will be explained in detail in Chapter 18.*

CHANGING PROPERTIES USING GRIPS

You can also use the grips to change the properties of a single or multiple object. To change the properties of an object, select the object to display the grips and then right-click to display the shortcut menu. In the shortcut menu, choose the **Properties** option to display

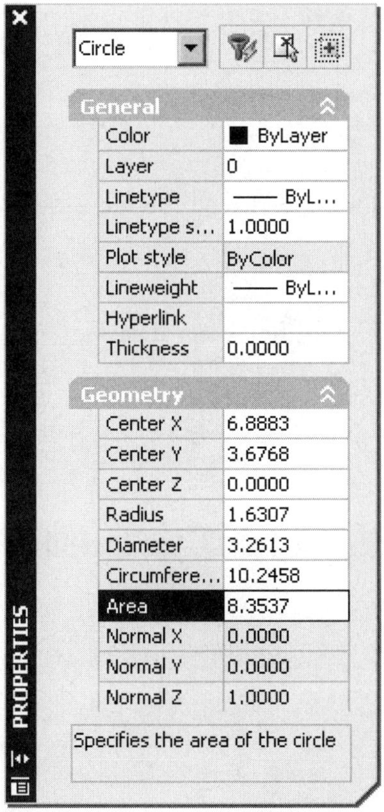

***Figure 6-16 PROPERTIES** palette for editing the properties of the objects*

the **PROPERTIES** palette. If you select a circle, AutoCAD will display **Circle** in the **No selection** drop-down list available on the upper left corner of the **PROPERTIES** palette. Similarly, if you select text, **Text** is displayed in the drop-down list. If you select several objects, AutoCAD will display all the objects in the selection drop-down list of the **PROPERTIES** palette. You can use this palette to change the properties (color, layer, linetype, linetypes scale, lineweight, thickness, and so on) of the gripped objects.

MATCHING PROPERTIES OF THE SKETCHED OBJECTS

Toolbar:	Standard > Match Properties
Menu:	Modify > Match Properties
Command:	MATCHPROP

The **MATCHPROP** command can be used to change some properties like color, layer, linetype, and linetype scale of the selected objects. However, in this case you need a source object whose properties will be forced on the destination objects. When you invoke this command, AutoCAD will prompt you to select the source object and then the destination objects. The properties of the destination objects will be changed to that of the source object. This command is a transparent command and can be used inside another command. The prompt sequence that will follow when you choose the **Match Properties** button from the **Standard** toolbar is given next.

Select Source Object: *Select the source object.*
Current active settings: Color Layer Ltype Ltscale Lineweight Thickness PlotStyle Text Dim Hatch Polyline Viewport
Select destination object(s) or [Settings]:

If you select the destination object in the **Select destination object(s) or [Settings]** prompt, the properties of the source object will be forced on it. If you select the **Settings** option, AutoCAD displays the **Property Settings** dialog box (Figure 6-17). The properties displayed in this dialog box are those of the source object. You can use this dialog box to edit the properties that are copied from source to destination objects.

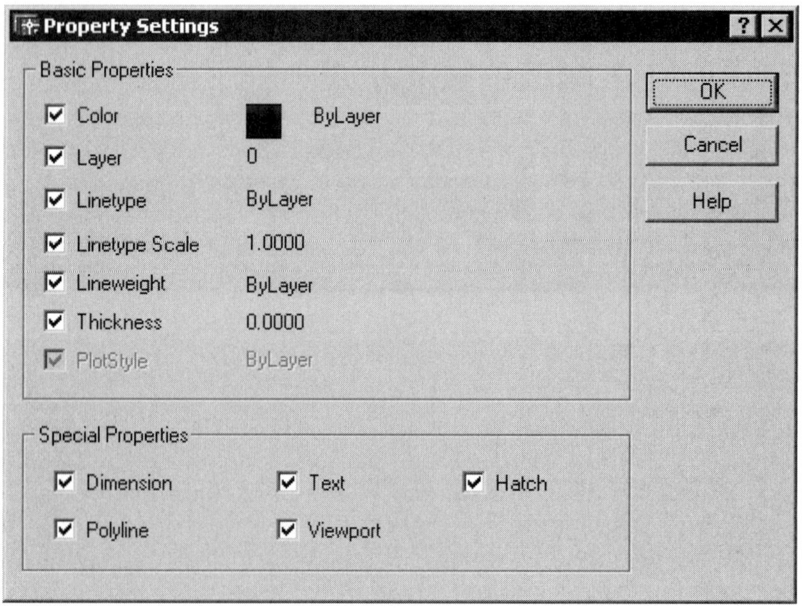

***Figure 6-17 Property Settings** dialog box*

The table shown next lists the properties associated with different types of objects.

Object	Color & layer	Linetype & Linetype Scale	Line-weight	Thick-ness	Text	Dime-nsion	Hatch	Plot style
3dface	x	x	x					x
Arc	x	x	x	x				x
AttDef	x			x	x			x
Body	x	x						x
Circle	x	x	x	x				x
Dimension	x	x				x		x
Ellipse	x	x	x					x
Hatch	x		x				x	x
Image	x	x						x
Insert	x	x	x					x
Leader	x	x	x			x		x
Line	x	x	x	x				x
Mtext	x				x			x
OLE object								
Point	x		x	x				x
2D polyline	x	x	x		x		x	
3D polyline	x	x	x				x	
3D mesh	x	x	x				x	
Pface mesh	x	x	x				x	
Ray	x	x	x				x	
Region	x	x	x	x			x	
2D solid	x	x	x				x	
ACIS solid	x	x	x				x	
Spline	x	x	x				x	
Text	x	x		x	x			x
Tolerance	x	x	x			x		x
Trace	x	x	x	x			x	
Viewport	x		x					x
Xline	x	x	x				x	
Xref	x	x	x				x	
Zombie	x	x	x				x	

QUICK SELECTION OF THE SKETCHED OBJECTS

Menu:	Tools > Quick Select
Command:	QSELECT

The **QSELECT** command creates a new selection set that will either include or exclude all objects that match the specified object type and property criteria. The **QSELECT** command can be applied to the entire drawing or existing selection set. If a drawing is partially opened, **QSELECT** does not consider the objects that are not loaded. The **QSELECT** command can be invoked by choosing the **Quick Select** button in the **Properties** palette. In the shortcut menu, the **QSELECT** command can be invoked by choosing **Quick Select**. When you invoke this command, the **Quick Select** dialog box will be displayed, see Figure 6-18. The **Quick Select** dialog box specifies the object filtering criteria and creates a selection set from that criteria.

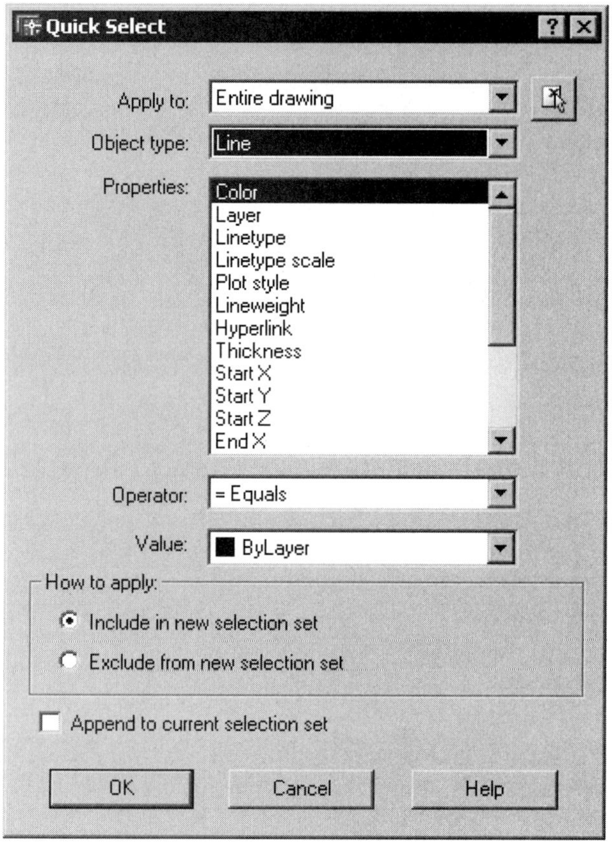

***Figure 6-18 Quick Select** dialog box*

Apply to

The **Apply to** drop-down list specifies whether to apply the filtering criteria to the entire drawing or to the current selection set. If there is an existing selection set, the **Current selection** is the default value. Otherwise the entire drawing is the default value. You can select the objects to create a selection set by choosing the **Select objects** button provided on the right side of this drop-down list. The **Quick Select** dialog box is temporarily closed when you choose this button and you will be prompted to select the objects. The dialog box will be redisplayed once a selection set is made.

Object type

This drop-down list specifies the type of object to be filtered. It lists all the available object types and if some objects are selected, it lists all the selected object types. **Multiple** is the default setting.

Properties

This list box displays the properties to be filtered. All the properties related to the object type will be displayed in this list box. The property selected from this list box will define the options that will be available in the **Operator** and **Value** drop-down list.

Operator

This drop-down list specifies the range of the filter for the chosen property. The filters that are available are given next.

- Equals =
- Not Equal < >
- Greater than >
- Less than <
- Select All
- Wildcard Match (For Hyperlink property)

Note
*The **Value** drop-down list will not be available when you select **Select All** from the **Operator** drop-down list.*

Value

This drop-down list specifies the property value of the filter. If the values are known, it becomes a list of the available values from which you can select a value. Otherwise you can enter a value.

How to apply Area

The options under this area are used to specify whether the filtered entities will be included or excluded from the new selection set. This area provides the following two radio buttons.

Include in new selection set

If this radio button is selected, the filtered entities will be included in the new selection set. If selected, this radio button creates a new selection set composed only of those objects that conform to the filtering criteria.

Exclude from new selection set

If this radio button is selected, the filtered entities will be excluded from the new selection set. This radio button creates a new selection set of objects that do not conform to the filtering criteria.

Append to current selection set

This creates a cumulative selection set by using multiple uses of Quick Select. It specifies whether the objects selected using the **QSELECT** command replace the current selection set or append the current selection set.

Tip

*Quick Select supports custom objects (objects that are created by some other applications) and their properties. If custom objects have other properties than AutoCAD, then the source application of the object should be running for the properties to be available by the **QSELECT**.*

MANAGING CONTENTS USING THE DESIGNCENTER*

Toolbar:	Standard > DesignCenter
Menu:	Tools > DesignCenter
Command:	ADCENTER

The **DESIGNCENTER** window is used to locate and organize drawing data, and to insert blocks, layers, external references, and other customized drawing content. These contents can be selected from either your own files, local drives, a network, or the Internet. You can even access and use the contents between files or from the Internet. You can use the **DESIGNCENTER** to conveniently drag and drop any information that has been previously created into a current drawing. This is a powerful tool that reduces repetitive tasks of again creating information that already exists. To invoke the **DESIGNCENTER** window, choose the **DesignCenter** button from the **Standard** toolbar. The **DESIGNCENTER** window is displayed, see Figure 6-19.

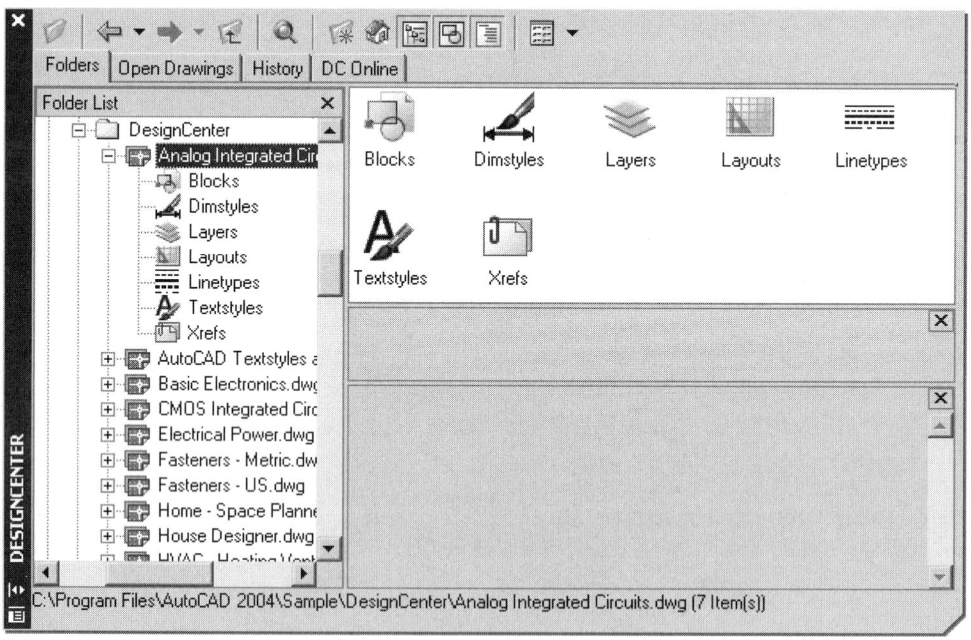

*Figure 6-19 **DESIGNCENTER** window*

This window can be moved to any location on the screen by picking and dragging it with the grab bar located on the left of the window. You can also resize it by clicking the borders and dragging them to the right or left. Right-clicking on the title bar of the window displays a shortcut menu that gives options to move, resize, close, dock, and hide the **DESIGNCENTER** window. The **Auto-Hide** button available on the grab bar acts as a toggle for hiding and displaying the **DESIGNCENTER**. Also, double-clicking on the title bar of the window docks the **DESIGNCENTER** window if it is undocked and vice versa. To use this option make sure that the **Allow Docking** option is selected from the shortcut menu that is displayed by right-clicking on the grab bar.

Note
The DESIGNCENTER can be turned on and off by pressing the CTRL+2 keys.

Figure 6-20 shows the **DESIGNCENTER** toolbar buttons. When you choose the **Tree View Toggle** button on the **DESIGNCENTER** toolbar, it displays the **Tree View** (Left Pane) with a tree view of the contents of the drives. If the tree view is not displayed, you can also right-click in the window and choose **Tree** from the shortcut menu that is displayed. Now, the window is divided into two parts, the **Tree View** (left pane) and the **Palette** (right pane). The Palette displays folders, files, objects in a drawing, images, Web-based content, and custom content. You can also resize both the **Tree View** and the **Palette** by clicking and dragging the bar between them to the right or the left.

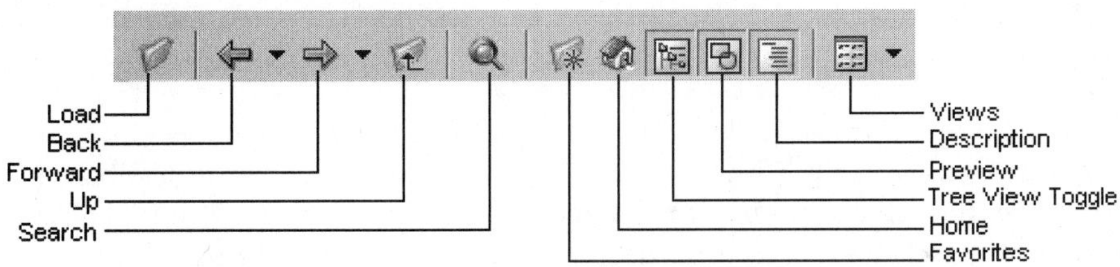

***Figure 6-20 DESIGNCENTER** toolbar buttons*

The DESIGNCENTER has four tabs provided below the DESIGNCENTER toolbar buttons. They are **Folders**, **Open drawings**, **History**, and **DC Online**. The description of these tabs is given next.

Folders Tab*

The **Folders** tab lists all the folders and files available in the local and network drives. When this tab is selected, the **Tree View** displays the tree view of the contents of the drives and the **Palette** displays the various folders, and files in a drawing, images, and the Web-based content available in the selected drive.

In the **Tree View** you can browse the contents of any folder by clicking on the plus sign (+) adjacent to it to expand the view. Further expanding the contents of a file displays the categories such as **Blocks**, **Dimstyles**, **Layers**, **Layouts**, **Linetypes**, **Textstyles**, and **Xrefs**. Clicking on any one of these categories in the **Tree View** displays the listing under the selected category in the Palette (Figure 6-21). Alternately, right-clicking a particular folder, file, or category of the file contents displays a shortcut menu. The **Explore** option in this shortcut menu also further expands the selected folder, file, or category of contents to display the

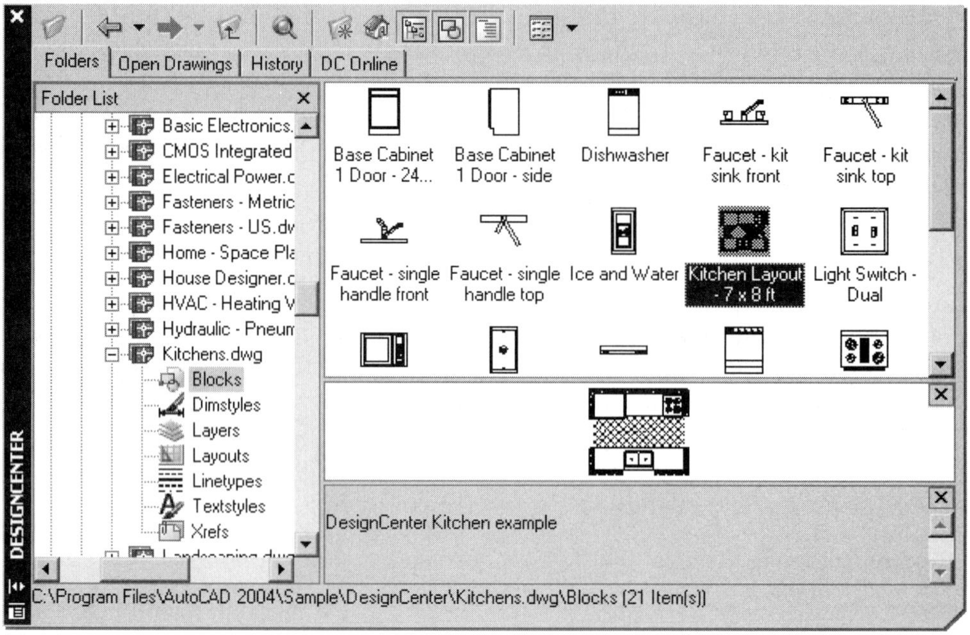

***Figure 6-21 DESIGNCENTER** displaying Tree pane, Palette, Preview pane, and the Description box*

listing of contents respectively. Choosing the **Preview** button from the toolbar displays an image of the selected object or file in a **Preview pane** below the Palette. Choosing the **Description** button displays a brief text description of the selected item if it has one in the **Description** box. When you click on a specific block name in the palette, its preview image and description that was defined earlier when creating the block are displayed in the Preview pane and the Description box respectively.

You can drag and drop any of the contents into the current drawing, or add them by double-clicking on the contents. These are then reused as part of the current drawing. When you double-click on specific Xrefs and blocks, AutoCAD displays the **External Reference** dialog box and the **Insert** dialog box respectively to help in attaching the external reference and inserting the block, respectively. Right-clicking a block displays the options of **Insert Block**, **Copy**, or **Create Tool Palette** and right-clicking an Xref displays the options of **Attach Xref** or **Copy** in the shortcut menus. Similarly, when you double-click a layer, text style, dimstyle, layout, or linetype style, they also get added to the current drawing. If any of these named objects already exist in the current drawing, duplicate definition is ignored and it is not added again. When you right-click on a specific linetype, layer, textstyle, layout, or dimstyle in the palette, a shortcut menu is displayed that gives you an option to **Add** or **Copy**. The **Add** option directly adds the selected named object to the current drawing. The **Copy** option copies the specific named object to the clipboard from where you can paste it into a particular drawing.

Note
You will learn more about inserting blocks in Chapter 14, Working with Blocks.

Right-clicking a particular folder or file in the **Tree View** displays a shortcut menu. The various options available in the shortcut menu besides those discussed earlier are **Add to Favorites**, **Organize Favorites**, **Create Tool Palette**, and **Set as Home**. **Add to Favorites** adds the selected file or folder to the **Favorites** folder, which contains the most often accessed files and folders. **Organize Favorites** allows you to reorganize the contents of the **Favorites** folder. When you select **Organize Favorites** from the shortcut menu, the **Autodesk** folder is opened in a window. **Create Tool Palette** adds the blocks of the selected file or folder to the **TOOL PALETTES** window, which contains the predefined blocks. **Set as Home** sets the selected file or folder as the **Home** folder. You will notice that when the **Design Center** command is invoked the next time, the file that was last set as the **Home** folder is displayed selected in the **DESIGNCENTER**.

Open Drawings Tab*

The **Open Drawings** tab lists all the drawings that are open, including the current drawing which is being worked on. When you select this tab, the **Tree View** (left pane) displays the tree view of all the drawings that are currently open and the **Palette** (right pane) displays the various contents in the selected drawing.

History Tab*

The **History** tab lists the most recent locations accessed through the **DESIGNCENTER**. When you select this tab, the **Tree View** (left pane) and the **Palette** (right pane) are replaced by a list box. Right-clicking a particular file displays a shortcut menu. The various options available in the shortcut menu are **Explore, Folders**, **Open Drawings**, **Delete**, and **Search**. The **Explore** option invokes the **Folders** tab of the DESIGNCENTER with the file selected in the **Tree View** and the contents available in the selected file displayed in the **Palette View**. The **Folders** option invokes the **Folders** tab of the **DESIGNCENTER**. The **Open Drawings** option invokes the **Open Drawings** tab of the **DESIGNCENTER**. The **Delete** option deletes the selected drawing from the History list. The **Search** option allows you to search for drawings or named objects such as blocks, textstyles, dimstyles, layers, layouts, external references, or linetypes. When you choose the **Search** option from the shortcut menu, the **Search** dialog box is displayed as shown in Figure 6-22.

You can select the type of object you want to search for from the **Look for** drop-down list. You can look for a drawing or a named object in your hard drives. If you are looking for a named object, you can first select the drive where you think it might be located from the **In** drop-down list or use the **Browse** button to locate a folder. The **Search subfolders** check box is selected by default. Then, you can enter the name of the named object you are looking for in the **Search for the word(s)** text box and then choose the **Search Now** button to start the search operation.

If you are looking for drawing files, three tabs are available in the **Search** dialog box that

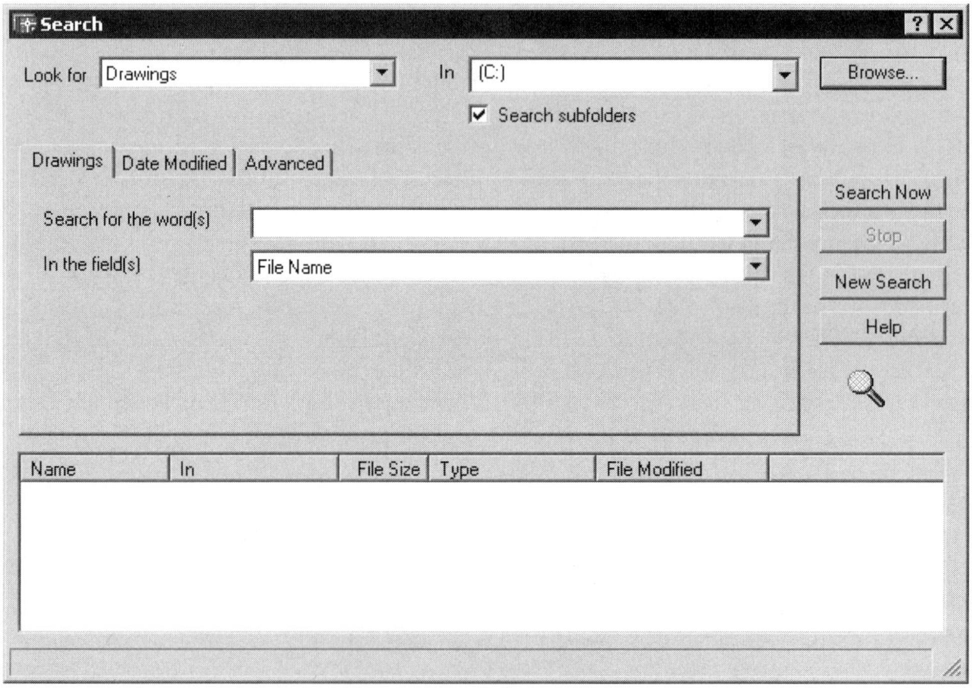

***Figure 6-22 Search** dialog box*

provide additional search criteria. You can conduct a search based on the following additional criteria.

1. Words that are part of the **File name, Title, Subject, Author's name,** and **keyword text** stored in the **summary tab** of the **Drawing Properties**. In the **Drawings** tab, enter the words in the **Search for the word(s)** edit box. The field in which the particular words are to be located can be selected from the **In the field(s)** drop-down list of the **Search** dialog box. You can enter text that might be part of the File name, title, subject, author, or description text of a particular drawing.

2. **Date last modified on**. You can choose the **Date Modified** tab in the **Search** dialog box and select the appropriate radio buttons there.

3. In the **Advanced** tab, you can select the category to look for from the **Containing** drop-down list of the **Search** dialog box. You can choose either of the following categories on which to base your search: **Block name**, **Block and drawing description**, **Attribute tag**, or **Attribute value**. If the named object or drawing you are searching for does not contain any blocks or attributes, and does not have any drawing description, you need not select any of these from the **Containing** drop-down list. In such a case, the Containing text edit box is not available. The text blocks, attributes, or drawing description can be entered in the **Containing text** edit box. You can also specify the **At least** or **At most** sizes of the file you are searching for in the **Size is** edit boxes to further narrow down the search.

After you have entered the appropriate criteria for searching the specific file, choose the **Search Now** button. The name, location, file size, type, and date last modified on are displayed at the bottom of the dialog box under the **Name**, **In**, **File Size**, **Type**, and **File Modified** columns of the list box respectively.

Now, when you double-click on the drawing name, it gets highlighted in the tree view of the **DESIGNCENTER**, and its contents are displayed in the palette, or it gets loaded in the palette. You can also right-click on the drawing name in the list box of the **Search** dialog box to display a shortcut menu. This shortcut menu provides you with the options of loading a drawing in the palette, inserting it as a block, attaching it as an external reference, opening it as a current drawing, or copying it to the pasteboard.

DC Online Tab*

The **DC Online** tab allows you to download the symbols, information regarding various manufacturer's products, and the online catalogs of various products from the **DesignCenter Online** window. To access the **DesignCenter Online** after establishing the Web connection choose the **Reconnect to DesignCenter** button. In the **DesignCenter Online** window the **Tree View** displays various folders under the **Standard Parts**, **Manufactures**, and the **Aggregators** heading. You can select the desired folder from the **Tree View** and the various contents available in the selected folder are displayed on the **Palette**. The preview and the description of the selected content are displayed in the Preview window. You can double-click or drag and drop the selected content from the Web page in the current drawing.

Choosing the **Back** button in the **DESIGNCENTER** toolbar displays the last item selected in the **DESIGNCENTER**. If you pick the down arrow available on the **Back** button, a list of the five recently visited items is displayed. You can view the desired item in the **DESIGNCENTER** by selecting it from the list. The **Forward** button is available only if you have chosen the **Back** button once. This button displays the same page as the current page before you choose the **Back** button. The **Up** button moves one level up in the tree structure from the current location. Choosing the **Favorites** button displays shortcuts to files and folders that are accessed frequently by you and are stored in the **Favorites** folder. This reduces the time you take to access these files or folders from their normal location. Choosing the **Tree View Toggle** button in the **DESIGNCENTER** toolbar displays or hides the tree pane with the tree view of the contents in a hierarchial form. Choosing the **Load** button displays the **Load** dialog box, as shown in Figure 6-23, whose options are similar to those of the standard **Select file** dialog box. When you select a file here and choose the **Open** button, AutoCAD displays the selected file and its contents in the **DESIGNCENTER**.

The **Views** button gives four display format options for the contents of the palette: **Large icons**, **Small icons**, **List**, and **Details**. The **List** option lists the contents in the palette while the **Details** option gives a detailed list of the contents in the palette with the name, file size, and type.

Right-clicking in the palette displays a shortcut menu with all the options provided in the

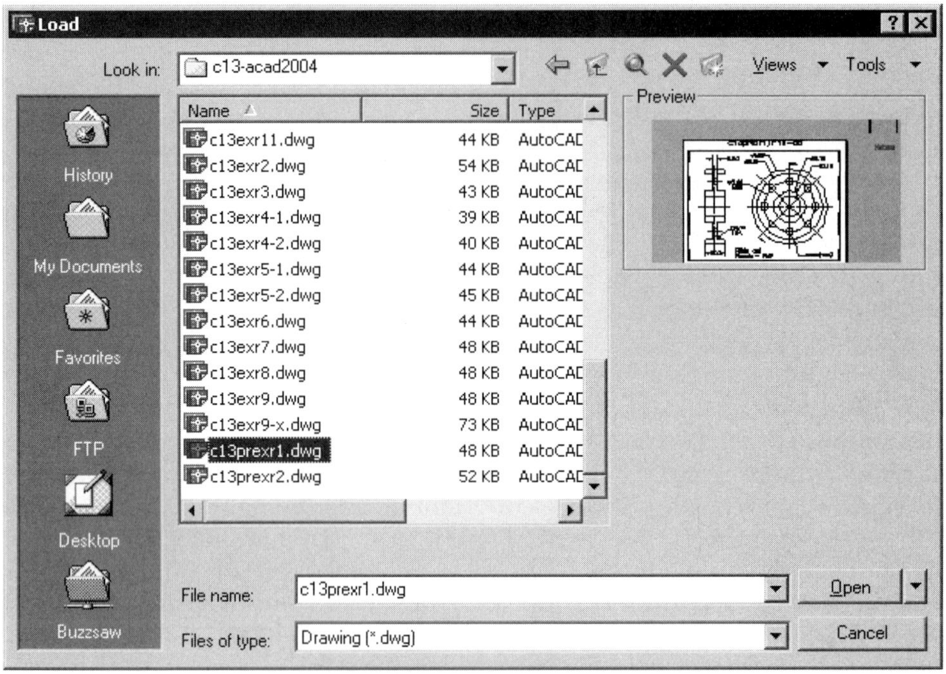

Figure 6-23 Load *dialog box*

DESIGNCENTER in addition to the **Add to Favorites**, **Organize favorites**, **Refresh**, and **Create Tool Palette of Blocks** options. The **Refresh** option refreshes the palette display if you have made any changes to it. The **Create Tool Palette of Blocks** option adds the drawings of the selected file or folder to the **TOOL PALETTES**, which contains the predefined blocks The following example will illustrate how to use the **DESIGNCENTER** to locate a drawing and then use its contents into a current drawing.

Example 1 *Architectural*

Use the **DESIGNCENTER** to locate and view contents of the drawing *Kitchens*. Also, use the **DESIGNCENTER** to insert a block from this drawing and also import a layer and textstyle from the *8th floor lighting.dwg* located in the **Sample** folder. Use these to make a drawing of a Kitchen plan (*MyKitchen.dwg*) and then add text to it as shown in Figure 6-24.

1. Open a new drawing using the **Start from Scratch** option. Make sure to select the **Imperial (feet and inches)** option in the **Create New Drawing** dialog box.

2. Change the units to **Architectural** using the **Drawing Units** dialog box. Increase the limits to 10',10'. Invoke the **ALL** option of the **ZOOM** command so that you are able to view the entire drawing.

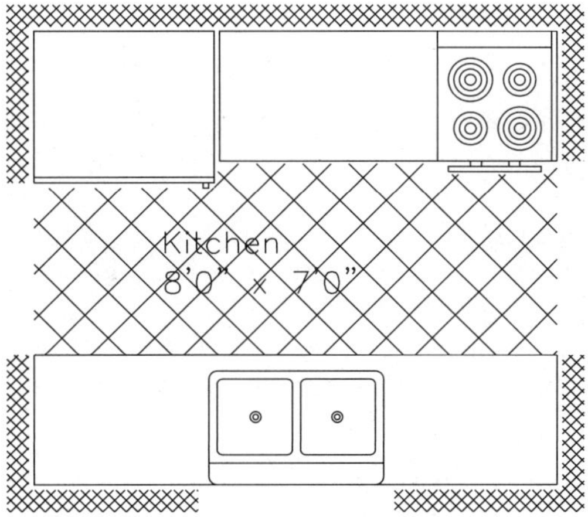

Figure 6-24 *Drawing for Example 1*

3. Choose the **DesignCenter** button from the **Standard** toolbar. The **DESIGNCENTER** window is displayed at its default location.

4. In the **DESIGNCENTER** toolbar, choose the **Tree View Toggle** button to display the **Tree View** and the **Palette** (if not already displayed). Also, choose the **Preview** button. You can resize the window, if need be, to view both the **Tree View** and the **Palette**, conveniently.

5. Choose the **Search** button in the **DESIGNCENTER** to display the **Search** dialog box. Here, select **Drawings** from the **Look for** drop-down list and **C:** from the **In** drop-down list. Select the **Search subfolders** check box. In the **Drawings** tab, type KITCHENS in the **Search for the word(s)** edit box and select **File Name** from the **In the field(s)** drop-down list. Now, choose the **Search Now** button to commence the search. After the drawing has been located, its details and path are displayed in a list box at the bottom of the dialog box.

6. Now, right-click on *Kitchens.dwg* in the list box of the **Search** dialog box and choose **Load into Content Area** from the shortcut menu. You will notice that the drawing and its contents are displayed in the **Tree view**.

7. Close the **Find** dialog box.

8. Double-click on *Kitchens.dwg* in the **Tree View** to display its contents, in case they are not displayed. You can also expand the contents by clicking on the plus sign adjacent to the File name in the **Tree view**.

9. Select **Blocks** in the **Tree View** to display the list of blocks in the drawing in the Palette. Using the left mouse button, drag and drop the block **Kitchen Layout-7x8 ft** in the current drawing.

10. Now, double-click on *8th floor lighting.dwg* located in the *Sample* folder in the same directory to display its contents in the Palette.

11. Select **Layers** in the **Tree View** to display the layers in the drawing. Drag and drop or double-click the layer **LOGO** from the **Palette** to the current drawing. Now, you can use this layer for placing the text in the current drawing after making it the current layer.

12. Select **Textstyles** to display the list of Textstyles in the **Palette**. Select **ROMANS** in the **Palette** and drag and drop it in the current drawing. You can use this textstyle for adding text to the current drawing.

13. Use the imported data to add text to the current drawing and complete it as shown in the Figure 6-24.

14. Save the current drawing as *MyKitchen.dwg*.

MAKING INQUIRIES ABOUT OBJECTS AND DRAWINGS

When you create a drawing or examine an existing one, you often need some information about the drawing. In manual drafting, you inquire about the drawing by performing measurements and calculations manually. Similarly, when drawing in an AutoCAD environment, you will need to make inquiries about data pertaining to your drawing. The inquiries can be about the distance from one location on the drawing to another, the area of an object like a polygon or circle, coordinates of a location on the drawing, and so on. AutoCAD keeps track of all the details pertaining to a drawing. Since inquiry commands are used to obtain information about the selected objects, these commands do not affect the drawings in any way. The following is the list of Inquiry commands:

AREA	**DIST**	**ID**	**LIST**	**DBLIST**
STATUS	**TIME**	**DWGPROPS**	**MASSPROP**	

Note

*The **MASSPROP** command will be discussed in Chapter 25.*

For most of the Inquiry commands, you are prompted to select objects; once the selection is complete, AutoCAD switches from graphics mode to text mode, and all the relevant information about the selected objects is displayed. For some commands information is displayed in the AutoCAD Text Window. The display of the text screen can be tailored to your requirements using a pointing device. Therefore, by moving the text screen to one side, you can view the drawing screen and the text screen simultaneously. If you select the **minimize** button or select the close button, you will return to the graphics screen. You can also return to the graphics screen by entering the **GRAPHSCR** command at the Command prompt. Similarly, you can return to the AutoCAD Text Window by entering **TEXTSCR** at the Command prompt.

Measuring the Area of the Objects

Toolbar:	Inquiry > Area
Menu:	Tools > Inquiry > Area
Command:	AREA

Finding the area of a shape or an object manually is time-consuming. In AutoCAD, the **AREA** command is used to automatically calculate the area of an object in square units. This command saves time when calculating the area of shapes, especially when the shapes are complicated or irregular.

You can use the default option of the **AREA** command to calculate the area and perimeter or circumference of the space enclosed by the sequence of specified points. For example, to find the area of an object (one which is not formed of a single object) you have created with the help of the **LINE** command (Figure 6-25), you need to select all the vertices of that object. By selecting the points, you define the shape of the object whose area is to be found. This is the default method for determining the area of an object. The only restriction is that all the points you specify should be in a plane parallel to the *XY* plane of the current UCS. You can make the best possible use of object snaps such as ENDpoint, INTersect, and TANgent, or even use running Osnaps, to help you select the vertices quickly and accurately. For AutoCAD to find the area of a shape, the shapes need not have been drawn with polylines; nor do the lines need to be closed. However, curves must be approximated with short straight segments. In such cases, AutoCAD computes the area by assuming that the first point and the last point are joined. The prompt sequence that will follow when you choose the **Area** button from the **Inquiry** toolbar is given next.

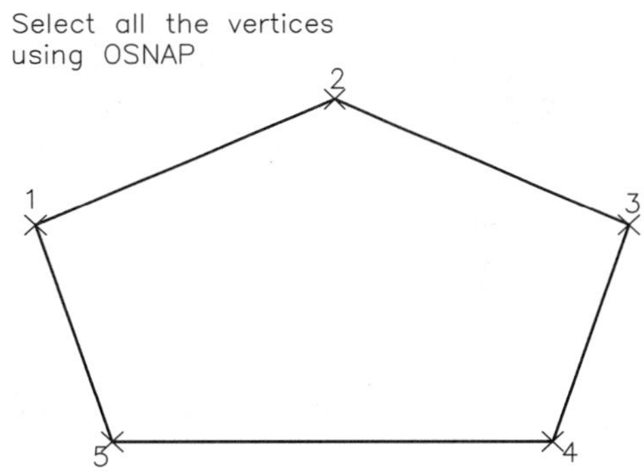

Figure 6-25 *Using the **AREA** command*

Specify first corner point or [Object/Add/Subtract]: *Specify first point.*
Specify next corner point or press ENTER for total: *Specify the second point.*
Specify next corner point or press ENTER for total: *Continue selecting until all the points enclosing the area have been selected.*
Specify next corner point or press ENTER for total: Enter
Area = X, Perimeter = Y

Here, X represents the numerical value of the area and Y represents the circumference/perimeter. It is not possible to accurately determine the area of a curved object, such as an arc, with the default (Point) option. However, the approximate area under an arc can be calculated by specifying several points on the given arc. If the object whose area you want to find is not closed (formed of independent segments) and has curved lines, you should use the following steps to determine the accurate area of such an object.

1. Convert all the segments in that object into polylines using the **PEDIT** command.
2. Join all the individual polylines into a single polyline. Once you have performed these operations, the object becomes closed and you can then use the **Object** option of the **AREA** command to determine the area.

If you specify two points on the screen, the **AREA** command will display the value of the area as 0.00; the perimeter value is the distance between the two points.

Object Option

You can use the **Object** option to find the area of objects such as polygons, circles, polylines, regions, solids, and splines. If the selected object is a polyline or polygon, AutoCAD displays the area and perimeter of the polyline. In case of open polylines, the area is calculated assuming that the last point is joined to the first point but the length of this segment is not added to the polyline length unlike the default option. If the selected object is a circle, ellipse, or planar closed spline curve, AutoCAD will provide information about its area and circumference. For a solid, the surface area is displayed. For a 3D polyline, all vertices must lie in a plane parallel to the *XY* plane of the current UCS. The extrusion direction of a 2D polyline whose area you want to determine should be parallel to the *Z* axis of the current UCS. In case of polylines which have a width, the area and length of the polyline are calculated using the centerline. If any of these conditions is violated, an error message is displayed on the screen. The following prompt sequence appears when you choose the **Area** button.

Specify first corner point or [Object/Add/Subtract]: O Enter
Select objects : Select an object Enter
Area = (X), Circumference = (Y)

X represents the numerical value of the area, and Y represents the circumference/perimeter.

Tip

In many cases, the easiest and most accurate way to find the area of an area enclosed by multiple objects is to use the ***BOUNDARY*** *command to create a polyline, and then use the* ***AREA Object*** *option.*

Add Option

Sometimes you want to add areas of different objects to determine a total area. For example, in the plan of a house, you need to add the areas of all the rooms to get the total floor area. In such cases, you can use the **Add** option. Once you invoke this option, AutoCAD activates the **Add** mode. By using the **First corner point** option at the **Specify first corner point or [Object/Subtract]:** prompt, you can calculate the area and perimeter by selecting points on the screen. Pressing ENTER after you have selected points defining the area that is to be added, calculates the total area, since the **Add** mode is on. The command prompt is as follows.

Specify next corner point or press ENTER for total [ADD mode]:

If the polygon whose area is to be added is not closed, the area and perimeter are calculated assuming that a line that connects the first point to the last point is added to close the polygon. The length of this area is added in the perimeter. The **Object** option adds the areas and perimeters of selected objects. While using this option, if you select an open polyline, the area is calculated considering the last point is joined to the first point but the perimeter does not consider the length of this assumed segment, unlike the **First corner point** option. When you select an object, the area of the selected object is displayed on the screen. At this time the total area is equal to the area of the selected object. When you select another object, AutoCAD displays the area of the selected object as well as the combined area (total area) of the previous object and the currently selected object. In this manner you can add areas of different objects. Until the **Add** mode is active, the string ADD mode is displayed along with all subsequent object selection prompts to remind you that the **Add** mode is active. When the **AREA** command is invoked, the total area is initialized to zero.

Subtract Option

The action of the **Subtract** option is the reverse of that of the **Add** option. Once you invoke this option, AutoCAD activates the **Subtract** mode. The **First corner point** and **Object** options work similar to the way they work in the ADD mode. When you select an object, the area of the selected object is displayed on the screen. At this time, the total area is equal to the area of the selected object. When you select another object, AutoCAD displays the area of the selected object as well as the area obtained by subtracting the area of the currently selected object from the area of the previous object. In this manner, you can subtract areas of objects from the total area. Until the **Subtract** mode is active, the string SUBTRACT mode is displayed along with all subsequent object selection prompts, to remind you that the **Subtract** mode is active. To exit the **AREA** command, press ENTER (null response) at the **Specify first corner point or [Object/Add/Subtract]** prompt. The prompt sequence for these two modes for Figure 6-26 is given next.

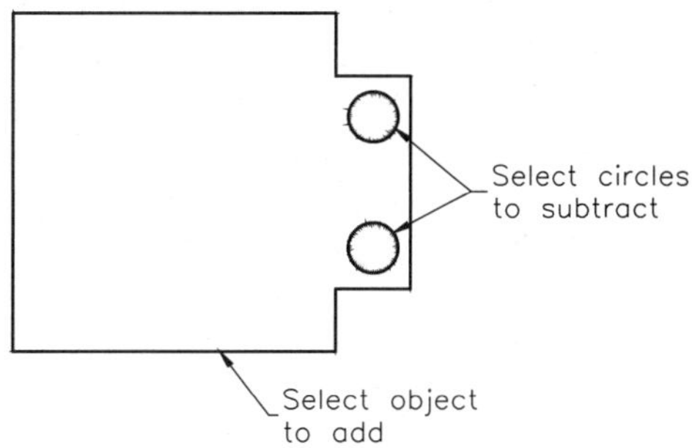

Figure 6-26 *Using the* ***Add*** *and* ***Subtract*** *options*

Specify first corner point or [Object/Add/Subtract]: A Enter
Specify first corner point or [Object/Subtract]: O Enter
(ADD mode) Select objects: *Select the polyline.*
Area = 2.4438, Perimeter = 6.4999
Total area = 2.4438
(ADD mode) Select objects: Enter
Specify first corner point or [Object/Subtract]: S Enter
Specify first corner point or [Object/Add]: O Enter
(SUBTRACT mode) Select object: *Select the circle.*
Area = 0.0495, Circumference = 0.7890
Total area = 2.3943
(SUBTRACT mode) Select objects: *Select the second circle.*

Area = 0.0495, Circumference = 0.7890
Total area = 2.3448
(SUBTRACT mode) Select object: Enter
Specify first corner point or [Object/Add]: Enter

The **AREA** and **PERIMETER** system variables hold the area and perimeter (or circumference in the case of circles) of the previously selected polyline (or circle). Whenever you use the **AREA** command, the **AREA** variable is reset to zero.

Tip
*If an architect wants to calculate the area of flooring and skirting in a room, the **Area** command provides you with the area and the perimeter of the room. You can use these parameters to calculate the skirting.*

Measuring the Distance Between Two Points

Toolbar:	Inquiry > Distance
Menu:	Tools > Inquiry > Distance
Command:	DIST

The **DIST** command is used to measure the distance between two selected points (Figure 6-27). The angles that the selected points make with the *X* axis and the *XY* plane are also displayed. The measurements are displayed in current units. Delta X (horizontal displacement), delta Y (vertical displacement), and delta Z are also displayed. The distance computed by the **DIST** command is saved in the **DISTANCE** variable. The prompt sequence that will follow when you choose the **Distance** button is given next.

Specify first point: *Specify a point.*
Specify second point: *Specify a point.*

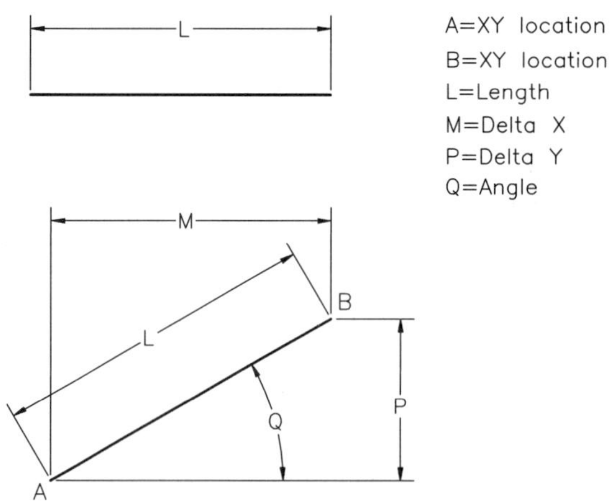

Figure 6-27 *Using the **DIST** command*

AutoCAD returns the following information.

Distance = *Calculated distance between the two points.*
Angle in XY plane = *Angle between the two points in the XY plane.*
Angle from XY plane = *Angle the specified points make with the XY plane.*
Delta X = *Change in X*, Delta Y = *Change in Y*, Delta Z = *Change in Z.*

If you enter a single number or fraction at the **Specify first point** prompt, AutoCAD will convert it into the current unit of measurement and display it in the command line.

Command: **DIST**
First point: 3-3/4 *(Enter a number or a fraction.)*
Distance = 3.7500

Note

The Z coordinate is used in 3D distances. If you do not specify the Z coordinates of the two points between which you want to know the distance, AutoCAD takes the current elevation as the Z coordinate value.

Identifying the Location of a Point on the Screen

Toolbar:	Inquiry > Locate Point
Menu:	Tools > Inquiry > ID Point
Command:	ID

The **ID** command is used to identify the position of a point you specify by displaying the *X*, *Y*, and *Z* coordinates of the point. The prompt sequence that will follow when you choose the **Locate Point** button from the **Inquiry** toolbar is given next.

Specify point: *Specify the point to be identified.*
X = X coordinate   Y = Y coordinate   Z = Z coordinate

AutoCAD takes the current elevation as the *Z* coordinate value. If an **Osnap** mode is used to snap to a 3D object in response to the **Specify point** prompt, the *Z* coordinate displayed will be that of the selected feature of the 3D object. You can also use the **ID** command to identify the location on the screen. This can be realized by entering the coordinate values you want to locate on the screen. AutoCAD identifies the point by drawing a blip mark at that location if the **BLIPMODE** system variable is on. For example, the following is the prompt sequence to find where the position X = 2.345, Y = 3.674, and Z = 1.0000 is located on the screen.

Specify point: 2.345,3.674,1.00 Enter
X = 2.345   Y = 3.674   Z = 1.0000

The coordinates of the point specified in the **ID** command are saved in the **LASTPOINT** system variable. You can locate a point with respect to the **ID** point by using the relative or polar coordinate system. You can also snap to this point by typing @ when AutoCAD prompts for a point.

Listing Information About Objects

Toolbar:	Inquiry > List
Menu:	Tools > Inquiry > List
Command:	LIST

The **LIST** command displays all the information pertaining to the selected objects. The information is displayed in the AutoCAD Text Window. The prompt sequence that follows when you choose the **List** button from the **Inquiry** toolbar is given next.

Select objects: *Select objects whose data you want to list.*
Select objects: Enter

Once you select the objects to be listed, AutoCAD shifts you from the graphics screen to the AutoCAD Text Window. The information displayed (listed) varies from object to object. The information on an object's type, its coordinate position with respect to the current UCS (user coordinate system), the name of the layer on which it is drawn, and whether the object is in model space or paper space is listed for all types of objects. If the color, lineweight, and the linetype are not BYLAYER, they are also listed. Also, if the thickness of the object is greater than 0, that is also displayed. The elevation value is displayed in the form of a *Z* coordinate (in the case of 3D objects). If an object has an extrusion direction different from the *Z* axis of the current UCS, the object's extrusion direction is also provided.

More information based on the objects in the drawing is also provided. For example, for a line the following information is displayed.

1. The coordinates of the endpoints of the line.
2. Its length (in 3D).
3. The angle made by the line with respect to the *X* axis of the current UCS.
4. The angle made by the line with respect to the *XY* plane of the current UCS.
5. Delta X, Delta Y, Delta Z: this is the change in each of the three coordinates from the start point to the endpoint.
6. The name of the layer in which the line was created.
7. Whether the line is drawn in Paper space or Model space.

The center point, radius, true area, and circumference of circles is displayed. For polylines, this command displays the coordinates. In addition, for a closed polyline, its true area and perimeter are also given. If the polyline is open, AutoCAD lists its length and also calculates the area by assuming a segment connecting the start point and endpoint of the polyline. In the case of wide polylines, all computation is done based on the centerlines of the wide segments. For a selected viewport, the **LIST** command displays whether the viewport is on and active, on and inactive, or off. Information is also displayed about the status of Hideplot and the scale relative to paper space. If you use the **LIST** command on a polygon mesh, the size of the mesh (in terms of M, X, N), the coordinate values of all the vertices in the mesh, and whether the mesh is closed or open in M and N directions are all displayed. As mentioned before, if all the information does not fit on a single screen, AutoCAD pauses to allow you to press ENTER to continue the listing.

Listing Information About All Objects in a Drawing

Command: DBLIST

The **DBLIST** command displays information pertaining to all the objects in the drawing. Once you invoke this command, information is displayed in the Command prompt. If you want to display the drawing information in the AutoCAD Text Window, press the F2 key on the keyboard. If the information does not fit on a single screen, AutoCAD pauses to allow you to press ENTER to continue the listing. To terminate the command, press ESC. To return to the graphics screen, close the AutoCAD Text Window. This command can be invoked by entering **DBLIST** at the Command prompt.

Checking the Time-Related Information

Menu: Tools > Inquiry > Time
Command: TIME

The time and date maintained by your system are used by AutoCAD to provide information about several time factors related to the drawings. Hence, you should be careful about setting the current date and time in your computer. The **TIME** command can be used to display information pertaining to time related to a drawing and the drawing session. The display obtained by invoking the **TIME** command is similar to the following.

Command: **TIME**
Current time: Monday, April 26 1999 at 11:22:35.069 AM
Times for this drawing:
Created: Thursday, April 15 1999 at 09:34:42.157 AM
Last updated: Friday, April 16 1999 at 02:54:25.700 PM
Total editing time: 0 days 04:05:00.440
Elapsed timer (on): 0 days 04:05:00.390
Next automatic save in: 0 days 01:57:38.181

Enter option [Display/ON/OFF/Reset]:

The foregoing display gives you information on the following.

Current Time

Provides today's date and the current time.

Created

Provides the date and time that the current drawing was created. The creation time for a drawing is set to the system time when the **NEW**, **WBLOCK**, or **SAVE** command is used to create that drawing file.

Last updated

Provides the most recent date and time you saved the current drawing. In the beginning, it is set to the drawing creation time, and it is modified every time you use the **QUIT** or **SAVE** command to save the drawing.

Total editing time

This specifies the total time spent on editing the current drawing since it was created. If you terminate the editing session without saving the drawing, the time you have spent on that editing session is not added to the total time spent on editing the drawing. Also, the last update time is not revised.

Elapsed timer

This timer operates while you are in AutoCAD. You can stop this timer by entering OFF at the **Enter an option [Display/ON/OFF/Reset]:** prompt. To activate the timer, enter ON. If you want to know how much time you have spent on the current drawing or part of the drawing in the current editing session, use the **Reset** option as soon as you start working on the drawing or part of the drawing. This resets the user-elapsed timer to zero. By default, this timer is ON. If you turn this timer OFF, the time accumulated in this timer up to the time you turned it OFF will be displayed.

Next automatic save in

As mentioned in the earlier chapters, AutoCAD automatically saves the drawing after the specified interval of time. The **Next automatic save in** option specifies when the next automatic save will be performed. The automatic save time interval can be set in the **Options** dialog box (**Open and Save** tab) or with the **SAVETIME** system variable. If the time interval has been set to zero, the **TIME** command displays the following message.

Next automatic time save in: <disabled>

If the time interval is not set to zero, and no editing has taken place since the previous save, the **TIME** command displays the following message.

Next automatic time save in: <no modification yet>

If the time interval is not set to zero, and editing has taken place since the previous save, the **TIME** command displays the following message.

Next automatic time save in: 0 days hh:mm:ss.msec

hh stands for hours
mm stands for minutes
ss stands for seconds
msec stands for milliseconds

At the end of the display of the **TIME** command, AutoCAD prompts as follows.

Enter an option [Display/ON/OFF/Reset]:

The information displayed by the **TIME** command is static. This means that the information is not updated dynamically on the screen. If you respond to the last prompt with DISPLAY (or D), the display obtained by invoking the **TIME** command is repeated. This display contains updated time values.

With the ON response, the user-elapsed timer is started, if it was off. As mentioned earlier, when you enter the drawing editor, by default the timer is on. The OFF response is just the opposite of the ON response and stops the user-elapsed time, if it is on. With the **Reset** option, you can set the user-elapsed time to zero.

Obtaining Drawing Status Information

Menu:	Tools > Inquiry > Status
Command:	STATUS

The **STATUS** command displays information about the prevalent settings of various drawing parameters, such as snap spacing, grid spacing, limits, current space, current layer, current color, and various memory parameters.

Command: **STATUS**

Once you enter this command, AutoCAD displays information similar to the following:

```
106 objects in Drawing.dwg
Model space limits are     X: 0.0000     Y: 0.0000(On)
                           X: 6.0000     Y: 4.4000
Model space uses           X:0.6335      Y:-0.2459   **Over
                           X:8.0497      Y: 4.9710   **Over
Display shows              X: 0.0000     Y:-0.2459
                           X: 8.0088     Y: 5.5266
Insertion base is          X: 0.0000     Y: 0.0000     Z: 0.0000
Snap resolution is         X: 0.2500     Y: 0.2500
Grid spacing is            X: 0.2500     Y: 0.2500

Current space:       Model space
Current layout:      Model
Current layer:       OBJ
Current color:       BYLAYER .... .... 7 (white)
Current linetype:    BYLAYER .... .... CONTINUOUS
Current lineweight: BYLAYER
Current elevation:   0.0000   thickness:   0.0000
Fill on  Grid on  Ortho off  Qtext off  Snap off  Tablet  on
Object snap modes: None
Free dwg disk space: 2047.7 MBytes
Free temp disk: 2047.7 MBytes
Free physical memory: 15.0 MBytes
Free swap file space: 1987.5 MBytes
```

All the values (coordinates and distances) on this screen are given in the format declared in the **UNITS** command. You will also notice **Over in the **Model space uses** or **Paper space uses** line. This signifies that the drawing is not confined within the drawing limits. The amount of memory available on the disk is given in the **Free dwg disk space:** line. Information on the name of the current layer, current color, current space, current linetype, current lineweight, current elevation, snap spacing (snap resolution), grid spacing, various tools that are on or off (such as Ortho, Snap, Fill, Tablet, Qtext), and which object snap modes are active is also provided by the display obtained by invoking the **STATUS** command.

Displaying Drawing Properties

Menu:	File > Drawing Properties
Command:	DWGPROPS

The **DWGPROPS** command displays information about the drawing properties. On choosing **Drawing Properties** from the **File** Menu, the **Drawing Properties** dialog box is displayed (Figure 6-28). This dialog box has four tabs under which information about the drawing is displayed. The information displayed in this dialog box helps you look for the drawing more easily. The tabs are as follows.

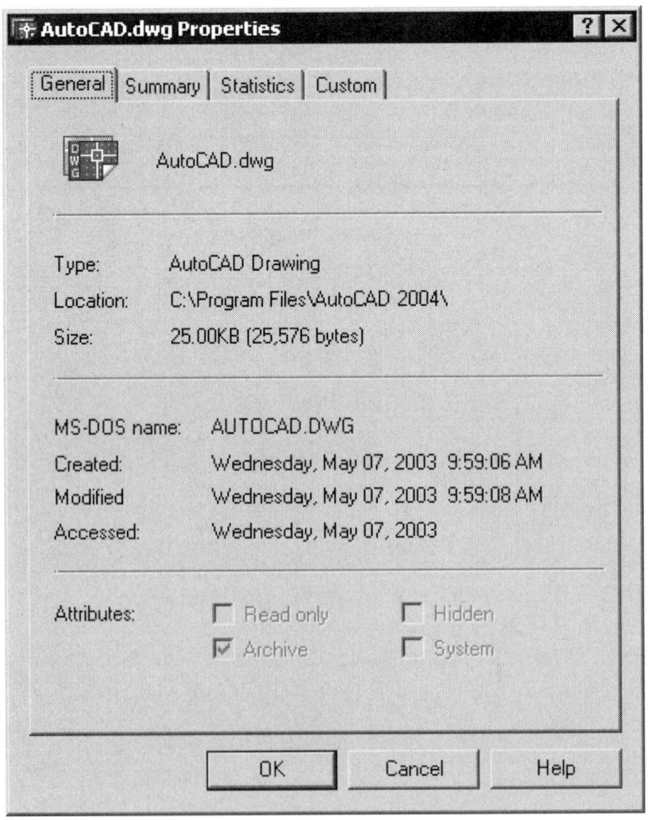

***Figure 6-28 Drawing Properties** dialog box*

General

This tab displays general properties about the drawing like the **Type**, **Size**, and **Location**.

Summary

The **Summary** tab displays predefined properties like the Author, title, and subject.

Statistics

This tab stores and displays data such as the file size and data such as the dates when the drawing was last saved on or modified on.

Custom

This tab displays custom file properties including values assigned by you.

Self-Evaluation Test

Answer the following questions, and then compare your answers to the correct answers given at the end of this chapter.

1. The number of grips depends on the selected object. (T/F)
2. You can use the **Options** dialog box to modify the grips parameters. (T/F)
3. You need at least one source object while using the **MATCHPROP** command. (T/F)
4. You cannot drag and drop the entities from the **DESIGNCENTER** window. (T/F)
5. A grip is a small square that is displayed on an object at its _________ points.
6. A line has _______ grip points and a polyline has _________ .
7. You can enable grips within a block by setting the system variable _________ to 1 (On).
8. The color of the unselected grips can also be changed by using the _________system variable.
9. You can access the Mirror mode by selecting a grip and then entering _________ or _________ from the keyboard or giving a null response by pressing the SPACEBAR four times.
10. The _________ drop-down list will not be available if you select **Select All** from the **Operator** drop-down list in the **Quick Select** dialog box.

Review Questions

Answer the following questions.

1. If you select a grip of an object, the grip becomes a hot grip. (T/F)

2. To cancel the grip, press the ESC key once. (T/F)

3. The Rotate mode allows you to rotate objects around the base point without changing their size. (T/F)

4. If you have already added hyperlink to the object, you can also use the grips to open a file associated with the hyperlink. (T/F)

5. Which system variable is used to modify the color of the selected grip?

   (a) **GRIPCOLOR** (b) **GRIPHOT**
   (c) **GRIPCOLD** (c) **GRIPBLOCK**

6. Which system variable is used to enable the display of the grips inside the blocks?

   (a) **GRIPCOLOR** (b) **GRIPHOT**
   (c) **GRIPCOLD** (c) **GRIPBLOCK**

7. Which system variable is used to modify the size of the grips?

   (a) **GRIPCOLOR** (b) **GRIPSIZE**
   (c) **GRIPCOLD** (c) **GRIPBLOCK**

8. By holding down which key you can select and make more than one grips hot?

   (a) SHIFT (b) CTRL
   (c) ESC (c) ALT

9. Which system variable is used to enable the grip mode?

   (a) **GRIPCOLOR** (b) **GRIPHOT**
   (c) **GRIPS** (c) **GRIPBLOCK**

10. When you double-click on a circle, the __________ palette is displayed.

11. The **GRIPSIZE** is defined in pixels, and its value can range from __________ to __________ pixels.

12. When you select a grip for editing, you are automatically in the __________ mode.

13. The __________ mode lets you move the selected objects to a new location.

14. The __________ mode allows you to scale the objects with respect to the base point without changing their orientation.

15. The Mirror mode allows you to mirror the objects across the __________ without changing the size of the objects.

Exercises

Exercise 1 — *General*

1. Use the **LINE** command to draw the shape as shown in Figure 6-29(a).
2. Use grips (Stretch mode) to get the shape as shown in Figure 6-29(b).
3. Use the Rotate and Stretch modes to get the copies as shown in Figure 6-29(c)

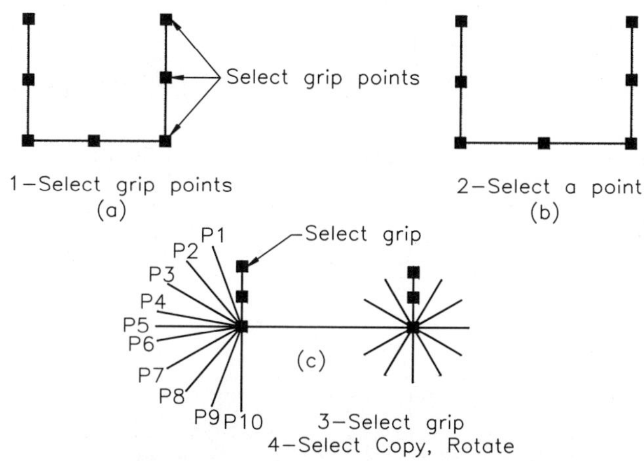

Figure 6-29 *Drawing for Exercise 1*

Exercise 2 — *Mechanical*

Use the draw and editing commands to make the drawing shown in Figure 6-30.

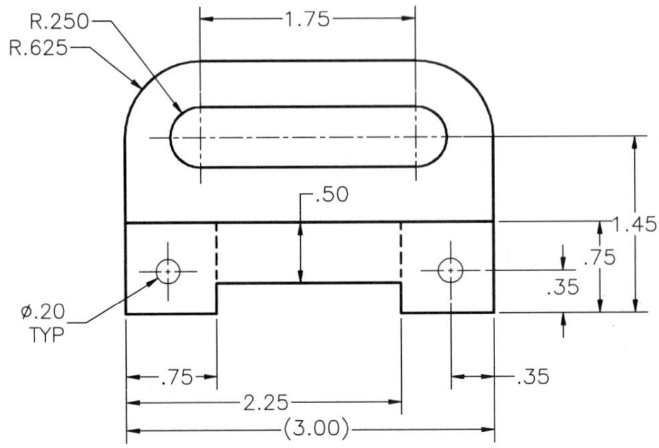

Figure 6-30 Drawing for Exercise 2

Exercise 3 *Mechanical*

Use the draw and editing commands to make the drawing shown in Figure 6-31.

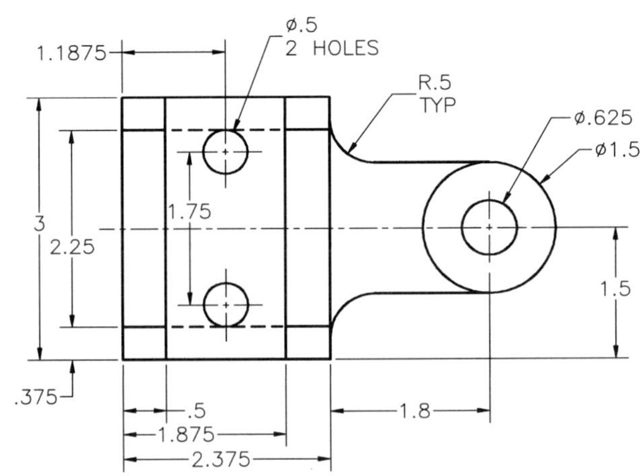

Figure 6-31 Drawing for Exercise 3

Problem Solving Exercise 1 *Mechanical*

Use the draw and editing commands to make the drawing shown in Figure 6-32.

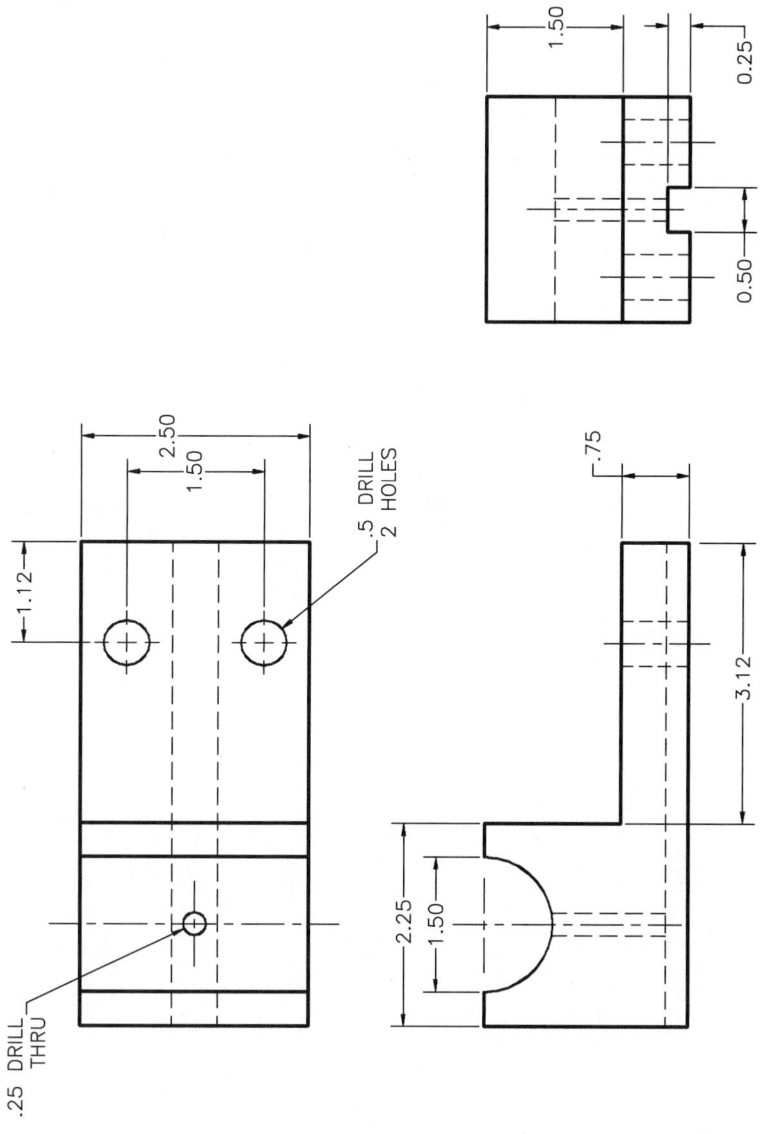

Figure 6-32 *Drawing for Problem Solving Exercise 1*

Problem Solving Exercise 2 *Mechanical*

Use the draw and editing commands to make the drawing shown in Figure 6-33. Assume the missing dimensions.

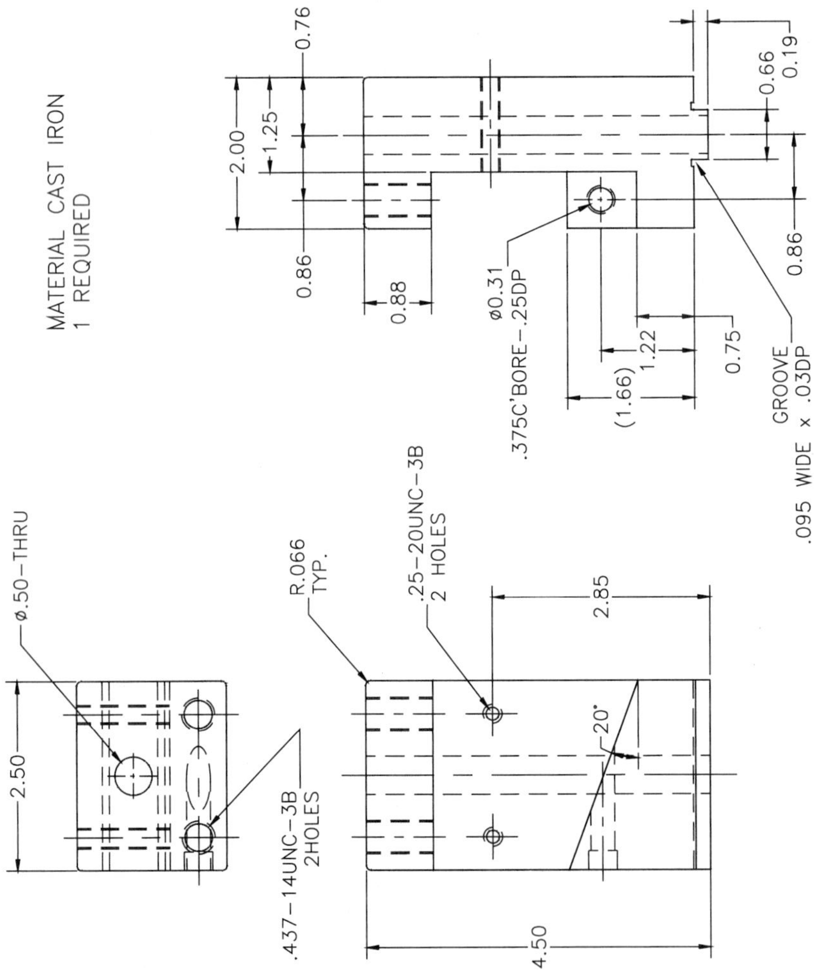

Figure 6-33 Drawing for Problem Solving Exercise 2

Problem Solving Exercise 3 *Mechanical*

Draw Figure 6-34 using draw and edit commands. Use the **MIRROR** command to mirror the shape 9 units across the *Y* axis so that the distance between two center points is 9 units. Mirror the shape across the *X* axis and then reduce the mirrored shape by 75 percent. Join the two ends to complete the shape of the open end spanner. Save the file. Assume the missing dimensions. Note that this is not a standard size spanner.

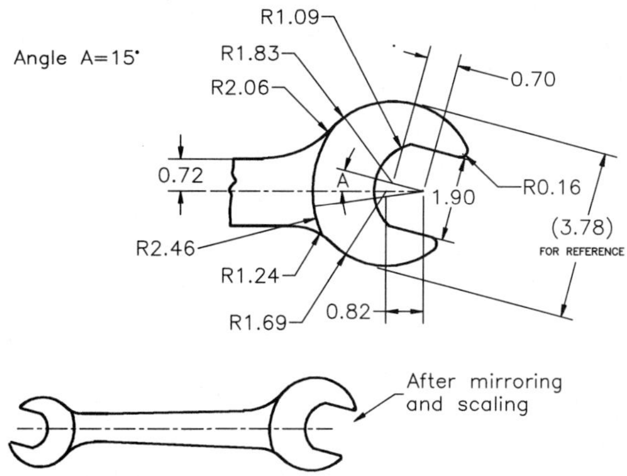

Figure 6-34 *Drawing for Problem Solving Exercise 3*

Problem Solving Exercise 4 *Architectural*

Draw the reception desk shown in Figure 6-35. To get the dimensions of the chairs, refer to Problem Solving Exercise 3 of Chapter 5.

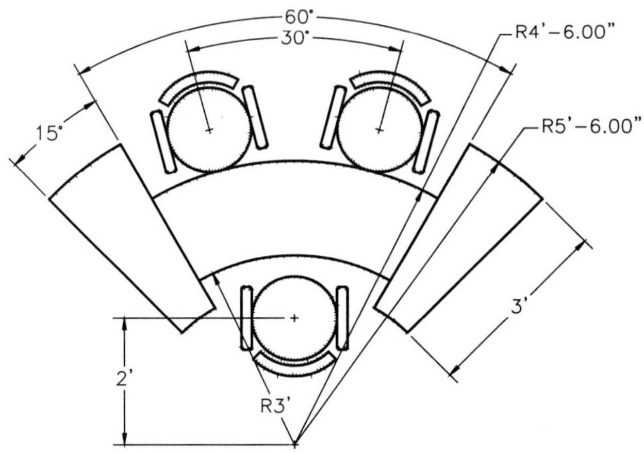

Figure 6-35 *Drawing for Problem Solving Exercise 4*

Answers to Self-Evaluation Test

1 - T, **2** - T, **3** - T, **4** - F, **5** - definition, **6** - three, two, **7** - **GRIPBLOCK**, **8** - **GRIPCOLOR**, **9** - **MIRROR**, **MI**, **10** - **Value**

Chapter 7

Controlling the Drawing Display and Creating Text

Learning Objectives

After completing this chapter you will be able to:

- *Use the* ***REDRAW*** *and* ***REGEN*** *commands.*
- *Use the* ***ZOOM*** *command and its options.*
- *Understand the* ***PAN*** *and* ***VIEW*** *commands.*
- *Understand the use of the* ***Aerial View*** *window.*
- *Draw text using the* ***TEXT*** *command.*
- *Create paragraph text using the* ***MTEXT*** *command.*
- *Edit text using the* ***DDEDIT*** *command.*
- *Use the* ***PROPERTIES*** *palette to change the properties of the text.*
- *Substitute fonts and specify alternate default fonts.*
- *Create text styles using the* ***STYLE*** *command.*
- *Determine text height.*
- *Check spellings and find and replace text.*

BASIC DISPLAY OPTIONS

Drawing in AutoCAD is much simpler than manual drafting in many ways. Sometimes while drawing, it is very difficult to see and alter minute details. In AutoCAD, you can overcome this problem by viewing only a specific portion of the drawing. For example, if you want to display a part of the drawing on a larger area, you can use the **ZOOM** command, which lets you enlarge or reduce the size of the drawing displayed on the screen. Similarly, you can use the **REGEN** command to regenerate the drawing and **REDRAW** to refresh the screen. In this chapter you will learn some of the drawing display commands, such as **REDRAW**, **REGEN**, **PAN**, **ZOOM**, and **VIEW**. These commands can also be used in the transparent mode. Transparent commands are commands that can be used while another command is in progress. Once you have completed the process involved with a transparent command, AutoCAD automatically returns you to the command with which you were working before you invoked the transparent command.

REDRAWING THE SCREEN

Menu:	View > Redraw
Command:	REDRAW

The **REDRAW** command (Figure 7-1) redraws the screen and is used to remove the small cross marks (blips) that appear when a point is specified on the screen when BLIPMODE is set to on. The blip mark is not treated as an element of the drawing. It also redraws the objects that do not display on the screen as a result of editing some other object. In AutoCAD, several commands redraw the screen automatically (for example, when a grid is turned off), but it is sometimes useful to redraw the screen explicitly. In AutoCAD, the **REDRAW** command can also be used in the transparent mode. This can be done by typing an apostrophe in front of the command. The apostrophe appended to a command indicates that the command is to be used as a transparent command (Command: **'REDRAW**). Use of the **REDRAW** command does not involve a prompt sequence; instead, the redrawing process takes place without any prompting for information.

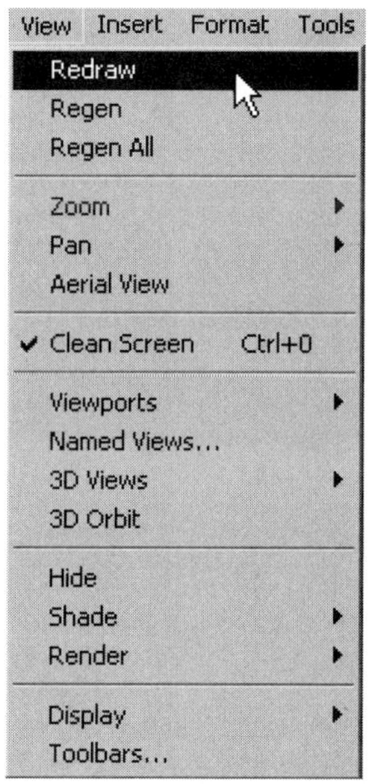

Figure 7-1 *Invoking the **REDRAW** command from the **View** menu*

The **REDRAW** command affects only the current viewport. If you have more than one viewport you can use the **REDRAWALL** command to redraw all the viewports. **Redraw** in the **View** menu is the **REDRAWALL** command.

Tip

*While working on complex drawings it may be better to set the **BLIPMODE** variable as **Off** (default) instead of using the **REDRAW** command to clear blips.*

REGENERATING THE DRAWINGS

Menu:	View > Regen
Command:	REGEN

The **REGEN** command makes AutoCAD regenerate the entire drawing to update it. The need for regeneration usually occurs when you change certain aspects of the drawing. All the objects in the drawing are recalculated and redrawn in the current viewport. One of the advantages of this command is that the drawing is refined by smoothing out circles and arcs. To use this command, enter **REGEN** at the Command prompt. AutoCAD displays the message **Regenerating model** while it regenerates the drawing. The **REGEN** command affects only the current viewport. If you have more than one viewport, you can use the **REGENALL** command to regenerate all the viewports. The **REGEN** command can be aborted by pressing ESC. This saves time if you are going to use another command that causes automatic regeneration.

Tip

*Under certain conditions, the **ZOOM** and **PAN** commands automatically regenerate the drawing. Some other commands also perform regenerations under certain conditions.*

ZOOMING THE DRAWINGS

Toolbar:	Zoom toolbar or Standard > Zoom flyout
Menu:	View > Zoom
Command:	ZOOM

Creating drawings on the screen would not be of much use if you could not magnify the drawing view to work on minute details. Getting close to or away from the drawing is the function of the **ZOOM** command. In other words, this command enlarges or reduces the view of the drawing on the screen, but it does not affect the actual size of the objects. In this way the **ZOOM** command functions like the zoom lens on a camera. When you magnify the apparent size of a section of the drawing, you see that area in greater detail. On the other hand, if you reduce the apparent size of the drawing, you see a larger area.

The ability to zoom in, or magnify, has been helpful in creating the minuscule circuits used in the electronics and computer industries. This is one of the most frequently used commands. Also, this command can be used transparently, which means that it can be used while working in other commands. The **ZOOM** command (Figure 7-2) can be invoked from the shortcut menu by right-clicking in the drawing area and choosing **Zoom** from the menu (Figure 7-3). This command has several options and can be used in a number of ways.

Command: **ZOOM** Enter
Specify corner of window, enter a scale factor (nX or nXP) , or
[All/Center/Dynamic/Extents/Previous/Scale/Window] <real time>:

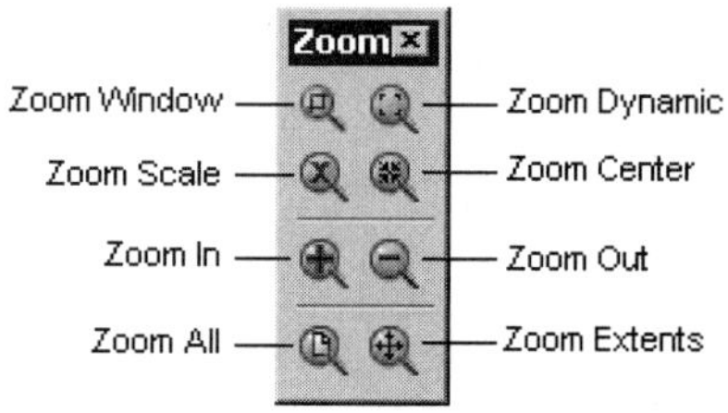

Figure 7-2 *Selecting* ***Zoom*** *options from the* ***Zoom*** *toolbar*

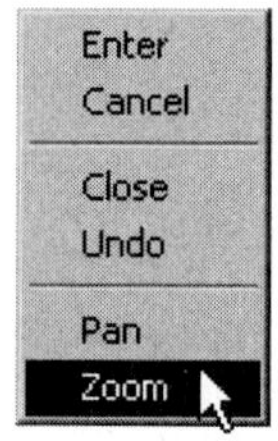

Figure 7-3 *Invoking the* ***ZOOM*** *command from the shortcut menu, when in the* ***LINE*** *command*

Realtime Zooming

You can use the **Realtime Zoom** to zoom in and zoom out interactively. To zoom in, invoke the command, then hold the pick button down and move the cursor up. If you want to zoom in further, release the pick button and bring the cursor down. Specify a point and move the cursor up again. Similarly, to zoom out, hold the pick button down and move the cursor down. If you move the cursor vertically up from the midpoint of the screen to the top of the window, the drawing is magnified by 100% (zoom in 2x magnification). Similarly, if you move the cursor vertically down from the midpoint of the screen to the bottom of the window, the drawing display is reduced 100% (zoom out 0.5x magnification). Realtime zoom is the default setting for the **ZOOM** command. Pressing ENTER after entering the **ZOOM** command automatically invokes the realtime zoom.

When you use the realtime zoom, the cursor becomes a magnifying glass and displays a plus sign (+) and a minus sign (–). When you reach the zoom out limit, AutoCAD does not display the minus sign (–) while dragging the cursor. Similarly, when you reach the zoom in limit, AutoCAD does not display the plus sign (+) while dragging the cursor. To exit the realtime zoom, press ENTER or ESC, or select **Exit** from the shortcut menu.

All Option

This option of the **ZOOM** command adjusts the display area on the basis of the drawing limits (Figure 7-4) or extents of the object, whichever is greater. Even if the objects are not within the limits, they are still included in the display. Hence, with the help of the **All** option, you can view the entire drawing in the current viewport (Figure 7-5).

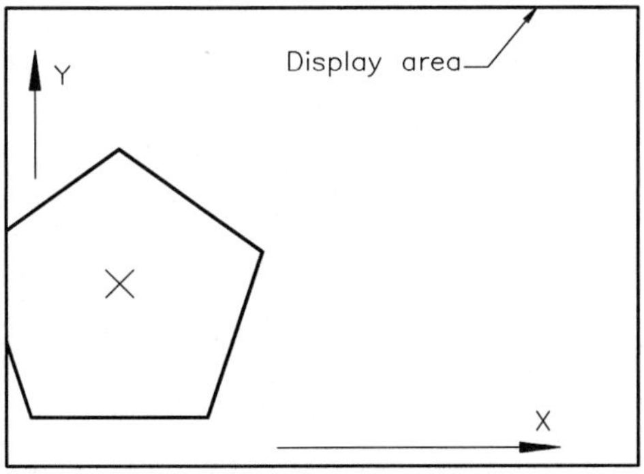

Figure 7-4 Drawing showing limits

*Figure 7-5 The **Zoom All** option*

Chapter 7

Center Option

This option lets you define a new display window by specifying its center point (Figures 7-6 and 7-7) and the magnification or height. Here, you are required to enter the **center** and the **height** of the subsequent screen display. If you press ENTER instead of entering a new center point, the center of the view will remain unchanged. Instead of entering a height, you can enter the **magnification factor** by typing a number. If you press ENTER at the height prompt, or if the height you enter is the same as the current height, magnification does not take place. For example, if the current height is 2.7645 and you press ENTER at the **magnification or height <2.7645>:** prompt, magnification will not take place. The smaller the value, the greater the enlargement of the image. You can also enter a number followed by **X**. This indicates the change in magnification, not as an absolute value, but as a value relative to the current screen. The prompt sequence is given next.

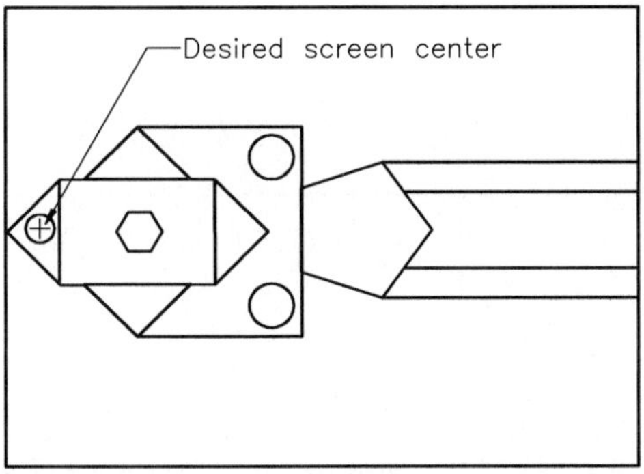

*Figure 7-6 Drawing before using the **ZOOM Center** option*

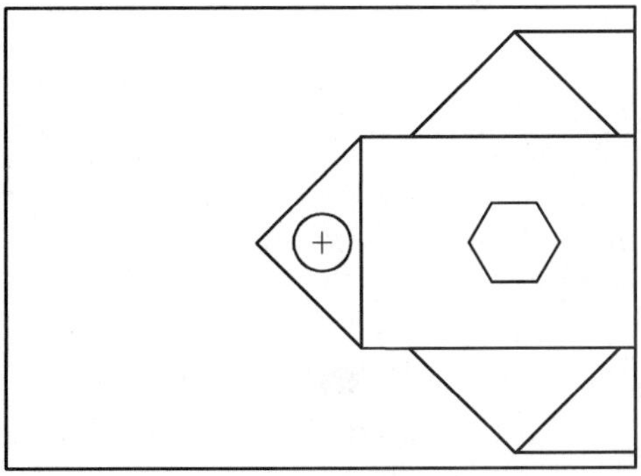

*Figure 7-7 Drawing after using the **ZOOM Center** option*

Command : **ZOOM** [Enter]
Specify corner of window, enter a scale factor (nX or nXP), or
[All/Center/Dynamic/Extents/Previous/Scale/Window] <real time>: **C** [Enter]

Specify center point : *Specify a center point.*
Enter magnification or height <current>: 5X Enter

In Figure 7-7, the current magnification height is 5X, which magnified the display five times. If you enter a value of 2, the size (height and width) of the zoom area changes to 2 X 2 around the specified center. In Figure 7-8, if you enter .12 as the height after specifying the center point as the circle's center, the circle will zoom to fit in the display area since its diameter is 0.12. The prompt sequence is given next.

Command: **ZOOM** Enter
Specify corner of window, enter a scale factor (nX or nXP), or
[All/Center/Dynamic/Extents/Previous/Scale/Window]<real time>: **C** Enter
Center point: *Select the center of the circle.*
Magnification or Height <5.0>: **0.12** Enter

Extents Option

As the name indicates, this option lets you zoom to the extents of the biggest object in the drawing. The extents of the drawing comprise the area that has the drawings in it. The rest of the empty area is neglected. With this option, all the objects in the drawing are magnified to the largest possible display (Figure 7-9).

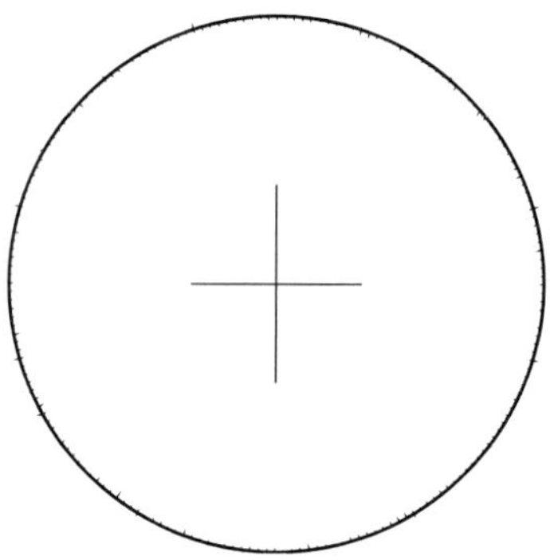

Figure 7-8 *Drawing after using the* ***ZOOM Center*** *option*

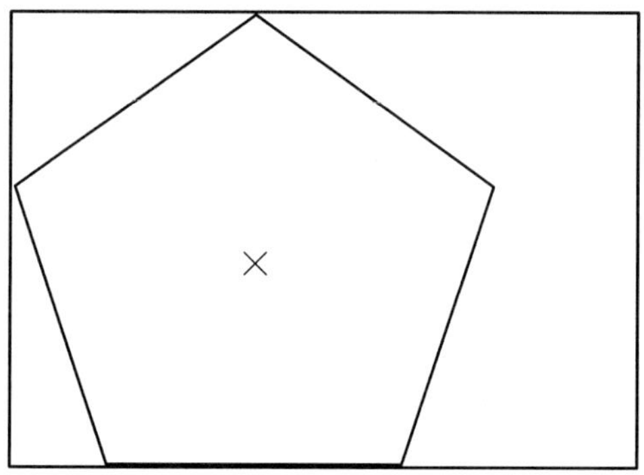

Figure 7-9 *The* ***ZOOM Extents*** *option*

Dynamic Option

This option displays the portion of the drawing that you have already specified. The prompt sequence for using this option of the **ZOOM** command is given next.

Command: **ZOOM** Enter
Specify corner of window, enter a scale factor (nX or nXP), or
[All/Center/Dynamic/Extents/Previous/Scale/Window]<real time>: **D** Enter

You can then specify the area you want to be displayed by manipulating a view box representing your viewport. This option lets you enlarge or shrink the view box and move it around. When you have the view box in the proper position and size, the current viewport is cleared by AutoCAD and a special view selection screen is displayed. This special screen comprises information regarding the current view as well as available views. In a color display, the different viewing windows are very easy to distinguish because of their different colors, but in a monochrome monitor, they can be distinguished by their shape.

Blue dashed box representing drawing extents

Drawing extents are represented by a dashed blue box (Figure 7-10), which constitutes the larger of the drawing limits or the actual area occupied by the drawing.

Green dashed box representing the current view

A green dashed box is formed to represent the area that the current viewport comprises when the **Dynamic** option of the **ZOOM** command is invoked (Figure 7-11).

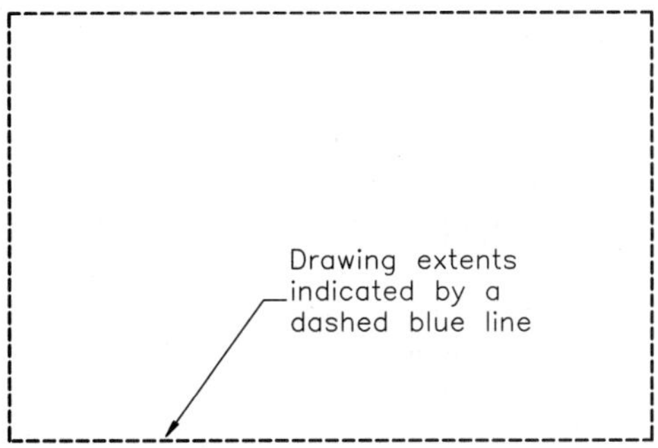

Figure 7-10 *Box representing drawing extents*

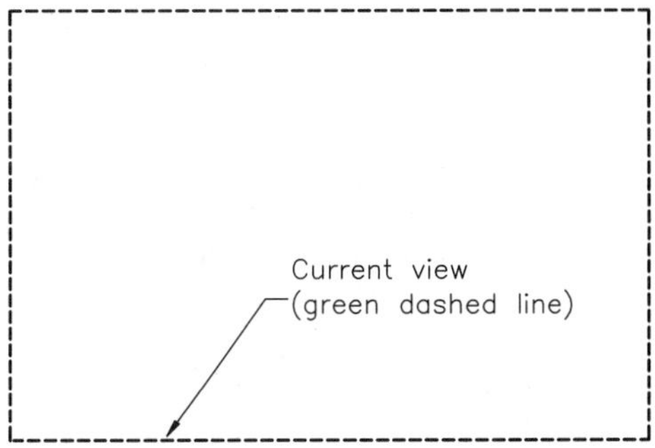

Figure 7-11 *Representation of the current view*

Panning view box (X in the center)

A view box initially of the same size as the current view box is displayed with an X in the center (Figure 7-12). You can move this box with the help of your pointing device. This box, known as the **panning view box**, helps you to find the center point of the zoomed display you want. When you have found the center, you press the pick button to make the zooming view box appear.

Zooming view box (arrow on the right side)

After you press the pick button in the center of the panning view box, the X in the center of the view box is replaced by an arrow pointing to the right edge of the box. This **zooming view box** (Figure 7-13) indicates the ZOOM mode. You can now increase or decrease the area of this box according to the area you want to zoom into. To shrink the box, move the pointer to the left; to increase it, move the pointer to the right. The top, right, and bottom sides of the zooming view box move as you move the pointer, but the left side remains fixed, with the zoom base point at the midpoint of the left side. You can slide it up or down along the left side. When you have the zooming view box in the desired size for your zoom display, press ENTER to complete the command and zoom into the desired area of the drawing. Before pressing ENTER, if you want to change the position of the zooming view box, click the pick

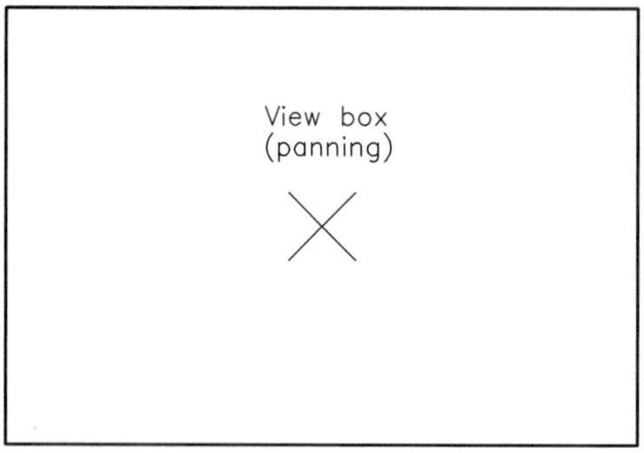

Figure 7-12 *The panning view box*

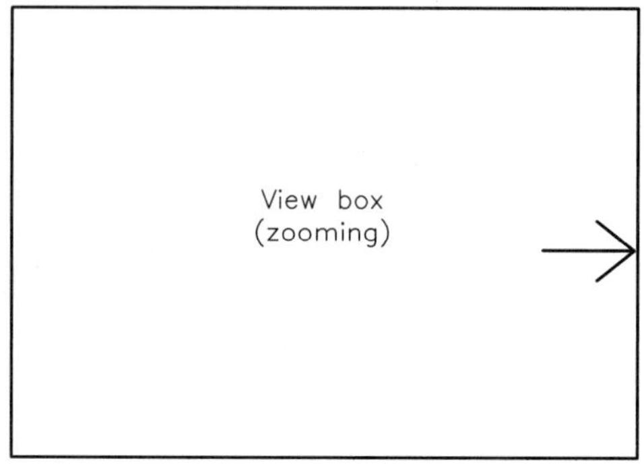

Figure 7-13 *The zooming view box*

button of your pointing device to make the panning view box reappear. After repositioning, press ENTER.

Previous Option

While working on a complex drawing, you may need to zoom in on a portion of the drawing to edit some minute details. Once the editing is over you may want to return to the previous view. This can be done using the **Previous** option of the **ZOOM** command. Without this option it would be very tedious to zoom back to previous views. AutoCAD saves the view specification of the current viewport whenever it is being altered by any of the ZOOM options or by the PAN, VIEW Restore, **DVIEW**, or **PLAN** commands (which are discussed later). Up to ten views are saved for each viewport. The prompt sequence for this option is given next.

Command: **ZOOM**
Specify corner of window, enter a scale factor (nX or nXP), or
[All/Center/Dynamic/Extents/Previous/Scale/Window]<real time>: **P**

Successive **ZOOM,** P commands can restore up to ten previous views. **VIEW** here refers to the area of the drawing defined by its display extents. If you erase some objects and then issue a **ZOOM** Previous command, the previous view is restored, but the erased objects are not.

Window Option

This is the most commonly used option of the **ZOOM** command. It lets you specify the area you want to zoom in on, by letting you specify two opposite corners of a rectangular window. The center of the specified window becomes the center of the new display screen. The area inside the window is magnified or reduced in size to fill the display as completely as possible. The points can be specified either by selecting them with the help of the pointing device or by entering their coordinates. The prompt sequence is given next.

Command: **ZOOM**

Specify corner of window, enter a scale factor (nX or nXP), or
[All/Center/Dynamic/Extents/Previous/Scale/Window]<real time>: *Specify a point.*
Specify opposite corner: *Specify another point.*

Whenever the **ZOOM** command is invoked, the window method is one of two default options. This is illustrated by the previous prompt sequence where you can specify the two corner points of the window without invoking any option of the **ZOOM** command. The **Window** option can also be used by entering **W**. In this case the prompt sequence is given next.

Command: **ZOOM** Enter
Specify corner of window, enter a scale factor (nX or nXP), or
[All/Center/Dynamic/Extents/Previous/Scale/Window]<real time>: **W** Enter
Specify first corner: *Specify a point.*
Specify opposite corner: *Specify another point.*

Scale Option

The **Scale** option of the **ZOOM** command is a very versatile option. It can be used in the following ways.

Scale: Relative to full view

This option of the **ZOOM** command lets you magnify or reduce the size of a drawing according to a scale factor (Figure 7-14). A scale factor equal to 1 displays an area equal in size to the area defined by the established limits. This may not display the entire drawing if the previous view was not centered on the limits or if you have drawn outside the limits. To get a magnification relative to the full view, you can enter any other number. For example, you can type 4 if you want the displayed image to be enlarged four times. If you want to decrease the magnification relative to the full view, you need to enter a number that is less than 1. In Figure 7-15, the image size decreased because the scale factor is less than 1. In other words, the image size is half of the full view because the scale factor is 0.5.

Figure 7-14 *Drawing before* ***ZOOM Scale*** *option*

Figure 7-15 *Drawing after* ***ZOOM Scale*** *option*

The prompt sequence is given next.

Command: **ZOOM** Enter
Specify corner of window, enter a scale factor (nX or nXP), or
[All/Center/Dynamic/Extents/Previous/Scale/Window]<real time>: **S** Enter
Enter a scale factor (nX or nXP) : **0.5** Enter

Scale: Relative to current view

The second way to scale is with respect to the current view (Figure 7-16). In this case, instead of entering only a number, enter a number followed by an **X**. The scale is calculated with reference to the current view. For example, if you enter **0.25X**, each object in the drawing will be displayed at one-fourth (1/4) of its current size. The following example increases the display magnification by a factor of 2 relative to its current value (Figure 7-17).

Command: **ZOOM** Enter
Specify corner of window, enter a scale factor (nX or nXP), or
[All/Center/Dynamic/Extents/Previous/Scale/Window]<real time>: **2X** Enter

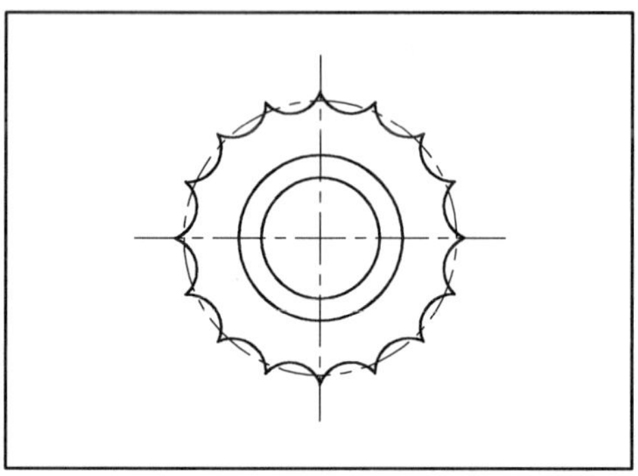

Figure 7-16 *Drawing before **ZOOM Scale** (X) option*

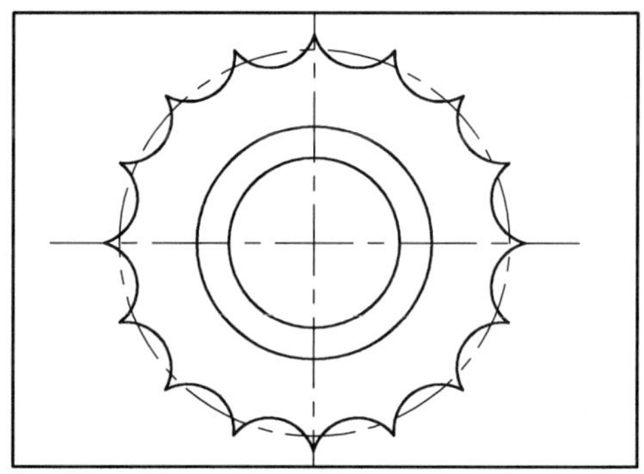

Figure 7-17 *Drawing after **ZOOM Scale** (X) option*

Scale: Relative to paper space units

The third method of scaling is with respect to paper space. You can use paper space in a variety of ways and for various reasons. For example, you can array and plot various views of your model in paper space. To scale each view relative to paper space units, you can use the **ZOOM XP** option. Each view can have an individual scale. The drawing view can be at any scale of your choice in a model space viewport. For example, to display a model space at one-fourth (1/4) the size of the paper space units, the prompt sequence is given next.

Command: **ZOOM** Enter
Specify corner of window, enter a scale factor (nX or nXP), or
[All/Center/Dynamic/Extents/Previous/Scale/Window]<real time>: **1/4XP** Enter

Note

For a better understanding of this topic, refer to "Model Space Viewports, Paper Space Viewports and Layouts" in Chapter 11.

Zoom In and Out

You can also zoom into the drawing using the **In** option, which doubles the image size.

Similarly, you can use the **Out** option to decrease the size of the image by half. To invoke these options from the command line, enter **ZOOM 2X** for the **In** option or **ZOOM .5X** for the **Out** option at the Command prompt. The center of the screen is taken as the reference point for enlarging or reducing the view of the drawing.

Exercise 1 *Mechanical*

Draw Figure 7-18 according to the given dimensions. Use the **ZOOM** command to get a bigger view of the drawing. Do not dimension the drawing.

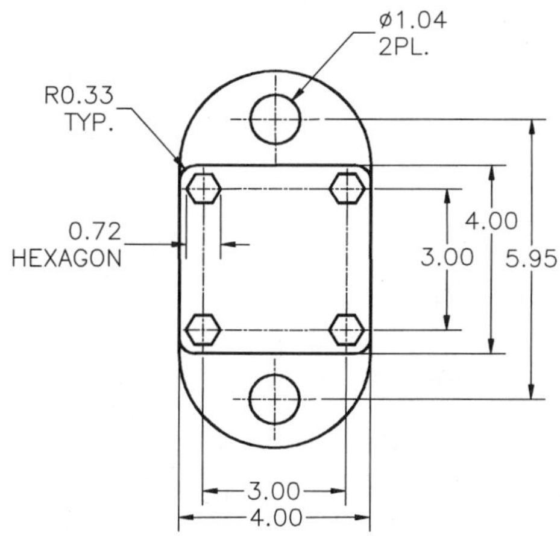

Figure 7-18 *Drawing for Exercise 1*

PANNING THE DRAWINGS

Toolbar:	Standard Toolbar > Pan
Menu:	View > Pan
Command:	PAN

You may want to view or draw on a particular area outside the current viewport. You can do this using the **PAN** command. If done manually, this would be like holding one corner of the drawing and dragging it across the screen. The **PAN** command allows you to bring into view portions of the drawing that are outside the display area of the current viewport. This is done without changing the magnification of the drawing. The effect of this command can be illustrated by imagining that you are looking at a big drawing through a window (**display window**) that allows you to slide the drawing right, left, up, and down to bring the part you want to view inside this window. You can invoke the **PAN** command from the shortcut menu also.

Panning in Realtime

You can use the **Realtime Pan** to pan the drawing interactively. To pan a drawing, invoke the command and then hold the pick button down and move the cursor in any direction. When you select the realtime pan, AutoCAD displays an image of a hand indicating that you are in PAN mode. **Realtime pan** is the default setting for the **PAN** command. Choosing the **Pan Realtime** button in the toolbar and entering **PAN** at the Command prompt automatically invokes the realtime pan. To exit the realtime pan, press ENTER or ESC, or choose **Exit** from the shortcut menu.

For the **PAN** command, there are various options that can be used to pan the drawing in a particular direction. These items can be invoked only from the menu (Figure 7-19).

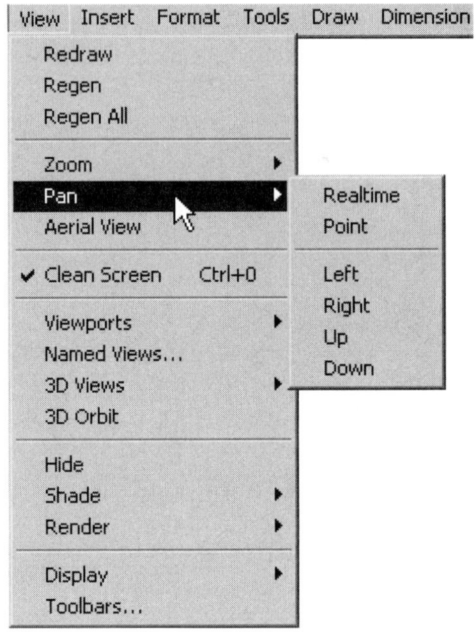

***Figure 7-19** Invoking the **PAN** command options from the **View** menu*

Point

In this option you are required to specify the displacement. To do this you need to specify in what direction to move the drawing and by what distance. You can give the displacement either by entering the coordinates of the points or by specifying the coordinates by using a pointing device. The coordinates can be entered in two ways. One way is to specify a single coordinate pair. In this case, AutoCAD takes it as a relative displacement of the drawing with respect to the screen. For example, in the following case the **PAN** command would shift the displayed portion of the drawing 2 units to the right and 2 units up.

Command: *Select the **Point** option from the **View** menu.*
Specify base point or displacement: **2,2** Enter
Specify second point: Enter

In the second case, you can specify two coordinate pairs. AutoCAD computes the displacement from the first point to the second. Here, displacement is calculated between point (3,3) and point (5,5).

Command: *Specify the Point item from the View menu.*
Specify base point or displacement: **3,3** Enter *(Or specify a point.)*
Specify second point: **5,5** Enter *(Or specify a point.)*

Left

Moves the drawing left so that some of the right portion of the drawing is brought into view.

Right
Moves the drawing right so that some of the left portion of the drawing is brought into view.

Up
Moves the drawing up so that some of the bottom portion of the drawing is brought into view.

Down
Moves the drawing down so that some of the top portion of the drawing is brought into view.

Tip
*You can use the scroll bars to pan the drawing vertically or horizontally. The scroll bars are located at the right side and the bottom of the drawing area. You can control the display of the scroll bars in the **Display** tab of the **Options** dialog box.*

CREATING VIEWS

Toolbar:	View> Named Views
Menu:	View > Named Views
Command:	View

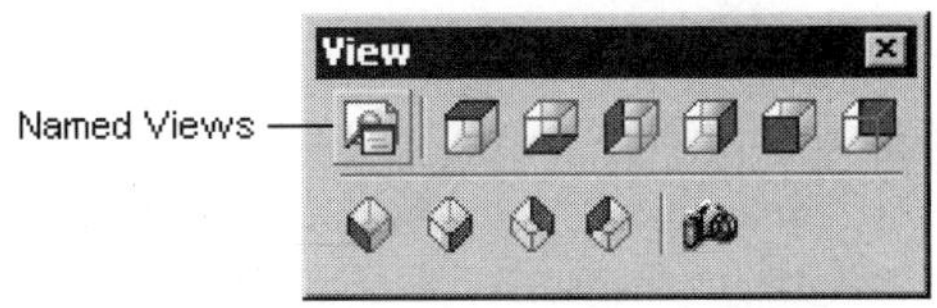

***Figure 7-20** Invoking **Named Views** from the **View** toolbar*

While working on a drawing, you may frequently be working with the **ZOOM** and **PAN** commands, and you may need to work on a particular drawing view (some portion of the drawing) more often than others. Instead of wasting time by recalling your zooms and pans and selecting the same area from the screen over and over again, you can store the view under a name and restore the view using the name you have given it. Choose the **Named Views** button available on the **View** toolbar (Figure 7-20) to invoke the **View** dialog box (Figure 7-21). This dialog box is used to save the current view under a name so that you can restore (display) it later. It does not save any drawing object data, only the view parameters needed to redisplay that portion of the drawing.

View Dialog Box
You can save and restore the views from the **View** dialog box (Figure 7-21). This dialog box is very useful when you are saving and restoring many view names. With this dialog box you can name the current view or restore some other view. The **Named Views** tab lists all the created named views . The **Orthographic and Isometric Views** tab lists all the preset views of the drawing, and allows you to set current any of those views. The following are the various options in the **Named Views** tab of the **View** dialog box.

Current View
The View list box displays a list of the named views in the drawing. The list appears with the names of all saved views and the space in which each was defined (Model space and paper space are discussed in Chapter 11).

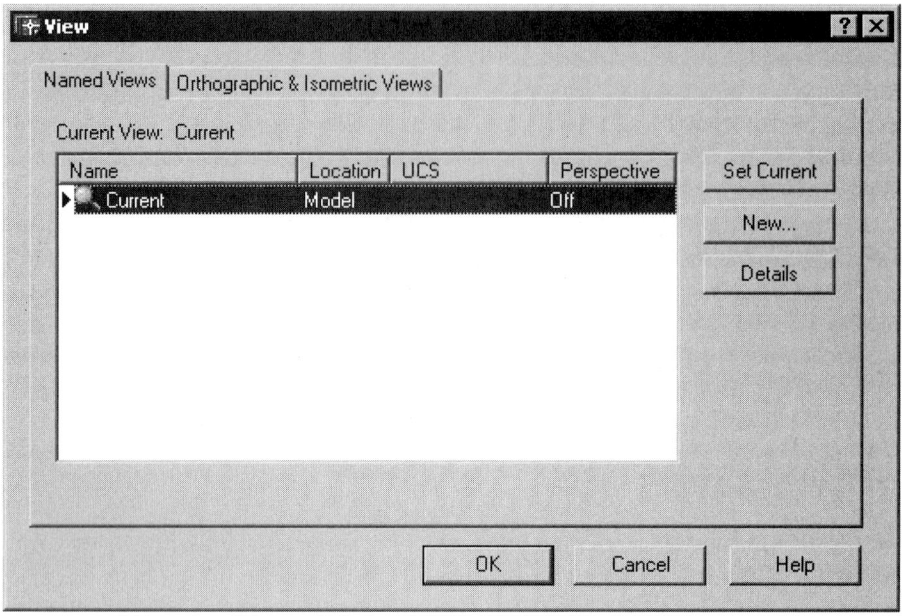

Figure 7-21 ***View*** *dialog box*

New

The **New** button allows you to create a new view and save it by giving it a name. When you choose the **New** button, the **New View** dialog box is displayed as shown in Figure 7-22.

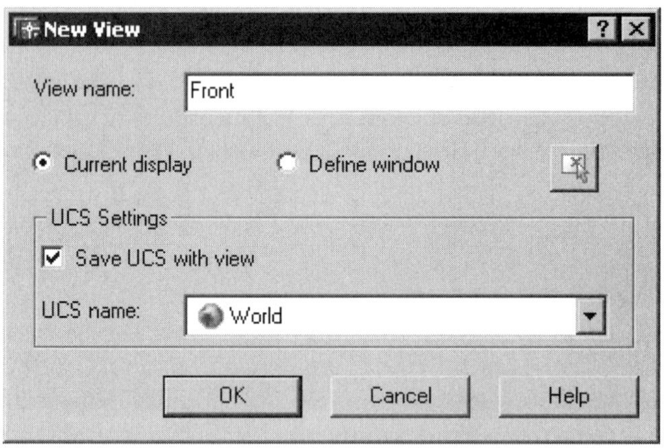

Figure 7-22 ***New View*** *dialog box*

Enter a name for the view in the **View name** edit box. If you want to save the current view, then activate the **Current display** radio button. If you want to save a rectangular portion of the current drawing as a view (without first zooming in on that area), then select the **Define window** radio button to activate it. You can specify two points on the screen to describe the window by choosing the **define view window** button or you can enter the *X* and *Y* coordinates in the **Specify first corner** and **Specify other corner** in the command lines. After specifying the view, choose the **Save View** button to save the named view.

Set Current

The **Set Current** button allows you to replace the current viewport by the view you specify. AutoCAD uses the center point and magnification of each saved view and executes a **ZOOM Center** with this information when a view is restored.

Note

If ***TILEMODE*** *is on, you cannot restore a Paper space view, refer to Chapter 11.*

Details

You can also see the description of the general parameters of a view by selecting the particular view and then choosing the **Details** button.

Tip

You can use the shortcut menu to rename and delete any named view in the dialog box. Highlight the named view and then right-click to display the shortcut menu.

Using the Command prompt

You can also use the **-VIEW** command to work with views at the Command prompt.

Command: **-VIEW** [Enter]
Enter an option [?/Orthographic/Delete/Restore/Save/Ucs/Window]:

You can use the various options to save, restore, delete, or list the views. If you try to restore a Model space while working in Paper space, AutoCAD automatically switches to floating model space. In this case AutoCAD will prompt you further as follows.

Enter view name to restore: *Select viewport for restoring*

You can select the viewport you want by selecting its border. This particular viewport must be on and active. The restored viewport also becomes the current one.

You can use the Window option to define the view. The prompt sequence is given next.

Command: **-VIEW** [Enter]
Enter an option [?/Orthographic/Delete/Restore/Save/UCS/Window]: **W** [Enter]
Enter view name to save: *Enter the name.*
Specify first corner: *Specify a point.*
Specify other corner: *Specify another point.*

Tip

*The various options of the **VIEW** command can be used transparently by entering **'-VIEW** at the Command prompt. The **VIEW** command that invokes the dialog box cannot be used transparently.*

AERIAL VIEW

Menu:	View > Aerial View
Command:	DSVIEWER

As a navigational tool, AutoCAD provides you with the option of opening another drawing display window along with the graphics screen window you are working on. This window, called the **Aerial View** window, can be used to view the entire drawing and select those portions you want to quickly zoom or pan. The AutoCAD graphics screen window resets itself to display the portion of the drawing you have selected for Zoom or Pan in the **Aerial View** window. You can keep the **Aerial View** window open as you work on the graphics screen, or minimize it so that it stays on the screen as a button which can be restored when required (Figure 7-23). As with all windows, you can resize or move it. The **Aerial View** window can also be invoked when you are in the midst of any command other than the **DVIEW** command.

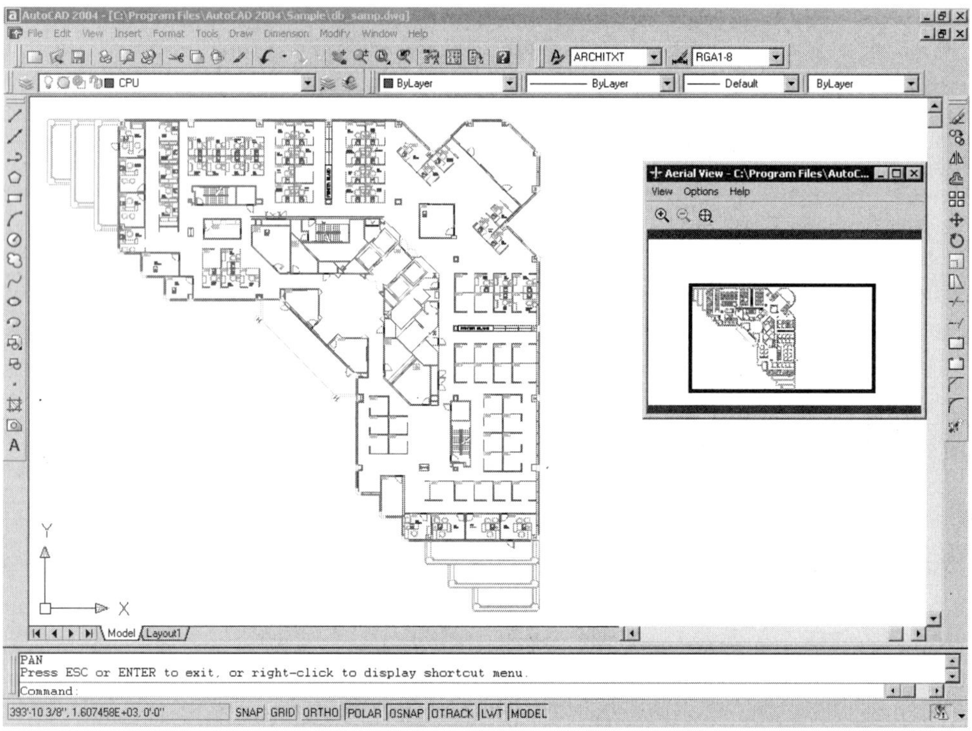

Figure 7-23 *AutoCAD graphics screen with the Aerial View window*

The **Aerial View** window has two menus, **View** and **Options**. It also has a toolbar containing zoom in, zoom out, and zoom global options. The following is a description of the available options in the **Aerial View**.

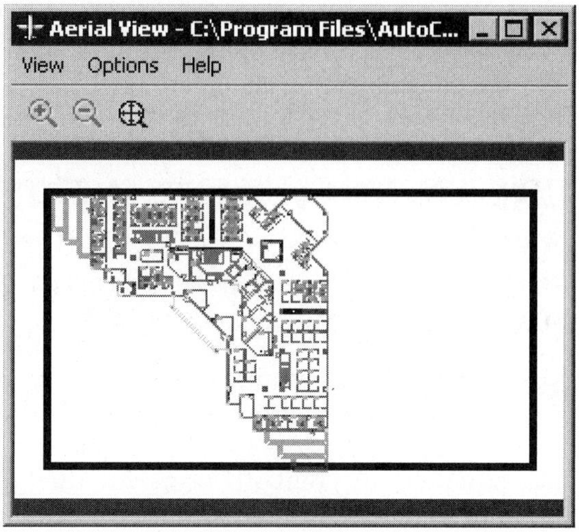

Figure 7-24 Aerial View window

Toolbar Buttons

The **Aerial View** window has three buttons: **Zoom In**, **Zoom Out**, and **Global** (Figure 7-24).

Zoom In

This option leads to magnification by a factor of **2** centered on the current view box.

Zoom Out

This option leads to reduction by **half** centered on the current view box.

Global

The **Global** option displays the complete drawing in the **Aerial View** window.

Menus

The menus available in the menu title bar are **View** and **Options**. The **View** menu contains **Zoom In**, **Zoom Out**, and **Global** options, which are the same as their respective buttons in the toolbar. The **Options** menu has the following options.

Auto Viewport

When you are working with viewports, you may need to change the view in the **Aerial View** window to display the current viewport view. You can achieve this by selecting the **Auto Viewport** option from the **Options** menu. If you are working in multiple viewports and you zoom or pan in the **Aerial View** window, only the view in the current viewport is affected.

Dynamic Update

When you make any changes in the current drawing, the view box is updated simultaneously in the **Aerial View** window if you have selected the **Dynamic Update** option. If **Dynamic Update** is unselected, the drawing is not updated simultaneously. The drawing in the **Aerial View** window is updated only when you move the cursor in the **Aerial View** window or invoke any of the menus or the toolbar buttons.

Realtime Zoom

If on, this option updates the drawing area when you are zooming in the **Aerial View** window in realtime.

CREATING TEXT

In manual drafting, lettering is accomplished by hand using a lettering device, pen, or pencil.

Chapter 7

This is a very time-consuming and tedious job. Computer-aided drafting has made this process extremely simple. Engineering drawings invoke certain standards to be followed in connection with the placement of a text in a drawing. In this section, you will learn how text can be added in a drawing by using **TEXT** and **MTEXT** commands.

CREATING SINGLE LINE TEXT

Toolbar:	Text > Single Line Text
Menu:	Draw > Text > Single Line Text
Command:	TEXT

The **TEXT** command lets you write text on a drawing and allows you to delete what has been typed by using the BACKSPACE key. You can enter multiple lines in one command. The **TEXT** command displays a line in the drawing area where you specified the start point after entering the height and the rotation angle. This line identifies the start point and the size of the text height entered. The characters appear on the screen as you enter them. When you press ENTER after typing a line, the cursor automatically places itself at the start of the next line and repeats the prompt **Enter text**. You can end the command by pressing the ENTER key or ESC key at the **Enter text** prompt.

The screen crosshairs can be moved irrespective of the cursor line for the text. If you specify a point, this command will complete the current line of the text and move the cursor line to the point you selected. This cursor line can be moved and placed anywhere on the screen; therefore, multiple lines of text can be entered at any desired location in the drawing area with a single **TEXT** command. By pressing BACKSPACE you can delete one character to the left of the current position of the cursor box. Even if you have entered several lines of text, you can use BACKSPACE and go on deleting until you reach the start point of the first line entered. Upon deletion of an entire line, **TEXT** displays a ***Deleted*** message in the Command prompt area.

This command can be used with most of the text alignment modes, although it is most useful in the case of left-justified texts. In the case of aligned texts, this command assigns a height appropriate for the width of the first line to every line of the text. Irrespective of the **Justify** option chosen, the text is first left-aligned at the selected point. After the **TEXT** command ends, the text is momentarily erased from the screen and regenerated with the requested alignment. For example, if you use the **Middle** option, the text will first appear on the screen as left-justified, starting at the point you designated as the middle point. After you end the command and press ENTER, it is regenerated with the proper alignment. The prompt sequence that follows when you choose this command is given next.

Specify start point of text or [Justify/Style]: *Specify the start point.*
Specify height <0.25>: **0.15** Enter
Specify rotation angle of text <0>: Enter
Enter text: *Enter first line of the text.*
Enter text: *Enter second line of the text.*
Enter text: Enter

Start Point Option

This is the default and the most commonly used option in the **TEXT** command. By specifying a start point, the text is left-justified along its baseline starting from the location of the starting point. Before AutoCAD draws the text, it needs the text height, the rotation angle for the baseline, and the text string to be drawn. It prompts you for all this information. The prompt sequence is given next.

Specify height <default>:
Specify rotation angle of text <default>
Enter text:

The **Specify height** prompt determines the distance by which the text extends above the baseline, measured by the capital letters. This distance is specified in drawing units. You can specify the text height by specifying two points or entering a value. In the case of a null response, the default height, that is, the height used for the previous text drawn in the same style, will be used.

The **Specify rotation angle of text** prompt determines the angle at which the text line will be drawn. The default value of the rotation angle is **0-degree** (3 o'clock, or east), and in this case the text is drawn horizontally from the specified start point. The rotation angle is measured in a counterclockwise direction. The last angle specified becomes the current rotation angle, and if you give a null response the last angle specified will be used as default. You can also specify the rotation angle by specifying two points. The text is drawn upside down if a point is specified at a location to the left of the start point.

As a response to the **Enter text:** prompt, enter the text string. Spaces are allowed between words. After entering the text, press ENTER.

Justify Option

AutoCAD offers various options to align the text. **Alignment** refers to the layout of the text. The main text alignment modes are **left, center,** and **right**. You can align a text using a combination of modes; for example, top/middle/baseline/bottom and left/center/right (Figure 7-25). **Top** refers to the line along which lie the top points of the capital letters; **Baseline** refers to the line along which lie their bases. Letters with descenders (such as p, g, y) dip below the baseline to the bottom.

When the **Justify** option is invoked, the user can place text in one of the fourteen various alignment types by selecting the desired alignment option. The orientation of the text style determines the command interaction for Text Justify. (Text styles and fonts are discussed later in this chapter). For now, assume that the text style orientation is horizontal. The prompt sequence that will follow when you choose the **Single Line Text** button to use this option is given next.

Specify start point of text or [Justify/style]: **J** Enter
Enter an option [Align/Fit/Center/Middle/Right/TL/TC/TR/ML/MC/MR/BL/BC/ BR]: *Select any of these options.*

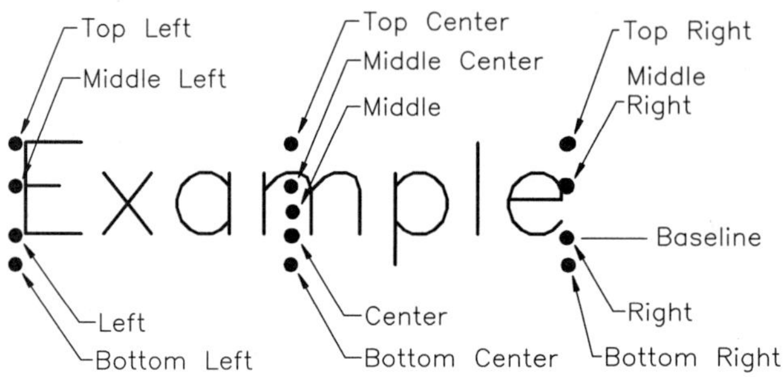

Figure 7-25 *Text alignment positions*

If the text style is vertically oriented (refer to the "Creating Text Styles" section later in this chapter), only four alignment options are available. The prompt sequence is as follows.

> Specify start point of text or [Justify/Style]: **J** Enter
> Enter an option [Align/Center/Middle/Right]:

If you know what justification you want, you can enter it directly at the **Specify start point of text or [Justify/Style]** prompt instead of first entering J to display the justification prompt. If you need to specify a style as well as a justification, you must specify the style first. The various alignment options are as follows.

Align Option

In this option the text string is written between two points (Figure 7-26). You must specify the two points that act as the endpoints of the baseline. The two points may be specified horizontally or at an angle. AutoCAD adjusts the text width (compresses or expands) so that it fits between the two points. The text height is also changed, depending on the distance between points and number of letters.

> Specify start point of text or [Justify/Style]: **J** Enter
> Enter an option [Align/Fit/Center/Middle/Right/TL/TC/TR/ML/MC/MR/BL/BC/BR]: **A** Enter
> Specify first endpoint of text baseline: *Specify a point.*
> Specify second endpoint of text baseline: *Specify a point.*
> Enter text: *Enter the text string.*

Fit Option

This option is very similar to the previous one. The only difference is that in this case you select the text height, and it does not vary according to the distance between the two points.

AutoCAD adjusts the letter width to fit the text between the two given points, but the height remains constant (Figure 7-26). The **Fit** option is not accessible for vertically oriented text styles. If you try the **Fit** option on the vertical text style, you will notice that the text string does not appear in the prompt. The prompt sequence is given next.

Enter an option [Align/Fit/Center/Middle/Right/TL/TC/TR/ML/MC/MR/BL/BC/BR]: **F**
Specify first endpoint of text baseline: *Specify a point.*
Specify second endpoint of text baseline: *Specify a point.*
Specify height<current>: *Enter the height.*
Enter text: *Enter the text.*

Note

You do not need to select the ***Justify*** *option (J) for selecting the text justification. You can enter the text justification by directly entering justification when AutoCAD prompts "Specify start point of text or [Justify/Style]:"*

Center Option

You can use this option to select the midpoint of the baseline for the text. This option can be invoked by entering **Justify** and then **Center** or **C**. After you select or specify the center point, you must enter the letter height and the rotation angle (Figure 7-26).

Specify start point of text or [Justify/Style]: C [Enter]
Specify center point of text: *Specify a point.*
Specify height<current>: **0.15** [Enter]
Specify rotation angle of text<0>: [Enter]
Enter text: **CENTER JUSTIFIED TEXT** [Enter]

Middle Option

Using this option you can center text not only horizontally, as with the previous option, but also vertically. In other words, you can specify the middle point of the text string (Figure 7-26). You can alter the text height and the angle of rotation to your requirement. The prompt sequence that will follow when you choose the **Single Line Text** button from the **Text** toolbar is given next.

Specify start point of text or [Justify/Style]: **M** [Enter]
Specify middle point of text : *Specify a point.*
Specify height<current>: **0.15** [Enter]
Specify rotation angle of text<0>: [Enter]
Enter text: **MIDDLE JUSTIFIED TEXT** [Enter]

Right Option

This option is similar to the default left-justified Start point option. The only difference is that the text string is aligned with the lower right corner (the endpoint you specify); that is,

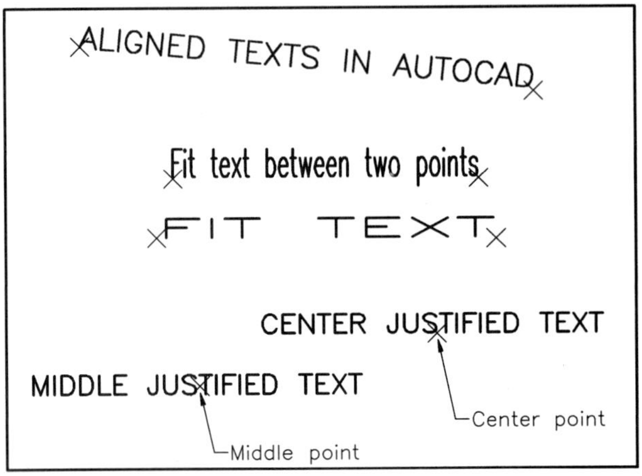

Figure 7-26 *Writing the text using* ***Align, Fit, Center,*** *and* ***Middle*** *options*

the text is **right-justified** (Figure 7-27). The prompt sequence that will follow when you choose this button is given next.

Specify start point of text or [Justify/Style]: R Enter
Specify right endpoint of text baseline: *Specify a point.*
Specify height<current>: **0.15** Enter
Specify rotation angle of text<0>: Enter
Enter text: **RIGHT JUSTIFIED TEXT** Enter

TL Option

In this option the text string is justified from the **top left** (Figure 7-27). The prompt sequence is given next.

Specify start point of text or [Justify/Style]: **TL** Enter
Specify top-left point of text: *Specify a point.*
Specify height<current>: **0.15** Enter
Specify rotation angle of text <0>: Enter
Enter text: **TOP/LEFT JUSTIFIED TEXT** Enter

Note

The rest of the text alignment options are similar to those just discussed, and you can try them on your own. The prompt sequence is almost the same as those given for the previous examples.

Style Option

With this option you can specify another existing text style. Different text styles can have different text fonts, heights, obliquing angles, and other features. This option can be invoked

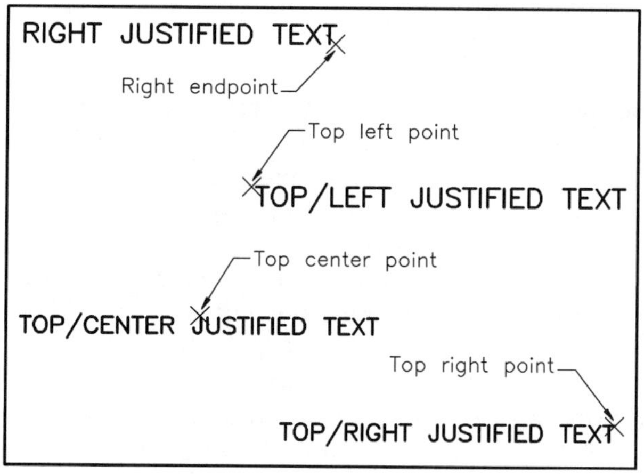

Figure 7-27 *Writing text using the Right, Top-Left, Top-Center, and Top-Right options*

by entering **TEXT** and then S at the next prompt. The prompt sequence that will follow when you choose the **Single Line Text** button for using this option is given next.

Specify start point of text or [Justify/Style]: **S** Enter
Enter style name or [?] <current>:

If you want to work in the previous text style, just press ENTER at the last prompt. If you want to activate another text style, enter the name of the style at the last prompt. You can also choose from a list of available text styles, which can be displayed by entering ?. After you enter ?, the next prompt is given next.

Enter text style(s) to list <*>:

Press ENTER to display the available text style names and the details of the current styles and commands in the **AutoCAD Text Window**.

Note

*With the help of the **Style** option of the **TEXT** command, you can select a text style from an existing list. If you want to create a new style, use the **STYLE** command, which is explained later in this chapter.*

DRAWING SPECIAL CHARACTERS

In almost all drafting applications, you need to draw **special characters** (symbols) in the normal text and in the dimension text. For example, you may want to draw the degree symbol (°) or the diameter symbol (ø), or you may want to underscore or overscore some text. This can be achieved with the appropriate sequence of control characters (control code). For each symbol, the control sequence starts with a percent sign written twice (%%). The character immediately following the double percent sign depicts the symbol. The control sequences for some of the symbols are given next.

Control sequence	**Special character**
%%c	Diameter symbol (ø)
%%d	Degree symbol (°)
%%p	Plus/minus tolerance symbol (±)
%%o	Toggle for overscore mode on/off
%%u	Toggle for underscore mode on/off
%%%	Single percent sign (%)

For example, if you want to draw **25° Celsius**, you need to enter 25%%dCelsius. If you enter **43.0%%c**, you get **43.0ø** on the drawing screen. To underscore (underline) text, use the **%%u** control sequence followed by the text to be underscored. For example, to underscore the text: **UNDERSCORED TEXT IN AUTOCAD**, enter **%%uUNDERSCORED TEXT IN AUTOCAD** at the prompt asking for text to be entered. To underscore and overscore a text string, include **%%u%%o** at the text string.

Note

*The special characters %%o and %%u act as toggles. For example, if you enter "**This %%utoggles%%u the underscore**", the word **toggles** will be underscored (toggles).*

None of these codes will be translated in the **TEXT** command until this command is complete. For example, to draw the degree symbol, you can enter **%%d**. As you are entering these symbols, they will appear as %%d on the screen. After you have completed the command and pressed ENTER, the code %%d will be replaced by the degree symbol (°).

You may be wondering why a percent sign should have a control sequence when a percent sign can easily be entered at the keyboard by pressing the percent (%) key. The reason is that sometimes a percent symbol is immediately followed (without a space) by a control sequence. In this case, the **%%%** control sequence is needed to draw a single percent symbol. To make the concept clear, assume you want to draw **67%±3.5.** Try drawing this text string by entering **67%%%p3.5.** The result will be **67%p3.5,** which is wrong. Now enter **67%%%%%p3.5** and notice the result on the screen. Here you obtain the correct text string on the screen, that is, **67%±3.5**. If there were a space between 67% and ±3.5, you could enter 67% %%p3.5 and the result would be **67% ±3.5**.

In addition to the control sequences shown earlier, you can use the %%nnn control sequence to draw special characters. The nnn can take a value in the range of 1 to 126. For example, to draw the & symbol, enter the text string **%%038**.

CREATING MULTILINE TEXT*

Toolbar:	Draw > Multiline Text
Menu:	Draw > Text > Multiline Text
Command:	MTEXT

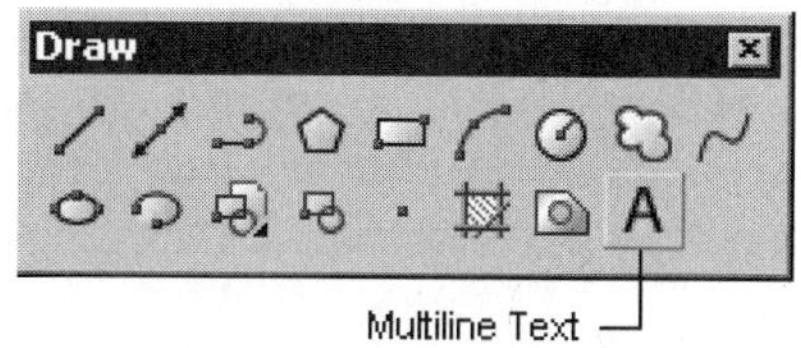

***Figure 7-28** Invoking **Multiline Text** from the **Draw** toolbar*

You can use the **MTEXT** command (Figure 7-28) to write a multiline text whose width can be specified by defining two corners of the text boundary or by entering a width, using coordinate entry. The text created by the **MTEXT** command is a single object regardless of the number of lines it contains.

Note

When you invoke the MTEXT command a sample text is attached to the cursor. By default the text "abc" is attached to the cursor. The MTJIGSTRING system variable stores the default contents of the sample text. You can specify a string of ten letters or numbers as the default sample text.

After specifying the width, you have to enter the text in the **Multiline Text Editor**. The following is the prompt sequence.

Command: **MTEXT** Enter
Current text style: "Standard". Text height: 0.2000
Specify first corner: *Select a point to specify first corner.*
Specify opposite corner or [Height/Justify/Line spacing/Rotation/Style/Width]: *Select an option or select a point to specify other corner.*

After selecting the first corner you can move the pointing device so that a box that shows the location and size of the paragraph text is formed. An arrow is displayed within the boundary, which indicates the direction of the text flow. When you define the text boundary, it does not mean that the text paragraph will fit within the defined boundary. AutoCAD only uses the width of the defined boundary as the width of the text paragraph. The height of the text boundary has no effect on the text paragraph. Once you have defined the boundary of the paragraph text, AutoCAD displays the **Multiline Text Editor** as shown in Figure 7-29.

Note

Although the box boundary that you specify controls the width of the paragraph, a single word is not broken to adjust inside the boundary limits. This means that if you write a single word whose width is more than the box boundary specified, AutoCAD will write the word irrespective of the box width, and therefore, will exceed the boundary limits.

The **Multiline Text Editor** consists of the **Text formatting toolbar**, **Text window** (with a ruler at the top), and **shortcut menu**. The following is the description of the options available in the **Text formatting** toolbar and the **shortcut menu**.

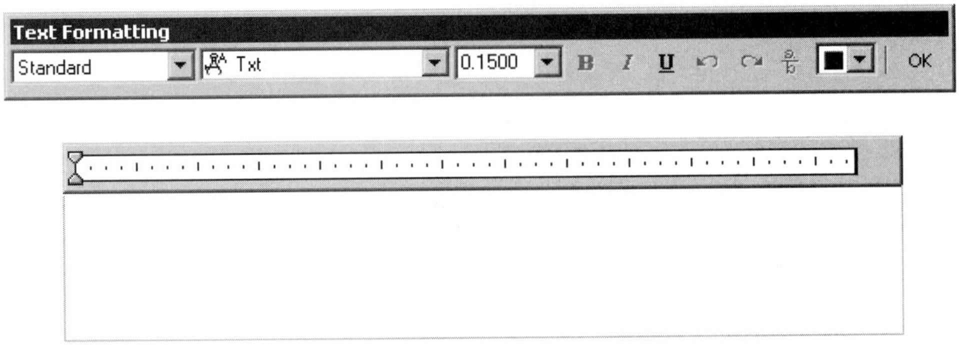

Figure 7-29 Multiline Text Editor

Text Formatting Toolbar

The options provided under this toolbar are as follows.

Style

The **Style** drop-down list is the first drop-down list available on the left of the **Text Formatting** toolbar. This drop-down list contains a list of all the text styles created in the current drawing. You can select the desired text style from this drop-down list. You can create a new text style using the **STYLE** command, which is explained later in this chapter.

Font

The **Font** drop-down list displays all the fonts available in AutoCAD. You can select the desired font from this drop-down list, see Figure 7-30.

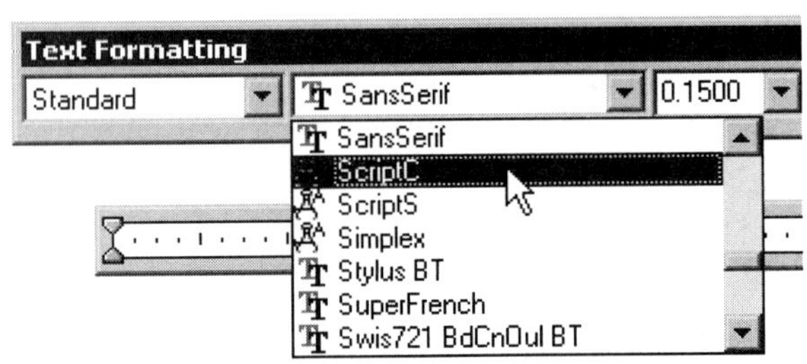

Figure 7-30 ***Font*** *drop-down list*

Note

Irrespective of the font assigned to a text style, you can assign a different font to that style for the current multiline text using the ***Font*** *drop-down list.*

Text Height

The **Text Height** edit box is used to specify the text height of the multiline text. The default value in this edit box is 0.2000. Once you modify the height, AutoCAD retains that value unless you change it. Remember that the **MTEXT** height does not affect the size specified for the **TEXT** command.

Bold, Italic, Underline

You can use the appropriate tool buttons located on the text box to make the selected text boldface, or italics, or create underlined text. Boldface and italics are not supported by SHX fonts and hence will not be available for the particular fonts. These three buttons toggle between on and off.

Undo

The **Undo** button allows you to undo the actions in the **Multiline Text Editor**. You can also press the CTRL+Z keys to undo the previous actions.

Redo

The **Redo** button allows you to redo the actions in the **Multiline Text Editor**. You can also press the CTRL+Y keys to redo the previous actions.

Stack

To create a fraction text you must use the stack button with special characters /, ^, and #. The character / stacks the text vertically with a line, and the character ^ stacks the text vertically without a line (tolerance stack). The character # stacks the text with a diagonal line. After you enter the text with the required special character between them, select the text, and then select the **Stack/Unstack** button. The stacked text is displayed equal to 70 percent of the actual height. If you enter two numbers separated by / or ^ and then press the Enter key, AutoCAD displays the **AutoStack Properties** dialog box (Figure 7-31). You can use this dialog box to control the stacking properties.

Figure 7-31 ***AutoStack Properties*** *dialog box*

Color

The **Color** drop-down list is used to set the color for the multiline text. You can also select the color from the **Select Color** dialog box that is displayed by selecting **Select Color** in the drop-down list.

Text Window

The **Text window** is used to enter the multiline text. The width of the active text area is determined by the width of the window that you specify when you invoke the **MTEXT** command. You can increase the size of the dialog box by dragging the right or the bottom edge of the box. You can also use the scroll bar to move up or down to display the text.

The ruler available on the top of the text window is used to specify the indentation of the current paragraph. The top slider of the ruler specifies the indentation of the first line of the paragraph while the bottom slider specifies the indentation of the other lines of the paragraph.

Shortcut Menu

In the **Text window**, right-click to display the shortcut menu shown in Figure 7-32. The shortcut menu provides various options to edit the multiline text. To edit the text, select it and then right-click. The shortcut menu is displayed. The **Undo** and the **Redo** options are used to undo or redo the last actions done in the Multiline Text editor. The **Cut** and **Copy** options can be used to move or copy the text from the text editor to any other application. Similarly using the **Paste** option, you can paste text from any windows text-based application to the **Multiline Text Editor**. The remaining options are discussed next.

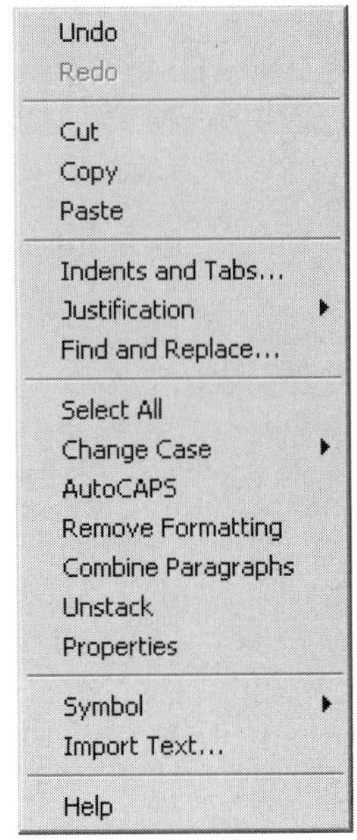

Figure 7-32 *Shortcut menu*

Indents and Tabs

The **Indents and Tabs** option allows you to set the indentation of the multiline text and the tab position. Note that as mentioned earlier, these options can also be set using the ruler and the sliders available on top of the drawing window. When you choose **Indents and Tabs** from the shortcut menu, the **Indents and Tabs** dialog box is displayed as shown in Figure 7-33. The various options available in the dialog box are discussed next.

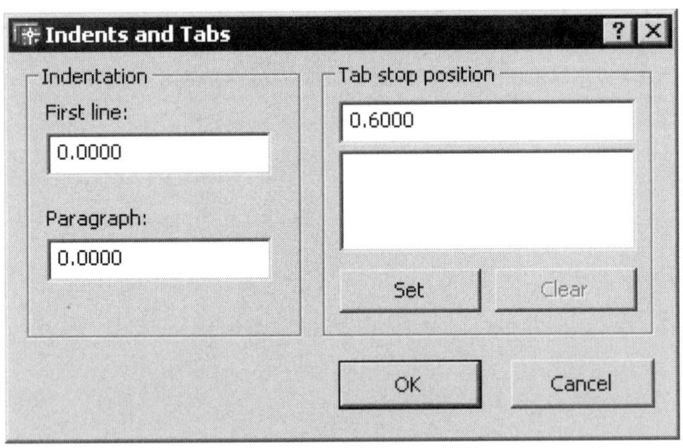

***Figure 7-33 Indents and Tabs** dialog* box

Indentation Area. You can set the indentation for the first line and the following lines of the multiline text by entering the values in the **First Line** and the **Paragraph** edit boxes.

Tab stop position Area. The **Tab stop position** area is used to set the position up to which the cursor will move when you press the TAB key once at the starting of a new paragraph. This position is called the tab position. To set the tab position, enter the value in the edit box available in this area and then choose the **Set** button. You will notice that the value appears in the list box available below this area. You can choose the **Clear** button to delete the selected tab position value. Notice that a mark appears in the ruler that defines the tab stop position.

Justification

In large complicated technical drawings, the **Justification** option is used to fit the text matter along a particular width. This option is used to control the justification and alignment of the text paragraph. For example, if the text justification is bottom-right (BR), the text paragraph will spill to the left and above the insertion point, regardless of how you define the width of the paragraph. When you select **Justification** from the shortcut menu, a cascading menu appears that displays the predefined text justifications. By default, the text is Top Left justified. You can choose the new justification from the cascading menu. The various justifications are TL, ML, BL, TC, MC, BC, TR, MR, BR. Figure 7-34 shows various text justifications for multiline text.

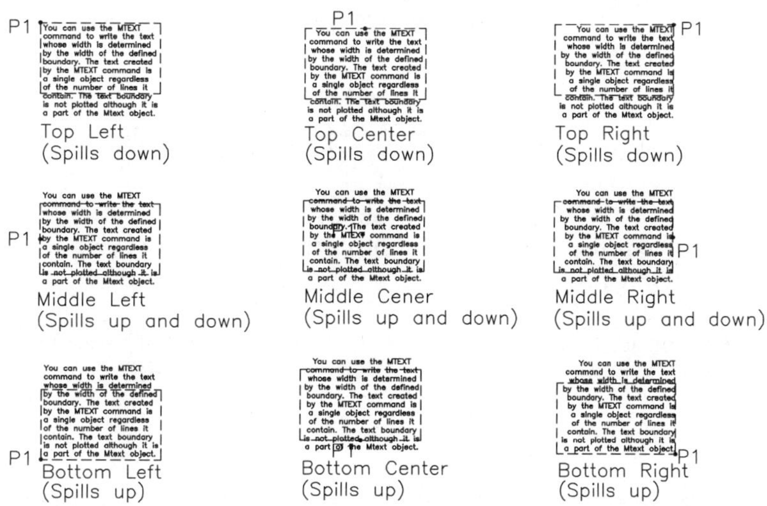

Figure 7-34 *Text justifications for **MTEXT** (P1 is the text insertion point)*

Find and Replace

When you choose this option, the **Replace** dialog box is displayed as shown in Figure 7-35. The various options available in this dialog box are discussed next.

Find what. You can define the text string that you need to find in this text box. You can locate a part of a word or the complete word mentioned in this text box.

Replace with. If you want some text to be replaced, enter new text for replacement in this edit box.

Match whole word only. If this check box is selected, AutoCAD will match the word in the text only if it is a single word identical to that mentioned in the **Find what** text box. For example, if you enter **and** in the **Find what** text box and select the **Match whole word only**

Figure 7-35 ***Replace*** *dialog box*

check box, AutoCAD will not match the string **and** in words like sand, land, band, and so on.

Match case. If this check box is selected, AutoCAD will match the word only if the case of all the characters in the word is identical to that of the word mentioned in the **Find what** text box.

Find Next. Choose this button if you want to continue the search for the text entered in the **Find what** box.

Replace. Choose this button if you want to replace the highlighted text with the text entered in the **Replace with** text box.

Replace All. If you choose this button, all the words in the current multiline text that match the word specified in the **Find what** text box will be replaced with the word entered in the **Replace with** box.

Select All

This option is used to select the complete text entered in the **Mutiline Text Editor**.

Change Case

This option is used to change the case of the selected text to uppercase or lowercase.

AutoCAPS

If you choose this option, the case of all the text written or imported after choosing this option will be changed to uppercase. However, the case of the text written before choosing this option is not changed.

Remove Formatting

This option is used to remove the formatting such as bold, italics, or underline from the selected text. To use this option, select the text whose formatting you need to change and then right-click to display the shortcut menu. In the menu, choose the **Remove Formatting** option. The formatting of the selected text will be removed.

Combine Paragraphs

This option is used to combine the selected paragraphs into a single paragraph. AutoCAD replaces the returns between all the paragraphs by a space. As a result, the lines in the resultant paragraph are in continuation.

Stack

The **Stack** option is used to stack the selected text if there are any stack characters (characters separated by /, #, or ^) available in the multiline text. Note that this option is available only if you select stack characters from the text.

Unstack

The **Unstack** option is used to unstack the selected stacked text. This option is available only if there are some stacked characters in the text.

Properties

This option is available only when you select a stacked text. When you choose this option, the **Stack Properties** dialog box is displayed as shown in Figure 7-36. This dialog box is used to edit the text and the appearance of the selected stacked text. The various options available in this dialog box are discussed next.

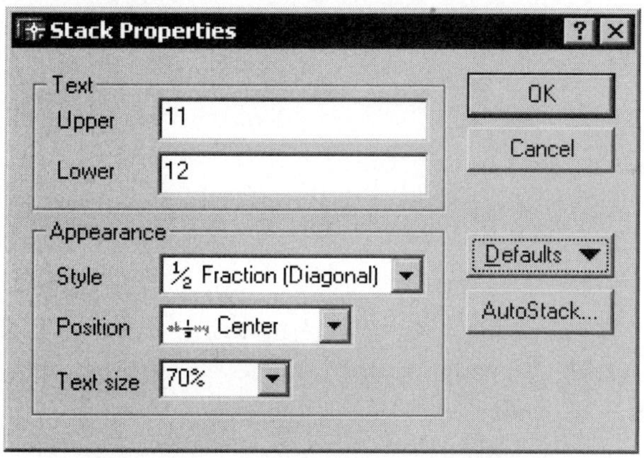

Figure 7-36 ***Stack Properties*** *dialog box*

Text Area. You can change the upper and the lower values of the stacked text by entering their values in the **Upper** and **Lower** text boxes respectively.

Appearance Area. You can change the style, position, and size of the stacked text by entering their values in the **Style**, **Position**, and **Text size** text boxes respectively.

Default. This option allows you to either restore the default values or save the new settings as the default settings for the selected stacked text.

AutoStack. When you choose this button, the **AutoStack Properties** dialog box is displayed as shown in Figure 7-31. The options in this dialog box were discussed under the **Stack** heading.

Symbol

This option is used to insert the special characters in the text. When you choose **Symbol** from the shortcut menu, a cascading menu appears that displays some predefined special characters. You can also choose **Other** from the cascading menu to display the **Character Map** dialog box. This dialog box has a number of other special characters that you can insert in multiline text. To insert the characters from the dialog box, select the character you want to copy and then choose the **Select** button. Once you have selected all the required special characters, choose the **Copy** button and then close the dialog box. Now, in the Text window of the **Multiline Text Editor**, position the cursor where you want to insert the special characters and right-click to display the shortcut menu. Choose **Paste** to insert the selected special character in the **Multiline Text Editor**.

Import Text

When you choose this option, AutoCAD displays the **Select File** dialog box. In this dialog box you can select any text file you want to import as the multiline text. The imported text is displayed in the text area. Note that only the ASCII or RTF files are interpreted properly.

Example 1 — *General*

In this example you will use the **Multiline Text Editor** dialog box to write the following text on the screen.

For long, complex entries, create multiline text using the MTEXT option. The angle is 10-degree. Diameter = 1/2" and Length = 32 1/2".

The font of the text is **Swis721 BT,** text height is 0.20, color red, and written at an angle of 10-degree with Middle-Left justification. Make the word "multiline" bold, underline the text "multiline text", and make the word "angle" italic. The line spacing type and line spacing between the lines are At least and 1.5x respectively. Use the symbol for degrees and replace the word "option" with "command".

1. The first step is to choose the **Multiline Text** button from the **Draw** toolbar. You can also invoke this command from the **Draw** menu or enter the command at the Command prompt. After invoking the command, specify the first corner on the screen to define the first corner of the paragraph text boundary. You need to specify the rotation angle of the text before specifying the second corner of the paragraph text boundary. The prompt sequence is given next.

   Current text style: STANDARD. Text height: 0.1000
   Specify first corner: *Select a point to specify first corner.*
   Specify opposite corner or [Height/Justify/Line spacing/Rotation/ Style/ Width]: *Enter R to specify the rotation angle.*

Specify rotation angle <0>: 10.
Specify opposite corner or [Height/Justify/Line spacing/Rotation/ Style/Width]: *Enter L to specify the line spacing.*
Enter line spacing type [At least/Exactly] <At least>: Enter
Enter line spacing factor or distance <1x>: 1.5x
Specify opposite corner or [Height/Justify/Line spacing/Rotation/Style/Width]: *Select another point to specify the other corner.*

The **Multiline Text Editor** is displayed.

2. Select **Swis721 BT** true type font from the **Font** drop-down list.

3. Enter **0.20** in the **Text height** edit box, if the value in this edit box is not 0.2.

4. Select **Red** from the **Color** drop-down list.

5. Now enter the text in the **Multiline Text Editor** as shown in Figure 7-37. To add the degrees symbol, right-click on the text window and choose **Symbol > Degrees** from the shortcut menu. When you type 1/2 after Dia = and then press the " key, AutoCAD displays the **AutoStack Properties** dialog box. Select **Convert it to a horizontal fraction** radio button, if it is not already selected. Also, make sure the **Enable AutoStacking** check box is checked. Now, close the dialog box.

   Similarly, when you type 1/2 after length = 32 1/2 and then press the " key, AutoCAD displays the **AutoStack Properties** dialog box. Select **Convert it to a diagonal fraction** radio button and make sure the **Enable AutoStacking** check box is checked. Close the dialog box.

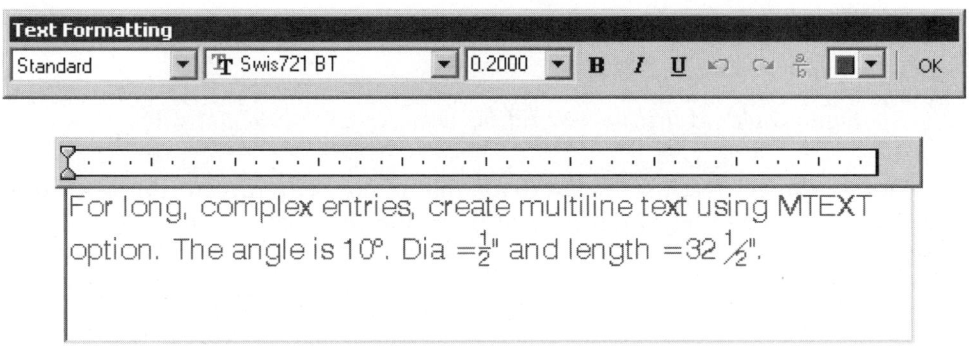

***Figure 7-37 Multiline Text Editor** dialog box*

6. Double-click on the word "multiline" to select it and then choose the **Bold** button to make it boldface. As it is already selected select the **Underline** button to underline it.

7. Similarly, highlight the word "text" and then select the **Underline** button to underline it.

8. Highlight the word "angle" by double-clicking on it (or pick and drag to select the text) and then choose the **Italic** button.

9. Right-click on the text window and choose **Justification > Middle Left** from the shortcut menu.

10. Click at the start of the multiline to move the cursor to the start. Now, right-click on the text window and choose **Find and Replace** from the shortcut menu. The **Replace** dialog box is displayed.

11. In the **Find what** edit box, enter **option** and in the **Replace with** edit box, enter **command**.

12. Choose the **Find Next** button. AutoCAD finds the word "option" and highlights it as shown in Figure 7-38. Choose the **Replace** button to replace **option** by **command**. The **AutoCAD** information box is displayed informing you that AutoCAD has finished searching for the word. Choose **OK** to close the information box.

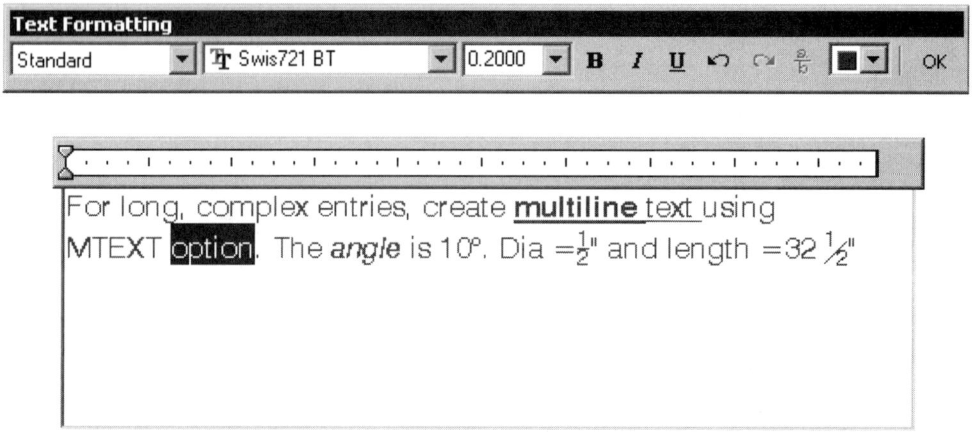

***Figure 7-38 Multiline Text Editor** with the selected word highlighted*

13. Now, choose **Cancel** to close the **Replace** dialog box to return to the **Multiline Text Editor**. Choose the **OK** button to exit the **Multiline Text Editor**. The text is displayed on the screen as shown in Figure 7-39.

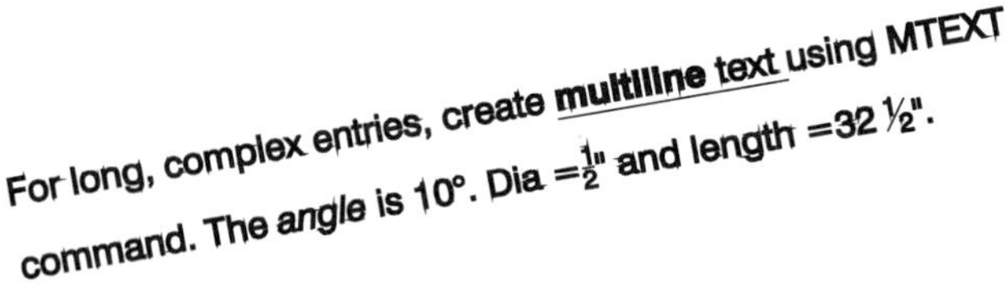

***Figure 7-39** Multiline text for Example 1*

Tip

*The text in the **Multiline Text Editor** can be selected by double-clicking on the word to select a word, by holding down the left mouse button of the pointing device and then dragging the cursor, or by triple-clicking on the text to select the entire line or paragraph.*

Exercise 2 *General*

Draw the text on the screen as shown in Figures 7-40 and 7-41. Use the special characters and text justification options shown in the drawing using the **TEXT** and **MTEXT** commands. The text height is 0.1 and 0.15 for Figures 7-40 and 7-41.

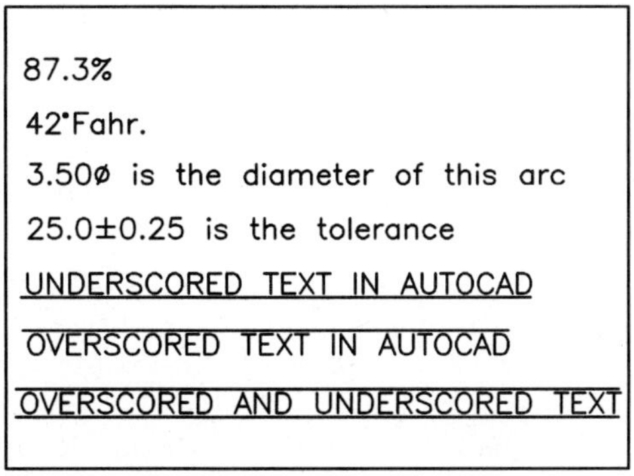

***Figure 7-40** Drawing with special characters*

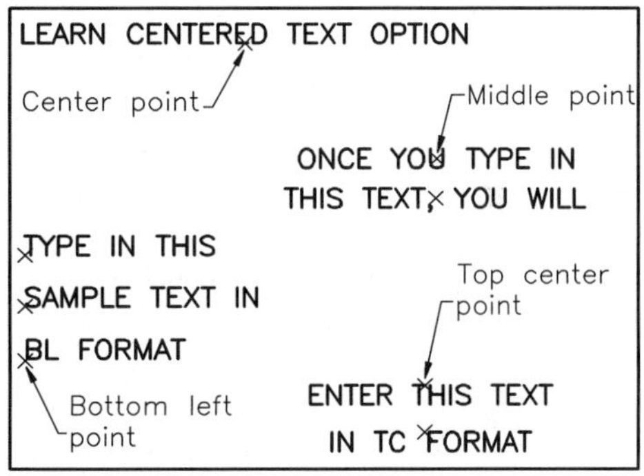

***Figure 7-41** Drawing for Exercise 2*

EDITING TEXT

The contents of **MTEXT** and **TEXT** object can be edited by using the **DDEDIT** and **PROPERTIES** commands. You can also use the AutoCAD editing commands, such as **MOVE**, **ERASE**, **ROTATE**, **COPY**, **MIRROR**, and **GRIPS** with any text object.

In addition to editing, you can also modify the text in AutoCAD. The modification that you can perform on the text include changing the scale and the justification of the text. The various editing and modifying operations are discussed next.

Editing Text Using DDEDIT Command

Toolbar:	Text > Edit Text
Menu:	Modify > Object > Text > Edit
Command:	DDEDIT

You can use the **DDEDIT** command to edit the text. If you select **TEXT** object, AutoCAD displays the **Edit Text** dialog box as shown in Figure 7-42 in which the selected text is displayed in the **Text** edit box. You can make the changes in the text string only and then choose the **OK** button. For the text object, you cannot modify any of its properties in the **Edit Text** dialog box. However, if you select multiline text created using the **MTEXT** command, AutoCAD displays the text in the **Multiline Text Editor**. You can make

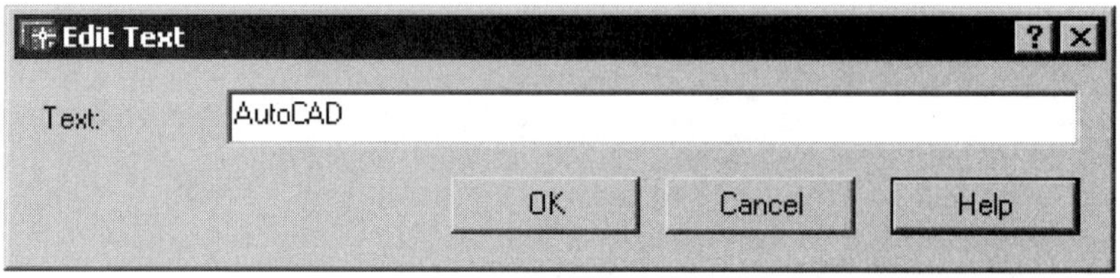

Figure 7-42 *Using the* ***Edit Text*** *dialog box to edit text*

the changes using the various options in the editor. Apart from changing the text string you are also allowed to change the properties of the paragraph text.

Command: **DDEDIT**
Select an annotation object or [Undo]: *Select a text object*

You can also invoke the **Edit Text** dialog box and the **Multiline Text Editor** from the shortcut menu. Select the text for editing and then right-click in the drawing area. A shortcut menu is displayed and depending on the text object you have selected, the **Text Edit** or **Mtext Edit** options are available in the menu. Selecting these options displays the respective dialog boxes.

Tip
You can also simply double-click on a text object or a multiline text object to display the ***Edit Text*** *dialog box or the* ***Multiline Text Editor*** *respectively without selecting or entering the* ***DDEDIT*** *command.*

Editing Text Using the PROPERTIES Palette

Using the **DDEDIT** command with the text object, you can only change the text string and not its properties such as the height, angle, and so on. In this case, you can use the **PROPERTIES** palette for changing the properties. Select the text and choose the **Properties** button in the **Standard** toolbar. AutoCAD displays the **PROPERTIES** palette with all the properties of the selected text as shown in Figure 7-43. Here you can change any value and also the text string. If you are editing a single line text, you can change the text string in the window. But for a multiline text you must choose the **Full editor** button in the **Content** edit box. AutoCAD automatically switches to the **Multiline Text Editor**, where you can make changes to the paragraph text.

Modifying the Scale of the Text

Toolbar:	Text > Scale Text
Menu:	Modify > Object > Text > Scale
Command:	SCALETEXT

You can modify the scale factor of the text using the **Scale Text** button in the **Text** toolbar. AutoCAD uses one of the justification options as the base point to scale the text. The prompt sequence that follows when you invoke this tool is given next.

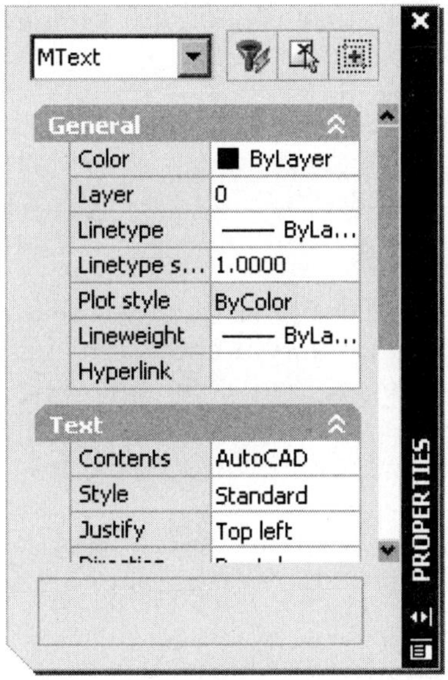

Figure 7-43 ***PROPERTIES*** *palette*

Select objects: *Select the text object to scale*
Select objects: [Enter]
Enter a base point option for scaling
[Existing/Left/Center/Middle/Right/TL/TC/TR/ML/MC/MR/BL/BC/BR] <current>:
Specify an option that will be used as the base point to scale the text.
Specify new height or [Match object/Scale factor] <0.2000>: *Enter the new height or select an option*

Match object

You can use the **Match object** option to select an existing text whose height will be used to scale the selected text.

Scale factor

You can use the **Scale factor** option to specify a scale factor to scale the text. You can also use the **Reference** option to specify the scale factor for the text.

Modifying the Justification of the Text

Toolbar:	Text > Justify Text
Menu:	Modify > Object > Text > Justify
Command:	JUSTIFYTEXT

You can modify the justification of the text using the **Justify Text** button in the **Text** toolbar. Note that even after modifying the justification using this command, the location of the text is not changed. The prompt sequence that follows when you invoke this tool is given next.

Select objects: *Select the text object whose justification needs to be changed*
Select objects: [Enter]
Enter a base point option for scaling
[Existing/Left/Center/Middle/Right/TL/TC/TR/ML/MC/MR/BL/BC/BR] <current>:
Specify the new justification for the text.

SUBSTITUTING FONTS

AutoCAD provides you the facility to designate the fonts that you want to substitute for other fonts used in the drawing. The information about font mapping is specified in the font mapping file (*acad.fmp*). The font mapping has the following advantages.

1. You can specify a font to be used when AutoCAD cannot find a font used in the drawing.

2. You can enforce the use of a particular font in your drawings. If you load a drawing that uses different fonts, you can use font mapping to substitute the desired font for the fonts used in the drawing.

3. You can use .shx fonts when creating or editing a drawing. When you are done and ready to plot the drawing, you can substitute other fonts for .shx fonts.

The font mapping file is an ASCII file with FMP extension containing one font mapping per line. The format on the line is given next.

Base name of the font file;Name of the substitute font with extension (ttf, shx, etc.)

For example, if you want to substitute the ROMANC font for SWISS.TTF; the entry is given next.

SWISS;ROMANC.SHX

You can enter this line in the *acad.fmp* file or create a new file. If you create a new font mapping file, you need to specify this new file. You can use the **Options** dialog box to specify the new font mapping file. To specify a font mapping table in the **Options** dialog box, choose **Options** from the **Tools** menu to display the **Options** dialog box. Choose the **Files** tab and click on the **plus** sign next to **Text Editor, Dictionary, and Font File Names**. Now, click on the plus sign next to **Font Mapping File** to display the path and the name of the font mapping file. Double-click on the file to display the **Select a file** dialog box. Select the new font mapping file and exit the **Options** dialog box. At the Command prompt enter **REGEN** to convert the existing text font to the font as specified in the new font mapping file. You can also use the **FONTMAP** system variable to specify the new font map file.

Command: **FONTMAP** [Enter]
Enter new value for FONTMAP, or . for none <"path and name of the current font mapping file">: *Enter the name of the new font mapping file.*

The following file is a partial listing of *acad.fmp* file with the new font mapping line added (swiss;romanc.shx).

swiss;romanc.shx
cibt;CITYB___.TTF
cobt;COUNB___.TTF
eur;EURR____.TTF
euro;EURRO___.TTF
par;PANROMAN.TTF
rom;ROMANTIC.TTF
romb;ROMAB___.TTF
romi;ROMAI___.TTF
sas;SANSS___.TTF
sasb;SANSSB__.TTF
sasbo;SANSSBO_.TTF
saso;SANSSO__.TTF

Note
The text styles that were created using the PostScript fonts are substituted with an equivalent TrueType font and plotted using the substituted font.

Specifying an Alternate Default Font

When you open a drawing file that specifies a font file that is not on your system or is not specified in the font mapping file, AutoCAD, by default, substitutes the *simplex.shx* font file. You can specify a different font file in the **Options** dialog box or do so by changing the **FONTALT** system variable.

Command: **FONTALT** Enter
New value for FONTALT, or . for none <“simplex.shx”>: *Enter the font file name.*

CREATING TEXT STYLES

Toolbar:	Text > Text Style
Menu:	Format > Text Style
Command:	STYLE

By default, whenever you write a text in AutoCAD, it is written using the default text style called **Standard**. This text style is assigned a default text font (TXT.SHX) and the default formatting. However, if you need to write a text using some other font and other parameters, you need to use the **Multiline Text Editor**. This is because only using this command you can change the formatting and font of the text.

However, it is a very tedious job to use the **Multiline Text Editor** overtone to write the text and change its properties. This is the reason AutoCAD provides you with an option of modifying the default text style or creating a new text style. After creating a new text, you can make it current. All the text written after making the new style current will use this style.

To create a new text style or to modify the default style, choose the **Text Style** button from the **Text** toolbar. The **Text Style** dialog box is displayed as shown in Figure 7-44.

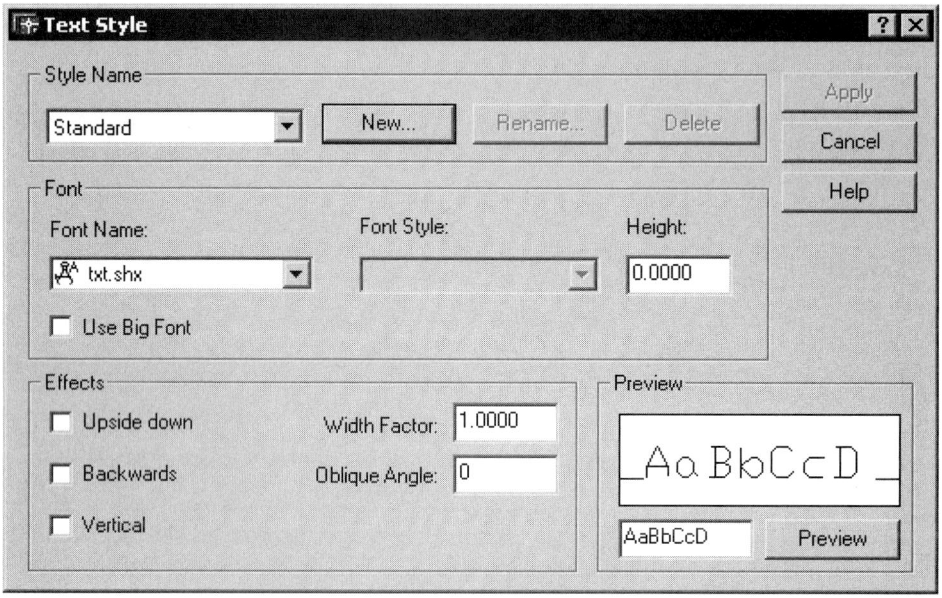

***Figure 7-44 Text Style** dialog box*

In the **Style Name** edit box the default style (Standard) will be displayed. For creating a new style, choose the **New** button to display the **New Text Style** dialog box (Figure 7-45) and enter the name of the style you want to create.

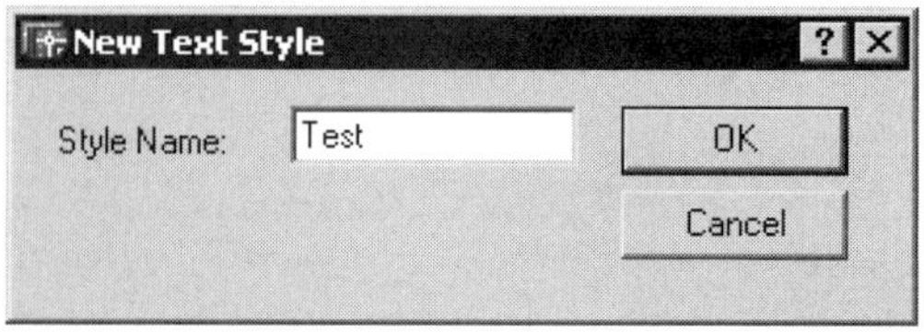

***Figure 7-45 New Text Style** dialog box*

A new style having the entered name and the properties present in the **Text Style** dialog box will be created. To modify this style, select the style name from the list box and then change the different settings by entering new values in the appropriate boxes. You can change the font by selecting a new font from the **Font Name** drop-down list. Similarly, you can change the text height, width, and oblique angle.

Remember that if you have already specified the height of the text in the **Text Style** dialog box, AutoCAD will not prompt you to enter the text height while writing the text using the **TEXT** command. The text will be created using the height specified in the text style. If you want AutoCAD to prompt you for the text height, specify 0 text height in the dialog box. For

Width Factor, 1 is the default. If you want the letters expanded, enter a width factor greater than 1. For compressed letters, enter a width factor less than 1. Similarly, for the **Oblique Angle**, 0 is the default. If you want the slant of the letters toward the right, the value should be greater than 0; to slant the letters toward the left, the value should be less than 0. You can also force the text to be written upside down, backwards, and vertically by checking their respective check boxes. As you make the changes, you can see their effect in the **Preview** box. After making the desired changes, choose the **Apply** button and the **Close** button to exit the dialog box. Figure 7-46 shows text objects with all these settings.

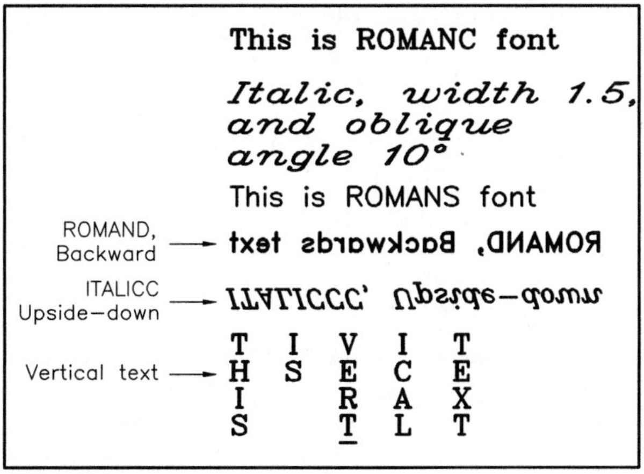

Figure 7-46 *Specifying different features to text style files*

Note

*You can also use the **-STYLE** command to define a new text style from the command line.*

DETERMINING TEXT HEIGHT

The actual text height is equal to the product of the **scale factor** and the **plotted text height**. Therefore, scale factors are important numbers for plotting the text at the correct height. This factor is a reciprocal of the drawing plot scale. For example, if you plot a drawing at a scale of ¼ = 1, you calculate the scale factor for text height as follows.

¼" = 1" (i.e., the scale factor is 4)

The scale factor for an architectural drawing that is to be plotted at a scale of ¼" = 1'0" is calculated as given next

¼" = 1'0", or ¼" = 12", or 1 = 48

Therefore, in this case, the scale factor is 48.

For a civil engineering drawing with a scale 1"= 50', the scale factor is shown next.

1" = 50', or 1" = 50X12", or 1 = 600

Therefore, the scale factor is 600.

Next, calculate the height of the AutoCAD text. If it is a full-scale drawing (1=1) and the text is to be plotted at 1/8" (0.125), it should be drawn at that height. However, in a civil engineering drawing a text drawn 1/8" high will look like a dot. This is because the scale for a civil engineering drawing is 1"= 50', which means that the drawing you are working on is 600 times larger. To draw a normal text height, multiply the text height by 600. Now the height will be as calculated below.

0.125" x 600 = 75

Similarly, in an architectural drawing, which has a scale factor of 48, a text that is to be 1/8" high on paper must be drawn 6 units high, as shown in the following calculation:

0.125 x 48 = 6.0

It is very important to evaluate scale factors and text heights before you begin a drawing. It would be even better to include the text height in your prototype drawing by assigning the value to the **TEXTSIZE** system variable.

CHECKING SPELLING

Menu:	Tools > Spelling
Command:	SPELL

You can check the spelling of text (text generated by the **TEXT** or **MTEXT** commands) by using the **SPELL** command. The prompt sequence is given next.

Command: **SPELL**
Select object: *Select the text for spell check or enter ALL to select all text objects.*

If no misspelled words are found in the selected text, then AutoCAD displays a message. If the spelling is incorrect for any word in the selected text, AutoCAD displays the **Check Spelling** dialog box as shown in Figure 7-47. The misspelled word is displayed under **Current word**, and correctly spelled alternate words are listed in the **Suggestions** box. The dialog box also displays the misspelled word with the surrounding text in the **Context** box. You may select a word from the list, ignore the correction, and continue with the spell check, or accept the change. To add the listed word in the custom dictionary, choose the **Add** button. The dictionary can be changed from the **Check Spelling** dialog box or by specifying the name in the **DCTMAIN** or **DCTCUST** system variables. Dictionary must be specified for spell check.

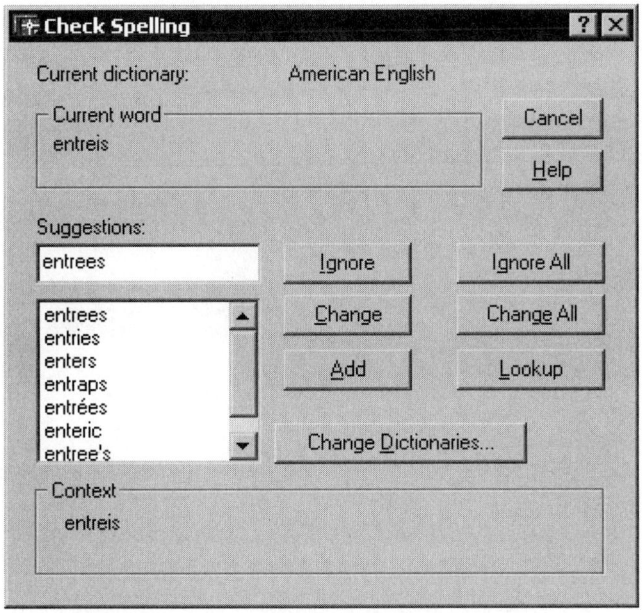

Figure 7-47 ***Check Spelling*** *dialog box*

Note

*You can also rename and use the Word 97 dictionary or any other dictionary. Choose the **Change Dictionaries** button to display the **Change Dictionaries** dialog box and enter the new dictionary name with the .cus extension.*

TEXT QUALITY AND TEXT FILL

AutoCAD supports **TrueType fonts**. You can use your own TrueType fonts by adding them to the Fonts directory. You can also keep your fonts in a separate directory, in which case you must specify the location of your fonts directory in the AutoCAD search path.

The resolution and text fill of the TrueType font text is controlled by the **TEXTFILL** and **TEXTQLTY** system variables. If **TEXTFILL** is set to 1, the text will be filled. If the value is set to 0, the text will not be filled. On the screen the text will appear filled, but when it is plotted the text will not be filled. The **TEXTQLTY** variable controls the quality of the TrueType font text. The value of this variable can range from 0 to 100. The default value is 50, which gives a resolution of 300 dpi (dots per inch). If the value is set to 100, the text will be drawn at 600 dpi. The higher the resolution, the more time it takes to regenerate or plot the drawing.

FINDING AND REPLACING TEXT

Toolbar:	Text > Find and Replace
Menu:	Edit > Find
Command:	FIND

You can use the **FIND** command to find and replace the text. The text could be a line text created by **TEXT** command, paragraph text created by **MTEXT** command, dimension annotation text, block attribute value, hyperlinks, or hyperlink description. When you invoke this command, AutoCAD displays the **Find and Replace** dialog box as shown in Figure 7-48. You can use this dialog box to perform the following functions.

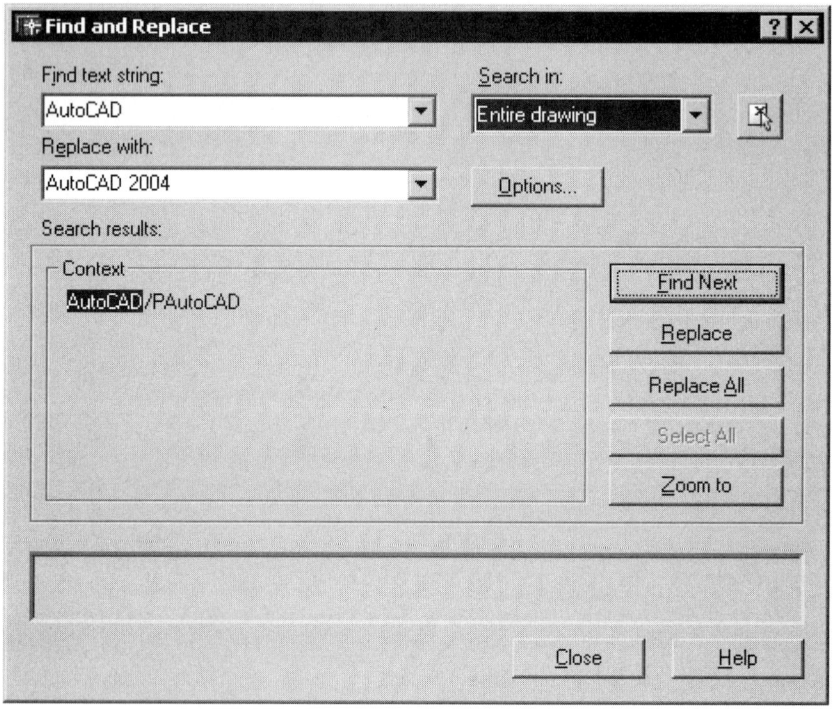

Figure 7-48 Find and Replace dialog box

Find Text

To find the text, enter the text you want to find in the **Find text string** edit box. You can search the entire drawing or confine the search to selected text. To select text, choose the **Select Objects** button. The **Find and Replace** dialog box is temporarily closed and AutoCAD switches to drawing window. Once you have selected the text, the **Find and Replace** dialog box is redisplayed. In the **Search in** drop-down list you can specify if you want to search the entire drawing or current selection. If you choose the Options button, AutoCAD displays the **Find and Replace Options** dialog box as shown in Figure 7-49. In this dialog box you can specify whether to find the whole word and whether to match the case of the specified text. To find the text, choose the **Find** button. AutoCAD displays the found text with the surrounding text in the Context area. To find the next occurrence of the text, choose the **Find Next** button.

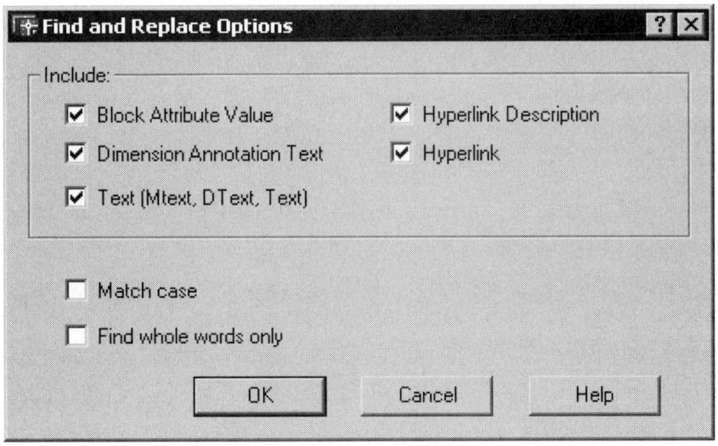

***Figure 7-49 Find and Replace Options** dialog box*

Replacing Text

If you want to replace the specified text with new text, enter the new text in the **Replace with** edit box. Now, if you choose the **Replace** button, only the found text will be replaced. If you choose the **Replace All** button, all occurrences of the specified text will be replaced with the new text.

Self-Evaluation Test

Answer the following questions and then compare your answers to the answers given at the end of this chapter.

1. Transparent commands cannot be used while another command is in progress. (T/F)

2. The **ALL** option of the **ZOOM** command displays the drawing limits or extents, whichever is greater. (T/F)

3. While using the ZOOM scale option with respect to the current view you have to enter a number followed by an **X**. (T/F)

4. You can use the scroll bars to pan the drawing in any direction. (T/F)

5. The **VIEW** command does not save any drawing object data, only the _________ parameters needed to redisplay that portion of the drawing.

6. Multiple lines of text can be entered at any desired location in the drawing area with a single _________ command.

7. With the _________ justification option of the **TEXT** command, AutoCAD adjusts the letter width to fit the text between the two given points, but the height remains constant.

8. In the **MTEXT** command, the height specified in the Multiline Text Editor does not affect the _________ system variable.

9. With the _________ command, if you are editing a single line text you can change the text string in the window, but for a multiline text you must choose the **Full editor** button in the **Content** edit box.

10. You can use the _________ system variable to specify the new font mapping file.

Review Questions

Answer the following questions.

1. After completion of a transparent command, AutoCAD returns you to the Command prompt. (T/F)

2. If the **BLIPMODE** variable is set to On, blip marks do not appear on the screen. (T/F)

3. With the **ZOOM** command, the actual size of the object changes. (T/F)

4. The **REDRAW** command can be used as a transparent command. (T/F)

5. The **TEXT** command does not allow you to see the text on the screen as you type it. (T/F)

6. Which command recalculates all the objects in a drawing and redraws the current viewport only?

   (a) **REDRAW** (b) **REDRAWALL**
   (c) **REGEN** (d) **REGENALL**

7. Which of the following commands cannot be used transparently?

   (a) **ZOOM** (b) **PAN**
   (c) **VIEW** (d) **REDRAW**

8. How many views are saved with the **Previous** option of the **ZOOM** command?

   (a) 6 (b) 8
   (c) 10 (d) 12

9. In the **Multiline Text Editor**, which of the following characters is used to stack the text with a diagonal line without using the **Autostack Properties** dialog box?

   (a) ^ (b) /
   (c) # (d) @

10. Which command can be used to create a new text style and modify the existing ones?

    (a) **TEXT** (b) **MTEXT**
    (c) **STYLE** (d) **SPELL**

11. The four main text alignment modes are __________, __________, __________, and __________.

12. You can use the __________ command to write a paragraph text whose width can be specified by defining the __________ of the text boundary.

13. When the **Justify** option is invoked, the user can place text in one of the __________ various alignment types by selecting the desired alignment option.

14. The text created by the __________ command is a single object regardless of the number of lines it contains.

15. Using the **MTEXT** command, the character __________ stacks the text vertically without a line (tolerance stack).

16. If you want to edit text, select it and then right-click such that the various editing options __________ in the menu become available.

17. You can view the entire drawing (even if it is beyond limits) with the help of the __________ option.

18. In the **ZOOM** Window option, the area inside the window is __________ to completely __________ the current viewport.

Exercises

Exercise 3 *General*

Draw the text on the screen as shown in Figure 7-50. Use the text justification that will produce the text as shown in the drawing. Assume a value for text height. Use the **PROPERTIES** palette to change the text as shown in Figure 7-51.

Figure 7-50 *Drawing for Exercise 3*

Figure 7-51 *Drawing for Exercise 3 (After changing the text)*

Exercise 4 *General*

Draw the text on the screen as shown in Figure 7-52. You must first define text style files using the **STYLE** command with the attributes as shown in the drawing. The text height is 0.25 units.

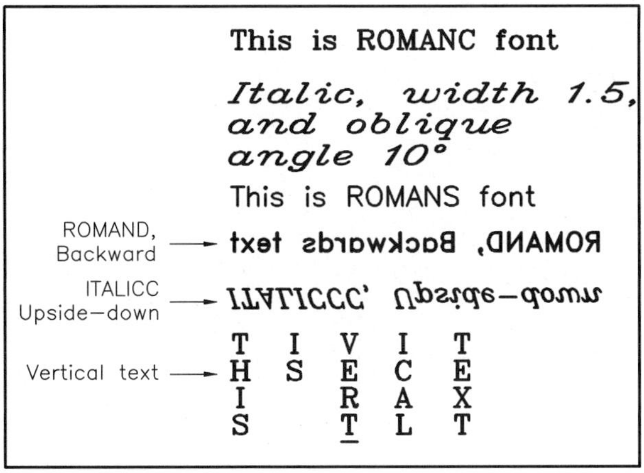

Figure 7-52 Drawing for Exercise 4

Exercise 5 *Mechanical*

Draw Figure 7-53 using the draw, edit, text, and display commands. Do not dimension the drawing.

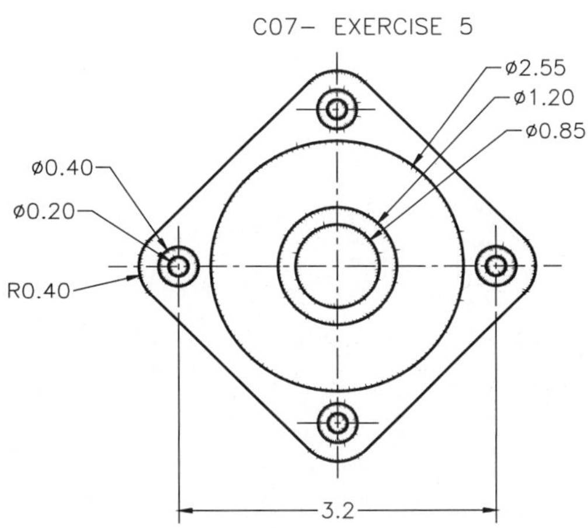

Figure 7-53 Drawing for Exercise 5

Exercise 6 *Mechanical*

Draw Figure 7-54 using the **MIRROR** command to duplicate the features that are identical. Also, add the text shown in the figure. Use the display commands to facilitate the process. Do not dimension the drawing.

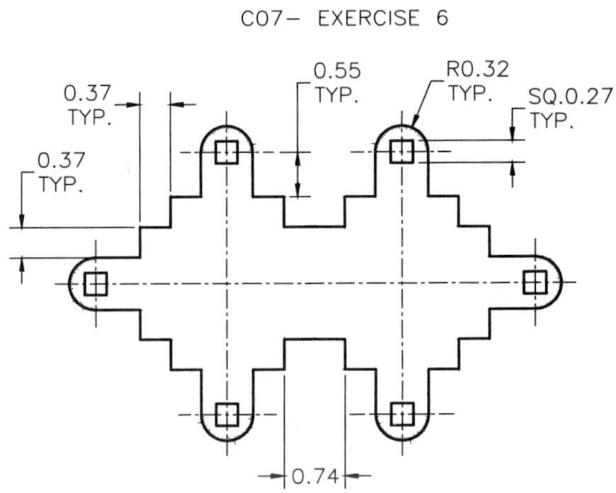

Figure 7-54 *Drawing for Exercise 6*

Exercise 7 *Mechanical*

Draw Figure 7-55 and also add the text shown in the figure. Use the display commands to facilitate the process. Do not dimension the drawing.

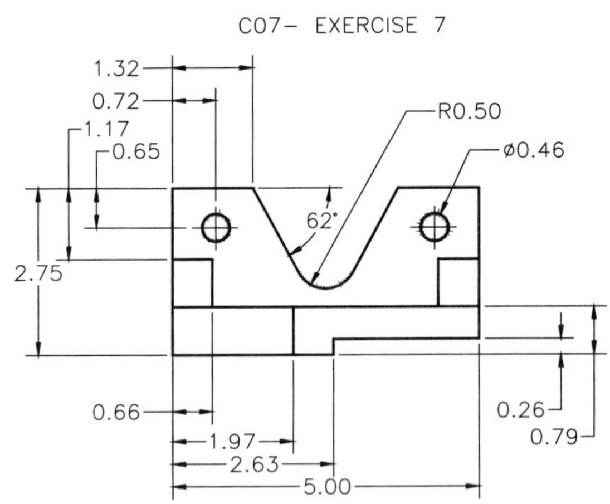

Figure 7-55 *Drawing for Exercise 7*

Exercise 8

Mechanical

Draw Figure 7-56 using the **MIRROR** command to duplicate the features that are identical. Use the display commands to facilitate the process. Add the text to the drawing but do not dimension the drawing.

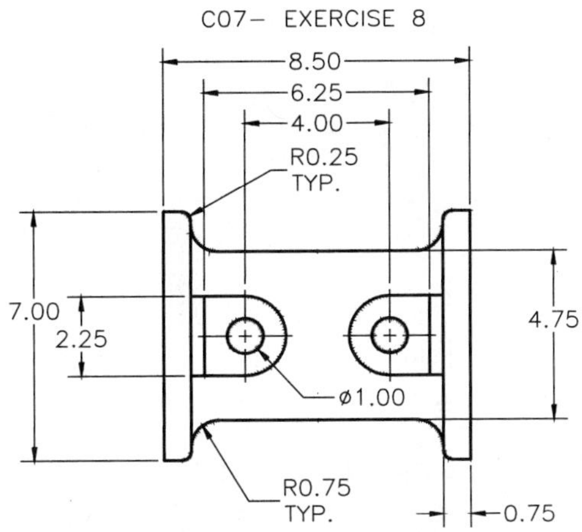

Figure 7-56 *Drawing for Exercise 8*

Problem Solving Exercise 1

Architectural

Draw Figure 7-57 using AutoCAD's draw, edit, and display commands. Also add the text to the drawing. Assume the missing dimensions. Do not dimension the drawing.

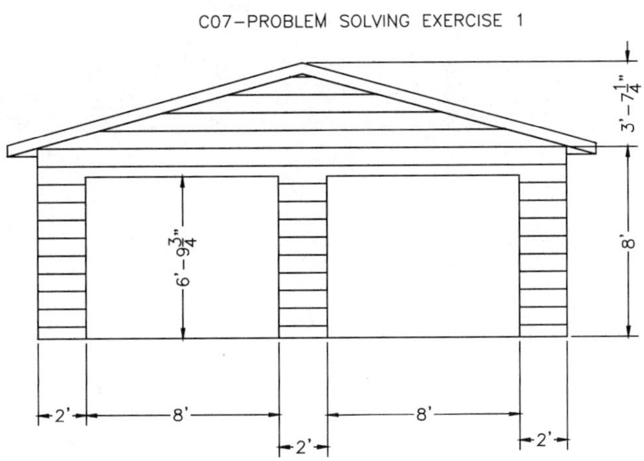

Figure 7-57 *Drawing for Problem Solving Exercise 1*

Problem Solving Exercise 2

Architectural

Draw Figure 7-58 using AutoCAD's draw, edit, and display commands. Also add the text to the drawing. Assume the missing dimensions. Do not dimension the drawing.

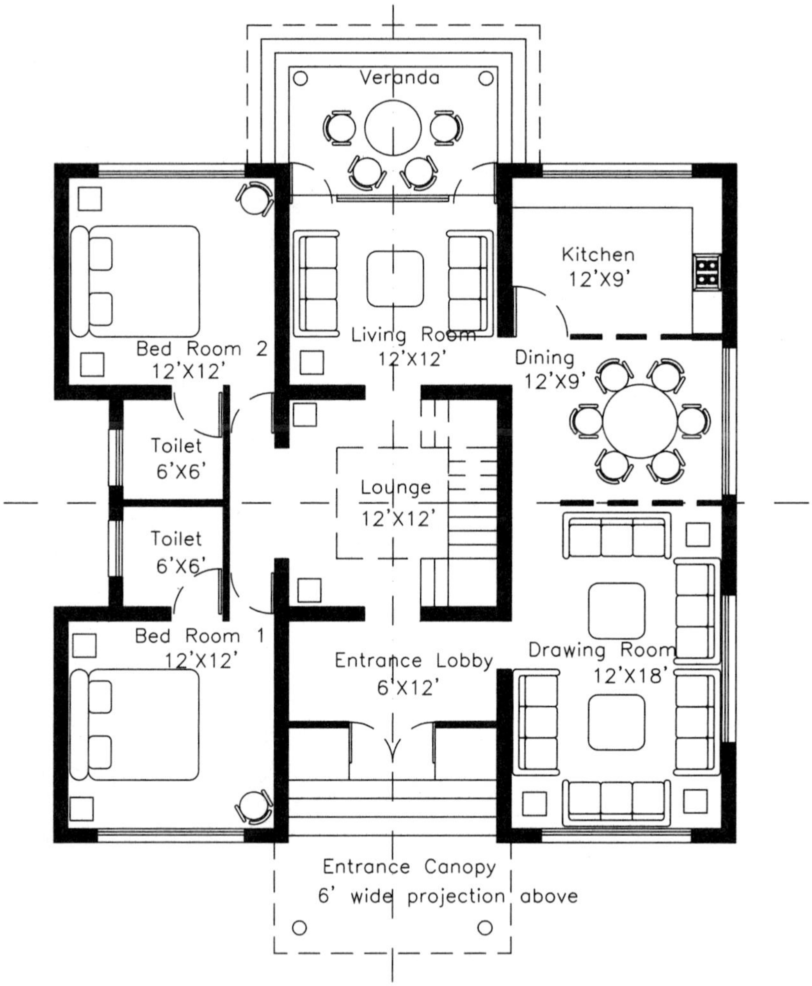

Figure 7-58 *Drawing for Problem Solving Exercise 2*

Answers to Self-Evaluation Test

1 - F, **2** - T , **3** - T, 4 - T , **5** - view , **6** - **TEXT**, **7** - Fit, **8** - TEXTSIZE, **9** - PROPERTIES, **10** - FONTMAP

Chapter 8

Basic Dimensioning, Geometric Dimensioning, and Tolerancing

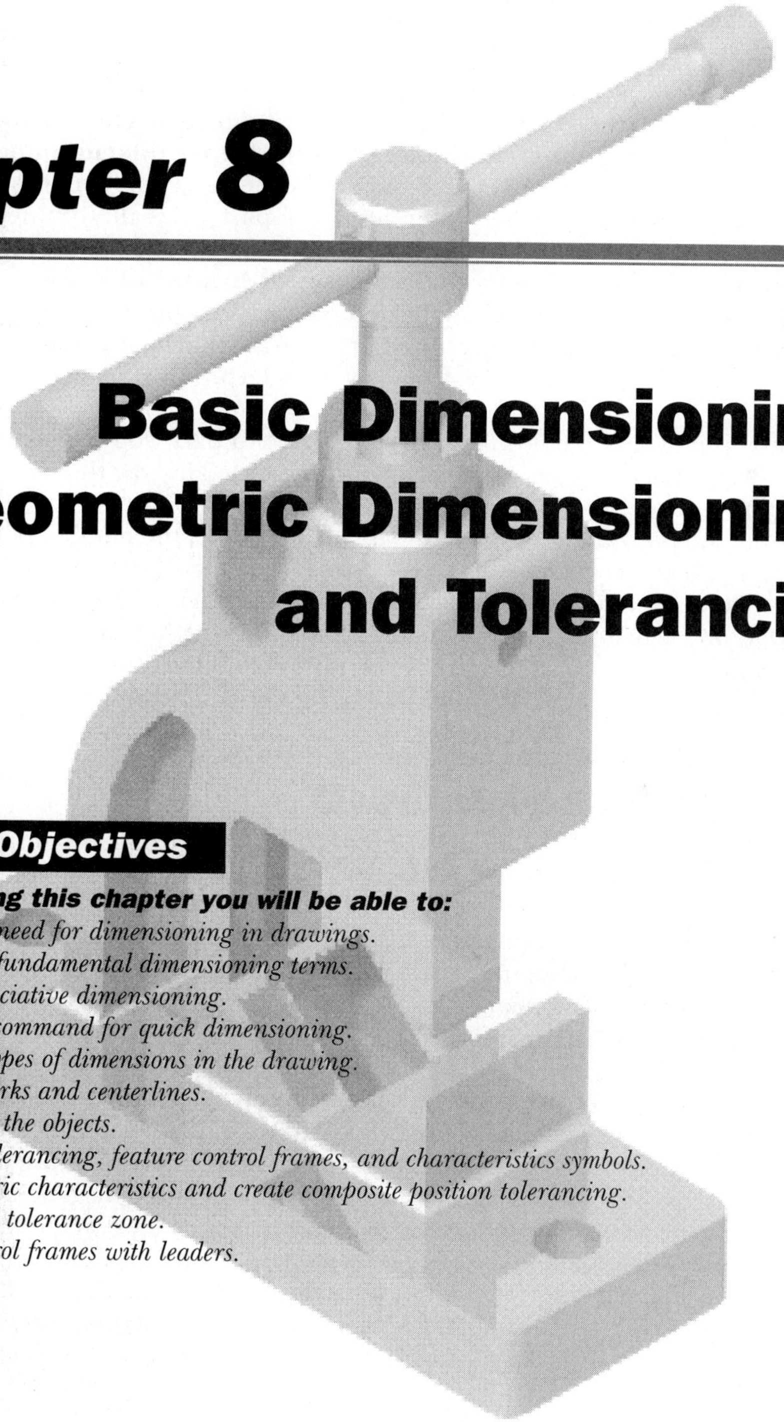

Learning Objectives

After completing this chapter you will be able to:

- *Understand the need for dimensioning in drawings.*
- *Understand the fundamental dimensioning terms.*
- *Understand associative dimensioning.*
- *Use the **QDIM** command for quick dimensioning.*
- *Create various types of dimensions in the drawing.*
- *Create center marks and centerlines.*
- *Attach leaders to the objects.*
- *Use geometric tolerancing, feature control frames, and characteristics symbols.*
- *Combine geometric characteristics and create composite position tolerancing.*
- *Use the projected tolerance zone.*
- *Use feature control frames with leaders.*

NEED FOR DIMENSIONING

To make designs more informative and practical, the drawing must convey more than just the graphic picture of the product. To manufacture an object, the drawing must contain size descriptions such as the length, width, height, angle, radius, diameter, and location of features. All this information is added to the drawing with the help of **dimensioning**. Some drawings also require information about tolerances with the size of features. This information conveyed through dimensioning is vital and often just as important as the drawing itself. With the advances in computer-aided design/drafting and computer-aided manufacturing, it has become mandatory to draw the part to actual size so that the dimensions reflect the actual size of the features. At times it may not be necessary to draw the object of the same size as the actual object would be when manufactured, but it is absolutely essential that the dimensions be accurate. Incorrect dimensions will lead to manufacturing errors.

By dimensioning, you are not only giving the size of a part, you are also giving a series of instructions to a machinist, an engineer, or an architect. The way the part is positioned in a machine, the sequence of machining operations, and the location of various features of the part depend on how you dimension the part. For example, the number of decimal places in a dimension (2.000) determines the type of machine that will be used to do that machining operation. The machining cost of such an operation is significantly higher than for a dimension that has only one digit after the decimal (2.0). Similarly, whether a part is to be forged or cast, the radii of the edges, and the tolerance you provide to these dimensions determine the cost of the product, the number of defective parts, and the number of parts you get from a single die.

DIMENSIONING IN AutoCAD

The objects that can be dimensioned in AutoCAD range from straight lines to arcs. The dimensioning commands provided by AutoCAD can be classified into four categories:

Dimension Drawing Commands
Dimension Style Commands
Dimension Editing Commands
Dimension Utility Commands

While dimensioning an object, AutoCAD automatically calculates the length of the object or the distance between two specified points. Also, settings such as the gap between the dimension text and the dimension line, the space between two consecutive dimension lines, arrow size, and text size are maintained and used when the dimensions are being generated for a particular drawing. The generation of arrows, lines (dimension lines, extension lines), and other objects that form a dimension is automatically performed by AutoCAD to save the user's time. This also results in uniform drawings. However, you can override the default measurements computed by AutoCAD and change the settings of various standard values. The modification of dimensioning standards can be achieved through the dimension variables.

The dimensioning functions offered by AutoCAD provide you with extreme flexibility in dimensioning by letting you dimension various objects in a variety of ways. This is of great help because different industries, such as architectural, mechanical, civil, or electrical, have different standards for the placement of dimensions.

FUNDAMENTAL DIMENSIONING TERMS

Before studying AutoCAD's dimensioning commands, it is important to know and understand various dimensioning terms that are common to linear, angular, radius, diameter, and ordinate dimensioning. Figures 8-1 and 8-2 show various dimensioning parameters.

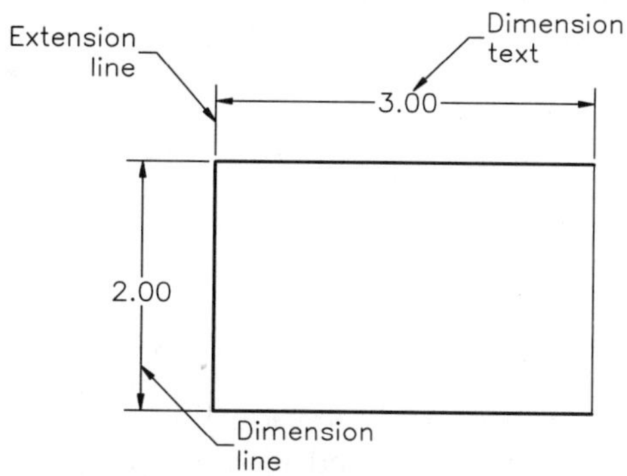

***Figure 8-1** Various dimension parameters*

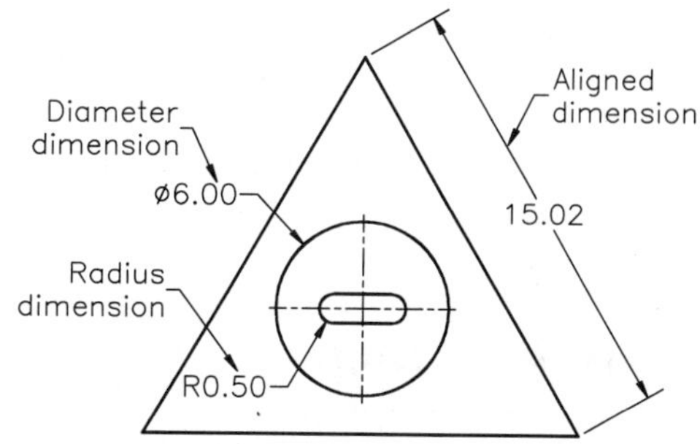

***Figure 8-2** Various dimension parameters*

Dimension Line

The **dimension line** indicates which distance or angle is being measured. Usually this line has arrows at both ends, and the dimension text is placed along the dimension line. By default the dimension line is drawn between the extension lines (Figure 8-1). If the dimension line does not fit inside, two short lines with arrows pointing inward are drawn outside the extension lines. The dimension line for angular dimensions (which are used to dimension angles) is an arc. You can control the positioning and various other features of the dimension lines by setting the parameters in the dimension styles. (The dimension styles are discussed in Chapter 10.)

Dimension Text

Dimension text is a text string that reflects the actual measurement (dimension value) between the selected points as calculated by AutoCAD. You can accept the value that AutoCAD returns or enter your own value. In case you use the default text, AutoCAD can be supplied with instructions to append the tolerances to it. Also, you can attach prefixes or suffixes of your choice to the dimension text.

Arrowheads

An **arrowhead** is a symbol used at the end of a dimension line (where dimension lines meet the extension lines). Arrowheads are also called **terminators** because they signify the end of the dimension line. Since the drafting standards differ from company to company, AutoCAD allows you to draw arrows, tick marks, closed arrows, open arrows, dots, right angle arrows, or user-defined blocks (Figure 8-3). The user-defined blocks at the two ends of the dimension line can be customized to your requirements. The size of the arrows, tick marks, user blocks, and so on can be regulated by using the dimension variables.

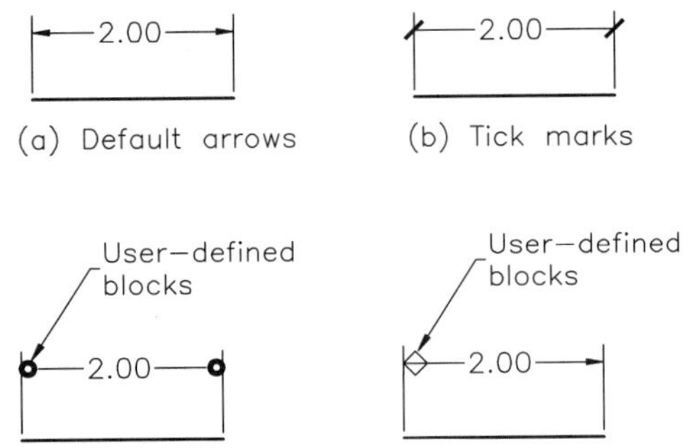

Figure 8-3 Using arrows, tick marks, and user-defined blocks

Extension Lines

Extension lines are drawn from the object measured to the dimension line (Figure 8-4). These lines are also called **witness lines.** Extension lines are used in linear and angular dimensioning. Generally, extension lines are drawn perpendicular to the dimension line. However, you can make extension lines incline at an angle by using the **DIMEDIT** command (Oblique option) or by selecting **Dimension Edit** from the **Dimension** toolbar. AutoCAD also allows you to suppress either one or both extension lines in a dimension (Figure 8-5). Other aspects of the extension line can be controlled by using the dimension variables (these variables are discussed in Chapter 10).

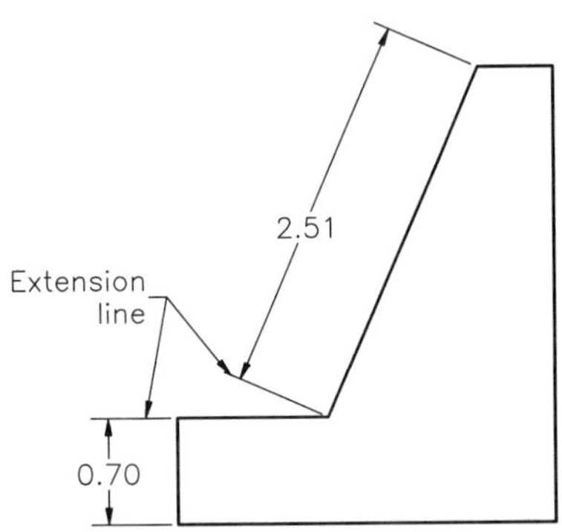

Figure 8-4 Extension lines

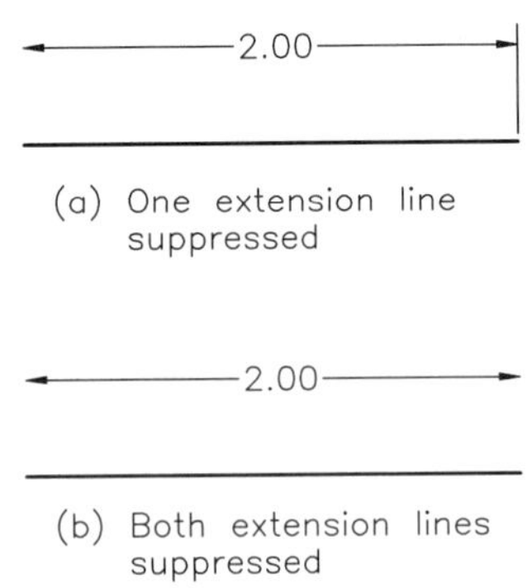

Figure 8-5 Extension line suppression

Leader

A **leader** is a line that stretches from the dimension text to the object being dimensioned. Sometimes the text for dimensioning and other annotations do not adjust properly near the object. In such cases, you can use a leader and place the text at the end of the leader line. For example, the circle shown in Figure 8-6 has a keyway slot that is too small to be dimensioned. In this situation, a leader can be drawn from the text to the keyway feature. Also, a leader can be used to attach annotations such as part numbers, notes, and instructions to an object.

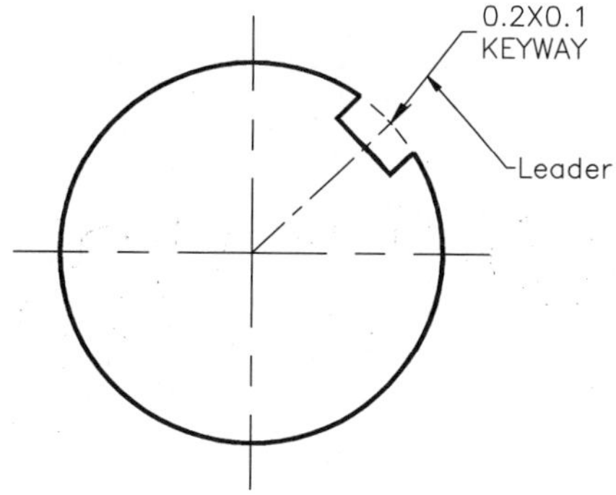

Figure 8-6 *Leader used to attach annotation*

Center Mark and Centerlines

The **center mark** is a cross mark that identifies the center point of a circle or an arc. Centerlines are mutually perpendicular lines passing through the center of the circle/arc and intersecting the circumference of the circle/arc. A center mark or the centerlines are automatically drawn when you dimension a circle or arc (see Figure 8-7). The length of the center mark and the extension of the centerline beyond the circumference of the circle is determined by the value assigned to the **DIMCEN** dimension variable or the value assigned to **Center Marks for Circles** in the **Dimension Style Manager** dialog box.

Alternate Units

With the help of **alternate units** you can generate dimensions for two systems of measurement at the same time (Figure 8-8). For example, if the dimensions are in inches, you can use the alternate units dimensioning facility to append metric dimensions to the dimensions (controlling the alternate units through the dimension variables is discussed in Chapter 10).

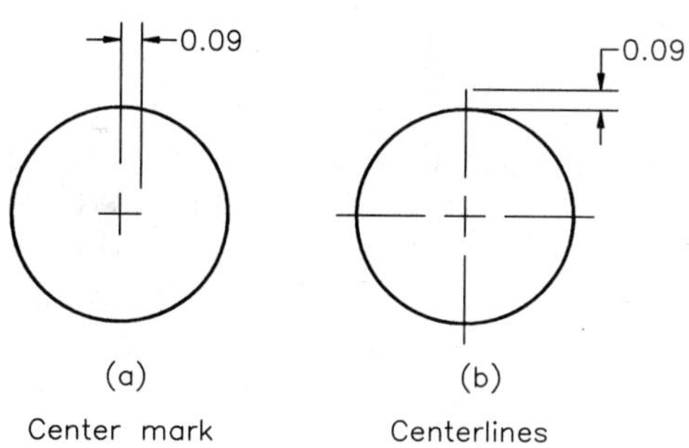

Figure 8-7 *Center mark and centerlines*

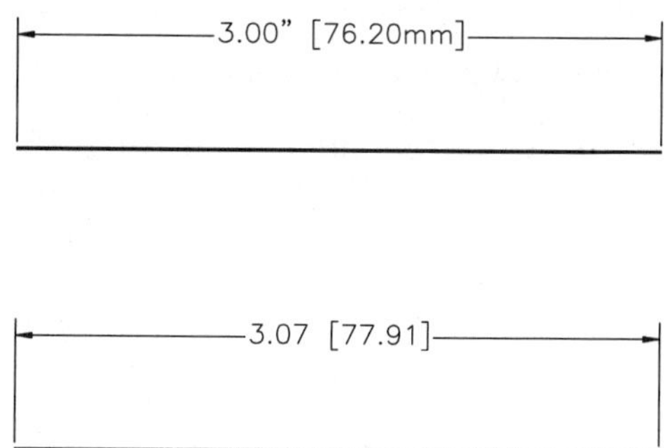

Figure 8-8 *Using alternate units dimensioning*

Tolerances

Tolerance is the amount by which the actual dimension can vary (Figure 8-9). AutoCAD can attach the plus/minus tolerances to the dimension text (actual measurement computed by AutoCAD). This is also known as **deviation tolerance**. The plus and minus tolerance that you specify can be the same or different. You can use the dimension variables to control the tolerance feature (these variables are discussed in Chapter 10).

Limits

Instead of appending the tolerances to the dimension text, you can apply the tolerances to the measurement itself (Figure 8-10). Once you define the tolerances, AutoCAD will automatically calculate the upper and lower **limits** of the dimension. These values are then displayed as a dimension text.

For example, if the actual dimension as computed by AutoCAD is 2.6105 units and the tolerance values are +0.025 and -0.015, the upper and lower limits are 2.6355 and 2.5955. After calculating the limits, AutoCAD will display them as dimension text, as shown in Figure 8-10. The dimension variables that control the limits are discussed in Chapter 10.

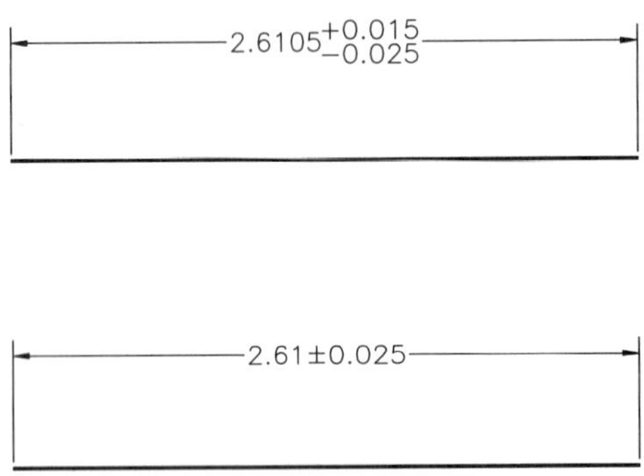

Figure 8-9 *Using tolerances with dimensions*

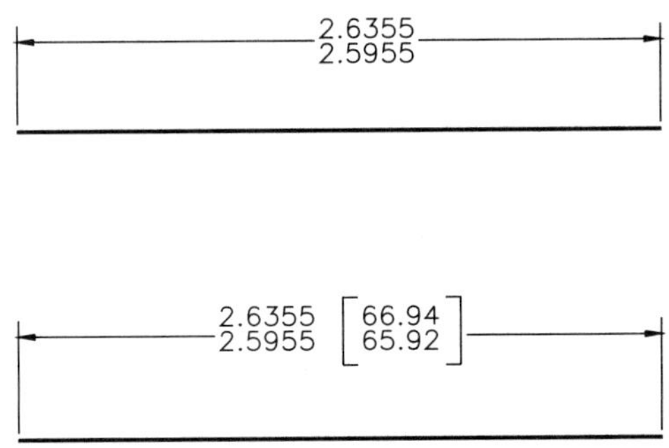

Figure 8-10 *Using limits dimensioning*

ASSOCIATIVE DIMENSIONS

Associative dimensioning is a method of dimensioning in which the dimension is associated with the object that is dimensioned. In other words, the dimension is influenced by the changes in the size of the object. In the previous releases of AutoCAD, the dimensions were not truly associative, but were related to the objects being dimensioned by definition points on the DEFPOINTS layer. To cause the dimension to be associatively modified, these definition points had to be adjusted along with the object being changed. If, for example, you use the **SCALE** command to change an object's size and select the object, the dimensions will not be modified. If you select the object and its defpoints (using the Crossing selection method), then the dimension will be modified. However, in AutoCAD 2002, a new concept called **true associative dimensioning** was introduced. This concept ensures that if the dimensions are associated to the object and the object changes its size, the dimensions will also change automatically. With the introduction of the true associative dimensions, there is no need to select the definition points along with the object. This eliminates the use of definition points for updating the dimensions.

The associative dimensions automatically update their values and location if the value or location of the object is modified. For example, if you edit an object using simple editing operations such as breaking using the **BREAK** command, then the true associative dimension will be modified automatically. The dimensions can be converted into the true associative dimensions using the **DIMREASSOCIATE** command. The association of the dimensions with the objects can be removed using the **DIMDISASSOCIATE** command. Both of these commands will be discussed later in this chapter.

The dimensioning variable **DIMASSOC** controls associativity of dimensions. The default value of this variable is **1**, which means turned on. When **DIMASSOC** is turned off then the dimension will be placed in the exploded format. This means that the dimensions will now be placed as a combination of individual arrowheads, dimension lines, extension lines, and text. Also, note that the exploded dimensions cannot be associated to any object.

DEFINITION POINTS

Definition points are the points drawn at the positions used to generate a dimension object. The definition points are used by the dimensions to control their updating and rescaling. AutoCAD draws these points on a special layer called **DEFPOINTS.** These points are not plotted by the plotter because AutoCAD does not plot any object on the **DEFPOINTS** layer. If you explode a dimension (which is as good as turning **DIMASSOC** off), the definition points are converted to point objects on the **DEFPOINTS** layer. In Figure 8-11, the small circles indicate the definition points for different objects.

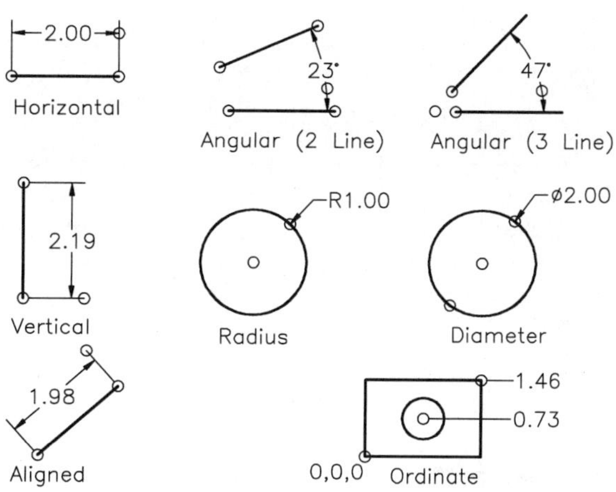

Figure 8-11 *Definition points of linear, angular, and ordinate dimensions*

The definition points for linear dimensions are the points used to specify the extension lines and the point of intersection of the first extension line and the dimension line. The definition points for the angular dimension are the endpoints of the lines used to specify the dimension and the point used to specify the dimension line arc. For example, for three-point angular dimension, the definition points are the extension line endpoints, the angle vertex, and the point used to specify the dimension line arc.

The definition points for the radius dimension are the center point of the circle or arc, and the point where the arrow touches the object. The definition points for the diameter dimension are the points where the arrows touch the circle. The definition points for the ordinate dimension are the UCS origin, the feature location, and the leader endpoint.

Note

In addition to the definition points just mentioned, the middle point of the dimension text serves as a definition point for all types of dimensions.

SELECTING DIMENSIONING COMMANDS

Using the Toolbar and the Dimension Menu

You can select the dimension commands from the **Dimension** toolbar by choosing the desired dimension button (Figure 8-12), or from the **Dimension** menu (Figure 8-13). The **Dimension** toolbar can also be displayed by right-clicking on any toolbar and choosing **Dimensions** from the shortcut menu.

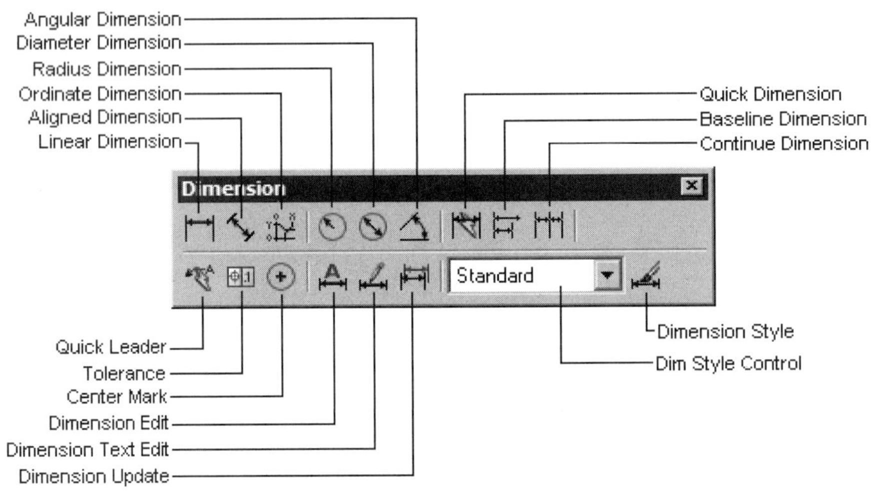

Figure 8-12 ***Dimension*** *toolbar*

Using the Command Line

You can directly enter a dimensioning command in the Command line or use the **DIM** or the **DIM1** commands to invoke the dimensioning commands. Both of these methods of using the command line are discussed next.

Using Dimensioning Commands

The first method of using the Command line is to directly enter the dimensioning command in the Command line. For example, if you want to draw the linear dimension, the **DIMLINEAR** command can be entered directly at the Command prompt.

Command: **DIMLINEAR**
Specify first extension line origin or <select object>: *Select a point or press ENTER.*
Specify second extension line origin: *Select second point.*
Specify dimension line location or
[Mtext/Text/Angle/Horizontal/Vertical/Rotated]: *Select a point to locate the position of the dimension.*
Command: *(After you have finished dimensioning, AutoCAD returns to the Command prompt.)*

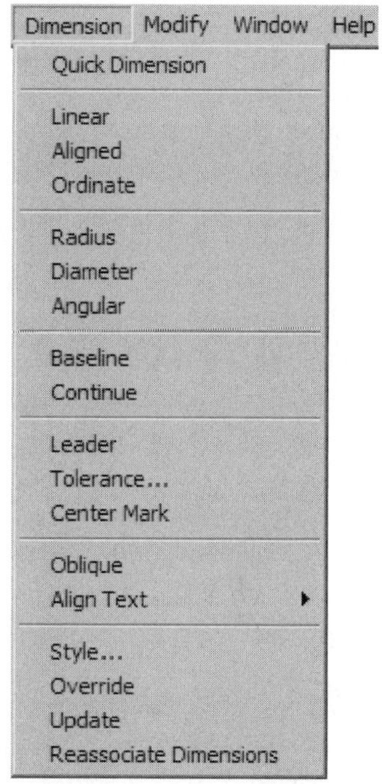

Figure 8-13 *Selecting dimensions from the* ***Dimension*** *menu*

DIM and DIM1 Commands

Since dimensioning has several options, it also has its own command mode. The **DIM** command keeps you in the dimension mode, and the **Dim:** prompt is repeated after each dimensioning command until you exit the dimension mode to return to the normal AutoCAD Command prompt. To exit the dimension mode, enter EXIT (or just E) at the **Dim:** prompt. You can also exit by pressing ESC. The previous command will be repeated if you press the SPACEBAR or ENTER at the **Dim:** prompt. In the dimension mode, it is not possible to execute the normal set of AutoCAD commands, except function keys, object snap overrides, control key combinations, transparent commands, dialog boxes, and menus.

Command: **DIM**
Dim: **Hor**
Specify first extension line origin or <select object>: *Select a point or press ENTER.*
Specify second extension line origin: *Select the second point.*
Specify dimension line location or [MText/Text/Angle]: *Select a point to locate the position of the dimension.*
Enter dimension text <default>: *Press ENTER to accept the default dimension.*
Dim: *(After you have finished dimensioning, AutoCAD returns to the* ***Dim:*** *prompt.)*

The **DIM1** command is similar to the **DIM** command. The only difference is that **DIM1** lets you execute a single dimension command and then automatically takes you back to the normal Command prompt.

Command: **DIM1**
Dim: **Hor**
Specify first extension line origin or <select object>: *Select a point or press ENTER.*

Specify second extension line origin: *Select the second point.*
Specify dimension line location or [MText/Text/Angle]: *Select a point to locate the position of the dimension.*
Enter dimension text <default>: *Press ENTER to accept the default dimension.*
Command: *(After you are done dimensioning, AutoCAD returns Command prompt.)*

AutoCAD has provided the following six fundamental dimensioning types.

Quick dimensioning	**Linear dimensioning**	**Diameter dimensioning**
Radius dimensioning	**Angular dimensioning**	**Ordinate dimensioning**

Figures 8-14 and 8-15 show the various fundamental dimension types.

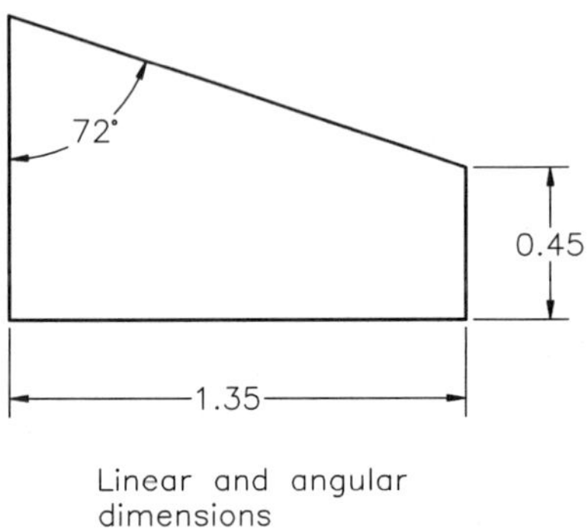

Figure 8-14 *Linear and angular dimensions*

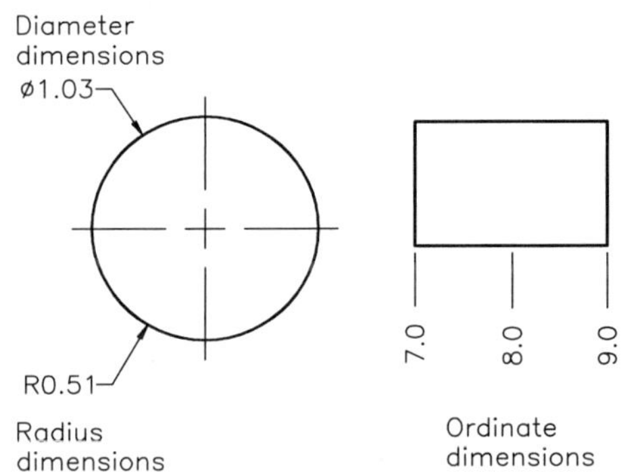

Figure 8-15 *Radius, diameter, and ordinate dimensions*

Note

*The **DIMDEC** variable sets the number of decimal places for the value of primary dimension and the **DIMADEC** variable for angular dimensions. For example, if **DIMDEC** is set to 3, AutoCAD will display the decimal dimension up to three decimal places (2.037).*

DIMENSIONING NUMBER OF OBJECTS TOGETHER

Toolbar:	Dimension > Quick Dimension
Menu:	Dimension > Quick Dimension
Command:	QDIM

The **QDIM** command allows you to dimension a number of objects at the same time. It also allows you to quickly edit dimension arrangements already existing in the drawing and also create new dimension arrangements. It is especially useful when creating a series of baseline or continuous dimensions. It also allows you to dimension multiple arcs and circles at the same time. When you are using the **QDIM** command, you can relocate the datum base point for baseline and ordinate dimensions. The prompt sequence that will follow when you choose this button is given next.

Select geometry to dimension: *Select the objects to be dimensioned and press ENTER.*
Select geometry to dimension: Enter
Specify dimension line position, or [Continuous/Staggered/Baseline/Ordinate/Radius/Diameter/datum point/Edit] <continuous>: *Press ENTER to accept default dimension arrangement and specify dimension line location or enter new dimension arrangement or Edit existing dimension arrangement.*

For example, you can dimension all the circles in a drawing (Figure 8-16) by using Quick Dimensioning as follows.

Select geometry to dimension: *Select all the circles.*
Select geometry to dimension: Enter
Specify dimension line position, or
[Continuous/Staggered/Baseline/Ordinate/Radius/Diameter/datumPoint/Edit/seTtings] <Continuous>: *Press D for diameter dimensioning and select a point where you want to position the radial dimension.*

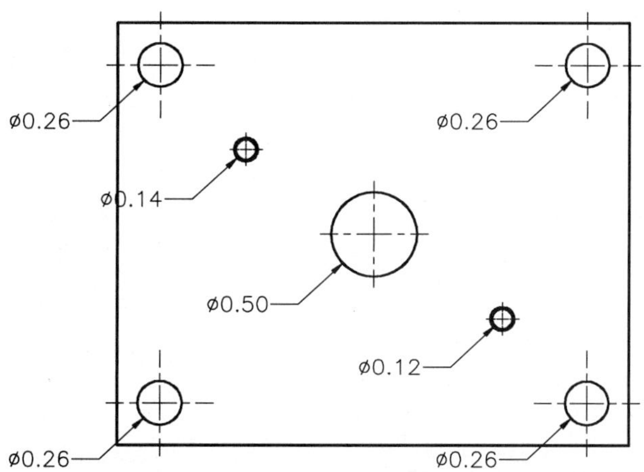

Figure 8-16 *Using the **QDIM** command to quickly dimension multiple circles*

CREATING LINEAR DIMENSIONS

Toolbar:	Dimension > Linear Dimension
Menu:	Dimension > Linear
Command:	DIMLIN or DIMLINEAR

Linear dimensioning applies to those dimensioning commands that measure the shortest distance between two points. You can directly select the object to dimension or select two points. The points can be any two points in the space, the endpoints of an arc or line, or any set of points that can be identified. To achieve accuracy, points must be selected with the help of object snaps or by selecting an object to dimension. In case the object selected is aligned then the linear dimensions will add **Horizontal** or **Vertical** dimensions to the object. The prompt sequence that will follow when you choose this button is given next.

Specify first extension line origin or <select object>: Enter
Select object to dimension: *Select the object.*
Specify dimension line location or
[Mtext/Text/Angle/Horizontal/Vertical/Rotated]: *Select a point to locate the position of the dimension.*

Instead of selecting the object, you can also select the two endpoints of the line that you want to dimension (Figure 8-17). Usually the points on the object are selected by using the **object snaps** (endpoints, intersection, center, etc.). The prompt sequence is as follows.

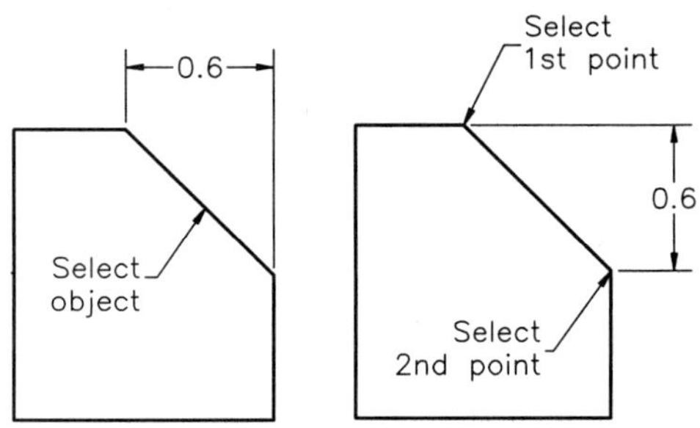

Figure 8-17 *Drawing linear dimension*

Specify first extension line origin or <select object>: *Select a point.*
Specify second extension line origin: *Select second point.*
Specify dimension line location or
[Mtext/Text/Angle/Horizontal/Vertical/Rotated]: *Select a point to locate the position of the dimension.*

When using the **DIMLINEAR** command, you can obtain the horizontal or vertical dimension by simply defining the appropriate dimension location point. If you select a point above or below the dimension, AutoCAD creates a horizontal dimension. If you select a point that is on the left or right of the dimension, AutoCAD creates a vertical dimension through that point.

DIMLINEAR Command Options

The options provided under this command are discussed next.

Mtext Option

The **Mtext** option allows you to override the default dimension text and also change the font, height, and so on, using the **Multiline Text Editor** dialog box. When you enter **M** at the **Specify dimension line location or [Mtext/Text/Angle/Horizontal/Vertical/Rotated]** prompt, the **Multiline Text Editor** is displayed. By default, it includes the <> code to represent the measured dimension text. You can change the text by entering a new text and deleting the <> code. You can also use the various options of the dialog box (explained in Chapter 7) and then choose **OK**. However, if you override the default dimensions, the dimensional associativity of the **dimension text** is lost. This means that if you modify the object using the definition points, AutoCAD will not recalculate the dimension text. Even if the dimension is a true associative dimension, the text will not be recalculated when the object is modified. The prompt sequence to invoke this option is given next.

Specify first extension line origin or <select object>: *Specify a point.*
Specify second extension line origin: *Specify second point.*
Specify dimension line location or
[Mtext/Text/Angle/Horizontal/Vertical/Rotated]: **M** *(Enter dimension text in* ***Mutiline Text Editor*** *dialog box and then choose* ***OK****.)*
Specify dimension line location or
[Mtext/Text/Angle/Horizontal/Vertical/Rotated]: *Specify the dimension location.*

Text Option

This option also allows you to override the default dimension. However, this option will prompt you to specify the new text value in the Command prompt itself, see Figure 8-18. The prompt sequence to invoke this option is given next.

Specify first extension line origin or <select object>: *Select a point.*
Specify second extension line origin: *Select second point.*
Specify dimension line location or
[Mtext/Text/Angle/Horizontal/Vertical/Rotated]: **T**
Enter dimension text <Current>: *Enter new text.*
Specify dimension line location or
[Mtext/Text/Angle/Horizontal/Vertical/Rotated]: *Specify the dimension location.*

Angle Option

This option lets you change the angle of the dimension text, see Figure 8-18.

Horizontal Option

This option lets you create a horizontal dimension regardless of where you specify the dimension location, see Figure 8-18.

Vertical Option

This option lets you create a vertical dimension regardless of where you specify the dimension location, see Figure 8-18.

Rotated Option

This option lets you create a dimension that is rotated at a specified angle, see Figure 8-18.

Note

If you override the default dimensions, the dimensional associativity of the dimension text is lost and AutoCAD will not recalculate the dimension when the object is scaled.

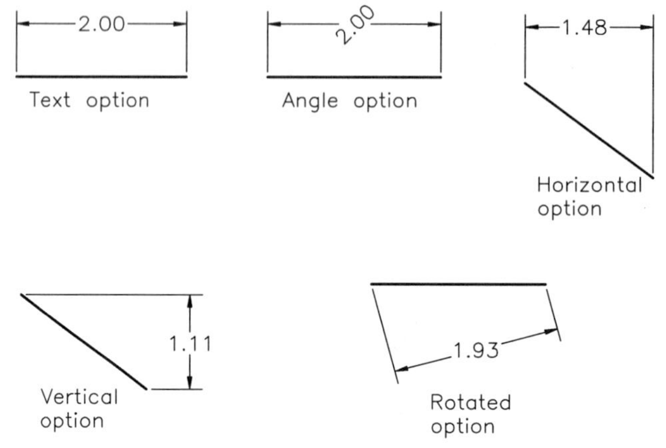

Figure 8-18 *Text, Angle, Horizontal, Vertical, and Rotated options*

Example 1 *General*

In this example, you will use linear dimensioning to dimension a horizontal line of 4 units length. The dimensioning will be done by selecting the object and by specifying the first and second extension line origins. Using the **Multiline Text Editor** modify the default text such that the dimension is underlined.

Selecting the Object

1. Choose the **Linear Dimension** button from the **Dimension** toolbar. The prompt sequence is as follows.

    Specify first extension line origin or <select object>: Enter
    Select object to dimension: *Select the line.*
    Specify dimension line location or
    [Mtext/Text/Angle/Horizontal/Vertical/Rotated]: **M**

    *The **Multiline Text Editor** will be displayed as shown in Figure 8-19. The default dimension value will be displayed in the **<>** code. Select this code and then choose the **Underline** button to underline the text.*

    Specify dimension line location or
    [Mtext/Text/Angle/Horizontal/Vertical/Rotated]: *Place the dimension.*
    Dimension text = 4.00

Specifying Extension Line Origins

2. Choose the **Linear Dimension** button from the **Dimension** toolbar. The prompt sequence is as follows.

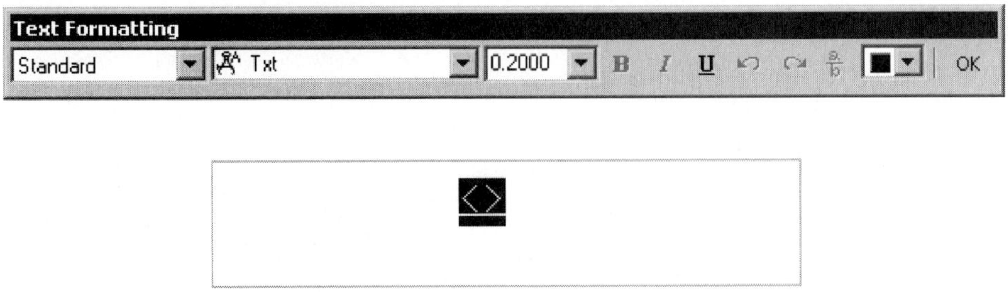

Figure 8-19 Multiline Text Editor

Specify first extension line origin or <select object>: *Select the first endpoint of the line using the Endpoint object snap, see Figure 8-20.*
Specify second extension line origin: *Select the second endpoint of the line using the Endpoint object snap, see Figure 8-20.*
Specify dimension line location or
[Mtext/Text/Angle/Horizontal/Vertical/Rotated]: **M**

*Select the code and then choose the **Underline** button to underline the text in the **Multiline Text Editor**.*

Specify dimension line location or
[Mtext/Text/Angle/Horizontal/Vertical/Rotated]: *Place the dimension.*
Dimension text = 4.00

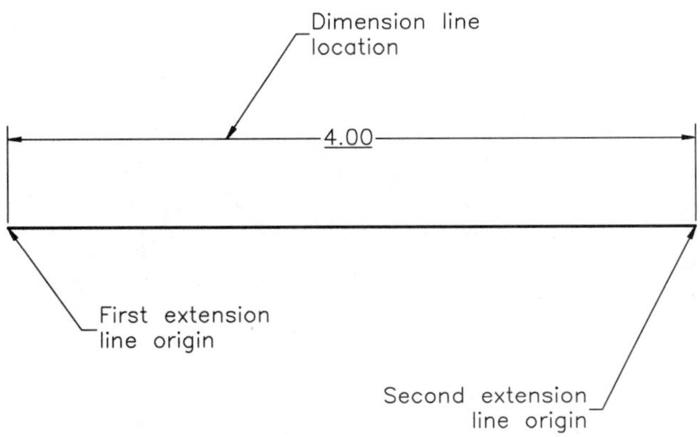

Figure 8-20 *Line for Example 1*

CREATING ALIGNED DIMENSIONS

Toolbar:	Dimension > Aligned Dimension
Menu:	Dimension > Aligned
Command:	DIMALIGNED

Generally, the drawing consists of various objects that are neither parallel to the *X* axis nor to the *Y* axis. Dimensioning of such objects can be done using **aligned dimensioning**. In horizontal or vertical dimensioning, you can only measure the shortest distance from the first extension line origin to the second extension line origin along the horizontal or vertical axis, respectively, whereas, with the help of aligned dimensioning, you can measure the true aligned distance between the two points. The working of the **ALIGNED** dimension command is similar to that of the other linear dimensioning commands. The dimension created with the **ALIGNED** command is **parallel to the object being dimensioned**. The prompt sequence that will follow when you choose this button is given next.

Specify first extension line origin or <select object>: *Specify the first point or press ENTER to select the object.*
Specify second extension line origin: *Specify second point.*
Specify dimension line location or
[Mtext/Text/Angle]: *Specify the location for the dimension line.*
Dimension text = Current.

The options provided under this command are similar to those under the **DIMLINEAR** command. Figure 8-21 illustrates aligned dimensioning.

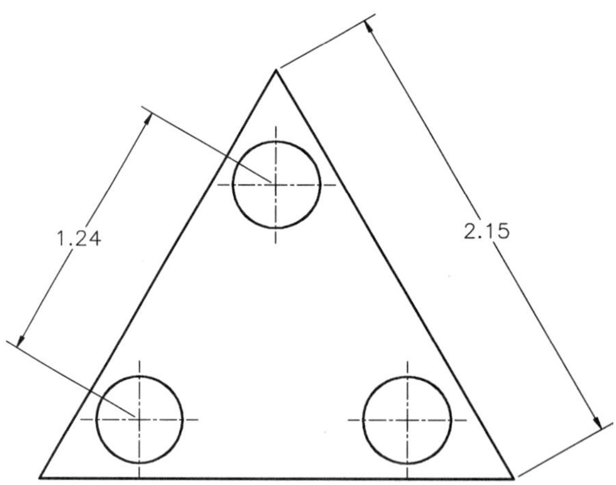

Figure 8-21 *Aligned dimensioning*

Exercise 1 — *Mechanical*

Draw the object shown in Figure 8-22 and then use linear and aligned dimensioning to dimension the part. The distance between the dotted lines is 0.5 units. The dimensions should be up to 2 decimal places. To get dimensions up to 2 decimal places, enter DIMDEC at the Command prompt and then enter 2. (There will be more information about dimension variable in Chapter 10.)

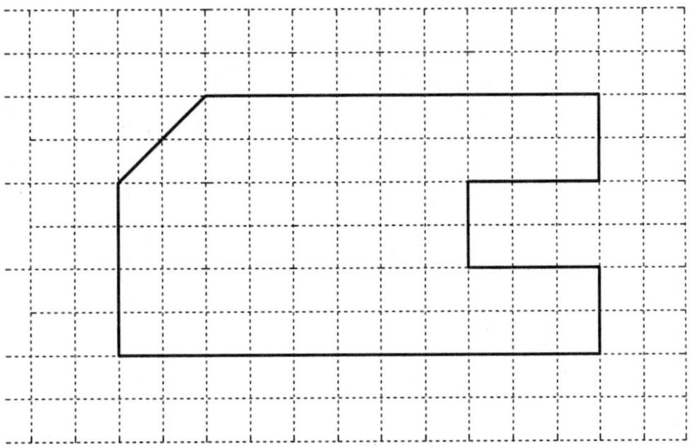

Figure 8-22 *Drawing for Exercise 1*

CREATING ROTATED DIMENSIONS

Rotated dimensioning is used when you want to place the dimension line at an angle (if you do not want to align the dimension line with the extension line origins selected) (Figure 8-23). The **ROTATED** dimension option will prompt you to specify the dimension line angle. You can invoke this command by using **DIMLINEAR** (Rotate option) or by entering **ROTATED** at the **Dim:** Command prompt. The prompt sequence is given next.

Dim: **ROTATED**
Specify angle of dimension line <0>: **110**
Specify first extension line origin or <select object>: *Select the lower right corner of the triangle.*
Specify second extension line origin: *Select the top corner.*
Specify dimension line location or [MText/Text/Angle]: *Select the location for the dimension line.*
Enter dimension text <2.0597>: Enter

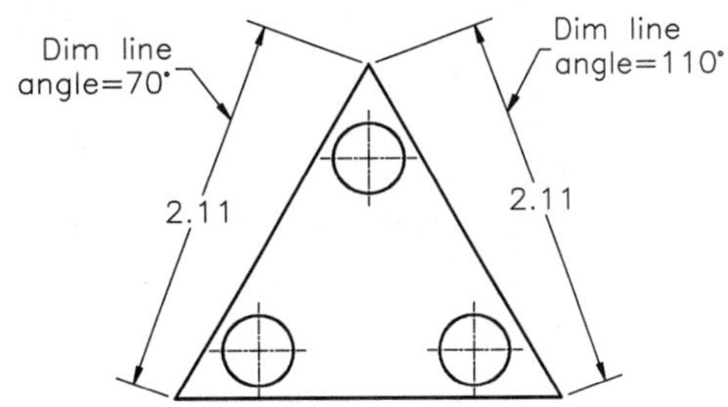

Figure 8-23 *Rotated dimensioning*

Tip

You can draw horizontal and vertical dimensioning by specifying the rotation angle of 0-degree for horizontal dimensioning and 90-degree for vertical dimensioning.

CREATING BASELINE DIMENSIONS

Toolbar:	Dimension > Baseline Dimension
Menu:	Dimension > Baseline
Command:	DIMBASE or DIMBASELINE

Sometimes in manufacturing, you may want to locate different points and features of a part with reference to a fixed point (base point or reference point). This can be accomplished by using **BASELINE dimensioning** (Figure 8-24). With this command you can continue a linear dimension from the first extension line origin of the first dimension. The new dimension line is automatically offset by a fixed amount to avoid overlapping of the dimension lines. This has to be kept in mind that **there must already exist a linear, ordinate, or angular associative dimension to use the Baseline dimensions.** When you choose the **Baseline Dimension** button, the last linear, ordinate, or angular dimension that was created will be selected and used as baseline. The prompt sequence that will follow when you choose this button is given next.

Specify a second extension line origin or [Undo/Select] <Select>: *Select the origin of the second extension line.*
Dimension text = current
Specify a second extension line origin or [Undo/Select] <Select>: *Select the origin of the second extension line.*
Dimension text = current
Specify a second extension line origin or [Undo/Select] <Select>: *Select the origin of the second extension line or press ENTER.*
Select base dimension: Enter

When you use the **DIMBASELINE** command, you cannot change the default dimension text. However, the **DIM** mode commands allows you to override the default dimension text.

Command: **DIM**
Dim: **HOR**
Specify first extension line origin or <select object>: *Select left corner (P1, Figure 8-24). (Use Endpoint object snap.)*
Specify second extension line origin: *Select the origin of the second extension line (P2).*
Specify dimension line location or [MText/Text/Angle]: **T**
Enter dimension text <1.0000>: **1.0**
Specify dimension line location or

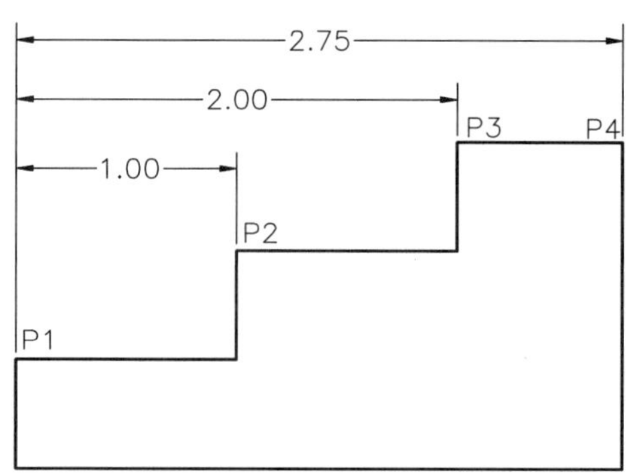

Figure 8-24 *Baseline dimensioning*

[MText/Text/Angle]: *Select the dimension line location.*
Dim: **BASELINE (or BAS)**
Specify a second extension line origin or [Select] <Select>: *Select the origin of the next extension line (P3).*
Enter dimension text <2.0000>: **2.0**
Dim: **BAS**
Specify a second extension line origin or [Select] <Select>: *Select the origin of the next extension line (P4).*
Enter dimension text <3.000>: **2.75**

The next dimension line is automatically spaced and drawn by AutoCAD.

CREATING CONTINUED DIMENSIONS

Toolbar:	Dimension > Continue Dimension
Menu:	Dimension > Continue
Command:	DIMCONT or DIMCONTINUE

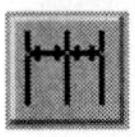

With this command you can continue a linear dimension from the second extension line of the previous dimension. This is also called as **chained** or **incremental dimensioning**. **Note that there must exist linear, ordinate, or angular associative dimension to use the Continue dimensions**. The prompt sequence that will follow when you choose this button is given next.

Specify a second extension line origin or [Undo/Select] <Select>: *Select the origin of the second extension line or press ENTER to select the existing dimension.*
Select continued dimension: *Select the dimension.*
Specify a second extension line origin or [Undo/Select] <Select>: *Specify the point on the origin of the second extension line.*
Dimension text = current
Specify a second extension line origin or [Undo/Select] <Select>: *Specify the point on the origin of the second extension line.*
Dimension text = current
Specify a second extension line origin or [Undo/Select] <Select>: Enter
Select continued dimension: Enter

In this case also the **DIM** command should be used if you want to change the default dimension text.

Command: **DIM**
Dim: **HOR**
Specify first extension line origin or <select object>: *Select left corner (P1, see Figure 8-25). (Use Endpoint object snap.)*
Specify second extension line origin: *Select the origin of the second extension line (P2, see Figure 8-25).*
Specify dimension line location or [MText/Text/Angle]: **T**

Enter dimension text <current>: **0.75**
Specify dimension line location or [MText/Text/Angle]: *Select the dimension line location.*
Dim: **CONTINUE**
Specify a second extension line origin or [Select] <Select>: *Select the origin of the next extension line (P3, see Figure 8-25).*
Enter dimension text <current>: Enter
Dim: **CONTINUE**
Specify a second extension line origin or [Select] <Select>: *Select the origin of next extension line (P4, see Figure 8-25).*
Enter dimension text <current>: Enter

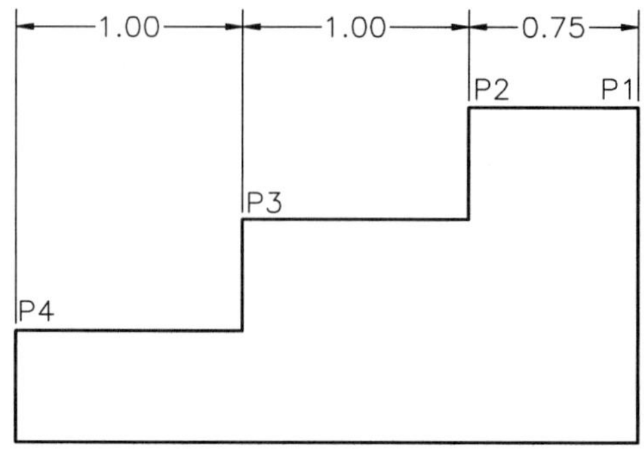

Figure 8-25 *Continue dimensioning*

The default base (first extension line) for the dimensions created with the **CONTINUE** command is the previous dimension's second extension line. You can override the default by pressing ENTER at the **Specify a second extension line origin or [Select] <Select>** prompt, and then specifying the other dimension. The extension line origin nearest to the selection point is used as the origin for the first extension line.

Tip
*You can use the **Select** option of the **DIMBASELINE** or the **DIMCONTINUE** command to select any other existing dimension to be used as baseline or continuous dimension.*

Exercise 2 *Mechanical*

Draw the object shown in Figure 8-26 and then use the **baseline** dimensioning to dimension the top half and **continue dimensioning** to dimension the bottom half. The distance between the dotted lines is 0.5 units.

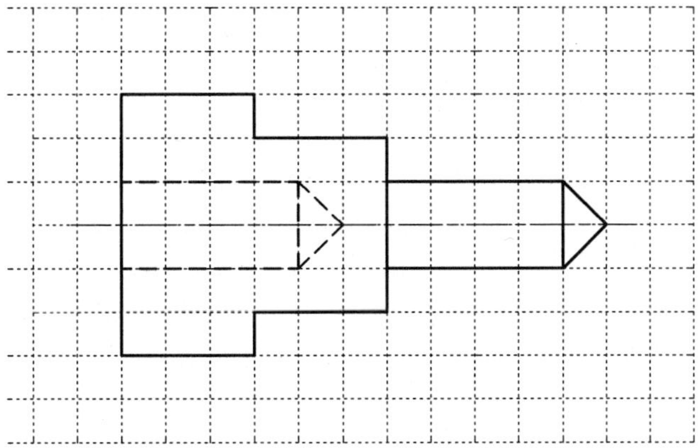

Figure 8-26 *Drawing for Exercise 2*

CREATING ANGULAR DIMENSIONS

Toolbar:	Dimension > Angular Dimension
Menu:	Dimension > Angular
Command:	DIMANGULAR or DIMANG

Angular dimensioning is used when you want to dimension an angle. This command generates a dimension arc (dimension line in the shape of an arc with arrowheads at both ends) to indicate the angle between two nonparallel lines. This command can also be used to dimension the vertex and two other points, a circle with another point, or the angle of an arc. For every set of points there exists one acute angle and one obtuse angle (inner and outer angles). If you specify the dimension arc location between the two points, you will get the acute angle; if you specify it outside the two points, you will get the obtuse angle. Figure 8-27 shows the four ways to dimension two nonparallel lines. The prompt sequence that will follow when you choose this button is given next.

Select arc, circle, line, or <specify vertex>: *Select the object or press ENTER to select a vertex point where two segments meet.*
Select second line: *Select the second object.*
Specify dimension arc line location or [Mtext/Text/Angle]: *Place the dimension or select an option.*
Dimension text = current

The methods of dimensioning various entities using this command are discussed next.

Dimensioning the Angle Between Two Nonparallel Lines

The angle between two nonparallel lines or two straight line segments of a polyline can be dimensioned with the **DIMANGULAR** dimensioning command. The vertex of the angle is taken as the point of intersection of the two lines. The location of the extension lines and

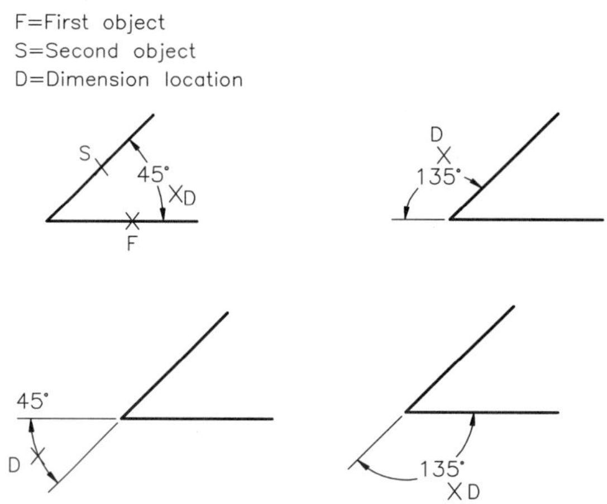

Figure 8-27 *Angular dimensioning between two nonparallel lines*

dimension arc is determined by how you specify the dimension arc location.

The following example illustrates the dimensioning of two nonparallel lines using the **DIMANGULAR** command invoked using the **Angular Dimension** button.

> Select arc, circle, line, or <specify vertex>: *Select the first line.*
> Select second line: *Select the second line.*
> Specify dimension arc line location or [Mtext/Text/Angle]: **M***(Enter the new value in the **Multiline Text Editor** dialog box.)*
> Specify dimension arc line location or [Mtext/Text/Angle]: *Specify the dimension arc location.*

Dimensioning the Angle of an Arc

Angular dimensioning can also be used to dimension the angle of an arc. In this case, the center point of the arc is taken as the vertex and the two endpoints of the arc are used as the extension line origin points for the extension lines (Figure 8-28). The following example illustrates the dimensioning of an arc using the **DIMANGULAR** command.

> Select arc, circle, line, or <specify vertex>: *Select the arc.*
> Specify dimension arc line location or [Mtext/Text/Angle]: *Specify a location for the arc line or select an option.*

Angular Dimensioning of Circles

The angular feature associated with the circle can be dimensioned by selecting a circle object at the **Select arc, circle, line, or <specify vertex>** prompt. The center of the selected circle is used as the vertex of the angle. The first point selected (when the circle is selected for angular dimensioning) is used as the origin of the first extension line. In a similar manner, the second point selected is taken as the origin of the second extension line (Figure 8-29).

The following is the prompt sequence for dimensioning a circle.

> Select arc, circle, line, or <specify vertex>: *Select the circle at the point where you want the first extension line.*
> Specify second angle endpoint: *Select the second point on or away from the circle.*
> Specify dimension arc line location or [Mtext/Text/Angle]: *Select the location for the dimension line.*

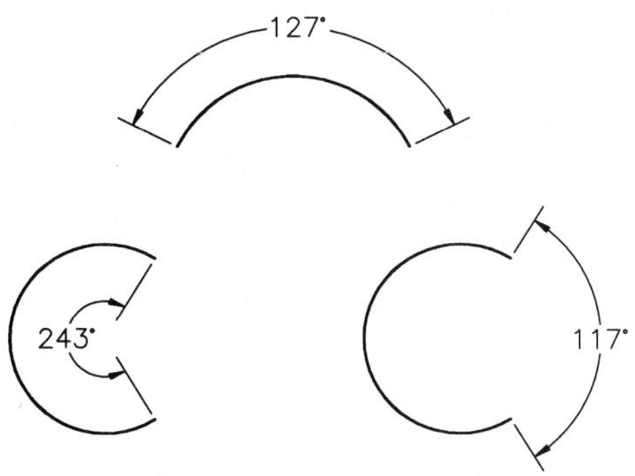

Figure 8-28 *Angular dimensioning of arcs*

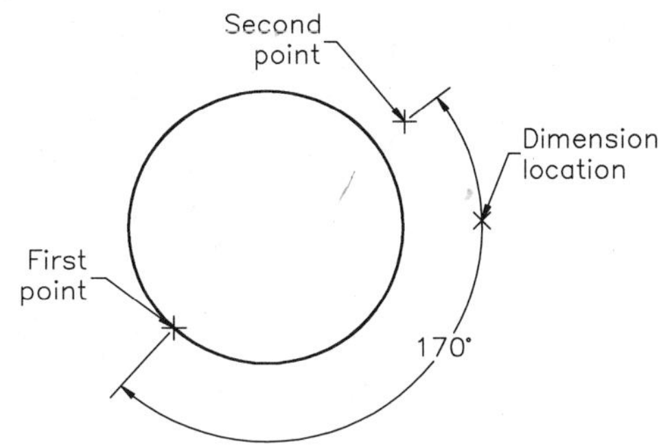

Figure 8-29 *Angular dimensioning of circles*

Angular Dimensioning Based on Three Points

If you press ENTER at the **Select arc, circle, line, or <specify vertex>** prompt, AutoCAD allows you to select three points to create an angular dimension. The first point is the vertex point, and the other two points are the first and second angle endpoints of the angle (Figure 8-30). The coordinate specifications of the first and the second angle endpoints must not be identical. However, the angle vertex coordinates and one of the angle endpoint coordinates can be identical. The following example illustrates angular dimensioning by defining three points.

> Select arc, circle, line, or <specify vertex>: Enter
> Specify angle vertex: *Specify the first point, vertex. This is the point where the two segments meet. If the two segments do not meet actually, use the* ***Apparent Intersection*** *object snap.*
> Specify first angle endpoint: *Specify the second point. This point will be the origin of the first extension line.*
> Specify second angle endpoint: *Specify the third point. This point will be the origin of the second extension line.*
> Specify dimension arc line location or [Mtext/Text/Angle]: *Select the location for the dimension line.*

Note

If you use the ***DIMANGULAR*** *command, you cannot specify the text location. With the* ***DIMANGULAR*** *command, AutoCAD positions the dimensioning text automatically. If you want to manually define the position of dimension text, use the* ***ANGULAR*** *option of the* ***DIM*** *command.*

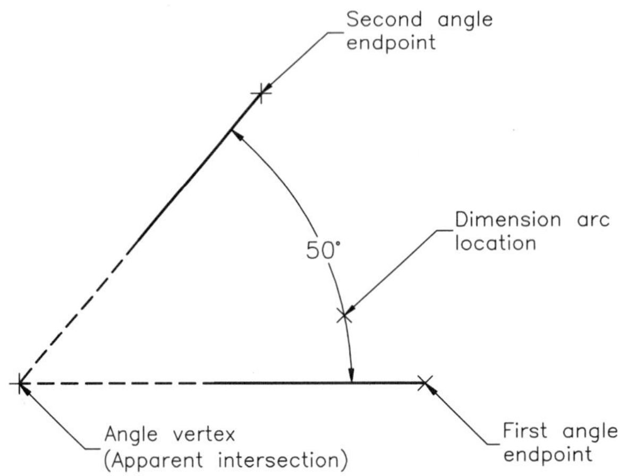

Figure 8-30 *Angular dimensioning for three points*

Exercise 3 *Mechanical*

Make the drawing shown in Figure 8-31 and then use **angular dimensioning to** dimension all angles of the part. The distance between the dotted lines is 0.5 units.

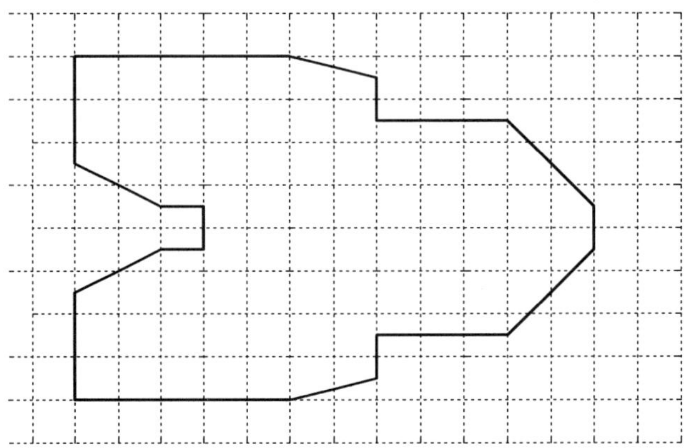

Figure 8-31 *Drawing for Exercise 3*

CREATING DIAMETER DIMENSIONS

Toolbar:	Dimension > Diameter Dimension
Menu:	Dimension > Diameter
Command:	DIMDIAMETER or DIMDIA

Diameter dimensioning is used to dimension a circle or an arc. Here, the measurement is done between two diametrically opposite points on the circumference of the circle or arc (Figure 8-32). The dimension text generated by AutoCAD commences with the Ø symbol, to indicate a diameter dimension. The prompt sequence that will follow when you choose this button is given next.

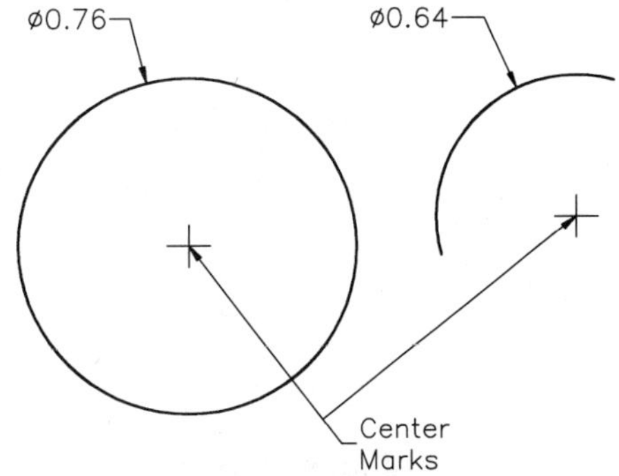

Figure 8-32 *Diameter dimensioning*

Select arc or circle: *Select an arc or circle by selecting a point anywhere on its circumference.*
Dimension text = Current
Specify dimension line location or [Mtext/Text/Angle]: *Specify a point to position the dimension.*

If you want to override the default value of the dimension text, use the Mtext or Text option. The control sequence %%C is used to obtain the diameter symbol ø. It is followed by the dimension text that you want to appear in the diameter dimension. For example, if you want to write a text that displays a value ø20, then enter %%c20 in the text prompt.

Tip
The control sequence %%d can be used to generate the degree symbol " ° " (45°).

CREATING RADIUS DIMENSIONS

Toolbar:	Dimension > Radius Dimension
Menu:	Dimension > Radius
Command:	DIMRADIUS or DIMRAD

Radius dimensioning is used to dimension a circle or an arc (Figure 8-33). Radius and diameter dimensioning are similar; the only difference is that instead of the diameter line, a radius line is drawn (half of the diameter line), which is measured from the center to any point on the circumference. The dimension text generated by AutoCAD is preceded by the letter **R** to indicate a radius dimension. If you want to use the default dimension text (dimension text generated automatically by AutoCAD), simply specify a point to position the dimension at the **Specify dimension line location or [Mtext/Text/Angle]** prompt. You

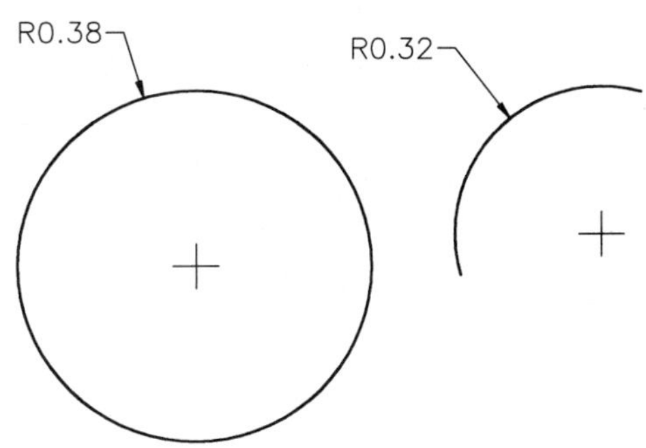

Figure 8-33 *Radius dimensioning*

Chapter 8

can also enter a new value or specify a prefix or suffix, or suppress the entire text by entering a blank space following the **Enter dimension text <current>** prompt. A center mark for the circle/arc is drawn automatically, provided the center mark value controlled by the **DIMCEN** variable is not 0. The prompt sequence that will follow when you choose this button is given next.

Select arc or circle: *Select the object you want to dimension.*
Dimension text = Current
Specify dimension line location or [Mtext/Text/Angle]: *Specify the dimension location.*

If you want to override the default value of the dimension text, use the **Text** or the **Mtext** option.

GENERATING CENTER MARKS AND CENTERLINES

Toolbar:	Dimension > Center Mark
Menu:	Dimension > Center Mark
Command:	DIMCENTER

When circles or arcs are dimensioned with the **DIMRADIUS or DIMDIAMETER** command, a small mark known as a center mark, or line known as centerline, may be drawn at the center of the circle/arc. Sometimes you may want to mark the center of a circle or an arc without using these dimensioning commands. This can be achieved with the help of the **DIMCENTER** command. You can invoke this command by choosing the **Center Mark** button from the **Dimension** toolbar or by entering **CENTER** (or **CEN**) at the **Dim:** prompt. When you invoke this command, you will be prompted to select the arc or the circle. The result of this command will depend upon the value of the **DIMCEN** variable. If the value of this variable is positive, center marks are drawn (Figure 8-34) and if the value is negative, centerlines are drawn (Figure 8-35).

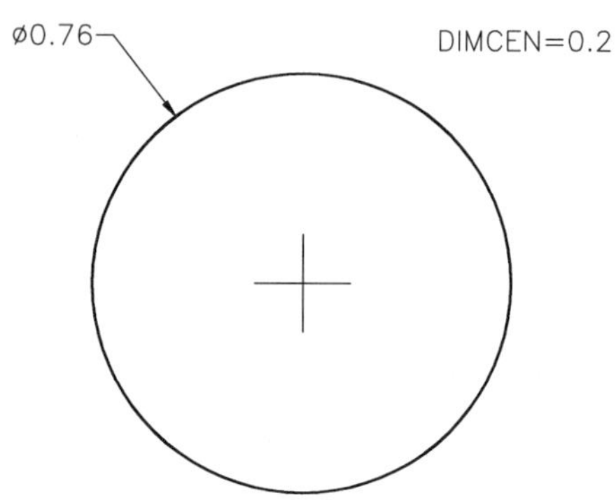

Figure 8-34** Using a positive value for **DIMCEN

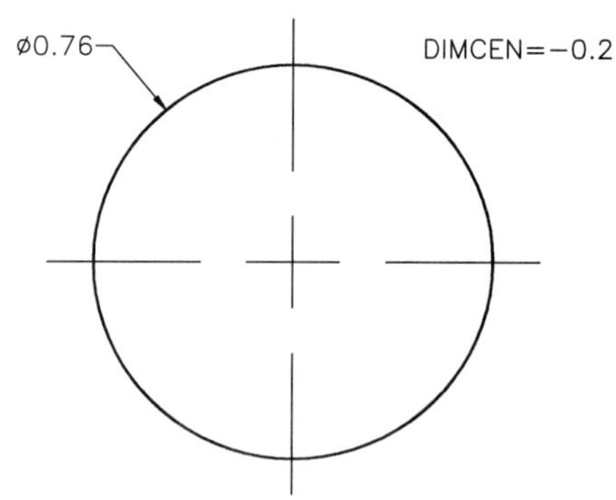

Figure 8-35** Using a negative value for **DIMCEN

Note

*The center marks created by **DIMCENTER** or **DIM, CENTER** are lines, not associative dimensioning objects, and they have an explicit linetype.*

Exercise 4 *Mechanical*

Draw the model shown in Figure 8-36 and then use the radius and diameter dimensioning commands to dimension the part. Use the **DIMCENTER** command to draw the centerlines through the circles. The distance between the dotted lines is 0.5 units.

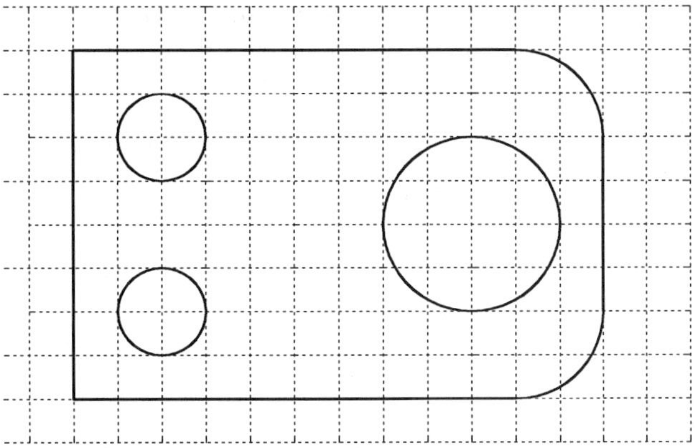

Figure 8-36 *Figure for Exercise 4*

CREATING ORDINATE DIMENSIONS

Toolbar:	Dimension > Ordinate
Menu:	Dimension > Ordinate
Command:	DIMORDINATE or DIMORD

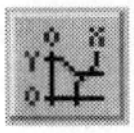

Ordinate dimensioning is used to dimension the *X* and *Y* coordinates of the selected point. This type of dimensioning is also known as **arrowless** dimensioning because no arrowheads are drawn in this type of dimensioning. Ordinate dimensioning is also called **datum dimensioning** because all dimensions are related to a common base point. The current UCS (user coordinate system) origin becomes the reference or the base point for ordinate dimensioning. With ordinate dimensioning you can determine the X or Y displacement of a selected point from the current UCS origin.

In ordinate dimensioning, AutoCAD automatically places the dimension text (*X* or *Y* coordinate value) and the leader line along the *X* or *Y* axis (Figure 8-37). Since ordinate dimensioning pertains to either the *X* coordinate or the *Y* coordinate, you should keep ORTHO on. When ORTHO is off, the leader line is automatically given a bend when you select the second leader line point that is offset from the first point. This allows you to generate offsets and avoid overlapping text on closely spaced dimensions. In ordinate dimensioning, only one extension line (leader line) is drawn.

The leader line for an *X* coordinate value will be drawn perpendicular to the *X* axis, and the leader line for a *Y* coordinate value will be drawn perpendicular to the *Y* axis. Since you cannot override this, the leader line drawn perpendicular to the *X* axis will have the dimension text aligned with the leader line. The dimension text is the X datum of the selected point. The leader line drawn perpendicular to the *Y* axis will have the dimension text, which is the Y datum of the selected point, aligned with the leader line. Any other alignment specification for the dimension text is nullified. Hence, changes in Text Alignment in the **Dimension Style Manager** dialog box (**DIMTIH** and **DIMTOH** variables) have no effect on the alignment of dimension text. You can specify the coordinate value you want to dimension at the **Specify leader endpoint or [Xdatum/Ydatum/MText/Text/Angle]** prompt.

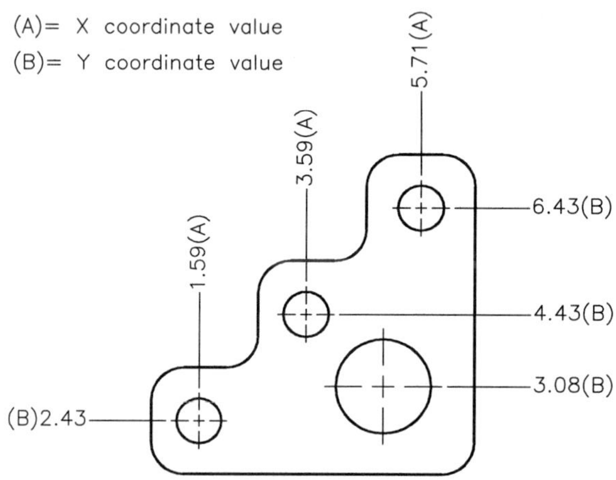

Figure 8-37 *Ordinate dimensioning*

If you select or enter a point, AutoCAD checks the difference between the feature location and the leader endpoint. If the difference between the *X* coordinates is greater, the dimension measures the *Y* coordinate; otherwise, the *X* coordinate is measured. In this manner AutoCAD determines whether it is an X or Y type of ordinate dimension. However, if you enter Y instead of specifying a point, AutoCAD will dimension the *Y* coordinate of the selected feature. Similarly, if you enter X, AutoCAD will dimension the *X* coordinate of the selected point. The prompt sequence that follows when you choose this button is given next.

Specify feature location: *Select a point on an object.*
Specify leader endpoint or [Xdatum/Ydatum/Mtext/Text/Angle]: *Enter the endpoint of the leader.*

You can override the default text with the help of the **Mtext** or the **Text** option. If you use the **Mtext** option, the **Multiline Text Editor** will be displayed. If you use the **Text** option, you will be prompted to specify the new text in the Command line itself.

Exercise 5 — *General*

Draw the model shown in Figure 8-38 and then use ordinate dimensioning to dimension the part. The distance between the dotted lines is 0.5 units.

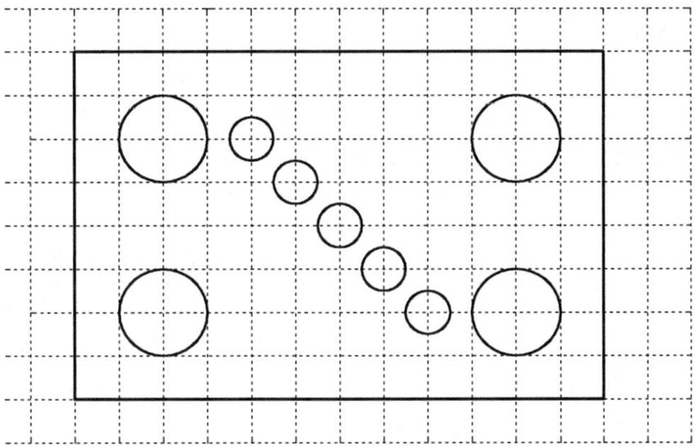

Figure 8-38 *Using the* ***DIMORDINATE*** *command to dimension the part*

CREATING TRUE ASSOCIATIVE DIMENSIONS

The true associative dimensions are the dimensions that are automatically modified when the objects to which they are associated are modified. As mentioned earlier, this concept was introduced in AutoCAD 2002. If the dimension attached to the object is a true associative dimension, then the dimension will be modified automatically when the object is modified. In this case you do not have to select the definition points of the dimensions. Any dimension can be converted into a true associative dimension with the help of the **DIMREASSOCIATE** command. This is discussed next.

Converting a Dimension into a True Associative Dimension

Menu:	Dimension > Reassociate Dimensions
Command:	DIMREASSOCIATE

The **DIMREASSOCIATE** command is used to create a true associative dimension by associating the selected dimension to the specified object. When you invoke this command, you will be prompted to select the objects. These objects are the dimensions to be associated. Once you select the dimensions to be associated, a cross is displayed and you are prompted to select the feature location. This cross implies that the dimension is not associated. You can define a new association point for the dimensions by selecting the objects or by using the object snaps. If you select a dimension that has already been associated to an object, the cross will be displayed inside a box. The prompt sequence that is displayed varies depending upon the type of dimension selected. In case of **linear**, **aligned**, **radius**, and **diameter** dimensions, you can directly select the object to associate the dimension. If the arcs or circles are assigned the angular dimensions using three points, then also you can select these arcs or circles directly for associating the dimensions. For rest of the dimension types, you can use the object snaps to specify the point to associate the dimensions.

REMOVING THE DIMENSION ASSOCIATIVITY (DIMDISASSOCIATE COMMAND)

Command:	DIMDISASSOCIATE

The **DIMREASSOCIATE** command is used to remove the associativity of the dimensions from the object to which they are associated using the **DIMREASSOCIATE** command. When you invoke this command, you will be prompted to select the dimensions to disassociate. The true association of the selected dimensions is automatically removed once you exit this command. The number of dimensions disassociated is displayed in the Command prompt.

DRAWING LEADERS

Toolbar:	Dimension > Quick Leader
Menu:	Dimension > Leader
Command:	QLEADER

The leader line is used to attach annotations to an object or when the user wants to show a dimension without using another dimensioning command. Sometimes leaders of the dimensions of circles or arcs are so complicated that you need to construct a leader of your own. The leaders can be created using the **QLEADER** command. The leaders drawn by using this command create the arrow and the leader lines as a single object. The text is created as a separate object. This command can create multiline annotations and offer several options such as copying existing annotations and so on. You can customize the leader and annotation by selecting the **Settings** option at the **Specify first leader point, or [Settings]<Settings>** prompt. The prompt sequence that will follow when you choose this button is given next.

Specify first leader point, or [Settings]<Settings>: *Specify the start point of the leader.*
Specify next point: *Specify endpoint of the leader.*
Specify next point: *Specify next point.*
Specify text width <current>: *Enter text width of multiline text.*
Enter first line of annotation text <MText>: *Press ENTER; AutoCAD displays the **Multiline Text Editor** dialog box. Enter text in the dialog box and then choose **OK** to exit.*

If you press ENTER at the **Specify first leader point, or [Settings]<Settings>** prompt then the **Leader Settings** dialog box is displayed. The **Leader Settings** dialog box gives you a number of options for the leader line and the text attached to it. It has the following tabs.

Annotation Tab

This tab provides you with various options to control annotation features, see Figure 8-39.

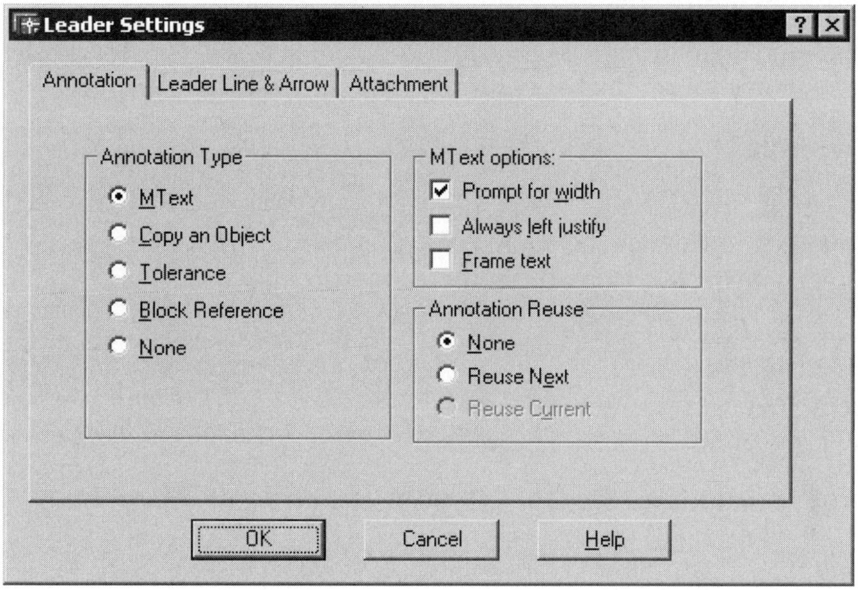

Figure 8-39 ***Leader Settings*** *dialog box,* ***Annotation*** *tab*

Annotation Type Area

Mtext. When selected, AutoCAD uses the **Multiline Text Editor** to create annotation. Three Mtext options are available.

Copy an Object. This option allows you to copy an existing annotation object (like mtext, single line text, tolerance, or block) and attach it at the end of the leader. For example, if you have a text string in the drawing that you want to place at the end of the leader, you can use the **Copy an Object** option to place it at the end of the leader.

Tolerance. When you select the **Tolerance** option, AutoCAD displays the **Geometric Tolerance** dialog box on the screen. Specify the tolerance in the dialog box and choose OK to exit. AutoCAD will place the specified Geometric Tolerance with feature control frame at the end of the leader (Figure 8-40).

Block Reference. The **Block Reference** option allows you to insert a predefined block at the end of the leader. When you select this option, AutoCAD will prompt you to enter the block name and insertion point.

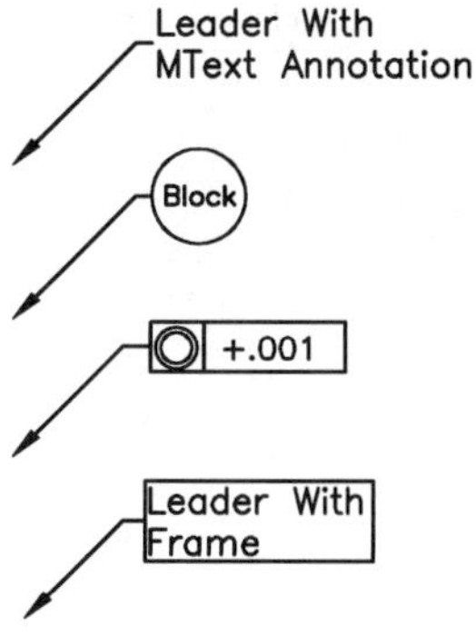

Figure 8-40 *Leader with MText, Block, Tolerance, and MText with Frame annotations*

None. This option creates a leader without placing any annotation at the end of the leader.

MText options Area

The options under this area will be available only if the **MText** radio button is selected from the **Annotation Type** area. This area provides you with the following options:

Prompt for width. Selecting this check box allows you to specify the width of the mtext annotation.

Always left justify. This option left justifies the mtext annotation in all situations. Selecting this check box makes the Prompt for width option unavailable.

Frame text. Selecting this check box draws a box around the mtext annotation.

Annotation Reuse Area

The options under this area allow you to reuse the annotation.

None. When selected, AutoCAD does not reuse the leader annotation.

Reuse Next. This option allows you to reuse the annotation that you are going to create next for all subsequent leaders.

Reuse Current. This option allows you to reuse the current annotation for all subsequent leaders.

Leader Line & Arrow Tab

The options under this area are related to the leader parameters, see Figure 8-41.

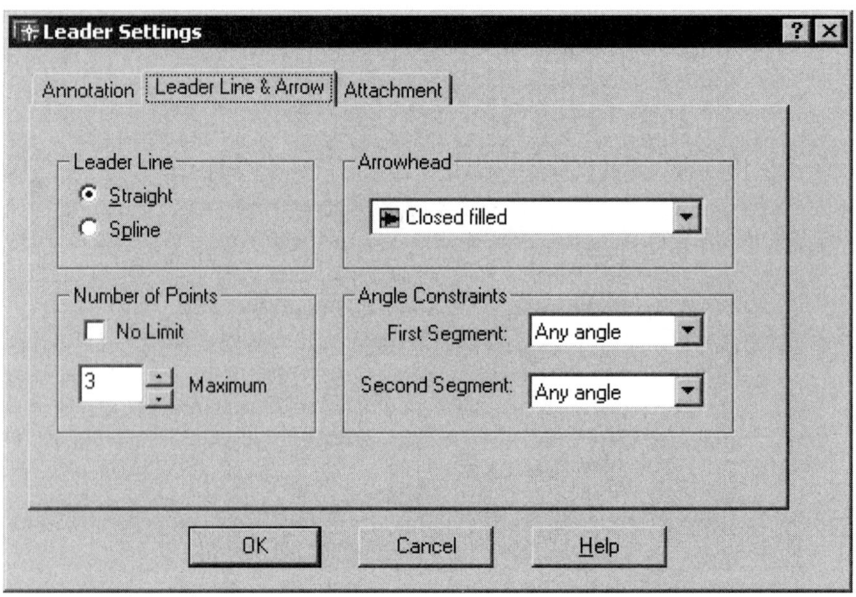

***Figure 8-41 Leader Settings** dialog box, **Leader Line & Arrow** tab*

Leader Line Area

This area gives the options for the **leader line type** such as straight or spline. The **Spline** option draws a spline through the specified leader points and the **Straight** option draws straight lines. Figure 8-42 shows straight and spline leader lines.

Number of Points Area

No Limit. If this check box is selected, you can define as many number of points as you want in the leader line. AutoCAD will keep prompting you for the next point until you press ENTER at this prompt.

Maximum. This spinner is used to specify the maximum number of points on the leader line. The default value is 3, which means that by default, there will be only three points in the leader line. You can specify the maximum number of points using this spinner. This has to be kept in mind that the start point of the leader is the first leader point. This spinner will be available only if the **No Limit** check box is cleared. Figure 8-42 shows a leader line with five number of points.

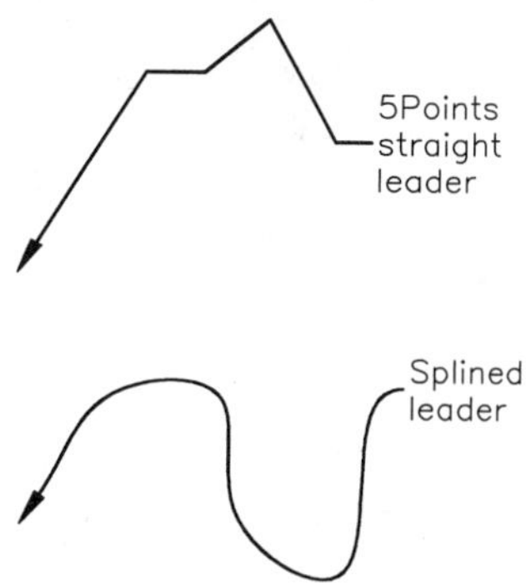

Figure 8-42 *Splined and straight leaders*

Arrowhead Area

The drop-down list provided under this area allows you to define a leader arrowhead. The arrowhead available here is the same as the one available for dimensioning. You can also use the user-defined arrows by selecting **User Arrow** from the drop-down list.

Angle Constraints Area

The options provided under this area are used to define the angle for the segments of the leader lines.

First Segment. This drop-down list is used to specify the angle at which the first leader line segment will be drawn. You can select the predefined values from this drop-down list.

Second Segment. This drop-down list is used to specify the angle at which the second leader line segment will be drawn.

Attachment Tab

The **Attachment** tab (Figure 8-43) will be available only if you have selected **MText** from the **Annotation Type** area of the **Annotation** tab. The options provided under this tab are used for attaching the Multiline Text to the leader. It has two columns: **Text on left side** and **Text on right side**. Both of these columns have five radio buttons below them. Each radio button corresponds to the option of attaching the multiline text. If you draw a leader from right to left, AutoCAD uses the settings under **Text on left side**. Similarly, if you draw a leader from left to right, AutoCAD uses the settings as specified under **Text on right side**. This area also provides you with the **Underline bottom line** check box. If this check box is selected then the last line of the multiline text will be underlined.

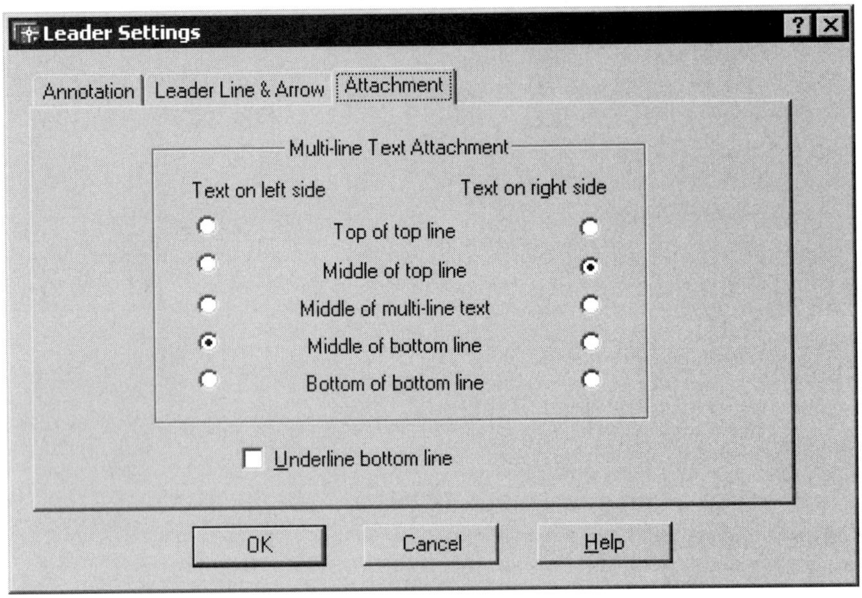

***Figure 8-43 Leader Settings** dialog box, Attachment tab*

Exercise 6 *Mechanical*

Make the drawing shown in Figure 8-44 and then use the **QLEADER** command to dimension the part as shown. The distance between the dotted lines is 0.5 units.

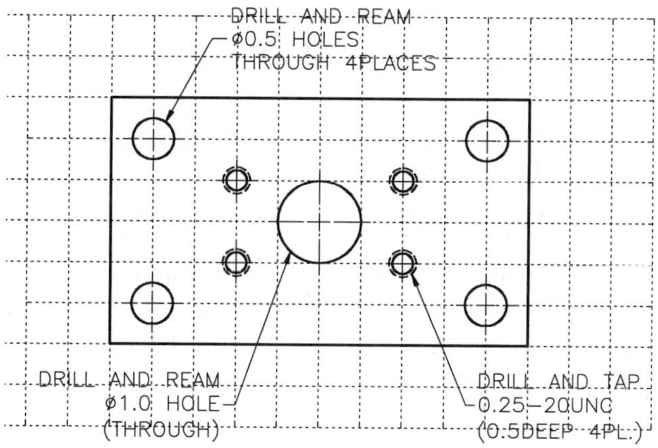

Figure 8-44 *Drawing for Exercise 6*

Note

You can use the Command line to create the leaders with the help of the ***LEADER*** *command. The options provided under this command are similar to those under the* ***QLEADER*** *command. The only difference is that the* ***LEADER*** *command uses the Command line.*

USING LEADER WITH THE DIM COMMAND

You can also draw a leader by using the **Leader** option of the **DIM** command. The **DIM, LEADER** command creates the arrow, leader lines, and the text as separate objects. This command has the feature of defaulting to the most recently measured dimension. Once you invoke the **DIM, LEADER** command and specify the first point, the prompt sequence is similar to that of the **LINE** command. The start point of the leader should be specified at the point closest to the object being dimensioned. After drawing the leader, enter a new dimension text or keep the default one.

Command: **DIM**
Leader (or **L**)
Leader start: *Specify the starting point of the leader.*
To point: *Specify the endpoint of the leader.*
To point: *Specify the next point.*
To point: Enter
Dimension text <current>: *Enter dimension text.*

The value between the angle brackets is the current value, that is, the measurement of the most recently dimensioned object. If you want to retain the default text, press ENTER at the **Dimension text <current>** prompt. You can enter text of your choice, specify a prefix/suffix, or suppress the text. The text can be suppressed by pressing the SPACEBAR, and then pressing ENTER at the **Dimension text <current>** prompt. An arrow is drawn at the start point of the leader segment if the length of the segment is greater than two arrow lengths. If the length of the line segment is less than or equal to two arrow lengths, only a line is drawn.

GEOMETRIC DIMENSIONING AND TOLERANCING

One of the most important parts of the design process is giving the dimensions and tolerances, since every part is manufactured from the dimensions given in the drawing. Therefore, every designer must understand and have a thorough knowledge of the standard practices used in industry to make sure that the information given on the drawing is correct and can be understood by other people. Tolerancing is equally important, especially in the assembled parts. Tolerances and fits determine how the parts will fit. Incorrect tolerances could result in a product that is not usable. In addition to dimensioning and tolerancing, the function and the relationship that exists between the mating parts is important if the part is to perform the way it was designed. This aspect of the design process is addressed by **geometric dimensioning and tolerancing**, generally known as **GDT**.

Geometric dimensioning and tolerancing is a means to design and manufacture parts with respect to actual function and the relationship that exists between different features of the same part or the features of the mating parts. Therefore, a good design is not achieved by just giving dimensions and tolerances. The designer has to go beyond dimensioning and think of the intended function of the part and how the features of the part are going to affect its function. For example, Figure 8-45 shows a part with the required dimensions and tolerances. In this drawing there is no mention of the relationship between the pin and the plate. Is the pin perpendicular to the plate? If it is, to what degree should it be perpendicular? Also, it does not mention on which surface the perpendicularity of the pin is to be measured. A design like this is open to individual interpretation based on intuition and experience. This is where geometric dimensioning and tolerancing play an important part in the product design process.

Figure 8-46 has been dimensioned using geometric dimensioning and tolerancing. The feature symbols define the datum (reference plane) and the permissible deviation in the perpendicularity of the pin with respect to the bottom surface. From a drawing like this, the chances of making a mistake are minimized.

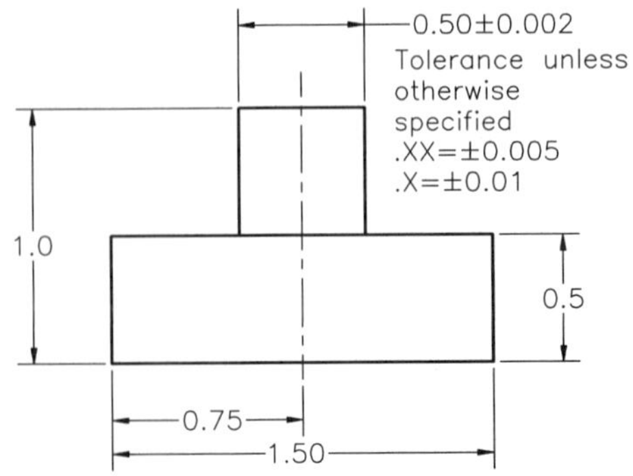

Figure 8-45 *Traditional dimensioning and tolerancing technique*

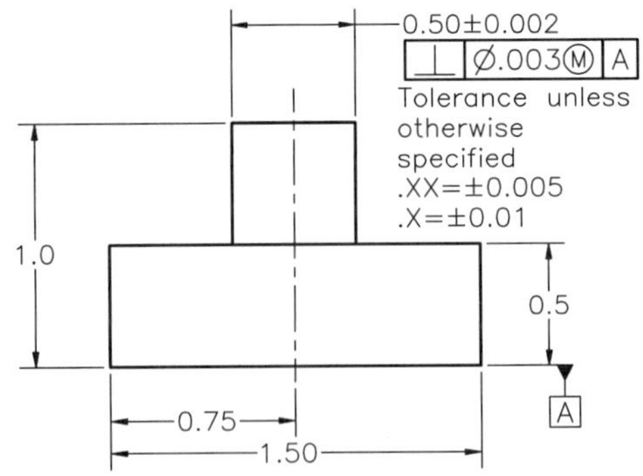

Figure 8-46 *Geometric dimensioning and tolerancing*

GEOMETRIC CHARACTERISTICS AND SYMBOLS

Before discussing the application of AutoCAD commands in geometric dimensioning and tolerancing, you need to understand the following feature symbols and tolerancing components. Figure 8-47 shows the geometric characteristics and symbols used in geometric dimensioning and tolerancing.

Kind of feature	Type of feature	Characteristics	
Related	Location	Position	⌖
		Concentricity or Coaxiality	◎
		Symmetry	⌯
	Orientation	Parallelism	//
		Perpendicularity	⊥
		Angularity	∠
Individual	Form	Cylindricity	⌭
		Flatness	⏥
		Circularity or Roundness	○
		Straightness	—
Individual or related	Profile	Surface Profile	⌓
		Line Profile	⌒
Related	Runout	Circular Runout	↗
		Total Runout	⌰

Figure 8-47 *Characteristics and symbols used in GTOL*

Note

These symbols are the building blocks of geometric dimensioning and tolerancing.

ADDING GEOMETRIC TOLERANCE

Toolbar: Dimension > Tolerance
Menu: Dimension > Tolerance
Command: TOLERANCE

Geometric tolerance displays the deviations of profile, orientation, form, location, and runout of a feature. In AutoCAD, geometrical tolerancing is displayed by feature control frames. The frames contain all the information about tolerances for a single dimension. To display feature control frames with the various tolerancing parameters, the specifications are entered in the **Geometric Tolerance** dialog box (Figure 8-48).

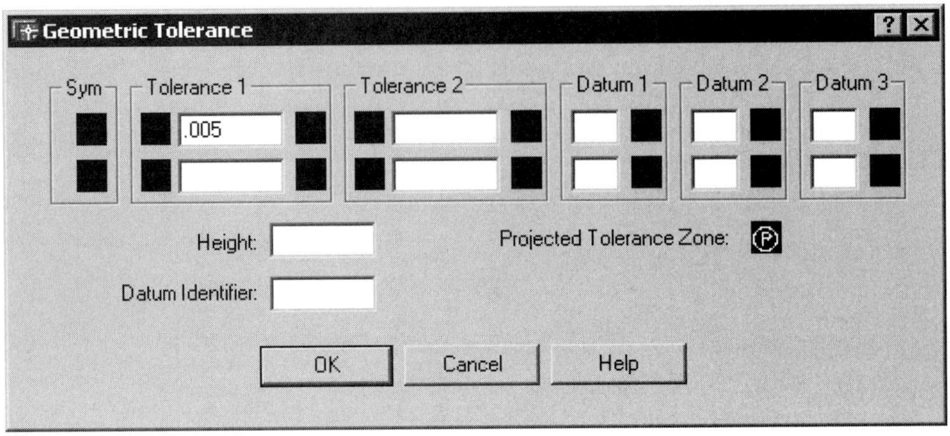

***Figure 8-48 Geometric Tolerance** dialog box*

The various components that constitute the GTOL are shown in Figures 8-49 and 8-50.

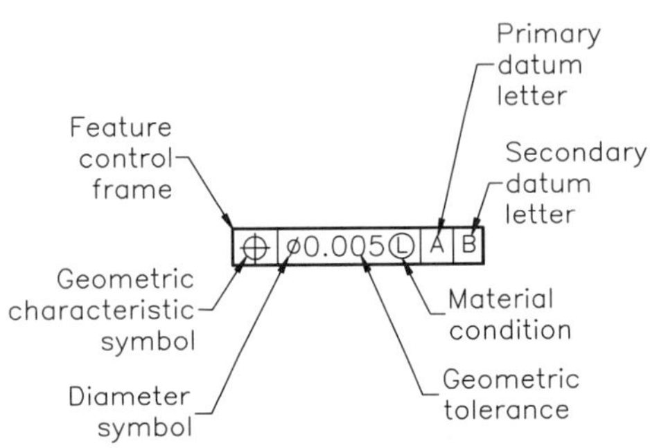

***Figure 8-49** Components of GTOL*

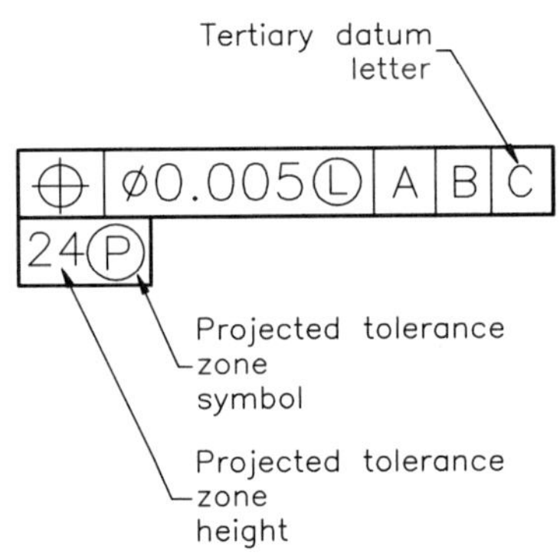

***Figure 8-50** Components of GTOL*

Feature Control Frame

The **feature control frame** is a rectangular box that contains the geometric characteristics symbols and tolerance definition. The box is automatically drawn to standard specifications; you do not need to specify its size. You can copy, move, erase, rotate, and scale the feature control frame. You can also snap to them using various Object snap modes. You can edit feature control frames using the **DDEDIT** command or you can also edit them using **GRIPS**. System variable **DIMCLRD** controls the color of the feature control frame. System variable **DIMGAP** controls the gap between the feature control frame and the text.

Geometric Characteristics Symbol

The geometric characteristics symbols indicate the characteristics of a feature like straightness, flatness, perpendicularity, and so on. You can select the required symbol from the **Symbol** dialog box (Figure 8-51). This dialog box is displayed by selecting the box provided in the **Sym** area of the **Geometric Tolerance** dialog box. To select the required symbol, just pick the symbol using the left mouse button. The symbol will now be displayed in the box under the **Sym** area.

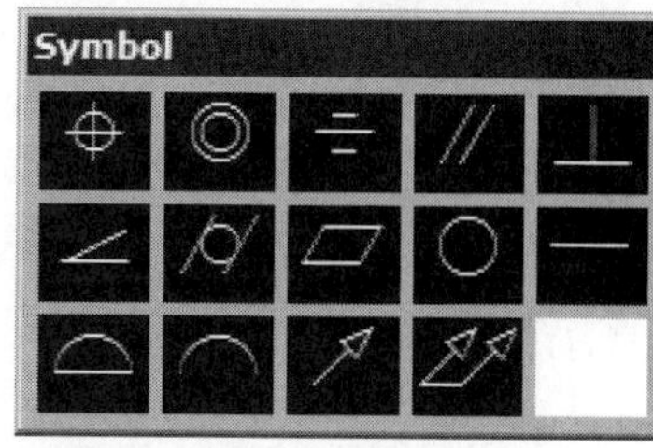

Figure 8-51 ***Symbol*** *dialog box*

Tolerance Value and Tolerance Zone Descriptor

The tolerance value specifies the tolerance on the feature as indicated by the tolerance zone descriptor. For example, a value of .003 indicates that the feature must be within a 0.003 tolerance zone. Similarly, ϕ.003 indicates that this feature must be located at true position within 0.003 diameter. The tolerance value can be entered in the edit box provided under the **Tolerance 1** or the **Tolerance 2** area of the **Geometric Tolerance** dialog box. The tolerance zone descriptor can be invoked by selecting the box, located to the left of the edit box. The system variable **DIMCLRT** controls the color of the tolerance text, variable **DIMTXT** controls the tolerance text size, and variable **DIMTXSTY** controls the style of the tolerance text. Using the **Projected Tolerance Zone**, inserts a projected tolerance zone symbol, which is an encircled P, after the projected tolerance zone value.

Material Condition Modifier

The **material condition modifier** specifies the material condition when the tolerance value takes effect. For example, ϕ.003(M) indicates that this feature must be located at true position within a 0.003 diameter at maximum material condition (MMC). The material condition modifier symbol can be selected from the **Material Condition** dialog box (Figure 8-52). This dialog box can be invoked by selecting the boxes located on the right side of the edit boxes under the **Tolerance 1**, **Tolerance 2**, **Datum 1**, **Datum 2**, and **Datum 3** areas of the **Geometric Tolerance** dialog box.

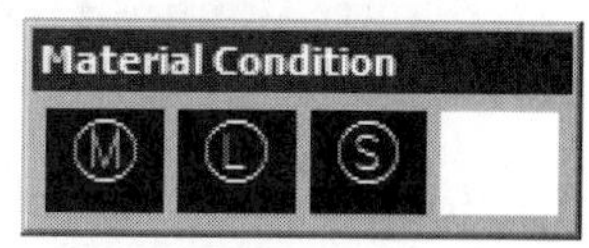

Figure 8-52 ***Material Condition*** *dialog box*

Datum

The datum is the origin, surface, or feature from which the measurements are made. The datum is also used to establish the geometric characteristics of a feature. The datum feature symbol consists of a reference character enclosed in a feature control frame. You can create the datum feature symbol by entering characters (like -A-) in the **Datum Identifier** edit box in the **Geometric Tolerance** dialog box and then selecting a point where you want to establish this datum.

You can also combine datum references with geometric characteristics. AutoCAD automatically positions the datum references on the right end of the feature control frame.

Example 2 *Mechanical*

In the following example, you will create a feature control frame to define perpendicularity specification (see Figure 8-53).

1. Chose the **Tolerance** button from the **Dimension** toolbar to display the **Geometric Tolerance** dialog box. Choose the upper box from the **Sym** area to display the **Symbol** dialog box. Select the **perpendicularity** symbol. It will now be displayed in the **Sym** area.

2. Select the box available on the left of the upper edit box under the **Tolerance 1** area. A diameter symbol will appear to denote a cylindrical tolerance zone.

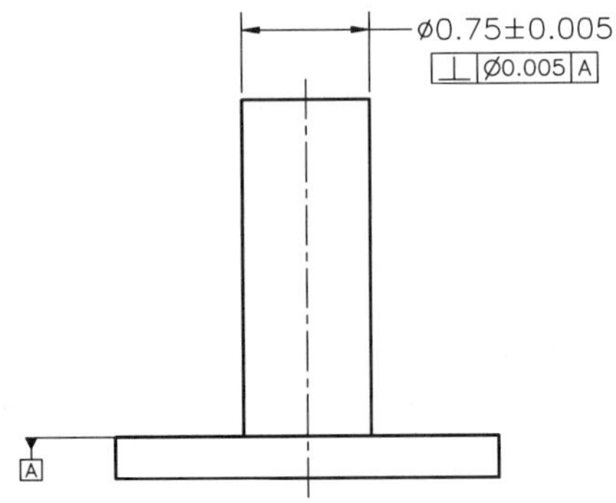

Figure 8-53 *Drawing for Example 2*

3. Enter **0.005** in the upper edit box under the **Tolerance 1** area.

4. Enter **A** in the edit box under the **Datum 1** area. Choose the **OK** button to accept the changes made in the **Geometric Tolerance** dialog box.

5. The **Enter tolerance location** prompt is displayed in the Command line area and the **Feature Control Frame** is attached to the cursor at its middle left point. Select a point to insert the frame.

6. To place the datum symbol, use the **TOLERANCE** command to display the **Geometric Tolerance** dialog box. In the **Datum Identifier** edit box, enter **A**.

7. Choose the **OK** button to accept the changes to the **Geometric Tolerance** dialog box, and then select a point to insert the frame.

COMPLEX FEATURE CONTROL FRAMES

Combining Geometric Characteristics

Sometimes it is not possible to specify all geometric characteristics in one frame. For example, Figure 8-54 shows the drawing of a plate with a hole in the center.

In this part, it is determined that surface C must be perpendicular to surfaces A and B within 0.002 and 0.004, respectively. Therefore, we need two frames to specify the geometric characteristics of surface C. The first frame specifies the allowable deviation in perpendicularity of surface C with respect to surface A. The second frame specifies the allowable deviation in perpendicularity of surface C with respect to surface B. In addition to these two frames, we need a third frame that identifies datum surface C.

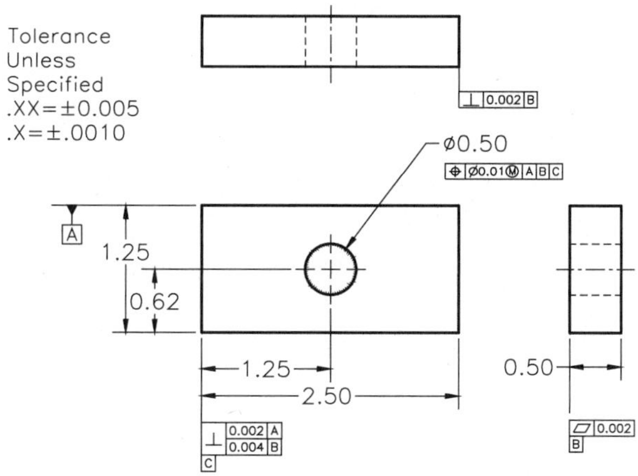

Figure 8-54 Combining feature control frames

All the three feature control frames can be defined in one instance of the **TOLERANCE** command.

1. Choose the **Tolerance** button to invoke the **Geometric Tolerance** dialog box. Select box under the **Sym** area to display the **Symbol** dialog box. Select the **perpendicular** symbol. AutoCAD will display the selected symbol in the first row of the **Sym** area.
2. Enter **0.002** in the first row edit box under the **Tolerance 1** area and enter **A** in the first row edit box under the **Datum 1** area.
3. Select the second row box under the **Sym** area to display the **Symbol** dialog box. Select the **perpendicular** symbol. AutoCAD will display the selected symbol in the second row box of the **Sym** area.
4. Enter **.004** in the second row edit box under the **Tolerance 1** area and enter **B** in the second row edit box under the **Datum 1** area.
5. In the **Datum Identifier** edit box enter **C**, and then choose the **OK** button to exit the dialog box.
6. In the graphics screen, select the position to place the frame.
7. Similarly, create the remaining feature control frames.

Composite Position Tolerancing

Sometimes the accuracy required within a pattern is more important than the location of the pattern with respect to the datum surfaces. To specify such a condition, composite position tolerancing may be used. For example, Figure 8-55 shows four holes (pattern) of diameter 0.15. The design allows a maximum tolerance of 0.025 with respect to datums A, B, and C at the maximum material condition (holes are smallest). The designer wants to maintain a closer positional tolerance (0.010 at MMC) between the holes within the pattern. To specify this requirement, the designer must insert the second frame. This is generally known as composite position tolerancing. AutoCAD provides the facility to create two composite position tolerance frames by means of the **Geometric Tolerance** dialog box. The composite tolerance frames can be created as follows:

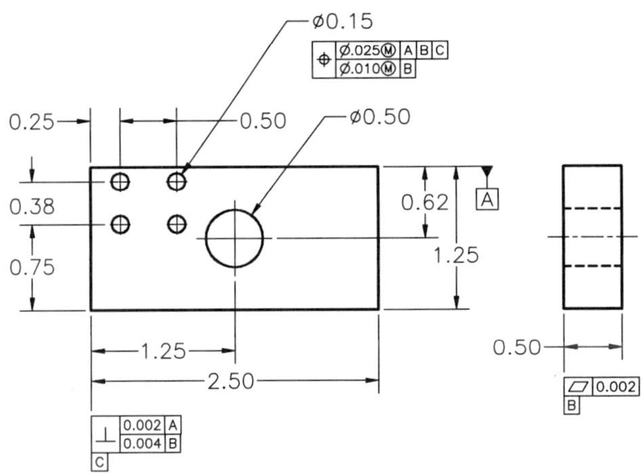

Figure 8-55 *Composite position tolerancing*

1. Invoke the **TOLERANCE** command to display the **Geometric Tolerance** dialog box. Select box under the **Sym** area to display the **Symbol** dialog box. Select the **position** symbol. AutoCAD will display the selected symbol in the first row of the **Sym** area.

2. In the first row of the **Geometric Tolerance** dialog box, enter the geometric characteristics and the datum references required for the first position tolerance frame.

3. In the second row of the **Geometric Tolerance** dialog box, enter the geometric characteristics and the datum references required for the second position tolerance frame.

4. When you have finished entering the values, choose the **OK** button in the **Geometric Tolerance** dialog box, and then select the point where you want to insert the frames. AutoCAD will create the two frames and automatically align them with the common position symbol, as shown in Figure 8-55.

USING FEATURE CONTROL FRAMES WITH LEADERS

The **Leader Settings** dialog box invoked using the **QLEADER** command has the Tolerance option that allows you to create the feature control frame and attach it to the end of the leader extension line, see Figure 8-56. The following is the prompt sequence for using the

QLEADER command with the Tolerance option:

> Specify first leader point, or [Settings] <Settings>: *Press ENTER to display the **Leader Settings** dialog box. Choose the **Annotation** tab. Now, select the **Tolerance** radio button from the **Annotation Type** area. Choose **OK**.*
> Specify first leader point, or [Settings] <Settings>: *Specify the start point of the leader line.*
> Specify next point: *Specify the second point.*
> Specify next point: *Specify the third point to display the **Geometric Tolerance** dialog box.*

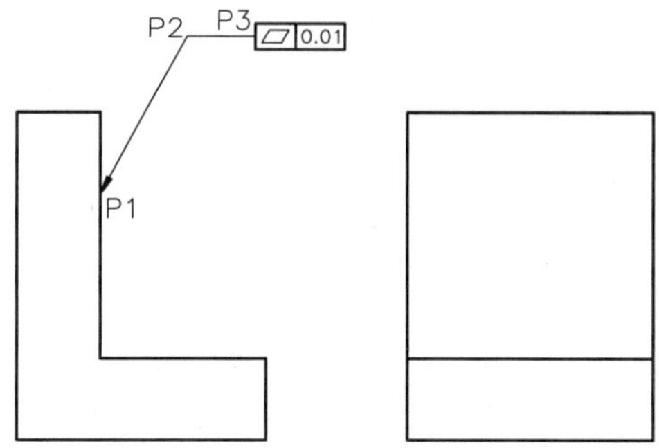

Figure 8-56 *Using the feature control frame with leaders*

PROJECTED TOLERANCE ZONE

Figure 8-57 shows two parts joined with a bolt. The lower part is threaded, and the top part has a drilled hole. When these two parts are joined, the bolt that is threaded in the lower part will have the orientation error that exists in the threaded hole. In other words, the error in the threaded hole will extend beyond the part thickness, which might cause interference, and the parts may not assemble. To avoid this problem, projected tolerance is used. The projected tolerance establishes a tolerance zone that extends above the surface. In Figure 8-57, the position tolerance for the threaded hole is 0.010, which extends 0.1 above the surface (datum A). By using projected tolerance, you can ensure that the bolt is within the tolerance zone up to the specified distance.

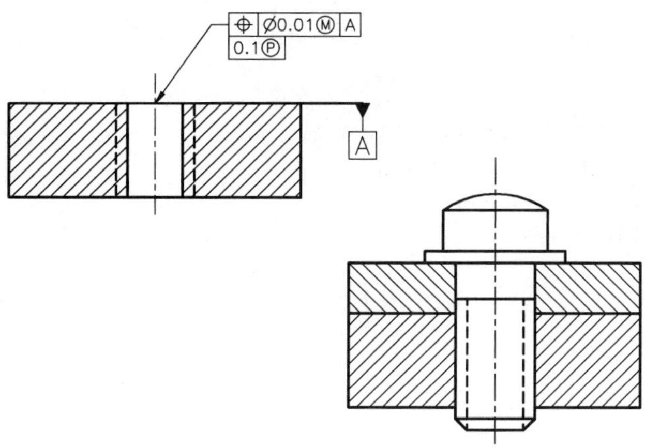

Figure 8-57 *Projected tolerance zone*

You can use the AutoCAD GDT feature to create feature control frames for the projected tolerance zone as follows:

1. Invoke the **QLEADER** command and then press ENTER at the **Specify first leader point, or [Settings] <Settings>** prompt to display the **Leader Settings** dialog box.

2. Choose the **Annotation** tab and then select the **Tolerance** radio button from the **Annotation Type** area. Choose **OK** to return to the Command line. Specify the first, second, and the third leader point as shown in Figure 8-57. As soon as you specify the third point, the **Geometric Tolerance** dialog box will be displayed. Choose the position symbol from the **Symbol** dialog box.

3. In the first row of the **Geometric Tolerance** dialog box, enter the geometric characteristics and the datum references required for the first position tolerance frame.

4. In the **Height** edit box, enter the height of the tolerance zone (0.1 for the given drawing) andselect the box on the right of Projected Tolerance Zone. The projected tolerance zone symbol will be displayed in the box.

5. Choose the **OK** button in the **Geometric Tolerance** dialog box. AutoCAD will create the two frames and automatically align them, as shown in Figure 8-57.

6. Similarly, attach the datum symbol also.

Example 3 *Mechanical*

In the following example, you will create a leader with a combination feature control frame to control runout and cylindricity (Figure 8-58).

1. Choose the **Quick Leader** button from the **Dimension** toolbar to invoke the **QLEADER** command. The prompt sequence is as follows.

   Specify first leader point, or [Settings] <Settings>: *Press ENTER to display the **Leader Settings** dialog box. Choose the **Annotation** tab. Select the **Tolerance** radio button from the **Annotation Type** area. Choose **OK**.*
   Specify first leader point, or [Settings] <Settings>: *Specify the leader start point as shown in Figure 8-58.*
   Specify next point: *Specify the second point of the leader.*
   Specify next point: *Specify the third point of the leader line to display the **Tolerance** dialog box.*

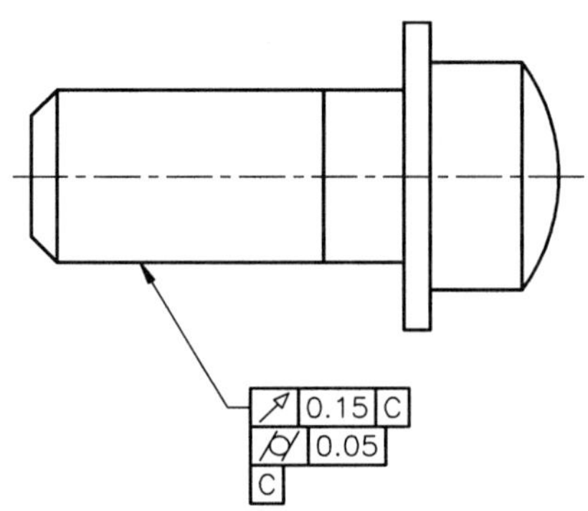

Figure 8-58 *Drawing for Example 3*

2. Choose the runout symbol from the **Symbol** dialog box. The **runout** symbol will be displayed on the first row of the **Sym** area. Enter **0.15** in the first row edit box under the Tolerance 1 area.

3. Enter **C** in the edit box under the **Datum 1** area.

4. Select the edit box on the second row of the **Sym** area and select the **cylindricity** symbol. The **cylindricity** symbol will be displayed in the second row of the **Sym** area.

5. Enter **0.05** in the second row edit box of the Tolerance 1 area.

6. Enter **C** in the **Datum Identifier** edit box.

7. Choose the **OK** button to accept the changes to the **Geometric Tolerance** dialog box. The control frames will be automatically attached at the end of the leader.

Self-Evaluation Test

Answer the following questions, and then compare your answers to the answers given at the end of this chapter.

1. You can specify dimension text of your own or accept the measured value computed by AutoCAD. (T/F)

2. The rotated dimension can be specified to get the effect of horizontal or vertical dimension. (T/F)

3. The center point of an arc/circle is taken as the vertex angle in the angular dimensioning of the arc/circle. (T/F)

4. You cannot combine GTOL with the leaders. (T/F)

5. The __________ command is used to dimension only the *X* coordinates of the selected object.

6. __________ symbols are the building blocks of geometric dimensioning and tolerancing.

7. In the __________ dimensions, the dimension text is, by default, aligned with the object being dimensioned.

8. The dimensions that are automatically updated when the object to which they are assigned change are called __________.

9. The _________ point is taken as the vertex point of the angular dimensions while dimensioning an arc or circle.

10. The feature control frame is _________ in shape.

Review Questions

Answer the following questions.

1. Only inner angles (acute angles) can be dimensioned with angular dimensioning. (T/F)

2. In addition to the most recently drawn dimension (the default base dimension), you can use any other linear dimension as the base dimension. (T/F)

3. In continued dimensions, the base for successive continued dimensions is the base dimension's first extension line. (T/F)

4. In rotated dimensioning, the dimension line is always aligned with the object being dimensioned. (T/F)

5. The dimensions can be converted into true associative dimensions using which command?

   (a) **DIMREASSOCIATE** (b) **DIMASSOCIATE**
   (c) **DIMDISASSOCIATE** (d) **None**

6. You can dimension more than one object in a single effort using which command?

   (a) **DIMLINEAR** (b) **DIMANGULAR**
   (c) **QDIM** (d) **None**

7. You can dimension different points and features of a part with reference to a fixed point using which command?

   (a) **DIMBASELINE** (b) **DIMANGULAR**
   (c) **DIMRADIUS** (d) **DIMALIGNED**

8. You can add geometric dimensions and tolerance to the current drawing using which command?

   (a) **GTOL** (b) **TOLERANCE**
   (c) **TEXT** (d) **None**

9. Which option of the **QLEADER** command allows you to make the settings for attaching tolerances to the leader?

   (a) **Settings** (b) **Set**
   (c) **Undo** (d) **None**

10. Geometric dimensioning and tolerancing is generally known as _________.

11. Give three examples of geometric characteristics that indicate the characteristics of a feature. ______________________________.

12. The three ways to return to the Command prompt from the **Dim:** prompt (dimensioning mode) are _________, _________, and _________.

13. The six fundamental dimensioning types provided by AutoCAD are ______________________________.

14. Horizontal dimensions measure displacement along the _________.

15. Vertical dimensions measure displacement along the _________.

Exercises

Exercise 7 *Mechanical*

Draw the object shown in Figure 8-59 and then dimension it. Save the drawing as a **DIMEXR7** drawing file.

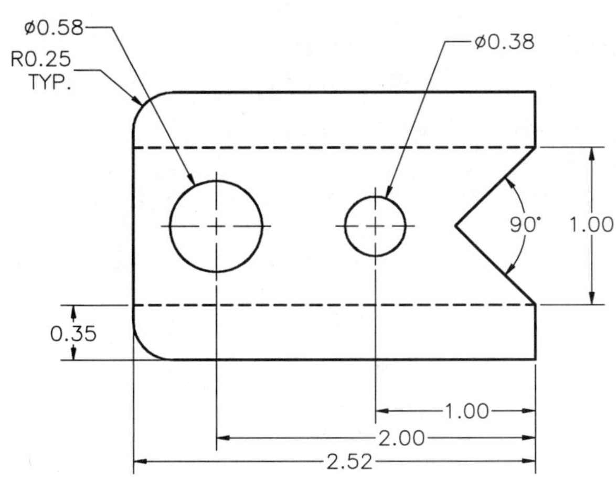

Figure 8-59 *Drawing for Exercise 7*

Exercise 8 *Mechanical*

Draw and dimension the object shown in Figure 8-60. Save the drawing as **DIMEXR8**.

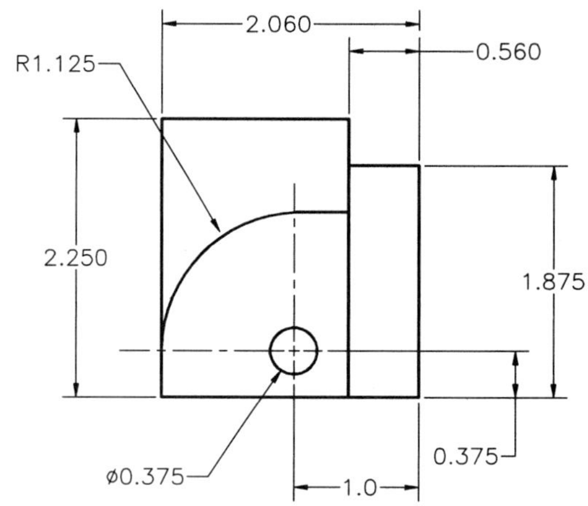

Figure 8-60 *Drawing for Exercise 8*

Exercise 9 *Mechanical*

Draw the object shown in Figure 8-61 and then dimension it. Save the drawing as **DIMEXR9**.

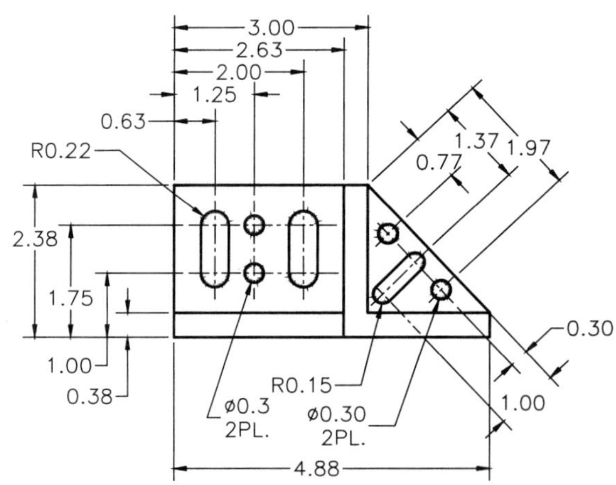

Figure 8-61 *Drawing for Exercise 9*

Exercise 10 *Mechanical*

Draw the object shown in Figure 8-62 and then dimension it. Save the drawing as **DIMEXR10**.

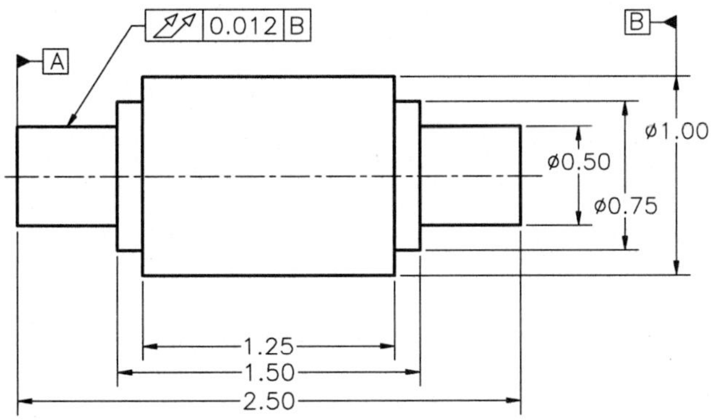

Figure 8-62 *Drawing for Exercise 10*

Chapter 8

Problem Solving Exercise 1 *Mechanical*

Draw Figure 8-63 and dimension it as shown in the drawing. The **FILLET** command should be used where needed. Save the drawing as **DIMPSE1**.

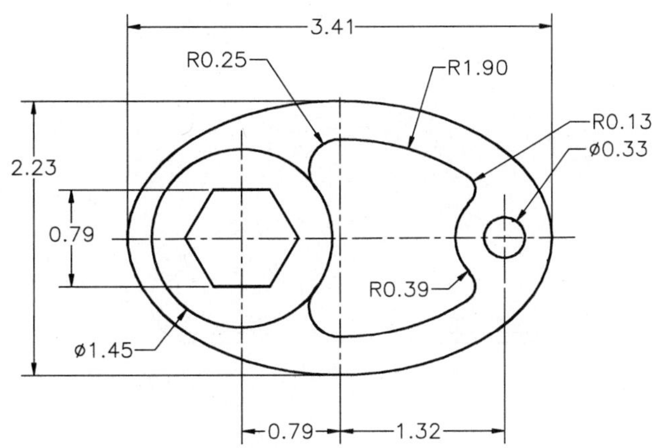

Figure 8-63 *Drawing for Problem Solving Exercise 1*

Problem Solving Exercise 2 *Mechanical*

Draw Figure 8-64 and then dimension it as shown in the drawing. Save the drawing as **DIMPSE2**.

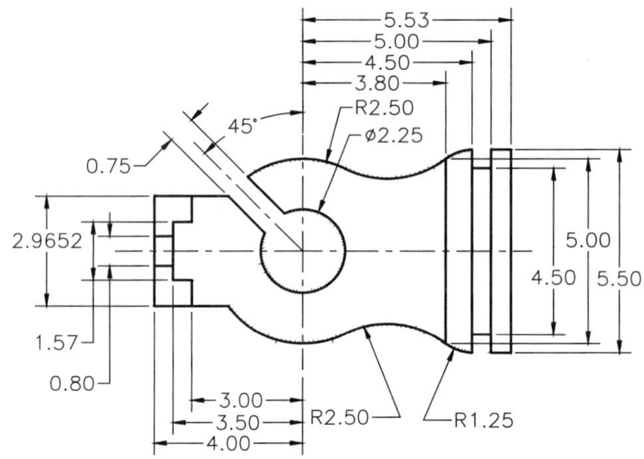

Figure 8-64 Drawing for Problem Solving Exercise 2

Problem Solving Exercise 3 *Mechanical*

Draw Figure 8-65 and then dimension it as shown in the drawing. Save the drawing as **DIMPSE3**.

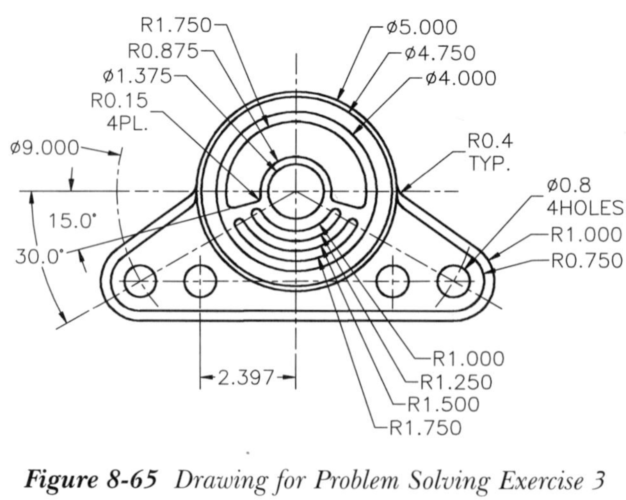

Figure 8-65 Drawing for Problem Solving Exercise 3

Problem Solving Exercise 4 *Mechanical*

Draw the three orthographic views of an object as shown in Figure 8-66 and then dimension it as shown in the drawing. Save the drawing as **DIMPSE4**.

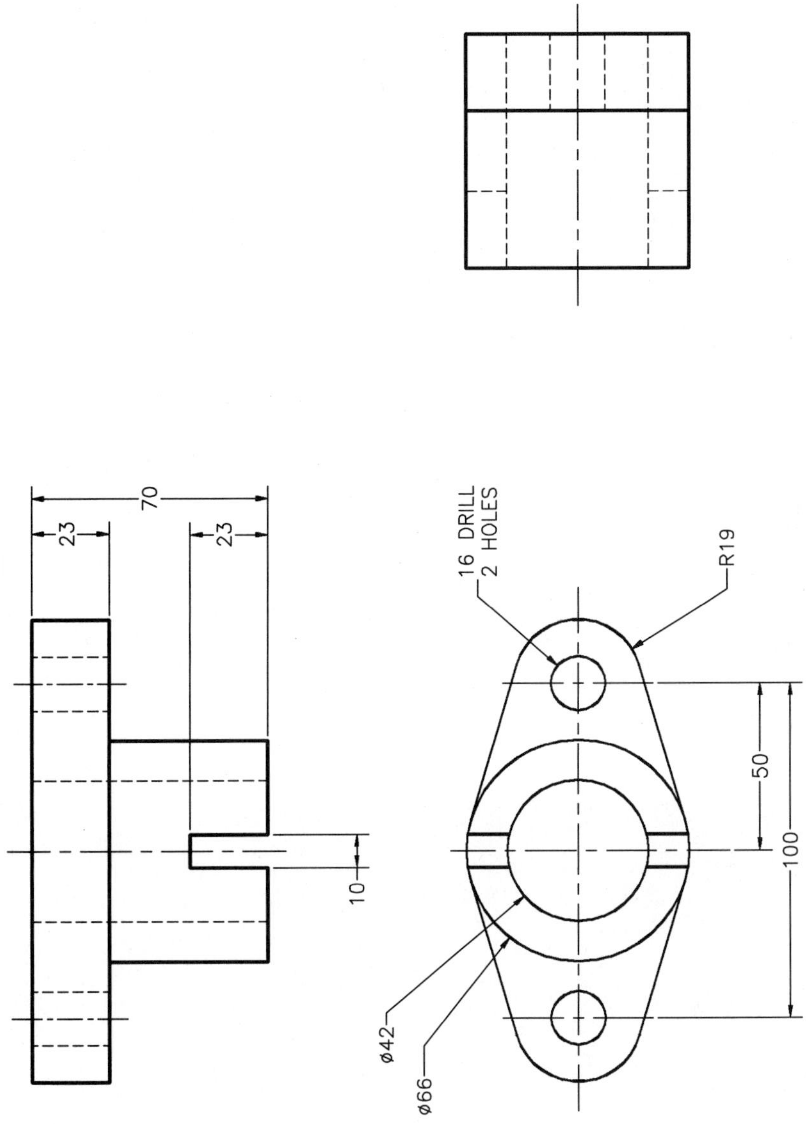

Figure 8-66 *Drawing for Problem Solving Exercise 4*

Answers to Self-Evaluation Test

1 - T, **2** - T, **3** - T, **4** - F, **5** -**DIMORDINATE**, **6** - geometric characteristics, **7** - aligned, **8** - true associative dimensions, **9** - center, **10** - rectangular

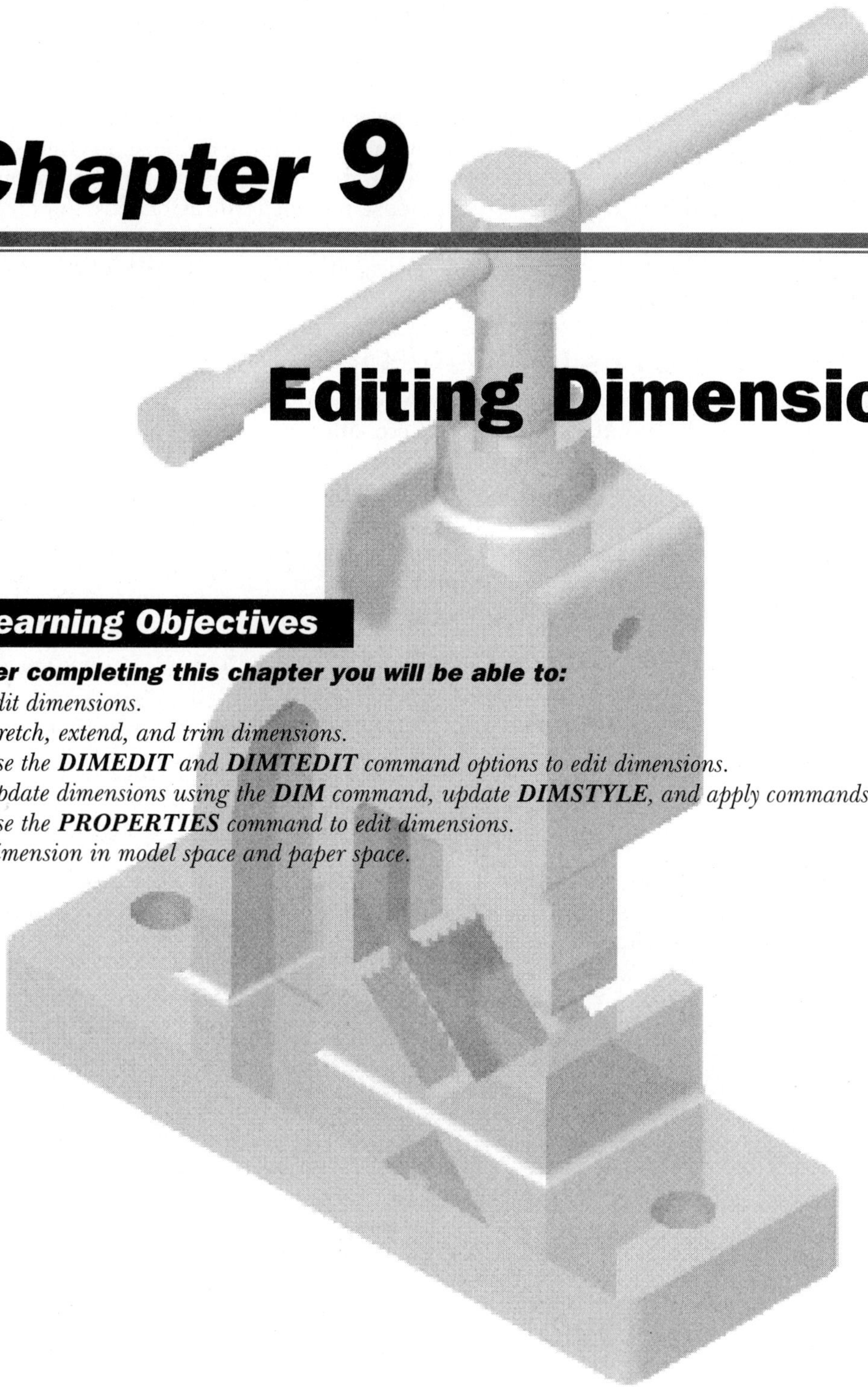

Chapter 9

Editing Dimensions

Learning Objectives

After completing this chapter you will be able to:

- *Edit dimensions.*
- *Stretch, extend, and trim dimensions.*
- *Use the **DIMEDIT** and **DIMTEDIT** command options to edit dimensions.*
- *Update dimensions using the **DIM** command, update **DIMSTYLE**, and apply commands.*
- *Use the **PROPERTIES** command to edit dimensions.*
- *Dimension in model space and paper space.*

EDITING DIMENSIONS USING EDITING TOOLS

For editing dimensions, AutoCAD has provided some special editing commands that work with dimensions. These editing commands can be used to define new dimension text, return to home text, create oblique dimensions, and rotate and update the dimension text. You can also use the **TRIM**, **STRETCH**, and **EXTEND** commands to edit the dimensions. In case the dimension assigned to the object is a true associative dimension, it will be automatically updated if the object is modified. However, if the dimension is not a true associative dimension, you will have to include the dimension along with the object in the edit selection set. The properties of the dimensioned objects can also be changed using the **PROPERTIES Palette** or the **Dimension Style Manager**.

Editing the Dimensions by Stretching

You can edit the dimension by stretching it. However, to stretch a dimension, appropriate definition points must be included in the selection crossing or window. As the middle point of the dimension text is a definition point for all types of dimensions, you can easily stretch and move the dimension text to any location you want. When you stretch the dimension text, the gap in the dimension line gets filled automatically. When editing, the definition points of the dimension being edited must be included in the selection crossing box. The dimension is automatically calculated when you stretch the dimensions.

Note

The dimension type remains the same after stretching. For example, the vertical dimension maintains itself as a vertical dimension and measures only the vertical distance even after the line it dimensions is modified and converted into an inclined line. The following example illustrates the stretching of object lines and dimensions.

Example 1 — *Mechanical*

In this example you will stretch the objects and dimensions shown in Figure 9-1 to a new location using grips. The new location of the lines and dimension is at a distance of 0.5 in the positive *Y* axis direction. See Figure 9-2.

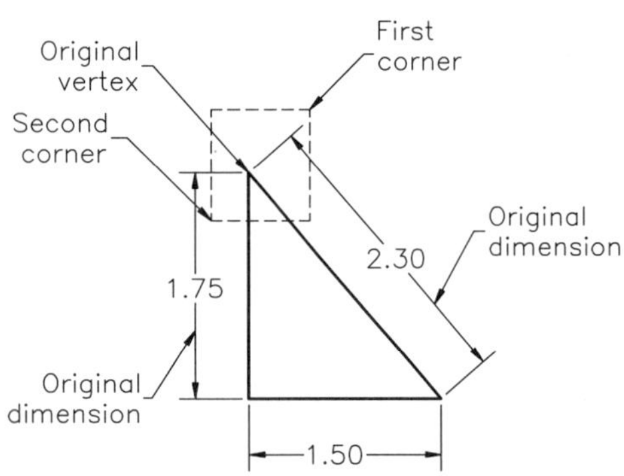

Figure 9-1 *Original location of lines and dimensions*

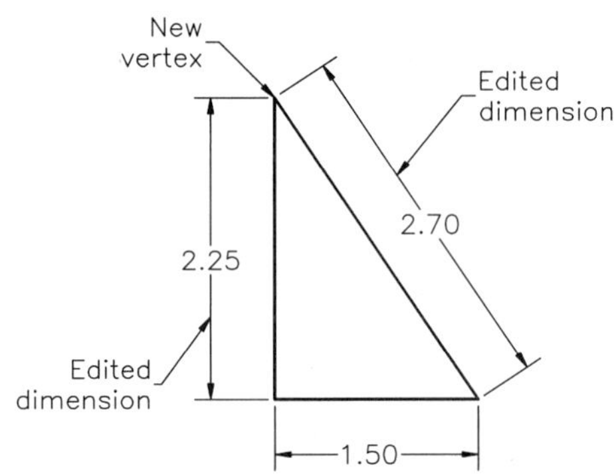

Figure 9-2 *New location of lines and dimensions*

1. Choose the **Stretch** button from the **Modify** toolbar. The prompt sequence is as follows:

   Select objects to stretch by crossing-window or crossing-polygon
   Select objects: Specify opposite corner: *Define a crossing window using the first and second corner as shown in Figure 9-1.*
   Select objects: Enter
   Specify base point or displacement: *Select original vertex using the osnaps as the base point.*
   Specify second point of displacement or <use first point as displacement>: **@0.5<90**

2. The selected entities will be stretched to the new location. The dimension that was initially 1.75 will become 2.25 and the dimension that was initially 2.30 will become 2.70, see Figure 9-2. Press ESC to remove the grip points from the objects.

Exercise 1 *General*

The two dimensions in Figure 9-3(a) are too close. Fix the drawing by stretching the dimension as shown in Figure 9-3(b).

1. Stretch the outer dimension to the right so that there is some distance between the two dimensions.
2. Stretch the dimension text of the outer dimension so that the dimension text is staggered (lower than the first dimension).

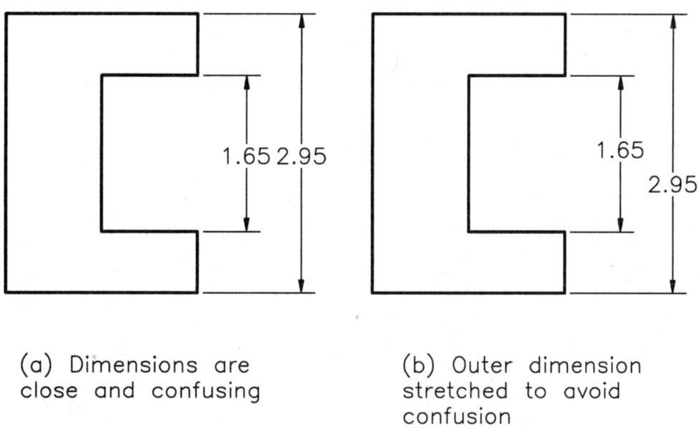

Figure 9-3 *Drawing for Exercise 1, stretching dimensions*

Editing Dimensions by Trimming and Extending

Trimming and extending operations can be carried out with all types of linear dimensions (horizontal, vertical, aligned, rotated) and the ordinate dimension. Even if the dimensions are true associative, you can trim and extend them. AutoCAD trims or extends a linear dimension between the extension line definition points and the object used as a boundary or trimming edge. To extend or trim an ordinate dimension, AutoCAD moves the feature location (location of the dimensioned coordinate) to the boundary edge. To retain the original ordinate value, the boundary edge to which the feature location point is moved should be orthogonal to the measured ordinate. In both cases, the imaginary line drawn between the two extension line definition points is trimmed or extended by AutoCAD, and the dimension is adjusted automatically. Figures 9-4 and 9-5 show the dimensions edited by trimming and extending.

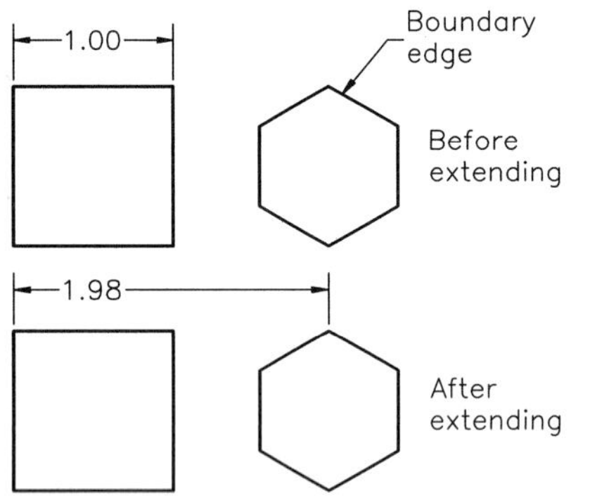

Figure 9-4 *Dimensions edited by extending*

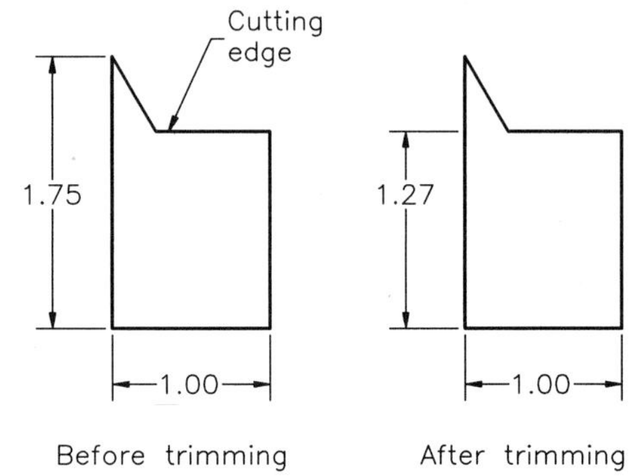

Figure 9-5 *Dimensions edited by trimming*

Exercise 2 *Mechanical*

Use the **Edgemode > Extend** option of the **TRIM** command to trim the dimension in Figure 9-6(a) so that it looks like Figure 9-6(b).

1. Make the drawing and dimension it as shown in Figure 9-6(a). Assume the dimensions where necessary.
2. Trim the dimensions by setting the **Edgemode** option of the **TRIM** command to **Extend**.

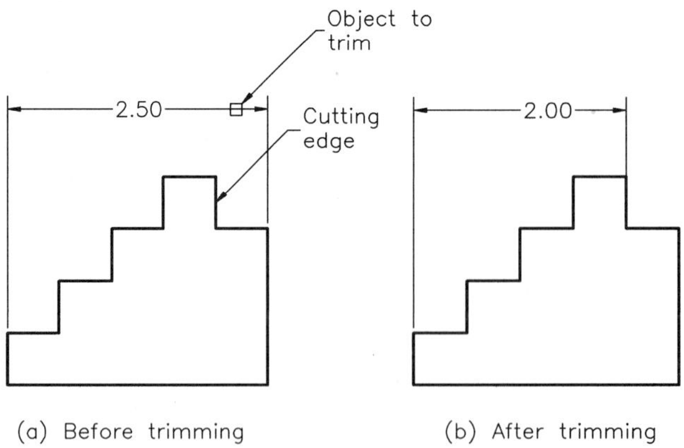

Figure 9-6 *Drawing for Exercise 2*

EDITING THE DIMENSIONS

Toolbar:	Dimension > Dimension Edit
Command:	DIMEDIT

The dimension text can be edited by using the **DIMEDIT** command (Figure 9-7). This command has four options: New, Rotate, Home, and Oblique. The prompt sequence that will follow when you choose this button is

Enter type of dimension editing (Home/New/Rotate/Oblique) <Home>: *Enter an option.*

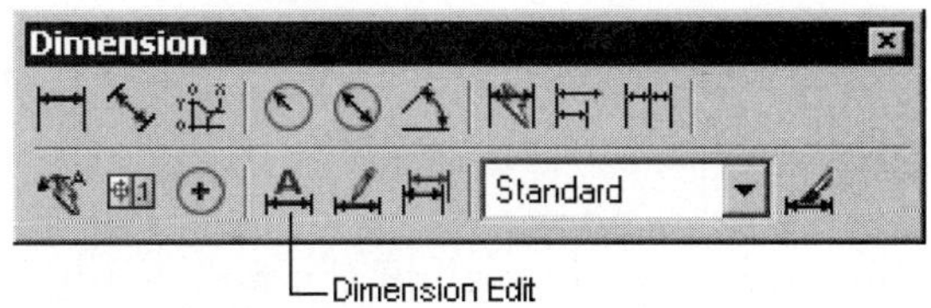

Figure 9-7 *Invoking the* ***DIMEDIT*** *command from the* ***Dimension*** *toolbar*

New

The **New** option is used to replace the existing dimension text with a new text string. When you invoke this option, the **Multiline Text Editor** will be displayed, see Figure 9-8. The default text will be displayed as angular brackets (<>). You can change the default text or add prefix or suffix to the text using this **Multiline Text Editor**. Once you have edited the dimension in the editor and chosen **OK**, you will be prompted to select the dimension to be changed. The <> prefix or suffix facility can also be used when you enter the new text string. In this case, the prefix/suffix is appended before or after the dimension measurement that is placed instead of the <> characters. For example, if you enter Ref-Dim before <> and mm after <> (Ref-Dim<>mm), AutoCAD will attach Ref-Dim as prefix and mm as suffix (Ref-Dim1.1257mm).

If information about the dimension style is available on the selected dimension, it is used to redraw the dimension, or the prevailing variable settings are used for the redrawing process.

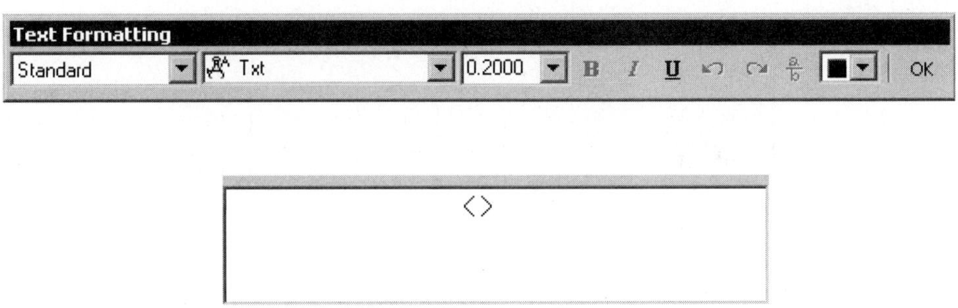

Figure 9-8 ***Multiline Text Editor*** *for editing the default text*

Rotate

The **Rotate** option is used to position the dimension text at a specified angle. With this option, you can change the orientation (angle) of the dimension text of any number of associative dimensions. The angle can be specified by entering its value at the **Specify angle**

for dimension text prompt or by specifying two points at the required angle. Once you have specified the angle, you will be prompted to select the dimension text to be rotated. You will notice that the text rotates around its middle point, see Figure 9-9.

Home

The Home option restores the text of a dimension to its original (home/default) location if the position of the text has been changed by stretching or editing, see Figure 9-9.

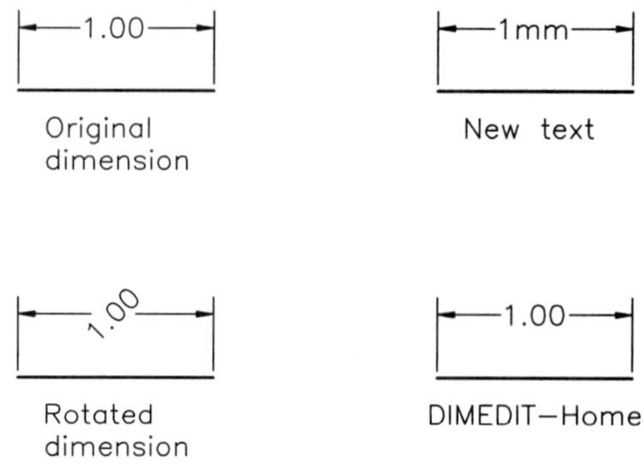

Figure 9-9 *Using the **DIMEDIT** command to edit dimensions*

Oblique

In the linear dimensions, extension lines are drawn perpendicular to the dimension line. The **Oblique** option bends the linear dimensions. It draws extension lines at an oblique angle (Figure 9-10). This option is particularly important to create isometric dimensions and can be used to resolve conflicting situations due to the overlapping of extension lines with other objects. Making an existing dimension oblique by specifying an angle oblique to it does not affect the generation of new linear dimensions. The oblique angle is maintained even after performing most editing operations. (See Chapter 21 for details about how to use this option.) When you invoke this option, you will be prompted to select the dimension to be edited. After selecting the dimensions you will be prompted to specify the obliquing angle. The extensions lines will be bent at the angle specified. You can also invoke this option by choosing **Oblique** from the **Dimension** menu.

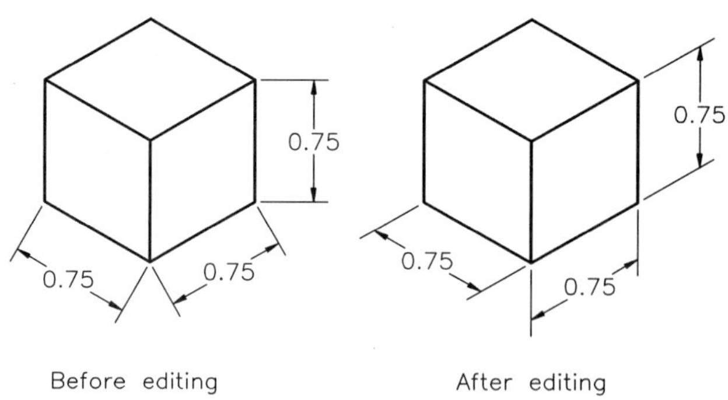

Figure 9-10 *Using the **Oblique** option to edit dimensions*

EDITING DIMENSION TEXT

Toolbar:	Dimension > Dimension Text Edit
Menu:	Dimension > Align Text
Command:	DIMTEDIT

The dimension text can also be edited by using the **DIMTEDIT** command. This command is used to edit the placement and orientation of a single existing dimension. You can apply this command, for example, in cases where dimension texts of two or more dimensions are too close together, creating confusion. In such cases, the **DIMTEDIT** command is invoked to move the dimension text to some other location so that there is no confusion. The prompt sequence that will follow when you choose this button is given next.

Select dimension: *Select the dimension to modify.*
Specify new location for dimension text or [Left/Right/Center/Home/Angle]:

Left

With this option, you can left-justify the dimension text along the dimension line. The vertical placement setting determines the position of the dimension text. The horizontally aligned text is moved to the left and the vertically aligned text is moved down, see Figure 9-11. This option can be used only with the linear, diameter, and radius dimensions.

Right

With this option, you can right-justify the dimension text along the dimension line. Similar to the Left option, the vertical placement setting determines the position of the dimension text. The horizontally aligned text is moved to the right, and the vertically aligned text is moved up, see Figure 9-11. This option can be used only with linear, diameter, and radius dimensions.

Center

With this option you can center-justify the dimension text for linear, and aligned dimensions, see Figure 9-11. The vertical setting controls the vertical position of the dimension text.

Home

The **Home** option is used to restore (move) the dimension text of a dimension to its original (home/default) location if the position of the text has been changed, see Figure 9-11.

Angle

With the Angle option, you can position the dimension text at the angle you specify, see Figure 9-11. The angle can be specified by entering its value at the **Specify angle for dimension text** prompt or by specifying two points at the required angle. You will notice that the text rotates around its middle point. If the dimension text alignment is set to Orient Text Horizontally, the dimension text is aligned with the dimension line. If information about the dimension style is available on the selected dimension, AutoCAD uses it to redraw the dimension, or the prevailing dimension variable settings are used for the redrawing process.

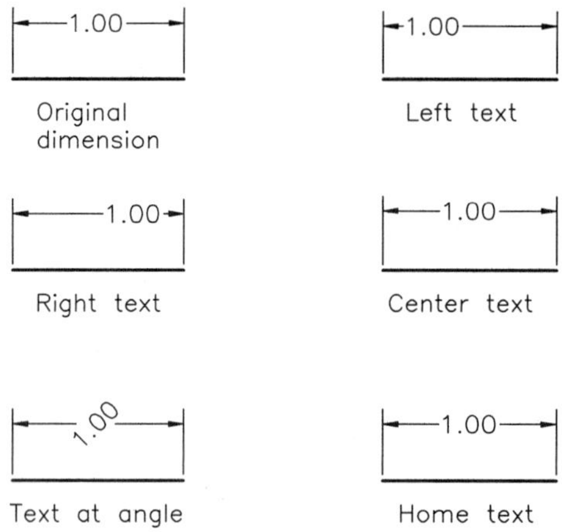

Figure 9-11 *Using **DIMTEDIT** command to edit dimensions*

Entering 0-degree angle changes the text to its default orientation.

UPDATING DIMENSIONS

Toolbar:	Dimension > Dimension Update
Menu:	Dimension > Update

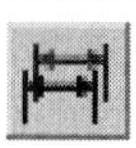

The **Update** option regenerates and updates the prevailing dimension entities (such as arrows heads and text height) using current settings for the dimension variables, dimension style, text style, and units. When you choose this button, you will be prompted to select the dimensions to be updated. You can select all the dimensions or specify the dimensions that should be updated.

EDITING DIMENSIONS WITH GRIPS

You can also edit dimensions by using GRIP editing modes. GRIP editing is the easiest and quickest way to edit dimensions. You can perform the following operations with GRIPS.

1. Position the text anywhere along the dimension line. Note that you cannot move the text and position it above or below the dimension line.
2. Stretch a dimension to change the spacing between the dimension line and the object line.
3. Stretch the dimension along the length. When you stretch a dimension, the dimension text automatically changes.
4. Move, rotate, copy, or mirror the dimensions.
5. Relocate a dimension origin.
6. Change properties such as color, layer, linetype, and linetype scale.
7. Load Web browser (if any *Universal Resource Locator* is associated with the object).

EDITING DIMENSIONS USING THE PROPERTIES PALETTE

Toolbar:	Standard > Properties
Menu:	Modify > Properties
Command:	PROPERTIES

You can also modify a dimension or leader by using the **PROPERTIES** palette. The **PROPERTIES** palette is displayed when you choose the **Properties** button from the **Standard** toolbar. The **PROPERTIES** palette can also be invoked by double-clicking on the dimension to be edited. All the properties of the selected object are displayed in the **PROPERTIES** palette (see Figure 9-12).

PROPERTIES Palette (Dimension)

You can use the **PROPERTIES** palette (Figure 9-12) to change the properties of a dimension, change the dimension text style, or change geometry, format, and annotation-related features of the selected dimension. The changes takes place dynamically in the drawing. The **PROPERTIES** palette provides various categories for modification of dimensions are as follows.

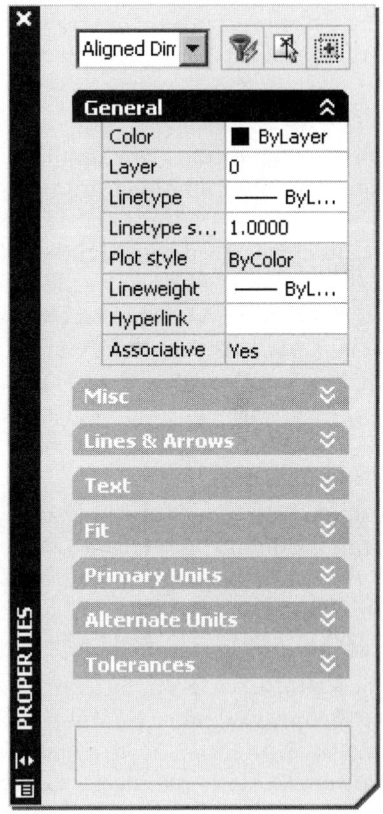

*Figure 9-12 **PROPERTIES** palette for dimensions*

General

In the general category, the various parameters displayed are **Color**, **Layer**, **Linetype**, **Linetype scale**, **Plot style**, **Lineweight**, **Hyperlink,** and **Associative** with their current values. For example, if you want to change the color of the selected object, then select Color property and then select the required color from the drop-down list. Similarly, Layer, Plot style, Linetype, and Lineweight can be changed from the respective drop-down lists. The Linetype scale can be changed manually at the corresponding cell.

Misc

This category displays the dimension style by name (for the **DIMSTYLE** system variable, use **SETVAR**). You can change the dimension style from the drop-down list for the selected dimension.

Lines & Arrows

The various parameters of the lines and arrows in the dimension such as arrowhead size, type, arrow lineweight, and so on can be changed in this category.

Text

The various parameters that control the text in the dimension object such as text color, text height, vertical position text offset, and so on can be changed in this category.

Fit

In the fit category, the various parameters are **Dim line forced**, **Dim line inside**, **Dim scale overall**, **Fit**, **Text inside**, and **Text movement**. All the parameters can be changed by the drop-down list except Dim scale overall (which can be changed manually).

Primary Units

In the primary units category, the various parameters displayed are **Decimal separator**, **Dim prefix**, **Dim suffix**, **Dim roundoff**, **Dim scale linear**, **Dim units**, **Suppress leading zeroes**, **Suppress trailing zeroes**, **Suppress zero feet**, **Suppress zero inches**, and **Precision**. Among these properties, Dim units, Suppress leading zeroes, Suppress trailing zeroes, Suppress zero feet, Suppress zero inches, and Precision properties can be changed with the corresponding drop-down lists and the other parameters can be changed manually.

Alternate Units

The alternate units are required when a drawing is to be read in two different units. For example, if an architectural drawing is to be read in both metric and feet-inches, you can turn the alternate units on. The primary units can be set to metric and the alternate units to architectural. As a result, the dimensions of the drawing will be displayed in metric units as well as in engineering. In the alternate unit category, there are various parameters for the alternate units. They can be changed only if the **Alt enabled** parameter is **on**. The parameters such as **Alt format**, **Alt precision**, **Alt suppress leading zeroes**, **Alt suppress trailing zeroes**, **Alt suppress zero feet**, and **Alt suppress zero inches** can be changed from the respective drop-down lists and others can be changed manually.

Tolerances

The parameters of this category can be changed only if the Tolerances display parameter has some mode of the tolerance selected. The various parameters are available and correspond to the mode of tolerance selected.

PROPERTIES Palette (Leader)

The **PROPERTIES** palette for **Leader** can be invoked by selecting a Leader and then choosing the **Properties** button from the **Standard** toolbar. You can also invoke the **PROPERTIES** palette (Figure 9-13) from the shortcut menu by right-clicking in the drawing area and choosing **Properties**. This palette can also be invoked by double-clicking in the leader to be edited. The various properties under the **PROPERTIES** palette (Leader) are described as follows.

General

The parameters in the general category are the same as those discussed in the previous section (**PROPERTIES** palette for dimensions).

Geometry
This category displays the coordinates of the Leader. The various parameters under this category are **Vertex**, **Vertex X**, **Vertex Y**, and **Vertex Z**. You can choose any vertex of the leader and change its coordinates.

Misc
This category displays the **Dim Style** and **Type** of the Leader. You can change the style name and the type of the Leader by using these properties.

Lines & Arrows
Lines and Arrows displays the various specifications of the arrowheads and linetypes for the Leader.

Fit
This category displays the **Dim scale overall** property that specifies the overall scale factor applied to size, distances, or the offsets of the Leader.

Text
Text category displays the **Text offset** to the dimension line and the vertical position (**Text pos vert**) of the dimension text and can be changed accordingly.

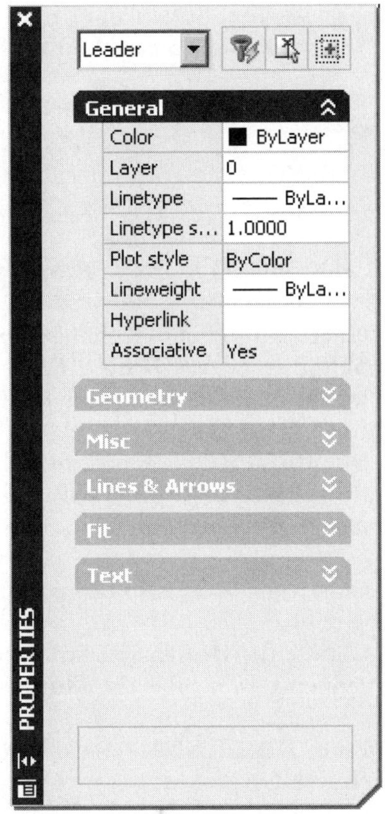

Figure 9-13 ***PROPERTIES*** *palette for leaders*

Example 2 *Mechanical*

In this example you will modify the dimensions in Figure 9-14 so that they match the dimensions given Figure 9-15.

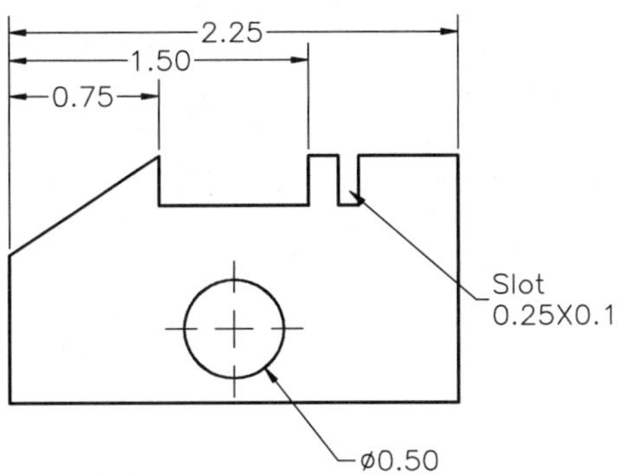

Figure 9-14 *Drawing for Example 2*

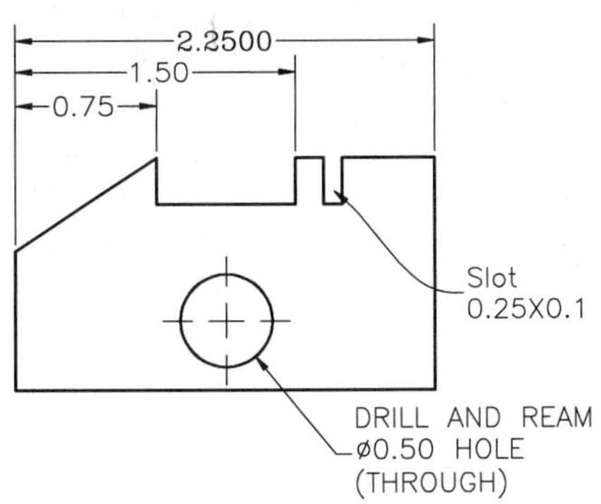

Figure 9-15 *Drawing after editing the dimensions*

1. Choose **Text Style** from the **Format** menu and create a style with the name **ROMANC**. Select **romanc.shx** as the font for the style.

2. Double-click on the dimension 2.25 to display the **PROPERTIES** palette.

3. In the **Text** category, select the **Text style** drop-down list and then select **ROMANC** style from this drop-down list. The changes will take place dynamically.

4. Select **0.0000** from the **Precision** drop-down list in the **Primary Units** area.

5. Once all the required changes are made in the linear dimension, choose the **Select Object** button in the **PROPERTIES** palette. You will be prompted to select the object. Select the leader line and then press ENTER.

6. The **PROPERTIES** palette will not display the leader options. Select **Spline with arrow** from the **Type** drop-down list in the **Misc** category. The straight line will be converted into a spline with an arrow dynamically.

7. Close the **PROPERTIES** palette.

8. Choose the **Dimension Edit** button from the **Dimension** toolbar. Enter **N** in the prompt sequence to display the **Multiline Text Editor**.

9. Enter **DRILL AND REAM %%C0.25 HOLE (THROUGH)** in the text editor and then choose **OK**.

10. You will be prompted to select the object to change. Select the diameter dimension and then press ENTER. The diameter dimension will be modified to the new value.

MODEL SPACE AND PAPER SPACE DIMENSIONING

Dimensioning objects can be drawn in model space or paper space. If the drawings are in model space, associative dimensions should also be created in model space. If the drawings are in model space and the associative dimensions are in paper space, the dimensions will not change when you perform such editing operations as stretching, trimming, and extending, or such display operations as zoom and pan in the model space viewport. The definition points of a dimension are located in the space where the drawing is drawn. You can select the **Scale dimensions to layout (paperspace)** check box under the **Scale for Dimension Features** area in the **Fit** tab of the **Modify**, **New**, or **Override** dialog boxes in the **Dimension Style Manager** dialog box, depending on whether you want to modify the present style or you want to create a new style (see Figure 9-16). Choose **OK/OK** to exit from both the dialog boxes. AutoCAD calculates a scale factor that is compatible with the model space and the paper space viewports. Choose **Update** from the **Dimension** menu and select the dimension objects for updating.

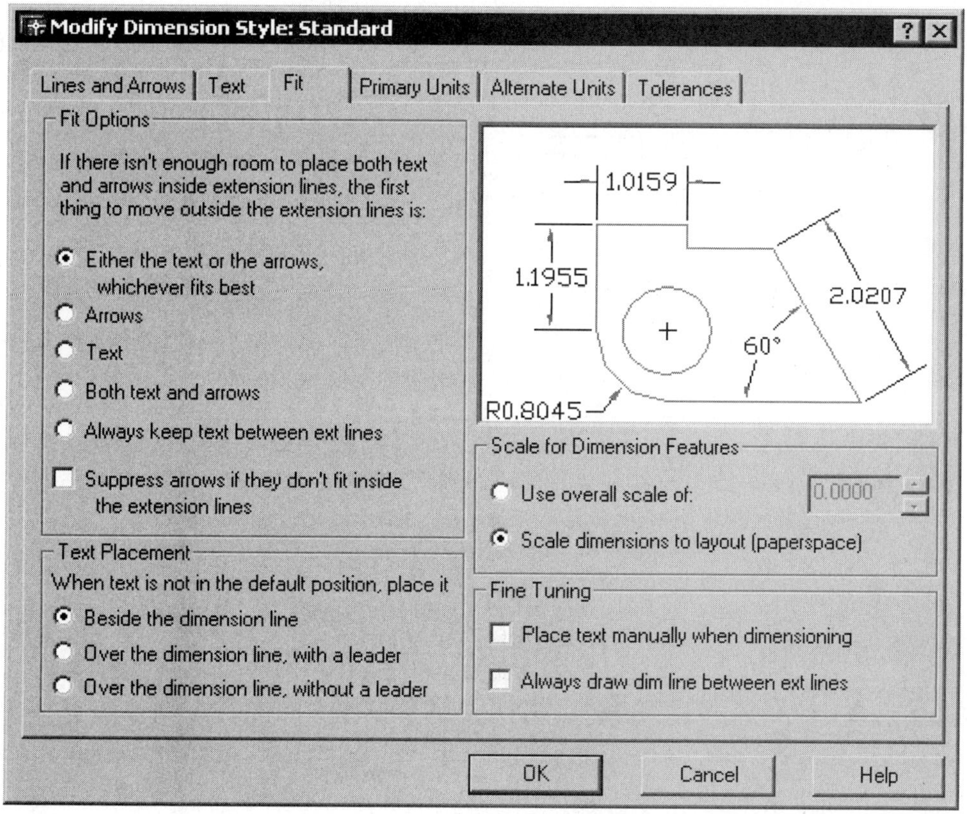

Figure 9-16 *Selecting paper space scaling in the* ***Modify Dimension Style*** *dialog box*

The drawing shown (Figure 9-17) uses paper space scaling. The main drawing and detail drawings are located in different floating viewports (paper space). The zoom scale factors for these viewports are different: 0.3XP, 1.0XP, and 0.5XP, respectively. When you use paper scaling, AutoCAD automatically calculates the scale factor for dimensioning so that the dimensions are uniform in all the floating viewports (model space viewports).

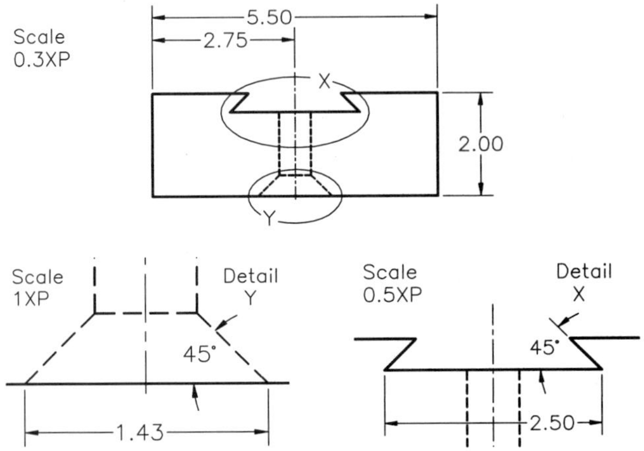

***Figure 9-17** Dimensioning in paper model space viewports using paper space scaling or setting **DIMSCALE** to 0*

Self-Evaluation Test

Answer the following questions and then compare your answers to the correct answers given at the end of this chapter.

1. In associative dimensioning, the items constituting a dimension (such as dimension lines, arrows, leaders, extension lines, and dimension text) are drawn as a **single object.** (T/F)

2. If the value of the variable **DIMDASSOC** is set to zero, the dimension lines, arrows, leaders, extension lines, and dimension text are drawn as **independent** objects. (T/F)

3. You cannot edit dimensions using the grips. (T/F)

4. The true associative dimensions cannot be trimmed or extended. (T/F)

5. You can use the ________ command to break the dimensions into individual entities.

6. The ________ option of the **DIMTEDIT** command is used to justify the dimension text toward the left side.

7. The ________ option of the **DIMTEDIT** command is used to justify the dimension text to the center of the dimension.

8. The ________ option of the **DIMEDIT** command is used to create a new text string.

9. The __________ option of the **DIMEDIT** command is used to bend the extension lines through the specified angle.

10. The __________ button from the **Dimension** toolbar is used to update the dimensions.

Review Questions

Answer the following questions.

1. The horizontal, vertical, aligned, and rotated dimensions cannot be edited using the grips.

2. Trimming and extending operations can be carried out with all types of linear (horizontal, vertical, aligned, and rotated) dimensions and with the ordinate dimension. (T/F)

3. To extend or trim an ordinate dimension, AutoCAD moves the feature location (location of the dimensioned coordinate) to the boundary edge. (T/F)

4. Once moved from the original location, the dimension text cannot be restored to its original position. (T/F)

5. With the __________ or __________ commands you can edit the dimension text.

6. The __________ command is particularly important for creating isometric dimensions and is applicable in resolving conflicting situations due to overlapping of extension lines with other objects.

7. The __________ command is used to edit the placement and orientation of a single existing dimension.

8. The __________ command regenerates (updates) prevailing associative dimension objects (like arrows and text height) using the current settings for the dimension variables, dimension style, text style, and units.

9. Explain when to use the **EXTEND** command and how it works with dimensions. ____.

10. Explain the use and working of the **PROPERTIES** command for editing dimensions. ____.

Exercises

Exercise 3 *General*

1. Create the drawing as shown in Figure 9-18. Assume the dimensions where necessary.
2. Dimension the drawing as shown in Figure 9-18.
3. Edit the dimensions so that they match the dimensions shown in Figure 9-19.

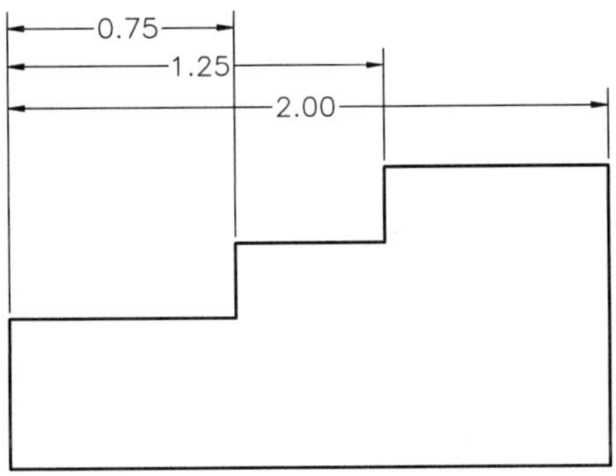

Figure 9-18 *Drawing for Exercise 3, before editing dimensions*

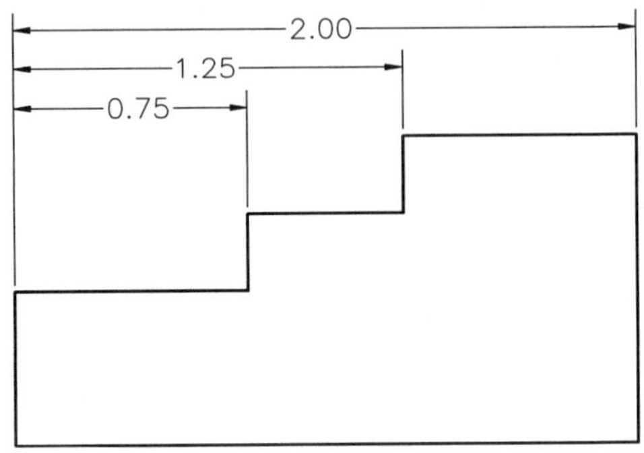

Figure 9-19 *Drawing for Exercise 3, after editing dimensions*

Exercise 4 *Mechanical*

1. Draw the object shown in Figure 9-20(a). Assume the dimensions where necessary.
2. Dimension the drawing and edit them as shown in Figure 9-20(b).

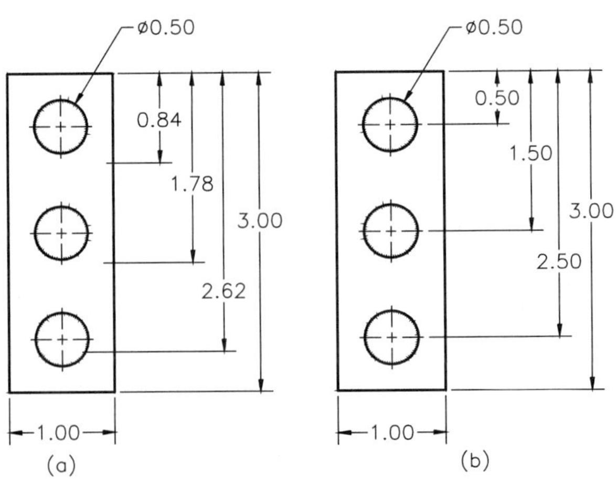

Figure 9-20 *Drawings for Exercise 4*

Exercise 5

Architectural

Create the drawing shown in Figure 9-21 and then dimension it. Assume the dimensions wherever necessary. After dimensioning the drawing, edit the dimensions so that they match the dimensions shown in Figure 9-21

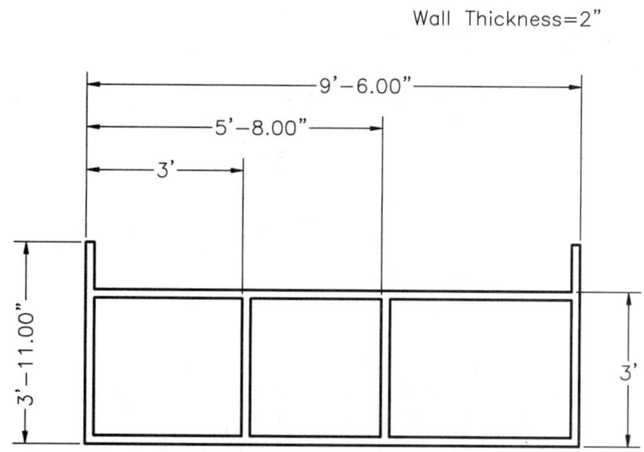

Figure 9-21 *Drawing for Exercise 5*

Problem Solving Exercise 1

Mechanical

Create the drawing in Figure 9-22 and then dimension it as shown. Edit the dimensions so that the dimensions are positioned as shown in the drawing. You may change the dimension text height and arrow size to 0.08 units.

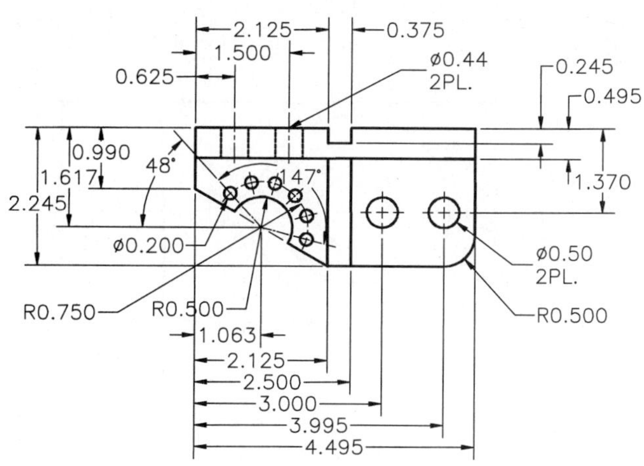

Figure 9-22 *Drawing for Problem Solving Exercise 1*

Problem Solving Exercise 2 *Mechanical*

Draw the front and side view of an object shown in Figure 9-23 and then dimension the two views. Edit the dimensions so that the dimensions are positioned as shown in the drawing. You may change the dimension text height and arrow size to 0.08 units.

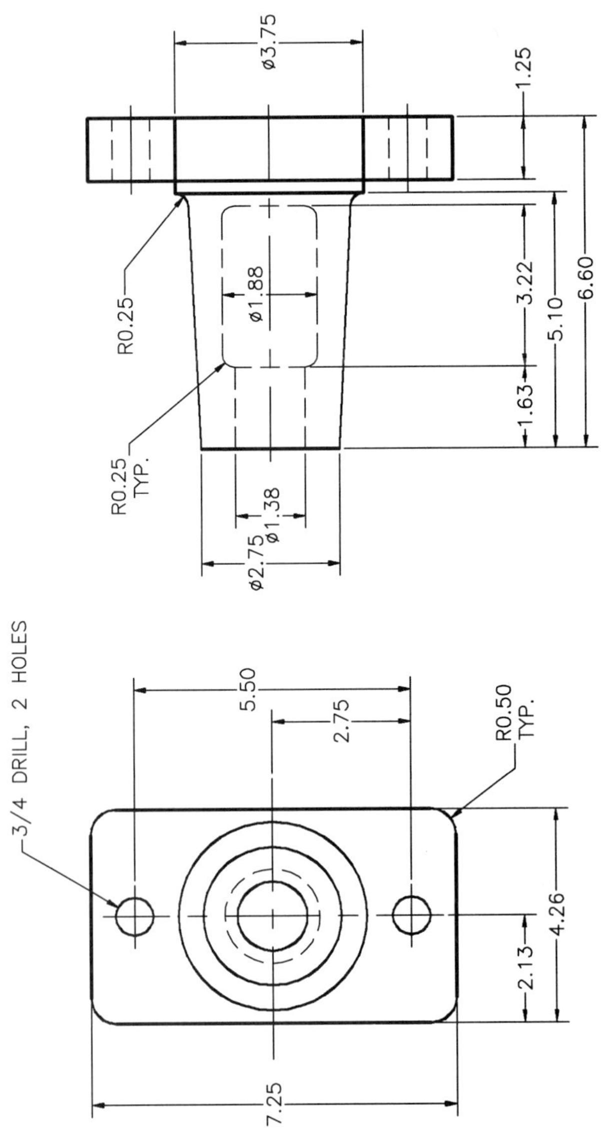

Figure 9-23 *Drawing for Problem Solving Exercise 2*

Problem Solving Exercise 3 — *Architectural*

Create the drawing of the floor plan shown in Figure 9-24 and then give the dimensions as shown. Edit the dimensions, if needed, so that they are positioned as shown in the drawing.

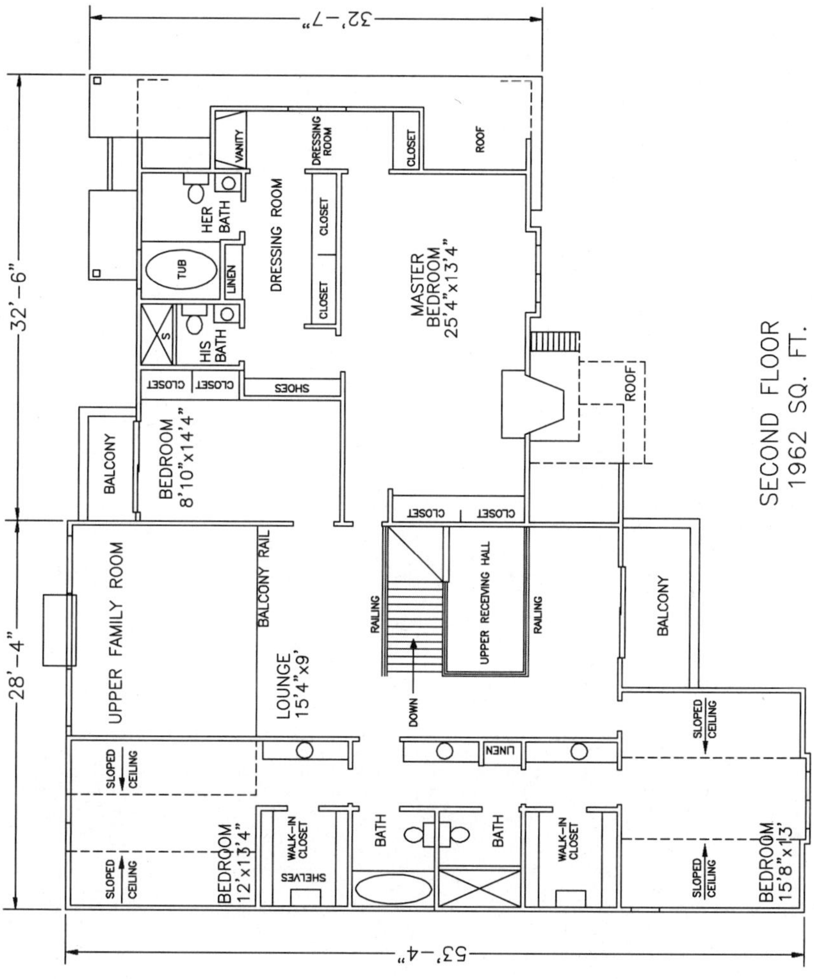

Figure 9-24 *Drawing for Problem Solving Exercise 3*

Answers to Self-Evaluation Test

1 - T, **2** - T, **3** - F, **4** - F, **5** - **EXPLODE**, **6** - **Left**, **7** - **Center**, **8** - **New**, **9** - **Oblique**, 10 - **Dimension Update**

Chapter 10

Dimension Styles and Dimensioning System Variables

Learning Objectives

After completing this chapter you will be able to:

- *Use styles and variables to control dimensions.*
- *Create dimensioning styles.*
- *Set dimension variables using the various tabs of the* ***New****,* ***Modify****, and* ***Override Dimension Style*** *dialog boxes.*
- *Set other dimension variables that are not in dialog boxes.*
- *Understand dimension style families and how to apply them in dimensioning.*
- *Use dimension style overrides.*
- *Compare and list dimension styles.*
- *Import externally referenced dimension styles.*

USING STYLES AND VARIABLES TO CONTROL DIMENSIONS

In AutoCAD, the appearance of dimensions on the drawing screen and the manner in which they are saved in the drawing database are controlled by a set of dimension variables. The dimensioning commands use these variables as arguments. The variables that control the appearance of the dimensions can be managed with dimension styles. You can use the **Dimension Style Manager** dialog box to control dimension styles and dimension variables through a set of dialog boxes.

CREATING AND RESTORING DIMENSION STYLES

Toolbar:	Dimension > Dimension Style
Menu:	Dimension > Style or Format > Dimension style
Command:	DIMSTYLE

The dimension styles control the appearance and positioning of dimensions and leaders in the drawing. If the default dimensioning style (STANDARD) does not meet your requirements, you can select another existing dimensioning style or create a new one that does. The default dimension style file name is **STANDARD**. Dimension styles can be created by using the **Dimension Style Manager** dialog box. Choose **Dimension > Dimension Style** (Figure 10-1) or **Dimension > Style** from the menu bar to invoke the **Dimension Style Manager** dialog box (Figure 10-2).

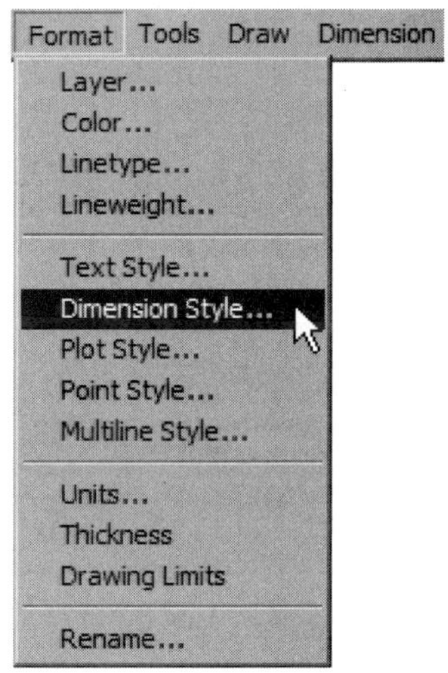

Figure 10-1 *Choosing* ***Dimension Style*** *from the* ***Format*** *menu*

In the **Dimension Style Manager** dialog box, choose the **New** button to display the **Create New Dimension Style** dialog box (Figure 10-3). Enter the dimension style name in the **New Style Name** text box and then select a style on which you want to base your style from the **Start With** drop-down list. The **Use for** drop-down list allows you to select the dimension type to which you want to apply the new dimension style. For example if you wish to use the new style for only the diameter dimension, you can select **Diameter dimension** from the **Use for** drop-down list. Choose the **Continue** button to display the **New Dimension Style** dialog box where you can define the new style.

In the **Dimension Style Manager** dialog box, the current dimension style name is shown in front of **Current dimstyle** and is also shown highlighted in the **Styles** list box. A brief description of the current style (its differences from the default settings) is also displayed in the **Description** area. The **Dimension Style Manager** dialog box also has a **Preview of** window that displays a preview of the current dimension style. A style can be made current (restored) by selecting the name of the dimension style you want to make current from the list of defined dimension styles and choosing the **Set Current** button. You can also make a style current by

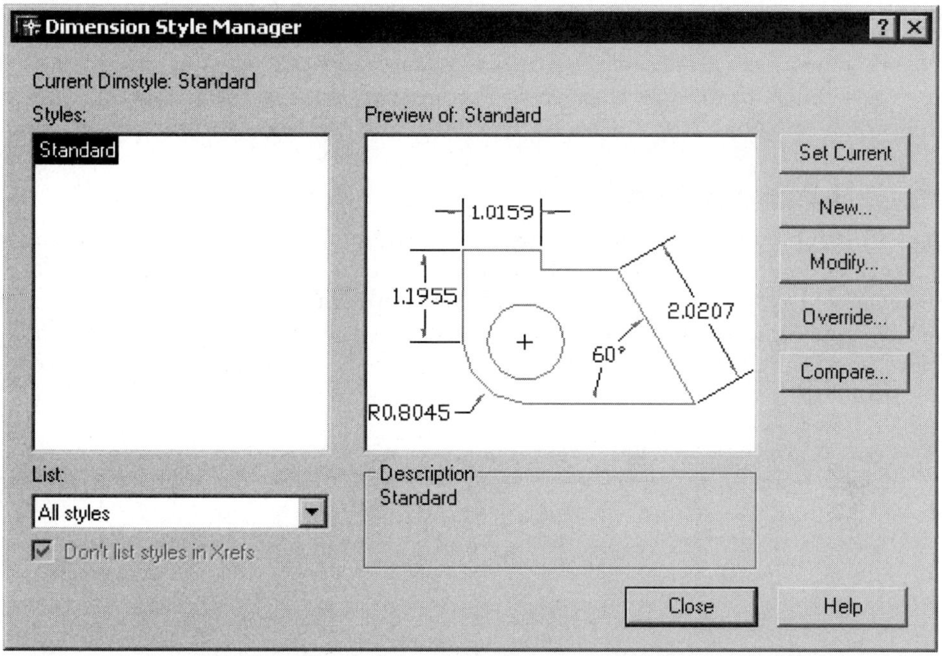

***Figure 10-2 Dimension Style Manager** dialog box*

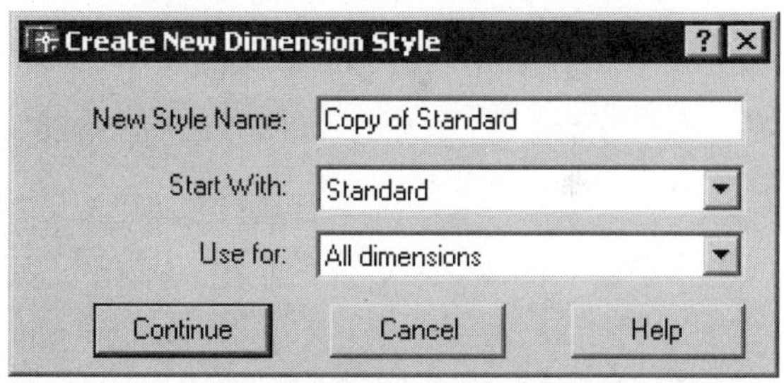

***Figure 10-3 Create New Dimension Style** dialog box*

double-clicking on the style name in the **Styles** list box. The drop-down list in the **Dimension** toolbar also displays the dimension styles. Selecting a Dimension style from this list also sets it current. The list of dimension styles displayed in the **Styles** list box is dependent on the option selected from the **List** drop down-list. If you select the **Styles in use** option, only the dimension styles in use will be listed in the **Style** list box. If you right-click on a style in the **Style** list box, a shortcut menu is displayed that provides you with the options to **Set current**, **Rename**, or **Delete** a dimension style. Selecting the **Don't list styles in Xrefs** check box does not list the names of Xref styles in the **Styles** list box. Choosing the **Modify** button displays the **Modify Dimension Style** dialog box where you can modify an existing style. Choosing the **Override** button displays the **Override Current Style** dialog box where you can define overrides to an existing style (discussed later in this chapter). Both these dialog boxes along with the **New Dimension Style** dialog box have identical properties. Choosing the **Compare** button displays the **Compare Dimension Styles** dialog box (also discussed later in this chapter) that allows you to compare two existing styles.

NEW DIMENSION STYLE DIALOG BOX

The **New Dimension Style** dialog box can be used to specify the dimensioning attributes (variables) that affect the various properties of the dimensions. The various tabs provided under the **New Dimension Style** dialog box are discussed next.

Lines and Arrows Tab

The options under the **Lines and Arrows** tab (Figure 10-4) of the **New Dimension Style** dialog box can be used to specify the dimensioning attributes (variables) that affect the format of the dimension lines and arrows. For example, the appearance and behavior of the dimension lines, extension lines, arrowheads, and center marks can be changed with this tab. If the settings of the dimension variables have not been altered in the current editing session, the settings displayed in the dialog box are the default settings.

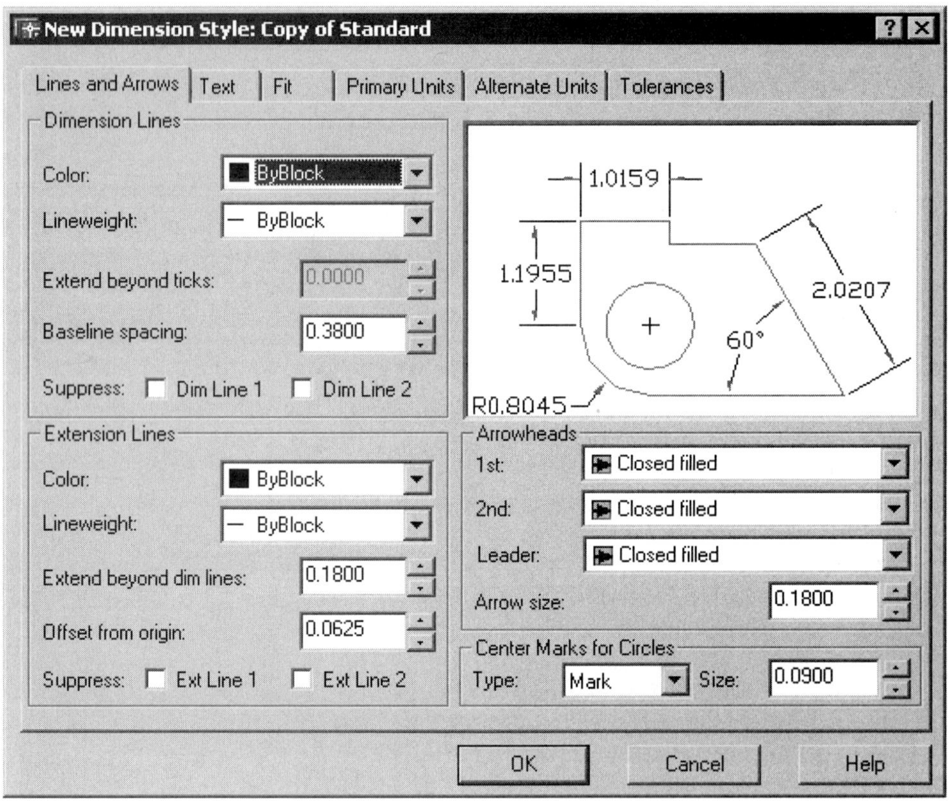

Figure 10-4 ***Lines and Arrow*** *tab of the* ***New Dimension Style*** *dialog box*

Dimension Lines Area

This area provides you with the options of controlling the display of the dimension lines and leader lines. These options are discussed next.

Color. This drop-down list is used to set the colors for the dimension lines and arrowheads. The dimension arrowheads have the same color as the dimension line because arrows constitute a part of the dimension line. The color you set here will also be assigned to the leader lines and arrows. The default color for the dimension lines and arrows is BYBLOCK. You can specify the color of the dimension line by selecting it from the **Color** drop-down list. You can also select **Other** from the **Color** drop-down list to display the **Select Color** dialog box where you can choose a specific color. The color number or the special color label is stored in the **DIMCLRD** variable.

Lineweight. This drop-down list is used to specify the lineweight for the dimension line. You can select the required lineweight by selecting it from this drop-down list. This value is also stored in the **DIMLWD** variable. The default value is BYBLOCK. Keep in mind that you cannot assign the lineweight to the arrowheads using this drop-down list.

Extend beyond ticks. The **Extend beyond ticks** (Oblique tick extension) spinner will be available only when you select the oblique, Architectural tick, or any such arrowhead type in the **1st** and **2nd** drop-down lists in the **Arrowheads** area. This spinner is used to specify the distance by which the dimension line will extend beyond the extension line. The extension value entered in the **Extend beyond ticks** edit box gets stored in the **DIMDLE** variable. By default, this edit box is disabled because the oblique arrowhead type is not selected.

Baseline spacing. The **Baseline spacing** (Baseline Increment) spinner is used to control the spacing between successive dimension lines drawn using baseline dimensioning, see Figure 10-5. You can specify the dimension line increment to your requirement by specifying the desired value using the **Baseline spacing** spinner. The default value displayed in the **Baseline spacing** spinner is 0.38 units. This spacing value is also stored in the **DIMDLI** variable.

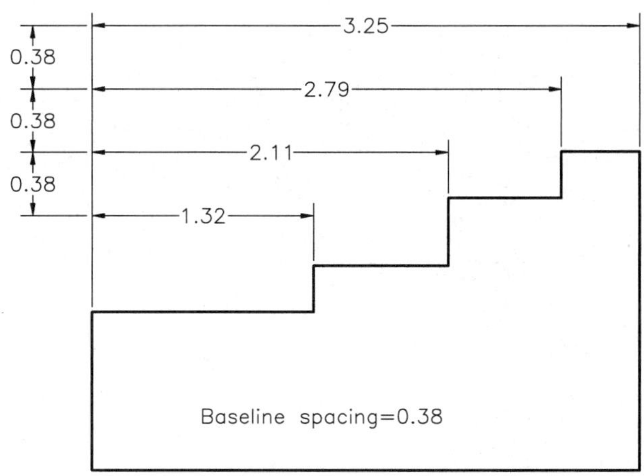

***Figure 10-5** Baseline increment*

Suppress. The **Suppress** check boxes control the display of the first and second dimension lines. By default, both dimension lines will be drawn. You can suppress one or both the dimension lines by selecting their corresponding check boxes. The values of these check boxes are stored in the **DIMSD1** and **DIMSD2** variables.

Note

The first and second dimension lines are determined by how you select the extension line origins. If the first extension line origin is on the right, the first dimension line is also on the right.

Extension Lines Area

Color. This drop-down list is used to control the color of the extension lines. The default extension line color is BYBLOCK. You can assign a new color to the extension lines by selecting it from this drop-down list. The color number or the color label is saved in the **DIMCLRE** variable.

Lineweight. This drop-down list is used to modify the lineweight of the extension lines. The default value is BYBLOCK. You can change the lineweight value by selecting a new value from this drop-down list. The value for lineweight is stored in the **DIMLWE** variable.

Extend beyond dim lines. It is the distance by which the extension lines extend past the dimension lines, see Figure 10-6. You can change the extension line offset using the **Extend beyond dim lines** spinner. This value is also stored in the **DIMEXE** variable. The default value for extension distance is 0.1800 units.

Offset from origin. It is the distance by which the extension line is offsetted from the point you specify as the origin of the extension line, see Figure 10-7. You may need to override this setting for specific dimensions when dimensioning curves and angled lines. You can specify an offset distance of your choice using this spinner. AutoCAD stores this value in the **DIMEXO** variable. The default value for this distance is 0.0625.

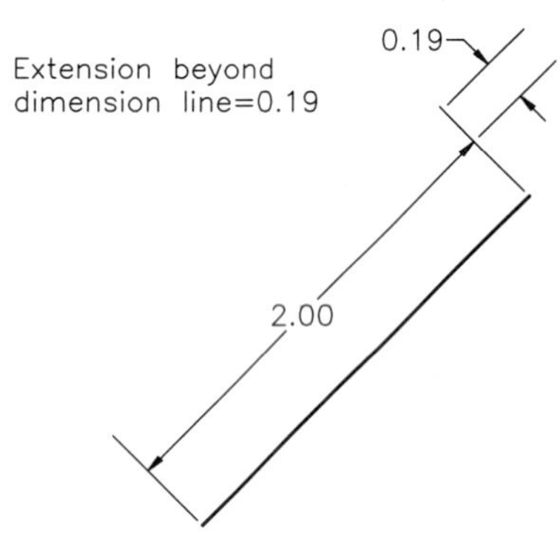

Figure 10-6 Extension beyond dimension lines

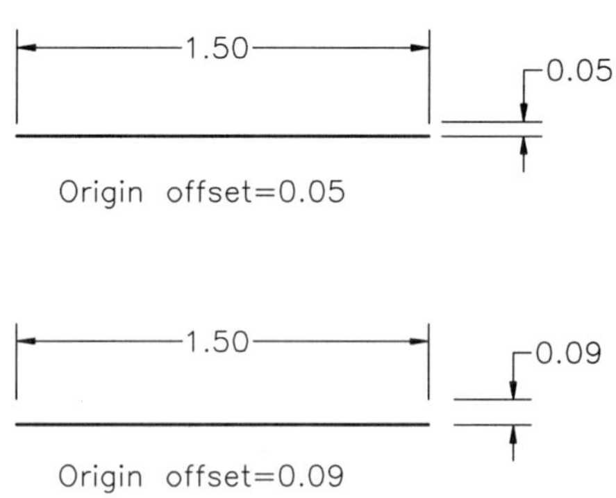

Figure 10-7 The origin offset

Suppress. The **Suppress** check boxes are used to control the display of the extension lines. By default, both extension lines will be drawn. You can suppress one or both extension lines by selecting the corresponding check boxes (Figure 10-8). The values of these check boxes are stored in the **DIMSE1** and **DIMSE2** variables.

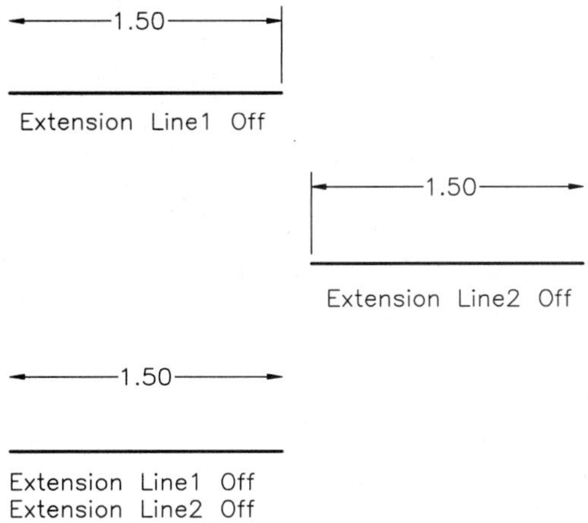

Figure 10-8 *Suppressing the extension lines*

Note

The first and second extension lines are determined by how you select the extension line origins. If the first extension line origin is on the right, the first extension line is also on the right.

Arrowheads Area

1st/2nd. When you create a dimension, AutoCAD draws the terminator symbols at the two ends of the dimension line. These terminator symbols, generally referred to as **arrowheads**, represent the beginning and end of a dimension. AutoCAD has provided nineteen standard termination symbols you can apply at each end of the dimension line. In addition to these, you can create your own arrows or terminator symbols. By default, the same arrowhead type is applied at both ends of the dimension line. If you select the first arrowhead, it is automatically applied to the second end by default. However, if you want to specify a different arrowhead at the second dimension line endpoint, you must select the desired arrowhead type from the **2nd** drop-down list. The first endpoint of the dimension line is the intersection point of the first extension line and the dimension line. The first extension line is determined by the first extension line origin. However, in angular dimensioning the second endpoint is located in a counterclockwise direction from the first point, regardless of how the points were selected when creating the angular dimension. The specified arrowhead types are selected from the **1st** and **2nd** drop-down lists. The first arrowhead type is saved in the **DIMBLK1** system variable, and the second arrowhead type is saved in the **DIMBLK2** system variable.

AutoCAD provides you with an option of specifying a user-defined arrowhead. To define a user-defined arrow, you must create one as a block. (See Chapter 14 for information regarding blocks.) Now, from the **1st** or the **2nd** drop-down list, select **User Arrow**. The **Select Custom Arrow Block** dialog box will be displayed, see Figure 10-9.

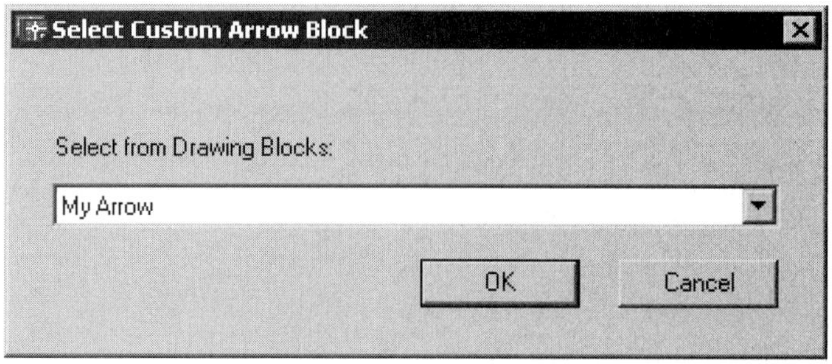

***Figure 10-9 Select Custom Arrow Block** dialog box*

All the blocks available in the current drawing will be available in the **Select from Drawing Blocks** drop-down list. You can select the desired block from this drop-down list and it will become the current arrowhead.

Creating an Arrowhead Block

1. To create a block for an arrowhead, you will use a 1 X 1 box, see Figure 10-10. AutoCAD automatically scales the block's X and Y scale factors to arrowhead size multiplied by the overall scale. The **DIMASZ** variable controls the length of the arrowhead. For example, if **DIMASZ** is set to 0.25, the length of the arrow will be 0.25 units. Also, if the length of the arrow is not 1 unit, it will leave a gap between the dimension line and the arrowhead block.

2. The arrowhead must be drawn as it would appear on the right side of the dimension line. Choose the **Make Block** button from the **Draw** toolbar to convert it into a block.

3. The insertion point of the arrowhead block must be the point that will coincide with the extension line, see Figure 10-10.

Leader. The **Leader** drop-down list displays arrowhead types for the Leader arrow. Here also you can either select the standard arrowheads from the drop-down list or select **User Arrow** that allows you to define and use a user-defined arrowhead type.

Arrow size. This spinner is used to define the size of the arrowhead, see Figure 10-11. The default value is 0.18 units. This value is stored in the **DIMASZ** system variable.

Center Marks for Circles Area

This area deals with options that control the appearance of center marks and centerlines in the radius and diameter dimensioning. However, keep in mind that the center marks or the centerlines will be drawn only when dimensions are placed outside the circle.

Type. The **Type** drop-down list provides you with an option of drawing a center mark or a centerline when you create radius or diameter dimensions. If you select **Mark**, a mark will be drawn at the center of the circle and if you select **Line**, centerlines will be drawn at the center of the circles. If you select **None**, no mark or line will be drawn at the center of the circle. If

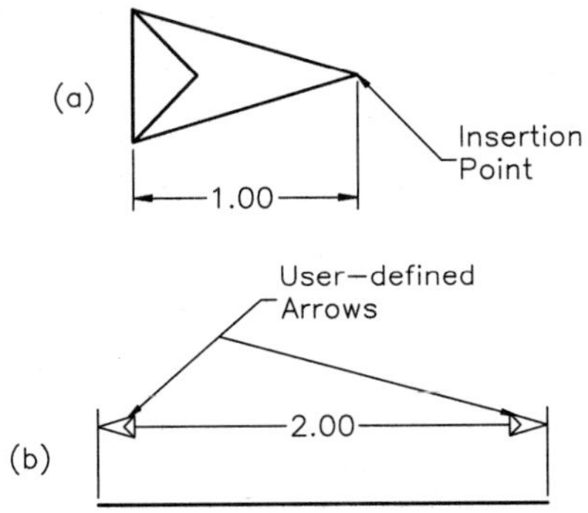

Figure 10-10 *Creating user-defined arrows*

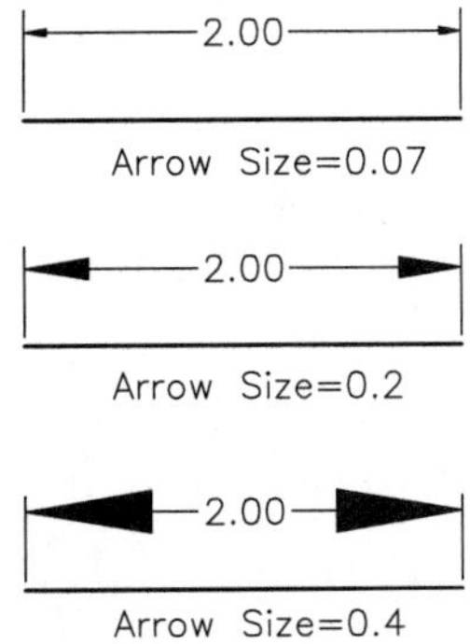

Figure 10-11 *Defining arrow sizes*

you use the **DIMCEN** command, a positive value will create a center mark, and a negative value will create a centerline. If the value is 0, AutoCAD does not create center marks or centerlines.

Size. This spinner is used to set the size of the center marks or centerlines. You can specify the required size of the center marks or centerlines using the **Size** spinner. This value is stored in the **DIMCEN** variable. The default value of the **DIMCEN** variable is 0.09.

Note

Unlike specifying a negative value for the ***DIMCEN*** *variable, you cannot enter a negative value in the* ***Size*** *spinner. Selecting* ***Line*** *from the* ***Type*** *drop-down list automatically treats the value in the* ***Size*** *spinner as the size for the centerlines and sets* ***DIMCEN*** *to the negative of the value shown.*

Exercise 1 — *Mechanical*

Draw Figure 10-12 and then set the values in the **Lines and Arrows** tab of the **New Dimension Style** dialog box to dimension the drawing as shown in this figure. (Baseline spacing = 0.25, Extension beyond dimension line = 0.10, Offset from origin = 0.05, Arrowhead size= 0.09.) Assume the missing dimensions.

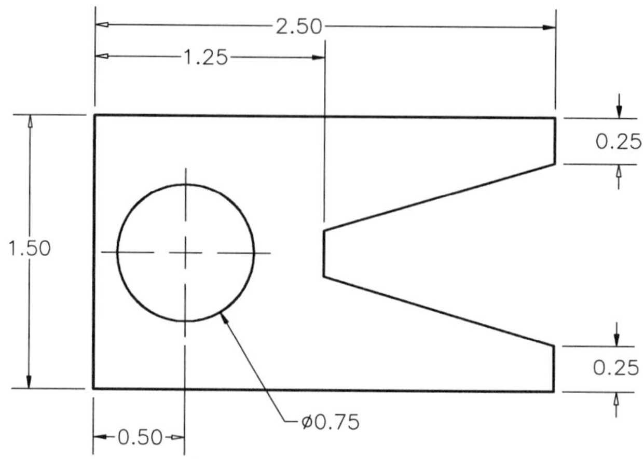

***Figure 10-12** Drawing for Exercise 1*

CONTROLLING DIMENSION TEXT FORMAT

Text tab

You can control the dimension text format through the **Text** tab of the **New Dimension Style** dialog box (Figure 10-13). In the **Text** tab, you can control the parameters such as the placement, appearance, horizontal and vertical alignment of dimension text, and so on. For example, you can force AutoCAD to align the dimension text along the dimension line. You can also force the dimension text to be displayed at the top of the dimension line. You can save the settings in a dimension style file for future use. The **New Dimension Style** dialog box has a Preview window that updates dynamically to display the text placement as the settings are changed. Individual items of the **Text** tab and the related dimension variables are described next.

Text Appearance Area

Text style. The **Text Style** drop-down list displays the names of the predefined text styles. From this list you can select the style name that you want to use for dimensioning. You must define the text style before you can use it in dimensioning (see "**Creating Text Styles**" in Chapter 7). Choosing the **[...]** button displays the **Text Style** dialog box that allows you to create a new or modify an existing text style. The value of this setting is stored in the **DIMTXSTY** system variable. The change in dimension text style does not affect the text style you are using to draw other text in the drawing.

Text color. This drop-down list is used to modify the color of the dimension text. The default color is BYBLOCK. If you choose the **Other** option from the **Text Color** drop-down list, the **Select Color** dialog box is displayed where you can choose a specific color. This color or color number is stored in the **DIMCLRT** variable.

Text height. This spinner is used to modify the height of the dimension text, see

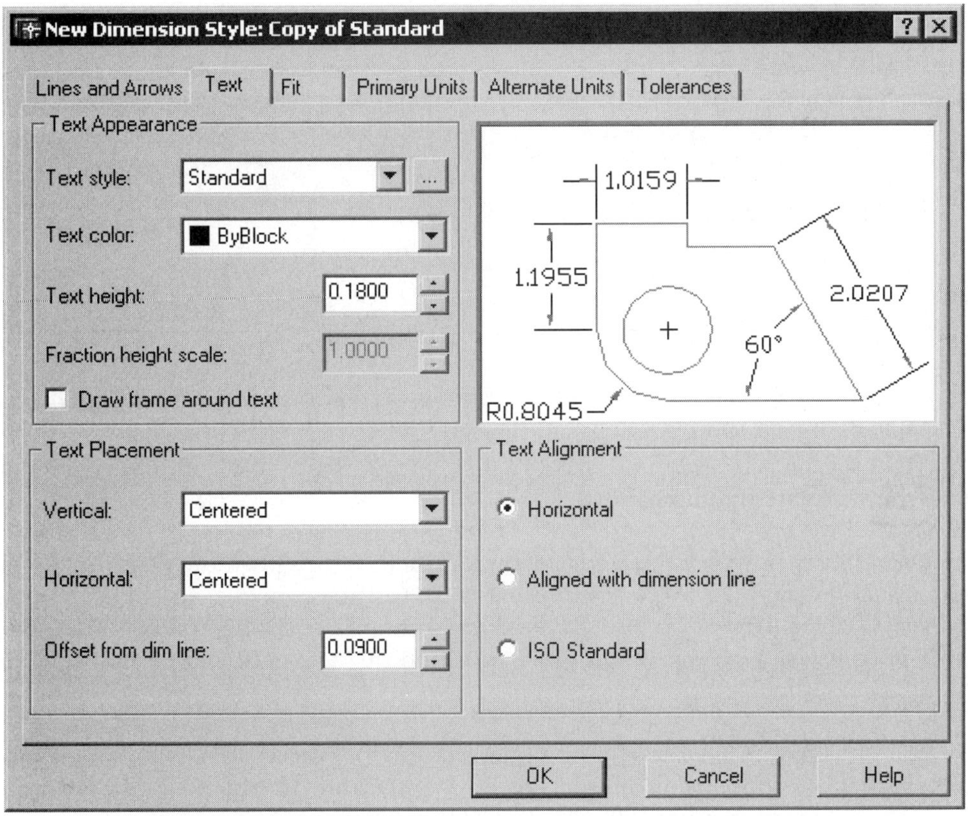

***Figure 10-13 Text** tab of the **New Dimension Style** dialog box*

Figure 10-14. You can change the dimension text height only when the current text style does not have a fixed height. In other words, the text height specified in the **STYLE** command should be zero. This is because a predefined text height (specified in the **STYLE** command) overrides any other setting for the dimension text height. This value is stored in the **DIMTXT** variable. The default text height is 0.1800 units.

Fraction height scale. This spinner is used to set the scale of the fractional units in relation to the dimension text height. This spinner will be available only when you select a format for the primary units in which you can define the values in fractions, such as architectural or fractional. This value is stored in the **DIMTFAC** variable.

Draw frame around text. When selected this check box draws a frame around the dimension text, see Figure 10-15. This value is stored as a negative value in the **DIMGAP** system variable.

Text Placement Area

Vertical. The **Vertical** drop-down list displays the options that control the vertical placement of the dimension text. The current setting is highlighted. Controlling the vertical placement of the dimension text is possible only when the dimension text is drawn in its normal (default)

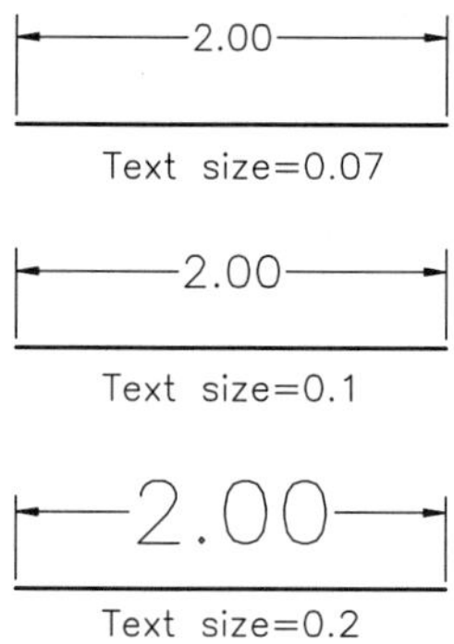

***Figure 10-14** Changing the dimension height*

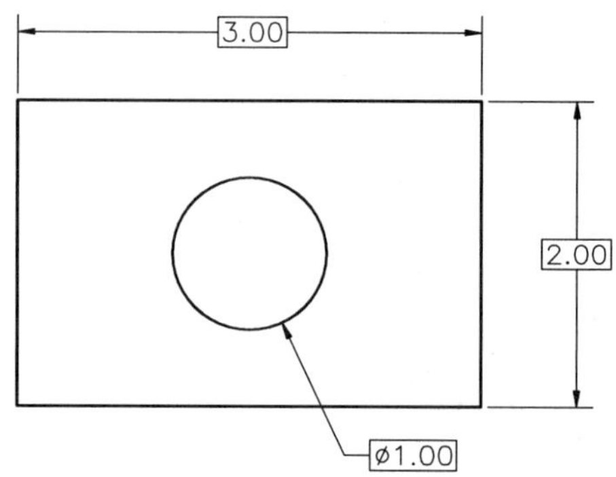

***Figure 10-15** Dimension text inside frame*

location. This setting is stored in the **DIMTAD** system variable. The options provided under this drop-down list are discussed next.

Centered. If this option is selected, the dimension text gets positioned on the dimension line in such a way that the dimension line is split to allow for placement of the text, see Figure 10-16. If the **1st** or **2nd Extension Line** option is selected in the **Horizontal** drop-down list, this centered setting will position the text on the extension line, not on the dimension line.

Above. If this option is selected, the dimension text is placed above the dimension line, except when the dimension line is not horizontal and the dimension text inside the extension lines is horizontal. The distance of the dimension text from the dimension line is controlled by the **DIMGAP** variable. This results in an unbroken solid dimension line being drawn under the dimension text, see Figure 10-16.

Outside. This option places the dimension text on the side of the dimension line.

JIS. This option lets you place the dimension text to conform to **JIS** (Japanese Industrial Standards) representation.

Note

The Horizontal and Vertical Placement options selected are reflected in the dimensions shown in the Preview window.

Horizontal. This drop-down list is used to control the horizontal placement of the dimension text. You can select the required horizontal placement from this list. However, remember that these options will be useful only when the **Place text manually when dimensioning** check box in the **Fine Tuning** area of the **Fit** tab is cleared. The options provided under this drop-down list are discussed next.

Centered. This option places the dimension text between the extension lines. This is the default option.

At Ext Line 1. This option is selected to place the text near the first extension line along the dimension line, see Figure 10-17.

At Ext Line 2. This option is selected to place the text near the second extension line along the dimension line, see Figure 10-17.

Over Ext Line 1. This option is selected to place the text over the first extension line and also along the first extension line, see Figure 10-17.

Over Ext Line 2. This option is selected to place the text over the second extension line and also along the second extension line, see Figure 10-17.

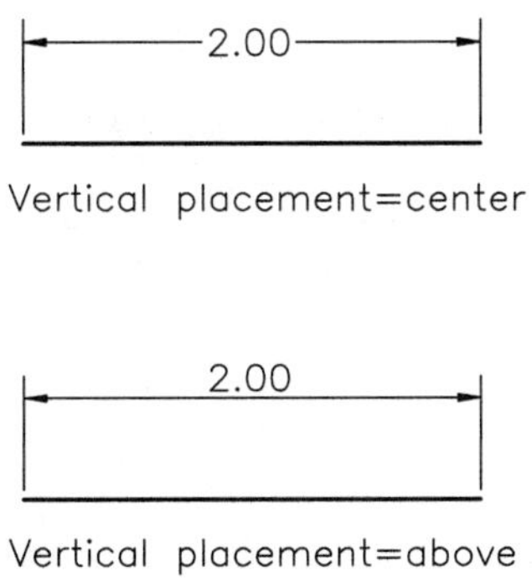

Figure 10-16 *Vertical text placement*

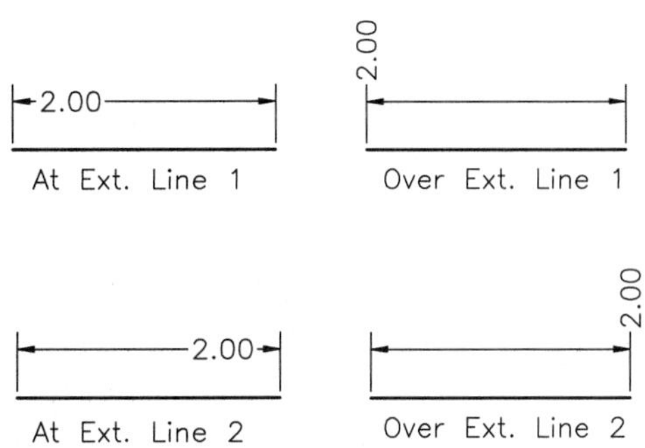

Figure 10-17 *Horizontal text placement*

Offset from dim line. This spinner is used to specify the distance between the dimension line and the dimension text (Figure 10-18). You can set the text gap you need using this spinner. The text gap value is also used as the measure of minimum length for the segments of the dimension line and in basic tolerance. The default value specified in this box is 0.09 units. The value of this setting is stored in the **DIMGAP** system variable.

Text Alignment Area

Horizontal. This is the default option and if selected, the dimension text is drawn horizontally with respect to the current UCS (user coordinate system). The alignment of the dimension line does not affect text alignment. Selecting this option turns both the **DIMTIH** and **DIMTOH** system variables **on**. The text is drawn horizontally even if the dimension line is at an angle.

Aligned with dimension line. Selecting this radio button aligns the text with the dimension line (Figure 10-19) and both the system variables **DIMTIH** and **DIMTOH** are off.

ISO Standard. If you select the **ISO Standard** radio button, the dimension text is aligned with the dimension line only when the dimension text is inside the extension lines. Selecting this option turns the system variable **DIMTOH** on; that is, the dimension text outside the extension line is horizontal regardless of the angle of the dimension line.

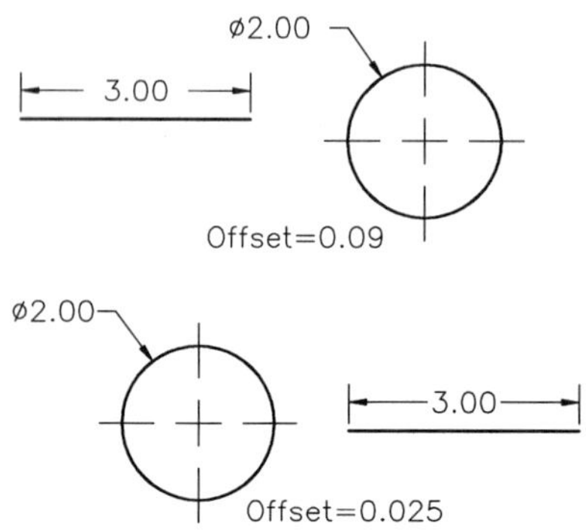

Figure 10-18 *Offset from dimension line*

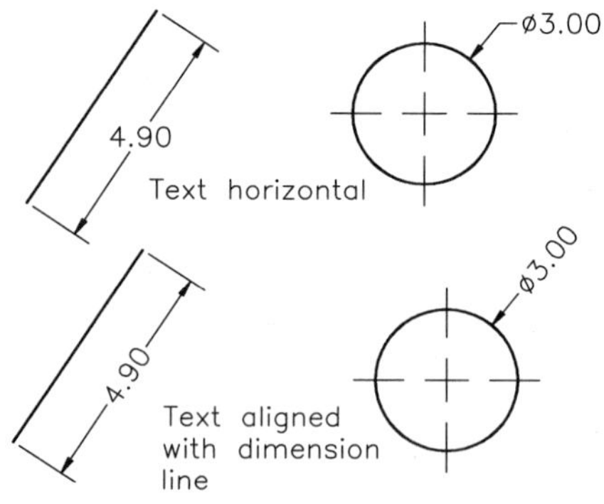

Figure 10-19 *Specifying test alignment*

Exercise 2 *Mechanical*

Draw Figure 10-20 and then set the values in the **Lines and Arrows** and **Text** tabs of the **New Dimension Style** dialog box to dimension the drawing as shown in the figure. (Baseline spacing = 0.25, Extension beyond dimension lines = 0.10, Offset from origin = 0.05, Arrow size = 0.09, Text height = 0.08.)

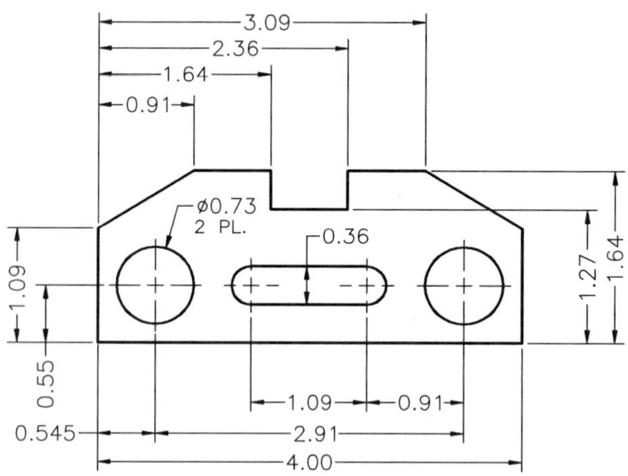

Figure 10-20 *Drawing for Exercise 2*

FITTING DIMENSION TEXT AND ARROWHEADS

Fit Tab

The **Fit** tab provides you with the options that control the placement of dimension lines, arrowheads, leader lines, text, and the overall dimension scale (Figure 10-21).

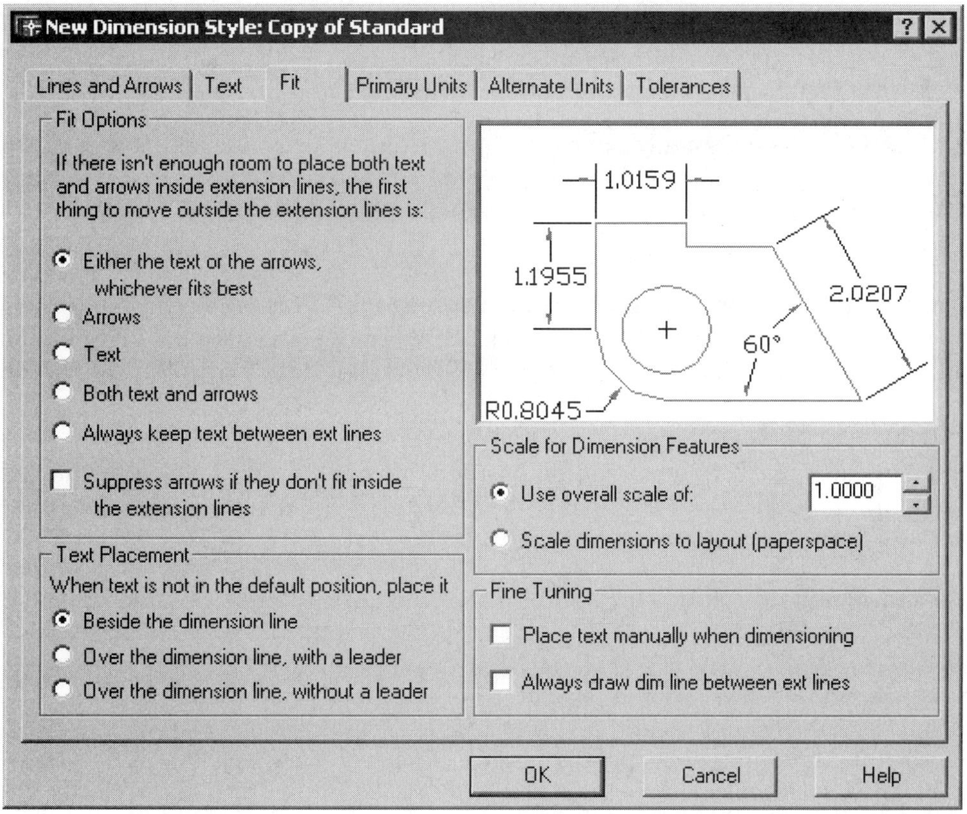

Figure 10-21 Fit *tab of the* ***New Dimension Style*** *dialog box*

Fit Options Area

These options are used to set the priorities for moving the text and arrowheads outside the extension lines, if the space between the extension lines is not enough to fit both of them.

Either the text or the arrows, whichever fits best. This is the default option. In this option, AutoCAD places the dimension where it fits best between the extension lines.

Arrows. When you select this option, AutoCAD places the text and arrowheads inside the extension lines if there is enough space to fit both. If space is not available for both the arrows and text, the arrows are moved outside the extension lines. If there is not enough space for text, both text and arrowheads are placed outside the extension lines.

Text. When you select this option, AutoCAD places the text and arrowheads inside the extension lines if there is enough space to fit both. If there is enough space to fit the arrows, the arrows will be placed inside the extension lines and the dimension text moves outside the extension lines. If there is not enough space for either text or arrowheads, both are placed outside the extension lines.

Both Text and Arrows. If you select this option, AutoCAD will place the arrows and dimension text between the extension lines if there is enough space available to fit both. Otherwise, both text and arrowheads are placed outside the extension lines.

Always keep text between ext lines. This option always keeps the text between extension lines even in cases where AutoCAD would not do so. Selecting this radio button does not affect radius and diameter dimensions. The value is stored in the **DIMTIX** variable and the default value is **off**.

Suppress arrows if they don't fit inside the extension lines. If you select this check box, the arrowheads are suppressed if the space between the extension lines is not enough to adjust them. The value is stored in the **DIMSOXD** variable and the default value is **off**.

Text Placement Area

This area provides you with the options to position the dimension text when it is moved from the default position. The value is stored in the **DIMTMOVE** variable. The options under this area are as follows.

Beside the dimension line. This option places the dimension text beside the dimension line.

Over the dimension line, with a leader. Selecting this option places the dimension text away from the dimension line and a leader line is created, which connects the text to the dimension line. But, if the dimension line is too close to the text, a leader is not drawn. The Horizontal placement decides whether the text is placed to the right or left of the leader.

Over the dimension line, without a leader. In this option, AutoCAD does not create a leader line if there is insufficient space to fit the dimension text between the extension lines. The dimension text can be moved freely, independent of the dimension line.

Scale for Dimension Features Area

The options under this area set the value for overall Dimension scale or scaling to paper space.

Use overall scale of. The current general scaling factor that pertains to all of the size-related dimension variables, such as text size, center mark size, and arrowhead size, is displayed in the **Use overall scale of** spinner. You can alter the scaling factor to your requirement by entering the scaling factor of your choice in this spinner. Altering the contents of this box alters the value of the **DIMSCALE** variable, since the current scaling factor is stored in it. The overall scale (**DIMSCALE**) is not applied to the measured lengths, coordinates, angles, or tolerance. The default value for this variable is 1.0. In this condition, the dimensioning variables assume their preset values and the drawing is plotted at full scale. The scale factor is the reciprocal of the drawing size and so the drawing is to be plotted at half size. The overall scale factor (**DIMSCALE**) will be the reciprocal of ½, which is 2.

Note

If you are in the middle of the dimensioning process and you change the ***DIMSCALE*** *value and save the changed setting in a dimension style file, the dimensions with that style will be updated.*

Tip

When you increase the limits of the drawing, increase the overall scale of the drawing using the ***Use overall scale of*** *spinner before dimensioning. This will save the time required in changing the individual scale factors of all the dimension parameters.*

Scale dimensions to layout (paperspace). If you select the **Scale dimensions to layout (paper space)** radio button, the scale factor between the current model space viewport and floating viewport (paper space) is computed automatically. Also, by selecting this radio button, you disable the **Use overall scale of** spinner (it is disabled in the dialog box) and the overall scale factor is set to 0. When the overall scale factor is assigned a value of 0, AutoCAD calculates an acceptable default value based on the scaling between the current model space viewport and paper space. If you are in paper space (**TILEMODE=0**), or are not using the **Scale dimensions to layout (paper space)** feature, AutoCAD sets the overall scale factor to 1; otherwise, AutoCAD calculates a scale factor that makes it possible to plot text sizes, arrow sizes, and other scaled distances at the values in which they have been previously set. (For further details regarding model space and layouts, refer to Chapter 11.)

Fine Tuning Area

The **Fine Tuning** area provides additional options governing placement of dimension text. The options are as follows.

Place text manually when dimensioning. When you dimension, AutoCAD places the dimension text in the middle of the dimension line (if there is enough space). If you select the **Place text manually when dimensioning** check box, you can position the dimension text anywhere along the dimension line. You will also notice that when you select this check box, the **Horizontal Justification** is ignored. This setting is saved in the **DIMUPT** system variable. The default value of this variable is **off**. Selecting this check box enables you to position the dimension text anywhere along the dimension line.

Always draw dim line between ext lines. This check box is selected when you want the dimension line to appear between the extension lines, even if the text and dimension lines are placed outside the extension lines. When you select this option in the radius and diameter dimensions (when default text placement is horizontal), the dimension line and arrows are drawn inside the circle or arc, and the text and leader are drawn outside. When you select the **Always draw dimension line between extension lines** check box, the **DIMTOFL** variable is set to on by AutoCAD. The default setting is off.

FORMATTING PRIMARY DIMENSIONS UNITS

Primary Units Tab

You can use the **Primary Units** tab of the **New Dimension Style** dialog box to control the dimension text format and precision values (Figure 10-22). You can use the options under this tab to control Units, Dimension Precision, and Zero Suppression for dimension measurements. AutoCAD lets you attach a user-defined prefix or suffix to the dimension text. For example, you can define the diameter symbol as a prefix by entering %%C in the **Prefix** edit box; AutoCAD will automatically attach the diameter symbol in front of the dimension text. Similarly, you can define a unit type, such as **mm**, as a suffix; AutoCAD will then attach **mm** at the end of every dimension text. This tab also enables you to define zero suppression, precision, and dimension text format.

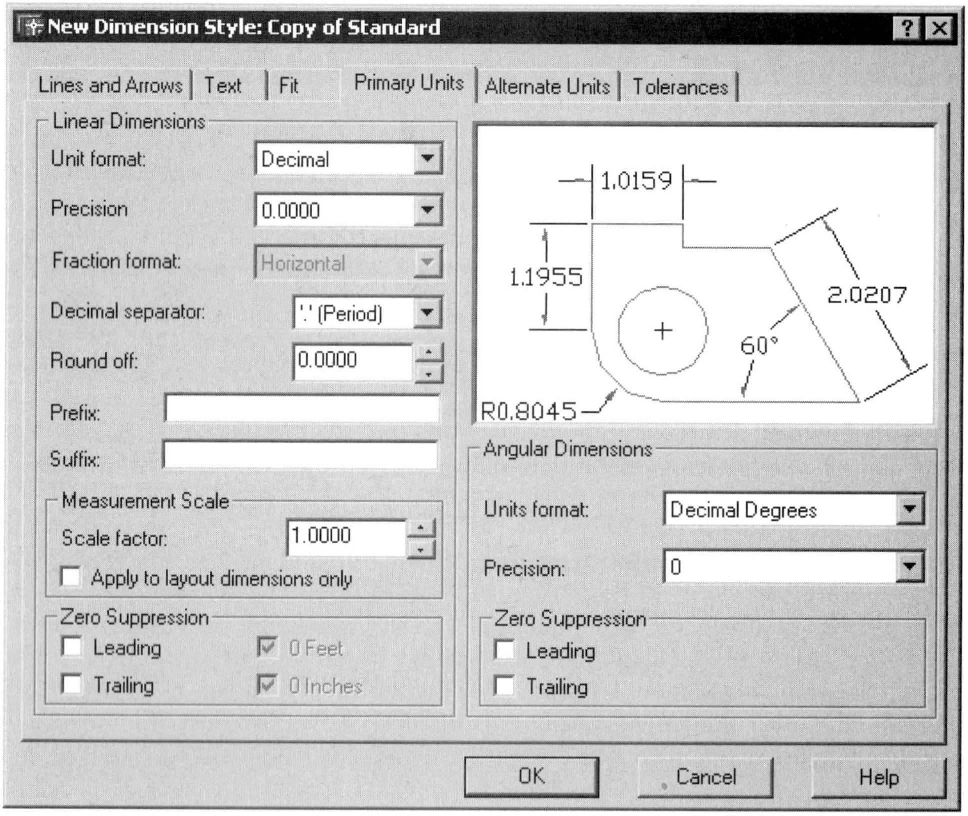

Figure 10-22 ***Primary Units*** *tab of the* ***New Dimension Style*** *dialog box*

Linear Dimensions Area

Unit format. This drop-down list provides you with the options of specifying the units for the primary dimensions. The formats that are available include Decimal, Scientific, Architectural, Engineering, Fraction, and Windows Desktop. Keep in mind that by selecting a dimension unit format, the drawing units (which you might have selected by using the **UNITS** command) are not affected. The unit setting for linear dimensions is stored in the **DIMLUNIT** system variable.

Precision. This drop-down list is used to control the number of decimal places for the primary units. The setting for precision (number of decimal places) is saved in the **DIMDEC** variable.

Fraction format. This drop-down list is used to set the fraction format. The options are Diagonal, Horizontal, and not stacked. This drop-down list will be available only when you select **Architectural** or **Fractional** from the **Unit format** drop-down list. The value is stored in the **DIMFRAC** variable.

Decimal Separator. This drop-down list is used to select an option that will be used as the decimal separator. For example, Period [.], Comma [,] or Space [ ]. If you have selected Windows desktop units in the **Unit Format:** drop-down list, AutoCAD uses the Decimal symbol settings. The value is stored in the **DIMDSEP** variable.

Round off. The **Round off** spinner sets the value for rounding off the dimension values. For example, if the **Round off** spinner is set to 0.05, a value of 0.06 will round off to 0.10. The number of decimal places of the round off value you enter in the edit box should be less than or equal to the value in the **Precision:** edit box. The value is stored in the **DIMRND** variable and the default value in the **Round Off:** edit box is 0. Also see Figure 10-23.

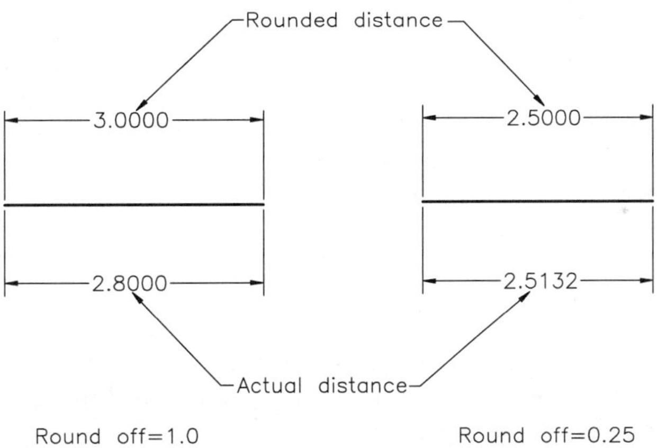

Figure 10-23 Rounding off the dimension measurements

Prefix. You can append a prefix to the dimension measurement by entering the prefix in this edit box. The dimension text is converted into **Prefix<dimension measurement>** format. For example, if you enter the text "Abs" in the Prefix edit box, "Abs" will be placed in front of the dimension text (Figure 10-24). The prefix string is saved in the **DIMPOST** system variable.

Note

*Once you specify a prefix, default prefixes such as **R** in radius dimensioning and Ø in diameter dimensioning are cancelled.*

Suffix. Just like appending a prefix, you can append a suffix to the dimension measurement by entering the desired suffix in this edit box. For example, if you enter the text mm in the

Suffix edit box, the dimension text will have <dimension measurement>mm format, see Figure 10-24. AutoCAD stores the suffix string in the **DIMPOST** variable.

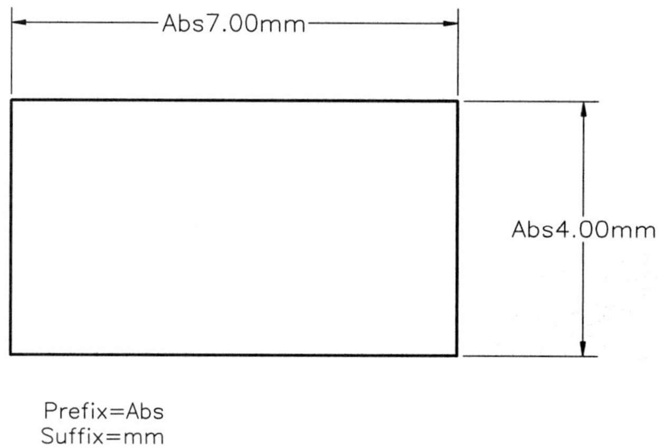

Figure 10-24 *Adding prefix and suffix to the dimensions*

Tip

*The **DIMPOST** variable is used to define both prefix and suffix to the dimension text. This variable takes a string value as its argument. For example, if you want to have a suffix for centimeters, set **DIMPOST** to cm. To establish a prefix to a dimension text, type the prefix text string and then "<>".*

Measurement Scale Area

Scale factor. You can specify a global scale factor for linear dimension measurements by setting the desired scale factor in the **Scale factor** spinner. All the linear distances measured by dimensions, which include radii, diameters, and coordinates, are multiplied by the existing value in this spinner. For example, if the value of the **Scale factor** spinner is set to 2, two unit segments will be dimensioned as 4 units (2 X 2). However, the angular dimensions are not affected. In this manner, the value of the linear scaling factor affects the contents of the default (original) dimension text (Figure 10-25). Default value for linear scaling is 1. With the default value, the dimension text generated is the actual measurement of the object being dimensioned. The linear scaling value is saved in the **DIMLFAC** variable.

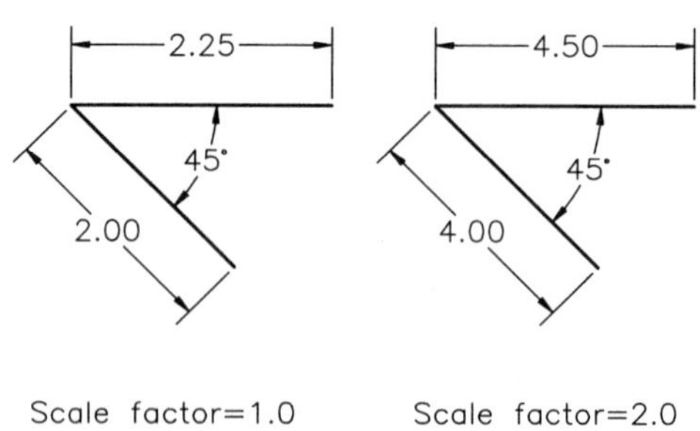

Figure 10-25 *Identical figures dimensioned using different scale factors*

Note

The linear scaling value is not exercised on rounding a value or on plus or minus tolerance values. Therefore, changing the linear scaling factor will not affect the tolerance values.

Apply to layout dimensions only. When you select the **Apply to layout dimensions only** check box, the scale factor value is applied only to the dimensions in the layout. The value is stored as a negative value in the **DIMLFAC** variable. If you change the **DIMLFAC** variable from the **Dim:** prompt, AutoCAD displays the viewport option to calculate the **DIMLFAC** variable. First, set the **TILEMODE** to 0 (paper space), and then invoke the **MVIEW** command to get the **Viewport** option.

Zero Suppression Area. The options under this area are used to suppress the leading or trailing zeros in the dimensioning. This area provides you with four check boxes. These check boxes can be selected to suppress the leading or trailing zeros or zeros in the feet and inches. The **0 Feet** and the **0 Inches** check boxes will be available only when you select **Engineering** or **Architectural** from the **Unit format** drop-down list. When Architectural units are being used, the **Leading** and **Trailing** check boxes are disabled. For example, if you select the **0 Feet** check box, the dimension text 0'-8 ¾" becomes 8 ¾". By default, the 0 Feet and 0 Inches value is suppressed. If you want to suppress the inches part of a feet-and-inches dimension when the distance in the feet portion is an integer value and the inches portion is zero, select the **0 Inches** check box. For example, if you select the **0 Inches** check box, the dimension text 3'-0" becomes 3'. Similarly, if you select the **Leading** check box, the dimension that was initially 0.53 will become .53. If you select the **Trailing** check box, the dimension that was initially 2.0 will become 2.

Angular Dimensions Area

This area provides you with the options to control the units format, precision, and zero suppression for Angular units.

Units format. The **Units format** drop-down list displays a list of unit formats for the angular dimensions. The default value in which the angular dimensions are displayed is **Decimal Degrees**. The value governing the unit setting for angular dimensions is stored in the **DIMAUNIT** variable.

Precision. You can select the number of decimal places for the angular dimensions from this drop-down list. This value is stored in the **DIMADEC** variable.

Zero Suppression. Similar to linear dimensions, you can suppress the **leading**, **trailing**, neither, or both zeros in the angular dimensions by selecting the respective check boxes in this area. The value is stored in the **DIMAZIN** variable.

FORMATTING ALTERNATE DIMENSION UNITS

Alternate Units tab

By default, the **Alternate Units** tab of the **New Dimension Style** dialog box is disabled and the value of the **DIMALT** variable is turned off. If you want to perform alternate units dimensioning, select the **Display Alternate Units** check box. By doing so, AutoCAD activates various options in this area (Figure 10-26). This tab sets the format, precision, angles, placement, scale, and so on for the alternate units in use. In this tab you can specify the values that will be applied to alternate dimensions.

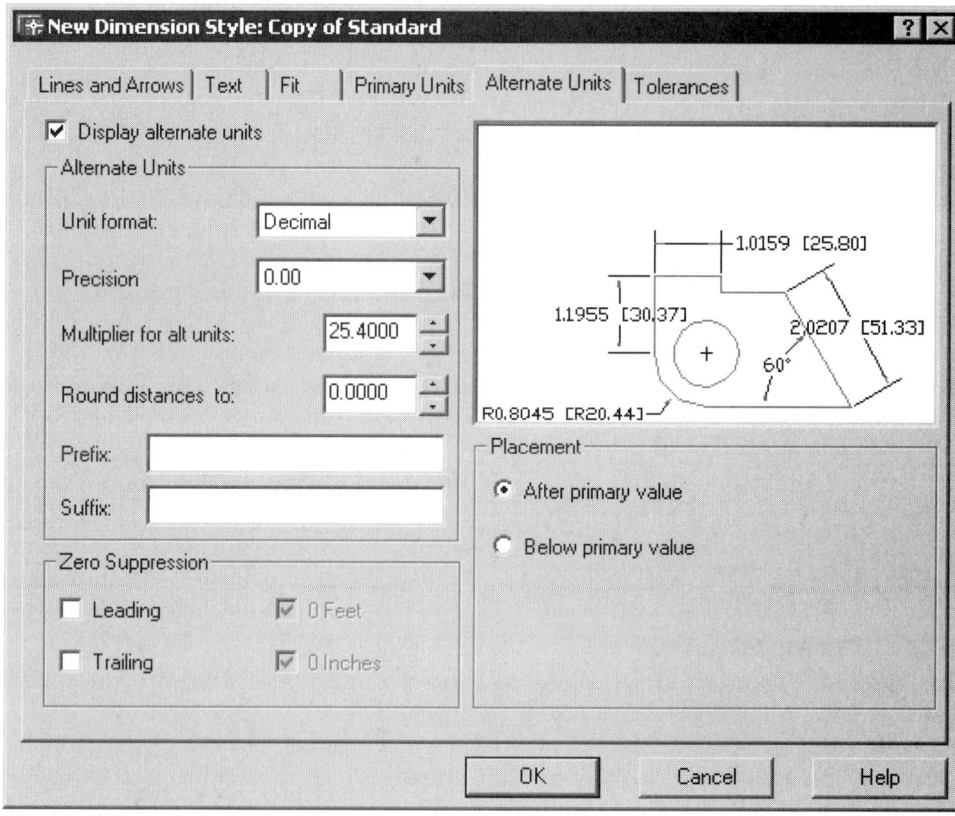

***Figure 10-26 Alternate Units** tab of the **New Dimension Style** dialog box*

Alternate Units Area

The options under this area are identical to those under the **Linear Dimensions** area of the **Primary Units** tab. This area provides you with the options to set the format for all dimension types except Angular.

Unit format. You can select a unit format to apply to the alternate dimensions from this drop-down list. The options under this drop-down list include Scientific, Decimal, Engineering, Architectural stacked, Fractional stacked, Architectural, Fractional, and Windows Desktop. The value is stored in the **DIMALTU** variable. The relative size of fractions is governed by the **DIMTFAC** variable.

Precision. You can select the number of decimal places for the alternate units from the **Precision** drop-down list. The value is stored in the **DIMALTD** variable.

Multiplier for alt units. To generate a value in the alternate system of measurement, you need a factor with which all the linear dimensions will be multiplied. The value for this factor can be set using the **Multiplier for alt units** spinner. The default value of 25.4 is for dimensioning in inches with alternate units in millimeters. This scaling value (contents of the **Multiplier for alt units** spinner) is stored in the **DIMALTF** variable.

Round distances to. This spinner is used to set a value to which you want all your measurements (made in alternate units) to be rounded off. This value is stored in the **DIMALTRND** system variable. For example, if you set the value of the **Round distances to** spinner to 0.25, all the alternate dimensions get rounded off to the nearest .25 unit.

Prefix/Suffix. The **Prefix** and **Suffix** edit boxes are similar to the edit boxes in the **Linear Dimensions** area of the **Primary Units** tab. You can enter the text or symbols that you want to precede or follow the alternate dimension text. The value is stored in the **DIMAPOST** variable. You can also use control codes and special characters to display special symbols.

Zero Suppression Area

This area allows you to suppress the leading or trailing zeros in decimal unit dimensions by selecting either, both, or none of the **Trailing** and **Leading** check boxes. Similarly, selecting the **0 Feet** check box suppresses the zeros in the feet area of the dimension, when the dimension value is less than a foot. Selecting the **0 inches** check box suppresses the zeros in the inches area of the dimension. For example, 1'-0" becomes 1'. The **DIMALTZ** variable controls the suppression of zeros for alternate unit dimension values. The values that are between 0 and 3 affect feet-and-inch dimensions only.

Placement Area

This area provides the options that control the positioning of the Alternate units. The value is stored in the **DIMAPOST** variable.

After primary value. Selecting the **After primary value** radio button places the alternate units dimension text after the primary units. This is the default option, see Figure 10-27.

Below primary value. Selecting the **Below primary value** radio button places the alternate units dimension text below the primary units, see Figure 10-27.

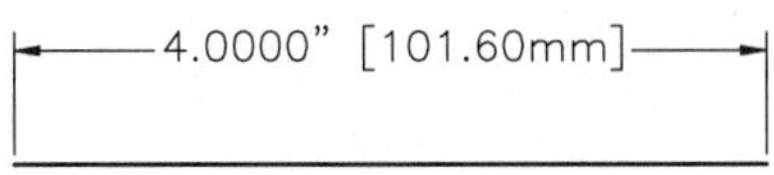

Alternate units after primary units

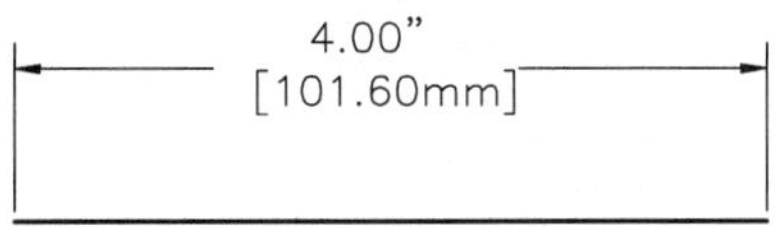

Alternate units below primary units

Figure 10-27 Placements of alternate units

Figure 10-27 illustrates the result of entering information in the **Alternate Units** tab. The

Chapter 10

decimal places get saved in the **DIMALTD** variable, the scaling value (contents of the **Multiplier for alt units** spinner) in the **DIMALTF** variable, and the suffix string (contents of the **Suffix** edit box) in the **DIMAPOST** variable. Similarly, the units format for alternate units are in **DIMALTU**, and suppression of zeros for alternate units decimal values are in **DIMALTZ**.

FORMATTING THE TOLERANCES

Tolerances Tab

The **Tolerances** tab (Figure 10-28) allows you to set the parameters for options that control the format and display of tolerance dimension text. These include the alternate unit tolerance dimension text.

Tolerance Format Area

The **Tolerance Format** area of the **Tolerances** tab (Figure 10-28) lets you specify the tolerance method, tolerance value, position of tolerance text, and precision and height of the tolerance text. For example, if you do not want a dimension to deviate more than plus 0.01 and minus 0.02, you can specify this by selecting **Deviation** from the **Method** drop-down list and then

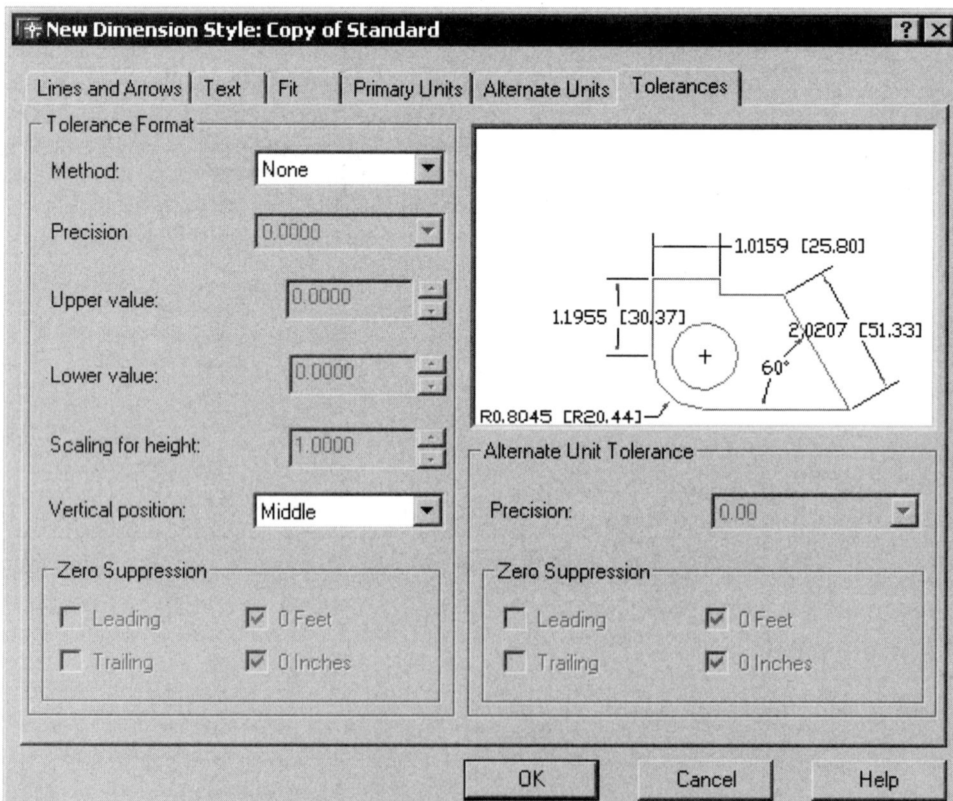

Figure 10-28 ***Tolerances*** *tab of the* ***New Dimension Style*** *dialog box*

specifying the plus and minus deviation in the **Upper Value** and the **Lower Value** edit boxes. When you dimension, AutoCAD will automatically append the tolerance to the dimension. The **DIMTP** variable sets the maximum (or upper) tolerance limit for the dimension text and **DIMTM** variable sets the minimum (or lower) tolerance limit for the dimension text. Different settings and their effects on relevant dimension variables are explained in the following sections.

Method. The **Method** drop-down list lets you select the tolerance method. The tolerance methods supported by AutoCAD are **Symmetrical**, **Deviation**, **Limits**, and **Basic**. These tolerance methods are described next.

None. Selecting the **None** option sets the **DIMTOL** variable to 0 and does not add tolerance values to the dimension text, that is, the **Tolerances** tab is disabled.

Symmetrical. This option is used to specify the symmetrical tolerances. When you select this option, the **Lower Value** spinner is disabled and the value specified in the **Upper Value** spinner is applied to both plus and minus tolerance. For example, if the value specified in the **Upper Value** spinner is 0.05, the tolerance appended to the dimension text is ±0.05, see Figure 10-29. The value of **DIMTOL** is set to 1 and the value of **DIMLIM** is set to 0.

Deviation. If you select the **Deviation** tolerance method, the values in the **Upper Value** and **Lower Value** spinners will be displayed as plus and minus dimension tolerances. If you enter values for the plus and minus tolerances, AutoCAD appends a plus sign (+) to the positive values of the tolerance and a negative sign (–) to the negative values of the tolerance. For example, if the upper value of the tolerance is 0.005 and the lower value of the tolerance is 0.002, the resulting dimension text generated will have a positive tolerance of 0.005 and a negative tolerance of 0.002 (Figure 10-29). Even if one of the tolerance values is 0, a sign is appended to it. On specifying the deviation tolerance, AutoCAD sets the **DIMTOL** variable value to 1 and the **DIMLIM** variable value to 0. The values in the **Upper Value** and **Lower Value** edit boxes are saved in the **DIMTP** and **DIMTM** system variables, respectively.

Limits. If you select the **Limits** tolerance method from the **Method** drop-down list, AutoCAD adds the upper value (contents of the **Upper Value** spinner) to the dimension text (actual measurement) and subtracts the lower value (contents of the **Lower Value** spinner) from the dimension text. The resulting values are displayed as the dimension text, see Figure 10-29. Selecting the **Limits** tolerance method results in setting the **DIMLIM** variable value to 1 and the **DIMTOL** variable value to 0. The numeral values in the **Upper Value** and **Lower Value** edit boxes are saved in the **DIMTP** and **DIMTM** system variables, respectively.

Basic. A basic dimension text is dimension text with a box drawn around it (Figure 10-29). The basic dimension is also called a reference dimension. Reference dimensions are used primarily in geometric dimensioning and tolerances. The basic dimension can be realized by selecting the basic tolerance method. The distance provided around the dimension text (distance between dimension text and the rectangular box) is stored as a negative value in the **DIMGAP** variable. The negative value signifies basic dimension. The default setting is off, resulting in the generation of dimensions without the box around the dimension text.

Precision. The **Precision** drop-down list is used to select the number of decimal places for the tolerance dimension text. The value is stored in **DIMTDEC** variable.

Upper value/Lower value. In the **Upper value** spinner the positive upper or maximum value is specified. If the method of tolerances is symmetrical, the same value is used as the lower value also. The value is stored in the **DIMTP** variable. In the **Lower** spinner the lower or minimum value is specified. The value is stored in the **DIMTM** variable.

Scaling for height. The **Scaling for height** spinner is used to specify the height of the dimension tolerance text relative to the dimension text height. The default value is 1, which means the height of the tolerance text is the same as the dimension text height. If you want the tolerance text to be 75 percent of the dimension height text, enter 0.75 in the **Scaling for height** edit box. The ratio of the tolerance height to the dimension text height is calculated by AutoCAD and then stored in the **DIMTFAC** variable. **DIMTFAC = Tolerance Height/Text Height.**

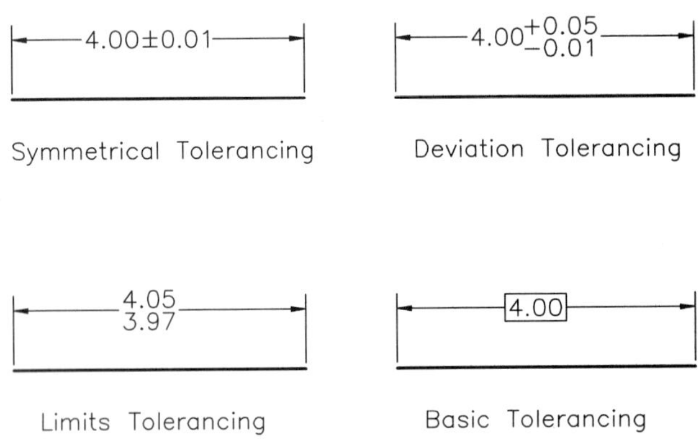

Figure 10-29 Various toleracing methods

Vertical position. This drop-down list allows you to specify the location of the tolerance text for deviation and symmetrical methods only. The three alignments that are possible are with the **Bottom**, **Middle**, or **Top** of the main dimension text. The settings are saved in the **DIMTOLJ** system variable (Bottom=0, Middle=1, and Top=2).

Zero Suppression Area

This area controls the zero suppression in the tolerance text depending on which one of the check boxes is selected. Selecting the **Leading** check box suppresses the leading zeros in all decimal tolerance text. For example, 0.2000 becomes .2000. Selecting the **Trailing** check box suppresses the trailing zeros in all decimal tolerance text. For example, 0.5000 becomes 0.5. Similarly, selecting both the boxes suppresses both the trailing and leading zeros and selecting none, suppresses none. If you select **0 Feet** check box, the zeros in the feet portion of the tolerance dimension text are suppressed if the dimension value is less than one foot. Similarly, selecting the **0 Inches** check box suppresses the zeros in the inches portion of the dimension text. The value is stored in the **DIMTZIN** variable.

Alternate Unit Tolerance Area

The options under this area define the precision and zero suppression settings for Alternate unit tolerance values. The options under this area will be available only when you are displaying the alternate units along with the primary units.

Precision. This drop-down list is used to set the number of decimal places to be displayed in the tolerance text of the alternate dimensions. This value is stored in the **DIMALTTD** variable.

Zero Suppression Area. Selecting the respective check boxes controls the suppression of the **Leading** and **Trailing** zeros in decimal values and the suppression of zeros in the Feet and Inches portions for dimensions in the feet and inches format. The value is stored in the **DIMALTTZ** variable.

Exercise 3 *Mechanical*

Draw Figure 10-30 and then set the values in the various tabs of the **New Dimension Style** dialog box to dimension it as shown. (Baseline spacing = 0.25, Extension beyond dim lines = 0.10, Offset from origin = 0.05, Arrowhead size =0.07, Text height = 0.08.) Assume the missing dimensions.

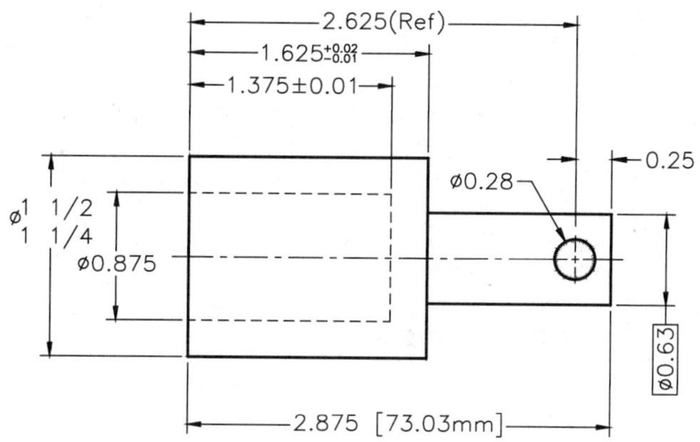

Figure 10-30 *Drawing for Exercise 3*

OTHER DIMENSIONING VARIABLES

Positioning Dimension Text (DIMTVP Variable)

You can position the dimension text with respect to the dimension line by using the **DIMTVP** system variable (Figure 10-31). In certain cases, **DIMTVP** is used with **DIMTAD** to control the vertical position of the dimension text. The **DIMTVP** value applies only when the **DIMTAD** is off. To select the vertical position of the dimension text to meet your requirement (over or under the dimension line), you must first calculate the numerical value by which you want to offset the text from the dimension line. The vertical placing of the text is done by offsetting the dimension text. The magnitude of the offset of dimension text is a product of text height and **DIMTVP** value. If the value of **DIMTVP** is 1, **DIMTVP** acts as **DIMTAD**. For example, if you want to position the text 0.25 units from the dimension line, the value of **DIMTVP** is calculated as follows.

DIMTVP=Relative Position value/Text Height value
DIMTVP=0.25/ 0.09=2.7778

The value 2.7778 is stored in the dimension variable **DIMTVP**. If the absolute value is less than 0.70, the dimension line is broken to accommodate the dimension text. Relative positioning is not effective on angular dimensions.

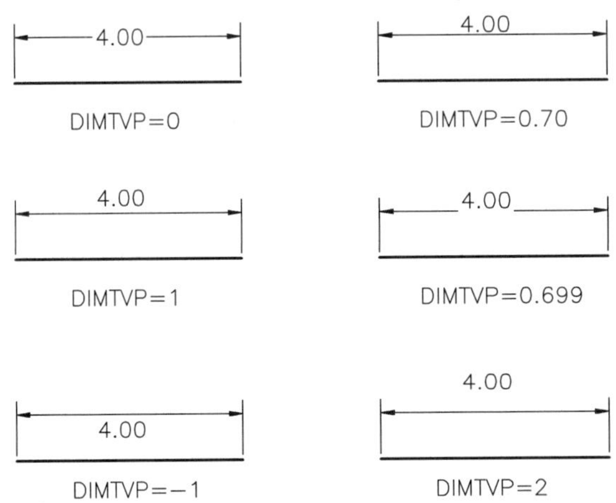

Figure 10-31 *Vertical positioning of dimension text*

DIMENSION STYLE FAMILIES

The dimension style feature of AutoCAD lets the user define a dimension style with values that are common to all dimensions. For example, the arrow size, dimension text height, or color of the dimension line are generally the same in all types of dimensioning such as linear, radial, diameter, and angular. These dimensioning types belong to the same family because they have some characteristics in common. In AutoCAD, this is called a **dimension style family,** and the values assigned to the family are called **dimension style family values.**

After you have defined the dimension style family values, you can specify variations on it for other types of dimensions such as radial and diameter. For example, if you want to limit the number of decimal places to two in radial dimensioning, you can specify that value for radial dimensioning. The other values will stay the same as the family values to which this dimension type belongs. When you use the radial dimension, AutoCAD automatically uses the style that was defined for radial dimensioning; otherwise, it creates a radial dimension with the values as defined for the family. After you have created a dimension style family, any changes in the parent style are applied to family members if the particular property is the same. Special suffix codes are appended to the dimension style family name that correspond to different dimension types. For example, if the dimension style family name is MYSTYLE and you define a diameter type of dimension, AutoCAD will append $4 at the end of the dimension style family name. The name of the diameter type of dimension will be MYSTYLE$4. The following are the suffix codes for different types of dimensioning.

Suffix Code	Dimension Type	Suffix Code	Dimension Type
0	Linear	2	Angular
3	Radius	4	Diameter
6	Ordinate	7	Leader

Example 1 *Mechanical*

The following example illustrates the concepts of dimension style families, Figure 10-32.

1. Specify the values for the dimension style family.
2. Specify the values for linear dimension.
3. Specify the values for diameter and radius dimensions.
4. After creating the dimension style, use it to dimension the given drawing.

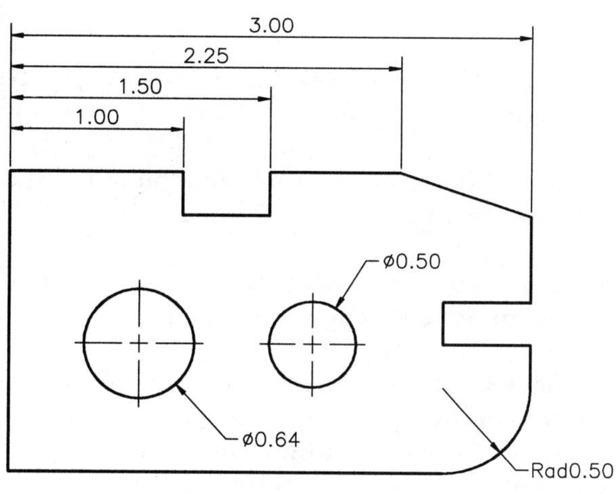

Figure 10-32 *Drawing for Example 1*

1. Open a new file and then draw the object shown in Figure 10-32.

2. Invoke the **Dimension Style Manager** dialog box by choosing the **Dimension Style** button from the **Dimension** toolbar. AutoCAD will display **Standard** in the **Styles** list box. Select the **Standard** style from the **Styles** list box.

3. Choose the **New** button to display the **Create New Dimension Style** dialog box. In this dialog box, enter **MyStyle** in the **New Style Name** edit box. Select **Standard** from the **Start With** drop-down list. Also select **All dimensions** from the **Use for** drop-down list. Now, choose the **Continue** button to display the **New Dimension Style: MyStyle** dialog box. In this dialog box, choose the **Lines and Arrows** tab and enter the following values.

   Baseline Spacing: 0.15 **Extension beyond dim line: 0.07**
   Offset from origin: 0.03 **Arrow size: 0.09**
   Center Mark for circle, Size: 0.05 **Center Mark for Circles Type: Line**

4. Choose the **Text** tab and change the following values:

   Text Height: 0.09 **Offset from dimension line: 0.03**

5. Choose the **Fit** tab and set the value of the **Use overall scale of** spinner in the **Scale for Dimension Features** area to **1**.

6. After entering the values, choose the **OK** button to return to the **Dimension Style Manager** dialog box. This dimension style contains the values that are common to all dimension types.

7. Now, choose the **New** button again in the **Dimension Style Manager** dialog box to display the **Create New Dimension Style** dialog box. AutoCAD displays **Copy of MyStyle** in the **New Style name** edit box. Select **MyStyle** from the **Start with** drop-down list if it is not already selected. From the **Use for** drop-down list, select **Linear dimensions**. Choose the **Continue** button to display the **New Dimension Style: MyStyle: Linear** dialog box and set the following values in the **Text** tab:

   a. Select the **Aligned with dimension line** radio button in the Text alignment area.
   b. In the **Text placement** area, from the **Vertical** drop-down list, select **Above**.

8. In the **Primary Units** tab, set the precision to two decimal places. Choose the **OK** button to return to the **Dimension Style Manager** dialog box.

9. Choose the **New** button again to display the **New Dimension Style** dialog box. Select **MyStyle** from the **Start with** drop-down list. Also select **Diameter dimension** type from the **Use for** drop-down list. Choose the **Continue** button to display the **New dimension Style: MyStyle: Diameter** dialog box.

10. Choose the **Primary units** tab and set the **Precision** to two decimal places. In the **Lines and Arrows** tab, select **Line** from the **Type** drop-down list in the **Center mark for circle** area. Choose the **OK** button to return to the **Dimension Style Manager** dialog box.

11. In this dialog box, choose the **New** button to display the **Create New Dimension Style** dialog box. Select **MyStyle** from the **Start with** drop-down list and **Radius dimensions** from the **Use for** drop-down list. Choose the **Continue** button to display the **New Dimension Style: MyStyle: Radial** dialog box.

12. Choose the **Primary Units** tab and set the precision to two decimal places. Enter **Rad** in the **Prefix** edit box.

13. In the **Fit** tab, select the **Text** radio button in the **Fit Options** area. Choose the **OK** button to return to the **Dimension Style Manager** dialog box.

14. Select **MyStyle** from the **Styles** list box and choose the **Set current** button. Choose the **Close** button to exit the dialog box.

15. Use the linear and baseline dimensioning to draw the linear dimensions as shown in Figure 10-32. You will notice that when you enter any linear dimensioning, AutoCAD automatically uses the values that were defined for the linear type of dimensioning.

16. Use the diameter dimensioning to dimension the circles as shown in Figure 10-32. Again, notice that the dimensions are drawn according to the values specified for the diameter type of dimensioning.

17. Now, use the radius dimensioning to dimension the fillet as shown in Figure 10-32.

USING DIMENSION STYLE OVERRIDES

Most of the dimension characteristics are common in a production drawing. The values that are common to different dimensioning types can be defined in the dimension style family. However, at times you might have different dimensions. For example, you may need two types of linear dimensioning: one with tolerance and one without. One way to draw these dimensions is to create two dimensioning styles. You can also use the dimension variable overrides to override the existing values. For example, you can define a dimension style (**MyStyle**) that draws dimensions without tolerance. Now, to draw a dimension with tolerance or update an existing dimension, you can override the previously defined value. You can override the values through the **Dimension Style Manager** dialog box or by setting the variable values at the Command prompt. The following example illustrates how to use the dimension style overrides.

Example 2 *Mechanical*

In this example, you will update the overall dimension (3.00) so that the tolerance is displayed with the dimension. You will also add two linear dimensions, as shown in Figure 10-33.

This problem can be solved by dimension style overrides as well as using the **Properties** palette. However, here only the dimension style overrides method will be discussed.

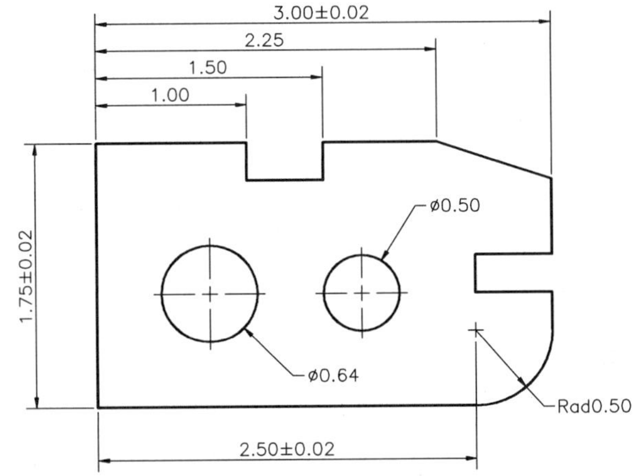

Figure 10-33 *Drawing for Example 2*

1. Invoke the **Dimension Style Manager** dialog box. Select **MyStyle** from the **Styles** list box and choose the **Override** button to display the **Override Current Style: My Style** dialog box. The options in this dialog box are identical to the **New Dimension Style** dialog box discussed earlier in the chapter.

2. Choose the **Tolerance** tab and select **Symmetrical** from the **Method** drop-down list.

3. Set the value of the **Precision** spinner to two decimal places. Set the value of the **Upper value** spinner to **0.02**. Choose the **OK** button to exit the dialog box (this does not save the style). You will notice that **<style overrides>** is displayed under **MyStyle** in the **Style** list box, indicating that the style overrides the **MyStyle** dimension style.

4. This **<style overrides>** is displayed until you save it under a new name or under the style it is displayed under, or until you delete it. Select **<style overrides>** and right-click to display the shortcut menu. Choose the **Save to current Style** option from the shortcut menu to save the overrides to the current style. Choosing the **Rename** option allows you to rename the style override and save it as a new style.

5. Choose **Update** from the **Dimension** menu and select the dimension that measures **3.00**. It will now display the symmetrical tolerance.

6. Draw the remaining two linear dimensions. They will automatically appear with the tolerances, see Figure 10-33.

Tip

*You can also use the **DIMOVERRIDE** command to apply the change to the existing dimensions. Apply the changes to the **DIMTOL, DIMTP,** and **DIMTM** variables.*

COMPARING AND LISTING DIMENSION STYLES

Choosing the **Compare** button in the **Dimension Style Manager** dialog box displays the **Compare Dimension Styles** dialog box where you can compare the settings of two dimensions styles or list all the settings of one of them (Figure 10-34).

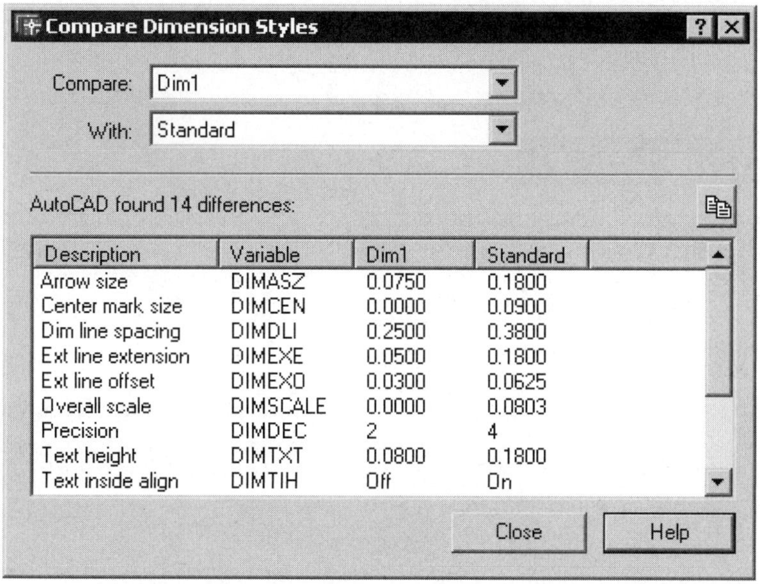

***Figure 10-34 Compare Dimension Styles** dialog*

The **Compare** and the **With** drop-down lists display the dimension styles in the current drawing and selecting dimension styles from the respective lists compare the two styles. In the **With** drop-down list, if you select **None** or the same style as selected from the **Compare** drop-down list, all the properties of the selected style are displayed. The comparison results are displayed under four headings: **Description** of the Dimension Style property, the **System Variable** controlling a particular setting, and the **values of the variable for both the dimension styles** which differ in the two styles in comparison. The number of differences between the selected dimension styles are displayed below the **With** drop-down list. The button provided in this dialog box prints the comparison results to the Windows clipboard from where they can be pasted to other Windows applications.

USING EXTERNALLY REFERENCED DIMENSION STYLES

The externally referenced dimensions cannot be used directly in the current drawing. When you Xref a drawing, the drawing name is appended to the style name and the two are separated by the vertical bar (|) symbol. It uses the same syntax as other externally dependent symbols. For example, if the drawing (FLOOR) has a dimension style called DECIMAL and you Xref this drawing in the current drawing, AutoCAD will rename the dimension style to FLOOR|DECIMAL. You cannot make this dimension style current, nor can you modify or override it. However, you can use it as a template to create a new style. To accomplish this, invoke the **Dimension Style Manager** dialog box. If the **Don't list styles in Xrefs** check box is selected, the styles in the Xref are not displayed. Clear this check box to display the Xref dimension styles and choose the **New...** button. In the **New Style Name:** edit box of the **New Dimension Style** dialog box, enter the name of the dimension style. AutoCAD will create a new dimension style with the same values as those of the externally referenced dimension style (FLOOR|DECIMAL).

Self-Evaluation Test

Answer the following questions, and then compare your answers to the correct answers given at the end of this chapter.

1. You can invoke the **Dimension Style Manager** dialog box using both the **Format** menu and the **Dimension** menu. (T/F)
2. The size of the arrow block is determined by the value stored in the **Arrow size** edit box. (T/F)
3. The default dimension style file name is **Drawing**. (T/F)
4. The size of the tolerance text with respect to the dimensions can be defined. (T/F)
5. The **DIMTVP** variable is used to control the __________ position of the dimension text.
6. When you select the **Arrows** option, AutoCAD places the text and arrowheads __________.
7. A basic dimension text is dimension text with a __________drawn around it.
8. The **Suppress** check boxes in the **Dimension Lines** area control the display of __________ and __________.
9. You can specify the tolerancing using __________ methods.
10. The __________ button in the **Dimension Style Manager** dialog box is used to override the current dimension style.

Review Questions

Answer the following questions.

1. You cannot replace the default arrowheads at the end of the dimension lines. (T/F)

2. When the **DIMTVP** variable has a negative value, the dimension text is placed below the dimension line. (T/F)

3. Dimension style name cannot be changed. (T/F)

4. The named dimension style associated with the dimension being updated by overriding is not updated. (T/F)

5. To use a dimension style for dimensioning you will have to first make it active using which button?

   (a) **Set Current** (b) **New**
   (c) **Override** (d) **Modify**

6. To add a suffix **mm** to the dimensions, which tab of the **Dimension Style Manager** dialog box will you use?

   (a) **Fit** (b) **Text**
   (c) **Primary Units** (d) **Alternate Units**

7. If you want to place the dimension text manually every time you create a dimension, which tab of the **Dimension Style Manager** dialog box will you use?

   (a) **Fit** (b) **Text**
   (c) **Primary Units** (d) **Alternate Units**

8. The size of the __________ is determined by the value stored in the **Arrow size** edit box.

9. When **DIMSCALE** is assigned a value of __________, AutoCAD calculates an acceptable default value based on the scaling between the current model space viewport and paper space.

10. If you use the **DIMCEN** command, a positive value will create a center mark, and a negative value will create a __________.

11. If you select the __________ check box, you can position the dimension text anywhere along the dimension line.

12. You can append a prefix to the dimension measurement by entering the desired prefix in the **Prefix** edit box of the __________ dialog box.

13. If you select the **Limits** tolerance method from the **Method** drop-down list, AutoCAD __________ the upper value to the dimension and __________ the lower value from the dimension text.

14. You can also use the __________ command to override a dimension value.

15. What is the dimension style family, and how does it help in dimensioning? ____________
____.

Exercises

Exercises 4 through 9 *Mechanical*

Make the drawings as shown in Figures 10-35 through 10-40. You must create dimension style files and specify values for different dimension types such as linear, radial, diameter, and ordinate. Assume the missing dimensions.

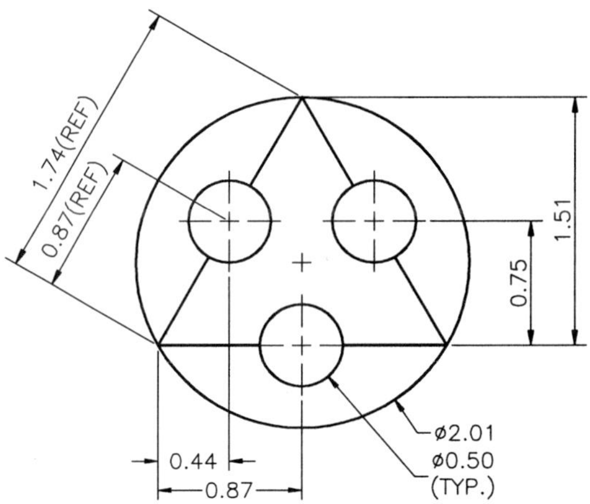

Figure 10-35 *Drawing for Exercise 4*

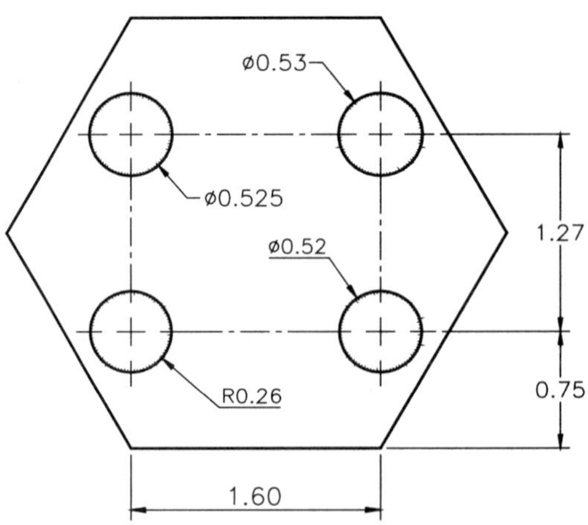

Figure 10-36 *Drawing for Exercise 5*

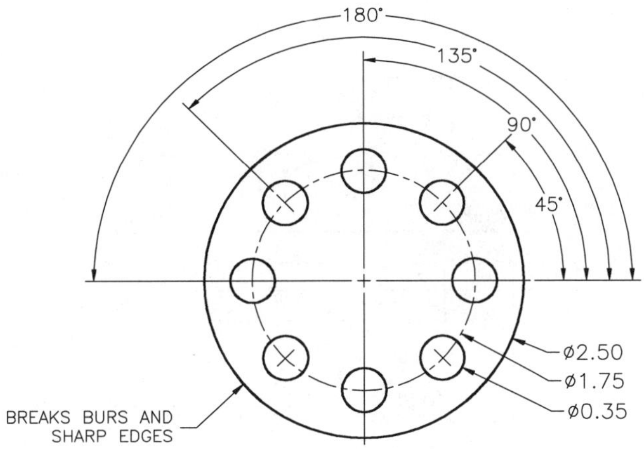

Figure 10-37 Drawing for Exercise 6

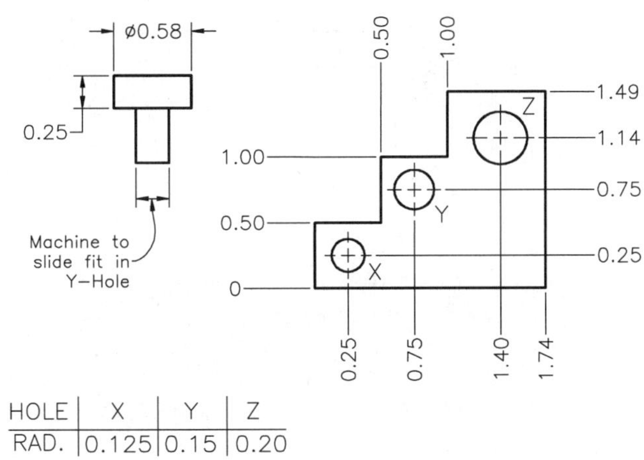

Figure 10-38 Drawing for Exercise 7

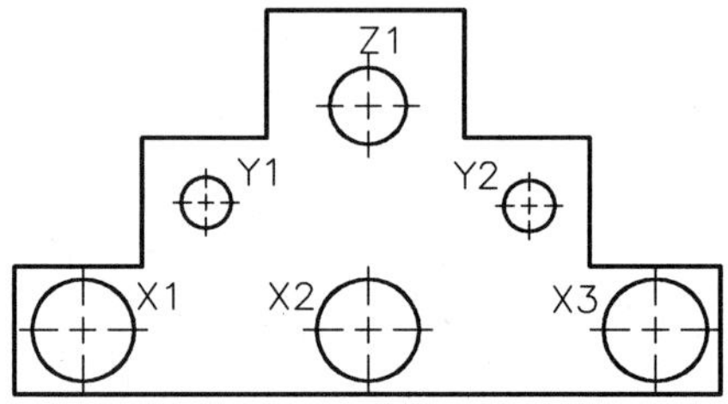

HOLE	X1	X2	X3	Y1	Y2	Z1
DIM.	R0.2	R0.2	R0.2	R0.1	R0.1	R0.15
QTY.	1	1	1	1	1	1
X	0.25	1.375	2.50	0.75	2.0	1.375
Y	0.25	0.25	0.25	0.75	0.75	1.125
Z	THRU	THRU	THRU	1.0	1.0	THRU

Figure 10-39 *Drawing for Exercise 8*

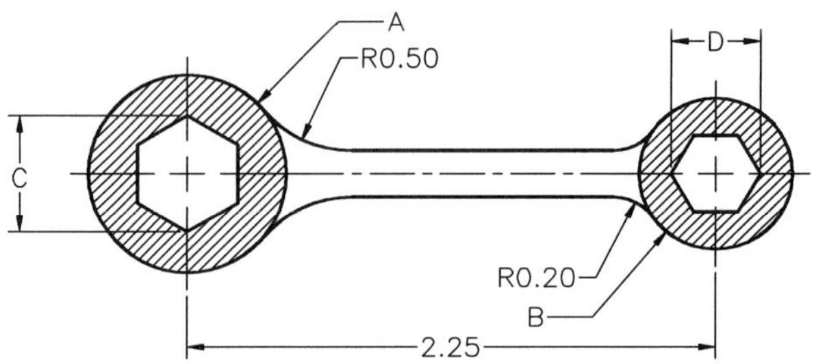

SPANNER NO.	A	B	C	D
S1	0.85	0.65	0.50	0.38
S2	1	0.75	0.59	0.44
S3	1.15	0.88	0.67	0.52
S4	1.25	0.95	0.74	0.56

Figure 10-40 *Drawing for Exercise 9*

Exercise 10 *Mechanical*

Draw the objects shown in Figure 10-41. You must create the dimension style and specify different dimensioning parameters. Also, suppress the leading and trailing zeros in the dimension style. Assume the missing dimensions.

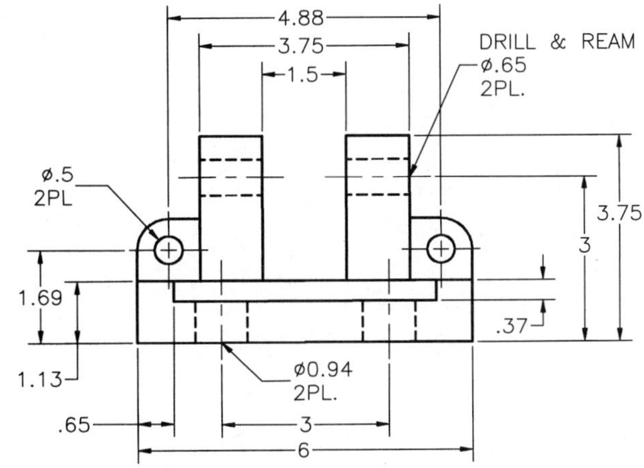

Figure 10-41 *Drawing for Exercise 10*

Exercise 11 *Mechanical*

Draw the objects shown in Figure 10-42. You must create the dimension style and specify different dimensioning parameters in the dimension style. Assume the missing dimensions.

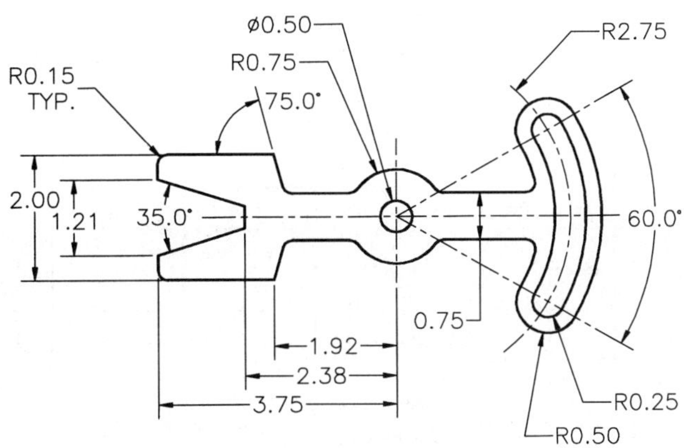

Figure 10-42 *Drawing for Exercise 11*

Exercise 12 *Mechanical*

Draw the objects shown in Figure 10-43. You must create the dimension style and specify different dimensioning parameters in the dimension style. Assume the missing dimensions.

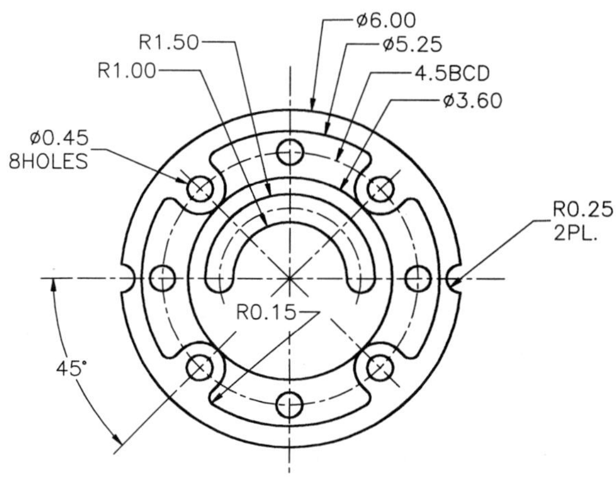

Figure 10-43 Drawing for Exercise 12

Problem Solving Exercise 1 *Mechanical*

Draw the objects shown in Figure 10-44. You must create the dimension style and specify different dimensioning parameters in the dimension style. Also, suppress the leading and trailing zeros in the dimension style. Assume the missing dimensions.

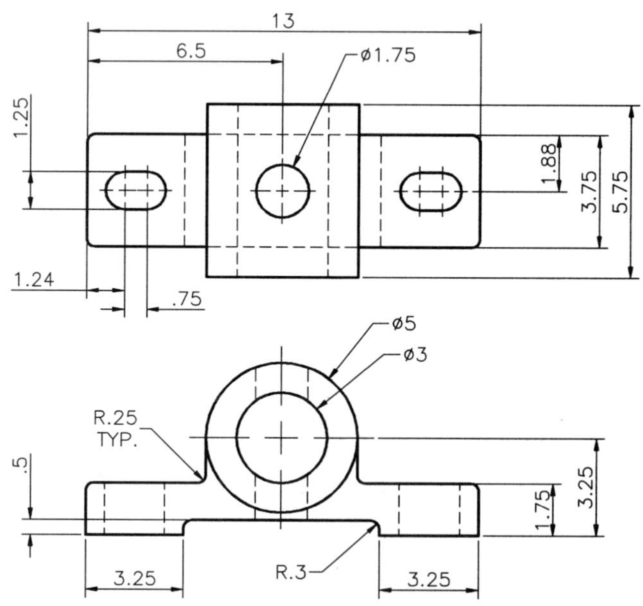

Figure 10-44 Drawing for Problem Solving Exercise 1

Problem Solving Exercise 2 *Mechanical*

Draw the shaft shown in Figure 10-45. You must create the dimension style and specify the dimensioning parameters based on the given drawing. Assume the missing dimensions.

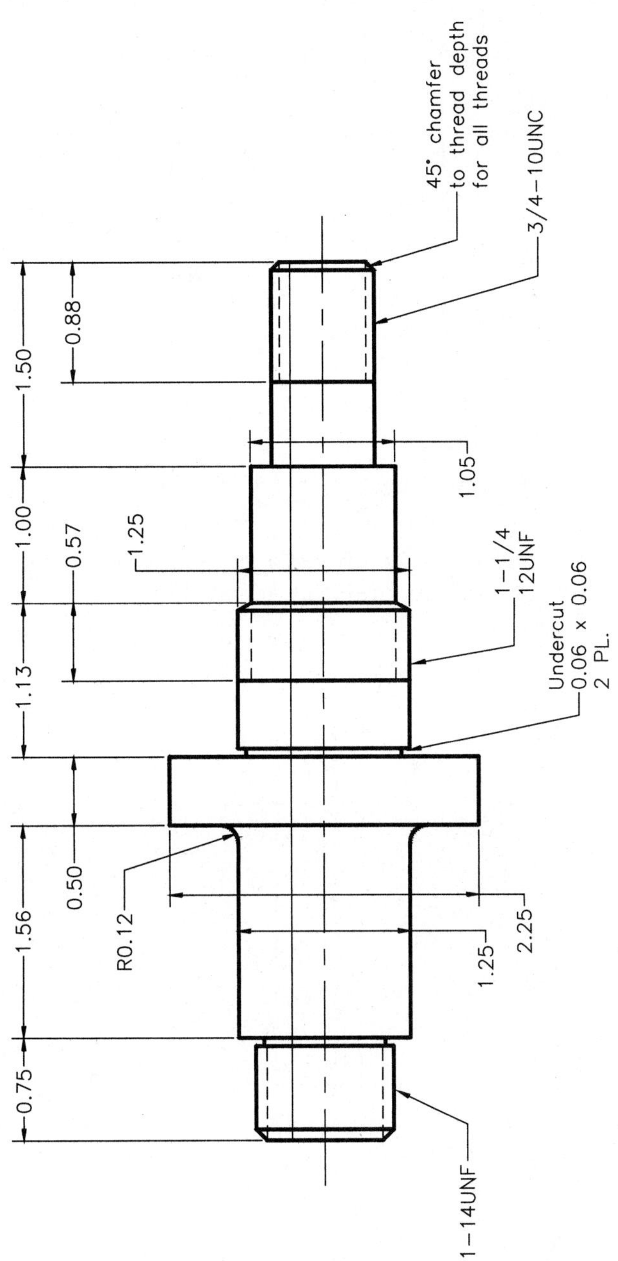

Figure 10-45 *Drawing for Problem Solving Exercise 2*

Problem Solving Exercise 3

Mechanical

Draw the connecting rod shown in Figure 10-46. You must create the dimension style and specify the dimensioning parameters based on the given drawing. Assume the missing dimensions.

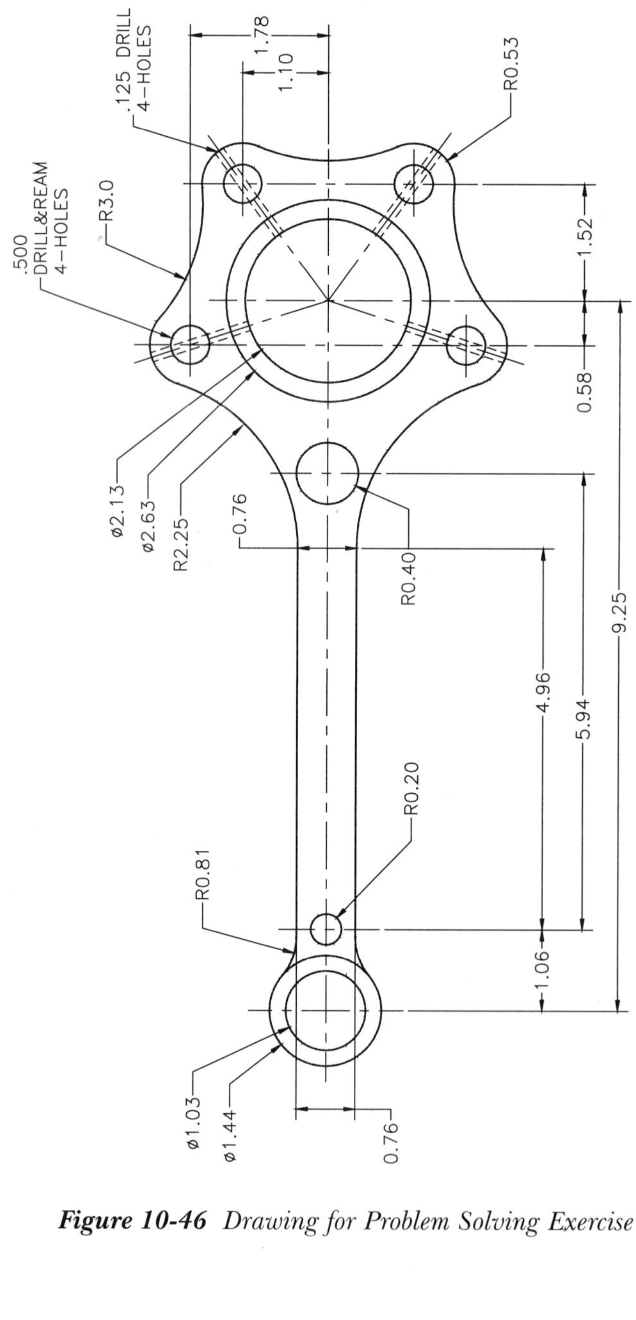

Figure 10-46 *Drawing for Problem Solving Exercise 3*

Problem Solving Exercise 4 *Architectural*

Draw the elevation of the house shown in Figure 10-47. Assume the missing dimensions.

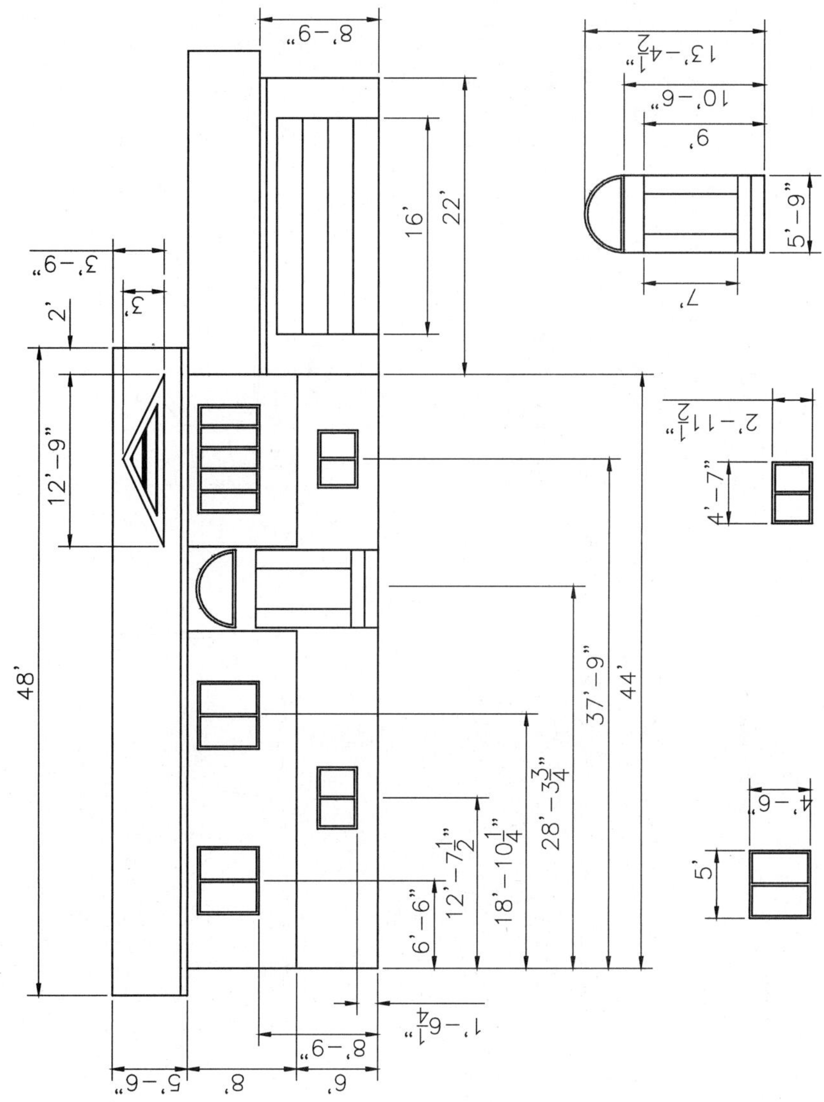

Figure 10-47 *Drawing for Problem Solving Exercise 4*

Problem Solving Exercise 5 *Architectural*

Create the drawing shown in Figure 10-48. Assume the missing dimensions.

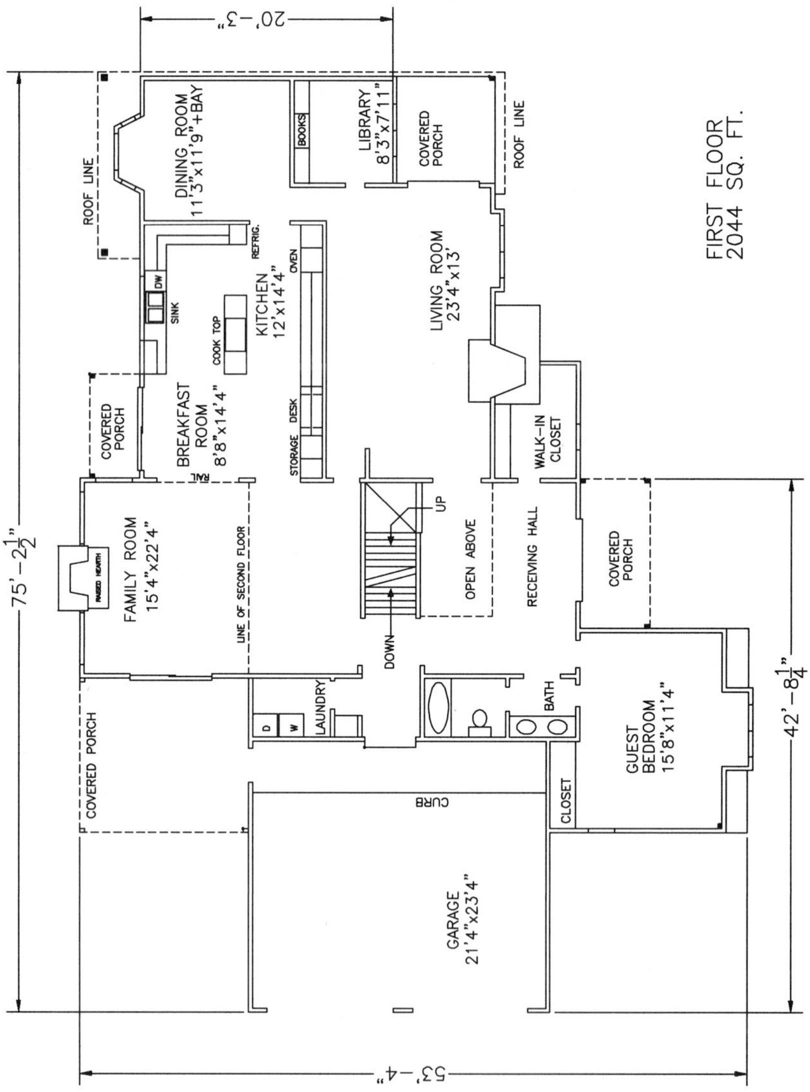

Figure 10-48 *Drawing for Problem Solving Exercise 5*

Problem Solving Exercise 6 — *Architectural*

Create the drawing shown in Figure 10-49. Assume the missing dimensions.

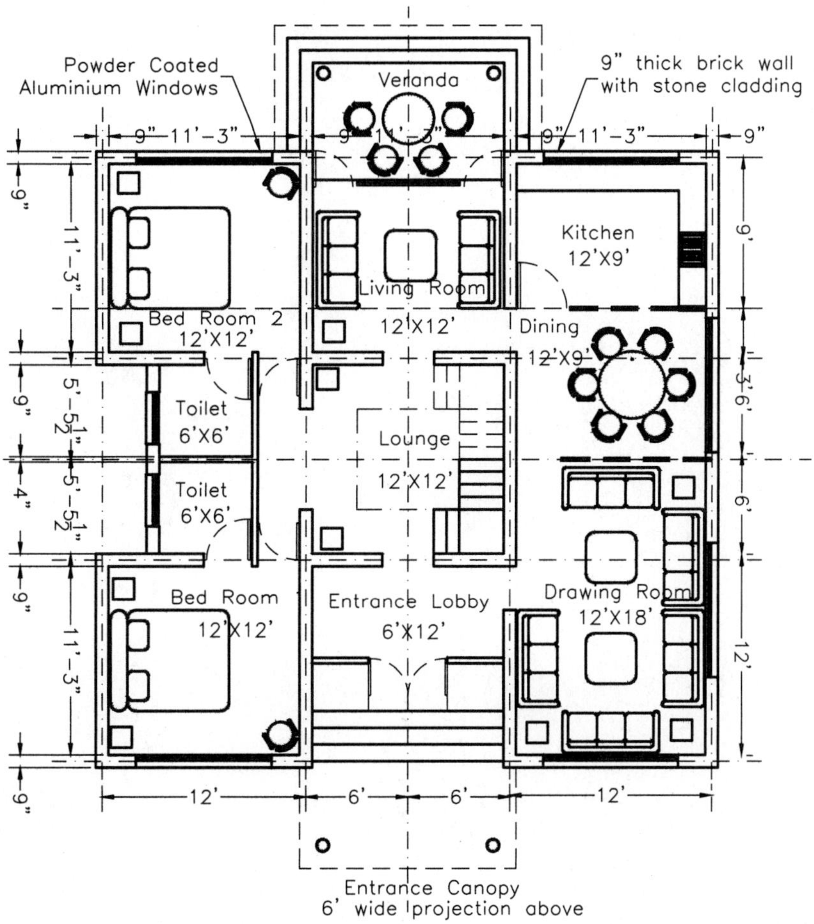

Figure 10-49 Drawing for Problem Solving Exercise 6

Answers to Self-Evaluation Test

1 - T **2** - T, **3** - **F**, **4** - T, **5** - vertical, **6** - inside, **7** - frame, **8** - first, second dimension lines, **9** - four, **10** - **Override**

Chapter 11

Model Space Viewports, Paper Space Viewports, and Layouts

Learning Objectives

After completing this chapter you will be able to:

- *Understand the concepts of model space and paper space.*
- *Create tiled viewports in the model space using various commands.*
- *Create floating viewports in layouts using various commands.*
- *Shift from paper space to model space using the* ***MSPACE*** *command.*
- *Shift from model space to paper space using the* ***PSPACE*** *command.*
- *Control the visibility of viewport layers with the* ***VPLAYER*** *command.*
- *Set linetype scaling in paper space using the* ***PSLTSCALE*** *command.*

MODEL SPACE AND PAPER SPACE/LAYOUTS

For ease in designing, AutoCAD provides two different types of environments. These environments are the model space and the paper space. The paper space is also called layout. The model space is basically used for the designing or the drafting work. This is the default environment that is active when you start using AutoCAD. Almost the entire design is created in the model space. The other environment is the paper space. This environment is used for plotting the drawings or generating the drawing views for the solid models. By default, the paper space provides you with two layouts. A layout can be considered a sheet of paper on which you can place the design created in the model space and then print it. You can also assign different plotting parameters to these layouts for plotting. Almost all the commands of the model space also work in the layouts. Note that you cannot select the drawing objects created in model space when they are displayed in the viewports in layouts. However, you can snap on to the different points of the drawing objects such as the endpoints, midpoints, center points, and so on, using the **OSNAP** options.

You can shift from one environment to the other environment by choosing the **Model** tab or the **Layout1**/**Layout2** tabs provided at the bottom of the drawing window. You can also shift from one environment to the other using the **TILEMODE** system variable. The default value of this variable is **1**. If the value of this system variable is set to **0**, you will be shifted to the layouts and if its value is set to **1**, you will be shifted to model space. The viewports created in model space are called **tiled viewports** and viewports in layouts are called **floating viewports**.

MODEL SPACE VIEWPORTS (TILED VIEWPORTS)

A viewport in the model space is defined as a rectangular area of the drawing window in which you can create the design. When you start AutoCAD, only one viewport is displayed in the model space. You can create multiple nonoverlapping viewports in the model space that display different views of the same object, see Figure 11-1. Each of the viewports will become the individual drawing area. This is generally used while creating the solid models. You can view the same solid model from different positions by creating the tiled viewports and define the distinct coordinate system configuration for each viewport. You can also use the **PAN** or **ZOOM** command to display different portions or different levels of detail of the drawing in each viewport. The tiled viewports can be created using the **VPORTS** command.

Creating Tiled Viewports

Toolbar:	Viewports > Display Viewports Dialog
Menu:	View > Viewports > New Viewports
Command:	VPORTS

As mentioned earlier, the display screen in the model space can be divided into multiple nonoverlapping tiled viewports. All these viewports are created only in the rectangular shape. The number of these viewports depend on the equipment and the operating system on which AutoCAD is running. Each tiled viewport contains a view of the drawing. The tiled viewports touch each other at the edges without overlapping. While using tiled viewports, you are not allowed to edit, rearrange, or turn individual viewports on or off. These viewports

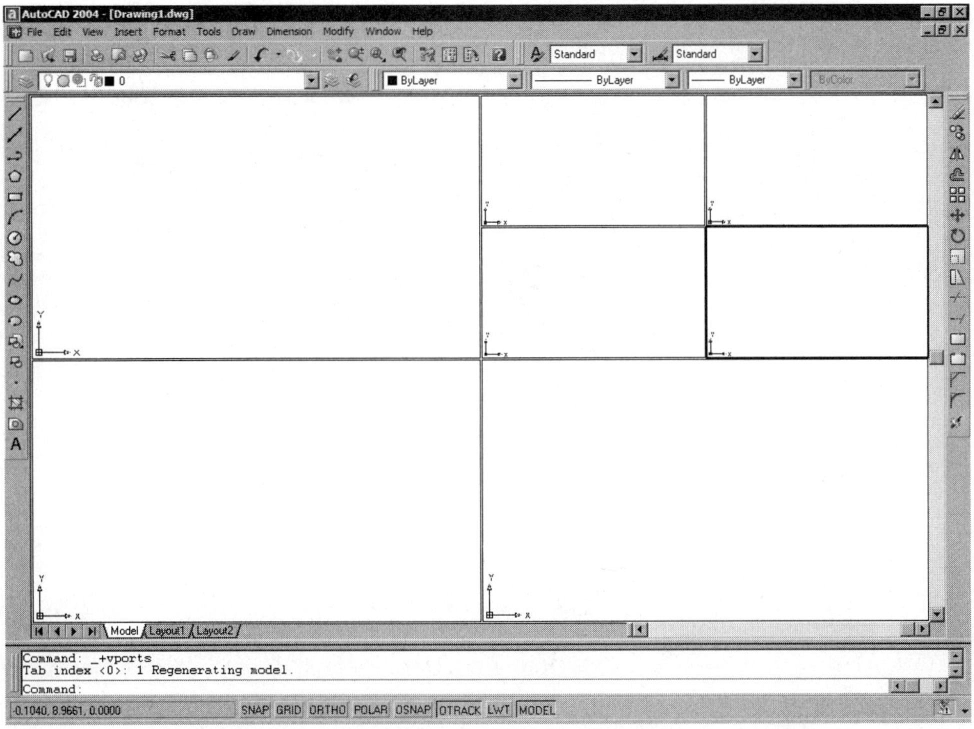

Figure 11-1 Screen display with multiple tiled viewports

are created using the **VPORTS** command when the system variable **TILEMODE** is set to 1 or the **Model** tab is active. When you choose the **Display Viewports Dialog** button (Figure 11-2), the **Viewports** dialog box is displayed. You can use this dialog box to create new viewport configurations and save them. The options under both the tabs of the **Viewports** dialog box are discussed next.

Figure 11-2 Displaying the ***Viewports*** *dialog box using the* ***Viewports*** *toolbar*

New Viewports Tab

The **New Viewports** tab of the **Viewports** dialog box (Figure 11-3) provides the options related to the standard viewport configurations that can be used. You can also save a user-defined configuration using this tab. The name for the new viewport configuration can be specified in the **New name** edit box. If you do not enter a name in this edit box, the viewport configuration you create is not saved and, therefore, cannot be used later. A list of standard viewport configurations is listed in the **Standard viewports** list box. This list also contains the ***Active Model Configuration***, which is the current viewport configuration. From the **Standard viewports** list, you can select and apply any one of the listed standard viewport configurations. A preview image of the selected configuration is displayed in the

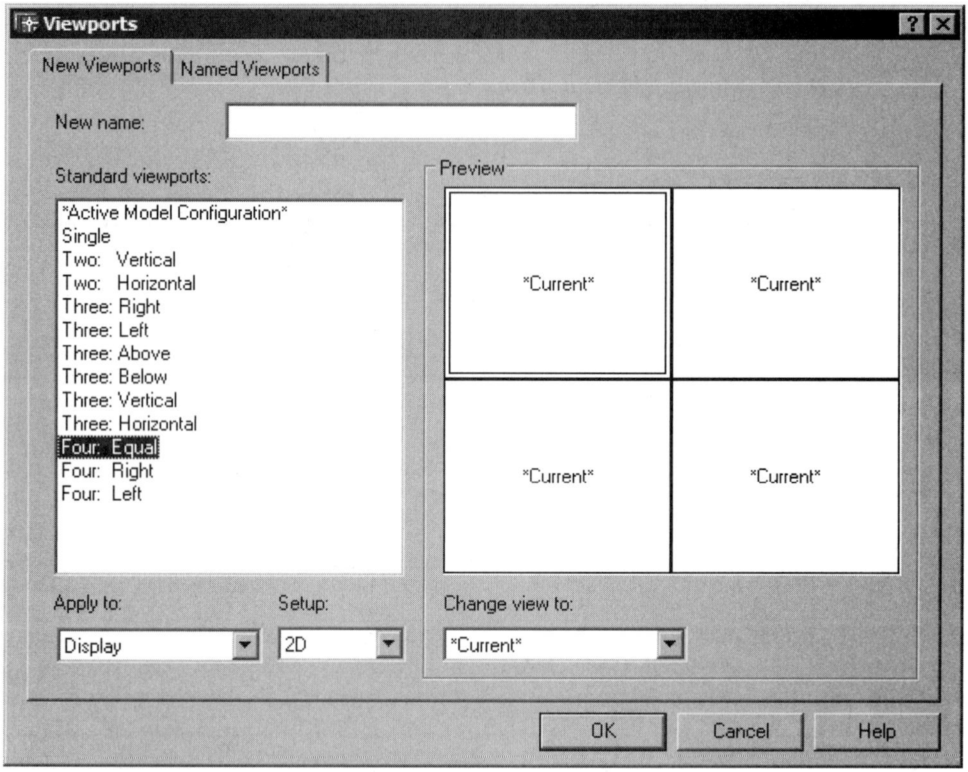

***Figure 11-3** The **New Viewports** tab of the **Viewports** dialog box*

Preview window. The **Apply to** drop-down list has the **Display** and **Current viewport** options. Selecting the **Display** option applies the selected viewport configuration to the entire display and selecting the **Current viewport** option applies the selected viewport configuration to only the current viewport. With this option you can create more viewports inside the existing viewports. The changes will be applied to the current viewport and the new viewports will be created inside the current viewport.

From the **Setup** drop-down list, you can select **2D** or **3D**. When you select the **2D** option, it creates the new viewport configuration with the current view of the drawing in all the viewports initially. Using the **3D** option applies the standard orthogonal and isometric views to the viewports. For example, if a configuration has four viewports, they are assigned the **Top**, **Front**, **Right**, and **South East Isometric** views respectively. You can also modify these standard orthogonal and isometric views by selecting from the **Change view to** drop-down list and replacing the existing view in the selected viewport. For example, you can select the viewport that is assigned the **Top view** and then choose **Bottom view** from the **Change view to** drop-down list to replace it. The Preview image in the **Preview** window reflects the changes you make. If you use the **2D** option, you can select a named viewport configuration to replace the selected one. Choose **OK** to exit the dialog box and apply the created or selected configuration to the current display in the drawing. When you save a new viewport configuration, it saves information about the number and position of viewports, the viewing direction and zoom factor, and the grid, snap, coordinate system, and UCS icon settings of the viewports.

Named Viewports Tab

The **Named Viewports** tab of the **Viewports** dialog box (Figure 11-4) displays the name of the current viewport next to **Current name**. The names of all the saved viewport configurations in a drawing are displayed in the **Named viewports** list box. You can select any one of the named viewport configurations and apply it to the current display. A preview image of the selected configuration is displayed in the **Preview** window. Choose **OK** to exit the dialog box and apply the selected viewport configuration to the current display. In the **Named viewports** list box, you can select a name and right-click to display a shortcut menu. Choosing **Delete** deletes the selected viewport configuration and choosing **Rename** allows you to rename the selected viewport configuration.

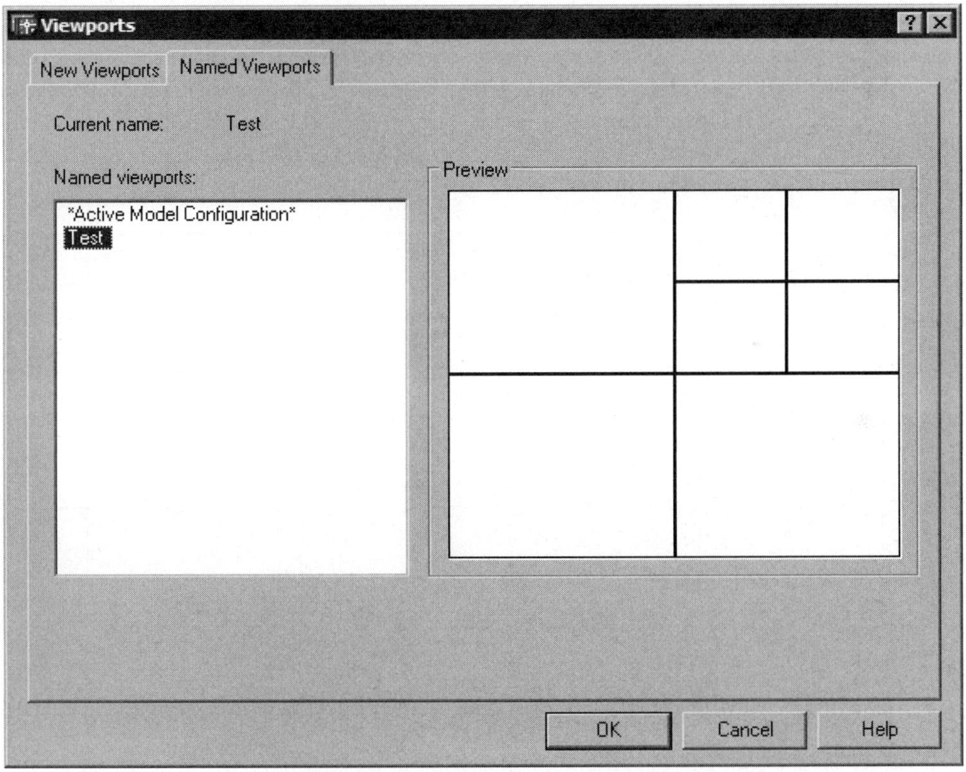

*Figure 11-4 The **Named Viewports** tab of the **Viewports** dialog box*

MAKING A VIEWPORT CURRENT

The viewport you are working in is called the current viewport. You can display several model space viewports on the screen, but you can work in only one of them at a time. You can switch from one viewport to another even when you are in the middle of a command. For example, you can specify the start point of the line in one viewport and the endpoint of the line in the other viewport. The current viewport is indicated by a border that is heavy compared to the borders of the other viewports. Also, the graphics cursor appears as a drawing cursor (screen crosshairs) only when it is within the current viewport. Outside the current viewport this cursor appears as an arrow cursor. You can enter points and select objects only from the current viewport. To make a viewport current, you can select it with the pointing device. Another method of making a viewport current is by assigning its identification number to the **CVPORT** system variable. The identification numbers of the named viewport configurations are not listed in the display.

JOINING TWO ADJACENT VIEWPORTS

AutoCAD provides you with an option of joining two adjacent viewports. However, remember that the viewports you wish to join should result in a rectangular-shaped viewport only. As mentioned earlier, the viewports in model space can only be in rectangular shape. Therefore, you will not be able to join two viewports in case they do not result in a rectangular shape. The viewports can be joined by choosing **View > Viewports > Join** from the menu bar. Upon choosing this, you will be prompted to select the dominant viewport. A dominant viewport is the one whose display will be retained after joining. After selecting the dominant viewport, you will be prompted to select the viewport to join. Figure 11-5 shows a viewport configuration before joining and Figure 11-6 shows a viewport configuration after joining.

Figure 11-5 *Selecting the dominant viewport and the viewport to join*

Tip

*You can also use the Command line to create, save, restore, delete, or join the viewport configurations. This is done using the **-VPORTS** command.*

Figure 11-6 *Viewports after joining*

PAPER SPACE VIEWPORTS (FLOATING VIEWPORTS)

As mentioned earlier, the viewports created in the layouts are called floating viewports. The viewports in the layouts are called floating viewports because unlike model space, the viewports in the layouts can be overlapping and can be of any shape. In layouts there is no restriction of the shape of the viewports. You can even convert a closed object into a viewport in the layouts. Figure 11-7 shows a layout with floating viewports.

The method of creating the floating viewports is discussed next.

Creating Floating Viewports

Toolbar:	Viewports > Display Viewports Dialog
Menu:	View > Viewports > New Viewports
Command:	VPORTS

This command is used for creating the floating viewports in layouts. However, when you invoke this command in the layouts, the dialog box displayed is slightly modified. Instead of the **Apply to** drop-down list in the **New Viewports** tab, the **Viewport**

Figure 11-7 *Screen display with multiple floating viewports*

Spacing spinner is displayed, see Figure 11-8. This spinner is used to set the spacing between the adjacent viewports. The rest of the options in both the **New Viewports** and the **Named Viewports** tabs of the **Viewports** dialog box are the same as those discussed under the tiled viewports. When you select a viewport configuration and choose **OK**, you will be prompted to specify the first and the second corner of a box that will act as a reference for placing the viewports. You will also be given an option of **Fit**. This option fits the configuration of viewports such that they fit exactly in the current display.

Tip

*You can also use the **+VPORTS** command to display the **Viewports** dialog box. When you invoke this command, you will be prompted to specify the **Tab Index**. Enter **0** to display the **New Viewports** tab and enter **1** to display the **Named Viewports** tab.*

Creating Polygonal Viewports

As mentioned earlier, you can create floating viewports with any closed shape. To create a polygonal viewport, choose **View > Viewports > Polygonal Viewport** from the menu bar. The viewports thus created can even be self-intersecting in shape. The prompt sequence that will follow when you choose this is given next.

Specify start point: *Specify the start point of the viewport.*

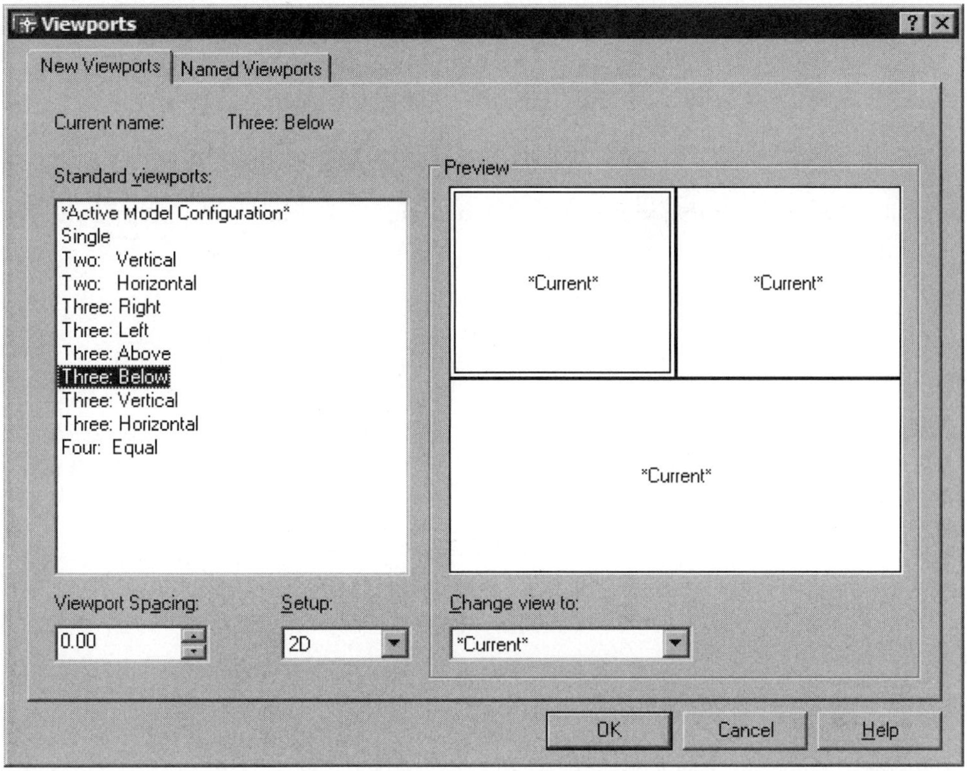

Figure 11-8 *The **New Viewports** tab of the **Viewports** dialog box displayed in layouts*

Specify next point or [Arc/Length/Undo]: *Specify the next point or select an option.*
Specify next point or [Arc/Close/Length/Undo]: *Specify the next point or select an option.*

The various options available in the prompt sequence are discussed next.

Arc

This option is used to switch to the arc mode for creating the viewports. When you invoke this option, the various options for creating the arcs will be displayed. You can switch back to the line mode by choosing the **Line** option.

Length

This option is used to specify the length of the next line of the viewport. The line will be drawn in the direction of the last drawn line segment. In case the last drawn segment was an arc, the line will be drawn tangent to the arc.

Undo

This option is used to undo the last drawn segment of the polygonal viewport.

Close

This option is used to close the polygon and create the viewport. The last entity that will be used to close the polygon will depend upon whether you were in the arc mode or in the line mode. If you were in the line mode, the last entity will be a line. If you were in the arc mode, the last entity will be an arc.

Figure 11-9 shows a polygonal viewport created using a combination of lines and arcs.

Figure 11-9 *A polygonal viewport*

Converting an Existing Closed Object into a Viewport

One of the major enhancements in the recent releases of AutoCAD is the option that allows you to convert an existing closed object into a viewport. However, remember that the object selected should be a single entity. The objects that can be converted into a viewport include polygons drawn using the **POLYGON** command, rectangles drawn using the **RECTANG** command, closed polylines, circles, ellipses, closed splines, or regions. To convert any of these objects into a viewport, choose **View > Viewports > Objects** from the menu bar. When you choose this, you will be prompted to select the object that will be converted into a viewport. Figure 11-10 shows a viewport created using a polygon of nine sides and Figure 11-11 shows a viewport created using a closed spline.

Note

*When you shift to the layouts, the **Page Setup** dialog box is displayed for printing and a rectangular viewport is created that fits the drawing area. If you want, you can retain this viewport or you can also delete this viewport using the **ERASE** command.*

TEMPORARY MODEL SPACE

Sometimes, when you create a floating viewport in the layout, the drawing is not displayed completely inside it, see Figures 11-10 and 11-11. In such cases you need to zoom or pan the drawings to fit them in the viewport. But when you invoke any of the **ZOOM** or the **PAN**

Figure 11-10 *A viewport created from a polygon*

Figure 11-11 *A viewport created from a closed spline*

commands in the layouts, the display of the entire layout is modified instead of the display inside the viewport. Now, to change the display of the viewports, you will have to switch to the temporary model space. The temporary model space is defined as a state when the model space is activated in the layouts. The temporary model space is exactly similar to the actual model space and you can make any kind of modifications in the drawing from temporary model space. Therefore, the main reason for invoking the temporary model space is that you can modify the display of the drawing. The temporary model space can be invoked by choosing the **Paper** button from the status bar. You can also switch to the temporary model space by double-clicking inside the viewports. You will see that the model space UCS icon is automatically displayed when you switch to the temporary model space. When you switch to the temporary model space, the extents of the viewport become the extents of the drawing. You can use the **ZOOM** and **PAN** commands to fit the model inside the viewport. The temporary model space can also be invoked using the **MSPACE** command.

Once you have modified the display of the drawing in the temporary model space, you have to switch back to the paper space. This is done by choosing the **Model** button from the status bar. You can also switch back to the paper space by double-clicking anywhere in the layout outside the viewport. This can also be done using the **PSPACE** command.

Example 1 *Mechanical*

In this example you will draw the object shown in Figure 11-12 and then create a floating viewport of the shape shown in Figure 11-13 to display the object in the layouts. The dimensions of the viewport are in the paper space. Do not dimension the object.

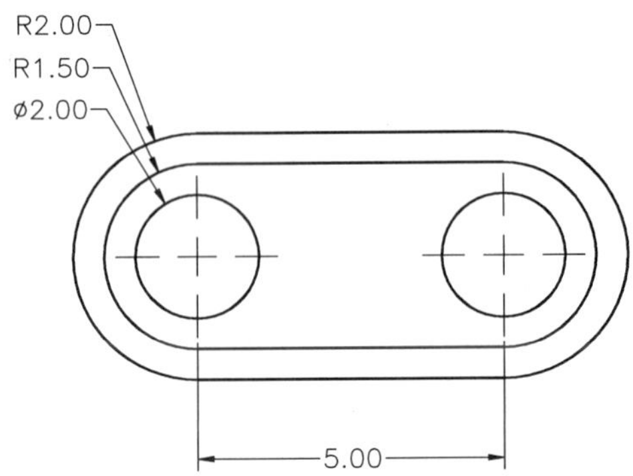

Figure 11-12 *Model for Example 1*

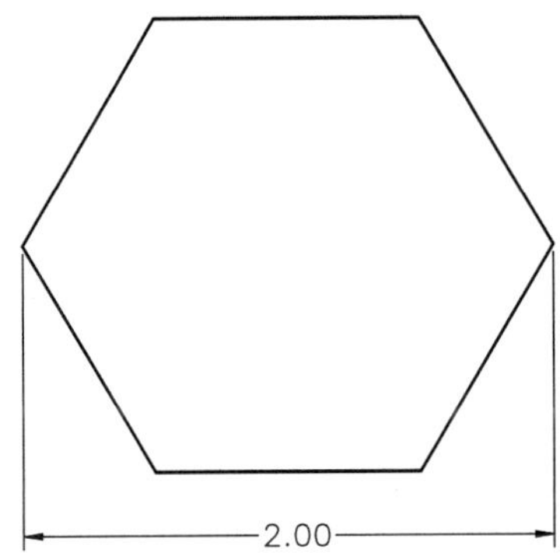

Figure 11-13 *Shape for the viewport*

1. Open a new drawing and then draw the object shown in Figure 11-12.

2. Choose the **Layout1** tab to switch to the layouts. The **Page Setup** dialog box for plotting the drawing will be displayed. Choose the **OK** button to exit this dialog box.

3. A rectangular viewport will be automatically created in this layout. Choose the **Erase** button from the **Modify** toolbar. You will be prompted to select the object. Enter **L** in this prompt to delete the last object, that is, the viewport in this case.

4. Choose the **Polygon** button from the **Draw** toolbar and create the required hexagon.

5. Choose **View > Viewports > Object** from the menu bar. You will be prompted to select the object. Select the hexagon and it will be converted into a viewport.

6. You will see that the complete object is not displayed inside the viewport. Therefore, you have to modify the display of this viewport. Double-click inside the viewport to switch to the temporary model space.

7. The viewport border will become thick, suggesting that you have switched to the temporary model space. Now, using the **ZOOM** and the **PAN** commands, fit the drawing inside the viewport.

8. Choose the **Model** button from the status bar to switch back to the paper space. The drawing is now displayed fully inside the viewport, see Figure 11-14.

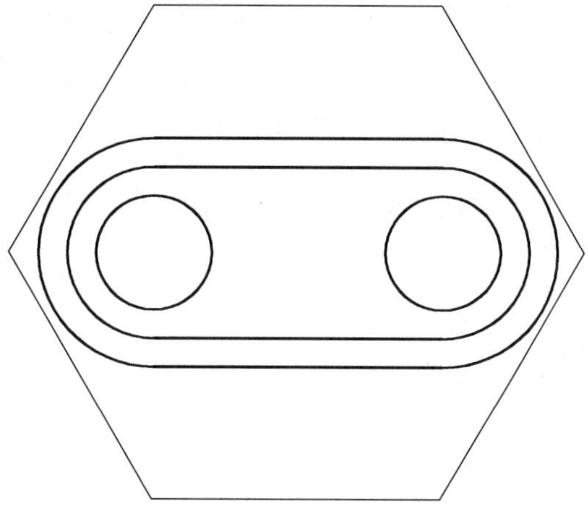

Figure 11-14 *Displaying the drawing inside polygonal viewport*

EDITING THE VIEWPORTS

You can perform various editing operations on the viewports. For example, you can control the visibility of the objects in the viewports, lock the display of the viewports, clip the existing viewports using an object, and so on. All these editing operations are discussed next.

Controlling the Display of the Objects in the Viewports

The display of the objects in the viewports can be turned on or off. If the display is turned off, the objects will not be displayed in the viewport. However, the object in the model space is not affected by this editing operation. To control the display of the objects, select the viewport entity in the layout and right-click to display the shortcut menu. In this menu choose **Display Viewport Objects > No**. If there is only one viewport, you will be prompted to confirm whether you really want to turn off all active viewports. Enter Y to turn off the display. However, if there are more than one viewports, the visibility of the selected viewports will be automatically turned off when you choose **Display Viewport Objects > No** from the shortcut menu. Similarly, you can again turn on the display of the objects by choosing **Display Viewport Objects > On** from the shortcut menu. This shortcut menu is displayed upon selecting the viewport and right-clicking. This editing operation can also be done using the **OFF** option of the **MVIEW** or the **-VPORTS** command.

Locking the Display in the Viewports

To avoid the accidental modification in the display of objects in the viewports, you can lock their display. If the display of a viewport is locked, the commands such as **ZOOM** and **PAN** do not work in it. Also, you cannot modify the view in the locked viewport. For example, if the display of a viewport is locked you cannot zoom or pan the display in that viewport or change the view in that viewport even if you switch to the temporary model space. To lock the display of the viewports, select the viewport and right-click on it to display the shortcut menu. In this menu choose **Display Locked > Yes**. Now, the display of this viewport will not be modified. However, you can draw objects or delete objects in this viewport by switching to the temporary model space. Similarly, you can unlock the display of the objects in the viewports by choosing **Display Locked > No** from the shortcut menu. You can also lock or unlock the display of the viewports using the **Lock** option of the **MVIEW** command or the **-VPORTS** command.

Controlling the Display of the Hidden Lines in the Viewports

While working with three-dimensional solid or surface models, there are a number of occasions where you have to plot the solid models such that the hidden lines are not displayed. If you are plotting the solid models in the model space (Tilemode=1), it can be easily done by selecting the **Hidden** option from the **Shade plot** drop-down list in the **Shaded viewport options** area of the **Plot Settings tab** of the **Plot** dialog box. But if you are plotting the drawings in the layouts, this option is not available. In this case you will have to control the display of the hidden lines in the viewports. To control the display of hidden lines, select the viewport and right-click to display the shortcut menu. In this menu choose **Shade Plot > Hidden**. Although the hidden lines will be displayed in the viewports, now they will not be displayed in the printouts. The display of the hidden lines can also be controlled using the **Shadeplot** option of the **MVIEW** command or the **-VPORTS** command.

Note

*You can also use the other options in the **Shade Plot** shortcut menu. These options are explained in detail in Chapter 12 (Plotting Drawings).*

Tip

Apart from the previously mentioned editing operations, you can also move, copy, rotate, stretch, scale, or trim the viewports using the respective commands. You can also use the grips to edit the viewports using the grips.

Clipping the Existing Viewports

Toolbar:	Viewports > Clip Existing Viewport
Command:	VPCLIP

You can modify the shape of the existing viewport by clipping it using an object or by defining the clipping boundary. The viewports can be clipped using the **VPCLIP** command. This command can also be invoked by choosing **Viewport Clip** from the **Viewports** toolbar. The prompt sequence that will follow when you invoke this command is given next.

Select viewport to clip: *Select the viewport to be clipped.*
Select clipping object or [Polygonal/Delete] <Polygonal>: *Select an object for clipping the viewport or specify an option.*

Select clipping object Option

This option is used to clip the viewport using a selected closed loop. The objects that can be used for clipping the viewports include circles, ellipses, closed polylines, closed splines, and regions. As soon as you select the clipping object, the original viewport will be deleted and the selected object will be converted into a viewport. The portion of the display that was common to both the original viewport and the object selected will be displayed. You can, however, change the display of the viewport using the **ZOOM** command or the **PAN** command in the viewports. Figure 11-15 shows a viewport and an object that will be used to clip the viewport and Figure 11-16 shows the new viewport created after clipping.

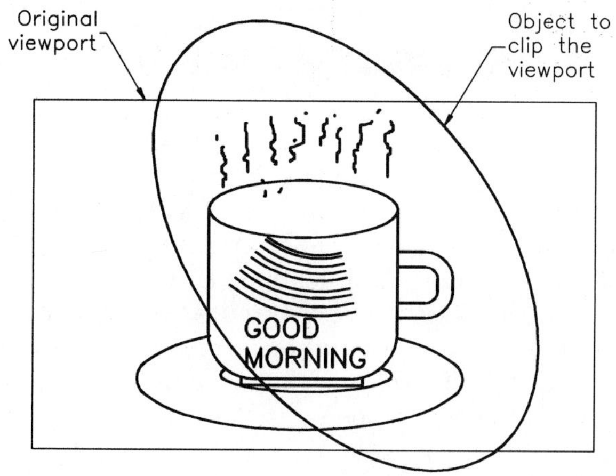

Figure 11-15 *Selecting the object for clipping the viewport*

Figure 11-16 *New viewport created after clipping*

Polygonal

This option is used to create a polygonal boundary for clipping the viewports. When you invoke this option, various options for creating a polygonal boundary will be displayed. You can draw a polygonal boundary for clipping the viewport using these options.

Delete

This option is used to delete the new clipping boundary created using an object or using the **Polygonal** option. The original viewport is restored when you invoke this option. Keep in mind that this option will be available only if the viewport has been clipped at least once. If the viewport is clipped more than once, you can restore only the last viewport clipping boundary using this option.

MANIPULATING THE VISIBILITY OF VIEWPORT LAYERS

Command: VPLAYER

You can control the visibility of layers inside a floating viewport with the **VPLAYER** or **LAYER** command. The **On/Off** or **Freeze/Thaw** option of the **LAYER** command controls the visibility of layers globally, including the viewports. However, with the **VPLAYER** command you can control the visibility of layers in individual floating viewports. For example, you can use the **VPLAYER** command to freeze a layer in the selected viewport. The contents of this layer will not be displayed in the selected viewport, although in the other viewports, the contents are displayed. This command can be used from either temporary model space or paper space. The only restriction is that **TILEMODE** be set to 0 (Off); that is, you can use this command only in the **Layout** tab.

Command: **VPLAYER** Enter
Enter an option [?/Freeze/Thaw/Reset/Newfrz/Vpvisdflt]:

? Option

You can use this option to obtain a listing of the frozen layers in the selected viewport. When you enter ?, you will be prompted to select the viewport. The AutoCAD text window will be displayed showing all the layers that are frozen in the current layer. If you invoke this option when you are in temporary model space, AutoCAD will temporarily shift you to paper space to let you select the viewport.

Freeze Option

The **Freeze** option is used to freeze a layer (or layers) in one or more viewports. When you select this option, you will be prompted to specify the name(s) of the layer(s) you want to freeze. If you want to specify more than one layer, the layer names must be separated by commas. You can also select an object whose layer you want to freeze. Once you have specified the name of the layer(s), AutoCAD prompts you to select the viewport(s) in which you want to freeze the specified layer(s). You can select one or all the viewports. The layers will be frozen after you exit this command.

Thaw Option

With this option, you can thaw the layers that have been frozen in viewports using the **VPLAYER** command or the **LAYER** command. Layers that have been frozen, thawed, turned on, or turned off globally are not affected by **VPLAYER Thaw**. For example, if a layer has been frozen, the objects on the frozen layer are not regenerated on any viewport even if **VPLAYER Thaw** is used to thaw that layer in any viewport. If you want to thaw more than one layer, they must be separated by commas. You can thaw the specified layers in the current, selected, or all the viewports.

Reset Option

With the **Reset** option, you can set the visibility of layer(s) in the specified viewports to their current default setting. The visibility defaults of a layer can be set by using the **Vpvisdflt** option of the **VPLAYER** command. When you invoke this option, you will be prompted to specify the names of the layers to be reset. You can reset the layers in the current viewport, in selected viewports, or in all the viewports.

Newfrz (New Freeze) Option

With this option, you can create new layers that are frozen in all the viewports. This option is used mainly where you need a layer that is visible only in one viewport. This can be accomplished by creating the layer with the **Newfrz** option and then thawing that particular layer in the viewport where you want to make the layer visible. When you invoke this option, you will be prompted to specify the name(s) of the new layer(s) that will be frozen in all the viewports. If you want to specify more than one layer, separate the layer names with commas. After you specify the name(s) of the layer(s), AutoCAD creates frozen layers in all viewports. Also, the default visibility setting of the new layer(s) is set to Frozen; therefore, if you create any new viewports, the layers created with **VPLAYER Newfrz** are also frozen in new viewports.

Vpvisdflt (Viewport Visibility Default) Option

With this option, you can set a default for the visibility of layer(s) in the subsequent new viewports. When a new viewport is created, the frozen/thawed status of any layer depends on the **Vpvisdflt** setting for that particular layer. When you invoke this option, you will be prompted to specify the names of the layer(s) whose visibility is to be changed. After specifying the name(s), you will be prompted to specify whether the layers should be frozen or thawed in the new viewports.

CONTROLLING THE LAYERS IN THE VIEWPORTS USING THE LAYER PROPERTIES MANAGER DIALOG BOX

You can use the **Layer Properties Manager** dialog box (Figure 11-17) to perform certain functions of the **VPLAYER** command, such as freezing/thawing layers in viewports.

Current VP Freeze

When the **TILEMODE** is turned off, you can freeze or thaw the selected layers in the current

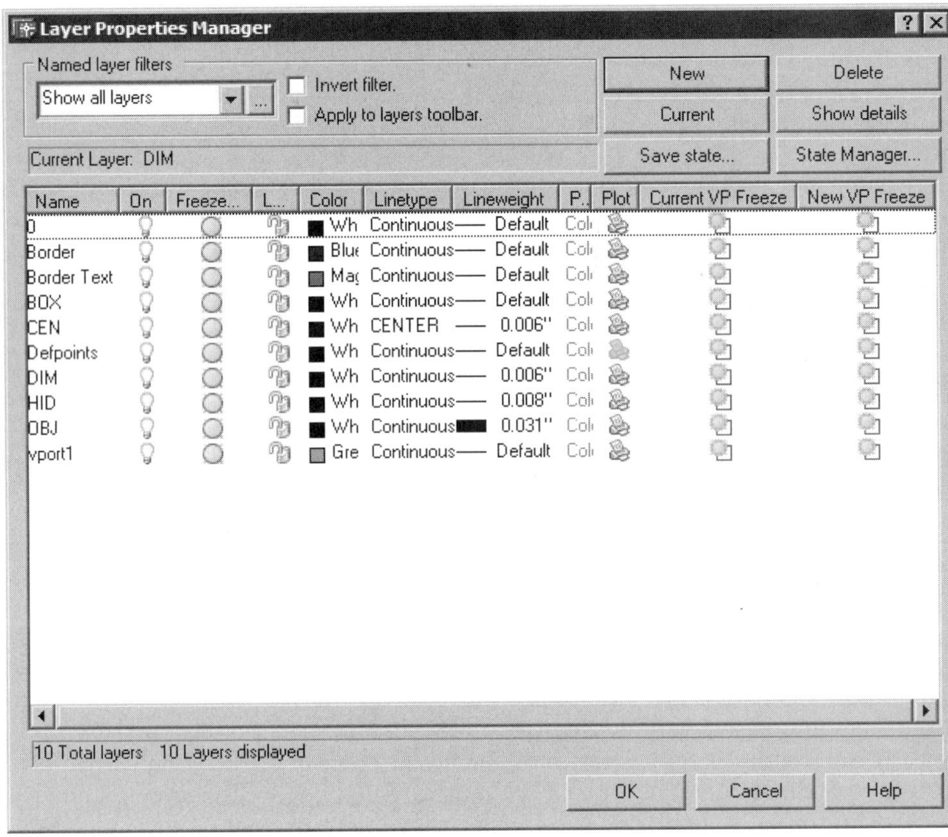

Figure 11-17 *Controlling the visibility of layers in viewports using the **Layer Properties Manager** dialog box*

floating viewport by selecting the **Current viewport Freeze** option. The frozen layers will still be visible in other viewports.

New VP Freeze

If you want to freeze some layers in the new floating viewports, then select the **New VP Freeze** toggle icon. AutoCAD will freeze the layers in subsequently created new viewports without affecting the viewports that already exist. If you start drawing on the frozen layer, objects drawn in the new viewport will not be displayed in the new viewport; however, in other viewports, the objects drawn in the new viewport will appear.

Note

*For more information about the **Layer Properties Manager** dialog box, see Chapter 4.*

PAPER SPACE LINETYPE SCALING (PSLTSCALE SYSTEM VARIABLE)

By default, the linetype scaling is controlled by the **LTSCALE** system variable. Therefore, the display size of the dashes depends on the **LTSCALE** factor, on the drawing limits, or on the drawing units. If you have different viewports with different zoom (XP) factors, the size of the dashes will be different for these viewports. Figure 11-18 shows three viewports with different sizes and different zoom (XP) factors. You will notice that the dash length is different in each of these three viewports.

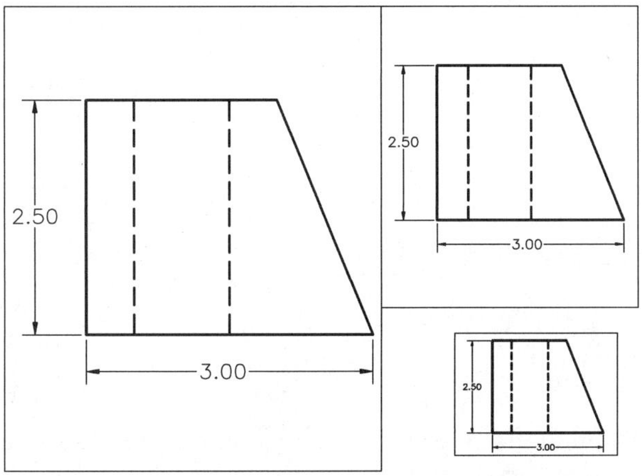

Figure 11-18 *Varying sizes of dashed lines with* ***PSLTSCALE*** *= 0*

Generally, it is desirable to have identical line spacing in all viewports. This can be achieved with the **PSLTSCALE** system variable. By default, **PSLTSCALE** is set to 0. In this case, the size of the dashes depends on the **LTSCALE** system variable and on the zoom (XP) factor of the viewport where the objects have been drawn. If you set **PSLTSCALE** to 1 and **TILEMODE** to 0, the size of the dashes for objects in the model space are scaled to match the **LTSCALE** of objects in the paper space viewport, regardless of the zoom scale of the viewports. In other words, if **PSLTSCALE** is set to 1, even if the viewports are zoomed to different sizes, the length of the dashes will be identical in all viewports. Figure 11-19 shows three viewports with different sizes. Notice that the dash length is identical in all three viewports.

Tip

You can control the scale factor for displaying the objects in the viewports using the ***Viewport Scale Control*** *drop-down list. This drop-down list is available in the* ***Viewports*** *toolbar.*

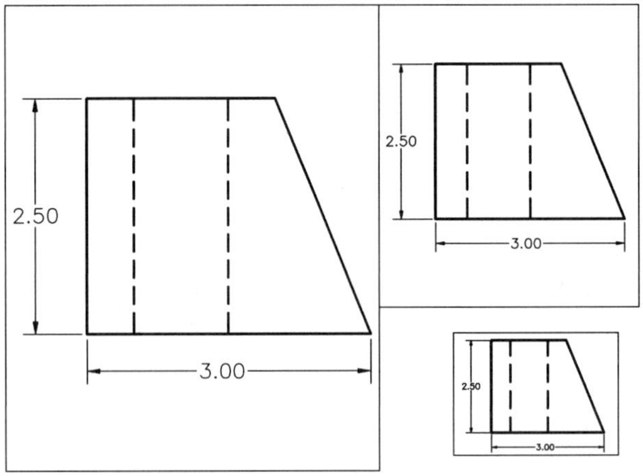

*Figure 11-19 Similar sizes of dashed lines with **PSLTSCALE** = 1*

INSERTING THE LAYOUTS

Toolbar:	Layouts > New Layout
Menu:	Insert > Layout > New Layout
Command:	LAYOUT

The **LAYOUT** command is used to create a new layout. It also allows you to rename, copy, save, and delete existing layouts. A drawing designed in the **Model** tab can be composed for plotting in the **Layout** tab. The prompt sequence is as follows.

Enter layout option [Copy/Delete/New/Template/Rename/SAveas/Set/?]<Set>:

The various options available in the prompt sequence are discussed next.

New Option

This option is used to create a new layout. When you invoke this command, you will be prompted to specify the name of the new layout. A new tab with the new layout name appears in the drawing.

Copy Option

This option copies a layout. When you invoke this option, you will be prompted to specify the layout that has to be copied. Upon specifying the layout to be copied, you will be prompted to specify the name of the new layout. If you do not enter a name for the new copied layout, the name of the copied layout is assumed with an incremental number in brackets next to it. For example, Layout 1 is copied as Layout1 (2). The name of the new layout appears as a new tab next to the copied **layout** tab.

Delete Option

This option deletes an exiting layout. When you invoke this option, you will be prompted to specify the name of the layout to be deleted. The current layout is the default layout for deleting. Remember that the **Model** tab cannot be deleted.

Template Option

This option creates a new template based on an existing layout template in either *.dwg* or *.dwt* files. This option invokes the **Select Template From File** dialog box, see Figure 11-20.

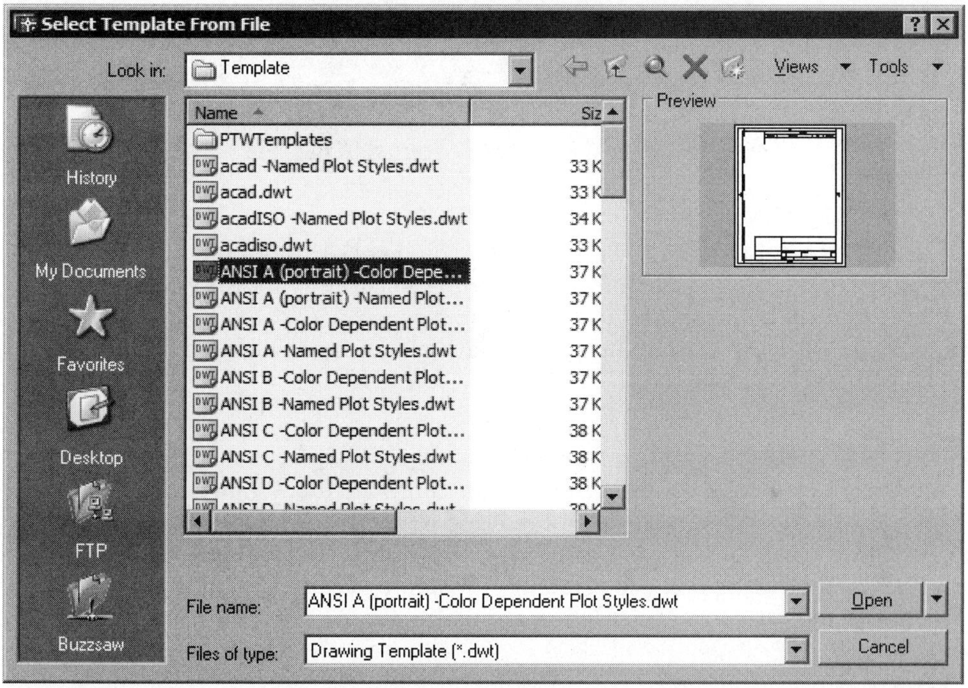

Figure 11-20 *Selecting the template file for creating the layout*

The layout and geometry from the specified template or drawing file is inserted into the current drawing. After the *dwt* or *dwg* file is selected, the **Insert Layout(s)** dialog box is displayed, as shown in Figure 11-21.

Choosing the **Layout from Template** button from the **Layouts** toolbar creates a layout using an existing template or drawing file. You can also choose **Layout** > **Layout from Template** from the **Insert** menu to directly invoke this option.

Note

If you insert a template that has a title block, the title block will be inserted as a block and all the text in the title block will be inserted as attributes. You will learn more about blocks and attributes in later chapters.

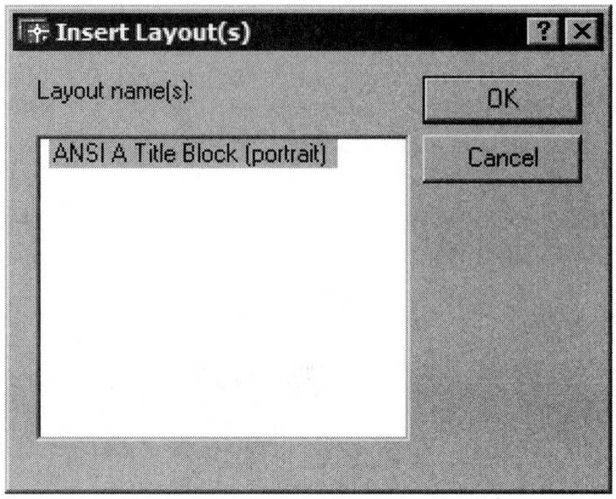

***Figure 11-21 Insert Layout(s)** dialog box*

Rename Option

This option allows you to rename a layout. When you invoke this option, you will be prompted to specify the name of the layout to be renamed. Upon specifying the name of the layout to be renamed, you will be prompted to specify the new name of the layout. The layout names have to be unique and can contain up to 255 characters, out of which only 32 are displayed in the tab. The characters in the name are not case sensitive.

SAveas Option

This option is used to save a layout in the drawing template file. When you invoke this option, you will be prompted to specify the layout that has to be saved. When you specify the name of the layout that has to be saved, the **Create Drawing File** dialog box is displayed, see Figure 11-22. In this dialog box you can enter the name of the template in the **File name** edit box. The layout templates can be saved in the *.dwg*, *.dwt*, or the *.dxf* format.

Set Option

Sets a layout as the current layout. When you invoke this option, you will be prompted to specify the name of the layout that has to be made current.

? Option

This option lists all the layouts that are available in the current drawing. The list will be displayed in the Command line. You can open the AutoCAD Text Window to view the list by pressing the F2 key.

Tip

You can right-click on the layout tabs to display the shortcut menu. This shortcut menu can be used for creating a layout, creating a layout using template, deleting, renaming, plotting, moving, or copying the layouts.

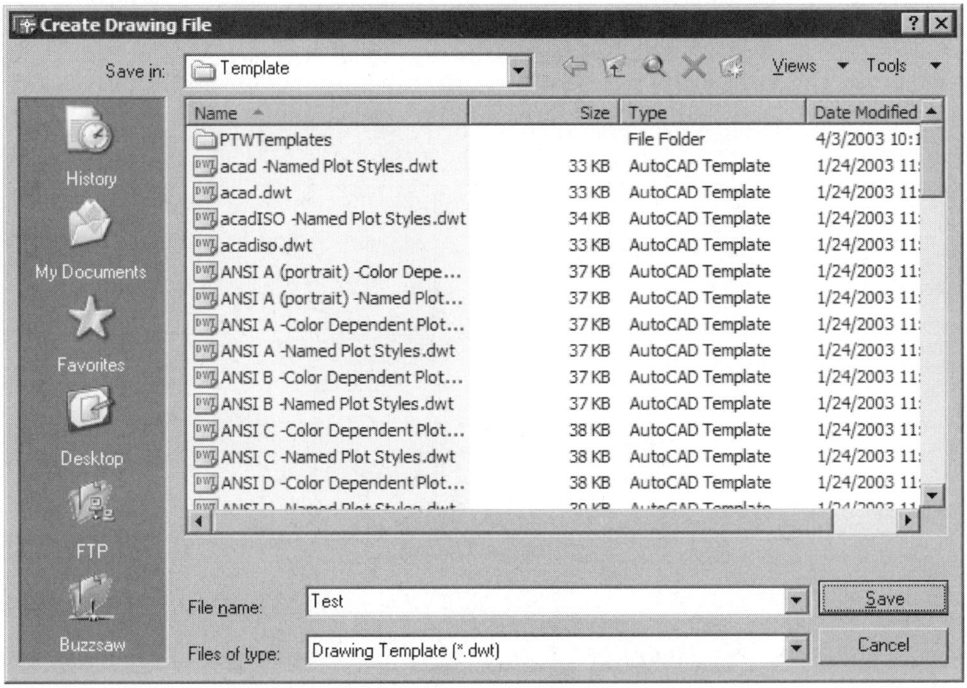

Figure 11-22 ***Create Drawing File*** *dialog box*

INSERTING A LAYOUT USING THE WIZARD

Menu:	Insert > Layouts > Layout Wizard
	Tools > Wizards > Create Layout
Command:	LAYOUTWIZARD

This command displays the **Layout Wizard** that guides you step-by-step through the process of creating a new layout.

DEFINING THE PAGE SETTINGS

Toolbar:	Layouts > Page Setup
Menu:	File > Page Setup
Command:	PAGESETUP

This command allows you to specify the Layout and Plot device settings for each new layout. You can also right-click the **Model** or any of the **Layout** tabs and choose **Page Setup** from the shortcut menu to invoke this command. When you invoke this command, the **Page Setup** dialog box is displayed, which will be discussed in Chapter 12.

Every time you select a new **Layout** tab, the **Page Setup** dialog box is displayed automatically. These default settings are controlled in the **Layout elements** area of the **Display** tab of the **Options** dialog box.

Example 2 *Mechanical*

In this example, you will learn how to create a drawing in the model space and then use the paper space to plot the drawing. The drawing to be used is shown in Figure 11-23.

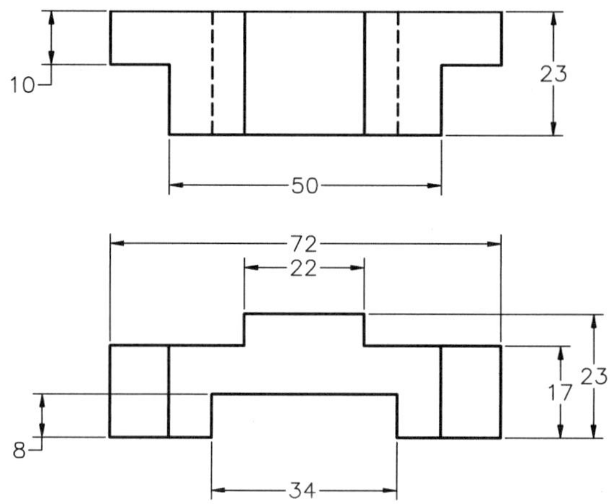

***Figure 11-23** Drawing for Example 2*

1. Based on the drawing size, calculate the limits. Since the drawing is 72 X 23, the limits should be about 75 X 75.

2. Set the limits and layers, and then draw the plan and front view as shown in Figure 11-23.

3. Choose the **Layout1** tab; AutoCAD displays the **Page Setup - Layout1** dialog box. Select the **Plot Device** tab and then select the printer or plotter that you want to use from the **Name** drop-down list. In this example, **HP Lasejet4000** is used. Choose the **Layout Settings** tab and then select the paper size that is supported by your plotting device from the **Paper size** drop-down list. In this example, the paper size is **A4 (210 x 297mm)**. Choose the **OK** button to accept the settings and exit the dialog box. AutoCAD displays **Layout1** on the screen with the default viewport. Delete this viewport.

4. Choose the **Single Viewport** button from the **Viewports** toolbar. The prompt sequence is as follows:

   Specify corner of viewport or
   [ON/OFF/Fit/Shadeplot/Lock/Object/Polygonal/Restore/2/3/4] <Fit>: *Specify the first corner of the viewport near the bottom left corner of the drawing window.*
   Specify opposite corner: **@297,210**
   Regenerating model.

5. Use the **ZOOM** command in the paper space to zoom to the extents of the viewport.

6. Double-click in the viewport to switch to the temporary model space and use the **ZOOM** command to zoom the drawing to 2XP. In this example, it is assumed that the scale factor is 2:1; therefore, the zoom factor is 2XP.

7. Create the dimension style with the text height of 1.5 and arrowhead height of 1.25. Define all the other parameters based on the text and arrowhead heights and then select the **Scale dimensions to layout (paperspace)** radio button from the **Scale for Dimension Features** area of the **Fit** tab of the **New Dimension Style** dialog box.

8. Using the new dimension style, dimension the drawing. Make sure that you do not change the scale factor. You can use the **PAN** command to adjust the display.

9. Double-click in the paper space to switch back to the paper space. Choose the **Plot** (CTRL+P) button from the **Standard Toolbar** to display the **Plot** dialog box.

10. Choose the **Window** button from the **Plot area** of the **Plot Settings** tab. The dialog box will be temporarily removed from the screen and you will be prompted to specify the first and the second corner of the window. Define a window such that the viewport is not included in the window.

11. As soon as you define both the corners of the window, the **Plot** dialog box will be redisplayed on the screen. Select **1:1** from the **Scale** drop-down list of the **Plot scale** area.

12. Select the **Center the plot** check box from the **Plot offset** area.

13. Choose the **Full Preview** button to display the plot preview. You can make any adjustments, if required, by redefining the window.

14. After you are satisfied with the preview, right-click and choose **Plot** from the shortcut menu. The drawing will be printed with the scale of 2:1. This means that two plotted units will be equal to one actual unit. Save this drawing with the name *Example2.dwg*.

WORKING WITH MVSETUP COMMAND

The **MVSETUP** command is a very versatile command and can be used both on the **Model** tab as well as on the **Layout** tab. This means that this command can be used when the **Tilemode** is set to **1** or when it is set to **0**. The uses of this command in both the drawing environments are discussed next.

Using the MVSETUP Command on the Model Tab

On the **Model** tab, this command is used to set the units, scale factor for the drawing, and the size of the paper. When you invoke this command, you will be prompted to specify whether or not you want to enable the paper space. Enter **NO** at this prompt to use this command on the **Model** tab. The prompt sequence that will follow when you enter **NO** is given next.

Enter units type [Scientific/Decimal/Engineering/Architectural/Metric]: *Specify the unit. In this case **Architectural** is selected.*

Architectural Scales
=======================

(480) 1/40"=1'
(240) 1/20"=1'
(192) 1/16"=1'
(96) 1/8"=1'
(48) 1/4"=1'

(24) 1/2"=1'
(16) 3/4"=1'
(12) 1"=1'
(4) 3"=1'
(2) 6"=1'
(1) FULL

Enter the scale factor: *Specify the scale factor for displaying the drawing.*
Enter the paper width: *Specify the width of the paper.*
Enter the paper height: *Specify the height of the paper.*

A box of the specified width and height will be drawn and the drawing will be displayed inside the box. The display of the drawing will depend upon the scale factor you have specified.

Using the MVSETUP Command on the Layout Tab

The way this command works on the **Layout** tab is entirely different from that on the **Model** tab. On the **Layout** tab it is used to insert a title block, create an array of viewports, align the objects in the viewports, and so on. The prompt sequence that will follow when you invoke this command on the **Layout** tab is given next.

[Align/Create/Scale viewports/Options/Title block/Undo]: *Select an option.*

Align Option

This option is used to align the objects in one of the viewports with that of another viewport. You can specify horizontal, vertical, angled, or rotated alignment. After you have selected the type of alignment, you will be prompted to select the base point. This base point is the point which will be used as the reference point for moving the objects in the specified viewport. Once you have specified the base point, you will be prompted to specify the point in the viewport that will be panned. This point will be aligned with the base point and thus the objects in this viewport will be moved.

Create Option

This option is used to create the viewports in the layouts. When you invoke this option, you will be prompted to specify whether you want to create a viewport, delete the objects in the viewport, or undo the changes of this option. The default option is **Create**. If this option is selected, the AutoCAD Text Window is displayed. This window provides you with four options. These options are discussed next.

0: **None**
1: **Single**
2: **Std. Engineering**
3: **Array of Viewports**

If you enter **0**, the viewports will not be created and the previous prompt will be displayed again. If you enter **1**, you will be prompted to specify two points for the bounding box of the viewport. A new viewport will be created inside the specified bounding box. If you enter **2**, you will be prompted to specify two corners of the bounding box. Inside the bounding box you have specified, four standard engineering viewports will be created. The first viewport will display the model from the top view, the second will display the model from the SE isometric view, the third will display the front view of the model (*ZX* plane of the current UCS), and the fourth will display the right-side view of the model (*YZ* plane of the current UCS). This option is generally used while working with 3D models. You will also be prompted to specify the distances between the viewports in the X direction and in the Y direction.

Note

The details about the UCS and the different views will be discussed in Chapters 22 (The User Coordinate System) and 23 (Getting Started with 3D), respectively.

If you enter **3**, you will be allowed to create an array of viewports along the *X* axis and the *Y* axis. You will be prompted to define a bounding box in which the viewports will be created. After you have defined the bounding box, you will be prompted to specify the number of viewports in the X and Y directions. Then you will be prompted to specify the distances between the viewports in the X and Y directions. Figure 11-24 shows an array of viewports in the layout with the distance between the viewports in the *X* axis and *Y* axis directions set to zero.

Scale viewports Option

This option is used to modify the scale factors of the viewports. This scale factor is defined by the ratio of the paper space units to model space units. You can select one or more than one viewport to modify the scale factors. If you select more than one viewport, you will be prompted to specify whether you want to define a uniform scale factor for the viewports or define individual scale factors. You will be first prompted to specify the number of the paper space units in the viewport and then the model space units in the viewport. The display in the viewports will be automatically modified once you have defined both the values.

Options Option

This option is used to set the parameters for inserting the title block. The parameters that can be set include layers, limits, units, and external reference options. Using the **Layers** option you can predefine the layer into which the title block will be inserted. Using the

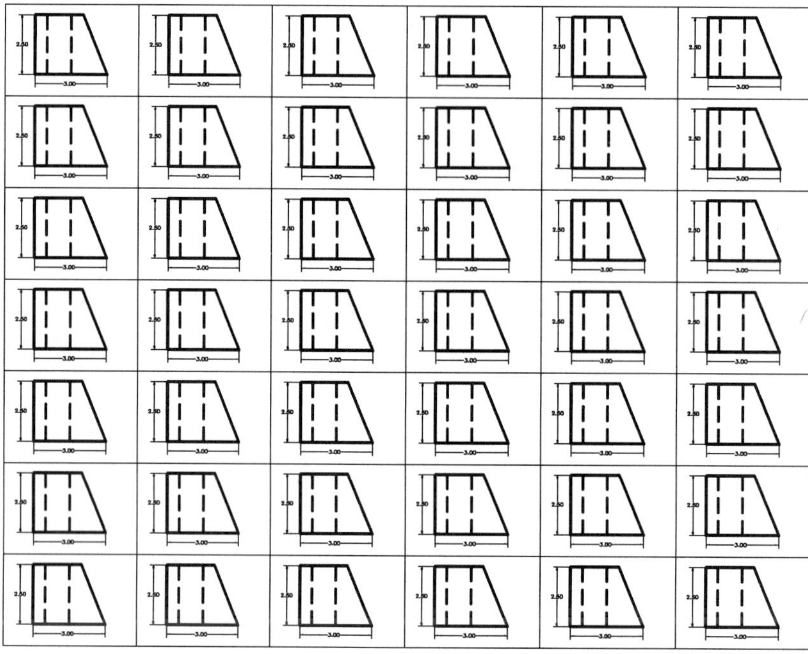

Figure 11-24 *Array of viewports*

Limits option you can specify whether or not the limits should be reset to the extents of the title block after it is inserted. Using the **Units** option you can set the units in which the size of the title block will be translated. Using the **Xref** option you can set whether the title block after inserting will become the entity of the current drawing or remain as an Xref object.

Title block Option

This option is used to insert the title block of the desired size in the current layout, delete the entities from the current layout, or reset the origin of the current layout. The default option is that for inserting a title block. If you select this option, the AutoCAD Text Window will be opened displaying different sizes and formats of the title blocks that can be inserted. The various options that will be displayed are given next.

0: None
1: ISO A4 Size(mm)
2: ISO A3 Size(mm)
3: ISO A2 Size(mm)
4: ISO A1 Size(mm)
5: ISO A0 Size(mm)
6: ANSI-V Size(in)
7: ANSI-A Size(in)
8: ANSI-B Size(in)
9: ANSI-C Size(in)

10: ANSI-D Size(in)
11: ANSI-E Size(in)
12: Arch/Engineering (24 x 36in)
13: Generic D size Sheet (24 x 36in)

You can enter the number corresponding to the desired title block and it will be inserted in the current layout. The properties will depend on the parameters set using the **Options** option. You can also add or delete the title blocks in this default list. Figure 11-25 shows a layout with an A4 size title block and a viewport created in it.

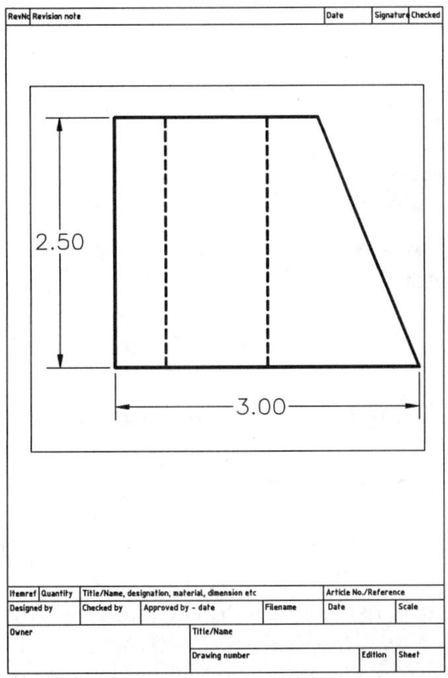

Figure 11-25 *Layout with the title block*

CONVERT DISTANCE BETWEEN MODEL SPACE AND PAPER SPACE

Toolbar:	Text > Convert distance between spaces
Command:	SPACETRANS

While working with drawings in layouts, you may need to find a distance value that is equivalent to a specific distance in the model space. For example, you may need to write text whose height should be equal to a similar text written in the model space. To convert these distances between the model space and layouts, AutoCAD provides the **SPACETRANS** command. Note that this command will not work in the model space. This command will work only in the layouts or in the temporary model space invoked from the layouts. When you invoke this command in the paper space, AutoCAD prompts you to specify the model space distance. Enter the original distance value that was measured in the model space. AutoCAD displays the paper space equivalent of the specified distance.

Similarly, when you invoke this command in the temporary model space, AutoCAD prompts you to specify the paper space distance. Enter the distance value measured in paper space. AutoCAD displays the model space equivalent of the specified distance.

Self-Evaluation Test

Answer the following questions, and then compare your answers to the correct answers given at the end of this chapter.

1. The viewports in the model space can be of any shape. (T/F)
2. The viewports in the model space can overlap each other. (T/F)
3. Two different tiled viewports can be joined. (T/F)
4. You cannot insert any additional layout in the current drawing. (T/F)
5. The _________ command is used to create the tiled viewports.
6. The viewports in the layouts are called _________ viewports.
7. The two working environments provided by AutoCAD are _________ and _________.
8. The _________ command is used to insert a title block in the current layout.
9. When you join two adjacent viewports, the resultant viewport is _________ in shape.
10. The default viewport that is created in the new layout is _________ in shape.

Review Questions

Answer the following questions.

1. Only the viewports created in the layouts can be polygonal in shape. (T/F)

2. You cannot lock the display of a floating viewport. (T/F)

3. An existing closed loop can be converted into a viewport in the model space. (T/F)

4. You can create an array of viewports using the **MVSETUP** command in the model space. (T/F)

5. Which command can be used to control the display of the objects in the viewports?

   (a) **MVIEW** (b) **DVIEW**
   (c) **LAYOUT** (d) **MSPACE**

6. Which command can be used to switch to the temporary model space?

   (a) **MVIEW** (b) **DVIEW**
   (c) **LAYOUT** (d) **MSPACE**

7. Which option of the **MVIEW** command in the paper space can be used to hide the hidden lines of the solid models in the printing?

   (a) **Hide** (b) **Shadeplot**
   (c) **Create** (d) **None**

8. Which command can be used for clipping an existing floating viewport?

   (a) **MVIEW** (b) **DVIEW**
   (c) **VPCLIP** (d) **MSPACE**

9. Which option of the **VPLAYER** command is used to create a layer that will be frozen in all the viewports?

   (a) **Freeze** (b) **Thaw**
   (c) **Newfrz** (d) **Reset**

10. The _________ option of the **MVSETUP** command can be used to set the parameters related to the inserting of the title blocks.

11. You can work only in the _________ tiled viewport.

12. The _________ command can be used to set similar linetype scale for all the viewports.

13. The _________ command is used to switch back to paper space from the temporary model space.

14. The _________ dialog box is used to save a viewport configuration in the model space.

15. Layers that have been frozen, thawed, switched on, or switched off globally are not affected by the _________ command.

Exercises

Exercise 1 *Mechanical*

In this exercise, you will perform the following operations.

a. In the model space make the drawing of the shaft shown in Figure 11-26.
b. Create three tiled viewports in the model space and then display the drawing in all three tiled viewports.
c. Create a new layout with the name **Title Block** and insert a title block of ANSI A size in this layout.
d. Create two viewports, one for the drawing and one for the detail "A". See Figure 11-26 for approximate size and location. The dimensions in detail "A" viewport must not show up in the other viewport. Also adjust the LTSCALE factor for hidden and center lines.
e. Plot the drawing.

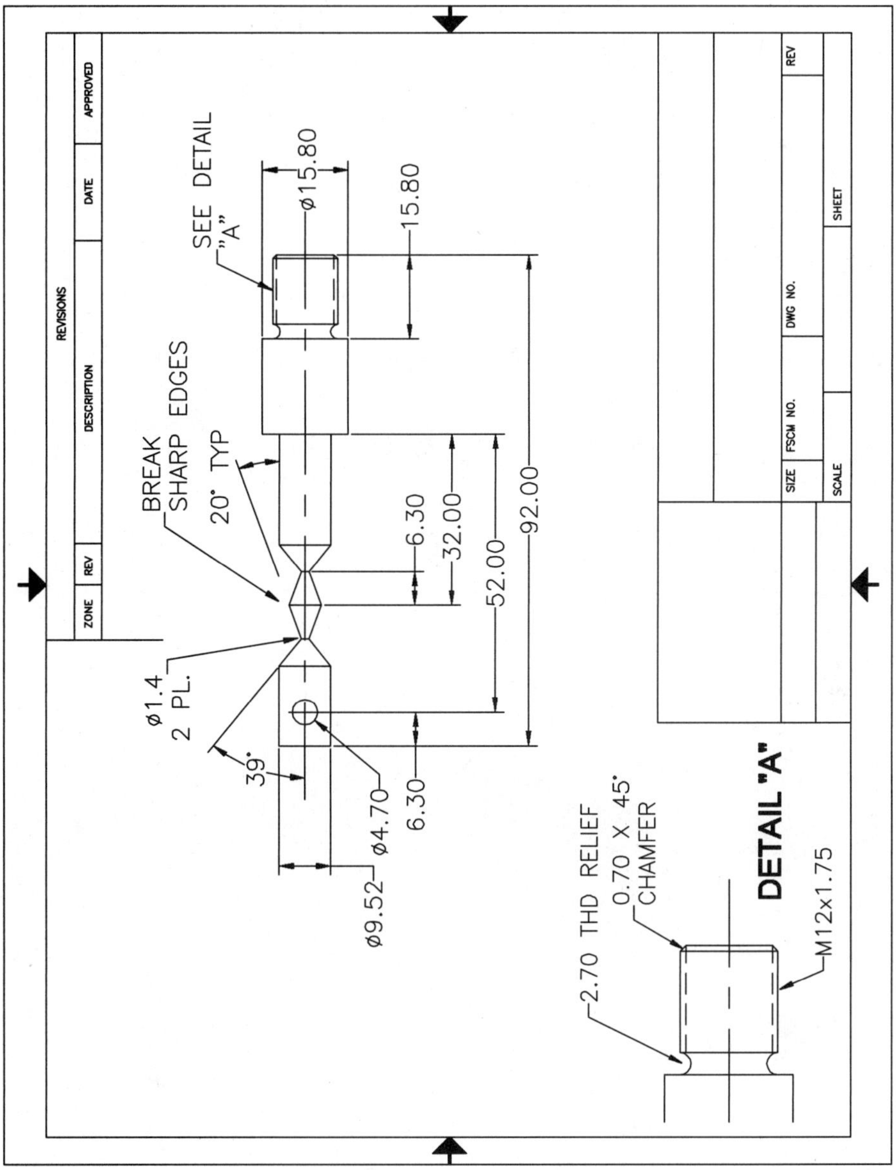

Figure 11-26 *Drawing for Exercise 1*

Answers to Self-Evaluation Test

1 - F, **2** - **F**, **3** - T, **4** - F, **5** - **VPORTS**, **6** - floating, **7** - model space and paper space/layouts, **8** - **MVSETUP**, **9** - rectangular, **10** - rectangular

Chapter 12

Plotting Drawings

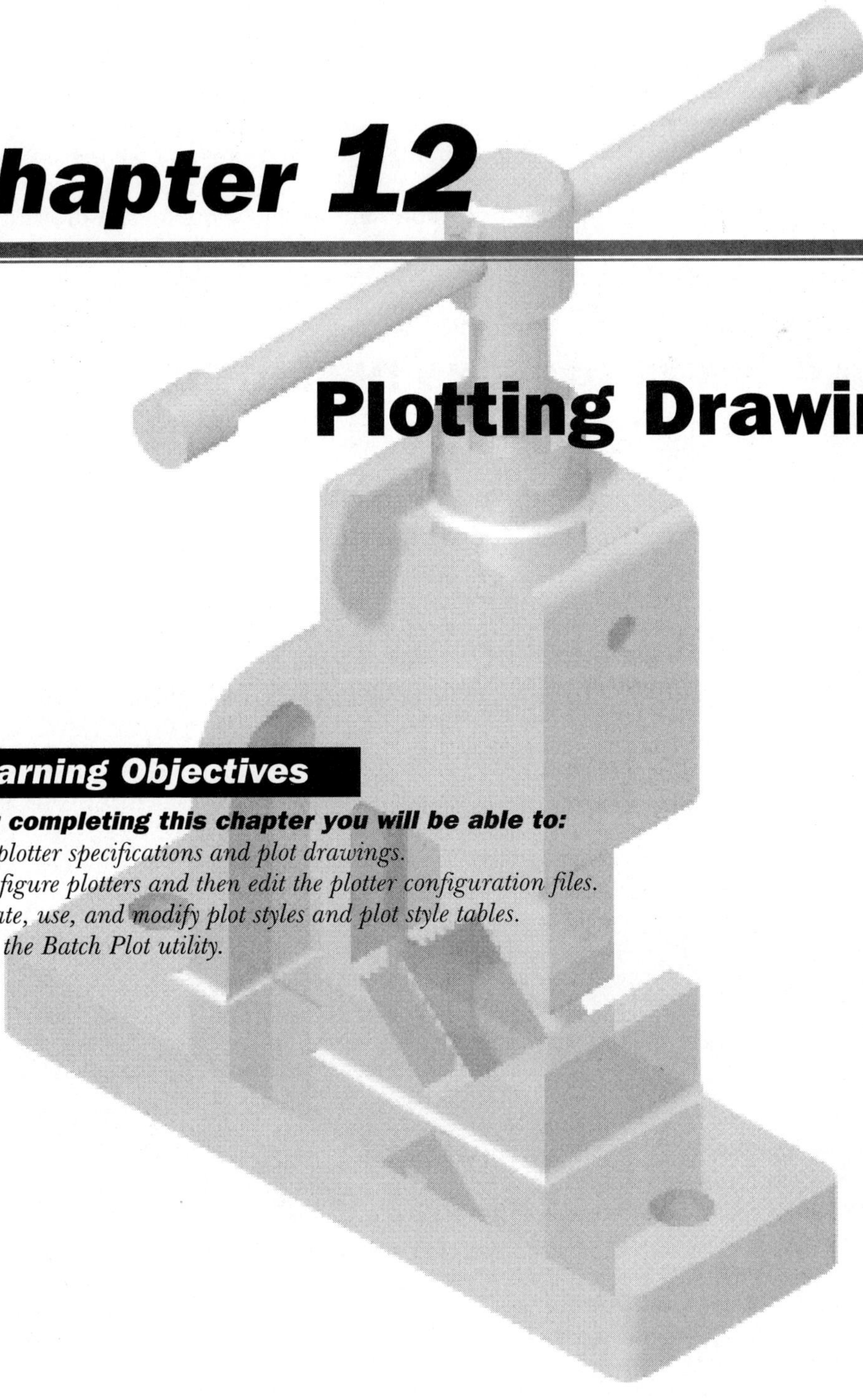

Learning Objectives

After completing this chapter you will be able to:

- *Set plotter specifications and plot drawings.*
- *Configure plotters and then edit the plotter configuration files.*
- *Create, use, and modify plot styles and plot style tables.*
- *Use the Batch Plot utility.*

PLOTTING DRAWINGS IN AutoCAD

When you are done with a drawing, you can store it on the computer storage device such as the hard drive or diskettes. However, to get a hard copy of the drawing, you should plot the drawing on a sheet of paper using a plotter or printer. A hard copy is a handy reference for professionals working on site. With the help of pen plotters, you can obtain a high-resolution drawing. Basic plotting has already been discussed in Chapter 2, Getting Started with AutoCAD. You can plot drawings in the **Model** tab or any of the layout tabs. A drawing has a **Model** and two layout tabs (**Layout 1**, **Layout 2**) by default. Each of these tabs has its own settings and can be used to create different plots. You can also create new layout tabs using the **LAYOUT** command. This is discussed in Chapter 11.

PLOTTING DRAWINGS USING THE PLOT DIALOG BOX

Toolbar:	Standard > Plot
Menu:	File > Plot
Command:	PLOT

The **PLOT** command is used to plot a drawing. When you invoke this command, the **Plot** dialog box is displayed as shown in Figure 12-1. You can also right-click on the **Model** tab or any of the layout tabs to display the shortcut menu and choose **Plot** to invoke the **Plot** dialog box.

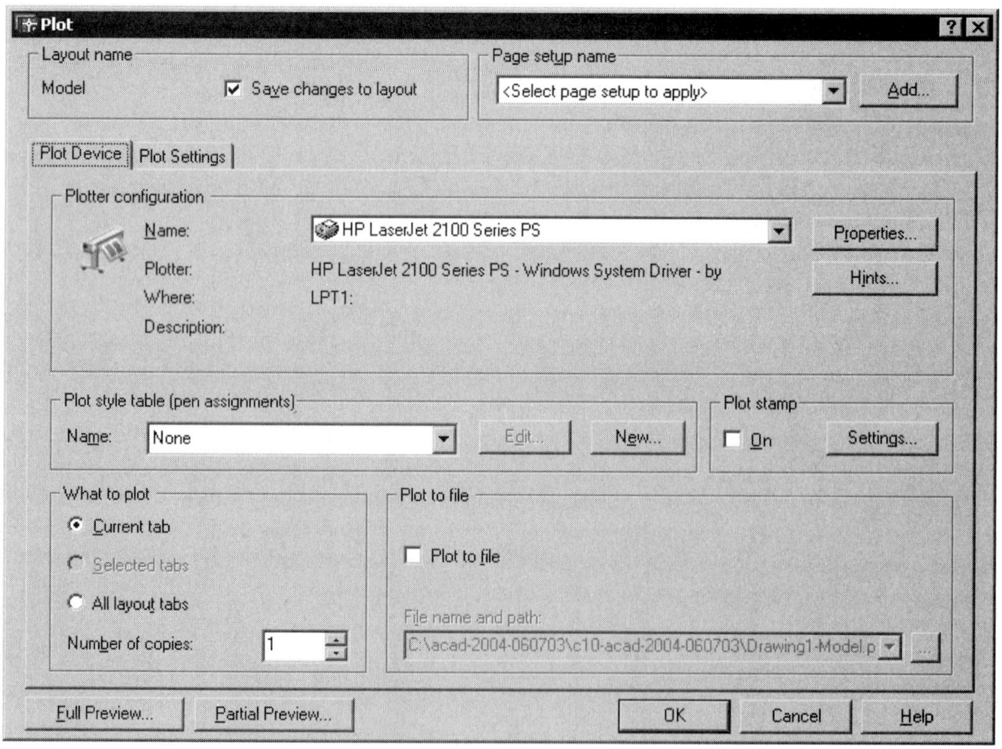

Figure 12-1 ***Plot Device*** *tab of the* ***Plot*** *dialog box*

Some values in this dialog box were set when AutoCAD was first configured. You can examine these values and if they conform to your requirements, you can start plotting directly. If you want to alter the plot specifications, you can do so through the options provided in the **Plot** dialog box. The available plot options are described next.

Layout name Area

***Figure 12-2 Layout name** area of the **Plot** dialog box*

This area (Figure 12-2) displays the current layout name or displays **Selected layouts** if multiple tabs have been selected. For example, if the **Model** tab is currently selected and you invoke the **PLOT** command, the **Layout name** area displays **Model**. A layout can be defined as the way in which the drawing has been arranged for plotting. A drawing can have many layouts, represented by different tabs at the bottom of the drawing area and each of these layouts can have different plotting settings. When you choose a layout tab for the first time, a **Page Setup** dialog box is displayed. This dialog box is the same as the **Plot** dialog box, except that it does not allow you to preview the plotted drawing. This dialog box is used to create layout settings, which can be saved as page setups and can be used later for plotting drawings. Page setups will be discussed later in this chapter. At this stage you can choose **OK** or **Cancel** to exit the dialog box. This area also provides the **Save changes to layout** check box. This check box is selected by default. If this check box is selected, any changes made to the layout such as zooming and so on will be reflected even after the **Plot** command is exited. If you select more than one layout for plotting, this option is not available.

Page setup name Area

The **Page setup name** drop-down list displays all the saved and named page setups. A page setup is the way a drawing is laid out on a sheet of paper to create a layout. It consists of all settings related to the plotting of a drawing and include the plot device being used, the scale, the pen settings, and so on. These settings can be saved as a named page setup, which can be later selected from this drop-down list and then be used for plotting a drawing. If you select **Previous Plot** from the drop-down list, the settings used for the last drawing plotted are applied to the current drawing. You can choose the base for the current page setup on a named page setup, or you can add a new named page setup by choosing the **Add** button that is located next to the drop-down list. When you choose this button, AutoCAD displays the **User Defined Page Setups** dialog box, see Figure 12-3. In this dialog box, you can create, delete, rename, or import page setups. All the settings applied to the current layout are saved in the user-defined page setup file.

The options under **User Defined Page Setups** dialog box are discussed next.

New page setup name

You can enter a name for the new user-defined page setup in this box. The current page setup can then be based on this named page setup.

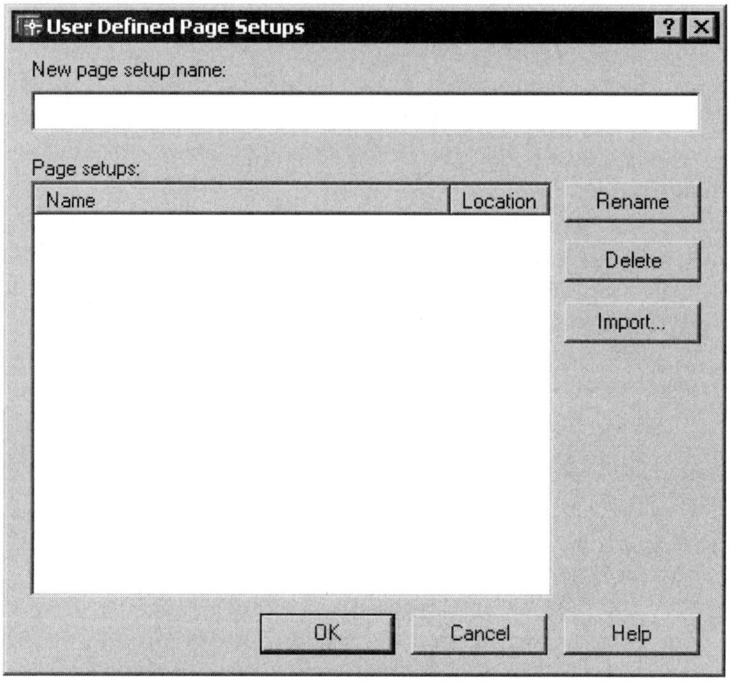

Figure 12-3 ***User Defined Page Setups*** *dialog box*

Page setups

This list box lists all the named and saved user-defined page setups. You can select a named page setup from the list box and use it for the current layout.

Rename

You can select any user-defined page setup from the list box and choose the **Rename** button to rename the page setup.

Delete

You can select a user-defined page setup from the list box and choose the **Delete** button to delete it.

Import

You can import a user-defined page setup from another drawing and apply it to a layout in the current drawing. When you choose the **Import** button, AutoCAD displays the standard **Select Page Setup From File** dialog box, where you can browse through the files and select a file for the page setup selection. After you have selected a file, choose the **Open** button. The **Import user defined page setup(s)** dialog box is displayed, see Figure 12-4. The **Page setups** list box in this dialog box displays all the page setups saved in the selected drawing. You can select any one of the page setups listed there and choose **OK** to return to the **User Defined Page Setups** dialog box. You will notice that the page setup you imported is listed in the

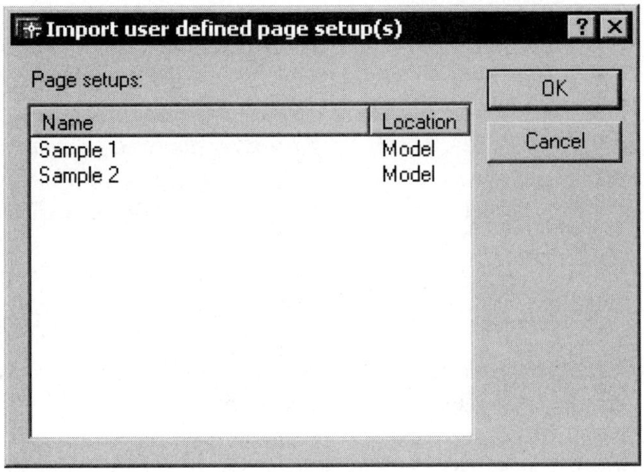

Figure 12-4 Import user defined page setup(s) *dialog box*

current **Page setups** list box. You can select it here and choose **OK** to get back to the **Plot** dialog box. The imported page setup is now listed in the **Page setup name** drop-down list and can be selected and used for the current plot.

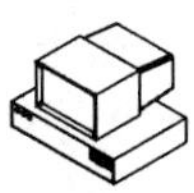

Tip

Select an existing page setup from the ***Page setup name*** *drop-down list and make modifications in it and then choose the* ***Add*** *button to create a new page setup based on an existing one.*

The **Plot** dialog box has two tabs. They are: **Plot Device** and **Plot Settings**. Both these tabs and the options available in them are discussed next.

Plot Device Tab (Plot Dialog Box)

In this tab (see Figure 12-1), information about the current configured plotter, plot style table, the layout or layouts to be plotted, and information about plotting to a file is displayed. All the options and areas in this tab are discussed next.

Plotter configuration Area

This area displays all the information about the configured printers and plotters currently selected from the **Name** drop-down list. It displays the plotter driver being used and the printer port being used. It also displays some description text about the selected plotter or printer. All the plotters that are currently configured are displayed in the **Name** drop-down list.

Note

To add plotters and printers to the ***Name*** *drop-down list, choose* ***Plotter Manager*** *from the* ***File*** *menu to display the* ***Plotters*** *window. Double-click on the* ***Add-A-Plotter Wizard*** *icon in this window to display the* ***Add Plotter*** *wizard. You can use this wizard to add a plotter to the list of configured plotters and a plotter configuration file (PC3) for the plotter is created. This file consists of all the settings needed by the specific plotter to plot. The* ***Plotters*** *window will be discussed later in this chapter in the section "****PLOTTERMANAGER*** *Command".*

Properties. If you want to check information about a configured printer or plotter, choose the **Properties** button. When you choose this button, the **Plotter Configuration Editor** is displayed. This dialog box lists all the details of the selected plotter under three tabs: **General**, **Ports**, and **Device and Document Settings**. The **Plotter Configuration Editor** will be discussed later in the "**Editing Plotter Configuration**" section of this chapter.

Hints. When you choose the **Hints** button, AutoCAD displays a help topic that provides some details about plotting drawings in AutoCAD. It also displays the criteria to select a plotter driver, and to select a system or nonsystem printer driver.

Plot style table (pen assignments) Area

This area in the **Plot** dialog box allows you to view and select a plot style table, edit the current plot style table, or create a new plot style table. A plot style table is a collection of plot styles. A plot style is a group of pen settings that are assigned to an object or layer and that determine the color, thickness, line ending, and the fill style of drawing objects when they are plotted. It is a named file that allows you to control the pen settings for a plotted drawing. You can select **None** from the **Name** drop-down list in this area of the dialog box if you want to plot a drawing without using any plot styles. You can assign different plot style tables to a drawing and plot the same drawing differently each time. The use of plot styles will be discussed later in this chapter.

Name. The **Name** drop-down list lists all of the available plot style tables. From this drop-down list, you can select a plot style table that is assigned to the current **Model** tab or layout tab. Also, if you have selected more than one layout tab, each with different plot style tables assigned to it, the **Name** edit box displays **Varies**.

Edit. You can edit a plot style table you have selected from the **Name** drop-down list by choosing the **Edit** button. This button is not available when you have selected **None** from the **Name** drop-down list. When you choose the **Edit** button, AutoCAD displays the **Plot Style Table Editor**, where you can edit the selected plot style table. This dialog box has three tabs: **General**, **Table View**, and **Form View**. The **Plot Style Table Editor** will be discussed later in the "**Using Plot Styles**" section of this chapter.

New. When you choose the **New** button, AutoCAD displays the **Add Named Plot Style Table** wizard, which is used to create a new plot style table. The use of this wizard will be discussed later in this chapter.

Plot stamp Area

The options under this area are used to set the plot stamp. Plot stamp is user-defined information that will be displayed on the sheet after plotting. You can set the plot stamping when you use the **On** check box. You can set the parameters for the plot stamp by using the **Plot Stamp** dialog box that is displayed when you choose the **Settings** button.

What to plot Area

This area in the **Plot** dialog box defines what you want to plot. You can decide to plot the **Model** tab, a layout tab, or multiple layout tabs. In this area of the dialog box you can also determine the number of copies of the drawing to be plotted. The options available are as follows.

Current tab. If this radio button is selected, AutoCAD plots the current **Model** or **Layout** tab. If multiple tabs are selected in the drawing, the tab that displays the viewing area gets plotted. You can select multiple tabs by holding down the CTRL key and then choosing the tabs. The selected tabs are highlighted.

Selected tabs. This radio button is not available if only the **Model** or only one of the layout tabs is selected. If you hold down the CTRL key and select multiple tabs, this option becomes available. You can then select it if you want to plot the multiple tabs.

All layout tabs. This option in the **Plot** dialog box plots all layout tabs, irrespective of the tabs that have been selected.

Number of copies. You can enter the number of copies that you want plotted in this edit box. You can also use the spinners to change the value in this edit box and plot the desired number of plots. If multiple layouts and copies are selected and some of the layouts are set for plotting to a file or AutoSpool, they will produce a single plot. Autospool allows you to send a file for plotting while you are working on another program.

Plot to file Area

The options in this area are used for plotting a drawing to a file instead of the plotter. By default, the options in this area are not available unless you select the **Plot to file** check box.

Plot to file. If you select this check box, AutoCAD plots the output to a file rather than to the plotter. Depending on the plotter selected, the file can be plotted in the *.dwf*, *.plt*, *.jpg*, or *.png* format. The remaining options in the **Plot to file** area will be available only when you select this check box.

File name and path. You can specify the file name and the location of the plot file in this field. By default, the plot file name is the drawing name with the tab name, separated by a hyphen, and it has a *.plt* file extension. The default location of the plot file is the directory where the drawing file resides. The last six locations that have been used to store the plot files previously are saved in this drop-down list. This way, whenever you need to store another plot file in a folder where you have stored an earlier plot file, you can select the path from the drop-down list, instead of typing the path or locating the folder again.

[...]. If you choose the [...] button, AutoCAD displays a standard **Browse for Plot File** dialog box. You can choose a location of the directory where you want to store the plot file using this dialog box.

Plot Settings Tab (Plot Dialog Box)

The **Plot Settings** tab in the **Plot** dialog box (Figure 12-5) controls the paper size, orientation, scale of plotting, plot offset, and other plotting options.

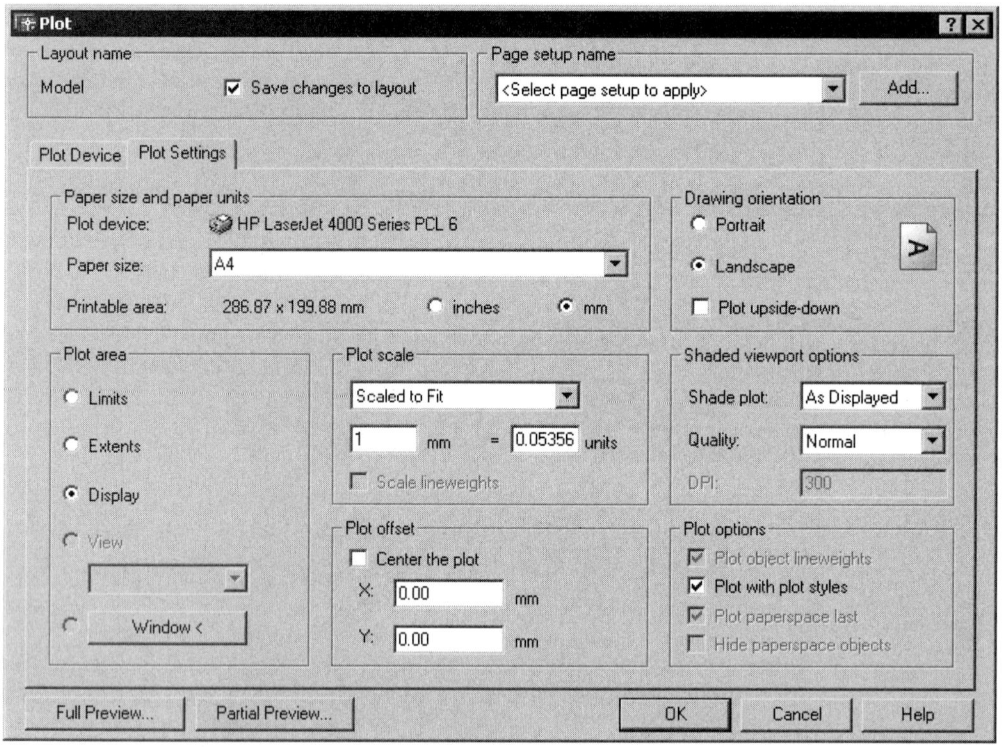

***Figure 12-5 Plot Settings** tab of the **Plot** dialog box*

The description of the options available in this tab is as follows.

Paper size and paper units Area

This area provides options that help you determine the paper size on which plotting takes place. The options available in this area are discussed next.

Plot device. The device that was selected for plotting in the **Plot Device** tab is displayed here.

Paper size. The **Paper size** drop-down list displays all the available standard paper sizes for the selected plotting device. You can select any size from the list to make it current. If **None** has been selected currently from the **Name** drop-down list in the **Plotter configuration** area of the **Plot Device** tab, AutoCAD displays the entire standard paper size list to select from.

Printable area. This is the actual dimension of the area on the paper that is used for plotting. It is based on the current paper size selected and is indicated by the width (in the *X*-axis direction) and the height (in *Y*-axis direction), when the default drawing orientation **Landscape** is selected in the **Drawing orientation** area.

inches or mm. These radio buttons specify the units to be used for plotting.

Note

*The paper size selected here is saved with the layout and overrides the PC3 file settings created by the **Add Plotter** wizard (discussed later in the "**PLOTTERMANAGER** Command" section of this chapter).*

Drawing orientation Area

This area provides options that help you specify the orientation of the drawing on the paper for the plotters that support landscape or portrait orientation. You can change the drawing orientation by selecting the **Portrait** or **Landscape** radio buttons, with or without selecting the **Plot Upside-Down** check box. The paper icon displayed on the right side of this area indicates the media orientation of the selected paper and the letter icon (A) on it indicates the orientation of the drawing on the page. The **Landscape** radio button is selected by default for AutoCAD drawings and orients the length of the paper along the *X* axis, that is horizontally. If we assume this orientation to be at a rotation angle of 0-degree, when selecting the **Portrait** radio button, the plot is oriented with the width along the *X* axis, that is equivalent to the plot being rotated through a rotation angle of 90-degree. Similarly, if you select both the **Landscape** radio button and the **Plot upside-down** check box at the same time, the plot gets rotated through a rotation angle of 180-degree and if you select both the **Portrait** radio button and the **Plot upside-down** check box at the same time, the plot gets rotated through a rotation angle of 270-degree. The AutoCAD screen conforms to the landscape orientation by default.

Plot Area

In this area you can specify the portion of the drawing to be plotted. You can also control the way the plotting will be carried out. The five options available in this area are described next.

Limits. When you select the **Limits** radio button, AutoCAD plots everything within the margins of the specified paper size. In the case of layouts, the origin is calculated at 0,0. If you are plotting from the **Model** tab, the complete area defined within the drawing limits is plotted.

Extents. If you select this option, the section of drawing that currently contains the objects is plotted. In this respect, this option resembles the **Extents** option of the **ZOOM** command. If you add objects to the drawing, they are also included in the plot and the extents of the drawing are recalculated. If you reduce the drawing extents by erasing, moving, or scaling the objects, the extents of the drawing are again recalculated. You can use the **Extents** option of the **ZOOM command** to determine what objects shall be plotted. If you use the **Extents** option when the perspective view is on and the position of the camera is not outside the drawing extents, the following message is displayed: **Plot of perspective view has been scaled to fit available area**. When no objects are drawn in the drawing area, this option is not available.

Display. If you select this radio button the portion of the drawing that is currently being displayed on the screen is plotted.

Note

*To be able to clearly view the differences between the three previously listed plotting options, it may be a good idea to make sure that the default scale options have been selected. If not, select **Scaled to Fit** from the **Scale** drop-down list in the **Plot scale** area of the dialog box if you are in the **Model** tab, and select 1:1 if you are working in any one of the **layouts** tabs.*

View. Selecting the **View** radio button in the **Plot area** of the dialog box enables you to plot a view that was created with the **VIEW** command. The view must be defined in the current drawing. If no view has been created, the **View** radio button and the drop-down list are not available. The **View** radio button and the drop-down list can be activated by creating a view in the current drawing. Once views have been created in the current drawing, you can select a view for plotting from the **View** drop-down list (Figure 12-6) and then choose **OK** in the **Plot** dialog box. When using the **View** option, the specifications of the plot depend on the specifications of the named view.

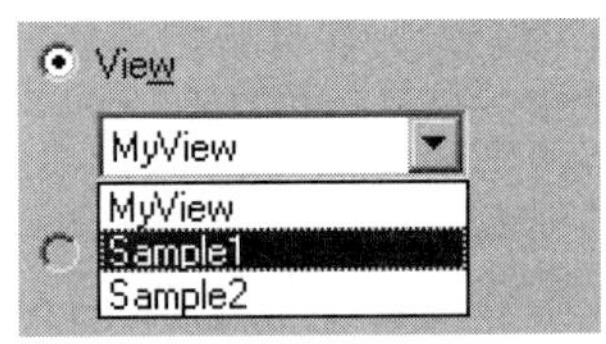

***Figure 12-6** Selecting a named view for plotting*

Window. With this option, you can specify the section of the drawing to be plotted by defining a window. The section of the drawing contained within the window defined by selecting a lower left corner and an upper right corner is plotted. To define a window, choose the **Window** button that is adjacent to the **Window** radio button. The **Window** radio button is not available until you have defined a window by choosing the **Window** button. The **Plot** dialog box is temporarily closed and you are allowed to specify two points on the screen that define a window, the area within which shall be plotted. Once you have defined the window, the **Plot** dialog box reappears on the screen. You will notice that the **Window** radio button is displayed as selected now. You can choose the **OK** button in the dialog box if you want to plot the drawing.

Note

*Sometimes, when using the **Window** option, the area you have selected may appear clipped off. This may happen because the objects are too close to the window you have defined on the screen. You need to redefine the window in this situation. Such errors can be avoided by using the preview options discussed later.*

Plot scale Area

This area controls the drawing scale of the plot area. The **Scale** drop-down list has thirty-one architectural and decimal scales apart from **Custom** and **Scaled to Fit** options. The default scale setting is **1:1** when you are plotting a layout. But if you are plotting in a **Model** tab, the default setting is **Scaled to Fit**. The **Scaled to Fit** option allows you to automatically fit the entire drawing on the paper. It is useful when you have to print a large drawing using a printer that uses a smaller size of paper or when you want to plot the drawing on a small sheet.

Whenever you select a standard scale from the drop-down list, the scale is displayed in the **Custom** edit boxes as a ratio of the plotted units to the drawing units. You can also change the scale factor manually in the **Custom** edit boxes. When you do so, the **Scale** edit box displays **Custom**. For example, for an architectural drawing, that is to be plotted at the scale 1/4"=1'-0", you can either enter 1/4"=1'-0" or 1=48 in the **Custom** edit boxes.

Note

*The **PSLTSCALE** system variable controls the paper space linetype scaling and has a default value of 1. This implies that irrespective of the zoom scale of the viewports, the linetype scale of the objects in the viewports remains the same. If you want linetype scale of objects in different viewports with different magnification factors to appear different, you should set the value of the **PSLTSCALE** variable to 0. This has been discussed in detail in Chapter 11 (Model Space Viewports, Paper Space Viewports, and Layouts).*

The **Scale lineweights** check box is available only if you are plotting in a layout tab. This option is not available in the **Model** tab. If you select the **Scale lineweights** check box, you can scale lineweights in proportion to the plot scale. Lineweights generally specify the linewidth of the printable objects and are plotted with the original linewidth size regardless of the plot scale.

Plot offset Area

This area of the **Plot Settings** tab allows you to specify an offset of the plotting area from the lower left corner of the paper. The lower left corner of a specified plot area is positioned at the lower left margin of the paper by default. If you select the **Center the plot** check box, AutoCAD automatically centers the plot on the paper by calculating the X and Y offset values. This check box is not available when you are working in one of the layout tabs and the **Layout** radio button is selected in the **Plot area** of the dialog box. You can specify an offset from the origin by entering positive or negative values in the **X** and **Y** edit boxes. For example, if you want the drawing to be plotted 4 units to the right and 4 units above the origin point, enter 4 in both the **X** and **Y** edit boxes. Depending on the units you had specified in the **Paper size and paper units** area of the dialog box, the offset values are either in inches or millimeters.

Shaded viewport options Area

The options in this area are used to print a shaded or a rendered image. These options are discussed next.

Shade plot. This drop-down list is used to select the technique that will be used to plot the drawings. If you select **As Displayed** from this drop-down list, the drawing will be plotted as it is displayed on the screen. If the drawing is hidden, shaded, or rendered, it will be printed as it is. Hidden geometry consists of objects that lie behind the facing geometry and displays the object as it would be seen in reality. If you select **Wireframe** option, the model will be printed in wireframe displaying all the hidden geometries even if it is shaded in the drawing. Selecting the **Hidden** option plots the drawing with hidden lines suppressed. Similarly, selecting the **Rendered** option plots the rendered image of the drawing.

Quality. This drop-down list is used to select printing quality in terms of dots per inch (dpi) for the printed drawing. The **Draft** option prints the drawing with 0 dpi, which results in the wireframe printout. The **Preview** option prints the drawing at 150 dpi, the **Normal** option prints the drawing at 300 dpi, the **Presentation** option prints the drawing at 600 dpi, the **Maximum** option prints the drawing at the selected plotting device's maximum dpi. You can also specify a custom dpi by selecting the **Custom** option from this drop-down list. The custom value of dpi can be specified in the **DPI** drop-down list that is enabled below the **Quality** drop-down list when you select the **Custom** option.

Note

*Selecting the **Draft** option from the **Quality** drop-down list prints the drawing in wireframe even if you select **Rendered** from the **Shade plot** drop-down list. Also, to plot the drawings in layouts with hidden lines suppressed, you need to use the **Shade Plot** option discussed in Chapter 11, Model Space Viewports, Paper Space Viewports, and Layouts.*

You will learn more about wireframe, hidden, shaded, and rendered models in later chapters.

Plot options Area

This area displays four additional plotting options that can be selected as per the plot requirements. They are described next.

Plot object lineweights. This check box is not available if the **Plot with plot styles** check box is selected. To activate this option, clear the **Plot with plot styles** check box. This check box is selected by default and AutoCAD plots the drawing with the specified lineweights. To plot the drawing without the specified lineweights clear this check box.

Plot with plot styles. When you select the **Plot with plot styles** check box, AutoCAD plots using the plot styles applied to objects in the drawing and defined in the plot style table. The different property characteristics associated with the different style definitions are stored in the plot style tables and can be easily attached to geometry. This setting replaces pen mapping used in earlier versions of AutoCAD.

Plot paperspace last. This check box is not available when you are in the **Model** tab because no paper space objects are present in the **Model** tab. When you are working in a layout tab, this option is available. By selecting the **Plot paperspace last** check box, you get an option of plotting model space geometry before paper space objects. Usually paper space geometry is plotted before model space geometry. This option is also useful when there are multiple tabs selected for plotting and you want to plot the model space geometry before the layout tabs.

Hide paperspace objects. This check box is used to specify whether or not the objects drawn in the layouts will be hidden or not while plotting. If this check box is selected, the objects created in the layouts will be hidden.

Plot Preview

You can view the plot on the specified paper size before actually plotting it by selecting the **Full Preview** or **Partial Preview** buttons in the **Plot** dialog box. Previews help you find errors and correct them prior to plotting, thus saving stationery and time. AutoCAD provides two types of plot previews. They are partial and full.

Full Preview

When you choose the **Full Preview** button AutoCAD displays the drawing on the screen just as it would be plotted on the paper. Full preview takes more time than the partial preview because regeneration of the drawing takes place. A smaller file size results in a faster regeneration. Once regeneration is performed, the dialog boxes on the screen are removed temporarily, and an outline of the paper size is shown. In the plot preview (Figure 12-7), the cursor is replaced with the Zoom Realtime icon. This Realtime icon can be used to zoom in and out interactively by holding the pick button down and then moving the pointing device. You can right-click to display a shortcut menu and then choose **Exit** to exit the preview or press the ENTER or ESC keys to return to the dialog box. You can also choose **Plot** to plot the drawing right away or choose the other zooming options available.

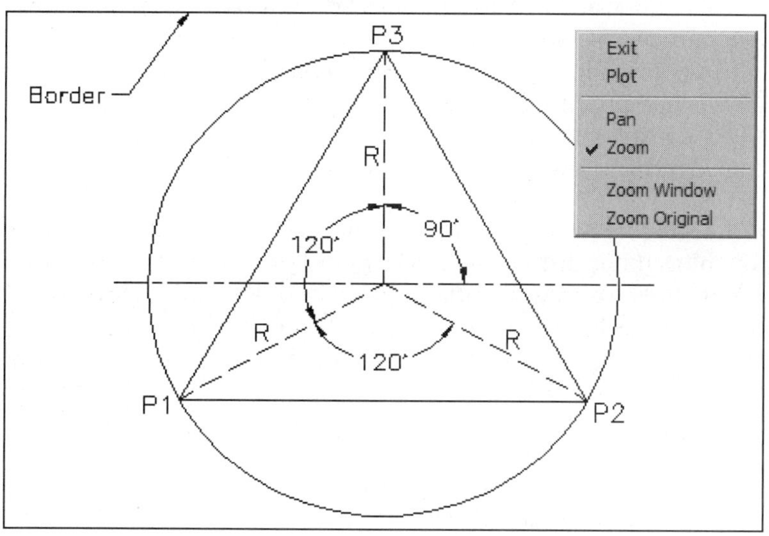

Figure 12-7 *Full plot preview with the shortcut menu*

Partial Preview

To generate a partial preview of a plot, choose the **Partial Preview** button. The **Partial Plot Preview** dialog box is displayed, see Figure 12-8. The paper size is graphically represented by the white paper icon. The dashed rectangle shows the printable area. The blue hatched

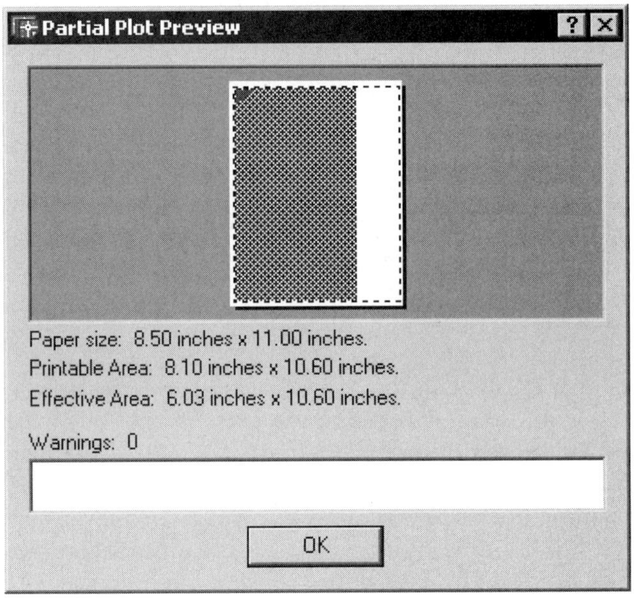

***Figure 12-8 Partial Plot Preview** dialog box*

rectangle is the section in the paper that is used by the image. This area is also known as the **effective area**. The dimensions of the paper, printable area, and effective area are also displayed numerically. If you define the origin of the plot so that the effective area is not accommodated in the graphic area of the dialog box, AutoCAD displays a warning. Hence, with the help of a partial preview, you can accurately see how the plot will fit on the paper. If there is something wrong with the specifications of the plot, AutoCAD provides warning messages so that corrections can be made before the actual plotting takes place.

The red triangle icon located at one of the corners of the effective area represents the orientation of the plot. If it is displayed in the upper left corner of the effective area, it means the plot has the default landscape orientation. In case of Portrait orientation, the red triangle icon is located at the lower left corner of the effective area. Similarly, the icon is displayed at the lower right corner in case of upside-down landscape orientation and is displayed at the upper right corner of the effective area in case of an upside-down portrait orientation.

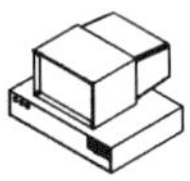

Tip

*You can also choose **Plot Preview** from the **File** menu to see a full preview of the plot, and at the same time, bypass the plot dialog box.*

After you have finished with all the settings and other parameters, if you choose the **OK** button in the **Plot** dialog box, AutoCAD starts plotting the drawing in the file or plotters as specified. AutoCAD displays the **Plot Progress** dialog box (Figure 12-9), where you can view the actual progress in plotting.

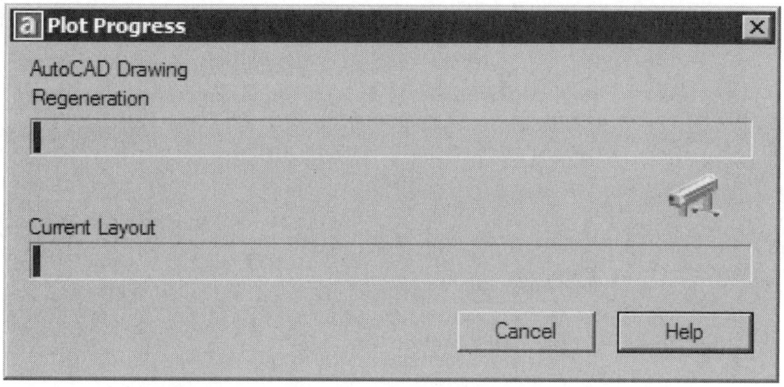

Figure 12-9 ***Plot Progress*** *dialog box*

ADDING PLOTTERS

In AutoCAD, the plotters can be added using the **PLOTTERMANAGER** command. This command is discussed next.

PLOTTERMANAGER Command

Menu:	File > Plotter Manager
Command:	PLOTTERMANAGER

When you invoke the **PLOTTERMANAGER** command, AutoCAD will display the **Plotters** window, see Figure 12-10.

The **Plotters** window is basically a Windows Explorer window. It displays all the configured plotters and the **Add-A-Plotter-Wizard** icon. You can right-click on any one of the icons belonging to plotters that already have been configured to display a shortcut menu. You can choose **Delete** from the shortcut menu to remove a plotter from the list of available plotters in the **Name** drop-down list in the **Plot/Page Setup** dialog box. You can also choose **Rename** from the shortcut menu to rename the name of the plotter configuration file or choose **Properties** to view the properties of the configured device.

Add-A-Plotter-Wizard

If you double-click on the **Add-A-Plotter-Wizard** icon in the **Plotters** window, AutoCAD guides you to configure a nonsystem plotter for plotting your drawing files. AutoCAD stores all the information of a configured plotter in configured plot (PC3) files. The **PC3** files are stored in *AutoCAD 2004\Plotters* folder by default. The steps for configuring a new plotter using the **Add-A-Plotter-Wizard** are as follows:

1. Open the **Plotters** window by choosing **Plotter Manager** from the **File** menu.

2. In the **Plotters** window, double-click on the **Add-A-Plotter-Wizard** icon.

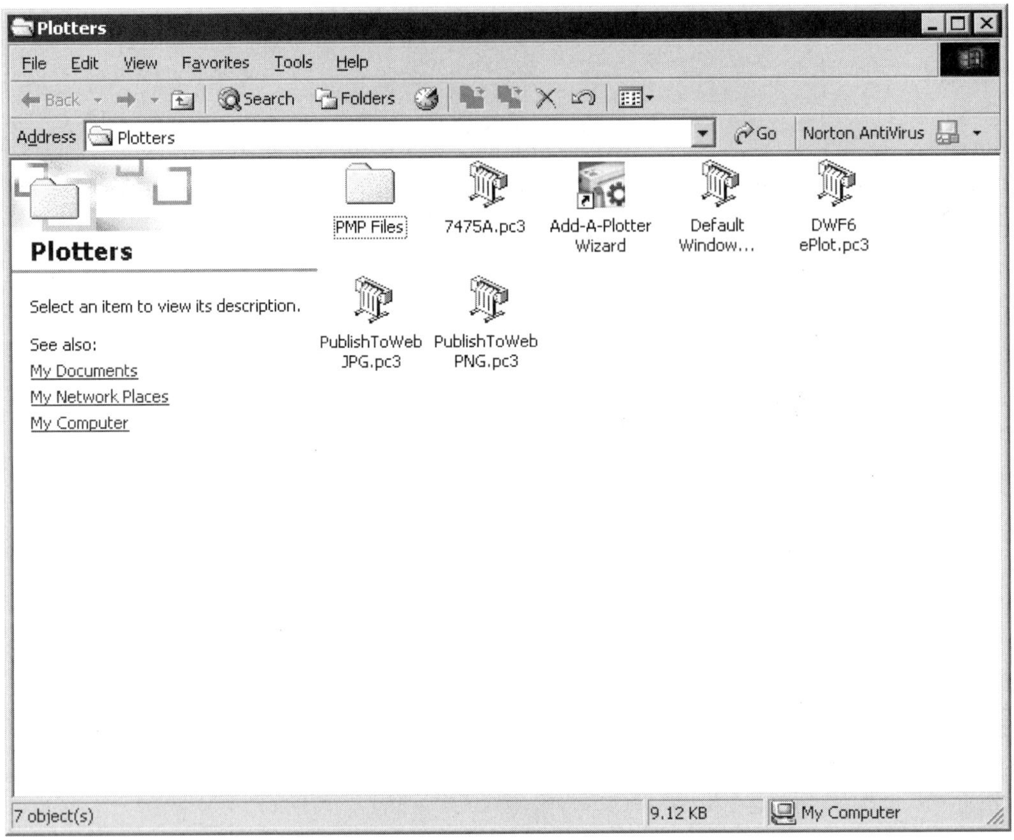

***Figure 12-10 Plotters** window*

3. In the **Add Plotter** wizard, carefully read the **Introduction** page, and then choose the **Next** button to advance to the **Add Plotter - Begin** page.

4. On the **Add Plotter - Begin** page, the **My Computer** radio button is selected by default. Choose the **Next** button. The **Add Plotter - Plotter Model** page is displayed.

5. On this page, select a manufacturer and model of your nonsystem plotter from the **Manufacturers** and **Models** list boxes respectively. Now, choose the **Next** button. The **Add Plotter - Import Pcp or Pc2** page is displayed.

   If your plotter is not present in the list of available plotters, and you have a driver disk for your plotter, choose the **Have Disk** button to locate the **HIF** file from the driver disk, and install the driver supplied with your plotter.

6. In the **Add Plotter - Import Pcp or Pc2** page, if you want to import configuring information from a **PCP** or a **PC2** file created with a previous version of AutoCAD, you

can choose the **Import File** button and select the file. Otherwise, simply choose the **Next** button to advance to the next page.

7. On the **Add Plotter** - **Ports** page, select the port from the list to use when plotting and choose **Next**.

8. On the **Add Plotter** - **Plotter Name** page, you can specify the name of the currently configured plotter or the default name will be entered automatically. Choose **Next**.

9. When you reach the **Add Plotter** - **Finish** page, you can choose the **Finish** button to exit the **Add-A-Plotter-Wizard**.

   You can also choose the **Edit Plotter Configuration** button to display the **Plotter Configuration Editor** where you can edit the current plotter's configuration. Also, in this page you can choose the **Calibrate Plotter** button to display the **Calibrate Plotter** wizard. This wizard allows you to calibrate your plotter by setting up a test measurement. After the test plotting it compares the plot measurements with the actual measurements and computes a correction factor.

Once you have chosen **Finish** to exit the wizard, a PC3 file for the newly configured plotter will be displayed in the **Plotters** window. This PC3 file contains all the settings needed by the plotter to plot. Also the newly configured plotter name is added to the **Name** drop-down list in the **Plotter configuration** area of the **Plot Device** tab in the **Plot** dialog box. You can now use the plotter for plotting.

EDITING PLOTTER CONFIGURATION

You can modify the properties of a selected plot device by using the **Plotter Configuration Editor**. This dialog can be invoked in several ways. As discussed earlier, when using the **Plot/ Page Setup** dialog boxes, you can choose the **Properties** button in the **Plotter configuration** area to display the **Plotter Configuration Editor**, see Figure 12-11. You can modify the default settings for a plotter while configuring it, by choosing the **Edit Plotter Configuration** button on the **Add Plotter** - **Finish** page of the **Add Plotters** wizard. You can also select the PC3 file for editing in the **Plotters** window using Windows Explorer (by default, PC3 files are stored in the *AutoCAD 2004\Plotters* folder) and double-click the file or right-click the file and choose **Open** from the shortcut menu. The three tabs available in the **Plotter Configuration Editor** are discussed next.

General tab

This tab contains basic information about the configured plotter or the PC3 file. You can make changes only in the **Description** area. The rest of the information in the tab is read only. This tab contains information of the configured plotter file name, plotter driver type, HDI driver file version number, name of the system printer (if any), and the location and name of the PMP file (if any calibration file is attached to the PC3 file).

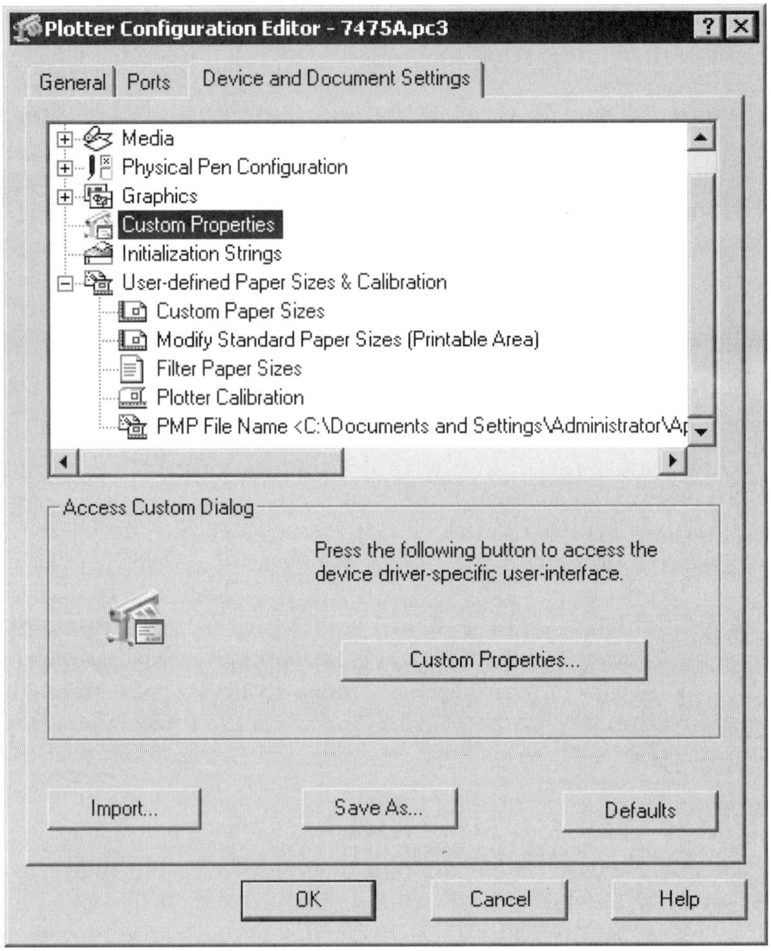

Figure 12-11 Plotter Configuration Editor

Ports Tab

This tab contains information about the communication between the plotting device and your computer. You can choose between a serial (local), parallel (local), or network port. The default settings for parallel ports is **LPT1** and **COM1** for serial ports. You can also change the port name if your device is connected to a different port. You can also select the **Plot to File** radio button, if you want to save the plot as a file. You can select **Autospool**, if you want plotting to occur automatically while you continue work on another application.

Device and Document Settings Tab

This tab contains plotting options specific to the selected plotter, displayed as a tree view in a window. For example, if you configure a nonsystem plotter you have the option to modify the pen characteristics. You can select any plotter properties from the tree view displayed in the window to change the values as required. Whenever you select an icon from the tree view in the window, corresponding information is displayed in an area below. For example, if you select **PMP File Name <None>** in the tree view in the window, a **PMP file** area is displayed below. This area contains the current settings and options to modify it. Information that is displayed within brackets (<>) can be modified. By default, **Custom properties** is displayed as highlighted in the window, since it contains properties that are modified commonly and the **Access Custom Dialog** area is displayed below the window. Choosing the **Custom Properties** button in this area displays a properties dialog box specific to the selected plotter. This properties dialog box has several properties of the selected plotter grouped and displayed under various tabs and can be modified here.

Once you have made the desired changes, choose **OK** to exit the properties dialog box and then choose **Save As** in the **Plotter Configuration Editor** if you want to save the changes you just made to the PC3 file. You can also import old plot configuration files from previous releases of AutoCAD (Release 13-PCP and Release 14-PC2) using the **Import** button and save some of the information from these settings as a new PC3 file. When you choose the **Import** button, the **Plotting Components** dialog box is displayed. This dialog box displays what to use to import AutoCAD 14 information into an AutoCAD 2004 drawing. For example, it tells you that to import PCP or PC2 file settings into a drawing in AutoCAD 2004 you should use the **Import PCP or PC2 Plot Settings Wizard**. Choose **OK** to exit this dialog box; the **Import** dialog box is displayed. Here you can select the file to import and then choose the **Import** button.

Tip

It is better to create a new PC3 file for a plotter and keep the original file as it is so that you encounter no error while using the specific printer later. The PC3 files determine the proper function of a plotter and any modifications may lead to errors.

IMPORTING PCP/PC2 CONFIGURATION FILES

If you want to import a PCP or PC2 configuration file or plot settings created by previous releases of AutoCAD into the **Model** tab or current layout for the drawing, you can also use the **PCINWIZARD** command to display the **Import PCP or PC2 Plot Settings Wizard**. All the information from a PCP or PC2 file regarding plot area, rotation, plot offset, plot optimization, plot to file, paper size, plot scale, and pen mapping can be imported. Read the **Introduction** page of the wizard that is displayed carefully and then choose the **Next** button. The Browse File Name page is displayed. Here you can either enter the name of the PCP or PC2 file directly in the **PC2 or PCP file name** edit box or then choose the **Browse** button to display the **Import** dialog box, where you can select the file to import. After you specify the file for importing, choose **Import** to return to the wizard. Choose the **Next** button to display the **Finish** page. After importing the files you can modify the rest of the plot settings for the current layout.

SETTING THE PLOT PARAMETERS

Before starting with the drawing, you can set various plotting parameters in the **Model** tab or in the **Layouts**. The plot parameters that can be set include the plotter to be used, plot style table, the paper size, units, and so on. All these parameters can be set using the **PAGESETUP** command discussed next.

PAGESETUP Command

Toolbar:	Layouts > Page Setup
Menu:	File > Page Setup
Command:	PAGESETUP

As discussed earlier, page setup contains settings required to plot a drawing. Each layout as well as the **Model** tab can have a unique page setup attached to it. You can use the **PAGESETUP** command to create named page setups that can be used later. A page setup consists of specifications for the layout page, plotting device, paper size, and settings for the layouts to be plotted. The **PAGESETUP** command can also be invoked from the shortcut menu by right-clicking in the current **Model** or any of the **Layout** tabs and choosing **Page Setup**. Remember that if you have not invoked the layouts even once in the current AutoCAD file, the **Page Setup** option will not be enabled when you right-click on the **Layout** tab.

When you invoke the **PAGESETUP** command, AutoCAD displays the **Page Setup** dialog box, see Figure 12-12.

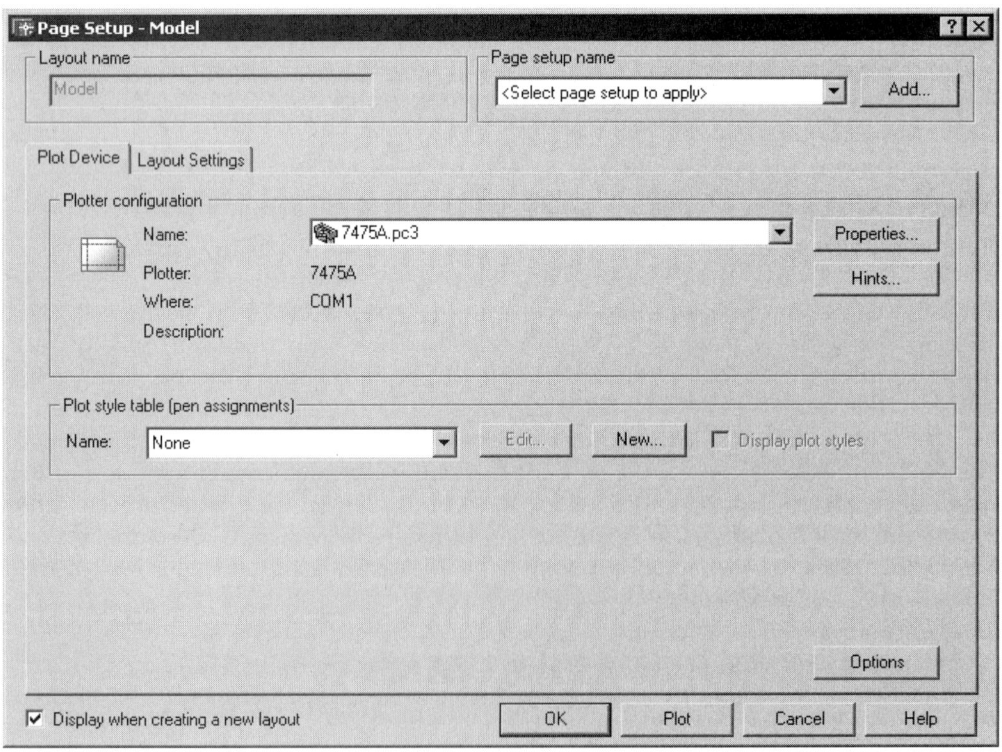

Figure 12-12 ***Page Setup*** *dialog box when displayed in the* ***Model*** *tab*

The **Page Setup** dialog box is similar to the **Plot** dialog box that has been described earlier except that it does not have the preview buttons and the **What to plot** and **Plot to file** areas. The first time you choose a layout tab in a drawing session, the **Page Setup** dialog box is automatically displayed and specifies the page layout and current plotting device settings. The layout settings are stored with the layout. The current Layout name is displayed in the **Layout Name** box, and in the **Page Setup Name** area, a list of all the named and saved page setups are available. You can choose to base the current page setup on a named page setup or you can add a new named page setup by choosing the **Add** button. AutoCAD displays the **User Defined Page Setups** dialog box (discussed earlier in this chapter in "Plotting Drawings Using the Plot Dialog Box"). The **Page Setup** dialog box has two tabs that are described next.

Tip

*When you invoke the **Page Setup** dialog box in the layouts, you can enter a name in the **Layout name** text box. The name that you enter will be made the name of the current layout.*

Plot Device Tab

This tab specifies the current plotting device for plotting the layout. This tab is similar to the **Plot Device** tab of the **Plot** dialog box, excluding the **What to Plot** and **Plot to file** areas. The additional feature in this tab is the **Options** button. If you choose the **Options** button, AutoCAD displays the **Plotting** tab of the **Options** dialog box (Figure 12-13). This tab has three areas that are discussed next.

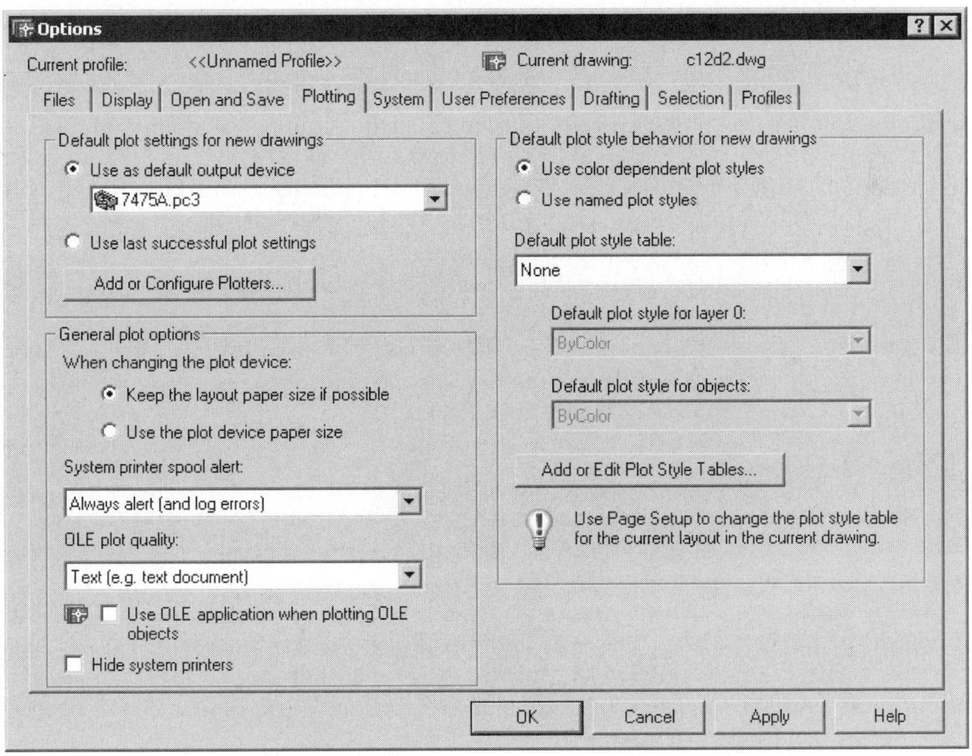

Figure 12-13 ***Plotting** tab of the **Options** dialog box*

Default plot settings for new drawings area. This area provides options to set the default plot settings. You can either decide to use the settings that were used successfully for the last plot created by selecting the **Use last successful plot settings** radio button or you can select a device from the **Use as default output device** drop-down list and use it as the default device every time you create a new plot. You can also choose the **Add or Configure Plotters** button to display the **Plotters** window and use the **Add-A-Plotter-Wizard** to configure a new plotter and then use it as the default plotter.

General plot options area. This area provides options that control the general plotting environment. The **Keep the layout paper size if possible** radio button is selected by default and allows you to keep plotting on the paper size selected from the **Paper size** drop-down list in the **Layout Settings** tab of the **Page Setup** dialog box irrespective of the plotting device selected as long as possible. When no longer possible, AutoCAD will display a warning saying that the paper size is not supported by the current device and will use the paper size that is specified in the PC3 file of the plotter or the default printer settings of the printer. This is also controlled by the system variable **PAPERUPDATE**. When this option is selected the value of this variable is set to 0. If you select the **Use the plot device paper size** radio button, only the paper size specified in the plotter configuration file (PC3) of the selected device is used and the value of the **PAPERUPDATE** variable is set to 1.

The **System printer spool alert** drop-down list displays options that determine whether AutoCAD will display an alert message if there is an error (due to conflict between the output or input port) and the drawing is spooled through the system printer and when it will do so. The options also determine if AutoCAD will log the error.

The **OLE plot quality** drop-down list provides options that control the quality of OLE objects, when they are plotted. The **OLEQUALITY** system variable also controls this option. OLE stands for Object Linking and Embedding and has been discussed later in Chapter 19. Selecting the **Use OLE application when plotting OLE objects** check box loads applications used to create OLE objects and is useful when you want to improve the quality of OLE objects. This option is also controlled by the **OLESTARTUP** system variable.

Default plot style behavior for new drawings area. This area provides options that allow you to set the default plot style mode and table for new drawings. These options shall be discussed later in this chapter, in "**Using Plot Styles**".

Layout Settings Tab

This tab specifies the layout settings of the drawing such as paper size, drawing orientation, plot area, plot scale, plot offset, and other plotting options. This tab is similar to the **Plot Settings** tab of the **Plot** dialog box (discussed earlier in the chapter). The settings specified in the **Page Setup** dialog box are stored with the layout. You can plot the current layout from the **Page Setup** dialog box by choosing the **Plot** button after specifying all the necessary parameters in the dialog box. You can also control whether the plotting should include lineweights and whether the lineweights are in proportion to the plot scale.

Importing a Page Setup

Command: PSETUPIN

As discussed earlier, it is possible for you to import a user-defined page setup from an existing drawing and use it in the current drawing or base the current page setup for the drawing on it. This option is available through the **Plot/Page Setup** dialog box. It is also possible to bypass the **Plot/Page Setup** dialog box and directly import a page setup from an existing drawing into a new drawing layout by using the **PSETUPIN** command. This command facilitates importing a saved and named page setup from a drawing into a new drawing. The settings of the named page setup can be applied to layouts in the new drawing. When you invoke this command, a standard file selection dialog box is displayed, where you can locate a drawing (*.dwg*) file whose pagesetups have to be imported. After you select a drawing file, AutoCAD displays the **Import User Defined Page Setups** dialog box (Figure 12-14). You can also enter **-PSETUPIN** at the Command prompt to display prompts at the command line.

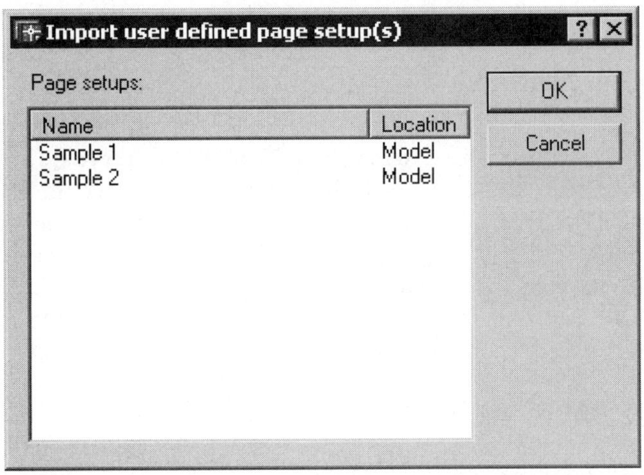

Figure 12-14 Import user defined page setup(s) *dialog box*

Note

If a page setup with the same name already exists in the current file, the ***AutoCAD Alert*** *box is displayed and you will be informed that a page setup with the same name already exists in the current file, do you want to redefine it? If you choose* ***Yes*** *in this dialog box, the current page setup will be redefined.*

USING PLOT STYLES

The plot styles can change the complete look of a plotted drawing. You can use this feature to override a drawing object's color, linetype, and lineweight. For example, if an object is drawn on a layer that is assigned the color red and no plot style is assigned to it, it will be plotted as red. But, if you have assigned a plot style to the object with the color blue, the object will be plotted as blue irrespective of the layer color it was drawn on. Similarly, you can change the end, join, and fill styles of the drawing, and also change the output effects like dithering, gray scales, pen assignments, and screening. Basically, you can use **Plot Styles** effectively to plot the same drawing in various ways.

Every object and layer in the drawing has a plot style property. The plot style characteristics are defined in the plot style tables attached to the **Model** tab, layouts, and viewports within the layouts. You can attach and detach different plot style tables to get different looks for your plots. Generally, there are two plot style modes. They are **Color-Dependent** and **Named**. The **Color-dependent** plot styles are based on object color and there are **255** color-dependent plot styles. It is possible to assign each color in the plot style a value for the different plotting properties and these settings are then saved in a color-dependent plot style table file that has a *.ctb* extension. Similarly, **Named** plot styles are independent of object color and you can assign any plot style to any object regardless of that object's color. These settings are saved in a named plot style table file that has a *.stb* extension. Every drawing in AutoCAD 2004 is in either one of the plot style modes.

Adding a Plot Style

Menu:	File > Plot Style Manager
Command:	STYLESMANAGER

All plot styles are saved in the **AutoCAD 2004\Plot Styles** folder. If you enter **STYLESMANAGER** at the Command prompt, AutoCAD displays the **Plot Styles Manager** window, see Figure 12-15. This window displays icons for all the available plot styles in addition to the **Add-A-Plot Style Table Wizard** icon. You can double-click on any of the plot style icons to display the **Plot Style Table Editor** dialog box and edit the selected plot style. When you double-click on the **Add-A-Plot Style Table Wizard** icon, the **Add Plot Style Table** wizard is displayed and you can use it to create a new plot style.

Add-A-Plot Style Table Wizard

If you want to add a new plot style table to your drawing, double-click on the **Add-A-Plot Style Table Wizard** in the **Plot Styles Manager** window to display the **Add Plot Style Table** wizard. You can also invoke the wizard by choosing **Add Plot Style Table** from the **Tools > Wizards** menu. The following are the steps for creating a new plot style table using the wizard:

1. Read the introduction page carefully and choose the **Next** button.

2. In the **Begin** page, select the **Start from scratch** radio button and choose **Next**. Selecting this option creates an absolutely new plot style table and does not use an existing file. Therefore, the **Browse File** page is not available.

   In addition to the **Start from scratch** option, this page has three more options. They are **Use an existing plot style table**, **Use My R14 Plotter Configuration (CFG)**, and **Use a PCP or PC2 file**. When you use the **Use an existing plot style table** option, an existing plot style table is used as a base for the new plot style table you are creating. In such a situation, the **Table Type** page of the wizard is not available and is not displayed because

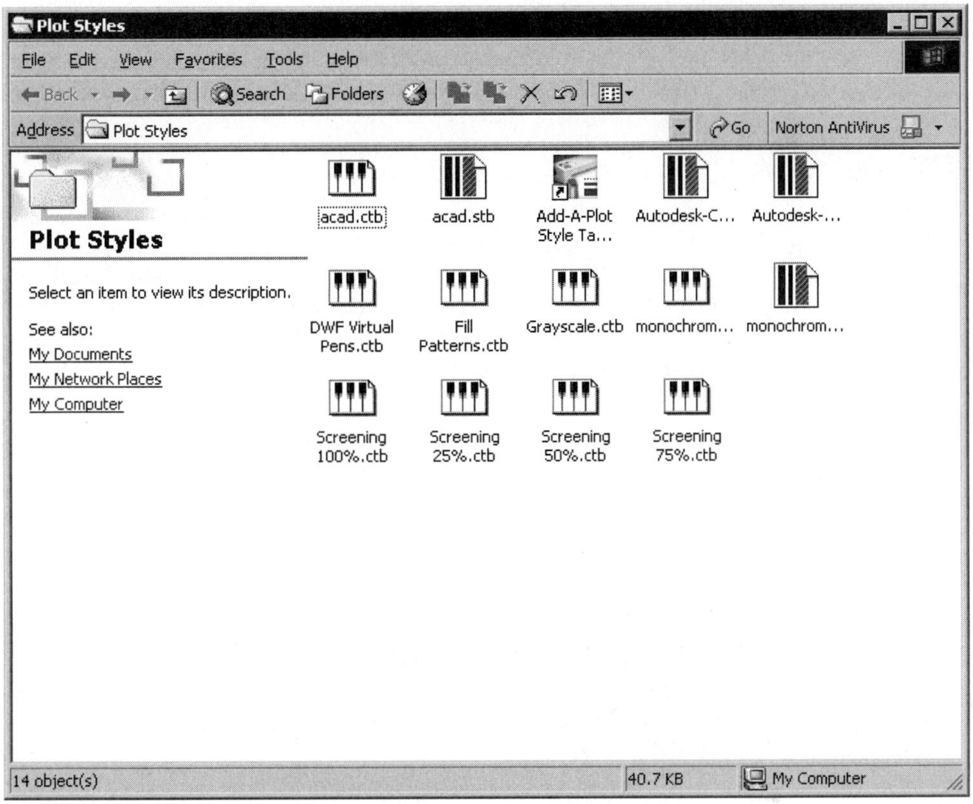

Figure 12-15 ***Plot Styles*** *window*

the table type will be based on the existing plot style table you are using to create a new one. With the **Use My R14 Plotter Configuration (CFG)** option, the pen assignments from the *acad2004.cfg* file are used as a base for the new table you are creating. If you are using the **Use a PCP or PC2** option, pen assignments saved earlier in a Release 14 PCP or PC2 file are used to create the new plot style.

3. In the **Pick Plot Style Table** page, select the **Named Plot Style Table** or the **Color-Dependent Plot Style Table** according to your requirement. Select the **Color-Dependent Plot Style Table** radio button and then choose the **Next** button.

4. Since, you have selected the **Start from scratch** radio button in the **Begin** page, the **Browse File** page is not available and the **File name** page is displayed. However, if you had selected any of the other three options available on the **Begin** page, the **Browse File** page is displayed. You can select an existing file from the drop-down list available in this page or choose the **Browse** button to display the **Select File** dialog box. You can then browse and select a file from a specific folder and choose **Select** to return to the wizard. You can also enter the name of the existing plot style table you want to base the new plot style table on, directly in the edit box. After you have specified the file name, choose

Chapter 12

Next to display the **File name** page of the wizard. In the **File name** page, enter a file name for the new plot style table and choose **Next**.

Note

*If you are using the pen assignments from the Release 14 acad.cfg file to define the new plot style table, you also have to specify the printer or plotter to use from the drop-down list available in the **Browse File** page of the wizard.*

5. The **Finish** page is displayed. This page gives you the option of choosing the **Plot Style Table Editor** button to display the **Plot Style Table Editor** and then edit the plot style table you have created. If you select the **Use this plot style table for new and pre-AutoCAD 2004 drawings** check box in this page of the wizard, the plot style table that you have created will become the default plot style table for all the drawings you create. This check box is available only if the plot style mode you have selected in the wizard is the same as the one you have specified as the default plot style mode in the **Default plot style behavior for new drawings** area in the **Plotting** tab of the **Options** dialog box. Choose **Finish** in the **Finish** page of the wizard to exit the wizard. A new plot style table gets added to the **Plot Styles** window and can be used for plotting.

Note

*You can also choose **Add Named Plot Style Table/ Add Color-Dependent Plot Style Table** from the **Tools > Wizards** menu to display wizards that are similar to the **Add Plot Style Table** wizard except that in the **Begin** page, the **Use an existing plot style table** option is not available. Also the **Table Type** page is not there and the **Finish** page has an additional option to use the new plot style table for the current drawing.*

Plot Style Table Editor

When you double-click on any of the plot style table icons available in the **Plot Styles** window, the **Plot Style Table Editor** is displayed where you can edit the particular plot style table. You can also choose the **Plot Style Table Editor** button in the **Finish** page to display the **Plot Style Table Editor** and choose the **Edit** button adjacent to the **Name** drop-down list in the **Plot style table (pen assignments)** area in the **Plot Device** tab of the **Plot** or **Page Setup** dialog box to display the **Plot Style Table Editor**.

The **Plot Style Table Editor** has three tabs: **General**, **Table View**, and **Form View**. You can edit all the properties of an existing plot style table using the different tabs. The description of the tabs are as follows.

General Tab

This tab provides the information about the file name, location of the file, version, and scale factor. All the information except the description are read only. You can enter a description about the plot style table in the **Description** text box here. If you select the **Apply global scale factor to non-ISO linetypes** check box, all the non-ISO linetypes in a drawing are scaled by the scale factor specified in the **Scale factor** edit box placed below the check box. If this check box is cleared (by default), the **Scale factor** edit box is not available.

Table View Tab

This tab displays all the plot styles, available with their properties in tabular form, that can be edited individually here, see Figure 12-16. In the case of a named plot style table, you can edit the existing styles or add new styles by choosing the **Add Styles** button. A new column with default style name **Style1** will be added in the table. You can change the style name if you want to. You can edit the various properties in the table by selecting a particular value you want to modify and a corresponding drop-down list is displayed. You can select a value from this drop-down list. This manner of editing is similar to the one we use when editing properties of objects using the **PROPERTIES** palette. The **Normal** plot style table is not available and therefore cannot be edited. This plot style is assigned to layers by default.

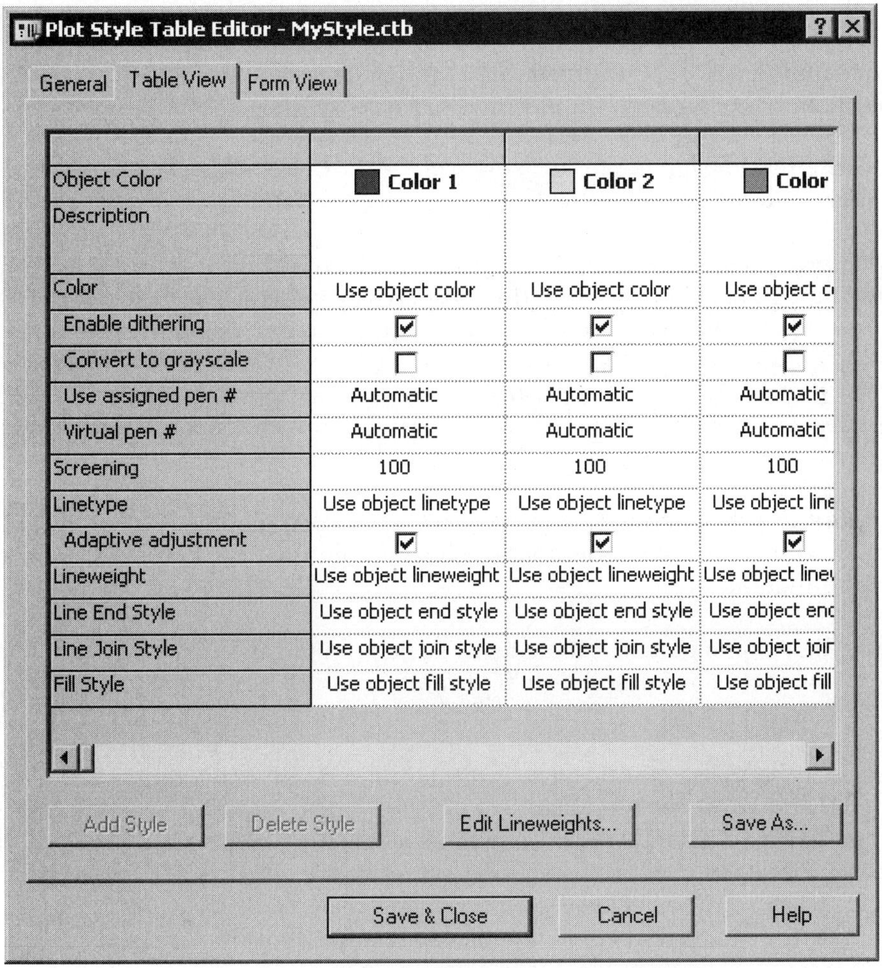

***Figure 12-16 Plot Style Table Editor (Table View** tab)*

You can select a particular plot style by clicking on the gray bar placed above the column and the entire column gets highlighted. You can select several plot styles by pressing the SHIFT key and selecting more plot styles. All the plot styles that are selected are highlighted. If you choose the **Delete Style** button now, the selected plot styles are removed from the table.

Choosing the **Edit Lineweights** button displays the **Edit Lineweights** dialog box. You can select the units for specifying the lineweights in the Units for listing area of the dialog box, and can use either **Millimeters** or **Inches**. You can also edit the value of a lineweight by selecting it in the **Lineweights** list box and choosing the **Edit Lineweight** button. After you have edited lineweights, choose the **Sort Lineweights** button to rearrange the lineweight values in the list box. Choose **OK** to exit the dialog box and return to **Plot Style Table Editor**.

With a color-dependent plot style table, the **Table View** tab displays all 255 plot styles, one for each color, and the properties can be edited in the table in the same way as discussed for named plot styles. The only difference is that you cannot add a new plot style or delete an existing one and therefore, the **Delete Style** and **Add Style** buttons are not available. The properties that can be defined in a plot style are discussed next.

Color. The color you assign to plot style overrides the color of the object in the drawing. The default value is **Use object color**. The other properties related to color are as follows.

Enable dithering. Dithering is described as a mixing of various colored dots to produce a new color. You can enable or disable this property by selecting or clearing the check box available in this field. Dithering is enabled by default and is independent of the color selected.

Convert to grayscale. If you are using plotter that supports grayscaling, selecting this option applies grayscale to the objects color. By default, this check box is not selected and the object's color is used.

Use assigned pen #. This property is applied only to pen plotters. The pens range from 1 to 32. The default value is **0** or **Automatic**, which implies that the pen used will be based on the plotter configuration.

Virtual pen #. Nonpen plotters can behave like pen plotters using virtual pens. The value of this property lies between 1 and 255. The default value is **0** or **Automatic**. This implies that AutoCAD will assign a virtual pen automatically from the AutoCAD Color Index (ACI).

Screening. This property of a plot style indicates the amount of ink used while plotting. The value ranges between 0 and 100. The default value is **100**, which creates the plot in its full intensity. Similarly, a value of **0** produces white color. This may be useful when plotting on a colored background.

Linetype. Like the property of color, the linetype assigned to a plotstyle overrides the object linetype on plotting. The default value is **Use object linetype**.

Adaptive adjustment. This property is applied by default and implies that the linetype scale of a linetype is applied such that on plotting, the linetype pattern will be completed.

Lineweight. The lineweight value assigned here overrides the value of the object lineweight on plotting. The default value is **Use object lineweight**.

Line End Style. This determines the manner in which a plotted line ends. The effect of this property is more noticeable when the thickness of the line is substantial. The line can end in a **Butt**, **Square**, **Round**, or **Diamond** shape. The default value is **Use object end style**.

Line Join Style. You can select the manner in which two lines join in a plotted drawing. The available options are **Miter**, **Bevel**, **Round**, and **Diamond**. The default value is **Use object join style**.

Fill Style. The fill style assigned to a plot style overrides the objects fill style, when plotted. The available options are **Solid**, **Checkerboard**, **Crosshatch**, **Diamonds**, **Horizontal Bars**, **Slant Left**, **Slant Right**, **Square Dots**, and **Vertical Bars**.

Form View Tab

This tab displays all the properties in a form, see Figure 12-17. All the available plot styles are displayed in the **Plot Styles** list box. You can select any style in the list box and then edit its properties in the **Properties** area.

While creating a named plot styles table (**.stb**), if you want to add a new plot style, choose the **Add Style** button and AutoCAD will display the **Add Plot Style** dialog box with default plot style name **Style 1**. You can change the name of the plot style in the Plot Style edit box, see Figure 12-18. Now, when you choose the **OK** button, the new style will be added in the **Plot Styles** list box and you can select it for editing. If you want to delete a style, select a style and then choose the **Delete Style** button. With a color-dependent plot styles the **Form View** tab also does not provide options that allow you to add or delete plot styles.

After editing, choose the **Save & Close** button to save and return to the **Plot Styles** window. You can also choose the **Save As** button to display the **Save As** dialog box and save a plot style table with another name. If you want to remove the changes you made to a plot style table, choose the **Cancel** button.

Applying Plot Styles

The **Model**, or any of the layout tabs, can be assigned a plot style table. The plot style table can be either named or color-dependent as discussed earlier. These plot style modes for a new drawing can be determined in the **Default plot style behavior for new drawings** area of the **Plotting** tab of the **Options** dialog box (see Figure 12-13). The **Use color dependent plot styles** option is selected by default and the drawings are assigned a color-dependent plot style. If you select the **Use named plot styles** radio button, the new drawings (not the current drawing) will be assigned a named plot style. The **PSTYLEPOLICY** system variable also controls the default plot style modes of the new drawings. A value of 0 implies a named plot style mode and a value of 1 implies a color-dependent plot style mode.

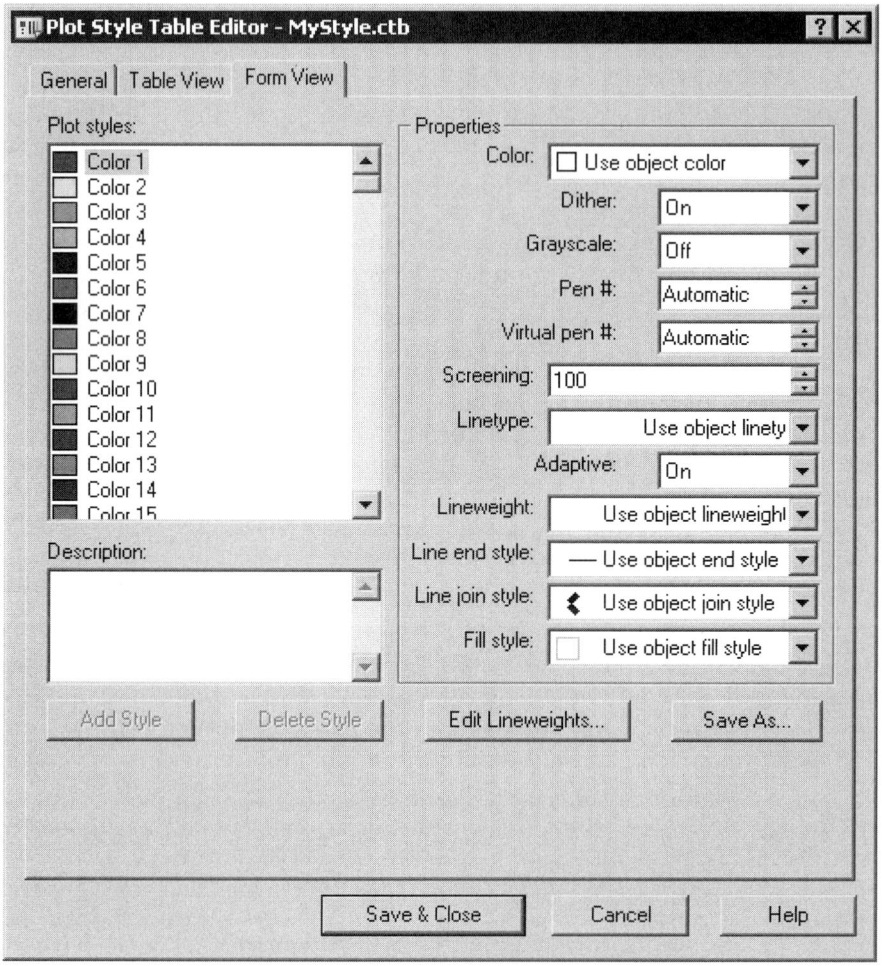

Figure 12-17 ***Plot Style Table Editor*** *(**Form View** tab)*

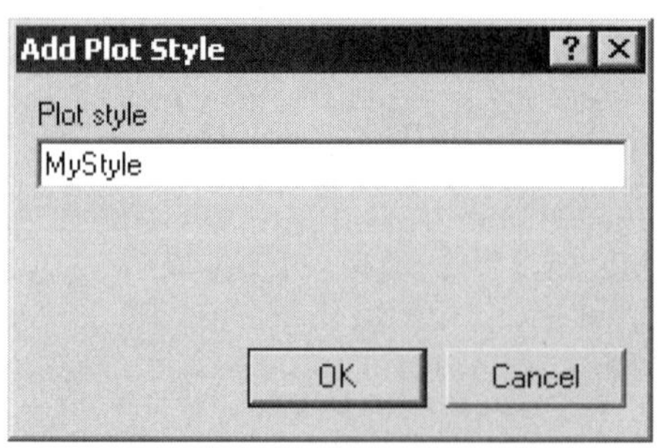

Figure 12-18 Add Plot Style *dialog box*

You can select a plot style table that you want to use as a default for the drawings from the **Default plot style table** drop-down list in the **Default plot style behavior for new drawings** area of the **Plotting** tab of the **Options** dialog box. If you select **None**, the drawing is plotted with the object properties as displayed on the screen. In this area of the **Options** dialog box, only when you select the **Use named plot styles** radio button, are the **Default plot style for layer 0** and the **Default plot style for objects** drop-down lists available. You can select the default plot styles that you want to assign to Layer 0 and to the objects in a drawing from these drop-down lists respectively. If you choose the **Add or Edit Plot Style Table** button in this area of the **Plotting** tab of the **Options** dialog box, the **Plot Styles** window is displayed. Here, you can double-click on any plot style table icon available and edit it using the **Plot Style Table Editor** that is displayed.

To change the plot style table for a current layout, you have to invoke the **Page Setup** or **Plot** dialog box and then select a plot style table from the **Name** drop-down list in the **Plot style table (pen assignments)** area of the **Plot Device** tab of the dialog box. A color-dependent plot style table can be selected and applied to a tab only if the default plot style mode already has been set to color dependent. Similarly, if you want to apply a named plot style table to a tab, the **Use named plot styles** radio button should have been selected in the **Plotting** tab of the **Options** dialog box.

A color-dependent plot style cannot be applied to objects or layers and therefore the **Plot Style Control** drop-down list in the **Properties** toolbar is not available when a drawing has a color-dependent plot style mode. The plot styles also appear grayed out in the **Layer Properties Manager** dialog box and cannot be selected and changed. But, named plot styles can be applied to objects and layers. A plot style applied to an object overrides the plot style applied to the layer on which the object is drawn. To apply a plot style to a layer, invoke the **Layer Properties Manager** dialog box where all the layers in the selected tab are displayed. Select a layer to which you want to apply a plot style and select the default plot style (Normal) currently applied to the layer. The **Select Plot Style** dialog box is displayed, see Figure 12-19. The **Plot styles** list box in this dialog box displays all the plot styles present in the plot style table attached to the current tab. You can select another plot style table to attach to the current tab from the **Active plot style table** drop-down list. You will notice that all the plot styles in the selected plot style table are displayed in the list box now. You can also choose the **Editor** button to display the **Plot Style Table Editor** to edit plot style tables, as discussed earlier. The **Select Plot Style** dialog box also displays the name of the original plot style assigned to the object adjacent to **Original**. Also, the new plot style to be assigned to the selected object is displayed next to **New**.

You can apply a named plot style to an object using the **Plot Style Control** drop-down list in the **Properties** toolbar or the **Properties** palette. The process is the same as that applied for layers, colors, linetypes, and lineweights. This named plot style is applied to an object irrespective of the tab on which it is drawn. If the plot style assigned to an object is present in the plot style table of the tab in which it is present, the object is plotted with the specified plot style. But, if the plot style assigned to the object is not present in the plot style table assigned to the tab on which it is drawn, the object will be plotted with the properties that are displayed on the screen. The default plot style assigned to an object is **Normal** and the default plot style assigned to a layer is **ByLayer**.

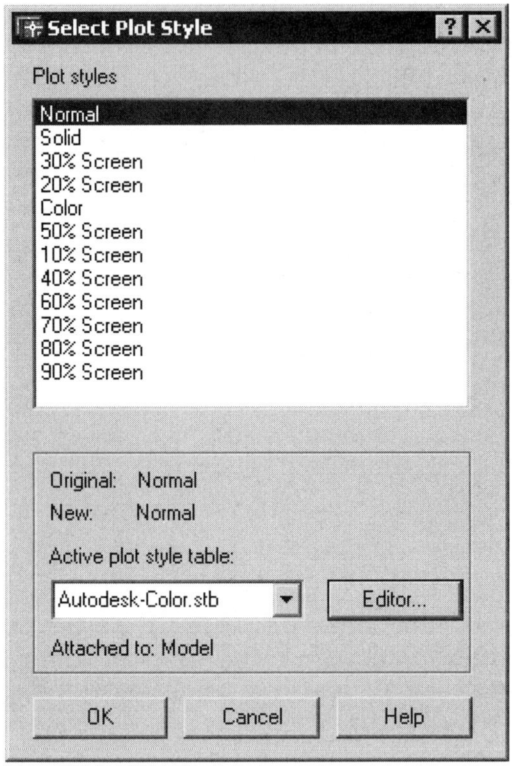

Figure 12-19 ***Select Plot Style*** *dialog box*

Setting the Current Plot Style

You can use the **PLOTSTYLE** command to set the current plot style for new objects or of selected objects. When you enter **PLOTSTYLE** at the Command prompt, and if no object has been selected in the drawing, AutoCAD displays the **Current Plot Style** dialog box, see Figure 12-20. But if any object selection is there in the drawing, then AutoCAD displays the **Select Plot Style** dialog box (Figure 12-19), which has been discussed earlier. You can select a plot style from the list box and choose **OK** to assign it to the selected objects in the drawing. All the plot styles present in the current plot style table that are assigned to the current tab are displayed in the list box in the **Current Plot Style** dialog box. You can select any one of these plot styles and choose **OK**. Now, when you create new objects, they will have the plot style that you had set current in the **Current Plot Style** dialog box. The parameters of this dialog box are described next.

Current plot style

The name of the current plot style is displayed adjacent to this label.

Plot style list box

This list box lists all the available plot styles that can be assigned to an object, including the default plot style, **Normal**.

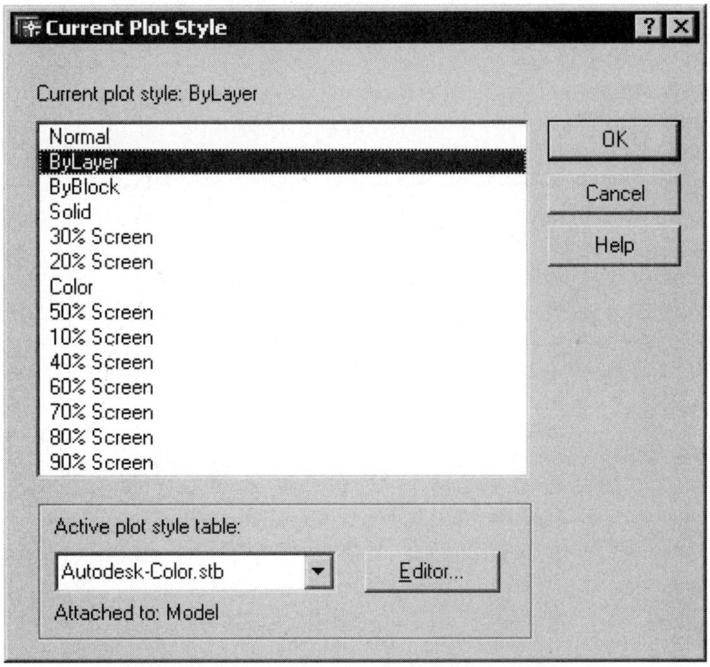

***Figure 12-20 Current Plot Style** dialog box*

Active plot style table
This drop-down list displays the names of all the available plot style tables. The current plot style table attached to the current layout or viewport is displayed in the edit box.

Editor
If you choose the **Editor** button, adjacent to the drop-down list, AutoCAD displays the **Plot Style Table Editor** to edit the selected plot style table.

Attached to
The tab to which the selected plot style table is attached, **Model** or any one of the layout tabs, is displayed next to this label.

Note
*You can also assign a current plot style using the **-PLOTSTYLE** command.*

Exercise 1 *Mechanical*

Make the drawing shown in Figure 12-21. Create a named plot style table *My Named Table.stb* with three plot styles: Style1, Style2, and Style3, in addition to the Normal plot style. The Normal plot style is used for plotting the object lines. These three styles have the following specifications.

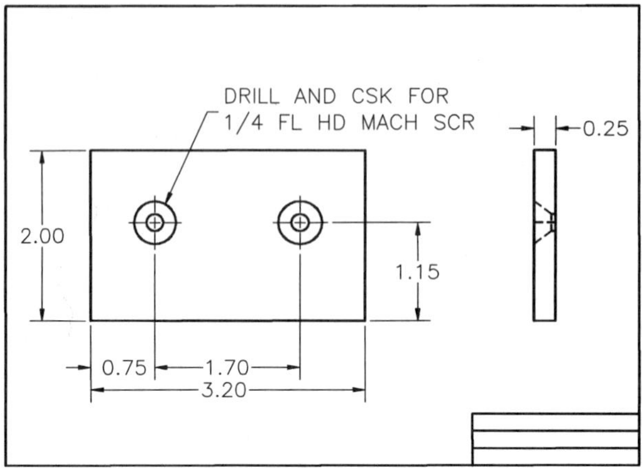

Figure 12-21 *Drawing for Exercise 1*

Style 1. This style has a value of Screening = 50. The dimensions, dimension lines, and the text in the drawing must be plotted with this style.

Style 2. This style has a value of Lineweight = 0.800. The border and title block must be plotted with this style.

Style 3. This style has a linetype of Medium Dash. The centerlines must be plotted with this plot style.

BATCH PLOTTING

Batch plotting is used to plot several drawings with or without their associated configuration files (PC3 files). Batch plot is not an AutoCAD command and cannot be run within AutoCAD. It is a Visual Basic batch plotting utility that can be executed at the system prompt. In Windows 95, 98, or NT operating systems, you can choose **Run**, and then select *Batchplt.exe* from *AutoCAD 2004* directory, see Figure 12-22.

Batch plot allows you to create a list of files that you want to plot. You can also assign a PC3 file to the drawing name. You can save the list in a batch plot list (BP3) file. Batch plot allows you to change the layout for plotting, change plotting parameters, or change a plotting device. When you choose **Batch Plot Utility** by the above-mentioned methods, the **AutoCAD Batch Plot Utility** window is displayed (Figure 12-23).

The following columns are displayed for a drawing file in this window.

Drawing File. The file name of the drawing to be plotted is displayed here.

Path. The full path for the drawing file is displayed.

*Figure 12-22 Invoking the **AutoCAD Batch Plot Utility** from the **Run** dialog box*

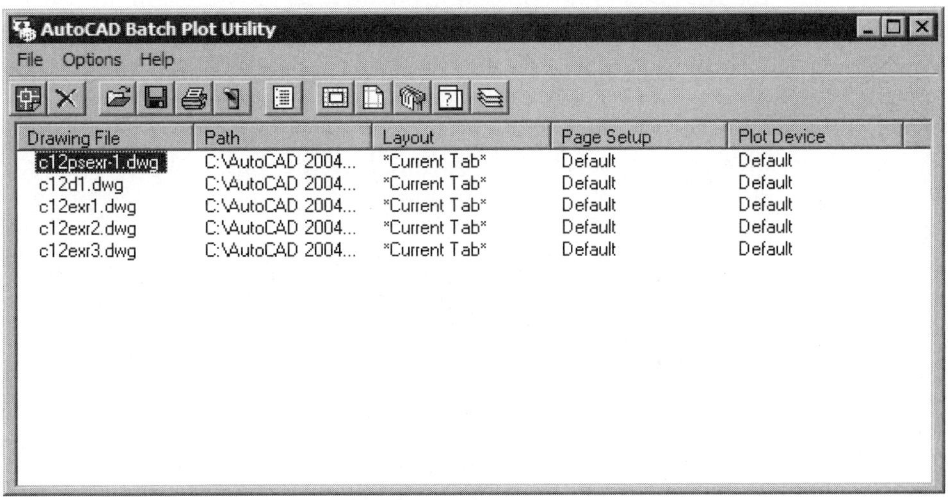

*Figure 12-23 **AutoCAD Batch Plot Utility** window*

Layout. The name of the layout of the drawing file that you want to plot is displayed.

Page Setup. The name of the page setup for plotting this drawing is displayed.

Plot Device. The name of the PC3 file for plotting the drawing is displayed.

The options available in the toolbar and the various menus provided in the **AutoCAD Batch Plot Utility** window are discussed next.

File Menu (AutoCAD Batch Plot Utility Window)

The following options are available in the **File** menu of the **AutoCAD Batch Plot Utility** window. These options are also available through the toolbar buttons in the dialog box, see Figure 12-24, or through the shortcut menu that is displayed when you select a drawing in the window and right-click.

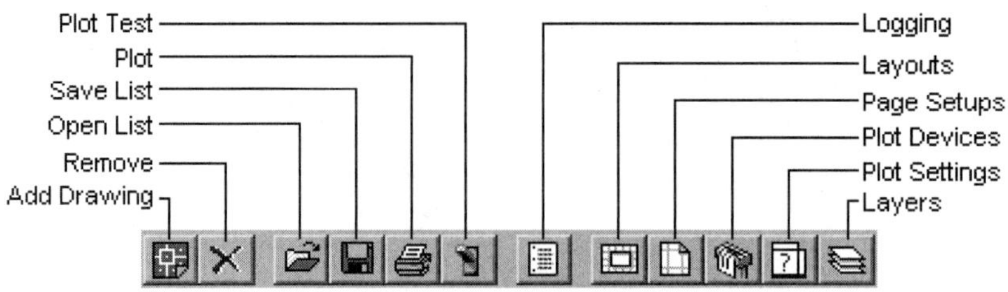

Figure 12-24 AutoCAD Batch Plot Utility *window toolbar*

Add Drawing

This option displays the **Add Drawing File** dialog box to select the drawing files you want to add to the batch plot file list. Select the files you want to add and then choose the **Open** button. The drawing will now be displayed in the **AutoCAD Batch Plot Utility** window.

Remove

You can use this option to remove a selected drawing from the batch plot list.

New List

You can use this option to open a new batch plot list window. If you have some unsaved changes in the current list, AutoCAD will prompt you to save the list. After you have saved the changes, the **AutoCAD Batch Plot Utility** window is displayed, where you can create a new list. This option is available only through the **File** menu in the batch plot utility window.

Open List

This option displays the **Open Batch Plot List** dialog box to select and load a saved Batch Plot List (BP3) file. The new file replaces the current file listing.

Append List

When you choose this option from the **File** menu in the batch plot utility window, the **Append Batch Plot List File** dialog box is displayed. You can use this dialog box to select a BP3 file. Once you have selected a BP3 file, choose **Open** and the files saved in this Batch Plot List (BP3) file get appended to the current file list. This option is not available in the toolbar.

Save List

This option saves the existing Batch Plot List and PC3 information in a BP3 file.

Plot

This option plots the drawing files listed in the list box according to the settings as defined in the corresponding PC3 file. The AutoCAD window comes to the front and the drawing currently plotted will be displayed. If the drawing is plotted successfully, a check mark will appear next to the drawing name and if it is not plotted properly, an X mark will appear next to the drawing name.

Plot Test

AutoCAD will load but does not plot each drawing in the batch. You can check for missing xrefs, fonts, or shapes. You can use the **Plot Test** option to correct any problems that will affect the drawings during plotting. The results of the test are displayed in the **Plot Test Results** window. You can choose **OK** to exit this dialog box. You can also save the plot results as a text file that can be viewed later by choosing the **Save As** button. The **Save Plot Test Results** dialog box is displayed where you can select a folder to save the text file in and then choose **Save** to return to the **Save Plot Test Results** dialog box. If you choose the **Append to log** button, the test results are added, especially if errors are encountered, to a log.

Logging

When you choose the **Logging** button, the **Logging** dialog box is displayed as shown in Figure 12-25.

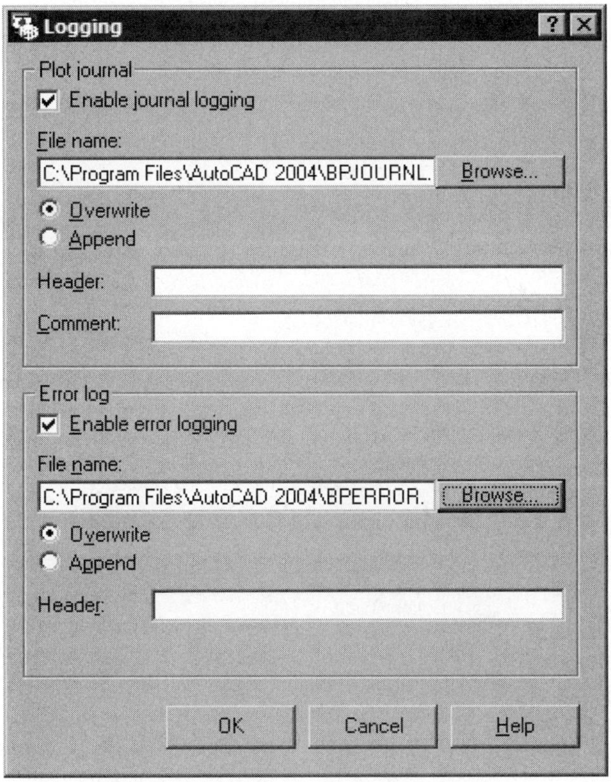

***Figure 12-25 Logging** dialog box*

You can create both plot journal logging and error logging by default. The **Enable journal logging** and **Enable error logging** check boxes are selected by default in the **Plot journal** and **Error log** areas of the **Logging** dialog box respectively. **Plot journal logging** creates a log about when and who plotted each drawing while **Error logging** shows whether any errors were encountered during plotting. The default *bpjournl.log* and *bperror.log* files are saved in the *AutoCAD 2004\Support\Batchplt* folder. These file names are displayed in the **File name** edit boxes in the dialog box by default. You can choose the **Browse** button adjacent to the **File Name** edit box in the **Plot journal** and **Error log** areas. The **Journal Filename** and **Error Log Filename** dialog boxes are displayed respectively and can be used to locate another folder where you can save the log files with different names if you want to.

The **Overwrite** button is selected by default in both the **Plot journal** and **Error log** areas of the **Logging** dialog box. This implies that the new batch plot log files shall overwrite the last log files that were saved. If you select the **Append** radio button, the current log file created will be also saved be in addition to the previous log file. You select this option when you need to keep the old log. You can add a line of text to the current log file by entering this text in the **Header** edit boxes. The text you enter in the **Comment** edit box gets added for each drawing in a file.

Exit

Exit closes the **AutoCAD Batch Plot Utility** window. The existing list of files and the corresponding plot configurations will be lost if they have not been saved.

Options Menu (AutoCAD Batch Plot Utility Window)

The following options are available in the **Options** menu of the **AutoCAD Batch Plot Utility** window.

Layouts

This option displays the **Layouts** dialog box, see Figure 12-26 (for single or multiple drawings), depending on the number of drawings selected.

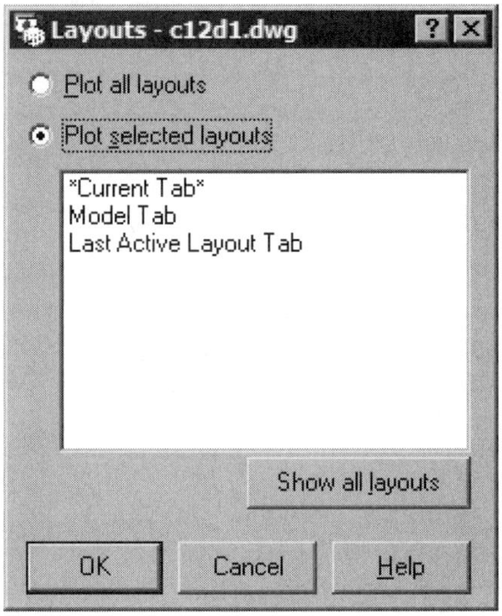

***Figure 12-26 Layouts** dialog box*

You can use this dialog box to change the layouts to plot. The default layout is **Current Tab** and it is the tab in which the drawing was saved last. You can choose the **Show all layouts** button to display all the layouts in the selected drawing. You can either select the **Plot all layouts** radio button to plot all the available layouts or select the **Plot selected layouts** radio button and select any layouts from the list box. Once you have selected the layouts in the list box, choose **OK** to exit the dialog box. The selected layouts are displayed in the **AutoCAD Batch Plot Utility** window under the **Layout** column.

Page Setups

This option displays the **Page Setups** dialog box, see Figure 12-27. This dialog box lists all the available setups. You can also import new setups from other drawings. Choose the [...] button to display the **Select Drawing** dialog box that is used to select a drawing whose page setup you want to import and choose **Open**. The selected drawing whose page setup is to be used is displayed in the **Load page setups from drawing or template area** of the dialog box. You can select a setup by selecting it in the list box to highlight it and then choosing the **OK** button to exit the dialog box. The selected page setup name is displayed in the **Page Setup** column of the **AutoCAD Batch Plot Utility** window.

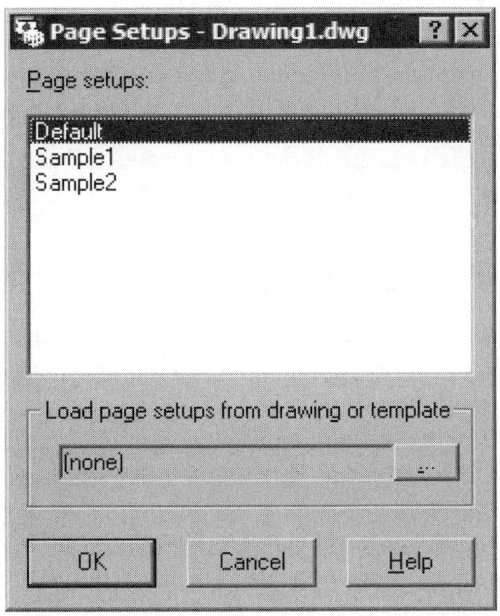

***Figure 12-27 Page Setups** dialog box*

Plot Devices

This option displays the **Plot Devices** dialog box, see Figure 12-28. This dialog box lists all the available plotting devices. You can specify a plot device for each drawing that will be used while plotting. The **Show plot device description** check box is selected by default and displays a description of the device selected from the **Plot devices** list box in the **Description** area. You can choose the **Browse** button to display the **Open Plot Configuration File** dialog box. You can select another plot configuration file (apart from the ones listed in the list box) using

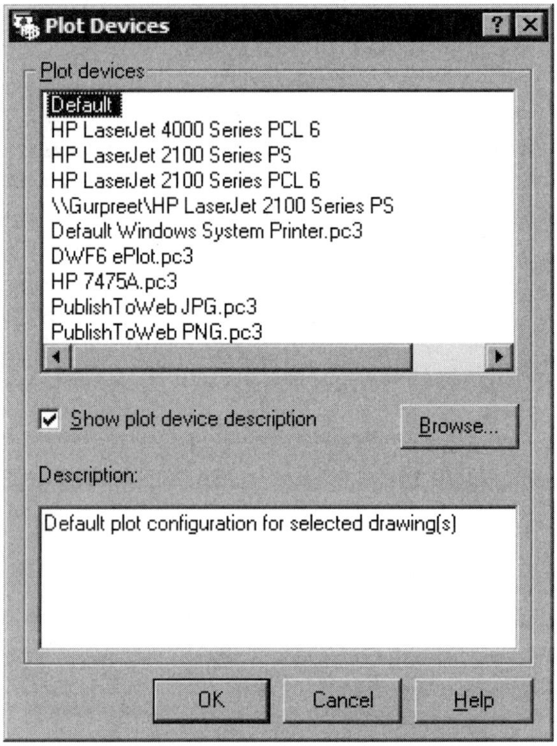

***Figure 12-28 Plot Devices** dialog box*

this dialog box and then choose the **Open** button to return to the **Plot Devices** dialog box. The selected plot configuration file gets added to the **Plot devices** list box. You can now select a plot configuration file from the list box to highlight it and choose the **OK** button to exit the **Plot Devices** dialog box. The selected device is displayed in the **Plot Device** column in the **AutoCAD Batch Plot Utility** window.

Plot Settings

This option displays the **Plot Setting** tab of the **Plot Settings** dialog box (Figure 12-29) where you can specify the area to plot, refer to scaling, and find other information related to plotting to a file for each of the drawings or all the drawings. You can set plot setting options using this dialog box. You can decide the area that you want to plot by selecting one of the radio buttons in the **Plot area**. These options are also available in the **Plot/Page Setup** dialog box, except the **Window** option. The scale at which the drawing is to be plotted is displayed in the **Plot scale** area. You can modify the scale that was selected earlier in the **Page setup/Plot** dialog box for plotting by selecting another scale from the **Scale** drop-down list in the **Plot scale** area of the **Plot Settings** dialog box. You can also select the **scale lineweights** check box in the **Plot scale** area, if you want to make sure that the lineweights in the drawing are to be plotted after they have been scaled as per the plot scale. You can plot a drawing to a file by specifying its name in the **File name** edit box. Its location can be specified by choosing the [**...**] button to display the **Plot-To-File** dialog box. Select the folder where you want to save the

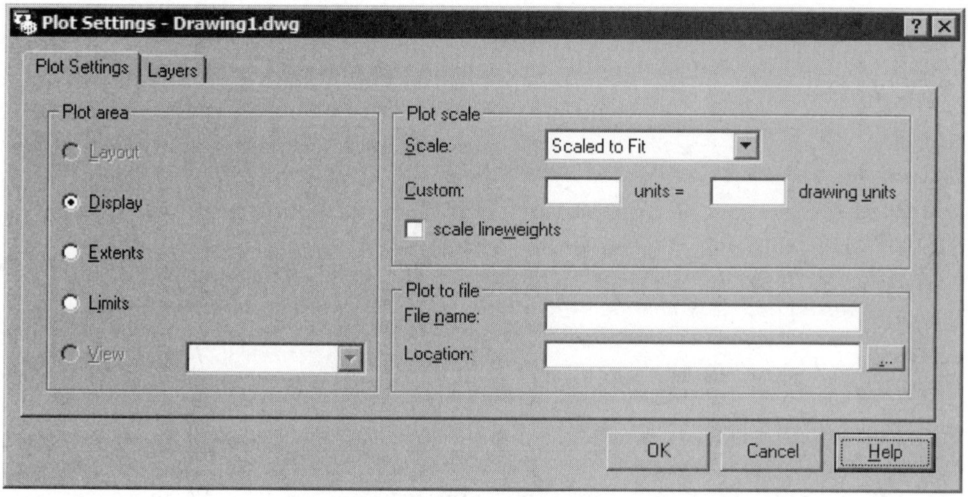

Figure 12-29 ***Plot Settings*** *dialog box*

plot using the dialog box and choose the **Save** button to return to the **Plot Settings** dialog box. Once all plot settings are specified in the dialog box, choose **OK** to exit it and return to the **AutoCAD Batch Plot Utility** window.

Layers

This option displays the **Layers** tab of the **Plot Settings** dialog box (Figure 12-30) where you can specify the layers to plot. You can select a layer or multiple layers in this list box and change their status from on to off or from off to on by choosing the **Off** or **On** buttons respectively. You can use the SHIFT or CTRL keys to select multiple layers in the list box. For

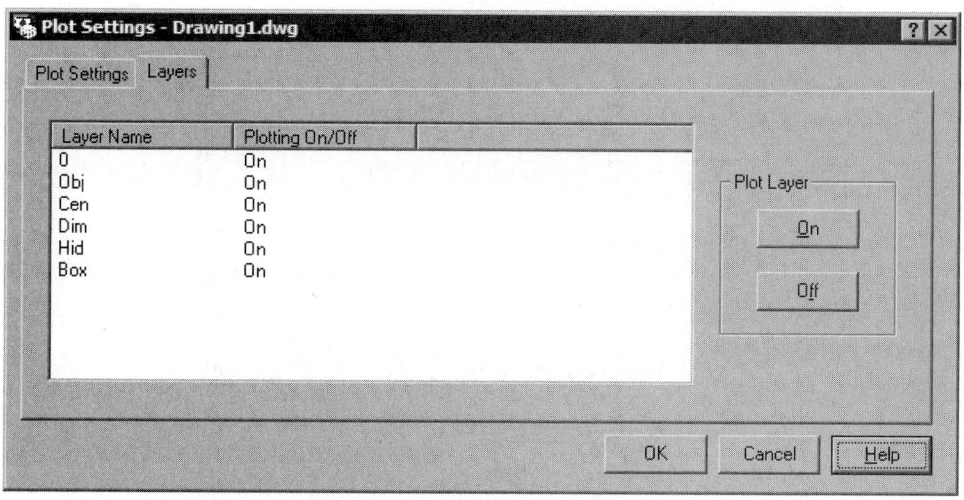

Figure 12-30 ***Plot Settings*** *dialog box (**Layers** tab)*

example, if you select a layer whose status is on and you want to turn it off, select the layer and choose the **Off** button. Once you have determined whether you want to turn off or turn on the plotting of layers in the list box, choose the **OK** button to return to the **AutoCAD Batch Plot Utility** dialog box.

Tip

PC3 files are the plot configuration files that contain the information about the plotter and all of its settings, as well as the paper or other output media. The PC3 files are completely drawing-independent and can be shared among AutoCAD users. This makes it possible to use batch plotting to plot the drawings on different plotters and printers.

Self-Evaluation Test

Answer the following questions, and then compare your answers to the correct answers given at the end of this chapter.

1. All the settings about a plotter are saved in the .PC3 file. (T/F)

2. Different objects in the same drawing can be plotted in different colors, with different linetypes and line widths. (T/F)

3. Full preview takes less time than partial preview. (T/F)

4. The **PSLTSCALE** system variable controls the paper space linetype scaling and has a default value of 1. (T/F)

5. The size of a plot can be specified by selecting any paper size from the __________ drop-down list in the **Plot** dialog box.

6. If you want to store the plot in a file and not have it printed directly on a plotter, select the __________ check box in the **Plot Device** tab of the **Plot** dialog box.

7. The scale for the plot can also be specified in the __________ edit boxes in the **Plot** dialog box.

8. If you select the __________ radio button in the **Plot** dialog box, the portion of the drawing that is in the current display is plotted.

9. In the **Partial Plot Preview** dialog box, the __________ is graphically represented by a white rectangle. The blue hatched rectangle is the section in the paper that is used by the __________ .

10. You can use the __________ window to create a list of drawing files that are plotted and then you can go back to working on other programs.

Review Questions

Answer the following questions.

1. When you choose a layout tab for the first time in a drawing, the **Page Setup** dialog box is displayed. (T/F)

2. By selecting the **View** radio button in the **Plot area** of the **Page Setup/Plot** dialog box, you can plot a view that was created with the **VIEW** command in the current drawing. (T/F)

3. If you do not want the hidden lines of a 3D object created in the **Model** tab, select the **Hide Objects** check box in the **Plot options** area of the **Plot Settings** tab of the **Plot** dialog box. (T/F)

4. The orientation of the drawing can be changed using the **Plot** dialog box. (T/F)

5. Which check box in the **Plot options area** of the **Plot/Page Setup** dialog box is available only if you are plotting in a layout tab?

   (a) **Plot paper space last** (b) **Hide objects**
   (c) **Plot with plot styles** (d) **None**

6. Which command when invoked displays the **Plotters** window?

   (a) **PLOTTER** (b) **PLOTTERMANAGER**
   (c) **PLOTSTYLE** (d) **None**

7. With which command is it possible to bypass the **Plot/Page Setup** dialog box and directly import a page setup from an existing drawing into a new drawing layout?

   (a) **PSETUPIN** (b) **PLOTTERMANAGER**
   (c) **PLOTSTYLE** (d) **None**

8. Which command is used to create a new plot style?

   (a) **STYLE** (b) **STYLESMANAGER**
   (c) **PLOTSTYLE** (d) **None**

9. Which command can be used to import a PCP file or PC2 files?

   (a) **PCINWIZARD** (b) **PCIN**
   (c) **PLOTSTYLE** (d) **None**

10. You can view the plot on the specified paper size before actually plotting it by selecting the __________ button in the **Plot** dialog box.

11. In the **Partial Plot Preview** dialog box, the red triangle icon located at one of the corners of the effective area (hatched blue rectangle) represents the __________ of the plot.

12. You can modify the properties of a selected plotting device by using the __________ .

13. The plot style modes available are __________ and __________ .

14. The **Plot Style Table Editor** has three tabs: __________ , __________ , and __________ .

15. Batch plotting allows you to create a list of files that you want to plot. This list can be saved as a __________ file.

Create the drawing shown in Figure 12-31 and plot it according to the following specifications.

Exercises

Exercise 2 *Mechanical*

Create and use a plot style table with the specified plot styles.

1. The drawing is to be plotted on 10 X 8 inch paper.

2. The object lines must be plotted with a plot style Style 1. Style 1 must have a value of lineweight = 0.800 mm.

3. The dimension lines must be plotted with plot style Style 2. Style 2 must have a value of screening = 50.

4. The centerlines must be plotted with plot style Style 3. Style 3 must have a linetype of Medium Dash and screening = 50.

5. The border and title block must be plotted with plot style Style 4. The value of lineweight =0.25 mm.

Exercise 3 *Mechanical*

Create the drawing shown in Figure 12-32 and plot the drawing according to your specifications.

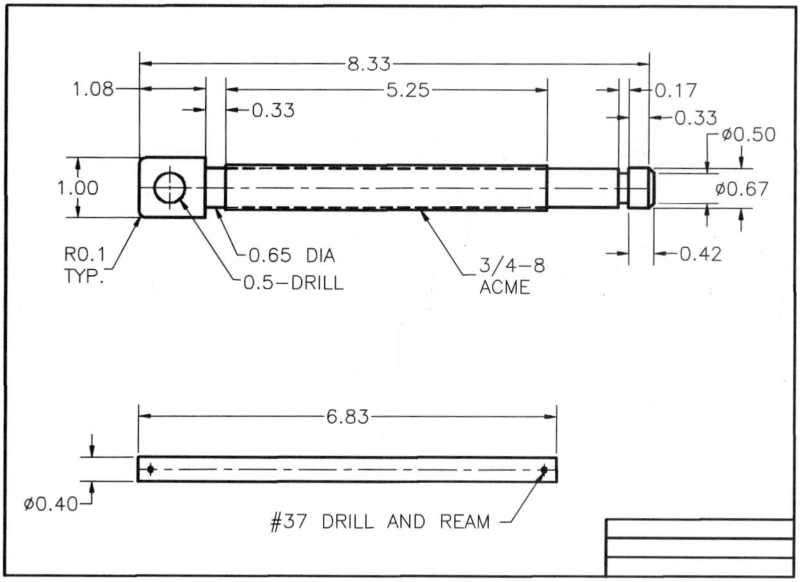

Figure 12-31 *Drawing for Exercise 2*

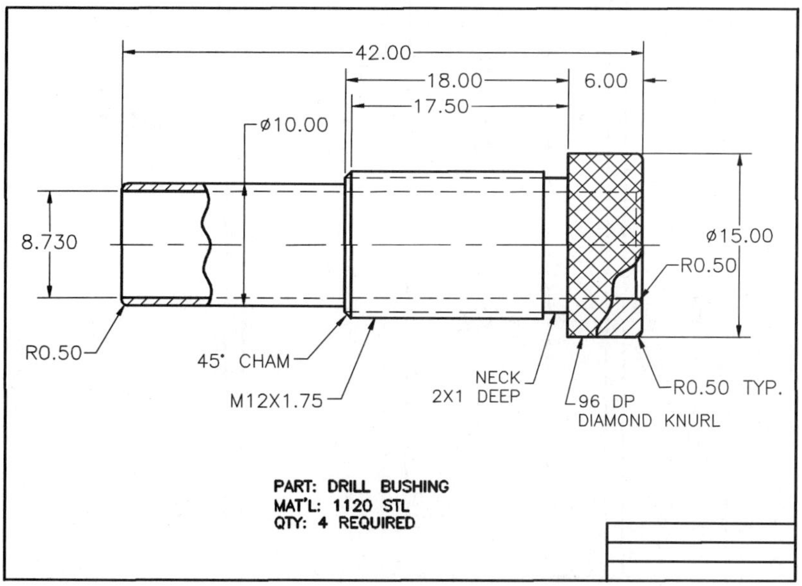

Figure 12-32 *Drawing for Exercise 3*

Problem Solving Exercise 1 *Mechanical*

Make the drawing shown in Figure 12-33 and plot it according to your specifications.

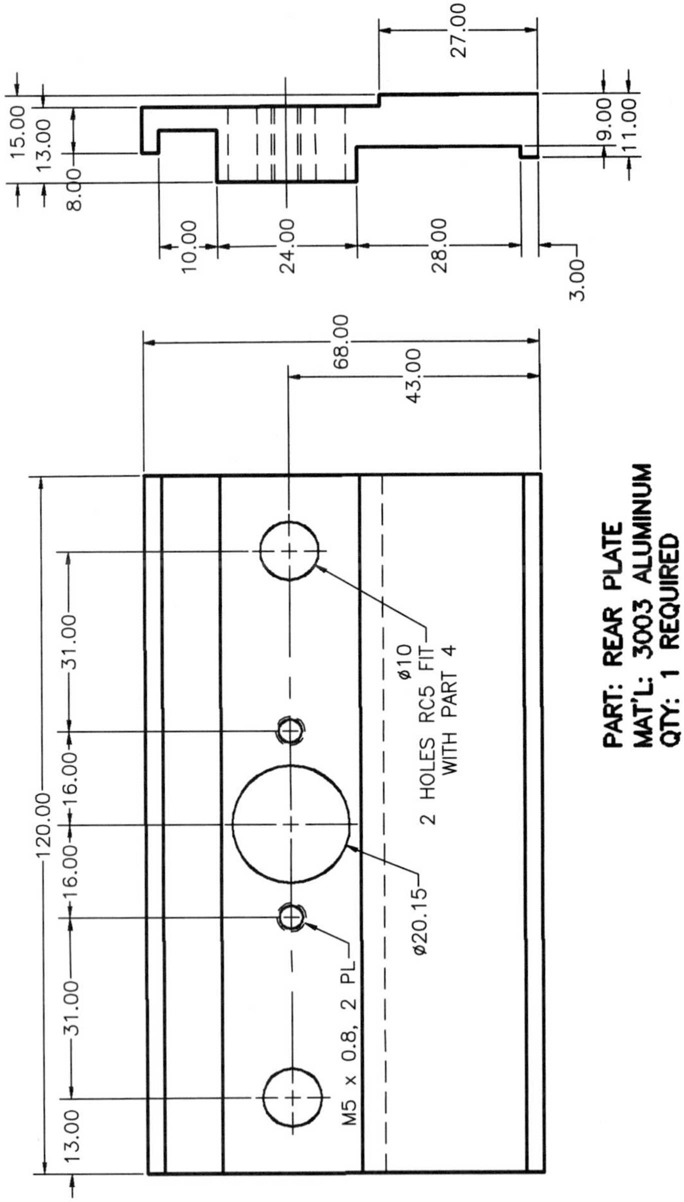

Figure 12-33 *Drawing for Problem Solving Exercise 1*

Answers to Self-Evaluation Test

1 - T, **2** - T, **3** - F, **4** - T, **5** - **Paper size**, **6** - **Plot to file**, **7** - **Custom**, **8** - **Display**, **9** - paper size, image, **10** - **AutoCAD Batch Plot Utility**

Chapter 13

Hatching Drawings

Learning Objectives

After completing this chapter you will be able to:

- *Use the **BHATCH** command to hatch an area using various patterns.*
- *Use boundary hatch with predefined, user-defined, and custom hatch patterns as options.*
- *Specify pattern properties.*
- *Preview and apply hatching.*
- *Use advanced hatching options and Ray Casting.*
- *Edit associative hatch and hatch boundary using **HATCHEDIT** command, **PROPERTIES** palette, and Grips.*
- *Hatch inserted blocks.*
- *Align hatch lines in adjacent hatch areas.*
- *Hatch by using the **HATCH** command at the Command prompt.*

- *Hatch inserted blocks.*
- *Align hatch lines in adjacent hatch areas.*
- *Hatch by using the **HATCH** command at the Command prompt.*

HATCHING

In many drawings, such as sections of solids, the sectioned area needs to be filled with some pattern. Different filling patterns make it possible to distinguish between different parts or components of an object. Also, the material the object is made of can be indicated by the filling pattern. You can also use these filling patterns in graphics for rendering architectural elevations of buildings, or indicating the different levels in terrain and contour maps. Filling objects with a pattern is known as hatching (Figure 13-1). This hatching process can be accomplished by using the **BHATCH** or **HATCH** command or by using the **TOOL PALETTES**.

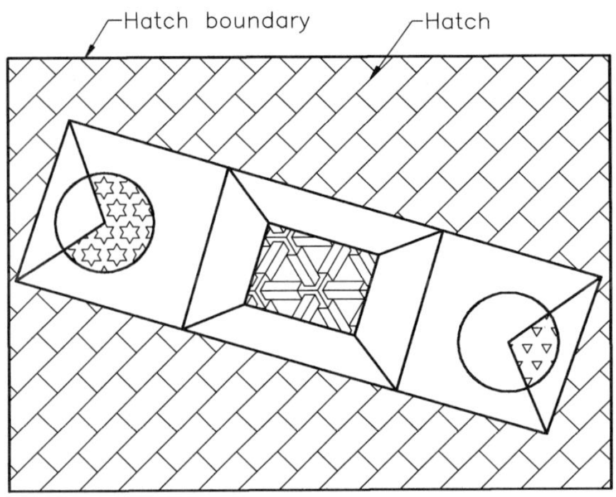

Figure 13-1 *Illustration of hatching*

Before using the **BHATCH** and **HATCH** commands, you need to understand some terms that are used when hatching. The following subsection describes some of the terms.

Hatch Patterns

AutoCAD supports a variety of hatch patterns (Figure 13-2). Every hatch pattern is composed of one or more hatch lines or a solid fill. The lines are placed at specified angles and spacing. You can change the angle and the spacing between the hatch lines. These lines may be broken into dots and dashes, or may be continuous, as required. The hatch pattern is trimmed or repeated, as required, to fill exactly the specified area. The lines comprising the hatch are drawn in the current drawing plane. The basic mechanism behind hatching is that the line objects of the pattern you have specified are generated and incorporated in the desired area in the drawing. Although a hatch can contain many lines, AutoCAD normally groups them together into an internally generated object and treats them as such for all practical purposes. For example, if you want to perform an editing operation, such as erasing the hatch, all you need to do is select any point on the hatch and press ENTER, the entire pattern gets deleted. If you want to break a pattern into individual lines to edit individual lines, you can use AutoCAD's **EXPLODE** command.

Figure 13-2 Some hatch patterns

Hatch Boundary

Hatching can be used on parts of a drawing enclosed by a boundary. This boundary may be lines, circles, arcs, polylines, 3D faces, or other objects, and at least part of each bounding object must be displayed within the active viewport. The **BHATCH** command automatically defines the boundary, whereas in the case of the **HATCH** command you have to define the boundary by selecting the objects that form the boundary of the hatch area.

HATCHING DRAWINGS USING THE BOUNDARY HATCH AND FILL DIALOG BOX*

Toolbar:	Draw > Hatch
Menu:	Draw > Hatch
Command:	BHATCH

The **BHATCH** (boundary hatch) command allows you to hatch a region enclosed within a boundary (closed area) by selecting a point inside the boundary or by selecting the objects to be hatched. This command automatically designates a boundary and ignores any other objects (whole or partial) that may not be a part of this boundary. When you invoke the **BHATCH** command, the **Boundary Hatch and Fill** dialog box is displayed (Figure 13-3) through which you can perform the hatching operation.

One of the advantages of this command is that you do not have to select each object comprising the boundary of the area you want to hatch as in the case of the **HATCH** command (discussed later). This is because this command defines a boundary by creating a polyline from all the objects comprising the boundary. By default, this polyline gets deleted; however, if you want to retain it, you can specify that. This command also allows you to preview the hatch before actually applying it.

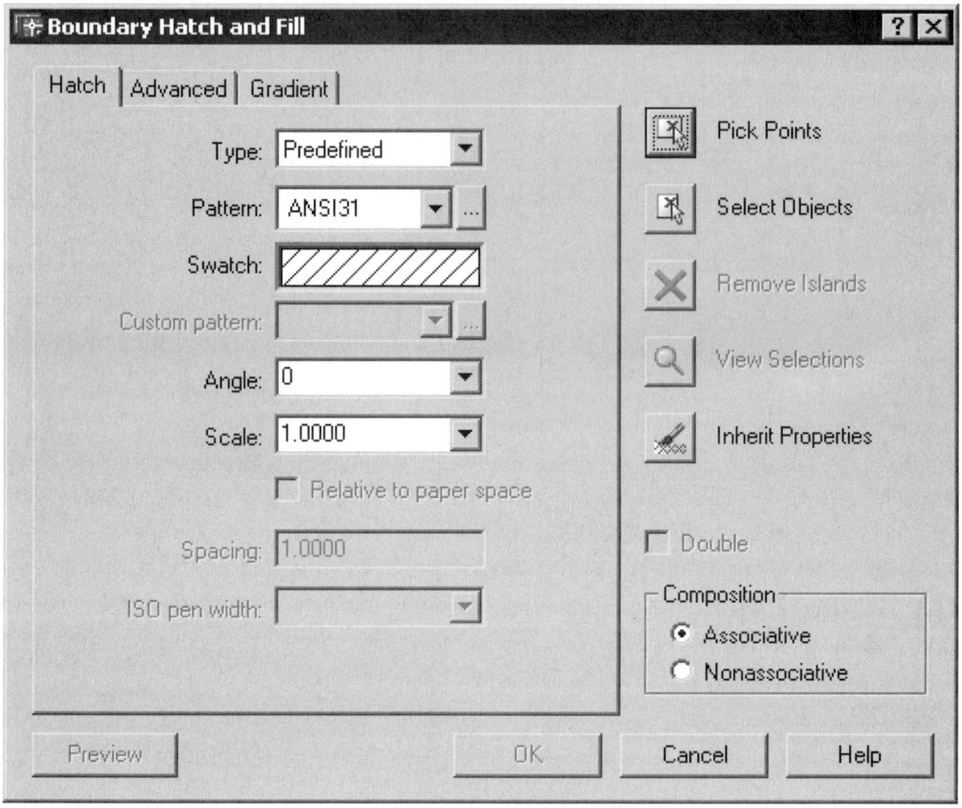

***Figure 13-3 Boundary Hatch and Fill** dialog box*

Example 1 — *General*

In this example, you will hatch a circle using the default hatch settings. Later in the chapter you will learn how to change the settings to get a desired hatch pattern.

1. Invoke the **Boundary Hatch and Fill** dialog box by choosing the **Hatch** button on the **Draw** toolbar.

2. Choose the **Pick Points** button in the **Boundary Hatch and Fill** dialog box. The dialog box temporarily closes.

3. Select a point inside the circle (P1) (Figure 13-4) and press ENTER. The **Boundary Hatch and Fill** dialog box is displayed again. You can now preview the hatching by choosing the **Preview** button. After specifying an internal point, you can also right-click to display a shortcut menu and select the **Preview** option.

4. After checking the preview, press ESC to complete the point specification; the dialog box reappears. You can also right-click to apply the hatch without invoking the dialog box.

5. Choose **OK** to apply the hatch to the selected object, Figure 13-5.

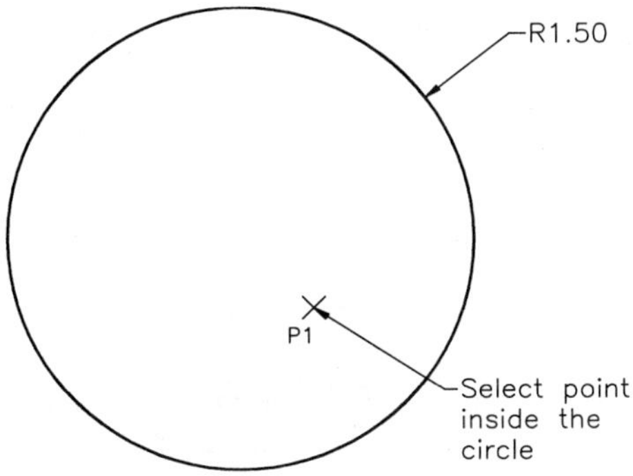

Figure 13-4 *Specifying a point to hatch the circle*

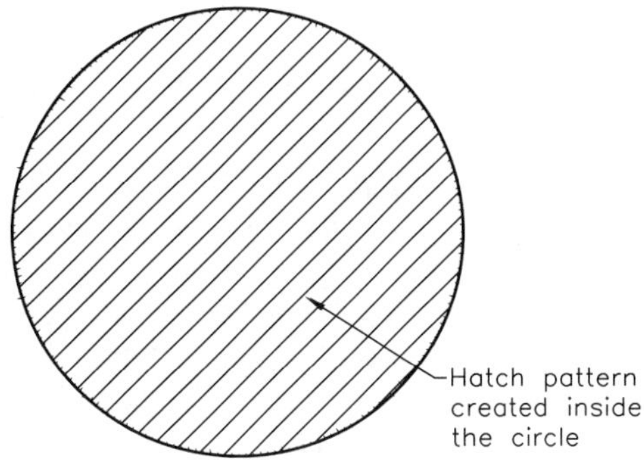

Figure 13-5 *Drawing after hatching*

Tip

*The **FILLMODE** system variable is set to On by default and hence the hatch patterns are displayed. In case of large hatching areas you can set the **FILLMODE** to Off, so that the hatch pattern is not displayed and regeneration time is saved.*

You can also hatch a selected face of a solid model. However, to hatch a face of the solid model, you need to align the UCS at that face. You will learn more about solid models and UCS in later chapters.

BOUNDARY HATCH AND FILL DIALOG BOX OPTIONS

The **Boundary Hatch and Fill** dialog box has several options that let you control various aspects of hatching, such as pattern type, scale, angle, boundary parameters, associativity, and so on. The following is a description of these options. The options have been grouped by the tab name in which they are found in the **Boundary Hatch and Fill** dialog box. The **Boundary Hatch and Fill** dialog box has three tabs: **Hatch**, **Advanced**, and **Gradient**. The description of these tabs follows.

Hatch Tab

This tab provides options that define the appearance of the hatching patterns that can be applied (Figure 13-3). For quick hatching of an object this tab can be used easily.

Type

The **Type** drop-down list (Figure 13-6) displays the types of patterns that can be used for hatching drawing objects. This list lets you choose the type of hatch pattern you want. The three types of hatch patterns available are **Predefined**, **User defined**, and **Custom**. The predefined type of patterns come with AutoCAD and are stored in the *acad.pat* and *acadiso.pat* files. The predefined type of hatch pattern

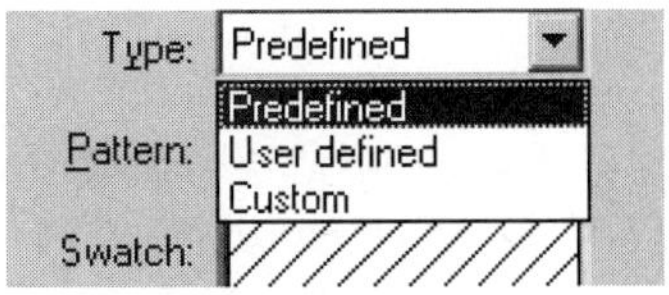

Figure 13-6 *The **Type** drop-down list displaying hatch pattern types*

is the default type. When you select **User defined** from the **Type** drop-down list a pattern of lines that is based on the current linetype is created. You can set the angle of the lines in the pattern and also the spacing between them. When you select **Custom** from the **Type** drop-down list, you can select a pattern that has been defined in a custom PAT file.

Note

*When using the patterns defined in custom PAT files, make sure that their search path has been added in the **Files** tab of the **Options** dialog box.*

Pattern

The **Pattern** drop-down list (Figure 13-7) displays the names of all the predefined patterns available in AutoCAD. You can select any pattern from this list and the selected pattern is displayed in the **Swatch** box. The selected pattern is stored in the **HPNAME** system variable. ANSI31 is the default pattern in the **HPNAME** system variable. The **Pattern** drop-down list is available only if you have selected **Predefined** from the **Type** drop-down list. The six most recently used hatch patterns are displayed at the top of the **Pattern** drop-down list.

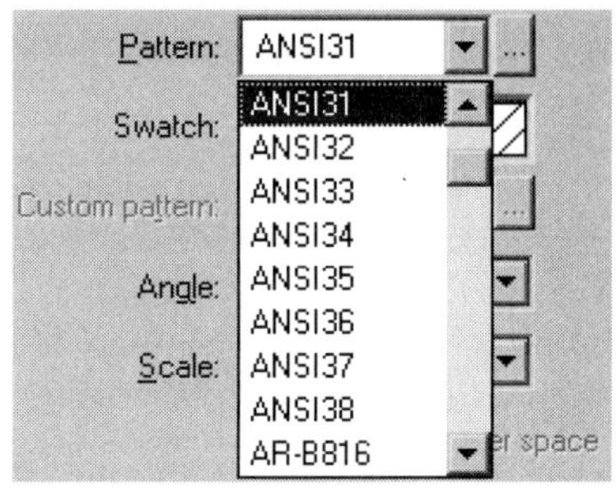

***Figure 13-7 Pattern** drop-down list*

To select a particular hatch pattern, you can also choose the [...] button that displays the **Hatch Pattern Palette** dialog box (Figure 13-8). This dialog box shows the images and names of the available hatch patterns. This dialog box is also displayed by clicking on the pattern displayed in the **Swatch** box. The **Hatch Pattern Palette** dialog box has four tabs. They are described as follows.

ANSI. This tab displays all the ANSI (American National Standards Institute)-defined hatch patterns that come with AutoCAD. ANSI31 is the default pattern.

ISO. This tab displays all the ISO (International Standard Organization)-defined hatch patterns that come with AutoCAD.

Other Predefined. This tab displays all the predefined hatch patterns other than ANSI- or ISO-defined patterns.

Custom. This tab displays all custom patterns described in any custom PAT file that is added in the search path of AutoCAD 2004.

Tip

*Using the **Hatch Pattern Palette** dialog box for selecting a hatch pattern is more convenient since you can view all the pattern names and the corresponding patterns simultaneously.*

Swatch

This box displays the selected pattern. When you click in the **Swatch** box, the **Hatch Pattern**

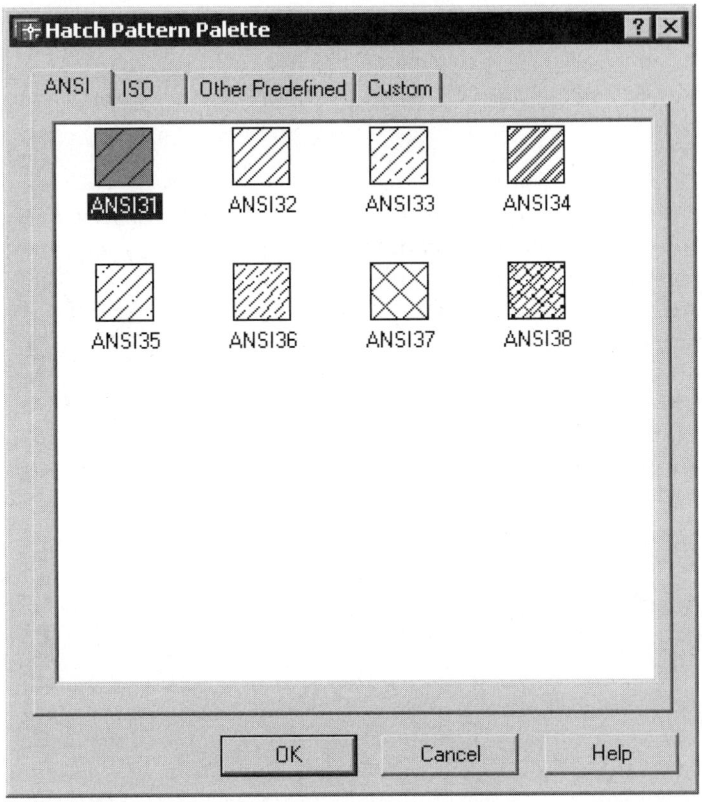

***Figure 13-8 Hatch Pattern Palette** dialog box (**ANSI** tab)*

Palette dialog box is displayed.

Custom Pattern

This drop-down list is available only if you have selected **Custom** from the **Type** drop-down list. It displays all the available custom hatch patterns defined in the custom PAT files that have been added to the AutoCAD search path. The recently used custom hatch patterns in a current drawing session appear in the list. The selected pattern is stored in the **HPNAME** system variable. You can also choose the **[...]** button; the **Hatch Pattern Palette** dialog box (**Custom** tab) appears. This tab displays the custom pattern paths.

Angle

The **Angle** drop-down list (Figure 13-9) displays angles (at an increment of 15-degree) by which you can rotate the hatch pattern with respect to the *X* axis of the current UCS. You can select any angle from this drop-down list or you can also enter an angle of rotation of your choice in the Angle edit box. The angle value is stored in the **HPANG** system variable. The angle of hatch lines of a particular hatch pattern is governed by the values specified

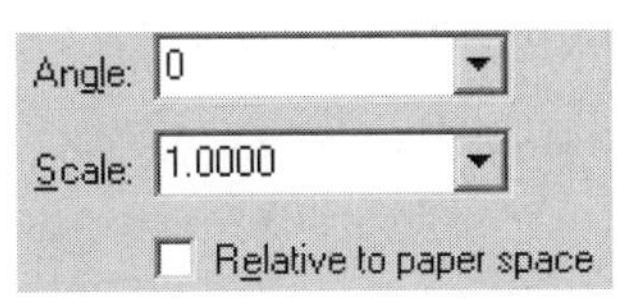

***Figure 13-9** The **Angle** and **Scale** drop-down lists*

in the hatch definition. For example, in the **ANSI31** hatch pattern definition, the specified angle of hatch lines is 45-degree. If you select an angle of 0, the angle of hatch lines will be 45-degree. If you enter an angle of 45-degree, the angle of the hatch lines will be 90-degree.

Scale

The **Scale** drop-down list (Figure 13-9) displays scale factors by which you can expand or contract the selected hatch pattern. The scale factors that are displayed range between 0.25 to 2 at increments of 0.25. You can also enter the scale factor of your choice in the **Scale** edit box. The **Scale** drop-down list is not available when you are using user-defined hatch patterns. The scale value is stored in the **HPSCALE** system variable. A value of 1 does not mean that the distance between the hatch lines is 1 unit. The distance between the hatch lines and other parameters of a hatch pattern is governed by the values specified in the hatch definition. For example, in the ANSI31 hatch pattern definition, the specified distance between the hatch lines is 0.125. If you select a scale factor of 1, the distance between the lines will be 0.125. If you enter a scale factor of 0.5, the distance between the hatch lines will be 0.5 X 0.125 = 0.0625.

Relative to Paper Space

If this check box (Figure 13-9) is selected then AutoCAD 2004 will automatically scale the hatched pattern relative to the paper space units. This option can be used to display the hatch pattern at a scale that is appropriate for your layout. This option is available only in a layout.

Spacing

The **Spacing** edit box is available only when you have selected **User defined** from the **Type** drop-down list. The value you enter in the **Spacing** edit box sets the spacing between the lines in a user-defined hatch pattern. This spacing is stored in the **HPSPACE** system variable. The default spacing value is 1.0000.

ISO Pen Width

The **ISO pen width** drop-down list is available only for ISO hatch patterns. You can select the desired pen width value from the **ISO pen width** drop-down list (Figure 13-10). The value selected specifies the ISO-related pattern scaling.

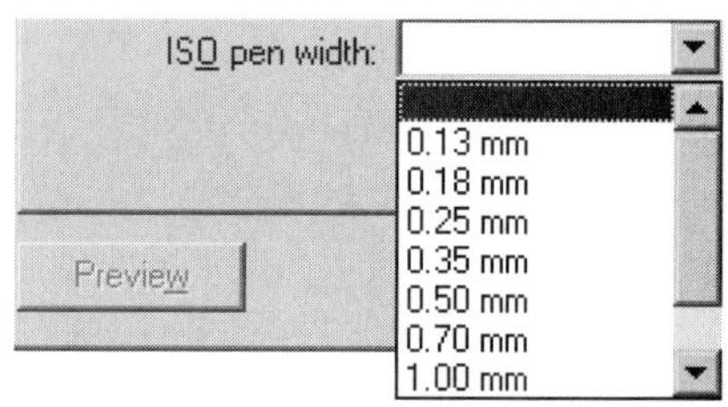

***Figure 13-10** The **ISO pen width** drop-down list*

User defined Hatch Patterns

To define a simple pattern, you can select **User defined** from the **Type** drop-down list. You will notice that the **Angle** and **Spacing** edit boxes and the **Double** check box are available in the dialog box. The **Scale** drop-down list is not available.

Angle

Just as with predefined patterns, you can specify in the **Angle** edit box an angle with respect to the *X* axis of the current UCS through which the hatch pattern is rotated. This value is stored in the **HPANG** system variable.

Spacing

This edit box is available only for user-defined patterns and allows you to specify the space between the hatch lines in the pattern. The entered spacing value is stored in the **HPSPACE** system variable.

Double

This check box is available only for user-defined patterns. When you select this check box, AutoCAD doubles the original pattern by drawing a second set of lines at right angles to the original lines in the hatch pattern. For example, if you have a parallel set of lines as a user-defined pattern and if you select the **Double** check box, the resulting pattern has two sets of lines intersecting at 90-degree. You can notice the effect of selecting the **Double** check box in Figure 13-11. If the **Double** check box is selected, the **HPDOUBLE** system variable is set to 1.

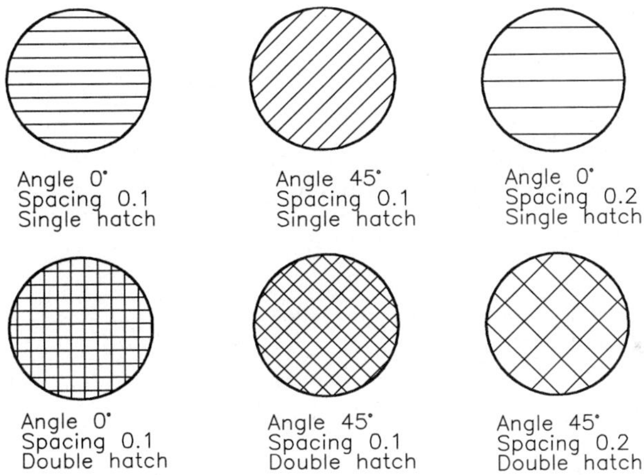

***Figure 13-11** Specifying angle and spacing for user-defined hatch patterns*

Custom Hatch Patterns

When you select **Custom** from the **Type** drop-down list in the **Boundary Hatch and Fill** dialog box, you will notice that the **Custom pattern** drop-down list is available. The **Angle** and **Scale** drop-down lists are also available. See Figure 13-12.

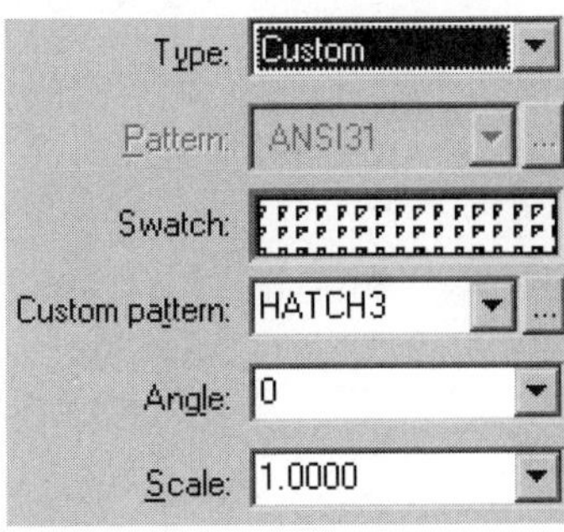

***Figure 13-12** Selecting a Custom pattern*

Custom Pattern

In AutoCAD, hatch patterns are normally stored in the file named *acad.pat*. You can select a hatch pattern from an individual file by selecting **Custom** from the **Type** drop-down list (Figure 13-12). The **Custom pattern** drop-down list is available now. You can choose the [...] button to display the **Custom** tab of the **Hatch Pattern Palette** dialog box. Here, the list box on the left side of the dialog box displays the paths of the custom hatch patterns. When you select a path, the corresponding pattern is displayed in a preview window on the left. Choose **OK** to return to the **Boundary Hatch and Fill** dialog box. The corresponding pattern is displayed in the **Swatch** window. You can also select a pattern from the **Custom pattern** drop-down list. You can also enter the name of the custom pattern in the field if you know the name of the hatch pattern. The name is held in the **HPNAME** system variable. If AutoCAD does not locate the entered pattern in the *acad.pat* file, it searches for it in a file with the same name as the pattern. You can specify an angle and scale for the custom pattern using the **Angle** and **Scale** drop-down lists as in the case of predefined patterns.

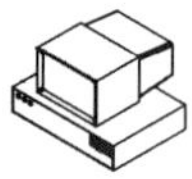

Tip

When you are hatching large areas, it is better to use a larger scale factor for the hatch pattern since you save time regenerating and plotting.

Also, when using a larger scale factor for the hatch pattern, the drawing appears to be neater and not too cluttered.

Exercise 1 *Mechanical*

In this exercise, you will hatch the given drawing using the hatch pattern named STEEL. Set the scale and the angle to match the drawing shown in Figure 13-13.

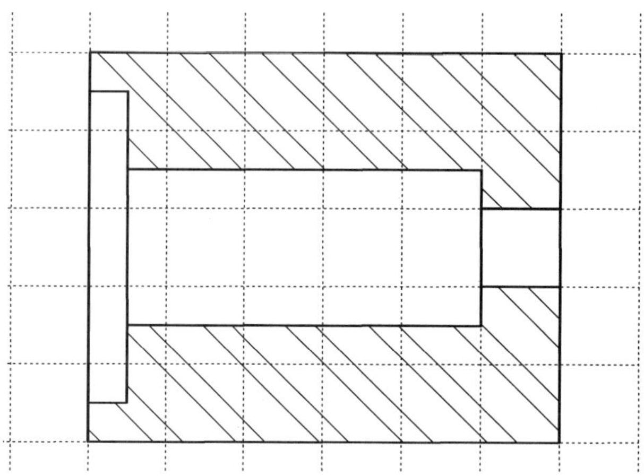

Figure 13-13 *Drawing for Exercise 1*

Selecting Hatch Boundary

The boundary selection options available in the **Boundary Hatch and Fill** dialog box allow you to define the hatch boundary. This is done by selecting a point inside a closed area to be hatched or by selecting the objects. There are other options also available here for removing islands, viewing the selection, using properties from an existing hatched object, and controlling associativity of the hatch pattern.

Pick Points

When you choose this button, AutoCAD automatically constructs a boundary from the objects that form a closed area. This is the easiest method to specify a region for hatching. When you choose the **Pick Points** button, the dialog box disappears temporarily to allow you to select a point within the closed area to be hatched. Select a point inside the object and a boundary is defined around the selected point. The following prompts appear when you choose the **Pick Points** button in the **Boundary Hatch and Fill** dialog box.

Select internal point: *Select a point inside the object to hatch.*
Selecting everything visible...
Analyzing the selected data...
Analyzing internal islands...
Select internal point: *Select another internal point or press ENTER to end selection.*

Once the internal points have been selected, press ENTER and the dialog box reappears. Choose **OK** to apply the hatch pattern to the region. If you want to hatch an object and leave another object contained in it without hatching, select a point inside the object you want to hatch. By default, when using the **BHATCH** command, selecting a point creates multiple boundaries; a boundary of the internal object is also created, and the region between these two boundaries is selected for hatching. In Figure 13-14, the **Pick Points** button has been used to hatch the square but not the triangle simply by selecting the point inside the square but outside the triangle (using the default settings of the dialog box).

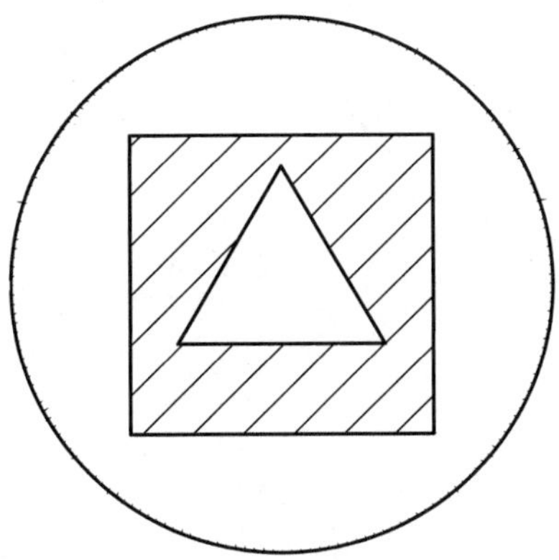

Figure 13-14 *Defining multiple hatch boundaries by selecting a point*

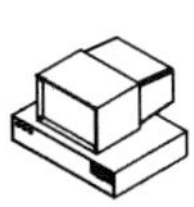

Tip

*When you have selected a point within an area to hatch and you realize you have selected the wrong area, you can enter **U** or **UNDO** at the Command prompt to undo the last selection. You can also undo a hatch pattern you have already applied to an area by entering **UNDO** at the Command prompt.*

Boundary Definition Error. This dialog box (Figure 13-15) is displayed if AutoCAD finds that the selected point is not inside the boundary or that the boundary is not closed. Choose **OK** and select another point inside the boundary or ensure that the boundary is closed.

***Figure 13-15 Boundary Definition Error** dialog box*

Note

*When selecting internal points for hatching, you should select only one point for one region. If you select another internal point in the same hatch region, a **Boundary Definition Error** dialog box is displayed with the message: **Boundary duplicates an existing boundary**. Choose **OK** to close the dialog box.*

Select Objects

This option in the **Boundary Hatch and Fill** dialog box lets you select objects that form the boundary for hatching. It is useful when you have to hatch an object and disregard objects that lie inside it or intersect with it. When you select this option, AutoCAD will prompt you to select objects. You can select the objects individually or use other object selection methods. In Figure 13-16, the **Select Objects** button is used to select the triangle. This uses the triangle as the hatch boundary, and everything inside it. The text inside the triangle, in this case, also gets hatched. To avoid hatching of such internal objects, select them at the next **Select Objects** prompt. In Figure 13-17, the text is selected to exclude it from hatching. When using the **Pick Points** option, the text automatically gets selected to be excluded from the hatching.

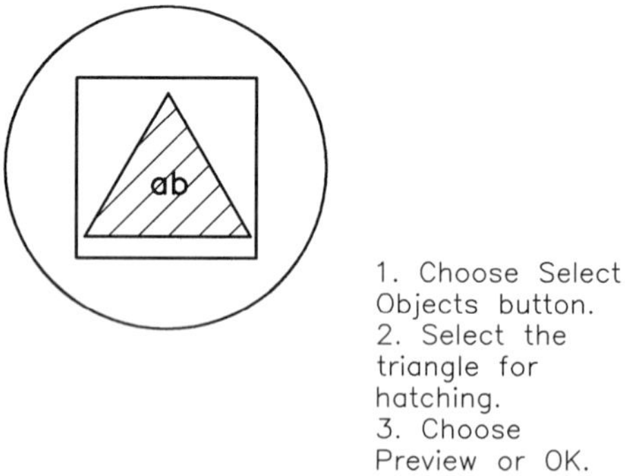

***Figure 13-16** Using the **Select Objects** button to specify the hatching boundary*

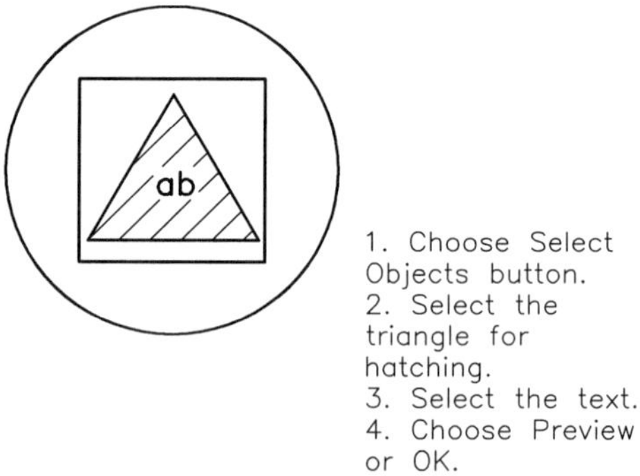

***Figure 13-17** Using the **Select Objects** button to exclude the text from hatching*

Tip

If you have an area to be hatched that has many objects intersecting with it, it is easier to select the entire object to be hatched rather than choosing internal points within each of the smaller regions created by the intersection.

You can also turn off layers that contain text or lines that make it difficult for you to select hatch boundaries.

Remove Islands

This option is used to remove any islands from the hatching area. Boundaries inside another boundary are known as islands. If you choose the **Pick Points** button to hatch an area, islands are not hatched by default. But, if you want to hatch the islands, you can choose the **Remove Islands** button in the **Boundary Hatch and Fill** dialog box. This applies, for example, if you have a rectangle with circles within, as shown in Figure 13-18. You can use the **Pick Points** option to select a point inside the rectangle; AutoCAD will select both the circles and the rectangle. To remove the circles (islands), you can use the **Remove Islands** option. When you select this option, AutoCAD will prompt you to select the islands to remove. Select the islands to remove and press ENTER to return to the dialog box. Choose **OK** to apply the hatch.

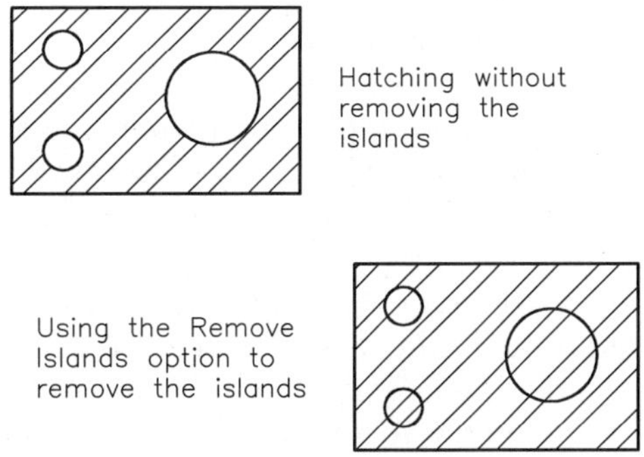

Figure 13-18 *Using the* ***Remove Islands*** *option to remove islands from the hatch area*

Tip

It may be a good idea to use the ***Select Objects*** *option to select the object containing islands if you want to remove the islands from the hatching area.*

View Selections

This option lets you view the selected boundaries before you apply the hatch pattern. This option is not available when you have not selected any points or boundaries. When you choose the **View Selections** button, the dialog box closes temporarily and the selected boundaries in the drawing are displayed as highlighted.

Inherit Properties

This option hatches the specified boundaries using the hatch properties of an existing object. When you invoke the **BHATCH** command and choose the **Inherit Properties** button, the following command sequence appears. Note that this prompt sequence does not appear when you select the hatch boundary first and then choose the **Inherit Properties** button.

Select associative hatch object: *Select the object from which you have to inherit the hatch pattern.*
Inherited Properties: Name <current>, Scale <current>, Angle <current>
Select Internal point: *Select a point inside the object to be hatched.*

When you press ENTER, the **Boundary Hatch and Fill** dialog box reappears with the name of the inherited hatch pattern in the **Pattern** edit box. The inherited hatch pattern is also displayed in the **Swatch** box. The selected pattern now becomes the current hatch pattern. If you then want to adjust the hatch properties, such as the angle or scale of the pattern, you can do so.

Composition

The **Composition** area of the dialog box provides options that control whether the hatch is associative or nonassociative with the hatch boundary. By default, the **Associative** radio button is selected, which implies that when you modify the boundary of a hatch object, the hatch patterns will be automatically updated to fill up the new area. If you select the **Nonassociative** radio button, the hatch pattern becomes independent of the boundary, which means that when you edit the boundary of a hatched object, the hatch pattern does not change with it. One of the major advantages with the associative hatch feature is that you can edit the hatch pattern or edit the geometry that is hatched without having to modify the associated pattern or boundary separately. After editing, AutoCAD automatically regenerates the hatch and the hatch geometry to reflect the changes. The hatch pattern can be edited by using the **HATCHEDIT** command, and the hatch geometry can be edited by using grips or AutoCAD editing commands.

Preview

You can use this option after you have selected the area to be hatched to preview the hatching before it is actually applied. This option is available only after you have selected the area to be hatched. When you choose the **Preview** button, the dialog box is temporarily closed and the object selected for hatching is temporarily filled with the specified hatch pattern. Previewing can also be done by right-clicking in the drawing area to display the shortcut menu, from where you can choose **Preview**. Pick a point on the screen or press ESC to redisplay the **Boundary Hatch and Fill** dialog box so that you can make modifications in the hatching operation. You can also press ENTER or right-click after previewing the drawing to accept the hatching and apply it to the selected boundary. Since you are accepting the current hatch settings by pressing ENTER or by right-clicking, the **BHATCH** command is completed and the **Boundary Hatch and Fill** dialog box is not displayed.

OK

By choosing the **OK** button, you can apply hatching in the area specified by the hatch boundary. This option works whether or not you have previewed your hatch. This button is not available if the hatch boundary is not defined.

Exercise 2 *Mechanical*

In this exercise, you will hatch the front view section of the drawing in Figure 13-19 using the hatch pattern for brass. Two views, top and front, are shown. In the top view, the cutting plane line indicates how the section is cut and the front view shows the full section view of the object. The section lines must be drawn only where the material is actually cut.

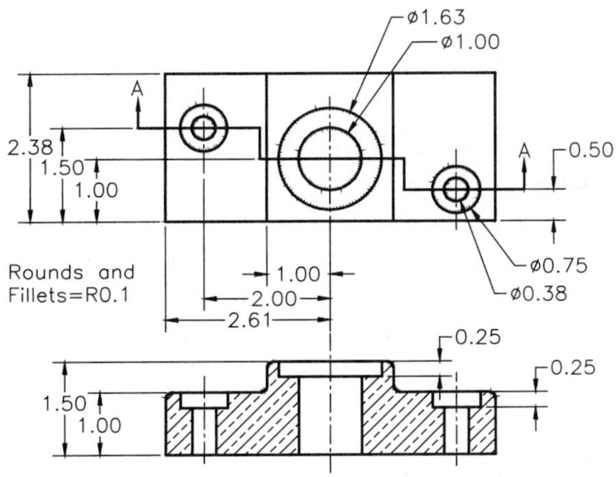

Figure 13-19 Drawing for Exercise 2

Advanced Tab

You can choose the **Advanced** tab of the **Boundary Hatch and Fill** dialog box (Figure 13-20) to specify options that improve the hatching speed. Defining a boundary by specifying an

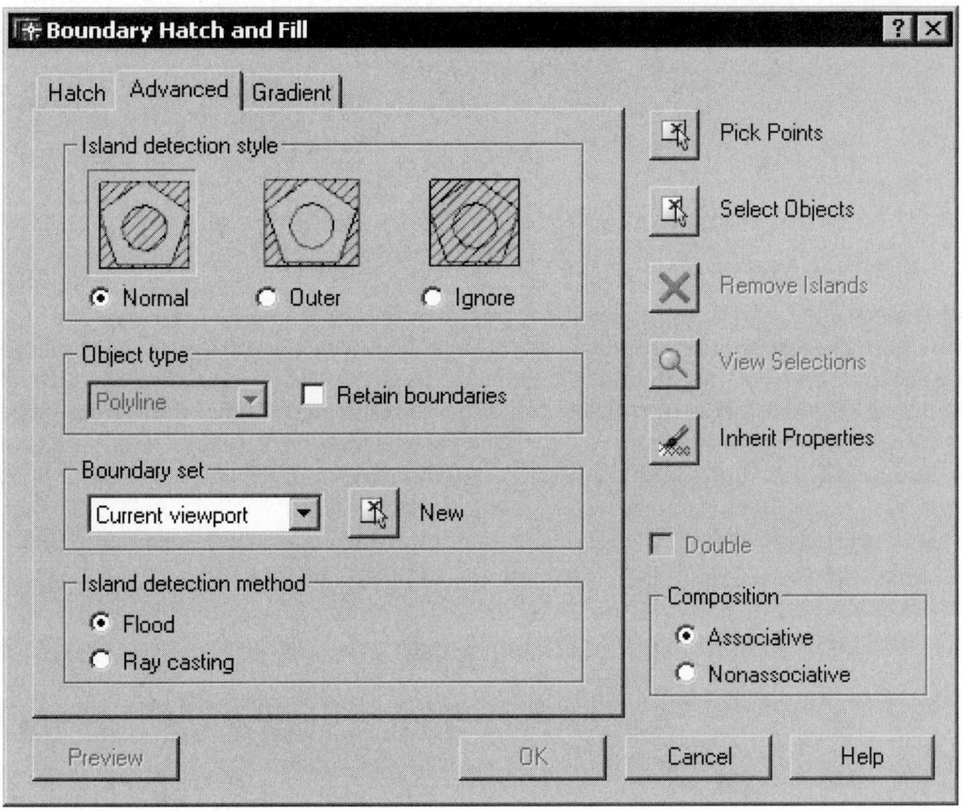

***Figure 13-20 Advanced** tab of the **Boundary Hatch and Fill** dialog box*

internal point is quite simple in the case of small and less complicated drawings. It may take more time in the case of large, complicated drawings because AutoCAD examines everything that is visible in the current viewport. Thus, the larger and more complicated the drawing, the more time it takes to locate the boundary edges. In such cases, setting parameters in the **Advanced** tab of the **Boundary Hatch and Fill** dialog box improves the speed of hatching.

Island detection style Area

The **Island detection style** area provides options to select the style of island detection during hatching (Figure 13-21). There are three styles available: **Normal**, **Outer**, and **Ignore**. To select a particular style, you can select the corresponding radio button. The effect of using the particular style is displayed in the form of an illustration in the image tile placed above the particular radio button. You can also choose the image tiles to select a particular island detection style. The results of using each of these styles is discussed next (Figure 13-22).

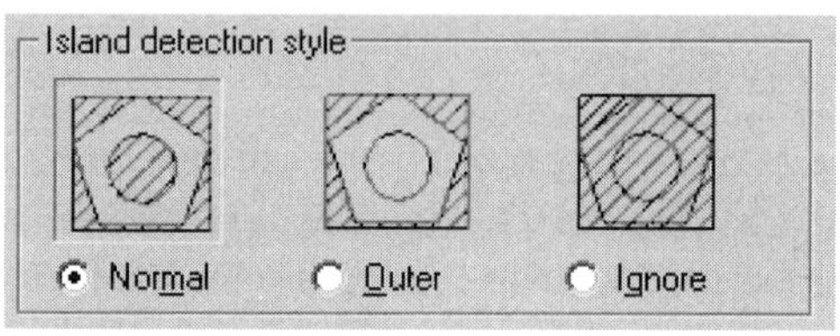

Figure 13-21 Island detection styles

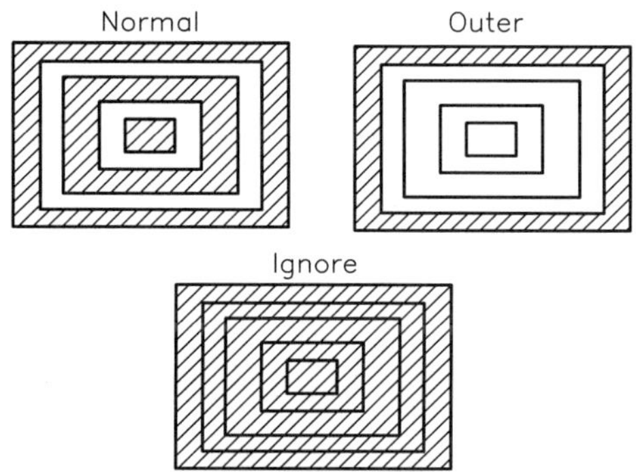

Figure 13-22 Using hatching styles

Note

The selection of an island detection style carries meaning only if the objects to be hatched are nested (that is, one or more selected boundaries is within another boundary).

Normal. The Normal style is selected by default. This style hatches inward starting at the outermost boundary. If it encounters an internal boundary, it turns off the hatching. An internal boundary causes the hatching to turn off until another boundary is encountered. In this manner, alternate areas of the selected object are hatched, starting with the outermost area. Thus, areas separated from the outside of the hatched area by an odd number of boundaries are hatched, while those separated by an even number of boundaries are not.

Outer. This particular option also lets you hatch inward from an outermost boundary, but the hatching is turned off if an internal boundary is encountered. Unlike the previous case, it does not turn the hatching on again. The hatching process in this case starts from both ends of each hatch line; only the outermost level of the structure is hatched, hence the name **Outer**.

Ignore. In this option, all areas bounded by the outermost boundary are hatched. The option ignores any hatch boundaries that are within the outer boundary. All islands are completely ignored and everything within the selected boundary is hatched.

Tip

It is also possible to set the pattern and the island detection style at the same time by using the ***HPNAME*** *system variable. AutoCAD stores Normal style code by adding* ***N*** *to the pattern name. Similarly,* ***O*** *is added for Outer style, and* ***I*** *for the Ignore style to the value of the* ***HPNAME*** *system variable. For example, if you want to apply the BOX pattern using the outer style of island detection, you should enter the* ***HPNAME*** *value to be* ***BOX, O****. Now, when you apply the hatch pattern, the BOX pattern is applied using the outer style of hatching.*

Note

The ***Normal, Outer,*** *and* ***Ignore*** *options are also available in the shortcut menu when you select an internal point in the drawing object and right-click in the drawing area. The shortcut menu shall be discussed later.*

Object type Area

When you select an internal point in a region to be hatched, a boundary is created around it. By default, these boundaries are removed as soon as the hatch pattern is applied. The **Object type** area provides options that allow you to specify whether these boundaries are to be retained as objects or not. You can also specify the type of object it can be saved as.

Retain Boundaries check box. The **Retain boundaries** check box is cleared by default, which implies that the hatch boundaries are not saved. If you select this check box, you can retain the defined boundary. When a hatching is successful and you want to keep the boundary as a polyline or region so that you can use it again, you can select the **Retain boundaries** check box. When you select this check box, the **Object type** drop-down list is available.

Object type drop-down list. The **Object type** drop-down list is available when you select the **Retain boundaries** check box. From this list you can select the type of object AutoCAD will create when you define a boundary for hatching. It has two options: **Polyline** and **Region**. **Polyline** is selected by default and the boundary created around the hatch area is a polyline. Similarly, when you select **Region**, the boundary of the hatch area is a region. Regions are two-dimensional areas that can be created from closed shapes or loops.

Boundary set Area

When hatching takes place normally, AutoCAD evaluates the entire drawing visible on the screen to determine the boundary. For larger drawings this may take a lot of time and therefore, slow down the process of hatching. The **Boundary set** area provides options that allow you to specify what is considered when hatching. You can either select an option from the drop-down list or you can create a new boundary set. The boundary set comprises the objects that the **BHATCH** command uses when constructing the boundary. The default boundary set is **Current viewport**, which comprises everything that is visible in the current viewport. Boundary hatching is made faster by specifying a boundary set because, in this case, AutoCAD does not have to examine everything on screen.

New. This option allows you to define a boundary area so that only a specific portion of the drawing is considered for hatching. You use this option to create a new boundary set. When you choose the **New** button, the dialog box is cleared from the screen temporarily to allow selection of objects to be included in the new boundary set. While constructing the boundary set, AutoCAD uses only those objects that you select and that are hatchable. If a boundary set already exists, it is replaced by the new one. If you do not select any hatchable objects, no new boundary set is created. AutoCAD retains the current set if there is one. Once you have selected the objects to form the boundary set, press ENTER and the dialog box reappears on the screen. You will notice that **Existing Set** gets added to the drop-down list. When you invoke the **BHATCH** command and you have not formed a boundary set, there is only one option, **Current Viewport**, available in the drop-down list. The benefit of creating a selection set is that when you select a point or select the objects to define the hatch boundary, AutoCAD will search only for the objects that are in the selection set. By confining the search to the objects in the selection set, the hatching process is faster. If you select an object that is not a part of the selection set, AutoCAD ignores them. When a boundary set is formed, it becomes the default for hatching until you exit the **BHATCH** command or select the **Current Viewport** option from the drop-down list.

Tip

To improve hatching speed in large drawings, you should zoom into the area to be hatched, so that defining the boundary to be hatched is easier. Also, since AutoCAD does not have to search the entire drawing to find hatch boundaries, the hatch process gets faster.

Island detection method Area

The **Island detection method** area of the **Boundary Hatch and Fill** dialog box allows you to select a method of island detection. The two methods available are **Flood** and **Ray casting**. These methods determine whether to include the objects that lie within the outermost boundary, referred to as islands, as boundary objects or not.

Flood. This method of defining the hatch boundary includes the islands as boundary objects. This is the default method for island detection.

Ray casting. This method casts rays in all directions or a particular direction from the point specified to the nearest object and then traces the boundary in a counterclockwise direction, thus excluding islands as boundary objects. Ray casting is discussed next.

RAY CASTING

You can use the **ray casting** technique to define a hatch boundary. In ray casting, AutoCAD casts an imaginary ray in all directions or in a particular specified direction and locates the object that is nearest to the point you have selected. After locating the object, it takes a left turn to trace the boundary and the boundary gets highlighted. If the point you have specified lies within the highlighted boundary, the islands within are ignored and the entire object is hatched. But, if the point lies outside the highlighted boundary, an error message is displayed and you are asked to select another point. Ray casting has five options: **Nearest**, **+X**, **-X**, **+Y**, and **-Y**, and **Angle**. All these options are available when you use the **-BHATCH** command. But, when you are using the **BHATCH** command, only the **Nearest** option can be used.

Using the BHATCH Command

When you invoke the **BHATCH** command, the **Boundary Hatch and Fill** dialog box is displayed. Choose the **Advanced** tab in the dialog box and select the **Ray casting** radio button. Choose the **Pick Points** button. The dialog box closes temporarily to allow you to specify a point for ray casting. In Figure 13-23, depending upon the location of the specified point, a boundary object is selected. If you select a point close to the outer circle, the circle gets highlighted as the hatching boundary. This is because imaginary rays are cast from the specified point and the circle is the nearest closed boundary available. When you press ENTER, the dialog box returns; choose the **Preview** button to see the preview of the hatching. You will notice that the entire circle is hatched and the inner objects (rectangle and triangle) have been ignored (Figure 13-24).

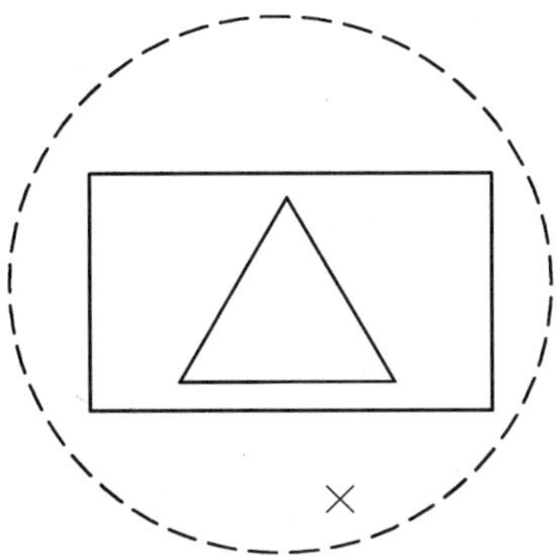

Figure 13-23 *Specifying a point closer to the outer circle during hatching using ray casting*

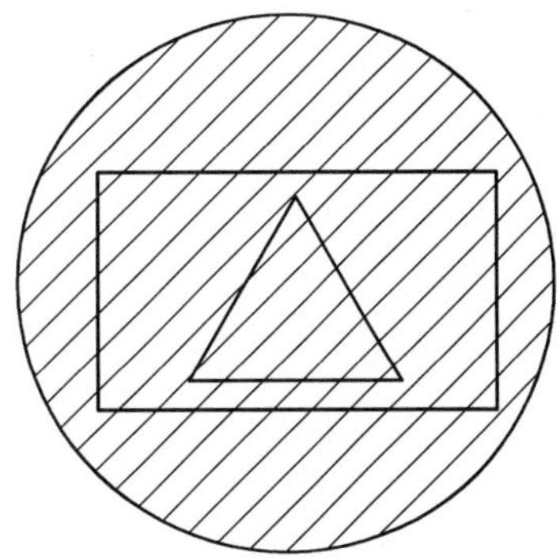

Figure 13-24 *The result of using the **Nearest** ray casting option*

Now, for the next time, select a point inside the circle but closer to the inner rectangle. The inner rectangle gets highlighted as it is nearest to the specified point and a **Boundary Definition Error** dialog box (Figure 13-25) is displayed suggesting the specified point is not within the highlighted boundary. The **Boundary Definition Error** dialog box has two buttons: **OK** and **Look at it**. If you choose **OK**, the dialog box is closed and you can select another point. If you choose **Look at it**, AutoCAD displays the highlighted boundary and the position of the specified point is shown marked with a (X) marker. An **AutoCAD Message** box is displayed with the message: **Press ENTER to continue**. You can press ENTER or choose **OK** to continue and specify another point.

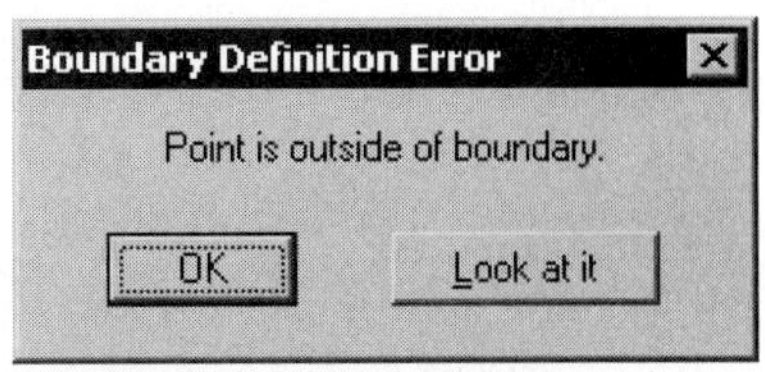

Figure 13-25 *The **Boundary Definition Error** dialog box*

Using the -BHATCH Command

You can also create hatch patterns from the command line using the **-BHATCH** command. The **-BHATCH** command is especially useful when you want to avoid boundary definition errors, by forcing the ray casting in a particular direction, or want to use other ray casting options that are not available through the dialog box. These options control the direction in which AutoCAD casts the ray to form a hatch boundary. When using the **-BHATCH** command, the dialog box is bypassed and the available options are **Nearest**, **+X**, **-X**, **+Y**, and **-Y**. The prompt sequence is given next.

Command: **-BHATCH**
Current hatch Pattern: ANSI31
Specify internal point or [Properties/Select/Remove islands/Advanced]: **A**
Enter an option [Boundary set/Retain boundary/Island detection/Style/Associativity]: **I**
Do you want island detection? [Yes/No] <Y>: **N**
Enter type of ray casting [Nearest/+X/-X/+Y/-Y/Angle] <Nearest>: *Specify an option.*
Enter an option [Boundary set/Retain boundary/Island detection/Style/Associativity]: Enter
Current hatch pattern: ANSI31
Specify internal point or [Properties/Select/Remove islands/Advanced]: *Select a point.*

Nearest. The **Nearest** option is the default option. When this option is used, AutoCAD casts an imaginary ray from the specified point to the nearest hatchable object, then takes a turn to the left, and continues the tracing process in an effort to form a boundary around the internal point. To make this process work properly, the point you select inside the boundary should be closer to the boundary than any other object that is visible on the screen.

+X, -X, +Y, and -Y options. These options can be explained better by studying Figure 13-26. Figure 13-26(a) shows that the points that can be selected are those that are nearest to the right edge of the circumference of the circle because the ray casting takes place in the direction of the positive X. If you select a point on the left of the rectangle that is placed inside the circle, no valid boundary is detected. This is because when you have selected this point, rays are cast horizontally from it in the positive X direction and the rectangle is encountered, which is invalid, since this point is not within the highlighted boundary.

Figure 13-26(b) shows that the points that can be used for selection are those that are nearest to the left edge of the circle because the ray casting takes place in the negative X direction. Any point selected on the right of the rectangle casts rays toward the negative X direction and encounters the rectangle, which is again an invalid boundary for the selected point.

In Figure 13-26(c), the ray casting takes place in the positive Y direction, and so the points that can be used for selection are the ones closest to the upper edge of the circle. If a point is selected below the rectangle, which is placed within the circle, rays are cast in the positive direction and a rectangle is encountered, which is an invalid boundary for the selected point.

In Figure 13-26(d), the ray casting takes place in the negative Y direction, and so the internal points that can be selected are the ones closest to the lower edge of the circle. When a point is selected above the rectangle, rays are cast in the negative Y direction and the rectangle is encountered, which is an invalid boundary for the selected point.

The valid areas for selecting points for the **+X**, **-X**, **+Y**, and **-Y** options of ray casting are

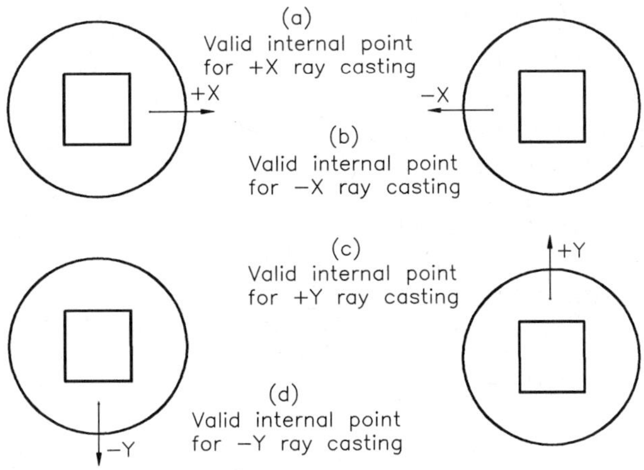

Figure 13-26 *Valid points for casting a ray in different directions to select the circle as boundary*

shown in Figures 13-27 and 13-28 as shaded. Points selected within the unshaded portions result in boundary errors.

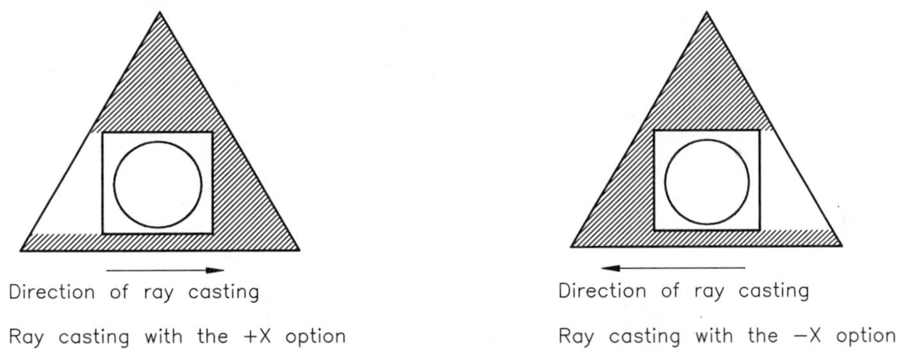

Figure 13-27 *Valid areas (shaded) for selecting points for the* ***-X*** *and* ***+X*** *options of ray casting*

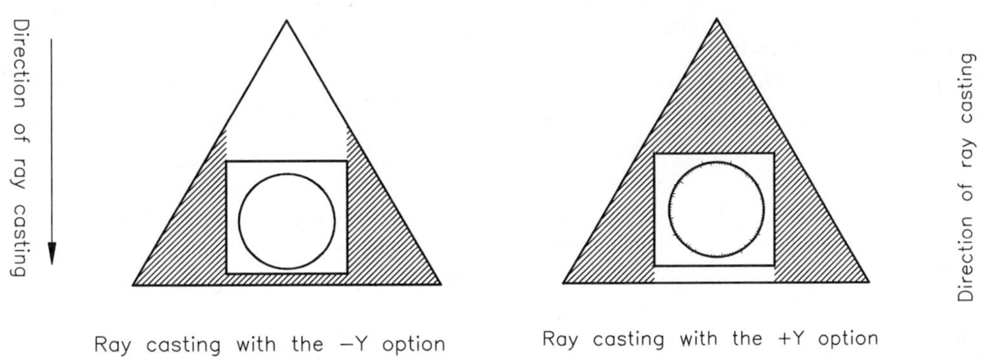

Figure 13-28 *Valid areas (shaded) for selecting points for the* ***-Y*** *and* ***+Y*** *options of ray casting*

The BHATCH Shortcut Menu

After invoking the **BHATCH** command and selecting a point that lies in the interior of the object to be hatched, or the object to be hatched itself, right-click to display the shortcut menu (Figure 13-29). This shortcut menu displays options that are also available through the **Boundary Hatch and Fill** dialog box. If you choose **Preview**, you can preview the hatching before applying it. If you choose **Enter**, the dialog box reappears. Choosing **Clear All** removes all the previous selections made. You can also select another island detection mode and also change from one method of hatching to another.

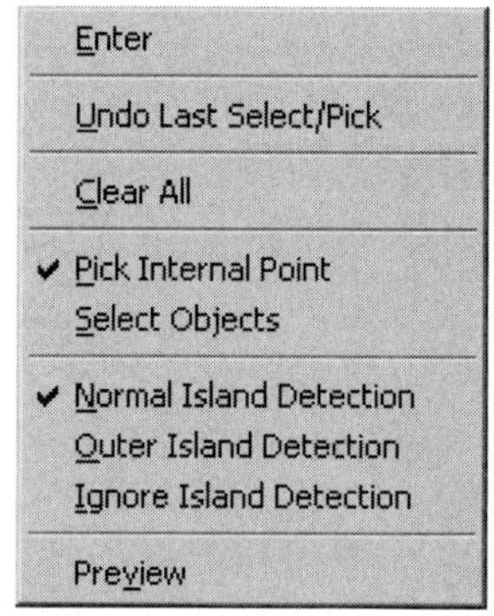

***Figure 13-29** The **BHATCH** shortcut menu*

Gradient Tab*

The **Gradient** tab of the **Boundary Hatch and Fill** dialog box (Figure 13-30) is used to specify the appearance of the gradient fill that can be applied to the selected objects.

One Color

The **One Color** radio button is used to specify the gradient fill comprising different shades and tints of a single color. By default, this radio button is selected in the dialog box. You can select the desired color by choosing the **Color Swatch** available on the right of the **One Color** radio button. Also, you can change the shade and tint of the selected color from the **Shade and Tint** slider available below the **Color Swatch**.

Two Color

The **Two Color** radio button is used to specify the gradient fill comprising different shades and tints of two colors. When you select this radio button, AutoCAD displays the color swatch for color 1 and color 2. Note that in this case, the **Shade and Tint** slider is not available.

Color Swatch

The **Color Swatch** is used to specify the color of the gradient fill. You can choose the **[...]** button available on the right of the color swatch to display the **Select Color** dialog box. You can select the desired color using this dialog box.

Shade and Tint Slider

The **Shade and Tint Slider** is used to specify the shade and tint of the color selected for the gradient fill.

Gradient Patterns

You can specify the desired display pattern of the gradient fill by selecting any one of the nine fixed patterns available below the **Shade and Tint Slider**.

Tip

*You can specify the gradient pattern by entering its value in the **GFNAME** system variable. The default value of this variable is 1. As a result, the first gradient pattern is selected.*

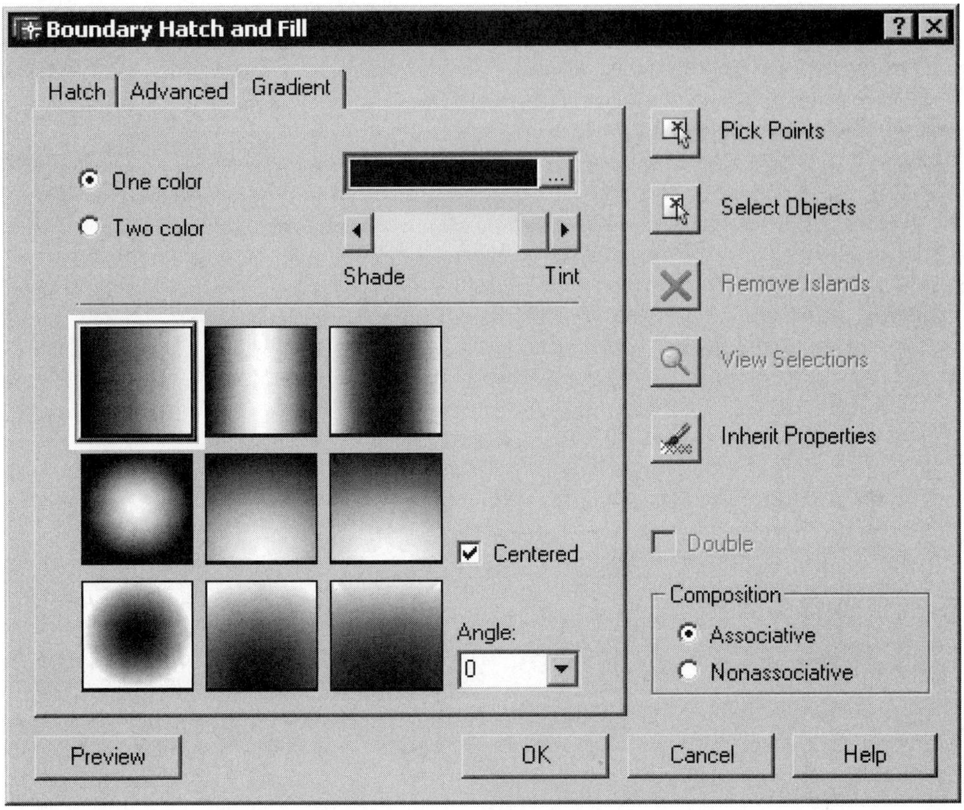

***Figure 13-30 Gradient** tab of the **Boundary Hatch and Fill** dialog box*

Centered

The **Centered** radio button is used to specify a symmetrical gradient fill. If this check box is not selected then the gradient fill is moved up and to the left. When you clear this check box the display is automatically updated in the gradient patterns.

Angle

The **Angle** drop-down list displays angles (at an increment of 15-degree) by which you can rotate the gradient fill with respect to the *X* axis of the current UCS. You can select any angle from this drop-down list or you can also enter an angle of your choice in the Angle edit box. The angle value is stored in the **GFANG** system variable.

Tip

Using the shortcut menu is more convenient and saves time.

HATCHING THE DRAWINGS USING THE TOOL PALETTES*

Toolbar:	Standard > Tool Palettes
Menu:	Tools > Tool Palettes Window
Command:	TOOLPALETTES

You can use the **TOOL PALETTES** window (Figure 13-31) to insert predefined hatch patterns and blocks in the drawings. By default, AutoCAD displays **TOOL PALETTES** as a window on the right of the drawing area. By default, the **TOOL PALETTES** window has three tabs: **Sample Office project**, **Imperial Hatches**, and **ISO Hatches**. In this chapter you will learn how to insert hatch patterns using the **Imperial Hatches** and **ISO Hatches** tabs of the **TOOL PALETTES**. Also, you will learn the insertion of predefined blocks using the **Sample Office project** tab of the **TOOL PALETTES** in the next chapter.

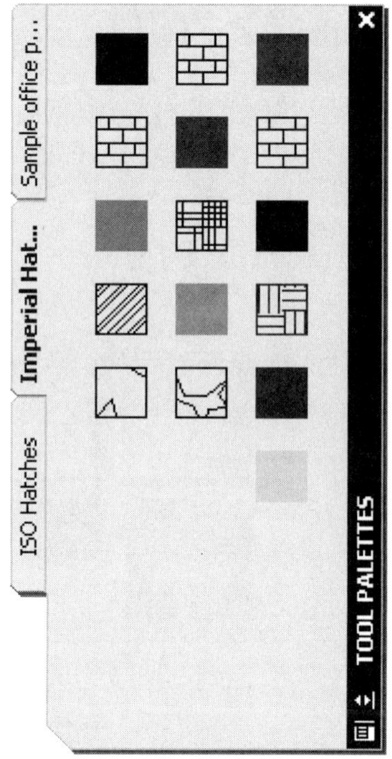

Figure 13-31 TOOL PALETTES window

The **Imperial Hatches** tab of the **TOOL PALETTES** provides you an option to insert the hatch patterns that are created using the Imperial units. The **ISO Hatches** tab of the **TOOL PALETTES** provides you an option to insert the hatch patterns that are created using the Metric units. You will notice that the hatch patterns provided in **Imperial Hatches** and **ISO Hatches** tabs of the **TOOL PALETTES** are similar. The basic difference between the two tabs is scale factor of the predefined hatch patterns.

Tip
*The **TOOL PALETTES** window can be turned on and off by pressing the CTRL+3 keys.*

Inserting Predefined Patterns in the Drawing

AutoCAD provides two methods to insert the predefined hatch patterns from the **TOOL PALETTES**: **Drag and Drop** method and **Select and Place** method. Both these methods of inserting the predefined hatch patterns using the **TOOL PALETTES** are discussed next.

Drag and Drop Method

To insert the predefined hatch pattern from the **TOOL PALETTES** in the drawing using this method, move the cursor over the desired predefined pattern in the **TOOL PALETTES**. You will notice that as you move the cursor over the hatch pattern, the hatch icon gets converted into a 3D icon. Also, a tooltip is displayed that shows the name and description of the block. Press and hold the left mouse button and drag the cursor within the area to be hatched. Release the left mouse button, and you will notice that the selected predefined hatch pattern is added to the drawing.

Select and Place Method

You can also add the predefined hatch patterns to the drawings using the select and place method. To add the hatch pattern, move the cursor over the desired pattern in the **TOOL PALETTE**. The pattern icon is changed to a 3D icon. Press the left mouse button, the selected hatch pattern is attached to the cursor and you are prompted to specify the insertion point of the hatch pattern. Move the cursor within the area to be hatched and press the left mouse button. The selected hatch pattern is inserted at the specified location.

Modifying the Properties of the Predefined Patterns Available in the TOOL PALETTES Window

To modify the properties of the predefined hatch patterns, move the cursor over the hatch pattern in the **TOOL PALETTES** and right-click on it to display the shortcut menu. Using the options available in this shortcut menu, you can cut or copy the desired hatch pattern available in one tab of **TOOL PALETTES** and paste it on the other tab. You can also delete and rename the selected hatch pattern using the **Delete Tool** and **Rename** options respectively. To modify the properties of the selected hatch pattern, choose **Properties** from the shortcut menu as shown in Figure 13-32. The **Tool Properties** dialog box is displayed (Figure 13-33).

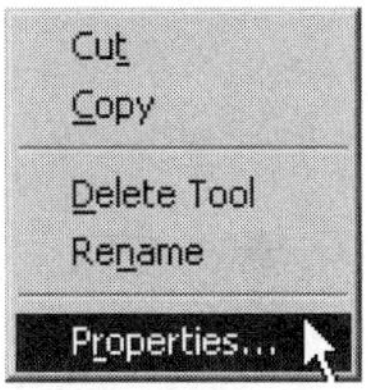

***Figure 13-32** Choosing **Properties** from the shortcut menu*

In the **Tool Properties** dialog box, the name of the selected hatch pattern is displayed in the **Name** edit box. You can also rename the hatch pattern by entering the new name in the **Name** edit box. The **Image** area available on the left of the **Name** edit box displays the image of the selected hatch pattern. If you enter a description of the hatch pattern in the **Description** text box, it is stored with the hatch definition in the **TOOL PALETTES**. Now, when you move the cursor over the hatch pattern in **TOOL PALETTES** and pause for a second, the description of the hatch pattern appears along with its name in the tooltip. The **Tool Properties** dialog box displays the properties of the selected hatch pattern under the following categories.

Pattern

In this category, you can change the Pattern Type, Pattern name, Angle, and Scale of the selected pattern. You can modify the spacing of the **User defined** patterns in the **Spacing** text box. Also, the ISO pen width of the ISO patterns can be redefined in the **ISO pen width** text box. The **Double** drop-down list is available only for the **User-defined** hatch patterns. You can select Yes or No from the **Double** drop-down list to determine the hatch pattern to double at right angles to the original pattern or not. When you choose the **[...]** button in the **Type** property field, AutoCAD displays the **Hatch Pattern Type** dialog box. You can select the type of hatch pattern from the **Pattern Type** drop-down list. If the **Predefined** pattern type is selected, you can specify the pattern name by either selecting it from the **Pattern** drop-down list or from the **Hatch Pattern Palette** dialog box displayed on choosing the **Pattern** button. Similarly, if you select the **Custom** pattern type, you can enter the name of the custom pattern in the **Custom Pattern** edit box. If you select the **User defined** pattern type, both the **Pattern** drop-down list and the **Custom Pattern** edit box are not available.

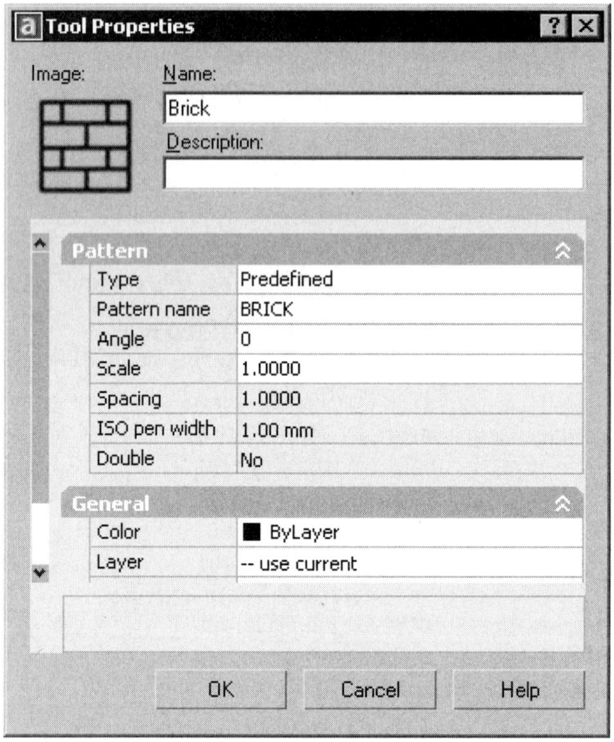

Figure 13-33 *The* ***Tool Properties*** *dialog box*

General

In this category, you can specify the general properties of the hatch pattern such as **Color**, **Layer**, **Linetype**, **Plot style**, and **Line weight** for the selected hatch pattern. The properties of a particular field can be modified from the drop-down list available on selecting that field.

When you select a particular field in the **Tool Properties** dialog box, its function is displayed in the display box available at the bottom of the dialog box. Choose **OK** to apply the changes and close the dialog box.

Tip

You can cut, copy, or rename the selected hatch pattern available on ***TOOL PALETTES*** *using the shortcut menu.*

HATCHING AROUND TEXT, DIMENSIONS, AND ATTRIBUTES

When you select a point within a boundary to be hatched and if it contains text, dimensions, and attributes, the hatch lines do not pass through the text, dimensions, and attributes present in the object being hatched by default. AutoCAD places an imaginary box around these objects that does not allow the hatch lines to pass through it. Remember that if you are using the select objects option to select objects to hatch, you must select the text/attribute/shape along with the object in which it is placed when defining the hatch boundary. If multiple line text is to be written, the **MTEXT** command is used. You can also select both the boundary and the text when using the window selection method. Figure 13-34 shows you how hatching takes place around text and traces.

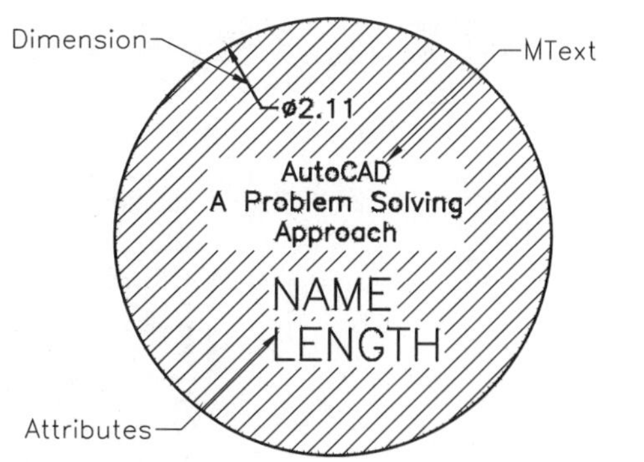

Figure 13-34 *Hatching around dimension, text, and attributes*

Note

In case of traces, the hatching stops exactly at the outline of the trace object unlike the case of text where the box formed around the text allows a small gap between the text and the hatch pattern. This gap makes the text easily readable.

EDITING HATCH PATTERNS

Using the HATCHEDIT Command

Toolbar:	Modify II > Edit Hatch
Menu:	Modify > Object > Hatch
Command:	HATCHEDIT

Figure 13-35 *Choosing* ***Edit Hatch*** *from the* ***Modify II*** *toolbar*

The **HATCHEDIT** command (Figure 13-35) can be used to edit the hatch pattern. When you invoke this command and select the hatch for editing, the **Hatch Edit** dialog box (Figure 13-36) is displayed on the screen. The **Hatch Edit** dialog box can also be invoked from the shortcut menu or by double-clicking on the hatch pattern. After selecting a hatched object, right-click in the drawing area, and choose **Hatch Edit** from the shortcut menu. The command sequence is as follows.

Command: **HATCHEDIT**
Select associative hatch objects: *Select a hatch object to edit.*

AutoCAD will display the **Hatch Edit** dialog box, where the hatch pattern can be changed or modified accordingly (Figure 13-36). This dialog box is the same as the **Boundary Hatch and Fill** dialog box except that only the options that control the hatch pattern are available. These available options work in the same way as they do in the **Boundary Hatch and Fill** dialog box.

The **Hatch Edit** dialog box has three tabs just like the **Boundary Hatch and Fill** dialog box.

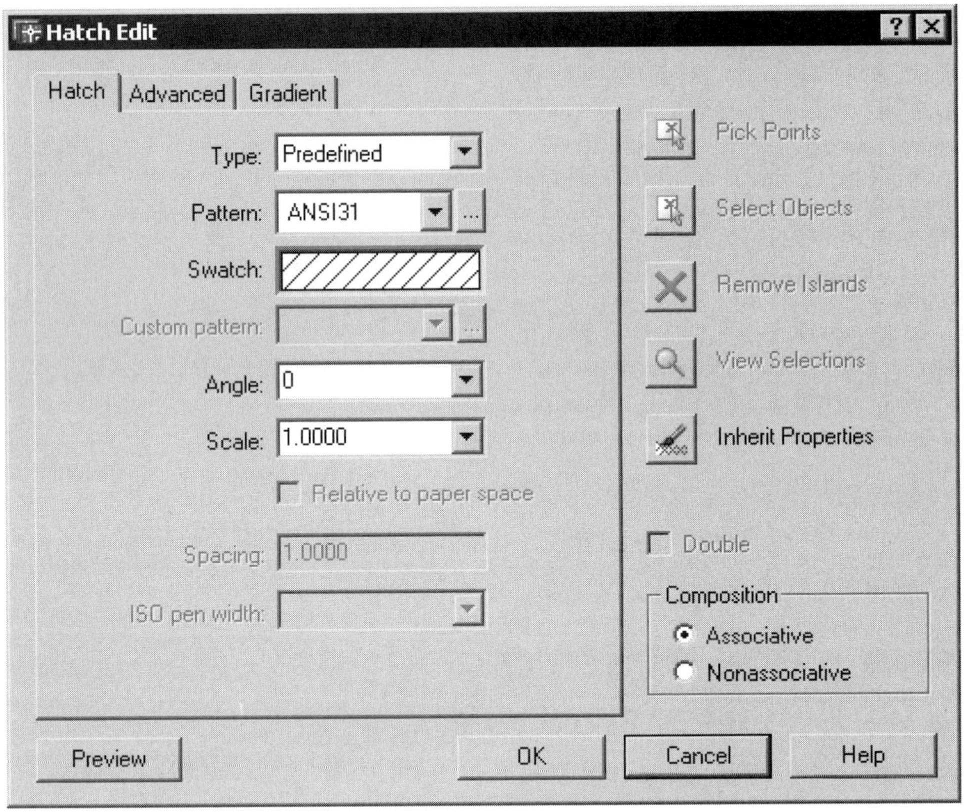

*Figure 13-36 The **Hatch** tab of the **Hatch Edit** dialog box*

They are **Hatch**, **Advanced**, and **Gradient**. In the **Hatch** tab, you can redefine the type of hatch pattern by selecting another type from the **Type** drop-down list. If you are using the **Predefined** pattern, you can select a new hatch pattern name from the **Pattern** drop-down list. You can also change the scale or angle by entering new values in the **Scale** or **Angle** edit box. When using **User defined** pattern, you can redefine the spacing between the lines in the hatch pattern by entering a new value in the **Spacing** edit box. If you are using **Custom** pattern type, you can select another pattern from the **Custom pattern** drop-down list. You can also redefine the island detection style by selecting either of the **Normal**, **Outer**, or **Ignore** styles in the **Advanced** tab of the dialog box. If you want to copy the properties from an existing hatch pattern, choose the **Inherit Properties** button, and then select the hatch whose properties you want to inherit. If the **Associative** radio button in the **Composition** area of the dialog box is selected, the pattern is associative. This implies that whenever you modify the hatch boundary, the hatch pattern is automatically updated. The appearance of the gradient fill can be specified using the **Gradient** tab of the dialog box. You can specify the gradient fill comprising different shades and tints of a single color or double colors by selecting the **One Color** and the **Two Color** radio buttons respectively. You can also specify the color of the gradient fill or the shade and tint of a single color using the color swatch and the **Shade and Tint Slider**. If the Centered button is selected, AutoCAD will apply a symmetrical gradient fill. Also, AutoCAD provides you with an option to select the desired display pattern of the gradient fill by selecting any one of the nine fixed patterns for the gradient fill.

Note

If a hatch pattern is associative, the hatch boundary can be edited using grips and editing commands and the associated pattern is modified accordingly. This is discussed later.

In Figure 13-37, the object is hatched using the ANSI31 hatch pattern. With the **HATCHEDIT** command, you can edit the hatch using the **Hatch Edit** dialog box to change the hatch to the one shown in Figure 13-38. You can also edit an existing hatch through the command line by entering **-HATCHEDIT** at the Command prompt.

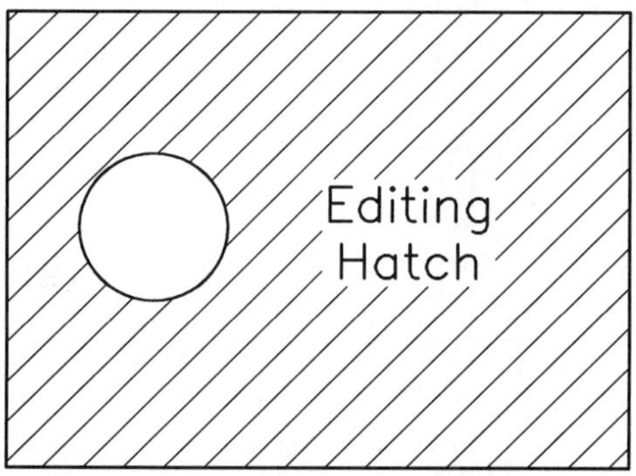

Figure 13-37 ANSI31 hatch pattern

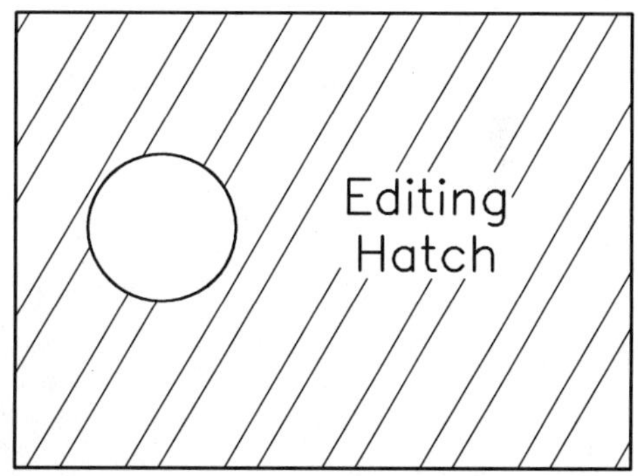

*Figure 13-38 Using the **HATCHEDIT** command to edit the hatch pattern*

Using the PROPERTIES Command

Toolbar:	Standard > Properties
Menu:	Modify, Tools > Properties
Command:	PROPERTIES

You can also use the **PROPERTIES** command to edit the hatch pattern. When you select a hatch pattern for editing, and invoke the **PROPERTIES** command, AutoCAD displays the **PROPERTIES** palette (Hatch) (Figure 13-39). You can also invoke the **PROPERTIES** palette by selecting the hatch pattern and right-clicking to display a shortcut menu. Choose **Properties** here; the **PROPERTIES** palette is displayed. This palette displays the properties of the selected pattern under the following categories.

General

In this category, you can change the general pattern properties like Color, Layer, Linetype, Linetype scale, Lineweight, and so on. When you click on the field corresponding to a particular property, a drop-down list is displayed from where you can select an option or value. Whenever a drop-down list does not get displayed, you can enter a value in the field itself.

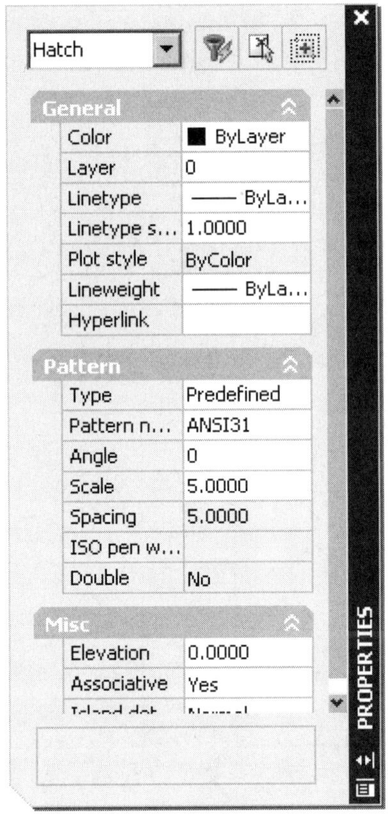

Figure 13-39 ***PROPERTIES*** *palette (Hatch)*

Pattern

In this category you can change the Pattern Type, Pattern name, Angle, and Scale of the pattern. In the case of ISO patterns, you can redefine the ISO pen width of the pattern and you can also modify the spacing of **User defined** patterns. You can also determine whether you want to have a **User defined** pattern as double or not. When you choose the **[...]** button in the **Type** property field, AutoCAD displays the **Hatch Pattern Type** dialog box (Figure 13-40). Here, you can select the type of hatch pattern from the **Pattern Type** drop-down list. If you have selected a **Predefined** pattern type, you can select a pattern name from the **Pattern** drop-down list or choose the **Pattern** button to display the **Hatch Pattern Palette** dialog box. You can then select a pattern here in one of the tabs and then choose **OK** to exit the dialog box. Now, choose **OK** in the **Hatch Pattern Type** dialog box to return to the **PROPERTIES** palette. You will notice that the pattern name you have selected

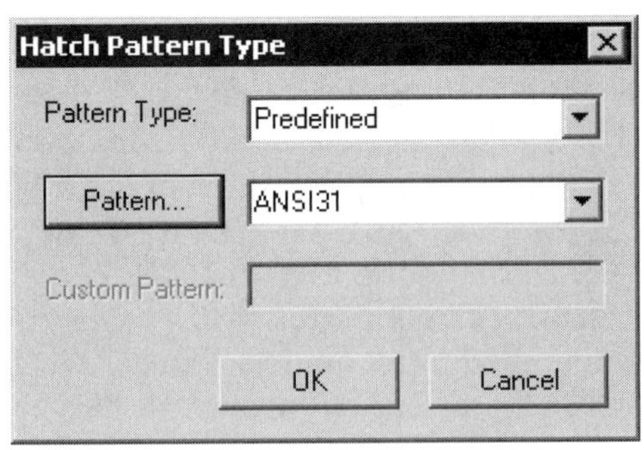

Figure 13-40 ***Hatch Pattern Type*** *dialog box*

is displayed in the **Pattern** name field. Similarly, if you select the **Custom** pattern type, you can enter the name of the custom pattern in the **Custom Pattern** edit box. If you select a **User defined** pattern type, both the **Pattern Type** drop-down list and the **Custom Pattern** edit box are not available. When you click in the **Pattern name** property field, the **[...]** button is displayed. When you choose the **[...]** button, the **Hatch Pattern Palette** dialog box is displayed with the default pattern ANSI31 selected. The values of the angle, scale, and spacing properties can be entered in the corresponding fields of the **PROPERTIES** palette. A value for the ISO pen width can be selected from the **ISO pen width** drop-down list. You can select **Yes** or **No** from the **Double** drop-down list to determine if you want the hatch pattern to double at right angles to the original pattern or not.

Tip

It is more convenient to choose the [...] button in the ***Pattern Type*** *edit box to display the* ***Hatch Pattern Type*** *dialog box and enter both the type and name of the pattern at the same time.*

Misc

In this category, the elevation, associativity, and island detection style properties of the hatch pattern can be changed. A new value of elevation for the hatch pattern can be entered in the **Elevation** field. You can select **Yes** or **No** from the **Associative** drop-down list to determine if a pattern is associative or not. Similarly, you can select an island detection style from the **Island detection style** drop-down list.

EDITING HATCH BOUNDARY

Using Grips

One of the ways you can edit the hatch boundary is by using grips. You can select the hatch pattern or the hatch boundaries. If you select the hatch pattern, the hatch highlights and a grip is displayed at the centroid of the hatch. A centroid for a region that is coplanar with the *XY* plane is the center of that particular area. However, if you select an object that defines the hatch boundary, the object grips are displayed at the defining points of the selected object. Once you change the boundary definition, and if the hatch pattern is associative, AutoCAD will reevaluate the hatch boundary and then hatch the new area. When you edit the hatch boundary, make sure that there are no open spaces in the hatch boundary. AutoCAD will not create a hatch if the outer boundary is not closed. Figure 13-41 shows the result of moving the circle and text, and shortening the bottom edge of the hatch boundary of the object shown in Figure 13-38. The objects were edited by using grips and since the pattern is associative, when modifications are made to the boundary object, the pattern automatically

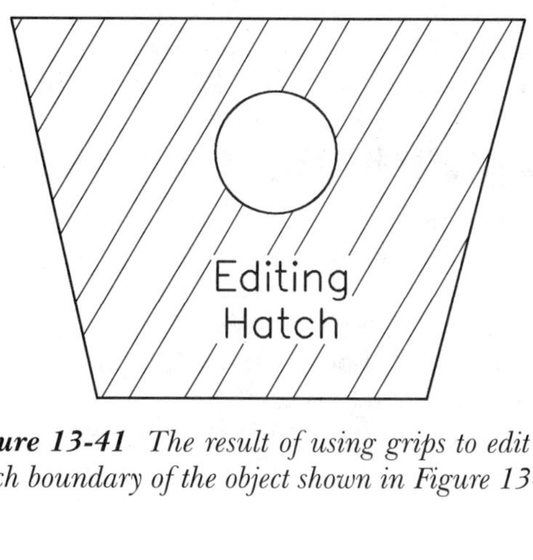

Figure 13-41 *The result of using grips to edit the hatch boundary of the object shown in Figure 13-38*

fills up the new area. If you select the hatch pattern, AutoCAD will display the hatch object grip at the centroid of the hatch. You can select the grip to make it hot and then edit the hatch object. You can stretch, move, scale, mirror, or rotate the hatch pattern. Once an editing operation takes place on the hatch pattern, it may be moved, rotated, or scaled; then, the associativity is lost and AutoCAD displays the message: "**Hatch boundary associativity removed**."

The **Retain boundaries** check box in the **Boundary Hatch and Fill** dialog box is cleared when a hatch is created by default. Hence, all objects selected or found by the selection or point selection processes during hatch creation are associated with the hatch as boundary objects. Editing any of them may affect the hatch. If, however, the **Retain boundaries** check box was selected when the hatch was created, only the retained polyline boundary created by **BHATCH** is associated with the hatch as its boundary object. Editing it will affect the hatch, and editing the objects selected or found by selection or point selection processes during the hatch created will have no effect on the hatch. You can, of course, select and simultaneously edit the objects selected or found by the selection or point selection processes, as well as the retained polyline boundary created by the **BHATCH** command.

Using AutoCAD Editing Commands

When you use the editing commands such as **MOVE**, **ROTATE**, **SCALE**, and **STRETCH**, associativity is maintained, provided all objects that define the boundary are selected for editing. If any of the boundary-defining objects is missing, the associativity will be lost and AutoCAD will display the message **Hatch boundary associativity removed**. When you rotate or scale an associative hatch, the new rotation angle and the scale factor are saved with the hatch object data. This data is then used to update the hatch. When using the **ARRAY**, **COPY**, and **MIRROR** commands, you can array, copy, or mirror just the hatch boundary, without the hatch pattern. Similarly, you can erase just the hatch boundary, using the **ERASE** command and associativity is removed. If you explode an associative hatch pattern, the associativity between the hatch pattern and the defining boundary is removed. Also, when the hatch object is exploded, each line in the hatch pattern becomes a separate object.

When the original boundary objects are being edited, the associated hatch gets updated, but new boundary objects do not have any effect. For example, when you have a square island within a circular boundary and then you erase the square, the hatch pattern is updated to fill up the entire circle. But, once the island is removed, another island cannot be added, since it was not calculated as a part of the original set of boundary objects.

HATCHING BLOCKS AND XREF DRAWINGS

When you apply hatching to inserted blocks and xref drawings, their internal structure is treated as if the block or xref drawing were composed of independent objects. This means that if you have inserted a block that consists of a circle within a rectangle and you want the internal circle to be hatched, all you have to do is invoke the **Boundary Hatch and Fill** dialog box and then choose the **Pick Points** button to select a point within the circle to generate the desired hatch, see Figure 13-42. If you want to hatch a block using the **Select Objects** option, the objects inside the blocks should be parallel to the current UCS. If the block is composed of objects such as arcs, circles, or polyline arc segments, they need not be uniformly scaled for hatching, see Figure 13-42.

When you xref a drawing, you can hatch any part of the drawing that is visible. Also, if you hatch an xref drawing and then use the **XCLIP** command to clip the xref drawing, the hatch pattern is not clipped, although the hatch boundary associativity is removed. Similarly, when you detach the xref drawing, the hatch pattern and its boundaries are not detached, although the hatch boundary associativity is removed.

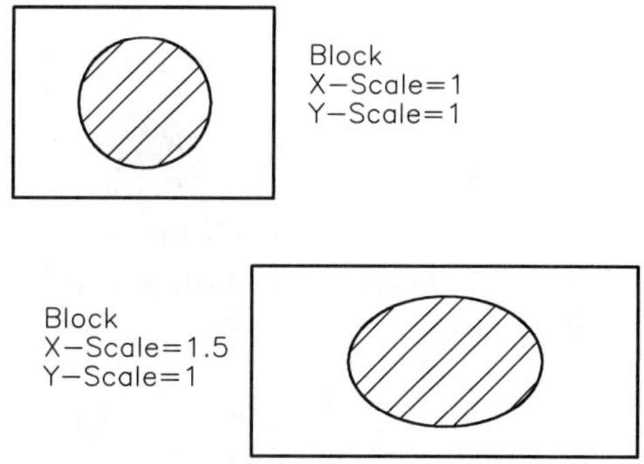

Figure 13-42 *Hatching inserted blocks*

PATTERN ALIGNMENT DURING HATCHING

Pattern alignment is an important feature of hatching, since on many occasions you need to hatch adjacent areas with similar or sometimes identical hatch patterns while keeping the adjacent hatch patterns properly aligned. Proper alignment of hatch patterns is taken care of automatically by generating all lines of every hatch pattern from the same reference point. The reference point is normally at the origin point (0,0). Figure 13-43 shows two adjacent hatch areas. The area on the right is hatched using the pattern ANSI32 at an angle of 0-degree and the area on the left is hatched using the same pattern at an angle of 90-degree. When you hatch these areas, the hatch lines may not be aligned, as shown in Figure 13-43(a). To align them, you can redefine the value of the **SNAPBASE** variable before you apply the hatch to the area on the left, so that the hatch lines are aligned, as shown in Figure 13-43(b). The reference point for hatching can be changed using the system variable **SNAPBASE**. The command prompt sequence is given next.

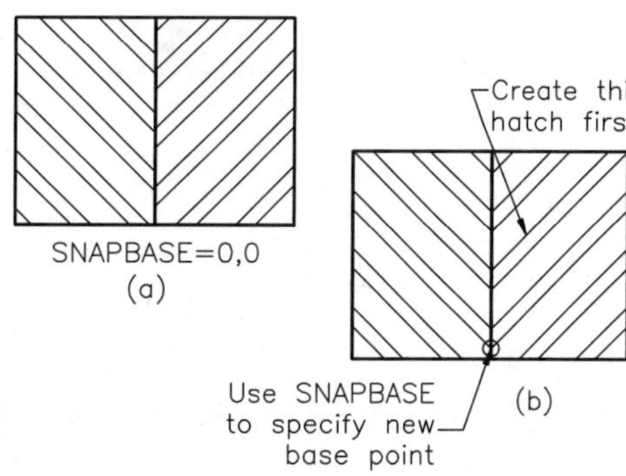

Figure 13-43 *Aligning hatch patterns using the* ***SNAPBASE*** *variable*

Command: **SNAPBASE**
Enter new value for SNAPBASE <0.0000, 0.0000>: *Enter the new reference point.*

In this case, the endpoint of the hatch in the area on the right is selected as the new reference point for hatching the area on the left. See Figure 13-43(b).

Note

The value of the ***SNAPBASE*** *variable can also be set using the* ***X base*** *and* ***Y base*** *edit boxes in the* ***Snap*** *area of the* ***Drafting Settings*** *dialog box (****Snap and Grid*** *tab).*

Tip

You should set the base point back to (0,0) when you are done with hatching the object; otherwise, the most recent value gets stored and is used the next time you apply a hatch pattern.

Exercise 3 *Mechanical*

In this exercise, you will hatch the given drawing using the hatch pattern ANSI31. Use the **SNAPBASE** variable to align the hatch lines shown in the drawing (Figure 13-44).

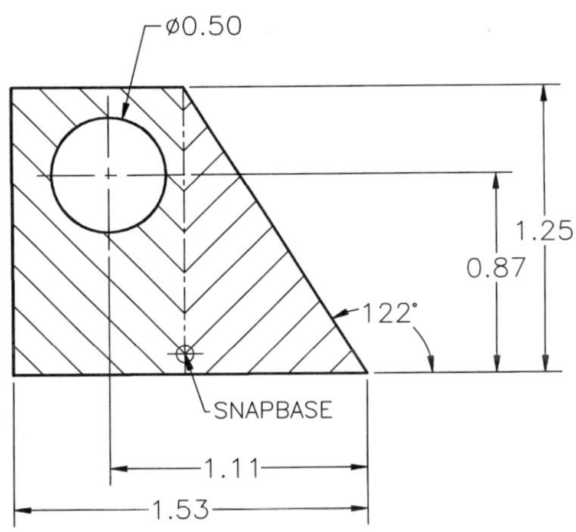

Figure 13-44 *Drawing for Exercise 3*

CREATING A BOUNDARY USING CLOSED LOOPS

Menu:	Draw > Boundary
Command:	BOUNDARY

The **BOUNDARY** command is used for creating a polyline or region around a selected point within a closed area, in a manner similar to the one used for defining a hatch boundary. This command works exactly like the **BHATCH** command, when the **Retain boundaries** check box is selected, except that the boundary object created is not hatched. When this command is entered, AutoCAD displays the **Boundary Creation** dialog box, shown in Figure 13-45.

The options in the **Boundary Creation** dialog box are similar to the options provided in the **Advanced** tab of the **Boundary Hatch and Fill** dialog box discussed earlier, although, only the options related to boundary selection and not hatch patterns are available. The **Pick Points** button in the **Boundary Creation** dialog box is used to create a boundary by selecting a point inside the object, whose boundary you want to create.

When you select the **Pick Points** option, the dialog box is cleared from the screen temporarily and the following prompt is displayed.

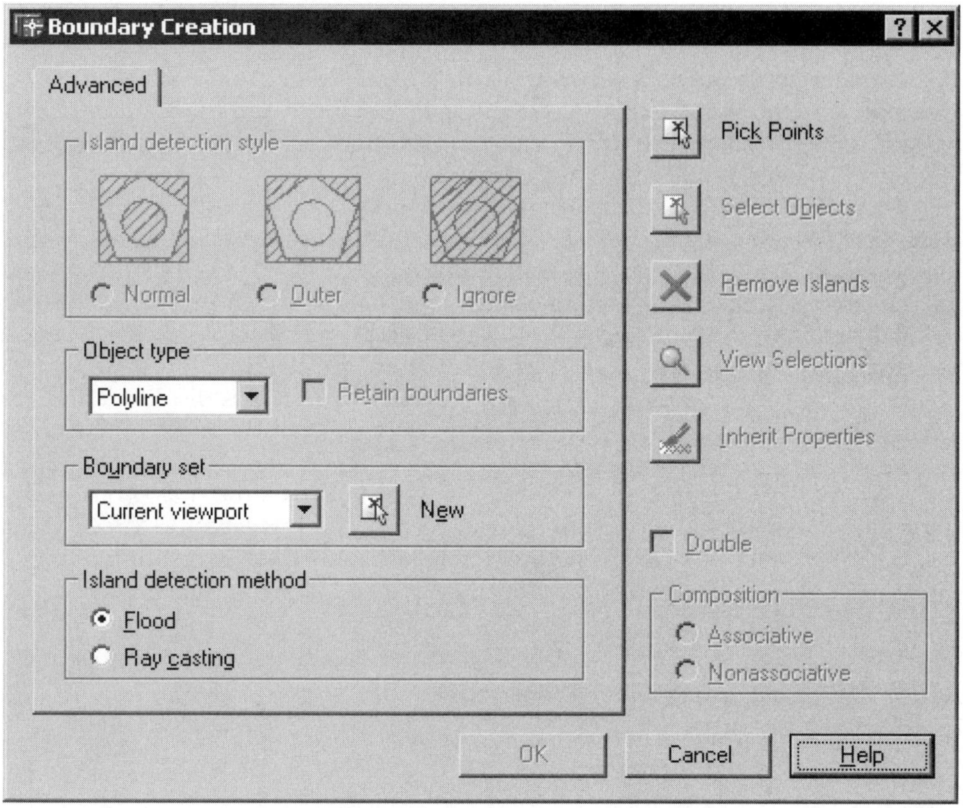

*Figure 13-45 The **Boundary Creation** dialog box*

Select internal point: *Select a point that lies within the boundary of the object.*

Once you select an internal point, a polyline or a region is formed around the boundary (Figure 13-46). To end this process, press ENTER at the **Select internal point** prompt. Whether the boundary created is a polyline or region is determined by the option you have selected from the drop-down list in the **Object Type** area of the **Boundary Creation** dialog box. **Polyline** is the default option. You can edit the boundary that has been created with the editing commands. The boundary is selected by using the last object selection option.

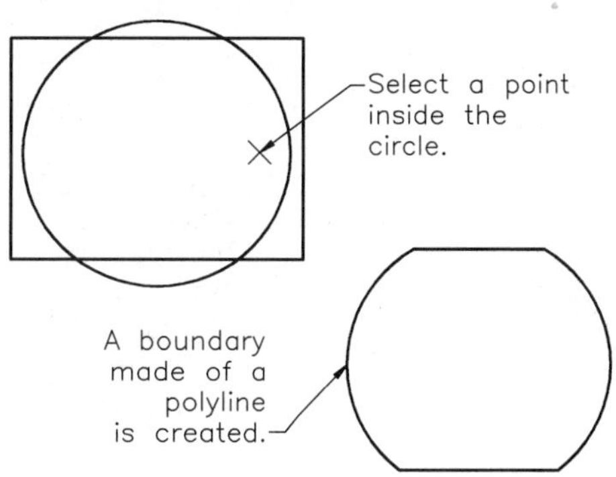

*Figure 13-46 Using **BOUNDARY** command to create a polyline boundary*

Just like when you used the **BHATCH** command, you can also define a boundary set here by choosing the **New** button in the **Boundary set** area of the dialog box. The dialog box will temporarily close to allow you to select objects to be used to create a boundary. The default boundary set is **current viewport**, which means everything visible in the current viewport consists of the boundary set. As discussed earlier, the boundary set option is especially useful when you are working with large and complex drawings, where examining everything on the screen becomes a time-consuming process.

The **Boundary Creation** dialog box also provides you the option of determining the island detection method. While **Flood** is the default option and islands are included as boundary objects, you can also select **Ray casting**, where rays shall be cast from the selected internal point in all directions (by default, using the dialog box). The nearest closed object it encounters is used for boundary creation. If the selected point does not lie within the encountered boundary, a **Boundary Definition Error** dialog box is displayed and you have to select another internal point.

The **-BOUNDARY** command can be used if you want to bypass the dialog box and use the command line. When using the command line, all the options (**Nearest**, **+X**, **-X**, **+Y**, and **-Y**) of ray casting are available. When using the dialog box, only the **Nearest** option is available. The different options of ray casting have been described earlier in the chapter.

Note

*The **HPBOUND** system variable controls the type of boundary object created using the **BOUNDARY** command. Its default value of 1 creates a polyline and if the value is 0, the object created is a region.*

Exercise 4 *Mechanical*

First see Figure 13-47 and then make the drawing shown in Figure 13-48 using the **BOUNDARY** command to create a hatch boundary. Copy the boundary from the drawing shown in Figure 13-48, and then hatch.

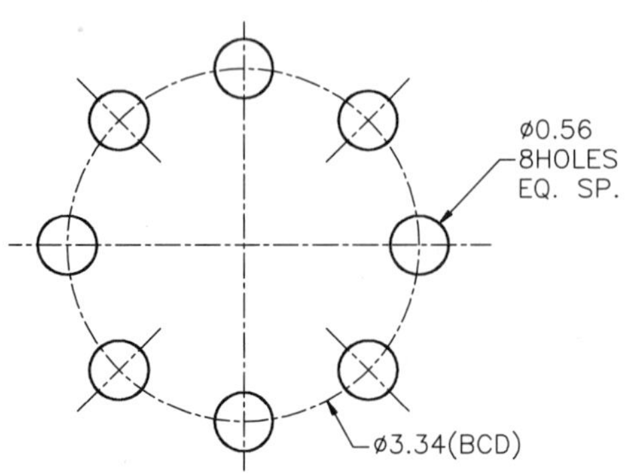

Figure 13-47 *Drawing for Exercise 4*

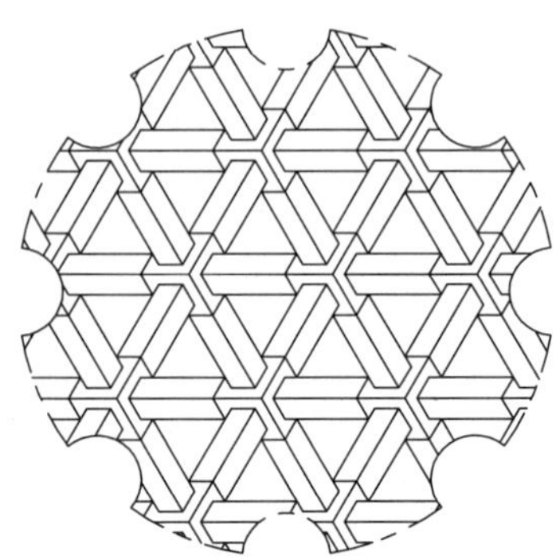

Figure 13-48 *Final drawing for Exercise 4*

OTHER FEATURES OF HATCHING

1. In AutoCAD, the hatch patterns are separate objects. The objects store the information about the hatch pattern boundary with reference to the geometry that defines the pattern for each hatch object.

2. When you save a drawing, the hatch patterns are stored with the drawing. Therefore, the hatch patterns can be updated even if the hatch pattern file that contains the definitions of hatch patterns is not available.

3. If the system variable **FILLMODE** is 0 (Off), the hatch patterns are not displayed. To see the effect of a changed **FILLMODE** value, you must use the **REGEN** command to regenerate the drawing after the value has been changed.

4. You can edit the boundary of a hatch pattern even when the hatch pattern is not visible (**FILLMODE**=0).

5. The hatch patterns created with earlier releases are automatically converted to AutoCAD 2004 hatch objects when the hatch pattern is edited.

6. When you save an AutoCAD 2004 drawing in Release 12 format (DXF), the hatch objects are automatically converted to Release 12 hatch blocks.

7. You can use the **CONVERT** command to change the pre-Release 14 hatch patterns to AutoCAD 2004 objects. You can also use this command to change the pre-Release 14 polylines to AutoCAD 2004 optimized format to save memory and disk space.

HATCHING USING THE HATCH COMMAND

Hatches can also be created with the **HATCH** command. However, such hatches are not associated with the boundary. Also, the **HATCH** command does not create its own boundary and hence, you have to select objects that will form the boundary for hatching. You can invoke this command only by entering **HATCH** at the AutoCAD Command prompt.

Note

*Although, the **Draw** menu displays **Hatch** and also the **Draw** toolbar displays the **Hatch** button, choosing them invokes the **BHATCH** command and not the **HATCH** command.*

When you enter the **HATCH** command, it does not display a dialog box as in the case of the **BHATCH** command. The prompt sequence is given next.

Command: **HATCH**
Enter a pattern name or [?/Solid/User defined] <ANSI31>: *Enter a pattern name or ?*

If you know the name of the pattern, enter it or you can use the ? option. You can also use the following options in response to this prompt:

? Option

If you enter **?** as the response to the **Enter a pattern name or [?/Solid/User defined] <ANSI31>:** prompt, the following prompt is displayed.

Enter pattern(s) to list <*>: Enter

When you press ENTER at the last prompt, the names and descriptions of all the defined hatch patterns are displayed in the AutoCAD Text window. If you are not sure about the name of the pattern you want to use, you can select this option before entering a name. You can also list a particular pattern by specifying it at the prompt.

Solid Option

This option is used to fill the selected boundary with solid. If you enter **S** as the response to the **Enter a pattern name or [?/Solid/User defined] <ANSI31>:** prompt, the following prompts are displayed.

Select objects to define hatch boundary or <direct hatch>,
Select objects: *Select a boundary for filling or press ENTER for direct hatch.*

If you select a boundary object for filling, it gets highlighted and when you press ENTER, the selected area is solid filled. But, if you press ENTER at the **Select objects:** prompt, you can use the direct hatch option and you will be allowed to create a boundary for filling. The prompt sequence for the direct hatch option is given next.

Select objects to define hatch boundary or <direct hatch>,
Select objects: Enter
Retain polyline boundary [Yes/No] <N>: *Enter* **Y** *to retain the polyline,* **N** *to lose it.*

When using the direct hatch option, a polyline boundary is created around the area to be hatched. You can retain this polyline if you enter Y at the previous prompt and if you do not want to retain the polyline boundary, enter N at the previous prompt. The prompt sequence that follows is similar to that of the **PLINE** command.

Specify start point: *Select the first point of the boundary.*
Specify next point or [Arc/Length/Undo]: *Select the next point of the boundary.*
Specify next point or [Arc/Close/Length/Undo]: *Select the next point of the boundary.*
Specify next point or [Arc/Close/Length/Undo]: **C** (*Enter* **C** *to close the boundary.)*
Specify start point for new boundary or <apply hatch>: Enter

The created boundary will be solid-filled. You can also specify another boundary by selecting another set of points and not pressing ENTER at the last command prompt.

Note

A solid-filled or hatched area using the direct hatch option need not be drawn within an existing area. You can create new solid-filled/hatched areas wherever you want. This is not possible using the ***BHATCH*** *command.*

Tip

The ***HATCH*** *command allows you to create a polyline hatch boundary, even if the objects selected do not form a closed area. This is useful when you need to hatch areas that do not form a closed area.*

User defined Option

This option is used for creating a user-defined array or double-hatched grid of straight lines. To invoke this option, enter **U** at the **Enter pattern name or [?/Solid/User defined] <ANSI31>**: prompt. After you enter **U**, the following prompts are displayed.

Specify angle for crosshatch lines <0>: *Specify the angle for hatching.*
Specify spacing between the lines <1.0000>: *Specify the spacing between hatch lines.*
Double hatch area? [Yes/No] <N> *Press ENTER if you do not want double hatching or enter Y.*

At the **Specify angle for crosshatch lines <0>**: prompt, you need to specify the angle at which you want the hatch to be drawn. This angle is measured with respect to the *X* axis of the current UCS. You can also specify the angle by selecting two points on the screen. This value is stored in the **HPANG** system variable.

At the second prompt, **Specify spacing between the lines <1.0000>**, you can specify the space between the hatch lines. The default value for spacing is 1.0000. The specified spacing value is stored in the **HPSPACE** system variable.

At the third prompt, **Double hatch area? [Yes/No] <N>**, you can specify if you want double hatching. The default is no (**N**), and in this case, double hatching is disabled. You can enter Yes (**Y**) if you want double hatching. In that case, AutoCAD draws a second set of lines at right angles to the original lines. If the double hatch is on, AutoCAD sets the **HPDOUBLE** system variable to 1. The prompts that follow are the same as mentioned earlier in the case of solid- filled areas.

Hatching by Specifying the Pattern Name

When you invoke the **HATCH** command, you can also specify the hatch pattern name at the **Enter a pattern name or [?/Solid/User defined] <ANSI31>:** prompt. With the hatch name you can also specify the hatch style by adding style option to the pattern name. For example, for the pattern ANSI32, which needs to be applied with Outer style, the pattern shall be entered as ANSI32,O. The format is as follows.

Command: **HATCH**
Enter a pattern name or [?/Solid/User defined] <ANSI31>: **Stars, I** *(Enter a pattern name proceeded by a comma and the style.)*

Specify a scale for the pattern <1.0000>: *Specify a scale.*
Specify an angle for the pattern <0>: *Specify the angle for hatching.*

By specifying **Stars, I** you are specifying the hatch pattern as Stars and the hatch style as Ignore. There are three hatch styles: **Normal** (N), **Outer** (O), and **Ignore** (I). They have been discussed earlier in this chapter under "Advanced Tab".

All hatch patterns drawn using hatch names are drawn as blocks, which means that the entire hatch pattern acts as a single object and you cannot erase the lines in the pattern individually. If you want to edit the lines in the pattern individually, you can type an asterisk (*) before the hatch pattern name. For example, you can enter *ANSI31 at the **Enter a pattern name or [?/Solid/User defined] <ANSI31>:** prompt to edit the lines in the pattern individually.

All the stored hatch patterns are assigned an initial scale (size) and a rotation of 0-degree with respect to the positive *X* axis. You can expand or contract according to your needs. You can scale the pattern by specifying a scale factor. You can also rotate the pattern with respect to the positive direction of the current UCS by entering a desired angle in response to the **Specify an angle for the pattern <0>:** prompt. You can even specify the angle manually by selecting or entering two points. For example, Figure 13-49 illustrates the hatch pattern ANSI31 with different scales and angles. The scale and angle are shown with the figures. Once you are finished with specifying the pattern, AutoCAD prompts you to select objects for hatching or to use the direct hatch option (discussed earlier).

Select objects to define hatch boundary or <direct hatch>,
Select objects: *Select a boundary for hatching or press ENTER for direct hatch.*

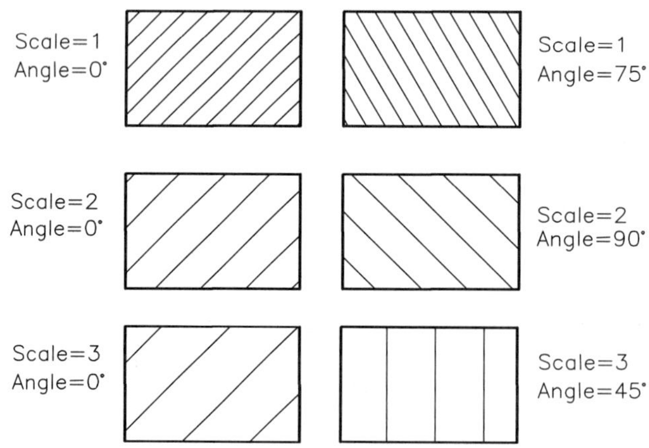

Figure 13-49 *Using different values for scale and angle*

Note

You can also reference the scale for the pattern to paper space. This is useful when plotting drawings, since the scale factor is then based on the plotted scale of the drawing. Enter XP after the scale in the ***Specify a scale for the pattern <1.000>:*** *prompt. For example, if you enter 1XP as the scale factor, AutoCAD will calculate the actual scale in Model space such that it matches a value of 1 in paper space and this model space equivalent value is stored as the default value for the next time you use the* ***HATCH*** *command.*

Defining the boundaries in the **HATCH** command is a bit tedious compared with defining them with the **BHATCH** command. In the **BHATCH** command, the boundaries are automatically defined, whereas with the **HATCH** command you need to define the boundaries by selecting the objects. You can use any object selection method to select the objects that define the boundary of the area to be hatched and the objects that you want to exclude from hatching. If the boundary objects you select do not form a closed boundary, depending upon the angle and type of hatch pattern, the resulting hatch varies. For example, in Figure 13-50(a), hatching does not take place. Here, only two lines lie completely within the window and the boundary is not closed. The pattern you have selected starts getting applied from one end of one of the lines to another line, if it is able to find it using its current angle and direction. When AutoCAD is unable to find another object to end the pattern, it displays the following message: **Unable to hatch the boundary**. When you change the angle of the pattern (ANSI31) from **0**-degree (default) to **135**-degree, see Figure 13-50(b), the hatch pattern starts getting applied from one end and continues until another end is being found. Some patterns, like TRIANG, need more than two edges to be completely drawn and therefore, are drawn incomplete in the absence of the required boundary edges.

In Figure 13-51(a), the boundary is selected by selecting objects. In this illustration, the lines LQ and MR extend beyond the hatch boundary, which makes the hatching begin outside the intended hatch area. You must break the top and bottom line at points Q and R so that lines LQ, QR, RM, and ML define the hatch boundary. Now, if you select the objects, AutoCAD will hatch the area bound by the four selected objects, as shown in Figure 13-51(b).

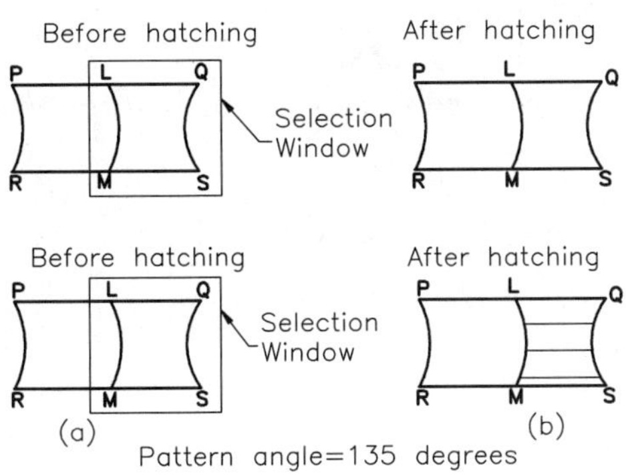

Figure 13-50 *Selecting objects using the* ***HATCH*** *command*

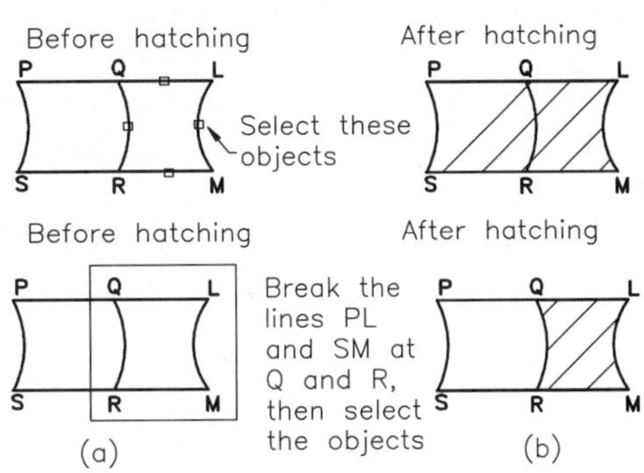

Figure 13-51 *Selecting objects using the* ***HATCH*** *command*

Note

*Patterns drawn using the **HATCH** command are nonassociative. This implies that whenever the hatch boundary is modified, the hatch pattern is not updated automatically.*

Self-Evaluation Test

Answer the following questions, and then compare your answers to the correct answers given at the end of this chapter.

1. Different filling patterns make it possible to distinguish between different parts or components of an object. (T/F)

2. If AutoCAD does not locate the entered pattern in the *acad.pat* file, it searches for it in a file with the same name as the pattern. (T/F)

3. When a boundary set is formed, it does not become the default for hatching. (T/F)

4. You can edit the hatch boundary and the associated hatch pattern by using grips. (T/F)

5. The __________ command automatically defines the hatch boundary, whereas in the case of the __________ command you have to define the hatch boundary by selecting the objects that form the boundary of the hatch area.

6. One of the ways to specify a hatch pattern from the group of stored hatch patterns is by selecting one from the __________ drop-down list.

7. The value you enter in the __________ edit box lets you rotate the hatch pattern with respect to the *X* axis of the current UCS.

8. The __________ option in the **Boundary Hatch and Fill** dialog box lets you select objects that form the boundary for hatching.

9. You can use the __________ option after you have selected the area to be hatched to see the hatching before it is actually applied.

10. The selection of a hatch style carries meaning only if the objects to be hatched are __________.

Review Questions

Answer the following questions.

1. The **BHATCH** command does not allow you to hatch a region enclosed within a boundary (closed area) by selecting a point inside the boundary. (T/F)

2. A **Boundary Definition Error** dialog box is displayed if AutoCAD finds that the selected point is not inside the boundary or that the boundary is not closed. (T/F)

3. You can improve the speed of hatching by setting parameters in the **Advanced** tab of the **Boundary Hatch and Fill** dialog box. (T/F)
4. When you use editing commands such as **MOVE**, **SCALE**, **STRETCH**, and **ROTATE**, associativity is lost, even if all the boundary objects are selected. (T/F)

5. The hatching procedure in AutoCAD does not work on inserted blocks. (T/F)

6. Patterns drawn using the **HATCH** command are associative. (T/F)

7. If the **Double** hatch box is selected, which system variable is set to 1?

   (a) **HPSPACE** (b) **HPDOUBLE**
   (c) **HPANG** (d) **HPSCALE**

8. The **ISO Pen Width** drop-down list is available only for which hatch patterns?

   (a) **ANSI** (b) **Predefined**
   (c) **Custom** (d) **ISO**

9. If you want to have the same hatching pattern, style, and properties as that of an existing hatch on the screen, which button should you choose?

   (a) **Inherit Properties** (b) **Select Objects**
   (b) **Pick Points** (d) **Remove Islands**

10. Which variable can you use to align the hatches in adjacent hatch areas?

    (a) **SNAPBASE** (b) **FILLMODE**
    (c) **SNAPANG** (d) **HPANG**

11. The specified hatch spacing value is stored in which system variable?

    (a) **HPDOUBLE** (b) **HPANG**
    (c) **HPSCALE** (d) **HPSPACE**

12. One of the advantages of using the __________ command is that you don't have to select each object comprising the boundary of the area you want to hatch, as in the case of the __________ command.

13. To select a custom hatch pattern, first select **Custom** from the **Type** drop-down list and then select the name of a previously stored hatch pattern from the __________ drop-down list.

14. There are three hatching styles from which you can choose: __________, __________, and __________.

15. If you select the __________ style, all areas bounded by the outermost boundary are hatched, ignoring any hatch boundaries that lie within the outer boundary.

Exercises

Exercise 5 *Mechanical*

Hatch the drawings in Figures 10-52 and 10-53 using the hatch pattern to match.

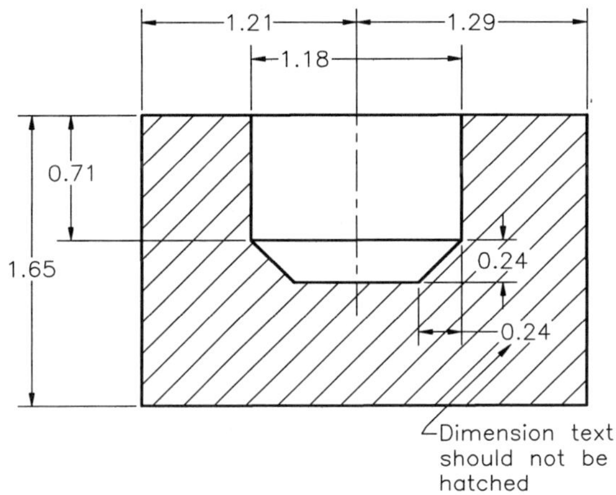

Figure 13-52 *Drawing for Exercise 5*

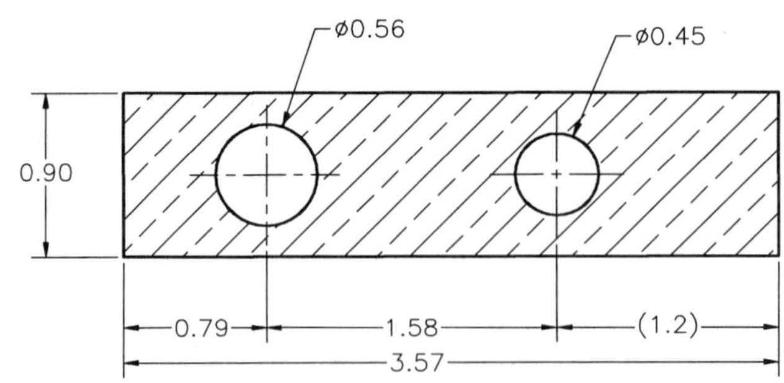

Figure 13-53 *Drawing for Exercise 5*

Exercise 6

General

Hatch the drawing in Figure 13-54 using the hatch pattern ANSI31. Use the **SNAPBASE** variable to align the hatch lines as shown in the drawing.

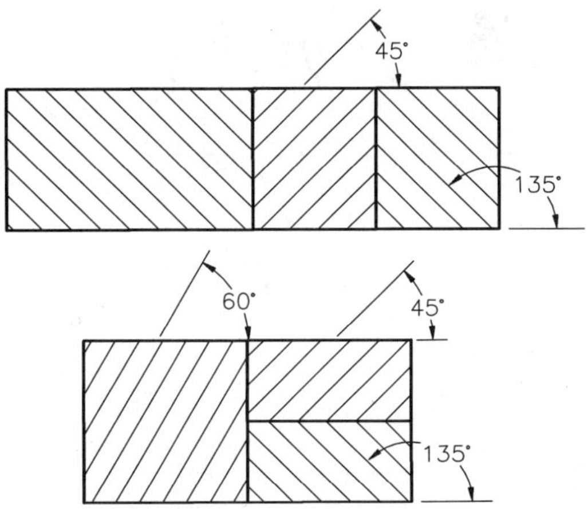

Figure 13-54 *Drawing for Exercise 6*

Exercise 7

Mechanical

Figure 13-55 shows the top and front views of an object. It also shows the cutting plane line. Based on the cutting plane line, hatch the front views in section. Use the hatch pattern of your choice.

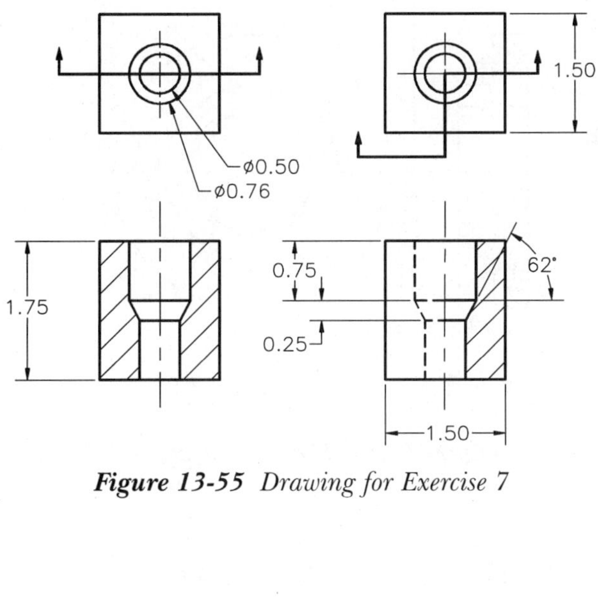

Figure 13-55 *Drawing for Exercise 7*

Exercise 8 *Mechanical*

Figure 13-56 shows the top and front views of an object. Hatch the front view in full section. Use the hatch pattern of your choice.

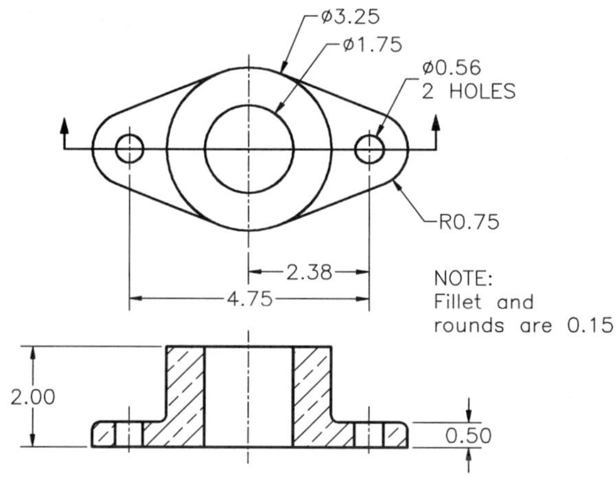

Figure 13-56 *Drawing for Exercise 8*

Exercise 9 *Mechanical*

Figure 13-57 shows the front and side views of an object. Hatch the side view in half section. Use the hatch pattern of your choice.

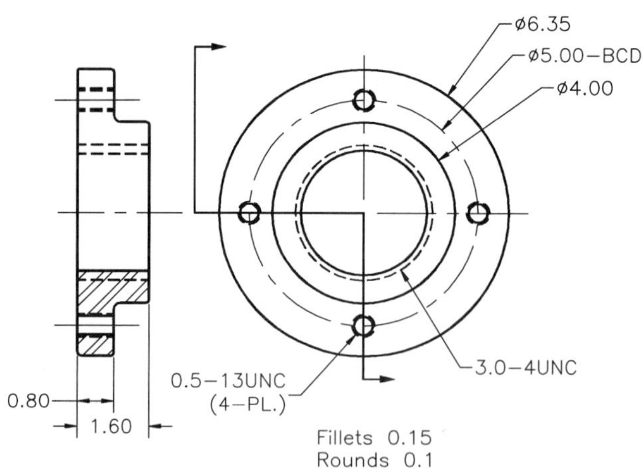

Figure 13-57 *Drawing for Exercise 9*

Exercise 10

Mechanical

Figure 13-58 shows the front view with broken section and top views of an object. Hatch the front view as shown. Use the hatch pattern of your choice.

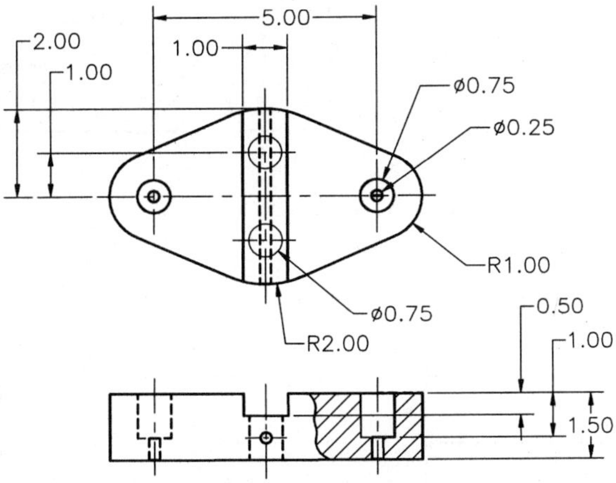

Figure 13-58 *Drawing for Exercise 10*

Exercise 11

Mechanical

Figure 13-59 shows the front, top, and side views of an object. Hatch the side view in section using the hatch pattern of your choice.

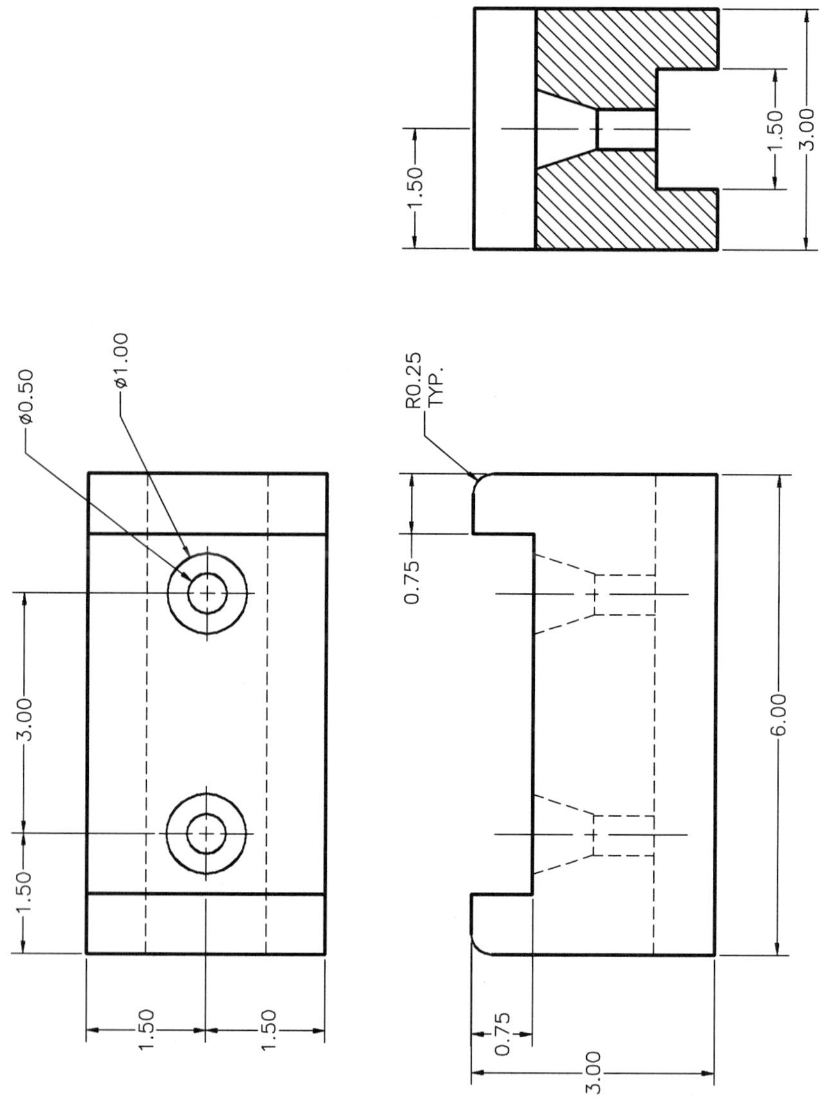

Figure 13-59 *Drawing for Exercise 11*

Problem Solving Exercise 1 *Mechanical*

Figure 13-60 shows an object with front and side views with aligned section. Hatch the side view using the hatch pattern of your choice.

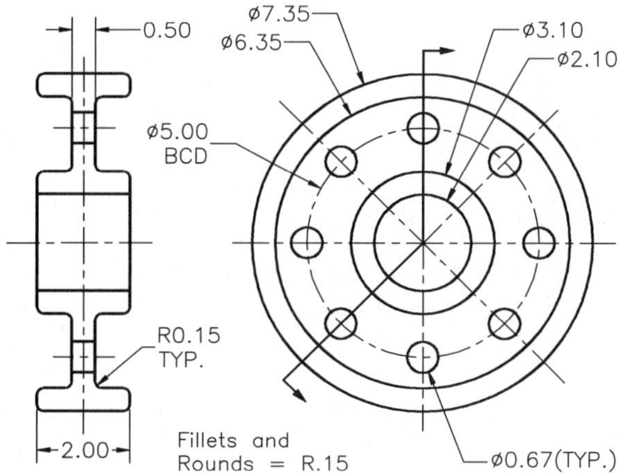

Figure 13-60 Drawing for Problem Solving Exercise 1

Problem Solving Exercise 2

Mechanical

Figure 13-61 shows the front view, side view, and the detail "A" of an object. Hatch the side view and draw the detail drawing as shown.

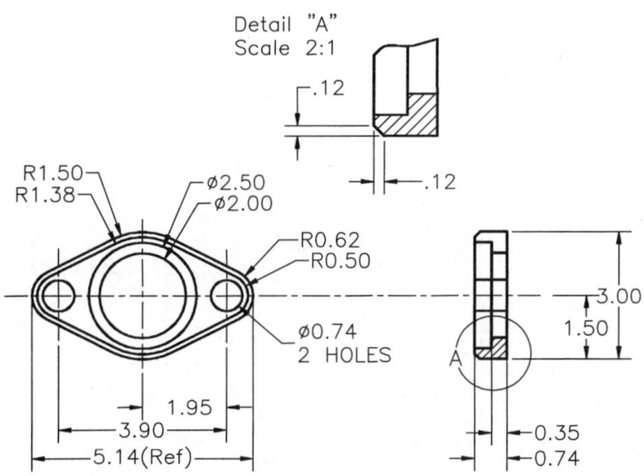

Figure 13-61 Drawing for Problem Solving Exercise 2

Problem Solving Exercise 3

Mechanical

Create the drawing shown in Figure 13-62. Assume the missing dimensions.

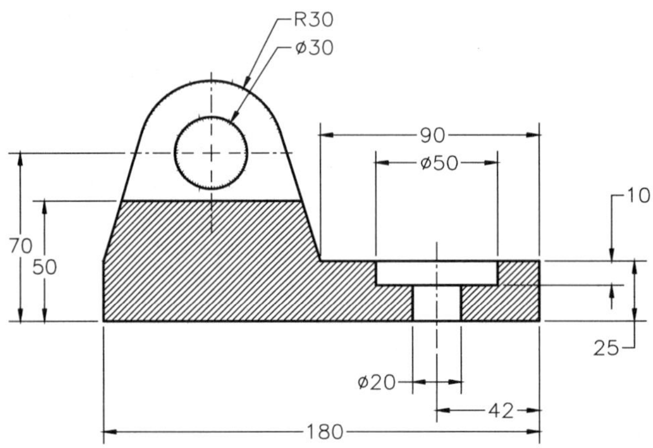

Figure 13-62 Drawing for Problem Solving Exercise 3

Problem Solving Exercise 4 *Mechanical*

Create the drawing shown in Figure 13-63. Assume the missing dimensions.

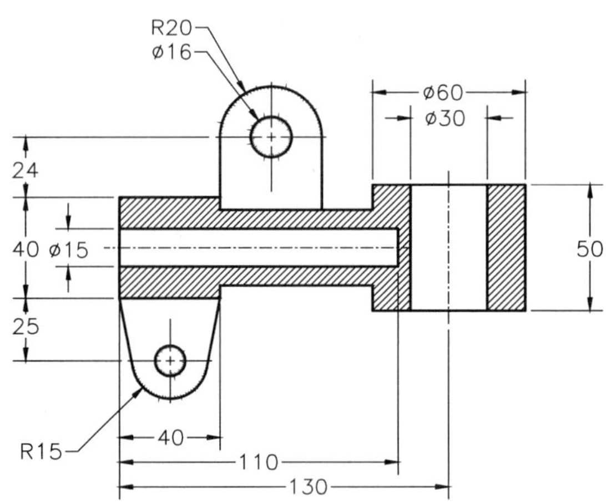

Figure 13-63 Drawing for Problem Solving Exercise 4

Answers to Self-Evaluation Test

1 - T, **2** - T, **3** - F, **4** - T, **5** - **BHATCH**, **HATCH**, **6** - **Pattern**, **7** - **Angle**, **8** - **Select Objects**, **9** - **Preview**, **10** - nested

Chapter 14

Working with Blocks

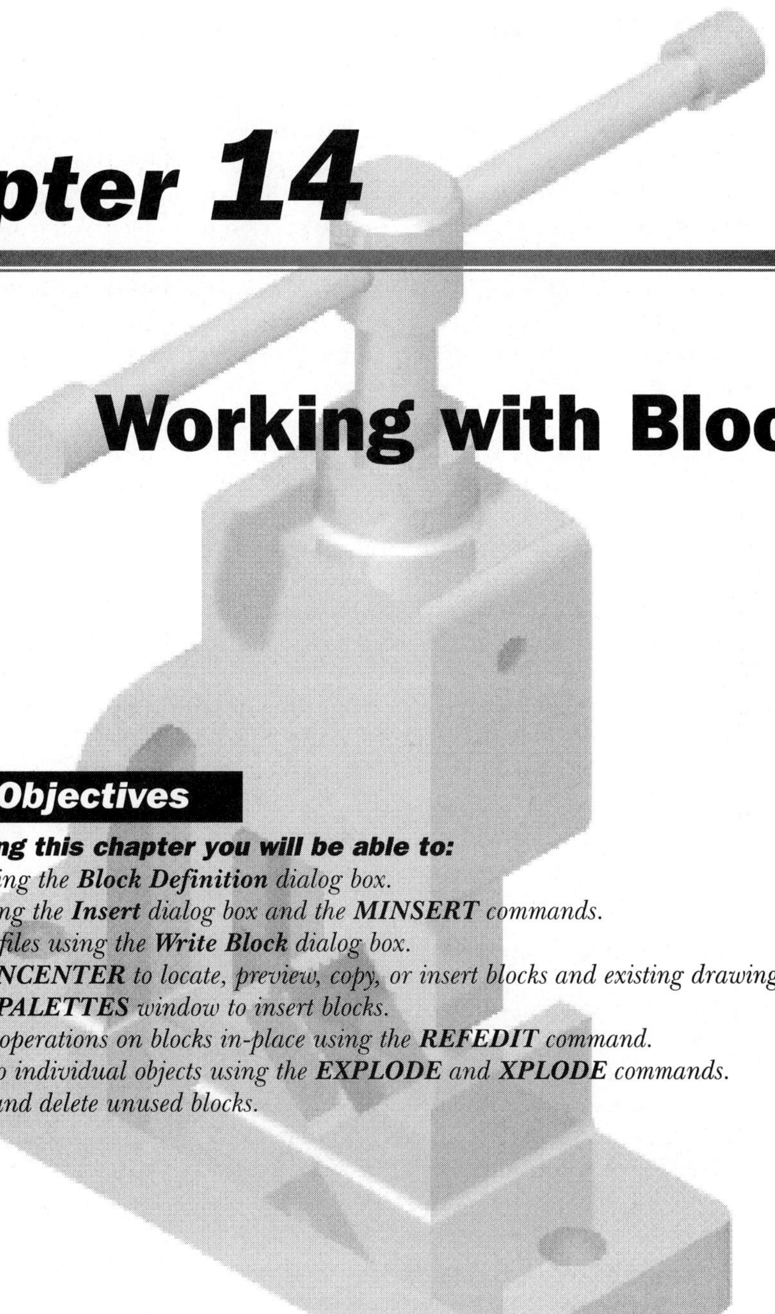

Learning Objectives

After completing this chapter you will be able to:

- *Create blocks using the* ***Block Definition*** *dialog box.*
- *Insert blocks using the* ***Insert*** *dialog box and the* ***MINSERT*** *commands.*
- *Create drawing files using the* ***Write Block*** *dialog box.*
- *Use the* ***DESIGNCENTER*** *to locate, preview, copy, or insert blocks and existing drawings.*
- *Use the* ***TOOL PALETTES*** *window to insert blocks.*
- *Perform editing operations on blocks in-place using the* ***REFEDIT*** *command.*
- *Split a block into individual objects using the* ***EXPLODE*** *and* ***XPLODE*** *commands.*
- *Rename blocks and delete unused blocks.*

THE CONCEPT OF BLOCKS

The ability to store parts of a drawing, or the entire drawing, so that they need not be redrawn when required in the same drawing or another drawing is a great benefit to the user. These parts of a drawing, entire drawings, or symbols (also known as **blocks**) can be placed (inserted) in a drawing at the location of your choice, in the desired orientation, and with the desired scale factor. The block is given a name (block name) and the block is referenced (inserted) by its name. All the objects within a block are treated as a single object. You can **MOVE**, **ERASE**, or **LIST** the block as a single object, that is, you can select the entire block simply by selecting anywhere on it. As for the edit and inquiry commands, the internal structure of a block is immaterial, since a block is treated as a primitive object, like a polygon. If a block definition is changed, all references to the block in the drawing are updated to incorporate the changes made to the block.

A block can be created using the **BLOCK** command. You can also save objects in a drawing, or an entire drawing as a drawing file using the **WBLOCK** command. The main difference between the two is that a wblock can be inserted in any other drawing, but a block can be inserted only in the drawing file in which it was created.

Another feature of AutoCAD is that instead of inserting a symbol as a block (which results in adding the content of the referenced drawing to the drawing in which it is inserted), you can reference the other drawings (Xref) in the current file. This means that the contents of the referenced drawing are not added to the current drawing file, although they become part of that drawing on the screen. This is explained in detail in Chapter 16.

Advantages of Using Blocks

Blocks offer many advantages. Here are some of them.

1. Drawings often have some repetitive features. Instead of drawing the same feature again and again, you can create a block of this feature and insert it wherever required. This helps you to reduce drawing time and better organize your work.
2. Another advantage of using blocks is that they can be drawn and stored for future use. You can thus create a custom library of objects required for different applications. For example, if your drawings are concerned with gears, you could create blocks of gears and then integrate these blocks with custom menus (see Chapter 32, for writing menus). In this manner, you could create an application environment of your own in AutoCAD.
3. The size of a drawing file increases as you add objects to it. AutoCAD keeps track of information about the size and position of each object in the drawing; for example, the points, scale factors, radii, and so on. If you combine several objects into a single object by forming a block with the **BLOCK** command, there will be a single scale factor, rotation angle, position, and so on, for all objects in the block, thereby saving storage space. Each object repeated in multiple block insertions needs to be defined only once in the block definition. Ten insertions of a gear made of forty-three lines and forty-one arcs require only ninety-four objects (10+43+41), while ten individually drawn gears require 840 objects [10 X (43+41)].
4. If the specification for an object changes, the drawing needs to be modified. This is a

very tedious task if you need to detect each location where the change is to be made and edit it individually. But if this object has been defined as a block, you can redefine it, and everywhere the object appears, it is revised automatically. This has been dealt with in Chapter 15, Defining Block Attributes.

5. Attributes (textual information) can be included in blocks. Different attribute values can be specified in each insertion of a block.
6. You can create symbols and then store them as blocks with the **BLOCK** command. Later on, with the **INSERT** command, you can insert the blocks in the drawing in which they were defined. There is no limit to the number of times you can insert a block in a drawing.
7. You can store symbols as drawing files using the **WBLOCK** command and later insert these symbols into any other drawing using the **INSERT** command.
8. Blocks can have varying X, Y, and Z scale factors and rotation angles from one insertion to another (Figure 14-1).

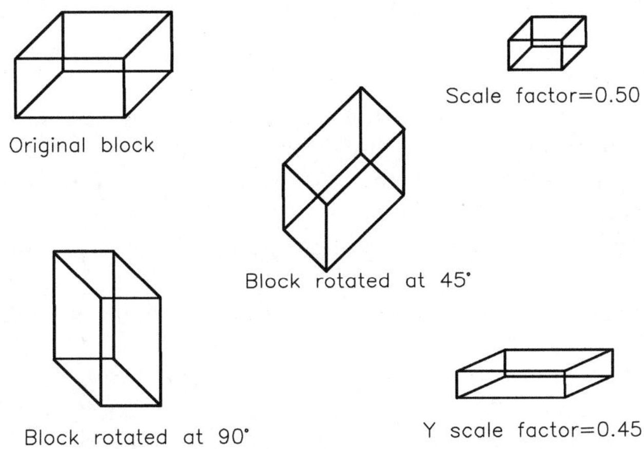

***Figure 14-1** Block with different specifications*

9. You can insert multiple copies of a block about a specified path using the **DIVIDE** or the **MEASURE** command.

FORMATION OF BLOCKS

Drawing Objects for Blocks

The first step in creating blocks is to draw the object(s) to be converted into a block. You can consider any symbol, shape, or view that may be used more than once for conversion into a block (Figure 14-2). Even a drawing that is to be used more than once can be inserted as a block.

Tip

Examine the sketch of the drawing you are about to begin and examine it carefully to look for any shapes, symbols, and components that are to be used more than once. They can then be drawn once and be stored as blocks.

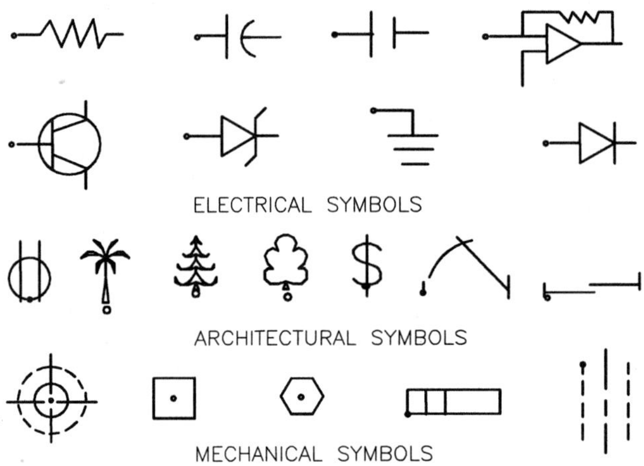

Figure 14-2 *Some common drafting symbols created to store in a library*

Checking Layers

You should be particularly careful about the layers on which the objects to be converted into a block are drawn. The objects must be drawn on layer 0 if you want the block to inherit the linetype and color of the layer on which it is inserted. If you want the block to retain the linetype and color of the layer on which it was drawn, draw the objects defining the block on that layer. For example, if you want the block to have the linetype and color of layer OBJ, draw the objects comprising the block on layer OBJ. Now, even if you insert the block on any other layer, the linetype and color of the block will be those of layer OBJ.

If the objects in a block are drawn with the color, linetype, and lineweight as ByBlock, this block when inserted assumes the properties of the current layer. To set the color to ByBlock, select **ByBlock** from the **Color Control** drop-down list of the **Properties** toolbar. You can also choose **Color** from the **Format** menu to display the **Select Color** dialog box and then choose the **ByBlock** button in the dialog box. Choose **OK** to close the dialog box. To set the linetype to ByBlock, select **ByBlock** from the **Linetype Control** drop-down list of the **Properties** toolbar. You can also select **ByBlock** in the **Linetype Manager** dialog box that is displayed when you choose **Linetype** from the **Format** menu and then choose the **Current** button. The current linetype is now set to **ByBlock**. Similarly, you can specify the lineweight as **ByBlock**.

Note

*As discussed previously, before you make a block you should check the layers on which the objects are made because these layers will determine the color and linetype of the block when inserted. To change the layer associated with an object, use the **Properties** toolbar, **PROPERTIES** palette, **Properties** option of the **CHANGE** command, or the **CHPROP** command.*

CONVERTING OBJECTS INTO A BLOCK

You can convert objects into a block either using the **Block Definition** dialog box or by using the command line. Both these options are discussed next.

Converting Objects Into Blocks Using the Block Definition Dialog Box

Toolbar:	Draw > Make Block
Menu:	Draw > Block > Make
Command:	BLOCK

When you invoke the **BLOCK** command (Figure 14-3), the **Block Definition** dialog box (Figure 14-4) is displayed. You can use the **Block Definition** dialog box to save any object or objects as a block.

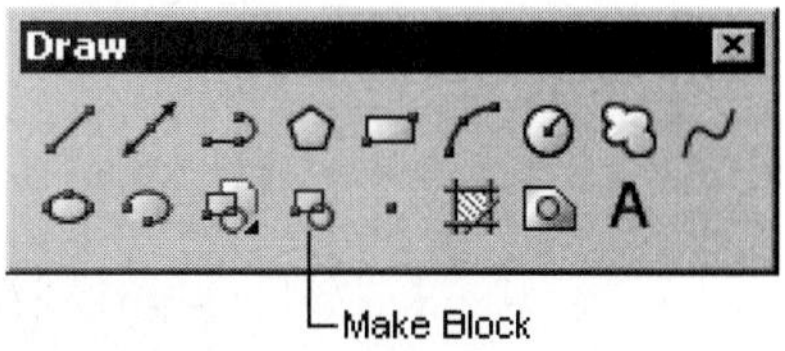

*Figure 14-3 **Make Block** button in the **Draw** toolbar*

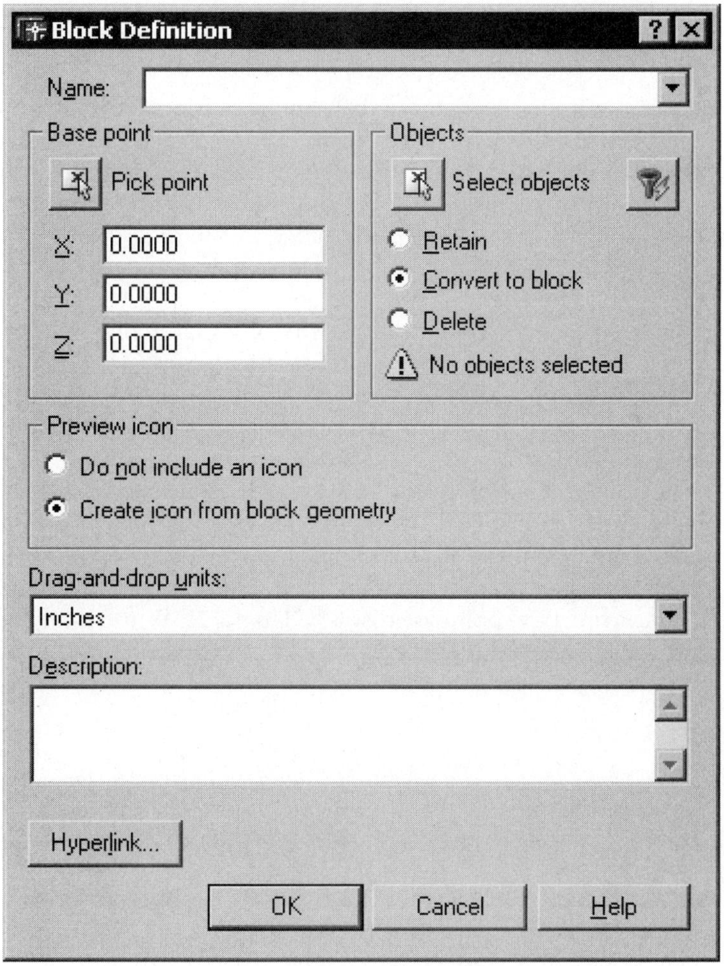

*Figure 14-4 **Block Definition** dialog box*

In the **Block Definition** dialog box, you can enter the name of the block you want to create in the **Name** edit box. All the block names present in a current drawing are displayed in the **Name** drop-down list in an alphabetical and numerical order. If a drawing contains more than six blocks, a scroll bar appears on the right side of the **Name** drop-down list, which can be moved up and down to view all the blocks in the drawing. This way you can verify whether a block you have defined has been saved. By default, the block name can have up to 255 characters. The name can contain letters, digits, blank spaces as well as any special characters including the $ (dollar sign), - (hyphen), and _ (underscore), provided they are not being used for any other purpose by AutoCAD or Windows.

Note

The block name is controlled by the ***EXTNAMES*** *system variable and the default value is 1. If the value is set to 0, the block name will be only thirty-one characters long and will not include spaces or any other special characters apart from a $ (dollar sign), - (hyphen), or _ (underscore).*

If a block already exists with the block name that you have specified in the **Name** edit box, an AutoCAD dialog box, warning that the block name is already defined, is displayed when you choose the **OK** button in the **Block Definition** dialog box. In this dialog box, you can either redefine the existing block by choosing the **Yes** button or you can exit it by choosing the **No** button. You can then use another name for the block in the dialog box for the block.

After you have specified a block name, you are required to specify the insertion base point. This point is used as a reference point to insert the block. Usually, either the center of the block or the lower left corner of the block is defined as the insertion base point. Later on, when you insert the block, you will notice that the block appears at the insertion point, and you can insert the block with reference to this point. The point you specify as the insertion base point is taken as the origin point of the block's coordinate system. You can specify the insertion point by choosing the **Pick point** button in the **Base point** area of the dialog box. The dialog box is temporarily removed, and you can select a base point on the screen. You can also enter the coordinates in the **X:**, **Y:**, and **Z:** edit boxes instead of selecting a point on screen. Once the insertion base point is specified, the dialog box reappears and the *X*, *Y*, and *Z* coordinates of the insertion base point are displayed in the **X**, **Y**, and **Z** edit boxes respectively. If no insertion base point is selected, AutoCAD assumes the insertion point coordinates to be 0,0,0, which are the default coordinates.

Tip

In drawings where the insertion point of a block is important, for example, in the case of inserting the block of a door symbol in architectural plans, it may be a good idea to first specify the requisite scale factors and rotation angles and then the insertion point. This way, you are able to preview the block symbol as it is to be inserted.

After specifying the insertion base point, you are required to select the objects that will constitute the block. Until the objects are selected, AutoCAD displays a warning: **No objects selected**, at the bottom of the **Objects** area. Choose the **Select objects** button. The dialog box is temporarily removed from the screen. You can select the objects on the screen using any selection method. After completing the selection process, right-click or press the ENTER key to return to the dialog box. The number of objects selected is displayed at the bottom of the **Objects** area of the dialog box.

The **Objects** area of the **Block Definition** dialog box also has a **QuickSelect** button. If you choose this button, the **Quick Select** dialog box is displayed, which allows you to define a selection set based on properties of objects. **Quick Select** is used in cases where the drawings are very large and complex. The **Quick Select** dialog box has been discussed earlier in Chapter 6, Editing Sketched Objects-II.

In the **Objects** area, if you select the **Retain** radio button it retains the selected objects that form the block as individual objects and does not convert them into a block after the block has been defined. If you select the **Convert to block** radio button (default option) it converts the selected objects into a block after you have defined the block. If you select the **Delete** radio button it deletes the selected objects from the drawing after the block has been defined.

Tip

*When you select the **Delete** radio button in the **Objects** area when defining a block, the selected objects are removed from the drawing. If you want them back, you can use the **OOPS** command. Using **U** at the Command prompt, or choosing the **Undo** button from the **Standard** toolbar will undo the block definition from the drawing.*

The **Preview icon** area gives you the following options. Selecting the **Do not include an icon** radio button does not create a preview icon and whenever a block name is selected from the name drop-down list, the **Preview Image box** is not displayed at all in this area of the dialog box. If you select the **Create icon from block geometry** radio button a **Preview icon** is created, which is saved with the block definition in the drawing. So, whenever you select a block name from the **Name** drop-down list of the **Block Definition** dialog box or the **DESIGNCENTER**, the preview image is displayed. The **DESIGNCENTER** has been discussed earlier in Chapter 6, Editing Sketched Objects-II. Similarly, in the **Description** text box, if you enter a textual description of the block, it is stored with the block definition and displayed on selecting the block name from the **Name** drop-down list of the **Block Definition** dialog box or the **DESIGNCENTER**.

Tip

Saving a preview image and the description text with the block definition makes it easier to identify a block and is an easy reference.

The **Drag-and-drop units** drop-down list displays the units the **DESIGNCENTER** will use when inserting the particular block. The default value is Inches. Choose the **OK** button to complete defining the block.

Tip

It is a good idea to create a unit block, especially for drawings where the same block is to be inserted with various scale factors. A unit block can be defined as a block that is created within a unit square area (1 unit x 1 unit). This way, every time you specify a scale factor, the block simply gets enlarged or reduced by the said amount directly, since the multiplication factor is 1.

Converting Objects Into a Block Using the Command Line

You can use the **-BLOCK** command to create blocks from the command line. The following prompt sequence saves a symbol as a block named SYMBOLX.

Command: **-BLOCK**
Enter block name or [?]: **SYMBOLX**
Specify insertion base point: *Select a point.*
Select objects: *Select all the objects comprising the block.*
Select objects: *Press ENTER when selection is complete.*

AutoCAD acknowledges the creation of the block by removing the objects defining the block from the screen. You can get the objects back at the same position by entering **OOPS**. You can use the **?** option to list all the blocks created in the drawing. The prompt sequence is given next.

Command: **-BLOCK**
Enter block name or [?]: ?
Enter block(s) to list <*>: Enter

Information similar to the following is displayed in the AutoCAD Text Window.

Defined blocks
"BLK1"
"BLK2"
"SYMBOLX"

User Blocks	*External References*	*Dependent Blocks*	*Unnamed Blocks*
n	*n*	*n*	*n*

Here **n** denotes a numeric value.

Defined blocks. The names of all the blocks defined in the drawing.

User Blocks. The number of blocks created by the user.

External References. The number of external references. These are the drawings that are referenced with the **XREF** command and are discussed later in Chapter 16.

Dependent Blocks. The number of externally dependent blocks. These are the blocks that are present in a referenced drawing.

Unnamed blocks. The number of unnamed blocks in the drawing. These are objects such as associative dimensions and hatch patterns.

Exercise 1 *General*

Draw a unit radius circle, and use the **BLOCK** command to create a block. Name the block CIRCLE. Figure 14-5 shows the steps involved in creating a block. Select the center as the insertion point. Verify if the block CIRCLE was saved.

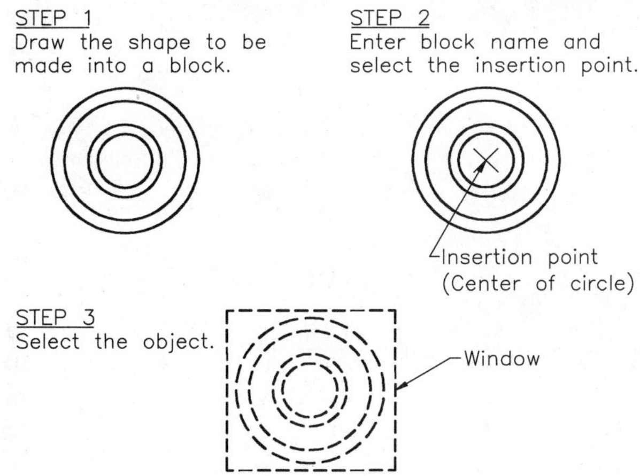

Figure 14-5 *Using the **BLOCK** command*

INSERTING THE BLOCKS USING THE INSERT DIALOG BOX

Toolbar:	Insert > Insert Block
	Draw > Insert
Menu:	Insert > Block
Command:	INSERT

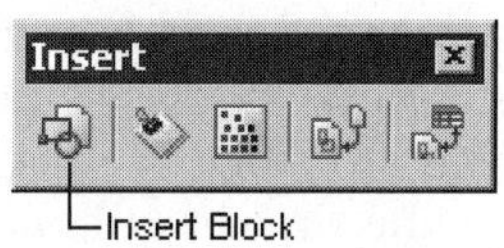

Figure 14-6 *Invoking the **INSERT** command from the **Insert** toolbar*

Insertion of a predefined block is possible with the **INSERT** command (Figure 14-6). An inserted block is called a block reference. You should determine the layer on which you want to insert the block, the location where you want to insert it, its size, and the angle by which you want the block to be rotated, prior to insertion. If the layer on which you want to insert the block is not current, you can use the **Layer Control** drop-down list in the **Layers** toolbar or choose the **Current** button in the **Layer Properties Manager** dialog box to make it current. When you invoke the **INSERT** command, the **Insert** dialog box (Figure 14-7) is displayed. You can specify the different parameters of the block or external file to be inserted in the **Insert** dialog box.

Name

This drop-down is used to specify the name of the block to be inserted. Select the name of the block to be inserted from this drop-down list. You can also enter the name of the block to be inserted. All the blocks created in the current drawing are available in the **Name** drop-down list.

Note

*The last block name selected from the **Name** drop-down list and inserted becomes the default name for the next insertion and is displayed in this drop-down list.*

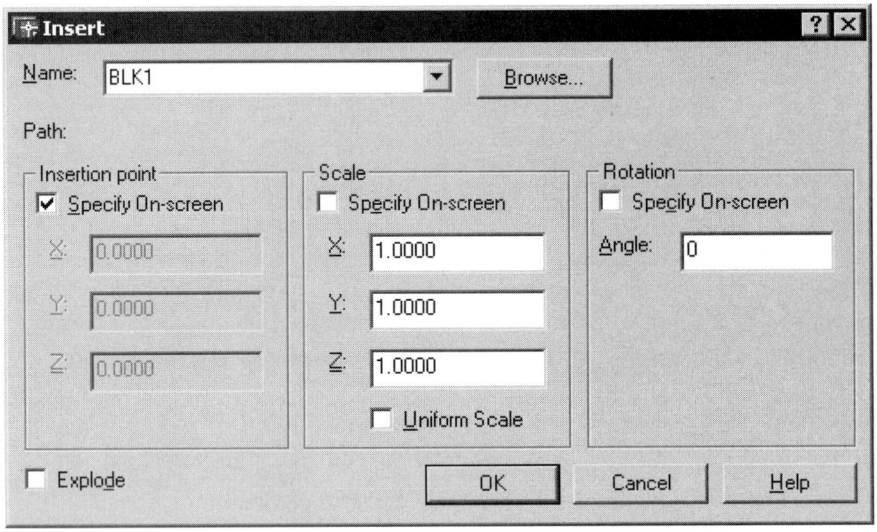

***Figure 14-7 Insert** dialog box*

The **Browse** button is used to insert external files. When you choose this button, the **Select Drawing File** dialog box (Figure 14-8) is displayed. This dialog box works in a manner similar to the standard **Select File** dialog box, which has been discussed earlier in Chapter 1. You can select a drawing file from the listed drawing files existing in a current directory, and can change the directory by selecting the desired directory in the **Look in** drop-down list. Once you select the drawing file in this dialog box and choose the **Open** button, the drawing file name is displayed in the **Name** drop-down list of the **Insert** dialog box. **Path** displays the path of the external file selected to be inserted. Now, if you want to change the block name, just change the name in the **Name** drop-down list. In this manner, the drawing can be inserted with a different name. Changing the original drawing does not affect the inserted drawing.

Note

*If the name you have specified in the **Name** edit box does not exist as a block in the current drawing, AutoCAD will search the drives and directories on the path (specified in the **Options** dialog box) for a drawing of the same name and if it finds it, it will insert it.*

*Also, suppose you have inserted a block in a drawing and then you want to insert a drawing with the same name as the block, AutoCAD will display a message saying that the specified block already exists and asks if you want to redefine it. If you choose **Yes**, the block in the drawing gets replaced by the drawing with the same name, that is, the block gets redefined.*

Tip

*You can create a block in a current drawing from an existing drawing file. This saves time by avoiding redrawing the object as a block. Locate and select an existing file using the **Browse** button and rename it with the name that you want to assign to the block in the **Name** edit box. When you choose **OK**, you can insert the block in the drawing or press the ESC key. The new block gets added to the current drawing.*

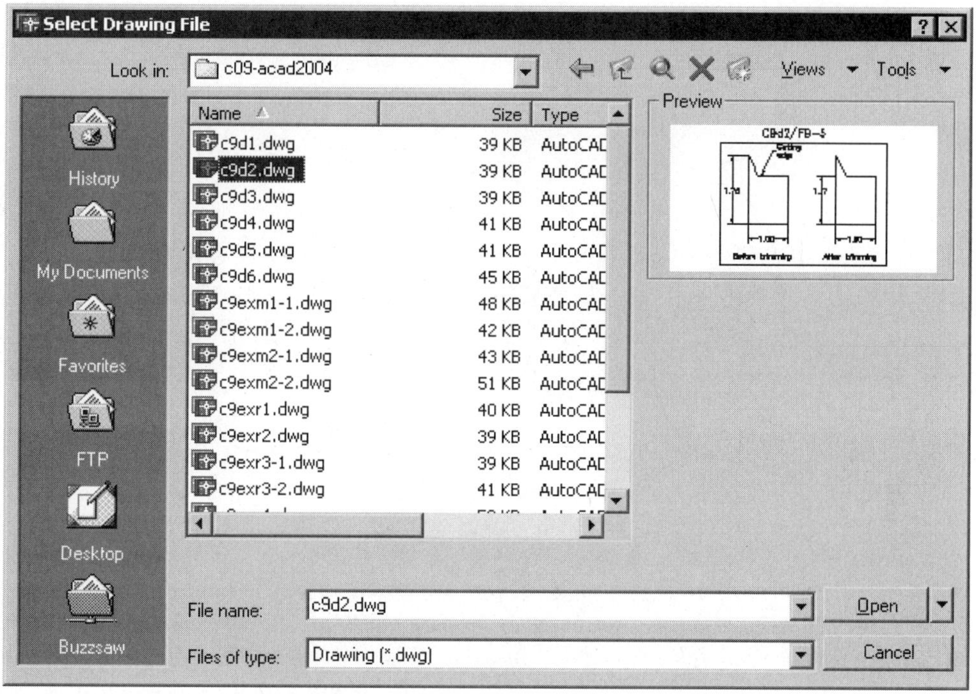

***Figure 14-8 Select Drawing File** dialog box*

Insertion point

When a block is inserted, the coordinate system of the block is aligned parallel to the current UCS. In the **Insertion point** area, you can specify the *X*, *Y*, and *Z* coordinate locations of the block insertion point in the **X**, **Y**, and **Z** edit boxes respectively. If you select the **Specify On-screen** check box, the **X**, **Y**, and **Z** edit boxes are not available and you can specify the insertion point on the screen. By default, the **Specify On-screen** check box is selected and hence, you can specify the insertion point on the screen.

Scale

In this area, you can specify the X, Y, and Z scale factors of the block to be inserted in the **X**, **Y**, and **Z** edit boxes respectively. By selecting the **Specify On-screen** check box, you can specify the scale of the block at which it has to be inserted on the screen. The **Specify On-screen** check box is cleared by default and the block is inserted with the scale factors of 1 along the three axes. Also, if the **Uniform Scale** check box is selected, the X scale factor value is assumed for the Y and Z scale factors also. This means that if this check box is selected, you need to specify only the X scale factor. All the dimensions in the block are multiplied by the X, Y, and Z scale factors you specify. These scale factors allow you to stretch or compress a block along the *X* and *Y* axes, respectively, according to your needs. You can also insert 3D objects into a drawing by specifying the third scale factor (since 3D objects have three dimensions), the Z scale factor. Figure 14-9 shows variations of the X and Y scale factors for the block, CIRCLE, during insertion.

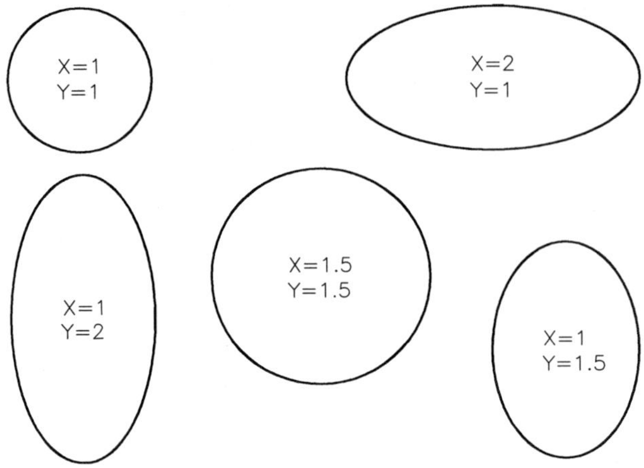

***Figure 14-9** Block inserted with different scale factors*

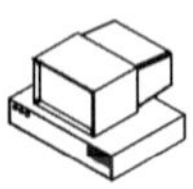

Tip

By specifying negative scale factors, you can insert a mirror image of a block along the particular axis. A negative scale factor for both the X and Y axes is the same as rotating the block reference through 180-degree, since it mirrors the block reference in the opposite quadrant of the coordinate system. In Figure 14-10, the effect of negative scale factor on block (DOOR) can be marked by a change in position of insertion point marked with an (X).

Also, specifying a scale factor of less than 1 inserts the block reference smaller than the original size and a scale factor greater than 1 inserts the block reference larger than its original size.

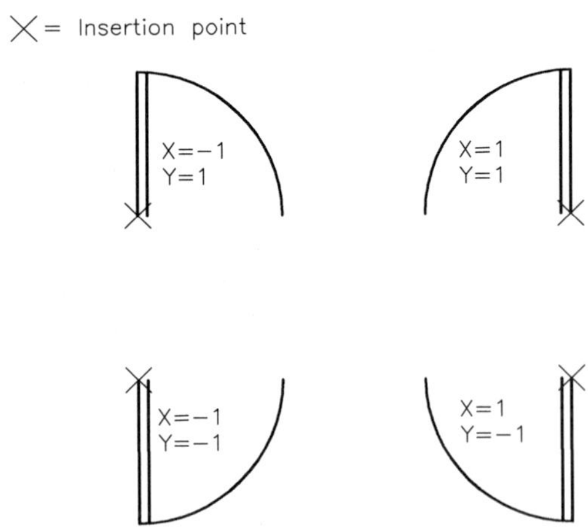

***Figure 14-10** Block inserted using negative scale factors*

Rotation

You can enter the angle of rotation for the block to be inserted in the **Angle** edit box. The insertion point is taken as the location about which rotation takes place. Selecting the **Specify**

On-screen check box allows you to specify the angle of rotation on the screen. This check box is cleared by default and the block is inserted at an angle of zero-degree.

Explode Check Box

By selecting this check box, the inserted block is inserted as a collection of individual objects. The function of the **Explode** check box is identical to that of the **EXPLODE** command. Once a block is exploded, the X, Y, and Z scale factors are identical. Hence, you are provided access to only one scale factor edit box (**X** edit box), the **Y** and **Z** edit boxes are not available, and the **Uniform Scale** check box also gets selected. The X scale factor is assigned to the Y and Z scale factors too. You must enter a positive scale factor value.

Note

*If you select the **Explode** check box before insertion, the block reference is exploded and the objects are inserted with their original properties such as layer, color, and linetype.*

Once you have entered the relevant information in the dialog box, choose the **OK** button. Once you do so, the **Insert** dialog box is removed from the screen and if you have selected the **Specify On-Screen** check boxes, you can specify the insertion point, scale, and angle of rotation with a pointing device. Whenever the insertion point is to be specified on screen (by default), you can specify scale factors and rotation angle values that will override the values specified in the dialog box, using the command line. Before you specify the insertion point on screen, you can right-click to display the shortcut menu, which has all the options available through the command line. However, if you have already specified an insertion point in the dialog box and have selected the **Specify on-screen** check boxes in the **Scale** or **Rotation** areas of the **Insert** dialog box, you are allowed to specify only the scale factors or the rotation angle through the command line. You can also use the command line where the following prompts appear.

Specify insertion point or [Scale/X/Y/Z/Rotate/PScale/PX/PY/PZ/PRotate]:
Enter X scale factor, specify opposite corner, or [Corner/XYZ] <1>:
Enter Y scale factor <use X scale factor>:
Specify rotation angle <0>:

Tip

Using the command line allows you to override scale factors and rotation angles already specified in the dialog box.

The command line options available are discussed next.

Scale	**X**	**Y**	**Z**	**Rotate**
PScale	**PX**	**PY**	**PZ**	**PRotate**

Scale

If you enter S at the **Specify Insertion point** prompt, you are asked to enter the scale factor. After entering the scale factor, the block assumes the specified scale factor, and AutoCAD lets you drag the block until you locate the insertion point on the screen. The *X*, *Y*, and *Z* axes are uniformly scaled by the specified scale factor. The prompt sequence for this option is given next.

Specify insertion point or [Scale/X/Y/Z/Rotate/PScale/PX/PY/PZ/PRotate]: **S**
Specify scale factor for XYZ axes: *Enter a value to preset general scale factor.*
Specify insertion point: *Select the insertion point.*
Specify rotation angle <0>: *Specify angle of rotation.*

X, Y, Z

With X, Y, or Z as the response to the **Specify insertion point** prompt, you can specify the X, Y, or Z scale factors before specifying the insertion point. The prompt sequence when you use the **X** option is given next.

Specify insertion point or [Scale/X/Y/Z/Rotate/PScale/PX/PY/PZ/PRotate]: **X**
Specify X scale factor: *Enter a value for the X scale factor.*
Specify insertion point: *Select the insertion point.*
Specify rotation angle <0>: *Specify angle of rotation.*

Rotate

If you enter R at the **Specify insertion point** prompt, you are asked to specify the rotation angle. You can enter the angle of rotation or specify the angle by specifying two points on the screen. Dragging is resumed only when you specify the angle, and the block assumes the specified rotation angle. The prompt sequence for this option is given next.

Specify insertion point or [Scale/X/Y/Z/Rotate/PScale/PX/PY/PZ/PRotate]: **R**
Specify rotation angle: *Enter a value for the rotation angle.*
Specify insertion point: *Select the insertion point.*
Enter X scale factor, specify opposite corner, or [Corner/XYZ] <1>: *Specify the X scale factor.*
Enter Y scale factor <use X scale factor>: *Specify the Y scale factor.*

PScale

The **PScale** option is similar to the **Scale** option. The difference is that in the case of the **PScale** option, the scale factor specified reflects only in the display of the block while it is dragged to the desired position. This implies that you can preview the effect of this scale factor on the block before actually applying it. After you have specified the insertion point for the block, you are again prompted for a scale factor. If you do not enter the scale factor, the block is drawn exactly as it was when created. The prompt sequence for this option is given next.

Specify insertion point or [Scale/X/Y/Z/Rotate/PScale/PX/PY/PZ/PRotate]: **PS**
Specify preview scale factor for XYZ axes: *Enter a value to preview the general scale factor.*
Specify insertion point: *Specify the insertion point for the block.*
Enter X scale factor, specify opposite corner, or [Corner/XYZ] <1>: *Specify the X scale factor.*
Enter Y scale factor <use X scale factor>: *Specify the Y scale factor.*
Specify rotation angle <0>: *Specify angle of rotation.*

PX, PY, PZ

These options are similar to the **PScale** option. The only difference is that the X, Y, or Z scale factors specified are reflected only in the display of the block while it is dragged to the desired position. After you have specified the insertion point for the block, you are again prompted for the X scale factor and the Y scale factor. If you do not enter the scale factor, the block is drawn exactly as it was when created.

Specify insertion point or [Scale/X/Y/Z/Rotate/PScale/PX/PY/PZ/PRotate]: **PX**
Specify preview X scale factor: *Enter a value to preview the X scale factor.*
Specify insertion point: *Specify the insertion point for the block.*
Enter X scale factor, specify opposite corner, or [Corner/XYZ] <1>: *Specify the X scale factor.*
Enter Y scale factor <use X scale factor>: *Specify the Y scale factor.*
Specify rotation angle <0>: *Specify angle of rotation.*

Note

*If you have already specified scale factors in the **X, Y,** and **Z** edit boxes of the **Scale** area of the **Insert** dialog box, the **PScale, PX, PY,** and **PZ** options work equivalent to the **Scale, X, Y,** and **Z** options respectively and allow you to insert the block with the PScale, PX, PY, or PZ values you specified.*

PRotate

The **PRotate** option is similar to the **Rotate** option. The difference is that the rotation angle specified reflects only in the display of the block while it is dragged to the desired position. After you have specified the insertion point for the block, you are again prompted for an angle of rotation. If you do not enter the rotation angle, the block is drawn exactly as it was when created (with the same angle of rotation as specified on the creation of the block). The prompt sequence is given next.

Specify insertion point or [Scale/X/Y/Z/Rotate/PScale/PX/PY/PZ/PRotate]: **PR**
Specify preview rotation angle: *Specify the preview value for rotation.*
Specify insertion point: Specify the insertion point for the block.
Enter X scale factor, specify opposite corner, or [Corner/XYZ]: *Specify the X scale factor.*
Enter Y scale factor < use X scale factor>: *Specify the Y scale factor.*
Specify rotation angle <0>: *Specify angle of rotation.*

Note

*If you have already specified rotation angle in the **Angle** edit box of the **Rotation** area of the **Insert** dialog box, the **PRotate** option works equivalent to the **Rotate** option and allows you to insert the block with the PRotate value you specified.*

Tip

The preview options (PScale, PX, PY, PZ and PRotate) that are preceded with a P allow you to view the effect of the scale factor or rotation angle specified before actually applying it to the block reference.

The XYZ and Corner Options

When you are specifying the insertion point and scale factors on screen, the following prompt appears after you have selected an insertion point.

> Enter X scale factor, specify opposite corner or [Corner/XYZ]: *Enter X scale factor, specify opposite corner or select an option.*

Here, the **XYZ** option can be used to enter a 3D block reference and the successive prompts allow you to specify the X, Y, and Z scale factors individually.

With the **Corner** option, you can specify both X and Y scale factors at the same time. When you invoke this option, and **DRAGMODE** is turned off, you are prompted to specify the other corner (the first corner of the box is the insertion point). You can also enter a coordinate value instead of moving the cursor. The length and breadth of the box are taken as the X and Y scale factors for the block. For example, if the X and Y dimensions (length and width) of the box are the same as that of the block, the block will be drawn without any change. Points selected should be above and to the right of the insertion block. Mirror images will be produced if points selected are below or to the left of the insertion point. The prompt sequence is given next.

> Enter X scale factor, specify opposite corner or [Corner/XYZ]: **C**
> Specify opposite corner: *Select a point as the corner.*

The **Corner** option also allows you to use a dynamic scaling technique when **DRAGMODE** is turned to Auto (default). You can also move the cursor at the **Enter X scale factor** prompt to change the block size dynamically and, when the size meets your requirements, select a point.

Note

*You should try to avoid using the **Corner** option if you want to have identical X and Y scale factors because it is difficult to select a point whose X distance equals its Y distance, unless the **SNAP** mode is on. If you use the **Corner** option, it is better to specify the X and Y scale factors explicitly or select the corner by entering coordinates.*

INSERTING BLOCKS USING THE COMMAND LINE

Predefined blocks can also be inserted in a drawing with the **-INSERT** command, using the command line. When using the **-INSERT** command, the dialog box is not displayed. To insert a drawing file as a block, enter the file name at the **Enter block name or [?]** prompt. You can enter ? if you do not know the name of the blocks in the drawing. To create a block with a different name from the drawing file, enter **blockname= file name** at the **Enter block name or[?]** prompt. If you enter **(~)** tilde character at the **Enter block name or [?]** prompt, the **Select Drawing File** dialog box is displayed on the screen. If you enter an **(*)** asterisk before entering the block name, the block is automatically exploded upon insertion. All the command line options discussed earlier are available through the **-INSERT** command. Also, if you make changes to a block and want to update the block in the existing drawing, enter **blockname=** at the **Specify insertion point** prompt. AutoCAD will display the prompt:

Block "current" already exists. Redefine it? [Yes/No] <No>: *Enter y to redefine it.*

Note

If you create a template file that has blocks created and stored in it and use this template file to create a new drawing, the blocks are available in the new drawing also.

Also when you insert a drawing into a current drawing, all the blocks belonging to the inserted drawing are also brought into the current drawing.

Exercise 1 — *General*

Using the dialog box, insert a block SQUARE into a drawing. The block insertion point is (1,2). The X scale factor is 2 units, the Y scale factor is 2 units, and the angle of rotation is 35-degree. It is assumed that the block SQUARE is already defined in the current drawing.

Choose the **Insert Block** button from the **Draw** toolbar. The **Insert** dialog box is displayed on the screen. In the **Name** edit box, enter SQUARE or select the Block SQUARE from the **Name** drop-down list. Clear the **Specify On-screen** check box. Now, in the **Insertion point** area, enter 1 in the **X** edit box and 2 in the **Y** edit box. In the **Scale** area, enter 2 in the **X** edit box and 2 in the **Y** edit box. In the **Rotation** area, enter 35 in the **Angle** edit box. Choose the **OK** button to insert the block on the screen.

Exercise 2 — *General*

a. Insert the block CIRCLE created in Exercise 1. Use different X and Y scale factors to get different shapes after inserting this block.
b. Insert the block CIRCLE created in Exercise 1. Use the **Corner** option to specify the scale factor.

Exercise 3 — *General*

a. Construct a triangle and form a block of it. Name the block TRIANGLE. Now, set the Y scale factor of the inserted block as 2.
b. Insert the block TRIANGLE with a rotation angle of 45-degree. After defining the insertion point, enter an X and a Y scale factor of 2.

USING DESIGNCENTER TO INSERT BLOCKS

You can use the **DESIGNCENTER** to locate, preview, copy, or insert blocks or existing drawings into a current drawing. On the **Standard** toolbar, choose the **DesignCenter** button to display the **DESIGNCENTER** window (Figure 14-11). In this window, choose the **Tree View Toggle** button to display the **tree pane** on the left side, if not already displayed. By default, the tree pane is displayed. Expand **My Computer** to display the *C:/Program Files/ACAD2004/Sample/DesignCenter* folder, by clicking on the plus (+) signs on the left of the respective folders. Click on the plus sign adjacent to the folder to display its contents. Select a drawing file you wish to use to insert blocks from and then click on the plus sign adjacent to the drawing

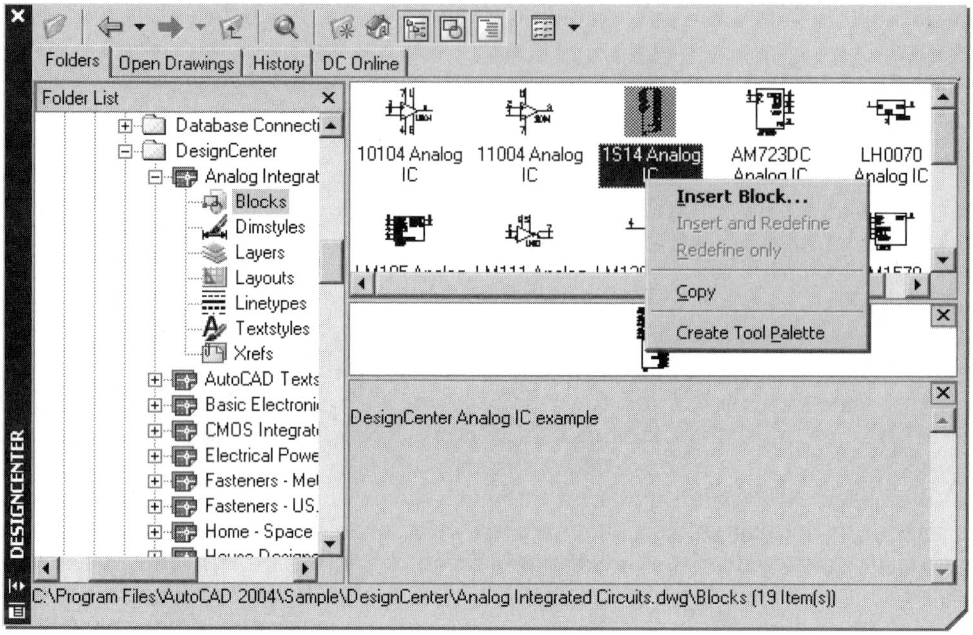

*Figure 14-11 Using the **DESIGNCENTER** to insert blocks*

again. All the icons depicting the different types of content in the selected drawing are displayed, that is, Blocks, Dimstylcs, Layers, Linetypes, Textstyles, and so on. Select **Blocks** by clicking on it and a list of blocks in the drawing is displayed in the palette. Select the block you wish to insert, and drag and drop the block into the current drawing. Later, you can move it in the drawing to a desired location. You can also right-click on the block name to display a shortcut menu. Select **Insert Block** from the shortcut menu. The **Insert** dialog box is displayed. Here, you can specify the **Insertion point**, **scale**, and **rotation angle** in the respective edit boxes. If you select the **Specify On-Screen** check boxes in the **Insert** dialog box, you are allowed to specify these parameters on the screen. The selected block is inserted into the current drawing based on the type of units specified when the block was created. For example, if you had created a block of 1" X 1", and specified feet as units at that time, the block when inserted using the DESIGNCENTER, will be 12" X 12". You can also choose **Copy** from the shortcut menu; the selected block is copied to the clipboard. Now, you can right-click in the drawing area to display a shortcut menu and choose **Paste**. The selected block is displayed attached to the cursor and as you move the cursor, the block also moves with it. You can now select a point where you want to paste the block. The advantage of this option over drag and drop is that you can select an insertion point at the time of pasting. In the case of a drag and drop operation, you need to move the inserted block to a specific point later and it is an additional step.

If you want to insert an entire drawing, select the folder in which the drawing resides by double-clicking on the folder. All the contents in the folder are displayed in the palette. Drag and drop the desired drawing into the current drawing. You can also select the drawing and right-click to display a shortcut menu. Choose **Insert as Block** from the shortcut menu. The

Insert dialog box is displayed, where you can specify the Insertion point, scale factors, and rotation angle, if required. You can also select the **Specify On-screen** check boxes to select everything on the screen.

Note
For a detailed explanation of the DESIGNCENTER, see Chapter 6, Editing Sketched Objects-II.

USING TOOL PALETTES TO INSERT BLOCKS*

You can use the **TOOL PALETTES** window shown in Figure 14-12 to insert predefined blocks in the current drawing. The **TOOL PALETTES** window has three tabs: **Sample Office project**, **Imperial Hatches**, and **ISO Hatches**. In this chapter, you will learn how to insert blocks using the **Sample Office project** tab of the **TOOL PALETTES**.

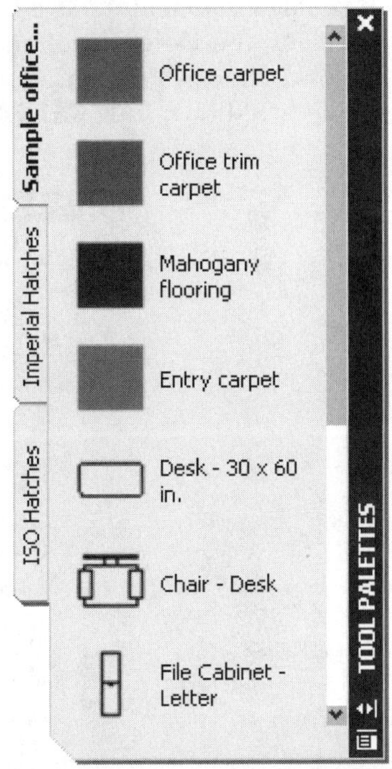

Figure 14-12 TOOL PALETTES window

Inserting Blocks in the Drawing

AutoCAD provides two methods to insert blocks from the **TOOL PALETTES**: **Drag and Drop** method and **Select and Place** method. Both these methods of inserting blocks using the **TOOL PALETTES** are discussed next.

Drag and Drop Method

To insert blocks from the **TOOL PALETTES** in the drawing using this method, move the cursor over the desired predefined block in the **TOOL PALETTES**. You will notice that as you move the cursor over the block, the block icon gets converted into a 3D icon. Also, a tooltip is displayed that shows the name and description of the block. Press and hold the left mouse button and drag the cursor to the drawing area. Release the left mouse button, and you will notice that the selected block is inserted in the drawing. You may need to modify the drawing display area to view the block.

Select and Place Method

You can also insert the desired block in the drawings using the select and place method. To insert the block, move the cursor over the desired block in the **TOOL PALETTES**. The block icon is changed to a 3D icon. Press the left mouse button, the selected block is attached to the cursor and you are prompted to specify the insertion point of the block. Move the cursor to the required location in the drawing area and press the left mouse button. The selected block is inserted at the specified location.

Remember that when a block is inserted from the **TOOL PALETTES**, you are not prompted to specify the rotation angle or the scale of the block. The blocks are automatically inserted with their default scale factor and rotation angle. The method of changing the default scale factor and rotation angle of the blocks in the **TOOL PALETTES** is discussed next.

Modifying the Properties of the Blocks Available in the TOOL PALETTES

To modify the properties of the blocks, move the cursor over the block in the **TOOL PALETTES** and right-click on it to display the shortcut menu. Using the options available in this shortcut menu, you can cut or copy the desired block available in one tab of **TOOL PALETTES** and paste it on the other tab. You can also delete and rename the selected block using the **Delete Tool** and **Rename** options, respectively. To modify the properties of the block, choose **Properties** from the shortcut menu. The **Tool Properties** dialog box is displayed as shown in Figure 14-13.

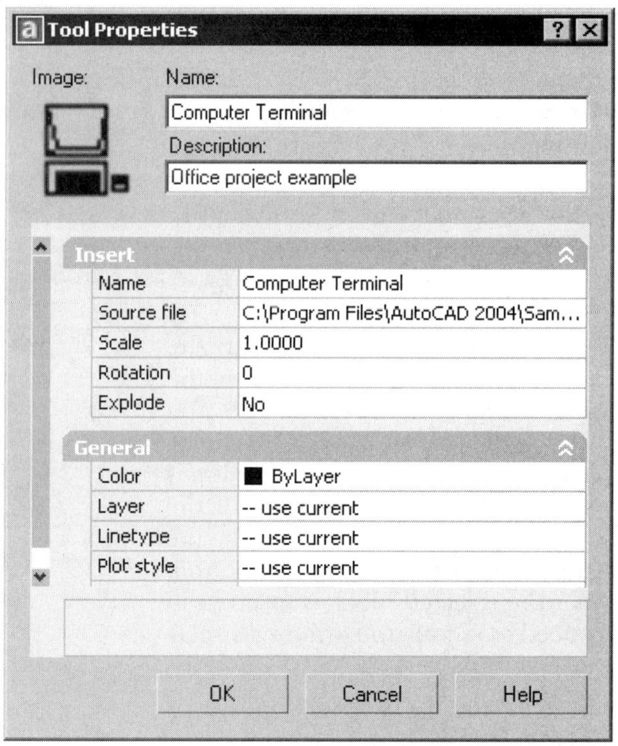

***Figure 14-13** The **Tool Properties** dialog box*

In the **Tool Properties** dialog box, the name of the selected block is displayed in the **Name** edit box. You can rename the block by entering the new name in the **Name** edit box. The **Image** area available on the left of the **Name** edit box displays the image of the selected block. If you enter a description of the block in the **Description** text box, it is stored with the block definition in the **TOOL PALETTES**. Now, when you move the cursor over the block in **TOOL PALETTES** and pause for a second, the description of the block appears along with its name in the tooltip. Note that when you select a particular field in the **Tool Properties** dialog box, its function is displayed in the description box available at the bottom of the dialog box. The **Tool Properties** dialog box displays the properties of the selected block under the following categories.

Insert

In this category, you can specify the insertion properties of the selected block such as its name, original location of the block file, scale, and rotation angle. The **Name** edit box specifies the name of the block. The **Source File** edit box displays the location of the file in which the selected block is created. When you choose the **[...]** button in the **Source File** edit box, AutoCAD displays the location of the file in the **Select Linked Drawing** dialog box. The **Scale** edit box is used to specify the scale factor of the block. The block will be inserted in the drawing according to the scale factor specified in this edit box. You can enter the angle of rotation for the block to be inserted in the **Rotation** edit box. The **Explode** edit box is used to specify whether the block will be exploded while inserting or will be inserted as a single entity.

General

In this category, you can specify the general properties of the block such as the **Color**, **Layer**, **Linetype**, **Plot style**, and **Lineweight** for the selected block. The properties of a particular field can be modified from the drop-down list available on selecting that field.

ADDING BLOCKS IN THE TOOL PALETTES*

By default, the **TOOL PALETTES** window displays the predefined blocks available in AutoCAD. You can also add the desired block and the drawing file to the **TOOL PALETTES** window. This is done using the **DESIGNCENTER**. AutoCAD provides two methods for adding the blocks from the **DESIGNCENTER** to the **TOOL PALETTES**: **Drag and Drop** method and **Shortcut menu**. These two methods are discussed next.

Drag and Drop Method

To add the blocks from the **DESIGNCENTER** in the **TOOL PALETTES**, move the cursor over the desired block in the **DESIGNCENTER**. Press and hold the left mouse button and drag the cursor to the desired location in the **TOOL PALETTES** window. As you drag the cursor in the **TOOL PALETTES** window, you will notice that a box with a + sign is attached to the cursor and a black line appears on the **TOOL PALETTES** window as shown in Figure 14-14. If you move the cursor up and down in the **TOOL PALETTES**, the black line also moves between the two consecutive blocks. This line is used to define the position of the block to be inserted in the **TOOL PALETTES** window. Release the left mouse button and you will notice that the selected block is added to the location specified by the black line in the **TOOL PALETTES**.

Shortcut Menu

You can also add the desired block from the **DESIGNCENTER** to the **TOOL PALETTES** using the shortcut menu. To add the block, move the cursor over the desired block in the

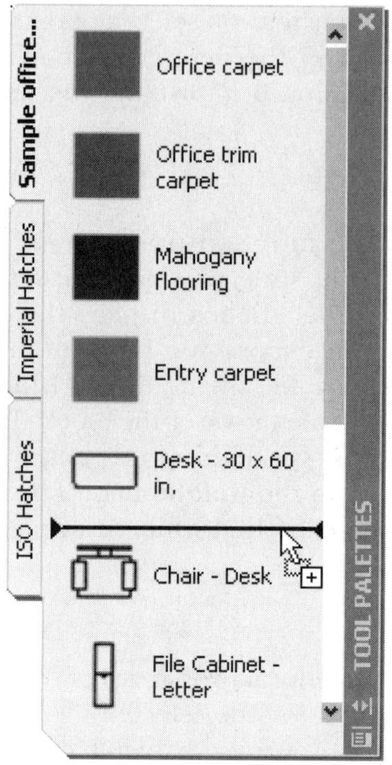

***Figure 14-14** Specifying the location of the block to be inserted in the **TOOL PALETTES** window*

DESIGNCENTER and right-click on it to display a shortcut menu. Choose **Create Tool Palette** from the shortcut menu. You will notice that a new tab with the name **New Tool Palette** is added to the **TOOL PALETTES** and the block is added in the new tab of the **TOOL PALETTES**. Also, a text box appears that displays the current name of the tab. You can change the name of the tab by entering a new name in this text box.

You can also add a number of blocks available in a drawing simultaneously to the **TOOL PALETTES** using the following two methods.

Right-click on the **Palette** area of the **DESIGNCENTER** to display the shortcut menu. Choose **Create Tool Palette** from the shortcut menu, a new tab is added to the **TOOL PALETTES** with the same name as that of the drawing selected in the **DESIGNCENTER**. This new tab contains all the blocks that were available in the drawing that you selected from the **DESIGNCENTER**.

You can also add the blocks available in the drawing simultaneously by right-clicking on the drawing in the **Tree View** of the **DESIGNCENTER**. A shortcut menu is displayed. Choose **Create Tool Palette** from the shortcut menu and you will notice that a new tab is added to the **TOOL PALETTES**, which contains all the blocks that were available in the selected drawing. You will also notice that the new tab has the same name as that of the selected drawing.

MODIFYING THE EXISTING BLOCKS IN THE TOOL PALETTES*

If you modify an existing block that was added to the **TOOL PALETTES**, and then insert it using the **TOOL PALETTES** in the same or a new drawing, you will notice that the modified block is inserted and not the original block. But note that if you insert the modified block from the **TOOL PALETTES** in a drawing in which the original block was already inserted, AutoCAD inserts the original block and not the modified block. This is because the file already has a block of the same name in its memory.

To insert the modified block, you first need to delete the original block from the current drawing using the **ERASE** command. Next, you need to delete the block from the memory of the current drawing. The unused block can be deleted from memory of the current drawing using the **PURGE** command. To invoke this command, enter **PURGE** at the Command prompt. The **Purge** dialog box is displayed. Choose the (+) sign located on the left of **Blocks** in the tree view available in the **Items not used in drawing** area. You will notice that a list of blocks available in the drawing is shown. Select the original block to be deleted from the memory of the current drawing and then choose the **Purge** button. The **Confirm Purge** dialog box is displayed that confirms the purging of the selected item. Choose **Yes** in this dialog box and then choose the **Close** button to exit the **Purge** dialog box. Next, when you insert the block using the **TOOL PALETTES**, the modified block is inserted in the drawing.

Note

*You will learn more about the **PURGE** command in Chapter18, Grouping and Advanced Editing of Sketched Objects. Deleting unused blocks using the command line is however, discussed later in this chapter.*

*If you create a block with the name that is defined in the **TOOL PALETTES**, the block in the **TOOL PALETTES** is redefined in the current drawing. However, when you open a new drawing and insert the block using the **TOOL PALETTES**, the original block will be inserted.*

LAYERS, COLORS, LINETYPES, AND LINEWEIGHTS FOR BLOCKS

A block possesses the properties of the layer on which it is drawn. The block may be composed of objects drawn on several different layers, with different colors, linetypes, and lineweights. All this information is preserved in the block. At the time of insertion, each object in the block is drawn on its original layer with the original linetype, lineweight, and color, irrespective of the current drawing layer, object color, object linetype, and object lineweights. You may want all instances of a block to have identical layers, linetype properties, lineweight, and color. This can be achieved by allocating all the properties explicitly to the objects forming the block. On the other hand, if you want the linetype and color of each instance of a block to be set according to the linetype and color of the layer on which it is inserted, draw all the objects forming the block on layer 0 and set the color, lineweight, and linetype to BYLAYER.

Objects with a BYLAYER color, linetype, and lineweight can have their colors, linetypes, and lineweights changed after insertion by changing the layer settings. If you want the linetype, lineweight, and color of each instance of a block to be set according to the current explicit linetype, lineweight, and color at the time of insertion, set the color, lineweight, and linetype of its objects to BYBLOCK. You can use the **PROPERTIES**, **CHPROP**, or **CHANGE** command to change some of the characteristics associated with a block (such as layer).

Note

The block is inserted on the layer that is current, but the objects comprising the block are drawn on the layers on which they were drawn when the block was being defined.

For example, assume block B1 includes a square and a triangle that were originally drawn on layer X and layer Y, respectively. Let the color assigned to layer X be red and to layer Y be green. Also, let the linetype assigned to layer X be continuous and for layer Y be hidden. Now, if we insert B1 on layer L1 with color yellow and linetype dot, block B1 will be on layer L1, but the square will be drawn on layer X with color red and linetype continuous, and the triangle will be drawn on layer Y with color green and linetype hidden.

The **BYLAYER** option instructs AutoCAD to assign objects within the block the color and linetype of the layers on which they were created. There are three exceptions:

1. If objects are drawn on a special layer (layer 0), they are inserted on the current layer. These objects assume the characteristics of the current layer (the layer on which the block is inserted) at the time of insertion, and can be modified after insertion by changing that layer's settings.

2. Objects created with the special color BYBLOCK are generated with the color that is current at the time of insertion of the block. This color may be explicit or BYLAYER. You are thus allowed to construct blocks that assume the current object color.

3. Objects created with the special linetype BYBLOCK are generated with the linetype that is prevalent at the time the block is inserted. Blocks are thus constructed with the current object linetype, which may be BYLAYER or explicit.

Note

If a block is created on a layer that is frozen at the time of insertion, it is not shown on the screen.

Tip

If you are providing drawing files to others for their use, using only BYLAYER settings provide the greatest compatibility with varying office standards for layer/color/linetype/lineweight, because they can be changed more easily after insertion.

NESTING OF BLOCKS

The concept of having one block within another block is known as the **nesting of blocks**. For example, you can insert several blocks by selecting them and then with the **BLOCK** command, create another block. Similarly, if you use the **INSERT** command to insert a drawing that contains several blocks into the current drawing, it creates a block containing nested blocks in the current drawing. There is no limit to the degree of nesting. The only limitation in nesting of blocks is that blocks that reference themselves cannot be inserted. The nested blocks must have different block names. Nesting of blocks affects layers, colors, and linetypes. The general rule is as follows.

If an inner block has objects on layer 0, or objects with linetype or color BYBLOCK, these objects may be said to behave like fluids. They "float up" through the nested block structure until they find an outer block with fixed color, layer, or linetype. These objects then assume the characteristics of the fixed layer. If a fixed layer is not found in the outer blocks, then the objects with color or linetype BYBLOCK are formed; that is, they assume the color white and the linetype CONTINUOUS.

Example 2 *General*

To make the concept of nested blocks clear, consider the following example.

1. Draw a rectangle on layer 0, and form a block of it named X.
2. Change the current layer to OBJ, and set its color to red and linetype to hidden.
3. Draw a circle on OBJ layer.
4. Insert the block X in the OBJ layer.
5. Combine the circle with the block X (rectangle) to form a block Y.
6. Now, insert block Y in any layer (say, layer CEN) with color green and linetype continuous.

You will notice that block Y is generated in color red and linetype hidden. Normally, block X, which is nested in block Y and created on layer 0, should have been generated in the color (green) and linetype (continuous) of the layer CEN. This is because the object (rectangle) on layer 0 floated up through the nested block structure and assumed the color and linetype of the first outer block (Y) with a fixed color (red), layer (OBJ), and linetype (hidden). If both the blocks (X and Y) were on layer 0, the objects in block Y would assume the color and linetype of the layer on which the block was inserted.

Example 3 *General*

1. Change the color of layer 0 to red.
2. Draw a circle and form a block of it, B1, with color BYBLOCK. It appears white because the color is set to BYBLOCK (Figure 14-15).
3. Set the color to BYLAYER and draw a square. The color of the rectangle is red.
4. Insert block B1. Notice that the block B1 (circle) assumes red color.
5. Create another block B2 consisting of Block B1 (circle) and rectangle.
6. Create a layer L1 with green color and hidden linetype. Insert block B2 in layer L1.
7. Explode block B2. Notice the change.
8. Explode block B1, circle. You will notice that the circle changes to white because it was drawn with color set to BYBLOCK.

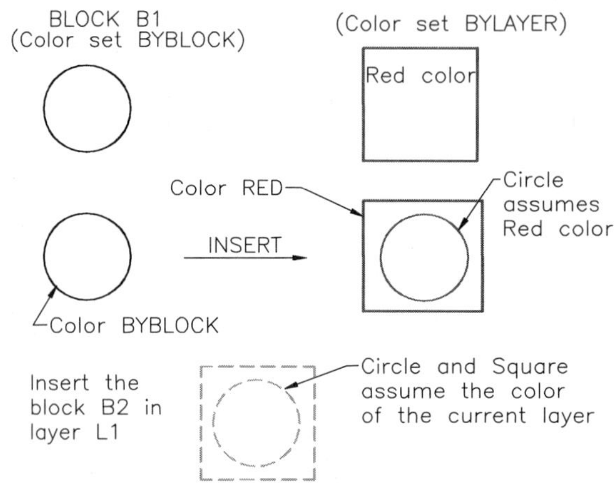

Figure 14-15 *Blocks versus layers and colors*

Exercise 4 — *General*

Part A

1. Draw a unit square on layer 0 and make it a block named B1.
2. Make a circle with radius 0.5 and change it into block B2.
3. Insert block B1 into a drawing with an X scale factor of 3 and a Y scale factor of 4.
4. Now, insert the block B2 in the drawing and position it at the top of B1.
5. Make a block of the entire drawing and name it PLATE.
6. Insert the block Plate in the current layer.
7. Create a new layer with different colors and linetypes and insert blocks B1, B2, and Plate.

Keep in mind the layers on which the individual blocks and the inserted block were made.

Part B

Try nesting the blocks drawn on different layers and with different linetypes.

Part C

Change the layers and colors of the different blocks you have drawn so far.

INSERTING MULTIPLE BLOCKS

Command: MINSERT

The **MINSERT** (multiple insert) command is used for multiple insertion of a block. This command comprises features of the **INSERT** and **ARRAY** commands. **MINSERT** is similar to the **-INSERT** command because it is used to insert blocks at the command level. The difference between these two commands is that the **MINSERT** command inserts more than one copy of the block in a rectangular fashion, similar to the command **ARRAY,** and blocks inserted by

MINSERT cannot be exploded. Preceding the block name with an asterisk does not explode it on insertion. With the **MINSERT** command, only one block reference is created. But, in addition to the standard features of a block definition (insert point, X/Y scaling, rotation angle, and so on), this block has repeated row and column counts. In this manner, using this command saves time and disk space. The prompt sequence is very similar to that of the **-INSERT** and **-ARRAY** commands. Consider an example in which you have to array a block named BENCH with the following specifications: number of rows = 4, number of columns = 3, unit cell distance between rows = 0.50, and unit cell distance between columns = 2.0. The prompt sequence for creating the arrangement of benches in an auditorium (Figure 14-16) is given next.

Command: **MINSERT**
Enter block name or [?] <current>: **BENCH**
Specify insertion point or [Scale/X/Y/Z/Rotate/PScale/PX/PY/PZ/PRotate]: *Specify the insertion point or enter an option.*
Enter X scale factor, specify opposite corner, or [Corner/XYZ] <1>: *Press ENTER, enter a number, or specify a point.*
Enter Y scale factor <use X scale factor>: *Press ENTER, enter a number or specify a point.*
Specify rotation angle <0>: Enter
Enter number of rows (---)<1>: **4**
Enter number of columns (|||) <1>: **3**
Enter distance between rows or specify unit cell (---): **0.50**
Specify distance between columns (|||): **2.0**

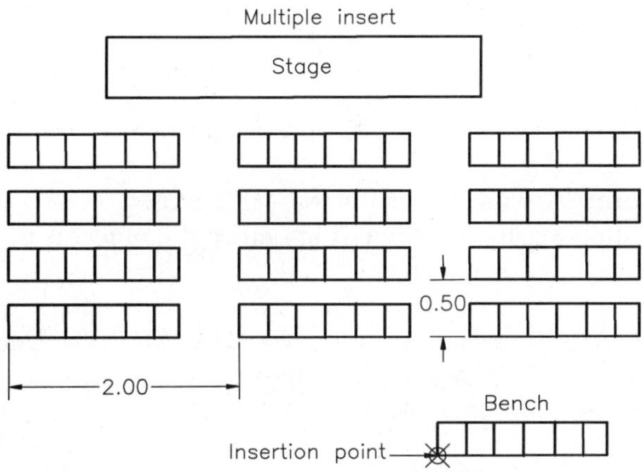

Figure 14-16 *Arrangement of benches in an auditorium*

The prompts pertaining to the number of rows and the number of columns require a positive nonzero integer. For the number of rows exceeding 1, you are prompted to specify the distance between the rows and similarly, if the number of columns exceeds 1, you need to specify the distance between columns. The distance between rows and columns can be a positive or a negative value. While specifying the distance between rows and columns, you can also specify a unit cell by selecting two points on screen that form a box whose width and height represent the distance between columns and rows. While inserting a block, if you specify block rotation, each copy of the inserted block is rotated through that angle. If you specify the rotation angle in the **MINSERT** command, the whole **MINSERT** block (array) is rotated by the specified rotation angle. If both the block and the **MINSERT** block (array) are rotated through the same angle, the block does not seem rotated in the **MINSERT** pattern. This is the same as inserting an array with the rotation angle equal to 0-degree. If you want to rotate individual blocks within the array, you should first generate a rotated block and then **MINSERT** it.

All the options for the **MINSERT** command are similar to the ones discussed earlier for the **-INSERT** command. The **?** option of the **Enter block name or [?]:** prompt lists the names of the blocks. When you specify the insertion point, AutoCAD prompts you for the scale factors. The options are X scale factor, Corner, and XYZ. The X scale factor is the default option and pressing ENTER at the **Enter X scale factor, specify opposite corner or [Corner/XYZ] <1>:** prompt prompts you to enter the Y scale factor. By default, the Y scale factor uses the X scale factor. Using the **Corner** option sets the scale factor by using the insertion point as the starting point and the opposite corner can be specified on screen. The **XYZ** option sets the X, Y, and Z scale factors. It prompts for the X, Y, Z scale factors separately.

Also, similar to the **INSERT** or **-INSERT** commands, when you use the **MINSERT** command, the scale factors and the angle of the block that is to be inserted can be preset and previewed prior to insertion. While using options that have P as a prefix, the preset values can be negated when finally inserting the block. The options available at the **Specify insertion point:** prompt are discussed next.

Scale. Sets the scale factor for the *X*, *Y*, and *Z* axes.
X, Y, or Z. Sets the X, Y, or Z scale factor.
Rotate. Sets the angle for the rotation of the array as well as the individual inserts.
PScale. Presets the preview scale for the *X*, *Y*, and *Z* axes before they are inserted into the specified position.
PX/PY/PZ. Presets the preview scale factor for the *X*/*Y*/*Z* axis.
PRotate. Presets the rotation angle for previewing before the object is inserted into position.

Note

In the array of blocks generated with the use of the ***MINSERT*** *command, you cannot alter the number of rows or columns or the spacing between them. The whole* ***MINSERT*** *pattern is considered as one object that cannot be exploded.*

When you right-click while specifying the insertion point at the ***Specify insertion point*** *prompt, a shortcut menu is displayed where all the command line options are available.*

Exercise 5 — *General*

1. Create a triangle with each side equal to 3 units.
2. Generate a block of the triangle.

3. Use the **MINSERT** command to insert the block to create a 3 by 3 array. The distance between the rows is 1 unit, and the distance between columns is 2 units.
4. Again, use the **MINSERT** command to insert the block to create a 3 by 3 array. The distance between the rows is 2 units and between the columns is 3 units, and the array is rotated 15-degree.

CREATING DRAWING FILES USING THE WRITE BLOCK DIALOG BOX

Command: WBLOCK

The blocks or symbols created by the **BLOCK** command can be used only in the drawing in which they were created. This is a shortcoming because you may need to use a particular block in different drawings. The **WBLOCK** command is used to export symbols by writing them to new drawing files that can then be inserted in any drawing. With the **WBLOCK** command, you can create a drawing file (*.dwg* extension) of specified blocks, selected objects in the current drawing, or the entire drawing. All the used named objects (linetypes, layers, styles, and system variables) of the current drawing are inherited by the new drawing created with the **WBLOCK** command. This block can then be inserted in any drawing.

When the **WBLOCK** command is invoked, the **Write Block** dialog box is displayed (Figure 14-17). This dialog box converts blocks into drawing files and also saves objects as drawing files. You can also save the entire current drawing as a new drawing file.

The **Write Block** dialog box has two main areas: **Source** and **Destination**. The **Source** area allows you to select objects and blocks, specify insertion base points and convert them into drawing files. In this area of the dialog box, depending on the selection you make, different default settings are displayed. By default, the **Objects** radio button is shown as selected and in the **Destination** area, the **File name and path** edit box displays *new block.dwg* as the new file name and its location. The **Block** drop-down list is also not available. Now, you can select objects in a drawing and save them as a wblock, and can enter a name and a path for the file. If the current drawing consists of blocks and if you want to save a block as a wblock, you can select the **Block** radio button which is available. When the **Block** radio button is selected, the **Block** drop-down list is available. The **Block** drop-down list displays all the block names in the current drawing and you can select a name of the block that you want to convert into a wblock. The **Base point** and **Objects** areas are not available, since the insertion points and objects have already been saved with the block definition. Also, you will notice that in the **Destination** area, the **File name and path** edit box displays the name of the selected block by default. This means that you can keep the name of the wblock the same as the selected block or you can change the name if you want to. Selecting the **Entire drawing** radio button selects the current drawing as a block and saves it as a new file. When you use this option, the **Base point** and **Objects** areas are not available.

The **Base point** area allows you to specify the base point of a wblock, which it uses as an insertion point. You can either enter values in the **X**, **Y**, and **Z** edit boxes or choose the **Pick point** button to select it on the screen. The default value is 0,0,0. The **Objects** area allows you to

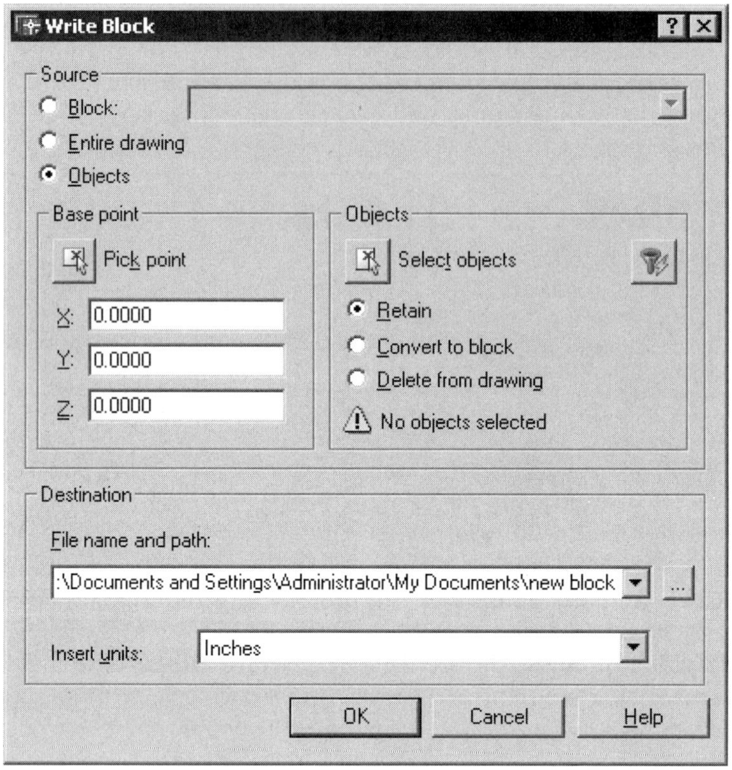

***Figure 14-17 Write Block** dialog box*

select objects to save as a file. You can use the **Select objects** button to select objects or use the **QuickSelect** button to set parameters in the **Quick Select** dialog box to select objects in the current drawing. The number of objects selected is displayed at the bottom of the **Objects** area. If the **Retain** radio button is selected in the **Objects** area, the selected objects in the current drawing are kept as such after they have been saved as a new file. If the **Convert to block** radio button is selected, the selected objects in the current drawing are converted into a block with the same name as the new file after being saved as a new file. Selecting the **Delete** radio button deletes the selected objects from the current drawing after they have been saved as a file.

Note

*Both the **Base point** and **Objects** areas are available in the **Write Block** dialog box only when the **Objects** radio button is selected in the **Source** area of the dialog box.*

The **Destination** area sets the File name, location, and units of the new file in which the selected objects are saved. In the **File name and path** edit box you can specify the file name and the path of the block or the selected objects. You can choose the **[...]** button to display the **Browse for folder** window, where you can specify a path for where the new file will be saved. From the **Insert Units** drop-down list, you can select the units the new file will use when inserted as a block. The settings for units are stored in the **INSUNITS** system variable and the default option Inches has a value of 1. On specifying the required information in the dialog box, choose **OK**. The objects or the block is saved as a new file in the path specified by you. A **WBLOCK Preview** window with the new file contents is displayed. This preview image is stored and displayed in the DESIGNCENTER, when using it to insert drawings and blocks.

Note

Whenever a drawing is inserted into a current drawing, it acts as a single object. It cannot be edited unless exploded.

Tip

*Using the **Entire drawing** radio button in the **Write Block** dialog box to save the current drawing as a new drawing is a good way to reduce the drawing file size, since all unused blocks, layers, linetypes, text styles, dimension styles, multiline styles, shapes, and so on are removed from the drawing. The new drawing does not contain any information that is no longer needed. This **Entire drawing** option is faster than the **-PURGE** command (which also removes unused named objects from a drawing file) and can be used whenever you have completed a drawing and want to save it. The **-PURGE** command has been discussed later in this chapter.*

CREATING DRAWING FILES USING THE COMMAND LINE

The **-WBLOCK** command can be used to create drawing files when you want to use the command line. When this command is invoked, AutoCAD displays the **Create Drawing File** dialog box, except when the system variable **FILEDIA** is set to 0. This dialog box displays a list of all the drawing files in the current directory. Enter the name of the output drawing file in the **File name** edit box and then choose the **Save** button. If the file name you specify is identical to some other file in the same directory path, AutoCAD displays a message saying that a file with the same name already exists and asks if you want to replace it. After choosing the **Save** button, the dialog box is cleared from the screen, and AutoCAD issues the following prompt asking you for the name of an existing block that you want to convert into a permanent symbol.

> Enter name of existing block or
> [=(block=output file)/*(whole drawing)] <define new drawing>: *Enter the name of a predefined block.*

The drawing file with the name specified in the dialog box is created from the predefined block.

If you want to assign the same name to the output drawing file as that of the existing block, enter an (=) equal sign as a response to the **Enter name of existing block** prompt and enter the same file name in the **File name** edit box of the **Create Drawing File** dialog box. If a block by the name you have already assigned to the output file does not exist in the current drawing, AutoCAD gives an appropriate message that no block by that name exists in the current drawing, and the **Enter name of existing block or** prompt is repeated. This option is the same as the **Block** option of the **WBLOCK** command.

If you have not yet created a block, but want to create an output drawing file of the objects in the current drawing, press ENTER at the **Enter name of existing block** prompt. You are then required to select an insertion point and select the objects to be incorporated in the drawing file. This option is the same as the **Objects** option of the **WBLOCK** command.

You can also store an entire drawing as a wblock. In other words, the current file is copied into a new one specified in the **File name** edit box. To do so, respond to the prompt **Enter name of existing block** with an asterisk (*). The entire drawing is saved in the file in the directory you have specified. This option is the same as the **Entire drawing** option of the **WBLOCK** command.

Export Data Dialog Box

You can also use the **Export Data** dialog box (Figure 14-18) to create drawing files. This dialog box can be invoked by choosing **Export** from the **File** menu, or by entering **EXPORT** at the Command prompt. In this dialog box, you can select **Block (*.dwg)** from the **Save as type** drop-down list, enter the name of the drawing file in the **File name** edit box, and then choose the **Save** button. The dialog box is cleared from the screen and AutoCAD issues the same prompts as the **-WBLOCK** command discussed earlier.

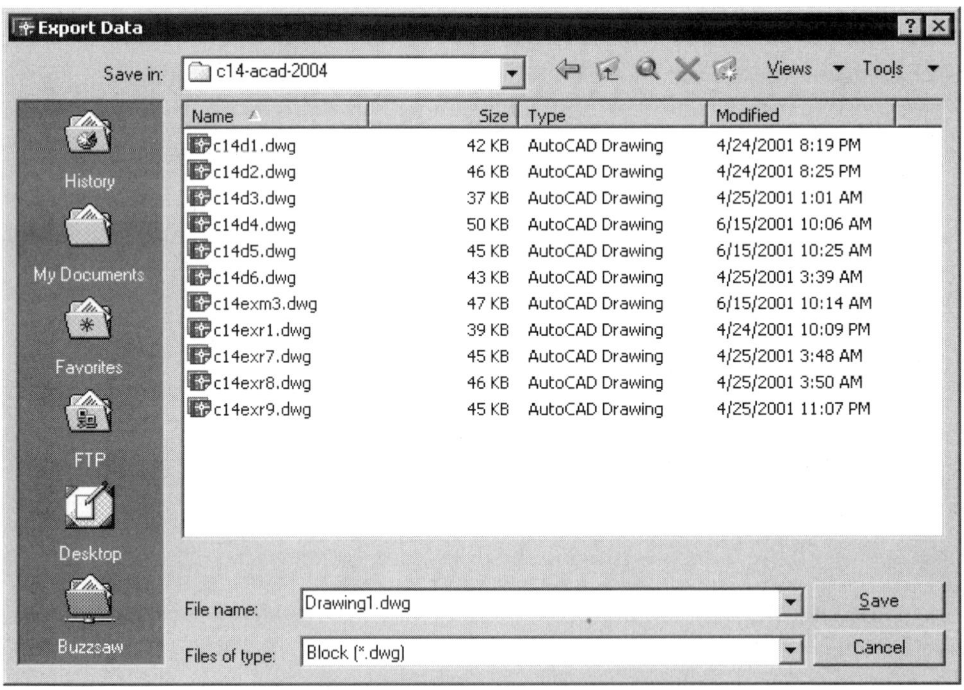

Figure 14-18 ***Export Data*** *dialog box*

DEFINING THE INSERTION BASE POINT

Menu:	Draw > Block > Base
Command:	BASE

The **BASE** command lets you set the insertion base point for a drawing just as you set the base insertion point using the **BLOCK** command. This base point is defined so that when you insert the drawing into some other drawing, the specified base point is placed on the insertion point. By default, the base point is at the origin (0,0,0). When a drawing is inserted on a current layer, it does not inherit the color, linetype, or thickness properties of the current layer. When you invoke the **BASE** command, the prompt sequence is given next.

Enter base point <0.0000,0.0000,0.0000>: *Specify a base point or press ENTER to accept the default.*

Tip

You can insert a drawing that you want to refer to for checking dimensions or certain features into the current drawing for reference. Then later you can use the ***UNDO*** *command to delete the inserted drawing. This way you can save stationery and time spent in printing.*

Note

Drawings inserted into a current drawing become part of the current drawing and increase the file size. Also, when the drawing is modified it does not get updated in the drawing; it has been inserted automatically. Hence, it is better to xref a drawing into the current drawing instead. External references is discussed in Chapter 16, Understanding External References.

Exercise 6 *General*

a. Create a drawing file named CHAIR using the **WBLOCK** command. Get a listing of your *.dwg* files and make sure that *CHAIR.dwg* is listed. Quit the drawing editor.
b. Begin a new drawing and insert the drawing file into the drawing. Save the drawing.

EDITING BLOCKS

Toolbar:	Refedit > Edit block or Xref
Menu:	Modify > Xref and Block Editing > Edit Reference In-Place
Command:	REFEDIT

You can edit blocks by breaking them into parts and then making modifications and redefining the block or by editing them in place. Both these methods have been discussed here.

Editing Blocks in Place*

When using AutoCAD, you can edit blocks in the current drawing by using the **REFEDIT** command, referred to as the **in-place reference editing**. This feature of AutoCAD allows you to make minor changes to blocks, wblocks, or drawings that have been inserted in the current drawing without having to break them up into component parts or having to open the original drawing and redefine them. This command saves valuable time of going back and forth between drawings and redefining blocks.

Note

*You cannot use the **REFEDIT** command on blocks that have been inserted using the **MINSERT** command. This command is discussed in reference to xrefs in Chapter 16, Understanding External References.*

You can invoke the **REFEDIT** command by choosing the **Edit block or Xref** button from the **Refedit** toolbar (Figure 14-19), choosing **In-place Xref and Block edit** > **Edit Reference** from the **Modify** menu, or by entering **REFEDIT** at the Command prompt. You can display the **Refedit** toolbar by right-clicking on any displayed toolbar and then selecting **Refedit** from the shortcut menu that is displayed. When you invoke the **REFEDIT** command, the following prompt is displayed.

Select Reference: *Select the block to edit.*

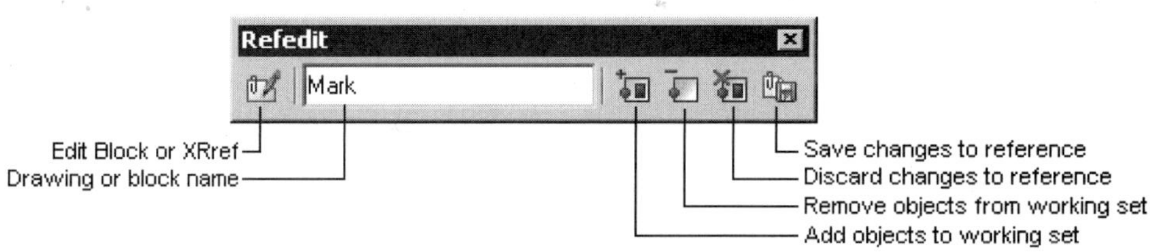

***Figure 14-19** The **Refedit** toolbar*

Once you have selected the block reference to be edited, the **Reference Edit** dialog box (Figure 14-20) is displayed. The **Reference Edit** dialog box has two tabs: **Identify Reference** and **Settings**. The description of these tabs follows.

Identify Reference Tab*

The **Identify Reference** tab (**Reference Edit** dialog box) provides information for identifying the reference edit, selecting the reference, and editing the references.

Reference name: The **Reference name** list box displays the name of the selected reference. A plus sign (+) next to the name implies that the selected block reference contains nested references. When you click on the plus sign, the nested reference is displayed in a tree view (Figure 14-21). If a block does not contain any nested blocks, this plus sign is not displayed next to the reference name in the list box. In the figure, the block MARK contains a nested block ELL. An image of the selected block is displayed in the **Preview** box. If you want to see a preview of the nested block, select its name in the list box and the corresponding preview image is displayed in the Preview box.

Path: It displays the location of the selected reference. Note that if the selected reference is a block, no path is displayed.

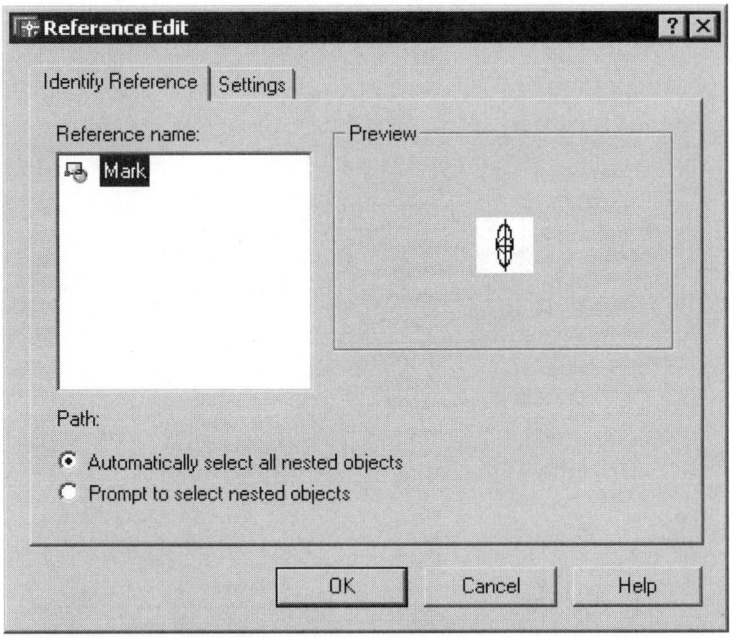

***Figure 14-20** The **Identify Reference** tab of the **Reference Edit** dialog box*

The **Automatically select all nested objects** radio button is selected by default and allows you to automatically include all the selected objects in the reference editing session. The selected objects also include the nested objects.

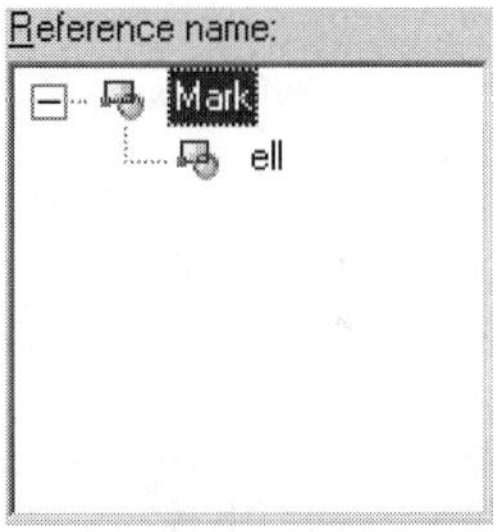

***Figure 14-21** The **Reference name** list box displaying nested references*

The **Prompt to select nested objects** radio button, when selected, allows you to individually select the nested objects in the reference editing session. If this radio button is selected and you choose the **OK** button, the **Reference Edit** dialog box is closed and you are prompted to select the nested objects in the reference that you want to edit.

Settings Tab*

The **Settings** tab of the **Reference Edit** dialog box (Figure 14-22) provides options for editing the references.

The **Create Unique layer, style, and block names** check box is selected by default and allows you to control layer and symbol names of objects that are extracted from the selected reference. If you clear this check box, the names of the layers remain the same as in the reference drawing. Similarly, the **Display attribute definitions for editing** check box, when selected, allows you to edit attributes and attribute definitions associated with block references. Modified attribute definitions are effective for future insertions, and no changes are made in

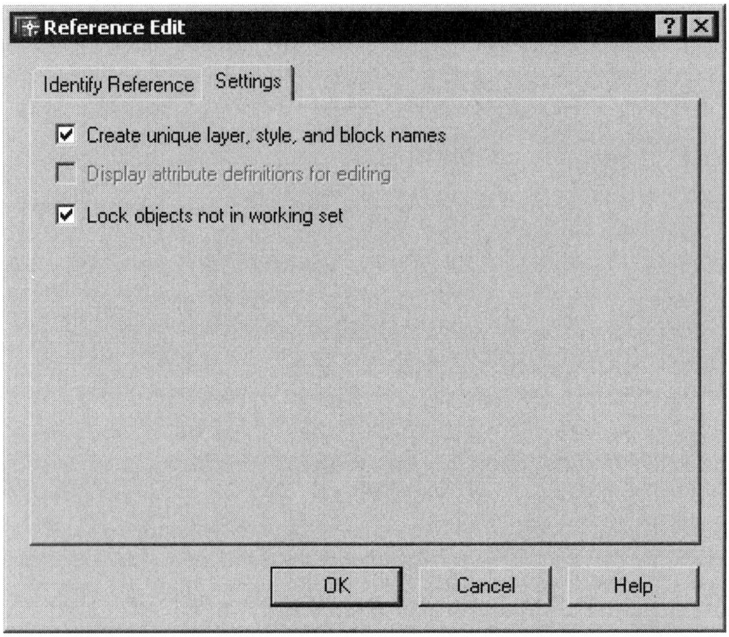

***Figure 14-22** The **Settings** tab of the **Reference Edit** dialog box*

the current insertions. Attributes are explained in detail in the next chapter, Chapter 15, Defining Block Attributes. The **Lock objects not in working set** check box, when selected, allows you to lock all the objects not in the working set.

After you have selected the required block that you want to edit, choose **OK** to exit the dialog box. The next prompt displayed is given next.

Select nested objects: *Select objects within the block that you want to modify.*
n entities added
Select nested objects: Enter
n items selected
Use REFCLOSE or the Refedit toolbar to end reference editing session.

Once you have selected the objects to be edited and pressed ENTER, the total number of objects selected is displayed. If you had not already invoked the **Refedit** toolbar, it shall be displayed now after you have selected the objects in the block. Also, you will notice that except the objects that you have selected for editing, the rest of the objects appear faded out. The percentage of fading can be controlled in the **Reference Edit fading intensity** area of the **Display** tab of the **Options** dialog box, see Figure 14-23. In this area, you can either use the slider bar to increase or decrease the fading intensity or you can enter a percentage in the edit

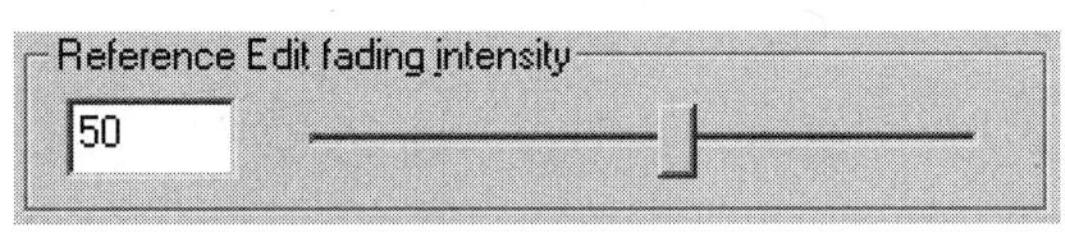

***Figure 14-23** The **Reference Edit fading intensity** area of the **Display** tab of the **Options** dialog box*

box. By default, the fading intensity value is set at 50. This value of the fading intensity is stored in the **XFADECTL** variable.

The objects that have been selected for editing are referred to as the **working set**. You can use any drawing and editing commands to make modifications to this working set as required. You can add or remove objects to the existing set and then choose the **Save back changes to reference** button in the **Refedit** toolbar. An AutoCAD warning box is displayed (Figure 14-24) saying that all reference edits will be saved and asking if you want to continue or cancel the command. Choose **OK** if you want to continue saving the modifications or choose **Cancel** if you want to cancel the command. When you choose **OK**, the modifications are made for the current block as well as for all future insertions.

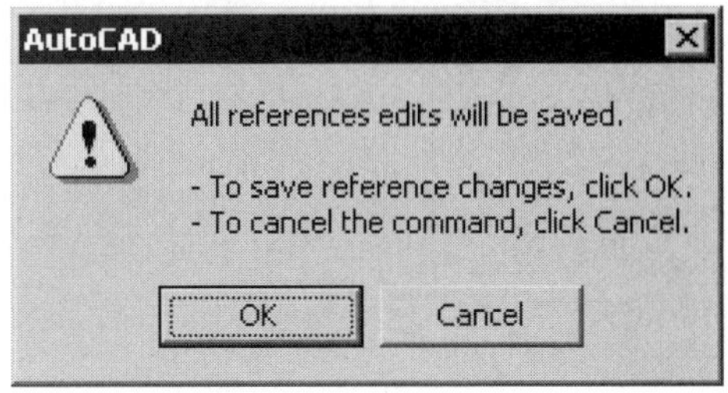

***Figure 14-24** AutoCAD message box displayed on choosing the **Save back changes to reference** button*

Note

The Automatic Save feature of AutoCAD is disabled during reference editing.

Sometimes, you may want to add another external object from the current drawing to the block. To do so, choose the **Add objects to working set** button in the **Refedit** toolbar. You can then select the objects you want to add to the working set. Press ENTER when the selection is made. These objects will then get deleted from the current drawing. Similarly, you can remove objects from a working set by choosing the **Remove objects from working set** button in the **Refedit** toolbar. You are then allowed to select objects in the working set that you want to remove from there. On completing the selection press ENTER. The objects that have been removed from the working set appear faded and get added back to the current drawing. You can also use the **REFSET** command to add or remove objects from the working set. But, this command can be used only when a reference has already been selected using the **REFEDIT** command.

If you want to exit reference editing without saving any changes to the block reference, choose the **Discard changes to reference** button in the **Refedit** toolbar. An AutoCAD alert message box is displayed saying that all changes made will be discarded (Figure 14-25). If you want to continue choose **OK** and if you want to get back to reference editing, choose **Cancel**. The **REFCLOSE** command can also be used to save or discard changes made during reference editing and then exit reference editing. Similar to the **REFSET** command, the **REFCLOSE** command can only be used when a block reference is selected for editing.

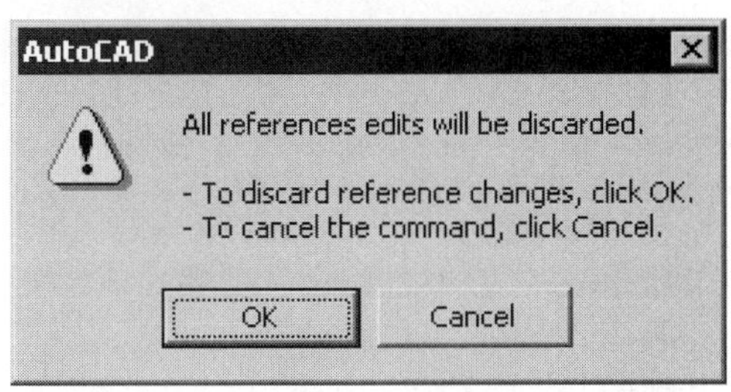

***Figure 14-25** AutoCAD message box displayed on choosing the **Discard changes to reference** button in the **Refedit** toolbar*

You can also use the **-REFEDIT** command to

display prompts on the command line. The prompts are as follows.

Command: **-REFEDIT**
Select reference: *Select the block reference that is to be edited.*
Select nesting level [Ok/Next] <Next>: *Press ENTER if the block reference has nested blocks and you want to edit them or enter O and press ENTER.*
Enter object selection method [All/Nested] <All>: *Enter A if you want to select all the selected objects in the reference editing session. Enter N if you want to individually select the nested objects in the reference editing session and press ENTER.*
Select nested objects: *Select the objects within the block for editing.*
n entities added
Select nested objects: Enter
n items selected
Display attribute definitions [Yes/No] <No>: *Press ENTER if you do not want to display attribute definitions or enter Y.*
Use REFCLOSE or the Refedit toolbar to end reference editing session.

You can now make changes in the selected objects using any editing and draw commands. Once the changes have been made, you can enter the **REFCLOSE** command to save or discard changes and then exit reference editing.

Breaking Up a Block

A block may be composed of different basic objects such as lines, arcs, polylines, and circles. All these objects are grouped together in the block and are treated as a single object. To edit any particular object of a block, the block needs to be "exploded," or split into independent parts. This is especially useful when an entire view or drawing has been inserted and a small detail needs to be corrected. This can be done in the following ways.

Selecting the Explode Check Box in the Insert Dialog Box

As discussed earlier, individual objects within a block cannot be edited unless the block is broken up (exploded). You can explode a block at the time of insertion by simply selecting the **Explode** check box before you choose **OK** in the **Insert** dialog box. Now, when the block is inserted, it would have been already exploded into its component objects.

Entering * as the Prefix of Block Name

To insert a block as a collection of individual objects (exploded), you can also enter an asterisk before its name, when using the command line. The prompt sequence for exploding the block BLOCKPLAN while inserting is given next.

Command: **-INSERT**
Enter block name or [?] <current>: ***BLOCKPLAN**
Specify insertion point or [Scale/X/Y/Z/Rotate/PScale/PX/PY/PZ/PRotate]: *Select the insertion point.*
Scale factor <1>: Enter
Specify rotation angle <0>: Enter

The inserted object will no longer be treated as a block. The figure consists of various objects that can be edited separately.

Note

Blocks with different X, Y, and Z values or nonuniformly scaled blocks can also be exploded. This was not possible in releases prior to the Release 13 of AutoCAD.

Using the EXPLODE Command

Toolbar:	Modify > Explode
Menu:	Modify > Explode
Command:	EXPLODE

As mentioned earlier, the other method to break a block into individual objects is by using the **EXPLODE** command (Figure 14-26). When you invoke the **EXPLODE** command you are prompted to select the block you want to explode.

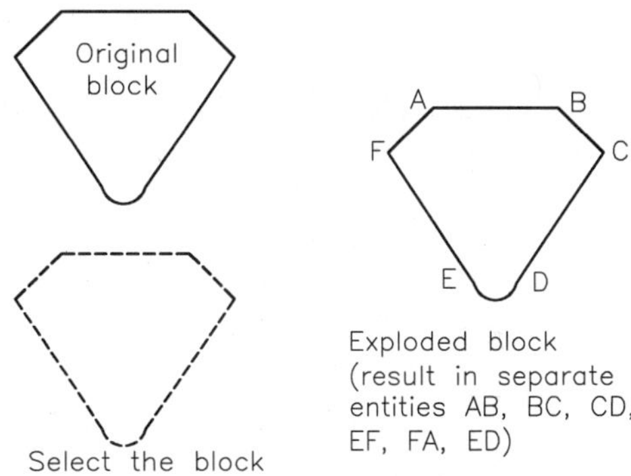

***Figure 14-26** Using the **EXPLODE** command*

Once you have selected the block you want to explode, press ENTER. The **EXPLODE** command explodes a block into its component objects regardless of the scale factors. It also does not have control over the properties such as layer, linetype, lineweight, and color of the component objects. When a block is exploded, there may be no visible change in the drawing. The drawing is identical except that the color and linetype may have changed because of floating layers, colors, or linetypes. The exploded block is now a group of objects that can be edited individually. To check whether the breaking of the block has taken place, select any object that was formerly a part of the block; only that particular object should be highlighted. After a block is exploded, the block definition continues to be in the block symbol table. After exploding a block you can modify it and then redefine it by using the **BLOCK** command. The **Block Definition** dialog box is displayed where you can select the name of the block you wish to redefine from the **Name** drop-down list. Then, use the dialog box options and redefine it. You can also use the **-BLOCK** command to redefine a block. Once you redefine a block, it automatically gets updated in the current drawing and for future use.

Note

*Blocks inserted with **MINSERT** and External references cannot be exploded.*

When redefining a block if you do not explode the block into its components prior to redefining, a warning message saying that the particular block references themselves are displayed by AutoCAD.

Using the XPLODE Command

Command:	XPLODE

With the **XPLODE** command, you can explode a block or blocks into component objects and simultaneously control their properties such as layer, linetype, color, and lineweight. The scale factor of the object to be exploded should be equal. Note that if the scale factor of the objects to be exploded is not equal, you need to change the value of the **EXPLMODE** system variable to 1. The command prompts are as follows.

Command: **XPLODE**
Select objects to XPlode
Select objects: Use *any object selection method and select objects and then press ENTER.*

On pressing ENTER, AutoCAD reports the total number of objects selected and also the number of objects that cannot be exploded. If you select multiple object to explode, AutoCAD further prompts you to specify whether the changes in the properties of the component objects should be made individually or globally. The prompt is given next.

XPlode Individually/<Globally>: *Enter i, g, or press ENTER to accept the default option.*

If you enter **i** at the above prompt, AutoCAD will modify each object individually, one at a time. The next prompt is given next.

Enter an option [All/Color/LAyer/LType/LWeight/Inherit from parent block/Explode] <Explode>: *Select an option.*

All the options available are discussed next.

All. This option sets all the properties such as color, layer, linetype, and lineweight of selected objects after exploding them. AutoCAD prompts you to enter new color, linetype, lineweight, and layer name for the exploded component objects.

Color. This option sets the color of the exploded objects. The prompt is given next.

New color [Truecolor/COlorbook]<BYLAYER>: *Enter a color option or press ENTER.*

When you enter BYLAYER, the component objects take on the color of the exploded object's layer and when you enter BYBLOCK, they take on the color of the exploded object.

Layer. This option sets the layer of the exploded objects. The default option is inheriting the current layer. The command prompt is as follows.

Enter new layer name for exploded objects <current>: *Enter an existing layer name or press ENTER.*

LType. This option sets the linetype of the components of the exploded object. The command prompt is given next.

Enter new linetype name for exploded objects <BYLAYER>: *Enter a linetype name or press ENTER to accept the default.*

LWeight. This option sets the lineweight of the components of the exploded object. The command prompt is given next.

Enter new lineweight <BYLAYER>: *Enter a lineweight or press ENTER to accept the default.*

Inherit from parent block. This option sets the properties of the component objects to that of the exploded parent object, provided the component objects are drawn on layer 0 and the color, lineweight, and linetype are BYBLOCK.

Explode. This option explodes the selected object exactly as in the **EXPLODE** command.

Selecting the **Globally** option, applies changes to all the selected objects at the same time and the options are similar to the ones discussed in the **Individually** option.

Tip

*Using the **XPLODE** command is better than the **EXPLODE** command since it gives you an added advantage of determining the properties of the exploded components of the block.*

RENAMING BLOCKS

Menu:	Format > Rename
Command:	RENAME

Blocks can be renamed with the **RENAME** command. AutoCAD displays the **Rename** dialog box, see Figure 14-27. This dialog box allows you to modify the name of an existing block. In the **Rename** dialog box, the **Named Objects** list box displays the categories of object types that can be renamed, such as blocks, layers, dimension styles, linetypes, and text styles, UCSs, views, and viewports. You can rename all of these except layer 0 and continuous linetype. When you select **Blocks** from the **Named Objects** list, the **Items** list box displays all the block names in the current drawing. When you select a block name you want to rename from the **Items** list box, it is displayed in the **Old Name** edit box. Enter the new name to be assigned to the block in the **Rename To** edit box. Choosing the **Rename To** button applies the change in name to the old name. Choose **OK** to exit the dialog box. For example, if you want to rename a block named Drawing1 to Drawing2, select Drawing1 from the Items list box. It is displayed in the **Old Name** edit box. Enter Drawing2 in the **Rename To** edit box and choose the **Rename To** button. Drawing2 appears in the **Items** list box. Now choose **OK** to exit the dialog box.

Note

*The layer 0 and Continuous linetype cannot be renamed and therefore do not appear in the **Items** list box, when Layers and Linetypes are selected in the named **Objects** list box.*

*You can also rename a block using the command line, if you enter **-RENAME** at the Command prompt.*

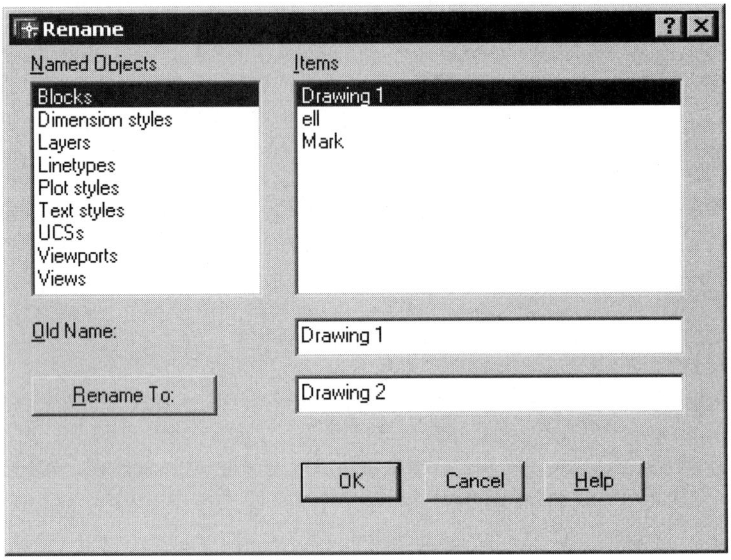

***Figure 14-27 Rename** dialog box*

DELETING UNUSED BLOCKS

Sometimes after completing a drawing, you may notice that the drawing contains several named objects, such as dimstyles, textstyles, layers, blocks, and so on that are not being used. Since these unused named objects unnecessarily occupy disk space, you may want to remove them. Unused blocks can be deleted with the **-PURGE** command. For example, if you want to delete an unused block named Drawing2, the prompt sequence is given next.

Command: **-PURGE**
Enter type of unused objects to purge
[Blocks/Dimstyles/LAyers/LTypes/Plotstyles/SHapes/textSTyles/Mlinestyles/All]: **B**
Enter names to purge <*>: Drawing2
Verify each name to be purged? [Yes/No] <Y>: Enter
Purge block "Drawing2"? <N>: **Y**

If there are no objects to remove, AutoCAD displays a message that there are no unreferenced objects to purge.

Note

*The unused blocks can also be deleted using the **PURGE** command discussed in Chapter 18 (Object Grouping and Editing Commands)*

*Preceding the name of the wblock with an (*) asterisk when entering the name of the wblock while using the **-WBLOCK** command, or selecting the **Entire drawing** radio button in the **Write Drawing** dialog box when creating a wblock using the **WBLOCK** command has the same effect as using the **-PURGE** command. But, the **WBLOCK** command is faster and deletes the unused named objects automatically, while the **-PURGE** command allows you to select the type of named objects you want to delete, and it also gives you an option to verify the objects before deletion occurs.*

Self-Evaluation Test

Answer the following questions, and then compare your answers with the correct answers given at the end of this chapter.

1. Individual objects in a block can be erased. (T/F)
2. A block can be mirrored by providing a scale factor of -1 for X. (T/F)
3. Blocks created by the **BLOCK** command can be used in any drawing. (T/F)
4. An existing block cannot be redefined. (T/F)
5. The ____________ command lets you create a drawing file (*.dwg* extension) of a block defined in the current drawing. (T/F)
6. The ____________ command can be used to change the name of a drawing file. (T/F)
7. The _________ command is used to place a previously created block in a drawing.
8. You can delete unreferenced blocks with the _________ command.

9. You can use the __________ to locate, preview, copy, and insert blocks or existing drawings into the current drawing.

10. The __________ command is used for in-place reference editing.

Review Questions

Answer the following questions.

1. An entire drawing can be converted into a block. (T/F)

2. The objects in a block possess the properties of the layer on which they are drawn, such as color and linetype. (T/F)

3. If the objects forming a block were drawn on layer 0 with color and linetype BYLAYER, then at the time of the insertion, each object that makes up a block is drawn on the current layer with the current linetype and color. (T/F)

4. Objects created with the special color BYBLOCK are generated with the color that is current at the time the block was inserted. (T/F)

5. In the array generated with the **MINSERT** command, there is no way to change the number of rows or columns or the spacing between them after insertion. The whole **MINSERT** pattern is considered as one object that cannot be exploded. (T/F)

6. The block name is controlled by which system variable?

   (a) **EXTNAMES** (b) **EXTMIN**
   (c) **EXTMAX** (d) **SAVENAME**

7. Which command should you use if you want to get back the objects that consist of the block and have been removed from the drawing?

   (a) **OOPS** (b) **BLIPS**
   (c) **BLOCK** (d) **UNDO**

8. By what amount is a block rotated if the values of both the X and Y scale factors is -1?

   (a) 90 (b) 180
   (c) 270 (d) 360

9. When you insert a drawing into a current drawing, how many of the blocks belonging to the inserted drawing are brought into the current drawing?

   (a) One (b) None
   (c) All (d) Two

10. What command is used to create a rectangular array of a block?

    (a) **ARRAY** (b) **INSERT**
    (c) **MINSERT** (d) **3DARRAY**

11. **The WBLOCK**-asterisk method or the **Entire drawing** option of the **WBLOCK** command have the same effect as the **PURGE** command. The only difference is that with the **PURGE** command, ____.

12. Using the __________ command has the added advantage of being able to control the properties of the exploded components of the block, besides being able to break up the block into its component objects.

13. The Automatic Save feature is __________ during in-place reference editing.

14. You cannot use the **REFEDIT** command on blocks inserted using the __________ command.

15. The Layer 0 and the Continuous linetype __________ be renamed using the **RENAME** command.

Exercises

Exercise 7 *Mechanical*

Draw part (a) of Figure 14-28 and define it as a block named A. Then, using the block insert command, insert the block in the plate as shown.

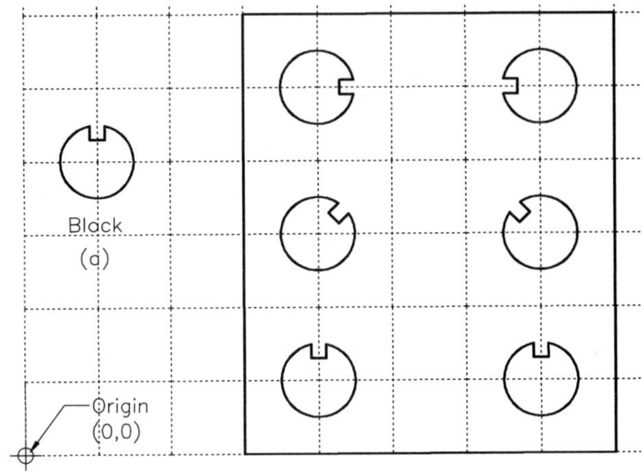

Figure 14-28 *Drawing for Exercise 7*

Exercise 8 *Piping*

Draw the diagrams in Figure 14-29 using blocks.
a. Create a block for the valve, Figure 14-29(a).
b. Use a thick polyline for the flow lines.

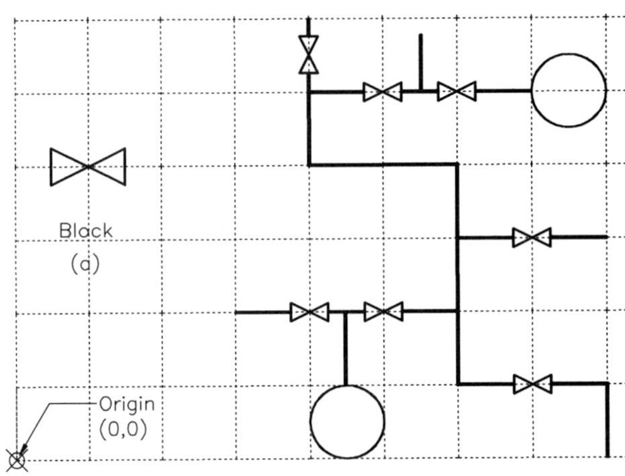

Figure 14-29 *Drawing for Exercise 8*

Exercise 9 *General*

Draw part (a) of Figure 14-30 and define it as a block named B. Then, using the relevant insertion method, generate the pattern as shown. Note that the pattern is rotated at 30-degree.

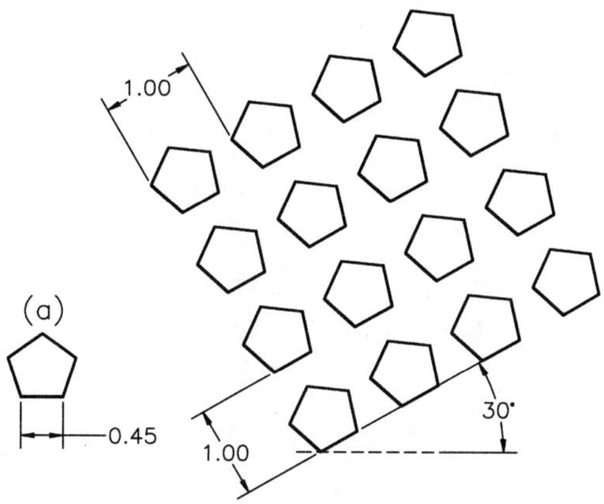

Figure 14-30 *Drawing for Exercise 9*

Exercise 10 *General*

Draw part (a) of Figure 14-31 and define it as a block named Chair. The dimensions of the chair can be referred from the Problem Solving Exercise 3 of Chapter 5. Then, using the block insert command, insert the chair around the table as shown.

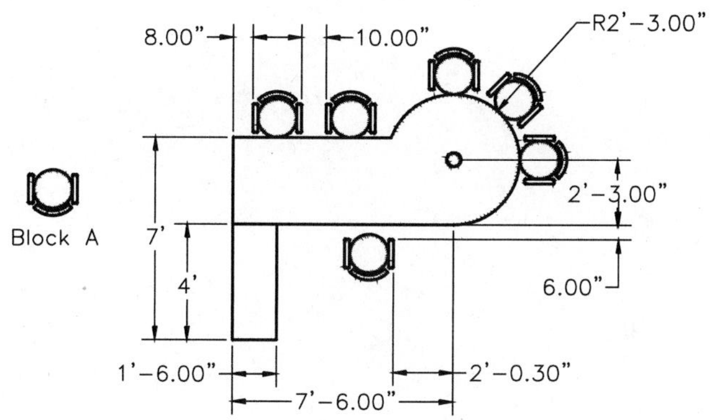

Figure 14-31 *Drawing for Exercise 10*

Answers to Self-Evaluation Test

1 - F, 2 - T, 3 - F, 4 - F, 5 - **WBLOCK**, 6 - **RENAME**, 7 - **INSERT**, 8 - **PURGE**, 9 - **DESIGNCENTER**, 10 - **REFEDIT**

AutoCAD
Part II

Author's Web sites

For Faculty: Please contact the author at **stickoo@calumet.purdue.edu** or **tickoo@cadcim.com** to access the Web site that contains the following:

1. **PowerPoint presentations, programs, and drawings used in this textbook.**
2. **Syllabus, chapter objectives and hints, and questions with answers for every chapter.**

For Students: You can download drawing-exercises, tutorials, programs, and special topics by accessing author's Web site at **www.cadcim.com** or **http://technology.calumet.purdue.edu/met/tickoo/students/students.htm**.

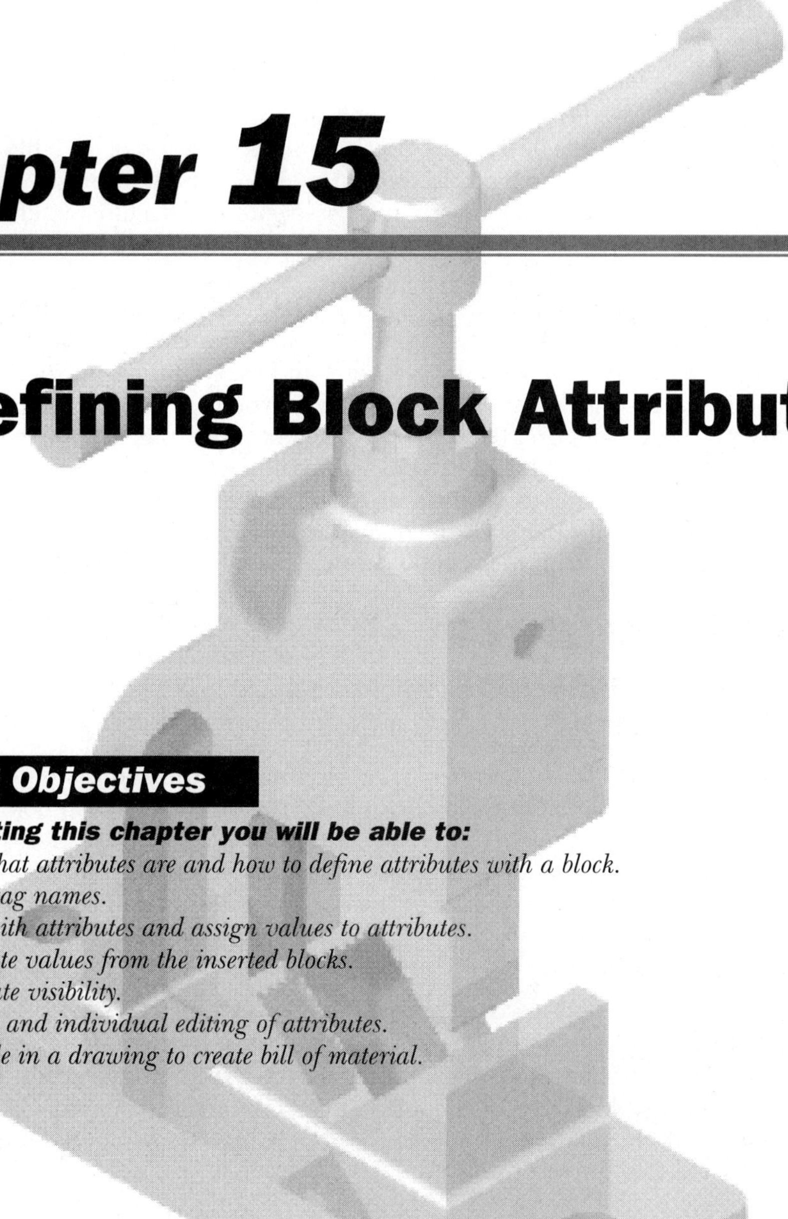

Chapter 15

Defining Block Attributes

Learning Objectives

After completing this chapter you will be able to:

- *Understand what attributes are and how to define attributes with a block.*
- *Edit attribute tag names.*
- *Insert blocks with attributes and assign values to attributes.*
- *Extract attribute values from the inserted blocks.*
- *Control attribute visibility.*
- *Perform global and individual editing of attributes.*
- *Insert a text file in a drawing to create bill of material.*

UNDERSTANDING ATTRIBUTES

AutoCAD has provided a facility that allows the user to attach information to blocks. This information can then be retrieved and processed by other programs for various purposes. For example, you can use this information to create a bill of material for a project, find the total number of computers in a building, or determine the location of each block in a drawing. Attributes can also be used to create blocks (such as title blocks) with prompted or preformatted text, to control text placement. The information associated with a block is known as **attribute value** or simply **attribute**. AutoCAD references the attributes with a block through tag names.

Before assigning attributes to a block, you must create an attribute definition by using the **ATTDEF** command. The attribute definition describes the characteristics of the attribute. You can define several attribute definitions (tags) and include them in the block definition. Each time you insert the block, AutoCAD will prompt you to enter the value of the attribute. The attribute value automatically replaces the attribute tag name. The information (attribute values) assigned to a block can be extracted and written to a file by using AutoCAD's **EATTEXT** command. This file can then be inserted in the drawing as a table or processed by other programs to analyze the data. The attribute values can be edited by using the **EATTEDIT** command. The display of attributes can be controlled with the **ATTDISP** command.

DEFINING ATTRIBUTES

Menu:	Draw > Block > Define Attributes
Command:	ATTDEF

When you invoke the **ATTDEF** command, the **Attribute Definition** dialog box is displayed, see Figure 15-1. The block attributes can be defined through this dialog box. When creating an attribute definition, you must define the mode, attributes, insertion point, and text information for each attribute. All this information can be entered in the dialog box. The following is the description of each area of the **Attribute Definition** dialog box.

Mode Area

The **Mode** area of the **Attribute Definition** dialog box has four check boxes: **Invisible**, **Constant**, **Verify**, and **Preset**, see Figure 15-2. These options determine the display and edit features of the block attributes. For example, if you select the **Invisible** check box, the attribute becomes invisible; that is, it is not displayed on the screen. Similarly, if the **Constant** check box is selected, the attribute becomes constant. This means that its value is predefined and cannot be changed. These options are described next.

Invisible

This option lets you create an attribute that is not visible on the screen, by default. Clear this check box if you want the attribute to be visible.

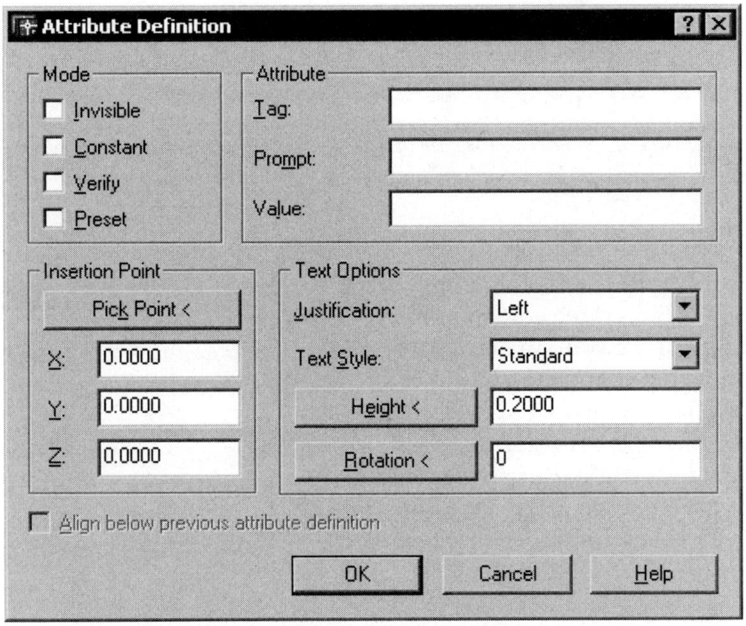

***Figure 15-1 Attribute Definition** dialog box*

Tip
The Invisible mode is especially useful when you do not want the attribute values to be displayed on the screen to avoid cluttering the drawing. Also, if the attributes are invisible, it takes less time to regenerate the drawing.

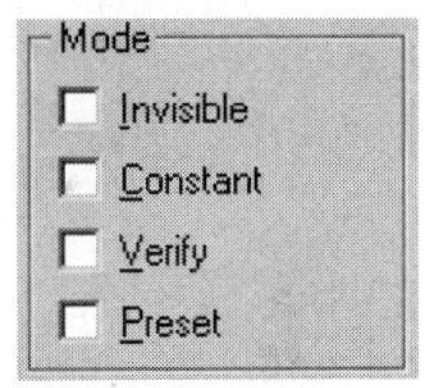

***Figure 15-2 Mode** area of the **Attribute Definition** dialog box*

You can make the invisible attribute visible by using the **ATTDISP** command discussed later in this chapter, in the section "Controlling Attribute Visibility".

Constant

This option lets you create an attribute that has a fixed value and cannot be changed after block insertion. When you select this mode, the **Prompt** edit box and the **Verify** and **Preset** check boxes are disabled. Since the value is constant, there is no need to be prompted for new values. This check box is cleared by default and you can use different attribute values for the blocks.

Verify

This option allows you to verify the attribute value you have entered when inserting a block by asking you twice for the data. If the value is incorrect, you can correct it by entering the new value. If this check box is cleared, you are not prompted for verification of the attribute values.

Preset

This option allows you to create an attribute that is automatically set to default value. The attribute values are not requested when you insert a block and default values are used. But unlike a constant attribute, the preset attribute value can be edited later.

Note

*Not selecting any of the check boxes in the **Mode** area displays all the prompts at the command line and the values will be visible on the screen. This is also referred to as the **Normal** mode.*

Attribute Area

The **Attribute** area (Figure 15-3) of the **Attribute Definition** dialog box has three edit boxes: **Tag**, **Prompt**, and **Value**, where you can enter values. You can enter up to 256 characters in these edit boxes. If the first character to be entered in any one of these edit boxes is a space, you should start with a backslash (\). But if the first character is a backslash (\), you should start the value to be entered with two backslashes (\\). The three edit boxes have been described next.

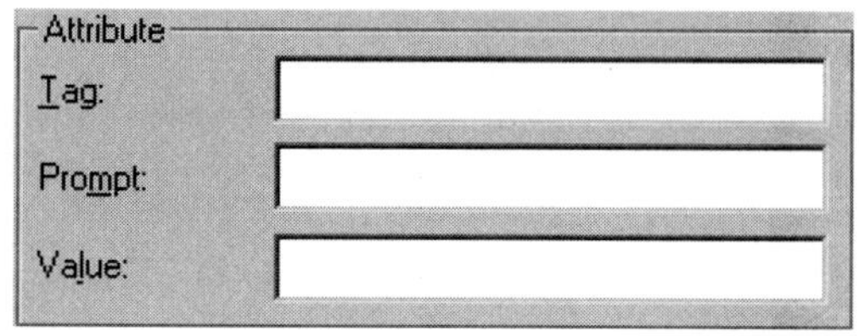

***Figure 15-3** Attribute area of the **Attribute Definition** dialog box*

Tag

This is like a label that is used to identify an attribute. For example, the tag name COMPUTER can be used to identify an item. Here you can enter the tag names as uppercase, lowercase, or both, but all lowercase letters are automatically converted into uppercase when displayed. The tag name cannot be null. Also, the tag name must not contain any blank spaces.

Tip

It is advisable to specify a tag name that reflects the contents of the item being tagged. For example, the tag name COMP or COMPUTER is an appropriate name for labeling computers.

Prompt

The text that you enter in the **Prompt** edit box is used as a prompt when you insert a block that contains the defined attribute. For example, if COMPUTER is the tag, you can enter WHAT IS THE MEMORY? or ENTER MEMORY: in the **Prompt** edit box. AutoCAD will then prompt you with this same statement when you insert the block with which the attribute is defined. If you have selected the **Constant** check box in the **Mode** area, the **Prompt** edit box is not available because no prompt is required if the attribute is constant. If you do not enter anything in the **Prompt** edit box, the entry made in the **Tag** edit box is used as the prompt.

Value

The entry in the **Value** edit box defines the default value of the specified attribute. If you do not enter a value, it is used as the value for the attribute. The entry of a value is optional.

Insertion Point Area

The **Insertion Point** area of the **Attribute Definition** dialog box (Figure 15-4) lets you define the insertion point of block attribute text. You can define the insertion point by entering the values in the **X**, **Y**, and **Z** edit boxes or by choosing the **Pick Point** button. When you choose this button, the dialog box is temporarily closed and you can select an insertion point on the screen or enter the *X*, *Y*, and *Z* values of the insertion point at the command line. Once you have specified the insertion point, the **Attribute Definition** dialog box is redisplayed.

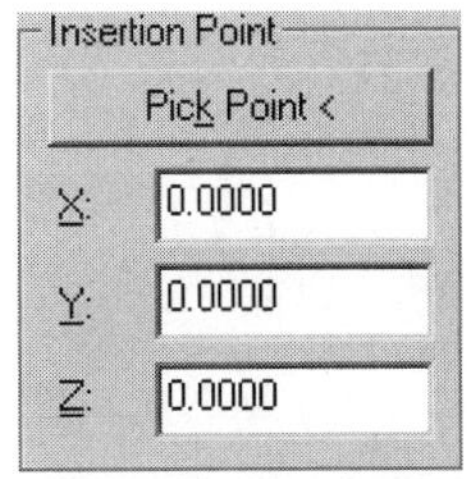

***Figure 15-4 Insertion Point** area of the **Attribute Definition** dialog box*

Just below the **Insertion Point** area of the dialog box is a check box labeled **Align below previous attribute definition**. This check box is not available when you use the **Attribute Definition** dialog box for the first time. After you have defined an attribute and when you press ENTER to display the **Attribute Definition** dialog box again, this check box is available. You can select this check box to place the subsequent attribute text just below the previously defined attribute automatically. When you select this check box, the **Insertion Point** area and the **Text Options** areas of the dialog box are not available and AutoCAD assumes previously defined values for text such as text height, text style, text justification, and text rotation. The text is automatically placed on the following line.

Text Options Area

The **Text Options** area of the **Attribute Definition** dialog box (Figure 15-5) lets you define the justification, text style, height, and rotation of the attribute text. To set the text justification, select a justification type from the **Justification** drop-down list. The default option is **Left**. Similarly, you can use the **Text Style** drop-down list to select a text style. All the text styles defined in the current drawing are displayed in the **Text Style** drop-down list. The default text style is **Standard**. You can specify the text height and text rotation in the **Height** and **Rotation** edit boxes. You can also define the text height by choosing the **Height** button. When you choose this button, AutoCAD temporarily exits the dialog box and lets you enter the height value by selecting points on the screen or from the command line. Once you have defined the height on the screen, the dialog box reappears and the defined text height is displayed in the edit box. Similarly, you can define the text rotation by choosing the **Rotation** button and then selecting points on the screen or by entering the rotation angle at the command line.

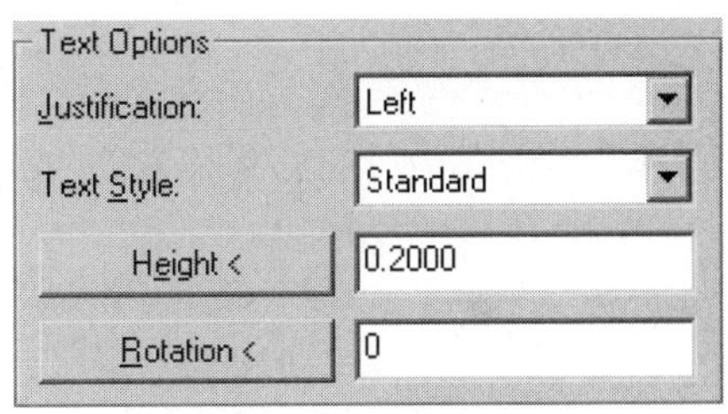

***Figure 15-5 Text Options** area of the **Attribute Definition** dialog box*

Note

The text style must be defined before it can be used to specify the text style.

*If you select a style that has the height predefined, AutoCAD automatically disables the **Height** edit box.*

*If you have selected the **Align** option from the **Justification** drop-down list, the **Height** and **Rotation** edit boxes are disabled.*

*If you have selected the **Fit** option from the **Justification** drop-down list, the **Insertion Point** area and the **Rotation** edit box is disabled. To specify the location of the attribute, choose **OK** from the dialog box. AutoCAD prompts you to specify first and second endpoint of text baseline. The text will be fit between the two endpoints that you specify.*

After you complete the settings in the **Attribute Definition** dialog box and choose **OK**, the attribute tag text is inserted in the drawing at the specified insertion point. Now, you can use the **BLOCK** or **WBLOCK** commands to select all the objects and attributes to define a block.

Note

*You can use the **-ATTDEF** command to display prompts on the command line. All the options available in the **Attribute Definition** dialog box are available through the command line too.*

Example 1

General

In this example, you will define the following attributes for a computer and then create a block using the **BLOCK** command. The name of the block is COMP.

Mode	Tag name	Prompt	Default value
Constant	ITEM		Computer
Preset, Verify	MAKE	Enter make:	CAD-CIM
Verify	PROCESSOR	Enter processor type:	Unknown
Verify	HD	Enter Hard-Drive size:	40 GB
Invisible, Verify	RAM	Enter RAM:	256 MB

1. Draw the computer as shown in Figure 15-6. Assume the dimensions, or measure the dimensions of the computer you are using for AutoCAD.

2. Invoke the **ATTDEF** command. The **Attribute Definition** dialog box is displayed.

3. Define the first attribute as shown in the preceding table. Select **Constant** check box in the Mode area because the mode

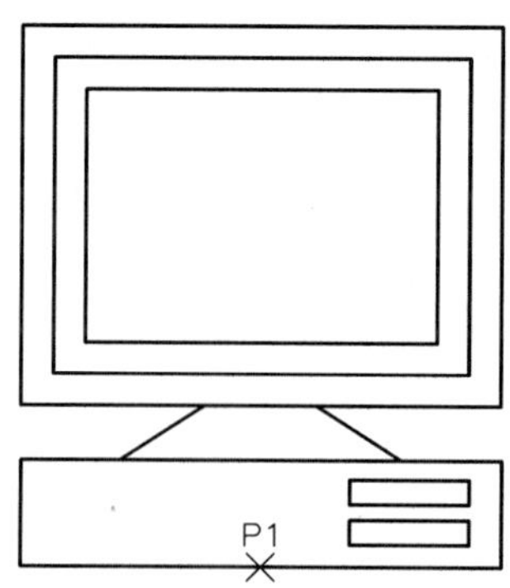

Figure 15-6 *Drawing for Example 1*

of the first attribute is constant. In the **Tag** edit box, enter the tag name, ITEM. Similarly, enter COMPUTER in the **Value** edit box. Note that the **Prompt** edit box is not available because the mode is constant.

4. In the **Insertion Point** area, choose the **Pick Point** button to define the text insertion point. Select a point below the insertion base point (P1) of the computer to place the text.

5. In the **Text Options** area, specify the justification, style, height, and rotation of the text.

6. Choose the **OK** button once you have entered information in the **Attribute Definition** dialog box.

7. Press ENTER to invoke the **Attribute Definition** dialog box again. Enter the mode and attribute information for the second attribute as shown in the table at the beginning of Example 1. You need not define the insertion point and text options again. Select the **Align below previous attribute definition** check box that is located just below the **Insertion Point** area. You will notice that when you select this check box, the **Insertion Point** and **Text Options** areas are not available. Now, choose the **OK** button. AutoCAD places the attribute text just below the previous attribute text.

8. Similarly, define the remaining attributes also (Figure 15-7).

9. Now, use the **BLOCK** command to create a block. The name of the block is COMP, and the insertion point of the block is P1, midpoint of the base. When you select the objects for the block, make sure you also select the attributes.

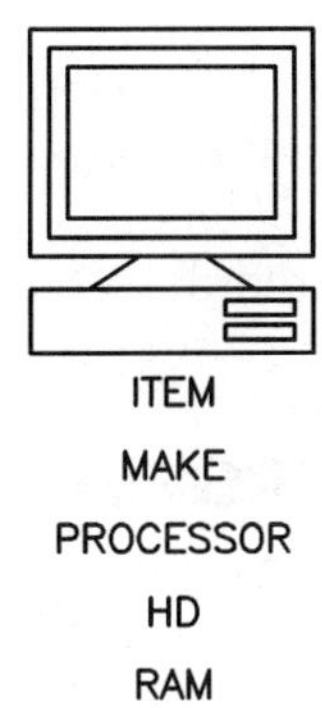

Figure 15-7 *Define attributes below the computer drawing*

Note

The order of prompts is the same as the order of attributes selection.

EDITING ATTRIBUTE DEFINITION

Toolbar:	Text > Edit Text
Menu:	Modify > Object > Text > Edit
Command:	DDEDIT

Using the **DDEDIT** command, you can edit text and attribute definitions, before you define the block. After invoking this command, AutoCAD will prompt you to **select an annotation object or [Undo]**. If you select an attribute created using the **Attribute Definition** dialog box, the **Edit Attribute Definition** dialog box is displayed and lists the tag name, prompt, and default value of the attribute (Figure 15-8).

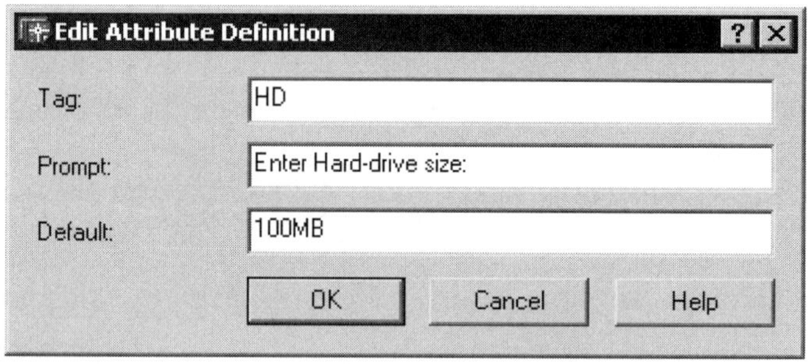

***Figure 15-8 Edit Attribute Definition** dialog box*

Tip

*When you invoke the **DDEDIT** command, the **Edit Attribute Definition** dialog box appears bigger in size than shown in Figure 15-8. To reduce the size of the dialog box, move the cursor toward the top or toward the bottom of the dialog box until the cursor is replaced by a two-sided arrow. Press the left mouse button and the size of the dialog box is reduced as shown in Figure 15-8.*

You can enter the new values in the respective edit boxes. Once you have entered the changed values, choose the **OK** button in the dialog box. After you exit the dialog box, AutoCAD will continue to prompt you to select another text or attribute object (Attribute tag). If you have finished editing and do not want to select another attribute object to edit, press ENTER to return to the Command prompt.

Using the PROPERTIES Palette

The **PROPERTIES** command has been already discussed in Chapter 4, Working with Drawing Aids. It can also be used to edit defined attributes. Select the attribute to modify and right-click to display a shortcut menu. Choose **Properties** here and the **PROPERTIES** palette is displayed, see Figure 15-9. You will notice that **Attribute** is displayed in the text box located at the top of the palette. Also, under the **Categorized** tab you will find that all the properties of the selected attribute are displayed under four headings. They are **General**, **Text**, **Geometry**, and **Misc**. You can change these values in their corresponding fields. For example, you can modify the color, layer, linetype, thickness, linetype scale, and so on, of the selected attribute under the **General** head. Similarly, you can modify the tag name, prompt, and value of the selected attribute in the **Tag**, **Prompt**, and **Value** fields under the **Text** heading. Under the **Text** head, you can also modify the text style, justification, text height, rotation angle, width factor, and obliquing angle values of the selected attribute. Under the **Geometry** heading you can redefine the insertion point of the selected attribute by choosing the button with an arrow icon, which is displayed when you select the **Position X**, **Position Y**, or **Position Z** fields. When you choose the button with the arrow icon, you are allowed to reposition an existing insertion point by selecting on the screen. You can also determine if you want the attribute text to appear upside down or backwards or not under the **Misc** heading. Here, you can also modify the attribute modes, which have been already defined. When you select a particular mode, say **Invisible**, a drop-down list is available in the

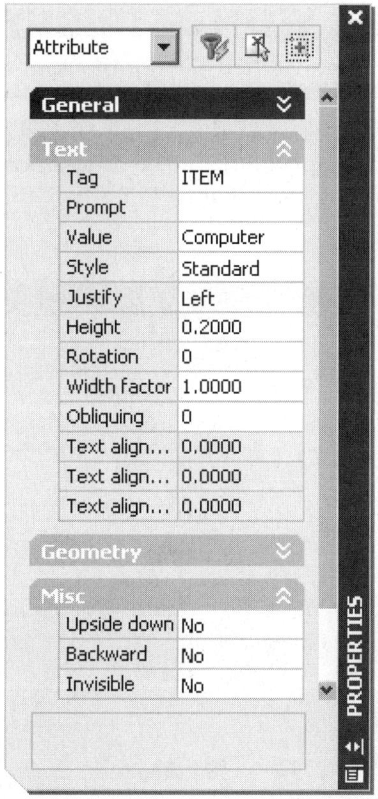

Figure 15-9 *Using the* ***PROPERTIES*** *palette to modify attributes*

corresponding field. This list displays two options, **Yes** and **No**. If you select **Yes**, the is made invisible and if you select **No**, the attribute is not made invisible. To select a particular mode, you should choose **Yes** from their corresponding drop-down lists.

Note

Remember that if you change the justification of the first attribute below which you have aligned the remaining attributes, the remaining attributes are not updated automatically. You will have to change the justifications of the remaining attributes individually to align them below the first attribute.

If you change the justification of the attributes from align to some other, the size of the text is also changed.

You can also use the ***CHANGE*** *command to edit the attribute objects using the Command line.*

INSERTING BLOCKS WITH ATTRIBUTES

The value of the attributes can be specified during block insertion, either at the command line or in the **Enter Attributes** dialog box. When you use the **INSERT** command or the **-INSERT** command to insert a block in a drawing (discussed earlier in Chapter 14, Working with Blocks), and after you have specified the insertion point, scale factors, and rotation angle, the **Enter Attributes** dialog box (Figure 15-10) is displayed if the system variable **ATTDIA** is set to **1**.

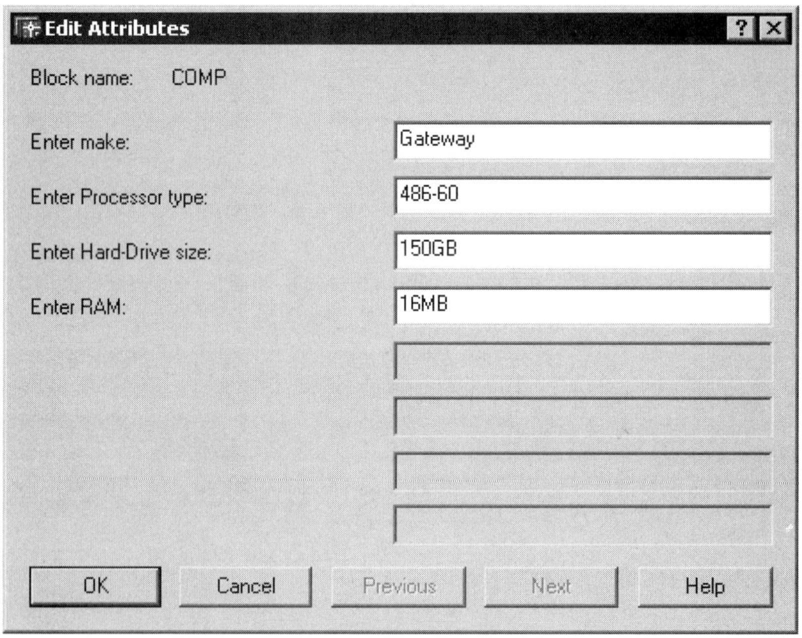

Figure 15-10 *The* ***Enter Attributes*** *dialog box*

The default value for **ATTDIA** is **0**, which disables the dialog box. The prompts and their default values, which you had specified with the attribute definition, are then displayed on the command line after you have specified the insertion point, scale, and rotation angle for the block to be inserted.

In the **Enter Attributes** dialog box, the prompts that were entered at the time of attribute definition in the dialog box are displayed with their default values in corresponding fields. If an attribute has been defined with the **Constant** mode, it is not displayed in the dialog box because a constant attribute value cannot be edited. You can enter the attribute values in the fields located next to the attribute prompt. If no new values are specified, the default values are displayed. Eight attribute values are displayed at a time in the dialog box. If there are more attributes, they can be accessed by using the **Next** or **Previous** buttons. The block name is displayed at the top of the dialog box. After entering the new attribute values, choose the **OK** button. AutoCAD will place these attribute values at the specified location.

Tip

*It is more convenient to use the **Edit Attributes** dialog box, since you can view all the attribute values at a glance and can correct them here itself, before placing. Therefore, it is a good idea to set the **ATTDIA** value to 1, before you insert a block with attributes.*

*Attributes can also be defined from the command line by setting system variable **ATTDIA** to 0 (default value). Now, when you use **INSERT** command, AutoCAD does not display the **Enter Attributes** dialog box. Instead, AutoCAD will prompt you to enter the attribute values for various attributes that have been defined in the block at the command line.*

Example 2 *General*

In this example, you will insert the block (COMP) that was defined in Example 1. The following is the list of the attribute values for computers.

ITEM	MAKE	PROCESSOR	HD	RAM
Computer	Gateway	486-60	150 MB	16 MB
Computer	Zenith	486-30	100 MB	32 MB
Computer	IBM	386-30	80 MB	8 MB
Computer	Del	586-60	450 MB	64 MB
Computer	CAD-CIM	Pentium-90	100 Min	32 MB
Computer	CAD-CIM	Unknown	600 MB	Standard

1. Make the floor plan drawing as shown in Figure 15-11 (assume the dimensions).

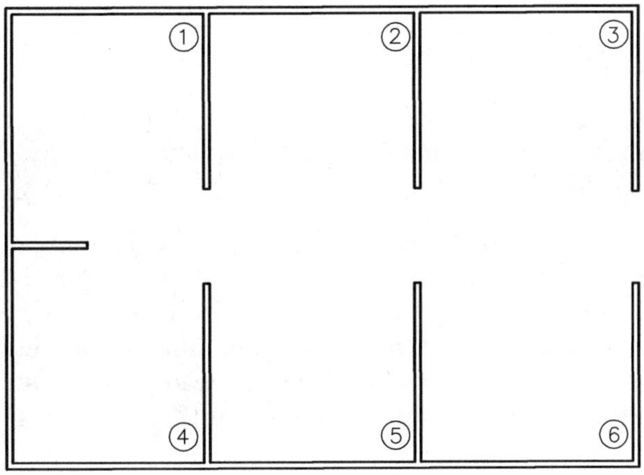

Figure 15-11 *Floor plan drawing for Example 2*

2. Set the system variable **ATTDIA** to 1. Use the **INSERT** command to insert the blocks. When you invoke the **INSERT** command, the **Insert** dialog box is displayed. Enter COMP in the **Name** edit box and choose **OK** to exit the dialog box. Select an insertion point on

screen to insert the block. After you specify the insertion point, the **Enter Attributes** dialog box is displayed, where you can change the attribute values, if you need to, in the different edit boxes.

3. Repeat the **INSERT** command to insert other blocks, and define their attribute values as shown in Figure 15-12.

4. Save the drawing for further use.

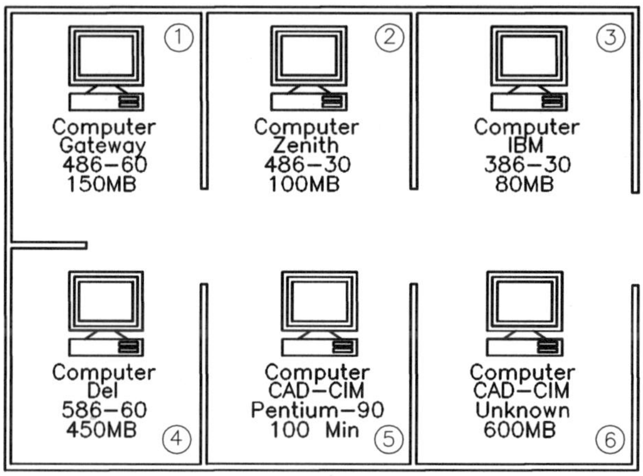

Figure 15-12 *The floor plan after inserting blocks and defining their attributes*

MANAGING ATTRIBUTES

Toolbar:	Modify II > Block Attribute Manager
Menu:	Modify > Menu > Attribute > Block Attribute Manager
Command:	BATTMAN

The **BATTMAN** command allows you to manage the attribute definitions for blocks in the current drawing through the **Block Attribute Manager** dialog box. This dialog box is shown in Figure 15-13. By default, any attribute changes you make are applied to all existing block references in the current drawing. **Immediate changes in the default value of the attribute in the existing drawing can only be seen if the attribute that is edited is in Constant mode**. If the mode is other than constant, the changes in the attribute can only be seen for new block insertions.

When you invoke this command, AutoCAD displays the number of blocks that exist in the current drawing, below the attribute list in the dialog box. The block can be in the model space or in the layout. The names of the existing blocks in the current drawing are displayed in the **Block** drop-down list. Attributes of the selected block are displayed in the attribute list. By default, **Tag**, **Prompt**, **Default**, and **Modes** attribute properties are displayed in the

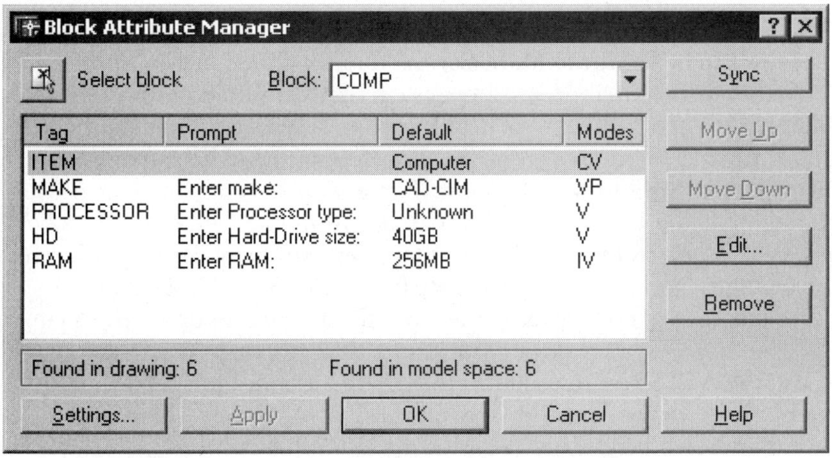

***Figure 15-13 Block Attribute Manager** dialog box*

attribute list. You can edit the attribute definitions in blocks, remove attributes from blocks, and change the order in which you were prompted for attribute values while inserting the block.

Select block

This button allows you to select a block from the drawing area whose attributes you want to edit. When you choose the **Select block** button, the dialog box closes temporarily until you select block from the drawing or cancel by pressing ESC. Once you have selected the block, the dialog box is redisplayed. The name of the block you have selected is displayed in the **Block** drop-down list. If you modify attributes of a block and then select a new block before saving the attribute changes you made, you are prompted to save the changes before selecting another block by displaying an alert message box (Figure 15-14).

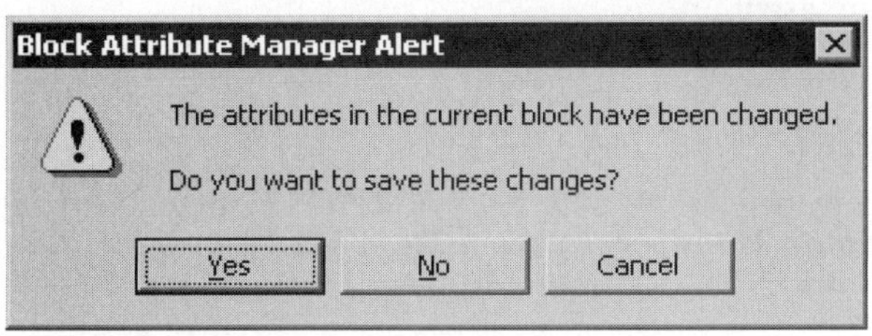

***Figure 15-14 Block Attribute Manager Alert** box*

Block Drop-down List

This list in the **Block Attribute Manager** dialog box allows you to choose the block whose attributes you want to modify. It lists all the blocks in the current drawing that have attributes.

Sync

This button of the **Block Attribute Manager** dialog box allows you to update all instances of the selected block with the attribute properties currently defined. This does not affect any of the values assigned to attributes in each block when attributes were defined. By using this button, you can update attributes in all block references in the current drawing with the changes you made to the block definition. For example, you may have used the **Block Attribute Manager** dialog box to modify attribute properties of several block definitions in your drawing. But you do not want it to automatically update existing block references when you made the changes. Now, when you are satisfied with the attribute changes you made, you can apply those changes to all blocks in the current drawing by using the **Sync** button.

You can also use the **ATTSYNC** command to update attribute properties in block references to match their block definition. This command can also be invoked by choosing the **Synchronize Attributes** button from the **Modify II** toolbar.

Note

Changing the attribute properties of an existing block does not affect the values assigned to the block. For example, in a block containing an attribute whose ***Tag*** *is Cost and whose value is 200, the value 200 remains unaffected if the* ***Tag*** *is changed to Unit Cost.*

Move Up

This button moves the selected attribute tag earlier in the prompt sequence. Now, when you insert the block, the attribute you have just moved up will appear earlier in the prompt sequence. The **Move Up** button is not available when a constant attribute is selected and if the attribute is at the topmost position in the attribute list.

Move Down

This button moves the selected attribute tag later in the prompt sequence. The **Move Down** button is not available when a constant attribute is selected and when the attribute is at the bottom position in the attribute list.

Tip

When you define a block, the order in which you select the attributes determines the order in which you are prompted for attribute information when you insert the block. You can use the ***Block Attribute Manager*** *to change the order of prompts that request attribute values.*

Edit

This button of the **Block Attribute Manager** allows you to modify attribute properties. To edit the attribute definition in blocks, select the attribute and then choose **Edit** from the dialog box. You can also double-click on the attribute to display the **Edit Attribute** dialog box (Figure 15-15). In the dialog box, **Active Block** displays the name of the block that you have selected and whose properties have to be edited.

The options in the **Edit Attribute** dialog box are discussed next.

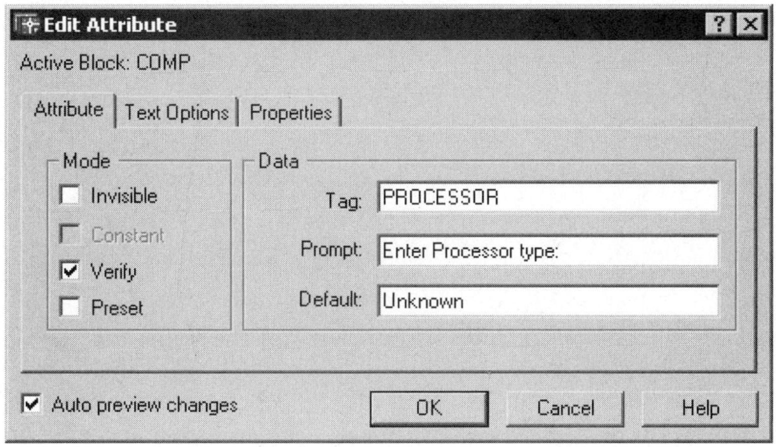

Figure 15-15 ***Edit Attribute*** *dialog box*

Attribute Tab

The options under the **Attribute** tab of the **Edit Attribute** dialog box are used to modify the mode and the data of the attributes (see Figure 15-15). The options under the **Mode** area can be selected for the block. The **Data** area of the dialog box has three edit boxes: **Tag**, **Prompt**, and **Default**, where you can enter the data to be changed.

Text Options Tab

The options under this tab (Figure 15-16) are used to set the properties that define the way an attribute's text is displayed in the drawing. The **Text Options** are discussed next.

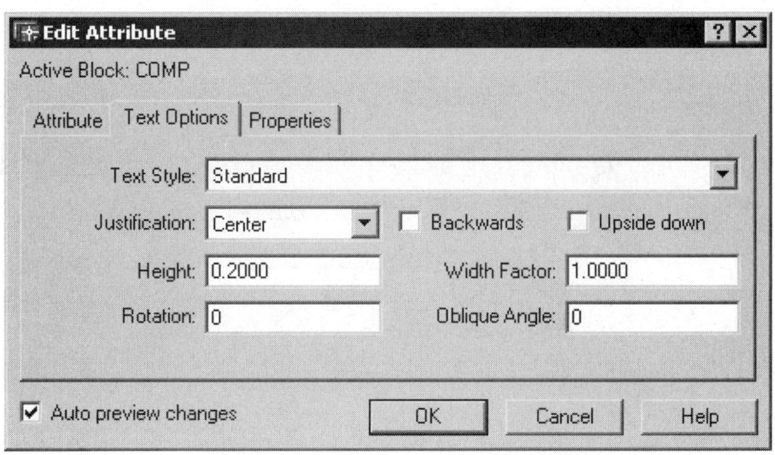

Figure 15-16 ***Text Options*** *tab of the* ***Edit Attribute*** *dialog box*

Text Style. This drop-down list specifies the text style for attribute text. Default values for this text style are assigned to the text properties displayed in this dialog box.

Justification. This drop-down list specifies how attribute text is justified.

Height. This edit box specifies the height of the attribute text. If the height is defined in the current text style, this edit box is not available.

Rotation. This edit box specifies the rotation angle of the attribute text.

Backwards. This check box specifies whether or not the text is displayed backwards. Select this box if you want to display the text backwards.

Upside Down. This check box specifies whether or not the text is displayed upside down. Select this box if you want to display the text upside down.

Width Factor. This edit box sets the character spacing for attribute text. Entering a value less than 1.0 condenses the text and a value greater than 1.0 expands it.

Oblique Angle. In this edit box, the angle at which the attribute text is slanted away from its vertical axis is specified.

Note

If you change the properties mentioned above of a ***Constant*** *attribute, you may need to regenerate the drawing to view the effects.*

Properties

Under this tab of the **Block Attribute Manager** (Figure 15-17), you can define the layer that the attribute is on and the color, lineweight, and linetype for the attribute's line. Also, if the drawing uses plot styles, you can assign a plot style to the attribute.
The **Properties** tab has the following options.

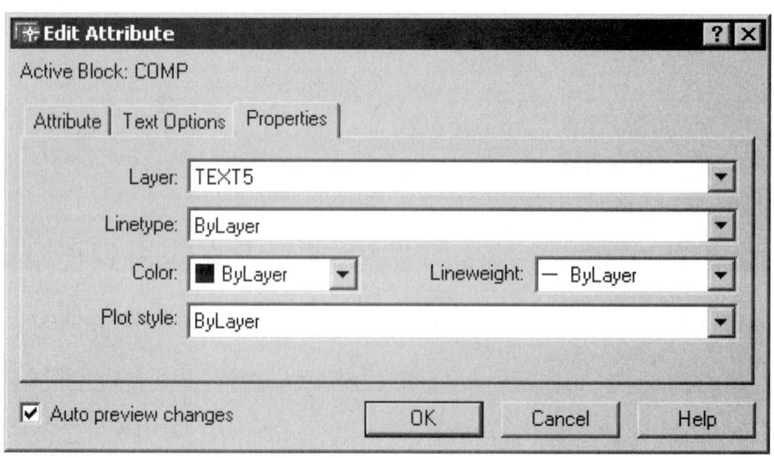

Figure 15-17 ***Properties*** *tab of the* ***Edit Attribute*** *dialog box*

Layer. This drop-down list specifies the layer that the attribute is on. The layer on which the attribute was defined is displayed here in this drop-down list. This cannot be changed using the dialog box.

Linetype. This drop-down list specifies the linetype of attribute text.

Color. This drop-down list specifies the attribute's text color.

Lineweight. This drop-down list specifies the lineweight of attribute text. The changes you make to this option do not come into effect if the **LWDISPLAY** system variable is off.

Plot Style. This drop-down list specifies the plot style of the attribute. If the current drawing uses color-dependent plot styles, the **Plot style** list in this dialog box is not available.

Auto preview changes

This check box controls whether or not the drawing area is immediately updated to display any visible attribute changes you make. If the **Auto preview changes** check box is selected, changes are immediately visible. If this check box is cleared, changes are not immediately visible.

Tip

__Move Up__, __Move Down__, __Edit__, __Remove__, and __Settings__ can be invoked by the shortcut menu, which appears by right-clicking on the attributes in the attribute list in the __Block Attribute Manager__ dialog box.

Remove

This button removes the selected attribute from the block definition. If the **Apply changes to existing references** check box is selected in the **Settings** dialog box (discussed later in the chapter) before you choose the **Remove** button, the attribute is removed from all instances of the block in the current drawing. If this check box is cleared before choosing the **Remove** button, the attribute is removed only from the attribute list and the blocks that will be inserted henceforth. You can remove attributes from block definitions and from all existing block references in the current drawing. Attributes removed from existing block references do not disappear in the drawing area until you regenerate the drawing using the **REGEN** command. You cannot remove all attributes from the block, at least one attribute should remain. However, if you want to remove all the attributes, you have to redefine the block.

Note that the **Remove** button is not available for blocks with only one attribute.

Settings

This button of the **Block Attribute Manager** opens the **Settings** dialog box (Figure 15-18), where you can customize how attribute information is listed in the **Block Attribute Manager**.

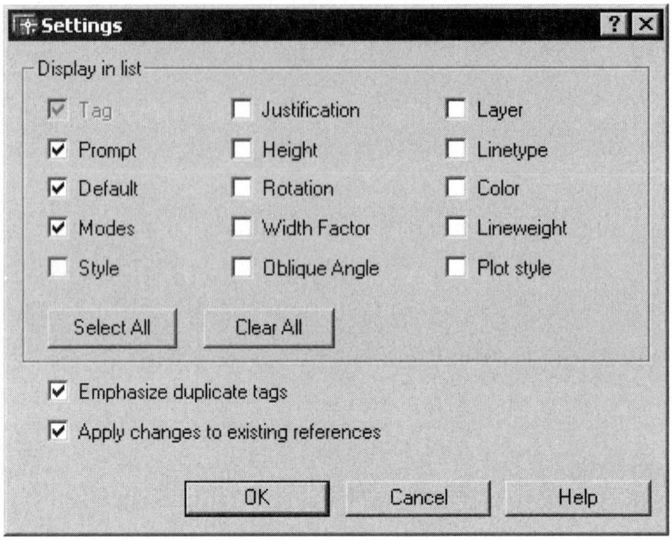

Figure 15-18 ***Settings*** *dialog box in the* ***Block Attribute Manager***

Display in list

You can specify the attribute properties you want to be displayed in the list by selecting the check boxes available in this area of the **Settings** dialog box. This area shows the check boxes of all the properties that are displayed in the attribute list. Only the check boxes of the properties selected are displayed in the attribute list. These properties can be viewed in the attribute list by scrolling the list. The **Tag** check box is always selected by default.

This area of the dialog box has two buttons, **Select all** and **Clear all**, which select and clear all the check boxes respectively.

Emphasize duplicate tags

This check box turns duplicate tag emphasis on and off. If this check box is selected, duplicate attribute tags are displayed in red type in the attribute list. For example, if a block is having PRICE as a tag more than once, the attribute PRICE will be displayed in red in the attribute list. If this check box is cleared, duplicate tags are not emphasized in the attribute list.

Apply changes to existing references

This check box of the dialog box specifies whether or not to update all existing instances of the block whose attributes you are modifying. Therefore, if this check box is selected, all instances of the block with the new attribute definitions are updated. If this check box is cleared, only new insertions of the block with the new attribute definitions are displayed.

Use **Sync** in the **Block Attribute Manager** to apply changes immediately to existing block instances. This temporarily overrides the **Apply changes to existing references** option.

Tip

If constant attributes or nested attributed blocks are affected by your changes, use the ***REGEN*** *command to update the display of those blocks in the drawing area.*

EXTRACTING ATTRIBUTES

Toolbar:	Modify II > Attribute Extract
Menu:	Tools > Attribute Extraction
Command:	EATTEXT

The **EATTEXT** command allows you to extract block attribute information from a drawing or from multiple drawings. This is saved to an external file in **Comma Delimited** ASCII text format, **Tab Delimited** ASCII text format, or in either **Microsoft Excel** or **Access** file format (if the application is installed on your system). When this command is invoked, the **Attribute Extraction** wizard is displayed on the screen. This wizard includes the following pages.

Select Drawing

The **Select Drawing** page (Figure 15-19) allows you to select the object either from a single drawing or from multiple drawings. The **Select Drawing** page has three radio buttons, namely **Select Objects**, **Current Drawing**, and **Select Drawings**.

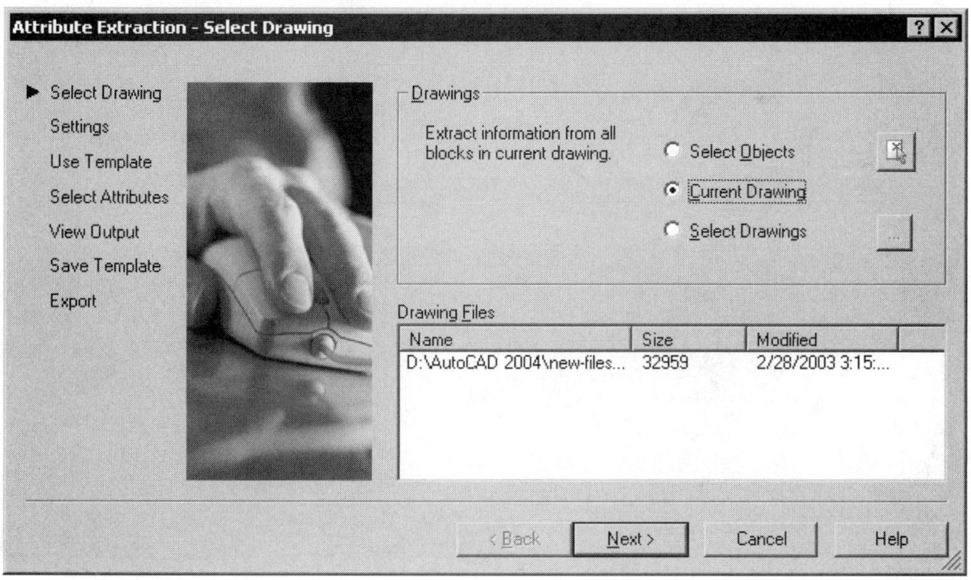

***Figure 15-19** Select Drawing page of the **Attribute Extraction** wizard*

Select Objects

If this radio button is selected, the **Select Objects** button is made available. Choose this button to select the blocks whose attribute values you want to extract. The wizard disappears from the screen temporarily until the block is selected or ESC is pressed. After the block is selected, the name of the drawing in which the selected block lies and its path address is displayed in the **Drawing Files** area under **Name** heading.

Current Drawing

This radio button allows you to extract the information from all the blocks in the current drawing. The drawing name with full path is displayed in the **Drawing Files** area under **Name** heading.

Select Drawings

If this radio button is selected, it makes the **[...]** button available. This button is chosen to display the **Select Files** dialog box. The dialog box allows you to select multiple drawings to extract block and attribute information. The names of the drawings selected are displayed in the **Drawing Files** area under **Name** heading with full path.

Settings

This page (Figure 15-20) has two check boxes, **Include xrefs** and **Include nested blocks**. Both of these check boxes can be selected or either of them or both of them can be cleared. These two check boxes allow you to specify whether you want to extract block attribute information from the external reference drawings or from nested blocks or from both.

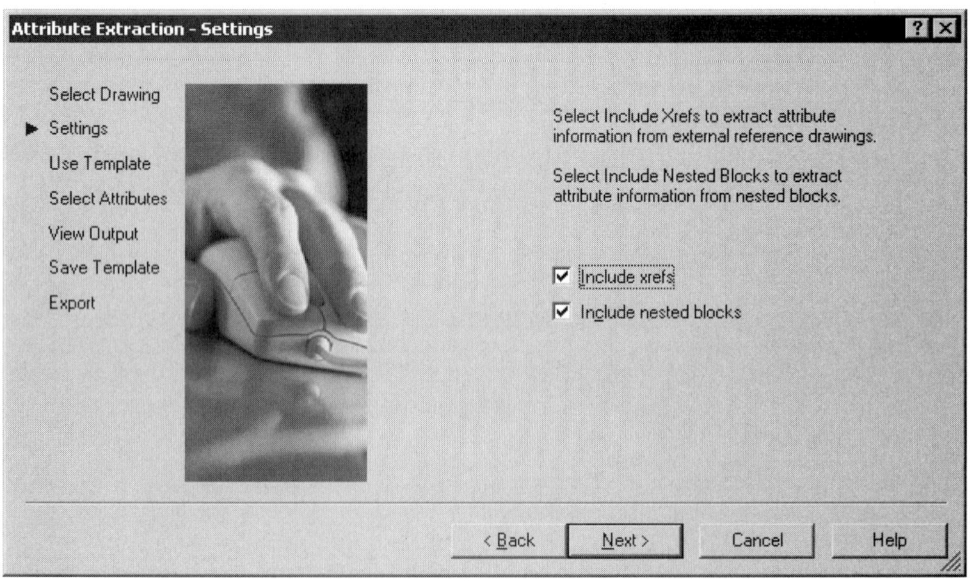

***Figure 15-20 Settings** page of the **Attribute Extraction** wizard*

Use Template

The **Use Template** page (Figure 15-21) allows you to use the block attribute settings from

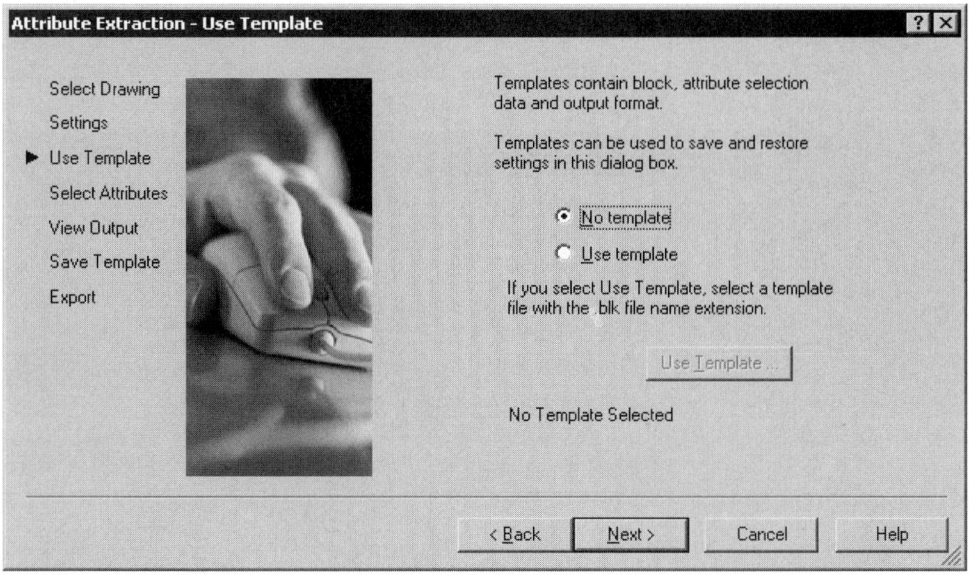

***Figure 15-21 Use Template** page of the **Attribute Extraction** wizard*

ones that are saved in the template file with an extension *.blk*. This page of the wizard has two radio buttons: **No template** and **Use template**. If you select the **Use template** button, you are allowed to select a template file with an extension *.blk*. A template file is a text file that allows you to specify the attribute values you want to extract and the information you want to retrieve about the block. It also lets you format the display of the extracted data. The file can be created by using any text editor, such as Notepad, Windows Write, or WordPad. You can also use a word processor or a database program to write the file.

Select Attributes

This page (Figure 15-22) allows you to select the blocks, attributes, and their information. In the **Blocks** area, select the blocks that have attributes you want to extract. In the **Attributes for block** area, select the attributes you want to extract. If you want to assign block or attribute aliases, use the corresponding column to enter the aliases. You can select the individual boxes to get this information in the exported file (output file), which is in text format.

View Output

The **View Output** page allows to preview the block attributes to be extracted. This is shown in the list (Figure 15-23). The list displays the attributes currently selected for extraction. This page shows two different views of the information listed. This can be seen by selecting the **Alternate View** button. First view displays all attributes of the block with block name and alternate view displays the values of each block attribute with block name. In exported files, blocks with the same name but different attributes are renamed, which are found in multiple drawings or in xrefs. After renaming, the block name is suffixed with a tilde character (~).

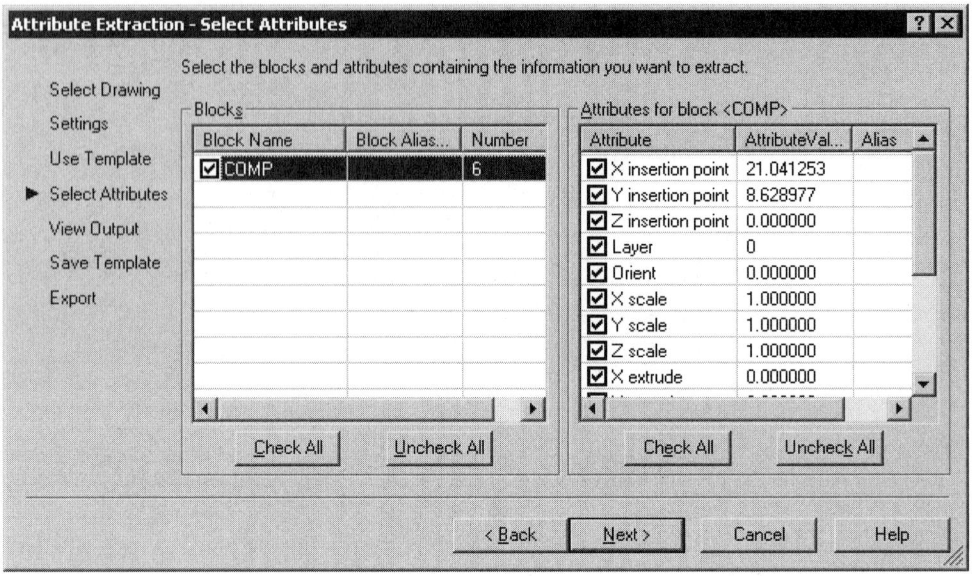

Figure 15-22 ***Select Attributes*** *page of the* ***Attribute Extraction*** *wizard*

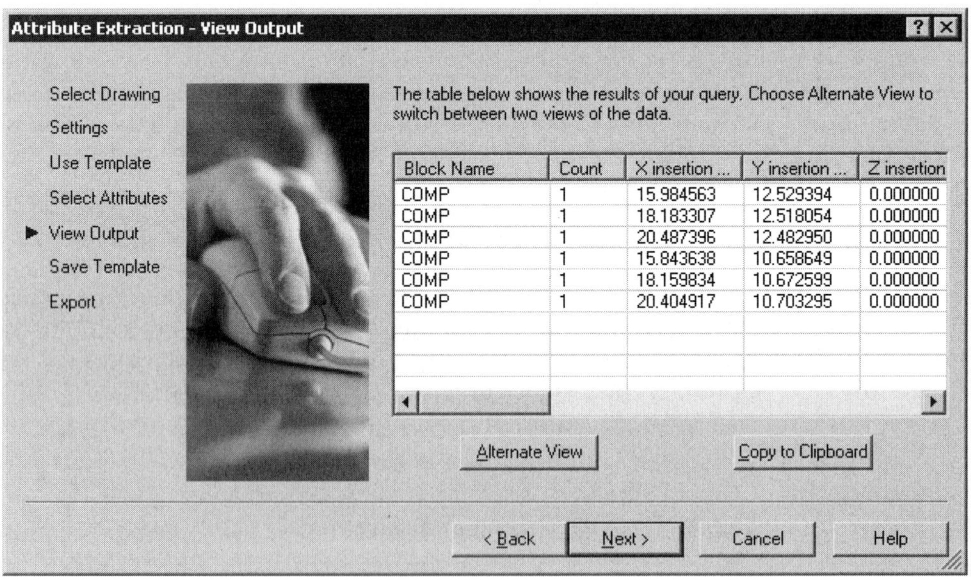

Figure 15-23 ***View Output*** *page of the* ***Attribute Extraction*** *wizard*

Save Template

This page of the wizard (Figure 15-24) allows you to save the attribute extraction settings that were made to a template file with *.blk* extension. This command can be used later as a template when you again invoke this command.

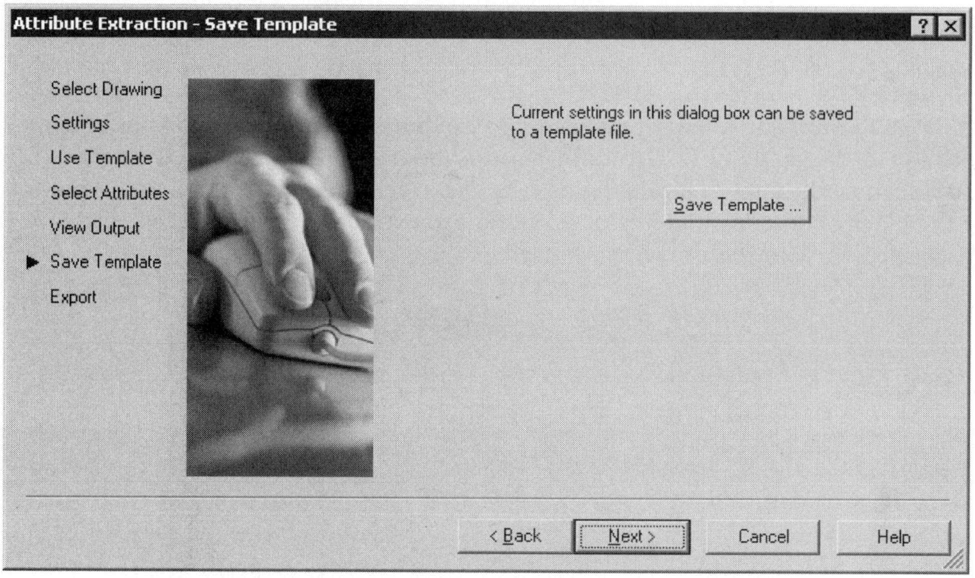

Figure 15-24 ***Save Template*** *page of the* ***Attribute Extraction*** *wizard*

Export

This page (Figure 15-25) allows you to give a name to the attribute extraction file and export the attribute information to the specified file. The file formats available in the **File Type** drop-down list are **CSV (Comma delimited) (*.csv)**, **Tab Delimited File (*.txt)**, and **Microsoft Excel (*.xls)**.

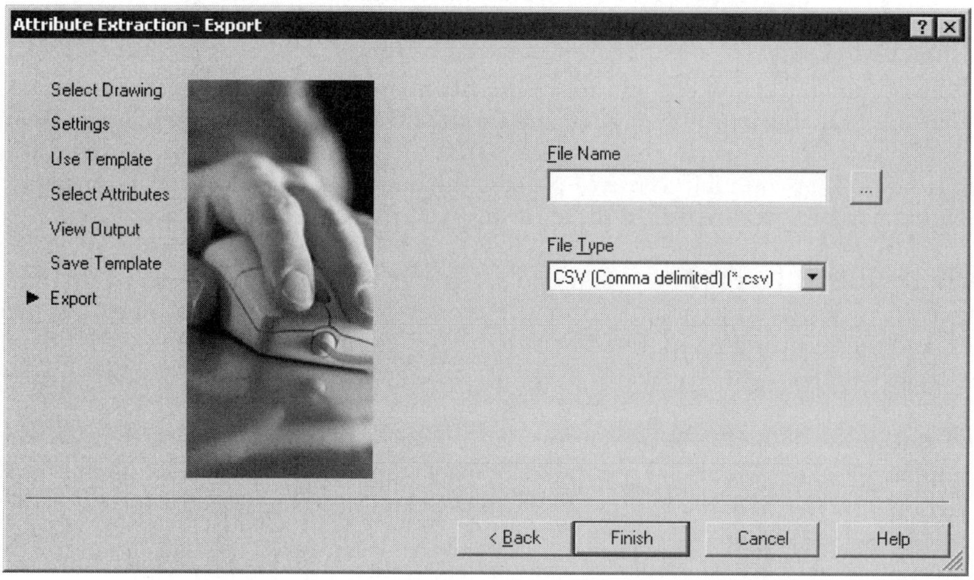

Figure 15-25 ***Export*** *page of the* ***Attribute Extraction*** *wizard*

Chapter 15

ATTEXT Command for Attribute Extraction

Command:	ATTEXT

The **ATTEXT** command allows you to use the **Attribute Extraction** dialog box (Figure 15-26) for extracting the attributes. The information about the **File format**, **Template file**, and **Output** file must be entered in the dialog box to extract the defined attribute. Also, you must select the blocks whose attribute values you want to extract. If you do not specify a particular block, all the blocks in the drawing are used.

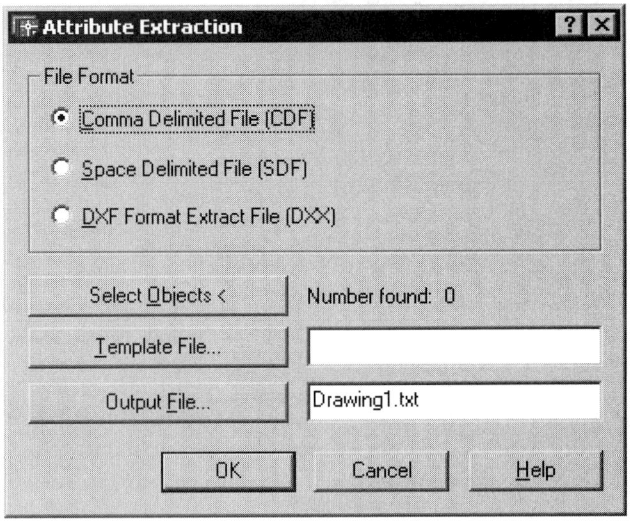

***Figure 15-26 Attribute Extraction** dialog box*

File Format Area

This area of the dialog box lets you select the file format. You can select either of the three radio buttons available in this area. They are **Comma Delimited File (CDF)**, **Space Delimited File (SDF)**, and **DXF Format Extract File (DXX)**. The format selection is determined by the application that you plan to use to process the data. Both CDF and SDF formats can be used with database software. All of these formats are printable.

Comma Delimited File (CDF). When you select this radio button, the extracted attribute information is displayed in a CDF format. Here, each character field is enclosed in single quotes, and the records are separated by a delimiter (comma by default). A CDF file is a text file with the extension *.txt*. Of the three formats available, this is the most cumbersome.

Space Delimited File (SDF). In SDF format, the records are of fixed width as specified in the template file. The records are separated by spaces and the character fields are not enclosed in single quotes. The SDF file is a text file with the extension *.txt*. This file format is the most convenient and easy to use.

DXF Format Extract File (DXX). If you select this file format, you will notice that the **Template**

File button and edit box in the **Attribute Extraction** dialog box are not available. This is because extraction in this file format does not require any template. The file created by this option contains only block references, attribute values, and end-of-sequence objects. This is the most complex of the three file formats available and is related to programming. The extension of these files is *.dxx*.

Select Objects

Select the blocks with attributes whose attribute information you want to extract. You can use any object selection method. Once you have selected the blocks that you want to use for attribute extraction, right-click or press ENTER. The **Attribute Extraction** dialog box is redisplayed and the number of objects you have selected is displayed adjacent to **Number found**.

Template File

When you choose the **Template File** button, the **Template File** dialog box is displayed, where you are allowed to select a template file that has been defined already. After you have selected a template file, choose the **Open** button to return to the **Attribute Extraction** dialog box. The name of the selected file is displayed in the **Template File** edit box. The template file is saved with the extension of the file as *.txt*. The following are the fields that you can specify in a template file (the comments given on the right are for explanation only; they must not be entered with the field description):

```
BL:LEVEL        Nwww000    (Block nesting level)
BL:NAME         Cwww000    (Block name)
BL:X            Nwwwddd    (X coordinate of block insertion point)
BL:Y            Nwwwddd    (Y coordinate of block insertion point)
BL:Z            Nwwwddd    (Z coordinate of block insertion point)
BL:NUMBER       Nwww000    (Block counter)
BL:HANDLE       Cwww000    (Block's handle)
BL:LAYER        Cwww000    (Block insertion layer name)
BL:ORIENT       Nwwwddd    (Block rotation angle)
BL:XSCALE       Nwwwddd    (X scale factor of block)
BL:YSCALE       Nwwwddd    (Y scale factor of block)
BL:ZSCALE       Nwwwddd    (Z scale factor of block)
BL:XEXTRUDE     Nwwwddd    (X component of block's extrusion direction)
BL:YEXTRUDE     Nwwwddd    (Y component of block's extrusion direction)
BL:ZEXTRUDE     Nwwwddd    (Z component of block's extrusion direction)
Attribute tag              (The tag name of the block attribute)
```

The extract file may contain several fields. For example, the first field might be the item name and the second field might be the price of the item. Each line in the template file specifies one field in the extract file. Any line in a template file consists of the name of the field, the width of the field in characters, and its numerical precision (if applicable). For example:

```
ITEM        N015002
BL:NAME     C015000
```

Where **BL:NAME** ------ Field name
Blankspaces --- Blank spaces (must not include the tab character)
C ------------------ Designates a character field
N ------------------ Designates a numerical field
015 --------------- Width of field in characters
002 --------------- Numerical precision

BL:NAME or **ITEM** — Indicates the field names; can be of any length.

C — Designates a character field; that is, the field contains characters or it starts with characters. If the file contains numbers or starts with numbers, then C will be replaced by N. For example, **N015002**.

015 — Designates a field that is fifteen characters long.

002 — Designates the numerical precision. In this example, the numerical precision is 2, or two places following the decimal. The decimal point and the two digits following the decimal are **included in the field width**. In the next example, (000), the numerical precision, is not applicable because the field does not have any numerical value (the field contains letters only).

After creating a template file, when you choose the **Template File** button, the **Template File** dialog box is displayed, where you can browse and select a template file (Figure 15-27).

Note

You can put any number of spaces between the field name and the character C or N (ITEM N015002). However, you must not use the tab characters. Any alignment in the fields must be done by inserting spaces after the field name.

In the template file, a field name must not appear more than once. The template file name and the output file name must be different.

The template file must contain at least one field with an attribute tag name because the tag names determine which attribute values are to be extracted and from which blocks. If several blocks have different block names but the same attribute tag, AutoCAD will extract attribute values from all selected blocks. For example, if there are two blocks in the drawing with the attribute tag PRICE, then when you extract the attribute values, AutoCAD will extract the value from both blocks (if both blocks were selected). To extract the value of an attribute, the tag name must match the field name specified in the template file. AutoCAD automatically converts the tag names and the field names to uppercase letters before making a comparison.

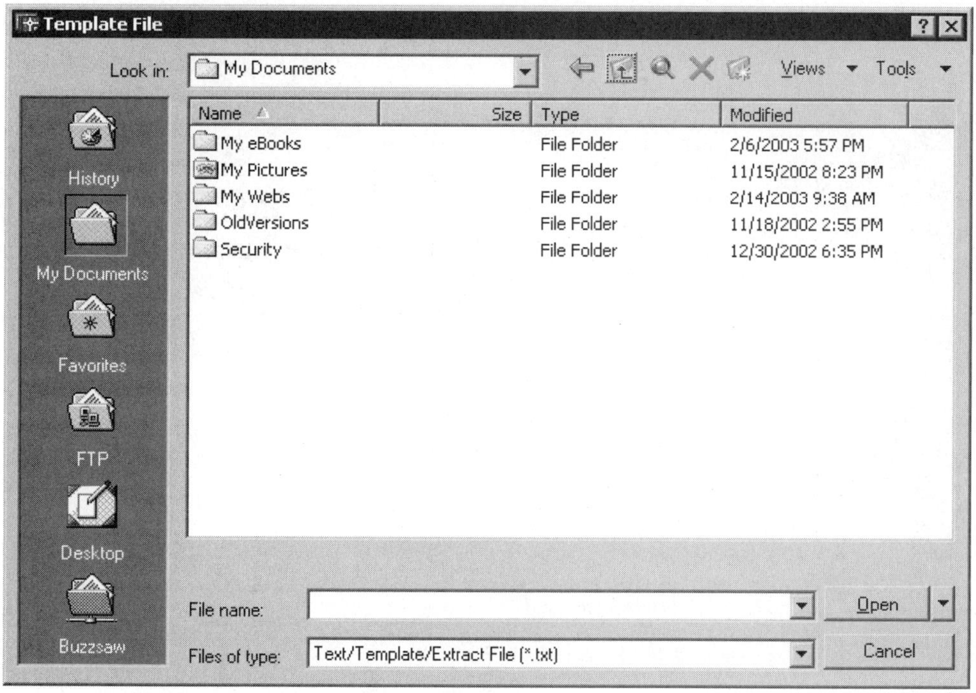

***Figure 15-27 Template File** dialog box to select a template file*

Output File

When you choose the **Output File** button, the **Output File** dialog box is displayed. You can select an existing file here, if you want the extracted or output file to be saved as an existing file. You can enter a name in the **File Name** edit box in the **Output File** dialog box (Figure 15-28) and then choose the **Save** button, if you want to save the output file as a new file. By default, the output file has the same name as the drawing name. For example, a drawing named *Drawing1.dwg* will have an output file by the name *Drawing1.txt* by default. Once a name for the output file is specified, it is displayed in the **Output File** edit box in the **Attribute Extraction** dialog box. You can also enter the file name in this edit box. As discussed earlier, AutoCAD appends *.txt* file extension for CDF or SDF files and *.dxx* file extension for DXF files.

Note

*You can also use the **-ATTEXT** command to extract attributes using the command line. Here, you are prompted to specify a file format for the extract information and you can also select specific blocks to extract their attribute information. You can specify a template file using the **Select Template File** dialog box and an extract file using the **Create extract file** dialog box.*

Example 3 *General*

In this example, you will extract the attribute values that were defined in Example 2. Extract the values of MAKE, PROCESSOR, HD, and RAM. These attribute values must be saved in a **Tab Delimited File** named TEST and arranged as shown in the following table.

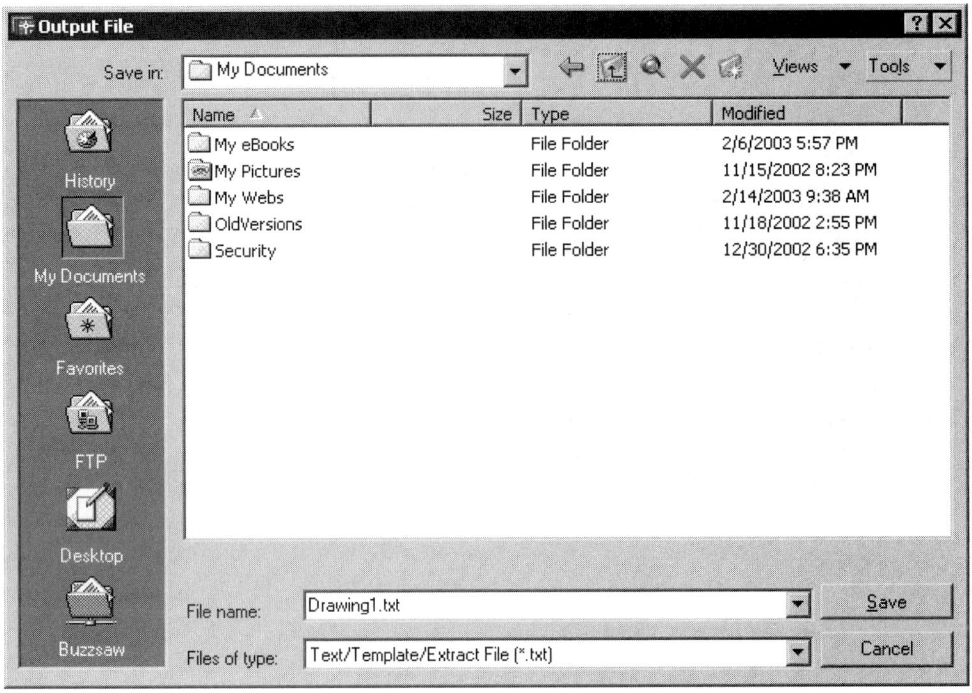

Figure 15-28 ***Output File*** *dialog box to save an output file*

COMP	1	Gateway	486-60	150 MB	16 MB	Computer
COMP	1	Zenith	486-30	100 MB	32 MB	Computer
COMP	1	IBM	386-30	80 MB	8 MB	Computer
COMP	1	Del	586-60	450 MB	64 MB	Computer
COMP	1	CAD-CIM	Pentium-90	100 Min	32 MB	Computer
COMP	1	CAD-CIM	Unknown	600 MB	Standard	Computer

1. Open the drawing you saved in Example 2.

2. Use the **EATTEXT** command to invoke the **Attribute Extraction** wizard. Select the **Select Objects** radio button from the **Select Drawing** page. Choose **Select Objects** button and select the objects in the current drawing whose attributes have to be extracted. After selecting, choose the **Next** button.

3. In the **Settings** page, clear both the check boxes of Xrefs and Nested blocks since there are no external references and nested blocks. Choose the **Next** button.

4. In the **Use Template** page, select the **No template** radio button and choose **Next**.

5. In the **Select Attributes** page, select the check box named **COMP** from the **Blocks** area. In the attribute list, clear all the check boxes and then select the **MAKE**, **PROCESSOR**, **HD**, **RAM**, and **ITEM** check boxes. Choose the **Next** button.

6. The **View output** page shows the result of your query. Here choose the view you want for the data to be displayed in an exported file. Choose the **Next** button.

7. In the **Save template** page, choose the **Next** button.

8. In the **Export** page, enter the name of the output file, **TEST**, and save it in **Tab Delimited File** format. Choose the **Finish** button.

9. You can view the resultant output file in the Notepad. Open the Notepad using the **Start** menu and then specify the location of the *Test.txt* file. The output file will be similar to the file shown at the beginning of Example 3.

CONTROLLING ATTRIBUTE VISIBILITY

Menu:	View > Display > Attribute Display
Command:	ATTDISP

The **ATTDISP** command allows you to change the visibility of all attribute values. Normally, the attributes are visible unless they are defined invisible by using the **Invisible** mode. The invisible attributes are not displayed, but they are a part of the block definition. You can select any one of the options of the **ATTDISP** command to turn the display of the attributes completely on or off. You can also select the **Normal** option, where the attributes that were created using the **Invisible** mode continue to be invisible. The options can be selected from the **View** > **Display** > **Attribute Display** cascading menu or can be entered in the command line. The prompt sequence for the command is given next.

Command: **ATTDISP**
Enter attribute visibility setting [Normal/ON/OFF] <Normal>: *Specify an option and press ENTER.*

When you select **ON**, all attribute values will be displayed, including the attributes that are defined with the **Invisible** mode. If you select **OFF**, all attribute values will become invisible. Similarly, if you select **N** (**Normal**), AutoCAD will display the attribute values the way they are defined, that is, the attributes that were defined invisible will stay invisible and the attributes that were defined visible are visible.

In Example 2, the RAM attribute was defined with the Invisible mode. Therefore, the RAM values are not displayed with the block. If you want to make the RAM attribute values visible (Figure 15-29), choose **On** from the **View > Display > Attribute Display** menu.

Tip

After you have defined the attribute values and saved them with the block definition, it may be a good idea to use the ***Off*** *option of the* ***ATTDISP*** *command. By doing this, the drawing is simplified and also regeneration time is reduced.*

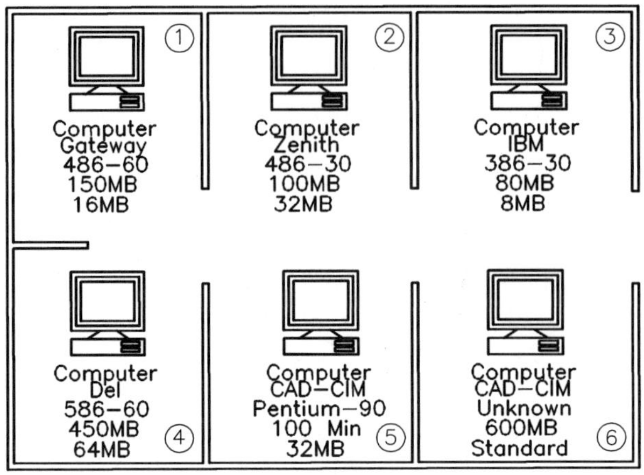

Figure 15-29 *Using the **ATTDISP** command to make the RAM attribute values visible*

EDITING BLOCK ATTRIBUTES

Toolbar:	Modify II > Edit Attribute
Menu:	Modify > Object > Attribute > Single
Command:	EATTEDIT

The block attributes can be edited using the **Enhance Attribute Editor** or the **ATTEDIT** command.

Editing Attributes Using Enhanced Attribute Editor

The **EATTEDIT** command allows you to edit the block attribute values through the **Enhanced Attribute Editor** dialog box (Figure 15-30). When you invoke this command, AutoCAD prompts you to select the block whose values you want to edit. After selecting the block, the **Enhanced Attribute Editor** dialog box is displayed. This dialog box allows you to change the **Attributes, Text Options**, and **Properties** of the block attribute. This dialog box can also be invoked by double-clicking on the block or the attribute in the current drawing. The **Enhanced Attribute Editor** dialog box displays all the attributes of the selected block and their properties. You can change the properties of the attribute you want. If an attribute has been defined with **Constant** mode, it is not displayed in the dialog box because a constant attribute value cannot be edited. The dialog box displays the following tabs.

Attribute Tab

The options under this tab are used to change the value of the attribute. Also, here the **Mode** of the attribute is not displayed. The other attributes of the block such as **Tag**, **Prompt**, and **Value** are displayed as shown in Figure 15-30.

Text Options Tab

Under this tab, the options related to the text of the attributes can be modified. This includes

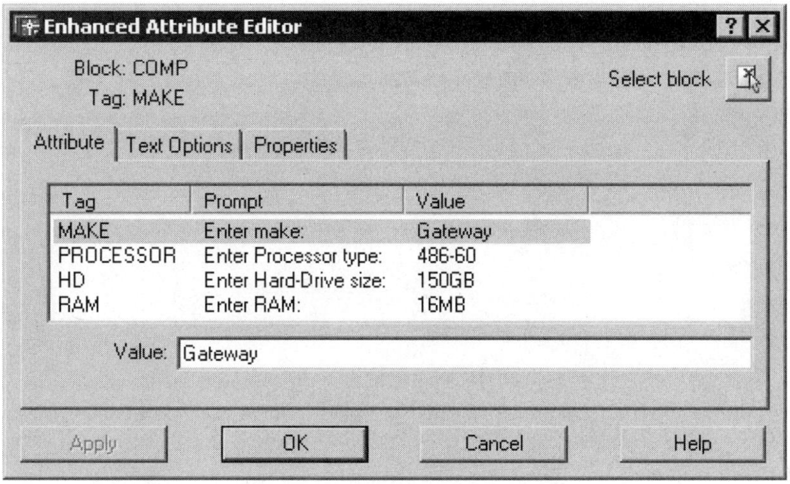

***Figure 15-30 Enhanced Attribute Editor** dialog box*

Text Style, **Justification**, **Height**, **Rotation**, **Backwards**, **Upside down**, and **Width Factor** as shown in Figure 15-31.

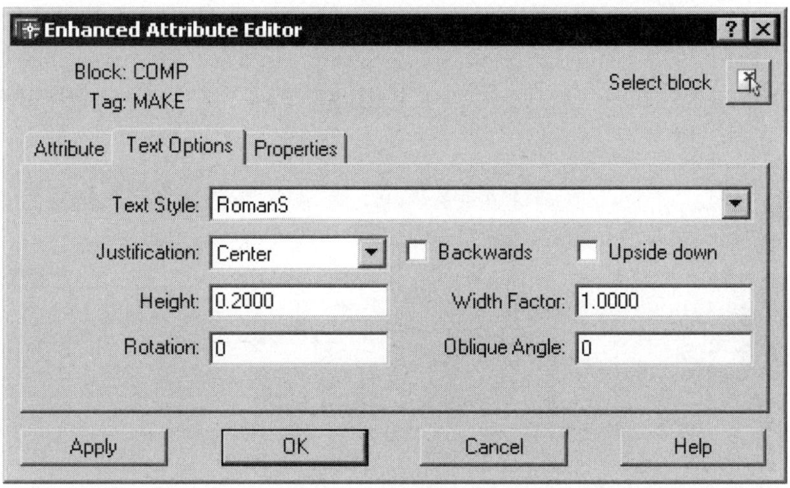

Figure 15-31 Text Options** in **Enhance Attribute Editor

Properties Tab

This tab can be used to edit the properties of the attributes such as **Layer**, **Linetype**, **Color**, **Lineweight**, and **Plotstyle**. Choose the **Properties** button of **Enhanced Attribute Editor** to display the dialog box as shown in Figure 15-32.

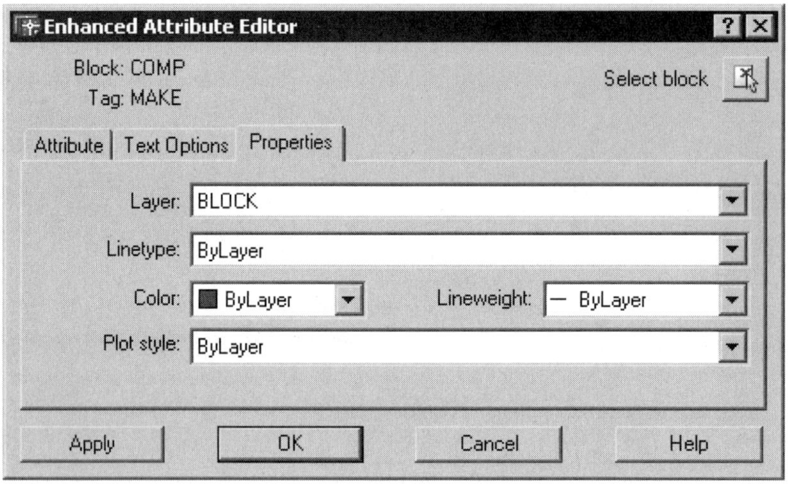

Figure 15-32 ***Properties*** *under* ***Enhanced Attribute Editor***

Editing Attributes Using the Edit Attributes Dialog Box

Command: ATTEDIT

The **ATTEDIT** command allows you to edit the block attribute values through the **Edit Attributes** dialog box. When you invoke this command, AutoCAD prompts you to select the block whose values you want to edit. After selecting the block, the **Edit Attributes** dialog box (Figure 15-33) is displayed.

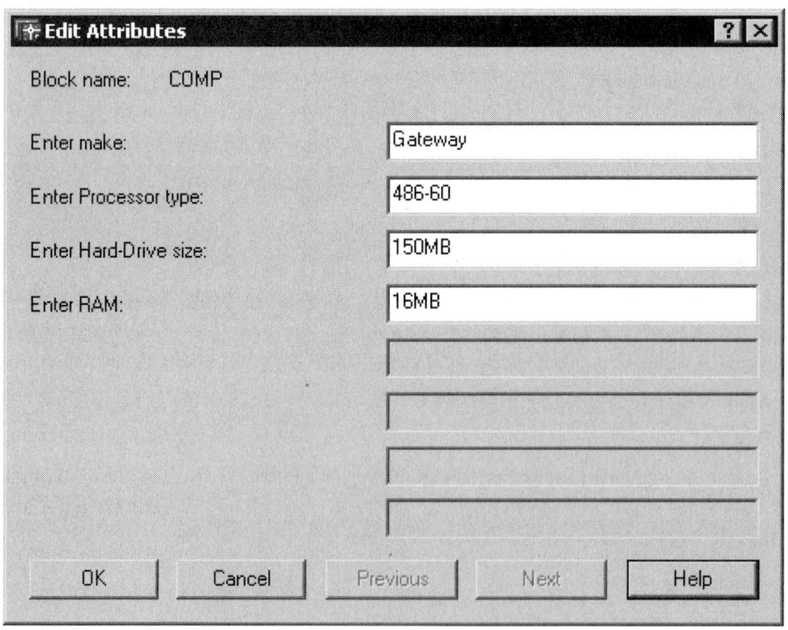

Figure 15-33 *Editing attribute values using the* ***Edit Attributes*** *dialog box*

The dialog box is similar to the **Enter Attributes** dialog box and shows the prompts and the attribute values of the selected block. If an attribute has been defined with the **Constant** mode, it is not displayed in the dialog box because a constant attribute value cannot be edited. To make any changes, select the existing value and enter a new value in the corresponding edit box. After you have made the modifications, choose the **OK** button. The attribute values are updated in the selected block.

If a selected block has no attributes, AutoCAD will display the alert message **That block has no editable attributes**. Similarly, if the selected object is not a block, AutoCAD again displays the alert message **That object is not a block**.

Note

*You cannot use the **ATTEDIT** command to perform global editing of attribute values, or to modify position, height, or style of the attribute value. The global editing can be performed using the **-ATTEDIT** command, which is discussed in the next section.*

Example 4 *General*

In this example, you will use the **EATTEDIT** command to change the attribute of the first computer (150 MB to 2.1 GB), which is located in Room-1.

1. Open the drawing that was created in Example 2. The drawing has six blocks with attributes. The name of the block is COMP, and it has five defined attributes, one of them invisible. Zoom in so that the first computer is displayed on the screen (Figure 15-34).

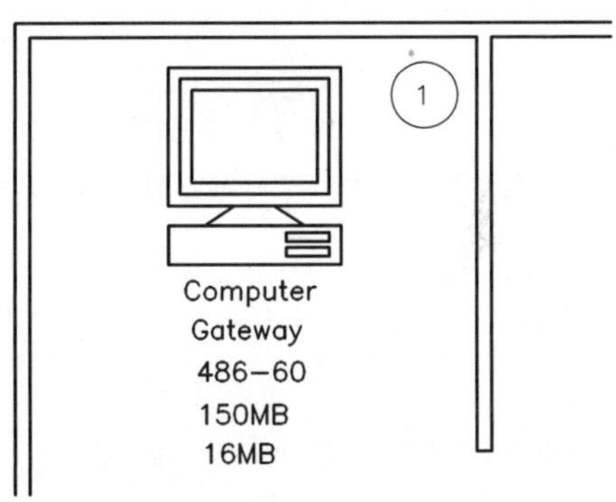

Figure 15-34 *Zoomed view of the first computer*

2. Invoke the **EATTEDIT** command. AutoCAD will prompt you to select a block. Select the computer located in Room-1.

   AutoCAD will display the **Enhanced Attribute Editor** dialog box (Figure 15-30) that shows the attribute prompts and the attribute values.

3. Edit the values, select **Apply**, and choose the **OK** button in the dialog box. When you exit the dialog box, the attribute values are updated.

Tip

*You can also edit attributes with the **FIND** command. This command can be invoked by choosing the **Find and Replace** button in the **Standard** toolbar or by entering **FIND** at the Command prompt. When you invoke this command, the **Find and Replace** dialog box is displayed.*

*When you know that a set of attributes in a drawing may have to be changed in the future, you can make a group out of them by using the **GROUP** command. Then later you can choose the **Select objects** button in the **Find and Replace** dialog box and enter G at the **Select objects** prompt and enter the name of the group. All the objects in the group will get selected.*

Global Editing of Attributes

Menu:	Modify > Object > Attribute >Global
Command:	-ATTEDIT

The **-ATTEDIT** command allows you to edit the attribute values independently of the blocks that contain the attribute reference. For example, if there are two blocks, COMPUTER and TABLE, with the attribute value PRICE, you can globally edit this value (PRICE) independently of the block that references these values. You can also edit the attribute values one at a time. For example, you can edit the attribute value (PRICE) of the block TABLE without affecting the value of the other block, COMPUTER. When you enter the **-ATTEDIT** command, AutoCAD displays the following prompt:

Command: **-ATTEDIT**
Edit attributes one at a time? [Yes/No]: N
Performing global editing of attribute values

If you enter **N** at this prompt, it means that you want to do the global editing of the attributes. However, you can restrict the editing of attributes by block names, tag names, attribute values, and visibility of attributes on the screen.

Editing Visible Attributes Only

After you select global editing, AutoCAD will display the following prompt:

Edit only attributes visible on screen? [Yes/No] <Y>: **Y**

If you enter Y at this prompt, AutoCAD will edit only those attributes that are visible and displayed on the screen. The attributes might not have been defined with the Invisible mode, but if they are not displayed on the screen, they are not visible for editing. For example, if you zoom in, some of the attributes may not be displayed on the screen. Since the attributes are not displayed on the screen, they are invisible and cannot be selected for editing.

Editing All Attribute

If you enter N at the previously mentioned prompt, AutoCAD flips from graphics to text screen and displays the following message on the screen:

Drawing must be regenerated afterwards.

Now, AutoCAD will edit all attributes even if they are not visible or displayed on the screen. Also, changes that you make in the attribute values are not reflected immediately. Instead, the attribute values are updated and the drawing is regenerated after you are done with the command.

Editing Specific Blocks

Although you have selected global editing, you can confine the editing of attributes to specific blocks by entering the block name at the prompt. For example

Enter block name specification <*>: **COMP**

When you enter the name of the block, AutoCAD will edit the attributes that have the given block (COMP) reference. You can also use the wild-card characters to specify the block names. If you want to edit attributes in all blocks that have attributes defined, press ENTER.

Editing Attributes with Specific Attribute Tag Names

Like blocks, you can confine attribute editing to those attribute values that have the specified tag name. For example, if you want to edit the attribute values that have the tag name MAKE, enter the tag name at the following AutoCAD prompt.

Enter attribute tag specification <*>: **MAKE**

When you specify the tag name, AutoCAD will not edit attributes that have a different tag name, even if the values being edited are the same. You can also use the wild-card characters to specify the tag names. If you want to edit attributes with any tag name, press ENTER.

Editing Attributes with a Specific Attribute Value

Like blocks and attribute tag names, you can confine attribute editing to a specified attribute value. For example, if you want to edit the attribute values that have the value 100 MB, enter the value at the following AutoCAD prompt.

Enter attribute value specification <*>: **100MB**

When you specify the attribute value, AutoCAD will not edit attributes that have a different value, even if the tag name and block specification are the same. You can also use the wild-card characters to specify the attribute value. If you want to edit attributes with any value, press ENTER.

Sometimes the value of an attribute is null, and these values are not visible. If you want to select the null values for editing, make sure you have not restricted the global editing to visible attributes. To edit the null attributes, enter \ at the following prompt.

Enter attribute value specification <*>: \

After you enter this information, AutoCAD will prompt you to select the attributes. You can select the attributes by selecting individual attributes or by using one of the object selection options (Window, Crossing, or individually).

Select Attributes: *Select the attribute values parallel to the current UCS only.*

After you select the attributes, AutoCAD will prompt you to enter the string you want to change and the new string. A string is a sequence of consecutive characters. It could also be a portion of the text. AutoCAD will retrieve the attribute information, edit it, and then update the attribute values.

Enter string to change: *Enter the value which is to be modified.*
Enter new string: *Enter the new value.*

The following is the complete command prompt sequence of the **-ATTEDIT** command. It is assumed that the editing is global and for visible attributes only.

Command: **-ATTEDIT**
Edit attributes one at a time? [Yes/No] <Y>: N
Performing global editing of attribute values.
Edit only attributes visible on screen? [Yes/No] <Y>: N
Drawing must be regenerated afterwards.
Enter block name specification <*>: Enter
Enter attribute tag specification <*>: Enter
Enter attribute value specification <*>: Enter
Enter string to change: *Enter the value to be modified.*
Enter new string: *Enter the new value.*

Note

*AutoCAD regenerates the drawing at the end of the command automatically unless system variable **REGENAUTO** is off, which controls automatic regeneration of the drawing.*

*If you select an attribute defined with the **Constant** mode using the **-ATTEDIT** command, prompt displays 0 found, since attributes with constant mode is uneditable.*

Example 5 *General*

In this example, you will use the drawing from Example 2 to edit the attribute values that are **highlighted** in the following table. The tag names are given at the top of the table (ITEM, MAKE, PROCESSOR, HD, RAM). The RAM values are invisible in the drawing.

	ITEM	MAKE	PROCESSOR	HD	RAM
COMP	Computer	Gateway	486-60	150 MB	**16 MB**
COMP	Computer	Zenith	486-30	100 MB	**32 MB**
COMP	Computer	IBM	386-30	80 MB	**8 MB**
COMP	Computer	Del	586-60	450 MB	**64 MB**

COMP	Computer	**CAD-CIM**	Pentium-90	100 Min	**32 MB**
COMP	Computer	**CAD-CIM**	**Unknown**	600 MB	Standard

Make the following changes in the **highlighted** attribute values (Figure 15-35).

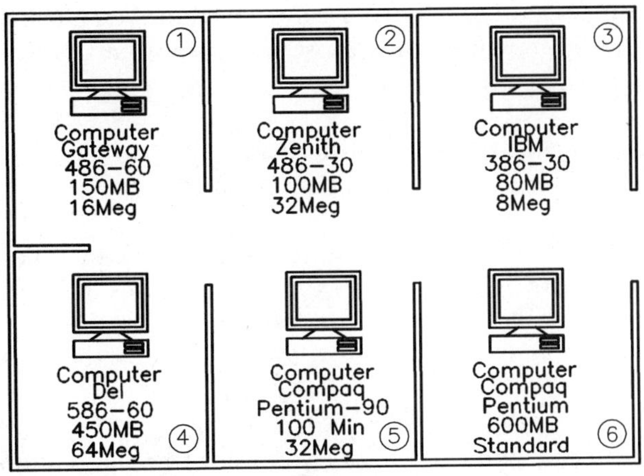

Figure 15-35 *Using* ***-ATTEDIT*** *to change the attribute values*

1. Change Unknown to Pentium.
2. Change CAD-CIM to Compaq.
3. Change MB to Meg for all attribute values that have the tag name RAM. (No changes should be made to the values that have the tag name HD.)

The following is the prompt sequence to change the attribute value from **Unknown** to **Pentium**.

1. Enter the **-ATTEDIT** command at the Command prompt. At the next prompt, enter **N** and press ENTER.

   Command: **-ATTEDIT**
   Edit attributes one at a time? [Yes/No] <Y>: **N**
   Performing global editing of attribute values.

2. We want to edit only those attributes that are visible on the screen, so press ENTER at the following prompt:

   Edit only attributes visible on screen? [Yes/No] <Y>: Enter

3. As shown in the table, the attributes belong to a single block, COMP. In a drawing, there could be more blocks. To confine the attribute editing to the COMP block only, enter the name of the block (COMP) at the next prompt.

   Enter block name specification <*>: **COMP**

4. At the next two prompts, enter the attribute tag name and the attribute value specification. When you enter these two values, only those attributes that have the specified tag name and attribute value will be edited.

   Enter attribute tag specification<*>: **Processor**
   Enter attribute value specification<*>: **Unknown**

5. Next, AutoCAD will prompt you to select attributes. Use any object selection option to select all the blocks. AutoCAD will search for the attributes that satisfy the given criteria (attributes belong to the block COMP, the attributes have the tag name Processor, and the attribute value is Unknown). Once AutoCAD locates such attributes, they will be highlighted.

6. At the next two prompts, enter the string you want to change, and then enter the new string.

   Enter string to change: **Unknown**
   Enter new string: **Pentium**

7. The following is the Command prompt sequence to change the make of the computers from **CAD-CIM** to **Compaq**.

   Command: **-ATTEDIT**
   Edit attributes one at a time? [Yes/No] <Y>: **N**
   Performing global editing of attribute values.
   Edit only attributes visible on screen? [Yes/No] <Y>: Enter
   Enter block name specification <*>: **COMP**
   Enter attribute tag specification <*>: **MAKE**
   Enter attribute value specification <*>: Enter
   Select Attributes: *Use any selection method to select the attributes.*
   n attributes selected.
   Select Attributes: Enter
   Enter string to change: **CAD-CIM**
   Enter new string: **Compaq**

8. The following is the Command prompt sequence to change **MB** to **Meg**.

   Command: **-ATTEDIT**
   Edit attributes one at a time? [Yes/No] <Y>: **N**
   Performing global editing of attribute values.

At the next prompt, you must enter **N** because the attributes you want to edit (tag name, RAM) are not visible on the screen.

Edit only attributes visible on screen? [Yes/No] <Y>: **N**
Drawing must be regenerated afterwards.
Enter block name specification <*>: **COMP**

At the next prompt, about the tag specification, you must specify the tag name because the text string MB also appears in the hard drive size (tag name, HD). If you do not enter the tag name, AutoCAD will change all MB attribute values to Meg.

Enter attribute tag specification <*>: **RAM**
Enter attribute value specification <*>:
n Attributes selected
Enter string to change: **MB**
Enter new string: **Meg**

9. Choose **On** from the **View > Display > Attribute Display** menu to display the invisible attributes on the screen.

Note
*You can also use the **-ATTEDIT** command to edit the attribute values individually. Attribute value, string, position, height, angle, layer, and color can be changed using this command.*

Example 6 *General*

In this example, you will use the drawing in Example 5 to edit the attributes individually (Figure 15-36). Make the following changes in the attribute values.

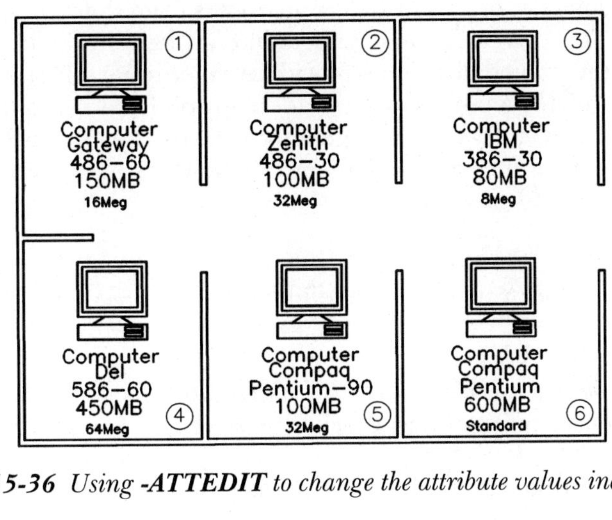

Figure 15-36 *Using **-ATTEDIT** to change the attribute values individually*

a. Change the attribute value 100 Min to 100 MB.

b. Change the height of all attributes with the tag name RAM to 0.075 units.

1. Load the drawing that you had saved in Example 5.
2. Double-click on the **100Min** attribute value in the drawing. The **Enhanced Attribute Editor** dialog box is displayed.
3. In the **Value** edit box, change **100 Min** to **100MB**. Choose **Apply** and then choose **OK**.
4. To change the height of attribute text, double-click on the value of **RAM** in the drawing. The **Enhanced Attribute Editor** dialog box is displayed.
5. In the dialog box, choose **Text Options** tab and change height to **0.075** in the **Height** edit box.
6. Choose **Apply** and then choose **OK**.
7. Repeat Steps 5 and 6 to change the height of other attribute values.

Tip

*When you are defining attributes and have certain attributes whose values are not known at that time, you should enter AAAA or something similar. Later on you can replace such text with the values that you have obtained, using the **ATTEDIT** or **-ATTEDIT** commands. This is easier than adding an attribute later after the block has already been defined.*

REDEFINING A BLOCK WITH ATTRIBUTES

If a block with attributes has been inserted a number of times in a drawing and you wish to now edit the block geometry, or add or delete attributes from them, it may take a lot of time. You can use the **ATTREDEF** command to redefine an existing block with attributes, and once redefined, all the copies of the block with attributes get updated automatically. But before using this command, you have to explode a copy of the block containing the attributes and make the changes or you may use new geometry. If you do not explode a block and try to use this command, an error message is displayed: **New block has no attributes**. The prompt sequence is as follows:

Command: **ATTREDEF**
Enter name of the block you wish to redefine: *Enter the name of the block to be redefined and press ENTER.*
Select objects for new Block...
Select objects: *Select the new block geometry with the attributes and press ENTER.*
Specify insertion base point of new Block: *Specify an insertion base point.*

After you select an insertion base point, the block is redefined and all the copies of the block in the drawing are updated. If you do not include an existing attribute when selecting objects for the new block, it will not be a part of the new block definition.

In-place Editing of Blocks with Attributes

Toolbar:	Refedit> Edit block or Xref
Menu:	Modify > Xref and Block Editing > Edit Reference In-Place
Command:	REFEDIT

The **REFEDIT** command has already been discussed earlier in Chapter 14, Working with Blocks. This command can also be used for in-place editing of blocks with attributes. After you invoke the **REFEDIT** command, select a block with attributes that you want to edit. The **Reference Edit** dialog box is displayed with the name of the block in the **Reference name** list box. To be able to edit attributes, you should select the **Display attribute definitions for editing** check box in the **Settings** tab of the dialog box and then choose **OK** to exit the dialog box. When you select this check box, it displays the attributes and makes them available for editing. At the next prompt, **Select nested objects:**, you can select any part of the block with attributes that need to be edited and press ENTER. If you want to edit some other objects that are part of the block, you can select them too. All the selected objects become a part of the working set. You will notice that when you select the block with attributes, the tags replace the assigned attribute values and objects in the drawing that were not selected appear faded.

You can now edit the block geometry or edit only the attributes. Enter the **DDEDIT** command at the Command prompt and select the attribute to be edited at the next prompt. The **Edit Attribute Definition** dialog box is displayed. Make changes in the values displayed in the **Tag**, **Prompt**, and **Default** edit boxes and choose **OK** to exit the dialog box. You can then select another attribute to edit and so on until you have made all the modifications in the attributes you intended and then press ENTER to exit the command. Now choose the **Save back changes to reference** button in the **Refedit** toolbar. An AutoCAD message box is displayed prompting you whether you want to save the changes made or cancel the command and discard the changes. Choose **OK** to save the completed changes and exit the **REFEDIT** command.

Once the block has been redefined, the blocks that already exist in the drawing are not updated, unless a change is made to a default value of an attribute that has been defined with the **Constant** mode. The modifications made using the **REFEDIT** command can be seen in the blocks inserted later.

Note

*You can also use the **-REFEDIT** command to change block attributes. The prompts are displayed on the command line. All the options available through the **REFEDIT** command are available here also. The **DDEDIT** command is used to make modifications in the attribute tags. Then, you can either choose the **Save back changes to reference** button in the **Refedit** toolbar or enter the **REFCLOSE** command to save the changes back to the block and exit the command.*

INSERTING TEXT FILES IN THE DRAWING

Toolbar:	Draw > Multiline Text
	Text > Multiline Text
Menu:	Draw > Text > Multiline Text
Command:	MTEXT

After you have extracted attribute information into a file, you may want to insert this text into a drawing. You can insert this text file in a drawing by using the **Import Text** option that is displayed when you right-click in the text window of the **Multiline Text Editor**. The **Multiline Text Editor** is displayed using the **MTEXT** command.

After you have invoked the **MTEXT** command, AutoCAD prompts you to enter the insertion point and other corner of the paragraph text box, within which the text file will be placed. After you specify these points, the **Multiline Text Editor** is displayed. To insert the text file *Test.txt* (created in Example 3), right-click in the text window of the **Multiline Text Editor** to display the shortcut menu. In this shortcut menu, choose **Import Text**. AutoCAD displays the **Select File** dialog box.

In the **Select File** dialog box, you can select the text file **Test** from the list box and then choose the **Open** button. The imported text is displayed in the text window of the **Multiline Text Editor** (Figure 15-37). Note that only ASCII files are properly interpreted. Now choose the **OK** button to get the imported text in the selected area on the screen (Figure 15-38).

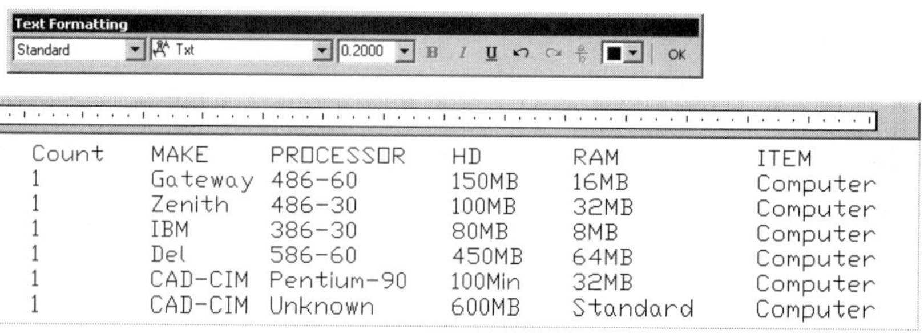

***Figure 15-37 Multiline Text Editor** displaying the imported text*

Block Name	Count	MAKE	PROCESSOR	HD	RAM	ITEM
COMP	1	Gateway	486-60	150MB	16MB	Computer
COMP	1	Zenith	486-30	100MB	32MB	Computer
COMP	1	IBM	386-30	80MB	8MB	Computer
COMP	1	Del	586-60	450MB	64MB	Computer
COMP	1	CAD-CIM	Pentium-90	100Min	32MB	Computer
COMP	1	CAD-CIM	Unknown	600MB	Standard	Computer

***Figure 15-38** Imported text file on the screen*

You can also use the **Multiline Text Editor** to change the text style, height, direction, width, rotation, line spacing, and attachment.

Self-Evaluation Test

Answer the following questions, and then compare your answers to the correct answers given at the end of this chapter.

1. Like a constant attribute, the **Preset** attribute cannot be edited. (T/F)
2. For tag names, any lowercase letters are automatically converted to uppercase. (T/F)
3. You can use the **EATTEDIT** command to modify the justification, height, or style of the attribute value. (T/F)
4. If you select the **On** option of the **ATTDISP** command, even the attributes defined with the **Invisible** mode are displayed. (T/F)
5. The entry in the **Value** edit box of the **Attribute Definition** dialog box defines the __________ of the specified attribute.
6. If you have selected the **Align** option from the **Justification** drop-down list, in the **Text Options** tab of the **Enhanced Attribute Editor** dialog box, the **Height** and **Rotation** edit boxes are __________.
7. You can use the __________, and __________ commands or the **PROPERTIES** palette to edit text or attribute definitions.
8. The default value of the **ATTDIA** variable is __________, which disables the dialog box.
9. In the __________ File, the records are not separated by a comma and the character fields are not enclosed in single quotes.
10. The __________ command is used to redefine an existing block with attributes in a drawing and once redefined, all the copies of the block in the drawing are updated.

Review Questions

Answer the following questions.

1. If you do not enter anything in the **Prompt** edit box, the entry made in the **Tag** edit box is used as the prompt. (T/F)
2. You can also use the **Find and Replace** dialog box to modify the attribute values. (T/F)
3. When using the **REFEDIT** command for in-place editing of attributes, the existing blocks in drawing do not get updated. Only subsequent insertions display the modified block. (T/F)

4. You can use ? and * in the string value. When these characters are used in string values, AutoCAD does not interpret them as wild-card characters. (T/F)

5. The template file name and the output file name can be the same. (T/F)

6. Not selecting any of the check boxes in the **Mode** area of the **Attribute Definition** dialog box displays all the prompts at the command line and the values will be visible on the screen. This is also referred to as which mode?

   (a) Formal (b) Normal
   (c) Abnormal (d) None of the above

7. Which of the following system variables when set to 0 will suppress the display of prompts for new values?

   (a) **ATTDIA** (b) **ATTREQ**
   (c) **ATTMODE** (d) **ATTDEF**

8. Selecting the **Off** option of which command turns off the visibility of all attribute values?

   (a) **ATTDISP** (b) **ATTEXT**
   (c) **ATTEDIT** (d) **ATTDEF**

9. AutoCAD regenerates the drawing at the end of the **-ATTEDIT** command, unless which of the following is turned off?

   (a) **AUTOSNAP** (b) **REGENAUTO**
   (c) **ATTDIA** (d) **ATTMODE**

10. If you need a leading blank space in the string to change, the character string must start with which one of the following characters?

    (a) space ( ) (b) backlash (\)
    (c) asterisk (*) (d) colon (:)

11. You can insert the text file in the drawing by choosing the __________ button in the **Multiline Text Editor** dialog box displayed when using the **MTEXT** command.

12. The function of the **Preset** option is ____.

13. If you select the **Constant** check box in the **Mode** area of the **Attribute Definition** dialog box, the **Prompt** edit box is __________.

14. You should select the __________ check box in the **Attribute Definition** dialog box to automatically place the subsequent attribute text just below the previously defined attribute.

15. The attribute extract file is saved as a _________ file.

Exercises

Exercise 1 *Electronics*

In this exercise, you will define the following attributes for a resistor and then create a block using the **BLOCK** command. The name of the block is RESIS.

Mode	Tag name	Prompt	Default value
Verify	RNAME	Enter name	RX
Verify	RVALUE	Enter resistance	XX
Verify, Invisible	RPRICE	Enter price	00

1. Draw the resistor as shown in Figure 15-39.

2. Enter **ATTDEF** at the AutoCAD Command prompt to invoke the **Attribute Definition** dialog box.

3. Define the attributes as shown in the preceding table, and position the attribute text as shown in Figure 15-39.

4. Use the **BLOCK** command to create a block. The name of the block is RESIS, and the insertion point of the block is at the left end of the resistor. When you select the objects for the block, make sure you also select the attributes.

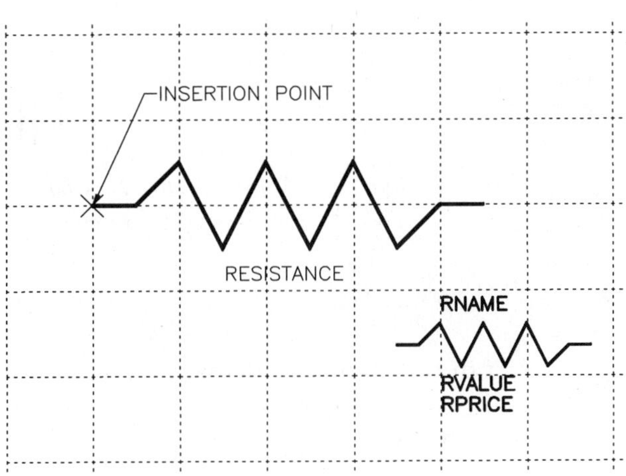

Figure 15-39 *Drawing of a resistor for Exercise 1*

Exercise 2 *Electronics*

In this exercise, you will use the **INSERT** command to insert the block that was defined in Exercise 1 (RESIS). The following is the list of the attribute values for the resistances in the electric circuit.

RNAME	RVALUE	RPRICE
R1	35	0.32
R2	27	0.25
R3	52	0.40
R4	8	0.21
RX	10	0.21

1. Draw the electric circuit diagram as shown in Figure 15-40 (assume the dimensions).
2. Set the system variable **ATTDIA** to 1. Use the **INSERT** command to insert the blocks, and define the attribute values in the **Enter Attributes** dialog box.
3. Repeat the **INSERT** command to insert other blocks, and define their attribute values as given in the table. Save the drawing as *attexr2.dwg* (Figure 15-41).

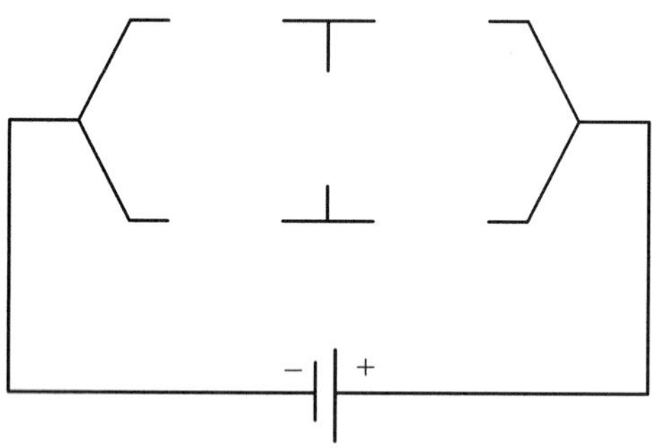

Figure 15-40 *Electric circuit diagram without resistors for Exercise 2*

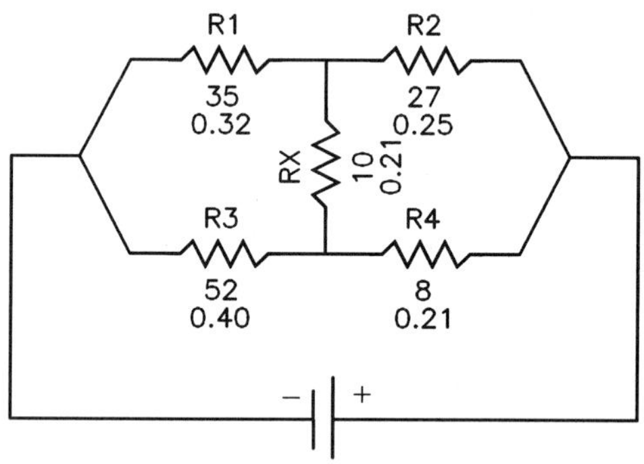

Figure 15-41 *Electric circuit diagram with resistors for Exercise 2*

Exercise 3 *Electronics*

In this example, you will extract the attribute values that were defined in Exercise 2. Extract the values of RNAME, RVALUE, and RPRICE. These attribute values must be saved in a Tab Delimited File format named **RESISLST** and arranged as shown in the following table.

RESIS	1	R1	35	0.32
RESIS	1	R2	27	0.25
RESIS	1	R3	52	0.40
RESIS	1	R4	8	0.21

RESIS	1	RX	10	0.21

1. Load the drawing **ATTEXR2** that you saved in Exercise 2.

2. Use the **EATTEXT** command to invoke the **Attribute Extraction** wizard, and select the **Select Objects** radio button. Now, choose **Select Objects** button and select the objects in the current drawing, whose attributes have to be extracted. Now, choose **Next** in the wizard.

3. On the **Settings** page, clear both the check boxes of **Xrefs** and **Nested blocks** and choose **Next**.

4. On the **Use Template** page, select the **No Template** radio button and choose **Next**.

5. On the **Select Attribute** page, under the Block Name heading, select the check box named **RESIS**. In the attribute list, clear all the check boxes and select **RNAME**, **RVALUE**, and **RPRICE** check boxes. Choose **Next**.

6. **View output** page shows the result of your query. Here choose the view you want the data to be displayed in exported file. Select the Next button.

7. Ignore the **Save template** page and choose **Next**.

8. On the **Export** page, enter the name of the output file, **RESISLST**, and save it in **Tab Delimited File** format. Choose **Finish**. Use Notepad to list the output file, **RESISLST**. The output file will be similar to the file shown at the beginning of Exercise 3.

Exercise 4 *Electronics*

In this exercise, you will use the **EATTEDIT** command to change the attributes of the resistances that are highlighted in the following table. You will also extract the attribute values and insert the text file in the drawing.

1. Load the drawing **ATTEXR2** that was created in Exercise 2. The drawing has five resistances with attributes. The name of the block is RESIS, and it has three defined attributes, one of them invisible.

2. Use the **EATTEDIT** command to edit the values that are **highlighted** in the following table.

RESIS	R1	**40**	0.32
RESIS	R2	**29**	0.25
RESIS	R3	52	**0.45**
RESIS	R4	8	**0.25**
RESIS	**R5**	10	0.21

3. Extract the attribute values, and write the values to a text file.

4. Use the **MTEXT** command to insert the text file in the drawing.

Exercise 5 *Electronics*

Use the information given in Exercise 3 to extract the attribute values, and write the data to the output file. The data in the output file should be Comma Delimited CDF. Use the **ATTEXT** and **-ATTEXT** commands to extract the attribute values.

Exercise 6 *Electronics*

In this exercise, you will create the blocks with the required attributes. Next, draw the circuit diagram shown in Figure 15-42 and then extract the attributes to create a bill of materials.

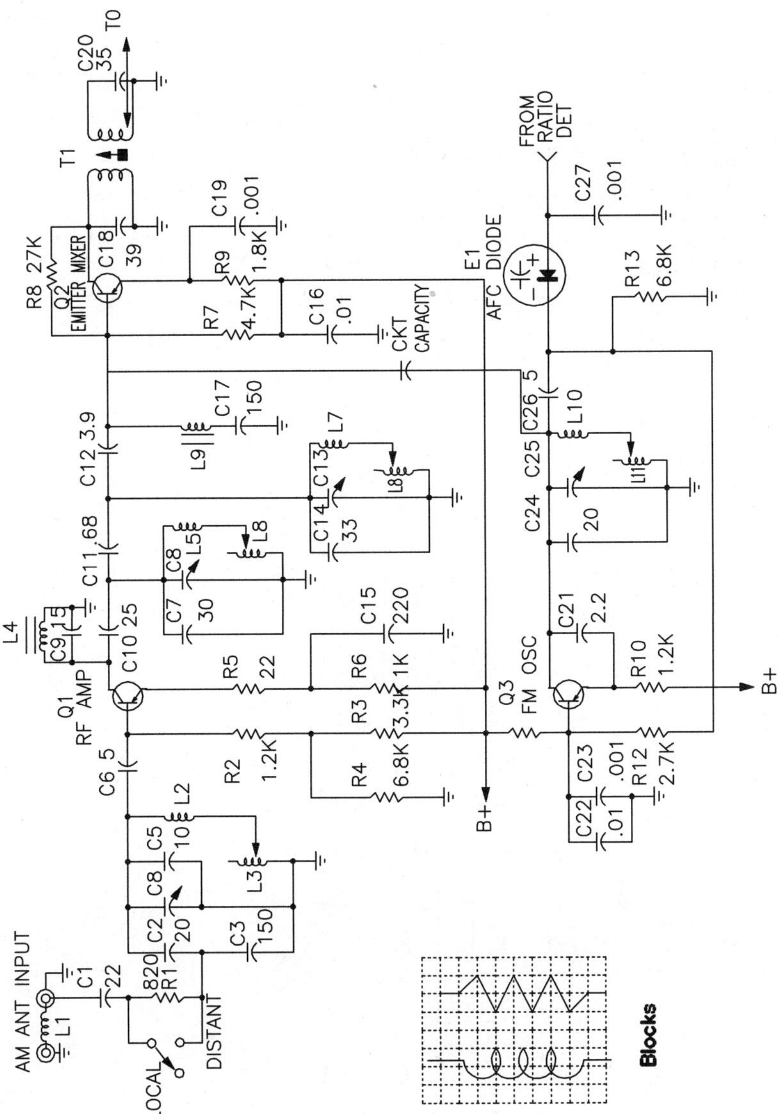

Figure 15-42 Drawing of the circuit diagram for Exercise 6

Exercise 7 — *Electronics*

In this exercise, you will create the blocks with the required attributes. Next, draw the circuit diagram shown in Figure 15-43 and then extract the attributes to create a bill of materials.

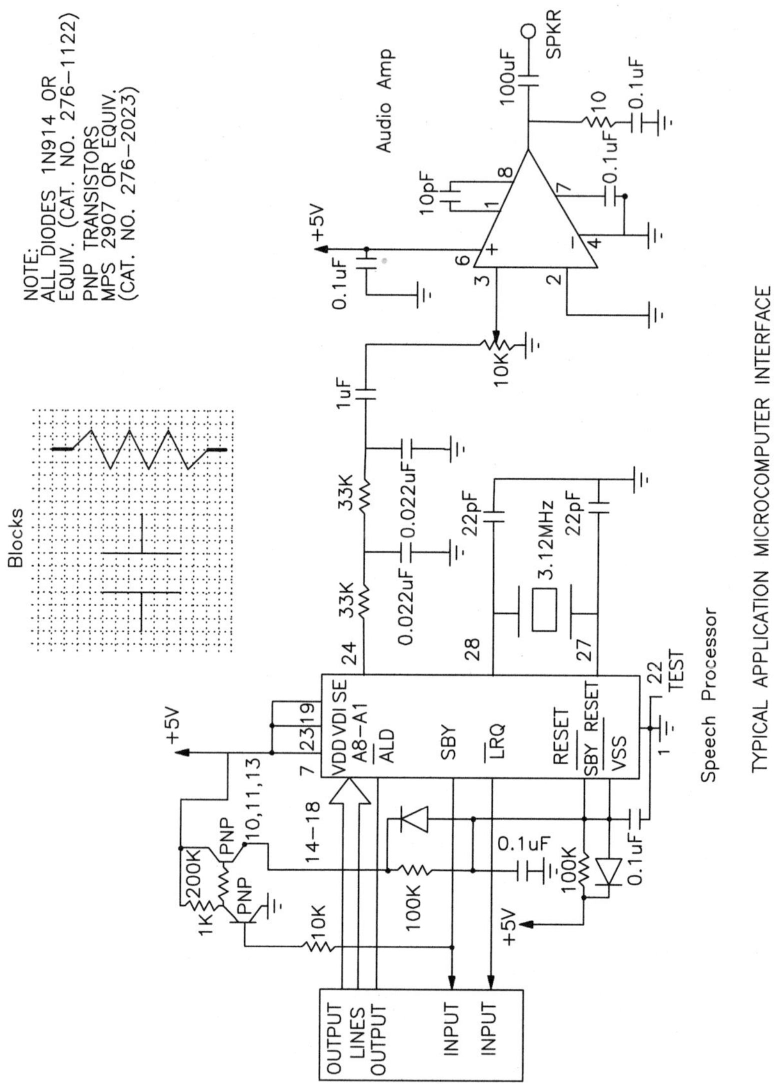

Figure 15-43 Drawing for Exercise 7

Answers to Self-Evaluation Test

1 - F, **2** - T, **3** - T, **4** - T, **5** - default value, **6** - disabled, **7** - **EATTEDIT**, **CHANGE**, **8** - 0, **9** - Space Delimited, **10** - **ATTREDEF**

Chapter 16

Understanding External References

Learning Objectives

After completing this chapter you will be able to:

- *Understand external references and their applications.*
- *Understand dependent symbols.*
- *Use the **XREF** command and its options.*
- *Use the **Attach**, **Unload**, **Reload**, **Detach**, and **Bind** options.*
- *Edit the path of an xref.*
- *Understand the difference between the **Overlay** and **Attachment** options and use the **XATTACH** command.*
- *Use the **XBIND** command to add dependent symbols.*
- *Use the **XCLIP** command to clip xref drawings.*
- *Understand demand loading.*
- *Use the **DESIGNCENTER** to attach a drawing as an xref.*
- *Use the **REFEDIT** command for In-place editing.*

EXTERNAL REFERENCES

The external reference feature allows you to reference an external drawing without making that drawing a permanent part of the existing drawing. For example, assume that we have an assembly drawing ASSEM1 that consists of two parts, SHAFT and BEARING. The SHAFT and BEARING are separate drawings created by two CAD operators or provided by two different vendors. We want to create an assembly drawing from these two parts. One way to create an assembly drawing is to insert these two drawings as blocks by using the **INSERT** command. Now assume that the design of BEARING has changed due to customer or product requirements. To update the assembly drawing, you have to make sure that you insert the BEARING drawing after the changes have been made. If you forget to update the assembly drawing, then the assembly drawing will not reflect the changes made in the piece part drawing. In a production environment, this could have serious consequences.

You can solve this problem by using the **external reference** facility, which lets you link the piece part drawings with the assembly drawing. If the xref drawings (piece part) get updated, the changes are automatically reflected in the assembly drawing. This way, the assembly drawing stays updated no matter when the changes were made in the piece part drawings. There is no limit to the number of drawings that you can reference. You can also have **nested references**. For example, the piece part drawing BEARING could be referenced in the SHAFT drawing, then the SHAFT drawing could be referenced in the assembly drawing ASSEM1. When you open or plot the assembly drawing, AutoCAD automatically loads the referenced drawing SHAFT and the nested drawing BEARING. When using external references, several people working on the same project can reference the same drawing and all the changes made are displayed everywhere the particular drawing is being used.

If you use the **INSERT** command to insert the piece parts, the piece parts become a permanent part of the drawing, and therefore, the drawing has a certain size. However, if you use the external reference feature to link the drawings, the piece part drawings are not saved with the assembly drawing. AutoCAD only saves the reference information with the assembly drawing; therefore, the size of the drawing is minimized. Like blocks, the xref drawings can be scaled, rotated, or positioned at any desired location, but they cannot be exploded. You can also use only a part of the Xref by making clipped boundary of Xrefs.

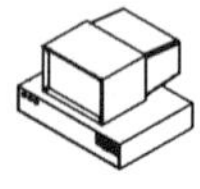

Tip

External Referenced drawings are useful for creating parts or subassemblies and then putting them together in one drawing to create the main assembly. You can also use it for laying out the contents of a drawing with multiple views before plotting.

DEPENDENT SYMBOLS

If you use the **INSERT** command to insert a drawing, the information about the named objects is lost if the names are duplicated. If they are unique, it is imported. The **named objects** are entries such as blocks, layers, text styles, and layers. For example, if the assembly drawing has a layer HIDDEN with green color and HIDDEN linetype, and the piece part BEARING has a layer HIDDEN with blue color and HIDDEN2 linetype, then when you insert the BEARING drawing in the assembly drawing, the values set in the assembly drawing will override the values of the inserted drawing (Figure 16-1). Therefore, in the assembly drawing, the layer HIDDEN will retain green color and HIDDEN linetype, ignoring the layer settings of the inserted drawing. Only those layers that have the same names are affected. Remaining layers that have different layer names are added to the current drawing.

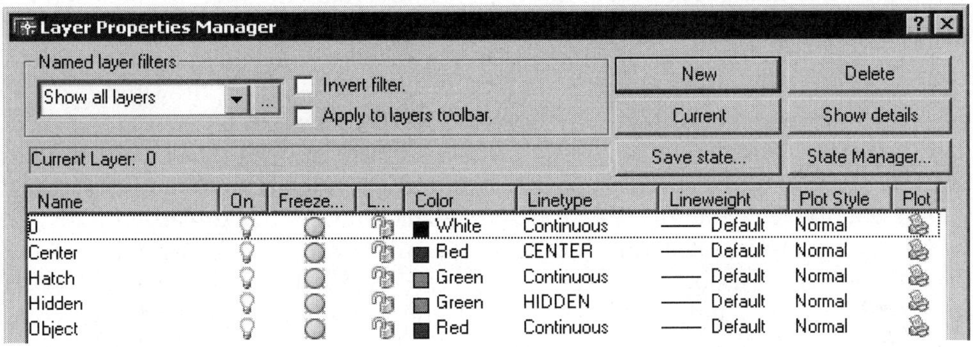

Figure 16-1 *Layer settings of the current drawing override the layers of the inserted drawing*

In the xref drawings, the information about named objects is not lost because AutoCAD will create additional named objects such as the specified layer settings as shown in Figure 16-2. For xref drawings, these named objects become dependent symbols (features such as layers, linetypes, object color, text style, and so on).

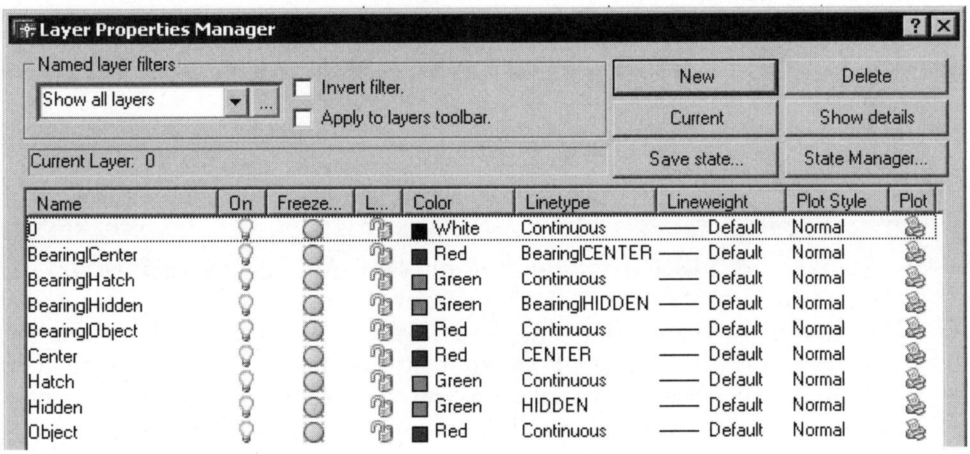

Figure 16-2 *Xref creates additional layers*

The layer HIDDEN of the xref drawing (BEARING) is appended with the name of the xref drawing BEARING, and the two are separated by the vertical bar symbol (|). The name of these layers appear in light gray color in the **Layer Control** drop-down list of the **Layers** toolbar. These layers can neither be selected nor be made current. The layer name HIDDEN changes to BEARING|HIDDEN. Similarly, CENTER is renamed BEARING|CENTER and

OBJECT is renamed BEARING|OBJECT (Figure 16-2). The information added to the current drawing is not permanent. It is added only when the xref drawing is loaded. If you detach the xref drawing, the dependent symbols are automatically erased from the current drawing.

When you xref a drawing, AutoCAD does not let you reference the symbols directly. For example, you cannot make the dependent layer, BEARING|HIDDEN, current. Therefore, you cannot add any objects to that layer. However, you can change the color, linetype, lineweight, plotstyle, or visibility (on/off, freeze/thaw) of the layer in the current drawing. If the **Retain changes to Xref layers** check box in the **External References (Xrefs)** area of the **Open and Save** tab of the **Options** dialog box is cleared, which also implies that the system variable **VISRETAIN** is set to 0, the settings are retained only for the current drawing session. This means that when you save and exit the drawing, the changes are discarded and the layer settings return to their default status. If this check box is selected (default), which also implies that the **VISRETAIN** variable is set to 1, layer settings such as color, linetype, on/off, and freeze/thaw are retained, and they are saved with the drawing and used when you open the drawing the next time. Whenever you open or plot a drawing, AutoCAD reloads each Xref in the drawing and as a result, the latest updated version of the drawing is loaded automatically.

Note

You cannot make the xref-dependent layers current in a drawing. Only when the xref drawing is bounded to the current drawing using the ***XBIND*** *command you can make the xref-dependent layers a permanent part of the current drawing and use them. The* ***XBIND*** *command will be discussed later in this chapter.*

MANAGING EXTERNAL REFERENCES IN A DRAWING

Toolbar:	Insert > External Reference
	Reference > External Reference
Menu:	Insert > Xref Manager
Command:	XREF

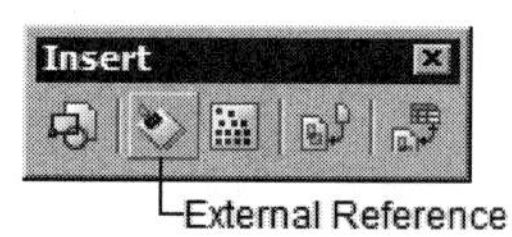

Figure 16-3 *The* ***Insert*** *toolbar*

When you invoke the **XREF** command (Figure 16-3), AutoCAD displays the **Xref Manager** dialog box (Figure 16-4).
The **Xref Manager** dialog box displays the status of each Xref in the current drawing and the relation between the various Xrefs. It allows you to attach a new xref, detach, unload, load an existing one, change an attachment to an overlay, or an overlay to an attachment. It also allows you to edit an xref's path and bind the xref definition to the drawing.

Apart from the methods mentioned in the command box, you can also invoke the **Xref Manager** dialog box by selecting an Xref in the current drawing and then right-clicking in the drawing area to display a shortcut menu. Now, choosing **Xref Manager** from the shortcut menu, the **Xref Manager** dialog box is displayed.

The upper left corner of the dialog box has two buttons: **List View** and **Tree View**.

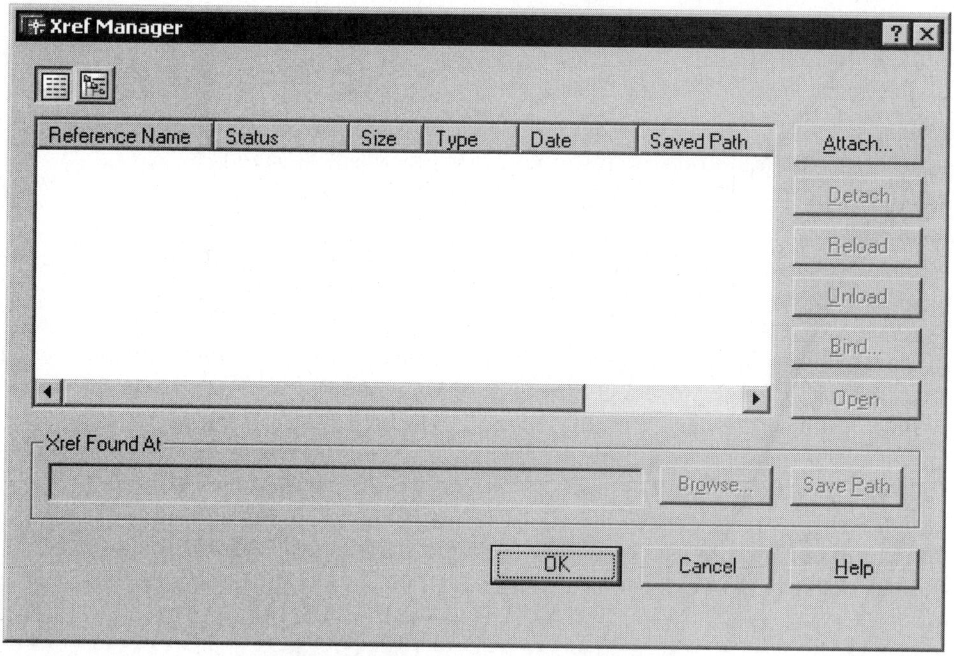

Figure 16-4 ***Xref Manager*** *dialog box*

List View

Choosing the **List View** button displays the xrefs present in the drawing in alphabetical order. This is the default view. The list view displays information about xrefs in the current drawing under the following headings.

Reference Name

This column lists the name of all existing references in the current drawing.

Status

This column lists the current status of each xref in the drawing. It lists whether an xref is loaded, unloaded, unreferenced, not found, orphaned, unresolved, or marked to be reloaded. A loaded xref implies that the xref is attached to the current drawing. You can then unload it and then reload it using the options in the dialog box (this will be discussed later). An xref selected to be unloaded or reloaded displays **Unload** and **Reload**, respectively, under the **Status** column. If the xref has nested references that cannot be found, the status is **Unreferenced**, and if the parent of the nested reference gets unloaded, or cannot be found, its status is described as **Orphaned**. An unreferenced xref will not be displayed. If the xref is not found in the search paths defined, its status is **Not Found**. A missing xref or one that cannot be found is **Unresolved**.

Size

The file size of each xref is listed here.

Type
This column lists whether the xref is an attachment or overlay.

Date
This column lists the date on which the xref drawing was last saved.

Saved Path
This column lists the path of the xref, that is, the route taken to locate the particular referenced drawing.

Choosing any of these headings, sorts and lists the Xrefs in the current drawing according to that particular title. For example, choosing **Reference Name**, sorts and lists the xrefs as per name. The column widths can be increased or decreased as per requirements. When you place your cursor at the edge of a column title button, the cursor changes to a horizontal resizing cursor. Now, press the pick button of your mouse and drag the column edge to increase or decrease its width. After you increase the column widths, it is possible that the width of the columns extend beyond the list box width. In such a case, a horizontal scroll bar appears at the bottom of the list box. You can use the scroll bar to view the columns that extend beyond the width of the list box.

Choosing the **Tree View** button, displays the xrefs in the drawing in a hierarchical tree view. It displays information on nested xrefs and their relationship with one another. Xrefs are indicated by an icon of a paper with a paper clip. This icon appears faded when the xref has been unloaded, and if there is a missing xref, a question mark appears. Similarly, an upward arrow indicates that the xref was reloaded and an arrow pointing downward indicates that the xref is unloaded. You can also choose **List View** and **Tree View** by pressing the F3 and F4 keys, respectively.

Attaching an Xref Drawing (Attach Option)
The **Attach** button of the **Xref Manager** dialog box is used to attach an xref drawing to the current drawing. This option can be invoked by choosing the **Attach** button in the **Xref Manager** dialog box. The following examples illustrate the process of attaching an xref to the current drawing. In this example, it is assumed that there are two drawings, SHAFT and BEARING. SHAFT is the current drawing that is loaded on the screen (Figure 16-5) and the BEARING drawing is saved on the disk. We want to xref the BEARING drawing in the SHAFT drawing.

1. The first step is to make sure that the SHAFT drawing is on the screen (draw the shaft drawing with assumed dimensions).

Tip
One of the drawings need not be on the screen. You could attach both drawings, BEARING and SHAFT, to an existing drawing, even if it is a blank drawing.

2. Invoke the **XREF** command to display the **Xref Manager** dialog box. In this dialog box, choose the **Attach** button. You can also choose **External Reference** from the **Insert** menu.

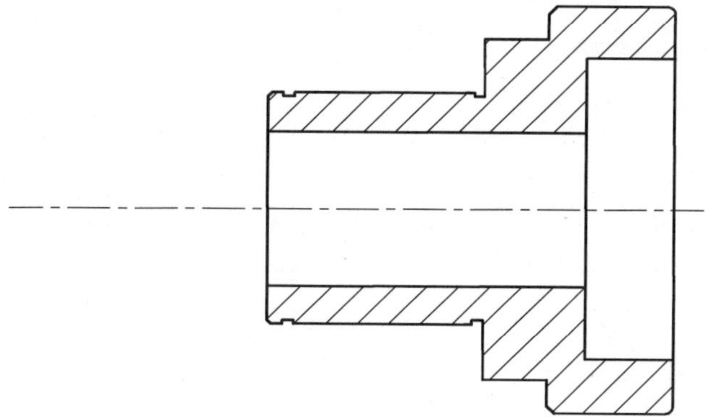

Figure 16-5 *Current drawing, SHAFT*

The **Select Reference File** dialog box (Figure 16-6) is displayed.

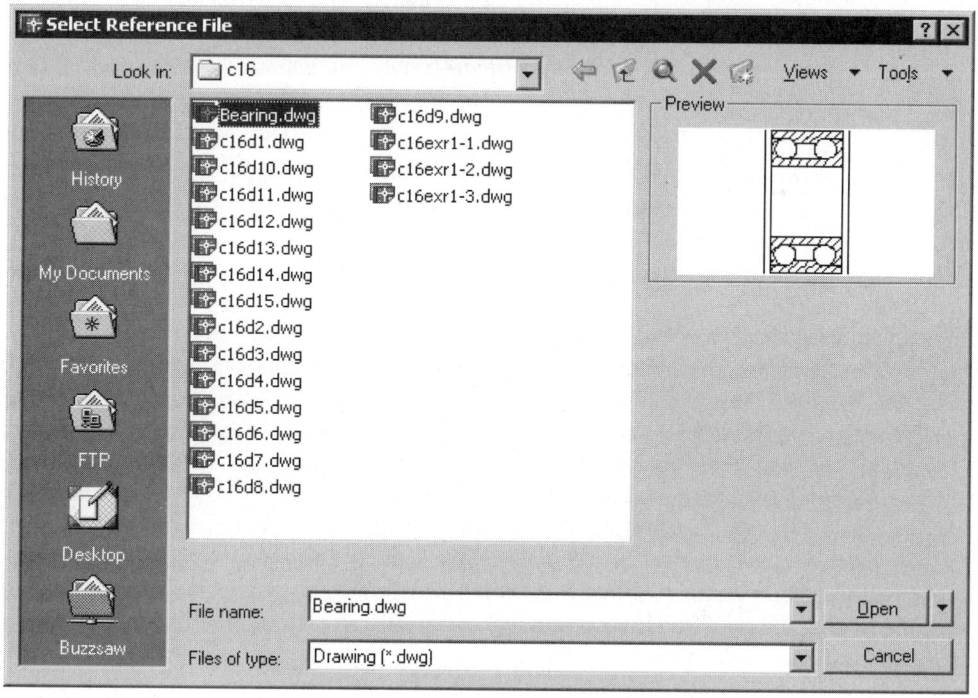

Figure 16-6 ***Select Reference File*** *dialog box*

Select the drawing that you want to attach (BEARING), and then choose the **Open** button. The **External Reference** dialog box is displayed on the screen (Figure 16-7). In the **External Reference** dialog box (Figure 16-7), the name of the file you have selected to be

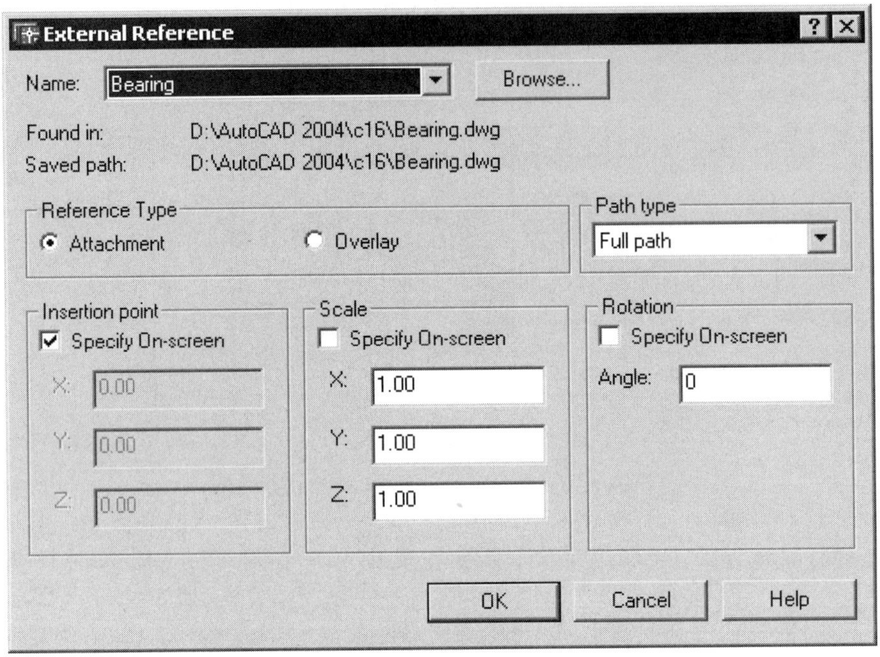

Figure 16-7 External Reference *dialog box*

attached to the current drawing as an xref is displayed in the **Name** edit box. You can also select a name of the file to attach from the **Name** drop-down list. The path of the file is displayed adjacent to **Found in**, located below the **Name** edit box. Also, the saved path of the file is displayed adjacent to **Saved Path**.

Note

*AutoCAD also searches for the xref file in the paths defined in the **Project Files Search Path** folder in the **Files** tab of the **Options** dialog box. This folder does not have any paths defined in it and displays **Empty** when the tree view is expanded. To define a search path, highlight **Empty** and choose the **Add** button in the dialog box. You can enter a project name here, if you want. Now, expand the tree view again and select **Empty** again and choose the **Browse** button. The **Browse for Folder** window is displayed. Select the folder that is to be searched for the file and choose OK. Then choose **Apply** and **OK** in the **Options** dialog box to exit it.*

In the **Reference Type** area, select the **Attachment** radio button if it is not already selected (default option). The **Overlay** option is discussed later in this chapter. The **Path type** drop-down list is used to specify whether you want to attach the drawing with Full path, Relative path, or No path. If you select the Full path option, the precise location of the xreffed drawing is saved. If you select the relative path, the position of the xreffed drawing with reference to the host drawing is saved. If you select the No path option, AutoCAD will search for the xreffed drawing in only that folder in which the host drawing is saved. You can either specify the insertion point, scale factors, and rotation angle in the respective **X**, **Y**, **Z** and **Angle** edit boxes or select the **Specify On-screen** check boxes to use the pointing device to specify them on the screen. By default, the X, Y, and Z scale factors are 1 and the rotation angle is 0. Accept the default values and choose the **OK** button in the **External Reference** dialog box. Specify the insertion point on the screen. After attaching the BEARING drawing as an xref, save the current drawing with the file name SHAFT (Figure 16-8).

Note

You can also use the ***-XREF*** *command to attach a drawing from the command line. All the options available in the dialog box are available through the command line too.*

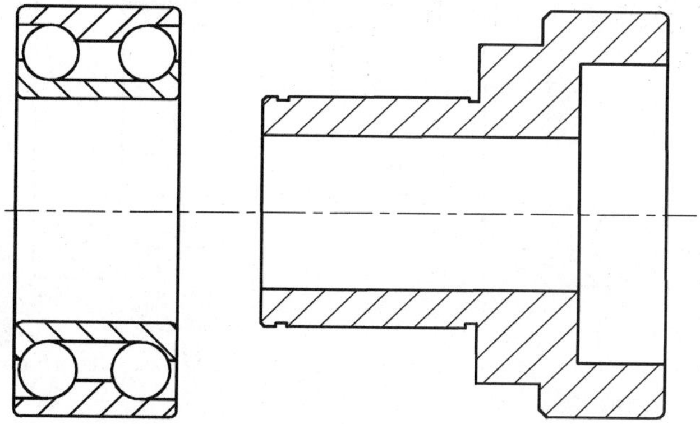

Figure 16-8 *Attaching xref drawing BEARING*

3. Load the drawing BEARING and make the changes shown in Figure 16-9 (draw polylines on the sides). Now, save the drawing with the file name BEARING.

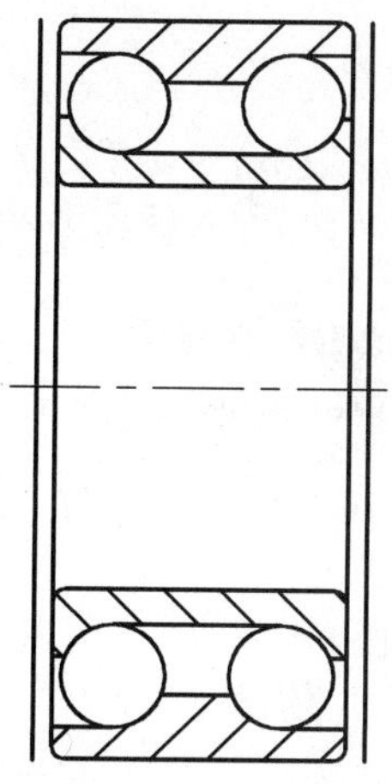

Figure 16-9 *Modifying the xref drawing BEARING*

4. Load the drawing SHAFT (Figure 16-10) on the screen. You will notice that the xref drawing BEARING is automatically updated. This is the most useful feature of the **XREF** command. You could also have inserted the BEARING drawing as a block, but if you had updated the BEARING drawing, the drawing in which it was inserted would not have been updated automatically.

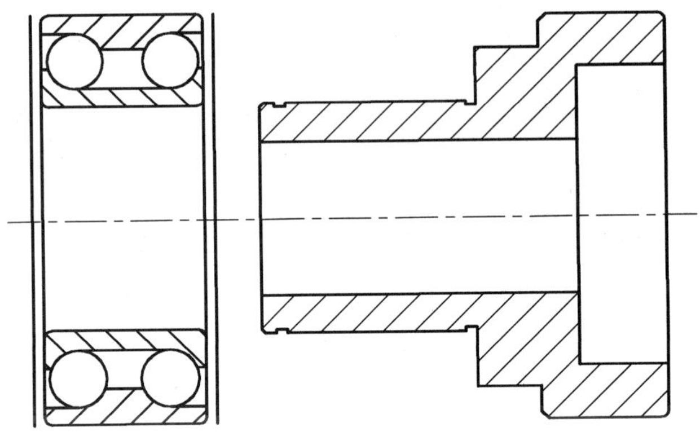

Figure 16-10 *After loading the drawing SHAFT, BEARING is automatically updated*

When you attach an xref drawing, AutoCAD remembers the name of the attached drawing. If you xref the drawing again, AutoCAD displays a message to that effect as given next.

Xref "BEARING" has already been defined.
Using existing definition.
Specify insertion point or [Scale/X/Y/Z/Rotate/PScale/PX/PY/PZ/PRotate]: *Specify where to place another copy of the xref.*

Note
If the xref drawing you want to attach is currently being edited, AutoCAD will attach the drawing that was last saved through the ***SAVE, WBLOCK,*** *or* ***QUIT*** *command.*

Points to Remember about Xref

1. When you enter the name of the xref drawing, AutoCAD checks for block names and xref names. If a block exists with the same name as the name of the xref drawing in the current drawing, the **XREF** command is terminated and an error message is displayed.

2. When you xref a drawing, the objects that are in the model space are attached. Any objects that are in the paper space are not attached to the current drawing.

3. The layer 0, DEFPOINTS, and the linetype CONTINUOUS are treated differently. The

current drawing layers 0, DEFPOINTS, and linetype CONTINUOUS will override the layers and linetypes of the xref drawing. For example, if the layer 0 of the current drawing is white and the layer 0 of the xref drawing is red, the white color will override the red.

4. The xref drawings can be nested. For example, if the BEARING drawing contains the reference INRACE and you xref the BEARING drawing to the current drawing, the INRACE drawing is automatically attached to the current drawing. If you detach the BEARING drawing, the INRACE drawing gets detached automatically.

5. You can rename an xref under the **Reference** column name in the list box of the **Xref Manager** dialog box by highlighting the xref and then clicking on it again. You can now enter a new name. An AutoCAD warning is displayed: **Caution! "XXXX" is an externally referenced block. Renaming it will also rename its dependent symbols**.

6. When you xref a drawing, AutoCAD stores the name and path of the drawing by default. If the name of the xref drawing or the path where the drawing was originally stored has changed or you cannot find it in the path specified in the **Options** dialog box, AutoCAD cannot load the drawing, plot it, or use the **Reload** option of the **XREF** command.

Detaching an Xref Drawing (Detach Option)

The **Detach** option can be used to detach or remove the xref drawings. If there are any nested xref drawings defined with the xref drawings, they are also detached. Once a drawing is detached, it is erased from the screen. To detach an xref drawing, select the file name in the **Xref Manager** dialog box list box to highlight it and then choose the **Detach** button. When you choose **OK** in the dialog box, the xref is completely removed from the current drawing. If you do not want to remove the specified xref from the current drawing, but have already selected the xref in the list box and chosen the **Detach** button, you can simply choose the **Cancel** button to cancel the detach operation.

You can also use the **-XREF** command to detach the xref drawings. When AutoCAD prompts for an xref file name to detach, you can enter the name of one xref drawing or the name of several drawings separated by commas. You can also enter * (asterisk), in which case all referenced drawings, including the nested drawings, will be detached.

Updating an Xref Drawing (Reload Option)

When you load a drawing, AutoCAD automatically loads the referenced drawings. The **Reload** option of the **XREF** command lets you update the xref drawings and nested xref drawings at any time. You do not need to exit the drawing editor and then reload the drawing. To reload the xref drawings, invoke the **Xref Manager**, select the drawings in the list box, and then choose the **Reload** button. AutoCAD will scan for the referenced drawings and the nested xref drawings and load the most recently saved version of the drawing.

The **Reload** option is generally used when the xref drawings are currently being edited and you want to load the updated drawings. The xref drawings are updated based on what is saved on the disk. Therefore, before reloading an xref drawing, you should make sure that the xref drawings that are being edited have been saved. If AutoCAD encounters an error while loading the referenced drawings, the **XREF** command is terminated, and the entire reload operation is canceled.

You can also reload the xref drawings by using the **-XREF** command. When you enter the **-XREF** command, AutoCAD will prompt you to enter the name of the xref drawing. You can enter the name of one xref drawing or the names of several drawings separated by commas. If you enter * (asterisk), AutoCAD will reload all xref and nested xref drawings.

Unloading an Xref Drawing (Unload Option)

The **Unload** option allows you to temporarily remove the definition of an xref drawing from a current drawing. However, AutoCAD retains the pointer to the xref drawings. When you unload the xref drawings, the drawings are not displayed on the screen. You can reload the xref drawings by using the **Reload** option.

Tip

It is recommended that you unload the referenced drawings if they are not being used. After unloading the xref drawings, the drawings load much faster and need less memory.

Adding an Xref Drawing (Bind Option)

The **Bind** option lets you convert the xref drawings to blocks in the current drawing. The bound drawings, including the nested xref drawings (that are no longer xrefs), become a permanent part of the current drawing. The bound drawing cannot be detached or reloaded. You can use this option when you want to send a copy of your drawing to a customer for review. Since all the xref drawings are a part of the existing drawing, you do not need to include the xref drawings or the path information. You can also use this option to safeguard the master drawing from accidental editing of the piece parts. To bind the xref drawings, select the file names in the **Xref Manager** dialog box list box and then choose the **Bind** button. The **Bind Xrefs** dialog box (Figure 16-11) is displayed. AutoCAD provides two methods to bind the xref drawing in the **Bind Type** area of the dialog box. These methods are discussed next.

***Figure 16-11** **Bind Xrefs** dialog box*

Bind

When you use the **Bind** option, AutoCAD binds the selected xref definition to the current drawing. All the xrefs are converted to blocks and the named objects are renamed. For example, if you xref the drawing BEARING with a layer named OBJECT, a new layer BEARING|OBJECT is created in the current drawing. When you bind this drawing, the xref dependent layer BEARING|OBJECT will become a locally defined layer BEARING0OBJECT (Figure16-12). If the BEARING0OBJECT layer already exits, AutoCAD will automatically increment the number, and the layer name becomes BEARING1OBJECT.

Insert

When you use the **Insert** option, AutoCAD inserts the xref drawing. The xrefs get converted into blocks. For example, if you xref the drawing SHAFT with a layer named OBJECT, a new layer SHAFT|OBJECT is created in the current drawing. If you use the **Insert** option to bind the xref drawing, the layer name SHAFT|OBJECT is renamed as OBJECT (Figure 16-12). If the object layer already exists, then the values set in the current drawing override the values of the inserted drawing.

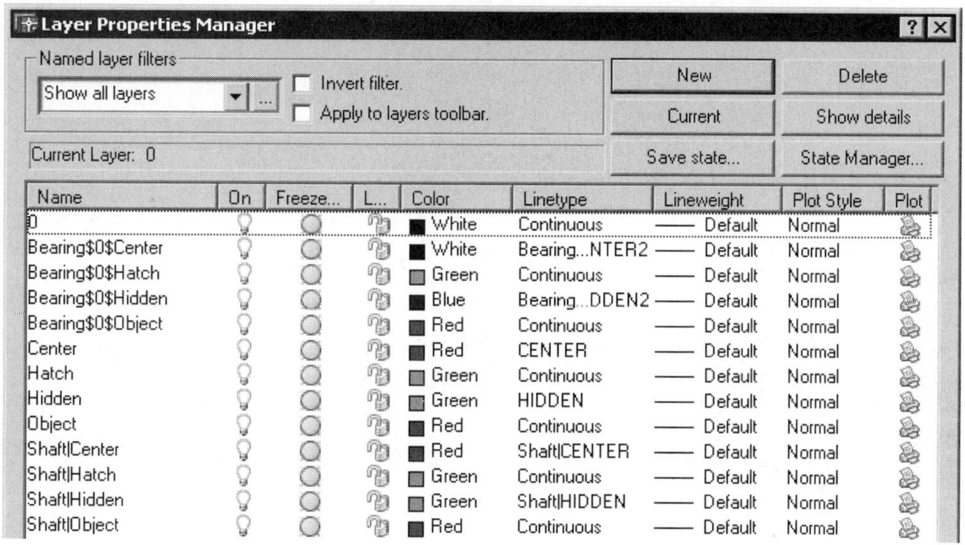

***Figure 16-12 Layer Properties Manager** dialog box*

Note

It is possible to bind only an individual or several xref-dependent named objects instead of the entire drawing into a current drawing. This will be discussed later in this chapter.

*You can also use the **-XREF** command to bind the xref drawings. Prompts are displayed on the command line.*

Editing an XREF's Path (Path Option)

By default, AutoCAD saves the path of the referenced drawing and displays it in the **Saved Path** column in the **Xref Manager** dialog box. As mentioned earlier, when AutoCAD loads the drawing containing a referenced file, and if it is not able to find the file at the location specified in the **Saved Path** column of the **Xref Manager** dialog box, it searches for the file in the current directory, and in the **Support File Search Path** locations specified in the **Files** tab of the **Options** dialog box. If a file with the same name is found here, it is loaded. Now, when you invoke the **Xref Manager** dialog box, you will notice that when you select an xref name in the list box to highlight it, the path displayed in the **Saved Path** column for the xref file is different from the one displayed in the **Xref Found At** edit box. To update the path of the xref file, choose the **Save Path** button. The new path is saved and displayed in the **Saved Path** column.

If AutoCAD cannot locate the specified file even in the directories specified in the **Files** tab of the **Options** dialog box, it will display an error message saying that it cannot find the specified file. The path of the file is displayed as a marker text in the current drawing. Now, when you invoke the **Xref Manager** dialog box, the status of the drawing is shown as **Not Found**. To specify a new path for the xref file, select the xref file name in the list box of the dialog box and choose the **Browse** button. The **Select new path** dialog box is displayed where you can locate the drawing to be used as xref. Once you have found the file, choose the **Open** button to return to the **Xref Manager** dialog box. The new path is displayed in the **Saved Path** column and the **Xref Found At** edit box. The specified xref file is reloaded and replaces the marker text in the drawing when you choose the **OK** button in the **Xref Manager** dialog box. If you remember the new location of the xref file, you can also enter it in the **Xref Found At** edit box. For example, if the drawing was originally in the C:\CAD\Proj1 subdirectory and the drawing has been moved to A:\Parts directory, the path must be edited so that AutoCAD is able to load the xref drawing.

You can also use the **-XREF** command to change the path using the prompts on the command line. When AutoCAD prompts you to enter the name of the xref whose path you want to edit, you can enter the name of one xref drawing or the names of several drawings separated by commas. You can also enter * (asterisk), in which case AutoCAD will prompt you for the path name of each xref drawing. The path name stays unchanged if you press ENTER when AutoCAD prompts for a new path name.

THE OVERLAY OPTION

As discussed earlier, when you are attaching an xref to a drawing, the **External Reference** dialog box is displayed. The **Reference Type** area of this dialog box has two radio buttons. They are **Attachment** and **Overlay**. The **Attachment** option is the default option. You can use any of these options to xref a drawing. The advantage of using the **Overlay** option is that you can access the desired drawing instead of the drawing along with its xreffed attachments. For example, consider three people working on three different drawings that are a part of the same project. The first designer is working on the layout of walls of a room, the second designer is working on the furniture layout of the room, and the third on the electrical layout of that room. The names of the drawings are WALLS, FURNITURE, and ELECTRICAL, respectively. Assume that the designer working on the furniture layout uses the **Attachment** option to xref the WALLS drawing so that he or she can check the furniture layout according to the wall layout. After insertion, the FURNITURE drawing will comprise the wall layout (xreffed drawing) along with the furniture layout (current drawing). Now, if the designer working on the electrical layout xrefs the FURNITURE drawing to check the location of electrical fittings with respect to the furniture, he/she will get the drawing that has the furniture layout as well as the wall layout. This is because the WALLS drawing was xreffed in the FURNITURE drawing using the **Attachment** option.

In the above example, the designer working on the ELECTRICAL drawing may not require the WALLS drawing. This is because at this stage, he/she is more interested in checking the electrical fittings with respect to the furniture. So the wall structure that is xreffed with the furniture layout needs to be avoided. This can be done using the **Overlay** option while xreffing the WALLS drawing in the FURNITURE drawing. This means that the designer working on the furniture layout needs to xref the wall layout using the **Overlay** option. Now, if the furniture layout is xreffed in some other drawing, the wall layout will not appear.

One of the problems with the **Attachment** option is that you cannot have circular reference. For example, assume you are designing the plant layout of a manufacturing unit. One person is working on the floor plan (see Figure 16-13), and the second person is working on the furniture layout in the offices (Figure 16-14). The names of the drawings are FLOORPLN and OFFICES, respectively. The person working on the office layout uses the **Attachment** option to insert the FLOORPLN drawing so that he or she has the latest floor plan drawing. The person who is working on the floor plan wants to reference the OFFICES drawing.

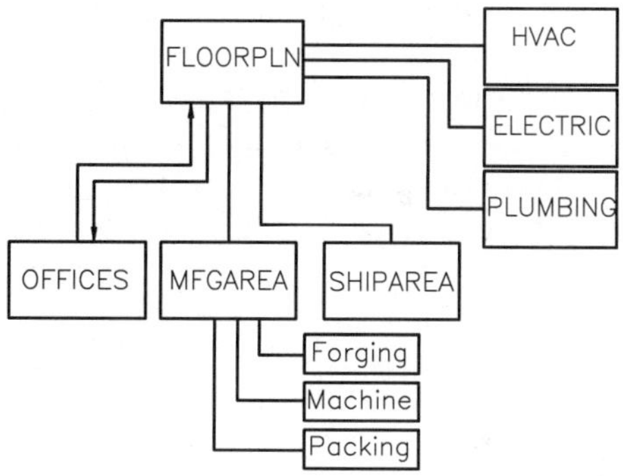

Figure 16-13 *Drawing files hierarchy*

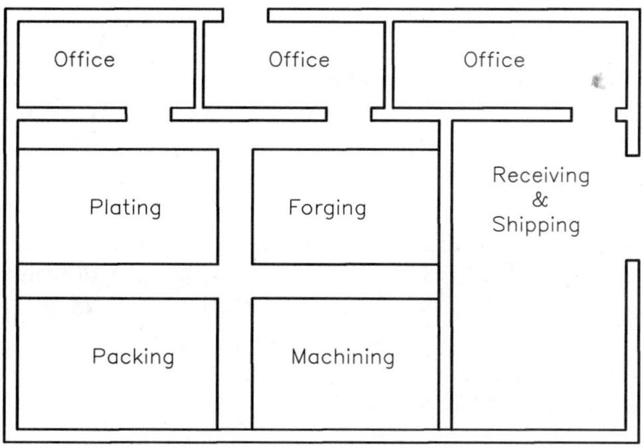

Figure 16-14 *Sample plant layout drawing*

Now, if the **Attachment** option is used to reference the drawings, AutoCAD displays an error message because by attaching the OFFICES drawing a circular reference is created (Figure 16-15). The AutoCAD message displayed is "**Circular references detected. Continue?**" If you choose the **No** button, the **XREF** command is canceled and no drawing is referenced. But, if you choose the **Yes** button, the following message is displayed **Breaking circular reference from "offices" to "current drawing"** and the particular file you wanted to reference is referenced.

***Figure 16-15 AutoCAD Alert** message box*

However, to overcome this problem of circular reference, you can use the **Overlay** option to overlay the OFFICES drawing. This is a very useful option because the **Overlay** option lets different operators avoid circular reference and share the drawing data without affecting the drawing. Overlaying allows you to view a referenced drawing without having to attach it to the current drawing. This option can be invoked by selecting the **Overlay** radio button in the **External Reference** dialog box, which is displayed after you have selected a drawing to reference. Also, when a drawing that has a nested overlay is overlaid, the nested overlay is not visible in the current drawing. This is another difference between attaching an xref and overlaying an xref to a drawing. This feature is especially useful when you want to reference a drawing that another user who is referencing your drawing does not need. Although the attachment will reference the nested reference too, the **overlay** option ignores nested references.

You can also use the **-XREF** command to display prompts on the command line and enter the **Overlay** option. Selecting the **Overlay** option, displays the **Enter Name of file to overlay** dialog box, where you can select the file you want to overlay.

Example 1 — *Architectural*

In this example, you will use the **Attachment** and **Overlay** options to attach and reference the drawings. Two drawings, PLAN and PLANFORG, are given. The PLAN drawing (Figure 16-16) consists of the floor plan layout, and the PLANFORG drawing (Figure 16-17) has the details of the forging section only. The CAD operator who is working on the PLANFORG drawing wants to xref the PLAN drawing for reference. Also, the CAD operator working on the PLAN drawing should be able to xref the PLANFORG drawing to complete the project. The following steps illustrate how to accomplish the defined task without creating a circular reference.

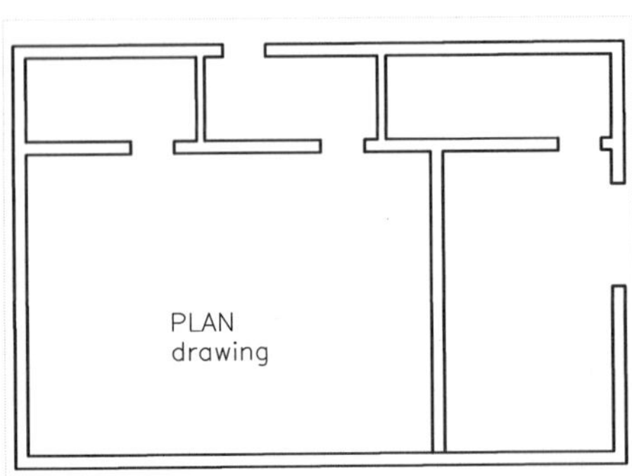

***Figure 16-16** PLAN drawing*

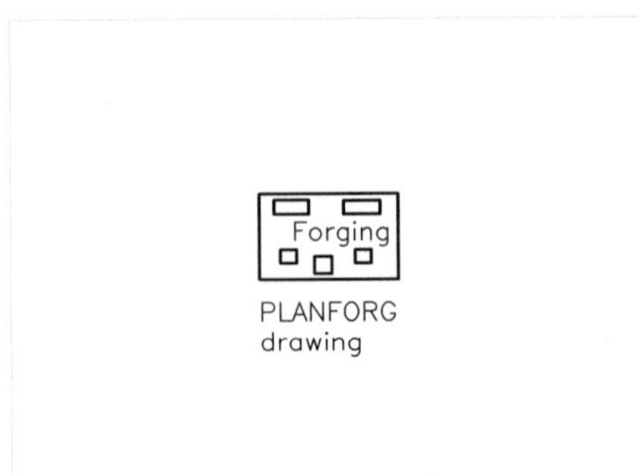

***Figure 16-17** PLANFORG drawing*

How circular reference is caused.

1. Load the drawing PLANFORG, invoke the **XREF** command, and choose the **Attach** button in the **Xref Manager** dialog box. Select the PLAN drawing in the list box of the **Select Reference File** dialog box and choose the **Open** button. The **External Reference** dialog box is displayed. The name of the PLAN drawing is displayed in the **Name** edit box and the **Attachment** radio button is selected by default in the **Reference Type** area of the dialog box. Choose **OK** to exit the dialog box and specify an insertion point on the screen. Now the drawing consists of PLANFORG and PLAN. Save the drawing.

2. Open the drawing file PLAN, and invoke the **XREF** command and attach the PLANFORG drawing using the same steps described in Step 1. AutoCAD will display the message that circular reference has been detected and ask you if you want to continue. If you choose to continue by choosing **Yes** in the AutoCAD message box, the circular reference is broken and you are allowed to reference the specific drawing.

   Another possible solution is for the operator working on the PLANFORG drawing to detach the PLAN drawing. This way, the PLANFORG drawing does not contain any reference to the PLAN drawing and would not cause any circular reference. The other solution is to use the **Overlay** option, as follows.

How to prevent circular reference:

3. Open the drawing PLANFORG (Figure 16-18) and select the **Overlay** radio button in the **External reference** dialog box, which is displayed after you have selected the PLAN drawing to reference. The PLAN drawing is overlaid on the PLANFORG drawing (Figure 16-19).

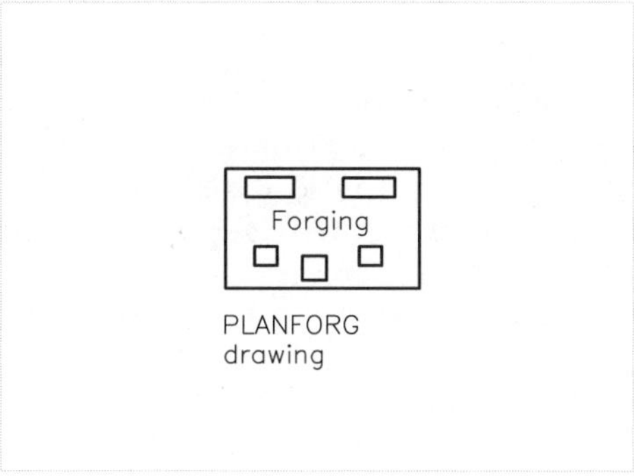

Figure 16-18 PLANFORG drawing

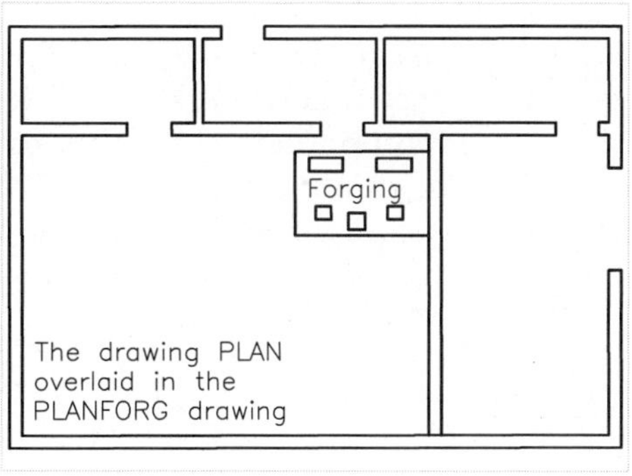

Figure 16-19 PLANFORG drawing after overlaying the PLAN drawing

4. Open the drawing file PLAN (Figure 16-20), and select the **Attachment** radio button in the **External References** dialog box, which is displayed when you have selected the PLANFORG drawing in the **Select Reference** dialog box to attach as an xref to the

PLAN drawing. You will notice that only the PLANFORG drawing is attached (Figure 16-21). The drawing that was overlaid in the PLANFORG drawing (PLAN) does not appear in the current drawing. This way, the CAD operator working on the PLANFORG drawing can overlay the PLAN drawing, and the CAD operator working on the PLAN drawing can attach the PLANFORG drawing, without causing a circular reference.

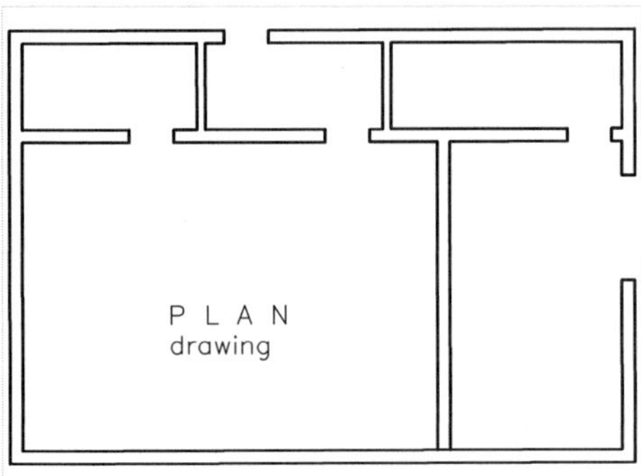

Figure 16-20 *PLAN drawing*

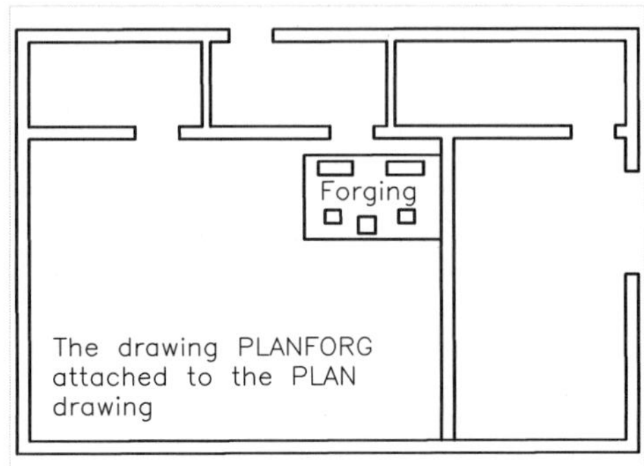

Figure 16-21 *PLAN drawing after attaching the PLANFORG drawing*

WORKING WITH THE XATTACH COMMAND

Toolbar:	Reference > External Reference Attach
Menu:	Insert > External Reference
Command:	XATTACH

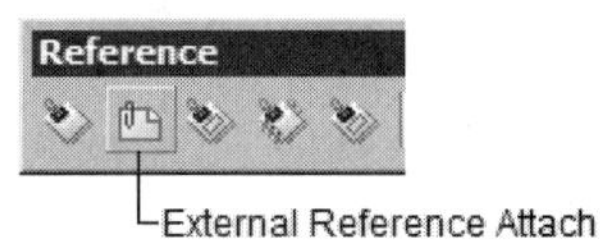

Figure 16-22 ***Reference*** *toolbar*

If you want to attach a drawing without invoking the **Xref Manager** dialog box, you can use the **XATTACH** command (Figure 16-22). When you invoke this command, AutoCAD displays the **Select Reference File** dialog box. This command makes it easier to attach a drawing, since most of the xref operations involve simply attaching a drawing file. After you have selected the drawing file to attach, choose the **Open** button. The **External Reference** dialog box is displayed. Select the **Attachment** radio button under the **Reference Type** area. You can specify the insertion point, scale, and rotation angle on screen or in the respective edit boxes.

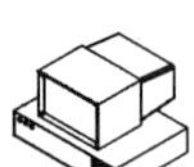

Tip

When you attach or reference a drawing that has a drawing order created by using the ***DRAWORDER*** *command, the drawing order is not maintained in the xref. To correct the drawing order, first open the xref drawing and specify the drawing order in it. Now, use the* ***WBLOCK*** *command to convert it into a drawing file and the* ***XATTACH*** *command to attach the newly created drawing file to the current drawing. This way the drawing order will be maintained.*

Note

AutoCAD maintains a log file (.xlg) for xref drawings if the ***XREFCTL*** *system variable is set to 1. This file lists information about the date and time of loading and other xref operations to be completed. This .xlg file is saved with the current drawing with the same name as the current drawing and is updated each time the drawing is loaded or any xref operations are carried out.*

OPENING AN XREFFED OBJECT IN A SEPARATE WINDOW*

If you are in the host drawing and you want to open a selected xreffed object in a separate window without using the **Select File** dialog box, you can use the **XOPEN** command. When you invoke this command, you are prompted to select the xref. Select the xreffed object that you want to open in a separate window. When you select the desired xref, AutoCAD opens the DWG file of the xreffed object in a separate window. You can now make the desired changes in the DWG file of the xreffed object. Save the changes and then close the drawing. When you open the host drawing, you will notice the **Communication Center** displays a message that the external reference file has changed and the xreffed drawing needs to be reloaded. Reload the xreffed drawing using thc **Xref Manager** dialog box and you will notice that the host drawing is updated.

Using the DESIGNCENTER to Attach a Drawing as Xref

The **DESIGNCENTER** can also be used to attach an xref to a drawing. Choose the **DesignCenter** button in the **Standard** toolbar to display the **DESIGNCENTER** window (Figure 16-23). In the **DESIGNCENTER** toolbar, choose the **Tree View Toggle** button, if not chosen already, to display the **tree pane**. Expand the **Tree view** and double-click the folder whose contents you want to view. The contents of the selected folder are displayed in the palette on the right side. From the list of drawings displayed in the palette, right-click the drawing you wish to attach as an xref. A shortcut menu is displayed. Choose **Attach as Xref**, the **External Reference** dialog box is displayed. You can also use the right button of your mouse to drag and drop the drawing into the current drawing. A shortcut menu is displayed again. Choose **Attach as Xref** and the **External Reference** dialog box is displayed.

The **Name** edit box displays the name of the selected file to be inserted as an xref and **Found in** displays the path of the file. Select the **Attachment** radio button in the **Reference type** area, if not already selected. Specify the **Insertion point**, **Scale**, and **Rotation** in the respective edit boxes or select the **Specify On-screen** check boxes to specify this information on the screen. Choose **OK** to exit the dialog box.

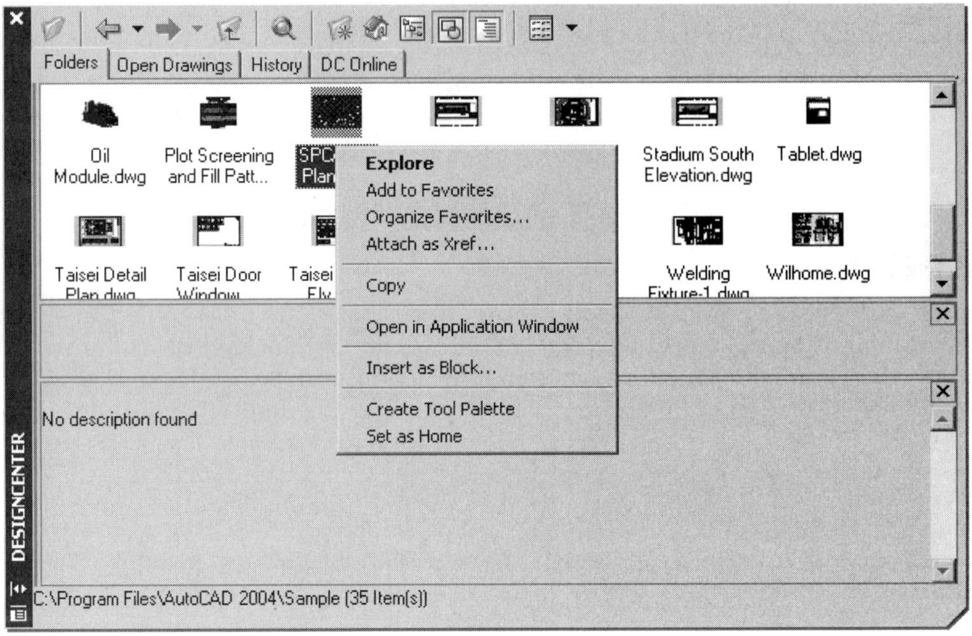

*Figure 16-23 **DESIGNCENTER** window*

ADDING DEPENDENT SYMBOLS TO A DRAWING

Toolbar:	Reference > External Reference Bind
Menu:	Modify > Object > External Reference > Bind
Command:	XBIND

You can use the **XBIND** command to add the selected dependent symbols of the xref drawing to the current drawing. The following example describes how to use the **XBIND** command (Figure 16-24).

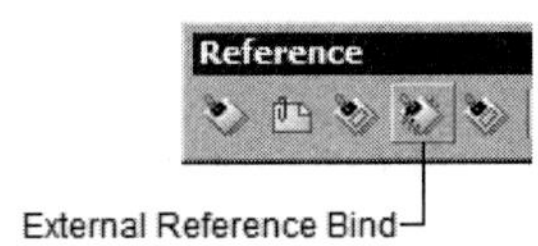

***Figure 16-24 Reference** toolbar*

1. Load the drawing BEARING that was created earlier when the **Attach** option of the **XREF** command was discussed. Make sure the drawing has the following layer setup. Otherwise, create the following layers, using the **Layer Properties Manager** dialog box.

Layer Name	Color	Linetype
0	White	Continuous
Object	Red	Continuous
Hidden	Blue	Hidden2
Center	White	Center2
Hatch	Green	Continuous

2. Draw a circle and use the **BLOCK** command to create a block of it. The name of the block is SIDE. Save the drawing as BEARING.

3. Start a new drawing with the following layer setup.

Layer Name	Color	Linetype
0	White	Continuous
Object	Red	Continuous
Hidden	Green	Hidden

4. Use the **XATTACH** command or the **Attach** option of the **XREF** command and attach the BEARING drawing to the current drawing. When you xref the drawing, the layers will be added to the current drawing, as discussed earlier in this chapter.

5. Now, invoke the **XBIND** command. The **Xbind** dialog box is displayed on the screen. This dialog box has two areas with list boxes. They are **Xrefs** and **Definitions to Bind**. If you want to bind the blocks defined in the xref drawing BEARING, first click on the plus sign adjacent to the xref BEARING. Icons for named objects in the drawing are displayed in a tree view (Figure 16-25). Click on the plus sign next to the Block icon. AutoCAD lists the blocks defined in the xref drawing (BEARING). Select the block BEARING|SIDE and then choose the **Add** button. The block name will be added to the **Definitions to Bind** area list box. Choose **OK** to exit the dialog box. AutoCAD will bind the block with the current drawing and a message at the command line is displayed: **1 Block(s) bound**. The name of the block will change to BEARING0SIDE. You can invoke the **Block Definition** dialog box using the **BLOCK** command and check the **Name** drop-down list to see if the block with the name BEARING0SIDE has been added to the drawing. If you want to insert the block, you must enter the new block name (BEARING0SIDE). You can also rename the block to another name that is easier to use.

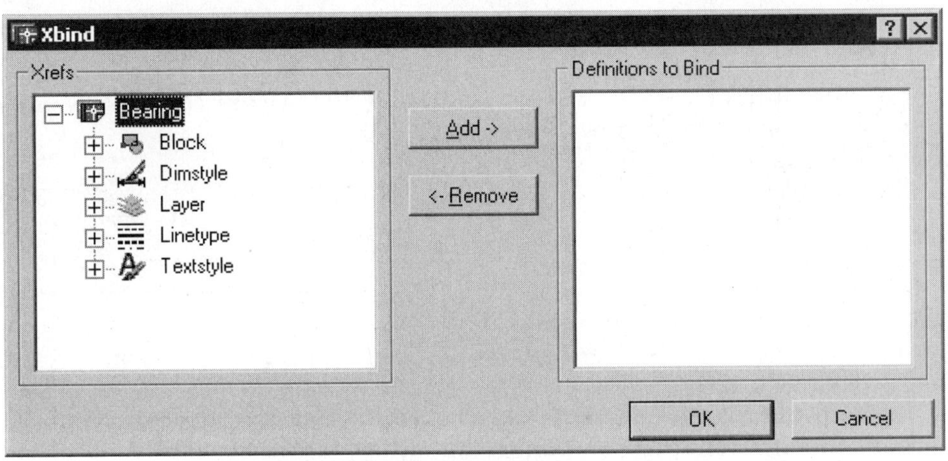

Figure 16-25 *The* ***Xbind*** *dialog box*

If the block contains a reference to another xref drawing, AutoCAD binds that xref drawing and all its dependent symbols to the current drawing also. Once you **XBIND** the dependent symbols, AutoCAD does not delete them, not even when the xref drawing is detached. For example, the block BEARING0SIDE will not be deleted even when you detach the xref drawing or end the drawing session.

You can also use the **-XBIND** command to bind the selected dependent symbols of the xref drawing, using the command line.

6. Similarly, you can bind the dependent symbols, BEARING|STANDARD (textstyle), BEARING|HIDDEN, and BEARING|OBJECT layers of the xref drawing. Click on the plus signs adjacent to the respective icons to display the contents and then select the layer or textstyle you want to bind and choose the **Add** button. The selected named objects are displayed in the **Definitions to Bind** area list box. If you have selected a named object that you do not want to bind, select it in the **Definitions to Bind** area list box and choose the **Remove** button. Once you have finished selecting the named objects that you want to bind to the current drawing, choose **OK**. A message indicating the number of named objects that are bound to the current drawing is displayed at the command line.

   Once bound, the layer names will change to BEARING0HIDDEN and BEARING0OBJECT. If the layer name BEARING0HIDDEN was already there, the layer will be named BEARING1HIDDEN. These two layers become a permanent part of the current drawing. Even if the xref drawing is detached or the current drawing session is closed, the layers are not discarded.

CLIPPING EXTERNAL REFERENCES

Toolbar:	Reference > External Reference Clip
Command:	XCLIP

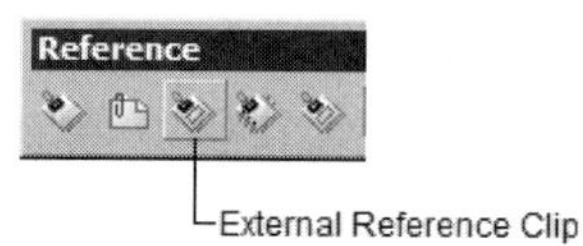

Figure 16-26 ***Reference*** *toolbar*

The **XCLIP** command (Figure 16-26) is used to trim an xref drawing after it has been attached to the drawing to display only a portion of the drawing (Figure 16-27). After you have attached an xref to a drawing, you can trim it to display only a portion of the drawing, using the **XCLIP** command (Figure 16-27). However, clipping the xref does not, in any way, modify the referenced drawing. You can also invoke the **XCLIP** command by selecting an xref and then right-clicking in the drawing area to display a shortcut menu and choosing **Xref Clip**. When you invoke this command, AutoCAD will prompt you to select the objects to clip. Select the xref objects and then define the clip boundary. The **XCLIP** command also defines a front and back clipping frame for 3D objects. The clipping boundary is specified

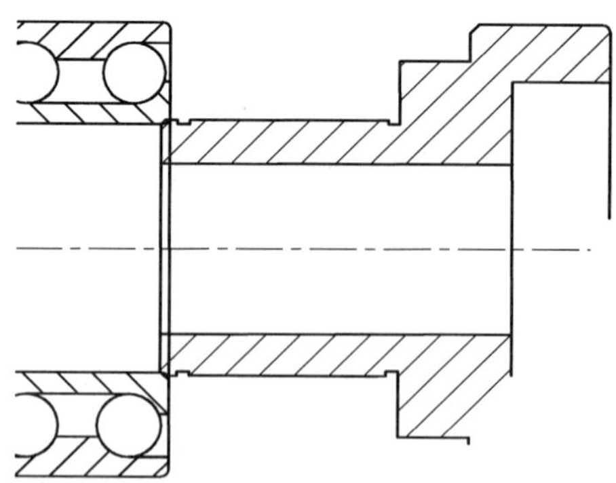

Figure 16-27 *Using* ***XCLIP*** *command to clip the xref drawings*

in the same plane as the object (Figures 16-28 and 16-29). The following is the prompt sequence.

Command: **XCLIP**
Select objects: *Select the xref that needs to be clipped.*
Enter clipping option
[ON/OFF/Clipdepth/Delete/generate Polyline/New boundary] <New>: *Press ENTER to specify a new clip boundary or enter an option.*
[Select polyline/Polygonal/Rectangular] <Rectangular>: *Press ENTER to select the **Rectangular** option, and then specify two corners of the rectangular boundary.*

You can also right-click in the drawing area at the **Enter clipping option** prompt to display a shortcut menu. This shortcut menu displays all the options available at the command line and can be selected here. All the clipping options available are discussed next.

ON/OFF

The **ON/OFF** option allows you to turn the clipping boundary on or off. When the clipping boundary is off, you can see the complete xref drawing and when it is on, the drawing that is within the clipping polygon is displayed.

Delete

The **Delete** option completely deletes the predefined clipping boundary and the entire xref gets displayed. The **ERASE** command cannot be used to delete the clipping boundary.

generate Polyline

This option displays a polyline coinciding with the boundary of the clipped xref drawing. The polyline boundary can be edited with no effect on the clipped drawing. For example, if you stretch the boundary, it does not affect the xref drawing. The edited boundary can be used later to specify a new clipping boundary.

Clipdepth

This option allows you to define the front and back clipping planes for 3D xref objects or blocks. The objects between the front and back planes will be displayed (Figures 16-28 and 16-29). You will be able to understand this option better after you have dealt with 3D drawings later in the book. In the figures that follow, the clipping boundary is defined aligned to the front face of the object, and therefore, the clipping planes will get defined parallel to it. The xref must contain a clip boundary before specifying a clip depth. If you use the **Clipdepth** option, it prompts you to specify a front clip point and a back clip point. Specifying these points, creates a clipping frame passing through them and parallel to the clipping boundary. If the front clipping plane is specified behind the back clipping plane, AutoCAD displays an error message and the clipdepth is also not applied. The **Distance** option for specifying the front or back clip points creates a clipping plane at a specified distance from the clipping boundary parallel to it. The **Remove** option removes both the front and back clipping planes and the entire object is visible.

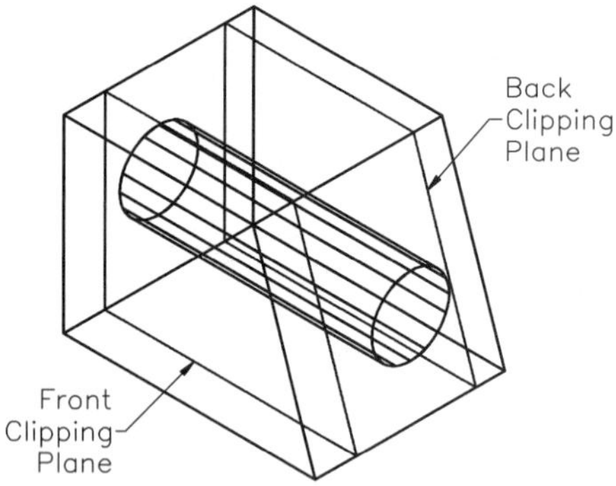

***Figure 16-28** 3D xref object before clipping*

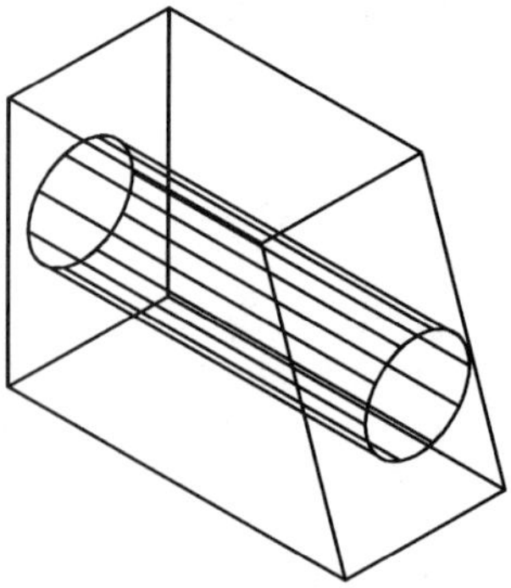

***Figure 16-29** 3D xref object after clipping*

The clipping boundary can be specified by using the **Rectangular**, **Polygonal**, or **Select polyline** options. The **Rectangular** option generates a rectangular boundary and **Polygonal** allows you to specify a boundary of any shape. You can also draw the boundary using **PLINE** or **POLYGON** commands and then use the **Select polyline** option to select this polyline as the clipping boundary. If a boundary already exists, AutoCAD will ask you if the old boundary should be deleted. You can enter YES if you want to delete the old boundary and define a new one. If you enter NO, the old boundary is retained and the **XCLIP** command ends.

DISPLAYING CLIPPING FRAME

You can use the **XCLIPFRAME** system variable to turn the clipping boundary on or off. The clipping boundary is invisible by default. This system variable can be set from the **Modify** menu (**Modify > Object > External Reference > Frame**), or by choosing the **External Reference Clip Frame** button from the **Reference** toolbar (Figure 16-30), or by entering **XCLIPFRAME** at the Command prompt. When the value is 0 (default), the clipping boundary is not displayed. When it is 1, the clipping boundary is displayed.

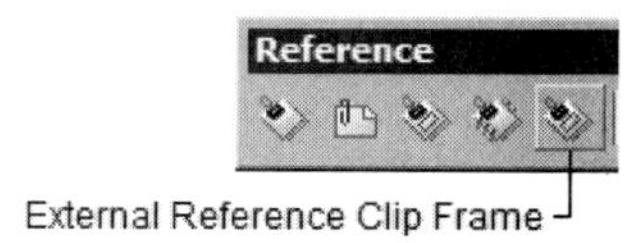

***Figure 16-30 Reference** toolbar*

Note

*The polyline generated when you use the **generate Polyline** option of the **XCLIP** command should not be confused with the clipping boundary. Although they both coincide with each other, you can select and edit only the polyline. When you select the clipping boundary, the xref object also gets selected.*

DEMAND LOADING

The demand loading feature loads only that part of the referenced drawing that is required in the existing drawing. For example, demand loading provides a mechanism by which objects on frozen layers are not loaded. Also, only the clipped portion of the referenced drawing can be loaded. This makes the xref operation more efficient since less disk space is used, especially when the drawing is reopened.

Demand loading is enabled by default. You can modify its settings in the **External References (XRefs)** area of the **Open and Save** tab of the **Options** dialog box (Figure 16-31). You can select any of the three settings available in the **Demand load Xrefs** drop-down list in this dialog box. They are **Disabled**, **Enabled**, and **Enabled with copy**. These options correspond to a value of **0**, **1**, and **2** of the **XLOADCTL** system variable, respectively, and are discussed next.

Setting	Value of XLOADCTL	Features
Disabled	0	1. Turns off demand loading. 2. Loads entire xref drawing file. 3. The file is available on the server and other users can edit the xref drawing.
Enabled	1	1. Turns on demand loading. 2. The referenced file is kept open. 3. Makes the referenced file read-only for other users.
Enabled with copy	2	1. Turns on demand loading with the copy option. 2. A copy of a referenced drawing is opened. 3. Other users can access and edit the original referenced drawing file.

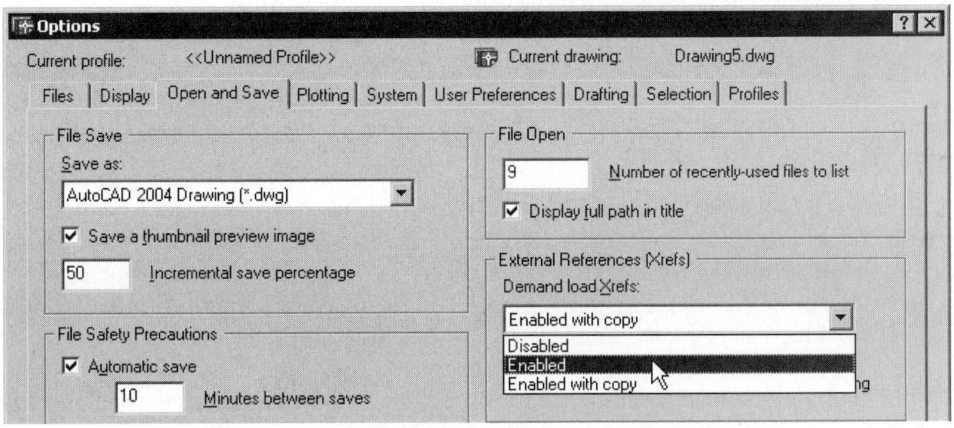

Figure 16-31 *Setting **XLOADCTL** using the **Options** dialog box*

You can also set the value of **XLOADCTL** at the command line. When you are using the **Enabled with copy** option of demand loading, the temporary copies of the xref are saved in the AutoCAD temporary files directory (defined in the **Temporary File Location** folder in the **Files** tab of the **Options** dialog box) or in a user-specified directory. The **XLOADPATH** system variable creates a temporary path to store demand loaded Xrefs.

Spatial and Layer Indexes

As mentioned earlier, the demand loading improves performance when the drawing contains referenced files. To make it work effectively and to take full advantage of demand loading, you must store a drawing with Layer and Spatial indexes. The layer index maintains a list of objects in different layers and the spatial index contains lists of objects based on their location in 3D space.

Layer and spatial indexes are created using the **Save Drawing As** dialog box. Choose the **Tools** button available on the upper right corner of the dialog box to display a shortcut menu. Choose the **Options** option from the shortcut menu to display the **Saveas Options** dialog box. Choose the **DWG Options** tab if it is not already chosen and select the type of index you want to save the file with from the **Index type** drop-down list. **None** is the default option, that is, no indexes are created. The other options available are: **Layer**, **Spatial**, and **Layer & Spatial**. Once you have selected the type of index to create, choose the **OK** button to exit the dialog box and choose the **Save** button in the **Save Drawing As** dialog box to save the drawing with the indexes. The **INDEXCTL** variable also controls the creation of layer index and spatial index and its value can be set using the command line. The following are the settings of the **INDEXCTL** system variable, and they correspond to the **Index Type** options available in the **DWG Options** tab of the **Saveas Options** dialog box.

Setting	Features	Index Type option
0	No index created.	None
1	Layer index created.	Layer
2	Spatial index created.	Spatial
3	Layer and spatial index created.	Layer and Spatial

EDITING REFERENCES IN PLACE

Menu:	Modify > Xref and Block editing > Edit Reference In-Place
Command:	REFEDIT

Often you have to make minor changes in the xref drawing if you want to save yourself from the trouble of going back and forth between drawings. In this situation, you can use **In-Place Reference editing** to select the xref, modify it, and then save it after modifications. This feature of AutoCAD has already been discussed in detail with reference to blocks in Chapter 14, Working with Blocks. The following steps explain the referencing editing process.

1. To edit In-place, choose **Xref and Block editing > Edit Reference In-Place** from the **Modify** menu. Select the xref you wish to edit. The **Reference Edit** dialog box is displayed.

2. In the **Reference name** list box of the **Reference Edit** dialog box, the selected reference and its nested references are listed. Select the specific xref you wish to edit from the list. You can cycle between the xrefs by using the **Next** button. A **Preview** window displays an image of the selected reference.

3. In the **Settings** tab of the dialog box, select both the **Create unique layer, style, and symbol names** and the **Lock objects not working in working set** check boxes. The **Display attribute definitions for editing** check box is selected and not available by default. In

the **Identify Reference** tab of the dialog box, select the **Prompt to select nested objects** radio button available below the **Path** area. Choose **OK** to return to the current drawing. At the **Select nested objects:** prompt, select the objects you want to modify and press ENTER. The **Refedit** toolbar (Figure 16-32) is displayed and the objects that have not been selected for modifications become faded. The fading is controlled by the **XFADECTL** system variable or by using the **Reference fading intensity** slider bar in the **Display** tab of the **Options** dialog box. You can add or remove objects to the working set by choosing the respective buttons in the **Refedit** toolbar or by using the **REFSET** command. Once you have defined a working set, all the standard AutoCAD commands can be used to modify them.

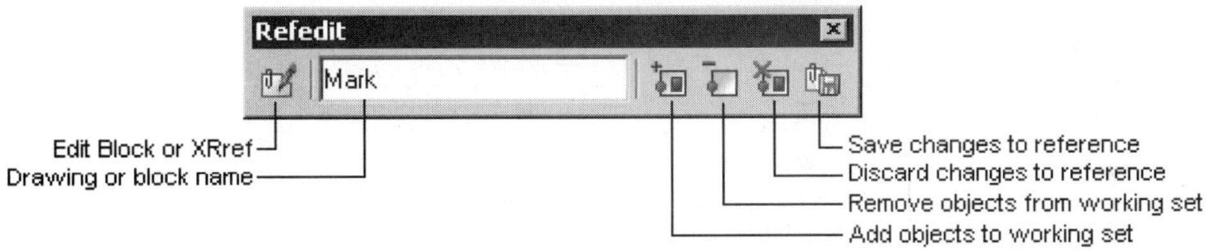

Figure 16-32 *The **Refedit** toolbar*

4. When you choose the **Save back changes to the reference** button in the **Refedit** toolbar or use the **REFSAVE** command, the modifications made are saved in the drawing used as reference as well as the current drawing. Choosing the **Discard changes to reference** button in the **Refedit** toolbar does not save the changes made. After you have saved the changes to the xref, AutoCAD displays a message: **All reference edits will be saved**. Choose **OK** to return to the drawing area. The current drawing is redisplayed and the fading is turned off.

Note

*In the **External References (Xrefs)** area of the **Open and Save** tab of the **Options** dialog box, if you select the **Allow other users to Refedit current drawing** check box, the current drawing can be edited in-place by other users even when it is open and is being referenced by another file. This option is selected by default and can also be controlled by the **XEDIT** system variable. The default value of **XEDIT** is 1 and can be changed using the command line.*

Self-Evaluation Test

Answer the following questions, and then compare your answers to the correct answers given at the end of this chapter.

1. If the assembly drawing has been created by inserting a drawing, the drawing will be updated automatically if a change is made in the drawing that was inserted. (T/F)

2. The external reference facility helps you keep the drawing updated no matter when the changes were made in the piece part drawings. (T/F)

3. Objects can be added to a dependent layer. (T/F)

4. While using the **Attachment** option during referencing a drawing, the drawing will reference the nested references too, while the **Overlay** option ignores nested references. (T/F)

5. The __________ are entries such as blocks, layers, and text styles.

6. If you use the **INSERT** command to insert a drawing, the information about the named objects is lost if the names are ____________ and if the names are ___________, the drawing is imported.

7. The ___________ button of the **Xref Manager** dialog box is used to attach an xref drawing to the current drawing.

8. The ________________ feature loads only that part of the referenced drawing that is required in the existing drawing.

9. The _________ option can be used to overcome the problem of circular reference.

10. If the **Retain changes to Xref layers** is __________, in the **External References (Xrefs)** area of the **Open and Save** tab of the **Options** dialog box, the layer settings such as color, linetype, on/off, and freeze/thaw are retained. The settings are saved with the drawing and are used when you xref the drawing the next time.

Review Questions

Answer the following questions.

1. If the xref drawings get updated, the changes are not automatically reflected in the assembly drawing when you open the assembly drawing. (T/F)

2. There is a limit to the number of drawings you can reference. (T/F)

3. It is not possible to have nested references. (T/F)

4. Like blocks, the xref drawings can be scaled, rotated, or positioned at any desired location. (T/F)

5. You can change the color, linetype, or visibility (on/off, freeze/thaw) of the dependent layer. (T/F)

6. Which of the following features lets you reference an external drawing without making this drawing a permanent part of the existing drawing?

   (a) demand loading (b) external reference
   (c) external clipping (d) insert drawing

7. If the xref has nested references that cannot be found, which of the following will be displayed under the status heading of the **List View** button in the **Xref Manager** dialog box?

   (a) Orphaned (b) Not found
   (c) Unreferenced (d) Unresolved

8. Which of the following commands can be used from the command line to attach a drawing?

   (a) **-XBIND** (b) **-XREF**
   (c) **XCLIP** (d) None of the above

9. Which of the following system variables, when set to 1, will allow AutoCAD to maintain a log file (*.xlg*) for xref drawings?

   (a) **XLOADCTL** (b) **XLOADPATH**
   (c) **XREFCTL** (d) **INDEXCTL**

10. Which of the following system variables when set to 0 will not allow the clipping boundary to be displayed?

   (a) **XCLIPFRAME** (b) **XLOADCTL**
   (c) **INDEXCTL** (d) **XREFCTL**

11. In the __________ drawings, the information regarding dependent symbols is not lost.

12. What is the function of the **XCLIPFRAME** system variable? Explain. ______________
    ____.

13. AutoCAD maintains a log file for xref drawings if the __________ variable is set to 1.

14. It is possible to edit xrefs in place using the __________ command.

15. You can use the __________ command to add selected dependent symbols from the xref drawing to the current drawing.

Exercises

Exercise 1 *Mechanical*

In this exercise, you will start a new drawing and xref the drawings Part-1 and Part-2. You will also edit one of the piece parts to correct the size and use the **XBIND** command to bind some of the dependent symbols to the current drawing. The following are detailed instructions for completing this exercise.

1. Start a new drawing, Part-1, and set up the following layers.

Layer Name	Color	Linetype
0	White	Continuous
Object	Red	Continuous
Hidden	Blue	Hidden2
Center	White	Center2
Dim-Part1	Green	Continuous

2. Draw Part-1 with dimensions as shown in Figure 16-33. Save the drawing as Part-1.

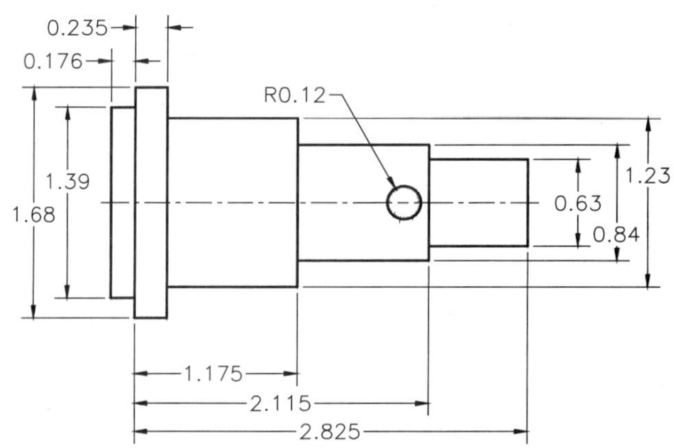

***Figure 16-33** Drawing of Part-1*

3. Start a new drawing, Part-2, and set up the following layers.

Layer Name	Color	Linetype
0	White	Continuous
Object	Red	Continuous
Hidden	Blue	Hidden
Center	White	Center
Dim-Part2	Green	Continuous
Hatch	Magenta	Continuous

4. Draw Part-2 with dimensions as shown in the Figure 16-34. Save the drawing as Part-2.

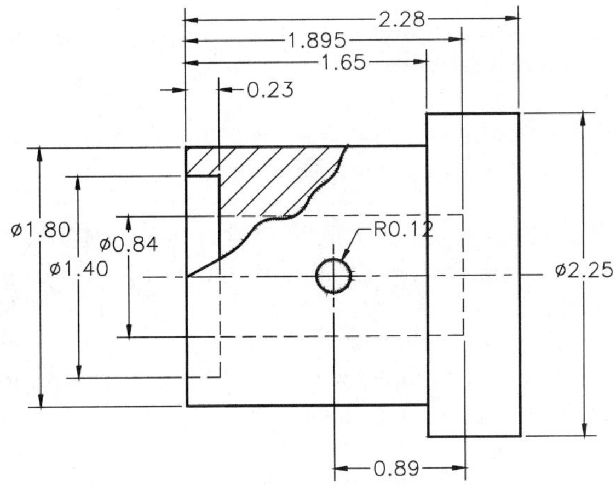

Figure 16-34 *Drawing of Part-2*

5. Start a new drawing, ASSEM1, and set up the following layers.

Layer Name	Color	Linetype
0	White	Continuous
Object	Blue	Continuous
Hidden	Yellow	Hidden

6. Xref the two drawings Part-1 and Part-2 so that the centers of the two drilled holes coincide. Notice the overlap as shown in Figure 16-35. Save the assembly drawing as ASSEM1.

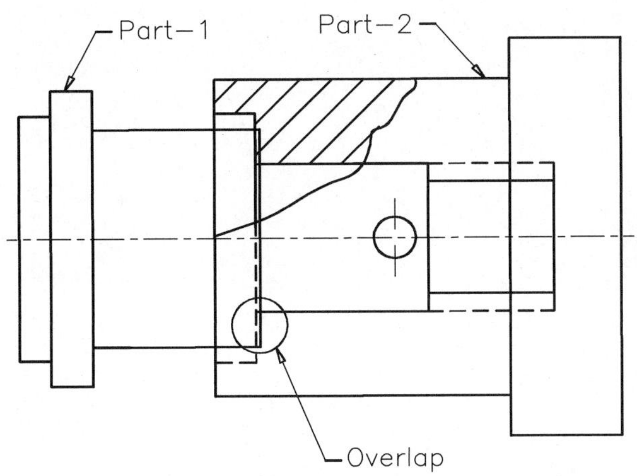

Figure 16-35 *Assembly drawing after attaching Part-1 and Part-2*

7. Open the drawing Part-1 and correct the mistake so that there is no overlap. You can do it by editing the line (1.175 dimension) so that the dimension is 1.160.

8. Open the assembly drawing ASSEM1 and notice the change in the overlap. The assembly drawing gets updated automatically.

9. Study the layers and notice how AutoCAD renames the layers and linetypes assigned to each layer. Check to see if you can make the layers belonging to Part-1 or Part-2 current.

10. Use the **XBIND** command to bind the OBJECT and HIDDEN layers that belong to drawing Part-1. Check again to see if you can make one of these layers current.

11. Use the **Detach** option to detach the xref drawing Part-1. Study the layer again, and notice that the layers that were edited with the **XBIND** command have not been erased. Other layers belonging to Part-1 are erased.

12. Use the **Bind** option of the **XREF** command to bind the xref drawing Part-2 with the assembly drawing ASSEM1. Open the xref drawing Part-1 and add a border or make any changes in the drawing. Now, open the assembly drawing ASSEM1 and check to see if the drawing is updated.

Answers to Self-Evaluation Test

1 - F, **2** - T, **3** - F, **4** - T, **5** - named objects, **6** - duplicated, unique, **7** - **Attach**, **8** - demand loading, **9** - **Overlay**, **10** - selected

Chapter 17

Working with Advanced Drawing Options

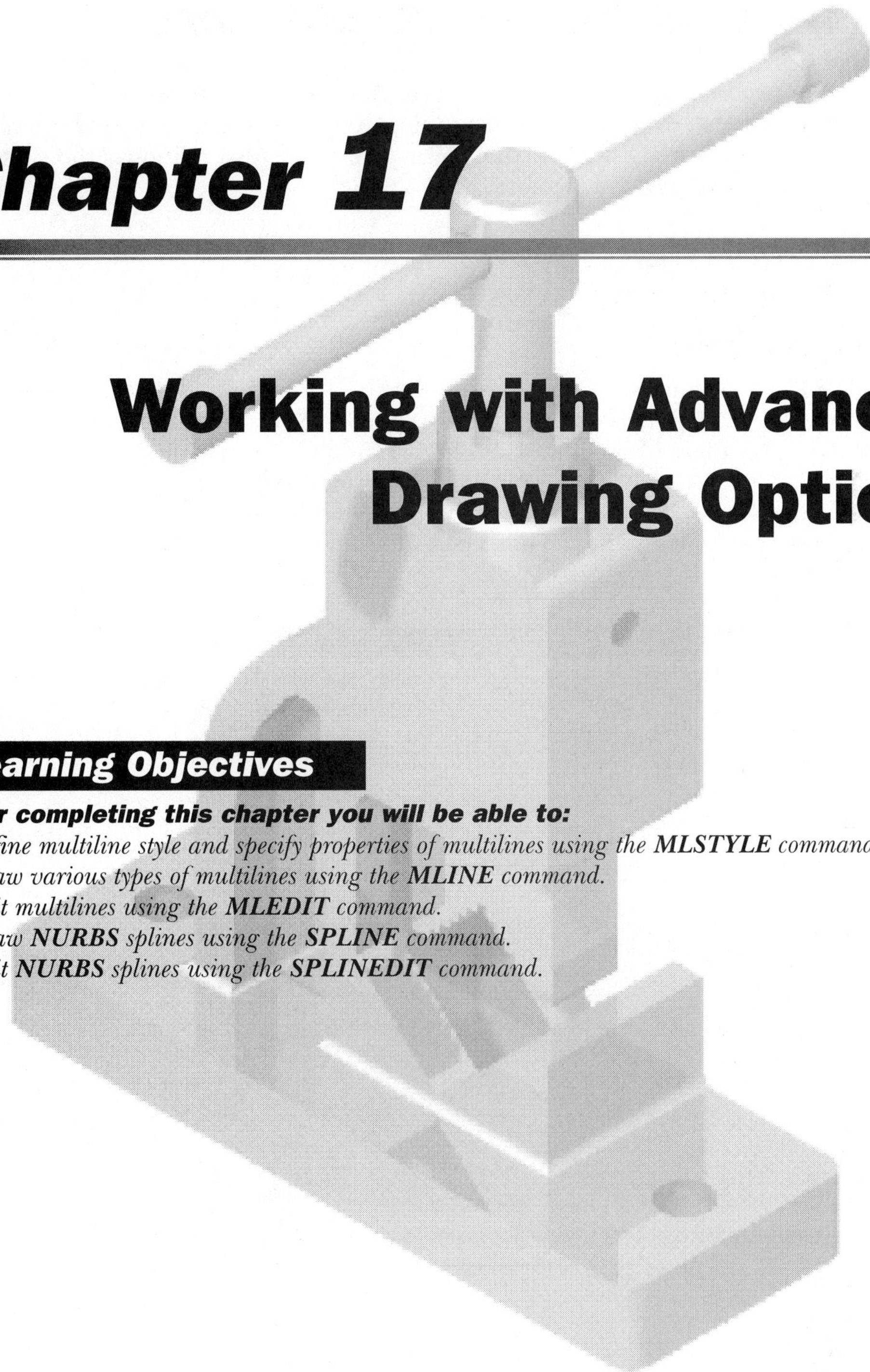

Learning Objectives

After completing this chapter you will be able to:

- *Define multiline style and specify properties of multilines using the* ***MLSTYLE*** *command.*
- *Draw various types of multilines using the* ***MLINE*** *command.*
- *Edit multilines using the* ***MLEDIT*** *command.*
- *Draw* ***NURBS*** *splines using the* ***SPLINE*** *command.*
- *Edit* ***NURBS*** *splines using the* ***SPLINEDIT*** *command.*

UNDERSTANDING THE USE OF MULTILINES

The **Multiline** feature allows you to draw composite lines that consist of multiple parallel lines. Multilines consist of between 1 and 16 parallel lines called **elements**. You can draw these lines with the **MLINE** command. Before drawing multilines, you need to set the multiline style. This can be accomplished using the **MLSTYLE** command. Also, editing of the multilines is made possible by the **MLEDIT** command.

DEFINING MULTILINE STYLE

Menu:	Format > Multiline Style
Command:	MLSTYLE

Using the **MLSTYLE** command, you can create the style of multilines. You can specify the number of elements in the multiline and the properties of each element. The style also controls the end caps, the end lines, and the color of multilines and fill. When you invoke this command, AutoCAD displays the **Multiline Styles** dialog box (Figure 17-1). With this dialog box, you can set the spacing between the parallel lines, linetype pattern, colors, solid fill, and capping arrangements. By default, the multiline style (STANDARD) has two lines that are offset at 0.5 and -0.5.

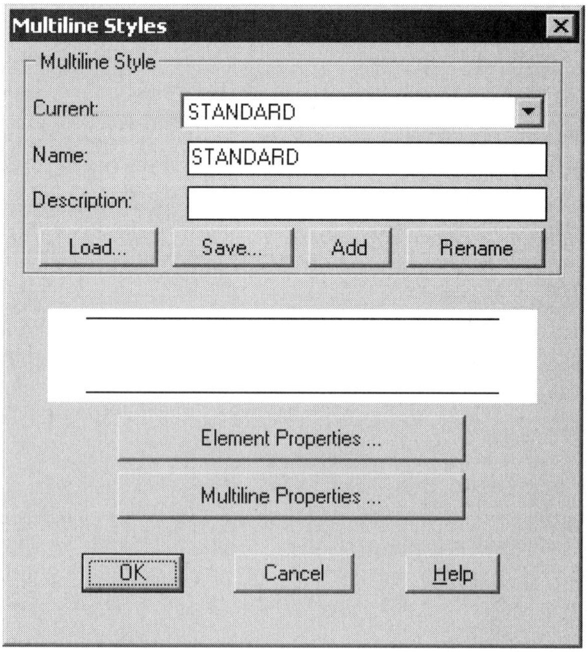

***Figure 17-1** The **Multiline Styles** dialog box*

Note

You cannot modify the style of a multiline that is used in the current drawing. You need to first delete the multiline from all the places where it is used and then only you can modify its style.

Multiline Style Area

The options provided under this area are used to create, load, save, and rename the multiline styles. This area has the following options.

Current

This edit box displays the name of the current multiline style. If several styles have been defined, you can select a style from the **Current** drop-down list to make it current. The current multiline style is saved in the **CMLSTYLE** system variable. By default this list displays the default style, **STANDARD**.

Name

This edit box lets you enter the name of the new multiline style you want to define. The name can have up to thirty-one characters and should not have spaces. By default, the current style name is displayed in this edit box. To create a new style, enter the name of the new style in this edit box and then choose the **Add** button.

Description

This edit box allows you to enter the description of the new multiline style. The description can be up to 255 characters long, including spaces. This is an optional edit box and you can leave it blank if you want.

Load

This button allows you to load a multiline style that was saved earlier, from an external multiline library file (*acad.mln*). When you choose this button, AutoCAD displays the **Load Multiline Styles** dialog box, see Figure 17-2. This dialog box displays all the multiline styles available in the current *mln* directory. If the style is saved in another directory or file, choose the **File** button to display the **Load Multiline Style From File** dialog box and then select the multiline file you want to load. Once the file is loaded, all the styles saved under it are displayed in the list box. You can select the style and then choose **OK** to make it current.

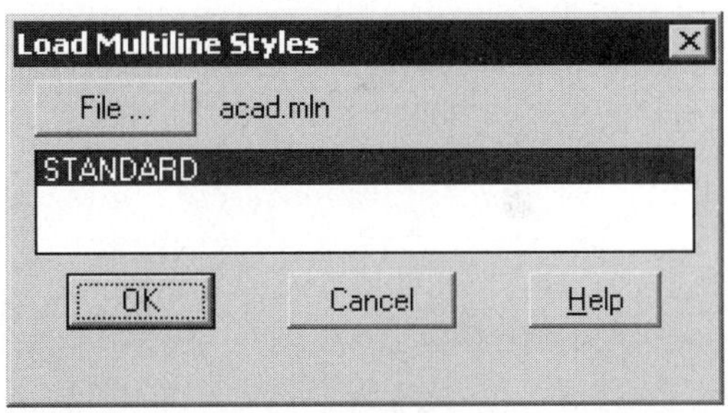

*Figure 17-2 The **Load Multiline Styles** dialog box*

Save

This button lets you save a multiline style to an external file (*.mln*). When you choose this button, AutoCAD displays the **Save Multiline Style** dialog box. This dialog box lists the names of the available predefined multiline style files (*.mln*). From the file listing, select, or enter the name of the file where you want to save the current multiline style and choose the **Save** button. By default, the multiline styles get saved in the *acad.mln* file in the *AutoCAD2004\Support* folder.

Add

This button allows you to add the new multiline style name displayed in the **Name** edit box to the **Current** drop-down list of available multiline styles. Enter the name in the **Name** edit box and then choose this button to create a new multiline style.

Rename

This button allows you to rename the current multiline style. The multiline style that is displayed in the **Current** drop-down list will be renamed using the name displayed in the **Name** edit box. You can even rename the default **STANDARD** style using this button.

Preview Area

The **Multiline Styles** dialog box also displays the current multiline configuration in the preview area. The panel will display the color, linetype, and relative spacing of the lines. As you select another multiline style and make it current, the image in the preview area also gets updated accordingly.

Element Properties

This button modifies the element properties of the multiline style. The element properties include the number of lines, color, and linetype of the lines comprising the multiline. When you choose this button, the **Element Properties** dialog box (Figure 17-3) will be displayed. This dialog box is used to define the element properties of a new multiline style. Before making any changes in this dialog box, you should enter a new multiline name in the **Name** edit box of the **Multiline Styles** dialog box, and then make it current using the **Add** button.

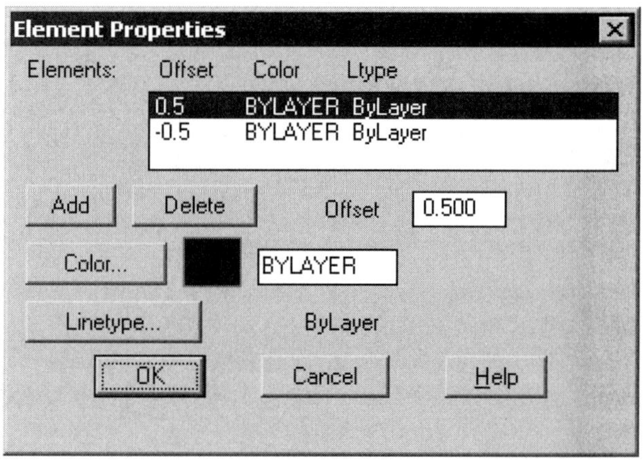

***Figure 17-3** The **Element Properties** dialog box*

This dialog box gives you the following options for setting the properties of the individual lines (elements) that constitute the multiline.

Elements

This list box displays the offset, color, and linetype of each line that constitutes the current multiline style. The lines are always listed in descending order based on the offset distance. For example, a line with a 0.5 offset will be listed first and a line with a 0.25 offset will be listed next. You can select an element to modify from the list box and the selected element is highlighted.

Add
This button allows you to add new lines to the current line style. The maximum number of lines you can add is 16. When you choose the **Add** button, AutoCAD inserts a line with the offset distance of 0.0, **Color** = BYLAYER, and **Ltype** = BYLAYER. The new element gets added between two existing elements. After the line is added, you can change its offset distance, color, or linetype. The offset distance can be entered in the **Offset** edit box. Similarly, you can choose the **Color** button to display the **Select Color** dialog box, where you can specify a color for the selected element of the multiline. Also, you can choose the **Linetype** button to display the **Select Linetype** dialog box. You can use this dialog box and its options to load and select the linetype you want to apply to the selected element of a multiline.

Delete
This button allows you to delete a selected element from the multiline style. The highlighted element is removed from the **Elements** list box.

Offset
This edit box allows you to change the offset distance of the selected line in the **Elements** list box. The offset distance is defined with respect to the origin (0,0). The offset distance can be a positive or a negative value, which enables you to center the lines.

Color
This button allows you to assign a color to the selected line. When you choose this button, AutoCAD displays the **Select Color** dialog box. You can select a color from the dialog box or enter a color number or name in the edit box located to the right of the color swatch. Choosing the color swatch also displays the **Select Color** dialog box.

Linetype
This button allows you to assign a linetype to the selected line. When you choose this button, AutoCAD displays the **Select Linetype** dialog box. After selecting a linetype in this dialog box, choose **OK** to exit. You can also choose the **Load** button in the **Select Linetype** dialog box to display the **Load or Reload Linetypes** dialog box, where you can select a linetype you want to use. The selected linetype is displayed in the **Select Linetype** dialog box list box. You can now select the specific linetype and apply it to a selected element of a multiline.

After you have added the elements to a multiline and set their properties, choose **OK** to exit the **Element Properties** dialog box. An image of the current multiline style is displayed in the line preview area in the **Multiline Styles** dialog box.

Multiline Properties
When you choose the **Multiline Properties** button in the **Multiline Styles** dialog box, AutoCAD displays the **Multiline Properties** dialog box, see Figure 17-4. You can use this dialog box to define multiline properties such as display joints, end caps, and background fill. The options provided under this dialog box are discussed next. It is advised that you first set the element properties of the multiline and then set these properties.

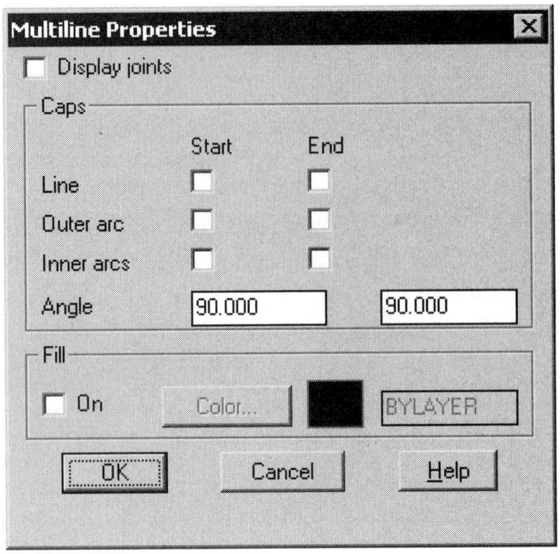

Figure 17-4 The ***Multiline Properties*** *dialog box*

Display Joints

If you select the **Display joints** check box, AutoCAD will draw a line across all elements of the multiline at the vertex points, where the multiline changes direction. The joints of the multiline are also called miter joints. If you draw only one multiline segment, no line is drawn at the miter joint.

Caps Area

The options under this area are used to set the type and location of the start and end caps of the multilines. The options under this area are discussed next.

Line. Selecting the corresponding **Start** and **End** check boxes of this option caps the multiline with a line at the start point and endpoint (Figure 17-5).

Outer Arc. Selecting the corresponding **Start** and **End** check boxes of this option draws a semicircular arc between the endpoints of the outermost lines (Figure 17-5).

Inner Arcs. This option draws a semicircular arc between the two innermost lines at the start point and endpoint of a multiline. The arc is drawn between the even-numbered inner lines (Figure 17-5). For example, if there are two inner lines, an arc will be drawn at the ends of these lines. On the other hand, if there are three inner lines, the two outer lines are capped with the arc and the middle line is not capped.

Angle. By default, the end caps of a multiline are drawn tangentially to the multiline elements, that is, the angle the cap makes with the multiline elements is 90-degree. You can control the cap angle at the start and end of a multiline by entering angles in the corresponding **Angle** edit boxes (Figure 17-5). This angle can be between 10-degree and 170-degree.

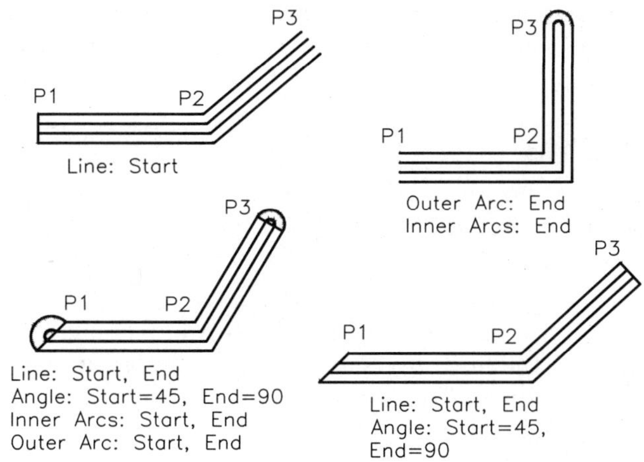

***Figure 17-5** Drawing multilines with different end cap specifications*

Fill Area

The options under this area are used to fill the multiline with a specified color. The **On** check box toggles the fill option on or off. By default, this check box is cleared. This means that there is no fill effect in the multiline. To specify a fill effect, select this check box. When you select this check box, the **Color** button, swatch, and edit box are made available. You can select the color of the fill by choosing the **Color** button. When you choose this button, the standard **Select Color** dialog box is displayed. You can also choose the color swatch to display the **Select Color** dialog box. After you have selected a color to apply to the fill, choose **OK** to exit the dialog box. You can also set the color by entering the color name or color number in the edit box located to the right of the color swatch box.

Once you have specified all the additional multiline properties in the **Multiline Properties** dialog box, choose **OK** to exit it with the modifications. The current modifications are visible in the preview area in the **Multiline Styles** dialog box.

DRAWING MULTILINES

Menu:	Draw > Multiline
Command:	MLINE

The **MLINE** command is used for drawing multilines. The following is the prompt sequence for the **MLINE** command.

Current settings: Justification = Top, Scale = 1.00, Style = STANDARD
Specify start point or [Justification/Scale/STyle]: *Select a start point or enter an option.*
Specify next point: *Select the second point.*
Specify next point or [Undo]: *Select next point or enter* **U** *for undo.*
Specify next point or [Close/Undo]: *Select next point, enter* **U**, *or enter* **C** *for close. You can also right-click to display the shortcut menu and choose* ***Enter***, ***Cancel***, ***Close***, ***Undo***, ***Pan***, *or* ***Zoom***.

When you invoke the **MLINE** command, it always displays the current status of the multiline justification, scale, and style name. Remember that all line segments created in a single **MLINE** command are a single entity. The options provided under this command are discussed next.

Justification Option

The justification determines how a multiline is drawn between the specified points. Three justifications are available for the **MLINE** command. They are **Top**, **Zero**, and **Bottom**.

Top

This justification produces a multiline in which the top line coincides with the points you selected on the screen. Since the line offsets in a multiline are arranged in descending order, the line with the largest positive offset will coincide with the selected points. This is the default justification option, see Figure 17-6.

Zero

This option will produce a multiline in which the zero offset position of the multiline coincides with selected points. Multilines will be centered if positive and negative offsets are equal, see Figure 17-6.

Bottom

This option will produce a multiline in which the bottom line (the line with the least offset distance) coincides with the selected point when the line is drawn from left to right, see Figure 17-6.

Scale Option

The **Scale** option allows you to change the scale of the multiline. For example, if the scale factor is 0.5, the distance between the lines (offset distance) will be reduced to half, as shown in Figure 17-7. Therefore, the width of the multiline will be half of what was defined in the multiline style. A negative scale factor will reverse the order of the offset lines. The multilines will now be drawn such that the line with the maximum negative offset distance is at the top and the line with the maximum positive offset distance is at the bottom. For example, if you enter a scale factor of -0.5, the order in which the lines are drawn will be reversed, and the offset distances will be reduced by half. Here it is assumed that the lines are drawn from left to right. If the lines are drawn from right to left, the offsets are reversed. Also, if the scale factor is 0, AutoCAD forces the multiline into a single line. However, the line will still possess the properties of a multiline. This scale factor does not affect the linetype scale factor (**LTSCALE**).

STyle Option

The **STyle** option allows you to change the current multiline style. The style must be defined before using the **STyle** option to change the style.

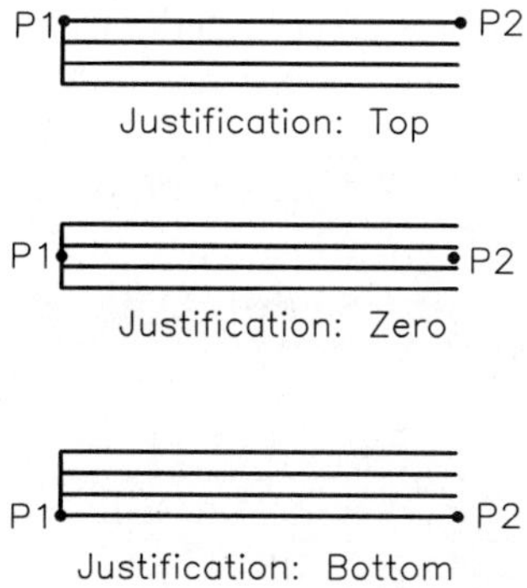

Figure 17-6 *Justifications in multilines*

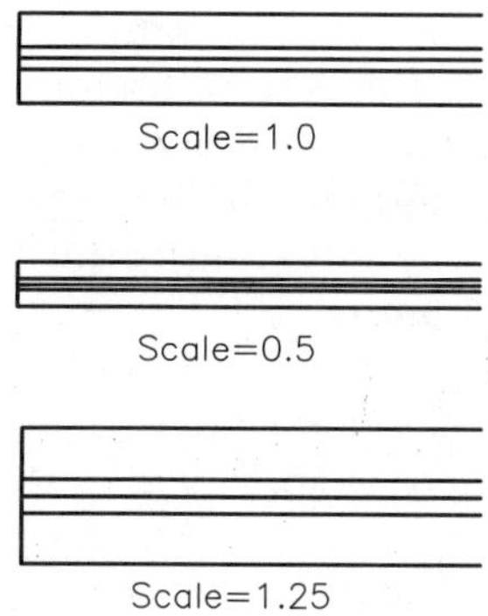

Figure 17-7 *Different multiline scales*

EDITING MULTILINES USING GRIPS

Multilines can be easily edited using grips (Figure 17-8). When you select a multiline, the grips appear at the endpoints based on the justification used when drawing multilines. For example, if the multilines are top-justified, the grips will be displayed at the endpoint of the first (top) line segment. Similarly, for zero and bottom-justified multilines, the grips are displayed on the centerline and bottom line, respectively.

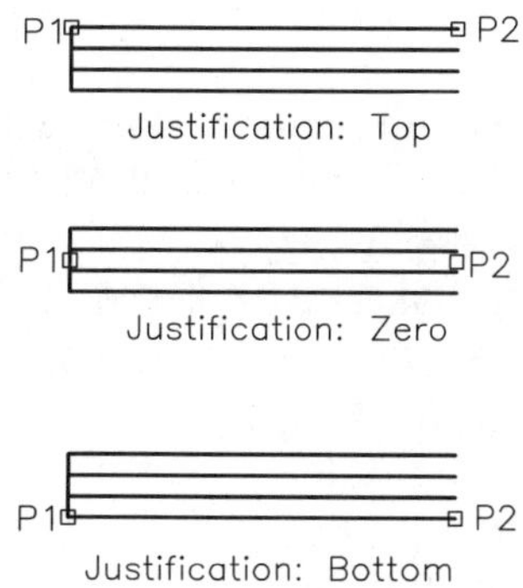

Figure 17-8 *Using grips to edit multilines*

Note

*Multilines do not support certain editing commands such as **BREAK, CHAMFER, FILLET, TRIM,** and **EXTEND**. However, commands such as **COPY, MOVE, MIRROR, STRETCH,** and **EXPLODE** and most object snap modes can be used with multilines.*

Tip

*You must use the **MLEDIT** command to edit multilines. The **MLEDIT** command has several options that make it easier to edit these lines.*

EDITING MULTILINES USING THE DIALOG BOX

Menu:	Modify > Multiline
Command:	MLEDIT

When you invoke the **MLEDIT** command, AutoCAD displays the **Multiline Edit Tools** dialog box, see Figure 17-9. This dialog box contains five basic editing tools. To edit a multiline, first select the editing operation you want to perform by double-clicking on the image tile or by selecting the image tile and then choosing **OK**. The name of the editing operation you select is displayed at the bottom left corner of the **Multiline Edit Tools** dialog box. Once you have selected the editing option, AutoCAD will prompt you to select the first and second multiline. After editing one set of multilines, you will be again prompted to select the first multiline or undo the last editing. Press ENTER after you have finished editing to exit this command. The following is the list of options for editing multilines.

Cross Intersection
Closed Cross, Open Cross, Merged Cross

Tee Intersection
Closed Tee, Open Tee, Merged Tee

Corner Joint

Adding and Deleting Vertices
Add Vertex, Delete Vertex

Cutting and Welding Multilines
Cut Single, Cut All, Weld All

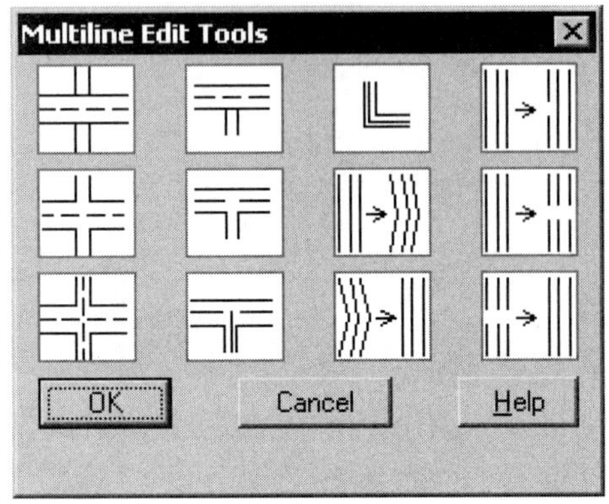

*Figure 17-9 The **Multiline Edit Tools** dialog box*

Cross Intersection

With the **MLEDIT** command options, you can create three types of cross intersections: closed, open, and merged. You must be careful about the order in which you select the multilines for editing. The order in which the multilines are selected determines the edited shape of a multiline (Figure 17-10). The cross intersection can belong to a self-intersecting multiline or to two different multilines.

When you invoke the **MLEDIT** command, the **Multiline Edit Tools** dialog box is displayed. Select the **Closed Cross** image tile and then choose the **OK** button. The dialog box is removed from the screen and the following prompts appear in the command line.

Select first mline: *Select the first multiline.*
Select second mline: *Select the second multiline.*
Select first mline or [Undo]: *Select another multiline, press ENTER to end the command, or enter U to undo the last operation.*

If you enter U, AutoCAD will undo the last editing operation and prompt you to select a first multiline. However, if you do not enter U and select another multiline, AutoCAD will prompt you to select the second multiline.

Tee Intersection

With the **MLEDIT** command options, you can create three types of tee-shaped intersections: closed, open, and merged. As with the cross intersection, you must be careful how you select

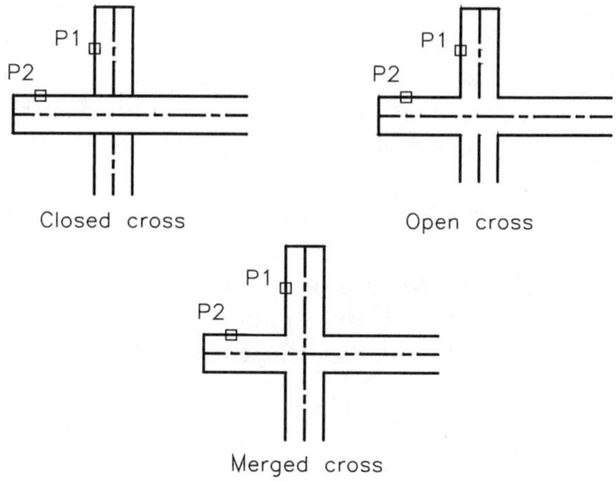

Figure 17-10 *Three types of cross intersections*

the objects because the order in which you select them determines the edited shape of a multiline (Figure 17-11). The prompt sequence for the tee intersection is the same as that for the cross intersection.

When you invoke the **MLEDIT** command, the **Multiline Edit Tools** dialog box is displayed. Select the **Closed Tee** image tile and then choose **OK**. The dialog box is removed from the screen, and the following prompts appear in the command line.

Select first mline: *Select the first multiline.*
Select second mline: *Select the intersecting multiline.*
Select first mline or [Undo]: *Select another multiline, enter **U** to undo the last operation, or press ENTER to end the command.*

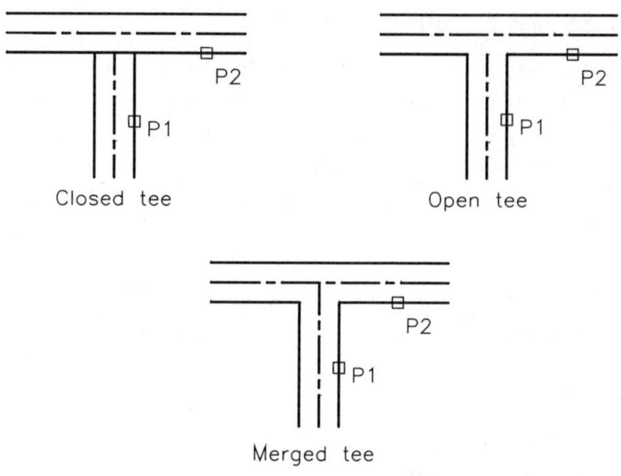

Figure 17-11 *Three types of tee intersections*

Corner Joint

The **Corner Joint** option creates a corner joint between the two selected multilines. The multilines must be two separate objects (multilines) or a self-intersecting multiline (Figure 17-12). When you specify the two multilines, AutoCAD trims or extends the first multiline to intersect with the second one.

The prompt sequence that follows when you invoke this option is given next.

Select first mline: *Select the multiline to trim or extend.*
Select second mline: *Select the intersecting multiline.*
Select first mline or [Undo]: *Select another multiline, enter **U** to undo the last operation, or press ENTER to end the command.*

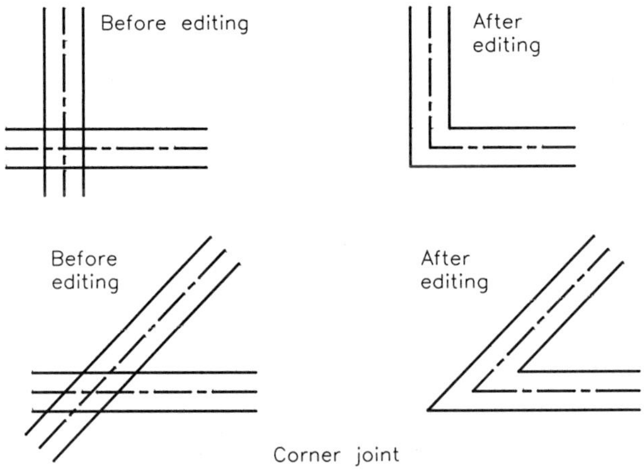

Figure 17-12 *Editing corner joints*

Adding and Deleting Vertices

You can use the **MLEDIT** command to add or delete the vertices of a multiline (Figure 17-13). When you select a multiline for adding a vertex, AutoCAD inserts a vertex point at the point where you clicked on the object while selecting it. Later, you can move the vertex by using grips. Similarly, you can also use the **MLEDIT** command to delete vertices by selecting the object whose vertex point you want to delete. AutoCAD removes the vertex that is nearest to the point where you click to select the multiline segment.

When you invoke the **MLEDIT** command, the **Multiline Edit Tools** dialog box is displayed. Select the **Add Vertex** image tile and then choose the **OK** button. The dialog box is removed from the screen, and the following prompts appear in the command line.

Select mline: *Select the multiline for adding vertex.*
Select mline or [Undo]: *Select another multiline for modifications, enter **U** to undo the last operation, or press ENTER to exit the command.*

11. Choose the **OK** button to accept the changes to the element properties and return to the **Multiline Styles** dialog box. The new multiline style will be displayed in the preview area.

12. Choose the **OK** button in the **Multiline Styles** dialog box to return to the drawing editor. To test the new multiline style, use the **MLINE** command and draw a series of lines (Figure 17-15).

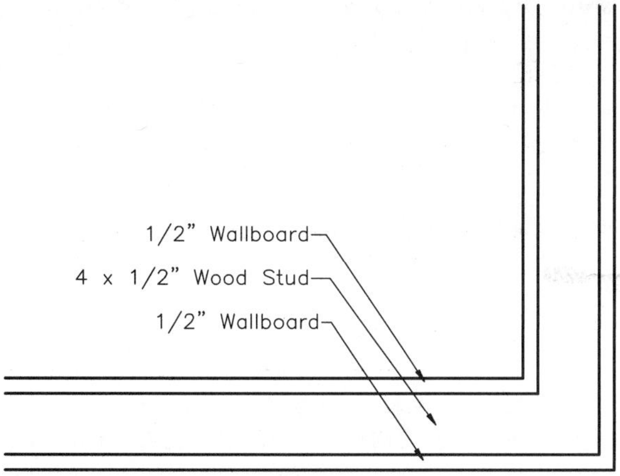

Figure 17-15 *Multiline style created for Example 1*

CREATING REVISION CLOUDS*

Toolbar:	Draw > Revcloud
Menu:	Draw > Revision Cloud
Command:	REVCLOUD

The **REVCLOUD** command is used to create a cloud shaped polyline. Figure 17-16 shows the use of the **REVCLOUD** command. This command can be used to highlight the details of a drawing. The prompt sequence that will follow when you choose the **Revcloud** button from the **Draw** toolbar is as follows.

Minimum arc length: 0.5000 Maximum arc length: 0.5000
Specify start point or [Arc length/Object] <Object>: *Specify the start point of the revision cloud.*
Guide crosshairs along cloud path...

As you move the cursor, different arcs of the cloud with varied radii are drawn. When the start point and endpoints meet, the revision cloud is completed and you get a message.

Revision Cloud Finished.

Using the **Arc length** option, you can define the length of the arcs to be drawn and all the arcs drawn are of constant length.

Using the **Object** option, you can convert a closed loop into a revision cloud. Note that the selected closed loop should be a single entity such as an ellipse, a circle, a rectangle, a polyline, and so on.

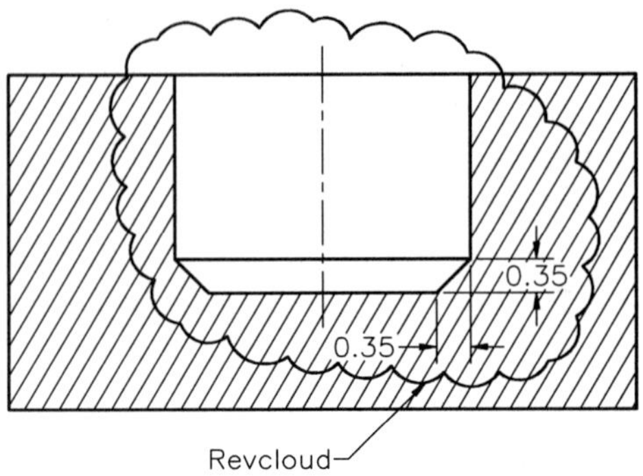

Figure 17-16 *Creating revclouds*

Note

REVCLOUD *stores the last used arc length in the system registry. The value is multiplied by* ***DIMSCALE*** *to provide consistency when the program is used with drawings that have different scale factors.*

CREATING WIPEOUTS*

Menu:	Draw > Wipeout
Command:	Wipeout

The **WIPEOUT** command creates a polygonal area to cover the existing objects with the current background color. The area defined by this command is governed by the wipeout frame. The frame can be turned on and off for editing and plotting the drawings, respectively. This command can be used to add notes and details to the drawing. The prompt sequence that will follow when you choose the **Wipeout** button from the **Draw** toolbar is as follows.

Specify first point or [Frames/Polyline] <Polyline>: *Specify the start point of the wipeout.*
Specify next point: *Specify the next point of the wipeout.*
Specify next point or [Undo]: *Specify the next point of the wipeout.*
Specify next point or [Close/Undo]: *Specify the next point of the wipeout.*

Figure 17-17 shows a drawing before creating wipeout and Figure 17-18 shows a drawing after creating wipeout. If you do not want the frame of the wipeout to be displayed, enter **F** at the Specify first point or [Frames/Polyline] <Polyline> prompt and turn the frame off. The next time when you draw a wipeout, the frames will not appear in it.

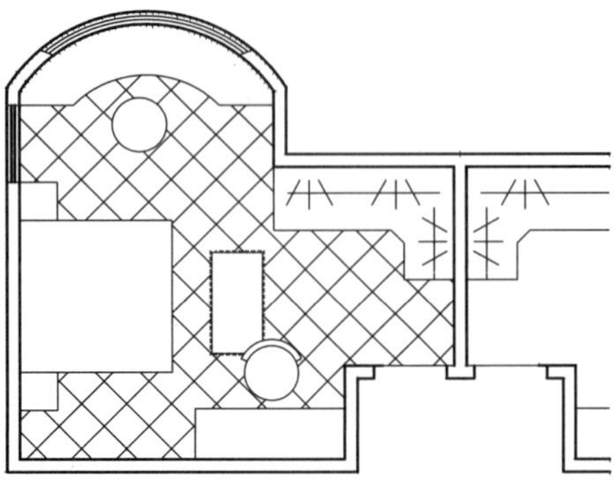
Figure 17-17 *Drawing before creating wipeout*

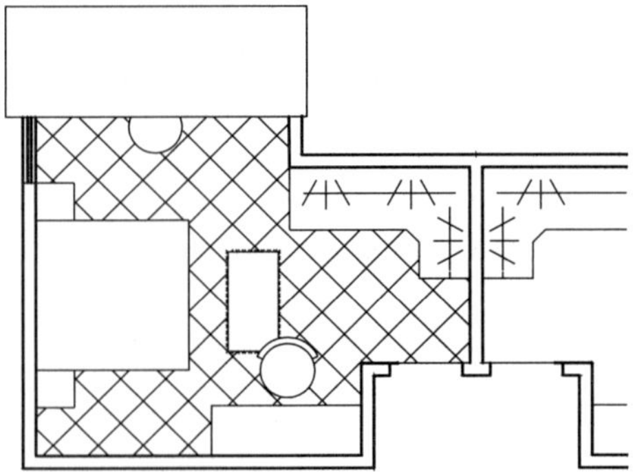
Figure 17-18 *Drawing after creating wipeout*

CREATING NURBS

Toolbar:	Draw > Spline
Menu:	Draw > Spline
Command:	SPLINE

The NURBS is an acronym for **NonUniform Rational Bezier-Spline**. These splines are considered true splines. In AutoCAD, you can create NURBS using the **SPLINE** command. The spline created with the **SPLINE** command (Figure 17-19) is different from the spline created using the **PLINE** command. The nonuniform aspect of the spline enables the spline to have sharp corners because the spacing between the spline elements that constitute a spline can be irregular. Rational means that irregular geometry such as arcs, circles, and ellipses can be combined with free-form curves. The Bezier-spline (B-spline) is the core that enables accurate fitting of curves to input data with Bezier's curve-fitting interface. Not only are spline curves more accurate compared to smooth polyline curves, but they also use less disk space. The following is the prompt sequence for creating the spline shown in Figure 17-20.

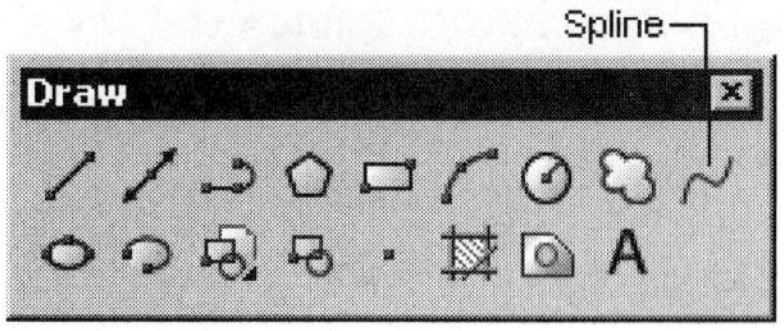

Figure 17-19 *Invoking the **SPLINE** command from the **Draw** toolbar*

Specify first point or [Object]: *Select point P1.*
Specify next point: *Select point P2.*
Specify next point or [Close/Fit tolerance] <start tangent>: *Select point P3.*
Specify next point or [Close/Fit tolerance] <start tangent>: *Select point P4.*
Specify next point or [Close/Fit tolerance] <start tangent>: *Select point P5.*
Specify next point or [Close/Fit tolerance] <start tangent>: *Select point P6.*
Specify next point or [Close/Fit tolerance] <start tangent>: *Press ENTER to end the process of point specification.*
Specify start tangent: *Press ENTER to accept the default start tangent.*
Specify end tangent: *Select point P7.*

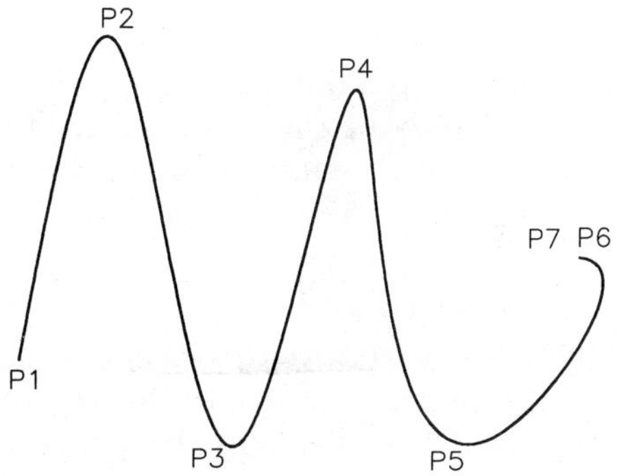

Figure 17-20 *Creating the spline*

You can specify the start and end tangents to change the direction in which the spline curve starts and ends at the **Specify start tangent:** and **Specify end tangent:** prompts. The **SPLINE** command options are discussed next.

Options for Creating the Splines

The options provided under this command for creating the splines are as follows.

Object

This option allows you to change a 2D or 3D splined polyline into a NURBS. The original splined polyline is deleted if the system variable **DELOBJ** is set to 1, which is the default value of the variable. You can change a polyline into a splined polyline using the **Spline** option of the **PEDIT** command.

Close

This option allows you to create closed NURBS. When you use this option, AutoCAD will automatically join the endpoint of the spline with the start point, and you will be prompted to define the start tangent only.

Fit Tolerance

This option allows you to control the fit of the spline between specified points. By default, this value is zero and so the spline passes through the points through which it is created. Using this option you can specify some tolerance value that will govern the spline creation, see Figure 17-21. The splines will be offsetted from the specified point by a distance equal to the tolerance value. The smaller the value of the tolerance, the closer the spline will be to the specified points.

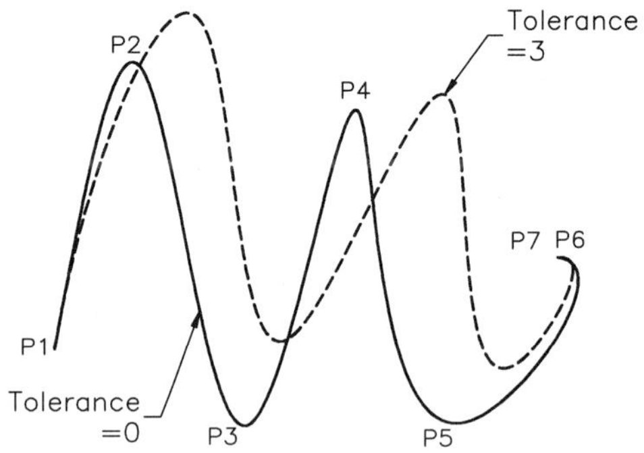

*Figure 17-21 Creating a spline with a **Fit Tolerance** of 0 and 3*

Start and End Tangents

This allows you to control the tangency of the spline at the start point and endpoint of the spline. If you press ENTER at these prompts, AutoCAD will use the default value. By default, the tangency is determined by the slope of the spline at the specified point.

EDITING SPLINES

Toolbar:	Modify II > Edit Spline
Menu:	Modify > Spline
Command:	SPLINEDIT

*Figure 17-22 Choosing the **Edit Spline** button from the **Modify II** toolbar*

The NURBS can be edited using the **SPLINEDIT** command (Figure 17-22). With this command, you can fit data in the selected spline, close or open the spline, move vertex points, and refine or reverse a spline. Apart from the ways mentioned in the preceding command box, you can also invoke the **SPLINEDIT** command by choosing **Spline Edit** from the shortcut menu that is displayed when you select a spline and right-click. The prompt sequence that will follow when you choose the **Edit Spline** button is given next.

Select spline: *Select the spline that is to be edited if not selected already, using the above-mentioned shortcut menu.*
Enter an option [Fit data/Close/Move vertex/Refine/rEverse/Undo]: *Select any one of the options.*

Options for Editing the Splines

The options provided under this command for editing the splines are described next.

Fit Data

When you draw a spline, the spline is fit to the specified points (data points). The **Fit Data** option allows you to edit these points. You can add, delete, or move the data points. These data points or control points are also referred to as fit points. For example, if you want to redefine the start and end tangents of a spline, select the **Fit Data** option, then select the **Tangents** option. The prompt sequence that will follow when you invoke this option is given next.

Select spline: *Select the spline that is to be edited.*
Enter an option [Fit data/Close/Move vertex/Refine/rEverse/Undo]: **F**
Enter a fit data option
[Add/Close/Delete/Move/Purge/Tangents/toLerance/eXit] <eXit>:

The start and end tangent points can be selected or their coordinates can be entered. The options available within the **Fit data** option are as follows.

Add. You can use this option to add new fit points to the spline. When you invoke this option, you will be prompted to specify the control point. This control point should be one of the existing control points on the spline. After selecting the existing control point, you will be prompted to specify the location of the new control points. The fit point you select and the next fit point appear as selected grips. You can now add a fit point between these two selected fit points as shown in Figure 17-23. If you select the start point or endpoints of the spline, only those points are highlighted. When you select the start point of the spline, you are prompted to specify whether you want to add the new fit point before or after the start point of the spline. AutoCAD will continue prompting for the location of new control points until you press ENTER at the **Specify new point <exit>** prompt.

Close/Open. This option allows you to close an open spline or open a closed spline, see Figure 17-24. If the spline is open, the **Close** option is available and if the spline is closed, the **Open** option is displayed.

Delete. This option allows you to delete a selected fit point from the spline. You can continue deleting fit points from a spline until only two fit points are left in the spline.

Move. You can move fit points by using this option. When you invoke this option, the start point of the spline is highlighted and the prompt sequence is as follows.

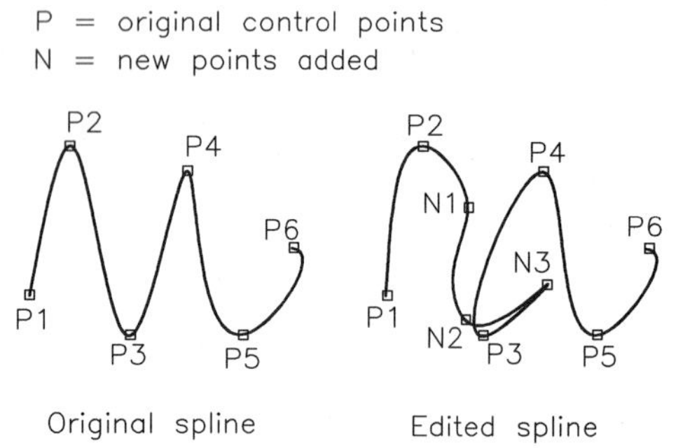

Figure 17-23 Original spline and edited spline after adding the data points

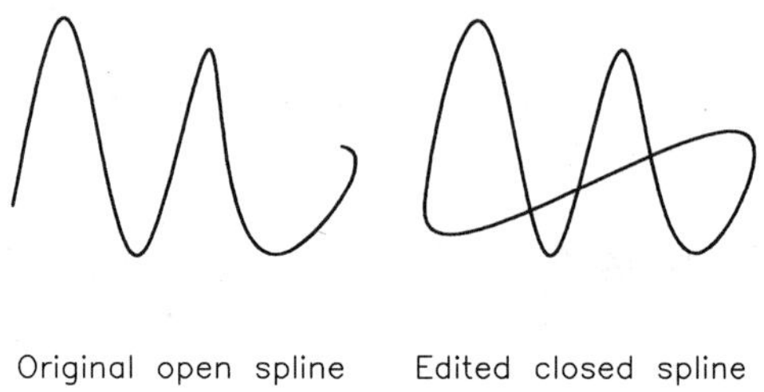

Figure 17-24 An open spline and a closed spline

Specify new location or [Next/Previous/Select point/eXit] <N>: *Select a new location for the start point using the mouse pick button or enter any one of these options.*

You can enter **N** if you want to select the next point, **P** if you want to select a previous point, or **S** if you want to select any other point. If you enter **X** at the preceding prompt, you can exit the command. Figure 17-25 shows the movement of data points in a spline. Remember that this option is used to move only the data points on the spline and not the control points on the Bezier control frame.

Purge. This option allows you to remove fit point data from a spline. This reduces the file size, which is useful when a drawing, for example, a landscape contour map, contains large number of splines. Purging simplifies the spline definition and the drawing file size. But, once fit point data has been removed from a spline, editing a spline gets difficult. Also, the **Fit Data** option is not displayed when you again use the **SPLINEDIT** command on the edited spline.

Tangents. This option allows you to modify the tangents of the start and endpoints of the selected spline, see Figure 17-26. When you invoke this option, you will be prompted to specify the start and end tangents for the spline. You can specify the start and end tangents or use the systems default tangents.

toLerance. This option allows you to modify the fit tolerance values of a selected spline. As discussed while creating the splines, different tolerance values produce different splines. A smaller tolerance value creates a spline that passes very closely through the definition points of a spline. When you invoke this option, you will be prompted to specify the tolerance value for the spline.

eXit. This option allows you to exit the **Fit data** option of editing the splines.

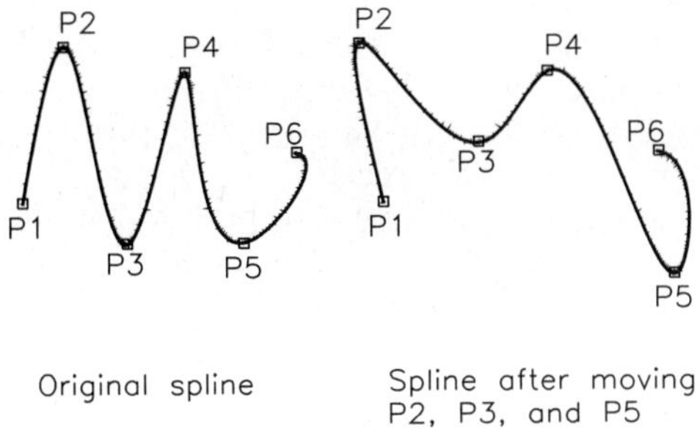

Figure 17-25 Original spline and edited spline after moving the data points

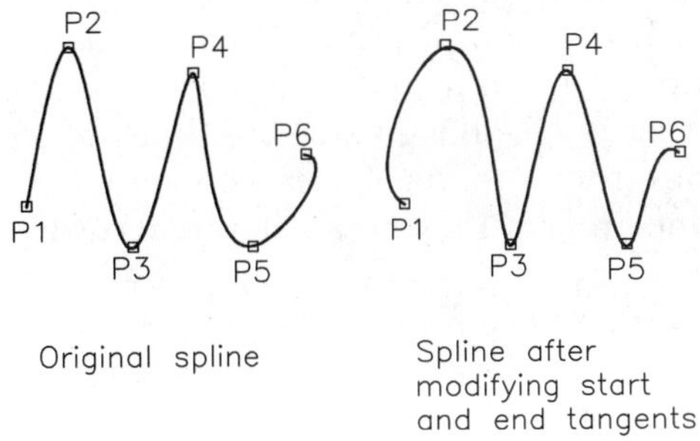

Figure 17-26 Original spline and the spline after modifying the start and end tangents

Close/Open

This option allows you to close an open spline or open a closed spline. When you select the **Close** option, AutoCAD lets you open, move the vertex, or refine or reverse the spline.

Move vertex

When you draw a spline, it is associated with the Bezier control frame. The **Move vertex** option allows you to move the vertices of the control frame. To display this frame with the spline, set the value of the **SPLFRAME** system variable to 1. The default value of this system variable is 0 and the frames are not displayed with the spline. The **Move vertex** option is similar to the **Move** option of the **Fit data** option of editing the splines. The prompt sequence that follows when you invoke this option is given next.

Enter an option [Fit data/Close/Move vertex/Refine/rEverse/Undo]: **M**
Specify new location or [Next/Previous/Select point/eXit] <N>:

Refine

This option allows you to refine a spline by adding more control points in the spline, elevating the order, or adding weight to vertex points. The prompt sequence that will follow when you invoke this option is given next.

Enter an option [Fit data/Close/Move vertex/Refine/rEverse/Undo]: **R**
Enter a refine option [Add control point/Elevate order/Weight/eXit] <eXit>:

Add control point. This option allows you to add more control points on the spline. When you invoke this option, you will be prompted to specify the location of the new point on the spline. You can directly specify the location of the new point on the spline using the left mouse button. A new control point will be added at the location you specified.

Elevate order. This option allows you to increase the order of a spline curve. An order of the curve can be defined as the highest power of the algebraic expression that defines the spline plus 1. For example, the order of a cubic spline will be 3 + 1 = 4. Using the **Elevate order** option, you can increase this order for a selected spline. This results in more control points on the curve and a greater possibility of controlling the spline. The value of the spline order varies from 4 to 26. You can only increase the order and not decrease it. For example, if a spline order is 18, you can elevate its order to any value greater than 18, but not less than 18.

Weight. You can also add weight to any of the vertices of the spline by using this option. When weight is added to a particular vertex, the spline gets pulled more toward it. Similarly, a lower value of weight of a particular spline vertex will result in the spline getting pulled less toward that particular vertex. In other words, adding weight to a particular point will force the selected point to maintain its tangency with the point. The more weight added to the point, the more the distance through which the spline will remain tangent to the point. By default, the spline gets pulled equally toward the vertices of the spline. The default value of weight provided to each control point is 1.0 and can have only positive values. Once you have added the weight to a point, you can proceed to the next point. You can also directly select the point to which the weight has to be added. The prompt sequence for using this option is given next.

Enter a refine option [Add control point/Elevate order/Weight/eXit] <eXit>: **W**
Enter new weight (current = 1.0000) or [Next/Previous/Select point/eXit] <N>:

rEverse

This option allows you to reverse the spline direction. This implies that when you apply this option to a spline, its start point becomes its endpoint and vice versa.

Undo

This option will undo the previous editing operation applied to a spline within the current session of the **SPLINEDIT** command. You can continue to use this option till you reach the spline as it was when you started to edit it.

Self-Evaluation Test

Answer the following questions, and then compare your answers to the correct answers given at the end of this chapter.

1. The name of the multiline style that you enter in the **Name** edit box of the **Multiline Styles** dialog box can have up to thirty-one characters and contain spaces. (T/F)

2. The **Zero** option will produce a multiline so that the zero offset position of the multiline coincides with the selected points. (T/F)

3. Multilines cannot be edited using grips. (T/F)

4. When using the **Top** justification option to draw a multiline, the line with the largest positive offset coincides with the points selected on the screen. (T/F)

5. When a multiline is cut, two separate multilines are created. (T/F)

6. The maximum number of lines you can add to a multiline style is up to _________.

7. When using the **Inner arcs** option for multilines, the capping arc is drawn between the _________ numbered inner lines.

8. The **Multiline Edit tools** dialog box contains _________ basic editing tools.

9. The _________ option of the **Fit data** option of the **SPLINEDIT** command reduces the file size by removing the fit data of the selected splines.

10. The _________ system variable stores the justification of the current multiline.

Review Questions

Answer the following questions.

1. By default, the *.mln* files are saved in the *AutoCAD2004/Support* folder. (T/F)

2. Once the **Purge** option of the **Fit Data** option of the **SPLINEDIT** command is used on a spline, editing it gets difficult and the **Fit Data** option of the **SPLINEDIT** command is no longer available for the particular spline. (T/F)

3. You can break or trim a multiline. (T/F)

4. You cannot change the lineweights of the lines that constitute the multiline. (T/F)

5. You cannot weld or join separate multilines. (T/F)

6. By default, the multiline styles are saved in which file?

   (a) *acad.mln* (b) *acad.pat*
   (c) *acad.mpr* (c) *acad.plt*

7. Which of the following is the default start and end angle for the multilines?

   (a) 90 (b) 180
   (c) 45 (d) 0

8. AutoCAD forces the multiline into a single line if the scale factor of the multiline is

   (a) 0 (b) 1
   (c) 2 (d) 5

9. Which of the following commands does not work on the multilines?

   (a) **MIRROR** (b) **FILLET**
   (c) **STRETCH** (d) **EXPLODE**

10. Which option can be used to reverse the direction of spline creation while editing the splines?

    (a) **Fit data** (b) **Refine**
    (c) **Reverse** (d) **Weight**

11. The current multiline style is saved in the __________ system variable and default value is STANDARD.

12. Three justifications are available for the **MLINE** command. They are __________ , __________ , and __________ .

13. Using the **MLEDIT** command options, you can create three types of cross intersections. They are __________ , __________ , and __________ .

14. If the Fit tolerance value of a spline is __________ , the spline passes exactly through the fit points of the spline.

15. To display the Bezier control frame with the spline, set the value of the __________ system variable to 1.

Exercises

Exercise 1 General

Create the drawing shown in Figure 17-27. Assume the missing dimensions.

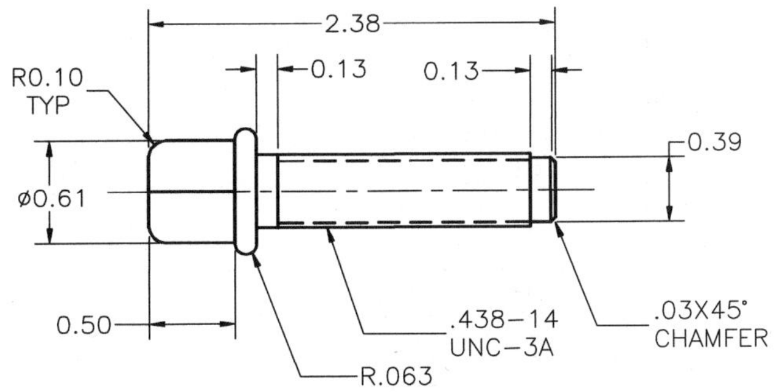

Figure 17-27 Drawing for Exercise 1

Exercise 2 Architectural

Create the drawing shown in Figure 17-28. Use the **MULTILINE** command to draw the walls. The wall thickness is 12 inches. Assume the missing dimensions.

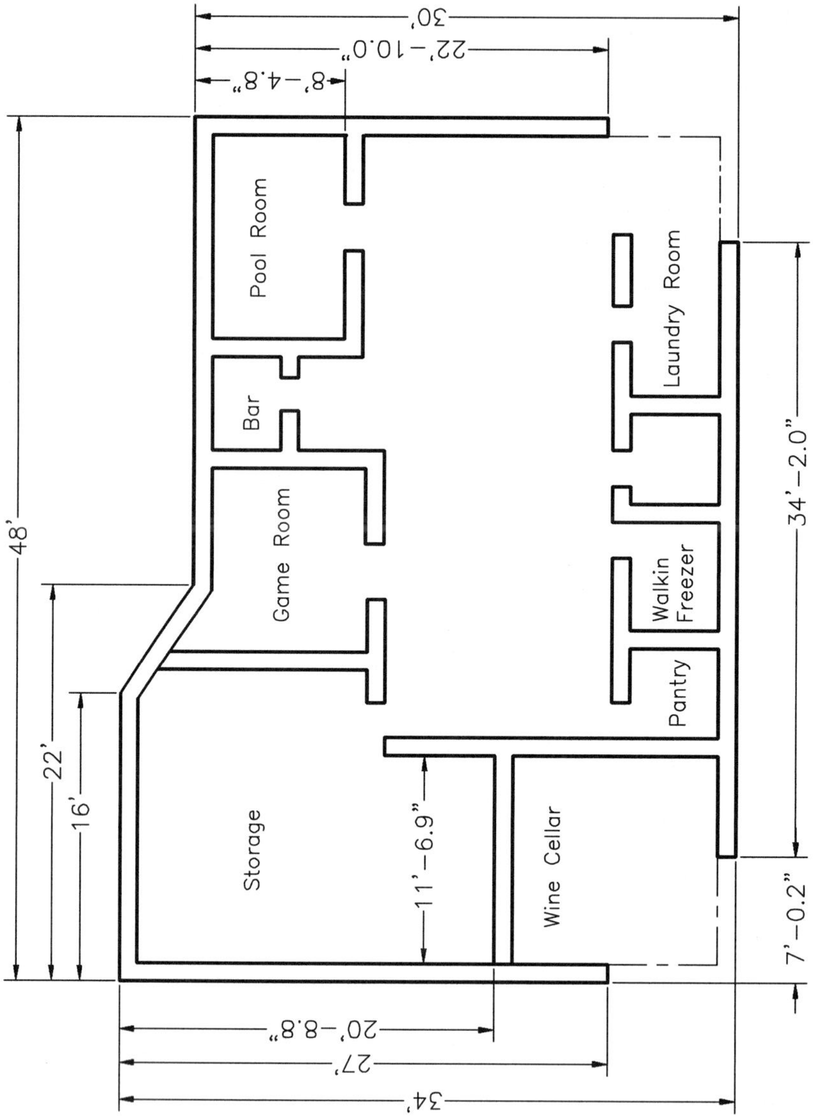

Figure 17-28 *Drawing for Exercise 2*

Problem Solving Exercise 1

General

Create the drawing shown in Figure 17-29. Some of the reference dimensions are given in the drawing. Assume the missing dimensions so that the drawing looks similar to the given drawing.

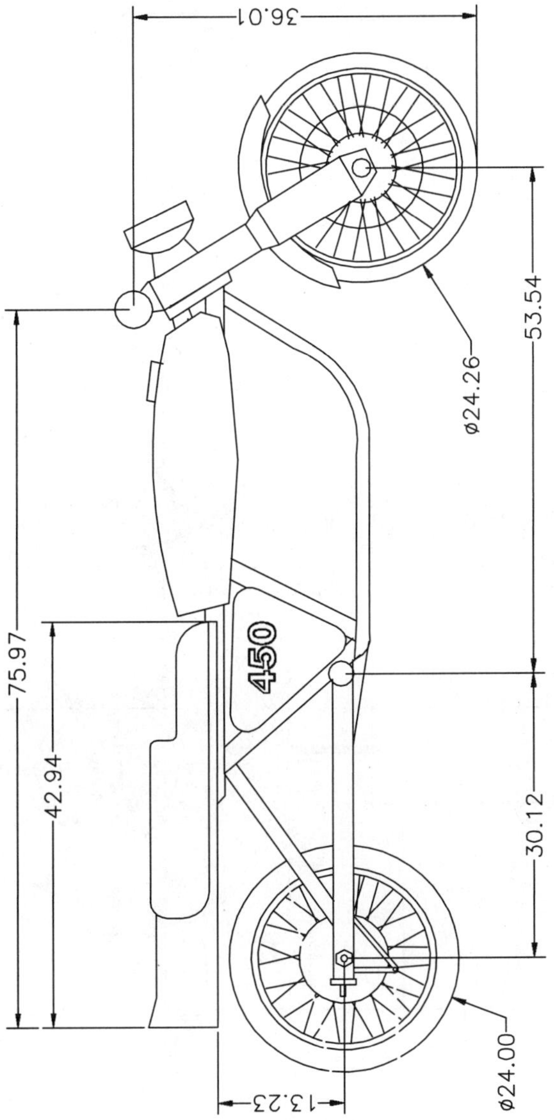

Figure 17-29 *Drawing for Problem Solving Exercise 1*

Problem Solving Exercise 2 *General*

Create the drawing shown in Figure 17-30. Some of the reference dimensions are given in the drawing. Assume the missing dimensions so that the drawing looks similar to the given drawing.

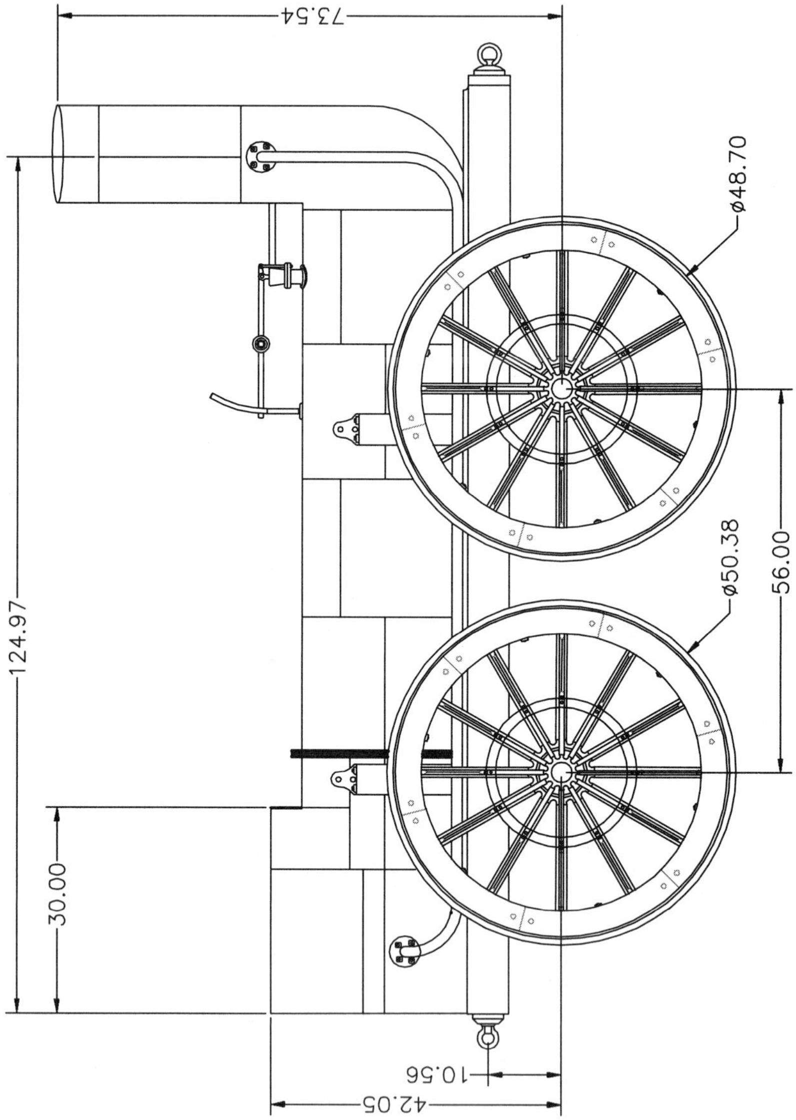

Figure 17-30 *Drawing for Problem Solving Exercise 2*

Problem Solving Exercise 3 *General*

Create the drawing shown in Figure 17-31. Some of the reference dimensions are given in the drawing. Assume the missing dimensions so that the drawing looks similar to the given drawing.

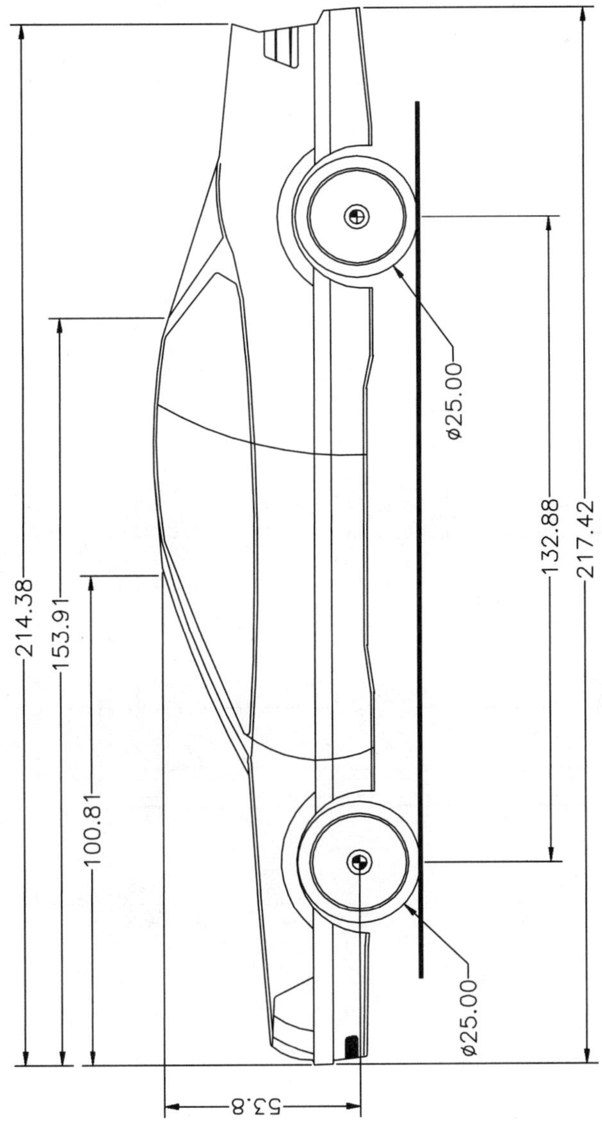

Figure 17-31 *Drawing for Problem Solving Exercise 3*

Problem Solving Exercise 4 *Architecture*

Create the drawing shown in Figure 17-32. Some of the reference dimensions are given in the drawing. Assume the missing dimensions so that the drawing looks similar to the given drawing.

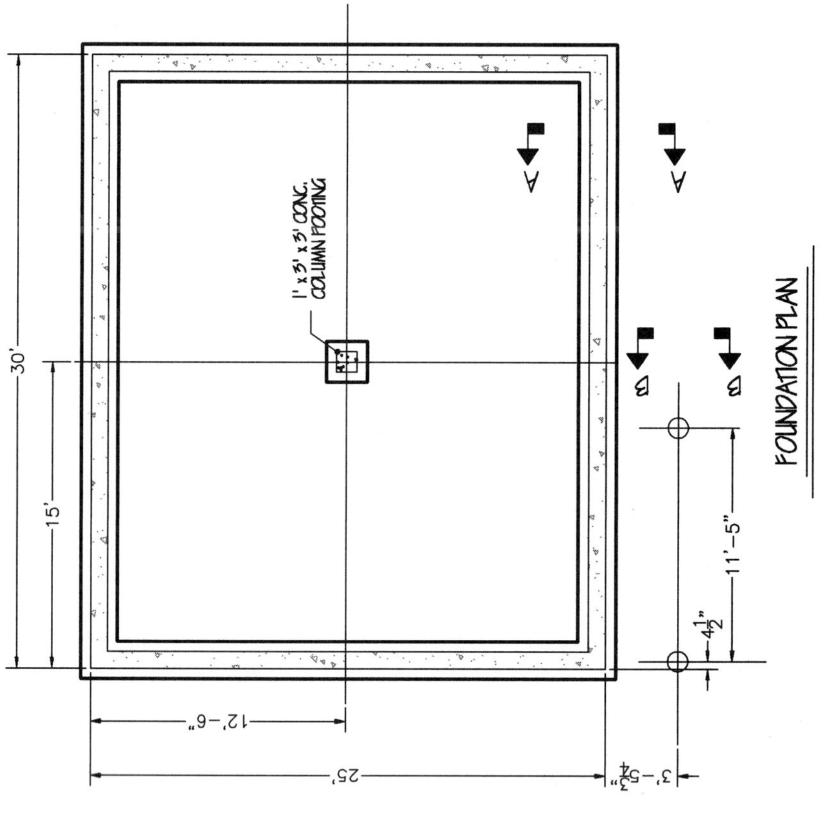

Figure 17-32 *Drawing for Problem Solving Exercise 4*

Problem Solving Exercise 5 *Architecture*

Create the drawing shown in Figure 17-33. Some of the reference dimensions are given in the drawing. Assume the missing dimensions so that the drawing looks similar to the given drawing.

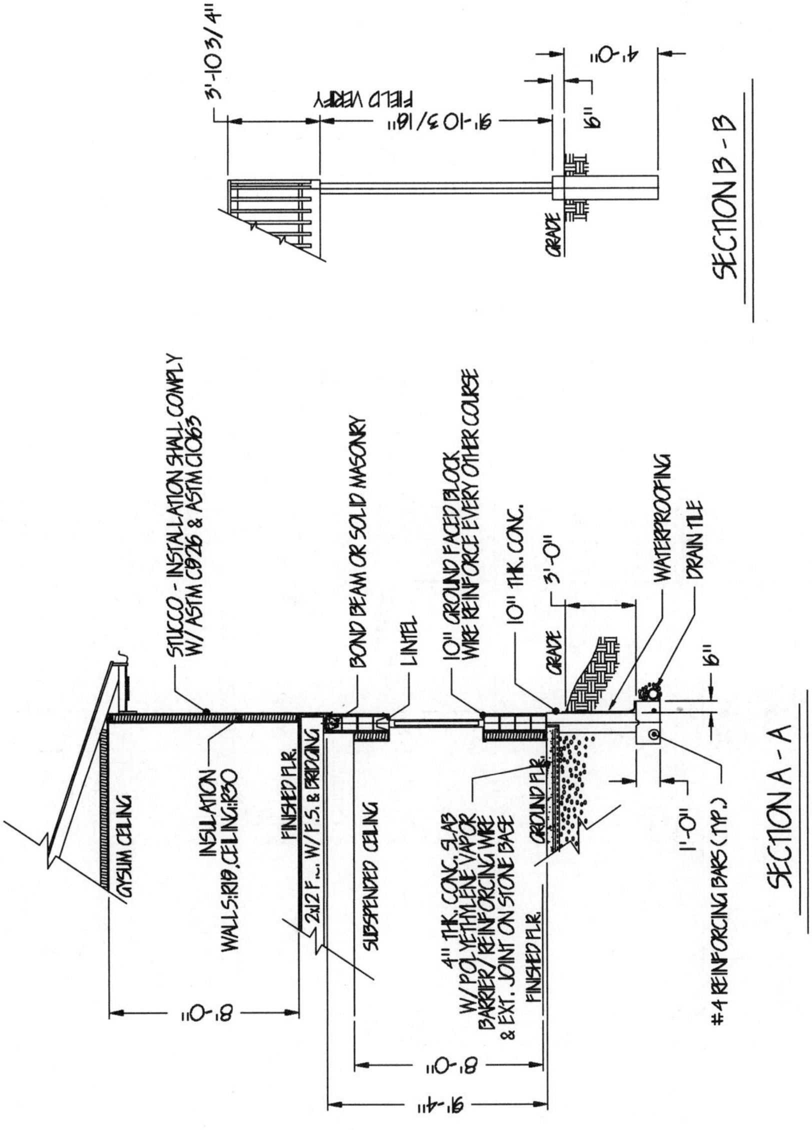

Figure 17-33 *Drawing for Problem Solving Exercise 5*

Problem Solving Exercise 6

Architecture

Create the drawing shown in Figure 17-34. Some of the reference dimensions are given in the drawing. Assume the missing dimensions so that the drawing looks similar to the given drawing.

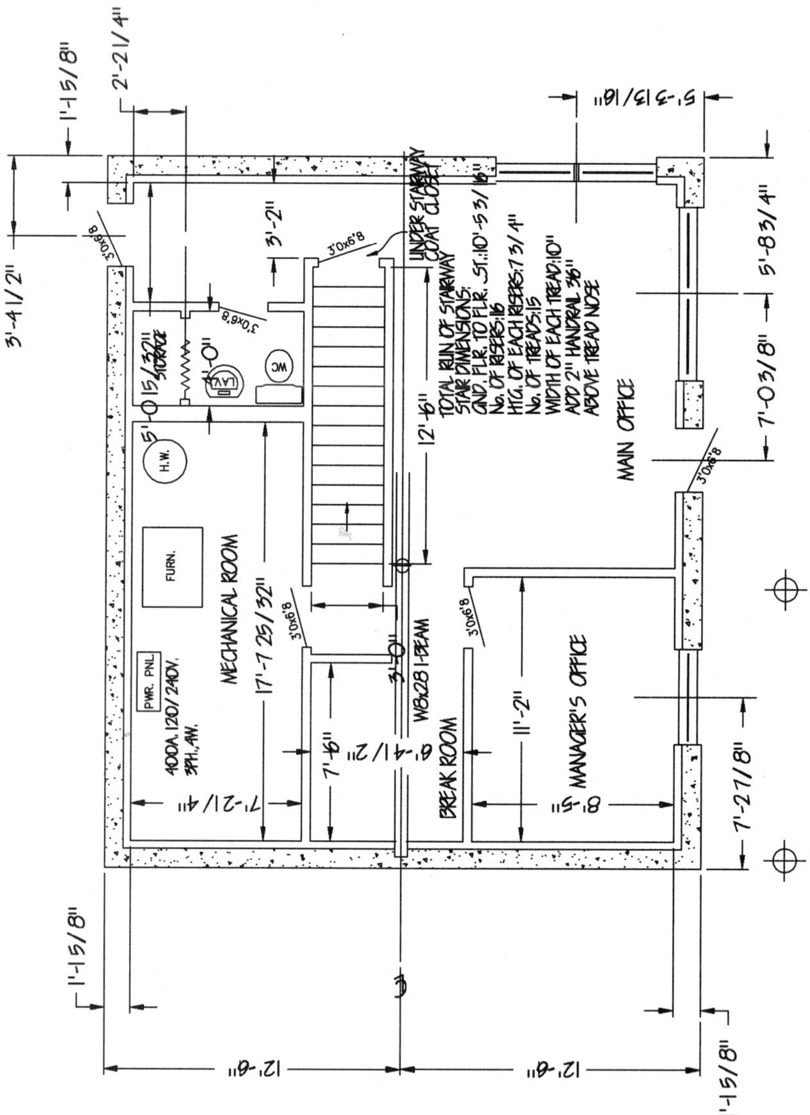

Figure 17-34 *Drawing for Problem Solving Exercise 6*

Problem Solving Exercise 7 — *Architecture*

Create the drawing shown in Figure 17-35. Some of the reference dimensions are given in the drawing. Assume the missing dimensions so that the drawing looks similar to the given drawing.

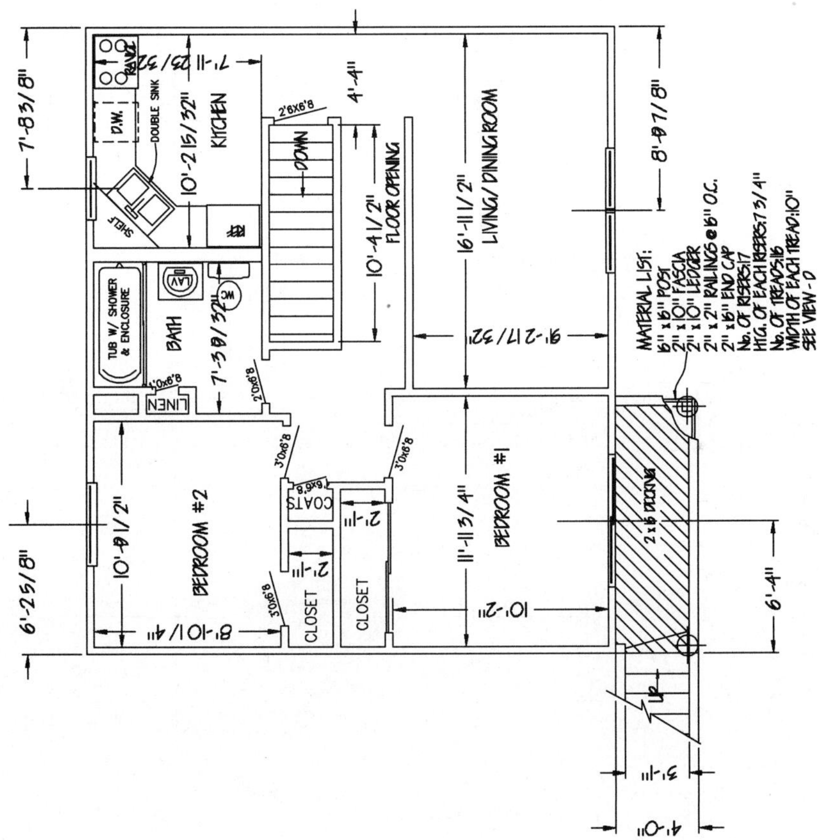

Figure 17-35 *Drawing for Problem Solving Exercise 7*

Problem Solving Exercise 8 *Architecture*

Create the drawing shown in Figure 17-36. Some of the reference dimensions are given in the drawing. Assume the missing dimensions so that the drawing looks similar to the given drawing.

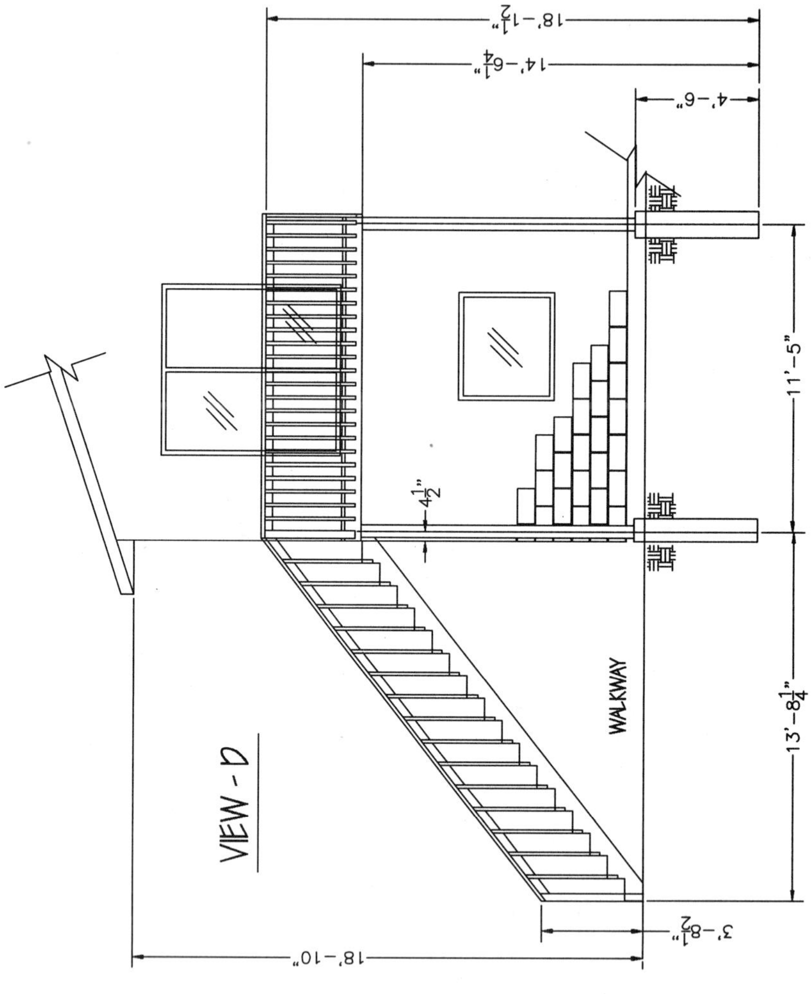

Figure 17-36 Drawing for Problem Solving Exercise 8

Answers to Self-Evaluation Test

1 - T, **2** - T, **3** - F, **4** - T, **5** - 16, **6** - even, **7** - five, **8** - **Purge**, **9** - smaller, **10** - **eXit**

Chapter 18

Grouping and Advanced Editing of Sketched Objects

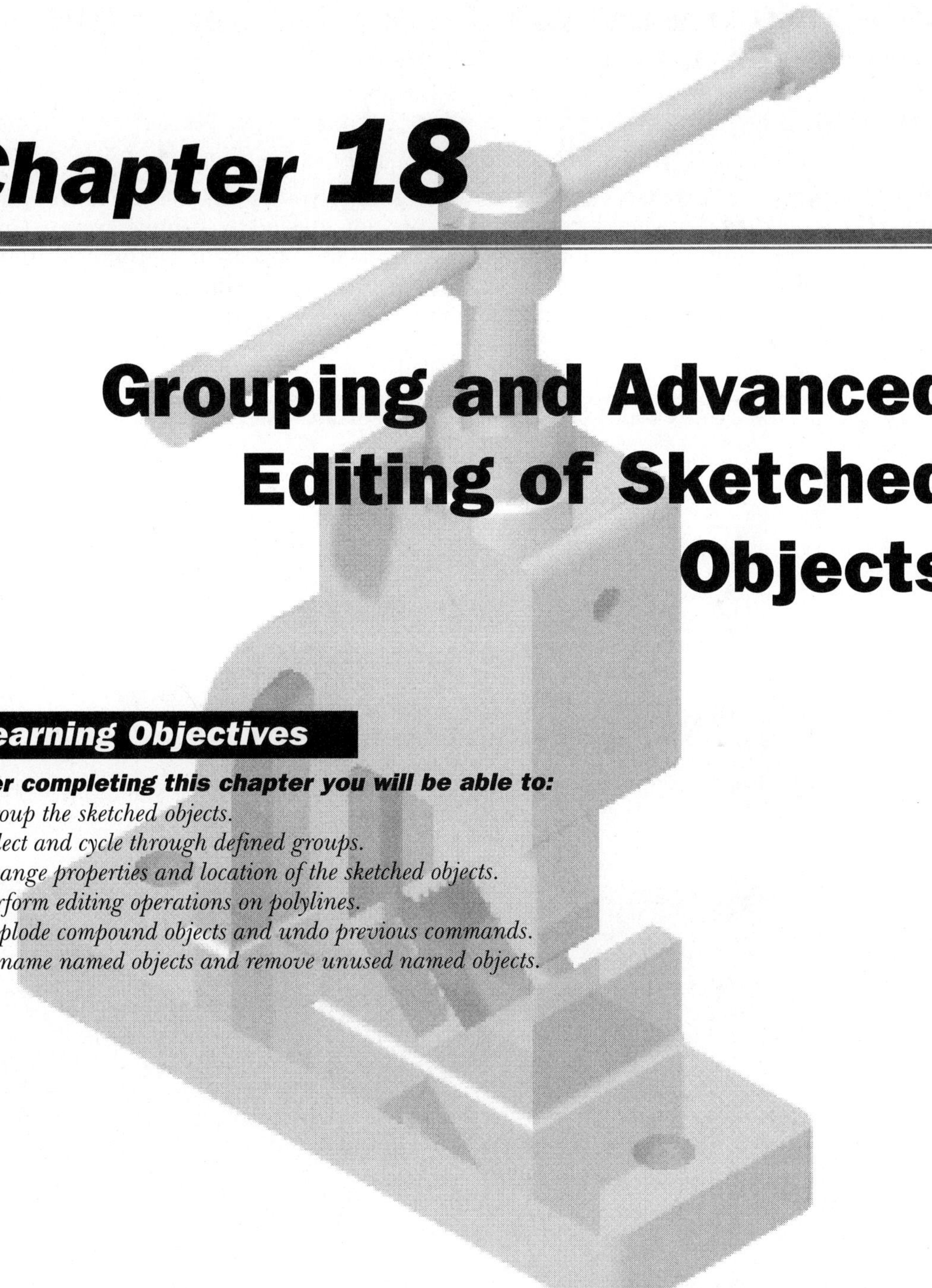

Learning Objectives

After completing this chapter you will be able to:

- *Group the sketched objects.*
- *Select and cycle through defined groups.*
- *Change properties and location of the sketched objects.*
- *Perform editing operations on polylines.*
- *Explode compound objects and undo previous commands.*
- *Rename named objects and remove unused named objects.*

GROUPING THE SKETCHED OBJECTS USING THE OBJECT GROUPING DIALOG BOX

Command: GROUP

You can use the **Object Grouping** dialog box (Figure 18-1) to group AutoCAD objects and assign a name to the group. Once you have created groups, you can select the objects by group name. The individual characteristics of an object are not affected by forming groups. Groups are simply a mechanism that enables you to select all the objects in a group together for editing. It makes the object selection process easier and faster. Objects can be members of several groups. Also, a larger group can contain another smaller group. This may be referred to as nested groups.

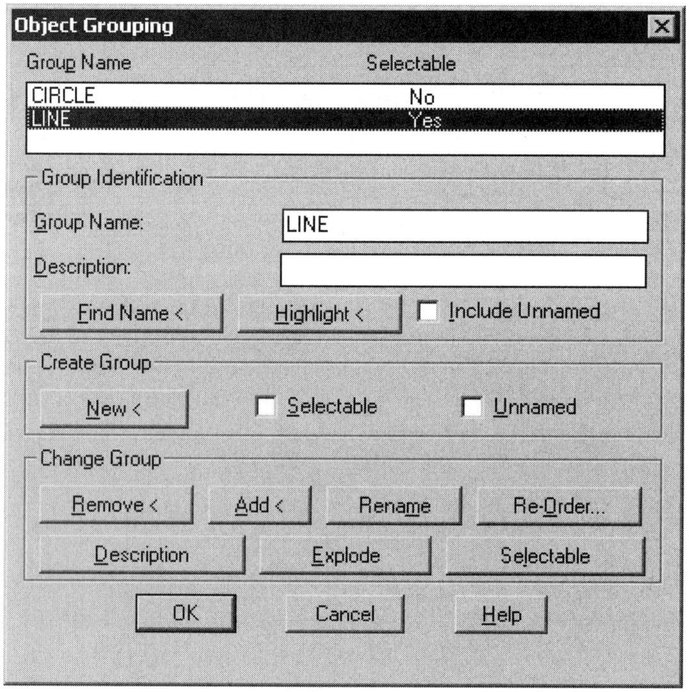

Figure 18-1 *The **Object Grouping** dialog box*

Although an object belongs to a group, you can still select an object as if it did not belong to any group. Groups can be selected by entering the group name or by selecting an object that belongs to the group. You can also highlight the objects in a group or sequentially highlight the groups to which an object belongs. The **Object Grouping** dialog box can be displayed by invoking the **GROUP** command.

Group Name List

The **Group Name** list in the **Object Grouping** dialog box displays the names of the existing groups. The list box also displays whether or not a group is selectable under the **Selectable** column. A selectable group is one in which all members are selected on selecting a single member. Members that are in frozen or locked layers are not selected. You can also make a group not selectable. This allows you to select the objects in the group individually.

Group Identification Area

The options under the **Group Identification** area are as follows.

Group Name

The **Group Name** edit box displays the name of the selected group. You can also use the **Group Name** edit box to enter the name of a new group. You can enter any name, but it is recommended that you use a name that reflects the type of objects in the group. For example, the group name WALLS can include all lines that form walls. Group names can be up to thirty-one characters long and can include letters, numbers, and special characters ($, _, and -). Group names cannot have spaces.

Description

The **Description** edit box displays the description of the selected group. It can be used to enter the description of a group. The description text can be up to sixty-four characters long, including spaces.

Find Name

The **Find Name** button is used to find the group name or names associated with an object. When you select this button, the dialog box temporarily disappears from the screen and AutoCAD prompts you to select an object of the group. Once you select the object, the **Group Member List** dialog box (Figure 18-2) will be displayed, showing the group name(s) to which the selected object belongs.

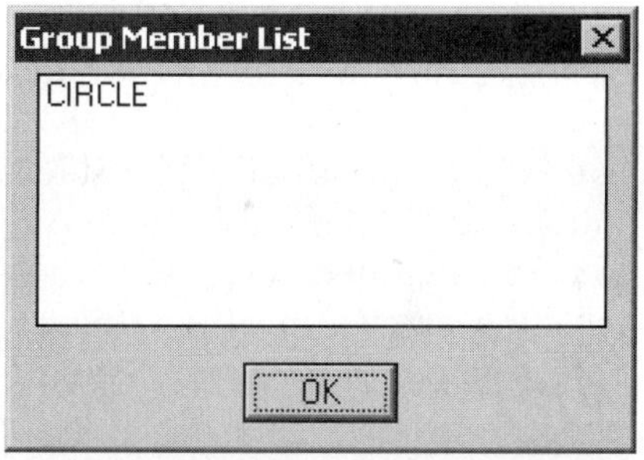

***Figure 18-2** The **Group Member List** dialog box*

Highlight

The **Highlight** button is used to highlight the objects in the selected group name. This button is available only when you select a group from the **Group Name** list box. When you choose the **Highlight** button, the objects that are members of the selected group are highlighted in the drawing and the **Object Grouping** dialog box (Figure 18-3) is displayed. Choose the **Continue** button or press ENTER to return to the **Object Grouping** dialog box.

***Figure 18-3** The **Object Grouping** dialog box*

Include Unnamed

The **Include Unnamed** check box is used to display the names of the unnamed groups in the **Object Grouping** dialog box. The unnamed groups are created when you copy the objects that belong to a group. AutoCAD automatically groups and assigns a name to the copied objects. The format of the name is A*n* (for example, *A1, *A2, *A3). If you select the **Include**

Unnamed check box, the unnamed group names (*A1, *A2, *A3, ...) will also be displayed in the **Group Name** list box of the dialog box. The unnamed groups can also be created by not assigning a name to the group (see Unnamed in the next section, "Create Group Area").

Create Group Area

The options under this area are as follows.

New

The **New** button is used to create a new group. Enter the name of the group in the **Group Name** edit box and then choose this button. Once you choose the **New** button, the **Object Grouping** dialog box will be temporarily closed and you will be prompted to select objects to include in the group. After selecting the objects, the dialog box will be redisplayed and a new group will be created from the selected objects. The new group name is now displayed in the **Group Name** list box of the dialog box. The group names in the **Group Name** list box are listed alphabetically.

Selectable

The **Selectable** check box allows you to define a group as selectable. A selectable group has a property that if you select any one object in the group, the entire group is selected. If you clear this check box while defining a group, the new group created will not be selectable. This allows you to select the members of the group individually and the entire group will not be selected upon selecting the individual entities of the group.

Even if a group is defined as selectable, and the **PICKSTYLE** system variable is set to 0, you will not be able to select the entire group by selecting one of its members. You will be able to individually select members of the group. You can also turn group selectability on or off by selecting or clearing the **Object Grouping** check box in the **Selection Modes** area of the **Selection** tab of the **Options** dialog box or by using the SHIFT+CTRL+A toggle keys.

Unnamed

When you create a group, you can assign it a name or leave it unnamed. If you select the **Unnamed** check box, AutoCAD will automatically assign a group name to the selected objects and these unnamed groups are listed in the **Group Name** list box. To view the name of the unnamed group the **Include Unnamed** check box should be selected. The format of the name is A*n* (*A1, *A2, *A3, ...), where *n* is a number incremented with each new unnamed group.

Change Group Area

All the options under this area, other than the **Re-order** button, will be available only when you highlight a group name by selecting it in the **Group Name** list box of the dialog box. This area of the **Object Grouping** dialog box modifies the properties of the existing group. The options are as follows.

Remove

The **Remove** button is used to remove objects from the selected group. Once you select the group name from the **Group Name** list box and then choose the **Remove** button, the dialog box will be temporarily closed and AutoCAD will display the following prompts at the command line.

Select objects to remove from group...
Remove objects: *(Select objects that you want to remove from the selected group.)*

When you are selecting objects to be removed on the screen, you will notice that all the objects belonging to the selected group are highlighted. Now, as soon as you select the objects to be removed from the group, they no longer appear highlighted. If you remove all objects from the group, the group name still exists, unless you use the **Explode** option to remove the group definition (discussed later in this section). If you remove objects and then add them later in the same drawing session, the objects retain their order in the group.

Add

The **Add** button is used to add objects to the selected group. When you select this option, AutoCAD will prompt you to select the objects you want to add to the selected group. The prompt sequence is similar to that for the **Remove** option. Similar to using the Remove option, all objects belonging to the selected group are highlighted and as soon as you select an object to be added to the group, it also gets highlighted.

Rename

The **Rename** button is used to rename the selected group. To rename a group, first select the group name from the **Group Name** list box, enter the new name in the **Group Name** edit box, and then choose the **Rename** button. AutoCAD will rename the specified group. You can use this option to rename unnamed groups that were discussed earlier.

Re-Order

The **Re-Order** option allows you to change the order of the objects in the selected group. The objects are numbered in the order in which you select them when selecting objects for the group. Sometimes when creating a tool path, you may want to change the order of these objects to get a continuous tool motion. You can do this using the **Re-Order** option. This option is also useful in some batch operations where one object needs to be on top of another object for display reasons. When you choose the **Re-Order** button, AutoCAD displays the **Order Group** dialog box (Figure 18-4) where you can change the order of the group members. The following example explains the use of the **Re-Order** option using the **Order Group** dialog box.

Example 1 *General*

Figure 18-5 shows four objects that are in group G1. Determine the order of these objects and then reorder them in a clockwise direction.

1. Invoke the **Object Grouping** dialog box, and make the group (G1) current by selecting it in the **Group Name** list box.

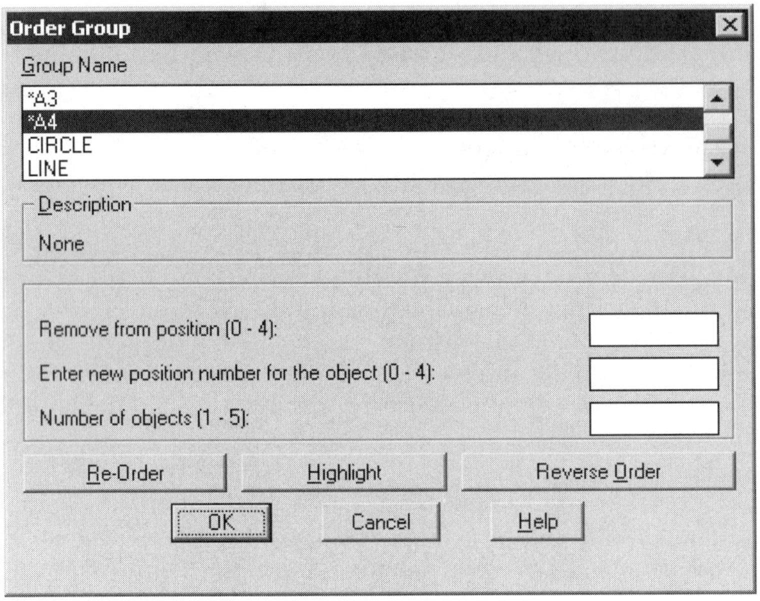

*Figure 18-4 The **Order Group** dialog box*

2. Choose the **Re-Order** button in the dialog box. The **Order Group** dialog box is displayed on the screen. Select group G1 in the **Group Name** list box, if not already selected, to highlight it. All the group names present in the current drawing are displayed in this list box. The **Description** area displays the corresponding description text that was entered in the **Description** edit box of the **Object Grouping** dialog box when defining the particular group.

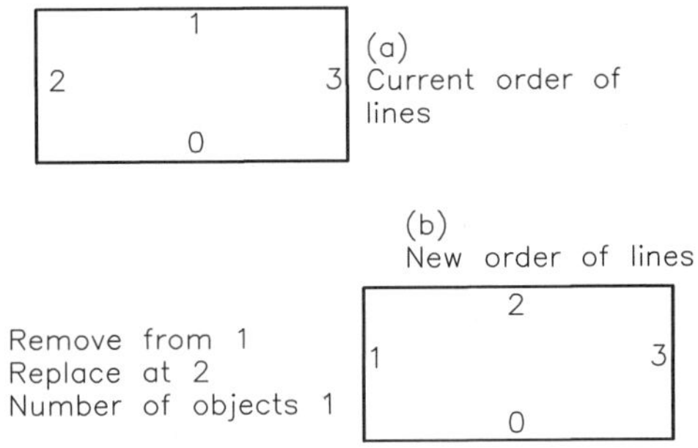

Figure 18-5 Changing the order of group objects

3. Use the **Highlight** button to highlight the current order of individual objects in the selected group. When you choose the **Highlight** button, the **Object Grouping** dialog box (Figure 18-6) is displayed. You can use the **Next** and **Previous** buttons to cycle through the group objects. Figure 18-5(a) shows the order of the objects in group G1. To create a tool path so that the tool moves in a clockwise direction, you need to reorder the objects as shown in Figure 18-5(b).

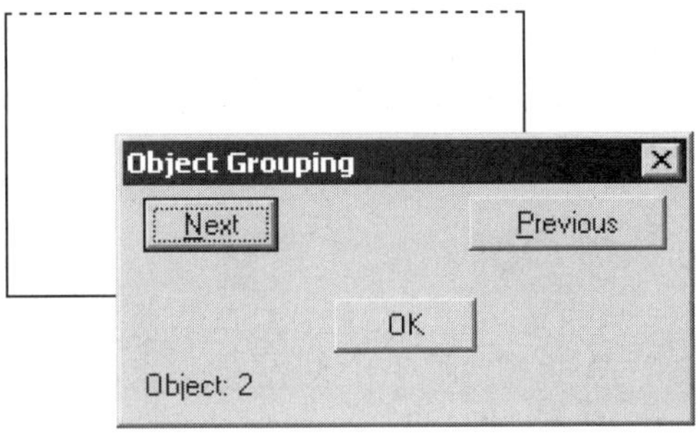

*Figure 18-6 Using the **Highlight** option to view the order of the group objects*

4. To get a clockwise tool path, you must switch object numbers 1 and 2. You can do this by entering the necessary information in the **Order Group** dialog box. Enter 1 in the **Remove from position** {0-(n-1)} edit box and 2 in the **Enter new position number for the object:** {0-(n-1)} edit box. Enter 1 in the **Number of objects (1-n)** edit box, since there is only one object to be replaced. Here, **n** is the number of objects in the group.

5. After entering the information, choose the **Re-Order** button to define the new order of the objects. You can confirm the change by choosing the **Highlight** button again and cycling through the objects.

Description

The **Description** button lets you change the description of the selected group. To change the description of a group, first select the group name from the **Group Name** list box, enter the new description in the **Description** edit box, and then choose the **Description** button. AutoCAD will update the description of the specified group.

Explode

The **Explode** option deletes group definition of the selected group. The objects that were in the group become regular objects without a group reference. If you had made copies of the group that was exploded in a drawing, they shall remain as unnamed groups in the drawing. You can select the **Include Unnamed** check box in the **Group Identification** area of the **Object Grouping** dialog box to view the names of the unnamed groups in the **Group Name** list box. You can then select them in the list box and explode them or rename them as required.

Selectable

The **Selectable** button changes the selectable status of the selected group. To change the selectable status of the group, first select the group name from the **Group Name** list box, and then choose the **Selectable** button. This button acts like a toggle key between the options of selectable and not selectable. If the selectable status for a group is displayed as "**Yes**", you can choose the **Selectable** button to change it to "**No**". Selectable implies that the entire group gets selected when one object in the group is selected. If the selectable status is "**No**", you cannot select the entire group by selecting one object in the group.

SELECTING GROUPS

You can select a group by name by entering G at the **Select objects** prompt. For example, if you have to move a particular group, choose the **Move** button from the **Modify** toolbar and the prompt sequence will be as follows.

Select objects: **G**
Enter group name: *Enter the groups name.*
n found
Select objects: Enter

Instead of entering G at the **Select Objects:** prompt, if you select any member of a selectable group, all the group members get selected. Make sure that the **PICKSTYLE** system variable is set to 1 or 3. Also, the group selection can be turned on or off by pressing SHIFT+CTRL+A. If the group selection is off and you want to erase a group from a drawing, invoke the **Erase** command from the **Modify** toolbar. The prompt sequence is as follows:

> Select objects: SHIFT+CTRL+A
> <Groups on> *Select an object that belongs to a group. (If the group has been defined as selectable, all objects belonging to that group will be selected.)*

If the group has not been defined as selectable, you cannot select all objects in the group, even if you turn the group selection on by pressing the SHIFT+CTRL+A keys. This setting can also be changed in the **Selection** tab of the **Options** dialog box. You can select or clear the **Object Grouping** check box in the **Selection Modes** area of the dialog box to turn group selectability on or off.

Tip
The combination of SHIFT, CTRL, and A keys is used as a toggle to turn the group selection on or off.

CYCLING THROUGH GROUPS

When you use the **GROUP** command to form object groups, AutoCAD lets you sequentially highlight the groups of which the selected object is a member. For example, assume that an object belongs to two different groups, and you want to highlight the objects in those groups. To accomplish this, press CTRL at the **Select objects** prompt of any command and select the object that belongs to different groups. AutoCAD will highlight the objects in one of the groups. Press the pick button of your pointing device to cycle through the groups (Figure 18-7). Choose the **Erase** command; the prompt sequence is given next.

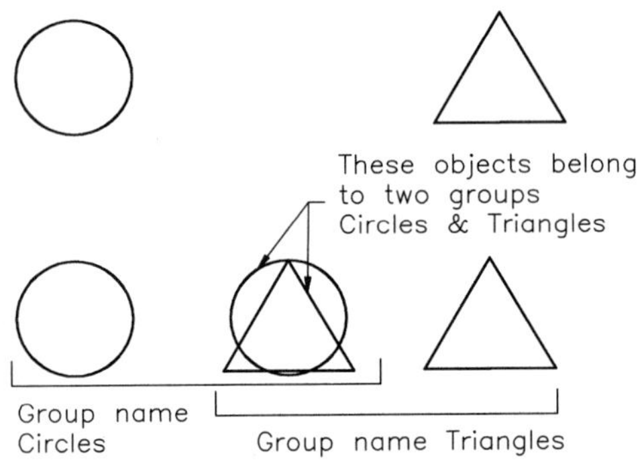

Figure 18-7 *Cycling through objects of different groups*

> Select objects: *Hold down CTRL.*
> <Cycle on> *Select the object that belongs to different groups. (Press the pick button on your pointing device to cycle through the groups.)*
> Select objects: Enter

Also, if objects are very close or directly on top of one another, such that they lie within the same selection pickbox, you can press CTRL at the **Select objects** prompt and repeatedly click your pointing device until the object you want is highlighted. Press ENTER to select it.

Note

You can also use the command line form of the ***GROUP*** *command by entering* ***-GROUP*** *at the Command prompt.*

CHANGING PROPERTIES AND LOCATION OF AN OBJECT

Using the PROPERTIES Palette

Toolbar:	Standard > Properties
Menu:	Tool > Properties
Command:	PROPERTIES

As discussed earlier in Chapter 4, Working with Drawing Aids, and Chapter 6, Editing Commands-II, invoking the **PROPERTIES** command displays the **PROPERTIES** palette that can be used for changing properties of objects. Depending on the type of object selected, various categories are displayed. The **General** category displays the general properties of objects, such as color, layer, linetype, linetype scale, plot style, lineweight, hyperlink, and thickness. How to change the general properties of objects using the **PROPERTIES** palette has already been discussed earlier in Chapter 4, Working with Drawing Aids. Here, the **Geometry** category will be discussed that contains properties that control the geometry of an object. Depending on the type of object selected, the **Geometry** category will contain a set of different properties. If you have selected many types of objects in a drawing, the edit box at the top of the **PROPERTIES** palette displays **All** and only the **General** category is available in the palette. You can select a type of object from the drop-down list and the corresponding categories of object properties are displayed in the palette.

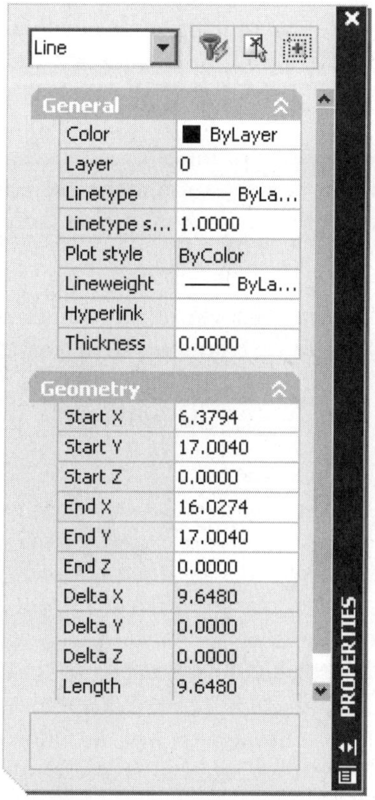

Figure 18-8 *The* ***PROPERTIES*** *palette*

Selecting Line

If you have selected a line in the drawing, the **PROPERTIES** palette appears similar to what is shown in Figure 18-8, where you can change the properties of the selected object.

Start X/Start Y/Start Z. The **Start X/Start Y/Start Z** fields in the **Geometry** category display the start point coordinates of the selected line. You can enter new values in these fields or use the **Pick Point** button that appears in the field you click in. When you choose the **Pick Point** button, a rubber-band line gets attached to the cursor on its start point and the endpoint of the line is fixed. You can now move the cursor and specify a new start point for the line.

End X/ End Y/ End Z. Similar to the start point you can also modify the location of the endpoint of a line by entering new coordinate values in the **End X**, **End Y**, **End Z** fields. You can also choose the **Pick Point** button in the respective fields and select a new location for the endpoint of the selected line. This **Pick Point** button appears in the fields when you click in them.

Note

*The **Delta X, Delta Y, Delta Z, Length,** and **Angle** fields are not available for change but the values in these fields get updated as the start and endpoints of the line are changed.*

Selecting Circle

When you select a circle, the **Geometry** category displays the categories as displayed in Figure 18-9. You can modify the values in these fields and the effects of modifying the values are discussed next.

Center X/Center Y/Center Z. Here under **Center X/ Center Y/Center Z** fields of Geometry category, you can specify a new location of the center point of the circle by entering new coordinate values in the **Center X**, **Center Y**, and **Center Z** fields. When you click in any of these fields, the **Pick Point** button is displayed. You can choose this button to locate a new location of the center point of the circle. The circle will be drawn at the new location with the same radius as before. It is as if the circle has moved from the old location to a new one.

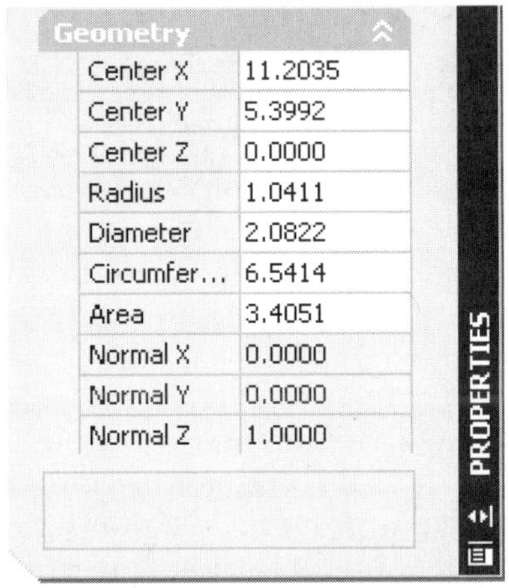

*Figure 18-9 The **Geometry** category of the **PROPERTIES** palette for a circle*

Radius/Diameter. You can enter a new value for the radius or diameter of the circle in the **Radius** or **Diameter** fields respectively. The circle's radius or diameter is modified as per the new values you have entered. As you modify the values of the radius or the diameter, the values in the **Circumference** and **Area** fields get modified accordingly.

Note

*You can also enter new values of the circumference or area in the **Circumference** and **Area** fields respectively if you want to create a circle with a given value of circumference or area. You will notice that the radius and diameter values are modified accordingly. The **Normal X, Normal Y,** and **Normal Z** fields are not available for modifications.*

Selecting MText

If you select mtext written in a drawing and invoke the **PROPERTIES** palette, apart from the **General** category, two more categories of properties are displayed: **Text** and **Geometry**. The properties under these two categories (Figure 18-10) and how they can be modified are discussed next.

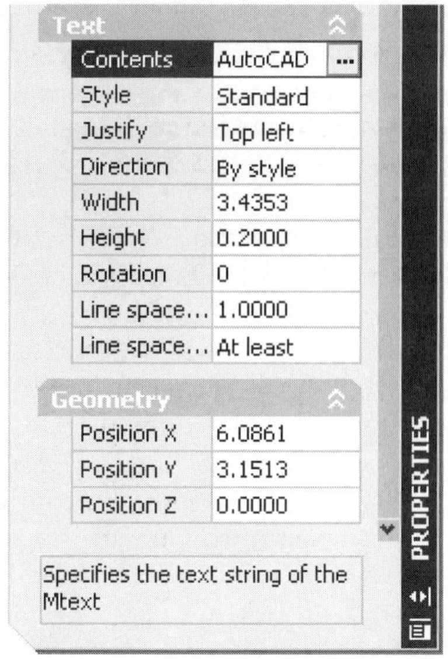

*Figure 18-10 The **Geometry** and **Text** categories of the **PROPERTIES** palette for text*

Contents. The current contents of the text are displayed in the **Contents** field under the **Text** category of the **PROPERTIES** palette. To modify the contents click in the field and the **[...]** button is displayed at the right corner of the field. You can then choose the **[...]** button to display the **Multiline Text Editor** where you can make the necessary modifications using the various options available in the dialog box. Once you have made the changes choose **OK** to exit the text editing and return to the **PROPERTIES** palette. The changes you made are reflected in the value of the **Contents** field of the **PROPERTIES** palette and in the drawing.

Style. When you click in the **Style** field in the **PROPERTIES** palette, a drop-down list of the defined text styles is displayed. By default only the Standard style is displayed. If you had defined more text styles in the drawing, they can be selected from the drop-down list and be applied to the current text in the drawing.

Justify. All the text justification options are available in the **Justify** drop-down list. This drop-down list is available when you click in the **Justify** field. By default, the Top left justification is applied to the text. You can select a text justification from the drop-down list to be applied to the selected text in the drawing.

Direction. This drop-down list displays the possible directions for the text. This option is generally available for the multiline text objects.

Width. In the **Width** field you can enter a new value for the paragraph text width. The changes are evident when you have multiline text in the drawing. You can increase or decrease the value of the paragraph text.

Height. You can enter a new value for the selected text in the drawing. The default text height is 0.2. But, you can make the text smaller or larger in the drawing by simply entering a new value in the **Height** field of the **PROPERTIES** palette.

Rotation. This field displays the current angle of the selected text and is **0**-degree, by default. You can change the angle of rotation of the text by entering an angle by which you want to rotate the selected text in the **Rotation** field and then pressing ENTER. The effect is immediately visible in the drawing.

Line space factor. The effect of changing the value of this field is more evident when you have selected paragraph or multiline text. This value controls the spacing between lines in a paragraph or multiline text. A scale factor of 1 is the default value. You can increase or decrease the spacing between the lines of the selected paragraph text by entering a new scale factor of your choice and pressing ENTER. The result is immediately visible in the drawing.

Line space distance. This field is used to modify the distance between the spacing of each line of the mtext. Click in this field and enter the new value in it.

Line space style. This field when selected displays the two line spacing options in the corresponding drop-down list. They are **Exactly** and **At least**. **At least** is the default option.

Position X/Position Y/Position Z. You can change the location of the insertion point of the selected text by entering new *X/Y/Z* coordinates for the text in the **Position X/Position Y/ Position Z** fields under the **Geometry** category in the **PROPERTIES** palette. You can also use the **Pick Points** button that is displayed in these fields when you click in them. When you choose the **Pick Points** button, a rubber-band line is attached between the cursor and the selected text and you can move the cursor and specify a new location for the text on the screen. The values of the three fields get updated automatically.

Selecting Block Reference

When you select a block in the drawing and invoke the **PROPERTIES** palette, apart from the **General** category, two more categories containing the properties of the block are available. They are **Geometry** and **Misc**. The properties under these categories are discussed next.

Position X/Position Y/Position Z. Here under the **Position X/Position Y/Position Z** fields of the **Geometry** category you can enter new *X*, *Y*, and *Z* coordinates for the selected block. The block reference shall move to the new location as you specify. You can also click in any one of the fields to display the **Pick Points** button and choose it to specify a new location for the block on the screen.

Scale X/Scale Y/Scale Z. You can specify new X, Y, and Z scale factors for the selected block in the **Scale X**, **Scale Y**, and **Scale Z** fields respectively. The current scale factors for the block are displayed in the **Scale X**, **Scale Y**, and **Scale Z** fields and as you change them, the modifications are reflected in the drawing.

Name. This field under the **Misc** category displays the name of the selected block reference and is not available for modifications.

Rotation. This field specifies the current angle of rotation of the selected block. You can specify a new angle of rotation for the selected block in this field. After specifying an angle, when you press ENTER, the selected block in the drawing is rotated through the specified angle.

Selecting Attribute

When you select an attribute before converting it into a block, apart from the **General** category, the other three categories under which the properties of an attribute are displayed are **Text**, **Geometry**, and **Misc**. How to use the **PROPERTIES** palette to edit an attribute has already been discussed in Chapter 15, Defining Block Attributes.

Tag. You can modify the text of the tag by entering a new value in the **Tag** field under the **Text** category. After you have modified the tag text and pressed ENTER, the old attribute text in the drawing is replaced by the new text.

Prompt. You can also change the text of the prompt and enter a new prompt value in the **Prompt** field.

Value. This field displays the default value of the selected attribute. You can change this default value by entering a new value in the **Value** field.

Style. When you click in this field, the **Style** drop-down list is available. You can select a style for the attribute text from this drop-down list. Only text styles that already have been defined are available in this drop-down list and can be selected.

Justify. You can select an attribute text justification from the **Justification** drop-down list.

Height. You can change the height of the selected attribute text by entering a new value in this field.

Rotation. You can also rotate the selected attribute text through an angle that you specify in the **Rotation** field.

Width Factor. You can modify the width factor of the attribute text by entering a new value in the **Width Factor** field of the **PROPERTIES** palette.

Obliquing. The attribute text can also be made slanting by entering an obliquing angle in the **Obliquing** field of the **PROPERTIES** palette.

Text alignment X/Text alignment Y/Text alignment Z. The **Text alignment X/Text alignment Y/Text alignment Z** fields of the **Geometry** category display the *X*, *Y*, and *Z* coordinates of the alignment point of the selected attribute. Note that these fields will be available only if you modify the properties such as the text alignment of the attributes while defining them. If these fields are available and you click on them, the **Pick Point** button appears in the field. You can use this button to specify the new location of the attribute.

Note

*If you modify the text alignment of the first attribute and define the remaining attributes below the previous, the remaining attributes will also have the **Text Alignment** fields active.*

Position X/Position Y/Position Z. You can enter new *X*, *Y*, and *Z* coordinates in the **Position X**, **Position Y**, and **Position Z** fields of the **Geometry** category to relocate the attribute text. You can also choose the **Pick Points** button that is displayed in these fields when you click in them to relocate the selected attribute text.

Misc. As discussed earlier, under the **Misc** category, you can redefine the attribute modes that have been defined at the time of the attribute definition. You can also make the attribute text appear upside-down or backwards by selecting **Yes** from the **Upside-down** drop-down list and the **Backward** drop-down list respectively.

Using the CHANGE Command

Command: CHANGE

With the help of the **CHANGE** command, you can change some of the characteristics associated with an object, such as location, color, layer, lineweight, and linetype. You can also change the geometry of the object using this command. The **CHANGE** command has two options, **Change point** and **Properties**, which are discussed next.

Change Point Option

You can change various features and the location of an object with the **Change point** option of the **CHANGE** command. For example, to change the endpoint of a line or a group of lines, the prompt sequence is given next.

Command: **CHANGE**
Select objects: *Select the line(s).*
Select objects: Enter
Select Change point or [Properties]: *Specify a point to be used as a new endpoint.*

Tip
*When the **ORTHO** mode is on, you can use the **CHANGE** command to change the endpoint of a line or group of lines to make them parallel to either the X axis or the Y axis.*

To change various features associated with the text, the prompt sequence is given next.

Command: **CHANGE**
Select objects: *Select the text.*
Select objects: Enter
Specify change point or [Properties]: Enter
Specify new text insertion point <no change>: *Specify the new text insertion point (location of text) or press ENTER for no change.*
Enter new text style <current>: *Enter the name of the new text style or press ENTER to accept the current style.*
Specify new height <current>: *Specify the new text height or press ENTER to accept the current value.*
Specify new rotation angle <current>: *Specify the new rotation angle or press ENTER to accept the current value.*
Enter new text <current>: *Enter the new text or press ENTER to accept the current text.*

The properties of an attribute definition text can also be changed just as you change the properties of text. The prompt sequence for changing the properties of an attribute definition text is as follows.

Command: **CHANGE**
Select objects: *Select the attribute definition text.*
Select objects: Enter
Specify change point or [Properties]: Enter
Specify new text insertion point <no change>: *Specify the new attribute definition text insertion point (location) or press ENTER if you do not want to modify the location of the attribute text.*
Enter new text style <current>: *Enter the name of the new text style or press ENTER to accept the current style.*
Specify new height <current>: *Specify the new attribute definition text height or press ENTER to accept the current value.*
Specify new rotation angle <current>: *Specify the new rotation angle or press ENTER to accept the current value.*
Enter new tag <current>: *Enter a new tag or press ENTER to accept the current value of tag.*
Enter new prompt <current>: *Enter a new prompt or press ENTER to accept the current value.*
Enter new default value <current>: *Enter the new default value or press ENTER to accept the current value.*

You can also change the position of an existing block and specify a new rotation angle to it using the **CHANGE** command. The prompt sequence is as follows.

Command: **CHANGE**
Select objects: *Select the block.*
Select objects: Enter
Specify change point or [Properties]: *Specify the new block insertion point or press ENTER.*
Specify new block insertion point <no change>: *Specify the new block insertion point or press ENTER to accept the current location of the block.*
Specify new block rotation angle <current>: *Specify the new rotation angle for the block or press ENTER to accept the current rotation angle.*

The radius of a circle can also be changed with the **Change point** option of the **CHANGE** command by specifying the new radius in the case of the circle. To change the radius of a circle, the prompt sequence is given next.

Command: **CHANGE**
Select objects: *Select the circle.*
Select objects: Enter
Specify change point or [Properties]: *Select a point to specify the radius of the circle.*

If more than one circle is selected, the same prompts are repeated for the next circle after the first circle is modified.

The Properties Option

The **Properties** option can be used to change the characteristics associated with an object.

Changing the layer of an object. If you want to change the layer on which an object exists and other characteristics associated with layers, you can use the **LAyer** option of the **Properties** option. The prompt sequence is given next.

Command: **CHANGE**
Select objects: *Select the object whose layer you want to change.*
Select objects: *If you have finished selection, press ENTER.*
Specify change point or [Properties]: P
Enter property to change [Color/Elev/LAyer/LType/ltScale/LWeight/Thickness/PLotstyle]: LA
Enter new layer name <0>: *Enter a new layer name.*
Enter property to change [Color/Elev/LAyer/LType/ltScale/LWeight/Thickness]: Enter

The selected object is now placed on the desired layer.

Changing the color of an object. Similarly, you can change the color of the selected object with the **Color** option of the **CHANGE** command. If you want to change the color and linetype to match the layer, you can use the **Properties** option of the **CHANGE** command and then change the color or linetype by entering BYLAYER at the **Enter new color:** prompt. The command prompt sequence for this option is similar to the one displayed previously.

Note

A block originally created on layer 0 assume the color of the new layer. Otherwise (if it was not created on layer 0), it will retain the color of the layer on which it was created.

Changing the thickness of an object. You can also change the thickness of the selected object in a similar manner. After you have invoked the **CHANGE** command and selected the object to change, use the **Properties** option and enter T at the **Enter property to change** prompt. You are then prompted to enter the new thickness you want to assign to the selected object.

Changing the lineweight of an object. You can similarly also change the lineweight of the selected object by using the **LWeight** option of the **CHANGE** command. The lineweight values are predefined and if you enter a value that is not predefined, a predefined value of lineweight, which comes closest to the specified value, is assigned to the selected object.

Tip

*You can see the effect of changing the thickness of an object in a 3D view. Choose **SE Isometric** from the **View > 3D Views** menu to view the effect of the changed thickness value of an object. To get back to the original view, simply choose the **Undo** button in the **Standard** toolbar.*

Changing the linetype/Linetype scale of an object. Similarly, you can change the linetype and linetype scale of the selected objects by entering the appropriate option at the **Enter property to change** prompt.

Changing the elevation of an object. You can also change the *Z* axis elevation of a selected object by entering E at the **Enter property to change** prompt, provided all the points in the particular object have the same *Z* value. After changing the elevation value of an object, you can view the changes in a 3D view.

Using the CHPROP Command

Command: CHPROP

You can also use the **CHPROP** command to change the properties of an object using the command line. The prompt sequence is as follows:

> Command: **CHPROP**
> Select objects: *Select the object whose properties you want to change.*
> Select objects: Enter
> Enter property to change [Color/LAyer/LType/ltScale/LWeight/Thickness]: *Specify the property to be modified.*

Note

*Apart from the commands discussed, you can also use the **Properties** toolbar to change some of the properties of the selected objects such as layer, color, linetype, lineweight, and plot style. These options have been discussed in Chapter 4, Working with Drawing Aids.*

Exercise 1 — *General*

Draw a hexagon on layer OBJ in red color. Let the linetype be hidden. Now, use the **PROPERTIES** command to change the layer to some other existing layer, the color to yellow, and the linetype to continuous. Also, in the **PROPERTIES** palette, under the **Geometry** category, specify a vertex in the **Vertex** field or select one using the arrow (Next or Previous) buttons and then relocate it. The values of the **Vertex X** and **Vertex Y** fields change as the coordinate values of the vertices change. Also, notice the change in the value in the **Area** field. You can also assign a start and end lineweight to each of the hexagon segments between the specified vertices. Use the **LIST** command to verify that the changes have taken place.

EXPLODING COMPOUND OBJECTS

Toolbar: Modify > Explode
Menu: Modify > Explode
Command: EXPLODE

The **EXPLODE** command is used to split compound objects such as blocks, polylines, regions, polyface meshes, polygon meshes, multilines, 3D solids, 3D meshes, bodies, or dimensions into the basic objects that make them up (Figure 18-11). For example, if you explode a polyline or a 3D polyline, the result will be

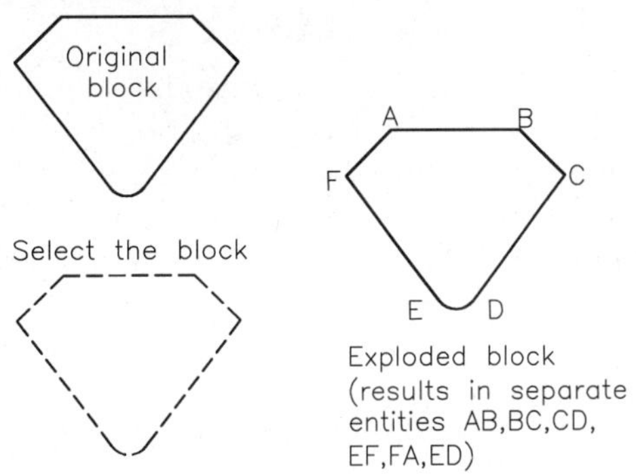

Figure 18-11 *Using the **EXPLODE** command*

ordinary lines or arcs (tangent specification and width are not considered). When a 3D polygon mesh is exploded, the result is 3D faces. Polyface meshes are turned into 3D faces, points, and lines. Upon exploding 3D solids, the planar surfaces of the 3D solid turn into regions, and nonplanar surfaces turn into bodies. Multilines are changed to lines. Regions turn into lines, ellipses, splines, or arcs. 2D polylines lose their width and tangent specifications and 3D polylines explode into lines. When a body is exploded, it changes into single-surface bodies, curves, or regions. When a leader is exploded, the components are lines, splines, solids, block inserts, text or Mtext, tolerance objects, and so on. Mtext explodes into single line text. This command is especially useful when you have inserted an entire drawing and you need to alter a small detail. After you invoke the **EXPLODE** command, you are prompted to select the objects you want to explode. After selecting the objects, press ENTER or right-click to explode the selected objects and then end the command.

When a block or dimension is exploded, there is no visible change in the drawing. The drawing remains the same except that the color and linetype may have changed because of floating layers, colors, or linetypes. The exploded block is turned into a group of objects that can be modified separately. To check whether the explosion of the block has taken place, select any object that was a part of the block. If the block has been exploded, only that particular object will be highlighted. With the **EXPLODE** command, only one nesting level is exploded at a time. Hence, if there is a nested block or a polyline in a block and you explode it, the inner block or the polyline will not be exploded. Attribute values are deleted when a block is exploded, and the attribute definitions are redisplayed.

Note

*Remember that the blocks inserted using the **MINSERT** command and x-refs cannot be exploded. This command was discussed in relation to the blocks in Chapter 14.*

Tip

*If you want to insert a block in the form of its separate components, while using the **INSERT** command select the **Explode** check box in the **Insert** dialog box. Also, if you are using the **-INSERT** command, type * in front of the block name at the **Enter block name or [?]:** prompt. The block will be inserted in the drawing as separate component objects and not as an entire block.*

EDITING POLYLINES

A polyline can assume various characteristics such as width, linetype, joined polyline, and closed polyline. You can edit polylines, polygons, or rectangles to attain the desired characteristics using the **PEDIT** command. In this section, we will be discussing how to edit simple 2D polylines. The following are the editing operations that can be performed on an existing polyline using the **PEDIT** command. They are discussed in detail later in this chapter.

1. A polyline of varying widths can be converted to a polyline of uniform width.

2. An open polyline can be closed and a closed one can be opened.

3. You can remove bends and curved segments between two vertices to make a straight polyline.

4. A polyline can be split up into two and individual polylines or polyarcs connected to one another can be joined into a single polyline.

5. If you select an entity that is not a polyline to edit using the **PEDIT** command, you are prompted to specify whether you want the entity to be converted into a polyline. You can avoid this prompt by setting the value of the **PEDITACCEPT** variable to 1.

6. Multiple polylines can be edited.

7. The appearance of a polyline can be changed by moving and adding vertices.

8. Curves of arcs and B-spline curves can be fitted to all vertices in a polyline, with the specification of the tangent of each vertex being optional.

9. The linetype generation at the vertices of a polyline can be controlled.

10. Multiple polylines can be joined together to form a single polyline.

Editing Single Polyline

Toolbar:	Modify II > Edit Polyline
Menu:	Modify > Object > Polyline
Command:	PEDIT

Apart from the methods displayed in the **PEDIT** command box as shown, you can also invoke the **PEDIT** command by choosing **Polyline Edit** from the shortcut menu that is displayed when you select a polyline and right-click. You can use the **PEDIT** command to edit any type of polyline. When you invoke this command you are prompted to select a polyline. If the selected line is an arc or a line (not a polyline), AutoCAD issues the following prompt.

Select polyline or [Multiple]:

If the selected entity is not a polyline, the message that is displayed is given next.

Object selected is not a polyline.
Do you want to turn it into one? <Y>:

If you want to turn the object into a polyline, respond by entering a Y and pressing ENTER or by simply pressing ENTER. To let the object be as it is, enter N. AutoCAD will then prompt you to select another polyline or object to edit. As mentioned earlier, you can avoid this prompt by setting the value of the **PEDITACCEPT** variable to **1**. The subsequent prompts and editing options depend on the type of polyline that has been selected. AutoCAD provides you the option of either selecting a single polyline or multiple polylines. In the case a single 2D polyline is selected, the next prompt displayed is given next.

Enter an option [Close/Join/Width/Edit vertex/Fit/Spline/Decurve/Ltype gen/Undo]: *Enter an option or press ENTER to end command.*

Note

Depending on the type of polyline selected, the ***PEDIT*** *command options change. In this chapter, the 2D polyline options have been discussed. The 3D polylines and 3D Polygon Mesh editing options will be discussed in Chapter 23, Getting Started with 3D.*

The options available in **Single** polyline selection method are discussed next:

Close (C) Option

This option is used to close an open polyline. **Close** creates the segment that connects the last segment of the polyline to the first. You will get this option only if the polyline is not closed. Figure 18-12 illustrates this option.

Open (O) Option

If the selected polyline is closed, the **Close** option is replaced by the **Open** option. Entering **O**, for open, removes the closing segment, see Figure 18-12.

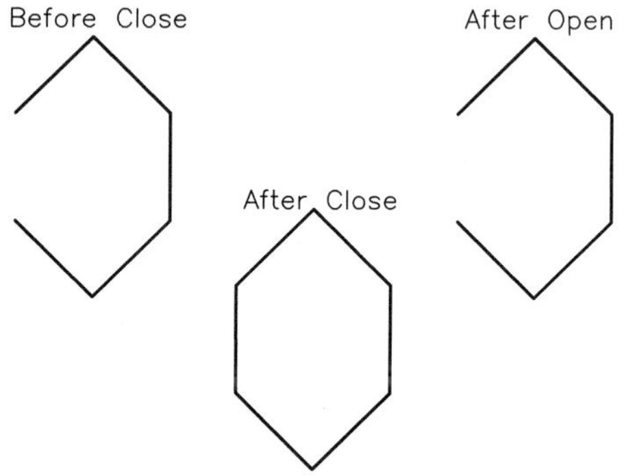

Figure 18-12 The ***Close*** *and the* ***Open*** *options*

Join (J) Option

This option appends lines, polylines, or arcs whose endpoints meet a selected polyline at any of its endpoints and adds (joins) them to it (Figure 18-13). This option can be used only if a polyline is open. After this option has been selected, AutoCAD asks you to select objects. Once you have chosen the objects to be joined to the original polyline, AutoCAD examines them to determine whether any of them has an endpoint in common with the current polyline, and joins such an object with the original polyline. The search is then repeated using new endpoints. They will not join if the endpoint of the object does not exactly meet the polyline. The line touching a polyline at its endpoint to form a T will not be joined. If two lines meet a polyline in a Y shape, only one of them will be selected, and this selection is unpredictable. To verify which lines have been added to the polyline, use the **LIST** command or select a part of the object. All the segments that are joined to the polyline will be highlighted.

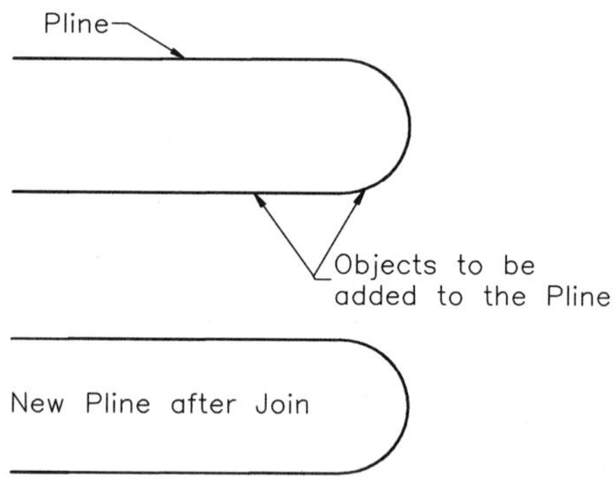

Figure 18-13 Using the ***Join*** *option*

Width (W) Option

The **W** option allows you to define a new, unvarying width for all segments of a polyline

(Figure 18-14). It changes the width of a polyline with a constant or a varying width. The desired new width can be specified either by entering the width at the keyboard or by specifying the width as the distance between two specified points. Once the width has been specified, the polyline assumes it. Here, you will change the width of the given polyline in the figure from 0.02 to 0.05 (Figure 18-15). After you invoke the **PEDIT** command, the next prompts are as follows.

Select polyline or [Multiple]: *Select the polyline.*
Enter an option [Close/Join/Width/Edit vertex/Fit/Spline/Decurve/Ltype gen/Undo]: **W**
Specify new width for all segments: **0.05**

Note
*Circles drawn using the **CIRCLE** command cannot be changed to polylines. However, polycircles can be drawn using the **Pline Arc** option (by drawing two semicircular polyarcs) or by using the **DONUT** command (you can modify the thickness using the **PEDIT Width** option).*

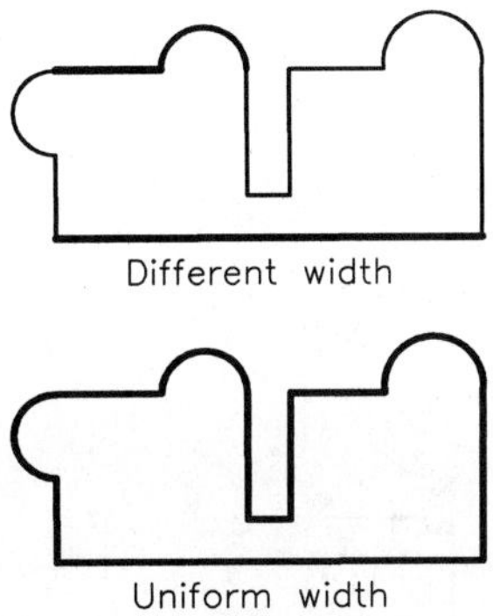

Figure 18-14 *Making the width of a polyline uniform*

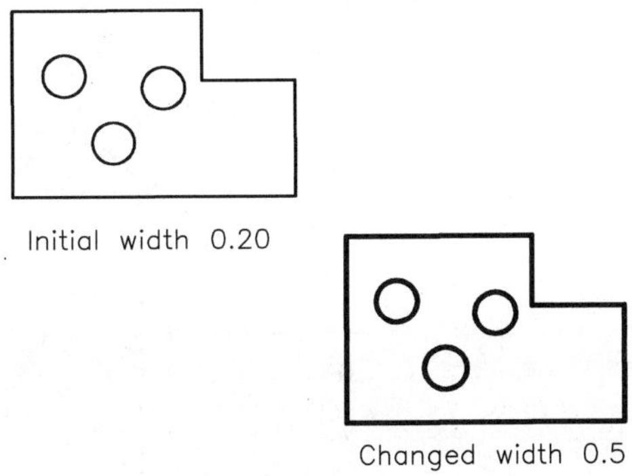

Figure 18-15 *Entering a new width for all segments*

Edit vertex (E) Option

The **Edit vertex** option lets you select a vertex of a polyline and perform different editing operations on the vertex and the segments following it. A polyline segment has two vertices. The first one is at the start point of the polyline segment; the other one is at the endpoint of the segment. When you invoke this option, an X marker appears on the screen at the first vertex of the selected polyline. If a tangent direction has been specified for this particular vertex, an arrow is generated in that direction. After this option has been selected, the next prompt appears with a list of options for this prompt. The prompt sequence after you invoke the **PEDIT** command is given next.

Select polyline or [Multiple]: *Select the polyline to be edited.*
Enter an option [Close/Join/Width/Edit vertex/Fit/Spline/Decurve/Ltype gen/Undo]: **E**
Enter a vertex editing option
[Next/Previous/Break/Insert/Move/Regen/Straighten/Tangent/Width/eXit]<N>: *Enter an editing option or press ENTER to accept default.*

All the options available for the **Edit vertex** option are discussed next.

Next and Previous options. These options move the X marker to the next or the previous vertex of a polyline. The default value in the **Edit vertex** option is one of these two options. The option that is selected as default is the one you chose last. In this manner, the **Next** and **Previous** options help you to move the X marker to any vertex of the polyline by selecting one of these two options, and then pressing ENTER to reach the desired vertex. These options cycle back and forth between first and last vertices, but cannot move past the first or last vertices, even if the polyline is closed (Figure 18-16).

The prompt sequence for using this option after you invoke the **PEDIT** command is given next.

Select polyline or [Multiple]: *Select the polyline to be edited.*
Enter an option [Close/Join/Width/Edit vertex/Fit/Spline/Decurve/Ltype gen/Undo]: **E**
Enter a vertex editing option
[Next/Previous/Break/Insert/Move/Regen/Straighten/Tangent/Width/eXit] <N>: *Enter N or P to move to the next or previous vertices respectively.*

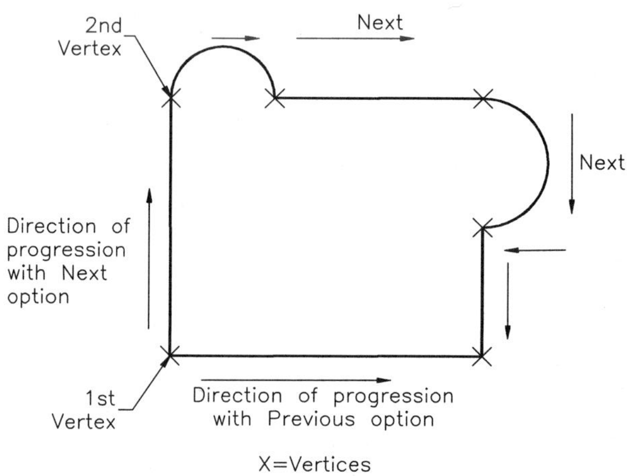

***Figure 18-16** The **Next** and **Previous** options*

Break option. With the **Break** option, you can remove a portion of the polyline as shown in Figure 18-17 or break it at a single point. The breaking of the polyline can be specified between any two vertices. By specifying two different vertices, all the polyline segments and vertices between the specified vertices are erased. If one of the selected vertices is at the endpoint of the polyline, the **Break** option will erase all the segments between the first vertex and the endpoint of the polyline. The exception to this is that AutoCAD does not erase the entire polyline if you specify the first vertex at the start point (first vertex) of the polyline and the second vertex at the endpoint (last vertex) of the polyline. If both vertices are at the endpoint of the polyline, or only one vertex is specified and its location is at the endpoint of the polyline, no change is made to the polyline. The last two selections of vertices are treated as invalid by AutoCAD, which acknowledges this by displaying the message ***Invalid***.

To use the **Break** option, first you need to move the marker to the first vertex where you want the split to start. Placement of the marker can be achieved with the help of the **Next** and **Previous** options. Once you have selected the first vertex to be used in the **Break** operation, invoke the **Break** option by entering **B** at the Command prompt. AutoCAD takes the vertex where the marker (X) is placed as the first point of breakup. The next prompt asks you to specify the position of the next vertex for breakup. You can enter **GO** at this prompt if you want to split the polyline at one vertex only. Otherwise, use the **Next** or **Previous** option to specify the position of the next vertex and then enter GO. The polyline segments between the two selected vertices is erased. The prompt sequence after you have invoked the **PEDIT** command is given next.

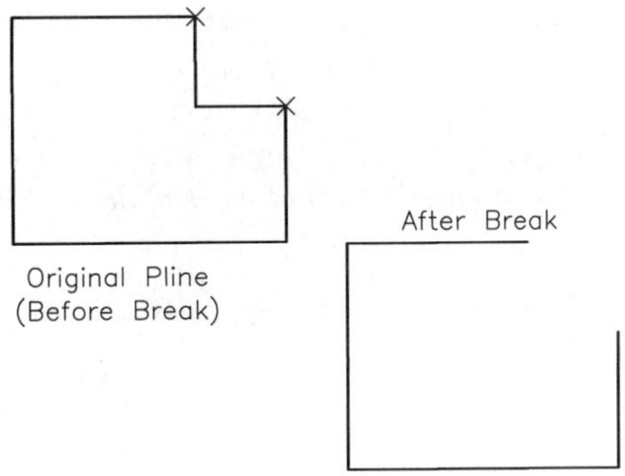

Figure 18-17 *Using the* ***Break*** *option*

Select polyline or [Multiple]: *Select the polyline to be edited.*
Enter an option [Close/Join/Width/Edit vertex/Fit/Spline/Decurve/Ltype gen/Undo]: **E**
Enter a vertex editing option
[Next/Previous/Break/Insert/Move/Regen/Straighten/Tangent/Width/eXit]<N>: *Enter N or P to locate the first vertex for the* ***Break*** *option.*
Enter a vertex editing option
[Next/Previous/Break/Insert/Move/Regen/Straighten/Tangent/Width/eXit]<N>: **B**

Once you invoke the **Break** option, AutoCAD treats the vertex where the marker (X) is displayed as the first point for splitting the polyline. The next prompt issued is given next.

Enter an option [Next/Previous/Go/eXit] <N>: *Enter G if you want to split the polyline at one vertex only or move the X marker using the* ***Next*** *or* ***Previous*** *option to specify the position of the next vertex for breakup.*

After you have specified the next position of the X marker using the **Next** and **Previous** options, entering **Go** deletes the polyline segment between the two markers specified. Now, exit the **Enter a vertex editing option...** prompt using the **eXit** option.

Insert option. The **Insert** option allows you to define a new vertex and add it to the polyline (Figure 18-18). You can invoke this option by entering I for Insert. You should invoke this option only after you have moved the marker (X) to the vertex after which the new vertex is to be added. The new vertex is inserted immediately after the vertex with the X mark. After you have invoked the **PEDIT** command, the prompt sequence for using the **Insert** option is as follows.

Select polyline or [Multiple]: *Select the polyline to be edited.*
Enter an option [Close/Join/Width/Edit vertex/Fit/Spline/Decurve/Ltype gen/Undo]: **E**
Enter a vertex editing option

[Next/Previous/Break/Insert/Move/Regen/Straighten/Tangent/Width/eXit]<N>: *Move the marker to the vertex after which the new vertex is to be inserted.*
Enter a vertex editing option
[Next/Previous/Break/Insert/Move/Regen/Straighten/Tangent/Width/eXit]<N>: **I**
Specify location for new vertex: *Move the cursor and select to specify the location of the new vertex or enter the coordinates of the new location.*

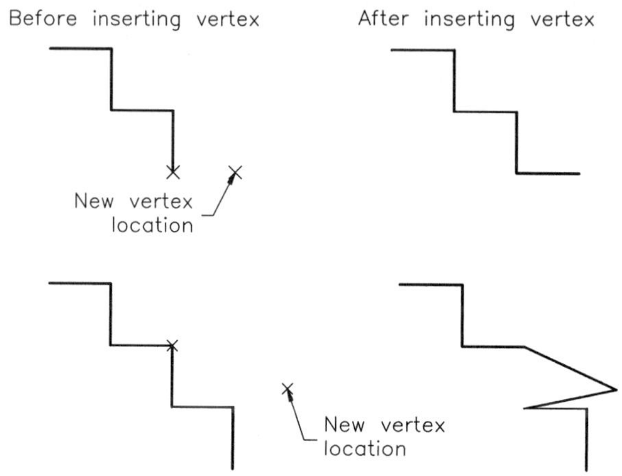

Figure 18-18 *Using the* ***Insert*** *option to define new vertex points*

Move option. This option lets you move the X-marked vertex to a new position (Figure 18-19). Before invoking the **Move** option, you must move the X marker to the vertex you want to relocate by selecting the **Next** or **Previous** option. The prompt sequence for relocating a vertex after you invoke the **PEDIT** command is given next.

Select polyline or [Multiple]: *Select the polyline to be edited.*
Enter an option [Close/Join/Width/Edit vertex/Fit/Spline/Decurve/Ltype gen/Undo]: **E**
Enter a vertex editing option
[Next/Previous/Break/Insert/Move/Regen/Straighten/Tangent/Width/eXit]<N>: *Enter N or P to move the X marker to the vertex you want to relocate.*
Enter a vertex editing option
[Next/Previous/Break/Insert/Move/Regen/Straighten/Tangent/Width/eXit]<N>: **M**
Specify new location for marked vertex: *Specify the new location for the selected vertex by selecting a new location using the pick button of your pointing device or by entering its coordinate values.*
Enter a vertex editing option
[Next/Previous/Break/Insert/Move/Regen/Straighten/Tangent/Width/eXit]<N>: *Enter an option or enter X to exit.*

Regen option. The **Regen** option regenerates the polyline to display the effects of edits you have made, without having to exit the vertex mode editing. It is used most often with the **Width** option.

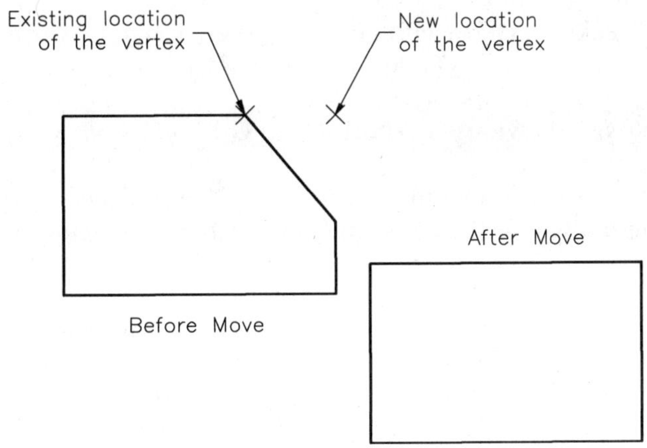

***Figure 18-19** The **Move** option*

Straighten option. The **Straighten** option can be used to straighten polyline segments or arcs between specified vertices (Figure 18-20). It deletes the arcs, line segments, or vertices between the two specified vertices and substitutes them with one polyline segment.

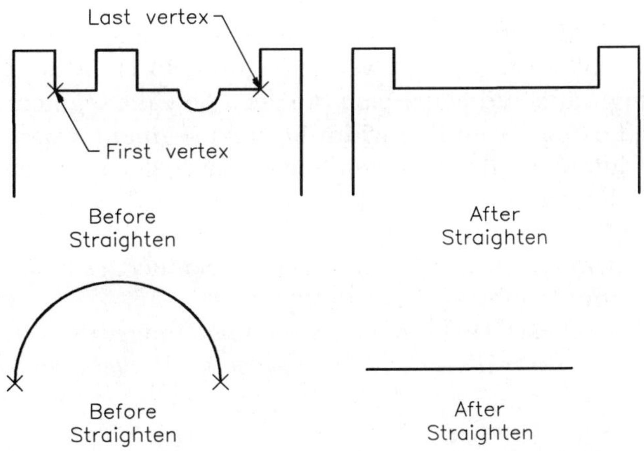

***Figure 18-20** Using the **Straighten** option to straighten polylines*

The prompt sequence to use this option after you have invoked the **PEDIT** command is given next.

Select polyline or [Multiple]: *Select the polyline to be edited.*
Enter an option [Close/Join/Width/Edit vertex/Fit/Spline/Decurve/Ltype gen/Undo]: **E**
Enter a vertex editing option
[Next/Previous/Break/Insert/Move/Regen/Straighten/Tangent/Width/eXit]<N>: *Move the marker to the desired vertex from where you want to start applying the **Straighten** option with the **Next** or **Previous** option.*

Enter a vertex editing option
[Next/Previous/Break/Insert/Move/Regen/Straighten/Tangent/Width/eXit]<N>: **S**
Enter an option [Next/Previous/Go/eXit] <N>: *Move the marker to the next desired vertex, until you reach the vertex you want to straighten.*
Enter an option [Next/Previous/Go/eXit] <N>: **G**

The polyline segments between two marker locations are replaced by a single straight line segment. If you specify a single vertex, the segment following the specified vertex is straightened, if it is curved.

Tangent option. The **Tangent** option is used to associate a tangent direction to the current vertex (marked by X). The tangent direction is used in curve fitting or the **Fit** option of the **PEDIT** command. This option is discussed in detail in the subsequent section on curve fitting. The prompt issued on using the **Tangent** option is given next.

Specify direction of vertex tangent: *Specify a point or enter an angle.*

You can specify the direction of the vertex tangent by entering an angle at the previous prompt or by selecting a point to express the direction with respect to the current vertex. You can then move the marker to another vertex using the **Next** or **Previous** option and change its direction of tangent or enter X to exit the **Enter a vertex editing option** prompt.

Width option. The **Width** option lets you change the starting and the ending widths of a polyline segment that follows the current vertex (Figure 18-21). By default, the ending width is equal to the starting width and hence, you can get a polyline segment of uniform width by pressing ENTER at the **Specify ending width for next segment <starting width>** prompt. You can also specify different starting and ending widths to get a varying-width polyline. The prompt sequence is given next.

Enter an option [Close/Join/Width/Edit vertex/Fit/Spline/Decurve/Ltype gen/Undo]: E
Enter a vertex editing option
[Next/Previous/Break/Insert/Move/Regen/Straighten/Tangent/Width/eXit]<N>: *Move the marker to the starting vertex of the segment whose width is to be altered, using the* ***Next*** *or* ***Previous*** *option.*
Enter a vertex editing option
[Next/Previous/Break/Insert/Move/Regen/Straighten/Tangent/Width/eXit]<N>: W
Specify starting width for next segment <current>: *Enter the revised starting width.*
Specify ending width for next segment <starting width>: *Enter the revised ending width or press ENTER to accept the default option of keeping the ending width of the segment equal to the starting width.*

If no difference is noticed in the appearance of the polyline, you may need to use the **Regen** option. The segment with the revised widths is redrawn after invoking the **Regen** option or when you exit the vertex mode editing.

eXit option. This option lets you exit the vertex mode editing and return to the **PEDIT** prompt.

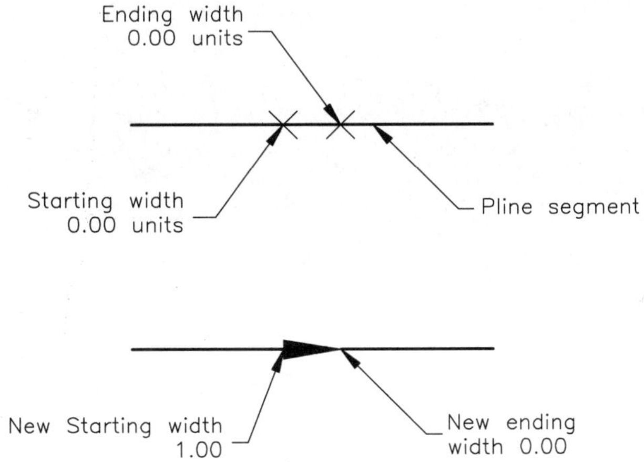

***Figure 18-21** Using the **Width** option to change the width of a polyline*

Fit Option

The **Fit** or **Fit curve** option generates a curve that passes through all the corners (vertices) of the polyline, using the tangent directions of the vertices (Figure 18-22). The curve is composed of a series of arcs passing through the corners (vertices) of the polyline. This option is used when you draw a polyline with sharp corners and need to convert it into a series of smooth curves. An example of this is a graph. In a graph, we need to show a curve by joining a series of plotted points. The process involved is called curve fitting; therefore, the name of this option. The vertices of the polyline are also known as the control points. The closer together these control points are, the smoother the curve. Therefore, if the **Fit** option does not give optimum results, insert more vertices into the polyline or edit the tangent directions of vertices and then use the **Fit** option on the polyline. Before using this option you may give each vertex a tangent direction. The curve is then constructed, according to the tangent directions you have specified. The following prompt sequence illustrates the **Fit** option after you invoke the **PEDIT** command.

Select polyline or [Multiple]: *Select the polyline to be edited.*
Enter an option [Close/Join/Width/Edit vertex/Fit/Spline/Decurve/Ltype gen/Undo]: **F**

If the tangent directions need to be edited, use the **Edit vertex** option of the **PEDIT** command. Move the X marker to each of the vertices that need to be changed. Now, you can invoke the **Tangent** option, and either enter the tangent direction in degrees or select points. The chosen direction is expressed by an arrow placed at the vertex. The prompt sequence after you invoke the **PEDIT** command is given next.

Select polyline or [Multiple]: *Select the polyline to be edited.*
Enter an option [Close/Join/Width/Edit vertex/Fit/Spline/Decurve/Ltype gen/Undo]: **E**
Enter a vertex editing option

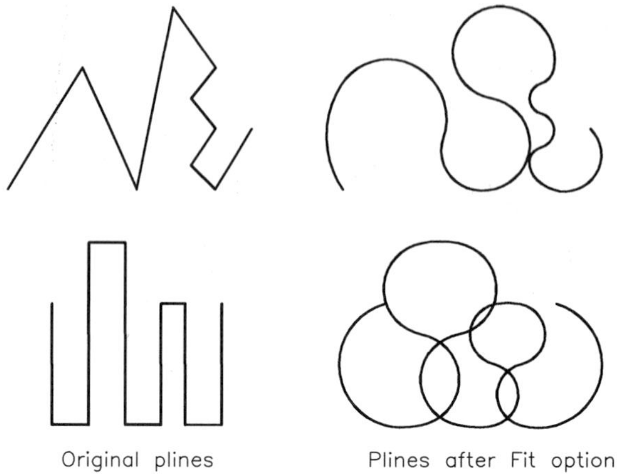

Figure 18-22 *The **Fit** option*

[Next/Previous/Break/Insert/Move/Regen/Straighten/Tangent/Width/eXit]<N>: **T**
Specify direction of vertex tangent: *Specify a direction in + or - degrees, or select a point in the desired direction. Press ENTER.*

Once you specify the tangent direction, use the **eXit** option to return to the previous prompt and use its **Fit** option.

Spline Option

The **Spline** option (Figure 18-23) also smoothens the corners of a straight segment polyline, as does the **Fit** option, but the curve passes through only the first and the last control points (vertices), except in the case of a closed polyline. The spline curve is stretched toward the other control points (vertices) but does not pass through them, as in the case of the **Fit** option. The greater the number of control points, the greater the force with which the curve is stretched toward them. The prompt sequence after you invoke the **PEDIT** command is as follows:

Select polyline or [Multiple]: *Select the polyline.*
Enter an option [Close/Join/Width/Edit vertex/Fit/Spline/Decurve/Ltype gen/Undo]: **S**

The generated curve is a B-spline curve. The **frame** is the original polyline without any curves in it. If the original polyline has arc segments, these segments are straightened when the spline's frame is formed. A frame that has width produces a spline curve that tapers smoothly from the width of the first vertex to that of the last. Any other width specification between the first width specification and the last is neglected. When a spline is formed from a polyline, the frame is displayed as a polyline with zero width and continuous linetype. Also, AutoCAD saves its frame so that it may be restored to its original form. Tangent specifications on control point vertices do not affect spline construction.

By default, the spline frames are not shown on screen, but you may want them displayed for

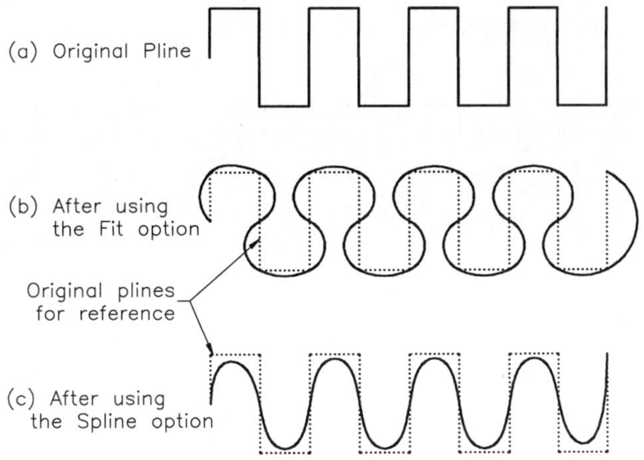

***Figure 18-23** The **Spline** option*

reference. In this case, the system variable **SPLFRAME** needs to be manipulated. The default value for this variable is zero. If you want to see the spline frame as well, set it to 1.

Now, whenever the **Spline** option is used on a polyline, the frame will also be displayed. Most editing commands such as **MOVE**, **ERASE**, **COPY**, **MIRROR**, **ROTATE**, and **SCALE** work similarly for both the Fit curves and Spline curves. They work on both the curve and its frame, whether the frame is visible or not. The **EXTEND** command changes the frame by adding a vertex at the point the last segment intersects with the boundary. If you use any of the previously mentioned commands and then use the **Decurve** option to decurve the spline curve and later use the **Spline** option again, the same spline curve is generated. The **BREAK**, **TRIM**, and **EXPLODE** commands delete the frame. The **DIVIDE**, **MEASURE**, **FILLET**, **CHAMFER**, **AREA**, and **HATCH** commands recognize only the spline curve and do not consider the frame. The **STRETCH** command first stretches the frame and then fits the spline curve to it.

When you use the **Join** option of the **PEDIT** command, the spline curve is decurved and the original spline information is lost. The **Next** and **Previous** options of the **Edit vertex** option of the **PEDIT** command move the marker only to points on the frame whether visible or not. The **Break** option discards the spline curve. **Insert**, **Move**, **Straighten**, and **Width** options refit the spline curve. Object Snaps consider the spline curve and not the frame; therefore, if you wish to snap to the frame control points, restore the original frame.

There are two types of spline curves:
Quadratic B-spline
Cubic B-spline

Both of them pass through the first and the last control points, which is characteristic of the spline curve. Cubic curves are very smooth. A cubic curve passes through the first and last control points, and the curve is closer to the other control points. Quadratic curves are not as smooth as the cubic ones, but they are smoother than the curves produced by the **Fit curve** option. A quadratic curve passes through the first and last control points, and the rest of the curve is tangent to the polyline segments between the remaining control points (Figure 18-24).

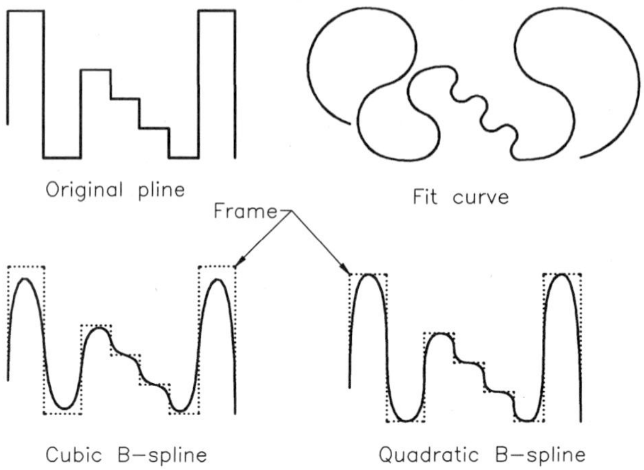

***Figure 18-24** Comparison of fit curve, quadratic B-spline, and cubic B-spline*

Generation of different types of spline curves. If you want to edit a polyline into a B-spline curve, you are first required to enter a relevant value in the **SPLINETYPE** system variable. A value of 5 produces the quadratic curve, whereas a value of 6 produces a cubic curve. The default value is 6, which implies that when we use the **Spline** option of the **PEDIT** command, a cubic curve is produced by default. You can change the value of the **SPLINETYPE** variable using the command line.

You can also set the value of the variable by selecting the **Polyvars** option in the **PEDIT** screen menu. The screen menu can be invoked using the Display screen menu option available on the **Display** tab of the **Options** dialog box. On doing so, the **Set Spline Fit Variables** dialog box (Figure 18-25) is displayed. You can select any type of curve from the list available. The **SPLINETYPE** variable is automatically set to the value of the type you have chosen from the dialog box. In case you are editing 3D polygon meshes, the smoothness of the meshes is controlled by the **SURFTYPE** variable. The value of the variable can be set on the command line. You can select the type of polygon mesh you want when you are using the **Smooth Surface** option of the **PEDIT** command from the **Set Spline Fit Variables** dialog box.

SPLINESEGS. The system variable **SPLINESEGS** governs the number of line segments used to construct the spline curves, and so you can use this variable to control the smoothness of the curve. The default value for this variable is 8. With this value, a reasonably smooth curve that does not need much regeneration time is generated. The greater the value of this variable, the smoother the curve, but greater regeneration time, and more space occupied by the drawing file.

Figure 18-26 shows cubic curves with different values for the **SPLINESEGS** parameter.

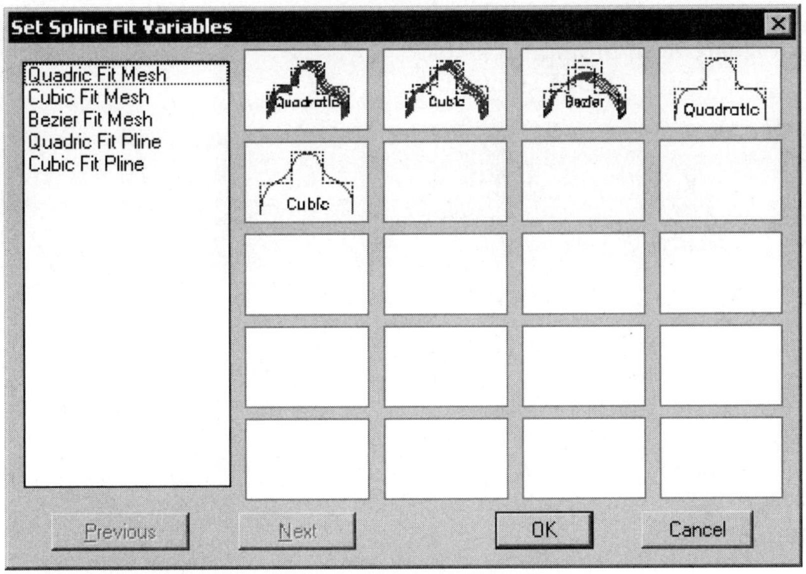

Figure 18-25 The ***Set Spline Fit Variables*** *dialog box*

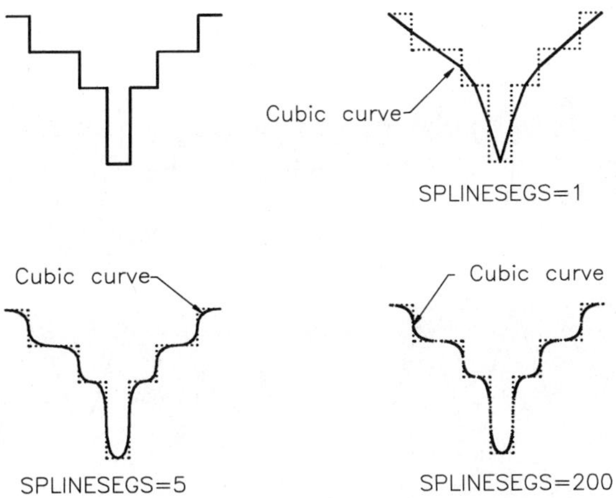

Figure 18-26 Using the ***SPLINESEGS*** *variable*

Decurve Option

The **Decurve** option straightens the curves generated after using the **Fit** or **Spline** option on a polyline. They return to their original shape (Figure 18-27). The polyline segments are straightened using the **Decurve** option. The vertices inserted after using the **Fit** or **Spline** option are also removed. Information entered for tangent reference is retained for use in future fit curve operations. You can also use this command to straighten out any curve drawn with the help of the **Arc** option of the **PLINE** command. Enter **D** at the **Enter an option**

[Close/Join/Width/Edit vertex/Fit/Spline/Decurve/Ltype gen/Undo] prompt to invoke the **Decurve** option of the **PEDIT** command.

Note

*If you use edit commands such as the **BREAK** or **TRIM** commands on spline curves, the **Decurve** option cannot be used.*

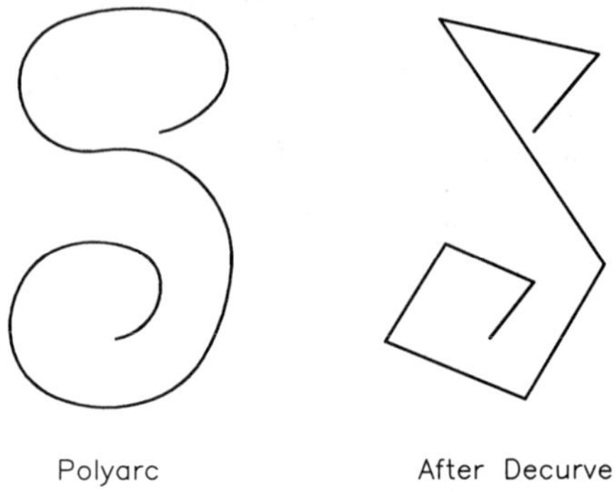

Figure 18-27 *The **Decurve** option*

Ltype gen Option

You can use this option to control the linetype pattern generation for linetypes other than Continuous with respect to the vertices of the polyline. This option has two modes ON and OFF. If turned off, the break in the noncontinuous linetypes will be avoided at the vertices of the polyline and a continuous segment will be displayed at the vertices (Figure 18-28). If turned on, this option generates the linetype in a continuous pattern such that the gaps may be displayed at the vertices. This option is not applicable to polylines with tapered segments. The command prompt displayed when you select a polyline and invoke the **Ltype gen** option of the **PEDIT** command is as follows.

Enter polyline linetype generation option [ON/OFF] <Off>: *Enter ON or press ENTER to accept the default value.*

The linetype generation for new polylines can also be controlled with the help of the **PLINEGEN** system variable, which acts as a toggle. The default value is 0 (off). Entering a value of 1 turns on the **Ltype gen** option.

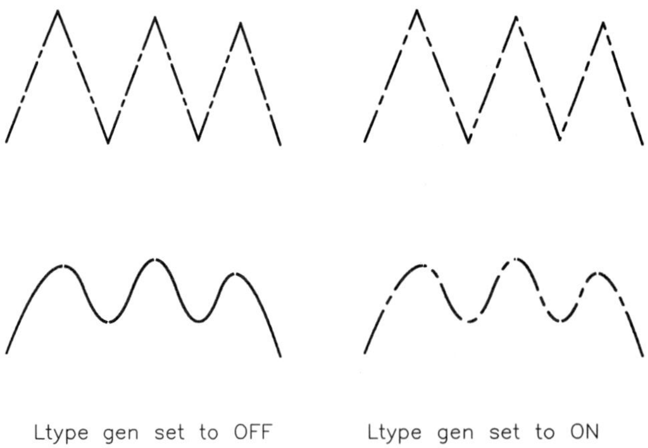

Figure 18-28 *Using the **Ltype gen** option*

Undo (U) Option

The **Undo** option negates the effect of the most recent **PEDIT** operation. You can go back as far as you need to in the current **PEDIT** session by using the **Undo** option repeatedly until you get the desired screen. If you started editing by converting an object into a polyline, and you want to change the polyline back to the object from which it was created, the **Undo** option of the **PEDIT** command will not work. In this case, you will have to exit to the Command prompt and use the **UNDO** command to undo the operation.

Editing Multiple Polylines

Selecting the **Multiple** option of the **PEDIT** command allows you to select more than one polyline for editing. You can select the polylines using any of the objects selection techniques. After the objects to be edited are selected, press ENTER or right-click to proceed with the command. The prompt sequence that will follow is given next.

Enter an option [Close/Open/Join/Width/Fit/Spline/Decurve/Ltype gen/Undo]:

Join

This option is used to join more than one polylines that may or may not be in contact with each other. Even the polylines that are not coincident can be joined using this option. The polylines are joined using a specified distance value called the **Fuzz distance**. After you select the **Join** option, the sequence of prompts is as follows.

Join Type = Extend
Enter fuzz distance or [Jointype] <current> : *Enter a distance or **J** for changing the **Jointype**.*

The prompt sequence to follow when you enter **J** at the above prompt is given next.

Enter join type [Extend/Add/Both] <Extend>:

Extend. This option extends or trims the selected polylines at the endpoints that are nearest to each other, see Figure 18-29 . Keep in mind that the segments of selected polylines that are nearest to each other should not be parallel. This means that the segments that are nearest to each other should intersect at some point when extended.

Add. This option joins the selected polylines by drawing a straight line between the nearest endpoints, see Figure 18-30. The fuzz distance should be greater than the actual distance between the two endpoints to be joined.

Both. This option joins the endpoints by extending or trimming if possible; otherwise, it adds a line segment between the two endpoints.

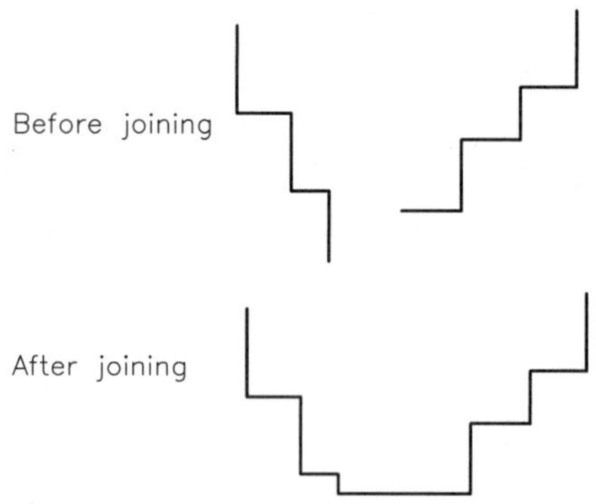

*Figure 18-29 Joining two polylines using the **Extend** option*

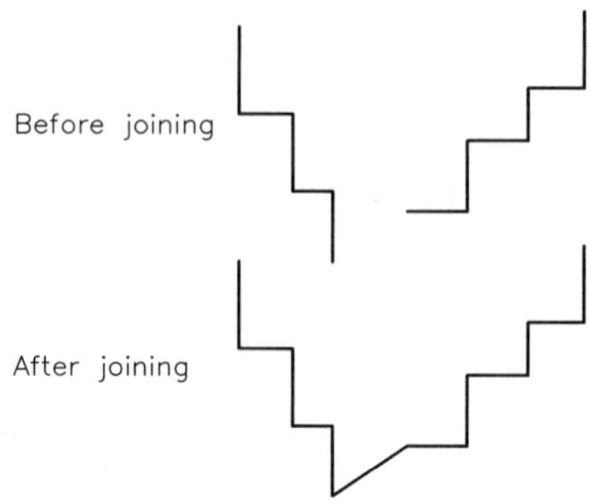

*Figure 18-30 Joining two polylines using the **Add** option*

Exercise 2 — *General*

a. Draw a line from point (0,0) to point (6,6). Convert the line into a polyline with starting width 0.30 and ending width 0.00.

b. Draw different polylines of varying width segments and then use the **Join** option to join all the segments into one polyline. After joining the different segments, make the width of the segments uniform, with the **Width** option.

Exercise 3 — *General*

a. Draw a staircase-shaped polyline, then use different options of the **PEDIT** command to generate a fit curve, a quadratic B-spline, and a cubic B-spline, and then convert the curves back to the original polyline.

b. Draw square-wave-shaped polylines, and use different options of the **Edit vertex** option to navigate around the polyline, split the polyline into two at the third vertex, insert more vertices at different locations in the original polyline, and convert the square-wave-shaped polyline into a straight-line polyline.

UNDOING THE PREVIOUS COMMANDS

Command: UNDO

The **LINE** and **PLINE** commands have the **Undo** option, which can be used to undo (nullify) the changes made within these commands. The **UNDO** command is used to undo a previous command or to undo more than one command at one time. This command can be invoked by entering **UNDO** at the Command prompt. **Undo** is also available from the **Edit** menu and the **Standard** toolbar, but this can only undo the previous command and only one command at a time. If you right-click in the drawing area a shortcut menu is displayed. Choose **Undo**. It is equivalent to **UNDO 1** or to enter **U** at the Command prompt. The **U** command can undo only one operation at a time. You can use **U** as many times as you want until AutoCAD displays the message: **Nothing to undo**. When an operation cannot be undone, the message name is displayed but no action takes place. External commands such as **PLOT** cannot be undone. The prompt sequence for the **UNDO** command is given next.

Command: **UNDO**
Enter the number of operations to undo or [Auto/Control/BEgin/End/Mark/Back] <1>: *Enter a positive number, an option, or press ENTER to undo a single operation.*

The various options of this command are discussed next.

Number (N) Option

This is the default option. This number represents the number of previous command sequences to be deleted. For example, if the number entered is 1, the previous command is undone; if the number entered is 4, the previous four commands are undone, and so on. This is identical to invoking the **U** command four times, except that only one regeneration takes place. AutoCAD lets you know which commands were undone by displaying a message after you press ENTER. The prompt sequence is given next.

Command: **UNDO**
Enter the number of operations to undo or [Auto/Control/BEgin/End/Mark/Back] <1>: 3
PLINE LINE CIRCLE

Control (C) Option

This option lets you determine how many of the options are active in the **UNDO** command. You can disable the options you do not need. With this option you can even disable the **UNDO** command. To access this option, type C. You will get the following prompt.

Enter an UNDO control option [ALL/None/One] <All>: *Enter an option or press ENTER to accept the default.*

ALL

The **All** option activates all the features (options) of the **UNDO** command.

None (N)

This option turns off **UNDO** and the **U** command. If you have used the **BEgin** and **End** options or **Mark** and **Back** options to create **UNDO** information, all of that information is undone. The prompt sequence for invoking the **Control** option is as follows.

Command: **UNDO**
Enter the number of operations to undo [Auto/Control/BEgin/End/Mark/Back] <1>: **C**
Enter an UNDO control option [ALL/None/One] <All>: **N**

If you try to use the **U** command now, while the **UNDO** command has been disabled, AutoCAD gives you the following message.

Command: **U**
U command disabled. Use **UNDO** command to turn it on.

The prompt sequence for the **UNDO** command after issuing the **None** option is given next.

Command: **UNDO**
Enter an UNDO control option [All/None/One] <All>: *Enter O or press ENTER to accept the default.*

To enable the **UNDO** options again, you must enter the **All** or **One** (one mode) option.

One (O)

This option restrains the **UNDO** command to a single operation. All **UNDO** information saved earlier during editing is scrapped. The prompt sequence is as follows.

Enter the number of operations to undo or [Auto/Control/BEgin/End/Mark/Back] <1>: **C**
Enter an UNDO control option [All/None/One] <All>: **O**

If you then enter the **UNDO** command, you will get the following prompt.

Command: **UNDO**
Enter an option [Control] <1>: *Press ENTER to undo the last operation or enter C to select another control option.*

In response to this last prompt, you can now either press ENTER to undo only the previous command, or go into the **Control** options by entering C. AutoCAD acknowledges undoing the previous command with messages such as given next.

Command: **UNDO**
Enter an option [Control] <1>: Enter
CIRCLE
Everything has been undone

Tip

*AutoCAD stores all information about all the **UNDO** operations taking place in a drawing session. When you use the **None** control option of the **UNDO** command, this information is removed and valuable disk space is made available. In case of limited disk space availability, it may be a good idea to use the **One** suboption of the **UNDO Control** command. This allows you to use both the **U** and **UNDO** commands and at the same time makes the disk space available.*

Auto (A) Option

Enter A to invoke this option. The following prompt is displayed.

Enter UNDO Auto mode [ON/OFF] <current>: *Select ON or OFF or press ENTER to accept the default.*

This option is On by default and every menu item is a single operation. When you use the **Begin** and **End** options to group a series of commands as a single operation, entering **U** at the Command prompt removes the objects individually and not as a group, if the Auto option is On. But if the **Auto** option is Off, entering **U** at the Command prompt will remove the entire group of objects you had grouped using the **Begin** and **End** options in one single operation.

Also, if you have put a marker in between the begin and end operations and if **Auto** is On, the group of commands till the marker is undone altogether on using the **Back** option. This group of commands is considered to be a single group and is undone as a single command, although, entering **U** at the Command prompt removes the objects individually until the marker is encountered. But, if the **Auto** option is off, entering **U** at the Command prompt or using the **Back** option of the **UNDO** command undoes the entire group of operations irrespective of the marker being there.

BEgin (BE) and End (E) Options

A group of commands is treated as one command for the **U** and **UNDO** commands (provided **Auto** option is Off) by embedding the commands between the **BEgin** and **End** options of the **UNDO** command. If you anticipate the removal of a group of successive commands later in a drawing, you can use this option, since all of the commands after the **BEgin** option and before the **End** option are treated as a single command by the **U** command (if the **Auto** option is Off). For example, the following sequence illustrates the possibility of removal of two commands.

Command: **UNDO**
Enter the number of operations to undo or [Auto/Control/BEgin/End/Mark/Back] <1>: **BE**
Command: **CIRCLE**
Specify center point for Circle or [3P/2P/Ttr (tan tan radius)]: *Specify the center.*
Specify radius of circle or [Diameter] <current>: *Specify the radius of the circle.*
Command: **PLINE**
Specify start point or [Arc/Close/Halfwidth/Length/Undo/Width]: *Select first point.*
Specify next point: *Select the next point.*
Specify start point or [Arc/Close/Halfwidth/Length/Undo/Width]: *Press ENTER.*
Command: **UNDO**
Enter the number of operations to UNDO or [Auto/Control/BEgin/End/Mark/Back]: **E**
Command: **U**

To start the next group once you have finished specifying the current group, use the **End** option to end this group. Another method is to enter the **BEgin** option to start the next group while the current group is active. This is equivalent to issuing the **End** option followed

Chapter 18

by the **BEgin** option. The group is complete only when the **End** option is invoked to match a **BEgin** option. If **U** or the **UNDO** command is issued after the **BEgin** option has been invoked and before the **End** option has been issued, only one command is undone at a time until it reaches the juncture where the **BEgin** option has been entered. If you want to undo the commands issued before the **BEgin** option was invoked, you must enter the **End** option so that the group is complete. This is demonstrated in the following example.

Example 2 — *General*

Enter the following commands in the same sequence as given, and notice the changes that take place on screen.

CIRCLE
POLYGON
UNDO BEgin (Make sure that the Auto is OFF)
PLINE
TRACE
U
DONUT
UNDO End
TEXT
U
U
U
U

The first **U** command will undo the **TRACE** command. If you repeat the **U** command, the **PLINE** command will be undone. Any further invoking of the **U** command will not undo any previously drawn object (**POLYGON** and **CIRCLE**, in this case), because after the **PLINE** is undone you have an **UNDO BEgin**. Only after you enter **UNDO End** can you undo the **POLYGON** and the **CIRCLE**. In the example, the second **U** command will undo the **TEXT** command, the third **U** command will undo the **DONUT** and **PLINE** commands (these are enclosed in the group), the fourth **U** command will undo the **POLYGON** command, and the fifth **U** command will undo the **CIRCLE** command. Whenever the commands in a group are undone, the name of each command or operation is not displayed as it is undone. Only the name, **GROUP**, is displayed.

Tip
*You can use the **BEgin** and **End** options only when the **UNDO Control** is set to **All** and the **Auto** option is Off.*

Mark (M) and Back Options

The **Mark** option installs a marker in the Undo file. The **Back** option lets you undo all the operations until the mark. In other words, the **Back** option returns the drawing to the point where the previous mark was inserted. For example, if you have completed a portion of your drawing and do not want anything up to this point to be deleted, you insert a marker and then proceed. Then, even if you use the **UNDO Back** option, it will work only until the marker. You can insert multiple markers, and with the help of the **Back** option you can return to the successive mark points. The following prompt sequence is used to introduce a mark point.

Command: **UNDO**
Enter the number of operations to undo or [Auto/Control/BEgin/End/Mark/Back]: **M**

The prompt sequence for using the **Back** option is given next.

Command: **UNDO**
Enter the number of operations to undo or [Auto/Control/BEgin/End/Mark/Back]: **B**

Once all the marks have been exhausted with the successive **Back** options, any further invoking of the **Back** option displays the message: **This will undo everything. OK? <Y>**. If you enter Y (Yes) at this prompt, all the operations carried out since you entered the current drawing session will be undone. If you enter N (No) at this prompt, the **Back** option will be disregarded.

You cannot undo certain commands and system variables, for example, **DIST**, **LIST**, **DELAY**, **NEW**, **OPEN**, **QUIT**, **AREA**, **HELP**, **PLOT**, **QSAVE** and **SAVE**, among many more. Actually, these commands have no effect that can be undone. Commands that change operating modes (**GRID**, **UNITS**, **SNAP**, **ORTHO**) can be undone, though the effect may not be apparent at first. This is the reason why AutoCAD displays the command names as they are undone.

REVERSING THE UNDO OPERATION

Toolbar:	Standard > Redo
Menu:	Edit > Redo
Command:	REDO

If you right-click in the drawing area, a shortcut menu is displayed. Choose **Redo** to invoke the **REDO** command. The **REDO** command brings back the process you removed previously using the **U** and **UNDO** commands. This command undoes the last **UNDO** command, but it must be entered immediately after the **UNDO** command and the objects previously undone reappear on the screen.

RENAMING NAMED OBJECTS

Menu:	Format > Rename
Command:	RENAME

You can edit the names of named objects such as blocks, dimension styles, layers, linetypes, styles, UCS, views, and viewports using the **Rename** dialog box. You can select the type of named object from the list provided in the **Named Objects** area of the dialog box. Corresponding names of all the objects of the specified type that can be renamed are displayed in the **Items** area. For example, if you want to rename the layer named LOCKED to HID, the process will be as follows.

1. Select **Layers** in the **Named Objects** list box. All the layer names in the current drawing that can be renamed are displayed in the **Items** list box.

2. Select LOCKED in the **Items** list box to highlight it. LOCKED is displayed in the **Old Name** edit box.

3. Enter HID in the **Rename To** edit box (Figure 18-31), and choose the **OK** button.

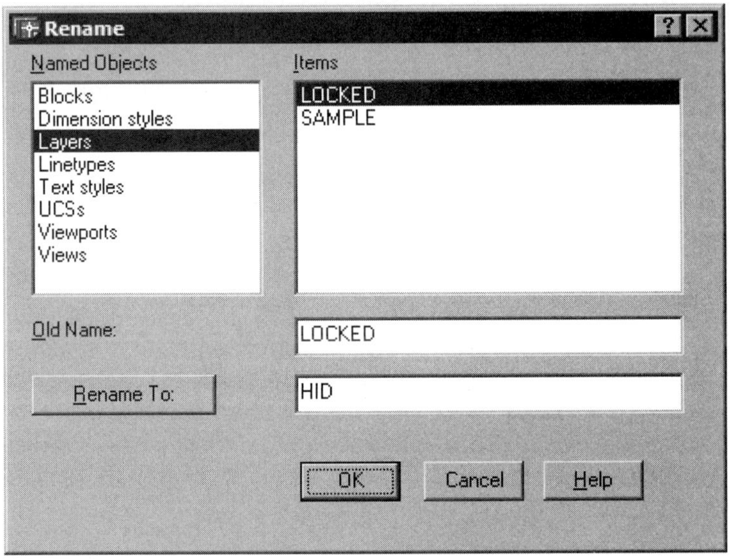

***Figure 18-31 Rename** dialog box*

Now the layer named LOCKED is renamed to HID. You can invoke the **Rename** dialog box again, view the **Layer Control** drop-down list in the **Layers** toolbar, or invoke the **Layer Properties Manager** dialog box to notice this change. You can rename blocks, dimension styles, linetypes, styles, UCS, views, and viewports in the same way.

Note

*The **RENAME** command cannot be used to rename drawing files created using the **WBLOCK** command. You can change the name of a drawing file by using the **Rename** option of your Operating System.*

*The Layer **0** and the **Continuous** linetype cannot be renamed and therefore do not appear in the **Items** list box of the **Rename** dialog box.*

REMOVING UNUSED NAMED OBJECTS

Menu:	File > Drawing Utilities > Purge
Command:	PURGE

This is another editing operation used for deletion and it was discussed earlier in relation to blocks. You can delete unused named objects such as blocks, layers, dimension styles, linetypes, text styles, and shapes with the help of the **PURGE** command. When you create a new drawing or open an existing one, AutoCAD records the named objects in that drawing and notes other drawings that reference the named objects. Usually only a few of the named objects in the drawing (such as layers, linetypes, and blocks) are used. For example, when you create a new drawing, the prototype drawing settings may contain various text styles, blocks, and layers which you do not want to use. Also, you may want to delete particular unused named objects such as unused blocks in an existing drawing. Deleting inactive named objects is important and useful because doing so reduces the space occupied by the drawing. With the **PURGE** command, you can select the named objects you want to delete. You can use this command at any time in the drawing session. When you invoke the **Purge** command, the **Purge** dialog box is displayed (Figure 18-32).

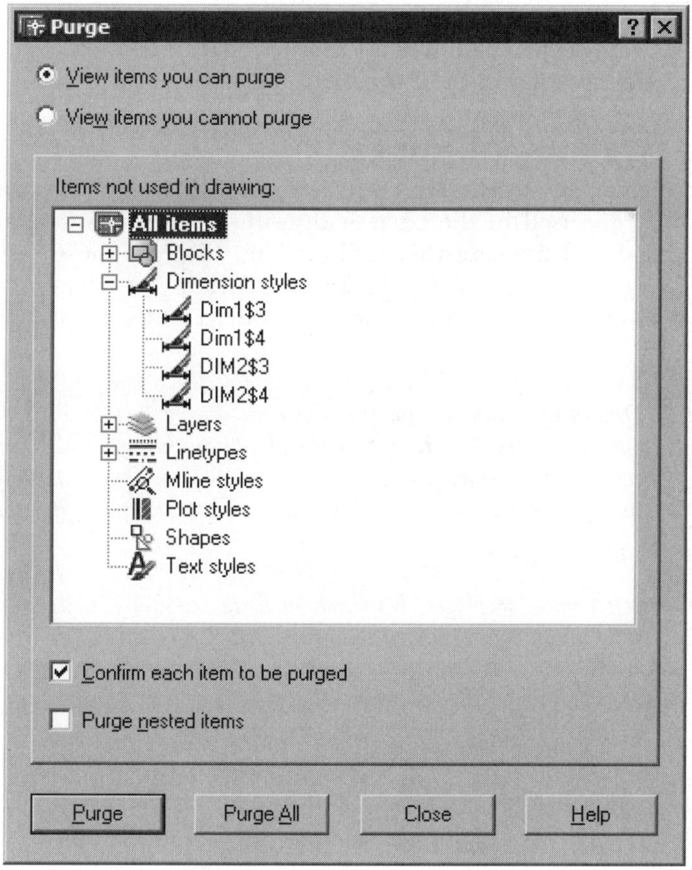

Figure 18-32 ***Purge*** *dialog box with items that can be purged*

Chapter 18

View items you can purge

When this radio button is selected, AutoCAD displays the tree view of all the named objects that are in the current drawing and those can be purged. When this radio button is chosen, **Items not used in drawing** area, and two check boxes, **Confirm each item to be purged** and **Purge nested items**, are displayed in the dialog box. These are as discussed as follows.

Items not used in drawing

This area lists all the named objects that can be purged. You can list the items of any object type by choosing the plus (+) sign or by double-clicking on the named object in the tree view. Select them to purge. You can select more than one item by holding down the SHIFT or the CTRL key and selecting the items.

Confirm each item to be purged

This check box is selected to confirm before purging the selected item. The **Confirm Purge** dialog box is displayed when items are selected to purge and after the **Purge** or **Purge All** buttons are chosen. You can confirm or cancel the items to be purged in this dialog box.

Purge nested items

This check box, when selected, removes all the nested items not in use. This removes the nested items only when you select the **Purge All** button or select blocks.

View items you cannot purge

This radio button is selected to display the tree view of the items you cannot purge. These items are those that are used in the current drawing or are default items that cannot be removed (Figure 18-33). Choosing this radio button displays the tree view in the **Items currently used in drawing** area of the dialog box. When you select an object in this tree view, information about why you cannot purge this object is displayed below the tree view.

Note

*The **Entire Drawing** option of the **Write Block** dialog box that is invoked on using the **WBLOCK** command or the **-WBLOCK asterisk** option that have been discussed earlier have the same effect as the **PURGE All** command. The only difference is that the unused named objects are removed automatically without any verifications, though these methods are much faster.*

*Standard objects created by AutoCAD (such as layer 0, STANDARD text style, and linetype CONTINUOUS) cannot be removed by the **PURGE** command, even if these objects are not used.*

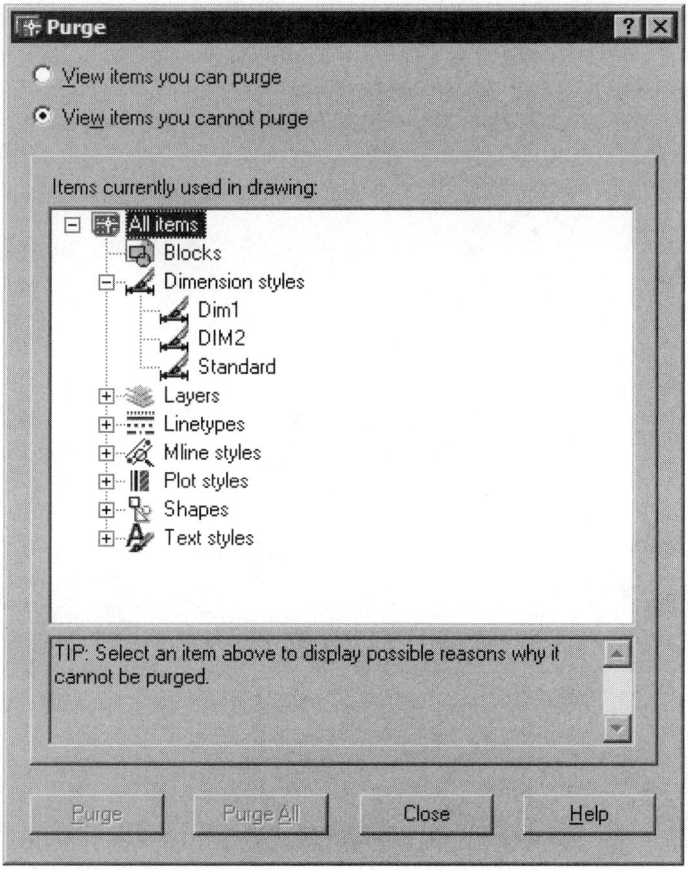

***Figure 18-33 Purge** dialog box with items that cannot be removed*

SETTING SELECTION MODES USING THE OPTIONS DIALOG BOX

Menu:	Tools > Options
Command:	OPTIONS

When you select a number of objects, the selected objects form a **selection set**. Selection of the objects is controlled in the **Options** dialog box (Figure 18-34) that is invoked by the **OPTIONS** command. Six selection modes are provided in the **Selection** tab of this dialog box. You can select any one of these modes or a combination of various modes.

Noun/verb selection

By selecting the **Noun/verb selection** check box, you can select the objects (noun) first and then specify the operation (verb) (command) to be performed on the selection set. This mode is active by default. For example, if you want to move some objects when the mode

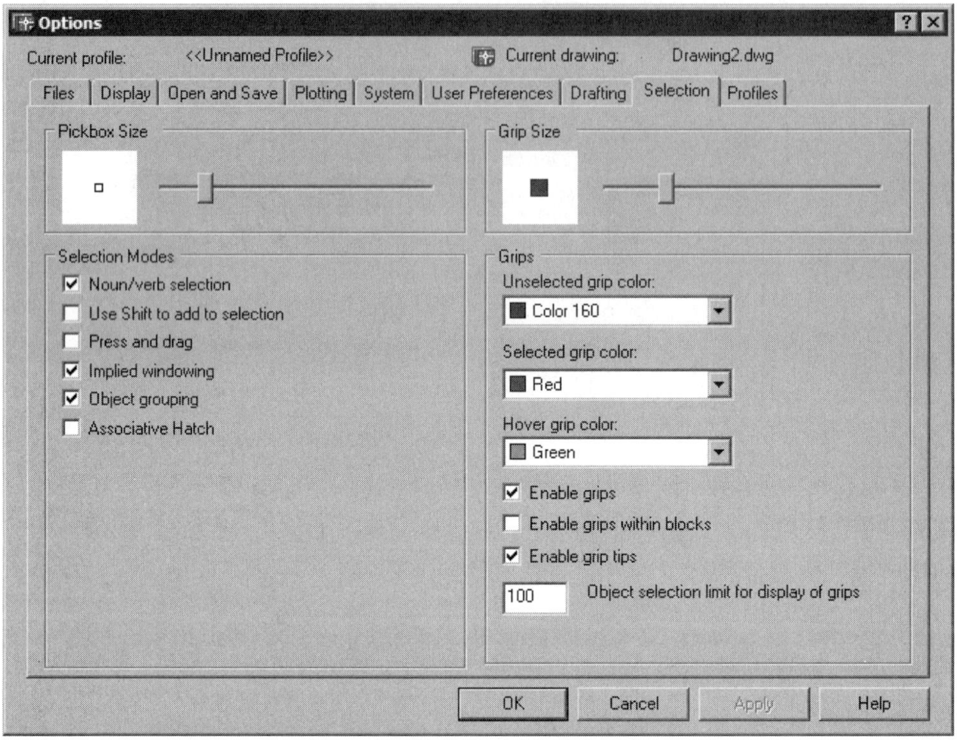

***Figure 18-34** The **Selection** tab of the **Options** dialog box*

Noun/verb selection is enabled, first select the objects to be moved, and then invoke the **MOVE** command. The objects selected are highlighted automatically when the **MOVE** command is invoked, and AutoCAD does not issue any **Select objects** prompt. The following are some of the commands that can be used on the selected objects when the **Noun/verb selection** mode is active.

ARRAY	**BLOCK**	**CHANGE**	**CHPROP**	**COPY**
DVIEW	**EXPLODE**	**ERASE**	**LIST**	**MIRROR**
MOVE	**PROPERTIES**	**ROTATE**	**SCALE**	**WBLOCK**

The following are some of the commands that are not affected by the **Noun/verb selection** mode. You are required to specify the objects (noun) on which an operation (command/verb) is to be performed, after specifying the command (verb).

BREAK	**CHAMFER**	**DIVIDE**	**EDGESURF**	**EXTEND**
FILLET	**MEASURE**	**OFFSET**	**REVSURF**	**RULESURF**
TABSURF	**TRIM**			

When the **Noun/Verb selection** mode is active, the **PICKFIRST** system variable is set to 1 (On). In other words, you can also activate the **Noun/verb selection** mode by setting the **PICKFIRST** system variable to 1 (On).

Use Shift to add to selection Option

The next option in the **Selection Modes** area of the **Selection** tab of the **Options** dialog box is **Use Shift to add to selection**. Selecting this option establishes the additive selection mode, which is the normal method of most Windows programs. In this mode, you have to hold down the SHIFT key when you want to add objects to the selection set. For example, suppose X, Y, and Z are three objects on the screen and you have selected the **Use Shift to add to selection** check box in the **Selection Modes** area of the **Options** dialog box. Select object X. It is highlighted and put in the selection set. After selecting X, and while selecting object Y, if you do not hold down the SHIFT key, only object Y is highlighted and it replaces object X in the selection set. On the other hand, if you hold down the SHIFT key while selecting Y, it is added to the selection set (which contains X), and the resulting selection set contains both X and Y. Also, both X and Y are highlighted. To summarize the concept, objects are added to the selection set only when the SHIFT key is held down while objects are selected. Objects can be discarded from the selection set by reselecting these objects while the SHIFT key is held down. If you want to clear an entire selection set quickly, draw a blank selection window anywhere in a blank drawing area. You can also right-click to display a shortcut menu and choose **Deselect All**. All selected objects in the selection set are discarded from it.

When the **Use Shift to add to selection** mode is active, the **PICKADD** system variable is set to 0 (Off). In other words, you can activate the **Use Shift to add to selection** mode by setting the **PICKADD** system variable to Off.

Press and drag

This selection mode is used to govern the way you can define a selection window or crossing window. When you select this option, you can create the window by pressing the pick button to select one corner of the window and continuing to hold down the pick button and dragging the cursor to define the other diagonal point of the window. When you have the window you want, release the pick button. If the **Press and drag** mode is not active, you have to select twice to specify the two diagonal corners of the window to be defined. This mode is not active by default. This implies that to define a selection window or crossing window, you have to select twice to define their two opposite corner points.

When the **Press and drag** mode is active, the **PICKDRAG** system variable is set to 1 (On). In other words, you can activate the **Press and drag** mode by setting **PICKDRAG** to On.

Implied windowing

By selecting this option, you can automatically create a Window or Crossing selection when the **Select objects** prompt is issued. The selection window or crossing window in this case is created in the following manner: At the **Select objects** prompt, select a point in empty space on the screen. This becomes the first corner point of the selection window. After this, AutoCAD asks you to specify the other corner point of the selection window. If the first corner point is to the right of the second corner point, a Crossing selection is defined; if the first corner point is to the left of the second corner point, a Window selection is defined. The **Implied windowing** check box is selected by default.

If this option is not active, or if you need to select the first corner in a crowded area where selecting would select an object, you need to specify Window or Crossing at the **Select objects** prompt, depending on your requirement.

When the **Implied windowing** mode is active, the **PICKAUTO** system variable is set to 1 (On). In other words, you can activate the **Implied windowing** mode by setting the value of the **PICKAUTO** system variable to 1 (On).

Object grouping

This turns the automatic group selection on and off. When this option is on and you select a member of a group, the whole group is selected. You can also activate this option by setting the value of the **PICKSTYLE** system variable to 1. (Groups were discussed earlier in this chapter.)

Tip

*You can clear the **Object grouping** check box in the **Selection Modes** area of the **Selection** tab of the **Options** dialog box to be able to select the objects of a group individually for editing without having to explode the group.*

Associative Hatch

If the **Associative Hatch** check box is selected in the **Selection modes** area of the **Selection** tab of the **Options** dialog box, the boundary object is also selected when an associative hatch is selected. You can also select this option by setting the value of the **PICKSTYLE** system variable to 2 or 3. Hatching and boundaries have been discussed in Chapter 13, Hatching Drawings. It is recommended that you select the **Associative Hatch** check box for most drawings.

Pickbox Size

The **Pickbox** slider bar controls the size of the pickbox. The size ranges from 0 to 20. The default size is 3. You can also use the **PICKBOX** system variable.

Self-Evaluation Test

Answer the following questions, and then compare your answers to the answers given at the end of this chapter.

1. Even if a group is defined as selectable, if the **PICKSTYLE** system variable is set to 0, you will not be able to select the entire group by selecting one of its members. (T/F)

2. Only the **CHANGE** command can be used to change the properties associated with an object. (T/F)

3. The **PROPERTIES** palette can be used to modify the geometry of objects apart from their general properties. (T/F)

4. The **RENAME** command can be used to change the name of a drawing file created by the **WBLOCK** command. (T/F)

5. A group of commands is treated as one command for the **U** and **UNDO** commands by embedding them between the __________ and __________ options of the **UNDO** command.

6. If the **Press and Drag** mode is not active, you have to select _________ to specify the two diagonal points of a selection window.

7. The _________ option of the **PEDIT** command's main prompt can be used to change the starting and ending widths of a polyline separately to a desired value.

8. The __________ option of the **PEDIT** command allows you to select more than one polyline.

9. You can move past the first and last vertices in a closed polyline by using either the ______ option or the __________ option.

10. While using the _______ command, if you select a line or arc, AutoCAD provides you with the option of converting them into a polyline first.

Chapter 18

Review Questions

Answer the following questions.

1. When you use the **GROUP** command to form object groups, AutoCAD lets you sequentially highlight the groups of which the selected object is a member. (T/F)

2. The color, linetype, lineweight, and layer on which a block is drawn can be changed with the help of the **CHANGE** command. (T/F)

3. After exploding an object, the object remains identical except that the color and linetype may change because of floating layers, colors, or linetypes. (T/F)

4. The **PURGE** command has the same effect as the **WBLOCK Entire drawing** or the **-WBLOCK asterisk** method. The only difference is that with the **PURGE** command, deletion takes place automatically. (T/F)

5. If you have made a copy of a group without naming the copy then its name is displayed in which of the following notations in the **Group Name** list box?

   (a) $A1 (b) @A1
   (c) %A1 (d) *A1

6. Which of the following option deletes the selected group from the drawing?

   (a) **Remove** (b) **Reorder**
   (c) **Explode** (c) **Rename**

7. Which of the following system variable controls the smoothness of the curve?

   (a) **SPLINETYPE** (b) **SPLINESEGS**
   (b) **PLINEGEN** (c) **PICKFIRST**

8. Which option of the **PEDIT** command can straighten a curve drawn with the help of **PLINE** command?

   (a) **Join** (b) **Close**
   (c) **Decurve** (d) None of the above

9. Which of the following properties cannot be changed using the **CHANGE** command?

   (a) **Lineweight** (b) **Color**
   (c) **Plotstyle** (d) **Thickness**

10. The ________ command can be used to change the name of a block.

11. If the __________ option of the **UNDO** command is off, any group of commands grouped together using the **Begin** and **End** options of the **UNDO** command are undone together.

12. Circles drawn using the **CIRCLE** command can be changed to _________.

13. The __________ option in the **Group** command lets you change the order of the objects in the selected group.

14. A group can be selected by entering ________ at the AutoCAD **Select objects**: prompt.

15. The ________ option of the **UNDO** command disables the **UNDO** and **U** commands entirely.

Exercises

Exercise 4 *Graphics*

Draw part (a) in Figure 18-35 and then, using the **PROPERTIES** and relevant **PEDIT** command options, convert it into parts (b), (c), and (d). The linetype used in part (d) is HIDDEN.

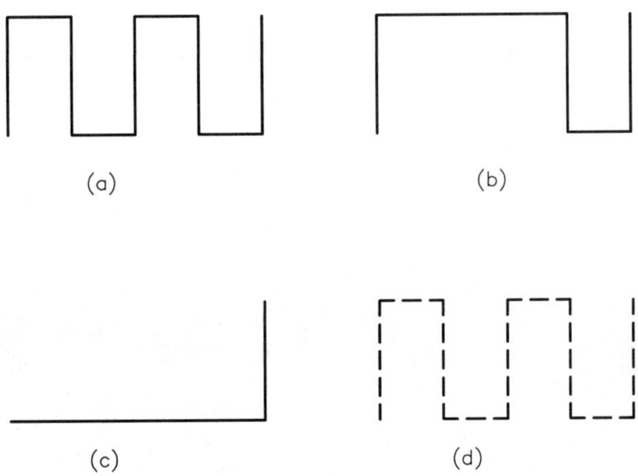

Figure 18-35 *Drawing for Exercise 4*

Exercise 5 *Mechanical*

Draw the object shown in Figure 18-36 using the **LINE** command. Change the object to a polyline with a width of 0.05.

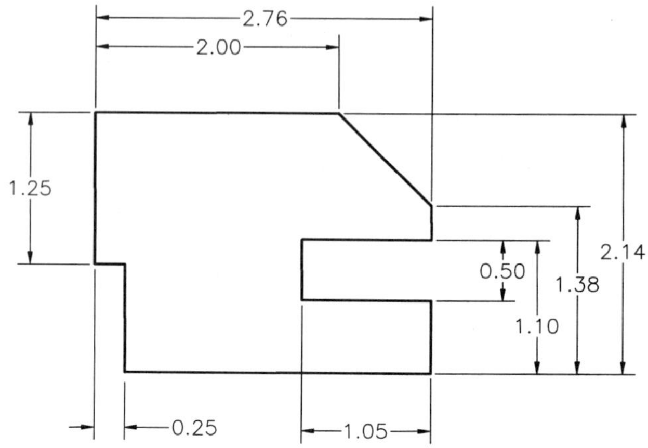

Figure 18-36 *Drawing for Exercise 5*

Exercise 6 *Graphics*

Draw part (a) in Figure 18-37; then, using the relevant **PEDIT** command options, convert it into drawing (b).

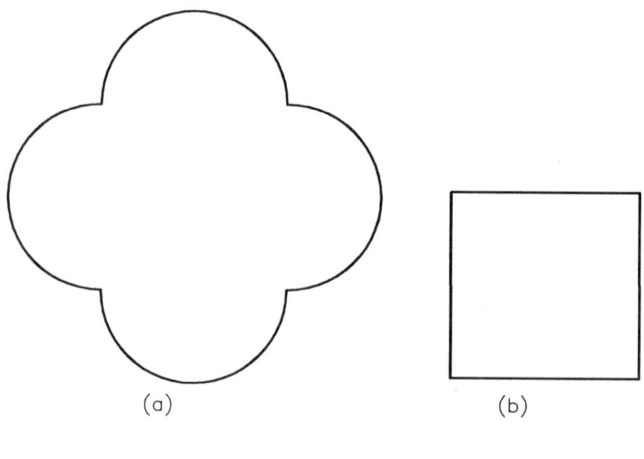

Figure 18-37 *Drawing for Exercise 6*

Exercise 7 *Graphics*

Draw part (a) in Figure 18-38; then, using the relevant **PEDIT** options, convert it into drawings (b), (c), and (d). Identify the types of curves.

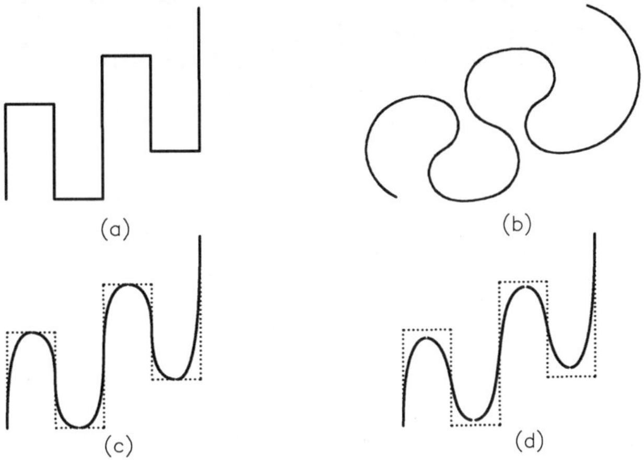

Figure 18-38 *Drawing for Exercise 7*

Exercise 8 *Graphics*

Draw part (a) in Figure 18-39; then, using the relevant **PEDIT** options, convert it into drawing (b).

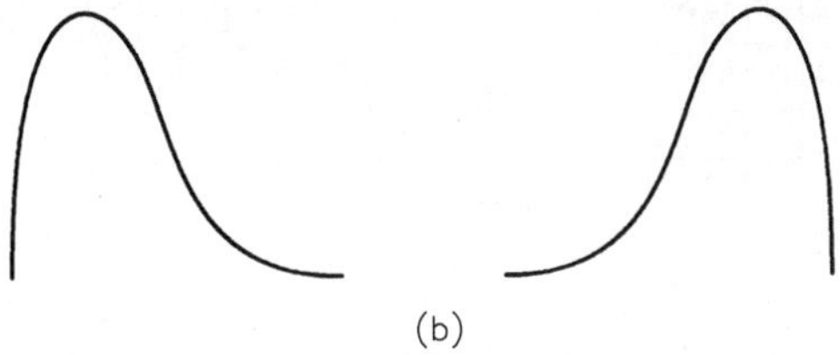

Figure 18-39 *Drawing for Exercise 8*

Exercise 9 *Graphics*

Draw part (a) in Figure 18-40; then, using the relevant **PEDIT** options, convert it into part (b) of the drawing.

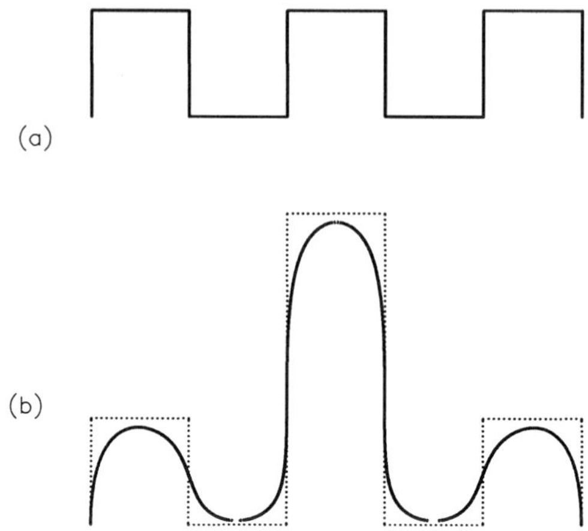

Figure 18-40 *Drawing for Exercise 9*

Exercise 10 *Mechanical*

Draw the illustration shown at the top of Figure 18-41; then use the required commands to obtain the illustration shown at the bottom. You can also use grips to obtain this illustration. Assume the missing dimensions.

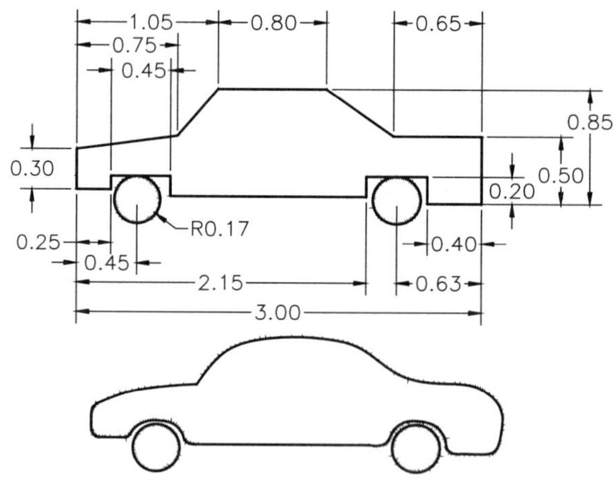

Figure 18-41 *Drawing for Exercise 10*

Exercise 11 *Mechanical*

Make the drawing shown in Figure 18-42. Use the splined polylines to draw the break lines. Dimension the drawing as shown.

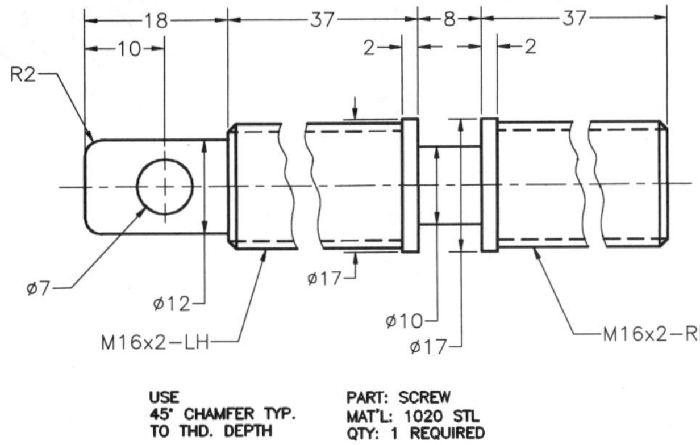

Figure 18-42 *Drawing for Exercise 11*

Answers to Self-Evaluation Test

1 - T, **2** - **F**, **3** - T, **4** - F, **5** - **Begin**, **End**, **6** - Twice, **7** - Width, **8** - **Multiple**, **9** - **Next**, **Previous**, **10** - **PEDIT**

Chapter 19

Working with Data Exchange and Object Linking and Embedding

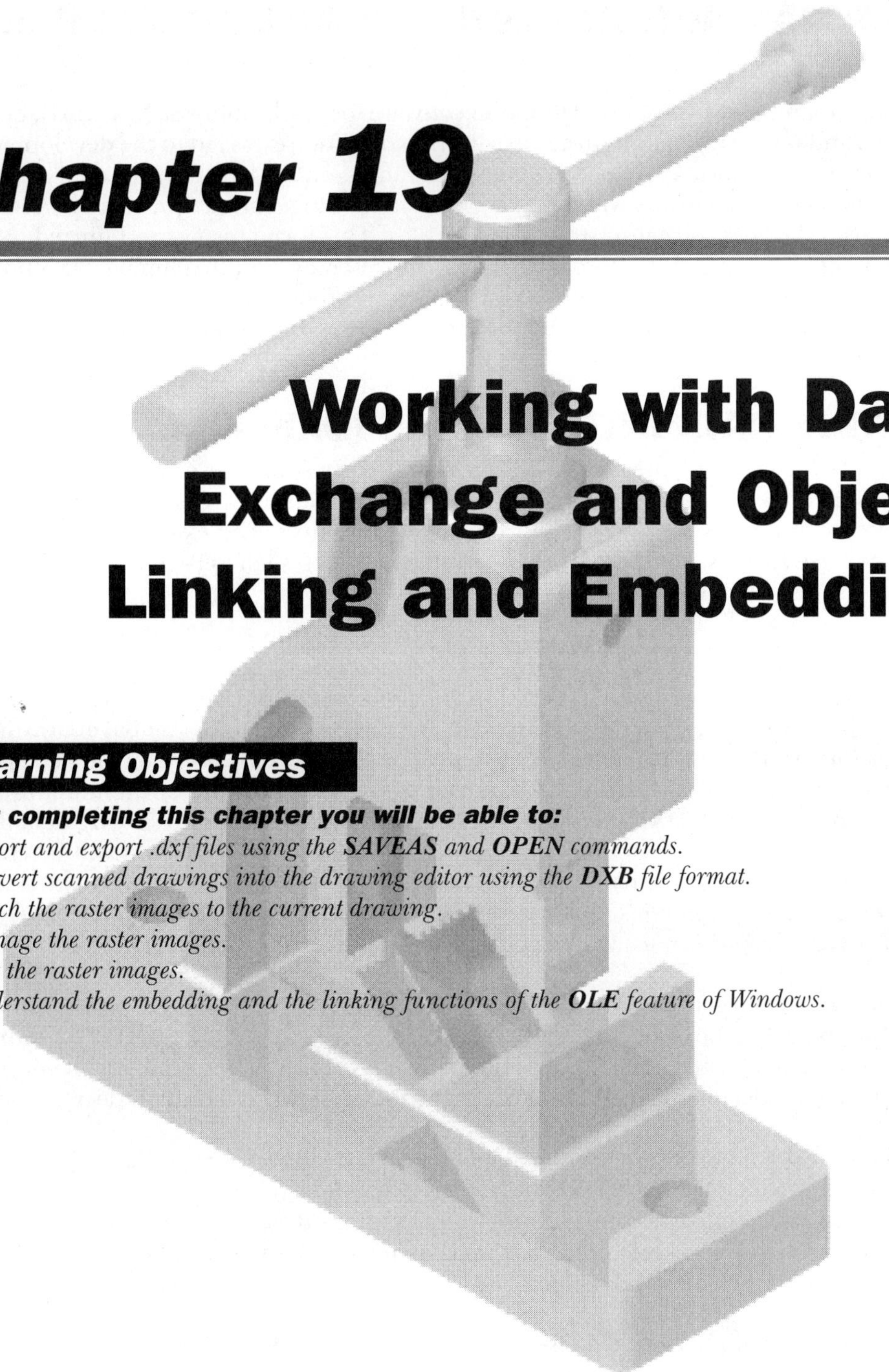

Learning Objectives

After completing this chapter you will be able to:

- *Import and export .dxf files using the **SAVEAS** and **OPEN** commands.*
- *Convert scanned drawings into the drawing editor using the **DXB** file format.*
- *Attach the raster images to the current drawing.*
- *Manage the raster images.*
- *Edit the raster images.*
- *Understand the embedding and the linking functions of the **OLE** feature of Windows.*

UNDERSTANDING THE CONCEPT OF DATA EXCHANGE IN AutoCAD

Different companies have developed different software for applications such as CAD, desktop publishing, and rendering. This nonstandardization of software has led to the development of various data exchange formats that enable transfer (translation) of data from one data processing software to another. This chapter will cover various data exchange formats provided in AutoCAD. AutoCAD uses the *.dwg* format to store drawing files. This format is not recognized by most other CAD software such as Intergraph, CADKEY, and MicroStation. To eliminate this problem so that files created in AutoCAD can be transferred to other CAD software for further use, AutoCAD provides various data exchange formats such as DXF (data interchange file) and DXB (binary drawing interchange).

CREATING DATA INTERCHANGE (DXF) FILES

The DXF file format generates a text file in ASCII code from the original drawing. This allows any computer system to manipulate (read/write) data in a DXF file. Usually, DXF format is used for CAD packages based on microcomputers. For example, packages like SmartCAM use DXF files. Some desktop publishing packages, such as PageMaker and Ventura Publisher, also use DXF files.

Creating a Data Interchange File

The **SAVE** or **SAVEAS** commands are used to create an ASCII file with a *.dxf* extension from an AutoCAD drawing file. Once you invoke any of these commands, the **Save Drawing As** dialog box (Figure 19-1) is displayed. You can also use the **-WBLOCK** command. The **Create Drawing File** dialog box is displayed, where you can enter the name of the file in the **File name** edit box and select **DXF [*dxf]** from the **File of type** drop-down list. By default, the DXF file to be created assumes the name of the drawing file from which it will be created. However, you can specify a file name of your choice for the DXF file by typing the desired file name in the **File name** edit box. Select the extension as **DXF [*dxf]**. This can be observed in the **Files of type** drop-down list where you can select the output file format.

Choose the **Tools** button to display the flyout. From the flyout, choose **Options** to display the **Saveas Options** dialog box, see Figure 19-2. In this dialog box, choose the **DXF Options** tab and enter the degree of accuracy for the numeric values. The default value for the degree of accuracy is sixteen decimal places. You can enter a value between 0 and 16 decimal places.

In this dialog box, you can also select the **Select objects** check box, which allows you to specify objects you want to include in the DXF file. In this case, the definitions of named objects such as block definitions, text styles, and so on, are not exported. Selecting the **Save thumbnail preview image** check box, saves a preview image with the file that can be previewed in the **Preview** window of the **Select File** dialog box.

Select the **ASCII** radio button. Choose **OK** to return to the **Save Drawing As** dialog box. Choose the **Save** button here. Now an ASCII file with a *.dxf* extension has been created, and

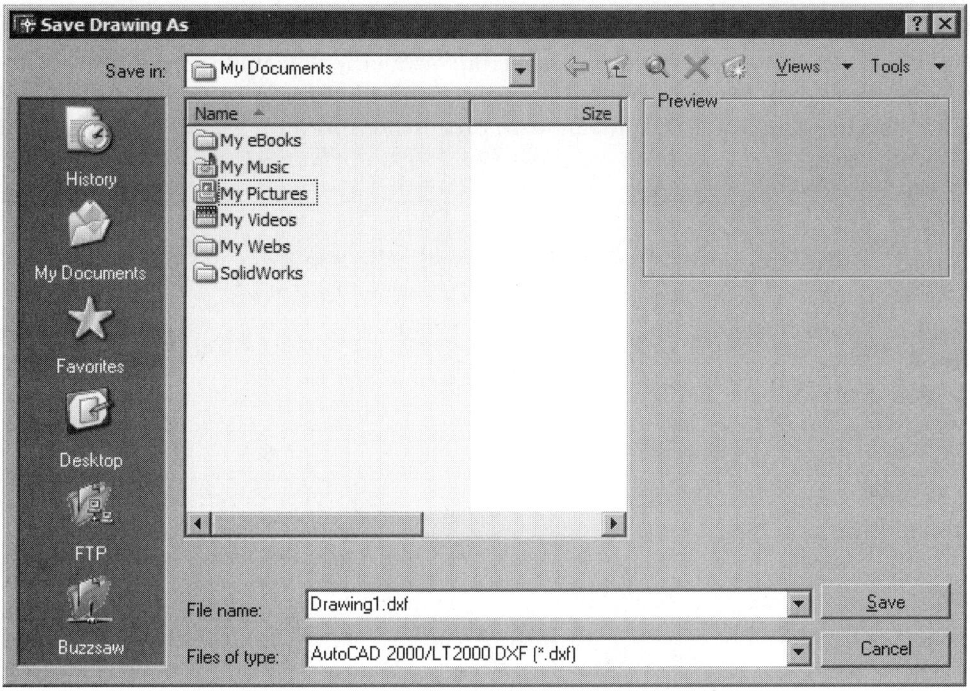

Figure 19-1 ***Save Drawing As*** *dialog box*

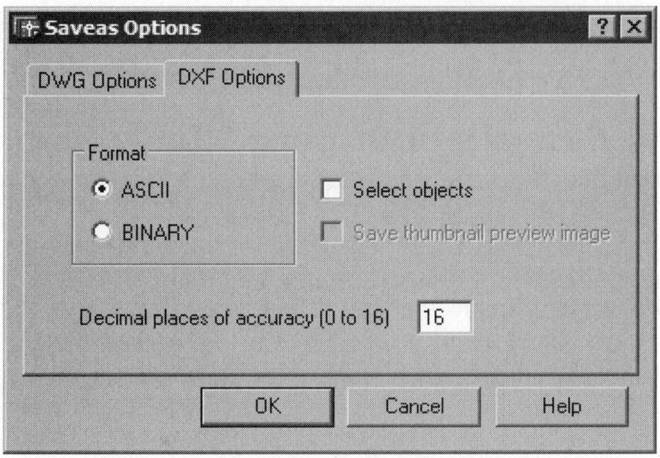

Figure 19-2 ***Saveas Options*** *dialog box*

this file can be accessed by other CAD systems. This file contains data on the objects specified. By default, DXF files are created in ASCII format. However, you can also create binary format files by selecting the **BINARY** radio button in the **Saveas Options** dialog box. Binary DXF files are more efficient and occupy only 75 percent of the ASCII DXF file. You can access a file in binary format more quickly than the same file in ASCII format.

Information in a DXF File

The DXF file contains data on the objects specified using the **Select objects** option in the **Save Drawing As** dialog box. You can change the data in this file to your requirement. To examine the data in this file, load the ASCII file in word processing software. A DXF file is composed of the following parts.

Header

In this part of the drawing database, all the variables in the drawing and their values are displayed.

Classes

This section deals with the internal database.

Tables

All the named objects such as linetypes, layers, blocks, text styles, dimension styles, and views are listed in this part.

Blocks

The objects that define blocks and their respective values are displayed in this part.

Entities

All the entities in the drawing, such as lines, circles, and so on, are listed in this part.

Objects

Objects in the drawing are listed in this part.

Converting DXF Files into Drawing Files

You can import a DXF file into a new AutoCAD drawing file with the **OPEN** command. After you invoke the **OPEN** command, the **Select File** dialog box (Figure 19-3) is displayed. From the **Files of type** drop-down list, select **DXF [*.dxf]**. In the **File name** edit box enter the name of the file you want to import into AutoCAD or select the file from the list. Choose the **Open** button. Once this is done, the specified DXF file is converted into a standard DWG file, regeneration is carried out, and the file is inserted into the new drawing. Now you can perform different operations on this file just as with other drawing files.

Tip

*A data interchange file (DXF) can also be inserted in the current drawing using the **INSERT** command.*

OTHER DATA EXCHANGE FORMATS

The other formats that can be used to exchange the data from one data processing format to the other are discussed next.

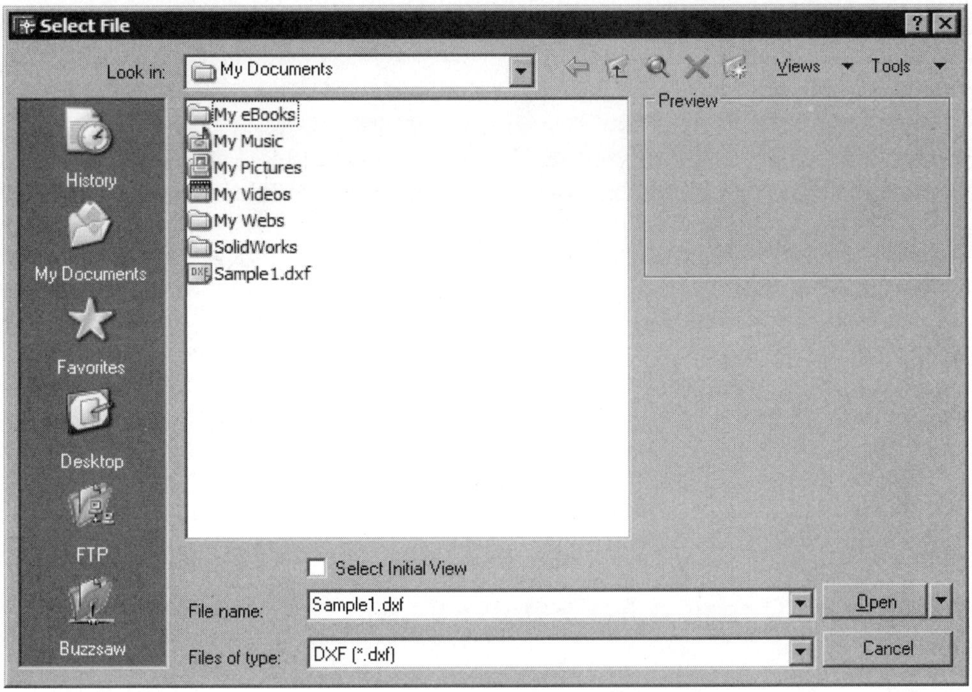

Figure 19-3 ***Select File*** *dialog box*

Importing Scanned Files into the Drawing Editor
DXB File Format

Menu:	Insert > Drawing Exchange Binary
Command:	DXBIN

AutoCAD offers another file format for data interchange, DXB. This format is much more compressed than the binary DXF format and is used when you want to translate large amount of data from one CAD system to another. For example, when a drawing is scanned, a DXB file is created. This file has huge amount of data in it. You can import a DXB file by choosing **Drawing Exchange Binary** from the **Insert** menu. When you choose this, the **Select DXB File** dialog box is displayed. In the **File name** edit box, enter the name of the file (in DXB format) you want to import into AutoCAD. Choose the **Open** button. Once this is done, the specified DXB file is converted into a standard DWG file and is inserted into the current drawing.

Note

Before importing a DXB file, you must create a new drawing file. No editing or drawing setup (limits, units, and so on) can be performed in this file. This is because if you import a DXB file into an old drawing, the settings (definitions) of layers, blocks, and so on, of the file being imported are overruled by the settings of the file into which you are importing the DXB file.

Creating and Using an ACIS file

Menu:	File > Export
Command:	EXPORT, ACISOUT

Trimmed Nurbs surfaces, regions, and solids can be exported to an ACIS file with an ASCII format. This can be accomplished by choosing **Export** from the **File** menu. The **Export Data** dialog box is displayed, enter a file name in the **File name** edit box. From the **Files of type** drop-down list, select **ACIS [*.sat]** and then choose the **Save** button. You will be prompted to select the solids, regions, or ACIS bodies you want to add to the file. Select the objects you wish to save as an ACIS file and press ENTER. AutoCAD appends the file extension *.sat* to the file name. You can also use the **ACISOUT** command to perform this function. It prompts you to select objects before the **Create ACIS File** dialog box is displayed on the screen. Keep in mind that only solids, regions, or ACIS bodies can be selected to add to the *.sat* file.

The **ACISIN** command displays the **Select ACIS File** dialog box. Select a file to import and choose the **Open** button. This command can also be invoked by choosing **ACIS File** from the **Insert** menu.

Creating and Using a 3D Studio File

Menu:	Tools > Export
Command:	EXPORT, 3DSOUT

3D geometry, views, lights, and objects with surface characteristics can be saved in the 3D Studio format. In the **Export Data** dialog box, enter the file name in the **File Name:** edit box and select **3D Studio [*.3ds]** from the **Files of type:** drop-down list. Choose **Save**. At the **Select objects:** prompt, select the objects you want to save in the 3D Studio format. Press ENTER; the file extension *.3ds* is automatically added to the file name. You can also use the **3DSOUT** command. AutoCAD prompts you to select the objects first. When you press ENTER, the **3D Studio Output File** dialog box is displayed. Enter the file name in the edit box and choose the **Save** button.

The **3DSIN** command displays the **3D Studio File Import** dialog box. Select the file you wish to import and choose the **Open** button. You can also invoke this command by choosing **3DStudio** from the **Insert** menu.

Creating and Using a Windows Metafile

Menu:	File > Export
Command:	EXPORT, WMFOUT

The Windows Metafile format (WMF) file contains screen vector and raster graphics format. In the **Export Data** dialog box or the **Create WMF File** dialog box, enter the file name. Select the objects you want to save in this file format. Select **Metafile [*.wmf]** from the **Files of type** drop-down list. The extension *.wmf* is appended to the file name.

The **WMFIN** command displays the **Import WMF** dialog box. Window metafiles are imported as blocks in AutoCAD. Select the *.wmf* file you want to import and choose the **Open** button. Specify an insertion point, rotation angle, and scale factor. Specify scaling by entering a **scale factor**, using the **corner** option to specify an imaginary box whose dimensions correspond to the scale factor or entering **xyz** to specify 3D scale factors. You can also invoke the **WMFIN** command by choosing **Windows Metafile** from the **Insert** menu.

Creating a BMP File

Menu:	File > Export
Command:	EXPORT, BMPOUT

This is used to create bitmap images of the objects in your drawing. In the **Export Data** dialog box, enter the name of the file, select **Bitmap [*.bmp]** from the **Files of type** drop-down list and then choose **Save**. Select the objects you wish to save as bitmap and press ENTER. Entering **BMPOUT** displays the **Create Raster File** dialog box. Enter the file name and choose **Save**. Select the objects to be saved as bitmap. The file extension *.bmp* is appended to the file name.

RASTER IMAGES

A raster image consists of small square-shaped dots known as pixels. In a colored image, the color is determined by the color of pixels. The raster images can be moved, copied, or clipped, and used as a cutting edge with the **TRIM** command. They can also be modified by using grips. You can also control the image contrast, transparency, and quality of the image. AutoCAD stores images in a special temporary image swap file whose default location is the Windows **Temp** directory. You can change the location of this file by modifying it under **Temporary File Location** in the **Files** tab of the **Options** dialog box.

The images can be 8-bit gray, 8-bit color, 24-bit color, or bitonal. When image transparency is set to On, the image file formats with transparent pixels is recognized by AutoCAD and transparency is allowed. The transparent images can be in color or grayscale. AutoCAD supports the following file formats.

Image Type	File Extension	Description
BMP	*.bmp, .dib, .rle*	Windows and OS/2 Bitmap Format
CALS-I	*.gp4, .mil, .rst*	Mil-R-Raster I
FLIC	*.flc, .fli*	Flic Autodesk Animator Animation
GEOSPOT	*.bil*	GeoSPOT (BIL files must be accompanied with HDR and PAL files with connection data in the same directory.)
IG4	*.ig4*	Image Systems group 4
IGS	*.igs*	Image Systems Grayscale
JFIF or JPEG	*.jpg, jpeg*	Joint Photographics Expert group
PCX	*.pcx*	Picture PC Paintbrush Picture
PICT	*.pct*	Picture Macintosh Picture
PNG	*.png*	Portable Network Graphic
RLC	*.rlc*	Run-length Compressed
TARGA	*.tga*	True Vision Raster based Data format
TIFF/LZW .tif	Tagged image file format	

When you store images as Tiled Images, that is, in the Tagged Image File Format (TIFF), you can edit or modify any portion of the image; only the modified portion is regenerated thus saving time. Tiled images load much faster compared to nontiled images.

Attaching Raster Images

Toolbar:	Reference > Image Attach
Menu:	Insert > Raster Image
Command:	IMAGEATTACH

Attaching Raster images do not make them part of the drawing. You can also drag and drop images from other files into the current drawing using the **DESIGNCENTER** window. When you invoke the **IMAGEATTACH** command, AutoCAD displays the **Select Image File** dialog box (Figure 19-4). Enter the name of the image file you want to attach to the current drawing in the **File name:** edit box or select the image from the list box. The name can be up to 255 characters long and can include spaces, numbers, and special characters that are not being used by AutoCAD or Windows. A preview image of the selected image is displayed in the **Preview** window of the dialog box. If you choose the **Hide Preview** button, the preview image of the selected image is not displayed in the **Preview** window. Choose the **Open** button.

AutoCAD displays the **Image** dialog box (Figure 19-5). The name of the selected image is displayed in the **Name** drop-down list. You can select another file by using the **Browse** button. The **Name** drop-down list displays names of all the images in the current drawing. The **Path** is also displayed just below this drop-down list. Select the **Specify on-screen** check boxes to specify the **Insertion point**, **Scale**, and **Rotation** Angle on screen. Alternately you can clear

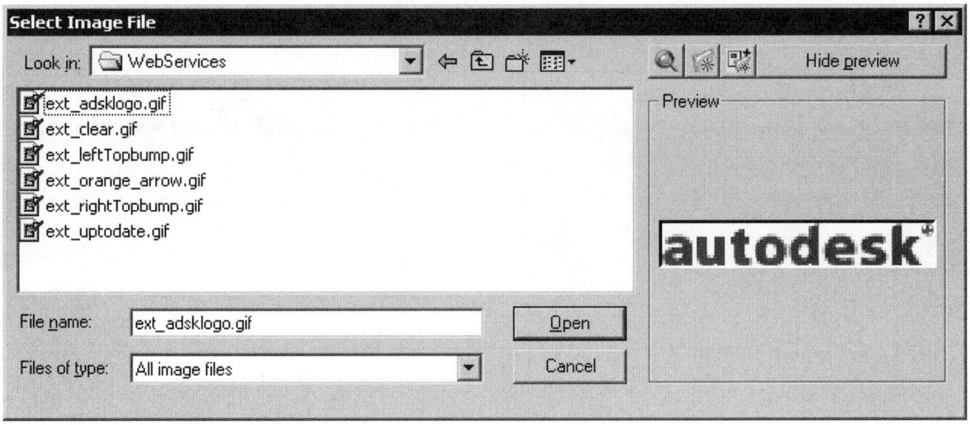

***Figure 19-4 Select Image File** dialog box*

these check boxes and enter values in the respective edit boxes. Choosing the **Details** button expands the **Image** dialog box and provides the image information such as the Horizontal and Vertical resolution values, current AutoCAD unit, and Image size in pixels and units. Choose **OK** to return to the drawing screen.

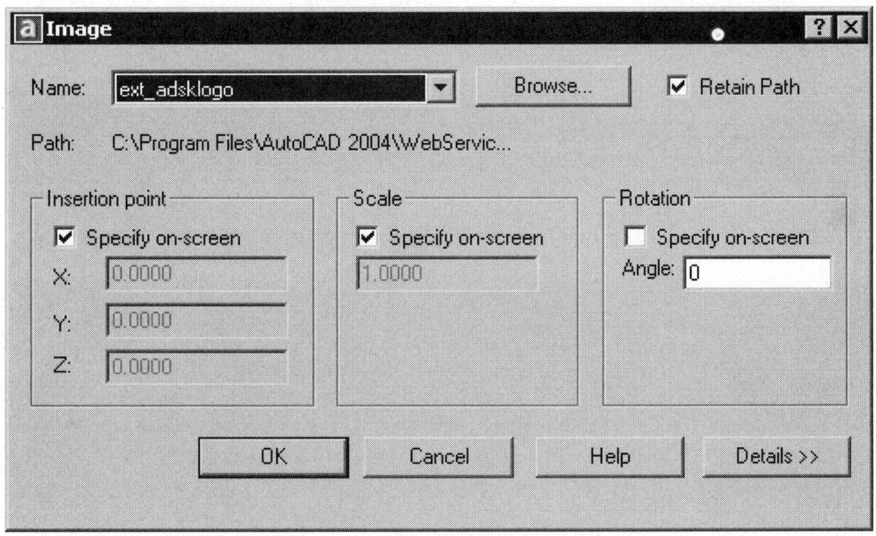

***Figure 19-5 Image** dialog box*

You can also attach and scale an image from the Internet. In the **Select Image File** dialog box, choose the **Search the Web** button. Once you have located the image file you wish to attach to the current drawing, enter the URL address in the **Look In** drop-down list and the Image file name in the **Name or URL** edit box. Choose the **Open** button. Specify the **Insertion point**, **scale** and **Rotation** angle in the **Image** dialog box. You can also right-click the image you wish to attach to the current drawing and choose **Properties** from the shortcut menu to display the **Properties** window. From here you can select, cut, and paste the URL address to the **File name:** edit box in the **Select File** dialog box.

Managing Raster Images (Image Manager Dialog Box)

Toolbar:	Reference > Image
	Insert > Image
Menu:	Insert > Image Manager
Command:	IMAGE

When you choose this button, AutoCAD displays the **Image Manager** dialog box (Figure 19-6). You can also invoke the **Image Manager** dialog box by selecting an image and right-clicking to display a shortcut menu. Choose **Image > Image Manager**. You can view image information either as a list or as a tree view by choosing the respective buttons located at the upper left corner of the **Image Manager** dialog box. The **List View** displays the names of all the images in the drawing, its loading status, size, date last modified on, and its search path. The **Tree View** displays the images in a hierarchy which shows its nesting levels within blocks and Xrefs. The **Tree View** does not display the status, size, or any other information about the image file. You can rename an image file in this dialog box. The various options in the **Image Manager** dialog box are as follows.

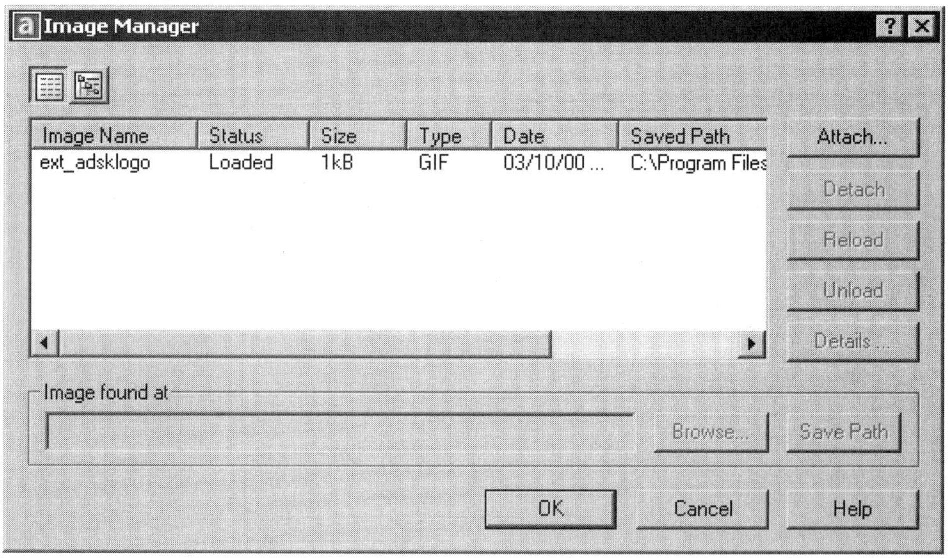

***Figure 19-6 Image Manager** dialog box*

Attach

If you want to attach an image file, choose the **Attach** button. The **Select Image File** dialog box is displayed. Select the file that you want to attach to the drawing and the selected image is displayed in the **Preview** box. Choose the **Open** button. The **Image** dialog box is displayed. The name of the file and its path and extension are displayed in the **Image** dialog box. You can use the **Browse** button to select another file. Choose the **OK** button in the **Image** dialog box and specify a point where you want to insert the image. Next, AutoCAD will prompt you to enter the scale factor. You can enter the scale factor or specify a point on the screen.

Detach

This option detaches the selected and all associated raster images from the current drawing. Once the image is detached, no information about the image is retained in the drawing. Note that the detaching does not take place dynamically. The image will be removed from the screen only after you choose the **OK** button in the **Image Manager** dialog box.

Reload

Reload simply reloads an image. The changes made to the image since the last insert will be loaded on the screen. You can change the status of the image file in the **Image Manager** dialog box by double-clicking on the current status. It changes from **Unload** to **Reload** and vice versa.

Unload

The **Unload** option unloads an image. An image is unloaded when it is not needed in the current drawing session. When you unload an image, AutoCAD retains information about the location and size of the image in the database of the current file. Unloading a file does not unlink the file from the drawing and the image frame is displayed. If you reload the image, the image will appear at the same point and in the same size as the image was before unloading. Unloading the raster images enhances AutoCAD performance. Also, the unloaded images are not plotted. If multiple images are to be loaded and the memory is insufficient, AutoCAD automatically unloads them.

Details

Details displays the **Image File Details** box as shown in Figure 19-7. This dialog box lists information about the image such as file name, saved path, file creation date, file size, file type, color and color path, pixel width and height, resolution, and default size. It also displays the image in the preview box.

Image found at Area

The edit box in this area displays the path of the selected image. You can edit this path and choose the **Save Path** button to save the new path. If you have changed the path of an image, choose the **Browse** button to display the **Select Image File** dialog box. In this dialog box, locate the image file and then choose the **Open** button. The new path is displayed in this edit box. Choose the **Save Path** button to save this new path. This path is also displayed in the **Saved Path** column of the dialog box.

Browse

This option displays the **Select Image File** dialog box to select the image file.

Save Path

This option saves the current path of the image.

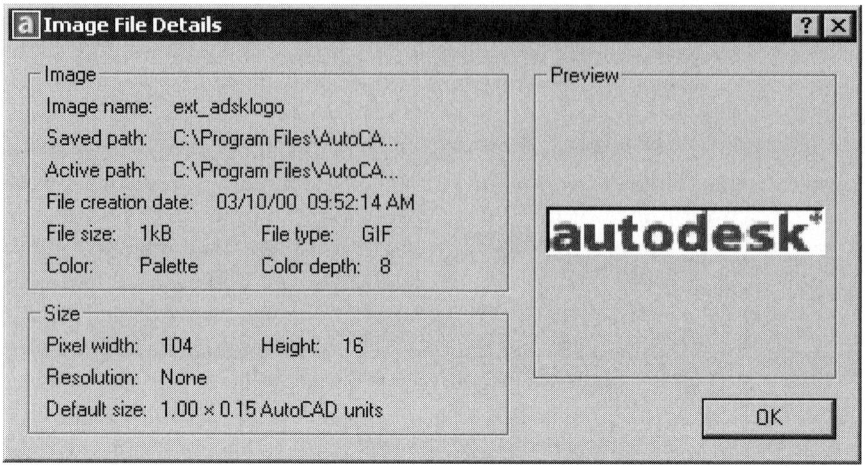

***Figure 19-7 Image File Details** dialog box*

Tip

*You can also use the Command line for managing the image files using the **-IMAGE** command. All the options of the **Image Manager** dialog box will be available in the Command line.*

EDITING RASTER IMAGE FILES

Clipping Raster Images

Toolbar:	Reference > Image Clip
Menu:	Modify > Clip > Image
Command:	IMAGECLIP

This command is used to clip the boundaries of the images and provides a desired shape to an image. You can also invoke this command by selecting an image to clip and by right-clicking to display the shortcut menu. Choose **Image > Clip** to invoke the **IMAGECLIP** command. When you invoke the command, AutoCAD will prompt you to select the image to clip. Select the raster image by selecting the boundary edge of the image. The image boundary must be visible to select the image. The clipping boundary must be specified in the same plane as the image or a plane that is parallel to it. The following prompt sequence is displayed when you choose the **Image Clip** button.

Select image to clip: *Select the image.*
Enter image clipping option [ON/OFF/Delete/New boundary] <New>:

New boundary

This option is used to define a boundary for clipping the image. When you invoke this option, you will be prompted to specify whether or not you want to delete the old boundary, if a boundary already exists. If you enter **No** at the **Delete old boundary? [No/Yes] <Yes>** prompt, this command will end. You can enter **Yes** to specify the new clipping boundary. The boundary can be defined in a rectangular shape or a polygonal shape.

ON

This option is used to turn the image clipping on. The image will be clipped using the last boundary.

OFF

This option is used to turn the image clipping off. If you have a clipped image, you can redisplay the entire image.

Delete

This option is used to delete the existing clipping boundary.

One such clipping is shown in Figures 19-8 and 19-9. Figure 19-8 shows the actual image and Figure 19-9 shows the clipped image where the clipping boundary becomes the midpoints of all four sides of the image.

Figure 19-8 *Image before clipping*

Figure 19-9 *Image after clipping*

Tip

You can improve the performance by turning on or off the image highlighting that is visible when you select an image. This can be done by selecting or clearing the ***Highlight raster image frame only*** *check box in the* ***Display*** *tab of the* ***Options*** *dialog box. You can also use the* ***IMAGEHLT*** *system variable. By default,* ***IMAGEHLT*** *is set to 0, that is, only the raster image frame will be highlighted.*

Adjusting the Raster Images

Toolbar:	Reference > Image Adjust
Menu:	Modify > Object > Image > Adjust
Command:	IMAGEADJUST

The **IMAGEADJUST** command allows you to adjust the brightness, contrast, and fading of the raster image. When you choose an image and right-click, a shortcut menu is displayed. Choose **Image > Adjust** to invoke the **IMAGEADJUST** command. When you invoke this command, AutoCAD prompts you to select image(s). After selecting the image(s), AutoCAD displays the **Image Adjust** dialog box (Figure 19-10) that you can use to adjust the image. As you adjust the brightness, contrast, or fade, the changes are dynamically made in the image. If you choose the **Reset** button, the values are returned to default values (Brightness= 50, Contrast= 50, and Fade= 0). You can also choose **Properties** in the shortcut menu to display the **PROPERTIES** window, where you can enter new values of **Brightness**, **Contrast**, and **Fade**.

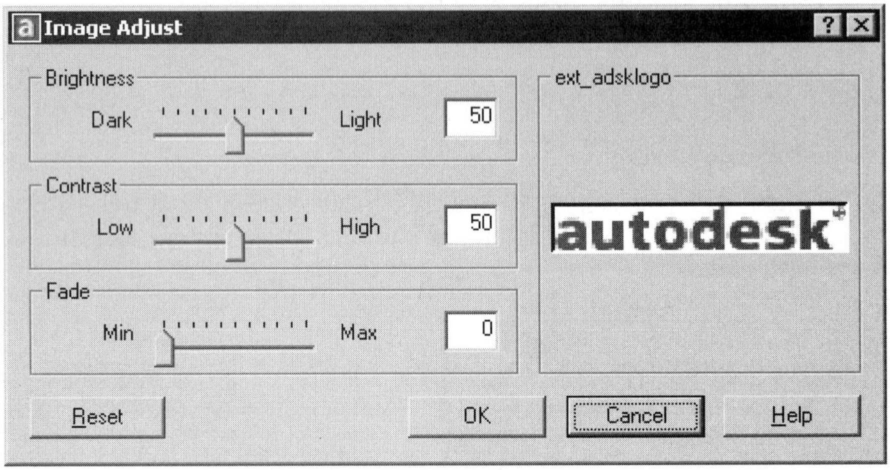

***Figure 19-10 Image Adjust** dialog box*

Modifying the Image Quality

Toolbar:	Reference > Image Quality
Menu:	Modify > Object > Image > Quality
Command:	IMAGEQUALITY

The **IMAGEQUALITY** command allows you to control the quality of the image that affects the display performance. A high-quality image takes a longer time to display. When you change the quality, the display changes immediately without causing a **REGEN**. The images are always plotted using a high-quality display. Although draft quality images appear more grainy, they are displayed more quickly.

Modifying the Transparency of an Image

Toolbar:	Reference > Image Transparency
Menu:	Modify > Object > Image > Transparency
Command:	TRANSPARENCY

The **TRANSPARENCY** command is used with bitonal images to turn the transparency of the image background on or off. A bitonal image is the one that consists of only a foreground and a background color. When you attach a bitonal image, the image assumes the color of the current layer. Also, the bitonal image and the bitonal image boundaries are always in the same color. If you turn on the transparency of an image, the background of the image will become transparent. Figure 19-11 shows two images one over the other with the transparency of the top image turned off. Figure 19-12 shows the figure with the transparency of the top image turned on. You can also right-click an image and choose **Properties** from the shortcut menu, which is displayed. In the **Object Properties** window under the **Misc** list, select an option from the **Transparency** drop-down list.

Figure 19-11 *Image before turning the transparency on*

Figure 19-12 *Image after turning the transparency on*

Note

You have to choose images in such a way that they can give the effect of transparency. Dull and opaque images will not give the effect of transparency.

Controlling the Display of Image Frames

Toolbar:	Reference > Image Frame
Menu:	Modify > Object > Image > Frame
Command:	IMAGEFRAME

The **IMAGEFRAME** command is used to turn the image boundary on or off. If the image boundary is off, the image cannot be selected with the pointing device, and therefore, cannot be accidentally moved or modified.

Changing the Display Order

Toolbar:	Modify II > Draworder
Menu:	Tools > Display Order
Command:	DRAWORDER

The **DRAWORDER** command is used to changes the display order of images and other objects. When you invoke this command, you will be prompted to select the object. The prompt sequence that will follow when you select the object is as follows.

Enter object ordering option [Above object/Under object/Front/Back] <Back>:

If you specify **Above object**, the selected object will be moved above the referenced object. If you select the **Under object**, the selected object will be moved under the referenced object. If you select **Front**, the selected object will be moved to the top of the draworder. If you select **Back**, the selected object will be moved to the bottom of the draworder.

Other Editing Operations

You can also perform other editing operations such as copy, move, and stretch to edit the raster image. Remember that you can also use the image as the trimming edge for trimming objects. However, you cannot trim an image. You can insert the raster image several times or make multiple copies of it. Each copy can have a different clipping boundary. You can also edit the image using grips. You can use the **PROPERTIES** window to change the image layer, boundary linetype, and color, and perform other editing commands such as changing the scale, rotation, width, and height by entering new values in the respective edit boxes. You can also display an image or turn off the display by selecting options from the **Show Image** drop-down list in the **Misc** field of the **PROPERTIES** window. If you select **Yes**, the image is displayed and if you select **No**, the display of the image is turned off. You can turn off the display of the image when you do not need it, thus improving the performance.

Scaling Raster Images

The scale of the inserted image is determined by the actual size of the image and the unit of measurement (inches, feet, and so on). For example, if the image is 1" X 1.26" and you insert this image with a scale factor of 1, the size of the image on the screen will be 1 AutoCAD unit by 1.26 AutoCAD units. If the scale factor is 5, the image will be five times larger. The image that you want to insert must contain the resolution information (DPI). If the image does not contain this information, AutoCAD treats the width of the image as one unit.

POSTSCRIPT FILES

PostScript is a page description language developed by Adobe Systems. It is used mostly in DTP (desktop publishing) applications. AutoCAD allows you to work with PostScript files. You can create and export PostScript files from AutoCAD so that these files can be used for DTP applications. PostScript images have higher resolution than raster images. The extension for these files is *.eps* (Encapsulated PostScript).

Creating the PostScript Files

Menu:	File > Export
Command:	EXPORT, PSOUT

As just mentioned, any AutoCAD drawing file can be converted into a PostScript file. This can be accomplished with the **PSOUT** command. When you invoke this command, the **Create PostScript File** dialog box is displayed (Figure 19-13). You can also use the **EXPORT** command to display the **Export Data** dialog box. The options are similar in both the dialog boxes.

In the **File name** edit box, enter the name of the PostScript (EPS) file you want to create. When you use the **EXPORT** command, in the **Export Data** dialog box, select **Encapsulated PS (*.eps)** from the **Files of type** drop-down list. Then you can choose the **Save** button to accept the default setting and create the PostScript file. You can also choose the

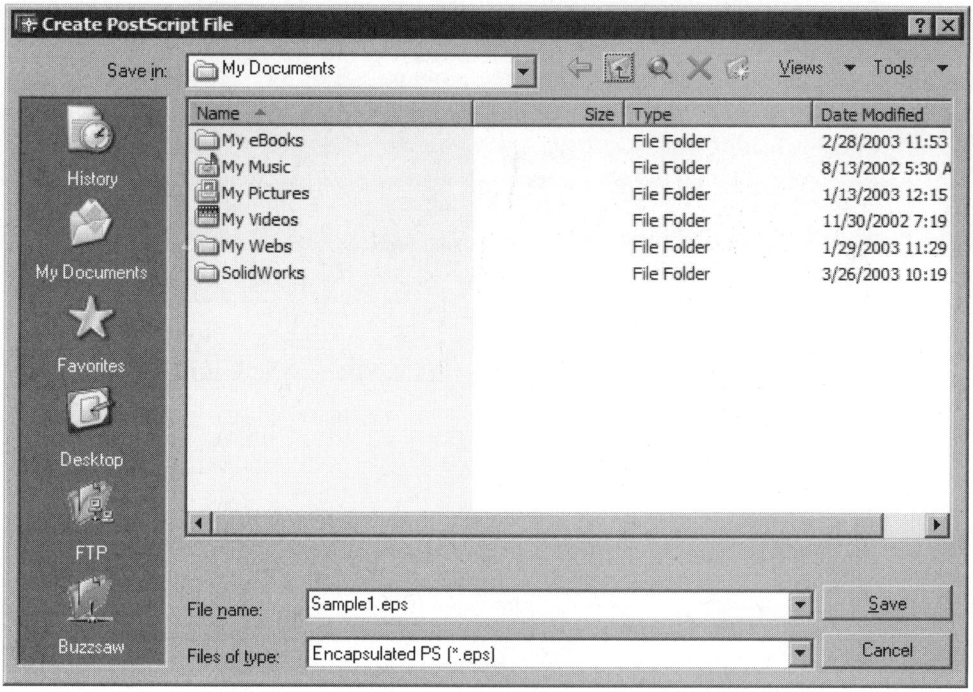

***Figure 19-13 Create PostScript File** dialog box*

Tools > Options in this dialog box to change the settings through the **PostScript Out Options** dialog box (Figure 19-14) and then save the file. The **PostScript Out Options** dialog box provides the following options.

Prolog Section Name

In this edit box, you can assign a name for a prolog section to be read from the *acad.psf* file.

What to plot Area

The **What to plot** area of the dialog box has the following options.

Display. If you specify this option when you are in model space, the display in the current viewport is saved in the specified EPS file. Similarly, if you are in layouts, the current view is saved in the specified EPS file.

Extents. If you use this option, the PostScript file created will contain the section of the AutoCAD drawing that currently holds objects. In this way, this option is similar to the **ZOOM Extents** option. If you add objects to the drawing, they are also included in the PostScript file to be created because the extents of the drawing are also altered. If you reduce the drawing extents by erasing, moving, or scaling objects, then you must use the **ZOOM Extents** or **ZOOM All** option. Only then does the **Extents** option of the **PSOUT** command understand the extents of the drawing to be exported. If you are in model space, the PostScript file is

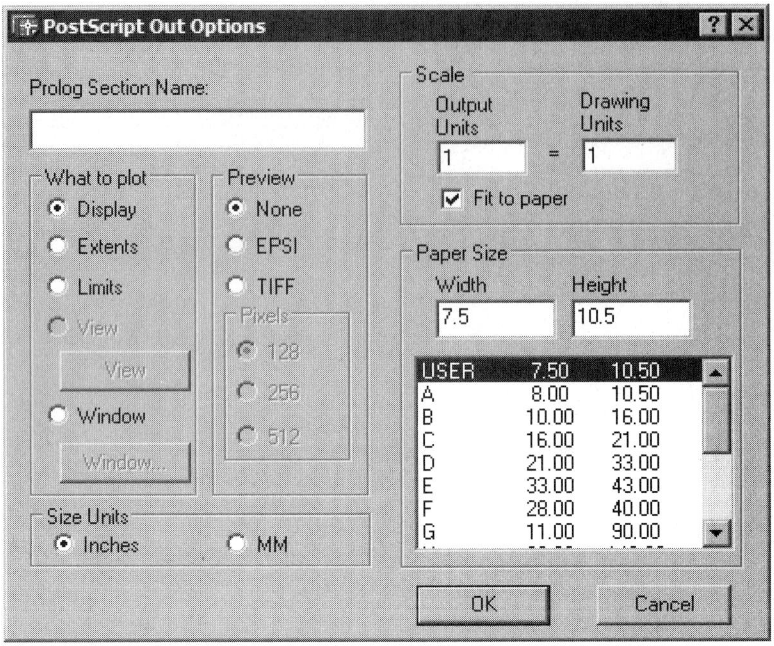

***Figure 19-14 PostScript Out Options** dialog box*

created in relation to the model space extents; if you are in paper space, the PostScript file is created in relation to the paper space extents.

Limits. With this option, you can export the whole area specified by the drawing limits. If the current view is not the plan view [viewpoint (0,0,1)], the **Limits** option exports the area just as the **Extents** option would.

View. Any view created with the **VIEW** command can be exported with this option. This option will be available only when you have created a view using the **VIEW** command. Choose the **View** button to display the **View Name** dialog box from where you can select the view.

Window. In this option, you need to specify the area to be exported with the help of a window. When this radio button is selected, the **Window** button is also available. Choose the **Window** button to display the **Window Selection** dialog box where you can select the **Pick** button and then specify the two corners of the window on the screen. You can also enter the coordinates of the two corners in the **Window Selection** dialog box.

Preview Area

The **Preview** area of the dialog box has two types of formats for preview images: **EPSI** and **TIFF**. If you want a preview image with no format, select the **None** radio button. If you select **TIFF** or **EPSI**, you are required to enter the pixel resolution of the screen preview in the **Pixels** area. You can select a preview image size of 128 X 128, 256 X 256, or 512 X 512.

Size Units Area
In this area, you can set the paper size units to **Inches** or **Millimeters** by selecting their corresponding radio buttons.

Scale Area
In this area, you can set an explicit scale by specifying how many drawing units are to be output per unit. You can select the **Fit to paper** check box so that the view to be exported is made as large as possible for the specified paper size.

Paper Size Area
You can select a size from the list or enter a new size in the **Width** and **Height** edit boxes to specify a paper size for the exported PostScript image.

OBJECT LINKING AND EMBEDDING (OLE)
With Windows, it is possible to work with different Windows-based applications by transferring information between them. You can edit and modify the information in the original Windows application, and then update this information in other applications. This is made possible by creating links between different applications and then updating those links, which in turn updates or modifies the information in the corresponding applications. This linking is a function of the OLE feature of Microsoft Windows. The OLE feature can also join together separate pieces of information from different applications into a single document. AutoCAD and other Windows-based applications such as Microsoft Word, Notepad, and Windows WordPad support the Windows OLE feature.

For the OLE feature, you should have a source document where the actual object is created in the form of a drawing or a document. This document is created in an application called a **server** application. AutoCAD for Windows and Paintbrush can be used as server applications. Now this source document is to be linked to (or embedded in) the **compound** (destination) document, which is created in a different application, known as the **container** application. AutoCAD for Windows, Microsoft Word, and Windows WordPad can be used as container applications.

Clipboard
The transfer of a drawing from one Windows application to another is performed by copying the drawing or the document from the server application to the Clipboard. The drawing or document is then pasted in the container application from the Clipboard. Hence, a Clipboard is used as a medium for storing the documents while transferring them from one Windows application to another. The drawing or the document on the Clipboard stays there until you copy a new drawing, which overwrites the previous one, or until you exit Windows. You can save the information present on the Clipboard with the *.clp* extension.

Object Embedding
You can use the embedding function of the OLE feature when you want to ensure that there is no effect on the source document even if the destination document has been changed through

the server application. Once a document is embedded, it has no connection with the source. Although editing is always done in the server application, the source document remains unchanged. Embedding can be accomplished by means of the following steps. In this example, AutoCAD for Windows is the server application and MS Word is the container application.

1. Create a drawing in the server application (AutoCAD).

2. Open MS Word (container application) by double-clicking on its shortcut icon at the desktop of the computer.

3. It is preferable to arrange both the container and the server windows so that both are visible as shown in Figure 19-15.

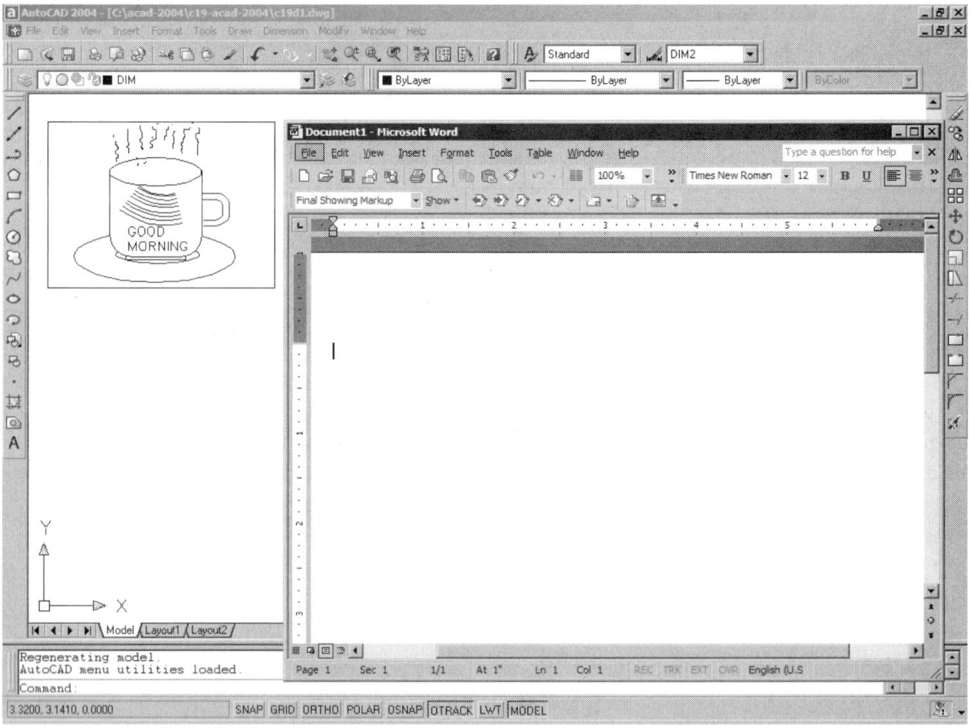

***Figure 19-15** AutoCAD graphics screen with the MS Word window*

4. In the AutoCAD graphics screen, choose the **Copy to Clipboard (Ctrl+C)** button from the **Standard** toolbar to invoke the **COPYCLIP** command. This command can be used in AutoCAD for embedding the drawings. This command can also be invoked from the **Edit** menu (Choose **Copy**), or by entering **COPYCLIP** at the command line. The next prompt, **Select objects**, allows you to select the entities you want to transfer. You can either select the full drawing by entering **ALL** or select some of the entities by selecting them. You can use any of the object selection methods for selecting the objects. With this command the selected objects are automatically copied to the Windows Clipboard.

5. After the objects are copied to the Clipboard, make the MS Word window active. To get the drawing from the Clipboard to the MS Word application (client), select the **Paste** in the Word application. Choose **Paste** from the **Edit** menu in MS Word (Figure 19-16). You can also use **Paste Special** from the **Edit** menu, which will display the **Paste Special** dialog box (Figure 19-17). In this dialog box, select the **Paste** radio button (default) for embedding, and then choose **OK**. The drawing is now embedded in the MS Word window.

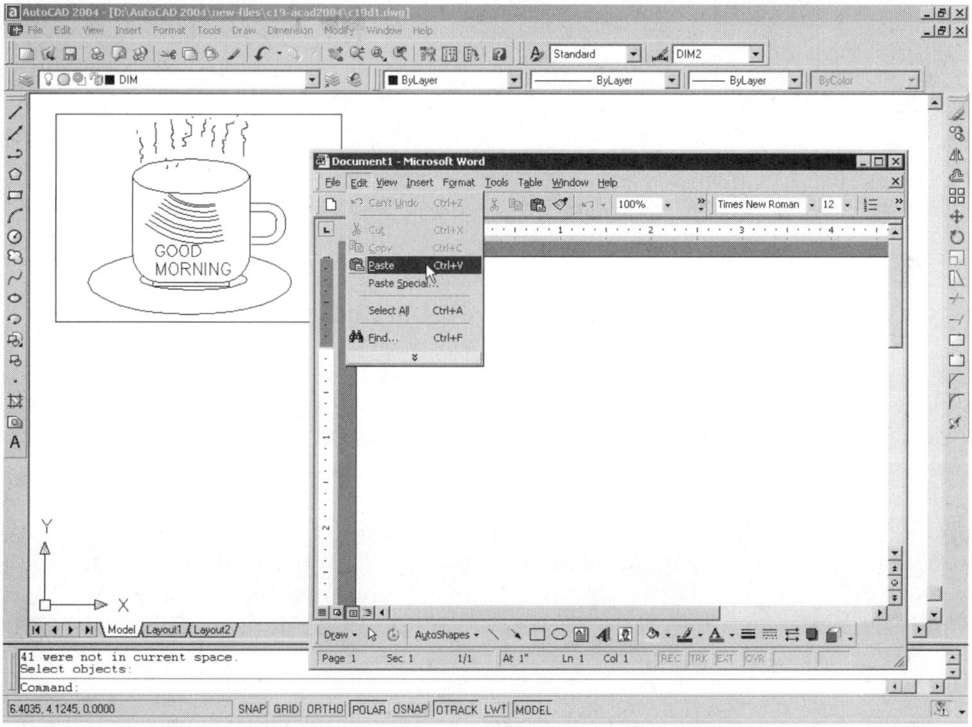

Figure 19-16 *Pasting a drawing to the MS Word application*

6. Your drawing is now displayed in the Write window, but it may not be displayed at the proper position. You can get the drawing in the current viewport by moving the scroll bar up or down in the MS Word window. You can also save your embedded drawing by choosing the **Save** button. It displays a **Save As** dialog box where you can enter a file name. You can now exit AutoCAD.

7. You can now edit your embedded drawing. Editing is performed in the server application, which in this case is AutoCAD for Windows. You can get the embedded drawing into the server application (AutoCAD) directly from the container application (MS Word) by double-clicking on the drawing in MS Word. The other method is by choosing **Edit > Drawing Object** in the **Edit** menu after selecting the drawing. (This menu item has replaced **Object**, which was present before pasting the drawing.)

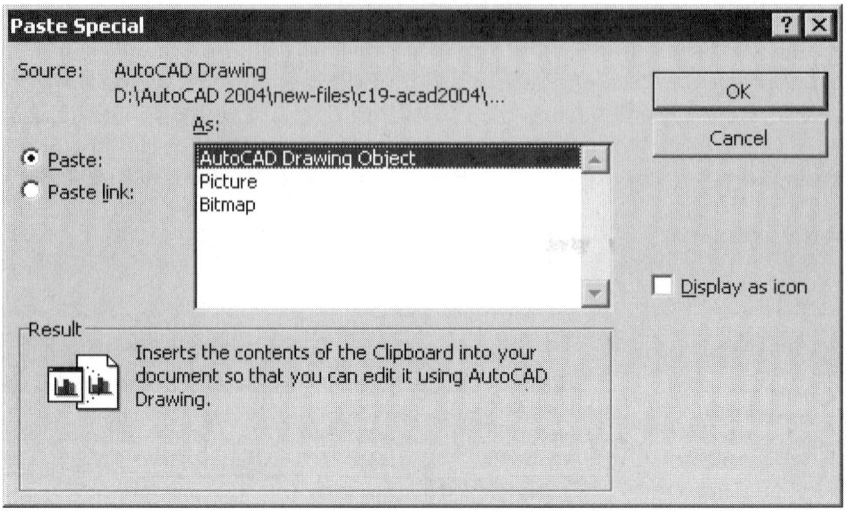

***Figure 19-17 Paste Special** dialog box*

8. Now you are in AutoCAD, with your embedded drawing displayed on the screen, but as a temporary file with a file name such as [Drawing in Document]. Here you can edit the drawing by changing the color and linetype or by adding and deleting text, entities, and so on. In Figure 19-18, the cup and plate have been hatched.

9. After you have finished modifying your drawing, choose **Update Microsoft Word** from the **File** menu in the server (AutoCAD). This menu item has replaced the previous **Save** menu item. When you choose **Update**, AutoCAD automatically updates the drawing in MS Word (container application). Now you can exit this temporary file in AutoCAD.

Note

Do not zoom or pan the drawing in the temporary file. If you do so, this will be included in the updating and the new file will display only that portion of the drawing that lies inside the original area.

10. This completes the embedding function, so you can exit the container application. While exiting, a dialog box that asks whether or not to save changes in MS Word is displayed.

Linking Objects

The linking function of OLE is similar to the embedding function. The only difference is that here a link is created between the source document and the destination document. If you edit the source, you can simply update the link, which automatically updates the client. This allows you to place the same document in a number of applications, and if you make a change in the source document, the clients will also change by simply updating the corresponding links. Consider AutoCAD for Windows to be the server application and MS Word to be the container application. Linking can be performed by means of the following.

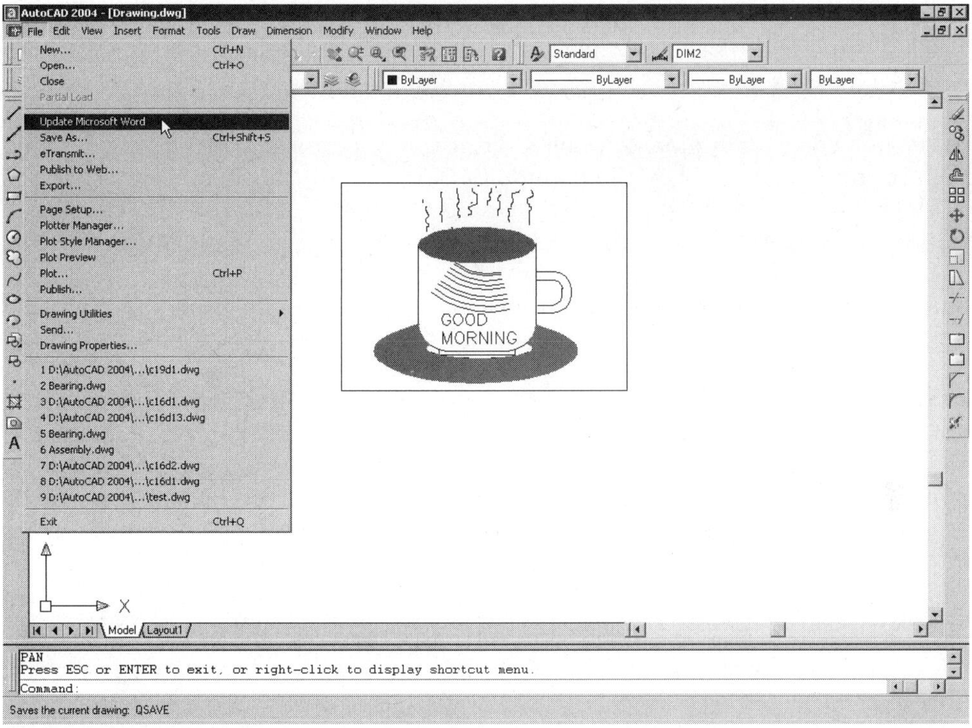

Figure 19-18 *Updating the drawing in AutoCAD*

1. Open a drawing in the server application (AutoCAD). If you have created a new drawing, then you must save the drawing before you can link it with the container application.

2. Open MS Word (container application) by double-clicking on its shortcut icon at the desktop of the computer.

3. It is preferable to arrange both the container and the server windows so that both are visible.

4. In the AutoCAD graphics screen, choose **Copy Link** from the **Edit** menu to invoke the **COPYLINK** command. This command can be used in AutoCAD for linking the drawing. This command copies all the objects that are displayed in the current viewport directly to the Clipboard. Here, you cannot select the objects for linking. If you want only a portion of the drawing to be linked, you can zoom into that view so that it is displayed in the current viewport prior to invoking the **COPYLINK** command. This command also creates a new view of the drawing having a name OLE1.

5. Make the MS Word window active. To get the drawing from the Clipboard to the Write (container) application, choose **Paste Special** from the **Edit** menu. This will display the **Paste Special** dialog box. In this dialog box, select the **Paste Link** radio button for linking. Choose **OK**. Note that the **Paste link** radio button in the **Paste Special** dialog box will not be available if you have not saved the drawing.

Chapter 19

6. The drawing is now displayed in the MS Word window and is linked to the original drawing. You can also save your linked drawing by choosing **Save** from the **File** menu. It displays a **Save As** dialog box where you can enter a file name.

7. You can now edit your original drawing. Editing can be performed in the server application, which in this case is AutoCAD for Windows. You can edit the drawing by changing the color and linetype or by adding and deleting text, entities, and so on. Then save your drawing in AutoCAD by using the **QSAVE** command. You can now exit AutoCAD.

8. You will notice that the drawing is automatically updated, and the changes made in the source drawing are present in the destination drawing also. This automatic updating is dependent on the selection of the **Automatic** radio button (default) in the **Links** dialog box (Figure 19-19). The **Links** dialog box can be invoked by choosing **Links** from the **Edit** menu. For updating manually, you can select the **Manual** radio button in the dialog box. In the manual case, after making changes in the source document and saving it, you need to invoke the **Links** dialog box and then choose the **Update Now** button. This will update the drawing in the container application and display the updated drawing in the MS Word document.

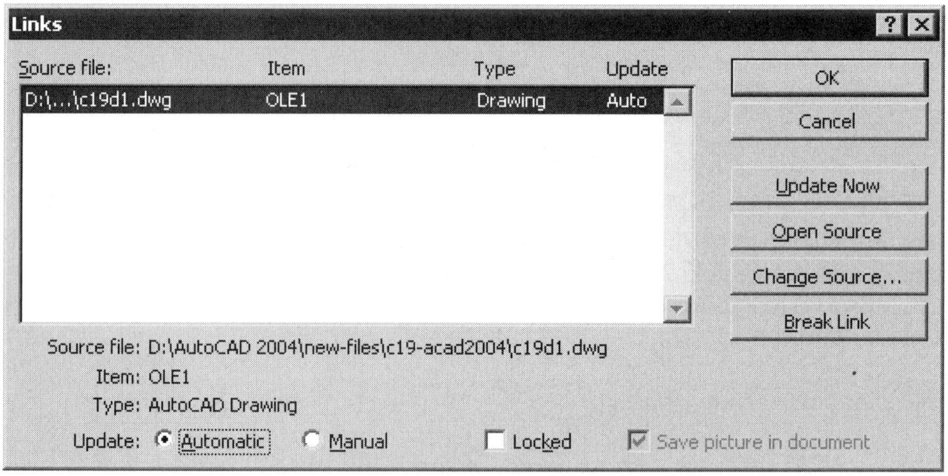

Figure 19-19 ***Links*** *dialog box*

9. Exit the container application after saving the updated file.

Linking Information into AutoCAD

Similarly, you can also embed and link information from a server application into an AutoCAD drawing. You can also drag selected OLE objects from another application into AutoCAD, provided this application supports Microsoft Activex and the application is running and visible on the screen. Dragging and dropping is like cutting and pasting. If you press the CTRL key while you drag the object it is copied to AutoCAD. Dragging and dropping an OLE object into AutoCAD embeds it into AutoCAD.

Linking Objects into AutoCAD

Start any server application such as the MS Word and open a document in it. Select the information you wish to use in AutoCAD with your pointing device and choose the **Copy** button to copy this data to the Clipboard. Open the AutoCAD drawing you wish to link this data to. Choose **Paste Special** from the **Edit** menu or use the **PASTESPEC** command. The **Paste Special** dialog box is displayed (Figure 19-20). In the **As** list box, select the data format you wish to use. For example, for a MS Word document, select **Microsoft Word Document**. Picture format uses a Metafile format. Select the **Paste Link** radio button to paste the contents of the Clipboard to the current drawing. If you select the **Paste** radio button, the data is embedded and not linked. Choose **OK** to exit the dialog box. The data is displayed in the drawing and can be positioned as needed.

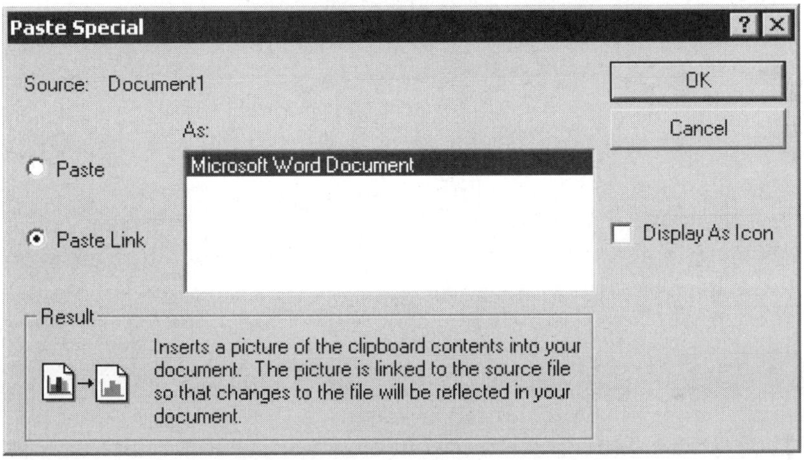

***Figure 19-20 Paste Special** dialog box*

You can also use the **INSERTOBJ** command by entering **INSERTOBJ** at the Command prompt, by choosing **OLE Object** from the **Insert** menu, or by choosing the **OLE Object** button in the **Insert** toolbar. This command links an entire file to a drawing from within AutoCAD. Using this command displays the **Insert Object** dialog box, as shown in Figure 19-21.

Select the **Create from File** radio button. Also select the **Link** check box. Choosing the **Browse** button, displays the **Browse** dialog box. Select a file you want to link from the list box or enter a name in the **File name** edit box and choose the **Open** button. The path of the file is displayed in the **File** edit box of the **Insert Object** dialog box. If you select the **Display As Icon** check box, an icon is also displayed in the dialog box. Choose **OK** to exit the dialog box, the selected file is linked to the AutoCAD drawing.

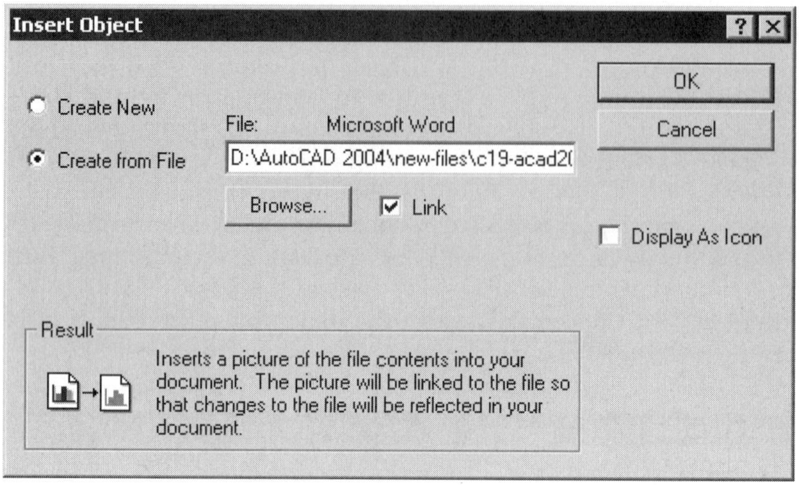

Figure 19-21 ***Insert Object*** *dialog box*

AutoCAD updates the links automatically by default, whenever the server document changes, but you can use the **OLELINKS** command to display the **Links** dialog box (Figure 19-22) where you can change these settings. This dialog box can also be displayed by choosing **OLE Links** from the **Edit** menu. In the **Links** dialog box, select the link you want to update and then choose the **Update Now** button. Then choose the **Close** button. If the server file location changes or if it is renamed, you can choose the **Change Source** button in the **Links** dialog box to display the **Change Source** dialog box. In this dialog box, locate the server file name and location and choose the **Open** button. You can also choose the **Break Link** button in the **Links** dialog box to disconnect the inserted information from the server application. This is done when the linking between the two is no longer required.

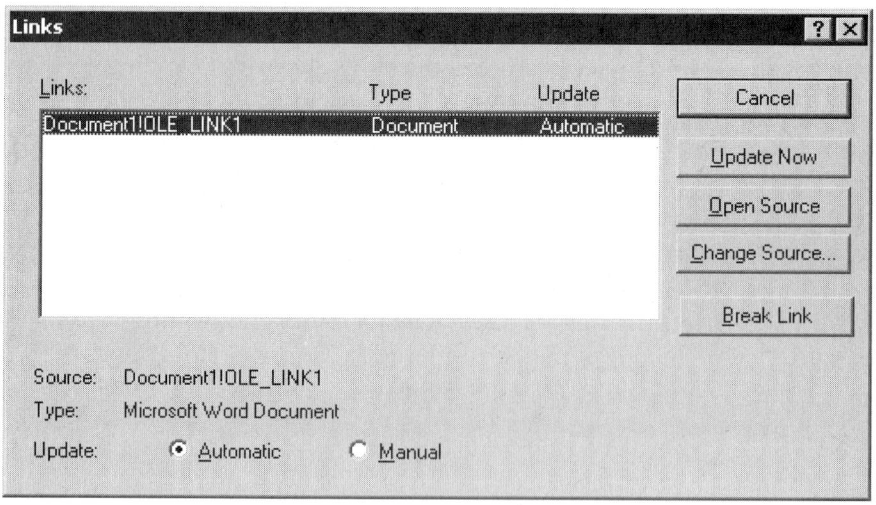

Figure 19-22 ***Links*** *dialog box*

Embedding Objects into AutoCAD

Open the server application and select the data you want to embed into the AutoCAD drawing. Copy this data to the Clipboard by choosing the **Copy** button in the toolbar or choosing **Copy** from the **Edit** menu. Open the AutoCAD drawing and choose **Paste** from the AutoCAD **Edit** menu. You can use the **PASTECLIP** command also. The selected information is embedded into the AutoCAD drawing.

You can also create and embed an object into an AutoCAD drawing starting from AutoCAD itself using the **INSERTOBJ** command. You can choose **OLE Object** from the **Insert** menu or choose the **OLE Object** button from the **Insert** toolbar, and the **Insert Object** dialog box is displayed (Figure 19-23). In this dialog box, select the **Create New** radio button and select the application you wish to use from the **Object Type** list box. Choose **OK**. The selected application opens. Now, you can create the information you wish to insert in the AutoCAD drawing and save it before closing the application. The **OLE Properties** dialog box is displayed (Figure 19-24) where you can resize and rescale the inserted objects. This dialog box is displayed by default. If you do not want to display this dialog box, clear the **Display dialog when pasting new OLE objects** check box in the **OLE Properties** dialog box. You can also clear the **Display OLE properties dialog** check box in the **System** tab of the **Options** dialog box. You can edit information embedded in the AutoCAD drawing by opening the server application by double-clicking on the inserted OLE object. You can also select the object and right-click to display a shortcut menu. Choose **Object** > **Edit**. After editing the data in the server application, choose **Update** from the **File** menu to reflect the modifications in the AutoCAD drawing.

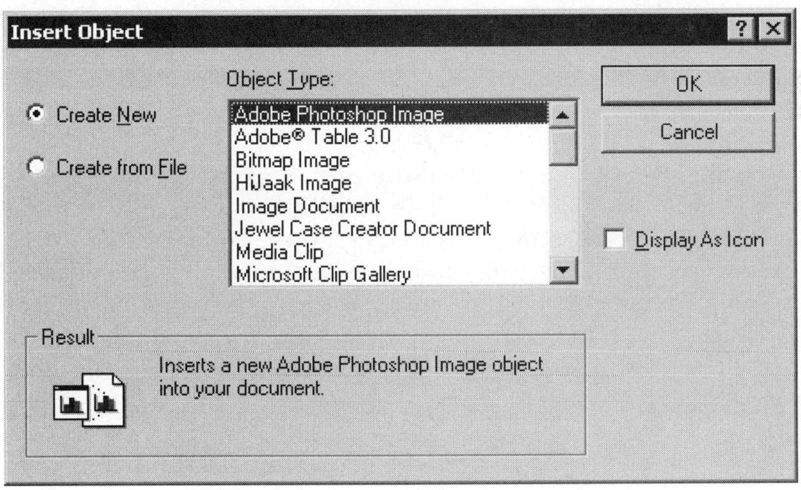

Figure 19-23 Insert Object *dialog box*

Working with OLE Objects

Select an OLE object and right-click to display a shortcut menu; choose **Properties**. The **OLE Properties** dialog box is displayed. You can also select the OLE object and use the **OLESCALE*** command. Specify a new height and a new width in the **Height** and **Width** edit

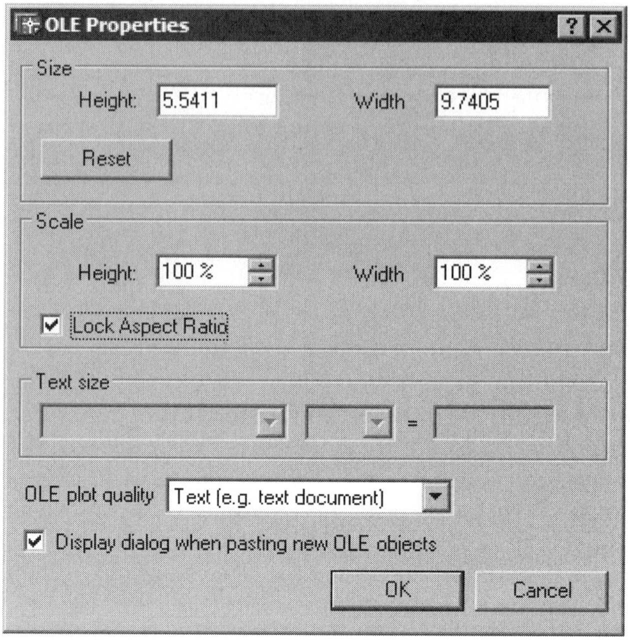

***Figure 19-24 OLE Properties** dialog box*

boxes in the **Size** area or under **Scale**. Enter a value in percentage of the current values in the **Height** and **Width** edit boxes. Here, if you select the **Lock Aspect Ratio** check box, whenever you change either the height or the width under **Scale** or **Size**, the respective width or height changes automatically to maintain the aspect ratio. If you want to change only the height or only the width, clear this check box. Choose **OK** to apply changes.

Choosing the **Reset** button restores the selected OLE objects to their original size, that is, the size they were when inserted. If the AutoCAD drawing contains an OLE object with text with different fonts and you wish to select and modify specific text, you can select a particular font and point size from the drop-down lists under the **Text size** area and enter the value in drawing units in the box after the equal (=) sign. For example, if you wish to select text in the Times Roman font, of point size 10 and modify it to size 0.5 drawing units, select Times Roman and 10 point size from the drop-down lists and in the text box after the equal (=) sign enter .5. The text that was in Times Roman and was of point size 10 will change to 0.5 drawing units in height.

The pointing device can also be used to modify and scale an OLE object in the AutoCAD drawing. Selecting the object displays the object frame and the move cursor. The move cursor allows you to select and drag the object to a new location. The middle handles allows you to select the frame and stretch it. It does not scale objects proportionately. The corner handles scale the object proportionately.

Select an OLE object and right-click to display the shortcut menu. Choosing **Cut** removes the object from the drawing and pastes it on the Clipboard, **Copy** places a copy of the selected object on the clipboard and **Clear** removes the object from the drawing and does not place it on the clipboard. Choosing **Object** displays the **Convert**, **Open**, and **Edit** options. Choosing **Convert** displays the **Convert** dialog box where you can convert objects from one type to another and **Edit** opens the object in the Server application where you can edit it and update it in the current drawing. **Undo** cancels the last action. **Bring to Front** and **Send to Back** options place the OLE objects in the front of or back of the AutoCAD objects. The **Selectable** option turns the selection of the OLE object on or off. If the **Selectable** option is on, the object frame is visible and the object is selected.

If you want to change the layer of an OLE object, select the object and right-click to display the shortcut menu; choose **Cut**. The selected object is placed on the clipboard. Change the current layer to the one you want to change the OLE object's layer to using the **Layer Properties Manager** dialog box. Now, choose **Paste** from the **Edit** menu to paste the contents of the clipboard in the AutoCAD drawing. The OLE object is pasted in the new layer, in its original size.

The **OLEHIDE** system variable controls the display of OLE objects in AutoCAD. The default value is 0, that is, all the OLE objects are visible. The different values and their effects are as follows.

0 All OLE objects are visible
1 OLE objects are visible in paper space only
2 OLE objects are visible in model space only
3 No OLE objects are visible

The **OLEHIDE** system variable affects both screen display and printing.

Self-Evaluation Test

Answer the following questions, and then compare your answers to the answers given at the end of this chapter.

1. The DXF file format generates a text file in ASCII code. (T/F)

2. You can directly open a DXF file in AutoCAD using the **OPEN** command. (T/F)

3. An image attached to AutoCAD cannot be detached. (T/F)

4. The file in the container application is automatically modified only when a link is maintained between the server application and container application. (T/F)

5. You can import a DXB file into an AutoCAD drawing file with the _________ command.

6. The _________ system variable controls the display of OLE objects in AutoCAD.

7. If you import a DXB file into an old drawing, the settings of layers, blocks, and so on, of the file being imported are __________ by the settings of the file into which you are importing the DXB file.

8. The __________ is used as a medium for storing the documents while transferring them from one Windows application to another.

9. You can get an embedded drawing into the server application directly from the container application by __________ on the drawing.

10. The **COPYLINK** command copies a drawing in the __________ to the Clipboard.

Review Questions

Answer the following questions.

1. The boundaries of the images can be clipped and the desired shape can be provided to them. (T/F)

2. The scanned files can be imported into the current session of AutoCAD using the **DXBIN** command. (T/F)

3. You can export an AutoCAD drawing into 3D Studio MAX. (T/F)

4. Raster images can be attached using the **IMAGEATTACH** command as well as **IMAGE** command. (T/F)

5. Which command can be used to open a DXF format file?

   (a) **OPEN** (b) **NEW**
   (c) **START** (d) **None**

6. Which command is used to detach an attached image file?

   (a) **IMAGE** (b) **IMAGEATTACH**
   (c) **IMAGECLIP** (d) **IMAGEADJUST**

7. Which command is used to modify the brightness of the image file?

   (a) **IMAGE** (b) **IMAGEATTACH**
   (c) **IMAGECLIP** (d) **IMAGEADJUST**

8. Which command is used to modify the frames by retaining only the desired portion of the images?

   (a) **IMAGE** (b) **IMAGEATTACH**
   (c) **IMAGECLIP** (d) **IMAGEADJUST**

9. The __________ command is used to create an ASCII format file with the *.dxf* extension from AutoCAD drawing files.

10. Binary DXF files are __________ efficient and occupy only 75 percent of the ASCII DXF file.

11. File access for files in binary format is __________ than for the same file in ASCII format.

12. In a __________ file, information is stored in the form of a dot pattern on the screen. This bit pattern is also known as __________ .

13. When you select an image file to attach, the __________ dialog box is displayed where you can define the insertion point, scale factor, and rotation angle for the image.

14. The clipping boundary for the raster images can be of __________ or the __________ shape.

Exercises

Exercise 1 *General*

In this exercise, you will create a cup and a plate as shown in Figure 19-15. Below this cup and plate, write the text in MS Word and then using OLE, paste it in the current drawing. The text to be written is given below.

These objects are drawn in AutoCAD 2004.

Answers to Self-Evaluation Test

1 - T, **2** - T, **3** - F, **4** - T, **5** - **DXBIN**, **6** - **OLEHIDE**, **7** - overruled, **8** - Clipboard, **9** - double-clicking, **10** - current display

Chapter 20

Technical Drawing with AutoCAD

Learning Objectives

After completing this chapter you will be able to:

- *Understand the concepts of multiview drawings.*
- *Understand X, Y, and Z axes; XY, and YZ, XZ planes; and parallel planes.*
- *Draw orthographic projections and position the views.*
- *Dimension a drawing.*
- *Understand the basic dimensioning rules.*
- *Draw sectional views using different types of sections.*
- *Hatch sectioned surfaces.*
- *Understand how to use auxiliary views and how to draw them.*

MULTIVIEW DRAWINGS

When designers design a product, they visualize the shape of the product in their minds. To represent that shape on paper or to communicate the idea to other people, they must draw a picture of the product or its orthographic views. Pictorial drawings, such as isometric drawings, convey the shape of the object, but it is difficult to show all of its features and dimensions in an isometric drawing. Therefore, in industry, multiview drawings are the accepted standard for representing products. Multiview drawings are also known as **orthographic projection drawings**. To draw different views of an object, it is very important to visualize the shape of the product. The same is true when you are looking at different views of an object to determine its shape. To facilitate visualizing the shapes, you must picture the object in 3D space with reference to the *X*, *Y*, and *Z* axes. These reference axes can then be used to project the image into different planes. This process of visualizing objects with reference to different axes is, to some extent, natural in human beings. You might have noticed that sometimes, when looking at objects that are at an angle, people tilt their heads. This is a natural reaction, an effort to position the object with respect to an imaginary reference frame (*X*, *Y*, *Z* axes).

UNDERSTANDING X, Y, AND Z AXES

To understand the *X*, *Y*, and *Z* axes, imagine a flat sheet of paper on the table. The horizontal edge represents the positive *X* axis and the other edge, the edge along the width of the sheet, represents the positive *Y* axis. The point where these two axes intersect is the origin. Now, if you draw a line perpendicular to the sheet passing through the origin, the line defines the positive *Z* axis (Figure 20-1). If you project the *X*, *Y*, and *Z* axes in the opposite direction beyond the origin, you will get the negative *X*, *Y*, and *Z* axes (Figure 20-2).

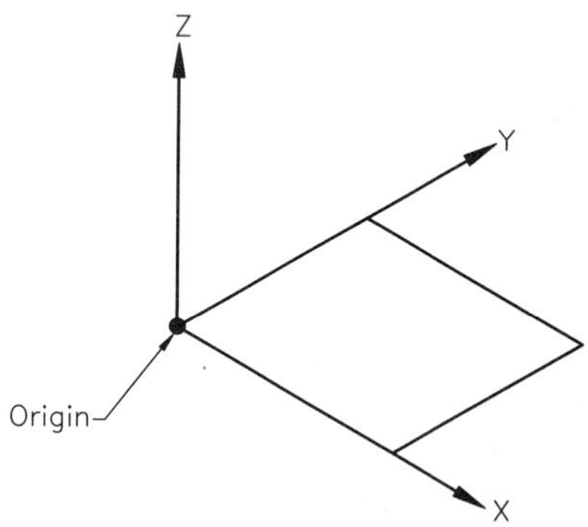

Figure 20-1 *X, Y, Z axes*

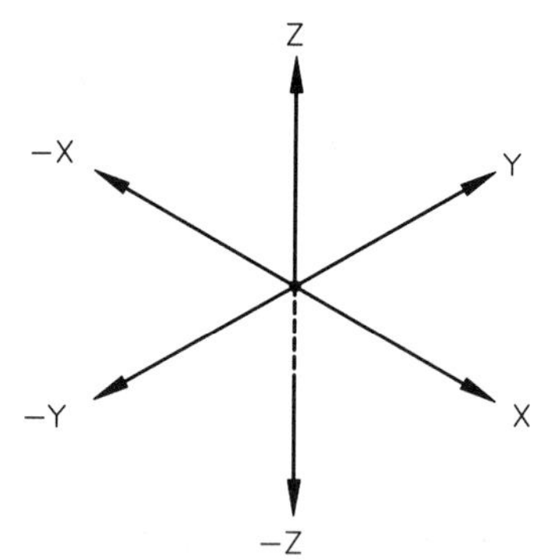

Figure 20-2 *Positive and negative axes*

The space between the *X* and *Y* axes is called the *XY* plane. Similarly, the space between the *Y* and *Z* axes is called the *YZ* plane, and the space between the *X* and *Z* axes is called the *XZ* plane (Figure 20-3). A plane that is parallel to these planes is called a parallel plane (Figure 20-4).

ORTHOGRAPHIC PROJECTIONS

The first step in drawing an orthographic projection is to position the object along the

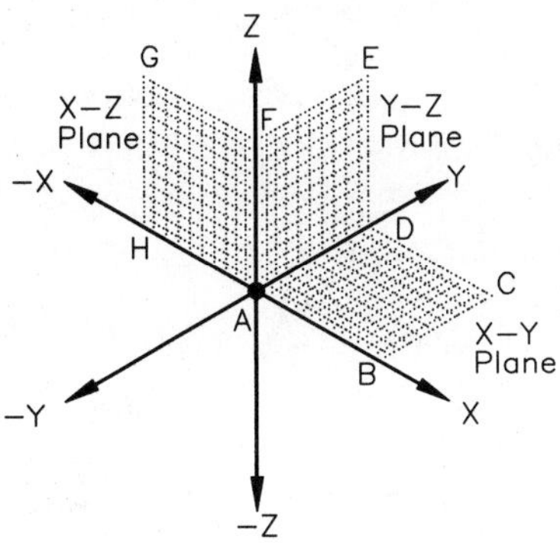

Figure 20-3 XY, YZ, and XZ planes

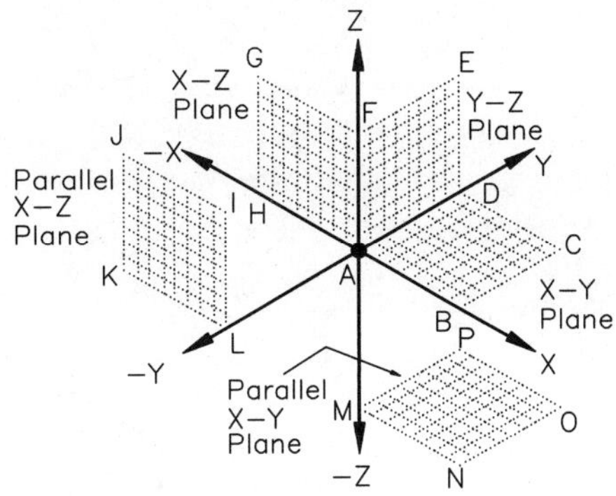

Figure 20-4 Parallel planes

imaginary *X*, *Y*, and *Z* axes. For example, if you want to draw orthographic projections of the step block shown in Figure 20-5, position the block so that the far left corner coincides with the origin, and then align the block with the *X*, *Y*, and *Z* axes.

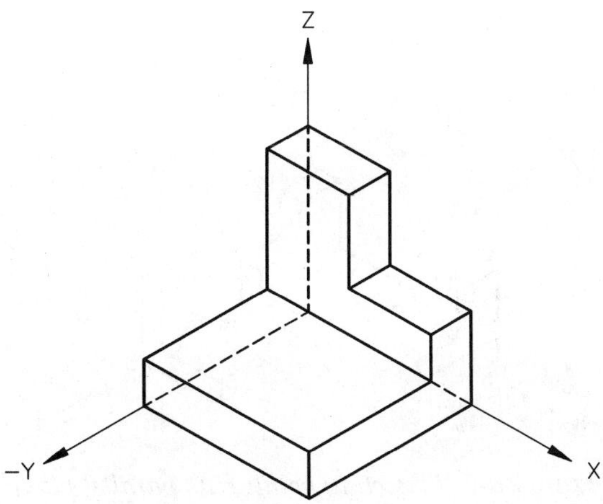

Figure 20-5 *Aligning the object with the X, Y, and Z axes*

Now you can look at the object from different directions. Looking at the object along the negative *Y* axis and toward the origin is called the front view. Similarly, looking at the object from the positive X direction is called the right side view. To get the top view, you look at the object from the positive *Z* axis. See Figure 20-6.

To draw the front, side, and top views, project the points onto the parallel planes. For example, if you want to draw the front view of the step block, imagine a plane parallel to the *XZ* plane located at a certain distance in front of the object. Now, project the points from the object onto the parallel plane (Figure 20-7), and join the points to complete the front view.

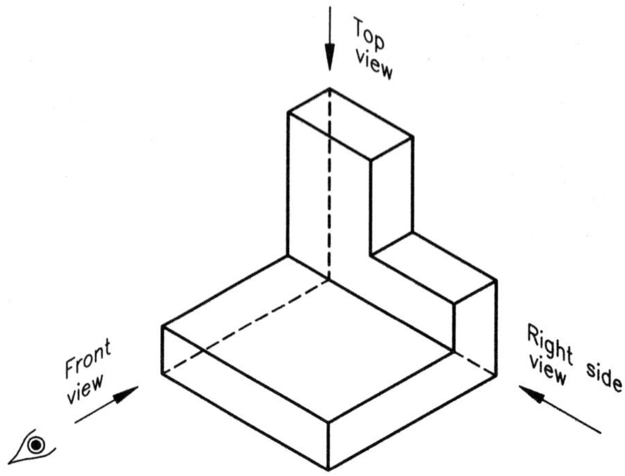

Figure 20-6 *Viewing directions for front, side, and top views*

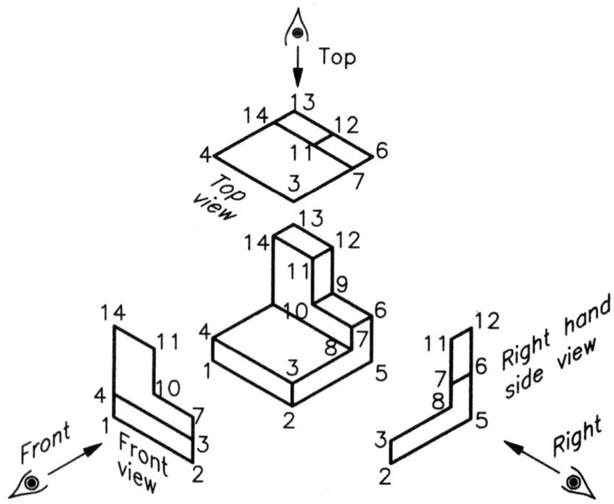

Figure 20-7 *Projecting points onto parallel planes*

Repeat the same process for the side and top views. To represent these views on paper, position the views as shown in Figure 20-8.

Another way of visualizing different views is to imagine the object enclosed in a glass box (Figure 20-9).

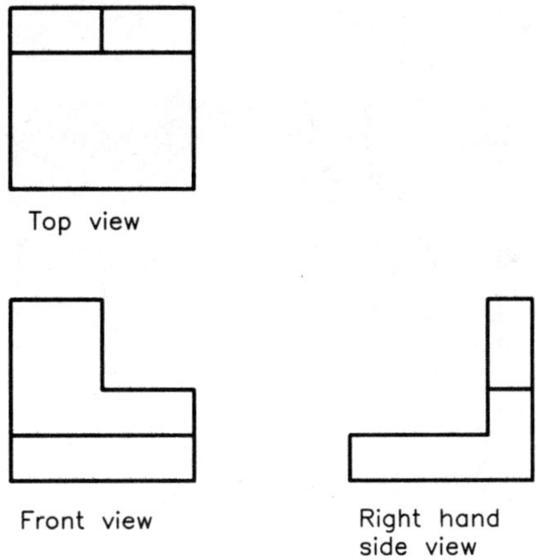

Figure 20-8 *Representing views on paper*

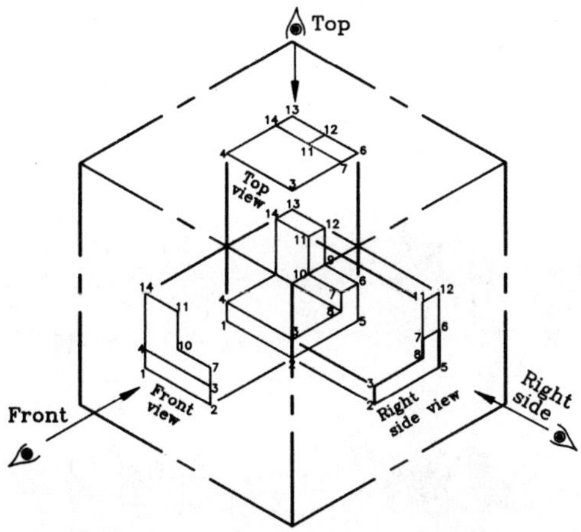

Figure 20-9 *Objects inside a glass box*

Now, look at the object along the negative *Y* axis and draw the front view on the front glass panel. Repeat the process by looking along the positive *X* and *Z* axes, and draw the views on the right side and the top panel of the box (Figure 20-10).

To represent the front, side, and top views on paper, open the side and the top panel of the glass box (Figure 20-11). The front panel is assumed to be stationary.

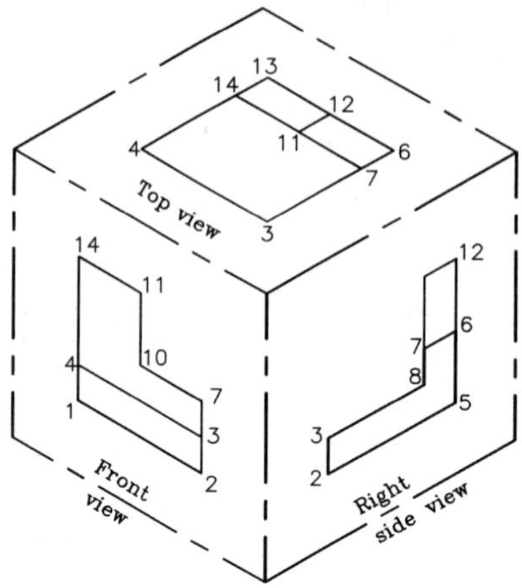

Figure 20-10 *Front, top, and side views*

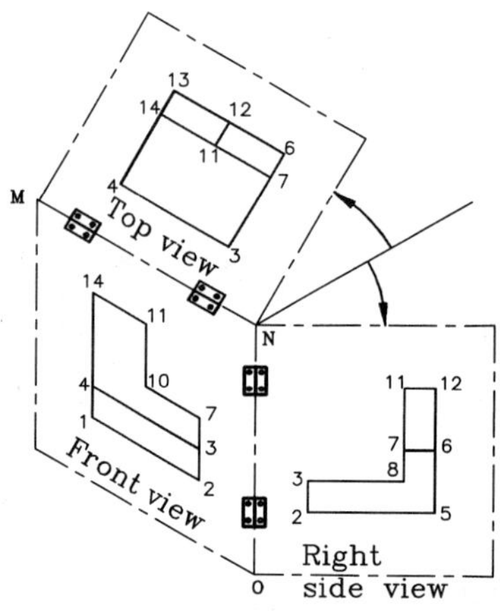

Figure 20-11 *Open the side and the top panel*

After opening the panels through 90-degree, the orthographic views will appear as shown in Figure 20-12.

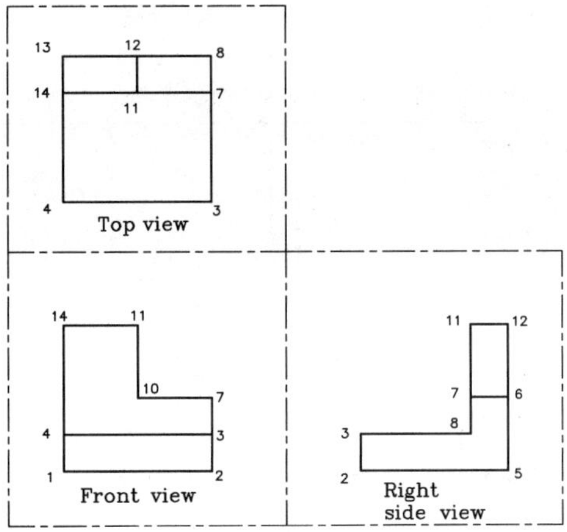

Figure 20-12 *Views after opening the box*

POSITIONING ORTHOGRAPHIC VIEWS

Orthographic views must be positioned as shown in Figure 20-13.

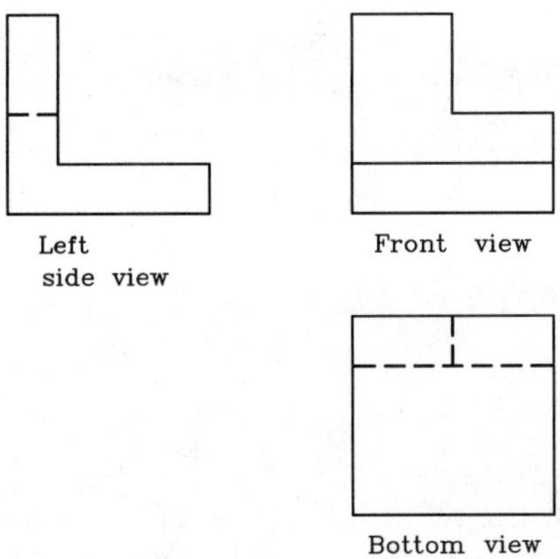

Figure 20-13 *Positioning orthographic views*

The right side view must be positioned directly on the right side of the front view. Similarly, the top view must be directly above the front view. If the object requires additional views, they must be positioned as shown in Figure 20-14.

The different views of the step block are shown in Figure 20-15.

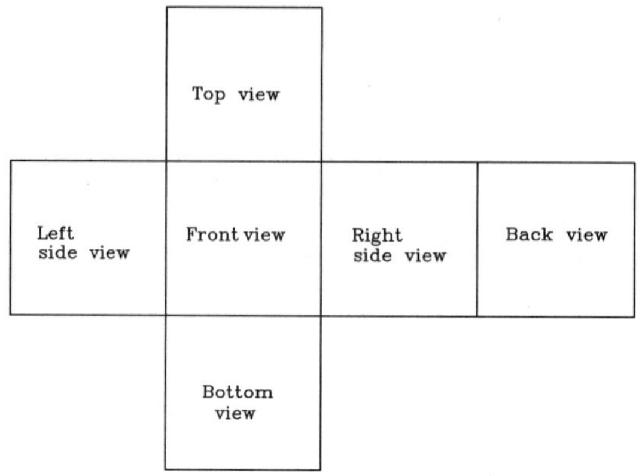

Figure 20-14 *Standard placement of orthographic views*

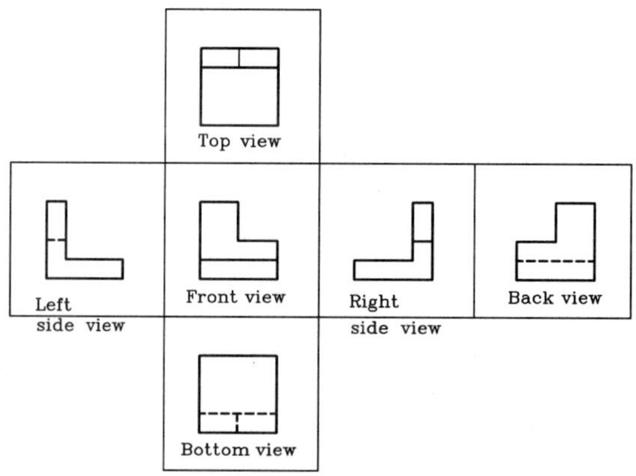

Figure 20-15 *Different views of the step block*

Example 1 *Mechanical*

In this example, you will draw the required orthographic views of the object in Figure 20-16. Assume the dimensions of the drawing.

Drawing the orthographic views of an object involves the following steps.

1. Look at the object, and determine the number of views required to show all of its features. For example, the object in Figure 20-16 will require three views only (front, side, and top).

2. Based on the shape of the object, select the side you want to show as the front view. Generally the front view is the one that shows the maximum number of features or that gives a better idea about the shape of the object. Sometimes the front view is determined by how the part will be positioned in an assembly.

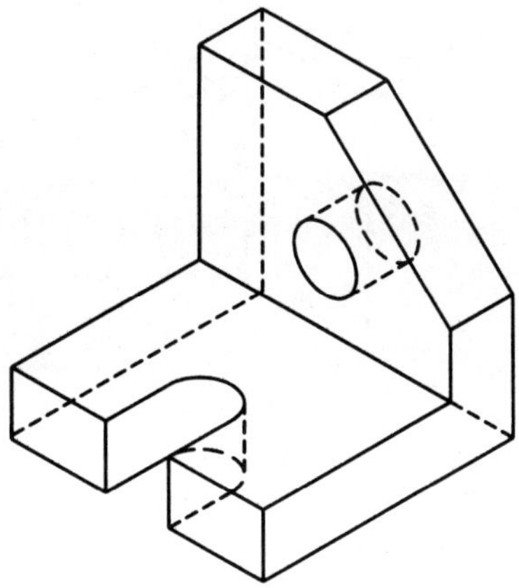

Figure 20-16 *Step block with hole and slot*

3. Picture the object in your mind, and align it along the imaginary *X*, *Y*, and *Z* axes (Figure 20-17).

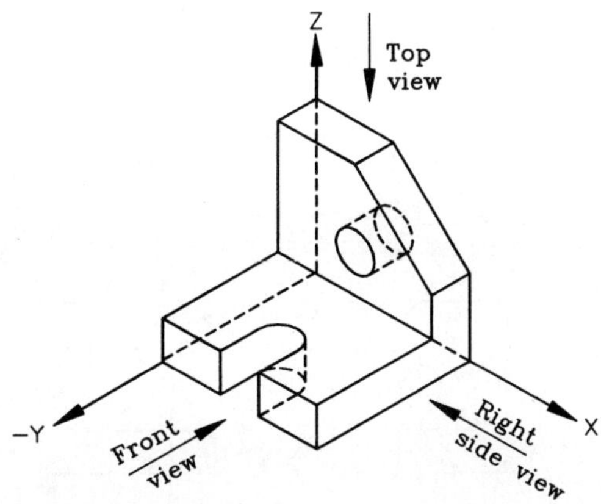

Figure 20-17 *Align the object with the imaginary X, Y, and Z axes*

4. Look at the object along the negative *Y* axis, and project the image on the imaginary *XZ* parallel plane, Figure 20-18.

5. Draw the front view of the object. If there are any hidden features, they must be drawn with hidden lines. The holes and slots must be shown with the centerlines.

6. To draw the right side view, look at the object along the positive *X* axis and project the image onto the imaginary *YZ* parallel plane.

7. Draw the right side view of the object. If there are any hidden features, they must be drawn with hidden lines. The holes and slots, when shown in side view, must have one centerline.

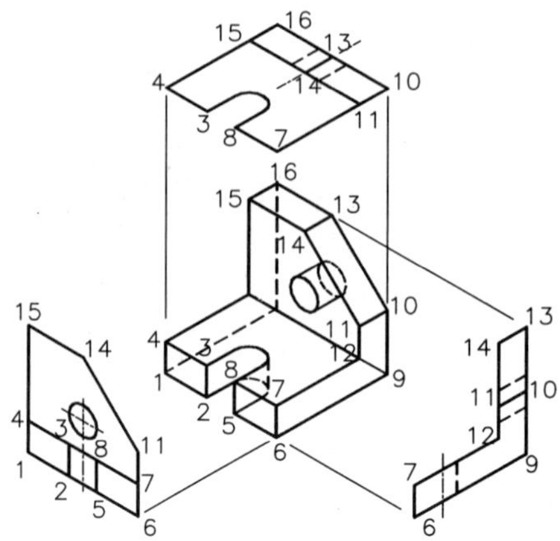

Figure 20-18 *Project the image onto the parallel planes*

8. Similarly, draw the top view to complete the drawing. Figure 20-19 shows different views of the given object.

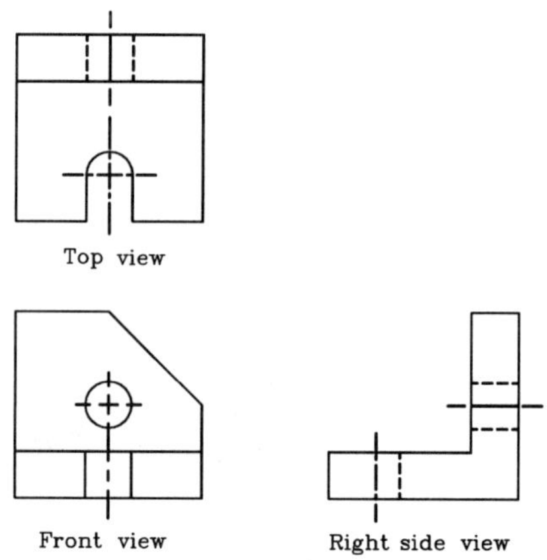

Figure 20-19 *Front, side, and top views*

Exercises 1 through 4 *Mechanical*

Draw the required orthographic views of the following objects in Figure 20-20 through Figure 20-23. The distance between the dotted lines is 0.5 units.

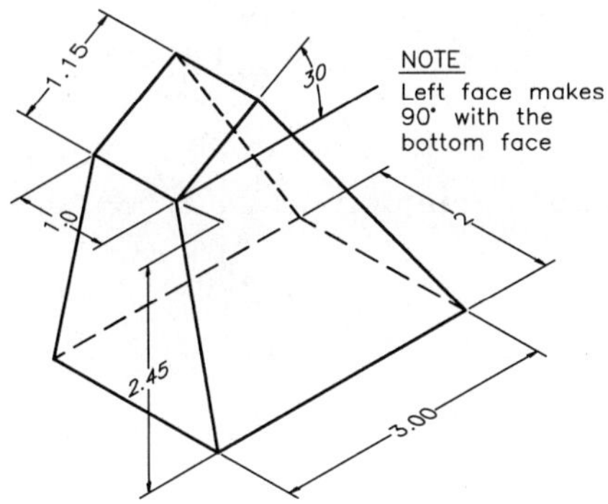

Figure 20-20 *Drawing for Exercise 1 (the object is shown as a surfaced wireframe model)*

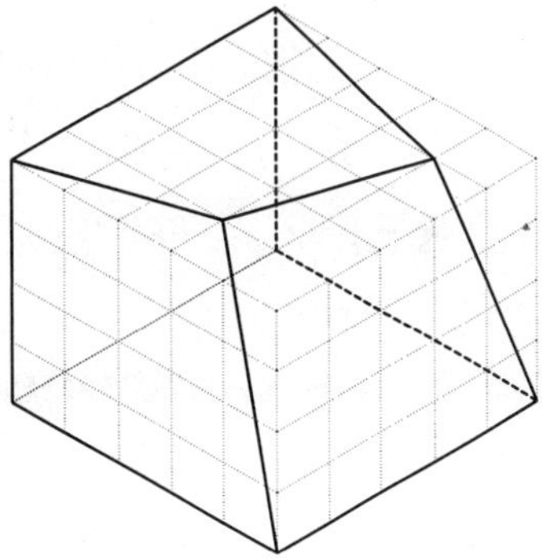

Figure 20-21 *Drawing for Exercise 2*

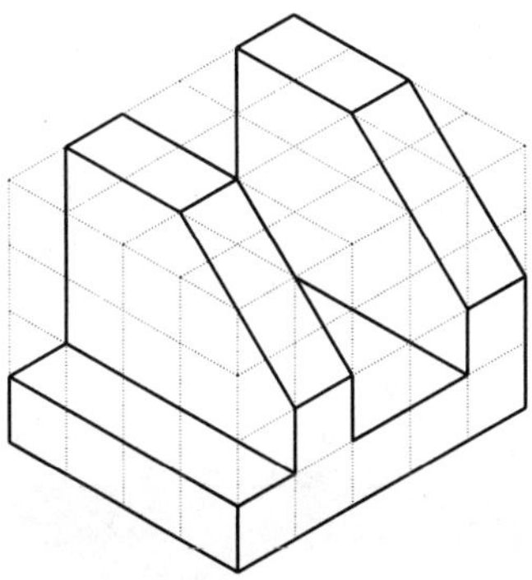

Figure 20-22 *Drawing for Exercise 3*

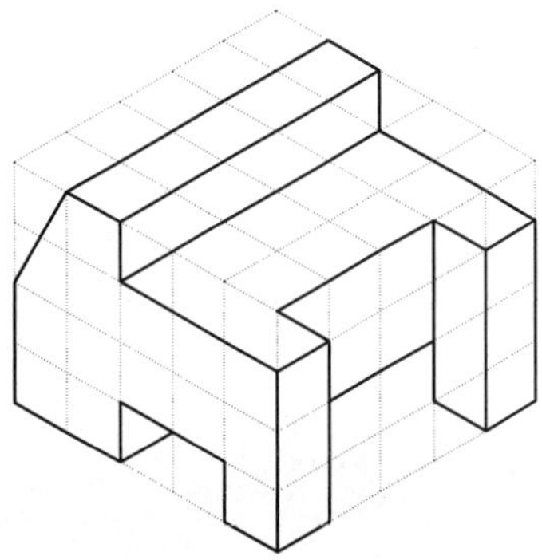

Figure 20-23 *Drawing for Exercise 4*

DIMENSIONING

Dimensioning is one of the most important elements in a drawing. When you dimension, you not only give the size of a part, you give a series of instructions to a machinist, an engineer, or an architect. The way the part is positioned in a machine, the sequence of machining operations, and the location of different features of the part depend on how you dimension it. For example, the number of decimal places in a dimension (2.000) determines the type of machine this will be used to do that machining operation. The machining cost of such an operation is significantly higher than a dimension that has only one digit after the decimal (2.0). If you are using a computer numerical control (CNC) machine, locating a feature may not be a problem, but the number of pieces you can machine without changing the tool depends on the tolerance assigned to a dimension. A closer tolerance (+.0001 -.0005) will definitely increase the tooling cost and ultimately the cost of the product. Similarly, if a part is to be forged or cast, the radius of the edges and the tolerance you provide to these dimensions determine the cost of the product, the number of defective parts, and the number of parts you get from the die.

When dimensioning, you must consider the manufacturing process involved in making a part and the relationships that exist among different parts in an assembly. If you are not familiar with any operation, get help. You must not assume things when dimensioning or making a piece part drawing. The success of a product, to a large extent, depends on the way you dimension a part. Therefore, never underestimate the importance of dimensioning in a drawing.

Dimensioning Components

A dimension consists of the following components (Figure 20-24):

Extension line
Arrows or tick marks
Dimension line
Leader lines
Dimension text

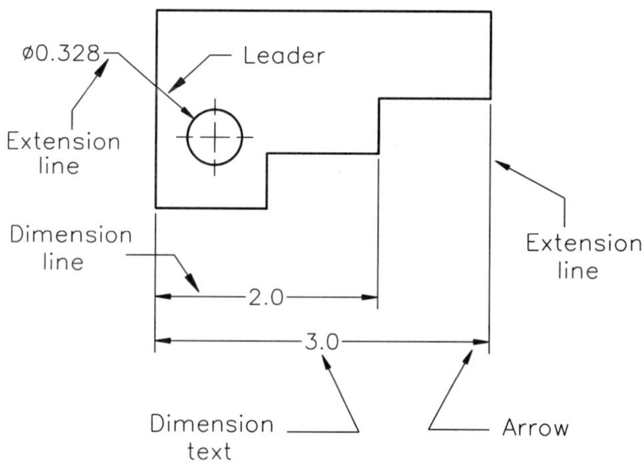

Figure 20-24 *Dimensioning components*

The **extension lines** are drawn to extend the points that are dimensioned. The length of the extension lines is determined by the number of dimensions and the placement of the dimension lines. These lines are generally drawn perpendicular to the surface. The **dimension lines** are drawn between the extension lines, at a specified distance from the object lines. The **dimension text** is a numerical value that represents the distance between the points. The dimension text can also consist of a variable (A, B, X, Y, Z12,...), in which case the value assigned to the variable is defined in a separate table. The dimension text can be centered around the dimension line or at the top of the dimension line. **Arrows or tick marks** are drawn at the end of the dimension line to indicate the start and end of the dimension. **Leader lines** are used when dimensioning a circle, arc, or any nonlinear element of a drawing. They are also used to attach a note to a feature or to give the part numbers in an assembly drawing.

Basic Dimensioning Rules

1. You should make the dimensions in a separate layer/layers. This makes it easy to edit or control the display of dimensions (freeze, thaw, lock, and unlock). Also, the dimension layer/layers should be assigned a unique color so that at the time of plotting you can assign the desired pen to plot the dimensions. This helps to control the line width and contrast of dimensions at the time of plotting.

2. The distance of the first dimension line should at least 0.375 units (10 units for metric drawing) from the object line. In CAD drawing, this distance may be 0.75 to 1.0 units (19 to 25 units for metric drawings). Once you decide on the spacing, it should be maintained throughout the drawing.

3. The distance between the first dimension line and the second dimension line must be at least 0.25 units. In CAD drawings, this distance may be 0.25 to 0.5 units (6 to 12 units for metric drawings). If there are more dimension lines (parallel dimensions), the distances between them must be same (0.25 to 0.5 units). Once you decide on the spacing (0.25 to 0.5), the same spacing should be maintained throughout the drawing. An appropriate snap setting is useful for maintaining this spacing. If you are using baseline dimensioning, you can use AutoCAD's **DIMDLI** variable to set the spacing. You must present the dimensions so that they are not crowded, especially when there is not much space (Figure 20-25).

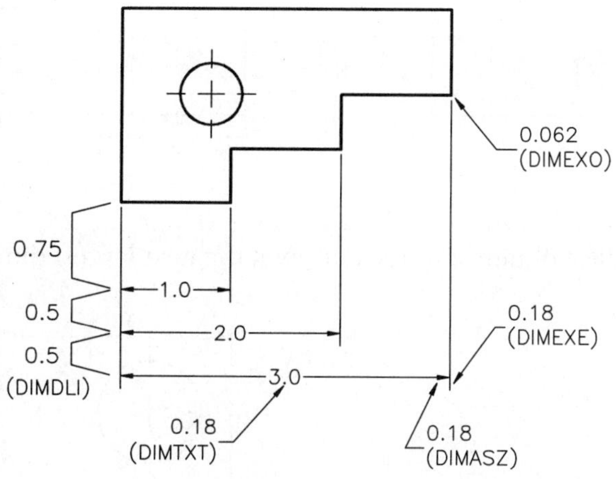

Figure 20-25 *Arrow size, text height, and spacing between dimension lines*

4. For parallel dimension lines, the dimension text can be staggered if there is not enough room between the dimension lines to place the dimension text. You can use the AutoCAD Object Grips feature or the **DIMTEDIT** command to stagger the dimension text (Figure 20-26).

Chapter 20

Note

You can change the grid, snap or UCS origin, and snap increment to make it easier to place the dimensions. You can also add the following lines to the AutoCAD menu file or toolbar buttons.

SNAP;R;\0;SNAP;0.25;GRID;0.25
SNAP;R;0,0;0;SNAP;0.25;GRID;0.5

The first line sets the snap to 0.25 units and allows the user to define the new origin of snap and grid display. The second line sets the grid to 0.5 and snap to 0.25 units. It also sets the origin for grid and snap to (0,0).

5. All dimensions should be given outside the view. However, the dimensions can be shown inside the view if they can be easily understood there and cause no confusion with other dimensions or details. (Figure 20-27.)

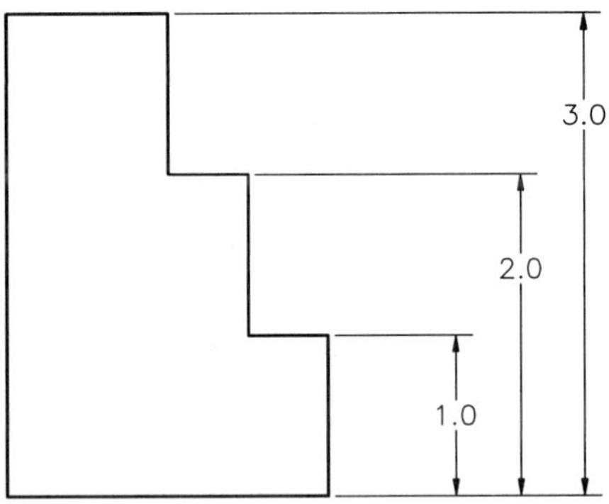

***Figure 20-26** Staggered dimensions*

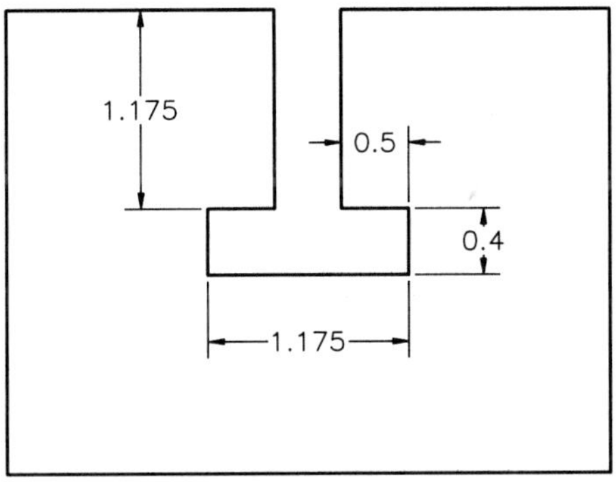

***Figure 20-27** Dimensions inside the view*

6. Dimension lines should not cross extension lines (Figure 20-28). You can accomplish this by giving the smallest dimension first and then the next largest dimension (Figure 20-29).

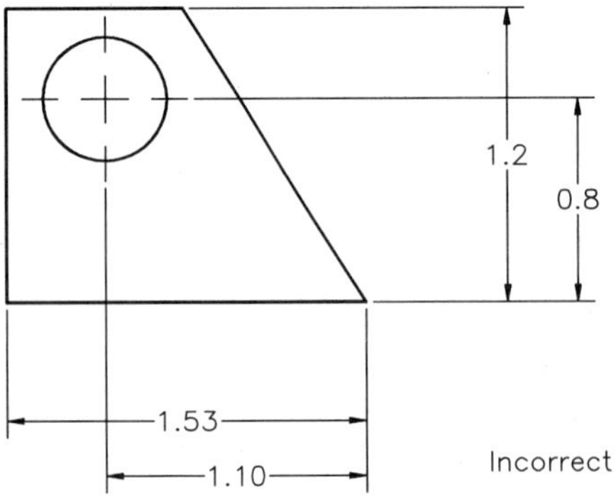

***Figure 20-28** Dimension lines should not cross extension lines*

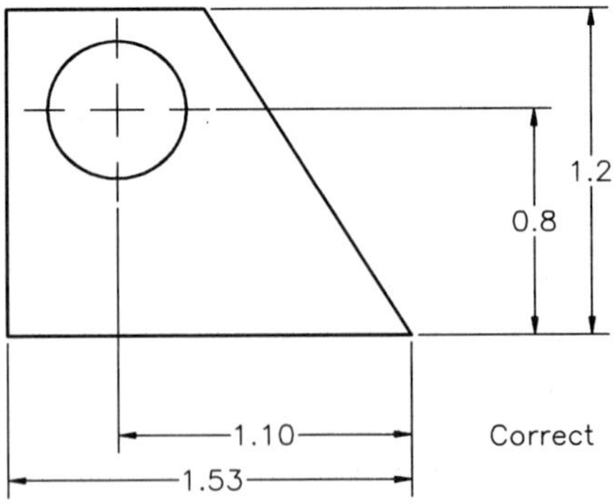

***Figure 20-29** Smallest dimension must be given first*

7. If you decide to have the dimension text aligned with the dimension line, then all dimension text in the drawing must be aligned (Figure 20-30). Similarly, if you decide to have the dimension text horizontal or above the dimension line, then to maintain uniformity in the drawing, all dimension text must be horizontal (Figure 20-31) or above the dimension line (Figure 20-32).

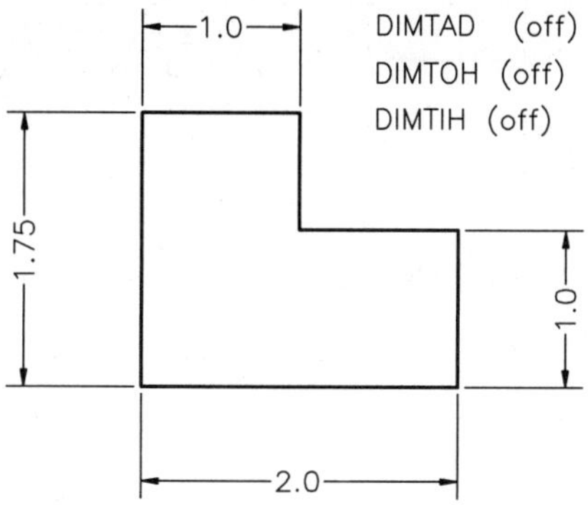

Figure 20-30 Dimension text aligned

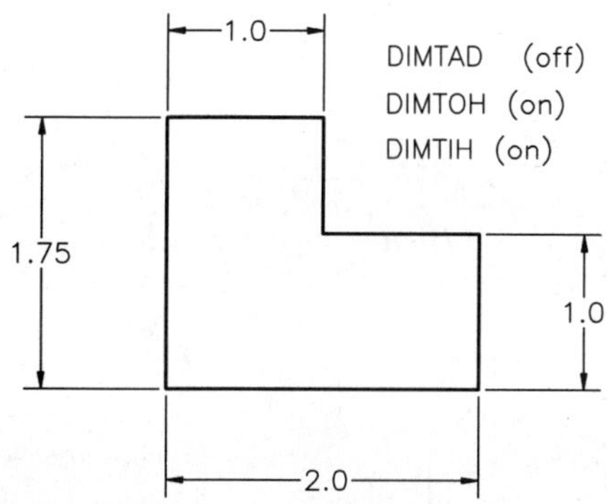

Figure 20-31 Dimension text horizontal

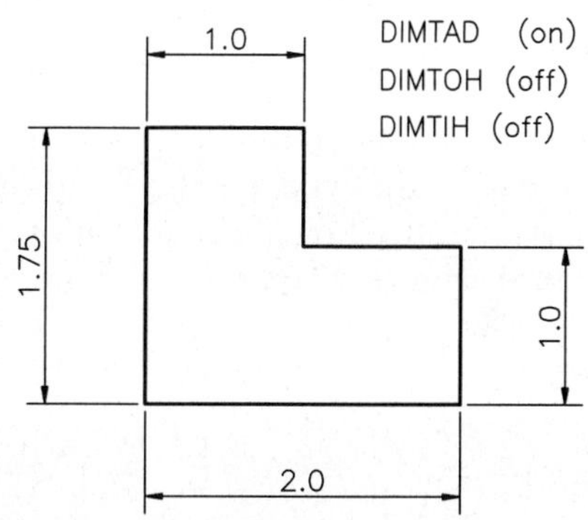

Figure 20-32 Dimension text above dimension line

8. If you have a series of continuous dimensions, they should be placed in a continuous line (Figure 20-33). Sometimes you may not be able to give the dimensions in a continuous line even after adjusting the dimension variables. In that case, give dimensions that are parallel (Figure 20-34).

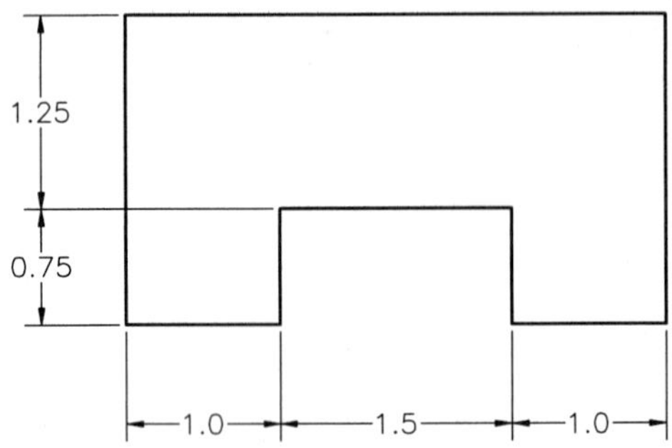

Figure 20-33 *Dimensions should be continuous*

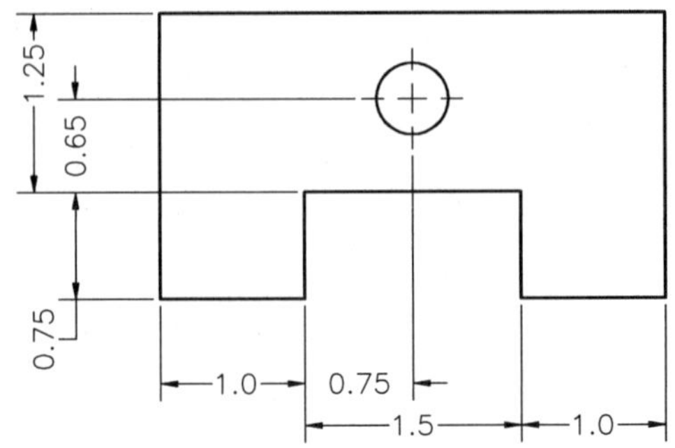

Figure 20-34 *Parallel dimensions*

9. You should not dimension to hidden lines. The dimension should be given where the feature is visible (Figure 20-35). However, in some complicated drawings you might be justified to dimension a detail with a hidden line.

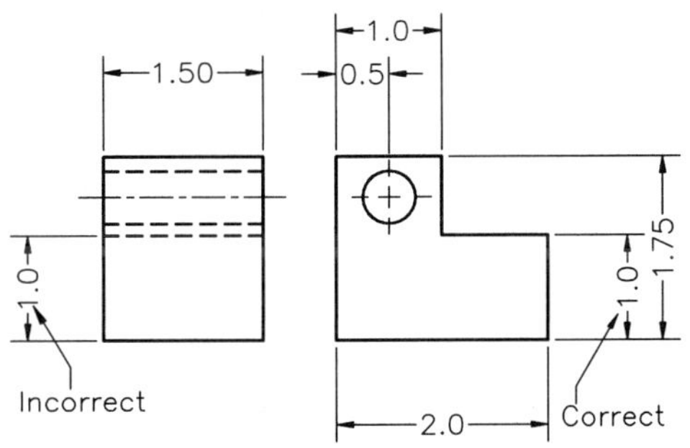

Figure 20-35 *Do not dimension with hidden lines*

10. The dimensions must be given where the feature that you are dimensioning is obvious and shows the contour of the feature (Figure 20-36).

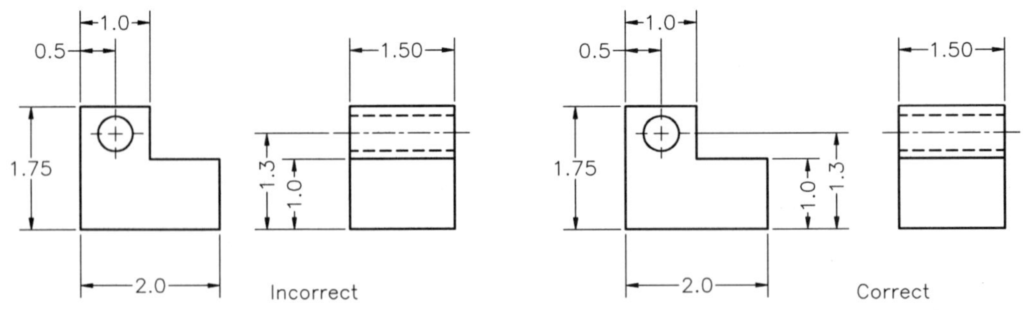

Figure 20-36 *Dimensions should be given where they are obvious*

11. The dimensions must not be repeated; this makes it difficult to update a dimension, and sometimes the dimensions might get confusing (Figure 20-37).

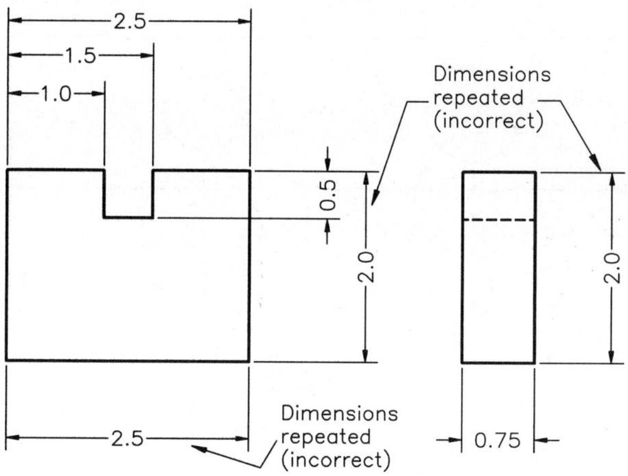

Figure 20-37 *Dimension must not be repeated*

12. The dimensions must be given depending on how the part will be machined and the relationship that exists between different features of the part (Figure 20-38).

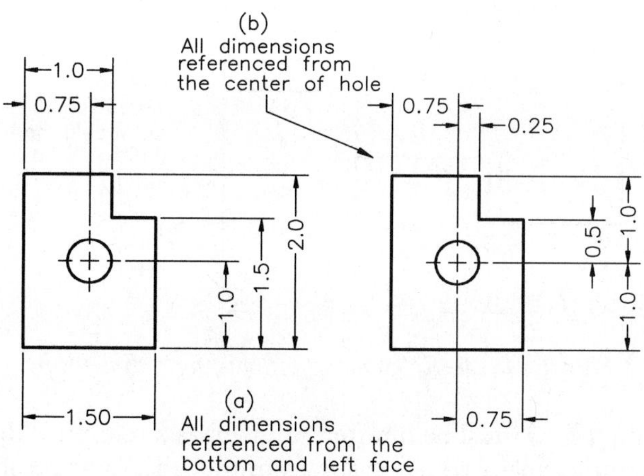

Figure 20-38 *When dimensioning, consider the machining processes involved*

13. When a dimension is not required but you want to give it for reference, it must be a reference dimension. The reference dimension must be enclosed in parentheses (Figure 20-39).

14. If you give continuous (incremental) dimensions for dimensioning various features of a part, the overall dimension must be omitted or given as a reference dimension (Figure 20-40). Similarly, if you give the overall dimension, one of the continuous

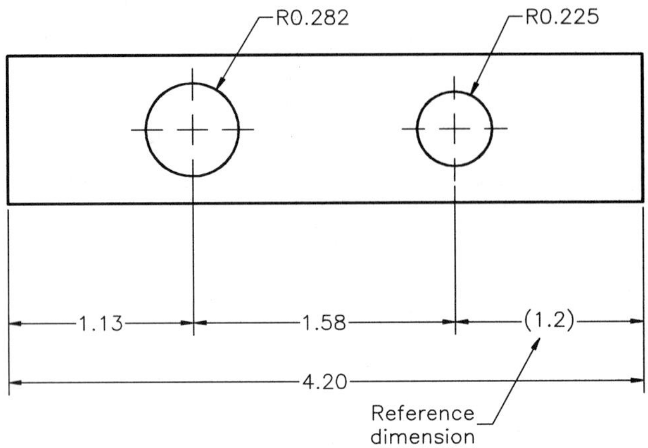

Figure 20-39 *Reference dimensions*

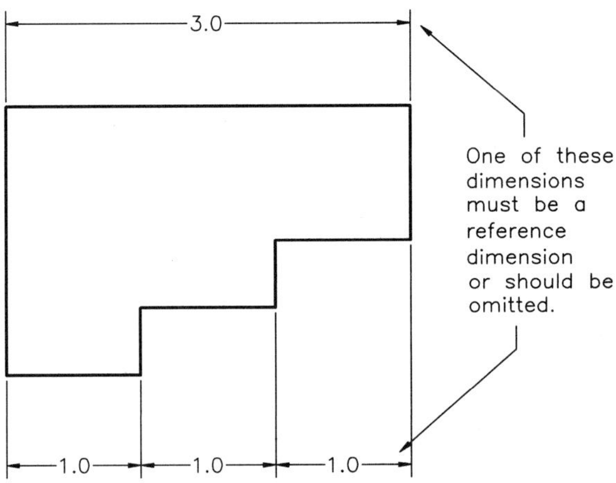

Figure 20-40 *Referencing or omitting a dimension*

(incremental) dimensions must be omitted or given as a reference dimension. Otherwise, there will be a conflict in tolerances. For example, the total positive tolerance on the three incremental dimensions shown in Figure 20-40 is 0.06. Therefore, the maximum size based on the incremental dimensions is (1 + 0.02) + (1 + 0.02) + (1 + 0.02) = 3.06. Also, the positive tolerance on the overall 3.0 dimension is 0.02. Based on this dimension, the overall size of the part must not exceed 3.02. This causes a conflict in tolerances: with incremental dimensions, the total tolerance is 0.06, whereas with the overall dimension the total tolerance is only 0.02.

15. If the dimension of a feature appears in a section view, you must not hatch the dimension text (Figure 20-41). You can accomplish it by selecting the dimension object when defining the hatch boundary. You can also accomplish this by drawing a rectangle around the dimension text and then hatching the area after excluding the rectangle from the hatch

boundary. (You can also use the **EXPLODE** command to explode the dimension and then exclude the dimension text from hatching. This is not recommended because by exploding a dimension, the associativity of the selected dimension is lost.)

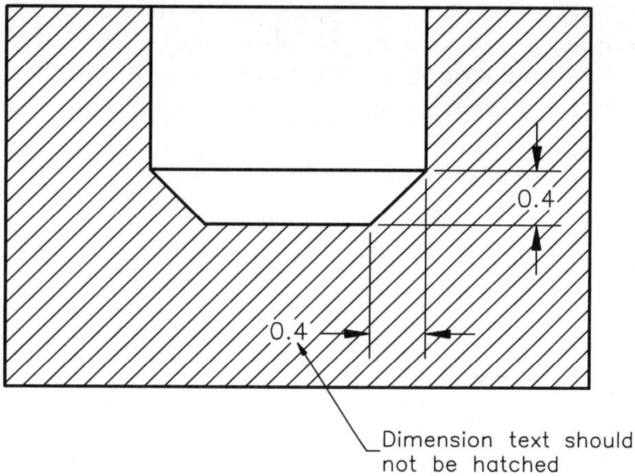

Figure 20-41 *Dimension text should not be hatched*

16. When dimensioning a circle, the diameter should be preceded by the diameter symbol (Figure 20-42). AutoCAD automatically puts the diameter symbol in front of the diameter value. However, if you override the default diameter value, you can use %%c followed by the value of the diameter (%%c1.25) to put the diameter symbol in front of the diameter dimension.

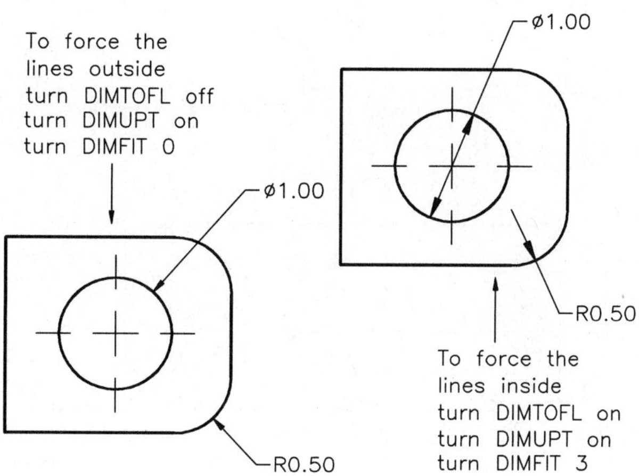

Figure 20-42 *Diameter should be preceded by the diameter symbol*

17. The circle must be dimensioned as a diameter, never as a radius. The dimension of an arc must be preceded by the abbreviation R (R1.25), and the center of the arc should be indicated

by drawing a small cross. You can use the AutoCAD **DIMCEN** variable to control the size of the cross. If the value of this variable is 0, AutoCAD does not draw the cross in the center when you dimension an arc or a circle. You can also use the **DIMCENTER** command or the **CENTER** option of the **DIM** command to draw a cross at the center of the arc or circle.

18. When dimensioning an arc or a circle, the dimension line (leader) must be radial. Also, you should place the dimension text horizontally (Figure 20-43).

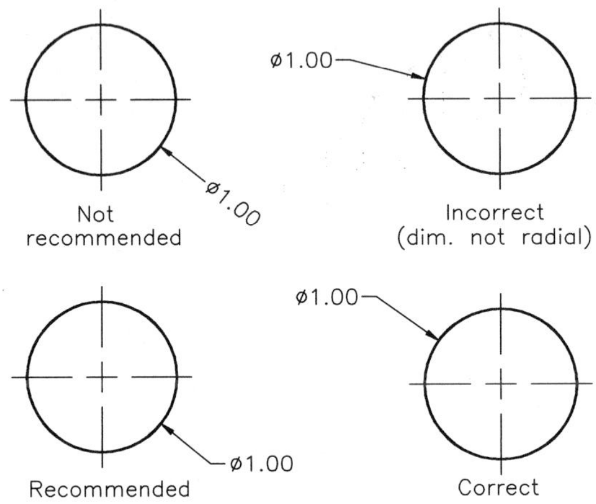

Figure 20-43 *Dimensioning a circle*

19. A chamfer can be dimensioned by specifying the chamfer angle and the distance of the chamfer from the edge. A chamfer can also be dimensioned by specifying the distances as shown in Figure 20-44.

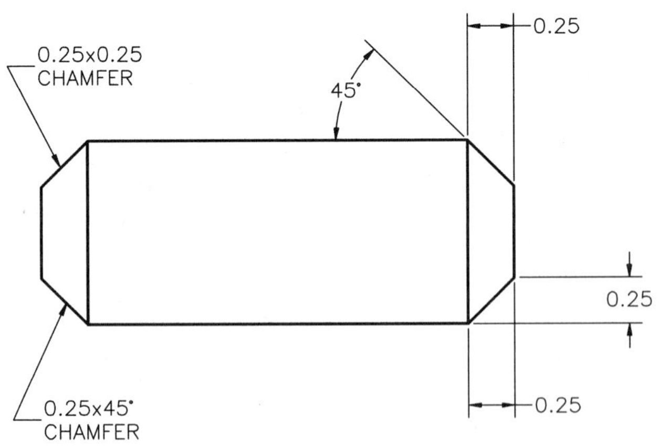

Figure 20-44 *Different ways of specifying chamfer*

20. A dimension that is not to scale should be indicated by drawing a straight line under the dimension text (Figure 20-45). You can draw this line by using the **DDEDIT** command. When you invoke this command, AutoCAD will prompt you to select an annotation object. When you select the object, the **Multiline Text Editor** dialog box is displayed. Select the text (< >), and then choose the **U** (underline) button.

21. A bolt circle should be dimensioned by specifying the diameter of the bolt circle, the diameter of the holes, and the number of holes in the bolt circle (Figure 20-46).

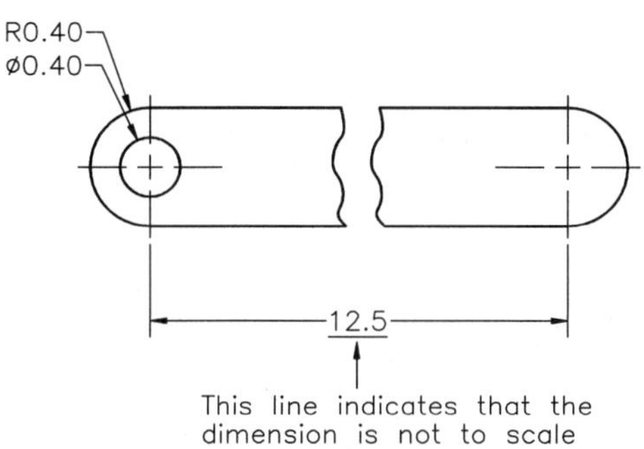

Figure 20-45 *Specifying dimensions that are not to scale*

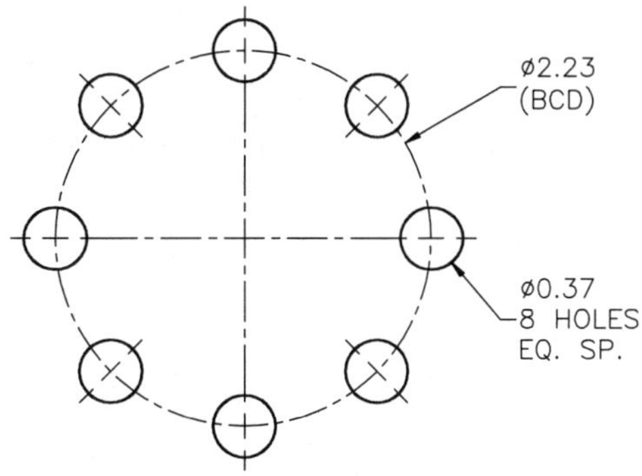

Figure 20-46 *Dimensioning a bolt circle*

Exercises 5 through 10 *Mechanical*

Draw the required orthographic views of the following objects, and then give the dimensions (refer to Figures 20-47 through 20-52). The distance between the grid lines is 0.5 units.

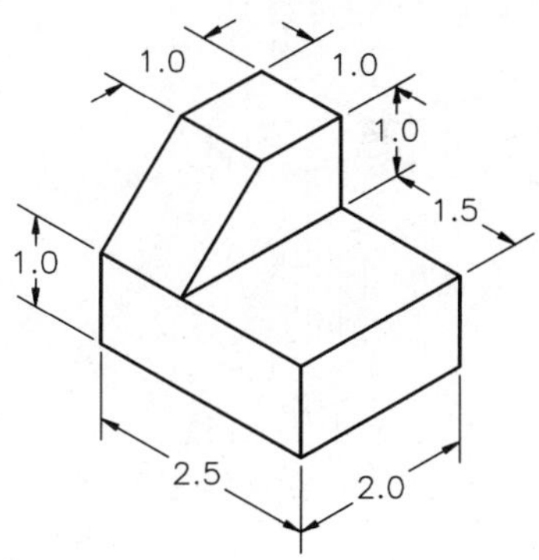

Figure 20-47 *Drawing for Exercise 5*

Chapter 20

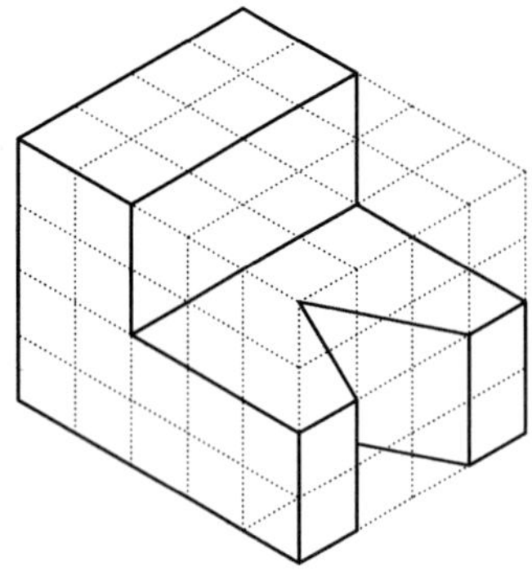

Figure 20-48 *Drawing for Exercise 6*

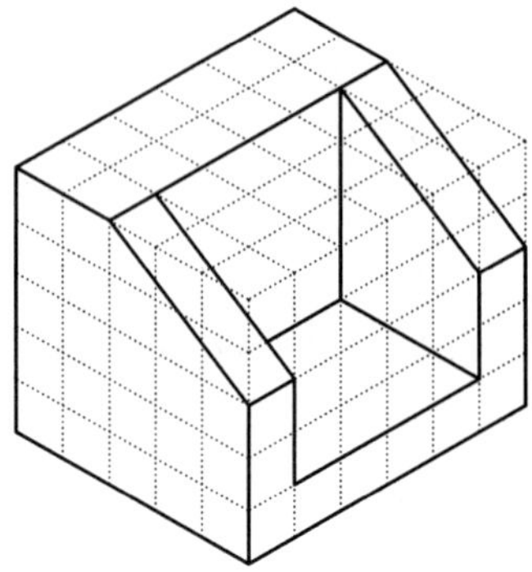

Figure 20-49 *Drawing for Exercise 7*

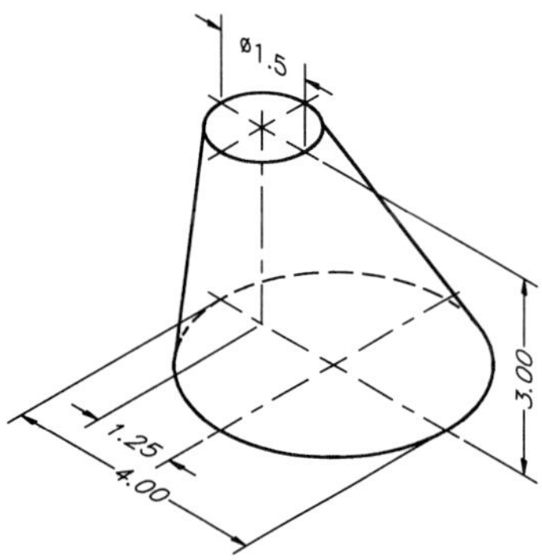

Figure 20-50 *Drawing for Exercise 8*

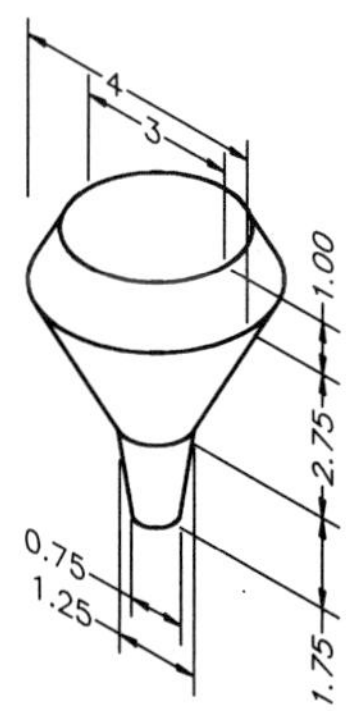

Figure 20-51 *Drawing for Exercise 9*

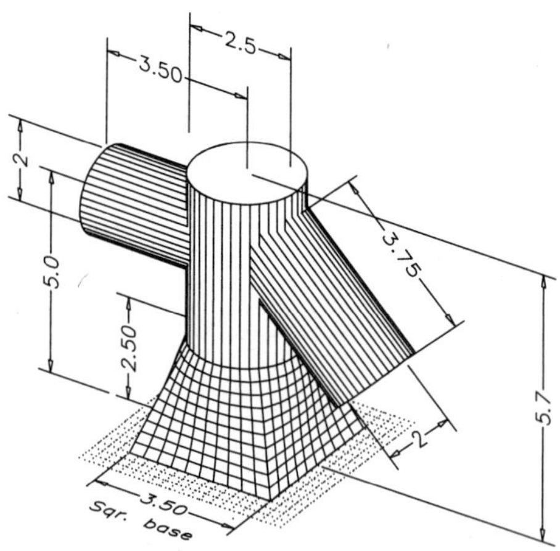

Figure 20-52 *Drawing for Exercise 10 (assume the missing dimensions)*

SECTIONAL VIEWS

In the principal orthographic views, the hidden features are generally shown by hidden lines. In some objects, the hidden lines may not be sufficient to represent the actual shape of the hidden feature. In such situations, sectional views can be used to show the features of the object that are not visible from outside. The location of the section and the direction of sight depend on the shape of the object and the features that need to be shown. Several ways to cut a section in the object are discussed next.

Full Section

Figure 20-53 shows an object that has a drilled hole, a counterbore, and a taper. In the orthographic views, these features will be shown by hidden lines (Figure 20-54).

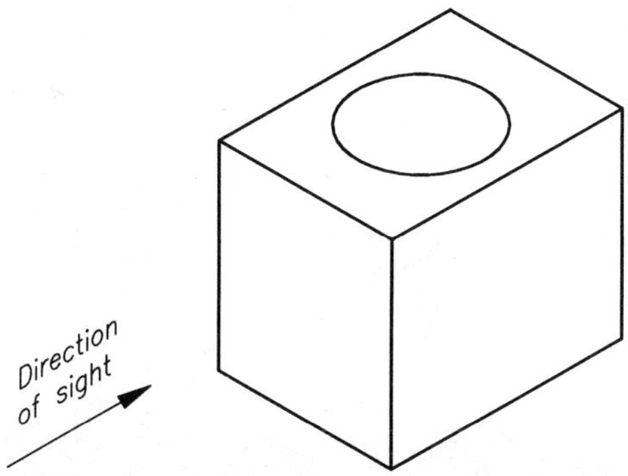

Figure 20-53 *Rectangular object with hole*

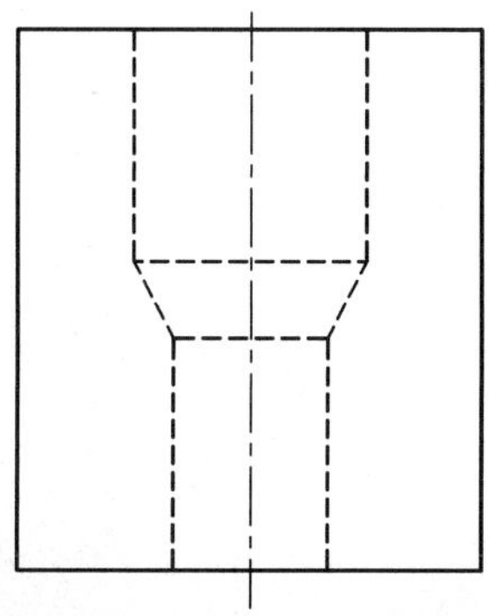

Figure 20-54 *Front view without section*

To better represent the hidden features, the object must be cut so the hidden features are visible. In the **full section**, the object is cut along the entire length of the object. To get a better idea of a full section, imagine that the object is cut into two halves along the centerline, as shown in Figure 20-55. Now remove the left half of the object and look at the right half in the direction that is perpendicular to the sectioned surface. The view you get after cutting the section is called a **full section** view (Figure 20-56).

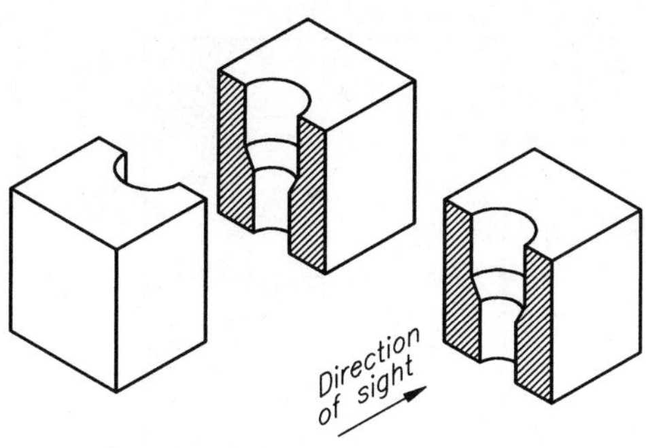

Figure 20-55 *One-half of the object removed*

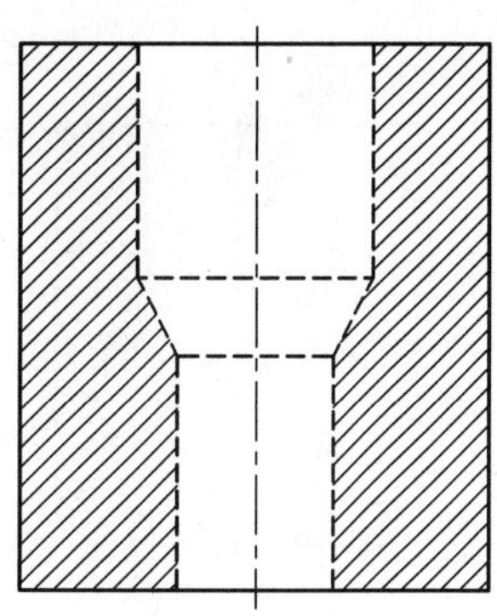

Figure 20-56 *Front view in full section*

In this section view, the features that would be hidden in a normal orthographic view are visible. Also, the part of the object where the material is actually cut is indicated by section lines. If the material is not cut, the section lines are not drawn. For example, if there is a hole, no material is cut when the part is sectioned, and so the section lines must not be drawn through that area of the section view.

Half Section

If the object is symmetrical, it is not necessary to draw a full section view. For example, in Figure 20-57 the object is symmetrical with respect to the centerline of the hole, so a full section is not required. Also, in some objects it may help to understand and visualize the shape of the hidden details better to draw the view in half section. In half section, one-quarter of the object is cut, as shown in Figure 20-58. To draw the view in half section, imagine one-quarter of the object removed, and then look in the direction that is perpendicular to the sectioned surface.

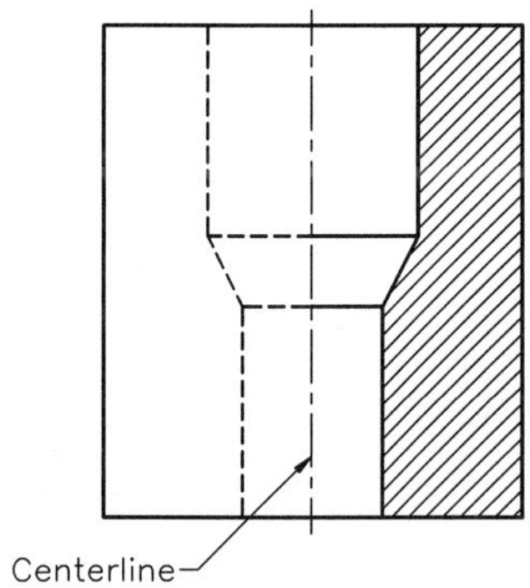

Figure 20-57 *Front view in half section*

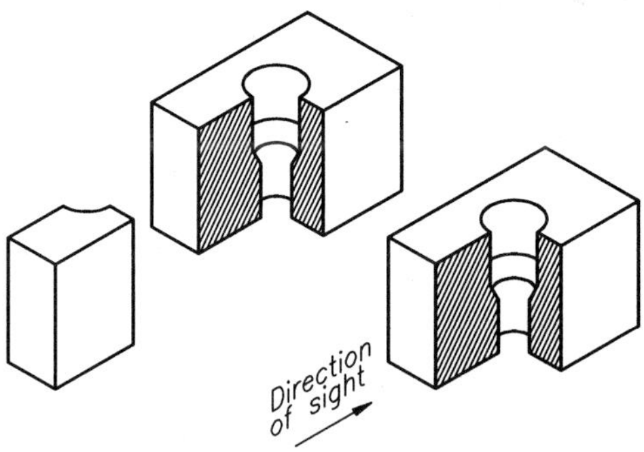

Figure 20-58 *One-quarter of the object removed*

You can also show the front view with a solid line in the middle, as shown in Figure 20-59. Sometimes the hidden lines representing the remaining part of the hidden feature are not drawn, as in Figure 20-60.

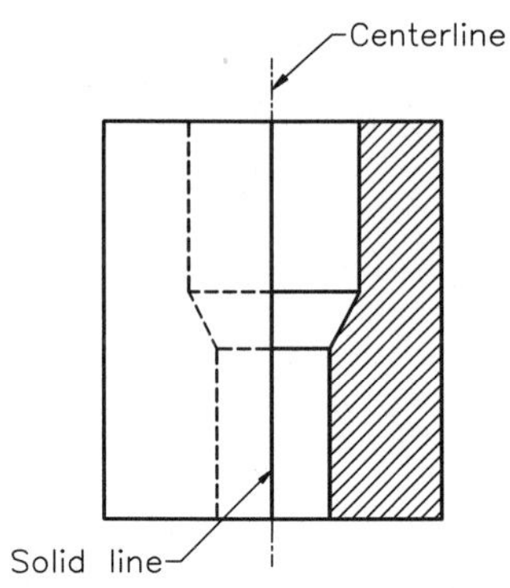

Figure 20-59 *Front view in half section*

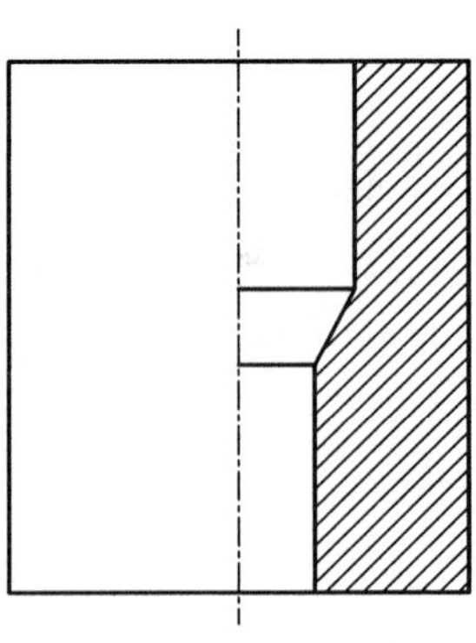

Figure 20-60 *Front view in half section*

Broken Section

In the **broken section**, only a small portion of the object is cut to expose the features that need to be drawn in section. The broken section is designated by drawing a thick zigzag line in the section view (Figure 20-61).

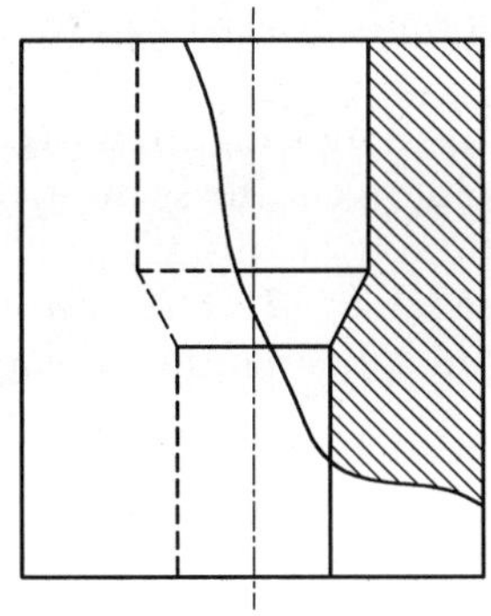

Figure 20-61 *Front view with broken section*

Revolved Section

The **revolved section** is used to show the true shape of the object at the point where the section is cut. The revolved section is used when it is not possible to show the features clearly in any principal view. For example, for the object in Figure 20-62, it is not possible to show the actual shape of the middle section in the front, side, or top view. Therefore, a revolved section is required to show the shape of the middle section.

The revolved section involves cutting an imaginary section through the object and then looking at the sectioned surface in a direction that is perpendicular to it. To represent the shape, the view is revolved 90-degree and drawn in the plane of the paper, as shown Figure 20-62. Depending on the shape of the object, and for clarity, it is recommended to provide a break in the object so that the object lines do not interfere with the revolved section.

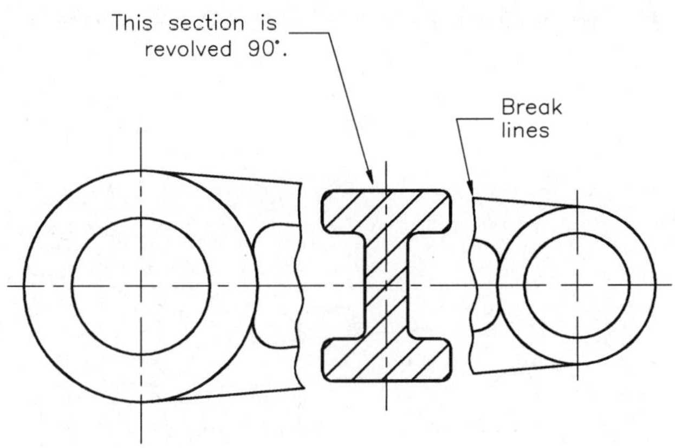

Figure 20-62 *Front view with revolved section*

Removed Section

The **removed section** is similar to the revolved section, except that it is shown outside the object. The removed section is recommended when there is not enough space in the view to show the revolved section or if the scale of the section is different from the parent object. The removed section can be shown by drawing a line through the object at the point where the revolved section is desired and then drawing the shape of the section, as in Figure 20-63.

The other way of showing a removed section is to draw a cutting plane line through the object where you want to cut the section. The arrows should point in the direction in which you are looking at the sectioned surface. The section can then be drawn at a convenient place in the drawing. The removed section must be labeled as shown in Figure 20-64. If the scale has been changed, it must be mentioned with the view description.

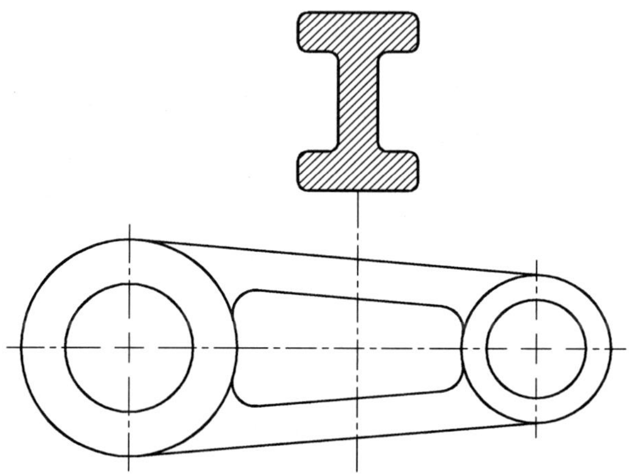

Figure 20-63 *Front view with removed section*

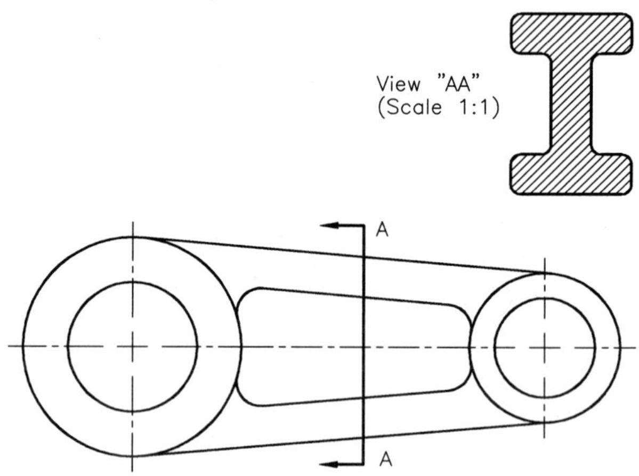

Figure 20-64 *Front view with removed section*

Offset Section

The **offset section** is used when the features of the object that you want to section are not in one plane. The offset section is designated by drawing a cutting plane line that is offset through the center of the features that need to be shown in section (Figure 20-65). The arrows indicate the direction in which the section is viewed.

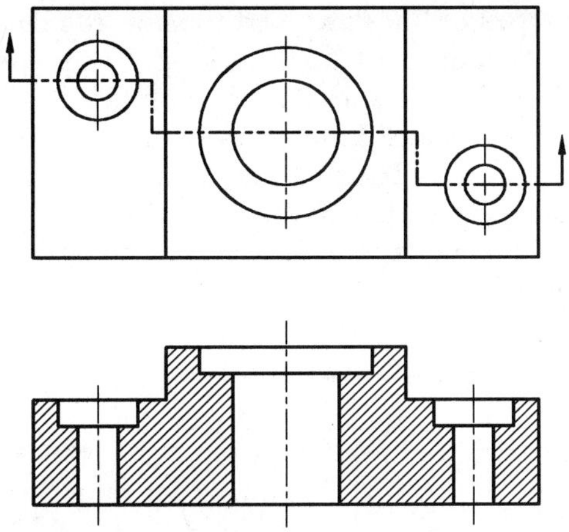

Figure 20-65 *Front view with offset section*

Aligned Section

In some objects, cutting a straight section might cause confusion in visualizing the shape of the section. Therefore, the aligned section is used to represent the shape along the cutting plane (Figure 20-66). Such sections are widely used in circular objects that have spokes, ribs, or holes.

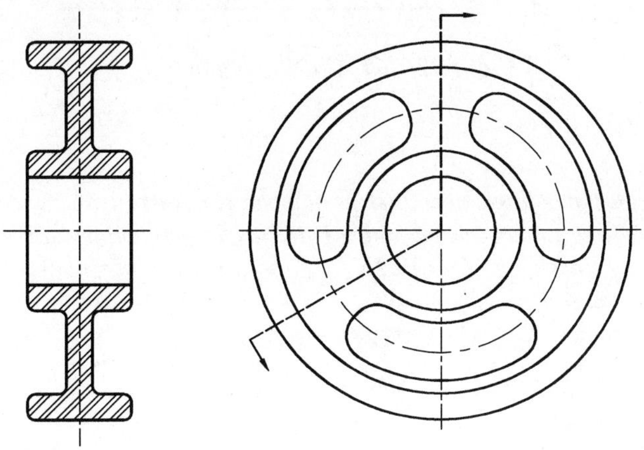

Figure 20-66 *Side view in section (aligned section)*

Cutting Plane Lines

Cutting plane lines are thicker than object lines (Figure 20-67). You can use the **PLINE** command to draw the polylines of desired width, generally 0.005 to 0.01. However, for drawings that need to be plotted, you should assign a unique color to the cutting plane lines and then assign that color to the slot of the plotter that carries a pen of the required tip width. (For details, see Chapter 12, Plotting Drawings.)

In industry, generally three types of lines are used to show the cutting plane for sectioning. The first line consists of a series of dashes 0.25 units long. The second type consists of a series of long dashes separated by two short dashes (Figure 20-68). The length of the long dash can vary from 0.75 to 1.5 units, and the short dashes are about 0.12 units long. The space between the dashes should be about 0.03 units. Third, sometimes the cutting plane lines might clutter the drawing or cause confusion with other lines in the drawing. To avoid this problem, you can show the cutting plane by drawing a short line at the end of the section (Figure 20-68 and Figure 20-69). The line should be about 0.5 units long.

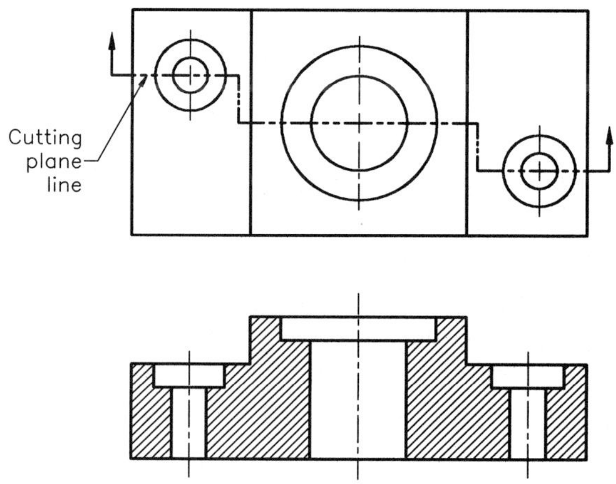

Figure 20-67 *Cutting plane line*

Note

In AutoCAD, you can define a new linetype that you can use to draw the cutting plane lines. See Chapter 31, Creating Linetypes and Hatch Patterns, for more information on defining linetypes. Add the following lines to the ***ACLT.LIN*** *file, and then load the linetypes before assigning it to an object or a layer.*

```
*CPLANE1,___ ___ ___ ___
A,0.25,-0.03
*CPLANE2,___ ___ ___ ___
A,1.0,-0.03,0.12,-0.03,0.12,-0.03
```

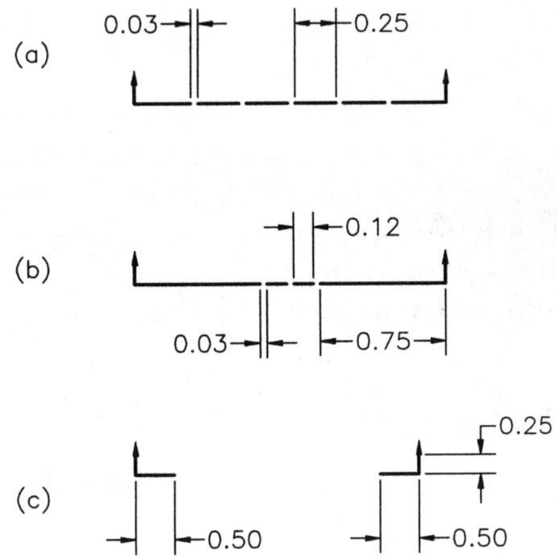

Figure 20-68 Cutting plane lines

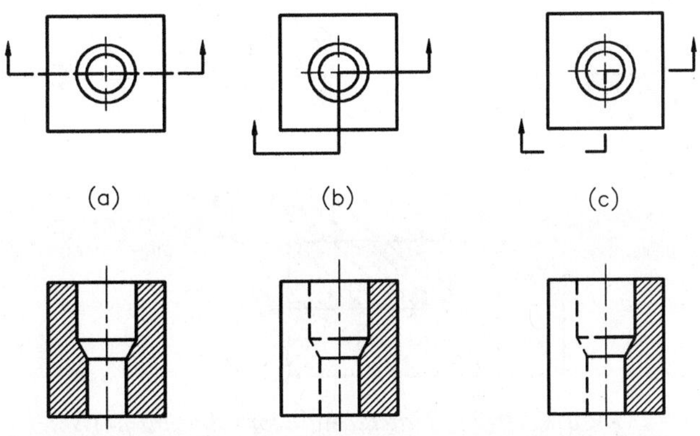

Figure 20-69 Application of cutting plane lines

Spacing for Hatch Lines

The spacing between the hatch (section) lines is determined by the space that is being hatched (Figure 20-70). If the hatch area is small, the spacing between the hatch lines should be smaller compared with a large hatch area.

In AutoCAD, you can control the spacing between the hatch lines by specifying the **scale factor** at the time of hatching. If the scale factor is 1, the spacing between the hatch lines is the same as defined in the hatch pattern file for that particular hatch. For example, in the following hatch pattern definition, the distance between the lines is 0.125.

***ANSI31, ANSI Iron, Brick, Stone masonry**
45, 0, 0, 0, .125

When the hatch scale factor is 1, the line spacing will be 0.125; if the scale factor is 2, the spacing between the lines will be 0.125 x 2 = 0.25.

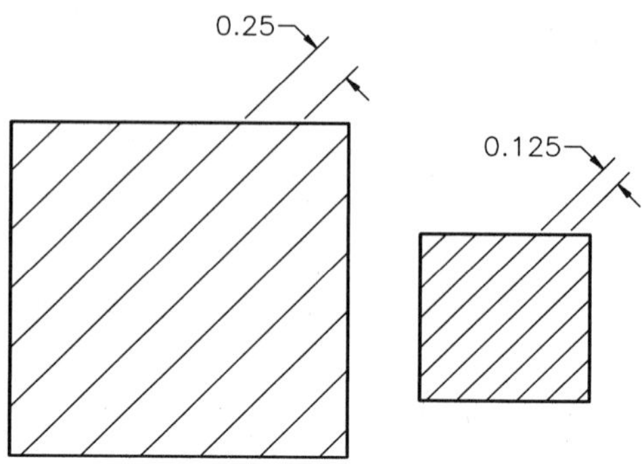

Figure 20-70 Hatch line spacing

Direction of Hatch Lines

The angle for the hatch lines should be 45-degree. However, if there are two or more hatch areas next to one another representing different parts, the hatch angle must be changed so that the hatched areas look different (Figure 20-71).

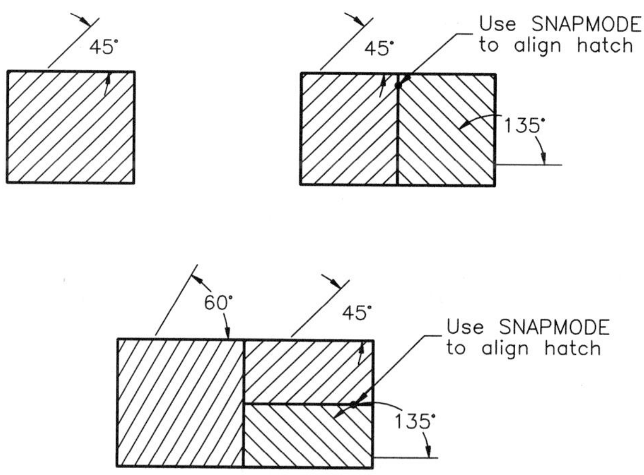

Figure 20-71 Hatch angle for adjacent parts

Also, if the hatch lines fall parallel to any edge of the hatch area, the hatch angle should be changed so that the lines are not parallel or perpendicular to any object line (Figure 20-72).

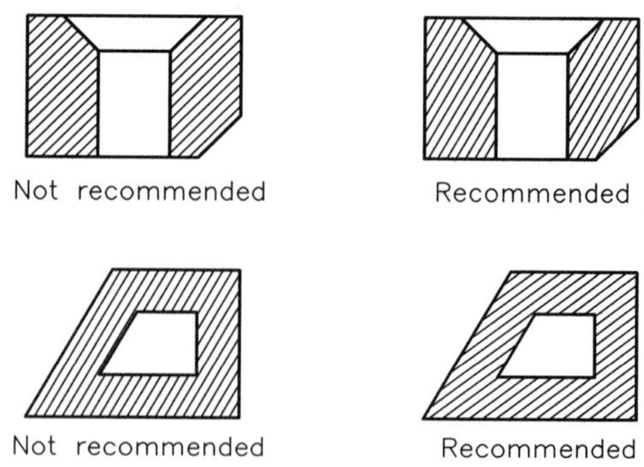

Figure 20-72 Hatch angle

Points to Remember

1. Some parts, such as bolts, nuts, shafts, ball bearings, fasteners, ribs, spokes, keys, and other similar items that do not show any important feature, if sectioned, should not be shown in section.

2. Hidden details should not be shown in the section view unless the hidden lines represent an important detail or help the viewer to understand the shape of the object.

3. The section lines (hatch lines) must be thinner than the object lines. You can accomplish this by assigning a unique color to hatch lines and then assigning the color to that slot on the plotter that carries a pen with a thinner tip.

4. The section lines must be drawn on a separate layer for display and editing purposes.

Exercises 11 and 12 *Mechanical*

In the following drawings (Figure 20-73 Figure and 20-74), the views have been drawn without a section. Draw these views in section as indicated by the cutting plane lines in each object.

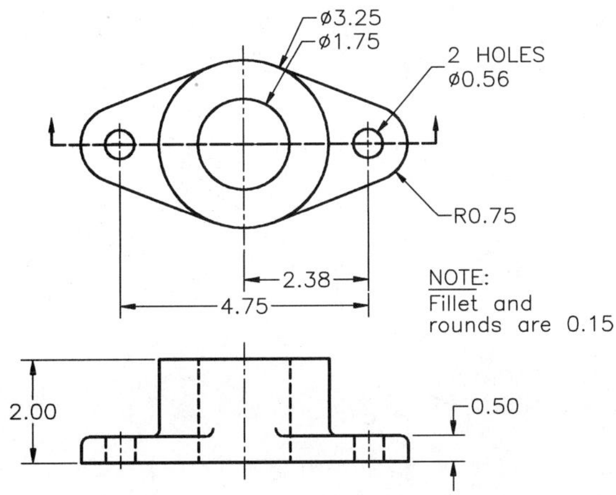

Figure 20-73 *Drawing for Exercise 11: draw the front view in section*

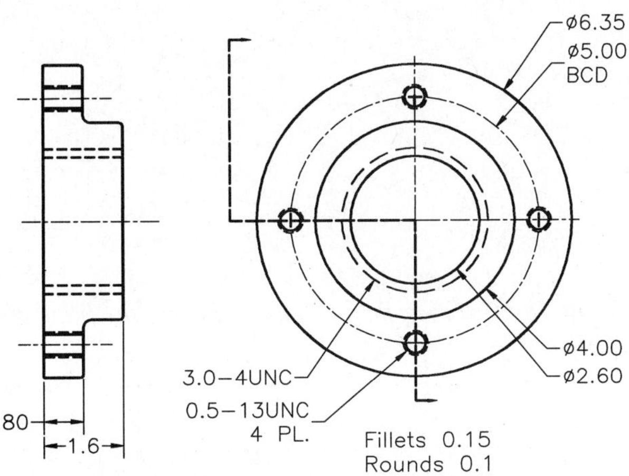

Figure 20-74 *Drawing for Exercise 12: draw the left side view in section*

Exercises 13 and 14 *Mechanical*

Draw the required orthographic views for the following objects in Figure 20-75 and Figure 20-76. Show the front view in section when the object is cut so that the cutting plane passes through the holes. Also, draw the cutting plane lines in the top view.

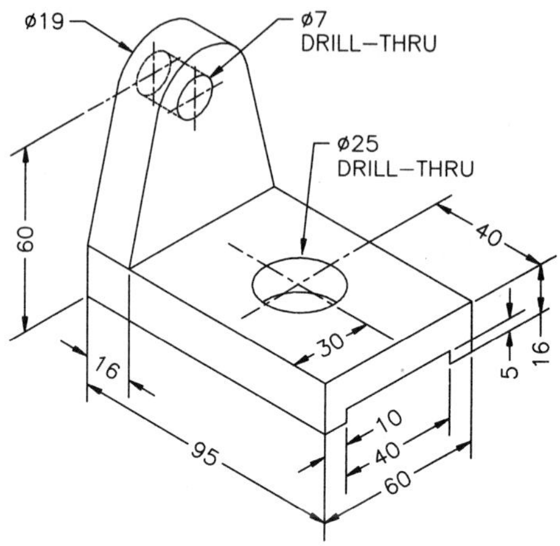

Figure 20-75 *Drawing for Exercise 13: draw the front view in full section*

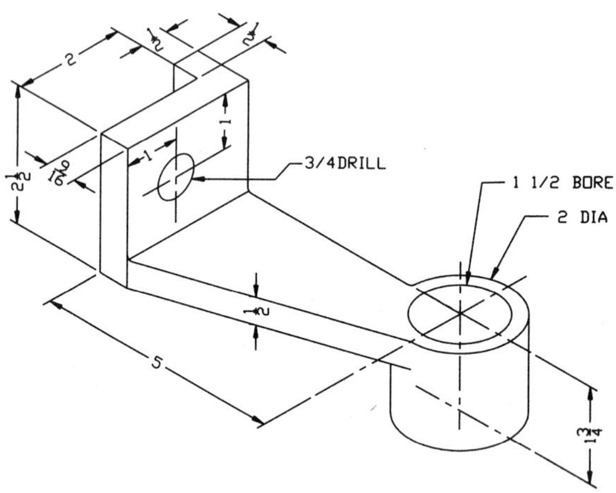

Figure 20-76 *Drawing for Exercise 14: draw the front view with offset section*

Exercise 15 *Mechanical*

Draw the required orthographic views for the objects in Figures 20-77 and 20-78 with the front view in section. Also, draw the cutting plane lines in the top view to show the cutting plane. The material thickness is 0.25 units. (The object has been drawn as a surfaced 3D wiremesh model.)

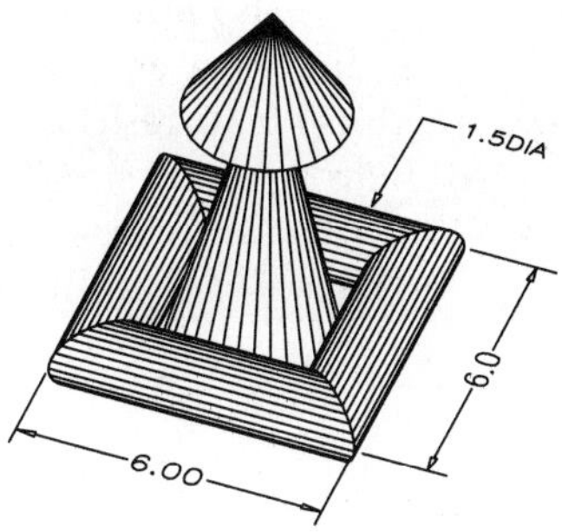

Figure 20-77 *Draw the front view in half section*

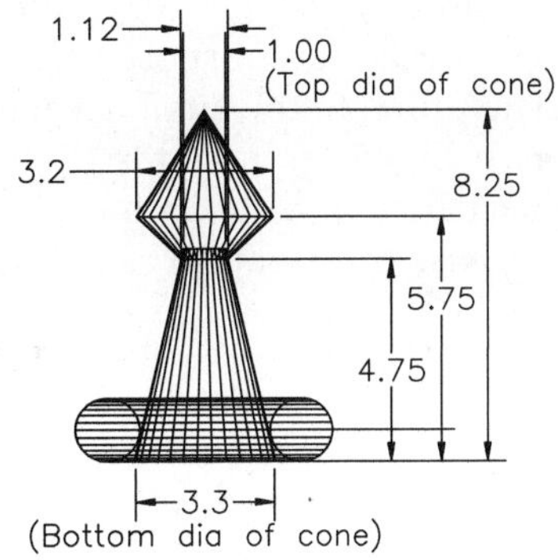

Figure 20-78 *Draw the front view in half section*

AUXILIARY VIEWS

As discussed earlier, most objects generally require three principal views (front view, side view, and top view) to show all features of the object. Round objects may require just two views. Some objects have inclined surfaces. It may not be possible to show the actual shape of the inclined surface in one of the principal views. To get the true view of the inclined surface, you must look at the surface in a direction that is perpendicular to the inclined surface. Then you can project the points onto the imaginary auxiliary plane that is parallel to the inclined surface. The view you get after projecting the points is called the **auxiliary view**, as shown in Figure 20-79 and Figure 20-80.

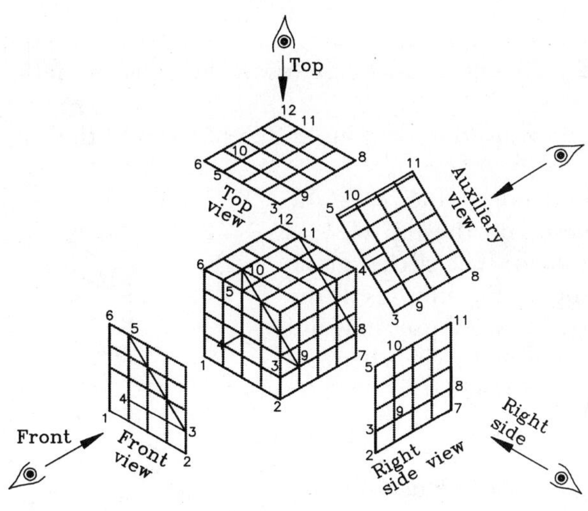

Figure 20-79 *Project points onto the auxiliary plane*

The auxiliary view in Figure 20-80 shows all features of the object as seen from the auxiliary view direction. For example, the bottom left edge is shown as a hidden line. Similarly, the lower right and upper left edges are shown as continuous lines. Although these lines are technically correct, the purpose of the auxiliary view is to show the features of the inclined surface. Therefore, in the auxiliary plane you should draw only those features that are on the inclined face, as shown in Figure 20-81. Other details that will help to understand the shape of the object may also be included in the auxiliary view. The remaining lines should be ignored because they tend to cause confusion in visualizing the shape of the object.

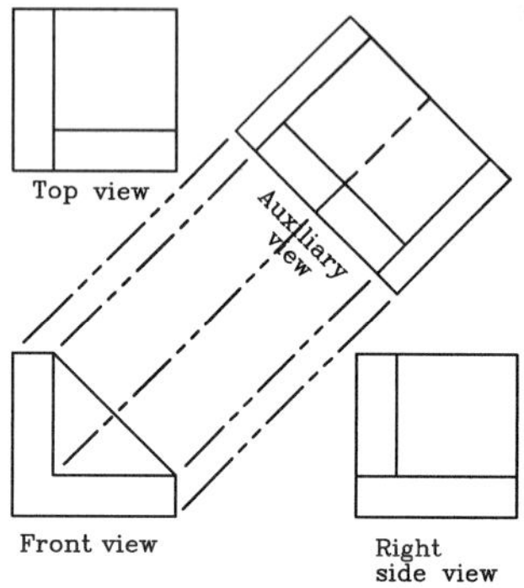

Figure 20-80 *Auxiliary, front, side, and top views*

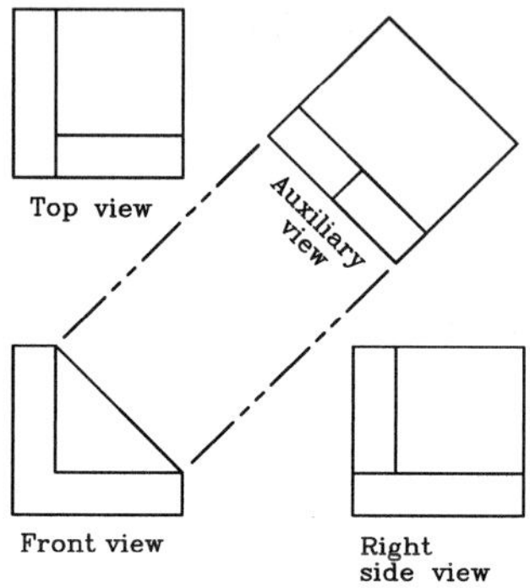

Figure 20-81 *Auxiliary, front, side, and top views*

How to Draw Auxiliary Views

The following example illustrates how to use AutoCAD to generate an auxiliary view.

Example 2 *Mechanical*

Draw the required views of the hollow triangular block with a hole in the inclined face as shown in Figure 20-82. (The block has been drawn as a solid model.)

The following steps are involved in drawing different views of this object.

1. Draw the required orthographic views: the front view, side view, and the top view as shown in Figure 20-83(a). The circles on the inclined surface appear like ellipses in the front and top views. These ellipses may not be shown in the orthographic views because they tend to clutter the views [Figure 20-83(b)].

2. Determine the angle of the inclined surface. In this example, the angle is 45-degree. Use the **SNAP** command to rotate the snap by

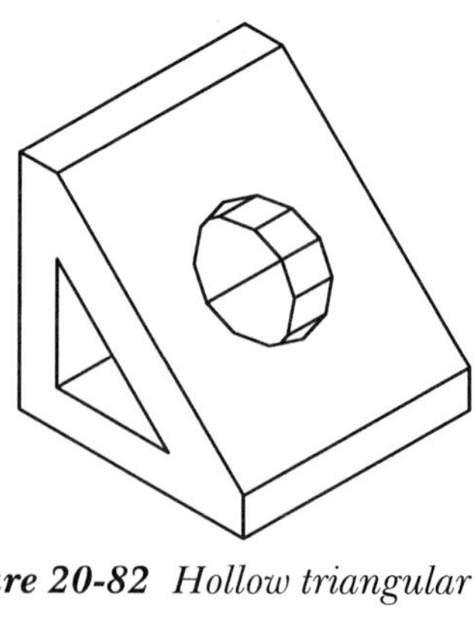

Figure 20-82 *Hollow triangular block*

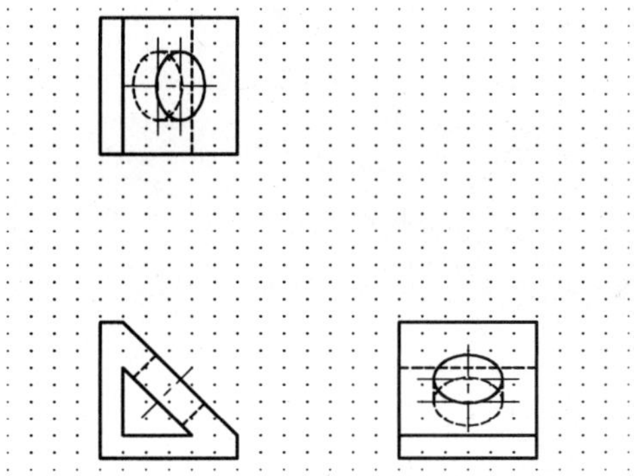

Figure 20-83(a) Front, side, and top views

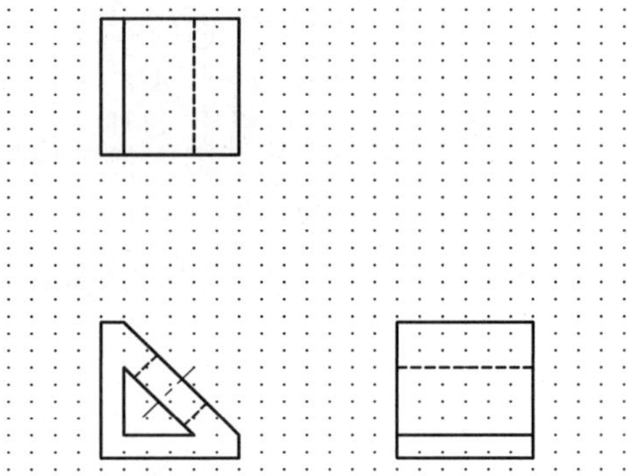

Figure 20-83(b) The ellipses may not be shown in the orthographic views

45-degree (Figure 20-84) or rotate the UCS by 45-degree around the Z axis.

Command: **SNAP** [Enter]
Specify snap spacing or [ON/OFF/Aspect/Rotate/Style/Type] <0.5000>: **R** [Enter]
Specify base point <0.0000,0.0000>: *Select P1.*
Specify rotation angle <0>: **45** [Enter]

Using the rotate option, the snap will be rotated by 45-degree and the grid lines will also be displayed at 45-degree (if the **GRID** is on). Also, one of the grid lines will pass through point P1 because it was defined as the base point.

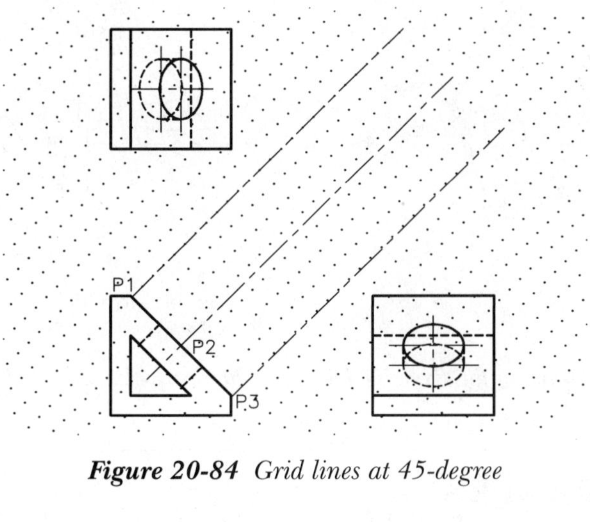

Figure 20-84 Grid lines at 45-degree

3. Turn **ORTHO** on, and project points P1, P2, and P3 from the front view onto the auxiliary plane. Now you can complete the auxiliary view and give dimensions. The projection lines can be erased after the auxiliary view is drawn (Figure 20-85).

Chapter 20

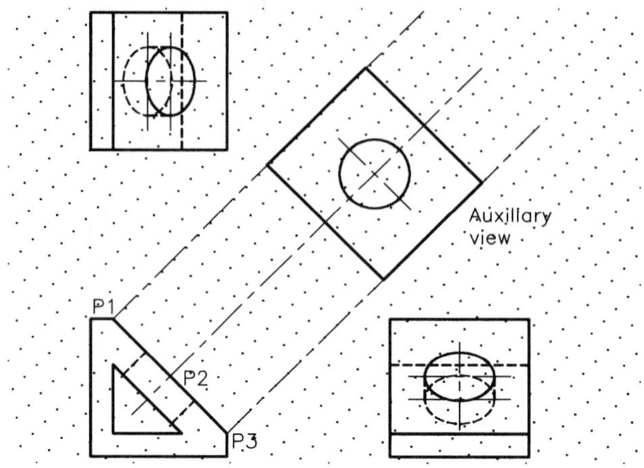

Figure 20-85 *Auxiliary, front, side, and top views*

Exercise 16 *General*

Draw the required orthographic and auxiliary views for the object in Figure 20-86. The object is drawn as a surfaced 3D wiremesh model.

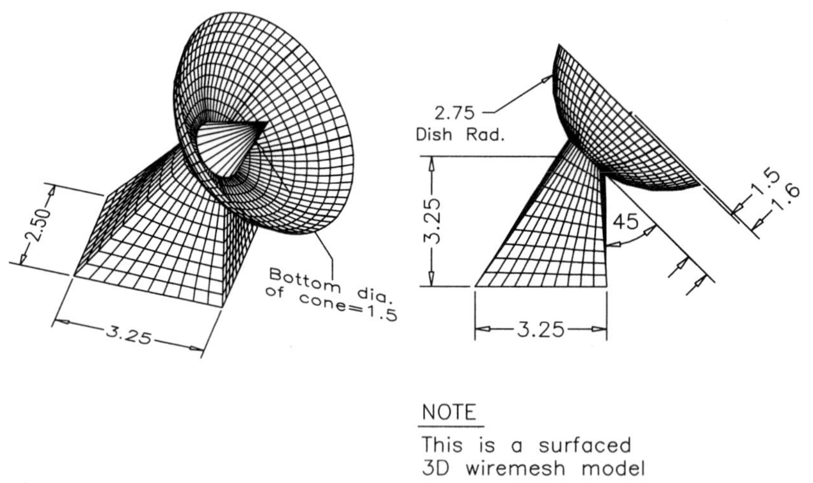

Figure 20-86 *Drawing for Exercise 16: draw the required orthographic and auxiliary views*

Exercises 17 and 18 *General*

Draw the required orthographic and auxiliary views for the following objects in Figure 20-87 and Figure 20-88. The objects are drawn as 3D solid models and are shown at different angles (viewpoints are different).

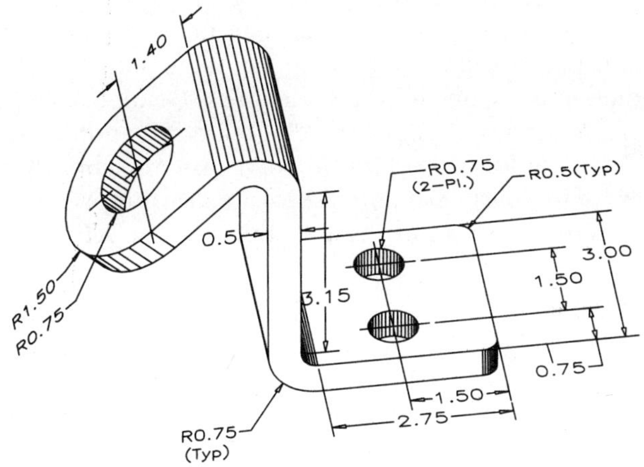

Figure 20-87 *Drawing for Exercise 17: draw the required orthographic and auxiliary views*

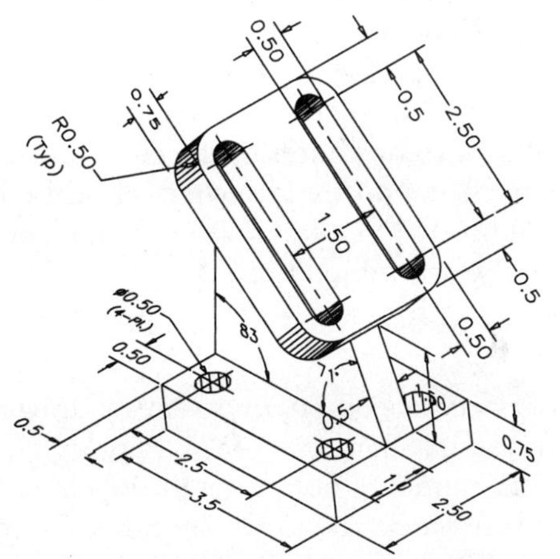

Figure 20-88 *Drawing for Exercise 18: draw the required orthographic and auxiliary views*

DETAIL DRAWING, ASSEMBLY DRAWING, AND BILL OF MATERIALS

Detail Drawing

A detail drawing is a drawing of an individual component that is a part of the assembled product. Detail drawings are also called **piece part drawings**. Each detail drawing must be drawn and dimensioned to completely describe the size and shape of the part. It should also contain information that might be needed in manufacturing the part. The finished surfaces should be indicated by using symbols or notes and all necessary operations on the drawing. The material of which the part is made and the number of parts that are required for production of the assembled product must be given in the title block. Detail drawings should also contain part numbers. This information is used in the bill of materials and the assembly drawing. The part numbers make it easier to locate the drawing of a part. You should make a detail drawing of each part, regardless of the part's size, on a separate drawing. When required, these detail drawings can be inserted in the assembly drawing by using the **XREF** command.

Assembly Drawing

The **assembly drawing** is used to show the parts and their relative positions in an assembled product or a machine unit. The assembly drawing should be drawn so that all parts can be shown in one drawing. This is generally called the main view. The main view may be drawn in full section so that the assembly drawing shows nearly all the individual parts and their locations. Additional views should be drawn only when some of the parts cannot be seen in the main view. The hidden lines, as far as possible, should be omitted from the assembly drawing because they clutter the drawing and might cause confusion. However, a hidden line may be drawn if it helps to understand the product. Only assembly dimensions should be shown in the assembly drawing. Each part should be identified on the assembly drawing by the number used in the detail drawing and in the bill of materials. The part numbers should be given as shown in Figure 20-89. It consists of a text string for the detail number, a circle (balloon), a leader line, and an arrow or dot. The text should be made at least 0.2 inches (5 mm) high and enclosed in a 0.4 inch (10 mm) circle (balloon). The center of the circle must be located not less than 0.8 inches (20 mm) from the nearest line on the drawing. Also, the leader line should be radial with respect to the circle (balloon). The assembly drawing may also contain an exploded isometric or isometric view of the assembled unit.

Bill of Materials

A **bill of materials** is a list of parts placed on an assembly drawing just above the title block (Figure 20-89). The bill of materials contains the part number, part description, material, quantity required, and drawing numbers of the detail drawings (Figure 20-90). If the bill of materials is placed above the title block, the parts should be listed in ascending order so that the first part is at the bottom of the table (Figure 20-91). The bill of materials may also be placed at the top of the drawing. In that case, the parts must be listed in descending order with the first part at the top of the table. This structure allows room for any additional items that may be added to the list.

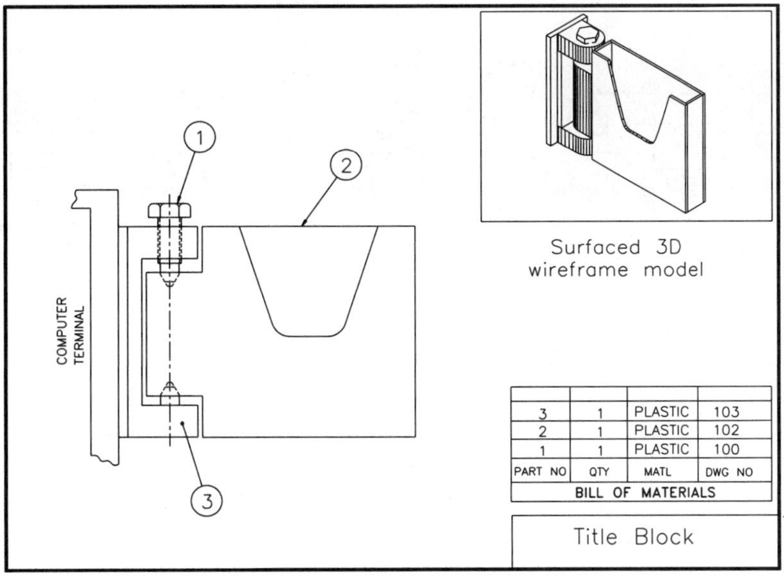

Figure 20-89 *Assembly drawing with title block, bill of materials, and surfaced 3D wireframe model*

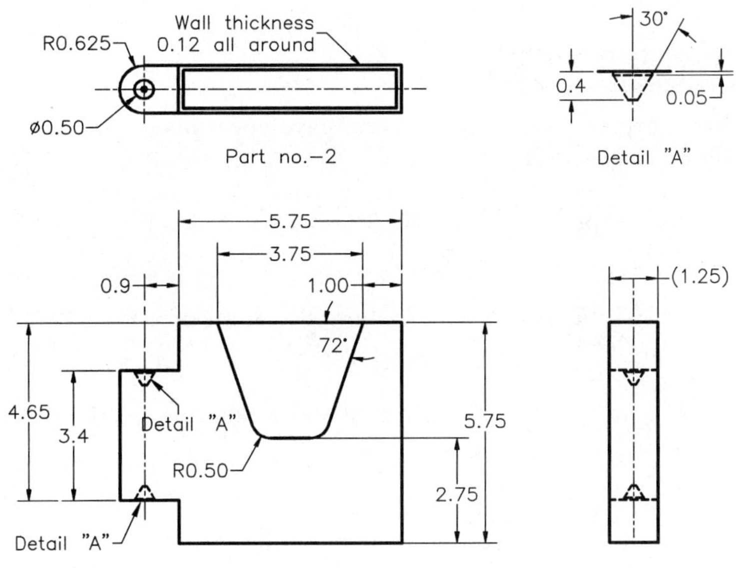

Figure 20-90 *Detail drawing (piece part drawing) of Part Number 2*

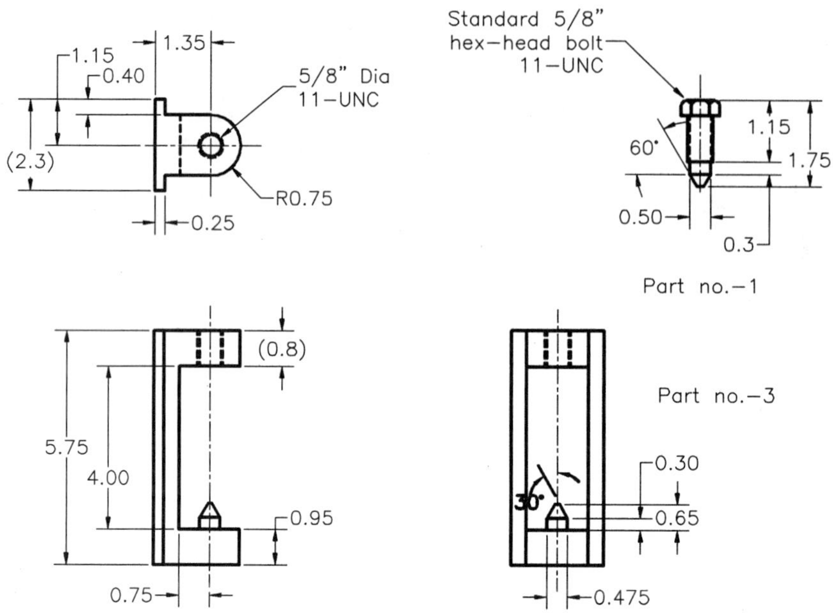

Figure 20-91 Detail drawings of Part Numbers 1 and 3

Self-Evaluation Test

Answer the following questions, and then compare your answers to the correct answers given at the end of this chapter.

1. If you draw a line perpendicular to the *X* and *Y* axes, this line defines the __________ axis.

2. The front view shows the maximum number of features or gives a better idea about the shape of the object. (T/F)

3. The number of decimal places in a dimension (for example, 2.000) determines the type of machine that will be used to do that machining operation. (T/F)

4. The dimension layer/layers should be assigned a unique color so that at the time of plotting you can assign a desired pen to plot the dimensions. (T/F)

5. All dimensions should be given inside the view. (T/F)

6. The dimensions can be shown inside the view if the dimensions can be easily understood there and cause no confusion with other object lines. (T/F)

7. The circle must be dimensioned as a(n) _________, never as a radius.

8. If you give continuous (incremental) dimensioning for dimensioning various features of a part, the overall dimension must be omitted or given as a(n) _________ dimension.

9. If the object is symmetrical, it is not necessary to draw a full section view. (T/F)

10. The removed section is similar to the _________section, except that it is shown outside the object.

11. In AutoCAD you can control the spacing between hatch lines by specifying the _________ at the time of hatching.

12. When dimensioning, you must consider the manufacturing process involved in making a part and the relationship that exists between different parts in an assembly. (T/F)

13. The distance between the first dimension line and the second dimension line must be _________ to _________units.

14. The reference dimension must be enclosed in _________.

15. A dimension must be given where the feature is visible. However, in some complicated drawings you might be justified to dimension a detail with a hidden line. (T/F)

Review Questions

Answer the following questions.

1. Multiview drawings are also known as _________ drawings.

2. The space between the *X* and *Y* axes is called the *XY* plane. (T/F)

3. A plane that is parallel to the *XY* plane is called a parallel plane. (T/F)

4. If you look at the object along the negative *Y* axis and toward the origin, you will get the side view. (T/F)

5. The top view must be directly below the front view. (T/F)

6. Before drawing orthographic views, you must look at the object and determine the number of views required to show all features of the object. (T/F)

7. By dimensioning, you are not only giving the size of a part, you are giving a series of instructions to a machinist, an engineer, or an architect. (T/F)

8. What are the components of a dimension? ______________________________ ______________________________.

9. Why should you make the dimensions in a separate layer/layers?

10. The distance of the first dimension line should be ________ to ________ units from the object line.

11. You can change grid, snap or UCS origin, and snap increments to make it easier to place the dimensions. (T/F)

12. For parallel dimension lines, the dimension text can be ________ if there is not enough room between the dimension lines to place the dimension text.

13. You should not dimension with hidden lines. (T/F)

14. A dimension that is not to scale should be indicated by drawing a straight line under the dimension text. (T/F)

15. The dimensions must be given where the feature that you are dimensioning is obvious and shows the contour of the feature. (T/F)

16. When dimensioning a circle, the diameter should be preceded by the ________.

17. When dimensioning an arc or a circle, the dimension line (leader) must be ________.

18. In radial dimensioning, you should place the dimension text vertically. (T/F)

19. If the dimension of a feature appears in a section view, you must hatch the dimension text. (T/F)

20. The dimensions must not be repeated; this makes it difficult to update a dimension, and the dimensions might get confusing. (T/F)

21. A bolt circle should be dimensioned by specifying the ________ of the bolt circle, the ________ of the holes, and the ________ of the holes in the bolt circle.

22. In the ________ section, the object is cut along the entire length of the object.

23. The part of the object where the material is actually cut is indicated by drawing ________ lines.

24. The _________ section is used to show the true shape of the object at the point where the section is cut.

25. The _________ section is used when the features of the object that you want to section are not in one plane.

26. Cutting plane lines are thinner than object lines. You can use the AutoCAD PLINE command to draw the polylines of desired width. (T/F)

27. The spacing between the hatch (section) lines is determined by the space being hatched. (T/F)

28. The angle for the hatch lines should be 45-degree. However, if there are two or more hatch areas next to one another, representing different parts, the hatch angle must be changed so that the hatched areas look different. (T/F)

29. Some parts, such as bolts, nuts, shafts, ball bearings, fasteners, ribs, spokes, keys, and other similar items that do not show any important feature, if sectioned, must be shown in section. (T/F)

30. Section lines (hatch lines) must be thicker than object lines. You can accomplish this by assigning a unique color to hatch lines and then assigning the color to that slot on the plotter carrying a pen with a thicker tip. (T/F)

31. The section lines must be drawn on a separate layer for _________ and _________ purposes.

32. In a broken section, only a _________ of the object is cut to _________ the features that need to be drawn in section.

33. The assembly drawing is used to show the _________ and their _________ in an assembled product or a machine unit.

34. Why shouldn't hidden lines be shown in an assembly drawing? ____________________ ____.

Exercises

Exercises 19 through 24 — *Mechanical*

Draw the required orthographic views of the following objects (the isometric view of each object is given in Figures 20-92 through 20-97). The dimensions can be determined by counting the number of grid lines. The distance between the isometric grid lines is assumed to be 0.5 units. Also dimension the drawings.

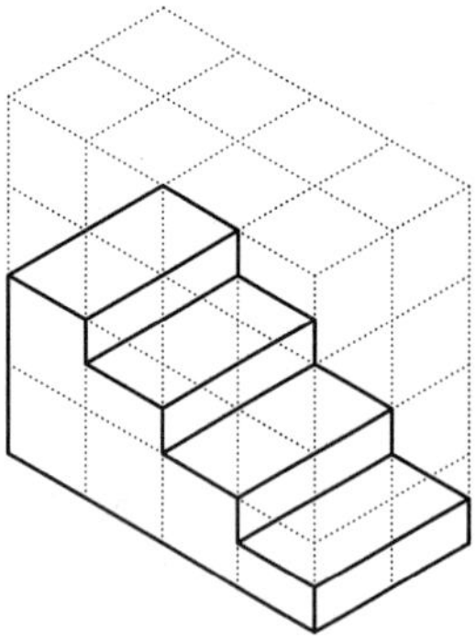

Figure 20-92 *Drawing for Exercise 19*

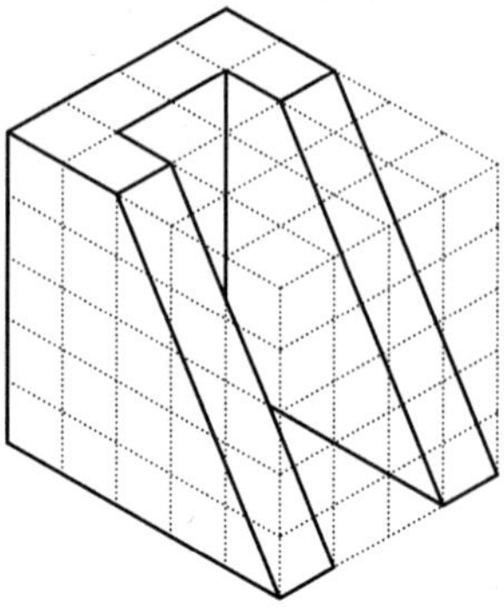

Figure 20-93 *Drawing for Exercise 20*

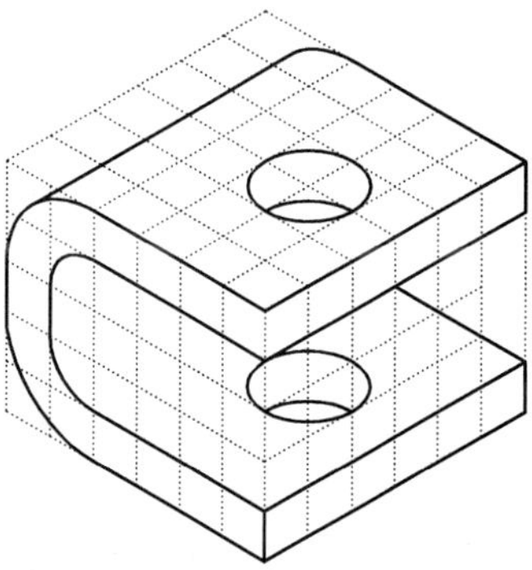

Figure 20-94 *Drawing for Exercise 21*

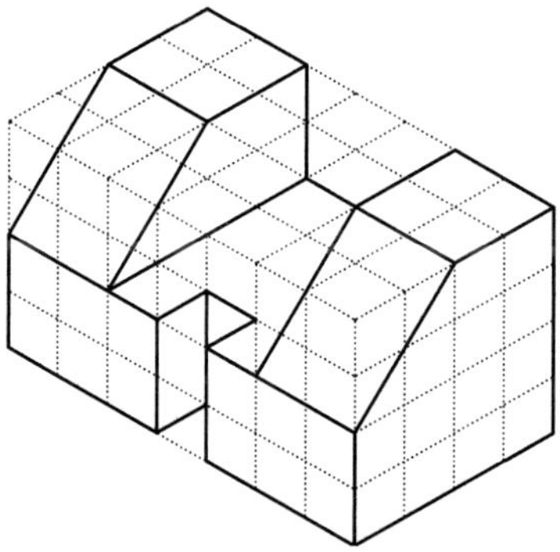

Figure 20-95 *Drawing for Exercise 22*

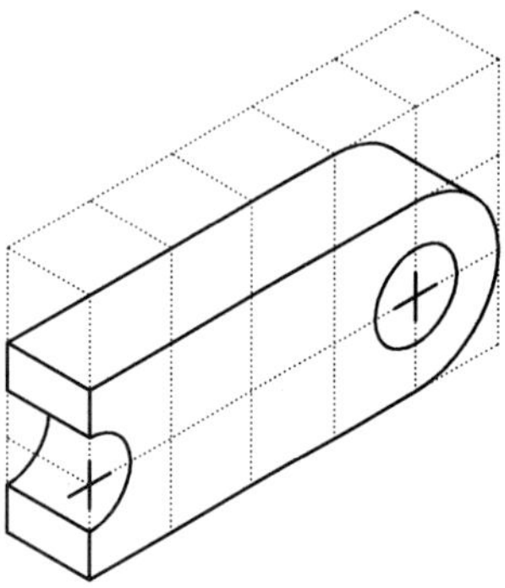

Figure 20-96 *Drawing for Exercise 23*

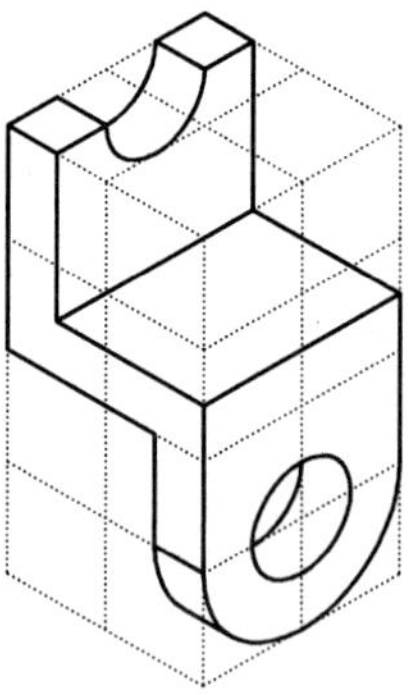

Figure 20-97 *Drawing for Exercise 24*

Exercise 25 *Electronics*

Draw the required orthographic views of the following object (Figures 20-98) and then give the dimensions. Fillet radius is 5mm.

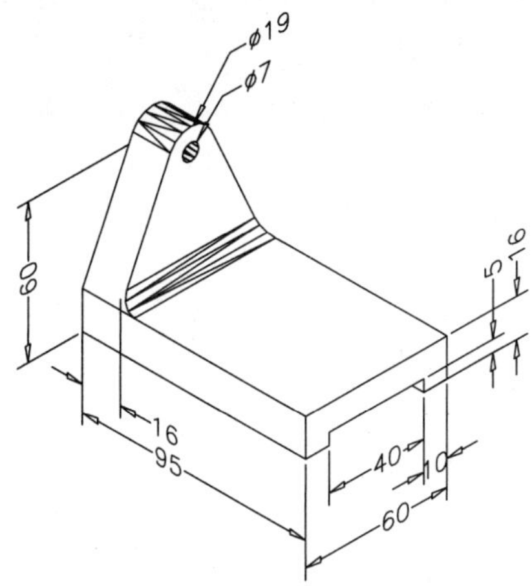

Figure 20-98 *Drawing for Exercise 25*

Exercises 26 *Mechanical*

Draw the required orthographic views of the following object (Figures 20-99) and then dimension the views. The fillet radius is 0.13.

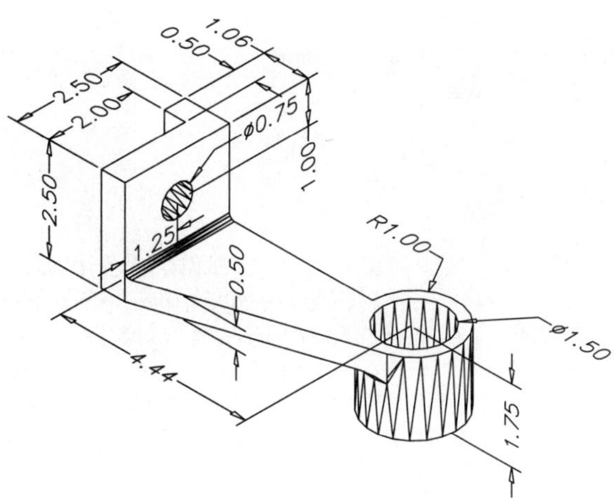

Figure 20-99 *Drawing for Exercise 26*

Answers to Self-Evaluation Test

1 - Z, **2** - T, **3** - T, **4** - T, **5** - F, **6** - T, **7** - diameter, **8** - reference, **9** - T, **10** - revolved, **11** - scale factor, **12** - T, **13** - 0.25 to 0.5, **14** - parentheses, **15** - T

Chapter 21

Isometric Drawing

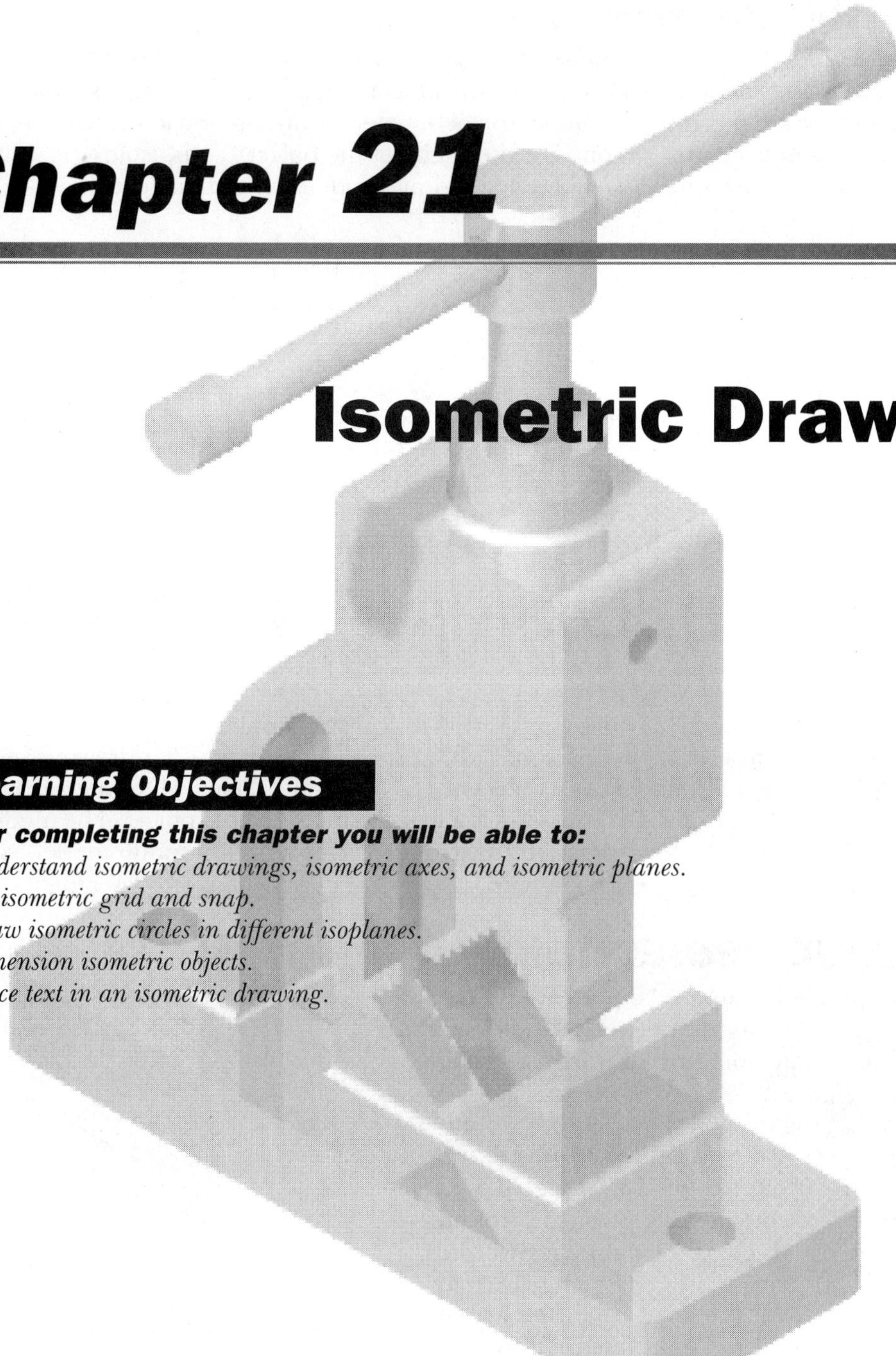

Learning Objectives

After completing this chapter you will be able to:

- *Understand isometric drawings, isometric axes, and isometric planes.*
- *Set isometric grid and snap.*
- *Draw isometric circles in different isoplanes.*
- *Dimension isometric objects.*
- *Place text in an isometric drawing.*

ISOMETRIC DRAWINGS

Isometric drawings are generally used to help visualize the shape of an object. For example, if you are given the orthographic views of an object (Figure 21-1), it takes time to put information together to visualize the shape. However, if an isometric drawing is given (Figure 21-2), it is much easier to conceive the shape of the object. Thus, isometric drawings are widely used in industry to help in understanding products and their features.

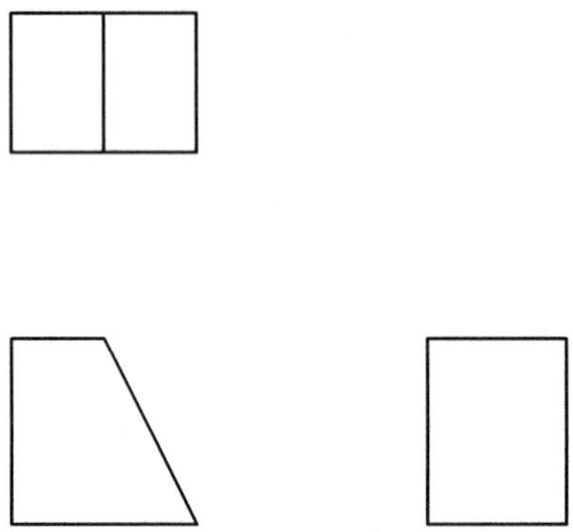

Figure 21-1 *Orthographic views of an object*

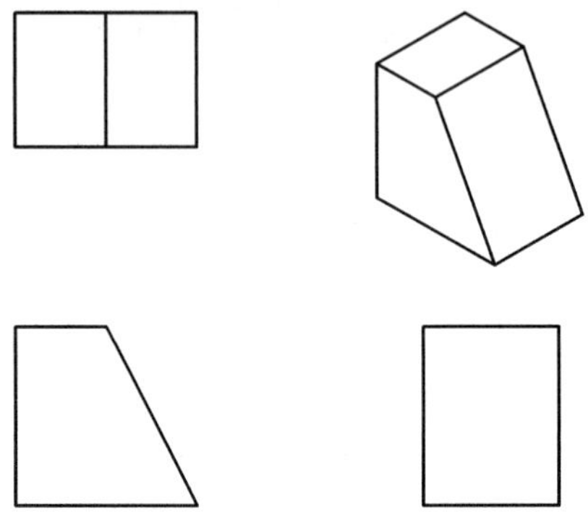

Figure 21-2 *Orthographic views with isometric drawing*

An isometric drawing should not be confused with a three-dimensional (3D) drawing. An isometric drawing is just a two-dimensional (2D) representation of a 3D drawing in a 2D plane. A 3D drawing is a 3D model of an object on the *X*, *Y*, and *Z* axes. An isometric drawing is a 2D drawing on a 2D plane. A 3D drawing is a true 3D model of the object. The model can be rotated and viewed from any direction. A 3D model can be a wireframe model, surface model, or solid model.

ISOMETRIC PROJECTIONS

The word isometric means "**equal measure**" because the three angles between the three principal axes of an isometric drawing are each 120-degree (Figure 21-3). An isometric view is obtained by rotating the object 45-degrees around the imaginary vertical axis, and then tilting the object forward through a 35°16' angle. If you project the points and the edges on the frontal plane, the projected length of the edges will be approximately 81 percent (isometric length/actual length = 9/11), which is shorter than the actual length of the edges. However, isometric drawings are always drawn to full scale because their purpose is to help the user visualize the shape of the object.

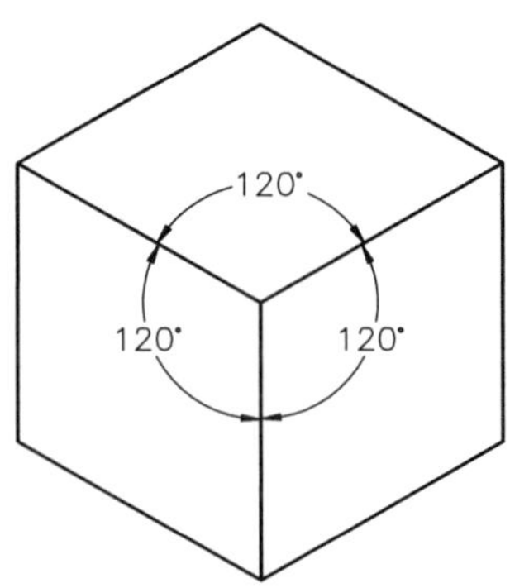

Figure 21-3 *Principal axes of an isometric drawing*

Isometric drawings are not meant to describe the actual size of the object. The actual dimensions, tolerances, and feature symbols must be shown in the orthographic views. Also, you should avoid showing any hidden lines in the isometric drawings, unless they show an important feature of the object or help in understanding the shape of the object.

ISOMETRIC AXES AND PLANES

Isometric drawings have three axes: **right horizontal axis** (P0,P1), **vertical axis** (P0,P2), and **left horizontal axis** (P0,P3). The two horizontal axes are inclined at 30-degree to the horizontal, or *X* axis (X1,X2). The vertical axis is at 90-degrees (Figure 21-4).

When you draw an isometric drawing, the horizontal object lines are drawn along or parallel to the horizontal axis. Similarly, the vertical lines are drawn along or parallel to the vertical axis. For example, if you want to make an isometric drawing of a rectangular block, the vertical edges of the block are drawn parallel to the vertical axis. The horizontal edges on the right side of the block are drawn parallel to the right horizontal axis (P0,P1), and the horizontal edges on the left side of the block are drawn parallel to the left horizontal axis (P0,P3). It is important to remember that the **angles do not appear true** in isometric drawings. Therefore, the edges or surfaces that are at an angle are drawn by locating the endpoints of the edges. The lines that are parallel to the isometric axes are called **isometric lines**. The lines that are not parallel to the isometric axes are called **nonisometric lines**.

Similarly, the planes can be **isometric planes** or **nonisometric planes**.

Isometric drawings have three principal planes, **isoplane right**, **isoplane top**, and **isoplane left**, as shown in Figure 21-5. The isoplane right (P0,P4,P10,P6) is defined by the vertical axis and the right horizontal axis. The isoplane top (P6,P10,P9,P7) is defined by the right and left horizontal axes. Similarly, the isoplane left (P0,P6,P7,P8) is defined by the vertical axis and the left horizontal axis.

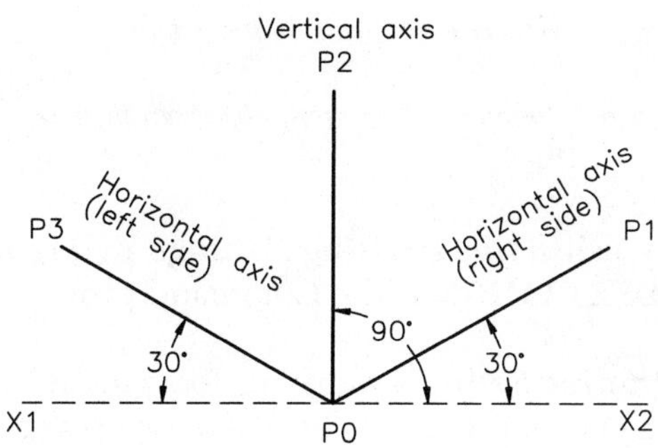

Figure 21-4 Isometric axes

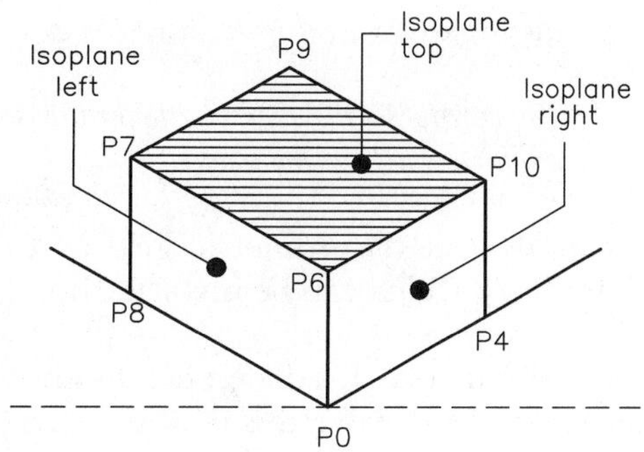

Figure 21-5 Isometric planes

SETTING THE ISOMETRIC GRID AND SNAP

You can use the **SNAP** command to set the isometric grid and snap. The isometric grid lines are displayed at 30-degrees to the horizontal axis. Also, the distance between the grid lines is determined by the vertical spacing, which can be specified by using the **GRID** or **SNAP** command. The grid lines coincide with the three isometric axes, which makes it easier to create isometric drawings. The following command sequence illustrates the use of the **SNAP** command to set the isometric grid and snap of 0.5 units (Figure 21-6).

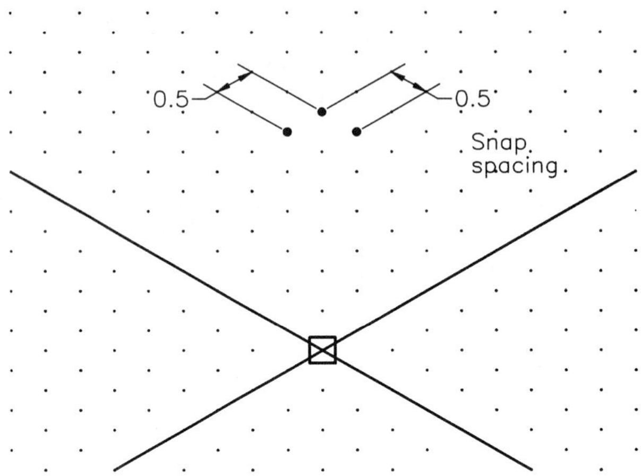

Figure 21-6 *Setting isometric grid and snap*

Command: **SNAP**
Specify snap spacing or [ON/OFF/Aspect/Rotate/Style/Type] <0.5000>: **S**
Enter snap grid style [Standard/Isometric] <S>: **I**
Specify vertical spacing <0.5000>: *Enter new snap distance.*

Note

*When you use the **SNAP** command to set the isometric grid, the grid lines may not be displayed. To display the grid lines, turn the grid on using the **GRID** command or press F7.*

You cannot set the aspect ratio for the isometric grid. Therefore, the spacing between the isometric grid lines will be the same.

You can also set the isometric grid and snap by using the **Drafting Settings** dialog box (Figure 21-7), which can be invoked by entering **DSETTINGS** at the Command prompt.

You can also invoke this dialog box by choosing **Drafting Settings** from the **Tools** menu. The other method to invoke this dialog box is by right-clicking **Snap**, **Grid**, **Polar,** or **Osnap**, on the status bar and choosing **Settings** from the shortcut menu.

The isometric snap and grid can be turned on/off by choosing the **On** box located in the **Snap and Grid/Object Snap** tabs of the **Drafting Settings** dialog box. The **Snap and Grid** tab also contains the radio buttons to set the snap type and style. To display the grid on the screen, make sure the grid is turned on.

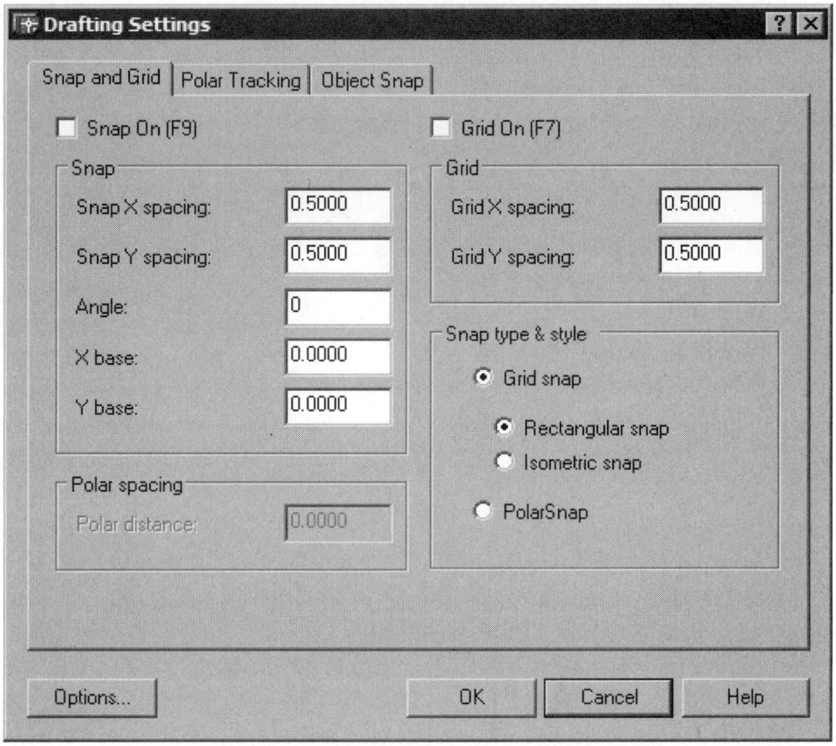

***Figure 21-7 Drafting Settings** dialog box*

When you set the isometric grid, the display of the crosshairs also changes. The crosshairs are displayed at an isometric angle, and their orientation depends on the current isoplane. You can toggle among isoplane right, isoplane left, and isoplane top by pressing the CTRL and E keys simultaneously (CTRL+E) or using the function key F5. You can also toggle among different isoplanes by using the **Drafting Settings** dialog box or by entering the **ISOPLANE** command at the Command prompt:

Command line: **ISOPLANE**
Enter isometric plane setting [Left/Top/Right] <Top>: **T**
Current Isoplane: **Top**

The Ortho mode is often useful when drawing in Isometric mode. In Isometric mode, Ortho aligns with the axes of the current isoplane.

Example 1 *Mechanical*

In this example, you will create the isometric drawing shown in Figure 21-8.

1. Use the **SNAP** command to set the isometric grid and snap. The snap value is 0.5 units.

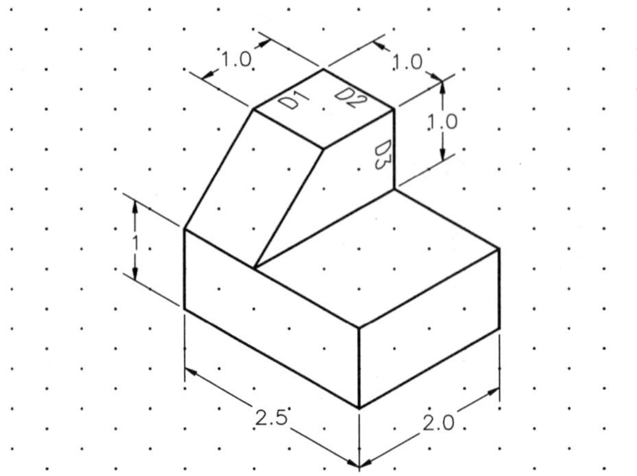

Figure 21-8 *Isometric drawing for Example 1*

Command: **SNAP**
Specify snap spacing or [ON/OFF/Aspect/Rotate/Style/Type] <0.5000>: **S**
Enter snap grid style [Standard/Isometric] <S>: **I**
Specify vertical spacing <0.5000>: **0.5** *(or press ENTER.)*

2. Change the isoplane to isoplane left by pressing the F5 key. Enter the **LINE** command and draw lines between points P1, P2, P3, P4, and P1 as shown in Figure 21-9.

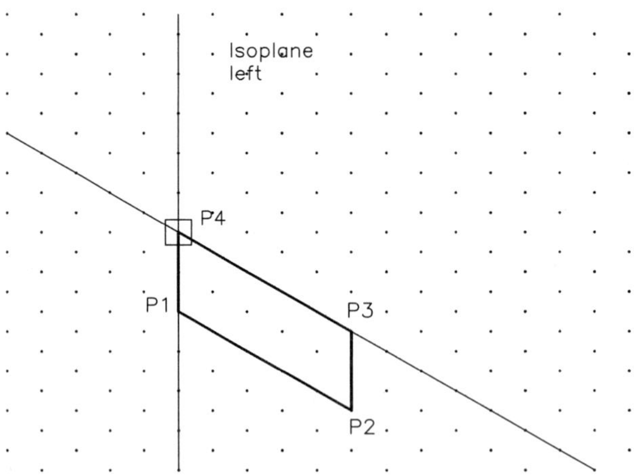

Figure 21-9 *Drawing the bottom left face*

3. Change the isoplane to isoplane right by pressing the F5 key. Invoke the **LINE** command and draw the lines as shown in Figure 21-10.

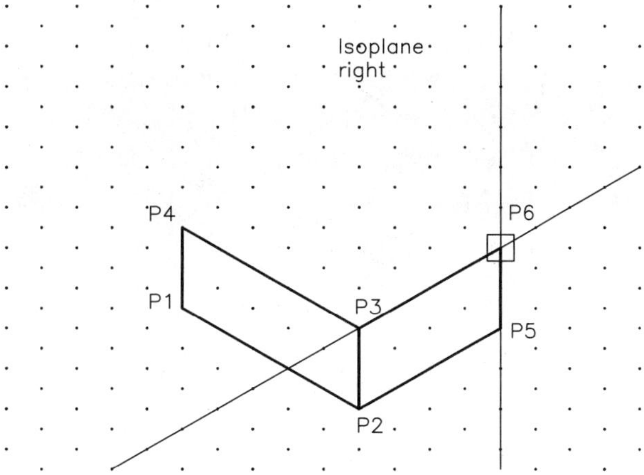

Figure 21-10 Drawing the bottom right face

4. Change the isoplane to isoplane top by pressing the F5 key. Invoke the **LINE** command and draw the lines as shown in Figure 21-11.

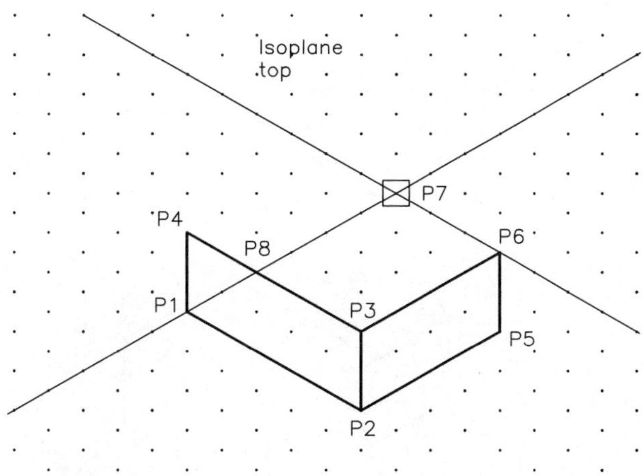

Figure 21-11 Drawing the top face

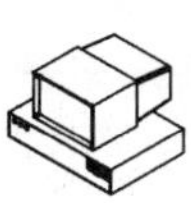

Tip

*You can increase the size of the crosshairs using the **Crosshair size** slider bar in the **Display** tab of the **Options** dialog box.*

5. Similarly, draw the remaining lines as shown in Figure 21-12.

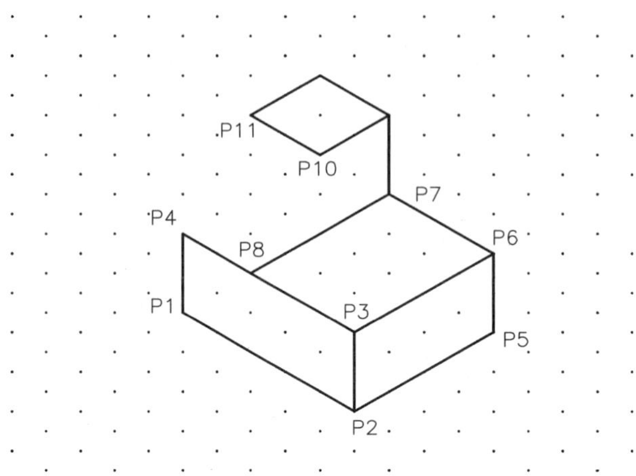

Figure 21-12 *Drawing the remaining lines*

6. The front left end of the object is tapered at an angle. In isometric drawings, oblique surfaces (surfaces at an angle to the isometric axis) cannot be drawn like other lines. You must first locate the endpoints of the lines that define the oblique surface and then draw lines between those points. To complete the drawing shown Figure 21-8, draw a line from P10 to P8 and from P11 to P4, see Figure 21-13.

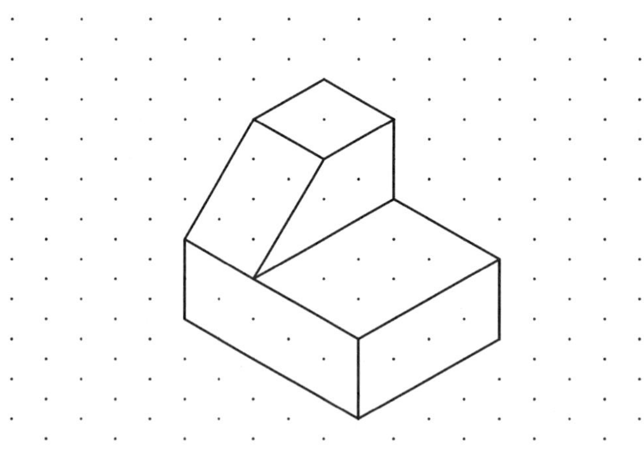

Figure 21-13 *Isometric drawing with the tapered face*

DRAWING ISOMETRIC CIRCLES

Isometric circles are drawn by using the **ELLIPSE** command and then selecting the **Isocircle** option. Before entering the radius or diameter of the isometric circle, you must make sure that you are in the required isoplane. For example, if you want to draw a circle in the right isoplane, you must toggle through the isoplanes until the required isoplane (right isoplane) is displayed. You can also set the required isoplane current before entering the **ELLIPSE** command. The crosshairs and the shape of the isometric circle will automatically change as you toggle through different isoplanes. As you enter the radius or diameter of the circle, AutoCAD draws the isometric circle in the selected plane, see Figure 21-14.

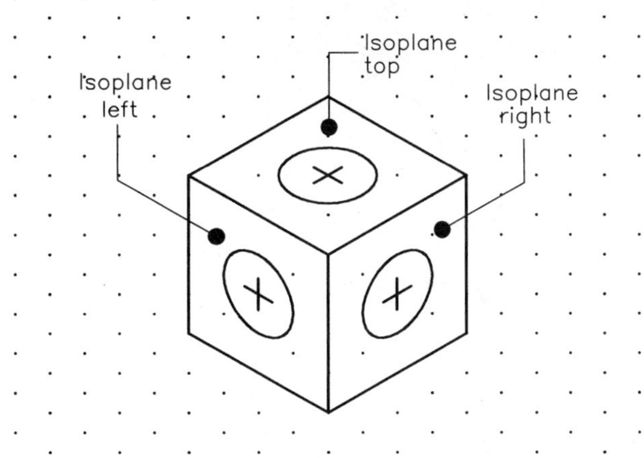

Figure 21-14 *Drawing isometric circles*

The prompt sequence to draw an isometric circle is:

Command: **ELLIPSE**
Specify axis endpoint of ellipse or [Arc/Center/Isocircle]: **I**
Specify center of isocircle: *Select a point.*
Specify radius of isocircle or [Diameter]: *Enter circle radius.*

Note

You must have the isometric snap on for the ***ELLIPSE*** *command to display the* ***Isocircle*** *option with the* ***ELLIPSE*** *command. If the isometric snap is not on, you cannot draw an isometric circle.*

Creating Fillets in the Isometric Drawings

To create fillets in isometric drawings, you need to first create an isometric circle and then trim the unwanted portion of the isometric circle. Remember that there is no method to directly create an isometric fillet.

DIMENSIONING ISOMETRIC OBJECTS

Isometric dimensioning involves two steps: (1) dimensioning the drawing using the standard dimensioning commands; (2) editing the dimensions to change them to oblique dimensions.

The following example illustrates the process involved in dimensioning an isometric drawing.

Example 2 *Mechanical*

In this example, you will dimension the isometric drawing created in Example 1.

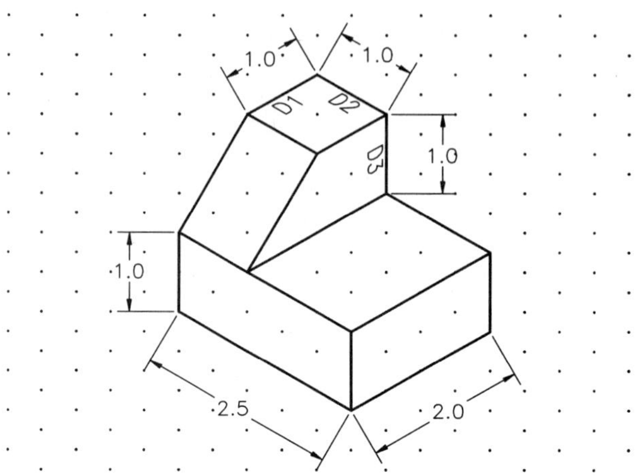

***Figure 21-15** Dimensioning an isometric drawing, before obliquing*

1. Dimension the drawing of Example 1 as shown in Figure 21-15. You can use the aligned or linear dimensions to dimension the drawing. Remember that when you select the points, you must use the Intersection or Endpoint object snap to snap the endpoints of the object you are dimensioning. AutoCAD automatically leaves a gap between the object line and the extension line as specified by the **DIMGAP** variable.

2. The next step is to edit the dimensions. First, enter the **DIM** command, and then enter **OBLIQUE** at the **Dim:** prompt or use the **Oblique** option of the **DIMEDIT** command. After selecting the dimension you want to edit, you are prompted to enter the obliquing angle. The obliquing angle is determined by the angle the extension line of the isometric dimension makes with the positive *X* axis.

   Command: **DIM**
   Dim: **OBLIQUE**
   Select object: *Select the dimension (D1).*
   Select object: *Press ENTER.*
   Enter obliquing angle (Press ENTER for none): **30**

For example, the extension line of the dimension labeled D1 makes a 150-degree angle with the positive *X* axis [Figure 21-16(a)]; therefore, the oblique angle is 150-degree. Similarly, the extension lines of the dimension labeled D2 make a 30-degrees angle with the positive *X* axis [Figure 21-16(b)]; therefore, the oblique angle is 30-degree. After you edit all dimensions, the drawing should appear as shown in Figure 21-17.

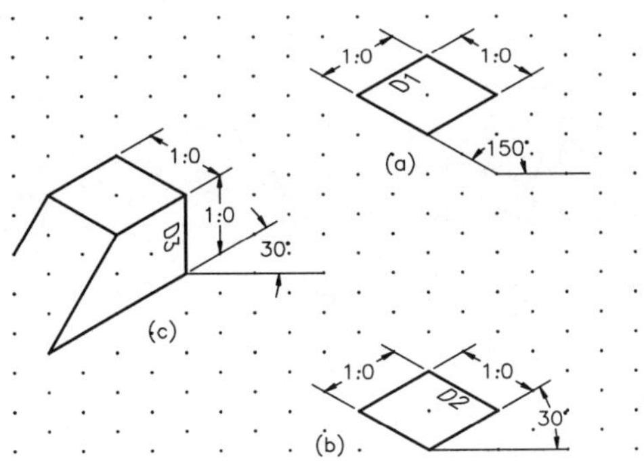

Figure 21-16 *Determining the oblique angle*

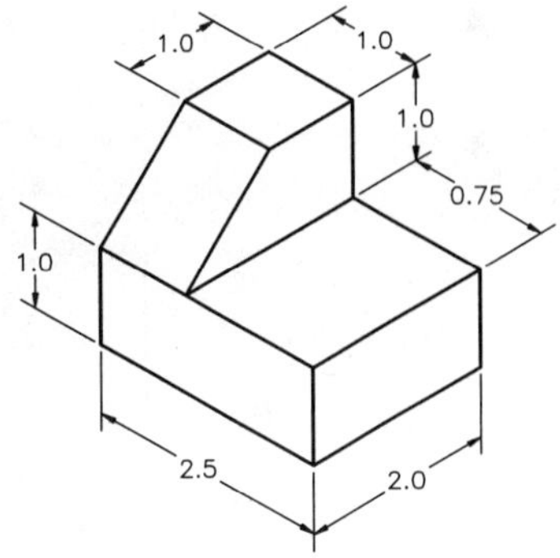

Figure 21-17 *Object with isometric dimensions*

ISOMETRIC TEXT

You cannot use regular text when placing text in an isometric drawing because the text in an isometric drawing is obliqued at positive or negative 30-degree. Therefore, you must create two text styles with oblique angles of 30-degree and negative 30-degree. You can use the **-STYLE** command or the **Text Style** dialog box to create a new text style as described here. (For more details, refer to "Creating Text Styles" in Chapter 7.)

Command: **-STYLE**
Enter name of text style or [?] <Standard>: **ISOTEXT1**
Specify full font name or font filename (TTF or SHX) <txt>: **ROMANS**
Specify height of text <0.0000>: **0.075**
Specify width factor <1.0000>: *Press ENTER.*
Specify obliquing angle <0>: **30**
Display text backwards? [Yes/No] <N>: **N**
Display text upside-down? [Yes/No] <N>: **N**
Vertical? <N> **N**

Similarly, you can create another text style, **ISOTEXT2**, with a negative 30-degree oblique angle. When you place the text in an isometric drawing, you must also specify the rotation angle for the text. The text style you use and the text rotation angle depend on the placement of the text in the isometric drawing, as shown in Figure 21-18.

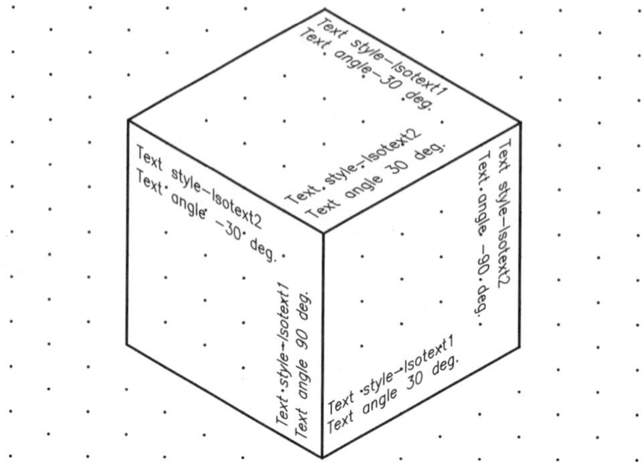

Figure 21-18 *Text style and rotation angle for isometric text*

Self-Evaluation Test

Answer the following questions, and then compare your answers with the correct answers given at the end of this chapter.

1. The word isometric means "__________" because the three angles between the three principal axes of an isometric drawing are each __________ degrees.

2. The ratio of isometric length to actual length in an isometric drawing is approximately __________.

3. The angle between the right isometric horizontal axis and the *X* axis is __________ degrees.

4. Isometric drawings have three principal planes: isoplane right, isoplane top, and __________.

5. What key combination or function key can you use to toggle among isoplane right, isoplane left, and isoplane top? __________.

6. You can only use the aligned dimension option to dimension an isometric drawing. (T/F)

7. Must the isometric snap be on to display the **Isocircle** option with the **ELLIPSE** command? (Y/N)

8. Do you need to specify the rotation angle when placing text in an isometric drawing? (Y/N). If yes, what are the possible angles? _________.

9. The lines that are not parallel to the isometric axes are called _________.

10. You should avoid showing any hidden lines in isometric drawings. (T/F)

Review Questions

Answer the following questions.

1. Isometric drawings are generally used to help in _________ the shape of an object.

2. An isometric view is obtained by rotating the object _________ degree around the imaginary vertical axis, and then tilting the object forward through a _________ angle.

3. If you project the points and the edges onto the frontal plane, the projected length of the edges will be approximately _________ percent shorter than the actual length of the edges.

4. When should hidden lines be shown in an isometric drawing?

5. Isometric drawings have three axes: right horizontal axis, vertical axis, and _________.

6. The angles do not appear true in isometric drawings. (T/F)

7. What are the lines called that are parallel to the isometric axis? _________.

8. What commands can you use to set the isometric grid and snap? _________.

9. Isometric grid lines are displayed at _________ degrees to the horizontal axis.

10. It is possible to set the aspect ratio for the isometric grid. (T/F)

11. You can also set the isometric grid and snap by using the **Drafting Settings** dialog box, which can be invoked by entering _________ at the Command prompt.

12. Isometric circles are drawn by using the **ELLIPSE** command and then selecting the _________ option.

13. Can you draw an isometric circle without turning the isometric snap on? (Yes/No)

14. Only aligned dimensions can be edited to change them to oblique dimensions. (T/F)

15. To place the text in an isometric drawing, you must create two text styles with oblique angles of __________ degrees and negative __________ degrees.

Exercises

Exercises 1 through 6 — *General*

Draw the following isometric drawings (Figures 21-19 through 21-24). The dimensions can be determined by counting the number of grid lines. The distance between the isometric grid lines is assumed to be 0.5 units. Dimension the odd-numbered drawings.

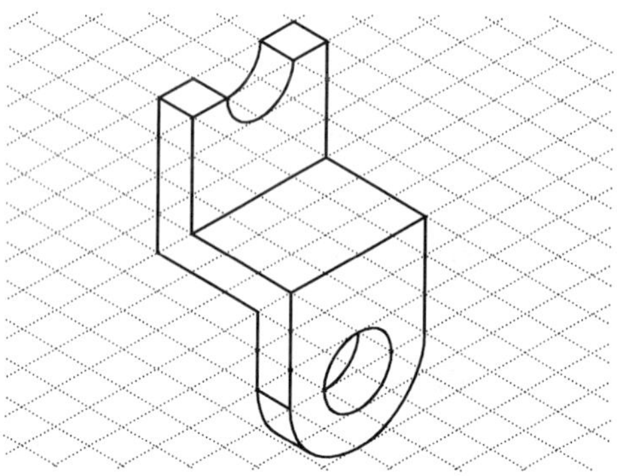

Figure 21-19 *Drawing for Exercise 1*

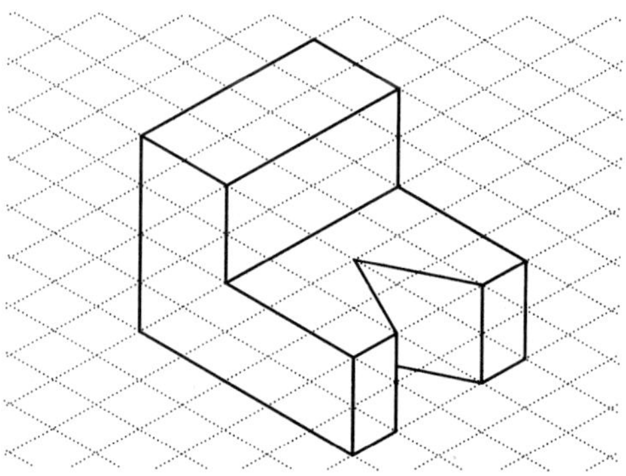

Figure 21-20 *Drawing for Exercise 2*

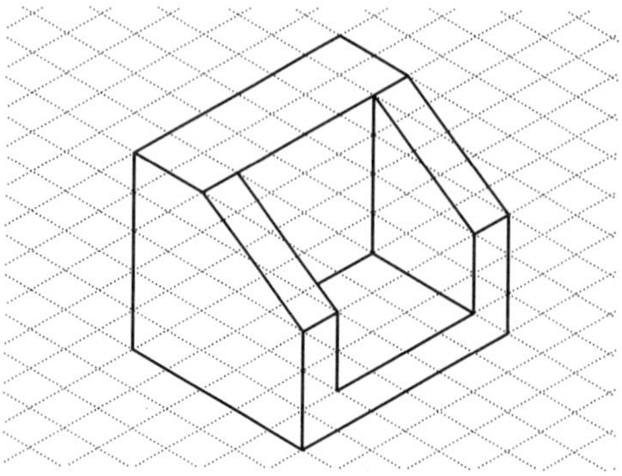

Figure 21-21 *Drawing for Exercise 3*

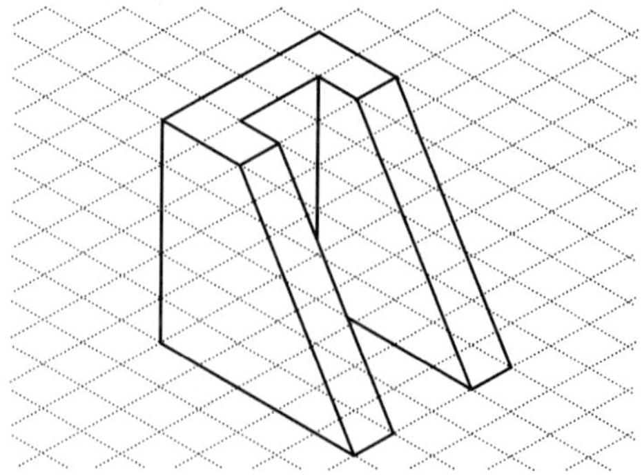

Figure 21-22 *Drawing for Exercise 4*

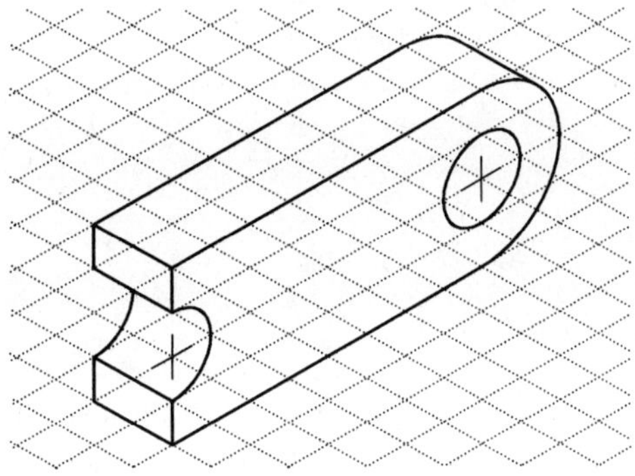

Figure 21-23 Drawing for Exercise 5

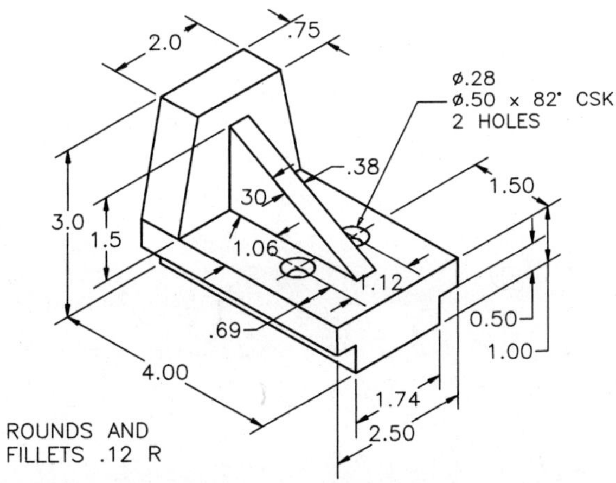

Figure 21-24 Drawing for Exercise 6

Answers to Self-Evaluation Test

1 - equal measure, 120, **2** - 9/11, **3** - 30, **4** - isoplane left, **5** - CTRL+E, or F5, **6** - F, **7** - Yes, **8** - Yes, 30-degree, -30-degree, 90-degree, etc., **9** - nonisometric lines, **10** - T

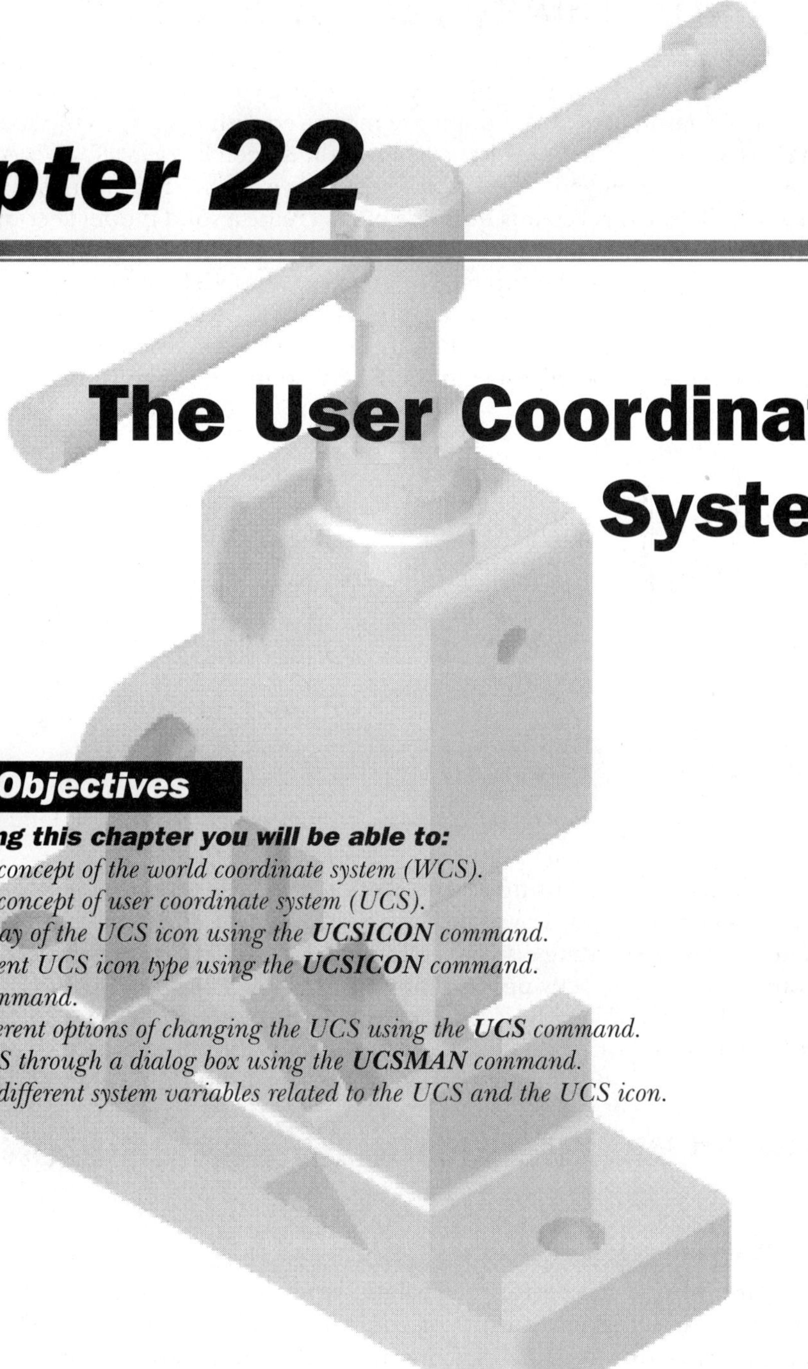

Chapter 22

The User Coordinate System

Learning Objectives

After completing this chapter you will be able to:
- *Understand the concept of the world coordinate system (WCS).*
- *Understand the concept of user coordinate system (UCS).*
- *Control the display of the UCS icon using the* ***UCSICON*** *command.*
- *Change the current UCS icon type using the* ***UCSICON*** *command.*
- *Use the* ***UCS*** *command.*
- *Understand different options of changing the UCS using the* ***UCS*** *command.*
- *Manage the UCS through a dialog box using the* ***UCSMAN*** *command.*
- *Understand the different system variables related to the UCS and the UCS icon.*

THE WORLD COORDINATE SYSTEM (WCS)

As discussed earlier, when you start a new AutoCAD drawing, by default, the world coordinate system (WCS) is established. The objects you have drawn until now use the WCS. In the WCS, the *X*, *Y*, and *Z* coordinates of any point are measured with respect to the fixed origin (0,0,0). This origin is located at the lower left corner of the screen by default. This coordinate system is fixed and cannot be moved. The WCS is generally used in 2D drawings, wireframe models, and surface models. However, it is not possible to create a solid model keeping the origin and the orientation of the *X*, *Y*, and *Z* axes at the same place. The reason for this is that in case you want to create a feature on the top face of an existing model, you can do it easily by shifting the working plane using the **Elevation** option of the **ELEV** command (Figure 22-1). But in case you want to create a feature on the faces other than the top and the bottom faces of an existing model, it is not possible using the **ELEV** command (Figure 22-2).

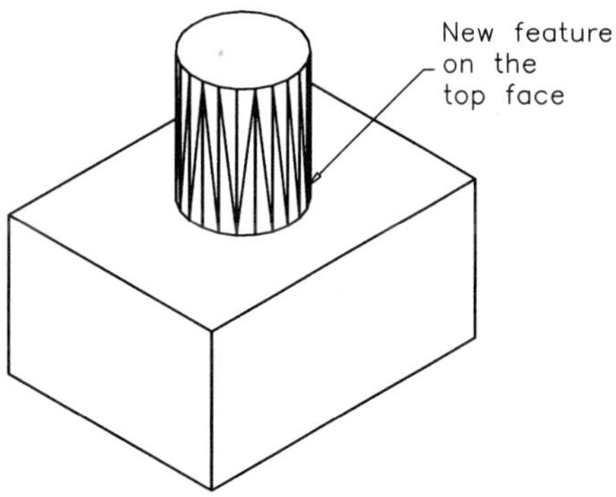

Figure 22-1 *Creating a feature on the top face*

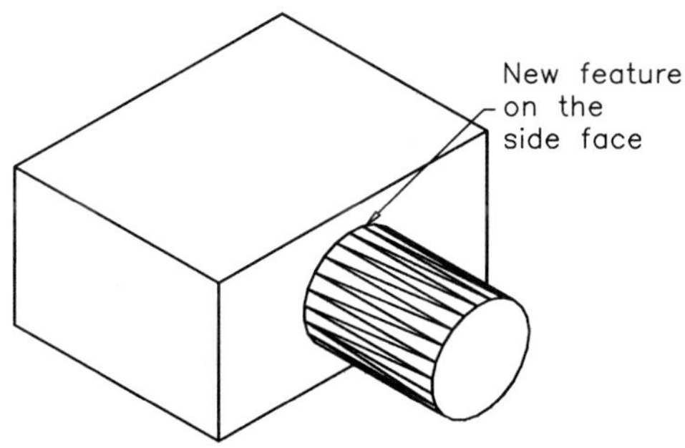

Figure 22-2 *Creating a feature on the side face*

This problem can be solved using a concept called the user coordinate system (UCS). Using the UCS, you can relocate and reorient the origin and *X*, *Y*, and *Z* axes and establish your own coordinate system, depending on your requirement. The UCS is mostly used in 3D drawings, where you may need to specify points that vary from each other along the *X*, *Y*, and *Z* axes. It is also useful for relocating the origin or rotating the *X* and *Y* axes in 2D work, such as ordinate dimensioning, drawing auxiliary views, or controlling hatch alignment. The UCS and its icon can be modified using the **UCSICON** and the **UCS** commands.

CONTROLLING THE VISIBILITY OF UCS ICON

Menu:	View > Display > UCS Icon
Command:	UCSICON

This command is used to control the visibility and the location of the UCS icon. The UCS icon is a geometric representation of the directions of the current *X*, *Y*, and *Z* axes. AutoCAD displays different UCS icons in model space and paper space as shown in Figures 22-3 and 22-4. By default, the UCS icon is displayed near the bottom left corner of the drawing area. You can change the location and visibility of this icon using the **UCSICON** command.

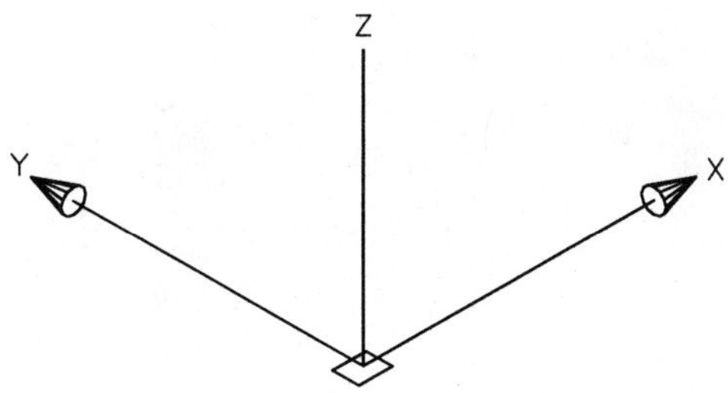

Figure 22-3 Model space UCS icon

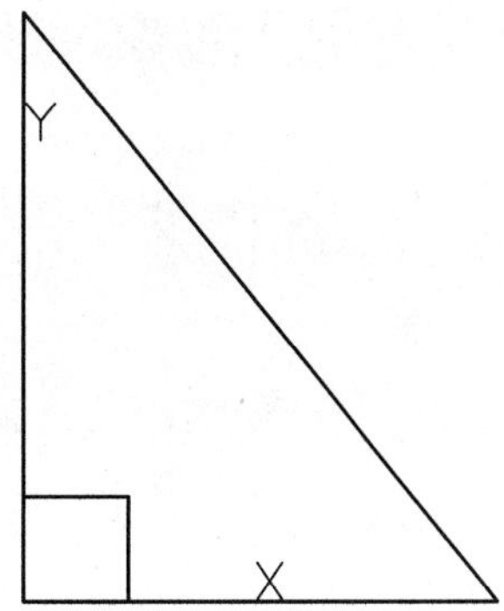

Figure 22-4 Paper space UCS icon

The following prompt sequence is issued when you invoke this command.

Enter an option [ON/OFF/All/Noorigin/ORigin/Properties] <ON>:

ON

This option is used to display the UCS icon on the screen.

OFF

This option is used to make the UCS icon invisible from the screen. When you invoke this option, the UCS icon will no longer be displayed on the screen. You can again turn on the display of the UCS icon using the **On** option of the **UCSICON** command.

All

This option is used to apply the changes to UCS icons in all the active viewports. If this option is not used, the changes will be applied only to the current viewport.

Noorigin

This option is used to display the UCS icon at the lower left corner of the viewport, irrespective of the actual location of the origin of the current UCS.

ORigin

This option is used to place the UCS icon at the origin of the current UCS.

Properties

When you invoke this command, the **UCS Icon** dialog box will be displayed as shown in Figure 22-5. The options provided in this dialog box are discussed next.

UCS icon style Area

2D. If this radio button is selected, the 2D UCS icon will be displayed on the screen instead of the 3D UCS icon, see Figure 22-6.

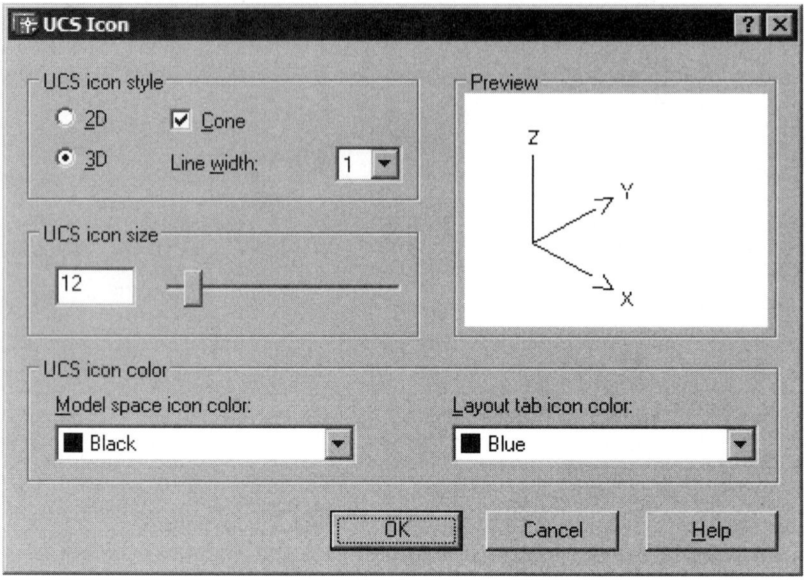

Figure 22-5 The ***UCS Icon*** *dialog box*

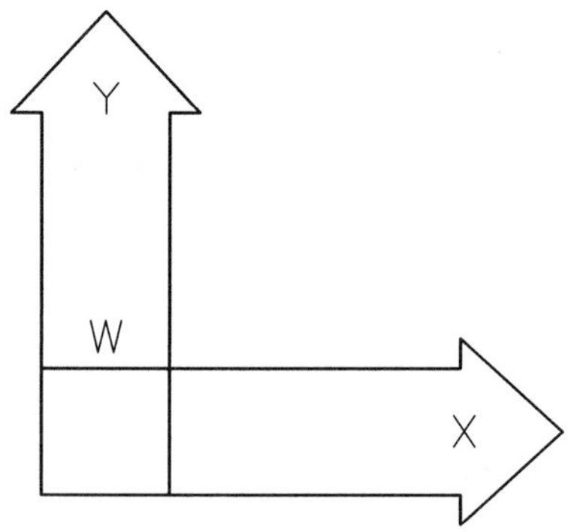

Figure 22-6 2D UCS icon at the World position

3D. If this radio button is selected, the 3D UCS icon will be displayed on the screen. This is the default option and AutoCAD displays the 3D UCS icon by default.

Cone. If this check box is cleared, the cones at the end of the *X* and *Y* axes of the 3D UCS icon will not be displayed. Instead, the arrows will be displayed. This option is not available if you select the **2D** radio button.

Line width. This drop-down list provides the width values that can be assigned to the 3D UCS icon. The default value for the line width is **1**. This drop-down list will not be available if the **2D** radio button is selected.

UCS icon size Area

The slider bar provided in this area is used to increase the size of the UCS icon. You can also enter the value of the size of the UCS icon directly in the edit box, but the value of the UCS icon size can vary between **5** and **95**. The default value of the size of the UCS icon is **12**.

UCS icon color Area

This area provides the drop-down lists that are used to change the color of the UCS icon in the **Model** space as well in the **Layout** tab. By default the color in the **Model** space is black and in the **Layout** is blue. You can assign any desired colors to the UCS icon. By default there are seven colors that are available in these drop-down lists. However, you can also select the desired color from the **Select Color** dialog box. This dialog box will be displayed if you select **More** from the **Model space icon color** or the **Layout tab icon color** drop-down list.

DEFINING NEW UCS

Toolbar:	UCS
Menu:	Tools > New UCS
Command:	UCS

The **UCS** command is used to set a new coordinate system by shifting the working plane (*XY* plane) to the desired location, see Figure 22-7. For certain views of the drawing it is better to have the origin of measurements at some other point on or relative to your drawing objects. This makes locating the features and dimensioning the objects easier. The change in the UCS can be viewed by the change in the position and orientation of the UCS icon, which is by default placed at the lower left corner of the drawing window. The origin and orientation of a coordinate system can be redefined using the **UCS** command. The prompt sequence is as follows.

Figure 22-7 *The* ***UCS*** *toolbar*

Command: **UCS**
Current ucs name: *WORLD*
Enter an option [New/Move/orthoGraphic/Prev/Restore/Save/Del/Apply/?/World]<World>: *Select an option.*

If the **UCSFOLLOW** system variable is set to 0, any change in the UCS does not affect the drawing view.

W (World) Option

With this option, you can set the current UCS back to the WCS, which is the default position. When the UCS is placed at the world position, a small rectangle is displayed at the point where all three axes meet in the UCS icon, see Figure 22-8. If the UCS is moved from its default position, this rectangle is no longer displayed, indicating that the UCS is not at the world position as shown in Figure 22-9.

Tip

In case you have selected the 2D UCS instead of the 3D UCS icon, the W will not be displayed if the UCS is not at the world position.

N (New) Option

The **New** option is used to define a new UCS by using various sub-options provided under this option. The prompt sequence is given next.

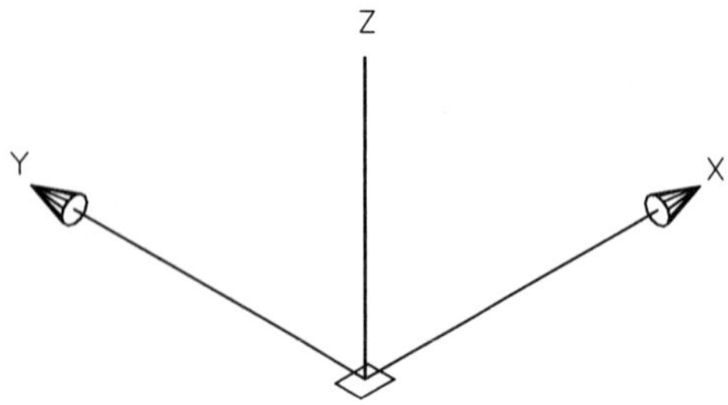

Figure 22-8 *UCS at the world position*

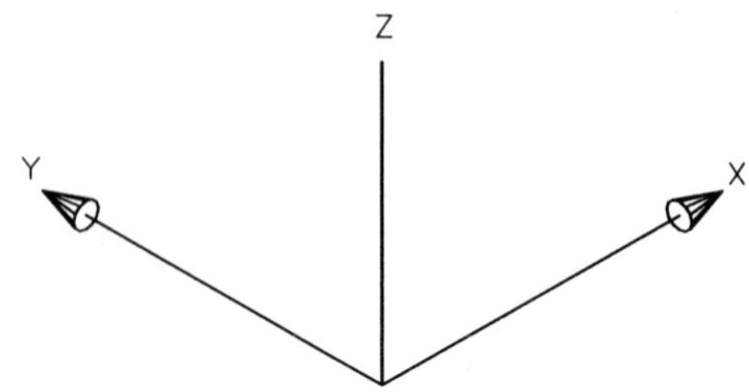

Figure 22-9 *UCS not at the world position*

Command: **UCS**
Enter an option [New/Move/orthoGraphic/Prev/Restore/Save/Del/Apply/?/World]<World>: **N**
Specify origin of new UCS or [ZAxis/3point/OBject/Face/View/X/Y/Z] <0,0,0>:

O (Origin) Option

This option is used to define a new UCS by changing the origin of the current UCS, see Figures 22-10 and 22-11. The directions of the *X*, *Y*, and *Z* axes remain unaltered. The new point defined will now be the origin point (0,0,0) for all the coordinate entries from this point on, until the origin is changed again. You can specify the coordinates of the new origin or simply pick it on the screen using the pointing device. The prompt sequence that will follow when you choose this button is given next.

Specify new origin point <0,0,0>: *Specify the origin point as shown in Figure 22-10.*

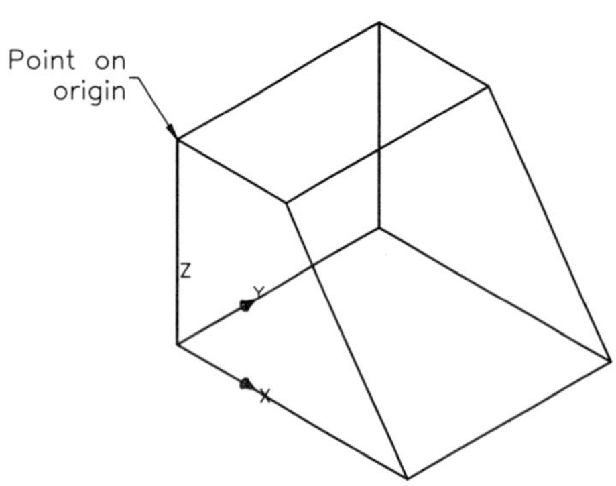

Figure 22-10 *Defining a new origin for the UCS*

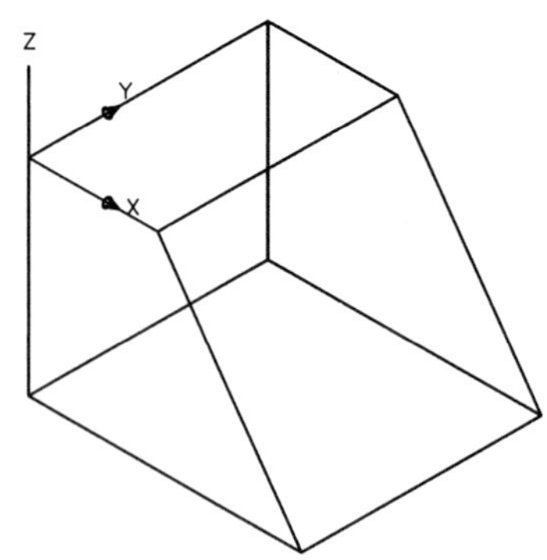

Figure 22-11 *Relocating UCS to the new origin*

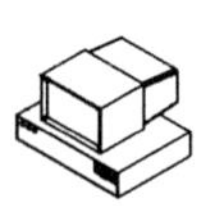

Tip

If you do not provide a Z coordinate for the origin, the Z coordinate is assigned the current elevation value.

Note

*The **Origin**, **ZAxis**, **3points**, **OBject**, **Face**, **View**, **X**, **Y**, and **Z** options can also be invoked by entering **N** at the **Enter an option [New/Move/orthoGraphic/Prev/Restore/Save/Del/Apply/ ?/World] <World>:** prompt. You can also invoke these options by directly entering them at the prompt sequence mentioned earlier.*

ZA (ZAxis) Option

This option is used to change the coordinate system by selecting the origin point of the *XY* plane and a point on the positive *Z* axis. After you specify a point on the *Z* axis, AutoCAD determines the *X* and *Y* axes of the new coordinate system accordingly. The prompt sequence that will follow when you choose this button is given next.

Specify new origin point <0,0,0>: *Specify the origin point as shown in Figure 22-12.*
Specify point on positive portion of Z-axis <default>: **@0,-1,0**

The front face of the model will now become the new work plane (Figure 22-13) and all the new objects will be oriented accordingly. If you give a null response for the **Specify point on the positive portion of Z axis <current>** prompt, the *Z* axis of the new coordinate system will be parallel to (in the same direction as) the *Z* axis of the previous coordinate system. Null responses to the origin point and the point on the positive *Z* axis establish a new coordinate system in which the direction of the *Z* axis is identical to that of the previous coordinate system; however, the *X* and *Y* axes may be rotated around the *Z* axis. The positive *Z* axis direction is also known as the extrusion direction.

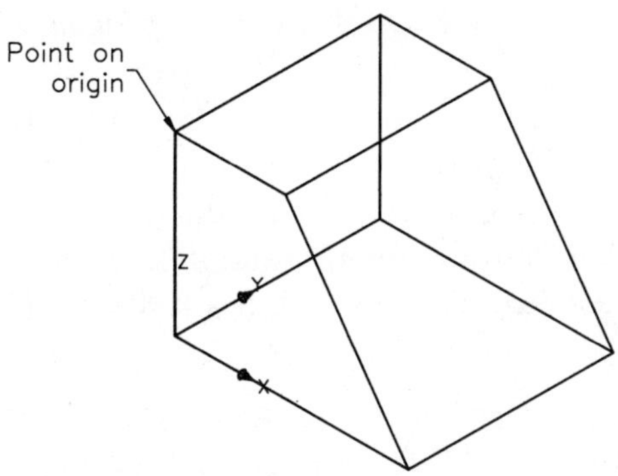

***Figure 22-12** Specifying point on origin*

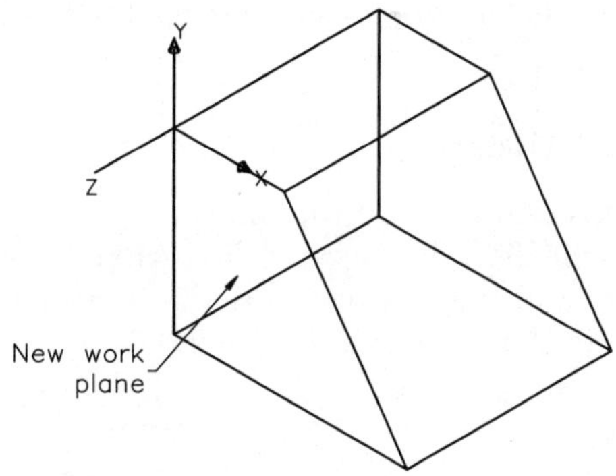

***Figure 22-13** Relocating the UCS using the **ZA** option*

3 (3point) Option

With this option you can establish a new coordinate system by specifying a new origin point, a point on the positive side of the new *X* axis, and a point on the positive side of the new *Y* axis (Figure 22-14). The direction of the *Z* axis is determined by applying the right-hand rule, about which you will learn in the next chapter. The **3point** option of the **UCS** command changes the orientation of the UCS to any angled surface. The prompt sequence that will follow when you choose this button is given next.

Specify new origin point <0,0,0>: *Specify the origin point of the new UCS.*
Specify point on positive portion of X-axis <1.0000,0.0000,0.0000>: *Specify a point on the positive portion of the X axis.*
Specify point on positive-Y portion of the UCS XY plane <0.0000,1.0000,0.0000>: *Specify a point on the positive portion of the Y axis.*

A null response to the **Specify new origin point <0,0,0>** prompt will lead to a coordinate system in which the origin of the new UCS is identical to that of the previous UCS. Similarly, null responses to the point on the *X* or *Y* axis prompt will lead to a coordinate system in which the *X* or *Y* axis of the new UCS is parallel to that of the previous UCS. In Figure 22-15, the UCS has been relocated by specifying three points (the origin point, a point on the positive portion of the *X* axis, and a point on the positive portion of the *Y* axis).

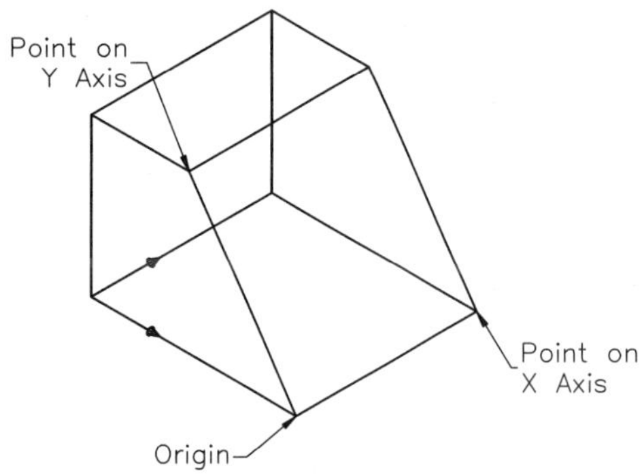

Figure 22-14 *Relocating the UCS using 3 points*

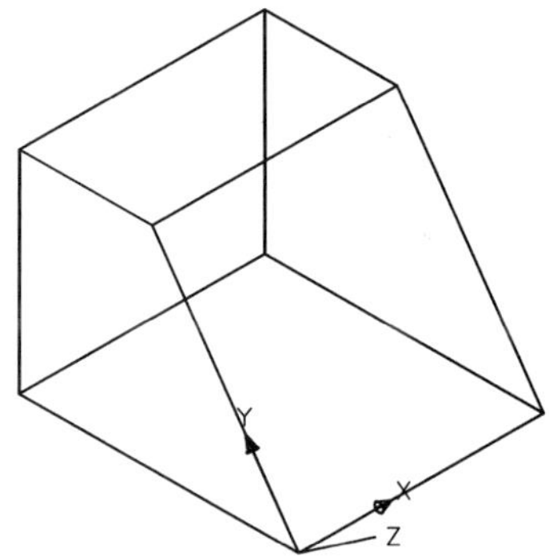

Figure 22-15 *UCS at new position*

Example 1

In this example you will draw a tapered rectangular block. After drawing the tapered block you will align the UCS on the inclined face of the block using the **3point** option of the **UCS** command. Then you will draw a circle on the inclined face. The dimensions for the block and the circle are given in Figure 22-16.

This example can be divided into the following steps.

- Draw the edges on the bottom face of the tapered block.
- Draw the edges on the top face of the tapered block.
- Draw the remaining edges of the block.
- Align the UCS on the inclined face.
- Draw the circle on the inclined face.

These steps are discussed next.

1. Open a new drawing and choose the **Line** button from the **Draw** toolbar. The prompt sequence is as follows:

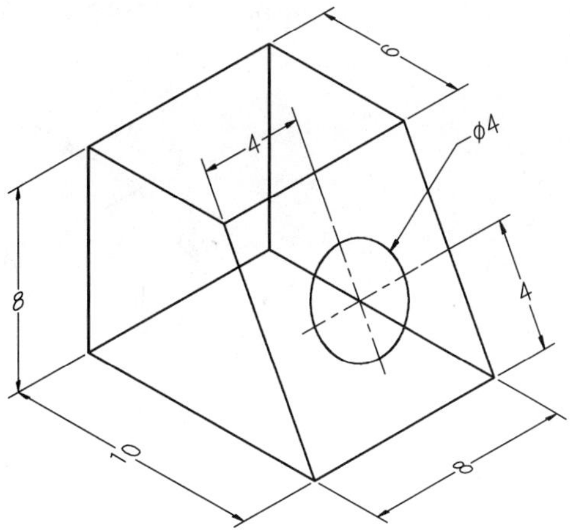

Figure 22-16 *Model for Example 1*

Specify first point: **1,1**
Specify next point or [Undo]: **@10,0**
Specify next point or [Undo]: **@0,8**
Specify next point or [Close/Undo]: **@-10,0**
Specify next point or [Close/Undo]: **C**

2. The top face of the model is at a distance of 8 units from the bottom face, in the positive *Z* direction. Therefore, you will have to define a UCS at a distance of 8 units to create the top face. This can be done by choosing the **Origin UCS** button from the **UCS** toolbar. The prompt sequence is as follows:

   Specify new origin point <0,0,0>: **0,0,8**

3. Choose the **Line** button from the **Draw** toolbar. The prompt sequence is as follows:

   Specify first point: **1,1**
   Specify next point or [Undo]: **@6,0**
   Specify next point or [Undo]: **@0,8**
   Specify next point or [Close/Undo]: **@-6,0**
   Specify next point or [Close/Undo]: **C**

4. When you open a new drawing, by default you view the model from the top view. Therefore, when viewing from the top, the three edges of the top side overlap with the corresponding bottom edges. Hence, you will not be able to see them. However, you can change the viewpoint to clearly view the model in 3D. Choose the **SE Isometric View** button from the **View** toolbar to proceed to the SE isometric view. You can now see the 3D shape of the objects, see Figure 22-17.

5. Join the remaining edges of the model using the **Line** button. The model after joining all the edges should look similar to the one shown in Figure 22-18.

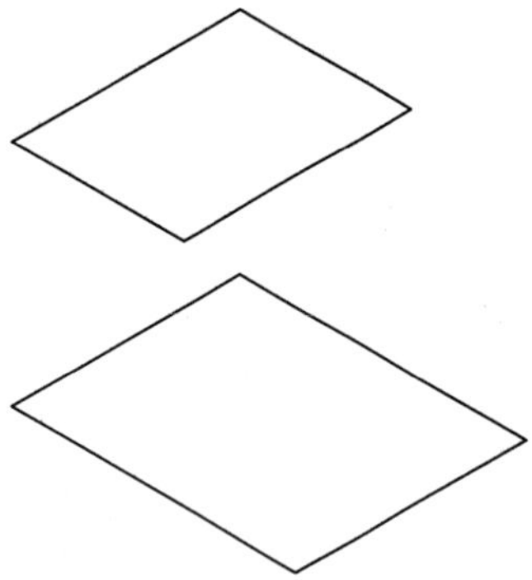

Figure 22-17 *3D view of the objects*

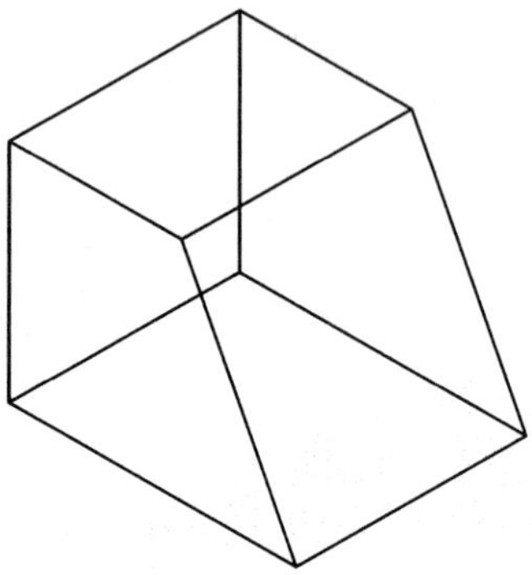

Figure 22-18 *Model after joining all the edges*

6. Now, you have to draw the circle at the inclined face. But to draw the circle at the inclined face, you will have to make the inclined face as the current working plane. Therefore, choose the **3 Point UCS** button from the **UCS** toolbar. The prompt sequence is as follows.

   Specify new origin point <0,0,0>: *Select point P1 as shown in Figure 22-19.*
   Specify point on positive portion of X-axis <default>: *Select point P2 as shown in Figure 22-19.*
   Specify point on positive-Y portion of the UCS XY plane <default>: *Select point P3 as shown in Figure 22-19.*

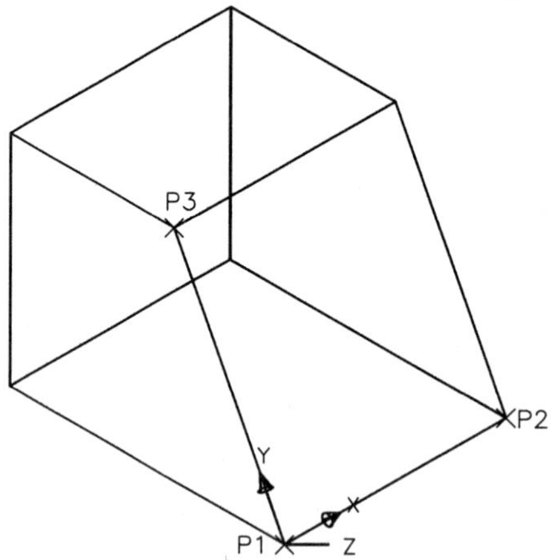

Figure 22-19 *Model after aligning the UCS on the inclined face*

7. Choose the **Circle** button from the **Draw** toolbar and draw the circle on the inclined face.

8. The final model should look similar to the one shown in Figure 22-20.

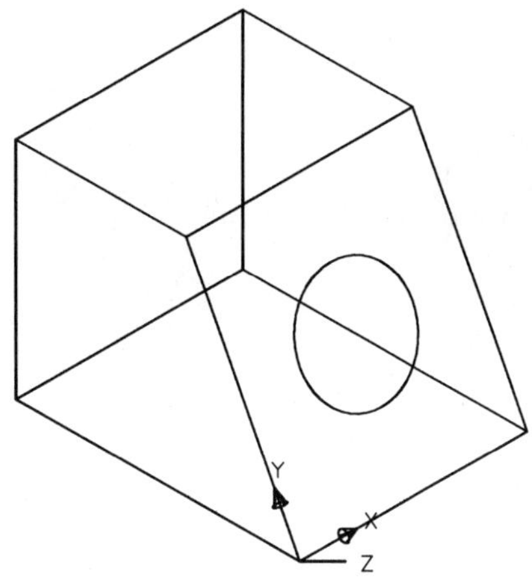

Figure 22-20 *Final model for Example 1*

OB (OBject) Option

With the **OB** (**OBject**) option of the **UCS** command, you can establish a new coordinate system by pointing to any object in an AutoCAD drawing. However, the objects that cannot be used as an object for defining a UCS are a 3D polyline, 3D mesh, viewport object, or xline. The positive *Z* axis of the new UCS is in the same direction as the positive *Z* axis of the object selected. If the *X* and *Z* axes are given, the new *Y* axis is determined by the right-hand rule. The prompt sequence that will follow when you choose this button is given next.

Select object to align UCS: *Select the object to align the UCS.*

In Figure 22-21, the UCS is relocated using the **OBject** option and is aligned to the circle. The origin and the *X* axis of the new UCS are determined by the following rules.

Arc. When you select an arc, the center of the arc becomes the origin for the new UCS. The *X* axis passes through the endpoint of the arc that is closest to the point selected on the object.

Circle/Cylinder/Ellipse. The center of the circle becomes the origin for the new UCS, and the *X* axis passes through the point selected on the object (Figure 22-22).

Line/Mline/Ray/Leader. The new UCS origin is the endpoint of the line nearest to the point selected on the line. The *X* axis is defined so that the line lies on the *XY* plane of the new UCS. Therefore, in the new UCS the *Y* coordinate of the second endpoint of the line is 0.

Tip

The linear edges of the solid models or regions are considered as individual lines when selected to align a UCS.

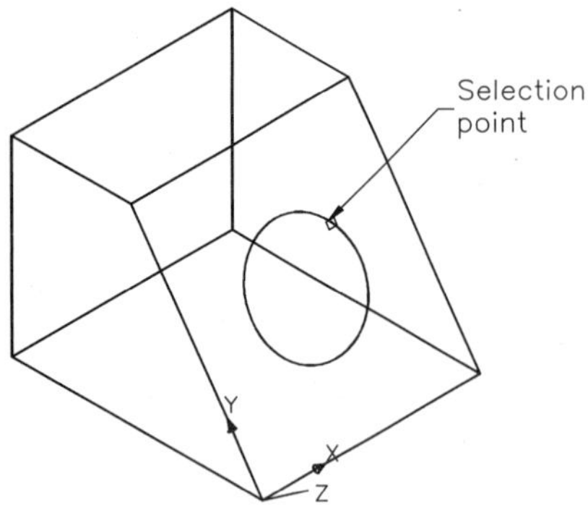

Figure 22-21 Relocating the UCS using a circle

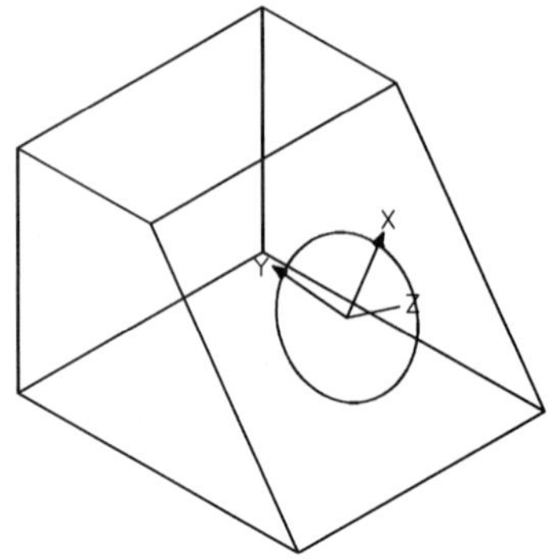

Figure 22-22 UCS at new location

Spline. The origin of the new UCS is the endpoint of the spline that is nearest to the point selected on the spline. An imaginary line will be drawn between the two endpoints of the spline and the *X* axis will be aligned along this imaginary line.

Trace. The origin of the new UCS is the "**start point**" of the trace. The new *X* axis lies along the direction of the selected trace.

Dimension. The middle point of the dimension text becomes the new origin. The *X* axis direction is identical to the direction of the *X* axis of the UCS that existed when the dimension was drawn.

Point. The position of the point is the origin of the new UCS. The directions of the *X*, *Y*, and *Z* axes will be same as those of the previous UCS.

Solid. The origin of the new UCS is the first point of the solid. The *X* axis of the new UCS lies along the line between the first and second points of the solid.

2D Polyline. The start point of the polyline or polyarc is treated as the new UCS origin. The *X* axis extends from the start point to the next vertex.

3D Face. The first point of the 3D face determines the new UCS origin. The *X* axis is determined from the first two points, and the positive side of the *Y* axis is determined from the first and fourth points. The *Z* axis is determined by applying the right-hand rule.

Shape/Text/Insert/Attribute/Attribute Definition. The insertion point of the object becomes the new UCS origin. The new *X* axis is defined by the rotation of the object around its positive *Z* axis. Therefore, the object you select will have a rotation angle of zero in the new UCS.

Tip

Except for 3D faces, the XY plane of the new UCS will be parallel to the XY plane existing when the object was drawn; however, X and Y axes may be rotated.

F (Face) Option

This option aligns the new UCS with the selected face of the solid object. The prompt sequence that will follow when you choose this button is given next.

Select face of solid object: *Select the face to align the UCS.*
Enter an option [Next/Xflip/Yflip] <accept>:

The **Next** option locates the new UCS on the next adjacent face or the back face of the selected edge. **Xflip** rotates the new UCS by 180-degree about the *X* axis and **Yflip** rotates it about the *Y* axis. Pressing ENTER at the **Enter an option [Next/Xflip/Yflip] <accept>** accepts the location of the new UCS as specified.

V (View) Option

The **V** (**View**) option of the **UCS** command allows you define a new UCS whose *XY* plane is parallel to the current viewing plane. The current viewing plane in this case is the screen of the monitor. Therefore, a new UCS is defined that is parallel to the screen of the monitor. The origin of the UCS defined in this option remains unaltered. This option is used mostly to view a drawing from an oblique viewing direction or for writing the text for the objects on the screen. As soon as you choose this button, a new UCS is defined parallel to the screen of the monitor.

X/Y/Z Options

With these options, you can rotate the current UCS around a desired axis. You can specify the angle by entering the angle value at the required prompt or by selecting two points on the screen with the help of a pointing device. You can specify a positive or a negative angle. The new angle is taken relative to the *X* axis of the existing UCS. The **UCSAXISANG** system variable stores the default angle by which the UCS is rotated around the specified axis, by using the **X**/ **Y**/ **Z** options of the **New** option of the **UCS** command. The right-hand thumb rule is used to determine the positive direction of rotation of selected axis.

X Option

In Figure 22-23, the UCS is relocated using the **X** option by specifying an angle about the *X* axis. The first model shows the UCS setting before the UCS was relocated and the second model shows the relocated UCS. The prompt sequence that will follow when you choose this button is given next.

Specify rotation angle about X axis <90>: *Specify the rotation angle.*

Y Option

In Figure 22-24, the UCS is relocated using the **Y** option by specifying an angle about the *Y* axis. The first model shows the UCS setting before the UCS was relocated and the second model shows the relocated UCS. The prompt sequence that will follow when you choose this button is given next.

Specify rotation angle about Y axis <90>: *Specify the angle.*

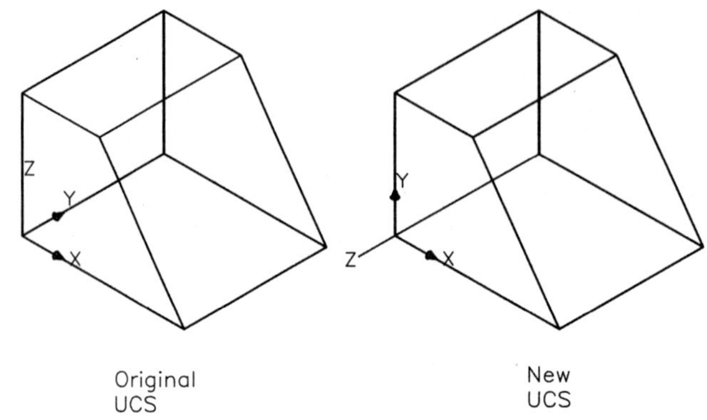

Figure 22-23 *Rotating the UCS about the X axis*

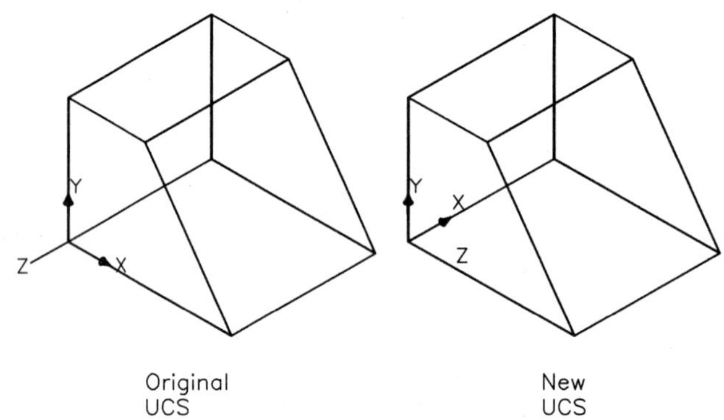

Figure 22-24 *Rotating the UCS about the Y axis*

Z Option

In Figure 22-25, the UCS is relocated using the **Z** option by specifying an angle about the *Z* axis. The first model shows the UCS setting before the UCS was relocated and the second model shows the relocated UCS. The prompt sequence that will follow when you choose this button is given next.

Specify rotation angle about Z axis <90>: *Specify the angle.*

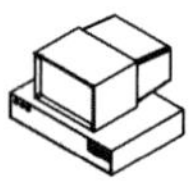

Tip

Rotation about the Z axis can be used as an alternative to using the ***SNAP*** *command to rotate the snap and grid.*

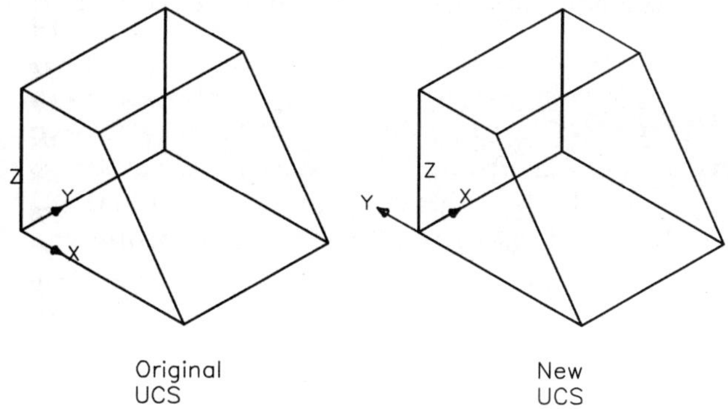

Figure 22-25 *Figure showing UCS rotated about the Z axis*

M (Move) Option

This option redefines the UCS by moving the UCS Origin in the positive or negative direction along the *Z* axis with respect to the current UCS Origin. The orientation of the *XY* plane remains the same. Remember that this button is not available in the **UCS I** toolbar. Instead, this button is available in the **UCS II** toolbar. The prompt sequence that will follow when you choose this button is given next.

Specify new origin point or [Zdepth]<0,0,0>: *Specify the point or the depth.*

If you enter a point at the previously mentioned prompt, then the origin of the current UCS is moved to the specified point. The prompt sequence that will follow when you enter **Z** at this prompt is:

Specify Zdepth<0>: *Specify the depth in the Z axis direction.*

The **Zdepth** option specifies the distance by which the UCS origin moves along the *Z* axis. When you use the **Move UCS** option, in case there are multiple viewports in the drawing, the change in the UCS is applied only to the current viewport. The **Move** option is not added to the **Previous** list.

G (Orthographic) Option

This option allows you to set current any one of the six orthographic UCSs provided with AutoCAD. They are **Top**, **Bottom**, **Front**, **Back**, **Left**, and **Right**. You can specify any one of these UCSs by choosing from the **UCS II** toolbar drop-down list of predefined UCSs (Figure 22-26). The orthographic UCSs are used normally when one is viewing and editing models.

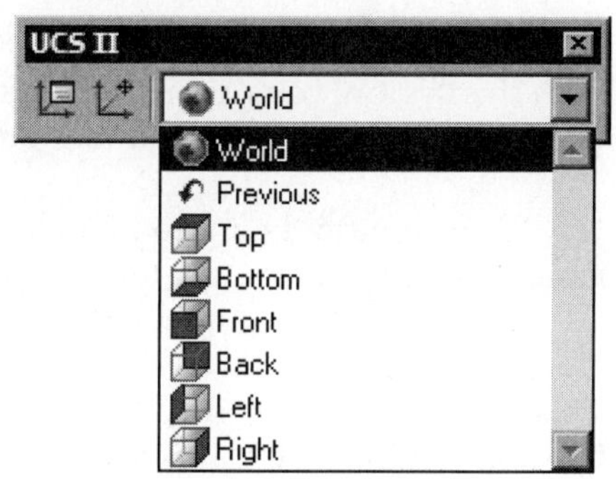

Figure 22-26 *Selecting orthographic UCS from the **UCS II** toolbar*

The orthographic UCSs are set with respect to the WCS. The **UCSBASE** system variable controls the UCS upon which the orthographic settings are based, and the initial value is the WCS.

P (Previous) Option

The **P** (**Previous**) option restores the current UCS settings to the previous UCS settings. The last ten UCS settings are saved by AutoCAD. You can go back to the previous ten UCS settings in the current space using the **Previous** option. If **TILEMODE** is off, the last ten coordinate systems in paper space and in model space are saved. When you choose this button, the previous UCS settings are automatically restored.

R (Restore) Option

With this option of the **UCS** command, you can restore a previously saved named UCS. Once a saved UCS is restored, it becomes the current UCS. The viewing direction of the saved UCS is not restored. You can also restore a named UCS by selecting it from the **UCS II** toolbar drop-down list. Because this option does not have a button for it, you have to invoke this option using the **UCS** button in the **UCS** toolbar. The prompt sequence that will follow when you choose the **UCS** button is given next.

Current ucs name: *WORLD*
Enter an option [New/Move/orthoGraphic/Prev/Restore/Save/Del/Apply/?/World] <World>: **R**
Enter name of UCS to restore or [?]:

You can specify the name of the UCS to be restored or list the UCSs that can be restored by pressing ENTER at the previous prompt. The prompt sequence that will be followed is given next.

Enter UCS name(s) to list <*>: *Specify the name of the UCS to list or give a null response to list all the available UCSs.*

If you give a null response at the previously mentioned prompt, then the AutoCAD text window will be opened listing all the available UCSs.

S (Save) Option

With this option, you can name and save the current UCS settings. When you are naming the UCS, the following points should be kept in mind.

1. The name can be up to 255 characters long.
2. The name can contain letters, digits, blank spaces and the special characters $ (dollar), - (hyphen), and _ (underscore).

This option also does not have a button. Therefore, this option will be invoked using the **UCS** button in the **UCS** toolbar. The prompt sequence that will follow when you choose the **UCS** button is given next.

Current ucs name: *WORLD*
Enter an option [New/Move/orthoGraphic/Prev/Restore/Save/Del/Apply/?/World] <World>: **S**
Enter name to save current UCS or [?]:

Enter a valid name for the UCS at this prompt. AutoCAD saves it as a UCS. You can also list the previously saved UCSs by pressing ENTER at this prompt. The next prompt sequence is:

Enter UCS name(s) to list <*>:

Enter the name of the UCS to list or give a null response to list all the available UCSs.

D (Delete) Option

The **D** (**Delete**) option is used to delete the selected UCS from the list of saved coordinate systems. This option also does not have a button. Therefore, this option will be invoked using the **UCS** button in the **UCS** toolbar. The prompt sequence that will follow when you choose the **UCS** button is given next.

Current ucs name: *WORLD*
Enter an option [New/Move/orthoGraphic/Prev/Restore/Save/Del/Apply/?/World] <World>: **D**
Enter UCS name(s) to delete <none>: *Specify the name of the UCS to delete.*

The UCS name you enter at this prompt is deleted. You can delete more than one UCS by entering the UCS names separated with commas or by using wild cards. If you delete a UCS that is current, it is renamed to **UCS Unnamed**.

A (Apply) Option

The **Apply** option applies the current UCS settings to a specified viewport or to all the active viewports in a drawing session. If the **UCSVP** system variable is set to 1, each viewport saves its UCS settings. The prompt sequence that will follow when you choose this button is:

Pick viewport to apply current UCS or [All]<current>: *Pick inside the viewport or enter A to apply the current UCS settings to all the viewports.*

? Option

By invoking this option, you can list the name of the specified UCS. This option gives you the name, origin, and *X*, *Y*, and *Z* axes of all the coordinate systems relative to the existing UCS. If the current UCS has no name, it is listed as WORLD or UNNAMED. The choice between these two names depends on whether the current UCS is the same as the WCS. Choose the **UCS** button to invoke this option. The prompt sequence that will follow is given next.

Current ucs name: *WORLD*
Enter an option [New/Move/orthoGraphic/Prev/Restore/Save/Del/Apply/?/World] <World>: **?**
Enter UCS name(s) to list <*>:

MANAGING UCS THROUGH DIALOG BOX

Toolbar:	UCS > Display UCS Dialog
	UCS II > Display UCS Dialog
Menu:	Tools > Named UCS
Command:	UCSMAN

The **UCSMAN** command displays the **UCS** dialog box (Figure 22-27). This dialog box can be used to restore the saved and orthographic UCSs, specify UCS icon settings, and rename UCSs. This dialog box has three tabs: the **Named UCSs** tab, the **Orthographic UCSs** tab, and the **Settings** tab.

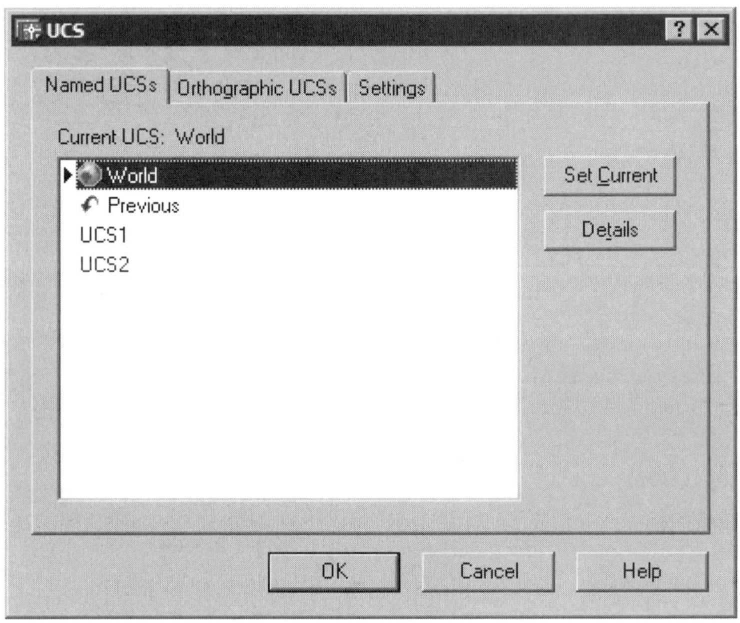

Figure 22-27 ***Named UCSs*** *tab of the* ***UCS*** *dialog box*

Named UCSs Tab

The list of all the coordinate systems defined (saved) on your system is displayed in the list box of this tab. The first entry in this list is always **World**, which means the WCS. The next entry is **Previous**. If you have defined any other coordinate systems in the current editing session, then it will also be displayed in the list box. Selecting the **Previous** entry and then choosing the **OK** button repeatedly allows you to go backward through the coordinate systems defined in the current editing session. **Unnamed** is the next entry in the list if you have not named the current coordinate system. If there are a number of viewports and unnamed settings, then only the current viewport UCS name is displayed in the list. The current coordinate system is indicated by a small pointer icon to the left of the coordinate system name. The current UCS name is also displayed next to **Current UCS**. If you want to make some other coordinate system current, select that coordinate system name in the UCS Names list, and then choose the **Set Current** button. To delete a coordinate system, select that coordinate system name, and then right-click to display a shortcut menu. The options are **Set Current**, **Rename**, **Delete**, and **Details**. Choose the **Delete** button to delete the selected UCS name. To rename a coordinate system, select that coordinate system name, and then right-click to display the shortcut menu and choose **Rename**. Now, enter the desired new name. You can also double-click the name to be modified and then change the name.

Tip
*All the changes and updating of the UCS information in the drawing are carried out only after you choose the **OK** button.*

If you want to check the current coordinate system's origin and *X*, *Y*, and *Z* axis values, select a UCS from the list and then choose the **Details** button. The **UCS Details** dialog box (Figure 22-28) containing that information is then displayed. You can also choose **Details** from the shortcut menu displayed on right-clicking a specific UCS name in the list box.

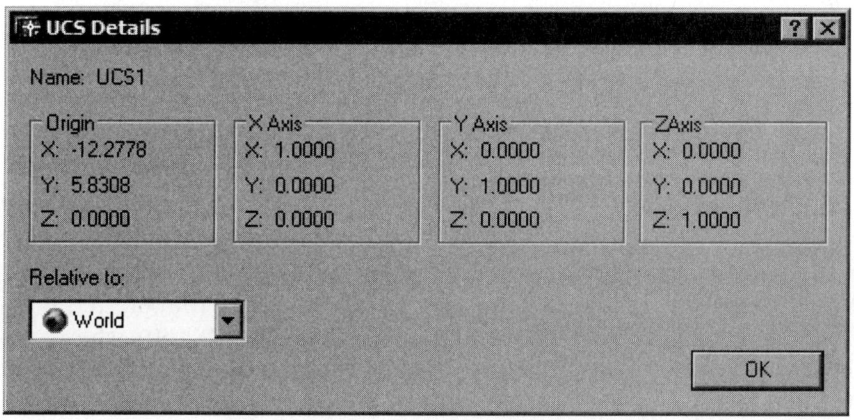

Figure 22-28 ***UCS Details*** *dialog box*

Setting UCS to Preset Orthographic UCSs Using the Orthographic UCSs Tab

The **UCS** dialog box with the **Orthographic UCSs** tab is displayed also on choosing **Orthographic UCS** > **Preset** from the **Tools** menu. You can select any one of the preset orthographic UCSs from the list in the **Orthographic UCSs** tab of the **UCS** dialog box (Figure 22-29). Selecting **Top** and choosing the **Set Current** button results in creation of the UCS icon in the top view, also known as the plan view. You can double-click on the UCS name you want to set as current. You can also right-click on the specific orthographic UCS name in the list and choose **Set Current** from the shortcut menu that is displayed. The **Current UCS** displays the name of the current UCS name. If the current settings have not been saved and named, the **Current UCS** name displayed is **Unnamed**.

The **Depth** field in the list box of this dialog box displays the distance between the *XY* plane

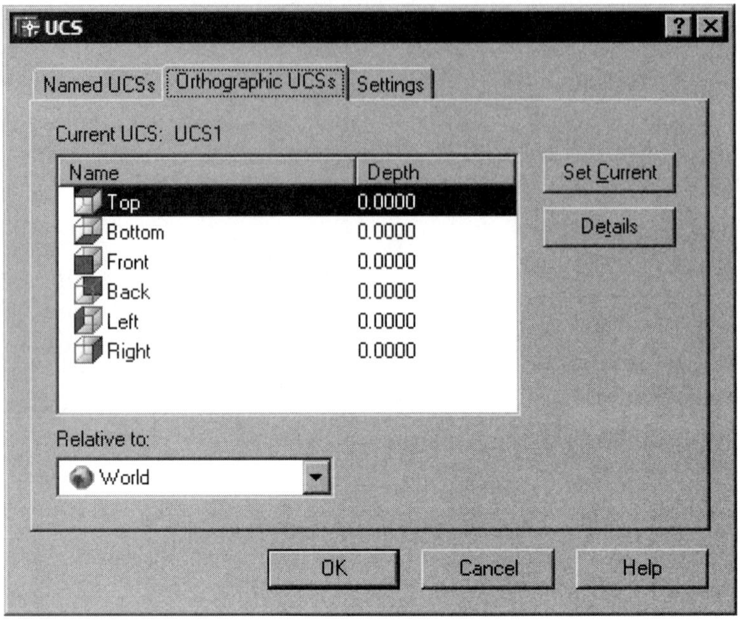

Figure 22-29 ***Orthographic UCSs*** *tab of the* ***UCS*** *dialog box*

of the selected orthographic UCS setting and a parallel plane passing through the origin of the UCS base setting. The system variable **UCSBASE** stores the name of the UCS which is considered the base setting; that is, it defines the origin and orientation. You can enter or modify values of **Depth** by double-clicking on the depth value of the selected UCS in the list box to display the **Orthographic UCS depth** dialog box (Figure 22-30), where you can enter new depth values. You can also right-click a specific orthographic UCS in the list box to display a shortcut menu. Choose **Depth** to display the **Orthographic UCS depth** dialog box. Enter a value in the **Top Depth** edit box or choose the **Select new origin** button to specify a new origin or depth on the screen.

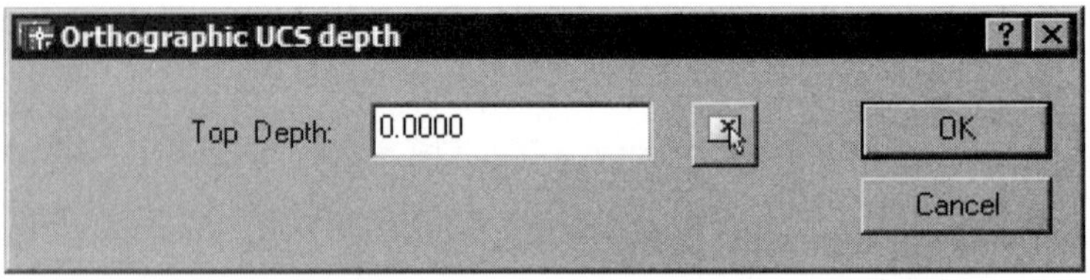

Figure 22-30 ***Orthographic UCS depth*** *dialog box*

In the **Orthographic UCSs** tab of the **UCS** dialog box, you are given the choice of establishing a new UCS relative to a named UCS or to the WCS. This choice is offered from the **Relative to** drop-down list. This list lists all the named UCSs in the current drawing. By default **World** is the base coordinate system. If you set current any one of the preset orthographic UCSs and then select a UCS other than **World** in the **Relative to:** drop-down list, the current UCS changes to **Unnamed**. Choosing the **Details** button displays the **UCS Details** dialog box with the origin and the *X*, *Y*, *Z* coordinate values of the selected UCS. You can also choose **Details** from the shortcut menu that is displayed on right-clicking on a UCS in the list box. The shortcut menu also has a **Reset** option. Choosing **Reset** from the shortcut menu restores the origin of the selected orthographic UCS to its default location (0,0,0).

Settings Tab

The **Settings** tab of the **UCS** dialog box (Figure 22-31) displays and modifies UCS and UCS icon settings of a specified viewport.

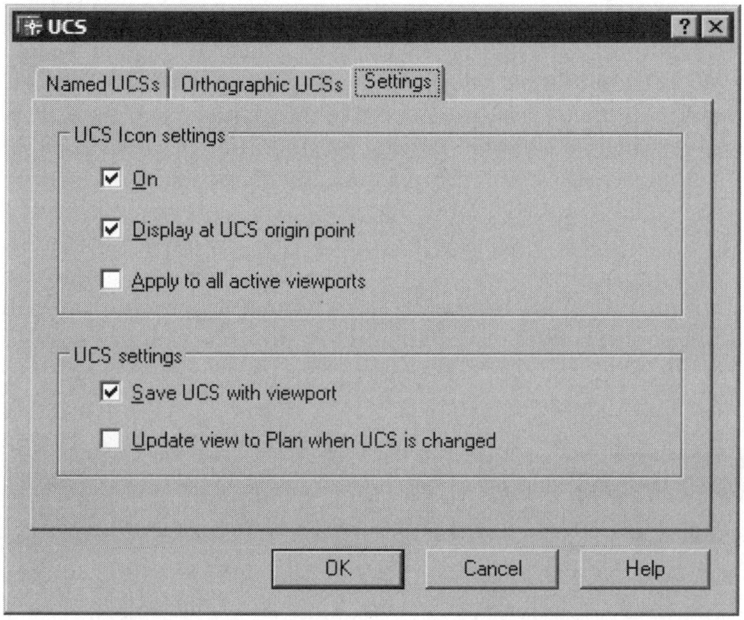

***Figure 22-31** **Settings** tab of the **UCS** dialog box*

UCS Icon settings Area

Selecting the **On** check box displays the UCS icon in the current viewport. It is similar to using the **UCSICON** command to set the display of the UCS icon to on or off. If you select the **Display at UCS origin point** check box, the UCS icon is displayed at the origin point of the coordinate system in use in the current viewport. If the origin point is not visible in the current viewport or if the check box is cleared, the UCS icon is displayed in the lower left corner of the viewport. Selecting the **Apply to all active viewports** check box applies the current UCS icon settings to all active viewports in the current drawing.

UCS settings Area

This area of the **UCS** dialog box **Settings** tab specifies UCS settings for the current viewport. Selecting the **Save UCS with viewport** check box saves the UCS with the viewport. The value is stored in the **UCSVP** system variable. If you clear this check box, the **UCSVP** variable is set to 0 and the UCS of the viewport reflects the UCS of the current viewport. When you select the **Update view to plan** check box, the plan view is restored when the UCS in the viewport is changed. Also, when the selected UCS is restored, the plan view is restored. The value is stored in the **UCSFOLLOW** system variable.

SYSTEM VARIABLES

The coordinate value of the origin of the current UCS is held in the **UCSORG** system variable. The *X* and *Y* axis directions of the current UCS are held in the **UCSXDIR** and **UCSYDIR** system variables, respectively. The name of the current UCS is held in the **UCSNAME** variable. All these variables are read-only. If the current UCS is identical to the WCS, the **WORLDUCS** system variable is set to 1; otherwise, it holds the value 0. The current UCS icon setting can be examined and manipulated with the help of the **UCSICON** system variable. This variable holds the UCS icon setting of the current viewport. If more than one viewport is active, each one can have a different value for the **UCSICON** variable. If you are in paper space, the **UCSICON** variable will contain the setting for the UCS icon of the paper space. The **UCSFOLLOW** system variable controls the automatic display of a plan view when you switch from one UCS to another. If **UCSFOLLOW** is set to 1, a plan view is automatically displayed when you switch from one UCS to another. The **UCSAXISANG** variable stores the default angle value for the *X*, *Y*, or *Z* axis around which the UCS is rotated using the ***X***, ***Y***, ***Z*** options of the **New** option of the UCS command. The **UCSBASE** variable stores the name of the UCS that acts as the base; that is, it defines the origin and orientation of the orthographic UCS setting. The **UCSVP** variable decides whether the UCS settings are stored with the viewport or not.

Self-Evaluation Test

Answer the following questions and then compare your answers with the correct answers given at the end of this chapter.

1. The world coordinate system can be moved from its position. (T/F)

2. The **View** option of the **New** option of the **UCS** command allows you to define a new UCS whose *XY* plane is parallel to the current viewing plane. (T/F)

3. The ellipse can not be used as an object for defining a new UCS. (T/F)

4. If you do not specify the *Z* coordinates of the new point while moving the UCS, the previous *Z* coordinates will be taken for the new value. (T/F)

5. By default the __________ is established as the coordinate system in the AutoCAD environment.

6. The __________ coordinate system can be moved and rotated to any desired location.

7. Once a saved UCS is restored, it becomes the __________ UCS.

8. The _________ command controls the display of the UCS icon.

9. You can change the 3D UCS icon to a 2D UCS icon using the _________ option of the **UCSICON** command.

10. The _________ point of the 3D face defines the origin of the new UCS.

Review Questions

Answer the following questions.

1. The **All** option is used to apply the changes in the UCS icon in all the active viewports. (T/F)

2. The line width of the UCS icon can be changed. (T/F)

3. The size of the UCS icon can vary between 5 and 100. (T/F)

4. The name of the current UCS is stored in the **UCSNAME** variable. (T/F)

5. Which option is used to restore the previous UCS settings?

   (a) ZAxis (b) Restore
   (c) Previous (d) Save

6. Which option is used to list the names of all the saved UCSs?

   (a) ZAxis (b) Restore
   (c) ? (d) Save

7. Which variable is used to control the automatic display of the plan view when you switch to the new UCS?

   (a) **UCSORG** (b) **UCSFOLLOW**
   (c) **UCS** (d) **UCSBASE**

8. Which variable stores the coordinates of the origin of the current UCS?

   (a) **UCSICON** (b) **UCSORG**
   (c) **UCSVP** (d) **UCSFOLLOW**

9. Which command is used to manage the UCS using a dialog box?

   (a) **UCS** (b) **UCSMAN**
   (c) **UCSICON** (d) **UCSBASE**

10. The __________ option of the **UCS** command is used to change the origin of the current UCS.

11. The __________ direction is known as the extrusion direction.

12. The lineweight of the UCS icon can be changed using the __________ option of the **UCSICON** command.

13. The __________ variable is used to control the *X* axis direction of the current UCS.

14. The __________ variable is used to control the *Y* axis direction of the current UCS.

15. The __________ variable is used to store the default angle value of the by which the UCS will be rotated using the **X**, **Y,** and **Z** option of the **New** option of the **UCS** command.

Exercises

Exercise 1 *Mechanical*

In this exercise you will draw the model shown in Figure 22-32. Assume the missing dimensions for the model.

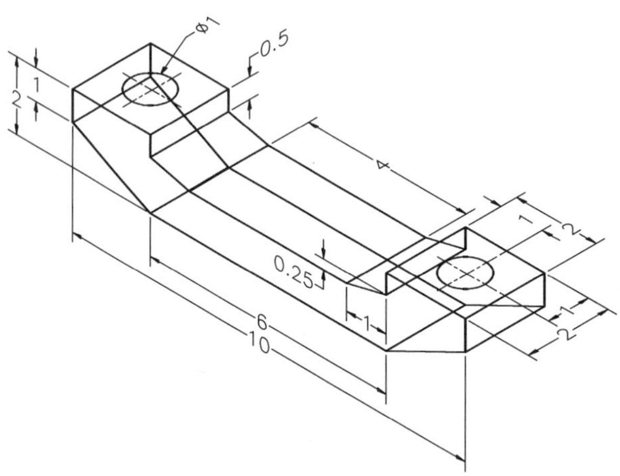

Figure 22-32 *Model for Exercise 1*

Exercises 2 and 3 *Mechanical*

Draw the objects shown in Figures 22-33 and 22-34. Assume the missing dimensions for the model. Use the **UCS** command to align the ucsicon and then draw the objects.

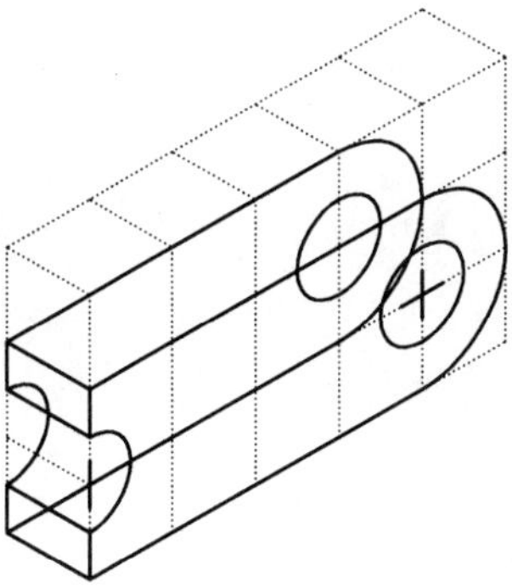

Figure 22-33 Drawing for Exercise 2

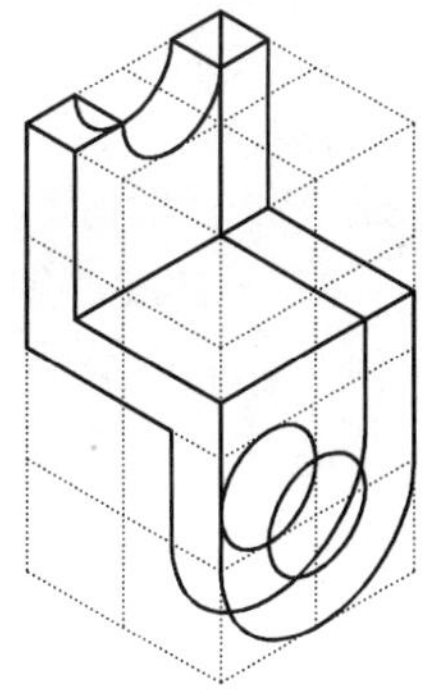

Figure 22-34 Drawing for Exercise 3

Exercises 4 and 5 *Mechanical*

Make the drawings shown in Figures 22-35 and 22-36. Assume the missing dimensions for the model. Use the **UCS** command to align the ucsicon and then draw the objects.

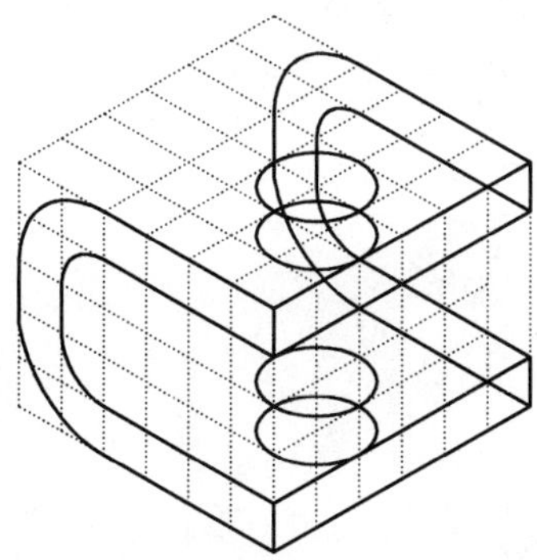

Figure 22-35 Drawing for Exercise 4

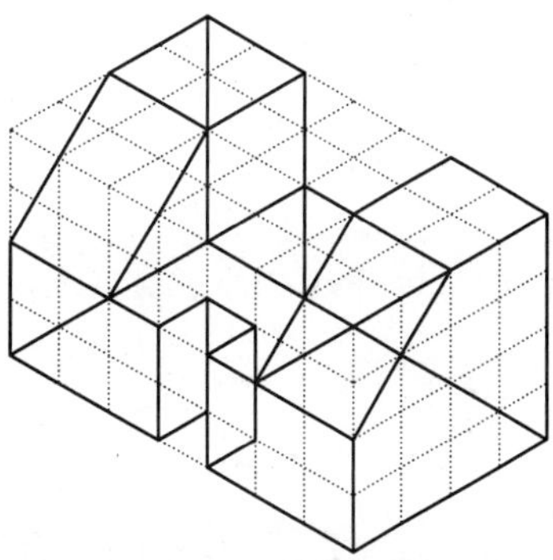

Figure 22-36 Drawing for Exercise 5

Exercise 6 *General*

Make the drawings shown in Figure 22-37. Assume the missing dimensions for the model. Use the **UCS** command to align the ucsicon and then draw the objects. The left side of the transition makes 90-degree angle with the bottom.

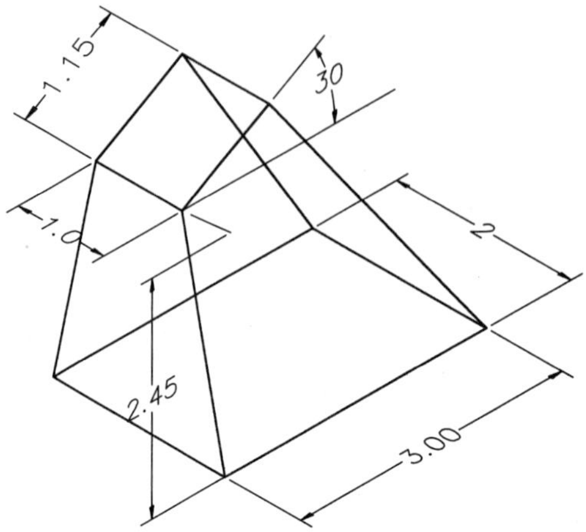

Figure 22-37 *Model for Exercise 6*

Exercise 7 *General*

Make the drawings shown in Figure 22-38. Assume the missing dimensions for the model. Use the **UCS** command to position the ucsicon and then draw the objects. The center of the top polygon is offset 0.75 units from the center of the bottom polygon.

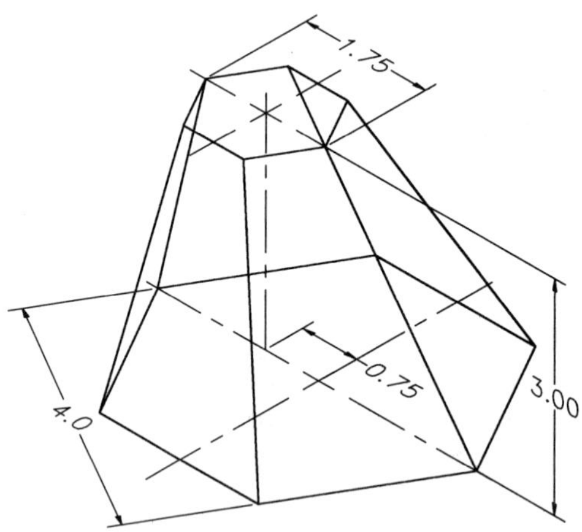

Figure 22-38 *Model for Exercise 7*

Answers to Self-Evaluation Test

1 - F, **2** - T, **3** - F, **4** - T, **5** - world coordinate system, **6** - user, **7** - current, **8** - **UCSICON**, **9** - **Properties**, **10** - first

Chapter 23

Getting Started with 3D

Learning Objectives

After completing this chapter you will be able to:

- *Understand the use of three-dimensional (3D) drawing.*
- *Understand various types of 3D models.*
- *Understand the conventions in AutoCAD.*
- *Use the **DDVPOINT** command to view the objects in the 3D space.*
- *Use the **VPOINT** command to view the objects in the 3D space.*
- *Understand various types of 3D coordinate systems.*
- *Create surface models using various options.*
- *Set the thickness and elevation for the objects.*

USE OF THREE-DIMENSIONAL DRAWING

The real world and literally whatever you see around in it is three-dimensional (3D). Each and every design idea you think of is 3D. However, until some time back, it was not possible to materialize these ideas in the design industry due to the lack of the third dimension since the drawings were done on the drawing sheets that are two dimensional objects with only *X* and *Y* coordinates. Therefore, it was not possible to draw the 3D objects. Since prototyping is a very long and costly affair, therefore, the designers had to suppress their ideas and convert the 3D designs into 2D drawings.

The use of computers and the CADD Technology has brought a significant improvement in materializing the design ideas and creating the engineering drawings. You can create a proper 3D object in the computer using the CADD Technology. This technology allows you to create models with the third coordinate called the *Z* coordinate in addition to the *X* and *Y* coordinates. Apart from materializing the design ideas, the 3D models have various other advantages. Some of them are given next.

1. **To generate the drawing views**. Once you have created the 3D model, its 2D drawing views can be automatically generated.

2. **To provide realistic effects**. You can provide realistic effects to the 3D models by assigning material and providing light effects to them. For example, an architectural drawing can be made more realistic and presentable by assigning the material to the walls and interiors and adding the lights to it. You can also create its bitmap image and use it in presentations.

3. **To create the assemblies and check them for interference**. You can assemble various 3D models and create the assemblies. Once the components are assembled, you can check them for interference to reduce the errors and material loss during manufacturing. You can also generate the 2D drawing views of the assemblies.

4. **To create the animation of the assemblies**. You can animate the assemblies and view the animation to provide the clear display of the mating parts in the assembly.

5. **To apply Boolean operations**. You can apply the Boolean operations to the 3D models.

6. **To calculate the mass properties**. You can calculate the mass properties of the 3D models.

7. **Cut Sections**. You can cut sections through the solid models to view the shape at that crosssection.

TYPES OF 3D MODELS

In AutoCAD, depending on their characteristics, the 3D models are divided into three categories. They are discussed next.

Wireframe Models

These models are created using the simple AutoCAD entities such as lines, polylines, rectangles, polygons, or some other entities such as 3D faces, and so on. To understand the wireframe models, consider a 3D model made up of match sticks or wires. These models consist of only the edges and you can see through these models. You cannot apply the Boolean operations on these models and also cannot calculate the mass properties of these models. Wireframe models are generally used in the bodybuilding of the vehicles. Figure 23-1 shows a wireframe model.

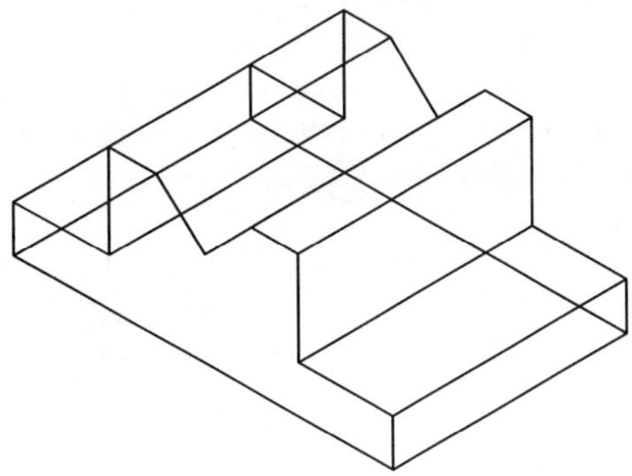

Figure 23-1 *A wireframe model*

Surface Models

These models are made up of one or more surfaces. They have a negligible wall thickness and are hollow inside. To understand these models, consider a wireframe model with a cloth wrapped around it. You cannot see through these models. These types of model are used in the plastic molding industry, shoe manufacturing, utensils manufacturing, and so on.

You can directly create a surface model or create a wireframe model and then convert it into a surface model. Remember that you cannot perform Boolean operations in the surface models also. Figure 23-2 shows a surface model created using a single surface and Figure 23-3 shows a surface model created using a combination of surfaces.

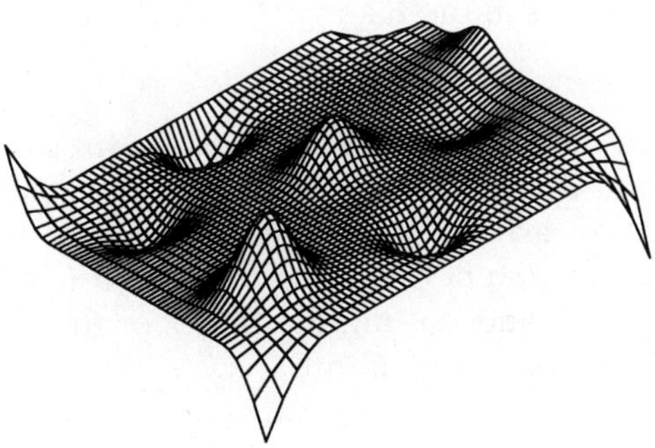

Figure 23-2 *A surface model created using a single surface*

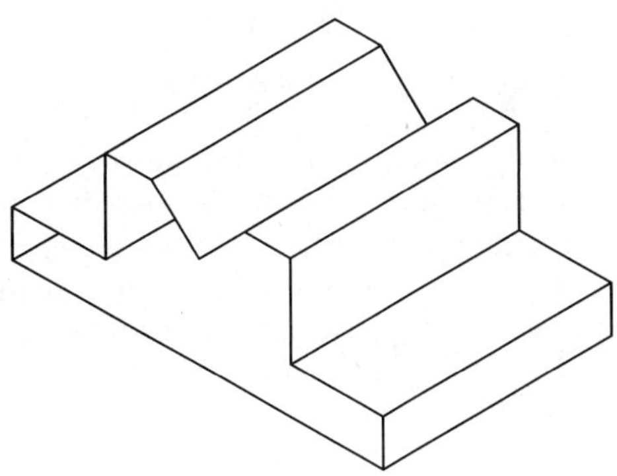

Figure 23-3 *A semi-finished surface model created using more than one surface*

Solid Models

These are the solid filled models having mass properties. To understand a solid model, consider a model made up of metal or wood. You can perform Boolean operations on these models, cut a hole through them, or even cut them into slices (Figure 23-4).

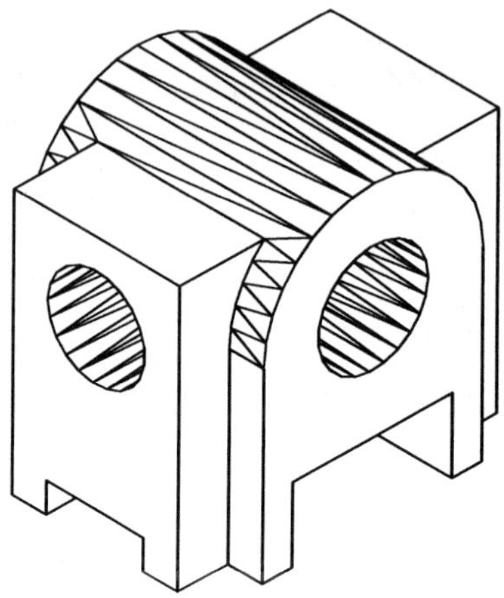

Figure 23-4 *A solid model*

CONVENTIONS IN AUTOCAD

It is important for you to know the conventions in AutoCAD before you proceed with the 3D. AutoCAD follows these three conventions:

1. Whenever you open a new drawing in AutoCAD, it is in the Plan view and whatever you draw, it will be in the world *XY* plane.

2. The right-hand rule is followed in AutoCAD to identify the *X*, *Y*, and *Z* axis directions. The right-hand rule states that if you keep the thumb, index finger, and the middle finger of the right hand mutually perpendicular to each other as shown in Figure 23-5(a), then the thumb of the right hand displays the direction of the positive *X* axis, the index finger displays the direction of the positive *Y* axis, and the middle finger of the right hand displays the direction of the positive *Z* axis, see Figure 23-5(b).

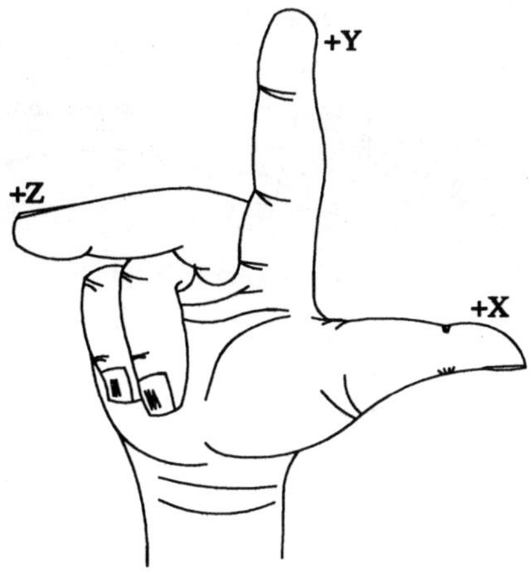

Figure 23-5(a) The orientation of the fingers as per the right-hand rule

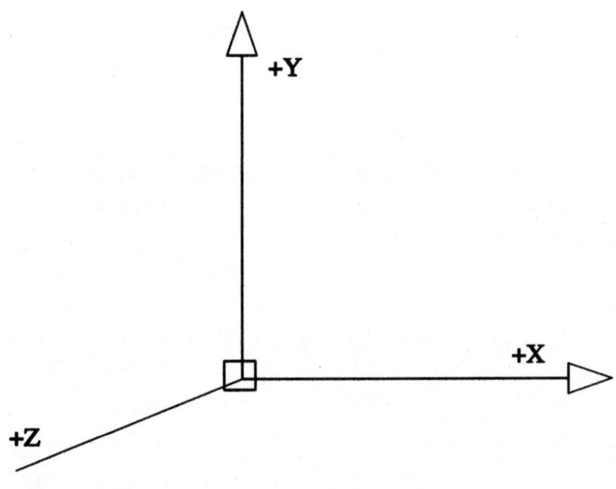

Figure 23-5(b) Orientation of the X, Y, and Z axes

3. The right-hand thumb rule is followed in AutoCAD to determine the direction of rotation or revolution in the 3D space. It is very difficult to specify the clockwise or the counterclockwise direction in the 3D space, and therefore, the right-hand thumb rule is followed. The right-hand thumb rule states that if the thumb of the right hand displays the direction of the axis of rotation, then the direction of the curled fingers will define the positive direction of rotation, see Figure 23-5(c).

This rule will be used during the rotation of 3D models or UCS and also during the creation of revolved surfaces and revolved solids.

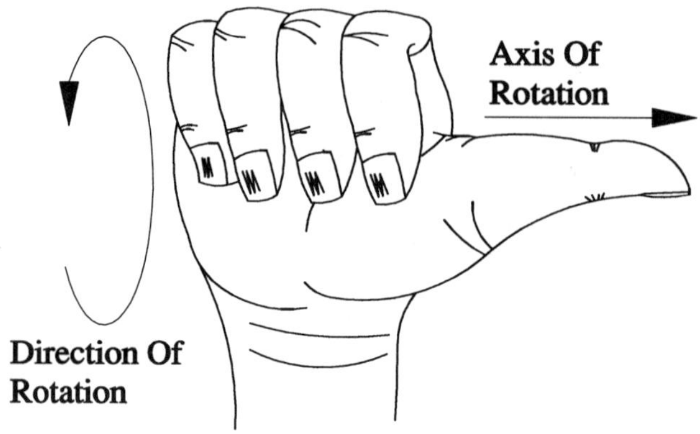

Figure 23-5(c) *Right-hand thumb rule showing the axis and direction of rotation*

CHANGING THE VIEWPOINT TO VIEW THE 3D MODELS

Until now, you have been drawing only the 2D entities in the *XY* plane and in the Plan view (see the preceding section "Conventions in AutoCAD"). But in the Plan view, it is very difficult to find out whether the object displayed is a 2D entity or a 3D model. For example, see Figure 23-6. In this view, it appears as though the object displayed is a rectangle. The reason for this is that by default you view the objects in the Plan view from the direction of the positive *Z* axis. This is a very confusing situation that has to be avoided. This situation can be avoided if the object is viewed from a direction that also displays the *Z* axis. In order to do that, you need to change the viewpoint so that the object is displayed in the space with all three axes as shown in Figure 23-7. In this view, you can also see the *Z* axis of the model along with the *X* and *Y* axes, and so it is clear that the original object is not a rectangle but a 3D model. The viewpoint can be changed by the **DDVPOINT** command, the **VPOINT** command, or using the **View** toolbar.

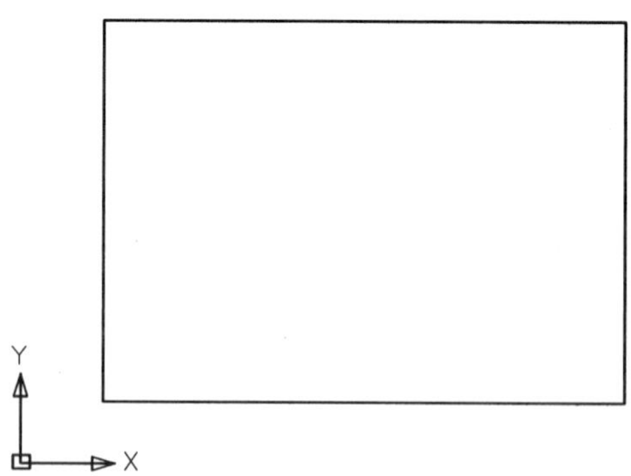

Figure 23-6 *Figure looking like a rectangle in the Plan view*

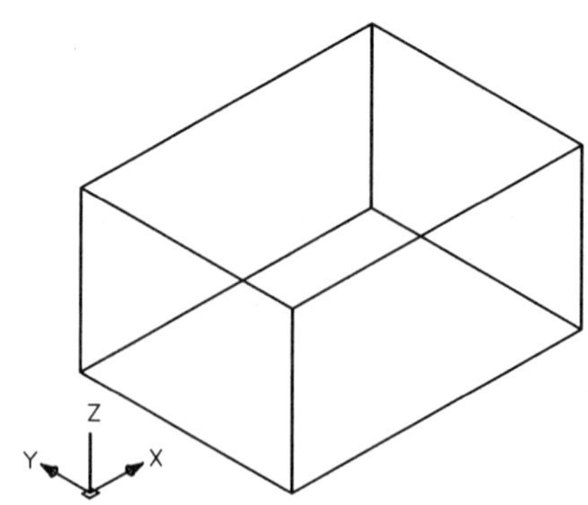

Figure 23-7 *Viewing the 3D model after changing the viewpoint*

Tip

Remember that changing the viewpoint does not affect the dimensions or the location of the model. When you change the viewpoint, the model is not moved from its original location. Changing the viewpoint only changes the direction from which you are viewing the model.

Changing the Viewpoint Using the Viewpoint Presets Dialog Box

Menu:	View > 3D Views > Viewpoint Presets
Command:	DDVPOINT

The **DDVPOINT** command is used to invoke the **Viewpoint Presets** dialog box, which can be used to set the viewpoint to view the 3D models. The dialog box sets the viewpoint with respect to the **angle in the XY plane** and the **angle from the XY plane**. Figure 23-8 shows different view direction parameters.

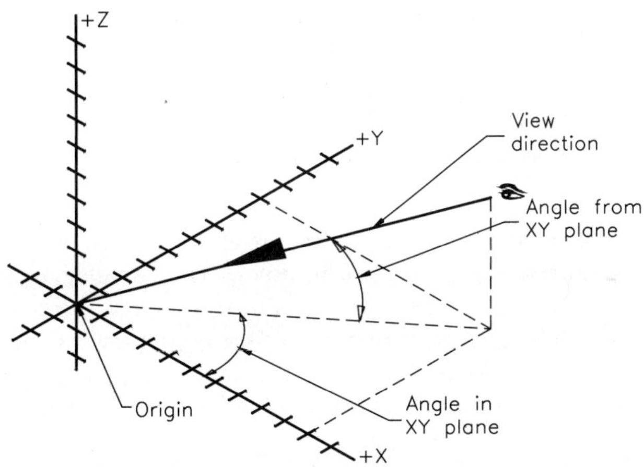

Figure 23-8 *View direction parameters*

Viewpoint Presets Dialog Box Options

The options available in the **Viewpoint Presets** dialog box (Figure 23-9) are discussed next.

Absolute to WCS. This radio button is selected to set the viewpoint with respect to the world coordinate system (WCS).

Relative to UCS. This radio button is selected to set the viewpoint with respect to the current user coordinate system (UCS).

Note

The WCS and the UCS has already been discussed in Chapter 22.

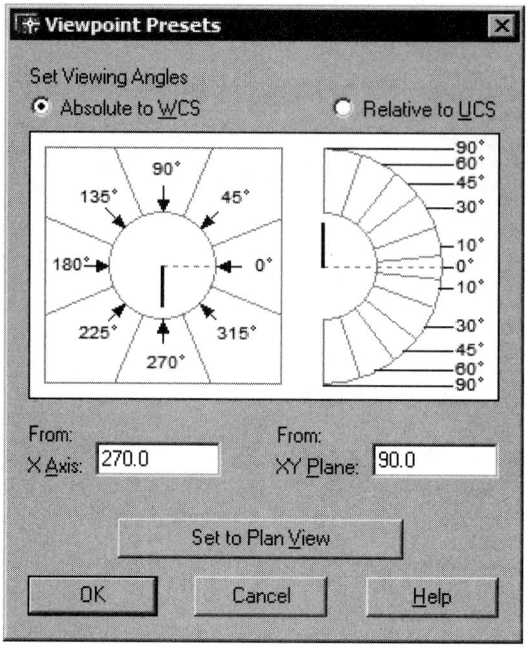

Figure 23-9 Viewpoint Presets *dialog box*

From: X Axis. This edit box is used to specify the angle in the *XY* plane from the *X* axis. You can directly enter the required angle in this edit box or you can select the desired angle from the image tile (Figure 23-10). There are two arms in this image tile and by default, they are placed one over the other. The gray arm displays the current viewing angle and the black one displays the new viewing angle. When you select a new angle from the image tile, the new angle is automatically displayed in the **X Axis** edit box. This value can vary between 0 and 359.9.

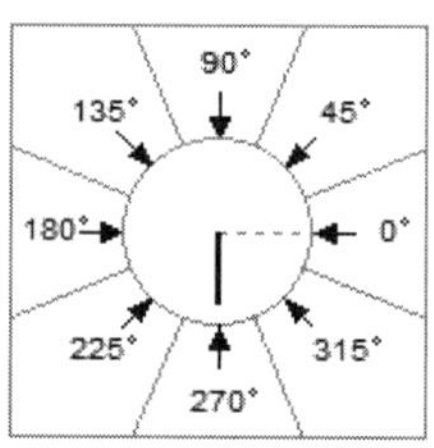

Figure 23-10 *The image tile for selecting the angle in the XY plane*

XY Plane. This edit box is used to specify the angle from the *XY* plane. You can directly enter the desired angle in this edit box, or can select the angle from the image tile as shown in Figure 23-11. This value can vary from -90 to +90.

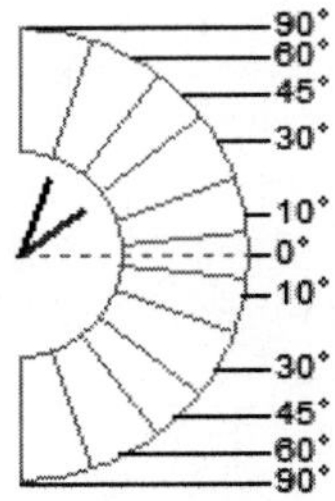

Figure 23-11 *The image tile for selecting the angle from the XY plane*

Set to Plan View. This button is chosen to set the viewpoint to the Plan view of the WCS or the UCS. If the **Absolute to WCS** radio button is selected, the viewpoint is set to the Plane view of the WCS and if the **Relative to UCS** radio button is selected, the viewpoint is set to the Plan view of the current UCS. Figures 23-12 and 23-13 show a 3D model from different viewing angles.

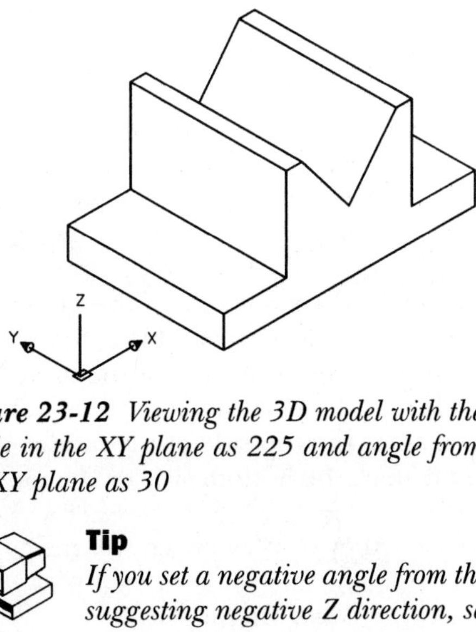

Figure 23-12 *Viewing the 3D model with the angle in the XY plane as 225 and angle from the XY plane as 30*

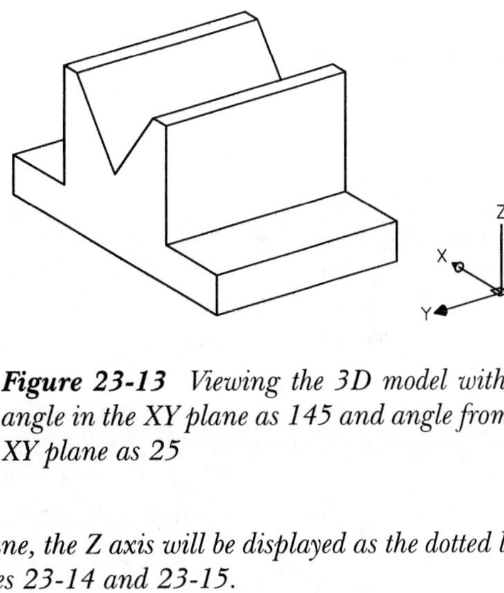

Figure 23-13 *Viewing the 3D model with the angle in the XY plane as 145 and angle from the XY plane as 25*

Tip

If you set a negative angle from the XY plane, the Z axis will be displayed as the dotted line, suggesting negative Z direction, see Figures 23-14 and 23-15.

In case of confusion in identifying the direction of the X, Y, and Z axes, use the right-hand rule.

Changing the Viewpoint Using the Command Line

Menu:	View > 3D Views > VPOINT
Command:	VPOINT

The **VPOINT** command is also used to set the viewpoint for viewing the 3D models. Using this command, the user can specify a point in the 3D space that will act as the viewpoint. The prompt sequence that will follow when you invoke this command is given next.

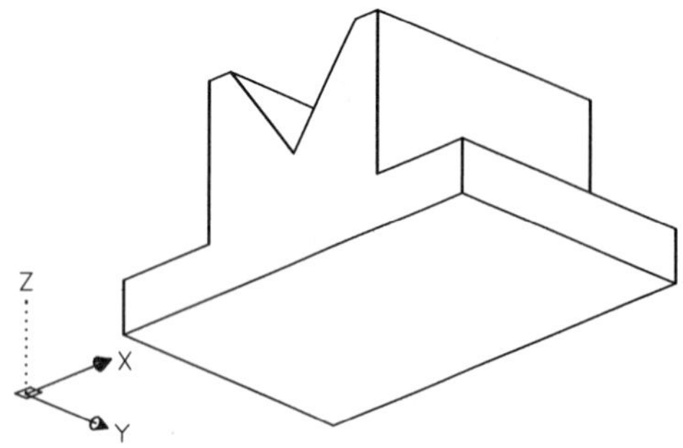

Figure 23-14 *Viewing the 3D model with the angle in the XY plane as 315 and angle from the XY plane as -25*

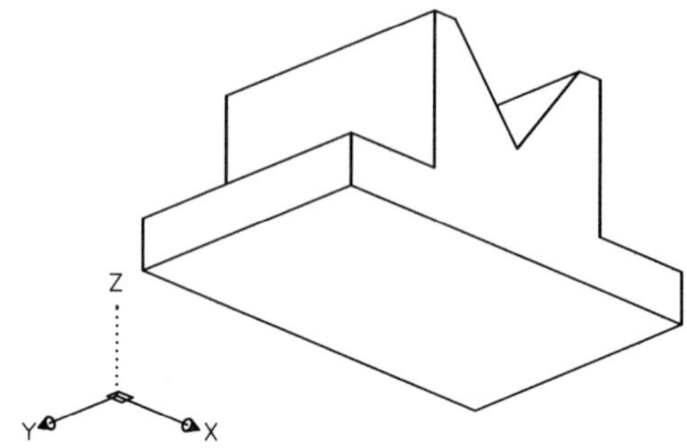

Figure 23-15 *Viewing the 3D model with the angle in the XY plane as 225 and angle from the XY plane as -25*

Current view direction: VIEWDIR=0.0000,0.0000,1.0000
Specify a viewpoint or [Rotate] <display compass and tripod>:

Specifying a Viewpoint

This is the default option and is used to set a viewpoint by specifying the location of the viewpoint (viewer) using the *X*, *Y*, and *Z* coordinates of that particular point. AutoCAD follows a convention of the sides of the 3D model for specifying the viewpoint. The convention states that if the UCS is at the World position (default position), then

1. The side at the positive *X* axis direction will be taken as the right side of the model.
2. The side at the negative *X* axis direction will be taken as the left side of the model.
3. The side at the negative *Y* axis direction will be taken as the front side of the model.
4. The side at the positive *Y* axis direction will be taken as the back side of the model.
5. The side at the positive *Z* axis direction will be taken as the top side of the model.
6. The side at the negative *Z* axis direction will be taken as the bottom side of the model.

Some of the standard viewpoint coordinates and the view they display are shown next.

Vpoint Value	View Displayed
1,0,0	Right side view
-1,0,0	Left side view
0,1,0	Back view
0,-1,0	Front view
0,0,1	Top view
0,0,-1	Bottom view
1,1,1	Right, Back, Top view
-1,1,1	Left, Back, Top view
1,-1,1	Right, Front, Top view
-1,-1,1	Left, Front, Top view
1,1,-1	Right, Back, Bottom view

-1,1,-1	Left, Back, Bottom view
1,-1,-1	Right, Front, Bottom view
-1,-1,-1	Left, Front, Bottom view

You can also enter the values in decimal, but the resultant views will not be the standard views. Figures 23-16 through 23-19 show the 3D model from different viewpoints.

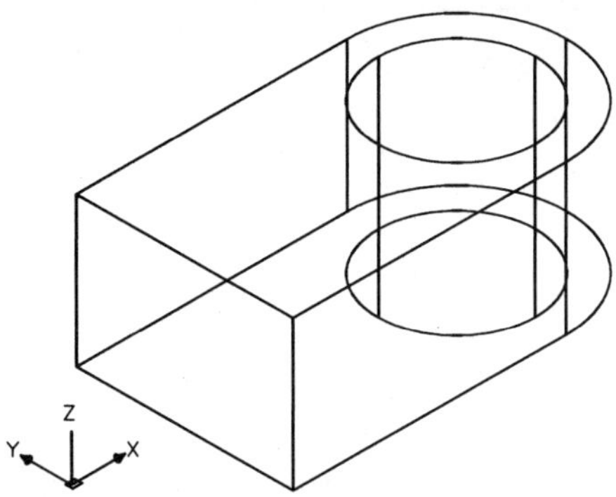

Figure 23-16 *Viewing the model with the viewpoint -1,-1,1*

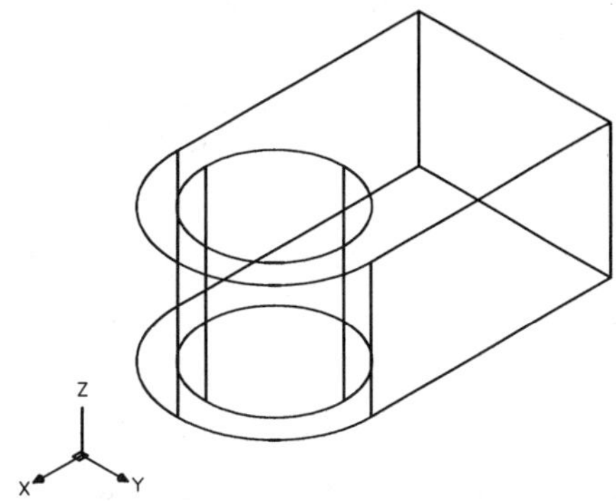

Figure 23-17 *Viewing the model with the viewpoint 1,1,1*

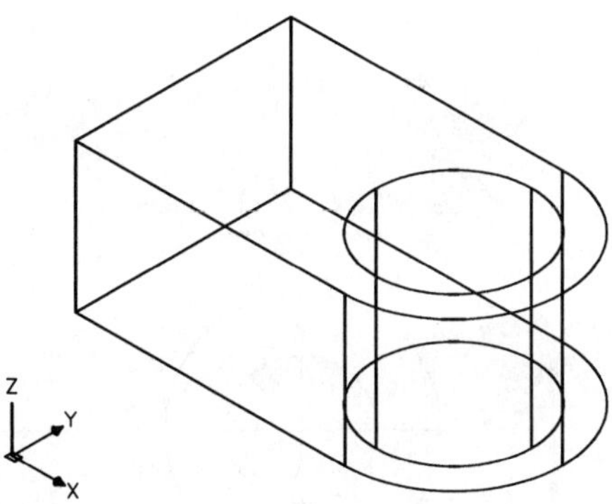

Figure 23-18 *Viewing the model with the viewpoint 1,-1,1*

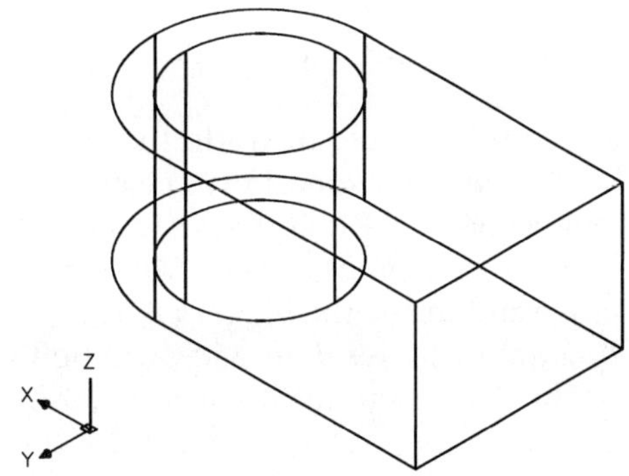

Figure 23-19 *Viewing the model with the viewpoint -1,1,1*

Compass and Tripod

When you press ENTER at the **Specify a viewpoint or [Rotate] <display compass and tripod>** prompt, a compass and an axis tripod is displayed as shown in Figure 23-20. You can directly set the viewpoint using this compass. The crosshairs appears on the compass and you can select any point on this compass to specify the viewpoint. The compass consists of two circles, a smaller circle and a bigger circle. Both of these circles are divided into four quadrants: first, second, third, and fourth. The resultant view will depend upon the quadrant and the circle on which you select the point. In the first quadrant, both the *X* and *Y* axes are positive; in the

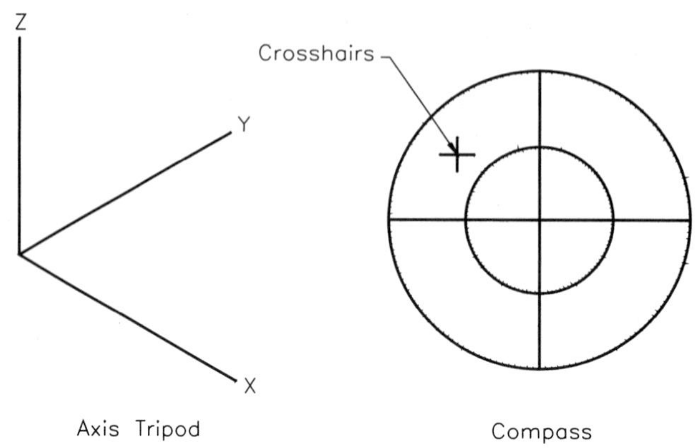

Figure 23-20 *The Axis tripod and the compass*

second quadrant, the *X* axis is negative and the *Y* axis is positive; in the third quadrant, both the *X* and *Y* axes are negative; and in the fourth quadrant, the *X* axis is positive and the *Y* axis is negative. Now, if you select the point inside the inner circle, it will be in the positive *Z* axis direction and if you select the point outside the inner circle and inside the outer circle, it will be in the negative *Z* axis direction. Therefore, if the previously mentioned statements are added, the following is concluded.

1. If you select a point inside the smaller circle in the first quadrant, the resultant view will be the Right, Back, Top view. If you select a point outside the smaller circle and inside the bigger circle in this quadrant, the resultant view will be the Right, Back, Bottom view (Figure 23-21).

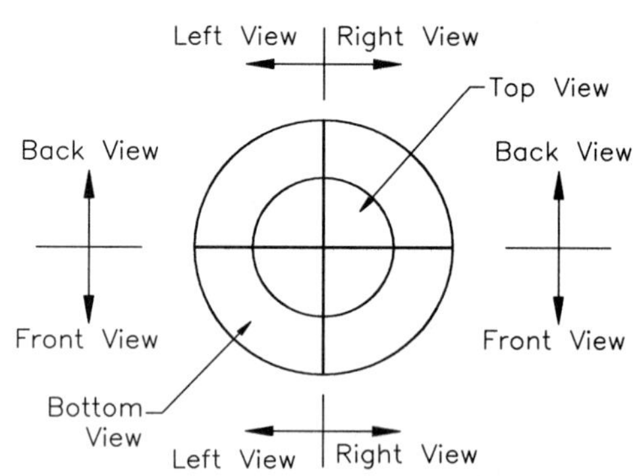

Figure 23-21 *The directions for the compass*

2. If you select a point inside the smaller circle in the second quadrant, the resultant view will be the Left, Back, Top view. If you select a point outside the smaller circle and inside the bigger circle in this quadrant, the resultant view will be the Left, Back, Bottom view (Figure 23-21).

3. If you select a point inside the smaller circle in the third quadrant, the resultant view will be the Left, Front, Top view. If you select a point outside the smaller circle and inside the bigger circle in this quadrant, the resultant view will be the Left, Front, Bottom view (Figure 23-21).

4. If you pick a select inside the smaller circle in the fourth quadrant, the resultant view will be the Right, Front, Top view. If you select a point outside the smaller circle and inside the bigger circle in this quadrant, the resultant view will be the Right, Front, Bottom view (Figure 23-21).

Rotate

This option is similar to setting the viewpoint in the **Viewpoint Presets** dialog box displayed by invoking the **DDVPOINT** command. When you select this option, you will be prompted to specify the angle in the *XY* plane and the angle from the *XY* plane. The angle in the *XY* plane will be taken from the *X* axis.

Changing the Viewpoint Using the View Toolbar

Apart from the **Viewpoint Presets** dialog box and the **VPOINT** commands, you can also set the viewpoint using the **View** toolbar (Figure 23-22).

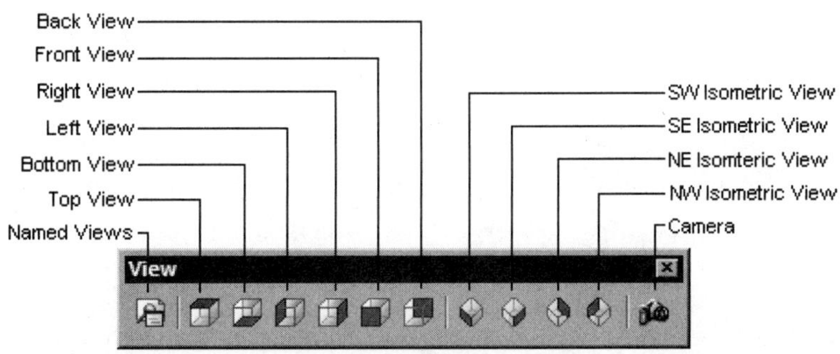

***Figure 23-22** View toolbar*

This toolbar consists of six preset orthographic views: Top View, Bottom View, Left View, Right View, Front View, and Back View. Apart from these orthographic views, this toolbar also has four preset isometric views: Southwest Isometric View, Southeast Isometric View, Northeast Isometric View, and Northwest Isometric View.

Tip

*When you choose any of the preset orthographic views from the **View** toolbar, the UCS is also aligned to that view.*

3D COORDINATE SYSTEMS

Similar to the 2D coordinate systems, there are two main types of coordinate systems. They are as follows.

Absolute Coordinate System

This type of coordinate system is similar to the 2D absolute coordinate system, in which the coordinates of the point are calculated from the origin (0,0). The only difference is that here along with the *X* and *Y* coordinates, the *Z* coordinates are also included. For example, if you want to draw a line in 3D space starting from the origin to a point say 10,6,6, the procedure to be followed is given next.

Command: **LINE**
Specify first point: **0,0,0**
Specify next point or [Undo]: **10,6,6**
Specify next point or [Undo]: Enter

Figure 23-23 shows the model drawn using the absolute coordinate systems.

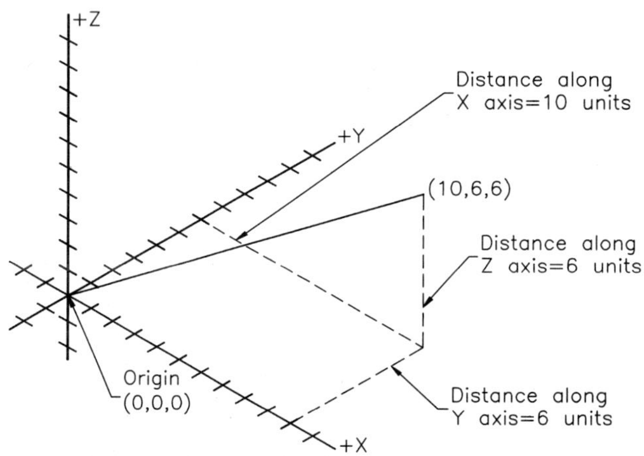

Figure 23-23 *Drawing a line from origin to 10,6,6*

Example 1

In this example you will draw the 3D wireframe model shown in Figure 23-24.

1. Choose the **Line** button from the **Draw** toolbar. The prompt sequence is given next.

   Specify first point: **2,2,0**
   Specify next point or [Undo]: **8,2,0**
   Specify next point or [Undo]: **8,6,0**
   Specify next point or [Close/Undo]: **2,6,0**
   Specify next point or [Close/Undo]: **C**

2. Choose **3D Views > Viewpoint Presets** from the **View** menu to display the **Viewpoint Presets** dialog box (Figure 23-25).

3. Set the value of the **X Axis** edit box to **225**.

4. Set the value of the **XY Plane** edit box to **38**. Choose **OK**.

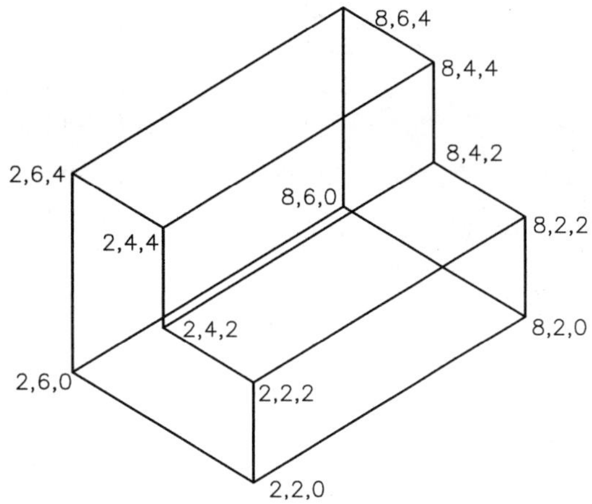

Figure 23-24 *3D wireframe model for Example 1*

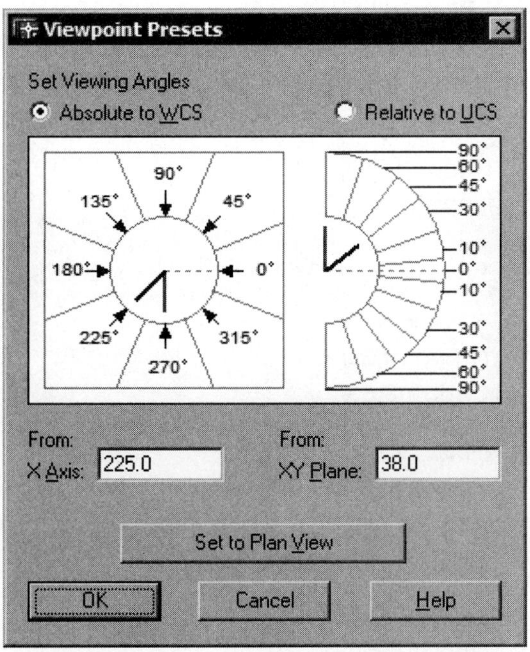

Figure 23-25 Viewpoint Presets *dialog box*

5. Choose the **Line** button from the **Draw** toolbar. The prompt sequence is as given next.

Specify first point: **2,2,0**
Specify next point or [Undo]: **2,2,2**
Specify next point or [Undo]: **2,4,2**
Specify next point or [Close/Undo]: **8,4,2**

Specify next point or [Close/Undo]: **8,2,2**
Specify next point or [Close/Undo]: **2,2,2**
Specify next point or [Close/Undo]: Enter

6. Choose the **Line** button from the **Draw** toolbar. The prompt sequence is given next.

Specify first point: **2,4,2**
Specify next point or [Undo]: **2,4,4**
Specify next point or [Undo]: **2,6,4**
Specify next point or [Close/Undo]: **2,6,0**
Specify next point or [Close/Undo]: Enter

7. Choose **3D Views > Viewpoint Presets** from the **View** menu to display the **Viewpoint Presets** dialog box.

8. Set the value of the **X Axis** edit box to **335**.

9. Set the value of the **XY Plane** edit box to **35**. Choose **OK**.

10. Choose the **Line** button from the **Draw** toolbar. The prompt sequence is given next.

Specify first point: **8,2,0**
Specify next point or [Undo]: **8,2,2**
Specify next point or [Undo]: Enter

11. Choose the **Line** button from the **Draw** toolbar. The prompt sequence is given next.

Specify first point: **8,4,2**
Specify next point or [Undo]: **8,4,4**
Specify next point or [Undo]: **8,6,4**
Specify next point or [Close/Undo]: **8,6,0**
Specify next point or [Close/Undo]: Enter

12. Choose the **Line** button from the **Draw** toolbar. The prompt sequence is given next.

Specify first point: **8,6,4**
Specify next point or [Undo]: **2,6,4**
Specify next point or [Undo]: Enter

13. Choose the **Line** button from the **Draw** toolbar. The prompt sequence is given next.

Specify first point: **8,4,4**
Specify next point or [Undo]: **2,4,4**
Specify next point or [Undo]: Enter

14. Enter **VPOINT** at the Command prompt. The prompt sequence is given next.

Current view direction: VIEWDIR=0.7424,-0.3462,0.5736
Specify a viewpoint or [Rotate] <display compass and tripod>: **-1,-1,1**

15. The final model should look similar to the one shown in Figure 23-26.

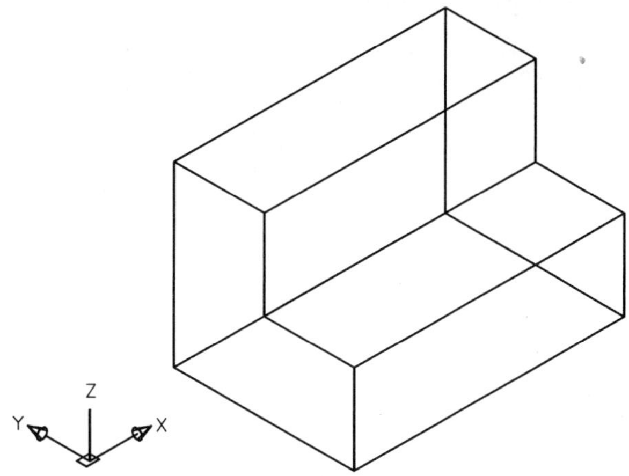

Figure 23-26 *Final 3D model for Example 1*

Relative Coordinate System

In the 3D, you are provided with three types of relative coordinate systems. They are as follows.

Relative Rectangular Coordinate System

This coordinate system is similar to the relative rectangular coordinate system, except that in 3D you have to enter the *Z* coordinate also along with the *X* and *Y* coordinates. The syntax of the relative rectangular system for the 3D is **@X coordinate,Y coordinate,Z coordinate**.

Relative Cylindrical Coordinate System

In this type of coordinate system, you can locate the point by specifying the distance of the point from the reference point, the angle in the *XY* plane, and its distance from the *XY* plane. The syntax of the relative cylindrical coordinate system is **@Distance from the reference point in the XY plane<Angle in the XY plane from the X axis, Distance from the XY plane along the Z axis**. Figure 23-27 shows various components of the relative cylindrical coordinate system.

Relative Spherical Coordinate System

In this type of coordinate system, you can locate the point by specifying the distance of the point from the reference point, the angle in the *XY* plane, and the angle from the *XY* plane. The syntax of the relative spherical coordinate system is **@Length of the line from the reference point<Angle in the XY plane from the X axis<Angle from the XY plane**. Figure 23-28 shows various components involved in the relative spherical coordinate system.

Tip

The major difference between the relative cylindrical and relative spherical coordinate system is that in the relative cylindrical coordinate system, the distance you specify is the distance from the reference point in the XY plane. On the other hand, in the relative spherical coordinate system, the distance you specify is the total length of the line in the 3D space.

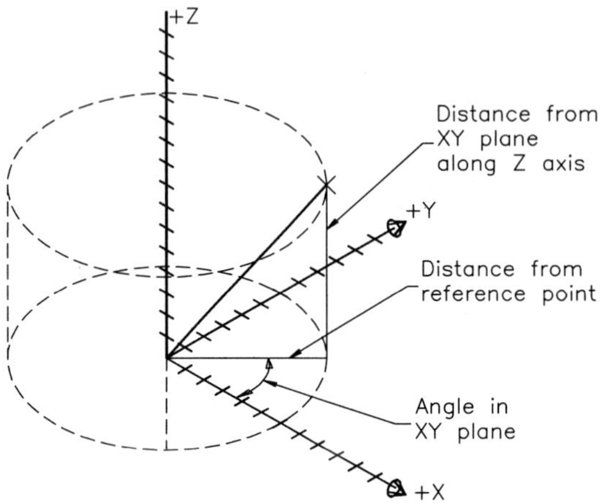

Figure 23-27 *Various components of the relative cylindrical coordinate system*

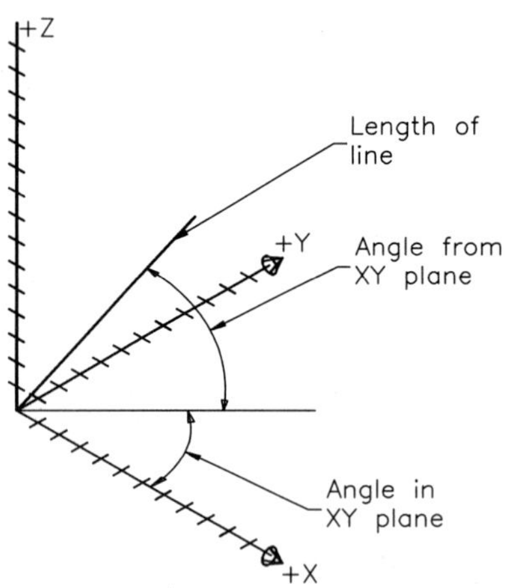

Figure 23-28 *Various components of the relative spherical coordinate system*

Example 2

In this example, you will draw the 3D wireframe model shown in Figure 23-29. The dimensions of the model are given in Figures 23-30 and 23-31.

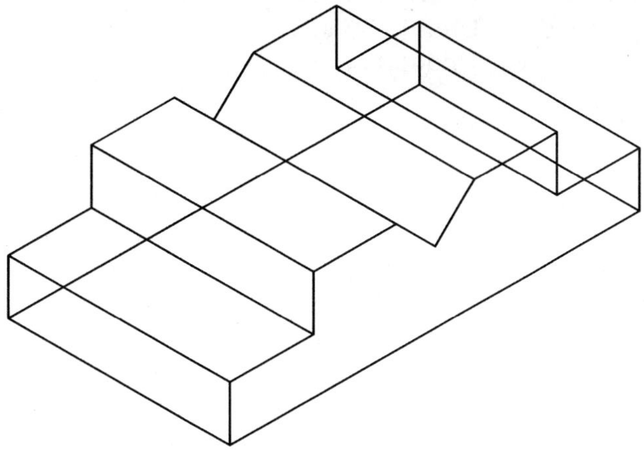

Figure 23-29 *Wireframe model for Example 2*

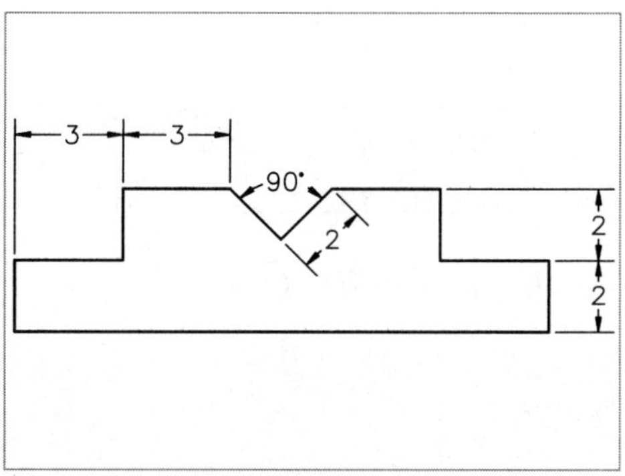

Figure 23-30 *Front view of the model*

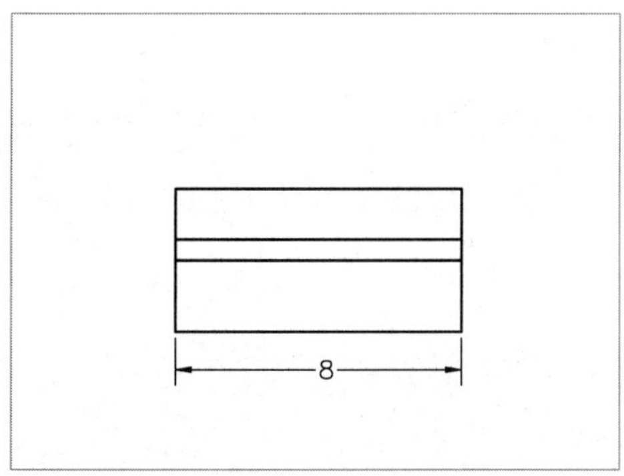

Figure 23-31 *Side view of the model*

1. Choose the **SW Isometric View** button from the **View** toolbar to shift to the southwest isometric view.

2. As the start point of the model is not given, you can start the model from any point. Choose the **Line** button from the **Draw** toolbar to invoke the **LINE** command. The prompt sequence is as follows:

   Specify first point: **4,2,0**
   Specify next point or [Undo]: **@0,0,2**
   Specify next point or [Undo]: **@3,0,0**
   Specify next point or [Close/Undo]: **@0,0,2**

Specify next point or [Close/Undo]: **@3,0,0**
Specify next point or [Close/Undo]: **@2<0<315**
Specify next point or [Close/Undo]: **@2<0<45**
Specify next point or [Close/Undo]: **@3,0,0**
Specify next point or [Close/Undo]: **@0,0,-2**
Specify next point or [Close/Undo]: **@3,0,0**
Specify next point or [Close/Undo]: **@0,0,-2**
Specify next point or [Close/Undo]: **C**

3. Choose the **Line** button from the **Draw** toolbar. The prompt sequence is as follows.

Specify first point: **4,2,0**
Specify next point or [Undo]: **@0,8,0**
Specify next point or [Undo]: **@0,0,2**
Specify next point or [Close/Undo]: **@3,0,0**
Specify next point or [Close/Undo]: **@0,0,2**
Specify next point or [Close/Undo]: **@3,0,0**
Specify next point or [Close/Undo]: **@2<0<315**
Specify next point or [Close/Undo]: **@2<0<45**
Specify next point or [Close/Undo]: **@3,0,0**
Specify next point or [Close/Undo]: **@0,0,-2**
Specify next point or [Close/Undo]: **@3,0,0**
Specify next point or [Close/Undo]: **@0,0,-2**
Specify next point or [Close/Undo]: **@0,-8,0**
Specify next point or [Close/Undo]: Enter

4. Complete the model by joining the remaining edges using the **LINE** command.

5. The final 3D model should look similar to the one shown in Figure 23-32.

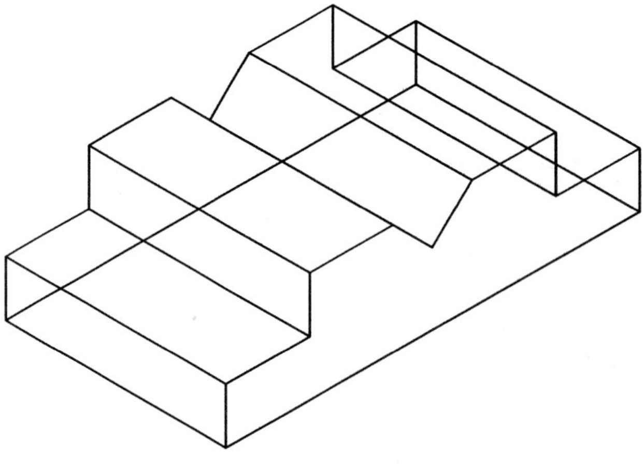

Figure 23-32 *Final 3D wireframe model for Example 2*

Exercise 1 *General*

In this exercise, you will draw the 3D wireframe model shown in Figure 23-33. The dimensions for the model are given in the figure itself. You can start the model from any point.

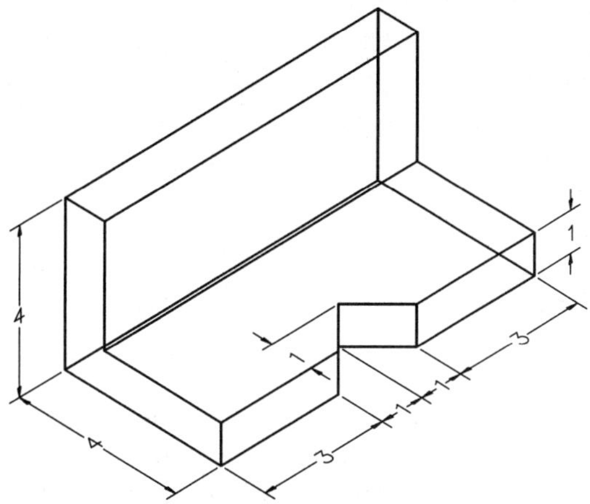

Figure 23-33 *Wireframe model for Exercise 1*

TRIM, EXTEND, AND FILLET COMMANDS IN 3D

As discussed in earlier chapters, you can use various options of these commands in 3D space. You can also use the **PROJMODE** and **EDGEMODE** variables. The **PROJMODE** variable sets the projection mode for trimming and extending. A value of 0 implies **True 3D mode**, that is, no projection. In this case, the objects must intersect in 3D space to trim or extend them. It is similar to using the **None** option of the **Project** option of the **EXTEND** and **TRIM** commands. A value of 1 projects onto the *XY* plane of the current UCS and is similar to using the **UCS** option of the **Project** option of the **TRIM** or **EXTEND** commands. A value of 2 projects onto the current view plane and is like using the **View** option of the **Project** option of the **TRIM** or **EXTEND** commands. The value of the **EDGEMODE** system variable controls how the cutting or trimming boundaries are decided. A value of 0 considers the selected edge with no extension. You can also use the **No Extend** option of the **Edge** option of the **TRIM** or **EXTEND** commands. A value of 1 considers an imaginary extension of the selected edge. This is similar to using the **Extend** option of the **Edge** option of the **TRIM** or **EXTEND** commands.

You can fillet coplanar objects whose extrusion directions are not parallel to the *Z* axis of the current UCS by using the **FILLET** command. The fillet arc exists on the same plane and has the same extrusion direction as the coplanar objects. If the coplanar objects to be filleted exist in opposite directions, the fillet arc will be on the same plane but will be inclined toward the positive *Z* axis.

SETTING THICKNESS AND ELEVATION FOR THE NEW OBJECTS

You can create the objects with a preset elevation or thickness using the **ELEV** command. This command is discussed next.

ELEV Command

Command: ELEV

This command is a transparent command and is used to set the elevation and thickness for new objects. The following prompt sequence is displayed when you invoke this command.

Specify new default elevation <0.0000>: *Enter the new elevation value.*
Specify new default thickness <0.0000>: *Enter the new thickness value.*

Elevation

This option is used to specify the elevation value for new objects. Setting the elevation is nothing but moving the working plane from its default position. By default, the working plane is on the world *XY* plane. You can move this working plane using the **Elevation** option of the **ELEV** command. However, remember that the working plane can be moved only along the *Z* axis (Figure 23-34). The default elevation value is 0.0 and you can set any positive or negative value for elevation. All the objects that will be drawn hereafter will be with the specified elevation. The **Elevation** option sets the elevation only for the new objects, and the existing objects are not modified using this option.

Thickness

This option is used to specify the thickness values for the new objects. It can be considered as another method of creating the surface models. Specifying the thickness creates the extruded surface models. The thickness is always taken along the *Z* axis direction (Figure 23-35). The 3D faces will be automatically applied on the vertical faces of the objects drawn with thickness. The **Thickness** option sets the thickness only for the new objects and the existing objects are not modified using this option.

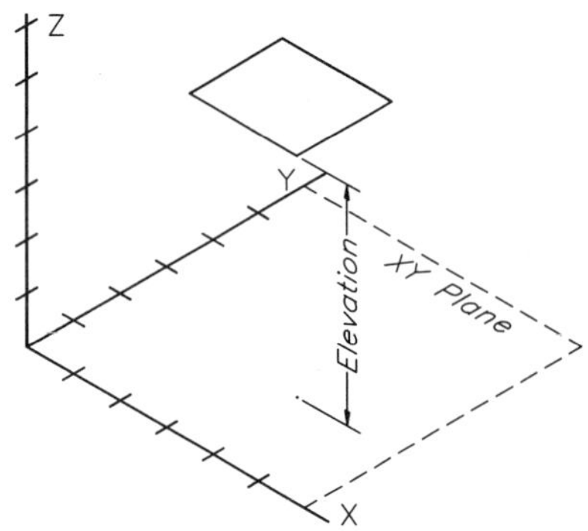

Figure 23-34 *Object drawn with elevation*

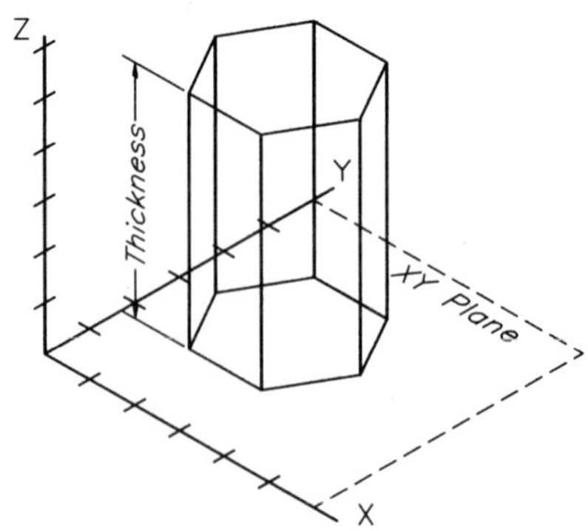

Figure 23-35 *Object drawn with thickness*

Note

You can suppress the hidden edges in the model by entering ***HIDE*** *at the command prompt. The* ***HIDE*** *command is discussed in the next section.*

Tip

The elevation value will be reset to 0.0 when you change the viewpoint using the preset orthographic views in the ***View*** *toolbar.*

The rectangles drawn using the ***RECTANG*** *command do not consider the thickness value set using the* ***ELEV*** *command. To draw a rectangle with thickness, use the* ***Thickness*** *option of the* ***RECTANG*** *command (Figure 23-36). To draw an ellipse with thickness, first set the value of the* ***PELLIPSE*** *system variable to* ***1****.*

To write a text with thickness, first write simple text and then change its thickness using any of the commands that modify the properties (Figure 23-37).

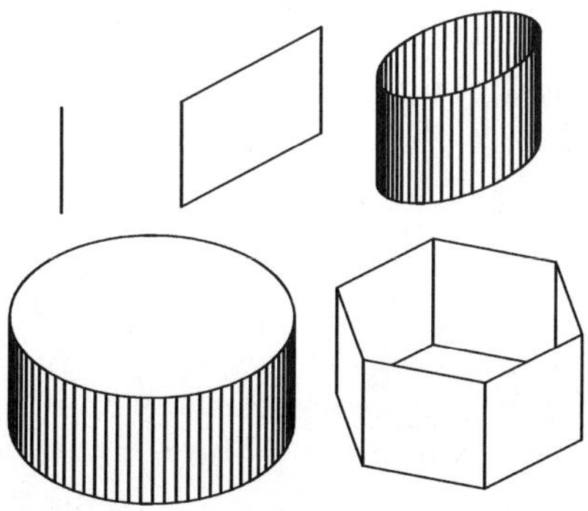

Figure 23-36 *A point, line, ellipse, circle, and a polygon drawn with a thickness of 5 units*

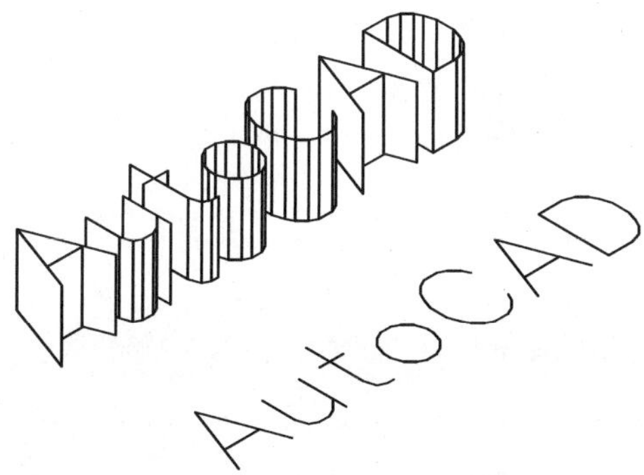

Figure 23-37 *Text with and without thickness*

SUPPRESSING THE HIDDEN EDGES

Toolbar:	Render > Hide
Menu:	View > Hide
Command:	HIDE

Whenever you create a surface or a solid model, the edges that lie at the back (also called the hidden edges) are also visible. As a result, the model appears like a wireframe model. You need to manually suppress the hidden lines in the 3D model using the **HIDE** command. When you invoke this command, all the objects on the screen are regenerated and the 3D models are displayed with the hidden edges suppressed. If the value of the **DISPSILH** system variable is set to **1**, the model is displayed only with the silhouette edges and the internal edges that have facets will not be displayed. The hidden lines are again included in the drawing when the next regeneration takes place. Figure 23-38 shows the surface model displaying the hidden edges and Figure 23-39 shows the surface model without the hidden edges.

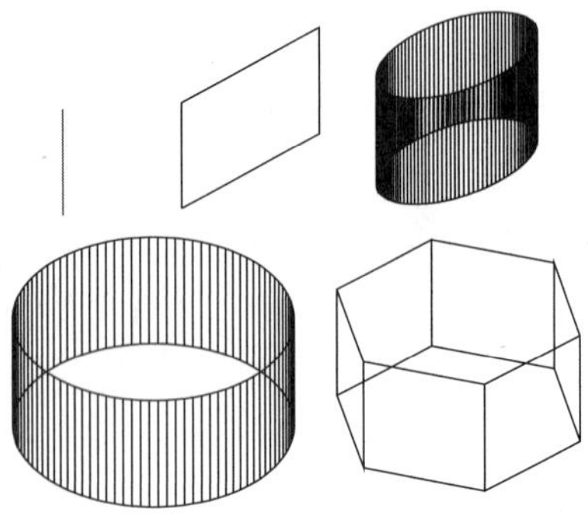

Figure 23-38 *3D models displaying all the edges*

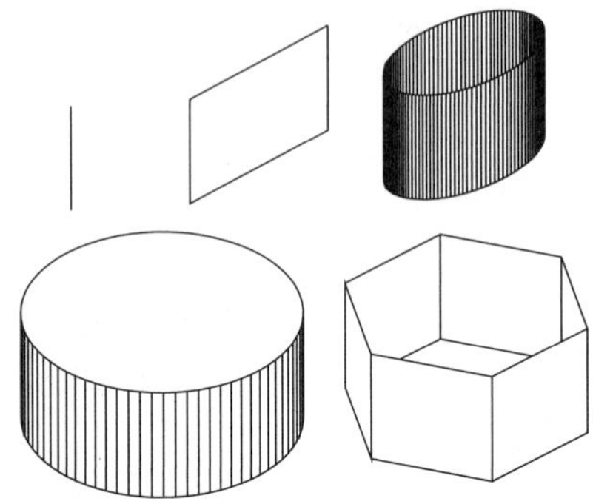

Figure 23-39 *Models with hidden edges suppressed*

Tip

*The **HIDE** command considers Circles, Solids, Traces, Polylines with width, Regions, and 3D Faces as opaque surfaces that will hide objects.*

CREATING 3D POLYLINES

Menu:	Draw > 3D Polyline
Command:	3DPOLY

The **3DPOLY** command is used to draw straight polylines in a 2D plane or in 3D space. This command is similar to the **POLYLINE** command except that using this command you can draw the polylines in the plane other then the *XY* plane also. However, this command does not provide the **Width** or the **Arc** option, and therefore you cannot create a 3D polyline with width or arc. The **Close** and the **Undo** options of this command are similar to those of the **POLYLINE** command.

Tip

*You can fit a spline curve about the 3D polyline using the **PEDIT** command. Use the **Spline curve** option to fit the spline curve and the **Decurve** option to remove the spline curve.*

CONVERTING WIREFRAME MODELS INTO SURFACE MODELS

Once you have created a wireframe model, you can convert it into a surface model using the **3DFACE** command and the **PFACE** command. These commands for converting a wireframe model into a surface model are discussed next.

Creating 3D Faces

Toolbar:	Surfaces > 3D Face
Menu:	Draw > Surfaces > 3D Face
Command:	3DFACE

This command is used to create 3D faces in the space. You can create three-sided or four-sided faces using this command. You can specify the same or a different *Z* coordinate value for each point of the face. When you invoke this command, you will be prompted to specify the first, second, third, and the fourth point of the 3D face. Once you have specified the first four points, it will again prompt you to specify the third and the fourth point. In this case, it will take the previous third and fourth point as the first and second point, respectively. This process will continue until you press ENTER in the **Specify third point or [Invisible] <exit>** prompt. Keep in mind that the points must be specified in the natural clockwise or the counterclockwise direction to create proper 3D face, see Figure 23-40.

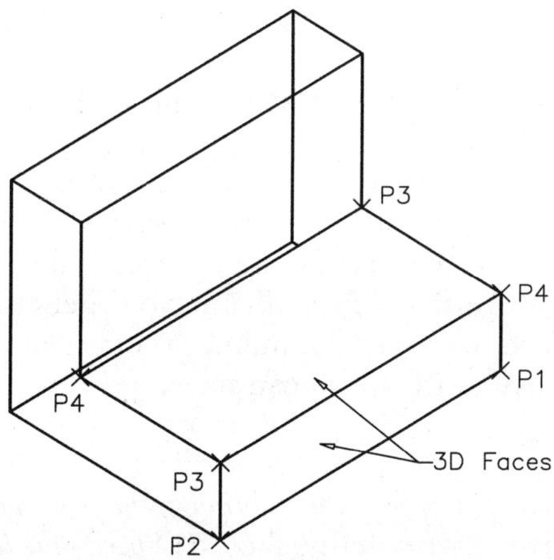

Figure 23-40 *3D faces applied to the wireframe model*

Figure 23-40 shows a simple 3D wireframe model. However, in case of complex wireframe models (Figure 23-41), it is not easy to apply the 3D faces. The reason for this is that when you apply a 3D face, it generates edges about all the points that you specify. These edges will be displayed on the wireframe model as shown in Figure 23-35. You can avoid these unwanted visible edges by applying the 3D faces with invisible edges. This is done by simply using the **Invisible** option. It is very important to mention here that the edge that will be created after you enter **I** at the **Specify next point or [Invisible] <exit>** prompt will not be the invisible edge, but the edge that will be created after this edge will be the invisible edge. Therefore, to make the P3,P4 edge invisible, specify P1 as first point, specify P2 as second point and before specifying P3 as third point, enter **I** at the **Specify third point or [Invisible] <exit>** prompt. Similarly, follow the same procedure for creating the other invisible edges (Figure 23-42).

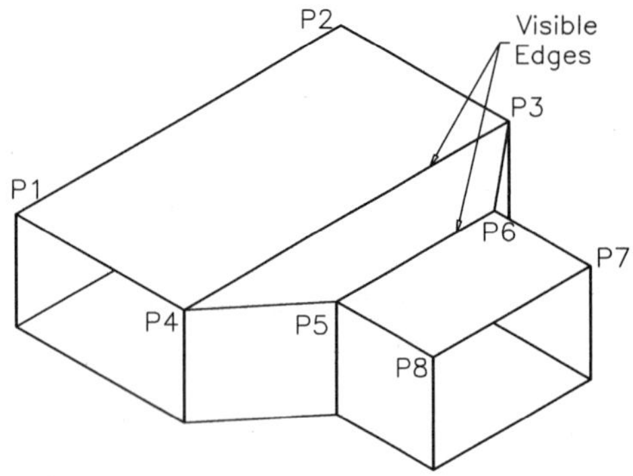

Figure 23-41 *Wireframe model with the edges*

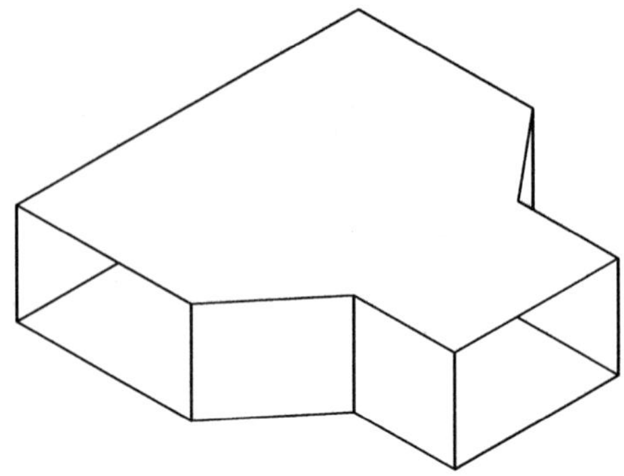

Figure 23-42 *Wireframe model without the edges*

Creating Polyface Meshes

Command: PFACE

The **PFACE** command is similar to the **3DFACE** command. This command also allows you to create a mesh of any desired surface shape by specifying the coordinates of the vertices and assigning the vertices to the faces in the mesh. The difference between them is that when you use the **3DFACE** command, you do not select the vertices that join another face twice. With the **PFACE** command, you need to select all the vertices of a face, even if they are coincident with vertices of another face. In this way, you can avoid generating unrelated 3D faces that have coincident vertices. Also, with this command, there is no restriction on the number of faces and the number of vertices the mesh can have.

Command: **PFACE**
Specify location for vertex 1: *Specify the location of the first vertex.*
Specify location for vertex 2 or <define faces>: *Specify the location of the second vertex.*
Specify location for vertex 3 or <define faces>: *Specify the location of the third vertex.*
Specify location for vertex 4 or <define faces>: *Specify the location of the fourth vertex.*
Specify location for vertex 5 or <define faces>: *Specify the location of the fifth vertex.*
Specify location for vertex 6 or <define faces>: *Specify the location of the sixth vertex.*
Specify location for vertex 7 or <define faces>: *Specify the location of the seventh vertex.*
Specify location for vertex 8 or <define faces>: *Specify the location of the eighth vertex.*
Specify location for vertex 9 or <define faces>: Enter

After defining the locations of all the vertices, press ENTER and you will assign vertices to the first face.

Face 1, vertex 1:
Enter a vertex number or [Color/Layer]: 1
Face 1, vertex 2:
Enter a vertex number or [Color/Layer] <next face>: 2

Face 1, vertex 3:
Enter a vertex number or [Color/Layer] <next face>: 3
Face 1, vertex 4:
Enter a vertex number or [Color/Layer] <next face>: 4

Once you have assigned the vertices to the first face, give a null response at the next prompt.

Face 1, vertex 5:
Enter a vertex number or [Color/Layer] <next face>:

You can also change the color and layer of the first face with in the **PFACE** command. For example, if you want to change the color of the first face to red, the following is the prompt sequence.

Face 1, vertex 5:
Enter a vertex number or [Color/Layer] <next face>: COLOR
Enter a color number or standard color name <BYLAYER>: RED

Now the first face will be red.

After assigning the vertices, give a null response at the next prompt. Now, AutoCAD provides you with the following prompt sequence in which you have to assign vertices to the second face.

Face 2, vertex 1:
Enter a vertex number or [Color/Layer]: 1
Face 2, vertex 2:
Enter a vertex number or [Color/Layer]: 4
Face 2, vertex 3:
Enter a vertex number or [Color/Layer]: 5
Face 2, vertex 4:
Enter a vertex number or [Color/Layer]: 4

Once you have assigned the vertices to the second face, give a null response at the next prompt.

Face 2, vertex 5:
Enter a vertex number or [Color/Layer]:
Face 3, vertex 1:
Enter a vertex number or [Color/Layer]:

Similarly, assign vertices to all the faces (Figure 23-43). If you want to make an edge invisible, specify a negative number for the first vertex of the edge. The display of the invisible edges of 3D solid surfaces is governed by the **SPLFRAME** system variable. If **SPLFRAME** is set to 0, invisible edges are not displayed. If this variable is set to a number other than 0, all invisible edges are displayed.

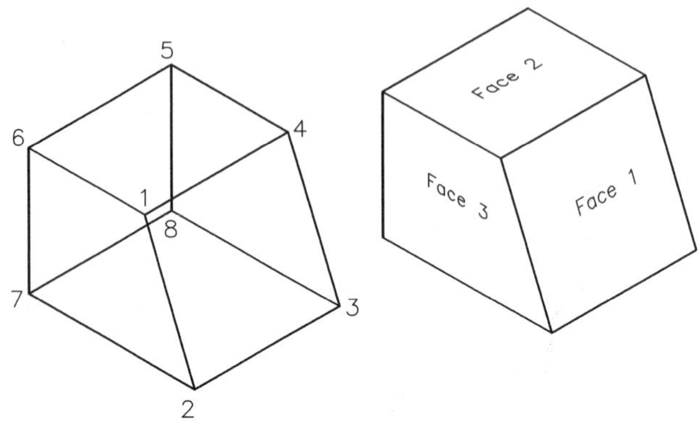

Figure 23-43 *Using the* ***PFACE*** *command*

Controlling the Visibility of the 3D Face Edges

Toolbar:	Surfaces > Edge
Menu:	Draw > Surfaces > Edge
Command:	EDGE

This command is used to control the visibility of the selected edges of the 3D faces. You can hide the edges of the 3D faces or display them using this command (Figures 23-44 and 23-45). When you invoke this command, you will be prompted to specify the 3D face edge to toggle the visibility or display. To hide an edge, select the edge, and to display the edges, enter **D** at the **Specify edge of 3dface to toggle visibility or [Display]** prompt. You can display all the invisible 3D face edges or the selected edges using this option.

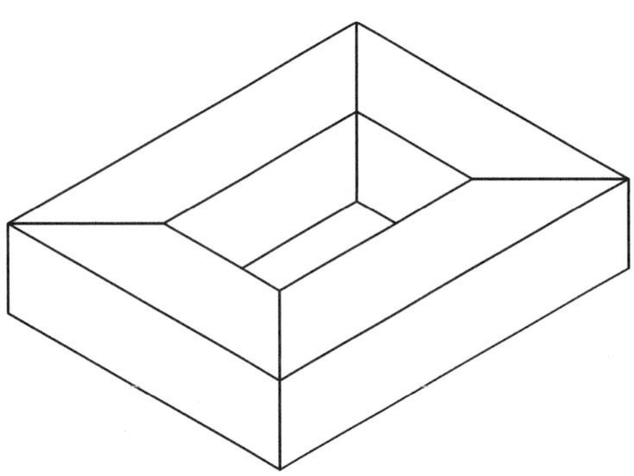

Figure 23-44 *Model with visible 3D face edges*

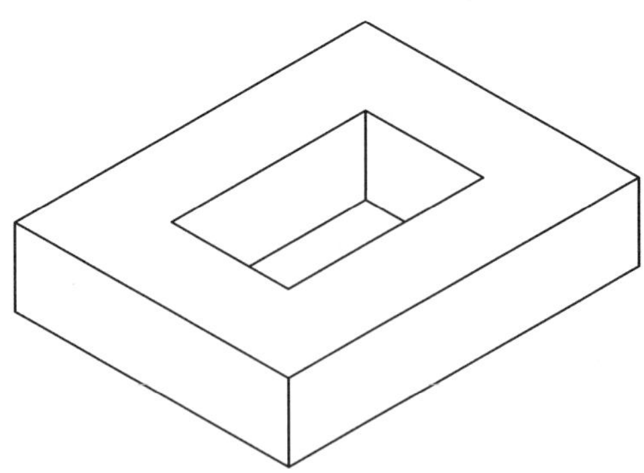

Figure 23-45 *Model after making the edges invisible*

Exercise 2 *General*

In this exercise, you will apply the 3D faces to the 3D wireframe model created in Example 2.

DIRECTLY CREATING THE SURFACE MODELS

The surface models created using the 3D faces consist only of straight edges. The surfaces with the curved edges cannot be created using the 3D faces. However, in AutoCAD, you can directly create various types of surface models consisting of straight edges or the curved edges. All of these surface models along with the commands used to create them are discussed next.

Creating Ruled Surfaces

Toolbar:	Surfaces > Ruled Surface
Menu:	Draw > Surfaces > Ruled Surface
Command:	RULESURF

The **RULESURF** command is used to create a ruled surface between two defining entities. You can create a ruled surface between two closed entities, two open entities, a point and a closed entity, and a point and an open entity. However, you cannot create a ruled surface between a closed and an open entity. Also, two points do not define a surface, therefore, you cannot create a ruled surface between two points. You can select lines, polylines, splines, circles, ellipses, arcs, or points for creating the ruled surfaces.

If the two defining entities are closed, the selection point does not matter. In case of circles, the start point of the surface will be determined by the combination of the direction of the current *X* axis and the current value of the **SNAPANG** system variable. In case of closed polylines, the ruled surface starts from the last vertex of the polyline and proceeds backwards along the polyline segment.

If the two defining entities are open, the selection points matter. If you select the open entities at the same end, it creates a straight polygon mesh, and if you select the open entities at the opposite ends, it creates a self-intersecting polygon mesh as shown in Figures 23-46 and 23-47.

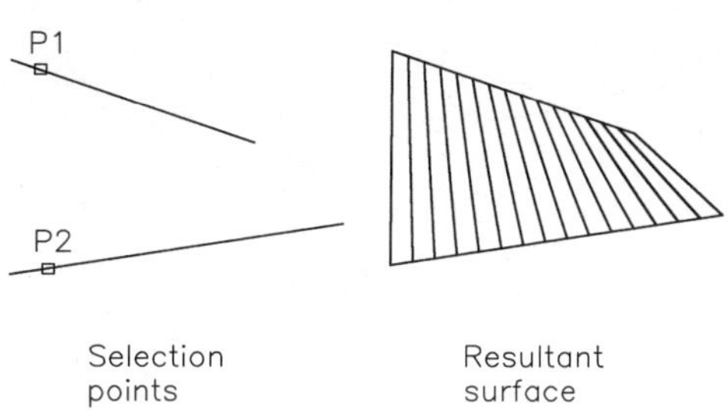

Figure 23-46 *Selecting objects at same ends*

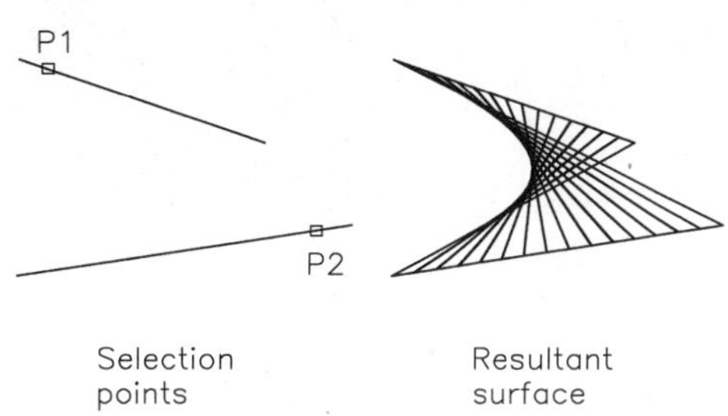

Figure 23-47 *Selecting objects at opposite ends*

The smoothness of the ruled surface can be increased using the **SURFTAB1** system variable, see Figures 23-48 and 23-49. The default value of this system variable is 6 and it excepts any integer between 2 and 32766 as its value. Remember, this value must be increased before creating the surface.

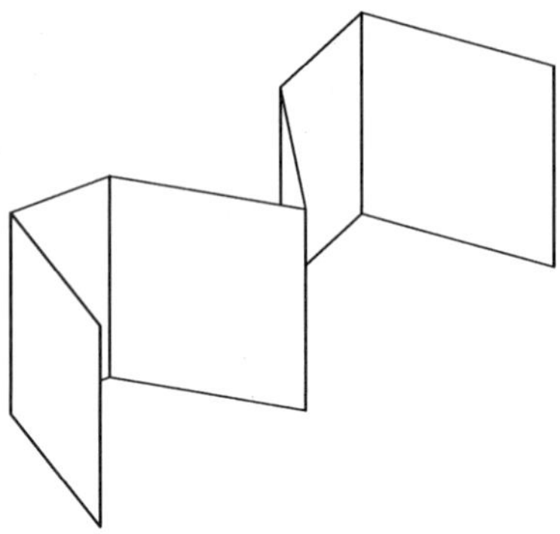

Figure 23-48 *Ruled surface created keeping the value of the SURFTAB1 system variable as 6*

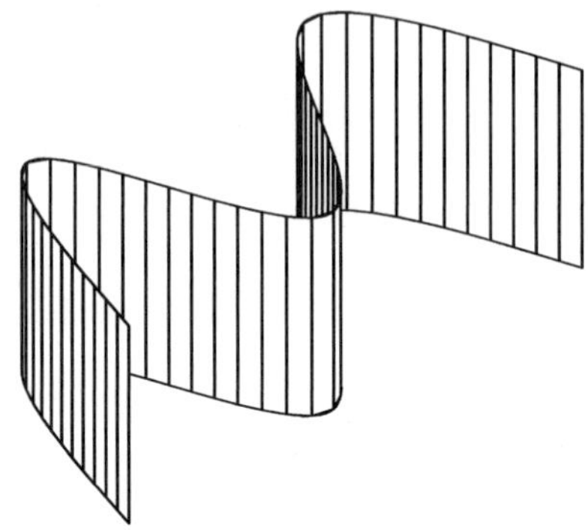

Figure 23-49 *Ruled surface created keeping the value of the SURFTAB1 system variable as 50*

Creating Tabulated Surfaces

Toolbar:	Surfaces > Tabulated Surface
Menu:	Draw > Surfaces > Tabulated Surface
Command:	TABSURF

The **TABSURF** command is used to create tabulated surfaces using a path curve along the direction defined by a direction vector. The entities that can be used as base curves are lines, circles, arcs, splines, 2D polylines, 3D polylines, or ellipses. The length of the tabulated surface will be equal to the length of the direction vector, and the point used to select the direction vector will determine the direction of the resultant surface, see Figures 23-50 and 23-51.

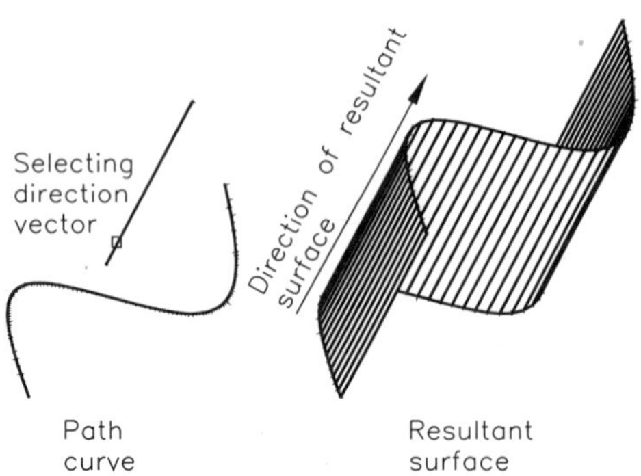

Figure 23-50 *Direction of tabulated surface*

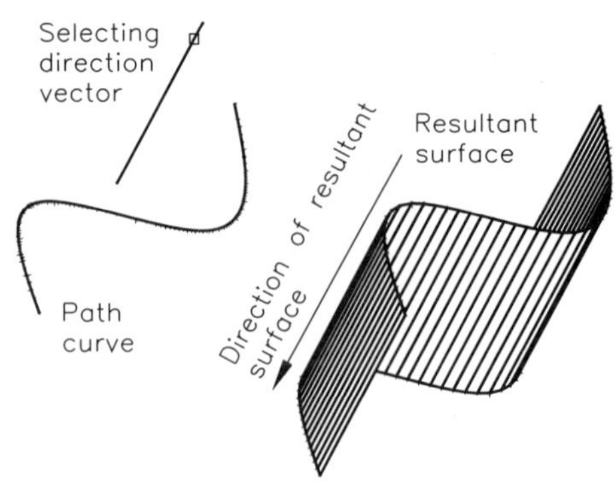

Figure 23-51 *Direction of tabulated surface*

If the direction vector consists of more than one segment, then all the intermediate points will be ignored and an imaginary vector will be drawn between the first and the last point on the polyline. This imaginary vector is considered as the direction vector and the tabulated surface is created along this imaginary vector, as shown in Figure 23-52.

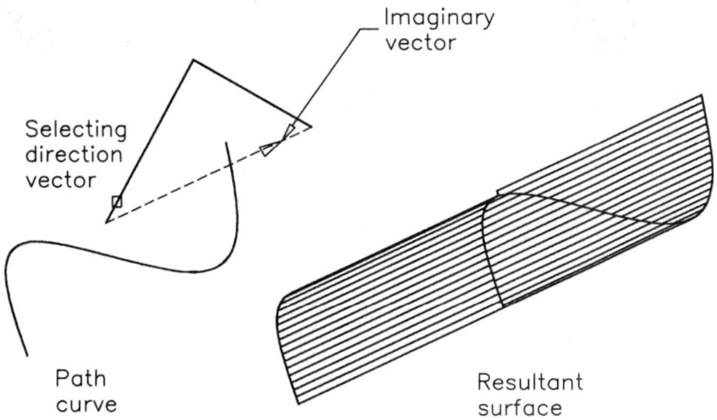

Figure 23-52 *Direction of tabulated surface*

Tip

You cannot select a curved entity as the direction vector for defining the direction of the tabulated surface.

*You can increase the smoothness of the tabulated surface by increasing the value of the **SURFTAB1** system variable.*

Creating Revolved Surfaces

Toolbar:	Surfaces > Revolved Surface
Menu:	Draw > Surfaces > Revolved Surface
Command:	REVSURF

The **REVSURF** command is used to create revolved surface by revolving a path curve about a specified axis of rotation. You can select an open entity or a closed entity as the path curve, however, the path curve should be a single entity. The direction of the revolution depends upon the side from which you select the revolution axis. Since it is not possible to define the clockwise and counterclockwise directions in the 3D space, therefore, AutoCAD uses the right-hand thumb rule (see the section "Conventions in AutoCAD") to determine the direction of revolution. Figure 23-53 shows a path curve and a revolution axis and Figure 23-54 shows the revolved surface.

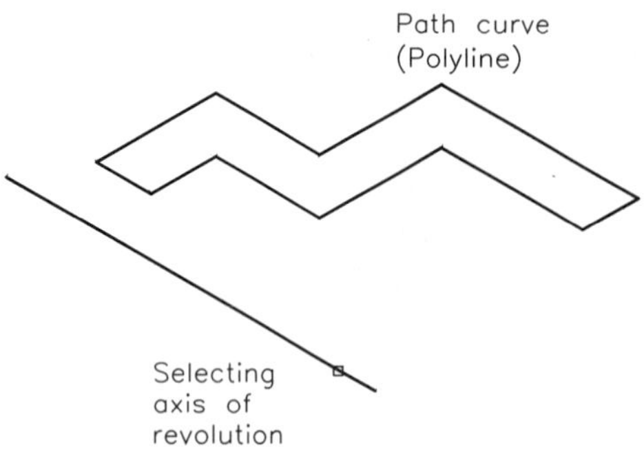

Figure 23-53 Selecting the object to revolve and the axis of revolution

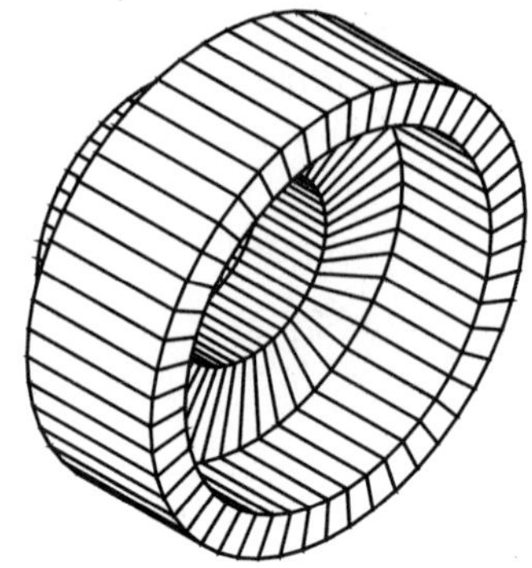

Figure 23-54 The resultant revolved surface

You can specify the start angle and the included angle for the revolved surface, see Figures 23-55 and 23-56.

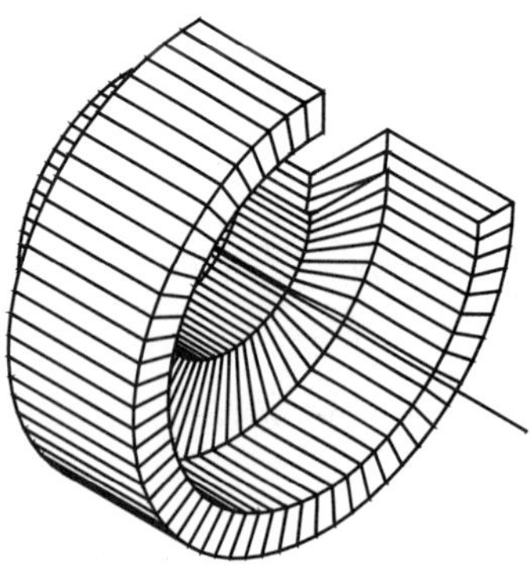

Figure 23-55 Revolved surface with start angle as 0 and included angle as 270

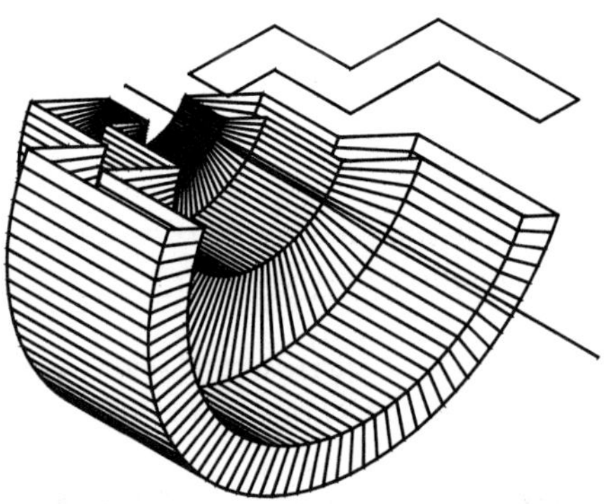

Figure 23-56 Revolved surface with start angle as 25 and included angle as 180

The smoothness of the tabulated surface depends upon the value of the **SURFTAB1** system variable. AutoCAD draws tabulated lines in the direction of the revolution and the number of tabulated lines is specified by the value of the **SURFTAB1** system variable. If the revolved surface is created using the path curve that is a line, arc, circle, spline, or a spline fit polyline, then another set of lines are drawn to divide it into equal-sized surfaces. The number of surfaces is defined using the **SURFTAB2** system variable (Figures 23-57).

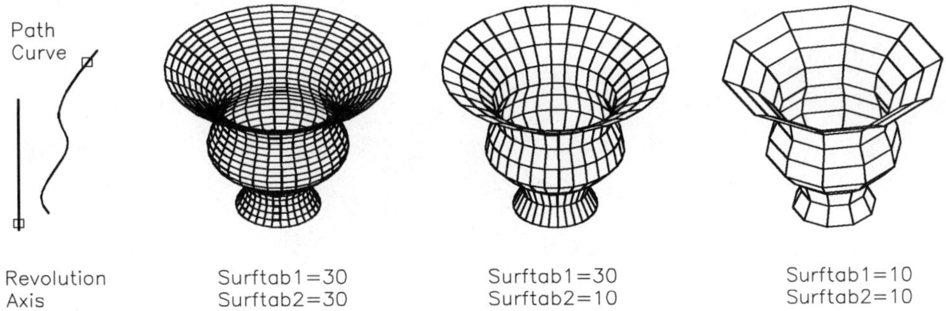

Figure 23-57 *Figure showing different values of SURFTAB1 and SURFTAB2 system variable*

Creating Four-sided Polygon Mesh

Toolbar:	Surfaces > Edge Surface
Menu:	Draw > Surfaces > Edge Surface
Command:	EDGESURF

This command is used to create a 3D polygon mesh using four adjoining edges forming a topologically closed loop. The entities that can be selected as the edges are lines, arcs, splines, open 2D polylines, or open 3D polylines. You can select the edges in any sequence. The smoothness of the edge surface will depend on the values of the **SURFTAB1** and **SURFTAB2** system variables. The edge selected first will define the M direction (**SURFTAB1** direction) and the other two edges that start from the endpoint of the first edge will define the N direction (**SURFTAB2** direction), see Figures 23-58 and 23-59.

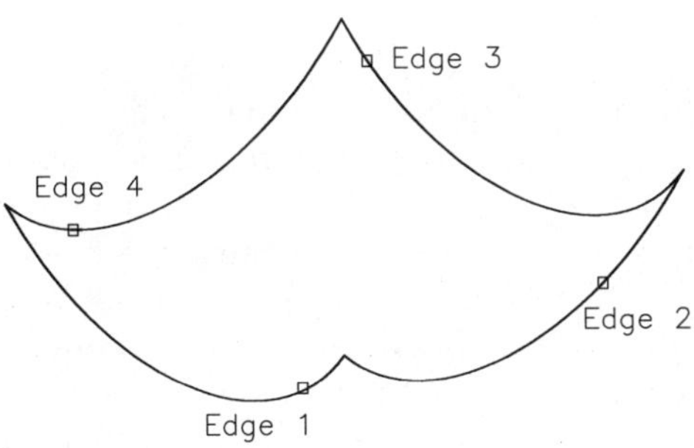

Figure 23-58 *Selecting edges for edge surface*

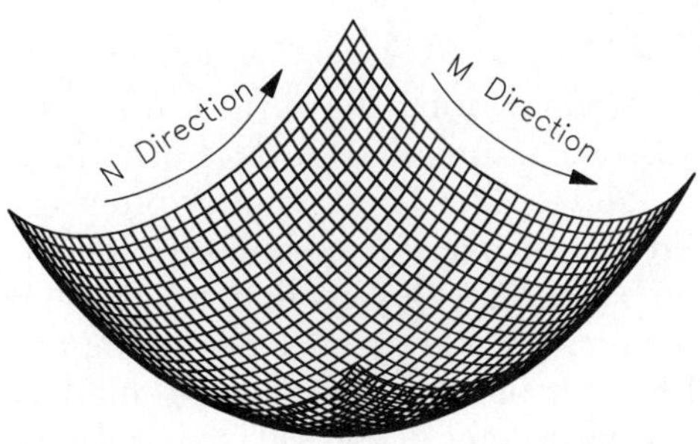

Figure 23-59 *Resultant edge surface*

Exercise 3 *Architectural*

In this exercise, you will create the table shown in Figure 23-60 using various types of surfaces. Assume the dimensions for the table.

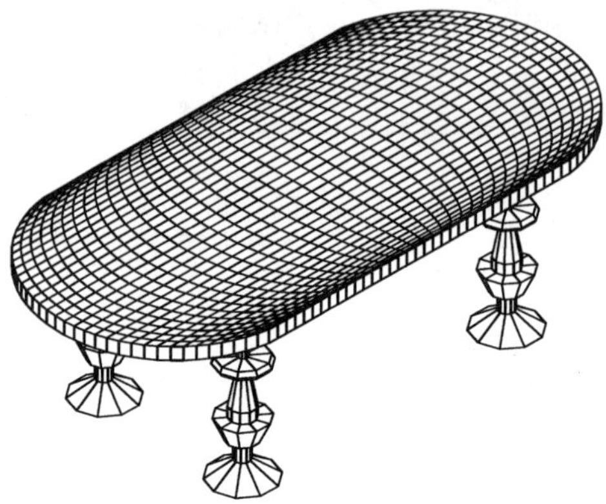

Figure 23-60 *Model for Exercise 3*

Hints

*1. Draw the top face and the bottom face of the table using the **EDGESURF** command.*
*2. Draw the side faces of the table using the **TABSURF** command.*
*3. Draw the legs of the table using the **REVSURF** command.*

DRAWING STANDARD SURFACE PRIMITIVES

Toolbar:	Surfaces
Menu:	Draw > Surfaces > 3D Surface
Command:	3D

Apart from creating the surface using the previously mentioned commands, you can also create some standard 3D surface primitives like box, cone, dish dome, pyramid, torus, sphere, mesh, and wedge. These standard surface primitives can be created using the **3D** command. When you invoke this command by choosing **Draw > Surfaces > 3D Surfaces** from the menu bar, the **3D Objects** dialog box will be displayed as shown in Figure 23-61. Double-click on the image tile of the surface you want to create. Remember that if you enter this command at the command line, it will be executed using the command line. In this case, the dialog box will not be displayed. You will have to select the required surface using the Command line only.

All the 3D surface primitives that can be created using this command can also be chosen directly from the **Surfaces** toolbar.

Creating a Box-Shaped Surface Mesh

This option is used to create a 3D polygon mesh in the shape of the box, see Figure 23-62. When you invoke this option, you will be prompted to specify the corner

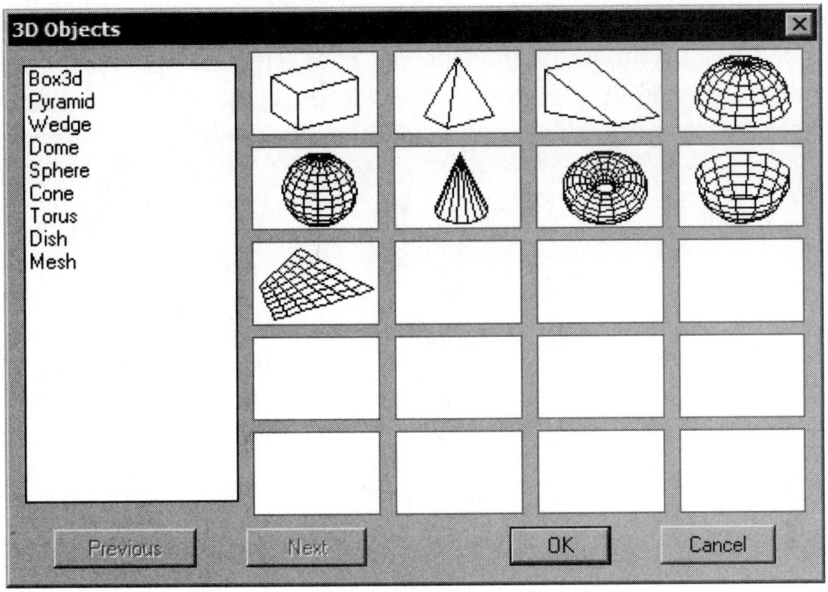

Figure 23-61 ***3D Objects*** *dialog box*

of the box. Once you specify the corner of the box, you will be prompted to specify the length, width, and the height of the box. In case you want to draw a cubical surface mesh, enter **C** at the **Specify width of box or [Cube]** prompt. The length you have specified will be taken as the length of the cubical surface mesh. Note that the length of the box will be taken along the *X* axis of the current UCS, the width along the *Y* axis, and the height along the *Z* axis of the current UCS. After specifying the length, width, and height, you will be prompted to specify the rotation angle of the box. You can also specify the rotation angle using the **Reference** option. The base point of the rotation in this case will be the corner of the box that you had specified.

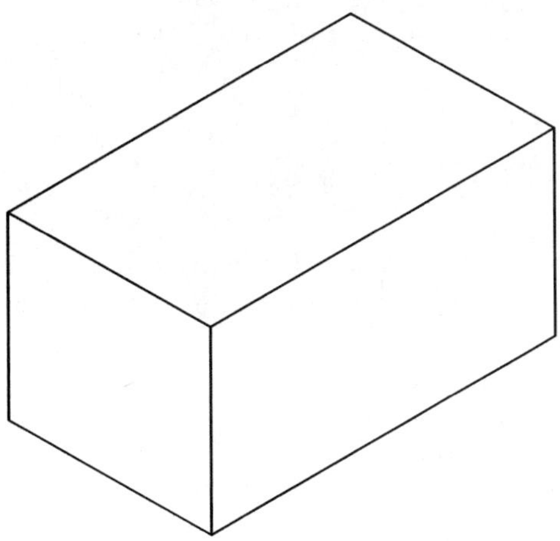

Figure 23-62 *3D box-shaped surface mesh*

Creating a Cone-Shaped Surface Mesh

You can create a surface mesh model of a cone (Figure 23-63) by using any one of the methods for creating surface mesh models. If you use the **3D** command, the prompt sequence will be as follows.

Command: **3D**
Enter an option
[Box/Cone/DIsh/DOme/Mesh/Pyramid/
Sphere/Torus/Wedge]: **CONE**

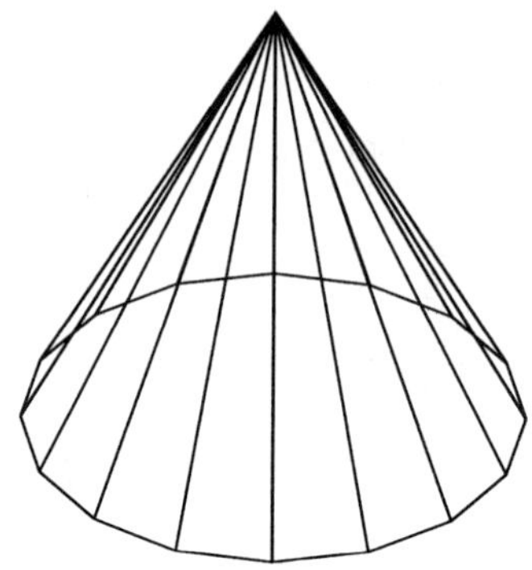

Figure 23-63 *3D cone-shaped surface mesh*

You can draw a truncated cone or a pointed cone.

Pointed Cone. Three parameters are required to draw a pointed cone:

1. Center point of the base of the cone.
2. Diameter/radius of the base of the cone.
3. Height of the cone.

Choose the **Cone** button in the **Surfaces** toolbar. The prompt sequence for drawing a pointed cone is given next.

Command: **AI_CONE**
Specify center point for base of cone: *Specify the center of base.*
Specify radius for base of cone or [Diameter]: *Specify the radius of the base.*

The next prompt asks you for the diameter/radius of the top. The default is zero (0), which results in a pointed cone.

Specify radius for top of cone or [Diameter] <0>: Enter
Specify height of cone: *Specify the height of the cone.*
Enter number of segments for surface of cone <16>: *Specify the number of segments for the cone.*

Note

The value of segments for the surface of cone can vary from 2 to 32767.

Truncated Cone. A truncated cone is a cone whose top side (pointed side) has been truncated (Figure 23-64). To draw a truncated cone, you require the following four parameters.

1. Center point of the base of the cone.
2. Diameter/radius of the base of the cone.
3. Diameter/radius of the top of the cone.
4. Height of the cone.

The prompt sequence for drawing a truncated cone is given next.

Command: **AI_CONE**
Specify center point for base of cone: *Specify the center of base.*
Specify radius for base of cone or [Diameter]: *Specify the radius of the base.*
Specify radius for top of cone or [Diameter] <0>: *Specify the radius of the top.*
Specify height of cone: *Specify the height of the cone.*
Enter number of segments for surface of cone <16>: *Specify the number of segments for the cone.*

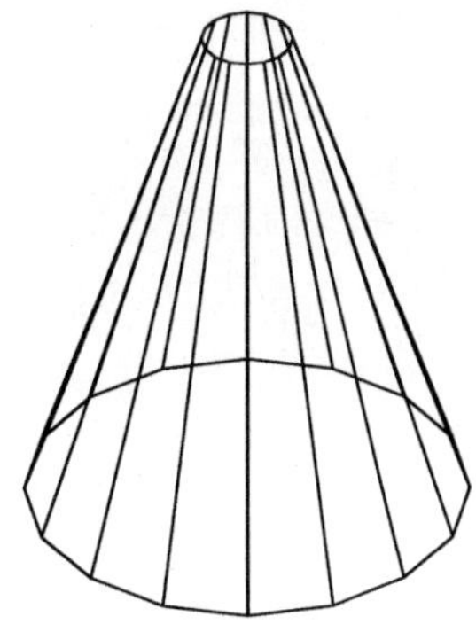

Figure 23-64 *Truncated 3D cone-shaped surface*

Tip
To draw a cylindrical surface mesh, draw a truncated cone with identical base and top diameters or radii.

Creating a Surface Mesh Dish

A dish is a hemisphere with the convex surface facing toward the bottom. Just as in the case of the dome, you can use any one of the methods described previously for drawing surface mesh models to construct a dish (Figure 23-65).

Command: **3D**
Enter an option
[Box/Cone/DIsh/DOme/Mesh/Pyramid/Sphere/Torus/Wedge]: DI
Specify center point of dish: *Specify the location of the center of the dish.*
Specify radius of dish or [Diameter]: *Specify the radius of the dish or enter D for the diameter.*
Enter number of longitudinal segments for surface of dish <16>: *Specify the number of longitudinal (vertical) segments for the dish.*
Enter number of latitudinal segments for surface of dish <8>: *Specify the number of latitudinal (horizontal) segments for the dish.*

Creating a Surface Mesh Dome

A dome is a hemisphere with the convex surface facing up (toward the top). You can create a surface mesh model of a dome (Figure 23-66) by using any one of the methods for drawing surface mesh models.

Command: **3D**
Enter an option
[Box/Cone/DIsh/DOme/Mesh/Pyramid/Sphere/Torus/Wedge]: DO
Specify center point of dome: *Specify the location of the center of the dome.*
Specify radius of dome or [Diameter]: *Specify the radius of the dome.*
Enter number of longitudinal segments for surface of dome <16>: *Specify the number of longitudinal (vertical) segments for the dome.*

Enter number of latitudinal segments for surface of dome <8>: *Specify the number of latitudinal (horizontal) segments for the dome.*

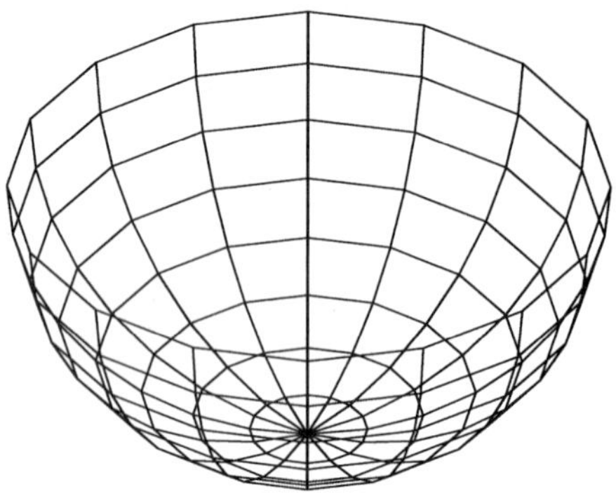

Figure 23-65 *3D surface dish*

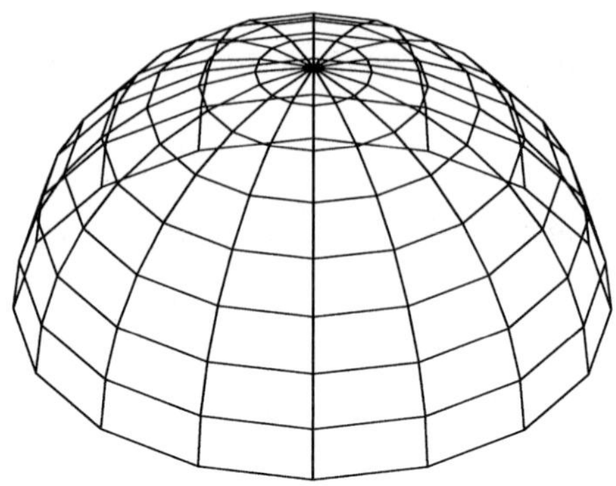

Figure 23-66 *3D surface dome*

Creating a Planar Mesh

You can create a planar model of a mesh (Figure 23-67) using any one of the methods for drawing surface mesh objects. The type of mesh obtained is also known as planar mesh because the mesh is generated in a single plane. The prompt sequence is given next.

Command: **3D**
Enter an option
[Box/Cone/DIsh/DOme/Mesh/Pyramid/
Sphere/Torus/Wedge]: **M**

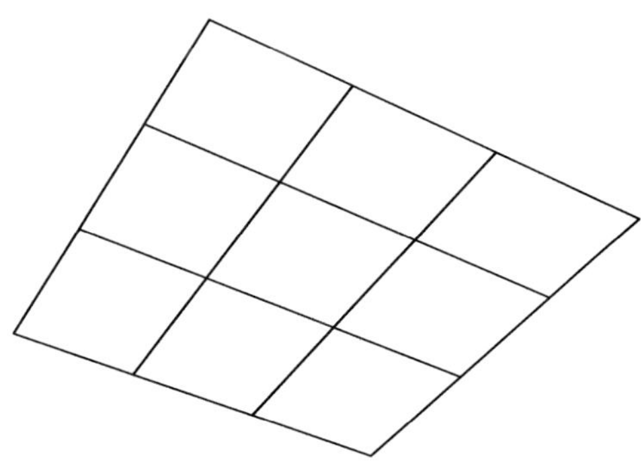

Figure 23-67 *3D Planar surface mesh*

In the next four prompts, AutoCAD asks you to specify the four corner points of the mesh.

Specify first corner point of mesh: *Specify the first corner point.*
Specify second corner point of mesh: *Specify the second corner point.*
Specify third corner point of mesh: *Specify the third corner point.*
Specify fourth corner point of mesh: *Specify the fourth corner point.*

Once you have specified the four corner points, AutoCAD prompts you to specify the number of rows (M) and the number of columns (N) of vertices. The number of vertices in both directions should be in the range of 2 to 256; that is, M and N should lie in the range of 2 to 256.

Enter mesh size in the M direction: *Enter the number of rows in the mesh.*
Enter mesh size in the N direction: *Enter the number of columns in the mesh.*

Creating a Pyramid Surface Mesh

You can create a surface mesh model of a pyramid by using any one of the methods for drawing surface meshes. Choose the **Pyramid** button in the **Surfaces** toolbar. You can enter AI_PYRAMID at the Command prompt also.

Command: **AI_PYRAMID**
Specify first corner point for base of pyramid: *Specify the location of the first base point.*
Specify second corner point for base of pyramid: *Specify the location of the second base point.*
Specify third corner point for base of pyramid: *Specify the location of the third base point.*

Once you have located the three base points on the screen, the next prompt issued is **Specify fourth corner point for base of pyramid or [Tetrahedron]**. At this prompt, you are given the choice of specifying the fourth base point of the pyramid or constructing a tetrahedron. Actually, a pyramid is any polyhedron with any polygonal base (from 3 to n base sides) and all other sides being triangles that all meet at the top. Therefore, a tetrahedron or an Egyptian-style pyramid is just a special case of pyramids, which never can have a flattened top in true geometrical terms. A more correct description of the flattened top case would be that AutoCAD enables you to create a truncated pyramid tetrahedron with a flattened top. The **Ridge** option simply contradicts the definition of a pyramid; it may be created with the Pyramid command, but it really is not a true pyramid.

Specify fourth corner point for base of pyramid or [Tetrahedron]: *Specify the location of the fourth base point.*

The next prompt issued is given below.

Specify apex point of pyramid or [Ridge/Top]:

At this prompt, you can select any one of the following options.

Pyramid with an Apex: Apex Point Option. The default option is the **Apex point** (Figure 23-68). With this option, you can specify the location of the apex of the pyramid. You must enter the *X*, *Y*, *Z* coordinates to specify the location of the apex or use the .X, .Y, .Z filters. If you simply select a point on the screen, the *X* and *Y* coordinates of the apex will assume the *X* and *Y* coordinates of the point selected, and the elevation, or the *Z* coordinate, will be the same as the *Z* coordinate of the base points. Hence, the apex of the pyramid will have the same elevation as the base of the pyramid. Only if you are in the 3D space [for example, your viewpoint is (1,-1,1)] you can specify the apex by selecting a point on the screen.

Pyramid with a Ridge: Ridge Option. This option (Figure 23-69) can be invoked by entering R at the **Specify apex point of pyramid or [Ridge/Top]** prompt. When you enter this option, the most recently drawn line of the base is highlighted. This indicates that the first point on the ridge will be perpendicular to the highlighted line. The next prompts are given below.

Specify first ridge end point of pyramid: *Specify a point inside the highlighted line.*
Specify second ridge end point of pyramid: *Specify a point inside the second highlighted line.*

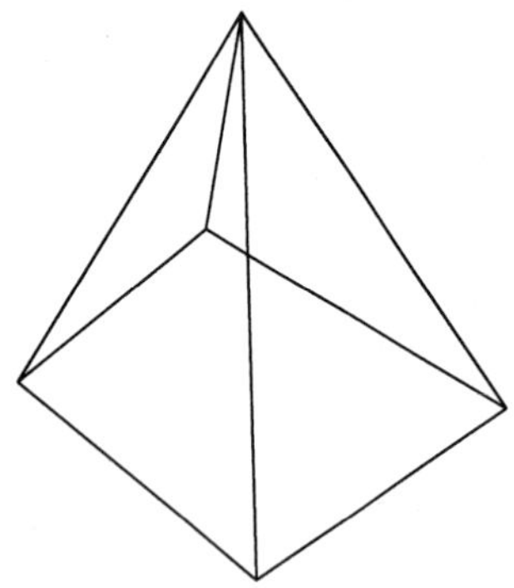

Figure 23-68 Surfaced pyramid with an apex

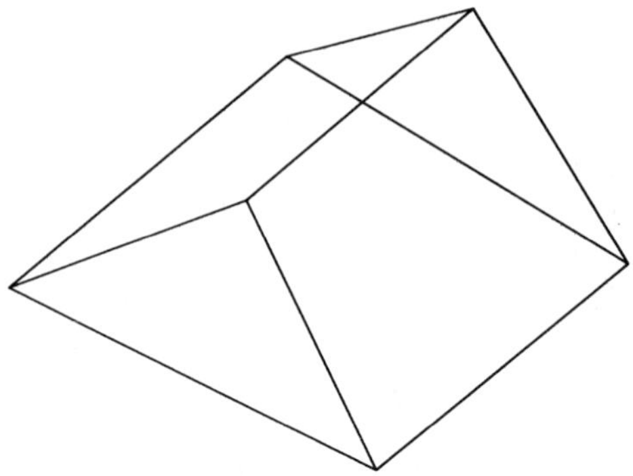

Figure 23-69 Surfaced pyramid with a ridge

Pyramid with a Top: Top Option. You can invoke this option (Figure 23-70) by entering T at the **Specify apex point of pyramid or [Ridge/Top]** prompt. When you invoke this option, a rubber-band line is attached to the first corner of the base. The rubber-band line stretches from the corner point to the crosshairs. This allows you to select the first top point with respect to the first corner, the second top point with respect to the second corner, the third top point with respect to the third corner, and the fourth top point with respect to the fourth corner. You should use filters to locate the points. If you specify the same location (coordinates) for all four top points, the result is a regular pyramid with its apex at the location where all the four top points are located.

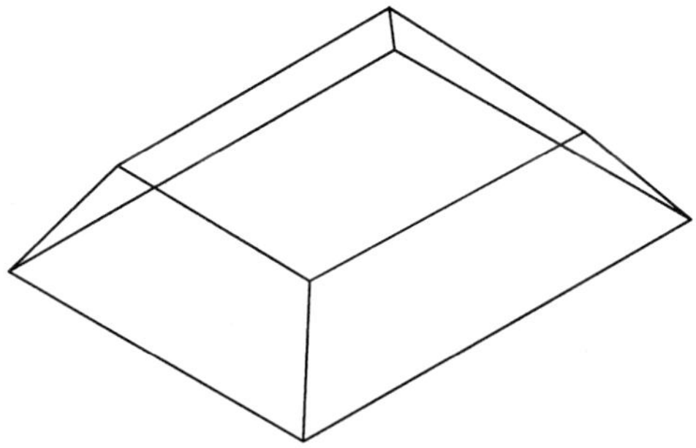

Figure 23-70 Surfaced pyramid with a top

Specify first corner point for top of pyramid: *Specify the location of the first top point.*
Specify second corner point for top of pyramid: *Specify the location of the second top point.*
Specify third corner point for top of pyramid: *Specify the location of the third top point.*
Specify fourth corner point for top of pyramid: *Specify the location of the fourth top point.*

Instead of entering the fourth base point for the pyramid, you can enter T for Tetrahedron at the **Specify fourth corner point for base of pyramid or [Tetrahedron]** prompt to draw a tetrahedron. You are going to construct two types of tetrahedrons. The first has four triangular sides and an apex, and the second has a flat top. You will first draw the tetrahedron with an apex and then a flat-topped tetrahedron.

Tetrahedron with an Apex: Apex Option. As mentioned before, with this option you can draw a tetrahedron that has four triangular sides and an apex (Figure 23-71). Choose the

Pyramid button in the **Surfaces** toolbar. The following is the prompt sequence to draw a tetrahedron with an apex.

Command: **AI_PYRAMID**
Specify first corner point for base of pyramid: *Specify the location of the first base point.*
Specify second corner point for base of pyramid: *Specify the location of the second base point.*
Specify third corner point for base of pyramid: *Specify the location of the third base point.*
Specify fourth corner point for base of pyramid or [Tetrahedron]: T
Specify apex point of tetrahedron or [Top]: *Specify the apex point.*

Flat-Topped Tetrahedron: Top Option. With this option, you can draw a tetrahedron that has a flat top (Figure 23-72). This option can be invoked by entering T at the **Specify apex point of tetrahedron or [Top]** prompt. The next prompts displayed are given next.

Specify apex point of tetrahedron or [Top]: T
Specify first corner point for top of tetrahedron: *Specify the location of the first top point.*
Specify second corner point for top of tetrahedron: *Specify the location of the second top point.*
Specify third corner point for top of tetrahedron: *Specify the location of the third top point.*

Tip
Specify the apex point and all the top points with the **.X**, **.Y**, **.Z** *filters, or enter the exact coordinate values.*

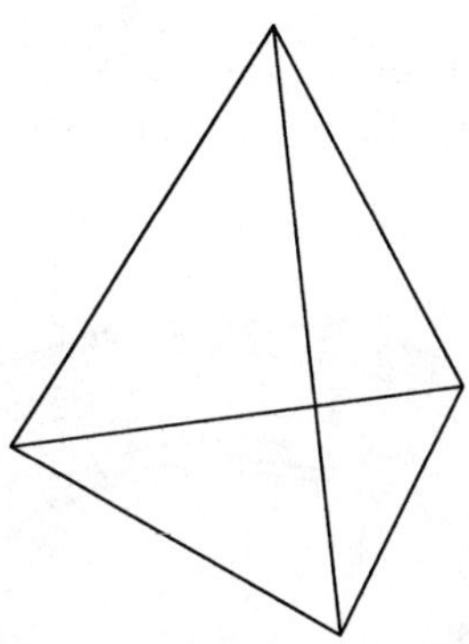

Figure 23-71 *Tetrahedron with an apex*

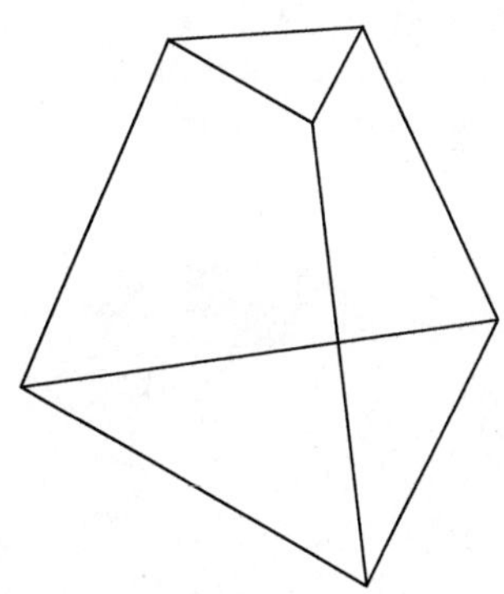

Figure 23-72 *Tetrahedron with a top*

Creating a Spherical Surface Mesh

The spherical surface mesh (Figure 23-73) has its central axis parallel to the *Z* axis, and the latitudinal lines are parallel to the *XY* plane of the current UCS. You can also append AI_ as a prefix to SPHERE and use it as a command. The prompt sequence is given next.

Command: **3D**
Enter an option
[Box/Cone/DIsh/DOme/Mesh/Pyramid/Sphere/Torus/Wedge]: S
Specify center point of sphere: *Specify the location of center point.*
Specify radius of sphere or [Diameter]: *Specify the diameter/radius of the sphere.*
Enter number of longitudinal segments for surface of sphere <16>: *Specify the number of longitudinal (vertical) segments for the sphere.*
Enter number of latitudinal segments for surface of sphere <16>: *Specify the number of latitudinal (horizontal) segments for the sphere.*

Creating a Surface Mesh Torus

A torus is like a thick, hollow ring. You can create a surface mesh model of a torus (Figure 23-74) by using any one of the methods for drawing surface mesh models.

Command: **3D**
Enter an option
[Box/Cone/DIsh/DOme/Mesh/Pyramid/Sphere/Torus/Wedge]: **T**
Specify center point of torus: *Specify the center of the torus.*
Specify radius of torus or [Diameter]: *Specify the radius of the torus.*
Specify radius of tube or [Diameter]: *Specify the diameter of the tube.*
Enter number of segments around tube circumference <16>: *Specify the number of segments for the tube.*
Enter number of segments around torus circumference <16>: *Specify the number of segments for the torus.*

The *XY* plane of the current UCS is parallel to the torus. If the center point of the torus has a zero *Z* coordinate value (the center point of the torus lies in the present construction plane), then the *XY* plane bisects it.

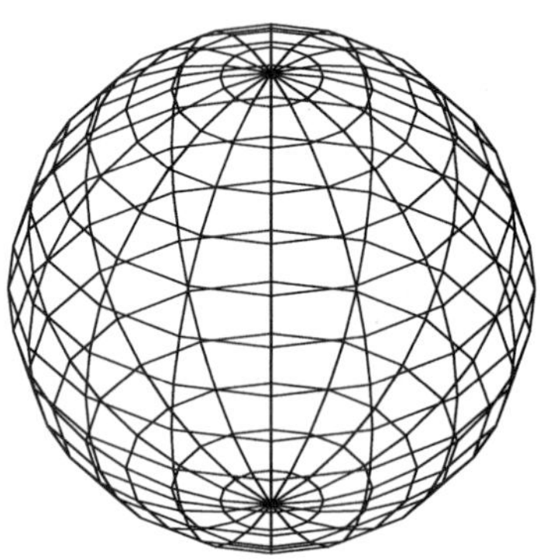

Figure 23-73 *Spherical surface mesh*

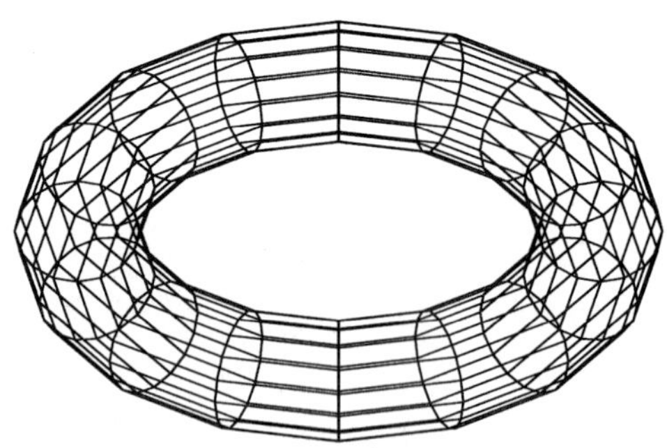

Figure 23-74 *Surfaced torus*

Creating a Surface Mesh Wedge

You can draw a surface mesh model of a right-angled wedge (Figure 23-75) by using any one of the methods for drawing surface mesh models. For example, you can initiate the creation of a 3D surface mesh model of a wedge by entering the **3D** command at the Command prompt and then entering WEDGE at the next prompt. You can also choose the **Wedge** button from the **Surfaces** toolbar. In the following example, you will draw a surface mesh model of a wedge using the **3D** command. Let the length of the wedge be 5 units, width 3 units, height 1 unit, and rotation 0-degree.

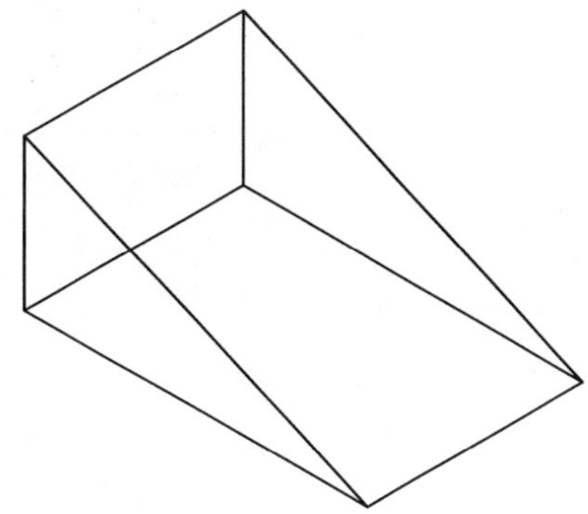

Figure 23-75 *3D wedge-shaped surface mesh*

Command: **3D**
Enter an option [Box/Cone/DIsh/DOme/Mesh/Pyramid/Sphere/Torus/Wedge]: W
Specify corner point of wedge: *Specify the corner point of the wedge.*
Specify length of wedge: **5**
Specify width of wedge: **3**
Specify height of wedge: **1**
Specify rotation angle of wedge about the Z axis: *Specify rotation around the Z axis.*

The sloped side of the wedge is always opposite the first corner specified at the **Specify corner point of wedge:** prompt. The base of the wedge is always parallel to the construction plane (*XY* plane) of the current UCS. The height is measured parallel to the *Z* axis.

Tip
To draw a wedge in which the triangular sides are not parallel, use the ***Ridge*** *option of* ***AI_PYRAMID*** *command.*

3DMESH Command

Toolbar:	Surfaces > 3D Mesh
Menu:	Draw > Surfaces > 3D Mesh
Command:	3DMESH

This command is used to create a free-form polygon mesh using a matrix of M X N size. When you invoke this command, you will be prompted to specify the size of the mesh in the M direction and the N direction, where M X N will be the total number of vertices in the mesh. Once you have specified the size of the mesh in the M and N directions, you will have to specify the coordinates of each and every vertex individually. For example, if the size of the mesh in the M and N directions is 12 X 12, then you will have to specify the coordinates of all the 144 vertices. You can specify any 2D or 3D coordinates for the vertices. Figure 23-76 shows a 4 X 4 free-form mesh. This command is mostly used for programming. For general drawing, use the **Mesh** option of the **3D** command. The 3D polygon mesh is always open in the M and N directions. You can edit this mesh or the mesh created using the **3D** command using the **PEDIT** command.

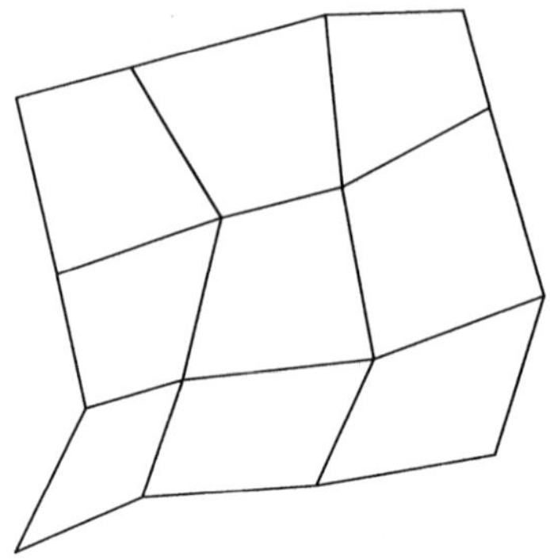

Figure 23-76 *Free-form 3D mesh*

EDITING THE SURFACE MESH

The surface meshes can be edited using the **PEDIT** command, which is discussed next.

PEDIT Command

Toolbar:	Modify II > Edit Polyline
Menu:	Modify > Polyline
Command:	PEDIT

You can also invoke this command using the shortcut menu displayed upon selecting the surface mesh and right-clicking on it. Choose the **Polyline Edit** option to invoke this command. The behavior of the **PEDIT** command differs, depending on whether the object selected is a line, polyline, or a surface mesh. The prompt sequence that will be followed when you choose this button is given next.

Select polyline or [Multiple]: *Select the surface mesh.*
Enter an option [Edit vertex/Smooth surface/Desmooth/Mclose/Nclose/Undo]:

Edit Vertex

This option is used for individual editing of the vertices of the mesh in the M direction or in the N direction. When you invoke this option, a cross will appear on the first point of the mesh and the following suboptions will be provided.

Next. This sub-option is used to move the cross to the next vertex.

Previous. This sub-option is used to move the cross to the previous vertex.

Left. This sub-option is used to move the cross to the previous vertex in the N direction of mesh.

Right. This sub-option is used to move the cross to the next vertex in the N direction of mesh.

Up. This sub-option is used to move the cross to the next vertex in the M direction of mesh.

Down. This sub-option is used to move the cross to the previous vertex in the M direction of the mesh.

Move. This sub-option is used to redefine the location of the current vertex. You can define the new location using any of the coordinate systems or you can directly select the location on the scree, see Figures 23-77 and 23-78.

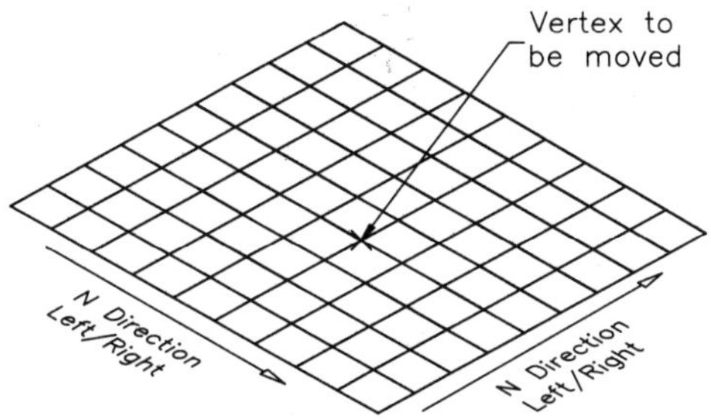

Figure 23-77 *Vertex before moving*

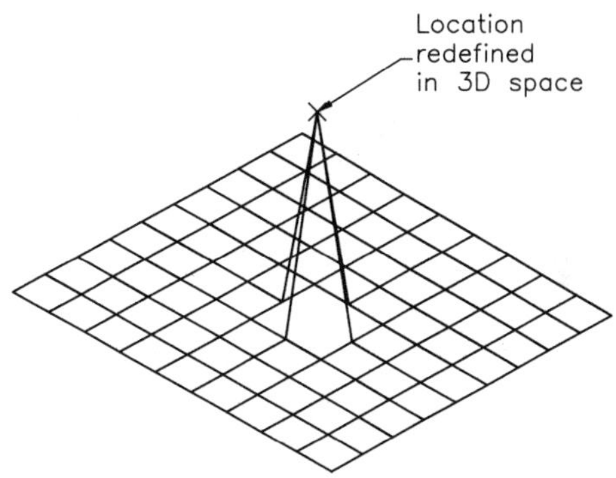

Figure 23-78 *Vertex after moving*

REgen. This sub-option is used to regenerate the 3D mesh.

eXit. This sub-option is used to exit the **Edit Vertex** option.

Smooth surface

This option is used to smoothen the surface of the 3D mesh by fitting a smooth surface, as shown in Figures 23-79 and 23-80. The smoothness of the surface will depend upon the **SURFU** and **SURFV** system variables. The value of these system variables can be set between 2 and 200.

The type of the smooth surface to be fit is controlled by the **SURFTYPE** system variable. If you set the value of the **SURFTYPE** system variable to **5**, it will fit a Quadratic B-spline surface, if you set the value to **6**, it will fit a Cubic B-spline surface, and if you set the value to **7**, it will fit a Bezier surface.

Desmooth

This option is used to remove the smooth surface that was fit on the 3D mesh using the **Smooth surface** option and restore to the original mesh.

Mclose

This option is used to close the 3D mesh in the M direction, as shown in Figures 23-81 and 23-82.

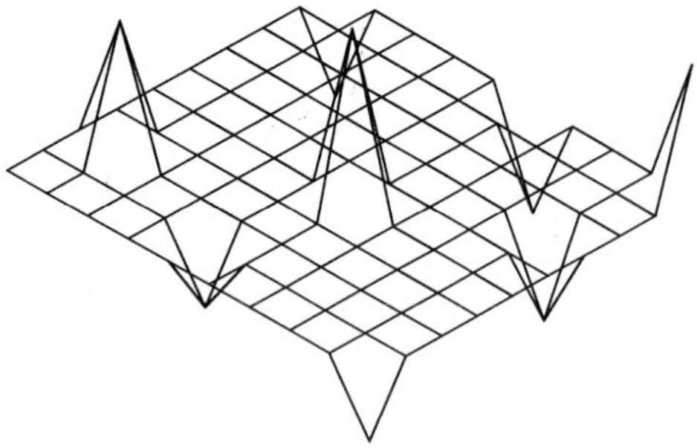

Figure 23-79 *3D mesh before fitting the smooth surface*

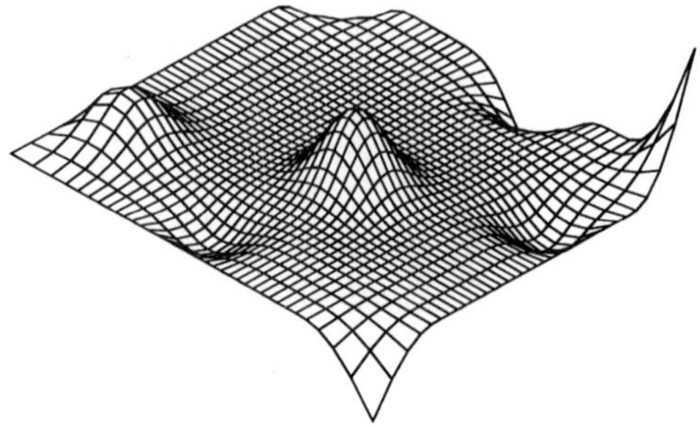

Figure 23-80 *3D mesh after fitting the smooth surface*

Nclose

This option is used to close the 3D mesh in the N direction, as shown in Figures 23-81 and 23-82.

Mopen

This option will be available if the 3D mesh is closed in the M direction. It is used to open the 3D mesh in the M direction (Figures 23-81 and 23-82).

Nopen

This option will be available if the 3D mesh is closed in the N direction. It is used to open the 3D mesh in the N direction (Figures 23-81 and 23-82).

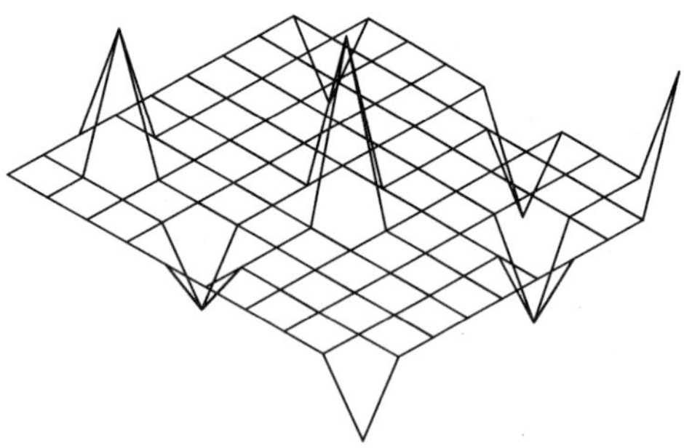

Figure 23-81 *3D mesh open in the M and N directions*

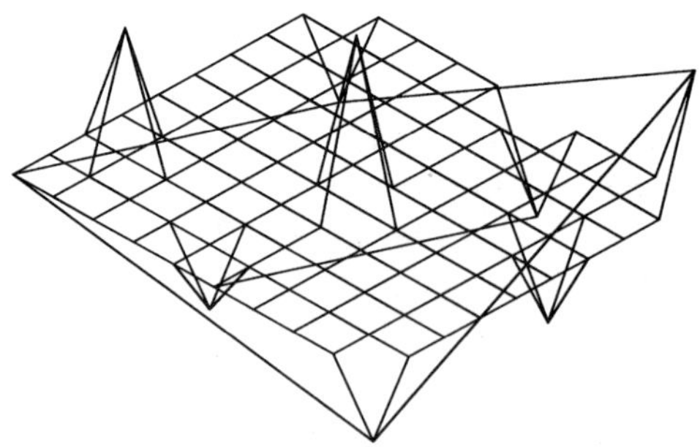

Figure 23-82 *3D mesh closed in the M and N directions*

Undo
This option is used to undo all the operations to the start of the **PEDIT** command.

eXit
This option is used to exit the **PEDIT** command.

MODIFYING THE PROPERTIES OF THE HIDDEN LINES
When you invoke the **HIDE** command, by default the hidden lines are suppressed and are not displayed in the surface or solid models. However, with AutoCAD, you can modify the hidden lines such that when you invoke the **HIDE** command, they appear as dotted lines and with different color. This makes it extremely easy to visualize the object. The properties of the hidden lines can be modified using the **Hidden Lines Settings** dialog box. To invoke this dialog box, right-click in the drawing area and choose **Options** from the shortcut menu. The **Options** dialog box appears. Choose the **User Preference** tab and then choose the **Hidden Line Settings** button to invoke the **Hidden Line Settings** dialog box shown in Figure 23-83.

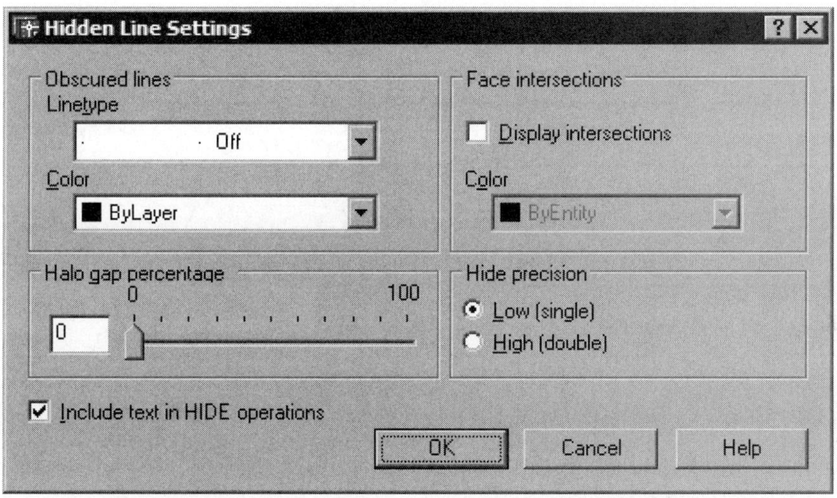

Figure 23-83* *Hidden Line Settings *dialog box*

Obscured lines Area
The options in this area are used to set the linetype and color of the hidden lines (also called obscured lines).

Linetype
The **Linetype** drop-down list is used to set the linetype for the obscured lines. You can select the desired linetype from this drop-down list. The linetype of the hidden lines can also be modified using the **OBSCUREDLTYPE** system variable. The default value of this variable is 0. As a result, the hidden lines are suppressed when you invoke the **HIDE** command. You can set any value between 0 and 11 for this system variable. The following table gives the details

of different values of this system variable and their corresponding linetypes that will be assigned to the hidden lines.

Value	Linetype	Sample
0	None	None
1	Solid	________________________
2	Dashed	– – – – – – – – – – – – – –
3	Dotted	....
4	Short Dash	– – – – – –
5	Medium Dash	— — — — — —
6	Long Dash	___ ___ ___ ___ ___ ___
7	Double Short Dash	___ ___ ___
8	Double Medium Dash	_____ _____ _____
9	Double Long Dash	______ ______ ______
10	Medium Long Dash	___ ___ ___ ___ ___
11	Sparse Dot	▪ ▪ ▪ ▪ ▪ ▪

Figure 23-84 shows a model with hidden linetype changed to dashed and Figure 23-85 shows the same model with hidden lines changed to dotted.

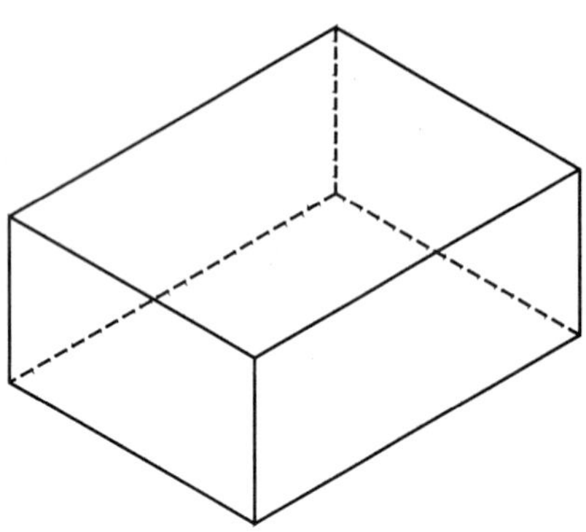

Figure 23-84 *Model with hidden lines changed to dashed*

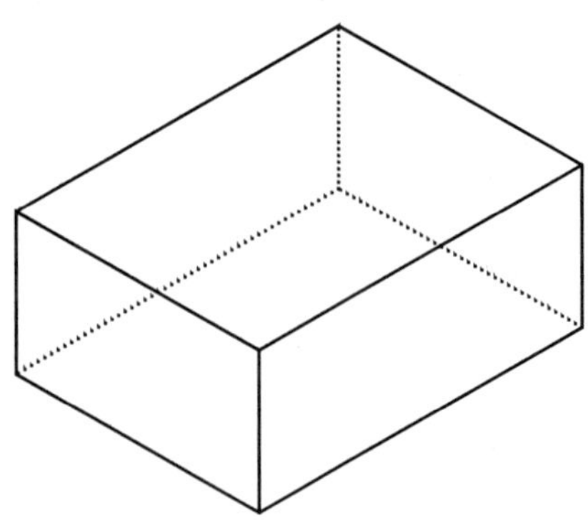

Figure 23-85 *Model with hidden lines changed to dotted*

Color

The **Color** drop-down list is used to define color for the obscured lines. If you define a separate color, the hidden lines will be displayed with that color when you invoke the **HIDE** command. You can select the required color from this drop-down list. This can also be done using the **OBSCUREDCOLOR** system variable. Using this variable, you can define the color that you want to assign to the hidden lines. The default value of this variable is 257. This value corresponds to the By Entity color. You can enter the number of any desired color at the sequence that will follow when you enter this system variable. For example, if you set the value of the **OBSCUREDLTYPE** variable to 2 and that of the **OBSCUREDCOLOR** to 1, the hidden lines will appear in red dashed lines when you invoke the **HIDE** command.

Note

*The linetype and color for the hidden lines defined using the previously mentioned variables are valid only when you invoke the **HIDE** command. They do not work if the model is regenerated.*

Face intersections Area

The options in the **Face intersections** area are used to display a 3D curve at the intersection of two surfaces. These options are discussed next.

Display intersections

If the **Display intersections** check box is selected, a 3D curve will be displayed that will define the intersecting portion of the 3D surfaces or solid models.

Color

The **Color** drop-down list is used to specify the color of the 3D curve that is displayed at the intersection of 3D surfaces or solid models.

Halo gap percentage Area

The **Halo gap percentage** area is used to specify in terms of percentage the distance by which the lines that lie behind a surface or solid model will be shortened, see Figures 23-86 and 23-87.

You can specify the distance in the edit box or using the slider.

Note

The obscured linetype in Figures 23-86 and 23-87 is changed to dashed.

Hide precision Area

The **Hide precision** area is used to define the precision of the hidden lines. You can select the **Low (single)** radio button to set the hide precision to single. Similarly, you can select the hide precision to double by selecting the **High (double)** radio button.

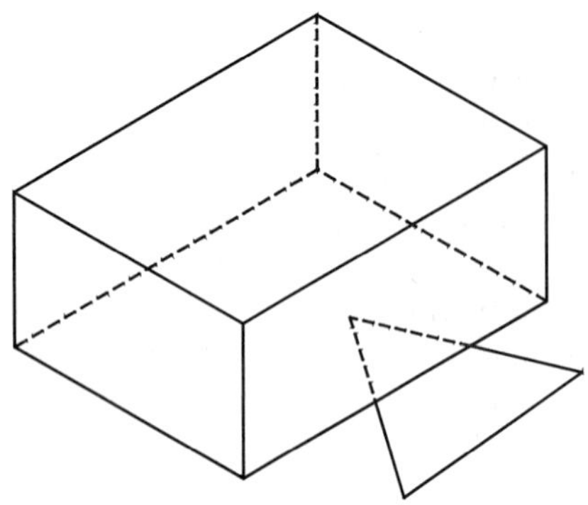

Figure 23-86 Hiding the lines with halo gap percentage as 0

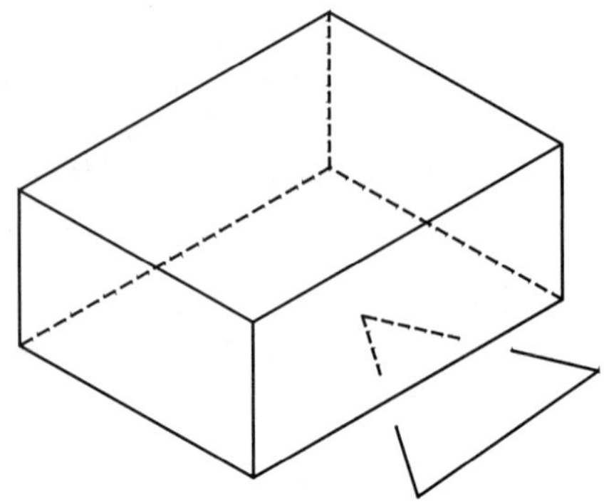

Figure 23-87 Hiding the lines with halo gap percentage as 40

Include text in HIDE operations

The **Include text in HIDE operations** check box is selected to hide the text that lies behind a surface or solid model when the **HIDE** command is invoked. If this check box is cleared, the text will not be hidden when you invoke the **HIDE** command, even if lies behind a surface or a solid model.

Figure 23-88 shows a solid model with a text. Notice that when you invoke the **HIDE** command, the text is also hidden. But as the obscured linetype is changed to dashed, the text appears in dashed lines. Figure 23-89 shows the same model with **Include text in HIDE operations** check box cleared. Notice that the text is not hidden and is displayed with continuous lines.

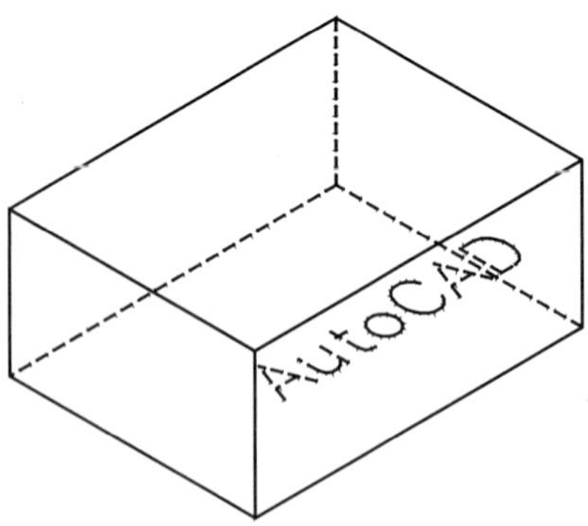

Figure 23-88 Text included while hiding the hidden lines

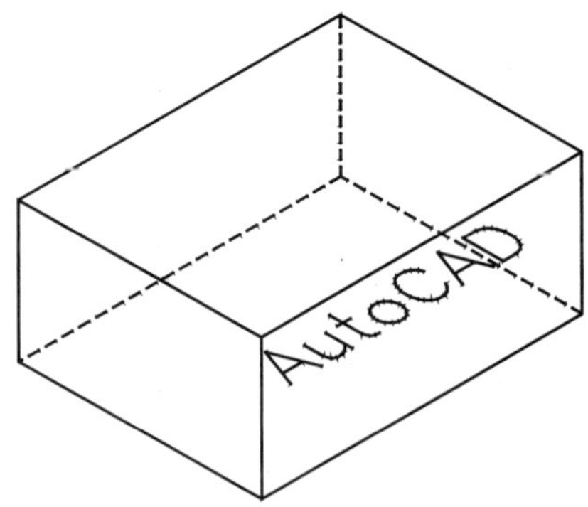

Figure 23-89 Text excluded while hiding the hidden lines

Self-Evaluation Test

Answer the following questions and then compare your answers to the correct answers given at the end of this chapter.

1. The **DDVPOINT** command is used to change the viewpoint to view the solid model. (T/F)
2. The wireframe models can be converted into surface models as well as solid models. (T/F)
3. The right-hand thumb rule is followed in AutoCAD to identify the *X*, *Y*, and *Z* axes direction. (T/F)
4. The right-hand rule is followed in AutoCAD to find out the direction of rotation of revolution in the 3D space. (T/F)
5. The _________ and the _________ commands can be used to change the viewpoint for viewing the models in the 3D space.
6. Changing the viewpoint moves the 3D model from its default position. (T/F)
7. Using the **DDVPOINT** command, you can set the viewpoint with respect to _________ and _________.
8. The various types of 3D coordinate systems are _________.
9. The _________ command is used to convert the wireframe model into a surface model.
10. The _________ command is used to draw standard 3D surface primitives.

Chapter 23

Review Questions

Answer the following questions.

1. You can perform Boolean operations only on the solid models. (T/F)
2. The **ELEV** command is a transparent command. (T/F)
3. You can directly write a text with thickness. (T/F)
4. You can draw a surface between a closed and an open entity using the **RULESURF** command. (T/F)

5. Which command is used to create a 3D polyline?

   (a) **POLYLINE** (b) **3DPOLY**
   (c) **3DPOLYLINE** (c) **POLY3D**

6. Which command is used for setting the elevation and thickness for new objects?

   (a) **ELEVATION** (b) **THICKNESS**
   (c) **ELEV** (d) **THICK**

7. Which command is used to suppress the hidden edges in the 3D model?

   (a) **SUPPRESS** (b) **HIDE**
   (c) **3DHIDE** (d) **EDGE**

8. When you open a new drawing, you are by default in which view?

   (a) SE Isometric View (b) SW Isometric View
   (c) Plan View (d) Bottom View

9. Which command is used to create a revolved surface?

   (a) **RULESURF** (b) **TABSURF**
   (c) **REVSURF** (d) **EDGESURF**

10. The __________ command is used to control the visibility of the edges of 3D faces.

11. To draw an ellipse with thickness, you have to set the value of the __________ system variable to __________.

12. The __________ system variable is used to increase the smoothness of the ruled surface.

13. The __________ option of the __________ command is used to fit a smooth surface on the 3D mesh.

14. The smoothness of the surface fitted on the 3D mesh is controlled using the __________ and the __________ system variables.

15. The __________ option of the **PEDIT** command is used to remove the smooth surface fitted on the 3D mesh.

Exercises

Exercises 4 through 6

General

In these exercises you will create the models shown in Figures 23-90 through 23-92. Assume the missing dimensions for the models.

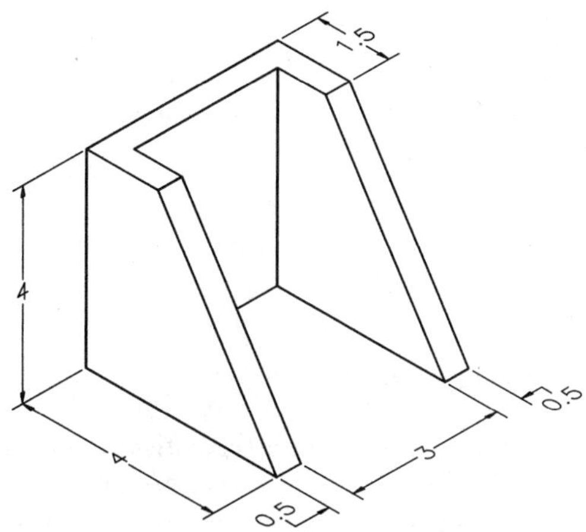

Figure 23-90 *Model for Exercise 4*

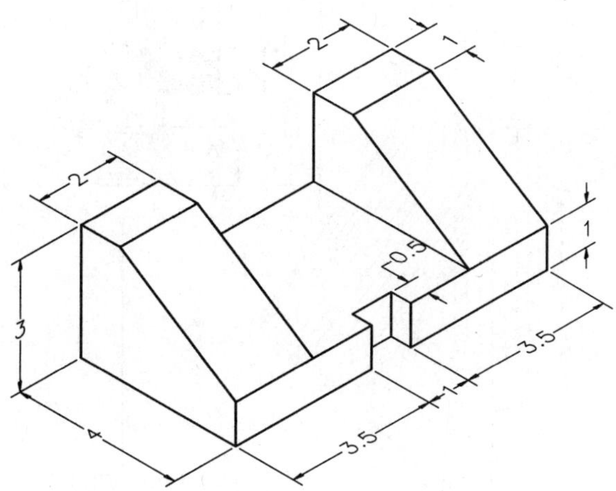

Figure 23-91 *Model for Exercise 5*

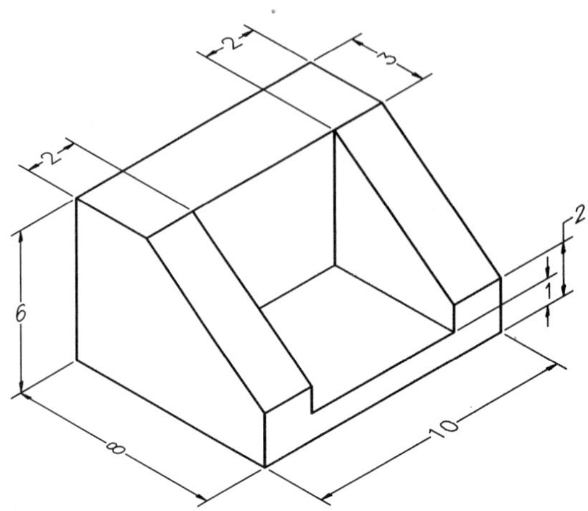

Figure 23-92 *Model for Exercise 6*

Problem Solving Exercise 1 — *General*

Create the drawing shown in Figure 23-93. The transition at the bottom can be drawn by using the **EDGESURF** command. The edges consist of the semicircle at the top, half rectangle at the bottom, and the two lines joining the endpoints of the semicircle to the endpoints of the bottom rectangle. The angle of the side tubes are 90- and 45-degrees.

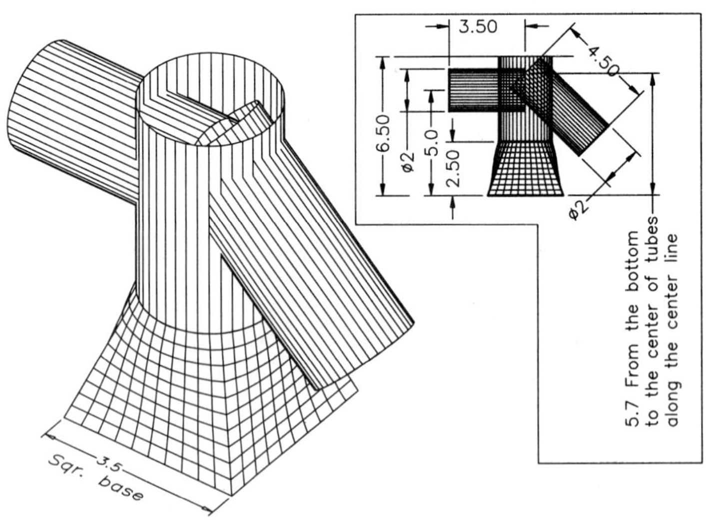

Figure 23-93 *Drawing for Problem Solving Exercise 1*

Problem Solving Exercise 2

General

Create the drawing shown in Figure 23-94. First, create one of the segments of the base and then use the **ARRAY** command to complete the base.

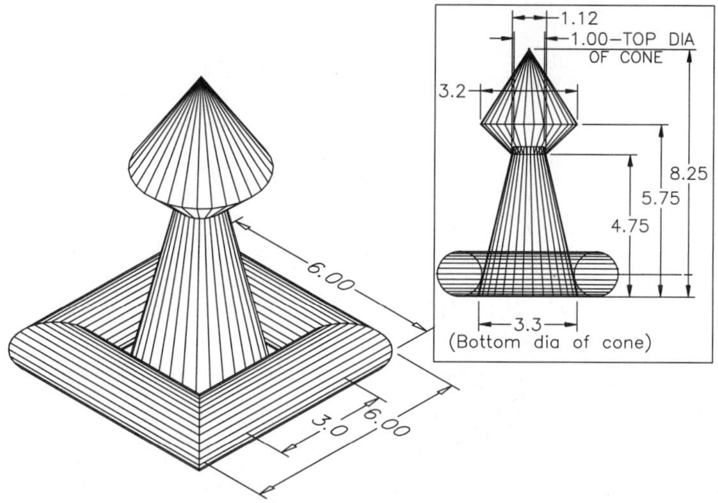

Figure 23-94 *Drawing for Problem Solving Exercise 2*

Problem Solving Exercise 3

General

Create the drawing shown in Figure 23-95. First, create the base unit (transition) and then the dish and cone.

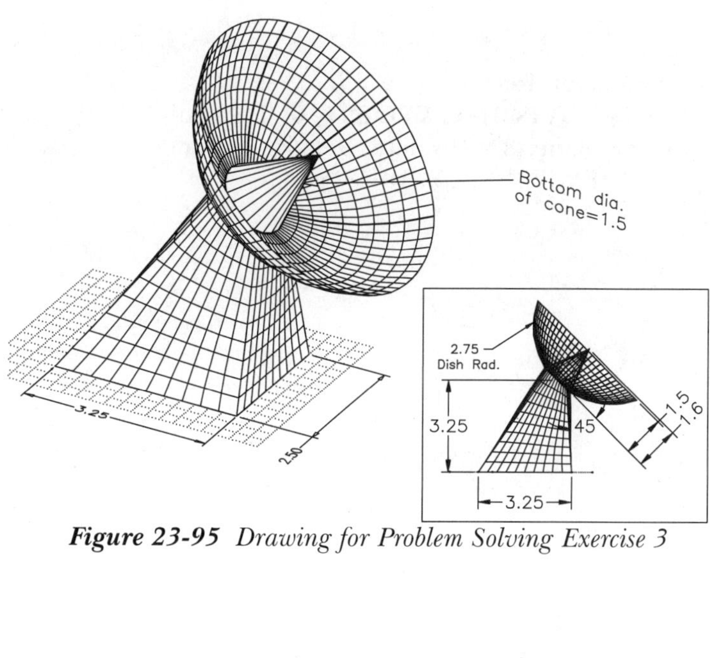

Figure 23-95 *Drawing for Problem Solving Exercise 3*

Problem Solving Exercise 4

General

Create the drawing shown in Figure 23-96. First, create one of the transitions and use the mirror command to get the second transition. To get the circular sections, draw circles and then create a surface between the circles.

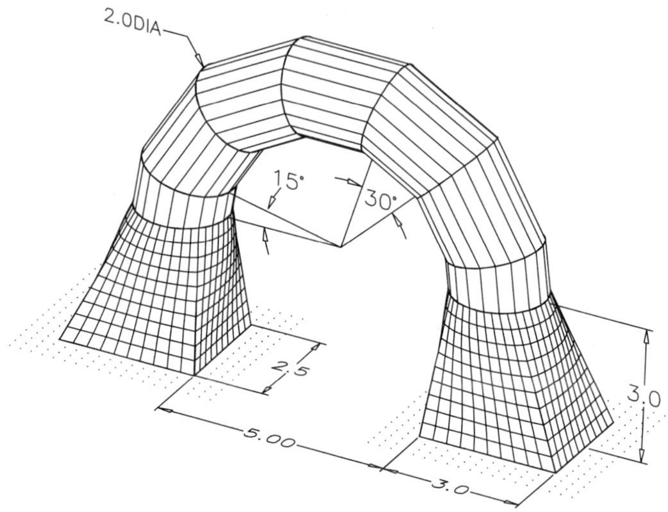

Figure 23-96 *Drawing for Problem Solving Exercise 4*

Answers to Self-Evaluation Test

1 - T, **2** - F, **3** - F, **4** - F, **5** - **DDVPOINT**, **VPOINT**, **6** - F, **7** - Angle in the *XY* plane from the *X* axis, Angle from the *XY* plane, **8** - Absolute, Relative rectangular, Relative cylindrical, and Relative spherical, **9** - **3DFACE**, **PFACE**, **10** - **3D**

Chapter 24

Creating Solid Models

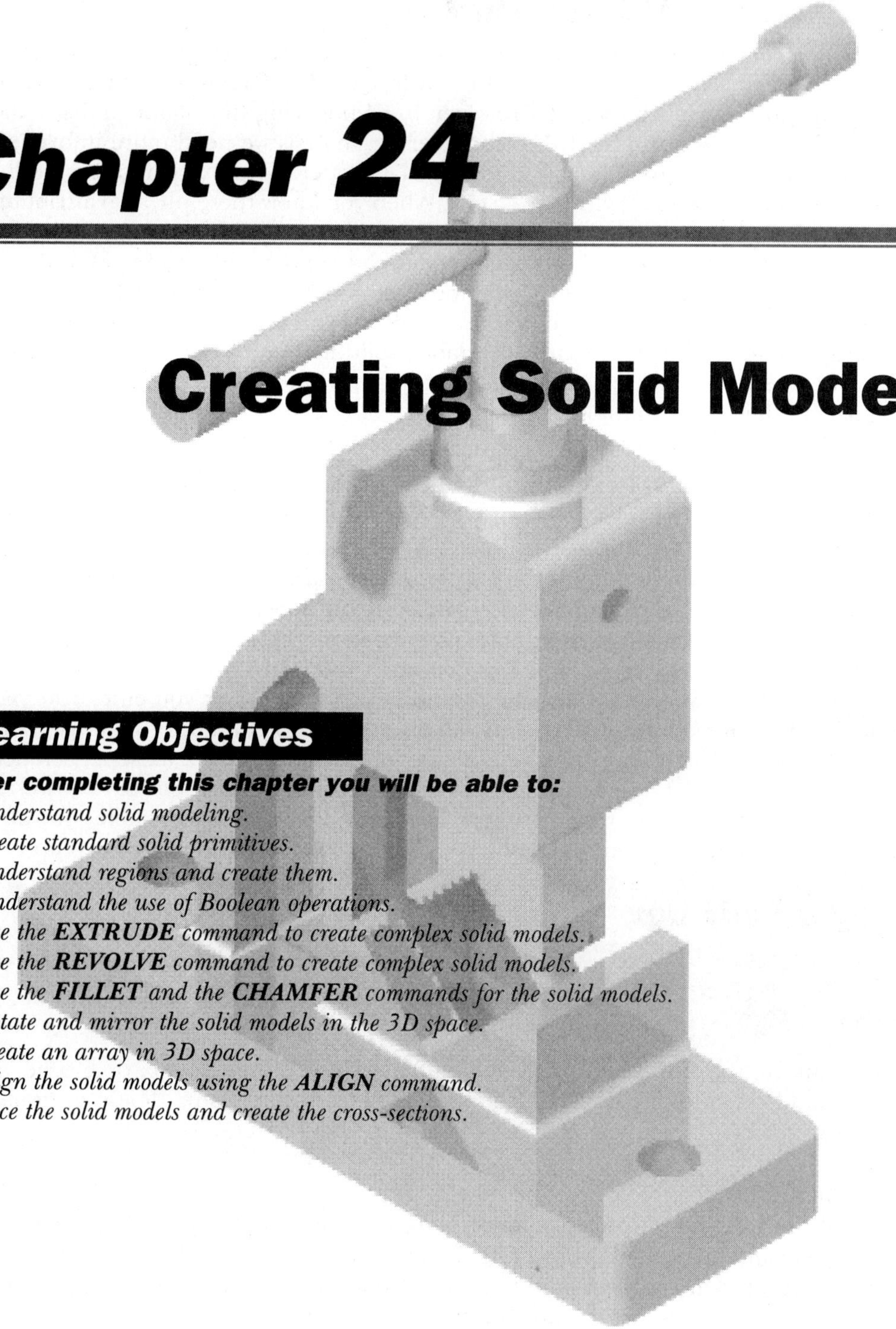

Learning Objectives

After completing this chapter you will be able to:

- *Understand solid modeling.*
- *Create standard solid primitives.*
- *Understand regions and create them.*
- *Understand the use of Boolean operations.*
- *Use the **EXTRUDE** command to create complex solid models.*
- *Use the **REVOLVE** command to create complex solid models.*
- *Use the **FILLET** and the **CHAMFER** commands for the solid models.*
- *Rotate and mirror the solid models in the 3D space.*
- *Create an array in 3D space.*
- *Align the solid models using the **ALIGN** command.*
- *Slice the solid models and create the cross-sections.*

WHAT IS SOLID MODELING?

Solid modeling is the process of building objects that have all the attributes of an actual solid object. For example, if you draw a wireframe or a surface model of a bushing, it is sufficient to define the shape and size of the object. However, in engineering, the shape and size alone are not enough to describe an object. For engineering analysis we need more information, such as volume, mass, moment of inertia, and material properties (density, Young's modulus, Poissons's ratio, thermal conductivity, and so on). When you know these physical attributes of an object, it can be subjected to various tests to make sure that it performs as required by the product specifications. It eliminates the need for building expensive prototypes and makes the product development cycle shorter. Solid models also make it easy to visualize the objects because we always think of and see the objects as solids. With computers getting faster and software getting more sophisticated and affordable, solid modeling will become the core of the manufacturing process. AutoCAD solid modeling is based on the ACIS solid modeler, which is a part of the core technology.

CREATING PREDEFINED SOLID PRIMITIVES

The solid primitives form the basic building blocks for a complex solid. ACIS has six predefined solid primitives that can be used to construct a solid model (box, wedge, cone, cylinder, sphere, and torus). The number of lines in a solid model representation is controlled by the value assigned to the **ISOLINES** variable. These lines are called tessellation lines. The number of lines determines the number of computations needed to generate a solid. If the value is high, it will take significantly more time to generate a solid on the screen. Therefore, the value you assign to the **ISOLINES** variable should be realistic. When you enter commands for creating solid primitives, AutoCAD Solids will prompt you to enter information about the part geometry. The height of the primitive is always along the positive *Z* axis, perpendicular to the construction plane. Just like surface meshes, solids are also displayed as wireframe models unless you hide, render, or shade them. The **FACETRES** system variable controls the smoothness in the shaded and rendered objects. The value of this variable can go up to 10.

Creating a Solid Box

Toolbar:	Solids > Box
Menu:	Draw > Solids > Box
Command:	BOX

You can use the **BOX** command to create a solid rectangular box or a cube. This command provides a number of options for creating the box. These options are discussed next.

Two Corner Option

This is the default option and using this option you can create a solid box by defining the first corner of the box and the other corner of the box (Figure 24-1). Note that the length of the box will always be taken along the *X* axis, the width along the *Y* axis and the height along the *Z* axis. Therefore, in this case when you specify the other corner, the value along the *X* axis will be taken as the length of the box, the value along the *Y* axis will be taken as the width of the box, and then you will be prompted to specify the height of the box. When you choose this button, the following prompt sequence will be issued.

Specify corner of box or [CEnter] <0,0,0>: **2,2,0**
Specify corner or [Cube/Length]: **@5,4,0** *(Length = 5, Width = 4)*
Specify height: **3**

Center-Length Option

The center of the box is the point where the center of gravity of the box lies. This option is used to create a box by specifying the center of the box, as well as the length, width, and height of the box (Figure 24-2). The following prompt sequence is issued when you choose the **Box** button.

Specify corner of box or [CEnter] <0,0,0>: **CE**
Specify center of box <0,0,0>: **4,4**
Specify corner or [Cube/Length]: **L**
Specify length: **8**
Specify width: **6**
Specify height: **3**

Tip

The Corner-Length option is similar to the Center-Length option except that in the Corner-Length option you will define the first corner of the box and the length, width, and height of the box.

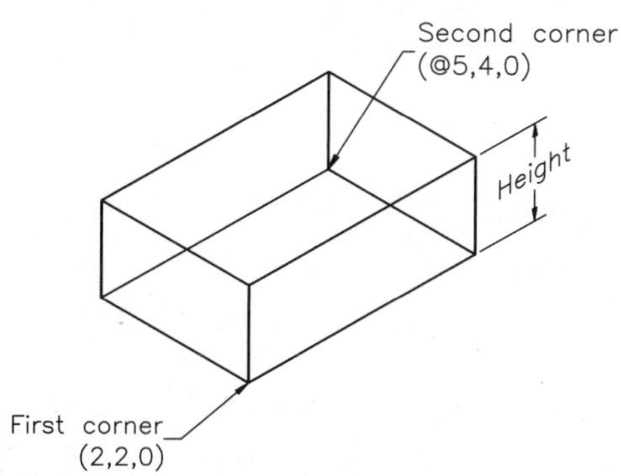

Figure 24-1 *Creating the box using the two corner option*

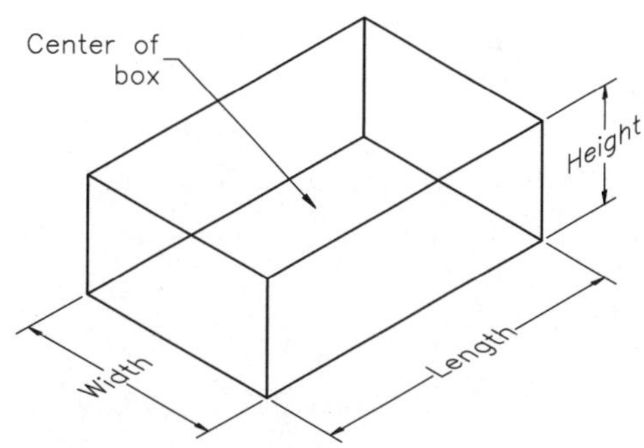

Figure 24-2 *Creating the box using the Corner-Length option*

Corner-Cube Option

This option is used to create a cube starting from a specified corner. Since you are creating a cube, it will prompt for only the length of the cube that will also be the width and the height (Figure 24-3). The prompt sequence for creating the cube using the Corner-Cube option is as follows.

Specify corner of box or [CEnter] <0,0,0>: **2,2**
Specify corner or [Cube/Length]: **C**
Specify length: **5**

Tip
The Center-Cube option is similar to the Corner-Cube option, except that in the Center-Cube option you will define the center and the length of the cube.

Creating a Solid Cone

Toolbar:	Solids > Cone
Menu:	Draw > Solids > Cone
Command:	CONE

The **CONE** command creates a solid cone with an elliptical or circular base. This command provides you with the option of defining the height of the cone or the location of the apex of the cone. Defining the location of the apex will also define the height of the cone and the orientation of the base of the cone from the *XY* plane. The methods of creating circular and elliptical cones are discussed next.

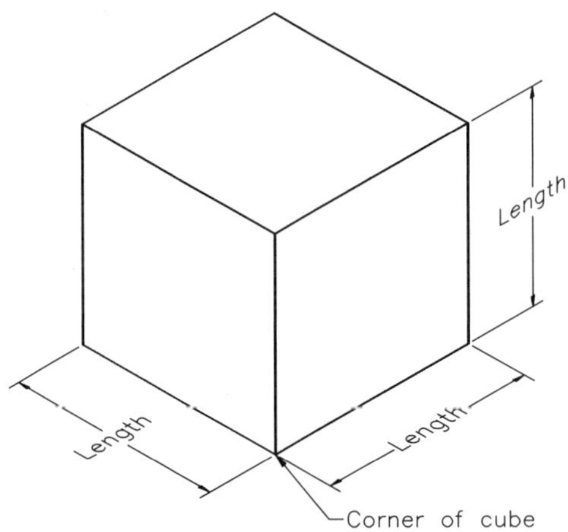

Figure 24-3 *Creating the cube using the Corner-Cube option*

Circular Cone

This method is used to create a cone with a circular base (Figure 24-4). The prompt sequence that will be issued when you choose the **Cone** button is as follows.

Current wire frame density: ISOLINES=4
Specify center point for base of cone or [Elliptical] <0,0,0>: *Specify the center of the base.*
Specify radius for base of cone or [Diameter]: *Specify radius or Enter **D** to specify the diameter of the cone.*
Specify height of cone or [Apex]: *Specify the height of the cone.*

To specify the apex (Figure 24-5), enter **A** at the **Specify height of cone or [Apex]** prompt. The prompt sequence that will follow is given next.

Specify apex point: *Specify the location of the apex of the cone.*

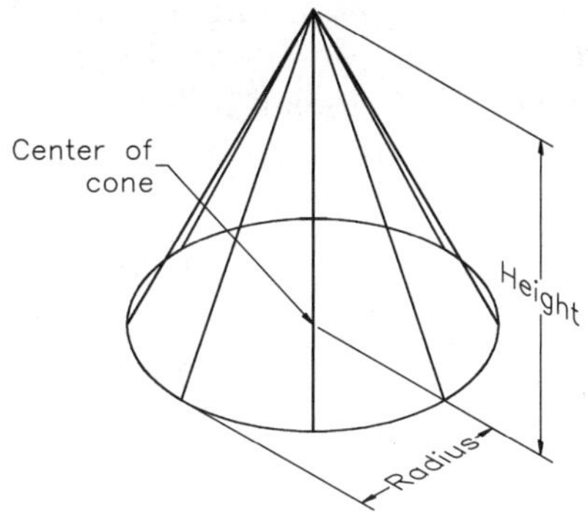

Figure 24-4 *Figure showing a circular cone*

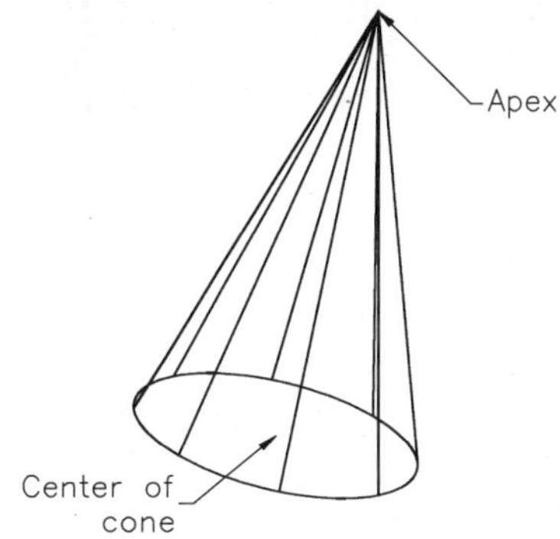

Figure 24-5 *Creating a cone using the **Apex** option*

Elliptical Cone

Using this method you can create a cone that has the elliptical base. To create an elliptical cone, enter **E** at the **Specify center point for base of cone or [Elliptical] <0,0,0>** prompt. You will be prompted to create the base ellipse for the cone. You can create the base ellipse using any of the methods for drawing the ellipse. However, you cannot specify the rotation in this case for defining the other axis. You will have to specify the length of the other axis. Once you have entered all these values, you will be prompted to specify the height of the cone or the apex of the cone. Figure 24-6 shows an elliptical cone.

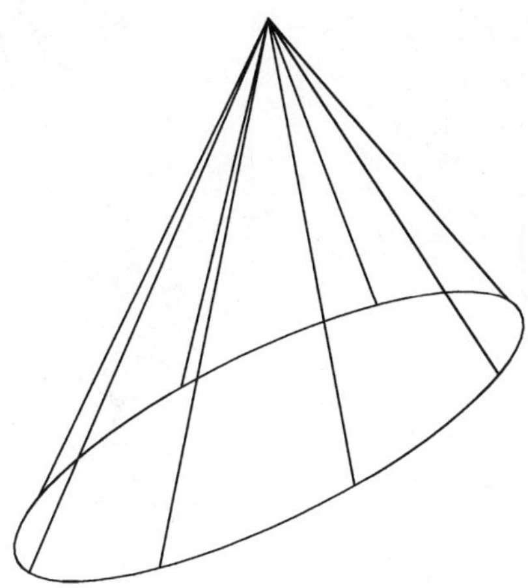

Figure 24-6 *An elliptical cone*

Creating a Solid Cylinder

Toolbar:	Solids > Cylinder
Menu:	Draw > Solids > Cylinder
Command:	CYLINDER

You can use the **CYLINDER** command to create a solid cylinder. Similar to the **CONE** command, this command also provides you with two options for creating the cylinder: **circular cylinder** and **elliptical cylinder**. This command allows you to define the height of the cylinder or the center of the other end for creating an inclined cylinder. The options for creating different types of cylinders are discussed next.

Circular Cylinder

This option is used to create a cylinder with circular base (Figure 24-7). The prompt sequence that follows when you choose the **Cylinder** button is as follows.

Current wire frame density: ISOLINES=4
Specify center point for base of cylinder or [Elliptical] <0,0,0>: *Specify the location of the center point.*
Specify radius for base of cylinder or [Diameter]: *Specify the radius.*
Specify height of cylinder or [Center of other end]: *Specify the height of the cylinder.*

You can also create an inclined cylinder by specifying the center of the other end (Figure 24-8). This is done by entering **C** at the **Specify height of cylinder or [Center of other end]** prompt. The prompt sequence that will follow is given next.

Specify center of other end of cylinder: *Specify the location of center of the other end.*

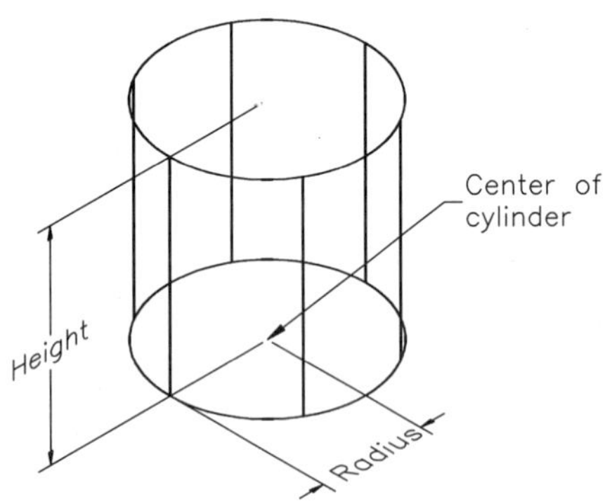

Figure 24-7 *Creating circular cylinder*

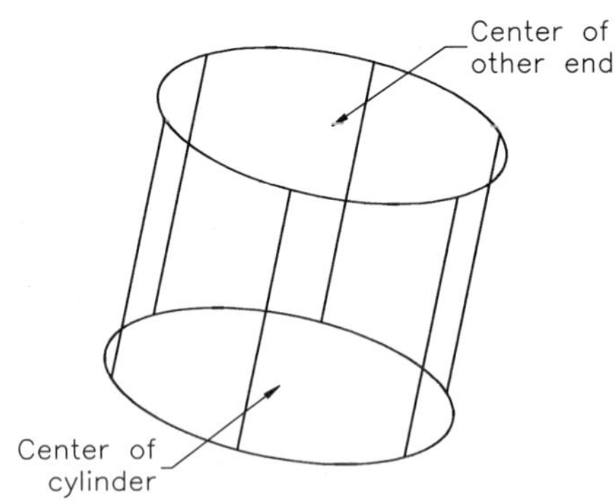

Figure 24-8 *Specifying the center of the other end*

Elliptical Cylinder

This option is used to create a cylinder with an elliptical base. The elliptical cylinder can be created by entering **E** at the **Specify center point for base of cylinder or [Elliptical] <0,0,0>**

prompt. You will be prompted to create the ellipse for the base. You can create the ellipse for the base using any of the methods for creating the ellipse. Note that you can not define the other axis using the rotation method. You can specify the height of the cylinder or the center of the other end of the cylinder, see Figures 24-9 and 24-10.

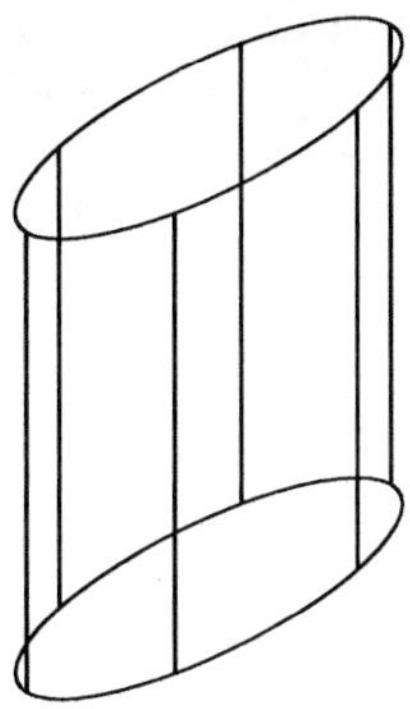

Figure 24-9 *Creating an elliptical cylinder*

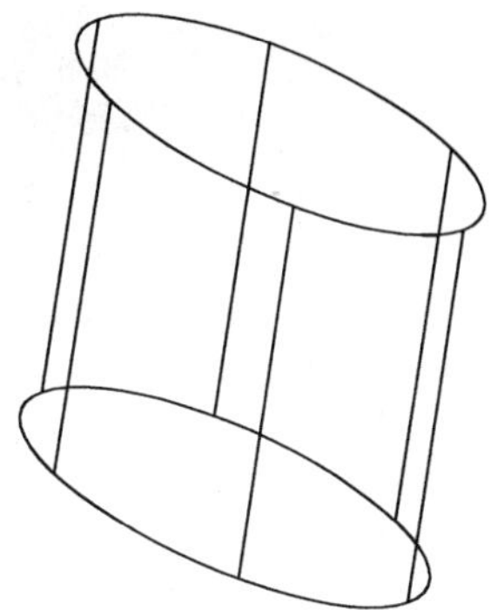

Figure 24-10 *Specifying the center of the other end*

Creating a Solid Sphere

Toolbar:	Solids > Sphere
Menu:	Draw > Solids > Sphere
Command:	SPHERE

The **SPHERE** command is used to create a solid sphere (Figure 24-11). When you choose the **Sphere** button, you will be asked to specify the center of the sphere. Once you have specified the center of the sphere, you can create the sphere by defining its radius or its diameter. The following prompt sequence will be issued when you choose the **Sphere** button.

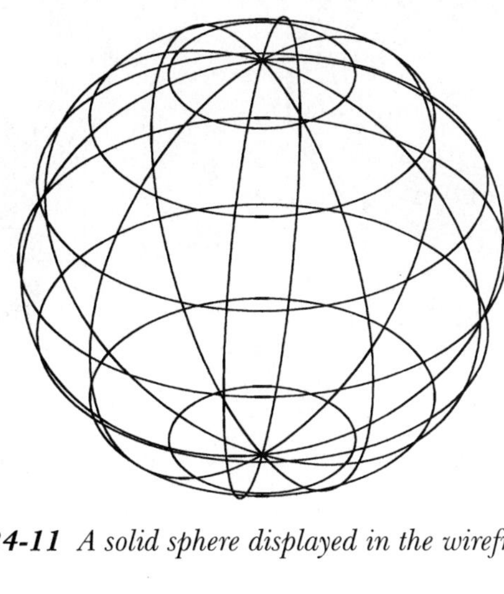

Figure 24-11 *A solid sphere displayed in the wireframe mode*

Current wire frame density: ISOLINES=4
Specify center of sphere<0,0,0>: *Specify the location of center of the sphere.*
Specify radius of sphere or [Diameter]: *Specify the radius.*

Creating a Solid Torus

Toolbar:	Solids > Torus
Menu:	Draw > Solids > Torus
Command:	TORUS

You can use the **TORUS** command to create a torus that is a tyre tube like shape, see Figure 24-12. When you select this command, AutoCAD will prompt you to enter the diameter or the radius of torus and the diameter or radius of tube (Figure 24-13). The radius of torus is the distance from the center of the torus to the centerline of the tube. This radius can have a positive or a negative value. If the value is negative, the torus has a rugby-ball like shape (Figure 24-14). A torus can be self-intersecting. If both the radii of the tube and the torus are positive and the radius of the tube is greater than the radius of the torus, the resulting solid looks like an apple (Figure 24-15). The torus is centered on the construction plane. The top half of the torus is above the construction plane and the other half is below the construction plane. The following prompt sequence is issued when you choose the **Torus** button.

Current wire frame density: ISOLINES=4
Specify center of torus <0,0,0>: *Specify the location of the center of the torus.*
Specify radius of torus or [Diameter]: *Specify the radius of the torus.*
Specify radius of tube or [Diameter]: *Specify the radius of the tube.*

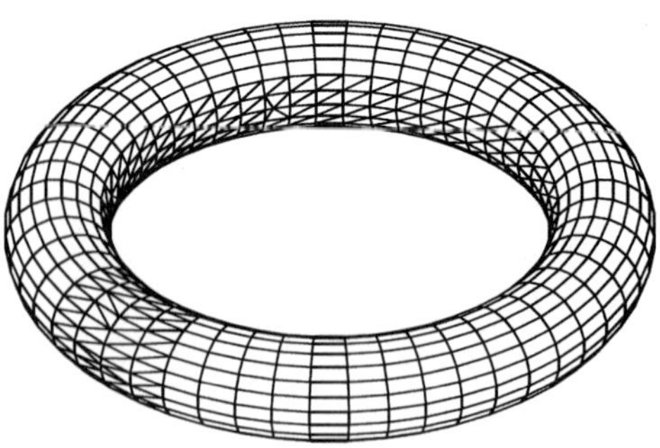

Figure 24-12 *Torus with hidden lines suppressed*

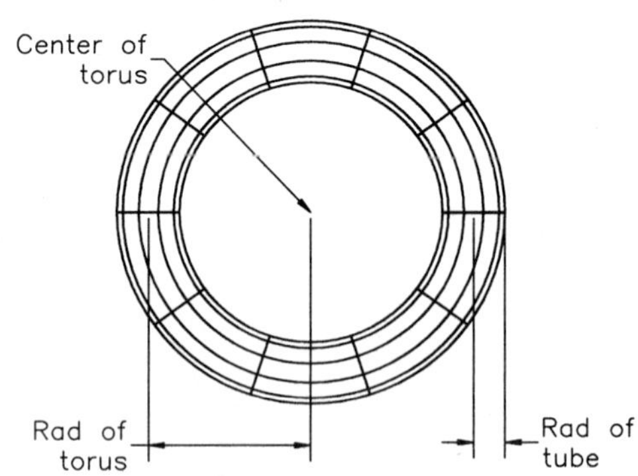

Figure 24-13 *Parameters associated with torus*

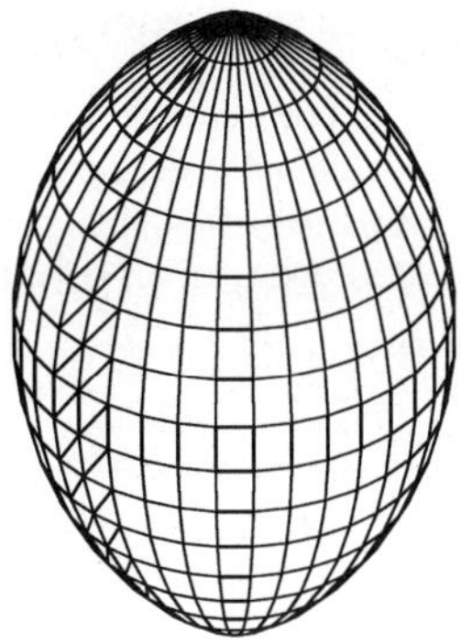

Figure 24-14 *Torus with negative value of radius of torus*

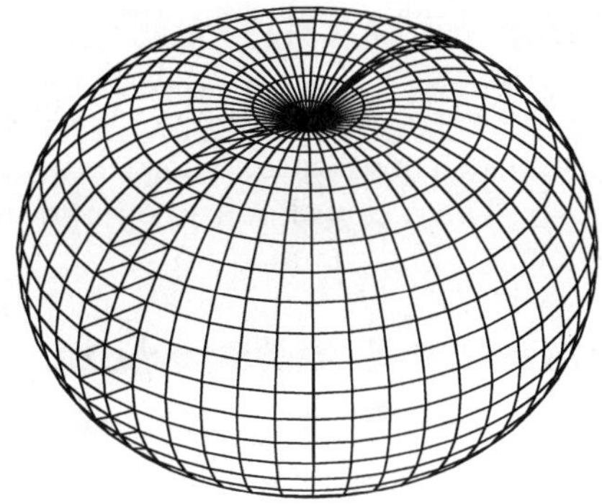

Figure 24-15 *Torus with radius of tube more than radius of torus*

Creating a Solid Wedge

Toolbar:	Solids > Wedge
Menu:	Draw > Solids > Wedge
Command:	WEDGE

This command is used to create a solid wedge and is similar to the **BOX** command. This means that this command provides you with the options of creating the wedge that are similar to those of the **BOX** command.

CREATING COMPLEX SOLID MODELS

Until now you have learned how to create simple solid models using the standard solid primitives. However, the realtime designs are not just the simple solid primitives, but complex solid models. These complex solid models can be created by modifying the standard solid primitives with the Boolean operations or directly by creating complex solid models by extruding or revolving the regions. All these options of creating complex solid models are discussed next.

CREATING REGIONS

Toolbar:	Draw > Region
Menu:	Draw > Region
Command:	REGION

The **Region** command is used to create regions from the selected loops or closed entities. Regions are the 2D entities with properties of the 3D solids. You can apply the Boolean operation on the regions and you can also calculate the mass properties of the regions. Bear in mind that the 2D entity you want to convert into a region should be a closed loop. Once you have created regions, the original object is deleted automatically. However, if the value of the **DELOBJ** system variable is set to **0**, the original object is retained. The valid selection set for creating the regions are closed polylines, lines, arcs, splines, circles, or ellipses. The current color, layer, linetype, and lineweight will be applied to the regions.

CREATING COMPLEX SOLID MODELS BY APPLYING THE BOOLEAN OPERATIONS

You can create complex solid models by applying the Boolean operations on the standard solid primitives. The various Boolean operations that can be performed are union, subtract, intersect, and interfere. The commands used to apply these Boolean operations are discussed next.

Combining Solid Models

Toolbar:	Solids Editing > Union
Menu:	Modify > Solids Editing > Union
Command:	UNION

The **UNION** command is used to apply the union Boolean operations on the selected set of solids or regions. You can create a composite solid or region by combining them using this command. You can combine any number of solids or regions. When you invoke this command, you will be asked to select the solids or regions to be added. Figure 24-17 shows the solid model created by uniting the solids shown in Figure 24-16.

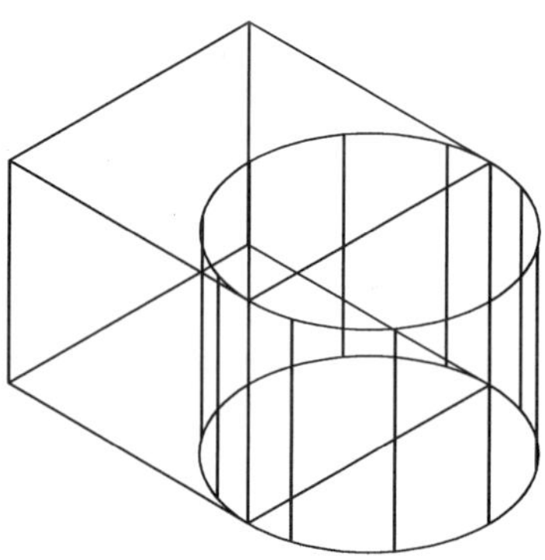

Figure 24-16 *Solid models before uniting*

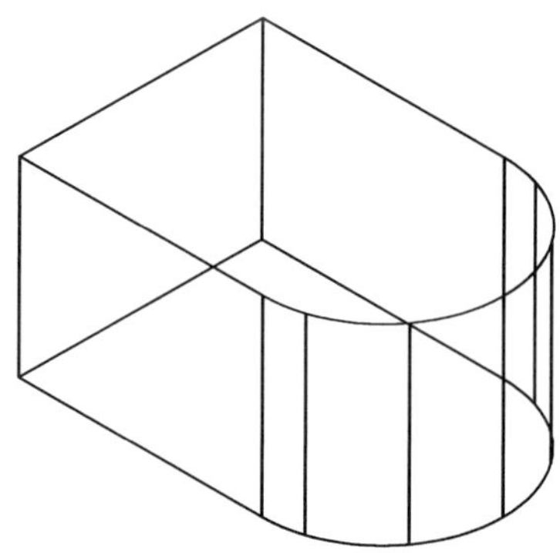

Figure 24-17 *Composite solid created after union*

Subtracting One Solid From the Other

Toolbar:	Solids Editing > Subtract
Menu:	Modify > Solids Editing > Subtract
Command:	SUBTRACT

This command is used to create a composite solid by removing the material common to the selected set of solids or regions. When you invoke this command, you will be prompted to select the set of solids or regions to subtract from. Once you have selected the set of solids or regions to subtract from, you will be prompted to select the solids or regions to subtract. The material common to the first selection set and the second selection set is removed from the first selection set. The resultant object will be a single composite solid, see Figures 24-18 and 24-19.

Note

If you subtract a single solid from two solids, the two solids are joined after the subtract operation.

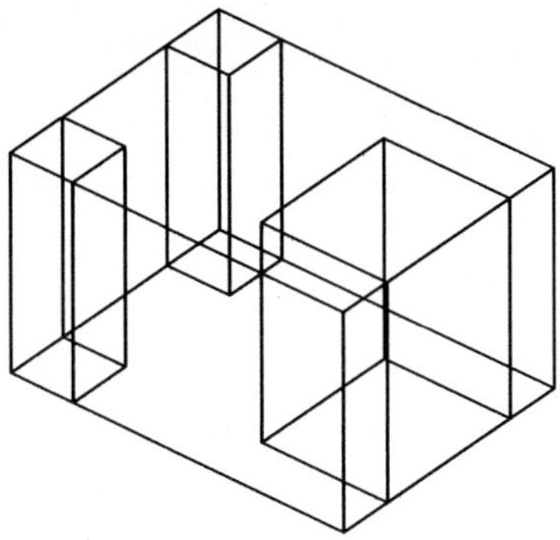

Figure 24-18 *Solid models before subtracting*

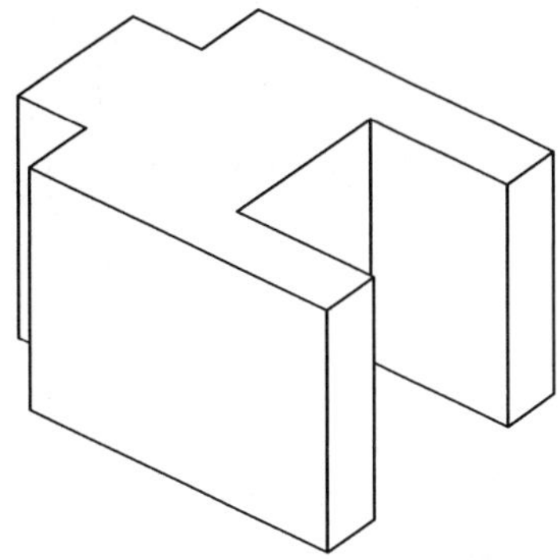

Figure 24-19 *Composite solid created after subtracting*

Intersecting Solid Models

Toolbar: Solids Editing > Intersect
Menu: Modify > Solids Editing > Intersect
Command: INTERSECT

The **INTERSECT** command is used to create a composite solid or region by retaining the material common to the selected set of solids or regions. When you invoke this command, you will be asked to select the solids or regions to intersect. The material common to all the selected solids or regions will be retained to create a new composite solid (Figures 24-20 and 24-21).

Checking Interference in Solids

Toolbar: Solids > Interfere
Menu: Draw > Solids > Interference
Command: INTERFERE

The **INTERFERE** command is used to create a composite solid model by retaining the material common to the selected sets of solids. The advantage of using this command is that the original objects are also retained. This command prompts you whether you want to create the interference solid or not. If you say yes, it will create the interference solid that can be moved out and used for analyzing the interference. This command is generally used for analyzing the interference between the mating parts of the assembly. Figure 24-22 shows two mating components of the assembly with interference between them and Figure 24-23 shows the interference solid created and moved out.

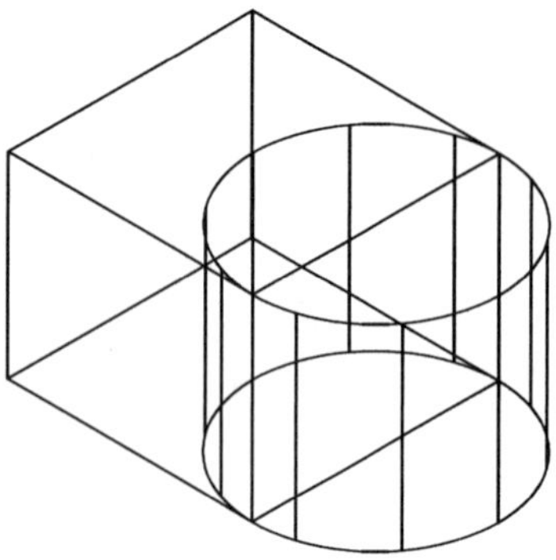

Figure 24-20 *Solid models before intersecting*

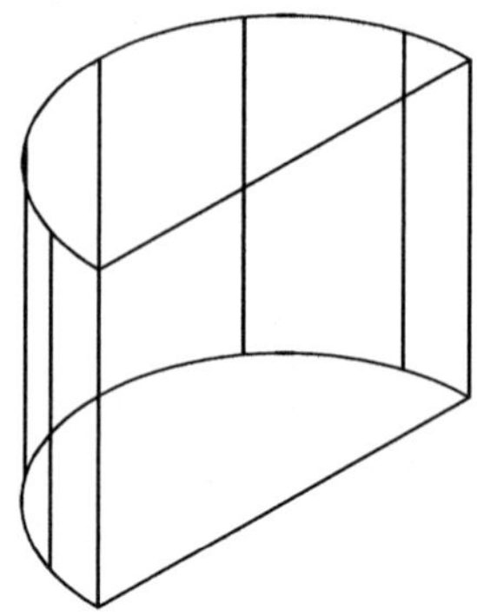

Figure 24-21 *Solid created after intersecting*

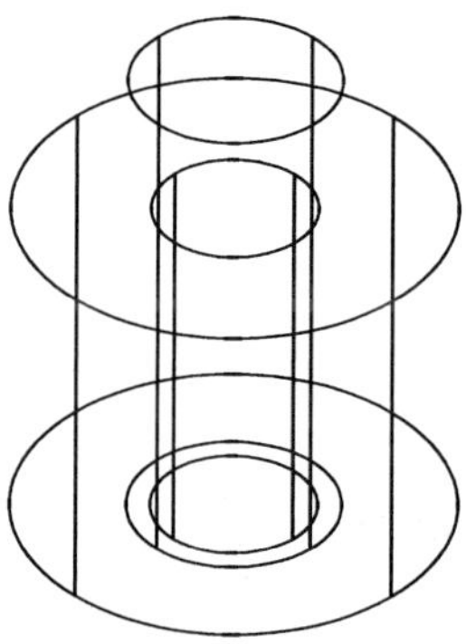

Figure 24-22 *Two mating components with interference*

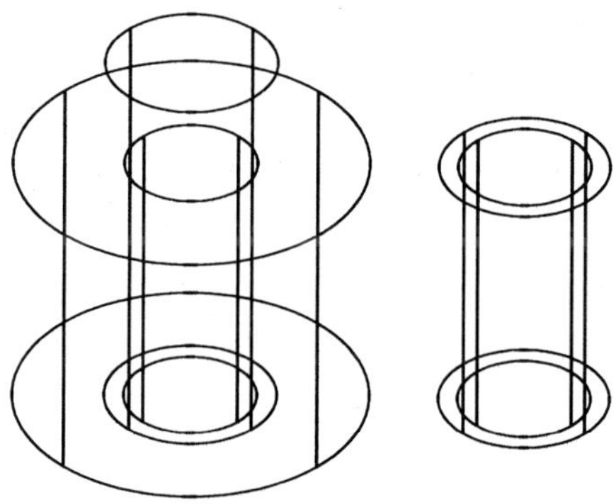

Figure 24-23 *Interference solid created using the* ***INTERFERENCE*** *command*

Example 1

In this example, you will create the solid model shown in Figure 24-24.

1. Increase the limits to 100,100. Zoom to the limits of drawing.

2. Choose the **SW Isometric View** button.

3. Enter **UCSICON** at the Command prompt. The prompt sequence is as follows:

   Enter an option [ON/OFF/All/Noorigin/ORigin/Properties] <ON>: **N**

4. Choose the **Box** button from the **Solids** toolbar. The prompt sequence is as follows.

   Specify corner of box or [CEnter] <0,0,0>: **10,10**
   Specify corner or [Cube/Length]: **L**

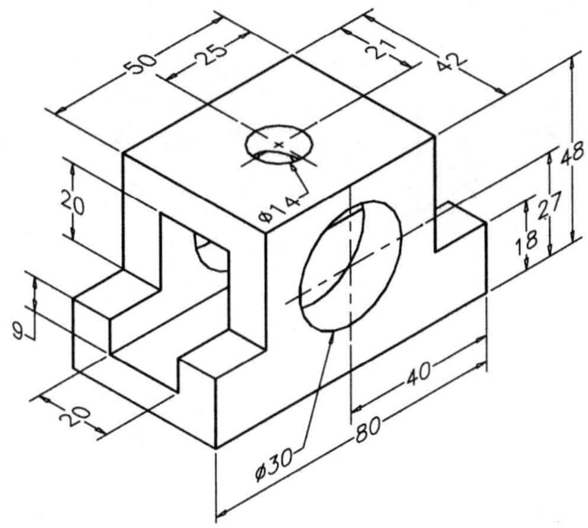

Figure 24-24 *Solid model for Example 1*

Specify length: **80**
Specify width: **42**
Specify height: **48**

5. Again choose the **Box** button from the **Solids** toolbar. The prompt sequence is as follows.

Specify corner of box or [CEnter] <0,0,0>: *Specify a point on the screen.*
Specify corner or [Cube/Length]: **L**
Specify length: **15**
Specify width: **42**
Specify height: **30**

6. Move the new box inside the old box using the midpoints of both the boxes. Copy the new box to the other side of the box (Figure 24-25).

7. Choose the **Subtract** button from the **Solids Editing** toolbar. The prompt sequence is as follows.

Select solids and regions to subtract from ..
Select objects: *Select the bigger box.*
Select objects: Enter
Select solids and regions to subtract ..
Select objects: *Select one of the smaller box.*
Select objects: *Select the other smaller box.*
Select objects: Enter

The model after subtracting the boxes and hiding the hidden lines should look similar to the one shown in Figure 24-26.

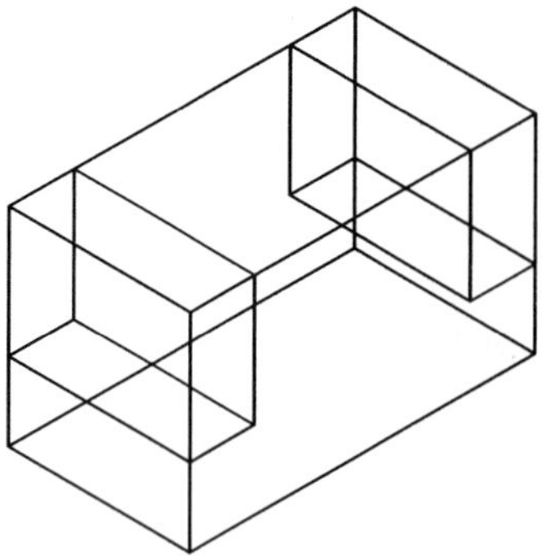

Figure 24-25 *Figure showing boxes moved inside the bigger box using the midpoints*

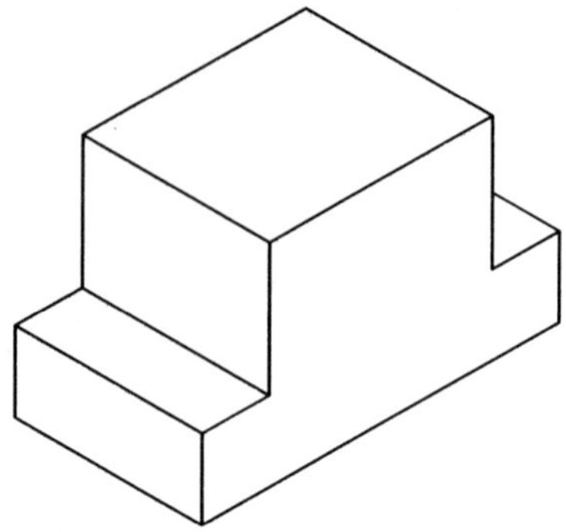

Figure 24-26 *Model after subtracting the smaller boxes and hiding hidden lines*

8. Again choose the **Box** button from the **Solids** toolbar. The prompt sequence is as follows.

   Specify corner of box or [CEnter] <0,0,0>: *Specify a point on the screen.*
   Specify corner or [Cube/Length]: **L**
   Specify length: **80**
   Specify width: **20**
   Specify height: **29**

9. Move the new box inside the existing model such that the midpoint of the left edge of the base of the box is 9 units above that of the model, see Figure 24-27. (Use the **From** option.)

10. Choose the **Subtract** button from the **Solids Editing** toolbar. The prompt sequence is as follows.

    Select solids and regions to subtract from ..
    Select objects: *Select the model.*
    Select objects: Enter
    Select solids and regions to subtract ..
    Select objects: *Select the box.*
    Select objects: Enter

The model after hiding the hidden lines should look similar to the one shown in Figure 24-28.

Tip
You can control the display of the silhouette lines in the solid model using the ***DISPSILH*** *system variable. Set the value of this variable to* ***1*** *to suppress the display of silhouette lines.*

11. Choose the **Cylinder** button from the **Solids** toolbar. The prompt sequence is as follows.

    Current wire frame density: ISOLINES=4
    Specify center point for base of cylinder or [Elliptical] <0,0,0>: *Specify the center of the*

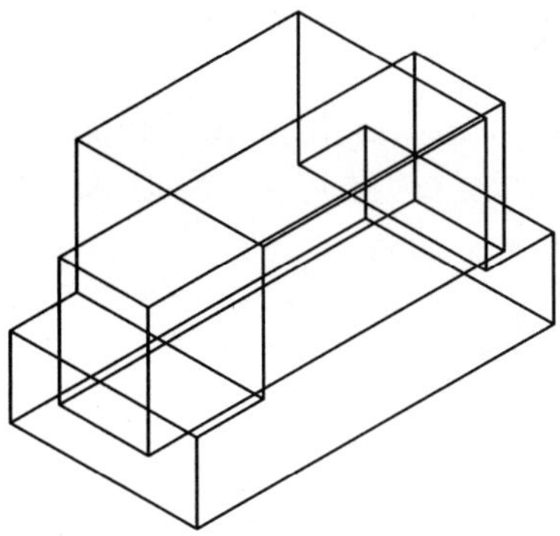

Figure 24-27 *Model before subtraction*

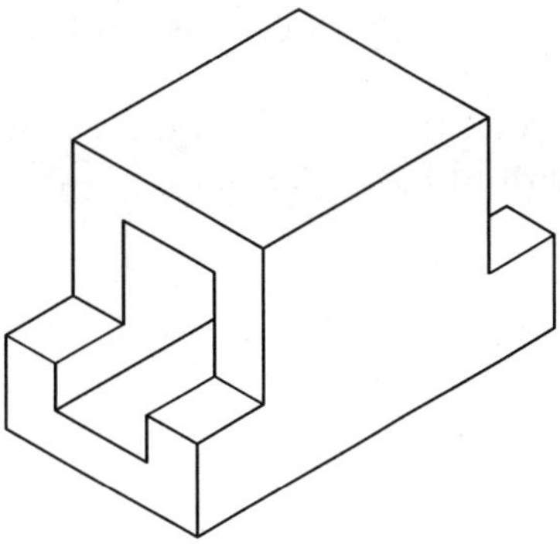

Figure 24-28 *Model after subtraction*

cylinder on the top face of the model using either ***Object Snap Tracking*** *or using the* ***From*** *option.*
Specify radius for base of cylinder or [Diameter]: **7**
Specify height of cylinder or [Center of other end]: **-10** *(The negative value of height creates the cylinder extruded in the negative Z direction, see Figure 24-29.)*

12. Subtract this cylinder from the existing model (Figure 24-30).

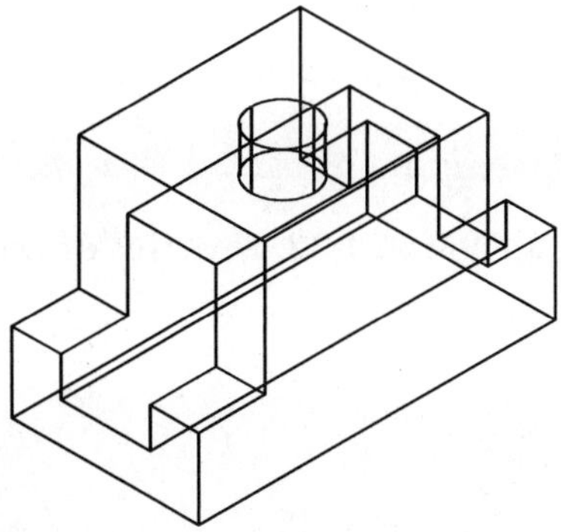

Figure 24-29 *Model before subtraction*

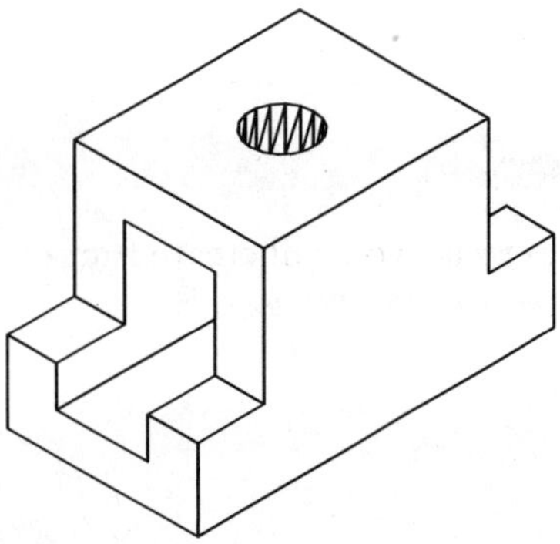

Figure 24-30 *Model after subtraction*

13. Choose the **X Axis Rotate UCS** button from the **UCS** toolbar. The prompt sequence is as follows.

Specify rotation angle about X axis <90>: Enter

14. Choose the **Cylinder** button from the **Solids** toolbar. The prompt sequence is as follows.

Current wire frame density: ISOLINES=4
Specify center point for base of cylinder or [Elliptical] <0,0,0>: *Use the* ***From*** *option to specify the center point of the cylinder at a distance of 27 units from the midpoint of the horizontal edge of the base.*

Specify radius for base of cylinder or [Diameter]: **15**
Specify height of cylinder or [Center of other end]: **-42**

15. Subtract this cylinder from the model. The final model should look similar to the one shown in Figure 24-31.

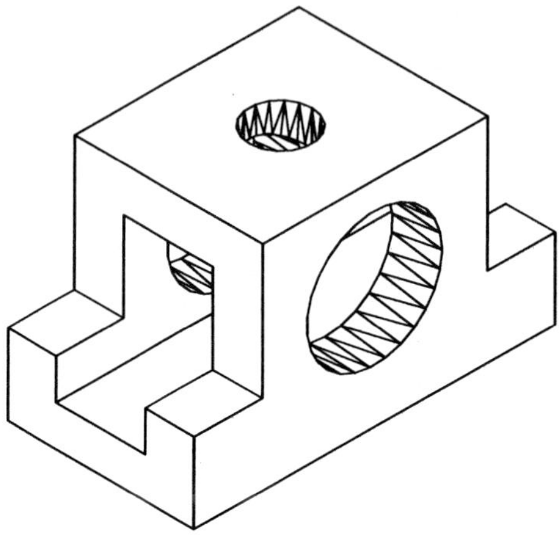

Figure 24-31 *Final solid model for Example 1*

Exercise 1 *Mechanical*

In this exercise you will create the solid model shown in Figure 24-32. Save this drawing with the name *\Ch-24\Exercise1.dwg*.

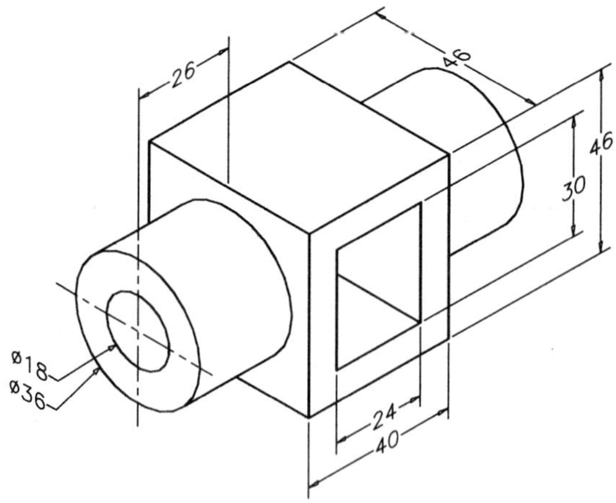

Figure 24-32 *Solid model for Exercise 1*

Sometimes the shape of the solid model is such that it cannot be created by just applying the Boolean operations on the standard solid primitives. For situations such as this, AutoCAD provides the commands called **EXTRUDE** and **REVOLVE**. Using these commands you can create the solid models of any complex shape. These commands are discussed next in detail.

CREATING EXTRUDED SOLIDS

Toolbar:	Solids > Extrude
Menu:	Draw > Solids > Extrude
Command:	EXTRUDE

The **EXTRUDE** command is used to create a complex solid model by extruding a closed 2D entity or a region along the *Z* axis direction or about a specified path. Remember that for extrusion, the original entity must be a closed loop or a region. The solid model can be created by extruding along the *Z* axis direction or about a specified path. Both of these options of creating the extruded solid are discussed next.

Extruding Along Z Axis

This is the default option and is used to create a solid model by extruding a closed 2D entity or a region along the *Z* axis direction. Figure 24-33 shows a region to be converted into an extruded solid and Figure 24-34 shows the solid created upon extruding the region. The prompt sequence that will be issued when you choose the **Extrude** button is given next.

Current wire frame density: ISOLINES=4
Select objects: *Select the region.*
Select objects: Enter
Specify height of extrusion or [Path]: **2**
Specify angle of taper for extrusion <0>: Enter

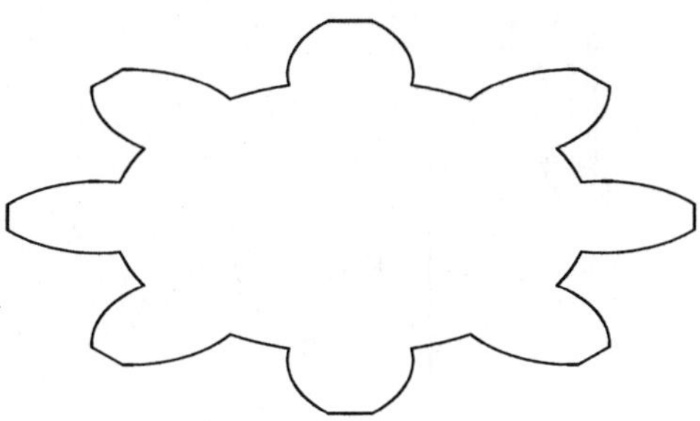

Figure 24-33 *Base object for extruding*

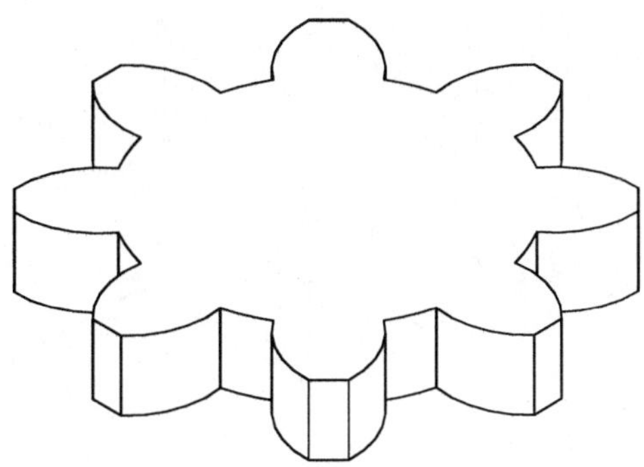

Figure 24-34 *Solid created upon extruding*

You can also specify the taper angle for the extruded solid. The positive value of the taper angle will taper in from the base object and the negative value will taper out of the base object (Figure 24-35).

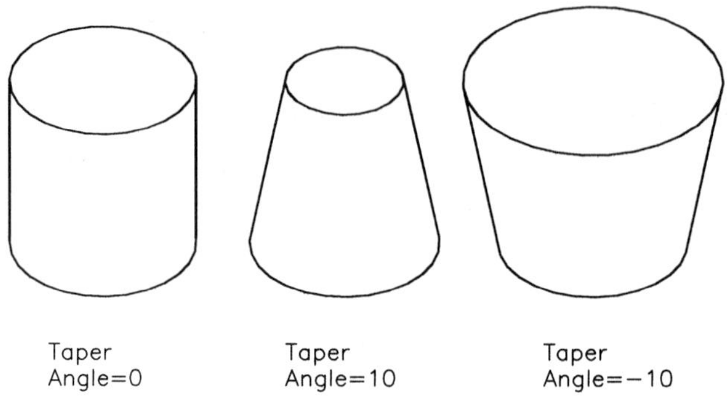

Figure 24-35 Results of various taper angles

Extruding About a Path

This option is used to create a solid by extruding a closed 2D entity or a region about a specified path. Bear in mind that the path of extrusion should be normal to the plane of the base object. If the path consists of more than one entity, all of them should be first joined using the **PEDIT** command so that the path remains a single entity. This option is generally used for creating complex pipelines and also by architects and interior designers for creating the beadings. The path used for extrusion can be a closed entity or an open entity and the valid entities that can be used as path are lines, circles, ellipses, polygons, arcs, polylines, or splines. You can not specify taper angle when you use a path. Figure 24-36 shows a base object and a path about which it has to be extruded and Figure 24-37 shows the solid created upon extruding about the specified path.

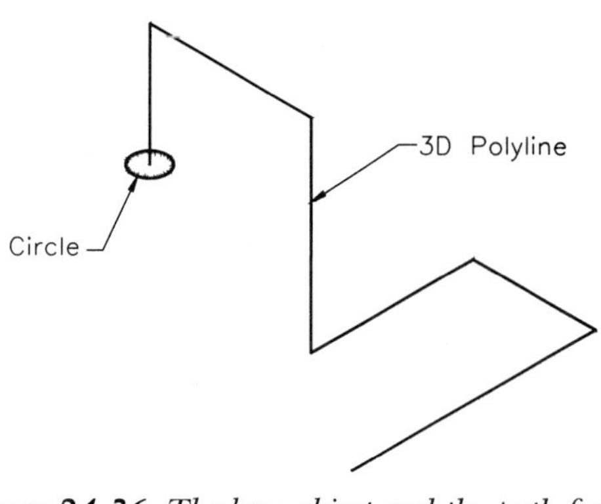

Figure 24-36 The base object and the path for extrusion

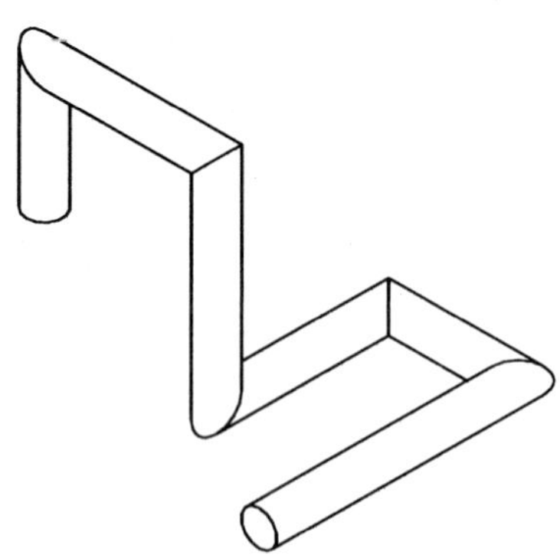

Figure 24-37 Solid model created upon extruding the base entity about the specified path

CREATING REVOLVED SOLIDS

Toolbar:	Solids > Revolve
Menu:	Draw > Solids > Revolve
Command:	REVOLVE

This command is used to create a complex solid by revolving closed 2D entities or regions about a specified revolution axis. The entire 2D entity should be on one side of the revolution axis. Self-intersecting and crossed entities cannot be revolved using this command. Remember that the direction of revolution is determined using the right-hand thumb rule. The axis of rotation can be defined by specifying two points, using the *X* or the *Y* axis of the current UCS, or using an existing object. The following prompt sequence will be issued when you choose this button.

Current wire frame density: ISOLINES=4
Select objects: *Select the object or region to revolve.*
Select objects: Enter
Specify start point for axis of revolution or
define axis by [Object/X (axis)/Y (axis)]:

Specifying the Start Point for Axis of Revolution

This option is used when you want to define the axis of revolution using two points: the start point and the endpoint of the axis of revolution. The positive direction of the axis will be from the start point to the endpoint and the direction of revolution will be defined using the right-hand thumb rule. Before revolving, make sure the complete 2D entity is on one side of the axis of revolution.

Object

This option is used to create a revolved solid by revolving a selected 2D entity or a region about a specified object. The valid entities that can be used as object for defining the axis of revolution are line, or a single segment of a polyline. If the polyline selected as the object consists of more than one entity, then AutoCAD draws an imaginary line from the start point of the first segment to the endpoint of the last segment. This imaginary line is then taken as the object for revolution.

X (axis)

This option uses the positive direction of the *X* axis of the current UCS for revolving the selected entity. If the selected entity is not completely on one side of the *X* axis, it will give you an error message that it cannot revolve the object.

Y (axis)

This option uses the positive direction of the *Y* axis of the current UCS as the axis of revolution for creating the revolved solid.

After setting the revolution axis option, you will be prompted to specify the angle of revolution. The default value is 360-degree. You can enter the required value at this prompt.

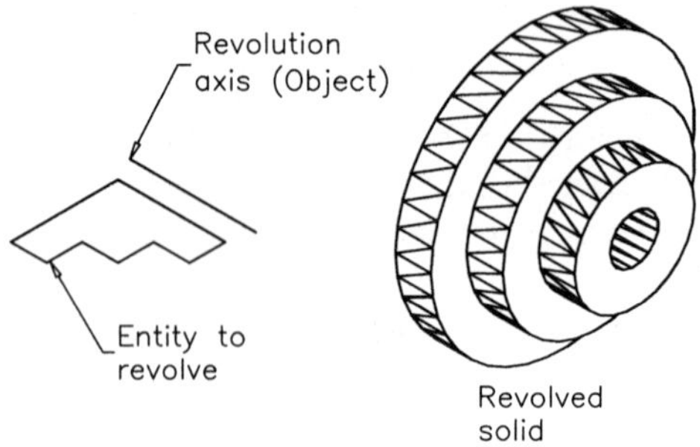

Figure 24-38 *Creating a revolved solid*

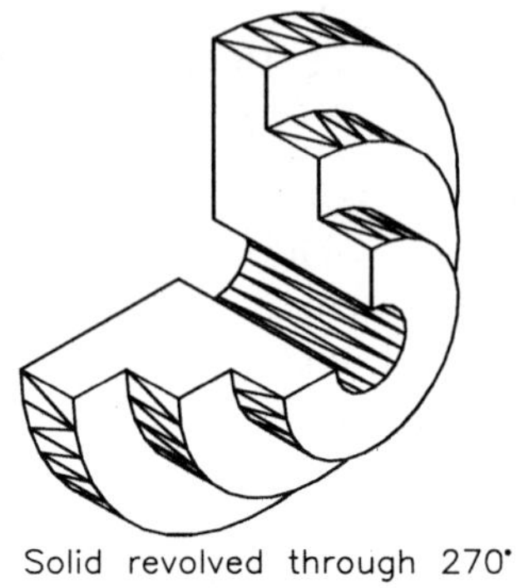

Figure 24-39 *Solid revolved to an angle*

Example 2

In this example you will create the solid model shown in Figure 24-40.

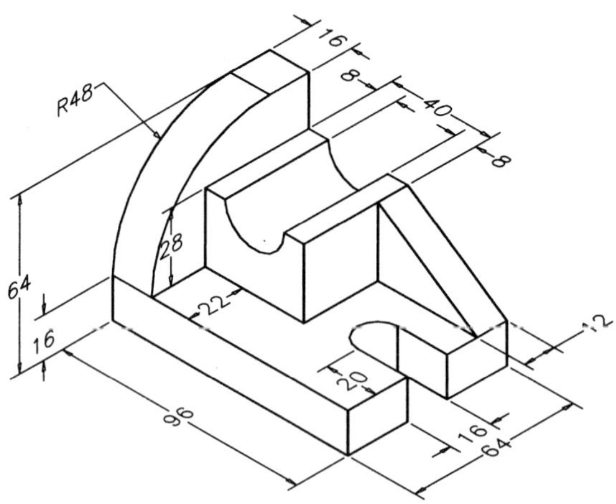

Figure 24-40 *Solid model for Example 2*

1. Increase the limits and then zoom to the limits of the drawing.

2. Using lines and arc, create the base of the model with the dimensions as shown in Figure 24-41. You can also use the **PLINE** command to create this sketch.

3. Choose the **Region** button from the **Draw** toolbar. The prompt sequence is as follows.

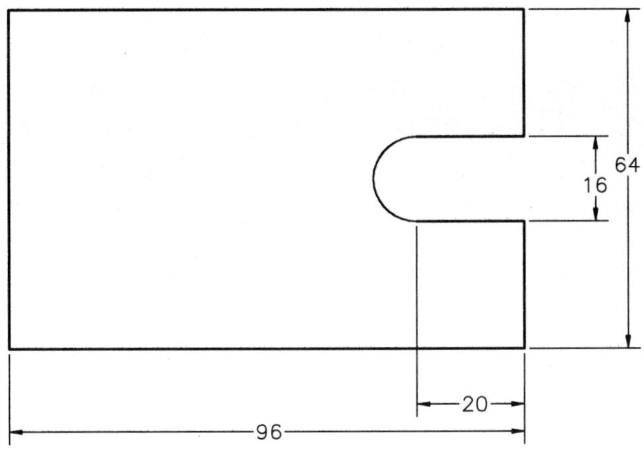

Figure 24-41 *Base for the solid model*

Select objects: *Select the complete base.*
Select objects: Enter
1 loop extracted.
1 Region created.

Note
If you created the sketch using the ***PLINE*** *command, you do not need to convert it into a region.*

4. Choose the **Extrude** button from the **Solids** toolbar. The prompt sequence is as follows.

   Select objects: *Select the region.*
   Select objects: Enter
   Specify height of extrusion or [Path]: **16**
   Specify angle of taper for extrusion <0>: Enter

5. Change the viewpoint to the SE Isometric view.

6. Choose the **Wedge** button from the **Solids** toolbar. The prompt sequence is as follows.

   Specify first corner of wedge or [CEnter] <0,0,0>: *Pick a point on the screen.*
   Specify corner or [Cube/Length]: **L**
   Specify length: **40**
   Specify width: **12**
   Specify height: **28**

7. Move this wedge using the endpoint so that it is properly aligned with the base, see Figure 24-42.

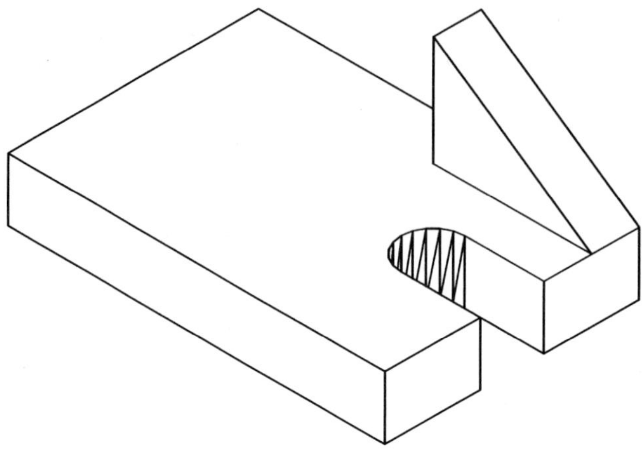

Figure 24-42 *Wedge aligned with the base*

8. Choose the **X Axis Rotate UCS** button from the **UCS** toolbar. The prompt sequence is as follows.

   Specify rotation angle about X axis <90>: Enter

9. Draw the object shown in Figure 24-43 and then convert it into a region.

10. Choose the **Extrude** button from the **Solids** toolbar. The prompt sequence is as follows.

    Current wire frame density: ISOLINES=4
    Select objects: *Select the region.*
    Select objects: Enter
    Specify height of extrusion or [Path]: **42**
    Specify angle of taper for extrusion <0>: Enter

11. Move it on the base using the endpoint as shown in Figure 24-44.

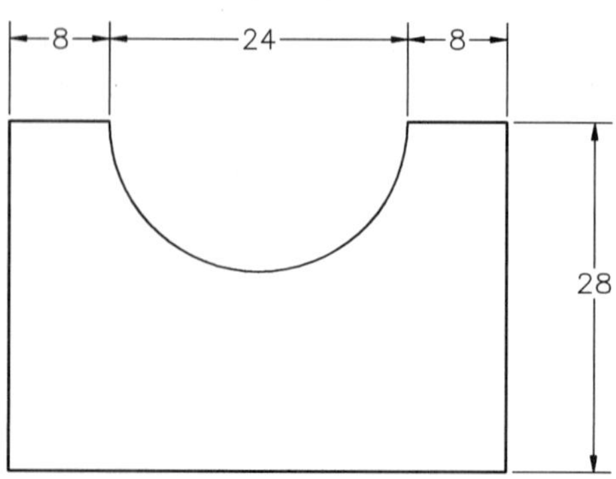

Figure 24-43 *The region to be extruded*

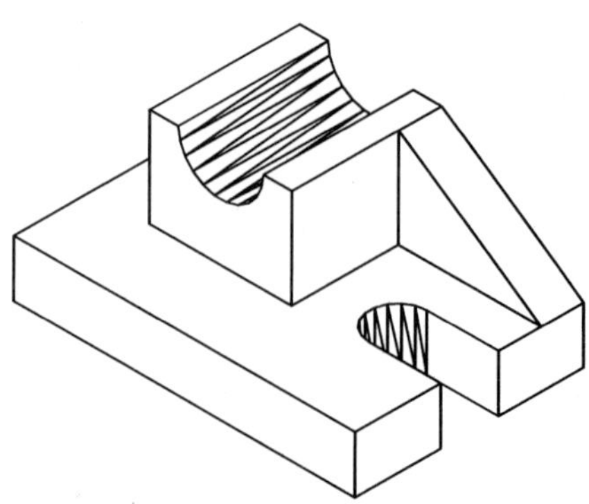

Figure 24-44 *Solid aligned with the base*

12. Choose the **Y Axis Rotation UCS** button from the **UCS** toolbar. The prompt sequence is as follows.

    Specify rotation angle about Y axis <90>: Enter

13. Draw the object shown in Figure 24-45 and then convert it into a region.

14. Extrude it and move it so that it is properly aligned with the base, see Figure 24-46.

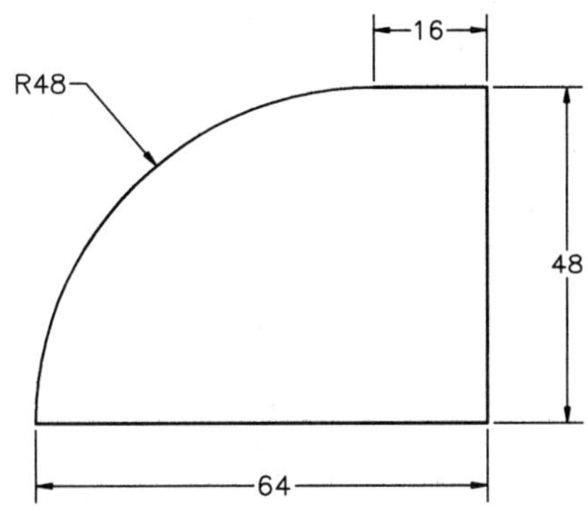

Figure 24-45 *The region before extruding*

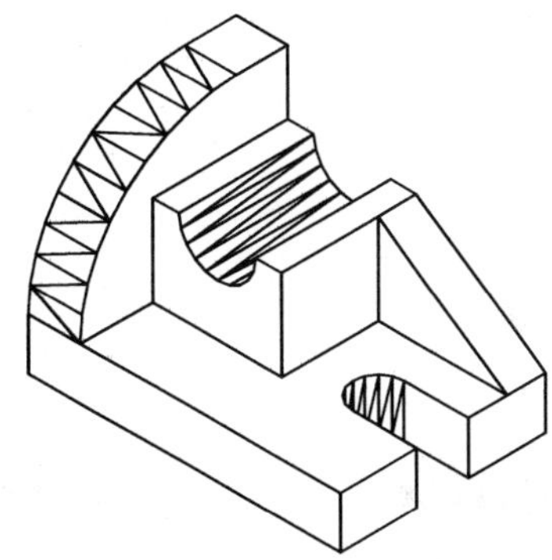

Figure 24-46 *Next solid aligned with the base*

15. Choose the **Union** button from the **Solids Editing** toolbar. The prompt sequence is as follows.

    Select objects: *Select all the objects.*
    Select objects: Enter

The final solid model should look similar to the one shown in Figure 24-47.

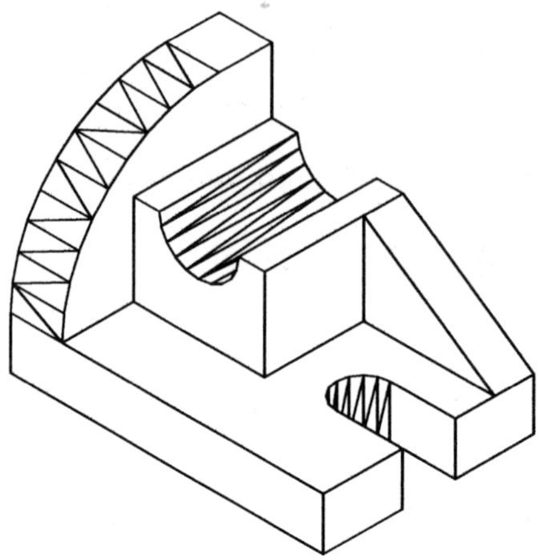

Figure 24-47 Solid model for Example 2

FILLETING SOLID MODELS

Toolbar:	Modify > Fillet
Menu:	Modify > Fillet
Command:	FILLET

As mentioned earlier, the **FILLET** command is used to round the edges or the corners of the models. This is generally done to reduce the stress concentration area in the model. The behavior of this command is different while working with 2D entities from the behavior while working with solid models. Therefore, it is very important for the user to first understand the use of this command to fillet the edges of the solid models. Figure 24-48 shows two lines that are selected to fillet. Now, as these lines are nothing but 2D entities, when you select these two lines to fillet, the result will be as shown in Figure 24-49.

Figure 24-48 *Lines before filleting*

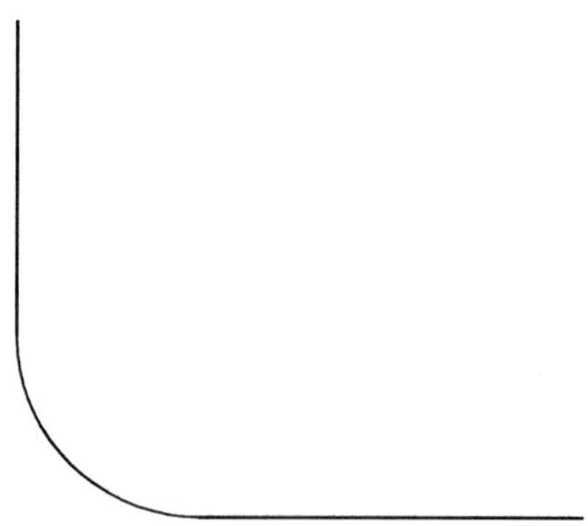

Figure 24-49 *Lines after filleting*

This shows that if actually there was a vertical edge at the corner of the two lines shown in Figure 24-48, then it would have been filleted. However, in the 3D models, you directly have the vertical edges and therefore, you have to select the vertical edge to fillet. Figure 24-50 shows a solid model. To fillet this model, you just need to select the vertical edges, see Figure 24-51.

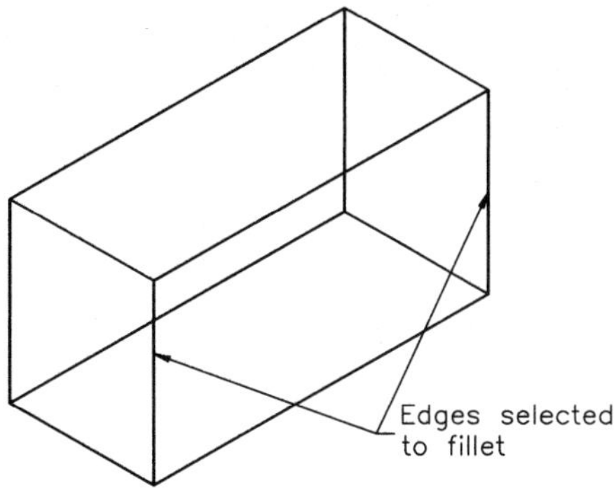

Figure 24-50 *Selecting the edges to fillet*

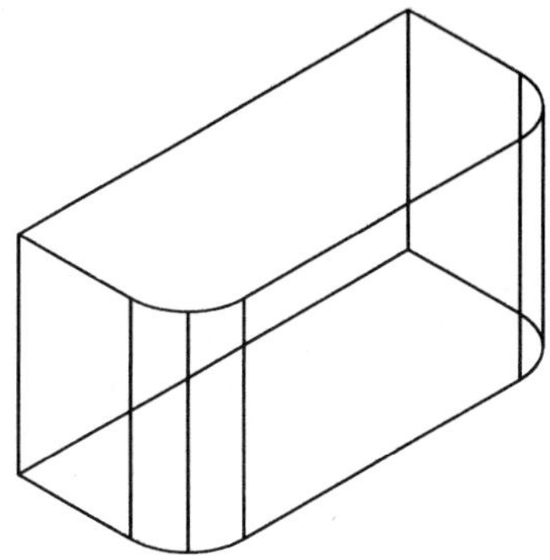

Figure 24-51 *Model after filleting the edges*

The prompt sequence that will be followed when you choose the **Fillet** button is given next.

Current settings: Mode = TRIM, Radius = 2.0000
Select first object or [Polyline/Radius/Trim/mUltiple]: *Select the edges to fillet. You will be allowed to select only one edge at this moment.*
Enter fillet radius <2.0000>: *Enter the required fillet radius.*
Select an edge or [Chain/Radius]: *Select the other edges to fillet or enter an option.*

Chain

This option is used to select the other tangential edges on the model. For example, you can select all the tangential edges on the top or the bottom face of the solid model shown in Figure 24-52 for filleting in just a single attempt using the **Chain** option.

Radius

This option is used to redefine the fillet radius.

CHAMFERING SOLID MODELS

Toolbar:	Modify > Chamfer
Menu:	Modify > Chamfer
Command:	CHAMFER

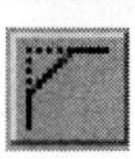

The **CHAMFER** command is used to bevel the edges of the solid models. This command is also used to reduce the area of the stress concentration in the solid models. The working of this command is also different while working with the solid models. The following prompt sequence is displayed when you choose the **Chamfer** button.

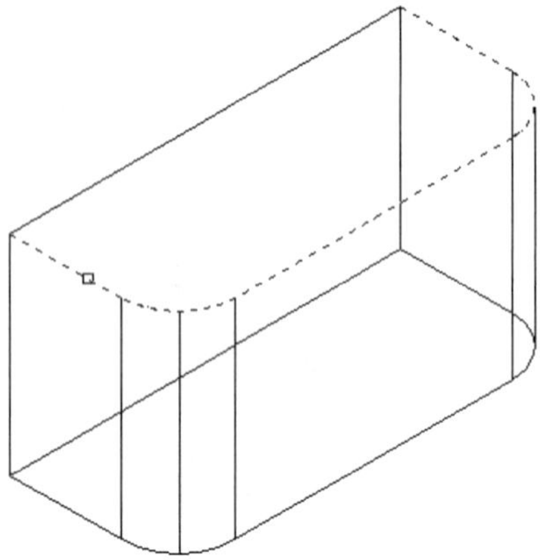

Figure 24-52 *Selecting the edges using the Chain option*

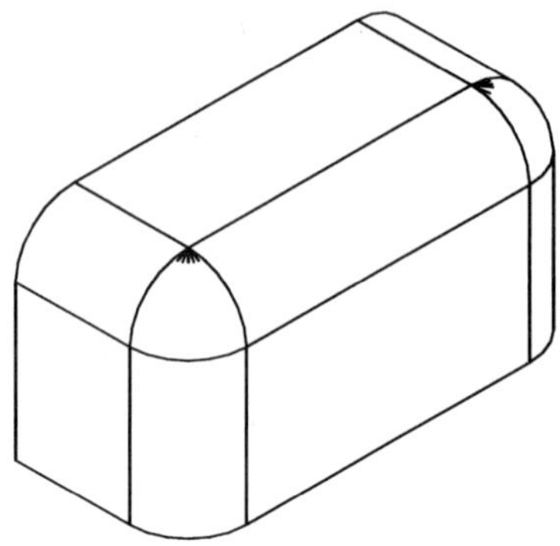

Figure 24-53 *The fillet created using the Chain option*

(TRIM mode) Current chamfer Dist1 = 1.0000, Dist2 = 0.5000
Select first line or [Polyline/Distance/Angle/Trim/Method/mUltiple]: *Select the edge to chamfer. One of the faces associated with the edge will be selected and highlighted.*
Base surface selection...
Enter surface selection option [Next/OK (current)] <OK>: *Give a null response if you want to make this face as the base surface. Otherwise enter **N** at this prompt.*
Specify base surface chamfer distance <default value>: *Specify the distance.*
Specify other surface chamfer distance <default value>: *Specify the distance.*
Select an edge or [Loop]: *Select the edge to fillet.*

Loop

This option is used to select all the edges composed of a loop on the selected face of the solid model. To use this option you will have to select any of the edges that comprise the loop at the **Select an edge loop or [Edge]** prompt.

ROTATING SOLID MODELS IN 3D SPACE

Menu:	Modify > 3D Operations > Rotate 3D
Command:	ROTATE3D

The **ROTATE3D** command is used to rotate the selected solid model in the 3D space about a specified axis. Once again the right-hand thumb rule will be used to determine the direction of rotation of solid model in the 3D space. The prompt sequence that will follow when you choose this command from the **Modify** menu is given next.

Current positive angle: ANGDIR=counterclockwise ANGBASE=0
Select objects: *Select the solid model.*
Select objects: Enter

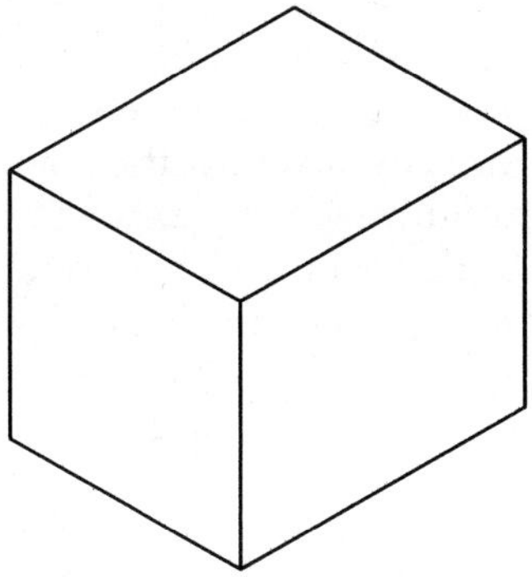

Figure 24-54 *Solid model before chamfering*

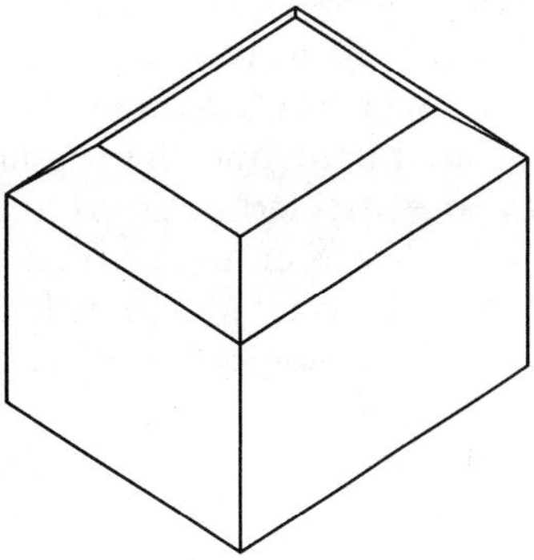

Figure 24-55 *Solid model after chamfering*

Specify first point on axis or define axis by
[Object/Last/View/Xaxis/Yaxis/Zaxis/2points]:

2points Option

This is the default option for rotating the solid models. This option allows you to rotate the solid model about an axis specified using two points. The direction of the axis will be from the first point to the second point. Using this direction of the axis you can calculate the direction of rotation of the solid model by applying the right-hand thumb rule. The prompt sequence that will follow when you invoke this option is given next.

[Object/Last/View/Xaxis/Yaxis/Zaxis/2points]: Enter
Specify first point on axis: *Specify the first point of the rotation axis. See Figure 24-56.*
Specify second point on axis: *Specify the second point of the rotation axis. See Figure 24-56.*
Specify rotation angle or [Reference]: *Specify the angle of rotation.*

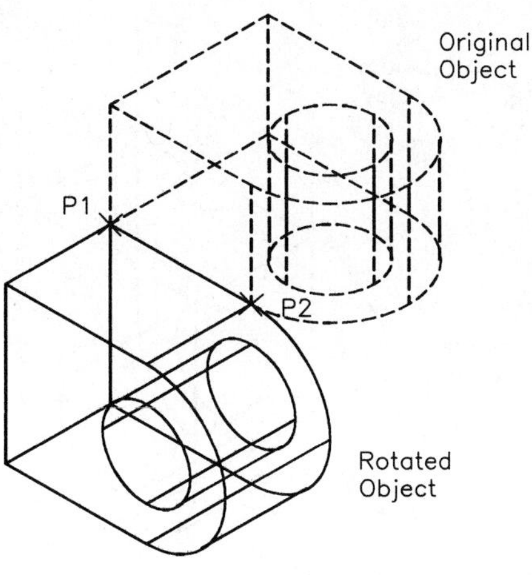

Figure 24-56 *Rotating the solid model using the 2points option*

Object Option

This option is used to rotate the solid model in the 3D space using a 2D entity. The 2D entities that can be used are lines, circles, arcs, or 2D polyline segments. If the selected entity is a line or a straight polyline segment, then it will be directly taken as the rotation axis. However, if the selected entity is an arc or a circle, then an imaginary axis will be drawn starting from the center and normal to the plane in which the arc or the circle is drawn. The object will then be rotated about this imaginary axis. The prompt sequence that will follow when you invoke this option is given next.

[Object/Last/View/Xaxis/Yaxis/Zaxis/2points]: **O**
Select a line, circle, arc, or 2D-polyline segment: *Select the 2D entity as shown in Figure 24-57.*
Specify rotation angle or [Reference]: *Specify the angle of rotation.*

Last Option

This option uses the same axis that was last selected to rotate the solid model.

View Option

This option is used to rotate the solid model about the viewing plane. In this case the viewing plane is the screen of the computer. This option draws an imaginary axis starting from the specified point and continues normal to the viewing plane. The model is then rotated about this axis. The prompt sequence that will follow when you invoke this option is given next.

Specify first point on axis or define axis by
[Object/Last/View/Xaxis/Yaxis/Zaxis/2points]: **V**
Specify a point on the view direction axis <0,0,0>: *Specify the point on the view plane as shown in Figure 24-58.*
Specify rotation angle or [Reference]: *Specify the angle of rotation.*

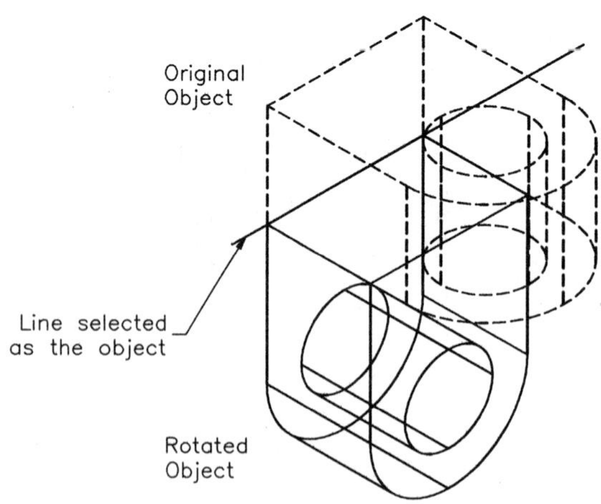

Figure 24-57 *Rotating the solid model using the Object option*

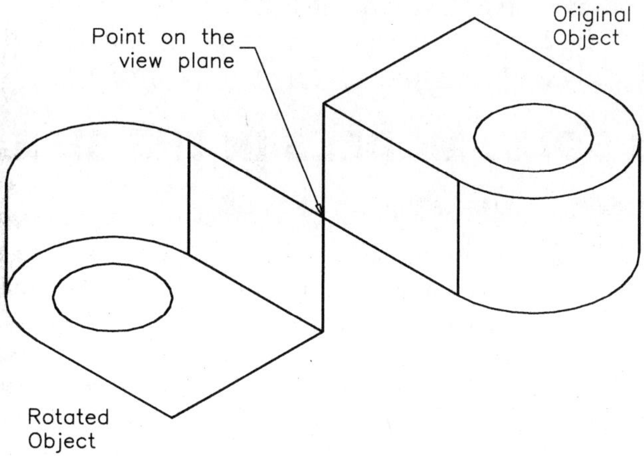

Figure 24-58 *Rotating the solid model using the View option by an angle of 180-degree*

Xaxis Option

This option is used to rotate the solid model about the positive *X* axis of the current UCS. When you invoke this option, you will be prompted to select a point on the *X* axis. The prompt sequence that will follow when you invoke this command is given next.

Specify first point on axis or define axis by
[Object/Last/View/Xaxis/Yaxis/Zaxis/2points]: **X**
Specify a point on the X axis <0,0,0>: *Specify the point on the X axis.*
Specify rotation angle or [Reference]: *Specify the angle of rotation.*

Yaxis Option

This option is used to rotate the solid model about the positive *Y* axis of the current UCS. When you invoke this option, you will be prompted to select a point on the *Y* axis. The prompt sequence that will follow when you invoke this command is given next.

Specify first point on axis or define axis by
[Object/Last/View/Xaxis/Yaxis/Zaxis/2points]: **Y**
Specify a point on the Y axis <0,0,0>: *Specify the point on the Y axis.*
Specify rotation angle or [Reference]: *Specify the angle of rotation.*

Zaxis Option

This option is used to rotate the solid model about the positive *Z* axis of the current UCS. When you invoke this option, you will be prompted to select a point on the *Z* axis. The prompt sequence that will follow when you invoke this command is given next.

Specify first point on axis or define axis by
[Object/Last/View/Xaxis/Yaxis/Zaxis/2points]: **Z**
Specify a point on the Z axis <0,0,0>: *Specify the point on the Z axis.*
Specify rotation angle or [Reference]: *Specify the angle of rotation.*

MIRRORING SOLID MODELS IN THE 3D SPACE

Menu:	Modify > 3D Operations > Mirror 3D
Command:	MIRROR3D

The **MIRROR3D** command is used to mirror the solid models about a specified plane in the space. The prompt sequence that will follow when you choose this command from the **Modify** menu is given next.

Select objects: *Select the solid model to be mirrored.*
Select objects: Enter
Specify first point of mirror plane (3 points) or
[Object/Last/Zaxis/View/XY/YZ/ZX/3points] <3points>:

3points Option

This is the default option for mirroring the solid models. As discussed earlier, a line can be defined by two points from which the line passes. Similarly, a plane can be defined by three points through which it passes. This option allows you to specify the three points from which the mirroring plane passes. The prompt sequence that follows when you invoke this command is given next.

Specify first point of mirror plane (3 points) or
[Object/Last/Zaxis/View/XY/YZ/ZX/3points] <3points>: Enter
Specify first point on mirror plane: *Specify the first point on the plane. See Figure 24-59.*
Specify second point on mirror plane: *Specify the second point on the plane. See Figure 24-59.*
Specify third point on mirror plane: *Specify the third point on the plane. See Figure 24-59.*
Delete source objects? [Yes/No] <N>: Enter

Object

This option is used to mirror the solid model using a 2D entity. The 2D entities that can be used to mirror the solids are circles, arcs, and 2D polyline segments. The prompt sequence that follows when you invoke this option is given next.

Specify first point of mirror plane (3 points) or
[Object/Last/Zaxis/View/XY/YZ/ZX/3points] <3points>: **O**
Select a circle, arc, or 2D-polyline segment: *Select the 2D entity as shown in Figure 24-60.*
Delete source objects? [Yes/No] <N>: Enter

Zaxis

This option allows you to define a mirroring plane using two points. The first point is the

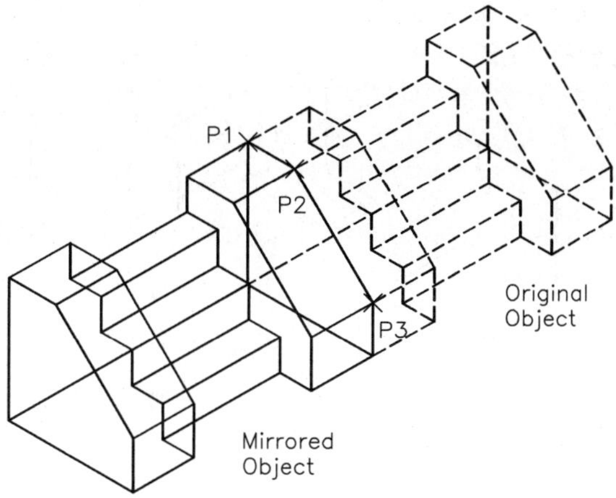

Figure 24-59 *Mirroring the solid model using the 3points option*

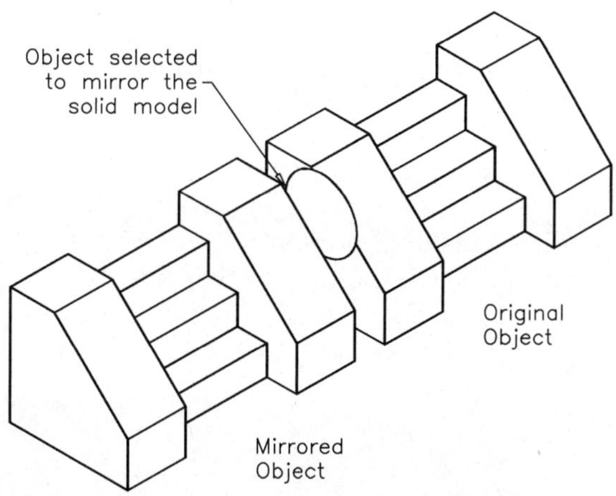

Figure 24-60 *Mirroring the solid model using the Object option*

point on the mirroring plane and the second point is a point on the positive direction of the *Z* axis of that plane. The prompt sequence that follows when you invoke this option is given next.

Specify first point of mirror plane (3 points) or
[Object/Last/Zaxis/View/XY/YZ/ZX/3points] <3points>: **Z**
Specify point on mirror plane: *Specify the point on the plane as shown in Figure 24-61.*
Specify point on Z-axis (normal) of mirror plane: *Specify the point on the Z direction of the plane as shown in Figure 24-61.*
Delete source objects? [Yes/No] <N>: Enter

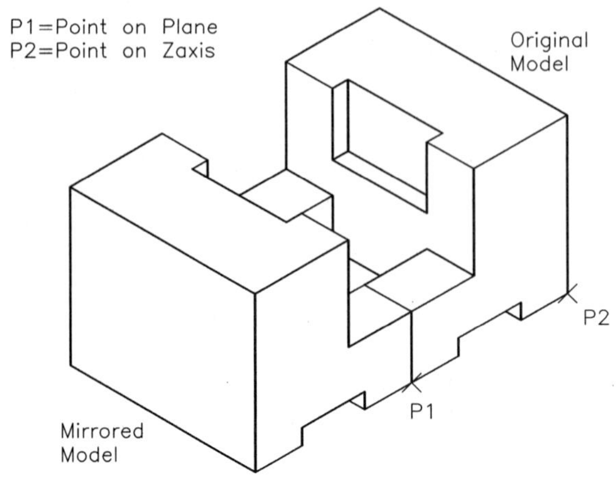

Figure 24-61 *Mirroring the solid model using the* ***Zaxis*** *option*

View Option

This is one of the most interesting options that is provided for mirroring the solid models. This option is used to mirror the selected solid model about the viewing plane. The viewing plane in this case is the screen of the monitor (Figure 24-62). The prompt sequence that will follow when you invoke this command is given next.

Specify first point of mirror plane (3 points) or
[Object/Last/Zaxis/View/XY/YZ/ZX/3points] <3points>: **V**
Specify point on view plane <0,0,0>: *Specify the point on the view plane.*
Delete source objects? [Yes/No] <N>: Enter

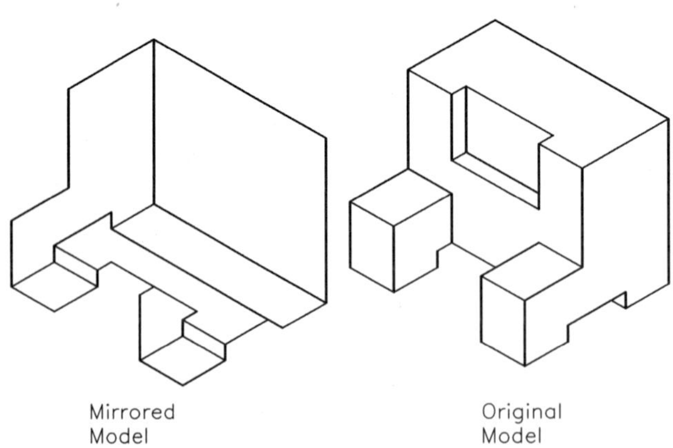

Figure 24-62 *A solid model mirrored using the* ***View*** *option and then isolated using the* ***MOVE*** *command*

Tip

Because the model is mirrored about the view plane, the new model will be placed over the original model when you view it from the current viewpoint. The best option to use to view the mirrored model is to hide the hidden lines using the **HIDE** *command. You can also move the new model away from the last model using the* **MOVE** *command or change the viewpoint for viewing both the models simultaneously.*

XY/YZ/ZX

These options are used to mirror the solid model about the *XY*, *YZ*, or *ZX* planes of the current UCS.

Tip

All these planes will be considered with reference to the current orientation of the UCS and not with the world position of the UCS. This means that when you select the **XY** *option to mirror the solid model, then the XY plane of the current UCS will be considered to mirror the model and not the XY plane of the world UCS.*

Example 3

In this example you will create the solid model shown in Figure 24-63. Assume the missing dimensions.

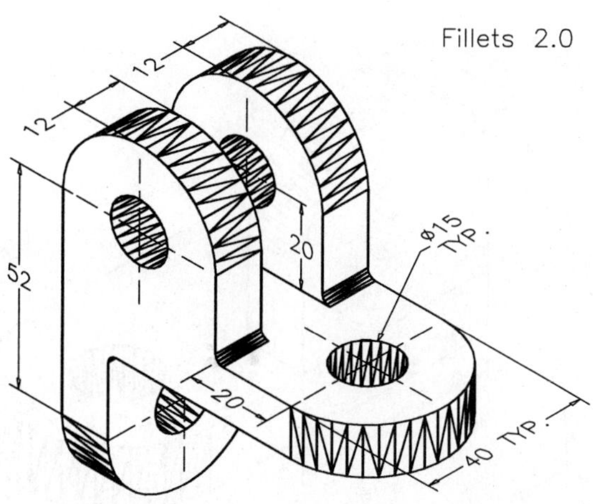

Figure 24-63 *Solid model for Example 3*

1. Open a new file and then set the limits to **100,100**. Change the overall scale factor to **10** using the **Dimension Style Manager** dialog box.

2. Create the base of the model and then change the viewpoint to **1,-1,1**, see Figure 24-64.

3. Change the UCS by rotating it around the *X* axis through an angle of 90-degree.

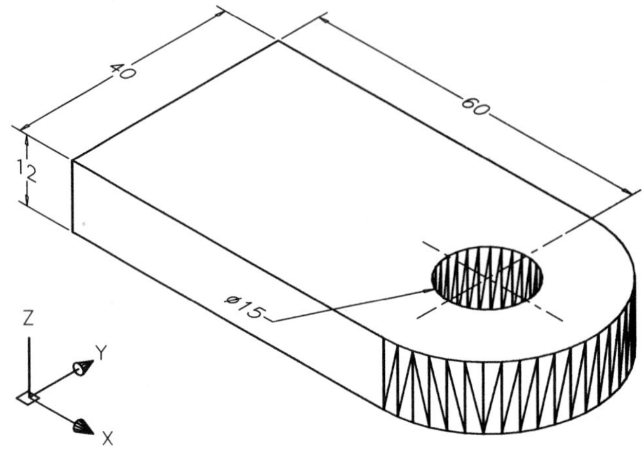

Figure 24-64 *Base for the model*

4. Change the viewpoint to the plan view of the current UCS using the **PLAN** command.

5. Create the next object and then move it using the endpoint to align with the base of the model as shown in Figure 24-65.

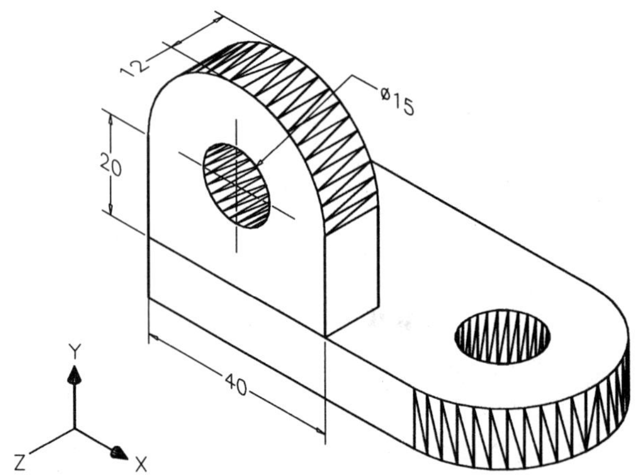

Figure 24-65 *Model after creating the next object and moving it*

6. Choose **Modify > 3D Operations > Mirror 3D** from the menu bar. The prompt sequence is as follows.

Select objects: *Select the last object.*
Select objects: Enter
Specify first point of mirror plane (3 points) or

[Object/Last/Zaxis/View/XY/YZ/ZX/3points] <3points>: **Z**
Specify point on mirror plane: *Select P1 as shown in Figure 24-66.*
Specify point on Z-axis (normal) of mirror plane: *Select P2 as shown in Figure 24-66.*
Delete source objects? [Yes/No] <N>: [Enter]

7. The object after mirroring and hiding the hidden lines should look similar to that shown in Figure 24-67.

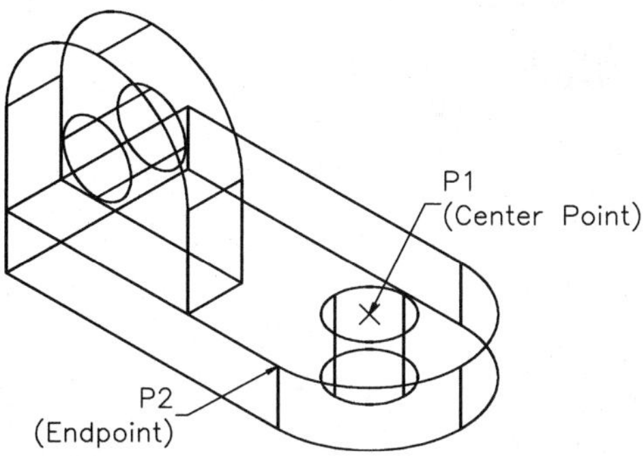

Figure 24-66 *Selecting the points to mirror*

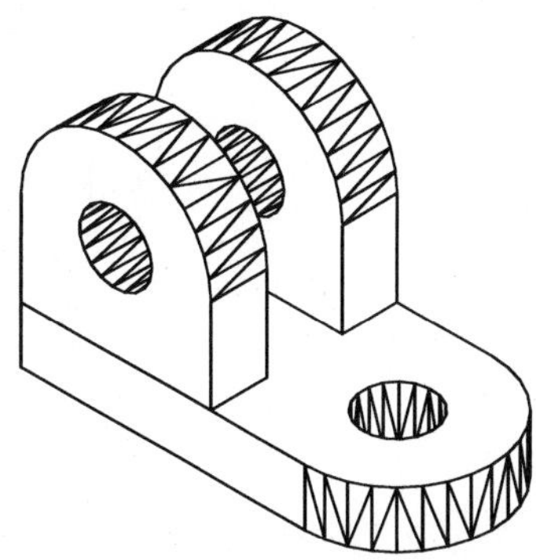

Figure 24-67 *Model after mirroring*

8. Change the UCS by rotating it around the *Y* axis through an angle of 90-degree.

9. Create the next object of the required dimensions and then move it using the endpoints as shown in Figure 24-68.

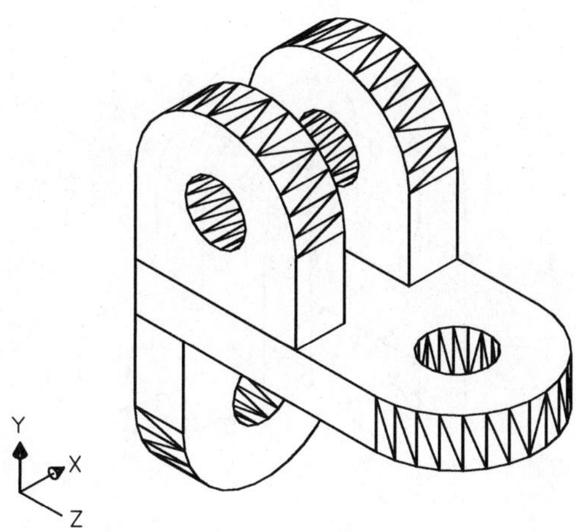

Figure 24-68 *Model after creating the next object and moving it*

10. Invoke the **UNION** command and then union all the objects, see Figure 24-69.

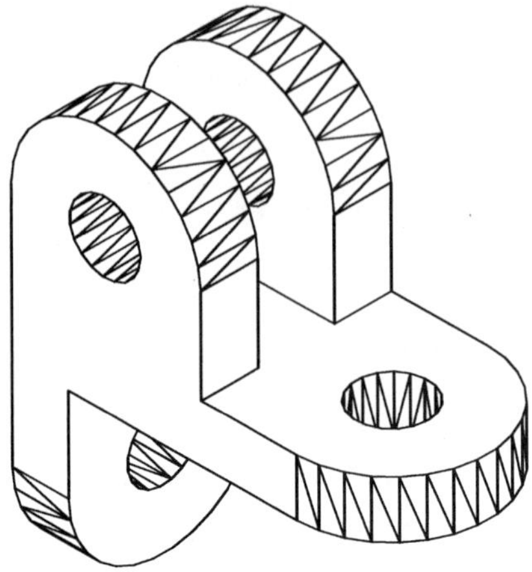

Figure 24-69 *Model after union*

11. Choose the **Fillet** button from the **Modify** toolbar. The following prompt sequence is displayed.

    Current settings: Mode = TRIM, Radius = 0.5000
    Select first object or [Polyline/Radius/Trim/mUltiple]: *Select first edge as shown in Figure 24-70.*
    Enter fillet radius <0.5000>: **2**
    Select an edge or [Chain/Radius]: *Select the second edge as shown in Figure 24-70.*
    Select an edge or [Chain/Radius]: *Select the third edge as shown in Figure 24-70.*
    Select an edge or [Chain/Radius]: Enter
    3 edge(s) selected for fillet.

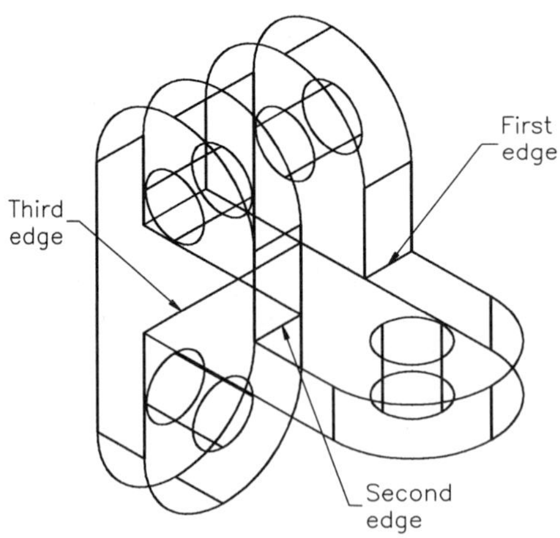

Figure 24-70 *Selecting the edges for filleting*

12. The final model for Example 3 should look similar to the one shown in Figure 24-71.

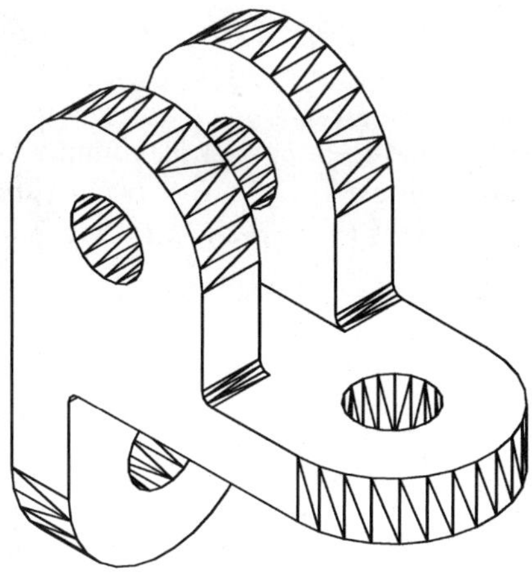

Figure 24-71 *The final model for Example 3*

CREATING ARRAYS IN THE 3D SPACE

Menu:	Modify > 3D Operations > 3D Array
Command:	3DARRAY

As mentioned in the previous chapters, the arrays are defined as the method of creating multiple copies of the selected object in a rectangular or a polar fashion. The 3D arrays can also be created similar to the 2D arrays. The only difference is that in 3D array, another factor called the *Z* axis is also taken into consideration. There are two types of 3D arrays. Both of these types are discussed next.

3D Rectangular Array

This is the method of arranging the solid model along the edges of a box. In this type of array you will have to specify three parameters. They are the rows (along the *X* axis), the columns (along the *Y* axis), and the levels (along the *Z* axis). You will also have to specify the distances between the rows, columns, and the levels. The 3D rectangular array can be easily understood by taking an example shown in Figure 24-72. This figure shows the two floors of a building. Initially, only one chair is placed on the ground floor. Now, if you create a 2D rectangular array then the chairs will be arranged only on the ground floor. However, when you create the 3D rectangular array, then the chairs will be arranged on the first floor (along the *Z* axis) as well as the ground floor, see Figure 24-73. In this example the number of rows is three, the number of columns is four, and the number of levels is two.

The prompt sequence that will follow when you choose this command from the **Modify** menu is given next.

Select objects: *Select the object to array.*
Select objects: Enter

Enter the type of array [Rectangular/Polar] <R>: Enter
Enter the number of rows (---) <1>: *Specify the number of rows along the X axis.*
Enter the number of columns (|||) <1>: *Specify the number of columns along the Y axis.*
Enter the number of levels (...) <1>: *Specify the number of levels along the Z axis.*
Specify the distance between rows (---): *Specify the distance between the rows.*
Specify the distance between columns (|||): *Specify the distance between the columns.*
Specify the distance between levels (...): *Specify the distance between the levels.*

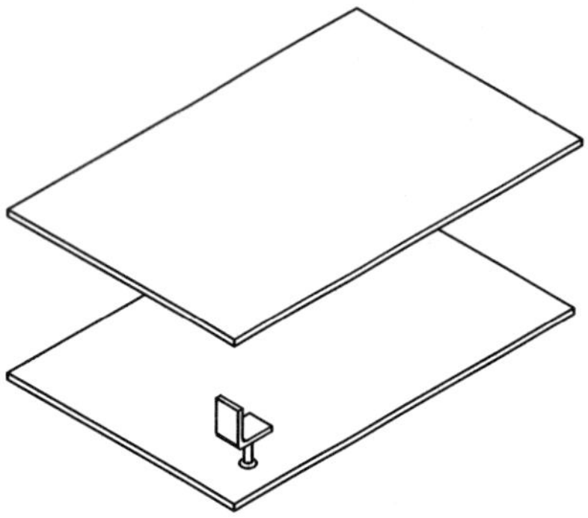

Figure 24-72 *The model before creating the rectangular array*

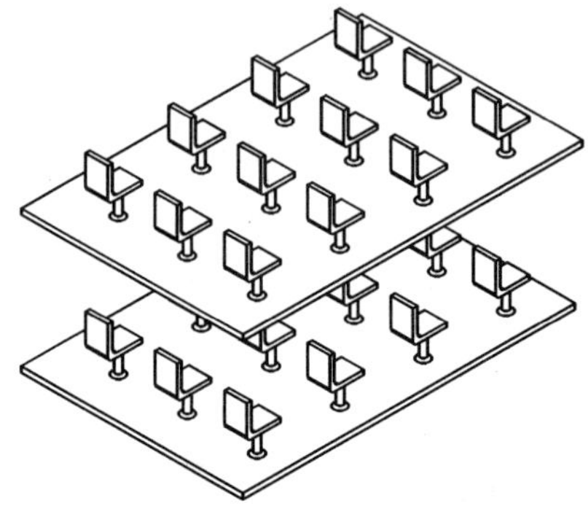

Figure 24-73 *The model after creating the array in the 3D space*

3D Polar Array

The 3D polar arrays are similar to the 2D polar arrays. The only difference is that in 3D you will have to specify an axis about which the solid models will be arranged. For example, consider the solid models shown in Figure 24-74. This figure shows a circular plate that has four holes. Now, to place the bolts in this plate, you can use the 3D polar array as shown in Figure 24-75. The axis for array is defined using the centers at the top and the bottom faces of the circular plate. The prompt sequence that will follow is given next.

Select objects: *Select the object to array.*
Select objects: Enter
Enter the type of array [Rectangular/Polar] <R>: **P**
Enter the number of items in the array: *Specify the number of items.*
Specify the angle to fill (+=ccw, -=cw) <360>: Enter
Rotate arrayed objects? [Yes/No] <Y>: Enter
Specify center point of array: *Specify the first point on the axis.*
Specify second point on axis of rotation: *Specify the second point on the axis.*

ALIGNING THE SOLID MODELS

Menu:	Modify > 3D Operations > Align
Command:	ALIGN

The **ALIGN** command is a very versatile and highly effective command. It is extensively used

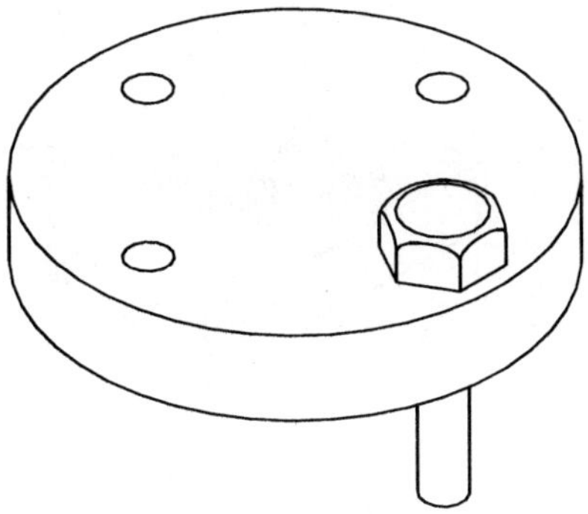

Figure 24-74 Model before creating the array

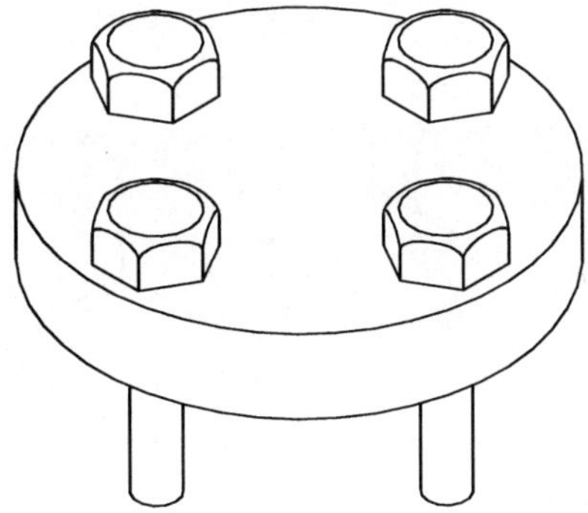

Figure 24-75 Model after creating the array

in solid modeling. As the name suggests, this command is used to align the selected solid model with another solid model. In addition to this, it can also be used to translate, rotate, and scale the selected solid model. This command uses pairs of source and destination points to align the solid model. The source point is a point with which you want to align the object. The destination point is a point on the destination object at which you want to place the source object. You can specify one, two, or three pairs of points to align the objects. However, the working of this command will be different for all the three cases. All three cases are discussed next individually.

Aligning the Objects Using One Pair of Points

When you align the object using just one pair of points, the working of this command will be similar to the **MOVE** command. That is, the source object will be as it is moved from its original location and will be placed on the destination object. Here, the source point will work as the first point of displacement and the destination point will work as the second point of displacement. A reference line will be drawn between the source and the destination points. This line will disappear once you exit the command. The prompt sequence is as follows.

Select objects: *Select the object to align.*
Select objects: Enter
Specify first source point: *Select S1 as shown in Figure 24-76.*
Specify first destination point: *Select D1 as shown in Figure 24-76.*
Specify second source point: Enter

Aligning the Objects Using Two Pairs of Points

The second case is to align the objects using two pairs of source and destination points. This method of aligning the objects forces the selected object to translate and rotate once. You are also provided with an option of scaling the source object to align with the destination object. The prompt sequence that follows is given next.

Select objects: *Select the object to align.*

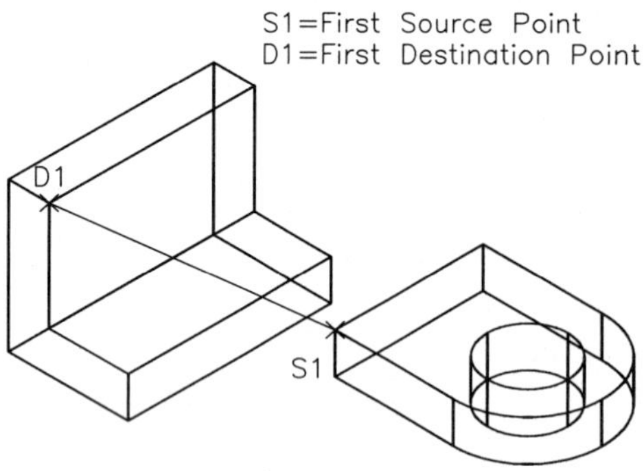

Figure 24-76 Models before aligning

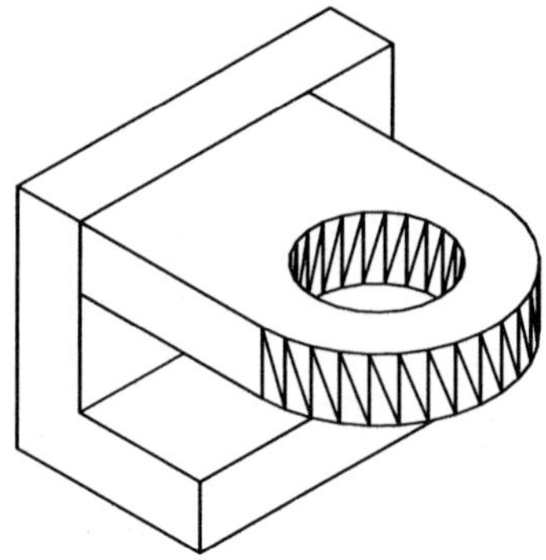

Figure 24-77 Models after aligning

Select objects: Enter
Specify first source point: *Select S1 as shown in Figure 24-78.*
Specify first destination point: *Select D1 as shown in Figure 24-78.*
Specify second source point: *Select S2 as shown in Figure 24-78.*
Specify second destination point: *Select D2 as shown in Figure 24-78.*
Specify third source point or <continue>: Enter
Scale objects based on alignment points? [Yes/No] <N>: **Y** *(You can also enter **N** at this prompt if you do not want to scale the object)*

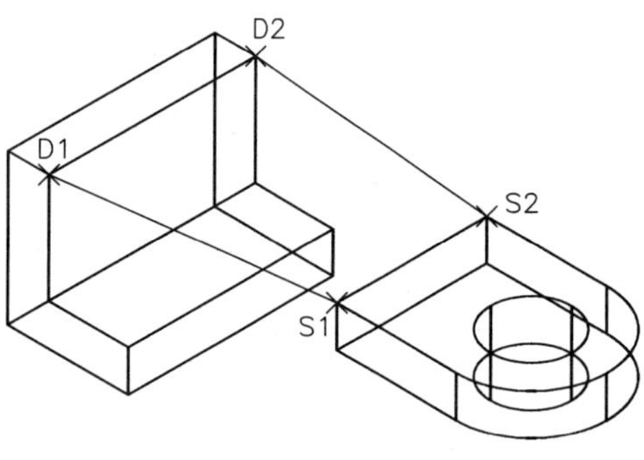

Figure 24-78 *Objects before aligning*

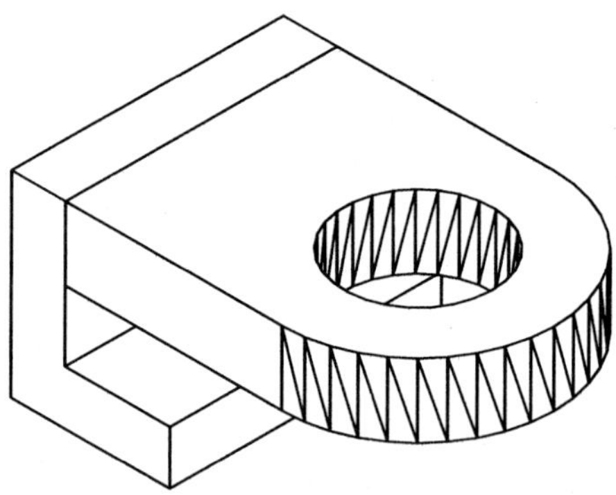

Figure 24-79 *Objects after aligning and scaling*

Aligning the Objects Using Three Pairs of Points

The third case is to align the objects using three pairs of points. This option forces the selected object to rotate twice and then translate, see Figures 24-80 and 24-81. The first pair of source and destination points is used to specify the base point of alignment, the second pair of source and destination points is used to specify the first rotation angle, and the third pair of source and destination points is used to specify the second rotation angle. In this case you will not be allowed to scale the object. The prompt sequence that follows is given next.

Select objects: *Select the object to align.*
Select objects: Enter

Specify first source point: *Select S1 as shown in Figure 24-80.*
Specify first destination point: *Select D1 as shown in Figure 24-80.*
Specify second source point: *Select S2 as shown in Figure 24-80.*
Specify second destination point: *Select D2 as shown in Figure 24-80.*
Specify third source point or <continue>: *Select S3 as shown in Figure 24-80.*
Specify third destination point: *Select D3 as shown in Figure 24-80.*

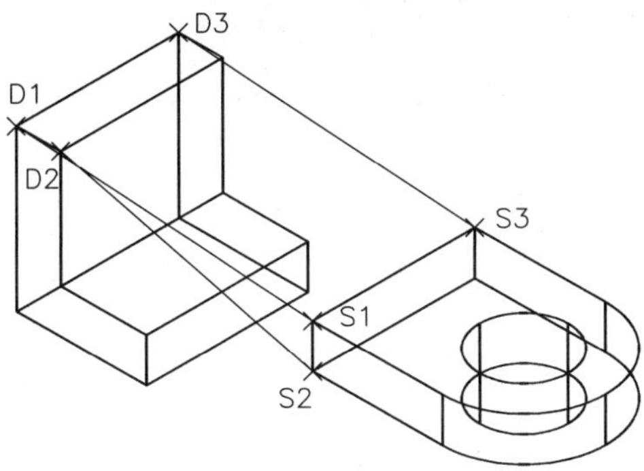

Figure 24-80 *Objects before aligning*

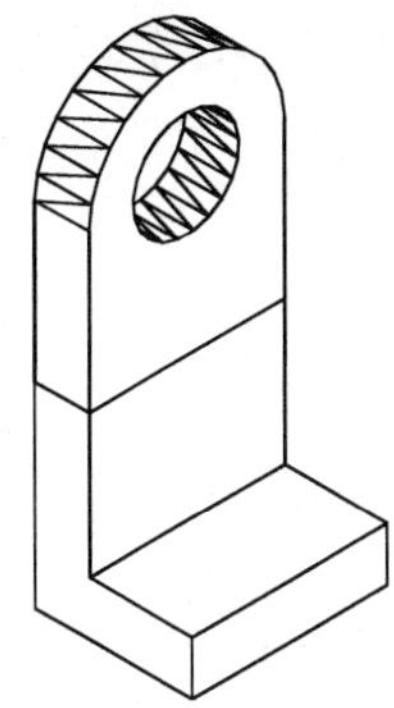

Figure 24-81 *Objects after aligning and rotating*

SLICING SOLID MODELS

Toolbar:	Solid > Slice
Menu:	Draw > Solids > Slice
Command:	SLICE

As the name suggests, the **SLICE** command is used to slice the selected solid with the help of a specified plane. You will be given an option to select the portion of the sliced solid that has to be retained. You can also retain both the portions of the sliced solids.

The prompt sequence that follows is given next.

Select objects: *Select the object to slice.*
Select objects: Enter
Specify first point on slicing plane by [Object/Zaxis/View/XY/YZ/ZX/3points] <3points>:

3points

This option is used to slice a solid using a plane defined by three points, see Figures 24-82 and 24-83. The prompt sequence that follows is given next.

Select objects: *Select the object to be sliced.*
Select objects: Enter
Specify first point on slicing plane by [Object/Zaxis/View/XY/YZ/ZX/3points] <3points>: Enter
Specify first point on plane: *Specify the point P1 on the slicing plane as shown in Figure 24-82.*

Specify second point on plane: *Specify the point P2 on the slicing plane as shown in Figure 24-82.*
Specify third point on plane: *Specify the point P3 on the slicing plane as shown in Figure 24-82.*
Specify a point on desired side of the plane or [keep Both sides]: *Select the portion of the solid to retain or enter B to retain both the portions of the sliced solid.*

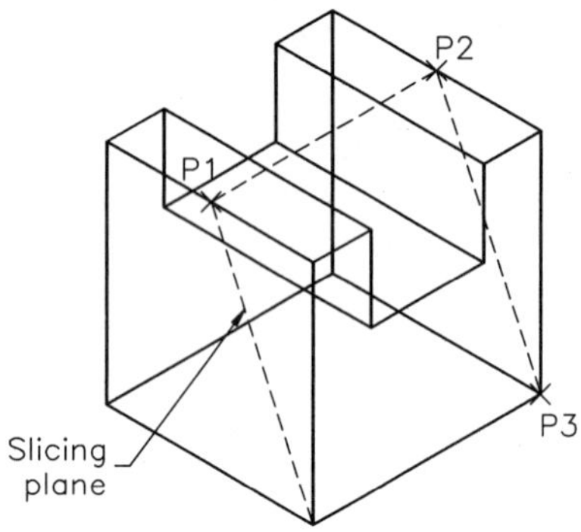

Figure 24-82 *Defining the slicing plane*

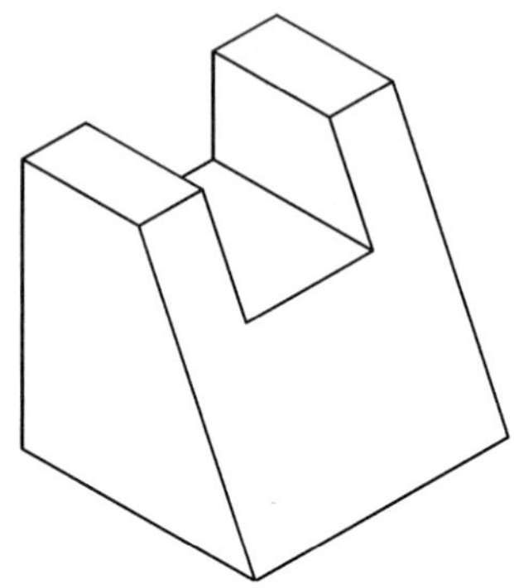

Figure 24-83 *Model after slicing*

Object

This option is used to slice the solid model using an object. The objects that can be used to slice the solid model include arcs, circles, ellipses, 2D polylines, and splines. The prompt sequence that will follow is given next.

Select objects: *Select the object to slice.*
Select objects: Enter
Specify first point on slicing plane by
[Object/Zaxis/View/XY/YZ/ZX/3points] <3points>: **O**
Select a circle, ellipse, arc, 2D-spline, or 2D-polyline: *Select the object to slice the solid.*
Specify a point on desired side of the plane or [keep Both sides]: *Specify the portion of the solid to retain.*

Zaxis

This option is used to slice the solid using a plane defined by two points. The first point is the point on the section plane and the second point is the point in the direction of the Z axis of the plane. The prompt sequence that follows is given next.

Select objects: *Select the solid to be sliced.*
Select objects: Enter
Specify first point on slicing plane by
[Object/Zaxis/View/XY/YZ/ZX/3points] <3points>: **Z**
Specify a point on the section plane: *Specify the point on the section plane.*
Specify a point on the Z-axis (normal) of the plane: *Specify the point in the direction of the Z axis of the section plane.*
Specify a point on desired side of the plane or [keep Both sides]: *Select the portion to retain.*

View

This option is used to slice the selected solid about the viewing plane, see Figures 24-84 and 24-85. The viewing plane in this case will be the screen of the monitor. You will be prompted to specify a point on the solid model through which the viewing plane will pass. The prompt sequence that will follow is given next.

Select objects: *Select the solid model to slice.*
Select objects: Enter
Specify first point on slicing plane by [Object/Zaxis/View/XY/YZ/ZX/3points] <3points>: **V**
Specify a point on the current view plane <0,0,0>: *Specify the point P1 as shown in Figure 24-84.*
Specify a point on the desired side of the plane or [keep Both sides]: *Specify the portion of the solid to retain.*

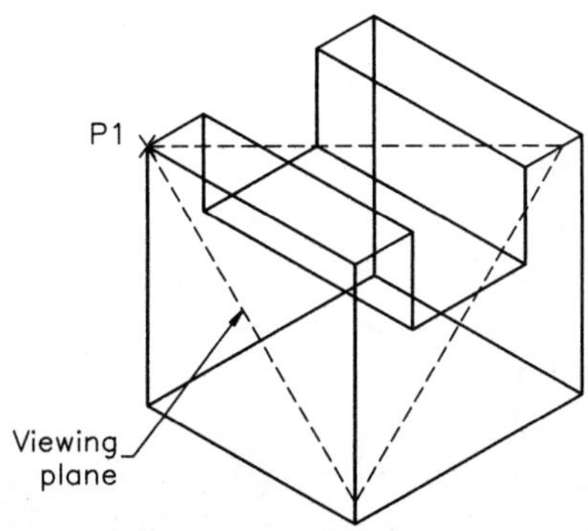

Figure 24-84 *Defining the slicing plane*

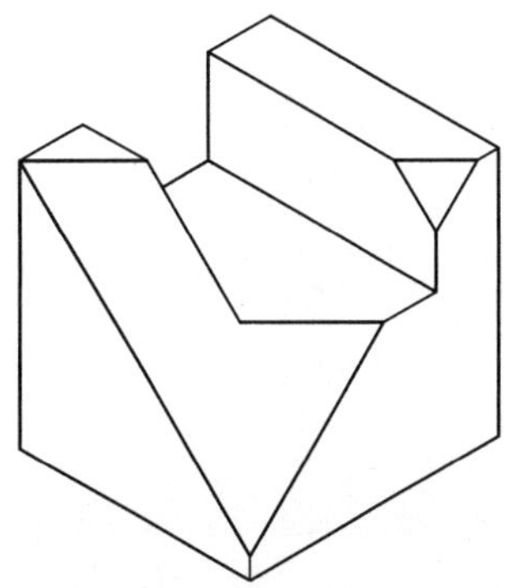

Figure 24-85 *Model after slicing*

XY, YZ, ZX

These options are used to slice the selected solid about the *XY*, *YZ*, or the *ZX* plane respectively. When you invoke this option, you will be prompted to select the point on the plane. The prompt sequence that will follow is given next.

Select objects: *Select the object to be sliced.*
Select objects: Enter
Specify first point on slicing plane by [Object/Zaxis/View/XY/YZ/ZX/3points] <3points>: *Select the XY, YZ, or the ZX plane.*
Specify a point on the XY-plane <0,0,0>: *Specify the point on the slicing plane.*
Specify a point on desired side of the plane or [keep Both sides]: *Specify the portion to retain.*

CREATING THE CROSS-SECTIONS OF THE SOLIDS

Toolbar:	Solid > Section
Menu:	Draw > Solids > Section
Command:	SECTION

The **SECTION** command is very similar to the **SLICE** command. The only difference is that this command does not chop the solid. Instead, this command creates a cross-section along the selected section plane. The cross-section thus created is a region. Note that the regions are created in the current layer and not in the layer in which the sectioned solid is stored. The prompt sequence that follows when you choose the **Section** button is given next.

Select objects: *Select the solid to section.*
Select objects: Enter
Specify first point on Section plane by [Object/Zaxis/View/XY/YZ/ZX/3points] <3points>:

3points

This option is used to define the section plane using three points.

Object

This option is used to specify the section plane using a planar object. The objects that can be used to create the sections are arcs, circles, ellipses, splines, or 2D polylines.

View

This option uses the current viewing plane to define the section plane. You will be prompted to specify the point on the current view plane. It will then automatically create a cross-section parallel to the current viewing plane and that passes through the specified point.

XY, YZ, ZX

These options are used to define the section planes that are parallel to the *XY*, *YZ*, or the *ZX* planes respectively. You will be prompted to specify the point through which the selected plane will pass, see Figures 24-86 and 24-87.

Tip

The number of regions created as cross-sections will be equal to the number of solids selected to create the cross-section.

*You can also hatch the cross-section created using the **SECTION** command. Select the entire region as the object for hatching. You may have to define a new UCS based on the section to hatch it.*

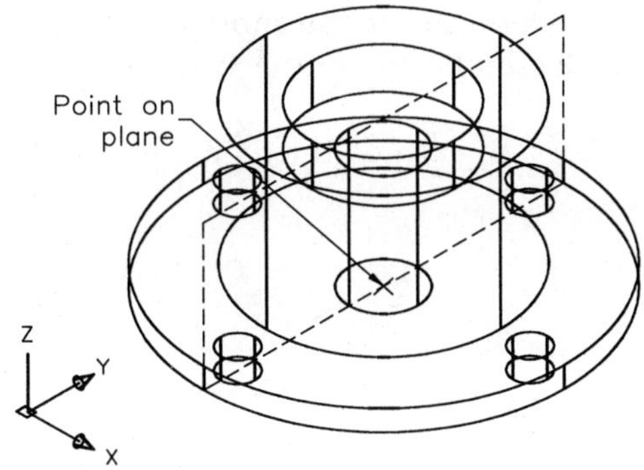

Figure 24-86 *Creating the cross-section along the YZ plane*

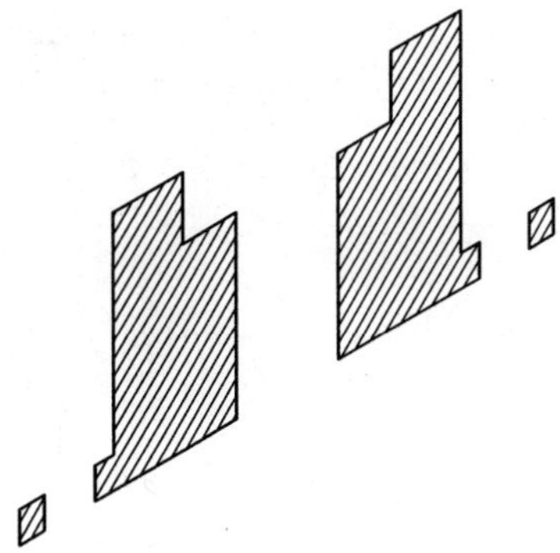

Figure 24-87 *Cross-section created, isolated, and hatched for clarity*

Example 4

In this example you will draw the solid model shown in Figure 24-88. The dimensions for the model are shown in Figures 24-89 and 24-91. After creating the model, slice it as shown in Figure 24-90.

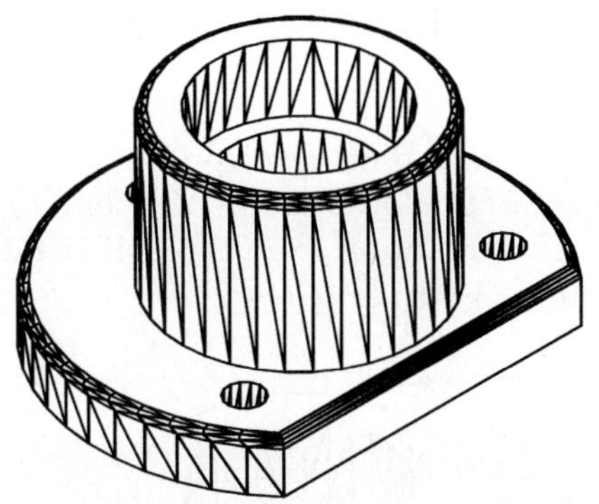

Figure 24-88 *Model for Example 4*

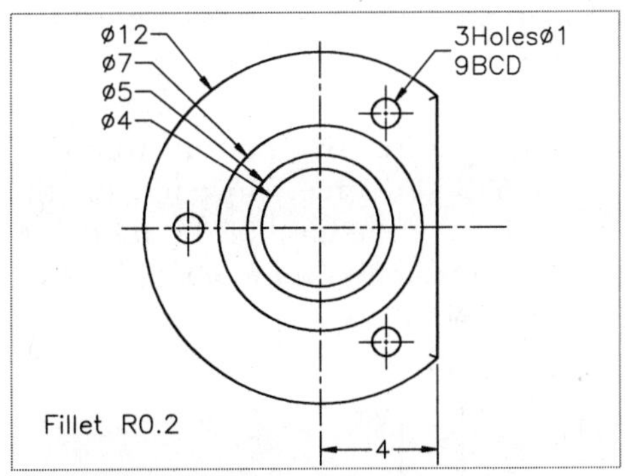

Figure 24-89 *Top view of the model*

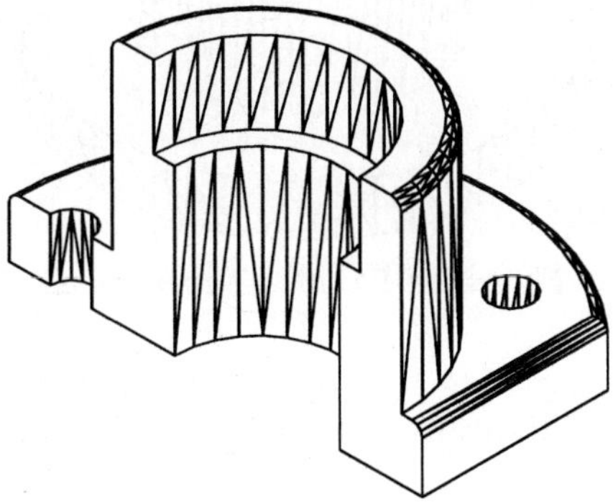

Figure 24-90 *Figure showing the sliced solid*

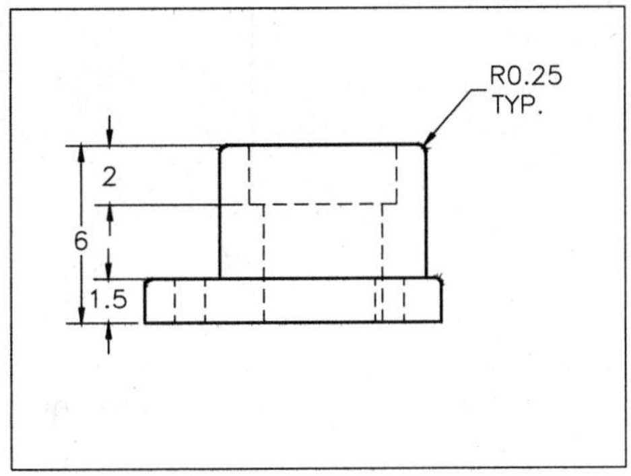

Figure 24-91 *Front view of the model*

1. Open a new file and then draw the sketch for the base of the model as shown in Figure 24-92. Convert it into a region using the **Region** button in the **Draw** toolbar.

2. Choose the **Extrude** button from the **Solids** toolbar and extrude the region to a distance of 1.5. See Figure 24-93.

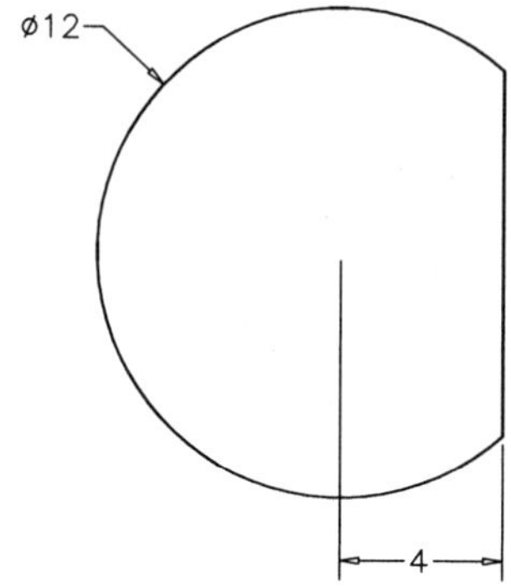

Figure 24-92 *Sketch for the base of the model*

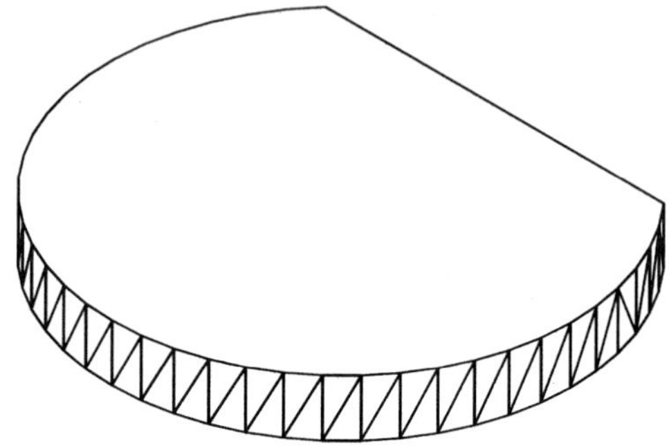

Figure 24-93 *Base of the model*

3. Rotate the UCS about the *X* axis through an angle of 90-degree. Draw the sketch for the next feature and convert into a region, see Figure 24-94.

4. Choose the **Revolve** button from the **Solids** toolbar and then revolve the sketch to an angle of 360-degree. Move it using the center point of the lower face to the center of the lower face of the base of the model. Choose the **Union** button from the **Solids Editing** toolbar and then union both the objects, see Figure 24-95.

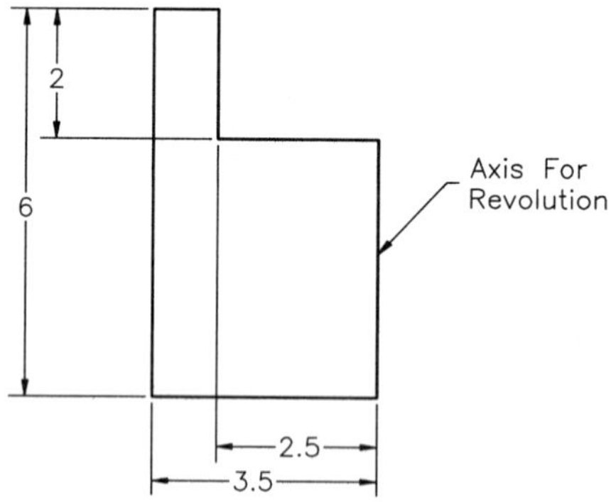

Figure 24-94 *Sketch for the next feature*

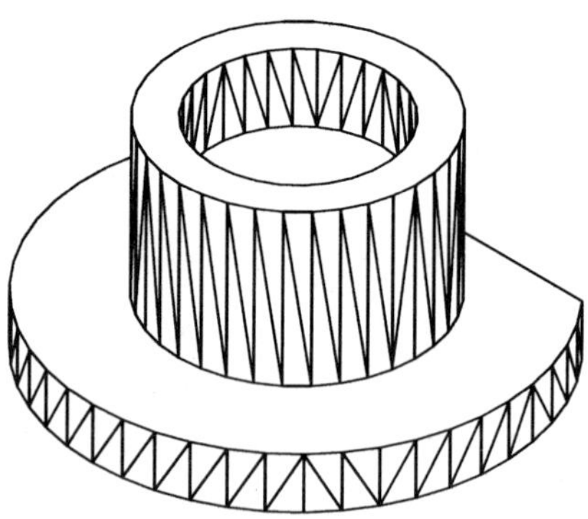

Figure 24-95 *Model after union*

5. Create a new cylinder of diameter 1 unit. The center of the base of the cylinder should lie at a distance of 1.5 units from the quadrant of the lower face of the base of the model.

6. Choose **3D Operation > 3D Array** from the **Modify** menu. The prompt sequence is as follows.

   Initializing... 3DARRAY loaded.
   Select objects: *Select the cylinder.*
   Select objects: Enter
   Enter the type of array [Rectangular/Polar] <R>: **P**
   Enter the number of items in the array: **3**
   Specify the angle to fill (+=ccw, -=cw) <360>: Enter
   Rotate arrayed objects? [Yes/No] <Y>: Enter
   Specify center point of array: *Select the center of the base of the model.*
   Specify second point on axis of rotation: *Select the center of the top face of the model.*

7. Subtract all three cylinders from the model using the **Subtract** button in the **Solids Editing** toolbar.

8. Create the central hole to be subtracted from the cylinder with the diameter 4 units and the height 6 units. Fillet the edges. The model should now look similar to the one shown in Figure 24-96.

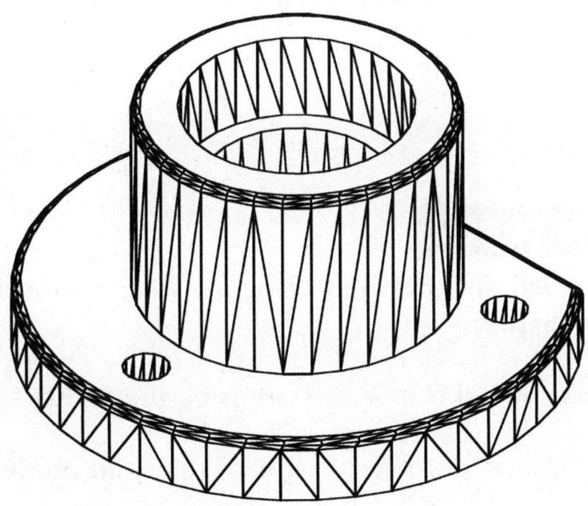

Figure 24-96 *Model after creating the fillet*

9. Relocate the UCS at the world position.

10. Choose the **Slice** button from the **Solids** toolbar. The prompt sequence is as follows.

    Select objects: *Select the model.*
    Select objects: Enter
    Specify first point on slicing plane by [Object/Zaxis/View/XY/YZ/ZX/3points] <3points>: **ZX**
    Specify a point on the ZX-plane <0,0,0>: *Select the center of the top face of the model.*

Specify a point on desired side of the plane or [keep Both sides]: *Specify a point on the back side of the model to retain it.*

11. The final model for Example 4 should look similar to the one shown in Figure 24-97.

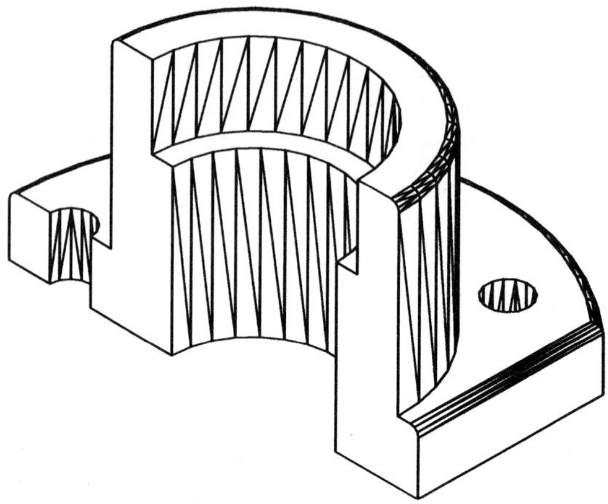

Figure 24-97 *Final model after slicing*

Self-Evaluation Test

Answer the following questions and then compare your answers to the correct answers given at the end of this chapter.

1. The edges of the solid model can be filleted using the **3DFILLET** command. (T/F)

2. You can select an ellipse as an object to rotate the solid model using the **ROTATE3D** command. (T/F)

3. The **CONE** command can be used to create a solid cone with a circular or elliptical base. (T/F)

4. Open entities can be converted into regions. (T/F)

5. The _________ option of the **CHAMFER** command is used to chamfer all the edges of the selected face of the solid model.

6. The _________ rule is used to determine the direction of rotation of the solid model in the 3D space.

7. The __________ command is used to create a rugby-ball like structure.

8. The __________ command is used to move, rotate, and scale the solid model in a single attempt.

9. The **SECTION** command creates a __________ along the plane of section.

10. You can extrude the selected region about a path using the __________ option of the **EXTRUDE** command.

Review Questions

Answer the following questions.

1. An open entity can be revolved. (T/F)

2. You cannot apply Boolean operations on the regions. (T/F)

3. The entity to be revolved should lie completely on one side of the revolution axis. (T/F)

4. You can select all the tangential edges of the selected solid model for filleting using the **Chain** option of the **FILLET** command. (T/F)

5. Which option of the **ROTATE3D** command is used to select a 2D entity as an object for rotating the solid models?

   (a) **2D** (b) **Last**
   (c) **Object** (d) **Entity**

6. Which value of the taper angle will taper the extruded model in from the base?

   (a) Positive (b) Negative
   (c) Zero (d) None

7. Which command is used to create a cube?

   (a) **BOX** (b) **CUBOID**
   (c) **POLYGON** (d) **CYLINDER**

8. Which command is used to check the interference between the selected solid models?

   (a) **INTERFERE** (b) **INTERSECT**
   (c) **INTERFERENCE** (d) **CHECK**

9. Which option of the **MIRROR3D** command is used to mirror the object about the view plane?

   (a) **Object** (b) **Last**
   (c) **View** (d) **Zaxis**

10. The __________ option of the **ROTATE3D** command is used to select the same axis that was last selected to rotate the solid model.

11. The __________ direction is known as the extrusion direction.

12. The __________ command is used to create an apple-like structure.

13. The __________ command is used to create a revolved solid.

14. In the **3DARRAY** command, the levels are arranged along the __________ axis.

15. The __________ option of the **REVOLVE** command is used to select a 2D entity as the revolution axis.

Exercises

Exercise 2

In this exercise, you will create the solid model shown in Figure 24-98. The dimensions for the model are shown in Figures 24-99 and 24-101. After creating the model slice it to get the model shown in Figure 24-100.

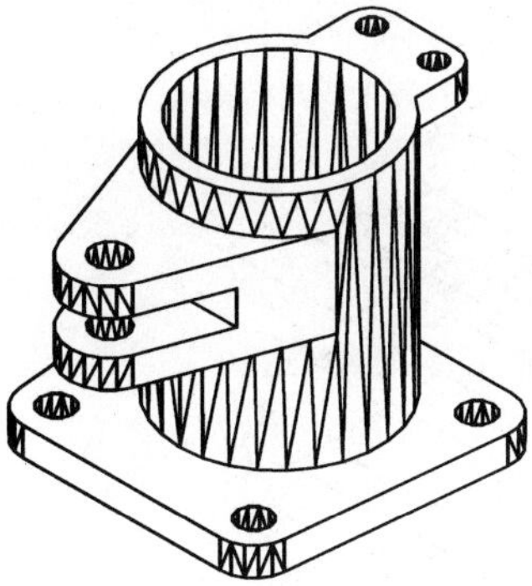

Figure 24-98 Model for Exercise 3

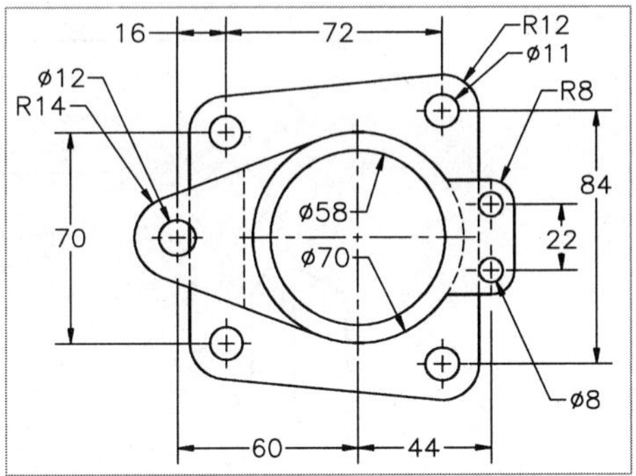

Figure 24-99 Top view of the model

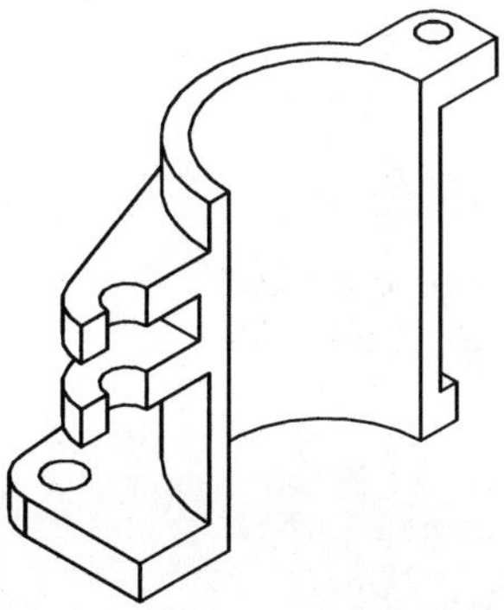

Figure 24-100 Model after slicing

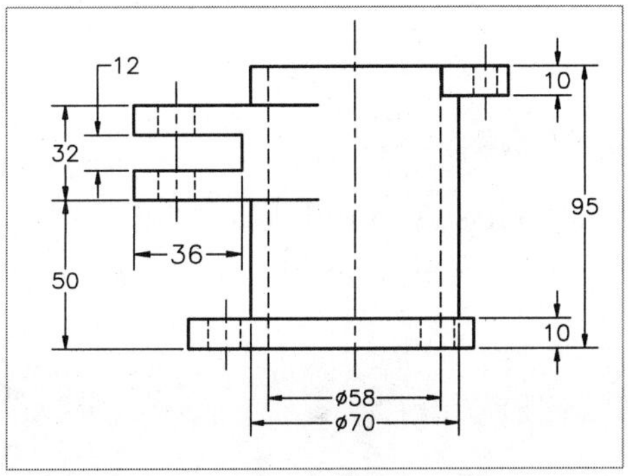

Figure 24-101 Front view of the model

Exercise 3

In this exercise, you will create the solid model shown in Figure 24-102. Assume the missing dimensions.

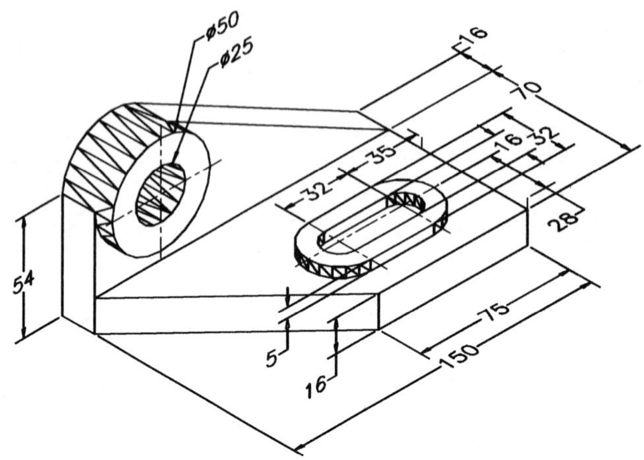

Figure 24-102 *Model for Exercise 3*

Exercise 4

In this exercise you will create the solid model shown in Figure 24-103. The dimensions for the model are given in the same figure. Assume the missing dimensions.

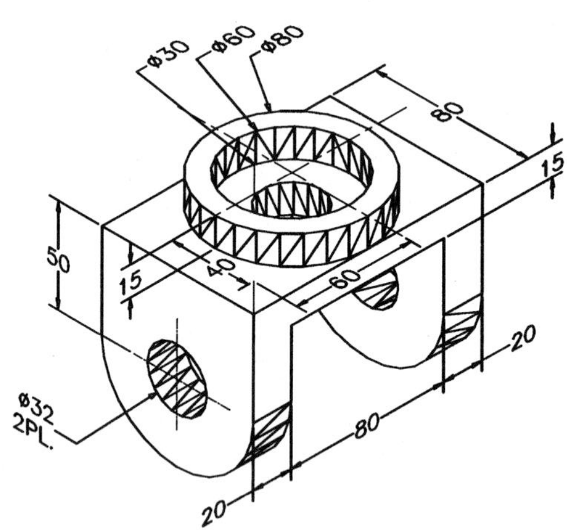

Exercise 5

In this exercise you will create the solid model shown in Figure 24-104. The dimensions are given in the drawing. The fillet radius is 5mm.

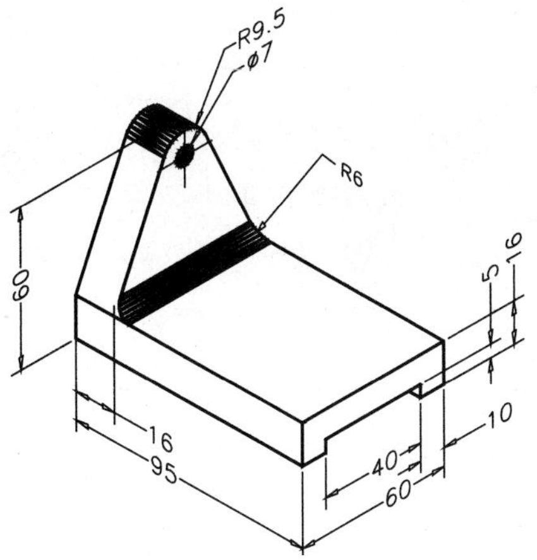

Figure 24-104 Solid model for Exercise 5

Exercise 6

In this exercise you will create the solid model shown in Figure 24-105. The dimensions for the model are given in the same figure. The fillet radius is 0.13 units.

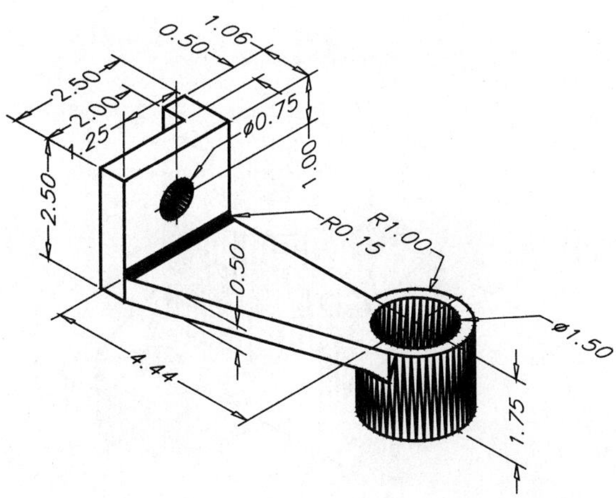

Figure 24-105 Solid model for Exercise 6

Exercise 7

In this exercise you will create the solid model shown in Figure 24-106. The dimensions for the model are given in the same figure. Assume the missing dimensions.

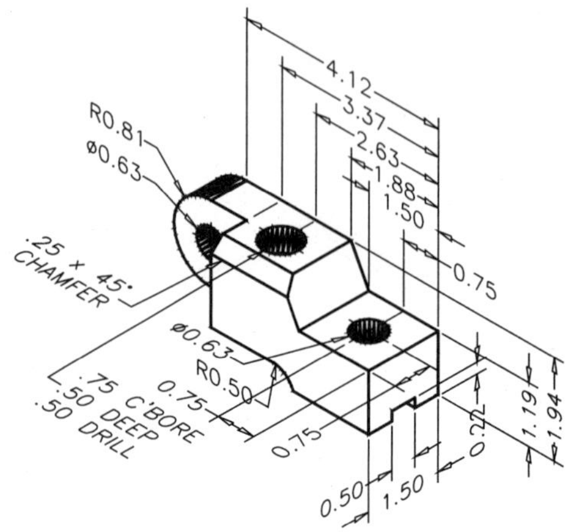

Figure 24-106 Solid model for Exercise 7

Problem Solving Exercise 1

In this exercise you will create the solid model with dimensions shown in Figure 24-107. Next, slice the object, create four viewports and arrange the views as shown in Figure 24-108.

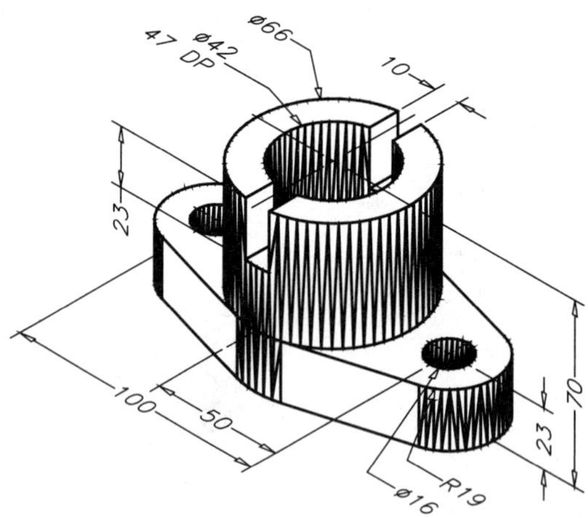

Figure 24-107 Solid model for Problem Solving Exercise 1

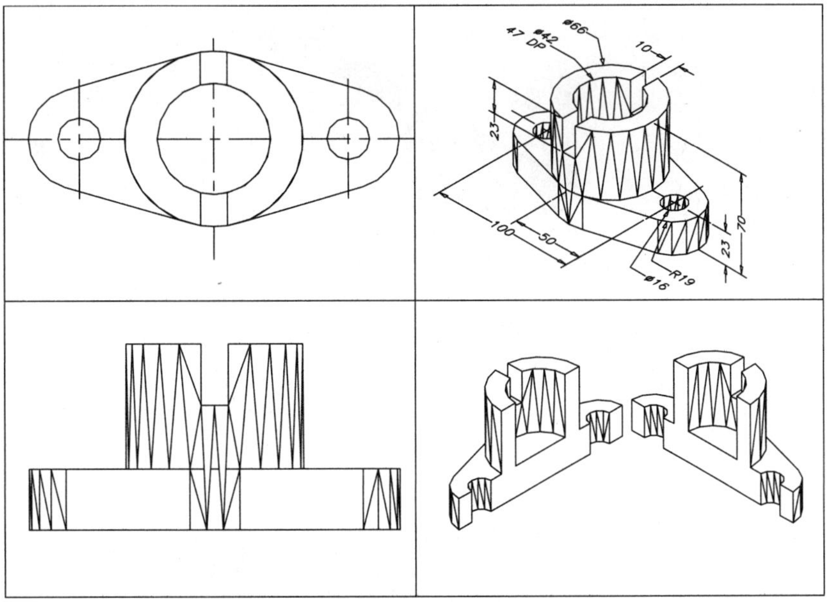

Figure 24-108 *Viewports for Problem Solving Exercise 1*

Problem Solving Exercise 2 — *Mechanical*

For the Gear Puller shown in Figure 24-109, draw the following.

Piece part drawings shown in Figures 24-110 through 24-114, assembly drawing with Bill of Materials shown in Figure 24-115, and the exploded view of the solid model shown in Figure 24-116.

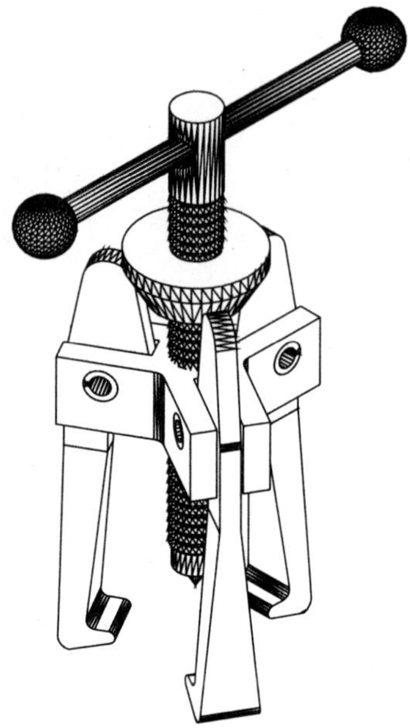

Figure 24-109 *Assembled solid model of the Gear Puller*

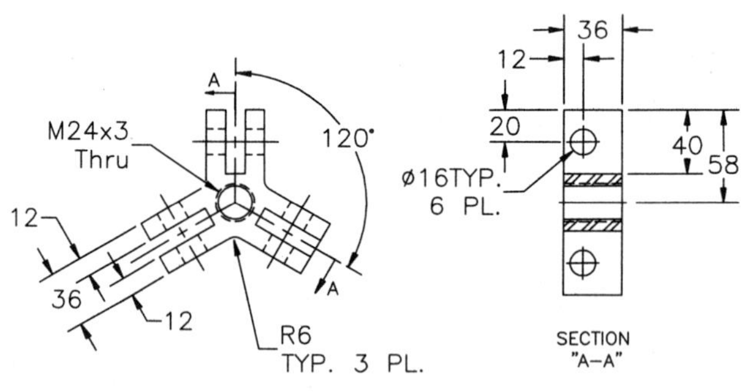

P1: Yoke

Figure 21-110 *Front and side views of the Yoke*

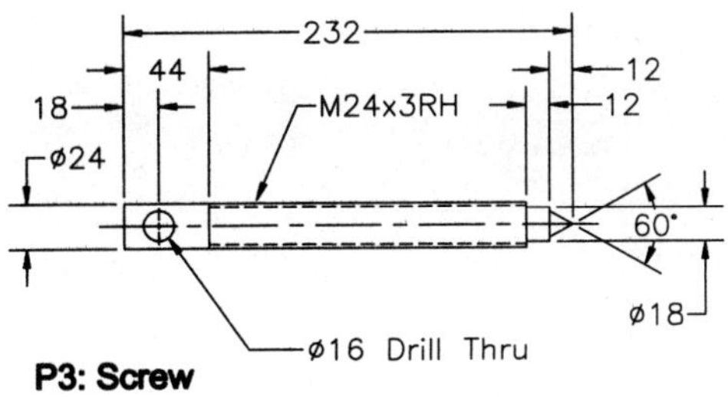

Figure 24-111 Front view of the Screw

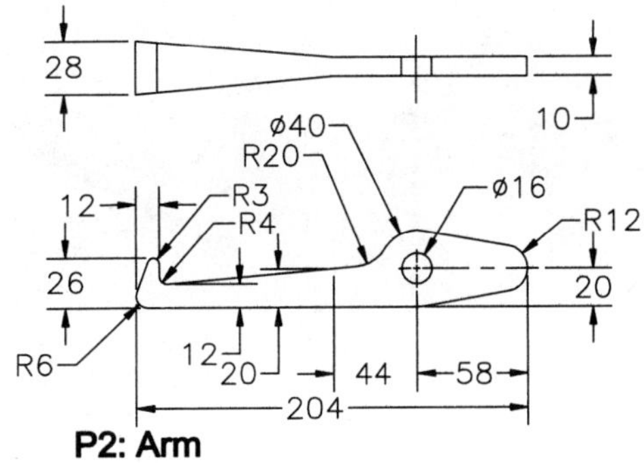

Figure 24-112 Front and top views of the Arm

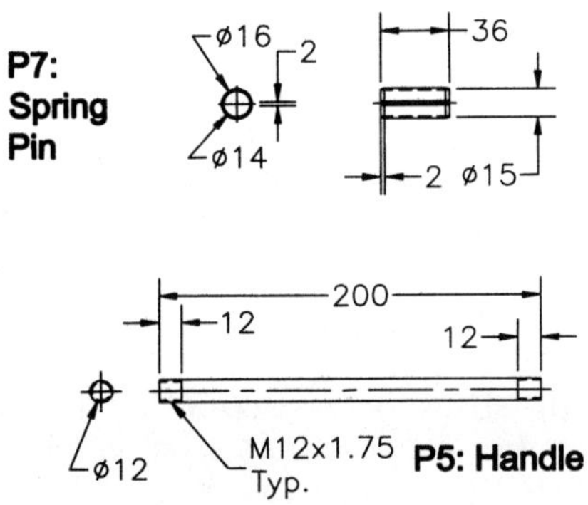

Figure 24-113 Views of Spring Pin and Handle

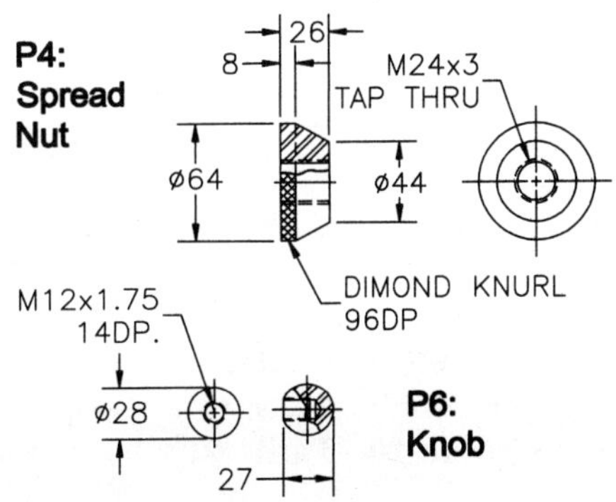

Figure 24-114 Views of Spread Nut and Knob

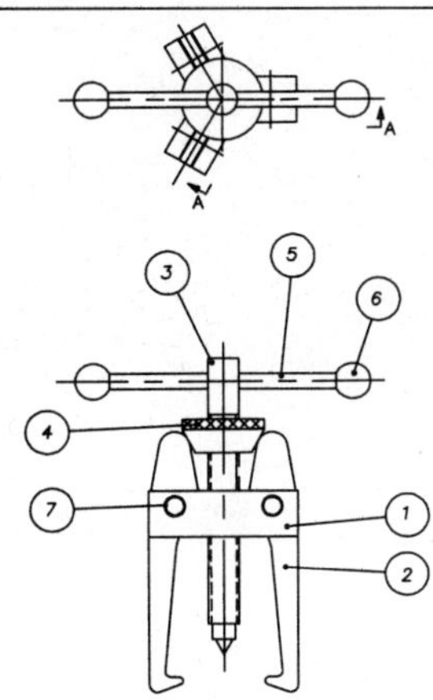

P.N.	Qty	Material	Description
1	1	1020 STL	Yoke
2	3	1020 STL	Arm
3	1	1020 STL	Screw
4	1	1020 STL	Spread nut
5	1	1020 STL	Handle
6	2	1020 STL	Knob
7	3	STL	Spring pin

Figure 24-114 Assembled view and BOM of the Gear Puller

Chapter 24

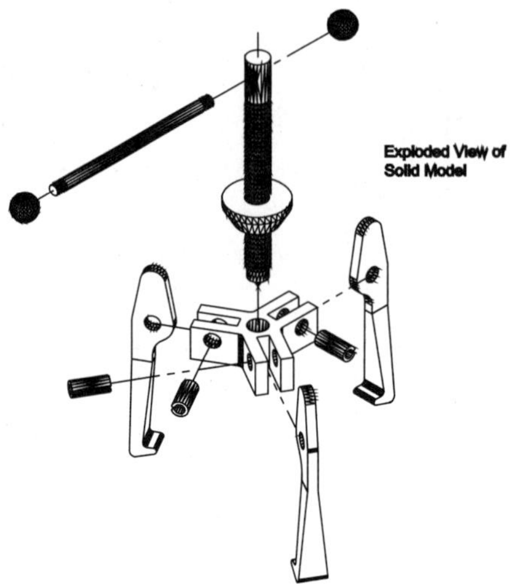

Figure 24-116 *Exploded solid model of the Gear Puller*

Answers to Self-Evaluation Test

1 - F, **2** - F, **3** - T, **4** - F, **5** - Loop, **6** - right-hand thumb, **7** - **TORUS**, **8** - **Align**, **9** - cross-section, **10** - **Path**

Chapter 25

Editing and Dynamic Viewing of 3D Objects

Learning Objectives

After completing this chapter you will be able to:

- *Edit the solid models using the* ***SOLIDEDIT*** *command.*
- *Generate the drawing views of the solid model using the* ***SOLVIEW*** *command.*
- *Generate the profiles and section in the drawing views using the* ***SOLDRAW*** *command.*
- *Create the profile images of the solid model using the* ***SOLPROF*** *command.*
- *Calculate the mass properties of the solid models using the* ***MASSPROP*** *command.*
- *Dynamically view 3D objects using the* ***3DORBIT*** *command.*
- *Use the* ***DVIEW*** *command for viewing the model.*

EDITING SOLID MODELS

Toolbar:	Solids Editing
Menu:	Modify > Solids Editing
Command:	SOLIDEDIT

One of the major enhancements in the recent releases of AutoCAD is the ability of users to edit the solid models. This has made the solid modeling in AutoCAD very user-friendly. In AutoCAD you can edit the solid models using the **SOLIDEDIT** command. You can select the options for editing the solid models from the **Solids Editing** toolbar. The prompt sequence that will follow when you invoke this command is given next.

Solids editing automatic checking: SOLIDCHECK=1
Enter a solids editing option [Face/Edge/Body/Undo/eXit] <eXit>:

Face Option

This option allows you to edit the faces of solid models. The suboptions provided under this option are **Extrude**, **Move**, **Rotate**, **Offset**, **Taper**, **Delete**, **Copy**, and **Color**. The next prompt is

Enter a face editing option
[Extrude/Move/Rotate/Offset/Taper/Delete/Copy/coLor/Undo/eXit]<eXit>: *Enter an option.*

Extrude

This option allows you to extrude selected faces of a solid model to a specific height or along a selected path (Figure 25-1). You can directly invoke this option by choosing the **Extrude Faces** button from the **Solids Editing** toolbar. The prompt sequence that will follow when you choose this button is given next.

Select faces or [Undo/Remove]: *Select faces to extrude or enter an option.*
Select faces or [Undo/Remove/ALL]: *Select another face, an option, or press ENTER.*

The **Undo** option cancels the selection of the most recent face you have selected. The **Remove** option allows you to remove a previously selected face from the selection set for extrusion. The **ALL** option selects all the faces of the specified solid. After you have selected a face for extrusion, the next prompt is given below.

Specify height of extrusion or [Path]: *Enter a height value or enter P to select a path.*

The Path option allows you to select a path for extrusion based on a specified line or curve. At the **Select extrusion path** prompt, select the line, circle, arc, ellipse, elliptical arc, polyline, or spline to be specified as the extrusion path. This path should not lie on the same plane as the selected face and should not have areas of high curvature.

A positive value for the height for extrusion extrudes the selected face outward, and a negative value extrudes the selected face inward. On specifying the height of extrusion, the next prompt is given below.

Specify angle of taper for extrusion <0>: *Specify a value between -90 and +90-degree or press ENTER to accept the default value of 0.*

A positive angle value tapers the selected face inward and a negative value tapers the selected face outward.

Move

This option allows you to move the selected faces from one location to another without changing the orientation of the solid (Figure 25-2). For example, you can move holes from one location to another without actually modifying the solid model. You can invoke this option by choosing the **Move Faces** button from the **Solids Editing** toolbar. The prompt sequence that will follow when you choose this button is given next.

Select faces or [Undo/Remove]: *Select one or more faces or enter an option.*
Select faces or [Undo/Remove/ALL]: *Select one or more faces, enter an option, or press ENTER.*
Specify a base point or displacement: *Specify a base point.*
Specify a second point of displacement: *Specify a point or press ENTER.*

The face moves to the specified location.

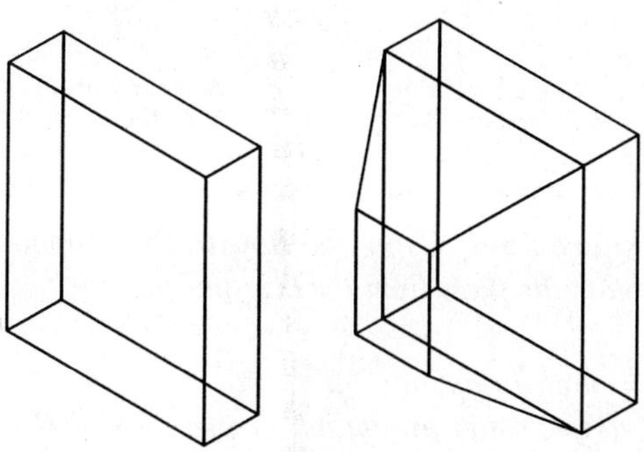

Figure 25-1 *Using the* ***Extrude Faces*** *option to extrude a solid face with a positive taper angle*

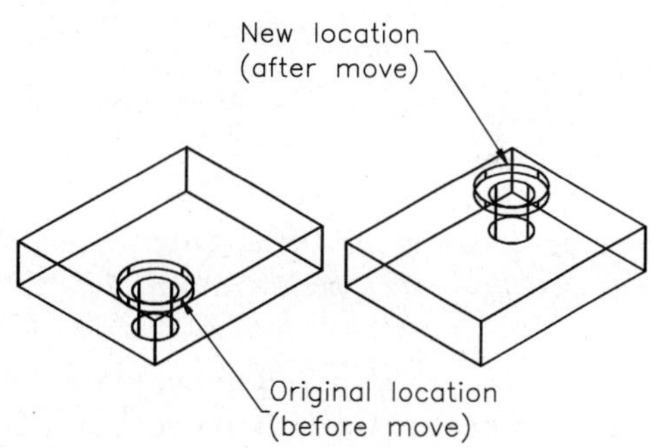

Figure 25-2 *Using the* ***Move Faces*** *option to move the hole from the original position*

Tip

If the feature you wish to move consists of more than one face, then you must select all the faces to move the entire feature.

Offset

The **Offset** option allows you to offset the selected faces of a solid model uniformly through a specified distance on the model. Offsetting takes place in the direction of the positive side of the face. For example, you can offset holes to a larger or smaller size in a 3D solid through a specified distance. In Figure 25-3, the hole on the solid has been offset to a bigger size. You can invoke this option by choosing the **Offset Faces** button from the **Solid Editing** toolbar. The following prompt sequence is displayed when you choose this button.

Select faces or [Undo/Remove]: *Select one or more faces or enter an option.*
Select faces or [Undo/Remove/ALL]: *Select one or more faces, enter an option, or press ENTER.*
Specify the offset distance: *Specify the offset distance.*

Rotate

This option allows you to rotate selected faces of a solid model through a specified angle. For example, you can rotate cuts or slots around a base point through an absolute or relative angle (Figure 25-4). The direction in which the rotation takes place is determined by the right-hand thumb rule. This option can be invoked by choosing the **Rotate Faces** button from the **Solids Editing toolbar**. The prompt sequence that will follow when you choose this button is given below.

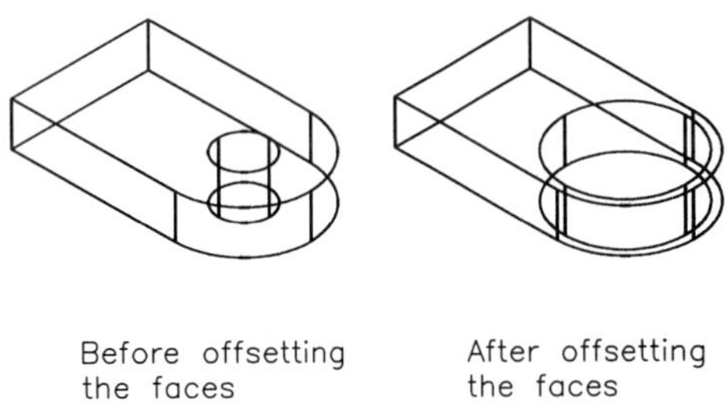

Figure 25-3 *Using the **Offset Faces** option to offset a solid face with a positive taper angle*

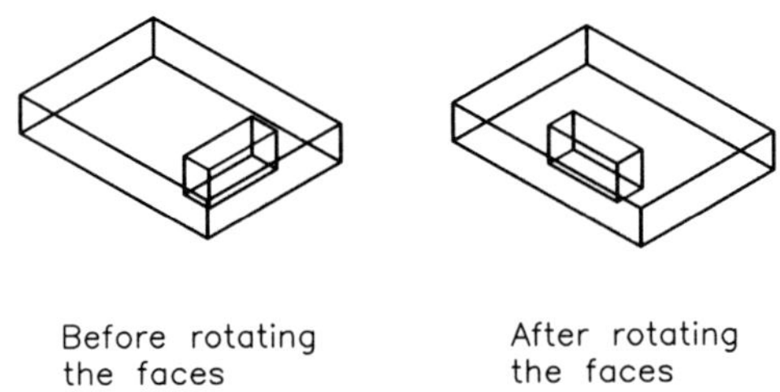

Figure 25-4 *Using the **Rotate Faces** option to rotate the faces from the original position*

Select faces or [Undo/Remove]: *Select face(s) or enter an option.*
Select faces or [Undo/Remove/ALL]: *Select face(s), or enter an option, or press ENTER.*
Specify an axis point or [Axis by object/View/Xaxis/Yaxis/Zaxis] <2points>: *Select a point, enter an option, or press ENTER.*

If you press ENTER to use the **2points** option, you will be prompted to select two points that will define the axis of rotation. The prompts sequence is given next.

Specify the first point on the rotation axis: *Specify the first point.*
Specify the second point on the rotation axis: *Specify the second point.*
Specify a rotation angle or [Reference]: *Specify the angle of rotation or enter R to use the reference option.*

The **Axis by object** option allows you to select an object to define the axis of rotation. The axis of rotation is aligned with the selected object; for example, if you select a circle, the axis of rotation is aligned along the axis that passes through the center of the circle and is normal to the plane of the circle. The selected objects can also be a line, ellipse, Arc, 2D, 3D, or LW polyline or even a spline. The prompts sequence is given next.

Select a curve to be used for the axis: *Select the object to be used to define axis of rotation.*
Specify a rotation angle or [Reference]: *Specify rotation angle or use the reference angle option.*

The **X**, **Y**, **Zaxis** options align the axis of rotation with the axis that passes through the specified point. The Command prompt is:

Specify the origin of rotation <0,0,0>: *Select a point through which the axis of rotation will pass.*
Specify a rotation angle or [Reference]: *Specify rotation angle or use the reference angle option.*

Taper

This option allows you to taper selected face(s) through a specified angle (Figure 25-5). The tapering takes place depending on the selection sequence of the base point and the second point along the selected face. It also depends on whether the taper angle is positive or negative. A positive value tapers the face inward and a negative value tapers the face outward. You can enter any value between -90-degree and +90-degree. The default value of 0-degree extrudes the face along the axis that is perpendicular to the selected face. If you have selected more than one face, they will all be tapered through the same angle. The prompt sequence is as follows.

Select faces or [Undo/Remove/ALL]: *Select one or more faces, enter an option, or press ENTER.*
Specify the base point: *Specify a base point.*
Specify another point along the axis of tapering: *Specify a point.*
Specify the taper angle: *Specify an angle between –90 and +90-degree.*

Delete

Removes selected face(s) from the specific 3D solid. It also removes chamfers and fillets (Figure 25-6). The prompt sequence is as follows.

Select faces or [Undo/Remove]: *Select the face to remove.*
Select faces or [Undo/Remove/ALL]: *Select one or more faces, enter an option, or press ENTER.*
Solid validation started.
Solid validation completed.

Copy

The **Copy** option allows you to copy a face as a body or region (Figure 25-7). The prompt sequence that follows when you choose this button is as follows.

Select faces or [Undo/Remove]: *Select the faces to copy.*
Select faces or [Undo/Remove/ALL]: *Select one or more faces, enter an option, or press ENTER.*

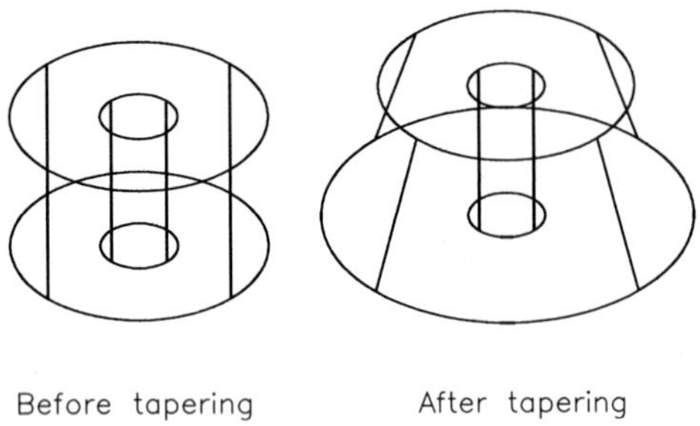

Figure 25-5 *Tapering the faces using the* ***Taper*** *option*

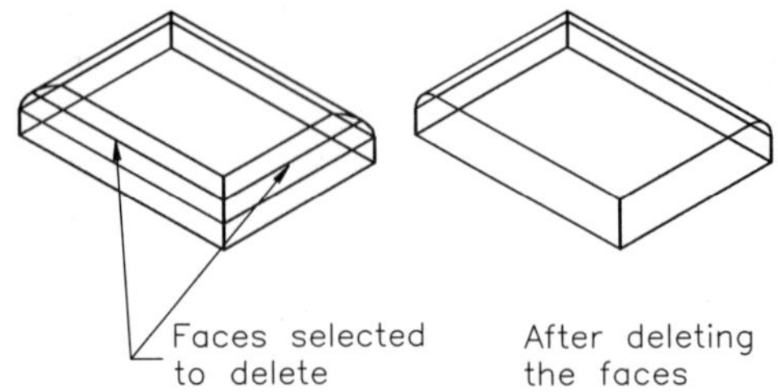

Figure 25-6 *Deleting the faces using the* ***Delete*** *option*

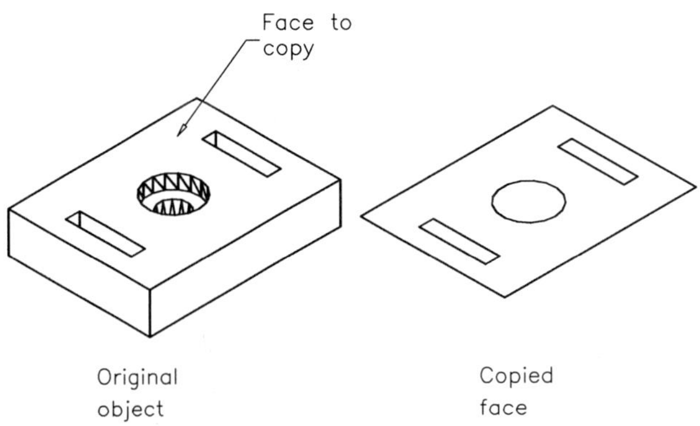

Figure 25-7 *Copying the faces using the* ***Copy*** *option*

Specify a base point or displacement: *Specify a base point.*
Specify a second point of displacement: *Specify the second point.*

Color

This option is used to change the color of the selected face. The prompt sequence that will follow when you choose this button is given next.

Select faces or [Undo/Remove/ALL]: *Select one or more faces, enter an option, or press ENTER.*

When you press ENTER at the previous prompt, the **Select Color** dialog box will be displayed. You can assign the color to the selected face using this dialog box.

Undo

This option allows you to cancel the changes made to the faces of the solid model during the **SOLIDEDIT** command. This option will be available only when you invoke the **SOLIDEDIT** command using the command line.

eXit

This option allows you to exit the face editing operations in the **SOLIDEDIT** command. This option will be available only when you invoke this command using the command line.

Edge Option

With the **Edit** option, you can modify properties of the edges of solid. You can copy and change the color of edges. This option provides the **Face**, **Edge**, **Body**, **Undo**, and **eXit** suboptions. These suboptions are discussed next.

Copy

This option allows you to copy individual edges of a solid model. The edges are copied as lines, arcs, circles, ellipses, or splines. You can also invoke this option by choosing the **Copy Edges** button from the **Solids Editing** toolbar. The prompt sequence that follows when you choose this button is given next.

Select edges or [Undo/Remove]: *Select the edges to be copied.*
Select edges or [Undo/Remove]: Enter
Specify a base point or displacement: *Specify the first point of displacement.*
Specify a second point of displacement: *Specify the second point of displacement.*

Color

You can use this option to change the color of a selected edge of a 3D solid. This option can be invoked by choosing the **Color Edges** button from the **Solids Editing** toolbar. The prompt sequence that follows when you choose this button is given next.

Select edges or [Undo/Remove]: *Select one or more edges or enter an option.*
Select edges or [Undo/Remove]: Enter

After you select edges and press ENTER, the **Select Color** dialog box is displayed. You can assign the color to the selected edges using this dialog box. You can also assign colors by entering their names or numbers. Individual edges can be assigned individual colors.

The **Undo** option cancels the modifications made during the **Edge** editing of the **SOLIDEDIT** command, and **eXit** option allows you to exit the **Edge** editing of this command.

Body Option

This option is used to edit the entire body of the solid model. The **Body** option provides the **Imprint**, **seParate solids**, **Shell**, **cLean**, **Check**, **Undo**, and **eXit** suboption. These suboptions are discussed next.

Imprint

This option allows you to imprint an object on the 3D solid object. Remember that the object to be imprinted should intersect one or more faces of the solid object. The objects that can be imprinted are arcs, circles, lines, 2D and 3D polylines, ellipses, splines, regions, bodies, and 3D solids. You can invoke this option by choosing the **Imprint** button from the **Solids Editing** toolbar. The prompt sequence that follows when you choose this button is given next.

Select a 3D solid: *Select an object.*
Select an object to imprint: *Select an object to be imprinted.*
Delete the source object [Yes/No] <N>: *Select an option or press ENTER.*
Select an object to imprint: *Select a new object or press ENTER.*

seParate solids

This option allows you to separate 3D solids with disjointed volumes into separate 3D solids. The solids with disjointed volumes are created by the union of two solids that are not in contact with each other. Keep in mind that the composite solids created using the Boolean operations that share the same volume cannot be separated using this option. You can invoke this option by choosing the **Separate** button from the **Solids Editing** toolbar. The prompt sequence that will follow when you choose this button is given next.

Select a 3D solid: *Select a solid object with multiple lumps.*

Shell

This is one of the most extensively used options of the **SOLIDEDIT** command. This option allows you to create a shell of a specified 3D solid. The shelling is defined as a process of scooping out the material from the solid model in such a manner that the walls with some thickness are left. One 3D object can have only one shell. You can exclude certain faces before the scooping of the material takes place. This means the selected face will not be included during the shelling, see Figure 25-8. The prompt sequence is as follows.

Select a 3D solid: *Select an object.*
Remove faces or [Undo/Add]: *Select one or more faces or enter an option.*
Remove faces or [Undo/Add/ALL]: *Select a face, enter an option, or press ENTER.*
Enter the shell offset distance: *Specify the wall thickness.*

Tip
If you have specified a positive distance, the shell that is created is to the inside perimeter of the solid, and in case it is negative, the shell is created to the outside perimeter of the 3D solid.

Clean

This option allows you to remove shared edges or vertices sharing the same surface or curve definition on either side of the edge or vertex. It removes all redundant edges and vertices, imprinted as well as unused geometry. It also merges faces that

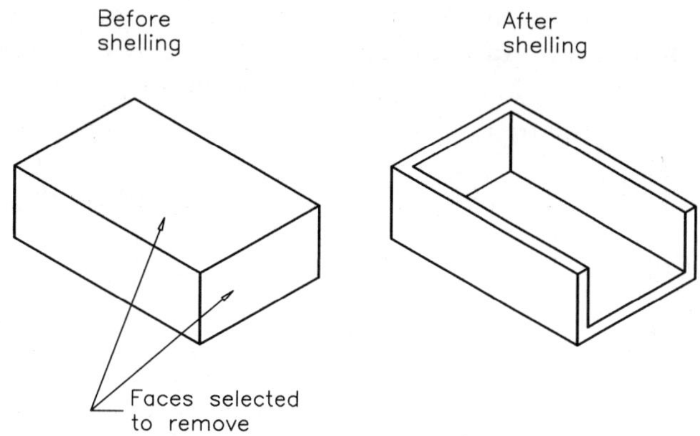

Figure 25-8 *Creating a shell in the solid model*

share the same surface. You can invoke this option by choosing the **Clean** button from the **Solids Editing** toolbar. The prompt sequence is as follows.

Select a 3D solid: *Select the solid model to clean.*

Check

This option can be used to check if a solid object created by you is a valid 3D solid object. This option is independent of the settings of the **SOLIDCHECK** variable. The **SOLIDCHECK** variable turns on or off solid validation for the current drawing session while using **SOLIDEDIT** options. The default value is 1; that is, solid validation is on. If the solid object is a valid 3D solid object, you can use solid editing options to modify it without AutoCAD displaying ACIS error messages. Solid editing can take place only on valid 3D solids. You can invoke this option by choosing the **Check** button from the **Solids Editing** toolbar. The prompt sequence is as follows.

Select a 3D solid: *Select a solid you want to check.*
This object is a valid ACIS solid.

The **Undo** option cancels the editing modifications made and the **eXit** option allows you to exit the body editing options.

Undo Option

This option allows you to cancel or undo the editing action taken.

Exit Option

This option allows you to exit the **SOLIDEDIT** command.

Example 1 *Mechanical*

In this example, you will create the solid model shown in Figure 25-9. After creating the model, use the options in the **Solids Editing** toolbar to modify it as shown in Figure 25-10.

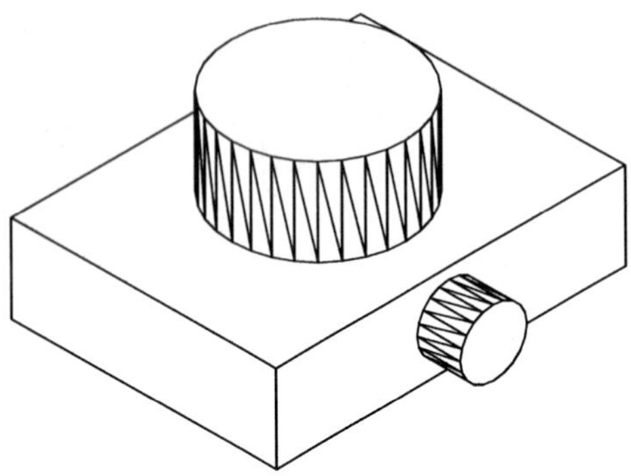

Figure 25-9 *Model for Example 1*

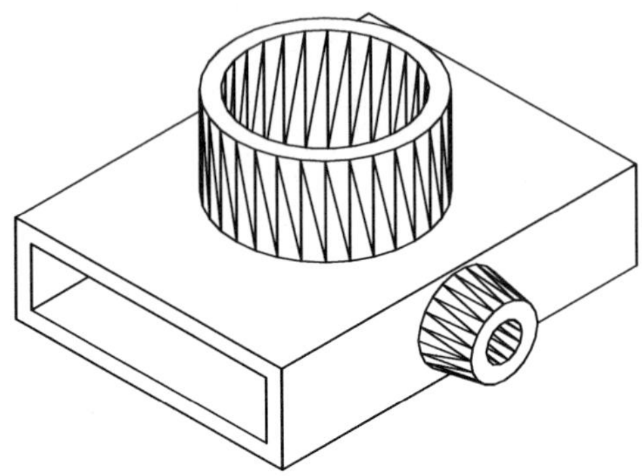

Figure 25-10 *Model after modifying*

1. Create the box and both the cylinders and move them so that they are placed at the appropriate position.

2. Union all the objects to make it a single solid model.

3. Choose the **Shell** button from the **Solids Editing** toolbar. The prompt sequence is as follows.

 Select a 3D solid: *Select the solid.*
 Remove faces or [Undo/Add/ALL]: *Select the top face of the bigger cylinder.*
 Remove faces or [Undo/Add/ALL]: *Select the front face of the smaller cylinder.*
 Remove faces or [Undo/Add/ALL]: *Select the left face of the box.*
 Remove faces or [Undo/Add/ALL]: Enter
 Enter the shell offset distance: **0.25**

 Solid validation started.
 Solid validation completed.
 Enter a body editing option
 [Imprint/seParate solids/Shell/cLean/Check/Undo/eXit] <eXit>: *Press ESC.*

4. Select the **Taper Faces** button from the **Solids Editing** toolbar. The prompt sequence is as follows.

 Solids editing automatic checking: SOLIDCHECK=1
 Enter a solids editing option [Face/Edge/Body/Undo/eXit] <eXit>: _face
 Enter a face editing option
 [Extrude/Move/Rotate/Offset/Taper/Delete/Copy/coLor/Undo/eXit] <eXit>: _taper
 Select faces or [Undo/Remove]: *Select the cylindrical faces of the smaller cylinder.*

Remove faces or [Undo/Add/ALL]: Enter
Specify the base point: *Select the center of the front face of the smaller cylinder.*
Specify another point along the axis of tapering: *Select the center at the back face of the smaller cylinder.*
Specify the taper angle: **-10**
Solid validation started.
Solid validation completed.
Enter a face editing option
[Extrude/Move/Rotate/Offset/Taper/Delete/Copy/coLor/Undo/eXit] <eXit>: *Press ESC.*

5. The final model should look similar to the one shown in Figure 25-10.

GENERATING THE DRAWING VIEWS OF THE SOLID MODEL

As mentioned earlier, one of the major advantages of working with the solid models is that the drawing views of the solid model are automatically generated once you have the solid model with you. All you have to do is to specify the type of drawing view required. This is done using the **SOLVIEW, SOLDRAW**, and **SOLPROF** commands. All these commands are discussed next.

SOLVIEW Command

Toolbar:	Solids > Setup View
Menu:	Draw > Solids > Setup > View
Command:	SOLVIEW

The **SOLVIEW** command is an externally defined ARX application (*solids.arx*). This command is used to create the documentation of the solid models in the form of the drawing views. This command guides you through the process of creating orthographic or auxiliary views. The **SOLVIEW** command creates floating viewports using orthographic projections to layout sectional view and multiview drawings of solids, while you are in the layout tab.

For the ease of documentation, this command automatically creates layers for visible lines, hidden lines, and section hatching for each view. **SOLVIEW** also creates layers for dimensioning that are visible in individual viewports. You can use these layers to draw dimensions in individual viewports. The following is the naming convention for layers.

Layer name	Object type
View name-VIS	Visible lines
View name-HID	Hidden lines
View name-DIM	Dimensions

View name-HAT Hatch patterns (for sections)

The view name is the user-defined layer name given to the view when created.

Note
The information stored on these layers is deleted and updated when you run SOLDRAW. Therefore, do not place any permanent drawing information on these layers.

The prompt sequence is as follows.

Command: **SOLVIEW**
Enter an option [Ucs/Ortho/Auxiliary/Section]: *Select an option to generate the drawing view.*

Ucs Option

This option allows you to create a profile view relative to a UCS. This is the first view that has to be generated. The reason for this is that the other options require existing viewports. The UCS can either be the current UCS or any other named UCS. The view generated using this option will be parallel to the *XY* plane of the current UCS. The next prompt sequence is

Enter an option [Named/World/?/Current] <Current>: *Enter an option or press ENTER to accept the current UCS.*

The **Named** option uses a named UCS to create the profile view. The prompts sequence is as follows.

Enter name of UCS to restore: *Enter the name of the UCS you want to use.*
Enter view scale <1.0>: *Enter a positive scale factor.*

This view scale value you enter is equivalent to zooming into your viewport by a factor relative to paper space. The default ratio is 1:1. The next prompt sequence is as follows.

Specify view center: *Specify the location of the view.*
Specify view center: Enter

This point is based on the Model space extents. The next prompt sequence is as follows.

Specify the first corner of viewport: *Specify a point.*
Specify the other corner of viewport: *Specify a point.*
Enter view name: *Enter a name for the view.*

The **World** option uses the *XY* plane of the WCS to create a profile view. The prompt sequences are the same as the **Named** option.

The **?** option lists all the named UCSs. The prompts sequence is given next.

Enter UCS names to list<*>: *Enter a name or press ENTER to list all the named UCSs.*

The **Current** option uses the *XY* plane of the current UCS to generate the profile view. This is also the default option.

Ortho Option

An orthographic view is generated by projecting the lines on to a plane normal to the current view. The **Ortho** option generates an orthographic view from the existing view. Keep in mind that the orthographic views created in AutoCAD are folded views and not the unfolded views. AutoCAD prompts you to specify the side of the viewport from where you want to project the view. The resultant view will depend upon the side of the viewport selected. The prompt sequence is as follows.

Specify side of viewport to project: *Select the edge of a viewport.*
Specify view center: *Specify the location of the orthographic view.*
Specify view center: Enter
Specify first corner of viewport: *Select first point.*
Specify opposite corner of viewport: *Select the second point.*
Enter view name: *Specify the view name.*

Auxiliary Option

This option creates an auxiliary view from an existing view. An auxiliary view is the one that is generated by projecting the lines from an existing view on to a plane that is inclined at a certain angle. AutoCAD prompts you to define the inclined plane used for the auxiliary projection. The prompt sequence is as follows.

Specify first point of inclined plane: *Specify the first point on the inclined plane.*
Specify second point of inclined plane: *Specify the second point on the inclined plane.*
Specify side to view from: *Specify a point.*
Specify view center: *Specify the location of the view.*
Specify view center: Enter
Specify first corner of viewport: *Specify the first corner of the viewport.*
Specify opposite corner of viewport: *Specify the second corner of the viewport.*
Enter view name: *Enter the view name.*

Section Option

This option is used to create the section views of the solid model. The type of the section view depends upon the alignment of the section plane and the side from which you view the model. Initially, the crosshatching is not displayed in the section view. To display the crosshatching, invoke the **SOLDRAW** command. AutoCAD prompts you to define the cutting plane, and the prompt sequence is as follows.

Specify first point of cutting plane: *Specify the first point on the cutting plane.*
Specify second point of cutting plane: *Specify the second point on the cutting plane.*

Once you have defined the cutting plane, you can specify the side to view from and the view scale, relative to paper space. The prompt sequence is given next.

Specify side to view from: *Specify the point in the direction from which the section is viewed.*
Enter view scale <current>: *Enter a positive number.*
Specify view center: *Specify the location for the view.*
Specify view center: Enter
Specify first corner of viewport: *Specify the first corner of the viewport.*
Specify opposite corner of viewport: *Specify the second corner of the viewport.*
Enter view name: *Enter the view name.*

SOLDRAW Command

Toolbar:	Solids > Setup Drawing
Menu:	Draw > Solids > Setup > Drawing
Command:	SOLDRAW

The **SOLDRAW** command generates profiles and sections in the viewports that were created with the **SOLVIEW** command. The basic function of this command is to clean up the viewports created by the **SOLVIEW** command. It creates the visible and hidden lines that represent the silhouettes and edges of solids in a viewport, and then projects them in a plane that is perpendicular to the viewing direction. When you use the **SOLDRAW** command with the sectional view, a temporary copy of the solid is created. This temporary copy is then chopped at the section plane defined by you. Then **SOLDRAW** creates the visible half as the section and discards the copy of the solid model. The crosshatching is automatically created using the current hatch pattern (HPNAME), hatch angle (HPANG), and hatch scale (HPSCALE). In this case, the layer **View name- HID** is frozen. If there are any existing profiles and sections in the selected viewports, they are deleted and replaced with new ones. The prompt sequence is as follows.

Select viewports to draw...
Select objects: *Select viewports to be drawn.*

Note

*To redraw drawings created with the **SOLVIEW** command in AutoCAD release 12 using the **SOLDRAW** command, first convert them from AME solids to AutoCAD 3D solids using the **AMECONVERT** command.*

SOLPROF Command

Toolbar:	Solids > Setup Profile
Menu:	Draw > Solids > Setup > Profile
Command:	SOLPROF

The **SOLPROF** command creates profile images of 3D solids by displaying only the edges and silhouettes of the curved surfaces of the solids in the current view. Note that before you invoke this command, you must be in the temporary model space of the layout. This means that the value of the **TILEMODE** variable should be set to **0**, but you should switch to the model space using the **MSPACE** command. The prompt sequence that follows when you choose the **Setup Profile** button is given next.

Select objects: *Select the object in the viewport.*
Select objects: Enter
Display hidden profile lines on separate layer? [Yes/No] <Y>: *Specify an option.*
Project profile lines onto a plane? [Yes/No] <Y>: *Specify an option.*
Delete tangential edges? [Yes/No] <Y>: *Specify an option.*
One solid selected.

Example 2 *Mechanical*

In this example, you will use the **SOLVIEW** and **SOLDRAW** commands to generate the top view and the sectioned front view of the solid model shown in Figure 25-11. Assume the dimensions for the model.

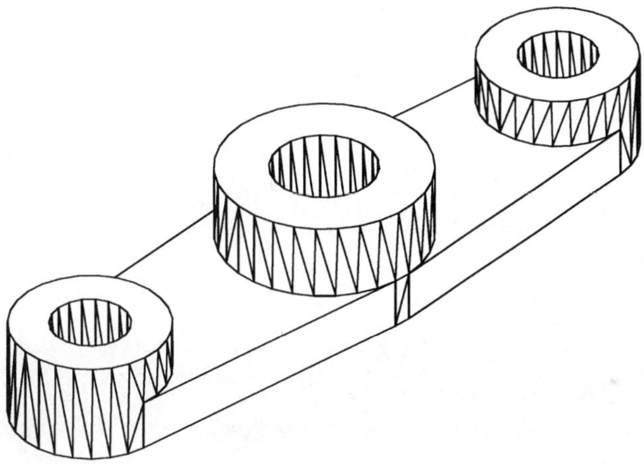

Figure 25-11 *Solid model to generate the drawing views*

1. Create the solid model by assuming the dimensions.

2. Switch to layout 1 and delete the default viewport.

3. Choose the **Setup View** button from the **Solids** toolbar. The prompt sequence is as follows.

   Enter an option [Ucs/Ortho/Auxiliary/Section]: **U**
   Enter an option [Named/World/?/Current] <Current>: **W**
   Enter view scale <1.0000>: Enter
   Specify view center: *Specify the view location close to the top left corner of the drawing window.*
   Specify view center <specify viewport>: Enter
   Specify first corner of viewport: *Specify the point of the viewport, see Figure 25-12.*
   Specify opposite corner of viewport: *Specify the second point, see Figure 25-12.*
   Enter view name: **Top**
   UCSVIEW = 1 UCS will be saved with view
   Enter an option [Ucs/Ortho/Auxiliary/Section]: Enter

4. Double-click in the viewport to switch to the temporary model space. Zoom the drawing to the extents using the **Extents** option of the **ZOOM** command. Now, switch back to the paper space using the **PSPACE** command. The layout after generating the top view should look similar to the one shown in Figure 25-12.

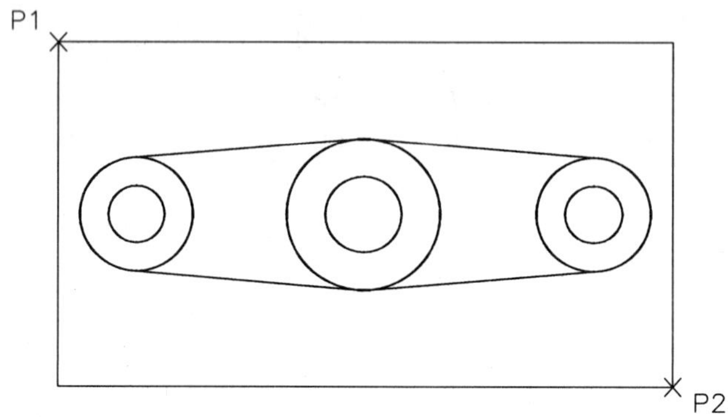

Figure 25-12 *Generating the top view*

5. Again choose the **Setup View** button from the **Solids** toolbar. The prompt sequence is as follows.

Enter an option [Ucs/Ortho/Auxiliary/Section]: **S**
Specify first point of cutting plane: *Select the left quadrant of the left cylindrical portion in the top view.*
Specify second point of cutting plane: *Select the right quadrant of the right cylindrical portion in the top view.*
Specify side to view from: *Select a point below the top view in the viewport.*
Enter view scale <current>: Enter
Specify view center: *Specify the location of the view below the top view.*
Specify view center <specify viewport>: Enter
Specify first corner of viewport: *Select first corner of the viewport.*
Specify opposite corner of viewport: *Select the second corner of the viewport.*
Enter view name: **Front**
UCSVIEW = 1 UCS will be saved with view
Enter an option [Ucs/Ortho/Auxiliary/Section]: Enter

6. Choose the **Setup Drawing** button from the **Solids** toolbar to invoke the **SOLDRAW** command. The prompt sequence is as follows.

Select viewports to draw..
Select objects: *Select the first viewport.*

Select objects: *Select the second viewport.*
Select objects: Enter

7. The drawing views should look similar to those shown in Figure 25-13.

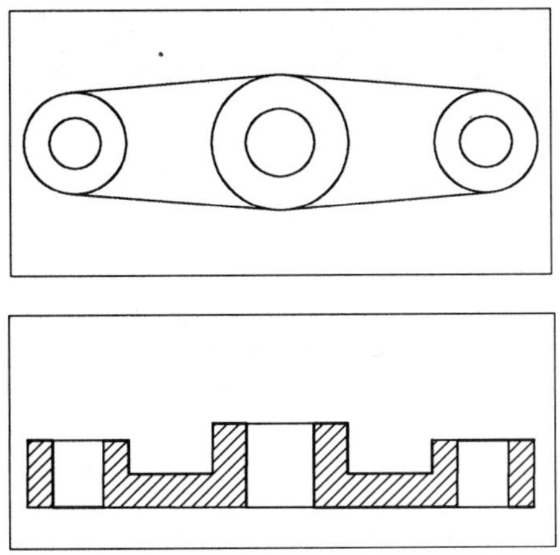

***Figure 25-13** Layout after generating the drawing views*

Tip

*You can change the scale and the type of hatch pattern using the properties window. You can also set different linetypes and colors to the different lines in the views by selecting their respective layers using the **LAYER** command.*

CALCULATING THE MASS PROPERTIES OF THE SOLID MODELS

Toolbar:	Inquiry > Region/Mass Properties
Menu:	Tools > Inquiry > Mass Properties
Command:	MASSPROP

The mass properties for the solids or the regions can be automatically calculated using the **MASSPROP** command. The various mass properties are displayed in the AutoCAD Text Window. You can also write these properties to a file. This file will be in the *.mpr* format. The properties thus calculated can be used in analyzing the solid models. Note that the mass properties are calculated by taking the density of the material of the solid model as one unit.

It is very important for you to first understand the various properties that will be calculated using this command. The properties are:

Mass. This property provides the measure of mass of a solid. This property will not be available for the regions.

Area. This property provides the measure of area of the region. This property will be available only for the regions.

Volume. This property tells you the measure of the space occupied by the solid. Since AutoCAD assigns a density of 1 to solids, the mass and volume of a solid are equal. This property will be available only for the solids.

Perimeter. This property provides the measure of perimeter of the region. This property will be available only for the regions.

Bounding Box. If the solid were to be enclosed in a 3D box, the coordinates of the diagonally opposite corners would be provided by this property. Similarly, for regions, the *X* and the *Y* coordinates of the bounding box are displayed.

Centroid. This property provides the coordinates of the center of mass for the selected solid. The density of a solid is assumed to be unvarying.

Moments of Inertia. This property provides the mass moments of inertia of a solid about the three axes. The equation used to calculate this value is given next.

mass_moments_of_inertia = object_mass * (radius of axis)2

The radius of axis is nothing but the radius of gyration. The values obtained are used to calculate the force required to rotate an object about the three axes.

Products of Inertia. The value obtained with this property helps you to determine the force resulting in the motion of the object. The equation used to calculate this value is

product _of_inertia YX,XZ = mass * dist centroid_to_YZ * dist centroid_to_XZ

Radii of Gyration. The equation used to calculate this value is given next.

gyration_radii = (moments_of_inertia/body_mass)1/2

Principal Moments and X-Y-Z Directions about Centroid. This property provides you with the highest, lowest, and middle value for the moment of inertia about an axis passing through the centroid of the object.

The prompt sequence that follows when you invoke the **MASSPROP** command is given next.

Select objects: *Select the solid model.*
Select objects: Enter

---------------- SOLIDS ----------------

Mass: 67278.5917

```
Volume:                      67278.5917
Bounding box:                X: 1.4012 -- 151.4012
              Y: 118.6406    -- 158.6406
              Z: 0.0000      -- 25.0000
Centroid:                    X: 76.4012
              Y: 138.6406
              Z: 9.2026
Moments of inertia:          X: 1308864825.5181
              Y: 507244127.3062
              Z: 1799135898.4724
Products of inertia:         XY: 712635884.7490
              YZ: 85837337.1138
              ZX: 47302745.3493
Radii of gyration:           X: 139.4790
              Y: 86.8301
              Z: 163.5285
Principal moments and X-Y-Z directions about centroid:
              I: 9991363.2928 along [1.0000 0.0000 0.0000]
              J: 108831209.8607 along [0.0000 1.0000 0.0000]
              K: 113244795.9294 along [0.0000 0.0000 1.0000]
Press ENTER to continue: Enter
```

Write analysis to a file? [Yes/No] <N>: *Specify an option.*

If you enter **Y** at the last prompt, the **Create Mass and Area Properties File** dialog box will be displayed. This dialog box is used to specify the name of the *.mpr* file. The mass properties will then be written in this file.

DYNAMIC VIEWING OF 3D OBJECTS

The advanced options for the dynamic viewing of solid models are one of the major enhancements in the recent releases of AutoCAD. These options make the entire process of the solid modeling very interesting. You can shade the solid models and dynamically rotate it on the screen to view it from different angles. You can also define clipping planes and then rotate the solid models such that whenever it passes through the cutting planes, it is sectioned just to be viewed. The original model can be restored as soon as you exit these commands. The various options and commands used to perform all these functions are discussed next.

Dynamically Rotating the View of the Model

Toolbar:	3D Orbit > 3D Orbit
Menu:	View > 3D Orbit
Command:	3DORBIT

The **3DORBIT** command allows you to visually maneuver around 3D objects to obtain different views. This is one of the most important commands for the advanced 3D viewing options. All the remaining commands can be invoked inside this command or using the **3D Orbit** toolbar, see Figure 25-14. This command activates a 3D Orbit view in the current viewport. You can click and drag your pointing device to view the 3D object from different angles. You can select individual objects or the entire drawing to view before you invoke the **3DORBIT** command. When you invoke the **3DORBIT** command, an arcball appears. This arcball is a circle with four smaller circles placed such that they divide the bigger circle into quadrants. The UCS icon is replaced by a shaded 3D UCS icon. In the 3D orbit viewing, the target is considered stationary and the camera location is considered to be moving around it. The target is the center of the arcball. **The 3DORBIT command is a transparent command and can be invoked inside any other command.**

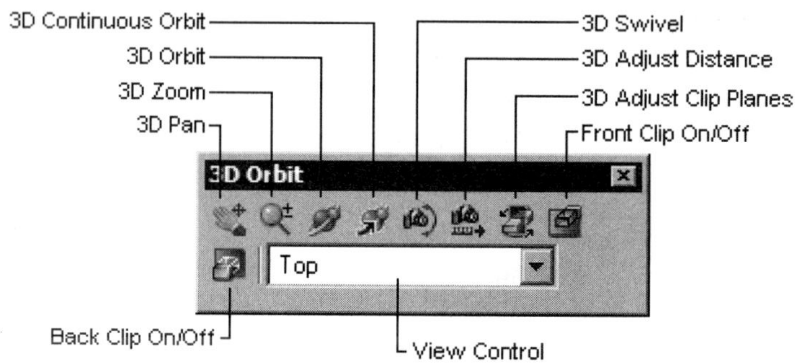

Figure 25-14 *The* ***3D Orbit*** *toolbar*

The cursor icon changes as you move the cursor in over the 3D Orbit view. The different icons indicate the different directions in which the view is being rotated. They are as follows.

Orbit mode. When you move the cursor within the arcball, the cursor looks like a sphere encircled by two lines. This icon is the **Orbit mode** cursor, and clicking and dragging the pointing device allows you to rotate the view freely as if your cursor was grabbing a sphere surrounding the objects and moving it around the target point. You can move horizontally, and vertically, as well as diagonally.

Roll mode. When you move the cursor outside the arcball, it changes to look like a sphere encircled by a circular arrow. When you click and drag, the view is rotated around an axis that extends through the center of the arcball, perpendicular to the screen.

Orbit Left-Right. When you move the cursor to the small circles placed on the left and right of the arcball, it changes into a sphere surrounded by a horizontal ellipse. When you click and drag, the view is rotated around the *Y* axis of the screen.

Orbit Up-down. When you move the cursor to the circles placed in the top or bottom of the arcball, it changes into a sphere surrounded by a vertical ellipse. Clicking and dragging the cursor rotates the view around the horizontal axis or the *X* axis of the screen, and passes it through the middle of the arcball.

Figure 25-15 shows a model being rotated using the **3DORBIT** command.

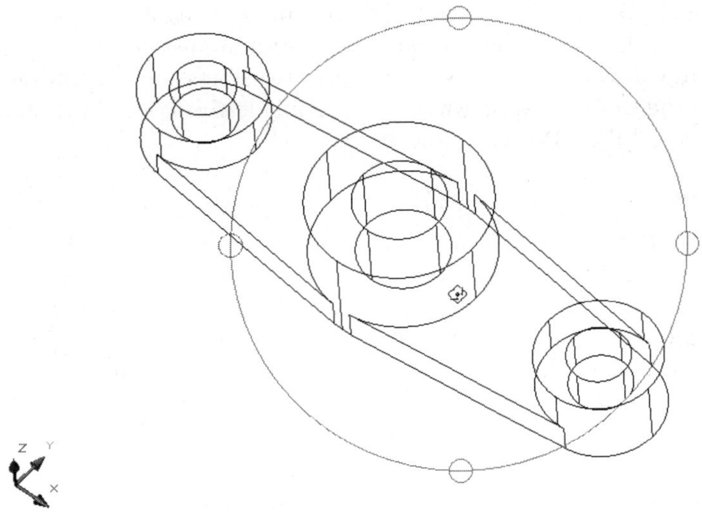

Figure 25-15 *3D orbit view, arcball with circular graphics, and the 3D shaded UCS icon*

You can invoke the other advanced viewing options using the shortcut menu that is displayed upon right-clicking in the drawing window when you are inside the **3DORBIT** command. The shortcut menu that is displayed is shown in Figure 25-16. The various options of this shortcut menu are discussed next.

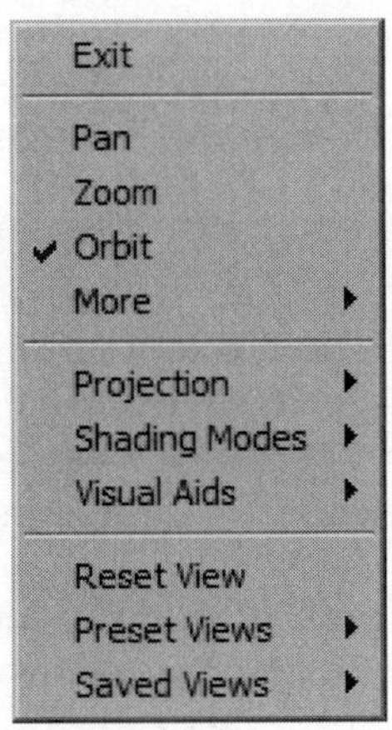

Figure 25-16 *The shortcut menu that is displayed upon right-clicking when you are inside the* ***3DORBIT*** *command*

Exit

This option is used to exit the **3DORBIT** command. You can also exit this command by pressing ESC.

Pan

The **Pan** option is the **3DPAN** command. This command can also be invoked by choosing the **3D Pan** button from the **3D Orbit** toolbar. This option allows you to drag the view to a new location in the drawing window. This command also starts the 3D interactive viewing. You can right-click in the drawing area, when you are inside this command, to display a shortcut menu that displays all the **3D Orbit** options.

Zoom

The **Zoom** option is the **3DZOOM** command. This command can also be invoked by choosing the **3D Zoom** button from the **3D Orbit** toolbar. Using this option is similar to using the camera's zoom lens. It makes the object appear closer or further away, but the position of the camera is not changed. You can invoke the other options of the **3DORBIT** command by displaying the shortcut menu. This shortcut menu is also displayed upon right-clicking in the drawing area when you are inside this command.

Orbit

This is the **3DORBIT** command. You can toggle between the **Pan**, **Zoom**, and **Orbit** options using this shortcut menu.

More

The suboptions provided by the **More** options are extensively used during the dynamic viewing of the solid model. All these suboptions are discussed next.

Adjust Distance. This suboption is the **3DDISTANCE** command and can also be invoked by choosing the **3D Adjust Distance** button from the **3D Orbit** toolbar. This suboption is similar to taking the camera closer to or farther away from the target object. This means that it makes the object appear closer or farther away. You can press the left button and drag the mouse to adjust the distance between the target and the camera. This option is generally used to reduce the distance between the camera and the object when the object rotates by a large angle in the **3DORBIT** command.

Swivel Camera. This suboption is the **3DSWIVEL** command and can also be invoked by choosing the **3D Swivel** button from the **3D Orbit** toolbar. It works similar to turning the camera on a tripod without changing the distance between the camera and the target. You can press the left button and drag the mouse to swivel the camera.

Continuous Orbit. This suboption is the **3DCORBIT** command. This suboption allows you to set the objects you select in a 3D view into continuous motion in a free-form orbit. You can also invoke this command by choosing the **3D Continuous Orbit** button from the **Solids** toolbar. While this suboption is being used, the cursor icon changes to the **continuous orbit** cursor, which is a sphere encircled by two continuous lines. Clicking in the drawing area and dragging the pointing device in any direction starts to move the object(s) in the direction you have dragged it. When you release the pointing device button, that is, stop clicking and dragging, the object continues to move in its orbit in the direction you specified. The speed at which the cursor is moved determines the speed at which the objects spin. You can click and drag again to specify another direction for rotation in the continuous orbit. While using this suboption you can right-click in the drawing area to display the shortcut menu and choose the other suboptions without exiting the command. Choosing **Pan**, **Zoom**, **Orbit, Adjust Clipping Plane** ends the continuous orbit.

Zoom Window/Extents. These suboptions are used to zoom the solid model using a window or to zoom the objects to the extents.

Orbit Maintains Z. This suboption is used to rotate the selected solid model in such a way that the *Z* axis orientation is kept constant while the object is being rotated.

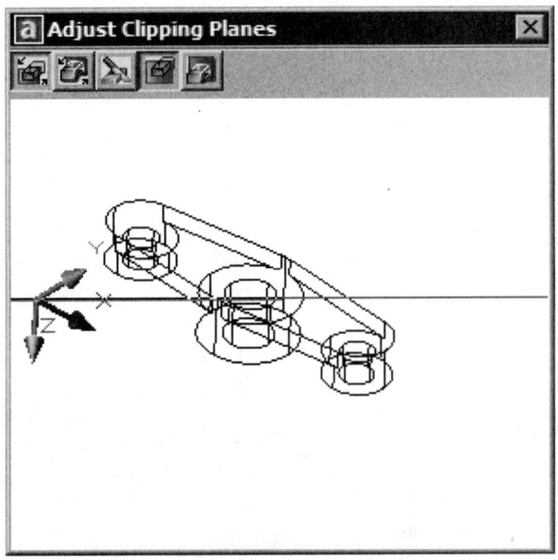

*Figure 25-17 The **Adjust Clipping Planes** window*

Adjust Clipping Planes. This suboption is the **3DClip** command and can also be invoked by choosing the **3D Adjust Clip Planes** button from the **Solids** toolbar. This suboption is used to adjust the clipping planes for sectioning the selected solid model. Note that the actual solid model is not modified by this command. This command is used only for the purpose of viewing and the original solid model will be restored when you exit this command. With this suboption you can adjust the location of the clipping planes. When you choose this suboption, the **Adjust Clipping Planes** window will be displayed as shown in Figure 25-17. In this window, the object appears rotated at 90-degree to the top of the current view in the window. This facilitates display of cutting planes. Setting the location of the front and back clipping planes in this window is reflected in the result in the current view. The various options of the **3DCLIP** command are displayed in the **Adjust Clipping Planes** toolbar in the window. There are five buttons provided in this window. The functions of these buttons are discussed next.

Adjust Front Clipping. This button is chosen to specify the location of the front clipping plane. When you move the line located near the bottom of the **Adjust Clipping Planes** window up or down, you can see the result in the 3D Orbit view.

Adjust Back Clipping. This button is chosen to specify the location of the back clipping plane. When you move the line located near the top of the **Adjust Clipping Planes** window, up or down, you can see the resulting view in the current 3D view.

Create Slice. Choosing this button, causes both the front and back clipping planes to move together. The slice is created between the two clipping planes and is displayed in the current 3D view. You can first adjust the front and the back clipping planes individually by choosing the respective options, and then by choosing the **Create Slice** button in the toolbar activate both the clipping planes simultaneously and display the result in the current view.

Front Clipping On. Turns on or off the sectioning using the front clipping plane, see Figure 25-18. You can toggle between front clipping on and off by choosing the **Front Clip On/Off** button in the **3D Orbit** toolbar.

Back Clipping On. Turns on or off the sectioning of the solid model using the back clipping plane. You can toggle between back clipping on and off by choosing the **Back Clip On/Off** button from the **3D Orbit** toolbar.

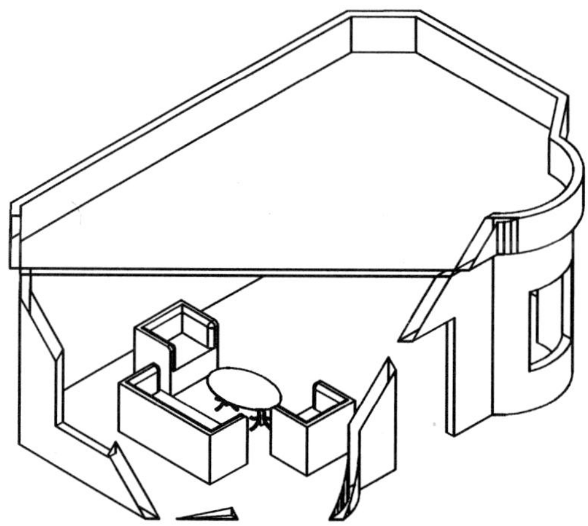

Figure 25-18 *The interior of a house displayed by setting the front clipping on and then hiding the hidden lines*

Projection

The suboptions provided under this option are used to specify the type of projection for the selected solid model.

Parallel. This suboption is chosen to display the selected model using the parallel projection method. It is the default method and ensures that no parallel lines in the model converge at any point.

Perspective. This suboption is chosen to use the one point perspective method to display the model. In this method, all the parallel lines in the model converge at a single point, thus providing the realistic view of the model when viewed with the naked eyes, see Figure 25-19.

Shading Modes

This option is the **SHADEMODE** command, and the suboptions provided under this option can also be invoked by choosing **Shade** from the **View** menu. All these suboptions are used to

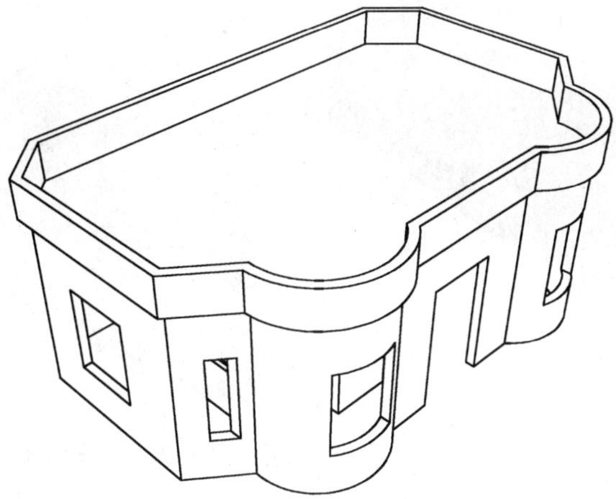

Figure 25-19 *Viewing the model using the perspective projection*

specify the mode of shading of the model while it is being rotated in the 3D orbit. Each of these suboptions used to shade the solid model is discussed next.

Wireframe. This suboption is used to display the solid model as a wireframe model. You can see through the solid model while the model is being rotated in the 3D orbit view.

Hidden. This suboption is used to display the solid model after hiding the hidden lines. You will not be able to see through the solid model.

Flat Shaded. This suboption applies the flat color on all faces of the solid model. The color of the shade will depend upon the color of the solid model. This mode of shading is not very effective as there is no smoothing of the faces (Figure 25-20). This is the reason why it is not used extensively to shade the solid model in the 3D orbit view.

Gouraud Shaded. This type of shading mode is similar to the **Flat Shading** with the only difference that the edges between the faces are very smooth (Figure 25-21). This type of shading is extensively used to shade the object in the 3D orbit view.

Flat Shaded, Edges On. This type of shading applies a flat shading on the faces of the solid model and at the same time the edges of the model are made visible, see Figure 25-22.

Gouraud Shaded, Edges On. This suboption applies the gouraud shading on the solid model and at the same time makes the edges of the model visible (Figure 25-23).

Visual Aids

The **3DORBIT** command provides you with three visual aids for the ease of visualizing the solid model in the 3D orbit view. These visual aids are **Compass**, **Grid**, and **UCS Icon**. A

Figure 25-20 Shading using flat shade

Figure 25-21 Shading using Gouraud shade

Figure 25-22 Model shaded using flat shading with the edges on

Figure 25-23 Model shaded using Gouraud shading with the edges on

check mark is displayed in front of the options you have selected. You can select none, one, two, or all three of the visual aids. Choosing **Compass** displays a sphere drawn within the arcball with three lines. These lines indicate the *X*, *Y*, and *Z* axis directions. Choosing the **Grid** displays a grid at the current *XY* plane. You can specify the height by using the **ELEVATION** system variable. The **GRID** system variable controls the display options of the **Grid** before using the **3DORBIT** command. The size of the grid will depend upon the limits of the drawing. Choosing the **UCS Icon** displays the UCS icon. If this option is not selected, then AutoCAD turns off the display of the UCS icon. While using any of the 3D Orbit options, interactive 3D viewing is on and a shaded 3D view UCS icon is displayed. The *X* axis is red, the *Y* axis is green, and the *Z* axis is blue or cyan. Figure 25-24 shows a Gouraud shaded solid model displaying the compass and grid, as well as the UCS icon.

Reset View

This option when chosen automatically restores the view that was current when you started

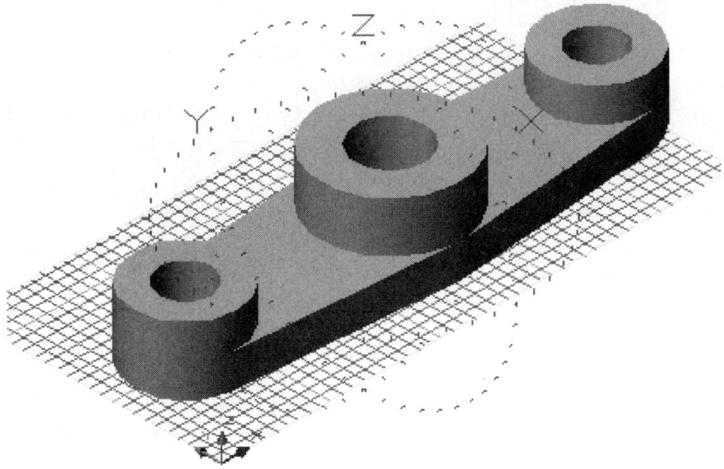

Figure 25-24 *Gouraud shaded solid model showing compass, grid, and the UCS icon*

rotating the solid model in the 3D orbit. Note that you will not exit the **3DORBIT** command if you choose this option.

Preset Views

This option is used to make current any of the six standard orthogonal views or the four isometric views.

Note

*If you have some views created using the **VIEW** command, then the **Saved Views** option will be added in the 3D Orbit shortcut menu. Use this option to activate the saved view.*

DVIEW Command

Command: DVIEW

The **DVIEW** command also allows you to visually maneuver around 3D objects to obtain different views. This command also offers you a choice between parallel viewing and perspective viewing. The difference between the two is that in parallel view, parallel lines in the 3D object remain parallel, whereas in perspective view, parallel lines meet at a vanishing point. Keep in mind that when the perspective view is current, the **ZOOM**, **PAN**, **'ZOOM**, **'PAN**, and **DSVIEWER** commands and scroll bars are not available. The **DVIEW** command uses the camera and target concept to visualize an object from any desired position in 3D space. The position from which you want to view the object (the viewer's eye) is the camera, and the focus point is the target. The line formed between these two points is the line of sight, also known as the viewing direction. To get different viewing directions, you can move the camera or target or both. Once you have the required viewing direction, you can change the distance between the camera and the target by moving the camera along the line of sight. The field of view can be changed by attaching a wide-angle lens or a telephoto lens to the camera. You can pan or twist (rotate) the image. The hidden lines in the object being viewed can be suppressed. You can clip those portions of the model that you do not want to see. All these options of the **DVIEW** command demonstrate how useful this command is. The **'ZOOM, DSVIEWER, PAN** commands and scroll bars are not available in the **DVIEW** command.

Command: **DVIEW** Enter
Select objects or <use DVIEWBLOCK>: *Select the objects you want to view dynamically or press ENTER to use the DVIEWBLOCK.*
Enter option
[CAmera/TArget/Distance/POints/PAn/Zoom/TWist/CLip/Hide/Off/Undo]: *Specify a point or enter an option.*

The point you specify is the start point for dragging. As you move the cursor, the viewing direction changes. The prompt sequence is given next.

Enter direction and magnitude angles: *Enter angles between 0-degree and 360-degree or specify on screen.*

Direction angle determines the front of the view and the magnitude angle determines the angle to which the view will roll. Both the angles should be specified in the same prompt separated by a comma. The options provided under this command are discussed next.

CAmera Option

With the **CAmera** option, you can rotate the camera about the target point. To invoke the CAmera option, enter CA at the **Enter option [CAmera/TArget/Distance/POints/PAn/Zoom/TWist/CLip/Hide/Off/Undo]** prompt. Once you have invoked this option, the drawing is still, but you can maneuver the camera up and down (above or below the target) or you can move it left or right (clockwise or counterclockwise around the target). Remember that when you are moving the camera, the target is stationary, and that when you are moving the target, the camera is stationary. The next prompt AutoCAD displays is as follows.

Specify camera location, or enter angle from XY plane,
or [Toggle (angle in)] <current>:

By default, this prompt asks you for the angle from the *XY* plane of the camera to the current UCS. This is nothing but the vertical (up or down) movement of the camera about the target. There are two ways to specify the angle.

1. You can specify the angle of rotation by moving the cursor in the graphics area until you attain the desired rotation and then choosing the pick button of your pointing device. You will notice that as you move the cursor in the graphics area, the angle of rotation is continuously displayed in the coordinates area of the status bar. Also, as you move the cursor, the object also moves dynamically. If you give a horizontal movement to the camera at the **Specify camera location, or enter angle from XY plane, or [Toggle (angle in)]**

<current> prompt, notice that the angle value displayed in the status bar does not change. This is because at **Specify camera location, or enter angle from XY plane, or [Toggle (angle in)] <current>** prompt, you are required to specify the vertical movement.

2. The other way to specify the angle of rotation is by entering the required angle of rotation value at the prompt. An angle of 90-degree (default value) from the *XY* plane makes the camera look at the object straight down from the top side of the object, providing you with the top view (plan view). In this case, the line of sight is perpendicular to the *XY* plane of the current UCS. An angle of negative 90-degree makes the camera look at the object straight up from the bottom side of the object. As you move the camera toward the top side of the object, the angle value increases. If you do not want to specify a vertical movement, press ENTER.

Once you have specified the angle from the *XY* plane, the following prompt sequence appears.

Specify camera location, or enter angle in XY plane from X axis,
or [Toggle (angle from)] <current>:

You will be prompted to specify the angle of the camera in the *XY* plane from the *X* axis. This is nothing but the horizontal (right or left) movement of the camera. You can enter the angle value in the range of -180 to 180-degree. You can move the camera toward the right side of the object (counterclockwise) by increasing the angle value. Similarly, you can move the camera toward the left side of the object (clockwise) by decreasing the angle value. You can also toggle between these two prompts using the **Toggle (angle from)** or **Toggle (angle in)** option in the two prompts. The use of the **toggle** option can be illustrated as follows.

Now, if you want to enter only the angle in the *XY* plane from the *X* axis, you can do it using two ways. The first is to give a null response to the **Specify camera location, or enter angle from XY plane, or [Toggle (angle in)] <default>** prompt. The other is by using the **Toggle (angle in)** option. Hence, to switch from the **Specify camera location, or enter angle from XY plane, or [Toggle (angle in)] <default>** prompt to the **Specify camera location, or enter angle in XY plane from X axis, or [Toggle (angle from)] <current>** prompt, you will have the following prompt sequence.

Specify camera location, or enter angle from XY plane,
or [Toggle (angle in)] <default>: T
Specify camera location, or enter angle in XY plane from X axis,
or [Toggle (angle from)] <current>:

In the same manner, if you want to switch from the **Specify camera location, or enter angle in XY plane from X axis, or [Toggle (angle from)] <current>** prompt to the **Specify camera location, or enter angle from XY plane, or [Toggle (angle in)] <default>** prompt, the following will be the prompt sequence.

Specify camera location, or enter angle in XY plane from X axis,
or [Toggle (angle from)] <current>: T

Specify camera location, or enter angle from XY plane,
or [Toggle (angle in)] <default>:

Note

*The **DVIEWBLOCK** is automatically displayed when you use the **DVIEW** command without selecting objects. You can define or redefine your own substitute **DVIEWBLOCK**, just like any other symbol block. To create a DVIEWBLOCK, create a 3D object and then convert it into a block. Enter the name of the block as **DVIEWBLOCK**. You will be prompted to redefine the **DVIEWBLOCK**. Redefine the block and the new object will now be displayed as the DVIEWBLOCK. Just be sure to define it to a 1X1X1 size so that it will be properly scaled in 3D space in the **DVIEW** command.*

TArget Option

With the **TArget** option, you can rotate the target point with respect to the camera. To invoke the **TArget** option, enter TA at the **Enter option [CAmera/ TArget/ Distance/ POints/ PAn/ Zoom/ TWist/ CLip/ Hide/ Off/ Undo]:** prompt. Once you have invoked this option, the camera is still, but you can maneuver the target point up or down or left or right about the camera. When you move the target, the camera is stationary. The prompt sequence for the **TArget** option is given next.

Command: **DVIEW**
Select objects or <use DVIEWBLOCK>: Enter
Enter option
[CAmera/TArget/Distance/POints/PAn/Zoom/TWist/CLip/Hide/Off/Undo]: **TA**
Specify camera location, or enter angle from XY plane,
or [Toggle (angle in)] <current>: *Specify the angle about which the target will lie above or below the camera.*
Specify camera location, or enter angle in XY plane from X axis,
or [Toggle (angle from)] <current>: *Specify the angle about which the target will lie left or right of the camera.*
Enter option
[CAmera/TArget/Distance/POints/PAn/Zoom/TWist/CLip/Hide/Off/Undo]: Enter

The prompt sequence does not reveal the difference between the **CAmera** option and the **TArget** option. The difference lies in the actual angle of view. For example, if you specify an angle of 90-degree in the **Specify camera location, or enter angle from XY plane, or [Toggle (angle in)] <current>** prompt of the **CAmera** option, your viewing direction is from the top of the object toward the bottom. If you specify the same angle in the same prompt for the **TArget** option, your viewing direction is from the bottom of the object toward the top.

Distance Option

As mentioned earlier, the line obtained by joining the camera position and the target position is known as the line of sight. The **Distance** option can be used to move the camera toward or away from the target along the line of sight. Invoking the **Distance** option enables perspective viewing. Since we have not used the **Distance** option until now, all the previous views were in parallel projection. In perspective display, the objects nearer to the camera appear bigger than objects that are farther away from the camera position. In other words, in perspective views, the parallel lines meet at a vanishing point. Another noticeable difference is that the regular coordinate system icon is replaced by the perspective icon. If the current UCS icon is the default 3D UCS icon, then the *X*, *Y*, and *Z* axes will be displayed in the dotted lines with no cone on the *X* and *Y* axes. If the current UCS icon is the 2D UCS icon, then a perspective box will be displayed instead of the UCS icon. This icon acts as a reminder that perspective viewing is enabled. The **distance** option can be invoked by entering D at the **Enter option [CAmera/TArget/Distance/ POints/PAn/Zoom/TWist/CLip/Hide/Off/Undo]** prompt. Once you invoke the **Distance** option, AutoCAD prompts you to specify the new distance between the camera and the target.

Specify new camera-target distance <1.0000>: *Enter the desired distance between the camera and the target.*

On the top side of the screen, a slider bar appears. It is marked from 0x to 16x. The current distance is represented by the 1x mark. This is verified by the fact that the slider bar moves right or left with respect to the 1x mark on the slider bar. As you move the slider bar toward the right, the distance between the camera and the target increases; as you move the slider bar toward the left, the distance between the camera and the target decreases. For example, when you move the slider bar to the 16x mark, the distance between the camera and the target increases 16 times, or in other terms, the camera moved away from the target on the line of sight 16 times the previous distance. The distance between the camera and the target is dynamically displayed in the status line. If you cannot display the entire object on the screen by moving the slider bar to the 16x mark, enter a larger distance at the keyboard. To revert to parallel viewing, invoke the **Off** option. If you want to magnify the drawing without turning the perspective viewing on, you can use the **Zoom** option. Figure 25-25 shows a model using the parallel projection and Figure 25-26 shows a model using the perspective projection.

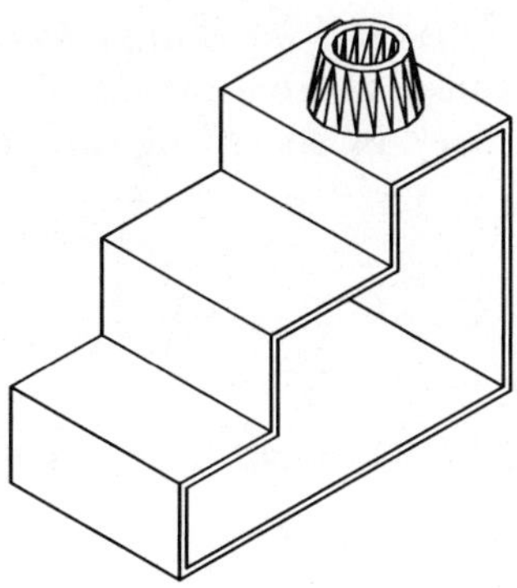

Figure 25-25 *Solid model in parallel projection*

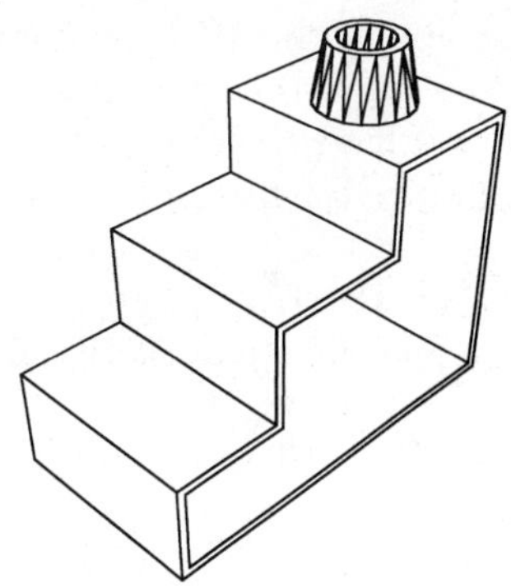

Figure 25-26 *Solid model in perspective projection*

POints Option

With the **POints** option, you can specify the camera and target positions (points) in the *X*, *Y*,

Z coordinates. In other words, this option allows you to define the line of sight using the target point and the camera point. You can specify the *X, Y, Z* coordinates of the point by any method used to specify points including Object snap and .X, .Y, .Z point filters. The *X, Y, Z* coordinate values are with respect to the current UCS. The prompt sequence to invoke this option is given next.

Command: **DVIEW**
Select objects or <use DVIEWBLOCK>: Enter
Enter option
[CAmera/TArget/Distance/POints/PAn/Zoom/TWist/CLip/Hide/Off/Undo]: **PO**

The target point needs to be specified first. A rubber-band line is drawn from the current target position to the drawing crosshairs. This is the line of sight.

Specify target point <current>: *Specify the location of the target or press ENTER.*

Once you have specified the target point, you are prompted to specify the camera point. A rubber-band line is drawn between the target point and the drawing crosshairs. This helps you to place the camera relative to the target.

Enter camera point <current>: *Specify the location of the camera press ENTER.*

Establishment of the new target point and camera point must be carried in parallel projection. If you specify these two points while the perspective projection is active, the perspective projection is temporarily turned off, until you specify the camera and target points. Once this is done, the object is again displayed in perspective. If the viewing direction is changed by the new target location and camera location, the preview image is regenerated to show the change.

PAn Option

The **PAn** option of the **DVIEW** command is similar to the **PAN** command. This option lets you shift the entire drawing with respect to the graphics display area. Just as with the **PAN** command, you have to specify the pan distance and direction by specifying two points. The prompt sequence for this option is given next.

Command: **DVIEW**
Select objects or <use DVIEWBLOCK>: Enter
Enter option
[CAmera/TArget/Distance/POints/PAn/Zoom/TWist/CLip/Hide/Off/Undo]: **PA**
Specify displacement base point: *Specify the first point.*
Specify second point: *Specify the second point.*

Zoom Option

The **Zoom** option of the **DVIEW** command is similar to the **ZOOM** command. With the help of this option, you can enlarge or reduce the drawing. This option can be invoked by entering Z at the following prompt.

Enter option
[CAmera/TArget/Distance/POints/PAn/Zoom/TWist/CLip/Hide/Off/Undo]: **Z**
Specify lens length <current>: *Specify the new lens length.*

Just as with the **Distance** option, in the **Zoom** option a slider bar, marked from 0x to 16x, is displayed on the top side of the screen. The default position on the slider bar is 1x. Two ways to zoom can now be specified. The first is when perspective is enabled. In this case, zooming is defined in terms of lens length. The 1x mark (default position) corresponds to 50.000 mm lens length. As you move the slider bar toward the right, the lens length increases; and as you move the slider bar toward the left, the lens length decreases. For example, when you move the slider bar to the 16x mark, the lens length increases 16 times, which is 16 X 50.000mm = 800.000mm. You can simulate the telephoto effect by increasing the lens length; by reducing the lens length, you can simulate the wide-angle effect. The lens length is dynamically displayed in the coordinates area of the status bar. If perspective is not enabled, zooming is defined in terms of the zoom scale factor. In this case, the **Zoom** option is similar to the **ZOOM Center** command, and the center point lies at the center of the current viewport. The 1x mark (default position) corresponds to a scale factor of 1. As you move the slider bar toward the right, the scale factor increases; as you move the slider bar toward the left, the scale factor decreases. For example, when you move the slider bar to the 16x mark, the scale factor increases 16 X 1 = 16 times. The scale factor is dynamically displayed on the status line.

Command: **DVIEW**
Select objects or <use DVIEWBLOCK>: *Select the objects.*
Enter option
[CAmera/TArget/Distance/POints/PAn/Zoom/TWist/CLip/Hide/Off/Undo]: **Z**
Specify zoom scale factor <1>: *Specify the scale factor.*

TWist Option

The **TWist** option allows you to rotate (twist) the view around the line of sight. In other words, you can say that the object on the screen is rotated around a point that is hinged to the center point of the screen. The display is always adjusted so that the target point is at the center of the screen. If you use a pointing device to specify the angle, the angle value is dynamically displayed in the coordinates area of the status line. A rubber-band line is drawn from the center (target point) to the crosshairs. As you move the crosshairs with a pointing device, the object on the screen is rotated around the line of sight. You can also enter the angle of twist using the keyboard. The angle of twist is measured in a counterclockwise direction, starting from the right side.

Command: **DVIEW**
Select objects or <use DVIEWBLOCK>: Enter
Enter option
[CAmera/TArget/Distance/POints/PAn/Zoom/TWist/CLip/Hide/Off/Undo]: **TW**
Specify view twist angle <0.00>: *Specify the angle of rotation (twist).*

CLip Option

The **CLip** option can be used to clip sections of the drawing. AutoCAD uses two invisible clipping planes to realize clipping. These clipping walls can be positioned anywhere on the screen and are perpendicular to the line of sight (line between target and camera). Once you position the clipping planes, AutoCAD conceals all the lines that are in front of the front clipping plane or behind the back clipping plane. The **CLip** option can be used in both parallel and perspective projections. When perspective is enabled, the front clipping plane is automatically enabled.

Note

*The clipping option of the **DVIEW** command is the same as that of the **3DORBIT** command. If you set the clipping using any one of the commands, the clipping in the other command is also set.*

If you have specified a positive distance, the clipping plane is placed between the target and the camera. If the distance you have specified is negative, the clipping plane is placed beyond the target.

The prompt sequence for the **CLip** option is given next.

Enter option
[CAmera/TArget/Distance/POints/PAn/Zoom/TWist/CLip/Hide/Off/Undo]: **CL**
Enter clipping option [Back/Front/Off]<Off>: *Select an option.*
Specify distance from target or [ON/OFF] <current>:

Once you have specified which clipping plane you want to set, a slider bar appears on the screen. As you move the pointer toward the right side of the slider bar, the negative distance between the target and the clipping plane increases. As you move toward the left side of the slider bar, the positive distance between the target and the clipping plane increases; hence, a greater portion of the drawing is clipped. The rightmost mark on the slider bar corresponds to a distance equal to the distance between the target and the farthermost point on the object you want to clip. After specifying the distance between one of the clipping planes (back clipping plane in this prompt sequence), you need to invoke the **CLip** option again and to specify the position of the front clipping plane in terms of the distance between the front clipping plane and the target. As you move the slider bar toward the right, the negative distance between the target and the front clipping plane increases. As the negative distance increases, a greater portion of the front side of the drawing is clipped. The rightmost mark on the slider bar corresponds to a distance equal to the distance between the target and the back clipping plane. Figure 25-27 shows a model with some portion clipped by setting the front clipping.

Hide Option

Sometimes, you may not want these hidden lines to show up on the screen, so you can use the **Hide** option. This option is used to suppress the hidden lines in the 3D model and is similar to the **HIDE** command.

Off Option

The **Off** option turns the perspective projection off. When the perspective projection is turned off, you will notice that the perspective icon is replaced by the regular UCS icon.

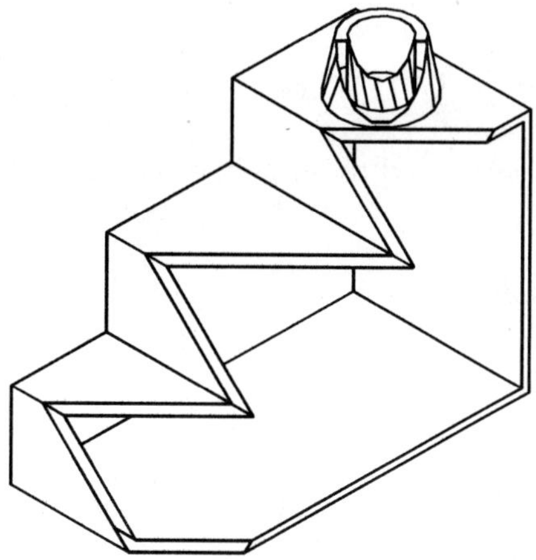

Figure 25-27 *A model clipped using the front clipping option*

Undo Option

The **Undo** option is similar to the **UNDO** command. The **Undo** option nullifies the result of the previous **DVIEW** operation. Just as in the case of the **UNDO** command, you can use this option a number of times to undo the results of multiple **DVIEW** operations.

Self Evaluation Test

Answer the following questions and then compare your answers with the answers given at the end of this chapter.

1. The solid models can be edited using the **EDITSOLID** command. (T/F)

2. You can create a shell in the solid model using the **Shell** option of the **SOLIDEDIT** command. (T/F)

3. Setting the clipping options in the **3DORBIT** command also sets them in the **DVIEW** command. (T/F)

4. The 3D model can be continuously rotated in the 3D orbit. (T/F)

5. The _________ command is used to generate the drawing views of the solid model.

6. The mass properties of the selected solid model can be written into a file of _________ format.

7. The __________ command is used to shade and dynamically rotate the selected solid model in the 3D orbit.

8. The __________ option of the **DVIEW** command is used to set the perspective projection.

9. The __________ command is used to directly rotate the solid model continuously in the 3D orbit.

10. The impression on the solid models can be removed using the __________ option of the **SOLIDEDIT** command.

Review Questions

Answer the following questions.

1. You can edit the wireframe models using the **SOLIDEDIT** command. (T/F)

2. You can move a selected face in the solid model from one location to the other. (T/F)

3. You can assign different colors to the different faces of the solid model. (T/F)

4. You can continuously rotate the selected solid model in the 3D orbit. (T/F)

5. Which command is used to generate the profiles and sections in the viewports created using the **SOLVIEW** command?

   (a) **SOLDRAW** (b) **SOLPROF**
   (c) **SOLEDIT** (d) **None**

6. Which command is used to set the shading mode for the 3D objects?

   (a) **SHADEMODE** (b) **SHADE**
   (c) **MODESHADE** (d) **None**

7. Which option of the **SOLIDEDIT** command is used to edit the faces of the solid model?

   (a) **Body** (b) **Face**
   (c) **Solid** (d) **None**

8. Which option of the **SOLVIEW** command is used to generate the drawing view relative the current UCS?

   (a) **New** (b) **Section**
   (c) **UCS** (d) **Ortho**

9. Which command is used to calculate the mass properties of the solid model?

   (a) **PROPERTIES** (b) **MASSPROP**
   (c) **CALCULATE** (d) **MASSPROPERTIES**

10. The __________ command is used to convert a 3D object into a block to be displayed as the DVIEWBLOCK in the **DVIEW** command.

11. When you select the distance option of the **DVIEW** command, the __________ viewing is enabled.

12. The __________ command creates profile images of 3D solids by displaying only the edges and silhouettes of the curved surfaces of the solids in the current view.

13. With the front clipping on in the **DVIEW** command, as you increase the negative distance, the __________ portion of the drawing is clipped.

14. The __________ command is used to adjust the clipping planes.

15. The __________ option in the shortcut menu of the **3DORBIT** command is used to define the type of projection for the solid model.

Exercises

Exercise 1 *Mechanical*

In this exercise, you will create the solid model shown in Figure 25-28. The dimensions for the solid model are shown in Figures 25-29 and 25-30. After creating the model, generate the top and sectioned front view of the model. Also, set the front clipping, and then shade the model using the Gouraud shading and dynamically rotate the model in the 3D orbit.

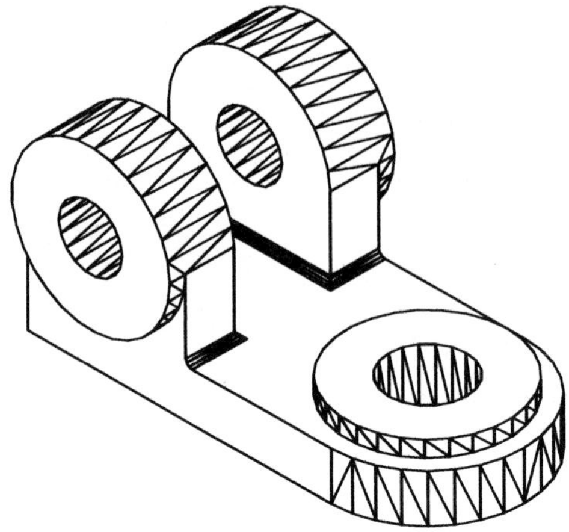

Figure 25-28 Solid model for Exercise 1

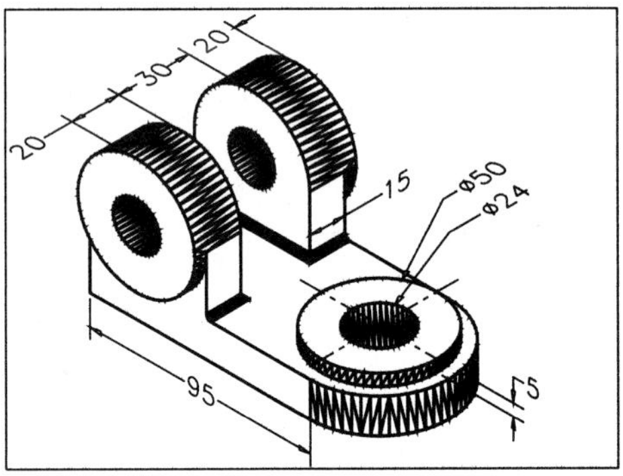

Figure 25-29 Dimensions for Exercise 1

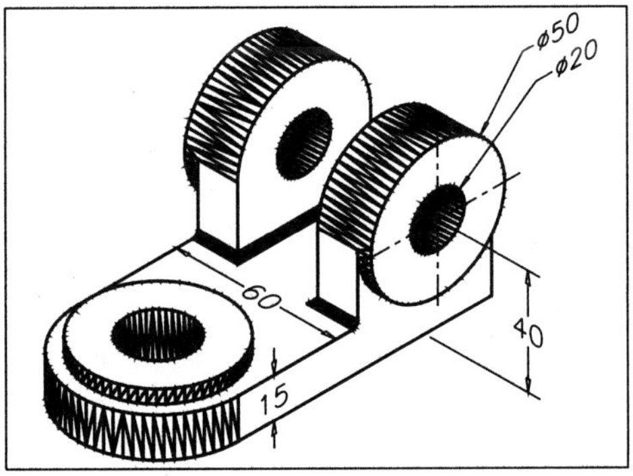

Figure 25-30 Dimensions for Exercise 1

Exercise 2 *Architectural*

In this exercise, you will create the solid model shown in Figure 25-31. After creating the model, shade and rotate it in the 3D orbit using the **3DORBIT** command.

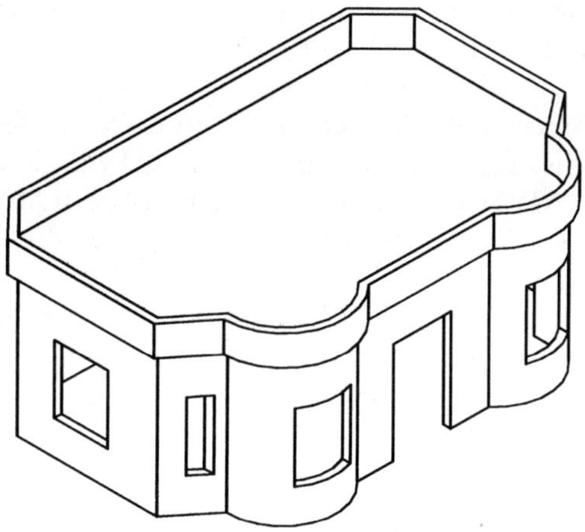

Figure 25-31 *Solid model for Exercise 2*

Problem Solving Exercise 1 *General*

a. Draw the computer as shown in Figure 25-32. Take the dimensions of the computer you are using. Construction of the given figure can be divided into the following steps:

1. Drawing the CPU box (central processing unit).
2. Drawing the base of the monitor.
3. Drawing the monitor.

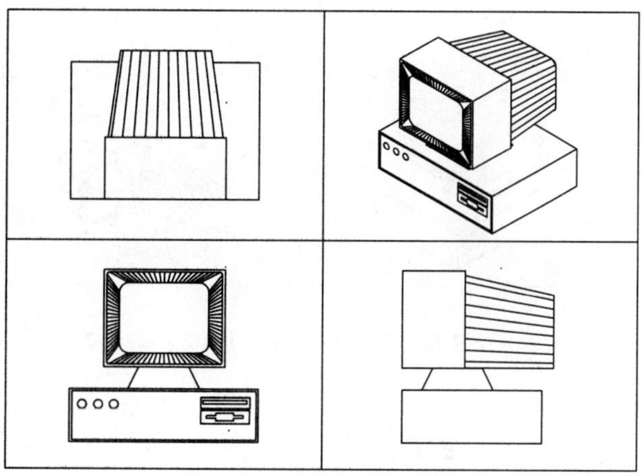

Figure 25-32 *Solid model for Problem Solving Exercise 1*

b. Once you have drawn the computer, create four viewports in the paper space and obtain different views of the computer as shown in the screen capture (Figure 25-32). You can use the **VPOINT** command to obtain the different views.

c. Next, use the **MVIEW** command to create another set of four viewports as shown in Figure 25-33. Center the figure of the computer in 3D view in all four viewports. Use the **CAmera** option of the **DVIEW** command to obtain the figure shown in the upper left viewport.

Use the **Distance** option of the **DVIEW** command to obtain the figure shown in the upper right viewport.

Use the **CLip** option of the **DVIEW** command to obtain the figure shown in the lower left viewport.

Use the **TWist** option of the **DVIEW** command to obtain the figure shown in the lower right viewport.

d. Notice the difference between the parallel projection and the perspective projection.

e. Use the **3DORBIT** command to view the object in 3D Orbit.

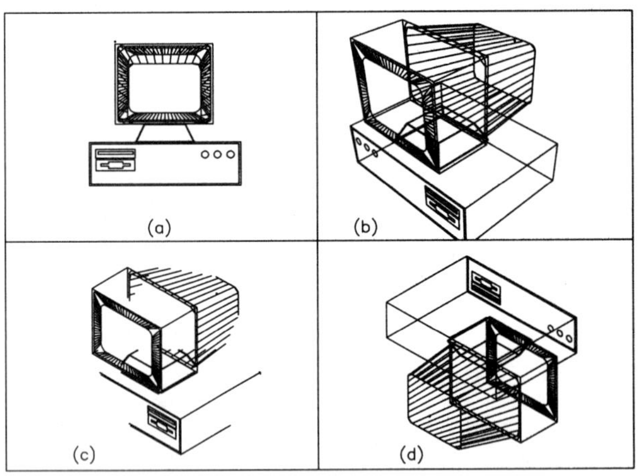

Figure 25-33 *Solid model for Exercise 1*

Problem Solving Exercise 2 — *Mechanical*

For the Tool Organizer shown in Figure 25-36, draw the following.

Piece part drawings shown in Figure 25-34, assembly drawing with Bill of Materials shown in Figure 25-35, solid model of the assembled Tool Organizer shown in Figure 25-36.

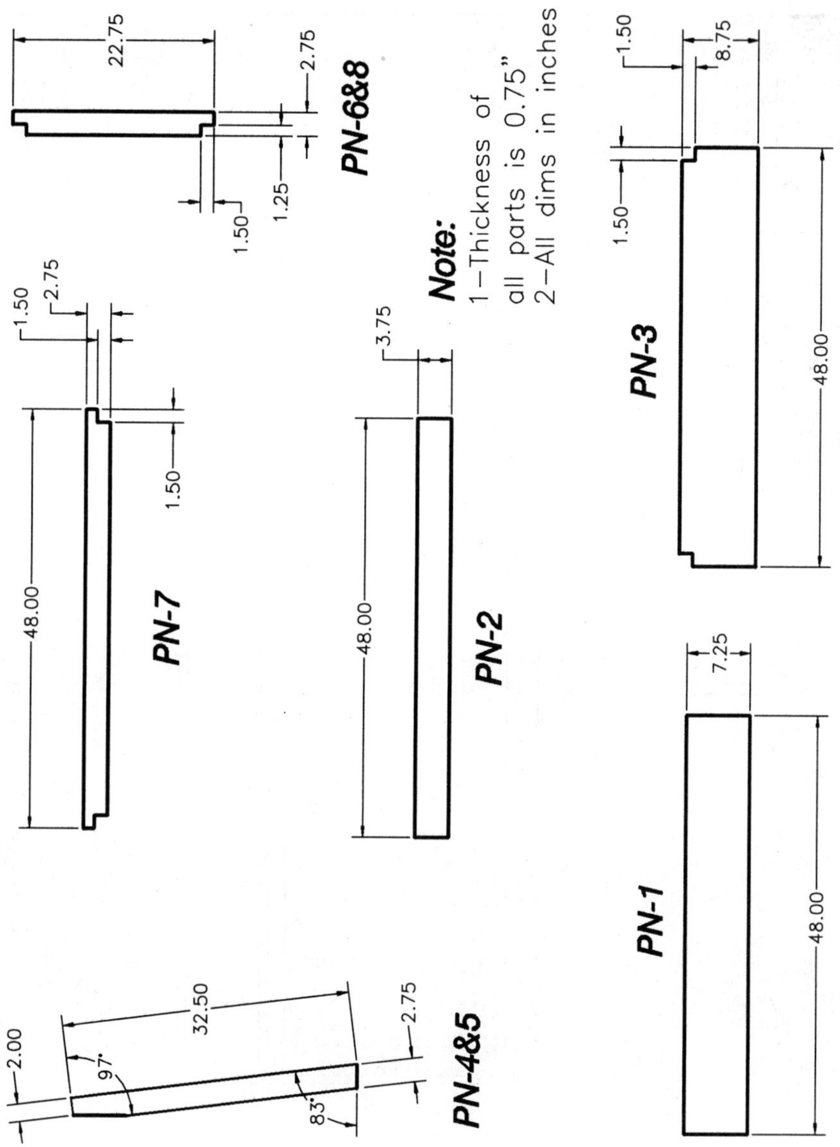

Figure 25-34 *Piece part drawings of Tool Organizer*

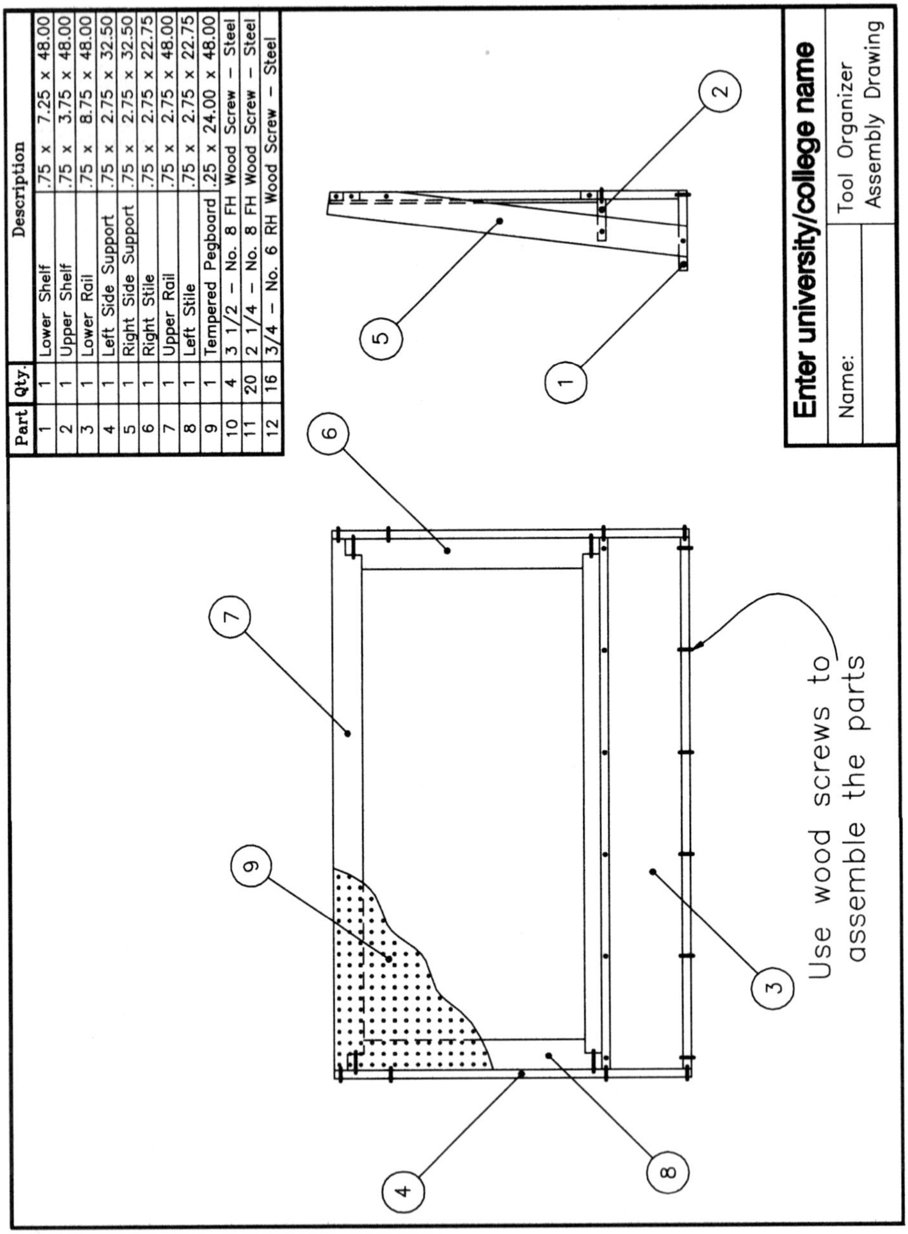

Figure 25-35 *Assembly drawing and BOM of Tool Organizer*

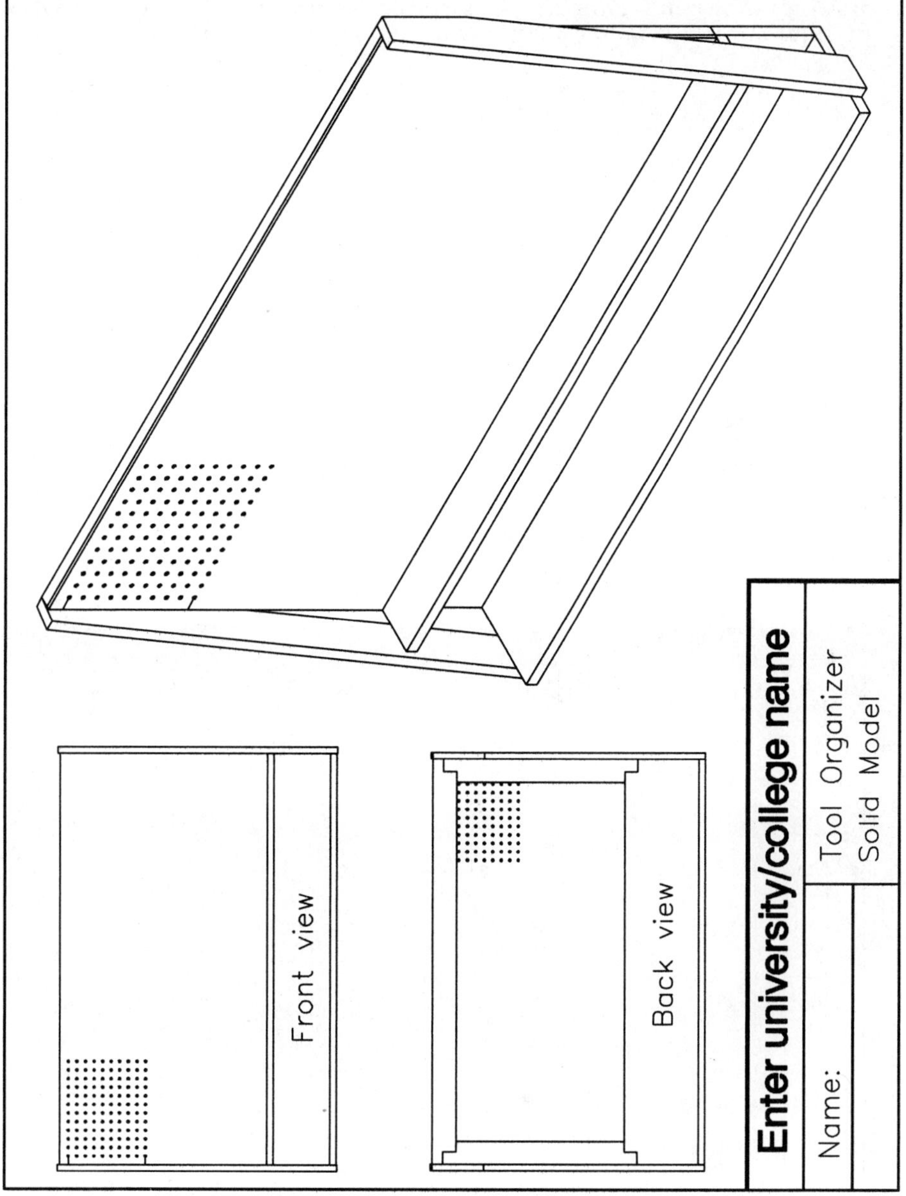

Figure 25-36 *Solid model of Tool Organizer*

Problem Solving Exercise 3 *Mechanical*

For the Work Bench shown in Figure 25-39, draw the following.

Piece part drawings shown in Figure 25-37, assembly drawing with Bill of Materials shown in Figure 25-38, solid model of the assembled Work Bench shown in Figure 25-39.

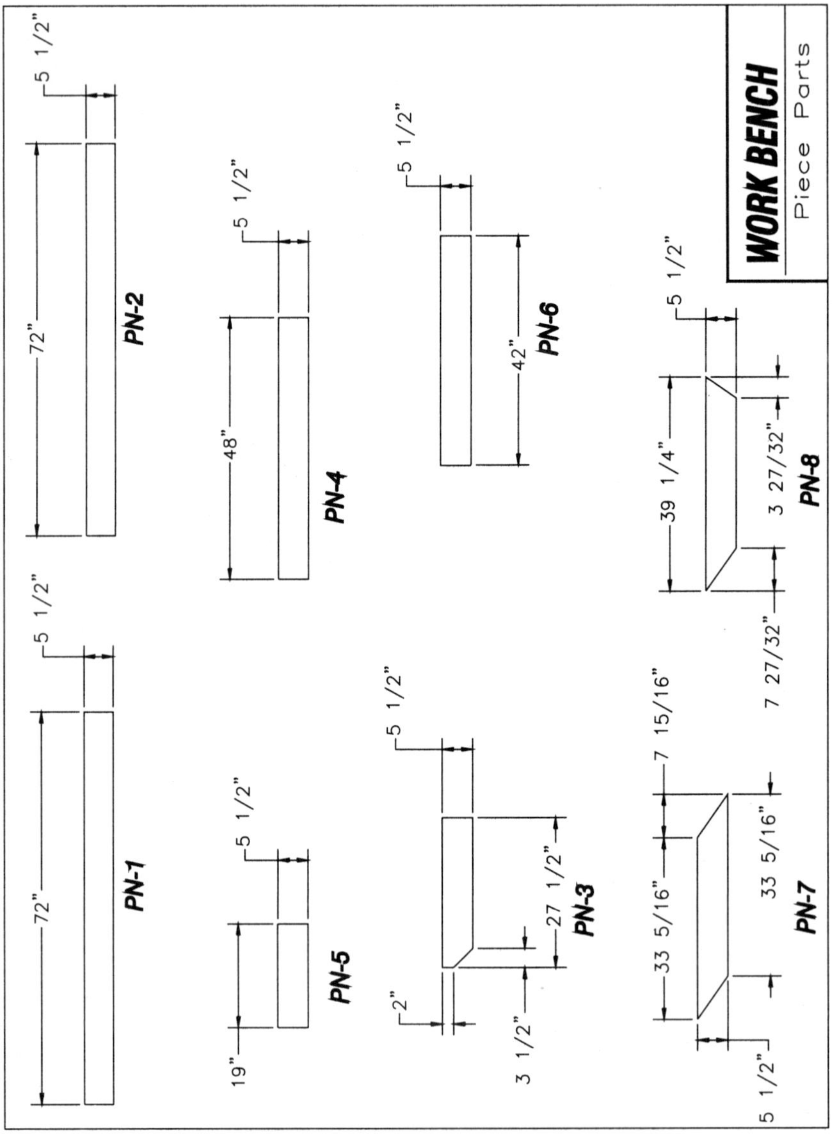

Figure 25-37 *Piece part drawings of Work Bench*

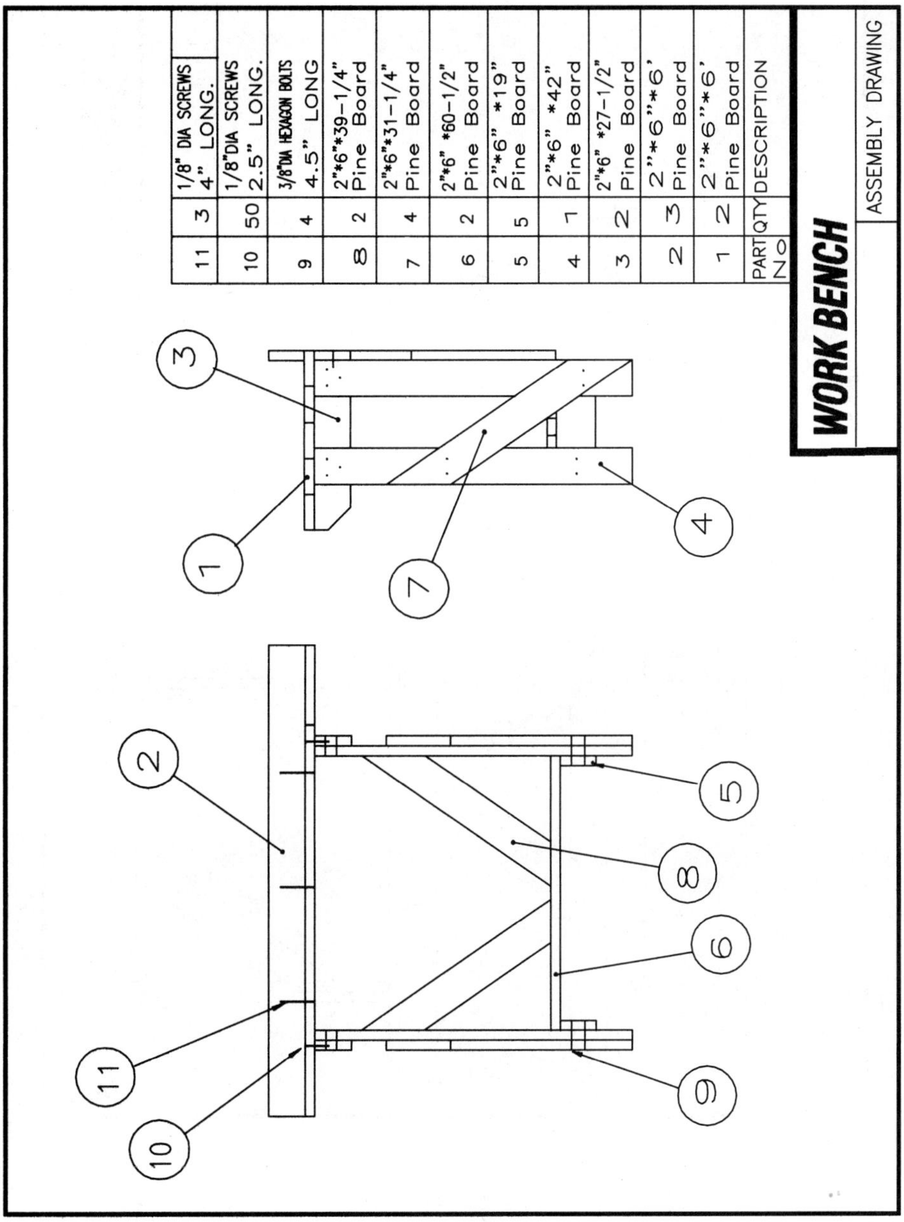

Figure 25-38 *Assembly drawing and BOM of Work Bench*

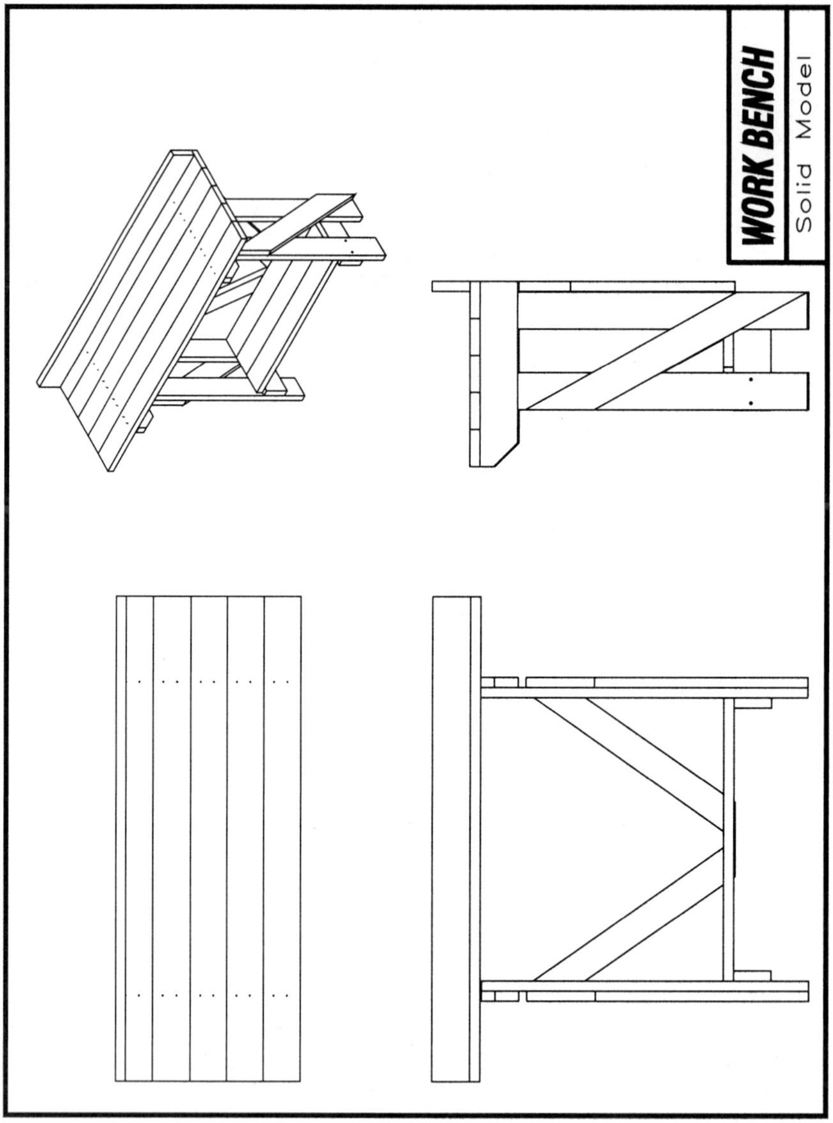

Figure 25-39 Solid model of Work Bench

Problem Solving Exercise 4 *Mechanical*

Create various components of the Pipe Vice assembly and then assemble them by moving, rotating, and aligning. The Pipe Vice assembly is shown in Figure 25-40. For your reference, the exploded view of the assembly is shown in Figure 25-41. The dimensions of the individual components are shown in Figures 25-42 through 25-47.

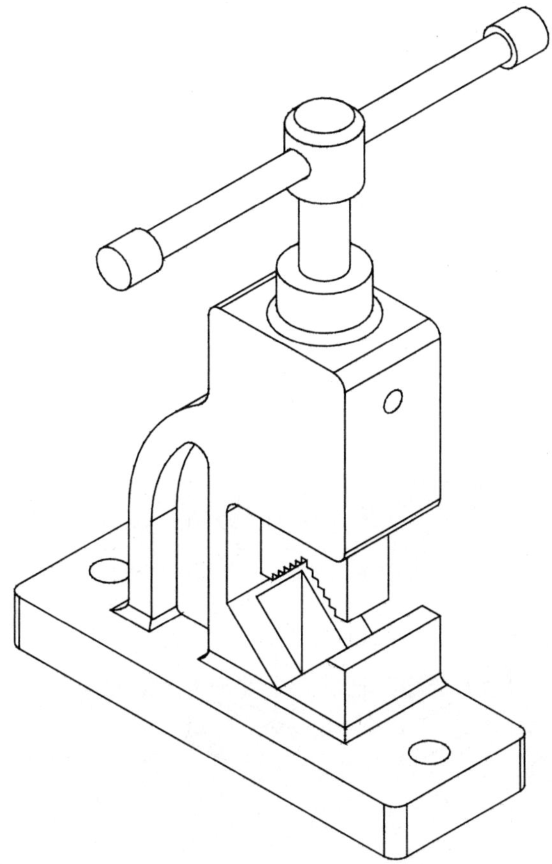

Figure 25-40 *Pipe Vice assembly*

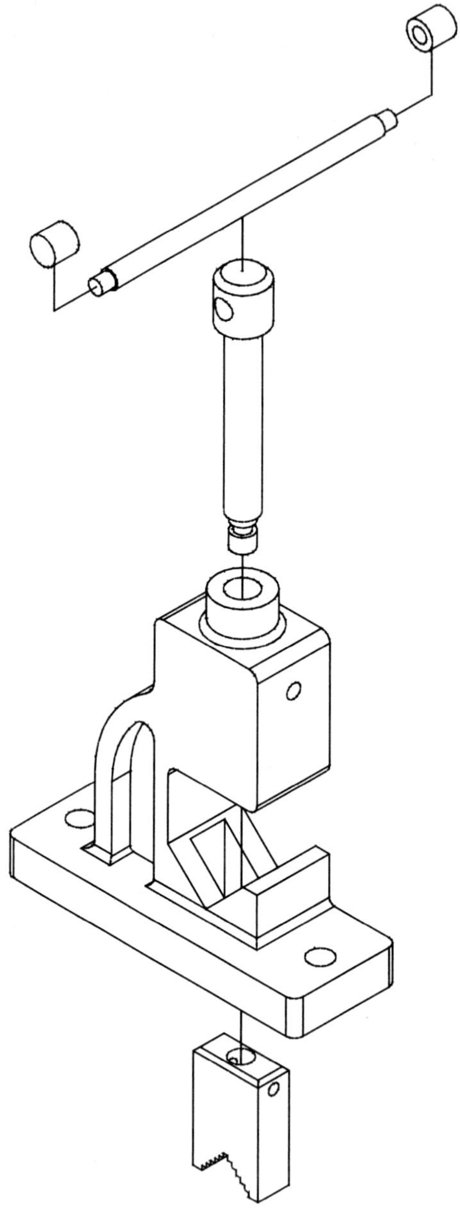

Figure 25-41 *Exploded view of the Pipe Vice assembly*

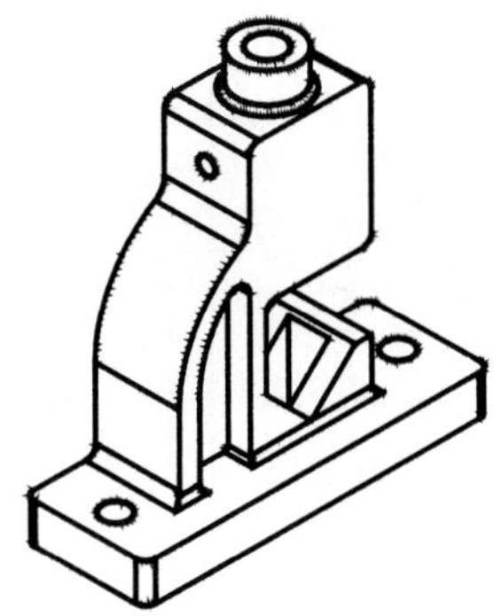

Figure 25-42 Isometric view of the Base

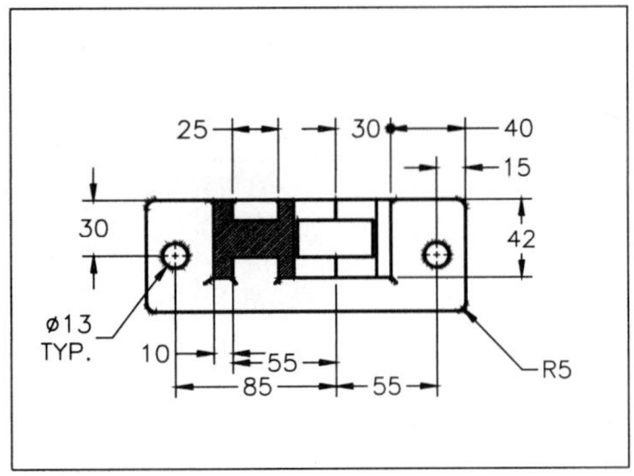

Figure 25-43 Sectioned top view of the Base

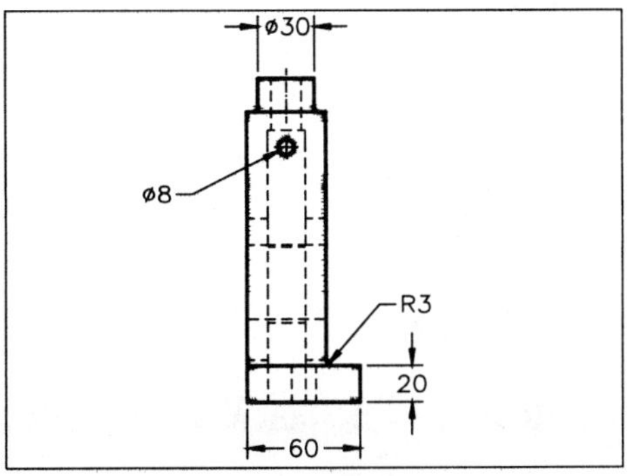

Figure 25-44 Side view of the Base

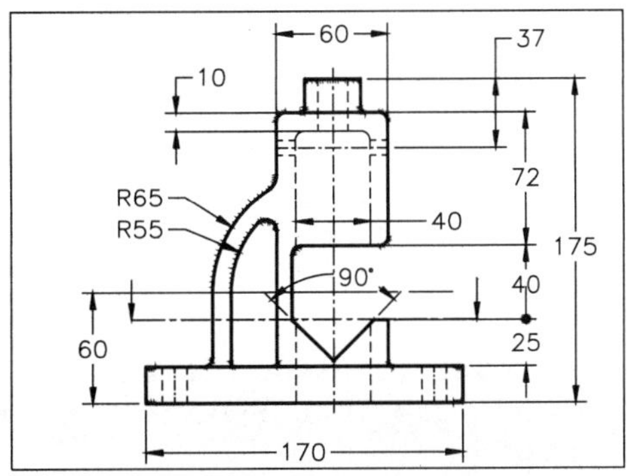

Figure 25-45 Front view of the Base

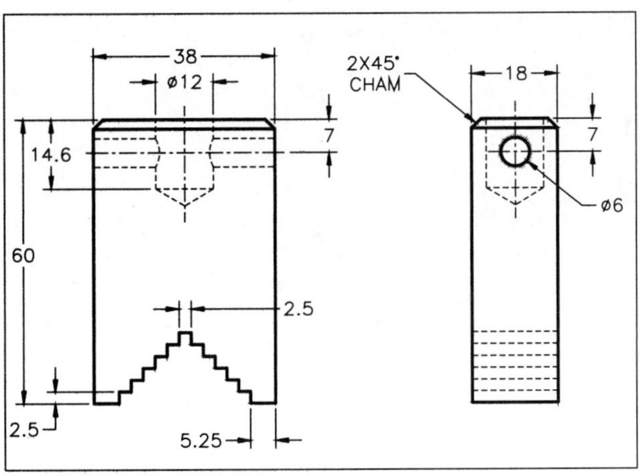

Figure 25-46 Dimensions of the Movable Jaw

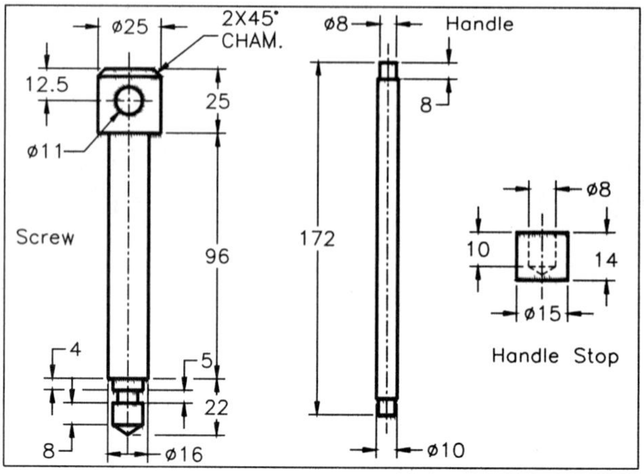

Figure 25-47 Dimensions of the Screw, Handle, and the Handle Stop

Answer to Self-Evaluation Test

1 - F, **2** - T, **3** - T, **4** - T, **5** - **SOLVIEW**, **6** - *.mpr*, **7** - **3DORBIT**, **8** - **Distance**, **9** - **3DCORBIT**, **10** - **Clean**.

Chapter 26

Rendering Designs

Learning Objectives

After completing this chapter you will be able to:

- *Understand rendering and the need to render the objects.*
- *Attach materials to the models and modify the materials.*
- *Define new materials and modify new materials.*
- *Insert and modify light sources.*
- *Select rendering type and render objects.*
- *Define, render, and modify scenes.*
- *Replay and print renderings.*
- *Define background and fog and apply background and fog.*
- *Define landscaping objects and attach them to the model.*
- *Configure, load, and unload AutoCAD Render.*

UNDERSTANDING THE CONCEPT OF RENDERING

A rendered image makes it easier to visualize the shape and size of a 3D object, compared to a wireframe image or a shaded image. A rendered object also makes it easier to express your design ideas to other people. For example, if you want to make a presentation of your project or a design, you do not need to build a prototype. You can use the rendered image to explain your design much more clearly because you will have a complete control over the shape, size, color, and surface material of the rendered image. Additionally, any required changes can be incorporated into the object, and the object can be rendered to check or demonstrate the effect of these changes. Thus, rendering is a very effective tool for communicating ideas and demonstrating the shape of an object.

Generally, the process of giving a realistic effect to the models (rendering) can be conveniently divided into four steps. These steps are given next.

1. Selecting the material to be assigned to the design.
2. Attaching the material to the design.
3. Adding the light effect to the model by assigning lights at different locations.
4. Using the **RENDER** command to render the objects.

SELECTING AND ATTACHING THE MATERIALS

AutoCAD provides you with number of materials that can be attached to the design. By default, these materials are stored in *render.mli* library. You can also create your own library for storing these materials. Once you have selected the material, you can assign materials directly to the objects. You can also assign the material to the blocks, layers, and to an AutoCAD color index (ACI). The commands that are used to select and attach the materials are discussed next.

Selecting the Material for the Design

Toolbar:	Render > Materials Library
Menu:	View > Render > Materials Library
Command:	MATLIB

To get the realistic rendered image of the object, it is necessary to assign materials to the objects. Therefore, once you have created the design, you now have to select the material to be assigned to the objects in the design. By default, global material is assigned to all the objects. The color of this material depends upon the color of the object. However, if you want to assign some other materials to the design, you will have to select them from the material library and import them in the current drawing. The material library is invoked by choosing the **Materials Library** button from the **Render** toolbar. When you choose this button, the **Materials Library** dialog box will be displayed as shown in Figure 26-1.

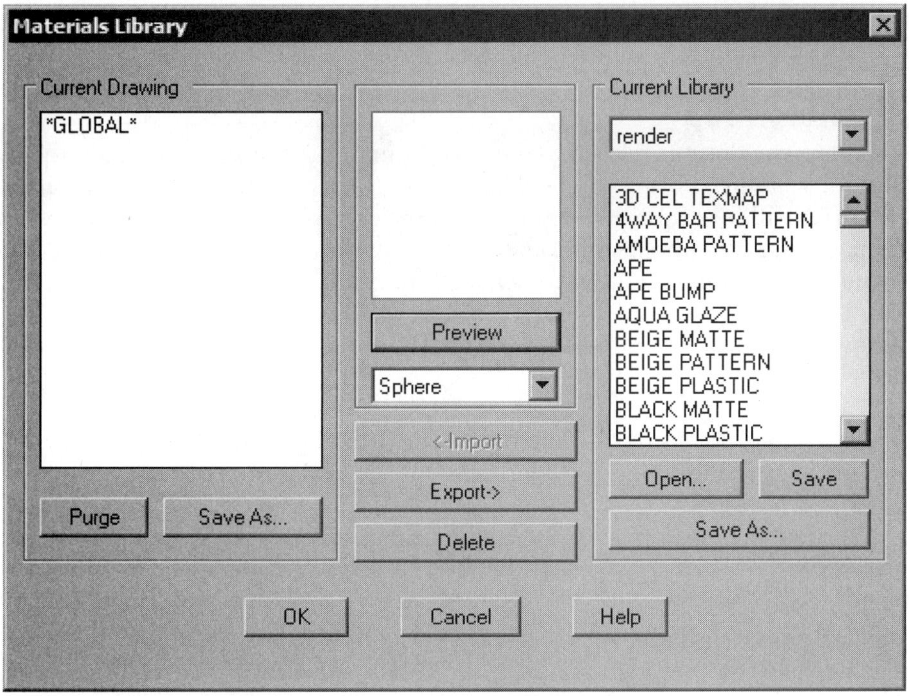

Figure 26-1 Materials Library *dialog box*

Current Drawing Area

This area provides you with the options related to the materials in the current drawing. The list box in the area initially displays only one material, *GLOBAL*. This material is by default assigned to all the new drawings. When you select a new material and import, it will be displayed in this list box.

Purge. This button is chosen to delete all the unused materials from the current drawing. If the list box displays any material that is not assigned to an object in the current drawing, it will be deleted when you choose this button. The **Confirm Purge** dialog box will be displayed to confirm the deletion of the unused material when you choose this button.

Save As. This button is chosen to save the list of the materials in the current drawing in a new material library. You can create your own library of the selected materials using this button. When you choose this button, the **Library File** dialog box will be displayed to save the library. The material library is saved in the *.mli* (materials library) format in the AutoCAD 2004 support directory.

Preview Area

This area displays a preview window, **Preview** button, and a drop-down list. You can preview the selected material in the preview window before importing it to the current drawing. Select the material from the list box under the **Current Library** or the **Current Drawing** area and then choose the **Preview** button to preview it. You can preview the material on a sphere or on a cube by selecting it from the drop-down list provided under this area.

Current Library Area

This area displays the options related to the current material library. By default, the materials saved under the *render.mli* library will be displayed in the list box of this area. All the available material libraries will be displayed in the drop-down list provided under this area. You can display the materials saved under some other library by selecting the library from the drop-down list. The material to be imported can be selected from the list box.

Open. Choose this button to open the **Library File** dialog box for selecting the material library file.

Save. Choose this button to save the changes made to the current material library. When you choose this button, the **Library File** dialog box will be displayed.

Save As. Choose this button to save the current material library with some other name.

Import

Choose this button to copy the selected material to the current drawing. You have to first select the material from the list box under the **Current Library** area and then choose this button. The material copied to the current drawing will be displayed in the list box of the current drawing area.

Export

Choose this button to add the material from the current drawing to the current library. If the selected material is already available in the current material library, the **Reconcile Exported Material Names** dialog box will be displayed as shown in Figure 26-2.

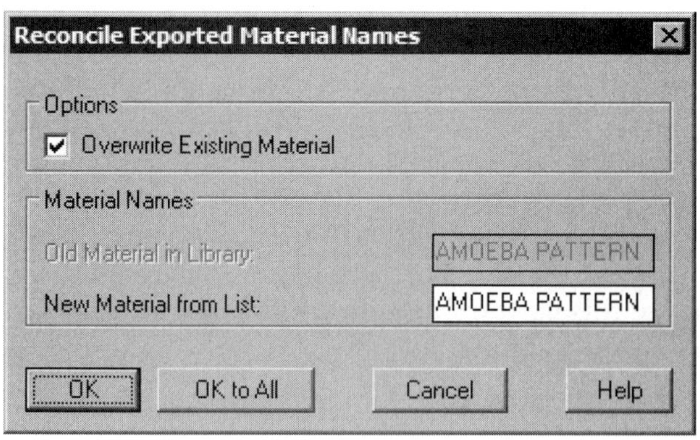

*Figure 26-2 **Reconcile Exported Material Names** dialog box*

Reconcile Exported Material Names Dialog Box Options

Overwrite Existing Material. This check box is selected to overwrite the existing material in the current material library. You can clear this check box to prevent the overwriting of the material.

Old Material in Library. This edit box displays the name of the old name of the material in the current material library that you are going to export. This edit box will be available only when you clear the **Overwrite Existing Material** check box.

New Material from List. This edit box is used to specify the name by which you want to save the selected material in the current library. Specifying a name in this edit box ensures that the current library now contains two similar materials but with different names.

Note
You cannot export the GLOBAL material that is displayed by default in the ***Current Drawing*** *area.*

Delete

This button deletes the selected material either from the current drawing or from the current library. If you delete a material from the current drawing that has been assigned to the objects in the drawing, a warning will be displayed that the selected material is attached in the current drawing.

Note
If you choose ***OK*** *without saving the changes made to the current material library, the* ***Library Modifications*** *dialog box will be displayed. This dialog box will inform you that the current material library has been changed. You will be given an option to save the changes in the current library, discard the changes, or cancel the closing of the* ***Materials Library*** *dialog box.*

Tip
After making the modifications in the current material library, choose the ***Save As*** *button from the* ***Current Library*** *to create another library. This way the materials in the current library are not modified.*

Attaching Selected Materials to the Objects

Toolbar:	Render > Materials
Menu:	View > Render > Materials
Command:	RMAT

Once you have selected the materials from the materials library, you have to attach them to the objects in the current drawing. This is done using the **RMAT** command. This command is also used to create a new material or modify the existing material. However, creating and editing material will be discussed later in this chapter.

This command can be invoked by choosing the **Materials** button from the **Render** toolbar. When you choose this button, the **Materials** dialog box will be displayed as shown in Figure 26-3.

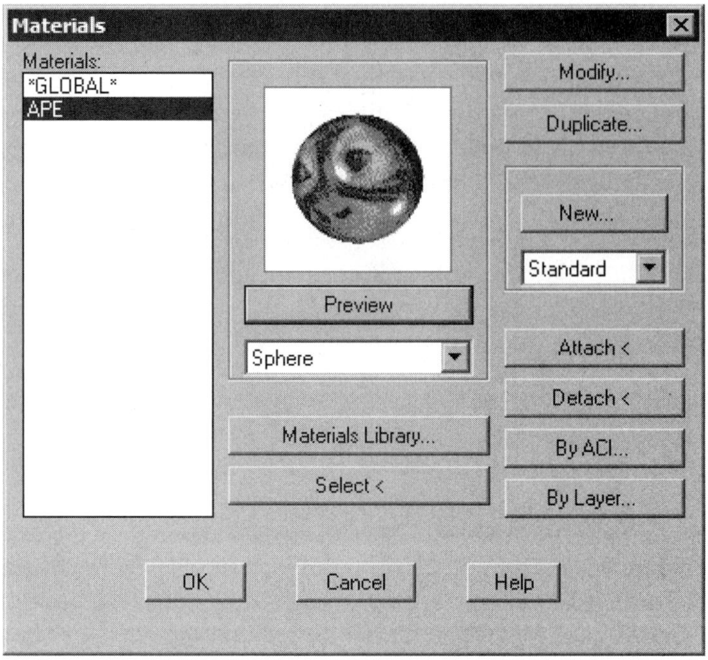

Figure 26-3 ***Materials*** *dialog box*

Materials Area

The list box provided under this area displays all the materials that are imported from the **Materials Library** dialog box. In addition to the imported materials, the default material is also displayed in this list box.

Preview Area

Using the options provided under this area you can preview the selected material. You have to first select the material from the list box provided under the **Materials** area and then choose the **Preview** button to preview the material. You can use a cube or a sphere to preview the material by selecting either of the two from the drop-down list provided under this area.

Attach

Choose this button to attach the selected material to the object in the current drawing. You have to first select the material from the **Materials** list box and then choose this button to attach the material. When you choose this button, the **Materials** dialog box is temporarily closed and you will be prompted to select the object from the drawing window to which the material has to be attached. Once you have selected all the objects to which the current material is attached, the **Materials** dialog box will be redisplayed.

Detach

This button detaches the selected material from the objects. You have to first select the material to be detached from the list box under the **Materials** area and then choose this button. The

Materials dialog box will be temporarily closed when you choose this button and you will be prompted to select the object from which you want to detach the selected material.

By ACI

Choose this button to assign the material using the AutoCAD Color Index numbers. This means that you can assign a particular material to an object of a specified color. Once a material has been assigned to a particular color, all the objects with that particular color will be rendered using the specified material. When you choose this button, the **Attach by AutoCAD Color Index** dialog box will be displayed, see Figure 26-4.

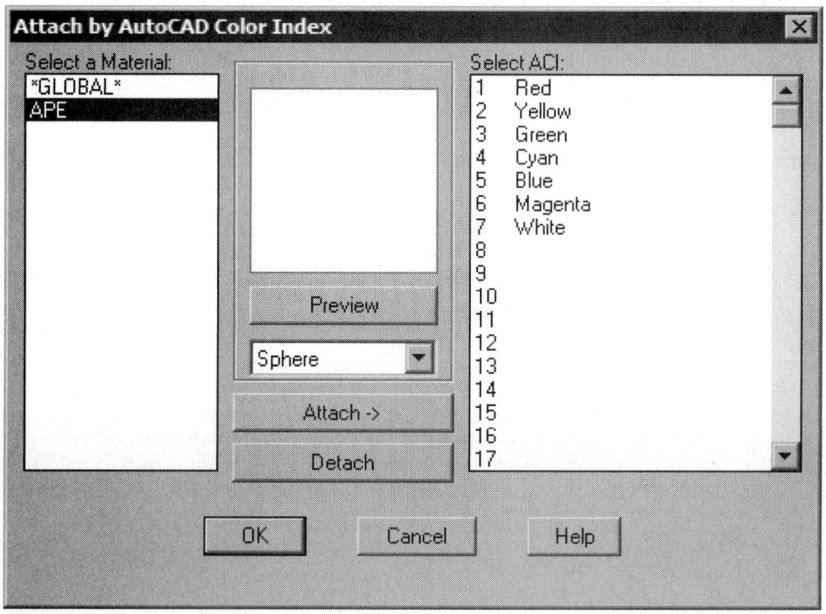

*Figure 26-4 **Attach by AutoCAD Color Index** dialog box*

This dialog box displays the materials in the current drawing under the **Select a Material** list box and all the 255 colors under the **Select ACI** list box. Select the material to be assigned from the **Select a Material** list box and the color to which the material has to be assigned from the **Select ACI** list box, and then choose the **Attach** button. The **Select ACI** list box will now display the color along with the material assigned to the color. Similarly, you can detach the material using the **Detach** button. You can also preview the material before attaching it to a color using the **Preview** area provided in this dialog box.

By Layer

This button is chosen to assign a particular material to a selected layer. This means that all the objects that are placed in the selected layer will be rendered using the specified material. When you choose this button, the **Attach by Layer** dialog box will be displayed as shown in Figure 26-5. This dialog box displays all the materials in the current drawing under the **Select**

a Material list box and the layers in the **Select Layer** list box. To assign a particular material to a layer, select the material from the **Select a Material** list box and the layer from the **Select Layer** list box, and then choose the **Attach** button. All the objects placed in the selected layer will now be rendered using the specified material. Similarly, you can detach the material from the selected layer. You can also preview the material before attaching it to the layer using the options provided under the **Preview** area.

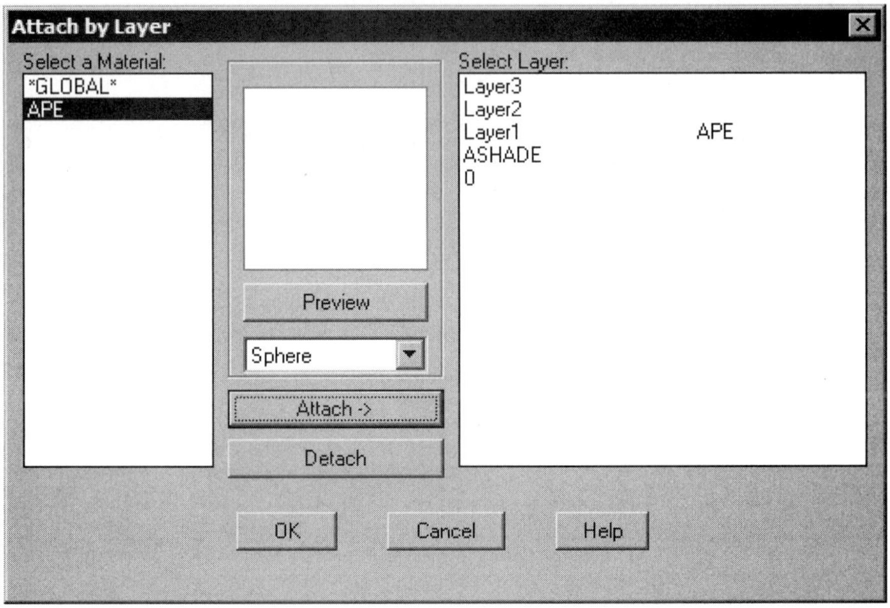

***Figure 26-5 Attach by Layer** dialog box*

Materials Library

Choose this button to invoke the **Materials Library** dialog box for importing the materials.

Tip

*You can invoke the **Materials Library** directly from the **Materials** dialog box by choosing the **Materials Library** button. This will save one step of invoking the **Materials Library** dialog box with the **MATLIB** command.*

Select

This button informs you of which material is attached to the selected object. When you choose this button, the **Materials dialog** box is temporarily closed and you will be prompted to select the object. As soon as you select the material, the **Materials dialog** box will be redisplayed and a message will be displayed below the **OK** button in this dialog box informing you about the material that is attached to the selected object.

Note
The remaining options for creating a material and editing the materials will be discussed later in this chapter.

UNDERSTANDING ELEMENTARY RENDERING

Toolbar:	Render > Render
Menu:	View > Render > Render
Command:	RENDER (RR)

In this section, you will learn about the elementary or the basic rendering. When you choose this button, the **Render** dialog box will be displayed, see Figure 26-6. There are basically three types of rendering. They are as follows.

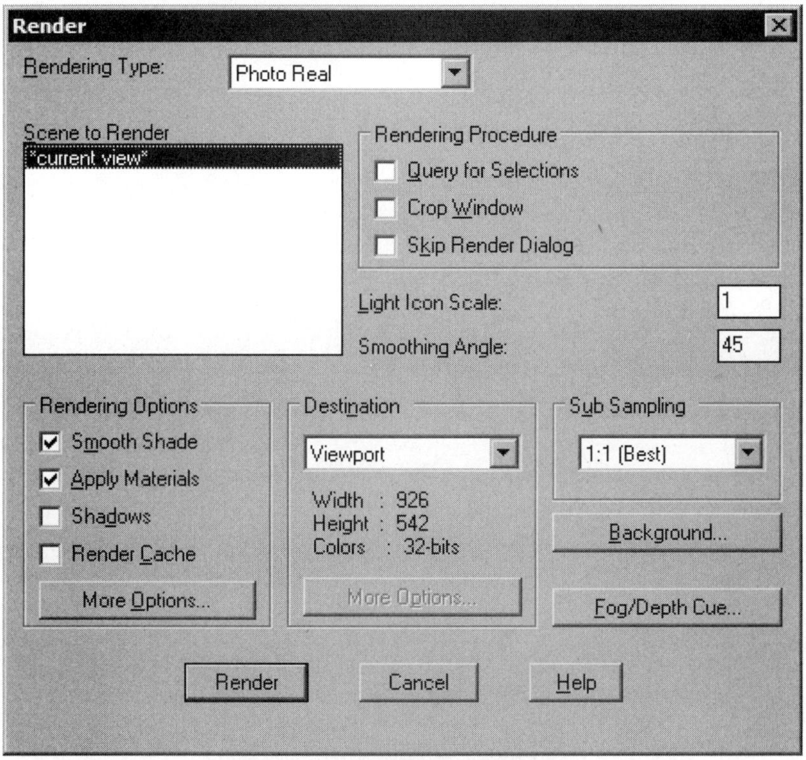

***Figure 26-6 Render** dialog box*

Render. This is the basic rendering option and renders the object without actually displaying the effect of the material. In this type of rendering, global material is attached to the model and rendered. The color of rendered image will depend on the color of the object. To render the design using this option, select it from the **Rendering Type** drop-down list in the **Render** dialog box and choose the **Render** button. Note that you cannot set the shadow option when you use this option and so it does not give a realistic effect to the model, see Figure 26-7.

Photo Real. This is the photorealistic rendering procedure. In this type of rendering, the specified material is attached to the model and rendered. You can view the bitmap images on the model and also view the transparent effect of the model if the material selected is transparent. If the lights are also added to the design, you can also view the shadows of the model. You can select this option from the **Rendering Type** drop-down list. Figure 26-8 shows a model to which material is applied, and then rendered using this option.

*Figure 26-7 Model rendered using the **Render** option*

*Figure 26-8 Model rendered using the **Photo Real** option displaying the effect of materials*

Photo Raytrace. This is also a photorealistic method of rendering. This method uses the ray tracing method for calculating the parameters such as the mirror effect, reflection, and refraction. If you have created your own material and provided the mirror effect to it, you can view the mirror effect only if you render using this option, see Figure 26-9.

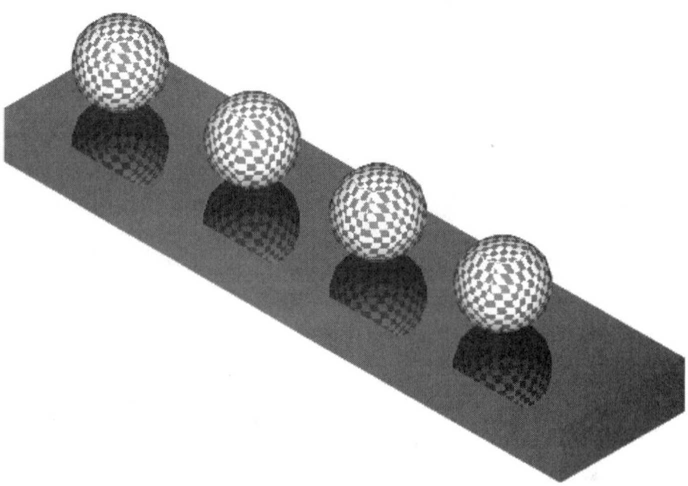

*Figure 26-9 Model rendered using the **Photo Raytrace** option showing the mirror effect*

Note

*The remaining options of the **Render** dialog box will be discussed in the **Understanding Advanced Rendering** section of this chapter.*

Example 1 *Architectural*

In this example, you will draw the staircase shown in Figure 26-10. You can assume your own dimensions. Attach the materials to the objects and then render the model.

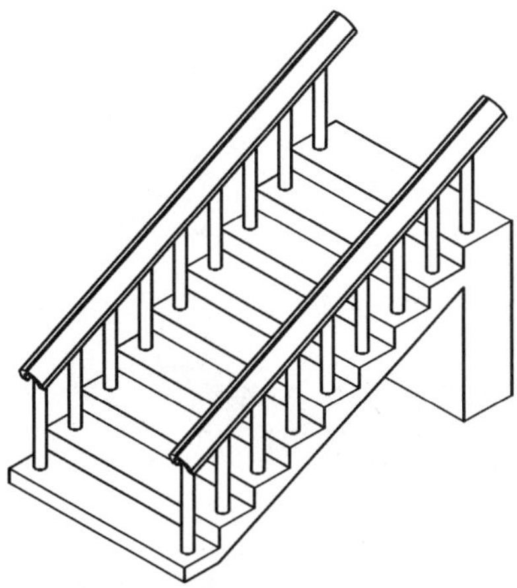

Figure 26-10 *Model for Example 1*

The materials that you have to assign to the model are given next.

Stairs	MARBLE PALE
Balusters	COPPER
Handrail	WOOD - WHITE ASH

1. Create the model shown in Figure 26-10. Assume your own dimensions.

2. Choose the **Materials** button from the **Render** toolbar to invoke the **Materials** dialog box. Choose the **Materials Library** button to invoke the **Materials Library** dialog box.

3. Select **MARBLE PALE** material from the **Current Library** list box and then choose the **Import** button. Similarly, import the other two materials. Choose **OK** to return to the **Materials** dialog box.

4. In the **Materials** dialog box, select the **MARBLE PALE** material from the **Materials** list box and then choose the **Attach** button. The **Materials** dialog box will be temporarily closed and you will be prompted to select the object to which you want to assign this material. Select the stairs and then press ENTER to redisplay the **Materials** dialog box.

Similarly, attach **COPPER** to all the balusters and **WOOD - WHITE ASH** to the handrails. Choose **OK** to exit the dialog box.

5. Choose the **Render** button from the **Render** toolbar to invoke the **Render** dialog box.

6. Select **Photo Real** from the **Rendering Type** drop-down list and then choose the **Render** button. The rendered image should look similar to the one shown in Figure 26-11.

Figure 26-11 *Rendered image of staircase*

ADDING LIGHTS TO THE DESIGN

Toolbar:	Render > Lights
Menu:	View > Render > Lights
Command:	LIGHT

Lights are vital to rendering a realistic image of an object. Without proper lighting, the rendered image may not show the features the way you expect. The lights can be added to the design using the **Lights** dialog box. This dialog box can be invoked by choosing the **Lights** button from the **Render** toolbar, see Figure 26-12.

AutoCAD render supports four light sources. All these light sources are discussed next.

Ambient Light

You can visualize ambient light as the natural light source that equally illuminates all surfaces of the objects, see Figure 26-13. Ambient light does not have a source, and therefore, has no location or direction. The ambient light is the default light that is applied to the design when you render it. You do not have to assign this light source to the design. However, you can modify the color and intensity of the ambient light using the options provided under the **Ambient Light** area of the **Lights** dialog box.

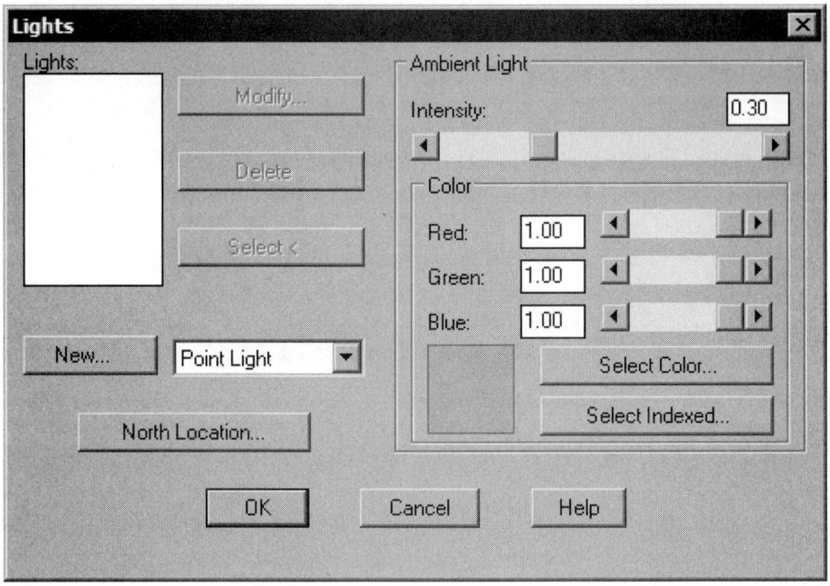

Figure 26-12 ***Lights*** *dialog box*

The slider bar provided under this area can be used to increase or decrease the intensity of the ambient light. By default, the color of the ambient light is white. You can also set the color of the light by selecting the color from the **Color** area. The two buttons provided under the **Color** area are used to create a custom color using the **Color** dialog box or select a standard color from AutoCAD Color Index. Normally, you should set the ambient light to a low value because high values gives a washed-out look to the image. If you want to create a dark room or a night scene, turn off the ambient light. With ambient light alone you cannot render a realistic image. Figures 26-14 and 26-15 show the same model rendered using different intensities of the ambient light.

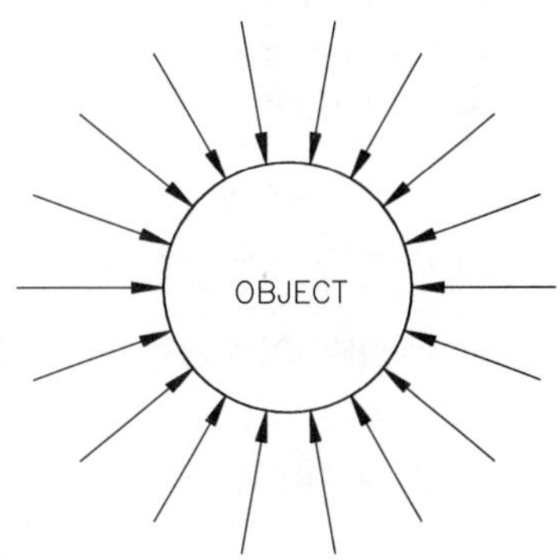

Figure 26-13 *Ambient light providing a constant illumination*

Point Light

A point light source emits light in all directions, and the intensity of the emitted light is uniform. You can visualize an electric bulb as a point light source. In AutoCAD render, if you

Figure 26-14 Model rendered with ambient light intensity = 0.3

Figure 26-15 Model rendered with ambient light intensity = 0.7

select to add a point light to the design, you can set the options of casting the shadow of the objects in the design. Figure 26-16 shows a point light source that radiates light uniformly in all directions.

As this is not a default light source, therefore, you will have to add this light source manually. This light source can be added by selecting **Point Light** from the drop-down list provided in front of the **New** button in the **Lights** dialog box and then choosing the **New** button. The **New Point Light** dialog box will be displayed, see Figure 26-17.

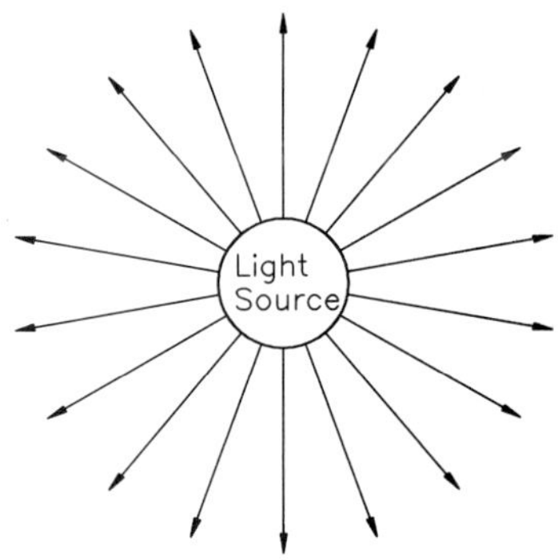

Figure 26-16 Point light source emitting light uniformly in all directions

New Point Light Dialog Box Options

Light Name

This edit box is used to specify the name of the point light. Specifying a light name is mandatory.

Intensity

This slider bar is used to specify the intensity of the point light. You can specify the intensity using the slider bar or directly entering the required value in the edit box.

Position Area

The options provided under this area are very important and are used to specify the location of the light source. This area provides you with the following two buttons:

Modify. Choose this button to specify the location of the point light source. The **New Point**

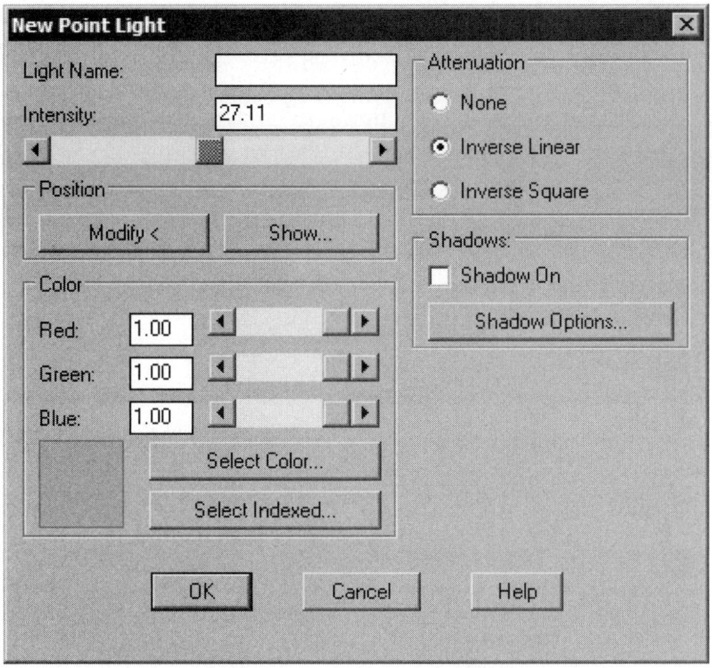

***Figure 26-17 New Point Light** dialog box*

Light dialog box will be temporarily closed and you will be prompted to specify the location of the light source. The **New Point Light** dialog box will be redisplayed once you specify the location.

Show. When you choose this button, the **Show Light Position** dialog box will be displayed as shown in Figure 26-18. This dialog box displays the *X*, *Y*, and *Z* coordinates of the light source and the target of the light. Since the point light source emits the light uniformly in all directions, there is no target for a point light.

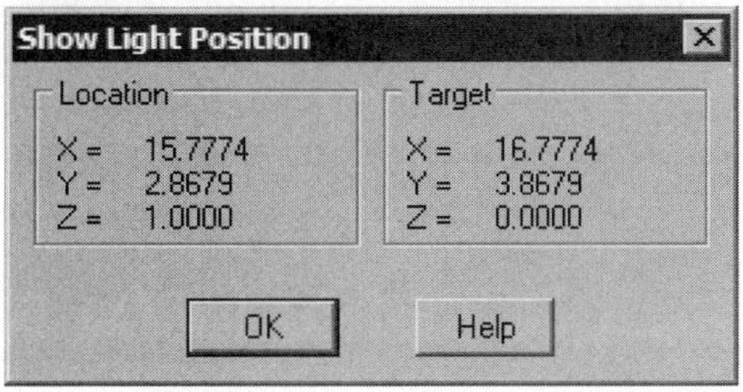

***Figure 26-18 Show Light Position** dialog box*

Note

A point light source emits light in all directions, so it has no fixed target. ***Target*** *area in the* ***Show Light Position*** *dialog box (Figure 26-18) is provided for a spotlight and distant light.*

Color Area

The options provided under this area are used to specify the color for the point light. You can define a custom color by using the RGB sliders provided under this area or by using the **Select Color** dialog box that is displayed when you choose the **Select Color** button. You can also select a color from the AutoCAD Color Index by choosing the **Select Indexed** button. The selected color will be displayed in the preview box provided under this area.

Attenuation Area

The light intensity is defined as the amount of light falling per unit of area. The intensity of light is inversely proportional to the distance between the light source and the object. Therefore, the intensity of light decreases as the distance increases. This phenomenon is called **Attenuation**. In AutoCAD, it occurs only with spotlights and point light.

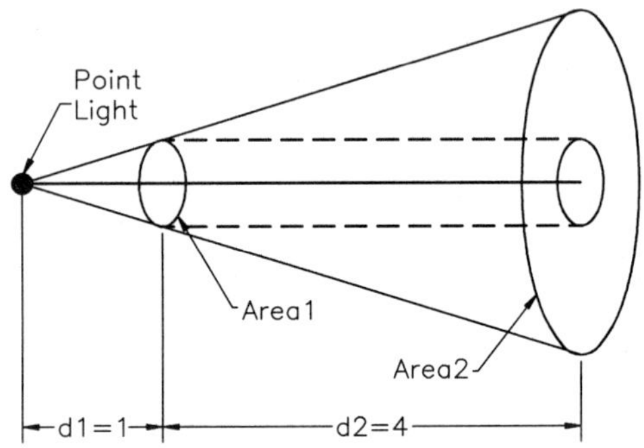

Figure 26-19 *The concept of attenuation*

In Figure 26-19, the light is emitted by a point source. Assume that the amount of light incident on Area1 is I. Therefore, the intensity of light on Area1 = I/Area. As the light travels farther from the source, it covers a larger area. The amount of light falling on Area2 is the same as on Area1, but the area is larger. Therefore, the intensity of light for Area2 is smaller (Intensity of light for Area2 = I/Area). Area1 will be brighter than Area2 because of higher light intensity. AutoCAD render provides you with the following three options for controlling the attenuation.

None. If you select the **None** option for light falloff, the brightness of objects is independent of distance. This means that objects that are far away from the point light source will be as bright as those close to the light source.

Inverse Linear. In this option, the light falling on the object (brightness) is inversely proportional to the distance of the object from the light source (Brightness = 1/Distance). As distance increases, brightness decreases. For example, assume that the intensity of the light source is **I** and the object is located at a distance of **2** units from the light source. Now, brightness or intensity = **I/2**. If the distance is 8 units, the intensity (light falling on the object per unit area) = **I/8**. The brightness is a linear function of the distance of the object from the light source.

Inverse Square. In this option, the light falling on the object (brightness) is inversely proportional to the square of the distance of the object from the light source (Brightness = 1/Distance^2). For example, assume that the intensity of the light source is **I** and the object is located at a distance of **2** units from the light source. Now, brightness or intensity = $\mathbf{I/(2)^2}$ = **I/4**. If the distance is 8 units, the intensity (light falling on the object per unit area) = $\mathbf{I/(8)^2}$ = **I/64**.

Shadows Area

The options provided under this area are used to control the display of the shadows of the objects after rendering. The shadows will be displayed only if the **Shadow On** check box provided under this area is selected. The other shadow related options can be controlled using the **Shadow Options** dialog box (Figure 26-20) displayed upon choosing the **Shadow Options** button provided under this area.

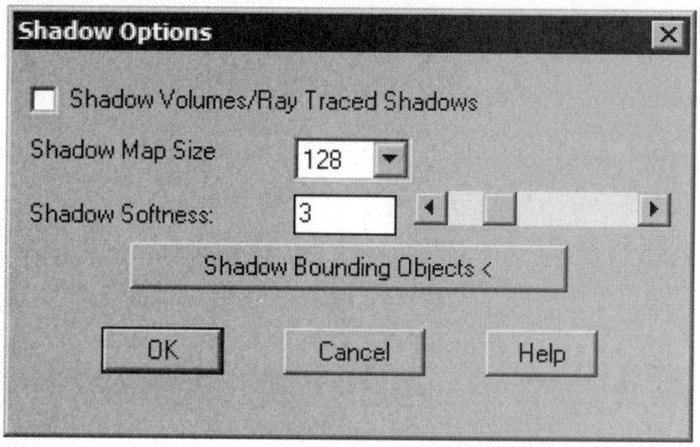

Figure 26-20 ***Shadow Options*** *dialog box*

Shadow Volumes/Ray Traced Shadows. If this check box is selected, the shadows that will be displayed will be volumetric and ray-traced for the photo real and photo ray-traced renders, respectively. Therefore, if this check box is selected, the remaining options of this dialog box will not be available.

Shadow Map Size. This drop-down list is used to select the map sizes for the shadows. The map sizes are calculated in terms of the pixels and the greater the number of pixels, the better the shadow will be. However, if the map size is large, the time taken to render the model will also increase. Figure 26-21 shows a model with shadow map size 128, and Figure 26-22 shows a model with shadow map size 1024.

Figure 26-21 *Shadow map size =128*

Figure 26-22 *Shadow map size =1024*

Shadow Softness. This slider bar is used to specify the softness for the shadow. If the value of this slider is increased, the shadow will be blurred.

Tip

You can view the best shadow effect by keeping the value of the shadow map size between 512 and 1024, and shadow softness between 3 and 4.

Shadow Bounding Objects. Choose this button to select the objects whose bounding boxes will be used to clip the shadows. When you choose this button, the **Shadow Options** dialog box will be temporarily closed and you will be prompted to select objects to clip the shadows.

Tip

You can remove the objects from the selection sets containing the objects for clipping the shadows by entering ***R*** *at the* ***Select the shadow bounding objects*** *prompt. You will then be prompted to select the objects to be removed.*

Example 2 — *Architectural*

In this example, you will apply the point light at the lintel of the window and at the lintel of the door of the model as shown in Figures 26-23 and 26-24. Choose your own materials to make the model look realistic and then render.

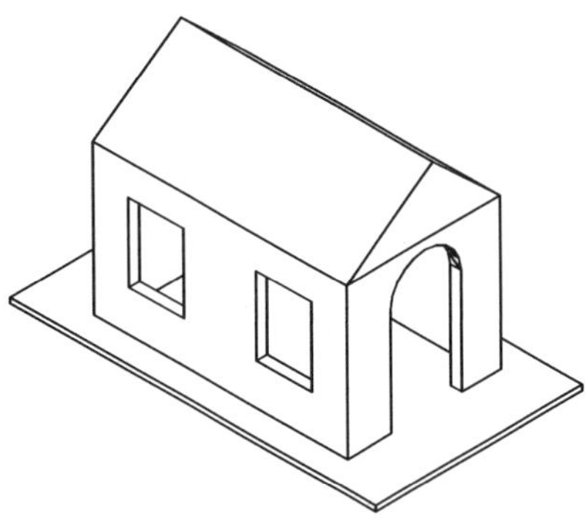

Figure 26-23 *Model for Example 2*

Figure 26-24 *Model after rendering*

1. Create the model as shown in Figure 26-23.

2. Choose your own materials and attach the materials to the model as described in Example 1. (Recommended: BROWN BRICK for Walls, SOUTH WEST PATTERN for flooring, and TILE WHITE for roof.)

3. Choose the **Lights** button from the **Render** toolbar to invoke the **Lights** dialog box.

4. Select **Point Light** from the drop-down list provided in front of the **New** button. Now, choose the **New** button to invoke the **New Point Light** dialog box.

5. Enter light name as **1** in the **Light Name** edit box. Select yellow color from the AutoCAD Color Index. Set the intensity based on the size of the objects in the drawing.

6. Choose the **Modify** button to set the position of the light **1**. Once you choose the **Modify** button, the dialog box will be exited and you will be prompted to specify the location of the light. Specify the location at the lintel of the window.

7. Select the **Shadows On** check box in the **Shadow** area. Choose **OK** to exit the dialog box.

8. Similarly, you can set the second light at the lintel of the door. Set the color of the light to green.

9. Choose the **Render** button in the **Render** toolbar to invoke the **Render** dialog box. Select **Photo Real** from the **Rendering Type** drop-down list. Select the **Shadows** check box from the **Rendering Options** area. The rendered image is shown in Figure 26-25.

Figure 26-25 *Rendered model of house*

Tip

*You can increase the smoothness of the rendered curved object by increasing the value of the **FACETRES** system variable. The default value of this variable is 0.5. Set this value close to 6.*

Spotlight

A spotlight emits light in the defined direction with a cone-shaped light beam. This light has a focused beam that starts from a point and is targeted at another point. This type of light is generally used to illuminate a specific point, such as illuminating a person on stage. The phenomenon of **attenuation** also applies to spotlights. This light can be applied by selecting **Spotlight** from the drop-down list provided in front of the **New** button and then choosing the **New** button. The **New Spotlight** dialog box (Figure 26-26) will be displayed when you choose this button.

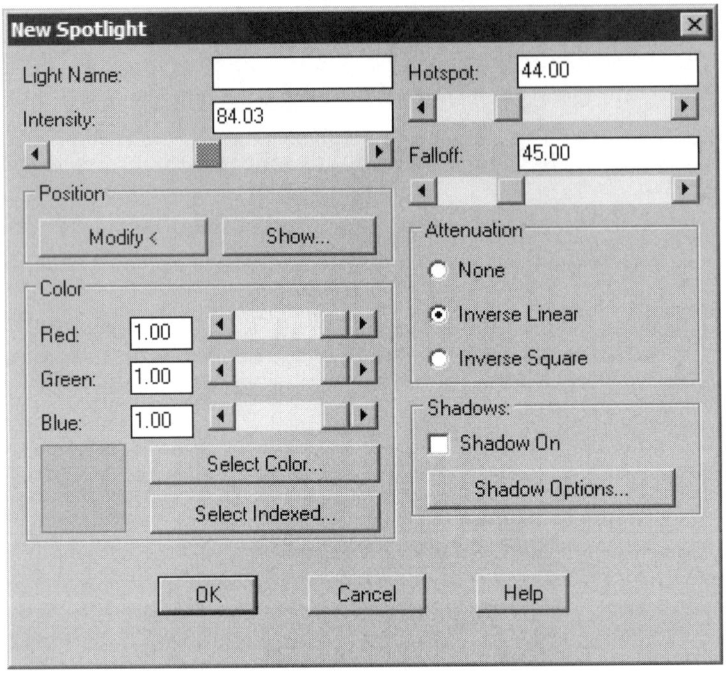

***Figure 26-26 New Spotlight** dialog box*

Most of the options under the **New Spotlight** dialog box are similar to those under the **New Point Light** dialog box. One of the differences is that when you choose the **Modify** button from the **Position** area, you will be prompted to enter two values. As already mentioned, this type of light source has a start point and a target point; therefore, you will be first prompted to specify the light target and then the light location.

Hotspot/Falloff

As mentioned earlier, the spotlight has a focused beam of light that is targeted at a particular point. Therefore, there are two cones that comprise the spotlight.

The hotspot is defined as the cone that carries the highest intensity light beam. In this cone, the light beam is most focused and is defined in terms of an angle, see Figure 26-27. The value of this angle can be adjusted with the help of the slider bar associated with it.

The other cone is called falloff and it specifies the full cone of light. It is the area around the hotspot, where the intensity of the beam of light is not very high as shown in Figures 26-27(a) and 26-27(b). It is also defined in terms of an angle that can be adjusted with the help of the slider bar associated with it. The value of the hotspot and the falloff can vary from 0 to 160.

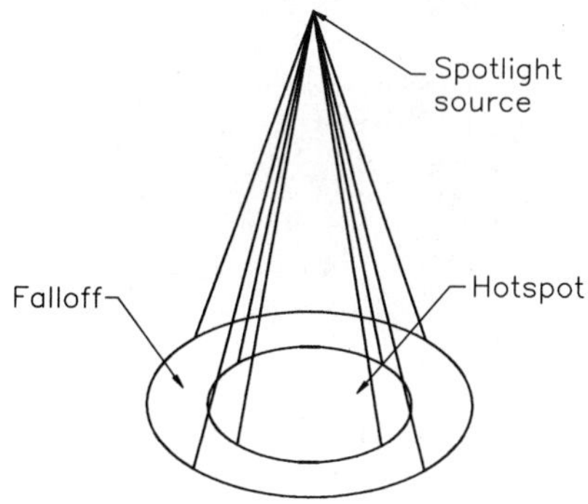

Figure 26-27(a) *The hotspot and falloff cones*

Figure 26-27(b) *The hotspot and falloff after rendering*

Tip

The value of hotspot should always be less than the falloff. If you have set the value of hotspot more than the value of falloff, then upon exiting the dialog box you will get an error message that ***Hotspot must be < or = Falloff****.*

Distant Light

A distant light source emits a uniform parallel beam of light in a single direction only (Figure 26-28). The intensity of the light beam does not decrease with the distance. It remains constant. For example, the sun rays can be assumed to be a distant light source because the light rays are parallel. When you use a distant light source in a drawing, the location of the light source does not matter; only the direction is critical. Distant light is used mostly to light objects or a backdrop uniformly and for getting the effect of sunlight. The distant light can be added by selecting **Distant Light** from the drop-down list provided in front of the **New** button and then choosing the **New** button.

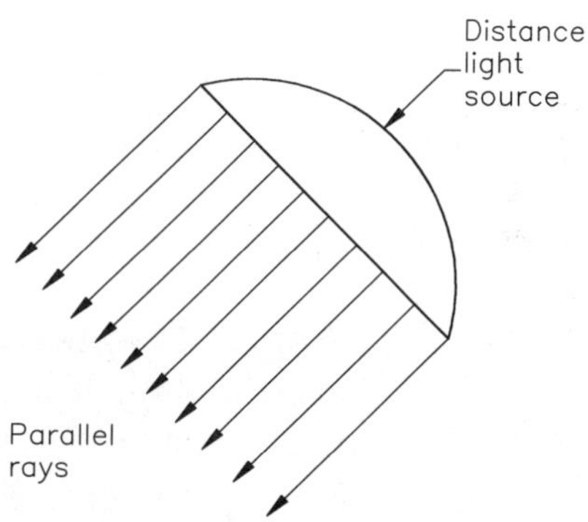

Figure 26-28 *Distant light source*

The **New Distant Light** dialog box will be displayed as shown in Figure 26-29. The options in this dialog box are similar to those under the **New Point Light** dialog box, except for a few additions. This dialog box allows you to define the distant light using the following two methods:

Using Azimuth and Altitude

You can set the **Azimuth** and **Altitude** of the distant light direction by moving the slider bars

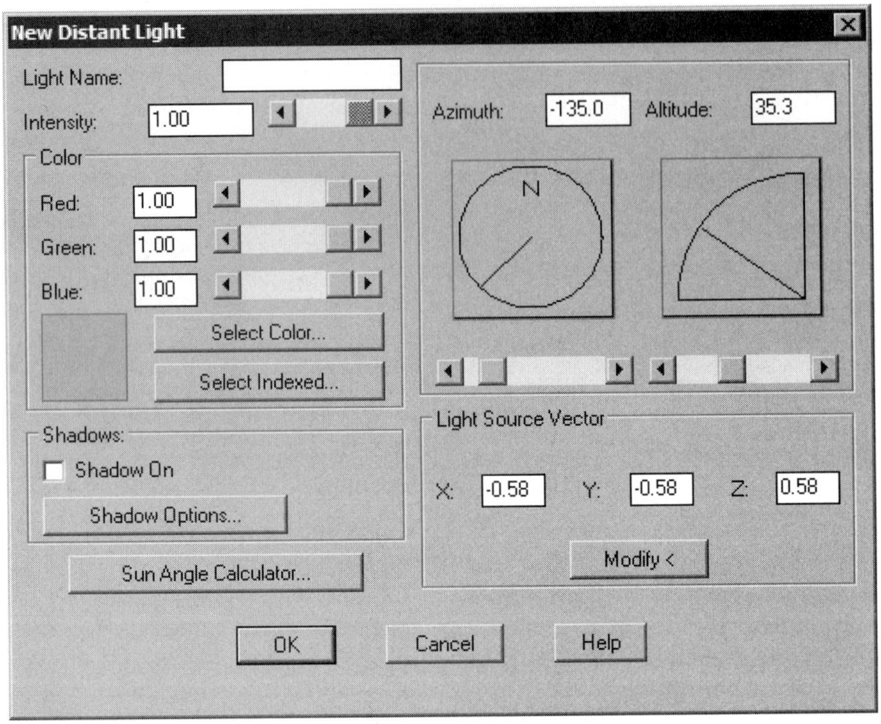

Figure 26-29 ***New Distant Light*** *dialog box*

associated with them. You can also set the angles by moving the arms of the angle in the **Azimuth** and **Altitude** image boxes. You can change the default light source vector by specifying the coordinates in the **X**, **Y**, and **Z** edit boxes under the **Light Source Vector** area.

Using the Sun Angle Calculator

This is the second and the simpler method of defining the distant light. With this method you can define a light anywhere in the world by defining its geographic location. You can also define the date and time at which you want to view the light effect. To use this method, choose the **Sun Angle Calculator** button from the **New Distant Light** dialog box to display the **Sun Angle Calculator** dialog box (Figure 26-30). This dialog box is divided into two areas. The left area describes the local settings. The right area describes the solar settings. Depending upon the local settings, solar setting will adjust itself.

In the local setting, you can set the date and clock time in the **Date** and **Clock Time** edit boxes, respectively. You can also set the latitude and longitude in the **Latitude** and **Longitude** edit boxes, respectively, and the directions as north, south, east, and west from **North** and **West** edit boxes. By default the *Y* axis is assumed north by AutoCAD.

You can also specify the geographic location of the place where you want to view the light effect. This is done by choosing the **Geographic Location** button in the **Sun Angle Calculator**

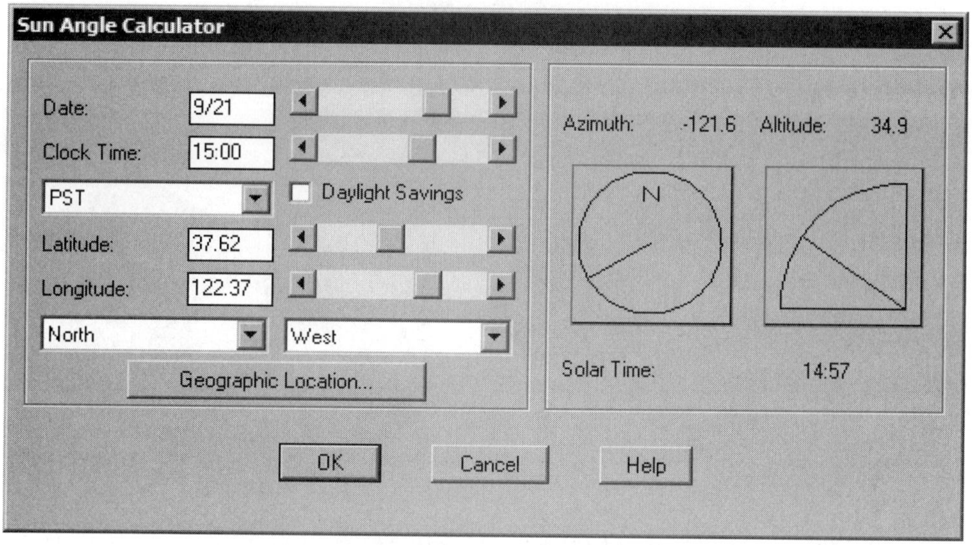

*Figure 26-30 **Sun Angle Calculator** dialog box*

dialog box. The **Geographic Location** dialog box will be displayed when you choose this button, see Figure 26-31.

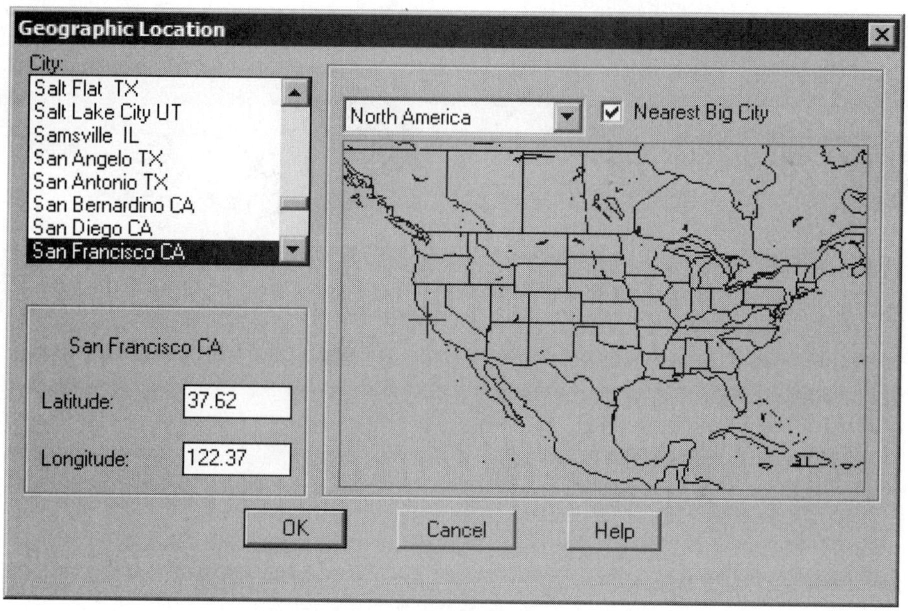

*Figure 26-31 **Geographic Location** dialog box*

You can select various locations from the drop-down list provided in this dialog box and the map of this location will be displayed. Also, all the nearest cities will be displayed in the **City** list box. You can also move the cursor in the world map area and it will show the city and the corresponding latitude and longitude. Once you have selected the desired city, you can go back to the **Sun Angle Calculator** dialog box and set the date and time. Figures 26-32 through 26-37 show the position of the shadows at different times.

Figure 26-32 *City: Chicago, Time: 11:00, Date: 1st May*

Figure 26-33 *City: Chicago, Time: 16:00, Date: 1st May*

Figure 26-34 *City: Tokyo, Time: 11:00, Date: 1st May*

Figure 26-35 *City: Tokyo, Time: 16:00, Date: 1st May*

Figure 26-36 *City: New Delhi, Time: 11:00, Date: 1st May*

Figure 26-37 *City: New Delhi, Time: 16:00, Date: 1st May*

Example 3 *Architectural*

In this example, you will attach distant light to the model shown in Figures 26-38 and 26-39. Choose your own materials. Set the distant light using the geographic location. The geographic location and other parameters that have to be set are given next.

City: New York, Date: 20 June, Time: 14:00.

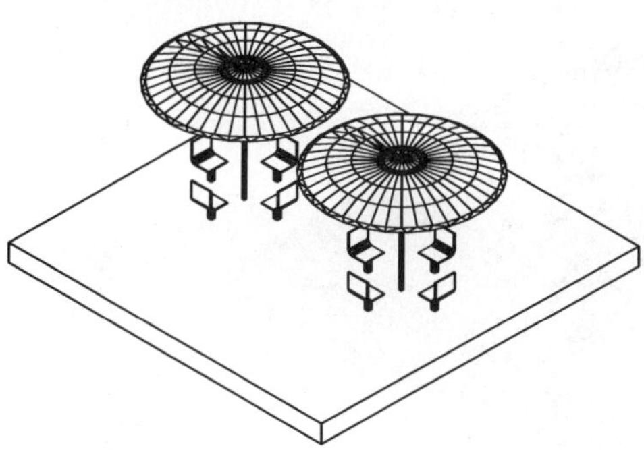

Figure 26-38 *Model for Example 3 before rendering*

Figure 26-39 *Model for Example 3 after rendering*

1. Create the model as shown in Figure 26-38.
2. Choose the materials and attach it to the model as described in Example 1.
3. Choose the **Light** button from the **Render** toolbar to invoke the **Lights** dialog box. Select **Distant Light** from the drop-down list and then choose the **New** button to invoke the **New Distant Light** dialog box.

4. Specify the name of the distant light as **D1** in the **Light Name** edit box. Select the **Shadow On** check box from the **Shadows** area. You can also specify the details of the shadow by choosing the **Shadow Options** button. Choose the **Sun Angle Calculator** button to invoke the **Sun Angle Calculator** dialog box.

5. In this dialog box, choose the **Geographic Location** button to invoke the **Geographic Location** dialog box. Select **North America** from the drop-down list provided above the map. Select **New York NY** from the **City** drop-down list. You can also select the city by clicking in the map of North America displayed in the **Geographic Location** dialog box. Choose **OK** to exit the dialog box.

6. In the **Sun Angle Calculator** dialog box, choose **MST** from the **PST** drop-down list. Set the date and clock time as **6/20** and **14:00** in the **Date** and **Clock Time** edit boxes, respectively. Choose **OK** to exit the **Sun Angle Calculator** dialog box. Choose **OK** to exit the **New Distant Light** dialog box. Once more, choose **OK** to exit the **Lights** dialog box.

7. Choose the **Render** button from the **Render** toolbar to invoke the **Render** dialog box. Select the **Photo Real** from the **Rendering Type** drop-down list. Select the **Shadows** check box from the **Rendering Options** area of the **Render** dialog box.

8. Choose the **Render** button to render the model and exit the dialog box. The final rendered model is shown in Figure 26-40.

Figure 26-40 *Final model for Example 3 after rendering*

MODIFYING LIGHTS

As mentioned earlier, the lights and lighting effects are important in rendering to create a realistic representation of an object. The sides of an object that face the light must appear brighter and the sides that are on the other side of the object must be darker. This smooth gradation of light produces a realistic image of the object. If the light intensity is uniform over the entire surface, the rendered object will not look realistic. For example, if you use the **SHADE** command to shade an object, that object does not look realistic because the displayed model lacks any gradation of light. Any number of lights can be added to the drawing. The color, location, and direction of all the lights can be specified individually. As mentioned before, you can specify attenuation for point lights and spotlights. AutoCAD also allows you to change the color, position, and intensity of any light source. The only limitation is that light types cannot be interchanged. For example, you cannot change a distant light into a point light. Any light can be modified by choosing the **Lights** button from the **Render** toolbar. The **Lights** list box will display all the lights that have been added to the drawing. Select the light you want to modify from this list box and then choose the **Modify** button. Depending upon the type of light selected, the dialog box will be displayed. For example, if you select a distant light for modifications, the **Modify Distant Light** dialog box will be displayed. One such **Modify Distant Light** dialog box is shown in Figure 26-41.

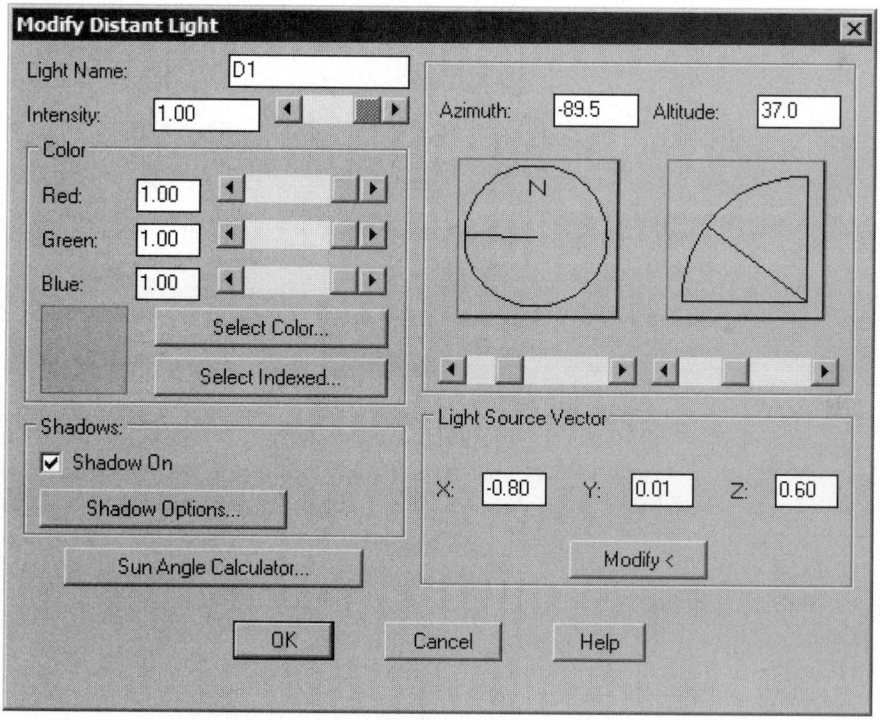

***Figure 26-41 Modify Distant Light** dialog box*

The options of the **Modify Light** dialog box are similar to those under the **New Light** dialog box.

CREATING NEW MATERIALS

AutoCAD allows you to create your own materials. This is done using the **Materials** dialog box. You can create four types of materials: **Standard**, **Granite**, **Marble**, and **Wood**. To create any of these materials, choose it from the drop-down list provided below the **New** button in the **Materials** dialog box and then choose the **New** button. The dialog box corresponding to the material that you have selected from the drop-down list will be displayed. Figure 26-42 shows a dialog box for creating a standard material.

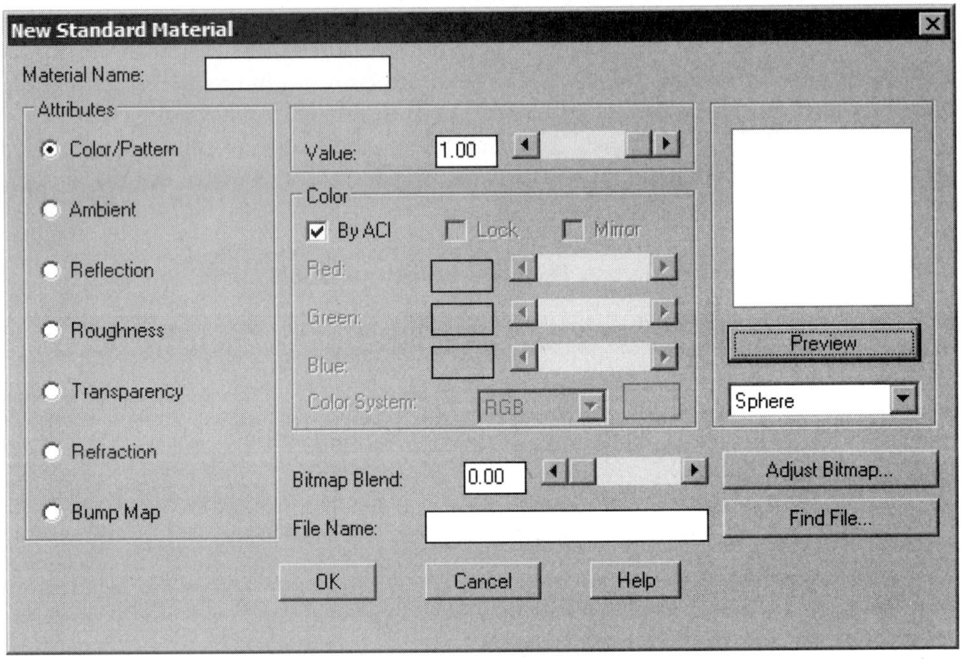

*Figure 26-42 **New Standard Material** dialog box*

You can define various parameters for the new material using this dialog box. To provide the mirror effect, select the **Mirror check** box that will be available when you select the **Reflection** radio button from the **Attributes** area. You can control the value of the reflection of the new material using the **Value** slider bar. You can also attach a bitmap image to the material by choosing the **Find File** button. When you choose this button, the **Bitmap File** dialog box will be displayed that allows you to select the bitmap file.

Tip

*AutoCAD has stored its bitmap images in the Textures subdirectory of the AutoCAD 2004 directory. These bitmap images are in the *.tga format. Therefore, to select any of these images, you will have to first select ***.tga** from the **Files of type** drop-down list in the **Bitmap File** dialog box.*

Note

*You can create a material similar to the original AutoCAD material. This is done by selecting the original material from the **Materials** area and then choosing the **Duplicate** button.*

MODIFYING THE MATERIAL

You can also modify AutoCAD materials using the **Materials** dialog box. To modify a material, select the material from the **Materials** area in the **Materials** dialog box and then choose the **Modify** button. When you choose this button, the **Modify Material** dialog box will be displayed. This dialog box will depend upon the type of material that is selected to be modified.

UNDERSTANDING ADVANCED RENDERING

Toolbar:	Render > Render
Menu:	View > Render > Render
Command:	RENDER

As mentioned earlier, render allows you to control the appearance of the objects in the design. This is done by defining the surface material and reflective quality of the surface and by adding lights to get the desired effects. The basic rendering has already been discussed earlier in this chapter, and now you will learn about the advanced rendering. You will also learn about the other options of the **Render** dialog box shown in Figure 26-43.

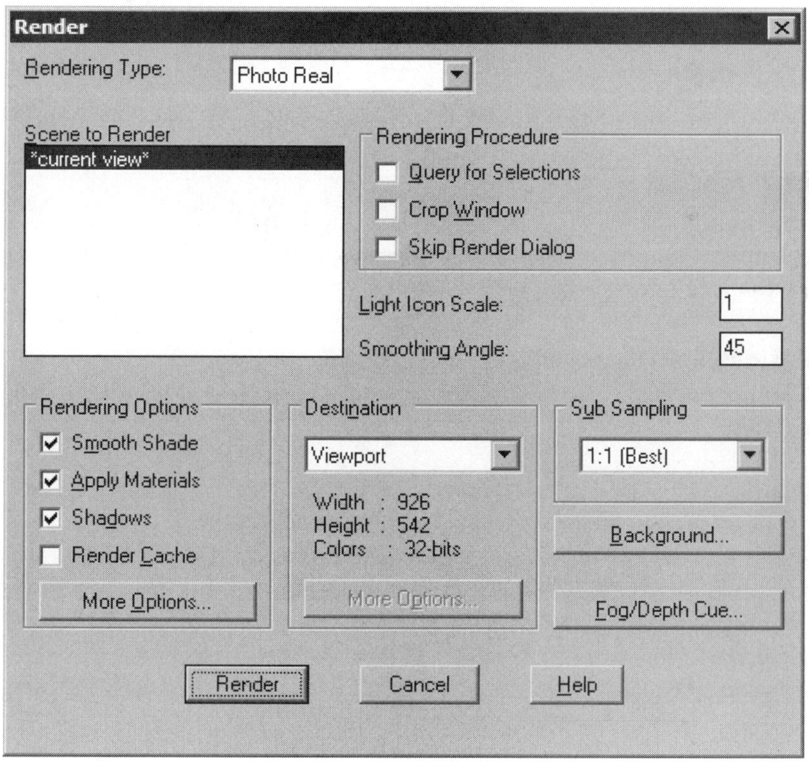

Figure 26-43 ***Render*** *dialog box*

Rendering Procedures Area

The **Rendering Procedure** area of the **Render** dialog box has the following options.

Query for Selections

If you select the **Query for Selections** check box, AutoCAD prompts you to select objects to render. Only the objects that you have selected will now be rendered. This way you can avoid rendering all the objects drawn on the screen and render only the desired objects. This reduces the time it takes to render large drawings.

Crop Window

If you select this check box, AutoCAD will prompt you to define a window. Only the portion of the design that lies completely inside the window will be rendered.

Skip Render Dialog

If you choose the **Skip Render Dialog** option, the current view is rendered without displaying the Render dialog box, and previous settings of render are taken into consideration. To invoke this dialog box again, when you invoke the **RENDER** command, you need to enter **RPREF** (Render preferences) at the Command prompt and clear this check box.

Light Icon Scale

The **Light Icon Scale** edit box can be used to set the size of the light blocks in the drawing.

Smoothing Angle

The **Smoothing Angle** option lets you specify the angle defined by two edges. The default value for smoothing an angle is 45-degree. Angles less than 45-degree make the edges smooth. Angles greater than 45-degree are taken as edges.

Rendering Options Area

Various rendering options are provided under the **Rendering Options** area of this dialog box. These are described next.

Smooth Shade

The **Smooth Shade** option allows you to smooth the rough edges. If this option is enabled, the rough-edged appearance of a multiface surface is smoothed. This option determines that the surface normals and colors across two or more adjoining faces are blended.

Apply Materials

The **Apply Materials** option allows you to assign surface materials to objects. If this check box is cleared, the objects in the drawing are assigned the ***GLOBAL*** material.

Shadows

When you select this option, AutoCAD generates shadows. The effect of shadows would not be visible if this check box is cleared even if the shadow option of the **Lights** dialog box is enabled. This option applies only to Photo Real and Photo Ray trace rendering.

Render Cache

Selecting this option results in writing rendering information to a temporary cache file that is stored in the memory of the current drawing. This file can then be used for the coming renderings. This eliminates the need for AutoCAD to recalculate all the objects for rendering. This saves time, especially when rendering solids.

More Options

If you choose the **More Options** button, the **Render Options** dialog box is displayed. The options in this dialog box depend on the type of rendering selected from the **Rendering Type** drop-down list. These dialog boxes provide the options to improve render quality.

Destination Area

The **Destination** area of the **Render** dialog box allows you to specify the destination for the rendered image output. The destination here refers to the kind of output of the rendering you desire. The kind of output can be selected from the drop-down list provided under this area. This drop-down list provides you with the following options.

Viewport

If you select the **Viewport** option, AutoCAD renders the object in the current viewport and the output is displayed in the drawing area. This is the default option.

Render Window

This option is selected to render the design to a window. When you select this option and choose the **Render** button, a window is opened displaying the rendered design. This option is generally used for plotting the rendered design. You can control the quality and size of the output using various buttons provided in this window.

File

The **File** option lets you output the rendered image to a file. When you select this option, the **More Options** button is activated. You can choose this button to specify the type and the size of output file. After setting all the options, when you choose the **Render** button, the **Rendering File** dialog box will be displayed. You can specify the name and the directory for the output file using this dialog box.

More Options

This button will be available only if you select **File** from the drop-down list provided under this area. You can use the **More Options** button to set the configuration for the output file through the **File Output Configuration** dialog box (Figure 26-44).

File Type Area. In this area, you can specify the type and rendering resolution of the output file. The file formats allowed are BMP, PCX, Postscript, TGA, and TIFF. The screen resolution can also be specified in this area.

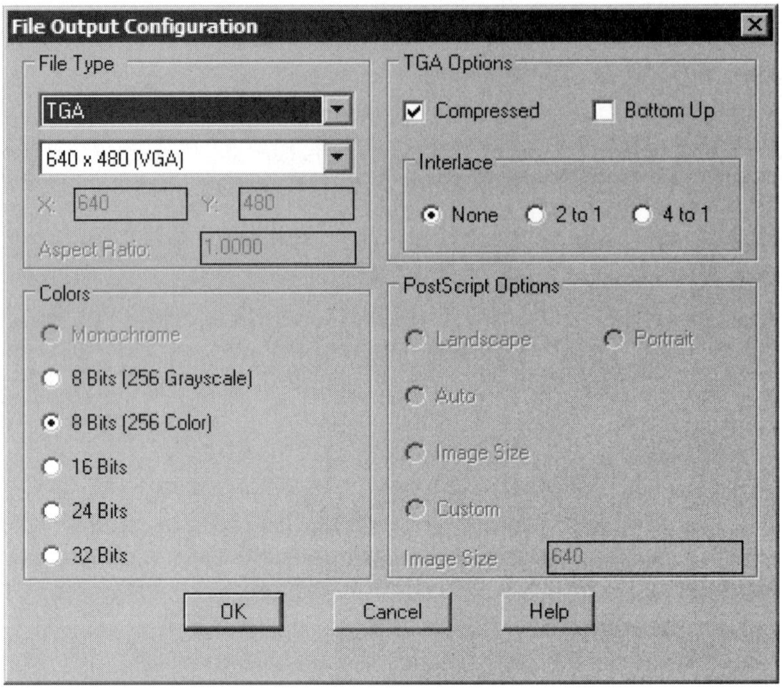

*Figure 26-44 **File Output Configuration** dialog box*

Colors Area. The colors in the output file can be specified in this area. The type of output file will determine the options that will be available in this area.

TGA Options Area. The **Compressed** option lets you specify compression for those file types that allow compression. The **Bottom Up** option lets you specify the scan line start point as bottom left instead of top left.

Interlace Area. Selecting the **None** option turns off line interlacing. Selecting the other two options turn on the interlacing.

PostScript Options Area. The **Landscape** and **Portrait** options in this area specify the orientation of the file. The **Auto** option automatically scales the image. The Custom option sets the image size in pixels. The **Image Size** edit box uses the explicit image size.

Sub Sampling

The **Sub Sampling** drop-down list controls the quality and time of rendering by reducing the number of pixels to be rendered. A **1:1** ratio (default) produces the best quality rendering but takes maximum time. If you are just testing a rendering, you can change the value to reduce rendering time. Figure 26-45 shows a model rendered with the ratio set to 1:1, and Figure 26-46 shows a model rendered with the ratio set to 6:1.

Figure 26-45 Model rendered keeping the ratio to 1:1

Figure 26-46 Model rendered keeping the ratio to 6:1

Background

You can choose the **Background** button in the **Render** dialog box to invoke the **Background** dialog box (Figure 26-47). This dialog box can also be displayed using the **BACKGROUND** command. The background option allows you to add a background to a rendering.

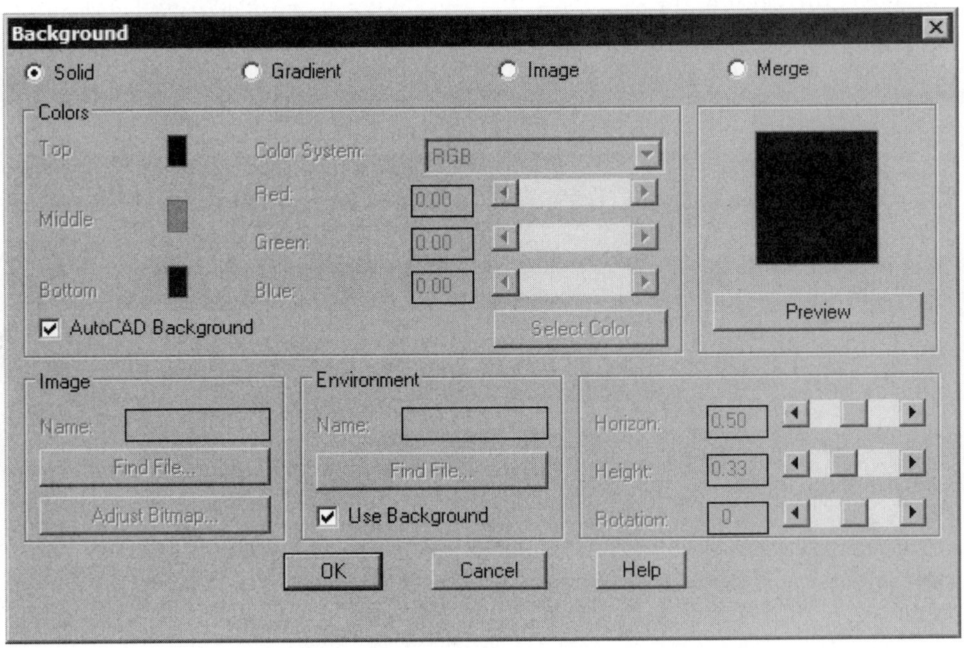

***Figure 26-47 Background** dialog box*

Solid

This option is selected to set a solid background for the rendering. By default, the solid AutoCAD background is selected. However, if you clear the **AutoCAD background** check box, the **Colors** area is available. In this area, you can specify a color for the background. Note that the background will display only a single color when you render using this option.

Gradient

Here you can specify background in the graded pattern of colors. You can specify colors from the **Colors** area by selecting different options. You can set the color using the **RGB** or **HLS** color patterns. You can also select the colors using the **Select Custom Color** button to assign it to the **Top**, **Middle**, and **Bottom** color swatch, and can see the preview by selecting the **Preview** button. Figure 26-48 shows a rendering with **Gradient** background.

Apart from the color area, there are three more buttons that are activated when you select this option. These buttons are **Horizon**, **Height**, and **Rotation**. The **Horizon** button represents the percentage of unrotated height. The **Height** button represents the percentage of the second color in a three-color pattern. The **Rotation** button will rotate the background pattern through the angle specified.

Image

This option is used to specify a bitmap file that will be used as a background for the rendering. When you select this button, the options under the **Image** area are activated. You can choose the **Find File** button to display the **Background Image** dialog box. This dialog box can be used to select the image file for the background. As mentioned earlier, AutoCAD image files are saved in the *Textures* subdirectory in the **.tga* format. You can select any of these files or other bitmap files using this dialog box, see Figure 26-49.

Figure 26-48 *Rendering with gradient background*

Figure 26-49 *Rendering with image as the background*

Merge

This option will merge the previous background settings with the current one. The previous background settings may be the solid, gradient, or image. To visualize it, draw a 3D model and then render it with the help of the solid, gradient, or image option of the **Background** dialog box. Without regenerating the model, now move this model to some distance and then use the **Merge** option of the **Background** dialog box to render the model again. You will see that the same model is visible in two different positions. This is because the **Merge** option has merged the previous rendering with the current one.

Environment Area

This option specifies an environment that creates more reflection and refraction effects on objects with reflective ray-traced materials. You can specify the environment with the help of the **Find File** button that displays the **Raytraced Environment Image** dialog box. Here you can specify the file for the environment.

Fog / Depth Cue

You can use the **Fog** option to add a misty effect to a rendering. You can also assign a color to the fog. When you choose the **Fog** button, the **Fog / Depth Cue** dialog box is displayed as shown in Figure 26-50. This dialog box can also be invoked using the **FOG** command. Select the **Enable Fog** check box to enable the fog. Enabling the fog does not affect the other

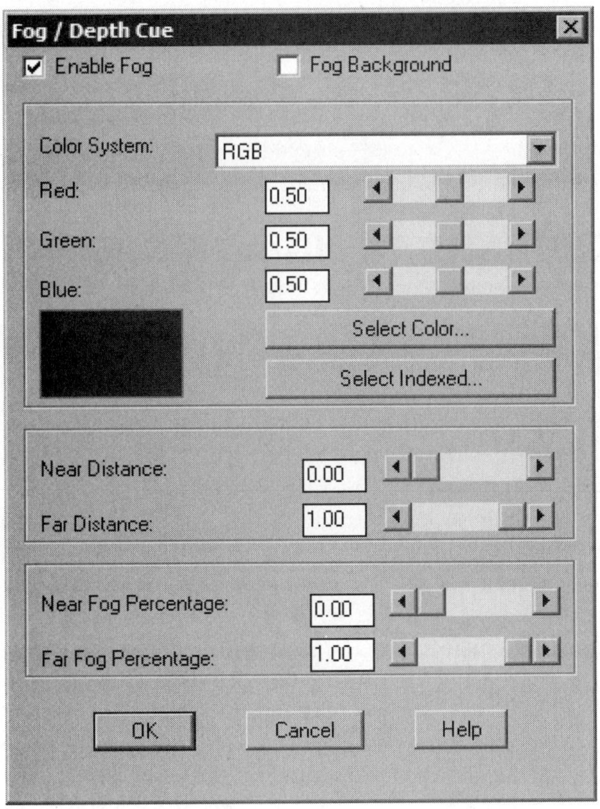

***Figure 26-50 Fog / Depth Cue** dialog box*

settings of the rendering. Choose the **Fog Background** check box to enable the fog to the model as well as background. With the help of the **Color System** you can assign color to the fog by various options specified in the dialog box. **Near Distance** and **Far Distance** options define where the fog begins and ends. The values specified in the **Near Distance** and **Far Distance** edit boxes are the percentage of the distance between the camera and the back clipping plane. The **Near Fog Percentage** and **Far Fog Percentage** edit boxes describe the percentage of fog at the near and far distances. Figure 26-51 shows a rendering displaying the fog effect.

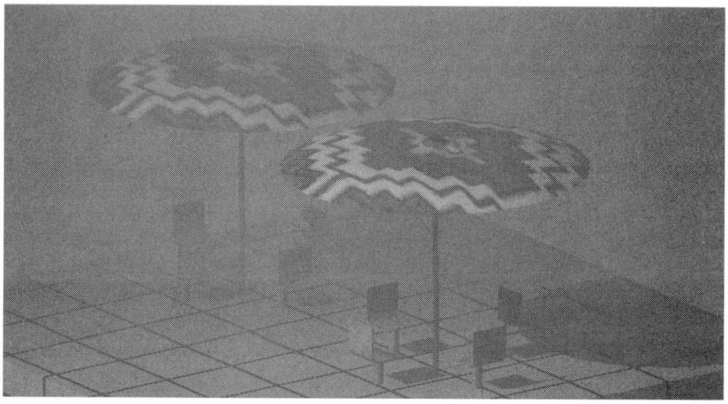

Figure 26-51 *A design rendered with the fog effect*

DEFINING AND RENDERING A SCENE

Toolbar:	Render > Scenes
Menu:	View > Render > Scene
Command:	SCENE

The rendering depends on the view that is current and the lights that are defined in the drawing. Sometimes the current view or the lighting setup may not be enough to show all features of an object. You might need different views of the object with a certain light configuration to show different features of the object. But when you change the view or define the lights for a rendering, the previous setup is lost. To avoid this, you can save the rendering information in a **scene**. For each scene, you can assign a view and the lights. When you render a particular scene, AutoCAD Render uses the view information and the lights that were assigned to that scene. It ignores the lights that were not defined in the scene. Defining scenes makes it convenient to render different views with the required lighting arrangement. The scenes that you create are displayed under the **Scene to Render** area of the **Render** dialog box. When you choose the **Scene** button from the **Render** toolbar, the **Scenes** dialog box will be displayed as shown in Figure 26-52. To create a new scene, choose the **New** button from the **Scenes** dialog box to display the **New Scene** dialog box as shown in Figure 26-53.

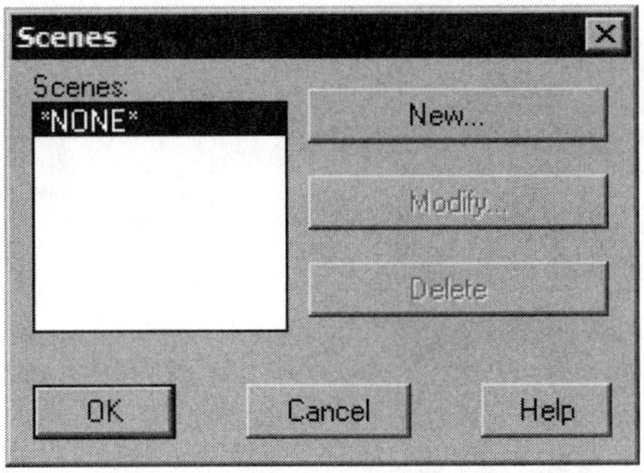

***Figure 26-52 Scenes** dialog box*

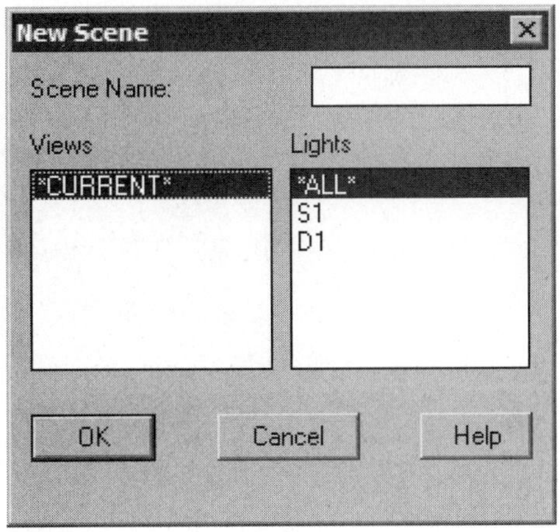

***Figure 26-53 New Scene** dialog box*

The **New Scene** dialog box will display all the views and lights available in the current drawing. You can specify the name of the current scene and select the view and lights that you want to display in the current scene. Now, choose **OK** from the **New Scene** dialog box to go back to the **Scenes** dialog box. This dialog box will now display the new scene in the **Scenes** area. Similarly, you can modify or delete an existing scene using this dialog box.

Example 4

In this example, you will open the model of Example 2 of this chapter and then create two scenes with the names **SCENE1** and **SCENE2**. The first scene should display the effect of light 1 and the second scene should display the effect of light 2.

1. Open the drawing of Example 2.

2. Choose the **Scenes** button from the **Render** toolbar to invoke the **Scenes** dialog box.

3. Choose the **New** button in the **Scenes** dialog box to invoke the **New Scene** dialog box.

4. Enter **SCENE1** in the **Scene Name** area and then select **1** from the **Lights** area. Choose **OK** to go back to the **Scenes** dialog box.

5. Again choose the **New** button to display the **New Scene** dialog box. Enter **SCENE2** in the **Scene Name** edit box. Select **2** from the **Lights** area. Now, choose **OK** to return to the **Scenes** dialog box. You can now see both scenes in the **Scene** area of the **Scenes** dialog box.

6. Choose **Render** button from the **Render** toolbar to invoke the **Render** dialog box. Select **SCENE1** from the **Scene to Render** area. Choose the **Render** button to render the model with the effect of light 1 only as shown in Figure 26-54.

Figure 26-54 *Rendering* ***SCENE1***

7. Similarly, render the drawing by selecting **SCENE2** from the **Scene to Render** area of the **Render** dialog box to view the effect of light 2, see Figure 26-55.

Figure 26-55 *Rendering* ***SCENE2***

Tip

If you want to see the effects of all the lights simultaneously on the model, choose the ***Current View*** *in the* ***Scene to Render*** *area in the* ***Render*** *dialog box.*

DEFINING AND RENDERING THE LANDSCAPING

Toolbar:	Render > Landscape New
Menu:	View > Render > Landscape New
Command:	LSNEW

The landscaping objects are the objects with a bitmap image attached to them. You can attach the landscaping objects to your model to give it a natural look. The display of the landscaping object depends on whether you choose one or two faces for it and whether it is view aligned. You make these choices according to your rendering requirements. When you choose this button, the **Landscape New** dialog box will be displayed as shown in Figure 26-56.

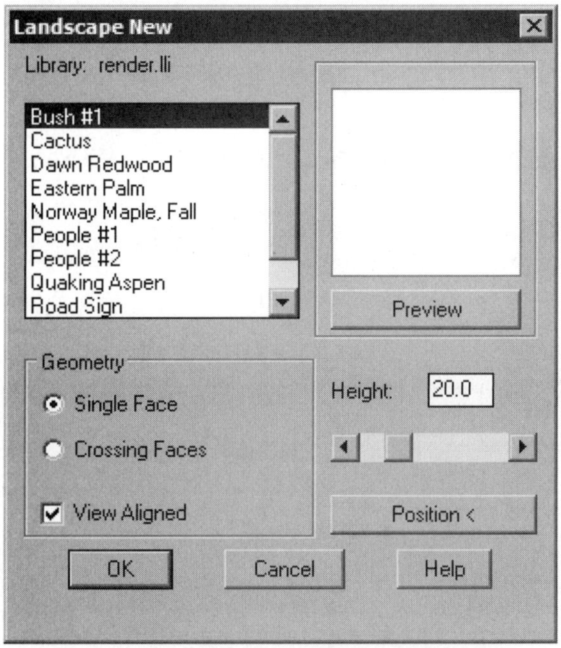

***Figure 26-56 Landscape New** dialog box*

This dialog box displays all the landscaping objects that are stored in the *render.lli* library. You can select the desired landscaping object from the list box of the **Landscape New** dialog box. To place a landscaping object, choose the **Position** button. The **Landscape New** dialog box will be temporarily closed and you will be prompted to specify the location for the object. Once you have defined the location, the dialog box will be redisplayed on the screen. The height of the landscaping object can be controlled using the **Height** spinner. You can define a single face landscaping object or a multiface landscaping object. You can also define a view aligned landscaping object. This object will be displayed as it is even if you change the viewpoint of the drawing. Once you have added the landscaping objects to the drawing, render it to view all the landscaping objects.

Figure 26-57 shows a rendering with the landscaping objects.

Figure 26-57 *Rendered scene showing the landscaping objects*

Example 5

In this example, you will attach the landscaping objects to the model of Example 3.

1. Open the drawing of Example 3. Using the **3DORBIT** command, change the orientation of the model and then zoom it, see Figure 26-58.

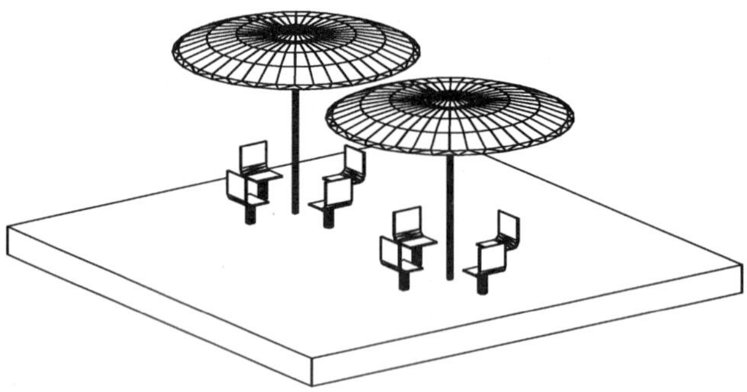

Figure 26-58 *Model after rotating and zooming*

2. Choose the **Landscape New** button from the **Render** toolbar to invoke the **Landscape New** dialog box.

3. In this dialog box, choose **Eastern Palm** from the list box. Choose the **Position** button to place it in the drawing. You will be prompted to choose the **Location of the base of the object**. Specify the location close to the left corner. The **Eastern Palm** landscape object will be placed there.

4. Similarly, place the **Eastern Palm** at the other two corners leaving the bottom left corner as shown in Figure 26-59.

5. Again, using the **Landscape New** dialog box, place **Sweetgum Summer** in the drawing as shown in Figure 26-60.

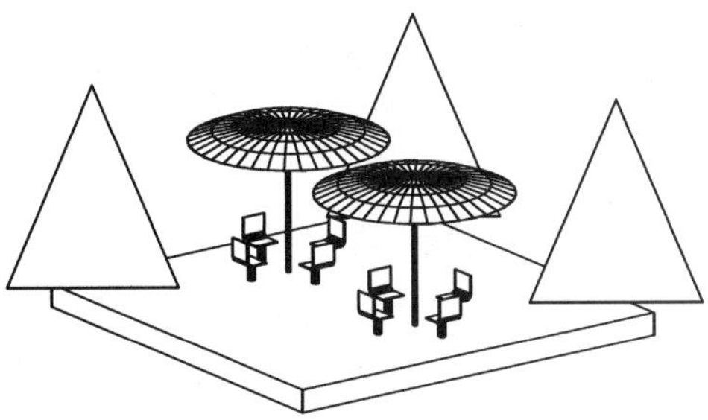

Figure 26-59 *Model after adding* ***Eastern Palm***

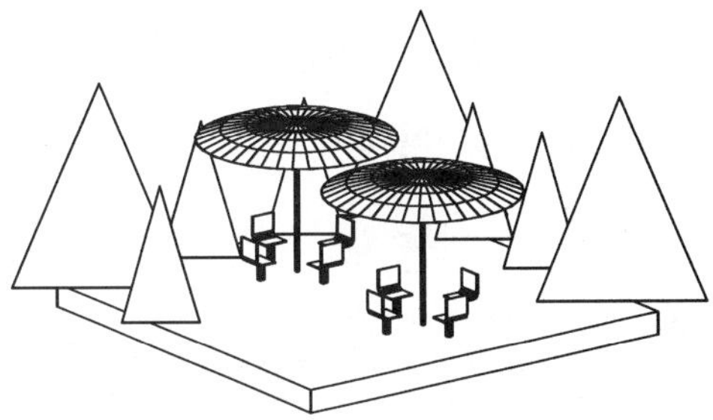

Figure 26-60 *Model after adding* ***Sweetgum Summer***

6. Render the drawing using the **Render** button. The drawing after rendering should look similar to the one shown in Figure 26-61.

Figure 26-61 *Model after rendering*

EDITING THE LANDSCAPING OBJECTS

Toolbar:	Render > Landscape Edit
Menu:	View > Render > Landscape Edit
Command:	LSEDIT

The landscaping objects placed in the drawings can also be edited. You can modify the height and location of the landscaping object or change the geometry. However, you cannot replace one landscaping object by the other. When you choose this button, you will be prompted to select the landscaping object to be edited. The **Landscape Edit** dialog box will be displayed when you select the object. The options under this dialog box are similar to those under the **Landscape New** dialog box.

CREATING USER-DEFINED LANDSCAPING OBJECTS

Toolbar:	Render > Landscape Library
Menu:	View > Render > Landscape Library
Command:	LSLIB

By default, *render.lli* contains the standard landscaping objects. In case you want to create your own landscaping objects, you will have to invoke the **Landscape Library** dialog box. This dialog box is invoked by choosing the **Landscape Library** button from the **Render** toolbar. The new landscaping object can be created by choosing the **New** button from the **Landscape Library** dialog box. You can save the new landscaping object in the same library or create a new library using the **Save** button.

MAPPING THE MATERIALS ON THE OBJECTS

Toolbar:	Render > Mapping
Menu:	View > Render > Mapping
Command:	SETUV

Mapping here is defined as the method of adjusting the coordinates and bitmap of the pattern of the material attached to the solid model. Mapping is generally required when the pattern of the material is not properly displayed on the object after rendering. This can be due to the sizes of the objects. For example, Figure 26-62 shows a model drawn within the limits of 12.00, 9.00. This model has **Brown Brick** attached to it. You can clearly see that the brick pattern is not clear after rendering. Figure 26-63 shows a model drawn with limits 100.00, 100.00 and with the same material. In this model, the brick pattern is clear. Therefore, to display the brick pattern in the smaller box, you need to adjust the coordinates and bitmap of the material.

The mapping is done with the help of the **Mapping** dialog box. To display this dialog box, choose the **Mapping** button from the **Render** toolbar. You will be prompted to select objects. Select the objects to be mapped to display the **Mapping** dialog box, see Figure 26-64.

This dialog box allows you to specify the projection of the material in planar, cylindrical,

Figure 26-62 Brick pattern on the model drawn within 12,9 limits

Figure 26-63 Brick pattern on the model drawn with limits 100,100

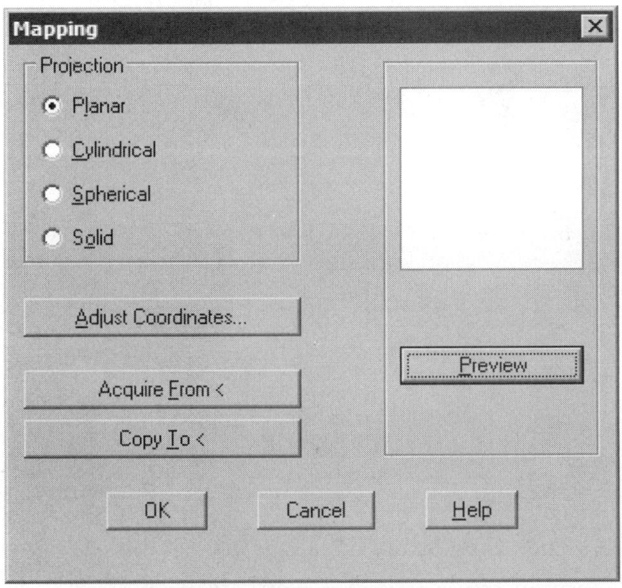

***Figure 26-64 Mapping** dialog box*

spherical, and solid forms. Figure 26-65 shows a model with planar projection and Figure 26-66 shows a model with cylindrical projection.

You can acquire the mapping from an existing object and copy on the selected object. You can preview the model in the preview window by choosing the **Preview** button.

You can also adjust the projection coordinates of the material with respect to the model by choosing the **Adjust Coordinates** button. When you choose this button, the **Adjust Coordinates**

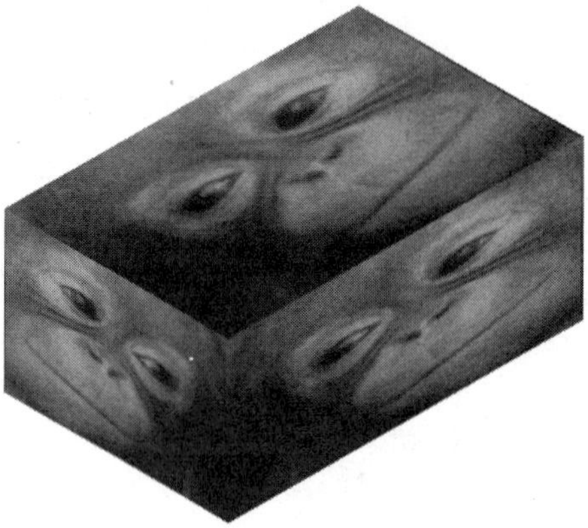

Figure 26-65 Model with planar projection

Figure 26-66 Model with cylindrical projection

dialog box will be displayed. This dialog box will depend upon the type of projection selected from the **Projection** area of the **Mapping** dialog box. Figure 26-67 shows the dialog box for the cylindrical projection type.

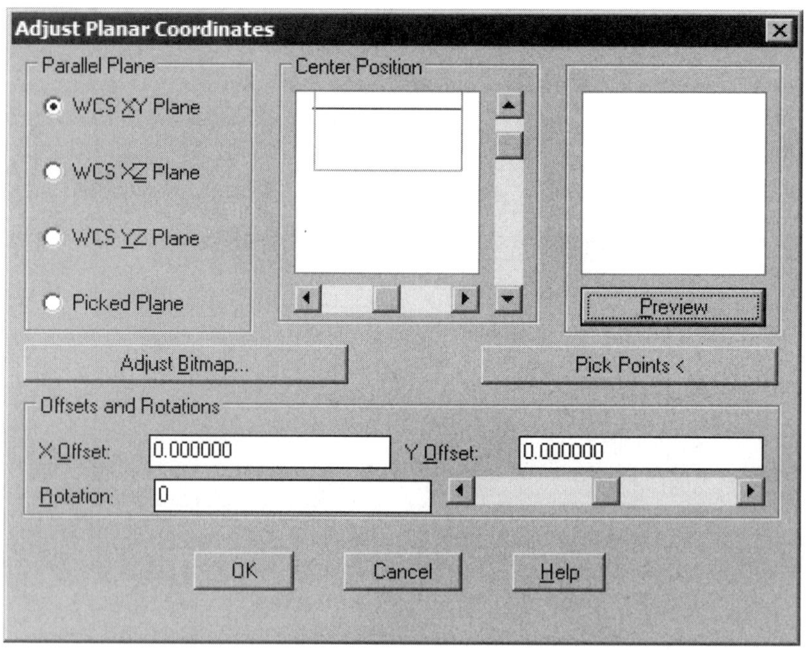

***Figure 26-67 Adjust Cylindrical Coordinates** dialog box*

The options under the **Adjust Coordinates** dialog box allow you to map the material with respect to the specified planes. You can also specify the offset and rotation for the pattern of the material along the specified axis. To adjust the bitmap image of the material, choose the

Adjust Bitmap button. When you choose this button, the **Adjust Object Bitmap Placement** dialog box will be displayed, see Figure 26-68. The options under this dialog box allow you to change the scale factor of the bitmap image of the material. You can also specify the offset values for the bitmap image of the material using this dialog box.

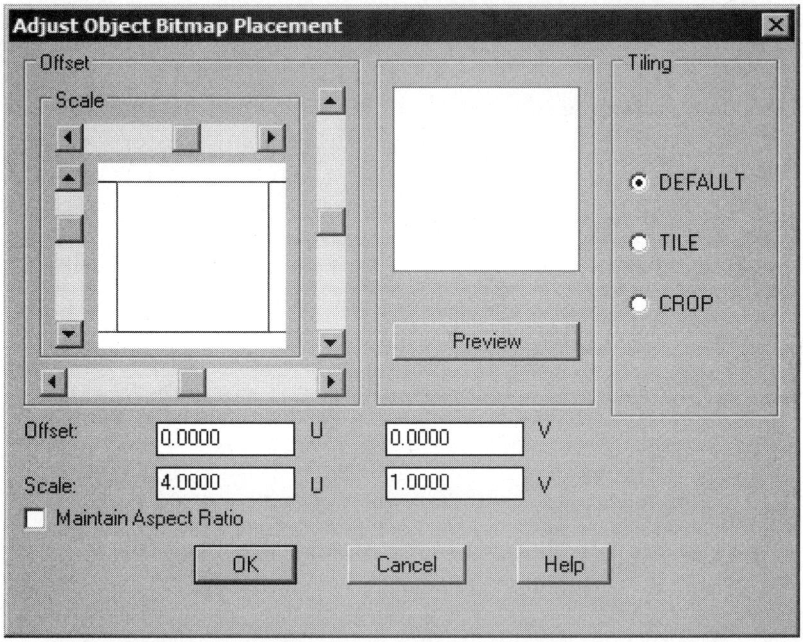

***Figure 26-68 Adjust Object Bitmap Placement** dialog box*

Figures 26-69 and 26-70 show the models rendered with different U and V scale factors set using the **Adjust Object Bitmap Placement** dialog box.

***Figure 26-69** Model rendered with U scale = 8, V scale = 2*

***Figure 26-70** Model rendered with U scale = 4, V scale = 1*

OBTAINING RENDERING INFORMATION

Toolbar:	Render > Statistics
Menu:	View > Render > Statistics
Command:	STATS

This command is used to obtain information about the last rendering. When you choose this button, the **Statistics** dialog box will be displayed as shown in Figure 26-71.

This dialog box provides the information about the name of the scene, the rendering type used, the time taken to produce the rendering, the number of faces processed by the rendering, and the number of triangles processed by the rendering. The information contained in the dialog box cannot be edited. However, the information can be saved to a file by selecting the **Save Statistics to File** check box. The name of the file can be specified in the edit box or you can use the **Find File** button. If a file by the specified name already exists, AutoCAD adds the present information to that file. You can use the **EDIT** function of any text editor to read the file.

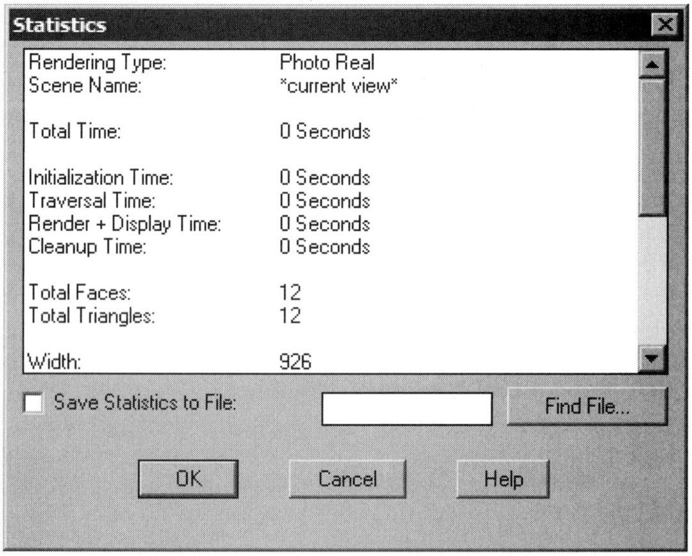

***Figure 26-71* Statistics** *dialog box*

SAVING A RENDERING

A rendered image can be saved by rendering to a file or by rendering to the screen and then saving the image. Redisplaying a saved rendering image requires very less time compared to the time involved in rendering. Various methods of saving the rendered image are discussed next.

Saving the Rendered Image to a File

You can save a rendered image directly to a file. One of the advantages of saving the rendered image is that you can redisplay the rendered image in less time as compared to again rendering the image. Another advantage is that when you render to the screen, the resolution of the rendering is limited by the resolution of your current display. However, if you render to a file, you can specify a higher resolution than that of your current display. Later you can display this rendered image that has a higher resolution. The rendered images can be saved in different formats, such as TGA, TIFF, BMP, PostScript, BMP, PCX, and IFF. The following steps explain the procedure of saving a rendering to a file.

1. Choose the **Render** button from the **Render** toolbar to invoke the **Render** dialog box.

2. In the **Destination** area, select **File** from the **Viewport** drop-down list.

3. Choose the **More Options** button from the **Destination** area. The **File Output Configuration** dialog box is displayed.

4. Specify the file type, rendering resolution, colors, and other options. Then choose the **OK** button to go back to the **Render** dialog box.

5. Choose the **Render** button in the **Render** dialog box. The **Rendering File** dialog box is displayed. Specify the name of the file to which you want to save the rendering and then choose the **OK** button. The rendered image will be saved.

Saving the Viewport Rendering

A rendered image in the viewport can be saved using the **SAVEIMG** command. This command can be invoked by choosing **Tools > Display Image > Save**. The **Save Image** dialog box will be displayed when you invoke this command, see Figure 26-72. This dialog box is used to specify the type, size, and offset for the output file. The valid output file formats are TGA, TIFF, and BMP. The following steps explain the procedure of saving a viewport rendering.

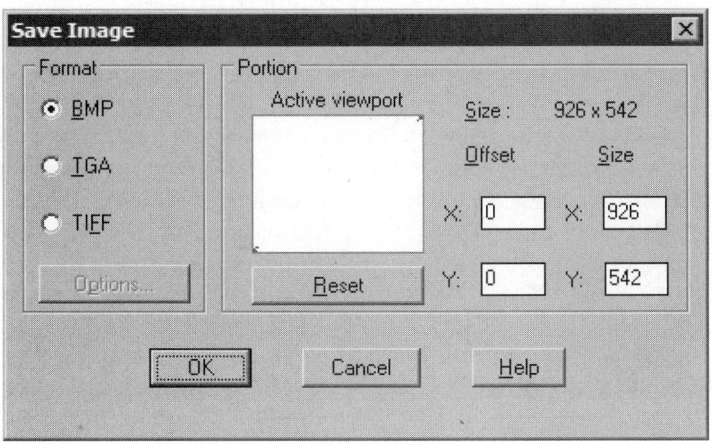

***Figure 26-72 Save Image** dialog box*

1. Invoke the **Render** dialog box and choose the **Render** button to render the object.

2. Choose **Tools > Display Image > Save** to invoke the **Save Image** dialog box.

3. Specify the file format (TGA, TIFF, or BMP), size, and offsets for the image, and then choose the **OK** button to display the **Image File** dialog box.

4. Specify the name of the file to which you want to save the rendering, and then choose the **OK** button. In this way, a rendered image in the viewport can be saved in the specified file format.

Saving a Render-Window Rendered Image

A rendered image in the **Render** window can be saved using the **Save** button available in the **Render** window. In this case, the rendered image will be saved in a **.bmp* format. The following steps explain the procedure for saving a render-window image.

1. Invoke the **Render** dialog box, and select the **Render Window** option in the **Destination** area. Choose the **Render** button to display the **Render** window. The rendered object is displayed in this window.

2. Choose the **Save** button in the **Render** window to save the rendered image. The **Save BMP** dialog box is displayed. Enter the file name in the **File Name** edit box, and then choose the **OK** button. You can set various options using the **Options** button in the **Render** window.

DISPLAYING A RENDERED IMAGE

Menu:	Tools > Display Image > View
Command:	REPLAY

The rendered images saved in the TGA, TIFF, or BMP format can be redisplayed on the screen using the **REPLAY** command. When you invoke this command, the **Replay** dialog box will be displayed as shown in Figure 26-73.

This dialog box is used to locate the saved image. You can select the type of file from the **Files of type** drop-down list and then select the image file. The **Image Specifications** dialog box will be displayed upon selecting the image file, see Figure 26-74. This dialog box is used to specify various image parameters. Specify the parameters and then choose **OK** to display the image on the screen. This image will be displayed as a raster image.

Note

*As the image displayed on the screen using the **Replay** command is a raster image, it will be removed from the display if any command that leads to regeneration is invoked.*

PLOTTING RENDERED IMAGES*

Potting the rendered design was one of the biggest problems in the previous releases of AutoCAD. You had to go through a number of steps to plot a rendered drawing. However, in

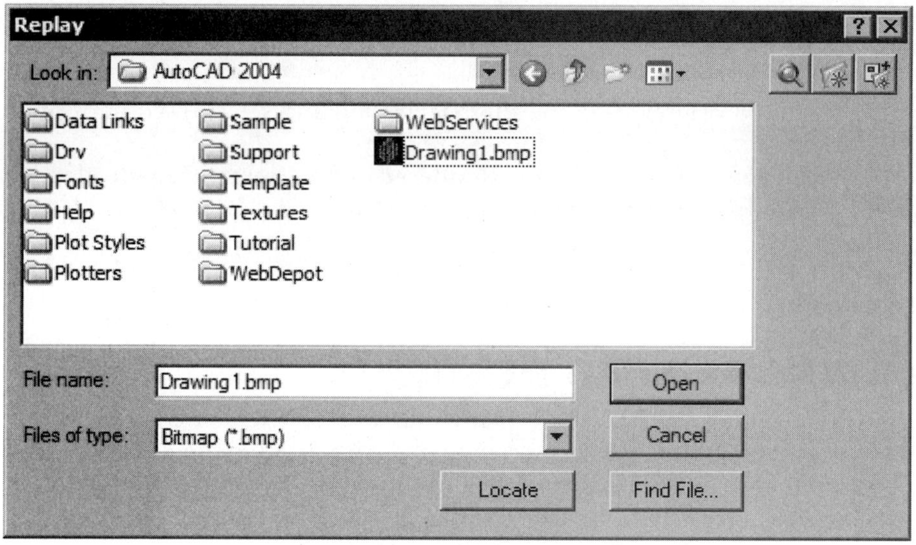

Figure 26-73 ***Replay*** *dialog box*

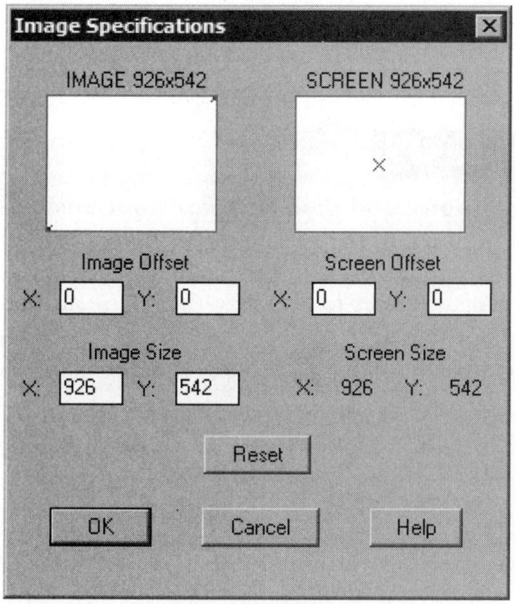

Figure 26-74 ***Image Specifications*** *dialog box*

this release of AutoCAD, plotting of rendered designs has become extremely easy. To plot a rendered design, follow the following steps.

1. Open the design to be plotted and then invoke the **Plot** dialog box. Remember that you

do not need to first render the design and then invoke the **Plot** dialog box. You can also render the design while plotting.

2. Select the required printer from the **Plot Device** tab and select the other options.

3. Invoke the **Plot Settings** tab. Select the **Rendered** option from the **Shade plot** drop-down list in the **Shaded viewport options** area.

4. Set the other options in this tab and then plot the design. The design will be first rendered and then plotted in the rendered form.

UNLOADING AutoCAD RENDER

When you select any AutoCAD **RENDER** command, AutoCAD Render is loaded automatically. If you do not need AutoCAD Render, you can unload it by entering the **ARX** at the Command prompt. You can reload AutoCAD Render by invoking the **RENDER** command or any other command associated with rendering (such as **SCENE**, **LIGHT**, and so on).

Command: **ARX**
Enter an option [?/Load/Unload/Commands/Options]: **U**
Enter ARX/DBX file name to unload: **ACRENDER**
ACRENDER successfully unloaded.

Self-Evaluation Test

Answer the following questions, and then compare your answers to the answers given at the end of this chapter.

1. A rendered image makes it easier to visualize the shape and size of a 3D object. (T/F)

2. You can unload AutoCAD Render by invoking the **ARX** command. (T/F)

3. Falloff occurs with a distant source of light. (T/F)

4. You can change the color of the ambient light. (T/F)

5. The __________ and __________ commands are used to import the material from the material library to the current drawing.

6. The __________ source emits a focused beam of light in the defined direction.

7. The default material library that is opened when you invoke the **MATLIB** command is __________.

8. The __________ command is used to insert the landscaping objects in the current drawing.

9. A __________ light source emits light in all directions, and the intensity of the emitted light is uniform.

10. In case you have saved the rendered image in a TGA, TIFF, or BMP format, you can use the ___________ command to display the image.

Review Questions

Answer the following questions.

1. You can also import the materials to the current drawing directly using the **RMAT** command. (T/F)

2. You can assign the materials to the object using the **AutoCAD Color Index**. (T/F)

3. The Photo Real rendering allows you to view the mirror effect of the material. (T/F)

4. You can increase or decrease the intensity of ambient light, but you cannot turn it completely off. (T/F)

5. Which light does not have a source, and hence, has no location or direction?

   (a) **Ambient** (b) **Point**
   (c) **Spot** (d) **Distant**

6. The intensity of light decreases as the distance increases. This phenomenon is called:

   (a) **Attenuation** (b) **Frequency**
   (c) **Light Effect** (d) **None**

7. Which light allows you to define the geographic location?

   (a) **Ambient** (b) **Point**
   (c) **Spot** (d) **Distant**

8. Apart from the **RENDER** command, which command can be used to change the background of rendering?

   (a) **FOG** (b) **ARX**
   (c) **LIGHT** (d) **BACKGROUND**

9. Which command is used to map the materials on the object?

   (a) **MAP** (b) **MATERIAL**
   (c) **SET** (d) **SETUV**

10. In the __________ source, the light falling on the object (brightness) is inversely proportional to the distance of the object from the light source.

11. The __________ command allows you to define the scenes for viewing the effect of the selected light sources.

12. A spotlight source emits a ____________ beam of light in one direction only.

13. Redisplaying a saved rendered image takes ____________ time as compared to the time involved in rendering.

14. You can assign the materials using the AutoCAD Color Index using the __________ dialog box.

15. Attenuation is defined as ____.

Exercises

Exercise 1 *General*

Create the 3D drawings as shown in Figure 26-75. Next, render the drawing after attaching the materials and inserting lights at appropriate locations so as to get a realistic model.

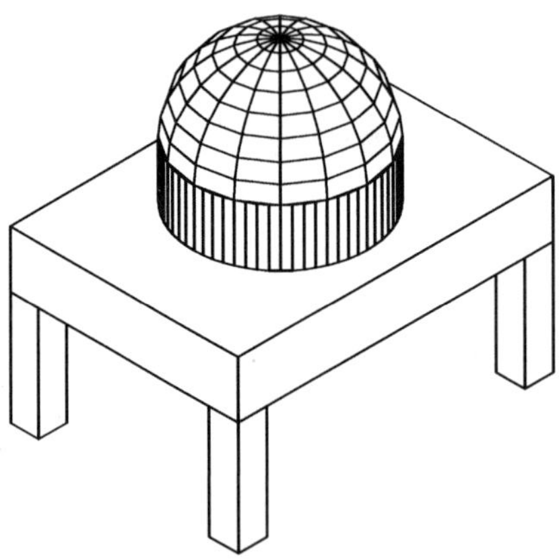

Figure 26-75 *Drawing for Exercise 1*

Exercise 2 *Architectural*

Create the model (Figure 26-76) and attach the materials. Choose your own materials. Apply the Point light and the Spotlight at the sill of the window and at bottom of the door. Also attach a background **VALLEY_1** and some landscaping objects to it and then render the model.

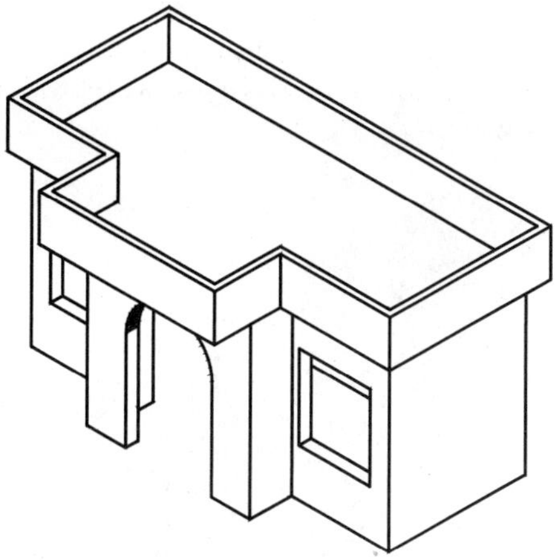

Figure 26-76 *Model for Exercise 2*

Exercise 3 — *General*

In this exercise, you will display the effect of fog on the model shown in Figure 26-77. Choose your own materials.

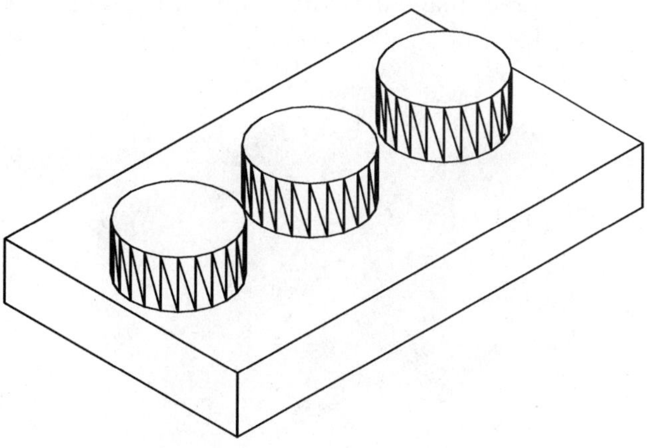

Figure 26-77 *Model for Exercise 3*

Tip

*The display of fog is based on the trial and error method. Use different combinations of options available in the **Fog** dialog box to get the desired effect. To clearly visualize the fog, use white color as the background of model.*

Exercise 4 *General*

Render the Tool Organizer of Chapter 25. Assign dark red wood material to the peg board and white ash to the remaining members. Also the wood texture of the two shelves should be in the opposite direction, Figure 26-78.

Figure 26-78 *Model for Exercise 4*

Exercise 5 *General*

Render the Work Bench of Chapter 25. Assign ash wood to all members. Adjust the mapping coordinates so that the rendered image looks as shown in Figure 26-79.

Figure 26-79 *Model for Exercise 5*

Answers to Self-Evaluation Test

1 - T, **2** - T, **3** - F, **4** - T, **5** - **MATLIB**, **RMAT**, **6** - spotlight, **7** - *render.mli*, **8** - **LSNEW**, **9** -distant, **10** - **REPLAY**

Chapter 27

Accessing External Database

Learning Objectives

After completing this chapter you will be able to:

- *Understand database and the database management system (DBMS).*
- *Understand the* ***AutoCAD database connectivity*** *feature.*
- ***Configure*** *external database.*
- *Access and edit a database using* ***DBCONNECT MANAGER****.*
- *Create* ***Links*** *with graphical objects.*
- *Create and display* ***Labels*** *in a drawing.*
- *Understand* ***AutoCAD SQL Environment*** *(ASE) and create queries using* ***Query Editor****.*
- *Form selection sets using* ***Link Select****.*
- *Convert ASE links into AutoCAD 2004 format.*

UNDERSTANDING DATABASE

Database

Database is a collection of data that is arranged in a logical order. In a database, the columns are known as **fields**, the individual rows are called **records**, and the entries in the database tables are known as **cells** that store data for a particular variable. For example, there are six computers in an office, and we want to keep a record of these computers on a sheet of paper. One of the ways of recording this information is to make a table with rows and columns as shown in Figure 27-1. Each column can have a heading that specifies a certain feature of the computer, such as COMP_CFG, CPU, HDRIVE, or RAM. Once the columns are labeled, the computer data can be placed in the columns. By doing this, you have created a database on a sheet of paper that contains information about the computers. The same information can be stored on a computer, known as a computerized database.

COMPUTER

COMP_CFG	CPU	HDRIVE	RAM	GRAPHICS	INPT_DEV
1	PENTIUM350	4300MB	64MB	SUPER VGA	DIGITIZER
2	PENTIUM233	2100MB	32MB	SVGA	MOUSE
3	MACIIC	40MB	2MB	STANDARD	MOUSE
4	386SX/16	80MB	4MB	VGA	MOUSE
5	386/33	300MB	6MB	VGA	MOUSE
6	SPARC2	600MB	16MB	STANDARD	MOUSE

Figure 27-1 *A table containing computer information*

Most of the database systems are extremely flexible and any modifications to or additions of fields or records can be done easily. Database systems also allow you to define relationships between multiple tables so that if the data of one table is altered, the corresponding values of another table with predefined relationships changes automatically.

Database Management System

The database management system (DBMS) is a program or a collection of programs (software) used to manage the database. For example, PARADOX, dBASE, INFORMIX, and ORACLE are database management systems.

Components of a Table

A database **table** is a two-dimensional data structure that consists of rows and columns as shown in Figures 27-2 and 27-3.

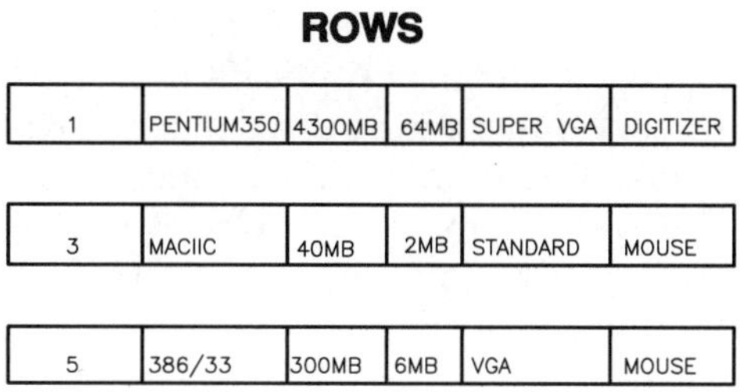

Figure 27-2 *Rows in a table (horizontal group)*

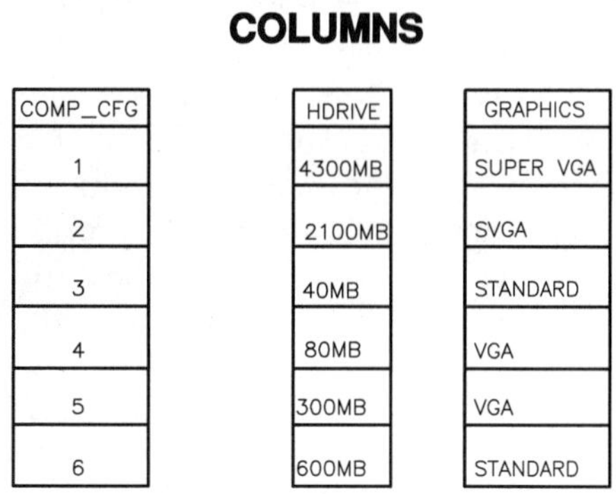

Figure 27-3 *Columns in a table (vertical group)*

Row

The horizontal group of data is called a **row**. For example, Figure 27-2 shows three rows of a table. Each value in a row defines an attribute of the item. For example, in Figure 27-2, the attributes assigned to COMP_CFG (1) include PENTIUM350, 4300MB, 64MB, and so on. These attributes are arranged in the first row of the table.

Column

A vertical group of data (attribute) is called a **column**. (See Figure 27-3.) HDRIVE is the column heading that represents a feature of a computer, and the HDRIVE attributes of each computer are placed vertically in this column.

AutoCAD DATABASE CONNECTIVITY

AutoCAD can be effectively used in associating data contained in an external database table with the AutoCAD graphical objects by linking. The **Links** are the pointers to the database tables from which the data can be referred. AutoCAD can also be used to attach **Labels** that will display data from the selected tables as text objects. AutoCAD database connectivity offers the following facilities.

1. A **DBCONNECT MANAGER** that can be used to associate links, labels, and queries with AutoCAD drawings.
2. An **External Configuration Utility** that enables AutoCAD to access the data from a database system.
3. A **Data View Window** that displays the records of a database table within the AutoCAD session.
4. A **Query Editor** that can be used to construct, store, and execute SQL queries. **SQL** is an acronym for **Structured Query Language**.
5. A **Migration Tool** that converts links and other displayable attributes of files created by earlier releases to AutoCAD 2004.
6. A **Link Select Operation** that creates iterative selection sets based on queries and graphical objects.

DATABASE CONFIGURATION

An External database can be accessed within AutoCAD only after configuring AutoCAD using Microsoft **ODBC** (Open Database Connectivity) and **OLE DB** (Object Linking and Embedding Database) programs. AutoCAD is capable of utilizing data from other applications, regardless of the format and the platform on which the file is stored. Configuration of a database involves creating a new **data source** that points to a collection of data and information about the required drivers to access it. A **data source** is an individual table or a collection of tables created and stored in an environment, catalog, or schema. Environments, catalogs, and schemas are the hierarchical database elements in most of the database management systems and they are analogous to Window-based directory structure in many ways. Schemas contain a collection of tables, while Catalogs contain subdirectories of schemas and Environment holds subdirectories of catalogs. The external applications supported by AutoCAD 2004 are **dBASE® V** and **III**, **Oracle® 8.0** and **7.3**, **Microsoft® Access®, PARADOX 7.0**, **Microsoft Visual FoxPro® 6.0**, **SQL Server 7.0** and **6.5**. The configuration process varies slightly from one database system to other.

DBCONNECT MANAGER

Menu:	Tools > dbConnect
Command:	DBCONNECT

When you invoke this command, the **DBCONNECT MANAGER** will be displayed as shown in Figure 27-4. Also, the **dbConnect** menu is added to the menu bar when you invoke this command. The **DBCONNECT MANAGER** enables you to access information from external database more effectively. The **DBCONNECT MANAGER** is dockable as well as resizable and contains a set of buttons and a tree view showing all the configured and available database. You can use **DBCONNECT MANAGER** for associating various database objects with an AutoCAD drawing. It contains two nodes in the tree view. The **Drawing nodes** displays all the open drawings and the associated database objects with the drawing. **Data Sources node** displays all the configured data on your system.

Figure 27-4 DBCONNECT MANAGER

AutoCAD contains several Microsoft Access sample database tables and a direct driver (*jet_dbsamples.udl*). The following example shows the procedure to configure a database with a drawing using the **DBCONNECT MANAGER**.

Example 1 — *General*

Configure a data source of Microsoft Access database with a diagram. Update and use the **jet_samples.udl** configuration file with new information.

1. Invoke the **DBCONNECT** command to display the **DBCONNECT MANAGER**. The **dbConnect** menu will be inserted between the **Modify** and the **Window** menus.

2. The **DBCONNECT MANAGER** will display **jet_dbsamples** under **Data Sources**. Right-click on **jet_dbsamples** to display the shortcut menu. In the shortcut menu choose **Configure** as shown in Figure 27-5.

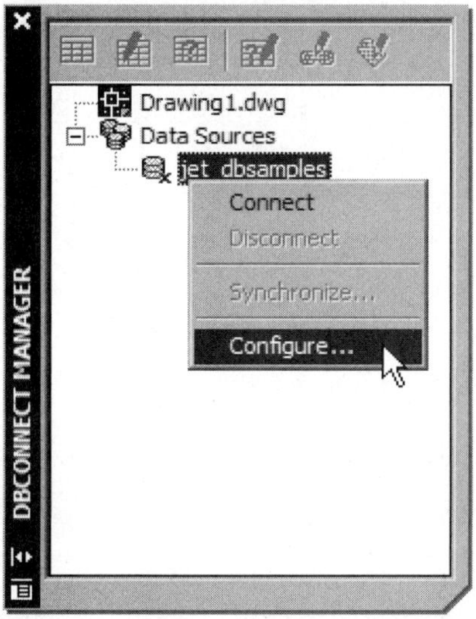

Figure 27-5 DBCONNECT MANAGER

3. The **Data Link Properties** dialog box will be displayed with the **Connection** tab as the current tab. (See Figure 27-6.)

4. In the **Data Link Properties** (**Connection** tab) dialog box, choose the [**...**] button adjacent to the **Select or enter a database name** text box. The **Select Access Database** dialog box is displayed as shown in Figure 27-7. Select **db_samples** (**.mdb file**) and choose the **Open** button.

Note

*If you are configuring a driver other than **Microsoft Jet** for database linking, you should consult the appropriate documentation for the configuring process.*

*The other tabs in the **Data Links Properties** dialog box are required for configuring various Database Providers supported by AutoCAD.*

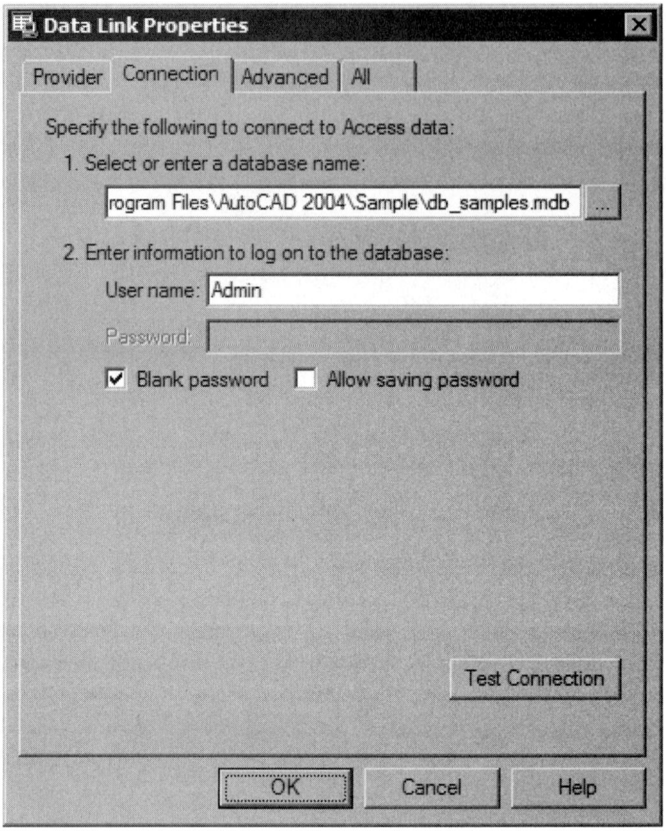

***Figure 27-6 Data Link Properties** dialog box*

5. Once the database name is selected, choose the **Test Connection** button to ensure that the database source has been configured correctly. If the source is configured correctly, the **Microsoft Data Link** message box is displayed, see Figure 27-8.

6. Choose the **OK** button to end the message and again choose the **OK** button to complete the configuration process.

7. After configuring a data source, double-click on **jet_dbsamples** in **DBCONNECT MANAGER**. All the sample tables will be displayed in the tree view and you can connect any table and link its records to your diagram.

VIEWING AND EDITING TABLE DATA FROM AutoCAD

After you have configured a data source using the **DBCONNECT MANAGER**, you can view as well as edit its tables within the AutoCAD session. The **Data View window** can be used for viewing and editing a database table. You can open tables in **Read-only** mode to view their content. But you cannot edit their records in **Read-only** mode. Some database systems may require a valid user-name and password before connecting to AutoCAD drawing files. The records of a database table can be edited by opening in **Edit** mode. The procedure to open a

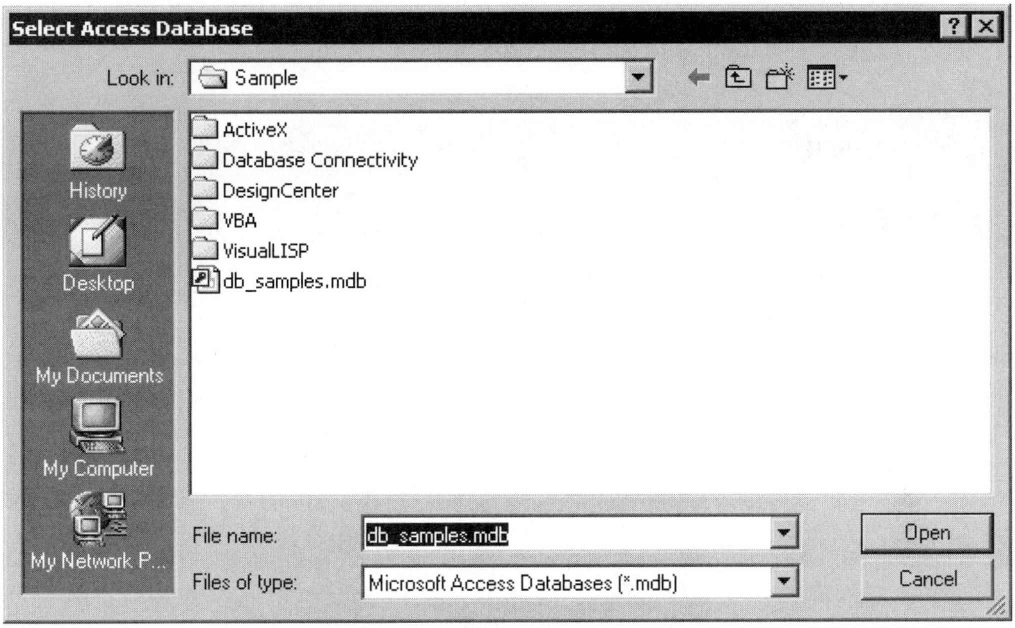

Figure 27-7 ***Select Access Database*** *dialog box*

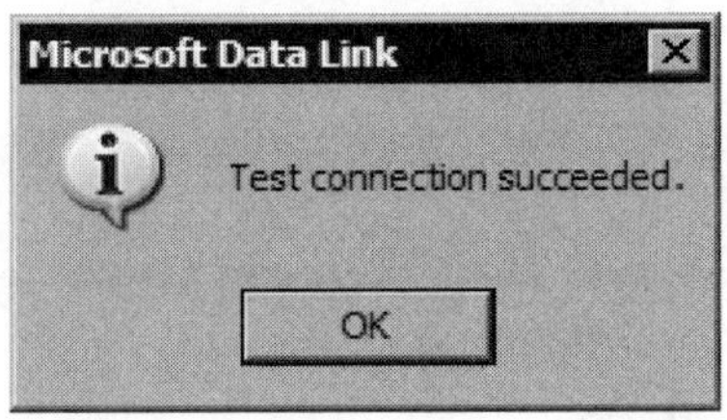

Figure 27-8 ***Microsoft Data Link*** *message box*

table in various modes are given next.

Read-only mode

The tables can be opened in the **Read-only** mode by selecting the table and then choosing the **View Table** button from the **DBCONNECT MANAGER**. You can also choose **View Table** from the shortcut menu displayed upon right-clicking on the selected table. This can also be done by choosing **View Data > View External Table** from the **dbConnect** menu. You will notice that the table that is displayed has gray background and you cannot edit the entries in it.

Edit mode

The tables can be opened in the **Edit** mode by choosing the **Edit Table** button from the **DBCONNECT MANAGER**. You can also choose **Edit Table** from the shortcut menu displayed upon right-clicking on the selected table. This can also be done by choosing **View Data > Edit External Table** from the **dbConnect** menu.

Note
The ***db_samples.mdb*** *file is available in the directory \AutoCAD 2004\Samples.*

Example 2 — *General*

In this example, you will select **Computer** table from the **jet_dbsamples** data source and edit rows in the table. Add a new computer and replace one by editing the table and save changes.

1. Select **Computer** table from the **jet_dbsamples** in the **DBCONNECT MANAGER** and right-click to invoke the shortcut menu. Choose **Edit Table** from the shortcut menu (Figure 27-9). You can also do the same by double clicking on the **Computer** table.

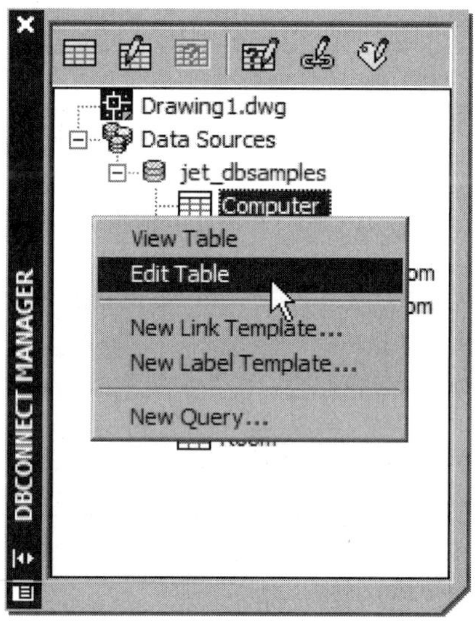

Figure 27-9 *Choosing* ***Edit Table*** *from the shortcut menu*

2. The **Data View** window is displayed with the **Computer** table in it. In the **Data View** window, you can resize, sort, hide, or freeze the columns according to your requirements.

3. To add a new item in the table, double-click on the first empty row. In **Tag_Number** column, type **24675**. Then type the following data into the columns.

Manufacturer:	**IBM**
Equipment_Description:	**PIII/450, 4500GL, NETX**
Item_Type:	**CPU**
Room:	**6035**

4. To edit the record for **Tag_Number 60298**, and display the following in **Data View** window

(Figure 27-10), double-click on each cell and type.

MANUFACTURER:	**CREATIVE**
Equipment_Description:	**INFRA 6000, 40XR**
Item_Type:	**CD DRIVE**
Room:	**6996**

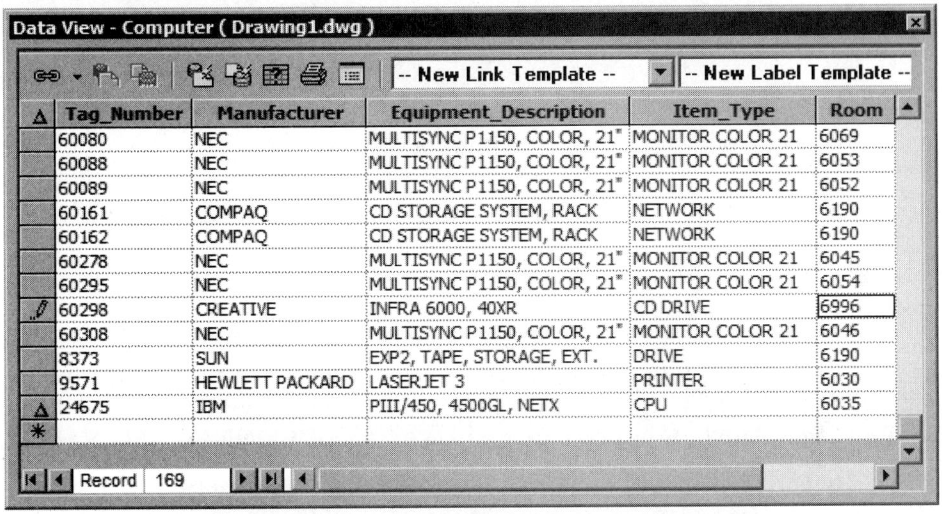

Figure 27-10 ***Data View*** *window after editing*

5. After you have made all the changes in the database table, you have to save the changes for further use. To save the changes in the table, right-click on the **Data View grid header**, a triangle mark available on the left of the **Tag_Number** column and choose **Commit** from the shortcut menu. The changes in the current table will be saved. The new record that you have entered does not necessarily get added at the end of the table, so if you close the table and then reopen it you might have to search for the new record in the table. Note that **if you quit Data View window without Committing, all the changes you have made during the editing session are automatically committed.**

Note

After making changes, if you do not want to save the changes, choose ***Restore*** *from the above shortcut menu.*

CREATING LINKS WITH GRAPHICAL OBJECTS

The main function of the database connectivity feature of AutoCAD is to associate data from external sources with its graphical objects. You can establish the association of the database table with the drawing objects by developing a **link**, which will make reference to one or more records from the table. But you **cannot** link nongraphical objects such as layers or linetypes with the external database. Links are very closely related with graphical objects and change with the change in graphical objects.

To develop links between database tables and graphical objects, you must create a **Link Template** that identifies the fields of the tables with which links are associated to share the template. For example, you can create a link template that uses the **Tag_number** from the **COMPUTER database table**. Link template also acts as a shortcut that points to the associated database tables. You can associate multiple links to a single graphical object using different link templates. This is useful in associating multiple database tables with a single drawing object. The following example will describe the procedure of linking using link template creation.

Example 3

General

Create a link template between **Computer** database table from **jet_dbsamples** and your drawing, and use **Tag_Number** as the **key field** for linking.

1. Open the drawing that has to be linked with the **Computer** database table.

2. Invoke the **DBCONNECT MANAGER**. Select **Computer** from the **jet_dbsamples** data source and right-click on it to invoke the shortcut menu.

3. Choose **New Link Template** from the shortcut menu to invoke the **New Link Template** dialog box (Figure 27-11). You can also select the table in the tree view and choose the **New Link Template** button from the toolbar in the **DBCONNECT MANAGER**.

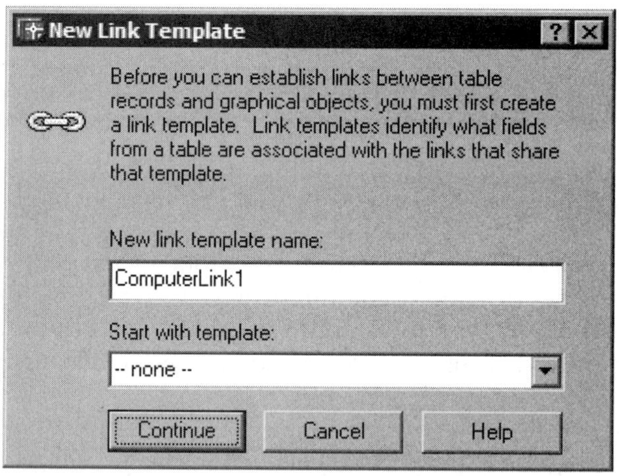

***Figure 27-11** New Link Template dialog box*

4. Choose the **Continue** button to accept default link template named **ComputerLink1**. The **Link Template** dialog box is displayed. Select the check box adjacent to **Tag_Number** to accept it as the **Key field** for associating the template with the block reference in the diagram. (See Figure 27-12.)

5. Choose the **OK** button to complete the new link template. The name of the link template will be displayed under the Drawing name node of the tree view in **DBCONNECT MANAGER**.

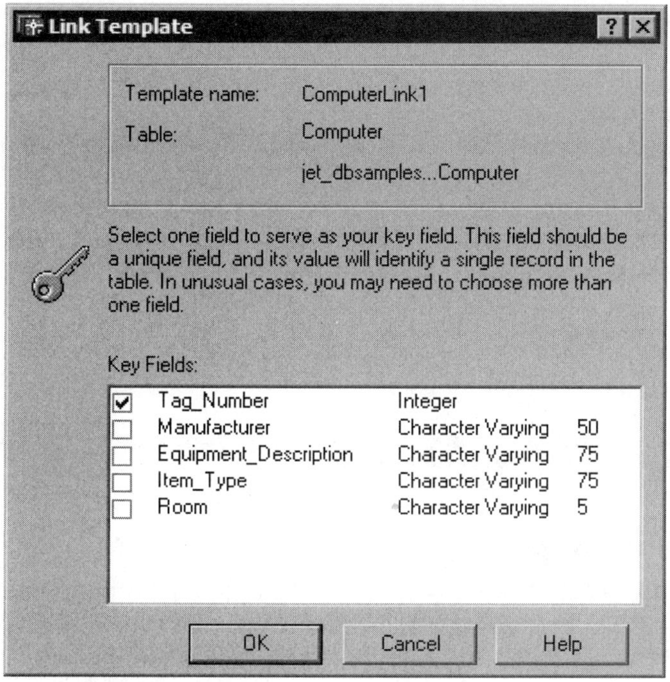

Figure 27-12 ***Link Template*** *dialog box*

6. To link a record, double-click on the **Computer** table to invoke the **Data View** window (**Edit** mode). In the table, go to Tag_Number **24675** and highlight the record (row).

7. Right-click on the row header and choose **Link!** from the shortcut menu (Figure 27-13). You can also link by choosing the **Link!** button from the toolbar in **Data View** window.

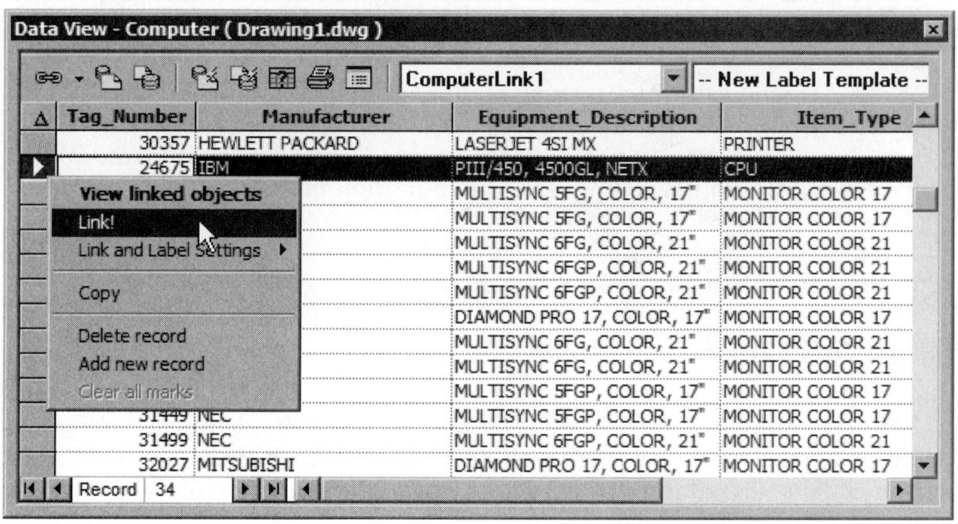

Figure 27-13 *Selecting* ***Link*** *from the shortcut menu*

8. You are prompted to select the objects. Select the required objects you want to link with the record. Repeat the process to link all the records to the corresponding block references in the diagram.

9. After linking, you can view the linked objects by choosing the **View Linked Objects in Drawing** button in the **Data View** window, and the linked objects will get highlighted in the drawing area.

Additional Link Viewing Settings

You can set a number of viewing options for linked graphical objects and linked records by using the **Data View and Query Options** dialog box (Figure 27-14). This dialog box can be invoked by choosing the **Data View and Query Options** button from the **Data View** window.

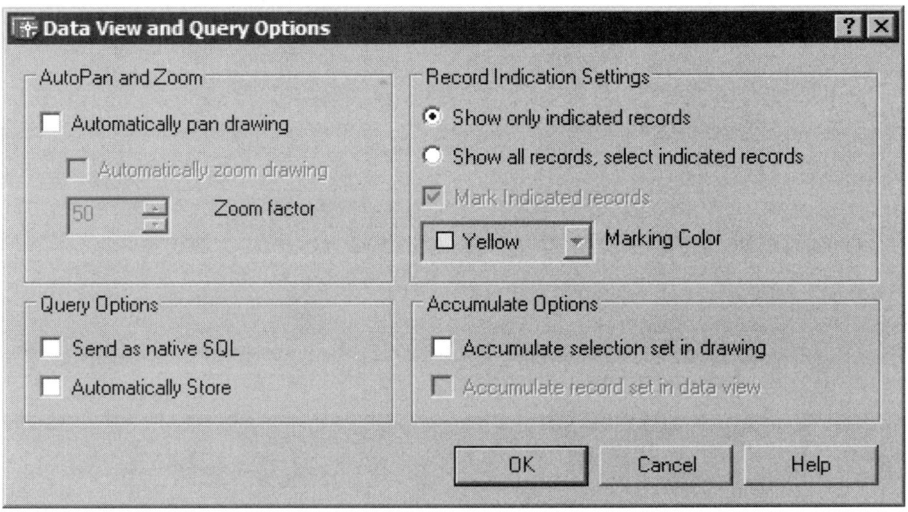

Figure 27-14 ***Data View and Query Options*** *dialog box*

You can set the **Automatically Pan Drawing** option so that the drawing is panned automatically to display objects linked with the current set of selected records in the **Data View** window. You can also set options for **Automatically zoom drawing** and the **Zoom factor**. You can also change the Record Indication settings and the indication marking color.

Note

Query Editor *dialog box is discussed later in this chapter.*

Editing Link Data

After linking data with the drawing objects, you may need to edit the data or update the **Key field** values. For example, you may need to reallocate the Tag-Number for the computer equipment or Room for each of the linked items. **Link Manager** can be used for changing the Key values. The next example describes the procedure of editing linked data.

Example 4 *General*

Use **Link Manager** to edit linked data from the Computer table and change the Key Value from **24352** to **24675**.

1. Open the diagram that was linked with the Computer table.

2. Select the diagram and choose **Links > Link Manager** from the shortcut menu. The **Link Manager** dialog box for the **Computer** table (Figure 27-15) is displayed.

***Figure 27-15 Link Manager** dialog box*

3. Select **24352** field in the **Value** column and choose the [**...**] button to invoke the **Column Values** dialog box.

4. Select **24675** from the list (Figure 27-16) and choose the **OK** button.

5. Again choose the **OK** button in **Link Manager** to accept the changes.

CREATING LABELS

You can use linking as a powerful mechanism to associate drawing objects with external database tables. You can directly access associated records in the database table by selecting linked objects in the drawing. But linking has some limitations. Consider that you want to include the associated external data with the drawing objects. Since during printing, the links are only the pointers to the external database table, they will not appear in the printed drawing. In such situations, a feature called **Labels** proves useful. The labels can be used for visible representation of external data in drawing.

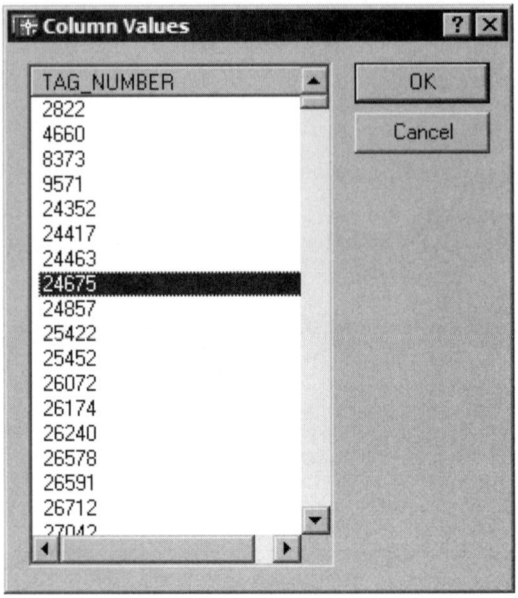

Figure 27-16 ***Column Values*** *dialog box*

Labels are the multiline text objects that display data from the selected fields in the AutoCAD drawing. The labels are of the following two types.

Freestanding labels

Freestanding labels exist in the AutoCAD drawing independent of the graphical objects. Their properties do not change with any change of graphical objects in the drawing.

Attached labels

The Attached labels are connected with the graphical objects they are associated with. If the graphical objects are moved, then the labels also move and if the objects are deleted, then the labels attached with them also get deleted.

Labels associated with the graphical objects in AutoCAD drawing are displayed with a leader. Labels are created and displayed by **Label Templates** and all the properties of label display can be controlled by using the **Label Template** dialog box. The next example will demonstrate the complete procedure of creating and displaying labels in AutoCAD drawing using the label template.

Example 5 *General*

Create a new label template in the **Computer** database table and use the following specifications for the display of labels in the drawing.

a. The label includes **Tag_Number**, **Manufacturer** and **Item_Type** fields.

b. The fields in the label are **0.25** in height, **Times New Roman** font, **black** (**Color 18**) in color and **Middle-left** justified.
c. The label offset starts with **Middle Center** justified and leader offset is **X=1.5** and **Y=1.5**.

1. Open the drawing where you want to attach a label with the drawing objects.

2. Choose **Tools** > **dbConnect** from the menu bar to invoke **DBCONNECT MANAGER**. Select **Computer** table from **jet_dbsamples** data source.

3. Right-click on **Computer** and choose **New Label Template** from the shortcut menu to invoke the **New Label Template** dialog box (Figure 27-17). You can also invoke the dialog box by choosing the **New Label Template** button from the **DBCONNECT MANAGER** toolbar.

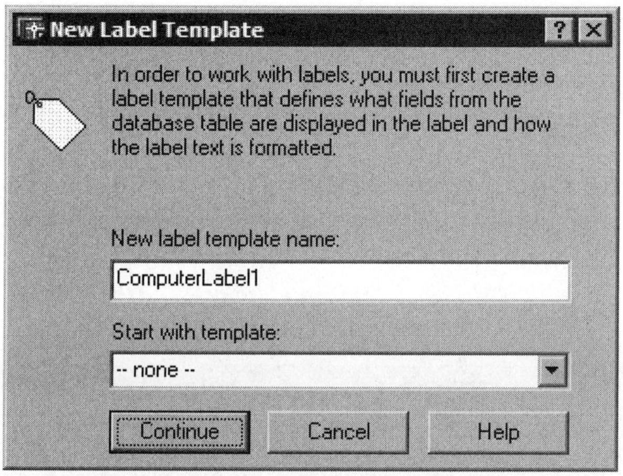

***Figure 27-17** **New Label Template** dialog box*

4. In the **New Label Template** dialog box, choose the **Continue** button to accept the default label template name **Computer Label1**. The **Label Template** dialog box is displayed.

5. In the **Label Template** dialog box, choose the **Label Fields** tab. From the **Field** drop-down list, one by one add **Tag_Number**, **Manufacturer**, **Item_Type** fields by using the **Add** button (Figure 27-18).

6. Highlight the field names by selecting them and choose the **Character** tab. In the **Character** tab, select the **Times New Roman** font and font height to **0.25**. Set the color to **Color 18**. Also select **Middle-left** justification in the **Properties** tab (Figure 27-19).

7. Choose the **Label Offset** tab and select **Middle-center** in the **Start**: drop-down list. Set the **Leader offset:** value to **X: 1.5** and **Y: 1.5** (Figure 27-20). Choose **OK**.

8. To display the label in the drawing, select the Computer table in the **DBCONNECT MANAGER**. Right-click on the table and choose **Edit Table** from the shortcut menu.

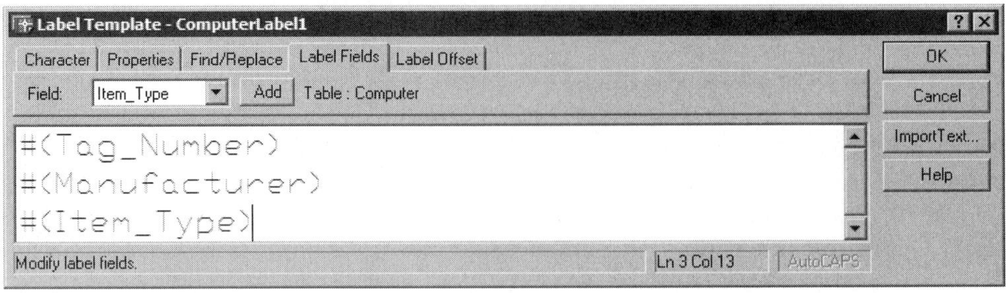

Figure 27-18 ***Label Template*** *dialog box (****Label Fields*** *tab)*

Figure 27-19 ***Label Template*** *dialog box (****Properties*** *tab)*

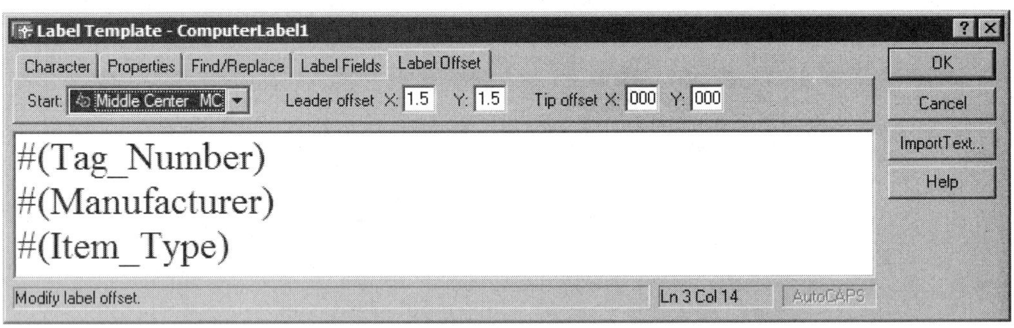

Figure 27-20 ***Label Template*** *dialog box (****Label Offset*** *tab)*

9. Select **Computer link1** (created in an earlier example) from the **link name** drop-down list and **Computer label1** from the **label name** drop-down list in the **Data View** window.

10. Select the record (row) you want to use as a label. Then choose the down arrow button provided on the right side of the **Links** button. Choose **Create Attached Labels** from the shortcut menu (Figure 27-21). Then choose the **Create Attached Label** button that replaces the **Links** button. Select the drawing object you want to label. The Label with given specifications is displayed in the drawing.

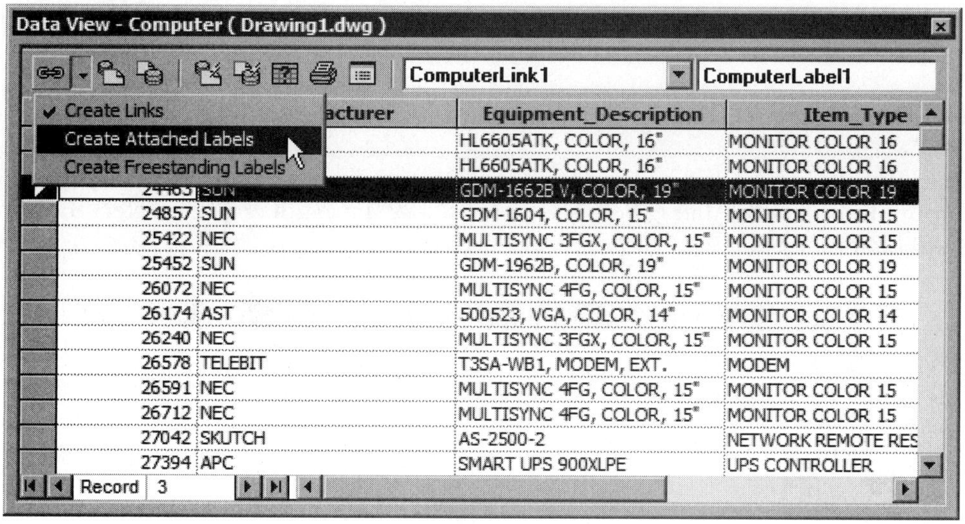

Figure 27-21 *Attaching the label to the record*

Note

You can create ***Freestanding Labels*** *in the same way by selecting* ***Create Freestanding Label*** *from the* ***Link and Label setting*** *menu in the* ***Data View*** *window.*

Updating Labels with New Database Values

You may have to change the data values in the database table after adding a label in the AutoCAD drawing. Therefore, you should update the labels in the drawing after making any alteration in the database table the drawing is linked with. The following is the procedure for updating all label values in the AutoCAD drawing.

1. After editing the database table, open the drawing that has to be updated. Select the diagram and choose **Label** > **Reload** from the shortcut menu.

2. The details in the label will be modified automatically and the changes will be reflected in the label attached to the diagram.

Importing and Exporting Link and Label Templates

You may want to use the link and label templates that have been developed by some other AutoCAD users. This is very useful when developing a set of common tools to be shared by all of the team members in a project. AutoCAD is capable of importing as well as exporting all the link and label templates that are associated with a drawing. The following is the procedure to export a set of templates from the current drawing.

1. From the **dbConnect** menu, choose **Templates** > **Export Template Set** to invoke the **Export Template Set** dialog box. In the dialog box in the **Save In list**, select the directory where you want to save the template set.

2. Under **File Name**, specify a name for the template set, and then choose the **Save** button to save the template in the specified directory.

The following is the procedure to import a set of templates into the current drawing.

1. From the **dbConnect** menu, choose **Templates** > **Import Template Set** to invoke the **Import Template Set** dialog box, see Figure 27-22. In the dialog box, select the template set to be imported.

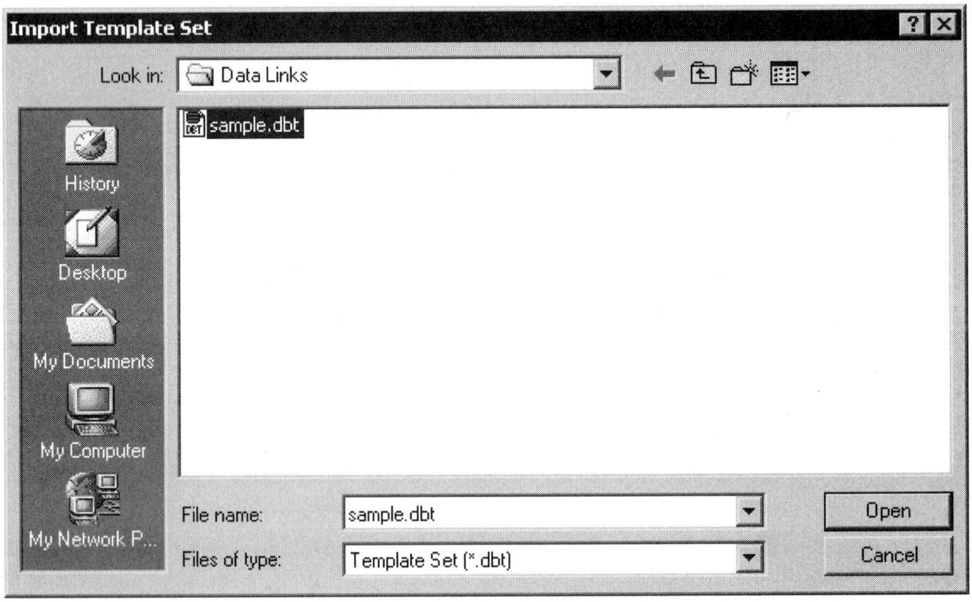

Figure 27-22 *Importing a template file*

2. Choose **Open** to import the template set into the current drawing.

AutoCAD displays an **Alert** box that can be used to provide a unique name for the template if there is a link or label template with the same name associated with the current drawing.

AutoCAD SQL ENVIRONMENT (ASE)

SQL is an acronym for **Structured Query Language**. It is often referred to as **Sequel**. SQL is a format in computer programming that lets the user ask questions about a database according to specific rules. The **AutoCAD SQL environment** (ASE) lets you access and manipulate the data that is stored in the external database table and link data from the database to objects in a drawing. Once you access the table, you can manipulate the data. The connection is made through a database management system (DBMS). The DBMS programs have their own methodology for working with database. However, the ASE commands work the same way regardless of the database being used. This is made possible by the ASE drivers that come with AutoCAD software. In AutoCAD 2004, SQL has been incorporated. For example, you want to prepare a report that lists all the computer equipment that costs more then $25. **AutoCAD Query Editor** can be used to easily construct a query that returns a subset of records or linked graphical objects that follow the previously mentioned criterion.

AutoCAD QUERY EDITOR

The AutoCAD **Query Editor** consists of four tabs that can be used to create new queries. The tabs are arranged in order of increasing the complexity. For example, if you are not familiar with **SQL** (Structured Query Language), you can start with **Quick Query** and **Range Query** initially to get familiar with query syntax.

You can start developing a query in one tab and subsequently add and refine the query conditions in the following tabs. For example, if you have created a query in the **Quick Query** tab and then decide to add an additional query using the **Query Builder** tab, when you choose the **Query Builder** tab, all the values initially selected in the previous tabs are displayed in this tab and additional conditions can be added to the query. But it is not possible to go backwards through the tabs once you have created queries with one of the advanced tabs. The reason for this is that the additional functions are not available in the simpler tabs. AutoCAD will prompt a warning indicating that the query will be reset to its default values if you attempt to move backward through the query tabs. The AutoCAD **Query Editor** has the following tabs for building queries.

Quick Query

This tab provides an environment where simple queries can be developed based on a single database field, single operator, and a single value. For example, you can find all records from the current table where the value of the **'Item_type'** field equals **'CPU'**.

Range Query

This tab provides an environment where a query can be developed to return all records that fall within a given range of values. For example, you can find all records from the current table where the value of the **'Room'** field is greater than or equal to 6050 and less than or equal to 6150.

Query Builder

This tab provides an environment where more complicated queries can be developed based on multiple search criteria. For example, you can find all records from the current table where the **'Item_type'** equals **'CPU'** and **'Room'** number is greater than **6050**.

SQL Query

This tab provides an environment where sophisticated queries can be developed that confirm with the SQL 92 protocol. For example, you can select * from **Item type.Room. Tag_Number** where.

Item_type = 'CPU' ; **Room >= 6050** and **<= 6150**
and
Tag_number > 26072

The following example describes the complete procedure of creating a new query using all the tabs of the **Query Editor**.

Example 6 *General*

Create a new Query for the **Computer** database table and use all the tabs of the **Query Editor** to prepare a SQL query.

1. Right-click on **Computer** in the **DBCONNECT MANAGER** window to display the shortcut menu. Choose **New Query** from the shortcut menu to invoke the **New Query** dialog box (Figure 27-23). You can also select **Computer** and choose the **New Query** button from the toolbar in the **DBCONNECT MANAGER** window to invoke this dialog box.

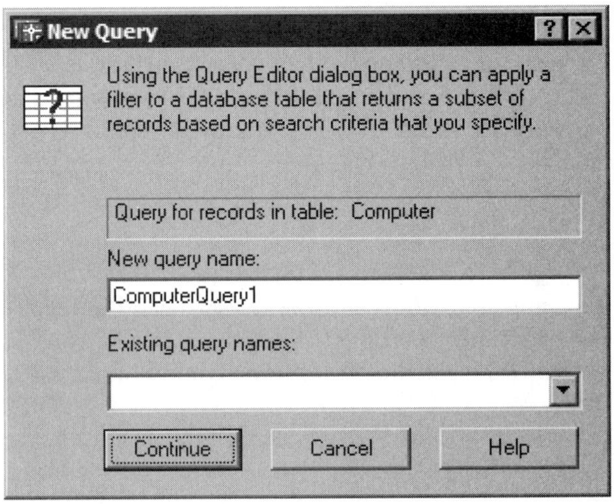

*Figure 27-23 **New Query** dialog box*

2. In the **New Query** dialog box, choose the **Continue** button to accept the default query name **ComputerQuery1**. The **Query Editor** is displayed.

3. In the **Quick Query** tab of **Query Editor** (Figure 27-24), select **'Item_Type'** from the **Field** list box, **'=Equal'** from the **Operator** drop-down list, and type **'CPU'** in the **Value** text box. You can also select **'CPU'** from the **Column Values** dialog box by choosing the **Look up values** button. Choose the **Store** button to save the query.

Note

*If you want to view the query, choose the **Execute** button from the **Query Editor**, however you will not be able to continue with the example using same **Query** dialog box.*

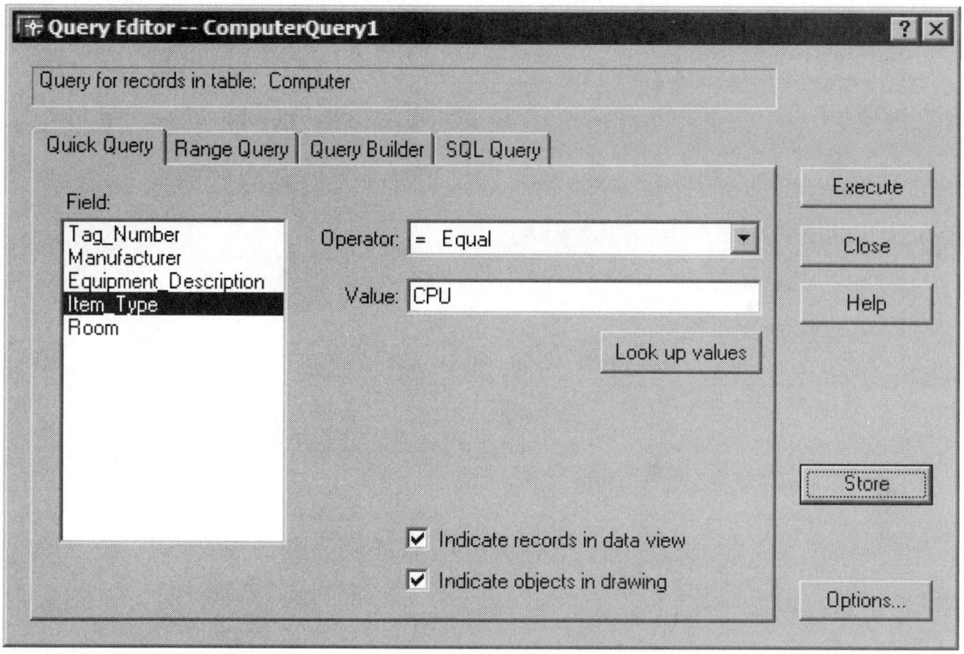

***Figure 27-24 Query Editor** (**Quick Query** tab)*

4. Choose the **Range Query** tab (Figure 27-25), and select **Room** from the **Field** list box. Enter **6050** in the **From** edit box and **6150** in the **Through** edit box. You can also select the values from the **Column Values** dialog box by choosing the **Look up values** button. Choose the **Store** button to save the query.

5. Choose the **Query Builder** tab, and add **Item_Type** in **Show fields** list box by selecting **Item_Type** from the **Fields in table:** list box and choosing the **Add** button. In the table area, change the entries of fields, Operator, Value, Logical, and Parenthetical grouping criteria as shown in Figure 27-26, by selecting each cell and selecting the values from the respective drop-down list or options lists. Choose the **Store** button to save the query.

6. Choose the **SQL Query** tab; the query conditions specified earlier will be carried to the tab automatically. Here you can create a query using multiple tables. But select **'Computer'** from the **Table** list box, **'Tag_Number'** from the **Fields** list box, **'>= Greater than or equal to'** from the **Operator** drop-down list and enter **26072** in the **Values** edit box. You

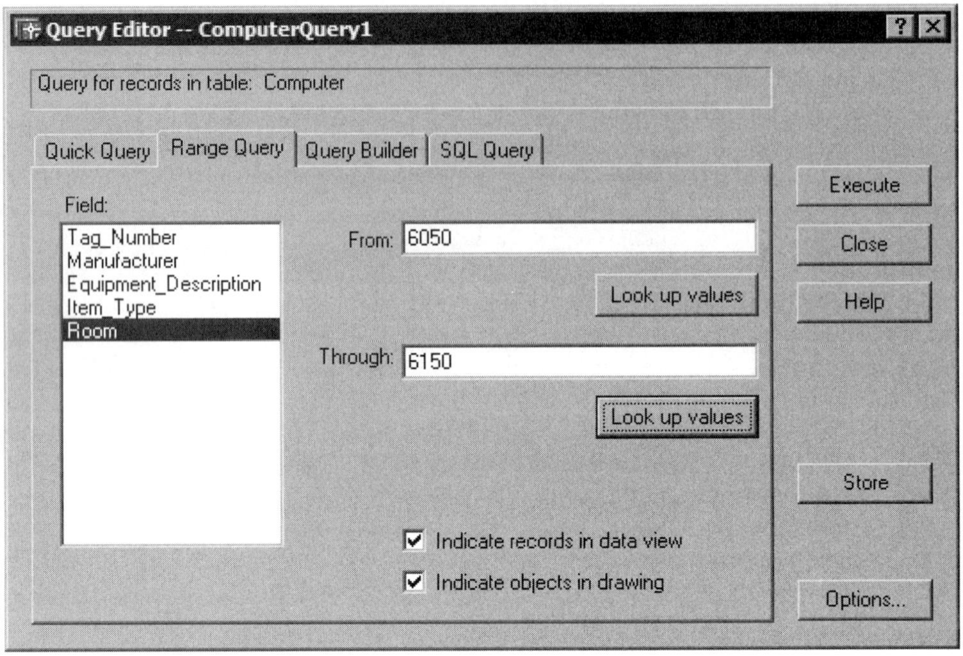

*Figure 27-25 **Query Editor** (**Range Query** tab)*

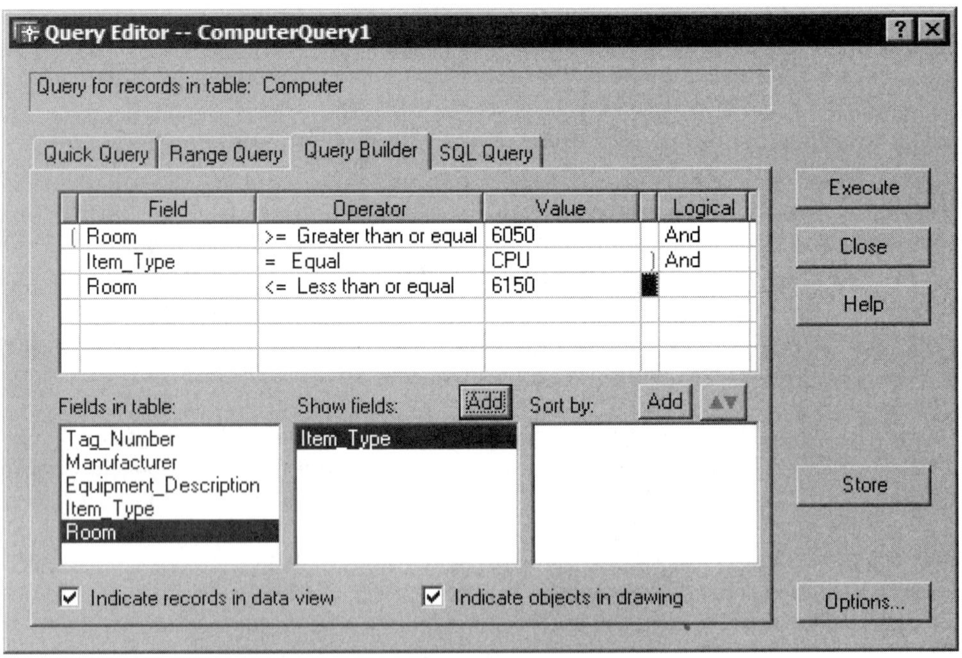

*Figure 27-26 **Query Editor** (**Query Builder** tab)*

can also select from the available values by choosing the [...] button (Figure 27-27). Choose the **Store** button to save the query.

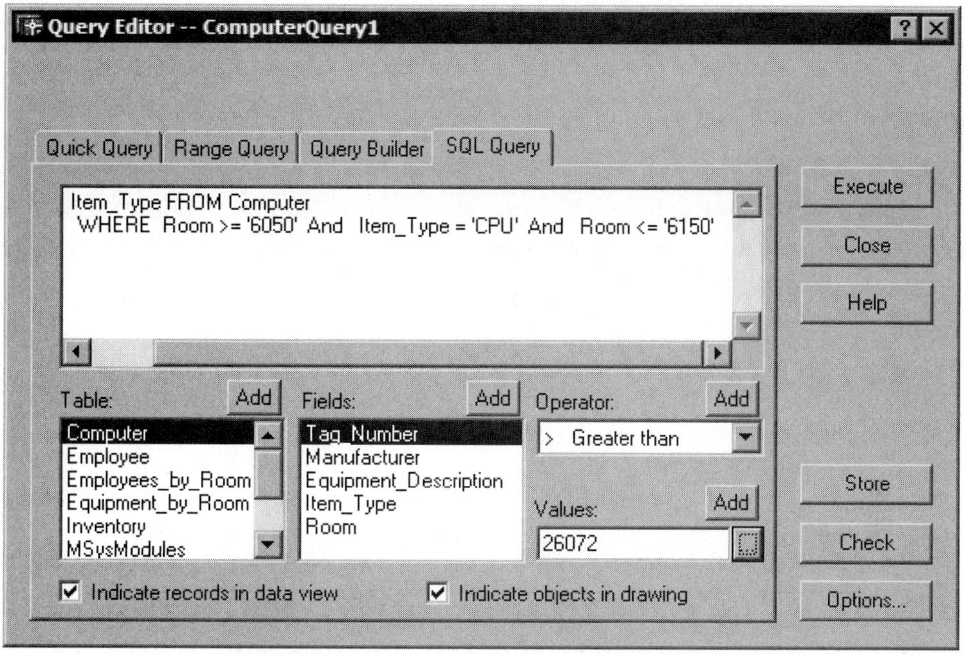

Figure 27-27 Query Editor (SQL Query tab)

7. To check whether the SQL syntax is correct, choose the **Check** button and AutoCAD will display the **Information box** to determine whether the syntax is correct. Choose the **Execute** button to display the **Data View** window showing a subset of records matching the specified query criteria.

Note

You can view the subset of records conforming to your criterion with any of the four tabs during the creation of queries.

Importing and Exporting SQL Queries

You may occasionally be required to use the queries made by some other user in your drawing or vice-versa. AutoCAD allows you to import or export stored queries. Sharing queries is very useful when developing common tools used by all team members on a project.

Note

*When you invoke the **DBCONNECT MANAGER** from the tools menu, the **dbConnect** menu is added to the menu bar. If it is not, enter **MENULOAD** at the Command prompt. From the **Menu Groups** tab of the **Menu Customization** dialog box, choose the **Browse** button and browse to the AutoCAD 2004 **Support** directory. Select the **dbcon.mnu** file. Now, choose the **Load** button in the **Menu Customization** dialog box. Now, choose the **Menu Bar** tab and select **dbConnect** from the **Menu Group** drop-down list. Choose **dbConnect** from the **Menus** list box and then choose the **Insert** button. The menu is added to the menu bar.*

The following is the procedure of exporting a set of queries from the current drawing.

1. From the **dbConnect** menu, choose **Queries** > **Export Query Set** to invoke the **Export Query Set** dialog box. In the dialog box, select the directory you want to save the query set to from the **Save In** list.

2. Under the **File Name** box, specify name for the query set, and then choose the **Save** button.

Following is the procedure of **importing a set of queries** into the current drawing:

1. From the **dbConnect** menu, choose **Queries** > **Import Query Set** to invoke the **Import Query Set** dialog box. In the dialog box, select the query set to be imported.

2. Choose the **Open** button to import the query set into the current drawing.

AutoCAD displays an alert box that you can use to provide a unique name for the query, if there is a query with the same name that is already associated with the current drawing.

FORMING SELECTION SETS USING THE LINK SELECT

It is possible to locate objects on the drawing on the basis of the linked nongraphic information. For example, you can locate the object that is linked to the first row of the **Computer** table or to the first and second rows of the **Computer** table. You can highlight specified objects or form a selection set of the selected objects. The **Link Select** is an advanced feature of the **Query Editor** that can be used to construct iterative selection sets of AutoCAD graphical objects and the database records. You can start constructing a query or selecting AutoCAD graphical objects for iterative selection process. The initial selection set is referred to as **set A**. Now you can select an additional set of queries or graphical objects to further refine your selection set. The second selection set is referred to as **set B**. To refine your final selection set you must establish a relation between set A and set B. The following is the list of available relationships or set operations (Figure 27-28).

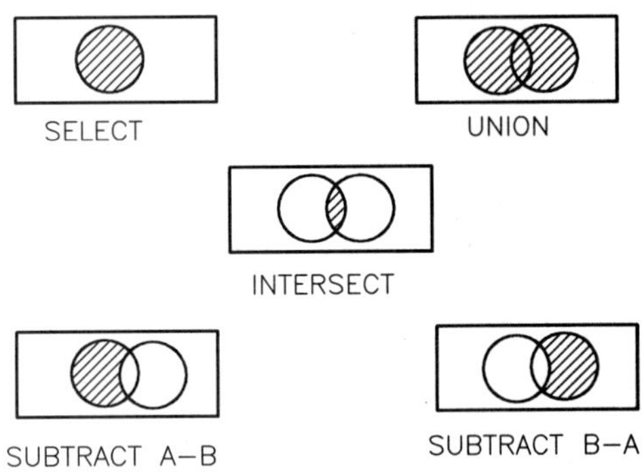

***Figure 27-28** Available relationships or Set operations*

Select

This creates an initial query or graphical objects selection set. This selection set can be further refined or modified using subsequent Link Select operations.

Union

This operation adds the outcome of a new selection set or query to the existing selection set. Union returns all the records that are members of set A **or** set B.

Intersect

This operation returns the intersection of the new selection set and the existing or running selection set. Intersection returns only the records that are common to both set A **and** set B.

Subtract A - B

This operation subtracts the result of the new selection set or query from the existing set.

Subtract B - A

This operation subtracts the result of the existing selection set or query from the new set.

After any of the Link Selection operation is executed, the result of the operation becomes the new running selection set and is assigned as set A. You can extend refining your selection set by creating a new set B and then continuing with the iterative process.

Following is the procedure to use **Link Select** for refining a selection set.

1. Choose **Links** > **Link Select** from the **dbConnect** menu to invoke the **Link Select** dialog box (Figure 27-29).

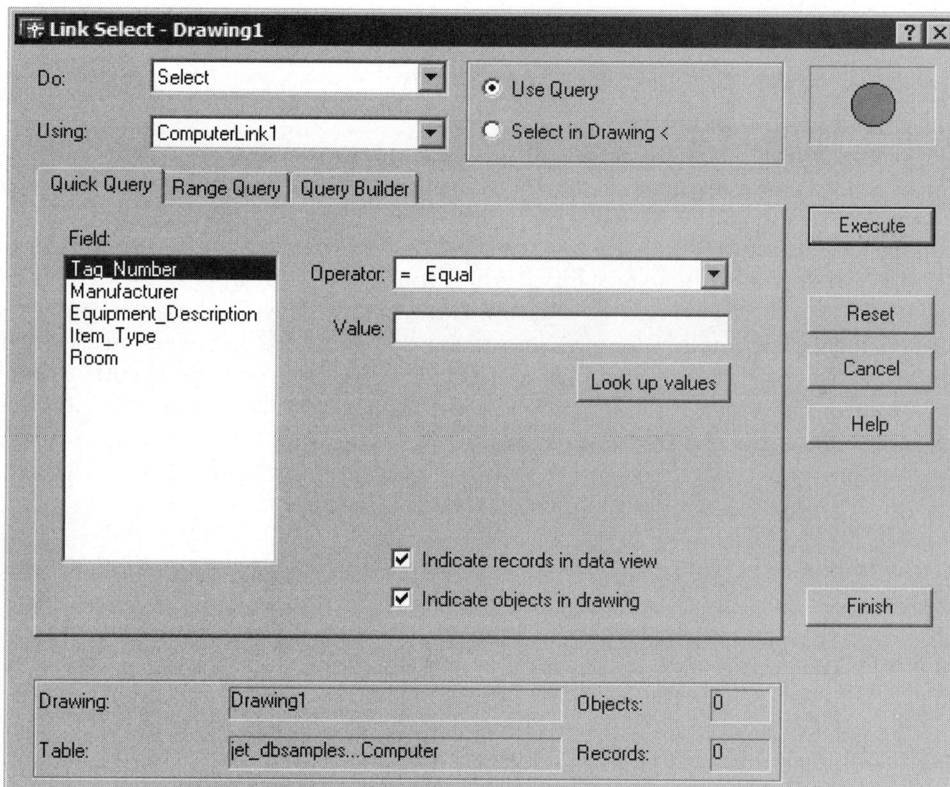

***Figure 27-29 Link Select** dialog box*

2. Select the **Select** option from the **Do** drop-down list for creating a new selection set. Also select a link template from the **Using** drop-down list.

3. Choose either the **Use Query** or **Select in Drawing <** option for creating a new query or drawing object selection set.

4. Choose the **Execute** button to execute the refining operation or **Select** to add your query or graphical object selection set.

5. Again choose any **Link Selection** operation from the **Do:** drop-down list.

6. Repeat steps 2 through 4 for creating a set B to the **Link Select** operation and then choose the **Finish** button to complete the operation.

Note

*You can choose the **Use Query** option to construct a new query or **Select in Drawing <** to select a graphical object from the drawing as a selection set.*

*If you select **Indicate Records in Data View**, then the Link Select operation result will be displayed in the **Data View** window and if you select **Indicate objects in drawing**, then AutoCAD displays a set of linked graphics objects in the drawing.*

CONVERSION OF ASE LINKS TO AutoCAD 2000 OR LATER FORMATS

From AutoCAD release 2000 or later, the links are stored in a different format. If you want to work with your links in AutoCAD 2004, you need to convert them to the AutoCAD 2000 or later format. You also need to create an AutoCAD 2004 configuration file that will point to the data source referred by the legacy links. When you open a drawing in AutoCAD 2004 that has links that were created by earlier releases, AutoCAD attempts to perform an automatic conversion of legacy information. But in some cases, where automatic conversion is not possible or a legacy data source does not match exactly the data source in AutoCAD 2004 format, you can use the **Link Conversion** dialog box. During the conversion process, AutoCAD writes the data source mapping information in **asi.ini** file and applies the information for further usage of the same data source. You can use the following steps to Convert AutoCAD R13 and R14 links to the AutoCAD 2000 or later formats.

1. Choose **Link Conversion** from the **dbConnect** menu to invoke the **Link Conversion** dialog box.

2. The **Link Conversion** dialog box will be displayed as shown in Figure 27-30. In the **Old Link Format** area, enter the following specifications.

   R13/R14 link format to be converted.
   Name of the Environment, Catalog, Schema, and Table.
   Link path name of the link that has to be formatted.

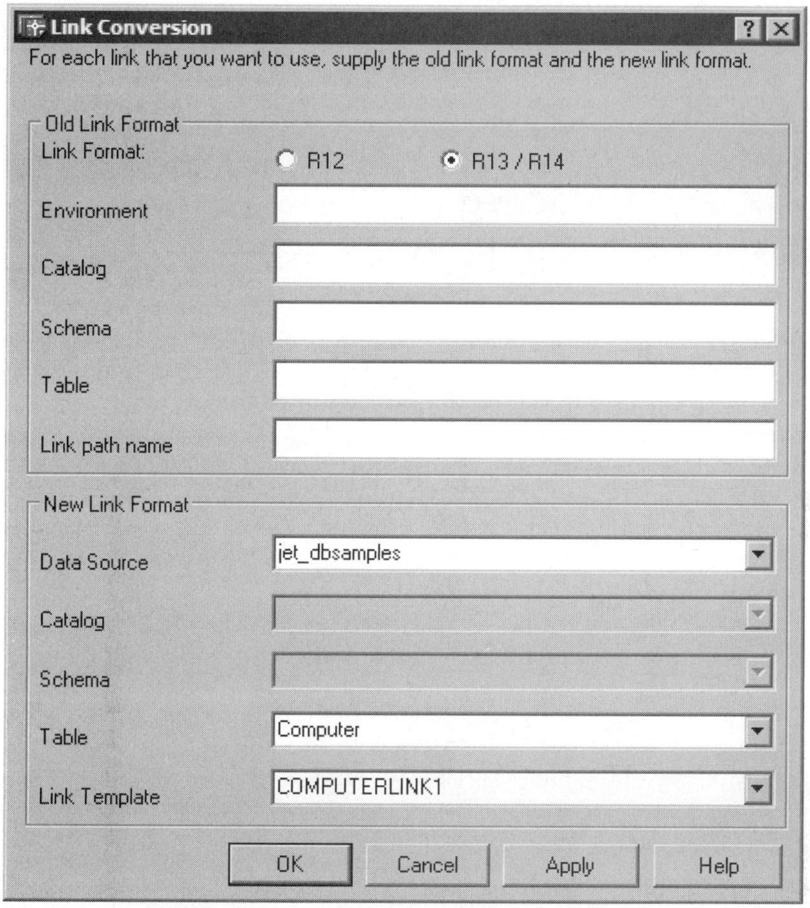

***Figure 27-30 Link Conversion** dialog box*

3. In the **New Link Format**, enter the following specifications.

   Name of the AutoCAD 2004 data source (for example, **jet_dbsamples**).
   Name of the new AutoCAD 2004 Environment, Catalog, Schema, and Table.
   Name of the AutoCAD 2004 link template.

4. Choose **Apply** and then the **OK** button to complete the conversion procedure.

5. Open the drawing you want to convert and save it in AutoCAD 2004 format drawing.

Saving AutoCAD 2004 Links to R13/R14 Formats

You can convert Links created in AutoCAD 2004 format as R13 or R14 formats, but you cannot save AutoCAD 2004 links in R12 format. When you select **Save As** for saving the links with drawings, select links to R13 or R14 formats and AutoCAD will automatically convert into the format of earlier release.

Self-Evaluation Test

Answer the following questions and then compare your answers with the answers given at the end of this chapter.

1. The horizontal group of data is called a __________.
2. A vertical group of data (attribute) is called a __________.
3. You can connect to an external database table by using __________ Manager.
4. You can edit an external database table within AutoCAD session. (T/F)
5. You can resize, dock, as well as hide the **Data View** window. (T/F)
6. Once you define a link, AutoCAD does not store that information with the drawing. (T/F)
7. To display the associated records with the drawing objects, AutoCAD provides__________.
8. Labels are of two types, __________ and __________.
9. It is possible to import as well as export Link and label templates. (T/F)
10. ASE stands for __________.

Review Questions

Answer the following questions.

1. The SQL statements let you search through the database and retrieve the information as specified in the SQL statements. (T/F)
2. What are the various components of a table?
3. What is a database?
4. What is a database management system (DBMS)?
5. A row is also referred to as a __________.
6. A column is sometimes called a __________.
7. The __________ acts like an identification tag for locating and linking a row.
8. Only one row can be manipulated at a time. This row is called the _________ .

9. **AutoCAD Query Editor** has four tabs, namely __________, __________, __________, and __________.

10. It is possible to go back to the previous query tab without resetting the values. (T/F)

11. You can execute query after any tab in **Query Editor.** (T/F)

12. Link select is an __________ implementation of AutoCAD **Query Editor**, which constructs selection sets of graphical objects or database records.

13. AutoCAD writes data source mapping information in the __________ file during the conversion process of links.

14. AutoCAD 2004 links __________ be converted into AutoCAD R12 format.

Exercises

Exercise 1 *General*

In this exercise, select **Employee** table from **jet_dbsamples** data source in **DBCONNECT MANAGER**, then edit the sixth row of the table (EMP_ID=1006, Keyser). Also, add a new row to the table, set the new row current, and then view it.

The row to be added has the following values.

EMP_ID	**1064**
LAST_NAME	**Joel**
FIRST_NAME	**Billy**
Gender	**M**
TITLE	**Marketing Executive**
Department	**Marketing**
ROOM	**6071**

Exercise 2 *General*

From the **Inventory** table in **jet_dbsamples** data source, build an SQL query step-by-step using all the tabs of the AutoCAD Query Editor. The conditions to be implemented are as follows.

1. Type of Item = **Furniture** and
2. Range of Cost = **200 to 650** and
3. Manufacturer = **Office master**

Answers to the Self-Evaluation Test

1 - Rows, **2** - Columns, **3** - **dbConnect**, **4** - T, **5** - T, **6** - F, **7** - Leaders, **8** - Freestanding, Attached, **9** - T, **10** - **AutoCAD SQL Environment**

Chapter 28

AutoCAD on the Internet

Learning Objectives

After completing this chapter, you will be able to:

- *Launch a Web browser from AutoCAD.*
- *Understand the importance of the uniform resource locator (URL).*
- *Open and save drawings to and from the Internet.*
- *Place hyperlinks in a drawing.*
- *Learn about the Autodesk Express Viewer plug-in and the DWF file format.*
- *View DWF files with a Web browser.*
- *Convert drawings to a DWF file format.*

INTRODUCTION

The Internet has become the most important and the fastest way to exchange information in the world. AutoCAD allows you to interact with the Internet in several ways. It can open and save files that are located on the Internet; it can launch a Web browser (AutoCAD 2004 includes a simple Web browser); and it can create Drawing Web Format (DWF) files for viewing as drawings on Web pages. Before you can use the Internet features in AutoCAD 2004, some components of Microsoft Internet Explorer must be present in your computer. If Internet Explorer version 5.01 (or later) or Netscape Navigator 4.7 (or later) is already installed on your computer, then you have these required components. If not, then the components are installed automatically during AutoCAD 2004's setup when you select: (1) the **Full Install**; or (2) the **Internet Tools** option during the **Custom** installation.

This chapter introduces the following Web-related commands.

Browser

Launches a Web browser from within AutoCAD.

Hyperlink

Attaches and removes a uniform resource locator (URL) to an object or an area in the drawing.

Hyperlinkfwd

Move to the next hyperlink (an undocumented command).

Hyperlinkback

Moves back to the previous hyperlink (an undocumented command).

Hyperlinkstop

Stops the hyperlink access action (an undocumented command).

Pasteashyperlink

Attaches a URL to an object in the drawing from text stored in the Windows Clipboard (an undocumented command).

Hyperlinkbase

A system variable for setting the path used for relative hyperlinks in the drawing.

In addition, all of AutoCAD 2004's file-related dialog boxes are "Web enabled."

CHANGES FROM AutoCAD RELEASE 14

The changes in Internet features of AutoCAD 2004 as compared to AutoCAD 14 are discussed next.

Attached URLs from R14

In AutoCAD R14, attached URLs were not active until the drawing was converted to a DWF file; in AutoCAD 2004, URLs are active in the drawing.

If you attached URLs to drawings in Release 14 (as well as LT 97 and LT 98), they are converted to AutoCAD 2004-style hyperlinks the first time you save the drawings in the AutoCAD 2004 DWG format.

CHANGED INTERNET COMMANDS

Many of AutoCAD's previous release Internet-related commands have been discontinued and replaced. The following summary describes the status of Internet commands that are used in the current release and the previous releases of AutoCAD.

BROWSER (launches a Web browser from within AutoCAD): continues to work in AutoCAD 2004; the default URL is now *http://www.autodesk.com*.

ATTACHURL (attaches a URL to an object or an area in the drawing): continues to work in AutoCAD 2004, but has been superceded by the **-HYPERLINK** command's **Insert** option.

SELECTURL (selects all objects with attached URLs): continues to work in AutoCAD 2004, but has been superceded by the **Quick Select** dialog box's **Hyperlink** option.

LISTURL (lists URLs embedded in the drawing): was removed from AutoCAD 2002 and hence is also not available in AutoCAD 2004. As a replacement, use the **Properties** dialog box's **Hyperlink** option.

DETACHURL (removes the URL from an object): continues to work in AutoCAD 2004, but has been superceded by the **-HYPERLINK** command's **Remove** option.

DWFOUT (exports the drawing and embedded URLs as a DWF file): continues to work in AutoCAD 2004, but has been superceded by the **Plot** dialog box's **ePlot** option.

DWFOUTD (exports the drawing in DWF format without a dialog box): was removed from AutoCAD 2002 and hence is also not available in AutoCAD 2004. As a replacement, use the **-PLOT** command's **ePlot** option.

INETCFG (configures AutoCAD for Internet access): was removed from AutoCAD 2002 and hence is also not available in AutoCAD 2004. It has been replaced by the **Internet** applet of the Windows Control Panel (choose the **Connection** tab).

INSERTURL (inserts a block from the Internet into the drawing): automatically executes the **INSERT** command and displays the **Insert** dialog box. You may type a URL for the block name.

OPENURL (opens a drawing from the Internet): automatically executes the **OPEN** command and displays the **Select File** dialog box. You may type a URL for the file name.

SAVEURL (saves the drawing to the Internet): automatically executes the **SAVE** command and displays the **Save Drawing As** dialog box. You may type a URL for the file name.

You are probably already familiar with the best-known uses of the Internet: e-mail (electronic mail) and the www (short for "World Wide Web"). E-mail allows the user to quickly exchange messages and data at very low cost. The Web brings together text, graphics, audio, and movies in an easy-to-use format. Other uses of the Internet include file transfer protocol (FTP) for effortless binary file transfer, Gopher (presents data in a structured, subdirectory-like format), and USENET, a collection of more than 10,000 news groups.

AutoCAD allows you to interact with the Internet in several ways. AutoCAD is able to launch a Web browser from within AutoCAD with the **BROWSER** command. Hyperlinks can be inserted in drawings with the **HYPERLINK** command, which lets you link the drawing with other documents on your computer and the Internet. With the **PLOT** command's **ePlot** option (short for "electronic plot), AutoCAD creates DWF files for viewing drawings in two-dimensional (2D) format on Web pages. AutoCAD can open, insert, and save drawings to and from the Internet through the **OPEN**, **INSERT**, and **SAVEAS** commands.

UNDERSTANDING URLS

The **uniform resource locator**, known as the URL, is the file naming system of the Internet. The URL system allows you to find any resource (a file) on the Internet. Example resources include a text file, a Web page, a program file, an audio or movie clip—in short, anything you might also find on your own computer. The primary difference is that these resources are located on somebody else's computer. A typical URL looks like the following examples.

ExampleURL	Meaning
http://www.autodesk.com	Autodesk Primary Web site
news://adesknews.autodesk.com	Autodesk News Server
ftp://ftp.autodesk.com	Autodesk FTP Server
http://www.autodeskpress.com	Autodesk Press Web Site

Note that the **http://** prefix is not required. Most of today's Web browsers automatically add in the *routing* prefix, which saves you a few keystrokes.

URLs can access several different kinds of resources such as Web sites, e-mail, and news groups, but they always take on the same general format as follows.

scheme://netloc

The scheme accesses the specific resource on the Internet including these.

Scheme	Meaning
file://	File is located on your computer's hard drive or local network
ftp://	File Transfer Protocol (used for downloading files)
http://	Hyper Text Transfer Protocol (the basis of Web sites)
mailto://	Electronic mail (e-mail)
news://	Usenet news (news groups)
telnet://	Telnet protocol
gopher://	Gopher protocol

The **://** characters indicate a network address. Autodesk recommends the following format for specifying URL-style file names with AutoCAD.

Resource	URL Format
Web Site	**http:**//*servername/pathname/filename*
FTP Site	**ftp:**//*servername/pathname/filename*
Local File	**file:**///*drive:/pathname/filename*
or	*drive:\pathname\filename*
or	**file:**///*drive | /pathname/filename*
or	**file:**//*localPC\pathname\filename*
or	**file:**////*localPC/pathname/filename*
Network File	**file:**//*localhost/drive*:*/pathname/filename*
or	*localhost\drive*:*\pathname\filename*
or	**file:**//*localhost/drive* **|** */pathname/filename*

The terminology can be confusing. The following definitions will help to clarify these terms.

Term	Meaning
servername	The name or location of a computer on the Internet, for example: *www.autodesk.com*
pathname	The same as a subdirectory or folder name
drive	The driver letter, such as C: or D:
localpc	A file located on your computer
localhost	The name of the network host computer

If you are not sure of the name of the network host computer, use Windows Explorer to check the Network Neighborhood for the network names of computers.

Launching a Web Browser

The **BROWSER** command lets you start a Web browser from within AutoCAD. Commonly used Web browsers include Netscape Navigator, Microsoft Internet Explorer, and Operasoft's Opera.

By default, the **BROWSER** command uses whatever brand of Web browser program is registered in your computer's Windows operating system. AutoCAD prompts you for the URL, such as *http://www.autodeskpress.com*. The **BROWSER** command can be used in scripts, toolbars or menu macros, and AutoLISP routines to automatically access the Internet.

Command: **BROWSER**
Enter Web location (URL) <http://www.autodesk.com>: *Enter a URL.*

The default URL is an HTML file added to your computer during AutoCAD's installation. After you type the URL and press ENTER, AutoCAD launches the Web browser and contacts the Web site. Figure 28-1 shows the popular Internet Explorer with the Autodesk Web site.

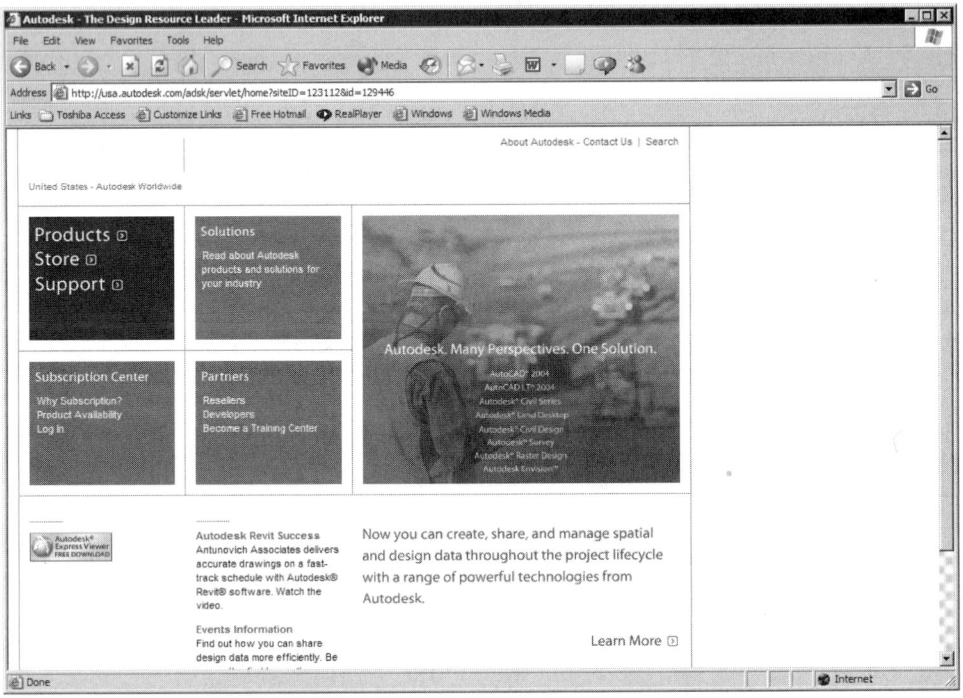

Figure 28-1 *Internet Explorer displaying the Autodesk Web site*

Changing the Default Web Site

To change the default Web page that your browser starts with from within AutoCAD, change the setting in the **INETLOCATION** system variable. The variable stores the URL used by the last executed **BROWSER** command and the **Browse the Web** dialog box. Make the change as follows.

Command: **INETLOCATION**
Enter new value for INETLOCATION <"http://www.autodesk.com">: *Type URL.*

DRAWINGS ON THE INTERNET

When a drawing is stored on the Internet, you access it from within AutoCAD 2004 using the standard **OPEN**, **INSERT**, and **SAVE** commands. (In Release 14, these commands were known as **OPENURL**, **INSERTURL**, and **SAVEURL**.) Instead of specifying the file's location with the usual drive-subdirectory-file name format such as *c:\acad2004\filename.dwg*, use the URL format. (Recall that the URL is the universal file-naming system used by the Internet to access any file located on any computer hooked up to the Internet.)

Opening Drawings from the Internet

The drawings from the Internet can be easily opened using the usual **Select File** dialog box. This dialog box can be displayed by invoking the **OPEN** command. Since the file is on the Web. Therefore, choose the **Search the Web** button from the **Select File** dialog box, see Figure 28-2.

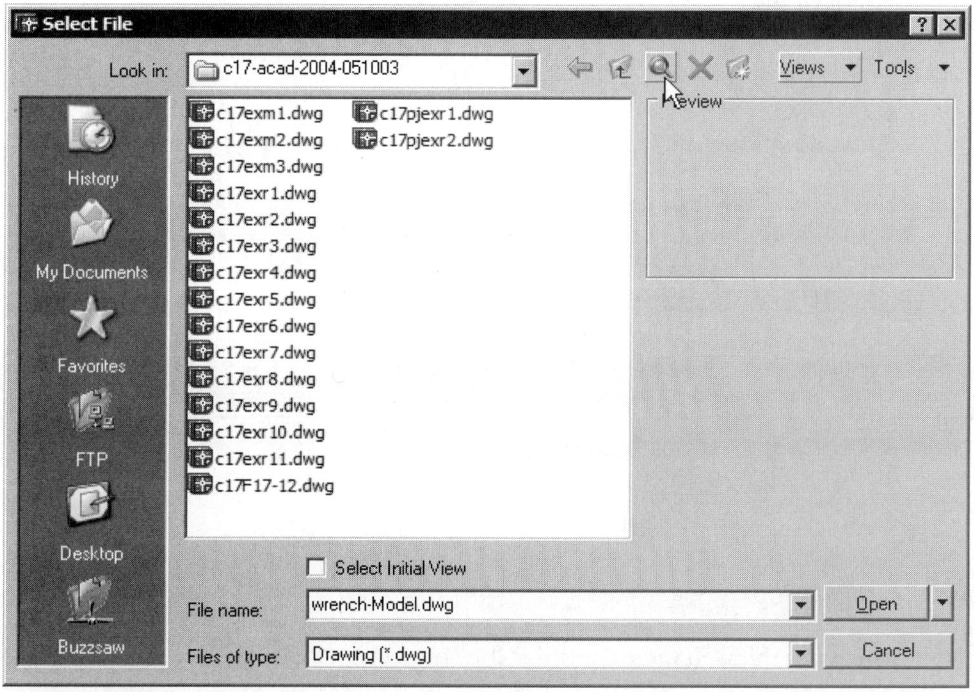

Figure 28-2 *Choosing the* ***Search the Web*** *button from the* ***Select File*** *dialog box*

When you choose the **Search the Web** button, AutoCAD opens the **Browse the Web** dialog box, see Figure 28-3. This dialog box is a simplified version of the Microsoft brand of Web browser. The purpose of this dialog box is to allow you to browse files at a Web site.

By default, the **Browse the Web** dialog box displays the contents of the URL stored in the **INETLOCATION** system variable. You can easily change this to another folder or Web site, by entering the required URL in the **Look in** edit box.

Along the top, the dialog box has six buttons. They are discussed next.

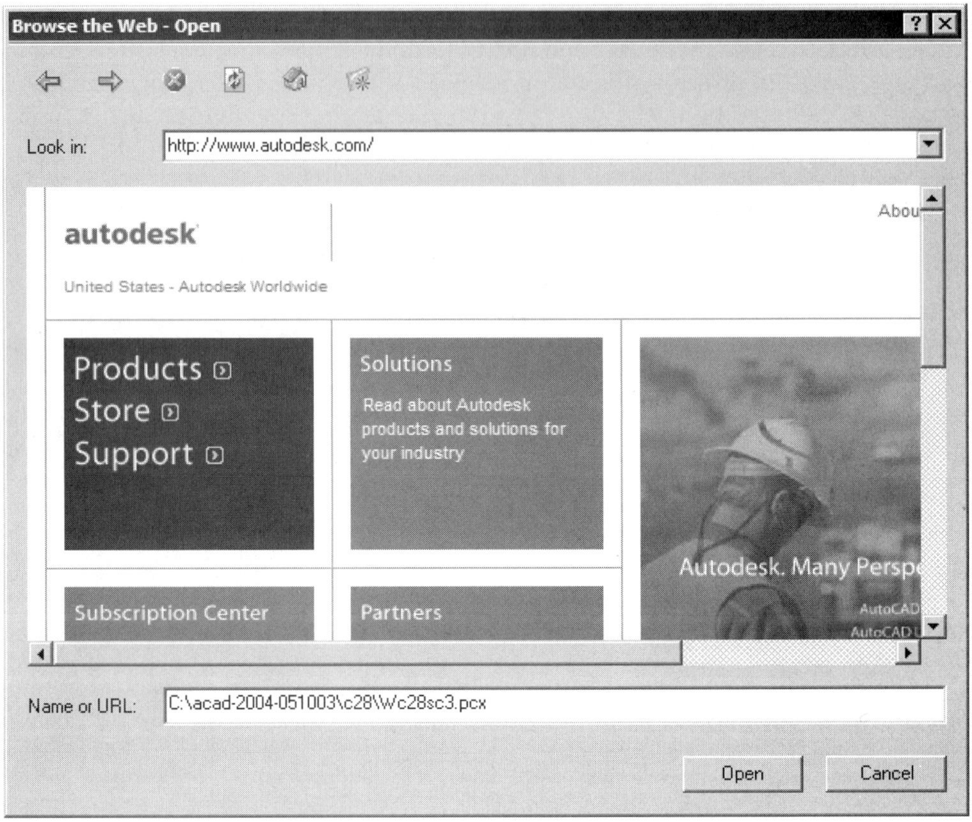

Figure 28-3 ***Browse the Web*** *dialog box*

Back. Go back to the previous URL.

Forward. Go forward to the next URL.

Stop. Halt displaying the Web page (useful if the connection is slow or the page is very large).

Refresh. Redisplay the current Web page.

Home. Return to the location specified by the **INETLOCATION** system variable.

Favorites. List stored URLs (hyperlinks) or bookmarks. If you have previously used Internet Explorer, you will find all your favorites listed here. Favorites are stored in the *\Windows\Favorites* folder on your computer.

The **Look in** field allows you to type the URL. Alternatively, click the down arrow to select a previous destination. If you have stored Web site addresses in the Favorites folder, then select a URL from that list.

You can either double-click a file name in the window, or type a URL in the **Name or URL** edit box. The following table gives templates for typing the URL to open a drawing file:

Drawing Location	Template URL
Web or HTTP Site	*http://servername/pathname/filename.dwg*
	http://practicewrench.autodeskpress.com/wrench.dwg
FTP Site	*ftp://servername/pathname/filename.dwg*
	ftp://ftp.autodesk.com
Local File	*drive:\pathname\filename.dwg*
	c:\acad 2004\sample\tablet2004.dwg
Network File	*\\localhost\drive:\pathname\filename.dwg*
	\\upstairs\d:\install\sample.dwg

When you open a drawing over the Internet, it will probably take much longer than opening a file found on your computer. During the file transfer, AutoCAD displays a dialog box to report the progress, see Figure 28-4. If your computer uses a 28.8 Kbps modem, you should allow about 5 to 10 min/MB of drawing file size. If your computer has access to a faster T1 connection to the Internet, you should expect a transfer speed of about 1 min/MB.

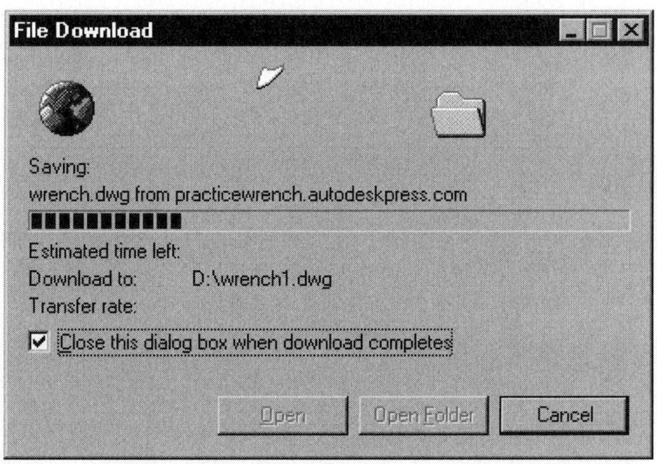

***Figure 28-4 File Download** dialog box*

It may be helpful to understand that the **OPEN** command does not copy the file from the Internet location directly into AutoCAD. Instead, it copies the file from the Internet to your computer's designated **Temporary** subdirectory such as *C:\Windows\Temp* (and then loads the drawing from the hard drive into AutoCAD). This is known as caching. It helps to speed up the processing of the drawing, since the drawing file is now located on your computer's fast hard drive, instead of the relatively slow Internet.

Note that the **Locate** and **Find File** buttons (in the **Select File** dialog box) do not work for locating files on the Internet.

Example 1 *General*

The author of this website has an area that allows you to practice using the Internet with AutoCAD. In this example, you will open a drawing file located at the author's Web site.

1. Start a new AutoCAD 2004 session.

2. Ensure that you have a live connection to the Internet. If you normally access the Internet via a telephone (modem) connection, dial your Internet service provider now.

3. Choose the **Open** button from the **Standard** toolbar. The **Select File** dialog box will be displayed.

4. Choose the **Search the Web** button to display the **Browse the Web** dialog box.

5. In the **Look in** field, type

   http://technology.calumet.purdue.edu/met/tickoo/students/acad-2000-usa/AutoCAD-Prob.htm

   Press ENTER. After a few seconds, the **Browse the Web** dialog box displays the Web page, see Figure 28-5.

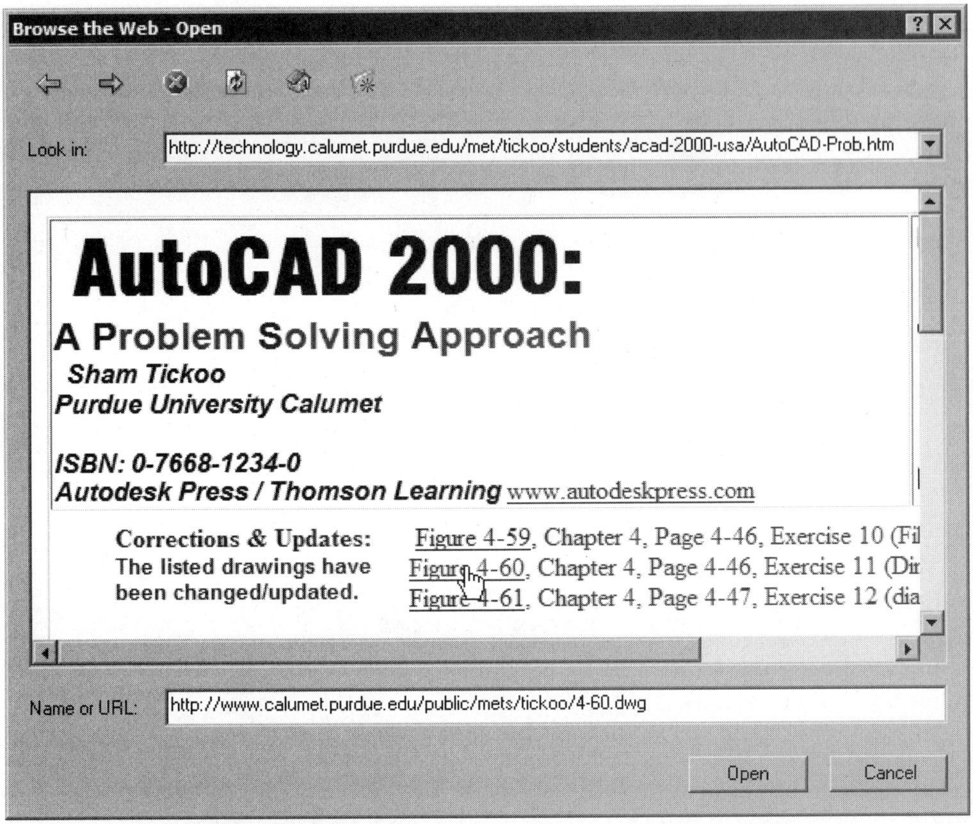

Figure 28-5 *Student's Web page of the author's Web site*

6. In the **Name or URL** edit box, type

   http://www.calumet.purdue.edu/public/mets/tickoo/4-60.dwg

   Press ENTER. AutoCAD begins transferring the file. Depending on the speed of your Internet connection, this will take between a couple of seconds and half minute. Once the file downloading is complete, the drawing of the wrench will be displayed in AutoCAD main window, see Figure 28-6.

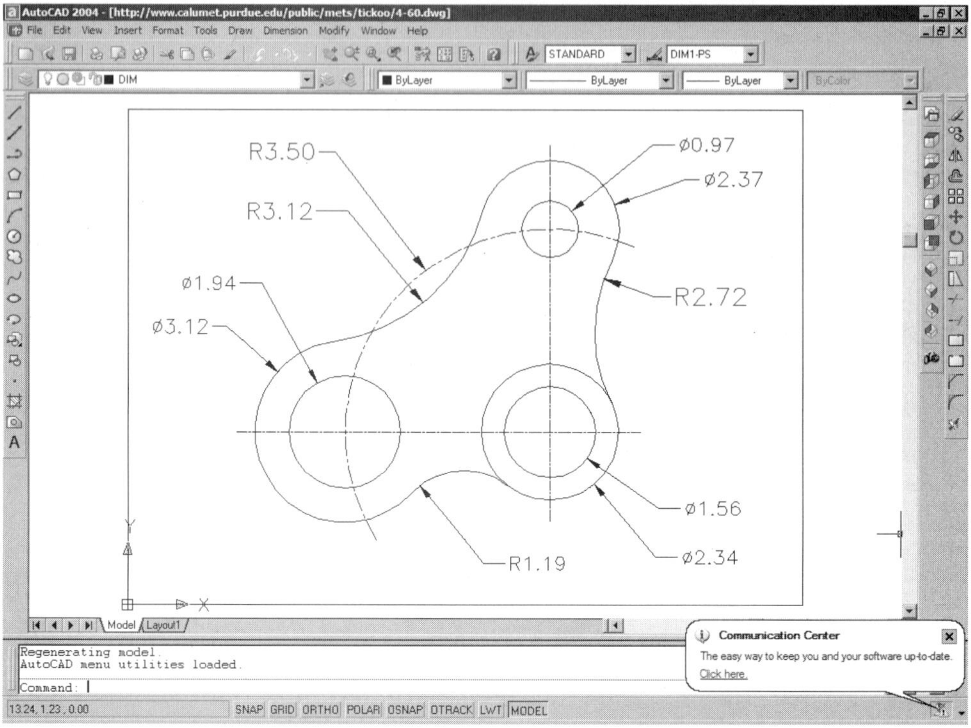

Figure 28-6 Drawing in AutoCAD

INSERTING A BLOCK FROM THE INTERNET

When a block (symbol) is stored on the Internet, you can access it from within AutoCAD using the **INSERT** command. When the **Insert** dialog box appears, choose the **Browse** button to display the **Select Drawing File** dialog box. This is identical to the dialog box discussed earlier.

After you select the file, AutoCAD downloads the file and continues with the prompt sequence of the **INSERT** command.

The process is identical for accessing external reference (xref) and raster image files. Other files that AutoCAD can access over the Internet include 3D Studio, SAT (ACIS solid modeling), DXB (drawing exchange binary), WMF (Windows metafile), and EPS (encapsulated PostScript). All of these options are found in the **Insert** menu on the menu bar.

ACCESSING OTHER FILES ON THE INTERNET

Most of the other file-related dialog boxes allow you to access files from the Internet or Intranet. This allows your firm or agency to have a central location that stores drawing standards. When you need to use a linetype or hatch pattern, for example, you access the

LIN or PAT file over the Internet. More than likely, you will have the location of these files stored in the Favorites list. Some examples include the following.

Linetypes. Choose **Linetype** from the **Format** menu. In the **Linetype Manager** dialog box, choose the **Load**, **File**, and **Look in Favorites** buttons.

Hatch Patterns. Use the Web browser to copy *.pat* files from a remote location to your computer.

Multiline Styles. Choose **Multiline Style** from the **Format** menu. In the **Multiline Styles** dialog box, choose the **Load**, **File**, and **Look in Favorites** buttons.

Layer Name. Choose the **Layers** button to display the **Layer Properties Manager** dialog box. In this dialog box, choose the **Restore State** button to display the **Layer State Manager** dialog box. In the **Layer Manager** dialog box, choose the **Import** and **Look in Favorites** buttons.

LISP and ARX Applications. Choose **Load Applications** from the **Tools** menu.

Scripts. Choose **Run Scripts** from the **Tools** menu.

Menus. Choose **Customize > Menus** from the **Tools** menu. In the **Menu Customization** dialog box, choose the **Browse** and **Look in Favorites** buttons.

Images. Choose **Display Image** > **View** from the **Tools** menu.

You cannot access text files, text fonts (SHX and TTF), color settings, lineweights, dimension styles, plot styles, OLE objects, or named UCSs over the Internet.

i-DROP

The i-drop plugin is installed on your system when you are installing AutoCAD 2004. If your system has i-drop installed, you can drag a drawing from a Web site and drop it in the AutoCAD main window. The drawing is displayed in AutoCAD. Using i-drop, you can insert blocks, symbols, and so on in your open drawing.

The i-drop can be used with earlier releases of AutoCAD and it can be downloaded for free from *www.autodesk.com/idrop*. A sample drawing of a chair is provided on this Web site, which you can drag and drop in the AutoCAD window.

Note

The drawings that are inserted using the i-drop are inserted as a block and the ***INSERT*** *command is executed automatically when you drop the drawing in the main window of AutoCAD.*

SAVING THE DRAWING TO THE INTERNET

When you are finished with editing a drawing in AutoCAD, you can save it to a file server on the Internet with the **SAVE** command. If you inserted the drawing from the Internet (using **INSERT**) into the default *Drawing.Dwg* drawing, AutoCAD insists you first save the drawing to your computer's hard drive.

When a drawing of the same name already exists at that URL, AutoCAD warns you, just as it does when you use the **SAVEAS** command. Recall from the **OPEN** command that AutoCAD uses your computer system's *Temp* subdirectory, therefore the reference to it in the dialog box.

ONLINE RESOURCES

To access the online resources, choose **Help > Online Resources** from the menu bar. The cascaded menu is displayed. The options that are available in the cascaded menu are **Product**, **Training**, **Customization**, and **Autodesk User Group International**. The functions of these options are discussed next.

Product Support

You can access this option only when you are connected to the Internet. When you choose this option, the Web page of the product support is displayed in the Web browser that is installed on your system. The information on the following can be obtained by selecting this option.

- Frequently asked questions.
- Product updates.
- Join discussion groups.
- Access available programs.

Note

*To get more information on the related topics in **Product Support** and the other **Online Resources** option, it is recommended to connect to the Internet and then choose the **Online Resources** option.*

Training

You can access this option only when you are connected to the Internet. When you choose this option, the Web page of training is displayed in the Web browser. The information on the following can be obtained by selecting this option.

- Information on Autodesk authorized training centers.
- General training centers.
- Autodesk Certification.
- Discussion groups.
- Learning tools.

Customization

You can access this option only when you are connected to the Internet. When you choose this option, the Web page is displayed in the Web browser that provides the following information.

- Answers to the questions that were asked by the users.
- Sample applications.
- Documentation.

Note

*When you choose any one of the **Online Resources** options, the **Live Update Status** Web browser window is displayed.*

Autodesk User Group International

When you choose this window, **Autodesk User Group International** Web browser window is displayed. This Web browser gives information about the AutoCAD user groups. There are some links in the Web page that can only be accessed when your system is connected to the Internet.

USING HYPERLINKS WITH AutoCAD

AutoCAD 2004 allows you to employ URLs in two ways:

1. Directly within an AutoCAD drawing.
2. Indirectly in DWF files displayed by a Web browser.

URLs are also known as hyperlinks, the term that is used throughout this book. Hyperlinks are created, edited, and removed with the **HYPERLINK** command. You can also use the command line with the help of the **-HYPERLINK** command.

HYPERLINK Command

Menu:	Insert > Hyperlink
Command:	HYPERLINK

When you invoke this command, you will be prompted to select the objects. Once the objects are selected, the **Insert Hyperlink** dialog box will be displayed, see Figure 28-7.

If you select an object that has a hyperlink already attached to it, the **Edit Hyperlink** dialog box is displayed as shown in Figure 28-8. The **Remove Link** button is also available in the dialog box. This button allows you to remove the hyperlink from the object.

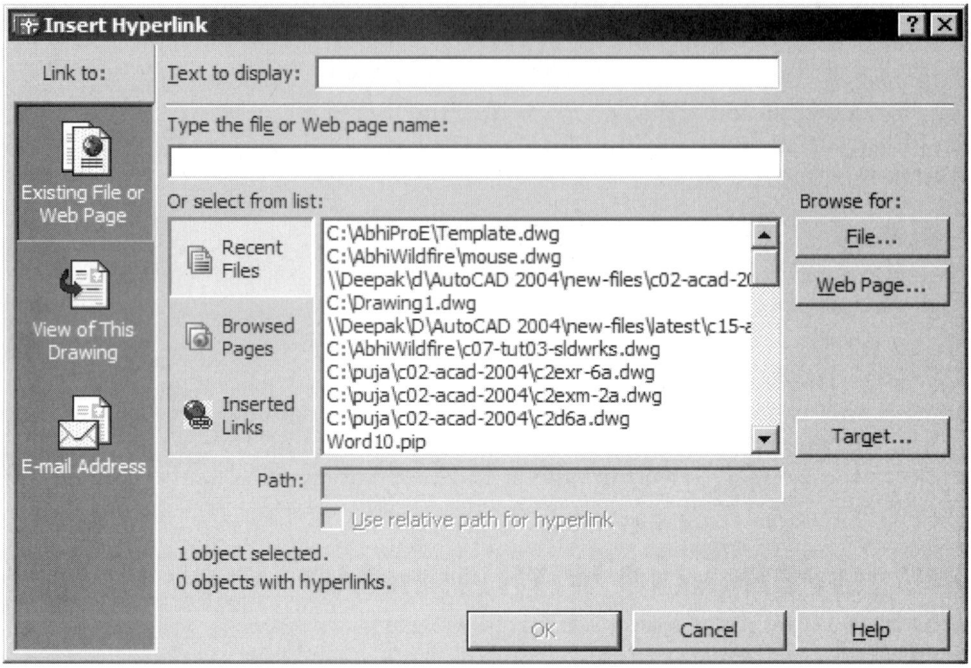

Figure 28-7 ***Insert Hyperlink*** *dialog box*

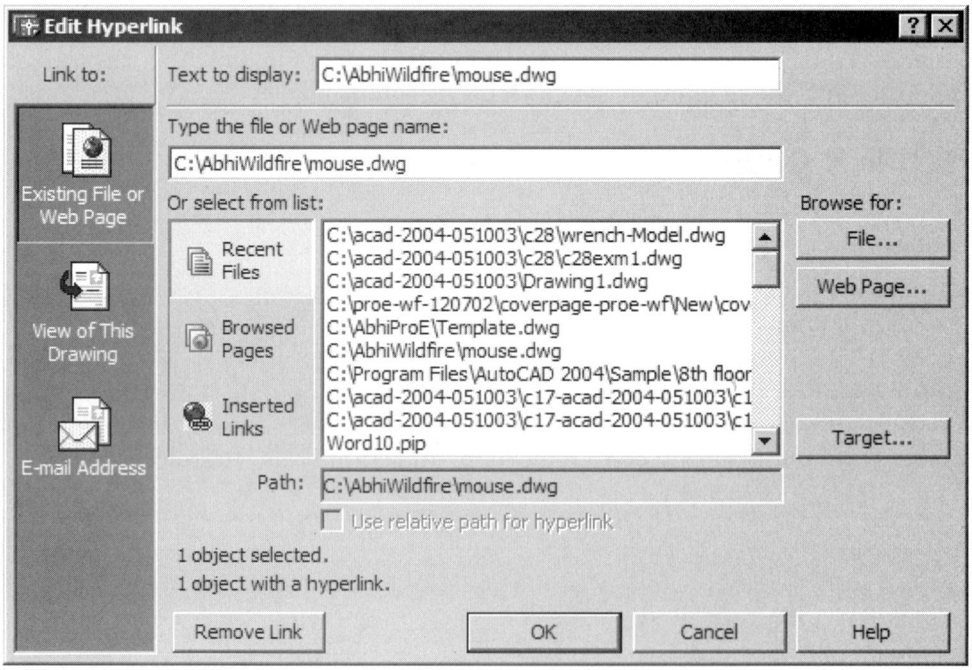

Figure 28-8 ***Edit Hyperlink*** *dialog box*

If you use the Command line for inserting the hyperlinks (**-HYPERLINK** command), you are also allowed to create hyperlink areas. The hyperlink area is a rectangular area that can be thought of as a 2D hyperlink (the dialog box-based **HYPERLINK** command does not create hyperlink areas). When you select the **Area** option, the rectangle is placed automatically on layer URLLAYER and colored red, see Figure 28-9.

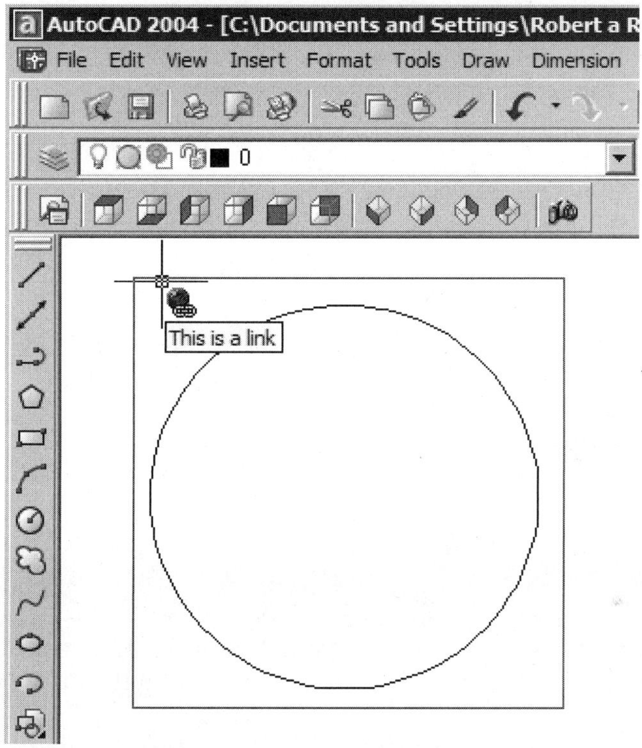

Figure 28-9 *A rectangular hyperlink area*

In the following sections, you learn how to apply and use hyperlinks in an AutoCAD drawing and in a Web browser through the dialog box-based **HYPERLINK** command.

Hyperlinks Inside AutoCAD

As mentioned earlier, AutoCAD allows you to add a hyperlink to any object in the drawing. An object is permitted just a single hyperlink. On the other hand, a single hyperlink may be applied to a selection set of objects.

You can determine whether or not an object has a hyperlink by passing the cursor over it. The cursor displays the "linked Earth" icon, as well as a tooltip describing the link, see Figure 28-10.

If, for some reason, you do not want to see the hyperlink cursor, you can turn it off. From the

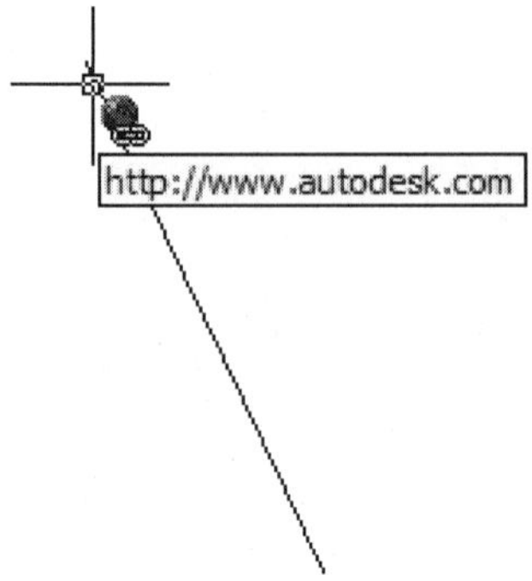

Figure 28-10 *The cursor reveals a hyperlink*

Tools menu, choose **Options**, and then choose the **User Preferences** tab. The **Display hyperlink cursor and shortcut menu** check box under the **Hyperlink** area toggles the display of the hyperlink cursor, as well as the **Hyperlink** tooltip option on the shortcut menu.

Attaching Hyperlinks

The following example explains attaching the hyperlinks.

Example 2 — *General*

In this example, you are given the drawing of a floorplan. To this drawing, you will add hyperlinks. These hyperlinks are another AutoCAD drawing. Hyperlinks must be attached to objects. For this reason, place some text in the drawing, and then attach the hyperlinks to the text.

1. Start the AutoCAD 2004 session.

2. Open the *8th floor plan.dwg* file found in the *AutoCAD 2004 \Sample* folder. Choose the **Model** tab to display the drawing in model space.

3. Save this file with the name *Example2.dwg*.

4. Using the **TEXT** command, write the text as shown in Figure 28-11.

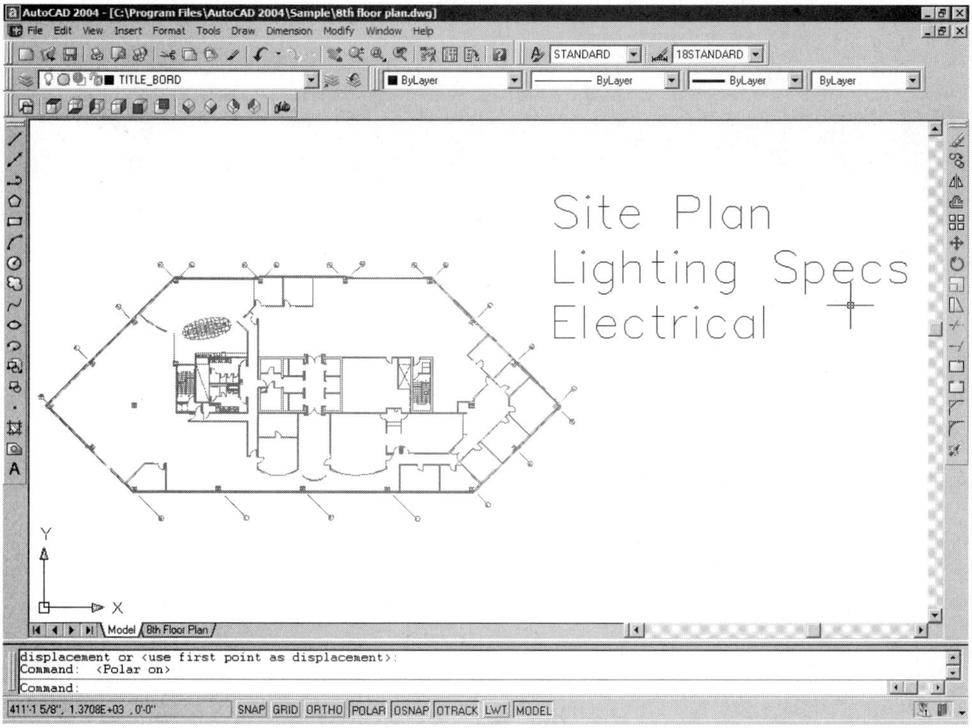

Figure 28-11 *Text placed in drawing*

5. Choose **Insert** > **Hyperlink** from the menu bar. You will be prompted to select the objects. Select the text **Site Plan** to display the **Insert Hyperlink** dialog box.

6. Choose the **File** button to display the **Browse the Web – Select Hyperlink** dialog box, see Figure 28-12.

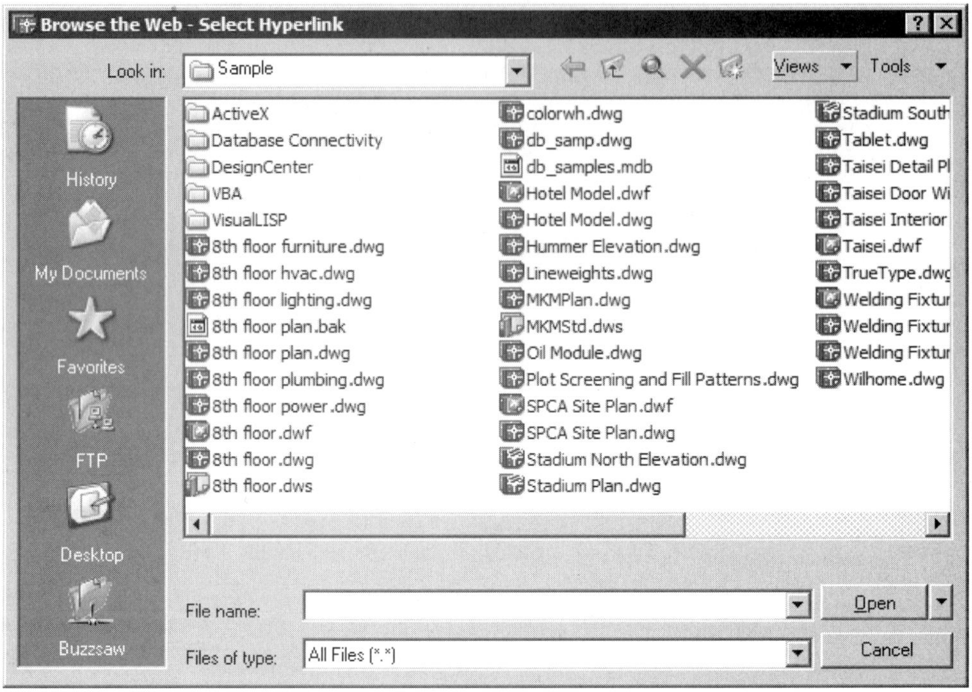

*Figure 28-12 **Browse the Web - Select Hyperlink** dialog box*

7. Go to AutoCAD 2004's *Sample* folder and select the *Hotel Model.dwg* file. Choose **Open**. AutoCAD does not open the drawing; rather, it copies the file's name to the **Insert Hyperlinks** dialog box. The name of the drawing you selected will be displayed in the **Type the file or web page name** edit box. The same name and path will also be displayed in the **Text to display** edit box. Remove the path in this edit box and retain only the file name.

8. Choose the **OK** button to close the dialog box. Move the cursor over the Site Plan text. Notice the display of the "linked Earth" icon; a moment later, the tooltip displays "Hotel Model.dwg", see Figure 28-13.

9. Connect to the Internet and then connect the URL *www.autodesk.com* to **Lighting Specs** using the **Web Page** button of the **Insert Hyperlink** dialog box.

10. You can directly open the file attached as the hyperlink to the object using the linked object. This can be done by selecting the object to which the file is linked and then right-clicking to display the shortcut menu. In the shortcut menu, choose **Hyperlink > Open (file name)**. In this case, the file name is *Hotel Model.dwg*. Therefore, right-click and select **Hyperlink** > **Open "Hotel Model.dwg"** from the shortcut menu, see Figure 28-14.

Figure 28-13 *The Hyperlink cursor and tooltip*

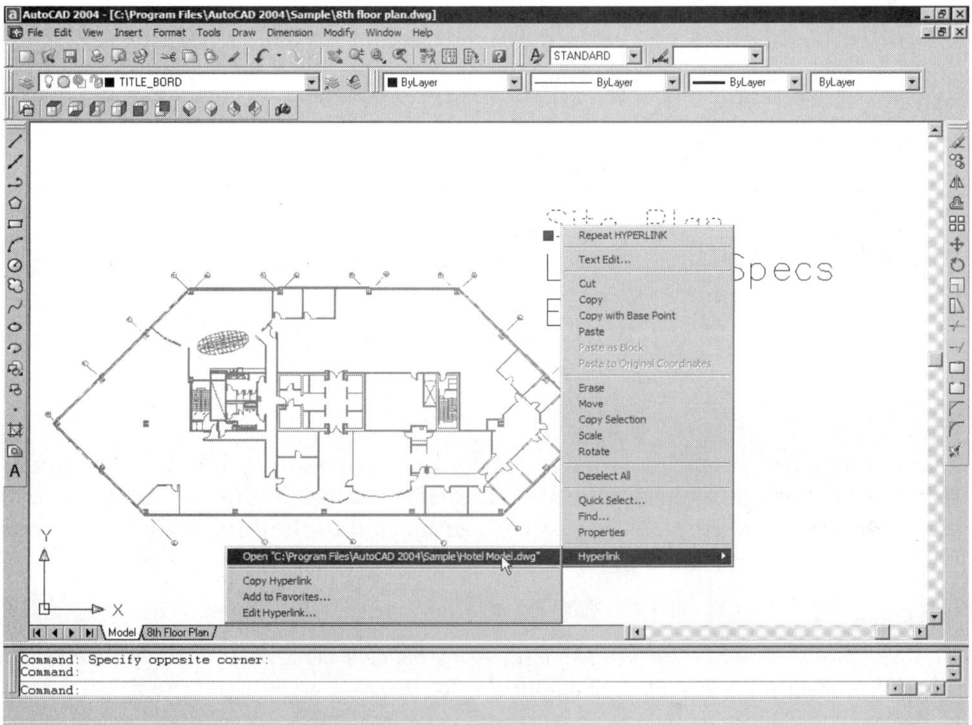

Figure 28-14 *Opening a hyperlink*

11. You can also see both drawings together. This is done by choosing **Tile Vertically** from the **Window** menu, see Figure 28-15.

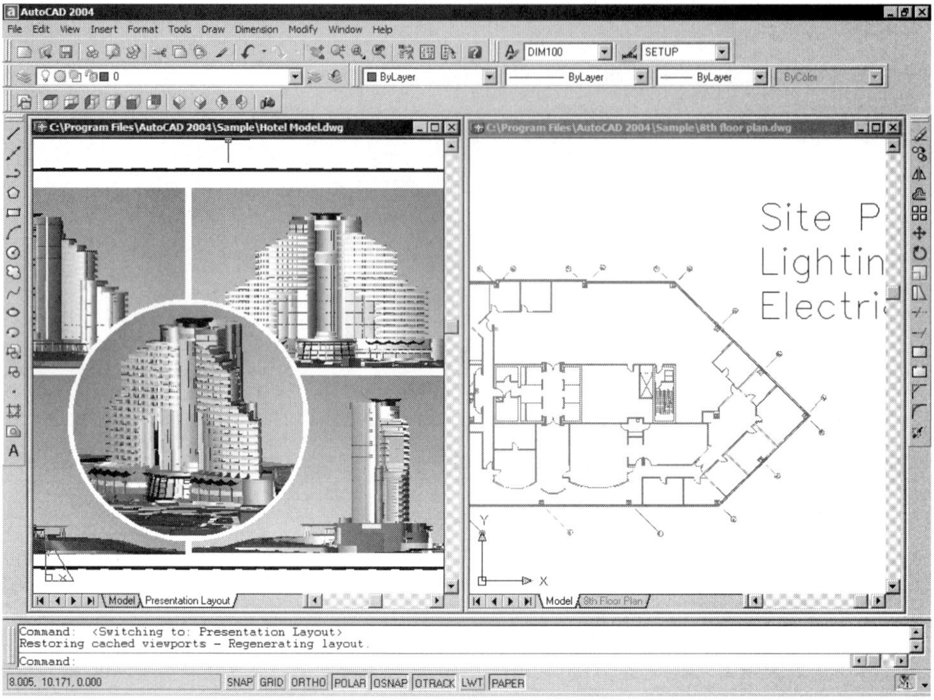

Figure 28-15 *Viewing both the drawings*

12. Select and then right-click on the "**Lighting Specs**" hyperlink. Select **Hyperlink > Open**. Notice that Windows starts your Web browser and opens the *www.autodesk.com* URL, see Figure 28-16. Make sure that you are connected to the Internet before opening this URL. If you are not connected to the Internet, this URL will not open.

Note

*If you have changed the default Web site using the **INTLOCATION** system variable, then your Web browser is opened with the modified URL.*

13. Save this drawing with the name *Example2.dwg*.

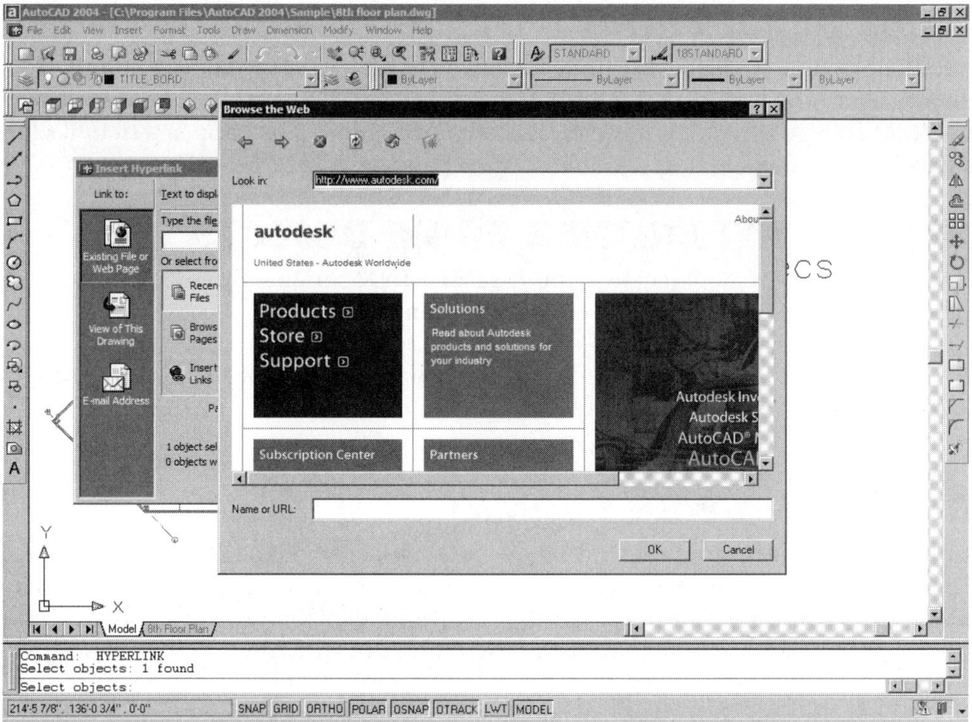

Figure 28-16 Web browser opens the URL that was hyperlinked to the selected text

PASTING AS HYPERLINK

AutoCAD 2004 has a shortcut method for pasting hyperlinks in the drawing. The hyperlink from one object can be copied and pasted to another object. This can be achieved by using the **PASTEASHYPERLINK** command.

1. In AutoCAD, select an object that has a hyperlink. Right-click to invoke the shortcut menu.

2. Choose the **Hyperlink** > **Copy Hyperlink** from the shortcut menu.

3. Choose **Edit** > **Paste as Hyperlink** from the menu bar. You are prompted to select the object to which the hyperlink is to be pasted.

4. Select the object and press ENTER. The hyperlink is pasted to the new object.

Tip

The ***MATCHPROP*** *command does not copy the hyperlinks from the source objects to the destination objects.*

EDITING HYPERLINKS

Once you know where the objects with hyperlinks are located, you can use the **HYPERLINK** command to edit the hyperlinks and related data. Select the hyperlinked object and invoke the **HYPERLINK** command. When the **Edit Hyperlink** dialog box appears (it looks identical to the **Insert Hyperlink** dialog box), make the changes and choose **OK**.

REMOVING HYPERLINKS FROM OBJECTS

To remove a URL from an object, use the **HYPERLINK** command on the object. When the **Edit Hyperlink** dialog box appears, choose the **Remove Link** button.

To remove a rectangular area hyperlink, you can simply use the **ERASE** command; select the rectangle and AutoCAD erases the rectangle. (Unlike in Release 14, AutoCAD no longer purges the URLLAYER layer.)

THE DRAWING WEB FORMAT

To display AutoCAD drawings on the Internet, Autodesk created a file format called drawing Web format (DWF). The DWF file has several benefits and some drawbacks over DWG files. The DWF file is compressed as much as eight times smaller than the original DWG drawing file. Therefore, it takes less time to transmit these files over the Internet, particularly with the relatively slow telephone modem connections. The DWF format is more secure, since the original drawing is not being displayed; another user cannot tamper with the original DWG file.

However, the DWF format has some drawbacks:

- You must go through the extra step of translating from DWG to DWF.
- DWF files cannot display rendered or shaded drawings.
- DWF is a flat 2D-file format; therefore, it does not preserve 3D data, although you can export a 3D view.
- AutoCAD itself cannot display DWF files.
- DWF files cannot be converted back to DWG format without using file translation software from a third-party vendor.
- Earlier versions of DWF did not handle paper space objects (version 2.x and earlier), or line widths, lineweights, and nonrectangular viewports (version 3.x and earlier).

To view a DWF file on the Internet, your Web browser needs to have a *plug-in* software extension called **Autodesk Express Viewer**. This viewer is installed on your system with AutoCAD 2004 installation. **Autodesk Express Viewer** allows Internet Explorer 5.01 (or later) to handle a variety of file formats. Autodesk makes this DWF plug-in freely available from its Web site at *http://www.autodesk.com*. It is a good idea to regularly check for updates to the DWF plug-in, which is updated frequently.

Autodesk Express Viewer is a stand-alone viewer that views and prints DWF files. The **Autodesk Express Viewer** is automatically installed on your computer when you install AutoCAD 2004.

Creating a DWF File

To create a DWF file from AutoCAD 2004, two methods can be used. The first method is to use the **PLOT** command and the second method is to use the **PUBLISH** command. The following steps explain you to create a DWF file using the **PLOT** command:

1. Open the file that has to be converted into a DWF file. Choose the **Plot** button from the **Standard** toolbar. The **Plot** dialog box is displayed as shown in Figure 28-17.

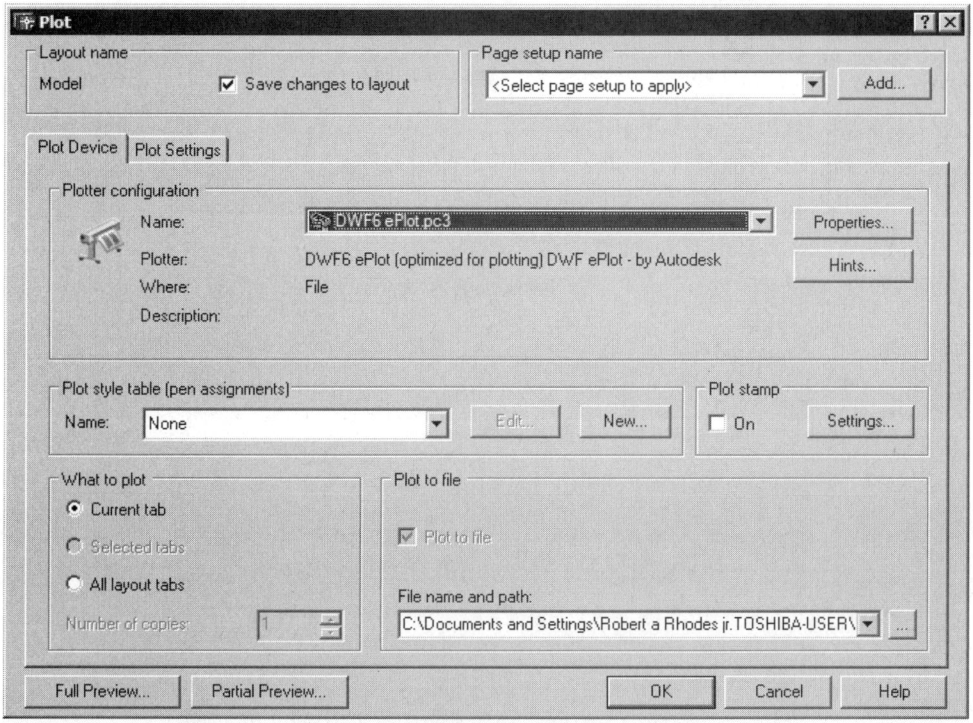

***Figure 28-17** **Plot** dialog box*

2. Select **DWF6 ePlot.pc3** from the **Name** drop-down list in the **Plotter Configuration** area.

3. Choose the **Properties** button to display the **Plotter Configuration Editor** dialog box, see Figure 28-18.

Chapter 28

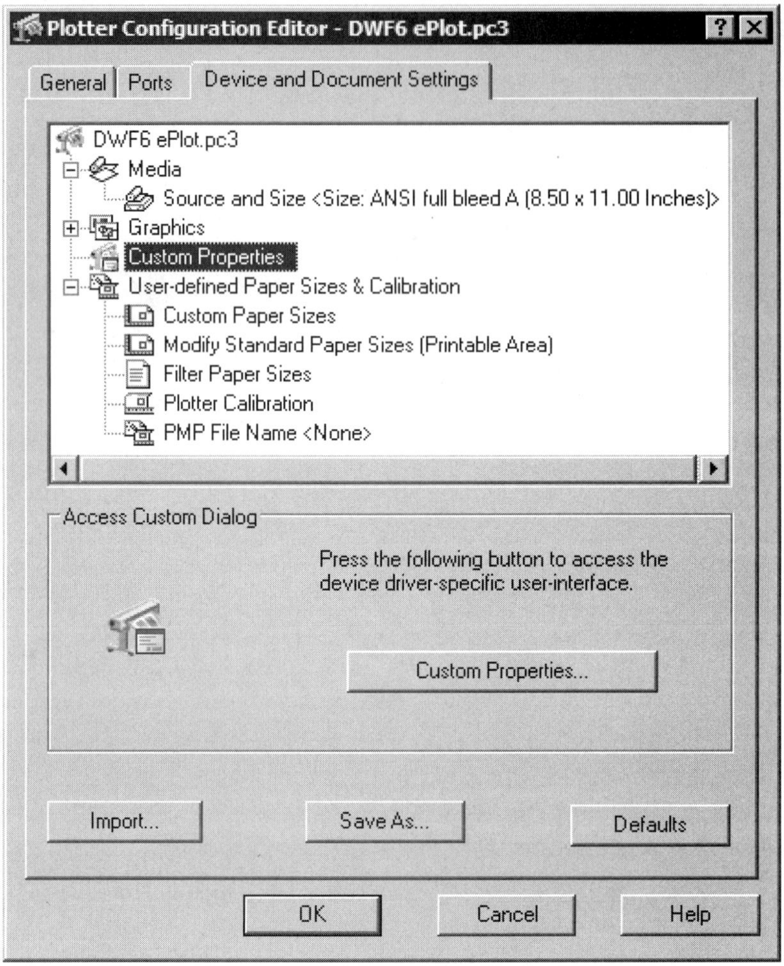

Figure 28-18 ***Plotter Configuration Editor*** *dialog box*

4. In the **Device and Documents Settings** tab, select **Custom Properties** in the tree view. The **Custom Properties** button will appear in the lower half of the dialog box.

5. Choose the **Custom Properties** button to display the **DWF6 eplot Properties** dialog box, see Figure 28-19. The options provided under this dialog box are discussed next.

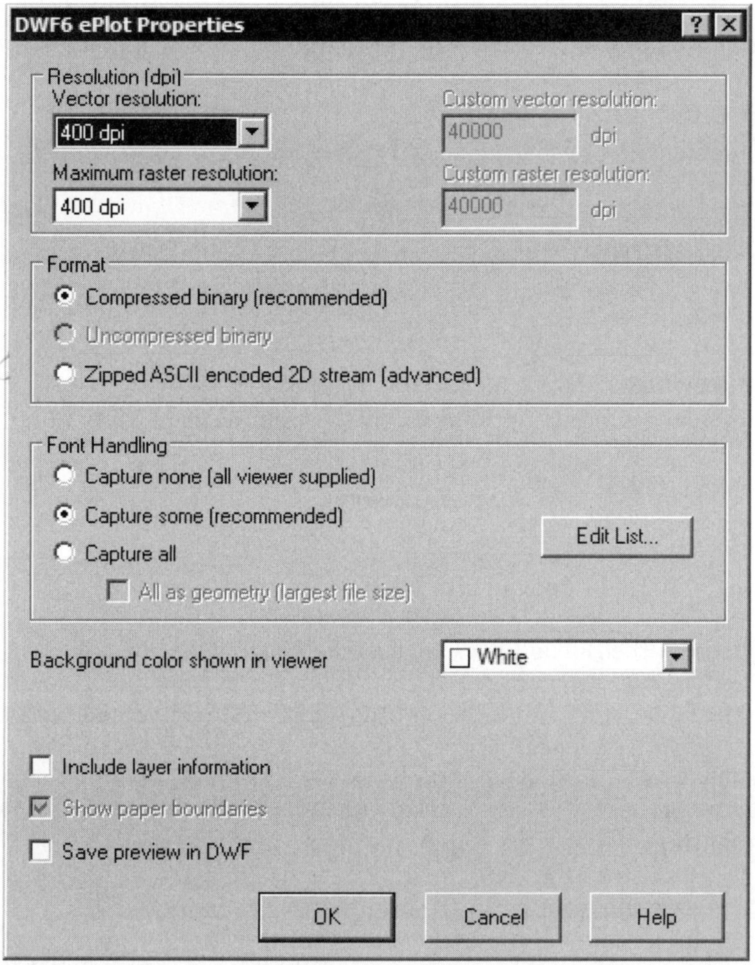

Figure 28-19 ***DWF6 ePlot Properties*** *dialog box*

Resolution area

Unlike AutoCAD DWG files that are based on real numbers, DWF files are saved using integer numbers. The higher the resolution, the better the quality of the DWF file. However, the size of the files also increases with the resolution.

Format area

Compression further reduces the size of the DWF file. You should always use compression, unless you know that another application cannot decompress the DWF file. Compressed binary format is seven times smaller than ASCII format. Again, this means the compressed DWF file transmits over the Internet seven times faster than a compressed ASCII DWF file.

Font Handling

This area of the **DWF6 ePlot Properties** dialog box is used to select the fonts that you need to include in the DWF file. You can choose the **Edit List** button from this area to display the list of the fonts. It should be noted that the fonts add to the size of the DWF file.

Background color shown in viewer

While white is probably the best background color, you may choose any of AutoCAD's 255 colors.

Include Layer information

This option includes layers, which allows you to toggle layers off and on when the drawing is viewed by the Web browser.

Show paper boundaries

Includes a rectangular boundary at the drawing's extents as displayed in the layouts.

Save preview in DWF

Saves the preview in DWF file.

6. In most cases, you would turn on all options.

7. Choose **OK** to exit the dialog boxes back to the **Plot** dialog box.

8. Accept the DWF file name and location listed in the **File name and path** text box in the **Plot to file** area, or type a new name. If necessary, change the location where the file will [be stored. The **[...]** button displays the **Browse for Plot File** dialog box.

9. In the **Plot Settings** tab, use the required options to create the DWF file.

10. Choose **OK** to save the drawing in DWF format.

The following is the procedure of how to create a DWF file using the **PUBLISH** command.

Choose the **Publish** button from the **Standard** toolbar. The **Publish Drawing Sheets** dialog box is displayed, see Figure 28-20.

Note

If you have not saved the drawing before invoking the ***PUBLISH*** *command, then the* ***Drawing Modified - Save Changes*** *dialog box is displayed. Choose* ***OK*** *to save the changes.*

The various options in this dialog box are discussed next.

List of drawing sheets area

This area lists the drawings that will be included for publishing. Right-click on any of the drawings to display a shortcut menu. In the shortcut menu, there are various options available that can be used on the drawings listed in this area.

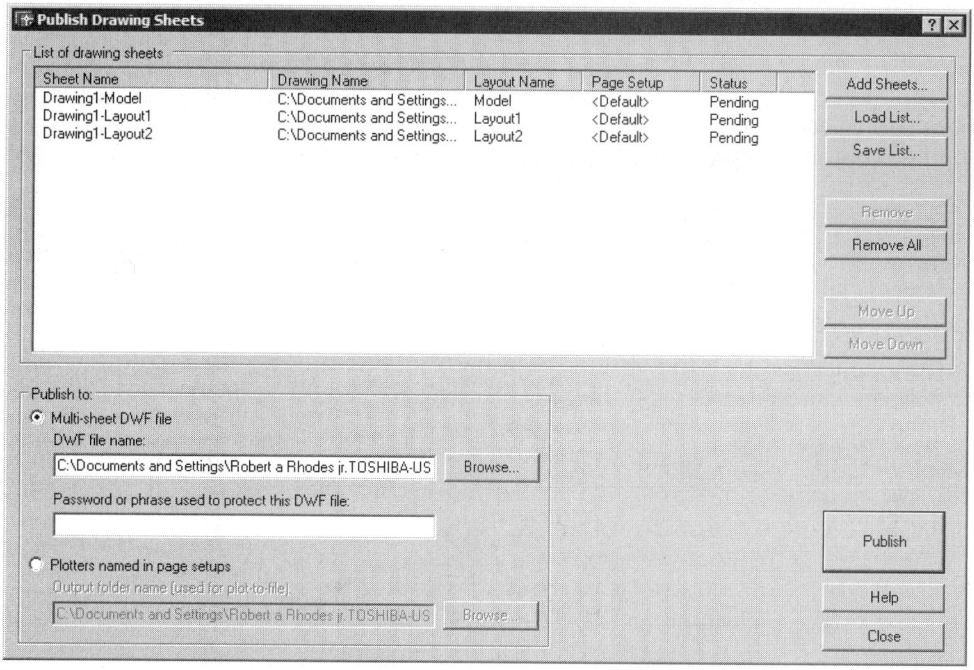

Figure 28-20 ***Publish Drawing Sheets*** *dialog box*

Add Sheets button

This button when selected displays the **Select Drawings** dialog box. Using this dialog box you can select the drawing files (*.dwg*). These files are then included in the creation of the DWF file. Remember that the model space and all the layouts of the drawing are added individually to the **Publish Drawing Sheets** dialog box as shown in Figure 28-20.

Load List button

After adding drawings in the list when you choose this button, the **List of Sheets Changed** dialog box is displayed. This dialog box prompts you to save the existing list of sheets that is listed in the **List of drawing sheets** area. The list of sheets is saved in the *.dsd* file format. When you choose to save the existing list, the **Save List As** dialog box is displayed. When you choose not to save the current list, the **Load List of Sheets** dialog box is displayed. You can select the *.dsd* or *.bp3* fie formats from this dialog box and select the list to load.

Save List button

This button when chosen displays the **Save List As** dialog box. This dialog box allows you to save the list of drawings or sheets.

Remove button

This button allows you to remove the selected sheet from the dialog box.

Remove All button

This button allows you to remove all the sheets from the dialog box.

Move Up button

This button allows you to move the selected drawing one position up in the list.

Move Down button

This button allows you to move the selected drawing one position down in the list.

Publish to area

This area provides you with the options of publishing the drawings to a multisheet DWF file or to a plotter. This area has two radio buttons that are discussed next.

Multi-sheet DWF file radio button

This radio button allows you to save the drawings in a DWF format and the path location of the file can be specified by using the **Browse** button.

Password or phrase used to protect this DWF file

This text box is used to enter the password for the DWF file in order to protect it. The password entered in this text box should be remembered because there is no way this password can be recovered.

Plotters named in page setups

This radio button is used to plot the file to the plotters that are set in the page setup of each layout.

Publish button

This button is used to start the process of publishing or creating the DWF file. If a drawing sheet fails to plot, the system continues to publish the remaining sheets.

Viewing DWF Files

One of the major enhancements in AutoCAD 2004 is the **Autodesk Express Viewer** that is automatically installed on your computer when you install AutoCAD 2004. This plug-in can be used to view the DWF files in addition to the other Web browsers. You can also use a Web browser with a special plug-in that allows the browser to correctly interpret the file for viewing the DWF files. Apart from viewing the DWF files, you can also view the DWG and DXF files using the **Autodesk Express Viewer**. Remember that you cannot view a DWF file with AutoCAD.

Autodesk updates the DWF plug-in approximately twice a year. Each update includes some new features. In summary, all versions of the DWF plug-in perform the following functions:

- Views DWF files created by AutoCAD within a browser.
- Right-clicking the DWF image displays a shortcut menu with commands.
- Real-time pan and zoom lets you change the view of the DWF file as quickly as a drawing file in AutoCAD.

- Embedded hyperlinks let you display other documents and files.
- File compression means that a DWF file appears in your Web browser much faster than the equivalent DWG drawing file would.
- Print the DWF file alone or along with the entire Web page.
- Works with Netscape Navigator or Microsoft Internet Explorer. A separate plug-in is required, depending upon which of the two browsers you use.
- Allows you to “drag and drop” a DWG file from a Web site into AutoCAD as a new drawing or as a block.
- Views a named view stored in the DWF file.
- Can specify a view using the *X* and *Y* coordinates.
- Toggles layers off and on.

You can use various buttons provided in the **Autodesk Express Viewer** for manipulating the view of the DWF file. However, keep in mind that the original objects of the DWF file cannot be modified. You can also right-click in the display screen to display a shortcut menu. The shortcut menu provides you with the options such as **Pan**, **Zoom**, **Layers**, and **Named Views**, see Figure 28-21. To select an option, choose it from the shortcut menu.

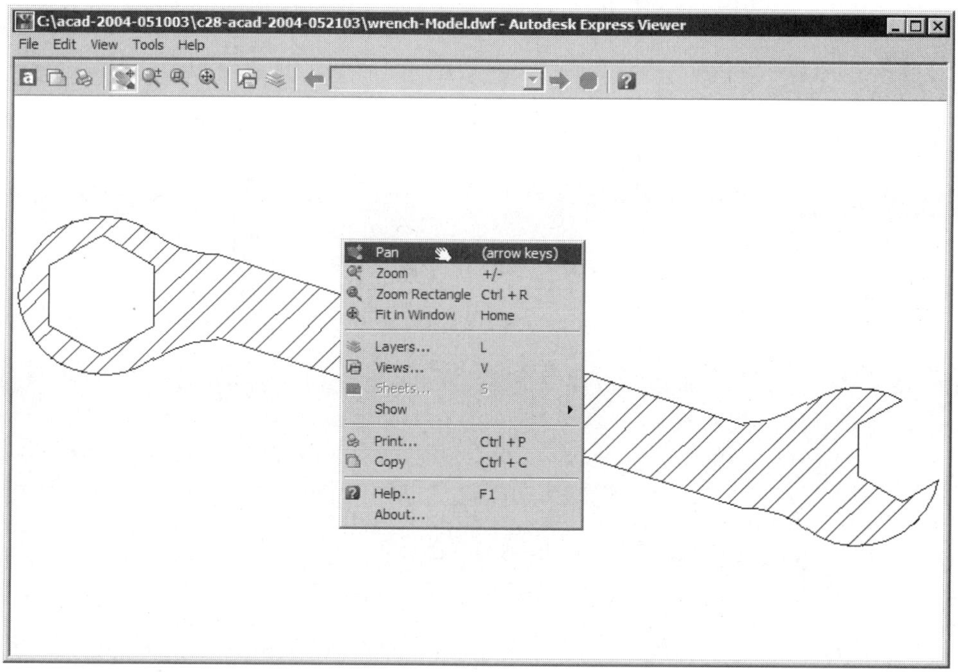

***Figure 28-21 Autodesk Express Viewer** with the shortcut menu*

Pan is the default option. Press the left mouse button and drag the cursor. The cursor changes to an open hand, signaling that you can pan the view around the drawing. This is exactly the same as realtime panning in AutoCAD.

The **Zoom** option works in the same way as in AutoCAD. This option displays a magnifying glass for zooming in and out of the current display.

Layers displays a nonmodal dialog box named **Layers**, see Figure 28-22 that lists all layers in the drawing. (A nonmodal dialog box remains on the screen; unlike AutoCAD's modal dialog boxes, you do not need to dismiss a nonmodal dialog box to continue working.) Select a layer name to toggle its visibility between on (yellow light bulb icon) and off (blue light bulb icon).

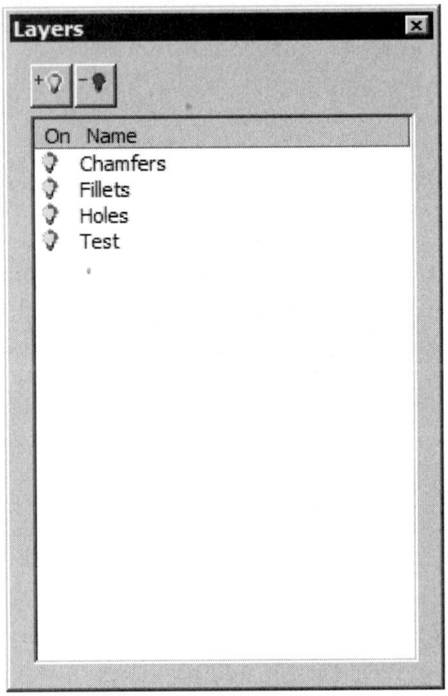

***Figure 28-22** Autodesk Express Viewer's **Layer** dialog box*

Views works only when the original DWG drawing file contained named views created with the **VIEW** command. Selecting **Views** displays a nonmodal dialog box named **Views** that allows you to select a named view, see Figure 28-23. Select a named view to see it; click the small **x** in the upper right corner to dismiss the dialog box.

Note

*To publish a sheet, the plotter in the page setup should be selected from the **Page Setup** dialog box. This plotter should be DWF6 eplot.pc3.*

EMBEDDING A DWF FILE

To let others view your DWF file on the Internet, you need to embed the DWF file in a Web page. Here are the steps to embed a DWF file in a Web page.

Figure 28-23 *Autodesk Express Viewer's* ***Views*** *dialog box*

1. The Hyper Text Markup Language (HTML) code is the most basic method of placing a DWF file in your Web page:

   <embed src=”filename.dwf”>

   The **<embed>** tag embeds an object in a Web page. The **src** option is short for “source.” Replace *filename.dwf* with the name of the DWF file. Remember to keep the quotation marks in place.

2. HTML normally displays an image as large as possible. To control the size of the DWF file, add the **Width** and **Height** options.

   <embed width=800 height=600 src=”filename.dwf”>

   The **Width** and **Height** values are measured in pixels. Replace 800 and 600 with any appropriate numbers such as 100 and 75 for a “thumbnail” image, or 300 and 200 for a small image.

3. To speed up the display of Web page, some users turn off the display of images. For this reason, it is useful to include a description, which is displayed in place of the image:

   <embed width=800 height=600 name=description src=”filename.dwf”>

The **Name** option displays a textual description of the image when the browser does not load images. You might replace description with the DWF filename.

4. When the original drawing contains named views created by the **VIEW** command, these are transferred to the DWF file. Specify the initial view for the DWF file:

```
<embed width=800 height=600 name=description namedview="viewname"
src="filename.dwf">
```

The **namedview** option specifies the name of the view to display upon loading. Replace **viewname** with the name of a valid view name. When the drawing contains named views, the user can right-click on the DWF image to get a list of all named views.

As an alternative, you can specify the 2D coordinates of the initial view:

```
<embed width=800 height=600 name=description view="0,0 9,12"
    src="filename.dwf">
```

The **View** option specifies the *X* and *Y* coordinates of the lower left and upper right corners of the initial view. Replace 0,0 9,12 with other coordinates. Since DWF is 2D only, you cannot specify a 3D viewpoint. You can use **View** or **Namedview**, but not both.

5. Before a DWF image can be displayed, the Web browser must have the DWF plug-in called "Autodesk Express viewer". For users of Netscape Communicator, you must include a description of where to get the Autodesk Express Viewer plug-in when the Web browser is lacking it.

```
<embed pluginspage=http://www.autodesk.com/products/express viewer
width=800 height=600 name=description view="0,0 9,12" src="filename.dwf">
```

The **pluginspage** option describes the page on the Autodesk Web site where the Autodesk Express Viewer plug-in can be downloaded.

The code previously listed works for Netscape Navigator. To provide for users of Internet Explorer, the following HTML code must be added:

```
<object classid ="clsid:B2BE75F3-9197-11CF-ABF4-08000996E931" codebase = "ftp:/
/    ftp.autodesk.com/pub/autocad/plugin/autodesk express viewer.cab#version=2,0,0,0"
width=800 height=600>
<param name="Filename" value="filename.dwf">
<param name="View" value="0,0 9,12">
<param name="Namedview" value="viewname">
<embed pluginspage=http://www.autodesk.com/products/express viewer
width=800 height=600 name=description view="0,0 9,12"
src="filename.dwf">
</object>
```

The two **<object>** and three **<param>** tags are ignored by Netscape Navigator; they are required for compatibility with Internet Explorer. The **classid** and **codebase** options tell Explorer where to find the plug-in. Remember that you can use **View** or **Namedview**, but not both.

6. Save the HTML file.

Self-Evaluation Test

Answer the following questions and then compare your answers to the answers given at the end of this chapter.

1. To access the online resources, choose **Help** > **Online Resources** from the menu bar. (T/F)
2. You cannot determine whether or not an object has a hyperlink. (T/F)
3. By default, the **Browse the Web** dialog box displays the contents of the URL stored in the **INETLOCATION** system variable. (T/F)
4. URLs are also known as hyperlinks. (T/F)
5. You cannot select text in the AutoCAD drawing to paste it as a hyperlink. (T/F)
6. Compression in the DWF file causes it to take ____________ time to transmit over the Internet.
7. Rectangular area hyperlinks are stored on __________ layer.
8. To see the location of hyperlinks in a drawing, use the __________ command.
9. The _________ is a HTML tag for embedding objects in a Web page.
10. When you attach a hyperlink to a block, the hyperlink data is __________ when you scale the block unevenly, stretch the block, or explode it.

Review Questions

Answer the following questions.

1. Can you launch a Web browser from within AutoCAD?
2. What does DWF mean?

3. What is the purpose of DWF files?

4. URL is an acronym for what?

5. Which of the following URLs are valid?
   (a) www.autodesk.com
   (b) http://www.autodesk.com
   (c) Both of the above.
   (d) None of the above.

6. FTP is an acronym for what?

7. What is a "local host"?

8. Are hyperlinks active in an AutoCAD 2004 drawing?

9. The purpose of URLs is to let you create _________ between files.

10. Can you can attach a URL to any object?

Answers to the Self-Evaluation Test

1 - T, **2** - F, **3** - T, **4** - T, **5** - T, **6** - less, **7** - URLLAYER, **8** - Hyperlink, **9** - <embed>, **10** - lost.

AutoCAD Part III (Customizing)

AutoCAD Part III (Customizing)

Author's Web Sites

For Faculty: Please contact the author at **stickoo@calumet.purdue.edu** or **tickoo@cadcim.com** to access the Web site that contains the following.

1. **PowerPoint presentations, programs, and drawings used in this textbook.**
2. **Syllabus, chapter objectives and hints, and questions with answers for every chapter.**

For Students:You can download drawing-exercises, tutorials, programs, and special topics by accessing author's web site at **www.cadcim.com** or **http://technology.calumet.purdue.edu/ /met/tickoo/students/students.htm**.

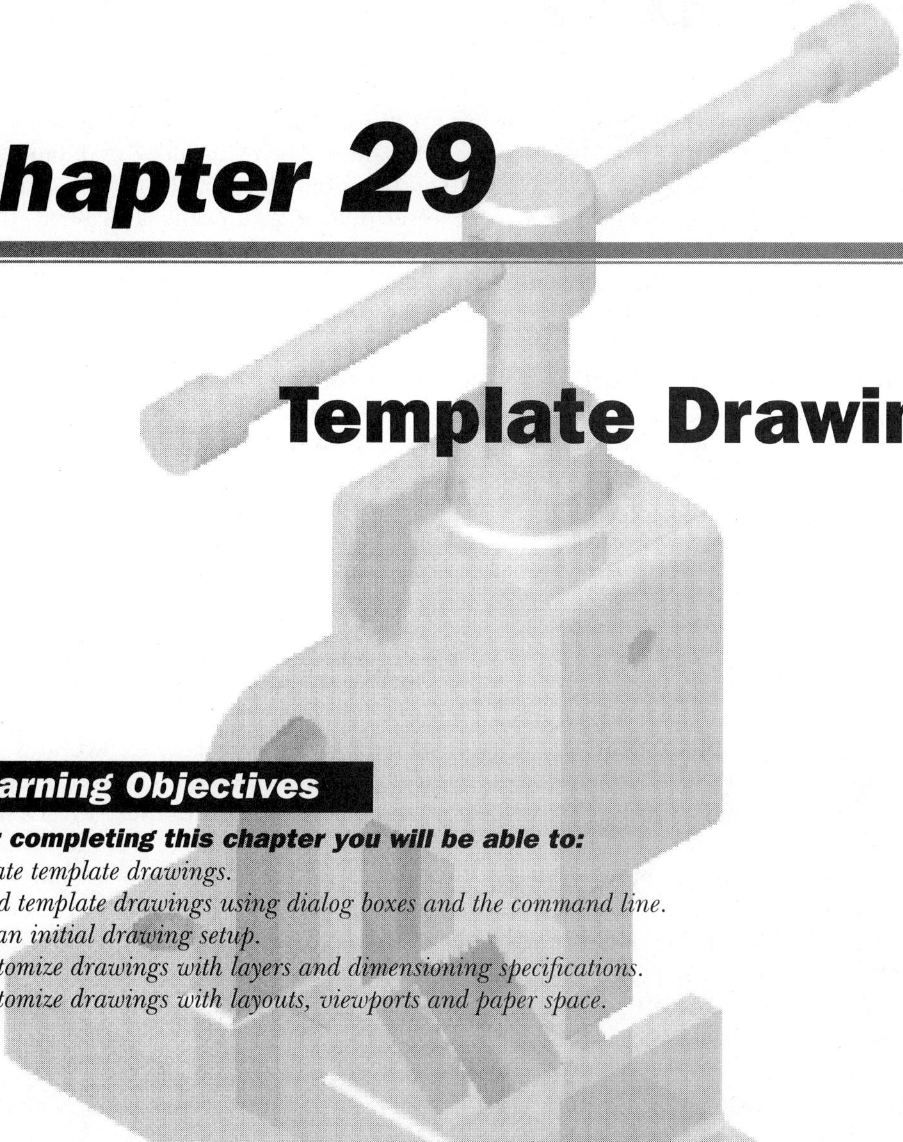

Chapter 29

Template Drawings

Learning Objectives

After completing this chapter you will be able to:

- *Create template drawings.*
- *Load template drawings using dialog boxes and the command line.*
- *Do an initial drawing setup.*
- *Customize drawings with layers and dimensioning specifications.*
- *Customize drawings with layouts, viewports and paper space.*

CREATING TEMPLATE DRAWINGS

One way to customize AutoCAD is to create template drawings that contain initial drawing setup information and if desired, visible objects and text. When the user starts a new drawing, the settings associated with the template drawing are automatically loaded. If you start a new drawing from scratch, AutoCAD loads default setup values. For example, the default limits are (0.0,0.0), (12.0,9.0) and the default layer is 0 with white color and continuous linetype. Generally, these default parameters need to be reset before generating a drawing on the computer using AutoCAD. A considerable amount of time is required to set up the layers, colors, linetypes, lineweights, limits, snaps, units, text height, dimensioning variables, and other parameters. Sometimes, border lines and a title block may also be needed.

In production drawings, most of the drawing setup values remain the same. For example, the company title block, border, layers, linetypes, dimension variables, text height, LTSCALE, and other drawing setup values do not change. You will save considerable time if you save these values and reload them when starting a new drawing. You can do this by creating template drawings that contain the initial drawing setup information configured according to the company specifications. They can also contain a border, title block, tolerance table, block definitions, floating viewports in the paper space, and perhaps some notes and instructions that are common to all drawings.

THE STANDARD TEMPLATE DRAWINGS

AutoCAD software package comes with standard template drawings like Acad.dwt, Acadiso.dwt, Ansi a (portrait) -color dependent plot styles.dwt, Din a1 -named plot styles.dwt, and so on. The ansi, din, and iso template drawings are based on the drawing standards developed by ANSI (American National Standards Institute), DIN (German), and ISO (International Organization for Standardization). When you start a new drawing and you are using the **Startup** dialog box, the **Create New Drawing** dialog box is displayed on the screen. To load the template drawing, select the **Use a Template** button and the list of standard template drawings is displayed. From this list you can select any template drawing according to your requirements. If you want to start a drawing with default settings, select the **Start from Scratch** button in the **Create New Drawing** dialog box. The following are some of the system variables, with the default values that are assigned to a new drawing.

System Variable Name	Default Value
CHAMFERA	0.5000
CHAMFERB	0.5000
COLOR	Bylayer
DIMALT	Off
DIMALTD	2
DIMALTF	25.4
DIMPOST	None
DIMASO	On
DIMASZ	0.18
FILLETRAD	0.5000
GRID	0.5000

GRIDMODE	0	
ISOPLANE	Left	
LIMMIN	0.0000,0.0000	
LIMMAX	12.0000,9.0000	
LTSCALE	1.0	
MIRRTEXT	1	(Text mirrored like other objects)
TILEMODE	1 (On)	

Example 1

Create a template drawing using **Advanced Setup** wizard of the **Create Drawings tab** with the following specifications and save it with the name *proto1.dwt*.

Units	Engineering with precision 0'-0.00"
Angle	Decimal degrees with precision 0.
Angle Direction	Counterclockwise
Area	144'x96'

Step 1

Select the option to show the **Startup** dialog box in the **System** tab of the **Options** dialog box. Choose **New** from the **File** menu to display the **Create New Drawing** dialog box. Choose the **Use a Wizard** button and select the **Advanced Setup** option as shown in Figure 29-1. Choose **OK**. The **Units** page of the **Advanced Setup** dialog box is displayed.

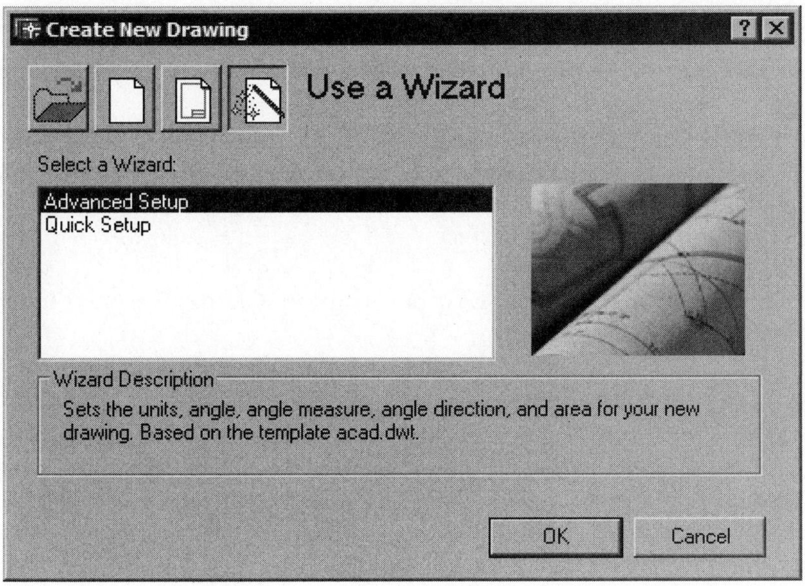

Figure 29-1 ***Advanced Setup*** *wizard option of the* ***Create New Drawing*** *dialog box*

Chapter 29

Step 2

Select the **Engineering** radio button. Select **0'-0.00"** precision from the **Precision** drop-down list as shown in Figure 29-2 and then choose the **Next** button. The **Angle** page of the **Advanced Setup** dialog box is displayed.

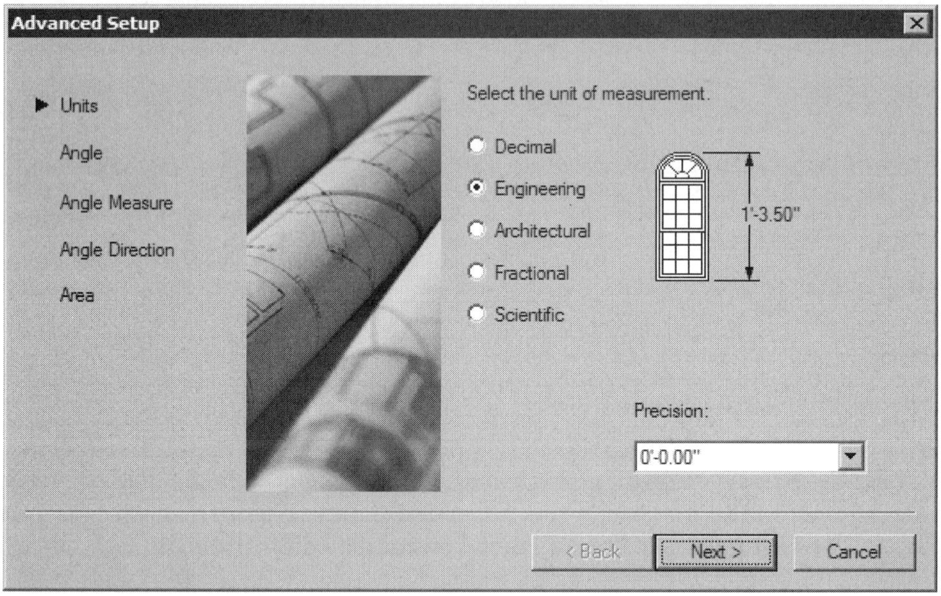

Figure 29-2 ***Units*** *page of the* ***Advanced Setup*** *dialog box*

Step 3

In the **Angle** page, select the **Decimal Degrees** radio button and select **0** from the **Precision** drop-down list as shown in Figure 29-3. Choose the **Next** button. The **Angle Measure** page of the **Advanced Setup** dialog box is displayed.

Step 4

In the **Angle Measure** page, select the **East** radio button. Choose the **Next** button to display the **Angle Direction** page.

Step 5

Select the **Counter-Clockwise** radio button and then choose the **Next** button. The **Area** page is displayed. Specify the area as 144' and 96' by entering the value of width and length as **144'** and **96'** in the **Width** and **Length** edit boxes and then choose the **Finish** button. Use the **All** option of the **ZOOM** command to display the new limits on the screen. Save the template drawing as *proto1.dwt.*

Note

*If you want to customize only units and area, you can use the **Quick Setup** in the **Create New Drawing** dialog box.*

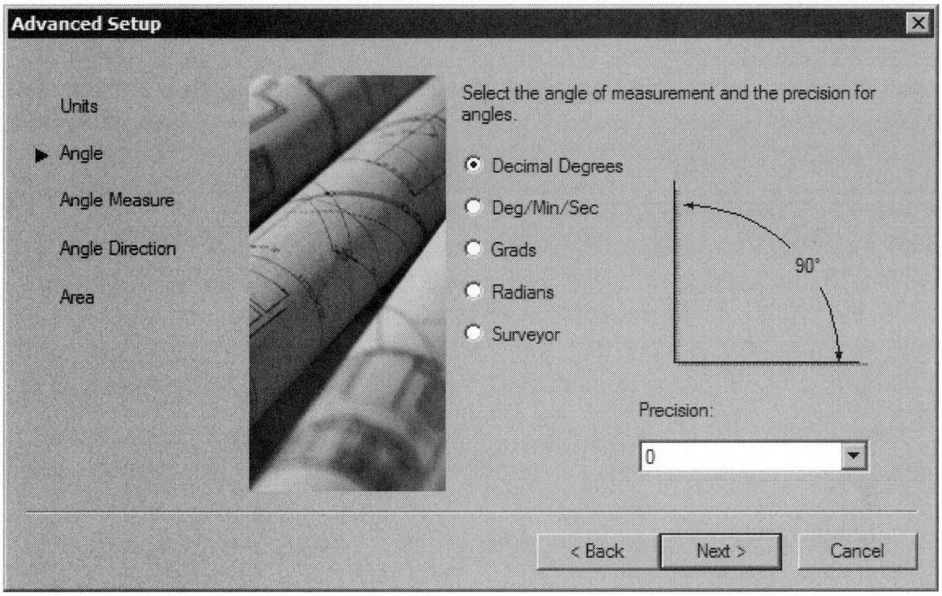

***Figure 29-3 Angle** page of the **Advanced Setup** dialog box*

Example 2

Create a template drawing with the following specifications. The template should be saved with the name *proto2.dwt*.

Limits	18.0,12.0
Snap	0.25
Grid	0.50
Text height	0.125
Units	3 digits to the right of decimal point
	Decimal degrees
	2 digits to the right of decimal point
	0 angle along positive *X* axis (east)
	Angle positive if measured counterclockwise

Step 1
Starting a new drawing

Start AutoCAD and choose **Start from Scratch** in the **Create New Drawing** dialog box. From the **Default Settings** area, select the **Imperial (feet and inches)** radio button as shown in Figure 29-4. Choose **OK** to open a new file.

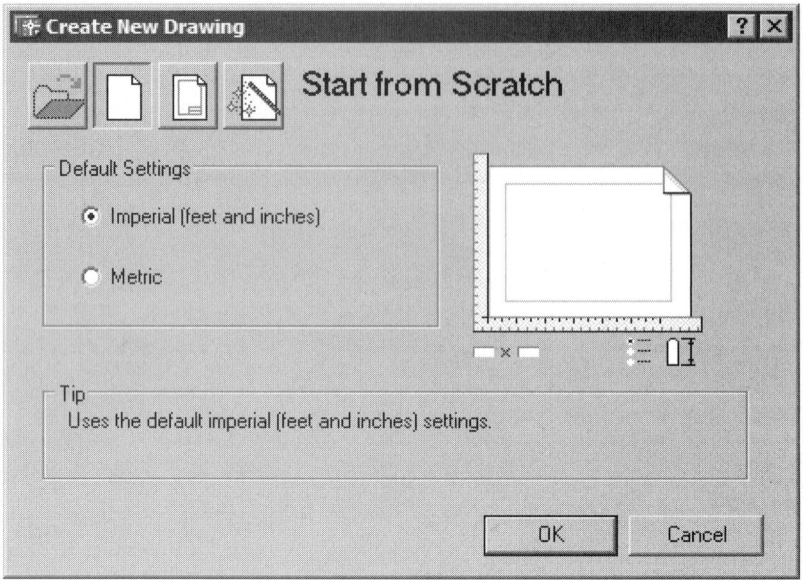

***Figure 29-4 Start from Scratch** option of the **Create New File** dialog box*

Step 2
Setting limits, snap, grid, and text size
The **LIMITS** command can be invoked by choosing **Drawing Limits** from the **Format** menu or by entering **LIMITS** at the Command prompt.

Command: **LIMITS**
Specify lower left corner or [ON/OFF] <0.00,0.00>: **0,0**
Specify upper right corner <12.0,9.0>: **18.0,12.0**

After setting the limits, the next step is to increase the drawing display area. Use the **ZOOM** command with the **All** option to display the new limits on the screen.

Now, right-click on the **Snap** or **Grid** button in the status bar to display the shortcut menu. Choose the **Settings** in the shortcut menu to display the **Drafting Settings** dialog box. You can also choose the **Object Snap Settings** button from the **Object Snap** toolbar to display the **Drafting Settings** dialog box. Choose the **SNAP** and **GRID** tab. Enter **0.25** and **0.25** in the **Snap X spacing** and **Snap Y spacing** edit boxes respectively. Enter **0.5** and **0.5** in the **Grid X spacing** and **Grid Y spacing** edit boxes respectively. Then choose **OK**.

Note
*You can also use **SNAP** and **GRID** commands to set these values.*

Size of the text can be changed by entering **TEXTSIZE** at the Command prompt.

Command: **TEXTSIZE**
Enter new value for TEXTSIZE <0.2000>: **0.125**

Step 3
Setting units

Choose **Format > Units** from the menu bar or enter **UNITS** at the Command prompt to invoke the **Drawing Units** dialog box shown in Figure 29-5. In the **Length** area, select **0.000** from the **Precision** drop-down list. In the **Angle** area, select **Decimal Degrees** from the **Type** drop-down list and **0.00** from the **Precision** drop-down list. Also make sure the **Clockwise** radio button from the **Angle** area is not selected.

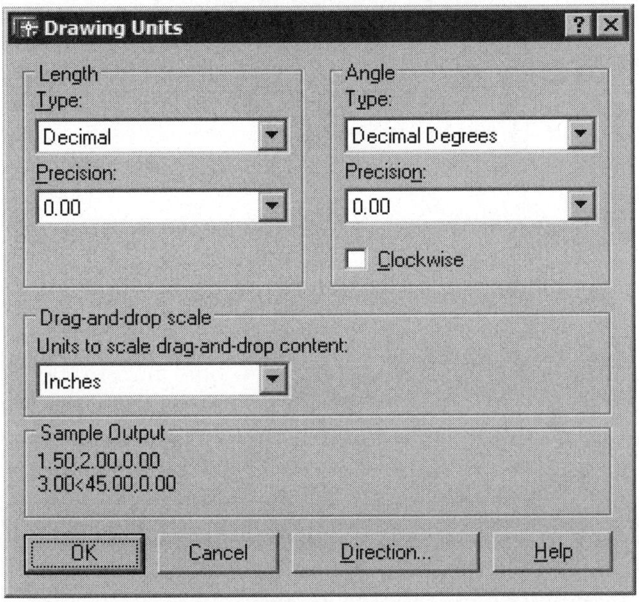

***Figure 29-5 Drawing Units** dialog box*

Choose the **Direction** button to display the **Direction Control** dialog box (Figure 29-6) and select the **East** radio button. Exit both the dialog boxes

Step 4

Now, save the drawing as *proto2.dwt* using the **SAVE** command. You must select **AutoCAD Drawing Template (*dwt)** from the **Files of type** drop-down list in the dialog box. This drawing is now saved as *proto2.dwt* on the default drive. You can also save this drawing on a diskette in drives A or B using the **Save Drawing As** dialog box.

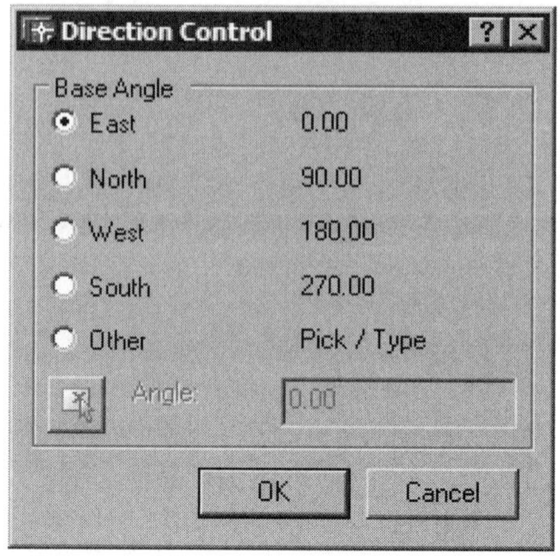

***Figure 29-6 Direction Control** dialog*

LOADING A TEMPLATE DRAWING

You can use the template drawing to start a new drawing. To use the preset values of the template drawing, start AutoCAD or select the **QNew** button from the **Standard** toolbar. The dialog box that appears will depend on whether you have selected the option to show **Startup** dialog box or not from the **Options** dialog box. If you have selected this option, the **Create New Drawing** dialog box appears. Choose the **Use a Template** option. All the templates that are saved in the default **Template** directory will be shown in the **Select a Template** list box, see Figure 29-7. If you have saved the template in any other file, choose the **Browse** button. The **Select a template file** dialog box is displayed. You can use this dialog box to browse the directory in which the template file is saved.

If you have selected the option of not showing the **Startup** dialog box, the **Select template** dialog box appears when you choose the **QNew** button. This dialog box also displays the default **Template** folder and all the template files saved in it, see Figure 29-8. You can use this dialog box to select the template file you want to open.

Using any of the previously mentioned dialog boxes, select *proto1.dwt* template drawing. AutoCAD will start a new drawing that will have the same setup as that of template drawing *proto1.dwt*.

You can have several template drawings, each with a different setup. For example, **PROTOB** for a 18" by 12" drawing, **PROTOC** for a 24" by 18" drawing. Each template drawing can be created according to user-defined specifications. You can then load any of these template drawings as discussed previously.

Note

*You can also use the command line to load a template drawing. Set the value of the **FILEDIA** system variable to **0** and then enter **QNEW** at the command line. You will be prompted to enter the name of the template file at the command line.*

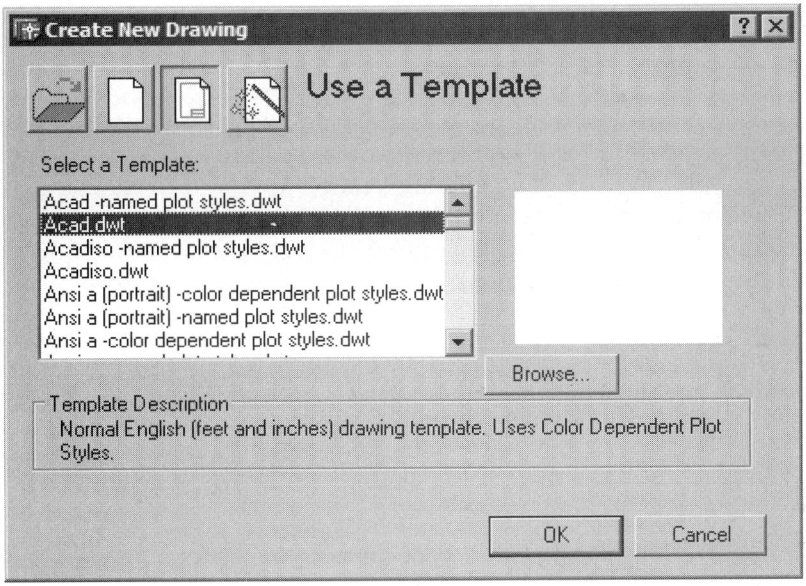

Figure 29-7 *Templates available in the default* ***Templates*** *directory*

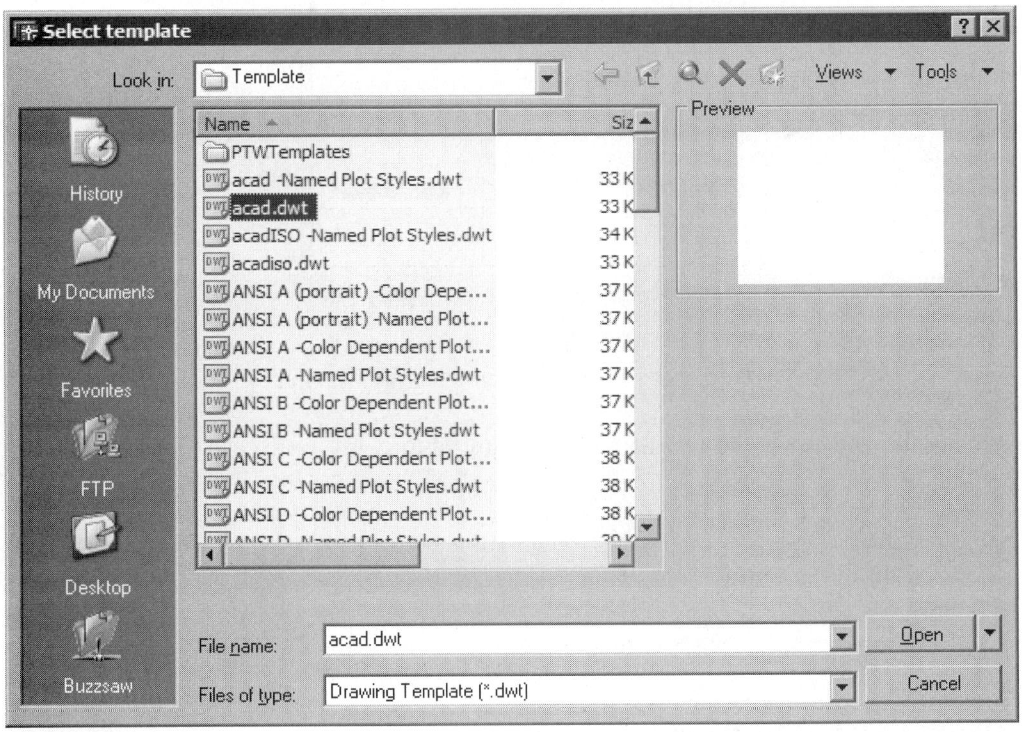

Figure 29-8 ***Select template*** *dialog box that appears if the option to show the* ***Startup*** *dialog box is not selected*

CUSTOMIZING DRAWINGS WITH LAYERS AND DIMENSIONING SPECIFICATIONS

Most production drawings need multiple layers for different groups of objects. In addition to layers, it is a good practice to assign different colors to different layers to control the line width at the time of plotting. You can generate a template drawing that contains the desired number of layers with linetypes and colors according to your company specifications. You can then use this template drawing to make a new drawing. The next example illustrates the procedure used for customizing a drawing with layers, linetypes, and colors.

Example 3

Create a template drawing *proto3.dwt* that has a border and the company's title block, as shown in Figure 29-9.

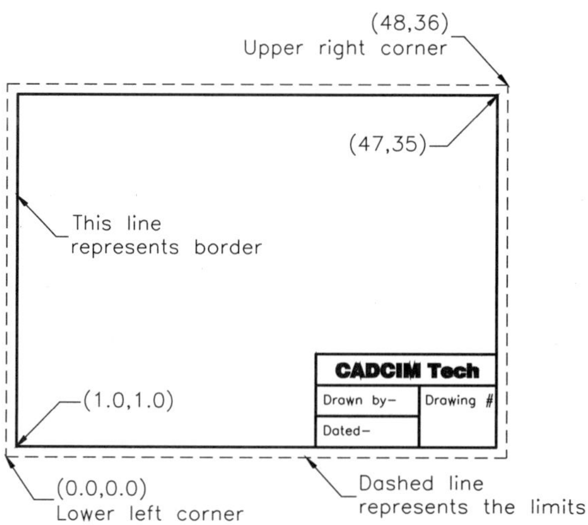

***Figure 29-9** Template drawing for Example 3*

This template drawing will have the following initial drawing setup.

Limits	48.0,36.0
Text height	0.25
Border line lineweight	0.012"
Ltscale	4.0

DIMENSIONS
Overall dimension scale factor 4.0
Dimension text above the extension line
Dimension text aligned with dimension line

LAYERS

Layer Names	Line Type	Color
0	Continuous	White
OBJ	Continuous	Red
CEN	Center	Yellow
HID	Hidden	Blue
DIM	Continuous	Green
BOR	Continuous	Magenta

Step 1
Setting limits, text size, polyline width, polyline and linetype scaling

Start a new drawing with default parameters by selecting the **Start from Scratch** option in the **Create New Drawings** dialog box. In the new drawing file, use the AutoCAD commands to set up the values as given for this example. Also, draw a border and a title block as shown in Figure 29-9. In this figure, the hidden lines indicate the drawing limits. The border lines are 1.0 units inside the drawing limits. For the border lines, increase the lineweight to a value of 0.012".

Use the following procedure to produce the prototype drawing for Example 3.

1. Invoke the **LIMITS** command by choosing **Drawing Limits** from the **Format** menu or by entering **LIMITS** at the Command prompt. The following is the prompt sequence

   Command: **LIMITS**
   Specify lower left corner or [ON/OFF] <0.00,0.00>: **0,0**
   Specify upper right corner <12.0,9.0>: **48.0,36.0**

2. Increase the drawing display area by invoking the **All** option of the **ZOOM** command

3. Enter **TEXTSIZE** at the Command prompt to change the text size.

   Command: **TEXTSIZE**
   Enter new value for TEXTSIZE <0.2000>: **0.25**

4. Next, you will draw the border using the **RECTANG** command. The prompt sequence to draw the rectangle is:

   Command: **RECTANG**
   Specify first corner point or [Chamfer/Elevation/Fillet/Thickness/Width]: **1.0,1.0**
   Specify other corner point or [Dimensions]: **47.0,35.0**

5. Now, select the rectangle and select **0.012"** from the **Lineweight Control** drop-down list in the **Properties** toolbar. Make sure the **Show/Hide Lineweight** button is chosen in the status bar.

6. Enter **LTSCALE** at the Command prompt to change the linetype scale.

   Command: **LTSCALE**
   Enter new linetype scale factor<Current>: **4.0**

Step 2
Setting dimensioning parameters

You can use the **Dimension Style Manager** dialog box to set the dimension variables. Choose the **Dimension Style** button from the **Dimension** toolbar or choose **Style** from the **Dimension** menu to invoke the **Dimension Style Manager** dialog box as shown in Figure 29-10.

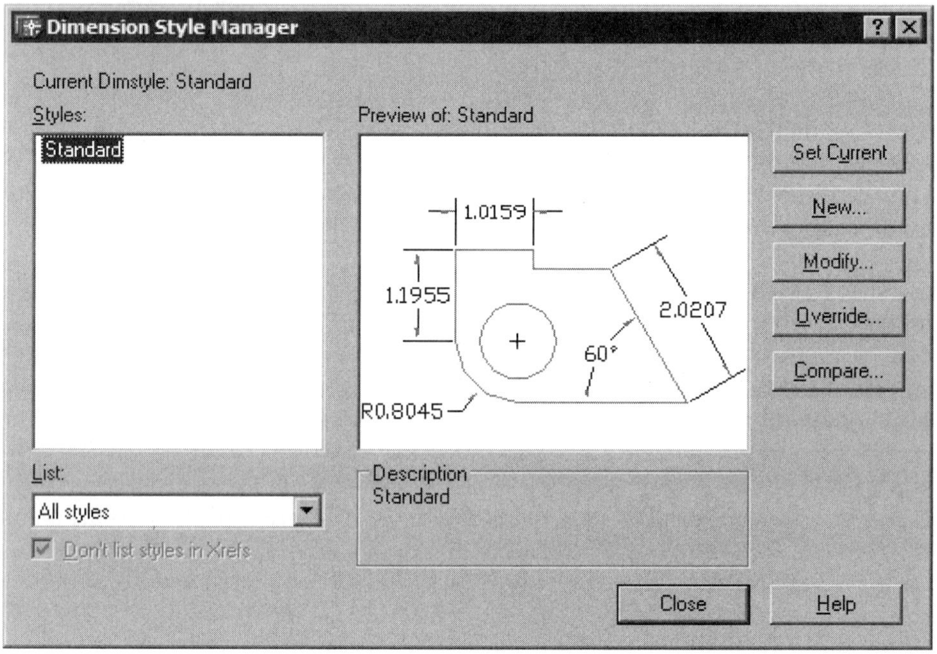

***Figure 29-10 Dimension Style Manager** dialog box*

You can also invoke this dialog box by entering **DIMSTYLE** at the Command prompt. Choose the **New** button from the **Dimension Style Manager** dialog box. The **Create New Dimension Style** dialog box is displayed. Specify new style name as **MYDIM1** in the **New Style Name** edit box as shown in the Figure 29-11 and then choose the **Continue** button. The **New Dimension Style:MYDIM1** dialog box is displayed.

Overall dimension scale factor

To specify dimension scale factor, choose the **Fit** tab of the **New Dimension Style:MYDIM1** dialog box. Set the value in the **Use overall scale of** as **4** in the **Scale for Dimension Features** area (Figure 29-12).

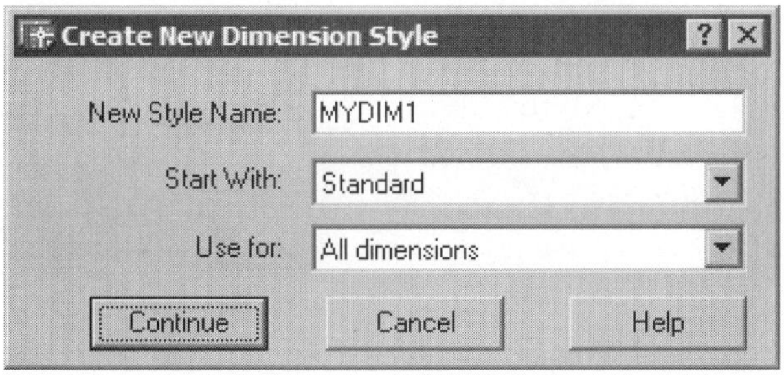

Figure 29-11 ***Create New Dimension Style*** *dialog*

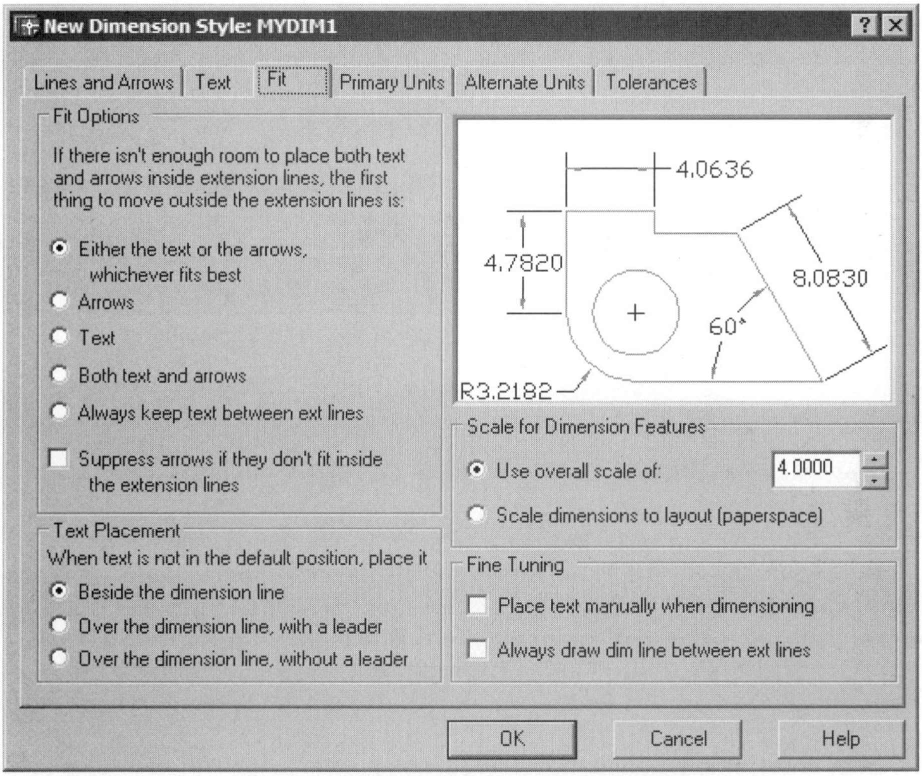

Figure 29-12 Fit *tab of the* ***New Dimension Style:MYDIM1*** *dialog box*

Dimension text over the dimension line

Choose the **Text** tab of the **New Dimension Style:MYDIM1** dialog box. Select the **Over the dimension line, with a leader** radio button from the **Text Placement** area.

Dimension text aligned with the dimension line

In the **Text Alignment** area of the **Text** tab, select the **Aligned with the dimension line** radio button and then choose **OK** to exit the **New Dimension Style:MYDIM1** dialog box.

Setting the new dimension style current

A new dimension style with the name **MYDIM1** is shown in the **Styles** area of the **Dimension Style Manager** dialog box. Select this dimension style and then choose the **Set Current** button to make it the current dimension style. Choose the **Close** button to exit this dialog box.

Step 3

Setting layers

Choose the **Layer Properties Manager** button from the **Layers** toolbar or choose **Layer** from the **Format** menu to invoke **Layer Properties Manager** dialog box as shown in Figure 29-13.

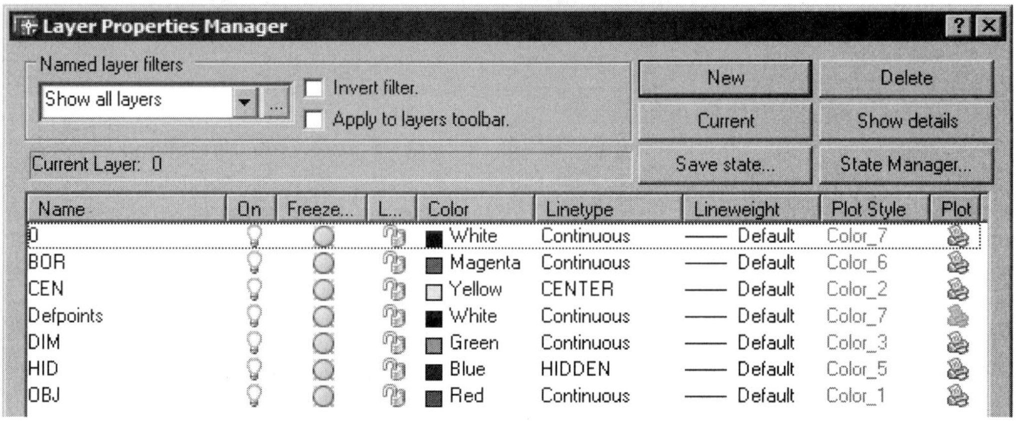

***Figure 29-13** Partial display of the **Layer Properties Manager** dialog box*

You can also invoke the **Layer Properties Manager** dialog box by entering **LAYER** at the Command prompt. Choose the **New** button in the **Layer Properties Manager** dialog box and rename **Layer1** as **OBJ**. Choose the color swatch of the **OBJ** layer to display the **Select Color** dialog box. Select the **Red** color and choose **OK**; the red color is assigned to the OBJ layer. Again choose the **New** button in the **Layer Properties Manager** dialog box and rename the **Layer1** as **CEN**. Choose the linetype swatch to display the **Select Linetype** dialog box.

If the different linetypes are not already loaded, choose the **Load** button to display the **Load or Reload Linetypes** dialog box. Select the **CENTER** linetype from the **Available Linetypes** area and choose **OK**. The **Select Linetype** dialog box will reappear. Select the **CENTER** linetype from the **Loaded linetypes** area and choose **OK**. Choose the color swatch to display the **Select Color** dialog box. Select the **Yellow** color and choose **OK**; the color yellow and linetype center is assigned to the layer CEN. Similarly different linetypes and different colors can be set for different layers mentioned in the example.

You can also use the **-LAYER** command to set the layers and linetypes from the Command prompt.

Step 4
Adding title block
Next, add the title block and the text as shown in Figure 29-8. After completing the drawing, save it as *proto3.dwt*. You have created a template drawing (PROTO3) that contains all the information given in Example 3.

CUSTOMIZING A DRAWING WITH LAYOUT

The Layout (paper space) provides a convenient way to plot multiple views of a three-dimensional (3D) drawing or multiple views of a regular two-dimensional (2D) drawing. It takes quite some time to set up the viewports in model space with different vpoints and scale factors. You can create prototype drawings that contain predefined viewport settings, with vpoint and other desired information. If you create a new drawing or insert a drawing, the views are automatically generated. The following example illustrates the procedure for generating a prototype drawing with paper space and model space viewports.

Example 4

Create a template drawing as shown in Figure 29-14 with four views in Layout3 (Paper space) that display front, top, side, and 3D views of the object. The plot size is 10.5 by 8 inches. The plot scale is 0.5 or 1/2" = 1". The paper space viewports should have the following vpoint setting:

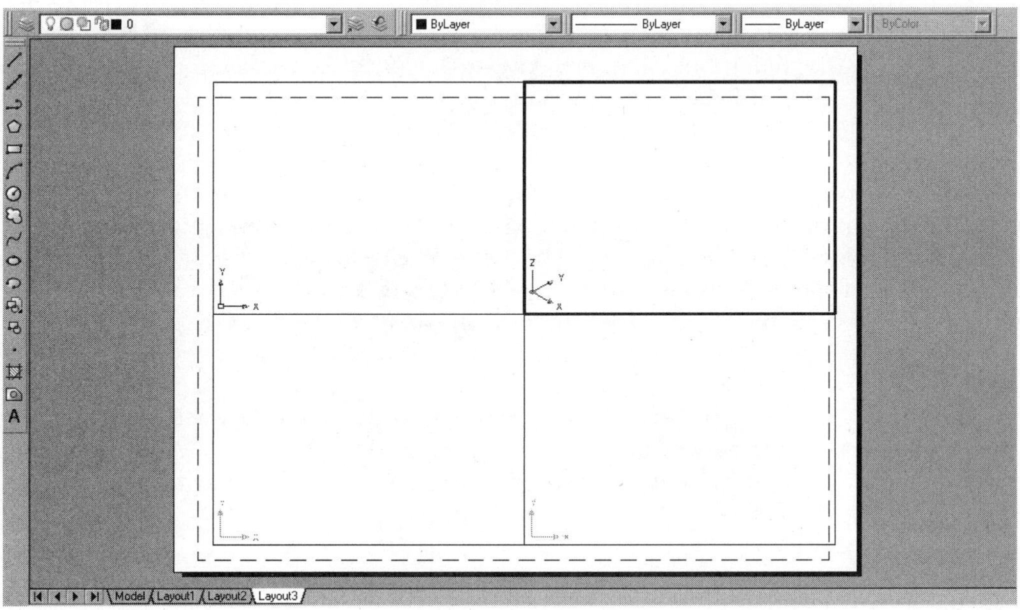

Figure 29-14 *Paper space with four viewports*

Viewports	Vpoint	View
Top right	1,-1,1	3D view

Top left	0,0,1	Top view
Lower right	1,0,0	Right side view
Lower left	0,-1,0	Front view

Start AutoCAD and create a new drawing. Use the following commands and options to set up various parameters.

Step 1

The first step is to create a new layout using the **LAYOUT** command. You can also right-click on **Model** or any **Layout** tab to display the shortcut menu. From the shortcut menu, choose **New layout**.

Command: **LAYOUT**
Enter layout option [Copy/Delete/New/Template/Rename/SAveas/Set/?] <set>: **N**
Enter new Layout name <Layout3>: **Layout3**

Step 2

Next, select the new layout (Layout 3) tab, the **Page Setup - Layout3** dialog box is displayed. The **Plot Device** tab is chosen by default. Select the printer or plotter that you want to use. In this example HP LaserJet4000 is used.

Step 3

Choose the **Layout Settings** tab and select the paper size that is supported by the selected plotting device. In this example the paper size is 8.5x11. Choose the **OK** button to accept the settings and exit the dialog box. The new layout (Layout3) is displayed on the screen with default viewport. Use the **ERASE** command to erase this viewport.

Step 4

Next, you need to set up a layer with the name VIEW for viewports and assign it green color. Invoke the **Layer Properties Manager** dialog box. Choose the **New** button and name the **Layer1** as **VIEW**. Choose the color swatch of the **VIEW** layer to display the **Select Color** dialog box. Select the color **Green** and choose the **OK** button. This color will be assigned to **View** layer. Also, make the **VIEW** layer current and then choose the **OK** button to exit.

Step 5

To create four viewports, use the **MVIEW** command. In order to invoke the **MVIEW** command choose **Viewports > 4Viewport** from the **View** menu or directly enter **MVIEW** command at the Command prompt. Then switch to model space to zoom the display to half the size.

Command: **MVIEW**
Specify corner of viewport or
[ON/OFF/Fit/Hideplot/Lock/Object/Polygonal/Restore/2/3/4] <Fit>:**4**
Specify first corner or [Fit] <Fit>: **0.25,0.25**
Specify opposite corner: **10.25,7.75**

Choose the **Paper** button in the status bar to activate the model space or enter **MSPACE** at the Command prompt.

Command: **MSPACE** (or **MS**)

Make the first viewport active by selecting a point in the viewport and then use the **ZOOM** command to specify the paper space scale factor to 0.5. The **ZOOM** command can be invoked by choosing **Zoom** > **Scale** from the **View** menu or by entering **ZOOM** at the Command prompt.

Command: **ZOOM**
Specify corner of window, enter a scale factor (nX or nXP), or
[All/Center/Dynamic/Extents/Previous/Scale/Window] <real time>: **0.5XP**

Now, make the next viewport active and specify the scale factor. Do the same for the remaining viewports.

Step 6

The next step is to change the viewpoints of different paper space viewports using the **VPOINT** command. To invoke this command, choose **3D Views** > **VPOINT** from the **View** menu or enter **VPOINT** at the Command prompt. The vpoint values for different viewports are shown in Example 5. To set the view point for the lower-left viewport the Command prompt sequence is as follows.

Command: **VPOINT**
Current view direction: VIEWDIR=0.0000,0.0000,1.0000
Specify a view point or [Rotate] <display compass and tripod>: **0,-1,0**

Similarly use the **VPOINT** command to set the vpoint of other viewports.

Step 7

Use the **Model** button in the status bar to change to paper space and then set a new layer PBORDER with yellow color. Make the PBORDER layer current, draw a border, and if needed a title block using the **PLINE** command. You can also change to paper space by entering **PSPACE** at the Command prompt.

The **PLINE** command can be invoked by choosing the **Polyline** button from the **Draw** toolbar or by choosing **Polyline** from the **Draw** menu. The **PLINE** command can also be invoked by entering **PLINE** at the Command prompt.

Command: **PLINE**
Specify start point: **0,0**
Current line-width is 0.0000
Specify next point or [Arc/Close/Halfwidth/Length/Undo/Width]: **0,8.0**
Specify next point or [Arc/Close/Halfwidth/Length/Undo/Width]: **10.5,8.0**

Specify next point or [Arc/Close/Halfwidth/Length/Undo/Width]: **10.5,0**
Specify next point or [Arc/Close/Halfwidth/Length/Undo/Width]: **C**

Step 8

The last step is to select the **Model** tab (or change the **TILEMODE** to 1) and save the prototype drawing. To test the layout that you just created, make the 3D drawing as shown in Figure 29-18 or make any 3D object. Switch to **Layout 3** tab, you will find four different views of the object (Figure 29-15). If the object views do not appear in the viewports, use the **PAN** commands to position the views in the viewports. You can freeze the VIEW layer so that the viewports do not appear on the drawing. You can plot this drawing from the Layout3 with a plot scale factor of 1:1 and the size of the plot will be exactly as specified.

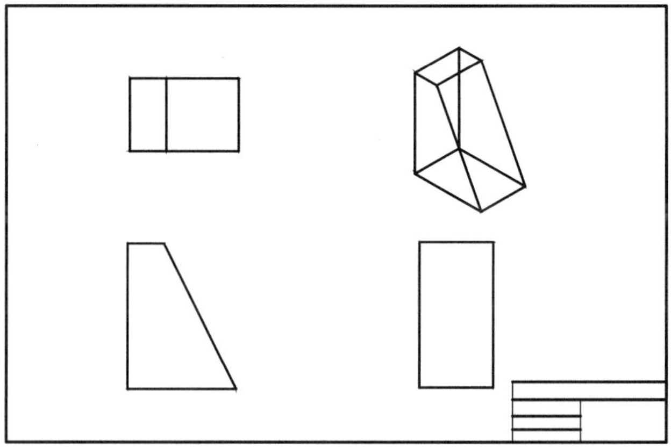

Figure 29-15 *Four views of a 3D object in paper space*

CUSTOMIZING DRAWINGS WITH VIEWPORTS

In certain applications you may need multiple model space viewport configurations to display different views of an object. This involves setting up the desired viewports and then changing the viewpoint for different viewports. You can create a prototype drawing that contains a required number of viewports and the viewpoint information. If you insert a 3D object in one of the viewports of the prototype drawing, you will automatically get different views of the object without setting viewports or viewpoints. The following example illustrates the procedure for creating a prototype drawing with a standard number (four) of viewports and viewpoints.

Example 5

Create a prototype drawing with four viewports, as shown in Figure 29-16.

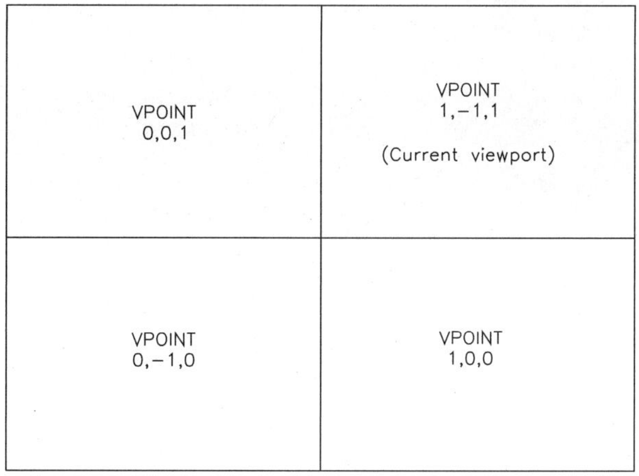

Figure 29-16 *Viewports with different viewpoints*

The viewports should have the following viewpoints.

Viewports	Vpoint	View
Top right	1,-1,1	3D view
Top left	0,0,1	Top view
Lower right	1,0,0	Right side view
Lower left	0,-1,0	Front view

Step 1

Start AutoCAD and create a new drawing from scratch.

Step 2

Setting viewports

Viewports and corresponding viewpoints can be set with the **VPORTS** command. You can also choose the **Display Viewports Dialog** button from the **Viewports** toolbar or choose **Viewports > New Viewports** from the **View** menu to display the **Viewports** dialog box as shown in Figure 29-17. Select **Four:Equal** from the **Standard Viewports** area. In the **Preview** area four equal viewports are displayed. Select **3D** from the **Setup** drop-down list. The four viewports with the different viewpoints will be displayed in the **Preview** area as Top, Front, Right and SE Isometric respectively. **Top** represents the viewpoints as (0,0,1), **Front** represents the viewpoints as (0,-1,0), **Right** represents the viewpoints as (1,0,0) and **SE Isometric** represents the viewpoints as (1,-1,1) respectively. Choose the **OK** button. Save the drawing as *proto5.dwt*.

Viewports and viewpoints can also be set by entering **-VPORTS** and **VPOINT** at the Command prompt respectively.

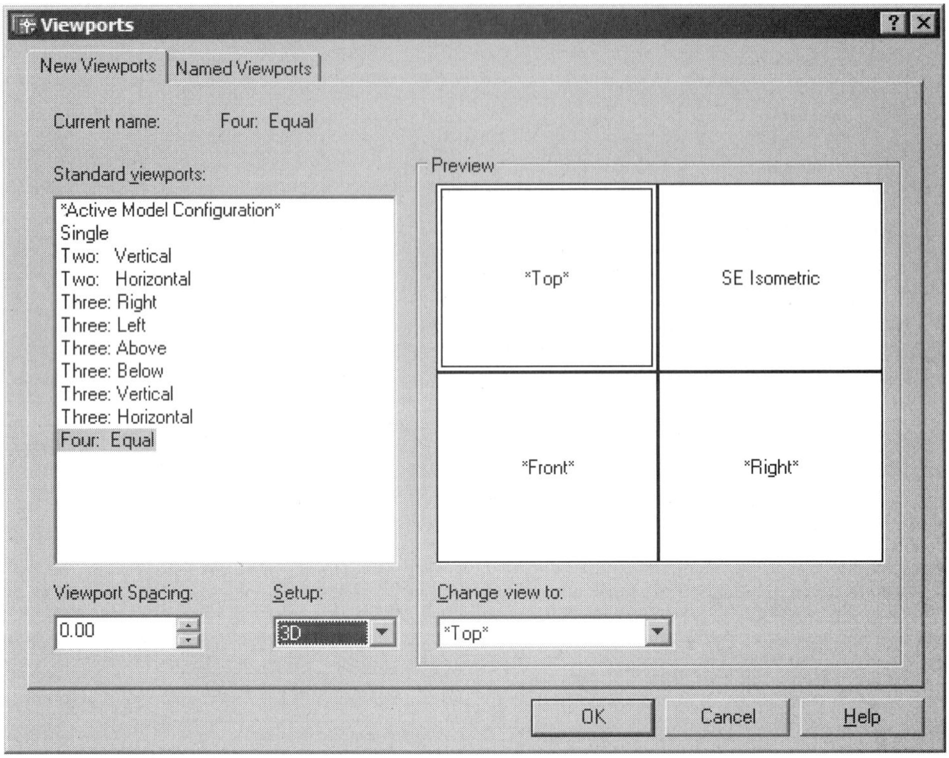

Figure 29-17 ***Viewports*** *dialog box*

Step 3

Start a new drawing and draw the 3D tapered block as shown in Figure 29-18.

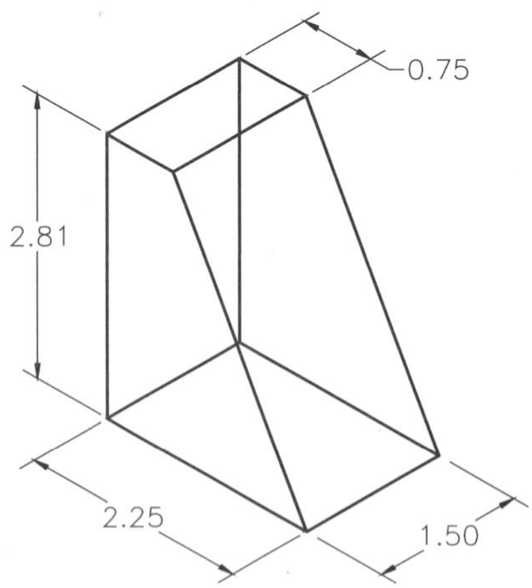

Figure 29-18 3D tapered block

Step 4

Again start a new drawing, TEST, using the prototype drawing *proto5.dwt*. Make the top right viewport current and insert or create a drawing shown in Figure 29-18. Four different views will be automatically displayed on the screen as shown in Figure 29-19.

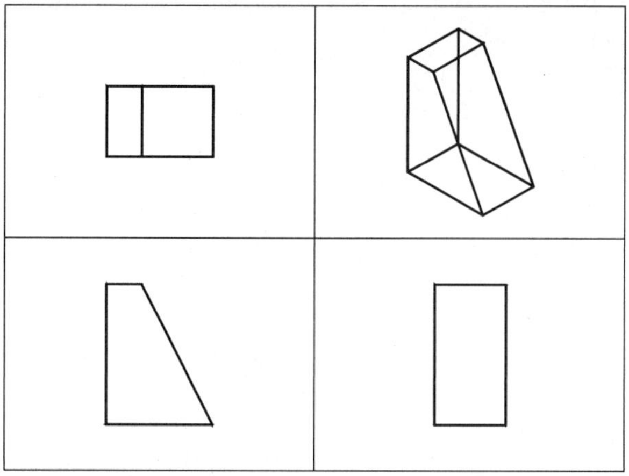

Figure 29-19 *Different views of 3D tapered block*

CUSTOMIZING DRAWINGS ACCORDING TO PLOT SIZE AND DRAWING SCALE

For controlling the plot area, it is recommended to use layouts. You can make the drawing of any size, use the layout to specify the sheet size, and then draw the border and title block. However you can also plot a drawing in the model space and set up the system variables so that the plotted drawing is to your specifications. You can generate a template drawing according to plot size and scale. For example, if the scale is 1/16" = 1' and the drawing is to be plotted on a 36" by 24" area, you can calculate drawing parameters like limits, **DIMSCALE**, and **LTSCALE** and save them in a template drawing. This will save considerable time in the initial drawing setup and provide uniformity in the drawings. The next example explains the procedure involved in customizing a drawing according to a certain plot size and scale.

Note

You can also use the paper space to specify the paper size and scale.

Example 6

Create a template drawing (**PROTO6**) with the following specifications.

Plotted sheet size	36" by 24" (Figure 29-20)
Scale	1/8" = 1.0'
Snap	3'

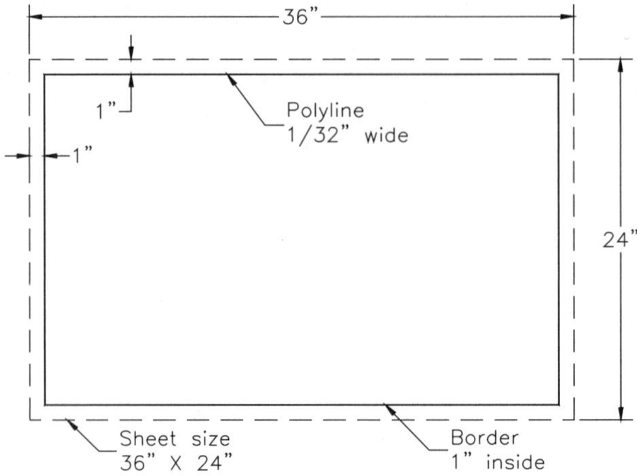

Figure 29-20 *Border of template drawing*

Grid	6'
Text height	1/4" on plotted drawing
Linetype scale	Calculate
Dimscale factor	Calculate
Units	Architectural Precision, 16-denominator of smallest fraction Angle in degrees/minutes/seconds Precision, 0d00' Direction control, base angle, east Angle positive if measured counterclockwise
Border	Border should be 1" inside the edges of the plotted drawing sheet, using PLINE 1/32" wide when plotted (Figure 29-20)

Step 1

Calculating limits, text height, linetype scale, dimension scale and polyline width

In this example, you need to calculate some values before you set the parameters. For example, the limits of the drawing depend on the plotted size of the drawing and the scale of the drawing. Similarly, **LTSCALE** and **DIMSCALE** depend on the plot scale of the drawing. The following calculations explain the procedure for finding the values of limits, ltscale, dimscale, and text height.

Limits

Given:

Sheet size 36" x 24"

Scale 1/8" = 1'

or 1" = 8'

Calculate:
X Limit
Y Limit
Since sheet size is 36" x 24" and scale is 1/8"=1'
Therefore, X Limit = 36 x 8' = 288'
Y Limit = 24 x 8' = 192'

Text height

Given:
Text height when plotted = 1/4"
Scale 1/8" = 1'
Calculate:
Text height
Since scale is 1/8" = 1'
or 1/8" = 12"
or 1" = 96"
Therefore, scale factor = 96
Text height = 1/4" x 96
= 24" = 2'

Linetype scale and dimension scale

Known:
Since scale is 1/8" = 1'
or 1/8" = 12"
or 1" = 96"

Calculate:
LTSCALE and DIMSCALE
Since scale factor = 96
Therefore, LTSCALE = Scale factor = 96
Similarly, DIMSCALE = 96
(All dimension variables, like DIMTXT and DIMASZ, will be multiplied by 96.)

Polyline Width

Given:
Scale is 1/8" = 1'
Calculate:
PLINE width
Since scale is 1/8" = 1'
or 1" = 8'
or 1" = 96"

Therefore,
PLINE width = 1/32 x 96
= 3"

After calculating the parameters, use the following AutoCAD commands to set up the drawing and save the drawing as *proto6.dwt.*

Step 2
Setting units
Start a new drawing and choose **Units** from the **Format** menu or enter **UNITS** at the Command prompt to display the **Drawing Units** dialog box. Choose **Architectural** from the **Type** drop-down list in the **Length** area. Choose **0'-01/16"** from the **Precision** drop-down list. Make sure the **Clockwise** radio button in the **Angle** area is not checked. Select **Deg/Min/Sec** from the **Type** drop-down list and select **0d00** from the **Precision** drop-down list in the **Angle** area. Now choose the **Direction** button to display the **Directional Control** dialog box. Choose the **East** radio button if it is not selected in the **Base Angle** area and then choose **OK**.

Step 3
Setting limits, snap and grid, textsize, linetype scale, dimension scale, dimension style and pline

To set the **LIMITS**, select **Drawing Limits** from the **Format** menu or enter **LIMITS** at the Command prompt.

> Command: **LIMITS**
> Specify lower left corner or [ON/OFF] <0'-0",0'-0">:**0,0**
> Specify upper right corner <1'-0",0'-9">: **288',192'**

Right-click on the **Snap** or **Grid** button in the status bar to invoke the shortcut menu. In the shortcut menu choose the **Settings** to display the **Drafting Settings** dialog box. You can also choose the **Object Snap Settings** button from the **Object Snap** toolbar to display the **Drafting Settings** dialog box. In the dialog box choose the **Snap and Grid** tab. Enter **3'** and **3'** in the **Snap X spacing** and **Snap Y spacing** edit boxes respectively. Enter **6'** and **6'** in the **Grid X spacing** and **Grid Y spacing** edit boxes respectively. Then choose **OK.**

You can also set these values by entering **SNAP** and **GRID** at the Command prompt.

The size of the text can be changed by entering **TEXTSIZE** at the Command prompt.

> Command: **TEXTSIZE**
> Enter new value for TEXTSIZE <current>: **2'**

To set the **LTSCALE**, choose the **Linetype** from the **Format** menu or enter **LINETYPE** at the Command prompt to invoke the **Linetype Manager** dialog box. Choose the **Show details** button. Specify the **Global scale factor** as **96** in the **Global scale factor** edit box.

You can also change the scale of the linetype by entering **LTSCALE** at the Command prompt.

To set the **DIMSTYLE**, choose the **Dimension Style** button from the **Dimension** toolbar or choose **Style** from the **Dimension** menu to invoke the **Dimension Style Manager** dialog box. Choose the **New** button from the **Dimension Style Manager** dialog box to invoke the **Create New Dimension** Style dialog box. Specify the new style name as **MYDIM2** in the **New Style Name** edit box and then choose the **Continue** button. The **New Dimension Style: MYDIM2** dialog box will be displayed. Choose the **Fit** tab and set the value in the **Use overall scale of** spinner to **96** in the **Scale for Dimension Features** area. Now choose the **OK** button to again display the **Dimension Style Manager** dialog box. Choose **Close** to exit the dialog box.

You can invoke **PLINE** command by choosing the **Polyline** button from the **Draw** toolbar or enter **PLINE** at the Command prompt.

Command: **PLINE**
Specify start point: **8',8'**
Current line-width is **0.0000**
Specify next point or [Arc/Close/Halfwidth/Length/Undo/Width]:**W**
Specify starting width<0.00>: **3**
Specify ending width<0'-3">: **ENTER**
Specify next point or [Arc/Close/Halfwidth/Length/Undo/Width]: **280',8'**
Specify next point or [Arc/Close/Halfwidth/Length/Undo/Width]: **280',184'**
Specify next point or [Arc/Close/Halfwidth/Length/Undo/Width]: **8',184'**
Specify next point or [Arc/Close/Halfwidth/Length/Undo/Width]: **C**

Now save the drawing as *proto6.dwt*.

Self- Evaluation Test

Answer the following questions and then compare your answers to the correct answers given at the end of this chapter.

1. The template drawings are stored in __________.

2. To use a template file, select the __________ option in the **Create Drawing** tab of __________ dialog box.

3. To start a drawing with default setup, select the __________ option in the **Create Drawing** tab of __________ dialog box.

4. If plot size is 36" x 24", and the scale is 1/2" = 1', then X Limit = __________ and Y Limit = __________.

5. You can use AutoCAD's __________ command to set up a viewport in paper space.

Review Questions

Answer the following questions.

1. The default value of **DIMSCALE** is __________.

2. The default value for **DIMTXT** is __________.

3. The default value for **SNAP** is __________.

4. Architectural units can be selected by using the __________ command or the __________ command.

5. Name three standard template drawings that come with AutoCAD software __________, __________, and __________.

6. If the plot size is 24" x 18", and the scale is 1 = 20, the X Limit = __________ and Y Limit = __________.

7. If the plot size is 200 x 150 and limits are (0.00,0.00) and (600.00,450.00), the **LTSCALE** factor = __________.

8. __________ provides a convenient way to plot multiple views of a 3D drawing or multiple views of a regular 2D drawing.

9. You can use the __________ command to change to paper space.

10. You can use AutoCAD's __________ command to change to model space.

11. The values that can be assigned to **TILEMODE** are __________ and __________.

12. In the model space, if you want to reduce the display size by half, the scale factor you enter in the **ZOOM**-Scale command is __________.

Exercises

Exercise 1 — *General*

Create a template drawing (*protoe1.dwt*) with the following specifications.

Units	Architectural with precision 0'-0 1/16
Angle	Decimal Degrees with precision 0.
Base angle	East.
Angle direction	Counterclockwise.
Limits	48' x 36'

Exercise 2 — *General*

Create a template drawing (*protoe2.dwt*) with the following specifications.

Limits	36.0,24.0
Snap	0.5
Grid	1.0
Text height	0.25
Units	Decimal
	Precision 0.00
	Decimal degrees
	Precision 0
	Base angle, East
	Angle positive if measured counterclockwise

Exercise 3 — *General*

Create a template drawing (*protoe3.dwt*) with the following specifications.

Limits	48.0,36.0
Text height	0.25
PLINE width	0.03
LTSCALE	4.0
DIMSCALE	4.0
Plot size	10.5 x 8

LAYERS

Layer Names	Line Type	Color
0	Continuous	White
OBJECT	Continuous	Green
CENTER	Center	Magenta
HIDDEN	Hidden	Blue
DIM	Continuous	Red
BORDER	Continuous	Cyan

Exercise 4 *General*

Create a prototype drawing with the following specifications (the name of the drawing is *protoe4.dwt*).

Limits	36.0,24,0
Border	35.0,23.0
Grid	1.0
Snap	0.5
Text height	0.15
Units	Decimal (up to 2 places)
LTSCALE	1
Current layer	Object

LAYERS

Layer Name	**Linetype**	**Color**
0	Continuous	White
Object	Continuous	Red
Hidden	Hidden	Yellow
Center	Center	Green
Dim	Continuous	Blue
Border	Continuous	Magenta
Notes	Continuous	White

This prototype drawing should have a border line and title block as shown in Figure 29-21.

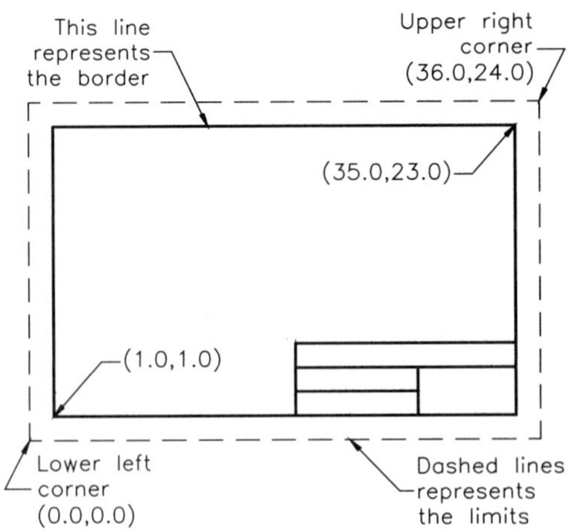

Figure 29-21 *Prototype drawing*

Exercise 5 General

Create a template drawing shown in Figure 29-22 with the following specifications and save it with the name *protoe5.dwt*.

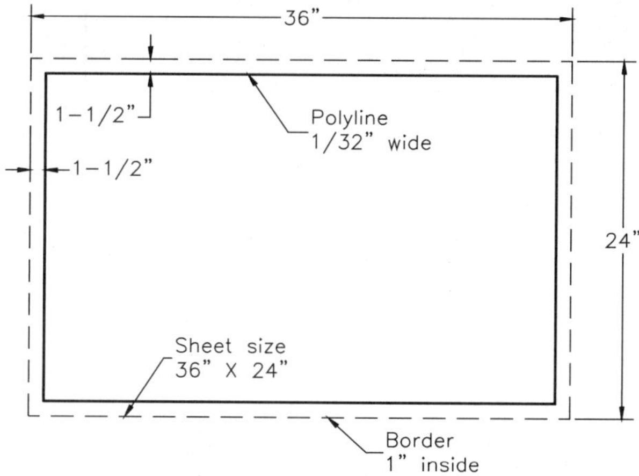

Figure 29-22 *Drawing for Exercise 5*

Plotted sheet size	36" x 24" (Figure 29-22)
Scale	1/2" = 1.0'
Text height	1/4" on plotted drawing
LTSCALE	24
DIMSCALE	24
Units	Architectural 32-denominator of smallest fraction to display Angle in degrees/minutes/seconds Precision 0d00"00" Angle positive if measured counterclockwise
Border	Border is 1-1/2" inside the edges of the plotted drawing sheet, using PLINE 1/32" wide when plotted.

Exercise 6 General

Create a prototype drawing with the following specifications (the name of the drawing is *protoe6.dwt*).

Plotted sheet size	24" x 18" (Figure 29-23)
Scale	1/2"=1.0'
Border	The border is 1" inside the edges of the plotted drawing sheet, using PLINE 0.05" wide when plotted (Figure 29-23)

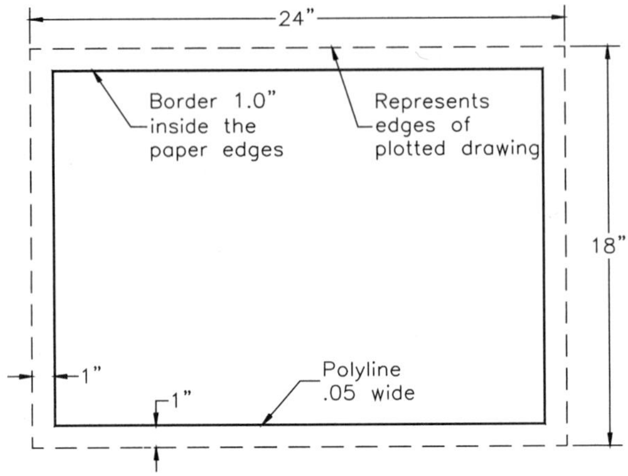

Figure 29-23 *Prototype drawing*

Dimension text over the dimension line
Dimensions aligned with the dimension line
Calculate overall dimension scale factor
Enable the display of alternate units
Dimensions to be associative.

Answers to Self-Evaluation Test

1 - *.dwt,* **2 - Use A Template, 3 - Start From Scratch, 4 -** 72"x48", **5 - MVIEW**.

Chapter 30

Script Files and Slide Shows

Learning Objectives

After completing this chapter you will be able to:

- *Write script files and use the* **SCRIPT** *command to run script files.*
- *Use the* **RSCRIPT** *and* **DELAY** *commands in script files.*
- *Invoke script files when loading AutoCAD.*
- *Create a slide show.*
- *Preload slides when running a slide show.*

WHAT ARE SCRIPT FILES?

AutoCAD has provided a facility called **script files** that allows you to combine different AutoCAD commands and execute them in a predetermined sequence. The commands can be written as a text file using any text editor like Notepad or AutoCAD's **EDIT** command (if the **ACAD.PGP** file is present and **EDIT** is defined in the file). These files, generally known as script files, have extension *.scr* (example: *plot1.scr*). A script file is executed with the AutoCAD **SCRIPT** command.

Script files can be used to generate a slide show, do the initial drawing setup, or plot a drawing to a predefined specification. They can also be used to automate certain command sequences that are used frequently in generating, editing, or viewing drawings. **Remember that the script files cannot access dialog boxes or menus. When commands that open dialog boxes are issued from a script file, AutoCAD runs the command line version of the command instead of opening the dialog box.**

Example 1

Write a script file that will perform the following initial setup for a drawing (file name *script1.scr*). It is assumed that the drawing will be plotted on 12x9 size paper (Scale factor for plotting = 4).

Ortho	On	Zoom	All
Grid	2.0	Text height	0.125
Snap	0.5	LTSCALE	4.0
Limits	0,0 48.0,36.0	DIMSCALE	4.0

Step 1: Understanding commands and prompt entries

Before writing a script file, you need to know the AutoCAD commands and the entries required in response to the Command prompts. To find out the sequence of the Command prompt entries, you can type the command and then respond to different Command prompts. The following is a list of AutoCAD commands and prompt entries for Example 1.

Command: **ORTHO**
Enter mode [ON/OFF] <OFF>: **ON**

Command: **GRID**
Specify grid spacing(X) or [ON/OFF/Snap/Aspect] <1.0>: **2.0**

Command: **SNAP**
Specify snap spacing or [ON/OFF/Aspect/Rotate/Style/Type] <1.0>: **0.5**

Command: **LIMITS**
Reset Model space limits:
Specify lower left corner or [ON/OFF] <0.0,0.0>: **0,0**
Specify upper right corner <12.0,9.0>: **48.0,36.0**

Command: **ZOOM**
Specify corner of window, enter a scale factor (nX or nXP), or
[All/Center/Dynamic/Extents/Previous/Scale/Window] <real time>: **A**

Command: **TEXTSIZE**
Enter new value for TEXTSIZE <0.02>: **0.125**

Command: **LTSCALE**
Enter new linetype scale factor <1.0000>: **4.0**

Command: **DIMSCALE**
Enter new value for DIMSCALE <1.0000>: **4.0**

Step 2: Writing the script file

Once you know the commands and the required prompt entries, you can write a script file using any text editor such as the Notepad.

You can also use the **EDIT** command. As you invoke the **EDIT** command, AutoCAD prompts you to enter the file to edit. Press ENTER in response to the prompt to display the **MS-DOS Editor**. Write the script file in the **MS-DOS Editor**. The following file is a listing of the script file for Example 1.

```
ORTHO
ON
GRID
2.0
SNAP
0.5
LIMITS
0,0
48.0,36.0
ZOOM
ALL
TEXTSIZE
0.125
LTSCALE
4.0
DIMSCALE 4.0
```

Notice that the commands and the prompt entries in this file are in the same sequence as mentioned earlier. You can also combine several statements in one line, as shown in the following list.

```
;This is my first script file, SCRIPT1.SCR
ORTHO ON
GRID 2.0
SNAP 0.5
LIMITS 0,0 48.0,36.0
ZOOM
ALL
TEXTSIZE 0.125
LTSCALE 4.0
DIMSCALE 4.0
```

Save the script file as *script1.scr* on A or C drive and exit the text editor. Remember that if you do not save the file in the *.scr* format, it will not work as a script file. Notice the space between the commands and the prompt entries. For example, between **ORTHO** command and **ON** there is a space. Similarly, there is a space between **GRID** and **2.0**.

Note

In the script file, a space is used to terminate a command or a prompt entry. Therefore, spaces are very important in these files. Make sure there are no extra spaces, unless they are required to press ENTER more than once.

After you change the limits, it is a good practice to use the ***ZOOM*** *command with the* ***All*** *option to increase the drawing display area.*

Tip

AutoCAD ignores and does not process any lines that begin with a semicolon (;). This allows you to put comments in the file.

RUNNING SCRIPT FILES

The **SCRIPT** command allows you to run a script file while you are in the drawing editor. Choose the **Run Script** button from the **Tools** toolbar to invoke the **Select Script File** dialog box as shown in the Figure 30-1. You can also invoke the **Select Script File** dialog box by entering **SCRIPT** at the Command prompt. You can enter the name of the script file or you can accept the default file name. The default script file name is the same as the drawing name. If you want to enter a new file name, type the name of the script file **without** the file extension (**.SCR**). (The file extension is assumed and need not be included with the file name.)

Step 3: Running the script file

To run the script file of Example 1, invoke the **SCRIPT** command, select the file **SCRIPT1**, and then choose the **Open** button in the **Select Script File** dialog box (Figure 30-1) . You will see the changes taking place on the screen as the script file commands are executed.

You can also enter the name of the script file at the Command prompt by setting **FILEDIA**=0. The sequence for invoking the script using the Command line is given next.

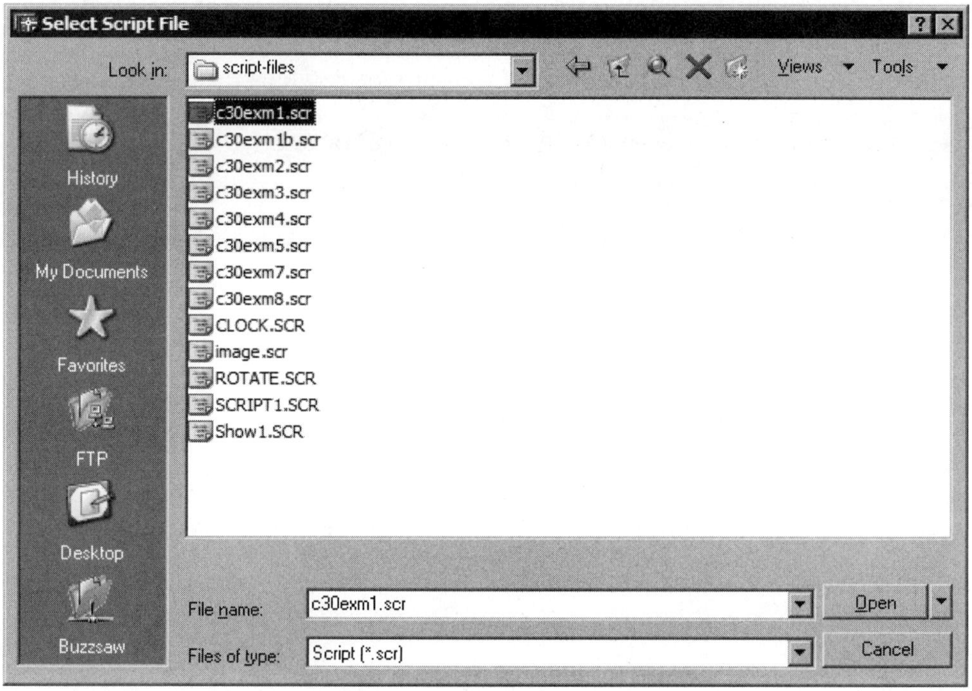

***Figure 30-1 Select Script File** dialog box*

Command: **FILEDIA**
Enter new value for FILEDIA <1>: **0**
Command: **SCRIPT**
Enter script file name <current>: *Script file name.*

Example 2

Write a script file that will set up the following layers with the given colors and linetypes (file name *script2.scr*).

Layer Names	Color	Linetype	Line Weight
Object	Red	Continuous	default
Center	Yellow	Center	default
Hidden	Blue	Hidden	default
Dimension	Green	Continuous	default
Border	Magenta	Continuous	default
Hatch	Cyan	Continuous	0.05

Step 1: Understanding commands and prompt entries

You need to know the AutoCAD commands and the required prompt entries before writing a script file. For Example 2, you need the following commands to create the layers with the given colors and linetypes.

Command: **-LAYER**
Enter an option
[?/Make/Set/New/ON/OFF/Color/Ltype/LWeight/Plot/PStyle/Freeze/Thaw/LOck/Unlock/stAte]: **N**
Enter name list for new layer(s): **OBJECT,CENTER,HIDDEN,DIM,BORDER, HATCH**

Enter an option
[?/Make/Set/New/ON/OFF/Color/Ltype/LWeight/Plot/PStyle/Freeze/Thaw/LOck/Unlock/stAte]: **L**
Enter loaded linetype name or [?] <Continuous>: **CENTER**
Enter name list of layer(s) for linetype "CENTER" <0>: **CENTER**

Enter an option
[?/Make/Set/New/ON/OFF/Color/Ltype/LWeight/Plot/PStyle/Freeze/Thaw/LOck/Unlock/stAte]: **L**
Enter loaded linetype name or [?] <Continuous>: **HIDDEN**
Enter name list of layer(s) for linetype "HIDDEN" <0>: **HIDDEN**

Enter an option
[?/Make/Set/New/ON/OFF/Color/Ltype/LWeight/Plot/PStyle/Freeze/Thaw/LOck/Unlock/stAte]: **C**
New color [Truecolor/COlorbook] <7 (white)>: **RED**
Enter name list of layer(s) for color 1 (red) <0>:**OBJECT**

Enter an option
[?/Make/Set/New/ON/OFF/Color/Ltype/LWeight/Plot/PStyle/Freeze/Thaw/LOck/Unlock/stAte]: **C**
New color [Truecolor/COlorbook] <7 (white)>: **YELLOW**
Enter name list of layer(s) for color 2 (yellow) <0>: **CENTER**

Enter an option
[?/Make/Set/New/ON/OFF/Color/Ltype/LWeight/Plot/PStyle/Freeze/Thaw/LOck/Unlock/stAte]: **C**
New color [Truecolor/COlorbook] <7 (white)>: **BLUE**
Enter name list of layer(s) for color 5 (blue)<0>: **HIDDEN**

Enter an option
[?/Make/Set/New/ON/OFF/Color/Ltype/LWeight/Plot/PStyle/Freeze/Thaw/LOck/Unlock/stAte]: **C**
New color [Truecolor/COlorbook] <7 (white)>: **GREEN**
Enter name list of layer(s) for color 3 (green)<0>: **DIM**

Enter an option
[?/Make/Set/New/ON/OFF/Color/Ltype/LWeight/Plot/PStyle/Freeze/Thaw/LOck/Unlock/stAte]: **C**

New color [Truecolor/COlorbook] <7 (white)>: **MAGENTA**
Enter name list of layer(s) for color 6 (magenta)<0>: **BORDER**

Enter an option
[?/Make/Set/New/ON/OFF/Color/Ltype/LWeight/Plot/PStyle/Freeze/Thaw/LOck/Unlock/stAte]: **C**
New color [Truecolor/COlorbook] <7 (white)>: **CYAN**
Enter name list of layer(s) for color 4 (cyan)<0>: **HATCH**

Enter an option
[?/Make/Set/New/ON/OFF/Color/Ltype/LWeight/Plot/PStyle/Freeze/Thaw/LOck/Unlock/stAte]: **LW**
Enter lineweight (0.0mm - 2.11mm):0.05
Enter name list of layers(s) for lineweight 0.05mm <0>:**HATCH**
[?/Make/Set/New/ON/OFF/Color/Ltype/LWeight/Plot/PStyle/Freeze/Thaw/LOck/Unlock/stAte]: Enter

Step 2: Writing the script file

The following file is a listing of the script file that creates different layers and assigns the given colors and linetypes to these layers.

```
;This script file will create new layers and
;assign different colors and linetypes to layers
LAYER
NEW
OBJECT,CENTER,HIDDEN,DIM,BORDER,HATCH
L
CENTER
CENTER
L
HIDDEN
HIDDEN
C
RED
OBJECT
C
YELLOW
CENTER
C
BLUE
HIDDEN
C
GREEN
DIM
C
MAGENTA
```

BORDER
C
CYAN
HATCH
(This is a blank line to terminate the **LAYER** command. End of script file.)

Save the script file as *script2.scr*.

Step 3: Running the script file
To run the script file of Example 2, choose the **Run Script** button from the **Tools** menu or enter **SCRIPT** at the Command prompt to invoke the **Select Script File** dialog box. Select **SCRIPT2.SCR** and then choose **Open**. You can also enter the **SCRIPT** command and the name of the script file at the Command prompt by setting **FILEDIA**=0.

Example 3

Write a script file that will rotate the circle and the line, as shown in Figure 30-2, around the lower endpoint of the line through 45-degree increments. The script file should be able to produce a continuous rotation of the given objects with a delay of two seconds after every 45-degree rotation (file name *script3.scr*). It is assumed that the line and circle are already drawn on the screen.

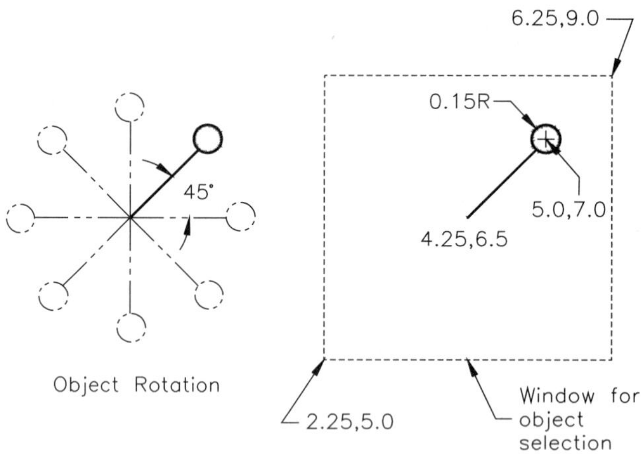

Figure 30-2 *Line and circle rotated through 45-degree increments*

Step 1: Understanding commands and prompt entries
Before writing the script file, enter the required commands and the prompt entries. Write down the exact sequence of the entries in which they have been entered to perform the given operations. The following is a list of the AutoCAD command sequence needed to rotate the circle and the line around the lower endpoint of the line:

Command: **ROTATE**
Current positive angle in UCS: ANGDIR=counterclockwise ANGBASE=0
Select objects: **W** *(Window option to select object)*
Specify first corner: **2.25, 5.0**
Specify opposite corner: **6.25, 9.0**
Select objects: Enter
Specify base point: **4.25,6.5**
Specify rotation angle or [Reference]: **45**

Step 2: Writing the script file

Once the AutoCAD commands, command options, and their sequences are known, you can write a script file. You can use any text editor to write a script file. The following is a listing of the script file that will create the required rotation of the circle and line of Example 3. The line numbers and *(Blank line for Return)* are not a part of the file. They are shown here for reference only.

```
1 ROTATE
2 W
3 2.25,5.0
4 6.25,9.0
5             (Blank line for Return.)
6 4.25,6.5
7 45
```

Line 1
ROTATE
In this line, **ROTATE** is an AutoCAD command that rotates the objects.

Line 2
W
In this line, W is the Window option for selecting the objects that need to be edited.

Line 3
2.25,5.0
In this line, 2.25 defines the *X* coordinate and 5.0 defines the *Y* coordinate of the lower left corner of the object selection window.

Line 4
6.25,9.0
In this line, 6.25 defines the *X* coordinate and 9.0 defines the *Y* coordinate of the upper right corner of the object selection window.

Line 5
Line 5 is a blank line that terminates the object selection process.

Line 6
4.25,6.5
In this line, 4.25 defines the *X* coordinate and 6.5 defines the *Y* coordinate of the base point for rotation.

Line 7
45
In this line, 45 is the incremental angle for rotation.

Note
One of the limitations of the script file is that all the information has to be contained within the file. These files do not let you enter information. For instance, in Example 3, if you want to use the Window option to select the objects, the Window option (W) and the two points that define this window must be contained within the script file. The same is true for the base point and all other information that goes in a script file. There is no way that a script file can prompt you to enter a particular piece of information and then resume the script file, unless you embed AutoLISP commands to prompt for user input.

Step 3: Saving the script file

Save the script file with the name *script3.scr*.

Step 4: Running the script file

Choose **Run Script** from the **Tools** menu or enter **SCRIPT** at the Command prompt to invoke the **Select Script File** dialog box. Select **SCRIPT3.SCR** and then choose **Open.** You will notice that the line and circle that were drawn on the screen are rotated once through an angle of 45-degree. However, there will be no continuous rotation of the sketched entities. The next section (Repeating Script Files) explains how to continue the steps mentioned in the script file. You will also learn how to add a time delay between the continuous cycles in later sections of this chapter.

REPEATING SCRIPT FILES

The **RSCRIPT** command allows the user to execute the script file indefinitely until canceled. It is a very desirable feature when the user wants to run the same file continuously. For example, in the case of a slide show for a product demonstration, the **RSCRIPT** command can be used to run the script file repeatedly until it is terminated by pressing the ESC (Escape) key. Similarly, in Example 3, the rotation command needs to be repeated indefinitely to create a continuous rotation of the objects. This can be accomplished by adding **RSCRIPT** at the end of the file, as shown in the following listing of the script file.

```
ROTATE
W
2.25,5.0
6.25,9.0
                (Blank line for Return.)
4.25,6.5
```

```
45
RSCRIPT
```

The **RSCRIPT** command on line 8 will repeat the commands from line 1 to line 7, and thus set the script file in an indefinite loop. If you run the *script3.scr* file now, you will notice that there is a continuous rotation of the line and circle around the specified base point. However, the speed at which the entities rotate makes it difficult to view the objects. As a result, you need to add time delay between every repetition. The script file can be stopped by pressing the ESC or the BACKSPACE key.

Note

You cannot provide conditional statements in a script file to terminate the file when a particular condition is satisfied unless you use the AutoLISP functions in the script file.

INTRODUCING TIME DELAY IN THE SCRIPT FILES

As mentioned earlier, some of the operations in the script files happen very quickly and make it difficult to see the operations taking place on the screen. It might be necessary to intentionally introduce a pause between certain operations in a script file. For example, in a slide show for a product demonstration, there must be a time delay between different slides so that the audience have enough time to see each slide. This is accomplished by using the **DELAY** command, which introduces a delay before the next command is executed. The general format of the **DELAY** command is.

Command: DELAY Time

Where **Command** ------ AutoCAD command prompt
DELAY ---------- **DELAY** command
Time ------------ Time in milliseconds

The **DELAY** command is to be followed by the delay time in milliseconds. For example, a delay of 2,000 milliseconds means that AutoCAD will pause for approximately two seconds before executing the next command. It is approximately two seconds because computer processing speeds vary. The maximum time delay you can enter is 32,767 milliseconds (about 33 seconds). In Example 3, a two-second delay can be introduced by inserting a **DELAY** command line between line 7 and line 8, as in the following file listing.

```
ROTATE
W
2.25,5.0
6.25,9.0
            (Blank line for Return.)
4.25,6.5
45
DELAY 2000
RSCRIPT
```

The first seven lines of this file rotate the objects through a 45-degree angle. Before the

RSCRIPT command on line 8 is executed, there is a delay of 2,000 milliseconds (about two seconds). The **RSCRIPT** command will repeat the script file that rotates the objects through another 45-degree angle. Thus, a slide show is created with a time delay of two seconds after every 45-degree increment.

RESUMING THE SCRIPT FILES

If you cancel a script file and then want to resume it, you can use the **RESUME** command.

Command: **RESUME**

The **RESUME** command can also be used if the script file has encountered an error that causes it to be suspended. The **RESUME** command will skip the command that caused the error and continue with the rest of the script file. If the error occurs when the command is in progress, use a leading apostrophe with the **RESUME** command (**'RESUME**) to invoke the **RESUME** command in transparent mode.

Command: **'RESUME**

COMMAND LINE SWITCHES

The command line switches can be used as arguments to the *acad.exe* file that launches AutoCAD. You can also use the **Options** dialog box to set the environment or by adding a set of environment variables in the *autoexec.bat* file. The command line switches and environment variables override the values set in the **Options** dialog box for the current session only. These switches do not alter the system registry. The following is the list of the command line switches.

Switch	Function
/c	Controls where AutoCAD stores and searches for the hardware configuration file. The default file is *acad 2004.cfg*.
/s	Specifies which directories to search for support files if they are not in the current directory
/b	Designates a script to run after AutoCAD starts
/t	Specifies a template to use when creating a new drawing
/nologo	Starts AutoCAD without first displaying the logo screen
/v	Designates a particular view of the drawing to be displayed on start-up of AutoCAD
/r	Reconfigures AutoCAD with the default device configuration settings
/p	Specifies the profile to use on start-up

INVOKING A SCRIPT FILE WHILE LOADING AutoCAD

The script files can also be run when loading AutoCAD, without getting into the drawing editor. The format of the command for running a script file when loading AutoCAD is.

"Drive**Program Files\AutoCAD2004\acad.exe" [existing-drawing] [/t template] [/v view] /b Script-file**

In the following example, AutoCAD will open the existing drawing (MYdwg1) and then run the script file (Setup) through the **Run** dialog box as shown in Figure 30-3.

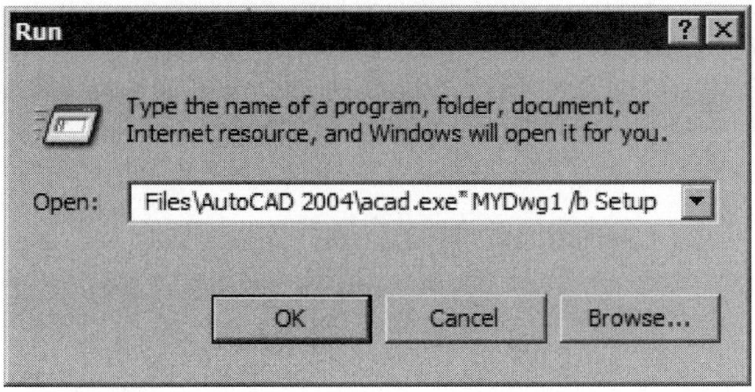

***Figure 30-3** Invoking script file when loading AutoCAD using the **Run** dialog box*

Example

"C:\Program Files\AutoCAD2004\acad.exe" MYdwg1 /b Setup

Where **AutoCAD2004** - AutoCAD2004 subdirectory containing AutoCAD system files
acad.exe -------- ACAD command to start AutoCAD
MYDwg1 -------- Existing drawing file name
Setup ------------- Name of the script file

In the following example, AutoCAD will start a new drawing with the default name (Drawing), using the template file temp1, and then run the script file (Setup).

Example

"C:\Program Files\AutoCAD2004\acad.exe" /t temp1 /b Setup

Where **temp1** ------------ Existing template file name
Setup ------------- Name of the script file

or

"C:\ProgramFiles\AutoCAD2004\acad.exe"/t temp1 "C:\MyFolder"/b Setup

Where C**Program Files\AutoCAD2004\acad.exe** Path name for acad.exe
C:**MyFolder ---** Path name for the Setup script file

In the following example, AutoCAD will start a new drawing with the default name (Drawing), and then run the script file (Setup).

Example

"C:\Program Files\AutoCAD 2004\acad.exe" /b Setup

Where **Setup** ------------- Name of the script file

Here, it is assumed that the AutoCAD system files are loaded in the AutoCAD 2004 directory.

Note

*To invoke a script file while loading AutoCAD, the drawing file or the template file specified in the command must exist in the search path. You cannot start a new drawing with a given name. You can also use any template drawing file that is found in the template directory to run a script file through the **Run** dialog box.*

Tip

You should avoid abbreviations to prevent any confusion. For example, a C can be used as a close option when you are drawing lines. It can also be used as a command alias for drawing a circle. If you use both of these in a script file, it might be confusing.

Example 4

Write a script file that can be invoked when loading AutoCAD and create a drawing with the following setup (filename *script4.scr*).

Grid	3.0
Snap	0.5
Limits	0,0
	36.0,24.0
Zoom	All
Text height	0.25
LTSCALE	3.0
DIMSCALE	3.0

Layers

Name	Color	Linetype
Obj	Red	Continuous
Cen	Yellow	Center
Hid	Blue	Hidden
Dim	Green	Continuous

Step 1: Writing the script file

Write a script file and save the file under the name *script4.scr*. The following is a listing of this script file that does the initial setup for a drawing.

```
GRID 3.0
SNAP 0.5
LIMITS 0,0 36.0,24.0 ZOOM ALL
TEXTSIZE 0.25
LTSCALE 3
DIMSCALE 3.0
LAYER NEW
OBJ,CEN,HID,DIM
```

L CENTER CEN
L HIDDEN HID
C RED OBJ
C YELLOW CEN
C BLUE HID
C GREEN DIM
(Blank line for ENTER.)

Step 2: Loading the script file through the run dialog box
After you have written and saved the file, quit the drawing editor. To run the script file, SCRIPT4, select **Start > Run** and then enter the following command line.

"C:\Program Files\AutoCAD 2004\acad.exe" /t EX4 /b SCRIPT4

Where **acad.exe** -------- ACAD to load AutoCAD
EX4 -------------- Drawing file name
SCRIPT4 ------- Name of the script file

Here it is assumed that the template file (EX4) and the script file (SCRIPT4) is on C drive. When you enter this line, AutoCAD is loaded and the file *ex4.dwt* is opened. The script file, SCRIPT4, is then automatically loaded and the commands defined in the file are executed.

In the following example, AutoCAD will start a new drawing with the default name (Drawing), and then run the script file (SCRIPT4) (Figure 30-4).

Example
"C:\Program Files\AutoCAD 2004\acad.exe" /b SCRIPT4
Where **SCRIPT4** ------- Name of the script file

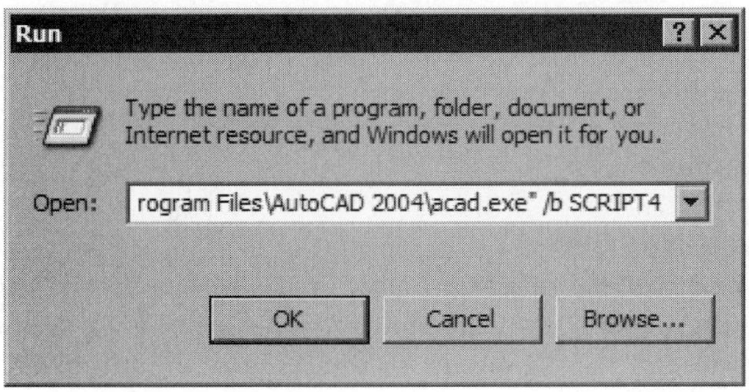

***Figure 30-4** Invoking script file when loading AutoCAD using the **Run** dialog box*

Here, it is assumed that the AutoCAD system files are loaded in the AutoCAD 2004 directory.

Example 5

Write a script file that will plot a 36" by 24" drawing to maximum plot size on a 8.5" by 11" paper, using your system printer/plotter. Use the **Window** option to select the drawing to be plotted.

Step 1: Understanding commands and prompt entries

Before writing a script file to plot a drawing, find out the plotter specifications that must be entered in the script file to obtain the desired output. To determine the prompt entries and their sequence to set up the plotter specifications, enter the **-PLOT** command. Note the entries you make and their sequence (the entries for your printer or plotter will probably be different). The following is a listing of the plotter specifications with the new entries.

Command: **-PLOT**
Detailed plot configuration? [Yes/No] <No>: **Yes**
Enter a layout name or [?] <Model>: Enter
Enter an output device name or [?] <HP LaserJet 4000 Series PCL 6>: Enter
Enter paper size or [?] <Letter (8 1/2 x 11 in)>: Enter
Enter paper units [Inches/Millimeters] <Inches>: **I**
Enter drawing orientation [Portrait/Landscape] <Landscape>: ***L***
Plot upside down? [Yes/No] <No>: **N**
Enter plot area [Display/Extents/Limits/View/Window] <Display>: **W**
Enter lower left corner of window <0.000000,0.000000>: **0,0**
Enter upper right corner of window <0.000000,0.000000>: **36,24**
Enter plot scale (Plotted Inches=Drawing Units) or [Fit] <Fit>: **F**
Enter plot offset (x,y) or [Center] <0.00,0.00>: **0,0**
Plot with plot styles? [Yes/No] <Yes>: **Yes**
Enter plot style table name or [?] (enter . for none) <>: .
Plot with lineweights? [Yes/No] <Yes>: Y
Enter shade plot setting [As displayed/Wireframe/Hidden/Rendered] <As displayed>: Enter
Write the plot to a file [Yes/No] <N>: N
Save changes to page setup? Or set shade plot quality? [Yes/No/Quality] <N>: Enter
Proceed with plot [Yes/No] <Y>: Y

Step 2: Writing the script file

Now you can write the script file by entering the responses to these prompts in the file. The following file is a listing of the script file that will plot a 36" by 24" drawing on 8.5" by 11" paper after making the necessary changes in the plot specifications. The comments on the right are not a part of the file.

Plot
y
(Blank line for ENTER, selects default layout.)
(Blank line for ENTER, selects default printer.)
(Blank line for ENTER, selects the default paper size.)

```
I
L
N
w
0,0
36,24
F
0,0
Y
.                    (Enter . for none)
Y
                     (Blank line for ENTER, plots as displayed.)
N
N
Y
```

Saving and running the script file for this example is the same as that described for previous examples. You can use a blank line to accept the default value for a prompt. A blank line in the script file will cause a Return. However, you must not accept the default plot specifications because the file may have been altered by another user or by another script file. Therefore, always enter the actual values in the file so that when you run a script file, it does not take the default values.

Exercise 1 *General*

Write a script file that will plot a 288' by 192' drawing on a 36" x 24" sheet of paper. The drawing scale is 1/8" = 1'. (The filename is *script9.scr*. In this exercise assume that AutoCAD is configured for the HPGL plotter and the plotter description is HPGL-Plotter.)

Example 6

Write a script file to animate a clock with continuous rotation of the second hand (longer needle) through 5-degree and the minutes hand (shorter needle) through 2 degree clockwise around the center of the clock, (Figure 30-5).

The specifications are given next.

Specifications for the rim made of donut.

Color of Donut	Blue
Inside diameter of Donut	8.0
Outside diameter of Donut	8.4
Center point of Donut	5,5

Specifications for the digit mark made of polyline.

Color of the digit mark	Green
Start point of Pline	5,8.5

Figure 30-5 Drawing for Example 6

Initial width of Pline	0.5
Final width of Pline	0.5
Height of Pline	0.5

Specification for second hand (long needle) made of polyline.

Color of the second hand	Red
Start point of Pline	5,5
Initial width of Pline	0.5
Final width of Pline	0.0
Length of Pline	3.5
Rotation of the second hand	5 degree clockwise

Specification for minute hand (shorter needle) made of polyline.

Color of the minute hand	Cyan
Start point of Pline	5,5
Initial width of Pline	0.35
Final width of Pline	0.0
Length of Pline	3.0
Rotation of the minute hand	2 degree clockwise

Step 1: Understanding the commands and prompt entries for creation of the clock
For this example you can create two script files and then link them. The first script file will demonstrate the creation of the clock on the screen. The next script file will demonstrate the rotation of the needles of the clock.

First write a script file to create the clock as follows and save the file under the name *clock.scr*. The following is the listing of this file.

Command: **-COLOR**
Enter default object color [Truecolor/COlorbook] <BYLAYER>: **Blue**
Command: **DONUT**
Specify inside diameter of donut<0.5>: **8.0**
Specify outside diameter of donut<0.5>: **8.4**
Specify center of donut or <exit>: **5,5**
Specify center of donut or <exit>: Enter
Command: **-COLOR**
Enter default object color [Truecolor/COlorbook] <BYLAYER>: **Green**
Command: **PLINE**
Specify start point: **5,8.5**
Specify next point or [Arc/Close/Halfwidth/Length/Undo/Width]: **Width**
Specify starting width<0.00>: **0.25**
Specify starting width<0.25>: **0.25**
Specify next point or [Arc/Close/Halfwidth/Length/Undo/Width]: **@0.25<270**
Specify next point or [Arc/Close/Halfwidth/Length/Undo/Width]: Enter
Command: **-ARRAY**
Select objects: **Last**
Select objects: Enter
Enter the type of array[Rectangular/Polar]<R>: **Polar**
Specify center point of array: **5,5**
Enter the number of items in the array: **12**
Specify the angle to fill(+= ccw, -=cw)<360>: **360**
Rotate arrayed objects ? [Yes/No]<Y>: **Y**
Command: **-COLOR**
Enter default object color [Truecolor/COlorbook] <BYLAYER>: **RED**
Command: **PLINE**
Specify start point: **5,5**
Specify next point or [Arc/Close/Halfwidth/Length/Undo/Width]: **Width**
Specify starting width<0.5>: **0.5**
Specify ending width<0.5>: **0**
Specify next point or [Arc/Close/Halfwidth/Length/Undo/Width]: **@3.5<0**
Specify next point or [Arc/Close/Halfwidth/Length/Undo/Width]: Enter
Command: **-COLOR**
Enter default object color [Truecolor/COlorbook] <BYLAYER>: **Cyan**
Command: **PLINE**
Specify start point: **5,5**
Specify next point or [Arc/Close/Halfwidth/Length/Undo/Width]: **Width**
Specify starting width<0.5>: **0.35**
Specify ending width<0.35>: **0**
Specify next point or [Arc/Close/Halfwidth/Length/Undo/Width]: **@3<90**
Specify next point or [Arc/Close/Halfwidth/Length/Undo/Width]: Enter
Command: **SCRIPT**
ROTATE.SCR

Now you can write the script file by entering the responses to these prompts in the file *color.scr*. Remember that while entering the commands in the script files, you do not need to add a hyphen (-) as a prefix to the command name to execute them from the command line. When a command is entered using the script file, the dialog box is not displayed and it is executed using the command line. For example, in this script file, the **COLOR** and the **ARRAY** command will be executed using the command line. Listing of the script file is as follows.

```
Color
Blue
Donut
8.0
8.4
5,5
                    Blank line for ENTER
Color
Green
Pline
5,8.5
W
0.25
0.25
@0.25<270
                    Blank line for ENTER
Array
L
                    Blank line for ENTER
P
5,5
12
360
Y
Color
Red
Pline
5,5
W
0.5
0
@3.5<0
                          Blank line for ENTER
Color
Cyan
Pline
5,5
w
```

```
0.35
0
@3<90
                              Blank line for ENTER
Script
ROTATE.SCR                    (Name of the script file that will cause rotation)
```

Save this file as *clock.scr* in a directory that is specified in the AutoCAD support file search path. It is recommended that the *rotate.scr* file should also be saved in the same directory. Remember that if the files are not saved in the directory that is specified in the AutoCAD support file search path using the **Options** dialog box, the linked script file (*rotate.scr*) may not run.

Step 2: Understanding the commands and sequences for rotation of the needles

The last line in the above script file is *rotate.scr*. This is the name of the script file that will rotate the clock hands. Before writing the script file, enter ROTATE command and respond to command prompts that will cause the desired rotation. The following is the listing of the AutoCAD command sequence needed to rotate the objects.

Command: **ROTATE**
Select objects: **L**
Select objects: [Enter]
Specify base point: 5,5
Specify rotation angle or [Reference]: -2
Command: **ROTATE**
Select objects: **C**
Specify first corner: 3,3
Specify other corner: 7,7
Select objects: **Remove**
Remove objects: **L**
Remove objects: [Enter]
Specify base point: 5,5
Specify rotation angle or [Reference]: -5

Now you can write the script file by entering the responses to these prompts in the file *rotate.scr*. The following is the listing of the script file that will rotate the clock hands.

```
Rotate
L
                              Blank line for ENTER
5,5
-2
Rotate
c
3,3
7,7
```

```
R
L
                         Blank line for ENTER
5,5
-5
Rscript
```

Save the above script file as *rotate.scr*. Now run the script file *clock.scr*. Since this file is linked with *rotate.scr*, it will automatically run *rotate.scr* after running *clock.scr*. Note that if the linked file is not saved in a directory specified in the AutoCAD support file search path, the last line of the *clock.scr* must include a fully-resolved path to *rotate.scr*, or AutoCAD would not be able to locate the file.

WHAT IS A SLIDE SHOW?

AutoCAD provides a facility using script files to combine the slides in a text file and display them in a predetermined sequence. In this way, you can generate a slide show for a slide presentation. You can also introduce a time delay in the display so that the viewer has enough time to view each slide.

A drawing or parts of a drawing can also be displayed using the AutoCAD display commands. For example, you can use **ZOOM**, **PAN**, or other commands to display the details you want to show. If the drawing is very complicated, it takes quite some time to display the desired information and it may not be possible to get the desired views in the right sequence. However, with slide shows you can arrange the slides in any order and present them in a definite sequence. In addition to saving time, this also helps to minimize the distraction that might be caused by constantly changing the drawing display. Also, some drawings are confidential in nature and you may not want to display some portions or views of them. You can send a slide show to a client without losing control of the drawings and the information that is contained in them.

WHAT ARE SLIDES?

A **slide** is the snapshot of a screen display; it is like taking a picture of a display with a camera. The slides do not contain any vector information like AutoCAD drawings, which means that the entities do not have any information associated with them. For example, the slides do not retain any information about the layers, colors, linetypes, start point, or endpoint of a line or viewpoint. Therefore, slides cannot be edited like drawings. If you want to make any changes in the slide, you need to edit the drawing and then make a new slide from the edited drawing.

CREATING SLIDES

In AutoCAD, slides are created using the **MSLIDE** command. If **FILEDIA** is set to 1, the **MSLIDE** command displays the **Create Slide File** dialog box (Figure 30-6) on the screen. You can enter the slide file name in this dialog box. If **FILEDIA** is set to 0, the command will prompt you to enter the slide file name.

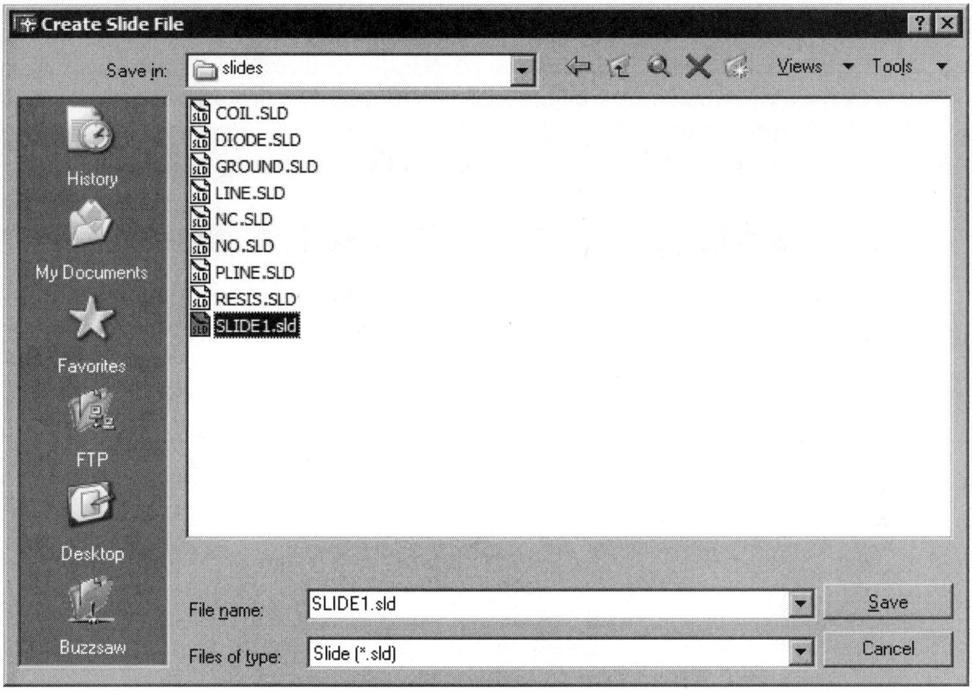

***Figure 30-6 Create Slide File** dialog box*

Command: **MSLIDE**
Enter name of slide file to create <Default>: *Slide file name.*

Example
Command: **MSLIDE**
Slide File: <Drawing1> **SLIDE1**
Where **Drawing1** ------- Default slide file name
SLIDE1 --------- Slide file name

In this example, AutoCAD will save the slide file as *slide1.sld*.

Note
*In model space, you can use the **MSLIDE** command to make a slide of the existing display in the current viewport. If you are in the paper space viewport, you can make a slide of the display in the paper space that includes any floating viewports.*

*When the viewports are not active, the **MSLIDE** command will make a slide of the current screen display.*

VIEWING SLIDES

To view a slide, use the **VSLIDE** command at the Command prompt. The **Select Slide File** dialog box is displayed as shown in the Figure 30-7. Choose the file you want to view and then choose **OK**. The corresponding slide will be displayed on the screen. If **FILEDIA** is 0, the slide that you want to view can be directly entered at the Command prompt.

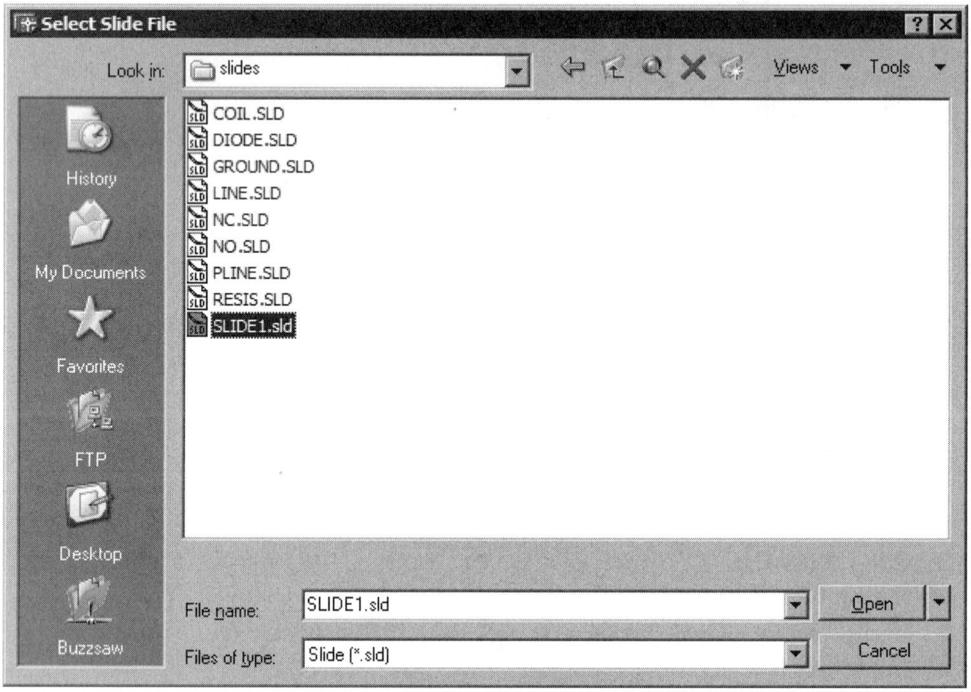

Figure 30-7 ***Select Slide Files*** *dialog box*

Command: **VSLIDE**
Enter name of slide file to create<Default>: *Name*.

Example

Command: **VSLIDE**
Slide file <Drawing1>: SLIDE1
Where **Drawing1** ------- Default slide file name
SLIDE1 --------- Name of slide file

Note

*After viewing a slide, you can use the **REDRAW** command, role the wheel on a wheel mouse, or pan with a wheel mouse to remove the slide display and return to the existing drawing on the screen.*

*Any command that is automatically followed by a **REDRAW** command will also display the existing drawing. For example, AutoCAD **GRID**, **ZOOM ALL**, and **REGEN** commands will automatically return to the existing drawing on the screen.*

You can view the slides on high-resolution or low-resolution monitors. Depending on the resolution of the monitor, AutoCAD automatically adjusts the image. However, if you are using a high-resolution monitor, it is better to make the slides using the same monitor to take full advantage of that monitor.

Example 7

Write a script file that will create a slide show of the following slide files, with a time delay of 15 seconds after every slide (Figure 30-8).

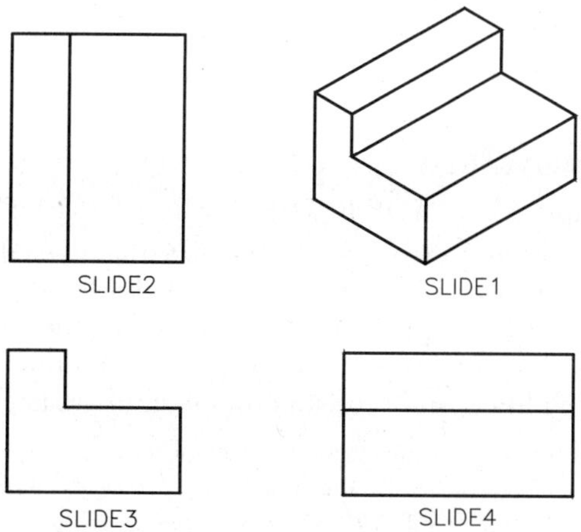

Figure 30-8 *Slides for slide show*

Step 1: Creating the slides

The first step in a slide show is to create the slides using the **MSLIDE** command. The **MSLIDE** command will invoke the **Create Slide File** dialog box. Enter the name of the slide as **SLIDE1** and choose the **Save** button to exit the dialog box. Similarly other slides can be created and saved. Figure 30-8 shows the drawings that have been saved as slide files **SLIDE1, SLIDE2, SLIDE3,** and **SLIDE4**. The slides must be saved to a directory in AutoCAD's search path or the script would not find the slides.

Step 2: Writing the script file

The second step is to find out the sequence in which you want these slides to be displayed, with the necessary time delay if any, between slides. Then you can use any text editor or the AutoCAD **EDIT** command (provided the *acad.pgp* file is present and **EDIT** is defined in the file) to write the script file with the extension *.scr*.

The following file is a listing of the script file that will create a slide show of the slides in Figure 30-8. The name of the script file is **SLDSHOW1**.

```
VSLIDE SLIDE1
DELAY 15000
VSLIDE SLIDE2
DELAY 15000
VSLIDE SLIDE3
DELAY 15000
VSLIDE SLIDE4
DELAY 15000
```

Step 3: Running the script file

To run this slide show, choose the **Run Script** from the **Tools** menu or enter **SCRIPT** at the Command prompt to invoke the **Select Script File** dialog box. Choose **SLDSHOW1** and choose **Open**. You can see the changes taking place on the screen.

PRELOADING SLIDES

In the script file of Example 7, VSLIDE SLIDE1 in line 1 loads the slide file, **SLIDE1**, and displays it on screen. After a pause of 15,000 milliseconds, it starts loading the second slide file, **SLIDE2**. Depending on the computer and the disk access time, you will notice that it takes some time to load the second slide file. The same is true for the other slides. To avoid the delay in loading the slide files, AutoCAD has provided a facility to preload a slide while viewing the previous slide. This is accomplished by placing an asterisk (*) in front of the slide file name.

```
VSLIDE SLIDE1        (View slide, SLIDE1.)
VSLIDE *SLIDE2       (Preload slide, SLIDE2.)
DELAY 15000          (Delay of 15 seconds.)
VSLIDE               (Display slide, SLIDE2.)
VSLIDE *SLIDE3       (Preload slide, SLIDE3.)
DELAY 15000          (Delay of 15 seconds.)
VSLIDE               (Display slide, SLIDE3.)
VSLIDE *SLIDE4
DELAY 15000
VSLIDE
DELAY 15000
RSCRIPT              (Restart the script file.)
```

Example 8

Write a script file to generate a continuous slide show of the following slide files, with a time delay of two seconds between slides SLD1, SLD2, SLD3

The slide files are located in different subdirectories, as shown in Figure 30-9.

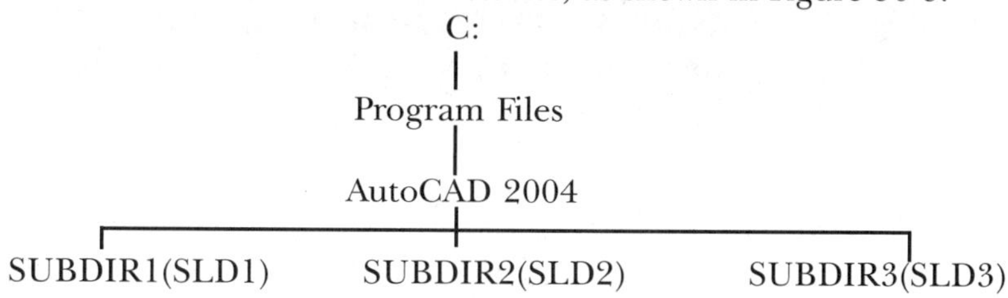

Figure 30-9 *Subdirectories of the C drive*

Where
C: -------------------- *Root directory.*
Program Files ------------ *Root directory.*
AutoCAD 2004 ----------- *Subdirectory where the AutoCAD files are loaded.*
SUBDIR1 ----------------- *Drawing subdirectory.*
SUBDIR2 ----------------- *Drawing subdirectory.*
SUBDIR3 ----------------- *Drawing subdirectory.*
SLD1 -------------------- *Slide file in SUBDIR1 subdirectory.*
SLD2 -------------------- *Slide file in SUBDIR2 subdirectory.*
SLD3 -------------------- *Slide file in SUBDIR3 subdirectory.*

The following is the listing of the script files that will generate a slide show for the slides in Example 8.

```
VSLIDE "C:/Program Files/AutoCAD 2004/SUBDIR1/SLD1.SLD"
DELAY 2000
VSLIDE "C:/Program Files/AutoCAD 2004/SUBDIR2/SLD2.SLD"
DELAY 2000
VSLIDE "C:/Program Files/AutoCAD 2004/SUBDIR3/SLD3.SLD"
DELAY 2000
RSCRIPT
```

Line 1
VSLIDE "C:/Program Files/AutoCAD 2004/SUBDIR1/SLD1.SLD"
In this line, the AutoCAD command **VSLIDE** loads the slide file **SLD1**. The path name is mentioned along with the command **VSLIDE**. If the path name directory contains spaces then the path name must be enclosed in quotes.

Line 2
DELAY 2000
This line uses the AutoCAD **DELAY** command to create a pause of approximately two seconds before the next slide is loaded.

Line 3
VSLIDE "C:/Program Files/**AutoCAD 2004/SUBDIR2/SLD2.SLD"**
In this line, the AutoCAD command **VSLIDE** loads the slide file **SLD2**, located in the subdirectory **SUBDIR2**. If the slide file is located in a different subdirectory, you need to define the path with the slide file.

Line 5
VSLIDE "C:/Program Files/**AutoCAD 2004/SUBDIR3/SLD3.SLD"**
In this line, the **VSLIDE** command loads the slide file SLD3, located in the subdirectory **SUBDIR3**.

Line 7
RSCRIPT
In this line, the **RSCRIPT** command executes the script file again and displays the slides on the screen. This process continues indefinitely until the script file is canceled by pressing the ESC key or the BACKSPACE key.

SLIDE LIBRARIES

AutoCAD provides a utility, SLIDELIB, which constructs a library of the slide files. The format of the SLIDELIB utility command is as follows.

SLIDELIB (Library filename) <(Slide list filename)

Example
SLIDELIB SLDLIB <SLDLIST
Where **SLIDELIB** ------ AutoCAD's SLIDELIB utility
SLDLIB --------- Slide library filename
SLDLIST ------- List of slide filenames

The SLIDELIB utility is supplied with the AutoCAD software package. You can find this utility (SLIDELIB.EXE) in the support subdirectory. The slide file list is a list of the slide filenames that you want in a slide show. It is a text file that can be written by using any text editor or AutoCAD's **EDIT** command (provided ACAD.PGP file is present and **EDIT** is defined in the file). The slide files in the slide file list should not contain any file extension (*.sld*). However, if you want to add a file extension it should be *.sld*.

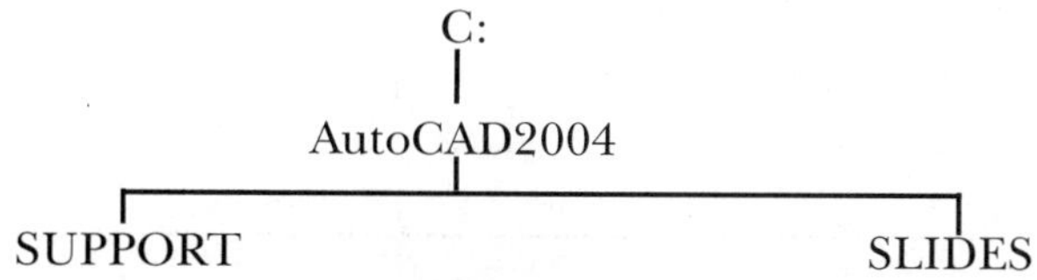

The slide file list can also be created by using the following command, if you have DOS version 5.0 or above. You can use the make directory (md) or change directory (cd) commands in the DOS mode while making or changing directories.

C:\AutoCAD2004\SLIDES>**DIR *.SLD/B>SLDLIST**

In this example assume that the name of the slide file list is **SLDLIST** and all slide files are in the SLIDES subdirectory. To use this command to create a slide file list, all slide files must be in the same directory.

When you use the SLIDELIB utility, it reads the slide file names from the file that is specified in the slide list and the file is then written to the file specified by the library. In the Example 9, the SLIDELIB utility reads the slide filenames from the file SLDLIST and writes them to the library file SLDLIB:

C:\>**SLIDELIB SLDLIB <SLDLIST**

Note

You ***cannot*** *edit a slide library file. If you want to change anything, you have to create a new list of the slide files and then use the SLIDELIB utility to create a new slide library.*

If you edit a slide while the slide is being displayed on the screen, the slide will not be edited. Instead, the current drawing that is behind the slide gets edited. Therefore, do not use any editing commands while you are viewing a slide. Use the ***VSLIDE*** *and* ***DELAY*** *commands only when viewing a slide.*

The path name is not saved in the slide library. This is the reason if you have more than one slide with the same name, even though they are in different subdirectories, only one slide will be saved in the slide library.

Example 9

Use AutoCAD's SLIDELIB utility to generate a continuous slide show of the following slide files with a time delay of 2.5 seconds between the slides. (The filenames are: SLDLIST for slide list file, SLDSHOW1 for slide library, SHOW1 for script file.)

front, top, rside, 3dview, isoview

The slide files are located in different subdirectories as shown in Figure 30-10.

C

Dwg-Files

Proj-A	***Proj-B***	***Slide-Files***
front	*3dview*	*sldlist*
top	*isoview*	*slideli.exe*
rside		

Where **C** ------------------ *(C drive.)*
Dwg-Files ------- *(Subdirectory where drawing files are located)*
Proj-A ------------ *(Drawing subdirectory.)*
Proj-B ------------ *(Drawing subdirectory.)*
Slide-Files ------ *(Directory where Slidelib.exe and sldlist are copied)*

Figure 30-10 *Drawing subdirectories of C drive*

Step 1

The first step is to create a list of the slide file names with the drive and the directory information. Assume that you are in the **Slide-Files** subdirectory. You can use a text editor or AutoCAD's EDIT function to create a list of the slide files that you want to include in the slide show. These files do not need a file extension. However, if you choose to give them a file extension, it should be .SLD. The following file is a listing of the file SLDLIST for Example 9.

```
c:\Dwg-Files\Proj-A\front
c:\Dwg-Files\Proj-A\top
c:\Dwg-Files\Proj-A\rside
c:\Dwg-Files\Proj-B\3dview
c:\Dwg-Files\Proj-B\isoview
```

Step 2

The second step is to use AutoCAD's SLIDELIB utility program to create the slide library. The name of the slide library is assumed to be **sldshow1** for this example. Before creating the slide library, copy the slide list file (SLDLIST) and the SLIDELIB utility from the support directory to the Slide-Files directory. This ensures that all the required files are in one directory. Enter **SHELL** command at AutoCAD Command prompt and then press the ENTER key at the OS Command prompt. The AutoCAD Shell Active dialog box will be displayed on the screen, see Figure 30-11. You can also use Windows DOS box by choosing **Programs > MS-DOS Prompt**.

Command: **SHELL**
OS Command: Enter

Now, enter the following to run the SLIDELIB utility to create the slide library. Here it is assumed that Slide-Files directory is the current directory.

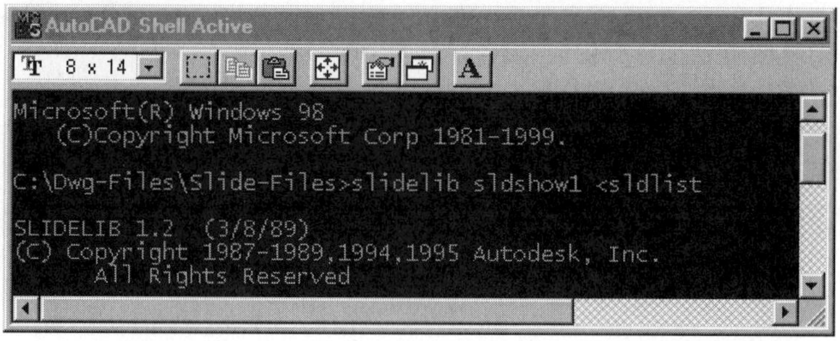

Figure 30-11 *AutoCAD Shell Active dialog box*

C:\Dwg-Files\Slide-Files>SLIDELIB sldshow1 <sldlist

Where **SLIDELIB** ------ AutoCAD's SLIDELIB utility
sldshow1 -------- Slide library
sldlist ------------ Slide file list

Step 3

Now you can write a script file for the slide show that will use the slides in the slide library. The name of the script file for this example is assumed to be SHOW1.

```
VSLIDE sldshow1(front)
DELAY 2500
VSLIDE sldshow1(top)
DELAY 2500
VSLIDE sldshow1(rside)
DELAY 2500
VSLIDE sldshow1(3dview)
DELAY 2500
VSLIDE sldshow1(isoview)
DELAY 2500
RSCRIPT
```

Step 4

Invoke the **Select Script File** (Figure 30-12) dialog box by choosing the **Run Script** button from the **Tools** menu or enter the **SCRIPT** command at the Command prompt. You can also enter the **SCRIPT** command at the Command prompt after setting the system variable FILEDIA to 0.

Command: **SCRIPT**
Enter script file name<default>: **SHOW1**

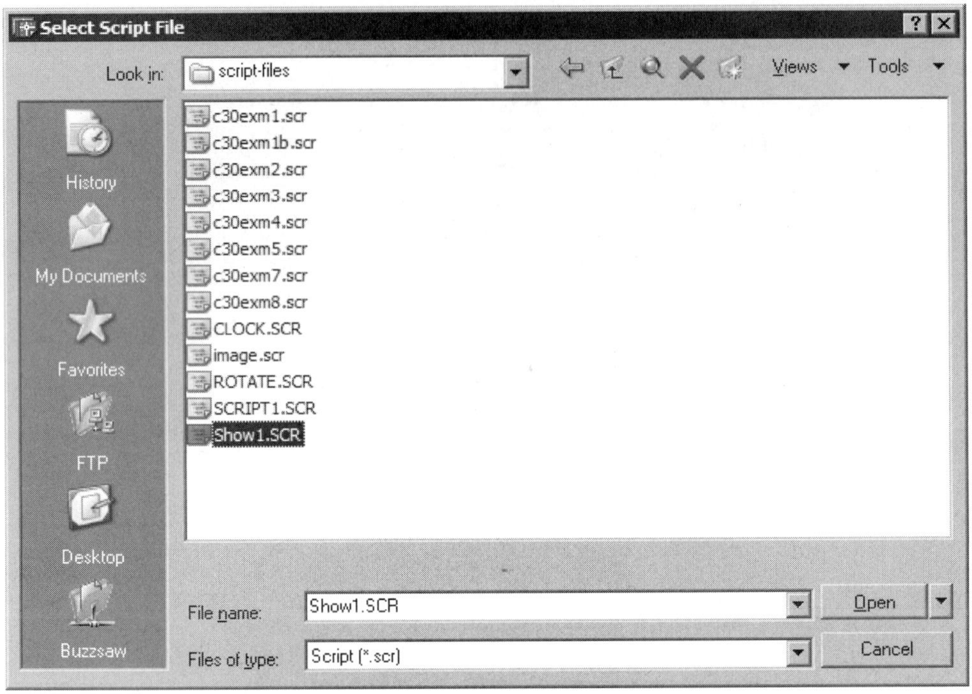

Figure 30-12 *Selecting script file from the* ***Select Script File*** *dialog box*

SLIDE SHOWS WITH RENDERED IMAGES

Slide shows can also contain rendered images. In fact, use of rendered images is highly recommended for a more dynamically pleasing, entertaining, and interesting show. AutoCAD provides the following AutoLISP function for using rendered images during a slide show presentation.

(C:REPLAY FILENAME TYPE [<XOFF> <YOFF> <XSIZE> <YSIZE>])

Where:

C:REPLAY ---------------- AutoLISP function calling AutoCAD's ***Replay*** command.

FILENAME --------------- Name of rendered file, without the extension. The file name should be enclosed in quotes. Include directory path and one forward slash (/) for directory structures.

TYPE -------------------- The rendered file type (e.g., Bmp, Tif, Tga). The file type should also be enclosed in quotes.

XOFF -------------------- The rendered file's "*X*" coordinate offset from 0,0.

YOFF -------------------- The rendered file's "Y" coordinate offset from 0,0.

XSIZE -------------------- The size of the rendered file to be displayed in the "X" (horizontal) direction measured in pixels from the offset position. This information can generally be obtained when creating a rendered file with the **SAVEIMG** command.

YSIZE -------------------- The size of the rendered file to be displayed in the "Y" (vertical) direction measured in pixels from the offset position. This information can generally be obtained when creating a rendered file with the **SAVEIMG** command.

AutoCAD has a file called BIGLAKE.TGA in its library as an example of a background texture map, see Figure 30-13.

Figure 30-13 *Displaying a rendered image using a script file*

But remember that from AutoCAD 2004 onwards, the texture maps are not located in the default directory. If the operating system on your computer is Windows 2000 or XP, the files are located in the *C:\Documents and Settings\Owner\Local Settings\Application Data\Autodesk\AutoCAD 2004\R16.0\enu\textures* directory. If the operating system is Windows 98, the files are located in the *C:\Windows\Local Settings\Application Data\Autodesk\AutoCAD 2004\R16.0\enu\textures* directory. To display this rendered image enter the following at the Command prompt. Modify the file path if you are working on Windows 98 operating system.

(C:REPLAY "C:\Documents and Settings\Owner\Local Settings\Application Data\Autodesk\AutoCAD 2004\R16.0\enu\textures/BIGLAKE" "TGA" 150 50 944 564)

Where:

C:REPLAY: ---------------- Function for AutoCAD to perform - issue the *Replay* command.
"C:/Doc../BIGLAKE" --- Name of the rendered image file, including drive and directory.
"TGA" -------------------- The rendered image file type.
150 -------------------- The "X" offset for the image displayed.
50 -------------------- The "Y" offset for the image displayed.
944 -------------------- The "X" size (horizontal) for the image displayed measured (in pixels) from the offset position.
564 -------------------- The "Y" size (vertical) for the image displayed measured (in pixels) from the offset position.

Combining Slides and Rendered Images

The following is an example of a script file (replay.scr) that combines the use of AutoCAD slides and rendered images (*.bmp files) into a seamless slide show presentation. The "F_" files are the wire FRAME images, while the "R_" files are the RENDERED images. It is assumed that the BMP image files are on A drive.

```
VSLIDE A:\SLIDES\F_ROLL
DELAY 3000
(C:REPLAY "A:/RENDERS/R_ROLL" "BMP" 0 0 944 564)
DELAY 3000
VSLIDE A:\SLIDES\F_BKCASE
DELAY 3000
(C:REPLAY "A:/RENDERS/R_BKCASE" "BMP" 0 0 944 564)
DELAY 3000
VSLIDE A:\SLIDES\F_MOUSE
DELAY 3000
(C:REPLAY "A:/RENDERS/R_MOUSE" "BMP" 0 0 944 564)
DELAY 3000
VSLIDE A:\SLIDES\F_TABLE2
DELAY 3000
(C:REPLAY "A:/RENDERS/R_TABLE2" "BMP" 0 0 944 564)
DELAY 3000
RSCRIPT
```

Self-Evaluation Test

Answer the following questions and then compare your answers to the correct answers given at the end of this chapter.

SCRIPT FILES

1. AutoCAD has provided a facility of __________ that allows you to combine different AutoCAD commands and execute them in a predetermined sequence.
2. Before writing a script file, you need to know the AutoCAD __________ and the __________ required in response to the command prompts.
3. The AutoCAD __________ command is used to run a script file.
4. In a script file, the __________ is used to terminate a command or a prompt entry.
5. The **DELAY** command is to be followed by __________ in milliseconds.

SLIDE SHOWS

6. Slides do not contain any __________ information, which means that the entities do not have any information associated with them.
7. Slides __________ edited like a drawing.
8. Slides can be created using the AutoCAD __________ command.
9. To view a slide, use the AutoCAD __________ command.
10. AutoCAD provides a utility that constructs a library of the slide files. This is done with AutoCAD's utility program called __________.

Review Questions

Answer the following questions.

SCRIPT FILES

1. The __________ files can be used to generate a slide show, do the initial drawing setup, or plot a drawing to a predefined specification.

2. In a script file, you can __________ several statements in one line.

3. When you run a script file, the default script file name is the same as the __________ name.

4. When you run a script file, type the name of the script file without the file __________.

5. One of the limitations of script files is that all the information has to be contained __________ the file.

6. The AutoCAD __________ command allows you to re-execute a script file indefinitely until the command is canceled.

7. You cannot provide a __________ statement in a script file to terminate the file when a particular condition is satisfied.

8. The AutoCAD __________ command introduces a delay before the next command is executed.

9. If the script file is canceled and you want to resume the script file, you need to use the ____________________ command.

SLIDE SHOWS

10. AutoCAD provides a facility through __________ files to combine the slides in a text file and display them in a predetermined sequence.

11. A __________ can also be introduced in the script file so that the viewer has enough time to view a slide.

12. Slides are the __________ of a screen display.

13. In model space, you can use the **MSLIDE** command to make a slide of the __________ display in the __________ viewport.

14. If you are in paper space, you can make a slide of the display in paper space that __________ any floating viewports.

15. If you want to make any change in the slide, you need to ____________________ the drawing, then make a new slide from the edited drawing.

16. If the slide is in the slide library and you want to view it, the slide library name has to be __________ with the slide filename.

17. You cannot __________ a slide library file. If you want to change anything, you have to create a new list of the slide files and then use the __________ utility to create a new slide library.

18. The path name __________ be saved in the slide library. Therefore, if you have more than one slide with the same name, although with different subdirectories, only one slide will be saved in the slide library.

Exercises

SCRIPT FILES

Exercise 2 — *General*

Write a script file that will do the following initial setup for a drawing:

Grid	2.0
Snap	0.5
Limits	0,0
	18.0,12.0
Zoom	All
Text height	0.25
LTSCALE	2.0

Overall dimension scale factor is 2
Aligned dimension text with the dimension line
Dimension text above the dimension line
Size of the center mark is 0.75

Exercise 3 — *General*

Write a script file that will set up the following layers with the given colors and linetypes (filename *scripte3.scr*).

Contour	Red	Continuous
SPipes	Yellow	Center
WPipes	Blue	Hidden
Power	Green	Continuous
Manholes	Magenta	Continuous
Trees	Cyan	Continuous

Exercise 4 *General*

Write a script file that will do the following initial setup for a new drawing.

Limits	0,0 24,18	
Grid	1.0	
Snap	0.25	
Ortho	On	
Snap	On	
Zoom	All	
Pline width	0.02	
PLine	0,0 24,0 24,18 0,18 0,0	
Ltscale	1.5	
Units	Decimal units	
	Precision 0.00	
	Decimal degrees	
	Precision 0	
	Base angle East (0.00)	
	Angle measured counterclockwise	
Layers		
Name	**Color**	**Linetype**
Obj	Red	Continuous
Cen	Yellow	Center
Hid	Blue	Hidden
Dim	Green	Continuous

Exercise 5 *General*

Write a script file that will plot a given drawing according to the following specifications. (Use the plotter for which your system is configured and adjust the values accordingly.)

Plot, using the Window option
Window size (0,0 24,18)
Do not write the plot to file
Size in inch units
Plot origin (0.0,0.0)
Maximum plot size (8.5,11 or the smallest size available on your printer/plotter)
90 degree plot rotation
No removal of hidden lines
Plotting scale (Fit)

Exercise 6 *General*

Write a script file that will continuously rotate a line in 10-degree increments around its midpoint (Figure 30-14). The time delay between increments is one second.

Exercise 7 *General*

Write a script file that will continuously rotate the following arrangements (Figure 30-15). One set of two circles and one line should rotate clockwise while the other set of two circles and one line should rotate counterclockwise. Assume the rotation to be 5-degree around the intersection of the lines for both sets of arrangements.

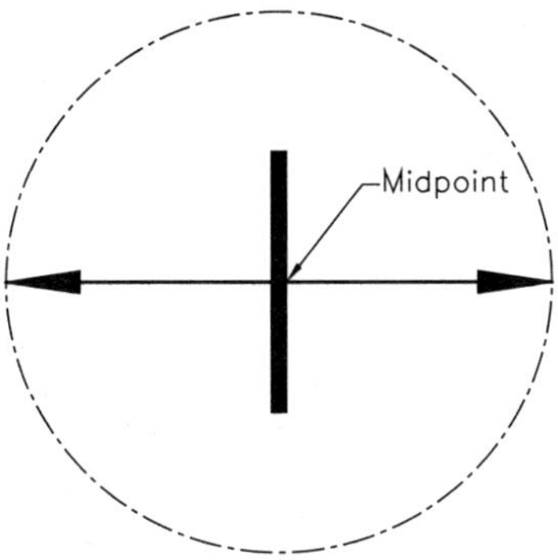

Figure 30-14 *Drawing for Exercise 6*

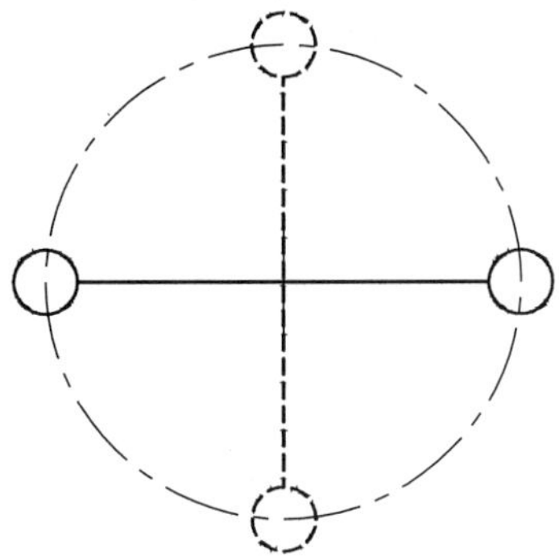

Figure 30-15 *Drawing for Exexrcise 7*

Specifications are given below.

Start point of the horizontal line	2,4
End point of the horizontal line	8,4
Center point of circle at the start point of horizontal line	2,4
Diameter of the circle	1.0
Center point of circle at the end point of horizontal line	8,4
Diameter of circle	1.0
Start point of the vertical line	5,1
End point of the vertical line	5,7
Center point of circle at the start point of the vertical line	5,1
Diameter of the circle	1.0
Center point of circle at the end point of the vertical line	5,7
Diameter of the circle	1.0

(Select one set of two circles and one line and create one group. Similarly select another set of two circles and one line and create another group. Rotate one group clockwise and another group counterclockwise.)

SLIDE SHOWS

Exercise 8 *General*

Make the slides shown in Figure 30-16 and write a script file for a continuous slide show. Provide a time delay of 5 seconds after every slide. (You do not need to use only the slides shown in Figure 30-16. You can use any slides of your choice.)

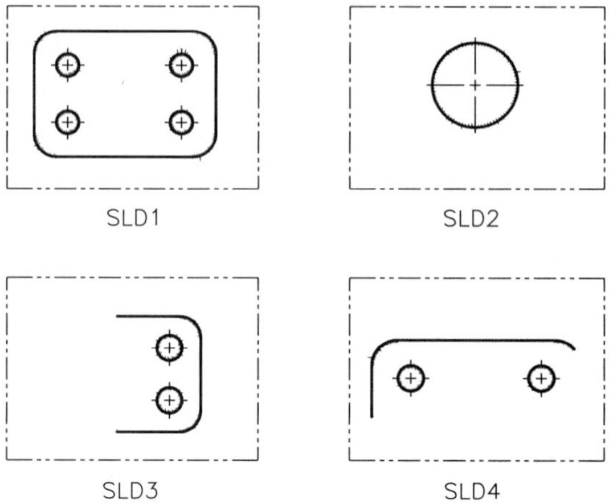

Figure 30-16 *Slides for slide show*

Exercise 9 *General*

List the slides in Exercise 8 in a file SLDLIST2 and create a slide library file SLDLIB2. Then write a script file SHOW2 using the slide library with a time delay of five seconds after every slide.

Answers to Self-Evaluation Test

1- **SCRIPT** files, **2**- Commands, options, **3**- **SCRIPT**, **4**- blank space, **5**- time, **6**- vector, **7**- cannot be, **8**- **MSLIDE**, **9**- **VSLIDE**, **10**- SLIDELIB

Chapter 31

Creating Linetypes and Hatch Patterns

Learning Objectives

After completing this chapter you will be able to:

Create Linetypes:
- *Write linetype definitions.*
- *Create different linetypes.*
- *Create linetype files.*
- *Determine **LTSCALE** for plotting the drawing to given specifications.*
- *Define alternate linetypes and modify existing linetypes.*
- *Create string and shape complex linetypes.*

Create Hatch Patterns:
- *Understand hatch pattern definition.*
- *Create new hatch patterns.*
- *Determine the effect of angle and scale factor on hatch.*
- *Create hatch patterns with multiple descriptors.*
- *Save hatch patterns in a separate file.*
- *Define custom hatch pattern file.*

STANDARD LINETYPES

The AutoCAD software package comes with a library of standard linetypes that has 38 different standard linetypes and seven complex linetypes, including ISO linetypes. These linetypes are saved in the *acad.lin* file. You can modify existing linetypes or create new ones.

LINETYPE DEFINITION

All linetype definitions consist of two parts: **header line and pattern line**.

Header Line

The **header line** consists of an asterisk (*) followed by the name of the linetype and the linetype description. The name and the linetype description should be separated by a comma. If there is no description, the comma that separates the linetype name and the description is not required.

The format of the header line is:

*** Linetype Name, Description**

Example

***HIDDENS,__ __ __ __ __ __**

Where * ------------------- Asterisk sign
HIDDENS ------ Linetype name
, -------------------- Comma
__ __ __ __ ------ Linetype description

All linetype definitions require a linetype name. When you want to load a linetype or assign a linetype to an object, AutoCAD recognizes the linetype by the name you have assigned to the linetype definition. The names of the linetype definition should be selected to help the user recognize the linetype by its name. For example, the linetype name LINEFCX does not give the user any idea about the type of line. However, a linetype name like DASHDOT gives a better idea about the type of line that a user can expect.

The linetype description is a textual representation of the line. This representation can be generated by using dashes, dots, and spaces at the keyboard. The graphic is used to display the linetypes on the screen using the **LINETYPE** command with the ? option or using the dialog box. The linetype description cannot exceed 47 characters.

Pattern Line

The **pattern line** contains the definition of the line pattern consisting of the alignment field specification and the linetype specification, separated by a comma.

The format of the pattern line is:

Alignment Field Specification, Linetype Specification

Example

A,.75,-.25,.75

Where **A** ----------------- Alignment field specification
, ------------------- Comma
.75,-.25,.75 ----- Linetype specification

The letter used for alignment field specification is A. This is the only alignment field supported by AutoCAD; therefore, the pattern line will always start with the letter A. The linetype specification defines the configuration of the dash-dot pattern to generate a line. The maximum number for dash length specification in the linetype is 12, provided the linetype pattern definition fits on one 80-character line.

ELEMENTS OF LINETYPE SPECIFICATION

All linetypes are created by combining the basic elements in a desired configuration. There are three basic elements that can be used to define a linetype specification.

Dash	(Pen down)
Dot	(Pen down, 0 length)
Space	(Pen up)

Example

______ . ______ . ______ . ______

Where **.** ------------------- Dot (pen down with 0 length)
Blank space -------------- Space (pen up)
______ ------------------- Dash (pen down with specified length)

The dashes are generated by defining a positive number. For example, .5 will generate a dash 0.5 units long. Similarly, spaces are generated by defining a negative number. For example, -.2 will generate a space 0.2 units long. The dot is generated by defining a 0 length.

Example

A,.5,-.2,0,-.2,.5

Where **0** ------------------ Dot (zero length)
-.2 ---------------- Length of space (pen up)
.5 ----------------- Length of dash (pen down)

CREATING LINETYPES

Before creating a linetype, you need to decide the type of line you want to generate. Draw the line on a piece of paper and measure the length of each element that constitutes the line. You need to define only one segment of the line, because the pattern is repeated when you draw a line. Linetypes can be created or modified by any one of the following methods.

Using a text editor like Notepad
Adding a new linetype in the *acad.lin* file
Using the **LINETYPE** command

The following example, Example 3, explains how to create a new linetype using the three method mentioned above.

Example 1

Create linetype DASH3DOT (Figure 31-1) with the following specifications.

Length of the first dash 0.5
Blank space 0.125
Dot
Blank space 0.125
Dot
Blank space 0.125
Dot
Blank space 0.125

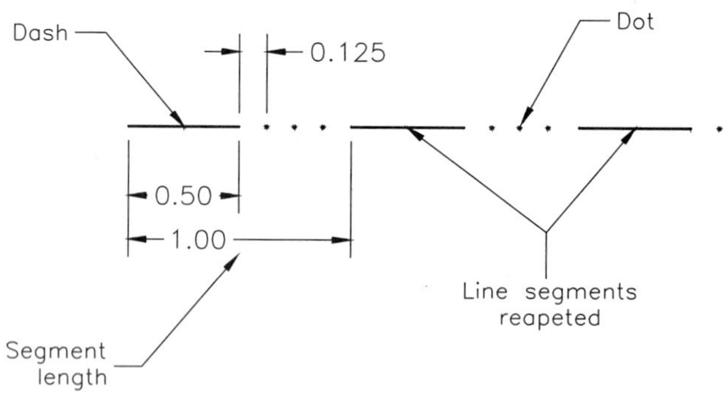

Figure 31-1 Linetype specifications of DASH3DOT

Using a Text Editor

Step 1: Writing definition of linetype

You can start a new linetype file and then add the line definitions to this file. To do this, use any text editor like Notepad to start a new file (*newlt.lin*) and then add the linetype definition of the DASH3DOT linetype. The name and the description must be separated by a comma (,). The description is optional. If you decide not to give one, omit the comma after the linetype name DASH3DOT.

***DASH3DOT,___ . . . ___ . . . ___**
A,.5,-.125,0,-.125,0,-.125,0,-.125

Save it as *newlt.lin* in AutoCAD's Support directory.

Step 2: Loading the linetype

To load this linetype, choose **Linetype** from the **Format** menu to display the **Linetype Manager** dialog box. Choose the **Load** button in the **Linetype Manager** dialog box to display the **Load or Reload Linetypes** dialog box. Choose the **File** button in the **Load or Reload Linetypes** dialog box to display the **Select Linetype File** dialog box as shown in Figure 31-2. Choose the *newlt.lin* file in the **Select Linetype File** dialog box and then choose **Open**. Again the **Load or Reload Linetypes** dialog box is displayed. Choose the **DASH3DOT** linetype in the **Available Linetypes** area and then choose **OK**. The **Linetype Manager** dialog box is displayed. Choose the **DASH3DOT** linetype and then choose the **Current** button to make the selected linetype current. Then choose **OK**.

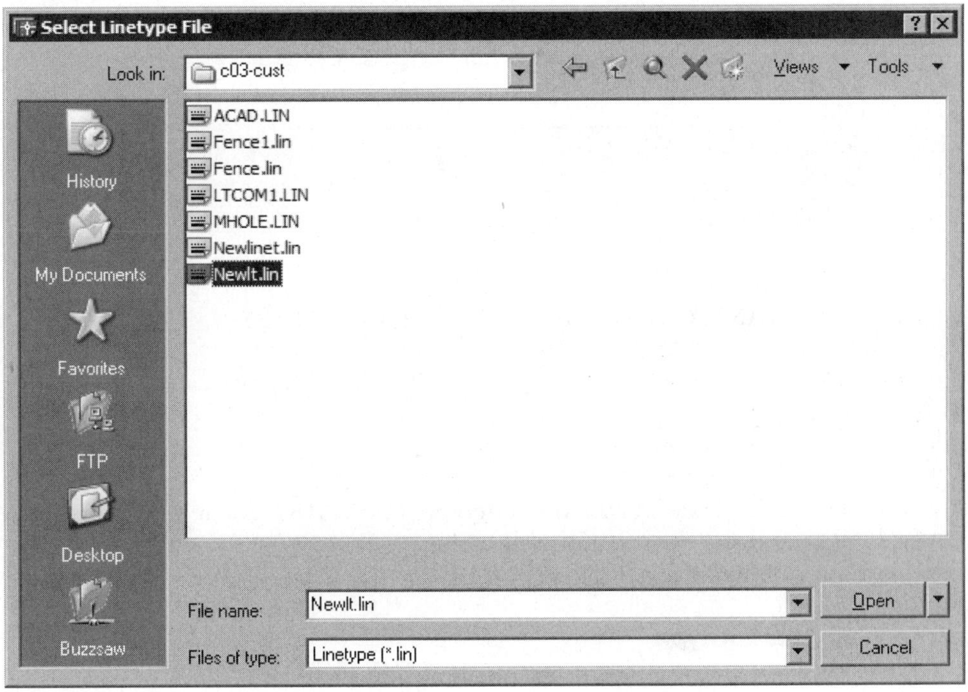

***Figure 31-2 Select Linetype File** dialog box*

Adding a New Linetype in the *acad.lin* File

Step 1: Adding a new linetype in the *acad.lin*

You can also use a text editor (like Notepad) to create a new linetype. Using the text editor, load the file and insert the lines that define the new linetype. The following file is a partial listing of the ***acad.lin*** file after adding a new linetype to the file.

```
*BORDER,__ __ . __ __ . __ __ . __ __ . __ __ . __ __ .
A,.5,-.25,.5,-.25,0,-.25
*BORDER2,__.__.__.__.__.__.__.__.__.__.__.__.__.__.
A,.25,-.125,.25,-.125,0,-.125
*BORDERX2,____ ____ . ____ ____ . ____ ____ .
A,1.0,-.5,1.0,-.5,0,-.5
*CENTERX2,________ __ ________ __ ________ __ __
A,2.5,-.5,.5,-.5
*DASHDOT,__ . __ . __ . __ . __ . __ . __ . __ .
A,.5,-.25,0,-.25
*DOTX2,. . . . . . . . . . . . . . . . .
A,.25,-.125
*HIDDEN2,_ _ _ _ _ _ _ _ _ _ _ _ _ _ _ _ _ _ _ _ _ _ _ _
A,.125,-.0625
*HIDDENX2,____ ____ ____ ____ ____ ____ ____ ____ ____
A,.5,-.25
*PHANTOM,______ __ __ ______ __ __ ______ __ __
A,1.25,-.25,.25,-.25,.25,-.25
*PHANTOMX2,___________ ____ ____ ___________
|
*GAS_LINE,Gas line ----GAS----GAS----GAS----GAS----GAS----GAS--
A,.5,-.2,["GAS",STANDARD,S=.1,R=0.0,X=-0.1,Y=-.05],-.25
*ZIGZAG,Zig zag /\/\/\/\/\/\/\/\/\/\/\/\/\/\/\/\/\/\/\/\/\
A,.0001,-.2,[ZIG,ltypeshp.shx,x=-.2,s=.2],-.4,[ZIG,ltypeshp.shx,r=180,x=.2,s=.2],-.2
*DASH3DOT,____ . . . ____ . . . ____
A,.5,-.125,0,-.125,0,-.125,0,-.125
```

The last two lines of this file define the new linetype, DASH3DOT. The first line contains the name DASH3DOT and the description of the line (___ . . . ___). The second line contains the alignment and the pattern definition.

Step 2: Loading the linetype

Save the file and then load the linetype using the **LINETYPE** command. The procedure of loading the linetype is the same as described earlier in this example. The lines and polylines that this linetype will generate are shown in Figure 31-3.

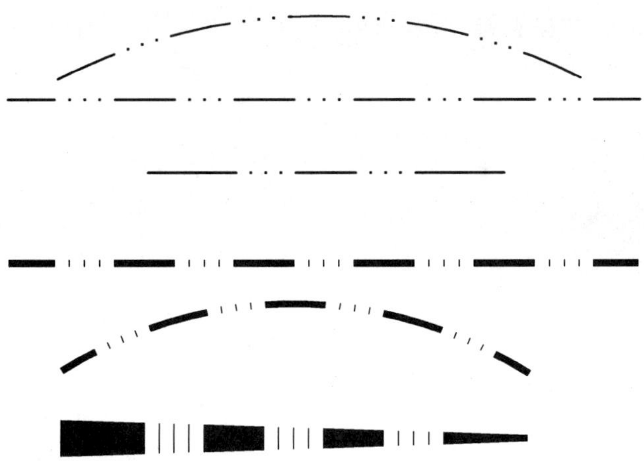

Figure 31-3 *Lines created by linetype DASH3DOT*

Note

If you change the LTSCALE factor, all lines in the drawing are affected by the new ratio.

Using the LINETYPE Command

Step 1: Creating a linetype

To create a linetype using the **LINETYPE** command, first make sure that you are in the drawing editor. Then enter the **-LINETYPE** command and select the **Create** option to create a linetype.

Command: **-LINETYPE**
Enter an option [?/Create/Load/Set]: **C**

Enter the name of the linetype and the name of the library file in which you want to store the definition of the new linetype.

Enter name of linetype to create: **DASH3DOT**

If **FILEDIA**=1, the **Create or Append Linetype File** dialog box (Figure 31-4) will appear on the screen. If **FILEDIA**=0, you are prompted to enter the name of the file.

Enter linetype file name for new linetype definition <default>: **Acad**

If the linetype already exists, the following message will be displayed on the screen.

Wait, checking if linetype already defined...
"Linetype" already exists in this file. Current definition is:
alignment, dash-1, dash-2, _____.
Overwrite?<N>

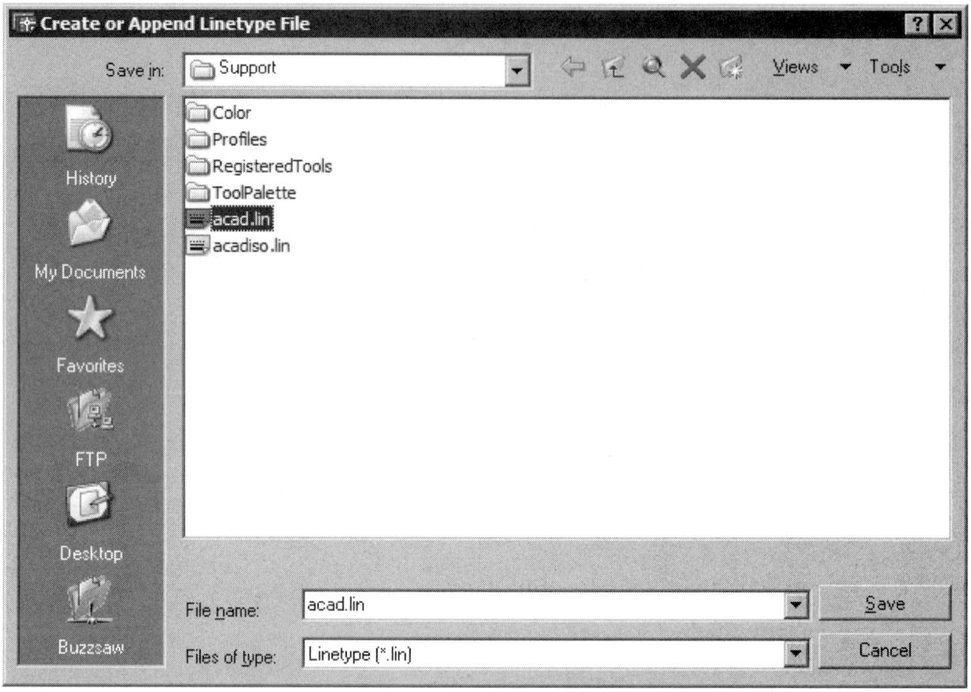

***Figure 31-4 Create or Append Linetype File** dialog box*

If you want to redefine the existing line style, enter Y. Otherwise, type N or press ENTER to choose the default value of N. You can then repeat the process with a different name of the linetype. After entering the name of the linetype and the library file name, you are prompted to enter the descriptive text and the pattern of the line.

Descriptive text: ***DASH3DOT,____ . . . ____ . . . ____**
Enter linetype pattern (on next line):
A,.5,-.125,0,-.125,0,-.125,0,-.125

Descriptive Text

***DASH3DOT,____ . . . ____ . . . ____**

For the descriptive text, you have to type an asterisk (*) followed by the name of the linetype. For Example 1, the name of the linetype is DASH3DOT. The name *DASH3DOT can be followed by the description of the linetype; the length of this description cannot exceed 47 characters. In this example, the description is dashes and dots ____ . . . ____. It could be any text or alphanumeric string. The description is displayed on the screen when you list the linetypes.

Pattern

A,.5,-.125,0,-.125,0,-.125,0,-.125

The line pattern should start with an alignment definition. By default, AutoCAD supports only one type of alignment—A. Therefore, it is displayed on the screen when you select the **LINETYPE** command with the **Create** option. After entering **A** for the pattern alignment, define the pen position. A positive number (.5 or 0.5) indicates a "pen-down" position, and a negative number (-.25 or -0.25) indicates a "pen-up" position. The length of the dash or the space is designated by the magnitude of the number. For example, 0.5 will draw a dash 0.5 units long, and -0.25 will leave a blank space of 0.25 units. A dash length of 0 will draw a dot (.). The following are the pattern definition elements for Example 1.

.5	**pen down**	**0.5 units long dash**
-.125	**pen up**	**.125 units blank space**
0	**pen down**	**dot**
-.125	**pen up**	**.125 units blank space**
0	**pen down**	**dot**
-.125	**pen up**	**.125 units blank space**
0	**pen down**	**dot**
-.125	**pen up**	**.125 units blank space**

After you enter the pattern definition, the linetype (DASH3DOT) is automatically saved in the *acad.lin* file.

Step 2: Loading the linetype

You can use the **LINETYPE** command to load the linetype or choose **Linetype** in the **Format** menu. The linetype (DASH3DOT) can also be loaded using the **-LINETYPE** command and selecting the **Load** option.

ALIGNMENT SPECIFICATION

As the name suggests, the alignment specifies the pattern alignment at the start and the end of the line, circle, or arc. In other words, the line always starts and ends with the dash (___). The alignment definition "A" requires the first element be a dash or dot (pen down), followed by a negative (pen up) segment. The minimum number of dash segments for alignment A is two. If there is not enough space for the line, a continuous line is drawn.

For example, in the linetype DASH3DOT of Example 1, the length of each line segment is 1.0 (.5 + .125 + .125 + .125 + .125 = 1.0). If the length of the line drawn is less than 1.00, a single line is drawn that looks like a continuous line, see Figure 31-5. If the length of the line is 1.00 or greater, the line will be drawn according to DASH3DOT linetype. AutoCAD automatically adjusts the length of the dashes and the line always starts and ends with a dash. The length of the starting and ending dashes is at least half the length of the dash as specified in the file. If the length of the dash as specified in the file is 0.5, the length of the starting and ending dashes is at least 0.25. To fit a line that starts and ends with a dash, the length of these dashes can also increase as shown in Figure 31-5.

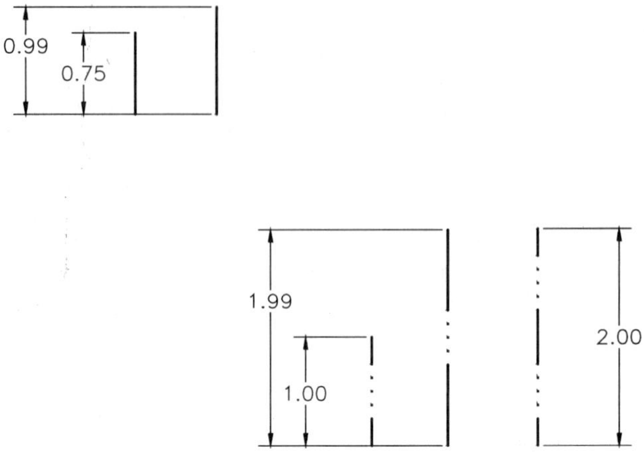

Figure 31-5 *Alignment of linetype DASH3DOT*

LTSCALE COMMAND

As mentioned earlier, the length of each line segment in the DASH3DOT linetype is 1.0 (.5 + .125 + .125 + .125 + .125 = 1.0). If you draw a line that is less than 1.0 units long, a single dash is drawn that looks like a continuous line, see Figure 31-6. This problem can be rectified by changing the linetype scale factor variable **LTSCALE** to a smaller value. This can be accomplished using the **LTSCALE** command.

Command: **LTSCALE**
Enter new linetype scale factor <default>: *New value.*

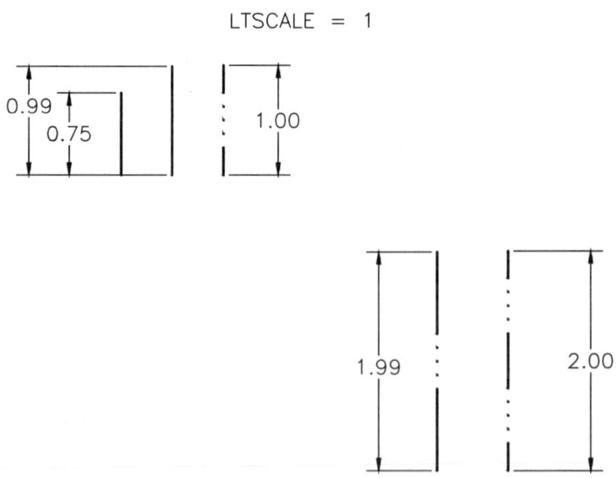

Figure 31-6 *Alignment when LTSCALE = 1*

The default value of the **LTSCALE** variable is **1.0**. If the LTSCALE is changed to 0.75, the length of each segment is reduced by 0.75 (1.0 x 0.75 = 0.75). Then, if you draw a line 0.75 units or longer, it will be drawn according to the definition of DASH3DOT (___ . . . ___) (Figures 31-7 and 31-8).

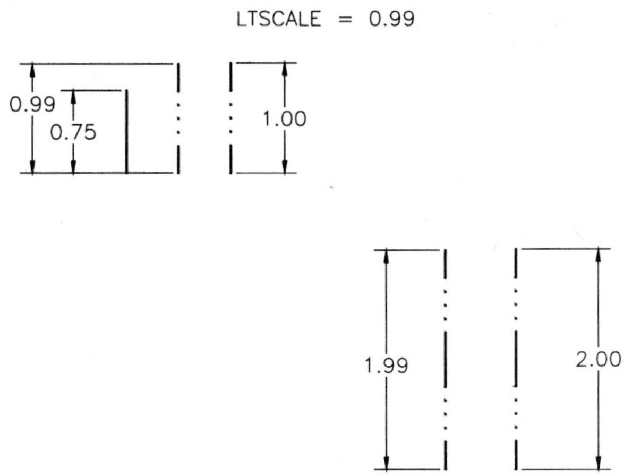

***Figure 31-7** Alignment when LTSCALE = 0.99*

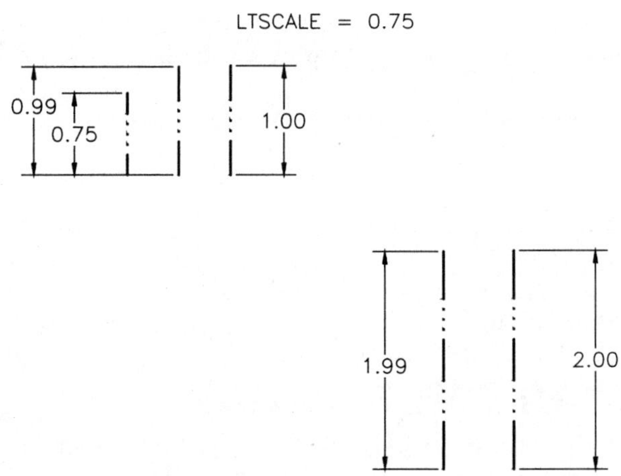

***Figure 31-8** Alignment when LTSCALE = 0.75*

The appearance of the lines is also affected by the limits of the drawing. Most of the AutoCAD linetypes work fine for drawings that have the limits 12,9. Figure 31-9 shows a line of linetype DASH3DOT that is four units long and the limits of the drawing are 12,9. If you increase the limits to 48,36 the lines will appear as continuous lines. If you want the line to appear the same as before **on the screen**, the LTSCALE needs to be changed. Since the limits of the drawing have increased four times, the LTSCALE should also be increased by the same amount.

If you change the scale factor to four, the line segments will also increase by a factor of four. As shown in Figure 31-9, the length of the starting and the ending dash has increased to one unit.

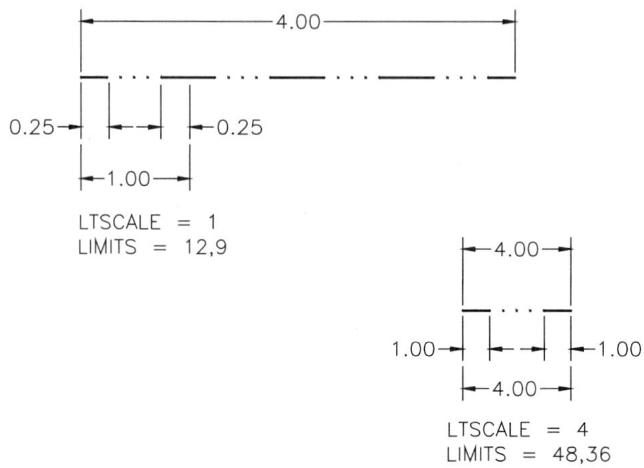

Figure 31-9 *Linetype DASH3DOT before and after changing the LTSCALE factor*

In general, the approximate LTSCALE factor for **screen display** can be obtained by dividing the X -limit of the drawing by the default X -limit (12.00). However, **it is recommended that the linetype scale must be set according to plot scale** discussed in the next section.

LTSCALE factor for SCREEN DISPLAY = X -limits of the drawing/12.00
Example
Drawing limits are 48,36
LTSCALE factor for screen display= 48/12 = 4

Drawing sheet size is 36,24 and scale is 1/4" = 1'
LTSCALE factor for screen display = 12 x 4 x (36 / 12) = 144

LTSCALE FACTOR FOR PLOTTING

The LTSCALE factor for plotting depends on the size of the sheet used to plot the drawing. For example, if the limits are 48 by 36, the drawing scale is 1:1, and you want to plot the drawing on a 48" by 36" size sheet, the LTSCALE factor is 1. If you check the specification of a hidden line in the *acad.lin* file, the length of each dash is 0.25. Therefore, when you plot a drawing with 1:1 scale, the length of each dash in a hidden line is 0.25.

However, if the drawing scale is 1/8" = 1' and you want to plot the drawing on a 48" by 36" paper, the LTSCALE factor must be 96 (8 x 12 = 96). If you increase the LTSCALE factor to 96, the length of each dash in the hidden line will increase by a factor of 96. As a result, the length of each dash will be 24 units (0.25 x 96 = 24). At the time of plotting, the scale factor must be 1:96 to plot the 384' by 288' drawing on a 48" by 36" size paper. Each dash of the hidden line that was 24" long on the drawing will be 0.25 (24/96 = 0.25) inch long when plotted. Similarly, if the desired text size on the paper is 1/8", the text height in the drawing must be 12" (1/8 x 96 = 12").

LTSCALE Factor for PLOTTING = Drawing Scale

Sometimes your plotter may not be able to plot a 48" by 36" drawing or you may like to decrease the size of the plot so that the drawing fits within a specified area. To get the correct dash lengths for hidden, center, or other lines, you must adjust the LTSCALE factor. For example, if you want to plot the previously mentioned drawing in a 45" by 34" area, the correction factor is.

Correction factor = 48/45
= 1.0666
New LTSCALE factor = LTSCALE factor x Correction factor
= 96 x 1.0666
= 102.4

New LTSCALE Factor for PLOTTING = Drawing Scale x Correction Factor

Note

If you change the LTSCALE factor, all lines in the drawing are affected by the new ratio.

CURRENT LINETYPE SCALING (CELTSCALE)

Like **LTSCALE**, the **CELTSCALE** system variable controls the linetype scaling. The difference is that **CELTSCALE** determines the current linetype scaling. For example, if you set the **CELTSCALE** to 0.5, all lines drawn after setting the new value for **CELTSCALE** will have the linetype scaling factor of 0.5. The value is retained in the **CELTSCALE** system variable. The first line (a) in Figure 31-10 is drawn with the **CELTSCALE** factor of 1 and the second line (b) is drawn with the **CELTSCALE** factor of 0.5. The length of the dashes is reduced by a factor of 0.5 when the **CELTSCALE** is 0.5.

The **LTSCALE** system variable controls the global scale factor. For example, if **LTSCALE** is set to 2, all lines in the drawing will be affected by a factor of 2. The net scale factor is equal to the product of **CELTSCALE** and **LTSCALE**. Figure 31-10(c) shows a line that is drawn with **LTSCALE** of 2 and **CELTSCALE** of 0.25. The net scale factor is = **LTSCALE** x **CELTSCALE** = 2 x 0.25 = 0.5.

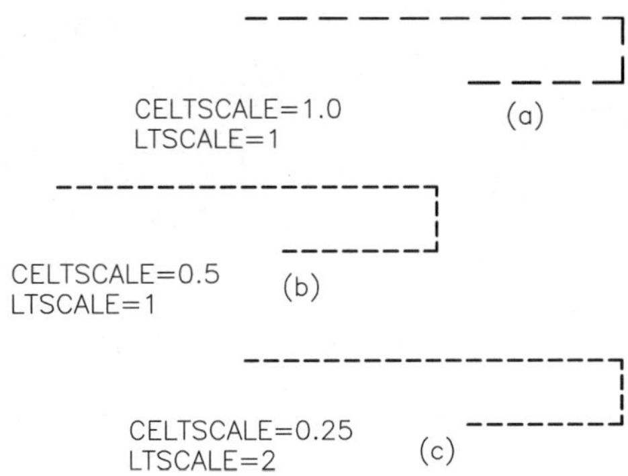

Figure 31-10 *Using* ***CELTSCALE*** *to control current linetype scaling*

Note

*You can change the current linetype scale factor of a line by using the **Properties** dialog box that can be invoked by choosing the **Properties** tool in the **Standard** toolbar. You can also use the **CHANGE** command and then select the **ltScale** option.*

ALTERNATE LINETYPES

One of the problems with the LTSCALE factor is that it affects all the lines in the drawing. As shown in Figure 31-11(a), the length of each segment in all DASH3DOT type lines is approximately equal, no matter how long the lines. You may want to have a small segment length if the lines are small and a longer segment length if the lines are long. You can accomplish this by using CELTSCALE (discussed later in this chapter) or by defining an alternate linetype with a different segment length. For example, you can define a linetype DASH3DOT and DASH3DOTX with different line pattern specifications.

```
*DASH3DOT,____ . . . ____ . . . ____ . . . ____
A,0.5,-.125,0,-.125,0,-.125,0,-.125
*DASH3DOTX,________ . . . __________
A,1.0,-.25,0,-.25,0,-.25,0,-.25
```

In the DASH3DOT linetype the segment length is one unit. Whereas, in the DASH3DOTX linetype the segment length is two units. You can have several alternate linetypes to produce the lines with different segment lengths. Figure 31-11(b) shows the lines generated by DASH3DOT and DASH3DOTX.

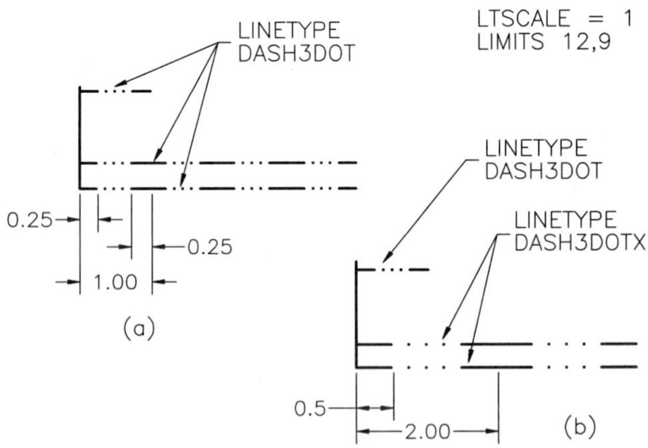

***Figure 31-11** Linetypes generated by DASH3DOT and DASH3DOTX*

Note

Although you may have used various linetypes with different segment lengths, the lines will be affected equally when you change the LTSCALE factor. For example, if the LTSCALE factor is 0.5, the segment length of DASH3DOT line will be 0.5 units and the segment length of DASH3DTX will be 1.0 units.

MODIFYING LINETYPES

You can also modify the linetypes that are defined in the *acad.lin* file. You must save a copy of the original *acad.lin* file before making any changes to it. You need a text editor, such as Notepad, to modify the linetype. You can also use the **EDIT** function of DOS, or the **EDIT** command (provided the *acad.pgp* file is present and **EDIT** is defined in the file). For example, if you want to change the dash length of the border linetype from 0.5 to 0.75, load the file, then edit the pattern line of the border linetype. The following file is a partial listing of the *acad.lin* file after changing the border and centerx2 linetypes.

```
;; AutoCAD Linetype Definition file, Version 2.0
;; Copyright 1991, 1992, 1993, 1994, 1996 by Autodesk, Inc.
;;
*BORDER,__ __ . __ __ . __ __ . __ __ . __ __ . __ __ . __ __ . __ __ .
A,.75,-.25,.75,-.25,0,-.25
*BORDER2,Border (.5x) __.__.__.__.__.__.__.__.__.__.__.
A,.25,-.125,.25,-.125,0,-.125
*BORDERX2,Border (2x) ____ ____ . ____ ____ . ___
A,1.0,-.5,1.0,-.5,0,-.5

*CENTER,Center ____ _ ____ _ ____ _ ____ _ ____ _ ____
A,1.25,-.25,.25,-.25
*CENTER2,Center (.5x) ___ _ ___ _ ___ _ ___ _ ___ _ ___
A,.75,-.125,.125,-.125
*CENTERX2,Center (2x) ________ __ ________ __ _____
A,3.5,-.5,.5,-.5

*DASHDOT,Dash dot __ . __ . __ . __ . __ . __ . __ . __
A,.5,-.25,0,-.25
*DASHDOT2,Dash dot (.5x) _._._._._._._._._._._._._._.
A,.25,-.125,0,-.125
*DASHDOTX2,Dash dot (2x) ____ . ____ . ____ . ___
A,1.0,-.5,0,-.5

*DASHED,Dashed __ __ __ __ __ __ __ __ __ __ __ __ __ _
A,.5,-.25
*DASHED2,Dashed (.5x) _ _ _ _ _ _ _ _ _ _ _ _ _ _ _ _ _ _
A,.25,-.125
*DASHEDX2,Dashed (2x) ____ ____ ____ ____ ____ ___
A,1.0,-.5
```

```
*DIVIDE,Divide ____ . . ____ . . ____ . . ____ . . ____
A,.5,-.25,0,-.25,0,-.25
*DIVIDE2,Divide (.5x) __..__..__..__..__..__..__..__.._
A,.25,-.125,0,-.125,0,-.125
*DIVIDEX2,Divide (2x) ________  . .  ________  . . _
A,1.0,-.5,0,-.5,0,-.5

*DOT,Dot . . . . . . . . . . . . . . . . . . . . . . . .
A,0,-.25
*DOT2,Dot (.5x) ........................................
A,0,-.125
*DOTX2,Dot (2x) . . . . . . . . . . . . . .
A,0,-.5

*HIDDEN,Hidden __ __ __ __ __ __ __ __ __ __ __ __ __ __ __
A,.25,-.125
*HIDDEN2,Hidden (.5x) _ _ _ _ _ _ _ _ _ _ _ _ _ _ _ _ _ _ _ _ _ _
A,.125,-.0625
*HIDDENX2,Hidden (2x) ____ ____ ____ ____ ____ ____ ____
A,.5,-.25

*PHANTOM,Phantom ______  __  __  ______  __  __  ______
A,1.25,-.25,.25,-.25,.25,-.25
*PHANTOM2,Phantom (.5x) ___ _ _ ___ _ _ ___ _ _ ___ _ _
A,.625,-.125,.125,-.125,.125,-.125
*PHANTOMX2,Phantom (2x) ____________    ____    ____   _
A,2.5,-.5,.5,-.5,.5,-.5
;; ISO 128 (ISO/DIS 12011) linetypes
;; The size of the line segments for each defined ISO line, is
;; defined for a usage with a pen width of 1 mm. To use them with
;; the other ISO predefined pen widths, the line has to be scaled
;; with the appropriate value (e.g. pen width 0,5 mm -> ltscale 0.5).
;;
*ACAD_ISO02W100,ISO dash __ __ __ __ __ __ __ __ __ __ __ __ __
A,12,-3
*ACAD_ISO04W100,ISO long-dash dot ____ . ____ . ____ . ____ . _
A,24,-3,.5,-3

*ACAD_ISO14W100,ISO dash triple-dot __ . . . __ . . . __ . . . _
A,12,-3,.5,-3,.5,-3,.5,-3
*ACAD_ISO15W100,ISO double-dash triple-dot __ __ . . . __ __ . .
A,12,-3,12,-3,.5,-3,.5,-3,.5,-3
;; Complex linetypes

;; Complex linetypes have been added to this file.
;; These linetypes were defined in ltypeshp.lin in
```

```
;; Release 13, and are incorporated in acad.lin in Release 14

;; These linetype definitions use ltypeshp.shx.
;;
*FENCELINE1,Fenceline circle ——0——0——0——0——0——0—
A,.25,-.1,[CIRC1,ltypeshp.shx,x=-.1,s=.1],-.1,1
*FENCELINE2,Fenceline square ——[]——[]——[]——[]——[]—-
A,.25,-.1,[BOX,ltypeshp.shx,x=-.1,s=.1],-.1,1
*TRACKS,Tracks -|-|-|-|-|-|-|-|-|-|-|-|-|-|-|-|-|-|-|-|-|-|-|-|-
A,.15,[TRACK1,ltypeshp.shx,s=.25],.15
*BATTING,Batting SSSSSSSSSSSSSSSSSSSSSSSSSSSSSSSSSSSSSSSSSSSS
A,.0001,-.1,[BAT,ltypeshp.shx,x=-.1,s=.1],-.2,[BAT,ltypeshp.shx,r=180,x=.1,s=.1],-.1
*HOT_WATER_SUPPLY,Hot water supply —— HW —— HW —— HW ——
A,.5,-.2,["HW",STANDARD,S=.1,R=0.0,X=-0.1,Y=-.05],-.2
*GAS_LINE,Gas line ——GAS——GAS——GAS——GAS——GAS——GAS—
A,.5,-.2,["GAS",STANDARD,S=.1,R=0.0,X=-0.1,Y=-.05],-.25
*ZIGZAG,Zig zag /\/\/\/\/\/\/\/\/\/\/\/\/\/\/\/\/\/\/\/\/\/\/\/\/
A,.0001,-.2,[ZIG,ltypeshp.shx,x=-.2,s=.2],-.4,[ZIG,ltypeshp.shx,r=180,x=.2,s=.2],-.2
```

Example 2

Create a new file, *newlint.lin*, and define a linetype VARDASH with the following specifications.

Length of first dash 1.0
Blank space 0.25
Length of second dash 0.75
Blank space 0.25
Length of third dash 0.5
Blank space 0.25
Dot
Blank space 0.25
Length of next dash 0.5
Blank space 0.25
Length of next dash 0.75

Step 1: Writing definition of linetype

Use a text editor and insert the following lines that define the new linetype **VARDASH**.

```
*VARDASH,———— ——— —— . —— ——— ————
A,1,-.25,.75,-.25,.5,-.25,0,-.25,.5,-.25,.75,-.25
```

Step 2: Loading the linetype

You can use the **LINETYPE** command to load the linetype or choose **Linetype** in the **Format** menu. The type of lines that this linetype will generate are shown in Figure 31-12.

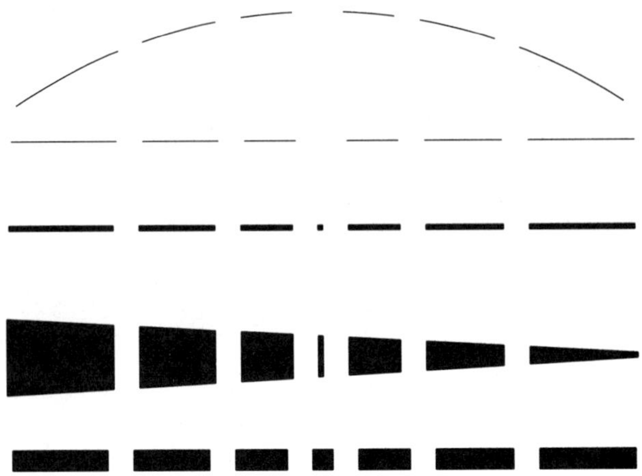

Figure 31-12 Lines generated by linetype VARDASH

COMPLEX LINETYPES

AutoCAD has provided a facility to create complex linetypes. The complex linetypes can be classified into two groups: string complex linetype and shape complex linetype. The difference between the two is that the string complex linetype has a text string inserted in the line, whereas the shape complex linetype has a shape inserted in the line. The facility of creating complex linetypes increases the functionality of lines. For example, if you want to draw a line around a building that indicates the fence line, you can do it by defining a complex linetype that will automatically give you the desired line with the text string (Fence). Similarly, you can define a complex linetype that will insert a shape (symbol) at predefined distances along the line.

CREATING A STRING COMPLEX LINETYPE

When writing the definition of a string complex linetype, the actual text and its attributes must be included in the linetype definition, refer to Figure 31-13. The format of the string complex linetype is as follows.

["String", Text Style, Text Height, Rotation, X-Offset, Y-Offset]

String. It is the actual text that you want to insert along the line. The text string must be enclosed in quotation marks (" ").

Text Style. This is the name of the text style file that you want to use for generating the text string. The text style must be predefined.

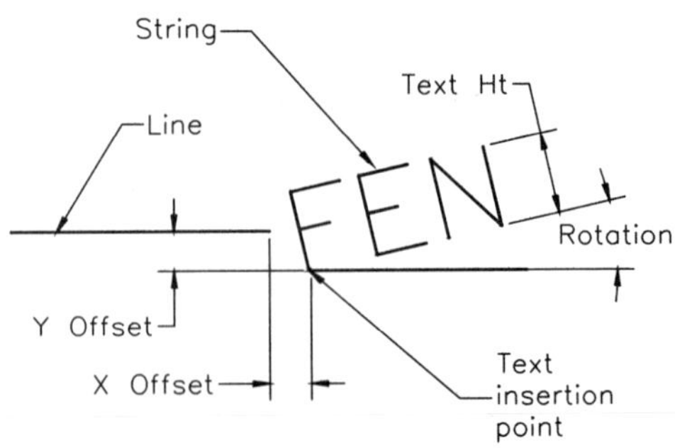

Figure 31-13 Attributes of a string complex linetype

Text Height. This is the actual height of the text, if the text height defined in the text style is 0. Otherwise, it acts as a scale factor for the text height specified in the text style. In Figure 31-13, the height of the text is 0.1 units.

Rotation. The rotation can be specified as an absolute or relative angle. In the absolute rotation the angle is always measured with respect to the positive *X* axis, no matter what AutoCAD's direction setting. The absolute angle is represented by letter "a".

In relative rotation the angle is always measured with respect to orientation of dashes in the linetype. The relative angle is represented by the letter "r". The angle can be specified in radians (r), grads (g), or degrees (d). The default is degrees.

X-Offset. This is the distance of the lower left corner of the text string from the endpoint of the line segment measured along the line. If the line is horizontal, then the X-Offset distance is measured along the *X* axis. In Figure 31-13, the X-Offset distance is 0.05.

Y-Offset. This is the distance of the lower left corner of the text string from the endpoint of the line segment measured perpendicular to the line. If the line is horizontal, then the Y-Offset distance is measured along the *Y* axis. In Figure 31-13, the Y-Offset distance is -0.05. The distance is negative because the start point of the text string is 0.05 units below the endpoint of the first line segment.

Example 3

In the following example, you will write the definition of a string complex linetype that consists of the text string "Fence" and line segments. The length of each line segment is 0.75. The height of the text string is 0.1 units, and the space between the end of the text string and the following line segment is 0.05, see Figure 31-14.

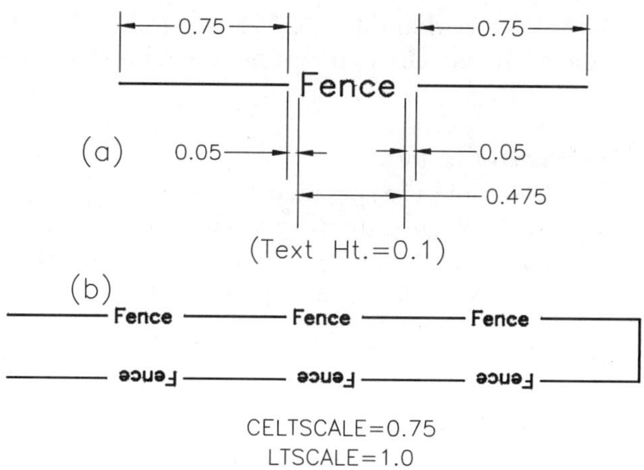

Figure 31-14 *The attributes of a string complex linetype and line specifications for Example 3*

Step 1: Determining the line specifications

Before writing the definition of a new linetype, it is important to determine the line specification. One of the ways this can be done is to actually draw the lines and the text the way you want them to appear in the drawing. Once you have drawn the line and the text to your satisfaction, measure the distances needed to define the string complex linetype. The values are given as follows.

Text string=	Fence
Text style=	Standard
Text height=	0.1
Text rotation=	0
X-Offset=	0.05
Y-Offset=	-0.05
Length of the first line segment=	0.75
Distance between the line segments=	0.575

Step 2: Writing the definition of string complex linetype

Use a text editor to write the definition of the string complex linetype. You can add the definition to the *acad.lin* file or create a separate file. The extension of the file must be *.lin*. The following file is the listing of the *fence.lin* file for Example 3. The name of the linetype is NEWFence1.

```
*NEWFence1,New fence boundary line
A,0.75,["Fence",Standard,S=0.1,A=0,X=0.05,Y=-0.05],-0.575
or
A,0.75,-0.05,["Fence",Standard,S=0.1,A=0,X=0,Y=-0.05],-0.525
```

Step 3: Loading the linetype

You can use the **LINETYPE** command to load the linetype or choose **Linetype** in the **Format** pulldown menu. Draw a line or any object to check if the line is drawn to the given specifications as shown in the Figure 31-15. Notice that the text is always drawn along the *X* axis. Also, when you draw a line at an angle, polyline, circle, or spline, the text string does not align with the object (Figure 31-15).

Step 4: Aligning the text with the line

In the NEWFence linetype definition, the specified angle is 0-degree (Absolute angle A = 0). Therefore, when you use the NEWFence linetype to draw a line, circle, polyline, or spline, the text string (Fence) will be at zero degrees. If you want the text string (Fence) to align with the polyline (Figure 31-16), spline, or circle, specify the angle as relative angle (R = 0) in the NEWFence linetype definition. The following is the linetype definition for NEWFence linetype with relative angle R = 0.

```
*NEWFence2,New fence boundary line
A,0.75,["Fence",Standard,S=0.1,R=0,X=0.05,Y=-0.05],-0.575
```

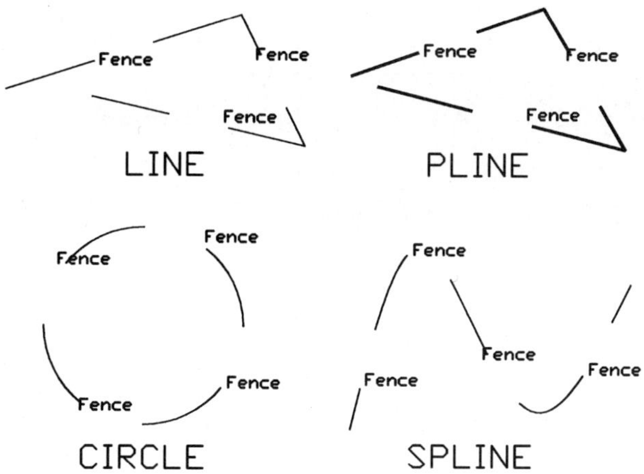

Figure 31-15 Using string complex linetype with angle A=0

Step 5: Aligning the midpoint of text with the line

In Figure 31-16, you will notice that the text string is not properly aligned with the circumference of the circle. This is because AutoCAD draws the text string in a direction that is tangent to the circle at the text insertion point.

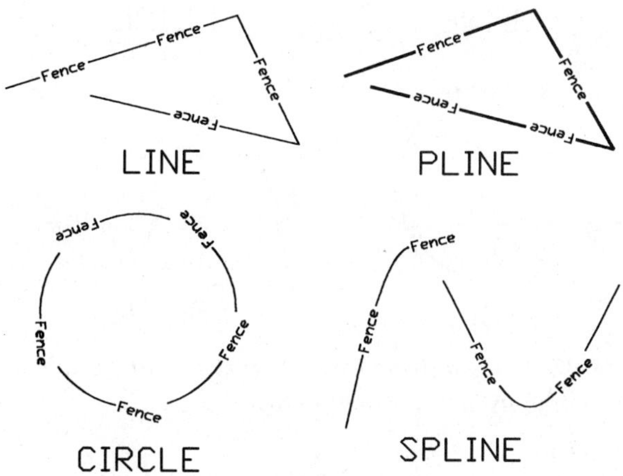

Figure 31-16 Using a string complex linetype with angle R = 0

To resolve this problem, you must define the middle point of the text string as the insertion point. Also, the line specifications should be measured accordingly. Figure 31-17 gives the measurements of the NEWFence linetype with the middle point of the text as the insertion point and Figure 31-18 shows the entities sketched with the selected linetype.

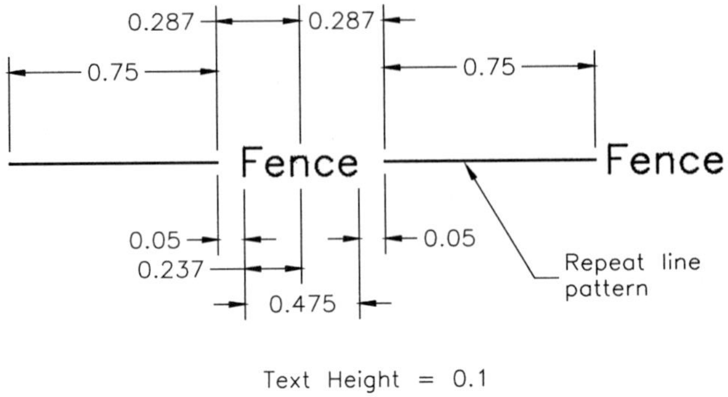

Figure 31-17 *Specifications of a string complex linetype with the middle point of the text string as the text insertion point*

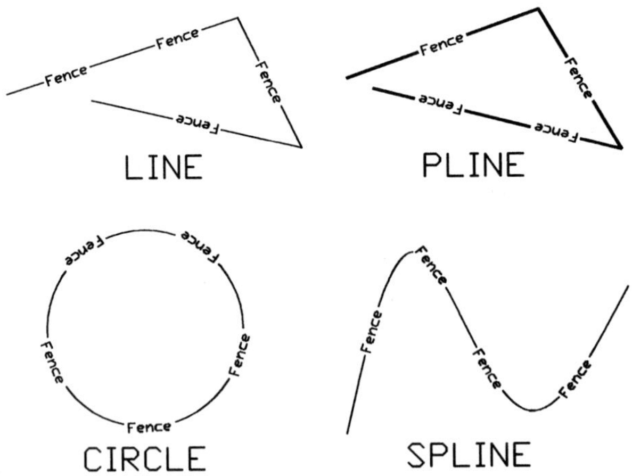

Figure 31-18 *Using a string complex linetype with the middle point of the text string as the text insertion point*

The following is the linetype definition for NEWFence linetype.

```
*NEWFence3,New fence boundary line
A,0.75,-0.287,["FENCE",Standard,S=0.1,X=-0.237,Y=-0.05],-0.287
```

Note

If no angle is defined in the line definition, it defaults to angle R = 0. Also, the text does not automatically insert to its midpoint like the regular text with MID justification.

Creating a Shape Complex Linetype

As with the string complex linetype, when you write the definition of a shape complex linetype, the name of the shape, the name of the shape file, and other shape attributes, like rotation, scale, X-Offset, and Y-Offset, must be included in the linetype definition. The format of the shape complex linetype is as follows.

[Shape Name, Shape File, Scale, Rotation, X-Offset, Y-Offset]

The following is the description of the attributes of Shape Complex Linetype (Figure 31-19).

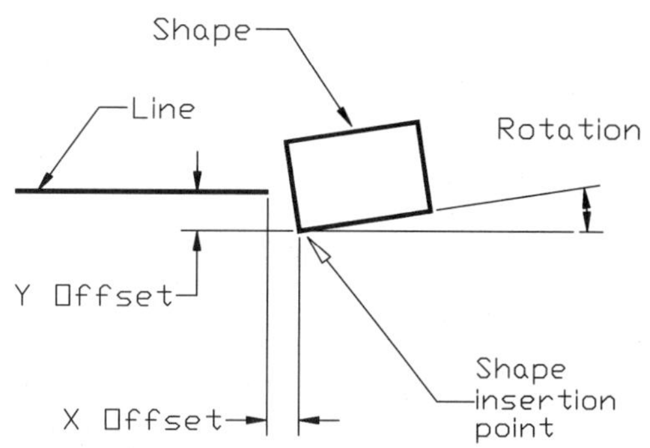

Figure 31-19 *The attributes of a shape complex linetype*

Shape Name. This is the name of the shape that you want to insert along the line. The shape name must exist; otherwise, no shape will be generated along the line.

Shape File. This is the name of the compiled shape file (*.shx*) that contains the definition of the shape being inserted in the line. The name of the subdirectory where the shape file is located must be in the ACAD search path. The shape files (*.shp*) must be compiled before using the **SHAPE** command to load the shape.

Scale. This is the scale factor by which the defined shape size is to be scaled. If the scale is 1, the size of the shape will be the same as defined in the shape definition (*.shp* file).

Rotation. The rotation can be specified as an absolute or relative angle. In absolute rotation, the angle is always measured with respect to the positive *X* axis, no matter what AutoCAD's direction setting. The absolute angle is represented by letter "a." In relative rotation, the angle is always measured with respect to the orientation of dashes in the linetype. The relative angle is represented by the letter "r." The angle can be specified in radians (r), grads (g), or degrees (d). The default is degrees.

X-Offset. This is the distance of the shape insertion point from the endpoint of the line segment measured along the line. If the line is horizontal, then the X-Offset distance is measured along the *X* axis. In Figure 31-19, the X-Offset distance is 0.2.

Y-Offset. This is the distance of the shape insertion point from the endpoint of the line segment measured perpendicular to the line. If the line is horizontal, then the Y-Offset distance is measured along the *Y* axis. In Figure 31-19, the Y-Offset distance is 0.

Example 4

Write the definition of a shape complex linetype that consists of the shape (Manhole; the name of the shape is MH) and a line. The scale of the shape is 0.1, the length of each line segment is 0.75, and the space between line segments is 0.2.

Step 1: Determining the line specifications

Before writing the definition of a new linetype, it is important to determine the line specifications. One of the ways this can be done is to actually draw the lines and the shape the way you want them to appear in the drawing (Figure 31-20). Once you have drawn the line and the shape to your satisfaction, measure the distances needed to define the shape complex linetype. In this example, the values are as follows.

Shape name MH
Shape file name *mhole.shx* (Name of the compiled shape file.)
Scale 0.1
Rotation 0
X-Offset 0.2
Y-Offset 0
Length of the first line segment = 0.75
Distance between the line segments = 0.2

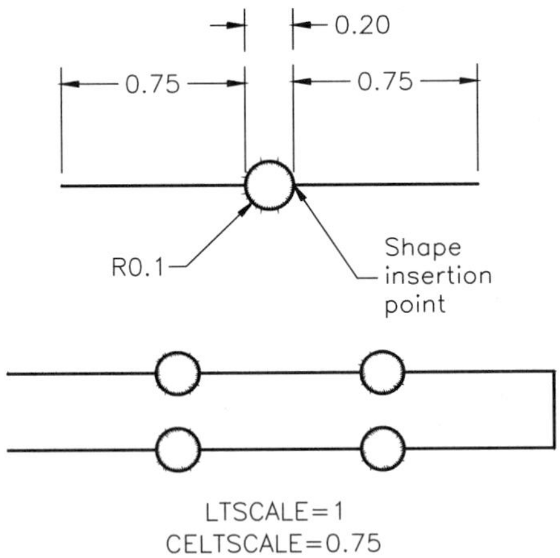

Figure 31-20 *The attributes of the shape complex linetype and line specifications for Example 4*

Step 2: Writing the definition of the shape

Use a text editor to write the definition of the shape file. The extension of the file must be *.shp*. The following file is the listing of the *mhole.shp* file for Example 4. The name of the shape is MH. (For details, see Chapter 11 of the Customizing AutoCAD 2004 book.")

```
*215,9,MH
001,10,(1,007),
001,10,(1,071),0
```

Step 3: Compiling the shape

Use the **COMPILE** command to compile the shape file (*.shp* file). When you use this command, the **Select Shape or Font File** dialog box will be displayed (Figure 31-21). If **FILEDIA** = 0, this command will be executed using the command line. The following is the command sequence for compiling the shape file.

Command: **COMPILE**
Enter shape (.SHP) or PostScript font (.PFB) file name: **MHOLE**

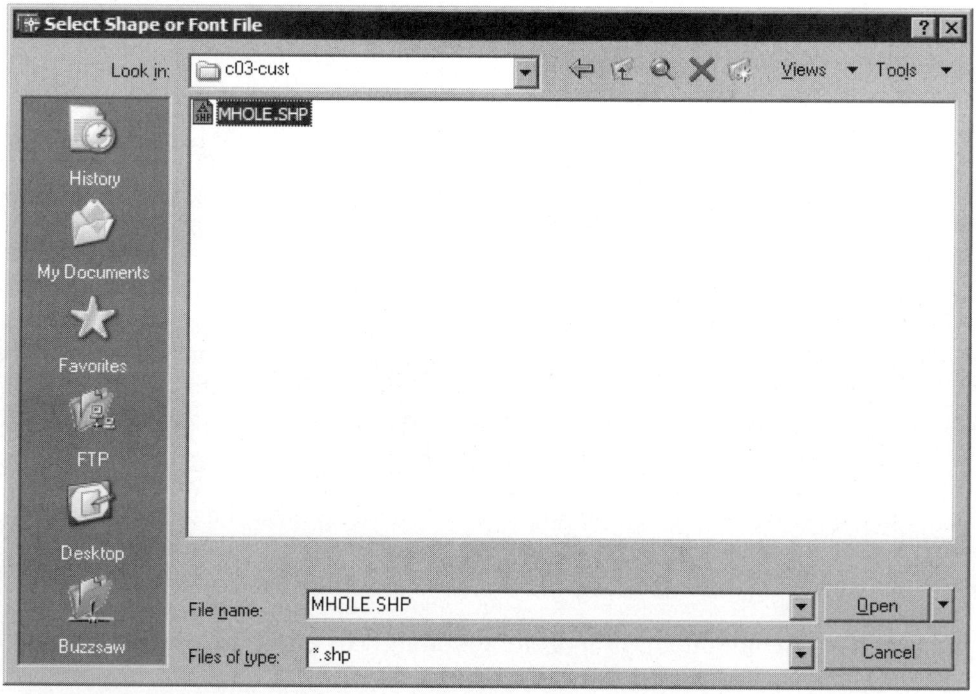

***Figure 31-21 Select Shape or Font File** dialog box*

Step 4: Writing the definition of the shape complex linetype

Use a text editor to write the definition of the shape complex linetype. You can add the definition to the *acad.lin* file or create a separate file. The extension of the file must be *.lin*. The following file is the listing of the *mhole.lin* file for Example 4. The name of the linetype is MHOLE.

```
*MHOLE,Line with Manholes
A,0.75,[MH,MHOLE.SHX,S=0.10,X=0.2,Y=0],-0.2
```

Step 5: Loading the linetype

To test the linetype, load the linetype using the **LINETYPE** command or choose **Linetype** in the **Format** pulldown menu. Assign the linetype to a layer. Draw a line or any object to check if the line is drawn to the given specifications. The shape is drawn upside down when you draw a line from right to left. Figure 31-22 shows the execution of the linetype *mhole.lin* using line, pline, and spline.

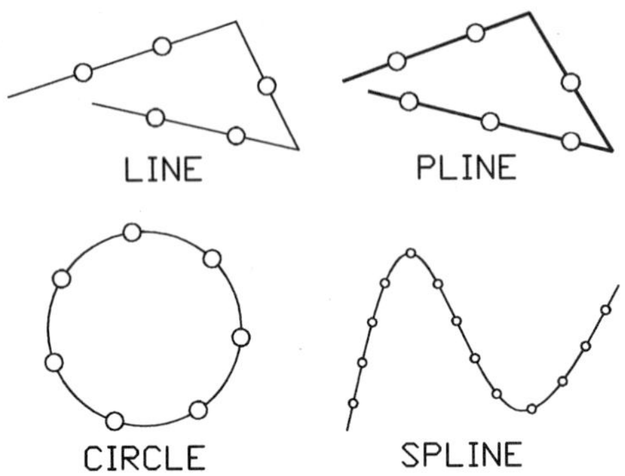

Figure 31-22 *Using a shape complex linetype*

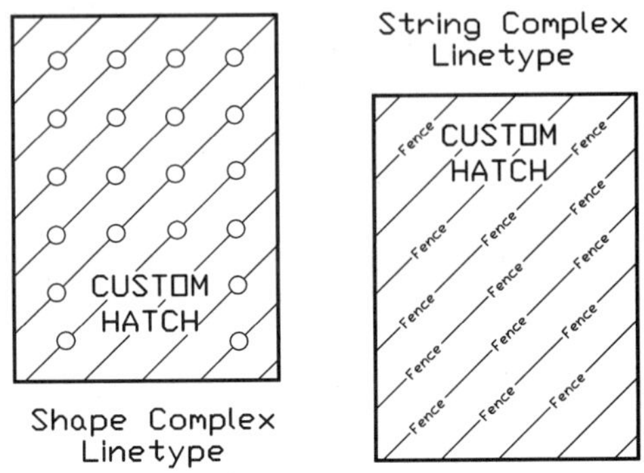

Figure 31-23 *Using shape and string complex linetypes to create custom hatch*

HATCH PATTERN DEFINITION

AutoCAD has a default hatch pattern library file, *acad.pat,* that contains 67 hatch patterns. Generally, you can hatch all the drawings using these default hatch patterns. However, if you need a different hatch pattern, AutoCAD lets you create your own hatch patterns. There is no limit to the number of hatch patterns you can define.

The hatch patterns you define can be added to the hatch pattern library file, *acad.pat*. You can also create a new hatch pattern library file, provided the file contains only one hatch pattern definition, and the name of the hatch is the same as the name of the file.

The hatch pattern definition consists of the following two parts: **Header Line and Hatch Descriptors**.

Header Line

The **header line** consists of an asterisk (*) followed by the name of the hatch pattern. The hatch name is the name used in the hatch command to hatch an area. After the hatch name comes the hatch description. Both are separated from each other by a comma (,). The general format of the header line is:

***HATCH Name [, Hatch Description]**

Where * ------------------------------ Asterisk
HATCH Name ----------- Name of hatch pattern
Hatch Description ------ Description of hatch pattern

The description can be any text that describes the hatch pattern. It can also be omitted, in which case, a comma should not follow the hatch pattern name.

Example
***DASH45, Dashed lines at 45-degree**
Where **DASH45** --------- Hatch name
Dashed lines at 45-degree ------- Hatch description

Hatch Descriptors

The **hatch descriptors** consist of one or more lines that contain the definition of the hatch lines. The general format of the hatch descriptor is:

Angle, X-origin, Y-origin, D1, D2 [,Dash Length.....]
Where **Angle** ------------ Angle of hatch lines
X-origin --------- *X* coordinate of hatch line
Y-origin --------- *Y* coordinate of hatch line
D1 ---------------- Displacement of second line (Delta-X)
D2 ---------------- Distance between hatch lines (Delta-Y)
Length ----------- Length of dashes and spaces (Pattern line definition)

Example
45,0,0,0,0.5,0.5,-0.125,0,-0.125
Where **45** ---------------- Angle of hatch line
0 ------------------- X-Origin
0 ------------------- Y-Origin
0 ------------------- Delta-X
0.5 ---------------- Delta-Y
0.5 ---------------- Dash (pen down)
-0.125 ------------ Space (pen up)
0 ------------------- Dot (pen down)
-0.125 ------------ Space (pen up)
0.5,-0.125,0,-0.125 Pattern line definition

Hatch Angle

X-origin and Y-origin. The hatch angle is the angle that the hatch lines make with the positive *X* axis. The angle is positive if measured counterclockwise (Figure 31-24), and negative if the angle is measured clockwise. When you draw a hatch pattern, the first hatch line starts from the point defined by X-origin and Y-origin.

The remaining lines are generated by offsetting the first hatch line by a distance specified by delta-X and delta-Y. In Figure 31-25(a), the first

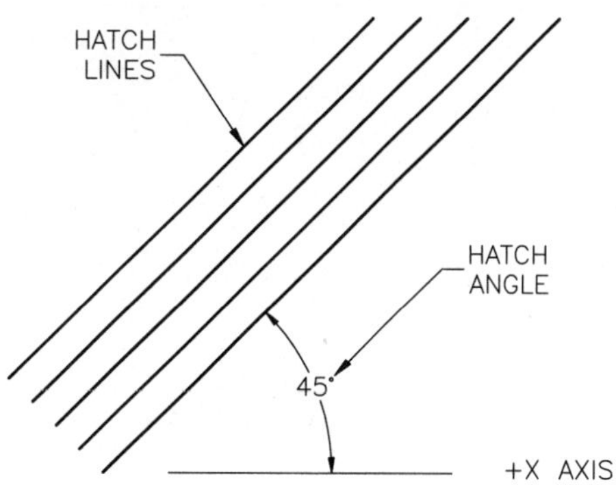

Figure 31-24 *Hatch angle*

hatch line starts from the point with the coordinates X = 0 and Y = 0. In Figure 31-25(b) the first line of hatch starts from a point with the coordinates X = 0 and Y = 0.25.

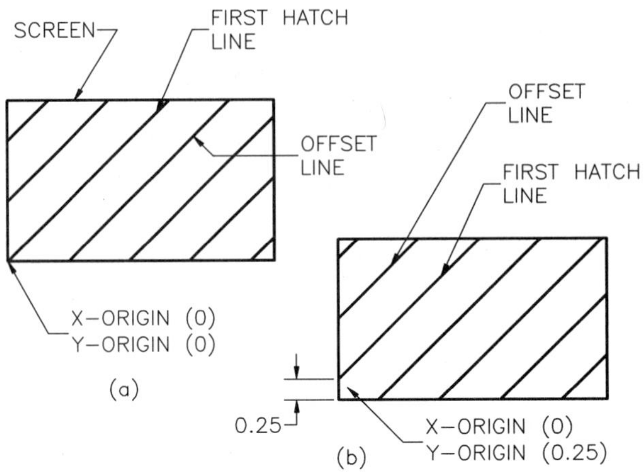

Figure 31-25 *X-origin and Y-origin of hatch lines*

Delta-X and Delta-Y. Delta-X is the displacement of the offset line in the direction in which the hatch lines are generated. For example, if the lines are drawn at a 0-degree angle and delta-X = 0.5, the offset line will be displaced by a distance delta-X (0.5) along the 0-angle direction. Similarly, if the hatch lines are drawn at a 45-degree angle, the offset line will be displaced by a distance delta-X (0.5) along a 45-degree direction (Figure 31-26). Delta-Y is the displacement of the offset lines measured perpendicular to the hatch lines. For example, if delta-Y = 1.0, the space between any two hatch lines will be 1.0 (Figure 31-26).

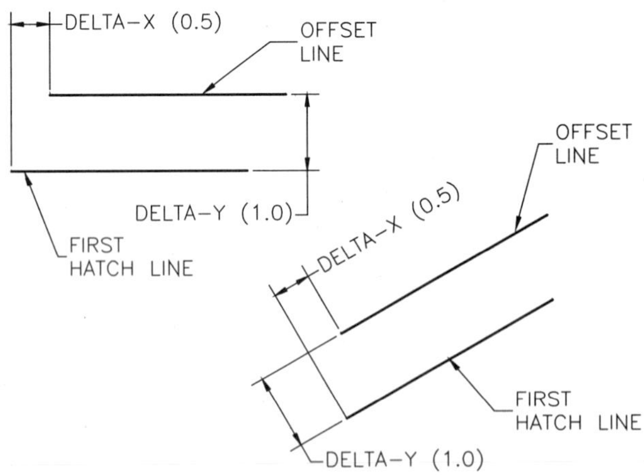

Figure 31-26 *Delta-X and delta-Y of hatch lines*

HOW HATCH WORKS

When you hatch an area, infinite number of hatch lines of infinite length are generated. The first hatch line always passes through the point specified by the X-origin and Y-origin. The remaining lines are generated by offsetting the first hatch line in both directions. The offset distance is determined by delta-X and delta-Y as shown in Figure 31-26. All selected entities that form the boundary of the hatch area are then checked for intersection with these lines. Any hatch lines found within the defined hatch boundaries are turned on, and the hatch lines outside the hatch boundary are turned off, as shown in Figure 31-27. Since the hatch lines are generated by offsetting, the hatch lines in all the areas of the drawing are automatically aligned relative to the drawing's snap origin. Figure 31-27(a) shows the hatch lines as computed by AutoCAD. These lines are not drawn on the screen; they are shown here for illustration only. Figure 31-27(b) shows the hatch lines generated in the circle that was defined as the hatch boundary.

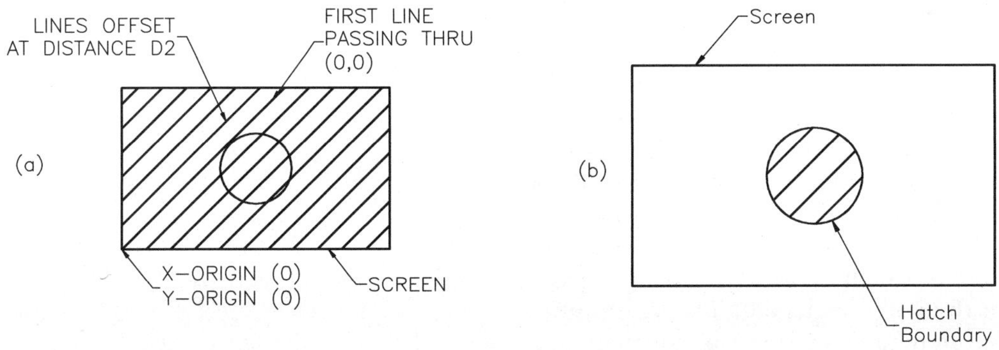

Figure 31-27 *Hatch lines outside the hatch boundary are turned off*

SIMPLE HATCH PATTERN

It is good practice to develop the hatch pattern specification before writing a hatch pattern definition. For simple hatch patterns it may not be that important, but for more complicated hatch patterns you should know the detailed specifications. Example 5 illustrates the procedure for developing a simple hatch pattern.

Example 5

Write a hatch pattern definition for the hatch pattern shown in Figure 31-28, with the following specifications.

Name of the hatch pattern =	HATCH1
X-Origin =	0
Y-Origin =	0
Distance between hatch lines =	0.5
Displacement of hatch lines =	0
Hatch line pattern =	Continuous

Step 1: Creating the hatch pattern file

This hatch pattern definition can be added to the existing *acad.pat* hatch file. You can use any text editor (like Notepad) to write the file. Load the *acad.pat* file that is located in **AutoCAD2004\SUPPORT** directory and insert the following two lines at the end of the file.

*HATCH1,Hatch Pattern for Example 5
45,0,0,0,.5

Where **45** ----------------- Hatch angle
0 ------------------- X-origin
0 ------------------- Y-origin
0 ------------------- Displacement of second hatch line
.5 ------------------ Distance between hatch lines

The first field of hatch descriptors contains the angle of the hatch lines. That angle is 45-degree with respect to the positive *X* axis. The second and third fields describe the *X* and *Y* coordinates of the first hatch line origin. The first line of the hatch pattern will pass through this point. If the values of the X-origin and Y-origin were 0.5 and 1.0, respectively, then the first line would pass through the point with the *X* coordinate of 0.5 and the *Y* coordinate of 1.0, with respect to the drawing origin 0,0. The remaining lines are generated by offsetting the first line, as shown in Figure 31-28.

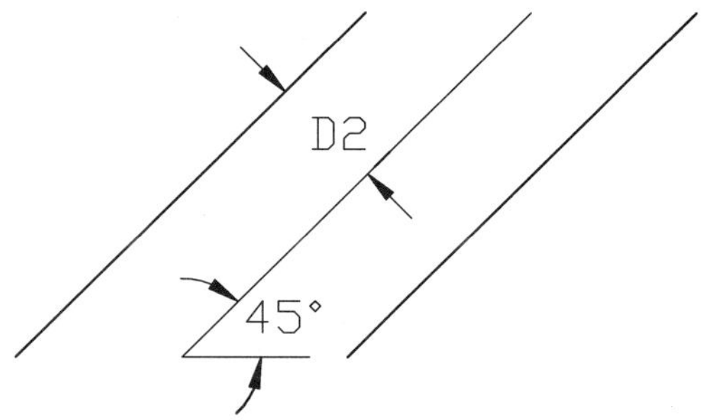

***Figure 31-28** Hatch pattern angle and offset distance*

Step 2: Loading the hatch pattern

Choose the **Hatch** button from the **Draw** toolbar or choose **Hatch** from the **Draw** menu to display the **Boundary Hatch and Fill** dialog box. Make sure **Predefined** is selected in the **Type:** edit box. Select the hatch pattern name from the drop-down list or choose the **[...]** button adjacent to the **Pattern:** drop-down list to display the **Hatch Pattern Palette** dialog box. Select the hatch pattern file displayed there. Then choose **OK** to display the **Boundary Hatch and Fill** dialog box again. Change the **Scale** and **Angle**, if needed, and then hatch an a circle to test the hatch pattern.

The **Boundary Hatch and Fill** dialog box can also be invoked by entering **BHATCH** at the Command prompt. Hatching can also be achieved by entering **-HATCH** at the Command prompt.

EFFECT OF ANGLE AND SCALE FACTOR ON HATCH

When you hatch an area, you can alter the angle and displacement of hatch lines you have specified in the hatch pattern definition to get a desired hatch spacing. You can do this by entering an appropriate value for angle and scale factor in the **HATCH** command.

To understand how the angle and the displacement can be changed, hatch an area with the hatch pattern HATCH1 in Example 5. You will notice that the hatch lines have been generated according to the definition of hatch pattern HATCH1. Notice the effect of hatch angle and scale factor on the hatch. Figure 31-29(a) shows a hatch with a 0-degree angle and a scale factor of 1.0. If the angle is 0, the hatch will be generated with the same angle as defined in the hatch pattern definition (45-degree in Example 5). Similarly, if the scale factor is 1.0, the distance between the hatch lines will be the same as defined in the hatch pattern definition. Figure 31-29(b) shows a hatch that is generated when the hatch scale factor is 0.5. If you measure the distance between the successive hatch lines, it will be 0.5 x 0.5 = 0.25. Figures 31-29(c) and (d) show the hatch when the angle is 45-degree and the scale factors are 1.0 and 0.5, respectively.

Scale and Angle can also be set by entering **-HATCH** at the Command prompt.

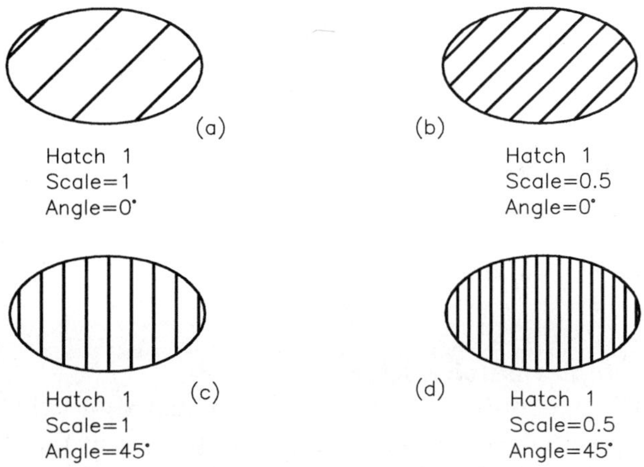

Figure 31-29 *Effect of angle and scale factor on hatch*

HATCH PATTERN WITH DASHES AND DOTS

The lines you can use in a hatch pattern definition are not restricted to continuous lines. You can define any line pattern to generate a hatch pattern. The lines can be a combination of dashes, dots, and spaces in any configuration. However, the maximum number of dashes you can specify in the line pattern definition of a hatch pattern is six. Example 6 uses a dash-dot line to create a hatch pattern.

Example 6

Write a hatch pattern definition for the hatch pattern shown in Figure 31-30, with the following specifications. Define a new path say C:\Program Files\Hatch1 and save the hatch pattern in that path.

Name of the hatch pattern	HATCH2
Hatch angle =	0

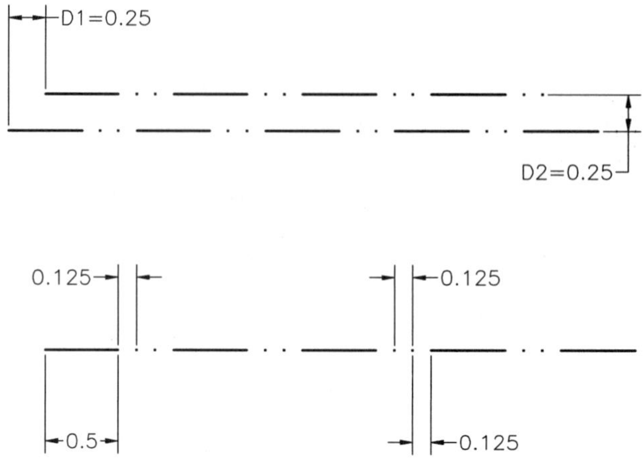

Figure 31-30 *Hatch lines made of dashes and*

X-origin =	0
Y-origin =	0
Displacement of lines (D1) =	0.25
Distance between lines (D2) =	0.25
Length of each dash =	0.5
Space between dashes and dots =	0.125
Space between dots =	0.125

Writing the definition of a hatch pattern

You can use the **EDIT** command or any text editor (Notepad) to edit the *acad.pat* file. The general format of the header line and the hatch descriptors is as follows.

***HATCH NAME, Hatch Description**
Angle, X-Origin, Y-Origin, D1, D2 [,Dash Length.....]

Substitute the values from Example 6 in the corresponding fields of the header line and field descriptor.

*HATCH2,Hatch with dashes and dots
0,0,0,0.25,0.25,0.5,-0.125,0,-0.125,0,-0.125

Where **0** ------------------- Angle
0 ------------------- X-origin
0 ------------------- Y-origin
0.25 --------------- Delta-X
0.25 --------------- Delta-Y
0.5 ---------------- Length of dash
-0.125 ------------ Space (pen up)
0 ------------------- Dot (pen down)
-0.125 ------------ Space (pen up)

0 ------------------- Dot
-0.125 ------------ Space

Specifying a New Path for Hatch Pattern Files

When you enter a hatch pattern name for hatching, AutoCAD looks for that file name in the **Support** directory or the directory paths specified in the support file search path. You can specify a new path and directory to store your hatch files.

Create a new folder **Hatch1** in C drive under the Program Files. Save the *acad.pat* file with hatch pattern **HATCH2** definition in the same subdirectory, Hatch1. Right-click in the drawing area to activate the shortcut menu. Choose **Options** from the shortcut menu to display the **Options** dialog box. The **Options** dialog box can also be invoked by choosing **Options** from the **Tools** menu or by directly entering **OPTIONS** at the Command prompt. Choose the **Files** tab in the **Options** dialog box to display the **Search paths, file names and file locations** area. Click on the **plus** sign of the **Support File Search Path** to display the different subdirectories of the **Support File Search Path** as shown in Figure 31-31. Now choose the **Add** button to display the space to add a new subdirectory. Enter the location of the new subdirectory, C:\Program Files\Hatch1 or click on the **Browse** button to specify the path. Choose the **Apply** button and then choose **OK** to exit the dialog box. You have created a subdirectory and specified the search path for the hatch files.

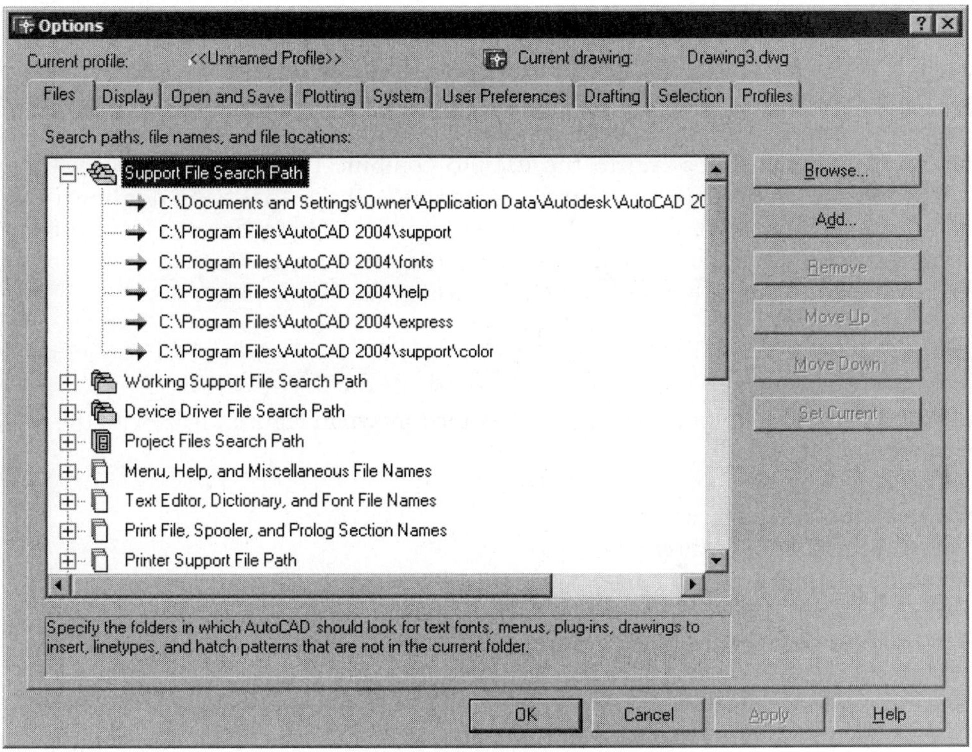

Figure 31-31 ***Options*** *dialog box*

Follow the procedure as described in Example 5 to activate the hatch pattern. The hatch thus generated is shown in Figure 31-32. Figure 31-32(a) shows the hatch with a 0-degree angle and a scale factor of 1.0. Figure 31-32(b) shows the hatch with a 45-degree angle and a scale factor of 0.5.

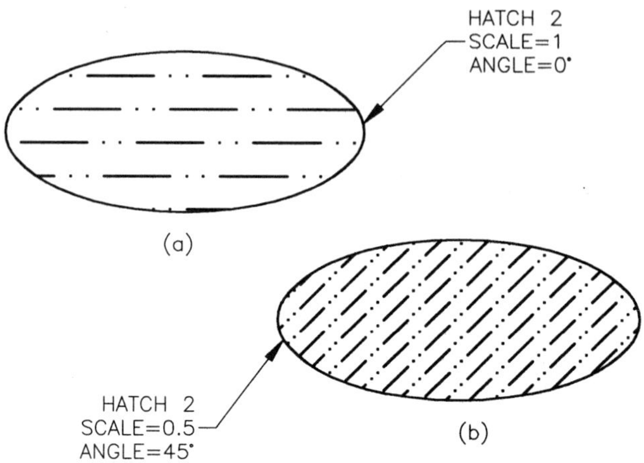

Figure 31-32 *Hatch pattern at different angles and scales*

HATCH WITH MULTIPLE DESCRIPTORS

Some hatch patterns require multiple lines to generate a shape. For example, if you want to create a hatch pattern of a brick wall, you need a hatch pattern that has four hatch descriptors to generate a rectangular shape. You can have any number of hatch descriptor lines in a hatch pattern definition. It is up to the user to combine them in any conceivable order. However, there are some shapes you cannot generate. A shape that has a nonlinear element, like an arc cannot be generated by hatch pattern definition. However, you can simulate an arc by defining short line segments because you can use only straight lines to generate a hatch pattern. Example 7 uses three lines to define a triangular hatch pattern.

Example 7

Write a hatch pattern definition for the hatch pattern shown in Figure 31-33, with the following specifications.

Name of the hatch pattern = HATCH3
Vertical height of the triangle = 0.5
Horizontal length of the triangle = 0.5
Vertical distance between the triangles = 0.5
Horizontal distance between the triangles = 0.5

Each triangle in this hatch pattern consists of the following three elements: a vertical line, a horizontal line, and a line inclined at 45-degree.

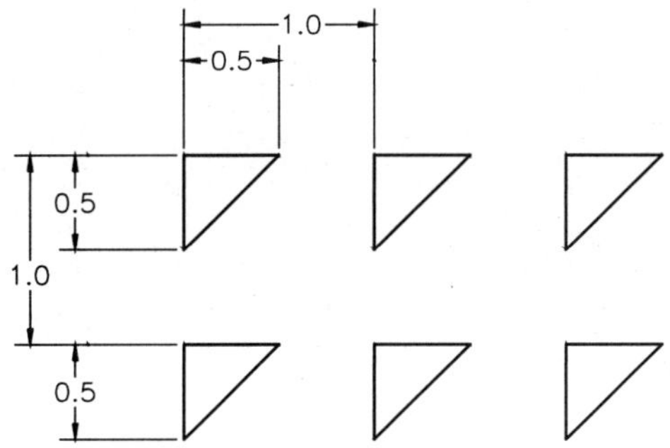

Figure 31-33 Triangle hatch pattern

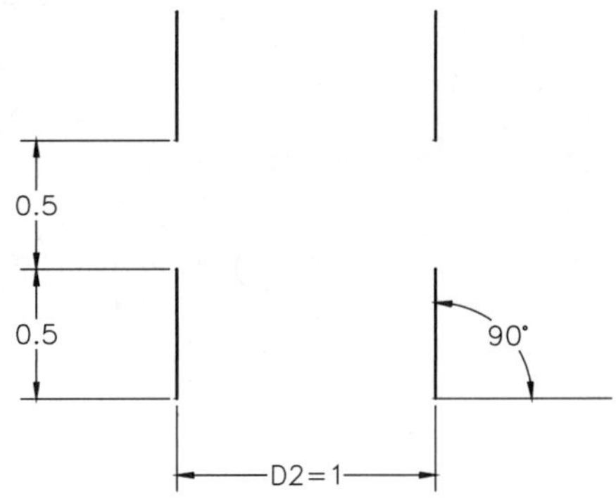

Figure 31-34 Vertical line

Step 1: Defining specifications for vertical line

For the vertical line, the specifications are (Figure 31-34).

Hatch angle =	90-degree
X-origin =	0
Y-origin =	0
Delta-X (D1) =	0
Delta-Y (D2) =	1.0
Dash length =	0.5
Space =	0.5

Substitute the values from the vertical line specification in various fields of the hatch descriptor to get the following line.

90,0,0,0,1,.5,-.5

Where **90** ----------------- Hatch angle
0 ------------------- X-origin
0 ------------------- Y-origin
0 ------------------- Delta-X
1 ------------------- Delta-Y
.5 ------------------ Dash (pen down)
-.5 ----------------- Space (pen up)

Step 2: Defining specifications of horizontal line

For the horizontal line (Figure 31-35), the specifications are given next.

Hatch angle =	0-degree
X-origin =	0
Y-origin =	0.5
Delta-X (D1) =	0
Delta-Y (D2) =	1.0
Dash length =	0.5
Space =	0.5

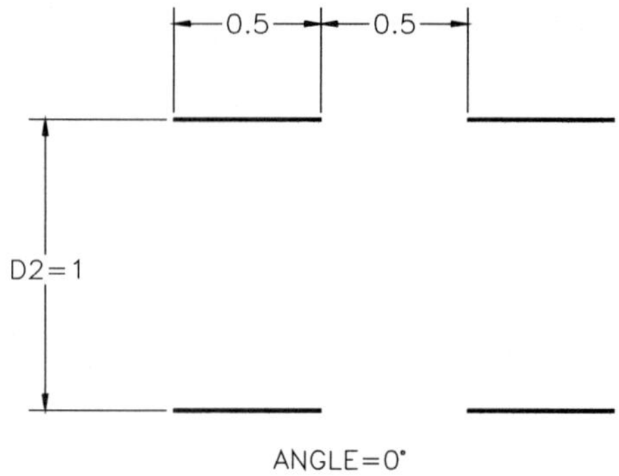

Figure 31-35 Horizontal line

The only difference between the vertical line and the horizontal line is the angle. For the horizontal line, the angle is 0-degree, whereas for the vertical line, the angle is 90-degree. Substitute the values from the vertical line specification to obtain the following line.

0,0,0.5,0,1,.5,-.5

Where **0** ------------------- Hatch angle
0 ------------------- X-origin
0.5 ---------------- Y-origin
0 ------------------- Delta-X
1 ------------------- Delta-Y
.5 ------------------ Dash (pen down)
-.5 ----------------- Space (pen up)

Step 3: Defining specifications of the inclined line

This line is at an angle; therefore, you need to calculate the distances delta-X (D1) and delta-Y (D2), the length of the dashed line, and the length of the blank space. Figure 31-36 shows the calculations to find these values.

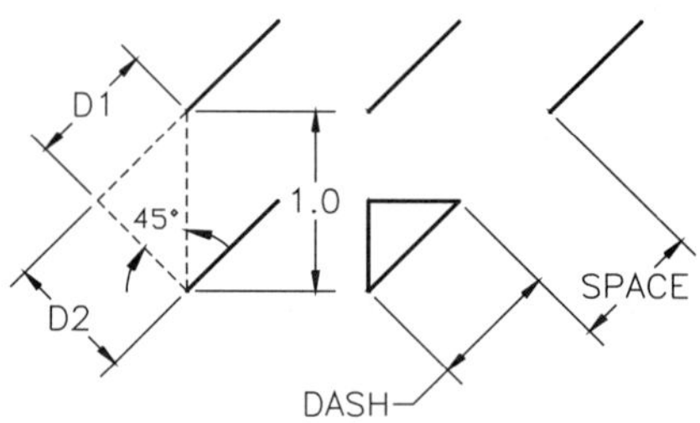

D1 = 1.0 x COS 45　　D2 = 1.0 x SIN 45
D1 = 0.7071　　D2 = 0.7071

DASH = SQRT(0.5^2 +0.5^2)
= .7071
SPACE = DASH = .7071

Figure 31-36 Line inclined at 45-degree

Hatch angle =	45-degree
X-Origin =	0
Y-Origin =	0
Delta-X (D1) =	0.7071
Delta-Y (D2) =	0.7071
Dash length =	0.7071
Space =	0.7071

After substituting the values in the general format of the hatch descriptor, you will obtain the following line.

45,0,0,.7071,.7071,.7071,-.7071

Where **45** ---------------- Hatch angle
0 ------------------ X-origin
0 ------------------ Y-origin
.7071 ------------- Delta-X
.7071 ------------- Delta-Y
.7071 ------------- Dash (pen down)
-.7071 ------------ Space (pen up)

Step 4: Loading the hatch pattern

Now you can combine the three lines and insert them at the end of the *acad.pat* file or you can enter the values in a separate hatch file and save it. You can also use the **EDIT** command to edit the file and insert the lines.

The following file is a partial listing of the *acad.pat* file, after adding the hatch pattern definitions from Examples 5, 6, and 7.

```
*SOLID, Solid fill
45, 0,0, 0,.125
*angle,Angle steel
0, 0,0, 0,.275, .2,-.075
90, 0,0, 0,.275, .2,-.075
*ansi31,ANSI Iron, Brick, Stone masonry
45, 0,0, 0,.125
*ansi32,ANSI Steel
45, 0,0, 0,.375
45, .176776695,0, 0,.375
*ansi33,ANSI Bronze, Brass, Copper
45, 0,0, 0,.25
45, .176776695,0, 0,.25, .125,-.0625
*ansi34,ANSI Plastic, Rubber
45, 0,0, 0,.75
45, .176776695,0, 0,.75
45, .353553391,0, 0,.75
45, .530330086,0, 0,.75
```

```
*ansi35,ANSI Fire brick, Refractory material
45, 0,0, 0,.25
45, .176776695,0, 0,.25, .3125,-.0625,0,-.0625
*ansi36,ANSI Marble, Slate, Glass
45, 0,0, .21875,.125, .3125,-.0625,0,-.0625
|
*swamp,Swampy area
0, 0,0, .5,.866025403, .125,-.875
90, .0625,0, .866025403,.5, .0625,-1.669550806
90, .078125,0, .866025403,.5, .05,-1.682050806
90, .046875,0, .866025403,.5, .05,-1.682050806
|
0, 0,0, .125,.125, .125,-.125
90, .125,0, .125,.125, .125,-.125
*HATCH1,Hatch at 45 Degree Angle
45,0,0,0,.5
*HATCH2,Hatch with Dashes & Dots:
0,0,0,.25,.25,0.5,-.125,0,-.125,0,-.125
*HATCH3,Triangle Hatch:
90,0,0,0,1,.5,-.5
0,0,0.5,0,1,.5,-.5
45,0,0,.7071,.7071,.7071,-.7071
```

Load the hatch pattern as described in Example 5 (Hatch3.pat) and test the hatch. Figure 31-37 shows the hatch pattern that will be generated by this hatch pattern (HATCH3). In Figure 31-37(a) the hatch pattern is at a 0-degree angle and the scale factor is 0.5. In Figure 31-37(b) the hatch pattern is at a -45-degree angle and the scale factor is 0.5.

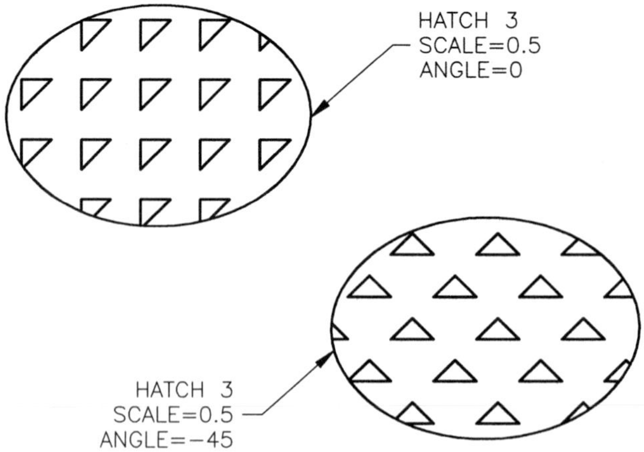

Figure 31-37 *Hatch generated by HATCH3 pattern*

SAVING HATCH PATTERNS IN A SEPARATE FILE

When you load a certain hatch pattern, AutoCAD looks for that definition in the *acad.pat* file. This is the reason, the hatch pattern definitions must be in that file. However, you can add the new pattern definition to a different file and then copy that file to *acad.pat*. Be sure to make a copy of the original *acad.pat* file so that you can copy that file back when needed. Assume the name of the file that contains your custom hatch pattern definitions is *customh.pat*.

1. Copy *acad.pat* file to *acadorg.pat*
2. Copy *customh.pat* to *acad.pat*

If you want to use the original hatch pattern file, copy the *acadorg.pat* file to *acad.pat*

CUSTOM HATCH PATTERN FILE

As mentioned earlier, you can add the new hatch pattern definitions to the **ACAD.PAT** file. There is no limit to the number of hatch pattern definitions you can add to this file. However, if you have only one hatch pattern definition, you can define a separate file. It has the following requirements:

1. The name of the file has to be the same as the hatch pattern name.
2. The file can contain only one hatch pattern definition.
3. The hatch pattern name and the hatch file name should be unique.
4. If you quite often use the hatch patterns saved on the A drive to hatch the drawings, you can add A drive to the AutoCAD search path using the **Options** dialog box. AutoCAD will automatically search the file on the A drive and will display it in the **Boundary Hatch and Fill** dialog box.

```
*HATCH3,Triangle Hatch:
90,0,0,0,1,.5,-.5
0,0,0.5,0,1,.5,-.5
45,0,0,.7071,.7071,.7071,-.7071
```

Note

The hatch lines can be edited after exploding the hatch with the ***EXPLODE*** *command. After exploding, each hatch line becomes a separate object.*

It is good practice not to explode a hatch because it increases the size of the drawing database. For example, if a hatch consists of 100 lines, save it as a single object. However, after you explode the hatch, every line becomes a separate object and you have 99 additional objects in the drawing.

Keep the hatch lines in a separate layer to facilitate editing of the hatch lines.

Assign a unique color to hatch lines so that you can control the width of the hatch lines at the time of plotting.

Tip

1. The file or the subdirectory in which hatch patterns have been saved must be defined in the ***Support File Search Path*** *in the* ***File*** *tab of the* ***Options*** *dialog box.*

2. The hatch patterns that you create automatically get added to AutoCAD's slide library as an integral part of AutoCAD 2004 and are displayed in the ***Preview Area*** *in the* ***Hatch Pattern Palette*** *dialog box under the* ***Boundary Hatch and Fill*** *dialog box. Hence there is no need to create a slide library.*

Self -Evaluation Test

Answer the following questions and then compare your answers to the correct answers given at the end of this chapter.

CREATING LINETYPES

1. The _________ command can be used to change the linetype scale factor.
2. The linetype description should not be more than _________ characters long.
3. A positive number denotes a pen _________ segment.
4. The segment length _________ generates a dot.
5. A negative number denotes a pen _________ segment.
6. The option_________ of the **LINETYPE** Command is used to generate a new linetype.
7. The description in the case of header line is _________. (optional/necessary)
8. The standard linetypes are stored in the file_________.
9. The_________ determines the current linetype scaling.

CREATING HATCH PATTERNS

10. The header line consists of an asterisk, the pattern name, and _________.
11. The *acad.pat* file contains _________ number of hatch pattern definitions.
12. The standard hatch patterns are stored in the file _________.

13. The first hatch line passes through a point whose coordinates are specified by __________ and __________.

Review Questions

CREATING LINETYPES

1. The __________ command can be used to create a new linetype.
2. The __________ command can be used to load a linetype.
3. In AutoCAD, the linetypes are saved in the __________ file.
4. AutoCAD supports only __________ alignment field specification.
5. A line pattern definition always starts with __________.
6. A header line definition always starts with __________.

CREATING HATCH PATTERNS

7. The perpendicular distance between the hatch lines in a hatch pattern definition is specified by __________.
8. The displacement of the second hatch line in a hatch pattern definition is specified by __________.
9. The maximum number of dash lengths that can be specified in the line pattern definition of a hatch pattern is __________.
10. The hatch lines in different areas of the drawing will automatically __________ since the hatch lines are generated by offsetting.
11. The hatch angle as defined in the hatch pattern definition can be changed further when you use the AutoCAD __________ command.
12. When you load a hatch pattern, AutoCAD looks for that hatch pattern in the __________ file.
13. The hatch lines can be edited after __________ the hatch by using the __________ command.

Exercises

CREATING LINETYPES

Exercise 1 *General*

Using the **LINETYPE** command, create a new linetype "DASH3DASH" with the following specifications.

Length of the first dash 0.75
Blank space 0.125
Dash length 0.25
Blank space 0.125
Dash length 0.25
Blank space 0.125
Dash length 0.25
Blank space 0.125

Exercise 2 *General*

Use a text editor to create a new file, *newlt2.lin*, and a new linetype, DASH2DASH, with the following specifications.

Length of the first dash 0.5
Blank space 0.1
Dash length 0.2
Blank space 0.1
Dash length 0.2
Blank space 0.1

Exercise 3 *General*

a. Write the definition of a string complex linetype (hot water line) as shown in Figure 31-38(a). To determine the length of the HW text string, you should first draw the text (HW) using any text command and then measure its length.

b. Write the definition of a string complex linetype (gas line) as shown in Figure 31-38(b). Determine the length of the text string as mentioned in part a.

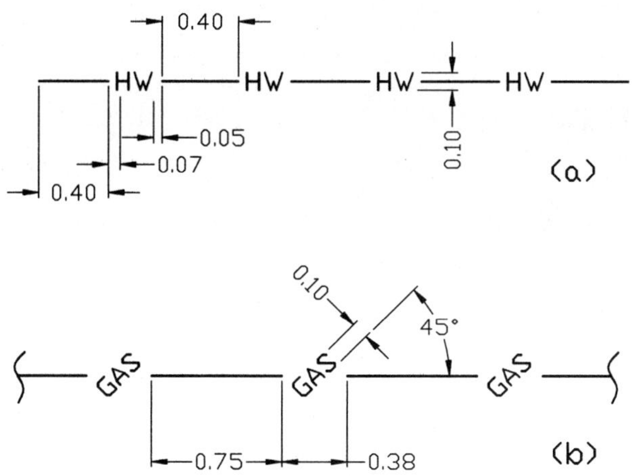

Figure 31-38 Specifications for string a complex linetype

Creating Hatch Patterns

Exercise 4 General

Determine the hatch pattern specifications and write a hatch pattern definition for the hatch pattern in Figure 31-39.

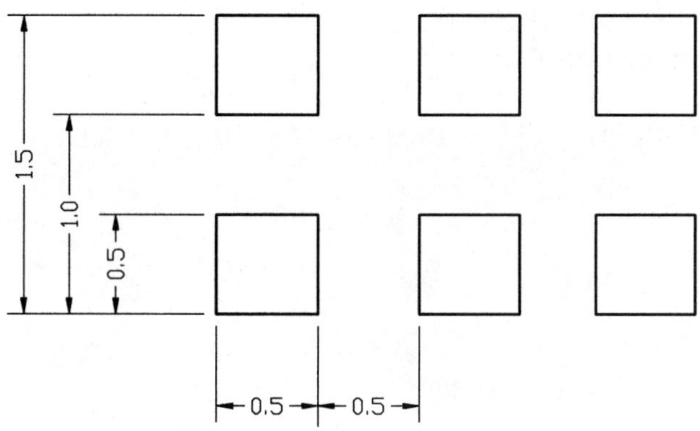

Figure 31-39 Drawing for Exercise 4

Exercise 5 *General*

Determine the hatch pattern specifications and write a hatch pattern definition for the hatch pattern in Figure 31-40.

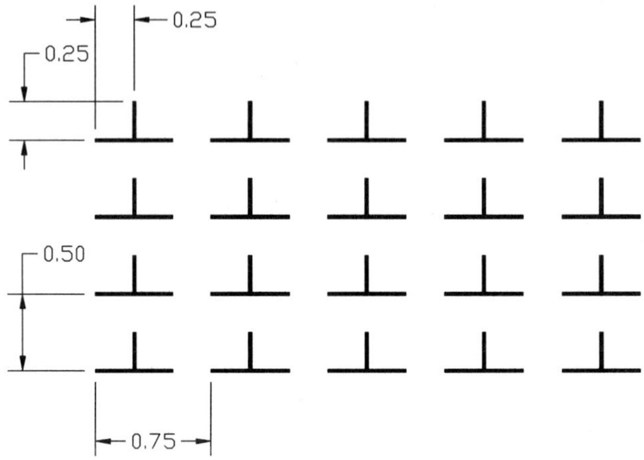

Figure 31-40 *Hatch pattern for Exercise 5*

Answers to Self-Evaluation Test

1. **LTSCALE**, **2.** 47, **3.** down, **4.** zero, **5.** up, **6.** **Create**, **7.** optional, **8.** *acad.lin*, **9.** **CELTSCALE**, **10.** Pattern Description, **11.** 67, **12.** *acad.pat*, **13.** X-origin, Y-origin.

Chapter 32

Pull-down, Shortcut, and Partial Menus and Customizing Toolbars

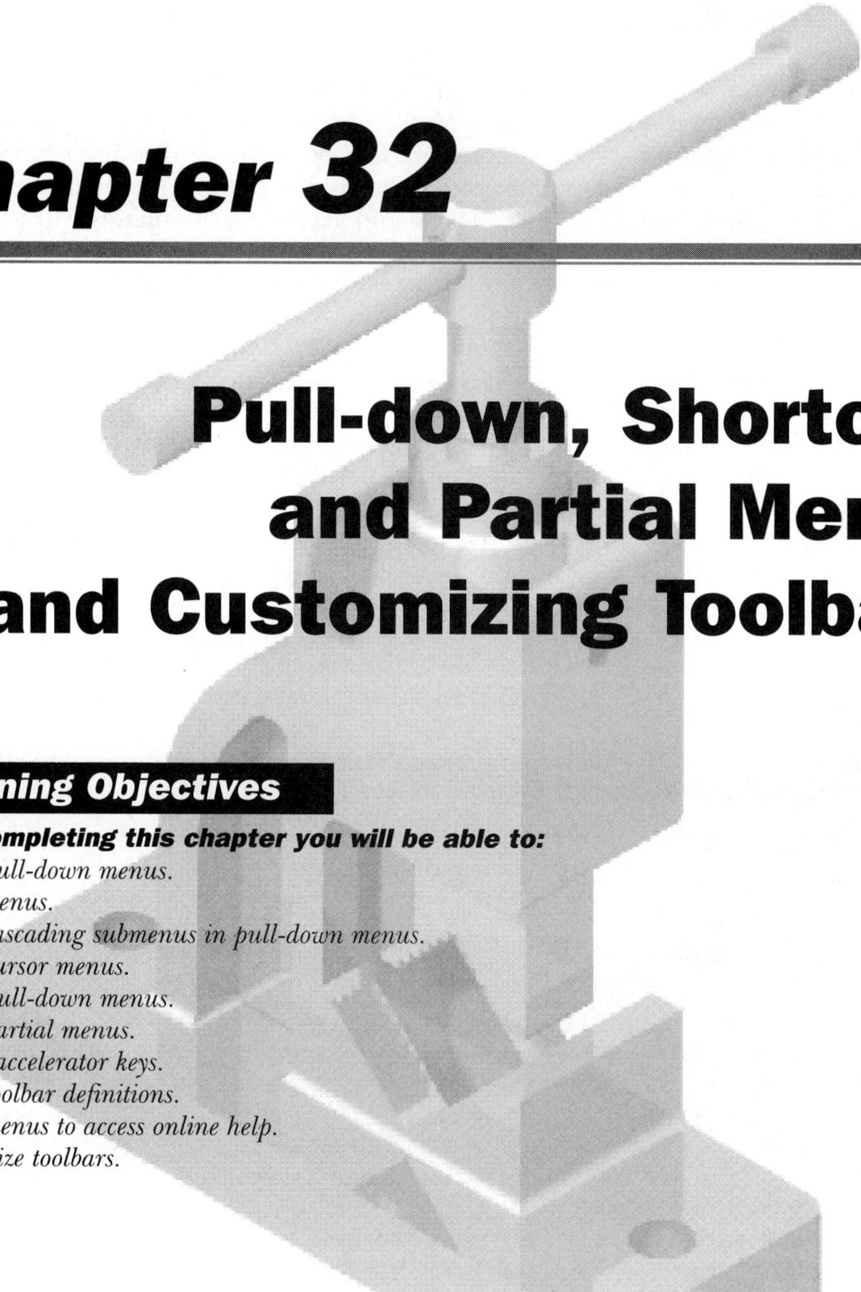

Learning Objectives

After completing this chapter you will be able to:

- *Write pull-down menus.*
- *Load menus.*
- *Write cascading submenus in pull-down menus.*
- *Write cursor menus.*
- *Swap pull-down menus.*
- *Write partial menus.*
- *Define accelerator keys.*
- *Write toolbar definitions.*
- *Write menus to access online help.*
- *Customize toolbars.*

AutoCAD MENU

The AutoCAD menu provides a powerful tool to customize AutoCAD. The AutoCAD software package comes with a standard menu file named *acad.mnu*. When you start AutoCAD, the menu file *acad.mnu* is automatically loaded. This file contains AutoCAD commands, separated under different headings for easy identification. For example, all drawing commands are under **Draw** and all editing commands are under the **Modify** menu. The headings are named and arranged to make it easier for you to locate and access the commands. However, there are some commands that you may never use. Also, some users might like to regroup and rearrange the commands so that it is easier to access the most frequently used commands.

AutoCAD lets the user eliminate rarely used commands from the menu file and define new ones. This is made possible by editing the existing *acad.mnu* file or writing a new menu file. There is no limit to the number of files you can write. You can have a separate menu file for each application. For example, you can have separate menu files for mechanical, electrical, and architectural drawings. You can load these menu files any time by using the AutoCAD **MENU** command. The menu files are text files with the extension *.mnu*. These files can be written by using any text editor like Wordpad or Notepad. The menu file can be divided into ten sections, each section identified by a section label. AutoCAD uses the following labels to identify different sections of the AutoCAD menu file.

*****SCREEN**	
*****TABLET(n)**	**n** is from 1 to 4
*****IMAGE**	
*****POP(n)**	**n** is from 1 to 499 (For Shortcut menus n=0 and 500 to 999)
*****BUTTONS(n)**	**n** is from 1 to 4
*****AUX(n)**	**n** is from 1 to 4
*****MENUGROUP**	
*****TOOLBARS**	
*****HELPSTRING**	
*****ACCELERATORS**	

The tablet menu can have up to four different sections. The **POP** menu (pull-down and cursor menu) can have up to 499 sections, and auxiliary and buttons menus can have up to four sections.

Tablet Menus	**Pull-Down and Cursor Menus**
***TABLET1	***POP0
***TABLET2	***POP1
***TABLET3	***POP2
***TABLET4	***POP3
	***POP4
BUTTONS Menus	***POP5
***BUTTONS1	***POP6
***BUTTONS2	***POP7

***BUTTONS3
***BUTTONS4

Auxiliary Menus
***AUX1
***AUX2
***AUX3
***AUX4

***POP8
|
***POP498
***POP499

STANDARD MENUS

The menu is a part of the AutoCAD standard menu file, *acad.mnu*. The *acad.mnu* file is automatically loaded when you start AutoCAD, unless the standard configuration of AutoCAD is not changed. The menus can be selected by moving the crosshairs to the top of the screen, on the menu bar area. If you move the pointing device sideways, different menu bar titles are highlighted and you can select the desired item by pressing the left button on your pointing device. Once the item is selected, the corresponding menu is displayed directly under the title (Figure 32-1). The menu can have 499 sections, named as POP1, POP2, POP3, . . ., POP499.

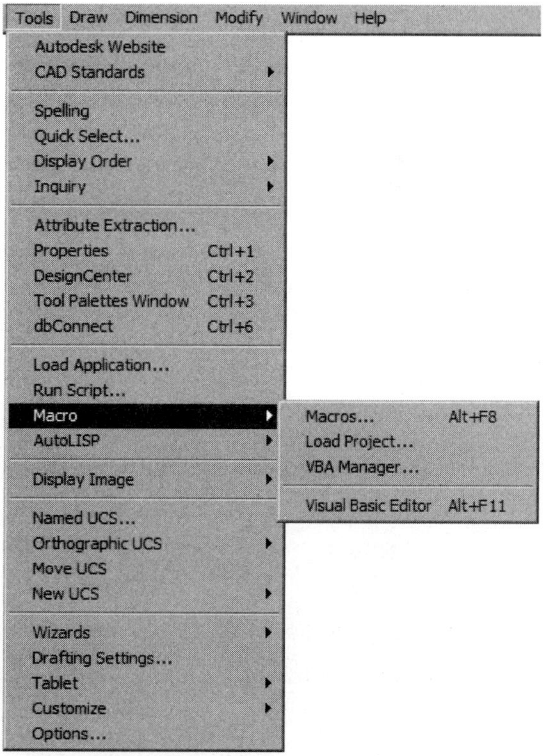

Figure 32-1 *Pull-down and cascading menus*

WRITING A MENU

Before you write a menu, you need to design a menu and arrange the commands the way you want them to appear on the screen. To design a menu, you should select and arrange the commands in a way that provides easy access to the most frequently used commands. A careful design saves a lot of time in the long run. Therefore, consider several possible designs with different command combinations, and then select the one best suited for the job. Suggestions from other CAD operators can prove very valuable.

The second important thing in developing a menu is to know the exact sequence of the commands and the prompts associated with each command. To better determine the prompt entries required in a command, you should enter all the commands and the prompt entries. The following is a description of some of the commands and the prompt entries required for Example 1.

LINE Command

Command: **LINE**

Specify first point:

LINE
<RETURN>

CIRCLE (C,R) Command

Command: **CIRCLE**
Specify center point for circle or [3P/2P/Ttr (tan tan radius)]: *Specify center point.*
Specify radius of circle or [Diameter]: *Enter radius.*

Notice the command and input sequence:

CIRCLE
<RETURN>
Center point
<RETURN>
Radius
<RETURN>

CIRCLE (C,D) Command

Command: **CIRCLE**
Specify center point for circle or [3P/2P/Ttr (tan tan radius)]: *Specify center point.*
Specify radius of circle or [Diameter]: **D**
Specify diameter of circle: *Enter diameter.*

Notice the command and input sequence:

CIRCLE
<RETURN>
Center Point
<RETURN>
D
<RETURN>
Diameter
<RETURN>

CIRCLE (2P) Command

Command: **CIRCLE**
Specify center point for circle or [3P/2P/Ttr (tan tan radius)]: **2P**

Specify first end point of circle's diameter: *Specify first point.*
Specify second end point of circle's diameter: *Specify second point.*

Notice the command and input sequence:

CIRCLE
<RETURN>
2P
<RETURN>
Select first point on diameter
<RETURN>
Select second point on diameter
<RETURN>

ERASE Command

Command: **ERASE**

Notice the command and input sequence:

ERASE
<RETURN>

MOVE Command

Command: **MOVE**

Notice the command and prompt entry sequence:

MOVE
<RETURN>

The difference between the **Center-Radius** and **Center-Diameter** options of the **CIRCLE** command is that in the first one, you are prompted to enter the radius value by default. Whereas in the second one, you need to enter **D** to use the **Diameter** option. This difference, although minor, is very important when writing a menu file. Similarly, the 2P (two-point) option of the **CIRCLE** command is different from the other two options. Therefore, it is important to know both the correct sequence of the AutoCAD commands and the entries made in response to the prompts associated with those commands.

You can use any text editor (like Notepad) to write the file. You can also use **EDIT** command to write the menu file. If you use the **EDIT** command, AutoCAD will prompt you to enter the file name you want to edit. The file name can be up to eight characters long, and the file extension must be *.mnu*. If the file name exists, it will automatically be loaded; otherwise a new file will be created. To understand the process of developing a pull-down menu, consider the following example.

Note

*If **EDIT** command does not work, check the acad.pgp file and make sure that this command is defined in the file.*

Example 1

Write a pull-down menu for the following AutoCAD commands and load them with the default menu items.

LINE	**ERASE**	**REDRAW**	**SAVE**
PLINE	**MOVE**	**REGEN**	**QUIT**
CIRCLE C,R	**COPY**	**ZOOM ALL**	**PLOT**
CIRCLE C,D	**STRETCH**	**ZOOM WIN**	
CIRCLE 2P	**EXTEND**	**ZOOM PRE**	
CIRCLE 3P	**OFFSET**		

Step 1: Designing the menu

The first step in writing any menu is to design the menu so that the commands are arranged in the desired configuration. Figure 32-2 shows one of the possible designs of the given menu. This menu has four different groups of commands; therefore, it will have four sections: POP1, POP2, POP3, and POP4, and each section will have a section label. The following file is a listing of the pull-down menu file for Example 1. **The line numbers are not a part of the file; they are shown here for reference only.**

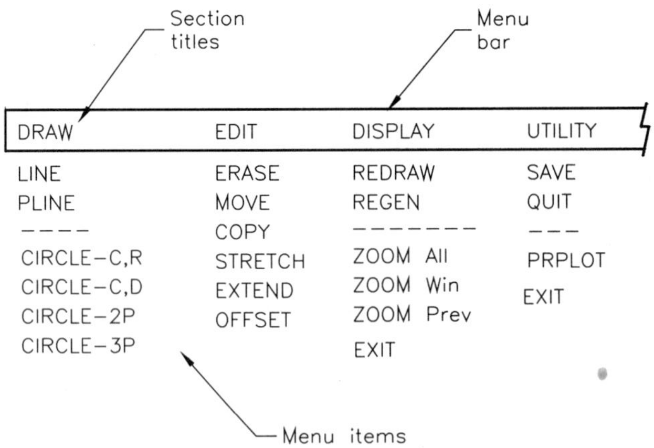

Figure 32-2 *Design of menu*

Step 2: Writing the menu file

```
1 ***POP1
2 [DRAW]
3 [LINE]*^C^CLINE
4 [PLINE]^C^CPLINE
```

```
5  [—]
6  [CIR-C,R]^C^CCIRCLE
7  [CIR-C,D]^C^CCIRCLE \D
8  [CIR-2P]^C^CCIRCLE 2P
9  [CIR-3P]^C^CCIRCLE 3P
10 ***POP2
11 [EDIT]
12 [ERASE]*^C^CERASE
13 [MOVE]^C^CMOVE
14 [COPY]^C^CCOPY
15 [STRETCH]^C^CSTRETCH;C
16 [EXTEND]^C^CEXTEND
17 [OFFSET]^C^COFFSET
18 ***POP3
19 [DISPLAY]
20 [REDRAW]'REDRAW
21 [REGEN]^C^CREGEN
22 [—]
23 [ZOOM-All]^C^C'ZOOM A
24 [ZOOM-Window]'ZOOM W
25 [ZOOM-Prev]'ZOOM PREV
26 [~Exit]^C
27 ***POP4
28 [UTILITY]
29 [SAVE]^C^CSAVE;
30 [QUIT]^C^CQUIT
31 [——]
32 [PLOT]^C^CPLOT
33 [EXIT]^C^CEXIT
```


Explanation

Line 1

*****POP1**

POP1 is the section label for the first pull-down menu. All section labels in the AutoCAD menu begin with three asterisks (***), followed by the section label name, such as POP1.

Line 2

[DRAW]

In this menu item **DRAW** is the menu bar title displayed when the cursor is moved in the menu bar area. The title names should be chosen so you can identify the type of commands you expect in that particular pull-down menu. In this example, all the drawing commands are under the title **DRAW** (Figure 32-3), all editing commands are under **EDIT**, and so on for other groups of items. The menu bar title can be of any length depending on the display resolution, display device, and the font size of the text. However, it is recommended to keep them short to accommodate other menu items. If a display device at a certain resolution provides a maximum of 80 characters; to have 16 sections in a single row in the menu bar, the length of each title should not exceed five characters. If the length of the menu bar items exceeds 80 characters, AutoCAD wraps the items that cannot be accommodated in 80 character space, and displays them in the next line. This results in two menu item lines in the menu bar.

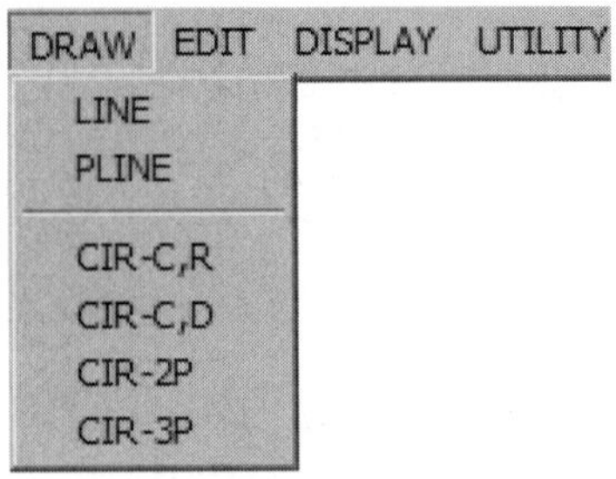

Figure 32-3 ***Draw*** *pull-down*

If the first line in a menu section is blank, the title of that section is not displayed in the menu bar area. Since the menu bar title is not displayed, you **cannot** access that menu. This allows you to turn off the menu section. For example, if you replace [**DRAW**] with a blank line, the **DRAW** section (POP1) of the menu will not be available; the second section (POP2) will be displayed in its place.

Example	
***POP1	Section label
	Blank line (turns off POP1)
[LINE:]^CLINE	Menu item
[PLINE:]^CPLINE	
[CIRCLE:]^CCIRCLE	

The menu bar titles are left-justified. If the first title is not displayed, the rest of the menu titles will be shifted to the left. In Example 1, if the **DRAW** title is not displayed in the menu bar area, then the **EDIT**, **DISPLAY**, and **FILE** sections of the menu will move to the left.

Line 3
***^C^CLINE**
In this menu item, the command definition starts with an asterisk (*). This feature allows the command to be repeated automatically until it is canceled by pressing ESCAPE, entering CTRL C, or by selecting another menu command. ^C^C cancels the existing command twice; **LINE** is an AutoCAD command that generates lines.

***^C^CLINE**

Where ***** ------------------- Repeats the menu item (command)
^C^C ---------- Cancels existing command twice
LINE ------------- AutoCAD's **LINE** command

Line 5
[--]
To separate two groups of commands in any section, you can use a menu item that consists of two or more hyphens (--). This line automatically expands to fill the entire width of the menu. You can use a blank line in a menu. If any section of a menu (**POP section) has a blank line, it is ignored.

Line 24
[ZOOM-Window]'ZOOM W
In this menu item the single quote (') preceding the **ZOOM** command makes the **ZOOM** Window command transparent. When a command is transparent, the existing command is not canceled. After the **ZOOM** Window command (Figure 32-4), AutoCAD will automatically resume the current operation.

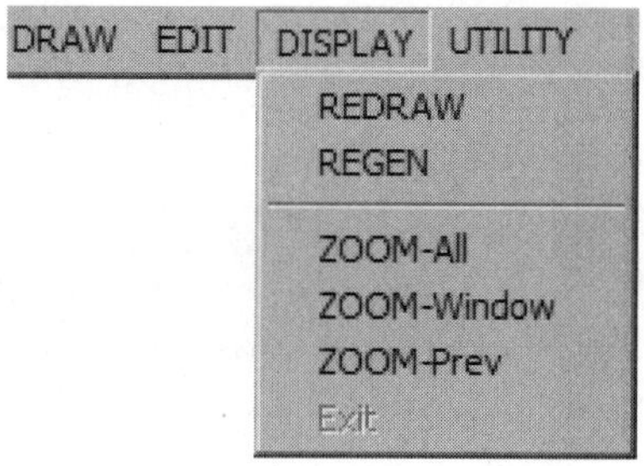

***Figure 32-4** Display menu*

[ZOOM-Window]'ZOOM W
Where **W** ----------------- Window option
ZOOM ----------- AutoCAD **ZOOM** command
' -------------------- Single quote makes **ZOOM** transparent

Line 26
[~Exit]^C
This menu item is specially meant for canceling the menu. Since this menu item has a tilde (~), the menu item is not available (displayed shaded), and if you select this item it will not cancel the menu. You can use this feature to disable a menu item or to indicate that the item is not a valid selection. If there is an instruction associated with the item, the instruction will not be executed when you select the item. For example, [~OSNAPS]^C^C$S=OSNAPS will not load the OSNAPS submenu on the screen.

Line 29
[SAVE]^C^CSAVE;
In this menu item, the semicolon (;) that follows the **SAVE** command enters RETURN. The semicolon is not required; the command will also work without a semicolon.

[SAVE]^C^CSAVE;
Where **SAVE** ------------- AutoCAD **SAVE** command
; ------------------- Semicolon enters RETURN

Line 31
[----]
This menu item has four hyphens. The line will extend to fill the width of the menu. If there is only one hyphen [-], AutoCAD gives a syntax error when you load the menu.

Line 33
[Exit]^C
In this menu item, **^C** command definition has been used to cancel the menu. This item provides you with one more option for canceling the menu. This is especially useful for new AutoCAD users who are not familiar with all AutoCAD features. The menu can also be canceled by any of the following actions:

1. Selecting a point.
2. Selecting an item in the screen menu area.
3. Selecting or typing another command.

4. Pressing ESC at the keyboard.
5. Selecting any menu title in the menu bar.

Note

For all menus, the menu items are displayed directly beneath the menu title and are left-justified. If any menu [for example, the rightmost menu (POP16)] does not have enough space to display the entire menu item, the menu will expand to the left to accommodate the entire length of the longest menu item.

You can use // (two forward slashes) for comment lines. AutoCAD ignores the lines that start with //.

The last line in the menu file must be terminated by ENTER or else it will be ignored.

From this example it is clear that every statement in the menu is based on the AutoCAD commands and the information that is needed to complete these commands. This forms the basis for creating a menu file and should be given consideration. Following is a summary of the AutoCAD commands used in Example 1 and their equivalents in the menu file.

AutoCAD Commands	Menu File
Command: **LINE**	**[LINE]^C^CLINE**
Command: **CIRCLE** Specify center point for circle or [3P/2P/Ttr (tan tan radius)]: Specify radius of circle or [Diameter]:	**CIR-C,R]^C^CCIRCLE**
Command: **CIRCLE** Specify center point for circle or [3P/2P/Ttr (tan tan radius)]: Specify radius of circle or [Diameter]: D Specify diameter of circle:	**[CIR-C,D]^C^CCIRCLE;\D**
Command: **CIRCLE** Specify center point for circle or [3P/2P/Ttr (tan tan radius)]: **2P** Specify first end point of circle's diameter: Specify second end point of circle's diameter:	**[CIR- 2P]^C^CCIRCLE;2P**
Command: **ERASE**	**[ERASE]^C^CERASE**
Command: **MOVE**	**[MOVE]^C^CMOVE**

LOADING MENUS

AutoCAD automatically loads the *acad.mnu* file when you get into the AutoCAD drawing editor, unless the *acad.mnu* file has been modified or unless a different menu file has been loaded. However, you can also load a different menu file by using the **MENU** command.

When you enter the **MENU** command at the Command prompt, AutoCAD displays the **Select Menu File** dialog box (Figure 32-5) on the screen. Select the menu file that you want to load and then choose the **Open** button.

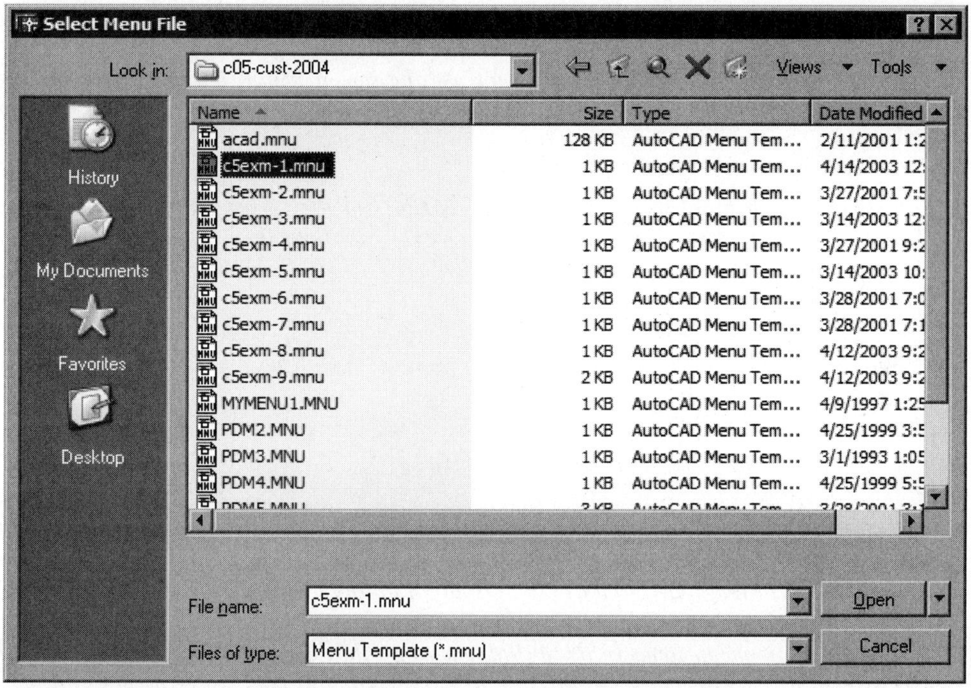

Figure 32-5 ***Select Menu File*** *dialog box*

You can also load the menu file from the command line, keeping **FILEDIA** as **0**.

Command: **FILEDIA**
Enter new value for FILEDIA <1>: **0**
Command: **MENU**

Enter menu file name or [. (for none)] <current menu>**: PDM1**
Where **PDM1** ------------ Name of menu file
<current menu> ------- Current menu file

After you enter the **MENU** command, AutoCAD will prompt for the file name. Enter the name of the menu file without the file extension (*.mnu*), since AutoCAD assumes the extension *.mnu*. AutoCAD will automatically compile the menu file into MNC and MNR files. When you load a menu file in windows, AutoCAD creates the following files

.mnc and **.mnr** files — When you load a menu file (*.mnu*), AutoCAD compiles the menu file and creates *.mnc* and *.mnr* files. The *.mnc* file is a compiled menu file. The *.mnr* file contains the bitmaps used by the menu.

.mns file — When you load the menu file, AutoCAD also creates an **.mns** file. This is an ASCII file that is the same as the *.mnu* file when you initially load the menu file. Each time you make a change in the contents of the file, AutoCAD changes the *.mns* file.

Note

After you load the new menu, you cannot use the screen menu, buttons menu, or digitizer because the original menu, acad.mnu, is not present and the new menu does not contain these menu areas.

*To activate the original menu again, enter **MENU** at the Command prompt to display the **Select Menu File** dialog box. Select the acad.mnu file from the Support directory and then choose the **Open** button. You can enter the values at the Command prompt if the **FILEDIA** system variable is 1.*

If you need to use input from a keyboard or a pointing device, use the backslash (\). The system will pause for you to enter data.

There should be no space after the backslash (\).

The menu items, menu labels, and command definition can be uppercase, lowercase, or mixed. You can introduce spaces between the menu items to improve the readability of the menu file. The blank lines are ignored and are not displayed on the screen.

If there are more menu items in the pull-down menu than the number of spaces available, the excess items are not displayed on the screen. For example, if the display device and the screen resolution limits the number of items to 21, items in excess of 21 will not be displayed on the screen and are therefore inaccessible.

If you are using a high-resolution graphics board, you can increase the number of lines that can be displayed on the screen. On some devices this is 80 lines.

RESTRICTIONS

The menus are easy to use and provide a quick access to frequently used AutoCAD commands. However, the options in the menus do not work during the following commands:

TEXT Command

After you assign the text height and the rotation angle to a **TEXT** command, the options in the menus will be displayed but they will not work.

SKETCH Command

The menus are made unavailable after you set the record increment in the **SKETCH** command.

Exercise 1 *General*

Write a menu for the following AutoCAD commands. The menu design is shown in Figure 32-6.

DRAW	EDIT	DISP/TEXT	UTILITY
LINE	FILLET0	TEXT,C	SAVE
PLINE	FILLET	TEXT,L	QUIT
ELLIPSE	CHAMFER	DTEXT,R	END
POLYGON	STRETCH	ZOOM WIN	DIR
DONUT	EXTEND	ZOOM PRE	PLOT
	OFFSET		

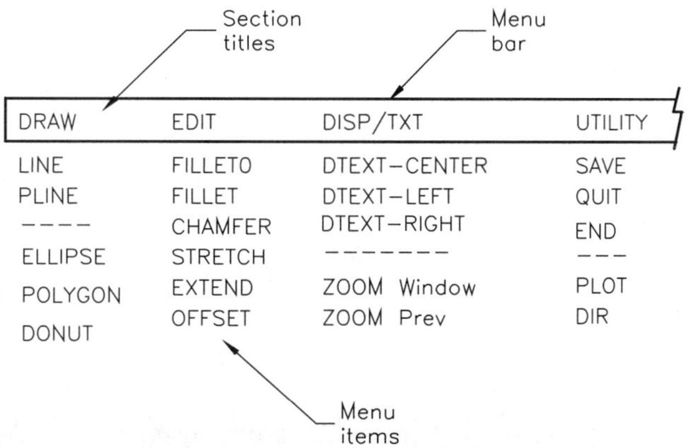

Figure 32-6 *Menu display for Exercise 1*

CASCADING SUBMENUS IN MENUS

The number of items in a menu or cursor menu can be very large, and sometimes they cannot all be accommodated on one screen. For example, consider a display device that can display a maximum of 21 menu items. If the pull-down menu or the cursor menu has more items than can be displayed, the excess menu items are not displayed on the screen and cannot be accessed. You can overcome this problem by using cascading menus that let you define smaller groups of items within a menu section. When an item is selected, it loads the cascading menu and displays the items, defined in the cascading menu, on the screen.

The cascading feature of AutoCAD allows pull-down and cursor menus to be displayed in a hierarchical order that makes it easier to select submenus. To use the cascading feature in pull-down and cursor menus, AutoCAD has provided some special characters. For example, -> defines a cascaded submenu and <- designates the last item in the menu. The following table lists some of the characters that can be used with the pull-down or cursor menus.

Character	**Character Description**
--	The item label consisting of two hyphens automatically expands to fill the entire width of the menu. Example: [**--**]
+	Used to continue the menu item to the next line. This character has to be the last character of the menu item. Example: [Triang:]^C^CLine;1,1;+3,1;2,2;
->	This label character defines a cascaded submenu; it must precede the name of the submenu. Example: [**->Draw**]
<-	This label character designates the last item of the cascaded pull-down or cursor menu. The character must precede the label item. Example: [**<-CIRCLE 3P**]**^C^CCIRCLE;3P**
<-<-...	This label character designates the last item of the pull-down or cursor menu and also terminates the parent menu. The character must precede the label item. Example: [**<-<-Center Mark**]**^C^C_dim;_center**
$(	This label character can be used with the pull-down and cursor menus to evaluate a DIESEL expression. The character must precede the label item. Example: **$**(if,**$**(getvar,orthomode),Ortho)
~	This item indicates the label item is not available (displayed shaded); the character must precede the item. Example: [~Application not available]
!.	When used as a prefix, it displays the item with a check mark.
&	When placed directly before a character, the character is displayed underscored. For example, [W&Block] is displayed as WBlock. It also specifies the character as a menu accelerator key in the pull-down or Shortcut menu.
/c	When placed directly before an item with a character, the character is displayed underscored. For example, [/BWBlock] is displayed as WBlock. It also specifies that the item has a menu accelerator key in the pull-down or Shortcut menu.

\t	The label text to the right of \t is displayed to the right side of the menu.

Consider a display device with a certain resolution and font size that provides space for a maximum of 80 characters. Therefore, if there are 10 menus, the length of each menu title should average eight characters. If the combined length of all menu bar titles exceeds 80 characters, AutoCAD automatically wraps the excess menu items and displays them on the next line in the menu bar. The following is a list of some additional features of the menu.

1. The section labels of the menus are ***POP1 through ***POP16. The menu bar titles are displayed in the menu bar.

2. The menus can be accessed by selecting the menu title from the menu bar at the top of the screen.

3. A maximum of 999 menu items can be defined in the menu. This includes the items that are defined in the submenus. The menu items in excess of 999 are ignored.

4. The number of menu items that can be displayed depends on the display device you are using. If the cursor or the menu contains more items than can be accommodated on the screen, the excess items are truncated. For example, if your system can display only 35 menu items, the menu items in excess of 35 are automatically truncated.

Example 2

Write a pull-down menu for the commands shown in Figure 32-7. The menu must use the AutoCAD cascading feature.

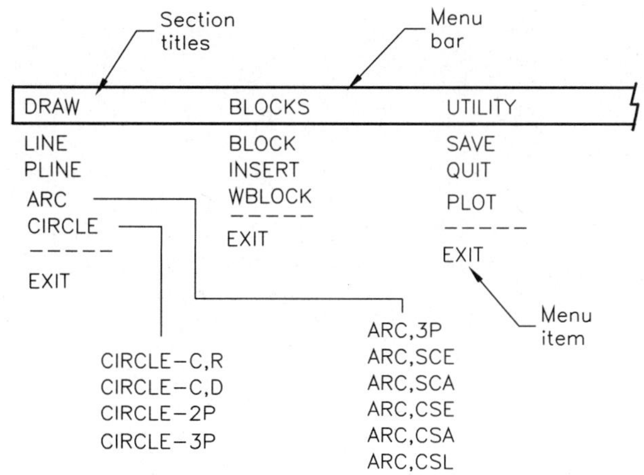

Figure 32-7 *Menu structure for Example 2*

Step 1: Writing the menu file

The following file is a listing of the menu for Example 2. **The line numbers are not a part of the menu; they are shown here for reference only.**

```
***POP1
[DRAW]
[LINE]^C^CLINE
[PLINE]^C^CPLINE
[->ARC]
  [ARC]^C^CARC
  [ARC,3P]^C^CARC
  [ARC,SCE]^C^CARC;\C
  [ARC,SCA]^C^CARC;\C;\A
  [ARC,CSE]^C^CARC;C
  [ARC,CSA]^C^CARC;C;\\A
  [<-ARC,CSL]^C^CARC;C;\\L
[->CIRCLE]
  [CIRCLE C,R]^C^CCIRCLE
  [CIRCLE C,D]^C^CCIRCLE;\D
  [CIRCLE 2P]^C^CCIRCLE;2P
  [<-CIRCLE 3P]^C^CCIRCLE;3P
[—]
[Exit]^C
***POP
[BLOCKS]
[BLOCK]^C^CBLOCK
[INSERT]*^C^CINSERT
[WBLOCK]^C^CWBLOCK
[—]
[Exit]^C
***POP3
[UTILITY]
[SAVE]^C^CSAVE
[QUIT]^C^CQUIT
[PLOT]^C^CPLOT
[—]
[Exit]^C
```

Explanation

Line 5

[->ARC]

In this menu item, **ARC** is the menu item label that is preceded by the special label characters **->**. These special characters indicate that the menu item has a submenu. The menu items that follow it (Lines 6-12) are the submenu items (Figure 32-8).

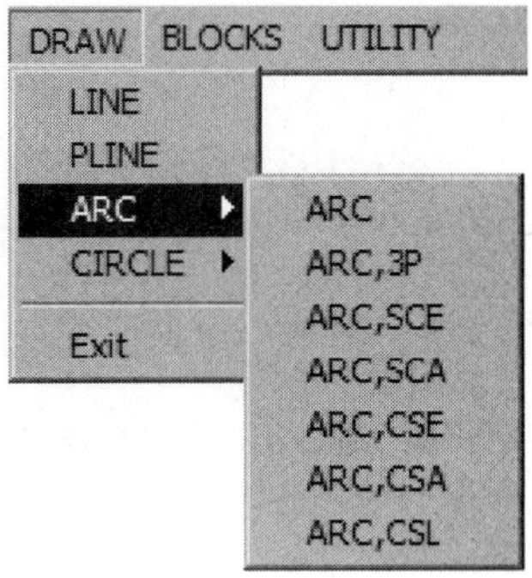

***Figure 32-8 Draw** menu with the **Arc** cascading menu*

Line 12

[<-ARC,CSL]^C^CARC;C;\\L

In this line, the menu item label **ARC,CSL** is preceded by another special label characters, **<-**, which indicates the end of the submenu. The item that contains these characters must be the last menu item of the submenu.

Lines 13 and 17

[->CIRCLE]

[<-CIRCLE 3P]^C^CCIRCLE;3P

The special characters -> in front of **CIRCLE** indicate that the menu item has a submenu; the characters <- in front of **CIRCLE 3P** indicate that this item is the last menu item in the submenu. When you select the menu item **CIRCLE** from the menu, it will automatically display the submenu on the side (Figure 32-9).

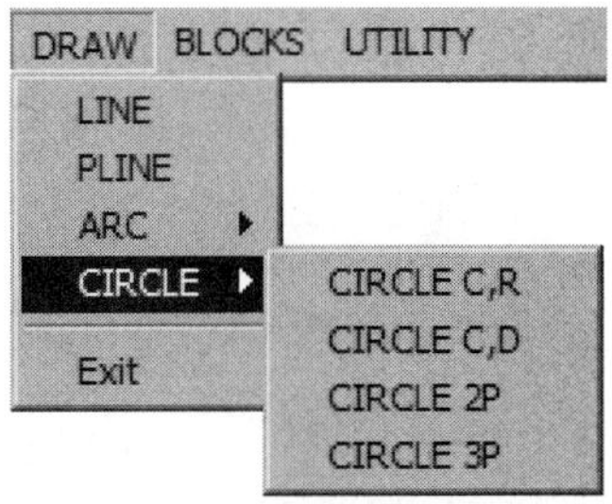

***Figure 32-9 Draw** menu with the **Circle** cascading menu*

Step 2: Saving and loading the menu file

Save the menu file with the name *exm2.mnu*. Use the **MENU** command to load this menu file. The previous menu is replaced by the menu that you select and the menu bar displays three items; **DRAW**, **BLOCKS**, and **UTILITY**.

Example 3

Write a menu that has the cascading submenus for the commands shown in Figure 32-10.

Writing the menu file

The following file is a listing of the menu for Example 3. **The line numbers are not a part of the menu; they are shown here for reference only.**

```
***POP1                              1
[CIR-DIM-DISP]                       2
[->CIRCLE]                           3
  [CIRCLE C,R]^C^C_CIRCLE            4
  [CIRCLE C,D]^C^C_CIRCLE;\_D        5
  [CIRCLE 2P]^C^C_CIRCLE;_2P         6
```

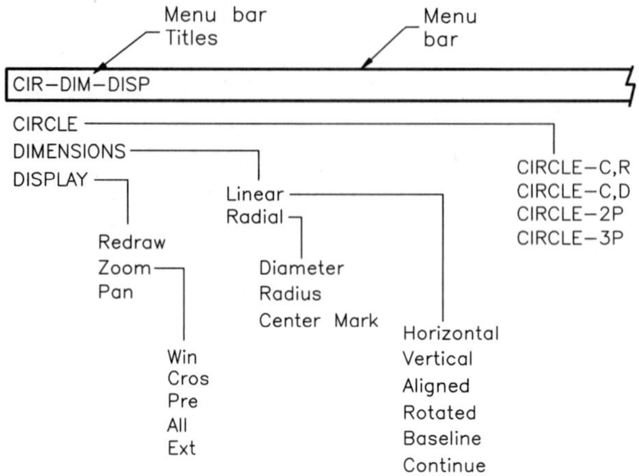

Figure 32-10 *Menu structure for Example 3*

```
7   [<-CIRCLE 3P]^C^C_CIRCLE;_3P
8   [->Dimensions]
9   [->Linear]
10  [Horizontal]^C^C_dimlinear
11  [Vertical]^C^C_dimlinear
12  [Aligned]^C^C_dimaligned
13  [Rotated]^C^C_dimrotated
14  [Baseline]^C^C_dimbaseline
15  [<-Continue]^C^C_dimcontinue
16  [->Radial]
17  [Diameter]^C^C_dimdiameter
18  [Radius]^C^C_dimradius
19  [<-<-Center Mark]^C^C_dimcenter
20  [->DISPLAY]
21  [REDRAW]^C^CREDRAW
22  [->ZOOM]
23  [...Win]^C^C_ZOOM;_W
24  [...Cros]^C^C_ZOOM;_C
25  [...Pre]^C^C_ZOOM;_P
26  [...All]^C^C_ZOOM;_A
27  [<-...Ext]^C^C_ZOOM;_E
28  [<-PAN]^C^C_Pan
```

Explanation

Lines 8 and 9

[->Dimensions]

[->Linear]

The special label characters **->** in front of the menu item **Dimensions** indicate that it has a submenu, and the characters -> in front of **Linear** indicate that there is another submenu. The second submenu **Linear** is within the first submenu Dimensions (Figure 32-11). The menu items on lines 10 to 15 are defined in the **Linear** submenu, and the menu items **Linear** and **Radial** are defined in the submenu Dimensions.

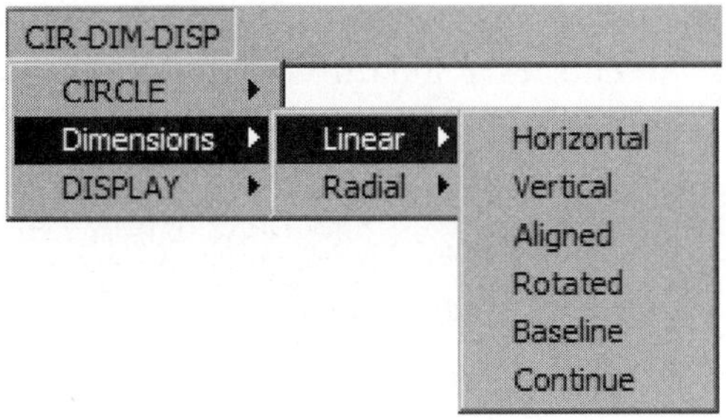

***Figure 32-11** **CIR-DIM-DISP** menu with the **Dimensions** cascading menus*

Line 16
[->Radial]
This menu item defines another submenu; the menu items on line numbers 17, 18, and 19 are part of this submenu.

Line 19
[<-<-Center Mark]^C^C_dim;_center
In this menu item the special label characters **<-<-** terminate the **Radial** and **Dimensions** (parent submenu) submenus.

Lines 27 and 28
[<-...Ext]^C^C_ZOOM;_E
[<-PAN]^C^C_Pan
The special characters **<-** in front of the menu item **...Ext** terminate the **ZOOM** submenu (Figure 32-12); the special character in front of the menu item **PAN** terminates the **DISPLAY** submenu.

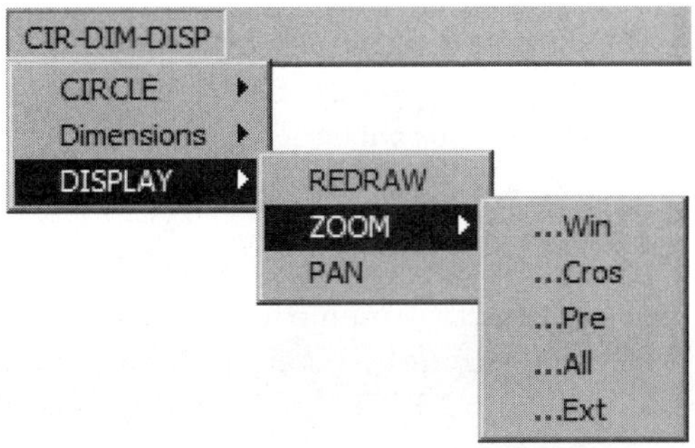

***Figure 32-12** **CIR-DIM-DISP** menu with the **Display** cascading menus*

SHORTCUT AND CONTEXT MENUS

The Shortcut menus are similar to the pull-down menus, except the Shortcut menu can contain only 499 menu items compared with 999 items in the pull-down menu. The section label of the Shortcut menu can be ***POP0 and ***POP500 to ***POP 999. The Shortcut menus are displayed near or at the cursor location. Therefore, they can be used to provide convenient and quick access to some of the frequently used commands. The Shortcut menus that are in the upper range are also referred to as Context menus. The following is a list of some of the features of the Shortcut menu.

1. The section label of the Shortcut menu are ***POP0 and ***POP500 to ***POP999. The menu bar title defined under this section label is not displayed in the menu bar.

2. On most systems, the menu bar title is not displayed at the top of the shortcut menu. However, for compatibility reasons you should give a dummy menu bar title.

3. The POP0 menu can be accessed through the **$P0=*** menu command. The shortcut menus POP500 through POP999 must be referenced by their alias names. The reserved alias names for AutoCAD use are **GRIPS**, **CMDEFAULT**, **CMEDIT**, and **CMCOMMAND**. For example, to reference POP500 for grips, use **GRIPS command line under POP500. This command can be issued by a menu item in another menu, such as the button menu, auxiliary menu, or the screen menu. The command can also be issued from an AutoLISP or ADS program.

4. A maximum of 499 menu items can be defined in the shortcut menu and items in excess of 499 are ignored. This includes the items that are defined in the shortcut submenus.

5. The number of menu items that can be displayed on the screen depends on the system you are using. If the shortcut or pull-down menu contains more items than your screen can accommodate, the excess items are truncated. For example, if your system displays 21 menu items, the menu items in excess of 21 are automatically truncated.

6. The system variable **SHORTCUTMENU** controls the availability of **Default**, **Edit**, and **Command** mode shortcut menus. If the value is 0, it restores R14 legacy behavior and makes the **Default**, **Edit**, and **Command** mode shortcut menus unavailable. The default value of this variable is 11.

Example 4

Write a Shortcut menu for the following AutoCAD commands using cascading submenus. The menu should be compatible with foreign language versions of AutoCAD. Use the third button of the **BUTTONS** menu to display the cursor menu.

Osnaps	Draw	DISPLAY
Center	Line	REDRAW
Endpoint	PLINE	ZOOM
Intersection	CIR C,R	...Win
Midpoint	CIR 2P	...Cen
Nearest	ARC SCE	...Prev
Perpendicular	ARC CSE	...All
Quadrant		...Ext
Tangent		PAN
None		

Writing the menu file

The following file is a listing of the menu file for Example 4. **The line numbers are not a part of the file; they are for reference only.**

```
***AUX1
;
$P0=*
***POP0
```

```
5 [Osnaps]
6 [Center]_Center
7 [End point]_Endp
8 [Intersection]_Int
9 [Midpoint]_Mid
10 [Nearest]_Nea
11 [Perpendicular]_Per
12 [Quadrant]_Qua
13 [Tangent]_Tan
14 [None]_Non
15 [—]
16 [->Draw]
17  [Line]^C^C_Line
18  [PLINE]^C^C_Pline
19  [CIR C,R]^C^C_Circle
20  [CIR 2P]^C^C_Circle;_2P
21  [ARC SCE]^C^C_ARC;\C
22  [<-ARC CSE]^C^C_Arc;C
23 [—]
24 [->DISPLAY]
25   [REDRAW]^C^_REDRAW
26   [->ZOOM]
27     [...Win]^C^C_ZOOM;_W
28     [...Cen]^C^C_ZOOM;_C
29     [...Prev]^C^C_ZOOM;_P
30     [...All]^C^C_ZOOM;_A
31     [<-...Ext]^C^C_ZOOM;_E
32   [<-PAN]^C^C_Pan
33 ***POP1
34 [SHORTCUTMENU]
35 [SHORTCUTMENU=0]^C^CSHORTCUTMENU;0
36 [SHORTCUTMENU=1]^C^CSHORTCUTMENU;1
```


Explanation

Line 1

*****AUX1**

AUX1 is the section label for the first auxiliary menu; *** designates the menu section. The menu items that follow it, until the second section label, are a part of this buttons menu.

Lines 2 and 3

;

**$P0=*

The semicolon (;) is assigned to the second button of the pointing device (the first button of the pointing device is the pick button); the special command **$P0=*** is assigned to the third button of the pointing device.

Lines 4 and 5
*****POP0**
[Osnaps]
The menu label **POP0** is the menu section label for the shortcut menu; **Osnaps** is the menu bar title. The menu bar title is not displayed, but is required. Otherwise, the first item will be interpreted as a title and will not be available.

Line 6
[Center]_Center
In this menu item, **_Center** is the center Object Snap mode. The menu files can be used with foreign language versions of AutoCAD, if AutoCAD commands and the command options are preceded by the underscore (_) character.

After loading the menu, if you press the third button of your pointing device, the shortcut menu (Figure 32-13) will be displayed at the cursor (screen crosshairs) location. If the cursor is close to the edges of the screen, the Shortcut menu will be displayed at a location that is closest to the cursor position. When you select a submenu, the items contained in the submenu will be displayed, even if the shortcut menu is touching the edges of the screen display area.

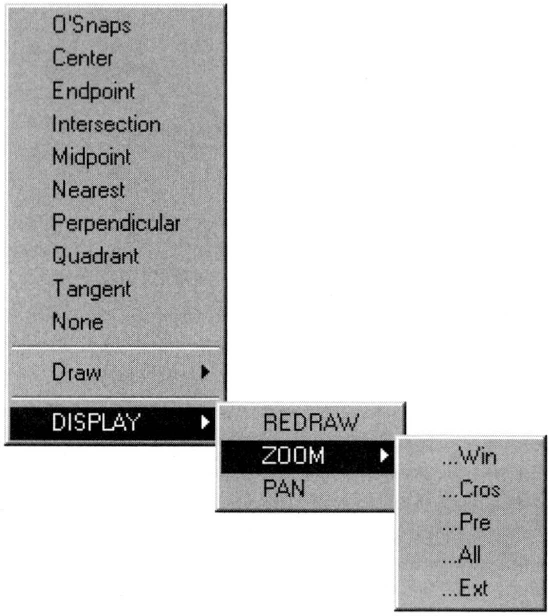

***Figure 32-13** Shortcut menu for Example 4*

Lines 33 and 34
*****POP1**
[Draw]
*****POP1** defines the first pull-down menu. If no POPn sections are defined or the status line is turned off, the Shortcut menu is automatically disabled.

Exercise 2 *General*

Write a menu for the following AutoCAD commands. Use a cascading menu for the **LINE** command options in the menu. (The layout of the menu is shown in Figure 32-14.)

LINE	**ZOOM All**	**TIME**
Continue	**ZOOM Win**	**LIST**
Close	**ZOOM Pre**	**DISTANCE**
Undo	**PAN**	**AREA**
CIRCLE	**DBLIST**	
ELLIPSE	**STATUS**	

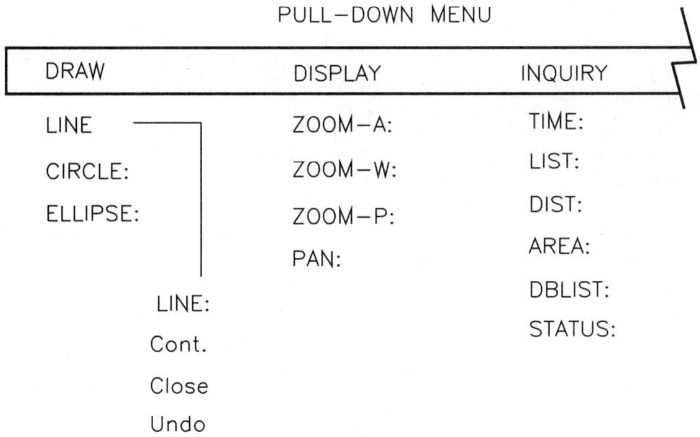

Figure 32-14 *Design of menu for Exercise 2*

SUBMENUS

The number of items in a pull-down menu or shortcut menu can be very large and sometimes they cannot all be accommodated on one screen. For example, consider a display device on which the maximum number of items that can be displayed is 21 items, the menu excess items are not displayed on the screen and cannot be accessed. You can overcome this problem by using submenus that let you define smaller groups of items within a menu section. When a submenu is selected, it loads the submenu items and displays them on the screen.

The menus that use AutoCAD's cascading feature are the most efficient and easy to write. The submenus follow a logical pattern that are easy to load and use without causing any confusion. It is strongly recommended to use the cascading menus whenever you need to write the pull-down or the shortcut menus. However, AutoCAD provides the option to swap the submenus in the menus. These menus can sometimes cause distraction because the original menu is completely replaced by the submenu when swapping the menus.

Submenu Definition

A submenu definition consists of two asterisk signs (**) followed by the name of the submenu. A menu can have any number of submenus and every submenu should have a unique name. The items that follow a submenu, up to the next section label or submenu label, belong to that submenu. Following is the format of a submenu label.

****Name**

Where ****** ----------------- Two asterisk signs (**) designate a submenu
Name ------------ Name of the submenu

Note

The submenu name can be up to 31 characters long.

The submenu name can consist of letters, digits, and the special characters like: $ (dollar), - (hyphen), and _ (underscore).

The submenu name cannot have any embedded blanks.

The submenu names should be unique in a menu file.

Submenu Reference

The submenu reference is used to reference or load a submenu. It consists of a dollar ($) sign followed by a letter that specifies the menu section. The letter that specifies a menu section is Pn, where n designates the number of the menu section. The menu section is followed by "=" sign and the name of the submenu that the user wants to activate. The submenu name should be without "**". Following is the format of a submenu reference.

$Section=Submenu

Where **$** ------------------- "$" sign
Section ---------- Menu section specifier
= ------------------ "=" sign
Submenu ------- Name of submenu

Example
$P1=P1A

Where **$P1** --------------- P1-Specifies pull-down menu section 1
P1A --------------- Name of submenu

Displaying a Submenu

When you load a submenu in a menu, the submenu items are not automatically displayed on the screen. For example, when you load a submenu **P1A** that has **DRAW-ARC** as the first item, the current title of POP1 will be replaced by the **DRAW-ARC**. But the items that are defined under DRAW-ARC are not displayed on the screen. To force the display of the new items on the screen, AutoCAD uses a special command $Pn=*.

$Pn=*

Where **P** ------------------ P for menu
Pn ----------------- Menu section number (1 to 10)
* ------------------- Asterisk sign (*)

LOADING SCREEN AND IMAGE TILE MENUS

From the menu, you can load any menu that is defined in the screen or image tile menu sections by using the appropriate load commands. It may not be needed in most of the applications, but if you want to, you can load the menus that are defined in other menu sections.

Loading Screen Menus

You can load any menu that is defined in the screen menu section from the menu by using the following load command.

$S=X $S=LINE

Where **S** ------------------- S specifies screen menu
X ------------------ Submenu name defined in screen menu section
LINE ------------- Submenu name defined in screen menu section

The first load command ($S=X) loads the submenu X that has been defined in the screen menu section of the menu file. The X submenu can contain 21 blank lines, so that when it is loaded it clears the screen menu. The second load command ($S=LINE) loads the submenu **LINE** that has also been defined in the screen menu section of the menu file.

Loading an Image Tile Menu

You can also load an image tile menu from the menu by using the following load command.

$I=IMAGE1 $I=*

Where **$I=Image1** ---- Load the submenu IMAGE1
$I=* ------------- Display the dialog box

This menu item consists of two load commands. The first load command $I=IMAGE1 loads the image tile submenu **IMAGE1** that has been defined in the image tile menu section of the file. The second load command $I=* displays the new dialog box on the screen.

Example 5

Write a pull-down menu for the following AutoCAD commands. Use submenus for the **ARC** and **CIRCLE** commands.

LINE	**BLOCK**	**QUIT**
PLINE	**INSERT**	**SAVE**
ARC	**WBLOCK**	**PLOT**
ARC 3P		
ARC SCE		
ARC SCA		
ARC CSE		
ARC CSA		
ARC CSL		
CIRCLE		
CIRCLE C,R		
CIRCLE C,D		
CIRCLE 2P		

Step 1: Designing the menu

The layout shown in Figure 32-15 is one of the possible designs for this menu. The **ARC** and **CIRCLE** commands are in separate groups that will be defined as submenus in the menu file.

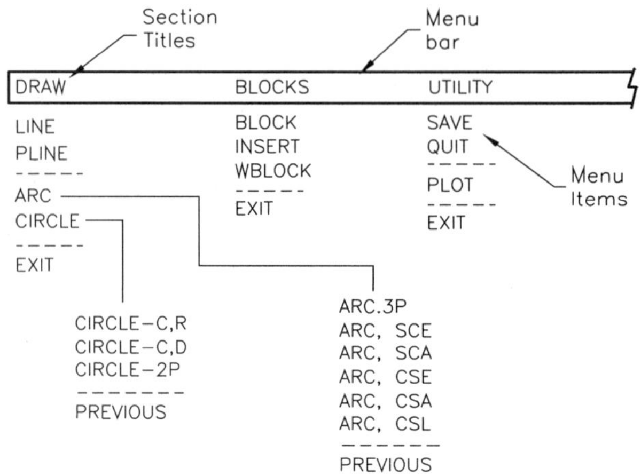

Figure 32-15 Design of menu for Example 5

Step 2: Writing the menu file

The following file is a listing of the menu for Example 5. **The line numbers are not a part of the menu file. They are given here for reference only.**

```
***POP1
**P1A
[DRAW]
[LINE]^C^CLINE
[PLINE]^C^CPLINE
[—]
[ARC]^C^C$P1=P1B $P1=*
[CIRCLE]^C^C$P1=P1C $P1=*
[—]
[Exit]^C

**P1B
[ARC]
[ARC,3P]^C^CARC
[ARC,SCE]^C^CARC \C
[ARC,SCA]^C^CARC \C \A
[ARC,CSE]^C^CARC C
[ARC,CSA]^C^CARC C \\A
[ARC,CSL]^C^CARC C \\L
[—]
[PREVIOUS]$P1=P1A $P1=*
```

```
22
23 **P1C
24 [CIRCLE]
25 [CIRCLE C,R]^C^CCIRCLE
26 [CIRCLE C,D]^C^CCIRCLE \D
27 [CIRCLE 2P]^C^CCIRCLE 2P
28 [—]
29 [PREVIOUS]$P1=P1A $P1=*
30
31 ***POP2
32 [BLOCKS]
33 [BLOCK]^C^CBLOCK
34 [INSERT]*^C^CINSERT
35 [WBLOCK]^C^CWBLOCK
36 [—]
37 [EXIT]$P1=P1A $P1=*
38
39 ***POP3
40 [UTILITY]
41 [SAVE]^C^CSAVE
42 [QUIT]^C^CQUIT
43 [~—]
44 [PLOT]^C^CPLOT
45 [~—]
46 [Exit]^C
47
```

Explanation

Line 2

****P1A**

**P1A defines the submenu P1A. All the submenus have two asterisk signs (**) followed by the name of the submenu. The submenu can have any valid name. In this example, P1A has been chosen because it is easy to identify the location of the submenu. P indicates that it is a menu, 1 indicates that it is in the first menu (POP1), and A indicates that it is the first submenu in that section.

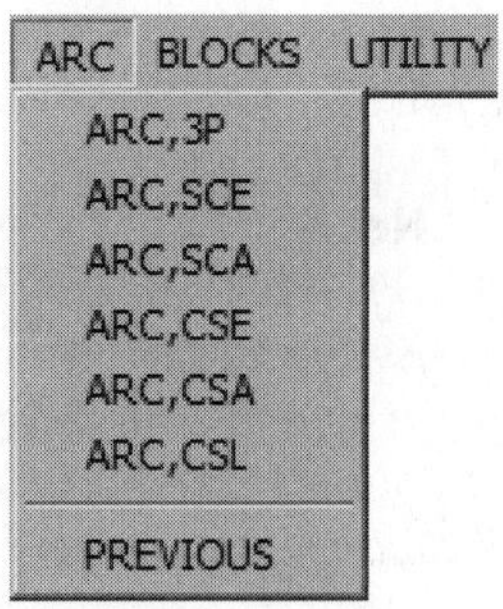

***Figure 32-16 Arc** pull-down menu replaces **Draw** menu*

Line 6

[--]

The two hyphens enclosed in the brackets will automatically expand to fill the entire width of the menu. This menu item cannot be used to define a command. If it does contain a command definition, the command is ignored. For example, if the menu item is [--]^C^CLINE, the command ^C^CLINE will be ignored.

Line 7
[ARC]^C^C$P1=P1B $P1=*
In this menu item, $P1=P1B loads the submenu P1B and assigns it to the first menu section (POP1), but the new menu is not displayed on the screen. $P1=* forces the display of the new menu on the screen.

For example, if you select **CIRCLE** from the first menu (POP1), the menu bar title **DRAW** will be replaced by **CIRCLE**, but the new menu is not displayed on the screen. Now, if you select **CIRCLE** from the menu bar, the command defined under the **CIRCLE** submenu will be displayed in the menu. To force the display of the menu that is currently assigned to POP1, you can use AutoCAD's special command **$P1=***. If you select **CIRCLE** from the first menu (POP1), the **CIRCLE** submenu will be loaded and automatically displayed on the screen.

Line 21
[EXIT]^C
When you select this menu item the current menu will be canceled. It will not return to the previous submenu (**DRAW**). If you check the menu bar, it will display **ARC** as the section title, not **DRAW**. Therefore, it is not a good practice to cancel a submenu. It is better to return to the first menu before canceling it or define a command that automatically loads the previous menu and then cancels the menu.

Example
[EXIT]$P1=P1A $P1=* ^C^C

Line 29
[PREVIOUS]$P1=P1A $P1=*
In this menu item $P1=P1A loads the submenu P1A, which happens to be the previous menu in this case. You can also use $P1= to load the previous menu. $P1=* forces the display of the submenu P1A.

Note
When you swap the pull-down menus, the menus will get incremented. For example, if a menu file loads POP1 through POP8 and then a partial menu or a LISP routine inserts a new menu at POP5, the existing pull-down menus will be incremented by one. Thus, $P7=P7A $P= will take the new P7A menu and use it to replace what used to be POP6, but has now been pushed over to POP7.*

Step 3: Loading the menu file
Save the menu file and load it using the **MENU** command as described in earlier examples.

PARTIAL MENUS

AutoCAD has a facility that allows users to write their own menus and then load them in the menu bar. For example, in Windows you can write partial menus, toolbars, and definitions for accelerator keys. After you write the menu, AutoCAD lets you load the menu and use it with the standard menu. For example, you could load a partial menu and use it like a menu.

You can also unload the menus that you do not want to use. These features make it convenient to use the menus that have been developed by AutoCAD users and developers.

Menu Section Labels

The following is a list of the additional menu section labels.

Section label	Description
***MENUGROUP	Menu file group name
***TOOLBARS	Toolbar definition
***HELPSTRING	Online help
***ACCELERATORS	Accelerator key definitions

Writing Partial Menus

The following example illustrates the procedure for writing a partial menu:

Example 6

In this example you will write a partial menu for Windows. The menu file has two menus, POP1 (**MyDraw**) and POP2 (**MyEdit**), as shown in Figure 32-17.

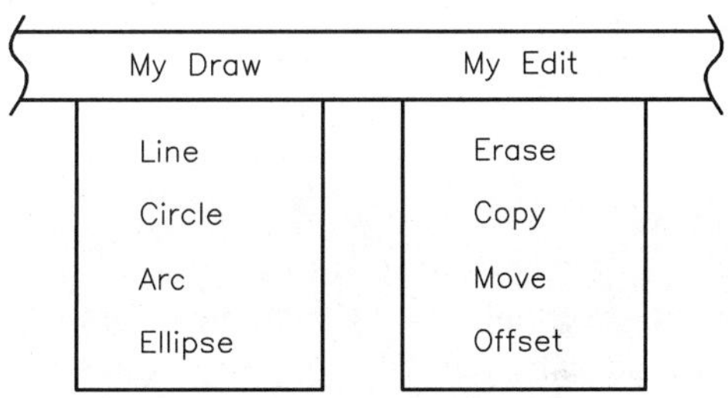

Figure 32-17 *Pull-down menus for Example 6*

Step 1: Writing the menu file

Use a text editor to write the following menu file. The name of the file is assumed to be *mymenu1.mnu*. The following is the listing of the menu file for this example.

```
***MENUGROUP=Menu1
***POP1
[/MMyDraw]
[/LLine]^C^CLine
[/CCircle]^C^CCircle
[/AArc]^C^CArc
[/EEllipse]^C^CEllipse
***POP2
[/EMyEdit]
[/EErase]^C^CErase
[/CCopy]^C^CCopy
[/VMove]^C^CMove
[/OOffset]^C^COffset
```

Explanation

Line 1

*****MENUGROUP=Menu1**

MENUGROUP is the section label and the Menu1 is the name tag for the menu group. The MENUGROUP label must precede all menu section definitions. The name of the MENUGROUP (Menu1) can be up to 32 characters long (alphanumeric), excluding spaces and punctuation marks. There is only one MENUGROUP in a menu file. All section labels must be preceded by *** (***MENUGROUP).

Line 2

*****POP1**

POP1 is the menu section label. The items on line numbers 3 through 7 belong to this section. Similarly, the items on line numbers 9 through 13 belong to the menu section **POP2**.

Line 3

[/MMyDraw]

/M defines the mnemonic key you can use to activate the menu item. For example, /M will display an underline under the letter M in the text string that follows it. If you enter letter M, AutoCAD will execute the command defined in that menu item. **MyDraw** is the menu item label. The text string inside the brackets [ ] , except /M, has no function. It is used for displaying the function name so that the user can recognize the command that will be executed by selecting that item.

Line 4

[/LLine]^C^CLine

In this line, the /L defines the mnemonic key, and the Line that is inside the brackets is the menu item label. ^C^C cancels the command twice, and the Line is the AutoCAD **LINE** command. The part of the menu item statement that is outside the brackets is executed when you select an item from the menu. When you select Line 4, the **LINE** command is executed.

Step 2: Loading the menu file

Save the file as *mymenu1.mnu*. Choose **Customize Menu** from the **Tools** menu or enter the **MENULOAD** command at the Command prompt to invoke the **Menu Customization** dialog box (Figure 32-18). To load the menu file, enter the name of the menu file, *mymenu1.mnu*, in the **File Name** edit box. You can also use the **Browse** button to invoke the **Select Menu File** dialog box. Select the name of the file, and then use the **OK** button to return to the **Menu Customization** dialog box. To load the selected menu file, choose the **LOAD** button. Choose **Yes** from the warning box that is displayed. The name of the menu group (**Menu1**) will be displayed in the **Menu Groups** list box.

You can also load the menu file from the command line, as follows:

Command: **FILEDIA**
Enter new value for FILEDIA <1>: **0** *(Disables the display of the file dialog boxes.)*
Command: **MENULOAD**
Enter name of menu file to load: **MYMENU1.MNU**

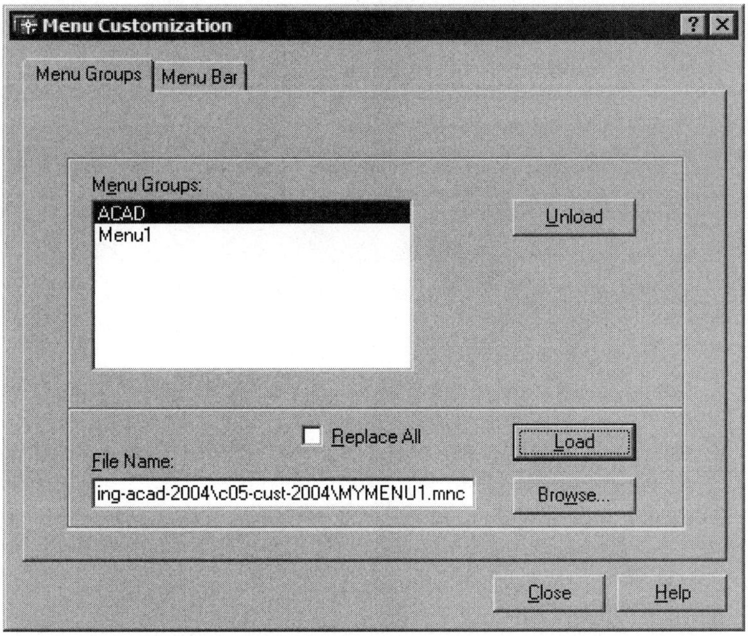

Figure 32-18 Menu Customization *dialog box (Menu Groups tab)*

You can also use the AutoLISP functions to set the **FILEDIA** system variable to 0 and then load the partial menu.

Command: **(SETVAR "FILEDIA" 0)**
Command: **(Command "MENULOAD" "MYMENU1")**

Step 3: Inserting menus in the Menu Bar

In the **Menu Customization** dialog box, choose the **Menu Bar** tab to display the menu bar options (Figure 32-19). In the **Menu Groups** list box select **Menu1**; the menus defined in the menu group (**Menu1**) will be displayed in the **Menus** list box. In the **Menus** list box select the menu (**MyDraw**) that you want to insert in the menu bar. In the **Menu Bar** list box select the position where you want to insert the new menu. For example, if you want to insert the new menu in front of Format, select the **Format** menu in the **Menu Bar** list box. Choose the **Insert** button to insert the selected menu (**MyDraw**) in the menu bar. The menu (**MyDraw**) is displayed in the menu bar located at the top of your screen.

You can also load the menu from the Command line. Once the menu is loaded, use the **MENUCMD** command (AutoLISP function) to display the partial menus.

Command: **(MENUCMD "P5=+Menu1.POP1")**
Command: **(MENUCMD "P6=+Menu1.POP2")**

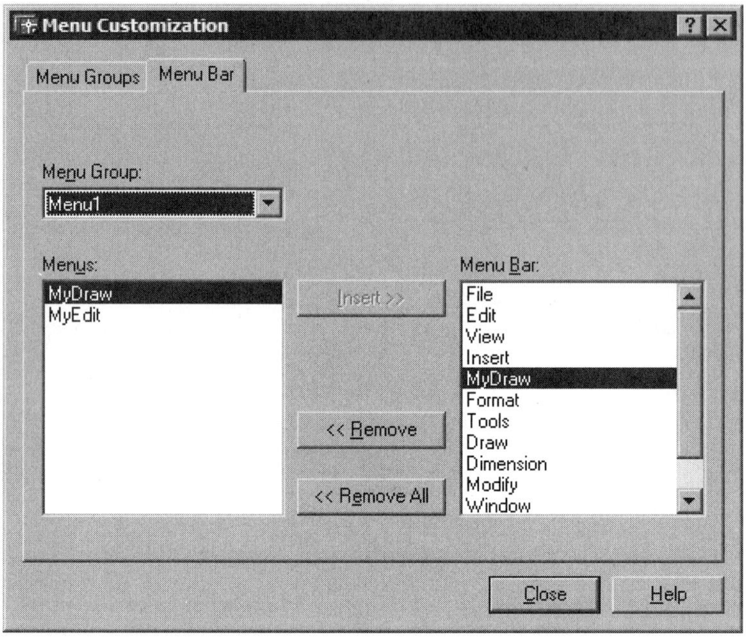

***Figure 32-19 Menu Customization** dialog box (**Menu Bar** tab)*

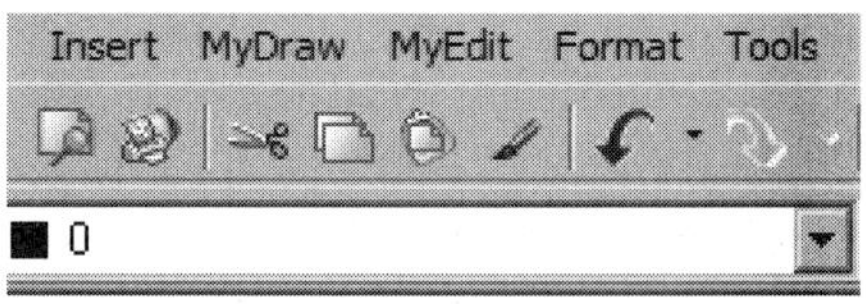

***Figure 32-20 MENUCMD** command places the menu titles in the menu bar*

After you enter these commands, AutoCAD will display the menu titles in the menu bar, as shown in Figure 32-20. If you select **MyDraw**, the corresponding menu as defined in the menu file will be displayed on the screen. Similarly, selecting **MyEdit** will display the corresponding edit menu.

MENUCMD is an AutoLISP function, and P5 determines where the POP1 menu will be displayed. In this example the POP1 (**MyDraw**) menu will be displayed as the fifth menu. Menu1 is the MENUGROUP name as defined in the menu file, and POP1 is the menu section label. The MENUGROUP name and the menu section label must be separated by a period (.).

Step 4: Unloading the menu

If you want to unload the menu (**MyDraw**), select the **Menu Bar** tab of the **Menu Customization** dialog box. In the **Menu Bar** list box select the menu item that you want to unload. Choose the **Remove** button to remove the selected item.

Step 5: Unloading the menugroup

You can also unload the menu groups (Menu1) using the **Menu Customization** dialog box. Choose the **Customize Menu** button from the **Tools** menu or enter **MENULOAD** or **MENUUNLOAD** at the Command prompt to invoke the **Menu Customization** dialog box. AutoCAD will display the names of the menu files in the **Menu Groups:** list box. Select the

Menu1 menu, and then choose the **Unload** button. AutoCAD will unload the menu group. Choose the **Close** button to exit the dialog box.

You can also unload the menu file from the command line, as follows.

Command: **FILEDIA**
Enter new value for FILEDIA <1>: **0**
Command: **MENUUNLOAD**
Enter the name of a MENUGROUP to unload: **MENU1.MNU**

The **MENUUNLOAD** command unloads the entire menu group. You can also unload an individual menu without unloading the entire menu group by using the following command.

Command: **(MENUCMD "P5=-)**

This command will unload the menu at position five (P5, **MyDraw** menu). The menu group is still loaded, but the P5 menu is not visible. The menu can also be reinitialized by using the **MENU** command to load the base menu *acad.mnu*. This will remove all partial menus and the tag definitions associated with the partial menus.

ACCELERATOR KEYS

AutoCAD for Windows also supports user-defined accelerator keys. For example, if you enter **C** at the Command prompt, AutoCAD draws a circle because it is a command alias for a circle as defined in the *acad.pgp* file. You cannot use the **C** key to enter the **COPY** command. To use the **C** key for entering the **COPY** command, you can define the accelerator keys. You can combine the SHIFT key with C in the menu file so that when you hold down the SHIFT key and then press the C key, AutoCAD will execute the **COPY** command. The following example illustrates the use of accelerator keys.

Example 7

In this example you will add the following accelerator keys to the partial menu of Example 6.

CONTROL+"E"	to draw an ellipse (**ELLIPSE** command)
SHIFT+"C"	to copy (**COPY** command)
[CONTROL+"Q"]	to quit (**QUIT** command)

Step 1: Writing the accelerators in a file

The following file is the listing of the partial menu file that uses the accelerator keys of Example 7.

```
***MENUGROUP=Menu1
***POP1
**Alias
[/MMyDraw]
[/LLine]^C^CLine
```

```
[/CCircle]^C^CCircle
[/AArc]^C^CArc
ID_Ellipse [/EEllipse]^C^CEllipse

***POP2
[/EMyEdit]
[/EErase]^C^CErase
ID_Copy [/CCopy]^C^CCopy
[/OOffset]^C^COffset
[/VMove]^C^CMov

***ACCELERATORS
ID_Ellipse [CONTROL+"E"]
ID_Copy [SHIFT+"C"]
[CONTROL"Q"]^C^CQuit
```

Explanation

This menu file defines three accelerator keys. The **ID_Copy [SHIFT+"C"]** accelerator key consists of two parts. The ID_Copy is the name tag, which must be the same as used earlier in the menu item definition. The SHIFT "C" is the label that contains the modifier (SHIFT) and the keyname (C). The keyname or the string, such as "ESCAPE," must be enclosed in quotation marks.

The accelerator keys can be defined in two ways. One way is to give the name tag followed by the label containing the modifier. The modifier is followed by a single character or a special virtual key enclosed in quotation marks [CONTROL+"E"] or ["ESCAPE"]. You can also use the plus sign (+) to concatenate the modifiers [SHIFT + CONTROL + "L"]. The other way of defining an accelerator key is to give the modifier and the key string, followed by a command sequence [CONTROL "Q"]^C^CQuit.

Step 2: Loading the accelerators

Save the file with the extension *.mnu*. Load the file using the **MENU** command. After you load the file, SHIFT+C will enter the **COPY** command and CTRL+E will draw an ellipse. Similarly, CTRL+Q will cancel the existing command and enter the **QUIT** command. If it does not work, you may have to unload the partial menu from Example 6 because it uses the same menugroup name.

Special Virtual Keys

The following are the special virtual keys. These keys must be enclosed in quotation marks when used in the menu file.

String	Description	String	Description
"F1"	F1 key	"NUMBERPAD0"	0 key
"F2"	F2 key	"NUMBERPAD1"	1 key
"F3"	F3 key	"NUMBERPAD2"	2 key
"F4"	F4 key	"NUMBERPAD3"	3 key
"F5"	F5 key	"NUMBERPAD4"	4 key

"F6"	F6 key	"NUMBERPAD5"	5 key
"F7"	F7 key	"NUMBERPAD6"	6 key
"F8"	F8 key	"NUMBERPAD7"	7 key
"F9"	F9 key	"NUMBERPAD8"	8 key
"F10"	F10 key	"NUMBERPAD9"	9 key
"F11"	F11 key	"UP"	UP-ARROW key
"F12"	F12 key	"DOWN"	DOWN-ARROW key
"HOME"	HOME key	"LEFT"	LEFT-ARROW key
"END"	END key	"RIGHT"	RIGHT-ARROW key
"INSERT"	INS key	"ESCAPE"	ESC key
"DELETE"	DEL key		

Valid Modifiers

The following are the valid modifiers:

String	Description
CONTROL	The CTRL key on the keyboard
SHIFT	The SHIFT key (left or right)
COMMAND	The Apple key on Macintosh keyboards
META	The meta key on UNIX keyboards

TOOLBARS

The contents of the toolbar and its default layout can be specified in the Toolbar section (***TOOLBARS) of the menu file. Each toolbar must be defined in a separate submenu.

Toolbar Definition. The following is the general format of the toolbar definition.

```
***TOOLBARS
**MYTOOLS1
TAG1 [Toolbar ("tbarname", orient, visible, xval, yval, rows)]
TAG2 [Button ("btnname", id_small, id_large)]macro
TAG3 [Flyout ("flyname", id_small, id_large, icon, alias)]macro
TAG4 [control (element)]
[—]
```

*****TOOLBARS** is the section label of the toolbar, and **MYTOOLS1** is the name of the submenu that contains the definition of a toolbar. Each toolbar can have five distinct items that control different elements of the toolbar: TAG1, TAG2, TAG3, TAG4, and separator ([—]). The first line of the toolbar (TAG1) defines the characteristics of the toolbar. In this line, **Toolbar** is the keyword, and it is followed by a series of options enclosed in parentheses. The following describes the available options.

tbarname This is a text string that names the toolbar. The tbarname text string must consist of alphanumeric characters with no punctuation other than a dash (-) or an underscore (_).

orient This determines the orientation of the toolbar. The acceptable values are Floating, Top, Bottom, Left, and Right. These values are not case-sensitive.

visible This determines the visibility of the toolbar. The acceptable values are Show and Hide. These values are not case-sensitive.

xval This is a numeric value that specifies the *X* ordinate in pixels. The *X* ordinate is measured from the left edge of the screen to the left side of the toolbar.

yval This is a numeric value that specifies the *Y* ordinate in pixels. The *Y* ordinate is measured from the top edge of the screen to the top of the toolbar.

rows This is a numeric value that specifies the number of rows.

The second line of the toolbar (TAG2) defines the button. In this line the **Button** is the key word and it is followed by a series of options enclosed in parentheses. The following is the description of the available options.

btnname This is a text string that names the button. The text string must consist of alphanumeric characters with no punctuation other than a dash (-) or an underscore (_). This text string is displayed as tooltip when you place the cursor over the button.

id_small This is a text string that names the ID string of the small image resource (16 by 16 bitmap). The text string must consist of alphanumeric characters with no punctuation other than a dash (-) or an underscore (_). The id_small text string can also specify a user-defined bitmap (Example: RCDATA_16_CIRCLE). The **bit map images must exist or you just get a question mark**.

id_big This is a text string that names the ID string of the large image resource (32 by 32 bitmap). The text string must consist of alphanumeric characters with no punctuation other than a dash (-) or an underscore (_). The id_big text string can also specify a user-defined bitmap (Example: RCDATA_32_CIRCLE).

macro The second line (TAG2) that defines a button, is followed by a command string (macro). For example, the macro can consist of ^C^CLine. It follows the same syntax as that of a standard menu item definition.

The third line of the toolbar (TAG3) defines the flyout control. In this line the **Flyout** is the key word, and it is followed by a series of options enclosed in parentheses. The following describes the available options.

flyname This is a text string that names the flyout. The text string must consist of alphanumeric characters with no punctuation other than a dash (-) or an underscore (_). This text string is displayed as tooltip when you place the cursor over the flyout button.

id_small This is a text string that names the ID string of the small image resource (16 by 16 bitmap). The text string must consist of alphanumeric characters with no punctuation other than a dash (-) or an underscore (_). The id_small text string can also specify a user-defined bitmap.

id_big This is a text string that names the ID string of the large image resource (32 by 32 bitmap). The text string must consist of alphanumeric characters with no punctuation other than a dash (-) or an underscore (_). The id_big text string can also specify a user-defined bitmap.

icon This is a Boolean key word that determines whether the button displays its own icon or the last icon selected. The acceptable values are **ownicon** and **othericon**. These values are not case-sensitive.

alias The alias specifies the name of the toolbar submenu that is defined with the standard ****aliasname** syntax.

macro The third line (TAG3), which defines a flyout control, is followed by a command string (macro). For example, the macro can consist of ^C^CCircle. It follows the same syntax as that of any standard menu item definition.

The fourth line of the toolbar (TAG4) defines a special control element. In this line the Control is the key word, and it is followed by the type of control element enclosed in parentheses. The following describes the available control element types.

element This parameter can have one of the following three values:
Layer: This specifies the layer control element.
Linetype: This specifies the linetype control element.
Color: This specifies the color control element.

The fifth line ([**--**]) defines a separator.

Example 8

In this example you will write a menu file for a toolbar for the **LINE**, **PLINE**, **CIRCLE**, **ELLIPSE**, and **ARC** commands. The name of the toolbar is **MyDraw1** (Figure 32-21).

Step 1: Writing the toolbars

Use any text editor to list the toolbars. The following is the listing of the menu file containing toolbars. In this file listing, ID specifies the name tag.

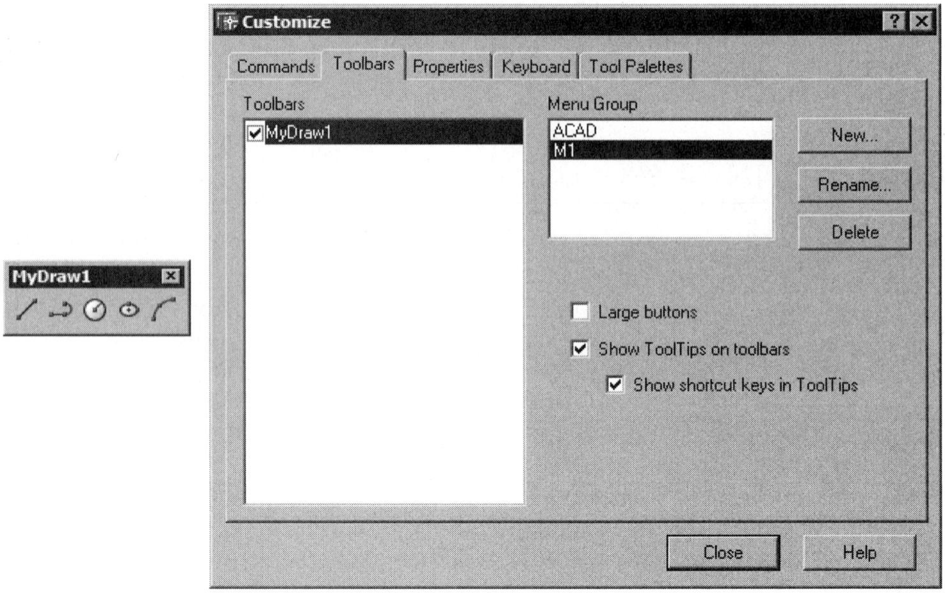

Figure 32-21 ***MyDraw1*** *toolbar for Example 8 and* ***Customize*** *dialog box*

```
MENUGROUP=M1
***TOOLBARS
**TB_MyDraw1
ID_MyDraw1[_Toolbar("MyDraw1", _Floating, _Hide, 10, 200, 1)]
ID_Line [_Button("Line", RCDATA_16_LINE, RCDATA_32_LINE)]^C^C_line
ID_Pline [_Button("Pline", RCDATA_16_PLine, RCDATA_32_PLine)]^C^C_PLine
ID_Circle[_Button("Circle", RCDATA_16_CirRAD, RCDATA_32_CirRAD)]^C^C_Circle
ID_ELLIPSE[_Button("Ellipse",RCDATA_16_EllCEN, RCDATA_32_EllCEN)]^C^C_ELLIPSE
ID_Arc[_Button("Arc 3Point", RCDATA_16_Arc3Pt, RCDATA_32_Arc3Pt)]^C^C_Arc
```

Step 2: Loading the menu file containing toolbars

Save the file with the extension *.mnu*. Use the **MENULOAD** command to load the **MyDraw1** menu group as discussed earlier while loading the partial menus. To display the new toolbar (**MyDraw1**) on the screen, select **Toolbars** from the **View** menu and then select **MyDraw1** in the **Menu Groups** list box. Turn the **MyDraw1** toolbar on in the **Toolbars** list box. MyDraw1 toolbar is displayed on the screen.

You can also load the new toolbar from the command line. After using the **MENULOAD** command to load the MyDraw1 menu group, use the **-TOOLBAR** command to display the MyDraw1 toolbar.

Command: **-TOOLBAR**
Enter toolbar name or [ALL]: **MYDRAW1**
Enter an option [Show/Hide/Left/Right/Top/Bottom/Float] <Show>: **S**

Note

Until Release 2002 of AutoCAD, you used to write ***ICON*** *in id_small and id_big. But from AutoCAD 2004, you need to replace* ***ICON*** *by* ***RCDATA****.*

Example 9

In this example you will write a menu file for a toolbar with a flyout. The name of the toolbar is **MyDraw2**, and it contains two buttons, **Circle** and **Arc**. When you select the **Circle** button, it should display a flyout with **Radius**, **Diameter**, **2P**, and **3P** buttons (Figure 32-22). Similarly, when you select the **Arc** button, it should display the **3Point**, **SCE**, and **SCA** buttons (Figure 32-22).

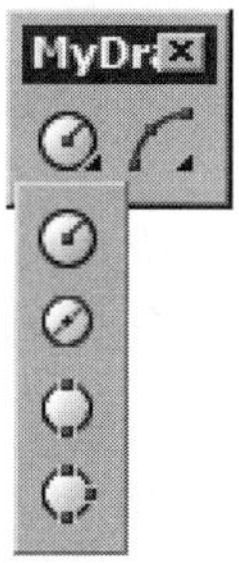

Figure 32-22 Circle *flyout in the* ***MyDraw2*** *toolbar*

Figure 32-23 Arc *flyout in the* ***MyDraw2*** *toolbar*

Step 1: Writing the toolbars menu file

Use any text editor to write the menu for the toolbars. The following is the menu file listing.

```
***Menugroup=M2
***TOOLBARS
**TB_MyDraw2
ID_MyDraw2[_Toolbar("MyDraw2", _Floating, _Show, 10, 100, 1)]
ID_TbCircle[_Flyout("Circle", RCDATA_16_Circle, RCDATA_32_Circle, _OtherIcon,
M2.TB_Circle)]
ID_TbArc[_Flyout("Arc", RCDATA_16_Arc, RCDATA_32_Arc, _OtherIcon, M2.TB_Arc)]
**TB_Circle
ID_TbCircle[_Toolbar("Circle", _Floating, _Hide, 10, 150, 1)]

ID_CirRAD[_Button("Circle C,R", RCDATA_16_CirRAD, RCDATA_32_CirRAD)]^C^C_Circle
ID_CirDIA[_Button("Circle C,D", RCDATA_16_CirDIA, RCDATA_32_CirDIA)]^C^C_Circle;\D
ID_Cir2Pt[_Button("Circle 2Pts", RCDATA_16_Cir2Pt, RCDATA_32_Cir2Pt)]^C^C_Circle;2P
ID_Cir3Pt[_Button("Circle 3Pts", RCDATA_16_Cir3Pt, RCDATA_32_Cir3Pt)]^C^C_Circle;3P

**TB_Arc
ID_TbArc[_Toolbar("Arc", _Floating, _Hide, 10, 150, 1)]
ID_Arc3PT[_Button("Arc,3Pts", RCDATA_16_Arc3PT, RCDATA_32_Arc3PT)]^C^C_Arc
```

ID_ArcSCE[_Button("Arc,SCE", RCDATA_16_ArcSCE, RCDATA_32_ArcSCE)]^C^C_Arc;\C
ID_ArcSCA[_Button("Arc,SCA", RCDATA_16_ArcSCA, RCDATA_32_ArcSCA)]^C^C_Arc;\C;\A

Explanation

ID_TbCircle[_Flyout("Circle", RCDATA**_16_Circle,** RCDATA**_32_Circle, _OtherIcon, M2.TB_Circle)]**

In this line M2 is the MENUGROUP name (***MENUGROUP=M2) and TB_Circle is the name of the toolbar submenu. **M2.TB_Circle** will load the submenu TB_Circle that has been defined in the M2 menugroup. If M2 is missing, AutoCAD will not display the flyout when you select the Circle button.

ID_CirDIA[_Button("Circle C,D", RCDATA**_16_CirDIA,** RCDATA**_32_CirDIA)] ^C^C_Circle;\D**

CirDIA is a user-defined bitmap that displays the **Circle-diameter** button. If you use any other name, AutoCAD will not display the desired button.

Step 2: Loading the menu file containing the toolbars
To load the toolbar, use the **MENULOAD** command to load the menu file. The **MyDraw1** toolbar will be displayed on the screen.

MENU-SPECIFIC HELP

AutoCAD for Windows allows access to online help. For example, if you want to define a helpstring for the **CIRCLE** and **ARC** commands, the syntax is as follows:

***HELPSTRINGS
ID_Copy *[This command will copy the selected object.]*
ID_Ellipse *[(This command will draw an ellipse.]*

The ***__HELPSTRING__** is the section label for the helpstring menu section. The lines defined in this section start with a name tag (ID_Copy) and are followed by the label enclosed in square brackets. The Helpstring tag names are limited to 12 characters. When a menu item is highlighted, AutoCAD searches for the name tag for the corresponding entry in the ***HELPSTRINGS section. If there is a match, the text string contained within the label is displayed in the status line.

CUSTOMIZING THE TOOLBARS

AutoCAD has provided several toolbars that should be sufficient for general use. However, sometimes you may need to customize the toolbars so that the commands that you use frequently are grouped in one toolbar. This saves time in selecting commands. It also saves the drawing space because you do not need to have several toolbars on the screen. The following example explains the process involved in creating and editing the toolbars.

Example 10

In this example you will create a new toolbar (**MyToolbar1**) that has **Line**, **Polyline**, **Circle** (**Center, Radius** option), **Arc** (**Center, Start, End** option), **Spline**, and **Paragraph Text** (**MTEXT**) commands. You will also change the image and tooltip of the Line button and perform other operations like deleting toolbars and buttons and copying buttons between toolbars.

Step1

Select the toolbars in the **View** menu. The **Toolbars** dialog box will appear on the screen. You can also invoke this dialog box by entering **TBCONFIG** at AutoCAD's Command prompt.

Step 2

Choose **New** button in the **Toolbars** dialog box; AutoCAD will display the **New Toolbar** dialog box at the top of the existing (**Toolbars**) dialog box.

Step 3

Enter the name of the toolbar as **MyToolbar1** in the **Toolbar name** edit box (Figure 32-24) and then choose the **OK** button to exit the dialog box. The name of the new toolbar (**MyToolbar1**) is displayed in the **Customize** dialog box.

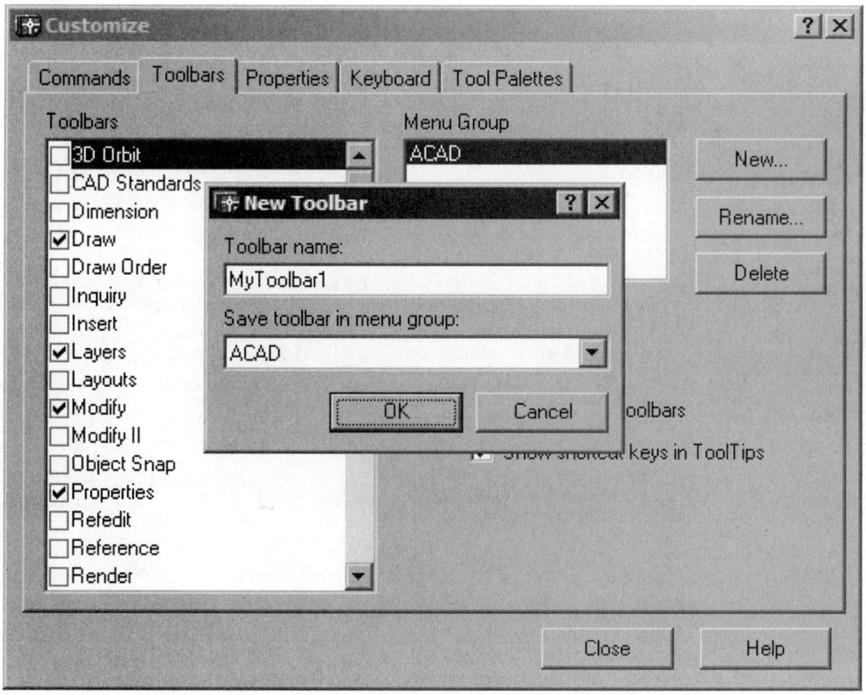

Figure 32-24 Customize and New Toolbar dialog box

Chapter 32

Step 4

Select the new toolbar so that it is highlighted, if it is not already selected, by selecting the box that is located just to the left of the toolbar name (**MyToolbar1**). The new toolbar (**MyToolbar1**) appears on the screen.

Step 5

Choose the **Commands** tab in the **Customize** dialog box (Figure 32-25). The **Categories** and **Commands** lists are displayed in the dialog box.

Step 6

In the **Categories** list box select the **Draw** item. The **Draw** commands are displayed in the Commands area of the **Customize** dialog box.

Step 7

Drag and drop the **Line** command in the **MyToolbar1** toolbar. Repeat the same for **Polyline**, **Circle** (**Center, Radius** option), **Arc** (**Center, Start, End** option), **Spline**, and **Text** (**Multiline text**) buttons.

Step 8

Now, select **Dimension** in the **Category** list box and then select and drag some of the dimensioning commands to **MyToolbar1**, see Figure 32-25. Similarly, you can open any command category and add commands to the new toolbar.

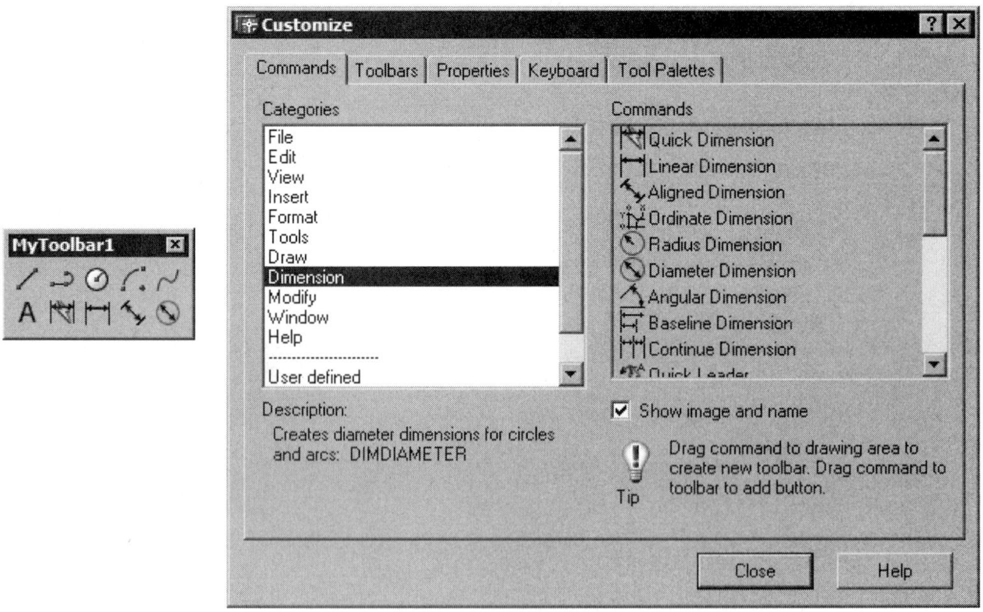

***Figure 32-25** MyToolbar1 toolbar and **Customize** dialog box*

Step 9

When you are done defining the commands for the new toolbar, choose the **Close** button in the **Customize Toolbars** dialog box to return to the AutoCAD screen.

Step 10

Test the buttons in the new toolbar (**MyToolbar1**).

Creating a New Image and Tooltip for a Button

Step 1

Make sure the button with the image you want to edit is displayed on the screen. In this example, you will edit the image of the **Line** button of **MyToolbar1**.

Step 2

Right-click on any toolbar and choose **Customize**; the **Customize** dialog box appears on the screen. After the dialog is displayed, click on the **Line** tool in the **MyToolbar1** toolbar; the button properties are displayed with the image of the existing **Line** tool (Figure 32-26).

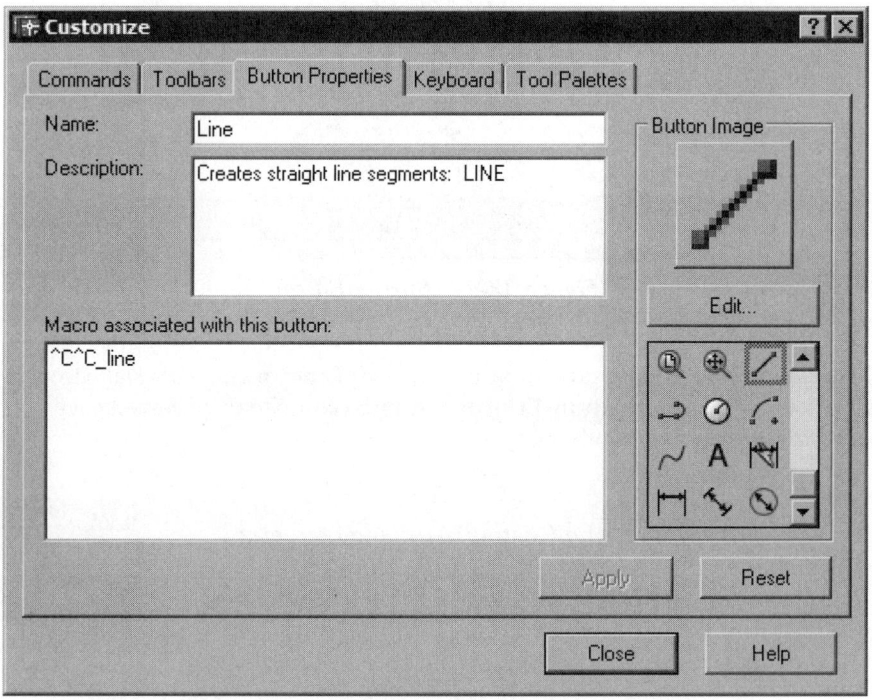

Figure 32-26 ***Button Properties*** *tab of the* ***Customize*** *dialog box*

Step 3

To edit the shape of the image, choose the **Edit** button to access the **Button Editor**. Select the Grid to display the grid lines. You can edit the shape by using different tools in the **Button**

Editor. You can draw a line by choosing the **Line** button and specifying two points. You can draw a circle or an ellipse by using the **Circle** button. The **Erase** button can be used to erase the image.

Step 4

Next, you will change the color of Line image. To accomplish this, erase the existing line, and then select the color and draw a line. Also, create the shape L in the lower right corner (Figure 32-27).

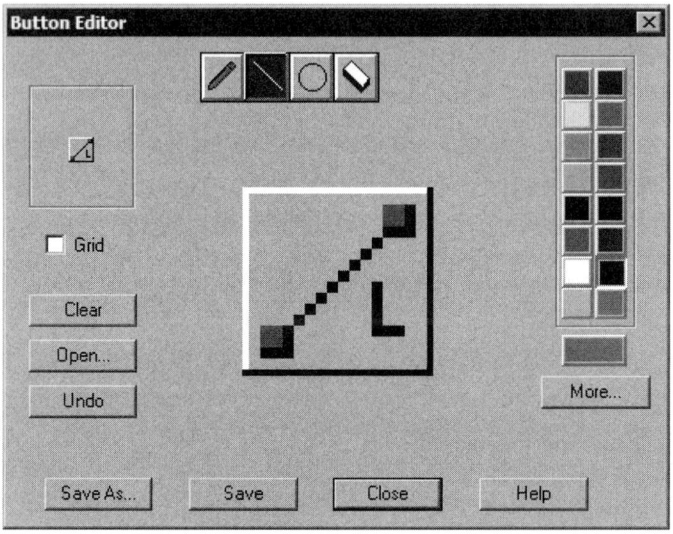

Figure 32-27 Button Editor

Step 5

Choose the **SaveAs** button and save the image as **MyLine** in the Tutorial directory. Choose the **Close** button to exit the **Button Editor**. Using **Save** instead of **SaveAs** will redefine the button image for all existing toolbars.

Step 6

In the **Button Properties** tab of the **Customize** dialog box, enter MyLine in the **Name** edit box. This changes the tooltip of this button. Now, choose the **Apply** button to apply the changes to the image. Close the dialog boxes to return to the AutoCAD screen. Notice the change in the button image and tooltip (Figure 32-28).

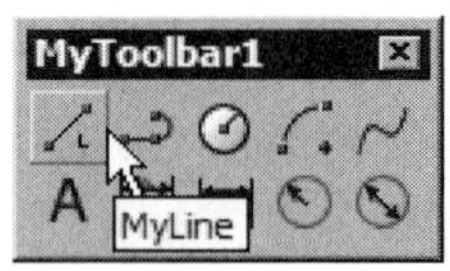

Figure 32-28 My Toolbar 1

Deleting a Button from the Toolbar

Step 1

Right-click on any toolbar to display the shortcut menu. Select **Customize** from the shortcut menu to display the **Customize** dialog box.

Step 2

Click and drag out the **Spline** button from the **MyToolbar1** toolbar. Choose **OK** from the warning box that is displayed. The button will be deleted from the toolbar. Repeat the above step to delete other buttons, if needed. Close the dialog boxes to return to the screen.

Deleting a Toolbar

Choose **Toolbars** from the **View** menu. In the **Toolbar** list box select, the toolbar that you want to delete and then choose the **Delete** button. The toolbar you selected is deleted.

Copying a Tool Button

Step 1

In this example you will copy the **Ordinate Dimension** button from the **Dimensioning** toolbar to **MyToolbar1**. Choose **Toolbars** from the **View** menu or right-click on any button in any toolbar to display the shortcut menu and then select **Customize** to display the **Customize** dialog box. In the **Customize** dialog box choose the **Commands** tab and then select the **Dimensions** in the **Categories** list box. The dimensioning commands are displayed in the **Commands** list box.

Step 2

Select and drag the **Ordinate Dimension** button from the **Commands** list box to **MyToolbar1** toolbar. The ordinate dimensioning button is copied to MyToolbar1.

Any changes made in the toolbars are saved in the *acad.mns* and *acad.mnr* files. The following is the partial listing of *acad.mns* file:

```
**MYTOOLBAR1
ID_MyToolbar1_0 [_Toolbar("MyToolbar1", _Floating, _Show, 512, 177, 1)]
ID_Line_0     [_Button("MyLine", "RCDATA.bmp", "RCDATA_24_LINE")]^C^C_line
ID_CircleCenterRadius_0 [_Button("Circle Center Radius", "RCDATA_16_CIRRAD",
"RCDATA_24_CIRRAD")]^C^C_circle
ID_Polyline_0 [_Button("Polyline", "RCDATA_16_PLINE", "RCDATA_24_PLINE")]^C^C_pline
ID_ArcCenterStartEnd_0 [_Button("Arc Center Start End", "RCDATA_16_ARCCSE",
"RCDATA_24_ARCCSE")]^C^C_arc _c
```

Creating Custom Toolbars with Flyout Icons

In this example you will create a custom toolbar with flyout buttons.

Step 1

Right-click on any toolbar, choose **Customize** from the shortcut menu to display the **Customize** dialog box on the screen. In the **Customize** dialog box, choose the **Toolbars** tab and choose the **New** button to display the **New Toolbar** dialog box. Enter the name of the new toolbar (for example, MYToolbar2) and then choose the **OK** button to exit the box.

Step 2

Choose the **Commands** tab and then select **Flyouts** from the **Categories** list box. The names of the flyouts are displayed in the **Commands** list box. Select and drag the **Draw** flyout button and drop it in the **MyToolbar2** toolbar. The flyout created this way contains all draw commands.

Step 3

Next, select **User defined** from the **Categories** list box. The names of the commands are displayed in the **Commands** list box, Figure 32-29. Select and drag the **User Defined Flyout** option and drop it in the **MyToolbar2** toolbar. Now, click on the **User defined Flyout** button in **MyToolbar2**; the AutoCAD message box appears on the screen. Choose the **OK** button to return to **Customize** dialog box. In the **Flyout Properties** tab, select **Inquiry** and then choose the **Apply** button. The **Inquiry** toolbar gets associated with the custom flyout. Similarly, you can add other user-defined flyouts to the toolbar, see Figure 32-29.

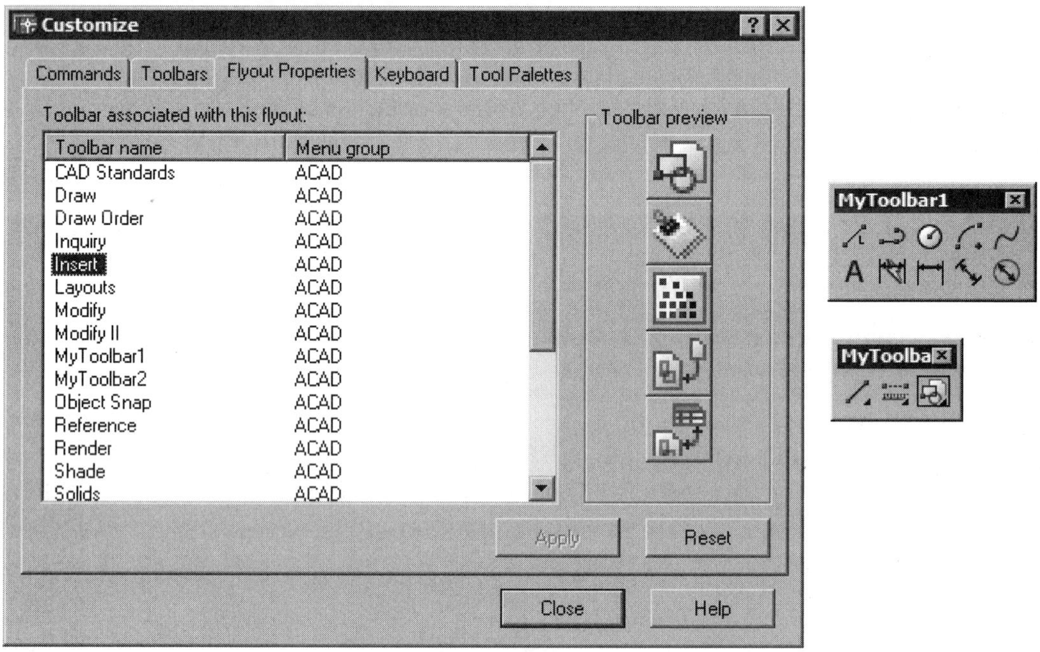

Figure 32-29 ***Customize*** *dialog box with the two custom toolbars*

Assigning Keyboard Shortcuts to Commands

Step 1

Choose **Toolbars** from the **View** menu to invoke the **Customize** dialog box. Choose the **Keyboards** tab and then select the **Draw Menu** in the **Categories** drop-down list. The draw commands are displayed in the **Commands** list box.

Step 2

In the **Press new shortcut key** edit box, specify a key combination to be used as the keyboard shortcut for the selected menu or toolbar item. To specify a value, simultaneously press CTRL and a letter on the keyboard. You can also simultaneously press CTRL+SHIFT and a letter.

In this example, hold the CTRL key down and press the L key. You can also click on the **Show All** button to display the shortcut keys.

Step 3

Choose the **Assign** button to assign the command to the shortcut key (Figure 32-30) and then click on the Close button to close the Customize dialog box. Now, if you hold the CTRL key down and press the L key from the keyboard, the LINE command is invoked.

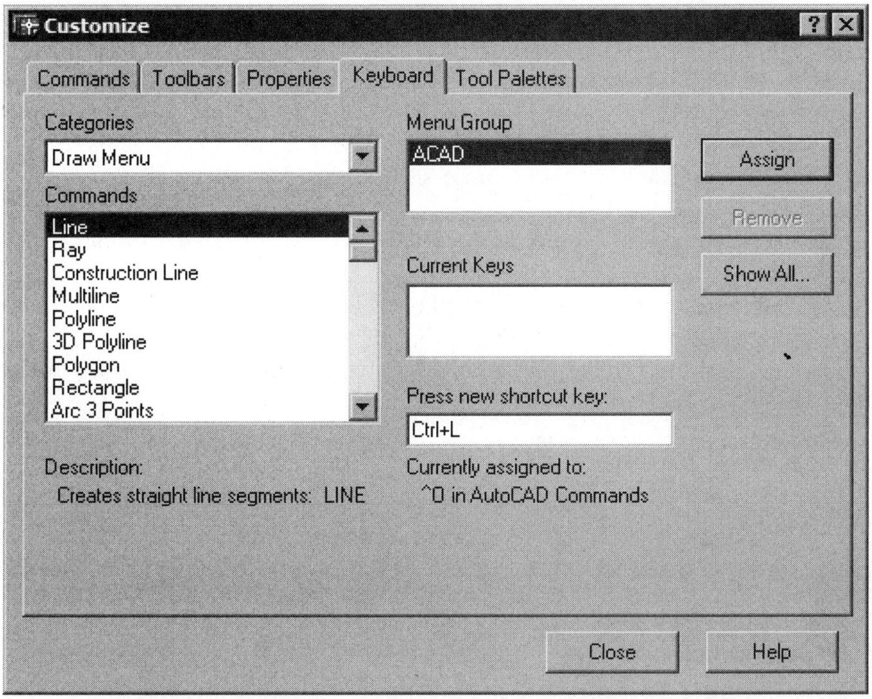

***Figure 32-30 Customize** dialog box with **Keyboard** tab*

Chapter 32

Note

You cannot reassign shortcut keys that are internally assigned to Windows, for example, F10, CTRL+F4, CTRL+F6, or CTRL+ALT+DEL. If you use an invalid key combination, AutoCAD does not show the combination in the edit box. Try using another key combination.

If the keyboard shortcut you specify is already assigned to another AutoCAD command, the "Currently assigned to" message is displayed.

Self-Evaluation Test

Answer the following questions and then compare your answers to the correct answers at the end of this chapter.

1. The name of the standard menu file is ________________ .

2. Menu files can be loaded by using the __________ commands.

3. Each section in the menu file is identified by a __________.

4. The pop menu can have up to __________.

5. The tablet menus can have up to __________ different sections.

6. __________ can be given if you need to use input from a keyboard or a pointing device.

7. The sign __________ defines a cascaded submenu while __________ designates the last item in the menu.

8. All section labels in the AutoCAD menu start with __________.

9. __________ designates the last item of the pull-down or cursor menu.

10. __________ when used as a prefix, displays the item with a check mark.

11. The system variable __________ controls the availability of the **Default**, **Edit**, and **Command** mode shortcut menus.

12. A submenu definition consists of __________.

13. The **Customize** dialog box can be invoked by either __________ or __________ command.

Review Questions

Answer the following questions.

1. A pull-down menu can have ______________ sections.

2. The length of the section title should not exceed ______________ characters.

3. The section titles in a pull-down menu are __________ justified.

4. In a pull-down menu, a line consisting of two hyphens ([--]) ______________ automatically to fill the ______________ of the menu.

5. If the menu item begins with a tilde (~), the items will be ______________ .

6. Every cascading menu in the menu file should have a ______________ name.

7. The cascading menu name can be ______________ characters long.

8. The cascading menu names should not have any ________________ blanks.

9. In Windows you can write partial menus, toolbars, and accelerator key definitions. (T/F)

10. A menu file can contain only one MENUGROUP. (T/F)

11. You can load the partial menu file by using the AutoCAD ____________ command.

12. Once the menu is loaded, you can use the ____________ (AutoLISP function) to display the partial menus.

Exercises

Exercise 3 *General*

Write a pull-down menu for the following AutoCAD commands. (The layout of the menu is shown in Figure 32-31.)

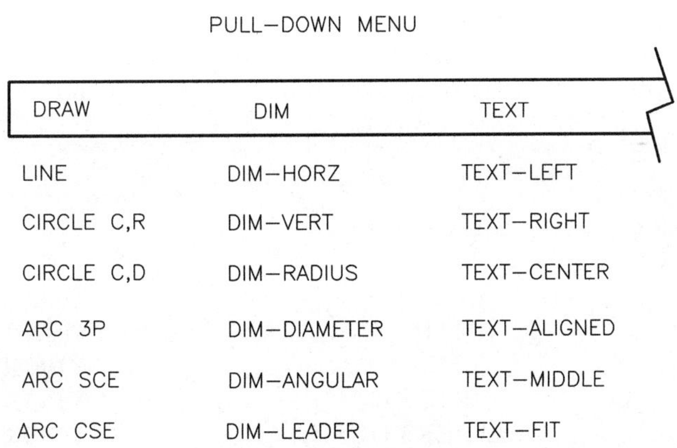

Figure 32-31 *Layout of menu*

LINE	**DIMLINEAR**	**TEXT LEFT**
CIRCLE C,R	**DIMALIGNED**	**TEXT RIGHT**
CIRCLE C,D	**DIMRADIUS**	**TEXT CENTER**
ARC 3P	**DIMDIAMETER**	**TEXT ALIGNED**
ARC SCE	**DIMANGULAR**	**TEXT MIDDLE**
ARC CSE	**QLEADER**	**TEXT FIT**

Exercise 4 *General*

Write a pull-down menu for the following AutoCAD commands.

LINE	**BLOCK**
PLINE	**WBLOCK**
CIRCLE C,R	**INSERT**
CIRCLE C,D	**BLOCK LIST**
ELLIPSE AXIS ENDPOINT	**ATTDEF**
ELLIPSE CENTER	**ATTEDIT**

Exercise 5 *General*

Write a partial menu for Windows. The menu file should have two menus, POP1 (**MyArc**) and POP2 (**MyDraw**). The **MyArc** menu should contain all **Arc** options and must be displayed at the sixth position. Similarly, the **MyDraw** menu should contain **LINE**, **CIRCLE**, **PLINE**, **TRACE**, **DTEXT**, and **MTEXT** commands and should occupy the ninth position.

Exercise 6 *General*

Write a menu file for a toolbar with a flyout. The name of the toolbar is **MyDrawX1**, and it contains two buttons, **Draw** and **Modify**. When you select the **Draw** button, it should display a flyout with **all draw** buttons (**Draw** commands). Similarly, when you choose the **Modify** button, it should display a flyout with **modify** buttons (Modify commands).

Exercise 7 *General*

Write a menu for the following AutoCAD commands.

LAYER NEW	**SNAP** 0.25	**UCS WORLD**
LAYER MAKE	**SNAP** 0.5	**UCS PREVIOUS**
LAYER SET	**GRID** 1.0	**VPORTS** 2
LAYER LIST	**DRID** 10.0	**VPORTS** 4
LAYER ON	**APERTURE** 5	**VPORTS SING**.
LAYER OFF	**PICKBOX** 5	

Answers to the Self-Evaluation Test

1. *acad.mnu*, **2**. **MENU** or the **MENULOAD, 3**. Section label, **4**. 499 sections, **5**. 4 sections, **6**. backslash, **7**. ->, <-, **8**. ***, **9**. <-<-, **10**. !, **11**. **SHORTCUTMENU**, **12**. **, **13**. **TBCONFIG**

Index

B

C

E

H

I

J

L

M

N

O

P

Q

R

S

V

W

X

Z